Volcanism in Antarctica:
200 Million Years of Subduction, Rifting and Continental Break-up

The Geological Society of London Books Editorial Committee

Chief Editor
Rick Law (USA)

Society Books Editors
Jim Griffiths (UK)
Dan Le Heron (Austria)
Mads Huuse (UK)
Rob Knipe (UK)
Phil Leat (UK)
Teresa Sabato Ceraldi (UK)
Lauren Simkins (US)
Randell Stephenson (UK)
Gabor Tari (Austria)
Mark Whiteman (UK)

Society Books Advisors
Kathakali Bhattacharyya (India)
Anne-Christine Da Silva (Belgium)
Xiumian Hu (China)
Spencer Lucas (USA)
Dolores Pereira (Spain)
Virginia Toy (Germany)
Georg Zellmer (New Zealand)

Geological Society books refereeing procedures

The scientific and production quality of the Geological Society's books matches that of its journals. Since 1997, all book proposals are reviewed by two individual experts and the Society's Books Editorial Committee. Proposals are only accepted once any identified weaknesses are addressed.

The Geological Society of London is signed up to the Committee on Publication Ethics (COPE) and follows the highest standards of publication ethics. Once a book has been accepted, the volume editors agree to follow the Society's Code of Publication Ethics and facilitate a peer review process involving two independent reviewers. This is overseen by the Society Book Editors who ensure these standards are adhered to.

Geological Society books are timely volumes in topics of current interest. Proposals are often devised by editors around a specific theme or they may arise from meetings. Irrespective of origin, editors seek additional contributions throughout the editing process to ensure that the volume is balanced and representative of the current state of the field.

Submitting a book proposal
More information about submitting a proposal and producing a book for the Society can be found at https://www.geolsoc.org.uk/proposals

It is recommended that reference to all or part of this book should be made in one of the following ways:

Smellie, J. L., Panter, K. S. and Geyer, A. (eds) 2021. *Volcanism in Antarctica: 200 Million Years of Subduction, Rifting and Continental Break-up*. Geological Society, London, Memoirs, **55**, https://doi.org/10.1144/M55

Quartini, E., Blankenship, D. D. and Young, D. A. 2021. Active subglacial volcanism in West Antarctica. *Geological Society, London, Memoirs*, **55**, 785–803, https://doi.org/10.1144/M55-2019-3

Geological Society Memoir No. 55

Volcanism in Antarctica:
200 Million Years of Subduction, Rifting and Continental Break-up

Edited by

J. L. Smellie
University of Leicester, UK

K. S. Panter
Bowling Green State University, USA

and

A. Geyer
Geosciences Barcelona – CSIC, Spain

2021
Published by
The Geological Society
London

The Geological Society of London

The Geological Society of London is a not-for-profit organisation, and a registered charity (no. 210161). Our aims are to improve knowledge and understanding of the Earth, to promote Earth science education and awareness, and to promote professional excellence and ethical standards in the work of Earth scientists, for the public good. Founded in 1807, we are the oldest geological society in the world. Today, we are a world-leading communicator of Earth science – through scholarly publishing, library and information services, cutting-edge scientific conferences, education activities and outreach to the general public. We also provide impartial scientific information and evidence to support policy-making and public debate about the challenges facing humanity. For more about the Society, please go to https://www.geolsoc.org.uk/

The Geological Society Publishing House (Bath, UK) produces the Society's international journals and books, and acts as European distributor for selected publications of the American Association of Petroleum Geologists (AAPG), the Geological Society of America (GSA), the Society for Sedimentary Geology (SEPM) and the Geologists' Association (GA). GSL Fellows may purchase these societies' publications at a discount. The Society's online bookshop is at https://www.geolsoc.org.uk/bookshop

To find out about joining the Society and benefiting from substantial discounts on publications of GSL and other Societies go to https://www.geolsoc.org.uk/membership or contact the Fellowship Department at: The Geological Society, Burlington House, Piccadilly, London W1J 0BG: Tel. +44 (0)20 7434 9944; Fax +44 (0)20 7439 8975; E-mail: enquiries@geolsoc.org.uk

For information about the Society's meetings, go to https://www.geolsoc.org.uk/events. To find out more about the Society's Corporate Patrons Scheme visit https://www.geolsoc.org.uk/patrons

Proposing a book
If you are interested in proposing a book then please visit: https://www.geolsoc.org.uk/proposals

Published by The Geological Society from:
The Geological Society Publishing House, Unit 7, Brassmill Enterprise Centre, Brassmill Lane, Bath BA1 3JN, UK

The Lyell Collection: www.lyellcollection.org
Online bookshop: www.geolsoc.org.uk/bookshop
Orders: Tel. +44 (0)1225 445046, Fax +44 (0)1225 442836

The publishers make no representation, express or implied, with regard to the accuracy of the information contained in this book and cannot accept any legal responsibility for any errors or omissions that may be made.

© The Geological Society 2021. Except as otherwise permitted under the Copyright, Designs and Patents Act, 1988, this publication may only be reproduced, stored or transmitted, in any form or by any other means, with the prior permission in writing of the publisher, or in the case of reprographic reproduction, in accordance with the terms of a licence issued by the Copyright Licensing Agency in the UK, or the Copyright Clearance Center in the USA. In particular, the Society permits the making of a single photocopy of an article from this issue (under Sections 29 and 38 of this Act) for an individual for the purposes of research or private study. Open access articles, which are published under a CC-BY licence, may be re-used without permission, but subject to acknowledgement.

Full information on the Society's permissions policy can be found at https://www.geolsoc.org.uk/permissions

British Library Cataloguing in Publication Data

A catalogue record for this book is available from the British Library.
ISBN 978-1-78620-536-0
ISSN 0435-4052

Distributors

For details of international agents and distributors see:
www.geolsoc.org.uk/agentsdistributors

Typeset by Nova Techset Private Limited, Bengaluru & Chennai, India
Printed and bound by CPI Group (UK) Ltd, Croydon CR0 4YY, UK

Contents

Acknowledgements — vii

Smellie, J. L., Panter, K. S. and Geyer, A. Introduction to volcanism in Antarctica: 200 million years of subduction, rifting and continental break-up — 1

Section 1: Tectonic history and overview — 7

Storey, B. C. and Granot, R. Chapter 1.1 Tectonic history of Antarctica over the past 200 million years — 9

Smellie, J. L. Chapter 1.2 Antarctic volcanism: volcanology and palaeoenvironmental overview — 19

Panter, K. S. Chapter 1.3 Antarctic volcanism: petrology and tectonomagmatic overview — 43

Geyer, A. Chapter 1.4 Antarctic volcanism: active volcanism overview — 55

Section 2: Gondwana break-up volcanism — 73

Elliot, D. H., White, J. D. L. and Fleming, T. H. Chapter 2.1a Ferrar Large Igneous Province: volcanology — 75

Elliot, D. H. and Fleming, T. H. Chapter 2.1b Ferrar Large Igneous Province: petrology — 93

Riley, T. R. and Leat, P. T. Chapter 2.2a Palmer Land and Graham Land volcanic groups (Antarctic Peninsula): volcanology — 121

Riley, T. R. and Leat, P. T. Chapter 2.2b Palmer Land and Graham Land volcanic groups (Antarctic Peninsula): petrology — 139

Luttinen, A.V. Chapter 2.3 Dronning Maud Land Jurassic volcanism: volcanology and petrology — 157

Section 3: Subduction-related volcanism — 183

Leat, P. T. and Riley, T. R. Chapter 3.1a Antarctic Peninsula and South Shetland Islands: volcanology — 185

Leat, P. T. and Riley, T. R. Chapter 3.1b Antarctic Peninsula and South Shetland Islands: petrology — 213

Smellie, J. L. Chapter 3.2a Bransfield Strait and James Ross Island: volcanology — 227

Haase, K. M. and Beier, C. Chapter 3.2b Bransfield Strait and James Ross Island: petrology — 285

Section 4: Post-subduction, slab-window volcanism — 303

Smellie, J. L. and Hole, M. J. Chapter 4.1a Antarctic Peninsula: volcanology — 305

Hole, M. J. Chapter 4.1b Antarctic Peninsula: petrology — 327

Section 5: Continental extension-related volcanism — 345

Smellie, J. L. and Rocchi, S. Chapter 5.1a Northern Victoria Land: volcanology — 347

Rocchi, S. and Smellie, J. L. Chapter 5.1b Northern Victoria Land: petrology — 383

Smellie, J. L. and Martin, A. P. Chapter 5.2a Erebus Volcanic Province: volcanology — 415

Martin, A. P., Cooper, A. F., Price, R. C., Kyle, P. R. and Gamble, J. A. Chapter 5.2b Erebus Volcanic Province: petrology — 447

Smellie, J. L., Panter, K. S. and Reindel, J. Chapter 5.3a Mount Early and Sheridan Bluff: volcanology — 491

Panter, K. S., Reindel, J. and Smellie, J. L. Chapter 5.3b Mount Early and Sheridan Bluff: petrology — 499

Wilch, T. I., McIntosh, W. C. and Panter, K. S. Chapter 5.4a Marie Byrd Land and Ellsworth Land: volcanology — 515

Panter, K. S., Wilch, T. I., Smellie, J. L., Kyle, P. R. and McIntosh, W. C. Chapter 5.4b Marie Byrd Land and Ellsworth Land: petrology 577

Smellie, J. L. and Collerson, K. D. Chapter 5.5 Gaussberg: volcanology and petrology 615

Section 6:   Tephra record 629

Di Roberto, A., Del Carlo, P. and Pompilio, M. Chapter 6.1 Marine record of Antarctic volcanism from drill cores 631

Narcisi, B. and Petit, J. R. Chapter 6.2 Englacial tephras of East Antarctica 649

Section 7: Active volcanoes 665

Geyer, A., Pedrazzi, D., Almendros, J., Berrocoso, M., López-Martínez, J., Maestro, A., Carmona, E., Álvarez-Valero, A. M. and de Gil, A. Chapter 7.1 Deception Island 667

Sims, K. W. W., Aster, R. C., Gaetani, G., Blichert-Toft, J., Phillips, E. H., Wallace, P. J., Mattioli, G. S., Rasmussen, D. and Boyd, E. S. Chapter 7.2 Mount Erebus 695

Gambino, S., Armienti, P., Cannata, A., Del Carlo, P., Giudice, G., Giuffrida, G., Liuzzo, M. and Pompilio, M. 741
Chapter 7.3 Mount Melbourne and Mount Rittmann

Dunbar, N. W., Iverson, N. A., Smellie, J. L., McIntosh, W. C., Zimmerer, M. J. and Kyle, P. R. 759
Chapter 7.4 Active volcanoes in Marie Byrd Land

Quartini, E., Blankenship, D. D. and Young, D. A. Chapter 7.5 Active subglacial volcanism in West Antarctica 785

Index 805

Acknowledgements

A volume such as this needs to be held to the highest standards. We are immeasurably grateful for the rigorous reviewing by numerous colleagues who gave unstintingly of their free time to review what were often long and very detailed chapters. We are thankful to all of them. They are as follows:

Meritxell Aulinas	Joan Martí
Marco Brenna	Adam Martin
Paola del Carlo	Nick Mortimer
Jim Cole	Patricia Mothes
Nelia Dunbar	David Murphy
David Elliot	Karoly Németh
Thomas Fleming	Robert Pankhurst
Michael Flowerdew	Dario Pedrazzi
Christophe Galerne	Chiara Petrone
John Gamble	Massimo Pompilio
Guido Giordano	Corina Risso
Roi Granot	Alessio di Roberto
Richard Hanson	Sergio Rocchi
Janet Hergt	Pierre-Simon Ross
Malcolm Hole	Christine Siddoway
Nels Iverson	Ian Skilling
Stefan Kraus	Rebecca Smith
Philip Kyle	Edmund Stump
Luis Lara	Christian Timm
Philip Leat	Dougal Townsend
Amanda Lough	Gerhard Wörner

Several additional reviewers also preferred to remain anonymous and we thank them too.

A volume of this size and complexity could not have been completed without the practical advice and behind-the-scenes support over three years by our Geological Society editorial handlers, particularly Bethan Phillips, to whom the editors offer their deepest thanks.

Last, but far from least, taking on a task with the immense scope of this Memoir and the time commitment required of us would have been impossible without the willing support of our families. To each of them, we offer our heartfelt apologies for all the times we were preoccupied and mentally elsewhere rather than being with them, together with our profound gratitude for their endless patience.

This volume is a contribution to the aims and objectives of the SCAR (Scientific Committee on Antarctic Research) Expert Group on Antarctic volcanism (AntVolc) (https://www.scar.org/science/antvolc/home/).

John Smellie (jls55@le.ac.uk)
Kurt Panter (kpanter@bgsu.edu)
Adelina Geyer (ageyer@geo3bcn.csic.es)

Introduction to volcanism in Antarctica: 200 million years of subduction, rifting and continental break-up

John L. Smellie*[1], Kurt S. Panter[2] and Adelina Geyer[3]

[1]Department of Geography, Geology and the Environment, University of Leicester, Leicester LE1 7RH, UK
[2]School of Earth, Environment and Society, Bowling Green State University, Bowling Green, OH 43403, USA
[3]Geosciences Barcelona, GEO3BCN – CSIC, Lluis Solé i Sabaris s/n, 08028 Barcelona, Spain

JLS, 0000-0001-5537-1763; KSP, 0000-0002-0990-5880; AG, 0000-0002-8803-6504

*Correspondence: jls55@leicester.ac.uk

Antarctica has undergone several important phases of volcanism throughout its long history. It was formerly at the heart of Gondwana but, from early Jurassic time (c. 200 Ma), it commenced the prolonged process of disintegration, which resulted in the dispersal and final disposition of the southern hemisphere continents that we are familiar with today (Veevers 2012; **Storey and Granot 2021**). As a consequence, volcanism has been particularly important in its construction and it is geographically widespread, although mainly located within West Antarctica (Fig. 1). Its effects have frequently been felt far outside of the continent. For example, it has been a driver of global mass extinctions (Burgess et al. 2015; Ernst and Youbi 2017) and it has potentially driven Antarctica climatically, and by implication the world, both into and out of glacials (Bay et al. 2006; McConnell et al. 2017). Conversely, Antarctica's volcanoes may have played a pivotal role in helping Life not only to survive multiple glacial episodes during the past few tens of millions of years but to undergo species diversification on the continent in spite of the dramatic climate variations (Fraser et al. 2014). Eruptions from Mount Erebus also represent a significant point source of gases and aerosols to the Austral polar troposphere, including affecting the ozone layer (Boichu et al. 2011; Zuev et al. 2015). Some of Antarctica's active volcanoes also have the potential to have a significant impact on southern hemisphere aviation (Geyer et al. 2017). Finally, Antarctica contains the world's largest and longest-lived glaciovolcanic province. The glaciovolcanic sequences contain a detailed record of the terrestrial Antarctic ice sheet going back to nearly 30 Ma and that record is now beginning to be tapped (e.g. Smellie et al. 2009; Smellie and Edwards 2016; **Wilch et al. 2021**). Despite these attributes, however, volcanism in Antarctica remains terra incognita to many Earth scientists, probably in large part because of its remoteness and inaccessibility.

The only previous volume to be devoted solely to Antarctic volcanism was published 30 years ago (i.e. LeMasurier and Thomson 1990). The scope of that volume was also considerably more restricted than in this Memoir: that is, to volcanoes <c. 30 myr old. That necessitated a main focus on the generally well-preserved volcanoes erupted into the West Antarctic Rift System (WARS), one of the world's major continental rift zones. The young remnants of formerly much more widespread arc-related volcanism, and intraplate volcanism in the sub-Antarctic region, were also included. Nevertheless, nothing else existed at the time and it was, and remains, a landmark publication that inspired a generation of Earth scientists to work in Antarctica, particularly volcanologists and petrologists. As a result, numerous new investigations rapidly followed and our knowledge of Antarctica's volcanism increased profoundly, with the generation of abundant new observations and many large datasets. However, as is always the case, much of the new knowledge has remained unpublished. A major intention of this volume is to capture that information and review it in a modern context after three decades of scientific advancement and enlightenment.

The Proterozoic and Paleozoic record of Antarctica's volcanism is patchy and comparatively poorly known (Riley et al. 2012; Goodge 2019). However, from the Early Jurassic (c. 200 Ma) onward, the record is much more complete and the multiple compositionally and tectonically diverse episodes of Antarctic volcanism have been intensively investigated. The volcanism can be divided into five categories (Smellie 2020): (1) Gondwana break-up volcanism (flood lavas and sills); (2) subduction-related continental margin arc, back-arc and marginal basin volcanism; (3) post-subduction slab-window basalts; (4) continental rift volcanism; and (5) intraplate volcanism of enigmatic origin. In this Memoir, the focus is on the better-preserved and much better-known record from c. 200 Ma, and each category is reviewed and assessed on a geographical basis in terms of its volcanology and eruptive palaeoenvironments; and petrology. Reviews of Antarctica's widely dispersed active volcanism, including tephrochronology (both onshore and offshore) and active subglacial volcanism, are also included. Overall, the objective of this Memoir is to review and assess the present state of knowledge of volcanism younger than c. 200 Ma right across Antarctica in all its aspects, and in the context of the interplay between the volcanism and the prevailing tectonic setting. Together with numerous geological maps, comprehensive tables of geochemical data and isotopic ages, and geophysical information, the intention is that this volume shall be the go-to resource for information on Antarctica's volcanism, *the* reference text for the coming decades. Our hope is that it shall act as a springboard for new proposals which, like the LeMasurier and Thomson volume, will revivify volcanic research in the region.

Volcanism in Antarctica: a brief overview

Gondwana break-up volcanism commencing at c. 190 Ma may have been driven by the effects of a large mantle plume (Storey 1995) and it is represented in Antarctica by two major voluminous volcanic provinces (Fig. 1a). One is a mafic large igneous province (LIP) that crops out throughout the Transantarctic Mountains and in Dronning Maud Land, with correlatives in South Africa, Tasmania, Australia and New Zealand (**Elliot et al. 2021; Elliot and Fleming 2021; Luttinen 2021**). The other consists of a series of felsic flare-ups that affected the entire Antarctic Peninsula and extended into southern South America (Chon Aike province: **Riley and Leat 2021a, b**). The mafic volcanism in Antarctica is

From: Smellie, J. L., Panter, K. S. and Geyer, A. (eds) 2021. *Volcanism in Antarctica: 200 Million Years of Subduction, Rifting and Continental Break-up*. Geological Society, London, Memoirs, **55**, 1–6,
First published online April 12, 2021, https://doi.org/10.1144/M55-2020-14
© 2021 The Author(s). This is an Open Access article distributed under the terms of the Creative Commons Attribution License (http://creativecommons.org/licenses/by/4.0/). Published by The Geological Society of London. Publishing disclaimer: www.geolsoc.org.uk/pub_ethics

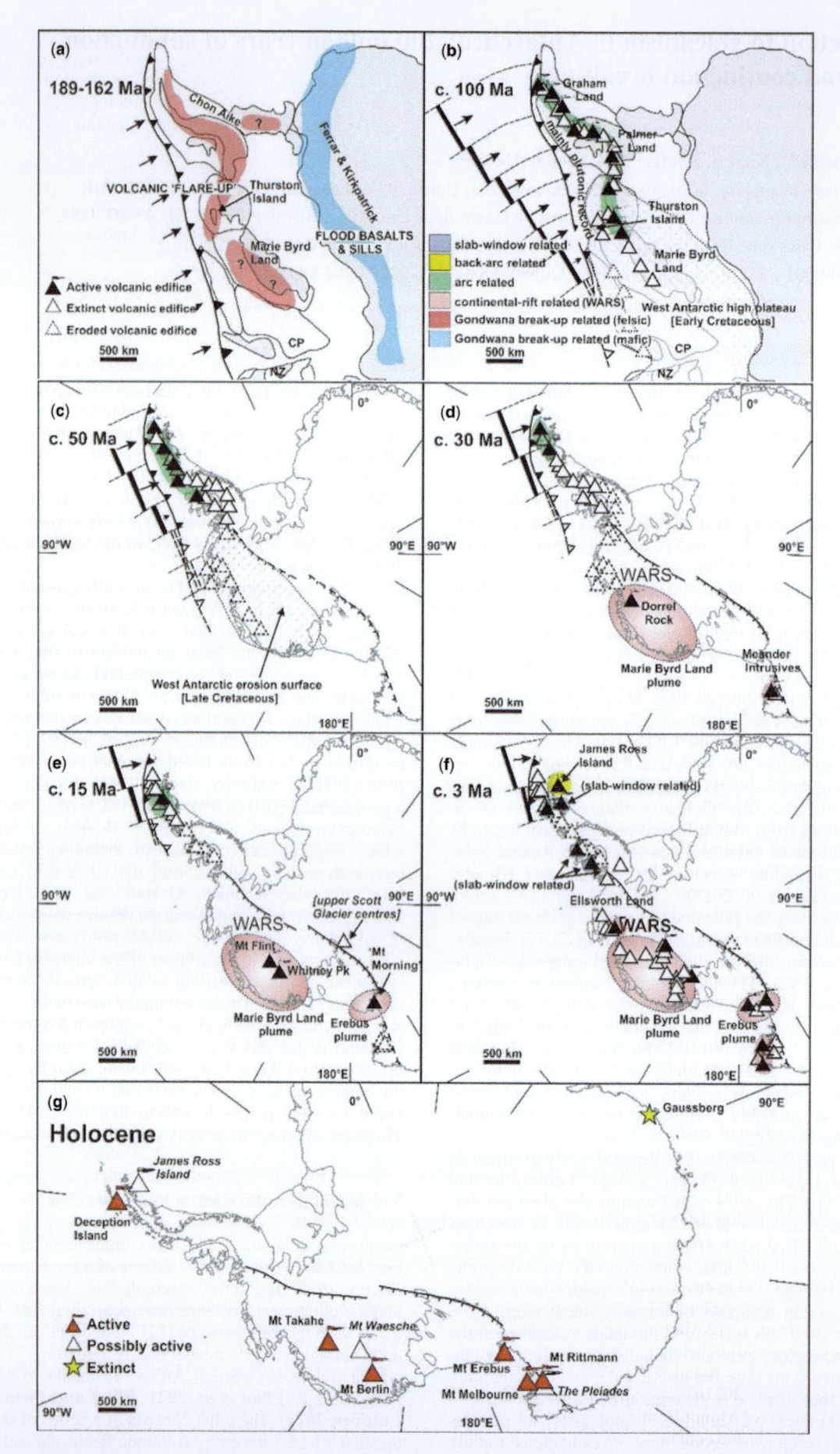

Fig. 1. Schematic time slices illustrating the different stages of volcanism that have affected Antarctica during the past 200 myr (modified after Smellie 2021*a*). The legends in diagrams (**a**) and (**b**) apply to diagrams (**a**)–(**f**), inclusive. In the Holocene depiction (**g**), only large volcanoes (≥*c.* 10 km basal diameter) are shown. The diagram also includes Gaussberg, a small, extinct, ultrapotassic volcanic centre 56 ka in age, to show its remarkable isolation from all other volcanic outcrops in Antarctica. NZ, New Zealand; CP, Campbell Plateau.

known as the Ferrar Supergroup but representatives in Dronning Maud Land show greater compositional affinities to coeval mafic lavas and sills in the Karoo of South Africa. The mafic LIP comprises voluminous flood lavas and thick dolerite sills. The event probably had a total volume of more than 0.5×10^6 km^3 but it may have been emplaced in a very short time, perhaps as little as 400 kyr (Marsh 2007; Burgess *et al.* 2015). Magma in the sills is estimated to have travelled *c.* 4000 km laterally, making it the longest interpreted magma flow on Earth (Elliot *et al.* 1999; Leat 2008). The earliest eruptive phase was characterized by phreatomagmatic pyroclastic eruptions from nested maar–diatreme vent complexes collectively called phreatocauldrons (White and McClintock 2001).

By contrast, the felsic Chon Aike province was erupted in three major pulses, individually *c.* 6–10 myr in duration, between 189 and 153 Ma (Pankhurst *et al.* 2000; Riley *et al.* 2001; **Riley and Leat 2021a, b**). Each corresponds to a predominantly explosive volcanic flare-up (*sensu* Paterson and Ducea 2015) resulting mainly in the eruption of rhyolite ignimbrites. A prominent Pacific-ward progression observed in the ages of the volcanism has been linked to the impact and sublithospheric melting effects of the large spreading head of the same plume responsible for the Ferrar–Karoo LIP. An arc-like (subduction-influenced) mantle source, rather than a plume, has also been postulated for the mafic LIP (Choi *et al.* 2019; **Elliot and Fleming 2021; Panter 2021**).

Coincident with the break-up magmatism, the Pacific margin of Gondwana was already the locus of a major long-lived continental magmatic arc, the products of which are today widely preserved throughout the Antarctic Peninsula (Fig. 1b–e) (**Leat and Riley 2021a, b**). The volcanism in the arc, and its plutonic equivalents, have been extensively dated, revealing that the activity probably took place as a series of magmatic flare-ups and lulls (Riley *et al.* 2018). This is particularly evident in the South Shetland Islands, which is the most intensively dated arc-related region (Haase *et al.* 2012; **Leat and Riley 2021a**). The locus of active arc volcanism in Palmer Land and much of Graham Land moved into the forearc region during the Late Cretaceous, probably due to orthogonal subduction coupled with enhanced time-integrated growth of an extensive accretionary complex. By contrast, the volcanism migrated obliquely away from the trench in the South Shetland Islands possibly due to a highly oblique subduction vector causing much more limited accretion and possibly forearc erosion (**Smellie 2021a**). The magmatic activity shut down progressively in a clockwise direction, commencing in Marie Byrd Land in the mid-Cretaceous (Larter *et al.* 2002). Subduction is active today (at a very slow rate) only at the northern tip of the Antarctic Peninsula, where a small ensialic marginal basin populated with numerous submarine volcanic centres opened up in response to plate boundary forces, including slab rollback (**Haase and Beier 2021; Smellie 2021b**). The marginal basin includes Deception Island, one of Antarctica's most active volcanoes (**Geyer *et al.* 2021**), which underwent a major caldera collapse eruption *c.* 4 kyr ago that dispersed ash more than 4000 km in an arcuate swathe across the Scotia Sea and much of East Antarctica (Antoniades *et al.* 2018). Coincidentally, from *c.* 12 Ma an extensive back-arc mafic alkaline volcanic field developed to the rear of Graham Land in the James Ross Island region (Fig. 1f) (**Haase and Beier 2021; Smellie 2021b**). It is dominated by the very large shield volcano of Mount Haddington, which is predominantly glaciovolcanic and contains an unrivalled record of the Antarctic Peninsula Ice Sheet (Smellie *et al.* 2008).

Following the progressive shut down of subduction, 'windows' opened up in the downgoing oceanic slab, allowing ingress and decompression melting of mantle unaffected by subduction metasomatism. As a consequence, several small monogenetic alkaline volcanic fields were constructed from *c.* 7.5 Ma, mainly along the flanks of the Antarctic Peninsula (Fig. 1f) (Hole *et al.* 1995). The outcrops are overwhelmingly glaciovolcanic and, like Mount Haddington, they also preserve a uniquely valuable record of the Antarctic Peninsula Ice Sheet, as well as influencing our understanding of glaciovolcanism (Smellie and Edwards 2016; **Smellie and Hole 2021**). New research suggests that melting of the downgoing slab may be involved in the generation of the slab-window basalts (**Hole 2021**).

Over a period of a few million years in the Cretaceous, Antarctica's magmatism shifted from arc-related to rift-related during the latest stage of Gondwana break-up that separated Zealandia from Antarctica (*c.* 90–83 Ma). Rare Late Cretaceous alkaline intraplate magmatism is found in southern Zealandia (Weaver and Smith 1989), including HIMU-like ocean island basalt (OIB) (Panter *et al.* 2006; Hoernle *et al.* 2020), and mafic dykes are present along the Ruppert and Hobbs coasts of Marie Byrd Land (Storey *et al.* 1999). However, the episode was largely amagmatic. The extension also probably caused widespread topographical lowering until much of the region in Marie Byrd Land (and coeval terrain in New Zealand) subsided down to, or close to, sea level (LeMasurier and Landis 1996).

Renewed extension during the Cenozoic resulted in the creation of the WARS, a very large continental rift characterized by widespread alkaline volcanism (Fig. 1d–f) (Siddoway 2008; **Smellie and Martin 2021; Smellie and Rocchi 2021; Wilch *et al.* 2021**). The rift contains numerous large and small volcanoes with basalt–trachyte, phonolite and rhyolite compositions (**Martin *et al.* 2021; Panter *et al.* 2021b; Rocchi and Smellie 2021**). Remote sensing studies have also suggested that numerous additional volcanic centres may be widespread beneath the West Antarctic Ice Sheet, including several that may be active (Behrendt *et al.* 1994; van Wyk de Vries *et al.* 2018; **Quartini *et al.* 2021**). The origins of the volcanism are disputed, and variable roles have been inferred for deep mantle plumes and shallow thermal anomalies with associated edge flow (LeMasurier and Landis 1996; Rocchi *et al.* 2005; **Martin *et al.* 2021; Panter *et al.* 2021b; Rocchi and Smellie 2021**). They are all linked within the broad concept of a diffuse alkaline magmatic province (DAMP), which also includes Late Cretaceous and Cenozoic volcanism in eastern Australia, Tasmania and Zealandia (Finn *et al.* 2005). The striking compositional commonality throughout the DAMP region has been explained by melting of a mantle source component inferred to underlie the entire region, with the characteristics of HIMU-like OIB. The mantle reservoir may have been emplaced as a large plume head, of late Cretaceous age or older, or lithosphere that has been metasomatized, or both sources exist (**Panter 2021**). Like alkaline volcanism in the Antarctic Peninsula, the volcanism in the WARS is also predominantly glaciovolcanic, and it has also provided uniquely important information on the development of the West and East Antarctic ice sheets (**Smellie and Rocchi 2021; Wilch *et al.* 2021**).

The tectonic setting of some of Antarctica's volcanoes is enigmatic, however. They include at least three small monogenetic phreatomagmatic edifices in the southern Transantarctic Mountains (upper Scott Glacier) and beneath the East Antarctic Ice Sheet, less than 300 km from the South Pole (Fig. 1e) (**Smellie *et al.* 2021**), and Gaussberg, which is an isolated pillow volcano on the East Antarctic coast far from any other expression of volcanism on the continent (Fig. 1g) (**Smellie and Collerson 2021**). The upper Scott Glacier and nearby subglacial centres are outside of the WARS and they are believed to have formed in response to the detachment and sinking of lithosphere into the convecting mantle beneath the East Antarctic Craton (**Panter et al. 2021a**). The composition of Gaussberg is unique in Antarctica and very rare worldwide, being formed of ultrapotassic lamproite. Its origin is probably linked to a small deep-sourced mantle plume, distinct from the large Kerguelen plume, that incorporated a component derived from ancient subducted sediment (Murphy et al. 2002).

Antarctica also contains several large active volcanoes (**Geyer 2021**) (Fig. 1g). Only two have been observed in eruption (Mount Erebus and Deception Island); both have been intensively investigated (Oppenheimer and Kyle 2008; **Geyer et al. 2021**; **Sims et al. 2021**). Mount Erebus also hosts the world's only semi-permanent phonolite lava lake. The presence of relict heat (Mount Berlin, Mount Melbourne and Mount Rittmann) and abundant englacial and marine tephras sourced in Mount Takahe, Mount Berlin, Mount Waesche, Mount Rittmann and, possibly, The Pleiades indicate that many others were active in recent geological time (<10 ka: Lee et al. 2019; **Dunbar et al. 2021**; **Gambino et al. 2021**; **Narcisi and Petit 2021**; **Di Roberto et al. 2021**). Three of the volcanoes are, or have been, monitored (Deception Island, Mount Erebus and Mount Melbourne) but only one has published hazard and risk assessments (Deception Island: Bartolini et al. 2014; Pedrazzi et al. 2018; **Geyer et al. 2021**).

Acknowledgements The authors gratefully acknowledge the collaborations and conversations with colleagues, too many to mention, over the many years that we have been actively engaged in Antarctic research. We are also grateful for the support provided by the various national funding agencies, without which none of our studies could have been implemented.

Author contributions JLS: conceptualization (lead), writing – original draft (lead); KSP: conceptualization (supporting); AG: conceptualization (supporting).

Funding This research received no specific grant from any funding agency in the public, commercial, or not-for-profit sectors.

Data availability Data sharing is not applicable to this article as no datasets were generated or analysed during the current study.

References

Antoniades, D., Giralt, S. et al. 2018. The timing and widespread effects of the largest Holocene volcanic eruption in Antarctica. *Scientific Reports*, **8**, 17279, https://doi.org/10.1038/s41598-018-35460-x

Bartolini, S., Geyer, A., Martí, J., Pedrazzi, D. and Aguirre-Díaz, G. 2014. Volcanic hazard on Deception Island (South Shetland Islands, Antarctica). *Journal of Volcanology and Geothermal Research*, **285**, 150–168, https://doi.org/10.1016/j.jvolgeores.2014.08.009

Bay, R.C., Bramall, N.E., Price, P.B., Clow, G.D., Hawley, R.L., Udisti, R. and Castellano, E. 2006. Globally synchronous ice core volcanic tracers and abrupt cooling during the last glacial period. *Journal of Geophysical Research: Atmospheres*, **111**, D11108, https://doi.org/10.1029/2005JD006306

Behrendt, J.C., Blankenship, D.D., Finn, C.A., Bell, R.E., Sweeney, R.E., Hodge, S.R. and Brozena, J.M. 1994. CASERTZ aeromagnetic data reveal late Cenozoic flood basalts(?) in the West Antarctic rift system. *Geology*, **22**, 527–530, https://doi.org/10.1130/0091-7613(1994)022<0527:CADRLC>2.3.CO;2

Boichu, M., Oppenheimer, C., Roberts, T.J., Tsanev, V. and Kyle, P. 2011. On bromine, nitrogen oxides and ozone depletion in the tropospheric plume of Erebus volcano (Antarctica). *Atmospheric Environment*, **45**, 3856–3866, https://doi.org/10.1016/j.atmosenv.2011.03.027

Burgess, S.D., Bowring, S.A., Fleming, T.H. and Elliot, D.H. 2015. High precision geochronology links the Ferrar Large Igneous Province with early Jurassic ocean anoxia and biotic crisis. *Earth and Planetary Science Letters*, **415**, 90–99, https://doi.org/10.1016/j.epsl.2015.01.037

Choi, S.H., Mukasa, S.B., Ravizza, G., Fleming, T.H., Marsh, B.D. and Bédard, J.H.J. 2019. Fossil subduction zone origin for magmas in the Ferrar Large Igneous Province, Antarctica: evidence from PGE and OS isotope systematics in the Basement Sill of the McMurdo Dry Valleys. *Earth and Planetary Science Letters*, **506**, 507–519, https://doi.org/10.1016/j.epsl.2018.11.027

Di Roberto, A., del Carlo, P. and Pompilio, M. 2021. Marine record of Antarctic volcanism from drill cores. *Geological Society, London, Memoirs*, **55**, https://doi.org/10.1144/M55-2018-49

Dunbar, N.W., Iverson, N.A., Smellie, J.L., McIntosh, W.C., Zimmerer, M.J. and Kyle, P.R. 2021. Active volcanoes in Marie Byrd Land. *Geological Society, London, Memoirs*, **55**, https://doi.org/10.1144/M55-2019-29

Elliot, D.H. and Fleming, T.H. 2021. Ferrar Large Igneous Province: petrology. *Geological Society, London, Memoirs*, **55**, https://doi.org/10.1144/M55-2018-39

Elliot, D.H., Fleming, T.W., Kyle, P.R. and Foland, K.A. 1999. Long-distance transport of magmatism the Jurassic Ferrar Large Igneous Province, Antarctica. *Earth and Planetary Science Letters*, **167**, 89–104, https://doi.org/10.1016/S0012-821X(99)00023-0

Elliot, D.H., White, J.D.L. and Fleming, T.H. 2021. Ferrar Large Igneous Province: volcanology. *Geological Society, London, Memoirs*, **55**, https://doi.org/10.1144/M55-2018-44

Ernst, R.E. and Youbi, N. 2017. How large igneous provinces affect global climate, sometimes cause mass extinctions, and represent natural markers in the geological record. *Palaeogeography, Palaeoclimatology, Palaeoecology*, **478**, 30–52, https://doi.org/10.1016/j.palaeo.2017.03.014

Finn, C.A., Müller, R.D. and Panter, K.S. 2005. A Cenozoic diffuse alkaline magmatic province (DAMP) in the southwest Pacific without rift or plume origin. *Geochemistry, Geophysics, Geosystems*, **6**, Q02005, https://doi.org/10.1029/2004GC000723

Fraser, C.I., Terauds, A., Smellie, J., Convey, P. and Chown, S. 2014. Geothermal activity helps life survive glacial cycles. *Proceedings of the National Academy of Sciences of the United States of America*, **111**, 5634–5639, https://doi.org/10.1073/pnas.1321437111

Gambino, S., Armienti, P. et al. 2021. Mount Melbourne and Mount Rittmann. *Geological Society, London, Memoirs*, **55**, https://doi.org/10.1144/M55-2018-43

Geyer, A. 2021. Antarctic volcanism: active volcanism overview. *Geological Society, London, Memoirs*, **55**, https://doi.org/10.1144/M55-2020-12

Geyer, A., Marti, A., Giralt, S. and Folch, A. 2017. Potential ash impact from Antarctic volcanoes: Insights from Deception Island's most recent eruption. *Scientific Reports*, **7**, 16534, https://doi.org/10.1038/s41598-017-16630-9

Geyer, A., Pedrazzi, D. et al. 2021. Deception Island. *Geological Society, London, Memoirs*, **55**, https://doi.org/10.1144/M55-2018-56

Goodge, J.W. 2019. Geological and tectonic evolution of the Transantarctic Mountains, from ancient craton to recent enigma. *Gondwana Research*, **80**, 50–122, https://doi.org/10.1016/j.gr.2019.11.001

Haase, K.M. and Beier, C. 2021. Bransfield Strait and James Ross Island: petrology. *Geological Society, London, Memoirs*, **55**, https://doi.org/10.1144/M55-2018-37

Haase, K.M., Beier, C., Fretzdorff, S., Smellie, J.L. and Garbe-Schönberg, D. 2012. Magmatic evolution of the South Shetland Islands, Antarctica, and implications for continental crust formation. *Contributions to Mineralogy and Petrology*, **163**, 1103–1119, https://doi.org/10.1007/s00410-012-0719-7

Hoernle, K., Timm, C. et al. 2020. Late Cretaceous (99–69 Ma) basaltic intraplate volcanism on and around Zealandia: Tracing upper mantle geodynamics from Hikurangi Plateau collision to Gondwana breakup and beyond. *Earth and Planetary Science Letters*, **529**, 115864, https://doi.org/10.1016/j.epsl.2019.115864

Hole, M.J. 2021. Antarctic Peninsula: petrology. *Geological Society, London, Memoirs*, **55**, https://doi.org/10.1144/M55-2018-40

Hole, M.J., Saunders, A.D., Rogers, G. and Sykes, M.A. 1995. The relationship between alkaline magmatism, lithospheric extension and slab window formation along continental destructive plate margins. *Geological Society, London, Special Publications*, **81**, 265–285, https://doi.org/10.1144/GSL.SP.1994.081.01.15

Larter, R.D., Cunningham, A.P., Barker, P.F., Gohl, K. and Nitsche, F.O. 2002. Tectonic evolution of the Pacific margin of Antarctica 1. Late Cretaceous tectonic reconstructions. *Journal of Geophysical Research: Solid Earth*, **107**, 2345, https://doi.org/10.1029/2000JB000052

Leat, P.T. 2008. On the long-distance transport of Ferrar magmas. *Geological Society, London, Special Publications*, **302**, 45–61, https://doi.org/10.1144/SP302.4

Leat, P.T. and Riley, T.R. 2021a. Antarctic Peninsula and South Shetland Islands: volcanology. *Geological Society, London, Memoirs*, **55**, https://doi.org/10.1144/M55-2018-52

Leat, P.T. and Riley, T.R. 2021b. Antarctic Peninsula and South Shetland Islands: petrology. *Geological Society, London, Memoirs*, **55**, https://doi.org/10.1144/M55-2018-68

Lee, M.J., Kyle, P.R., Iverson, N.A., Lee, J.I. and Han, Y. 2019. Rittmann volcano, Antarctica as the source of a widespread 1252 ± 2 CE tephra layer in Antarctica ice. *Earth and Planetary Science Letters*, **521**, 169–176, https://doi.org/10.1016/j.epsl.2019.06.002

LeMasurier, W.E. and Landis, C.A. 1996. Mantle-plume activity recorded by low-relief erosion surfaces in West Antarctica and New Zealand. *Geological Society of America Bulletin*, **108**, 1450–1466, https://doi.org/10.1130/0016-7606(1996)108<1450:MPARBL>2.3.CO;2

LeMasurier, W.E. and Thomson, J.W. (eds) 1990. *Volcanism of the Antarctic Plate and Southern Oceans*. American Geophysical Union Antarctic Research Series, **48**.

Luttinen, A.V. 2021. Dronning Maud Land Jurassic volcanism: volcanology and petrology. *Geological Society, London, Memoirs*, **55**, https://doi.org/10.1144/M55-2018-89

Marsh, B.D. 2007. Magmatism, magma, and magma chambers. *Treatise on Geophysics*, **6**, 276–333.

Martin, A.P., Cooper, A.F., Price, R.C., Kyle, P.R. and Gamble, J.A. 2021. Erebus Volcanic Province: petrology. *Geological Society, London, Memoirs*, **55**, https://doi.org/10.1144/M55-2018-80

McConnell, J.R., Burke, A. *et al.* 2017. Synchronous volcanic eruptions and abrupt climate change similar to 17.7 ka plausibly linked by stratospheric ozone depletion. *Proceedings of the National Academy of Sciences of the United States of America*, **114**, 10 035–10 040, https://doi.org/10.1073/pnas.1705595114

Murphy, D.T., Collerson, K.D. and Kamber, B.S. 2002. Lamproites from Gaussberg, Antarctica: possible transition zone melts of Archaean subducted sediments. *Journal of Petrology*, **43**, 981–1001, https://doi.org/10.1093/petrology/43.6.981

Narcisi, B. and Petit, J.R. 2021. Englacial tephras of East Antarctica. *Geological Society, London, Memoirs*, **55**, https://doi.org/10.1144/M55-2018-86

Oppenheimer, C. and Kyle, P.R. (eds) 2008. Volcanology of Erebus volcano, Antarctica. *Journal of Volcanology and Geothermal Research*, **177**, 531–754, https://doi.org/10.1016/j.jvolgeores.2008.08.016

Pankhurst, R.J., Riley, T.R., Fanning, C.M. and Kelley, S.P. 2000. Episodic silicic volcanism in Patagonia and the Antarctic Peninsula: Chronology of magmatism associated with the break-up of Gondwana. *Journal of Petrology*, **41**, 605–625, https://doi.org/10.1093/petrology/41.5.605

Panter, K.S. 2021. Antarctic volcanism: petrology and tectonomagmatic overview. *Geological Society, London, Memoirs*, **55**, https://doi.org/10.1144/M55-2020-10

Panter, K.S., Blusztajn, J., Hart, S.R., Kyle, P.R., Esser, R. and McIntosh, W.C. 2006. The origin of HIMU in the SW Pacific: Evidence from intraplate volcanism in southern New Zealand and subantarctic islands. *Journal of Petrology*, **47**, 1673–1704, https://doi.org/10.1093/petrology/egl024

Panter, K.S., Reindel, J. and Smellie, J.L. 2021a. Mount Early and Sheridan Bluff: petrology. *Geological Society, London, Memoirs*, **55**, https://doi.org/10.1144/M55-2019-2

Panter, K.S., Wilch, T.I., Smellie, J.L., Kyle, P.R. and McIntosh, W.C. 2021b. Marie Byrd Land and Ellsworth Land: petrology. *Geological Society, London, Memoirs*, **55**, https://doi.org/10.1144/M55-2019-50

Paterson, S.R. and Ducea, M.N. 2015. Arc magmatic tempos: gathering the evidence. *Elements*, **11**, 91–98, https://doi.org/10.2113/gselements.11.2.91

Pedrazzi, D., Németh, K., Geyer, A., Álvarez-Valero, A.M., Aguirre-Díaz, G. and Bartolini, S. 2018. Historic hydrovolcanism at Deception Island (Antarctica): implications for eruption hazards. *Bulletin of Volcanology*, **80**, 11, https://doi.org/10.1007/s00445-017-1186-9

Quartini, E., Blankenship, D.D. and Young, D.A. 2021. Active subglacial volcanism in West Antarctica. *Geological Society, London, Memoirs*, **55**, https://doi.org/10.1144/M55-2019-3

Riley, T.R. and Leat, P.T. 2021a. Palmer Land and Graham Land volcanic groups (Antarctic Peninsula): volcanology. *Geological Society, London, Memoirs*, **55**, https://doi.org/10.1144/M55-2018-36

Riley, T.R. and Leat, P.T. 2021b. Palmer Land and Graham Land volcanic groups (Antarctic Peninsula): petrology. *Geological Society, London, Memoirs*, **55**, https://doi.org/10.1144/M55-2018-51

Riley, T.R., Leat, P.T., Pankhurst, R.J. and Harris, C. 2001. Origins of large volume rhyolitic volcanism in the Antarctic Peninsula and Patagonia by crustal melting. *Journal of Petrology*, **42**, 1043–1065, https://doi.org/10.1093/petrology/42.6.1043

Riley, T.R., Flowerdew, M.J. and Whitehouse, M.J. 2012. U–Pb ion-microprobe zircon geochronology from the basement inliers of eastern Graham Land, Antarctic Peninsula. *Journal of the Geological Society, London*, **169**, 381–393, https://doi.org/10.1144/0016-76492011-142

Riley, T.R., Burton-Johnson, A., Flowerdew, M.J. and Whitehouse, M.J. 2018. Episodicity within a mid-Cretaceous magmatic flare-up in West Antarctica: U–Pb ages of the Lassiter Coast intrusive suite, Antarctic Peninsula, and correlations along the Gondwana margin. *Geological Society of America Bulletin*, **130**, 1177–1196, https://doi.org/10.1130/B31800.1

Rocchi, S. and Smellie, J.L. 2021. Northern Victoria Land: petrology. *Geological Society, London, Memoirs*, **55**, https://doi.org/10.1144/M55-2019-19

Rocchi, S., Armienti, P. and Di Vincenzo, G. 2005. No plume, no rift magmatism in the West Antarctic Rift. *United States Geological Survey Professional Papers*, **388**, 435–447.

Siddoway, C. 2008. Tectonics of the West Antarctic rift system: New light on the history and dynamics of distributed intracontinental extension (invited paper). *In*: Cooper, A.K., Barrett, P.J. *et al.* (eds) *Antarctica: A Keystone in a Changing World. Proceedings of the 10th International Symposium on Antarctic Earth Sciences*. The National Academies Press, Washington, DC, 91–114.

Sims, K.W.W., Aster, R. *et al.* 2021. Mount Erebus. *Geological Society, London, Memoirs*, **55**, https://doi.org/10.1144/M55-2019-8

Smellie, J.L. 2020. The role of volcanism in the making of Antarctica. *In*: Oliva, M. and Ruiz-Fernández, J. (eds) *Past Antarctica. Paleoclimatology and Climate Change*. Academic Press, London, 69–88.

Smellie, J.L. 2021a. Antarctic volcanism: volcanology and palaeoenvironmental overview. *Geological Society, London, Memoirs*, **55**, https://doi.org/10.1144/M55-2020-1

Smellie, J.L. 2021b. Bransfield Strait and James Ross Island: volcanology. *Geological Society, London, Memoirs*, **55**, https://doi.org/10.1144/M55-2018-58

Smellie, J.L. and Collerson, K.D. 2021. Gaussberg: volcanology and petrology. *Geological Society, London, Memoirs*, **55**, https://doi.org/10.1144/M55-2018-85

Smellie, J.L. and Edwards, B.E. 2016. *Glaciovolcanism on Earth and Mars: Products, Processes and Palaeoenvironmental Significance*. Cambridge University Press, Cambridge, UK.

Smellie, J.L. and Hole, M.J. 2021. Antarctic Peninsula: volcanology. *Geological Society, London, Memoirs*, **55**, https://doi.org/10.1144/M55-2018-59

Smellie, J.L. and Martin, A.P. 2021. Erebus Volcanic Province: volcanology. *Geological Society, London, Memoirs*, **55**, https://doi.org/10.1144/M55-2018-62

Smellie, J.L. and Rocchi, S. 2021. Northern Victoria Land: volcanology. *Geological Society, London, Memoirs*, **55**, https://doi.org/10.1144/M55-2018-60

Smellie, J.L., Johnson, J.S., McIntosh, W.C., Esser, R., Gudmundsson, M.T., Hambrey, M.J. and van Wyk de Vries, B. 2008. Six million years of glacial history recorded in the James Ross Island Volcanic Group, Antarctic Peninsula. *Palaeogeography, Palaeoclimatology, Palaeoecology*, **260**, 122–148, https://doi.org/10.1016/j.palaeo.2007.08.011

Smellie, J.L., Haywood, A.M., Hillenbrand, C.-D., Lunt, D.J. and Valdes, P.J. 2009. Nature of the Antarctic Peninsula Ice Sheet during the Pliocene: Geological evidence and modelling results compared. *Earth-Science Reviews*, **94**, 79–94, https://doi.org/10.1016/j.earscirev.2009.03.005

Smellie, J.L., Panter, K.S. and Reindel, J. 2021. Mount Early and Sheridan Bluff: volcanology. *Geological Society, London, Memoirs*, **55**, https://doi.org/10.1144/M55-2018-61

Storey, B.C. 1995. The role of mantle plumes in continental breakup: case histories from Gondwanaland. *Nature*, **377**, 301–308, https://doi.org/10.1038/377301a0

Storey, B.C. and Granot, R. 2021. Tectonic history of Antarctica over the past 200 million years. *Geological Society, London, Memoirs*, **55**, https://doi.org/10.1144/M55-2018-38

Storey, B.C., Leat, P.T., Weaver, S.D., Pankhurst, R.J., Bradshaw, J.D. and Kelley, S. 1999. Mantle plumes and Antarctica–New Zealand rifting: evidence from mid-Cretaceous mafic dykes. *Journal of the Geological Society, London*, **156**, 659–671, https://doi.org/10.1144/gsjgs.156.4.0659

van Wyk de Vries, M., Bingham, R.G. and Hein, A.S. 2018. A new volcanic province: an inventory of subglacial volcanoes in West Antarctica. *Geological Society, London, Special Publications*, **461**, 231–248, https://doi.org/10.1144/SP461.7

Veevers, J.J. 2012. Reconstructions before rifting and drifting reveal the geological connections between Antarctica and its conjugates in Gondwanaland. *Earth-Science Reviews*, **111**, 249–318, https://doi.org/10.1016/j.earscirev.2011.11.009

Weaver, S.D. and Smith, I.E.M. 1989. New Zealand intraplate volcanism. *In*: Johnson, R.W., Knutson, J. and Taylor, S.R. (eds) *Intraplate Volcanism in Eastern Australia and New Zealand*. Cambridge University Press, Cambridge, UK, 157–188.

White, J.D.L. and McClintock, M.K. 2001. Immense vent complex marks flood-basalt eruption in a wet, failed rift: Coombs Hills, Antarctica. *Geology*, **29**, 935–938, https://doi.org/10.1130/0091-7613(2001)029<0935:IVCMFB>2.0.CO;2

Wilch, T.I., McIntosh, W.C. and Panter, K.S. 2021. Marie Byrd Land and Ellsworth Land: volcanology. *Geological Society, London, Memoirs*, **55**, https://doi.org/10.1144/M55-2019-39

Zuev, V.V., Zueva, N.E., Savelieva, E.S. and Gerasimov, V.V. 2015. The Antarctic ozone depletion caused by Erebus volcano gas emissions. *Atmospheric Environment*, **122**, 393–399, https://doi.org/10.1016/j.atmosenv.2015.10.005

Section 1

Tectonic history and overview

Pliocene glaciovolcanic sequence intruded by scoria cone vent at Cape Roget, northern Victoria Land. The image illustrates the superb clean exposure of many of Antarctica's volcanoes in coastal regions.

Photograph by J. L. Smellie

Chapter 1.1

Tectonic history of Antarctica over the past 200 million years

Bryan C. Storey[1]* and Roi Granot[2]

[1]Gateway Antarctica, University of Canterbury, Christchurch 8041, New Zealand

[2]Department of Geological and Environmental Sciences, Ben-Gurion University of the Negev, Beer-Sheva 84105, Israel

RG, 0000-0001-5366-188X

*Correspondence: b.storey@xtra.co.nz

Abstract: The tectonic evolution of Antarctica in the Mesozoic and Cenozoic eras was marked by igneous activity that formed as a result of simultaneous continental rifting and subduction processes acting during the final stages of the southward drift of Gondwana towards the South Pole. For the most part, continental rifting resulted in the progressive disintegration of the Gondwana supercontinent from Middle Jurassic times to the final isolation of Antarctica at the South Pole following the Cenozoic opening of the surrounding ocean basins, and the separation of Antarctica from South America and Australia. The initial rifting into East and West Gondwana was proceeded by emplacement of large igneous provinces preserved in present-day South America, Africa and Antarctica. Continued rifting within Antarctica did not lead to continental separation but to the development of the West Antarctic Rift System, dividing the continent into the East and West Antarctic plates, and uplift of the Transantarctic Mountains. Motion between East and West Antarctica has been accommodated by a series of discrete rifting pulses with a westward shift and concentration of the motion throughout the Cenozoic leading to crustal thinning, subsidence, elevated heat flow conditions and rift-related magmatic activity. Contemporaneous with the disintegration of Gondwana and the isolation of Antarctica, subduction processes were active along the palaeo-Pacific margin of Antarctica recorded by magmatic arcs, accretionary complexes, and forearc and back-arc basin sequences. A low in magmatic activity between 156 and 142 Ma suggests that subduction may have ceased during this time. Today, following the gradual cessation of the Antarctic rifting and surrounding subduction, the Antarctic continent is situated close to the centre of a large Antarctic Plate which, with the exception of an active margin on the northern tip of the Antarctic Peninsula, is surrounded by active spreading ridges.

At the start of the Mesozoic Era, Antarctica was the centre piece or keystone to the Gondwana supercontinent which had remained stable for almost 350 myr. During that time, Gondwana drifted southwards from a more equatorial position (Torsvik *et al.* 2012). The slow southward drift was temporally disrupted at *c.* 250 Ma as Gondwana voyaged north but headed south again at *c.* 200 Ma (Torsvik and Cox 2013). In middle Jurassic times, the progressive disintegration of the supercontinent changed the global continental configuration, leading to the opening of major ocean gateways and the isolation of Antarctica at the South Pole. Today, the tectonic Antarctic Plate is bordered by six different tectonic plates and is almost entirely surrounded by spreading ridges with Cenozoic isolation upon the South Pole.

The Antarctic continent can be divided into two physiographical provinces, East and West Antarctica, separated by a spectacular mountain range, the Transantarctic Mountains (TAM), that stretch from north Victoria Land bordering the western Ross Sea to the Weddell Sea (Fig. 1). Cratonic East Antarctica comprises Archean and Proterozoic–Cambrian terranes amalgamated during Precambrian and Cambrian times (Fitzsimons 2000). In contrast, West Antarctica comprises a collage of five tectonic blocks separated by rifts and topographical depressions (Fig. 1): the Antarctic Peninsula, Thurston Island, the Ellsworth Whitmore Mountains (EWM), Haag Nunataks and Marie Byrd Land (Dalziel and Elliot 1982). The Antarctic Peninsula has generally been considered as a near-complete Mesozoic–Cenozoic continental arc system formed above an eastward-dipping palaeo-Pacific subduction zone (Suarez 1976; Burton-Johnson and Riley 2015). However, Vaughan and Storey (2000) suggested that the Antarctic Peninsula may have consisted of three fault-bounded terranes that amalgamated in Late Cretaceous time (Albian). The Ellsworth Whitmore Mountains block is a displaced fragment of the Permo-Triassic Gondwanide Fold Belt that was originally located in the Natal Embayment off South Africa in Gondwana before undergoing 90° counter-clockwise rotation prior to or during the Gondwana break-up (Randall and Mac Niocaill 2004). Haag Nunataks is a small fragment of

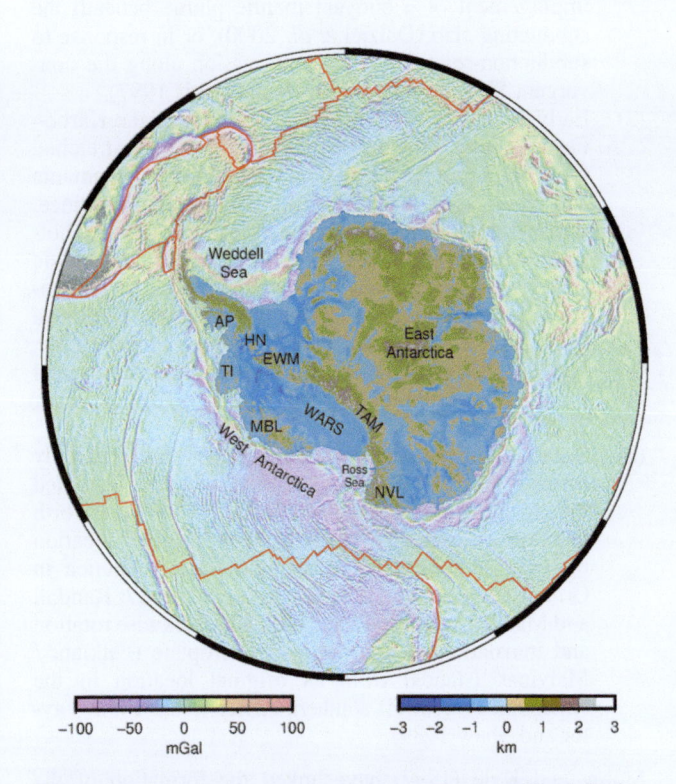

Fig. 1. Tectonic map of Antarctica superimposed on a satellite-derived free-air gravity field (offshore: Sandwell *et al.* 2014) and sub-ice topography (Fretwell *et al.* 2013) showing the Transantarctic Mountains (TAMts), the crustal blocks of West Antarctica and the West Antarctic Rift System (WARS). AP, Antarctic Peninsula; EWM, Ellsworth Whitmore Mountains; HN, Haag Nunataks; MBL, Marie Byrd Land; TI, Thurston Island.

Neoproterozoic craton similar to parts of East Antarctica (Millar and Pankhurst 1987). In contrast, the Thurston Island (Pankhurst *et al.* 1993; Riley *et al.* 2016) and Marie Byrd Land blocks (Mukasa and Dalziel 2000) contain Mesozoic subduction-related magmatic rocks. In parallel to these processes, a long rift, the West Antarctic Rift System, has evolved, leading to the dismembering of the plate into the East and West Antarctic plates.

This paper reviews the tectonic history of Antarctica during the Mesozoic and Cenozoic eras which provides the backdrop for the volcanic and magmatic evolution of the continent. The magmatic evolution in itself provides valuable insights into the lithospheric and tectonic processes that shaped the Antarctic continent. Three interacting tectonic processes affected Antarctica during the last 200 myr: continental break-up, rifting and subduction.

Continental break-up

The initial fragmentation of the Gondwana supercontinent was preceded by several major tectonic and igneous events prior to earliest seafloor spreading in the Jurassic (Dalziel *et al.* 2013):

(1) Latest Paleozoic–early Mesozoic Gondwanide orogenesis and formation of the Gondwanian Fold Belt that extended from the Sierra de la Ventana of Argentina, through the Cape Fold Belt in southern Africa to the Pensacola Mountains along the Transantarctic margin of East Antarctica (Du Toit 1937). The enigmatic Gondwanide folding may have developed in response to either flat-slab subduction (Lock 1980), perhaps due to the impingement of a buoyant mantle plume beneath the subducting slab (Dalziel *et al.* 2000), or in response to subduction-related dextral compression along the convergent SW margin of Gondwana (Curtis 1997).

(2) Early Jurassic (*c.* 183 Ma) emplacement of the Karoo–Ferrar Large Igneous Province (LIP) that stretched from southern Africa through the TAM to Tasmania and New Zealand (Cox 1992). This basaltic province, which is generally linked to a mantle plume (Bouvet hotspot), was in part synchronous with Middle Jurassic extrusion of voluminous silicic volcanic rocks in Patagonia (Chon Aike province) and the Antarctic Peninsula (Pankhurst and Rapela 1995; Pankhurst *et al.* 1998a, 2000) attributed to melting of continental crust proximal to the mantle-plume thermal anomaly (Bryan *et al.* 2002).

(3) Rotation and translation of microplates that originally formed part of the Gondwanian Fold Belt; these included the 90° counter-clockwise rotation of the Ellsworth Whitmore Mountains block from its original location between southernmost Africa and East Antarctica in Gondwana (Schopf 1969; Grunow *et al.* 1987; Randall and Mac Niocaill 2004), and the 180° clockwise rotation and translation of the Lafonian microplate (Falkland/Malvinas Islands) from its original location in the Natal embayment off southern Africa (Adie 1952; Taylor and Shaw 1989).

Storey and Kyle (1997) have linked the formation of the Gondwana LIP that formed the Karoo–Ferrar–Chon Aike provinces and the rotation and translation of the microplates to a large thermal anomaly centred in the Weddell Sea region that ultimately resulted in, or at least contributed to, the break-up of Gondwana. However, the timing and exact geodynamic processes that occurred during initial rifting remain unclear and no adequate dynamic model exists to explain these events. In addition, geophysical interpretations have provided little support for major motion and rotation of crustal blocks during Jurassic extension in the Weddell Sea region (Jordan *et al.* 2017). Jordan *et al.* (2017) proposed an alternative model that predicts *c.* 500 km of movement of the Haag and the Ellsworth microplates with 30° of block rotation during crustal extension in a Weddell Sea rift zone. In this model, the Weddell Sea rift zone would have formed in response to distributed crustal extension within a broad plate boundary region between East and West Antarctica. To reconcile the geological and palaeomagnetic data from the crustal blocks, they suggest that 60° of rotation that is unaccounted for by the geophysically imaged Jurassic extension may have occurred earlier during the Gondwanian orogenesis. This is possible within the transpressional tectonic regime suggested by Curtis (1997) for Gondwanian events.

The final break-up and separation of East and West of Gondwana was initiated (at *c.* 167 Ma) along a rift zone which comprised the Somali and Mozambique basins, the southern Africa–East Antarctica (Dronning Maud Land) conjugate margins, and the Weddell Sea embayment, with seafloor spreading commencing about 160–165 myr before propagating clockwise around Antarctica (Ghidella *et al.* 2002; Konig and Jokat 2006; see the review by Torsvik *et al.* 2008). Early Africa–Antarctic spreading offshore Dronning Maud Land has been dated as magnetic anomaly M24 (*c.* 150 Ma: Roeser *et al.* 1996; Jokat *et al.* 2003; Malinverno *et al.* 2012), with the earliest seafloor in the Weddell Sea dated at 147 Ma (Konig and Jokat 2006). A model for the early Indian–Antarctic spreading system places the onset of seafloor spreading in the Enderby Basin at anomaly M9 (130 Ma: Gaina *et al.* 2007), consistent with the opening history between India and Australia (Williams *et al.* 2013). Although volcanism preceded the initial break-up of Gondwana in Middle Jurassic times, the separation of India from Antarctica *c.* 130 Ma was followed by volcanic activity, with the earliest magmatic activity in the Kerguelen area dated to *c.* 118 Ma (Frey *et al.* 2000; Nicolaysen *et al.* 2001). Early Australia–Antarctic spreading has been identified by a Late Cretaceous ridge system slightly older than anomaly 34, at *c.* 90 Ma (Tikku and Cande 1999). In the South Tasman Sea, between eastern Australia and the Lord Howe Rise and New Zealand, seafloor spreading began in the Late Cretaceous (*c.* 83 Ma) (Gaina *et al.* 1998). The TAM were exhumed at this time with Early and Late Cretaceous and Cenozoic stages of uplift and exhumation (Fitzgerald and Gleadow 1988; Busetti *et al.* 1999; Fitzgerald 2002; Lisker and Laufer 2013). In the middle–late Eocene the final detachment of Australia from Antarctica led to the opening of the first circum-Antarctic oceanic gateway south of Tasmania causing radical changes in oceanic circulation patterns (Brown *et al.* 2006). Seafloor spreading in the Drake Passage and Scotia Sea region is generally considered to have commenced before 26 Ma (Barker 2001) or *c.* 30 Ma (Eagles and Livermore 2002), resulting in the development of the circumpolar current and the final isolation of Antarctica (Barker and Thomas 2004) (Fig. 2). However, Eagles and Jokat (2014) have suggested that the Drake Passage developed as an intermediate-depth ocean gateway through a sequence of extensional basins (50–30 Ma) that were succeeded by seafloor spreading with deep ocean flow forming from 30 to 6 Ma.

West Antarctic Rift System

A broad region of extended continental crust between the TAM and the Pacific margin (Fig. 1) is known as the West Antarctic Rift System (LeMasurier 1990; Behrendt *et al.* 1991; Busetti *et al.* 1999; Wilson and Luyendyk 2009; Chaput *et al.* 2014) (WARS in Fig. 1). The incipient motion between

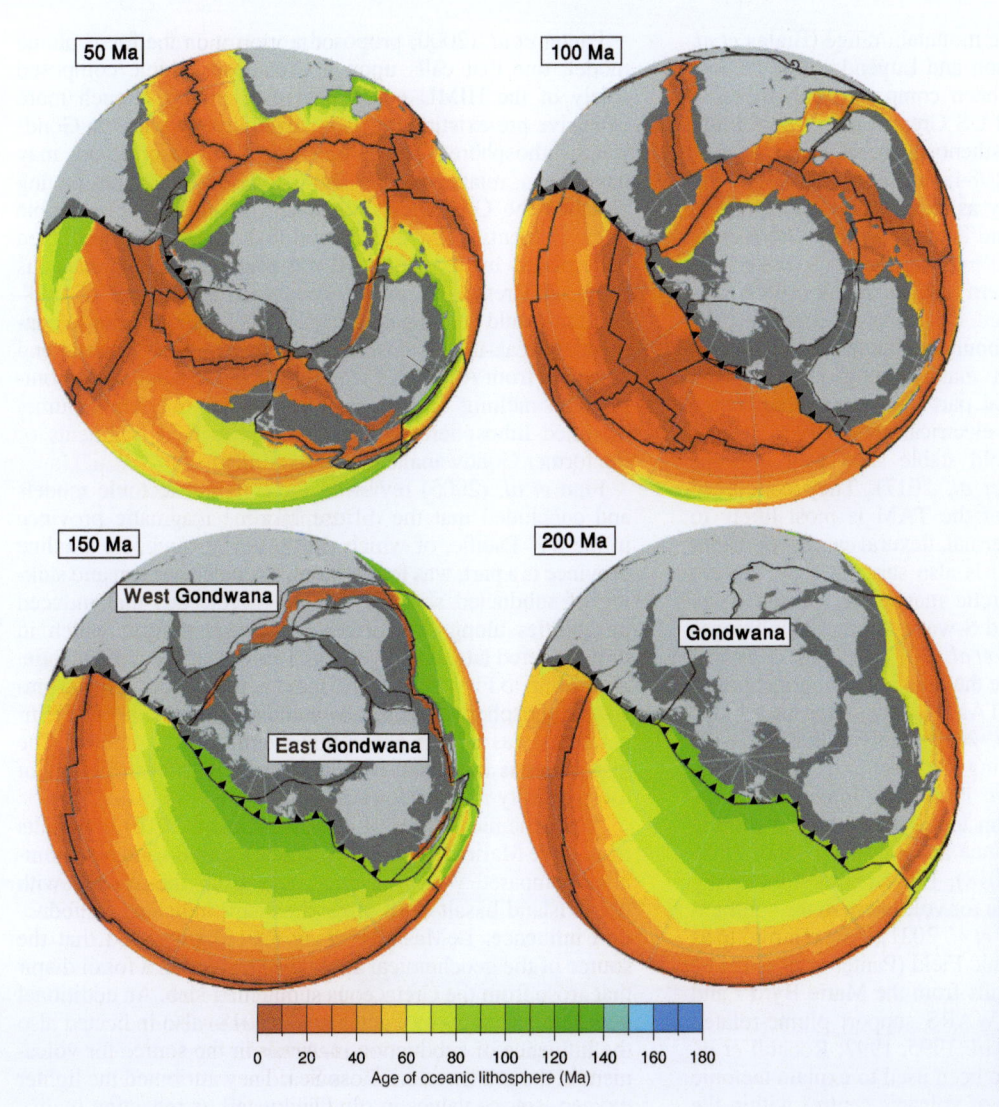

Fig. 2. Plate reconstructions from 200 to 50 Ma in 50 myr time intervals. The base map shows the age of oceanic lithosphere at the time of formation. Triangles denote subduction zones; black lines denote divergent margins and transform faults. The reconstructions and plate boundaries are based on the compilation by Müller *et al.* 2019.

East and West Antarctica and the formation of the WARS started in the late Mesozoic. During the Cenozoic the relative motion between East and West Antarctica has progressively shifted and concentrated along the western Ross Sea sector. It is broadly associated with a belt of Cenozoic alkaline magmatic rocks (48 Ma to presently active) described separately in this Memoir. West Antarctica may have been an orogenic highland in the Early Cretaceous that subsided over late Cretaceous and Cenozoic time with up to 200 km of extension between 68 and 46 Ma (Wilson and Luyendyk 2009). In the Eocene and Oligocene (43–26 Ma: Cande *et al.* 2000) a pulse of motion resulted in the formation of seafloor spreading in the northwestern end of the Ross Sea, along the Adare Trough and Northern Basin (Davey *et al.* 2016). Motion between East and West Antarctica has progressively slowed in the Neogene and lasted until *c.* 11 myr ago (Granot and Dyment 2018). Interestingly, the rotation poles that describe these motions (Granot *et al.* 2013; Granot and Dyment 2018) were located close to the centre of the rift system, suggesting that while the Ross Sea sector has undergone extensional motion, the central part went through minimal dextral transcurrent motion and the motions in the eastern parts were predominantly oblique convergence.

The WARS is marked by a topographical trough 750–1000 km wide and 3000 km long, running from near the Ellsworth–Whitmore Mountains to the Ross Embayment–northern Victoria Land (LeMasurier and Thomson 1990; Behrendt *et al.* 1991, 1992). The rift is characterized by thin crust approximately 20–30 km thick, deep rift basins, and an anomalous mantle within the Ross Sea region and including Marie Byrd Land (Behrendt *et al.* 1991; Chaput *et al.* 2014; O'Donnell *et al.* 2017; Ramirez *et al.* 2017; Shen *et al.* 2018). The WARS is similar in size to the East African Rift System, and to the Basin and Range province of the western USA (Tessensohn and Wörner 1991), and is geometrically asymmetrical. Marie Byrd Land was long considered to be the eastern flank of the rift system (Studinger *et al.* 2006) but it is now known that the crust of Marie Byrd Land is extended throughout (Spiegel *et al.* 2016). Thermochronological data (Spiegel *et al.* 2016) suggested that rifting in this sector occurred in two episodes: the earliest event between *c.* 100 and 60 Ma led to widespread tectonic denudation and basin-and-range style block faulting with about 3 km of uplift in the central part (LeMasurier and Rex 1989); the later episode started during the Early Oligocene and was confined to the eastern part of Marie Byrd Land. Uplift of the Marie Byrd Land dome may have started at *c.* 20 Ma (Spiegel *et al.* 2016). The opposite flank of the rift system in northern Victoria Land consists of the TAM, the uplifted roots of the early Paleozoic Ross orogen (Stump 1995). Although the TAM block is the world's longest rift shoulder, the source of its high elevation is still not fully resolved. Four competing models have been suggested by Wannamaker *et al.* (2017) and Shen *et al.* (2018). Firstly, a crustal root below the mountains, suggested to be residual from regional extensional collapse of a high elevation plateau with thicker than normal crust centred on West Antarctica, could contribute to uplift of the mountain range (Bialas *et al.* 2007; Block *et al.* 2009; Wilson and Luyendyk 2009). Secondly, uplift processes have been compared with those of the margins of the extensional US Great Basin, where lithospheric replacement by hot asthenosphere of lower density occurred (Smith and Drewry 1984; Rocholl *et al.* 1995) and incorporated thermal buoyancy as an essential uplift load in addition to erosion, ice load and flexural rigidity. Stern *et al.* (2005) attributed as much as 50% of peak height to the effects of glacial erosion. Thirdly, Stern and ten Brink (1989) proposed an elegant model based on a cantilevered flexural upwarp involving a regional boundary fault. Recent results of a 550 km long-transect of magnetotelluric geophysical soundings spanning the central part of the mountain range revealed a lithosphere of high electrical resistivity to at least 150 km depth, implying a cold stable state well into the upper mantle (Wannamaker *et al.* 2017). They concluded that at least the central part of the TAM is most likely to have been elevated by a non-thermal, flexural cantilever mechanism. A flexural uplift model is also supported by P-wave speed variations for the Antarctic mantle by Hansen *et al.* (2014). In contrast, new P- and S-wave tomographic images have been interpreted by Brenn *et al.* (2017) to suggest thermal buoyancy and flexural uplift are the principle components that lead to uplift of the northern TAM. Lastly, a recent seismic tomography study (Shen *et al.* 2018) showed that the wide southern TAM is undergoing lithospheric foundering, whereby the lower lithosphere is sinking into the mantle. They attributed this mechanism to the convergence motion that prevailed in this region since the Eocene (Granot *et al.* 2013; Granot and Dyment 2018). Lithospheric foundering has been considered responsible for volcanism in the southern TAM (Licht *et al.* 2018; Panter *et al.* 2021), now referred to as the Upper Scott Glacier Volcanic Field (Panter *et al.* 2021).

Geochemical studies of basalts from the Marie Byrd Land and Ross Sea sectors of the WARS support plume-related sources for volcanism (Hart *et al.* 1995, 1997; Rocholl *et al.* 1995). Plume models have also been used to explain tectonic doming and the spatial pattern of volcanic centres within the Marie Byrd Land province (LeMasurier and Rex 1989; Hole and LeMasurier 1994; LeMasurier and Landis 1996). Some models appeal to a single young 'active' plume concurrent with the onset of volcanism 28–35 myr ago (Kyle *et al.* 1994; LeMasurier and Landis 1996), while others favour a passive model involving a 'fossilized' plume head fixed at the base of the lithosphere (Rocholl *et al.* 1995; Hart *et al.* 1997). In the fossil plume model, the arrival of a plume head prior to the mid-Cretaceous break-up of New Zealand from Antarctica may explain the extremely broad distribution (over 5000 km) and significant age span (100 myr) of HIMU (high $^{238}U/^{204}Pb = \mu$)-type alkaline volcanism found throughout the continental borderlands of the SW Pacific (Lanyon *et al.* 1993; Hart *et al.* 1997; Panter *et al.* 1997). Magmatism associated with either Jurassic (184 Ma: Encarnación *et al.* 1996) or Cretaceous (100 Ma: Weaver *et al.* 1994) rifting events may signal early plume–lithosphere interaction. Seismic tomography (Spasojevic *et al.* 2010) and subsidence data (Sutherland *et al.* 2010) support evidence for a very long-lived deep mantle thermal anomaly off West Antarctica, probably triggered by cessation of subduction along the Gondwana margin. An analysis of P-wave velocities indicate a deep-seated low-velocity zone beneath Marie Byrd Land, which has been interpreted by Hansen *et al.* (2014) as a deep mantle plume ponded below the 660 km discontinuity that would thermally perturb the overlying mantle. The presence of low-viscosity mantle beneath a portion of coastal West Antarctica is consistent with the presence of lithospheric mantle that is of elevated temperature or altered composition (Lloyd *et al.* 2015; Barletta *et al.* 2018).

Panter *et al.* (2000) proposed a variant on the fossil plume model; one that calls upon a Cretaceous plume composed solely of the HIMU component overlain by a much more extensive pre-existing metasomatized layer within the Gondwana lithosphere. The plume-driven metasomatism may have been related to the Jurassic Bouvet plume having enriched the Gondwana lithosphere in highly incompatible trace elements. The plume-modified lithosphere was then underplated by a smaller HIMU plume in mid-Cretaceous times. A Cretaceous plume head with a diameter of 600–800 km would encompass these HIMU localities prior to continental break-up. Extension and rifting of New Zealand and Australia from Antarctica would have led to adiabatic decompression melting of the fossil plume and overlying plume-modified lithosphere within widely dispersed fragments of the former Gondwanaland supercontinent.

Finn *et al.* (2005) reviewed the different tectonic models and concluded that the diffuse alkaline magmatic province in the SW Pacific, of which the Antarctic Cenozoic alkaline province is a part, was formed by sudden detachment and sinking of subducted slabs in the late Cretaceous that induced instabilities along the former Gondwana margin which in turn triggered lateral and vertical flow of warm Pacific mantle. According to Finn *et al.* (2005), the combination of metasomatized lithosphere underlain by mantle at slightly elevated temperatures was key to generating Cenozoic magmatism. The hypothesis is borne out by the discovery of local regions of low-viscosity mantle (Barletta *et al.* 2018).

A geochemical study (LeMasurier *et al.* 2016) of basalts from three Marie Byrd Land volcanoes with Ba and Nd anomalies compared with volcanic rocks from the WARS with ocean island basalt (OIB)-like chemistry indicated a subduction influence. LeMasurier *et al.* (2016) suggested that the source of the geochemical anomalies resided in a fossil diapir that arose from the Cretaceous subducting slab. An additional geochemical study by Panter *et al.* (2018) also indicated also the influence of subduction materials in the source for volcanism in the northwestern Ross Sea. They attributed the lighter oxygen isotope values in olivine crystals as reflecting hydrothermally altered oceanic lithosphere (at high temperature) that was introduced into the upper mantle by subduction along the proto-Pacific margin of Gondwana or longer-term recycling from ancient oceanic lithosphere.

Alternative interpretations to both plume-driven and passive rifting have been proposed by Rocchi *et al.* (2002, 2003, 2005), which suggest that magma genesis and emplacement in north Victoria Land and Ross Embayment is due to the reactivation of pre-existing NW–SE trans-lithospheric faults (Salvini *et al.* 1997), which promoted local decompression melting of an enriched mantle previously veined during a Late Cretaceous amagmatic extensional rift phase.

Subduction processes

For much, if not all, of the Mesozoic and part of the Cenozoic, subduction of the proto-Pacific ocean floor took place on the Panthalassic margin of Gondwana (Barker *et al.* 1991). Evidence for subduction is provided by the magmatic record as documented by igneous rocks, and volcanoclastic-derived sedimentary successions, together with structurally deformed sequences interpreted as accretionary complexes formed in forearc regions on the Antarctic Peninsula (Storey and Garrett 1985), and magmatic rocks on Thurston Island (Leat *et al.* 1993) and on Marie Byrd Land (Mukasa and Dalziel 2000). For much of the Mesozoic, the subduction record is in part preserved in Zealandia (Mortimer *et al.* 2018), which was located outboard of Marie Byrd Land up to the Mid Cretaceous.

Antarctic Peninsula

The Mesozoic geology of the Antarctic Peninsula has traditionally been interpreted as a complete Andean-type arc–trench system (Suarez 1976; Smellie 1981; Storey and Garrett 1985). Subduction is interpreted to have been active before, during and after partial separation of the Antarctic Peninsula from Gondwana by seafloor spreading in the Weddell Sea. The main tectonic elements are accretion–subduction complexes on the western Pacific margin of the peninsula, a magmatic arc active from about 240 to 10 Ma, represented by the Antarctic Peninsula Batholith (Leat *et al.* 1995), and thick back-arc and retro-arc basin sequences (Macdonald and Butterworth 1990) on the eastern Weddell Sea side. The polarity of the system is consistent with east-directed subduction of proto-Pacific ocean floor. However, based on the identification of a major ductile shear zone, the eastern Palmer Land Shear Zone, Vaughan and Storey (2000) presented a testable hypothesis that at least three terranes of parautochthonous or allochthonous origin may have formed the Antarctic Peninsula. The terrane model has been tested recently by Burton-Johnson and Riley (2015) in light of recent data; the authors have returned to the earlier interpretation for the peninsula as having evolved as an *in situ* Andean-type continental arc (see Burton-Johnson and Riley 2015 for a full review) from Paleozoic to Cenozoic times based on stratigraphic correlations, new geochronological dating of Paleozoic basement and the similarity of Nd isotopic signatures across different domains. They conclude that;

- continental margin magmatism was active along the Antarctic Peninsula during the Carboniferous–Jurassic period prior to and during the initial stages of Gondwana break-up;
- subduction may have ceased between 156 and 142 Ma (Leat *et al.* 1995), as indicated by a low in magmatic activity;
- renewed subduction resulted in extensive arc, forearc and back-arc sequences in an extensional setting;
- magmatism peaked between *c.* 120 and 90 Ma, coinciding with the Palmer Land transpressional event and formation of the East Palmer Land Shear Zone, the origin of which remains unclear;
- magmatic activity began to wane following the Palmer Land deformational event, although there were local peaks in magmatic activity in the latest Cretaceous–Eocene.

Subduction ceased following a series of sequential ridge crest–trench collisions (Barker 1982; Larter and Barker 1991) where segments of the Pacific Phoenix spreading ridge jammed the subduction zone on the western margin of the peninsula. Following the collisions, magmatism waned until the production of scattered intra-plate alkaline volcanism from 6.5 to 0.1 Ma (Hole and Larter 1993). Subduction continues on one remaining segment on the western margin of the Antarctic Peninsula where the Drake Plate is subducting beneath the South Shetland Plate with the opening of the Bransfield Strait in a back-arc setting (Barker *et al.* 1991; Lawver *et al.* 1996; Christeson *et al.* 2003).

Thurston Island

The Thurston Island block, which includes Thurston Island and the adjacent Eights Coast and the Jones mountains, records only Pacific-margin magmatism dated from Carboniferous to Late Cretaceous times (Pankhurst *et al.* 1993), with no associated exposed sedimentary successions. The igneous rocks form a uniformly calc-alkaline, high-alumina dominantly metaluminous suite typical of subduction settings (Leat *et al.* 1993; Riley *et al.* 2016). The magmatic record on Thurston Island itself was dominated by Late Jurassic (152–142 Ma) and Early Cretaceous (125–110 Ma) bimodal suites with Triassic and Middle Jurassic magmatism (Riley *et al.* 2016). Volcanism in the Jones Mountains became predominantly silicic between 100 and 90 Ma prior to the cessation of subduction along this part of the margin.

Marie Byrd Land

Similar to Thurston Island, Marie Byrd Land preserves a scattered record of subduction-related magmatic activity during the Paleozic and Mesozoic, with peaks of activity during the Carboniferous, Permian (Pankhurst *et al.* 1998*b*) and Cretaceous; related forearc basin and accretionary complexes occur in present-day New Zealand which was outboard of Marie Byrd Land prior to mid-Cretaceous rifting. In Marie Byrd Land a widespread group of Cretaceous (124–95 Ma) calc-alkaline I-type granodiorite plutons intruded into basement rocks. In the western sector the plutons are of Early Cretaceous age (124–108 Ma), whereas they extend to slightly younger ages in eastern Marie Byrd Land (Weaver *et al.* 1994). The Cretaceous magmatic rocks record an important change in the tectonic regime from a subducting to an extensional margin prior to separation of Zealandia from Marie Byrd Land. Subduction ceased at about 105 Ma either just prior to (Luyendyk 1995) or immediately following collision of the spreading ridge with the trench (Bradshaw 1989). The age pattern suggests that subduction ceased first along the western sector at *c.* 108 Ma and persisted until about 95 Ma in the east. Prior to seafloor spreading a voluminous suite of mafic dykes and sills (dated 107 ± 5 Ma: Storey *et al.* 1999) and anorogenic silicic rocks, including syenites and peralkaline granitoids (95–102 Ma), were emplaced in Marie Byrd Land during a rifting event (Weaver *et al.* 1994). A migmatite-cored gneiss dome in the Fosdick Mountains was exhumed from mid- to lower-middle crustal depths during this rifting event which was the incipient stage of the WARS mentioned above. Prior to and during exhumation, major crustal melting and deformation included transfer and emplacement of voluminous granitic material and mantle-derived diorite dykes (McFadden *et al.* 2015) during a transition from wrench to oblique extension (Saito *et al.* 2013).

Concluding remarks

In contrast to the Paleozoic evolution of Antarctica within Gondwana, the Mesozoic and Cenozoic evolution was dominated by igneous activity related to continental rifting and subduction processes. Although these two processes operated simultaneously for much of this time period, it is not clear whether these processes were causally related or whether they operated independently of each other. Ultimately, these processes led to the disintegration of Gondwana and the current isolation of Antarctica at the South Pole and in the centre of an Antarctic plate surrounded by spreading ridges. With the exception of magmatic activity related to the opening of the Bransfield Strait on the northern tip of the Antarctic Peninsula and the final relaxation stages of magmatism within the WARS, the continent is amagmatic and surrounded by passive margins. This is in marked contrast to the Mesozoic and much of the Cenozoic where thermal anomalies within the mantle coupled with active subduction and rifting resulted in the wide range of igneous activity, particularly within West Antarctica.

Acknowledgements We are very grateful to the referees and volume editors for their constructive comments and for drawing attention to some key references that had been omitted.

Author contributions BCS: writing – original draft (lead), writing – review & editing (lead); RG: writing – review & editing (supporting).

Funding This research received no specific grant from any funding agency in the public, commercial, or not-for-profit sectors.

Data availability Data sharing is not applicable to this article as no datasets were generated or analysed during the current study.

References

Adie, R.J. 1952. The position of the Falkland Islands in a reconstruction of Gondwanaland. *Geological Magazine*, **89**, 401–410, https://doi.org/10.1017/S0016756800068102

Barker, P.F. 1982. The Cenozoic subduction history of the pacific margin of the Antarctic Peninsula: ridge crest–trench interactions. *Journal of the Geological Society, London*, **139**, 787–801, https://doi.org/10.1144/gsjgs.139.6.0787

Barker, P.F. 2001. Scotia Sea regional tectonic evolution: implications for mantle flow and palaoecirculation. *Earth-Sciences Review*, **55**, 1–39, https://doi.org/10.1016/S0012-8252(01)00055-1

Barker, P.F. and Thomas, E. 2004. Origin, signature and palaeoclimatic influence of the Antarctic Circumpolar Current. *Earth-Science Reviews*, **66**, 143–162, https://doi.org/10.1016/j.earscirev.2003.10.003

Barker, P.F., Dalziel, I.W.D. and Storey, B.C. 1991. Tectonic development of the Scotia Arc region. *Oxford Monographs on Geology and Geophysics*, **17**, 215–244.

Barletta, V.R., Bevis, M. *et al.* 2018. Observed rapid bedrock uplift in Amundsen Sea Embayment promotes ice-sheet stability. *Science*, **360**, 1335–1339, https://doi.org/10.1126/science.aao1447

Behrendt, J.C., LeMasurier, W.E., Cooper, A.K., Tessensohn, F., Tréhu, A. and Damaske, D. 1991. Geophysical studies of the West Antarctic rift system. *Tectonics*, **10**, 1257–1273, https://doi.org/10.1029/91TC00868

Behrendt, J.C., LeMasurier, W. and Cooper, A.K. 1992. The West Antarctic Rift System – A propagating rift captured by a mantle plume? *In*: Yoshida, K., Kaminuma, K. and Shiraishi, K. (eds) *Recent Progress in Antarctic Earth Science*. Terra Science, Tokyo, 315–322.

Bialas, R.W., Buck, W.R., Studinger, M. and Fitzgerald, P. 2007. Plateau collapse model for the Transantarctic Mountains West Antarctic Rift System: Insights from numerical experiments. *Geology*, **35**, 687–690, https://doi.org/10.1130/G23825A.1

Block, A.E., Bell, R. and Studinger, M. 2009. Antarctic crustal thickness from satellite gravity: implications for the Transantarctic and Gamburtsev subglacial mountains. *Earth and Planetary Science Letters*, **288**, 194–203, https://doi.org/10.1016/j.epsl.2009.09.022

Bradshaw, J.D. 1989. Cretaceous geotectonic patterns in the New Zealand region. *Tectonics*, **8**, 803–820, https://doi.org/10.1029/TC008i004p00803

Brenn, G.R., Hansen, S.E. and Park, Y. 2017. Variable tectonic loading and flexural uplift along the Transantarctic Mountains, Antarctica. *Geology*, **45**, 463–466, https://doi.org/10.1130/G38784.1

Brown, B., Gaina, C. and Muller, R.D. 2006. Circum-Antarctic palaeobathymetry: Illustrated examples from Cenozoic to recent times. *Palaeogeography, Palaeoclimatology and Palaeoecology*, **231**, 158–168, https://doi.org/10.1016/j.palaeo.2005.07.033

Bryan, S.E., Riley, T.R., Jerram, D.A., Stephens, D.J. and Leat, P.T. 2002. Silicic volcanism: An undervalued component of large igneous provinces and volcanic rifted margins. *Geological Society of America Special Papers*, **362**, 99–120, https://doi.org/10.1130/0-8137-2362-0.97

Burton-Johnson, A. and Riley, T.R. 2015. Autochthonous v. accreted terrane development of continental margins: a revised *in situ* tectonic history of the Antarctic Peninsula. *Journal of the Geological Society, London*, **172**, 822–835, https://doi.org/10.1144/jgs2014-110

Busetti, M., Spadini, G., van der Wateren, F.M., Cloetingh, S.A.P.L. and Zanolla, C. 1999. Thermo-mechanical modelling of the West Antarctic Rift System, Ross Sea Antarctica. *Global and Planetary Change*, **23**, 79–103, https://doi.org/10.1016/S0921-8181(99)00052-1

Cande, S.C., Stock, J., Muller, R.D. and Ishihara, T. 2000. Cenozoic motion between East and West Antarctica. *Nature*, **404**, 145–150, https://doi.org/10.1038/35004501

Chaput, J., Aster, R.C. *et al.* 2014. The crustal thickness of West Antarctica. *Journal of Geophysical Research: Solid Earth*, **119**, 378–395, https://doi.org/10.1002/2013JB010642

Christeson, G.L., Barker, D.H.N., Austin, J.A. and Dalziel, I.W.D. 2003. Deep crustal structure of Bransfield Strait: Initiation of a back-arc basin by rift reactivation and propagation. *Journal of Geophysical Research: Solid Earth*, **108**, 2492, https://doi.org/10.1029/2003JB002468

Cox, K.G. 1992. Karoo igneous activity, and the early stages of the break-up of Gondwanaland. *Geological Society, London, Special Publications*, **68**, 137–148, https://doi.org/10.1144/GSL.SP.1992.068.01.09

Curtis, M.L. 1997. Gondwana age dextral transpression and spatial kinematic partitioning within the Heritage Range, Ellsworth Mountains, West Antarctica. *Tectonics*, **16**, 172–181, https://doi.org/10.1029/96TC01418

Dalziel, I.W.D. and Elliot, D.H. 1982. West Antarctica: Problem child of Antarctica. *Tectonics*, **1**, 3–19, https://doi.org/10.1029/TC001i001p00003

Dalziel, I.W.D., Lawver, L.A. and Murphy, J.B. 2000. Plumes, orogenesis, and supercontinental fragmentation. *Earth and Planetary Science Letters*, **178**, 1–11, https://doi.org/10.1016/S0012-821X(00)00061-3

Dalziel, I.W.D., Lawver, L., Norton, I.O. and Gahagan, L.M. 2013. The Scotia Arc: genesis, evolution, global significance. *Annual Review of Earth and Planetary Sciences*, **41**, 767–793, https://doi.org/10.1146/annurev-earth-050212-124155

Davey, F.J., Granot, R., Cande, S.C., Stock, J.M., Selvans, M. and Ferraccioli, F. 2016. Synchronous oceanic spreading and continental rifting in West Antarctica. *Geophysical Research Letters*, **43**, 6162–6169, https://doi.org/10.1002/2016GL069087

Du Toit, A.L. 1937. *Our Wandering Continents*. Oliver & Boyd, Edinburgh.

Eagles, G. and Jokat, W. 2014. Tectonic reconstructions for paleobathymetry in Drake Passage. *Tectonophysics*, **611**, 28–50, https://doi.org/10.1016/j.tecto.2013.11.021

Eagles, G. and Livermore, R.A. 2002. Opening history of Powell Basin, Antarctic Peninsula. *Marine Geology*, **185**, 195–205, https://doi.org/10.1016/S0025-3227(02)00191-3

Encarnación, J., Fleming, T.H., Elliot, D.H. and Eales, H.V. 1996. Synchronous emplacement of Ferrar and Karoo dolerites and the early breakup of Gondwana. *Geology*, **24**, 535–538, https://doi.org/10.1130/0091-7613(1996)024<0535:SEOFAK>2.3.CO;2

Finn, C.A., Muller, R.D. and Panter, K.S. 2005. A Cenozoic diffuse alkaline magmatic province (DAMP) in the southwest Pacific without rift or plume origin. *Geochemistry, Geophysics, Geosystems*, **6**, Q02005, https://doi.org/10.1029/2004GC000723

Fitzgerald, P.G. 2002. Tectonics and landscape evolution of the Antarctic plate since the breakup of Gondwana, with an emphasis on the West Antarctic Rift System and the Trans Antarctic Mountains. *Royal Society of New Zealand Bulletin*, **35**, 453–469.

Fitzgerald, P.G. and Gleadow, A.J.W. 1988. Fission track geochronology, tectonics and structure of the Transantarctic Mountains in Northern Victoria Land, Antarctica. *Isotope Geoscience*, **73**, 169–198, https://doi.org/10.1016/0168-9622(88)90014-0

Fitzsimons, I.C.W. 2000. A review of tectonic events in the East Antarctic Shield and their implications for Gondwana and earlier supercontinents. *Journal of African Earth Sciences*, **31**, 3–23, https://doi.org/10.1016/S0899-5362(00)00069-5

Fretwell, P., Pritchard, H.D. *et al.* 2013. Bedmap2: improved ice bed, surface and thickness datasets for Antarctica. *The Cryosphere*, **7**, 375–393, https://doi.org/10.5194/tc-7-375-2013

Frey, F.A., Coffin, M.F. *et al.* 2000. Origin and evolution of a submarine large igneous province; the Kerguelen Plateau and Broken Ridge, southern Indian Ocean. *Earth and Planetary Science Letters*, **176**, 73–89, https://doi.org/10.1016/S0012-821X(99)00315-5

Gaina, C., Müller, R.D., Royer, J.-Y., Stock, J., Hardebeck, J. and Symonds, P. 1998. The tectonic history of the Tasman Sea: A puzzle with 13 pieces. *Journal of Geophysical Research: Solid Earth*, **103**, 12 413–12 433, https://doi.org/10.1029/98JB00386

Gaina, C., Müller, R.D., Brown, B., Ishihara, T. and Ivanov, K.S. 2007. Breakup and early seafloor spreading between India and Antarctica. *Geophysical Journal International*, **170**, 151–169, https://doi.org/10.1111/j.1365-246X.2007.03450.x

Ghidella, M.E., Yanez, G. and LaBreque, J. 2002. Raised tectonic implications for the magnetic anomalies of the western Weddell Sea. *Tectonophysics*, **347**, 65–86, https://doi.org/10.1016/S0040-1951(01)00238-4

Granot, R. and Dyment, J. 2018. Late Cenozoic unification of East and West Antarctica. *Nature Communications*, **9**, 3189, https://doi.org/10.1038/s41467-018-05270-w

Granot, R., Cande, S.C., Stock, J.M. and Damaske, D. 2013. Revised Eocene–Oligocene kinematics for the West Antarctic rift system. *Geophysical Research Letters*, **40**, 279–284, https://doi.org/10.1029/2012GL054181

Grunow, A.M., Kent, D.V. and Dalziel, I.W.D. 1987. Evolution of the Weddell Sea basin: new palaeomagnetic constraints. *Earth and Planetary Science Letters*, **86**, 16–26, https://doi.org/10.1016/0012-821X(87)90184-1

Hansen, S., Graw, J.H. *et al.* 2014. Imaging the Antarctic mantle using adaptively parameterized P-wave tomography: Evidence for heterogeneous structure beneath West Antarctica. *Earth and Planetary Science Letters*, **408**, 66–78, https://doi.org/10.1016/j.epsl.2014.09.043

Hart, S.R., Blusztajn, J. and Craddock, C. 1995. Cenozoic volcanism in Antarctica: Jones Mountains and Peter I Island. *Geochimica et Cosmochimica Acta*, **59**, 3379–3388, https://doi.org/10.1016/0016-7037(95)00212-I

Hart, S.R., Blusztajn, J., LeMasurier, W.E., Rex, W.C., Hawkesworth, C.E. and Arndt, N.T.E. 1997. Hobbs Coast Cenozoic volcanism: Implications for the West Antarctic rift system. *Chemical Geology*, **139**, 223–248, https://doi.org/10.1016/S0009-2541(97)00037-5

Hole, M.J. and Larter, R.D. 1993. Trench-proximal volcanism following ridge crest–trench collision along the Antarctic Peninsula. *Tectonics*, **12**, 897–910, https://doi.org/10.1029/93TC00669

Hole, M.J. and LeMasurier, W.E. 1994. Tectonic controls on the geochemical composition of Cenozoic mafic alkaline volcanic rocks from West Antarctica. *Contributions to Mineralogy and Petrology*, **117**, 187–202, https://doi.org/10.1007/BF00286842

Jokat, W., Boebel, T., Konig, M. and Meyer, U. 2003. Timing and geometry of early Gondwana breakup. *Journal of Geophysical Research: Solid Earth*, **108**, 2428, https://doi.org/10.1029/2002JB001802

Jordan, T.A., Ferraccioli, F. and Leat, P.T. 2017. New geophysical compilations link crustal block motion to Jurassic extension and strike-slip faulting in the Weddell Sea Rift System of West Antarctica. *Gondwana Research*, **42**, 29–48, https://doi.org/10.1016/j.gr.2016.09.009

Konig, M. and Jokat, W. 2006. The Mesozoic break-up of the Weddell Sea. *Journal of Geophysical Research: Solid Earth*, **111**, B12102, https://doi.org/10.1029/2005JB004035

Kyle, P.R., Pankhurst, R., Mukasa, S., Panter, K., Smellie, J. and McIntosh, W. 1994. Sr, Nd and Pb isotopic variations in the Marie Byrd Plume, West Antarctica. *US Geological Survey Circular*, **1107**, 184.

Lanyon, R., Varne, R. and Crawford, A.J. 1993. Tasmanian Tertiary basalts, the Balleny Plume, and opening of the Tasman Sea (southwest Pacific Ocean). *Geology*, **21**, 555–558, https://doi.org/10.1130/0091-7613(1993)021<0555:TTBTBP>2.3.CO;2

Larter, R.D. and Barker, P.F. 1991. Effects of ridge crest–trench interactions on Antarctic–Phoenix Spreading: Forces on a young subducting plate. *Journal of Geophysical Research: Solid Earth*, **96**, 19 583–19 607, https://doi.org/10.1029/91JB02053

Lawver, L.A., Sloan, B.J. *et al.* 1996. Distributed, active extension in Bransfield Basin, Antarctic Peninsula: evidence from multibeam bathymetry. *GSA Today*, **6**, 1–6.

Leat, P.T., Storey, B.C. and Pankhurst, R.J. 1993. Geochemistry of Palaeozoic–Mesozoic Pacific rim orogenic magmatism. Thurston Island area, West Antarctica. *Antarctic Science*, **5**, 281–296, https://doi.org/10.1017/S0954102093000380

Leat, P.T., Scarrow, J.H. and Millar, I.L. 1995. On the Antarctic Peninsula batholith. *Geological Magazine*, **132**, 399–412, https://doi.org/10.1017/S0016756800021464

LeMasurier, W.E. 1990. Late Cenozoic volcanism on the Antarctic Plate: An overview. *American Geophysical Union Antarctic Research Series*, **48**, 1–18.

LeMasurier, W.E. and Landis, C.A. 1996. Mantle-plume activity recorded by low-relief erosion surfaces in West Antarctica and New Zealand. *Bulletin Geological Society of America*, **108**, 1450–1466, https://doi.org/10.1130/0016-7606(1996)108<1450:MPARBL>2.3.CO;2

LeMasurier, W.E. and Rex, D.C. 1989. Evolution of linear volcanic ranges in Marie Byrd Land, West Antarctica. *Journal of Geophysical Research: Solid Earth*, **94**, 7223–7236, https://doi.org/10.1029/JB094iB06p07223

LeMasurier, W.E. and Thomson, J.W. (eds) 1990. *Volcanoes of the Antarctic Plate and Southern Oceans*. American Geophysical Union Antarctic Research Series, **48**.

LeMasurier, W.E., Choi, S.H., Hart, S.R., Mukasa, S. and Rogers, N. 2016. Reconciling the shadow of a subduction signature with rift geochemistry and tectonic environment in Eastern Marie Byrd Land, Antarctica. *Lithos*, **260**, 134–153, https://doi.org/10.1016/j.lithos.2016.05.018

Licht, K.J., Groth, T., Townsend, J.P., Hennessy, A.J., Hemming, S.R., Flood, T.P. and Studinger, M. 2018. Evidence for extending anomalous Miocene volcanism at the edge of the East Antarctic craton. *Geophysical Research Letters*, **45**, 3009–3016, https://doi.org/10.1002/2018GL077237

Lisker, F. and Laufer, A.L. 2013. The Mesozoic Victoria Basin: Vanished link between Antarctica and Australia. *Geology*, **41**, 1043–1046, https://doi.org/10.1130/G33409.1

Lloyd, A., Wiens, D.A. *et al.* 2015. A seismic transect across West Antarctica: Evidence for mantle thermal anomalies beneath the Bentley Subglacial Trench and the Marie Byrd Land Dome. *Journal of Geophysical Research: Solid Earth*, **120**, 8439–8460, https://doi.org/10.1002/2015JB012455

Lock, B.E. 1980. Flat-plate subduction and the Cape Fold Belt of South Africa. *Geology*, **8**, 35–39, https://doi.org/10.1130/0091-7613(1980)8<35:FSATCF>2.0.CO;2

Luyendyk, B.P. 1995. Hypothesis for Cretaceous rifting of East Gondwana caused by subducted slab capture. *Geology*, **23**, 373–376, https://doi.org/10.1130/0091-7613(1995)023<0373:HFCROE>2.3.CO;2

Macdonald, D.I.M. and Butterworth, P.J. 1990. The stratigraphy, setting and hydrocarbon potential of the Mesozoic sedimentary basins of the Antarctic Peninsula. *AAPG Studies in Geology*, **31**, 100–105.

Malinverno, A., Hildebrandt, J., Tominaga, M. and Channell, J.E.T. 2012. M-sequence geomagnetic polarity time scale (MHTC12) that steadies global spreading rates and incorporates astrochronology constraints. *Journal of Geophysical Research: Solid Earth*, **117**, B06104, https://doi.org/10.1029/2012JB009260

McFadden, R.R., Teyssier, C., Siddoway, C.S., Cosca, M.A. and Fanning, C.M. 2015. Mid-Cretaceous oblique rifting of West Antarctica: Emplacement and rapid cooling of the Fosdick Mountains migmatite-cored gneiss dome. *Lithos*, **232**, 306–318, https://doi.org/10.1016/j.lithos.2015.07.005

Millar, I.l. and Pankhurst, R.J. 1987. Rb–Sr geochronology of the region between the Antarctic peninsula and the Transantarctic Mountains: Haag Nunataks and Mesozoic granitiods. *American Geophysical Union Geophysical Monograph Series*, **40**, 151–160.

Mortimer, N., Gans, P.B. et al. 2018. Regional volcanism of northern Zealandia: post-Gondwana break-up magmatism on an extended, submerged continent. *Geological Society, London, Special Publications*, **463**, 199–226, https://doi.org/10.1144/SP463.9

Mukasa, S.B. and Dalziel, I.W.D. 2000. Marie Byrd Land, West Antarctica: Evolution of Gondwana's Pacific margin constrained by zircon U–Pb geochronology and feldspar common-Pb isotopic compositions. *Geological Society of American Bulletin*, **112**, 611–627, https://doi.org/10.1130/0016-7606(2000)112<611:MBLWAE>2.0.CO;2

Müller, D.R., Zahirovic, S. et al. 2019. A global plate model including lithospheric deformation along major rifts and orogens since the Triassic. *Tectonics*, **38**, 1884–1907, https://doi.org/10.1029/2018TC005462

Nicolaysen, K., Bowring, S., Frey, F., Weis, D., Ingle, S., Pringle, M.S. and Coffin, M.S. 2001. Provenance of Proterozoic garnet–biotite gneiss recovered from Elan Bank, Kerguelen Plateau, southern Indian Ocean. *Geology*, **29**, 235–238, https://doi.org/10.1130/0091-7613(2001)029<0235:POPGBG>2.0.CO;2

O'Donnell, J.P., Selway, K. et al. 2017. The uppermost mantle seismic velocity and viscosity structure of central West Antarctica. *Earth and Planetary Science Letters*, **472**, 38–49, https://doi.org/10.1016/j.epsl.2017.05.016

Pankhurst, R.J. and Rapela, C.R. 1995. Production of Jurassic rhyolites by anatexis of the lower crust of Patagonia. *Earth and Planetary Science Letters*, **134**, 23–36, https://doi.org/10.1016/0012-821X(95)00103-J

Pankhurst, R.J., Millar, I.L., Grunow, A.M. and Storey, B.C. 1993. The pre-Cenozoic magmatic history of the Thurston Island crustal block, West Antarctica. *Journal of Geophysical Research: Solid Earth*, **98**, 11 835–11 849, https://doi.org/10.1029/93JB01157

Pankhurst, R.J., Leat, P.T., Sruoga, P., Rapela, C.W., Márquez, M., Storey, B.C. and Riley, T.R. 1998a. The Chon-Aike silicic igneous province of Patagonia and related rocks in Antarctica: a silicic large igneous province. *Journal of Volcanology and Geothermal Research*, **81**, 113–136, https://doi.org/10.1016/S0377-0273(97)00070-X

Pankhurst, R.J., Weaver, S.D., Bradshaw, J.D., Storey, B.C. and Ireland, T.R. 1998b. Geochronology and geochemistry of pre-Jurassic superterranes in Marie Byrd Land, Antarctica. *Journal of Geophysical Research: Solid Earth*, **103**, 2529–2547, https://doi.org/10.1029/97JB02605

Pankhurst, R.J., Riley, T.R., Fanning, C.M. and Kelley, S.P. 2000. Episodic silicic volcanism in Patagonia and the Antarctic Peninsula: chronology of magmatism associated with break-up of Gondwana. *Journal of Petrology*, **41**, 605–625, https://doi.org/10.1093/petrology/41.5.605

Panter, K., Blusztajn, J., Hart, S.R. and Kyle, P. 1997. Late Cretaceous–Neogene basalts from Chatham Island: Implications for HIMU mantle beneath continental borderlands of the Southwest Pacific. *In*: Seventh Annual V. M. Goldschmidt Conference, June 2–6, 1997, Tucson, Arizona. LPI Contribution, **921**, 156–157.

Panter, K.S., Hart, S.R., Kyle, P., Blusztajn, J. and Wilch, T. 2000. Geochemistry of Late Cenozoic basalts from the Crary Mountains: Characterization of mantle sources in Marie Byrd Land, Antarctica. *Chemical Geology*, **165**, 215–241, https://doi.org/10.1016/S0009-2541(99)00171-0

Panter, K.S., Castillo, P. et al. 2018. Melt origin across a rifted continental margin: a case for subduction-related metasomatic agents in the lithospheric source of alkaline basalt, northwest Ross Sea, Antarctica. *Journal of Petrology*, **59**, 517–558, https://doi.org/10.1093/petrology/egy036

Panter, K.S., Reindel, J. and Smellie, J.L. 2021. Mount Early and Sheridan Bluff: petrology. *Geological Society, London, Memoirs*, **55**, https://doi.org/10.1144/M55-2019-2

Ramirez, C., Nyblade, A. et al. 2017. Crustal structure of the Transantarctic Mountains, Ellsworth Mountains and Marie Byrd Land, Antarctica: constraints on shear wave velocities, Poisson's ratios and Moho depths. *Geophysical Journal International*, **211**, 1328–1340, https://doi.org/10.1093/gji/ggx333

Randall, D.E. and Mac Niocaill, C. 2004. Cambrian palaeomagnetic data confirm a Natal embayment location for the Ellsworth–Whitmore Mountains, Antarctica, in Gondwana reconstructions. *Geophysical Journal International*, **157**, 105–116, https://doi.org/10.1111/j.1365-246X.2004.02192.x

Riley, T.R., Flowerdew, M.J., Pankhurst, R.J., Leat, P.T., Millar, I.L., Fanning, C.M. and Whitehouse, M.J. 2016. A revised geochronology of Thurston Island, West Antarctica, and correlations along the proto-Pacific margin of Gondwana. *Antarctic Science*, **29**, 47–60, https://doi.org/10.1017/S0954102016000341

Rocchi, S., Armienti, P., D'Orazio, M., Tonarini, S., Wijbrans, J.R. and Di Vincenzo, G. 2002. Cenozoic magmatism in the western Ross Embayment: Role of mantle plume v. plate dynamics in the development of the West Antarctic Rift System. *Journal Geophysical Research*, **107**, 2195, https://doi.org/10.1029/2001JB000515

Rocchi, S., Storti, F., Di Vincenzo, G. and Rossetti, F. 2003. Intraplate strike-slip tectonics as an alternative to mantle plume activity for the Cenozoic rift magmatism in the Ross Sea region, Antarctica. *Geological Society, London, Special Publications*, **210**, 145–158, https://doi.org/10.1144/GSL.SP.2003.210.01.09

Rocchi, S., Di Vincenzo, G. and Armienti, P. 2005. No plume, no rift magmatism in the West Antarctic rift. *Geological Society of America Special Papers*, **388**, 435–447.

Rocholl, A., Stein, M., Molzahn, M., Hart, S.R. and Worner, G. 1995. Geochemical evolution of rift magmas by progressive tapping of a stratified mantle source beneath the Ross Sea Rift, northern Victoria Land, Antarctica. *Earth and Planetary Science Letters*, **131**, 207–224, https://doi.org/10.1016/0012-821X(95)00024-7

Roeser, H.A., Fritsch, J. and Hinz, K. 1996. The development of the crust off Dronning Maud Land, East Antarctica. *Geological Society, London, Special Publications*, **108**, 243–264, https://doi.org/10.1144/GSL.SP.1996.108.01.18

Saito, S., Brown, M., Korhonen, F.J., McFadden, R.R. and Siddoway, C.S. 2013. Petrogenesis of Cretaceous mafic intrusive rocks, Fosdick Mountains, West Antarctica: Melting of the subcontinental arc mantle along the Gondwana margin. *Gondwana Research*, **23**, 1567–1580, https://doi.org/10.1016/j.gr.2012.08.002

Salvini, F., Brancolini, G., Busetti, M., Storti, G., Mazzarini, F. and Coren, F. 1997. Cenozoic geodynamics of the Ross Sea region, Antarctica: Crustal extension, intraplate strike-slip faulting, and tectonic inheritance. *Journal Geophysical Research*, **102**, 24 669–24 696, https://doi.org/10.1029/97JB01643

Sandwell, D.T., Müller, R.D., Smith, W.H.F., Garcia, E. and Francis, R. 2014. New global marine gravity model from CryoSat-2 and Jason-1 reveals buried tectonic structure. *Science*, **346**, 65–67, https://doi.org/10.1126/science.1258213

Schopf, J.M. 1969. Ellsworth Mountains: position in West Antarctica due to sea floor spreading. *Science*, **164**, 63–66, https://doi.org/10.1126/science.164.3875.63

Shen, W., Wiens, D.A. et al. 2018. The crust and upper mantle structure of central and West Antarctica from Bayesian inversion of Rayleigh wave and receiver functions. *Journal of Geophysical Research: Solid Earth*, **123**, 7824–7849, https://doi.org/10.1029/2017JB015346

Smellie, J.L. 1981. A complete arc–trench system recognized in Gondwana sequences of the Antarctic Peninsula region. *Geological Magazine*, **118**, 139–159, https://doi.org/10.1017/S001675680003435X

Smith, A.J. and Drewry, D.J. 1984. Delayed phase change due to hot asthenosphere causes Transantarctic Mountains uplift? *Nature*, **309**, 536–538, https://doi.org/10.1038/309536a0

Spasojevic, S., Gurnis, M. and Sutherland, R. 2010. Inferring mantle properties with an evolving dynamic model of the Antarctica–New Zealand region from the Late Cretaceous. *Journal of*

Geophysical Research: Solid Earth, **115**, B05402, https://doi.org/10.1029/2009JB006612

Spiegel, C., Lindow, J. *et al.* 2016. Tectonomorphic evolution of Marie Byrd Land – Implications for Cenozoic rifting activity and onset of West Antarctic glaciation. *Global and Planetary Change*, **145**, 98–115, https://doi.org/10.1016/j.gloplacha.2016.08.013

Stern, T.A. and ten Brink, U.S. 1989. Flexural uplift of the Transantarctic Mountains. *Journal of Geophysical Research: Solid Earth*, **94**, 10 315–10 330, https://doi.org/10.1029/JB094iB08p10315

Stern, T.A., Baxter, A.K. and Barrett, P.J. 2005. Isostatic rebound due to glacial erosion within the Transantarctic Mountains. *Geology*, **33**, 221–224, https://doi.org/10.1130/G21068.1

Storey, B.C. and Garrett, S. 1985. Crustal growth of the Antarctic Peninsula by accretion, magmatism and extension. *Geological Magazine*, **122**, 5–14, https://doi.org/10.1017/S0016756800034038

Storey, B.C. and Kyle, P.R. 1997. An active mantle mechanism for Gondwana breakup. *South African Journal of Geology*, **100**, 283–290.

Storey, B.C., Leat, P.T., Weaver, S.D., Pankhurst, R.J., Bradshaw, J.D. and Kelley, S. 1999. Mantle plumes and Antarctica–New Zealand rifting: evidence from Mid-Cretaceous mafic dykes. *Journal of the Geological Society, London*, **156**, 659–671, https://doi.org/10.1144/gsjgs.156.4.0659

Studinger, M., Bell, R.E., Fitzgerald, P.G. and Buck, W.R. 2006. Crustal architecture of the Transantarctic Mountains between the Scott and Reedy Glacier region and South Pole from aerogeophysical data. *Earth and Planetary Science Letters*, **250**, 182–199, https://doi.org/10.1016/j.epsl.2006.07.035

Stump, E. 1995. *The Ross Orogen of the Transantarctic Mountains*. Cambridge University Press, Cambridge.

Suarez, M. 1976. Plate-tectonic model for southern Antarctic Peninsula and its relation to southern Andes. *Geology*, **4**, 211–214, https://doi.org/10.1130/0091-7613(1976)4<211:PMFSAP>2.0.CO;2

Sutherland, R., Spasojevic, S. and Gurnis, M. 2010. Mantle upwelling after Gondwana subduction death explains anomalous topography and subsidence histories of eastern New Zealand and West Antarctica. *Geology*, **38**, 155–158, https://doi.org/10.1130/G30613.1

Taylor, R.K. and Shaw, J. 1989. The Falkland Islands: new paleomagnetic data and their origin as a displaced terrane from southern Africa. *American Geophysical Union Geophysical Monograph Series*, **50**, 59–72.

Tessensohn, F. and Wörner, G. 1991. The Ross Sea Rift System, Antarctica: Structure, evolution and analogues. *In*: Thomson, M.R.A., Crame, J.A. and Thomson, J.W. (eds) *Geological Evolution of Antarctica*. Cambridge University Press, New York, 273–278.

Tikku, A.A. and Cande, S.C. 1999. The oldest magnetic anomalies in the Australian–Antarctic Basin: Are they isochrones? *Journal of Geophysical Research: Solid Earth*, **104**, 661–677, https://doi.org/10.1029/1998JB900034

Torsvik, T.H. and Cox, L.R.M. 2013. Gondwana from top to base in space and time. *Gondwana Research*, **24**, 999–1030, https://doi.org/10.1016/j.gr.2013.06.012

Torsvik, T.H., Gaina, C. and Redfield, T.F. 2008. Antarctica and global paleogeography: From Rodinia, through Gondwanaland and Pangea, to the birth of the southern ocean and the opening of gateways. *In*: Cooper, A.K., Barrett, P.J., Stagg, H., Storey, B., Stump, E. and Wise, W. (eds) *Antarctica, A Keystone in a Changing World, Proceedings of the 10th International Symposium on Antarctic Earth Sciences*. The National Academies Press, Washington, DC, 125–140.

Torsvik, T.H., Van der Voo, R. *et al.* 2012. Phanerozoic polar wander, palaeogeography and dynamics. *Earth-Science Reviews*, **114**, 325–368, https://doi.org/10.1016/j.earscirev.2012.06.007

Vaughan, A.P.M. and Storey, B.C. 2000. The eastern Palmer Land shear zone: a new terrane accretion model for the Mesozoic development of the Antarctic Peninsula. *Journal of the Geological Society, London*, **157**, 1243–1256, https://doi.org/10.1144/jgs.157.6.1243

Wannamaker, P., Hill, G. *et al.* 2017. Uplift of the Central Transantarctic Mountains. *Nature Communications*, **8**, 1588, https://doi.org/10.1038/s41467-017-01577-2

Weaver, S.D., Storey, B.C., Pankhurst, R.J., Mukasa, S.B., DiVenere, V.J. and Bradshaw, J.D. 1994. Antarctica–New Zealand rifting and Marie Byrd Land lithospheric magmatism linked to ridge subduction and mantle plume activity. *Geology*, **22**, 811–814, https://doi.org/10.1130/0091-7613(1994)022<0811:ANZRAM>2.3.CO;2

Williams, S.E., Whittaker, J.M., Granot, R. and Müller, D.R. 2013. Early India–Australia spreading history revealed by newly detected Mesozoic magnetic anomalies in the Perth Abyssal Plain. *Journal of Geophysical Research: Solid Earth*, **118**, 3275–3284, https://doi.org/10.1002/jgrb.50239

Wilson, D.S. and Luyendyk, B.P. 2009, West Antarctic paleotopography estimated at the Eocene–Oligocene climate transition. *Geophysical Research Letters*, **36**, L16302, https://doi.org/10.1029/2009GL039297

Chapter 1.2

Antarctic volcanism: volcanology and palaeoenvironmental overview

John L. Smellie

Department of Geography, Geology and the Environment, University of Leicester, Leicester LE1 7RH, UK

0000-0001-5537-1763

jls55@leicester.ac.uk

Abstract: Since Jurassic time (c. 200 Ma), Antarctica has had a greater diversity of volcanism than other southern continents. It includes: (1) voluminous mafic and felsic volcanism associated with the break-up of Gondwana; (2) a long-lived continental margin volcanic arc, including back-arc alkaline volcanism linked to slab rollback; (3) small-volume mafic alkaline volcanism associated with slab-window formation; and (4) one of Earth's major continental rift zones, the West Antarctic Rift System (WARS), with its numerous large alkaline central volcanoes. Several of Antarctica's volcanoes are still active. This chapter is a review of the major volcanic episodes and their principal characteristics, in their tectonic, volcanological and palaeoenvironmental contexts. Jurassic Gondwana break-up was associated with large-scale volcanism that caused global environmental changes and associated mass extinctions. The volcanic arc was a major extensional arc characterized by alternating volcanic flare-ups and lulls. The Neogene rift-related alkaline volcanism is dominated by effusive glaciovolcanic eruptions, overwhelmingly as both pāhoehoe- and 'a'ā-sourced lava-fed deltas. The rift is conspicuously poor in pyroclastic rocks due to the advection and removal of tephra erupted during glacial intervals. Volcanological investigations of the Neogene volcanism have also significantly increased our knowledge of the critical parameters and development of the Antarctic Ice Sheet.

The products of volcanism are widespread in Antarctica (Figs 1 & 2). Although this Memoir extends back to 200 Ma, older volcanism is also present but the outcrops are scattered, generally small, and their tectonic setting and volcanological and palaeoenvironmental significance are much more poorly known (e.g. Goodge 2019 and references therein). By contrast, the evidence for volcanism since Jurassic time is widespread and generally well exposed (Smellie 2020b). It is linked to an unusually diverse range of tectonic settings, which vary from large-scale mafic effusive outpourings and associated felsic explosive activity connected with supercontinent break-up, to small-volume mafic monogenetic edifices and volcanic fields associated with subducted-slab windows. A very long-lived (>200 myr) continental margin volcanic arc is also present, with related flood lava volcanism in a back-arc setting and numerous small, mainly submarine centres in a Quaternary ensialic marginal basin. Neogene alkaline volcanism is also widespread within the West Antarctic Rift System (WARS), one of the world's largest continental rifts. The tectonic diversity of Antarctica's volcanism is unmatched by any other southern hemisphere continent. It is used here as a convenient basis for this overview of Antarctica's volcanism and eruptive palaeoenvironmental record.

Early Jurassic break-up of Gondwana: flood lavas, phreatocauldrons and flare-ups

The fragmentation of Gondwana during the Early Jurassic is generally linked with the impact of a mantle plume (e.g. Storey 1995; Storey and Kyle 1997), although shallow, plate-driven processes are becoming seen as a plausible alternative hypothesis (e.g. Hastie et al. 2014; Peace et al. 2020) (Fig. 3). The break-up was associated with widespread and voluminous volcanism and associated hypabyssal intrusions. In addition to outcrops in Antarctica, the episode is also represented in South Africa, southern South America, Australia and New Zealand. Cumulatively, it represents a large igneous province (LIP) sometimes informally called the Gondwana LIP (Storey and Kyle 1997), which is one of the largest continental LIPs on Earth (Sensarma et al. 2018). The volcanism has two parts: mafic and felsic. The earliest-formed volcanic units are mafic tholeiitic dolerite sills and coeval flood lavas. The outcrops in Antarctica (Transantarctic Mountains; excluding Dronning Maud Land) are called the Ferrar Dolerite Formation (intrusions) and Kirkpatrick Basalt Formation (mainly lavas), and they are collectively called the Ferrar Supergroup (Elliot et al. 2021). The age of the Ferrar Supergroup is 183–182 Ma and it may have been intruded over less than 0.4 myr, similar to estimates for the coeval Karoo volcanism in South Africa (Svensen et al. 2007, 2012; Burgess et al. 2015). The total volume of mafic magma involved is >0.5 × 10^6 km^3 (Elliot et al. 2021; Luttinen 2021) and an even greater volume can be inferred if extensive geophysically detected subsurface outcrops under the East Antarctic Ice Sheet and offshore, interpreted as early break-up magmas, are taken into account (Studinger et al. 2004; Leat 2008; Ferraccioli et al. 2009; Paxman et al. 2019). The magmatism probably had a global environmental impact as it was coeval with the significant Toarcian mass extinction event (Burgess et al. 2015; Ernst and Youbi 2017; Elliot et al. 2021). It probably influenced extinctions by mediating global climate and environmental change: for example, perturbations in pCO$_2$, CH$_4$, SO$_2$ and halogens, sea-level changes, oceanic anoxia, calcification crises, evolutionary radiations, the release of gas hydrates, and rapid global warming and cooling crises (Svensen et al. 2007; Storey et al. 2013).

The Ferrar sills are typically 100–300 m thick. Associated dykes are much thinner (<2 m) and inconspicuous. The sills probably have a local cumulative thickness of c. 1500 m and a total volume of c. 1.7 × 10^5–2.0 × 10^5 km^3, larger than sills in the Karoo (3000 km^3: Svensen et al. 2007). Some sills are extremely extensive. For example, the Basement Sill in Victoria Land has an estimated outcrop area of c. 10 000 km^2 (Marsh 2007) and the Peneplain Sill may be even larger (c. 19 000 km^3: Gunn and Warren 1962). Two distinctive compositional types of sills and lavas are recognized in the Transantarctic Mountains, and their compositional uniformity over large distances has been used to suggest that the magma flowed laterally for c. 4000 km from their source, which was assumed to be a large plume-head centre situated between Dronning Maud Land and South Africa (e.g. Leat 2008). This is the longest interpreted lateral flow of magma on Earth. The sills intruded the Beacon Supergroup, an unusually long-lived intermontane succession of continental sedimentary rocks with ages

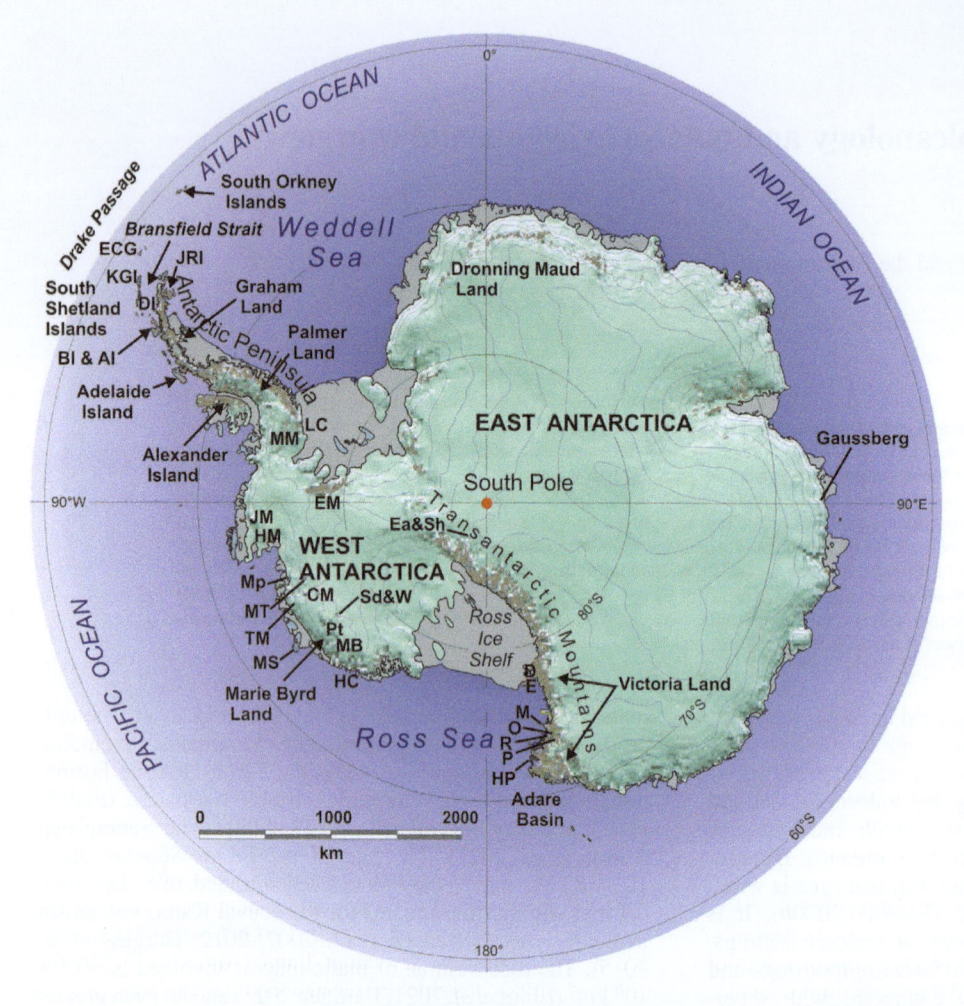

Fig. 1. Map showing the locations of places mentioned in the text. Abbreviations: AI, Anvers Island; BI, Brabant Island; CM, Crary Mountains; D, Mount Discovery; DI, Deception Island; E, Mount Erebus; Ea, Mount Early; ECG, Elephant and Clarence Islands Group; EM, Ellsworth Mountains; HC, Hobbs Coast; HM, Hudson Mountains; HP, Hallett Peninsula; JRI, James Ross Island; JM, Jones Mountains; KGI, King George Island; LC, Lassiter Coast; M, Mount Melbourne; MB, Mount Berlin; MM, Merrick Mountains; Mp, Mount Murphy; MS, Mount Siple; MT, Mount Takahe; O, Mount Overlord; P, The Pleiades; Pt, Mount Petras; R, Mount Rittmann; Sd, Mount Sidley; Sh, Sheridan Bluff; TM, Toney Mountain; W, Mount Waesche.

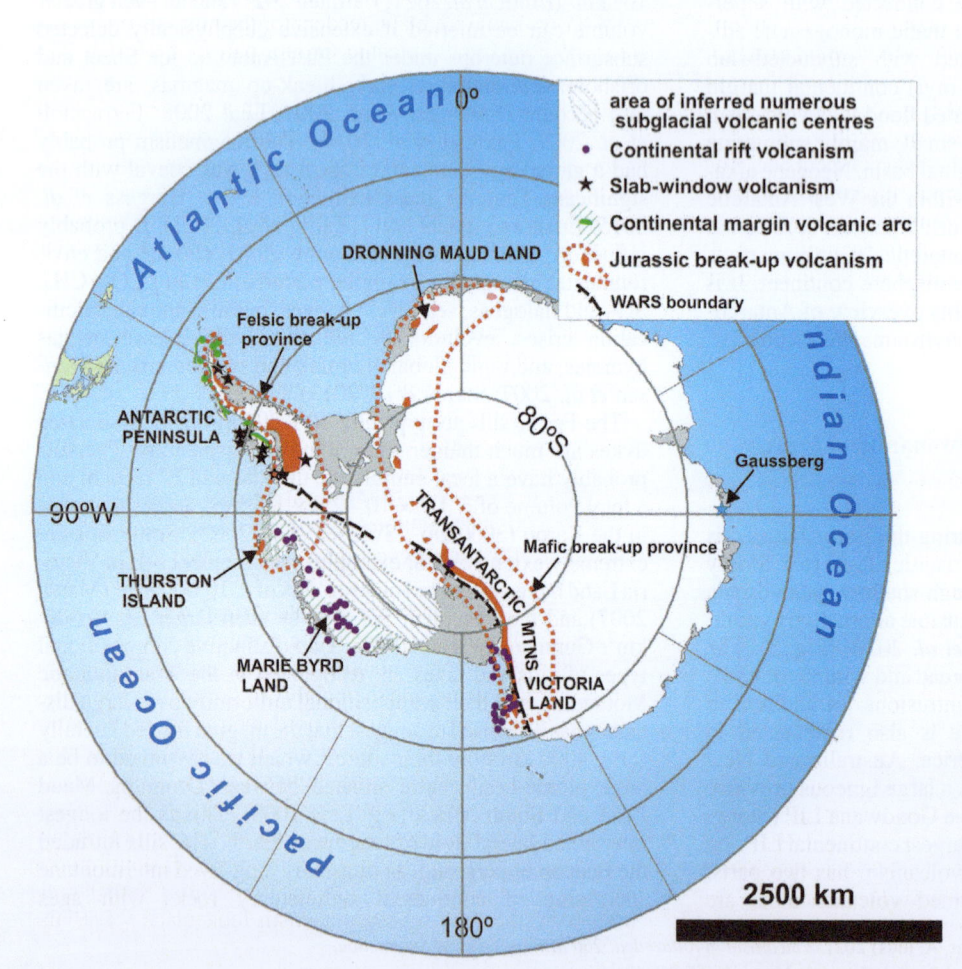

Fig. 2. Map showing the distribution of volcanic outcrops in Antarctica, going back 200 myr. The West Antarctic Rift System (WARS) boundary is after LeMasurier (2008).

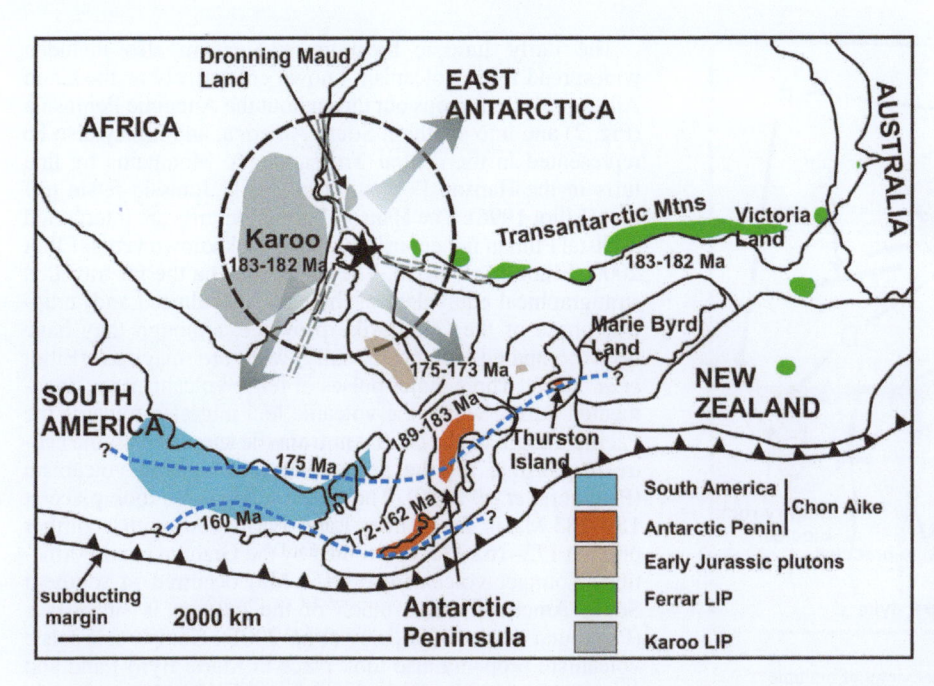

Fig. 3. Reconstructed Gondwana continent showing the mantle plume thought to be responsible for subsequent break-up, and the distribution and ages of associated volcanism forming the Ferrar and Chon Aike LIPs (adapted from Smellie 2020*b*).

spanning the Devonian–Triassic (Barrett 1991; Isbell 1999). Intrusion and inflation of the sills may have contributed to the final physiographical closure of the Beacon Basin(s), which may already have been shallowing due to progressive crustal thickening (Bialas *et al.* 2007; but see Lisker and Läufer 2013 for an alternative view).

Some sills merge and others terminate in dykes but no feeder dyke swarm has been identified in the Transantarctic Mountains, or any Strombolian deposits representing possible vent locations (cf. Pedersen *et al.* 2017). Volcanic plugs at a few localities in Victoria Land are potential feeders, but on a very small scale (Ross *et al.* 2008). However, the Ferrar LIP has been interpreted as a possible example of a flood lava system characterized by sill-driven feeder dykes in an absence of significant regional tectonic stresses (Muirhead *et al.* 2014). Known as the 'cracked lid' model, the emplacement and inflation of the Ferrar sill network generated fractures in the overlying country rock that were intruded by a plethora of narrow cross-cutting dykes with widely varying orientations (Fig. 4). Conversely, sills are uncommon in Dronning Maud Land and the dykes present there, although small (typically <2 m wide, although sometimes a few tens of metres thick: Riley *et al.* 2005; A. Luttinen pers. comm.), are presumed to have acted as feeders (Luttinen 2021). The Dronning Maud Land occurrences are linked genetically to the Karoo province in South Africa. The Karoo also includes feeder dyke swarms, although on a larger scale and generally wider than those in Dronning Maud Land (typically 18 m wide: Le Gall *et al.* 2002; Hastie *et al.* 2014). Dykes in Palmer Land (Antarctic Peninsula), interpreted as possible Ferrar-age equivalents by Vaughan *et al.* (1999), may be mafic feeders for the bimodal but mainly felsic Palmer Land Volcanic Group, part of the Chon Aike province (see later).

The Ferrar sills were associated with a widespread coeval cover of subaerial flood lavas (Kirkpatrick Basalt Formation). However, the earliest volcanic events, represented by the Prebble and Mawson formations and Exposure Hill rocks, include large-scale chaotic assemblages of phreatomagmatic pyroclastic deposits (mainly products of pyroclastic density currents) with rafts of lava and Beacon sedimentary rocks, intimately intruded by Ferrar dolerite sills and dykes. Those outcrops are interpreted as the infills of multiple coalesced maar–diatreme vent complexes, collectively called phreatocauldrons in recognition of their likely cumulative negative relief (Fig. 5). They represent explosive volcanic activity prior to effusion of the flood lava sequences. Similar explosively-generated deposits are a less well-known but important component of the basal stages of several flood lava provinces elsewhere (White and McClintock 2001; Ukstins Peate *et al.* 2003; Ross *et al.* 2005, 2008; McClintock and White 2006; Ross and White 2006; McClintock *et al.* 2008). They may have played a significant role in triggering climate change and global mass extinctions because of their mafic compositions, which are associated with higher volatile contents, especially CO_2, F, Cl and sulfur species, and their ability to send ash and volatiles directly into the stratosphere (Thordarson *et al.* 2003; Jolley and Widdowson 2005; Ross *et al.* 2005; McClintock *et al.* 2008). Water clearly played a significant part in the pyroclastic eruptions of the Ferrar province. However, the source was stored and recycled groundwater, and they were not erupted subaqueously, thus providing possible indirect confirmation that the precursor Beacon sedimentary basin(s) were largely infilled by the time Ferrar volcanism took place (cf. Bialas *et al.* 2007).

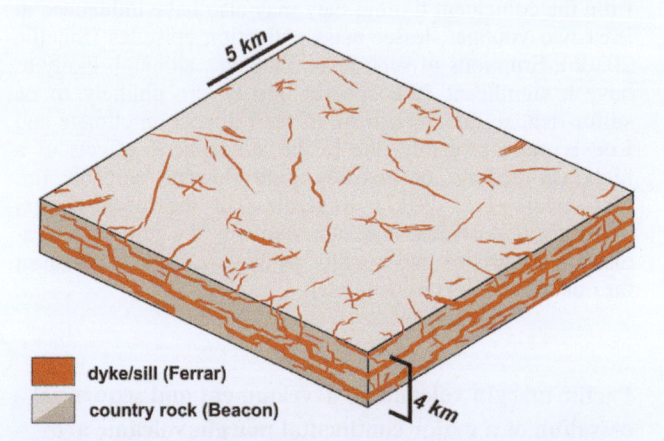

Fig. 4. Schematic depiction of the 'cracked lid' model of multiple small dykes genetically linked to inflation of subjacent Ferrar sills intruding Beacon Supergroup sedimentary strata. The dykes, whose widths are exaggerated in the diagram for clarity, are envisaged feeding flood lava effusion in the overlying coeval Kirkpatrick Basalt Formation. The essentially decussate pattern of dyke intrusion suggests that regional stresses were negligible. After Muirhead *et al.* (2014).

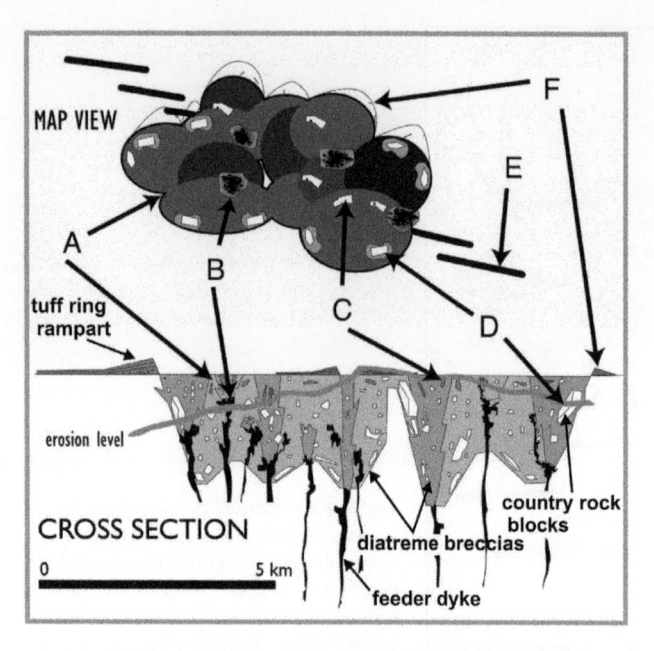

Fig. 5. Sketch showing map and cross-sectional views of multiple coalesced maar–diatreme volcanoes in the Coombs Hills area of the Transantarctic Mountains envisaged by White and McClintock (2001; modified). Coalescence of the centres led to a wide area of negative relief, termed a phreatocauldron.

The succeeding Kirkpatrick lavas may have had a minimum volume of 7000 km^3 but they are restricted by erosion to relatively small outcrops in Victoria Land, central Transantarctic Mountains and equivalents in Dronning Maud Land (Fleming et al. 1995; Elliot et al. 2021; Luttinen 2021). The individual lavas are pāhoehoe, commonly 500 m in cumulative thickness but locally exceeding 750 m, and they reach 2500 m in the Karoo-related province of Dronning Maud Land (Luttinen 2021). The individual lava lobes range in thickness from decimetres to several metres, and have widths of a few metres to several hundred metres. A few are associated with substantial thicknesses of 'hyaloclastite' (probably tuff breccia, *sensu* White and Houghton 2006) and pillow lava, suggesting that some of the lavas may have migrated into lakes as lava-fed deltas, similar to Paleocene flood lavas in Greenland (Pedersen et al. 2018). However, the proportion of volcaniclastic rocks in the Kirkpatrick Basalt Formation generally is small (c. 60 km^3: Elliot et al. 2021). Tabular sheet lavas several tens of metres to >100 m in thickness and tens of kilometres across are also present (Luttinen 2021). In Dronning Maud Land they have been divided up into flow fields based on contact relationships, associated interbeds, type of flow unit, colour and geochemical composition (Luttinen 2021); most are compound-braided pāhoehoe types (*sensu* Jerram 2002). Overall, the characteristics of the pāhoehoe lavas are those of inflated pāhoehoe (cf. Self et al. 1996; Keszthelyi et al. 2006). 'A'ā lavas have not been unequivocally identified but some lavas have brecciated surfaces and may be rubbly pāhoehoe (Luttinen et al. 2010; cf. Rowland and Walker 1990). The lava outcrops are comparatively small relicts of what was probably once a very extensive volcanic field.

The Kirkpatrick Basalt Formation also contains epiclastic interbeds sometimes associated with pillow lava and generally assigned a lacustrine origin (Elliot et al. 2021; Luttinen 2021). Additionally, red and yellowish-green to grey-green boles are present and probably represent palaeosols formed under a variety of oxidizing and reducing surface weathering conditions during substantial breaks in activity (e.g. Retallack 2001).

The Early Jurassic break-up magmatism also included widespread felsic volcanism known collectively as the Chon Aike province. It crops out throughout the Antarctic Peninsula (Fig. 2) and into southern South America, and it may also be represented in the central Transantarctic Mountains by fine tuffs in the Hanson Formation, an Early Jurassic felsic tuff unit (Elliot 1996). The Hanson Formation tuffs are interpreted as distal Plinian fall deposits and they lack known vents (Elliot 2000; Elliot et al. 2017). They may thus be the far-travelled stratigraphical equivalent of the earliest (Palmer Land) eruptive phase of the Chon Aike province, although they have been compared compositionally with arc magmas (Elliot et al. 2017). Three major pulses of felsic volcanism are documented during which the volcanic loci migrated towards the Pacific margin of the continent from the supposed plume centre responsible for the Ferrar and Karoo mafic volcanism (Pankhurst et al. 2000). The oldest felsic pulse took place at 189–183 Ma in the Palmer Land Formation, with a further pulse at 172–162 Ma represented by the Graham Land Formation. Younger volcanism (to 153 Ma) occurred in southern South America, where much of the outcrop is subsurface (Pankhurst et al. 2000; Riley et al. 2001). Comparable felsic volcanism probably also took place in Marie Byrd Land and may be hinted at by Early Jurassic (175–173 Ma) plutons in flanking areas (Jones Mountains and Ellsworth Mountains: Storey et al. 1988; Riley et al. 2017b). Any evidence is either obscured by the extensive ice cover or else has been removed by erosion.

The Chon Aike is a felsic province regarded as a volcanic flare-up (i.e. a period of unusually high magma flux: Paterson and Ducea 2015). It is linked to crustal anatexis caused by the migration of the Ferrar–Karoo plume head (Pankhurst et al. 2000) (Fig. 3). The presence of a coeval subduction zone along the Pacific margin probably acted as a self-limiting factor on the geographical spread of the plume and which intensified the effects of sublithospheric heating. The estimated volume of magma erupted, c. 0.5×10^6 km^3, is comparable with the volume of mafic magma involved in the Ferrar LIP (Pankhurst et al. 2000). The Chon Aike was a bimodal volcanic episode with a total thickness of c. 1 km of volcanic products, mainly rhyolitic ignimbrites and fewer rhyolite lavas, with minor mafic lavas (Hunter et al. 2006; Riley et al. 2010; Riley and Leat 2021). Eruption probably took place from multiple calderas associated with large stratovolcanoes. The felsic flare-ups probably contributed an atmospheric loading similar to that associated with the mafic (Ferrar) LIP and, from the coincident timing, they may also have influenced at least two younger, lesser mass extinction episodes (Smellie 2020b). Eruptions of such felsic magmas, although likely to have a significant stratospheric impact, are unlikely to be sulfur-rich, so the mechanism of their impact on climate and Life is more probably due to the atmospheric effects of a high ash-loading, or possibly high fluorine aerosols (cf. McConnell et al. 2017). Regardless of the mechanism(s) involved, it is evident that, cumulatively, the Jurassic volcanism in Antarctica had a major palaeoenvironmental impact far outside of the continent itself.

Pacific margin volcanism: development and sequential cessation of a major continental margin volcanic arc

The timing of the initiation of subduction-related magmatism along the Pacific margin of Gondwana is uncertain but preceded 200 Ma (see Smellie 2020a, b for a synthesis). Moreover, whether volcanic activity in the arc was subsequently continuous or episodic is unclear. Early Jurassic plutons in the central and southern Antarctic Peninsula may be subduction-related (Riley et al. 2017a) and could have fed coeval volcanism. Alternatively, the Early Jurassic magmatism may have been related to extension, and subduction (or, perhaps, its latest phase?) did not commence until Late Jurassic time, with the arc becoming a prominent established feature in the Early Cretaceous (Leat and Riley 2021b). However, if the Hanson Formation is indeed arc-derived and it overlapped in age with the Kirkpatrick Basalt Formation (Elliot et al. 2017), then it provides evidence for Early Jurassic subduction. The arc could have been a prominent orographical feature extending along the length of the Pacific margin of Gondwana (Fig. 2), including South America, the Antarctic Peninsula, Marie Byrd Land and New Zealand. Early Cretaceous granitoids in the Lassiter Coast, southern Palmer Land, have been interpreted as evidence for high magma flux in the arc represented by three discrete episodes of pluton emplacement, at 130–126, 118–113 and 108–102 Ma (Riley et al. 2017a, 2018). The plutons are mainly tonalite to granodiorite in composition (i.e. relatively felsic) and any associated volcanism is likely to have been mainly pyroclastic rather than effusive, with eruptions probably from large calderas similar to those associated with the Chon Aike. Together, they have been interpreted as a succession of volcanic flare-ups. However, unlike the Chon Aike, the impact on mass extinctions of the three Early Cretaceous volcanic episodes is much less obvious and may not be significant (Smellie 2020b).

Following the mid-Cretaceous separation of New Zealand from Marie Byrd Land at c. 90–80 Ma (Lawver and Gahagan 1994), offset sections of the Phoenix–Antarctic spreading centre collided with the coeval trench and subduction shut down progressively in a northerly direction (Larter et al. 2002). The collision ages are best defined for the Antarctic Peninsula where they began c. 50 myr ago offshore of southern Alexander Island (Hole et al. 1995). Subduction ceased following each collision and the arc-related magmatism stopped c. 10–20 myr ahead of each ridge collision (Barker 1982; McCarron and Larter 1998). The most recent collision took place opposite northern Graham Land at c. 3 Ma. However, it did not occur at the South Shetland Trench, and subduction is still taking place there at a very slow rate (Larter 1991; Maldonado et al. 1994). In addition, the axis of active arc magmatism migrated trenchwards from Late Cretaceous time (c. 70 Ma) and relocated to Alexander Island and Adelaide Island, but whether a similar relocation occurred in the South Shetland Islands is less certain (Leat and Riley 2021a; see also Smellie 2020a). The reason for the migration is unknown but the width of the forearc is substantially greater opposite Alexander Island and Adelaide Island compared with that opposite the South Shetland Islands, which may imply that the active axis of volcanism, linked as it is to slab depth, simply followed the oceanward migration of the trench (and subjacent slab) because of time-integrated growth of the accretionary prism. Moreover, essentially orthogonal subduction and associated accretion may have characterized much of the region south of the South Shetland Islands. Because of the oroclinal bend of the Antarctic Peninsula, the subduction trajectory was more oblique opposite the South Shetland Islands (cf. McCarron and Larter 1998). It may thus have triggered tectonic erosion of the forearc there (cf. Maldonado et al. 1994), thus counteracting any potential trenchward migration (also see below).

The most extensive outcrops of subduction-related volcanic rocks are present in the Antarctic Peninsula. However, most are pervasively hydrothermally altered by coeval intrusions of the Antarctic Peninsula batholith, contact metamorphism, and regional metamorphism to zeolite and prehnite–pumpellyite facies (Burn 1981; Smellie et al. 1984; Leat and Riley 2021b). The alteration obscures many volcanological details, and makes tracing and matching volcanic units exceedingly difficult. Although the outcrops are often well exposed in coastal cliffs, access is challenging or impossible because of the scarcity of beaches and the lack of sea ice in the region during the austral summer. Inland, the snow-covered alpine topography can also make overland access problematic. Conversely, beaches are common in the South Shetland Islands and the topography is much more subdued. Together with a lack of pervasive alteration in most of the outcrops, the South Shetland Islands contain the most intensively investigated outcrops of subduction-related volcanic rocks in Antarctica (e.g. Smellie et al. 1984; Birkenmajer 2001; Machado et al. 2005; Haase et al. 2012; Leat and Riley 2021a).

With its well-developed forearc and back-arc basins (Macdonald and Butterworth 1990), subdued elevation (mostly <2 km), lack of a back-arc fold-and-thrust belt, and overall relatively thin crust, the Antarctic Peninsula most resembles an extensional continental arc (Ducea et al. 2015). However, the volcanism in the South Shetland Islands is overwhelmingly mafic–intermediate and occurrences of felsic rocks (rhyolites and dacites) are rare (Smellie et al. 1984; Smellie 2020a; Leat and Riley 2021b). Continental margin arcs are characteristically poor in mafic volcanic products (Ducea et al. 2015). The mafic, partly tholeiitic, compositions in the South Shetland Islands are characteristic of arcs founded on thin crust (Smellie 2020a). Indeed, their petrological characteristics, especially the lack of crustal contamination, have led to some authors referring to the South Shetland Islands as typical island arc rocks (Birkenmajer et al. 1990; Machado et al. 2005). Geophysical studies suggest that crustal thicknesses vary from 36–42 km beneath Graham Land to 25–30 km beneath the South Shetland Islands (Grad et al. 2002). The volcanism in the South Shetland Islands may thus correspond to a so-called transitional continental arc (*sensu* Ducea et al. 2015).

The subduction-related outcrops are grouped together as the geographically extensive feature called the Antarctic Peninsula volcanic arc (Leat and Riley 2021b). Because of the relatively easy access, stratigraphic details of volcanic rocks on Alexander Island, Adelaide Island and the South Shetland Islands are comparatively well described (summarized by Leat and Riley 2021a). Because of some distinctive lava compositions corresponding to high-magnesian andesites formed by the melting of young hot lithosphere associated with ridge subduction (McCarron and Smellie 1998) and an incorrect assumption that coeval arc volcanism was present in Palmer Land, the Alexander Island Volcanic Group was interpreted as genetically distinct from the rest of the Antarctic Peninsula volcanic arc. However, such unusual compositions are occasionally found in arcs under certain circumstances (e.g. Kelemen 1995; Goss et al. 2013). The Alexander Island volcanism is now considered to be a consequence of the trenchward migration of the active arc volcanism away from Palmer Land during the Late Cretaceous (Leat and Riley 2021a). Alexander Island volcanism is predominantly effusive, with mainly mafic–intermediate lava compositions (rhyolitic and dacitic ignimbrites are common in the Colbert Mountains: McCarron and Millar 1997). Ages range from c. 80 to c. 46 Ma, and there is a pronounced northerly migration of the volcanism that was linked to the effects of the progressive collision and subduction of three offset spreading ridge segments (McCarron and Larter 1998; McCarron and Smellie 1998). Adelaide Island, 200 km to the north of Alexander Island, also contains substantial thicknesses (up to c. 2 km) of volcanic rocks in three formations that vary from basaltic andesite and andesite lavas, to rhyolitic tuffs and ignimbrites (Riley et al. 2012). As on Alexander Island (Burn 1981), all are intensely hydrothermally altered and they have latest Cretaceous ages (c. 75–67 Ma) overlapping with volcanism on Alexander Island. However there is no age migration, probably because subduction was continuous (and orthogonal) offshore of Adelaide Island until only a single long Phoenix–Antarctic ridge segment collided at the adjacent trench, rather than multiple offset ridge segments arriving at different times.

The Cretaceous–Miocene volcanic stratigraphy of the South Shetland Islands is the most frequently described part of the magmatic arc. From studies particularly by Polish geologists (e.g. Birkenmajer 2001), King George Island has the most complicated stratigraphy in Antarctica, despite encompassing a period just 30 myr in duration, and for an island just 90 km long and 25 km wide (Fig. 6). It illustrates how the easier access and generally lesser alteration has enabled much more frequent detailed stratigraphic investigations of the volcanic rocks there than elsewhere in Antarctica. King George Island also contains the only exposed Paleogene terrestrial environmental record, with an internationally important glacial–interglacial stratigraphy (Fig. 7). At least five glacial periods have been described, together with four interglacials (Birkenmajer 1996); although, of the putative Eocene glacials, one (Krakow) has been disproved (Dingle and Lavelle 1998) and the other (Ezcurra: Birkenmajer et al. 2005) has yet to be substantiated. The two best-known and verified glacial episodes (Polonez and Melville) are dominated by marine sedimentary rocks, but all the other units are volcanic. The volcanic sequences are well exposed and have been subject to several detailed investigations but few included volcanological–palaeoenvironmental aspects (Porebski and Gradzinski 1987, 1990; Smellie et al. 1998; Troedson and Smellie 2002).

Like Alexander Island, the South Shetland Islands show a prominent along-arc migration of the volcanism (and plutonism), with the inferred younging to the NE interpreted to indicate the progressive shutting down of volcanic centres rather than recent tilting and enhanced erosion of westerly areas (Pankhurst and Smellie 1983; Smellie et al. 1984). Since these studies, the number of K–Ar and (increasingly) $^{40}Ar/^{39}Ar$ isotopic ages determined on the volcanic rocks has expanded substantially; it is the largest dating dataset for any part of the magmatic arc (see Leat and Riley 2021a for a compilation). When the entire dataset is considered, it is now evident that the published migration model was simplistic (see also Willan and Kelley 1999). It is clear that the main volcanic axes associated with each eruptive phase have prominent northeasterly orientations and there is an overall migration to the SE, a trend that is continued into the marginal basin where most of the youngest volcanism is present (Fig. 8) (cf. Leat and Riley 2021a). The width of the different-aged volcanic zones, c. 20 km, is similar to typical widths of active magmatism in continental arcs at any time, which is caused by magma focusing (Ducea et al. 2015). However, the presence of, admittedly rare, inliers of Cretaceous rocks among the Paleogene outcrops suggests that the distribution of Cretaceous volcanic centres, at least, may have been geographically more widespread. The Cretaceous volcanic zones, particularly the Early Cretaceous one but not the Paleogene zone, also show a prominent geographical coincidence with the Pacific Margin Anomaly (PMA). The PMA is a prominent linear belt of positive long-wavelength magnetic anomalies that runs down the west side of the Antarctic Peninsula and is attributed to Cretaceous mafic plutons in the Antarctic Peninsula Batholith (Garrett et al. 1987; Parra et al. 1988; Soloviev et al. 2018). Its displacement to the NW of the islands emphasizes that a southeasterly migration of the volcanic axis, *away from* the South Shetland Trench, occurred during the Paleogene. Younger volcanism, of Oligocene age, is confined to southeastern King George Island (Smellie et al. 1984, 1998; Birkenmajer et al. 1986; Troedson and Smellie 2002). The youngest arc

KING GEORGE ISLAND SUPERGROUP

FILDES BLOCK/TERRANE

FILDES PENINSULA GROUP
Winkel Point Fmn (59-39.5 Ma)
Schneider Bay Fmn (57.7-42.9 Ma)

BARTON HORST
CARDOZO COVE GROUP
Admiralen Peak Fmn (43.7 Ma)
Znosko Glacier Fmn (60.4-56.8 Ma)

DUFAYEL ISLAND GROUP
Dalmor Bank Fmn (51.9 Ma)
Gdynia Point Fmn

MARTEL INLET GROUP
Goetel Glacier Fmn
Ullman Spur Fmn
Domeyko Glacier Fmn
Visca Anchorage Fmn (66.7 Ma)
Keller Peninsula Fmn

WARSAWA BLOCK/TERRANE

POINT HENNEQUIN GROUP
Mt Wawel Fmn (28.3-24.5 Ma)
Vieville Glacier Fmn (43.9 Ma)

POLONIA GLACIER GROUP
Sukiennice Hills Fmn
Lions Cove Fmn (42.1 Ma)

EZCURRA INLET GROUP
Point Thomas Fmn (37.4 Ma)
Arctowski Cove Fmn (66.7 Ma)

BARANOWSKI GLACIER GROUP
Zamek Fmn
Llano Point Fmn (77 Ma)

PARADISE COVE GROUP
Demay Point Fmn
Creeping Slope Fmn
Uchatka Point Fmn (67.7 Ma)

MAGDA NUNATAK COMPLEX (49.4 Ma)

KRAKOW ICEFIELD SUPERGROUP

KRAKOW BLOCK/TERRANE

LEGRU BAY GROUP
Vaureal Peak Fmn
Martins Head Fmn (25.7 Ma)
Harnasie Hill Fmn (>21.9 Ma)
Dunikowski Ridge Fmn (30.8-29.5 Ma)

CHOPIN RIDGE GROUP
Wesele Cove Fmn
Boy Point Fmn (>22.4-22.3 Ma)
Polonez Cove Fmn (32-30 Ma)
Mazurek Point Fmn (74.1 Ma; 37.6-34.4 Ma)

MELVILLE BLOCK

MOBY DICK GROUP
Cape Melville Fmn (>20.1-19.9 Ma)
Destruction Bay Fmn (23.6 Ma)
Sherratt Bay Fmn

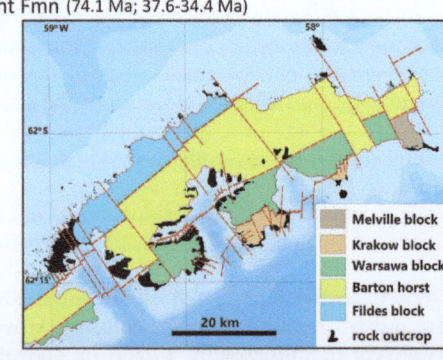

Fig. 6. Stratigraphy of King George Island constructed by Birkenmajer (2001). The ages shown are by K–Ar and have high errors (1–5 myr) and some ages have been disproved. Despite covering a period of just 30 myr, this is the most complicated stratigraphy in Antarctica, and was facilitated by the relatively easy access and generally low alteration of the rocks. The inset shows the distribution of structural blocks (also called 'terranes': Birkenmajer 2001) used as a framework for the stratigraphy. Note that many of the faults shown are conjectural but the main NE-trending faults have some topographical and geological support.

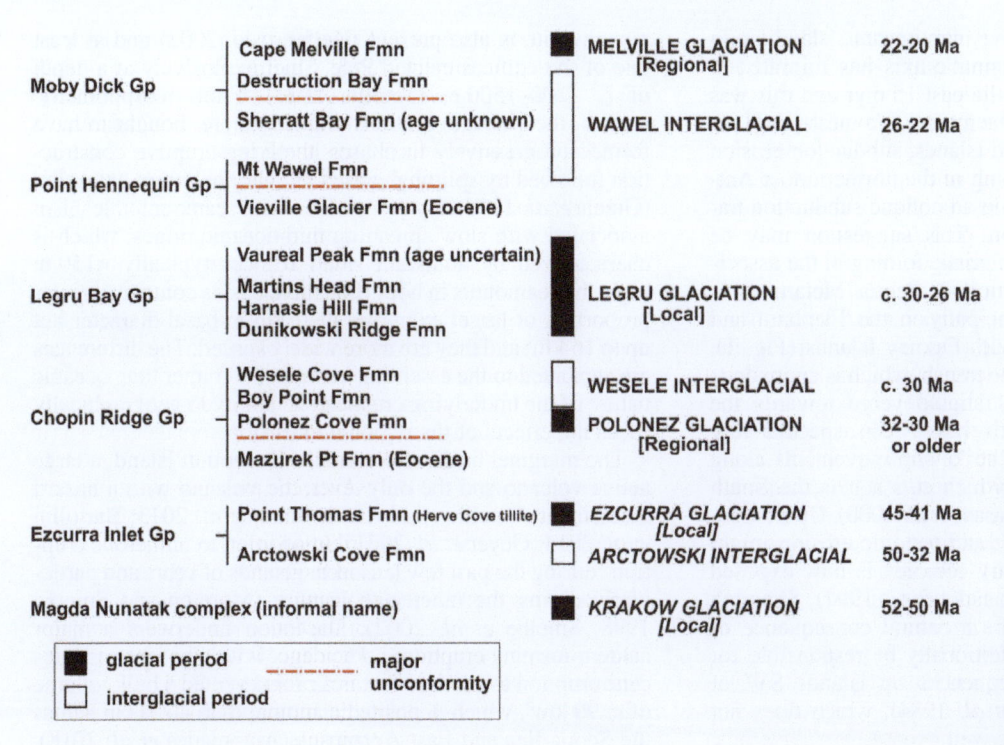

Fig. 7. Summary of glacial and interglacial periods on King George Island recognized by Birkenmajer (1996) and local stratigraphical names. The text in italics indicates environmental episodes either disproven ('*Krakow Glaciation*', '*Arctowski Interglacial*') or unconfirmed ('*Ezcurra Glaciation*').

volcanism may be represented by a small basalt volcano on King George Island at Melville Peak (<300 ka) and a submerged basaltic andesite–andesite volcano with a very young age (less than a few million years?) in the Eastern Basin of the Bransfield Strait (Smellie 2021). However, in general, evidence for arc volcanism in the South Shetland Islands is essentially absent after *c.* 20 Ma. This is probably due either to subduction of young buoyant oceanic crust switching off arc volcanism, as happened further south in the Antarctic Peninsula (Barker 1982), or to later subsidence of any arc volcanics in Bransfield Strait (Smellie 1990; Fretzdorff *et al.* 2004).

Migrations of the active volcanic axis, both toward and away from the trench, are well known in magmatic arcs and are usually ascribed to changes in slab dip (Ducea *et al.* 2015). There is currently no published explanation for why the slab dip should have changed over time beneath the South Shetland Islands. In another example, in the sub-Antarctic

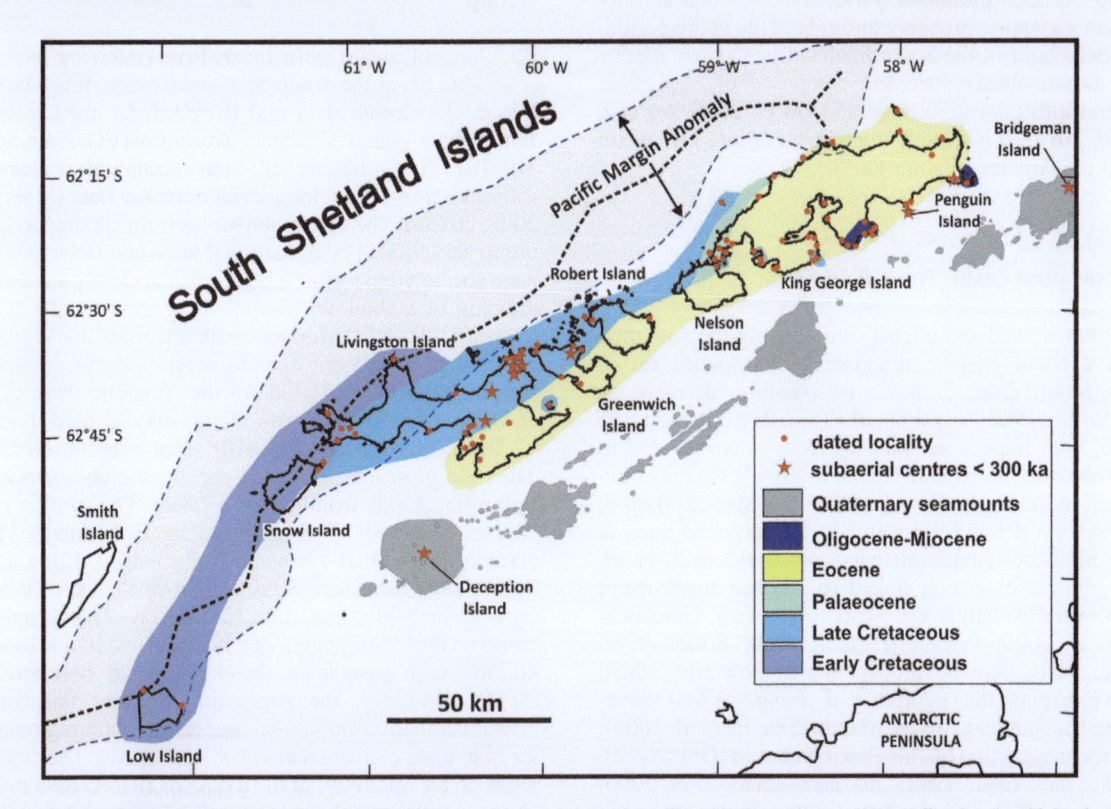

Fig. 8. Map of the South Shetland Islands showing the distribution of ages of volcanism. The ages form discrete NE-trending bands which migrate (get younger) in a southeasterly direction. The youngest volcanism is associated with the opening of Bransfield Strait as a young ensialic marginal basin (seamount centres shown). The location of the Pacific Margin Anomaly is also indicated, with its axis shown as a black dashed line (after Parra *et al.* 1988). It appears to coincide mainly with early Cretaceous-age volcanic outcrops.

South Sandwich Islands, an active intra-oceanic island arc in the eastern Scotia Sea, the magmatic axis has migrated c. 70 km away from the trench in the past 15 myr and this was attributed to tectonic erosion of the forearc (Vanneste and Larter 2002). For the South Shetland Islands, subduction erosion may be linked to oroclinal bending of the northernmost Antarctic Peninsula, which resulted in an oblique subduction trajectory and frontal compression. This suggestion may be supported by the presence of enigmatic folding in the associated Mesozoic–Cenozoic accretionary Scotia Metamorphic Complex, which crops out principally in the Elephant and Clarence Islands Group and South Orkney Islands (Fig. 1). The folding verges away from the trench, which is anomalous in an accretionary prism (they should verge towards the trench), although the structures have been speculatively explained as a result of sinistral strike-slip movements along the Shackleton Fracture Zone, which cuts across the South Shetland Trench at its NE end (Trouw et al. 2000). Uplift associated with oblique convergence and tectonic erosion might also explain why the accretionary complex is now exposed above sea-level (see also Maldonado et al. 1994). Shearing and strike-slip faulting would be a natural consequence of oblique subduction and may additionally be responsible for open folding of Cretaceous sequences on islands SW of King George Island (Smellie et al. 1984), which does not affect younger rocks.

Finally, four major pulses (i.e. flare-ups) of arc volcanism have been identified in the South Shetland Islands, at 130–110, 90–70, 60–40 and 30–20 Ma (Willan and Kelley 1999; Fretzdorff et al. 2004; Haase et al. 2012; Smellie 2020a; Leat and Riley 2021a). Haase et al. (2012) suggested that, within the limitations of the dataset of published ages, each flare-up may have lasted for c. 10 myr and might have been even shorter (e.g. just 2–3 myr) in some areas. The erupted products of each episode probably dominated the mass budget of the South Shetlands arc. However, the presence of isolated ages falling between the major pulses indicates that activity continued sporadically, probably throughout the entire period. Thus, the behaviour of the South Shetland Islands arc can be described as non-steady state and composed of alternating flare-ups and lulls (sensu Paterson and Ducea 2015; see also Ducea et al. 2015). It is a likely microcosm for arc volcanism throughout the Antarctic Peninsula.

Ensialic marginal basin: Bransfield Strait

As subduction slowed and almost ceased along the Antarctic Peninsula, a small ensialic marginal basin opened up in Bransfield Strait, from c. 4 Ma or possibly as early as c. 22 Ma (Barker 1982; Birkenmajer 1992). Its tectonic setting is disputed, with hypotheses ranging from extension due to rollback at the South Shetland Trench following the cessation of spreading in the Drake Passage (after c. 4 Ma; or possibly 6.7 Ma based on a rapid reduction in convergence rates at the South Shetland Trench at that time: Maldonado et al. 1994), to oblique extension linked to sinistral transcurrent movement of the Antarctic and Scotia plates (e.g. González-Casado et al. 2000; Fretzdorff et al. 2004; Solari et al. 2008). The basin may be propagating southwards, which could help to explain the occurrence of young (<3 Ma) volcanism on the Brabant and Anvers islands (Smellie et al. 2006), but northerly propagation has also been proposed (Gràcia et al. 1996; Barker and Austin 1998; González-Casado et al. 2000; Christeson et al. 2003; Fretzdorff et al. 2004). Swath studies in Bransfield Strait have shown that the marginal basin floor is characterized by numerous small submarine volcanic ridges and seamounts. They are mainly mafic pillow mounds but rare rhyolite is also present (Keller et al. 2002) and at least one of the edifices ejected Pele's hair explosively at a depth of c. 1200–1500 m (Smellie 2021). From morphometric studies, the volcanic mounds and ridges are thought to have formed progressively in phases involving eruptive construction followed by splitting caused by marginal basin extension (Gràcia et al. 1996). By comparison with seamount volcanism associated with slow spreading mid-oceanic ridges, which is characterized by abundant small edifices typically <150 m high, the seamounts in both Bransfield basins contain a greater proportion of larger examples (2–7 km in basal diameter but up to 16 km) and they are more widely spaced. The differences are attributed to the essentially continental rather than oceanic nature of the underlying crust and its ability to geographically focus the effects of the regional extension.

The marginal basin also contains Deception Island, a large active volcano and the only Antarctic volcano with a hazard assessment (Smellie et al. 2002; Martí et al. 2013; Bartolini et al. 2014; Geyer et al. 2021). In addition to numerous eruptions during the past few tens of thousands of years and particularly during the nineteenth century (Moreton and Smellie 1998; Smellie et al. 2002), Deception underwent a major caldera-forming eruption c. 4 kyr ago. It was the largest Holocene eruption to occur in Antarctica and vented a bulk volume of c. 90 km^3, which deposited ash more than 4000 km across the Scotia Sea and East Antarctica (Antoniades et al. 2018). Although the well-documented 1969 eruption of Deception was glaciovolcanic (but through very thin ice: Smellie 2002), and ice still covers half of the island during the current peak interglacial conditions, the deposits preserved on the island show no obvious evidence for glaciovolcanism.

Voluminous back-arc alkaline volcanism and Neogene cryosphere evolution: the James Ross Island Volcanic Group

Back-arc volcanism in the James Ross Island region is situated c. 250 km SE of the South Shetland Trench. It is a large mafic (basaltic) volcanic field and the products are known as the James Ross Island Volcanic Group (JRIVG: Smellie 2021). The JRIVG is probably still active, although it characteristically exists in a very long-lived dormant state (Smellie et al. 2008, 2013a). The volcanism varies from alkaline to less commonly tholeiitic in composition (Haase and Beier 2021). It has been attributed to rollback effects of the slowly subducted slab drawing in a shallow 'plume' of pristine mantle (i.e. unaffected by slab-related metasomatism) from the Weddell Sea, which then underwent upwelling and associated decompression melting on the SE side of the Antarctic Peninsula (Hole et al. 1995). JRIVG volcanism may extend back to c. 12.5 Ma (Marenssi et al. 2010). That age considerably predates the most usually stated age for the commencement of slab rollback (i.e. from c. 4 or 6.7 Ma). The conflict in timing may be resolved if the proposal by Birkenmajer (1992) is correct: that is, that extension (reflected by rifting and basin formation in Bransfield Strait) might date back to latest Oligocene–earliest Miocene time (c. 22 Ma). That proposal was based on the identification of a 'thermal event' and extensional faulting with associated dykes, dated at between 26 and 20 Ma. However, the suggestion remains unsubstantiated given that there is no known association with regional tectonics (e.g. plate convergence rates; see below). The conflict also relies on the reliability of the dating of JRIVG basalts found as clasts in tills, which were mostly determined by the K–Ar method (including ages of 12.4 and 7.13 Ma: Sykes 1988; Marenssi et al. 2010); however, a relatively old age of 9.2 Ma was determined (also on a lava clast in till) using

^{40}Ar/^{39}Ar and may be more reliable (Jonkers et al. 2002). In situ rocks date back to only 6.2 Ma, consistent with an age for rollback commencing at c. 6.7 Ma, although the presence of a deep 'keel' of likely JRIVG rock beneath Mount Haddington implies an even older volcanic history (Smellie et al. 2008; Jordan et al. 2009). Notably, neither the compositional nature nor the eruptive periodicity of the JRIVG volcanism changed across the transition from active to passive subduction (i.e. associated with slab rollback or transtensional effects). The evolution of the Cenozoic convergence rate of the Phoenix/Aluk slab is relatively well constrained. Two early stages of slowing are known, at 52.3 and 47.3 Ma. Both could have had profound effects on the overriding plate and might have been associated with slab rollback (McCarron and Larter 1998), although they had no apparent effect on the axis of active arc volcanism in the South Shetland Islands (Fig. 8). A further slowing of the convergence rate at c. 13 Ma could have created sufficient slab pull to initiate or enhance corner flow of pristine Weddell Sea mantle into an arc-rear 'thinspot', as postulated by Hole et al. (1995). However, the history of plate convergence does not show any rapid slowing until c. 6.7 Ma (from 40–60 to <10 mm a^{-1}), which is the earliest time that slab rollback is thought to have been initiated (Maldonado et al. 1994).

Although a significant role for the resulting magmatism, involving the melting of the subducted slab, is now being considered (Hole 2021), it is unclear yet how the new explanation might help to resolve either the timing of the JRIVG or the mechanism of its occurrence. Perhaps significantly, tholeiitic–alkaline mafic volcanism is also present in a similar arc-rear position in southern South America. Volcanic activity in the extra-Andean Patagonian basalt province mostly took place between 30 and 20 Ma but recurred in the Plio-Pleistocene. It consists mainly of 'plateau lavas', which are voluminous eruptions of basalt that cover a wide area, but includes numerous small volcanic centres in the southernmost, much younger outcrops (in the Pali Aike Volcanic Field: D'Orazio et al. 2000; Muñoz et al. 2000; de Ignacio et al. 2001; Ross et al. 2011). Although a connection with slab windows has been postulated for the volcanism, other workers prefer a slab rollback model linked to plate convergence processes, which initiated variable slab dips and caused corner flow of pristine mantle. However, both models may be correct. The field relationships and ages of the volcanism indicate that the much more widespread plateau basalt volcanism is older (Oligocene–Miocene), whereas the younger volcanism (Plio-Pleistocene) consists of numerous small-volume monogenetic volcanic centres (tuff cones, maars and scoria cones). Ridge–trench collision also did not occur until 14 Ma, placing a maximum age on the formation of slab windows. Thus, the older plateau basalt volcanism is plausibly related to a combination of slab rollback and shallow asthenospheric upwelling of pristine mantle (associated with more rapid plate convergence rates, the opposite to the situation in the Antarctic Peninsula), whilst the younger much smaller-volume volcanism can be linked to slab-window formation. The similarities with the contrasting models for volcanism in the JRIVG (flood lava eruptions (Smellie 2021); i.e. corner flow in the mantle wedge driven by slab rollback) and further south in the Antarctic Peninsula (i.e. small-volume monogenetic volcanic fields (Smellie 1999; Hole et al. 1995; Smellie and Hole 2021); slab-window-related) are self-evident but have not been highlighted previously.

An unusual feature of JRIVG volcanism is that it remained basaltic throughout its c. 12 myr duration (Smellie 1987; Košler et al. 2009; Haase and Beier 2021). However, the Antarctic Plate has been stationary since Late Cretaceous times. Under a model involving long-lived shallow diapiric upwelling and decompression melting of Weddell Sea mantle, the static conditions (i.e. no plate migration) would have favoured the construction of a substantial edifice (i.e. Mount Haddington, probably Antarctica's largest volcano). This is similar to the way very large volcanoes were constructed on Mars, a planet lacking plate tectonics and characterized by very long-lived stationary mantle plumes, whether internally generated or impact induced (e.g. Carr 1974; Reese et al. 2004). What triggered the repeated episodes of melting and eruption of large volumes of (only) mafic magma in the JRIVG is uncertain but two possible options are: (a) regional plate-tectonic effects (plate 'jostling') triggering short-lived diapiric upwelling episodes in the mantle; or (b) melting in response to glacial–interglacial fluctuations in the thickness of ice overburden. Both suggestions are speculative but, on a geological timescale, the periods between eruptions (i.e. few tens of thousands of years to >100 ka: Smellie et al. 2008) are probably rather too short to be related to large-scale plate effects. The other suggestion is a climate modulation similar to that documented for volcanic eruptions in Iceland, which are also situated above upwelling mantle (e.g. Jull and McKenzie 1996). The characteristically long repose periods between JRIVG eruptions more closely resemble a climate modulation as they are on a scale that resembles Milankovitch cyclicity (i.e. peaks at 26, 41 and c. 125 ka). Further (and more precise) dating of eruptions may help to verify this suggestion.

The JRIVG covers an area of 7000 km^2 and has an estimated volume of erupted products of >4500 km^3 (Smellie et al. 2013a). It is dominated by the huge volcanic shield of Mount Haddington, with a basal diameter of 60–80 km. Despite a summit elevation of just 1600 m, the volcano has a substantial root extending to 5 km with a volume below sea-level of c. 3000 km^3. Volcano spreading (i.e. settling and accompanying lateral displacement) has deformed the soft underlying Cretaceous sediments (Oehler et al. 2005; Jordan et al. 2009). The resulting deformation is particularly evident on the west side of the island facing Graham Land, where beds are steeply dipping on the coast, with thrust faults and anticlines, and give way inland to less steeply dipping younger Cretaceous strata. The Mio-Pliocene volcano-related deformation may be an alternative explanation for the deformation of the Cretaceous strata, which are currently interpreted using a model of progressive syndepositional basin subsidence (Whitham and Marshall 1988; Hathway 2000).

Mount Haddington and most of the satellite centres in the JRIVG are constructed principally of the products of multiple pāhoehoe lava-fed deltas, many of which were very voluminous (a few tens of cubic kilometres to 100 km^3) and correspond to flood lavas (Smellie et al. 2013a). By comparison with the most voluminous historical effusive flood lava eruption, Laki (Iceland) in 1873–74, which produced c. 15 km^3 of magma over 14 months, the individual and much larger effusive eruptions on Mount Haddington must have taken years to decades to be emplaced. This is comparable with flood basalt eruptions in LIPs and, similarly, the Mount Haddington eruptions were followed by substantial periods of repose (cf. Self et al. 1997; Muirhead et al. 2014), although the JRIVG is not a LIP.

Most of the JRIVG eruptions occurred in a glacial setting, and their features have been used to significantly improve our knowledge of glaciovolcanic edifice construction and eruptive processes (Skilling 1994, 2002; Smellie 2006; Smellie et al. 2008; Calabozo et al. 2015; Nehyba and Nývlt 2015). The JRIVG has also been the principal source of information about critical parameters of the northern part of the terrestrial Antarctic Peninsula Ice Sheet (APIS) from 6.2 Ma. The evidence indicates that the ice cover was comparatively thin (a few hundred metres) and wet-based (polythermal) throughout its history, apart from during the last glacial when it was much thicker and probably cold-based (Hambrey *et al.* 2008; Smellie *et al.* 2008, 2009; Nelson *et al.* 2009; Davies *et al.* 2012, 2013). Nývlt *et al.* (2011) suggested that the evidence for wet-based conditions may be an artefact caused by thermal activity of the JRIVG. However, active volcanism is only likely to change the climate-based thermal regime of a glacial cover in a comparatively narrow region on the summit of a volcano, within a zone approximately three times the width of any associated magma chamber (Smellie and Edwards 2016; Smellie 2018). Mount Haddington lacks a crustal magma chamber (Jordan *et al.* 2009), so the precise width of the thermally affected zone is uncertain but unlikely to be unusually large. Moreover, the evidence for wet-based conditions comes from volcanic outcrops situated variably 20–40 km from the Mount Haddington summit (i.e. well beyond any plausible volcano-related thermal effects). Therefore, it is highly unlikely that the thermal regime of the glacial cover on James Ross Island (away from the summit region) was influenced by the volcanism.

The thickest ice conditions (*c.* 700–800 m) appear to have occurred during the late Quaternary. The thickest ice was probably associated with the Last Glacial Maximum (LGM), as indicated by a glaciovolcanic lava-fed delta >650 m thick at <0.08 Ma and erratics on the 600 m asl (above sea-level) summit of Terrapin Hill, a Quaternary (*c.* 0.66 Ma) tuff cone on the north side of James Ross Island. Such an ice sheet would have been capable of largely drowning topography in the northern Antarctic Peninsula (Smellie *et al.* 2008, 2009), although it was much thinner than the 'giant', 2 km-thick APIS envisaged by others (Denton *et al.* 1991; Denton and Hughes 2002; cf. Nývlt *et al.* 2011). Moreover, the ice was probably comparatively resistant to global warming, up to mean global temperatures 4.5° above present (Smellie *et al.* 2009). Thus, the Peninsula region was ice-poor rather than ice-free during interglacials (Johnson *et al.* 2009; Salzmann *et al.* 2011). This is probably because, although the APIS comprises just 3% of the grounded ice-sheet area of the Antarctic continent, precipitation is three–four times greater than elsewhere and it receives *c.* 13% of the total mass input (Drewry and Morris 1992; van Lipzig *et al.* 2004).

Monogenetic glaciovolcanism associated with post-subduction slab-window formation

Following the progressive collision of offset sections of the Phoenix–Antarctic spreading centre with the Antarctic Peninsula trench from Eocene time, subduction ceased but a northerly propagating slab window opened up due to the continued sinking of the leading oceanic plate (the Phoenix or Aluk slab: Hole *et al.* 1995). This has been considered as the trigger for the uprise and decompression melting of pristine mantle drawn up from beneath the slab (see also Hole 2021). The resulting small-volume melts were erupted at the surface as numerous monogenetic volcanoes forming extensive volcanic fields (Fig. 9).

Eruptions took place from 7.7 Ma (Smellie 1999) and rapidly became widespread in southern areas. They were joined by volcanism in northern Antarctic Peninsula from *c.* 4 Ma, with a volcano-free gap of *c.* 500 km between the two outcrop regions. Despite the opening up of slab windows at progressively younger times in a northerly direction, there is no similar progression of eruptive ages seen in the slab-window-related volcanism, an enigma that still requires an explanation (Hole 2021; Smellie and Hole 2021). The edifices vary from a few hundred metres to *c.* 7 km in diameter and are *c.* 100–800 m high. They range from rare isolated scoria cones to (mainly) glaciovolcanic tuyas and a few sheet-like sequences (Smellie and Hole 2021). Most crop out in monogenetic volcanic fields with areas ranging from 1400 to >2000 km^2 (possibly >7000 km^2 in one example: Smellie 1999).

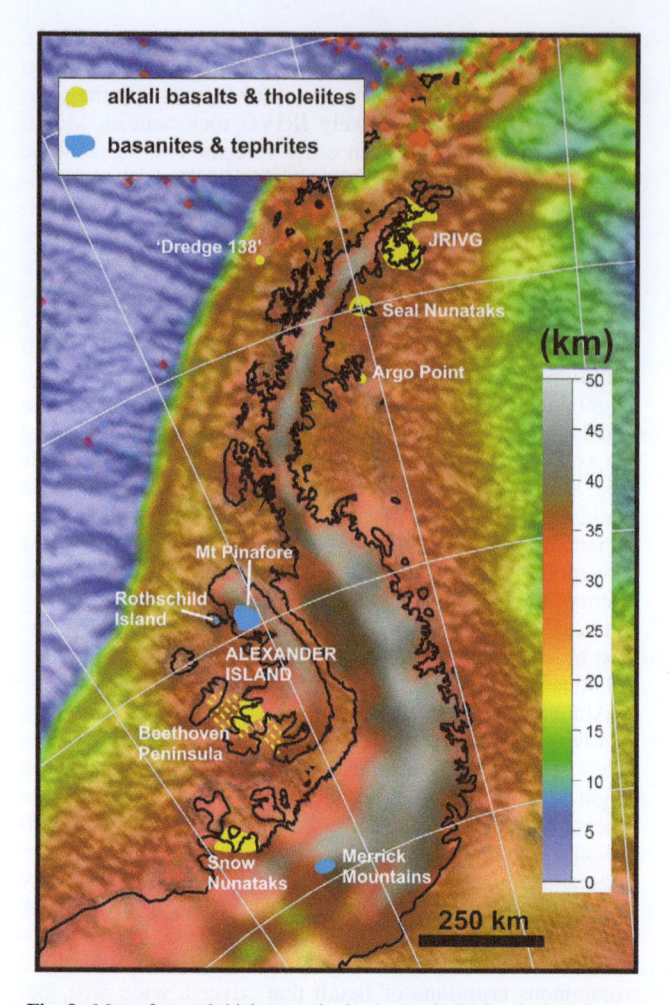

Fig. 9. Map of crustal thicknesses in the Antarctic Peninsula, based on gravity inversion (image: Nick Kusznir, reproduced with permission). The locations of alkaline slab-window basalts and similar-age compositionally indistinguishable back-arc basalts of the JRIVG are also shown. Note the prominent correlation between crustal thickness and compositions of the basalts. 'Dredge 138' indicates the location of a small submarine volcano formed of slab-window basalt (Hole and Larter 1993).

From the distribution of the outcrops, it is clear that, apart from two tiny isolated outcrops in Merrick Mountains (central southern Palmer Land), all of the Neogene alkaline volcanic rocks, including the compositionally similar JRIVG, occur on the flanks of the Antarctic Peninsula where crustal thicknesses can be expected to be relatively thin (Fig. 9) (Renner *et al.* 1985; Kusznir *et al.* 2018; Pappa *et al.* 2019). The implication is that greater crustal thicknesses generally prevented the alkaline magmas from being erupted. There is also a significant correlation between the erupted compositions and crustal thicknesses: viz. the more undersaturated, more alkaline lavas in northern Alexander Island and Merrick Mountains, composed of tephrites and basanites with locally abundant ultramafic nodules, crop out above thicker crust; the more common outcrops of lavas with less undersaturated compositions, including the JRIVG, composed of olivine basalts, alkali basalts and tholeiites generally lacking nodules, crop out above much thinner crust (Fig. 9). The more alkaline group was also formed by lower degrees of melting, consistent with its smaller erupted volumes (Hole 1988, 1990; Smellie 1999). The Moho would have acted as a density barrier where the magmas were temporarily stored and underwent the small degrees of fractionation observed. Moreover, the alkali basalts, olivine basalts and tholeiites consistently have slightly higher $^{87}Sr/^{86}Sr$ ratios and lower $^{143}Nd/^{144}Nd$ ratios than the basanites, interpreted as the incorporation of a variable subduction-zone component in the mantle wedge traversed by the magmas (Hole 1988, 1990). Moho depth differs by c. 10–15 km between the occurrences of the two magma types (Fig. 9) and the incorporation of the subduction component would be enhanced by migration over a greater vertical distance. Thus, greater degrees of melting (and erupted volumes) for the alkali/olivine basalt and tholeiitic occurrences may be explained by the greater vertical distance travelled to reach the Moho, and the correspondingly shallower equilibration pressures. It is therefore suggested that the high-relief sub-crustal topography of the Antarctic Peninsula exerted an important control on the location, compositions and erupted volumes of the Neogene slab-window volcanism.

Practically all of the slab-window-related centres were erupted in association with a glacial cover. Together with erupted units in the JRIVG, they have been used to advance our knowledge of glaciovolcanism generally, including: the relationship between sequence types and ice thickness (Smellie and Skilling 1994); the characteristics of sheet-like sequences and their mode of formation (Smellie et al. 1993; Smellie 2008); improved lithofacies-based models for tuya (Surtseyan) edifice construction (Skilling 1994, 2002; Smellie and Hole 1997); and the hydraulics of tuya eruptions (Smellie 2006). Palaeoenvironmentally, the outcrops have also been a major source of information on the critical parameters and evolution of the terrestrial APIS. As a result, more is known about the history and evolution of the terrestrial APIS from Late Miocene time (<7.7 Ma) than is known for most other pre-Quaternary ice sheets on Earth (e.g. Smellie et al. 2006, 2008, 2009; Hambrey et al. 2008; Nelson et al. 2009). The studies have demonstrated that a relatively thin (<c. 400 m), wet-based draping ice cover was present throughout the Antarctic Peninsula most of the time. The APIS therefore mainly resembled an ice field, although there were also occasional episodes of relatively thick ice (e.g. at c. 5.5 and 3 Ma, and LGM). It is also noticeable that typical ice thicknesses in the Antarctic Peninsula were significantly thicker (mainly 200–400 m, and at times much thicker (up to 750–800 m): Smellie et al. 2009) than inferred for the East Antarctic Ice Sheet in northern Victoria Land during the Late Miocene–Pliocene (mostly <200 m: Smellie et al. 2011b). This is probably due to the warmer and wetter conditions causing greater precipitation in the narrow high peninsula, with its more northerly position flanked by large oceanic masses, and greater exposure to large numbers of cyclonic weather systems tracking across it (Turner et al. 1998).

Alkaline continental rift volcanism (West Antarctic Rift System): influence of the West Antarctic Ice Sheet

Antarctica is also host to one of the world's great continental rift zones, known as the West Antarctic Rift System (WARS). It is c. 3000 km long and 750–1000 km wide, comparable with the East African Rift and the Basin and Range province. Conversely, its floor is substantially lower and large areas of the WARS are well below sea-level (down to c. 2555 m bsl (below sea-level): LeMasurier 2008). Although the earliest age of volcanism in the rift (Eocene: Wilch and McIntosh 2000; Smellie and Rocchi 2021; Wilch et al. 2021) might suggest that the WARS formed primarily during Cenozoic rifting, much of West Antarctica was affected by a much more significant precursor episode of extension and crustal thinning associated with the separation of New Zealand from Marie Byrd Land during the Late Cretaceous (from c. 105 Ma: Lawver and Gahagan 1994; Siddoway 2008). Perhaps as a result, the West Antarctic Plateau, a postulated high topographical area that may have formed the Pacific flank of the long-lived (Devonian–Triassic) Beacon Supergroup sedimentary basin(s), subsided close to or somewhat below sea-level (Bialas et al. 2007; Fitzgerald et al. 2007; but see also Elliot 2013). A regionally extensive, generally low-relief erosion surface, known as the West Antarctic Erosion Surface (WAES), was created in Marie Byrd Land and Ellsworth Land (Jones Mountains), principally between c. 85 and 75 Ma (LeMasurier and Landis 1996). A likely temporal equivalent is also present in New Zealand (the Waipounamu Erosion Surface). The extensive crustal thinning during this period is often described as amagmatic but alkaline dykes and plutons with ages of 107–95 Ma are present in Marie Byrd Land and may have fed volcanic activity, now removed by erosion associated with the formation of the WAES. The Cretaceous alkaline magmatism has been assigned a plume origin linked to the break-up episode (Storey et al. 1999). The presence of a weak marine mass extinction at the Cenomanian–Turonian boundary (c. 90 Ma: Ernst and Youbi 2017) provides possible support for suggesting that volcanic activity was associated with the Late Cretaceous intraplate magmatism in Marie Byrd Land (and Zealandia: Hoernle et al. 2020), since volcanic emissions of carbon dioxide and associated climate change have been identified as likely culprits. However, the Antarctic magmatism also broadly coincided with several LIP events (e.g. Caribbean–Colombian, Kerguelen Plateau and Madagascar: Storey et al. 2013), and they probably played the leading role.

Extension and rifting were renewed between c. 50 Ma and present, although the episode involved significantly less extension than during the Late Cretaceous and was largely confined to the western Ross Sea, adjacent to the Transantarctic Mountains. The earliest alkaline magmatism associated with the Cenozoic extension consisted of alkaline plutons and dykes of mainly Eocene age. They included a single pluton in eastern Marie Byrd Land (the Dorrel Rock gabbro close to Mount Murphy: 35–34 Ma); volcanism also occurred at 37 Ma (at Mount Petras: Rocchi et al. 2006; Wilch et al. 2021). However, the episode is best represented by numerous plutons and dykes in northern Victoria Land known as the Meander Intrusive Group (52–18 Ma: Rocchi et al. 2002; Ross et al. 2002; Smellie and Rocchi 2021). Emplacement of the Meander Intrusive Group coincided in time with oceanic crust formation in the Adare Basin and continental rifting in the Northern Basin offshore of northern Victoria Land (c. 43–26 Ma: Cande and Stock 2006), which implies a genetic link with tectonism outside of the WARS (i.e. seafloor spreading between Australia and Antarctica: Davey et al. 2016). It is generally unknown if the Meander Intrusive Group plutonic episode also fed volcanic activity as erosion has removed virtually the entire record. However, possible evidence for an associated volcanic suite is present at one locality in northern Victoria Land, Vulcan Hills (between Mount Melbourne and Mount Overlord: Fig. 1), and more widespread associated volcanic activity is inferred (Smellie and Rocchi 2021). The intrusive activity overlapped in time with alkaline volcanism, which began at c. 37 Ma in Marie Byrd Land and became voluminous after c. 14 Ma, and which constructed numerous large, shield-like, composite central volcanoes throughout the WARS, (Hamilton 1972; LeMasurier and Thomson 1990; Smellie et al. 2011a, b; LeMasurier 2013; Smellie and Rocchi 2021; Smellie and Martin 2021; Wilch et al. 2021). The unusual terminology used to describe these large polygenetic volcanoes (discussed by Wilch et al. 2021) reflects the fact that, despite being formed of alternating massive (effusive) and fragmental products and developing evolved compositions similar to stratovolcanoes (i.e. composite volcanoes), the individual volcanoes generally have relatively low gradients (c. 5°–10°) like shield volcanoes. The low gradients reflect a predominantly effusive origin, but with emplacement overwhelmingly as lava-fed deltas, particularly in Victoria Land (Smellie *et al.* 2011*a*, 2014; and unpublished information of the author). The same probably also applies to volcanoes in Marie Byrd Land (at Mount Murphy and Crary Mountains, at least: Wilch *et al.* 2021; and unpublished information of the author) but the erosional dissection of most of the volcanoes is minor and their internal structure is largely unknown. The lava-fed delta sequences are dominated by alternating massive subaerial lava and subaqueous fragmental (non-explosively-formed) tuff breccia lithofacies, in cogenetic couplets (e.g. LeMasurier 2002; Smellie *et al.* 2011*a, b*, 2013*b*). By contrast, some of the central volcanoes in Victoria Land are simple stratovolcanoes, with steep flank gradients (e.g. Mount Erebus, Mount Discovery, Mount Melbourne, Mount Overlord, Mount Lubbock and Mount Harcourt); most of those in northern Victoria Land are unusually small.

Volcanism in the WARS is widespread throughout Marie Byrd Land and Victoria Land. It is likely that a significant amount of the volcanic activity in Marie Byrd Land is hidden by the West Antarctic Ice Sheet, as numerous subglacial volcanoes have been tentatively identified aerogeophysically (Fig. 2) (e.g. Behrendt *et al.* 1994, 2002; van Wyk de Vries *et al.* 2018; but see also Quartini *et al.* 2021). Behrendt *et al.* (1994) speculated that the volcanism in the WARS, which they considered included comparatively large centres (2 km thick and >10 km in basal diameter), potentially had an erupted volume of 10^6 km^3, which is similar to areas of flood basalts (e.g. in LIPs). However, given a lifetime of >30 myr for the volcanism, the time-integrated magma discharge rate is far lower than in LIPs. Moreover, the absence of basalt clasts of appropriate age in subglacial sediments from the WARS suggests that subglacial volcanism is either not widespread or is covered by a sedimentary drape in the region sampled (Vogel *et al.* 2006). The Victoria Land volcanism is included in the McMurdo Volcanic Group, which is divided into several volcanic provinces and volcanic fields (Smellie and Rocchi 2021) (Fig. 10), whereas that in Marie Byrd Land and Ellsworth Land is included in the Marie Byrd Land Volcanic Group, with two subdivisions (Marie Byrd Land and Thurston Island volcanic provinces: Wilch *et al.* 2021). There are also several active (dormant) and potentially active volcanoes within the WARS, including Mount Takahe, Mount Berlin and Mount Waesche in Marie Byrd Land; Mount Erebus in southern Victoria Land; and Mount Melbourne, Mount Rittmann and The Pleiades in northern Victoria Land, together with candidate subglacial volcanoes inferred to be active (Smellie and Rocchi 2021; Dunbar *et al.* 2021; Quartini *et al.* 2021; Sims *et al.* 2021; Smellie and Martin 2021; see also Geyer *et al.* 2021).

Two principal contrasting origins are inferred for volcanoes in the WARS (Panter 2021; Panter *et al.* 2021*b*). Those in Marie Byrd Land (Marie Byrd Land Volcanic Group: Wilch *et al.* 2021) and in southern Victoria Land (Erebus Volcanic Province: Smellie and Martin 2021) may be related to mantle plumes. Many of the individual volcanoes are very large. They include Mount Erebus, which is 40 km in basal diameter and rises to 3794 m, and Mount Sidley, the highest volcano in Antarctica, rising to 4181 m. Additionally, Toney Mountain may be Antarctica's tallest volcano (c. 6595 m; it has a summit elevation of 3595 m asl and rests on basement at 3000 m bsl: LeMasurier 2013). The plume model is supported by: (a) a radial distribution of volcano ages and associated synvolcanic updoming in Marie Byrd Land; (b) the three-fold radial symmetry of major edifices in the Erebus Volcanic Province; (c) seismic tomography studies that show distinct low-velocity shear-wave anomalies below Ross Island and Marie Byrd Land extending down to 800–1200 km (Fig. 11); and (d)

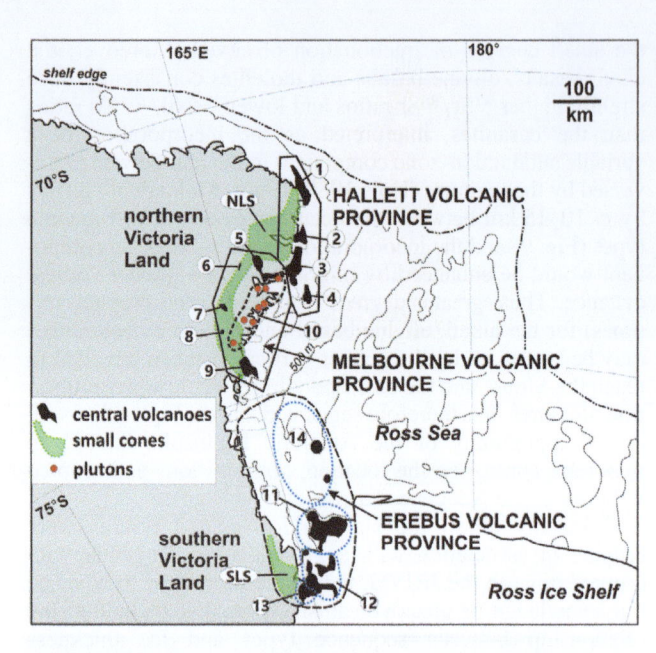

Fig. 10. Map showing the distribution of the McMurdo Volcanic Group (excluding the Scott Glacier Volcanic Field of the southern Transantarctic Mountains). The plutons shown are contained in the Meander Intrusive Group. Note also the widespread distribution of small pyroclastic cones (SLS, Southern Local Suite (volcanic field); NLS, Northern Local Suite (volcanic field)). See Smellie and Martin (2021) and Smellie and Rocchi (2021) for descriptions of the numbered volcanic fields.

spatially variable high heat flow; as well as petrological studies that have documented ocean island basalt compositional affinities and high U/Pb ratios (HIMU (high ^{238}U/^{204}Pb = μ): LeMasurier and Rex 1982, 1989; Hansen *et al.* 2014; Schroeder *et al.* 2014; Brenn *et al.* 2017; Seroussi *et al.* 2017; Phillips *et al.* 2018; Panter 2021; Martin *et al.* 2021; Panter *et al.* 2021*b*; Quartini *et al.* 2021). By contrast and despite essentially identical compositions throughout the WARS, it has been suggested that the origin of volcanism in northern Victoria Land is better explained by melting and craton-directed edge flow of a prominent shallow mantle thermal anomaly (Fig. 11) (Graw *et al.* 2016; Panter *et al.* 2018; Rocchi and Smellie 2021). The low-wave-speed shallow anomaly is one of several identified in the Balleny–Tasman Belt, a strike-slip zone extending between Victoria Land and Australia that may be an incipient (diffuse) intraplate boundary (Danesi and Morelli 2000; Storti *et al.* 2007). The northern Victoria Land volcanoes are further distinguished into two types: inland stratovolcanoes, which are few, relatively small and little eroded, including Mount Overlord which is only 7 km in basal diameter and rises just 800 m above the surrounding land surface; and numerous large coalesced shield volcanoes that form a prominent linear zone along the coast. Individual volcanoes in the latter are commonly 25–>35 km in original diameter (although much modified, mainly by marine erosion) and rise to 2 km (Smellie and Rocchi 2021). Although stratovolcanoes are also present in the coastal zone, they are rare and small, and just two are known (Mount Lubbock and Mount Harcourt, both with basal diameters of c. 15 km). The coastal volcanism probably erupted from north–south-trending faults that are conjugate with a series of prominent NW–SE-trending strike-slip faults genetically linked to the Balleny–Tasman strike-slip belt (Salvini *et al.* 1997; Rossetti *et al.* 2006; Rocchi and Smellie 2021). The coastal volcanism has also been modelled to show that Cenozoic extension of crust previously thinned during Cretaceous extension would cause necking in a narrow zone at the junction between thick East Antarctic crust and much thinner

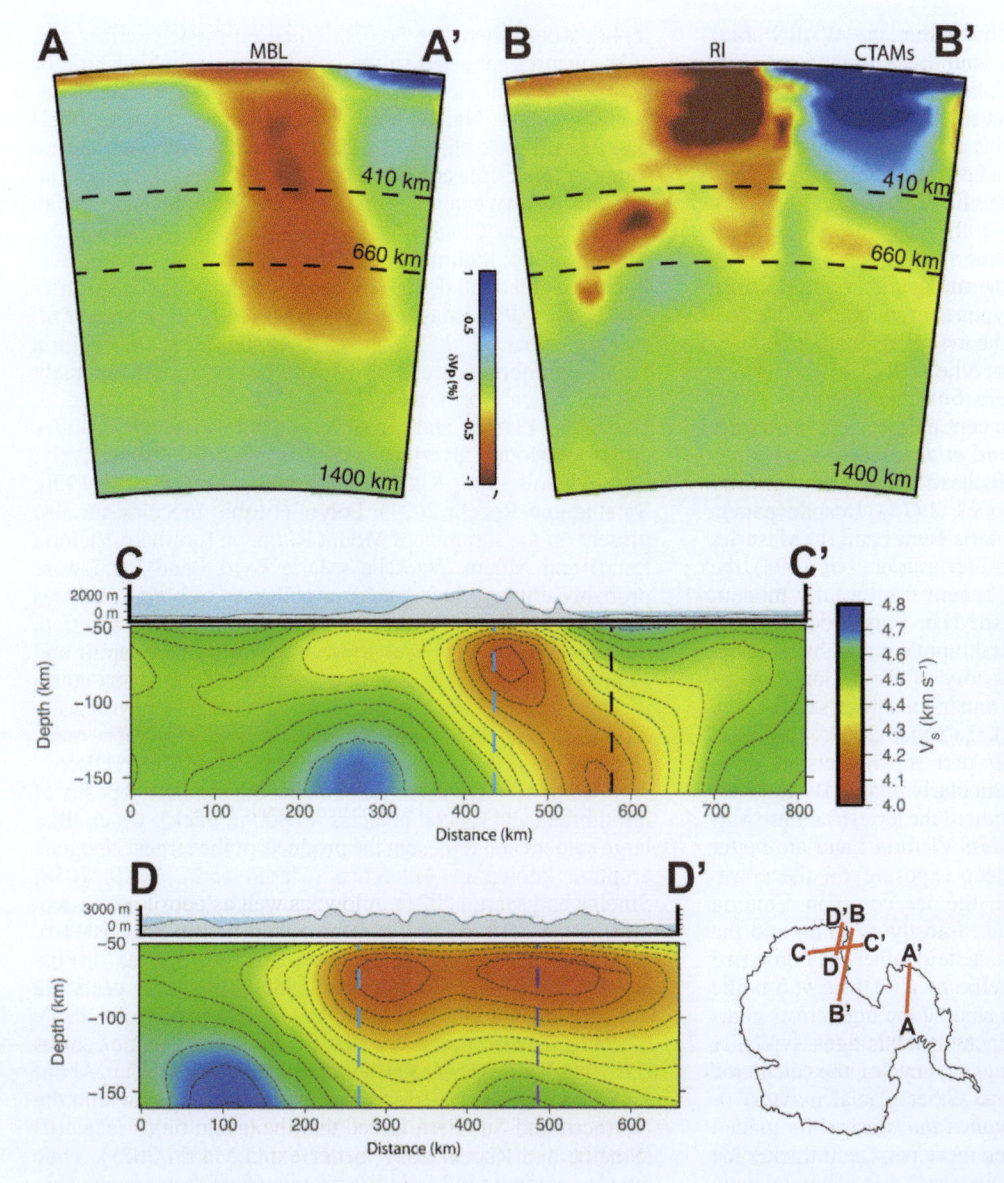

Fig. 11. Seismic tomographic profiles through Marie Byrd Land (MBL: A–A′), Ross Island (RI: B–B′) and northern Victoria Land (C–C′ and D–D′) showing the shear-wave structure. Zones with low velocity (orange and yellow colours) are interpreted as being due to mantle upwelling with associated melting. Note the different vertical scales between sections A–A′ and B–B′ v. C–C′ and D–D′. The basal boundary for upwelling is much shallower beneath northern Victoria Land compared with southern Victoria Land and Marie Byrd Land. Diagrams are from Hansen *et al.* (2014) and Graw *et al.* (2016). See also Phillips *et al.* (2018). CTAMs, central Transantarctic Mountains.

West Antarctic crust (Huerta and Harry 2007), and in which melting became focused (Rocchi *et al.* 2002; Panter *et al.* 2018). The coastal volcanic belt thus marks the junction between thick and thin crust, and is the structural margin of East Antarctica.

Volcanism in the coastal belt of northern Victoria Land is much more voluminous than that inland and it differs in broad geochemical characteristics. It is dominated by mafic compositions, whereas the inland volcanoes include a greater proportion of differentiated compositions. The latter also evolved to high-silica rhyolites that mostly reached peralkalinity consistent with the presence of substantial cooling and fractionating crustal magma chambers (Rocchi and Smellie 2021). The differences may be ascribed to variable magmatic fluxes, with the coastal volcanism experiencing a much higher flux than the inland volcanism. Thus, magma flux along the coast kept the crust hot and the magmas remained as gabbros for a much longer period, whereas the lower flux inland led to isolated magma chambers, more rapid cooling and fractionation. The coastal volcanic sequences have also been likened broadly to proto-seaward-dipping reflectors formed by volcanism at a 'non-volcanic' rifted margin (*sensu* Menzies *et al.* 2002), but which stalled essentially at a pre-rift stage (Rocchi and Smellie 2021). An oceanic basin did not develop offshore of northern Victoria Land, although there is evidence for continental rupture that may predate much of the onshore volcanism (Davey *et al.* 2016). Thus, the volcanism remains in a subaerial position and did not subside below sea-level.

Inland volcanoes within the WARS are largely uneroded and they typically show only the subaerial products of the latest eruptions (Hamilton 1972; Smellie and Rocchi 2021; Wilch *et al.* 2021). By contrast, the coastal volcanoes are often deeply eroded and contain very detailed environmental histories, which are the focus of recent and ongoing investigations (Wilch and McIntosh 2002; Smellie *et al.* 2011*a, b*, 2014; Smellie and Rocchi 2021; Smellie and Martin 2021; Wilch *et al.* 2021). The volcanism coincided with the inception and development of the Antarctic Ice Sheet and glaciovolcanic sequences are prominent. The information the outcrops contain provides important environmental data for a period well beyond instrumental records, including the growth, decay and critical parameters of the ice sheets in Antarctica; interglacial environments are much less documented. It may be that the impingement of a mantle plume below Marie Byrd Land from *c.* 37 Ma (at least), causing uplift that raised large areas a few hundred metres above sea-level, facilitated the initiation and growth of the West Antarctic Ice Sheet (Smellie 2020*b*; cf. Wilson *et al.* 2013). Moreover, geothermal heat currently associated with the volcanism in Marie Byrd Land may be having an influence on drawdown of the West Antarctic Ice Sheet (Schroeder *et al.* 2014; Seroussi *et al.* 2017; Loose *et al.* 2018; Quartini *et al.* 2021). Some eruptions from Antarctic volcanoes (overwhelmingly found within the WARS) may also influence regional climate and have the potential to accelerate Antarctica out of glacials (McConnell et al. 2017).

The large polygenetic centres in the WARS are overwhelmingly effusive and exposed sections are dominated by multiple superimposed glaciovolcanic 'a'ā lava-fed deltas, which gives them their shield-like profiles (Smellie et al. 2011a, b, 2013b). Although mainly mafic (basanite), they include rare evolved examples (tephriphonolite and trachyte: Wilch et al. 2021; and unpublished information of the author and Sergio Rocchi). Other glaciovolcanic sequence types are rare but they include sheet-like sequences, including the first plausible felsic sheet-like sequences to be described (Smellie et al. 2011a). The volcanic sequences in the WARS were the first to demonstrate the existence of an extensive pre-Quaternary ice sheet in Antarctica (Craddock et al. 1964; Rutford et al. 1968, 1972; Hamilton 1972; LeMasurier 1972), well ahead of the earliest drilling discoveries by DSDP (Hayes et al. 1975). Despite papers describing the glacial setting in Marie Byrd Land (LeMasurier 1972; LeMasurier and Rex 1982; LeMasurier et al. 1994), the investigations were too early to benefit from a fully modern understanding of glaciovolcanism (i.e. post-2000; see the summary of palaeoenvironmental applications by Smellie 2018), and the environmental history of the region remains generally poorly described and understood (but see Smellie 2001; Wilch and McIntosh 2000, 2002, 2007; Haywood et al. 2009; Wilch et al. 2021). In part, this reflects the often very poor exposure available, particularly inland. By contrast, critical parameters and the evolution of the terrestrial East Antarctic Ice Sheet (EAIS) in northern Victoria Land are better known due to the often excellent exposure on the coast. Between c. 12 Ma and present, the ice cover in Victoria Land comprised relatively thin ice (mostly <200 m) and the region would have resembled an ice field rather than drowned by a much thicker ice sheet (Smellie et al. 2011a, b; Smellie and Rocchi 2021). However, because of the high errors characteristic of the published $^{40}Ar/^{39}Ar$ isotopic ages (typically 40–80 ka, i.e. much more than a glacial cycle), the calculated thicknesses cannot be assigned to either glacial maxima or minima. They are regarded as 'typical thicknesses' for glacial conditions during the period since more persistent thicker ice would have left a prominent and unmistakable glaciovolcanic record. Much thicker (and wet-based) ice must have existed at times: for example, to deposit abraded basement erratics on top of volcanic deposits at Harrow Peaks (400 m asl: Smellie et al. 2018); up to an elevation of 500 m asl on northern Adare Peninsula (Hamilton 1972; Johnson et al. 2008); and up to c. 900 m asl on Minna Hook (unpublished information of the author). Moreover, the distribution and variable ages of localities showing evidence for warm- and cold-based ice coeval with eruptions during the late Miocene and Pliocene suggest that the thermal regime of the EAIS was polythermal overall: that is, a patchwork mosaic of wet-based ice and ice frozen to its bed (Smellie et al. 2014). The prevailing paradigm for EAIS evolution that has existed since the early 1980s states that the thermal regime underwent a change from wet-based dynamic ice to cold-based and stable in a single unidirectional step either at 14.5 Ma or c. 2.5–3 Ma (Webb and Harwood 1991; Wilson 1995; Lewis et al. 2007; Barrett 2013). However, judging ice thermal regime from sedimentary records is extremely challenging (Hambrey and Glasser 2012), and only drill-core material is available since appropriate-age outcrops are absent onshore in Victoria Land. By contrast, thermal regime is often straightforward to determine from glaciovolcanic sequences (Smellie 2018). Thus, the new paradigm involving polythermal ice, based on the volcanic evidence, is a significant change.

Pyroclastic products are uncommon in the WARS, probably because most eruptions coincided with an ice cover. Any tephra would therefore be distributed on the ice surface and subsequently advected to the ocean, leaving behind little or no onshore geological record apart from in ice cores (see Dunbar et al. 2021; Narcisi and Petit 2021). In most cases, their presence in rock outcrop probably signifies eruptions during ice-poor or ice-free conditions, presumably during interglacial periods. Known examples in Marie Byrd Land include Plinian fall deposits at Chang Peak (close to Mount Waesche) and Mount Sidley; ignimbrites on Mount Sidley and Mount Berlin; and welded fall deposits at several volcanoes (LeMasurier and Rex 1989; LeMasurier and Kawachi 1990; Panter et al. 1994; Dunbar et al. 2021; Wilch et al. 2021). In Victoria Land, ignimbrites occur on Mount Overlord and, uniquely in Antarctica, form an extensive plateau-like outcrop at Deception Plateau and possibly at Malta Plateau (localities close to Mount Overlord and The Pleiades, respectively: Fig. 1) (Noll 1985; Kyle 1990; Schmidt-Thomé et al. 1990; Smellie and Rocchi 2021). Polymict lithic breccias are also present on the summits of Mount Rittmann (northern Victoria Land) and Mount Waesche (Marie Byrd Land), and were probably erupted during caldera collapses at both volcanoes (Smellie and Rocchi 2021; Dunbar et al. 2021). In addition, the presence of numerous loose trachyte pumice lapilli and large broken bombs scattered on several of the scoria cones in The Pleiades attests to a relatively recent felsic pyroclastic eruption (Kyle 1982; Smellie and Rocchi 2021). The mid-Miocene (c. 12 Ma) Mason Spur volcano (near Mount Discovery, southern Victoria Land) contains the thickest deposits of ignimbrites and related breccias (c. 800 m thick), which fill a large caldera and represent the products of the largest Neogene eruption known in Antarctica (Martin et al. 2010, 2018; Smellie and Martin 2021). A block and ash deposit is also present on the northern Hallett Peninsula (Smellie et al. 2011a). Englacial pyroclastic layers (cryotephras) erupted in the WARS during the last few hundreds of thousand years are described by Dunbar et al. (2021) and Narcisi and Petit (2021).

There are also numerous small mafic monogenetic centres (i.e. small volcanoes, sensu White and Ross 2011). About 130 are known in Victoria Land, and are grouped within the Northern and Southern Local Suite volcanic fields (Fig. 10) (Smellie and Rocchi 2021; Smellie and Martin 2021). They are also common in Marie Byrd Land where they are parasitic or flank vents on the numerous large central volcanoes (LeMasurier 1972). They overwhelmingly comprise small scoria cones but rare tuff cones have been described at Mount Petras, Mount Murphy, Mount Siple and on the Hobbs Coast (Marie Byrd Land: LeMasurier et al. 1994; Wilch and McIntosh 2000, 2007; Smellie 2001; Wilch et al. 2021), and a few examples are present in the Mount Melbourne Volcanic Field (northern Victoria Land: Wörner and Viereck 1989; Giordano et al. 2012). The latter includes a small tuff cone north of Mount Melbourne, which erupted during the Marine Isotope Stage 16 (MIS) glacial (c. 640 ka) and was interpreted as glaciovolcanic. It erupted under relatively thin, cold-based ice (Smellie et al. 2018). In addition, a few tuyas have been postulated near Mount Murphy, Mount Takahe, Mount Berlin and the Hobbs Coast (Marie Byrd Land); in the Hudson Mountains (Ellsworth Land); and at Shield Nunatak (southern flank of Mount Melbourne, northern Victoria Land), but they are rare generally in the WARS (Wörner and Viereck 1989; Smellie 2001; Giordano et al. 2012; Wilch et al. 2021). The rarity of tuyas in the WARS contrasts with their abundance in the Antarctic Peninsula (cf. Skilling 1994; Smellie and Hole 1997).

Finally, the radiating age trends of magmatism in the WARS together with the reconstructed morphology of the subvolcanic West Antarctic erosion surface have been used to infer domical uplift of the central WARS in Marie Byrd Land, usually attributed to the influence of an underlying large mantle plume during the Neogene (LeMasurier and Rex 1989; LeMasurier and Landis 1996; LeMasurier 2008; but see Paulsen and Wilson 2010 for an alternative view involving transmitted far-field plate-tectonic effects). It is less well known, however, that the magmatism can also be used to broadly constrain vertical movements of the WARS rift margin in the Transantarctic Mountains of Victoria Land. For example, using reasoning similar to Rocchi et al. (2006) for the Dorrel Rock gabbro in Marie Byrd Land, it is inferred that, for plutons in the Meander Intrusive Group to be exposed at the surface in northern Victoria Land, at least 3 km of overburden has to be removed since Eocene–Oligocene time, presumably with accompanying uplift of similar magnitude. Possible support for this is provided by thermochronological studies in northern Victoria Land. For example, Prenzel et al. (2018) suggested uplift and exhumation of c. 3.5 km (ranging from <2 to c. 4.8 km) caused by erosion since Oligocene time ($\leq c$. 35 Ma), possibly within a period of just 5 myr, and Olivetti et al. (2016) identified an episode of rapid uplift at 26 Ma. The additional presence of younger subaerially erupted volcanic rocks (i.e. scoria cones and glaciovolcanic 'aʻā lava-fed deltas) at sea-level throughout the coastal volcanic belt, both in northern Victoria Land and in the Erebus Volcanic Province of southern Victoria Land, implies that no uplift or possibly some currently unquantified collapse of the rift shoulder has occurred during the last c. 12 myr, at least (Smellie et al. 2011b; Rocchi and Smellie 2021; Smellie and Martin 2021). However, whilst the observation is consistent with an absence of a plume below northern Victoria Land, it is apparently at odds with the supposed presence of a plume (and associated doming effects) below the Erebus Volcanic Province. However, Ross Island has a 'root' extending to c. 2.2 km caused by gravitational subsidence of the volcanic pile into weak Mio-Pliocene strata of the Victoria Land Basin (Aitken et al. 2012), which may have compensated for plume-related uplift. Similar reasoning may apply to other volcanoes in the Erebus Volcanic Province. Gravitational settling is unlikely to have occurred below the northern Victoria Land volcanoes as they are founded on strong Paleozoic basement rocks (Hamilton 1972).

Isolated within-plate volcanism of enigmatic origin

Small, Early Miocene (c. 25–19 Ma) mafic volcanic centres with tholeiitic and alkali basalt compositions are present inboard of the WARS and high on its southern flank, at Mount Early, Sheridan Bluff and a subglacial site 200 km from the South Pole (Fig. 1) (Licht et al. 2018; Panter et al. 2021a; Smellie et al. 2021). They are the most southerly volcanic occurrences on Earth. Mount Early is a small glaciovolcanic tuff cone and pillow mound, whereas Sheridan Bluff is a small shield volcano with an original basal diameter of c. 6 km (Smellie et al. 2021). Studies underway are providing significant palaeoenvironmental information for the Early Miocene. These isolated volcanoes may have an origin related to extensional stresses caused by the WARS, or to flexural bending linked to motion between East and West Antarctica (Granot and Dyment 2018). However, formation of the magmas is probably distinct from those in the WARS, and may be a result of the detachment and sinking of lithosphere into the convecting mantle beneath the East Antarctic Craton (Panter et al. 2021a).

Elsewhere, volcanism with an enigmatic origin is also present at Gaussberg, a small isolated outcrop 370 m high on the coast of East Antarctica at 66° 47′ S, 89° 18′ E (Fig. 1) (Tingey et al. 1983; Smellie and Collerson 2021). The nearest exposed rock is c. 150 km away and consists of Precambrian basement, but aeromagnetic investigations suggest that Gaussberg may be one of a cluster of several similar-sized subglacial volcanic edifices present within a 30 km radius. Gaussberg is a pile of ultrapotassic lamproite pillow lava c. 1200 m high with a possible original diameter of c. 10 km (identified magnetically). Other lithofacies are minor. The nunatak represents a relatively small pillow volcano, which erupted subglacially 56 kyr ago, during the last glacial when the local ice cover was much thicker than present (c. 1300 m). Because of the confining effects of ice, glaciovolcanic pillow mounds might be expected to have higher aspect ratios than analogous mounds emplaced subaqueously. Published data on glaciovolcanic pillow mounds seem to contradict this but are too few to be definitive (Smellie 2013). Gaussberg was probably constructed by multiple overlapping vents, which created a broad, low-profile shield-like edifice under the ice. As in Victoria Land, the presence of erratics and striations on Gaussberg indicate that it was overridden by wet-based ice at LGM, similar to several localities known in Victoria Land. Because of its isolated location and unusual lamproite composition, the importance of Gaussberg environmentally and petrologically is out of all proportion to its small size. The genesis of the Gaussberg lamproites has been ascribed to melting of a sediment-contaminated deep mantle source followed by entrainment in a plume, although not the plume responsible for volcanism in the Kerguelen Plateau nearby (Smellie and Collerson 2021).

Summary

Antarctica has been affected by volcanic processes on a range of scales and under variable tectonic conditions (i.e. Gondwana-break-up, subducting continental margin, and slab-window and continental rifting). They differ widely in the duration, volume and geographical distribution of the erupted products. Jurassic volcanism associated with Gondwana break-up was short lived but is represented by widely dispersed, astonishingly high-volume, mafic and felsic flare-ups that included the longest interpreted lateral flow of (mafic) magma on Earth, and had a major environmental impact on Life on Earth that potentially triggered several mass extinctions.

Long-lived subduction, including during the period prior to that considered in this Memoir, created a continental margin arc. It was formerly a major physiographical feature that dominated the Pacific margin of Antarctica, from Marie Byrd Land to the Antarctic Peninsula. Subduction now persists only at the northern tip, at the South Shetland Trench. It was an extensional arc, with transitional arc characteristics in the South Shetland Islands due to the relatively thin crustal thicknesses there (c. 25–30 km). Its early (pre-200 Ma) history is very poorly known but from the Jurassic onwards it was probably characterized by alternating episodes of high magma-flux flare-ups and magmatic lulls, which have been identified in the South Shetland Islands and Alexander Island, and for plutons in southern Palmer Land. The reasons for the episodic nature of the magmatism are uncertain. Like other continental margin arcs, they probably relate to fluctuations in subduction parameters caused by transient regional stresses or far-field changes in plate motions (i.e. plate 'jostling'), which triggered extension-related flare-ups and compression-related lulls. The arc volcanism migrated trenchwards from the Late Cretaceous in the Antarctic Peninsula, probably because of enhanced growth of the coeval accretionary complex caused by an orthogonal subduction trajectory and the cuspate (concave to the west) shape of the peninsula helping to trap and preferentially accumulate subducted sediments. This caused the subducted slab to migrate trenchwards, taking the volcanic axis with it. By contrast, subduction was much more oblique opposite the northern Antarctic Peninsula, in the South Shetland Islands, which probably elicited tectonic erosion of the forearc. As a result, the volcanic axis there migrated *away* from the trench. Associated important features of the Antarctic continental margin arc include voluminous (flood lava) alkaline back-arc volcanism in the James Ross Island region, which includes Antarctica's largest volcano (Mount Haddington). A small intra-arc marginal basin also opened up in Bransfield Strait, where the largest Holocene eruption took place during caldera formation on Deception Island. The progressive stepwise cessation of subduction in a northerly direction created a series of slab windows that triggered small-degree alkaline mafic melts, which were erupted as numerous small-volume monogenetic volcanoes mainly on the flanks of the peninsula. The relative alkalinity and erupted volumes of the slab-window magmas show a clear correlation with crustal thickness. The back-arc and slab-window basalts also interacted with the Antarctic Peninsula Ice Sheet, and its critical parameters and history are now relatively well known as a result of several focused volcanological–palaeoenvironmental investigations since the 1980s.

Neogene alkaline rift-related volcanism in Antarctica in the West Antarctic Rift System (WARS: and also, to a large extent, alkaline back-arc- and slab-window-related volcanism in the Antarctic Peninsula) is overwhelmingly effusive. As a consequence, prior to year 2000, most published maps were based on geochemical investigations of the lavas, and fragmental rocks were poorly represented. The absence of substantial pyroclastic deposits is probably because the WARS volcanism coincided with the inception and development of the Antarctic Ice Sheet. Tephra produced by any pyroclastic eruptions during glacials would have fallen on the surrounding ice and was advected to the Southern Ocean, thus removing any onshore evidence. However, several caldera rim exposures preserve pyroclastic deposits, mainly in Marie Byrd Land. Their presence confirms that explosive eruptions also occurred, and may have been relatively common (cf. the Pleistocene–Holocene ice-core record). Pyroclastic deposits are possibly only preserved on the cratered summits of volcanoes in the WARS because, when they were active, the focused high heat flow may have prevented significant thicknesses of ice from accumulating. This is the case for Mount Erebus and Mount Melbourne today. Because of the volcanothermal effects, the crater regions of large volcanoes in the WARS are thus unlikely locations in which to derive unambiguous information on the prevailing climate. However, because of their broad (up to 12 km) basin-like shape and more dispersed heat flow, volcano calderas will permit the widespread build-up of ice, not only during glacials but probably also in interglacials. This is demonstrated by some caldera volcanoes considered to be active in the WARS today (e.g. Mount Berlin, Mount Takahe and Mount Rittmann), which have an extensive ice cover. It is also true of active caldera volcanoes in Iceland (e.g. Grimsvötn, Eyjafjallajökull and Katla), which support prominent summit ice caps. Thus, evidence for the prevailing climatic regime may be preserved in the summit regions of caldera volcanoes, although not for the thermal regime of any ice (as explained earlier for the James Ross Island Volcanic Group (JRIVG)). In general, during periods of much reduced ice cover in prominent interglacials, tephra can be preserved more widely on ice-poor or ice-free volcano flanks unless removed subsequently by glacial erosion (e.g. tephras preserved on Mount Sidley (Marie Byrd Land) and possibly Deception Plateau (northern Victoria Land)).

There is a conspicuous volcanological difference between the alkaline volcanism in the West Antarctic Rift region and the Antarctic Peninsula. The WARS is an 'a'ā province dominated by multistorey 'a'ā lava-fed deltas in numerous central volcanoes. It is the only 'a'ā-dominated glaciovolcanic province currently known. This is probably the reason why Marie Byrd Land (and the WARS generally) has very few examples of pillow lavas and all are small outcrops. By contrast, the Antarctic Peninsula (comprising back-arc and slab-window volcanism) is a pāhoehoe province characterized by numerous glaciovolcanic tuyas in monogenetic volcanic fields, and pāhoehoe lava-fed deltas that largely constructed the Mount Haddington shield volcano and associated satellite centres. Thus, pāhoehoe lava-fed deltas are the predominant eruptive unit in the Antarctic Peninsula. Iceland and British Columbia, the other large glaciovolcanic regions on Earth are also dominated by pāhoehoe. The overall differences between the two Antarctic regional provinces are probably related to compositional and possibly tectonic contrasts, the latter influencing magma volumes and thus discharge rates. The Antarctic Peninsula volcanism is typically more weakly alkaline and includes tholeiites. In the WARS, the volcanism is generally more alkaline, plume-related and much greater volumes were erupted.

A single occurrence of ultrapotassic lamproite magma is also present at Gaussberg, a small isolated outcrop on the otherwise almost rock-free coast of East Antarctica. The outcrop is a compositionally unusual volcanic centre, constructed of subglacially-erupted pillow lava. It is also unusually important for petrological studies and for palaeoenvironmental information linked to the last glacial in an onshore region largely devoid of other environmental evidence.

Following several investigations of volcanism in the WARS in recent years, the East Antarctic Ice Sheet in Victoria Land is becoming increasingly well documented and understood. The characteristics and evolution of the West Antarctic Ice Sheet are less well known, however, due to the remoteness of the outcrops and (especially) the frequent lack of suitable exposures. Characteristics of the interglacial periods are infrequently described in any area. With the exception of scoria cones, which require subaerial conditions, and lapilli tuffs, which can form in a variety of eruptive settings (subaerial and glaciovolcanic), the presence of pyroclastic deposits in the WARS generally requires non-glacial conditions for their preservation since they otherwise get removed quickly by ice advection. Their occurrence, therefore, is broadly diagnostic of interglacial periods and they are a focus of several current volcanological–palaeoenvironmental investigations in the region.

Acknowledgements The author gratefully acknowledges the support of his research over several decades by numerous Antarctic organizations and their logistical staff, in particular the British Antarctic Survey, but including Programma Nazionale di Ricerche in Antartide (PNRA; Italy), National Science Foundation (USA), Antarctica New Zealand and Programa Antártico Brasileiro (PROANTAR). The number of colleagues who have helped the author in his research is too large to mention everybody but particular thanks go to the following (in no order): Mike Thomson, Alan Haywood, Jo Johnson, Sergio Rocchi, Mike Hambrey, Rob Larter, Peter Fretwell, Mike Tabecki, Ian Skilling, Malcolm Hole, Tom Jordan, Kurt Panter, Bill McIntosh, Gianfranco di Vincenzo, Adam Martin, Philip Kyle, Stefan Kraus, Wes LeMasurier and Adelina Geyer. Finally, the author is also very grateful to Nick Kusznir for permission to reproduce his figure showing Antarctic Peninsula crustal thicknesses; and to David Elliot, Kurt Panter and Adelina Geyer for their careful and constructive comments, which helped to improve this chapter.

Author contributions JLS: conceptualization (lead), formal analysis (lead), investigation (lead), writing – original draft (lead).

Funding This research was supported by grants from the Transantarctic Association, which were particularly helpful.

Data availability Data sharing is not applicable to this article as no datasets were generated during the current study.

References

Aitken, A.R.A., Wilson, G.S., Jordan, T., Tinto, K. and Blakenmore, H. 2012. Flexural controls on late Neogene basin evolution in southern McMurdo Sound, Antarctica. *Global and Planetary Change*, **80–81**, 99–112, https://doi.org/10.1016/j.gloplacha.2011.08.003

Antoniades, D., Giralt, S. *et al.* 2018. The timing and widespread effects of the largest Holocene volcanic eruption in Antarctica. *Nature Scientific Reports*, **8**, 17279, https://doi.org/10.1038/s41598-018-35460-x

Barker, D.H.N. and Austin, J.A. 1998. Rift propagation, detachment faulting, and associated magmatism in Bransfield Strait, Antarctic Peninsula. *Journal of Geophysical Research*, **103**, 24 017–24 043, https://doi.org/10.1029/98JB01117

Barker, P.F. 1982. The subduction history of the Pacific margin of the Antarctic Peninsula region: ridge crest–trench interactions. *Journal of the Geological Society, London*, **139**, 787–801, https://doi.org/10.1144/gsjgs.139.6.0787

Barrett, P.J. 1991. The Devonian to Triassic Beacon Supergroup of the Transantarctic Mountains and correlatives in other parts of Antarctica. *In*: Tingey, R.J. (ed.) *The Geology of Antarctica*. Oxford University Press, Oxford, Uk, 120–152.

Barrett, P.J. 2013. Resolving views on Antarctic Neogene glacial history – the Sirius debate. *Earth and Environmental Science Transactions of the Royal Society of Edinburgh*, **104**, 31–53, https://doi.org/10.1017/S175569101300008X

Bartolini, S., Geyer, A., Martí, J., Pedrazzi, D. and Aguirre-Díaz, G. 2014. Volcanic hazard on Deception Island (South Shetland Islands, Antarctica). *Journal of Volcanology and Geothermal Research*, **285**, 150–168, https://doi.org/10.1016/j.jvolgeores.2014.08.009

Behrendt, J.C., Blankenship, D.D., Finn, C.A., Bell, R.E., Sweeney, R.E., Hodge, S.R. and Brozena, J.M. 1994. Evidence for late Cenozoic flood basalts (?) in the West Antarctic rift system revealed by the CASERTZ aeromagnetic survey. *Geology*, **22**, 527–530, https://doi.org/10.1130/0091-7613(1994)022<0527:CADRLC>2.3.CO;2

Behrendt, J.C., Blankenship, D.D., Morse, D.L., Finn, C.A. and Bell, R.E. 2002. Subglacial volcanic features beneath the West Antarctic Ice Sheet interpreted from aeromagnetic and radar ice sounding. *Geological Society, London, Special Publications*, **202**, 337–355, https://doi.org/10.1144/GSL.SP.2002.202.01.17

Bialas, R.W., Buck, W.R., Studinger, M. and Fitzgerald, P.G. 2007. Plateau collapse model for the Transantarctic Mountains–West Antarctic rift system: Insights from numerical experiments. *Geology*, **35**, 687–690, https://doi.org/10.1130/G23825A.1

Birkenmajer, K. 1992. Evolution of the Bransfield Basin and rift, West Antarctica. *In*: Yoshida, Y., Kaminuma, K. and Shiraishi, K. (eds) *Recent Progress in Antarctic Earth Science*. Terrapub, Tokyo, 405–410.

Birkenmajer, K. 1996. Tertiary glacial/interglacial palaeoenvironments and sea-level changes, King George Island, West Antarctica. An overview. *Bulletin of the Polish Academy of Sciences, Earth Sciences*, **44**, 157–181.

Birkenmajer, K. 2001. Mesozoic and Cenozoic stratigraphic units in parts of the South Shetland Islands and northern Antarctic Peninsula (as used by the Polish Antarctic Programmes). *Studia Geologica Polonica*, **118**, 1–188.

Birkenmajer, K., Delitala, M.C., Narębski, W., Nicoletti, M. and Petrucciani, C. 1986. Geochronology of Tertiary island-arc volcanics and glacigenic deposits, King George Island, South Shetland Islands (West Antarctica). *Bulletin of the Polish Academy of Sciences, Earth Sciences*, **34**, 257–273.

Birkenmajer, K., Soliani, E. and Kawashika, K. 1990. Reliability of potassium-argon dating of Cretaceous–Tertiary island-arc volcanic suites of King George Island, South Shetland Islands (West Antarctica). *Neues Jahrbuch für Geologie und Paläontologie*, **1**, 127–140.

Birkenmajer, K., Gaździcki, A., Krajewski, K.P., Przybycin, A., Solecki, A., Tatur, A. and Yoon, H.I. 2005. First Cenozoic glaciers in West Antarctica. *Polish Polar Research*, **26**, 3–12.

Brenn, G.R., Hansen, S.E. and Park, Y. 2017. Variable thermal loading and flexural uplift along the Transantarctic Mountains, Antarctica. *Geology*, **45**, 463–466, https://doi.org/10.1130/G38784.1

Burgess, S.D., Bowring, S.A., Fleming, T.H. and Elliot, D.H. 2015. High precision geochronology links the Ferrar Large Igneous Province with early Jurassic ocean anoxia and biotic crisis. *Earth and Planetary Science Letters*, **415**, 90–99, https://doi.org/10.1016/j.epsl.2015.01.037

Burn, R.W. 1981. Early Tertiary calc-alkaline volcanism on Alexander Island. *British Antarctic Survey Bulletin*, **53**, 175–193.

Calabozo, F.M., Strelin, J.A., Orihashi, Y., Sumino, H. and Keller, R.A. 2015. Volcano–ice–sea interaction in the Cerro Santa Marta area, northwest James Ross Island, Antarctic Peninsula. *Journal of Volcanology and Geothermal Research*, **297**, 89–108, https://doi.org/10.1016/j.jvolgeores.2015.03.011

Cande, S.C. and Stock, J.M. 2006. Constraints on the timing of extension in the Northern Basin, Ross Sea. *In*: Futterer, D.K., Damaske, D., Kleinschmidt, G., Miller, H. and Tessensohn, F. (eds) *Antarctica: Contributions to Global Earth Sciences*. Springer, Berlin, 319–326.

Carr, M.H. 1974. Tectonism and volcanism of the Tharsis region of Mars. *Journal of Geophysical Research*, **79**, 3943–3949, https://doi.org/10.1029/JB079i026p03943

Christeson, G.L., Barker, D.H.N., Austin, J.A. and Dalziel, I.W.D. 2003. Deep crustal structure of Bransfield Strait: initiation of a backarc basin by rift reactivation and propagation. *Journal of Geophysical Research*, **108**, 2492, https://doi.org/10.1029/2003JB002468

Craddock, C., Bastien, T.W. and Rutford, R.H. 1964. Geology of the Jones Mountains area. *In*: Adie, R.J. (ed.) *Antarctic Geology*. North-Holland, Amsterdam, 171–187.

Danesi, S. and Morelli, A. 2000. Group velocity of Rayleigh waves in the Antarctic region. *Physics of the Earth and Planetary Interiors*, **122**, 55–66, https://doi.org/10.1016/S0031-9201(00)00186-2

Davey, F.J., Granot, R., Cande, S.C., Stock, J.M., Selvans, M. and Ferraccioli, F. 2016. Synchronous oceanic spreading and continental rifting in West Antarctica. *Geophysical Research Letters*, **43**, 6162–6169, https://doi.org/10.1002/2016GL069087

Davies, B.J., Hambrey, M.J., Smellie, J.L., Carrivick, J.L. and Glasser, N.F. 2012. Antarctic Peninsula Ice Sheet evolution during the Cenozoic Era. *Quaternary Science Reviews*, **31**, 30–66, https://doi.org/10.1016/j.quascirev.2011.10.012

Davies, B.J., Glasser, N.F., Carrivick, J.L., Hambrey, M.J., Smellie, J.L. and Nývlt, D. 2013. Landscape evolution and ice-sheet behaviour in a semi-arid polar environment: James Ross Island, NE Antarctic Peninsula. *Geological Society, London, Special Publications*, **381**, 353–395, https://doi.org/10.1144/SP381.1

De Ignacio, C., López, I., Oyarzun, R. and Márquez, A. 2001. The northern Patagonia Somuncura plateau basalts: a product of slab-induced, shallow asthenospheric upwelling? *Terra Nova*, **13**, 117–121, https://doi.org/10.1046/j.1365-3121.2001.00326.x

Denton, G.H. and Hughes, T.J. 2002. Reconstructing the Antarctic Ice Sheet at the Last Glacial Maximum. *Quaternary Science Reviews*, **21**, 193–202, https://doi.org/10.1016/S0277-3791(01)00090-7

Denton, G.H., Prentice, M.L. and Burckle, L.H. 1991. Cainozoic history of the Antarctic Ice Sheet. *In*: Tingey, R.J. (ed.) *The Geology of Antarctica*. Oxford University Press, Oxford, UK, 365–433.

Dingle, R.V. and Lavelle, M. 1998. Antarctic Peninsular cryosphere: Early Oligocene (c. 30 Ma) initiation and a revised glacial chronology. *Journal of the Geological Society, London*, **155**, 433–437, https://doi.org/10.1144/gsjgs.155.3.0433

D'Orazio, M., Agostini, S., Mazzarini, F., Innocenti, F., Manetti, P., Haller, M.J. and Lahsen, A. 2000. The Pali Aike Volcanic Field, Patagonia: slab-window magmatism near the tip of South America. *Tectonophysics*, **321**, 407–427, https://doi.org/10.1016/S0040-1951(00)00082-2

Drewry, D.J. and Morris, E. 1992. The response of large ice sheets to climatic change. *Philosophical Transactions of the Royal Society B: Biological Sciences*, **338**, 235–242, https://doi.org/10.1098/rstb.1992.0143

Ducea, M.N., Saleeby, J.B. and Bergantz, G. 2015. The architecture, chemistry, and evolution of continental magmatic arcs. *Annual Review of Earth and Planetary Sciences*, **43**, 10.1–10.33, https://doi.org/10.1146/annurev-earth-060614-105049

Dunbar, N.W., Iverson, N.A., Smellie, J.L., McIntosh, W.C., Zimmerer, M.J. and Kyle, P.R. 2021. Active volcanoes in Marie Byrd Land. *Geological Society, London, Memoirs*, **55**, https://doi.org/10.1144/M55-2019-29

Elliot, D.H. 1996. The Hanson Formation: a new stratigraphical unit in the Transantarctic Mountains, Antarctica. *Antarctic Science*, **8**, 389–394, https://doi.org/10.1017/S0954102096000569

Elliot, D.H. 2000. Stratigraphy of Jurassic pyroclastic rocks in the Transantarctic Mountains. *Journal of African Earth Sciences*, **31**, 77–89, https://doi.org/10.1016/S0899-5362(00)00074-9

Elliot, D.H. 2013. The geological and tectonic evolution of the Transantarctic Mountains: a review. *Geological Society, London, Special Publications*, **381**, 7–35, https://doi.org/10.1144/SP381.14

Elliot, D.H., Larsen, D., Fanning, C.M., Fleming, T.H. and Vervoort, J.D. 2017. The Lower Jurassic Hanson Formation of the Transantarctic Mountains: implications for the Antarctic sector of the Gondwana plate margin. *Geological Magazine*, **154**, 777–803, https://doi.org/10.1017/S0016756816000388

Elliot, D.H., White, J.D.L. and Fleming, T.H. 2021. Ferrar Large Igneous Province: volcanology. *Geological Society, London, Memoirs*, **55**, https://doi.org/10.1144/M55-2018-44

Ernst, R.E. and Youbi, N. 2017. How large igneous provinces affect global climate, sometimes cause mass extinctions, and represent natural markers in the geological record. *Palaeogeography, Palaeoclimatology, Palaeoecology*, **478**, 30–52, https://doi.org/10.1016/j.palaeo.2017.03.014

Ferraccioli, F., Armadillo, E., Jordan, T., Bozzo, E. and Corr, H. 2009. Aeromagnetic exploration over the East Antarctic Ice Sheet: a new view of the Wilkes Subglacial Basin. *Tectonophysics*, **478**, 62–77, https://doi.org/10.1016/j.tecto.2009.03.013

Fitzgerald, P.G., Bialas, R.W., Buck, W.R. and Studinger, M. 2007. A plateau collapse model for the formation of the West Antarctic rift system/Transantarctic Mountains. *United States Geological Survey Open-File Report*, **2007-1047**, extended abstract 087.

Fleming, T.H., Foland, K.A. and Elliot, D.H. 1995. Isotopic and chemical constraints on the crustal evolution and source signature of Ferrar magmas, North Victoria Land, Antarctica. *Contributions to Mineralogy and Petrology*, **121**, 217–236, https://doi.org/10.1007/BF02688238

Fretzdorff, S., Worthington, T.J., Haase, K.M., Hekinian, R., Franz, L., Keller, R.A. and Stoffers, P. 2004. Magmatism in the Bransfield Basin: Rifting of the South Shetland Arc? *Journal of Geophysical Research*, **109**, B12208, https://doi.org/10.1029/2004JB003046

Garrett, S.W., Renner, R.G.B., Jones, J.A. and McGibbon, K.J. 1987. Continental magnetic anomalies and the evolution of the Scotia Sea. *Earth and Planetary Science Letters*, **81**, 273–281, https://doi.org/10.1016/0012-821X(87)90163-4

Geyer, A., Pedrazzi, D. et al. 2021. Deception Island. *Geological Society, London, Memoirs*, **55**, https://doi.org/10.1144/M55-2018-56

Giordano, G., Lucci, F., Phillips, D., Cozzupoli, D. and Runci, V. 2012. Stratigraphy, geochronology and evolution of the Mt. Melbourne volcanic field (North Victoria land, Antarctica). *Bulletin of Volcanology*, **74**, 1985–2005, https://doi.org/10.1007/s00445-012-0643-8

González-Casado, J.M., Giner Robles, J.L. and López-Martínez, J. 2000. Bransfield Basin, Antarctic Peninsula: Not a normal back-arc basin. *Geology*, **28**, 1043–1046, https://doi.org/10.1130/0091-7613(2000)28<1043:BBAPNA>2.0.CO;2

Goodge, J.W. 2019. Geological and tectonic evolution of the Transantarctic Mountains, from ancient craton to recent enigma. *Gondwana Research*, **80**, 50–122, https://doi.org/10.1016/j.gr.2019.11.001

Goss, A.R., Kay, S.M. and Mpodozis, C. 2013. Andean adakite-like high-Mg andesites on the northern margin of the Chilean–Pampean flat-slab (27–28.5°S) associated with frontal arc migration and fore-arc subduction erosion. *Journal of Petrology*, **54**, 2193–2234, https://doi.org/10.1093/petrology/egt044

Gràcia, E., Canals, M., Farràn, M., Prieto, M.J., Sorribas, J. and GEBRA Team. 1996. Morphostructural evolution of the Central and Eastern Bransfield Basins (NW Antarctic Peninsula). *Marine Geophysical Researches*, **18**, 429–448, https://doi.org/10.1007/BF00286088

Grad, M., Guterch, A., Janik, T. and Środa, P. 2002. Seismic characteristic of the crust in the transition zone from the Pacific Ocean to the northern Antarctic Peninsula, West Antarctica. *Royal Society of New Zealand Bulletin*, **35**, 493–498.

Granot, R. and Dyment, J. 2018. Late Cenozoic unification of East and West Antarctica. *Nature Communications*, **9**, 3189, https://doi.org/10.1038/s41467-018-05270-w

Graw, J.H., Adams, A.N., Hansen, S.E., Wiens, D.A., Hackworth, L. and Park, Y. 2016. Upper mantle shear wave velocity structure beneath northern Victoria Land, Antarctica: Volcanism and uplift in the northern Transantarctic Mountains. *Earth and Planetary Science Letters*, **449**, 48–60, https://doi.org/10.1016/j.epsl.2016.05.026

Gunn, B.M. and Warren, G. 1962. Geology of Victoria Land between the Mawson and Mulock Glaciers, Antarctica. *New Zealand Geological Survey Bulletin*, **71**, 1–157.

Haase, K.M. and Beier, C. 2021. Bransfield Strait and James Ross Island: petrology. *Geological Society, London, Memoirs*, **55**, https://doi.org/10.1144/M55-2018-37

Haase, K.M., Beier, C., Fretzdorff, S., Smellie, J.L. and Garbe-Schönberg, D. 2012. Magmatic evolution of the South Shetland Islands, Antarctica, and implications for continental crust formation. *Contributions to Mineralogy and Petrology*, **163**, 1103–1119, https://doi.org/10.1007/s00410-012-0719-7

Hambrey, M.J. and Glasser, N.F. 2012. Discriminating glacier thermal and dynamic regimes in the sedimentary record. *Sedimentary Geology*, **251**, 1–33, https://doi.org/10.1016/j.sedgeo.2012.01.008

Hambrey, M.J., Smellie, J.L., Nelson, A.E. and Johnson, J.S. 2008. Late Cenozoic glacier–volcano interaction on James Ross Island and adjacent areas, Antarctic Peninsula region. *American Geological Society Bulletin*, **120**, 709–731, https://doi.org/10.1130/B26242.1

Hamilton, W. 1972. *The Hallett Volcanic Province, Antarctica*. United States Geological Survey Professional Papers, **456-C**.

Hansen, S.E., Graw, J.H. et al. 2014. Imaging the Antarctic mantle using adaptively parameterized P-wave tomography: evidence for heterogeneous structure beneath West Antarctica. *Earth and Planetary Science Letters*, **408**, 66–78, https://doi.org/10.1016/j.epsl.2014.09.043

Hastie, W.W., Watkeys, M.K. and Aurbourg, C. 2014. Magma flow in dyke swarms of the Karoo LIP: implications for the mantle plume hypothesis. *Gondwana Research*, **25**, 736–755, https://doi.org/10.1016/j.gr.2013.08.010

Hathway, B. 2000. Continental rift to back-arc basin: Jurassic–Cretaceous stratigraphical and structural evolution of the Larsen Basin, Antarctic Peninsula. *Journal of the Geological Society, London*, **157**, 417–432, https://doi.org/10.1144/jgs.157.2.417

Hayes, D.E., Frakes, L.A. et al. 1975. Introduction. *Initial Reports of the Deep Sea Drilling Project*, **28**, 1–5.

Haywood, A.M., Smellie, J.L. et al. 2009. Middle Miocene to Pliocene history of Antarctica and the Southern Ocean. *Developments in Earth & Environmental Sciences*, **8**, 401–463.

Hoernle, K., Timm, C. et al. 2020. Late Cretaceous (99–69 Ma) basaltic intraplate volcanism on and around Zealandia: Tracing upper mantle geodynamics from Hikurangi Plateau collision to Gondwana breakup and beyond. *Earth and Planetary Science Letters*, **529**, 115864, https://doi.org/10.1016/j.epsl.2019.115864

Hole, M.J. 1988. Post-subduction alkaline volcanism along the Antarctic Peninsula. *Journal of the Geological Society, London*, **145**, 985–989, https://doi.org/10.1144/gsjgs.145.6.0985

Hole, M.J. 1990. Geochemical evolution of Pliocene–Recent post-subduction alkali basalts from Seal Nunataks, Antarctic Peninsula. *Journal of Volcanology and Geothermal Research*, **40**, 149–167, https://doi.org/10.1016/0377-0273(90)90118-Y

Hole, M.J. 2021. Antarctic Peninsula: petrology. *Geological Society, London, Memoirs*, **55**, https://doi.org/10.1144/M55-2018-40

Hole, M.J. and Larter, R.D. 1993. Trench-proximal volcanism following ridge crest-trench collision along the Antarctic Peninsula. *Tectonics*, **12**, 897–910, https://doi.org/10.1029/93TC00669

Hole, M.J., Saunders, A.D., Rogers, G. and Sykes, M.A. 1995. The relationship between alkaline magmatism, lithospheric extension and slab window formation along continental destructive plate margins. *Geological Society, London, Special Publications*, **81**, 265–285, https://doi.org/10.1144/GSL.SP.1994.081.01.15

Huerta, A.D. and Harry, D.L. 2007. The transition from diffuse to focused extension: Modeled evolution of the West Antarctic Rift system. *Earth and Planetary Science Letters*, **255**, 133–147, https://doi.org/10.1016/j.epsl.2006.12.011

Hunter, M.A., Riley, T.T., Cantrill, D.J., Flowerdew, M.J. and Millar, I.L. 2006. A new stratigraphy for the Latady Basin, Antarctic Peninsula: Part 1, Ellsworth Land Volcanic Group. *Geological Magazine*, **143**, 777–796, https://doi.org/10.1017/S0016756806002597

Isbell, J.L. 1999. The Kukri Erosion Surface; a reassessment of its relationship to rocks of the beacon Supergroup in the central Transantarctic Mountains, Antarctica. *Antarctic Science*, **11**, 228–238, https://doi.org/10.1017/S0954102099000292

Jerram, D.A. 2002. Volcanology and facies architecture of flood basalts. *Geological Society of America Special Papers*, **362**, 119–132.

Johnson, J.S., Hillenbrand, C.-D., Smellie, J.L. and Rocchi, S. 2008. The last deglaciation of Cape Adare, northern Victoria Land, Antarctica. *Antarctic Science*, **20**, 581–587, https://doi.org/10.1017/S0954102008001417

Johnson, J.S., Smellie, J.L., Nelson, A.E. and Stuart, F.M. 2009. History of the Antarctic Peninsula Ice Sheet since the early Pliocene – evidence from cosmogenic dating of Pliocene lavas on James Ross Island, Antarctica. *Global and Planetary Change*, **69**, 205–213, https://doi.org/10.1016/j.gloplacha.2009.09.001

Jolley, D.W. and Widdowson, M. 2005. Did Paleogene North Atlantic rift-related eruptions drive early Eocene climate cooling? *Lithos*, **79**, 355–366, https://doi.org/10.1016/j.lithos.2004.09.007

Jonkers, H.A., Lirio, J.M., del Valle, R.A. and Kelley, S.P. 2002. Age and environment of Miocene–Pliocene glaciomarine deposits, James Ross Island, Antarctica. *Geological Magazine*, **139**, 577–594, https://doi.org/10.1017/S0016756802006787

Jordan, T.A., Ferraccioli, F., Jones, P.C., Smellie, J.L., Ghidella, M. and Corr, H. 2009. Airborne gravity reveals interior of Antarctic volcano. *Physics of the Earth and Planetary Interiors*, **175**, 127–136, https://doi.org/10.1016/j.pepi.2009.03.004

Jull, M. and McKenzie, D. 1996. The effect of deglaciation on mantle melting beneath Iceland. *Journal of Geophysical Research: Solid Earth*, **101**, 21 815–21 828, https://doi.org/10.1029/96JB01308

Kelemen, P.B. 1995. Genesis of high Mg# andesites and the continental crust. *Contributions to Mineralogy and Petrology*, **120**, 1–19, https://doi.org/10.1007/BF00311004

Keller, R.A., Fisk, M.R., Smellie, J.L., Strelin, J.A. and Lawver, L.A. 2002. Geochemistry of back arc basin volcanism in Bransfield Strait, Antarctica: Subducted contributions and along-axis variations. *Journal of Geophysical Research*, **107**, https://doi.org/10.1029/2001JB000444

Keszthelyi, L., Self, S. and Thordarson, T. 2006. Flood basalts on Earth, Io and Mars. *Journal of the Geological Society, London*, **163**, 253–264, https://doi.org/10.1144/0016-764904-503

Košler, J., Magna, T., Mlčoch, B., Mixa, P., Nývlt, D. and Holub, F.V. 2009. Combined Sr, Nd, Pb and Li isotope geochemistry of alkaline lavas from northern James Ross Island (Antarctic Peninsula) and implications for back-arc magma formation. *Chemical Geology*, **258**, 207–218, https://doi.org/10.1016/j.chemgeo.2008.10.006

Kusznir, N., Ferraccioli, F. and Jordan, T. 2018. Refining Gondwana plate reconstructions using Antarctic and Southern Ocean crustal thickness mapping from gravity inversion. *Geophysical Research Abstracts*, **20**, EGU2018-19196.

Kyle, P.R. 1982. Volcanic geology of The Pleiades, northern Victoria Land, Antarctica. *In*: Craddock, C. (ed.) *Antarctic Geoscience*. University of Wisconsin, Madison, WI, 747–754.

Kyle, P.R. 1990. Melbourne Volcanic Province. *American Geophysical Union Antarctic Research Series*, **48**, 48–52.

Larter, R.D. 1991. Preliminary results of seismic reflection investigations and associated geophysical studies in the area of the Antarctic Peninsula. Discussion. *Antarctic Science*, **3**, 217–222, https://doi.org/10.1017/S0954102091210251

Larter, R.D., Cunningham, A.P. and Barker, P.F. 2002. Tectonic evolution of the Pacific margin of Antarctica. 1. Late Cretaceous tectonic reconstructions. *Journal of Geophysical Research*, **107**, 2345, https://doi.org/10.1029/2000JB000053

Lawver, L.A. and Gahagan, L.M. 1994. Constraints on timing and extension in the Ross Sea region. *Terra Antartica*, **1**, 545–552.

Leat, P.T. 2008. On the long-distance transport of Ferrar magmas. *Geological Society, London, Special Publications*, **302**, 45–61, https://doi.org/10.1144/SP302.4

Leat, P.T. and Riley, T.R. 2021a. Antarctic Peninsula and South Shetland Islands: volcanology. *Geological Society, London, Memoirs*, **55**, https://doi.org/10.1144/M55-2018-52

Leat, P.T. and Riley, T.R. 2021b. Antarctic Peninsula and South Shetland Islands: petrology. *Geological Society, London, Memoirs*, **55**, https://doi.org/10.1144/M55-2018-68

Le Gall, B., Tshoso, G. et al. 2002. ^{40}Ar/^{39}Ar geochronology and structural data from the giant Okavango and related mafic dyke swarms, Karoo igneous province, northern Botswana. *Earth and Planetary Science Letters*, **2002**, 595–606, https://doi.org/10.1016/S0012-821X(02)00763-X

LeMasurier, W.E. 1972. Volcanic record of Cenozoic glacial history of Marie Byrd Land. *In*: Adie, R.J. (ed.) *Antarctic Geology and Geophysics*. Universitetsforlaget, Oslo, 251–259.

LeMasurier, W.E. 2002. Architecture and evolution of hydrovolcanic deltas in Marie Byrd Land, Antarctica. *Geological Society, London, Special Publications*, **202**, 115–148, https://doi.org/10.1144/GSL.SP.2002.202.01.07

LeMasurier, W.E. 2008. Neogene extension and basin deepening in the West Antarctic rift inferred from comparisons with the East African rift and other analogs. *Geology*, **36**, 247–250, https://doi.org/10.1130/G24363A.1

LeMasurier, W.E. 2013. Shield volcanoes of Marie Byrd Land, West Antarctic rift: oceanic island similarities, continental signature, and tectonic controls. *Bulletin of Volcanology*, **75**, 726, https://doi.org/10.1007/s00445-013-0726-1

LeMasurier, W.E. and Kawachi, Y. 1990. Mount Waesche. *American Geophysical Union Antarctic Research Series*, **48**, 208–211.

LeMasurier, W.E. and Landis, C.A. 1996. Mantle plume activity recorded by low relief erosion surfaces in west Antarctica and New Zealand. *Geological Society of America Bulletin*, **108**, 1450–1466, https://doi.org/10.1130/0016-7606(1996)108<1450:MPARBL>2.3.CO;2

LeMasurier, W.E. and Rex, D.C. 1982. Volcanic record of Cenozoic glacial history in Marie Byrd Land and western Ellsworth Land:

revised chronology and evaluation of tectonic factors. *In*: Craddock, C. (ed.) *Antarctic Geoscience*. University of Wisconsin, Madison, WI, 725–734.

LeMasurier, W.E. and Rex, D.C. 1989. Evolution of linear volcanic ranges in Marie Byrd Land, West Antarctica. *Journal of Geophysical Research*, **94**, 7223–7236, https://doi.org/10.1029/JB094iB06p07223

LeMasurier, W.E. and Thomson, J.W. (eds) 1990. *Volcanoes of the Antarctic Plate and Southern Oceans*. American Geophysical Union Antarctic Research Series, **48**.

LeMasurier, W.E., Harwood, D.M. and Rex, D.C. 1994. Geology of Mount Murphy Volcano: An 8-m.y. history of interaction between a rift volcano and the West Antarctic ice sheet. *Geological Society of America Bulletin*, **106**, 265–280, https://doi.org/10.1130/0016-7606(1994)106<0265:GOMMVA>2.3.CO;2

Lewis, A.R., Marchant, D.R., Ashworth, A.C., Hemming, S.R. and Machlus, M.L. 2007. Major middle Miocene global climate change: Evidence from East Antarctica and the Transantarctic Mountains. *Geological Society of America Bulletin*, **119**, 1449–1461, https://doi.org/10.1130/0016-7606(2007)119[1449:MMMGCC]2.0.CO;2

Licht, K.J., Groth, T., Townsend, J.P., Hennessy, A.J., Hemming, S.R., Flood, T.P. and Studinger, M. 2018. Evidence for extending anomalous Miocene volcanism at the edge of the East Antarctic craton. *Geophysical Research Letters*, **45**, 3009–3016, https://doi.org/10.1002/2018GL077237

Lisker, F. and Läufer, L. 2013. The Mesozoic Victoria Basin: Vanished link between Antarctica and Australia. *Geology*, **41**, 1043–1046, https://doi.org/10.1130/G33409.1

Loose, B., Naveira Garabato, A.C., Schlosser, P., Jenkins, W.J., Vaughan, D. and Heywood, K.J. 2018. Evidence of an active volcanic heat source beneath the Pine Island Glacier. *Nature Communications*, **9**, 2431, https://doi.org/10.1038/s41467-018-04421-3

Luttinen, A.V. 2021. Droning Maud Land Jurassic volcanism: volcanology and petrology. *Geological Society, London, Memoirs*, **55**, https://doi.org/10.1144/M55-2018-89

Luttinen, A.V., Leat, P.T. and Furnes, H. 2010. Björnnutane and Sembberget basalt lavas and the geochemical provinciality of Karoo magmatism in western Dronning Maud Land, Antarctica. *Journal of Volcanology and Geothermal Research*, **198**, 1–18, https://doi.org/10.1016/j.jvolgeores.2010.07.011

Macdonald, D.I.M. and Butterworth, P.J. 1990. The stratigraphy, setting and hydrocarbon potential of the Mesozoic sedimentary basins of the Antarctic Peninsula. *AAPG Studies in Geology*, **31**, 101–125.

Machado, A., Chemale, F., Conceição, R.V., Kawashita, K., Morata, D., Oteíza, O. and Van Schmus, W.R. 2005. Modeling of subduction components in the genesis of the Meso-Cenozoic igneous rocks from the South Shetland arc, Antarctica. *Lithos*, **82**, 435–453, https://doi.org/10.1016/j.lithos.2004.09.026

Maldonado, A., Larter, R.D. and Aldaya, F. 1994. Forearc tectonic evolution of the South Shetland Margin, Antarctic Peninsula. *Tectonics*, **13**, 1345–1370, https://doi.org/10.1029/94TC01352

Marenssi, S.A., Casadío, S. and Santillana, S.N. 2010. Record of Late Miocene glacial deposits on Isla Marambio (Seymour Island), Antarctic Peninsula. *Antarctic Science*, **22**, 193–198, https://doi.org/10.1017/S0954102009990629

Marsh, B.D. 2007. Magmatism, magma, and magma chambers. *Treatise on Geophysics*, **6**, 276–333.

Martí, J., Geyer, A. and Aguirre-Diaz, G. 2013. Origin and evolution of the Deception Island caldera (South Shetland Islands, Antarctica). *Bulletin of Volcanology*, **75**, 732, https://doi.org/10.1007/s00445-013-0732-3

Martin, A.P., Cooper, A.F. and Dunlap, W.J. 2010. Geochronology of Mount Morning, Antarctica: two-phase evolution of a long-lived trachyte–basanite–phonolite eruptive center. *Bulletin of Volcanology*, **72**, 357–371, https://doi.org/10.1007/s00445-009-0319-1

Martin, A.P., Smellie, J.L., Cooper, A.F. and Townsend, D.B. 2018. Formation of a spatter-rich pyroclastic density current deposit in a Neogene sequence of trachytic–mafic igneous rocks at Mason Spur, Erebus volcanic province, Antarctica. *Bulletin of Volcanology*, **80**, 13, https://doi.org/10.1007/s00445-017-1188-7

Martin, A.P., Cooper, A.F., Price, R.C., Kyle, P.R. and Gamble, J.A. 2021. Erebus Volcanic Province: petrology. *Geological Society, London, Memoirs*, **55**, https://doi.org/10.1144/M55-2018-80

McCarron, J.J. and Larter, R.D. 1998. Late Cretaceous to early Tertiary subduction history of the Antarctic Peninsula. *Journal of the Geological Society, London*, **155**, 255–268, https://doi.org/10.1144/gsjgs.155.2.0255

McCarron, J.J. and Millar, I.L. 1997. The age and stratigraphy of forearc magmatism on Alexander Island, Antarctica. *Geological Magazine*, **134**, 507–522, https://doi.org/10.1017/S0016756897007437

McCarron, J.J. and Smellie, J.L. 1998. Tectonic implications of forearc magmatism and generation of high-magnesian andesites: Alexander Island, Antarctica. *Journal of the Geological Society, London*, **155**, 269–280, https://doi.org/10.1144/gsjgs.155.2.0269

McClintock, M. and White, J.D.L. 2006. Large-volume phreatomagmatic vent complex at Coombs Hills, Antarctica records wet, explosive initiation of flood basalt volcanism in the Ferrar LIP. *Bulletin of Volcanology*, **68**, 215–239, https://doi.org/10.1007/s00445-005-0001-1

McClintock, M., White, J.D.L., Houghton, B.F. and Skilling, I.P. 2008. Physical volcanology of a large crater-complex formed during the initial stages of Karoo flood basalt volcanism, Sterkspruit, Eastern Cape, South Africa. *Journal of Volcanology and Geothermal Research*, **172**, 93–111, https://doi.org/10.1016/j.jvolgeores.2005.11.012

McConnell, J.R., Burke, A. *et al.* 2017. Synchronous volcanic eruptions and abrupt climate change *c*. 17.7 ka plausibly linked by stratospheric ozone depletion. *Proceedings of the National Academy of Sciences of the United States of America*, **114**, 10 035–10 040, https://doi.org/10.1073/pnas.1705595114

Menzies, M.A., Klemperer, S.L., Ebinger, C.J. and Baker, J. 2002. Characteristics of volcanic rifted margins. *Geological Society of America Special Papers*, **362**, 1–14.

Moreton, S. and Smellie, J.L. 1998. Identification and correlation of distal tephra layers in deep sea sedimentary cores, Scotia Sea, Antarctica. *Annals of Glaciology*, **27**, 285–289, https://doi.org/10.3189/1998AoG27-1-285-289

Muirhead, J.D., Airoldi, G., White, J.D.L. and Rowland, J.V. 2014. Cracking the lid: sill-fed dikes are the likely feeders of flood basalt eruptions. *Earth and Planetary Science Letters*, **406**, 187–197, https://doi.org/10.1016/j.epsl.2014.08.036

Muñoz, J., Troncoso, R., Duhart, P., Crignola, P., Farmer, L. and Stern, C.R. 2000. The relation of the mid-Tertiary coastal magmatic belt in south-central Chile to the late Oligocene increase in plate convergence rate. *Revista Geológica de Chile*, **27**, 177–203, https://doi.org/10.4067/S0716-02082000000200003

Narcisi, B. and Petit, J.R. 2021. Englacial tephras of East Antarctica. *Geological Society, London, Memoirs*, **55**, https://doi.org/10.1144/M55-2018-86

Nehyba, S. and Nývlt, D. 2015. 'Bottomsets' of the lava-fed delta of James Ross Island Volcanic Group, Ulu Peninsula, James Ross Island, Antarctica. *Polish Polar Research*, **36**, 1–24, https://doi.org/10.1515/popore-2015-0002

Nelson, A.E., Smellie, J.L. *et al.* 2009. Neogene glacigenic debris flows on James Ross Island, northern Antarctic Peninsula, and their implications for regional climate history. *Quaternary Science Reviews*, **28**, 3138–3160, https://doi.org/10.1016/j.quascirev.2009.08.016

Noll, M.R. 1985. *Mount Overlord, Northern Victoria Land, Antarctica*. MSc thesis, New Mexico Institute of Mining and Technology.

Nývlt, D., Kosler, J., Mlčoch, B., MIxa, P., Lisá, L., Bubík, M. and Hendriks, B.W.H. 2011. The Mendel Formation: Evidence for Late Miocene climatic cyclicity at the northern tip of the Antarctic Peninsula. *Palaeogeography, Palaeoclimatology, Palaeoecology*, **299**, 363–384, https://doi.org/10.1016/j.palaeo.2010.11.017

Oehler, J.-F., van Wyk de Vries, B. and Labazuy, P. 2005. Landslides and spreading of oceanic hot-spot and arc shield volcanoes on Low Strength Layers (LSLs): an analogue modelling approach. *Journal of Volcanology and Geothermal Research*, **144**, 169–189, https://doi.org/10.1016/j.jvolgeores.2004.11.023

Olivetti, V., Rossetti, F. et al. 2016. Contrasting exhumation and deformation style along a rift shoulder: insights from the Transantarctic Mountains. *Rendiconti della Società Geologica Italiana*, **40**(Suppl. 1), 549.

Pankhurst, R.J. and Smellie, J.L. 1983. K–Ar geochronology of the South Shetland Islands, Lesser Antarctica: apparent lateral migration of Jurassic to Quaternary island arc volcanism. *Earth and Planetary Science Letters*, **66**, 214–222, https://doi.org/10.1016/0012-821X(83)90137-1

Pankhurst, R.J., Riley, T.R., Fanning, C.M. and Kelley, S.P. 2000. Episodic silicic volcanism in Patagonia and the Antarctic Peninsula: Chronology of magmatism associated with the break-up of Gondwana. *Journal of Petrology*, **41**, 605–625, https://doi.org/10.1093/petrology/41.5.605

Panter, K.S. 2021. Antarctic volcanism: petrology and tectonomagmatic overview. *Geological Society, London, Memoirs*, **55**, https://doi.org/10.1144/M55-2020-10

Panter, K.S., McIntosh, W.C. and Smellie, J.L. 1994. Volcanic history of Mount Sidley, a major alkaline volcano in Marie Byrd Land, Antarctica. *Bulletin of Volcanology*, **56**, 361–376, https://doi.org/10.1007/BF00326462

Panter, K.S., Castillo, P. et al. 2018. Melt origin across a rifted continental margin: a case for subduction-related metasomatic agents in the lithospheric source of alkaline basalt, NW Ross Sea, Antarctica. *Journal of Petrology*, **59**, 517–558, https://doi.org/10.1093/petrology/egy036

Panter, K.S., Reindel, J. and Smellie, J.L. 2021a. Mount Early and Sheridan Bluff: petrology. *Geological Society, London, Memoirs*, **55**, https://doi.org/10.1144/M55-2019-2

Panter, K.S., Wilch, T.I., Smellie, J.L., Kyle, P.R. and McIntosh, W.C. 2021b. Marie Byrd Land and Ellsworth Land: petrology. *Geological Society, London, Memoirs*, **55**, https://doi.org/10.1144/M55-2019-50

Pappa, F., Ebbing, J. and Ferraccioli, F. 2019. Moho depths of Antarctica: Comparison of seismic, gravity, and isostatic results. *Geochemistry, Geophysics, Geosystems*, **20**, 1629–1645, https://doi.org/10.1029/2018GC008111

Parra, J.C., Yáñez, G. and Grupo de Trabajo USAC. 1988. Reconocimiento aeromagnético en la peninsula Antártica y mares circundantes, integración de información obtenida a diferentes Alturas. *Serie Científica Instituto Antártico Chileno*, **38**, 117–131.

Paterson, S.R. and Ducea, M.N. 2015. Arc magmatic tempos: gathering the evidence. *Elements*, **11**, 91–98, https://doi.org/10.2113/gselements.11.2.91

Paulsen, T.S. and Wilson, T.J. 2010. Evolution of Neogene volcanism and stress patterns in the glaciated West Antarctic Rift, Marie Byrd Land, Antarctica. *Journal of the Geological Society, London*, **167**, 401–416, https://doi.org/10.1144/0016-76492009-044

Paxman, G.J.G., Jamieson, S.S.R. et al. 2019. Subglacial geology and geomorphology of the Pensacola–Pole Basin, East Antarctica. *Geochemistry, Geophysics, Geosystems*, **20**, 2786–2807, https://doi.org/10.1029/2018GC008126

Peace, A.L., Phethean, J.J.J. et al. 2020. A review of Pangaea dispersal and Large Igneous Provinces – in search of a causative mechanism. *Earth-Science Reviews*, **206**, 102902, https://doi.org/10.1016/j.earscirev.2019.102902

Pedersen, A.K., Larsen, L.M. and Pedersen, G.K. 2017. Lithostratigraphy, geology and geochemistry of the volcanic rocks of the Vaigat Formation on Disko and Nuussuaq, Paleocene of West Greenland. *Geological Survey of Denmark and Greenland Bulletin*, **39**, https://doi.org/10.34194/geusb.v39.4354

Pedersen, A.K., Larsen, L.M. and Pedersen, G.K. 2018. Lithostratigraphy, geology and geochemistry of the volcanic rocks of the Maligât Formation and associated intrusions on Disko and Nuussuaq, Paleocene of West Greenland. *Geological Survey of Denmark and Greenland Bulletin*, **40**.

Phillips, E.H., Sims, K.W.W. et al. 2018. The nature and evolution of mantle upwelling at Ross Island, Antarctica, with implications for the source of HIMU lavas. *Earth and Planetary Science Letters*, **498**, 38–53, https://doi.org/10.1016/j.epsl.2018.05.049

Porebski, S.J. and Gradzinski, R. 1987. Depositional history of the Polonez Cove Formation (Oligocene), King George Island, West Antarctica: a record of continental glaciation, shallow marine sedimentation and contemporaneous volcanism. *Studia Geologica Polonica*, **93**, 7–62.

Porebski, S.J. and Gradzinski, R. 1990. Lava fed Gilbert-type delta in the Polonez Cove Formation (Oligocene), King George Island, West Antarctica. *International Association of Sedimentologists Special Publications*, **10**, 335–351.

Prenzel, J., Lisker, F., Monsees, N., Balestrieri, M.L., Läufer, A. and Spiegel, C. 2018. Development and inversion of the Mesozoic Victoria Badin in the Terra Nova Bay (Transantarctic Mountains) derived from thermochronological data. *Gondwana Research*, **53**, 110–128, https://doi.org/10.1016/j.gr.2017.04.025

Quartini, E., Blankenship, D.D. and Young, D.A. 2021. Active subglacial volcanism in West Antarctica. *Geological Society, London, Memoirs*, **55**, https://doi.org/10.1144/M55-2019-3

Reese, C.C., Solomatov, V.S., Baumgardner, J.R., Stegman, D.R. and Vezolainen, A.V. 2004. Magmatic evolution of impact-induced Martian mantle plumes and the origin of Tharsis. *Journal of Geophysical Research*, **109**, E08009, https://doi.org/10.1029/2003JE002222

Renner, R.G.B., Sturgeon, L.J.S. and Garrett, S.W. 1985. *Reconnaissance gravity and aeromagnetic surveys of the Antarctic Peninsula*. British Antarctic Survey Scientific Reports, **110**.

Retallack, G.J. 2001. *Soils of the Past. An Introduction to Paleopedology*. 2nd edn. Blackwell Science, Oxford, UK.

Riley, T.R. and Leat, P.T. 2021. Palmer Land and Graham Land volcanic groups (Antarctic Peninsula): volcanology. *Geological Society, London, Memoirs*, **55**, https://doi.org/10.1144/M55-2018-36

Riley, T.R., Leat, P.T., Pankhurst, R.J. and Harris, C. 2001. Origins of large volume rhyolitic volcanism in the Antarctic Peninsula and Patagonia by crustal melting. *Journal of Petrology*, **42**, 1043–1065, https://doi.org/10.1093/petrology/42.6.1043

Riley, T.R., Leat, P.T., Curtis, M.L., Millar, I.L., Duncan, R.A. and Fazel, A. 2005. Early–Middle Jurassic dolerite dykes from western Dronning Maud Land (Antarctica): identifying mantle sources in the Karoo Large Igneous Province. *Journal of Petrology*, **46**, 1489–1524, https://doi.org/10.1093/petrology/egi023

Riley, T.R., Flowerdew, M.J., Hunter, M.A. and Whitehouse, M.J. 2010. Middle Jurassic rhyolite volcanism of eastern Graham Land, Antarctic Peninsula: age, correlations and stratigraphic relationships. *Geological Magazine*, **147**, 581–595, https://doi.org/10.1017/S0016756809990720

Riley, T.R., Flowerdew, M.J. and Whitehouse, M.J. 2012. Chrono- and lithostratigraphy of a Mesozoic–Tertiary fore- to intra-arc basin: Adelaide Island, Antarctic Peninsula. *Geological Magazine*, **149**, 768–782, https://doi.org/10.1017/S001675681100 1002

Riley, T.R., Flowerdew, M.J., Pankhurst, R.J., Curtis, M.L., Millar, I.L., Fanning, C.M. and Whitehouse, M.J. 2017a. Early Jurassic magmatism on the Antarctic Peninsula and potential correlation with the Subcordilleran plutonic belt of Patagonia. *Journal of the Geological Society, London*, **174**, 365–376, https://doi.org/10.1144/jgs2016-053

Riley, T.R., Flowerdew, M.J., Pankhurst, R.J. and Leat, P.T. 2017b. A revised chronology of Thurston Island, West Antarctica, and correlations along the proto-Pacific margin of Gondwana. *Antarctic Science*, **29**, 47–60, https://doi.org/10.1017/S0954102016000341

Riley, T.R., Burton-Johnson, A., Flowerdew, M.J. and Whitehouse, M.J. 2018. Episodicity within a mid-Cretaceous magmatic flare-up in West Antarctica: U–Pb ages of the Lassiter Coast intrusive suite, Antarctic Peninsula, and correlations along the

Gondwana margin. *Geological Society of America Bulletin*, **130**, 1177–1196, https://doi.org/10.1130/B31800.1

Rocchi, S. and Smellie, J.L. 2021. Northern Victoria Land: petrology. *Geological Society, London, Memoirs*, **55**, https://doi.org/10.1144/M55-2019-19

Rocchi, S., Armienti, P., D'Orazio, M., Tonarini, S., Wijbrans, J.R. and Di Vincenzo, G. 2002. Cenozoic magmatism in the western Ross Embayment: Role of mantle plume v. plate dynamics in the development of the West Antarctic Rift System. *Journal of Geophysical Research: Solid Earth*, **107**, ECV 5-1–ECV 5-22, https://doi.org/10.1029/2001JB000515

Rocchi, S., LeMasurier, W.E. and Di Vincenzo, G. 2006. Oligocene to Holocene erosion and glacial history in Marie Byrd Land, West Antarctica, inferred from exhumation of the Dorrel Rock intrusive complex and form volcano morphologies. *Geological Society of America Bulletin*, **118**, 991–1005, https://doi.org/10.1130/B25675.1

Ross, P.-S. and White, J.D.L. 2006. Debris jets in continental phreatomagmatic volcanoes: a field study of their subterranean deposits in the Coombs Hills vent complex, Antarctica. *Journal of Volcanology and Geothermal Research*, **149**, 62–84, https://doi.org/10.1016/j.jvolgeores.2005.06.007

Ross, P.-S., Ukstins Peate, I., McClintock, M.K., Xu, Y.G., Skilling, I.P., White, J.D.L. and Houghton, B.F. 2005. Mafic volcaniclastic deposits in flood basalt provinces: A review. *Journal of Volcanology and Geothermal Research*, **145**, 281–314, https://doi.org/10.1016/j.jvolgeores.2005.02.003

Ross, P.-S., McClintock, M.K. and White, J.D.L. 2008. Geological evolution of the Coombs–Allan Hills area, Ferrar large igneous province, Antarctica: debris avalanches, mafic pyroclastic density currents, phreatocauldrons. *Journal of Volcanology and Geothermal Research*, **172**, 38–60, https://doi.org/10.1016/j.jvolgeores.2005.11.011

Ross, P.-S., Delpit, S., Haller, M.J., Németh, K. and Corbella, H. 2011. Influence of the substrate on maar-diatreme volcanoes – An example of a mixed setting from the Pali Aike volcanic field, Argentina. *Journal of Volcanology and Geothermal Research*, **201**, 253–271, https://doi.org/10.1016/j.jvolgeores.2010.07.018

Ross, S., Fioretti, A.M. and Cavazzini, G. 2002. Petrography, geochemistry and geochronology of the Cenozoic Cape Crossfire, Cape King and No Ridge igneous complexes (northern Victoria Land, Antarctica). *Royal Society of New Zealand Bulletin*, **35**, 215–225.

Rossetti, F., Storti, F. *et al*. 2006. Eocene initiation of Ross Sea dextral faulting and implications for East Antarctic neotectonics. *Journal of the Geological Society, London*, **163**, 119–126, https://doi.org/10.1144/0016-764905-005

Rowland, S.K. and Walker, G.P.L. 1990. Pahoehoe and aa in Hawaii: volumetric flow rate controls the lava structure. *Bulletin of Volcanology*, **52**, 615–628, https://doi.org/10.1007/BF00301212

Rutford, R.H., Craddock, C. and Bastien, T.W. 1968. Late Tertiary glaciation and sea-level changes in Antarctica. *Palaeogeography, Palaeoclimatology, Palaeoecology*, **5**, 15–39, https://doi.org/10.1016/0031-0182(68)90058-8

Rutford, R.H., Craddock, C., White, C.M. and Armstrong, R.L. 1972. Tertiary glaciation in the Jones Mountains. *In*: Adie, R.J. (ed.) *Antarctic Geology and Geophysics*. Universitetsforlaget, Oslo, 239–243.

Salvini, F., Brancolini, G., Busetti, M., Storti, F., Mazzarini, F. and Coren, F. 1997. Cenozoic geodynamics of the Ross Sea Region, Antarctica: crustal extension, intraplate strike–slip faulting and tectonic inheritance. *Journal of Geophysical Research*, **102**, 24 669–24 696, https://doi.org/10.1029/97JB01643

Salzmann, U., Riding, J.B., Nelson, A.E. and Smellie, J.L. 2011. How likely was a green Antarctic Peninsula during warm Pliocene interglacials? A critical reassessment based on new palynofloras from James Ross Island. *Palaeogeography, Palaeoclimatology, Palaeoecology*, **309**, 73–82, https://doi.org/10.1016/j.palaeo.2011.01.028

Schmidt-Thomé, M., Mueller, P. and Tessensohn, F. 1990. Malta Plateau. *American Geophysical Union Antarctic Research Series*, **48**, 53–59.

Schroeder, D.M., Blankenship, D.D., Young, D.A. and Quartini, E. 2014. Evidence for elevated and spatially variable geothermal flux beneath the West Antarctic Ice Sheet. *Proceedings of the National Academy of Sciences of the United States of America*, **111**, 9070–9072, https://doi.org/10.1073/pnas.1405184111

Self, S., Thordarson, Th. *et al*. 1996. A new model for the emplacement of Columbia River basalts as large, inflated pahoehoe lava flow fields. *Geophysical Research Letters*, **23**, 2689–2692, https://doi.org/10.1029/96GL02450

Self, S., Thordarson, T. and Keszthelyi, L. 1997. Emplacement of continental flood basalt lava flows. *American Geophysical Union Geophysical Monograph Series*, **100**, 381–410.

Sensarma, S., Storey, B.C. and Malviya, V.P. 2018. Gondwana Large Igneous Provinces (LIPs): distribution, diversity and significance. *Geological Society, London, Special Publications*, **463**, 1–16, https://doi.org/10.1144/SP463.11

Seroussi, H., Ivins, E.R., Wiens, D.A. and Bondzio, J. 2017. Influence of a West Antarctic mantle plume on ice sheet basal conditions. *Journal of Geophysical Research*, **122**, 7127–7155, https://doi.org/10.1002/2017JB014423

Siddoway, C.S. 2008. Tectonics of the West Antarctic Rift System: New light on the history and dynamics of distributed intracontinental extension. *In*: Cooper, A.K., Barrett, P.J. *et al*. (eds) *Antarctica: A Keystone in a Changing World. Proceedings of the 10th International Symposium on Antarctic Earth Sciences*. The National Academies Press, Washington, DC, 91–114.

Sims, K.W.W., Aster, R. *et al*. 2021. Mount Erebus. *Geological Society, London, Memoirs*, **55**, https://doi.org/10.1144/M55-2019-8

Skilling, I.P. 1994. Evolution of an englacial volcano: Brown Bluff, Antarctica. *Bulletin of Volcanology*, **56**, 573–591, https://doi.org/10.1007/BF00302837

Skilling,, I.P. 2002. Basaltic pahoehoe lava-fed deltas: large-scale characteristics, clast generation, emplacement processes and environmental discrimination. *Geological Society, London, Special Publications*, **202**, 91–113, https://doi.org/10.1144/GSL.SP.2002.202.01.06

Smellie, J.L. 1987. Geochemistry and tectonic setting of alkaline volcanic rocks in the Antarctic Peninsula: a review. *Journal of Volcanology and Geothermal Research*, **32**, 269–285, https://doi.org/10.1016/0377-0273(87)90048-5

Smellie, J.L. 1990. Graham Land and South Shetland Islands. *American Geophysical Union Antarctic Research Series*, **48**, 302–312.

Smellie, J.L. 1999. Lithostratigraphy of Miocene–Recent, alkaline volcanic fields in the Antarctic Peninsula and eastern Ellsworth Land. *Antarctic Science*, **11**, 362–378, https://doi.org/10.1017/S0954102099000450

Smellie, J.L. 2001. Lithofacies architecture and construction of volcanoes in englacial lakes: Icefall Nunatak, Mount Murphy, eastern Marie Byrd Land, Antarctica. *International Association of Sedimentologists Special Publications*, **30**, 73–98.

Smellie, J.L. 2002. The 1969 subglacial eruption on Deception Island (Antarctica): events and processes during an eruption beneath a thin glacier and implications for volcanic hazards. *Geological Society, London, Special Publications*, **202**, 59–79, https://doi.org/10.1144/GSL.SP.2002.202.01.04

Smellie, J.L. 2006. The relative importance of supraglacial v. subglacial meltwater escape in basaltic subglacial tuya eruptions: An important unresolved conundrum. *Earth-Science Reviews*, **74**, 241–268, https://doi.org/10.1016/j.earscirev.2005.09.004

Smellie, J.L. 2008. Basaltic subglacial sheet-like sequences: evidence for two types with different implications for the inferred thickness of associated ice. *Earth-Science Reviews*, **88**, 60–88, https://doi.org/10.1016/j.earscirev.2008.01.004

Smellie, J.L. 2013. Quaternary vulcanism: subglacial landforms. *In*: Elias, S.A. (ed.) *The Encyclopedia of Quaternary Science*. 2nd edn. Elsevier, Amsterdam, 780–802.

Smellie, J.L. 2018. Glaciovolcanism – a 21st century proxy for palaeo-ice. *In*: Menzies, J. and van der Meer, J.J.M. (eds) *Past Glacial Environments (Sediments, Forms and Techniques)*. 2nd edn. Elsevier, Amsterdam, 335–375.

Smellie, J.L. 2020a. Antarctic Peninsula – geology and dynamic development. *In*: Kleinschmidt, G. (ed.) *Geology of the Antarctic Continent*. Gebrüder Borntraeger, Stuttgart, Germany, 18–86.

Smellie, J.L. 2020b. The role of volcanism in the making of Antarctica. *In*: Oliva, M. and Ruiz, J. (eds) *Past Antarctica. Paleoclimatology & Climate Change*. Academic Press, London, 69–88.

Smellie, J.L. 2021. Bransfield Strait and James Ross Island: volcanology. *Geological Society, London, Memoirs*, **55**, https://doi.org/10.1144/M55-2018-58

Smellie, J.L. and Collerson, K.D. 2021. Gaussberg: volcanology and petrology. *Geological Society, London, Memoirs*, **55**, https://doi.org/10.1144/M55-2018-85

Smellie, J.L. and Edwards, B.E. 2016. *Glaciovolcanism on Earth and Mars: Products, Processes and Palaeoenvironmental Significance*. Cambridge University Press, Cambridge, UK.

Smellie, J.L. and Hole, M.J. 1997. Products and processes in Pliocene–Recent, subaqueous to emergent volcanism in the Antarctic Peninsula: examples of englacial Surtseyan volcano construction. *Bulletin of Volcanology*, **58**, 628–646, https://doi.org/10.1007/s004450050167

Smellie, J.L. and Hole, M.J. 2021. Antarctic Peninsula: volcanology. *Geological Society, London, Memoirs*, **55**, https://doi.org/10.1144/M55-2018-59

Smellie, J.L. and Martin, A.P. 2021. Erebus Volcanic Province: volcanology. *Geological Society, London, Memoirs*, **55**, https://doi.org/10.1144/M55-2018-62

Smellie, J.L. and Rocchi, S. 2021. Northern Victoria Land: volcanology. *Geological Society, London, Memoirs*, **55**, https://doi.org/10.1144/M55-2018-60

Smellie, J.L. and Skilling, I.P. 1994. Products of subglacial eruptions under different ice thicknesses: two examples from Antarctica. *Sedimentary Geology*, **91**, 115–129, https://doi.org/10.1016/0037-0738(94)90125-2

Smellie, J.L., Pankhurst, R.J., Thomson, M.R.A. and Davies, R.E.S. 1984. *The Geology of the South Shetland Islands: VI. Stratigraphy, Geochemistry and Evolution*. British Antarctic Survey Scientific Reports, **87**.

Smellie, J.L., Hole, M.J. and Nell, P.A.R. 1993. Late Miocene valley-confined subglacial volcanism in northern Alexander Island, Antarctic Peninsula. *Bulletin of Volcanology*, **55**, 273–288, https://doi.org/10.1007/BF00624355

Smellie, J.L., Millar, I.L., Rex, D.C. and Butterworth, P.J. 1998. Subaqueous, basaltic lava dome and carapace breccia on King George Island, South Shetland Islands, Antarctica. *Bulletin of Volcanology*, **59**, 245–261, https://doi.org/10.1007/s004450050189

Smellie, J.L., López-Martínez, J. et al. 2002. *Geology and Geomorphology of Deception Island (1:25 000 Scale)*. BAS GEOMAP Series, Sheets 6-A and 6-B, supplementary text. British Antarctic Survey, Cambridge, UK.

Smellie, J.L., McArthur, J.M., McIntosh, W.C. and Esser, R. 2006. Late Neogene interglacial events in the James Ross Island region, northern Antarctic Peninsula, dated by Ar/Ar and Sr-isotope stratigraphy. *Palaeogeography, Palaeoclimatology, Palaeoecology*, **242**, 169–187, https://doi.org/10.1016/j.palaeo.2006.06.003

Smellie, J.L., Johnson, J.S., McIntosh, W.C., Esser, R., Gudmundsson, M.T., Hambrey, M.J. and van Wyk de Vries, B. 2008. Six million years of glacial history recorded in the James Ross Island Volcanic Group, Antarctic Peninsula. *Palaeogeography, Palaeoclimatology, Palaeoecology*, **260**, 122–148, https://doi.org/10.1016/j.palaeo.2007.08.011

Smellie, J.L., Haywood, A.M., Hillenbrand, C.-D., Lunt, D.J. and Valdes, P.J. 2009. Nature of the Antarctic Peninsula Ice Sheet during the Pliocene: Geological evidence and modelling results compared. *Earth-Science Reviews*, **94**, 79–94, https://doi.org/10.1016/j.earscirev.2009.03.005

Smellie, J.L., Rocchi, S. and Armienti, P. 2011a. Late Miocene volcanic sequences in northern Victoria Land, Antarctica: products of glaciovolcanic eruptions under different thermal regimes. *Bulletin of Volcanology*, **73**, 1–25, https://doi.org/10.1007/s00445-010-0399-y

Smellie, J.L., Rocchi, S., Gemelli, M., Di Vincenzo, G. and Armienti, P. 2011b. A thin predominantly cold-based Late Miocene East Antarctic ice sheet inferred from glaciovolcanic sequences in northern Victoria Land, Antarctica. *Palaeogeography, Palaeoclimatology, Palaeoecology*, **307**, 129–149, https://doi.org/10.1016/j.palaeo.2011.05.008

Smellie, J.L., Johnson, J.S. and Nelson, A.E. 2013a. *Geological Map of James Ross Island. 1. James Ross Island Volcanic Group (1:125 000 Scale)*. BAS GEOMAP 2 Series, Sheet 5. British Antarctic Survey, Cambridge, UK.

Smellie, J.L., Wilch, T. and Rocchi, A. 2013b. 'A'ā lava-fed deltas: a new reference tool in paleoenvironmental research. *Geology*, **41**, 403–406, https://doi.org/10.1130/G33631.1

Smellie, J.L., Rocchi, S. et al. 2014. Glaciovolcanic evidence for a polythermal Neogene East Antarctic Ice Sheet. *Geology*, **42**, 39–41, https://doi.org/10.1130/G34787.1

Smellie, J.L., Rocchi, S., Johnson, J.S., Di Vincenzo, G. and Schaefer, J.M. 2018. A tuff cone erupted under frozen-bed ice (northern Victoria Land, Antarctica): linking glaciovolcanic and cosmogenic nuclide data for ice sheet reconstruction. *Bulletin of Volcanology*, **80**, 12, https://doi.org/10.1007/s00445-017-1185-x

Smellie, J.L., Panter, K.S. and Reindel, J. 2021. Mount Early and Sheridan Bluff: volcanology. *Geological Society, London, Memoirs*, **55**, https://doi.org/10.1144/M55-2018-61

Solari, M.A., Hervé, F., Martinod, J., Le Roux, J.P., Ramírez, L.E. and Palacios, C. 2008. Geotectonic evolution of the Bransfield Basin, Antarctic Peninsula: insights from analogue models. *Antarctic Science*, **20**, 185–196, https://doi.org/10.1017/S095410200800093X

Soloviev, V.D., Bakhmutov, V.G., Korchagin, I.N. and Yegorova, T.P. 2018. New geophysical data about the Pacific Margin (West Antarctica) Magnetic Anomaly sources and origin. *Ukrainian Antarctic Journal*, **17**, 20–31, https://doi.org/10.33275/1727-7485.1(17).2018.28

Storey, B.C. 1995. The role of mantle plumes in continental breakup: case histories from Gondwanaland. *Nature*, **377**, 301–308, https://doi.org/10.1038/377301a0

Storey, B.C. and Kyle, P.R. 1997. An active mantle mechanism for Gondwana breakup. *South African Journal of Geology*, **100**, 283–290.

Storey, B.C., Hole, M.J., Pankhurst, R.J., Millar, I.L. and Vennum, W. 1988. Middle Jurassic within-plate granites in West Antarctica and their bearing on the break-up of Gondwanaland. *Journal of the Geological Society, London*, **145**, 999–1007, https://doi.org/10.1144/gsjgs.145.6.0999

Storey, B.C., Leat, P.T., Weaver, S.D., Pankhurst, R.J., Bradshaw, J.D. and Kelley, S. 1999. Mantle plumes and Antarctica–New Zealand rifting: evidence from mid-Cretaceous mafic dykes. *Journal of the Geological Society, London*, **156**, 659–671, https://doi.org/10.1144/gsjgs.156.4.0659

Storey, B.C., Vaughan, A.P.M. and Riley, T.R. 2013. The link between large igneous provinces, continental break-up and environmental change: evidence reviewed from Antarctica. *Earth and Environmental Science Transactions of the Royal Society of Edinburgh*, **104**, 1–14, https://doi.org/10.1017/S175569101300011X

Storti, F., Salvini, F., Rossetti, F. and Phipps Morgan, J. 2007. Intraplate termination of transform faulting within the Antarctic continent. *Earth and Planetary Science Letters*, **260**, 115–126, https://doi.org/10.1016/j.epsl.2007.05.020

Studinger, M., Bell, R.E., Blankenship, D.D., Buck, W.R. and Karner, G.D. 2004. Sub-ice geology inland of the Transantarctic Mountains in light of new aerogeophysical data. *Earth and Planetary Science Letters*, **220**, 391–408, https://doi.org/10.1016/S0012-821X(04)00066-4

Svensen, H., Planke, S., Chevallier, L., Malthe-Sørenssen, A., Corfu, F. and Jamtveit, B., 2007. Hydrothermal venting of greenhouse gases triggering Early Jurassic global warming. *Earth and Planetary Science Letters*, **256**, 554–566, https://doi.org/10.1016/j.epsl.2007.02.013

Svensen, H., Corfu, F., Polteau, S., Hammer, Ø. and Planke, S. 2012. Rapid magma 1508 emplacement in the Karoo Large Igneous

Province. *Earth and Planetary Science Letters*, **325–326**, 1–9, https://doi.org/10.1016/j.epsl.2012.01.015

Sykes, M.A. 1988. New K–Ar age determinations on the James Ross island Volcanic Group, north-east Graham Land, Antarctica. *British Antarctic Survey Bulletin*, **80**, 51–56.

Thordarson, T., Self, S., Miller, D.J., Larsen, G. and Vilmundardóttir, E.G. 2003. Sulphur release from flood lava eruptions in the Veidivötn, Grímsvötn and Katla volcanic systems, Iceland. *Geological Society, London, Special Publications*, **213**, 103–121, https://doi.org/10.1144/GSL.SP.2003.213.01.07

Tingey, R.J., MacDougall, I. and Gleadow, A.J.W. 1983. The age and mode of formation of Gaussberg, Antarctica. *Journal of the Geological Society of Australia*, **30**, 241–246, https://doi.org/10.1080/00167618308729251

Troedson, A.L. and Smellie, J.L. 2002. The Polonez Cove Formation of King George Island, West Antarctica: stratigraphy, facies and palaeoenvironmental implications. *Sedimentology*, **49**, 277–301, https://doi.org/10.1046/j.1365-3091.2002.00441.x

Trouw, R.A.J., Passchier, C.W., Valeriano, C.M., Simões, L.S.A., Paciullo, F.V.P. and Ribiero, A. 2000. Deformational evolution of a Cretaceous subduction complex: Elephant Island, South Shetland Islands, Antarctica. *Tectonophysics*, **319**, 93–110, https://doi.org/10.1016/S0040-1951(00)00021-4

Turner, J., Marshall, G.J. and Lachlan-Cope, T.A. 1998. Analysis of synoptic-scale low pressure systems within the Antarctic Peninsula sector of the circumpolar trough. *International Journal of Climatology*, **18**, 253–280, https://doi.org/10.1002/(SICI)1097-0088(19980315)18:3<253::AID-JOC248>3.0.CO;2-3

Ukstins Peate, I., Larsen, M. and Lesher, C.E. 2003. The transition from sedimentation to flood volcanism in the Kangerlussuaq Basin, East Greenland: basaltic pyroclastic volcanism during initial Palaeogene continental break-up. *Journal of the Geological Society, London*, **160**, 759–772, https://doi.org/10.1144/0016-764902-071

van Lipzig, N.P.M., King, J.C., Lachlan-Cope, T.A. and van den Broeke, M.R. 2004. Precipitation, sublimation and snow drift in the Antarctic Peninsula region from a regional atmospheric model. *Journal of Geophysical Research*, **109**, D24, https://doi.org/10.1029/2004JD004701

Vanneste, L.E. and Larter, R.D. 2002. Sediment subduction, subduction erosion, and strain regime in the northern South Sandwich forearc. *Journal of Geophysical Research*, **107**, https://doi.org/10.1029/2001JB000396

van Wyk de Vries, M., Bingham, R.G. and Hein, A.S. 2018. A new volcanic province: an inventory of subglacial volcanoes in West Antarctica. *Geological Society, London, Special Publications*, **461**, 231–248, https://doi.org/10.1144/SP461.7

Vaughan, A.P.M., Millar, I.L. and Thistlewood, L. 1999. The Auriga Nunataks shear zone: Mesozoic transfer faulting and arc deformation in northwest Palmer Land, Antarctica. *Tectonics*, **18**, 911–928, https://doi.org/10.1029/1999TC900008

Vogel, S.W., Tulaczyk, S., Carter, S., Renne, P., Turrin, B. and Grunow, A. 2006. Geologic constraints on the existence and distribution of West Antarctic subglacial volcanism. *Geophysical Research Letters*, **33**, L23501, https://doi.org/10.1029/2006GL027344

Webb, P.-N. and Harwood, D.M. 1991. Late Cenozoic glacial history of the Ross Embayment, Antarctica. *Quaternary Science Reviews*, **10**, 215–223, https://doi.org/10.1016/0277-3791(91)90020-U

White, J.D.L. and Houghton, B.F. 2006. Primary volcaniclastic rocks. *Geology*, **34**, 677–680, https://doi.org/10.1130/G22346.1

White, J.D.L. and McClintock, M.K. 2001. Immense vent complex marks flood-basalt eruption in a wet, failed rift: Coombs Hills, Antarctica. *Geology*, **29**, 935–938, https://doi.org/10.1130/0091-7613(2001)029<0935:IVCMFB>2.0.CO;2

White, J.D.L. and Ross, P.-S. 2011. Maar-diatreme volcanoes: A review. *Journal of Volcanology and Geothermal Research*, **201**, 1–29, https://doi.org/10.1016/j.jvolgeores.2011.01.010

Whitham, A.G. and Marshall, J.E.A. 1988. Syn-depositional deformation in a Cretaceous succession, James Ross Island, Antarctica. Evidence from vitrinite reflectivity. *Geological Magazine*, **125**, 583–591, https://doi.org/10.1017/S0016756800023402

Wilch, T.I. and McIntosh, W.C. 2000. Eocene and Oligocene volcanism at Mt. Petras, Marie Byrd Land: implication s for middle Cenozoic ice sheet reconstructions in West Antarctica, *Antarctic Science*, **12**, 477–491, https://doi.org/10.1017/S0954102000000560

Wilch, T.I. and McIntosh, W.C. 2002. Lithofacies analysis and $^{40}Ar/^{39}Ar$ geochronology of ice–volcano interactions at Mt. Murphy and the Crary Mountains, Marie Byrd Land, Antarctica. *Geological Society, London, Special Publications*, **2002**, 237–253, https://doi.org/10.1144/GSL.SP.2002.202.01.12

Wilch, T.I. and McIntosh, W.C. 2007. Miocene–Pliocene ice–volcano interactions at monogenetic volcanoes near Hobbs Coast, Marie Byrd Land, Antarctica. *United States Geological Survey Open-File Report*, **2007-1047**, extended abstract 074.

Wilch, T.I., McIntosh, W.C. and Panter, K.S. 2021. Marie Byrd Land and Ellsworth Land: volcanology. *Geological Society, London, Memoirs*, **55**, https://doi.org/10.1144/M55-2019-39

Willan, R.C.R. and Kelley, S.P. 1999. Mafic dike swarms in the South Shetland Islands volcanic arc: Unravelling multiepisodic magmatism related to subduction and continental rifting. *Journal of Geophysical Research*, **104**, 23 051–23 068, https://doi.org/10.1029/1999JB900180

Wilson, D.S., Pollard, D., DeConto, R.M., Jamieson, S.S.R. and Luyendyk, B.P. 2013. Initiation of the West Antarctic Ice Sheet and estimates of total Antarctic ice volume in the earliest Oligocene. *Geophysical Research Letters*, **40**, 4305–4309, https://doi.org/10.1002/grl.50797

Wilson, G.S. 1995. The Neogene East Antarctic Ice Sheet: A dynamic or stable feature? *Quaternary Science Reviews*, **14**, 101–123, https://doi.org/10.1016/0277-3791(95)00002-7

Wörner, G. and Viereck, L. 1989. The Mt. Melbourne Volcanic Field (Victoria Land, Antarctica) I. Field observations. *Geologisches Jahrbuch*, **E38**, 369–393.

Chapter 1.3

Antarctic volcanism: petrology and tectonomagmatic overview

Kurt Samuel Panter

School of Earth, Environment and Society, Bowling Green State University, Bowling Green, OH 43403, USA

0000-0002-0990-5880

kpanter@bgsu.edu

Abstract: Petrological investigations over the past 30 years have significantly advanced our knowledge of the origin and evolution of magmas emplaced within and erupted on top of the Antarctic Plate. Over the last 200 myr Antarctica has experienced: (1) several episodes of rifting, leading to the fragmentation of Gondwana and the formation by *c.* 83 Ma of the current Antarctica Plate; (2) long-lived subduction that shut down progressively eastwards along the Gondwana margin in the Late Cretaceous and is still active at the northernmost tip of the Antarctic Peninsula; and (3) broad extension across West Antarctica that produced one of the Earth's major continental rift systems. The dynamic tectonic history of Antarctica since the Triassic has led to a diversity of volcano types and igneous rock compositions with correspondingly diverse origins. Many intriguing questions remain about the petrology of mantle sources and the mechanisms for melting during each tectonomagmatic phase. For intraplate magmatism, the upwelling of deep mantle plumes is often evoked. Alternatively, subduction-related metasomatized mantle sources and melting by more passive means (e.g. edge-driven flow, translithospheric faulting, slab windows) are proposed. A brief review of these often competing models is provided in this chapter along with recommendations for ongoing petrological research in Antarctica.

Magmatism in Antarctica over the last 200 myr includes: (1) voluminous magmatic activity associated with continental break-up; (2) igneous activity connected with continental–ocean convergence and back-arc extension; (3) igneous activity that is transitional between active and passive margin tectonics; and (4) magmatism related to one of the Earth's major continental rift systems. In addition, some widely scattered volcanism, without any clear association with tectonic processes, occurs on both continental and oceanic portions of the Antarctic Plate. With the variety of tectonomagmatic settings comes a diversity of volcanic rock compositions ranging from basalt to rhyolite that encompass both alkaline and sub-alkaline (i.e. calc-alkalic and tholeiitic) magma series, and rare occurrences of ultrapotassic series rocks (e.g. lamproite). This geochemical diversity originates from the melting of different mantle sources, defined modally and compositionally, and from varied conditions and mechanisms responsible for melt generation. Melt origin is a subject of considerable debate among Antarctic scientists but also globally for petrologists studying intraplate (e.g. http://www.mantleplumes.org/) and post-subduction continental margin (e.g. Castillo 2008; Calmus *et al.* 2011; Seghedi *et al.* 2011; Rabayrol *et al.* 2019) volcanism.

Antarctica is uniquely suited to help disentangle many long-standing petrological problems given its progressive isolation from destructive plate boundaries over the past 90 myr and its relative geostationary position for most of that time. The dampening of near-field tectonic processes through the Cenozoic is beneficial in helping to constrain the relative contributions of lithospheric and sublithospheric mantle melt sources, and the influence of lithospheric processes and subcrustal architecture, on melting regimes.

The occurrences and petrology of igneous rocks, along with fragments of crust and mantle materials (i.e. xenoliths) that the magmas bring to the surface, also add to our understanding of the history and dynamics of Antarctica's ice sheets. Petrological contributions to heat-flux estimates are critical to our knowledge of the conditions on the bottom of ice sheets (i.e. wet-based v. cold-based) and their stability. Petrological studies provide information on mantle rheology used to constrain glacial isostatic adjustments. Furthermore, the age and geochemical fingerprinting of volcanic material recovered in ice cores provide critical stratigraphic links and add to our understanding ice-core timescales. Finally, it is worth noting that although the continent's remoteness, harsh environment and its mostly ice-covered terrain present significant obstacles to fieldwork, most of the rock exposed in its extremely dry climate is unfettered by vegetation and, on the whole, remarkably fresh.

The primary objective of this Memoir is to provide an up-to-date volcanological, geochronological and petrological information resource on Mesozoic and Cenozoic magmatism in Antarctica, with the goal of presenting new ideas, stimulating new discovery and reaffirming the important contributions that research on igneous rocks can provide to our overall knowledge of Antarctica and to the global perspective. This overview chapter is structured to first deliver a brief review of the tectonomagmatic history over the past 200 myr, highlighting the occurrences, compositional types and primary origins of igneous rocks that are discussed in detail within the chapters that follow. Next, emphasis is placed on several areas of controversy concerning the primary sources and the fundamental mechanism for melting of the mantle beneath Antarctica. The chapter ends with some priorities for the advancement of petrological research on Antarctic volcanism.

Tectonomagmatic history

The past 200 myr of volcanism in Antarctica is intimately linked to its dynamic tectonic history (Jordan *et al.* 2020; Storey and Granot 2021). Over this period, igneous activity resulted from simultaneous tectonic processes of continental fragmentation and subduction as the Gondwana supercontinent drifted southward towards the South Pole. Progressive disintegration of Gondwana changed the global continental configuration and eventually led to the creation of the Southern Ocean. By the Middle–Late Eocene Australia had separated from Antarctica and by the Middle Oligocene the opening of the Drake Passage established the Antarctic Circumpolar Current, isolating the continent of Antarctica from all other landmasses (Storey and Granot 2021). Today, the Antarctic Plate is bordered by six major tectonic plates and is surrounded by spreading ocean ridges. Physiographically, the continent itself is divided into East and West Antarctica,

From: Smellie, J. L., Panter, K. S. and Geyer, A. (eds) 2021. *Volcanism in Antarctica: 200 Million Years of Subduction, Rifting and Continental Break-up*. Geological Society, London, Memoirs, **55**, 43–53,

First published online January 15, 2021, https://doi.org/10.1144/M55-2020-10

© 2021 The Author(s). Published by The Geological Society of London. All rights reserved.

For permissions: http://www.geolsoc.org.uk/permissions. Publishing disclaimer: www.geolsoc.org.uk/pub_ethics

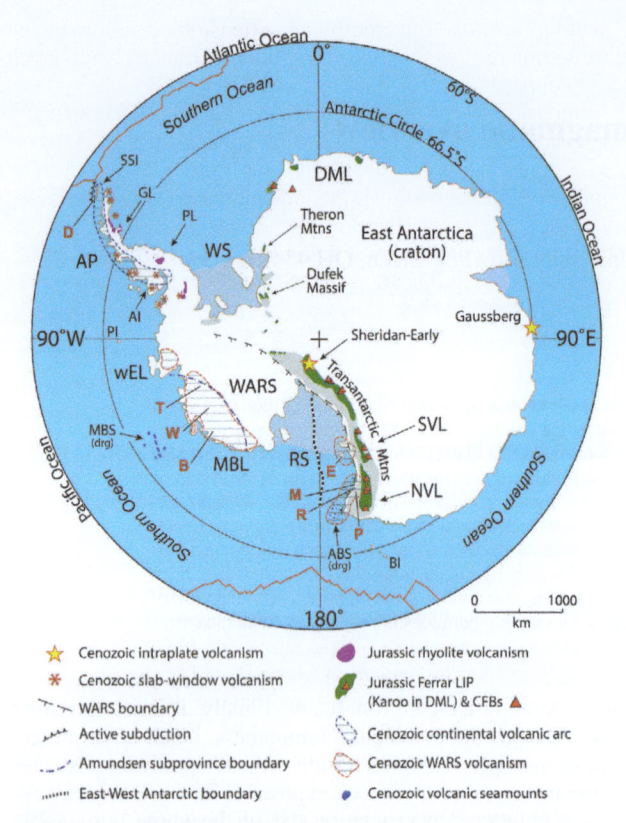

Fig. 1. Map of Antarctica showing the distribution of Mesozoic (Jurassic and Cretaceous) and Cenozoic tectonomagmatic provinces and volcanism. Distribution of Ferrar and Karoo large igneous provinces (LIPs) and associated continental flood basalts (CFBs) after Luttinen (2018) and Elliot and Fleming (2021). The locations of Jurassic rhyolite volcanism are from Riley and Leat (2021). The boundary of the Amundsen geotectonic subprovince of the West Antarctic Rift System (WARS) is after Jordan *et al.* (2020), and the geological boundary between East and West Antarctica is after Tinto *et al.* (2019) and Jordan *et al.* (2020). The locations of active volcanoes indicated by red letters are: B, Mount Berlin; D, Deception Island; E, Mount Erebus; M, Mount Melbourne; P, The Pleiades; R, Mount Rittmann; T, Mount Takahe; W, Mount Waesche. Other abbreviations: ABS, Adare Basin seamounts collected by dredging (drg: Panter *et al.* 2018); AI, Alexander Island; AP, Antarctic Peninsula; BI, Balleny Islands; DML, Dronning Maud Land; GL, Graham Land; MBL, Marie Byrd Land; MBS, Marie Byrd seamounts collected by dredging (drg: Kipf *et al.* 2014); NVL, northern Victoria Land; PL, Palmer Land; PI, Peter I Island; RS, Ross Sea; SSI, South Shetland Islands; SVL, southern Victoria Land; WS, Weddell Sea; wEL, western Ellsworth Land. Cenozoic WARS volcanism in MBL and wEL belong to the Marie Byrd Land Volcanic Group (Wilch *et al.* 2021), and Cenozoic volcanism in NVL and SVL belongs to the McMurdo Volcanic Group (Kyle 1990).

with the Transantarctic Mountains marking the boundary (Fig. 1). Geologically, the boundary between East and West Antarctica is poorly known, although recent geophysical studies suggest it may be beneath the middle of the Ross Sea (Fig. 1) (Tinto *et al.* 2019; Jordan *et al.* 2020). The older East Antarctic Craton is composed of sutured Archean–Cambrian terranes, and younger Paleozoic magmatic events and extensive sedimentary basins. The younger West Antarctic province consists of terranes amalgamated over an extended period beginning in the Neoproterozoic and culminating in the Late Cretaceous at the Antarctic Peninsula. Beginning in the Late Cretaceous and ending in the Late Miocene, the relative motion between East and West Antarctica caused thinning of continental crust between the Transantarctic Mountains and the Pacific margin in several stages (Storey and Granot 2021). This extensional episode created the West Antarctic Rift System (WARS) (Fig. 1). Elsewhere, back-arc extension on the northern segment of the Antarctic Peninsula, where the Phoenix Plate (also known as Drake Plate) is slowly subducting beneath the South Shetland Plate, has produced the Bransfield Strait marginal basin in a continental setting (Smellie 2021).

Jurassic–Early Cretaceous magmatism

The break-up of Gondwana was initiated in the early Jurassic by a mantle-plume-driven event that formed the Karoo and Ferrar large igneous provinces (LIPs) (Storey and Kyle 1997). Continental flood basalts and dyke swarms of the Karoo LIP occur in southern Africa and Dronning Maud Land, Antarctica (Luttinen 2018) (Fig. 1). The Ferrar LIP (*c.* 183 Ma: Elliot and Fleming 2021) is distributed along the Transantarctic Mountains as tholeiitic continental flood basalts (Kirkpatrick Basalt) and as intrusives (Dufek Massif and Ferrar Dolerite) (Fig. 1). Early–Middle Jurassic silicic volcanic rocks dominated by large-volume rhyolite-rhyodacite ignimbrites and crystal-rich tuffs (Riley and Leat 2021) are associated with the Chon Aike province of southern South America, and are exposed across the central and eastern Antarctic Peninsula (Fig. 1). These enormous and contemporaneous magmatic provinces are considered to have formed in a continental marginal back-arc setting that was underplated by a mantle plume, probably promoting extension and rifting, which ultimately caused the initial Late Jurassic break-up (*c.* 165–160 Ma) of Gondwana supercontinent on its western margin (Storey and Granot 2021). In contrast, the break-up of India from the East Gondwana margin in the Early Cretaceous (*c.* 135–130 Ma) was unlikely to have been initiated by mantle-plume activity, although a thermal influence associated with the emerging Kerguelen plume may have contributed (Gaina *et al.* 2007). Warmer material from the Kerguelen plume was likely to have been channelled from beneath the continental plate to the thinned lithosphere of the nascent Indian Ocean where decompression led to volcanism (*c.* 119 Ma) and the early construction of the Kerguelen Plateau (Olierook *et al.* 2019).

Along the proto-Pacific margin of Gondwana in the Middle Jurassic–Early Cretaceous ongoing convergence between oceanic and continental lithosphere produced forearc volcanism that consisted of a calc-alkaline bimodal (basalt–dacite/rhyolite) suite of lavas and volcanoclastics (Fossil Bluff Group) exposed on the eastern portion of Alexander Island near the base of the Antarctic Peninsula (Fig. 1) (Leat and Riley 2021) and correlative volcanoclastic deposits further north on Adelaide Island (Jordan *et al.* 2020). The rest of the Antarctic Peninsula volcanic arc was established by the Early Cretaceous and remained active with scattered volcanism along its length until the Early Miocene (Leat and Riley 2021).

Late Cretaceous–Paleocene volcanism

Late Cretaceous rifting related to the final stage of Gondwana break-up separated the continental fragments that became Zealandia. It did not produce any known volcanism in Antarctica, although alkaline dykes (i.e. intrusive equivalents of basalt, shoshonite, mugearite, hawaiite, potassic trachybasalt and basaltic andesite) were emplaced just prior to separation (107 Ma), and are found along the Ruppert and Hobbs coasts of Marie Byrd Land (Storey *et al.* 1999). Moreover, there was extensive intraplate volcanism in Zealandia that occurred prior to and just after break-up between 99 and 79 Ma (Hoernle *et al.* 2020 and references therein). As in the Jurassic, continental break-up in the Late Cretaceous may have also been associated with mantle-plume activity (Weaver et al. 1994; Storey et al. 1999; Hoernle et al. 2020). By the early Cenozoic, volcanism on the Amundsen Sea portion of the Antarctic Plate erupted alkali basalt–basaltic trachyandesite compositions between 65 and 56 Ma (Kipf et al. 2014), forming the Marie Byrd seamounts (Fig. 1). A mantle-plume underplate ('fossil plume') has been proposed as a source for the Marie Byrd seamounts (Kipf et al. 2014), as well as for the middle–late Cenozoic continental alkaline volcanism associated with the WARS (Rocholl et al. 1995; Hart et al. 1997; Panter et al. 2000).

Early Eocene–Early Miocene magmatism

Volcanism was generally small volume and widely scattered from the Eocene until the Early Miocene. In northern Victoria Land, Early Eocene–Early Miocene alkaline magmatism emplaced plutons and cogenetic dyke swarms (Meander Intrusive Group) that are part of the Western Ross Supergroup (Smellie and Martin 2021; Smellie et al. 2021). The intrusives, which are likely to be remnants of a subvolcanic system, are considered to be the result of transtensional stresses and directed mantle flow from beneath the thinned West Antarctic lithosphere and under the East Antarctic Craton (Rocchi and Smellie 2021). The earliest continental alkaline volcanism in the Cenozoic occurred during the Late Eocene–Early Miocene (37–20 Ma) in Marie Byrd Land (Wilch et al. 2021). The oldest volcanic rocks in Marie Byrd Land are found near the centre of a c. 1000 × 500 km structural dome that is located on the north flank of the WARS. The Marie Byrd Land dome has been interpreted as a topographical expression of a Cenozoic mantle plume (LeMasurier and Landis 1996).

Early Miocene alkaline intraplate activity in southern Victoria Land (i.e. the Erebus Volcanic Province of the McMurdo Volcanic Group) erupted mostly trachyte and rhyolite compositions at Mount Morning but is also evident from subsurface, mostly basaltic, materials recovered from drill cores taken within this region (Nyland et al. 2013; Martin et al. 2021). The fundamental cause of volcanism has been explained by two potentially competing hypotheses: (1) melting caused by edge-driven mantle flow; and (2) melting driven by a Cenozoic mantle plume (Martin et al. 2021). Approximately 1000 km to the south of the Erebus Volcanic Province and significantly inboard of the southern boundary of the WARS, Early Miocene volcanic activity (c. 20 Ma) produced alkaline and olivine tholeiite basalts at Mount Early and Sheridan Bluff (Fig. 1) of the Upper Scott Glacier Volcanic Field (Smellie 2021). The cause of this isolated activity is considered to be a response to the detachment and sinking of lithosphere into the convecting mantle beneath the East Antarctic Craton (Shen et al. 2017; Licht et al. 2018; Panter et al. 2021a).

Continental arc volcanism along the west coast of the Antarctic Peninsula, which may have initiated in the Early Cretaceous and was active up until the Early Miocene (c. 23 Ma), erupted magmas that are dominated by calc-alkaline series compositions ranging from basalt to rhyolite, with lesser amounts of adakite, high-Mg andesite, and intermediate to silicic (benmoreite and peralkaline trachyte) high-Zr rocks (Leat and Riley 2021). The origins of these compositions are explained by: (1) melting of variably depleted mantle-wedge material fluxed by hydrous fluids from the subducting slab (e.g. calc-alkaline series, South Shetland Islands); (2) partial melting of slab material and the equilibration of those melts under hydrous conditions within the mantle wedge prior to eruption (e.g. high-Mg andesite group, Alexander Island); (3) partial melting of garnet peridotite within the mantle wedge that incorporated melts of mafic slab material (e.g. adakitic group, South Shetland Islands); and (4) mantle partial melting that was triggered by arc extension or within a back-arc setting (e.g. high-Zr group, southern Graham Land–northern Palmer Land: Fig. 1) (Leat and Riley 2021).

Subduction terminated along the Antarctic Peninsula as segments of the Antarctic–Phoenix spreading centre collided with the trench northwards over time from c. 50 Ma at Alexander Island to c. 15 Ma at Adelaide Island to between 6 and 3 Ma SW of the Hero Fracture Zone (Hole 2021). Subduction continues to this day at a slow rate beneath the South Shetland Islands and the Bransfield Strait (Haase and Beier 2021; Leat and Riley 2021).

Middle Miocene–Holocene volcanism

Since the Middle Miocene, intraplate alkaline volcanism became more abundant and widespread across the extended crust of the WARS in Marie Byrd Land (Marie Byrd Land Volcanic Group) and within the western Ross Sea region (McMurdo Volcanic Group) of northern and southern Victoria Land (Fig. 1). The rift-related volcanism produced over 30 major polygenetic shield and stratovolcanoes (each ≥30–1800 km^3 of exposed volume above ice-sheet or sea level), with Mount Sidley located in Marie Byrd Land being the tallest at 4285 m above sea level and having the most complete petrogenetic record (Panter et al. 1994, 1997; Panter 1995). In addition to the central volcanoes there are numerous monogenetic volcanic fields, including seamounts in the southern Ross Sea (Aviado et al. 2015; Martin et al. 2021) and in the oceanic Adare Basin (Fig. 1) (Panter et al. 2018). Active volcanoes (i.e. erupted during the Holocene) include: Mount Berlin and Mount Takahe, and potentially Mount Waesche, within the Marie Byrd Land Volcanic Province (Dunbar et al. 2021); Mount Melbourne, Mount Rittmann and The Pleiades in the Melbourne Volcanic Province (Gambino et al. 2021); and Mount Erebus with its persistent phonolite lava lake located within the Erebus Volcanic Province (Fig. 1) (Sims et al. 2021). Overall, the volcanism consists of a full spectrum of alkaline series compositions ranging from basaltic types to intermediate to phonolite, trachyte, rhyolite and pantellerite (Martin et al. 2021; Panter et al. 2021b; Rocchi and Smellie 2021). Rare tholeiitic basalts associated with the WARS occur in the Fosdick Mountains in western Marie Byrd Land, and in the Hudson and Jones mountains in western Ellsworth Land (Panter et al. 2021b).

Existing models for magmatism within the WARS are grouped into two fundamental types: those that involve mantle plumes as a melt source and trigger for volcanism; and those that call upon more passive mechanisms to promote melting of pre-existing metasomatically veined lithosphere. The cause of metasomatism has been ascribed to several mechanisms, including enrichments from subduction-derived melts and fluids, plume-derived melts and fluids, and extension-related autometasomatism. The presence of active plumes is supported by geophysical evidence that includes high heat flow and slow seismic-wave velocity anomalies beneath Marie Byrd Land and Ross Island in south Victoria Land (e.g. Phillips et al. 2018; Lloyd et al. 2019). Edge-driven mantle convection along the boundary between the thinned lithosphere and the much thicker East Antarctic Craton in the western Ross Sea (also known as the western Ross Embayment) was established by the Eocene (Faccenna et al. 2008), and is considered a 'passive' mechanism to promote decompressive melting of the asthenosphere and the heating of the lithosphere at its base (Panter et al. 2018). The melt origins of rift-related West Antarctic volcanism will be discussed in more detail in a separate section.

Since the Middle–Late Miocene, volcanism became more widely distributed along the length of the Antarctic Peninsula

(Fig. 1). All igneous activity took place after, or nearly synchronous with, the cessation of subduction. In the southern portion of the Peninsula (Alexander Island and Palmer Land) post-subduction intraplate volcanism post-dates ridge–crest–trench collision by more than 40 myr (Hole 2021). In the northern part of the Antarctic Peninsula (Graham Land north of 66° S: Fig. 1) more recent and ongoing subduction has produced a more complex tectonic setting for volcanism, which includes activity developed on the thin crust within the Bransfield marginal basin, as well as continental back-arc and intraplate settings (Smellie 2021). Post-subduction volcanism produced extensive monogenetic volcanic fields, large polygenetic shield volcanoes and small isolated centres. The largest volcanic field on the Antarctic Peninsula is the James Ross Island Volcanic Group, covering an area of 7000 km^2 and a total volume estimated at over 4500 km^3, located near the northern end of the peninsula and includes Mount Haddington, which is considered the largest volcano in Antarctica (Smellie 2021). Deception Island, which is located within the Bransfield Strait, is the only volcano in the region that has been observed in eruption (Geyer et al. 2021). Compositionally, magmas forming the volcanic islands and seamounts of the Bransfield Strait are tholeiitic and range from basalt to rhyolite, whereas volcanism on the peninsula to the SE (i.e. James Ross Island, Paulet Island, Cape Purvis) consists predominantly of mafic alkaline magmas (Haase and Beier 2021). Post-subduction volcanism further to the south (Seal Nunataks, Alexander Island, Snow Nunataks, Merrick Mountains) is also dominated by mafic alkaline types (alkali basalt, basanite and hawaiite), although olivine tholeiite compositions occur in several places (Hole 2021). Camptonite, a variety of lamprophyre, occurs with other dykes/sills of alkali basalt, basanite and phonotephrite in the southeastern part of Alexander Island, with one dyke yielding a K–Ar whole-rock age of 15 ± 1 Ma (Rowley and Smellie 1990; Smellie 1999), which is the oldest date yet recorded for post-subduction alkaline magmatism on the Antarctic Peninsula.

Volcanism in the Bransfield Strait is considered to be the result of extension related to the rollback of the Phoenix Plate causing partial melting of shallow mantle, to a high degree, with variable input of materials and fluids from the subducting slab (Haase and Beier 2021). The origin of the rest of the post-subduction alkaline volcanism on the Antarctic Peninsula has been ascribed to back-arc extension or slab-window tectonics as a way to promote the decompression melting of asthenospheric mantle. As an alternative to the slab-window hypothesis, Hole (2021) proposes that melting of slab-hosted pyroxenite can produce the geochemical characteristics of the post-subduction alkaline volcanism and may account for the lack of age progression of volcanism, as well as the relatively short period of activity (i.e. predominantly ≤7 myr).

On the Wilhelm II Coast of East Antarctica, volcanism in the Late Quaternary produced an isolated nunatak that is located more than 2500 km from any other exposed volcano on the continent (Fig. 1). Gaussberg (−66.8° S, 89.2° E) is a small (c. 1 km diameter, 370 m high) glacially eroded centre that consists of pillow lavas and hyaloclastitic deposits of lamproite composition (Smellie and Collerson 2021). The fundamental cause for volcanism is unknown. Gaussberg is not located within an extensional setting and is not considered to be related to the Kerguelen plume (Murphy et al. 2002) – a mantle plume that is currently supplying melt to the Big Ben volcano on Heard Island located nearly 1750 km to the NW in the Indian Ocean. However, a separate plume-derived source is inferred based on geochemical signatures that indicate contributions from sediments that were subducted and recycled from the deep mantle (Zhang et al. 2019).

Deliberation on melt origins

Over the past 30 years there has been tremendous progress in our understanding of the primary origins of volcanism in Antarctica facilitated by more comprehensive petrological data coverage, along with better constraints on age as described within the chapters of this Memoir. Other contributions include higher-resolution architecture of the upper mantle and lithosphere, as well as finer-tuned plate reconstructions, all an outcome of extensive geological and geophysical campaigns (e.g. Lloyd et al. 2019; Pappa et al. 2019; Jordan et al. 2020; Storey and Granot 2021 and references therein). Despite these advances, considerable debate remains. Petrological studies on intraplate alkaline magmatism associated with the WARS deviate on whether primary melts were produced within the asthenosphere or lithospheric mantle and whether melting was facilitated by extension-enhanced edge-driven mantle flow or by mantle plumes. Debate also exists among petrologists who study intraplate magmatism closely associated with continental margin tectonics (i.e. Antarctic Peninsula and the Ferrar–Karoo LIPs). Compositions erupted in these areas have been explained by upwelling asthenosphere and plumes or mantle modified by subduction activity. An additional complexity to all petrological interpretations is the uncertainty as to whether the primary melt has been significantly modified during ascent through the continental plate (e.g. Panter et al. 2018) and/or has suffered post-emplacement alteration (e.g. Leat and Riley 2021).

Ferrar Large Igneous Province

Jurassic magmatism that produced the Ferrar LIP (i.e. Kirkpatrick Basalt and Ferrar Dolerite) is exposed within a narrow linear band over a distance of 3500 km (Fig. 1). The magmatism occurred over a very short time interval of 349 ± 0.49 kyr at c. 183 Ma (Elliot and Fleming 2021). There has been extensive study and debate on the origin of Ferrar magmatism stemming from its unique distribution and geochemistry relative to other LIPs. The long and narrow distribution of Ferrar outcrops is explained by two different scenarios: (1) magmas are sourced from a single region of mantle melting followed by long-distance lateral migration within the crust; and (2) magmas were supplied from multiple, roughly aligned melt regions (i.e. 'linear source') and emplaced within the crust with restricted lateral migration (Elliot and Fleming 2021). Overall, mafic compositions are relatively uniform province-wide and the least fractionated samples (MgO = 9–10 wt%), which are those collected from chilled margins of dolerite sills that contain forsteritic olivine, and possess mantle-like oxygen and osmium isotopic signatures. However, these olivine-bearing dolerites also have enriched initial Sr, Nd and Pb isotopic values which, along with trace-element patterns on normalized multi-element plots, suggest the addition of continental crust (Elliot and Fleming 2021). To explain both the geographical distribution and the coherent but unique geochemistry some researchers call upon a single mantle-plume source (i.e. the current Bouvet hotspot) and long-distance transport of magma at various depths within the crust (e.g. Storey and Kyle 1997; Elliot et al. 1999; Vaughan and Storey 2007), whereas others call for the linear melting of subduction-modified mantle sources along the length of the proto-Pacific Gondwana margin (e.g. Hergt et al. 1991; Molzahn et al. 1996; Ivanov et al. 2017). A recent study by Choi et al. (2019) on the Basement Sill, a Ferrar dolerite intrusion exposed in the McMurdo Dry Valleys, favours an arc-like mantle source rather than a plume source based on platinum group element abundances and Os isotopes. The authors suggest that rapid decompression of hydrated mantle-wedge materials parallel to the Gondwana margin subduction zone facilitated the large-volume and short-lived activity of Ferrar magmatism. Luttinen (2018) proposed that the melting which formed the Karoo LIP occurred under the influence of both active subduction and an active mantle-plume head. He called upon subduction-modified mantle sources to explain the Nb-depleted compositions found in the South Karoo LIP, Dronning Maud Land (Fig. 1) and, along with distribution of the coeval and similarly Nb-depleted compositions of the Ferrar LIP, reconstructed plume-influenced and subduction-influenced regions of the palaeo-Pacific region of Gondwana (refer to Luttinen 2018, fig. 5). Elliot and Fleming (2021) concur that the geochemical evidence supports a subduction-related source for Ferrar magmatism but that there is still significant uncertainty as to whether the primary melt was derived from a single region or multiple sites within the mantle. In their comprehensive review of Ferrar LIP petrology, the authors outline some of the major issues left to be resolved by new research.

West Antarctic Rift System

The origin of Cenozoic intraplate alkaline magmatism found within the broad region of thinned West Antarctic lithosphere has been explained by a variety of models that attempt to reconcile compositional characteristics within the framework of the last several hundred million years of Antarctica's tectonic history. As reviewed above, that history includes subduction and several stages of Gondwana break-up that may have been influenced by mantle-plume activity. A key point, however, is that most of the alkaline volcanism is much younger (<14 Ma). Furthermore, from the time of the Middle Miocene the mantle dynamics and tectonic development of the WARS is simple relative to other young continental rifts of comparable size such as the western United States (i.e. Basin and Range/Rio Grande Rift) and the East African Rift System (Fig. 2). Moreover, mafic alkaline compositions from the WARS have a relatively coherent geochemistry (Fig. 3) that is similar to ocean island basalt (OIB) with HIMU (high $^{238}U/^{204}Pb = \mu$)-like isotopic affinities ($^{206}Pb/^{204}Pb > 19.5$, $^{87}Sr/^{86}Sr < 0.7035$, $^{143}Nd/^{144}Nd > 0.5128$) (Martin *et al.* 2021; Panter *et al.* 2021*b*; Rocchi and Smellie 2021). Hence, sources and mechanisms for melting may be similar across the whole of the WARS. Furthermore, the magmatism has been linked compositionally and tectonically to a much broader region of long-lived, diffuse, intraplate alkaline magmatism (DAMP (diffuse alkaline magmatic province): Finn *et al.* 2005) found in Zealandia and eastern Australia (Panter *et al.* 2006; Timm *et al.* 2010; Price *et al.* 2014; van der Meer *et al.* 2017; Scott *et al.* 2020).

Models for the generation of intraplate alkaline magmas within the WARS are varied, and include active plumes, ancient plumes ('fossil plumes') and metasomatized lithosphere. The HIMU-like signature of basalts, along with their OIB-like geochemical and isotopic characteristics, constitute the petrological basis for mantle-plume source models. Alternatively, models that involve the melting of metasomatized lithosphere are also used to explain geochemical characteristics. Metasomatic enrichment of lithospheric mantle is based, in part, on the prevalence of negative K anomalies displayed by West Antarctic basalts on mantle-normalized multi-element diagrams (Fig. 3). The K anomalies are considered to be a result of incomplete melting and retention of hydrous potassic minerals (amphibole ± phlogopite) in their sources. Mantle amphiboles are stable at temperatures less than 1150°C (Mandler and Grove 2016) and therefore reside only within the lithosphere. The veined lithosphere model involves a multistage process (cf. Pilet *et al.* 2008) that begins with a low-degree melt from the asthenosphere (within either the garnet or the spinel stability fields) that penetrates the base of the lithosphere and forms hydrous-phase-bearing (pargasite amphibole and/or phlogopite) cumulates in veins. Later melting of these relatively small-volume, trace-element- and volatile-rich metasomatic veins at high degrees, and the reaction of this silica-undersaturated liquid with peridotite in the surrounding mantle during ascent, can reproduce the major- and trace-element and isotopic characteristics of the mafic alkaline magmas (Pilet 2015; Panter *et al.* 2018). Whether magmatism is sourced by metasomatism or a plume (or both), any petrological model must also take into account the tectonomagmatic history of the region, as well as evidence provided by mantle xenoliths and geophysical studies.

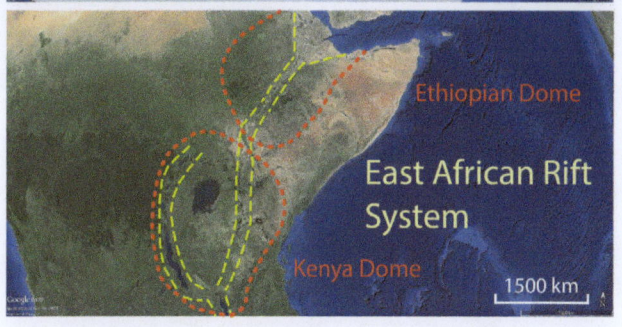

Fig. 2. Comparison of the Earth's three major Cenozoic continental rift systems and associated features: Basin and Range/Rio Grande Rift System of North America; West Antarctic Rift System; and East African Rift System. Google Earth Pro images are all at the same scale.

For the Marie Byrd Land Volcanic Group (Panter *et al.* 2021*b*), two principal types of mantle reservoirs appear to dominate as sources for volcanism: (1) HIMU plume; and (2) subduction-modified mantle. On the western and eastern extents of this region, melt sources for volcanism have been ascribed to lithosphere metasomatized by subduction (Hart *et al.* 1995; Gaffney and Siddoway 2007). Metasomatized lithosphere is also invoked as the source for volcanoes in the north-central and eastern portion of Marie Byrd Land (LeMasurier *et al.* 2016). Across the central portion of the region, Panter *et al.* (2000) proposed subduction-metasomatized lithosphere underlain by a HIMU plume, while Hart *et al.* (1997) preferred a stratified heterogeneous plume source for the volcanism. In both of these studies, the source of plume materials

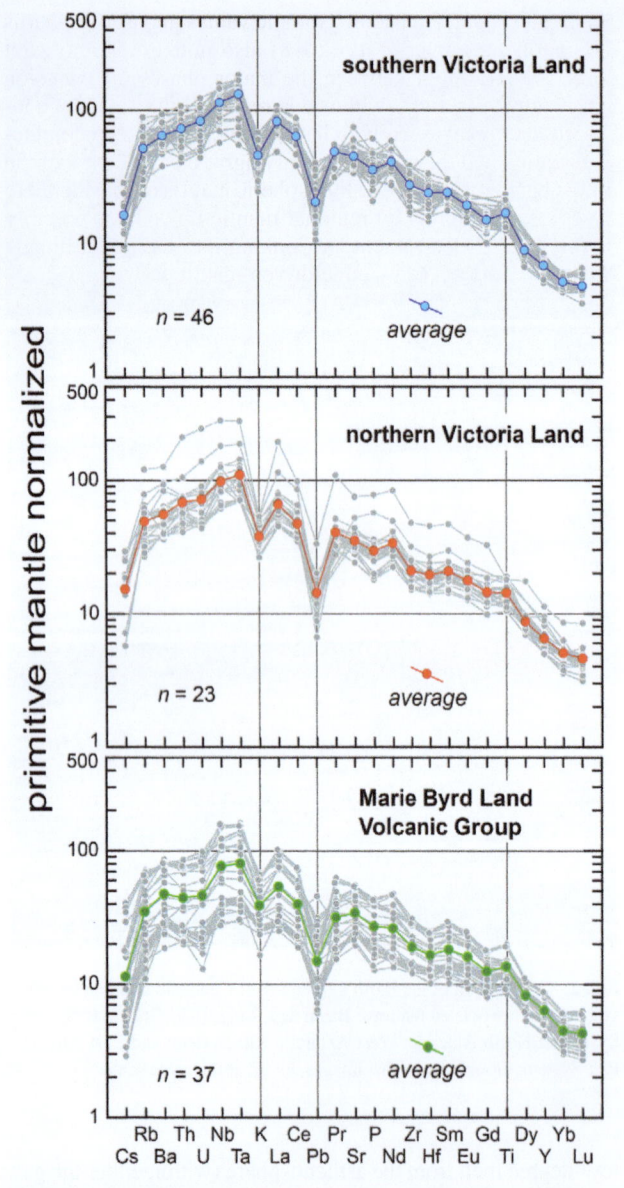

Fig. 3. Comparison of mafic basaltic compositions (MgO = 8–12 wt%) within the West Antarctic Rift System. Elemental concentrations are normalized to the primitive mantle values provided by McDonough and Sun (1995). Referenced datasets for southern Victoria Land (Erebus Volcanic Province) are from Martin *et al.* (2021), northern Victoria Land (Hallett and Melbourne volcanic provinces) from Rocchi and Smellie (2021), and Marie Byrd Land and western Ellsworth Land (Marie Byrd Land Volcanic Group (MBLVG) after Wilch *et al.* 2021) from Panter *et al.* (2021*b*).

is considered to have been emplaced before Gondwana break-up, and hence was 'fossilized' at the base of the lithosphere and tapped later by Cenozoic volcanism. Panter *et al.* (2021*b*) document a regional compositional gradient that is likely to be explained by mixing between metasomatized lithosphere and HIMU-like plume (^{206}Pb/^{204}Pb > 20) materials. An additional observation is that the current known distribution of the most radiogenic Pb signatures found in Marie Byrd Land reveals a relatively narrow and broadly linear feature. At *c.* 90 Ma this feature was roughly parallel with several other colinear features and may reflect continental strike-slip zones formed in response to oblique subduction prior to Gondwana break-up (Eagles *et al.* 2004). Recent geophysical studies (e.g. Heeszel *et al.* 2016; Lloyd *et al.* 2019) indicate a deep thermal anomaly beneath central Marie Byrd Land, suggesting a mantle-plume influence on volcanism and tectonism.

Translithospheric faults developed during the break-up may have helped to localize upwelling plume materials beneath the region (Riefstahl *et al.* 2020).

In southern Victoria Land, mantle sources and causes for volcanism (Erebus Volcanic Province of the McMurdo Volcanic Group) are also the subject of strong debate (Martin *et al.* 2021). The 'Erebus plume' was proposed by Kyle *et al.* (1992) and Esser *et al.* (2004) to explain the production of large volumes of phonolitic magmas from basanitic melts beneath the active Mount Erebus volcano. The authors suggest that uplift and crustal extension may have been enhanced by buoyancy from a plume, and may explain the radial pattern of volcanism on Ross Island and a similar pattern centred on Mount Discovery. More recent detailed geochemical and isotopic (Sr, Nd, Hf and Pb) studies of Mount Erebus and the rest of Ross Island by Sims *et al.* (2008) and Phillips *et al.* (2018) support a plume source. Phillips *et al.* (2018) used model mixing between a depleted MORB mantle (DMM) and HIMU sources to explain the Ross Island samples, and concluded that the HIMU isotopic signature originates from high time-integrated (Archean–Early Proterozoic) crustal materials that were recycled and upwelled from the deep mantle. They employed a seismic tomography model which suggested that a low-velocity zone exists to depth of *c.* 1200 km beneath Ross Island, supporting the deep plume hypothesis. In contrast, Martin *et al.* (2021) base interpretations on an extensive compilation of geochemical and isotopic data from basalts found across the Erebus Volcanic Province (i.e. including the Mount Discovery Volcanic Field, Mount Morning Volcanic Field and the Southern Local Suite, in addition to the Ross Island Volcanic Field). Trace-element and isotopic variations display a province-wide gradient which they interpret as being caused by the incorporation of an eclogitic crustal component in mantle melts to a greater degree eastwards from Mount Morning, which lies adjacent to the Transantarctic Mountains, to Cape Crozier at the eastern end of Ross Island. Their model favours subduction-derived sources that are being melted by edge-driven mantle flow rather than by a deep plume source. Other support for subduction-modified lithospheric mantle beneath the Erebus Volcanic Province comes from the study of peridotite and pyroxenite mantle xenoliths hosted by the basalts (Gamble *et al.* 1988; Martin *et al.* 2013, 2014; Aviado *et al.* 2015). In a recent petrological, geochemical and isotopic (He and Os) study by Day *et al.* (2019), glass- and amphibole-rich veins and replacement textures in peridotites (e.g. dunite and harzburgite) indicate extensive and multiple phases of melt–rock interactions in young (*c.* 250 Ma) mantle lithosphere beneath Hut Point Peninsula on Ross Island. The refractory nature of the lithosphere is suggested to be a consequence of subduction-processed mantle that was later metasomatized during Cretaceous subduction prior to Gondwana break-up. Day *et al.* (2019) concluded that modification of mantle lithosphere by subduction produced the HIMU reservoirs for the Cenozoic volcanism.

In northern Victoria Land, mantle sources and causes for igneous intrusion (Meander Intrusive Group) and younger volcanism (Melbourne Volcanic Province and Hallett Volcanic Province) are discussed by Rocchi and Smellie (2021). Rocholl *et al.* (1995) first proposed three mantle source components to explain the geochemistry and isotopic compositions of mafic volcanism in this region: depleted asthenosphere (DMM); fossilized plume head (HIMU); and enriched mantle lithosphere (EM). The authors envisage a pre-rift mantle that was stratified in these components (i.e. top down = EM→HIMU→DMM), and during rift development rising asthenosphere progressively replaced the overlying sources to supply DMM in greater proportion within the rift relative to rift shoulder volcanism. Alternatively, Rocchi *et al.* (2002, 2005) and Nardini *et al.* (2009) proposed a multistage model for the origin of the OIB HIMU-like magmatism without plume influence. The model first calls upon sublithospheric mantle enrichment in incompatible elements by small-degree asthenospheric melts (i.e. autometasomatism) during the amagmatic phase of Late Cretaceous Gondwana rifting. This stage was followed more than 30 myr later (Middle Eocene–Oligocene) by the melting of these metasomes within what has been transformed into subcontinental lithosphere. The cause of melting is considered to be the result of newly established edge-driven mantle flow developed at the margin between thinned West Antarctic lithosphere and the East Antarctic Craton along with localized decompression in response to strike-slip movement along pre-existing NW–SE-orientated translithospheric faults (Rocchi et al. 2002). This activity generated magmas of the Meander Intrusive Group. The same mechanisms caused melting of metasomatized lithosphere to supply magmas to the younger (Middle Miocene–present) volcanoes of the Melbourne and Hallett provinces, many of which are localized along rift boundary normal faults. Rocchi and Smellie (2021) review evidence from mafic igneous rocks and mantle xenoliths (e.g. Coltorti et al. 2004; Nardini et al. 2009; Melchiorre et al. 2011; Perinelli et al. 2011), and conclude that the metasomatic agents responsible for mantle enrichment were alkaline OIB-like melts. Furthermore, based on Pb and Os isotopic systematic, Rocchi and Smellie (2021) propose that the HIMU signature is an in-grown component that was initiated (i.e. U/Pb ratios increased) during the Cretaceous autometasomatic episode. In contrast, Broadley et al. (2016) considered metasomatism of the lithosphere beneath northern Victoria Land to have been derived from Paleozoic subduction based on the measurement of noble gases and halogens in peridotite and pyroxenite xenoliths. They also suggest that the HIMU-like geochemical signatures of xenoliths and magmas is an in-grown feature developed by the dehydration of hydrated metasomes, which may concentrate U, Th and Sr related to fluid mobile elements Pb and Rb, and hence produces source reservoirs with high U/Pb and Th/Pb ratios, and low Rb/Sr ratios (cf. Panter et al. 2006). Lastly, depleted MORB-type He and Ne isotopic signatures measured on hydrous and anhydrous peridotites by Correale et al. (2019) indicated metasomatism of the lithosphere beneath northern Victoria Land by asthenospheric melts that are unlikely to be plume-related.

The origin of seamount volcanism in the oceanic Adare Basin, northwestern Ross Sea (Fig. 1), was evaluated by Panter et al. (2018) using a suite of basalts erupted across the transition from oceanic lithosphere to thinned continental lithosphere of northern Victoria Land. The basalts across this ocean–continent transect display systematic variations in major- and trace-element concentrations and isotopic values (Sr, Nd, Pb and O). The authors maintain that the variations are not caused by crustal contamination or by changes in the degree of mantle partial melting but consider them to be a function of the thickness and age of the mantle lithosphere that they have traversed. The isotopic signature of the most silica-undersaturated and incompatible-element-enriched basalts have relatively low $^{87}Sr/^{86}Sr$ (≤ 0.7030) and $\delta^{18}O_{olivine}$ ($\leq 5.0‰$), with high $^{143}Nd/^{144}Nd$ (c. 0.5130) and $^{206}Pb/^{204}Pb$ (≥ 20) ratios, and are considered to have been derived from a sublithospheric source contaminated by carbonate-rich subducted materials. Small-degree melts of this source rose and froze within the cooler lithospheric mantle to produce amphibole-rich veins (Fig. 4). Later melting of the metasomes produced silica-undersaturated liquids that reacted with the surrounding peridotite (Fig. 4). This reaction occurred to a greater extent as melt traversed through thicker and older lithosphere, resulting in the progressive increase in SiO_2 concentration, and Sr and oxygen isotopic values, and decrease in Nb/Y ratios, and Nd and Pb isotopic values in basalts towards the continent. Panter et al. (2018) also noted the significant delay of 20–30 myr between the major phases of extension and magmatic activity within this region (Fig. 4). They concluded that this was likely to have been controlled by conductive heating and the rate of thermal migration at the base of the lithosphere. Heating of the lithospheric mantle is considered to have been facilitated by regional mantle upwelling, possibly driven by slab detachment and sinking into the lower mantle (Finn et al. 2005), or by edge-driven mantle flow.

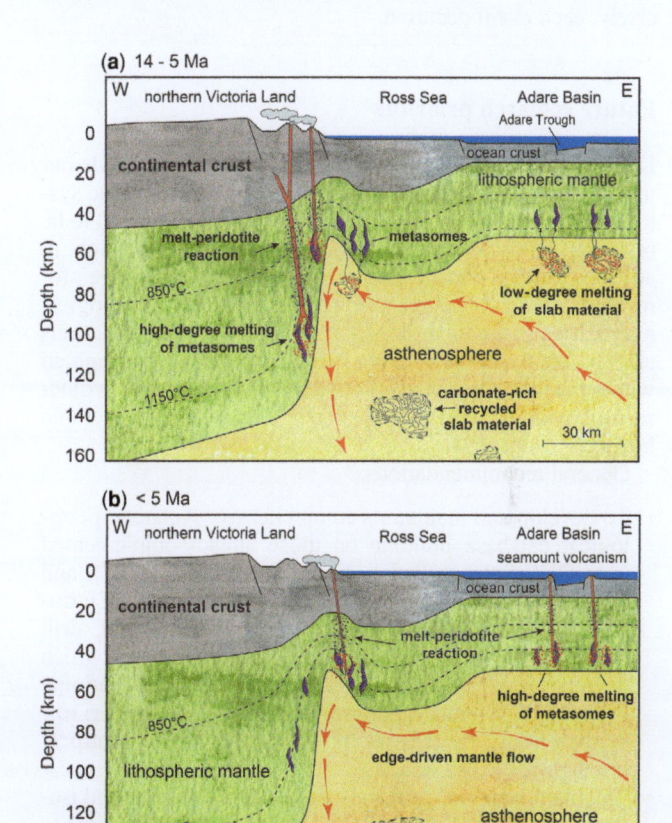

Fig. 4. Schematic model for the Late Oligocene–recent petrogenesis of alkaline volcanism in the northern Victoria Land (Hallett Volcanic Province) and the oceanic Adare Basin seamounts (Fig. 1) after Panter et al. (2018). Plate architecture from craton to ocean in the northern Ross Sea region is shown for two time frames: 14–5 and <5 Ma. Thinned lithosphere ('necked zone') beneath the rift boundary is considered to be a result of an earlier period (c. 80–40 Ma) of focused extension (Huerta and Harry 2007). The craton-directed edge-driven convective flow, depicted by the red arrows, is considered to have been established in the Eocene (Faccenna et al. 2008). To explain the geochemistry and isotopic compositions of basalts, and the time delay between rifting and volcanism, a multistage process is used. (a) Decompression melting of upwelling subduction-derived materials produced carbonate-rich silicate liquids that rose and froze within the cooler lithosphere to form amphibole-rich veins ('metasomes'). Conductive heating at the base of the lithosphere by edge-driven flow eventually (c. 25 myr after focused extension) reached temperatures to melt metasomes ($\geq 1150°C$) and produce silica-undersaturated liquid. The reaction of this liquid with the surrounding peridotite modified the melt composition as it traversed the thicker continental plate and erupted to form large, elongated shield volcanoes along the continental coastline. (b) The thermal evolution of the lithosphere oceanwards reached the melting temperature of the earlier formed metasomes and produced Pliocene–Pleistocene volcanic islands (e.g. Possession Islands) and the seamounts located on the continental shelf and within the oceanic Adare Basin.

In summary, the origin of melt produced beneath the WARS, along with the metasomatized mantle xenoliths that are commonly found within them, remains a subject of ongoing interest and vigorous debate. The long-standing elements of controversy centre on whether mantle-plume activity has played a role in rift development and magmatism, and whether subduction-derived agents have caused metasomatism. Intertwined within the various hypotheses are outstanding issues of whether the OIB HIMU-like geochemical signature found throughout the rift, and beyond, is ancient (i.e. recycled by plume) or relatively recent (i.e. in-grown by metasomatism over a few hundreds of millions of years) and when, more precisely, each event occurred.

Future research priorities

Despite the major advances that have been made towards our understanding of the cause and petrogenesis of magmatic systems in Antarctica, there are still many critical issues left to be resolved. This Memoir provides a platform on which to advance our knowledge and to formulate new objectives for research. Below I suggest some priorities for petrological research on magmatism in Antarctica. The recommendations are inspired by the gaps in our knowledge that are highlighted within the following chapters but also include input from other scientists with expertise in Antarctica's magmatic and tectonic history.

General recommendations:

- To develop and maintain a comprehensive Antarctic petrological database building on those already implemented (e.g. GNS Science Petlab, https://pet.gns.cri.nz/#/), and to fully integrate datasets with physical repositories of intrusive and extrusive rock samples, sediment, tephra, drill core, dredge materials, xenoliths, etc., several of which have already been established (e.g. GNS Science National Petrology Reference Collection, https://pet.gns.cri.nz/#/nprc; and the US Polar Rock Repository, http://research.bpcrc.osu.edu/rr/)
- To obtain more uniform and complete geochemical and isotopic (e.g. Sr, Nd, Pb, Hf, Os, O, etc.) datasets in order to facilitate comparison across Antarctica and beyond.
- To promote the application of novel and cutting-edge petrological tools such as high-precision micro-analytical techniques, and the study of halogen, noble gas and other volatiles from minerals and melt inclusions.
- To encourage and advocate for field campaigns, including drilling and dredging, designed not only to include targeted areas that have a strong potential for high-profile research outcomes but also to promote exploration of understudied areas, such as subglacial and seafloor environments.
- To foster close collaborations between petrological investigations and geophysical programmes (e.g. seismic and magnetotellurics studies) with the aim of integrating physical with geochemical models for magmatic systems from their source to the surface.

Overarching objectives (not exhaustive and not listed in order of priority):

- To provide a comprehensive geochemical and isotopic characterization of the HIMU mantle source component in East and West Antarctica.
- To constrain the age, extent, character and cause of metasomatism beneath West Antarctica.
- To develop rigorous petrological and geophysical criteria and to test for the presence or absence of deep mantle plumes in West Antarctica with specific focus on areas beneath Ross Island and the Marie Byrd Land dome.
- To determine the mantle origin of the Ferrar Large Igneous Province and whether melting was from a single source or multiple sources, with the corollary of designing a multidiscipline approach to assess magma transport and emplacement processes.
- To obtain a comprehensive chronology of post-subduction volcanism on the Antarctic Peninsula to better constrain the relationship between the end of subduction, slab-window formation and alkaline magma genesis.
- To apply up-to-date geochemical modelling and analytical techniques to better constrain the petrogenesis of peralkaline compositions that include the rare occurrences of pantellerite erupted within the West Antarctic Rift System.
- To investigate the underlying cause of small-volume tholeiite and lamproite magmas that occur within several Cenozoic intraplate alkaline volcanic fields in West Antarctica and in a few isolated deposits in East Antarctica.
- To evaluate the influence of ice-sheet loading and unloading on melt productivity, magma evolution and eruptability (cf. Nyland *et al.* 2013) across Antarctica.

Acknowledgements The author would like to thank the many people and institutions that have played a vital role in research on Antarctic volcanism over the years. Foremost I want to thank the US National Science Foundation and its private contractors, the people that 'made it happen', for field logistical support. I am also grateful to the collaborative interactions with scientists and facilities of the US Geological Survey, British Antarctic Survey, New Zealand Antarctic Research Programme, GNS Science and several universities in New Zealand (Victoria, Canterbury and Otago). I will always be grateful to Phil Kyle, Bill McIntosh, Nelia Dunbar, Thom Wilch, John Smellie and John Gamble for introducing me to the world of Antarctic volcanology, and for their friendship and mentorship over the years. I also wish to thank Stan Hart, Jurek Blusztajn, Pat Castillo, John Valley and Robert Pankhurst for their knowledge and generosity, as well as Sébastien Pilet for challenging my traditional views on the origin of intraplate alkaline magmas. I greatly appreciate my Italian colleagues Sergio Rocchi, Gianfranco Di Vincenzo, Paola Del Carlo, Massimo Tiepolo and Franco Talarico for their collaboration on the petrology of volcanic materials recovered from the ANDRILL AND-2A core. Furthermore, I must acknowledge the hard work of my students Joanne Antibus, Suvankar Charkraborty, Susan Krans, Yuyu Li, Rosie Nyland, Kari Odegaard, Ellen Redner, Jenna Reindel, Mary Scanlan, Dennis Wingrove and Brian Winter who contributed significant time and effort to advance research on Antarctic volcanism. Finally, this overview benefited from the constructive reviews by Phil Kyle and Adelina Geyer, with editorial oversight provided by John Smellie.

Author contributions KSP: conceptualization (lead), data curation (supporting), formal analysis (lead), funding acquisition (supporting), investigation (supporting), visualization (lead), writing – original draft (lead), writing – review & editing (lead).

Funding This research received no specific grant from any funding agency in the public, commercial, or not-for-profit sectors.

Data availability Data sharing is not applicable to this article as no datasets were generated or analysed during the current study.

References

Aviado, K.B., Rilling-Hall, S., Bryce, J.G. and Mukasa, S.B. 2015. Submarine and subaerial lavas in the West Antarctic Rift System: temporal record of shifting magma source components from the lithosphere and asthenosphere. *Geochemistry,*

Geophysics, Geosystems, **16**, 4344–4361, https://doi.org/10.1002/2015GC006076

Broadley, M.W., Ballentine, C.J., Chavrit, D., Dallai, L. and Burgess, R. 2016. Sedimentary halogens and noble gases within Western Antarctic xenoliths: implications of extensive volatile recycling to the sub continental lithospheric mantle. *Geochimica et Cosmochimica Acta*, **176**, 139–156, https://doi.org/10.1016/j.gca.2015.12.013

Calmus, T., Pallares, C., Maury, R.C., Aguillón-Robles, A., Bellon, H., Benoit, M. and Michaud, F. 2011. Volcanic markers of the post-subduction evolution of Baja California and Sonora, Mexico: slab tearing v. lithospheric rupture of the Gulf of California. *Pure and Applied Geophysics*, **168**, 1303–1330, https://doi.org/10.1007/s00024-010-0204-z

Castillo, P.R. 2008, Origin of the adakite–high-Nb basalt association and its implications for postsubduction magmatism in Baja California, Mexico. *Geological Society of America Bulletin*, **120**, 451–462, https://doi.org/10.1130/B26166.1

Choi, S.H., Mukasa, S.B., Ravizza, G., Fleming, T.H., Marsh, B.D. and Bédard, J.H.J. 2019. Fossil subduction zone origin for magmas in the Ferrar Large Igneous Province, Antarctica: Evidence from PGE and Os isotope systematics in the Basement Sill of the McMurdo Dry Valleys. *Earth and Planetary Science Letters*, **506**, 507–519, https://doi.org/10.1016/j.epsl.2018.11.027

Coltorti, M., Beccaluva, L., Bonadiman, C., Faccini, B., Ntaflos, T. and Siena, F. 2004. Amphibole genesis via metasomatic reaction with clinopyroxene in mantle xenoliths from Victoria Land, Antarctica. *Lithos*, **75**, 115–139, https://doi.org/10.1016/j.lithos.2003.12.021

Correale, A., Pelorosso, B., Rizzo, A.L., Coltorti, M., Italiano, F., Bonadiman, C. and Giacomoni, P.P. 2019. The nature of the West Antarctic Rift System as revealed by noble gases in mantle minerals. *Chemical Geology*, **524**, 104–118, https://doi.org/10.1016/j.chemgeo.2019.06.020

Day, J.M.D., Harvey, R.P. and Hilton, D.R. 2019. Melt-modified lithosphere beneath Ross Island and its role in the tectonomagmatic evolution of the West Antarctic rift system. *Chemical Geology*, **518**, 45–54, https://doi.org/10.1016/j.chemgeo.2019.04.012

Dunbar, N.W., Iverson, N.A., Smellie, J.L., McIntosh, W.C., Zimmerer, M.J. and Kyle, P.R. 2021. Active volcanoes in Marie Byrd Land. *Geological Society, London, Memoirs*, **55**, https://doi.org/10.1144/M55-2019-29

Eagles, G., Gohl, K. and Larter, R.D. 2004. High-resolution animated tectonic reconstruction of the South Pacific and West Antarctic margin. *Geochemistry, Geophysics, Geosystems*, **5**, Q07002, https://doi.org/10.1029/2003GC000657

Elliot, D.H. and Fleming, T.H. 2021. Ferrar Large Igneous Province: petrology. *Geological Society, London, Memoirs*, **55**, https://doi.org/10.1144/M55-2018-39

Elliot, D.H., Fleming, T.H., Kyle, P.R. and Foland, K.A. 1999. Long-distance transport of magmas in the Jurassic Ferrar Large Igneous Province, Antarctica. *Earth and Planetary Science Letters*, **167**, 89–104, https://doi.org/10.1016/S0012-821X(99)00023-0

Esser, R.P., Kyle, P.R. and McIntosh, W.C. 2004. $^{40}Ar/^{39}Ar$ dating of the eruptive history of Mount Erebus, Antarctica: volcano evolution. *Bulletin of Volcanology*, **66**, 671–686, https://doi.org/10.1007/s00445-004-0354-x

Faccenna, C., Rossetti, F., Becker, T.W., Danesi, S. and Morelli, A. 2008. Recent extension driven by mantle upwelling beneath the Admiralty Mountains (East Antarctica). *Tectonics*, **27**, TC4015, https://doi.org/10.1029/2007TC002197

Finn, C.A., Müller, R.D. and Panter, K.S. 2005. A Cenozoic diffuse alkaline magmatic province (DAMP) in the southwest Pacific without rift or plume origin. *Geochemistry, Geophysics, Geosystems*, **6**, Q02005, https://doi.org/10.1029/2004GC000723

Gaffney, A.M. and Siddoway, C.S. 2007. Heterogeneous sources for Pleistocene lavas of Marie Byrd Land, Antarctica: new data from the SW Pacific diffuse alkaline magmatic province. *United States Geological Survey Open-File Report*, **2007-1047**, extended abstract 063.

Gaina, C., Müller, R.D., Brown, B., Ishihara, T. and Ivanov, S. 2007. Breakup and early seafloor spreading between India and Antarctica. *Geophysical Journal International*, **170**, 151–169, https://doi.org/10.1111/j.1365-246X.2007.03450.x

Gambino, S., Armienti, P. *et al.* 2021. Mount Melbourne and Mount Rittmann. *Geological Society, London, Memoirs*, **55**, https://doi.org/10.1144/M55-2018-43

Gamble, J.A., McGibbon, F., Kyle, P.R., Menzies, M.A. and Kirsch, I. 1988. Metasomatised xenoliths from Foster Crater, Antarctica: Implications for lithospheric structure and processes beneath the Transantarctic Mountain front. *Journal of Petrology*, Special Volume, Issue 1, 109–138, https://doi.org/10.1093/petrology/Special_Volume.1.109

Geyer, A., Pedrazzi, D. *et al.* 2021. Deception Island. *Geological Society, London, Memoirs*, **55**, https://doi.org/10.1144/M55-2018-56

Hart, S.R., Blusztajn, J. and Craddock, C. 1995. Cenozoic volcanism in Antarctica; Jones Mountains and Peter I Island. *Geochimica et Cosmochimica Acta*, **59**, 3379–3388, https://doi.org/10.1016/0016-7037(95)00212-I

Hart, S.R., Blusztajn, J., LeMasurier, W.E. and Rex, D.C. 1997. Hobbs Coast Cenozoic volcanism: implications for the West Antarctic rift system. *Chemical Geology*, **139**, 223–248, https://doi.org/10.1016/S0009-2541(97)00037-5

Haase, K. and Beier, C. 2021. Bransfield Strait and James Ross Island: petrology. *Geological Society, London, Memoirs*, **55**, https://doi.org/10.1144/M55-2018-37

Heeszel, D.S., Wiens, D.A. *et al.* 2016. Upper mantle structure of central and West Antarctica from array analysis of Rayleigh wave phase velocities. *Journal of Geophysical Research: Solid Earth*, **121**, 1758–1775, https://doi.org/10.1002/2015jb012616

Hergt, J.M., Peate, D.W. and Hawkesworth, C.J. 1991. The petrogenesis of Mesozoic Gondwana low-Ti flood basalts. *Earth and Planetary Science Letters*, **105**, 134–148, https://doi.org/10.1016/0012-821X(91)90126-3

Hoernle, K., Timm, C. *et al.* 2020. Late Cretaceous (99–69 Ma) basaltic intraplate volcanism on and around Zealandia: tracing upper mantle geodynamics from Hikurangi Plateau collision to Gondwana breakup. *Earth and Planetary Science Letters*, **529**, https://doi.org/10.1016/j.epsl.2019.115864

Hole, M.J. 2021. Antarctic Peninsula: petrology. *Geological Society, London, Memoirs*, **55**, https://doi.org/10.1144/M55-2018-40

Huerta, A.D. and Harry, D.L. 2007. The transition from diffuse to focused extension: Modeled evolution of the West Antarctic Rift system. *Earth and Planetary Science Letters*, **255**, 133–147, https://doi.org/10.1016/j.epsl.2006.12.011

Ivanov, A.V., Meffre, S., Thompson, J., Corfu, F., Kamenetsky, V.S., Kamenetsky, M.B. and Demonterova, E.I. 2017. Timing and genesis of the Karoo-Ferrar large igneous province: new high precision U–Pb data for Tasmania confirm short duration of the major magmatic pulse. *Chemical Geology*, **455**, 32–43, https://doi.org/10.1016/j.chemgeo.2016.10.008

Jordan, T.A., Riley, T.R. and Siddoway, C.S. 2020. The geological history and evolution of West Antarctica. *Nature Reviews Earth and Environment*, **1**, 117–133, https://doi.org/10.1038/s43017-019-0013-6

Kipf, A., Hauff, F. *et al.* 2014. Seamounts off the West Antarctic margin: a case for non-hotspot driven intraplate volcanism. *Gondwana Research*, **25**, 1660–1679, https://doi.org/10.1016/j.gr.2013.06.013

Kyle, P.R. 1990. A. McMurdo Volcanic Group Western Ross Embayment. *American Geophysical Union, Antarctic Research Series*, **48**, 18–25.

Kyle, P.R., Moore, J.A. and Thirlwall, M.F. 1992. Petrologic evolution of anorthoclase phonolite lavas at Mount Erebus, Ross Island, Antarctica. *Journal of Petrology*, **33**, 849–875, https://doi.org/10.1093/petrology/33.4.849

Leat, P.T. and Riley, T.R. 2021. Antarctic Peninsula and South Shetland Islands: petrology. *Geological Society, London, Memoirs*, **55**, https://doi.org/10.1144/M55-2018-68

LeMasurier, W.E. and Landis, C.A. 1996. Mantle-plume activity recorded by low-relief erosion surfaces in West Antarctica and New Zealand. *Geological Society of America Bulletin*, **108**, 1450–1466, https://doi.org/10.1130/0016-7606(1996)108<1450:MPARBL>2.3.CO;2

LeMasurier, W.E., Choi, S.H., Hart, S.R., Mukasa, S.B. and Rogers, N.W. 2016. Reconciling the shadow of a subduction signature with rift geochemistry and tectonic environment in eastern Marie Byrd Land, Antarctica. *Lithos*, **260**, 134–153, https://doi.org/10.1016/j.lithos.2016.05.018

Licht, K.J., Groth, T., Townsend, J.P., Hennessy, A.J., Hemming, S.R., Flood, T.P. and Studinger, M. 2018. Evidence for extending anomalous Miocene volcanism at the edge of the East Antarctic craton. *Geophysical Research Letters*, **45**, 3009–3016, https://doi.org/10.1002/2018GL077237

Lloyd, A.J., Wiens, D.A. et al. 2019. Seismic Structure of the Antarctic upper mantle based on adjoint tomography. *Journal of Geophysical Research: Solid Earth*, **125**, https://doi.org/10.1029/2019JB017823

Luttinen, A.V. 2018. Bilateral geochemical asymmetry in the Karoo large igneous province. *Scientific Reports*, **8**, 5223, https://doi.org/10.1038/s41598-018-23661-3

Mandler, B.E. and Grove, T.L. 2016. Controls on the stability and composition of amphibole in the Earth's mantle. *Contributions to Mineralogy and Petrology*, **171**, 68, https://doi.org/10.1007/s00410-016-1281-5

Martin, A.P., Cooper, A.F. and Price, R.C. 2013. Petrogenesis of Cenozoic, alkalic volcanic lineages at Mount Morning, West Antarctica and their entrained lithospheric mantle xenoliths: lithospheric v. asthenospheric mantle sources. *Geochimica et Cosmochimica Acta*, **122**, 127–152, https://doi.org/10.1016/j.gca.2013.08.025

Martin, A.P., Cooper, A.F. and Price, R.C. 2014. Increased mantle heat flow with on-going rifting of the West Antarctic rift system inferred from characterisation of plagioclase peridotite in the shallow Antarctic mantle. *Lithos*, **190–191**, 173–190, https://doi.org/10.1016/j.lithos.2013.12.012

Martin, A.P., Cooper, A.F., Price, R.C., Kyle, P.R. and Gamble, J.A. 2021. Erebus Volcanic Province: petrology. *Geological Society, London, Memoirs*, **55**, https://doi.org/10.1144/M55-2018-80

McDonough, W.F. and Sun, S.S. 1995. The composition of the earth. *Chemical Geology*, **120**, 223–253, https://doi.org/10.1016/0009-2541(94)00140-4

Melchiorre, M., Coltorti, M., Bonadiman, C., Faccini, B., O'Reilly, S.Y. and Pearson, N.J. 2011. The role of eclogite in the rift-related metasomatism and Cenozoic magmatism of Northern Victoria Land, Antarctica. *Lithos*, **124**, 319–330, https://doi.org/10.1016/j.lithos.2010.11.012

Molzahn, M., Reisberg, L. and Wörner, G. 1996. Os, Sr, Nd, Pb, O isotope and trace element data from the Ferrar flood basalts, Antarctica: evidence for an enriched subcontinental lithospheric source. *Earth and Planetary Science Letters*, **144**, 529–546, https://doi.org/10.1016/S0012-821X(96)00178-1

Murphy, D.T., Collerson, K.D. and Kamber, B.S. 2002. Lamproites from Gaussberg, Antarctica: possible transition zone melts of Archaean subducted sediments. *Journal of Petrology*, **43**, 981–1001, https://doi.org/10.1093/petrology/43.6.981

Nardini, I., Armienti, P., Rocchi, S., Dallai, L. and Harrison, D. 2009. Sr–Nd–Pb–He–O isotope and geochemical constraints on the genesis of Cenozoic magmas from the West Antarctic Rift. *Journal of Petrology*, **50**, 1359–1375, https://doi.org/10.1093/petrology/egn082

Nyland, R.E., Panter, K.S. et al. 2013. Volcanic activity and its link to glaciation cycles: single-grain age and geochemistry of Early Miocene volcanic glass from ANDRILL AND-2A core, Antarctica. *Journal of Volcanology and Geothermal Research*, **250**, 106–128, https://doi.org/10.1016/j.jvolgeores.2012.11.008

Olierook, H.K.H., Jiang, Q., Jourdan, F. and Chiaradia, M. 2019. Greater Kerguelen large igneous province reveals no role for Kerguelen mantle plume in the continental breakup of eastern Gondwana. *Earth and Planetary Science Letters*, **511**, 244–255, https://doi.org/10.1016/j.epsl.2019.01.037

Panter, K.S. 1995. *Geology, Geochemistry and Petrogenesis of the Mount Sidley Volcano, Marie Byrd Land, Antarctica*. PhD thesis, New Mexico Institute of Mining and Technology, Socorro, New Mexico, USA.

Panter, K.S., McIntosh, W.C. and Smellie, J.L. 1994. Volcanic history of Mount Sidley, a major alkaline volcano in Marie Byrd Land, Antarctica. *Bulletin of Volcanology*, **56**, 361–376, https://doi.org/10.1007/BF00326462

Panter, K.S., Kyle, P.R. and Smellie, J.L. 1997. Petrogenesis of a phonolite–trachyte succession at Mount Sidley, Marie Byrd Land, Antarctica. *Journal of Petrology*, **38**, 1225–1253, https://doi.org/10.1093/petroj/38.9.1225

Panter, K.S., Hart, S.R., Kyle, P., Blusztanjn, J. and Wilch, T. 2000. Geochemistry of Late Cenozoic basalts from the Crary Mountains: characterization of mantle sources in Marie Byrd Land, Antarctica. *Chemical Geology*, **165**, 215–241, https://doi.org/10.1016/S0009-2541(99)00171-0

Panter, K.S., Blusztajn, J., Hart, S., Kyle, P., Esser, R. and McIntosh, W. 2006. The origin of HIMU in the SW Pacific: evidence from intraplate volcanism in southern New Zealand and Subantarctic Islands. *Journal of Petrology*, **47**, 1673–1704, https://doi.org/10.1093/petrology/egl024

Panter, K.S., Castillo, P. et al. 2018. Melt origin across a rifted continental margin: a case for subduction-related metasomatic agents in the lithospheric source of alkaline basalt, northwest Ross Sea, Antarctica. *Journal of Petrology*, **59**, 517–558, https://doi.org/10.1093/petrology/egy036

Panter, K.S., Reindel, J. and Smellie, J.L. 2021*a*. Mount Early and Sheridan Bluff: petrology. *Geological Society, London, Memoirs*, **55**, https://doi.org/10.1144/M55-2019-2

Panter, K.S., Wilch, T.I., Smellie, J.L., Kyle, P.R. and McIntosh, W.C. 2021*b*. Marie Byrd Land and Ellsworth Land: petrology. *Geological Society, London, Memoirs*, **55**, https://doi.org/10.1144/M55-2019-50

Pappa, F., Ebbing, J. and Ferraccioli, F. 2019. Moho depths of Antarctica: comparison of seismic, gravity, and isostatic results. *Geochemistry, Geophysics, Geosystems*, **20**, 1629–1645, https://doi.org/10.1029/2018GC008111

Perinelli, C., Armienti, P. and Dallai, L. 2011. Thermal evolution of the lithosphere in a rift environment as inferred from the geochemistry of mantle cumulates, northern Victoria Land, Antarctica. *Journal of Petrology*, **52**, 665–690, https://doi.org/10.1093/petrology/egq099

Phillips, E.H., Sims, K.W.W. et al. 2018. The nature and evolution of mantle upwelling at Ross Island, Antarctica, with implications for the source of HIMU lavas. *Earth and Planetary Science Letters*, **498**, 38–53, https://doi.org/10.1016/j.epsl.2018.05.049

Pilet, S. 2015. Generation of low-silica alkaline lavas: petrological constraints, models, and thermal implications. *Geological Society of America Special Papers*, **514**, 514–517, https://doi.org/10.1130/2015.2514(17)

Pilet, S., Baker, M.B. and Stolper, E.M. 2008. Metasomatized lithosphere and the origin of alkaline lavas. *Science*, **320**, 916–919, https://doi.org/10.1126/science.1156563

Price, R.C., Nicholls, I.A. and Day, A. 2014. Lithospheric influences on magma compositions of late Mesozoic and Cenozoic intraplate basalts (the Older Volcanics) of Victoria, south-eastern Australia. *Lithos*, **206–207**, 179–200, https://doi.org/10.1016/j.lithos.2014.07.027

Rabayrol, F., Hart, C.J.R. and Thorkelson, D.J. 2019. Temporal, spatial and geochemical evolution of late Cenozoic post-subduction magmatism in central and eastern Anatolia, Turkey. *Lithos*, **336–337**, 67–96, https://doi.org/10.1016/j.lithos.2019.03.022

Riefstahl, F., Gohl, K. et al. 2020. Cretaceous intracontinental rifting at the southern Chatham Rise margin and initialisation of seafloor spreading between Zealandia and Antarctica. *Tectonophysics*, **776**, 228–298, https://doi.org/10.1016/j.tecto.2019.228298

Riley, T. and Leat, P.T. 2021. Palmer Land and Graham Land volcanic groups (Antarctic Peninsula): petrology. *Geological Society, London, Memoirs*, **55**, https://doi.org/10.1144/M55-2018-51

Rocchi, S. and Smellie, J.L. 2021. Northern Victoria Land: petrology. *Geological Society, London, Memoirs*, **55**, https://doi.org/10.1144/M55-2019-19

Rocchi, S., Armienti, P., D'Orazio, M., Tonarini, S., Wijbrans, J.R. and Di Vincenzo, G. 2002. Cenozoic magmatism in the western Ross Embayment: role of mantle plume v. plate dynamics in the development of the West Antarctic Rift System. *Journal of Geophysical Research: Solid Earth*, **107**, ECV 5-1–ECV 5-22, https://doi.org/10.1029/2001JB000515

Rocchi, S., Armienti, P. and Di Vincenzo, G. 2005. No plume, no rift magmatism in the West Antarctic Rift. *Geological Society of America Special Papers*, **388**, 435–447, https://doi.org/10.1130/0-8137-2388-4.435

Rocholl, A., Stein, M., Molzahn, M., Hart, S.R. and Wörner, G. 1995. Geochemical evolution of rift magmas by progressive tapping of a stratified mantle source beneath the Ross Sea Rift, Northern Victoria Land, Antarctica. *Earth and Planetary Science Letters*, **131**, 207–224, https://doi.org/10.1016/0012-821X(95)00024-7

Rowley, P.D. and Smellie, J.L. 1990. Southwestern Alexander Island. *American Geophysical Union Antarctic Research Series*, **48**, 277–279.

Scott, J.M., Pontesilli, A., Brenna, M., White, J.D.L., Giacalone, E., Palin, M.J. and le Roux, P.J. 2020. The Dunedin Volcanic Group and a revised model for Zealandia's alkaline intraplate volcanism. *New Zealand Journal of Geology and Geophysics*, **63**, 510–529, https://doi.org/10.1080/00288306.2019.1707695

Seghedi, I., Matenco, L., Downes, H., Mason, P.R.D., Szakács, A. and Pécskay, Z. 2011. Tectonic significance of changes in post-subduction Pliocene–Quaternary magmatism in the south east part of the Carpathian–Pannonian region. *Tectonophysics*, **502**, 146–157, https://doi.org/10.1016/j.tecto.2009.12.003

Shen, W., Wiens, D.A. *et al.* 2017. Seismic evidence for lithospheric foundering beneath the southern Transantarctic Mountains, Antarctica. *Geology*, **46**, 71–74, https://doi.org/10.1130/G39555.1

Sims, K.W.W., Blichert-Toft, J. *et al.* 2008. A Sr, Nd, Hf, and Pb isotope perspective on the genesis and long-term evolution of alkaline magmas from Erebus volcano, Antarctica. *Journal of Volcanology and Geothermal Research*, **177**, 606–618, https://doi.org/10.1016/j.jvolgeores.2007.08.006

Sims, K.W.W., Aster, R. *et al.* 2021. Mount Erebus. *Geological Society, London, Memoirs*, **55**, https://doi.org/10.1144/M55-2019-8

Smellie, J.L. 1999. Lithostratigraphy of Miocene–Recent, alkaline volcanic fields in the Antarctic Peninsula and eastern Ellsworth Land. *Antarctic Science*, **11**, 352–378, https://doi.org/10.1017/S0954102099000450

Smellie, J.L. 2021. Bransfield Strait and James Ross Island: volcanology. *Geological Society, London, Memoirs*, **55**, https://doi.org/10.1144/M55-2018-58

Smellie, J.L. and Collerson, K.D. 2021. Gaussberg: volcanology and petrology. *Geological Society, London, Memoirs*, **55**, https://doi.org/10.1144/M55-2018-85

Smellie, J.L. and Martin, A.P. 2021. Erebus Volcanic Province: volcanology. *Geological Society, London, Memoirs*, **55**, https://doi.org/10.1144/M55-2018-62

Smellie, J.L., Panter, K.S. and Reindel, J. 2021. Mount Early and Sheridan Bluff: volcanology. *Geological Society, London, Memoirs*, **55**, https://doi.org/10.1144/M55-2018-61

Storey, B.C. and Kyle, P.R. 1997. An active mantle mechanism for Gondwana breakup. *South African Journal of Geology*, **100**, 283–290.

Storey, B.C. and Granot, R. 2021. Tectonic history of Antarctica over the past 200 million years. *Geological Society, London, Memoirs*, **55**, https://doi.org/10.1144/M55-2018-38

Storey, B.C., Leat, P.T., Weaver, S.D., Pankhurst, R.J., Bradshaw, J.D. and Kelly, S. 1999. Mantle plumes and Antarctica–New Zealand rifting: evidence from mid-Cretaceous mafic dykes. *Journal of the Geological Society, London*, **156**, 659–671, https://doi.org/10.1144/gsjgs.156.4.0659

Timm, C., Hoernle, K. *et al.* 2010. Temporal and geochemical evolution of the Cenozoic intraplate volcanism of Zealandia. *Earth-Science Reviews*, **98**, 38–64, https://doi.org/10.1016/j.earscirev.2009.10.002

Tinto, K.J., Padman, L. *et al.* 2019. Ross Ice Shelf response to climate driven by the tectonic imprint on seafloor bathymetry. *Nature Geoscience*, **12**, 441–449, https://doi.org/10.1038/s41561-019-0370-2

van der Meer, Q.H.A., Waight, T.E., Scott, J.M. and Münker, C. 2017. Variable sources for Cretaceous to recent HIMU and HIMU-like intraplate magmatism in New Zealand. *Earth and Planetary Science Letters*, **469**, 27–41, https://doi.org/10.1016/j.epsl.2017.03.037

Vaughan, A.P. and Storey, B.C. 2007. A new supercontinent self-destruct mechanism: evidence from the Late Triassic–Early Jurassic. *Journal of the Geological Society, London*, **164**, 383–392, https://doi.org/10.1144/0016-76492005-109

Weaver, S.D., Storey, B.C., Pankhurst, R.J., Mukasa, S.B., DiVenere, V.J. and Bradshaw, J.D. 1994. Antarctic–New Zealand rifting and Marie Byrd Land lithospheric magmatism linked to ridge subduction and mantle plume activity. *Geology*, **22**, 811–814, https://doi.org/10.1130/0091-7613(1994)022<0811:ANZRAM>2.3.CO;2

Wilch, T.I., McIntosh, W.C. and Panter, K.S. 2021. Marie Byrd Land and Ellsworth Land: volcanology. *Geological Society, London, Memoirs*, **55**, https://doi.org/10.1144/M55-2019-39

Zhang, Y., Wang, C., Zhu, L., Jin, Z. and Li, W. 2019. Partial melting of mixed sediment–peridotite mantle source and its implications. *Journal of Geophysical Research: Solid Earth*, **124**, 6490–6503, https://doi.org/10.1029/2019JB017470

Chapter 1.4

Antarctic volcanism: active volcanism overview

A. Geyer

Geosciences Barcelona, GEO3BCN – CSIC, Lluis Solé i Sabaris s/n, 08028 Barcelona, Spain

0000-0002-8803-6504

ageyer@geo3bcn.csic.es

Abstract: In the last two centuries, demographic expansion and extensive urbanization of volcanic areas have increased the exposure of our society to volcanic hazards. Antarctica is no exception. During the last decades, the permanent settlement and seasonal presence of scientists, technicians, tourists and logistical personnel close to active volcanoes in the south polar region have increased notably. This has led to an escalation in the number of people and the amount of infrastructure exposed to potential eruptions. This requires advancement of our knowledge of the volcanic and magmatic history of Antarctic active volcanoes, significant improvement of the monitoring networks, and development of long-term hazard assessments and vulnerability analyses to carry out the required mitigation actions, and to elaborate on the most appropriate response plans to reduce loss of life and infrastructure during a future volcanic crisis. This chapter provides a brief summary of the active volcanic systems in Antarctica, highlighting their main volcanological features, which monitoring systems are deployed (if any), and recent (i.e. Holocene and/or historical) eruptive activity or unrest episodes. To conclude, some notes about the volcanic hazard assessments carried out so far on south polar volcanoes are also included, along with recommendations for specific actions and ongoing research on active Antarctic volcanism.

Volcanic eruptions are among the most captivating, and yet most destructive, natural phenomena. For example, two recent volcanic events illustrate the inherent and often unexpected dangers of active volcanoes: the 2019 eruption of White Island, New Zealand, and the 2018 eruption of Anak Krakatoa, Indonesia. In both cases, human deaths were caused by primary volcanic explosivity and by the impact of volcano-induced tsunamis, respectively (Global Volcanism Program 2019, 2020). Over 1500 subaerial volcanoes have been active during the Holocene (i.e. over the past *c.* 10 kyr), and more than a third of these have erupted one or more times during recorded history (Siebert *et al.* 2011). During the last two centuries, demographic expansion and extensive urbanization of volcanic areas have increased the exposure of our society to volcanic hazards, with about 9–10% of the world's population living within proximity (<100 km) of active and/or potentially active volcanoes (Tilling 2005; Erfurt-Cooper 2010). In the last 20 years, about 790 eruptions have occurred worldwide causing over 1300 deaths, more than US$850 million in direct economic losses and several billion in indirect ones; the latter mainly related to the long-range disruption of air traffic due to the presence of ash particles in the atmosphere (e.g. the 2010 Eyjafjallajökull eruption in Iceland, the 2011 Grímsvötn eruption in Iceland and the 2015 Cordón Caulle eruption in Chile) (EM-DAT International Disaster Database, January 2016 data).

A volcano is considered to be active if it is currently erupting (or in unrest), or it has erupted during the Holocene (Siebert *et al.* 2011). Based on this definition, from the hundreds of volcanoes located in Antarctica and described in this Memoir, at least eight large volcanoes are either known to be, or are likely to be, active (Deception Island, Mount Erebus, Mount Melbourne, Mount Rittmann, Mount Takahe, Balleny Islands, The Pleiades and, possibly, Mount Waesche), whilst the remaining 12 are either likely extinct (e.g. large volcanoes such as Mount Siple, Toney Mountain) or else the Holocene activity comprised small monogenetic scoria cones that may not erupt again (Table 1; Fig. 1a). Of those considered active, five of them have reports of repeated volcanic activity in historical times (Table 1; Fig. 1a) (Global Volcanism Program: http://www.volcano.si.edu). However, despite the considerable number of active volcanic centres, the hazard posed by Antarctic eruptions has received little interest so far.

The lack of focus on Antarctic volcanic hazards can be explained by the isolation of the continent, and thus to the perceived lack of an exposed population. However, this overlooks the fact that tephra generated during a moderately to highly explosive eruption has the potential to affect not only nearby areas but also to have a regional, or even global, impact (Geyer *et al.* 2017). In addition, the scarcity of outcrops and the challenging conditions for mapping the volcanic deposits far from (or even relatively close to) the eruptive vents, make it difficult to reconstruct the eruptive history of some of these active volcanic systems, and to estimate the

Table 1. *List of Antarctic volcanoes and last eruptions*

Volcano name	Last eruption	Latitude (°)	Longitude (°)
Berlin, Mount	8350 BCE	−76.05	−136
Buckle Island	1899 CE	−66.76	163.25
Deception Island	1970 CE	−63.001	−60.652
Erebus, Mount	May 2020 CE (continuing)	−77.53	167.17
Hudson Mountains	207 BCE ± 240 years	−74.33	−99.42
James Ross Island	A few thousand years ago	−64.15	−57.75
Melbourne, Mount	1892 CE ± 30 years	−74.35	164.7
Melville Peak	A few thousand years ago?	−62.02	−57.67
Paulet Island	<1 kyr ago	−63.579	−55.78
Penguin Island	1905 CE	−62.1	−57.93
Pleiades, The	1050 BCE	−72.67	165.5
Rittmann, Mount	1252 CE ± 2 years	−73.45	165.5
Royal Society Range	Holocene?	−78.25	163.33
Seal Nunataks Group	Historical?	−65.03	−60.05
Siple, Mount	Holocene?	−73.43	−126.67
Takahe, Mount	5550 BCE	−76.28	−112.08
Toney Mountain	Holocene?	−75.8	−115.83
Waesche, Mount	Holocene?	−77.17	−126.88
Young Island	1839 CE	−66.42	162.47

According to the Global Volcanism Program (https://volcano.si.edu/, last accessed 28 May 2020).

From: Smellie, J. L., Panter, K. S. and Geyer, A. (eds) 2021. *Volcanism in Antarctica: 200 Million Years of Subduction, Rifting and Continental Break-up.* Geological Society, London, Memoirs, **55**, 55–72,

First published online January 15, 2021, https://doi.org/10.1144/M55-2020-12

© 2021 The Author(s). Published by The Geological Society of London. All rights reserved.

For permissions: http://www.geolsoc.org.uk/permissions. Publishing disclaimer: www.geolsoc.org.uk/pub_ethics

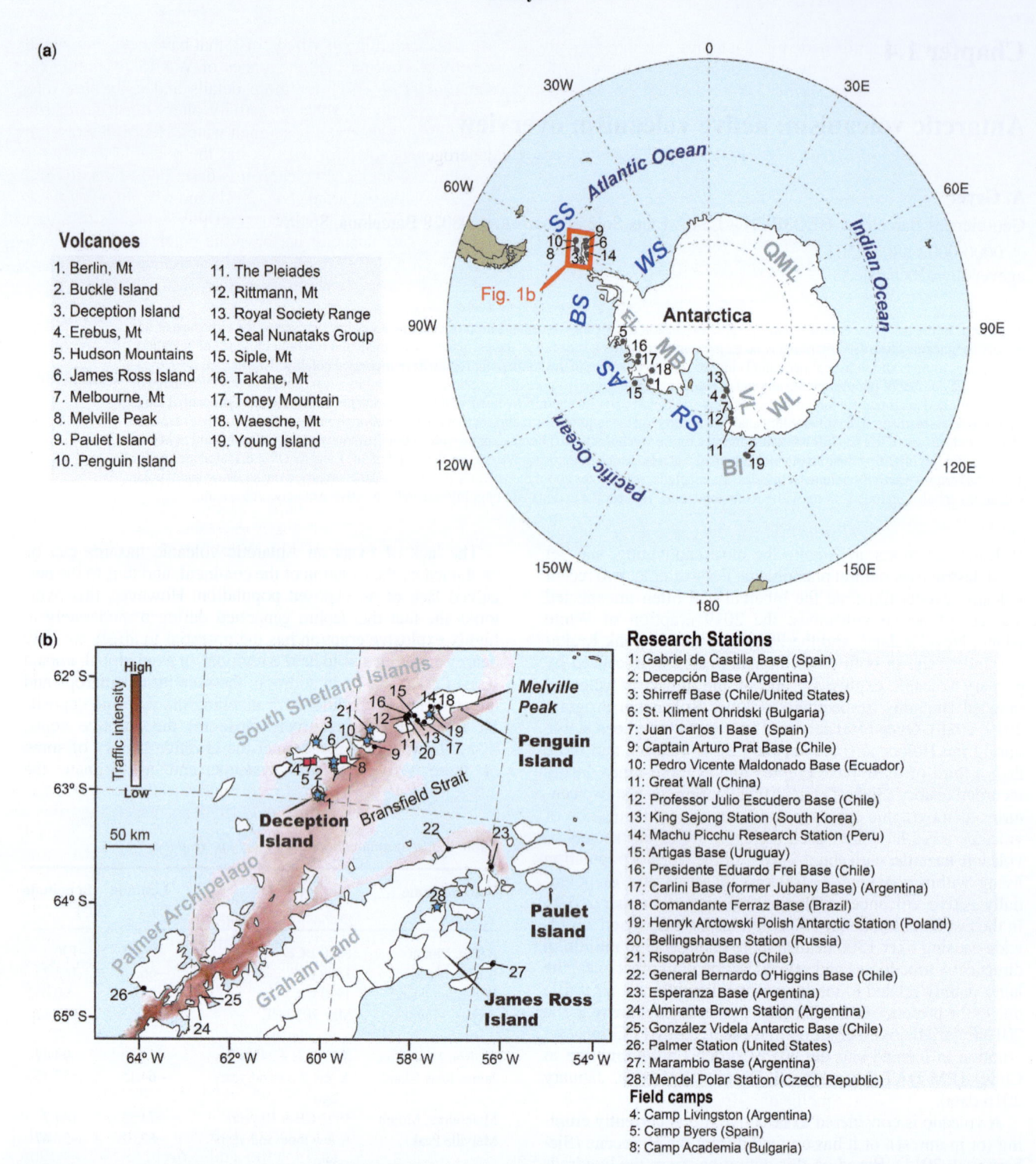

Fig. 1. (a) The locations of the Antarctic volcanoes listed in Table 1. AS, Amundsen Sea; BS, Bellingshausen Sea; QML, Queen Maud Land; MB, Marie Byrd Land; RS, Ross Sea; SS, Scotia Sea; VL, Victoria Land; WS, Weddell Sea; WL, Wilkes Land. (b) The locations of year-round (black dots) and temporary (only austral summer, blue stars) research stations in Graham Land, Palmer Archipelago and the South Shetland Islands. Red squares correspond to temporary field camps. The intensity of vessel traffic in the tourist season 2012–13 is also indicated (source: Bender *et al.* 2016).

magnitude and extent of their past eruptions – two fundamental aspects necessary to carry out a proper volcanic hazard assessment and to evaluate their potential future eruptive activity.

Nonetheless, the permanent settlement and seasonal presence of scientists, technicians, tourists and logistical personnel close to active Antarctic volcanoes has increased significantly during the last decades. For example, the Antarctic Peninsula region together with the South Shetland Islands hosts 22 research stations (both temporary and year-round) and three summer field camps, which are located inside (e.g. Gabriel de Castilla Base, Spain) or within a 150 km radius distance from the very active Deception Island volcano (Fig. 1b). Nearby, the Palmer Archipelago and the northwestern coast of the Antarctic Peninsula are both important tourist destinations exceeding 30 000 visitors per year (IAATO: https://iaato.org/) with a significant increase in vessel traffic during the tourist season (Fig. 1b). This escalation in the amount of exposed infrastructure and population to a future eruption of an Antarctic volcano urges the need to: (i) advance our knowledge of the volcanic and magmatic history of those south polar volcanoes with confirmed or suspected Holocene activity; (ii) provide the foundations for understanding the potential for upcoming eruptive activity; (iii) significantly improve the monitoring networks and our capacity for decoding monitoring data recorded during a volcanic crisis and, henceforth, the future eruption forecast capacity; (iv) develope vulnerability analyses and long-term volcanic hazard assessments; (v) carrying out the required mitigation strategies (e.g. fortifying buildings, implementing educational programmes addressed at enhancing public awareness); and (vi) establish effective early-warning systems, alert protocols, evacuation routes and communication strategies. All these actions will improve our response to future volcanic crises in Antarctica and thereby reduce the loss of life and infrastructure.

This Geological Society Memoir seeks to provide an up-to-date description of the current knowledge of Antarctic volcanism during the last 200 myr, covering tectonic, petrological, volcanological and palaeoenvironmental aspects. The main objective of this overview chapter is to offer a brief summary of the active volcanic systems in Antarctica, highlighting their main volcanological features, deployed monitoring systems (if any) and recent (i.e. Holocene and/ or historical) eruptive activity or unrest episodes. I will also present a short description of those volcanic areas for which (although not confirmed) there is a suspicion that they may have experienced activity over the last few thousand years. The reader is referred to the different chapters in this Memoir for more details on the described volcanoes and additional references. The chapter ends with some notes on the volcanic hazard assessments carried out so far on south polar volcanoes, along with recommendations for specific actions and ongoing research into active volcanism in Antarctica.

Active Antarctic volcanism

Active volcanism in Antarctica can be grouped as:

- Hotspot-related oceanic islands (e.g. Balleny Islands: Lanyon et al. 1993).
- Intraplate rift-related alkaline volcanism across the West Antarctic Rift System (WARS) in:
  (a) Marie Byrd Land (Marie Byrd Land Volcanic Province, e.g. Mount Berlin, Mount Takahe: Dunbar et al. 2021; Panter et al. 2021; Wilch et al. 2021);
  (b) Ellsworth Land (Hudson Mountains: Dunbar et al. 2021; Panter et al. 2021; Wilch et al. 2021);
  (c) northern Victoria Land (e.g. Mount Melbourne and Mount Rittmann: Gambino et al. 2021; Rocchi and Smellie 2021; Smellie and Rocchi 2021);
  (d) southern Victoria Land (e.g. Mount Erebus: Martin et al. 2021; Sims et al. 2021; Smellie and Martin 2021).
- Volcanism associated with the closing stages of very slow subduction close to the northeastern tip of the Antarctic Peninsula in the James Ross Island Group (e.g. James Ross Island) and post-subduction volcanism further south on the Antarctic Peninsula (e.g. Hole 2021; Smellie 2021b; Smellie and Hole 2021).
- Back-arc rifting volcanism related to the opening of the Bransfield Strait (e.g. Deception Island: Geyer et al. 2021).

In addition, during recent years the observations from geophysical surveys and from ice cores and subglacial sampling have revealed over 21 sites under the West Antarctic Ice Sheet (WAIS) that exhibit characteristics consistent with volcanic activity that has interacted with the current manifestation of the WAIS (Quartini et al. 2021). Active subglacial volcanism in West Antarctica, is mainly concentrated along crustal thickness gradients bounding the central WARS and in intraplate rift sites with thinned, rifted crust, that have been tectonically reactivated during multiple stages of WARS formation (see Quartini et al. 2021 for more details and additional references). Subglacial volcanic activity alters the ice's thermal structure and generates basal melt water. Hence, the resulting heterogeneous geothermal flux has the potential to affect ice dynamics. Subglacial volcanism is described in detail within this Memoir by Quartini et al. (2021) and will not be discussed further here. The main objective of this section is to present those active volcanoes that have the potential to be actively monitored and for which long-term volcanic hazard assessments have been, or could be, performed in the future.

Oceanic islands

Balleny Islands. Although, for practical reasons, volcanoes on the oceanic part of the Antarctic Plate are not included in this Memoir, volcanoes in the Balleny Islands are just 500–600 km from the nearest volcanic outcrops on the continent (Adare Peninsula, northern Victoria Land). They also qualify as active or potentially active (<10 ka in age).

The Balleny Islands are a 160 km-long chain of volcanic islands extending from 66.20° to 67.63° S and from 162.25° to 165.0° E. As described by Wright and Kyle (1990a), they are located at the southern end of a submarine ridge system that extends north to New Zealand but which is offset by the Indian–Antarctic Ridge (Fig. 1a). Although discovered in 1839, the geology has not yet been studied in detail (LeMasurier et al. 1990; Wright and Kyle 1990a). No scientific stations have been established and only a few landings have been made on any of the islands.

The archipelago is composed of three large islands (Young, Buckle and Sturge) and three small islands (Row, Borradaile and Sabrina). Volcanism in the Balleny Islands area is thought to be linked to the Balleny Fracture Zone (Falconer 1972; Kyle and Cole 1974; LeMasurier et al. 1990) and has been attributed to the Balleny hotspot (Lanyon et al. 1993). The volcanic deposits described to date consist of interbedded tuffs, agglomerate scoria and lava flows (Wright and Kyle 1990a). At least two of the islands, Young Island (66.42° S, 162.47° E; 1340 m asl (above sea level)) and Buckle Island (66.76° S, 163.25° E; 1239 m asl), are potentially still active given the dark eruption columns that were reported in 1839 and 1899, respectively (LeMasurier et al. 1990; Wright and Kyle 1990a).

Victoria Land

Active volcanism in Victoria Land is represented by the McMurdo Volcanic Group, formed as the result of continental rifting along the West Antarctic Rift (see Rocchi and Smellie 2021; Smellie 2021a; Smellie and Rocchi 2021 for more details and additional references). The McMurdo Volcanic Group is divided from north to south into the Hallett, Melbourne and Erebus volcanic provinces, the latter two hosting active volcanoes.

Melbourne Volcanic Province. The Melbourne Volcanic Province, located in the coastal area of northern Victoria Land, includes the active volcanic centres of Mount Melbourne, Mount Rittmann and The Pleiades (e.g. Gambino et al. 2021; Rocchi and Smellie 2021; Smellie and Rocchi 2021) (Figs 1a & 2a). Mount Melbourne stratovolcano (74.35° S, 164.70° E; 2732 m asl), part of the Mount Melbourne Volcanic Field, was discovered in 1841 by Sir J.C. Ross, a British Royal Navy officer and polar explorer. Mount Rittmann (73.45° S, 165.50° E; 2600 m asl), included in the Mount Overlord Volcanic Field, was not discovered until the 1988–89 campaign by the Italian National Antarctic

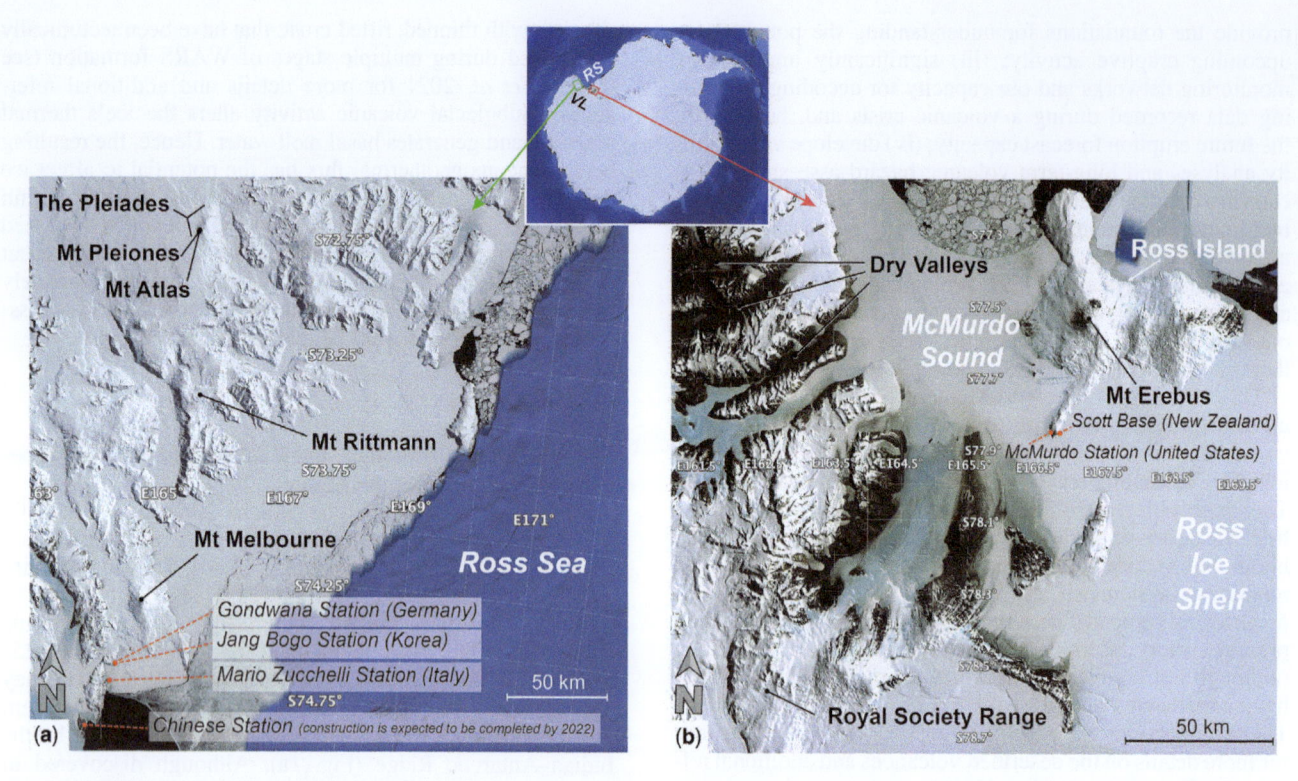

Fig. 2. (**a**) Map of the Melbourne Volcanic Province in north Victoria Land showing the location of Mount Melbourne, Mount Rittman and The Pleiades Volcanic Field. The names and locations of the scientific bases present in the area are also indicated. (**b**) Map of the Erebus Volcanic Province in south Victoria Land showing the locations of Mount Erebus and the Royal Society Range, as well as the McMurdo (USA) and Scott Base (New Zealand) Antarctic scientific stations. Image maps obtained from Google Earth Pro. Source: United States Geological Survey.

Programme (Armienti and Tripodo 1991; Gambino *et al.* 2021). The Pleiades (72.67° S, 165.5° E; 3040 m asl), comprising The Pleiades Volcanic Field, is situated 120 km inland from the coast of northern Victoria Land and is the furthest inland outcrop included in the Melbourne Volcanic Province with possible activity in the Holocene (Kyle 1990; Smellie 2001). Of the three active volcanic centres, Mount Melbourne is the one closest to scientific infrastructures, located at about 42 km from the Italian Mario Zucchelli Station, 33 km from the South Korean Jang Bogo Station and the German Gondwana Station, and approximately 65 km from the 5th Chinese Station (construction of the latter is expected to be completed by 2022: Fig. 2a). By contrast, Mount Rittman and The Pleiades are located about 135 and 215 km distance from the Korean scientific base, respectively (Fig. 2a). Parts of the summits of Mount Melbourne and Mount Rittmann, together with some areas of Mount Erebus, are protected under the Antarctic Treaty System (https://www.ats.aq), designated as Antarctic Specially Protected Area (ASPA) No. 175 (https://www.ats.aq/devph/en/apa-database, last accessed 28 May 2020). The purpose of this protected status is to safeguard their unique high-altitude geothermal environments, which due to their physical and chemical characteristics support biological communities that are not just regionally but globally unique. The summit of Mount Melbourne is also considered to be a separate ASPA (No. 118) due to its geothermally heated soils with a diverse and distinctive biological community (e.g. Broady *et al.* 1987; Nicolaus *et al.* 1996; Skotnicki *et al.* 2001).

Mount Melbourne and Mount Rittmann became active about 4.0 Ma (Mount Rittmann: Armienti and Tripodo 1991) and 2.7 Ma (Mount Melbourne: Wörner and Viereck 1989). Their magmatic activity comprises effusive eruptions of mafic compositions, such as alkali basalts and hawaiites, and explosive products of trachytic compositions producing Plinian eruptions (e.g. Armienti and Tripodo 1991; Giordano *et al.* 2012; Gambino *et al.* 2021; Rocchi and Smellie 2021). Both volcanoes show evidence of recent volcanic activity comprising fumarolic fields that are mainly located at or near their summits (Gambino *et al.* 2021 and references therein).

Mount Melbourne has been active on several occasions during the last 150 kyr. The most recent eruptive products, which are well exposed on the edifice, consist of trachytic to rhyolitic pumice-fall deposits which are likely to have originated from Plinian-style eruptions (Giordano *et al.* 2012). Based on the depth of burial of two tephra layers exposed in ice cliffs at Mount Melbourne and the determined snow accumulation rates (0.5–2.2 m a^{-1}) from collected snow sequences, the last major eruption occurred between 1862 and 1922 CE (Lyon 1986). In volcanic hazard terms, given that the last eruptions were explosive and associated with the most evolved magma compositions, Plinian explosive activity must be considered to be a potential eruptive scenario in the near future (Giordano *et al.* 2012; Gambino *et al.* 2021).

The extent of recent activity on Mount Rittman is still poorly known due to the scarcity of outcrops. Recently, tephrochronological studies have found evidence that Mount Rittman erupted in 1252 CE and deposited a tephra layer which spread over a distance of 2000 km (Del Carlo *et al.* 2015; Di Roberto *et al.* 2019; Lee *et al.* 2019). The ash-fallout deposits of this eruption have been identified in several ice cores from East and West Antarctica, making it the most widely correlated tephra layer in the Antarctic ice, an extremely valuable marker layer and one of the largest Holocene eruptions in Antarctica (Lee *et al.* 2019). Mount Rittmann is likely to have had a considerable eruptive history prior to 1252 CE. Further fieldwork and dating of the volcanic rocks will help to reveal the early volcanic history of this active volcano and to confirm the occurrence of the 1252 CE eruption. An eruption of similar characteristics today would form a long-lasting ash cloud, ash fall on nearby research stations, and disruption of air traffic to and from scientific stations on Ross Island, which lie nearly 500 km to the south (Lee *et al.* 2019) (Fig. 2b).

The Pleiades Volcanic Field is described as a group of at least 12 small, youthful-looking scoria cones with well-preserved craters and two domes (e.g. Rocchi and Smellie 2021; Smellie and Rocchi 2021). The highest points are Mount Atlas (*c.* 3040 m asl) and Mount Pleiones (*c.* 3020 m asl), located at the southern end of The Pleiades (Fig. 2a); together they are regarded as a small stratovolcano comprising several overlapping and nested scoria cones (Esser and Kyle 2002; Smellie and Rocchi 2021). The volcanic field ranges compositionally from basanites to tephriphonolites, and pumice lapilli are common on the surface of several of the centres, indicating the occurrence of a moderately explosive Holocene eruption (Rocchi and Smellie 2021; Smellie and Rocchi 2021). The Pleiades volcanic activity dates back over 847 kyr, and the presence of several cryptotephra layers recovered in ice cores, with compositions similar to rocks in The Pleiades, suggests the occurrence of several eruptions in the area during the last 123 kyr (e.g. Koeberl 1989; Esser and Kyle 2002; Dunbar 2003; Narcisi *et al.* 2016). Also, the youthful appearance of The Pleiades suggests that Holocene eruptions may have occurred and the volcanic group could be considered to be active (Kyle 1990; Smellie 2001), although the youngest reliable isotopic age is *c.* 20 ka (Smellie and Rocchi 2021).

Available monitoring data for Mount Melbourne and Mount Rittmann clearly indicate that both systems are currently active and characterized by: (i) active fumaroles fed by volcanic fluids; (ii) seismicity, including local seismic events and tremors related to the volcanoes' internal activity; and (iii) slow inflation/deflation ground deformation processes at the summit area of Mount Melbourne (see Gambino *et al.* 2021 for more details and additional references). Monitoring activities at Mount Melbourne were begun in the late 1980s by the Italian National Antarctic Programme (PRNA) with the set-up of a volcanological observatory and the installation of a global positioning system (GPS-tilt), and seismic networks on the volcano summit and flanks (Bonaccorso *et al.* 1997; Gambino *et al.* 2021). However, seismic and ground deformation data were mainly collected in the 1990s. In 2010–11, the Korean Polar Research Institute installed some broadband seismic stations, aimed at studying the structure beneath Mount Melbourne by using teleseismic data (see Gambino *et al.* 2021 and references therein). Recently, during the 2016–17 and 2017–18 Antarctic field seasons, new seismological, geochemical and volcanological research has been carried out on Mount Melbourne and Mount Rittmann in the framework of the ICE-VOLC project (Gambino *et al.* 2021). In 2016, the first gas monitoring measurements for Mount Melbourne concluded that the volcano's geothermal activity is still ongoing with CO_2 and CH_4 being the most common gases released at concentrations always above air concentrations. The isotopic compositions of these gases also confirmed their volcanic origin (Gambino *et al.* 2021). Due to the very recent discovery of the Mount Rittmann volcano, and the longer distance to the nearest Antarctic bases, very little monitoring information is available, although seismic measurements were collected in January 2017 as part of the ICE-VOLC project (see Gambino *et al.* 2021 for more details). At the time of writing, The Pleiades Volcanic Field is not being, and never has been, monitored (Lee *et al.* 2019).

Erebus Volcanic Province. The Erebus Volcanic Province, in southern Victoria Land, includes the active volcanic centre of Mount Erebus (77.53° S, 167.17° E; 3794 m asl) on Ross Island (southern Ross Sea), and the numerous small centres at the base of the inland mountain chain of the Royal Society Range (78.25° S, 163.33° E; 4025 m asl) and Dry Valleys (77.50° S, 162.00° E; max. 1400 m asl), located along the western shore of McMurdo Sound (Figs 1a & 2b). Mount Erebus is less than 40 km from McMurdo Station (USA) and Scott Base (New Zealand), and its summit area, which has fumarolic activity and high surface temperatures, is an ASPA (No. 130) and is thereby protected under the Antarctic Treaty.

Volcanic centres associated with the Royal Society Range and Dry Valleys are mainly Quaternary in age, and are composed of basanite lava flows and more than 50 basaltic vents, ranging from tiny amorphous scoria mounds to scoria cones up to 300 m high. Tephra layers in glaciers and geomorphic evidence may also suggest Holocene activity (LeMasurier *et al.* 1990; Wright and Kyle 1990*b*; Martin *et al.* 2021; Smellie and Martin 2021). Most of the volcanic outcrops have been mapped from a distance or from aerial photographs thanks to the distinctive colour contrast between volcanic and basement rocks, coupled with generally good exposures (Wright and Kyle 1990*b*). Detailed geological and volcanological descriptions of the different outcrops are scarce and very limited (see Wright and Kyle 1990*b*; Panter *et al.* 2003; Wingrove 2005; Smellie and Martin 2021 for more details).

Mount Erebus, the southernmost exposed active volcano on Earth, is a polygenetic, alkaline composite volcano. Its volcanic history can be divided into three distinct building phases (see Sims *et al.* 2021 for more details; also Smellie and Martin 2021): (i) Proto-Erebus shield (1.3–1.0 Ma); (ii) Proto-Erebus cone (from 1.0 Ma until the caldera collapse at about 750 ka); and (iii) a 'Modern' Erebus cone (250 ka–present), which forms the present-day summit area edifice. This last phase is characterized by two caldera-forming events (80–25 and 25–11 ka). Compositionally, the products of Mount Erebus have varied from basanites (Proto-Erebus shield-building phase) to phonotephrites (Proto-Erebus cone-building phase), and finally to tephriphonolitic–phonolitic lavas and minor trachytes (Modern Erebus phase) (Martin *et al.* 2021). Modern Erebus volcanic activity is largely confined to the older caldera structures and is dominated by lava flows. Over 20 eruptive events have occurred during the Holocene, and at least 10 during the last two centuries, as confirmed by historical observations (see Sims *et al.* 2021 for more details). Historical eruptive activity has been characterized by the existence of an active lava lake within the innermost crater, which has experienced significant elevation variations, and Strombolian explosions with bomb ejecta and minor ash (e.g. Kyle 1994; Sims *et al.* 2021 and references therein). The style, magnitude and frequency of these Strombolian eruptions have varied over the last few decades. Between 1973 and 1984 there were approximately six small episodes per day. In 1984, eruption frequency and intensity increased considerably. Starting in September 1984 until January 1985, there was a period of strong Strombolian activity (see Sims *et al.* 2021 for more details). In October 1993 powerful phreatic explosions near the rim of the innermost crater created a new subcrater (80 m in diameter). During 2001–04 Strombolian activity was much less frequent, with a notable pause during 2004. Since mid- 2005 there has been a return to more frequent Strombolian activity. The exact reasons for these variations in the lava lake eruptive style are not yet fully explained but are assumed to be linked to: (i) processes of gas accumulation at depth in the conduit system feeding the convecting lava lake; (ii) changes in the geometry or roughness of the vent/conduit system; and (iii) changes in the gas/magma supply from the deep system (e.g. through the size and flux of CO_2 slugs from depth: Aster *et al.* 2008; Sweeney *et al.* 2008; Sims *et al.* 2021).

Due to the proximity of Mount Erebus to McMurdo Station and Scott Base (Fig. 2b), monitoring activities have been facilitated and strongly promoted (e.g. Kyle 1994; Oppenheimer and Kyle 2008). Since the late 1970s there have been many studies measuring gas compositions and fluxes including: (i) aerosol sampling; (ii) ultraviolet spectroscopy for SO_2 flux measurements; (iii) open-path Fourier transform infrared spectroscopy of magma degassing; and (iv) infrared analysis of flank CO_2 abundance and flux (see Sims *et al.* 2021 for more details and additional references). As highlighted by Sims *et al.* (2021), gas measurements, especially SO_2 and CO_2 fluxes and ratios, are fundamental for volcano hazard assessment and, consequently, should be an integral part of any monitoring programme to evaluate the current state of activity of the volcano (cf. Sims *et al.* 2021). Mount Erebus's persistent lava lake is an open-conduit degassing magmatic system, and is a significant point source of gases and aerosols to the polar troposphere. Aerosol measurements have shown that the Mount Erebus volcano and its gas plume: (i) represent a potentially important point source for chemical species to a wide region of the Polar Plateau and are preserved in glacio-chemical records; (ii) provide around a third of the Antarctic sulfate budget observed at the South Pole; and (3) significantly contribute to the aerosol inventory observed in the deep interior Antarctic snow (see Sims *et al.* 2021 for more details and additional references).

Seismic and infrasonic monitoring observations at Mount Erebus were initiated in 1974 by New Mexico Tech (USA) and Victoria University (New Zealand) (see Sims *et al.* 2021 for more details). As described by Sims *et al.* (2021), the network was augmented in the early 2000s under the auspices of New Mexico Tech's Mount Erebus Volcano Observatory (MEVO) project by installation of a digital data acquisition system, which incorporated short-period sensors together with infrasound and intermediate-period (30 s natural period) seismometers in a continuous recording system. In 2016, the instruments were removed from the mountain at the conclusion of the MEVO project (see Sims *et al.* 2021 for more details). Currently, the volcano's summit activity is recorded by a temporary seismic network maintained by the Incorporated Research Institutions for Seismology (IRIS) Consortium. In addition, the United States Antarctic Program (USAP) is contemplating the installation of a multi-year network of five IRIS-supported seismographs and infrasonic sensors to continue monitoring activities at Mount Erebus (see Sims *et al.* 2021 for more details). Additionally, for many years, eruptive activity has been recorded by cameras deployed at the crater rim (see Sims *et al.* 2021 for more details).

During the last two decades, our understanding of Mount Erebus's internal structure and eruption dynamics has considerably improved thanks to more sophisticated seismic, photographic/video and infrasonic data and observations (Sims *et al.* 2021). For example, it is now well known that the eruptive rates of the lava lake vary from timescales of weeks/months to many tens or more strong events per day (e.g. Rowe *et al.* 2000). Also, according to current monitoring records, tremor-like signals are unusual at Mount Erebus, and large internal seismic events or swarms are not observed (see Sims *et al.* 2021 for more details). As described by Sims *et al.* (2021), tremor signals, when they occur, correlate to repetitive, stick–slip tabular-iceberg collisions or grounding events happening near Ross Island, which are observable at hundreds of kilometres from their source in Antarctica (MacAyeal *et al.* 2008; Martin *et al.* 2010). These observations are interpreted to be a consequence of the Mount Erebus open-vent conditions (Sims *et al.* 2021). Under these particular circumstances, deviatoric stresses or magmatic pressures cannot accumulate within the system and generate measurable seismicity (Sims *et al.* 2021).

In terms of volcanic hazards, McMurdo Station and Scott Base are highly exposed to the potential effects of volcanic activity at Mount Erebus, especially in the case of an explosive eruption. The scientific bases are indispensable to the USAP and the New Zealand Antarctic Research Programme (NZARP), respectively, and McMurdo also provides logistical and backup support to other international programmes (e.g. New Zealand, Italy and Australia). With the current knowledge of Mount Erebus's past eruptive history, the probability of a Plinian eruption from the summit, or even an effusive flank eruption, is assumed to be fairly small (see Sims *et al.* 2021 for more details). However, a phreatic or phreatomagmatic eruption, as happened in 1993, is more likely to occur in the near future, posing a significant hazard to anyone working on the volcano's summit or flanks. As a consequence, air traffic, which is critical for the USAP and NZARP science support and safety of live operations, could be potentially disrupted (Sims *et al.* 2021). In addition, energetic Strombolian events, like those which occurred in 1984–85, would also hinder any scientific activity on (or near) the flanks of Mount Erebus.

Marie Byrd Land and Ellsworth Land

Marie Byrd Land, with a very high percentage of land covered by glacial ice and a paucity of exposed outcrops, is among the least accessible and least frequently visited regions in Antarctica (Panter *et al.* 2021). However, the repository of tephra deposits in ice cores, and blue ice areas, offers a good record of its local explosive volcanism (Dunbar *et al.* 2021). Active volcanic activity in Marie Byrd Land is grouped within the Marie Byrd Land Volcanic Group, formally defined by Wilch *et al.* (2021) to include Marie Byrd Land and the Thurston Island volcanic provinces. Cenozoic volcanism in Marie Byrd Land and the western portion of Ellsworth Land (Jones and Hudson mountains) extends for nearly 1500 km along the northern shoulder of the WARS (Panter *et al.* 2021). This region lies between the subduction (to post-subduction)-related volcanism of the Antarctic Peninsula to the east and the rift-related volcanism of the western Ross Sea to the west (e.g. Hole 2021; Leat and Riley 2021*a*, *b*; Panter 2021; Rocchi and Smellie 2021; Smellie 2021*a*; Smellie and Rocchi 2021). The petrogenesis of the Marie Byrd Land Volcanic Group is still under debate and is discussed in detail in Panter *et al.* (2021). The oldest known volcanism in the Marie Byrd Land Volcanic Group is dated at 36.6 Ma but most of the documented volcanism has occurred since 13.4 Ma (LeMasurier *et al.* 1990; Wilch and McIntosh 2000; Wilch *et al.* 2021).

Active volcanic activity within the Marie Byrd Land Volcanic Province takes place in three volcanic fields (Fig. 3a): (i) Eastern Marie Byrd Land; (ii) Flood Range; and (iii) Executive Committee Range (Blankenship *et al.* 1993; Iverson *et al.* 2017; Dunbar *et al.* 2021; Martin *et al.* 2021; Quartini *et al.* 2021). Possible activity centred on Mount Siple, according to satellite observations (Global Volcanism Program 2012), has never been confirmed and there is no evidence of recent eruptions on Mount Siple itself (Wilch *et al.* 2021). It is therefore probably inactive. In western Ellsworth Land, subglacial volcanic activity has also been detected in the Thurston Island Volcanic Province within the Hudson Mountains Volcanic Field (Corr and Vaughan 2008; Quartini *et al.* 2021).

Eastern Marie Byrd Land Volcanic Field. The Eastern Marie Byrd Land Volcanic Field (Fig. 3a) includes several small volcanic centres and three isolated polygenetic central volcanoes of which Mount Takahe and Toney Mountain are considered to be active (see Panter *et al.* 2021; Wilch *et al.* 2021 for more details and additional references).

Mount Takahe (76.28° S, 112.08° W; 3460 m asl) (Fig. 3b) is a Late Quaternary isolated shield volcano. The non-dissected

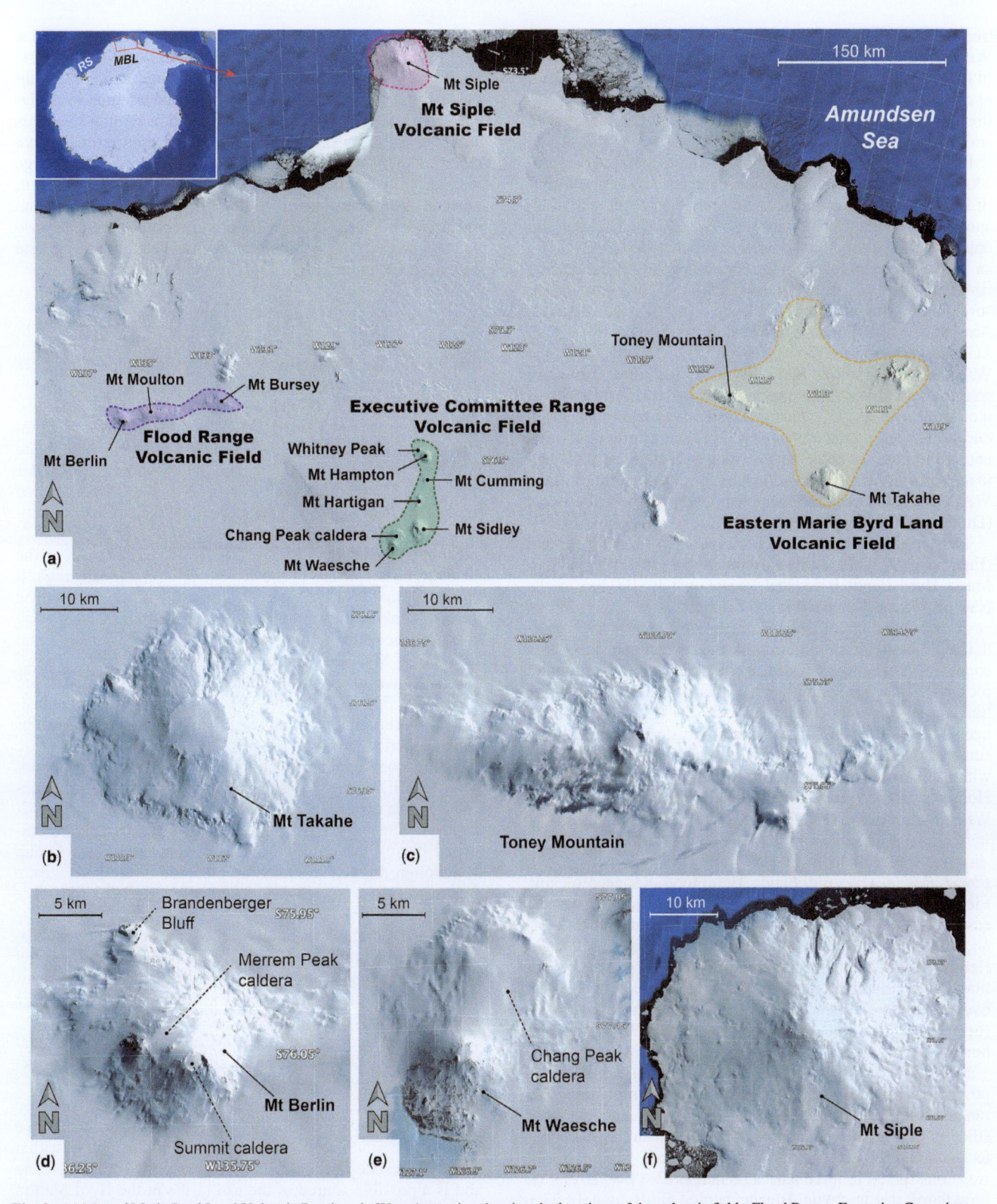

Fig. 3. (**a**) Map of Marie Byrd Land Volcanic Province in West Antarctica showing the locations of the volcanic fields: Flood Range, Executive Committee Range, Eastern Marie Byrd Land and Mount Siple. The names of the individual volcanoes are also included. The limits of the different volcanic fields are indicated in coloured dashed lines and are defined according to Wilch *et al.* (2021). Image map obtained from Google Earth Pro. Source: United States Geological Survey and PGC/NASA. (**b**)–(**f**) Google Earth Pro images (source: United States Geological Survey) of: (**b**) Mount Takahe; (**c**) Mount Toney; (**d**) Mount Berlin; (**e**) Mount Waesche; and (**f**) Mount Siple. Mount Siple is a well-formed young stratocone. It was previously regarded as potentially active but the evidence supporting recent activity is unreliable and it is more likely to be extinct.

symmetrical edifice has a flat, broad 8 km-diameter snow-filled summit caldera with limited outcrops that provide – together with limited flank outcrops – insights into the volcano's history. The age, distribution and elevation of hydroclastic deposits (including pillow lavas and breccias, as well as hyaloclastites and tuff cones) suggest ice-level fluctuations up to 400–575 m above the current ice-sheet level between 65.5 ± 5.2 and 23.7 ± 5.6 ka (Dunbar *et al.* 2021; Wilch *et al.* 2021). Outcrops around the caldera rim are mainly subaerial deposits, including pyroclastic deposits, lavas and hydrovolcanic tuffs (McIntosh *et al.* 1985; Dunbar *et al.* 2021; Wilch *et al.* 2021). These subaerial units suggest that the bulk of the eruptions occurred above the ice-sheet level (McIntosh *et al.* 1985; Dunbar *et al.* 2021). Mount Takahe is compositionally bimodal, consisting of basanite, hawaiite, mugearite and trachyte (Panter *et al.* 2021).

Based on observations from blue-ice areas and ice cores, the Mount Takahe eruptive record over the past 100 kyr consists of relatively sparse explosive eruptions that generated widespread tephra deposits (Dunbar *et al.* 2021). The volcano appears to have produced a large eruption around 8.2 ka BP, as indicated by an 8.2 ± 5.4 ka tephra deposit on the caldera rim (Dunbar *et al.* 2021; Wilch *et al.* 2021). The tephra corresponding to this eruption is widely distributed across West Antarctica, and is recognizable at Byrd Station (80.02° S, 119.52° W), Siple Dome (79.468° S, 112.086° W) and WAIS Divide (79.468° S, 112.086° W) ice cores, providing an important time-stratigraphic horizon in the climate archives (see Dunbar *et al.* 2021 for more details and additional references). According to Dunbar *et al.* (2021), this eruption, large enough to produce a visible tephra layer in blue-ice areas and in ice cores, is likely to have changed the albedo of much of the WAIS for some period of time following the eruptive event (Dunbar *et al.* 2021).

Toney Mountain (75.8° S, 115.83° W; 3595 m asl) (Fig. 3c), the least studied of the major volcanoes within Marie Byrd Land, is located approximately 100 km to the NW of Mount Takahe (Fig. 3a) (Panter *et al.* 2021; Wilch *et al.* 2021). It consists of a major central volcano with a well-defined 3 km-wide summit caldera and several parasitic scoria cones. Its geochemistry has been characterized by four whole-rock XRF analyses, comprising one each for hawaiite, latite, trachyte and rhyolite (Panter *et al.* 2021). It appears to have been active in Late Miocene and Pleistocene times, and some of the ash bands in the Byrd Station ice core deposited within the past 30 kyr may have been from Toney Mountain; Holocene activity is also possible (LeMasurier *et al.* 1990; Wilch *et al.* 2021).

Flood Range Volcanic Field. The Flood Range Volcanic Field is located in northwestern Marie Byrd Land, near the eastern coast of the Ross Sea. It consists of three east–west-aligned trachytic central shield volcanoes (Fig. 3a): Mount Berlin, Mount Moulton and Mount Bursey. Age data reveal that Flood Range volcanism migrated more than 100 km over a period of approximately 10 myr from east to west (between 2.77 ± 0.06 and <0.010 Ma) with Mount Berlin hosting the Holocene activity (Panter *et al.* 2021).

Mount Berlin (76.05° S, 136° W; 3478 m asl) consists of a polygenetic central volcano with a magma composition that ranges from basalts to felsic varieties which include both silica-undersaturated and silica-oversaturated trachytes (Panter *et al.* 2021). According to available data, Mount Berlin was constructed in three stages (Wilch *et al.* 1999, 2021): (i) growth of the trachytic Brandenberger Bluff tuya (2.77 ± 0.06 Ma), in which successive subglacial, phreatomagmatic and subaerial lithofacies record the rise of the volcano above a palaeo-ice-sheet surface; (ii) construction of the Merrem Peak shield volcano with eruptions from the Merrem Peak caldera (from 578 ± 9 to 143 ± 5.7 ka), mostly comprised of trachytic volcanic deposits dominated by pumiceous pyroclastic rocks; and (iii) growth of the volcanic edifice up to 3478 m asl and a southeastward shift of the vent area to the 2 km-diameter summit caldera.

Current volcanic activity of Mount Berlin is demonstrated by the presence of several ice-tower-topped steaming fumarolic caves (Wilch *et al.* 1999; Dunbar *et al.* 2021). Although only three young eruptions can be documented through investigation of surface outcrops on Mount Berlin (at 25.9 ± 2.0, 18.4 ± 5.8 and 10.4 ± 5.3 ka: Wilch *et al.* 1999, 2021; Dunbar *et al.* 2021), its explosive volcanic activity during the last two evolutionary stages is well documented in englacial tephra found at Mount Moulton and in several ice cores and blue-ice areas (Dunbar *et al.* 2021). Based on the tephra record, Mount Berlin appears to have produced a very large number of volcanic events over the past 100 kyr. Indeed, most of the tephra layers in the Byrd Station ice core, and most probably also in the WAIS Divide core, record extensive volcanic activity from Mount Berlin between around 18 and 24 kyr ago (Dunbar *et al.* 2021).

Executive Committee Range Volcanic Field. The Executive Committee Range Volcanic Field, in central Marie Byrd Land, consists of five north–south-aligned central volcanoes that are progressively younger towards the south (Fig. 3e) (Dunbar *et al.* 2021; Wilch *et al.* 2021): Whitney Peak and Mount Hampton (13.36 ± 0.05–8.6 ± 1.0 Ma), Mount Cumming (10.4 ± 1.0–10.0 ± 1.0 Ma, with a parasitic cone dated at 3.0 ± 0.1 Ma), Mount Hartigan (8.50 ± 0.66–6.02 ± 0.50 Ma), Mount Sidley (5.77 ± 0.12–4.24 ± 0.08 Ma), and the Mount Waesche and Chang Peak caldera volcanic complex (1.98 ± 0.05–<0.01 Ma).

Mount Waesche (77.17° S 126.88° W; 3292 m asl) is the southernmost volcano of the Executive Committee Range. It is a small and relatively young-looking volcano situated on the SW flank of the summit caldera of the older Chang Peak volcano (the latter ranging in age between 1.98 ± 0.05 and 1.08 ± 0.05 Ma: LeMasurier and Kawachi 1990; Panter 1995; Dunbar *et al.* 2021; Panter *et al.* 2021; Wilch *et al.* 2021) (Fig. 3e). The outcrops are largely composed of subaerial lavas classifiable in at least four compositional groups which describe two evolutionary lineages (see Dunbar *et al.* 2021 for more details): (i) Group 1 (trachytes) and 2 (phonolites); and (ii) Group 3 (basalts) and 4 (basanites–tephriphonolites). Scoria cones are also associated with the lava groups 2, 3 and 4. The 1.5 km-diameter Mount Waesche caldera collapsed during the eruptive episode that formed lava Group 4, although most of the material included in the latter group are post-caldera in age (see Dunbar *et al.* 2021 for more details).

Mount Waesche does not have the same degree of geochronological constraints as Mount Berlin or Mount Takahe but available isotopic dating results indicate activity at least since 1 ± 0.2 Ma up until 0.092 ± 0.026 Ma, with a large pulse of effusive magmatism occurring between 0.2 and 0.14 Ma (LeMasurier and Rex 1989; LeMasurier and Kawachi 1990; Panter 1995; Ackert *et al.* 2013). Nonetheless, the fact that the youngest lavas are too young to date by K–Ar and the existence of well-preserved volcanic deposits, both on the flanks of the volcano and in a large, tephra-bearing blue-ice field on the south side of the volcanic edifice, suggest that Mount Waesche may have been active during the Holocene and is a possible source of the ash layers in the Byrd Station ice core that were deposited over the past 30 kyr (see Dunbar *et al.* 2021 for more details and additional references). A future eruption from Mount Waesche would be most likely to form a mafic scoria cone and/or lava flow that would have an impact on the local environment but with a restricted geographical distribution (Dunbar *et al.* 2021).

Hudson Mountains Volcanic Field. The Hudson Mountains Volcanic Field (extending from 73.75° to 74.92° S and from 98.33° to 100.5° W) is located in western Ellsworth Land, along the eastern coast of the Amundsen Sea (Fig. 1a). The mountains consist of around 20 mostly snow- and ice-covered volcanic nunataks at 200–750 m asl. The majority of the outcrops consist of remnants of relatively small-volume subaerially erupted lavas, scoria and tuff cones, as well as subaqueous to emergent sequences that are possibly glaciovolcanic (Panter et al. 2021; Wilch et al. 2021 and references therein). Most of the volcanism is basaltic with minor occurrences of intermediate (i.e. latite) and felsic (i.e. trachyte) lavas, ranging in age from Middle Miocene to the end of the Pliocene (Panter et al. 2021; Wilch et al. 2021 and references therein). Current activity in the volcanic field has been suggested based on an unconfirmed observation of steam in 1974, and a possible eruption seen in satellite data in 1985, also unconfirmed (Rowley et al. 1990). More recently, subglacial to emergent volcanic activity near the Hudson Mountains was suggested by the presence of a bright local englacial tephra layer identified in radar data (Corr and Vaughan 2008). The source region for the tephra layer coincides with a subglacial topographical high adjacent to the Hudson Mountains (Golynsky et al. 2018) and the layer depth dates the eruption at 207 BCE ± 240 years, which matches well with the previously unattributed conductivity signals measured at two nearby Byrd Station and Siple Dome ice cores (Hammer et al. 1997; Kurbatov et al. 2006; Corr and Vaughan 2008; Quartini et al. 2021).

Antarctic Peninsula

On the Antarctic Peninsula, Neogene volcanism is mostly post-subduction, resulting from a series of collisions between segments of an oceanic spreading centre and the Antarctic Peninsula trench between *c.* 50 and 4 Ma, and the progressive opening of 'windows' in the downgoing oceanic slab (Hole 1988; Hole et al. 1991, 1994; Smellie and Hole 2021). The most recent volcanic activity, within the James Ross Island Volcanic Group, occurred on James Ross Island and Paulet Island, and in the Seal Nunataks Volcanic Field (Figs 1a & 4a). From 7.7 Ma until present, the magmas have erupted either: (i) in a back-arc position coeval with the final stages of subduction at the South Shetland Trench (Leat and Riley 2021*b*; Smellie and Hole 2021) leading to the formation of a large shield volcano, Mount Haddington, on James Ross Island; or (ii) as the result of decompression melting of mantle rising within slab windows, generating extensive monogenetic volcanic fields and small isolated centres, from sodic alkaline to tholeiitic composition (e.g. Seal Nunataks Volcanic Field and volcanic fields on Alexander Island and on southern Palmer Land: see Smellie and Hole 2021 for more details and additional references). Despite the difference in tectonic

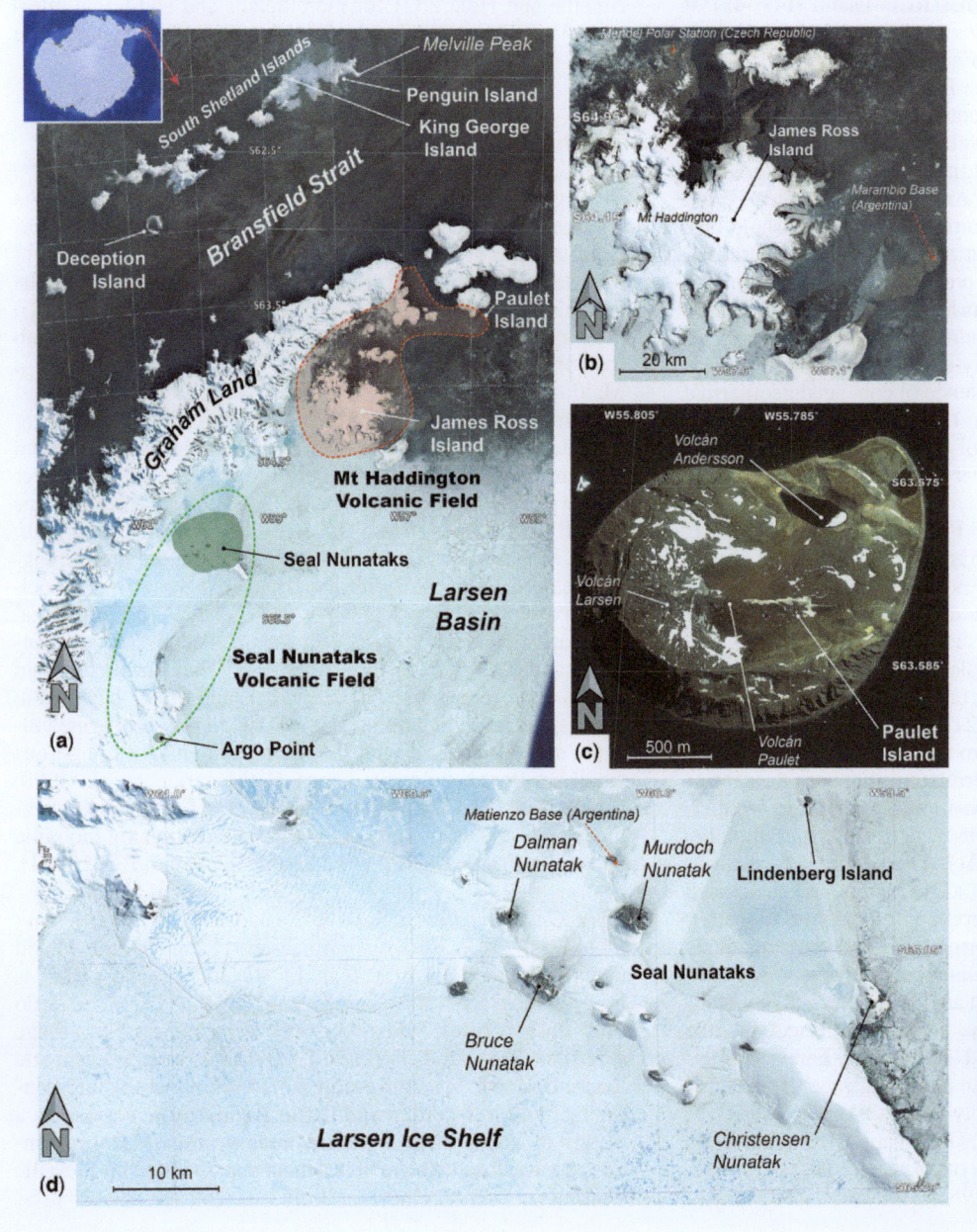

Fig. 4. (**a**) Map of Graham Land (Antarctic Peninsula), the Bransfield Strait and the South Shetland Islands. The limits of the Mount Haddington and Seal Nunataks volcanic fields are indicated by coloured dashed lines and are defined according to Smellie and Hole (2021). The image map was obtained from Google Earth Pro. Source: Landsat-Copernicus. (**b**) Image of James Ross Island showing the location of Mount Haddington, as well as the Mendel Polar Station (Czech Republic) and the Marambio Base (Argentina). Image obtained from Google Earth Pro. Source: Maxar Technologies. (**c**) Image of the Seal Nunataks showing the location of the Matienzo Base (Argentina). Image obtained from Google Earth Pro. Source: Landsat-Copernicus.

settings, the late Neogene volcanism occurrences throughout the Antarctic Peninsula have similar composition, lithofacies and eruptive palaeoenvironments (Smellie 2021b).

James Ross Island (64.14° S, 57.75° W; 1638 m asl) is located on the SE side of the northeastern extremity of the Antarctic Peninsula (Figs 1a & 4a). The island hosts the Czech Antarctic Base, Mendel Polar Station, and is less than 100 km from the Marambio (Argentina), Esperanza (Argentina) and General Bernardo O'Higgins (Chile) scientific bases (Fig. 1b). James Ross Island, the largest single outcrop of the James Ross Island Volcanic Group, is dominated by volcanic products related to the construction of Mount Haddington, a polygenetic shield volcano with associated (more youthful) flank tuff cones and other pyroclastic centres (Smellie 2021b) (Fig. 4b). The volcanic rocks of James Ross Island, which range compositionally from alkali basalts to hawaiites, consist of pāhoehoe lava, pillow lava and hyaloclastite, and scoria and tuff cones intruded by dykes and plugs (Košler et al. 2009; Calabozo et al. 2015; Haase and Beier 2021; Smellie 2021b). Well-preserved tuff cones and pyroclastic cones on the eastern flank below the summit ice cap are considered to be only a few thousand years old, suggesting that the system may still be active (Smellie et al. 2008; Smellie 2021b).

Paulet Island (63.59° S, 55.78° W; 351 m asl) is located about 120 km to the NE of James Ross Island (Fig. 4a). It has a very simple stratigraphy consisting of a basal lava sequence overlain by two steep coalesced pyroclastic cones (informally named by González-Ferrán 1995) (Fig. 4c): (i) Volcán Larsen, older and represented by a small and very subdued crater; and (ii) Volcán Paulet, a younger cone with a fresh-looking small crater. In addition, a water-filled possible maar, informally called Volcán Andersson (González-Ferrán 1995), may be present on the northern side of the island (see Smellie 2021b for more details and additional references). Compositionally, Paulet Island products consist of hawaiites and alkali basalts. The island has been considered potentially active due to the apparent residual high heat that may be responsible for the area's largely snow-free state (González-Ferrán 1995). However, the existence of enhanced geothermal warming has never been proven (Smellie 2021b). Nevertheless, the youthful appearance of the Paulet volcano scoria cone suggests that it may be very young (<1 kyr) (Smellie 1990b).

The Seal Nunataks Volcanic Field (65.13° S, 60.00° W; 368 m asl) is a small cluster of 16 isolated nunataks and islands situated on the east coast of the northern Antarctic Peninsula (Fig. 4a, d). The geology of the different nunataks has been extensively described and is summarized in Smellie and Hole (2021). Seal Nunataks Volcanic Field is formed from multiple small eruptive centres (ages ranging from 4 ± 1 to <0.1 Ma: Smellie and Hole 2021) with a very limited geographical extent of the erupted products, which hinders obtaining any potential stratigraphic correlations between the centres (Smellie and Hole 2021). Compositionally, volcanic products found in the different nunataks include alkali basalt, tholeiite and basaltic andesite (Hole 2021). The different formations described by Smellie and Hole (2021) include dykes, as well as subaerial (including lavas, lapilli tuffs and minor spatter: e.g. Christensen Nunatak Formation) and subaqueously erupted volcanic deposits (including pillow lavas: e.g. Bruce Nunatak Formation). The current activity of the Seal Nunataks Volcanic Field is under debate. Several historical records exist that report eruptions and fumaroles. Larsen (1894) described an eruption of black ash from Lindenberg (64.92° S, 59.70° W) and fumarolic activity at Christensen (65.10° S, 59.57° W) in December 1893. Observations of tephras on the ice surface, suggesting a twentieth century eruption, have been reported in 1968, 1982, 1985 and 2010 (Baker 1968; González-Ferrán 1985; Smellie 1990d; Kraus et al. 2013). However, these may have been due to aeolian tephra redistribution (Global Volcanism Program 1982). In January 1982, evidence of recent volcanism in the form of fresh-looking basaltic lavas was observed (Global Volcanism Program 1982; González-Ferrán 1983). These had apparently emerged from the central crater at Dallman volcano (65.02° S, 60.32° W) and flowed to its NW foot. In addition, the Larsen Ice Shelf, in the vicinity of Murdoch volcano (65.03° S, 60.03° W), had been widely covered by abundant basaltic lapilli and ash. Active fumaroles were also observed at the northern side of Dallman's summit and in a small scoria cone SE of Murdoch (Global Volcanism Program 1982). However, in 1988 a British expedition noted that tephra far from nunataks was found only in ice-cored moraines, suggesting a glacial rather than pyroclastic origin (Global Volcanism Program 1982; Smellie and Hole 2021). No fumarolic activity was detected at that time, although water vapour resulting from solar radiant heating of ice-cored moraines was observed. According to Smellie and Hole (2021), the observations by Larsen (1894) and González-Ferrán (1983) are inexplicable and should probably be discarded, implying that the Seal Nunataks Volcanic Field is either inactive or extinct (Smellie 1990d, 1999a; Smellie and Hole 2021). Nonetheless, the fact that some outcrops have been dated <0.1 Ma (see Smellie and Hole 2021 for more details and additional references) may be an indication that volcanic activity may resume in the future.

The Bransfield Strait

The Bransfield Strait is a c. 4 Ma back-arc rift developed in the continental lithosphere of the Antarctic Peninsula that has been the site of subduction of the Phoenix Plate throughout the Mesozoic (Barker 1982; McCarron and Larter 1998). Its tectonic evolution has been explained as being related to: (i) the passive subduction of the former Phoenix Plate and slab rollback of the South Shetland Trench (Smellie et al. 1984; Maldonado et al. 1994; Lawver et al. 1995, 1996); (ii) sinistral movement between the Antarctic and Scotia plates causing oblique extension along the Antarctic Peninsula's continental margin (Rey et al. 1995; Klepeis and Lawver 1996; Lawver et al. 1996; Gonzalez-Casado et al. 2000); or (iii) mechanisms (i) and (ii) occurring simultaneously (Galindo-Zaldívar et al. 2004; Maestro et al. 2007). Bransfield Basin, the morphological feature associated with the Bransfield Strait, has associated seismicity, volcanism and active hydrothermal circulation (Smellie 2021b). Volcanism is essentially concentrated at the central and eastern sectors of the Bransfield Basin and is represented by: (i) volcanic islands (e.g. Deception, Bridgeman and Penguin islands); (ii) six large submarine volcanic edifices aligned with the basin axis; and (iii) numerous small scattered volcanic cones (LeMasurier et al. 1990; Smellie 2021b). Compositionally, magmas forming the volcanic islands and seamounts of the Bransfield Strait are tholeiitic and range from basalt to rare rhyolite (Haase and Beier 2021).

King George Island. Melville Peak (62.02° S, 57.67° W; 549 m asl) is located on northeastern King George Island (South Shetland Islands: Figs 1a, 4a & 5a). The island is host to nine scientific stations and the Presidente Eduardo Frei Base (Chile), which operates as a permanent village (Fig. 1b). The island also has a 1300 m-long airstrip (Teniente Rodolfo Marsh Martin Aerodrome), with numerous intercontinental and intracontinental flights each season, serving as a means of transport to the many nearby bases. All the mentioned infrastructures are between 40 and 75 km from Melville Peak.

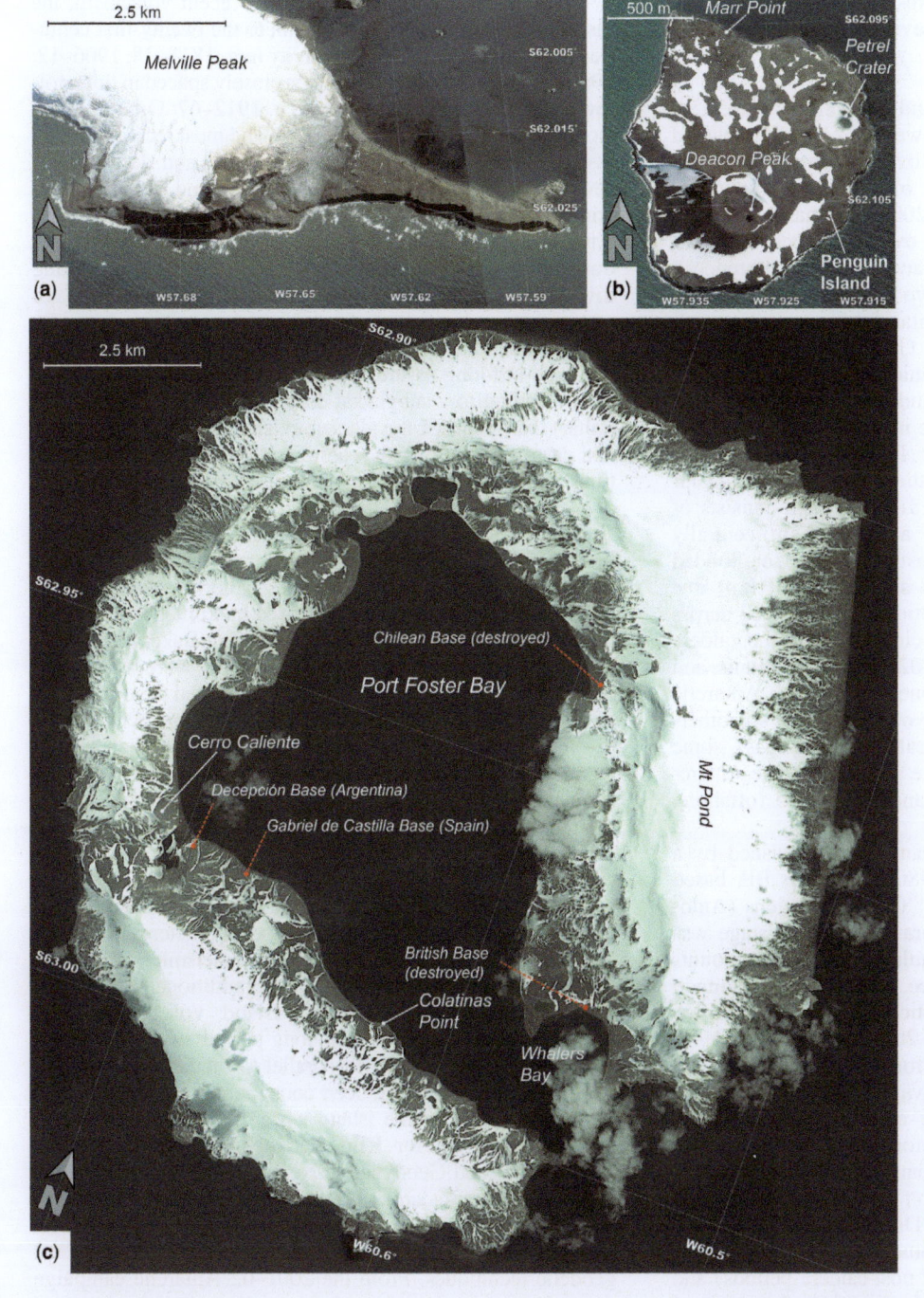

Fig. 5. (a) Image of Melville Peak on King George Island. (b) Image of Penguin Island. (c) Image of Deception Island showing the location of the current Spanish (Gabriel de Castilla Base) and Argentinian (Decepción Base) scientific bases. The positions of the destroyed Chilean and British bases are also indicated. Images obtained from Google Earth Pro. Source: Maxar Technologies.

Melville Peak consists of an eroded stratovolcano with a central basaltic plug and a summit crater (Barton 1965; Birkenmajer 1982; Smellie 1990a). Two principal phases of construction have been described for Melville Peak (Smellie 1990a, 2021b) (Fig. 5a): (i) a volumetrically dominant eruptive phase related to the construction of a small tuff cone or stratocone; and (ii) the growth of a small parasitic vent that erupted after the main eruptive phase, and after a period of erosion (possibly of glacial origin). Dating ages for lavas from the older stage of the volcano range from 296 ± 27 to 72 ± 15 ka (Birkenmajer and Keller 1990; Smellie 2021b). Smellie (2021b) suggests that the eruption was probably in an unconfined subaerial setting with magma interacting with seawater-replenished groundwater, causing predominantly phreatomagmatic eruptions. Despite the absence of any crater landform, an ash layer found 30 km away in a NE Bransfield Strait marine sediment core with compositional similarity to Melville Peak suggests possible Holocene activity within the past few thousand years (Keller et al. 2003; Kraus et al. 2013).

Penguin Island. Penguin Island (62.1° S, 57.92° W; 180 m asl) is located 2 km off the SE coast of King George Island, west of the axis of the Bransfield Rift (Figs 1a, 4a & 5b) (LeMasurier et al. 1990; Smellie 1990c, 2021b). Three geological units have been identified, all of mildly alkaline basaltic (Weaver et al. 1979) or low-potassium calc-alkaline (Pańczyk and Nawrocki 2011) compositions (see Smellie 2021b for more details and additional references): (i) Marr Point Formation (2.7 ± 0.2 Ma) (Pańczyk and Nawrocki 2011), a basal lava platform; (ii) Deacon Peak Formation (>340 years old: Birkenmajer 1980), a scoria cone with a few thin black basalt lavas at its base, which contains a smaller scoria cone (c. 174 years old: Birkenmajer 1980); and (iii) Petrel Crater Formation, which consists of maar deposits (c. 115 years old: Birkenmajer 1980) on the NE side of the island. Possible fumarolic activity was observed in 1966 within the maar's crater (González-Ferrán 1995), but this has not been confirmed. However, the youthful form of the Deacon Peak scoria cone may suggest that the fumarolic activity observed in the nineteenth century may correspond to Penguin Island and not to Bridgeman Island, as previously stated (González-Ferrán and Katsui 1970).

Deception Island. Deception Island (62.98° S, 60.65° W; 542 m asl) is located at the southwestern end of Bransfield Basin (Figs 1a, 4a & 5c). With over 30 eruptions during the Holocene, it is considered one of the most active volcanoes in Antarctica (Smellie *et al.* 2002), and has been intensively investigated and monitored over several decades (see Geyer *et al.* 2021 for more details and additional references). Today, Deception Island hosts two scientific bases, which operate every year during the Antarctic summer (Fig. 5c): Decepción Base (Argentina) and Gabriel de Castilla Base (Spain). In addition, the volcanic island is less than 100 km from another six bases and field camps (Fig. 1b). Deception Island is also one of the most popular tourist destinations in Antarctica, with over 15 000 visitors per year (data from International Association of Antarctica Tour Operators (IAATO) statistics, 2018). The island consists of a composite volcano, including a 8.5 × 10 km centrally located caldera (Fig. 5c). Port Foster bay, the sea-flooded part of the caldera depression, is smaller (6 × 10 km) due to the numerous post-caldera eruptions that occurred across a 1.5–2 km-wide zone around the circular rim of the caldera wall margins (see Geyer *et al.* 2021 for more details and additional references) (Fig. 5c). Due to its unique Antarctic flora and fauna, mainly associated with the island's geothermal activity (i.e. fumaroles and heated ground areas), some parts of inland Deception Island are protected by the Antarctic Treaty: ASPA 140 (onshore) and ASPA 145 (offshore, within Port Foster).

The evolution of Deception Island is distinguished by a caldera-forming event dated as 3980 ± 125 years BP, based on tephrochronology studies and ^{14}C determinations (Antoniades *et al.* 2018). The pre-caldera evolutionary stage was characterized by the formation of multiple shoaling seamounts and a main subaerial volcanic edifice, with magmas ranging in composition from basaltic to basaltic–andesitic and basaltic–trachyandesitic (see Geyer *et al.* 2021; Smellie 2021*b* for more details). During the caldera-forming eruption between 30 and 60 km^3 (dense rock equivalent (DRE)) of magma erupted in the form of dense basaltic–andesitic pyroclastic density current deposits, which unconformably covered the pre-caldera material with a pyroclastic mantle several tens of metres thick (Smellie 2001, 2002*a*; Martí *et al.* 2013). Since the caldera collapse, volcanic activity on Deception Island has been characterized by extensive eruptions occurring on the outer island slopes (mostly during early post-caldera periods) and small-volume eruptions (<0.1 km^3) located along the structural borders of the caldera and the interior of Port Foster bay with variable degrees of explosiveness depending on the water source (i.e. groundwater or surface water), amount and provenance (i.e. aquifer, sea, ice melting, etc.) that interacted with the rising or erupting magma (e.g. Baker *et al.* 1975; Smellie 2001, 2002*b*; Pedrazzi *et al.* 2014, 2018). However, Deception Island tephra are present in many lacustrine cores of neighbouring islands, and in marine sediments from the Bransfield Strait and the Scotia Sea (>800 km distant) (e.g. Hodgson *et al.* 1998; Moreton and Smellie 1998; Fretzdorff and Smellie 2002; Lee *et al.* 2007; Liu *et al.* 2016; Antoniades *et al.* 2018), and even in South Pole ice cores (e.g. Aristarain and Delmas 1998), suggesting that some recent post-caldera eruptions may have had significantly higher explosivity than those experienced during historical times. Post-caldera magmas outline a well-defined geochemical evolutionary trend showing the widest compositional range for eruptions on Deception Island, which vary from basalts to rhyolites (Aparicio *et al.* 1997; Smellie 2002*a*, 2021*b*; Geyer *et al.* 2019, 2021).

Regarding Deception Island's most recent volcanism, the historical record from the eighteenth to the twenty-first centuries indicates periods of high activity (e.g. 1818–28, 1906–12, 1967–70) with numerous eruptions closely spaced in time, followed by decades of dormancy (e.g. 1912–67: Orheim 1972; Roobol 1980, 1982; Smellie 2002*c*). The most recent volcanic events occurred in 1967, 1969 and 1970, and destroyed or severely damaged scientific bases (British and Chilean) operating at that time on the island. During the 1967 and 1970 eruptive episodes, several clustered vents opened simultaneously at the northwestern coast of Port Foster Bay, generating diverse types of volcanic landforms based on distinct levels of water–magma interaction (see Geyer *et al.* 2021 for more details and additional references). The 1969 eruption occurred when a 4 km-long fissure opened beneath glacial ice at Mount Pond, leading to a subglacial eruption that generated a flood (lahar) that modified the glacier, extended the local coastline, and destroyed the British scientific station and the infrastructures related to the whaling industry at Whalers Bay (Baker *et al.* 1975; Smellie 2002*b*).

Monitoring of volcanic activity at Deception Island began in the 1950s with the deployment of a seismometer at the Argentinian base. In 1965, a second seismometer was installed at the Chilean base and an improvised instrument was used at the British station during the 1969 eruption (Baker *et al.* 1975; Smellie 2002*b*). However, all instruments were abandoned once the 1967, 1969 and 1970 eruptions forced the evacuation of these bases. Monitoring of the local seismic activity was re-established in 1986 through field surveys carried out by Argentinian and Spanish researchers during the Antarctic summer (see Geyer *et al.* 2021 for more details and additional references). Since then, the volcanic island has been seismically monitored using: (i) short-period seismic stations (1986–90); (ii) small-aperture seismic arrays (1994–99); (iii) a combination of seismic arrays and a seismic network composed of five or six three-component seismometers (1999–present); and (iv) since 2008 a broadband seismic station (see Geyer *et al.* 2021 for more details and additional references). Local seismicity at Deception Island volcano includes volcano-tectonic earthquakes, long-period events and episodes of volcanic tremor. In parallel to the seismic network, Spanish researchers have been constructing a geodetic reference framework for the island since 1991–92. The REGID (Spanish acronym for REd Geodinámica Isla Decepción: Deception Island Geodynamic Network) network, composed of 15 benchmarks around Deception Island's interior bay (see Geyer *et al.* 2021 for more details), was designed with the aim of enabling geodynamic studies based on GNSS-GPS geodetic techniques. From the 2001–02 Antarctic campaign onwards, observations have been made during every austral summer over periods of 5–6 days (see Geyer *et al.* 2021 for more details and additional references).

Three episodes of unrest have been registered since the deployment of the monitoring network: 1992, 1999 and 2014–15. All three episodes have been interpreted as consequences of magma intrusions based on the seismological characteristics of the earthquakes, and the observations of simultaneous volcanological anomalies in gravity and magnetic data, surface deformation, high ground and water temperatures, and gas emissions. However, none of the intrusions produced a volcanic eruption (see Geyer *et al.* 2021 for more details and additional references).

In January 1992 a significant increase in the number and magnitude of seismic events was detected (Ortiz *et al.* 1997). Different geophysical observations, including small gravity irregularities and magnetic anomalies correlated with the successive seismic swarms, suggest that a magmatic intrusion took place during this unrest episode (Ortiz *et al.* 1992, 1997; García *et al.* 1997). Changes in fumarolic emissions, an increase in fumarole and groundwater temperatures, and possible ground deformation at the Argentinian base support the hypothesis of a magmatic intrusion (Ortiz *et al.* 1992, 1997). After a few weeks, the anomalous level of seismic activity started to decline and by the end of February it was back to pre-1992 levels.

During the 1998–99 summer campaign, seismic activity increased again. The two largest earthquakes in the series, with magnitudes of 2.8 and 3.4, occurred on 11 and 20 January 1999, and were felt by personnel from the Spanish scientific base Gabriel de Castilla. There were also a number of volcanic tremor episodes. Results show that the 1999 seismicity, although influenced by regional tectonics, was triggered by a shallow magmatic intrusion that perturbed the regional stress field (Carmona *et al.* 2010). Additional support for this hypothesis comes from changes in fumarolic emissions (Agusto *et al.* 2004; Caselli *et al.* 2004) and variations in the patterns of deformation, from radial extension and uplift to slow compression and subsidence (Fernandez-Ros *et al.* 2007; Berrocoso *et al.* 2008). In 2012–13 anomalous high soil and seawater temperatures were registered at Cerro Caliente and at Colatinas Point, respectively (Berrocoso *et al.* 2018). These were followed by a sharp increase in seismic activity (Almendros *et al.* 2018) and a period of volcano inflation that continued until 2015 (Berrocoso *et al.* 2018; Rosado *et al.* 2019). The similarities between this and the previous unrest events suggest a relationship to the stress changes induced by a magmatic intrusion at shallow depths (Almendros *et al.* 2018).

Volcanic hazard and risk management of Antarctic volcanoes

Past eruptions in Antarctica with a direct impact on infrastructure were only reported for Deception Island volcano during the 1967, 1969 and 1970 eruptions. These events severely damaged or destroyed the local scientific bases and slightly impacted the neighbouring stations distributed along the South Shetland Islands (e.g. King George Island: Baker *et al.* 1975; Roobol 1982). Nonetheless, an eruption occurring today in many of the active volcanoes described in the previous section could have a much larger impact. During the last few decades, the substantial number of scientific bases installed on the Antarctic continent and the Antarctic Peninsula, the expansion of touristic activities, and the elevated air and vessel traffic in the southern hemisphere in general has increased society's exposure to an eruption in the south polar region.

As for other volcanic areas worldwide, the impact of a future Antarctic eruption strongly depends on its geographical location, size, type and duration, which controls the types of hazards as well as their spatial and temporal extents. Due to their spread capacity, ash fallout would be the most worrying hazard derived from a south polar volcanic eruption. Ash fallout may lead to severe damage and roof collapse of infrastructures located close to the eruptive vent, and lead to short-term respiratory effects related to ash inhalation, including asthma and bronchitis attacks (Hansell *et al.* 2006; Horwell and Baxter 2006; Gudmundsson 2011). On a more regional scale, due to the strong winds and the low altitude (8–10 km) of the tropopause in certain areas of Antarctica (Smellie 1999*b*), ash-fallout deposits may rapidly disperse over a very wide area, far from the erupting volcano, as observed during the 1970 eruption at Deception Island (Baker *et al.* 1975). Regarding the influence of eruption size and vent location, lower ash plumes (<10 km height) from high-latitude (>70°) eruptive events are likely to be confined close to the South Pole due to wind patterns moderated by the encircling polar jet stream, while higher plumes have a greater potential for transcontinental ash dispersal (Geyer *et al.* 2017). However, contrary to this situation, ash from lower-latitude Antarctic volcanoes (e.g. Deception Island) is more likely to encircle the globe, even for moderate size eruptions. So, volcanic ash clouds could reach up to tropical latitudes, a vast part of the Atlantic coast of South America, South Africa and/or Oceania. Thus, a wider dispersion of volcanic particles than previously believed may impose significant consequences on global aviation safety (Geyer *et al.* 2017).

One aspect of Antarctic volcanism that can promote high eruptive plumes is the likelihood that magma will interact with external water sources, such as sea or ice meltwater. This fact enhances the risk of the occurrence of hydrovolcanism, during which magma–water interaction can turn small-volume eruptions into highly explosive ash-forming events (e.g. White and Houghton 2000). For example, direct observations of past eruptions on Deception Island indicate that hydrovolcanic activity from maars and tuff cones has been the major cause of volcanic hazards (e.g. Roobol 1982; Smellie 2002*c*; Pedrazzi *et al.* 2018). Also, past phreatic or phreatomagmatic eruptions at Mount Erebus (e.g. 1993 eruption) have demonstrated that these may pose a significant hazard to anyone working on the volcano's summit region or flanks, and could impact local/regional air traffic (Sims *et al.* 2021). Aside from ash fallout, other hazards related to hydrovolcanic events include ballistic impacts, subordinate dilute pyroclastic density currents and, for eruptions occurring in proximity to the coastline, tsunamis (Smellie 2002*c*).

Despite all these considerations, a detailed volcanic hazard assessment has been conducted only for Deception Island volcano (Bartolini *et al.* 2014) and, in a preliminary way, for Mount Erebus (Poirot 2002; Asher 2014). Nonetheless, long-term hazard assessments are crucial to help decision-makers allocate resources for hazard prevention and to design evacuation strategies aimed at reducing the loss of life. In the case of Deception Island, the first available volcanic hazard map was made by Roobol (1982), who mainly focused on assessing the zones threatened by lahars. Based on the impact of the most recent historical eruptions of 1842, 1967, 1969 and 1970, Smellie (2002*c*) produced a summary hazard map identifying potential areas affected by tsunamis, mudflows, dilute pyroclastic density currents, lava flows and/or tephra fallout. More recently, Bartolini *et al.* (2014) presented a long-term volcanic hazard assessment of Deception Island that considered both the temporal and spatial probabilities for the different eruptive scenarios in order to ultimately evaluate the hazard level on different parts of the island. In all three cases, and as for other Antarctic volcanoes, hazard assessment on Deception Island has always been (and still is) limited by the lack of a complete geological record of the volcano's past eruptive activity.

Mount Erebus and Deception Island have the largest and most comprehensive monitoring networks. However, to the best of my knowledge Deception Island is the only one with an implemented alert scheme reported as part of its Management Plan: Antarctic Specially Managed Area (ASMA) No. 4 (Deception Island Management Group: http://www.deceptionisland.aq, last accessed 25 May 2020; Management Plan for ASMA No. 4: https://www.ats.aq/devAS/Meetings/Documents/87, last accessed 25 May 2020). This alert scheme, together with the emergency evacuation plan, is kept up to date and is under continuous review. The current alert system at Deception Island is based on the continuous record of volcanic activity on the island during the approximately 4 months of the austral summer field season, coincident with the maximum human presence on the island. Captains of ships entering Port Foster, and pilots of aircraft or helicopters overflying the island, must request information on VHF Channel 16 Marine about the state of the island's volcanic activity monitored by the Gabriel de Castilla (Spain) and Decepción (Argentina) bases (Management Plan for ASMA No. 4: https://www.ats.aq/devAS/Meetings/Documents/87, last accessed 25 May 2020). To communicate this information, and similar to other active volcanoes worldwide, a traffic light system is used.

Future research priorities

Even if the threat presented by Antarctic volcanoes is rising due to an increase in human presence and newly constructed infrastructures, only two volcanoes are properly monitored and just one has an appropriate volcanic hazard assessment, early warning system and alert protocol. Hazard and risk management in volcanic environments is, in general, extremely challenging. The remoteness of Antarctica, the difficult terrain and adverse climate conditions prevent the deployment of proper monitoring networks and the scientific studies required to complete an accurate hazard assessment. In addition, and contrary to the majority of the active volcanic areas worldwide, Antarctic volcanoes are not located in specific countries, being under the umbrella of the Antarctic Treaty. As a consequence, the deployment (and maintenance) of monitoring networks relies on the participating countries, national research programmes or even individual research groups that are willing to deal with the economic cost and to invest human effort and infrastructure. Normally, this interest comes from the proximity of specific scientific bases to the different active volcanoes (e.g. McMurdo Station to Mount Erebus).

In parallel with volcano monitoring activities and volcanic hazard assessment, mitigation actions should also include education programmes for the people working and/or visiting these hazardous environments. The community exposed to the diverse volcanic hazards described above needs to be properly informed and prepared for any potential dangers related to volcanic emergencies. International guidelines should be prepared to ensure the safety of those working or visiting any Antarctic volcanic active regions. Antarctic tour guides need to be specially trained for emergencies and should have sufficient geological knowledge to be able to assess situations of imminent danger (cf. Erfurt-Cooper 2010). To be aware of the potential dangers in volcanic environments, visitors need to know beforehand how to prevent accidents and who is in charge and/or responsible in an emergency. In addition, to mitigate the potential risks, effective warning and rescue systems and community awareness-raising programmes need to be in place at the most visited active volcanoes (e.g. Deception Island and Mount Erebus), and the diverse stakeholders and decision-makers need to be aware of the current state of the active volcano.

From a scientific point of view, it is fundamental to work towards a complete geological record, including more geochronological data for the reconstruction of eruptive histories of the potentially active volcanoes. This information constitutes the basis for hazard assessments, hazard-zonation maps and long-term probability forecasts (Tilling 2005). Additionally, the most active volcanoes, especially those close to scientific stations and on tourist routes, need to be properly monitored. Geochemical and geophysical monitoring systems help to improve the knowledge of the volcanoes' internal structures and ongoing activity. It is necessary to work towards monitoring systems that allow year-round near real-time telemetered data acquisition, which is crucial to understanding a given volcano's evolution and to recognize changes in its state. The longer the monitoring period prior to a volcanic unrest, the earlier reliable detection of a measurable departure from a baseline dormant volcanic system (i.e. 'normal' behaviour) (Tilling 2005). As highlighted by Tilling (2005), only continuous volcano-monitoring data can diagnose a volcano's ongoing behaviour, which in turn constitutes the only scientific basis for short-term (hours to months in advance) forecasts of an impending eruption, or of mid-course changes to an active eruption. An efficient early warning system allows more time for scientists to inform decision-makers, and for them to prepare for a potential evacuation and to notify nearby bases, vessels and potentially affected airports of an impending eruption. As seen for many volcano observatories worldwide, optimum volcano monitoring is achieved through a combination of techniques, including, for example, seismicity, ground deformation and gas chemistry. Considering the remoteness of Antarctic volcanoes and their restricted accessibility, volcano-monitoring systems clearly benefit from the most recent and fast-evolving remote techniques that acquire, process and display data in real time or near-real time (e.g. satellite-based systems for monitoring ground deformation or gas monitoring). These techniques are likely to increasingly complement and, in some cases, supplant the conventional ground-based techniques currently used. In any case, simple monitoring approaches are better than no monitoring at all (Tilling 2005).

Finally, for sufficiently studied volcanoes, vulnerability analysis and long-term volcanic hazard assessments, including event trees and hazard maps, should be prepared in order to help decision-makers to carry out the required mitigation strategies (e.g. to fortify buildings), and to define evacuation routes and emergency management plans during future volcanic crises.

Acknowledgements The author is grateful for the logistical support of the Spanish Polar Programme, and to all the military staff of the Spanish Antarctic Base Gabriel de Castilla for their constant help and logistical support, without which this research would not have been possible. I thank Nelia Dunbar and the editors, John Smellie and Kurt Panter, for their constructive comments that have helped to improve this manuscript. The English editing was by Grant George Buffett.

Author contributions AG: conceptualization (lead), investigation (lead), visualization (lead), writing – original draft (lead).

Funding This research received funding from Ministerio de Economía, Industria y Competitividad, Gobierno de España grants (POSVOLDEC (CTM2016-79617-P) (AEI/FEDER, UE); PEVOLDEC (CTM2011-13578-E/ANT); and VOLCLIMA (CGL2015-72629-EXP)) and Ramón y Cajal contract (RYC-2012-11024). This research is part of POLARCSIC activities.

Data availability All data generated or analysed during this study are included in this published article (and its supplementary information files).

References

Ackert, R.P., Putnam, A.E., Mukhopadhyay, S., Pollard, D., DeConto, R.M., Kurz, M.D. and Borns, H.W. 2013. Controls on interior West Antarctic Ice Sheet Elevations: inferences from geologic constraints and ice sheet modeling. *Quaternary Science Reviews*, **65**, 26–38, https://doi.org/10.1016/J.QUASCIREV.2012.12.017

Agusto, M.R., Caselli, A.T. and Dos Santos Afonso, M. 2004. Manifestaciones de piritas framboidales en fumarolas de la Isla Decepción (Antártida): implicancias genéticas. *Revista de la Asociación Geológica Argentina*, **59**, 152–157.

Almendros, J., Carmona, E., Jiménez, V., Díaz-Moreno, A. and Lorenzo, F. 2018. Volcano-tectonic activity at Deception Island volcano following a seismic swarm in the Bransfield Rift (2014–2015). *Geophysical Research Letters*, **45**, 4788–4798, https://doi.org/10.1029/2018gl077490

Antoniades, D., Giralt, S. *et al.* 2018. The timing and widespread effects of the largest Holocene volcanic eruption in Antarctica. *Scientific Reports*, **8**, 17279, https://doi.org/10.1038/s41598-018-35460-x

Aparicio, A., Menegatti, N., Petrinovic, I., Risso, C. and Viramonte, J.G. 1997. El volcanismo de Isla Decepción (Península Antártida). *Boletin Geológico y Minero*, **108**, 235–258.

Aristarain, A.J. and Delmas, R.J. 1998. Ice record of a large eruption of Deception Island Volcano (Antarctica) in the XVIITH century. *Journal of Volcanology and Geothermal Research*, **80**, 17–25, https://doi.org/10.1016/s0377-0273(97)00040-1

Armienti, P. and Tripodo, A. 1991. Petrography and chemistry of lavas and comagmatic xenoliths of Mt. Rittmann, a volcano discovered during the IV Italian expedition in Northern Victoria Land (Antarctica). *Memoire della Società Geologica Italiana*, **46**, 427–451.

Asher, C. 2014. *Modeling Ash Fall and Debris Flow Hazards of Mt Erebus, Antarctica*. PhD thesis, University of Canterbury, Christchurch, New Zealand.

Aster, R., Zandomeneghi, D., Mah, S., McNamara, S., Henderson, D.B., Knox, H. and Jones, K. 2008. Moment tensor inversion of very long period seismic signals from Strombolian eruptions of Erebus Volcano. *Journal of Volcanology and Geothermal Research*, **177**, 635–647, https://doi.org/10.1016/j.jvolgeores.2008.08.013

Baker, P.E. 1968. Comparative volcanology and petrology of the atlantic island-arcs. *Bulletin Volcanologique*, **32**, 189, https://doi.org/10.1007/BF02596591

Baker, P.E., McReath, I., Harvey, M.R., Roobol, M.J. and Davies, T.G. 1975. *The Geology of the South Shetland Islands: Volcanic Evolution of Deception Island*. British Antarctic Survey Scientific Reports, **78**.

Barker, P.F. 1982. The Cenozoic subduction history of the Pacific margin of the Antarctic Peninsula: ridge crest–trench interactions. *Journal of the Geological Society, London*, **139**, 787–801, https://doi.org/10.1144/gsjgs.139.6.0787

Bartolini, S., Geyer, A., Martí, J., Pedrazzi, D. and Aguirre-Díaz, G. 2014. Volcanic hazard on Deception Island (South Shetland Islands, Antarctica). *Journal of Volcanology and Geothermal Research*, **285**, 150–168, https://doi.org/10.1016/j.jvolgeores.2014.08.009

Barton, C.M. 1965. *The Geology of the South Shetland Islands. III. The Stratigraphy of King George Island*. British Antarctic Survey Scientific Reports, **44**.

Bender, N.A., Crosbie, K. and Lynch, H.J. 2016. Patterns of tourism in the Antarctic Peninsula region: a 20-year analysis. *Antarctic Science*, **28**, 194–203, https://doi.org/10.1017/S0954102016000031

Berrocoso, M., Fernández-Ros, A. *et al.* 2008. Geodetic Research on Deception Island and its Environment (South Shetland Islands, Bransfield Sea and Antarctic Peninsula) During Spanish Antarctic Campaigns (1987–2007). *In*: Capra, A. and Dietrich, R. (eds) *Geodetic and Geophysical Observations in Antarctica: An Overview in the IPY Perspective*. Springer, Berlin, 97–124, https://doi.org/10.1007/978-3-540-74882-3_6

Berrocoso, M., Prates, G. *et al.* 2018. Caldera unrest detected with seawater temperature anomalies at Deception Island, Antarctic Peninsula. *Bulletin of Volcanology*, **80**, 41, https://doi.org/10.1007/s00445-018-1216-2

Birkenmajer, K. 1980. Age of the Penguin Island Volcano, South Shetland Islands (West Antarctica). *Bulletin of the Polish Academy of Sciences*, **27**, 69–76.

Birkenmajer, K. 1982. Structural evolution of the Melville Peak Volcano, King George Island (South Shetland Islands, West Antarctica). *Bulletin of the Polish Academy of Sciences*, **24**, 341–351.

Birkenmajer, K. and Keller, R.A. 1990. Pleistocene age of the Melville Peak volcano, King George 1497 Island, West Antarctica, by K–Ar dating. *Bulletin of the Polish Academy of Sciences*, **38**, 17–24.

Blankenship, D.D., Bell, R.E., Hodge, S.M., Brozena, J.M., Behrendt, J.C. and Finn, C.A. 1993. Active volcanism beneath the West Antarctic ice sheet and implications for ice-sheet stability. *Nature*, **361**, 526–529, https://doi.org/10.1038/361526a0

Bonaccorso, A., Gambino, S., Falzone, G. and Privitera, E. 1997. The volcanological observatory of the Mt. Melbourne (Northern Victoria Land, Antarctica). *In*: Ricci, C.A. (ed.) *The Antarctic Region: Geological Evolution and Processes*. Terra Antartica, Siena, Italy, 1083–1186.

Broady, P., Given, D., Greenfield, L. and Thompson, K. 1987. The biota and environment of fumaroles on Mt Melbourne, northern Victoria Land. *Polar Biology*, **7**, 97–113, https://doi.org/10.1007/BF00570447

Calabozo, F.M., Strelin, J.A., Orihashi, Y., Sumino, H. and Keller, R.A. 2015. Volcano–ice–sea interaction in the Cerro Santa Marta area, northwest James Ross Island, Antarctic Peninsula. *Journal of Volcanology and Geothermal Research*, **297**, 89–108, https://doi.org/10.1016/j.jvolgeores.2015.03.011

Carmona, E., Almendros, J., Peña, J.A. and Ibáñez, J.M. 2010. Characterization of fracture systems using precise array locations of earthquake multiplets: An example at Deception Island volcano, Antarctica. *Journal of Geophysical Research: Solid Earth*, **115**, B06309, https://doi.org/10.1029/2009jb006865

Caselli, A.T., dos Santos Afonso, M. and Agusto, M.R. 2004. Gases fumarólicos de la isla Decepción (Shetland del Sur, Antártida): variaciones químicas y depósitos vinculados a la crisis sísmica de 1999. *Revista de la Asociación Geológica Argentina*, **59**, 291–302.

Corr, H.F.J. and Vaughan, D.G. 2008. A recent volcanic eruption beneath the West Antarctic ice sheet. *Nature Geoscience*, **1**, 122–125, https://doi.org/10.1038/ngeo106

Del Carlo, P., Di Roberto, A. *et al.* 2015. Late Pleistocene–Holocene volcanic activity in northern Victoria Land recorded in Ross Sea (Antarctica) marine sediments. *Bulletin of Volcanology*, **77**, 1–17, https://doi.org/10.1007/s00445-015-0924-0

Di Roberto, A., Colizza, E., Del Carlo, P., Petrelli, M., Finocchiaro, F. and Kuhn, G. 2019. First marine cryptotephra in Antarctica found in sediments of the western Ross Sea correlates with englacial tephras and climate records. *Scientific Reports*, **9**, 1–10, https://doi.org/10.1038/s41598-019-47188-3

Dunbar, N.W. 2003. Tephra layers in the Siple Dome and Taylor Dome ice cores, Antarctica: Sources and correlations. *Journal of Geophysical Research*, **108**, 2374, https://doi.org/10.1029/2002jb002056

Dunbar, N.W., Iverson, N.A., Smellie, J.L., McIntosh, W.C., Zimmerer, M.J. and Kyle, P.R. 2021. Active volcanoes in Marie Byrd Land. *Geological Society, London, Memoirs*, **55**, https://doi.org/10.1144/M55-2019-29

Erfurt-Cooper, P. 2010. Introduction. *In*: Erfurt-Cooper, P. and Cooper, M. (eds) *Volcano and Geothermal Tourism*. Routledge, London, 3–31, https://doi.org/10.4324/9781849775182

Esser, R.P. and Kyle, P.R. 2002. 40Ar/39Ar chronology of the McMurdo Volcanic Group at The Pleiades, northern Victoria Land, Antarctica. *Royal Society of New Zealand Bulletin*, **35**, 415–418.

Falconer, R.K.H. 1972. The Indian–Antarctic–Pacific triple junction. *Earth and Planetary Science Letters*, **17**, 151–158, https://doi.org/10.1016/0012-821X(72)90270-1

Fernandez-Ros, A.M., Berrocoso, M. and Ramirez, M.E. 2007. Volcanic deformation models for Deception Island (South Shetland Islands, Antarctica). *United States Geological Survey Open-File Report*, **2007-1047**, extended abstract 094.

Fretzdorff, S. and Smellie, J.L. 2002. Electron microprobe characterization of ash layers in sediments from the central Bransfield basin (Antarctic Peninsula): evidence for at least two volcanic sources. *Antarctic Science*, **14**, 412–421, https://doi.org/10.1017/S0954102002000214

Galindo-Zaldívar, J., Gamboa, L., Maldonado, A., Nakao, S. and Bochu, Y. 2004. Tectonic development of the Bransfield Basin and its prolongation to the South Scotia Ridge, northern Antarctic Peninsula. *Marine Geology*, **206**, 267–282, https://doi.org/10.1016/j.margeo.2004.02.007

Gambino, S., Armienti, P. et al. 2021. Mount Melbourne and Mount Rittmann. *Geological Society, London, Memoirs*, **55**, https://doi.org/10.1144/M55-2018-43

García, A., Blanco, I., Torta, J.M., Astiz, M.M., Ibáñez, J. and Ortiz, R. 1997. A search for the volcanomagnetic signal at Deception volcano (South Shetland I., Antarctica). *Annals of Geophysics*, **40**, 319–327, https://doi.org/10.4401/ag-3914

Geyer, A., Marti, A., Giralt, S. and Folch, A. 2017. Potential ash impact from Antarctic volcanoes: Insights from Deception Island's most recent eruption. *Scientific Reports*, **7**, 16534, https://doi.org/10.1038/s41598-017-16630-9

Geyer, A., Álvarez-Valero, A.M., Gisbert, G., Aulinas, M., Hernández-Barreña, D., Lobo, A. and Marti, J. 2019. Deciphering the evolution of Deception Island's magmatic system. *Scientific Reports*, **9**, 373, https://doi.org/10.1038/s41598-018-36188-4

Geyer, A., Pedrazzi, D. et al. 2021. Deception Island. *Geological Society, London, Memoirs*, **55**, https://doi.org/10.1144/M55-2018-56

Giordano, G., Lucci, F., Phillips, D., Cozzupoli, D. and Runci, V. 2012. Stratigraphy, geochronology and evolution of the Mt. Melbourne volcanic field (North Victoria Land, Antarctica). *Bulletin of Volcanology*, **74**, 1985–2005, https://doi.org/10.1007/s00445-012-0643-8

Global Volcanism Program 1982. Report on Seal Nunataks Group (Antarctica). In: Institution, S. (ed.) *Scientific Event Alert Network Bulletin*, **7**:9. Smithsonian Institution, https://doi.org/10.5479/si.GVP.SEAN198209-390050

Global Volcanism Program 2012. Report on Siple (Antarctica). In: Sennert, S.K. (ed.) *Weekly Volcanic Activity Report, 20 June–26 June 2012*. Smithsonian Institution and United States Geological Survey, https://volcano.si.edu/showreport.cfm?doi=GVP.WVAR20120620-390025

Global Volcanism Program 2019. Report on Krakatau (Indonesia). In: Krippner, J.B. and Venzke, E. (eds) *Bulletin of the Global Volcanism Network*, **44**:3. Smithsonian Institution, https://doi.org/10.5479/si.GVP.BGVN201903-262000

Global Volcanism Program 2020. Report on Whakaari/White Island (New Zealand). In: Krippner, J.B. and Venzke, E. (eds) *Bulletin of the Global Volcanism Network*, **45**:2. Smithsonian Institution, https://volcano.si.edu/showreport.cfm?doi=10.5479/si.GVP.BGVN202002-241040

Golynsky, A.V., Ferraccioli, F. et al. 2018. New magnetic anomaly map of the Antarctic. *Geophysical Research Letters*, **45**, 6437–6449, https://doi.org/10.1029/2018GL078153

González-Ferrán, O. 1983. The Seal Nunataks: an active volcanic group on the Larsen Ice Shelf, West Antarctica. In: Oliver, R.L., James, P.R. and Jago, J.B. (eds) *Antarctic Earth Science*. Cambridge University Press, New York, 334–337.

Gonzalez-Casado, J.M., Giner-Robles, J.L. and Lopez-Martinez, J. 2000. Bransfield Basin, Antarctic Peninsula: Not a normal back-arc basin. *Geology*, **28**, 1043–1046, https://doi.org/10.1130/0091-7613(2000)28<1043:BBAPNA>2.0.CO;2

González-Ferrán, O. 1985. Volcanic and tectonic evolution of the Northern Antarctic Peninsula–Late Cenozoic to recent. *Tectonophysics*, **114**, 389–409, https://doi.org/10.1016/0040-1951(85)90023-X

González-Ferrán, O. 1995. *Volcanes de Chile*. Instituto Geografico Militar, Santiago.

González-Ferrán, O. and Katsui, Y. 1970. Estudio integral del volcanismo cenozoico superior de las Islas Shetland del Sur, Antartica. *Serie Científica Instituto Antártico Chileno*, **22**, 123–174.

Gudmundsson, G. 2011. Respiratory health effects of volcanic ash with special reference to Iceland. A review. *The Clinical Respiratory Journal*, **5**, 2–9, https://doi.org/10.1111/j.1752-699X.2010.00231.x

Haase, K.M. and Beier, C. 2021. Bransfield Strait and James Ross Island: petrology. *Geological Society, London, Memoirs*, **55**, https://doi.org/10.1144/M55-2018-37

Hammer, C.U., Clausen, H.B. and Langway, C.C. 1997. 50 000 years of recorded global volcanism. *Climatic Change*, **35**, 1–15, https://doi.org/10.1023/A:1005344225434

Hansell, A.L., Horwell, C.J. and Oppenheimer, C. 2006. The health hazards of volcanoes and geothermal areas. *Occupational and Environmental Medicine*, **63**, 149–156, https://doi.org/10.1136/oem.2005.022459

Hodgson, D.A., Dyson, C.L., Jones, V.J. and Smellie, J.L. 1998. Tephra analysis of sediments from Midge Lake (South Shetland Islands) and Sombre Lake (South Orkney Islands), Antarctica. *Antarctic Science*, **10**, 13–20, https://doi.org/10.1017/S0954102098000030

Hole, M.J. 1988. Post-subduction alkaline volcanism along the Antarctic Peninsula. *Journal of the Geological Society, London*, **145**, 985–998, https://doi.org/10.1144/gsjgs.145.6.0985

Hole, M.J. 2021. Antarctic Peninsula: petrology. *Geological Society, London, Memoirs*, **55**, https://doi.org/10.1144/M55-2018-40

Hole, M.J., Rogers, G., Saunders, A.D. and Storey, M. 1991. Relation between alkalic volcanism and slab-window formation. *Geology*, **19**, 657–660, https://doi.org/10.1130/0091-7613(1991)019<0657:RBAVAS>2.3.CO;2

Hole, M.J., Saunders, A.D., Rogers, G. and Sykes, M.A. 1994. The relationship between alkaline magmatism, lithospheric extension and slab window formation along continental destructive plate margins. *Geological Society, London, Special Publications*, **81**, 265, https://doi.org/10.1144/GSL.SP.1994.081.01.15

Horwell, C.J. and Baxter, P.J. 2006. The respiratory health hazards of volcanic ash: a review for volcanic risk mitigation. *Bulletin of Volcanology*, **69**, 1–24, https://doi.org/10.1007/s00445-006-0052-y

Iverson, N.A., Lieb-Lappen, R., Dunbar, N.W., Obbard, R., Kim, E. and Golden, E. 2017. The first physical evidence of subglacial volcanism under the West Antarctic Ice Sheet. *Scientific Reports*, **7**, 11457, https://doi.org/10.1038/s41598-017-11515-3

Keller, R.A., Domack, E. and Drake, A. 2003. Potential for tephrochronology of marine sediment cores from Bransfield Strait and the northwestern Weddell Sea. In: *XVI INQUA Congress Programs with Abstracts: July 23–30, 2003, Reno Hilton Hotel and Conference Center, Reno, Nevada, USA*. Desert Research Institute, Reno, NV, 236.

Klepeis, K.A. and Lawver, L.A. 1996. Tectonics of the Antarctic–Scotia plate boundary near Elephant and Clarence islands, West Antarctica. *Eos, Transactions of the American Geophysical Union*, **76**, 710.

Koeberl, C. 1989. Iridium enrichment in volcanic dust from blue ice fields, Antarctica, and possible relevance to the K/T boundary event. *Earth and Planetary Science Letters*, **92**, 317–322, https://doi.org/10.1016/0012-821X(89)90056-3

Košler, J., Magna, T., Mlčoch, B., Mixa, P., Nývlt, D. and Holub, F.V. 2009. Combined Sr, Nd, Pb and Li isotope geochemistry of alkaline lavas from northern James Ross Island (Antarctic Peninsula) and implications for back-arc magma formation. *Chemical Geology*, **258**, 207–218, https://doi.org/10.1016/j.chemgeo.2008.10.006

Kraus, S., Kurbatov, A. and Yates, M. 2013. Geochemical signatures of tephras from Quaternary Antarctic Peninsula volcanoes. *Andean Geology*, **40**, 1–40, https://doi.org/10.5027/andgeoV40n1-a01

Kurbatov, A.V., Zielinski, G.A., Dunbar, N.W., Mayewski, P.A., Meyerson, E.A., Sneed, S.B. and Taylor, K.C. 2006. A 12,000 year record of explosive volcanism in the Siple Dome Ice Core, West Antarctica. *Journal of Geophysical Research: Atmospheres*, **111**, https://doi.org/10.1029/2005JD006072

Kyle, P.R. 1990. The Pleiades. *American Geophysical Union, Antarctic Research Series*, **48**, 60–64.

Kyle, P.R. (ed.). 1994. *Volcanological and Environmental Studies of Mount Erebus, Antarctica*. American Geophysical Union Antarctic Research Series, https://doi.org/10.1029/AR066

Kyle, P.R. and Cole, J.W. 1974. Structural control of volcanism in the McMurdo Volcanic Group, Antarctica. *Bulletin Volcanologique*, **38**, 16–25, https://doi.org/10.1007/BF02597798

Lanyon, R., Varne, R. and Crawford, A.J. 1993. Tasmanian Tertiary basalts, the Balleny plume, and opening of the Tasman Sea (southwest Pacific Ocean). *Geology*, **21**, 555–558, https://doi.org/10.1130/0091-7613(1993)021<0555:ttbtbp>2.3.co;2

Larsen, C. 1894. The voyage of the Jason to the Antarctic regions. *Geographical Journal*, **4**, 333–344, https://doi.org/10.2307/1773537

Lawver, L.A., Keller, R.A., Fisk, M.R. and Strelin, J.A. 1995. Bransfield Strait, Antarctic Peninsula active extension behind a dead arc. *In*: Taylor, B. (ed.) *Backarc Basins*. Springer, New York, 315–342.

Lawver, L.A., Sloan, B.J. *et al.* 1996. Distributed, active extension in Bransfield Basin, Antarctic Peninsula: Evidence from multibeam bathymetry. *GSA Today*, **6**, 1–6.

Leat, P.T. and Riley, T.R. 2021*a*. Antarctic Peninsula and South Shetland Islands: volcanology. *Geological Society, London, Memoirs*, **55**, https://doi.org/10.1144/M55-2018-52

Leat, P.T. and Riley, T.R. 2021*b*. Antarctic Peninsula and South Shetland Islands: petrology. *Geological Society, London, Memoirs*, **55**, https://doi.org/10.1144/M55-2018-68

Lee, M.J., Kyle, P.R., Iverson, N.A., Lee, J.I. and Han, Y. 2019. Rittmann volcano, Antarctica as the source of a widespread 1252 ± 2 CE tephra layer in Antarctica ice. *Earth and Planetary Science Letters*, **521**, 169–176, https://doi.org/10.1016/j.epsl.2019.06.002

Lee, Y.I., Lim, H.S., Yoon, H.I. and Tatur, A. 2007. Characteristics of tephra in Holocene lake sediments on King George Island, West Antarctica: implications for deglaciation and paleoenvironment. *Quaternary Science Reviews*, **26**, 3167–3178, https://doi.org/10.1016/j.quascirev.2007.09.007

LeMasurier, W.E. and Kawachi, Y. 1990. Mount Waesche. *American Geophysical Union Antarctic Research Series*, **48**, 208–211.

LeMasurier, W.E. and Rex, D.C. 1989. Evolution of linear volcanic ranges in Marie Byrd Land, West Antarctica. *Journal of Geophysical Research: Solid Earth*, **94**, 7223–7236, https://doi.org/10.1029/JB094iB06p07223

LeMasurier, W.E., Thomson, J.W., Baker, P.E., Kyle, P.R., Rowley, P.D., Smellie, J.L. and Verwoerd, W.J. (eds) 1990. *Volcanoes of the Antarctic Plate and Southern Oceans. American Geophysical Union Antarctic Research Series*, **48**, https://doi.org/10.1029/AR048

Liu, E.J., Oliva, M. *et al.* 2016. Expanding the tephrostratigraphical framework for the South Shetland Islands, Antarctica, by combining compositional and textural tephra characterisation. *Sedimentary Geology*, **340**, 49–61, https://doi.org/10.1016/j.sedgeo.2015.08.002

Lyon, G.L. 1986. Stable isotope stratigraphy of ice cores and the age of the last eruption at Mount Melbourne, Antarctica. *New Zealand Journal of Geology and Geophysics*, **29**, 135–138, https://doi.org/10.1080/00288306.1986.10427528

MacAyeal, D.R., Okal, E.A., Aster, R.C. and Bassis, J.N. 2008. Seismic and hydroacoustic tremor generated by colliding icebergs. *Journal of Geophysical Research: Earth Surface*, **113**, https://doi.org/10.1029/2008JF001005

Maestro, A., Somoza, L., Rey, J., Martínez-Frías, J. and López-Martínez, J. 2007. Active tectonics, fault patterns, and stress field of Deception Island: A response to oblique convergence between the Pacific and Antarctic plates. *Journal of South American Earth Sciences*, **23**, 256–268, https://doi.org/10.1016/j.jsames.2006.09.023

Maldonado, A., Larter, R.D. and Aldaya, F. 1994. Forearc tectonic evolution of the South Shetland Margin, Antarctic Peninsula. *Tectonics*, **13**, 1345–1370, https://doi.org/10.1029/94tc01352

Martí, J., Geyer, A. and Aguirre-Diaz, G. 2013. Origin and evolution of the Deception Island caldera (South Shetland Islands, Antarctica). *Bulletin of Volcanology*, **75**, 1–18, https://doi.org/10.1007/s00445-013-0732-3

Martin, A.P., Cooper, A.F., Price, R.C., Kyle, P.R. and Gamble, J.A. 2021. Erebus Volcanic Province: petrology. *Geological Society, London, Memoirs*, **55**, https://doi.org/10.1144/M55-2018-80

Martin, S., Drucker, R., Aster, R., Davey, F., Okal, E., Scambos, T. and MacAyeal, D. 2010. Kinematic and seismic analysis of giant tabular iceberg breakup at Cape Adare, Antarctica. *Journal of Geophysical Research: Solid Earth*, **115**, https://doi.org/10.1029/2009JB006700

McCarron, J.J. and Larter, R.D. 1998. Late Cretaceous to early Tertiary subduction history of the Antarctic Peninsula. *Journal of the Geological Society, London*, **155**, 255–268, https://doi.org/10.1144/gsjgs.155.2.0255

McIntosh, W.C., LeMasurier, W.E., Ellerman, P.J. and Dunbar, N.W. 1985. A reinterpretation of glacio-volcanic interaction at Mount Takahe and Mount Murphy, Marie Byrd Land, Antarctica. *Antarctic Journal of the United States*, **20**, 57–59.

Moreton, S.G. and Smellie, J.L. 1998. Identification and correlation of distal tephra layers in deep-sea sediment cores, Scotia Sea, Antarctica. *Annals of Glaciology*, **27**, 285–289, https://doi.org/10.3189/1998AoG27-1-285-289

Narcisi, B., Petit, J.R., Langone, A. and Stenni, B. 2016. A new Eemian record of Antarctic tephra layers retrieved from the Talos Dome ice core (Northern Victoria Land). *Global and Planetary Change*, **137**, 69–78, https://doi.org/10.1016/j.gloplacha.2015.12.016

Nicolaus, B., Lama, L., Esposito, E., Manca, M.C., di Prisco, G. and Gambacorta, A. 1996. 'Bacillus thermoantarcticus' sp. nov., from Mount Melbourne, Antarctica: a novel thermophilic species. *Polar Biology*, **16**, 101–104, https://doi.org/10.1007/s003000050034

Oppenheimer, C. and Kyle, P. 2008. Volcanology of Erebus volcano, Antarctica. *Journal of Volcanology and Geothermal Research*, **177**, 531–754, https://doi.org/10.1016/j.jvolgeores.2008.10.006

Orheim, O. 1972. *A 200-Year Record of Glacier Mass Balance at Deception Island, Southwest Atlantic Ocean, and Its Bearing on Models of Global Climate Change*. Institute of Polar Studies Report, **42**.

Ortiz, R., Vila, J. *et al.* 1992. Geophysical features of Deception Island. *In*: Yoshida, Y., Kaminuma, K. and Shiraishi, K. (eds) *Recent Progress in Antarctic Earth Science*. Terra Scientific, Tokyo, 443–448.

Ortiz, R., García, A. *et al.* 1997. Monitoring of the volcanic activity of Deception Island, South Shetland Islands, Antarctica (1986–1995). *In*: Ricci, C.A. (ed.) *The Antarctic Region: Geological Evolution and Processes*. Terra Antartica, Siena, Italy, 1071–1076.

Pańczyk, M. and Nawrocki, J. 2011. Pliocene age of the oldest basaltic rocks of Penguin Island (South Shetland Islands, northern Antarctic Peninsula). *Geological Quarterly*, **55**, 335–344.

Panter, K.S. 1995. *Geology, Geochemistry and Petrogenesis of the Mount Sidley Volcano, Marie Byrd Land, Antarctica*. PhD thesis, New Mexico Institute of Mining and Technology, Socorro, New Mexico, USA.

Panter, K.S. 2021. Antarctic volcanism: petrology and tectonomagmatic overview. *Geological Society, London, Memoirs*, **55**, https://doi.org/10.1144/M55-2020-10

Panter, K.S., Blusztajn, J., Wingrove, D., Hart, S. and Mattey, D. 2003. Sr, Nd, Pb, Os, O isotope, Major and trace element data from basalts, South Victoria Land, Antarctica: Evidence for open-system processes in the evolution of mafic alkaline magmas. *General Assembly of the European Geosciences Union, Geophysical Research Abstracts*, **5**, 07583.

Panter, K.S., Wilch, T.I., Smellie, J.L., Kyle, P.R. and McIntosh, W.C. 2021. Marie Byrd Land and Ellsworth Land: petrology. *Geological Society, London, Memoirs*, **55**, https://doi.org/10.1144/M55-2019-50

Pedrazzi, D., Aguirre-Díaz, G., Bartolini, S., Martí, J. and Geyer, A. 2014. The 1970 eruption on Deception Island (Antarctica): eruptive dynamics and implications for volcanic hazards. *Journal of the Geological Society, London*, **171**, 765–778, https://doi.org/10.1144/jgs2014-015

Pedrazzi, D., Németh, K., Geyer, A., Álvarez-Valero, A.M., Aguirre-Díaz, G. and Bartolini, S. 2018. Historic hydrovolcanism at Deception Island (Antarctica): implications for eruption hazards. *Bulletin of Volcanology*, **80**, 11, https://doi.org/10.1007/s00445-017-1186-9

Poirot, C. 2002. *Volcanic Hazard Assessment of Mount Erebus, Ross Island, Antarctica*. PhD thesis, University of Canterbury, Christchurch, New Zealand.

Quartini, E., Blankenship, D.D. and Young, D.A. 2021. Active subglacial volcanism in Antarctica. *Geological Society, London, Memoirs*, **55**, https://doi.org/10.1144/M55-2019-3

Rey, J., Somoza, L. and Martínez-Frías, J. 1995. Tectonic, volcanic, and hydrothermal event sequence on Deception Island (Antarctica). *Geo-Marine Letters*, **15**, 1–8, https://doi.org/10.1007/bf01204491

Rocchi, S. and Smellie, J.L. 2021. Northern Victoria Land: petrology. *Geological Society, London, Memoirs*, **55**, https://doi.org/10.1144/M55-2019-19

Roobol, M.J. 1980. A model for the eruptive mechanism of Deception Island from 1820 to 1970. *British Antarctic Survey Bulletin*, **49**, 137–156.

Roobol, M.J. 1982. The volcanic hazard at Deception Island, South Shetland Islands. *British Antarctic Survey Bulletin*, **51**, 237–245.

Rosado, B., Fernández-Ros, A., Berrocoso, M., Prates, G., Gárate, J., de Gil, A. and Geyer, A. 2019. Volcano-tectonic dynamics of Deception Island (Antarctica): 27 years of GPS observations (1991–2018). *Journal of Volcanology and Geothermal Research*, **381**, 57–82, https://doi.org/10.1016/j.jvolgeores.2019.05.009

Rowe, C., Aster, R., Kyle, P., Dibble, R. and Schlue, J. 2000. Seismic and acoustic observations at Mount Erebus Volcano, Ross Island, Antarctica, 1994–1998. *Journal of Volcanology and Geothermal Research*, **101**, 105–128, https://doi.org/10.1016/S0377-0273(00)00170-0

Rowley, P.D., Laudon, T.S., La Prade, K.E. and LeMasurier, W.E. 1990. Hudson Mountains. *American Geophysical Union Antarctic Research Series*, **48**, 289–293.

Siebert, L., Simkin, T. and Kimberly, P. 2011. *Volcanoes of the World*, 3rd edn. University of California Press, Berkeley, CA.

Sims, K.W.W., Aster, R. et al. 2021. Mount Erebus. *Geological Society, London, Memoirs*, **55**, https://doi.org/10.1144/M55-2019-8

Skotnicki, M.L., Selkirk, P.M., Broady, P., Adam, K.D. and Ninham, J.A. 2001. Dispersal of the moss *Campylopus pyriformis* on geothermal ground near the summits of Mount Erebus and Mount Melbourne, Victoria Land, Antarctica. *Antarctic Science*, **13**, 280–285, https://doi.org/10.1017/S0954102001000396

Smellie, J.L. 1990a. Melville Peak. *American Geophysical Union Antarctic Research Series*, **48**, 331–333.

Smellie, J.L. 1990b. Paulet Island. *American Geophysical Union Antarctic Research Series*, **48**, 331–338.

Smellie, J.L. 1990c. Penguin Island. *American Geophysical Union Antarctic Research Series*, **48**, 322–324.

Smellie, J.L. 1990d. Seal Nunataks. *American Geophysical Union Antarctic Research Series*, **48**, 349–351.

Smellie, J.L. 1999a. Lithostratigraphy of Miocene–Recent, alkaline volcanic fields in the Antarctic Peninsula and eastern Ellsworth Land. *Antarctic Science*, **11**, 362–378, https://doi.org/10.1017/S0954102099000450

Smellie, J.L. 1999b. The upper Cenozoic tephra record in the south polar region: a review. *Global and Planetary Change*, **21**, 51–70, https://doi.org/10.1016/S0921-8181(99)00007-7

Smellie, J.L. 2001. Lithostratigraphy and volcanic evolution of Deception Island, South Shetland Islands. *Antarctic Science*, **13**, 188–209, https://doi.org/10.1017/S0954102001000281

Smellie, J.L. 2002a. Geology. *In*: Smellie, J.L., López-Martínez, J., Thomson, J.W. and Thomson, M.R.A. (eds) *Geology and Geomorphology of Deception Island (1:25 000 Scale)*. BAS GEOMAP Series, Sheets 6-A and 6-B, supplementary text. British Antarctic Survey, Cambridge, UK, 11–30.

Smellie, J.L. 2002b. The 1969 subglacial eruption on Deception Island (Antarctica): events and processes during an eruption beneath a thin glacier and implications for volcanic hazards. *Geological Society, London, Special Publications*, **202**, 59–79, https://doi.org/10.1144/GSL.SP.2002.202.01.04

Smellie, J.L. 2002c. Volcanic hazard. *In*: Smellie, J.L., López-Martínez, J., Thomson, J.W. and Thomson, M.R. A. (eds) *Geology and Geomorphology of Deception Island (1:25 000 Scale)*. BAS GEOMAP Series, Sheets 6-A and 6-B, supplementary text. British Antarctic Survey, Cambridge, UK, 47–53.

Smellie, J.L. 2021a. Antarctic volcanism: volcanology and palaeoenvironmental overview. *Geological Society, London, Memoirs*, **55**, https://doi.org/10.1144/M55-2020-1

Smellie, J.L. 2021b. Bransfield Strait and James Ross Island: volcanology. *Geological Society, London, Memoirs*, **55**, https://doi.org/10.1144/M55-2018-58

Smellie, J.L. and Hole, M.J. 2021. Antarctic Peninsula: volcanology. *Geological Society, London, Memoirs*, **55**, https://doi.org/10.1144/M55-2018-59

Smellie, J.L. and Martin, A.P. 2021. Erebus Volcanic Province: volcanology. *Geological Society, London, Memoirs*, **55**, https://doi.org/10.1144/M55-2018-62

Smellie, J.L. and Rocchi, S. 2021. Northern Victoria Land: volcanology. *Geological Society, London, Memoirs*, **55**, https://doi.org/10.1144/M55-2018-60

Smellie, J.L., Pankhurst, R.J., Thomson, M.R.A. and Davies, R.E.S. 1984. The geology of the south Shetland Islands: VI. Stratigraphy, geochemistry and evolution. *British Antarctic Survey Scientific Reports*, **87**, 2–83.

Smellie, J.L., López-Martínez, J. et al. (eds). 2002. *Geology and Geomorphology of Deception Island (1:25 000 Scale)*. BAS GEOMAP Series, Sheets 6-A and 6-B, supplementary text. British Antarctic Survey, Cambridge, UK.

Smellie, J.L., Johnson, J.S., McIntosh, W.C., Esser, R., Gudmundsson, M.T., Hambrey, M.J. and van Wyk de Vries, B. 2008. Six million years of glacial history recorded in volcanic lithofacies of the James Ross Island Volcanic Group, Antarctic Peninsula. *Palaeogeography, Palaeoclimatology, Palaeoecology*, **260**, 122–148, https://doi.org/10.1016/J.PALAEO.2007.08.011

Sweeney, D., Kyle, P.R. and Oppenheimer, C. 2008. Sulfur dioxide emissions and degassing behavior of Erebus volcano, Antarctica. *Journal of Volcanology and Geothermal Research*, **177**, 725–733, https://doi.org/10.1016/j.jvolgeores.2008.01.024

Tilling, R.I. 2005. Volcano hazards. *In*: Marti, J. and Ernst, G. (eds) *Volcanoes and the Environment*. Cambridge University Press, Cambridge, UK, 55–89.

Weaver, S.D., Saunders, A.D., Pankhurst, R.J. and Tarney, J. 1979. A geochemical study of magmatism associated with the initial stages of back-arc spreading. *Contributions to Mineralogy and Petrology*, **68**, 151–169, https://doi.org/10.1007/bf00371897

White, J.D.L. and Houghton, B. 2000. Surtseyan and related phreatomagmatic eruptions. *In*: Sigurdsson, H., Houghton, B.F., McNutt, S.R., Rymer, H. and Stix, J. (eds) *Encyclopedia of Volcanoes*. Academic Press, San Diego, CA, 495–511.

Wilch, T.I. and McIntosh, W.C. 2000. Eocene and Oligocene volcanism at Mount Petras, Marie Byrd Land: implications for middle Cenozoic ice sheet reconstructions in West Antarctica. *Antarctic Science*, **12**, 477–491, https://doi.org/10.1017/S0954102000000560

Wilch, T.I., McIntosh, W.C. and Dunbar, N.W. 1999. Late Quaternary volcanic activity in Marie Byrd Land: Potential ^{40}Ar/^{39}Ar-dated time horizons in West Antarctic ice and marine cores. *GSA Bulletin*, **111**, 1563–1580, https://doi.org/10.1130/0016-7606(1999)111<1563:lqvaim>2.3.co;2

Wilch, T.I., McIntosh, W.C. and Panter, K.S. 2021. Marie Byrd Land and Ellsworth Land: volcanology. *Geological Society, London, Memoirs*, **55**, https://doi.org/10.1144/M55-2019-39

Wingrove, D. 2005. *Early Mixing in the Evolution of Alkaline Magmas: Chemical and Oxygen Isotopic Evidence from Phenocrysts, Royal Society Range, Antarctica*. Master's thesis, Bowling Green State University, Bowling Green, Ohio, USA.

Wörner, G. and Viereck, L. 1989. The Mt. Melbourne Volcanic Field (Victoria Land, Antarctica) I: Field observations. *Geologisches Jahrbuch*, **E38**, 369–393.

Wright, A.C. and Kyle, P.R. 1990a. Balleny Islands. *In*: LeMasurier, W.E. and Thomson, J.W. (eds) *Volcanoes of the Antarctic Plate and Southern Oceans*. American Geophysical Union Antarctic Research Series, **48**, 449–451.

Wright, A.C. and Kyle, P.R. 1990b. Royal Society Range. *American Geophysical Union Antarctic Research Series*, **48**, 131–133.

Section 2

Gondwana break-up volcanism

View of Ferrar Dolerite sills cutting a Permian sedimentary sequence in the Royal Society Range, southern Victoria Land. The sills are the most visible part of the spectacularly voluminous Ferrar Large Igneous Province, erupted in response to continental break-up forces during the Early Jurassic. The most prominent sill seen in the view is $c.$ 200 m thick.

Photograph by J. L. Smellie

Chapter 2.1a

Ferrar Large Igneous Province: volcanology

David H. Elliot[1]*, **James D. L. White**[2] **and Thomas. H. Fleming**[3]

[1]School of Earth Sciences and Byrd Polar and Climate Research Center, Ohio State University, Columbus, OH 43210, USA

[2]Geology Department, University of Otago, PO Box 56, Dunedin, New Zealand

[3]Department of Earth Sciences, Southern Connecticut State University, New Haven, CT 06515, USA

DHE, 0000-0002-6111-0508; THF, 0000-0001-7091-7699

*Correspondence: elliot.1@osu.edu

Abstract: Preserved rocks in the Jurassic Ferrar Large Igneous Province consist mainly of intrusions, and extrusive rocks, the topic of this chapter, comprise the remaining small component. They crop out in a limited number of areas in the Transantarctic Mountains and southeastern Australia. They consist of thick sequences of lavas and sporadic occurrences of volcaniclastic rocks. The latter occur mainly beneath the lavas and represent the initial eruptive activity, but also are present within the lava sequence. The majority are basaltic phreatomagmatic deposits and in at least two locations form immense phreatocauldrons filled with structureless tuff breccias and lapilli tuffs with thicknesses of as much as 400 m. Stratified sequences of tuff breccias, lapilli tuffs and tuffs are up to 200 m thick. Thin tuff beds are sparsely distributed in the lava sequences. Lava successions are mainly 400–500 m thick, and comprise individual lavas ranging from 1 to 230 m thick, although most are in the range of 10–100 m. Well-defined colonnade and entablature are seldom displayed. Lava sequences were confined topographically and locally ponded. Water played a prominent role in eruptive activity, as exhibited by phreatomagmatism, hyaloclastites, pillow lava and quenching of lavas. Vents for lavas have yet to be identified.

The discovery of thick dolerite sills in south Victoria Land (Fig. 1) was made by the National Antarctic Expedition, 1901–04, the first of R.F. Scott's expeditions. The field setting of the rocks was described by the expedition geologist H.T. Ferrar (1907) and the petrography by G.T. Prior (1907). Dolerites, mainly erratics, were also collected by members of the British Antarctic Expedition, 1907–09, from the Beardmore Glacier and Ferrar Glacier regions, as well as the coastal region of Victoria Land as far north as the David Glacier (Benson 1916; Mawson 1916). Benson (1916) also noted an erratic with the petrography of a tholeiite and with interstices filled by skeletal feldspar in dark brown glass, a rock that today would be considered a lava. Campbell-Smith (1924) described the dolerites collected from Buckley Island and Mount Darwin at the head of the Beardmore Glacier and from south Victoria Land by the British Antarctic ('Terra Nova') Expedition, 1910–13. Several erratics from the Terra Nova Bay region were later reported (Campbell-Smith 1964) to have glassy mesostases and zeolite-filled amygdales, which, with one exception, were thought to be intrusive rocks rather than lavas. The Australasian Antarctic Expedition, 1911–14, collected dolerites from Horn Bluff, as well as dolerite erratics, and these were described by Browne (1923). L.M. Gould, geologist on the first Byrd Antarctic Expedition (1928–30), collected a diabase from Mount Fridtjof Nansen in the Queen Maud Mountains (Gould 1931, 1935), thus extending the known distribution of the dolerite sills yet farther along the Transantarctic Mountains.

Harrington (1958) suggested the name Ferrar Group for the dolerite sills and dykes that are so abundant in the Dry Valleys, and which are known to occur throughout much of the Transantarctic Mountains. During the International Geophysical Year (1957–58) extrusive equivalents were found *in situ* in the Allan Hills–Coombs Hills region and at Westhaven Nunatak, both in south Victoria Land (Fig. 1) (Gunn and Warren 1962). At the same time a layered basic intrusion was discovered at the Dufek Massif, Pensacola Mountains (Aughenbaugh 1961; Walker 1961), and dolerite sills were found in the Theron Mountains and Whichaway Nunataks (Stephenson 1966). Grindley (1963) broadened the name Ferrar Group to include basaltic lavas that cap the Devonian–Triassic Beacon succession into which the dolerite sills were intruded, and named the lavas the Kirkpatrick Basalt (although commonly referred to as basalts, strictly speaking the majority of the Ferrar rocks have a basaltic andesite composition). Thus, basaltic lavas and pyroclastic rocks first reported by Gunn and Warren (1962) were included in the Ferrar Group. Subsequent field investigations showed that igneous rocks assignable to the Ferrar Group are widespread in the Transantarctic Mountains, cropping out from the Theron Mountains near the Weddell Sea to Horn Bluff, NW of north Victoria Land (Fig. 1). Later, Ford (1976) correlated the Dufek intrusion with the Ferrar Group. Kyle *et al.* (1981) introduced the name Ferrar Supergroup for all the intrusive and extrusive rocks of Jurassic age, and subsequently Kyle (1998) used the name Ferrar Large Igneous Province (FLIP) for these tholeiitic sills, dykes and extrusive rocks. That name now encompasses those rocks as well as the Dufek intrusion and the tholeiites in southeastern Australia, Tasmania and New Zealand (Milnes *et al.* 1982; Hergt *et al.* 1991; Mortimer *et al.* 1995; Bromfield *et al.* 2007). It is proposed here that the name Ferrar Large Igneous Province be formally established for these rocks together with those in southeastern Australasia, all of which are characterized by distinctive chemistry (see Elliot and Fleming 2021). The name Ferrar Group is retained for the rocks belonging to the FLIP but restricted to outcrops in Antarctica.

The extant Ferrar Group is dominated by intrusive rocks, and the subordinate lavas and pyroclastic deposits are scattered in relatively small areas between the Grosvenor Mountains at the head of the Shackleton Glacier and the Litell Rocks situated in the lower reaches of the Rennick Glacier in north Victoria Land (Elliot and Fleming 2008, 2017). Details of the occurrence and distribution of the intrusive rocks are given in the chapter on the geochemistry of the Ferrar LIP (Elliot and Fleming 2021).

The Ferrar province has a limited exposed volume. The volume of dolerite sills is estimated to be about 1.7×10^5 km^3, assuming an outcrop belt 150 km wide. The lavas are estimated to have a volume of several thousand cubic kilometres assuming continuity within the principal areas of outcrop

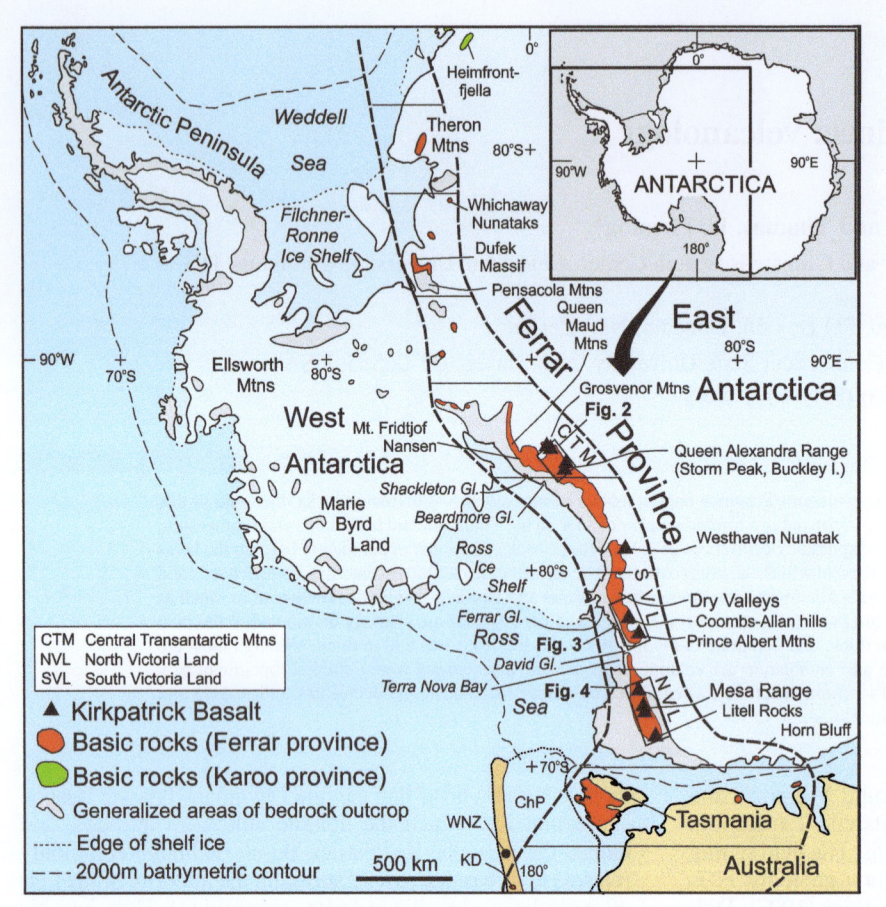

Fig. 1. Location map for the Ferrar Large Igneous Province. In a Gondwana reconstruction, New Zealand would have been off Tasmania. WNZ (solid outline), South Island, New Zealand, west of the Alpine fault. Kirwans Dolerite (KD) with Ferrar composition crops out in northwestern South Island. ChP, Challenger Plateau. Heavy dotted outline includes both WNZ and ChP.

(Fleming et al. 1995), but originally it must have been much greater. Although the geochemistry is described in the next chapter (Elliot and Fleming 2021), the division into two chemical types is noted here (Fleming et al. 1992, 1995). The Mount Fazio Chemical Type (MFCT) forms about 99% of the province and most of the analysed rocks. The Scarab Peak Chemical Type (SPCT) occurs only as the capping lava of most sequences and as a few sills in the Weddell Sea sector of the Ferrar province.

The age of the Ferrar rocks remained a little uncertain, other than Mesozoic, until the advent of radiometric age determinations when it was established that they are Jurassic in age. Initial results using the whole-rock K–Ar method were superseded by the analysis of plagioclase using the $^{40}Ar/^{39}Ar$ technique, but issues remained that were concerned mainly with monitor ages and the differences compared to U–Pb ages. The early U–Pb age determinations by multigrain zircon analysis (Encarnación et al. 1996; Minor and Mukasa 1997) have been overtaken by the single-crystal chemical-abrasion isotope-dilution thermal ionization mass spectrometry (CA-ID-TIMS) method (Table 1) (Burgess et al. 2015). Because of the sparsity of zircon in the extrusive rocks, U–Pb age determinations are primarily for dolerites. A restricted duration of emplacement (<0.4 Ma) is suggested for 14 Ferrar dolerite sills, with ages ranging from 182.78 ± 0.04 to 182.59 ± 0.08 Ma; two granophyre samples of the Dufek intrusion gave ages of 182.70 ± 0.05 and 182.63 ± 0.03 Ma (Burgess et al. 2015). A dolerite from Red Hill, Tasmania, part of the Ferrar LIP, gave an age of 182.54 ± 0.06 Ma (Burgess et al. 2015). Ivanov et al. (2017) reported three ID-TIMS U–Pb zircon ages for granophyres in Tasmanian dolerites, the ages ranging between 182.90 ± 0.21 and 182.65 ± 0.42 Ma. Kirkpatrick Basalt lavas from three different sections and forming the capping lava in each case yielded ages of 182.64 ± 0.08, 182.54 ± 0.20 and 182.43 ± 0.04 Ma

Table 1. *Single-grain and multigrain U–Pb zircon ages determined for the Ferrar Large Igneous Province*

Location	Sample no.	Rock type	Age (Ma)
CA-ID-TIMS (single grain), Antarctica (Burgess et al. 2015)			
Forrestal Range	PRR -8633	Granophyre	182.700 ± 0.045
Forrestal Range	PRR-09305	Granophyre	182.629 ± 0.029
Nilsen Plateau	96-65-11	Dolerite	182.590 ± 0.079
Roberts Massif	96-74-6	Dolerite	182.746 ± 0.054
Rougier Hill	96-51-67	Dolerite	182.753 ± 0.037
Mount Falla	90-53-12	Dolerite	182.85 ± 0.34
Wahl Glacier	85-6-16	Dolerite	182.753 ± 0.037
Mount Picciotto	85-4-4	Dolerite	182.616 ± 0.049
Mount Picciotto	85-4-18	Dolerite	182.633 ± 0.049
Dawson Peak	85-5-6	Dolerite	182.779 ± 0.033
Pandora Spire	A-236-A	Dolerite	182.689 ± 0.038
Pearse Valley	90-76-13	Dolerite	182.776 ± 0.059
Labyrinth	04-03-04	Dolerite	182.750 ± 0.048
Bull Pass	05-06-01	Dolerite	182.680 ± 0.038
Mount Bumstead	96-55-2	Lava	182.48 ± 0.20
Mount Bumstead	96-52-1	Lava	182.54 ± 0.20
Storm Peak	85-76-63	Lava	182.430 ± 0.036
Brimstone Peak	97-55-1	Lava	182.635 ± 0.077
CA-ID-TIMS (single grain), Tasmania (Burgess et al. 2015)			
Red Hill	97-17	Granophyre	182.540 ± 0.059
AA/CA-ID-TIMS (single grain), Tasmania (Ivanov et al. 2017)			
Northwest Bay	2013-289	Granophyre	182.90 ± 0.21
Northwest Bay	2013-288	Granophyre	182.65 ± 0.42
Cape Q Elizabeth	2013-290	Granophyre	182.75 ± 0.45
TIMS (multigrain), Antarctica (Encarnación et al. 1996)			
Dawson Peak	90-63-9	Dolerite	183.4 ± 1.4
Pearse Valley	90-76-12	Dolerite	183.8 ± 1.6
TIMS (multigrain), Antarctica (Minor and Mukasa 1997)			
Forrestal Range	93D-76	Granophyre	183.9 ± 0.3
Forrestal Range	93D-86	Granite dyke	182.7 ± 0.4

CA-ID-TIMS, chemical-abrasion isotope-dilution thermal ionization mass spectrometry; AA, air abrasion.

(Burgess et al. 2015). The latter suggests that at least part of the capping unit is permissibly slightly younger than the bulk of the Ferrar LIP. One lava from close to the base of the lava sequence gave a poorly constrained age of 182.48 ± 0.20 Ma, which is indistinguishable from the sill ages. The duration of Ferrar LIP magmatism was estimated to be 349 ± 0.49 ka (Burgess et al. 2015).

The Ferrar LIP, like the Karoo LIP of South Africa, exposes the supracrustal architecture of the plumbing system. The Ferrar LIP differs from the Karoo and many other LIPs in the very limited duration of emplacement (Burgess et al. 2015), a linear outcrop pattern along the Transantarctic Mountains, the dominance of a single set of chemical compositions and, perhaps most importantly, a distinctive isotopic signature (Elliot and Fleming 2008, 2017). The geochemistry is discussed in the next chapter.

Extrusive rocks (distribution, volumes and extent)

Extrusive rocks occur as isolated and limited outcrops in two distinct regions in the central Transantarctic Mountains (Fig. 2): at the head of the Shackleton Glacier in the Grosvenor Mountains and Otway Massif; and adjacent to the Beardmore Glacier in the Queen Alexandra Range (Barrett et al. 1986). In Victoria Land, outcrops, apart from an isolated occurrence at Westhaven Nunatak (Fig. 1) (Gunn and Warren 1962), are scattered over 600 km between the Allan–Coombs hills region (Kyle et al. 1983; Bradshaw 1987; Roland and Wörner 1996; Demarchi et al. 2001; Ross et al. 2008a) and Litell Rocks (Skinner et al. 1981) (Figs 3 & 4), with the only extensive outcrops found in the Mesa Range (Gair 1966; Elliot et al. 1986b; Brotzu et al. 1988; Hornig 1993; Hanemann and Viereck-Götte 2004; Viereck-Götte et al. 2007). Lavas comprise the bulk of the extrusive rocks, with a small proportion being volcaniclastic and formed by explosive eruptions, and an even smaller proportion being reworked volcaniclastic debris. The minimum volume of lavas, assuming continuity between outcrops within the extant areas, has been estimated to be c. 7000 km^3 (Fleming et al. 1995), and for the volcaniclastic rocks is estimated to be c. 60 km^3 (c. 37 km^3 in the Queen Alexandra Range and Otway Massif; c. 20 km^3 in south Victoria Land; c. 1.0 km^3 in the Prince Albert Mountains; and c. 0.2 km^3 in the southern and eastern Mesa Range region, north Victoria Land). The extrusive rocks are remnants of what must have been, at one time, extensive volcanic fields.

Volcaniclastic rocks

Distribution and thickness. Volcaniclastic rocks are assigned to the Prebble Formation in the central Transantarctic Mountains (Hanson and Elliot 1996; Elliot and Hanson 2001); to the Mawson Formation in south Victoria Land (Ballance and Watters 1971; Korsch 1984; Bradshaw 1987; White and McClintock 2001; Reubi et al. 2005; Ross and White 2005a; Elliot et al. 2006; McClintock and White 2006; Ross et al. 2008a) and in the Prince Albert Mountains (Elliot 2002); and are known as the Exposure Hill rocks (formerly Exposure Hill Formation: Elliot et al. 1986a) in north Victoria Land (Viereck-Götte et al. 2007). Volcaniclastic rocks typically underlie the lavas, but in a few places are found intercalated in the lower part of the lava sequence. Thicknesses of stratified volcaniclastic rocks range up to 200 m (Hanson and Elliot 1996), but unstratified accumulations infilling vent complexes comprising diatreme structures (called 'phreatocauldrons' by White and McClintock 2001) have a vertical extent of at least 370 m at the Otway Massif (Elliot and Hanson 2001) and more than 400 m in the Coombs–Allan hills area where the outcrop area is more than 30 km^2 (White and McClintock 2001). Similar features are known from the Karoo province (McClintock et al. 2008), but have not been widely identified in the massive volcaniclastic deposits of other flood basalt provinces (Ross et al. 2005).

Exposed stratigraphic contacts with underlying Jurassic or Triassic beds are few and far between. Contacts are present in the Queen Alexandra Range (Hanson and Elliot 1996), possibly at Shapeless Mountain and Coombs Hills, south Victoria Land (Korsch 1984; Elliot and Grimes 2011, respectively), and in the Deep Freeze Range and east of Gair Mesa, north Victoria Land (Viereck-Götte et al. 2007), where the basal part of the basaltic pyroclastic succession is interbedded with the upper part of the underlying silicic Shafer Peak Formation. More complex contact relationships characterize the northern part of Coombs Hills, where pyroclastic rocks and lava rafts (with an enclosed fossil tree stump: Garland et al. 2007) are closely associated with tilted and broken 'rafts' of Lashly Formation country rock of varying sizes, and all enclosed in rock mapped as a mixture of Ferrar Dolerite sills and dykes cutting the Lashly Formation (Grapes et al. 1974). White et al. (2009) interpreted these relationships to indicate that at the edge of the Coombs Hills phreatocauldrons the intruding sills had shoaled to the surface (cf. Muirhead et al. 2014). These sills engulfed broken blocks of the shallow

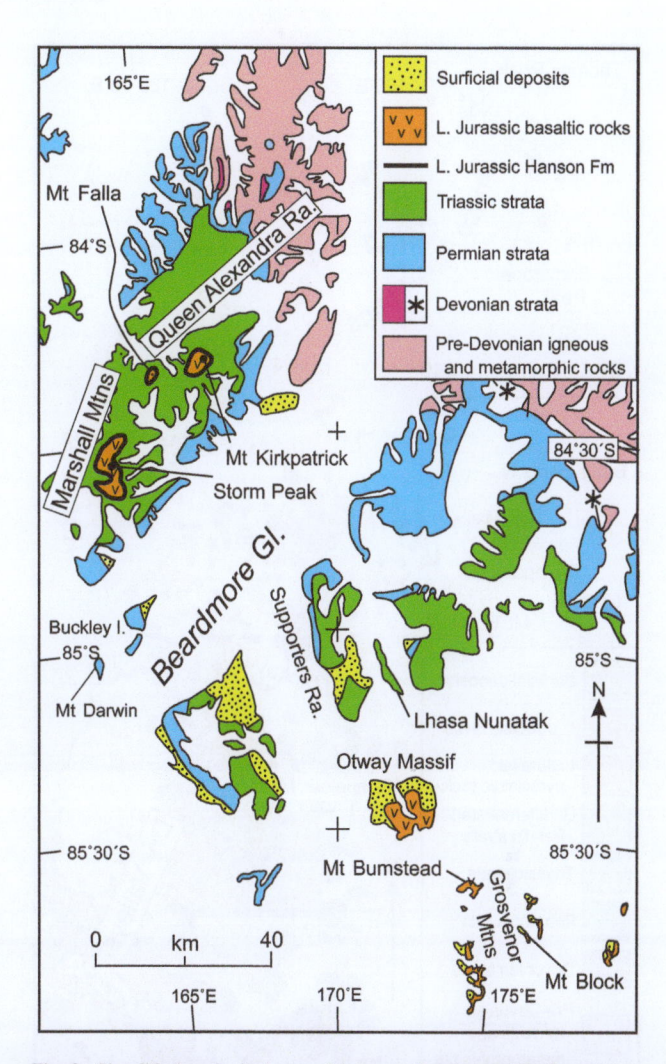

Fig. 2. Simplified geological map of the upper Beardmore Glacier region, central Transantarctic Mountains, illustrating the distribution of the Prebble Formation and Kirkpatrick Basalt lavas (combined as L. Jurassic basaltic rocks in the explanation). Ferrar Dolerite sills and dykes occur throughout the Devonian–Triassic Beacon strata.

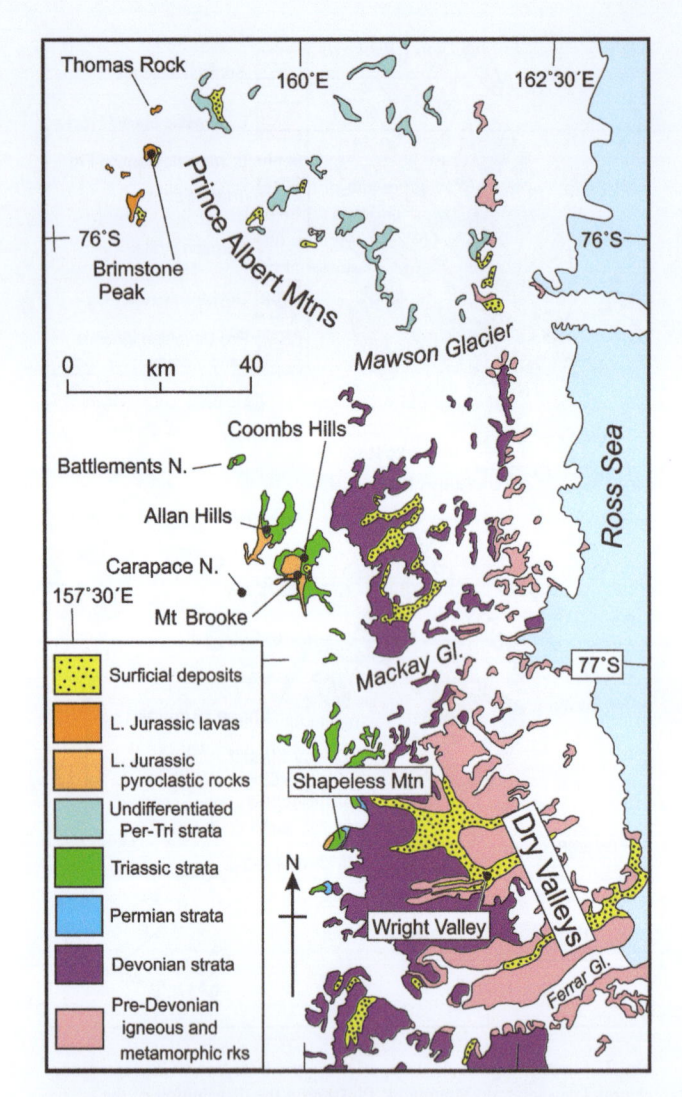

Fig. 3. Simplified geological map of south Victoria Land to show the distribution of the Mawson Formation and Kirkpatrick Basalt lavas. Ferrar Dolerite sills and dykes occur throughout the Devonian–Triassic Beacon strata; however, at Battlements Nunatak sills alone are present.

country rock beneath which magma was initially intruded, and enclose surface-emplaced pyroclastic deposits.

Facies and origin in south Victoria Land. Stratified Ferrar volcaniclastic rocks include both primary volcaniclastic deposits and others inferred to comprise debris redeposited in stream and lake environments. All of these are varieties of mafic volcaniclastic deposits (MVD) such as are known to be associated with many large igneous provinces (Ross *et al.* 2005). The lithofacies scheme used is shown in Table 2, with summary interpretations of the different lithofacies. Outcrops in the Allan and Coombs hills areas provide exceptional exposures of these rocks, which at all known localities mark the initiation of Ferrar volcanism.

At Allan Hills in South Victoria Land (Figs 1 & 3) prominent stratified deposits exposed over *c.* 6 km (Fig. 5a) include two different kinds of beds (Ballance and Watters 1971; Grapes *et al.* 1974; Ross and White 2005*a*; Ross *et al.* 2008*a*). Thick beds with associated accretionary lapilli are laterally persistent and contain small 'rags' of once-glassy basalt indicating inhomogeneous temperatures within the pyroclastic currents that deposited them. Their thickness and extent indicate emplacement from large-volume eruptions from a source beyond Allan Hills, inferred to have been in neighbouring Coombs Hills (Fig. 3). With metres-thick deposits of basaltic

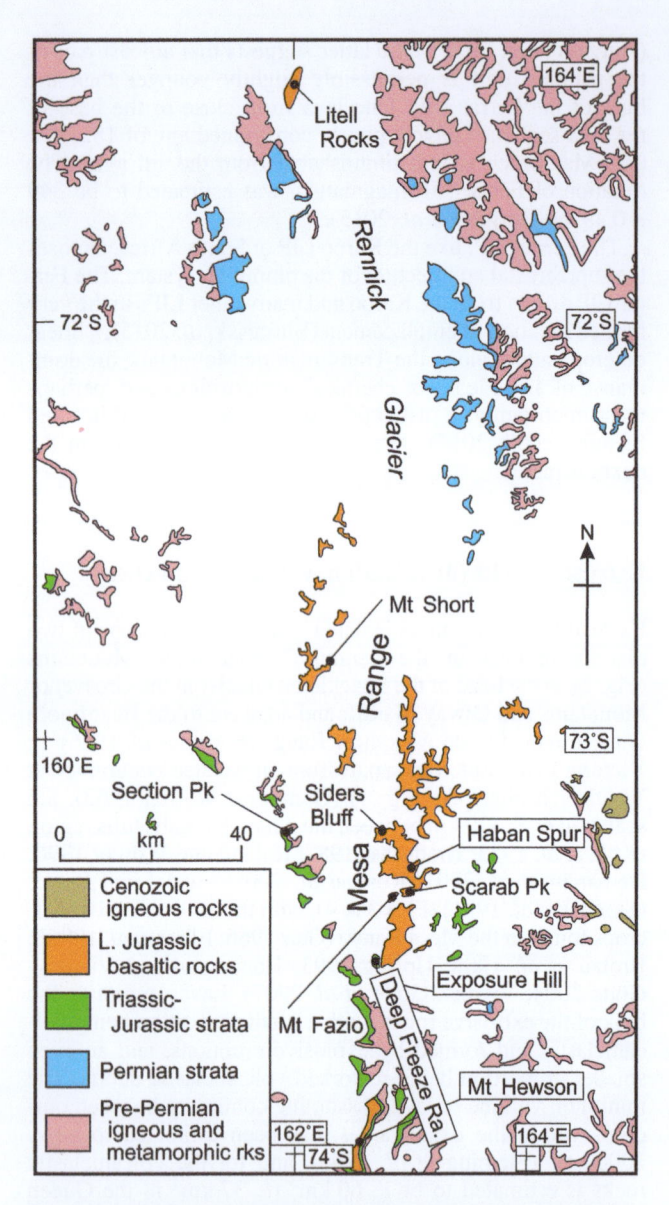

Fig. 4. Simplified geological map of north Victoria Land to illustrate the distribution of the Exposure Hill rocks and Kirkpatrick Basalt lavas. Ferrar Dolerite sills and dykes occur throughout the Permian–Triassic Beacon strata.

pyroclastic density currents (PDCs) covering areas of at least 100 km^2, the volumes produced by single eruptions were on the order of 1 km^3. The PDCs are inferred by Ross and White (2005*a*) to have been dilute, moist and turbulent, based on lack of welding, abundance of lithic particles and the presence of large (up to 4.5 cm) rim-type (Schumacher and Schmincke 1991) accretionary lapilli.

The thick beds overlie thin-bedded deposits with bedding features like those of small tuff rings (planar lamination, local dunes, bombs and blocks with bedding sags: Ross and White 2005*a*). The thin-bedded unit extends for more than 1 km without clear thinning or fining trends, and comprises numerous lenses overlapping along a single stratigraphic level. Multiple local sources, probably lying along a major fissure, are inferred from the lenticularity and large blocks (some exceeding 2 m). The layering, blocks and thickness variations are reminiscent of tuff ring deposits, but in a linear array rather than surrounding a single vent (cf. Sohn and Park 2005).

At Coombs Hills (Fig. 3), non-stratified volcaniclastic rocks are dominant over a large area (Bradshaw 1987; White and McClintock 2001; Ross and White 2006; White *et al.* 2009).

Table 2. *Facies descriptions and interpretations for rocks in the Mawson Formation, Coombs Hills and Allan Hills areas*

Facies	Observations	Interpretations
Heterolithological lapilli tuff (LTh and TBh)	1. Volumetrically dominant facies at Coombs Hills (locally tuff breccia in grain size) 2. At Coombs Hills lacks bedding planes for >300 m vertically; forms thick (up to 15 m) widespread layers at Allan Hills 3. Lateral variations in grain size and componentry at Coombs Hills, but no systematic vertical variations; various vertical variations in thick beds at Allan Hills 4. Comprises formerly glassy basalt fragments (mostly blocky ones), microcrystalline basalt fragments (at Allan Hills), sand-grade detrital quartz particles, Beacon fragments, composite clasts (recycled peperite) and rare granite fragments 5. Basaltic clasts are variably vesicular, mostly dense to incipiently vesicular (vesicularity index of Houghton and Wilson 1989) 6. Host for LTa and TBj zones 7. Contains rafts of Lashly Formation (sandstone to siltstone, plus silicic tuff and silicic tuffaceous sandstone) and rafts of layered, fine-grained, mafic volcaniclastic rocks (some with accretionary lapilli); most of these rafts dip steeply	• The clast assemblage suggests phreatomagmatic eruptions affecting the upper part of the Beacon sequence, occasionally as far down as the base of Victoria Group • At Coombs Hills, an origin as one or several lahars (Hanson and Elliot 1996) – or subaerial pyroclastic flows – filling a pre-existing topographical depression is not favoured because of observations 2, 3 and 7 • At Coombs Hills, emplacement in a vent complex by subterranean debris jets is inferred • Several cycles of eruption were probably necessary to reach proportions of formerly glassy basalt clasts observed in LTh (e.g. Bélanger and Ross 2018)
Beacon-rich lapilli tuff (LTa and TBa)	• Forms steep, pipe-like bodies cross-cutting LTh, with sharp contacts • Locally a tuff breccia • Outlines in map view are generally simple, elliptical in shape, with a long-axis length a few decimetres to a few hundreds of metres, aspect of ratio 0.2–0.8 • Same types of clasts as LTh, but with more abundant country-rock fragments	• Phreatomagmatic fragmentation, vent complex setting • Deposited by Beacon-rich subterranean debris jets • Jets originated when phreatomagmatic explosions occurred near the walls or floor of the vent complex, causing fragmentation of abundant Victoria Group material
Basalt-rich lapilli tuff and tuff breccia (LTj and TBj) – general	• Forms zones, metres to hundreds of metres-wide, cross-cutting LTh • Locally a lapilli tuff • Contains more vesicular basaltic clasts than in other facies (except TBhr) • No more than 5% Beacon clasts in the lapilli + block size fraction	• Vent complex setting
TBj – type 1	• Often relatively sharp contacts with host, relatively compact shapes • Same types of clasts as LTh, but with more abundant basaltic fragments; blocky basalt clasts present • Not strongly associated with basalt pods and/or peperite domains	• Phreatomagmatic fragmentation • Debris jets propelled by explosions taking place well away from country rocks • Material in the jets is richer in basalt than the surrounding LTh debris because of the addition of juvenile basalt
TBj – type 2	• Abundant in western Coombs Hills • Generally diffuse gradational contacts with the host rock, outlines can be very complex (e.g. octopus-like) • Contains fluidal basalt fragments and composite clasts, and few or no blocky clasts • Spatially associated with *in situ* peperite domains, and/or pods of glassy basalt	• Somewhat less violent origin than for LTa, LTh and type 1 TBj zones • Juvenile clast-forming processes inferred to be similar to those in peperite (mostly non-explosive: e.g. surface tension effects, magma-'sediment' density contrasts, instabilities in vapour films) • Mixing of juvenile fragments with surrounding volcaniclastic material (incorporation of quartz grains, etc.)
'Raggy' heterolithological tuff breccia (TBhr)	• Volumetrically minor at Coombs Hills • Tuff breccia version of LTh with abundant 'rags'; common at the top of thick beds at Allan Hills • Rags = relatively vesicular, glassy basaltic fragments, elongate, up to several decimetres long, with bent shapes, delicate ends that form spiral shapes and displaying accommodation of the surrounding clasts • Rags can be aligned in any orientation or be 'randomly' dispersed in unbedded Coombs Hills deposits; weak subhorizontal orientations at the top of Allan Hills thick beds	• Rags transported while still plastic (high temperature) • LTh-type material simultaneously transported with rags was probably cool (quenched, blocky basalt clasts and Beacon material) • Zones containing 'randomly' or subvertically aligned rags could have formed when phreatomagmatic explosions accelerated vesiculating melt not directly involved in the explosions • Zones containing subhorizontally aligned rags at Allan Hills reflect emplacement by pyroclastic density currents

After Ross and White (2005*a*, 2006), Ross *et al.* (2008*a, b, c*).

Within this area, domainal outcrop patterns are characteristic (Fig. 6), with steeply dipping, irregular and diffuse contacts separating different lithofacies defined primarily by differences in the relative abundance of juvenile particles v. country-rock clasts and country-rock-derived sediment grains (Fig. 7; Table 2). The most abundant lithofacies is a

Fig. 5. Layered Mawson deposits at Allan Hills. (**a**) Aerial view northwards of laterally extensive layering (arrows). (**b**) Avalanche deposits in the central Allan Hills. Lashly Formation strata, disrupted Ferrar magma intrusion, showing (lower left) broken coal and fine-grained beds enclosed in sandstone. Locally the deposits (not in image) contain domains of fluidally deformed basalt inferred to have triggered avalanching.

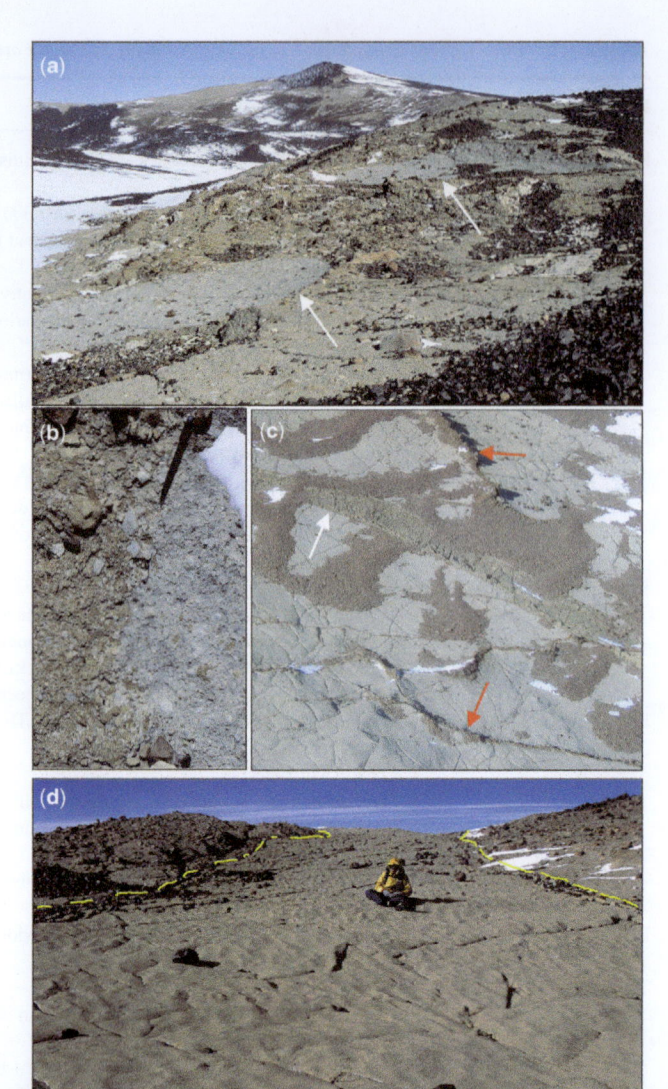

Fig. 6. Distinctive elements of Mawson vent-complex architecture at Coombs Hills. (**a**) Domainal outcrop interpreted as cross-cutting clastic deposits; pale domains in the midground (white arrows) are LTa, c. 10 m across. (**b**) Detail of a boundary between LTh and LTa domains. The pencil (upper edge) is for scale. (**c**) Regular thick clastic dyke (white arrow) and several irregular thin basaltic dykes (red arrows). Field of view is c. 100 m wide. (**d**) Clastic dyke c. 10 m wide (pale grey rock between the yellow dashed lines); the clast mixtures of the dyke are typically heterolithic like the host, but lack the host's coarser lapilli and blocks (dolerite clasts have accumulated on the surface of the host rocks).

heterogeneous lapilli tuff (locally coarser-grained tuff breccia) consisting of 70–80% glassy juvenile fragments and up to 30% Beacon sedimentary rock fragments. Other domains are much richer in sedimentary rock fragments (50–90%) or in juvenile material (up to 70–90%).

The non-stratified rocks locally contain tilted rafts of thin-bedded deposits including accretionary lapilli, interpreted to be remnants of tuff-ring deposits (White and McClintock 2001), and similar beds locally overlie the non-stratified deposits (McClintock and White 2006). Another very common component of the non-stratified deposits, also present in both thick and thin beds at Allan Hills, is a kind of composite clast (White and Houghton 2006). These comprise thin tendrils of glassy basalt mingled on centimetric to millimetric scales with clastic material, most commonly ash (Fig. 8), and are interpreted as fragments produced by disruption of the peperitic zones commonly formed where basalt magma intruded and mingled with previously deposited tephra (Ross and White 2006).

Clastic dykes, specifically pyroclastic ones (Fig. 6), are another distinctive type of deposit. These dykes cut both stratified and non-stratified rocks at Coombs Hills (Ross and White 2005b, 2006), and are also present at Allan Hills (Grapes et al. 1974; Ross and White 2005a). Some of these have exceptional width (up to 75 m).

A clastic deposit with a very low proportion of volcanic material at Allan Hills was emplaced as a large avalanche of Lashly Formation sandstone, shale and coal beds (Fig. 5b), into which basalt was injected before or during emplacement (Reubi et al. 2005; Lockett and White 2008). It comprises a chaotic assemblage of breccia domains and megablocks as large as 80 m, derived from underlying rocks of the Beacon Supergroup, produced by progressive, pervasive and relatively uniform fragmentation of initial megablocks during transport plus minor disruption and ingestion of the substrate. The avalanche was emplaced by northward flow into a pre-existing topographical depression carved into the Beacon sequence (Reubi et al. 2005).

At Carapace Nunatak (Fig. 3), pillow lavas (Ballance and Watters 1971), hyaloclastites and coarse tuffs with accretionary lapilli (Bradshaw 1987; Ross et al. 2008a) crop out. The tuffs are most probably primary eruption-fed deposits of distal lahar runout flows or possibly PDCs, perhaps including those that emplaced the thick beds at Allan Hills. More significant reworking by streams is a less-favoured possibility.

Thin, subvolcanic dolerite dykes and sills that pass into peperite are widely scattered, and best developed at Shapeless Mountain (Korsch 1984) and at Coombs Hills (Elliot and Grimes 2011). Phreatic explosion vent deposits, comprising

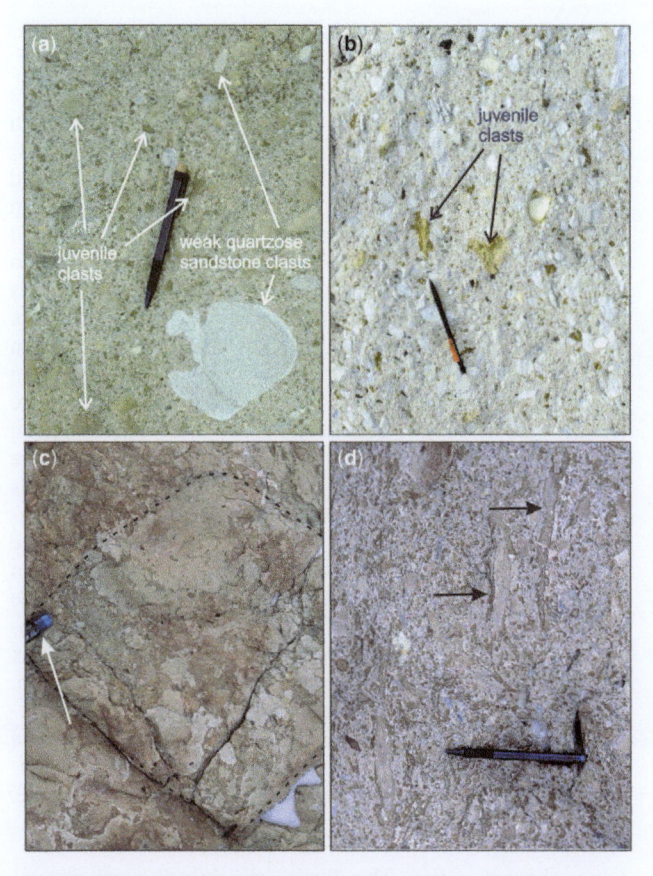

Fig. 7. Dominant clastic lithofacies of Ferrar primary volcaniclastic deposits, with examples illustrated from Coombs Hills. (**a**) Heterolithic lapilli tuff (LTh). (**b**) Accidental-rich lapilli tuff (LTa). (**c**) Juvenile-rich tuff breccia (TBj). The pencil head is for scale (white arrow). (**d**) Juvenile-rich lapilli tuff (LTj) with 'raggy' fluidal clasts (arrows).

Fig. 8. Distinctive Mawson composite clasts, lapilli (2–64 mm) and clasts >64 mm that are composite bombs and blocks; all are considered to be fragments of peperite. (**a**) Angular peperite block. Scale given by the pencil end (arrow). (**b**) Small-scale mingling of basalt (tan, smooth) with LTh lapilli tuff containing quartzo-feldspathic sand from Beacon country rock (darker, grainy). (**c**) The photomicrograph, with polarizer at 23° (1/4 crossed), shows once-glassy basalt enclosed in, and enclosing, sedimentary grains; field of view is 5 mm.

a mixture of fragments from the host strata in a matrix lacking basaltic clasts, have been identified at Coombs Hills where they cut Triassic Lashly Formation strata (Elliot and Grimes 2011).

The mafic volcaniclastic deposits of south Victoria Land have been interpreted in different ways. Gunn and Warren (1962) interpreted the thick-bedded to unbedded deposits comprising a mixture of clast sizes ranging from blocks or boulders to silt-grade fragments as a tillite of glacial origin. Ballance and Watters (1971) called the rocks in the Allan Hills a 'diamictite', identified possible vent deposits and inferred deposition of layered rocks by volcanic mudflows. Bradshaw (1987) further identified vent deposits at Coombs Hills, and noted that granite boulders contained in them had been carried upward, during eruption, through hundreds of metres of stratigraphy.

Interpreting both past work and new mapping results, White and McClintock (2001) recognized the grouping of 'vent deposits' at Coombs Hills as comprising a large vent complex, or 'phreatocauldron', formed by the same suite of processes that form diatremes (Fig. 9). Ross and White (2006) presented new mapping of the internal structure of this complex, and followed their field analysis with a series of experiments to investigate structures characteristic of subterranean vent-excavation processes (Ross *et al.* 2008*a*, *b*, 2013; McClintock *et al.* 2009; Valentine *et al.* 2015). White *et al.* (2009) and Muirhead *et al.* (2014) also discussed the role of sills in this setting, suggesting that an alternative interpretation would be that dykes feeding this complex were rooted in a sill(s) at a shallow level below the complex. Such a reconstruction would be consistent with apparent 'shoaling' of sills just north of the diatreme complex (White *et al.* 2009), and inferred processes of generation for

70 m-wide clastic dykes at Coombs Hills (Ross and White 2005*b*).

Facies and origin in other areas. At the Otway Massif several hundred metres of massive tuff breccias overlain by thin sequences of tuff beds and capped by lavas at the Otway Massif are interpreted as forming another phreatocauldron (Elliot and Hanson 2001; Elliot and Fleming 2008). These tuff breccias are less well exposed than those at Coombs Hills, but have many of the same characteristics: megablocks of Hanson Formation strata are enclosed in tuff breccia; accidental blocks of silicic tuff and dolerite are up to 10 m in length; areas within the tuff breccias show divergent particle orientations, and varying clast sizes and types; and clastic dykes are scattered widely. Non-bedded tuff breccias with cross-cutting relationships occur at Ambalada Peak, Prince Albert Mountains (Elliot 2002), and probably are another example. At Agate Peak and Exposure Hill, adjacent to the Mesa Range, tuff breccias are interpreted as vent fillings (Elliot *et al.* 1986*a*; Schöner *et al.* 2007).

Hanson and Elliot (1996) identified extensive deposits of stratified volcaniclastic rock underlying Kirkpatrick Basalt lavas in the Queen Alexandra Range, central Transantarctic Mountains (Fig. 2). Based on a particle population comprising a mixture of glassy basaltic (sideromelane) fragments with sand grains comminuted from country rock, tabular deposit geometries, local entrainment of surface debris into the basal layers and local fluvial channelling, these are interpreted as deposits of lahars. Associated deposits contain accretionary lapilli, and the lahars may have been fed directly from very water-rich

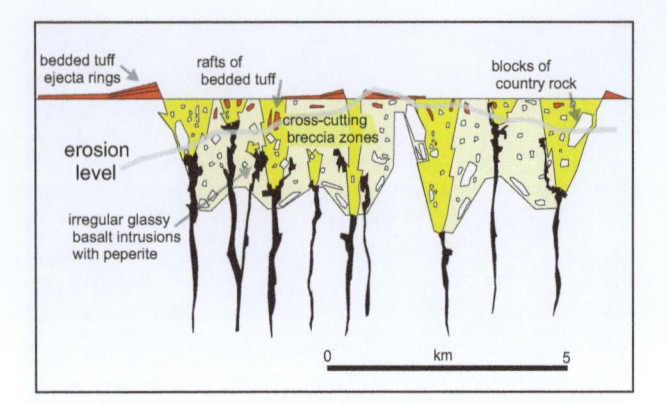

Fig. 9. Schematic cross-section of the Coombs Hills geology showing inferred subterranean development of non-bedded lithofacies in a 'phreatocauldron' or diatreme complex (after White and McClintock 2001) locally capped by, and containing tilted remnants of, bedded tuff and lapilli tuff from surficial tuff rings.

eruptions. Remnants of phreatomagmatic vents, along with small sills and dykes of pyroclastic material, attest to a depositional setting that included volcanic centres. This took place in a basin that was filling with thick Prebble Formation deposits, and beneath which magma interacted with weakly consolidated Falla Formation and older sandstones.

Stratified volcaniclastic rocks (tuff breccias, lapilli tuffs, tuffs and reworked debris) crop out in isolated areas from Mount Pratt, east of Mount Bumstead, to the Mesa Range, and represent remnants of volcanic constructs. Accretionary lapilli are not uncommon. Finer-grained tuffs are attributed to air-fall or base-surge deposition and interpreted as remnants of tuff rings and tephra cones. Most deposits can be attributed to phreatomagmatic activity in which the violent interaction between hot magma and water-rich sedimentary rocks produced mixtures of juvenile pyroclasts and disaggregated siliciclastic sandstones. In contrast, at one locality in the Prince Albert Mountains a c. 10 m-thick breccia, with blocky basalt clasts, spatter up to 2 m long, basalt shreds up to 30 cm long and carbonaceous sedimentary clasts, suggests a locally derived deposit. The overlying 20 m, which lack the number and size of large clasts although basalt shreds are ubiquitous, is succeeded by a few metres of tuff breccia with basalt and sedimentary clasts up to 3 m long (Elliot 2002). Together with stratigraphically associated massive and weakly bedded tuffs, these rocks suggest close proximity to a phreatomagmatic vent.

At Allan Hills and elsewhere in the Ferrar LIP, widespread layers of the same sorts of commonly heterolithological lapilli tuffs were emplaced in different ways: (1) by PDCs; (2) by lahars; and (3) by streamflow. Because they have the same components and similar textures, rocks with features indicating deposition by rivers may well include both primary volcaniclastic rocks from runout of eruption-fed lahars, and those produced by erosion of primary deposits and redeposition of their particles. Those derived from erosion of earlier deposits may be better termed volcaniclastic sandstones and conglomerates. It is important, for context, to appreciate that the variations in dimensions of both non-bedded (tens to hundreds of metres) and bedded (hundreds of metres to kilometres) deposits reflects processes affecting those particular sites, and that the exposed areas are very small relative to the scale of the Ferrar outcrop belt (hundreds to thousands of kilometres). Details of individual beds at different sites hold the record of which depositional process(es) were dominant from one area to the next, and a rich archive of information to be extracted in future research.

Effusive rocks

Distribution and thickness. Lava sequences (Fig. 10) are commonly 400–500 m thick but attain more than 750 m in the Mesa Range (Elliot *et al.* 1986*b*; Mensing *et al.* 1991; Elliot and Fleming 2008). The lower contact is locally exposed only at the Otway Massif, the Queen Alexandra Range and in Victoria Land, and not at all in the Grosvenor Mountains and the Prince Albert Mountains. Individual lavas range from 1 to 230 m in thickness; many are tens of metres thick, which leads to the layer-cake aspect evident in the field. Although it cannot be demonstrated in the field, lavas thicker than about 100 m are interpreted to be confined topographically in a rift system and locally ponded in the case of the 230 m-thick lava at the Otway Massif (Elliot and Fleming 2008). Thin lavas, in some sections, occur in packages and may occur anywhere in the sequence. Although the thin lavas might be the products of a local volcanic centre, at Storm Peak in the Queen Alexandra Range (Elliot and Fleming 2008, fig 10A) lavas 3–8 on the western ridge are represented by a single lava (designated X in that figure) on the adjacent ridge. This suggests that multiple thin lavas are most probably flow lobes of a much thicker single lava, and the similarity in chemistry of five of those six lavas at Storm Peak provides support for this interpretation. The Mesa Range in north Victoria Land has an almost continuous exposure of lavas over a distance of about 75 km. In a photogeological study of these rocks, Petri *et al.* (1997) distinguished five lava units that were traced throughout the range.

Lava characteristics. The facies classification scheme of Jerram (2002) can be applied to the Kirkpatrick Basalt lavas (Elliot and Fleming 2008). The lavas belong predominantly to the 'tabular-classic flow facies', and, being laterally extensive, many probably represent the sheet lobes of Self *et al.* (1997). These lavas range in thickness from several metres to more than 100 m. Within individual lavas a variety of fracture types commonly occur, but no consistent pattern has been observed (Fig. 10), and the classical colonnade and overlying entablature is seldom displayed. Perhaps one of the most remarkable lavas crops out in the Queen Alexandra Range: at Storm Peak it is as much as 135 m thick, and consists of a thin basal chilled-contact followed by a 3 m-thick colonnade (the latter not always present), 130 m of entablature and a 3-m-thick, slightly vesicular, upper crust. The entablature is entirely tachylitic and comprises downward and outward, diverging and branching, columns 10–20 cm across (Elliot and Fleming 2008, fig. 11). Although occurring as a tachylitic lava at all measured localities in the Marshall Mountains, on the north face of Mount Falla the tachylitic interval appears to thin and pinch out over a few hundred metres, and then reappear and pinch out again in the central part of the face (Fig. 11). The thickness changes in the tachylitic interval probably reflect patterns of quenching and alteration rather than real pinching and swelling of the lava. Stratigraphic sections measured on the NE and NW ridges of Mount Falla illustrate the challenges in lava correlation and in identification of lavas across and between outcrops (see Barrett *et al.* 1986, pl. 1c, secs 71 and 10). The sedimentary interbeds provide the only reliable framework for correlation within the central Transantarctic Mountains (Barrett *et al.* 1986) and in the Mesa Range (Elliot *et al.* 1986*b*), although detailed geochemical analysis of lavas might yield additional acceptable correlations. The second lava at Mount Kirkpatrick (Elliot and Fleming 2017, fig. 7b), given the MgO concentration and the presence of a thin colonnade and overlying thick black conchoidally fractured basalt, is permissibly correlative with the tachylitic lava at Storm Peak, Mount Falla and elsewhere in the Marshall Mountains. The much greater apparent thickness, half as much again as the

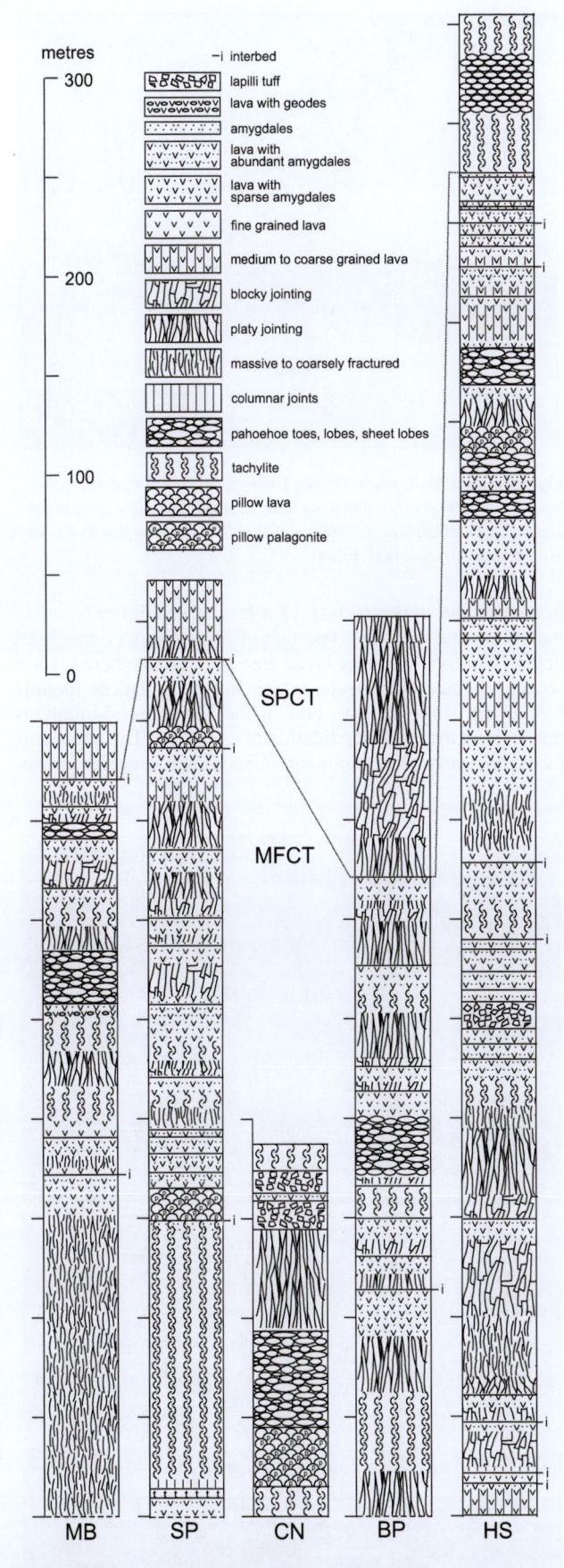

Fig. 10. Simplified stratigraphic columns for Kirkpatrick Basalt lava sequences in the five principal areas of outcrop. Mount Bumstead (MB), Storm Peak (SP), Carapace Nunatak (CN) and Brimstone Peak (BP) from Fleming (unpublished; see also Barrett *et al.* 1986 for MB and SP). Haban Spur (HS) from Elliot *et al.* (1986*b*). If the inferred correlation between Mount Bumstead and the Otway Massif (Barrett *et al.* 1986) is correct, then the lava sequence in that region is more than 550 m thick.

Fig. 11. Kirkpatrick Basalt lava sequence, 475 m thick, exposed on the north face of Mount Falla. (**a**) Aerial view from near the NE ridge. The black tachylitic lava is exposed on the NE ridge where it is 101 m thick, but this tachylitic interval pinches out westwards. It reappears as prominent cliffs just above the talus slope across most of the width of the mountain (white arrows) but pinches out farther to the west (out of view). Another prominent black tachylitic lava (68 m thick; indicated by black and white arrows) crops out about one-third of the way up the lava sequence along the NE ridge, and that tachylitic interval also thins and thickens westwards. (**b**) Aerial view from the NW. The arrows correspond with those in (a), and illustrate the lava sequence farther across the mountain front. (Images: D.H. Elliot.)

lava at Mount Falla, is attributed to a fault, down to the north, which is supported by a thin vertical zone of brecciated lava. In north Victoria Land the capping lava of the sequence at all examined localities is also tachylitic (Elliot and Fleming 2008, fig. 10B) and exhibits a variety of fracture patterns, but lacks any columnar jointing. Examples of relatively thin tachylitic lavas are present in many measured sections. A role for water in the rapid chilling and formation of tachylite is implied by the overlying lacustrine bed at Storm Peak and other outcrops in the Marshall Mountains, and is suggested by the lacustrine bed that underlies the capping lava in the Mesa Range; elsewhere such direct evidence is lacking for the tachylitic lavas. Quenched lavas of such thickness (>100 m) must record abrupt flooding of lava surfaces in a topographically confined setting, which has been interpreted as a rift (Elliot and Fleming 2008). The proximal cause could have been a rearrangement of drainage resulting from lava emplacement.

Regional ponding of lavas by topography, on a scale of at least tens of kilometres, is inferred for the very thick lavas (>100 m), and in particular for the basal lava at Mount Bumstead (>175 m) and the >230 m-thick correlative lava at adjacent Otway Massif (Barrett *et al.* 1986, pl. 1c). The palaeosol and tuffaceous beds at the Otway Massif (Elliot and Hanson 2001), which are conformable with the overlying lavas, suggest that the lavas are not simply filling a collapsed or partially collapsed phreatocauldron. All measured sections at the Otway Massif include one or more basal lavas more than 150 m thick.

Intervals of thin lavas are present in many sections, constitute the 'compound-braided flow facies' of Jerram (2002) and represent the flow lobes of Self *et al.* (1997). These sets of thin lavas are scattered among the sections, and are interpreted as either sets of pāhoehoe toes, lobes and sheet lobes marginal to, and break-outs from, thicker lavas of the 'tabular-classic flow facies' or, much less probably because of the absence of evidence for local vents, locally derived thin lavas. However, the lack of a clear relationship, in most instances, to thicker lavas makes their interpretation problematic, as does the absence of any sign of local vents. At Mount Bumstead, where there is an accessible, north-facing, snow-free outcrop, the irregular character of these sets of thin lava units is evident (Fig. 12). Marked differences in individual lava lobe thicknesses are dependent on the degree of inflation, and the commonly thicker, pod-shaped, dense bodies may represent lava tunnels.

The intra-lava characteristics are typical of tholeiitic flood lavas (Self *et al.* 1997). Pāhoehoe toes may be present at the base of a lava, in some instances stacked three or four high, and pass up into sheet lobes. Chilled margins of lobes and lava sheets are normally only a few tens of centimetres thick and are vesicular. Only rarely are there pipe vesicles that show stretching and inclination (Fig. 13), and spiracles are equally uncommon (Barrett *et al.* 1986, fig. 40). Nevertheless, at Mount Kirkpatrick small pipe vesicles in the lowest lava show a consistent west to SW flow direction, and in the Mesa Range inclined pipes suggest an overall westerly flow. Within the body of some thick lavas, vertical pipes (Fig. 14), a few to tens of centimetres in diameter, are filled by segregations of coarser-grained basalt; these pipes interconnect with similar horizontal pipes and sheets. Vesicular upper crusts are almost uniformly present, and in some cases are metres thick. Vesicles may be scattered or in distinct layers, reflecting pulses of inflation, and commonly are filled by secondary minerals. The upper surface of a lava is seldom exposed in map view and only one instance of a ropy surface has been noted. The tops of some lavas are strongly weathered, form weakly defined palaeosols, which in rare instances include root traces (Fig. 15a, b), and in the Grosvenor Mountains may exhibit the marked oxidation of a red bole. The upper surfaces may carry plant fragments. Geodes and irregular masses

Fig. 13. Siders Bluff, north Victoria Land. Inclined pipe vesicles in the base of a lava overlying a thin pyroclastic interbed filling low spots in the underlying lava (Elliot *et al.* 1986*b*; unit 35 at 571.5 m in section 81-2). Ice axe for scale. (Image: D.H. Elliot).

Fig. 12. Kirkpatrick Basalt lavas exposed on the north face of Mount Bumstead. Numerous thin strongly altered lava lobes occupy the interval between a prominent, *c.* 15 m-thick, black tachylitic interval (indicated by the white arrow; the top of this interval lies 9 m below the top of a 68 m-thick lava), which crosses the whole image, and the capping lava (*c.* 45 m thick on the right-hand edge of the image). (Image: D.H. Elliot.)

Fig. 14. Haban Spur, north Victoria Land. Pipes, up to 3–4 cm across, of coarser-grained basalt in the upper part of a lava, which here pass abruptly to scattered vesicles (Elliot *et al.* 1986*b*; unit 11 at 264 m in section 82-3). Hammer is for scale (arrow). (Image: D.H. Elliot.)

Fig. 15. Mount Block, Grosvenor Mountains. (**a**) Strongly oxidized and weathered upper part of a lava, forming a red bole. Amygdaloidal basalt fragments, partially destroyed by weathering processes, are scattered throughout the profile. Ice axe (80 cm long) is for scale. (**b**) Branching tube-like bodies interpreted as rootlets (scale in centimetres). (**c**) Tricuspate glass shards (*c.* 0.25 mm across) identified in the weathering profile. Mount Block, Grosvenor Mountains (Fig. 2) (section 61: Barrett *et al.* 1986; Elliot *et al.* 1991). (Images: D.H. Elliot.)

of secondary minerals occur between many pāhoehoe toes or lobes, between pillows and at some lava contacts.

Secondary minerals include quartz, chalcedony, calcite, zeolites, apophyllite, green phyllosilicate and gypsum. Zeolites recognized include stilbite, epistilbite, heulandite, chabazite, mordenite, clinoptilolite, analcite, natrolite, scolecite and erionite (Barrett *et al.* 1986; Vezzalini *et al.* 1994; Conaway *et al.* 2005). Crystallization of apophyllite in mid-Cretaceous time, with associated zeolite overgrowths, suggests elevated temperatures long after thermal perturbations associated with the Ferrar magmatic event (Fleming *et al.* 1999). At least some of the zeolites post-date apophyllite crystallization and, with their presence in lavas high in the sequence, suggest the possibility of a now-eroded thick lava sequence (Elliot 1970) and/or a middle Jurassic–early Cretaceous sedimentary overburden (Elliot and Fleming 2008), either of which could have been more than 1 km thick. Apatite fission-track data from Victoria Land have also been interpreted to suggest a now-eroded Mesozoic sedimentary sequence formerly overlying the Ferrar lavas (Lisker and Läufer 2013).

Thick intervals of hyaloclastite and pillow lava are prominent at the base of the sequence at Carapace Nunatak (south Victoria Land) and at Thomas Rock in the Prince Albert Mountains. At Mount Fazio in the Mesa Range a pillow lava stack about 50 m high is present (Fig. 16) and passes laterally into a large tachylite mound. Thinner intervals of hyaloclastite and pillow lava are also present higher in sections at the Mesa Range (Fig. 17) (see Elliot and Fleming 2008, fig. 14) and in the Marshall Mountains (Storm Peak). Such intervals indicate standing water in lakes or fluvial channels. The *c.* 100 m thickness of hyaloclastite and pillow lava at Thomas Rock (Fig. 18) must have originally extended laterally for more than 1000 m (the length of the outcrop), which implies a major lacustrine setting. Water played a significant role in the formation of the Ferrar extrusive rocks as shown by the phreatomagmatic deposits, hyaloclastite formation and quenching of lavas. The Thomas Rock hyaloclastite rocks in particular suggest eruption into an environment with significant local topography. Features advocated by Deschamps *et al.* (2014) to indicate subaqueous eruptions have not yet been observed.

Interbeds

The lava sequences are broken by interbeds, which are mainly lacustrine but also include tuff and lapilli tuff beds (a number

Fig. 16. Mount Fazio, north Victoria Land. Lavas are numbered: (1) lowest lava; (2) 15 m-thick lava with columnar jointing; (3) 3 m-thick lava consisting of small pāhoehoe toes and lobes, which is overlain by a thin volcaniclastic interbed; (4) *c.* 50 m-thick lava with a 5 m-thick columnar jointed interval at the base, followed by *c.* 45 m of tachylite; the tachylite passes laterally to small pillows within the white dashed lines, and then to larger pillows (above and to the left of the dashed white lines); the large pillows extend over the top of the tachylite interval; the irregular top of the lava is separated from the overlying lava (5) by patchy thin volcaniclastic beds; and (5) lava with crude columnar jointing (section 82-4: Elliot *et al.* 1986*b*).

Fig. 17. Mount Short, north Victoria Land. A 14 m-thick stack of lava pillows (Elliot *et al.* 1986*b*; 6 m above base of section 82-23). Pillows are from 0.5 m to tens of metres across; the upper 2 m consists of pillow breccia. Lava is overlain by 25 cm of bedded pyroclastic debris. Ice axe (80 cm long) in the lower left is for scale (arrow). (Image: D.H. Elliot.)

Fig. 18. At Thomas Rock, Prince Albert Mountains, Kirkpatrick Basalt hyaloclastite and pillow lava overlies Mawson Formation pyroclastic deposits (lower left). The hyaloclastite and pillow lava interval is about 100 m thick, and is overlain by lavas. Inset illustrates the right-hand end of the hyaloclastite and pillow lava outcrop (ice axe in lower left is for scale). (Image: D.H. Elliot.)

containing accretionary lapilli), redeposited tuffs, and palaeosols developed on lava crusts (e.g. Elliot *et al.* 1986*b*, 1991; Elliot and Hammer 1996). The occurrence of tricuspate shards in several weathering profiles in the Grosvenor Mountains (Fig. 15c) has been interpreted as the vertical mixing of silicic ash by vertisol processes (Elliot *et al.* 1991), and continuation at a low level of the silicic magmatism recorded in the underlying Hanson Formation (Elliot *et al.* 2016). Interbeds are common near the base of the lava sequences and characteristically separate the capping SPCT lava from the underlying MFCT lavas. The SPCT lava and the underlying interbed are a significant marker for correlation in the Kirkpatrick Basalt throughout the Transantarctic Mountains, and the lacustrine bed above the 135 m-thick lava at Storm Peak is similarly a widespread marker bed in the Queen Alexandra Range.

The interbeds have yielded a varied flora and fauna. Identifiable plant remains, which include ferns, cycads, conifers and plant microfossils, have been found at Carapace Nunatak and the Mesa Range region (Plumstead 1962; Townrow 1967; Ribecai 2007; Bomfleur *et al.* 2011; Heiger *et al.* 2015). Vertebrates are restricted to Pholidophoroid fish remains recovered from the interbeds at Storm Peak and elsewhere in the Marshall Mountains (Schaeffer 1972). Lake beds, from the Grosvenor Mountains to the Mesa Range, have yielded abundant invertebrates which are principally conchostracans but include ostracods, notostracans, syncarids, molluscs, beetle elytra and insects (Carpenter 1969; Ball *et al.* 1979; Tasch 1987; Shen 1994; Stigall *et al.* 2008; Bomfleur *et al.* 2011).

Vent locations and eruption rates

The locations of phreatocauldrons are clear and pyroclastic rocks, where exposed, always occur beneath the lavas, suggesting that numerous eruptive centres existed in the early stages of extrusive activity. Locally pyroclastic rocks (hyaloclastite with pillow basalt, tuff and lapilli tuff) may be intercalated higher in the lava successions. Co-location of vents for effusion of the lavas might be expected, but possible connections between a sill or a dyke and the lava sequences have not been observed. Northern Coombs Hills shows apparent 'shoaling' of a sill toward the surface, but an overlying flood lava sequence is not present at that site.

The Dry Valleys region (see the discussion of the sills in Chapter 2.1b, Elliot and Fleming 2021) was clearly the centre for emplacement of the Basement Sill if not the stratigraphically higher sills as well. Although three principal centres for magma intrusion and eruption (central Transantarctic Mountains, south Victoria Land and north Victoria Land) have been suggested (Elliot and Fleming 2008), this distribution may simply reflect extant intrusive and extrusive rock outcrop. The scattered outcrops and the nature of the volcaniclastic lithofacies in the central Transantarctic Mountains and Victoria Land point to multiple local centres for the initial stages of Ferrar extrusive activity. Intervals of volcaniclastic rocks within the basalt sequences also suggest continuing local sources. The phreatocauldrons demonstrate major eruptive centres lying within the present outcrop belt from at least north Victoria Land to the central Transantarctic Mountains, and suggest that vents for the lavas were probably also located within the outcrop belt. If correct, and given that the present narrow outcrop belt is only a fraction of its original extent, then magmas must have migrated laterally away from that linear system of vents.

Ross *et al.* (2008*a*) noted basalt plugs at Coombs Hills and estimated potential magma eruption rates. Their calculations for a conduit with a 10 m radius suggested that the rate of magma effusion would be sufficient for construction of a typical flood basalt field. No feeder dykes or dyke swarms, such as identified for the Columbia River Basalt (Tolan *et al.* 1989; Self *et al.* 1997), have been found. However, shallow dyke patterns are complex at some sites (Airoldi *et al.* 2011, 2012, 2016; Muirhead *et al.* 2012, 2014) and have been interpreted as possible feeders. Strombolian deposits interbedded with the lavas, such as observed in West Greenland and interpreted as possible vents (Pedersen *et al.* 2017), have not been observed.

Palaeoenvironments

Several lines of evidence point to groundwater and lakes exerting significant control on both the near-surface eruptive processes for the pyroclastic rocks and the flood lavas. The phreatomagmatic deposits already noted indicate abundant groundwater in the underlying Triassic–Jurassic strata, and probably played a crucial role for the generation of the phreatocauldrons in an overall rift setting. Hyaloclastite and pillow

Fig. 19. Mount Fazio, north Victoria Land. Tree (arrow) rooted in an interbed and engulfed by lava (Elliot *et al.* 1986*b*; unit 4, section 82.4). In Figure 16 this locality is out of sight just to the right of the tachylite bluff. Person is for scale at the base of the tree. (Image: D.H. Elliot.)

Fig. 20. Haban Spur, north Victoria Land. The rooted tree-stump diameter is approximately 80 cm (Elliot *et al.* 1986*b*; base of unit 23 at 524 m in section 82-3). Hammer is for scale. (Image: D.H. Elliot.)

lava intervals formed where lava entered standing water, which occurred regionally at the base of the lava stack in south Victoria Land (Carapace Nunatak and Thomas Rock), and later in the eruptive cycle at Storm Peak, Brimstone Peak and the Mesa Range. Lake beds are direct evidence of more protracted intervals of water ponding on the lava field surfaces. Conchostracans and the other invertebrates suggest relatively shallow water, and this shallowness is supported in the Marshall Mountains by a lacustrine bed that passes laterally into the weathered upper part of a lava. Plant debris associated with the weathering horizons, together with fossil logs (Storm Peak, Haban Spur and other sites) and tree stumps (Mount Fazio, Fig. 19; Haban Spur, Fig. 20), indicate flourishing vegetation between eruptive events. The flora and fauna recovered have been interpreted to suggest a temperate but strongly seasonal climate (Elliot and Hammer 1996; Garland *et al.* 2007).

Volcanic gas input to the Jurassic atmosphere from the phreatomagmatism at Allan and Coombs hills was unlikely to have been significant (Ross *et al.* 2008*a*), but no estimates have been made for Ferrar magmatism as a whole. Thordarson and Self (2003) have made such calculations for the historical Laki eruption in Iceland and Self *et al.* (2014) for the Roza eruption of the Columbia River Basalt province, and show the possible magnitude of such atmospheric perturbations. McElwain *et al.* (2005) speculated on the possibility that devolatilization of coals in the Gondwana sequence by Ferrar and Karoo magmatism might have been responsible for the Toarcian Oceanic Anoxic Event. High precision dating of the Ferrar province (Burgess *et al.* 2015) strengthens the possibility that Ferrar (and Karoo) magmatism is indeed related to, if not responsible for, the Toarcian event. The large sill complexes of the Ferrar LIP (and Karoo LIP) may have been important for that event, in an analogous way to the Siberian Traps and end-Permian extinction (Burgess *et al.* 2017), although coal beds in Antarctica were not extensive.

Future studies

Although the Ferrar LIP in Antarctica has been studied for more than 50 years, access is neither easy nor simple and much remains to be discovered. The following are suggested lines of future research that may prove particularly fruitful.

1. Can flow directions of lavas and sills be determined using AMS (as has been applied in various studies of dykes and sills) in order to establish the location of eruption centres, and whether linear or point sources?
2. What are the physical conditions and cooling patterns for thick lavas with tachylitic intervals?
3. Can lava correlations be established in the central Transantarctic Mountains and the Mesa Range that would enable the evaluation of the three-dimensional relationships of lavas?
4. Would detailed investigation of the lateral relationships in both thick lavas and sets of thin lavas confirm the applicability, to the Ferrar Province, of the general model of flood basalt field development (Self *et al.* 1997)?
5. There are many aspects of the primary volcaniclastic rocks (and deposits of material reworked from them) that merit further study. For example, what do the bedded deposits say about the distribution and nature of the volcanic province prior to eruption of the flood lavas?
6. The relationships of pyroclastic rocks at Coombs and Allan hills with country rock, sills, dykes and lavas are complex. They provide a window of fortuitous exposure that could allow assessment of near-surface (both closely below, and initial eruptive) processes taking place elsewhere in the Ferrar and other flood basalt provinces worldwide.
7. Different studies have come to strongly differing conclusions about the significance of intrusion geometry characterizing the Ferrar LIP. Do the apparently overwhelmingly predominant sills suggest a near-neutral stress regime, or is the dyke-swarm signature of an East Africa-like rift concealed in Antarctica? Were mid-crustal dykes formed which then fed upper-crustal sills and, if so, why?

Summary

The extrusive component of the Ferrar Large Igneous Province comprises an initial phreatomagmatic phase of activity followed by massive lava effusion. The early stages, involving large-scale magma–water interaction, created very large and unusual diatreme complexes (phreatocauldrons with vertical exposure of more than 300 m). Tuff rings and tuff cones that were built early were largely removed by later volcanic activity or by subsequent erosion. Locally, with diminished involvement of water, magmatic activity switched to Vulcanian and/or Strombolian. Only scattered remnants exist of any constructional volcanic topography. With exhaustion of subsurface water and/or increased magma supply rates, activity changed to the quiet effusion of flood lavas, with early but local deposition of thick hyaloclastites and pillow lavas in pre-existing topography. Lavas accumulated during a relatively short time interval to form sequences as much as 750 m thick and constructed from as many as 41 lavas, although commonly much fewer. Lavas in excess of 100 m in thickness are present in most measured sections. The internal features of the individual lavas are typical of flood lavas. Lacustrine interbeds and palaeosols mark more substantial breaks in eruptive activity.

Acknowledgements Reviews by Richard Hanson and Pierre-Simon Ross, and suggestions by the Editor, John Smellie, have greatly improved the manuscript. This is Byrd Polar and Climate Research Center Contribution No. 1580.

Author contributions DHE: writing – original draft (equal); **JDLW**: writing – original draft (equal), writing – review & editing (equal); **THF**: writing – original draft (equal), writing – review & editing (equal).

Funding DHE and THF acknowledge significant support over many years from the Office of Polar Programs, National Science Foundation, Washington, DC. JDLW was supported by Antarctica New Zealand for fieldwork in south Victoria Land and by University of Otago research grants.

Data availability All data presented in this review have been published previously in the cited literature.

References

Airoldi, G., Muirhead, J.D., White, J.D.L. and Rowland, J.V. 2011. Emplacement of magma at shallow depth and development of local vents: insights from field relationships at Allan Hills (South Victoria Land, East Antarctica). *Antarctic Science*, **23**, 281–296, https://doi.org/10.1017/S0954102011000095

Airoldi, G., Muirhead, J.D., Zanella, E. and White, J.D.L. 2012. Emplacement process of Ferrar Dolerite sheets at Allan Hills (South Victoria Land, Antarctica) inferred from magnetic fabric. *Geophysical Journal International*, **188**, 1046–1060, https://doi.org/10.1111/j.1365-246X.2011.05334.x

Airoldi, G.M., Muirhead, J.D., Long, S.M., Zanella, E. and White, J.D.L. 2016. Flow dynamics in mid-Jurassic dikes and sills of the Ferrar large igneous province and implications for long-distance magma transport. *Tectonophysics*, **683**, 182–199, https://doi.org/10.1016/j.tecto.2016.06.029

Aughenbaugh, N.B. 1961. Preliminary report on the geology of the Dufek Massif. *International Geophysical Year World Data Center A, Glaciology Report*, **4**, 155–193.

Ball, H.W., Borns, H.W., Hall, B.A., Brooks, H.K., Carpenter, F.M. and Delavoryas, T. 1979. Biota, age, and significance of lake deposits, Carapace Nunatak, Victoria Land, Antarctica. *In*: Laskar, B. and Raja Rao, C.S. (eds) *Fourth International Gondwana Symposium: Papers, Volume 1*. Hindustan Publishing Corporation, Delhi, 166–175.

Ballance, P. and Watters, W.A. 1971. The Mawson Diamictite and the Carapace Sandstone formations of the Ferrar Group at Allan Hills and Carapace Nunatak, Victoria Land, Antarctica. *New Zealand Journal of Geology and Geophysics*, **14**, 512–527, https://doi.org/10.1080/00288306.1971.10421945

Barrett, P.J., Elliot, D.H. and Lindsay, J.F. 1986. The Beacon Supergroup (Devonian–Triassic) and Ferrar Group (Jurassic) in the Beardmore Glacier area, Antarctica. *Antarctic Research Series American Geophysical Union*, **36**, 339–428.

Bélanger, C. and Ross, P.-S. 2018. Origin of non-bedded pyroclastic rocks in the Cathedral Cliff diatreme, Navajo volcanic field, New Mexico. *Bulletin of Volcanology*, **80**, https://doi.org/10.1007/s00445-018-1234-0

Benson, W.N. 1916. Report on the Petrology of the Dolerites Collected by the British Antarctic Expedition, 1907–09. *British Antarctic Expedition, 1907–09, Reports of Scientific Investigations, Geology*, **2**, part 9, 153–160.

Bomfleur, B., Schneider, J.W., Schöner, R., Viereck-Götte, L. and Kerp, H. 2011. Fossil sites in the continental Victoria and Ferrar Groups (Triassic–Jurassic) of north Victoria Land. *Polarforschung*, **80**, 88–99.

Bradshaw, M.A. 1987. Additional field interpretation of the Jurassic sequence at Carapace Nunatak and Coombs Hills, south Victoria Land, Antarctica. *New Zealand Journal of Geology and Geophysics*, **30**, 37–49, https://doi.org/10.1080/00288306.1987.10422192

Bromfield, K., Burrett, C.F., Leslie, R.A. and Meffre, S. 2007. Jurassic volcaniclastic–basaltic andesite–dolerite sequence in Tasmania: new age constraints for fossil plants from Lune River. *Australian Journal of Earth Sciences*, **54**, 965–974, https://doi.org/10.1080/08120090701488297

Brotzu, P., Capaldi, G., Civetta, L., Melluso, L. and Orsi, G. 1988. Jurassic Ferrar dolerites and Kirkpatrick basalts in northern Victoria Land (Antarctica): stratigraphy, geochronology and petrology. *Memorie della Societa Geologica Italiana*, **43**, 97–116.

Browne, W.R. 1923. The dolerites of King George Land and Adelie Land. *Australasian Antarctic Expedition, 1911–14, Scientific Reports Series A, Geology*, **3**, 245–258.

Burgess, S.D., Bowring, S.A., Fleming, T.H. and Elliot, D.H. 2015. High precision geochronology links the Ferrar Large Igneous Province with early Jurassic ocean anoxia and biotic crisis. *Earth and Planetary Science Letters*, **415**, 90–99, https://doi.org/10.1016/j.epsl.2015.01.037

Burgess, S.D., Muirhead, J.D. and Bowring, S.A. 2017. Initial pulse of Siberian Traps sills as the trigger of the end-Permian mass extinction. *Nature Communications*, **8**, 164, https://doi.org/10.1038/s41467-017-00083-9

Carpenter, F.M. 1969. Fossil insects from Antarctica. *Psyche*, **76**, 418–425, https://doi.org/10.1155/1969/17070

Campbell-Smith, W. 1924. The plutonic and hypabyssal rocks of South Victoria Land. *British Antarctic ('Terra Nova') Expedition, 1910–13, Natural History Reports, Geology*, **1**, 167–227.

Campbell-Smith, W. 1964. Volcanic rocks of Cape Adare and erratics from the Terra Nova Bay region, etc *British Antarctic ('Terra Nova') Expedition, 1910, Natural History Report, Geology*, **2**, 151–206.

Conaway, C.H., Fleming, T.H. and Elliot, D.H. 2005. Preliminary investigation of the secondary minerals in the Kirkpatrick Basalt, Prince Albert Mountains. *Antarctic Journal of the United States*, **33**, 344–347.

Deschamps, A., Grigné, C., Le Saout, M., Soule, S.A., Allemand, P., Van Vliet Lanoe, B. and Floc'h, F. 2014. Morphology and dynamics of inflated subaqueous basaltic lava flows. *Geochemistry, Geophysics, Geosystems*, **15**, 2128–2150, https://doi.org/10.1002/2014GC005274

Demarchi, G., Antonini, P., Piccirillo, E.M., Orsi, G., Civetta, L. and D'Antonio, M. 2001. Significance of orthopyroxene and major element constraints on the petrogenesis of Ferrar tholeiites from southern Prince Albert Mountains, Victoria land, Antarctica. *Contributions to Mineralogy and Petrology*, **142**, 127–146, https://doi.org/10.1007/s004100100287

Elliot, D.H. 1970. Jurassic tholeiites of the central Transantarctic Mountains, Antarctica. *In*: Gilmour, E.H. and Stradling, D. (eds) *Proceedings of the Second Columbia River Basalt Symposium, Cheney, Washington, March 1969*. Eastern Washington State College Press, Cheney, WA, 301–325.

Elliot, D.H. 2002. Paleovolcanological setting of the Mawson Formation: evidence from the Prince Albert Mountains, Victoria Land. *Royal Society of New Zealand Bulletin*, **35**, 185–192.

Elliot, D.H. and Fleming, T.H. 2008. Physical volcanology and geological relationships of the Ferrar Large Igneous Province, Antarctica. *Journal of Volcanology and Geothermal Research*, **172**, 20–37, https://doi.org/10.1016/j.jvolgeores.2006.02.016

Elliot, D.H. and Fleming, T.H. 2017. The Ferrar large Igneous Province: field and geochemical constraints on supra-crustal (high-level) emplacement of the magmatic system. *Geological Society, London, Special Publications*, **463**, 41–58, https://doi.org/10.1144/SP463.1

Elliot, D.H. and Fleming, T.H. 2021. Ferrar Large Igneous Province: petrology. *Geological Society, London, Memoirs*, **55**, https://doi.org/10.1144/M55-2018-39

Elliot, D.H. and Grimes, C.G. 2011. Triassic and Jurassic strata at Coombs Hills, south Victoria Land: stratigraphy, petrology and cross-cutting breccia pipes. *Antarctic Science*, **23**, 268–280, https://doi.org/10.1017/S0954102010000994

Elliot, D.H. and Hammer, W.R. 1996. Paleoclimatic indicators in Jurassic volcanic strata, Transantarctic Mountains, Antarctica. *In*: Mitra, N.D. (ed.) *Gondwana Nine*. Oxford and IBH Publishing, New Delhi, 895–907.

Elliot, D.H. and Hanson, R.E. 2001. Origin of widespread, exceptionally thick basaltic phreatomagmatic tuff breccia in the Middle Jurassic Prebble and Mawson formations, Antarctica. *Journal of Volcanology and Geothermal Research*, **111**, 183–201, https://doi.org/10.1016/S0377-0273(01)00226-8

Elliot, D.H., Haban, M.A. and Siders, M.A. 1986a. The Exposure Hill Formation, Mesa Range. *American Geophysical Union Antarctic Research Series*, **46**, 267–278.

Elliot, D.H., Siders, M.A. and Haban, M.A. 1986b. Jurassic tholeiites in the region of the upper Rennick Glacier, North Victoria Land. *American Geophysical Union Antarctic Research Series*, **46**, 249–265.

Elliot, D.H., Bigham, J. and Jones, F.S. 1991. Interbeds and weathering profiles inthe Jurassic basalt sequence, Beardmore Glacier region, Antarctica. *In*: Ulbrich, H. and Rocha Campos, A.C. (eds) *Gondwana Seven Proceedings. Papers presented at the Seventh International Gondwana Symposium, Sao Paulo, 1988*. Instituto de Geosciencias, Universidade de São Paulo, São Paulo, Brazil, 653–667.

Elliot, D.H., Fortner, E.H. and Grimes, C.B. 2006. Mawson breccias intrude Beacon strata at Allan Hills, south Victoria Land: regional implications. *In*: Fütterer, D.K., Kleinschmidt, G., Miller, H. and Tessensohn, F. (eds) *Antarctica: Contributions to Global Earth Sciences*. Springer, Berlin, 291–298.

Elliot, D.H., Larsen, D., Fanning, C.M., Fleming, T.H. and Vervoort, J.D. 2016. The Lower Jurassic Hanson Formation of the Transantarctic Mountains: implications for the Antarctic sector of the Gondwana Plate margin. *Geological Magazine*, **154**, 777–803, https://doi.org/10.1017/S0016756816000388

Encarnación, J., Fleming, T.H., Elliot, D.H. and Eales, J.V. 1996. Synchronous emplacement of Ferrar and Karoo dolerites and the early breakup of Gondwana. *Geology*, **24**, 535–538, https://doi.org/10.1130/0091-7613(1996)024<0535:SEOFAK>2.3.CO;2

Ferrar, H.T. 1907. Report on the field geology of the region explored during the 'Discovery' Antarctic Expedition, 1901–1904. *National Antarctic Expedition 1901–1904, Natural History, Geology (Field Geology, Petrography)*, **1**, 1–100.

Fleming, T.H., Elliot, D.H., Jones, L.M., Bowman, J.R. and Siders, M.A. 1992. Chemical and isotopic variations in an iron-rich lava flow from North Victoria Land, Antarctica: Implications for low-temperature alteration and the petrogenesis of Ferrar magmas. *Contributions to Mineralogy and Petrology*, **111**, 440–457, https://doi.org/10.1007/BF00320900

Fleming, T.H., Foland, K.A. and Elliot, D.H. 1995. Isotopic and chemical constraints on the crustal evolution and source signature of Ferrar magmas, North Victoria Land, Antarctica. *Contributions to Mineralogy and Petrology*, **121**, 217–236, https://doi.org/10.1007/BF02688238

Fleming, T.H., Foland, K.A. and Elliot, D.H. 1999. Apophyllite $^{40}Ar/^{39}Ar$ and Rb–Sr geochronology: potential utility and application to the timing of secondary mineralization of the Kirkpatrick Basalt, Antarctica. *Journal of Geophysical Research*, **104**, 20 081–20 095, https://doi.org/10.1029/1999JB900138

Ford, A.B. 1976. *Stratigraphy of the Layered Gabbroic Dufek Intrusion, Antarctica*. United States Geological Survey Bulletin, **1405-D**.

Gair, H.S. 1966. The geology from the upper Rennick Glacier to the coast, northern Victoria Land, Antarctica. *New Zealand Journal of Geology and Geophysics*, **10**, 309–344, https://doi.org/10.1080/00288306.1967.10426742

Garland, M.J., Bannister, J.M., Lee, D.E. and White, J.D.L. 2007. A coniferous tree stump of Middle Jurassic age from the Ferrar Basalt, Coombs Hills, southern Victoria Land, Antarctica. *New Zealand Journal of Geology and Geophysics*, **50**, 263–269 https://doi.org/10.1080/00288300709509836

Gould, L.M. 1931. Some geographical results of the Byrd Antarctic Expedition. *Geographical Review*, **21**, 177–200, https://doi.org/10.2307/209272

Gould, L.M. 1935. Structure of the Queen Maud Mountains, Antarctica. *Geological Society of America Bulletin*, **46**, 973–984, https://doi.org/10.1130/GSAB-46-973

Grapes, R.H., Reid, D.L. and McPherson, J.G. 1974. Shallow dolerite intrusion and phreatic eruption in the Allan Hills region, Antarctica. *New Zealand Journal of Geology and Geophysics*, **17**, 563–577, https://doi.org/10.1080/00288306.1973.10421581

Grindley, G.W. 1963. The geology of the Queen Alexandra Range, Beardmore Glacier, Ross Dependency, Antarctica; with notes on the correlation of Gondwana sequences. *New Zealand Journal of Geology and Geophysics*, **6**, 307–347, https://doi.org/10.1080/00288306.1963.10422067

Gunn, B.M. and Warren, G. 1962. *The Geology of Victoria Land between the Mawson and Mulock Glaciers, Ross Dependency, Antarctica*. New Zealand Geological Survey Bulletin, **71**.

Hanemann, R. and Viereck-Götte, L. 2004. Geochemistry of Jurassic Ferrar lava flows, sills and dikes sampled during the joint German–Italian Antarctic Expedition 1999–2000. *Terra Antartica*, **11**, 39–54.

Hanson, R.E. and Elliot, D.H. 1996. Rift-related Jurassic basaltic phreatomagmatic volcanism in the central Transantarctic Mountains: precursory stage to flood-basalt effusion. *Bulletin of Volcanology*, **58**, 327–347, https://doi.org/10.1007/s004450050143

Harrington, H.J. 1958. Nomenclature of rock units in the Ross Sea Region, Antarctica. *Nature*, **182**, 290–291, https://doi.org/10.1038/182290a0

Heiger, T.J., Serbet, R., Harper, C.J., Taylor, T.N., Taylor, E.L. and Gulbranson, E. 2015. Cheirolepidiaceous diversity: An anatomically preserved pollen cone from the Lower Jurassic of southern Victoria Land, Antarctica. *Review of Palaeobotany and Palynology*, **220**, 78–87, https://doi.org/10.1016/j.revpalbo.2015.05.003

Hergt, J.M., Peate, D.W. and Hawkesworth, C.J. 1991. The petrogenesis of Mesozoic Gondwana low-Ti flood basalts. *Earth and Planetary Science Letters*, **105**, 134–148, https://doi.org/10.1016/0012-821X(91)90126-3

Hornig, I. 1993. High-Ti and low-Ti tholeiites in the Jurassic Ferrar Group, Antarctica. *Geologisches Jahrbuch*, **E47**, 335–369.

Houghton, B. and Wilson, C.J.N. 1989. A vesicularity index for pyroclastic deposits. *Bulletin of Volcanology*, **51**, 451–462, https://doi.org/10.1007/BF01078811

Ivanov, A.V., Meffre, S., Thompson, J., Corfu, F., Kamenetsky, V.S., Kamenetsky, M.B. and Demonterova, W.I. 2017. Timing and genesis of the Karoo–Ferrar large igneous province: New high precision U–Pb data for Tasmania confirm short duration of the major magmatic pulse. *Chemical Geology*, **346**, 32–43, https://doi.org/10.1016/j.chemgeo.2016.10.008

Jerram, D.A. 2002. Volcanology and facies architecture of flood basalts. *Geological Society of America Special Papers*, **362**, 119–132, https://doi.org/10.1130/0-8137-2362-0.119

Korsch, R. 1984. The structure of Shapeless Mountain, Antarctica, and its relation to Jurassic igneous activity. *New Zealand Journal of Geology and Geophysics*, **27**, 487–504, https://doi.org/10.1080/00288306.1984.10422268

Kyle, P.R. 1998. Ferrar Dolerite Clasts from CRP-1 Drillcore. *Terra Antarctica*, **5**, 611–612.

Kyle, P.R., Elliot, D.H. and Sutter, J.F. 1981. Jurassic Ferrar Supergroup tholeiites from the Transantarctic Mountains, Antarctica, and their relationship to the initial fragmentation of Gondwana. *In*: Creswell, M.M. and Vella, P. (eds) *Gondwana Five*. Balkema, Rotterdam, The Netherlands, 283–287.

Kyle, P.R., Pankhurst, R.J. and Bowman, J.R. 1983. Isotopic and chemical variations in Kirkpatrick Basalt Group rocks from southern Victoria Land. *In*: Oliver, R.L., James, P.R. and Jago, J.B. (eds) *Antarctic Earth Science*. Australian Academy of Science, Canberra, 234–237.

Lisker, F. and Läufer, A.L. 2013. The Mesozoic Victoria Basin: vanished link between Antarctica and Australia. *Geology*, **41**, 1043–1046, https://doi.org/10.1130/G33409.1

Lockett, G.M. and White, J.D.L. 2008. Coal-fragment rank and contact relationships of debris avalanche and primary pyroclastic deposits in the Mawson Formation, Ferrar LIP, Allan Hills, Antarctica. *Journal of Volcanology and Geothermal Research*, **172**, 61–74, https://doi.org/10.1016/j.jvolgeores.2006.02.017

Mawson, D. 1916. Petrology of rock collections from the mainland of South Victoria Land. *British Antarctic Expedition, 1907–09, Reports of Scientific Investigations, Geology*, **2**, part 13, 201–234.

McClintock, M. and White, J.D.L. 2006. Large phreatomagmatic vent complex at Coombs hills, Antarctica: wet, explosive initiation of flood basalt volcanism in the Ferrar–Karoo LIP. *Bulletin of Volcanology*, **68**, 215–239, https://doi.org/10.1007/s00445-005-0001-1

McClintock, M., White, J.D.L., Houghton, B.F. and Skilling, I.P. 2008. Physical volcanology of a large crater-complex formed during the initial stages of Karoo flood basalt volcanism, Sterkspruit, Eastern Cape, South Africa. *Journal of Volcanology and Geothermal Research*, **172**, 93–111, https://doi.org/10.1016/j.jvolgeores.2005.11.012

McClintock, M., Ross, P.-S. and White, J.D.L. 2009. The importance of the transport system in shaping the growth and form of kimberlite volcanoes. *Lithos*, **112**, 465–472, https://doi.org/10.1016/j.lithos.2009.04.014

McElwain, J.C., Wade-Murphy, J. and Hesselbo, S.P. 2005. Changes in carbon dioxide during an oceanic anoxic event linked to intrusion into Gondwana coals. *Nature*, **435**, 479–482, https://doi.org/10.1038/nature03618

Mensing, T.M., Faure, G., Jones, L.M. and Hoefs, J. 1991. Stratigraphic correlation and magma evolution of the Kirkpatrick Basalt in the Mesa Range, northern Victoria Land, Antarctica. *In*: Ulbrich, H. and Rocha Campos, A.C. (eds) *Gondwana Seven Proceedings. Papers presented at the Seventh International Gondwana Symposium, Sao Paulo, 1988*. Instituto de Geosciencias, Universidade de São Paulo, São Paulo, Brazil, 653–667.

Milnes, A.R., Cooper, B. and Cooper, J.A. 1982. The Jurassic Wisanger basalt of Kangaroo Island, South Australia. *Royal Society of South Australia Transactions*, **106**, 1–13.

Minor, D. and Mukasa, S. 1997. Zircon U–Pb and hornblende $^{40}Ar-^{39}Ar$ ages for the Dufek layered mafic intrusion, Antarctica: Implications for the age of the Ferrar large igneous province. *Geochimica et Cosmochimica Acta*, **61**, 2497–2504, https://doi.org/10.1016/S0016-7037(97)00098-7

Mortimer, N., Parkinson, D., Raine, J.I., Adams, C.J., Graham, I.J., Oliver, P.J. and Palmer, K. 1995. Ferrar magmatic province rocks discovered in New Zealand: Implications for Mesozoic Gondwana geology. *Geology*, **23**, 185–188, https://doi.org/10.1130/0091-7613(1995)023<0185:FMPRDI>2.3.CO;2

Muirhead, J.D., Airoldi, J., Rowland, J.V. and White, J.D.L. 2012. Interconnected sills and inclined sheet intrusions control shallow magma transport in the Ferrar large igneous province, Antarctica. *Bulletin of the Geological Society of America*, **124**, 162–180, https://doi.org/10.1130/B30455.1

Muirhead, J.D., Airoldi, G., White, J.D.L. and Rowland, J.V. 2014. Cracking the lid: Sill-fed dikes are the likely feeders of flood basalt eruptions. *Earth and Planetary Science Letters*, **406**, 187–197, https://doi.org/10.1016/j.epsl.2014.08.036

Pedersen, A.K., Larsen, L.M. and Pedersen, G.K. 2017. Lithostratigraphy, Geology and Geochemistry of the Volcanic Rocks of the Vaigat Formation on Disko and Nuussuaq, Paleocene of West Greenland. Geological Survey of Denmark and Greenland, Bulletin, **39**.

Petri, A., Salvini, F. and Storti, F. 1997. Geology of the Ferrar Supergroup in the Mesa Range, northern Victoria Land, Antarctica: a photogeological study. *In*: Ricci, C.A. (ed.) *The Antarctic Region: Geological Evolution and Processes*. Terra Antartica, Siena, Italy, 305–312.

Plumstead, E.P. 1962. Fossil floras of Antarctica, with an appendix on Antarctic fossil wood by R. Krause. *Trans-Antarctic Expedition 1955-1958. Scientific Report, Geology*, **9**.

Prior, G.T. 1907. Report on the rock specimens collected during the 'Discovery' Antarctic Expedition, 1901–04. *National Antarctic Expedition, 1901-04, Natural History, Geology*, **1**, 101–140.

Reubi, O., Ross, P.-S. and White, J.D.L. 2005. Debris avalanche deposits associated with large igneous province volcanism: An example from the Mawson Formation, central Allan Hills, Antarctica. *Geological Society of America Bulletin*, **117**, 1615–1628, https://doi.org/10.1130/B25766.1

Ribecai, C. 2007. Early Jurassic miospores from Ferrar Group of Carapace Nunatak, South Victoria Land, Antarctica. *Reviews of Palaeobotany and Palynology*, **144**, 3–12, https://doi.org/10.1016/j.revpalbo.2005.09.005

Roland, N.W. and Wörner, G. 1996. Kirkpatrick flows and associated pyroclastics: new occurrences, definition, and aspects of a Jurassic Transantarctic Rift. *Geologisches Jahrbuch, Reihe B*, **89**, 97–121.

Ross, P.-S. and White, J.D.L. 2005a. Mafic, large-volume, pyroclastic density current deposits from phreatomagmatic eruptions in the Ferrar large igneous province. *Journal of Geology*, **113**, 627–649, https://doi.org/10.1086/449324

Ross, P.-S. and White, J.D.L. 2005b. Unusually large clastic dykes formed by elutriation of a poorly sorted, coarser-grained source. *Journal of the Geological Society, London*, **162**, 579–582, https://doi.org/10.1144/0016-764904-127

Ross, P.-S. and White, J.D.L. 2006. Debris jets in continental phreatomagmatic volcanoes: A field study of their subterranean deposits in the Coombs Hills vent complex, Antarctica. *Journal of Volcvanology and Geothermal Research*, **149**, 62–84, https://doi.org/10.1016/j.jvolgeores.2005.06.007

Ross, P.-S., Ukstins Peate, I., McClintock, M.K., Xu, Y.G., Skilling, I.P., White, J.D.L. and Houghton, B.F. 2005. Mafic volcaniclastic deposits in flood basalt provinces: A review. *Journal of Volcanology and Geothermal Research*, **145**, 281–314, https://doi.org/10.1016/j.jvolgeores.2005.02.003

Ross, P.-S., White, J.D.L. and McClintock, M.K. 2008a. Geological evolution of the Coombs–Allan Hills area, Ferrar large igneous province, Antarctica: debris avalanche, mafic pyroclastic density currents, phreatocauldrons. *Journal of Volcanology and Geothermal Research*, **172**, 38–60, https://doi.org/10.1016/j.jvolgeores.2005.11.011

Ross, P.-S., White, J., Zimanowski, B. and Büttner, R. 2008b. Rapid injection of particles and gas into non-fluidized granular material, and some volcanological implications. *Bulletin of Volcanology*, **70**, 1151–1168, https://doi.org/10.1007/s00445-008-0230-1

Ross, P.-S., White, J.D.L., Zimanowski, B. and Büttner, R., 2008c. Multiphase flow above explosion sites in debris-filled volcanic vents: Insights from analogue experiments. *Journal of Volcanology and Geothermal Research*, **178**, 104–112, https://doi.org/10.1016/j.jvolgeores.2008.01.013

Ross, P.-S., White, J.D.L., Valentine, G.A., Taddeucci, J., Sonder, I. and Andrews, R.G. 2013. Experimental birth of a maar–diatreme volcano. *Journal of Volcanology and Geothermal Research*, **260**, 1–12, https://doi.org/10.1016/j.jvolgeores.2013.05.005

Schaeffer, B. 1972. *A Jurassic Fish from Antarctica*. American Museum Novitates, **2495**.

Schöner, R., Viereck-Götte, L., Schneider, J. and Bomfleur, B. 2007. Triassic–Jurassic sediments and multiple volcanic events in North Victoria Land, Antarctica: A revised stratigraphic model. *United States Geological Survey Open-File Report*, **2007-1047**, Short Research Paper 102, https://doi.org/10.3133/ofr20071047srp102

Schumacher, R. and Schmincke, H.-U. 1991. Internal structure and occurrence of accretionary lapilli – a case study at Laacher See volcano. *Bulletin of Volcanology*, **53**, 612–634, https://doi.org/10.1007/BF00493689

Self, S., Thordarson, T. and Keszthelyi, L. 1997. Emplacement of continental flood basalt lava flows. *American Geophysical Union Geophysical Monograph Series*, **100**, 381–410.

Self, S., Schmidt, A. and Mather, T.A. 2014. Emplacement characteristics, time scales, and volcanic gas release rates of continental flood basalt eruptions on Earth. *Geological Society of America Special Papers*, **505**, 319–337, https://doi.org/10.1130/2014.2505(16)

Shen, Y. 1994. Jurassic conchostracans from Carapace Nunatak, southern Victoria Land, Antarctica. *Antarctic Science*, **6**, 105–113, https://doi.org/10.1017/S0954102094000131

Skinner, D.N.B., Tessensohn, F. and Vetter, U. 1981. Lavas in the Ferrar group of Litell Rocks, north Victoria Land. *Geologisches Jahrbuch, Reihe B*, **41**, 251–259.

Sohn, Y.K. and Park, K.H. 2005. Composite tuff ring/cone complexes in Jeju Island, Korea: possible consequences of substrate collapse and vent migration. *Journal of Volcanology and Geothermal Research*, **141**, 157–175, https://doi.org/10.1016/j.jvolgeores.2004.10.003

Stephenson, P.J. 1966. Geology 1. Theron Mountains, Shackleton Range, and Whichaway Nunataks *Trans-Antarctic Expedition, 1955–1958. Scientific Reports*, **8**.

Stigall, A.L., Babcock, L.E., Briggs, D.E.G. and Leslie, S.A. 2008. Taphonomy of lacustrine interbeds in the Kirkpatrick Basalt (Jurassic), Antarctica. *PALAIOS*, **23**, 344–355, https://doi.org/10.2110/palo.2007.p07-029r

Tasch, P. 1987. *Fossil Conchostraca of the Southern Hemisphere and Continental Drift: Paleontology, Biostratigraphy, and Dispersal*. Geological Society of America, Memoirs, **186**.

Thordarson, T. and Self, S. 2003. Atmospheric and environmental effects of the 1783–1784 Laki eruption: A review and reassessment. *Journal of Geophysical Research: Atmospheres*, **108**, 4011, https://doi.org/10.1029/2001JD002042

Tolan, T.L., Reidel, S.P., Beeson, M.H., Anderson, J.L., Fecht, K.R. and Swanson, D.A. 1989. Revisions to the estimates of the areal extent and volume of the Columbia River Basalt Group. *Geological Society of America Special Papers*, **239**, 1–20.

Townrow, J.A. 1967. Fossil plants from Allan and Carapce Nunataks, and from the upper Mill and Shackleton Glaciers, Antarctica. *New Zealand Journal of Geology and Geophysics*, **10**, 456–473, https://doi.org/10.1080/00288306.1967.10426750

Valentine, G.A., Graettinger, A.H., Macorps, É., Ross, P.-S., White, J.D.L., Döhring, E. and Sonder, I., 2015. Experiments with vertically and laterally migrating subsurface explosions with applications to the geology of phreatomagmatic and hydrothermal explosion craters and diatremes. *Bulletin of Volcanology*, **77**, 1–17, https://doi.org/10.1007/s00445-015-0901-7

Vezzalini, G., Quartieri, S., Rossi, A. and Alberti, A. 1994. Occurrence of zeolites from Northern Victoria Land (Antarctica). *Terra Antartica*, **1**, 96–99.

Viereck-Götte, L., Schöner, R., Bomfleur, B. and Schneider, J. 2007. Multiple shallow level sill intrusions coupled with hydromagmatic explosive eruptions marked the initial phase of Ferrar large igneous province magmatism in northern Victoria Land. *United States Geological Survey Open-File Report*, **2007-1047**, Short Research Paper 104, https://doi.org/10.3133/of2007-1047.srp104

Walker, P.T. 1961. Study of some rocks and minerals from the Dufek Massif, Antarctica. *International Geophysical Year World Data Center A, Glaciology Report*, **4**, 195–213.

White, J.D.L. and Houghton, B.F. 2006. Primary volcaniclastic rocks. *Geology*, **34**, 677–680, https://doi.org/10.1130/G22346.1

White, J.D.L. and McClintock, M.K. 2001. Immense vent complex marks flood-basalt eruption in a wet, failed rift: Coombs Hills, Antarctica. *Geology*, **29**, 935–938, https://doi.org/10.1130/0091-7613(2001)029<0935:IVCMFB>2.0.CO;2

White, J.D.L., Bryan, S.E., Ross, P.-S., Self, S. and Thordarson, T. 2009. Physical volcanology of large igneous provinces: update and review. *IAVCEI Special Publications*, **2**, 291–321.

Chapter 2.1b

Ferrar Large Igneous Province: petrology

David H. Elliot[1][*][2] and Thomas. H. Fleming[2]

[1]School of Earth Sciences and Byrd Polar and Climate Research Center, Ohio State University, Columbus, OH 43210, USA

[2]Department of Earth Sciences, Southern Connecticut State University, New Haven, CT 06515, USA

DHE, 0000-0002-6111-0508; THF, 0000-0001-7091-7699

*Correspondence: elliot.1@osu.edu

Abstract: The Lower Jurassic Ferrar Large Igneous Province consists predominantly of intrusive rocks, which crop out over a distance of 3500 km. In comparison, extrusive rocks are more restricted geographically. Geochemically, the province is divided into the Mount Fazio Chemical Type, forming more than 99% of the exposed province, and the Scarab Peak Chemical Type, which in the Ross Sea sector is restricted to the uppermost lava. The former exhibits a range of compositions (SiO_2 = 52–59%; MgO = 9.2–2.6%; Zr = 60–175 ppm; Sr_i = 0.7081–0.7138; ε_{Nd} = −6.0 to −3.8), whereas the latter has a restricted composition (SiO_2 = c. 58%; MgO = c. 2.3%; Zr = c. 230 ppm; Sr_i = 0.7090–0.7097; ε_{Nd} = −4.4 to −4.1). Both chemical types are characterized by enriched initial isotope compositions of neodymium and strontium, low abundances of high field strength elements, and crust-like trace element patterns. The most basic rocks, olivine-bearing dolerites, indicate that these geochemical characteristics were inherited from a mantle source modified by subduction processes, possibly the incorporation of sediment. In one model, magmas were derived from a linear source having multiple sites of generation each of which evolved to yield, in sum, the province-wide coherent geochemistry. The preferred interpretation is that the remarkably coherent geochemistry and short duration of emplacement demonstrate derivation from a single source inferred to have been located in the proto-Weddell Sea region. The spatial variation in geochemical characteristics of the lavas suggests distinct magma batches erupted at the surface, whereas no clear geographical pattern is evident for intrusive rocks.

An overview of the Ferrar Large Igneous Province (FLIP) is given in the introduction to the Ferrar LIP volcanology chapter in this Memoir (Elliot et al. 2021), and includes a summary of the existing age determinations. In brief, based on U–Pb zircon analyses, the duration of emplacement of the Ferrar Dolerite and Dufek intrusion is estimated to be 349 ± 0.49 kyr, with ages ranging from 182.78 ± 0.04 to 182.59 ± 0.08 Ma (Burgess et al. 2015). A granophyric dolerite and granophyres from Tasmania yielded ages within uncertainty of the Ferrar and Dufek results (Burgess et al. 2015; Ivanov et al. 2017). Ages for three Kirkpatrick Basalt lavas also lie within uncertainty, although one is permissibly slightly younger (Burgess et al. 2015).

Here, the distribution and thickness of the dolerite sills are summarized, the geochemistry of the intrusive and extrusive rocks is considered, the nature of the primary basalt magma and its source in the mantle are evaluated, and the mode of emplacement of the magmas is assessed.

Distribution of dolerite sills, dykes and large intrusions

Dolerite sills are the most widespread expression of the FLIP (Gunn and Warren 1962; summarized in Elliot and Fleming 2004) and crop out in a nearly continuous belt from Horn Bluff to the Theron Mountains (Fig. 1). Sills are commonly 100–300 m thick and cumulative thicknesses of 1500 m occur where the 2.0–2.5 km-thick Devonian–Triassic Beacon sequence is most extensively developed (Barrett 1991; Collinson et al. 1994; Bradshaw 2013). Locally in south Victoria Land, the Basement Sill thickens to as much as 700 m (Marsh and Zeig 1997) and a 1 km-thick sill forms the 15 km-long Warren Range (Grapes and Reid 1971). The Basement Sill is estimated to have occurred continuously over an area of 10 000 km² in the Dry Valleys region of south Victoria Land (Marsh 2007) and appears to have a large-scale lobe structure. In north Victoria Land, intrusions present in discontinuous outcrops at identical stratigraphic positions in Beacon strata were correlated over a distance of more than 200 km (Roland and Tessensohn 1987). Sills are predominantly near parallel to bedding (Fig. 2), but none have been described that have the saucer-shape recorded, for instance, from the Karoo of South Africa (Galerne et al. 2008, 2011; Coetzee and Kisters 2017, 2018; Sheth 2018, fig. 8–63). A sill may merge with another sill, or may appear to be concordant with another sill with or without slivers of sedimentary rock between them. North of Mackay Glacier, south Victoria Land, massive dolerite sills are reported to coalesce and attain a thickness of more than 1000 m (Pocknall et al. 1994). Inclined sheets are prominent in parts of south Victoria Land (e.g. Hamilton 1965; Morrison and Reay 1995), and are reported from Mount Howe, central Transantarctic Mountains (Fig. 1) (Doumani and Minshew 1965). Massive dolerite bodies up to a few kilometres in diameter and a possible laccolith have been reported by Gunn and Warren (1962), and dyke-like bodies as much as 30 km long, 1.5–3.0 km wide and at least 1500 m in vertical outcrop by Gunn (1963). Gunn (1966) discussed a dyke-like dolerite body up to 1.6 km (1 mile) wide and inferred to extend for 24 km (15 miles) SW from just south of the Mackay Glacier; it intersects three sills, and is the feeder for the uppermost sill in that region. Gunn and Warren (1962) described dykes feeding into sills, dykes tens of metres wide and swarms of thin dykes cutting Beacon strata in south Victoria Land (e.g. Allan Hills, see Muirhead et al. 2012; Terra Cotta Mountain, see Morrison and Reay 1995). At Mount Gran (Fig. 3) a massive dolerite plug cuts across, and may be connected to, a number of sills of varying thickness and forming a network. The regional distribution of sills and dykes in the Dry Valleys region is documented by McElroy and Rose (1987), Woolfe et al. (1989), Allibone et al. (1991), Pocknall et al. (1994), Turnbull et al. (1994), and Isaac et al. (1996), and the architecture of magma emplaced into supracrustal rocks is discussed by Marsh and co-workers (Marsh 2004, 2007; see the section entitled 'Magma emplacement at supracrustal

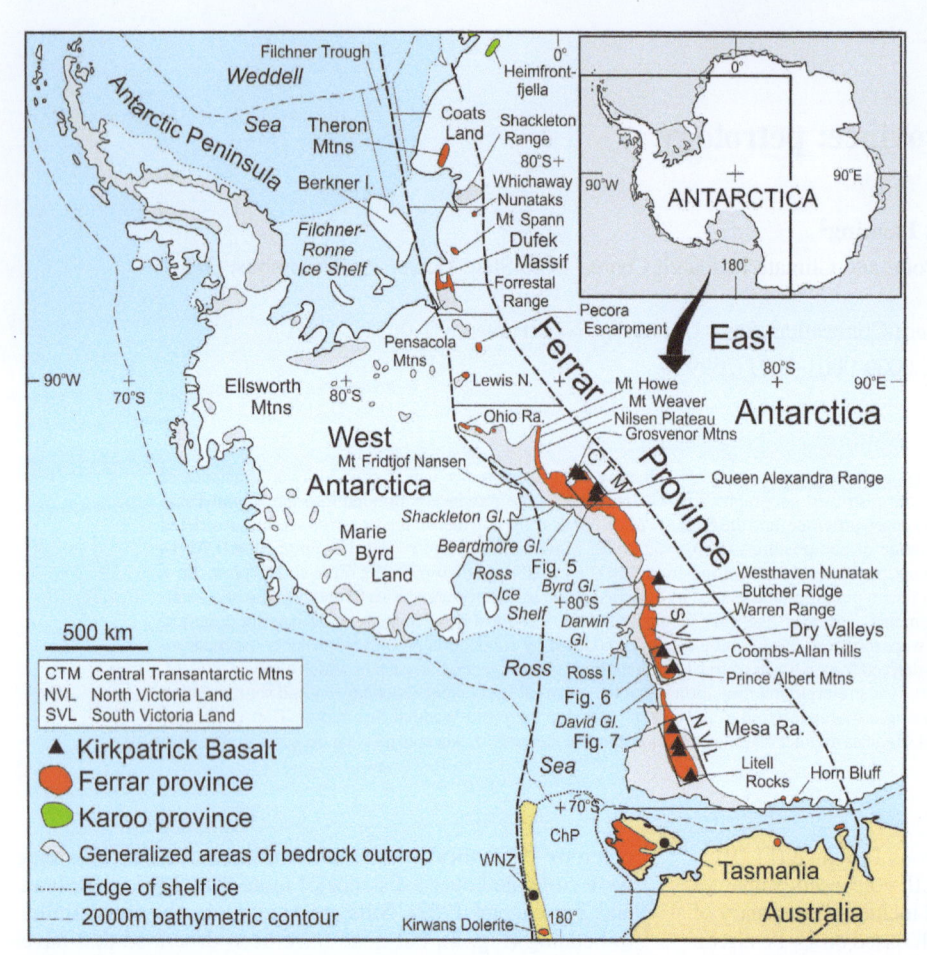

Fig. 1. Location map for the Ferrar Large Igneous Province of Antarctica and southeastern Australasia. WNZ, South Island of New Zealand west of the Alpine Fault, with the Kirwans Dolerite located. ChP, Challenger Plateau, which separates New Zealand from Australia in reconstructed Gondwana. Within the Ferrar province, the Scarab Peak Chemical Type (SPCT) has been identified in sills in the Theron Mountains and Whichaway Nunataks adjacent to the Filchner–Ronne Ice Shelf region and as the capping lava in the Transantarctic Mountains.

depths and evolution' later in this chapter). Elsewhere in the Transantarctic Mountains the numerous sills are mainly parallel to bedding, and steeply-dipping dykes are very sparsely scattered (summarized in Elliot and Fleming 2004, 2017). Dykes cutting across sills emplaced earlier are relatively rare but have been observed, mainly in the Dry Valleys region (e.g. Mount Feather, south Victoria Land, Fleming *et al.* 2005, 2012; McIntyre Promontory, central Transantarctic Mountains, Elliot and Fleming 2004). Sills may terminate in dykes, exchange stratigraphic positions, and locally form small dolerite masses and thin inclined sheets (Fig. 4). Kilometre-size dolerite masses occur locally, as in the Supporters Range and Lhasa Nunatak (Fig. 5), in the Warren Range (Fig. 1) (Grapes and Reid 1971) and Convoy Range (Fig. 6) (Pocknall *et al.* 1994), and at Butcher Ridge (Fig. 1). Aeromagnetic surveys over Butcher Ridge suggest that it is a gabbroic body about 3000 km^2 in area and with a minimum thickness of 1–2 km (Behrendt *et al.* 2002). The exposed part of the intrusion is remarkable for the inclined and contorted layers of interleaved andesite and rhyolite composition, which are cut by thin dolerite intrusions (Marshak *et al.* 1981; Shellhorn 1982; Nelson and Cottle 2016).

The layered basic Dufek intrusion (Fig. 1) (Ford and Boyd 1968; Ford 1976; Ford and Himmelberg 1991) was originally estimated, on the basis of aeromagnetic data, to have a volume of about 50 000 km^3 (Behrendt *et al.* 1981), but Ferris *et al.* (1998) argued that it comprises two much smaller bodies with a total volume of about 6600 km^3. Palaeomagnetic data have been interpreted to support the latter interpretation (Gee *et al.* 2013). More recently, Semenov *et al.* (2014) reviewed geophysical data for the Dufek intrusion, and supported the original size estimates and suggested that the smaller volume proposed by Ferris *et al.* (1998) is not consistent with petrographical observations and petrological models for layered basic intrusions.

Dolerite sills with Ferrar chemistry crop out in Tasmania (Hergt *et al.* 1989*b*) and New Zealand (Mortimer *et al.* 1995). Lavas with Ferrar chemistry are present in Tasmania (Bromfield *et al.* 2007), Kangaroo Island off South Australia (Milnes *et al.* 1982; Hergt *et al.* 1991) and in the subsurface in western Victoria (Hergt *et al.* 1991). The possibility of magma compositions similar to the Ferrar tholeiites in the Golden Gate lava sequence in the Karoo Province (Fig. 7) was suggested by Elliot and Fleming (2000), and argued for by Riley *et al.* (2006) for some of the Underberg dykes, but has yet to be confirmed. An extension of the Karoo province is present in Queen Maud Land as intrusive and extrusive rocks, and in the Theron Mountains where it forms some of the Lower Jurassic sills (Brewer *et al.* 1992; Leat 2008).

Subglacially, Ferrar tholeiites are inferred from geophysics to overlie East Antarctic basement rocks for about 500 km across the Wilkes Subglacial Basin from north Victoria Land (Ferraccioli *et al.* 2009), which is consistent with the isolated dolerite occurrences at, and to the east of, Horn Bluff. Dolerite is likewise inferred to occur for some 400 km inland from south Victoria Land (Studinger *et al.* 2004). However, Ferrar sills appear not to be present inland from the Scott Glacier toward the South Pole (Studinger *et al.* 2006). Nevertheless, it is probable that Ferrar rocks originally extended for a significant distance over the East Antarctic basement in the sector from the central Transantarctic Mountains to the Theron Mountains. Similarly, Ferrar rocks must have been present, but for a lesser distance, towards the Gondwana plate margin. Geophysical data for the Ross Ice Shelf region (Tinto *et al.* 2019) suggest that west of longitude 180° there is stretched crust related to the Lower Paleozoic Ross Orogen. If so, it is possible that Ferrar rocks extended, in the Ross embayment

Fig. 2. (a) Dolerite sills at spot height 3120 m, Nilsen Plateau, central Transantarctic Mountains (Fig. 1). Sills, indicated by numbers, were intruded into Permian Beacon strata. The height of the face is about 1000 m; view to the east. (Image: D.H. Elliot.) (b) Stack of sills (numbered) at Mount Joyce, Prince Albert Mountains (Fig. 6). Only slivers and short stratigraphic sections of undifferentiated Beacon strata separate the sills. View to the NW; the height from the ice surface to the summit is about 800 m. Grid references for images are given in Appendix A. (Image: T.H. Fleming.)

Fig. 3. (a) Ferrar intrusive rocks east of Mount Gran, south Victoria Land (Fig. 6). The dolerite plug cross-cuts the lowest thick sill and others higher up the rock face, and forms the feeder for other sills intruded into Devonian quartzose sandstones. View to the north; the height of the face is about 400 m. (Image: D.H. Elliot.) (b) Terra Cotta Mountain, south Victoria Land. Steeply inclined dolerite intrusions flank the mountain, thin dykes occur throughout the face and a climbing sill is present on the right-hand side. View to the south; the height of the face about 400 m. (Image: T.H. Fleming.)

sector, for some 200 km towards the Jurassic plate margin. Ferrar emplacement was probably limited by deformed plate margin rocks.

Prior geochemical studies

Prior (1907), in his study of the dolerites collected by the National Antarctic Expedition, 1901–04, many of which are glacial erratics from the Dry Valleys and other localities in south Victoria Land, provided the first geochemical analysis of these tholeiitic rocks. Subsequently, Benson (1916) published analyses of two tholeiite glacial erratics from Cape Royds, Ross Island, collected by the British Antarctic Expedition, 1907–09. The first analysis of an *in situ* dolerite, a rock sample from Horn Bluff that was collected by the Australasian Antarctic Expedition (1911–14), was reported by Browne (1923). The analysis of a dolerite (Stewart 1934) from Mount Fridtjof Nansen (Fig. 1) collected by L.M. Gould on the first Byrd Antarctic Expedition (1928–30) extended the distribution of the dolerite sills into the continental interior. Intense study of the dolerites began in the International Geophysical Year, with Gunn (1962, 1963, 1966) and Hamilton (1964, 1965) publishing the first modern investigations of sill geochemistry in south Victoria Land, and with Compston *et al.* (1968) reporting the unusually high initial Sr isotope ratio (high initial ratios of Sr had been reported previously for Tasmanian dolerites by Heier *et al.* 1965).

Subsequent studies on the chemistry of sills and lavas in south Victoria Land (the Dry Valleys region and the Prince Albert Mountains: Fig. 6), have been reported by Kyle *et al.* (1983), Morrison and Reay (1995), Wilhelm and Wörner (1996), Antonini *et al.* (1997, 1999), Demarchi *et al.* (2001), Ross *et al.* (2008) and Elliot and Fleming (2017). In particular, B.D. Marsh and collaborators (e.g. Marsh 2004, 2007; Bédard *et al.* 2007; Forsha and Zieg 2007; Zavala *et al.* 2011; Zieg and Marsh 2012) undertook a detailed investigation of the sills (Basement, Peneplain and Beacon sills) in the Dry Valleys of south Victoria Land. In north Victoria Land, north of the David Glacier (Fig. 8), the sills have been investigated by Brotzu *et al.* (1988), Hornig (1993), Antonini *et al.* (1997), Hanemann and Viereck-Götte (2004, 2007b), Melluso *et al.* (2014) and Elliot and Fleming (2017). Hornig (1993) reported on the sills at Scar Bluffs and Anxiety Nunataks, coastal George V Land (east of Horn Bluff). The Kirkpatrick Basalt lavas in south Victoria Land have been studied by Kyle *et al.* (1983), Wilhelm and Wörner (1996), Antonini *et al.* (1999), Demarchi *et al.* (2001) and Elliot and Fleming

Fig. 4. (**a**) Dolerite sill (a few tens of metres thick) abruptly thinning at a dyke and continuing as a thin sill from which a second dyke extends. The head of the LaPrade Valley is immediately east of Shenk Peak (the high point in the image), view to the west, Shackleton Glacier region (Fig. 5). (Image: D.H. Elliot.) (**b**) Small dolerite dykes and masses (arrows) intruded into Triassic Fremouw Formation strata (pale sandstones and grey, slope-forming fine-grained beds); vertical bluffs of thin-bedded strata in the lower part of the image grade laterally into the slope-forming beds. View to the NE; the height of the face is about 100 m. Dismal Buttress, Shackleton Glacier region (Fig. 5). (Image: D.H. Elliot.)

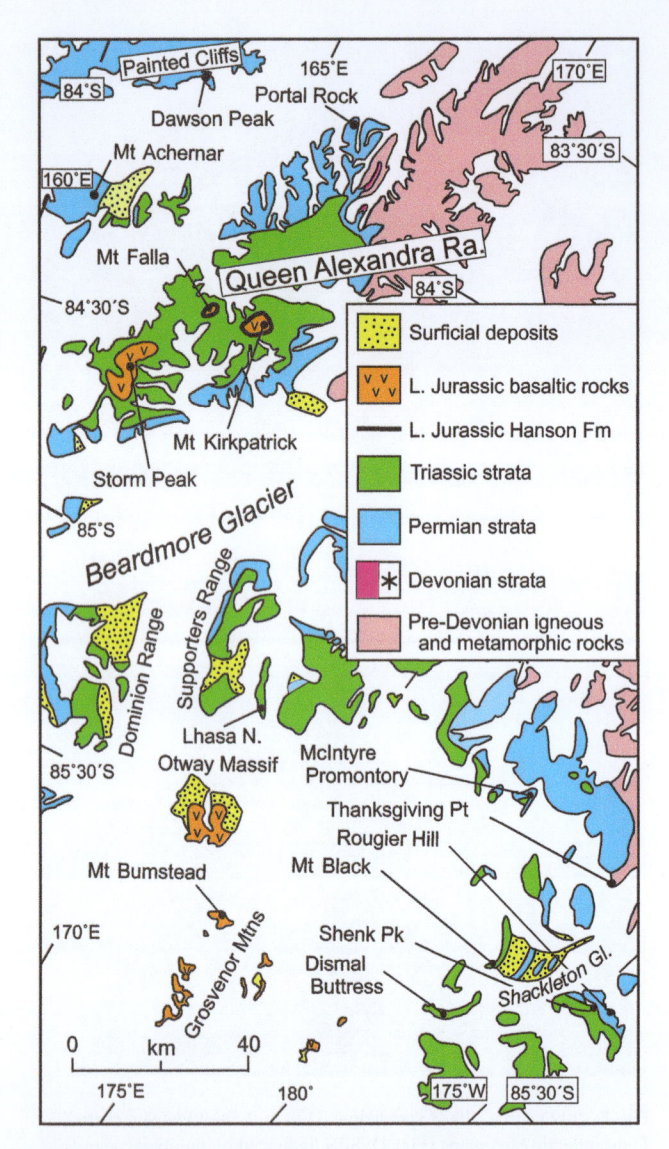

Fig. 5. Simplified geological map of the Queen Alexandra Range and Shackleton Glacier region, central Transantarctic Mountains. Ferrar Dolerite sills are co-extensive with Permian and Triassic strata. The map orientation is with north up (cf. the box in Fig. 1).

(2017). The geochemistry of the Kirkpatrick Basalt in north Victoria Land has been investigated by Mensing *et al.* (1984, 1991), Siders and Elliot (1985), Brotzu *et al.* (1988, 1992), Fleming *et al.* (1992, 1995) and Elliot *et al.* (1995). Chemical data from these studies document that the Ferrar Dolerite sills and Kirkpatrick Basalt lavas are predominantly basaltic andesite in composition. The intrusion at Butcher Ridge appears to be unique in the Ferrar province, in that it records a significant volume of silicic rocks and significant interaction with crustal rocks (Shellhorn 1982; Kyle *et al.* 1999; Nelson *et al.* 2014).

In the long stretch of the Transantarctic Mountains, from the Darwin Glacier region to the Theron Mountains (Fig. 1), the geochemistry of the lavas and sills in the central Transantarctic Mountains (Queen Alexandra Range and Shackleton Glacier region: Fig. 5) has been reported by Elliot (1970), Faure *et al.* (1974, 1991) and Elliot and Fleming (2017), for a sill at Portal Rock by Hergt *et al.* (1989*a*), and sills at Mount Achernar, central Transantarctic Mountains, and Roadend Nunatak, Darwin Glacier, by Faure *et al.* (1991). Brief information on sills at the Nilsen Plateau and the Ohio Range is given in Riley *et al.* (2020). Analyses of sills at Mount Schopf (Ohio Range), Lewis Nunatak (Thiel Mountains) and Pecora Escarpment, and dykes at Cordiner Peak (Pensacola Mountains) have been published by Ford and Kistler (1980), Vennum and Storey (1987), Leat (2008) and Harris (2014). Sills in the Theron Mountains and Whichaway Nunataks (Fig. 1) have been investigated by Stephenson (1966) and Brewer *et al.* (1992), and Ferrar dykes in the Shackleton Range reported by Stephenson (1966), Spaeth *et al.* (1995) and Leat (2008). Many of the sills in the Theron Mountains are not part of the FLIP, but geochemically are allied with the Karoo Large Igneous Province (Leat *et al.* 2006). The Dufek intrusion, a layered basic intrusion in the Pensacola Mountains (Ford and Boyd 1968, Ford and Himmelberg 1991), is part of the Ferrar province. Assuming it is a single intrusion, the lower part, the base of which is not exposed, crops out in the Dufek Massif, and the upper part, capped by a kilometre-thick granophyre, forms the Forrestal Range. A lamprophyric dyke in the Pensacola Mountains (Leat *et al.* 2000) has a similar age but is chemically very distinct and, strictly speaking, is not a Ferrar rock.

Petrography

The Dufek intrusion (Ford and Himmelberg 1991) and the thickest sills exhibit mineral layering with the associated cumulate textures. The mineralogy of the Dufek intrusion is

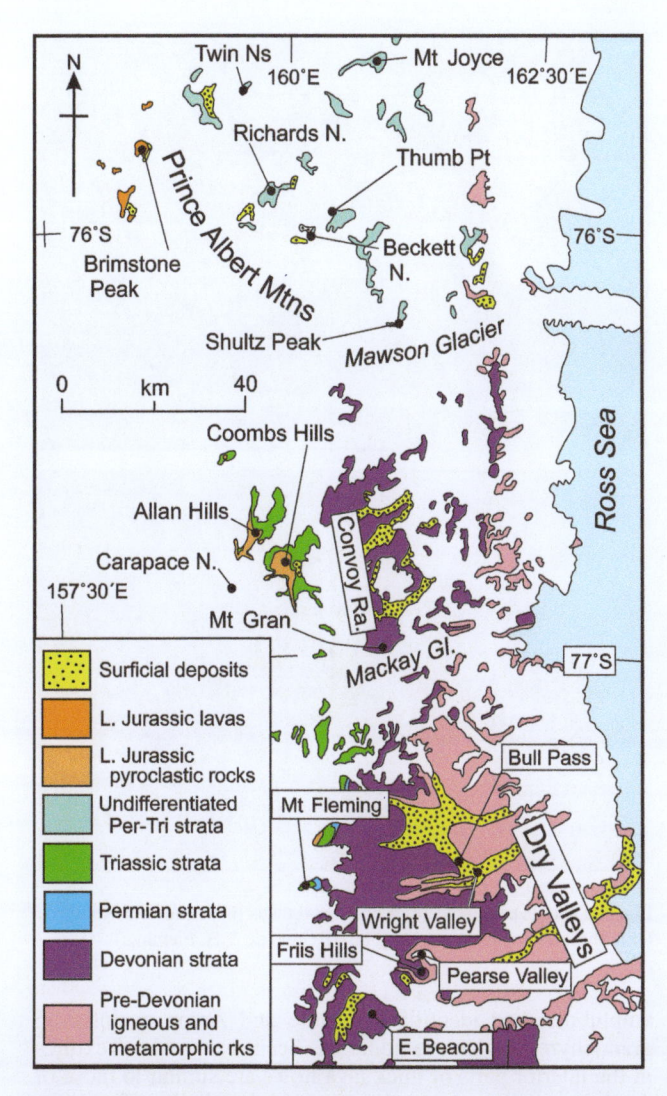

Fig. 6. Simplified geological map of south Victoria Land. Ferrar Dolerite sills are co-extensive with Devonian, Permian and Triassic strata. The map orientation is with north up (cf. the box in Fig. 1).

typical of layered basic intrusions, with plagioclase–two pyroxene (augite and pigeonite, with the latter commonly inverted to orthopyroxene) cumulates dominant, and interspersed thinner anorthosites and pyroxenites. This lower part is 1.7 km thick but the base is not exposed and might be as much as 2–3 km below the surface. This hidden basal part probably contains olivine and chromite cumulates, assuming that the Dufek is similar to other layered basic intrusions. An estimated 2–3 km-thick section is hidden beneath the snowfield between the Dufek Massif and the Forrestal Range. The upper (but not connected in outcrop) part of the intrusion (also about 1.7 km thick) in the Forrestal Range is dominated by plagioclase–two pyroxene cumulates but with more evolved compositions and significantly more iron–titanium oxides. A thick anorthosite interval occurs low in the Forrestal sequence, which is capped by a 300 m-thick granophyre.

The intensively studied Basement Sill in south Victoria Land (e.g. Bédard *et al.* 2007; Boudreau and Simon 2007; Hersum *et al.* 2007; Charrier 2010; Jerram *et al.* 2010; Zavala *et al.* 2011) includes a ponded lower zone, the Dais intrusion, in which websterite and anorthosite cumulate-textured layers up to 0.5 m thick extend continuously for several hundred metres. Although orthopyroxene and plagioclase dominate, they are accompanied by augite and inverted pigeonite, together with minor groundmass quartz, biotite and ilmenite. Layering has also been identified at Thumb Point (Fig. 6) by

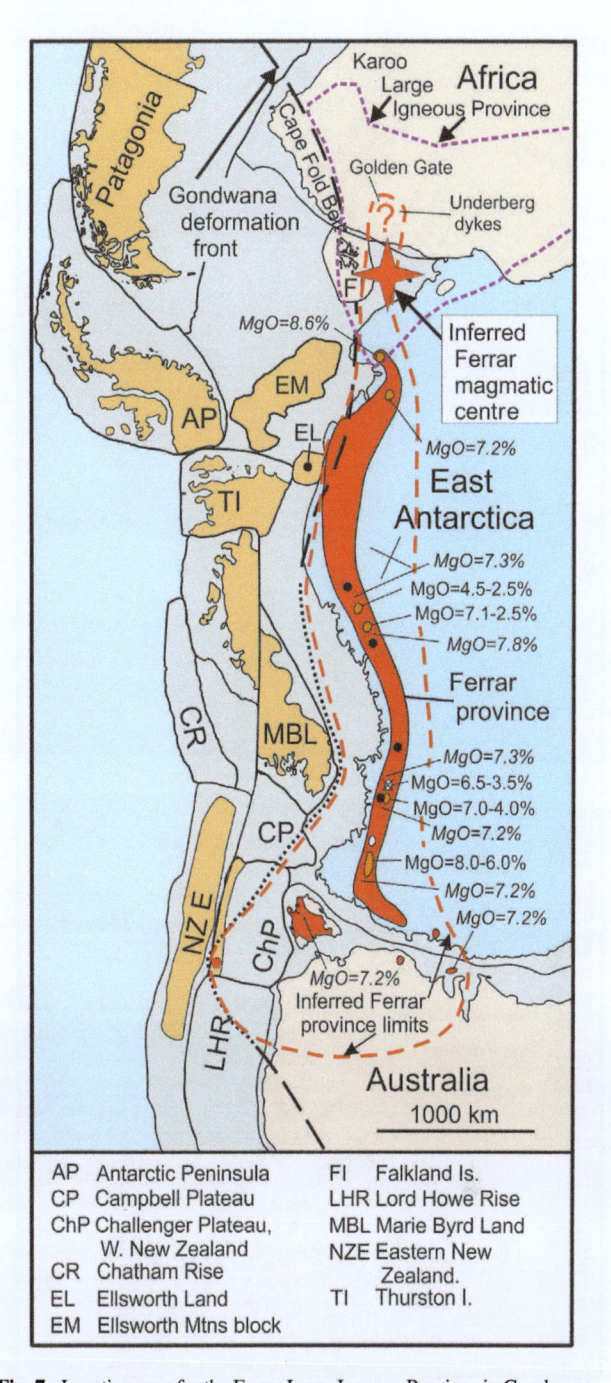

Fig. 7. Location map for the Ferrar Large Igneous Province in Gondwana (modified from a reconstruction provided by the PLATES Project at the Institute of Geophysics, University of Texas at Austin). Known outcrop areas of the SPCT composition rocks (in orange) are superimposed on the overall Ferrar distribution (SPCT compositions form sills in the Weddell Sea sector and cap the lava successions elsewhere; note there is no SPCT composition at the lava outcrop localities in white). Range of MgO compositions for lavas (excluding the SPCT composition) are linked to outcrop areas. Approximate locations of olivine-bearing dolerite sills (chilled margin MgO *c.* 9%) are marked by black dots. Sill chilled margins show no spatial pattern of compositions (highest sill MgO% in various regions is indicated in italics). The locations of possible Ferrar compositions within the Karoo Province are indicated (Golden Gate and Underberg dykes, both adjacent to Lesotho).

Ricker (1964), Gunn (1966) and Wilhelm and Wörner (1996), in several sills in the Dry Valleys by Gunn (1963), and at the Warren Range by Grapes and Reid (1971).

Chilled margins of sills may include microphenocrysts, but for the most part the sills show ophitic to doleritic textures. Excluding the Basement Sill (and possibly other unexamined

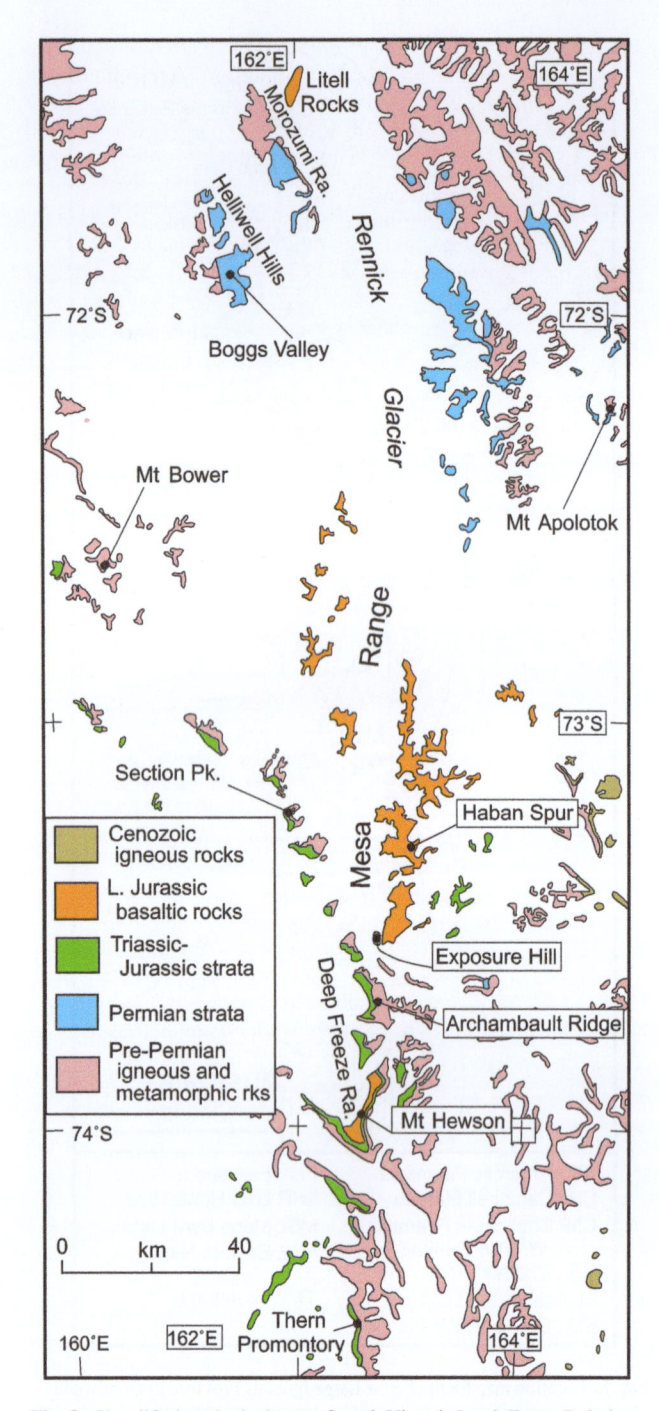

Fig. 8. Simplified geological map of north Victoria Land. Ferrar Dolerite sills are co-extensive with Permian, Triassic and Lower Jurassic (pre-Ferrar) strata. The map orientation is with north up (cf. the box in Fig. 1).

Fig. 9. Dolerite pegmatite. (**a**) A horizontal sheet. (**b**) A thin vein. Rougier Hill, Shackleton Glacier region (Fig. 5). (Images: T.H. Fleming.)

amphibole, and identifiable apatite and zircon enclosed in granophyre in which sanidine has been recognized. Textures in the interior parts of thick lava flows are similar to those of the dolerite sills and the mineralogy is also similar. The mesostasis in many lavas is quite variable, and quench textures are commonly exhibited. The latter comprise feldspar microlites with needle-like overgrowths, feather-like pyroxene aggregates and skeletal iron oxides, all enclosed in lightly to strongly oxidized glass. A flow at Carapace Nunatak has a mesostasis entirely of light brown glass. Quartz–K-feldspar intergrowths range between graphic intergrowths and cryptocrystalline. A variety of accessory minerals has been recorded in the more evolved rocks at Thern Promontory, north Victoria Land (Fig. 8) (Melluso *et al.* 2014), and include fayalite, amphibole, zircon and apatite, amongst others. The granophyres in the Red Hill intrusion in Tasmania contain other trace minerals (Melluso *et al.* 2014). The sequence of liquidus phases is olivine–orthopyroxene–pigeonite–augite plus plagioclase.

The crystal size distribution has been investigated in the Basement Sill (Marsh 1998; Jerram *et al.* 2010), the Beacon (Asgard) Sill (Zieg and Marsh 2012), and in the Thumb Point sill and lavas from Brimstone Peak, south Victoria Land (Wilhelm and Wörner 1996). Wilhelm and Wörner (1996) estimated nucleation rates and cooling histories, the latter yielding estimated times of *c.* 1500 years for the sill, *c.* 200 years for the *c.* 150 m-thick capping Scarab Peak Chemical Type (SPCT) flow and <100 years for a *c.* 100 m-thick Mount Fazio Chemical Type (MFCT) flow.

Geochemistry

Many of the studies of sill geochemistry have been directed at the internal evolution, but it is the compositional range of

thick sills), the dolerites exhibit a narrow range of mineralogy, which is principally plagioclase (labradorite), two pyroxenes (augite and pigeonite) and iron–titanium oxides. Orthopyroxene is present in some sills with more basic compositions, and also occurs in a number of chilled margins of lava flows where it is commonly rounded; it is commonly mantled by augite. Exsolution in pyroxenes is common. Sills with the most basic composition may carry forsteritic olivine or pseudomorphs thereafter. The interstitial groundmass in the sills is mainly a quartz–feldspar intergrowth, which may be granophyric; primary biotite has been recorded in a few sills. Secondary alteration of plagioclase and mafic minerals is common. In the more differentiated sills, dolerite pegmatite (Fig. 9) or schlieren may have clinopyroxene partially replaced by fine-grained lavas and chilled margins of thick lavas and sills, particularly the olivine-bearing dolerite sills, that provide the context for the province-wide evolution of magmas emplaced in the uppermost crust, in supracrustal strata or at the surface. These olivine-bearing dolerite sills have MgO = c. 9% and constitute the least-evolved Ferrar magma compositions, and thus are regarded as the starting point for assessing both the possible magma sources in the mantle and the subsequent evolution of those primary magmas. The term olivine-bearing dolerite is used here only for those sills with chilled margins having olivine crystals and MgO = 9–10%.

As noted in the volcanology chapter (Elliot *et al.* 2021), the Ferrar magmas fall into two chemical types (Siders and Elliot 1985), designated the Mount Fazio Chemical Type (MFCT) and the Scarab Peak Chemical Type (SPCT) (Fleming *et al.* 1992). Geographically, the two chemical types overlap each other, except that the SPCT has not been recognized in SE Australasia (Fig. 7). Stratigraphically, the SPCT always forms the youngest lavas, but in the Theron Mountains occurs as a sill that is presumed to be younger than the MFCT sills. The Dufek intrusion is grouped with the MFCT lavas and sills but, without an exposed chilled margin, some uncertainty remains. The adjacent Cordiner Peaks dykes with MFCT chemistry were considered to provide the best approximation to the composition of the original Dufek magma (Ford and Kistler 1980).

The olivine-bearing dolerite composition anchors the primitive end of the MFCT trend. Both chemical types are characterized by enriched Nd and Sr initial isotope ratios, crust-like trace element patterns, and depletions in high field strength elements (HFSEs) such as Ti and P. In brief, MFCT rocks (excluding the Dufek intrusion) have a range of compositions ($^{87}Sr/^{86}Sr_i = c$. 0.7081–0.7138; MgO = 9.2–2.6%; Zr = 60–175 ppm), whereas the SPCT has a restricted composition ($^{87}Sr/^{86}Sr_i = c$. 0.7095; MgO = c. 2.3%; Zr = c. 230 ppm). Comparisons between the two are discussed later. Although the timing of the Ferrar and Karoo provinces is nearly identical (Svensen *et al.* 2012; Sell *et al.* 2014), the geochemistry is quite distinct. Karoo basic magmas are much more diverse and, with a few exceptions, are geochemically distinct, having, for example, less enriched Sr, Nd and Pb isotopic compositions, and higher HFSE concentrations (Marsh *et al.* 1997; Jourdan *et al.* 2007; Neumann *et al.* 2011; Heinonen *et al.* 2014).

Previous studies of the Ferrar tholeiites have been cited in the introduction to this section. Those investigations acquired data in various analytical laboratories, and therefore comparisons between datasets are hampered by differences in precision and accuracy, by inter-laboratory biases, and, in the case of isotope measurements, by different and evolving standards and precisions. This concern is illustrated by the analysis of 10 SPCT samples distributed between the Mesa Range and the Grosvenor Mountains, which showed that the variations in concentrations, excluding the more mobile elements, fall within analytical precisions (Elliot *et al.* 1999; see also Fleming *et al.* 1992). Other authors (Hornig 1993; Molzahn *et al.* 1996; Antonini *et al.* 1997) have analysed SPCT samples, but results differ markedly for some elements, although are similar for others. Presentation of all the existing data leads to considerable analytical scatter and expanded fields, which, although broadly showing trends, lacks the clarity needed for an accurate portrayal of petrogenetic relationships. Further, alteration affects at least the more mobile elements, thus producing a 'geological' scatter of data, let alone the high-temperature hydrothermal exchange affecting Sr isotope compositions that may occur in chilled margins of sills. Finally, it is not always evident exactly where an analysed sample was collected in either a lava or a sill, and thus the possible effects of *in situ* differentiation are often unclear. The data presented here have been selected in an attempt to reduce these biases and uncertainties, and therefore have been drawn from the authors' studies of the Ferrar rocks extending geographically from the Mesa Range to the Nilsen Plateau (Fig. 1). Data are presented for samples that span the full range of magma compositions. In most cases this is illustrated using previously unpublished Ferrar mineral and whole-rock data from the central Transantarctic Mountains. Data sources are given in the figure captions.

Mineral chemistry

Early studies (e.g. Gunn 1966) reported mineral compositions determined by optical methods. Here, only mineral compositions determined by electron microprobe are considered. The Dufek intrusion is discussed separately.

Feldspar. MFCT plagioclase compositions typically range from calcic bytownite to calcic andesine, whereas the SPCT plagioclase has a more restricted range from calcic labradorite to sodic andesine (Table 1; Fig. 10) (Brotzu *et al.* 1992; Fleming 1995; Elliot *et al.* 1995; Antonini *et al.* 1999; Demarchi *et al.* 2001; Hanemann and Viereck-Götte 2004; Melluso *et al.* 2014). The more sodic plagioclase occurs as rims and groundmass grains. The groundmass may include alkali feldspar, which ranges from albite to orthoclase and even sanidine, with the K-feldspars occurring in granophyric intergrowths and granophyres (Barrett *et al.* 1986; Hornig 1993; Melluso *et al.* 2014). Anorthoclase has also been reported by the latter two authors. Plagioclase in the Dais layered body in the Basement Sill, south Victoria Land (Bédard *et al.* 2007) occurs in a variety of distinct textural settings (e.g. cumulate crystals, inclusions in other minerals, schlieren). The overwhelming composition is sodic bytownite in the chilled margin and lower gabbronorite, and calcic bytownite in the overlying rocks, but with some more sodic rim compositions; oligoclase is present in pegmatitic schlieren.

Pyroxene. Excluding the Basement Sill in the Dry Valleys and other sills with cumulates, orthopyroxene is present in a number of MFCT lavas and sills, and also occurs in chilled margins (Brotzu *et al.* 1992; Hornig 1993; Elliot *et al.* 1995; Fleming 1995; Antonini *et al.* 1999; Demarchi *et al.* 2001; Hanemann and Viereck-Götte 2004; Melluso *et al.* 2014). The most Mg-rich pigeonite (Mg# c. 84: Mg# = (Mg/[Mg + Fe] × 100) may co-exist with orthopyroxene (Mg# c. 85), and those and many other pyroxenes may exhibit marked Fe-enrichment both in rims and as core compositions in the more evolved tholeiites (Fig. 10). Augite similarly shows Fe-enrichment in grain rims (Mg# c. 20). The pyroxenes in lavas include compositions that bridge the gap between augite and pigeonite, which reflects quenching. Hedenbergitic pyroxenes (with Mg# <10) are present in the most-evolved MFCT compositions (Melluso *et al.* 2014). The SPCT pyroxene compositions (only pigeonite and augite) are relatively restricted (Table 2; Fig. 10), and also exhibit Fe-enriched rims. Based on the two-pyroxene geothermometer (Ishii 1975; Lindsley 1983; Lindsley and Andersen 1983), temperatures of crystallization for MFCT rocks range from 1200 to 1050°C for the lavas, and as low as 850°C for the more evolved rocks, and for SPCT lavas 1105–1070°C (Brotzu *et al.* 1992; Elliot *et al.* 1995; Fleming 1995; Melluso *et al.* 2014).

In the Dais intrusion (the basal part of the Basement Sill) the majority of primocryst orthopyroxenes have Mg# >80, with a scattering down to Mg# c. 65 (Bédard *et al.* 2007). Ca-rich clinopyroxenes in the lower part of the intrusion have Mg# >70

Table 1. *Plagioclase compositions determined by microprobe analysis for samples from Dawson Peak and Storm Peak*

Sample	85-71-1	85-71-1	85-75-1	85-75-9	85-75-13	85-76-39	85-76-39	85-76-54	85-76-54	85-76-60	85-76-60
Type	MFCT Sill	MFCT Sill	MFCT Lava	MFCT Lava	MFCT Lava	MFCT Lava	MFCT Lava	MFCT Lava	MFCT Lava	SPCT Lava	SPCT Lava
Region	CTM	CTM	CTM	CTM	CTM	CTM	CTM	CTM	CTM	CTM	CTM
SiO_2	45.87	50.09	48.69	51.24	50.08	48.79	52.38	50.52	53.64	53.37	55.49
Al_2O_3	33.77	31.12	32.11	30.36	30.98	32.16	28.92	31.04	28.27	29.04	27.43
FeO	0.45	0.76	0.48	0.83	0.88	0.58	0.8	1.06	1.08	1.02	0.72
MgO	0.15	0.40	0.29	0.17	0.09	0.21	0.13	0.05	0.07	0.01	0.05
CaO	17.60	13.71	15.55	13.94	13.94	16.17	12.44	14.18	11.49	11.83	9.65
Na_2O	1.24	3.03	2.44	3.21	3.08	2.16	4.09	3.01	4.54	4.57	5.33
K_2O	0.07	0.23	0.12	0.19	0.21	0.10	0.31	0.19	0.37	0.22	0.45
Total	99.14	99.33	99.67	99.94	99.27	100.17	99.07	100.06	99.48	100.07	99.11
Si^{4+}	2.132	2.300	2.238	2.339	2.304	2.234	2.405	2.310	2.449	2.423	2.523
Al^{3+}	1.850	1.684	1.739	1.633	1.680	1.736	1.565	1.672	1.512	1.554	1.470
Fe^{2+}	0.018	0.029	0.019	0.032	0.034	0.022	0.031	0.041	0.041	0.039	0.027
Mg^{2+}	0.010	0.027	0.020	0.011	0.006	0.014	0.009	0.003	0.005	0.001	0.004
Ca^{2+}	0.876	0.675	0.765	0.682	0.687	0.793	0.612	0.695	0.562	0.576	0.470
Na^+	0.111	0.270	0.217	0.284	0.275	0.191	0.364	0.267	0.402	0.403	0.470
K^+	0.004	0.013	0.007	0.011	0.013	0.006	0.018	0.011	0.022	0.013	0.026
Total	5.001	4.999	5.005	4.992	4.999	4.997	5.004	4.995	4.993	5.008	4.990
O^{2-}	8.000	8.000	8.000	8.000	8.000	8.000	8.000	8.000	8.000	8.000	8.000
An	88.4	70.5	77.3	69.8	70.5	80.1	61.6	71.4	57.0	58.1	48.7
Ab	11.2	28.2	21.9	29.1	28.2	19.3	36.6	27.5	40.8	40.6	48.6
Or	0.4	1.4	0.7	1.1	1.3	0.6	1.8	1.1	2.2	1.3	2.7

Analyses were performed at the Ohio State University Electron Microprobe Laboratory.

The analyses have been selected to show the range of compositions represented by the olivine-bearing dolerite sill at Dawson Peak and the lavas at Storm Peak. Analysis 85-75-1 is for the basal, thin basic flow. Samples 85-75-9 and 85-75-13 are from the 135 m-thick tachylitic flow 2. The two analyses for samples 85-76-39 (flow 13), 85-76-54 (flow 14) and 85-76-60 (flow 15) represent the range of compositions in those flows. Sample 85-76-60 is the SPCT flow. Data are from Fleming (1995).

with less Mg-rich augites more common in the upper part (Bédard *et al.* 2007).

Olivine. Olivine, or pseudomorphs after olivine, occurs principally in sills and has been recorded in south Victoria Land (Thumb Point, Gunn 1966; Skinner and Ricker 1968; Wilhelm and Wörner 1996: Roadend Nunatak, Hergt *et al.* 1989*a*), in the Queen Alexandra Range region (Painted Cliffs, Gunn 1966; Dawson Peak, Fleming 1995) and at Nilsen Plateau (McLelland 1967). In the olivine-bearing dolerite sills the composition range is Fo_{77}–Fo_{88} (Table 2; Fig. 10) (see Fleming 1995). Pseudomorphs after olivine have been noted only in one flow from north Victoria Land (Fleming *et al.* 1995). Melluso *et al.* (2014) reported fayalitic olivine as a rare groundmass phase in evolved rocks from Thern Promontory, north Victoria Land, and from granophyre at Red Hill, Tasmania (Fo_2–Fo_6).

Oxides. Titanomagnetite with exsolved ilmenite is a common accessory mineral in the lavas and sills (Brotzu *et al.* 1992; Hornig 1993; Elliot *et al.* 1995; Fleming 1995). Temperature of subsolidus re-equilibration was estimated to be *c.* 870°C. Titanomagnetite grains lacking exsolution lamellae lie in the range Usp_{56-76}. Independent ilmenite occurs in sills. Melluso *et al.* (2014) reported co-existing magnetite and ilmenite in evolved rocks, and calculated temperatures of crystallization of *c.* 750–820°C. Chromite occurs as inclusions in olivine and as independent grains in an olivine-bearing dolerite sill analysed by Fleming (1995), who reported that the independent grains are zoned (Cr# = 48–69: Cr# = $(Cr^3/[Cr^3 + Al^3] \times 100)$) and rimmed by titanomagnetite.

Dufek intrusion. Plagioclase cores in the Dufek intrusion have a limited compositional range (An_{50}–An_{79}) but individual grains exhibit little zoning (Ford and Himmelberg 1991), most probably due to annealing. The lower part of the intrusion is unexposed; however, orthopyroxene, with Mg# *c.* 70, present in the lowest exposed rock unit is replaced by pigeonite low in the section and is present up into the upper gabbros, becoming more Fe-rich (Mg# *c.* 40) (Himmelberg and Ford 1976). Ca-rich pyroxene is increasingly Fe-rich throughout

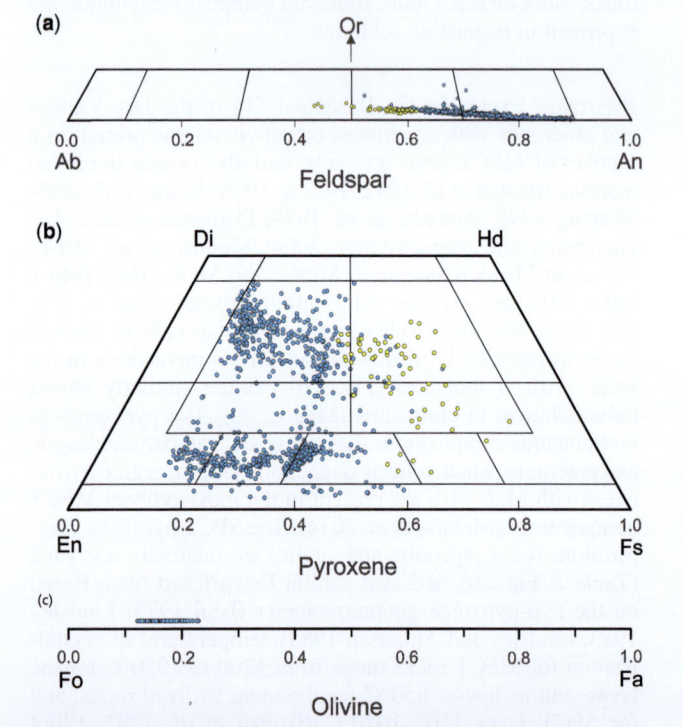

Fig. 10. Compositional variation preserved in minerals from sills at Dawson Peak and lavas at Storm Peak determined by electron microprobe analysis. (**a**) Plagioclase. (**b**) Pyroxene. (**c**) Olivine. Plagioclase from MFCT samples are in blue, SPCT samples are in yellow. Data source: Fleming (1995).

Table 2. *Pyroxene and olivine compositions determined by microprobe analysis for Ferrar rocks from Dawson Peak and Storm Peak*

Sample	85-71-1	85-71-1	85-75-1	85-75-1	85-75-9	85-75-9	85-75-9	85-75-36	85-75-36	85-76-63	85-76-63	85-71-2	85-71-2
Type	MFCT	MFCT	MFCT	MFCT	MFCT	MFCT	MFCT	MFCT	MFCT	SPCT	SPCT	MFCT	MFCT
	Sill	Sill	Lava	Lava	Lava	Lava	Lava	Lava	Lava	Lava	Lava	Sill	Sill
	pig	aug	pig	aug	pig	subcalcic	aug	aug	pig	pig	aug	ol	ol
Region	CTM	CTM	CTM	CTM	CTM	CTM	CTM	CTM	CTM	CTM	CTM	CTM	CTM
SiO_2	55.09	52.84	52.06	52.93	51.59	50.35	48.42	52.18	52.71	49.12	49.17	39.81	38.28
TiO_2	0.15	0.27	0.20	0.21	0.29	0.73	1.14	0.40	0.24	0.43	0.74		
Al_2O_3	2.40	2.79	0.68	1.18	0.93	1.91	3.26	1.38	0.78	0.82	1.48	0.03	0.03
FeO	8.58	7.14	20.32	8.51	22.51	19.87	17.83	11.57	17.96	30.79	20.58	13.05	20.89
MnO	0.25	0.17	0.41	0.32	0.48	0.43	0.38	0.19	0.46	0.62	0.45	0.24	0.32
MgO	27.94	18.48	20.25	18.62	18.6	15.08	12.60	16.06	22.31	12.99	10.55	46.10	39.44
CaO	4.66	17.75	4.97	17.36	4.98	10.51	15.06	17.85	4.47	4.07	15.51	0.21	0.09
Na_2O	0.04	0.11	0.08	0.08	0.03	0.14	0.14	0.15	0.05	0.02	0.07		
Cr_2O_3	1.07	0.24	0.02	0.06	0.04	0.01	0.11	0.11	0.05	0.00	0.02		
NiO												0.27	0.10
Total	100.17	99.79	98.99	99.27	99.45	99.03	98.95	99.88	99.02	98.86	98.57	99.71	99.14
Si^{4+}	1.945	1.932	1.971	1.958	1.964	1.936	1.877	1.950	1.968	1.960	1.940	0.996	0.999
Ti^{4+}	0.004	0.007	0.006	0.006	0.008	0.021	0.033	0.011	0.007	0.013	0.022		
Al^{3+}	0.100	0.120	0.030	0.052	0.042	0.087	0.149	0.061	0.034	0.039	0.069	0.001	0.001
Fe^{2+}	0.253	0.218	0.644	0.263	0.717	0.639	0.579	0.361	0.561	1.028	0.679	0.273	0.456
Mn^{2+}	0.007	0.005	0.013	0.010	0.016	0.014	0.013	0.006	0.014	0.021	0.015	0.005	0.007
Mg^{2+}	1.470	1.007	1.143	1.027	1.055	0.865	0.729	0.895	1.242	0.773	0.621	1.718	1.534
Ca^{2+}	0.176	0.695	0.197	0.688	0.203	0.433	0.625	0.715	0.179	0.174	0.656	0.006	0.003
Na^+	0.003	0.008	0.006	0.006	0.002	0.010	0.011	0.011	0.003	0.002	0.005		
Cr^{3+}	0.030	0.007	0.001	0.002	0.001	0.000	0.010	0.003	0.001	0.000	0.001		
NiO												0.005	0.002
Total	3.988	4.001	4.011	4.012	4.008	4.004	4.021	4.013	4.009	4.009	4.006	3.004	3.001
O^{2-}	6.000	6.000	6.000	6.000	6.000	6.000	6.000	6.000	6.000	6.000	6.000	4.000	4.000
Wo	9.3	36.2	10.1	34.8	10.3	22.4	32.4	36.3	9.0	8.8	33.5		
En	77.4	52.4	57.5	51.9	53.4	44.6	37.7	45.4	62.7	39.1	31.7		
Fs	13.3	11.4	32.4	13.3	36.3	33.0	29.9	18.3	28.3	52.0	34.7		
Fo												86.3	77.1
Fa												13.7	22.9

Analyses performed at the Ohio State University Electron Microprobe Laboratory.

The analyses represent the range of pyroxene and olivine compositions in the olivine-bearing dolerite sill at Dawson Peak and in the lavas at Storm Peak, central Transantarctic Mountains. Pyroxene compositions: olivine-bearing dolerite 85-71-1; analysis 85-75-1 is for the basal, thin basic flow; sample 85-75-9 is from the 135 m-thick tachylitic flow 2; the two analyses for samples 85-76-36 (flow 12) and 85-76-63 (flow 15) represent the range in compositions in those flows; sample 85-76-63 is the SPCT flow. Olivine compositions are from olivine-bearing dolerite sill sample 85-71-2. Data are from Fleming (1995).

the intrusion, ranging between Mg# *c.* 75 and Mg# *c.* 35. Temperatures of crystallization were estimated to be in the range 1180–1040°C, falling to about 800°C for the late stages. Forsteritic olivine does not occur in the lower part of the intrusion, indicating that a thick section is most likely to be concealed beneath the surface. The rare occurrence of fayalitic olivine in the upper part of the intrusion has been noted (Himmelberg and Ford 1976). The common oxide is titanomagnetite with ilmenite exsolution lamellae; independent ilmenite is rare (Himmelberg and Ford 1977). Preliminary results of ongoing investigations were reported in Grimes *et al.* (2008) and Carnes *et al.* (2011).

Major and trace element geochemistry

The Ferrar tholeiites (Ford and Kistler 1980; Kyle 1980; Mensing *et al.* 1984, 1991; Siders and Elliot 1985; Brotzu *et al.* 1988, 1992; Faure *et al.* 1991; Brewer *et al.* 1992; Fleming *et al.* 1992, 1995; Hornig 1993; Elliot *et al.* 1995; Fleming 1995; Morrison and Reay 1995; Wilhelm and Wörner 1996; Antonini *et al.* 1997, 1999; Demarchi *et al.* 2001; Hanemann and Viereck-Götte 2004, 2007*b*; Ross *et al.* 2008; Melluso *et al.* 2014; Elliot and Fleming 2017) constitute two distinct compositional groups: the bulk of the rocks belong to the Mount Fazio Chemical Type (MFCT) and the remaining *c.* 1% to the Scarab Peak Chemical Type (SPCT) (Fleming *et al.* 1992). Representative chemical analyses are presented in Table 3; in the Province overall, the SPCT has a markedly restricted composition (Mg# = 22–24), whereas the MFCT has a broad range of compositions (Mg# *c.* 11–69), with the olivine-bearing dolerite sill margins representing the most primitive liquid compositions and having the highest Mg numbers. Evolved tholeiite samples from Thern Promontory have the lowest Mg numbers. There is some uncertainty as to whether these Thern Promontory rocks represent a lava sequence or just two sills (see Brotzu *et al.* 1988, 1992; Lanza and Zanella 1993; Melluso *et al.* 2014; the authors, who have not visited the locality, prefer the sill interpretation, in which case the highly evolved compositions, with MgO concentrations as low as 0.6%, reflect *in situ* differentiation within a sill interior, a view supported by L. Viereck pers. comm. June 2018). The bulk of the lavas have MgO between 2.5 and 7.5%. High MgO (>7.5%) in some thicker lavas represents an accumulation of pyroxene and reflects *in situ* differentiation in those flows, which is supported by relatively high Cr and Ni. Coherent trends on variation diagrams for fine-grained rocks (Fig. 11) demonstrate the MFCT forms a related set of magma compositions. Large ion lithophile elements (LILEs) show marked scatter, which is greatly reduced if those analyses with high loss-on-ignition are excluded (e.g. Fleming *et al.* 1992). HFSEs show regular increases with decreasing Mg number. Incompatible elements plotted on a mid-ocean ridge basalt (MORB)-normalized diagram

Table 3. *Major and trace element analyses of a sill (85-72-2) at Dawson Peak, nine MFCT lavas and one SPCT lava at Storm Peak*

Sample	85-72-2	85-75-1	85-76-39	85-76-36	85-76-33	85-76-20	85-76-17	85-76-29	85-76-23	85-76-49	85-75-11	85-76-60
Type	Sill	Flow 1	Flow 13	Flow 12	Flow 11	Flow 8	Flow 7	Flow 10	Flow 9	Flow 14	Flow 2	Flow 15
SiO_2	52.68	54.26	55.46	57.25	57.99	59.10	57.89	58.90	59.41	58.66	59.71	57.85
TiO_2	0.49	0.66	0.95	1.25	1.37	1.43	1.55	1.53	1.56	1.62	1.52	2.00
Al_2O_3	16.22	15.49	14.26	13.58	13.34	13.14	12.99	12.79	12.91	12.87	12.56	12.07
Fe_2O_3	1.06	1.14	1.38	1.45	1.49	1.50	1.60	1.55	1.52	1.60	1.57	1.84
FeO	7.04	7.57	9.17	9.63	9.94	10.02	10.64	10.32	10.16	10.67	10.45	12.25
MnO	0.16	0.20	0.19	0.18	0.18	0.19	0.22	0.18	0.18	0.18	0.18	0.19
MgO	9.15	7.08	5.71	4.44	3.81	3.28	3.34	3.21	2.90	2.93	2.61	2.28
CaO	11.36	9.59	9.07	8.80	8.38	8.07	7.32	7.48	7.56	7.65	7.23	7.05
Na_2O	1.54	3.36	2.33	2.19	2.47	2.42	1.94	2.05	2.39	2.71	2.26	2.39
K_2O	0.23	0.58	1.36	1.07	0.86	0.68	2.34	1.81	1.24	0.92	1.72	1.83
P_2O_5	0.06	0.10	0.13	0.15	0.17	0.17	0.19	0.17	0.18	0.20	0.19	0.26
Total	100.00	100.00	100.00	100.00	100.00	100.00	100.00	100.00	100.00	100.00	100.00	100.00
LOI	1.13	1.89	1.46	0.83	0.97	1.03	3.15	1.84	0.82	1.55	0.90	0.36
Mg#	69.8	62.5	52.6	45.1	40.6	36.9	35.9	35.7	33.7	32.9	30.8	24.9
Trace elements by XRF (Ni, Cr, Sr, Zr) and ICP-MS												
Ni	119	66	49	32	23	13	13	13	11	12	6	8
Cr	398	109	47	43	32	28	25	27	24	17	18	15
Rb	6.9	11.7	42.2	59.0	65.6	73.2	53.0	65.0	68.5	78.1	75.6	67.1
Sr	111	101	132	135	134	137	66	136	141	131	136	124
Y	15.9	22.0	28.8	36.4	37.1	40.2	37.1	38.6	41.1	43.3	39.5	53.6
Zr	58	94	121	151	162	172	180	172	177	184	180	202
Nb	2.84	5.26	6.78	9.78	9.84	10.57	10.14	10.80	11.59	10.62	10.82	10.79
Cs	0.52	0.74	1.01	2.09	2.50	4.03	1.47	2.57	2.91	3.19	2.79	1.56
Ba	70	377	264	296	292	315	414	355	366	335	380	383
La	6.64	11.66	15.70	20.80	22.75	23.55	22.74	23.76	24.97	26.21	25.18	26.09
Ce	13.11	23.89	31.44	41.88	45.00	48.11	45.63	47.81	50.05	52.37	50.49	54.14
Pr	1.60	2.88	3.69	4.98	5.35	5.66	5.40	5.54	5.89	6.16	5.85	6.48
Nd	6.76	12.02	15.29	20.20	21.41	23.52	22.27	22.81	24.71	25.37	24.16	26.98
Sm	1.95	3.08	4.09	5.12	5.54	5.84	5.85	5.86	6.09	6.51	6.22	7.53
Eu	0.64	0.92	1.04	1.28	1.39	1.42	1.37	1.40	1.55	1.52	1.52	1.84
Gd	2.13	3.23	4.26	5.29	5.60	6.05	5.72	5.64	5.97	6.23	6.03	7.92
Tb	0.41	0.63	0.81	0.97	1.01	1.12	1.03	1.04	1.13	1.22	1.09	1.51
Dy	2.69	4.08	5.22	6.05	6.50	6.89	6.81	6.62	6.96	7.50	6.94	9.80
Ho	0.62	0.83	1.12	1.28	1.39	1.47	1.43	1.38	1.50	1.58	1.49	2.05
Er	1.79	2.49	3.35	3.78	3.98	4.24	4.19	4.06	4.26	4.78	4.49	6.10
Tm	0.25	0.36	0.47	0.55	0.57	0.61	0.58	0.58	0.62	0.68	0.63	0.85
Yb	1.60	2.27	2.85	3.45	3.49	3.82	3.69	3.68	3.83	4.12	3.98	5.41
Lu	0.26	0.36	0.48	0.52	0.57	0.61	0.61	0.59	0.62	0.65	0.61	0.83
Hf	1.40	2.37	3.19	3.89	4.38	4.58	4.64	4.66	4.87	5.01	4.93	5.70
Ta	0.22	0.34	0.46	0.58	0.65	0.68	0.70	0.69	0.72	0.75	0.74	0.86
Pb	2.86	5.69	7.33	9.16	10.31	17.14	11.60	10.98	11.83	11.62	12.08	11.68
Th	1.47	2.77	3.86	4.98	5.61	5.91	6.02	6.01	6.26	6.51	6.49	6.31
U	0.29	0.58	0.83	1.02	1.16	1.29	1.22	1.25	1.28	1.34	1.35	1.23

LOI, loss on ignition.
Analyses recalculated to 100%.
Iron partitioned: $Fe_2O_3/FeO = 0.15$.
Analyses were performed at the GeoAnalytical Laboratory at Washington State University.

(Fig. 12) highlight the crust-like patterns of the Ferrar rocks. Rare earth element (REE) diagrams (Fig. 13) illustrate patterns typical of continental tholeiites: enriched light REE but flat patterns for medium REE and heavy REE, with a negative Eu anomaly except in the olivine-bearing dolerite compositions, indicate the role of plagioclase during differentiation at crustal depths. Platinum group elements (PGEs) in north Victoria Land tholeiites (Hanemann and Viereck-Götte 2007*a*) show modest correlations with MgO (or Mg#). A detailed investigation of the PGEs in the Basement Sill in south Victoria Land revealed a positive correlation between Os and Ir at MgO less than 8%, and positive (convex-shaped) slopes between the Os–Ir–Ru group and the Pt–Pd–Rh group (Choi *et al.* 2019*a*). Preliminary results of investigations into PGEs in the Dufek intrusion have been reported by Mukasa *et al.* (2007) and Hanemann *et al.* (2009).

On a classical AFM diagram (Fig. 14), the MFCT compositions exhibit strong Fe-enrichment typical of tholeiitic rocks, with the Thern Promontory rocks (Melluso *et al.* 2014) and an interstitial glass from a Mesa Range lava (Elliot *et al.* 1995) showing the most extreme *in situ* tholeiitic magma evolution (although none is likely to be a liquidus composition). The extreme Fe-enrichment (Fe^t c. 14%) of the SPCT rocks is comparable to that of ferrobasalts from mid-ocean ridges.

Olivine-bearing dolerite sills are known to crop out in the central Transantarctic Mountains and south Victoria Land but not north Victoria Land; however, this may simply be an artefact of exposure or lack of discovery. The lavas, however, show regional variations in predominant major element compositions (Elliot and Fleming 2017). The majority of lavas in the central Transantarctic Mountains are evolved with only a few having Mg# >45, whereas in south Victoria Land lavas have Mg# = 40–65, and in north Victoria Land the Mesa Range lavas have a relatively restricted range of Mg# = 50–62. Excluding the olivine-bearing dolerites sills, chilled margins of sills in the Shackleton Glacier region have a Mg# range of 45–65, and in the Queen Alexandra Range region the range in Mg# is 56–65, which is in contrast to the lavas (Mg# <45). In south Victoria Land, the sills in the Dry Valleys have a Mg# range of 40–60, and in the Prince Albert Mountains the Mg# range is 55–62. In north Victoria Land the Mg# of the sills lies in the range 48–62. There is no spatial pattern in the geochemistry of the sills, but quite clearly the opposite is the case for the lavas and implies the eruption of regionally distinct batches of magma. This probably reflects differing residence times in crustal magma chambers prior to supracrustal emplacement. Further, it suggests the sills might be an episode of magma emplacement distinct from that of the extrusive rocks.

Isotope geochemistry

The unusually high initial Sr isotope composition of Ferrar tholeiites (Sr_i c. 0.711) from the Dry Valleys region of south Victoria Land was established by Compston et al. (1968), following on from investigation of the Tasmanian dolerites by Heier et al. (1965). These early results for the Ferrar dolerites were extended to the Kirkpatrick Basalt lavas in the Queen Alexandra Range (Table 4: all Sr and Nd isotope data have been recalculated to an age of 182.7 Ma), and the Sr isotope compositions (Sr_i = 0.7094–0.7133) were related to large-scale contamination of basaltic magma by granitic rocks (Faure et al. 1972, 1974, 1982).

Subsequent oxygen isotope studies on whole-rock lavas from all major outcrop regions (Hoefs et al. 1980; Kyle et al. 1983; Mensing et al. 1984, 1991) found a wide range of $\delta^{18}O$ values (6.0–9.3), and the weak correlations with initial $^{87}Sr/^{86}Sr$ were interpreted to support crustal assimilation. Later work (Fleming et al. 1992) revealed that much of the range of whole-rock oxygen isotope compositions could be found in a single chemically homogeneous lava flow ($\delta^{18}O$ = 5.8–8.1), with the plagioclase separates from that flow having a markedly limited range ($\delta^{18}O$ = 5.5–5.8), which approaches mantle-like values. The large range of previously published whole-rock compositions was reinterpreted to be the result of low-temperature interaction (alteration) of fine-grained and glassy components in the rocks with meteoric water. The Sr–O variations in the chilled margins of

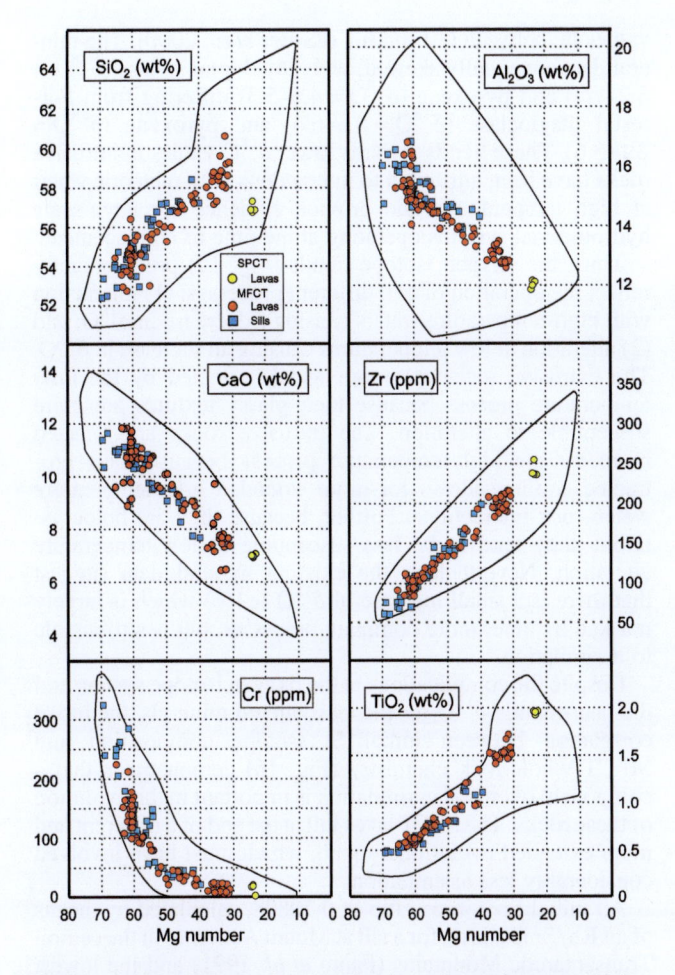

Fig. 11. Variation diagrams for selected major and trace elements v. Mg# for Ferrar Group lavas and sills to illustrate the geochemical coherence of the MFCT and the restricted and different composition of the SPCT. Plotted data include only fine-grained samples of lavas and chilled margins of sills from north Victoria Land, south Victoria Land and the central Transantarctic Mountains. Data sources: Elliot and Fleming (2017) and unpublished data. The outlined fields reflect data published in the literature over a period of more than 50 years, and which are compiled in the GEOROC database (Sarbas et al. 2017). The greater dispersion in those data is attributed to a combination of analytical issues, alteration and in situ differentiation in some larger magma bodies. Analyses reflecting cumulate compositions have been removed but it is more difficult to identify samples affected by in situ differentiation at the evolved end of the compositional spectrum (Mg# <30).

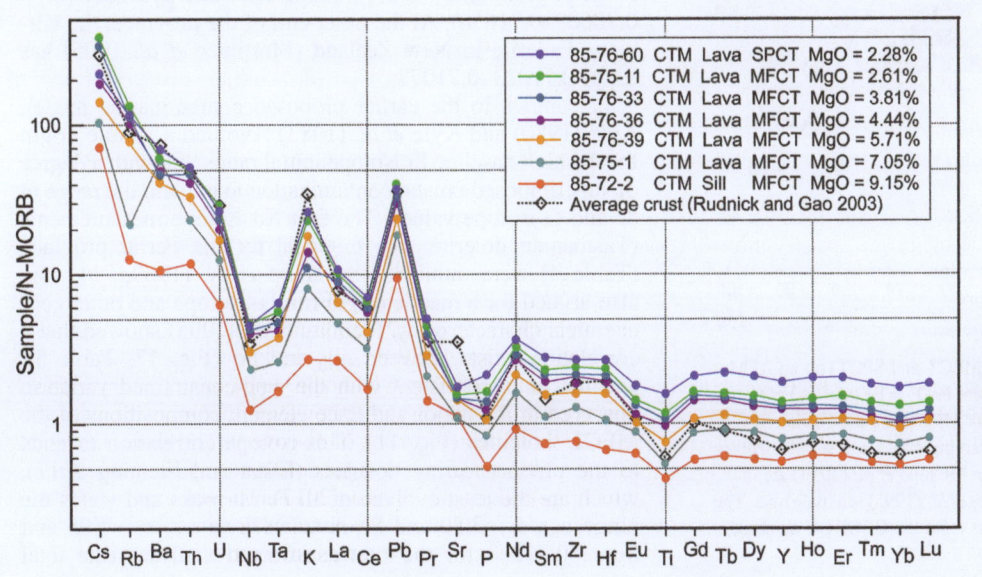

Fig. 12. MORB-normalized trace element diagram for samples (from Dawson Peak and Storm Peak) of the Ferrar Group selected to cover the entire range of MgO concentrations observed. The normalization factors are from Sun and McDonough (1989). Data source: Table 4 and Elliot and Fleming (2017 and supplementary data therein).

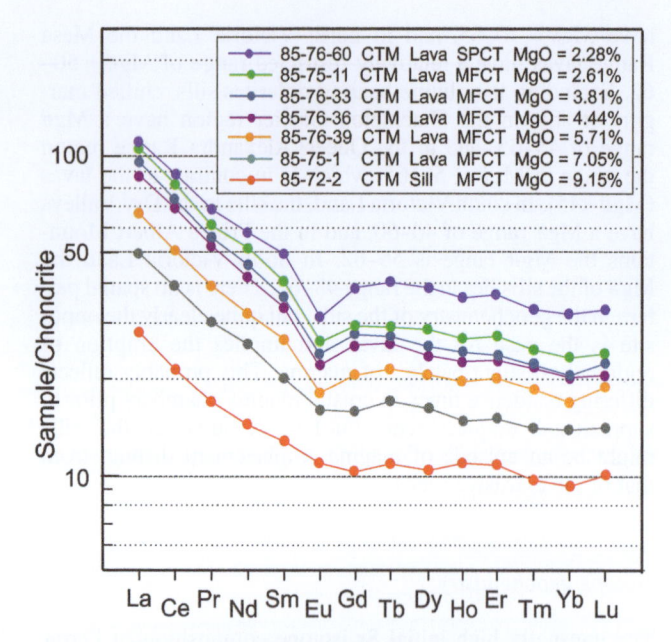

Fig. 13. Chondrite-normalized rare earth element diagram for Ferrar Group samples (from Dawson Peak and Storm Peak) illustrated in Figure 12. The normalization factors are from Sun and McDonough (1989).

Tasmanian sills and the sill at Portal Rock had also been interpreted as the result of meteoric water interactions (Hergt *et al.* 1989*a*, *b*). In contrast to the extrusive rocks, sills have been shown to have a range of compositions that trend towards very low values ($\delta^{18}O$ = 1.9–6.1) (Hergt *et al.* 1989*b*; Faure *et al.* 1991). For the Dufek intrusion, whole-rock $\delta^{18}O$ values for the lower section in the Dufek Massif are 5.0–6.9, but the Forrestal Range upper section is much more varied and $\delta^{18}O$ values lie between 0.0 and 6.1 (Kistler *et al.* 2000). The mineral data are equally skewed, with Dufek plagioclase ($\delta^{18}O$ = 6.2–7.7) and pyroxene ($\delta^{18}O$ = 4.6–5.3) differing from Forrestal plagioclase ($\delta^{18}O$ = 0.3–6.4) and pyroxene ($\delta^{18}O$ = 2.1–5.6). These trends towards lower $\delta^{18}O$ values in intrusive rocks have been attributed to interactions with meteoric water at high temperatures and provide evidence for large-scale hydrothermal systems operating at the time of emplacement.

Thus, the oxygen isotope trends (Fig. 15) predominantly reflect the operation of two different processes: (1) interaction with high-temperature waters causing a decrease in $\delta^{18}O$; and (2) alteration at low temperatures causing an increase in $\delta^{18}O$. The extrusive rocks are more widely affected by the low-temperature process because their glassy textures are more susceptible to alteration. The intrusive rocks are affected more by the high-temperature process because their protracted cooling allows for more extended high-temperature water–rock interactions. Further, because they are holocrystalline they tend to be less susceptible to low-temperature alteration. Nevertheless, the existing mineral data suggest that there is a small increase in $\delta^{18}O$ (<1%) which is largely masked by other more dominant processes, but is attributable to assimilation.

Despite the complications in the oxygen isotope system and the near-mantle $\delta^{18}O$ for the least altered minerals, the broad correlation between initial Sr isotope composition and MFCT whole-rock chemistry (Fig. 16) demonstrates that a component of crustal assimilation is important in the evolution of these rocks. The SPCT lavas fall at the end of a different and more extended evolutionary path, which must have involved considerably less assimilation.

An initial Sr isotope ratio of 0.70808, calculated by means of an Rb/Sr isochron for a sill at Mount Achernar in the central Transantarctic Mountains (Faure *et al.* 1991) and the lowest for a Ferrar sill or lava (excluding the Dufek intrusion), is here confirmed for the chilled margin of that sill (Table 5). The range in Sr isotope initial ratios for Ferrar rocks in the Ross Sea sector varies between 0.70710 and 0.71381 (Table 5), which is the result of secondary processes, as well as crustal assimilation. In the Weddell Sea sector, elevated initial Sr isotope ratios (Sr_i >0.710) were reported by Ford and Kistler (1980) for the Pecora Escarpment (Fig. 1), and by Brewer *et al.* (1992) and Leat (2008) for the Whichaway Nunataks and Theron Mountains (Sr_i >0.70819). The Dufek intrusion (including the capping granophyre) has an initial Sr isotope ratio range for whole-rock analyses of 0.70830–0.71541 (Ford *et al.* 1986; Kistler *et al.* 2000) but with pyroxene as low as 0.70763. Mukasa *et al.* (2003) reported a wider range of preliminary data for plagioclase and pyroxene (Sr_i = 0.70609–0.71656). At the other end of the province, the Kirwans Dolerite in New Zealand (Mortimer *et al.* 1995) has Sr_i of 0.71023–0.71073.

In contrast to the earlier proposed contamination model, Kyle (1980) and Kyle *et al.* (1983) favoured a mantle origin for the high baseline Sr isotope initial ratios, but with a degree of superimposed crustal contamination to explain the range in Sr and O isotope values. The first Nd isotope measurements (Tasmanian dolerite: ε_{Nd} c. −5.1) for the Ferrar province (Table 4) were published by Hergt *et al.* (1989*a*, *b*), who also argued for a mantle origin for the isotope and other geochemical characteristics. Fleming *et al.* (1995) showed that a correlation exists between ε_{Nd} and Sr_i (Fig. 17; Table 5), and that it is consistent with the well-constrained variation observed in the major and trace element compositions of the MFCT tholeiites (Fig. 11). This isotope correlation extends to the olivine-bearing dolerites (Elliot and Fleming 2017), which are the least-evolved of all Ferrar rocks and yields the range ε_{Nd} c. −3.80 and Sr_i c. 0.70878 to ε_{Nd} c. −5.95 and Sr_i c. 0.71288 for the best-constrained analyses (the total

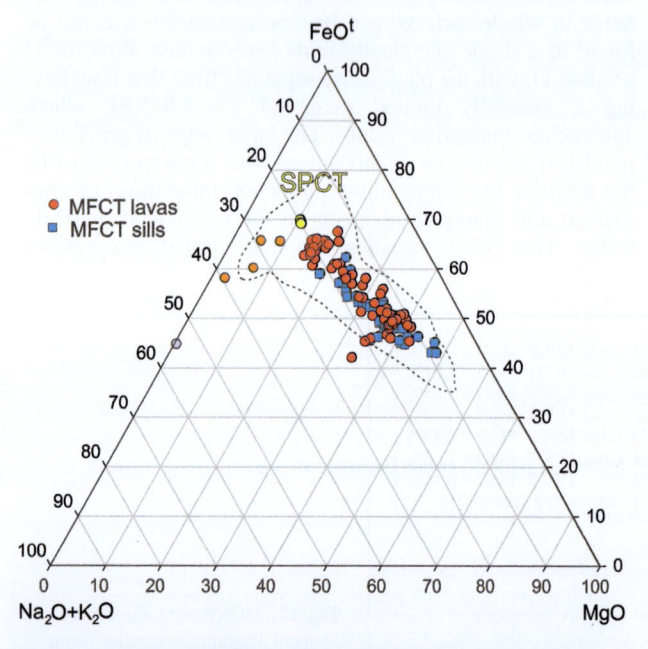

Fig. 14. Compositions of Ferrar rocks (MFCT and SPCT) on an AFM diagram ($Na_2O + K_2O$–FeO^T–MgO, where FeO^T = Fe total as FeO) illustrating the Fe-enrichment trend of the MFCT and the Fe-rich SPCT composition. Data sources: Elliot and Fleming (2017 and supplementary data therein); evolved compositions from Melluso *et al.* (2014) are in orange, and an interstitial glass from Elliot *et al.* (1995) is in pale blue. The field for the Ferrar province as a whole is from the GEOROC database (Sarbas *et al.* 2017).

Table 4. *Summary of Nd, Sr and O isotope data for Ferrar Large Igneous Province tholeiites*

Region	Location	Chemical type	ε_{Nd}	Sr$_i$	$\delta^{18}O$	Reference
Tasmania		MFCT wr sill*	−6.6 to −5.1	0.70934–0.71278	1.9–6.1	Hergt et al. (1989b)
New Zealand	Reefton	MFCT wr sill	−5.6 to −5.3	0.71023–0.71073		Mortimer et al. (1995)
North Victoria Land	Mesa Range	SPCT wr lavas		0.70863–0.70957	5.5–8.2	Fleming et al. (1992)
		SPCT wr lavas	−4.4 to −4.2	0.70858–0.70958		Fleming et al. (1995)
		MFCT wr lavas	−5.5 to −4.8	0.70872–0.71160		
		MFCT pyx.	−5.4	0.70951–0.70955		
		MFCT plag.		0.71061		
		SPCT wr lavas	−4.4 to −4.1	0.70954–0.70968		Elliot et al. (1999)
		SPCT plag.		0.70952		
	Thern Promontory	MFCT wr sill†	−5.9 to −5.1	0.71141–0.71304		Brotzu et al. (1992)
South Victoria Land	Prince Albert Mountains	MFCT wr sills, lavas	−5.6 to −3.3	0.71015–0.71198	4.8–8.0	Molzahn et al. (1996)
		MFCT pyx.	−5.3 to −4.4	0.70955–0.71201	5.2–6.2	
		MFCT plag.	−5.6 to −3.0	0.70763–0.71360	6.0–13.3	
		SPCT wr lavas	−3.5	0.70948	6	
		SPCT pyx.		0.70987	5.2	
		SPCT plag.		0.70987	18.3	
	Prince Albert Mountains	MFCT wr lavas†	−5.7 to −4.7	0.71028–0.71213		Antonini et al. (1999)
		SPCT wr lava	−3.8 to −3.3	0.70938–0.70973		
	Prince Albert Mountains	MFCT wr lavas	−5.4 to −5.1	0.70959–0.71381		Fleming (unpublished data)
	Prince Albert Mountains	SPCT wr lava	−4.4 to −4.3	0.70903–0.70929		Elliot et al. (1999)
		SPCT plag.		0.70949		
	Prince Albert Mountains	MFCT wr lavas		0.7098–7115	6.2–8.3	Kyle et al. (1983)
	Carapace Nunatak	MFCT wr lavas	−5.6 to −5.2	0.71063–0.71127		Fleming et al. (1998)
	Dry Valleys	MFCT wr sills	−5.7 to −5.2	0.71054–0.71191		
	Roadend Nunatak	MFCT wr sills		0.7091–0.7152	4.7–7.1	Faure et al. (1991)
Central Transantarctic Mountains	Storm Peak	MFCT wr lavas	−6.0 to −4.6	0.70970–0.71289		Fleming (1995)
		MFCT pyx.	−5.5 to −4.9	0.70982–0.71273	6.1–6.6	
		MFCT plag.		0.71024–0.71283	6.4–6.8	
		SPCT wr lava	−4.3 to −4.2	0.70957–0.70968		
		SPCT pyx.	−4.3 to −4.2	0.70962–0.71269	6.9	
		SPCT plag.		0.70963–0.70970	5.3–5.9	
	Dawson Peak	MFCT wr sill	−4.1 to −3.9	0.70987–0.71009		
		MFCT Ol-dol sill	−4.0 to −3.7	0.70768–0.70869		
		MFCT Ol-dol. plag		0.70877		
	Mount Achernar	MFCT wr sills		0.70710–0.71027	4.4–6.5	Faure et al. (1991)
	Queen Alexandra Range	MFCT wr sills	−5.4 to −3.8	0.70808–0.71264		Fleming (unpublished data)
	Portal Rock	MFCT wr sill‡	−6.0 to −5.2	0.70901–0.71082	1.9–6.1	Hergt et al. (1989a)
	Shackleton Glacier region	MFCT wr sills	−5.4 to −4.5	0.70859–0.71139		Fleming (unpublished data)
	Storm Peak	SPCT wr lavas	−4.3 to −4.2	0.70945–0.70949		Elliot et al. (1999)
	Grosvenor Mountains	SPCT wr lavas	−4.4 to −4.3	0.70946–0.70947		
	Nilsen Plateau	MFCT wr sills	−5.4 to −3.9	0.70971–0.71368		Fleming (unpublished data)
Theron Mountains		MFCT wr sills†	−5.0 to −3.7	0.70817–0.70955		Leat et al. (2006)
		SPCT wr sill†	−3.9 to −3.8	0.70878–0.70992		
Dufek Intrusion	Dufek Massif	MFCT wr		0.70828–0.71486	5.0–6.9	Kistler et al. (2000)
		MFCT pyx.		0.70743–0.70912	4.6–5.3	
		MFCT plag.		0.70896–0.70984	6.2–7.7	
	Forrestal Range	MFCT wr		0.70874–0.71200	0.1–6.2	
		MFCT pyx.		0.70816–0.71172	3.2–4.1	
		MFCT plag.		0.70932–0.71244	3.2–5.4	
	Dufek Massif	MFCT wr†		0.70609–0.71656		Mukasa et al. (2003)

Plag., plagioclase; pyx., pyroxene; wr, whole rock. Sr and Nd data are calculated to an age of 182.7 Ma.
*Hergt et al. (1989a, b) data renormalized to $^{146}Nd/^{144}Nd = 0.7219$ and adjusted to LaJolla Nd standard $^{143}Nd/^{144}Nd = 0.511843$.
†Data as reported.
‡Nd measurements at lower resolution and as reported.

range for all Ferrar province analysed samples, excluding the Dufek intrusion, is slightly greater: ε_{Nd} c. −3.0 to 6.6 and Sr$_i$ c. 0.70710–0.71381). The SPCT rocks have highly evolved major and trace element compositions, but isotopically are closer to the olivine-bearing dolerites.

Unfortunately, there are no published Nd isotope data to complement the low Sr$_i$ of pyroxene in the Dufek intrusion other than a reported initial Nd isotope ratio range of 0.51213–0.51233 for plagioclase and pyroxene (Mukasa et al. 2003). It should be noted that the full range of reported initial Sr and Nd isotope compositions is greater than that in Figure 17, and, as already noted, it is attributed to analysis in different laboratories at different times, and alteration effects.

There are few Pb isotope analyses for the Ferrar province as a whole. Hergt et al. (1989b), Mortimer et al. (1995) and Antonini et al. (1999) analysed whole-rock samples and provided initial ratio data (cf. Brewer et al. 1992; Kyle et al. 1987), whereas Molzahn et al. (1996) and Mukasa et al. (2003) analysed plagioclase, which requires little or no correction for *in situ* U and Th decay. The $^{207}Pb/^{204}Pb$ initial ratios of whole rocks and of plagioclase lie in the range 15.61–15.68, with

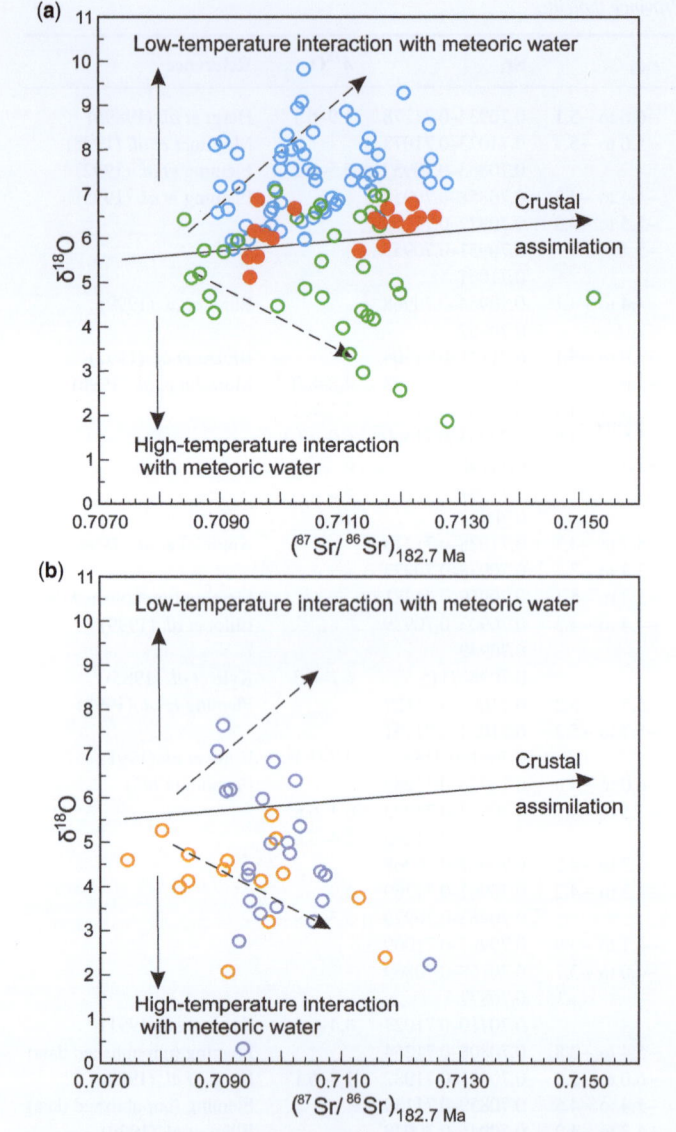

Fig. 15. ^{87}Sr/^{87}Sr$_i$ (at 182.7 Ma) v. δ^{18}O for Ferrar province rocks and minerals throughout the Transantarctic Mountains. (**a**) Whole-rock Kirkpatrick Basalt lavas (blue circles), Ferrar Dolerite sills (green circles), and pyroxene and plagioclase (red circles). See Table 4 for the data sources. (**b**) Pyroxene (orange circles) and plagioclase (purple circles) data for the Dufek intrusion (Kistler *et al.* 2000). Dashed arrows represent diagrammatic paths of evolution depending on high- or low-temperature alteration.

^{208}Pb/^{204}Pb ratios of 38.24–38.54 and ^{206}Pb/^{204}Pb ratios of 18.55–18.64. These ratios, plotting above the Northern Hemisphere Reference Line, reflect the high abundance of Pb and its crustal character. Osmium isotopes also have been measured (Molzahn *et al.* 1996; Brauns *et al.* 2000; Hergt and Brauns 2001; Mukasa *et al.* 2003). Os concentrations are quite low, leading to significant uncertainties, but initial ratios are consistent with a mantle origin (^{187}Os/^{188}Os$_i$ = 0.145 ± 0.049–0.194 ± 0.023). This has been confirmed by a detailed investigation of the Basement Sill in south Victoria Land (Choi *et al.* 2019*a*), which reported subchondritic Os/Ir ratios (<0.33) and a least radiogenic value of ^{187}Os/^{188}Os = 0.1609 ± 0.0003 (2σ), although the total range is quite extended (up to ^{187}Os/^{188}Os$_i$ = 8.100 ± 1.600).

Mensing *et al.* (1984, 1991) reported variable sulfur isotope compositions for Mesa Range tholeiites, which were attributed to outgassing under a range of oxygen fugacities; a conclusion also reached for the Kirkpatrick Basalt at Mount Falla, Queen Alexandra Range (Faure *et al.* 1984). Low sulfur saturation has been proposed to account for the PGE abundances in MFCT sills in north Victoria Land (Hanemann and Viereck-Götte 2007*a*).

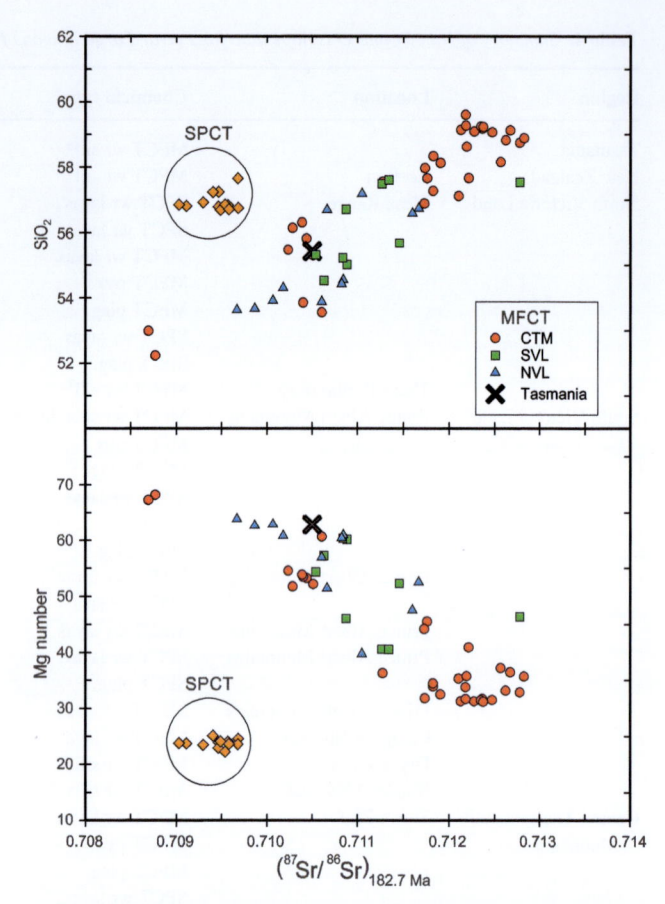

Fig. 16. ^{87}Sr/^{87}Sr$_i$ (at 182.7 Ma) v. SiO$_2$ and Mg number, illustrating MFCT correlations that reflect fractional crystallization and crustal assimilation. Data points represent the Ferrar province from north Victoria Land to the central Transantarctic Mountains. The SPCT composition must have followed a different evolutionary path. Data sources are the same as for Figures 11 and 15.

Several conclusions have been drawn from these results. First, Ferrar rocks have high initial strontium isotope ratios, which begin at a baseline value of 0.708 (with the majority >0.709), and low ε_{Nd} values (most are more negative than −3.7). Second, mineral separates confirm that both high and low whole-rock δ^{18}O values result from secondary processes. Third, there is a clear correlation between Sr and Nd isotope ratios, which, together with the whole-rock chemistry, points to a path of low-pressure evolution involving both fractional crystallization and assimilation of crustal material, from olivine-bearing dolerite to andesitic compositions.

Post-emplacement alteration and secondary mineralization

Fleming *et al.* (1989, 1992, 1993) proposed a mid-Cretaceous alteration event affecting the Kirkpatrick Basalt lavas based on a 103 Ma Rb/Sr array or 'errorchron' derived from SPCT samples. They attributed it to tectonism related to the break-up of Antarctica and Australia, and the development of associated hydrothermal systems, which caused mobility of Rb. This event is reflected by scattered K/Ar dates and anomalous palaeomagnetic pole positions determined for the lavas

Table 5. *MgO%, Rb, Sr, Sm and Nd concentrations, and $^{87}Sr/^{86}Sr$ and $^{143}Nd/^{144}Nd$ present-day and calculated initial ratios for selected tholeiites*

Sample	MgO %	Rb (ppm)	Sr (ppm)	$^{87}Rb/^{86}Sr$	$^{87}Sr/^{86}Sr$*	$^{87}Sr/^{86}Sr$†	ε_{Sr}‡	Sm (ppm)	Nd (ppm)	$^{147}Sm/^{144}Nd$	$^{143}Nd/^{144}Nd$*	$^{143}Nd/^{144}Nd$†	ε_{Nd}‡
85-72-2	9.14	6.3	115.6	0.1575	0.709173(10)	0.708764(10)	60.8	1.85	7.4	0.1510	0.512381(5)	0.512200(5)	−3.95
85-75-1	6.99	10.5	100.2	0.3044	0.711421(10)	0.710630(11)	87.3	2.94	12.4	0.1432	0.512310(6)	0.512139(6)	−5.16
85-76-42	6.15	41.5	123.0	0.9772	0.712707(10)	0.719169(16)	80.7	3.63	15.4	0.1423	0.512330(7)	0.512160(7)	−4.74
85-76-39	5.67	44.6	131.1	0.9792	0.713053(11)	0.710509(17)	85.6	4.03	17.1	0.1425	0.512317(7)	0.512147(7)	−5.00
85-76-36	4.44	57.4	136.3	1.2201	0.714786(9)	0.711617(19)	101.3	4.95	21.8	0.1374	0.512289(6)	0.512125(6)	−5.43
85-76-33	3.80	65.9	136.3	1.3991	0.715539(14)	0.711905(23)	105.4	5.30	23.5	0.1365	0.512277(6)	0.512114(6)	−5.64
85-76-20	3.26	74.1	139.0	1.5431	0.715782(9)	0.711773(23)	103.5	5.79	25.6	0.1367	0.512274(5)	0.512111(5)	−5.71
85-76-17	3.25	56.9	64.3	2.5632	0.717934(9)	0.711276(36)	96.5	5.82	26.1	0.1349	0.512284(6)	0.512123(6)	−5.47
85-76-29	3.18	61.5	131.9	1.3493	0.716273(8)	0.712768(20)	117.7	5.84	26.1	0.1352	0.512270(5)	0.512108(50	−5.75
85-76-49	2.89	73.4	134.2	1.5841	0.715850(9)	0.711735(23)	103.0	6.22	27.5	0.1364	0.512286(4)	0.512123(4)	−5.46
85-76-23	2.88	67.2	142.9	1.3630	0.716161(9)	0.712740(40)	117.2	5.83	26.1	0.1352	0.512265(7)	0.512109(7)	−5.89
85-75-11	2.58	76.8	137.6	1.5946	0.716270(10)	0.712128(24)	108.6	5.83	26.0	0.1355	0.512271(6)	0.512109(6)	−5.74
85-76-60	2.26	66.7	128.8	1.4979	0.713460(9)	0.709569(22)	72.2	7.42	31.3	0.1435	0.512353(6)	0.512181(6)	−4.32
11-1-3	6.03	22.8	169.5	0.3885	0.709088(9)	0.708078(10)	48.2	3.17	13.8	0.1395	0.512344(7)	0.512177(7)	−4.41

*Present-day measured isotopic ratios normalized with $^{87}Sr/^{86}Sr = 0/_{0}/119400$ or $^{143}Nd/^{144}Nd = 0/_{0}/121900$. 2σ mean within-run uncertainties in the last digits are given in parentheses. Mean values (and 1σ external reproducibilities) for standards measured during the same period are: SRM 987, $^{87}Sr/^{86}Sr = 0.710243$ (± 0.000010); and LaJolla Nd, $^{143}Nd/^{144}Nd = 511843$ (± 0.000005).

†Calculated model initial ratios at 182.7 Ma with decay constants of 1.42×10^{-11} (^{87}Rb) or 6.54×10^{-12} (^{147}Sm); uncertainties (in parentheses) provide for uncertainties in present-day measured isotopic ratios, parent/daughter ratios (0.5% for $^{87}Rb/^{86}Sr$ and 0.1% for $^{147}Sm/^{144}Nd$) and age (± 1.8 Ma).

‡Conventional ε notation for 182.7 Ma with reference values of $^{87}Rb/^{86}Sr = 0.085$, $^{87}Sr/^{86}Sr = 0.7047$, $^{147}Sm/^{144}Nd = 0.1966$ and $^{143}Nd/^{144}Nd = 0.512638$.

Samples were selected to illustrate the range of MgO contents and isotopic compositions in a single relatively restricted region. Ferrar tholeiites from the central Transantarctic Mountains: olivine-bearing dolerite (85-72-2) with the highest MgO content of all sills and lavas from near Dawson Peak; lavas from Peterson Ridge near Storm Peak (see also Table 3); and the sill (11-1-3), with the lowest Sr isotope initial ratio recorded for any lava or sill, from near Mount Achernar. Data are from Fleming (1995) and previously unpublished data (sample 11-1-3). Grid references are given in Appendix A.

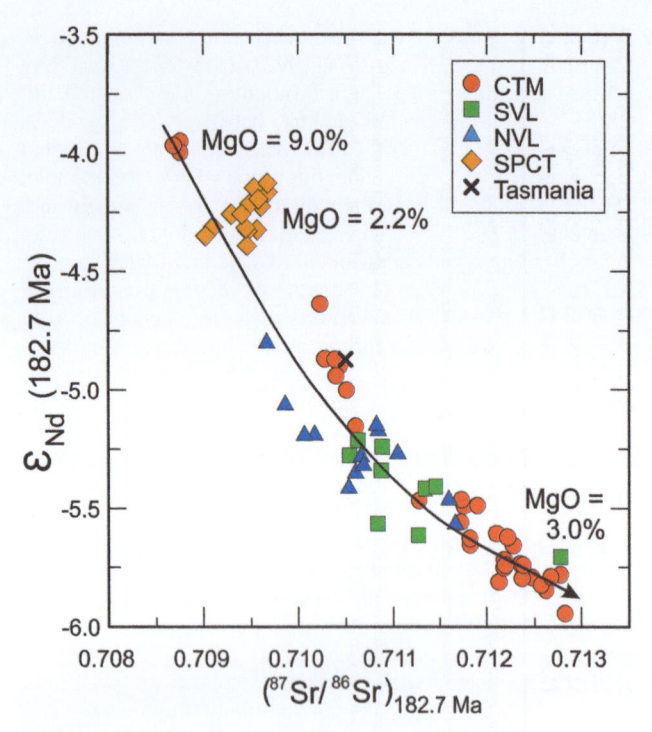

Fig. 17. ε_{Nd} v. $^{87}Sr/^{87}Sr_i$ (at 182.7 Ma) for Ferrar Large Igneous Province sills and lavas from north Victoria Land to the central Transantarctic Mountains. Data sources: for Tasmania, Hergt *et al.* (1989*b*); for Antarctica, Fleming (1995), Fleming *et al.* (1992, 1995) and Elliot *et al.* (1999); and unpublished data for south Victoria Land. Other data are not plotted: (1) because of analytical uncertainties (not given, large or highly variable) and interlaboratory biases; and (2) because whole-rock $^{87}Sr/^{87}Sr$ ratios for sill margins are affected by high-temperature alteration and, lacking measurement of Sr isotope ratios for plagioclase in the same rock, are subject to uncertainties.

(McIntosh *et al.* 1986; Delisle and Fromm 1989; Faure and Mensing 1993; Mensing and Faure 1996). Molzahn *et al.* (1999) dated apophyllite from vugs in the Kirkpatrick Basalt in the Prince Albert Mountains by the $^{40}Ar/^{39}Ar$ method and determined a crystallization age of 96.7 ± 0.6 Ma, which they interpreted in terms of an alteration event. A less well-defined apophylllite crystallization event was dated at 125–112 Ma. Age determinations of apophyllites by the $^{40}Ar/^{39}Ar$ and Rb/Sr methods (Fleming *et al.* 1999) extended those earlier results. Total $^{40}Ar/^{39}Ar$ gas ages vary from 133 to 114 Ma for the Queen Alexandra Range, 114–95 Ma for south Victoria Land and 100–76 Ma for north Victoria Land. Rb/Sr model ages range from 144 to 94 Ma; in some instances ages are concordant with the Ar total gas ages and in others are as much as 14 myr older. These data have been interpreted to record the early stage of uplift of the Transantarctic Mountains (Fleming *et al.* 1999). The differing patterns of age were attributed to the mountain range consisting of several major blocks which had different uplift histories (Fitzgerald 2002), let alone differing hydrological systems and thermal regimes. Further, the youngest apophyllite ages are broadly comparable to that of the metamorphic core complex in Marie Byrd Land (about 105 Ma), which marks separation of the New Zealand microcontinent from West Antarctica (Siddoway 2007) and the initiation of the West Antarctic Rift System.

The zeolite assemblage in the lava successions (see Elliot *et al.* 2021) suggests the possibility of a now eroded overburden of Ferrar lavas and/or Mesozoic sedimentary strata 1 km or more thick. Rather than recording the early stages of uplift, Lisker and Läufer (2013) have argued that a Jurassic–early Cretaceous sedimentary basin, overlying the Ferrar lavas but now eroded, better explains the apatite fission-track uplift data for the Transantarctic Mountains. A variety of thermal regimes and hydrological systems would also have existed in and beneath such a basin, thus leading to secondary mineralization and young Ar ages for lavas, and Cretaceous ages for apophyllites.

Magma emplacement at supracrustal depths and evolution

Building on the early work of Gunn (1962, 1966) on the sills in the Dry Valleys region, investigation of the Basement Sill has provided fundamental information on the mode of emplacement at upper-crustal depths (low pressure) and subsequent textural evolution of basic magmas (Bédard *et al.* 2007; Charrier 2010; Jerram *et al.* 2010; Charrier and Manochehri 2013; Petford and Mirhadizadeh 2017). The Basement Sill, which has been identified over an area of about 10 000 km^2 (Marsh 2007), is interpreted to be the result of injection of large batches of magma with an entrained tongue of orthopyroxene. Magmas spread outwards, as a series of lobes, from an inferred point of origin, which is postulated to be a vertical conduit connected at depth to the magma source (Marsh *et al.* 2005; Souter *et al.* 2006). The relatively fast cooling of the sill resulted in preservation of compaction and interstitial liquid segregation features, which are generally lost in more slowly cooled and thoroughly annealed layered basic intrusions (e.g. Dufek intrusion). Comparable injection of magma batches has also been proposed for the Beacon (Asgard) Sill in the Dry Valleys (Zieg and Marsh 2012). With this as a model, sills elsewhere may be interpreted as lateral injections principally into Beacon strata, and possibly from at least three principal centres spaced along the Transantarctic Mountains (Elliot and Fleming 2008). Based on their modelling results for the Basement Sill, Petford and Mirhadizadeh (2017) estimated lateral emplacement times. Assuming a constant viscosity of 33 Pa s, together with continuous and uniform flow in chemically coherent magma, lateral transport over 3000 km could be accomplished in about 1 year. At higher viscosities (e.g. 10^4 Pa s) a similar distance would take less than 2 × 10^5 years. Further, they estimated that the Basement Sill could have been filled in 10^5 years, provided viscosity and supply rate remained constant.

Ongoing studies of the mode of emplacement and accumulation in the lower part of the Dufek intrusion have been reported by Cheadle *et al.* (2007), Grimes *et al.* (2008), Carnes *et al.* (2011) and Gee *et al.* (2013). They suggest multiple magma injection events, as recorded by xenolith-rich layers and sharp contacts between modal units. Cheadle and Gee (2017) reported studies on mineral orientation and magnetic data aimed at assessing the physical processes operating in the development of cumulate rocks.

In situ geochemical evolution of magmas in sills is by fractional crystallization, with evidence for segregation of interstitial liquids shown in vertical pipes and schlieren of more evolved compositions (e.g. Zavala *et al.* 2011). Plagioclase cumulates, accompanied by migration of differentiates away from the site, were noted by Hergt *et al.* (1989*a*) for the Portal Rock sill in the Queen Alexandra Range.

The compositions of chilled margins of sills and fine-grained lavas reflect varying degrees of evolution of basic magmas (MgO *c.* 9%) at crustal depths by fractional crystallization (pyroxene–plagioclase–oxide) together with minor crustal assimilation (Menzies and Kyle 1990; Fleming 1995; Fleming *et al.* 1995; Antonini *et al.* 1999). The majority of the lavas are basaltic andesite and andesite in composition, but range from basalt to dacite (but to dacite only if the evolved Thern Promontory rocks are lavas and not evolved portions of sills, and discounting the contaminated Butcher Ridge rocks). Interstitial glass and minerals demonstrate the continued evolution at low pressures of dry tholeiitic magmas to silicic compositions with the crystallization of ferrohedenbergite, fayalite, quartz, alkali feldspar and a variety of trace minerals (e.g. monazite, allanite) (Melluso et al. 2014).

The Butcher Ridge igneous complex (Fig. 1) (Marshak et al. 1981; Shellhorn 1982; Kyle et al. 1999; Nelson et al. 2014) comprises rocks ranging from basalt to rhyolite, but the evolved components (high-K andesite to high-K rhyolite compositions) are interpreted to be the result of interaction with crustal materials, not the evolution of a magmatic system by simple fractional crystallization. Analysis of the vitrophyric rocks shows a high water content, and widespread hydration by snow- and ice-derived water (Nelson et al. 2018).

Separation, on the Sr–Nd isotope correlation diagram, of the olivine-bearing dolerites from the rest of the MFCT rocks suggests that they might form a separate but related intrusive event, in the same sense as that of the SPCT. The distinctive lava compositions from north Victoria Land (uniformly high MgO) compared with south Victoria Land (moderate MgO) and the Queen Alexandra Range and Grosvenor Mountains (almost uniformly low MgO) also suggest magma pulses and differing extents of evolution before eruption (the precision of age determinations does not yet allow a temporal evaluation of this possibility). Despite the strong correlation between Nd and Sr isotope compositions (Fig. 17), the Mg number, as an indicator of evolution, does not correlate quite as well with isotopic evolution. This may result from fractionation before and/or after crustal input, and indicates more complex evolutionary paths resulting from differing crustal-level residence times and episodic assimilation of crustal materials. These complexities are illustrated by the lavas at Storm Peak (Table 5) for which decreasing MgO is not accompanied by smoothly changing initial isotope ratios of strontium and ε_{Nd} values.

Origin

The geochemical characteristics of the Ferrar rocks, specifically the enriched initial isotope ratios of Sr, Nd and Pb but also mantle-like $\delta^{18}O$ and Os isotopes plus the low HFSE abundances (particularly Ti) and crust-like trace element patterns even in the most basic olivine-bearing dolerites (MgO c. 9%), have posed major questions for the understanding of the origin of the primary magmas in the mantle and their subsequent evolution to the least-evolved Ferrar rock. The presence of forsteritic olivine (Fo_{88}) in the olivine-bearing dolerites is consistent with equilibrium with the mantle, and the absence of a Eu anomaly indicates that the chemical composition of these most basic Ferrar rocks was little affected by low-pressure processes. Superimposed on this are the crustal evolution of the MFCT olivine-bearing dolerite composition to the most-evolved andesitic composition, and also the evolution of the primary magma to yield the SPCT magma type, which is highly evolved geochemically but, compared with the MFCT, less evolved isotopically.

The most distinctive characteristic of the SPCT, apart from its restricted composition, is evident in Figure 17, which shows the lack of isotopic evolution relative to the evolved MFCT rocks. In all other geochemical characteristics, it is similar to the MFCT and thus a Ferrar magma type. Fleming et al. (1995) suggested that it could have been derived from an olivine-bearing dolerite composition by fractional crystallization but with only very limited assimilation of crustal material.

A similar conclusion was also reached by Antonini et al. (1999). In contrast, Brotzu et al. (1992) suggested the relatively evolved tholeiites of the Thern Promontory and Archambault Ridge (Fig. 8) provided a link between the low-TiO_2 (MFCT) and high-TiO_2 (SPCT) rocks, although the Sr and Nd isotope data (Fig. 17) render this proposal most improbable. On the other hand, experimental studies by Hanemann and Viereck-Götte (2007b) suggested that the major and trace element differences can be attributed to different oxygen fugacities, activities of water and depths of magma evolution. In their model, the MFCT and SPCT rocks were generated from the same source but the former evolved at greater depths in the crust with higher oxygen fugacity (f_{O_2}) and activity of water (a_{H_2O}), and the latter at shallower crustal depths with lower f_{O_2} and a_{H_2O}. It should be noted that emplacement of the SPCT lavas and sills is a post-MFCT late-stage short-lived event in the Ferrar province.

A mantle origin for the geochemical characteristics, as opposed to crustal contamination of either basaltic magmas or an isotopically depleted mantle source, was first proposed by Kyle (1980), and later attributed to a source in the subcontinental mantle lithosphere enriched by crustal materials (Kyle et al. 1983). Hergt et al. (1989b), in a study of the Tasmanian dolerites, evaluated the lithospheric source proposal and pointed out that crustal contamination of mid-ocean ridge basalt (MORB), oceanic island basalt (OIB) and island arc tholeiite (IAT) type parental magmas is incompatible with the geochemistry of those tholeiites (which are part of the Ferrar province). Rather, the mantle source had assimilated a small proportion (<3%) of subducted sediment, thus giving enriched mantle characteristics somewhat similar to enriched MORB (i.e. E-MORB). Menzies and Kyle (1990) reviewed the possible alternative sites of generation in the lithosphere and/or the asthenosphere and advocated a Dupal-like mantle with $^{87}Sr/^{86}Sr = 0.704$–0.707 and a strong subduction zone signature suggesting crustal recycling. Fleming et al. (1995) proposed a somewhat different process, which involved a depleted mantle contaminated with Paleozoic-age crustal materials either by sediment subduction, tectonic erosion of continental crust or delamination of lower crustal materials. Partial melting followed by a melting–assimilation–storage–homogenization (MASH) process (Hildreth and Moorbath 1988) was proposed as the path to yield the most primitive Ferrar magmas, followed by assimilation–fractional crystallization processes in the upper crust to explain the observed range in isotopic and geochemical compositions. On the basis of isotope and trace element data, Molzahn et al. (1996) considered the Ferrar source to be subcontinental mantle lithosphere modified by crustal material. Antonini et al. (1999) noted that the Sr–Nd–Pb isotope signatures are consistent with the origin put forward by Hergt et al. (1989b), and they proposed that Ferrar magmas were generated by high degrees of partial melting of an enriched mantle (E-MORB type) which later interacted with crustal materials during assimilation–fractionation–crystallization processes. However, their contention that crustal-level interaction between mantle-derived magmas and lower continental crust (granulite) created the geochemical characteristics of their least-evolved Ferrar rock (MgO = 5.3%) has been questioned (Hergt 2000).

The geochemical characteristics of the olivine-bearing dolerites (MgO = 9%, together with low abundances of HFSEs, and crust-like trace element patterns and isotopes) compound the problem of the mantle origin. To help elucidate the mantle source, Molzahn et al. (1996) and Brauns et al. (2000) examined Os isotopes in Ferrar and Tasmanian tholeiites: the former consist of lavas and one sill sample, which has a high Mg# (71.9) but no petrographical evidence of olivine, whereas the latter consist of cumulates. The authors concluded that the mantle-like Os initial isotope ratios of whole rocks and minerals (Ferrar) and oxides (Tasmania) require assimilation of crustal materials prior to the generation of Ferrar magmas in the mantle. Hergt and Brauns (2001) evaluated the constraints on possible source compositions, whether it was a depleted subcontinental lithospheric mantle or a plume-related mantle modified by an enriched partial melt, and concluded that it was still unresolved. Alternatively, Mukasa et al. (2003) suggested a previously melted harzburgitic mantle later enriched by subduction processes as the Ferrar source. Subsequently, Mukasa et al. (2007) argued that the PGE abundances (extreme depletion in Os and Ir compared to Ru, Pt and Pd) are incompatible with a plume origin and proposed that the FLIP magmas originated by decompression melting in a subduction zone. Foden et al. (2012) advocated, on the basis of major elements, for the Ferrar magmas being derived from a mantle source more depleted than MORB. To account for the lithophile trace elements and isotopic compositions, they suggested melting of a depleted harzburgitic lithospheric source contaminated by a small fraction of upper crustal material.

With emphasis on the high SiO_2 of the Ferrar rocks, a model involving hydrous and anhydrous melting of fertile and depleted spinel lherzolites has been proposed by Demarchi et al. (2001). However, this model was put forward without consideration of isotope data. Further alternatives were proposed by Ivanov et al. (2017), who invoked either wet-sediment subduction and slab dehydration at the mantle transition or mantle melting followed by metasomatism involving subduction-derived fluids as mechanisms for generating the Ferrar geochemical characteristics.

Whatever the source, it had to have been enriched isotopically relative to E-MORB, and have HFSE element depletions greater than, and REE abundances lower than, E-MORB, yet carrying a crustal signature. Sediment subduction into the mantle appears to be mandated in order to generate the 'crustal' signature. Using PGE abundances and Os isotopic data, Choi et al. (2019a) argued that the Ferrar signature was acquired as a result of wet-sediment subduction and metasomatism of the overlying mantle wedge. The mantle wedge, converted to a hydrated peridotite–pyroxenite mix, underwent decompression melting in an extensional regime. This tectonic regime, initiated earlier in the Jurassic (Elliot et al. 2016), controlled the decompression and facilitated the rapid generation of magma, which led to the short duration of emplacement. Choi et al. (2019b) further suggested that decompression melting was a far-field effect of plume-related instabilities in the proto-Weddell Sea region.

Transport path

The linear distribution and geochemical coherence of Ferrar magmatic rocks has raised significant questions regarding the geographical location of their mantle source. Two alternatives have been presented: were the Ferrar magmas generated at a number of centres along the linear outcrop pattern, or were they generated at a point source and migrated laterally at depths, in some cases for thousands of kilometres?

Elliot (1976) and Cox (1978) advocated, and Storey and Alabaster (1991) similarly suggested, a line source for the Ferrar province magmas, in which magmas were generated from domains in the mantle directly underlying the region of magma emplacement and with minimal lateral transport. Given the linear geographical extent of the province, it was related to the Gondwana plate margin, which had been active for much of the Paleozoic Era and into Mesozoic (early Jurassic) time. The notion of a line source is not inconsistent with the models that require generation of an enriched lithospheric mantle source resulting from the incorporation of subducted sediment.

The linear model involves magma generation along a trend parallel or sub-parallel to the Antarctic basement boundary. That boundary delineates a substantial crustal thickness change (c. >35 km thick craton v. c. <25 km in West Antarctica: Chaput et al. 2014; Ramirez et al. 2017), and thus might have controlled the sites of Ferrar magma generation. Over a distance of 3500 km, the linear trend crosses several lithospheric provinces (Fig. 18) and magmas with a variety of geochemical and isotopic characteristics might be expected due to variations in source composition and extents of partial melting, in contrast to the geochemical coherence of the Ferrar magmas. A key might be the trend of the crustal thickness change, a fundamental property of the Antarctic Plate in that it marks the boundary between basement terrains and Phanerozoic orogenic belts (Elliot 2013). In this scenario, magma generation would be controlled by the trend of the early Paleozoic Ross Orogen. To generate the geochemically coherent Ferrar rocks spread over 3500 km there would have to have

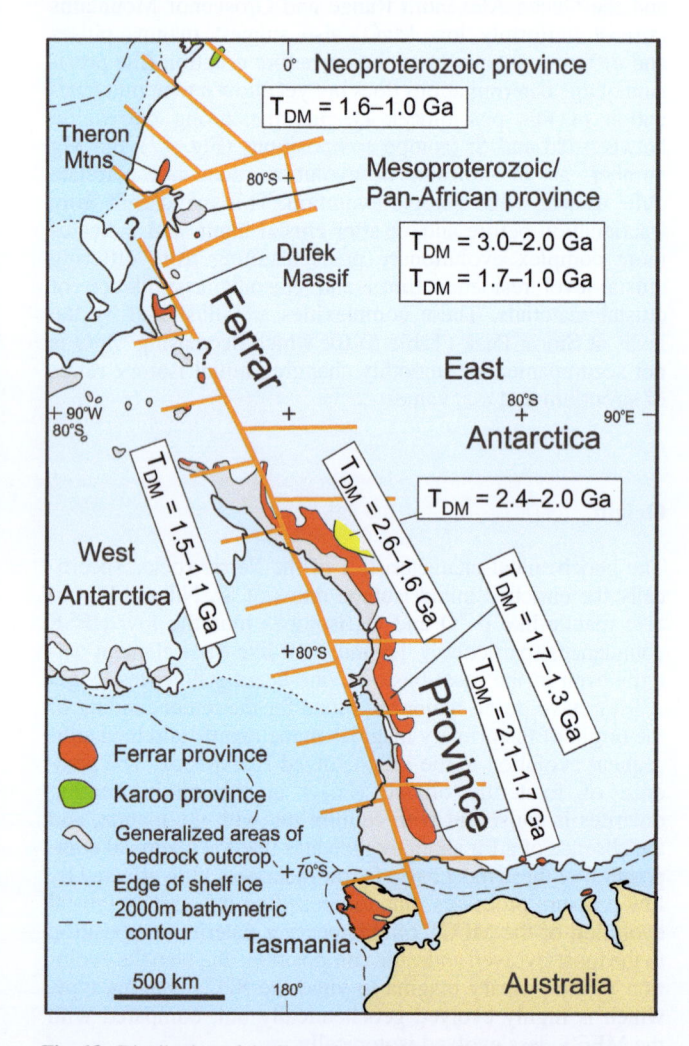

Fig. 18. Distribution of the Ferrar Large Igneous Province and lithospheric domains. The approximate domain margins are marked by solid orange lines and the inferred extent of the domains by orange 'ladders'. The Precambrian domain bordering the central Transantarctic Mountains is in yellow. Data sources for domains: Armienti et al. (1990), Borg et al. (1990), Cox et al. (2000), Leat et al. (2005), Black et al. (2010), Will et al. (2010), Loewy et al. (2011) and Goodge et al. (2012). The original lateral extent of the Ferrar province is largely speculative. T_{DM}, depleted mantle model age.

been a uniform mantle source reservoir modified by the incorporation of a consistent amount of subducted sediment of uniform composition, and uniformity in the composition of magmas generated. The creation of such a widespread uniform reservoir by subduction-related processes seems most unlikely.

An alternative (and preferred) hypothesis, a geographically restricted source combined with large-scale lateral transport, was first proposed by Fleming et al. (1997) because of the geochemical coherence of the Ferrar magmas, and by Storey and Kyle (1997), but the transport paths differed (see Elliot and Fleming 2017). The unique chemistry and tightly constrained composition of the SPCT led Elliot et al. (1999) to advocate long-distance transport of Ferrar magmas and suggested migration at various crustal depths. Strong support for a single source is given by the fact that, in the linear model, magmas from subjacent mantle sources would have traversed several different lithospheric provinces, as first noted by Leat (2008) (Fig. 18). However, this is not reflected in the coherent isotope characteristics of the Ferrar rocks and, in particular, the evolved but highly restricted SPCT composition, which would require identical magma generation and evolutionary processes, and identical end products over a linear distance of more than 3000 km. In this model, Ferrar magmas were generated in the lithospheric mantle, migrated into the crust and were then dispersed laterally at mid- to lower-crustal depths. The possibility of such long-distance transport is demonstrated by the Mackenzie dyke swarm (Baragar et al. 1996; Ernst and Buchan 1997), which was emplaced at mid-crustal depths and has been traced for 2500 km across the Canadian Shield.

Storey and Kyle (1997) argued for supracrustal transport through sills, and Ferris et al. (2003) further suggested that the Dufek intrusion formed the crustal magma chamber from which the Ferrar magmas migrated along the Transantarctic Mountains. Airoldi et al. (2016) and Magee et al. (2016) also advocated long-distance transport through sills. The contention that Mg# and MgO decrease from the point of origin along the length of the Transantarctic Mountains (Leat 2008; Magee et al. 2016, 2019) is misleading because, for the province as a whole, it is not supported by the geochemical data for the lavas nor for the sills. There is no spatial pattern with respect to the inferred proto-Weddell Sea source region (Elliot and Fleming 2017) (Fig. 7). In addition, long-distance sill transport throughout the province is regarded as improbable because it requires magmas to cross a pre-Devonian palaeotopographical high separating the central Transantarctic Mountains from south Victoria Land (the Ross High of Collinson et al. 1994), and another palaeotopographical high separating the south and north Victoria Land Beacon basins (Collinson et al. 1994) (Fig. 19). Magmas would also have to be transported to the Permo-Triassic basin of Tasmania (Veevers et al. 1994), the relationship of which to the north Victoria Land basin is uncertain because it is offset from, not along strike with, the north Victoria Land basin in a Gondwana reconstruction. In south Victoria Land, magmas would have had to burrow down hundreds of metres through the Taylor Group and penetrate basement granitic and gneissic rocks in the Dry Valleys region to form the very thick Basement Sill and its associated feeder (Fig. 20). That proposed feeder, on rising from depth, must have traversed basement rock. In addition, there are examples of dykes cutting the pre-Devonian basement in the Dry Valleys region and elsewhere in south Victoria Land (Darwin Glacier and Prince Albert Mountains regions: Haskell et al. 1965; Skinner and Ricker 1968, respectively). Dolerite intrusions, including thick sills, are also present in basement granitic rocks at the Nilsen Plateau (McLelland 1967), Mount Weaver (Doumani and Minshew 1965) and at Thanksgiving Point alongside the Shackleton Glacier (Figs 1 & 5). Thick dykes transecting basement rocks are few and widely scattered, but indicate transport of magmas from depth at those sites (Elliot and Fleming 2004), not transport through supracrustal sills. Dykes have been inferred geophysically to extend southwards from the Dufek intrusion (Ferris et al. 2003) and to occur at depth orientated parallel to major structures in the central Transantarctic Mountains (Goodge and Finn 2010); however, no major dyke swarms, such as occur in the Karoo of southern Africa (e.g. Coetzee and Kisters 2018), have been identified. Some support for transport in the lower crust is given by Ramirez et al. (2017), who suggested that geophysically interpreted mafic layering within or near the base of the crust of the Transantarctic Mountains may be related to the FLIP.

The actual path taken by the magmas at depth remains speculative (Fig. 21). In the Ross Sea sector of the Transantarctic Mountains, the outcrop distribution, the occurrence of dykes and the proposed Basement Sill feeder, and the phreatomagmatic centres all suggest that magmas migrated locally into supracrustal rocks to form sills and to the surface to be erupted as lavas and pyroclastic rocks. Although Karoo dolerite sills were not intruded into Cape Fold Belt deformed strata, the Dufek intrusion in the Weddell Sea sector was

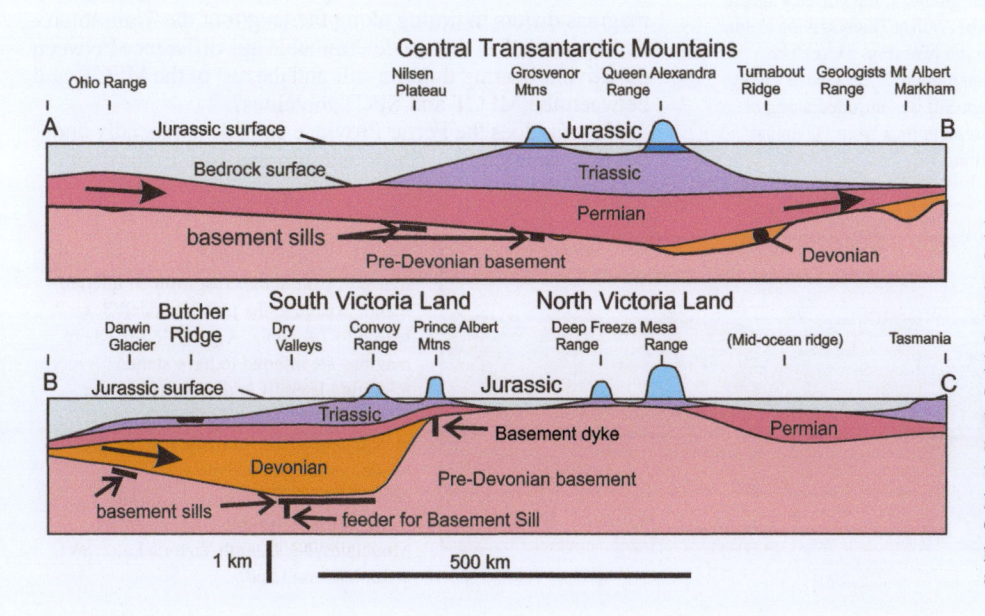

Fig. 19. Diagrammatic section from the Ohio Range to Tasmania (see Fig. 1) along the length of the Transantarctic Mountains to illustrate the current distribution of extrusive rocks, and the known distribution of sills and dykes cutting basement rocks (projected onto the line of section). Sills are present in all stratigraphic successions. Permian strata thin markedly, or are absent, over palaeogeographical highs. Heavy arrows denote magma paths if transport from the point of origin were through supracrustal sills.

emplaced by vertical magma migration into the folded Paleozoic strata of the Pensacola Mountains, the only such instance in the Ferrar province. The Ferrar extrusive rocks in the Ross Sea sector are interpreted to have been erupted into a rift valley system (Elliot and Fleming 2008), which is now located on the edge of the pre-Devonian basement and close to the lithospheric boundary between cratonic East Antarctica and the outboard Paleozoic orogenic belts that form the disrupted and displaced continental fragments making up West Antarctica (Dalziel and Lawver 2001). The geophysical interpretation of Ferrar rocks occupying a rift or rifts in the Wilkes Subglacial Basin in the hinterland of north Victoria Land (Ferraccioli *et al.* 2009) suggests other rift basins in the hinterland of the Transantarctic Mountains might have existed in Jurassic time.

Some outstanding issues

The generation of the mantle source composition: the most basic Ferrar magmas (MgO = 9–10%) have high SiO_2 (52%), enriched Sr and Nd isotope compositions, mantle oxygen and Os isotope compositions, and trace elements with low abundances but crustal patterns. These characteristics imply an unusual mantle source that is subduction-related rather than plume-related. What new studies might verify the proposal that partial melting of peridotitic material, metasomatized by subducted sediment, in the mantle wedge below the Gondwana margin generated Ferrar primary magmas?

If Ferrar magmas are subduction-related and have distributed sources along its outcrop length, why does the Ferrar province exhibit such geochemical coherence? Why is there so little magma diversity beyond the single MFCT trend and the restricted SPCT composition? In particular, what controlled the generation of the SPCT composition, which is identical for over 3000 km?

Assuming it is not simple vertical migration of magma from the mantle along the length of the Ferrar province, what is the transport path for crustal dispersal from the putative proto-Weddell Sea point source? Why do the Ferrar magmas show no spatial pattern of changing composition related to distance from the source?

What are the flow patterns in sills in the various regions? Would they show dispersion from central conduits, as is the case for the Basement Sill in the Dry Valleys region?

Would careful evaluation of sill geometry reveal saucer-shaped intrusions, such as are documented in the Karoo?

Can age determinations clarify if emplacement of the Ferrar magmas differs in timing along the length of the Transantarctic Mountains? Is there a determinable age difference between the olivine-bearing dolerite sills and the rest of the MFCT, and between the MFCT and SPCT tholeiites?

How far does the Ferrar Province extend subglacially under the East Antarctic Ice Sheet and into the Ross Embayment?

Fig. 20. Olivine-bearing dolerite sills emplaced in pre-Devonian basement rocks. (**a**) A sill cutting basement granite at the Nilsen Plateau, central Transantarctic Mountains (Fig. 1). (**b**) A sill at Thanksgiving Point, central Transantarctic Mountains. View to the NW (Fig. 5). (**c**) The Basement Sill intruding Cambro-Ordovician granitoid near Pearse Valley, south Victoria Land (Fig. 6). The Peneplain Sill was intruded along or close to the non-conformity separating basement rock from Devonian strata. View to the SE. (Images: T.H. Fleming.)

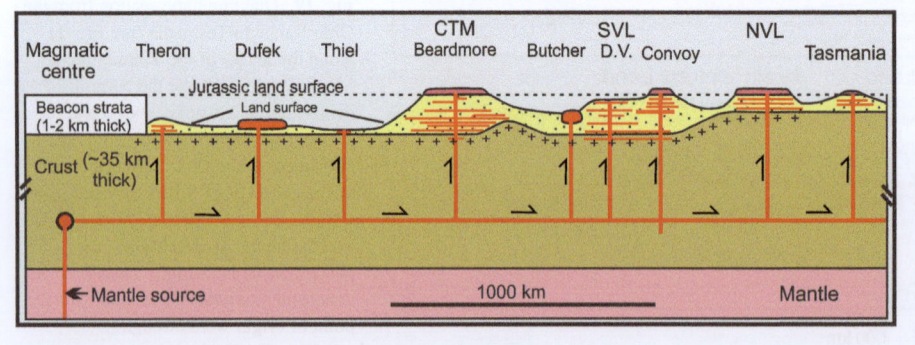

Fig. 21. Schematic model for a Ferrar magma transport path in the crust from an inferred mantle source in the proto-Weddell Sea region to Tasmania. A possible site where magmas are inferred to have started migrating laterally is represented by the circle, which is located at mid- to lower-crustal depths. The ultimate magma source resided in the mantle below the proto-Weddell Sea region. D.V., Dry Valleys; CTM, central Transantarctic Mountains; NVL, north Victoria Land; SVL, south Victoria Land.

Is the Dufek intrusion definitively one or two bodies?

Can a chilled margin composition be identified and/or liquid compositions be reconstructed for the Dufek intrusion?

Is there any clue to the lower hidden section of the Dufek intrusion in the sediments derived from it in the Filchner Trough region?

Are any of the inferred basaltic bodies in the Weddell Sea sector, such as Berkner Island and the dipping reflector sequences offshore Coats Land (Hunter et al. 1996; Jordan et al. 2017), part of the FLIP?

Do Ferrar compositions occur for certain in the Karoo Large Igneous Province? If so, are they confined to the region south and east of Lesotho?

Summary

The Ferrar Large Igneous Province (FLIP) differs from all other such provinces in that it has an extant linear outcrop pattern and its emplacement was probably controlled by lithospheric structure, which itself is defined by the boundary between the craton and Phanerozoic belts, and by the early Jurassic extensional tectonic regime. Geochemically, the province is unique among large igneous provinces (LIPs) in significant Sr, Nd and Pb isotope enrichment, and the low abundances of high field strength elements (HFSEs) and their crustal pattern even in the most basic olivine-bearing dolerites. The coherence of the province-wide geochemical data for the Mount Fazio Chemical Type (MFCT) compositions suggests a common origin in the mantle and similar evolutionary processes. Both models for the source – the single source and long-distance transport model, and the linear source model with multiple sites of mantle origin – have uncertainties. The highly evolved Scarab Peak Chemical Type (SPCT) composition strongly implies a single source region and evolution, and long-distance magma transport. The processes in the mantle source region that resulted in the Ferrar magmas, most probably involving assimilation of subducted material and then melting to produce the primary magma composition, remain somewhat uncertain.

Acknowledgements Reviews by Janet Hergt and Marco Brenna are much appreciated and have considerably improved the manuscript. In particular, the authors thank John Smellie for the invitation to contribute to this Memoir. This is Byrd Polar and Climate Research Center contribution No. 1581.

Author contributions DHE: data curation (supporting), formal analysis (equal), investigation (equal), writing – original draft (lead), writing – review & editing (equal); **THF**: data curation (lead), formal analysis (equal), investigation (equal), writing – original draft (supporting), writing – review & editing (equal).

Funding The authors acknowledge significant support over many years from the Office of Polar Programs, National Science Foundation, Washington, DC.

Data availability All data are either already published, included in the tables in this paper, or in the case of unpublished data, can be obtained, upon reasonable request, from the authors

Appendix A: Grid references for field photographs and samples in Tables 3 and 5

Location	Longitude	Latitude
Figure 2a: Point 3120, Nilsen Plateau	159° 15.5′ W	86° 28.4′ S
Figure 2b: Mount Joyce	160° 49′ E	75° 36′ S
Figure 3a: East of Mount Gran	161° 06.0′ E	76° 58.5′ S
Figure 3b: Terra Cotta Mountain	161° 15′ E	77° 54′ S
Figure 4a: Shenk Peak	174° 45′ W	85° 11′ S
Figure 4b: Dismal Buttress	178° 00′ W	85° 27′ S
Figure 8a: Rougier Hill, lowest sill	174° 33.4′ W	85° 09.5′ S
Figure 8b: Rougier Hill, lowest sill	174° 33.4′ W	85° 09.5′ S
Figure 18a: Cougar Cyn, Nilsen Plateau	160° 40.0′ W	86° 18.4′ S
Figure 18b: Thanksgiving Point	177° 00.0′ W	84° 56.7′ S
Figure 18c: SE of Pearse Valley	161° 34.7′ E	77° 45.0′ S
Sill near Dawson Peak	162° 25.2′ E	83° 50.5′ S
Lavas at Storm Peak (Peterson Ridge)	163° 55.7′ E	84° 34.1′ S
Sill near Mount Achernar	160° 53.9′ E	84° 11.3′ S

Coordinates without a decimal point from are the Gazetteer of Antarctic.
Place names with a decimal point are from USGS topographical sheets.

References

Airoldi, G.M., Muirhead, J.D., Long, S.M., Zanella, E. and White, J.D.L. 2016. Flow dynamics in mid-Jurassic dikes and sills of the Ferrar large igneous province and implications for long-distance transport. *Tectonophysics*, **682**, 182–199, https://doi.org/10.1016/j.tecto.2016.06.029

Allibone, A.H., Forsyth, P.J., Sewell, R.J., Turnbull, I.M. and Bradshaw, M.A. 1991. Geology of the Thundergut Area, Southern Victoria Land, Antarctica, Scale 1:50 000. New Zealand Geological Survey Miscellaneous Geological Map, **21**.

Antonini, P., Demarchi, G., Piccirillo, E.M. and Orsi, G. 1997. Distinct magma pulses in the Ferrar tholeiites of Thern Promontory (Victoria Land, Antarctica). *Terra Antartica*, **4**, 33–39.

Antonini, P., Piccirillo, E.M., Petrini, R., Civetta, L., D'Antonio, M. and Orsi, G. 1999. Enriched mantle – Dupal signature in the genesis of the Jurassic Ferrar tholeiites from Prince Albert Mountains (Victoria Land, Antarctica). *Contributions to Mineralogy and Petrology*, **136**, 1–19, https://doi.org/10.1007/s004100050520

Armienti, P., Ghezzo, C., Innocenti, F., Manetti, P., Rocchi, S. and Tonarini, S. 1990. Isotope geochemistry and petrology of granitoid suites from Granite Harbour Intrusives of the Wilson Terrane, north Victoria Land, Antarctica. *European Journal of Mineralogy*, **2**, 103–123, https://doi.org/10.1127/ejm/2/1/0103

Baragar, W.R.A., Ernst, R.E., Hulbert, L. and Peterson, T. 1996. Longitudinal petrochemical variation in the Mackenzie Dyke Swarm, northwestern Canadian Shield. *Journal of Petrology*, **37**, 317–359, https://doi.org/10.1093/petrology/37.2.317

Barrett, P.J. 1991. The Devonian to Triassic Beacon Supergroup of the Transantarctic Mountains and correlatives in other parts of Antarctica. *Oxford Monographs on Geology and Geophysics*, **17**, 120–152.

Barrett, P.J., Elliot, D.H. and Lindsay, J.F. 1986. The Beacon Supergroup (Devonian–Triassic) and Ferrar Group (Jurassic) in the Beardmore Glacier area, Antarctica. *Antarctic Research Series*, **36**. American Geophysical Union, Washington, DC, 339–428.

Bédard, J.J., Marsh, B.D., Hersum, T.G., Naslund, H.R. and Mukasa, S.B. 2007. Large-scale mechanical redistribution of orthopyroxene and plagioclase in the Basement Sill, Ferrar Dolerites, McMurdo Dry Valleys, Antarctica: Petrological, mineral–chemical and field evidence for channelized movement of crystals and melt. *Journal of Petrology*, **48**, 2289–2326, https://doi.org/10.1093/petrology/egm060

Behrendt, J.C., Drewry, D.J., Jankowski, E. and Grim, M.S. 1981. Aeromagnetic and radio echo ice-sounding measurements over the Dufek intrusion, Antarctica. *Journal of Geophysical Research*, **86**, 3014–3020, https://doi.org/10.1029/JB086iB04p03014

Behrendt, J.C., Damaske, D., Finn, C.A., Kyle, P.R. and Wilson, T.J. 2002. Draped aeromagnetic survey in Transantarctic Mountains over area of Butcher Ridge igneous complex showing extent of underlying mafic intrusion. *Journal of Geophysical Research: Solid Earth*, **107**, EPM 3-1–EPM 3-10, https://doi.org/10.1029/2001JB000376

Benson, W.N. 1916. Report on the Petrology of the Dolerites collected by the British Antarctic Expedition, 1907–09. *Reports of Scientific Investigations, Geology*, **II**, 153–160.

Black, L.P., Everard, J.L. *et al.* and 2010. Controls on Devonian–Carboniferous magmatism in Tasmania, based on inherited zircon age patterns, Sr, Nd and Pb isotopes, and major and trace element geochemistry. *Australian Journal of Earth Sciences*, **57**, 933–968, https://doi.org/10.1080/08120099.2010.509407

Borg, S.G., DePaolo, D.J. and Smith, B.M. 1990. Isotopic structure and tectonics of the central Transantarctic Mountains. *Journal of Geophysical Research*, **95**, 6647–6667.

Boudreau, A. and Simon, A. 2007. Crystallization and degassing in the Basement Sill, McMurdo Dry Valleys, Antarctica. *Journal of Petrology*, **48**, 1369–1386, https://doi.org/10.1093/petrology/egm022

Bradshaw, M.A. 2013. The Taylor Group (Beacon Supergroup): the Devonian sediments of Antarctica. *Geological Society, London, Special Publications*, **381**, 67–97, https://doi.org/10.1144/SP381.23

Brauns, C.M., Hergt, J.M., Woodhead, J.D. and Maas, R. 2000. Os isotopes and the origin of the Tasmanian Dolerites. *Journal of Petrology*, **41**, 905–918, https://doi.org/10.1093/petrology/41.7.905

Brewer, T.S., Hergt, J.M., Hawkesworth, C.J., Rex, D. and Storey, B.C. 1992. Coats Land dolerites and the generation of Antarctic continental flood basalts. *Geological Society, London, Special Publications*, **68**, 185–208, https://doi.org/10.1144/GSL.SP.1992.068.01.12

Bromfield, K., Burrett, C.F., Leslie, R.A. and Meffre, S. 2007. Jurassic volcaniclastic–basaltic andesite–dolerite sequence in Tasmania: new age constraints for fossil plants from Lune River. *Australian Journal of Earth Sciences*, **54**, 965–974, https://doi.org/10.1080/08120090701488297

Brotzu, P., Capaldi, G., Civetta, L., Melluso, L. and Orsi, G. 1988. Jurassic Ferrar dolerites and Kirkpatrick basalts in northern Victoria Land (Antarctica): stratigraphy, geochronology and petrology. *Memorie della Societa Geologica Italiana*, **43**, 97–116.

Brotzu, P., Capaldi, G., Civetta, L., Orsi, G., Gallo, G. and Melluso, L. 1992. Geochronology and geochemistry of Ferrar rocks from north Victoria Land, Antarctica. *European Journal of Mineralogy*, **4**, 605–617, https://doi.org/10.1127/ejm/4/3/0605

Browne, W.R. 1923. The dolerites of King George Land and Adelie Land. *Australasian Antarctic Expedition, 1911–14, Scientific Reports Series A, Geology*, **3**, 245–258.

Burgess, S.D., Bowring, S.A., Fleming, T.H. and Elliot, D.H. 2015. High precision geochronology links the Ferrar Large Igneous Province with early Jurassic ocean anoxia and biotic crisis. *Earth and Planetary Science Letters*, **415**, 90–99, https://doi.org/10.1016/j.epsl.2015.01.037

Carnes, J.D., Cheadle, M., Gee, J.S., Grimes, C.B. and Swapp, S.M. 2011. The magmatic and thermal history of the Dufek Complex, Antarctica. *Eos, Transactions of the American Geophysical Union*, Fall Meeting 2011, **92**, abstract V33C-2661.

Chaput, J., Aster, R.C. *et al.* 2014. The crustal thickness of West Antarctica. *Journal of Geophysical Research: Solid Earth*, **119**, 378–395, https://doi.org/10.1002/2013JB010642

Charrier, A.D. 2010. *Emplacement History of the Basement Sill, Antarctica: Injection Mechanics of Crystal-Laden Slurries*. PhD thesis, Johns Hopkins University, Baltimore, Maryland, USA.

Charrier, A.D. and Manochehri, S. 2013. The compaction of ultramafic cumulates in layered intrusions – time and length scales. *Eos, Transactions of the American Geophysical Union*, Fall Meeting 2013, **94**, abstract V54B-01.

Cheadle, M. and Gee, J.S. 2017. Quantitative textural insights into the formation of gabbro in mafic intrusions. *Elements*, **13**, 409–414, https://doi.org/10.2138/gselements.13.6.409

Cheadle, M., Meurer, W.P., Grimes, C.B., Gee, J.S. and McCullough, B.C. 2007. Understanding the magmatic construction of the Dufek complex, Antarctica. *Eos, Transactions of the American Geophysical Union*, **88**(52), Fall Meeting Supplement, abstract V53D-02.

Choi, S.H., Mukasa, S.B., Ravizza, G., Fleming, T.H., Marsh, B.D. and Bédard, J.H.J. 2019*a*. Fossil subduction zone origin for magmas in the Ferrar Large Igneous Province, Antarctica: Evidence from PGE and Os isotope systematics in the Basement Sill of the McMurdo Dry valleys. *Earth and Planetary Science Letters*, **506**, 507–519, https://doi.org/10.1016/j.epsl.2018.11.027

Choi, S.H., Mukasa, S.B., Ravizza, G., Fleming, T.H., Marsh, B.D. and Bédard, J.H.J. 2019*b*. Fossil subduction zone origin for magmas in the Ferrar Large Igneous Province, Antarctica: Evidence from PGE and Os isotope systematics in the Basement Sill of the McMurdo Dry valleys. *MantlePlumes.org*, http://www.mantleplumes.org/FerrarLIP.html

Coetzee, A. and Kisters, A.F.M. 2017. Dyke-sill relationships in Karoo dolerites as indicators of propagation and emplacement processes of mafic magmas in shallow crust. *Journal of Structural Geology*, **97**, 172–188, https://doi.org/10.1016/j.jsg.2017.03.002

Coetzee, A. and Kisters, A.F.M. 2018. The elusive feeders of the Karoo Large Igneous Province and their structural controls. *Tectonophysics*, **747–748**, 146–162, https://doi.org/10.1016/j.tecto.2018.09.007

Collinson, J.W., Elliot, D.H., Isbell, J.L. and Miller, J.M.G. 1994. Permian–Triassic Transantarctic Basin. *Geological Society of America Memoirs*, **184**, 173–222.

Compston, W., McDougall, I. and Heier, K.S. 1968. Geochemical comparison of rthe Mesozoic basaltic rocks of Antarctica, South Africa, South America and Tasmania. *Geochimica et Cosmochimica Acta*, **33**, 129–149, https://doi.org/10.1016/S0016-7037(68)80001-8

Cox, K.G. 1978. Flood basalts, subduction, and the break-up of Gondwanaland. *Nature*, **274**, 47–49, https://doi.org/10.1038/274047a0

Cox, S.C., Parkinson, D.L., Allibone, A.H. and Cooper, A.F. 2000. Isotopic character of Cambro-Ordovician plutonism, southern Victoria Land, Antarctica. *New Zealand Journal of Geology and Geophysics*, **43**, 501–520, https://doi.org/10.1080/00288306.2000.9514906

Dalziel, I.W.D. and Lawver, L.A. 2001. The lithospheric setting of the West Antarctic Ice Sheet. *American Geophysical Union Antarctic Research Series*, **77**, 29–44.

Delisle, G. and Fromm, K. 1989. Further evidence for a Cretaceous thermal event in north Victoria Land. *Geologisches Jahrbuch*, **38E**, 143–151.

Demarchi, G., Antonini, P., Piccirillo, E.M., Orsi, G., Civetta, L. and D'Antonio, M. 2001. Significance of orthopyroxene and major element constraints on the petrogenesis of Ferrar tholeiites from southern Prince Albert Mountains, Victoria land, Antarctica. *Contributions to Mineralogy and Petrology*, **142**, 127–146, https://doi.org/10.1007/s004100100287

Doumani, G.A. and Minshew, V.H. 1965. General geology of the Mount Weaver area, Queen Maud Mountains, Antarctica. *American Geophysical Union Antarctic Research Series*, **6**, 127–139.

Elliot, D.H. 1970. Jurassic tholeiites of the central Transantarctic Mountains, Antarctica. *In*: Gilmour, E.H. and Stradling, D. (eds) *Proceedings of the Second Columbia River Basalt Symposium, March 1969*. Eastern Washington State College Press, Cheney, WA, 301–325.

Elliot, D.H. 1976. Tectonic setting of the Jurassic Ferrar Group, Antarctica. *In*: Gonzalez, F.O. (ed.) *Proceedings of the Symposium* on *Andean and Antarctic Volcanology Problems*. IAVCEI Special Series, 357–372.

Elliot, D.H. 2013. The geological and tectonic evolution of the Transantarctic Mountains: a review. *Geological Society, London, Special Publications*, **381**, 7–35, https://doi.org/10.1144/SP381.14

Elliot, D.H. and Fleming, T.H. 2000. Weddell triple junction: The principal focus of Ferrar and Karoo magmatism during initial breakup of Gondwana. *Geology*, **28**, 539–542, https://doi.org/10.1130/0091-7613(2000)28<539:WTJTPF>2.0.CO;2

Elliot, D.H. and Fleming, T.H. 2004. Occurrence and dispersal of magmas in the Jurassic Ferrar Large Igneous Province, Antarctica. *Gondwana Research*, **7**, 223–237, https://doi.org/10.1016/S1342-937X(05)70322-1

Elliot, D.H. and Fleming, T.H. 2008. Physical volcanology and geological relationships of the Ferrar Large Igneous Province, Antarctica. *Journal of Volcanology and Geothermal Research*, **172**, 20–37, https://doi.org/10.1016/j.jvolgeores.2006.02.016

Elliot, D.H. and Fleming, T.H. 2017. The Ferrar large Igneous Province: field and geochemical constraints on supra-crustal (high-level) emplacement of the magmatic system. *Geological Society, London, Special Publications*, **463**, 41–58, https://doi.org/10.1144/SP463.1

Elliot, D.H., Fleming, T.H., Haban, M.A. and Siders, M.A. 1995. Petrology and mineralogy of the Kirkpatrick Basalt and Ferrar Dolerite, Mesa Range region, north Victoria Land, Antarctica. *American Geophysical Union Antarctic Research Series*, **67**, 103–141.

Elliot, D.H., Fleming, T.H., Kyle, P.R. and Foland, K.A. 1999. Long Distance Transport of Magmas in the Jurassic Ferrar Large Igneous Province, Antarctica. *Earth and Planetary Science Letters*, **167**, 87–104, https://doi.org/10.1016/S0012-821X(99)00023-0

Elliot, D.H., Larsen, D., Fanning, C.M., Fleming, T.H. and Vervoort, J.D. 2016. The Lower Jurassic Hanson Formation of the Transantarctic Mountains: implications for the Antarctic sector of the Gondwana Plate margin. *Geological Magazine*, **154**, 777–803, https://doi.org/10.1017/S0016756816000388

Elliot, D.H., White, J.D.L. and Fleming, T.H. 2021. Ferrar Large Igneous Province: volcanology. *Geological Society, London, Memoirs*, **55**, https://doi.org/10.1144/M55-2018-44

Ernst, R.E. and Buchan, K.L. 1997. Giant radiating dyke swarms: their use in identifying Pre-Mesozoic large igneous provinces and mantle plumes. *American Geophysical Union Geophysical Monograph Series*, **100**, 247–272.

Faure, G. and Mensing, T.M. 1993. K–Ar dates and paleomagnetic evidence for Cretaceous alteration of Mesozoic basaltic lava flows, Mesa Range, northern Victoria Land, Antarctica. *Chemical Geology*, **109**, 305–315, https://doi.org/10.1016/0009-2541(93)90077-V

Faure, G., Hill, R.L., Jones, L.M. and Elliot, D.H. 1972. Isotope composition of strontium and silica content of Mesozoic basalt and dolerite from Antarctica. *In*: Adie, R.J. (ed.) *Antarctic Geology and Geophysics*. Universitetsforlaget, Oslo, 617–624.

Faure, G., Bowman, J.R., Elliot, D.H. and Jones, L.M. 1974. Strontium isotope composition and petrogenesis of the Kirkpatrick Basalt, Queen Alexandra Range, Antarctica. *Contributions to Mineralogy and Petrology*, **48**, 153–169, https://doi.org/10.1007/BF00383353

Faure, G., Pace, K.K. and Elliot, D.H. 1982. Systematic variations of $^{87}Sr/^{86}Sr$ ratios and major element concentrations in the Kirkpatrick Basalt of Mt. Falla, Queen Alexandra Range, Transantarctic Mountains. *In*: Craddock, C. (ed.) *Antarctic Geoscience*. University of Wisconsin Press, Madison, WI, 715–723.

Faure, G., Hoefs, J. and Mensing, T.M. 1984. Effect of oxygen fugacity on sulfur isotope compositions and magnetite concentrations in the Kirkpatrick Basalt, Mount Falla, Queen Alexandra Range, Antarctica. *Chemical Geology*, **46**, 301–311, https://doi.org/10.1016/0009-2541(84)90173-6

Faure, G., Mensing, T.M., Jones, L.M., Hoefs, J. and Kibler, E.M. 1991. Isotopic and geochemical studies of Ferrar Dolerite sills in the Transantarctic Mountains. *In*: Ulbrich, H. and Rocha Campos, A.C. (eds) *Gondwana Seven Proceedings. Papers presented at the Seventh International Gondwana Symposium, Sao Paulo, 1988*. Instituto Geosciências, Universidade de São Paulo, São Paulo, Brazil, 669–683.

Ferraccioli, F., Armadillo, E., Jordan, T., Bozzo, E. and Corr, H. 2009. Aeromagnetic exploration over the East Antarctic Ice Sheet: A new view of the Wilkes Subglacial Basin. *Tectonophysics*, **478**, 62–77, https://doi.org/10.1016/j.tecto.2009.03.013

Ferris, J.K., Johnson, A. and Storey, B.C. 1998. Form and extent of the Dufek intrusion, Antarctica, from newly compiled aeromagnetic data. *Earth and Planetary Science Letters*, **154**, 185–202, https://doi.org/10.1016/S0012-821X(97)00165-9

Ferris, J.K., Storey, B.C., Vaughan, A.P.M., Kyle, P.R. and Jones, P.C. 2003. The Dufek and Forrestal intrusions, Antarctica: A centre for Ferrar Large Igneous Province dike emplacement? *Geophysical Research Letters*, **30**, 1348, https://doi.org/10.1029/2002GL016719

Fitzgerald, P.G. 2002. Tectonics and landscape evolution of the Antarctic Plate since the breakup of Gondwana, with an emphasis on the West Antarctic rift system and the Transantarctic Mountains. *Bulletin of the Royal Society of New Zealand*, **35**, 453–469.

Fleming, T.H. 1995. *Isotopic and Chemical Evolution of the Ferrar Group, Beardmore Glacier Region, Antarctica*. PhD thesis, Ohio State University, Columbus, Ohio, USA.

Fleming, T.H., Elliot, D.H., Jones, L.M. and Bowman, J.R. 1989. Secondary alteration or iron-rich tholeiitic rocks of the Kirkpatrick Basalt, northern Victoria Land. *Antarctic Journal of the United States*, **24**, 37–40.

Fleming, T.H., Foland, K.A., Elliot, D.H. and Miller, C.A. 1998. Isotopic and chemical constraints on the magmatic evolution of the Kirkpatrick Basalt, Carapace Nunatak, south Victoria Land, Antarctica. *Geological Society of America, Abstracts with Programs*, **30**(2), 17.

Fleming, T.H., Elliot, D.H., Jones, L.M., Bowman, J.R. and Siders, M.A. 1992. Chemical and isotopic variations in an iron-rich lava flow from North Victoria Land, Antarctica: Implications for low-temperature alteration and the petrogenesis of Ferrar magmas. *Contributions to Mineralogy and Petrology*, **111**, 440–457, https://doi.org/10.1007/BF00320900

Fleming, T.H., Elliot, D.H., Foland, K.A., Jones, L.M. and Bowman, J.R. 1993. Disturbance of Rb–Sr and K–Ar isotopic systems in the Kirkpatrick Basalt, North Victoria Land, Antarctica: Implications for mid-Cretaceous tectonism. *In*: Findlay, R.H., Banks, M.R., Veevers, J.J. and Unrug, R. (eds) *Gondwana 8 – Assembly, Evolution, and Dispersal*. Balkema, Rotterdam, The Netherlands, 411–424.

Fleming, T.H., Foland, K.A. and Elliot, D.H. 1995. Isotopic and chemical constraints on the crustal evolution and source signature of Ferrar magmas, North Victoria Land, Antarctica. *Contributions to Mineralogy and Petrology*, **121**, 217–236, https://doi.org/10.1007/BF02688238

Fleming, T.H., Heimann, A., Foland, K.A. and Elliot, D.H. 1997. $^{40}Ar/^{39}Ar$ geochronology of Ferrar Dolerite sills from the Transantarctic Mountains, Antarctica: Implications for the age and origin of the Ferrar Magmatic Province. *Bulletin of the Geological Society of America*, **109**, 533–546, https://doi.org/10.1130/0016-7606(1997)109<0533:AAGOFD>2.3.CO;2

Fleming, T.H., Foland, K.A. and Elliot, D.H. 1999. Apophyllite $^{40}Ar/^{39}Ar$ and Rb-Sr geochronology: potential utility and application to the timing of secondary mineralization of the Kirkpatrick Basalt, Antarctica. *Journal of Geophysical Research: Solid Earth*, **104**, 20 081–20 095, https://doi.org/10.1029/1999JB900138

Fleming, T.H., Elliot, D.H. and Calhoun, A. 2005. Geographic variations in chilled margin chemistry of Jurassic dolerite intrusions in the Dry Valley region of south Victoria Land, Antarctica. *Eos, Transactions of the American Geophysical Union*, **86**(52), Fall Meeting Supplement, V14C–V102.

Fleming, T.H., Burgess, S.D., Elliot, D.H. and Bowring, S. 2012. Space–time–geochemical constraints on the emplacement of the Ferrar Large Igneous Province in South Victoria Land, Antarctica. *Geological Society of America Abstracts with Programs*, **44**(7), 541.

Foden, J., Sossi, P., Segui, D., Robinson, F. and Tappert, R. 2012. The implications of mantle lithospheric delamination for the termination of Cambrian Pacific margin orogenesis: the creation of the source of the Tasmanian–Ferrar Jurassic large igneous province. In: *34th International Geological Congress (IGC) Australia 2012, Brisbane Convention and Exhibition Centre (BCEC), Queensland, Australia, 5–10 August 2012: Congress Handbook*. Australian Geoscience Council, Canberra, abstract 3818.

Ford, A.B. 1976. *Stratigraphy of the Layered Gabbroic Dufek Intrusion*. United States Geological Survey Bulletin, **1405-D**.

Ford, A.B. and Boyd, W.W. 1968. The Dufek intrusion – a major stratiform gabbroic body in the Pensacola Mountains. In: *Proceedings of the 23rd International Geological Congress, Prague, August 1968, Volume 2*, Academia, Prague, 13–28.

Ford, A.B. and Himmelberg, G.R. 1991. Geology and crystallization of the Dufek intrusion. *Oxford Monographs on Geology and Geophysics*, **17**, 175–214.

Ford, A.B. and Kistler, R.W. 1980. K–Ar age, composition, and origin of Mesozoic mafic rocks related to Ferrar Group, Pensacola Mountains, Antarctica. *New Zealand Journal of Geology and Geophysics*, **23**, 371–390, https://doi.org/10.1080/00288306.1980.10424146

Ford, A.B., Kistler, R.W. and White, L.D. 1986. Strontium and oxygen isotope study of the Dufek Intrusion. *Antarctic Journal of the United States*, **21**, 63–66.

Forsha, C.J. and Zieg, M.J. 2007. Mineralogy and Texture of the Peneplain sill, McMurdo Dry Valleys, Antarctica. *Eos, Transactions of the American Geophysical Union*, **88**(52), Fall Meeting Supplement, abstract V43A-1121.

Galerne, C.Y., Neumann, E-R. and Planke, S. 2008. Emplacement mechanisms of sill complexes: Information from the geochemical architecture of the Golden Valley Sill complex, South Africa. *Journal of Volcanology and Geothermal Research*, **177**, 425–440, https://doi.org/10.1016/j.jvolgeores.2008.06.004

Galerne, C.Y., Galland, O., Neumann, E. and Planke, S. 2011. 3D relationships between sills and their feeders: evidence from the Golden Valley Sill Complex (Karoo Basin) and experimental modelling. *Journal of Volcanology and Geothermal Research*, **202**, 189–199, https://doi.org/10.1016/j.jvolgeores.2011.02.006

Gee, J.S., Cheadle, M.J., Meurer, W.P. and Grimes, C.B. 2013. How big is the Dufek Intrusion? Paleomagnetic constraints on the cooling history of the Dufek layered intrusion. *Eos, Transactions of the American Geophysical Union*, Fall Meeting 2013, **94**, abstract GP34A-05.

Goodge, J.W. and Finn, C.A. 2010. Glimpses of East Antarctica: Aeromagnetic and satellite magnetic view from the central Transantarctic Mountains of West Antarctica. *Journal of Geophysical Research: Solid Earth*, **115**, https://doi.org/10.1029/2009JB006890

Goodge, J.W., Fanning, C.M., Norman, M.D. and Bennett, V.C. 2012. Temporal, isotopic and spatial relations of Early Paleozoic Gondwana-margin arc magmatism, central Transantarctic Mountains, Antarctica. *Journal of Petrology*, **53**, 2027–2065, https://doi.org/10.1093/petrology/egs043

Grapes, R.H. and Reid, D.L. 1971. Rythmic layering and multiple intrusion in the Ferrar Dolerite of South Victoria Land, Antarctica. *New Zealand Journal of Geology and Geophysics*, **14**, 600–604, https://doi.org/10.1080/00288306.1971.10421950

Grimes, C.B., Cheadle, M., Gee, J.S., Meurer, W.P., Swapp, S. and Lusk, M.W. 2008. The role of magma replenishment in the construction of the lower 500 m of the layered mafic Dufek intrusion. *Eos, Transactions of the American Geophysical Union*, **89**(53), Fall Meeting Supplement, abstract V13C-2134.

Gunn, B.M. 1962. Differentiation in Ferrar Dolerites, Antarctica. *New Zealand Journal of Geology and Geophysics*, **5**, 820–863, https://doi.org/10.1080/00288306.1962.10417641

Gunn, B.M. 1963. Layered intrusions in the Ferrar dolerites. *Mineralogical Society of America, Special Papers*, **1**, 124–133.

Gunn, B.M. 1966. Modal and element variation in Antarctic tholeiites. *Geochimica et Cosmochimica Acta*, **30**, 881–920, https://doi.org/10.1016/0016-7037(66)90026-3

Gunn, B.M. and Warren, G. 1962. *Geology of Victoria Land between the Mawson and Mulock Glaciers, Antarctica*. Bulletin of the Geological Survey of New Zealand, **71**.

Hamilton, W.B. 1964. Diabase sheets differentiated by liquid fractionation, Taylor Glacier region, south Victoria Land. In: Adie, R.J. (ed.) *Antarctic Geology*. North-Holland, Amsterdam, 442–454.

Hamilton, W.B. 1965. *Diabase Sheets of the Taylor Glacier Region, Victoria Land, Antarctica*. United States Geological Survey Professional Papers, **456-B**.

Hanemann, R. and Viereck-Götte, L. 2004. Geochemistry of Jurassic Ferrar lava flows, sills and dikes sampled during the joint German–Italian Antarctic Expedition 1999–2000. *Terra Antartica*, **11**, 39–54.

Hanemann, R. and Viereck-Götte, L. 2007a. Platinum-group elements in sills of the Jurassic Ferrar Large Igneous Province from northern Victoria Land, Antarctica. *United States Geological Survey Open-File Report*, **2007-1047**, Short Research Paper 032, https://doi.org/10.3133/ofr20071047srp032

Hanemann, R. and Viereck-Götte, L. 2007b. Evolution of low-Ti and high-Ti rocks of the Jurassic Ferrar Large Igneous Province, Antarctica: Constraints from crystallization experiments. *United States Geological Survey Open-File Report*, **2007-1047**, extended abstract 070.

Hanemann, R., Melcher, F., Mukasa, S.B., Viereck-Goette, L. and Abratis, M. 2009. PGE-enrichment with late-stage Fe–Ti oxide crystallization observed in the Dufek–Forrestal layered mafic intrusion, Antarctica. *Eos Transactions of the American Geophysical Union*, **90**(52), Fall Meeting Supplement, abstract V21A-1973.

Harris, M. 2014. *Geochemistry of some Ferrar Large Igneous Province Intrusive Rocks in the Transantarctic Mountains, Antarctica*. BS thesis, Ohio State University, Columbus, Ohio, USA.

Haskell, T.R., Kennett, J.P. and Prebble, W.M. 1965. Geology of the Brown Hills and Darwin Mountains, Antarctica. *Transactions of the Royal Society of New Zealand*, **2**, 231–248.

Heier, K.S., Compston, W. and McDougall, I. 1965. Thorium and uranium concentrations, and the isotope composition of strontium in the differentiated Tasmanian dolerites. *Geochimica et Cosmochimica Acta*, **29**, 643–659, https://doi.org/10.1016/0016-7037(65)90061-X

Heinonen, J.S., Carlson, R.W., Riley, T.R., Luttinen, A.V. and Horan, M.F. 2014. Subduction modified oceanic crust mixed with a depleted mantle reservoir in the sources of the Karoo continental flood basalt province. *Earth and Planetary Science Letters*, **394**, 229–241, https://doi.org/10.1016/j.epsl.2014.03.012

Hergt, J.M. 2000. Comment on: 'Enriched mantle – Dupal signature in the genesis of the Jurassic Ferrar tholeiites from Prince Albert Mountains (Victoria Land, Antarctica)' by Antonini P et al. (Contributions to Mineralogy and Petrology 136, 1–19; 1999). *Contributions to Mineralogy and Petrology*, **139**, 240–244, https://doi.org/10.1007/s004100000130

Hergt, J.M. and Brauns, C.M. 2001. On the origin of the Tasmanian dolerites. *Australian Journal of Earth Sciences*, **48**, 543–549, https://doi.org/10.1046/j.1440-0952.2001.00875.x

Hergt, J.M., Chappell, B.W., Faure, G. and Mensing, T.M. 1989a. The geochemistry of Jurassic dolerites from Portal Peak, Antarctica. *Contributions to Mineralogy and Petrology*, **102**, 298–305, https://doi.org/10.1007/BF00373722

Hergt, J.M., Chappell, B.W., McCulloch, T.M., McDougall, I. and Chivas, A.R. 1989b. Geochemical and isotopic constraints on the origin of the Jurassic dolerites of Tasmania. *Journal of Petrology*, **30**, 841–883, https://doi.org/10.1093/petrology/30.4.841

Hergt, J.M., Peate, D.W. and Hawksworth, C.J. 1991. The petrogenesis of Mesozoic Gondwana low-Ti flood basalts. *Earth and*

Planetary Science Letters, **105**, 134–148, https://doi.org/10.1016/0012-821X(91)90126-3

Hersum, G.T., Marsh, B.D. and Simon, C.A. 2007. Contact partial melting of granitic country rock, melt segregation, and re-injection as dikes into Ferrar Dolerite Sills, McMurdo Dry Valleys, Antarctica. *Journal of Petrology*, **48**, 2125–2148, https://doi.org/10.1093/petrology/egm054

Hildreth, W. and Moorbath, S. 1988. Crustal contributions to arc magmatism in the Andes of central Chile. *Contributions to Mineralogy and Petrology*, **98**, 455–489, https://doi.org/10.1007/BF00372365

Himmelberg, G.R. and Ford, A.B. 1976. Pyroxenes of the Dufek intrusion. *Journal of Petrology*, **17**, 219–243, https://doi.org/10.1093/petrology/17.2.219

Himmelberg, G.R. and Ford, A.B. 1977. Iron–titanium oxides of the Dufek intrusion. *American Mineralogist*, **62**, 623–633

Hoefs, J., Faure, G. and Elliot, D.H. 1980. Correlation of $\delta^{18}O$ and initial $^{87}Sr/^{86}Sr$ ratios in Kirkpatrick Basalt on Mt. Falla, Transantarctic Mountains. *Contributions to Mineralogy and Petrology*, **75**, 199–203, https://doi.org/10.1007/BF01166760

Hornig, I. 1993. High-Ti and low-Ti tholeiites in the Jurassic Ferrar Group, Antarctica. *Geologisches Jahrbuch*, **E47**, 335–369.

Hunter, S.J., Johnson, A.C. and Aleshkova, N.D. 1996. Aeromagnetic data from the southern Weddell Sea embayment and adjacent areas: synthesis and interpretation. *Geological Society, London, Special Publications*, **108**, 143–154, https://doi.org/10.1144/GSL.SP.1996.108.01.10

Isaac, M.J., Chinn, T.J., Edbrooke, S.W. and Forsyth, P.J. 1996. *Geology of the Olympus Range Area, Southern Victoria Land, Antarctica, Scale 1:50 000*. Institute of Geological and Nuclear Sciences Geological Map, **20**.

Ishii, T. 1975. The relations between temperature and composition of pigeonite in some lavas and their application to geothermometry. *Mineralogical Journal*, **8**, 48–57, https://doi.org/10.2465/minerj.8.48

Ivanov, A.V., Meffre, S., Thompson, J., Corfu, F., Kamenetsky, V.S., Kamenetsky, M.B. and Demonterova, E.I. 2017. Timing and genesis of the Karoo–Ferrar large igneous province: New high precision U–Pb data for Tasmania confirm short duration of the major magmatic pulse. *Chemical Geology*, **455**, 32–43, https://doi.org/10.1016/j.chemgeo.2016.10.008

Jerram, D.A., Davis, G.R., Mock, A., Charrier, A. and Marsh, B.D. 2010. Quantifying 3D crystal populations, packing and layering in shallow intrusions: A case study from the Basement Sill, Dry Valleys, Antarctica. *Geosphere*, **6**, 537–548, https://doi.org/10.1130/GES00538.1

Jordan, T.A., Ferraccioli, F. and Leat, P.T. 2017. New geophysical compilations link crustal block motion to Jurassic extension and strike-slip faulting in the Weddell Sea Rift System of West Antarctica. *Gondwana Research*, **42**, 29–48, https://doi.org/10.1016/j.gr.2016.09.009

Jourdan, F., Bertrand, H., Schörer, U., Blichert-Toft, J., Féraud, G. and Kampunzu, A.B. 2007. Major and trace element and Sr, Nd, Hf, and Pb isotope compositions of the Karoo Large Igneous Province, Botswana–Zimbabwe: lithosphere vs mantle plume combination. *Journal of Petrology*, **48**, 1043–1077, https://doi.org/10.1093/petrology/egm010

Kistler, R.W., White, L.D. and Ford, A.B. 2000. *Strontium and Oxygen Isotopic Data and Age for the Layered Gabbroic Dufek Intrusion, Antarctica*. United States Geological Survey Open-File Report, **00-133**.

Kyle, P.R. 1980. Development of heterogeneities in the subcontinental mantle: evidence from the Ferrar Group, Antarctica. *Contributions to Mineralogy and Petrology*, **73**, 89–104, https://doi.org/10.1007/BF00376262

Kyle, P.R., Pankhurst, R.J. and Bowman, J.R. 1983. Isotopic and chemical variations in Kirkpatrick Basalt Group rocks from southern Victoria Land. *In*: Oliver, R.L., James, P.R. and Jago, J.B. (eds) *Antarctic Earth Science*. Australian Academy of Science, Canberra, 234–237.

Kyle, P.R., Pankhurst, R.J., Bowman, J.R., Millar, I.L. and McGibbon, R. 1987. Enriched subcontinental lithospheric mantle along the Pacific margin of Gondwana: isotopic studies of Jurassic Ferrar Supergroup tholeiites. *In*: *Abstracts of the Fifth International Symposium on Antarctic Earth Sciences, Cambridge, August 1987*. Cambridge University Press, Cambridge, UK, 86.

Kyle, P.R., Pankhurst, R.J., Esser, R. and Shelhorn, M.A. 1999. Petrogenesis of the Jurassic Butcher Ridge Igensous Complex, Ferrar Large Igneous Province, Antarctica. *In*: *Proceedings of the 8th International symposium on Antarctic Earth Sciences, 5–9 July, 1999, Wellington, New Zealand, Programme and Abstracts*. Royal Society of New Zealand, Wellington, 178.

Lanza, R and Zanella, E. 1993. Paleomagnetism of the Ferrar dolerite in the northern Prince Albert Mountains (Victoria Land, Antarctica). *Geophysical Journal International*, **114**, 501–511, https://doi.org/10.1111/j.1365-246X.1993.tb06983.x

Leat, P.T. 2008. On the long-distance transport of Ferrar magmas. *Geological Society, London, Special Publications*, **302**, 45–61, https://doi.org/10.1144/SP302.4

Leat, P.T., Riley, T.R., Storey, B.C., Kelley, S.P. and Millar, I.L. 2000. Middle Jurassic ultramafic lamprophyre dyke within the Ferrar magmatic province, Pensacola Mountains, Antarctica. *Mineralogical Magazine*, **64**, 95–111, https://doi.org/10.1180/002646100549021

Leat, P.T., Dean, A.A., Millar, I.L., Kelley, S.P., Vaughan, A.P.M. and Riley, T.R. 2005. Lithospheric mantle domains beneath Antarctica. *Geological Society, London, Special Publications*, **246**, 359–380, https://doi.org/10.1144/GSL.SP.2005.246.01.15

Leat, P.T., Luttinen, A.V., Storey, B.C. and Millar, I.L. 2006. Sills of the Theron Mountains, Antarctica: evidence for long distance transport of mafic magmas during Gondwana break-up. *In*: Hanski, E., Mertanen, S., Rämö, T. and Vuollo, J. (eds) *Dyke Swarms: Markers of Crustal Evolution*. Taylor and Francis, Abingdon, UK, 183–199.

Lindsley, D.H. 1983. Pyroxene thermometry. *American Mineralogist*, **68**, 447–493.

Lindsley, D.H. and Andersen, D.J. 1983. A two-pyroxene thermometer. *Journal of Geophysical Research: Solid Earth*, **88**(S02), A887–A905, https://doi.org/10.1029/JB088iS02p0A887

Lisker, F. and Läufer, A.L. 2013. The Mesozoic Victoria Basin: Vanished link between Antarctica and Australia. *Geology*, **41**, 1043–1046, https://doi.org/10.1130/G33409.1

Loewy, S.L., Dalziel, I.W.D., Pisarevsky, S., Connelly, J.N., Tait, J., Hanson, R.E. and Bullen, D. 2011. Coats Land crustal block, East Antarctica: A tectonic tracer for Laurentia? *Geology*, **39**, 859–862, https://doi.org/10.1130/G32029.1

Magee, C., Muirhead, J.D. *et al*. 2016. Lateral flow in mafic sill complexes. *Geosphere*, **12**, 809–841, https://doi.org/10.1130/GES01256.1

Magee, C., Ernst, R.E., Muirhead, J.D., Phillips, T. and Jackson, C.A.L. 2019. Magma transport pathways in large igneous provinces: lessons from combining field observations and seismic reflection data. *In*: Srivastava, R.K., Ernst, R.E. and Peng, P. (eds) *Dyke Swarms of the World: A Modern Perspective*. Springer, 45–85, https://doi.org/10.1007/978-981-13-1666-1_2

Marsh, B.D. 1998. On the interpretation of crystal size distributions in magmatic systems. *Journal of Petrology*, **39**, 553–599, https://doi.org/10.1093/petroj/39.4.553

Marsh, D.B. 2004. A magmatic mush column Rosetta Stone: The McMurdo Dry Valleys of Antarctica. *Eos, Transactions of the American Geophysical Union*, **86**, 497–502, https://doi.org/10.1029/2004EO470001

Marsh, B D. 2007. Magmatism, magma, and magma chambers. *Treatise on Geophysics*, **6**, 273–323, https://doi.org/10.1016/B978-0-444-53802-4.00116-0

Marsh, D.B. and Zieg, M.J. 1997. The Dais layered intrusion: a new discovery in the Basement Sill of the McMurdo Dry Valleys. *Antarctic Journal of the United States*, **32**, 18–20.

Marsh, B.D., Hersum, T.G., Simon, A.C., Charrier, A.D. and Souter, B.J. 2005. Discovery of a funnel-like deep feeder zone for the Ferrar Dolerites, McMurdo Dry Valleys, Antarctica. *Eos, Transactions of American Geophysical Union*, **86**(52), Fall Meeting Supplement, abstract V14C-03.

Marsh, J.S., Hooper, P.R., Rehacek, J., Duncan, R.A. and Duncan, A.R. 1997. Stratigraphy and age of Karoo basalts of Lesotho and implications for correlations within the Karoo igneous province. *American Geophysical Union Geophysical Monograph Series*, **100**, 247–272.

Marshak, S., Kyle, P.R., McIntosh, W., Samsonov, V. and Shellhorn, M. 1981. Butcher Ridge igneous complex, Cook Mountains, Antarctica. *Antarctic Journal of the United States*, **16**, 54–55.

McElroy, C.T. and Rose, G. 1987. *Geology of the Beacon Heights Area, Southern Victoria Land, Antarctica, Scale 1:50 000*. New Zealand Geological Survey Miscellaneous Series Map, **15**.

McIntosh, W.C., Kyle, P.R. and Sutter, J.F. 1986. Paleomagnetic results from the Kirkpatrick Basalt group, Mesa Range, north Victoria Land, Antarctica. *American Geophysical Union Antarctic Research Series*, **46**, 289–303.

McLelland, D. 1967. *Geology of the Basement Complex, Thorvald Nilsen Mountains, Antarctica*. MS thesis, University of Nevada, Reno, Nevada, USA.

Melluso, L., Hergt, J.M. and Zanetti, A. 2014. The late crystallization stages of low-Ti, low-Fe tholeiitic magmas; insights from evolved Antarctic and Tasmanian rocks. *Lithos*, **188**, 72–83, https://doi.org/10.1016/j.lithos.2013.10.032

Mensing, T.M. and Faure, G. 1996. Cretaceous alteration of Jurassic volcanic rocks, Pain Mesa, northern Victoria Land, Antarctica. *Chemical Geology*, **129**, 153–161, https://doi.org/10.1016/0009-2541(95)00155-7

Mensing, T.M., Faure, G., Jones, L.M., Bowman, J.R. and Hoefs, J. 1984. Petrogenesis of the Kirkpatrick Basalt, Solo Nunatak, north Victoria Land, Antarctica. *Contributions to Mineralogy and Petrology*, **87**, 101–108, https://doi.org/10.1007/BF00376216

Mensing, T.M., Faure, G., Jones, L.M. and Hoefs, J. 1991. Stratigraphic correlation and magma evolution of the Kirkpatrick Basalt in the Mesa Range, northern Victoria Land, Antarctica. *In*: Ulbrich, H. and Rocha Campos, A.C. (eds) *Gondwana Seven Proceedings. Papers presented at the Seventh International Gondwana Symposium, Sao Paulo, 1988*. Institut Geosciências, Universidade de São Paulo, São Paulo, Brazil, 653–667.

Menzies, M.A. and Kyle, P.R. 1990. Continental volcanism: a crust–mantle probe. *Oxford Monographs on Geology and Geophysics*, **16**, 157–177.

Milnes, A.R., Cooper, B.J. and Cooper, J.A. 1982. The Jurassic Wisanger basalt of Kangaroo Island, South Australia. *Royal Society of South Australia Transactions*, **106**, 1–13.

Molzahn, M., Reisberg, L. and Wörner, G. 1996. Os, Sr, Nd, Pb, O isotope and trace element data from the Ferrar flood basalts, Antarctica: evidence for an enriched subcontinental lithospheric source. *Earth and Planetary Science Letters*, **144**, 529–546, https://doi.org/10.1016/S0012-821X(96)00178-1

Molzahn, M., Wörner, G., Henjes-Kunst, F. and Rocholl, A. 1999. Constraints on the Cretaceous thermal event in the Transantarctic Mountains from alteration processes in Ferrar flood basalts. *Global and Planetary Change*, **23**, 45–60, https://doi.org/10.1016/S0921-8181(99)00050-8

Morrison, A.D. and Reay, A. 1995. Geochemistry of Ferrar Dolerite sills and dykes at Terra Cotta Mountains, south Victoria Land, Antarctica. *Antarctic Science*, **7**, 73–85, https://doi.org/10.1017/S0954102095000113

Mortimer, N., Parkinson, D., Raine, J.I., Adams, C.J., Graham, I.J., Oliver, P.J. and Palmer, K. 1995. Ferrar magmatic province rocks discovered in New Zealand: Implications for Mesozoic Gondwana geology. *Geology*, **23**, 185–188, https://doi.org/10.1130/0091-7613(1995)023<0185:FMPRDI>2.3.CO;2

Muirhead, J.D., Airoldi, J., Rowland, J.V. and White, J.D.L. 2012. Interconnected sills and inclined sheet intrusions control shallow magma transport in the Ferrar large igneous province, Antarctica. *Bulletin of the Geological Society of America*, **124**, 162–180, https://doi.org/10.1130/B30455.1

Mukasa, S.B., Andronikov, A.V. and Carlson, R.W. 2003. Myth of the Dufek Plume: Nd, Sr, Pb and Os isotopic and trace element data in support of a subduction origin. *In*: *9th International Symposium on Antarctic Earth Sciences, Potsdam, Germany, 8–12 September 2003*. Programme and Abstracts. Terra Nostra, Potsdam, Germany, 238.

Mukasa, S.B., Ravizza, G., Bédard, J., Choi, S., Andronikov, A.V. and Fleming, T.H. 2007. Dufek layered mafic intrusion and Basement Sill, Antarctica: constraints on their magma sources based on PGE abundance patterns, Nd–Sr–Pb isotopic ratios and trace element modeling. *Eos Transactions of the American Geophysical Union*, **88**(52), Fall Meeting Supplement, abstract V53D-03.

Nelson, D.A. and Cottle, J.M. 2016. Formation of layering in a hypabyssal intrusion by shear-induced fracture, exsolution, and rapid devitrification. *Goldschmidt Conference Abstracts*, **2016**. 2262.

Nelson, D.A., Cottle, J.M., Barboni, M. and Schoene, B. 2014. Petrologic significance of silicic magmatism in the Ferrar Large Igneous Province: geochemistry and geochronology of the Butcher Ridge Igneous Complex, Antarctica. *Eos, Transactions of the American Geophysical Union*, Fall Meeting 2014, **95**, abstract V33A-4831.

Nelson, D.A., Cottle, J.M. and Bindeman, I. 2018. Jurassic volcanic glass in the Ferrar large Igneous Province of Antarctica preserves evidence for hydration by glacial meltwater. *Eos, Transactions of the American Geophysical Union*, Fall Meeting 2018, **99**, abstract V23J-0182.

Neumann, E.R., Svensen, H., Galerne, C.Y. and Planke, S. 2011. Multistage evolution of dolerites in the Karoo large igneous province, central South Africa. *Journal of Petrology*, **52**, 959–984, https://doi.org/10.1093/petrology/egr011

Petford, N. and Mirhadizadeh, S. 2017. Image-based modelling of lateral flow: the Basement Sill, Antarctica. *Royal Society Open Science*, **4**, 161083, https://doi.org/10.1098/rsos.161083

Pocknall, D.T., Chinn, T.J., Sykes, R. and Skinner, D.N.B. 1994. *Geology of the Convoy Range Area, Southern Victoria Land, Antarctica, Scale 1:50 000*. Institute of Geological and Nuclear Sciences Geological Map, **11**.

Prior, G.T. 1907. Report on the rock specimens collected during the 'Discovery' Antarctic Expedition, 1901–1904. *National Antarctic Expedition 1901–1904, Natural History, Geology (Field Geology, Petrography)*, **1**, 101–140.

Ramirez, C., Nyblade, A. *et al.* 2017. Crustal structure of the Transantarctic Mountains, Ellsworth Mountains and Marie Byrd Land, Antarctica: constraints on shear wave velocities, Poisson's ratios and Moho depths. *Geophysical Journal International*, **211**, 1328–1340, https://doi.org/10.1093/gji/ggx333

Ricker, J. 1964. Outline of the geology between the Mawson and Priestley Glaciers, Victoria Land. *In*: Adie, R.J. (ed.) *Antarctic Geology*. North-Holland, Amsterdam, 265–275.

Riley, T.R., Curtis, M.L., Leat, P.T., Watkeys, M.K., Duncan, R.A., Millar, I.L. and Owens, W.H. 2006. Overlap of Karoo and Ferrar magma types in KwaZulu–Natal, South Africa. *Journal of Petrology*, **47**, 541–566, https://doi.org/10.1093/petrology/egi085

Riley, J.R., Taylor, B.M. and Fleming, T.H. 2020. Geochemistry of Jurassic Dolerite Intrusions in the Ohio Range and Southern Queen Maud Mountains, Antarctica. *Geological Society of America Abstracts with Programs*, **52**(2), https://doi.org/10.1130/abs/2020SE-345188

Roland, N.W. and Tessensohn, F. 1987. Rennick faulting – An early phase of Ross Sea rifting. *Geologisches Jahrbuch*, **66B**, 203–229.

Ross, P.-S., White, J.D.L. and McClintock, M.K. 2008. Physical volcanology of mafic volcaniclastic deposits and lavas in the Coombs–Allan Hills area, Ferrar large igneous province, Antarctica. *Journal of Volcanology and Geothermal Research*, **172**, 38–60, https://doi.org/10.1016/j.jvolgeores.2005.11.011

Rudnick, R.L. and Gao, S. 2003. Composition of the continental crust. *Treatise on Geochemistry*, **3**, 1–64.

Sarbas, B., Jochum, K.P., Nohl, U. and Weis, U. 2017. The geochemical databases GEOROC and GeoReM – what's new? *Eos, Transactions of the American Geophysical Union*, Fall Meeting 2017, **98**, abstract V23D-2629.

Sell, B., Ovtcharova, M. *et al.* 2014. Evaluating the temporal link between the Karoo LIP and climatic–biologic events of the Toarcian Stage with high-precision U–Pb geochronology. *Earth and Planetary Science Letters*, **408**, 48–56, https://doi.org/10.1016/j.epsl.2014.10.008

Semenov, V.S., Mikhailov, V.M., Koptev-Dvornikov, E.V., Ford, A.B., Shulyatin, O.G., Semenov, S.V. and Tkacheva, D.A. 2014. Layered Jurassic intrusions in Antarctica. *Petrology*, **22**, 547–573, https://doi.org/10.1134/S0869591114060034 (original Russian text in *Petrologiya*, 2014, **22**(6), 592–619).

Shellhorn, M.A. 1982. *The Role of Crustal Contamination at the Butcher Ridge Igneous Complex, Antarctica.* MS thesis, New Mexico Institute of Mining and Technology, Socorro, New Mexico, USA.

Sheth, H. 2018. *A Photographic Atlas of Flood Basalt Volcanism.* Springer, Berlin.

Siddoway, C.S. 2007. Tectonics of the West Antarctic Rift System: New light on the history and dynamics of distributed intracontinental extension. *United States Geological Survey Open-File Report*, **2007-1047**, 91–114.

Siders, M.A. and. Elliot, D.H. 1985. Major and trace element geochemistry of the Kirkpatrick Basalt, Mesa Range, Antarctica. *Earth and Planetary Science Letters*, **72**, 54–64, https://doi.org/10.1016/0012-821X(85)90116-5

Skinner, D.N.B. and Ricker, J. 1968. The geology of the region between the Mawson and Priestley Glaciers, north Victoria Land. *New Zealand Journal of Geology and Geophysics*, **11**, 1041–1075, https://doi.org/10.1080/00288306.1968.10420768

Souter, B.J., Marsh, B., Malolepszy, Z. and Morin, P. 2006. 3D structure of the feeder zone of the McMurdo Dry valleys magmatic system, Antarctica. *Eos, Transactions of the American Geophysical Union*, **87**(36), Joint Assembly Supplement, abstract V41A-21.

Spaeth, G., Hotten, R., Peters, M. and Techmer, K. 1995. Mafic dykes in the Shackleton range, Antarctica. *Polarforschung*, **63**, 101–121.

Stephenson, P.J. 1966. Geology 1: Theron Mountains, Shackleton Range and Whichaway Nunataks. *Trans-Antarctic Expedition, 1955–1958. Scientific Reports*, **8**.

Stewart, D. 1934. The petrography of some Antarctic rocks. *American Mineralogist*, **19**, 150–160.

Storey, B.C. and Alabaster, T. 1991. Tectonomagmatic controls on Gondwana breakup models: evidence from the proto-Pacific margin of Antarctica. *Tectonics*, **10**, 1274–1288, https://doi.org/10.1029/91TC01122

Storey, B.C. and Kyle, P.R. 1997. An active mantle mechanism for Gondwana break-up. *South African Journal of Geology*, **100**, 283–290.

Studinger, M., Bell, R.E., Blankenship, D.D., Buck, W.R. and Karner, G.D. 2004. Sub-ice geology inland of the Transantarctic Mountains in light of new aerogeophysical data. *Earth and Planetary Science Letters*, **220**, 391–408, https://doi.org/10.1016/S0012-821X(04)00066-4

Studinger, M., Bell, R.E., Fitzgerald, P.G. and Buck, W.R. 2006. Crustal architecture of the Transantarctic Mountains between the Scott and Reedy Glacier region and the South Pole from aerogeophysical data. *Earth and Planetary Science Letters*, **250**, 182–199, https://doi.org/10.1016/j.epsl.2006.07.035

Sun, S-s. and McDonough, W.F. 1989. Chemical and isotopic systematics of oceanic basalts: implications for mantle composition and processes. *Geological Society, London, Special Publications*, **42**, 313–345, https://doi.org/10.1144/GSL.SP.1989.042.01.19

Svensen, H., Corfu, F., Polteau, S., Hammer, Ø. and Planke, S. 2012. Rapid magma emplacement in the Karoo Large Igneous Province. *Earth and Planetary Science Letters*, **325–326**, 1–9, https://doi.org/10.1016/j.epsl.2012.01.015

Tinto, K.J., Padman, L. *et al.* 2019. Ross Ice Shelf response to climate driven by the tectonic imprint on seafloor bathymetry. *Nature Geoscience*, **12**, 441–449, https://doi.org/10.1038/s41561-019-0370-2

Turnbull, I.M., Allibone, A.H., Forsyth, P.J. and Heron, D.W. 1994. *Geology of the Bull Pass–St Johns Range Area, Southern Victoria Land, Antarctica, Scale 1:50 000.* Institute of Geological and Nuclear Sciences Geological Map, **14**.

Veevers, J.J., Conaghan, P.J. and Powell, C.McA. 1994. Eastern Australia. *Geological Society of America Memoirs*, **184**, 11–171.

Vennum, W.R. and Storey, B.C. 1987. Correlation of gabbroic and diabasic rocks from the Ellsworth Mountains, Hart Hills, and Thiel Mountains, West Antarctica. *American Geophysical Union Geophysical Monograph Series*, **40**, 129–138.

Wilhelm, S. and Wörner, G. 1996. Crystal size distribution in Jurassic Ferrar flows and sills (Victoria Land, Antarctica): evidence for processes of cooling, nucleation and crystallization. *Contributions to Mineralogy and Petrology*, **125**, 1–15, https://doi.org/10.1007/s004100050202

Will, T.M., Frimmel, H.E., Zeh, A., Le Roux, P. and Schmädicke, E. 2010. Geochemical and isotopic constraints on the tectonic and crustal evolution of the Shackleton Range, East Antarctica, and correlation with other Gondwana crustal segments. *Precambrian Research*, **180**, 85–112, https://doi.org/10.1016/j.precamres.2010.03.005

Woolfe, K.J., Kirk, P.A. and Sherwood, A.M. 1989. *Geology of the Knobhead Area, Southern Victoria Land, Antarctica, Scale 1:50 000.* New Zealand Geological Survey Miscellaneous Series Map, **19**.

Zavala, K., Leitch, A.M. and Fisher, G.W. 2011. Silicic segregations of the Ferrar Dolerite sills, Antarctica. *Journal of Petrology*, **52**, 1927–1964, https://doi.org/10.1093/petrology/egr035

Zieg, M.J. and Marsh, D.B. 2012. Multiple reinjections and crystal-mush compaction in the Beacon sill, McMurdo Dry Valleys, Antarctica. *Journal of Petrology*, **53**, 2567–2591, https://doi.org/10.1093/petrology/egs059

Chapter 2.2a

Palmer Land and Graham Land volcanic groups (Antarctic Peninsula): volcanology

Teal R. Riley* and Philip T. Leat

British Antarctic Survey, High Cross, Madingley Road, Cambridge CB3 0ET, UK
TRR, 0000-0002-3333-5021; PTL, 0000-0003-3824-8557
*Correspondence: trr@bas.ac.uk

Abstract: The break-up of Gondwana during the Early–Middle Jurassic was associated with flood basalt volcanism in southern Africa and Antarctica (Karoo–Ferrar provinces), and formed one of the most extensive episodes of continental magmatism of the Phanerozoic. Contemporaneous felsic magmatism along the proto-Pacific margin of Gondwana has been referred to as a silicic large igneous province, and is exposed extensively in Patagonian South America, the Antarctic Peninsula and elsewhere in West Antarctica. Jurassic-age silicic volcanism in Patagonia is defined as the Chon Aike province and forms one of the most voluminous silicic provinces globally. The Chon Aike province is predominantly pyroclastic in origin, and is characterized by crystal tuffs and ignimbrite units of rhyolite composition. Silicic volcanic rocks of the once contiguous Antarctic Peninsula form a southward extension of the Chon Aike province and are also dominated by silicic ignimbrite units, with a total thickness exceeding 1 km. The ignimbrites include high-grade rheomorphic ignimbrites, as well as unwelded, lithic-rich ignimbrites. Rhyolite lava flows, air-fall horizons, debris-flow deposits and epiclastic deposits are volumetrically minor, occurring as interbedded units within the ignimbrite succession.

An episode of enhanced silicic magmatism during the Early–Middle Jurassic is recognized along the central and eastern Antarctic Peninsula and is not thought to be directly related to subduction processes (Riley *et al.* 2017*a*). This dominantly silicic volcanism has been interpreted to be closely associated with rifting and mantle-plume activity linked to Gondwana break-up (Fig. 1) and extension in the Weddell Sea (e.g. Storey *et al.* 2013). Subaerial silicic volcanic rocks of Early Jurassic (*c.* 183 Ma) and Middle Jurassic (*c.* 170 Ma) age crop out extensively in SE Palmer Land (southern Antarctic Peninsula) and eastern Graham Land (northern Antarctic Peninsula), respectively. Both episodes of volcanic activity are correlated with the extensive Chon Aike province of Patagonia (Pankhurst *et al.* 1998, 2000). The entire province has been identified as a silicic large igneous province (Bryan and Ferrari 2013) and is, in part, contemporaneous with the Karoo and Ferrar flood basalt provinces of southern Africa and Antarctica (Svensen *et al.* 2012; Burgess *et al.* 2015). This widespread silicic volcanism has been attributed to the extensive melting of subduction-modified mafic underplated Grenvillian-age lower crust (Riley *et al.* 2001).

The Jurassic-age silicic volcanic rocks of eastern Graham Land and SE Palmer Land are dominated by ignimbrites and crystal-lithic tuffs, and their extent and field characteristics are described here. Basaltic to intermediate volcanic rocks of the Chon Aike province and its correlatives on the Antarctic Peninsula are relatively rare, accounting for <10% of the entire province. Basaltic rocks are also described here from both the Early and Middle Jurassic volcanic episodes.

Geological setting

Antarctic Peninsula

The Antarctic Peninsula was interpreted by Suárez (1976) as an autochthonous continental arc of the West Gondwanan margin and was associated with long-lived subduction during the Mesozoic. The identification of a major shear zone in eastern Palmer Land from the southern Antarctic Peninsula lead to the composite geology of the Antarctic Peninsula being reinterpreted (Vaughan and Storey 2000) as a series of para-autochthonous and allochthonous terranes which had accreted onto the Gondwana margin. This hypothesis has been reinterpreted to a simpler, *in situ*, model of continental arc evolution (Burton-Johnson and Riley 2015).

The Middle Jurassic felsic volcanic rocks of the east coast of the Antarctic Peninsula (Riley and Leat 1999) are associated with tonalite, quartz diorite and granodiorite plutonic rocks also of Middle Jurassic age (Pankhurst *et al.* 2000), and sequences of sedimentary rocks deposited in a terrestrial setting (e.g. Hunter *et al.* 2005). These Mesozoic formations of the northeastern Antarctic Peninsula unconformably overlie Carboniferous–Triassic metasedimentary rocks of the Trinity Peninsula Group (Barbeau *et al.* 2010; Bradshaw *et al.* 2012). The Trinity Peninsula Group has an estimated thickness of *c.* 5 km and Hathway (2000) has interpreted the

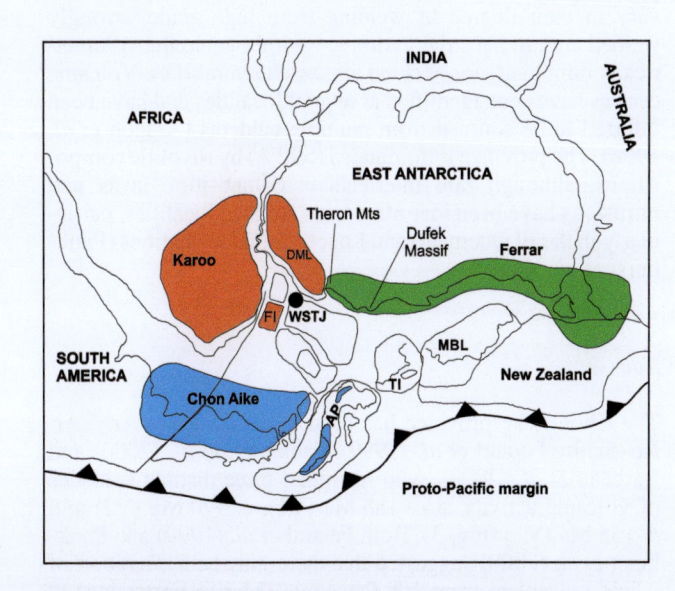

Fig. 1. Reconstruction of pre-break-up western Gondwana showing the major magmatic provinces (Karoo, Ferrar and Chon-Aike) (after Storey *et al.* 1992). DML, Dronning Maud Land; MBL, Marie Byrd Land; TI, Thurston Island; WSTJ, Weddell Sea triple junction.

From: Smellie, J. L., Panter, K. S. and Geyer, A. (eds) 2021. *Volcanism in Antarctica: 200 Million Years of Subduction, Rifting and Continental Break-up*. Geological Society, London, Memoirs, **55**, 121–138,
First published online January 19, 2021, https://doi.org/10.1144/M55-2018-36
© 2021 The Author(s). Published by The Geological Society of London. All rights reserved.
For permissions: http://www.geolsoc.org.uk/permissions. Publishing disclaimer: www.geolsoc.org.uk/pub_ethics

depositional environment as submarine fans along a passive continental margin. The Trinity Peninsula Group overlaps with and, in part, is likely to overlie Ordovician–Permian-age crystalline basement (e.g. Smellie and Millar 1995; Hervé et al. 1996; Bradshaw et al. 2012; Riley et al. 2012). The volcanic successions of eastern Graham Land also overlie or are, in part, interbedded with sedimentary rocks of the Jurassic-age Botany Bay Group (Hunter et al. 2005).

The Mesozoic volcanic sequences of Graham Land have all been subject to low- to medium-grade metamorphism and deformation, possibly during the Palmer Land deformation event (107–103 Ma: Vaughan et al. 2002) or during a Late Jurassic-age Peninsula deformation event (Storey et al. 1987). The Early–Middle Jurassic-age volcanic rocks of Palmer Land include the silicic Brennecke and Mount Poster formations (e.g. Hunter et al. 2006), which are associated with basaltic volcanic units of the Hjort and Swenney formations. The volcanic units are also associated with the widespread shallow-marine sedimentary rocks of the Latady Group (Hunter and Cantrill 2006). The sedimentary sequences of the Orville Coast of Palmer Land are known locally as the Mount Hill Formation and are interpreted to be a correlative of the Latady Formation. The granitoid plutons of Palmer Land are mid-Cretaceous in age (c. 125–102 Ma: Riley et al. 2018) and form part of a belt of Gondwana margin magmatism referred to locally as the Lassiter Coast Intrusive Suite, which intrudes the Jurassic volcanic and sedimentary sequences.

Chon Aike province

The Chon Aike province of Patagonian South America crops out in Argentina, Chile and offshore (Fig. 2) (Pankhurst et al. 1998), and is contiguous with the contemporaneous silicic volcanic rocks of the Antarctic Peninsula and Thurston Island (Riley et al. 2010, 2017b). The volcanic sequences exposed across the eastern region of the Chon Aike province are flat-lying, massive ignimbrite sheets, whereas the silicic volcanic rocks of the Andean Cordillera outcrop as linear outcrops which are locally deformed and strongly affected by hydrothermal alteration. The Chon Aike province is dominated by phenocryst-poor ignimbrites and crystal tuffs. The ignimbrites vary in their degree of welding from high-grade, strongly welded and, in part, rheomorphic ignimbrites to the volumetrically dominant, non-welded lithic-rich ignimbrites. Volcanic centres have been identified at several localities and have been inferred to be sourced from multiple calderas (Aragón et al. 1996). The province is dominated (>90%) by rhyolitic compositions, although rare intermediate composition lavas and intrusions have been identified from isolated localities, particularly in the El Quemado and Lonco-Tapial formations (Pankhurst et al. 1998).

Age

The Chon Aike province has been dated (U–Pb, $^{40}Ar/^{39}Ar$, Rb–Sr) by Féraud et al. (1999), Pankhurst et al. (2000) and Barbeau et al. (2009), who identified three distinct episodes of volcanic activity at c. 185 Ma (V1), c. 170 Ma (V2) and c. 155 Ma (V3) (Fig. 3). Both Féraud et al. (1999) and Pankhurst et al. (2000) suggested that there may be a migration of silicic volcanism from NE Patagonia (Marifil Formation) at 185 Ma (V1) to the SW (El Quemado Formation) at 155 Ma (V3), with the central area forming the Chon Aike Formation at 170 Ma (V2). However, provenance analysis by Barbeau et al. (2009) demonstrated that the straightforward age progression from the NE to the SW may not be valid after they identified all ages (V1–V3) present in felsic rocks from the southernmost Andes.

Both the V1 and V2 volcanic episodes have been identified from the Antarctic Peninsula, with V2-age (171–167 Ma) silicic tuffs and ignimbrites cropping out across large parts of eastern Graham Land (Pankhurst et al. 2000; Riley et al. 2010), whilst V1-age (185–181 Ma) rhyolitic ignimbrites

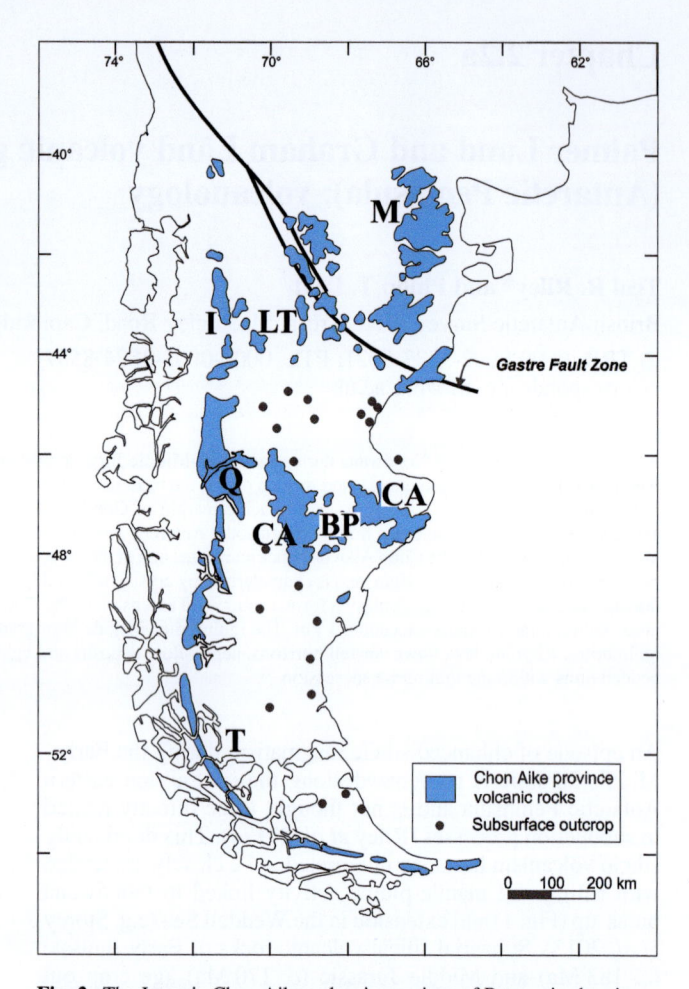

Fig. 2. The Jurassic Chon Aike volcanic province of Patagonia showing the extent of outcrop and subsurface exposures. Individual formations: M, Marifil; LT, Lonco Trapial; I, Ibañez; Q, El Quemado; CA, Chon-Aike; BP, Bajo Pobre; T, Tobífera (after Pankhurst et al. 1998).

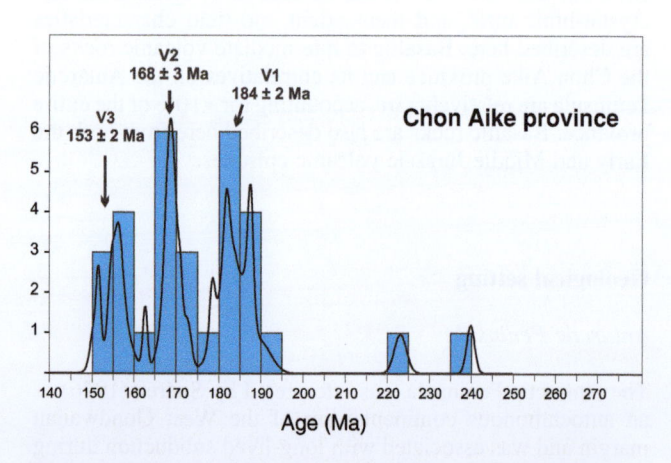

Fig. 3. Age histograms and cumulative probability curve for ages from the Chon Aike province (Pankhurst et al. 2000). The two Triassic ages are interpreted to be from inherited zircon grains and represent a period of Triassic magmatism, which was widespread in Patagonia and West Antarctica.

have been identified from large parts of central and eastern Palmer Land (Pankhurst et al. 2000; Hunter et al. 2006).

Graham Land Volcanic Group

General geology

The collective term, the Antarctic Peninsula Volcanic Group (APVG), was defined by Thomson and Pankhurst (1983) to describe presumed Mesozoic-age volcanic rocks of the Antarctic Peninsula. However, this term does not consider the tectonic setting, geochemistry or emplacement age, but instead groups together rocks of all compositions of Mesozoic age from a range of tectonic settings (fore-, intra-, back-arc or non-arc). In an attempt to introduce some better-defined constraints for the volcanic record of the Antarctic Peninsula, Riley et al. (2010) described the Graham Land Volcanic Group with the following key characteristics:

- Age: the volcanic rocks were erupted in the interval 177–162 Ma, with the main pulse of volcanism at 170–168 Ma.
- Lithology: the volcanic successions are dominated by ignimbrites, crystal and crystal-lithic tuffs, accretionary lapilli tuffs, and vitric tuffs.
- Composition: overall, the volcanic succession is bimodal, but >90% of the magmatism is silicic (rhyolite–rhyodacite) in composition, with rarer mafic units identified.
- Association: the silicic volcanic rocks are coeval with tonalities and granodiorites, which are interpreted to be the subvolcanic equivalents of the volcanic units.
- Geochemistry: isotopically the volcanic rocks have a restricted range ($^{87}Sr/^{86}Sr_{168}$ = 0.7065–0.7070; $\varepsilon_{Nd_{168}}$ = −3 to −2).
- Source: volcanic centres are difficult to identify, but are interpreted to be caldera-fed, successions, which are extensive over large areas.

The Graham Land Volcanic Group includes the Mapple Formation and its correlative units at Hope Bay (Kenney Glacier Formation: Birkenmajer 1993), Churchill Peninsula (Lyttleton Ridge), Joinville Island and across the entire area of Jason Peninsula. It also includes the intermediate volcanic rocks exposed at Botany Bay and Tower Peak (Mount Tucker Formation: Millar et al. 1990) and the basaltic rocks of Jason Peninsula (Riley et al. 2003) (Fig. 4). The Graham Land Volcanic Group is largely restricted to the north and east coasts of Graham Land, but may also include volcanic rocks exposed further south, potentially including the rhyolites (Marsh 1968) exposed at Cape Robinson (Fig. 4).

The volcanic successions defined by Riley and Leat (1999) and Riley et al. (2010) are correlated with the Chon Aike province of South America, and specifically with the V2 event of Pankhurst et al. (2000). These volcanic units were erupted along the continental margin of Gondwana in the early stages of break-up, and were the result of the melting of the lower crust and mafic underplate that have been modified by subduction-related fluids and melts (Riley et al. 2001).

The volcanic rocks of the Graham Land Volcanic Group are, in many respects, distinct from the volcanic successions which crop out along the west coast of the Antarctic Peninsula (Leat and Riley 2021), where volcanism is characterized by reworked volcanic sequences deposited in a forearc–intra-arc setting. Volcanism of the west coast of the Antarctic Peninsula is dominated by intermediate compositions, but rare rhyolitic centres have been identified in Palmer Land (Leat and Scarrow 1994) and southeastern Palmer Land (Flowerdew et al. 2005). These volcanic units are Cretaceous in age and are not the characteristic thick plateau, caldera-fed ignimbrites of the east coast, but are smaller, central volcano-fed outflow facies.

The silicic volcanic rocks of the east coast of the Antarctic Peninsula are associated with granitoid plutons of Early–Middle Jurassic age (Riley et al. 2017a) and thick sequences of terrestrial sedimentary rocks (Riley and Leat 1999). Jurassic-age silicic volcanic rocks have been identified from multiple sites across eastern Graham Land (Fig. 4). They are summarized here with details of their occurrence, physical characteristics and their age (radiometric/stratigraphic).

Oscar II Coast: Mapple Formation

Occurrence. Silicic volcanic rocks outcrop extensively along the Oscar II Coast of the eastern Antarctic Peninsula (Fleet 1968). Silicic volcanic rocks are interbedded with and overlie at least 600 m of sedimentary rocks of the Botany Bay Group (Hunter et al. 2005). Riley and Leat (1999) defined the Mapple Formation from the Mapple Glacier region (Fig. 5) where there is a maximum observed thickness of 1000 m of moderately welded, westward-dipping ignimbrites. Because of within-flow facies variations and significant faulting, it is difficult to correlate units, so a type stratigraphic section for the Mapple Formation was not presented.

Characteristics. In general terms, the Mapple Formation of the Oscar II Coast is dominated by ignimbrite units and crystal tuffs (c. 80%), which display significant variation in terms of their degree of welding and lithic content. The volcanic rocks are almost entirely subaerial, although localized subaqueous emplacement is evident in rare units and may have been in a lacustrine environment. The nature of the outcrop in a valley glacier terrain means that eruptive centres are difficult to identify, although proximity to potential source regions can be interpreted from mass-flow deposits and co-ignimbrite breccias. The ignimbrite and tuff units are typically phenocryst poor, but have a mineral assemblage characterized by quartz, plagioclase, K-feldspar, muscovite, biotite, apatite, magnetite, titanite and zircon. Individual ignimbrite flow units are typically in the range 5–10 m, but individual flow units up to 80 m have been identified (Riley and Leat 1999). The degree of welding varies from unwelded, lithic-rich ignimbrites through to higher-grade rheomorphic ignimbrites and lava-like tuffs with parataxitic textures. This variation in emplacement style implies a significant variation in eruptive temperature.

Ignimbrite units of the Mapple Formation are typically moderately to poorly welded, they have a low lithic content and are difficult to trace laterally over any distance. At the western extent of the Mapple Glacier an 800 m ignimbrite succession is exposed. The volcanic units are dominated by moderate to non-welded ignimbrite units. A lithic-rich basal layer has been identified from many of the ignimbrite units (Fig. 6a) and, less commonly, an associated fine-grained ash layer. The ignimbrites typically have oblate, often crystal-rich pumice, combined with lithic concentrations of up to 10%. Strongly welded ignimbrite units have a eutaxitic texture with pumice-flattening ratios of up to 6:1 (Fig. 6b).

Ignimbrite units, which have experienced secondary flow after agglutination, are defined as rheomorphic ignimbrites or rheoignimbrites (Wolff and Wright 1981). These high-grade units have only been identified at three localities throughout the Mapple Formation (Riley and Leat 1999). One unit, identified on the southern margin of the Stubb Glacier (Fig. 5), is 8 m in thickness, and is characterized by a pronounced parataxitic texture with rotated lithic fragments, boudinaged clasts, elongated vesicles and minor flow folds. The number of lithic fragments is unusually high for a rheomorphic ignimbrite, and larger fragments (5–10 cm) have often been rotated during secondary flow (Fig. 6c) with

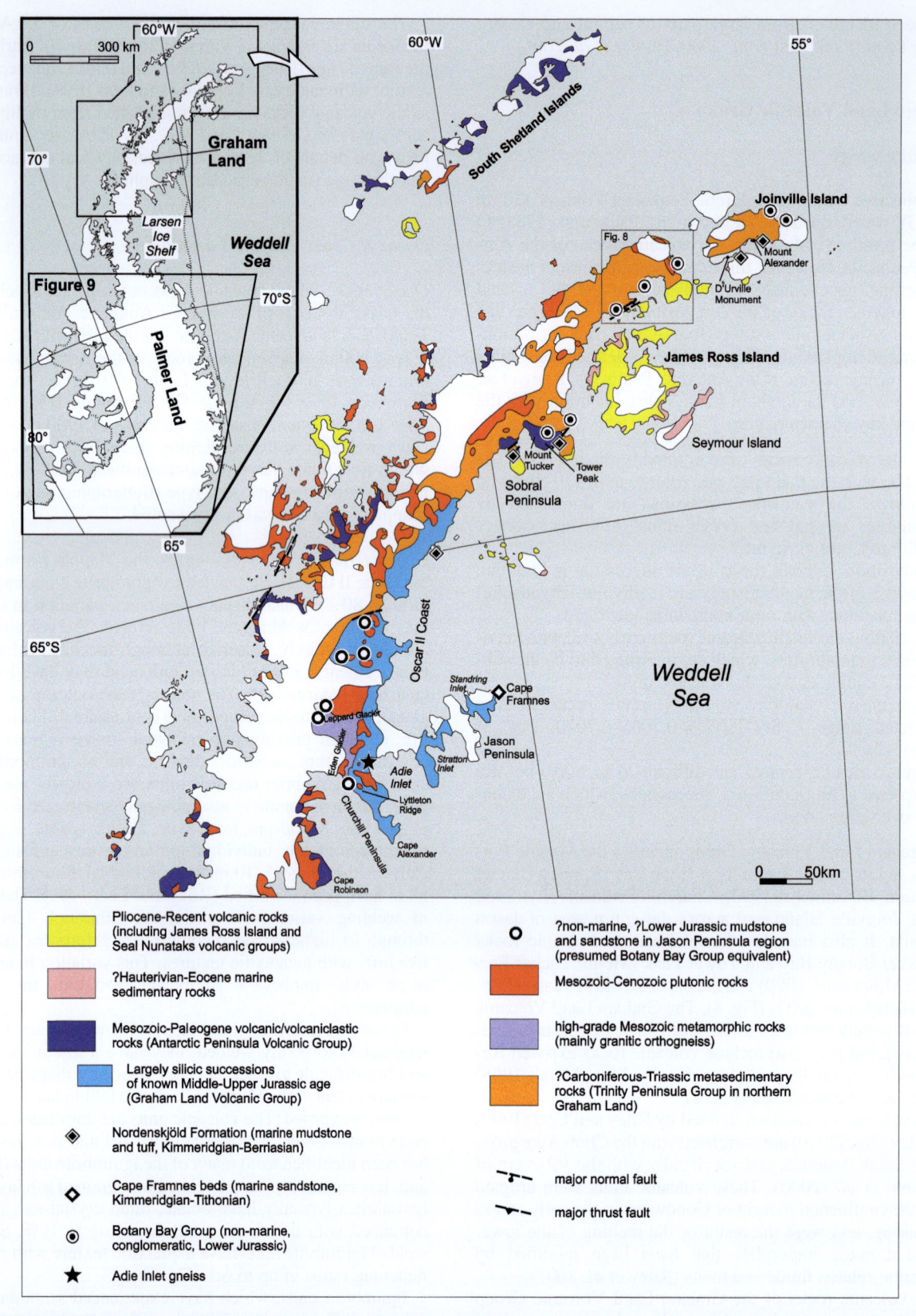

Fig. 4. Simplified geological map of the Graham Land (northern Antarctic Peninsula) showing the main outcrop area of Jurassic Antarctic Peninsula Volcanic Group rocks and other key units (from Hathway 2000; Riley *et al.* 2010).

minor flow folding developed in the lee of the fragments. Boudinaging of lithic fragments is also developed in the rheoignimbrite (Fig. 6d) and demonstrates the deformation of more competent material within a sheared ash-rich matrix. All of these deformation features are diagnostic of occurring in a rheomorphic state, after the agglutination of ash and lithic fragments (Riley and Leat 1999).

Rhyolite and rhyodacitic lava flows are rare in the Mapple Formation, and only account for *c.* 10% of the volcanic succession (Riley and Leat 1999). The lava flows identified in the Mapple Formation are typically fine grained and characterized by flow banding. Lava flows have been identified that are up to 30 m in thickness and are often associated with rhyolite feeder dykes. A flow-banded lava on the southern side of the

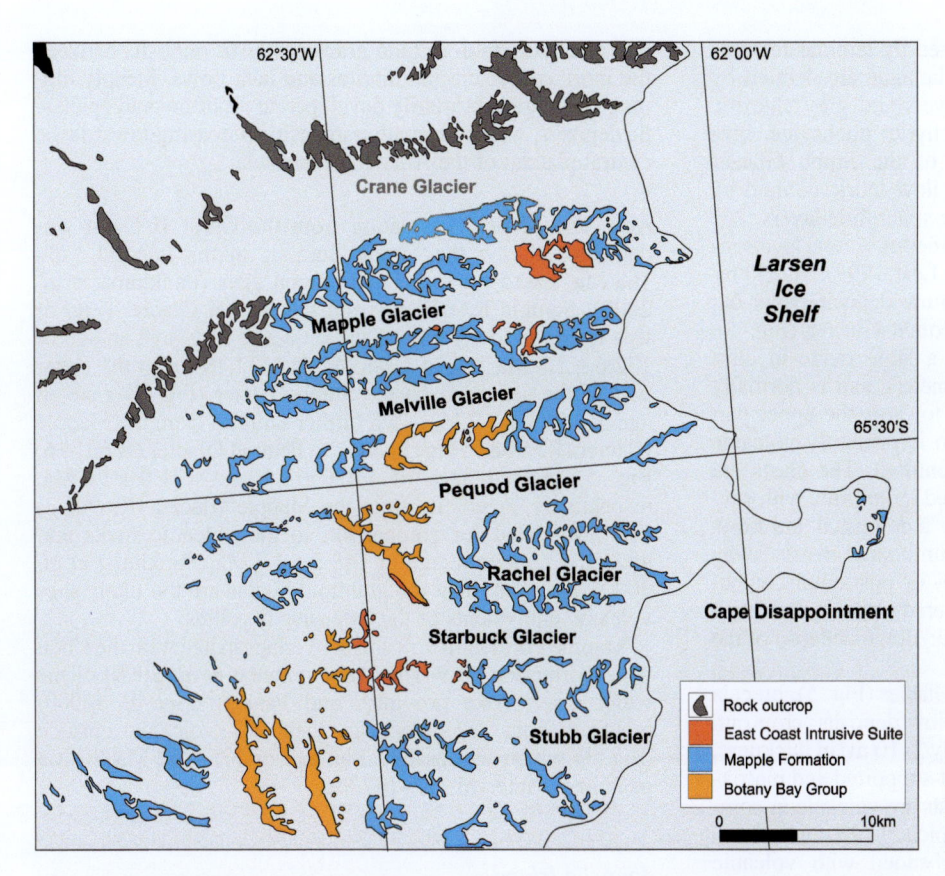

Fig. 5. Simplified geological outcrop map of the Oscar II Coast area (after Riley *et al.* 2010).

Fig. 6. (**a**) Lithic-rich basal layer of ignimbrite on the Mapple Glacier, Oscar II Coast. The notebook is 16 cm in length. (**b**) Eutaxitic texture in welded silicic ignimbrite from the Mapple Glacier, Oscar II Coast. The lens cap is 52 mm in diameter. Fiamme are marked. (**c**) Parataxitic texture and rotated lithic clasts in silicic rheoignimbrite from the Stubb Glacier, Oscar II Coast. The lens cap is 52 mm in diameter. (**d**) Boudinaged clasts in parataxitic, silicic rheoignimbrite from the Stubb Glacier, Oscar II Coast. The lens cap is 52 mm in diameter. (**e**) Flow-banded rhyolite dyke from the Stubb Glacier, Oscar II Coast. Banding is linear at the dyke margins and more chaotic towards the centre of the dyke. The lens cap is 52 mm in diameter. (**f**) Accretionary lapilli in a planar-bedded silicic ash-fall unit from the Stubb Glacier, Oscar II Coast. The lens cap is 52 mm in diameter. The lapilli zones are circled.

Mapple Glacier (Fig. 5) is characterized by laminar features and is strongly devitrified. The flow laminae are defined by alternating cream-coloured felsitic layers and grey chloritic layers. A c. 25 m-thick, feldspar-phyric pitchstone lava crops out at the eastern extremity of the Stubb Glacier (Fig. 5) and is flow folded, with the flow fabric defined by poorly preserved vitric bands and pale spherulitic layers.

Mass-flow deposits have been identified at several locations in the Mapple Formation (Riley and Leat 1999). A 250 m-thick, boulder-rich, subaerial debris-flow deposit crops out along the southern margin of the Pequod Glacier (Fig. 5). The deposit is clast-supported, with a large range in clast size (from <0.5 m up to 20 m in diameter), and is normally graded with the clast size decreasing towards the upper part of the deposit. The clast composition is primarily volcanic with only rare sedimentary clasts identified. The clasts are dominated by blocks of lava and welded ignimbrite, with several of the larger clasts surrounded by a deformed and flow-foliated matrix which grades into an unfoliated matrix away from the clast and may preserve syn- or post-emplacement flow. This mass-flow deposit is considered a proximal deposit and is interpreted as the result of a flank/dome collapse of the edifice (Riley and Leat 1999).

At the western end of the Pequod Glacier (Fig. 5), breccia units, interpreted as subaerial debris-flow deposits, crop out. The breccia beds are up to 40 m (mostly 2–10 m) in thickness, are weakly normally graded, with clast-supported and matrix-supported varieties identified. The clasts are volcanic in composition (tuff and lava), angular and typically <10 cm in diameter. The breccia deposits are interbedded with volcanic (ignimbrite, crystal tuff) and thin mud–sand grade sedimentary (volcaniclastic) units. Monomict and polymict variations are observed, with the monomict deposits potentially formed by the collapse of a lava dome or lava carapace, whereas the polymict deposits may indicate the collapse of a compound volcanic edifice.

Rhyolite dykes from the Mapple Formation are typically 2–3 m wide and characterized by flow folding (Riley and Leat 1999). They are associated with rhyolitic lava units and the dykes may represent feeder bodies. A petrogenetic link between the lava flows and rhyolite dykes is supported by their major element chemistry and petrography (Riley et al. 2001). Several dykes feed a 30 m-thick rhyolitic lava flow along the southern margin of the Stubb Glacier (Fig. 5). The dykes preserve a flow-banded fabric parallel to the dyke margin, with more chaotic flow folds toward the dyke centre (Fig. 6e).

A finely laminated, planar-bedded air-fall tuff unit characterized by numerous armoured accretionary lapilli has been identified on the southern side of the Stubb Glacier (Riley and Leat 1999) (Fig. 5). The air-fall unit is at least 1 m thick and is interbedded in a succession of ignimbrite flow units. The upper ignimbrite preserves a fine-grained basal layer and an overlying lithic-rich horizon, before passing into the main ignimbrite unit. The air-fall tuff is thinly bedded, containing multiple slightly flattened accretionary lapilli (Fig. 6f), and is interpreted to have been derived from a co-ignimbrite ash cloud (Riley and Leat 1999). The accretionary lapilli have a diameter up to 15 mm and have up to four concentric layers, and have an oblate morphology as a result of compaction. The lapilli are well sorted (Fig. 6f) and the flattening is parallel to bedding, features typical of co-ignimbrite air-fall tuffs (e.g. McPhie et al. 1993) or potentially surge deposits (Brown et al. 2010).

The Mapple Formation is metamorphosed up to greenschist facies and was deformed during the Cretaceous-age Palmer Land tectonic event (Vaughan et al. 2002). The deformation in eastern Graham was relatively weak, and generated low-angle faults and open folds. The low-angle faults localized movement on mud- to sand-grade epiclastic deposits between the more competent ignimbrites and lava flows. Steeply dipping cleavage is primarily developed in mudflow and epiclastic deposits, with deformation intensity increasing towards the central plateau of the Antarctic Peninsula.

Age. The Mapple Formation from the Oscar II Coast was deposited during the Middle Jurassic in the interval 173–168 Ma, based on four U–Pb zircon ages (Pankhurst et al. 2000). Sample R.6619.14 from the Rachel Glacier (Fig. 5) was from the lower part of the sequence and gave an age of 172.6 ± 1.8 Ma, whilst a sample (R.6632.10) from the upper part of the sequence on the Stubb Glacier (Fig. 5) gave an age of 168.3 ± 2.2 Ma. Two further samples from the Mapple Glacier (R.6908.7: Fig. 5) and the Pequod Glacier (R.6914.6: Fig. 5) yielded ages of 170.0 ± 1.4 and 171.0 ± 1.1 Ma, respectively. A granite from the Mapple Glacier (R.6906.3) was also dated for comparison to the volcanic rocks and yielded a concordia age of 168.5 ± 1.7 Ma (Pankhurst et al. 2000), indicating that the granitoid bodies are the likely subvolcanic equivalents of the extensive rhyolites.

Mapple Formation volcanism overlaps in age with the Chon Aike Formation (169–168 Ma: Pankhurst et al. 2000) of the wider Chon Aike province, and Pankhurst et al. (2000) included both formations in their V2 volcanic episode (Fig. 3) which was assigned an age of 173–162 Ma, with a peak in the interval 172–167 Ma.

Joinville Island

Occurrence. Joinville Island is located to the NE of the Antarctic Peninsula (Fig. 4) and is, geologically, a continuation of northern Graham Land, dominated by sedimentary successions of the Trinity Peninsula Group (Carboniferous–Triassic), the Botany Bay Group (Early–Middle Jurassic) and the Nordenskjöld Formation (Lower Jurassic–Cretaceous). Outcrops of subaerial volcanic rocks have been identified from southern Joinville Island (Elliot 1967; Riley et al. 2010), although any stratigraphic relationships are obscured by the extensive intrusion of granitoid plutons of likely Cretaceous age.

Characteristics. The volcanic rocks are best exposed at D'Urville Monument (Elliot 1967) and Mount Alexander (Riley et al. 2010). At D'Urville Monument (Fig. 4) rhyolite lava flows are the main lithology, and are characterized by flow banding and devitrification of the glassy bands. Rarer silicic tuffs also crop out, which are lithic rich in part. At Mount Alexander (Fig. 4), a 300 m volcanic succession, crops out and is dominated by moderately to strongly welded ignimbrites in association with fine-grained plagioclase- and amphibole-phyric rhyolitic lavas. A separate succession of unfossiliferous, but presumed Botany Bay Group, mudstones and siltstones are interbedded with silicic volcanic rocks, which include tuffs, hyaloclastite and flow-banded lavas (Fig. 7a, b), indicating contemporaneous sedimentation and volcanism.

Age. Two rhyolitic volcanic rocks (R.8308.1 and R.8309.1) from near Mount Alexander (Fig. 4) on the south coast of Joinville Island, which are interbedded with Botany Bay Group sedimentary rocks, have been dated (U–Pb) by Riley et al. (2010). Sample R.8308.1, from near the exposed base of the volcanic succession, is a rhyolitic, autoclastic breccia which is host to silicified quartz–plagioclase-phyric rhyolite fragments. R.8308.1 yielded a poorly constrained weighted mean age of 168 ± 11 Ma, which is considered to be the approximate eruptive age. Sample R.8309.1 is a flow-banded rhyolite lava cropping out higher in the succession than R.8308.1, although relatively close to the exposed base of the Botany Bay Group succession. The lithology is fine grained, and is plagioclase- and quartz-phyric; it is partly devitrified and displays chaotic flow banding. Sample R.8309.1 gave a weighted mean age of 162.4 ± 1.8 Ma, which was considered by Riley *et al.* (2010) to be the eruptive age.

Hope Bay: the Kenney Glacier Formation

Occurrence. The volcanic rocks at Hope Bay (Fig. 8) overlie and are interbedded with the Early–Middle Jurassic sedimentary rocks (Mount Flora Formation, Botany Bay Group: Elliot and Gracanin 1983) first described by Halle (1913), who assigned a Middle Jurassic age based on comparisons with the Jurassic flora of rocks from Yorkshire.

The volcanic rocks are dominated by silicic ignimbrites, lavas and breccia units, and have been described by Birkenmajer (1992, 1993), who defined the stratigraphy and referred to the succession as the Kenney Glacier Formation. An angular unconformity between the Kenney Glacier Formation and the underlying sedimentary Mount Flora Formation was described by Birkenmajer (1993), although Farquharson (1984) was less certain regarding the contact between the two units.

Ignimbrite flow units and lapilli tuff layers interbedded in the upper part of the sedimentary rocks of the Mount Flora Formation have been described by Elliot and Gracanin (1983) and Farquharson (1983). Such a relationship indicates a gradual transition from terrestrial shallow-water sedimentation to entirely volcanic conditions, with contemporaneous sedimentation and volcanism at some point.

Characteristics. The Kenney Glacier Formation overlies the Mount Flora Formation, exposed in the upper parts of Mount Flora at Hope Bay (Fig. 8); it also unconformably overlies the Hope Bay Formation (Trinity Peninsula Group: Birkenmajer 1992). The Kenney Glacier Formation succession is dominated by rhyolite–dacite lavas, ignimbrites, tuff units and agglomerates (Birkenmajer 1993). Farquharson (1984) identified fine-grained agglomerates and welded tuffs/ignimbrites from the Mount Flora Formation, including a 26 m-thick

Fig. 7. (a) Flow-banded rhyolite lava. Mount Alexander area, Joinville Island. The lens cap is 52 mm in diameter. (b) Rhyolitic lithic-rich tuff. Mount Alexander area, Joinville Island. The ice axe is 65 cm in length. (c) View of ignimbrite units south along Lyttleton Ridge on the Churchill Peninsula. (d) Flow-banded rhyolite lava/pitchstone with alternating glassy and cherty layers and hematite mineralization. Eden Glacier, Churchill Peninsula area. The lens cap is 58 mm in diameter. (e) Eutaxitic texture in welded silicic ignimbrite from the Churchill Peninsula, Mapple Formation. The lens cap is 58 mm in diameter. (f) Oblate, crystal-rich pumice in ignimbrite from Lyttleton Ridge on the Churchill Peninsula. The lens cap is 58 mm in diameter. (g) Almost spheroidal accretionary lapilli in a planar-bedded silicic ash-fall unit from the Churchill Peninsula. The lens cap is 58 mm in diameter. (h) Intense hydrothermal alteration causing rusty coloration in rhyolitic tuffs. Astro Cliffs, southern Churchill Peninsula. The cliff height is approximately 95 m.

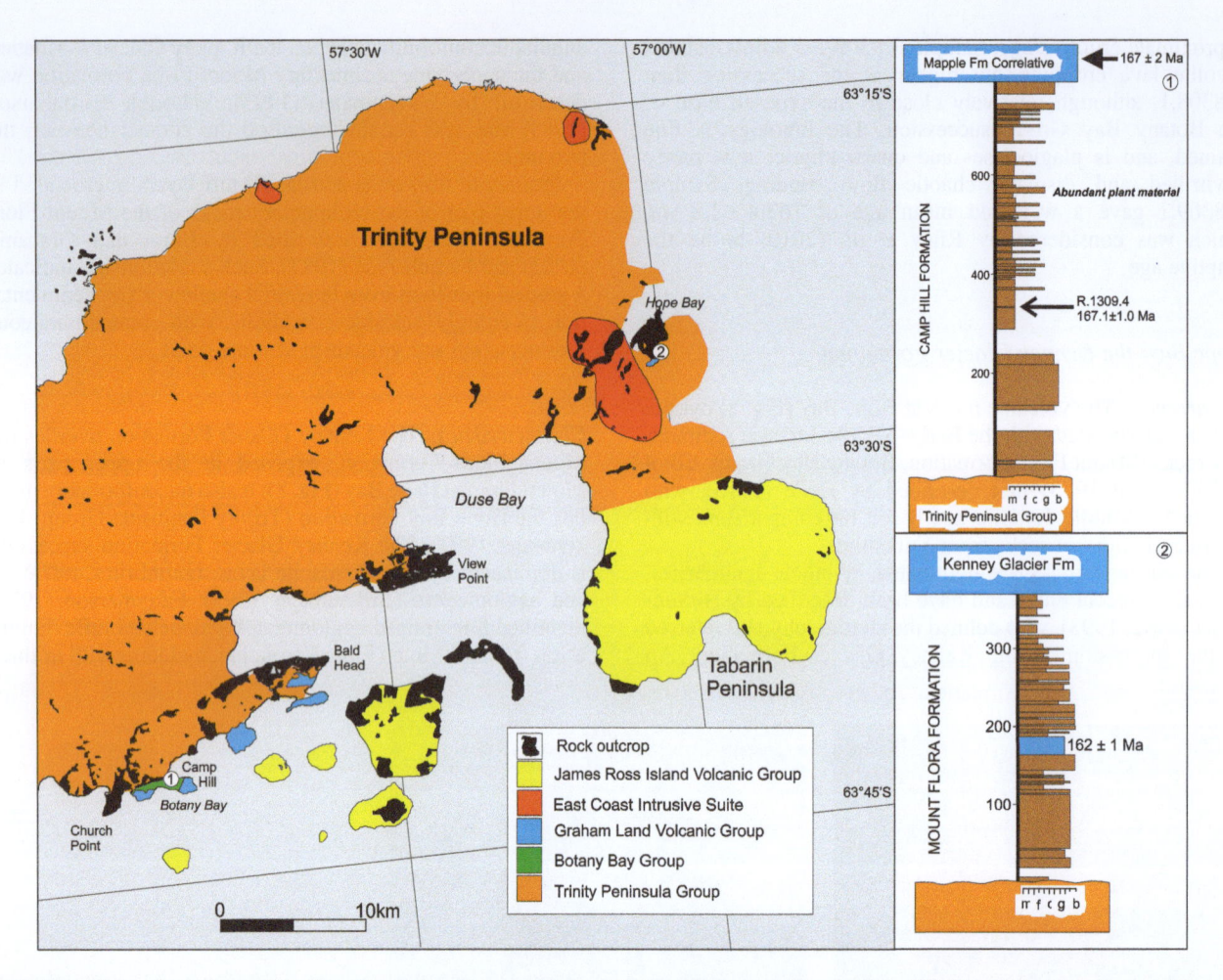

Fig. 8. Simplified geological map of the Hope Bay–Botany Bay area. The inset stratigraphic logs are from the Camp Hill (1) and Hope Bay (2) areas (Hunter *et al.* 2006).

ignimbrite unit which forms a prominent band on the north face of Mount Flora.

Age. Pankhurst *et al.* (2000) dated (U–Pb) an ignimbrite (R.609.1) from the Mount Flora Formation. The sample was taken from a volcanic unit interbedded with sedimentary rocks near the upper part of the Mount Flora Formation succession. Pankhurst *et al.* (2000) reported a concordant age of 162.2 ± 1.1 Ma, which also gives a maximum age of the overlying, but broadly contemporaneous, Kenney Glacier Formation and is contemporaneous with the volcanism reported from the adjacent successions on Joinville Island (Riley *et al.* 2010).

Camp Hill, Botany Bay and Bald Head

Occurrence. The volcanic succession of the Camp Hill–Botany Bay area crops out at Church Point (Fig. 8), to the west of Camp Hill and at Bald Head (Farquharson 1984). The volcanic units at Camp Hill are only 50 m in thickness and conformably overlie a 780 m sequence of terrestrial sedimentary rocks (fluviatile–deltaic) of the Botany Bay Group (Camp Hill Formation).

Characteristics. The minor succession of volcanic rocks exposed at Camp Hill are distinct from the Mapple Formation silicic units and their correlatives seen elsewhere in eastern Graham Land; they are overwhelmingly dominated by agglomerate near the base of the exposed section, and are overlain by a thick sequence of garnetiferous andesite lavas, tuffs and volcaniclastics near the upper part of the succession (Farquharson 1984). The sedimentary Camp Hill Formation, underlying the silicic volcanic rocks, itself lacks volcanic units, with the exception of a single accretionary lapilli horizon which occurs approximately half way up the section (Farquharson 1984). The lapilli are set in a matrix of devitrified glass shards, and provide the only evidence of contemporaneous volcanism and sedimentation at Botany Bay.

Age. Millar *et al.* (1990) dated (Sm–Nd) igneous garnet from an andesitic sill which cuts the base of the volcanic succession at Camp Hill, and produced an age of 152 ± 8 Ma. However, Millar *et al.* (1990) observed that the Sm and Nd concentrations were unusually high for garnet and may indicate that inclusions were present, and that the Sm–Nd age may not represent a reliable crystallization age. Pankhurst *et al.* (2000) dated (U–Pb) an andesite sill cutting the same Camp Hill succession dated by Millar *et al.* (1990) and determined an age of 166.9 ± 1.6 Ma, confirming that the Sm–Nd age does not precisely date the intrusion of the sills, or that magmatism continued until at least 152 Ma, or that the dated zircons are inherited. Hunter *et al.* (2005) dated detrital zircon grains from the single lapilli tuff horizon (described by Farquharson 1984) interbedded with the sedimentary Camp Hill Formation, and reported a maximum eruption age of 167.1 ± 1.1 Ma. The U–Pb geochronology of Pankhurst *et al.* (2000) and Hunter *et al.* (2005) indicates a Middle Jurassic age for the volcanic rocks at Botany Bay, which correlates with the Mapple Formation of eastern Graham Land. An Early Jurassic age suggested by Rees (1993), based on the fossil flora assemblage from the Camp Hill Formation, is therefore not considered reliable.

Mount Tucker and Tower Peak

Occurrence. The sedimentary Tower Peak Formation (Botany Bay Group), with a maximum thickness of 124 m (Farquharson 1984), predominantly consists of debris-flow-fed conglomerates. The upper half of the exposed sequence at Tower Peak and Mount Tucker (Fig. 4) is interbedded with several subaerial volcanic horizons. The Tower Peak Formation is conformably overlain by a sequence of agglomerates and garnet-bearing andesite lavas (c. 750 m thick: Aitkenhead 1975), akin to the succession observed at Camp Hill. The volcanic rocks are also in faulted contact with metasedimentary rocks of the Trinity Peninsula Group.

Characteristics. Detailed descriptions of the volcanic component of the Tower Peak Formation are lacking, but Farquharson (1984) does describe that the volcanic units of the upper part of the sequence include a 21 m-thick massive crystal-lithic tuff, a 6 m-thick altered tuff unit containing abundant accretionary lapilli and rare agglomerate horizons.

Age. Millar et al. (1990) dated (Sm–Nd) three garnet-bearing conglomerate samples which overlie the crystal-lithic tuff from the Tower Peak Formation. They did not yield a precise age, but instead fall on an errorchron yielding an age of 165 ± 57 Ma. The combined Tower Peak and Camp Hill Sm–Nd data yielded an age of 156 ± 6 Ma (Millar et al. 1990).

Hunter et al. (2005) dated (U–Pb) volcanic fragments from a conglomerate bed from the upper part of the Tower Peak Formation. The zircons yielded a concordia age of 168.9 ± 1.3 Ma, which was interpreted to date deposition of the sedimentary Tower Peak Formation and suggests contemporaneous sedimentation with volcanic activity of the Mapple Formation.

Jason Peninsula

Occurrence. The geology of the Jason Peninsula (Fig. 4) is dominated by rhyolitic lava flows and ignimbrites (Smellie 1991) which are petrographically akin to the silicic volcanic rocks of the Mapple Formation (Riley and Leat 1999). The rhyolitic volcanic rocks of the Jason Peninsula are associated, in part, with rare mafic lavas and sills (Saunders 1982; Riley et al. 2003, 2016) that crop out at Standring Inlet and Stratton Inlet (Fig. 4).

The silicic volcanic rocks are presumed to be overlain by fossiliferous sedimentary rocks (Riley et al. 1997), which are seen to crop out at the eastern extremity of the Jason Peninsula, at Cape Framnes (Fig. 4). The sedimentary rocks at Cape Framnes were interpreted to represent the northern extent of the Latady Group (Hunter and Cantrill 2006), which is exposed extensively in the southern Antarctic Peninsula. The fossil assemblage at Cape Framnes was assigned a Kimmeridgian–early Tithonian age (Riley et al. 1997).

Characteristics. A succession, at least 450 m thick (from sea level to summit), of silicic volcanic rocks is exposed on the Jason Peninsula, but the lithologies are intensely frost-shattered and any stratigraphical interpretations are not possible (Smellie 1991). The outcrops around the southeastern margins of the Jason Peninsula form an area of severely broken regolith and the lithologies are all hematite coated following the oxidation of finely disseminated Fe-sulfide. The rock types are all silicic volcanic lithologies and are dominated by quartz–feldspar porphyritic rhyolite lavas. Ignimbrite units are also present, and are characterized by the occurrence of fiamme and minor lithic fragments. Quartz (and amethyst) mineralization is prominent at all localities, often forming euhedral crystals in veins and cavities, and associated with agate development; the rhyolites have undergone significant silicification, with SiO_2 values up to 85 wt% (Riley et al. 2001). The ignimbrites locally contain carbonized wood (petrified tree trunks), which were identified in the upper parts of the sections from the western Jason Peninsula (del Valle et al. 1997). The tree trunks are 0.5 m in diameter and are not found in life position, but were interpreted to have been sheared by ash flows and transported a short distance; as a result, subaerial deposition is inferred (del Valle et al. 1997).

Mafic volcanic rocks have been identified from two areas (Standring Inlet and Stratton Inlet: Fig. 4) on the Jason Peninsula (Saunders 1982; Riley et al. 2003, 2016). At Standring Inlet (Fig. 4), mafic volcanic rocks and volcaniclastic sedimentary rocks crop out adjacent to outcrops of silicic crystal tuffs, but no stratigraphic relationship is observed between the two. The basaltic rocks are fine grained with individual lava units up to 50 m in thickness, which are associated with minor breccia units. The basaltic lavas are plagioclase-phyric with sparse accessory phases, including orthopyroxene, apatite and titanomagnetite. The basaltic rocks from the lowermost unit at Standring Inlet are typically aphyric, fine-grained lavas, whereas those from the upper flow unit are plagioclase porphyritic. The basaltic rocks from Stratton Inlet (Fig. 4) are also fine grained and typically aphyric.

Age. There are no reported U–Pb ages for any of the rhyolitic volcanic rocks from the Jason Peninsula. del Valle et al. (1997) compiled previously published K–Ar ages, which fall in the range 191–156 Ma; however, these K–Ar results do not accurately date an accurate eruption age.

Riley et al. (2003) reported $^{40}Ar/^{39}Ar$ ages from the basaltic rocks of the Jason Peninsula (Stratton Inlet and Standring Inlet), and dated both whole-rock and plagioclase phenocrysts to constrain the emplacement age. Riley et al. (2003) calculated ages in the range 175–168 Ma, overlapping with the age of the Mapple Formation (173–168 Ma: Pankhurst et al. 2000; Riley et al. 2010), indicating that the rhyolitic and basaltic magmatism of northern and eastern Graham Land were essentially coeval. Detrital zircon grains from the Kimmeridgian–early Tithonian sedimentary rocks at Cape Framnes (Riley et al. 1997) have been reported by Riley et al. (2010), and three distinct age populations were identified at 179 ± 6, 166 ± 4 and 147 ± 5 Ma. The age of the zircon populations accurately date the local volcanic events; the 166 ± 4 Ma zircon population is interpreted to be derived from the local silicic volcanic rocks on the Jason Peninsula, whilst the 179 ± 6 Ma population may record a volcanic episode correlating to the V1 event of Pankhurst et al. (2000) or from the Early Jurassic episode of granitoid emplacement (Riley et al. 2017a). The youngest detrital zircon age (147 ± 5 Ma) overlaps with the Kimmeridgian–early Tithonian age indicated by the fossil assemblage described by Riley et al. (1997), and indicates at least some degree of contemporaneous Late Jurassic sedimentation and volcanism.

Churchill Peninsula

Occurrence. At the Churchill Peninsula (Fig. 4) silicic volcanic rocks have been identified in association with granitoid plutons. There is no observed stratigraphic relationship with the sedimentary Botany Bay Group, although a small outcrop of hornfelsed terrestrial mudstones and siltstones is in faulted contact with the silicic volcanic rocks in the Eden Glacier region (Riley et al. 2010). The Jurassic-age volcanic rocks of the Churchill Peninsula region are, however, occasionally seen to unconformably overlie granitic and migmatitic gneisses of Ordovician–Permian age (Riley *et al.* 2012). This crystalline basement is the probable correlative of the Permian migmatites of Adie Inlet (Millar *et al.* 2002).

Characteristics. There are numerous localities in the Churchill Peninsula region where thick sequences of volcanic rocks crop out, such as Lyttleton Ridge on the eastern side of the Eden Glacier and west of Adie Inlet (Fig. 4). The most significant concentration and thickness of silicic volcanic rocks occurs to the west of Adie Inlet and the adjacent mountains on the eastern side of the Eden Glaicer. Also, Lyttleton Ridge preserves a sequence of volcanic rocks at least 450 m thick (Fig. 7c). Crystal and crystal-lithic tuffs are the predominant rock type in the region, with less significant thicknesses of ignimbrites, vitric tuffs and flow-banded rhyolite lavas with hematite mineralization (Fig. 7d). The sequence at Lyttleton Ridge is fairly typical of the region, and consists of weathered and fractured light grey crystal tuffs. They dip at 20°–30° to the NE and are bedded in rhythmic units typically a few metres thick. They have an 'apparent porphyritic' texture with clusters of feldspar phenocrysts evident throughout the sequence. Although the succession is dominated by crystal tuffs, pitchstone-like vitric tuffs also occur, as well as rare, strongly welded ignimbrites with well-developed eutaxitic textures (Fig. 7e) and also crystal-rich oblate pumice (Fig. 7f). Rare ash horizons have also been identified, which are often exploited by doleritic sills. Quartz porphyry and microgranite sheets are also present and cut the entire sequence of volcanic rocks; the quartz porphyry dykes can be several tens of metres in thickness. In the region to the west of Adie Inlet, accretionary lapilli-rich units (Fig. 7g) and associated ash horizons are more common, although they still only account for c. 1% of the exposed succession; they are often interbedded with leucocratic crystal-tuff units.

Overall, the sequence is altered by iron oxidation and has undergone silicification. The lithologies can appear quite siliceous with the alteration intense in areas, particularly at the southern extremity of the Churchill Peninsula at Cape Alexander (Fig. 7h) where the alteration is akin to that of the Jason Peninsula, with silicification, pyritization and oxidation of the iron sulfides. Also, in the Adie Inlet area, the crystal tuffs are characterized by distinct bands of intensely altered rocks and are due to intense hydrothermal activity.

The volcanic rocks were overwhelmingly deposited subaerially, although Marsh (1968) interpreted some successions as being deposited subaqueously. The general style of volcanism, with thick, monotonous, crystal tuffs, suggests an intracaldera-type succession, particularly in the area to the north of the Churchill Peninsula.

Age. Only one published age from the silicic volcanic rocks of the Churchill Peninsula area has been reported; Pankhurst (1982) calculated a five-point Rb–Sr isochron from a rhyolitic crystal tuff from Gulliver Nunatak (Fig. 9), which yielded an age of 174 ± 2 Ma, essentially contemporaneous with the Mapple Formation volcanic rocks farther north.

Granitoid magmatism coeval with the silicic volcanism has also been identified from the Churchill Peninsula area (Riley *et al.* 2010). The granitoids are isotopically identical to the silicic volcanic rocks (Riley *et al.* 2001) and fall within the same age interval. Granite and granodiorite plutons from the Churchill Peninsula–Leppard Glacier (Fig. 4) area have been dated by Pankhurst (1982: Rb–Sr) at 173 ± 6, 169 ± 3 and 167 ± 2 Ma, and by Riley *et al.* (2012: U–Pb) at 177 ± 3 and 173 ± 3 Ma. The ages indicate that the granitoids are likely to be the subvolcanic equivalents of the Mapple Formation volcanic rocks.

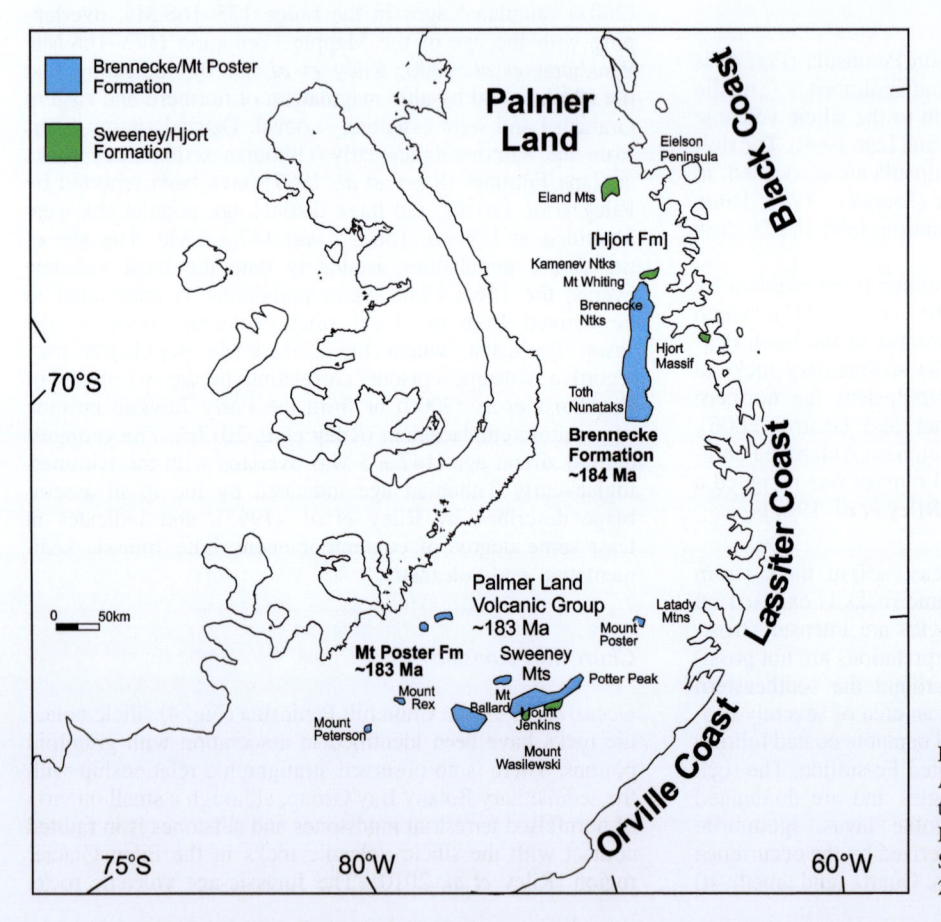

Fig. 9. Sketch map of Palmer Land (southern Antarctic Peninsula) showing the extent of the Early Jurassic Mount Poster and Brennecke formations, and the basaltic Sweeney and Hjort formations (Hunter *et al.* 2006; Riley *et al.* 2016).

Palmer Land Volcanic Group (previously Ellsworth Land Volcanic Group)

General geology

The Palmer Land Volcanic Group (Smellie 2020) replaces the previously defined Ellsworth Land Volcanic Group (Hunter et al. 2006) following the redefined boundaries for Palmer Land (southern Antarctic Peninsula).

The volcanic rocks of the Palmer Land Volcanic Group (Hunter et al. 2006; Smellie 2020) are exposed along the southeastern margin of the Antarctic Peninsula and are, in many respects, similar to the Graham Land Volcanic Group defined by Riley et al. (2010). Both groups are dominated by silicic pyroclastic rocks forming thick (>1 km) caldera-fed successions. The Palmer Land Volcanic Group rocks have an average eruption age of 183 Ma (Hunter et al. 2006) and are isotopically distinct from the Mapple Formation (Riley et al. 2001) and its correlatives, and are thought to relate to the V1 event from the northeastern region of the Chon Aike province (Pankhurst et al. 2000). The Palmer Land Volcanic Group (Smellie 2020) is formed of the Mount Poster Formation as defined by Rowley et al. (1982) and revised by Hunter et al. (2006), the Brennecke Formation (Wever and Storey 1992), the Hjort Formation (Wever and Storey 1992), the Sweeney Formation (Hunter et al. 2006), and related un-named units elsewhere in southeastern Palmer Land (Riley et al. 2016).

The Palmer Land Volcanic Group has the following key characteristics:

- Age: the volcanic rocks were erupted in the interval 189–167 Ma, with the main pulse of volcanism at 184–183 Ma.
- Lithology: the volcanic successions are dominated by moderately welded ignimbrites, crystal and lithic tuffs.
- Composition: overall, the volcanic succession is bimodal, but >80% of the magmatism is silicic (rhyodacite) in composition, with basaltic lava successions identified from several localities.
- Geochemistry: isotopically, the volcanic rocks have a broad range reflecting crustal input ($^{87}Sr/^{86}Sr_{168}$ = 0.7110–0.7213; $\varepsilon_{Nd_{168}}$ = −9 to −2).
- Source: overlapping and/or nested calderas for the silicic volcanic rocks and fissure/vent-fed pāhoehoe basaltic lavas.

Mount Poster Formation

Occurrence. The Mount Poster Formation was defined by Rowley et al. (1982), who described a 600 m succession of silicic volcanic rocks from Mount Poster, adjacent to the Latady Mountains (Fig. 9). The Mount Poster Formation was adjudged to continue further west to include the volcanic outcrops of the Orville Coast area (Fig. 9). The Mount Poster Formation volcanic rocks described by Rowley et al. (1982) overlie and are interbedded with fluvial or lacustrine sedimentary rocks of the Latady Formation (Williams et al. 1972). The volcanic outcrops at Mount Poster are metamorphosed and exhibit significant deformation which makes stratigraphic correlations difficult and geochemical analysis unreliable. Riley and Leat (1999) described the Mount Poster Formation from the Sweeney Mountains (Fig. 9) area where deformation is less pronounced. They interpreted the volcanic successions as intracaldera facies of rhyodacitic composition, dominated by poorly welded ignimbrites with an average thickness of at least 500 m, but locally, potentially reaching almost 1 km in thickness. Minor extracaldera facies volcanic rocks include ignimbrite flow units and basaltic lava units interbedded with sedimentary rocks of the Latady Formation.

Characteristics. The Mount Poster Formation of the Sweeney Mountains area is dominated by silicic poorly to moderately welded ignimbrites, which are crystal rich and weather to a characteristic purple/blue colour. Minor ash units and rare lavas are also present. Plagioclase clusters (up to 8 cm diameter) are a key characteristic of the ignimbrites, giving an 'apparent' porphyritic texture. The ignimbrites are characterized by lithic-rich beds (Fig. 10a, b), which include small fragments (<5 cm) of ignimbritic material and rounded quartzite/vein quartz clasts (typically <10 cm diameter). Microclasts (<1 cm) of red mudstone also occur in many of the ignimbrite beds. Finely disseminated, variably weathered pyrite accounts for much of the difference in colour between outcrops, whilst late-stage epidote mineralization and ubiquitous veins of quartz veining are common. In the strongly welded ignimbrite units, eutaxitic textures are developed (Fig. 10c), whilst the poorly welded ignimbrites are host to oblate, plagioclase-phyric pumices (Fig. 10d), which account for up to 20% of the rock. At rare localities, strongly welded ignimbrite units, with pumice flattening ratios of up to 10:1, occur in association with rheomorphic ignimbrites that have well-defined parataxitic texture.

An intracaldera setting for the Mount Poster Formation was suggested by Riley et al. (2001) based on the dense welding of most of the ignimbrites, the thickness of the succession (>500 m), combined with the low-grade alteration and faulting, and dyke emplacement at the caldera margins. Such features are characteristic of intracaldera ignimbrite-dominated successions from the western United States (Lipman 1984) and Alaska (Bacon et al. 1990).

The Mount Poster Formation silicic volcanic rocks are interpreted to be sourced from several separate volcanic centres, which may have resulted in a sequence of overlapping or nested calderas forming the present-day elongate outcrop pattern from west to east (Fig. 11). There is limited field evidence to support a pattern of overlapping calderas, although faulting and dyke emplacement at the contact with the presumably contemporaneous Sweeney Formation probably mark the caldera margin, and rare high aspect-ratio extracaldera ignimbrites exposed near Mount Ballard (Fig. 9) are interpreted to represent distal outflow units (Hunter et al. 2006).

Age. Fanning and Laudon (1999) dated (U–Pb sensitive high-resolution ion microprobe (SHRIMP)) three samples from the Mount Poster Formation, which gave ages of 189 ± 3 (Sweeney Mountains), 188 ± 3 (Mount Peterson: Fig. 9) and 167 ± 3 Ma (Mount Rex: Fig. 9). Hunter et al. (2006) dated (U–Pb SIMS) several silicic ignimbrites (Fig. 11) from the intracaldera succession of the Sweeney Mountains which gave an average age of 183.4 ± 1.4 Ma, overlapping with the nearby Brennecke Formation. The Early Jurassic ages from the Mount Poster Formation are akin to the nearby Brennecke Formation (c. 184 ± 2 Ma: Pankhurst et al. 2000), which has been considered to be part of the same Early Jurassic V1 event of Pankhurst et al. (2000). The 167 ± 3 Ma age reported from Mount Rex overlaps with the age of the Mapple Formation of eastern Graham Land (Pankhurst et al. 2000) and is distal to the main outcrop extent of the Mount Poster Formation, and is anomalous in comparison to all other ages from the Palmer Land Volcanic Group. The two Early Jurassic ages, coupled with a Middle and Late Jurassic deposition of the Latady Formation (e.g. Quilty 1983; Thomson 1983), indicate that the volcanic rocks of the Mount Poster Formation are from the base of the Latady Basin stratigraphy. A direct contact between the volcanic rocks and the Latady Group is not exposed but the presence of an air-fall tuff (Witte Nunataks Formation: Hunter et al. 2004) interbedded with sandstone and mudstone of the Latady Formation indicate that volcanism

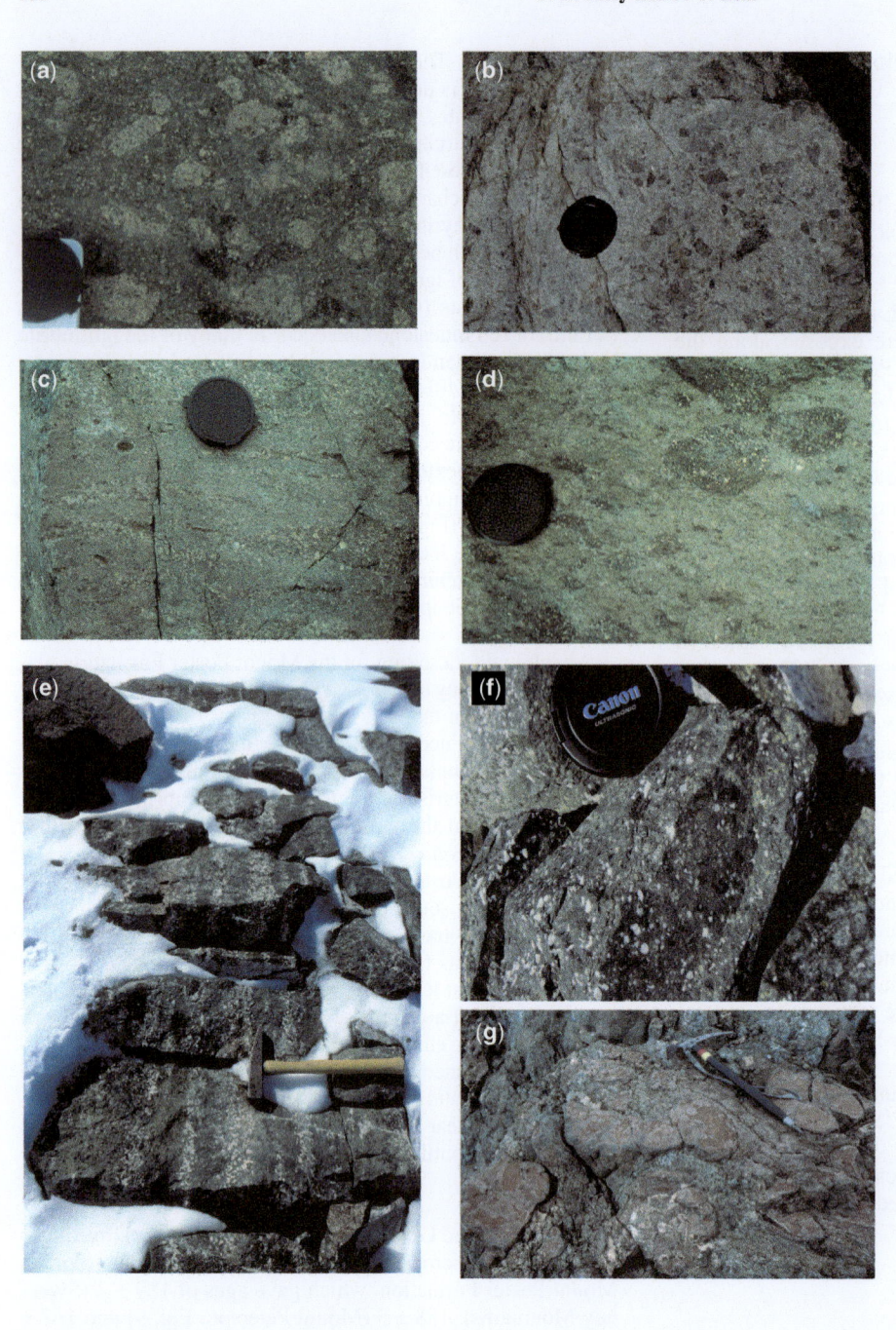

Fig. 10. (**a**) Lithic-rich Mount Poster Formation ignimbrite unit from Mount Jenkins, Sweeney Mountains. The lens cap is 52 mm in diameter. (**b**) Lithic-rich ignimbrite unit from Mount Wasilewski, southern Palmer Land. The lens cap is 54 mm in diameter. (**c**) Flattened pumice (fiamme) in Mount Poster Formation ignimbrite from SE of Mount Jenkins, Sweeney Mountains. The lens cap is 52 mm in diameter. (**d**) Oblate, crystal-rich pumice in the Mount Poster Formation ignimbrite, Potter Peak, Sweeney Mountains. The lens cap is 52 mm in diameter. (**e**) Linear pattern of vesicle zones in amygdaloidal basalts of the Sweeney Formation, southern Mount Jenkins, Sweeney Mountains. The length of the hammer is 35 cm. (**f**) Amygdaloidal pāhoehoe basalt from the Hjort Formation at Kamenev Nunataks, Black Coast. The lens cap is 60 mm in diameter. (**g**) Pillow basalts of the Sweeney Formation, northern Mount Jenkins, Sweeney Mountains. The length of the ice axe is 60 cm.

continued during deposition of the Latady Group as defined by Hunter and Cantrill (2006). Early Jurassic silicic magmatism is also contemporaneous with magmatism of the Karoo and Ferrar provinces of southern Africa and East Antarctica (Burgess *et al.* 2015).

Sweeney Formation

Previous workers assigned all sedimentary facies of southeastern Palmer Land to the Latady Formation, and all of the volcanic facies to the Mount Poster Formation (e.g. Laudon *et al.* 1983). However, no volcaniclastic rocks are found in the Latady Group (Hunter and Cantrill 2006), and the Mount Poster Formation (Rowley *et al.* 1982) was originally defined from the silicic intracaldera deposits at Mount Poster where there is no outcrop of basaltic material. Based on these criteria, Hunter *et al.* (2006) assigned the extracaldera facies, both volcanic and sedimentary, to a separate formation – the Sweeney Formation – as part of the Ellsworth Land Volcanic Group (now termed the Palmer Land Volcanic Group: Smellie 2020).

Occurrence. The Sweeney Formation crops out in the lower parts of the succession at Mount Edward, Potter Peak West, Mount Ballard, Mount Jenkins and to the east of Mount Wasilewski (Fig. 9). No contact between the Sweeney Formation and the Latady Group has been identified in the field, and the contact with the Mount Poster Formation is either faulted or intruded by dolerite dykes (Riley and Leat 1999). The most significant thickness of the Sweeney Formation crops out at the type section where approximately 300 m of black, finely laminated mudstone and sandstone are interbedded with at least 1 km of basaltic lava and silicic pyroclastic (distal ignimbrite) deposits. Elsewhere in the Sweeney Mountains the sedimentary facies are more typically only tens of metres thick. The ridges south and SE of Mount Ballard are underlain by at least 600 m of predominantly basaltic volcanic facies with minor interbedded fine-grained sedimentary rock.

Characteristics. The Sweeney Formation is dominated by volcanic lithologies with less-voluminous fine-grained sedimentary facies. The most distinctive facies is an amygdaloidal basalt with a melanocratic, fine-grained matrix and pale-

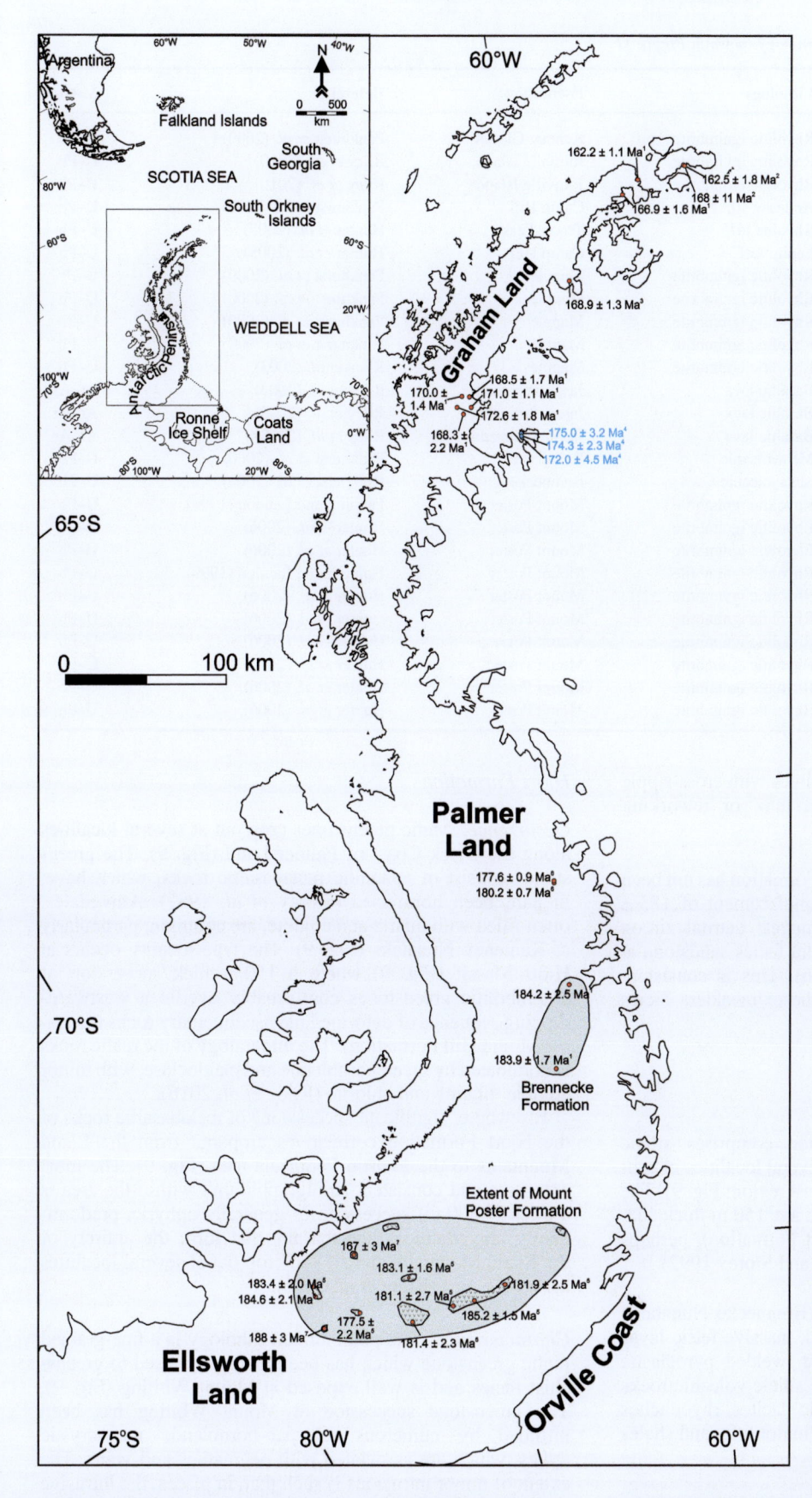

Fig. 11. The locations of dated (U–Pb) samples referenced in this study. Black text (U–Pb zircon); blue text (^{40}Ar/^{39}Ar). Data source and sample information: [1]Pankhurst et al. (2000); [2]Riley et al. (2010); [3]Hunter et al. (2005); [4]Riley et al. (2003); [5]Hunter et al. (2006); [6]Riley et al. (2005); and [7]Fanning and Laudon (1999). The full sample information is provided in Table 1.

coloured amygdales that have a linear arrangement, and may define vesicle zones in pāhoehoe lavas or a flow-unit carapace (Fig. 10e), similar to the Kamenev Formation (Storey et al. 1987) of the Black Coast (Fig. 10f). The amygdales vary in abundance and diameter (from <1 to up to 6 cm) between different lava units, with many amygdales being elongate. Other basaltic units are either weakly porphyritic, with phenocrysts of plagioclase and clinopyroxene, or aphyric. Vesicular lava units and minor pillow basalts have also been identified at both Mount Ballard and Mount Jenkins (Fig. 10g). Individual pillows are enclosed in a hyaloclastite matrix, and a hyaloclastic breccia has been identified to the south of Mount Jenkins. Other volcanic facies are either pyroclastic, including ash-rich and pebble-rich ignimbritic units or finely bedded ash-fall

Table 1. *Ages and data sources for the geochronology shown in Figure 11*

Sample ID	Age (Ma)	Lithology	Formation	Reference	Method
R.609.1	162.2 ± 1.1	Rhyolitic ignimbrite	Kenney Glacier	Pankhurst et al. (2000)	U–Pb
R.8309.1	162.5 ± 1.8	Rhyolitic ignimbrite	Joinville Island	Riley et al. (2010)	U–Pb
R.8308.1	168 ± 11	Rhyolitic ignimbrite	Joinville Island	Riley et al. (2010)	U–Pb
R.631.1	166.9 ± 1.6	Andesite sill	Camp Hill	Pankhurst et al. (2000)	U–Pb
D.9125.5	168.9 ± 1.3	Rhyolite tuff	Tower Peak	Hunter et al. (2005)	U–Pb
R.1309.4	167.1 ± 1.1	Lapilli tuff	Camp Hill	Hunter et al. (2005)	U–Pb
R.6906.3	168.5 ± 1.7	Rhyolitic ignimbrite	Mapple	Pankhurst et al. (2000)	U–Pb
R.6914.6	171.0 ± 1.1	Rhyolitic ignimbrite	Mapple	Pankhurst et al. (2000)	U–Pb
R.6908.7	170.0 ± 1.4	Rhyolitic ignimbrite	Mapple	Pankhurst et al. (2000)	U–Pb
R.6619.14	172.6 ± 1.8	Rhyolitic ignimbrite	Mapple	Pankhurst et al. (2000)	U–Pb
R.6632.10	168.3 ± 2.2	Rhyolitic ignimbrite	Mapple	Riley et al. (2003)	U–Pb
R.6642.6A	175.0 ± 3.2	Basaltic lava	Jason Peninsula	Riley et al. (2003)	Ar–Ar
R.6642.7	174.3 ± 2.3	Basaltic lava	Jason Peninsula	Riley et al. (2003)	Ar–Ar
R.6642.8	172.0 ± 4.5	Basaltic lava	Jason Peninsula	Riley et al. (2003)	Ar–Ar
R.4182.10	184.2 ± 2.5	Metavolcanic	Brennecke	Pankhurst et al. (2000)	U–Pb
R.4197.2	183.9 ± 1.7	Metavolcanic	Brennecke	Pankhurst et al. (2000)	U–Pb
unknown	167 ± 3	Rhyolitic ignimbrite	Mount Poster	Fanning and Laudon (1999)	U–Pb
R.8122.5	183.4 ± 2.0	Rhyolitic ignimbrite	Mount Poster	Hunter et al. (2006)	U–Pb
R.8125.1	184.6 ± 2.1	Rhyolitic ignimbrite	Mount Poster	Hunter et al. (2006)	U–Pb
unknown	188 ± 3	Rhyolitic ignimbrite	Mount Poster	Fanning and Laudon (1999)	U–Pb
R.8120.2	177.5 ± 2.2	Rhyolitic ignimbrite	Mount Poster	Hunter et al. (2006)	U–Pb
R.6878.1	183.1 ± 1.6	Rhyolitic ignimbrite	Mount Poster	Hunter et al. (2006)	U–Pb
R.6888.2	181.1 ± 2.7	Rhyolitic ignimbrite	Mount Poster	Hunter et al. (2006)	U–Pb
R.6871.3	181.4 ± 2.3	Rhyolitic ignimbrite	Mount Poster	Hunter et al. (2006)	U–Pb
R.7103.1	185.2 ± 1.5	Rhyolitic ignimbrite	Mount Poster	Hunter et al. (2006)	U–Pb
R.7108.2	181.9 ± 2.5	Rhyolitic ignimbrite	Mount Poster	Hunter et al. (2006)	U–Pb

horizons and graded lapilli tuffs, sometimes with cross-ripple lamination consistent with deposition into, or reworking by, water.

Age. An eruption age for the Sweeney Formation has not been determined, but a maximum age of emplacement of 183 ± 4 Ma has been inferred from the youngest detrital zircon peak age calculated from an extracaldera facies sandstone at Potter Peak West (Hunter et al. 2006). This is consistent with contemporaneous deposition of the extracaldera facies and intracaldera volcanism.

Brennecke Formation

Occurrence. The Brennecke Formation comprises silicic metavolcanic units which crop out at several localities in central and eastern Palmer Land (Black Coast region: Fig. 9). The rhyolites may be contemporaneous with a c. 150 m-thick succession of basalt–andesite lavas (Hjort Formation), perhaps forming a bimodal association (Wever and Storey 1992).

Characteristics. At the type locality (Brennecke Nunataks: Fig. 9) a succession of metamorphosed, massive felsic lavas are interbedded with strongly foliated, welded pyroclastic rocks and black shales. Elsewhere, the silicic volcanic rocks are generally buff-coloured, porphyritic dacites, rhyodacites and associated granophyric dykes with minor tuffs and shales (Wever and Storey 1992).

Age. Two samples were dated by Pankhurst et al. (2000) from the Brennecke Formation and gave consistent ages, indicating correlation with the V1 volcanic episode of the Chon Aike province and the Mount Poster Formation of southern Palmer Land. Sample R.4182.10 is a rhyodacitic ignimbrite from Brennecke Nunataks which gave a U–Pb concordia age of 184.2 ± 2.5 Ma, whilst a silicic ignimbrite from Toth Nunataks (R.4197.2) gave a concordia age of 183.9 ± 1.7 Ma.

Hjort Formation

Occurrence. Mafic greenstones crop out at several localities along the Black Coast of Palmer Land (Fig. 9). The greenstones consist of aphanitic metabasaltic rocks, which have, in part, been hornfelsed (Storey et al. 1987). Amygdales, often filled with quartz and epidote, are common, particularly at Kamenev Nunataks (Fig. 9). The type locality occurs at Hjort Massif (Fig. 9) where a 150 m-thick succession of amphibolitic greenstones conformably overlie a westward-dipping sequence of deformed metasedimentary rocks (probable Mount Hill Formation). The mineralogy of the mafic rocks is dominated by green hornblende and plagioclase, with minor epidote, titanite and chlorite (Riley et al. 2016).

Elsewhere, significant successions of metabasaltic rocks of the Hjort Formation correlatives crop out from the Eland Mountains to the Eielson Peninsula area (Fig. 9). The most extensive and consistent stratigraphic unit within the area is a thick (≥800 m) succession of generally aphyric, predominantly amygdaloidal, metabasalts that form the entirety of the Eland Mountains massif and crop out at several localities across the area (Riley et al. 2016).

Characteristics. The predominant lithology is a fine-grained mafic greenstone which has been metamorphosed to greenschist facies and is well exposed at Mount Whiting (Fig. 9). The greenstone succession at Mount Whiting has been intruded by numerous feldspar–hornblende megacrystic dykes, which are associated with pegmatites and veins. The extent of minor intrusions is such that, in places, the intrusive rocks can constitute >50% of the total outcrop.

Identifying individual units that comprise the Hjort Formation succession is difficult due to the nature of the isolated outcrops, the widespread deformation and extensive lower-greenschist-facies metamorphism. Pipe amygdales have been described from the lower parts of several units, whilst more circular (up to 2–3 cm diameter) amygdales of quartz and epidote composition occur towards the upper parts of the unit and can form concentrated horizons that are parallel to the other units (Fig. 10f). These units are likely to preserve pāhoehoe lavas, which are several metres in thickness.

The metavolcanic succession exposed across the Black Coast is punctuated by several aphyric, cream–grey-coloured, felsic units which have been described by Riley *et al.* (2016). The felsic units are up to 5 m thick, and are characterized by pervasive epidote mineralization and have also been devitrified. Globular inclusions of basalt have been described from the felsic units, which have cuspate margins and commonly chilled margins (Riley *et al.* 2016). These contact relationships, together with the presence of chilled margins within basaltic enclaves that have become entrained within the felsic unit, suggest that both basaltic and felsic units were co-magmatic at the time of emplacement.

Age. Two felsic samples from the dominantly basaltic succession at Eland Mountains were dated by Riley *et al.* (2016) using U–Pb zircon geochronology, and provide a crystallization age for the more extensive basalt successions of the Black Coast. The felsic units have a mean thickness of 2 m and are parallel to the stratification of the metabasalt succession. The contact relationships indicate that the felsic units are interbedded and therefore their age accurately reflects that of the major metabasaltic unit. Sample N10.2.1 comes from an interbedded silicic unit near the base of the basaltic pile and gave a concordia age of 180.2 ± 0.7 Ma, whereas sample N10.7.1 comes from an interbedded silicic unit close to the centre of the basaltic pile and yielded a concordia age of 177.6 ± 1.0 Ma (Riley *et al.* 2016). Therefore an age for the 800 m basaltic succession at Eland Mountains of *c.* 178 Ma was preferred, post-dating the main episode of silicic volcanism of the Brennecke and Mount Poster formations.

Summary

Graham Land Volcanic Group

The broad extent of Jurassic-age volcanic rocks exposed across the Antarctic Peninsula is well recognized, but compiling a clear regional picture and correlating key events has been less easy to establish. Pankhurst *et al.* (1998) attempted to correlate areas of Antarctic Peninsula silicic volcanism with the extensive Jurassic-age silicic volcanism of the Chon Aike province in Patagonia, whilst Riley *et al.* (2010) attempted correlations across the northern Antarctic Peninsula using geochronology and field correlations.

Age and lithological comparisons between the Mapple Formation (Oscar II Coast area: Fig. 4) and exposures at Hope Bay, Joinville Island, Botany Bay, Tower Peak, and the Churchill and Jason peninsulas (Riley *et al.* 2010) demonstrate that volcanism can be assigned to a common event, in the interval 172–162 Ma and with a peak at *c.* 168 Ma.

The field characteristics of the Mapple Formation described by Riley and Leat (1999) from the Cape Disappointment area are similar to the silicic volcanic rocks from the other localities, wherein the succession is dominated (typically >80%) by ash-flow tuffs in the form of crystal tuffs, crystal-lithic tuffs or ignimbrites. Individual units vary in thickness from <0.5 to up to 80 m, and there is a wide variation in the degree of welding (rare rheomorphic ignimbrites exist) and lithic content. The Mapple Formation and its correlatives are also characterized by rhyolitic, frequently flow-laminated, lava flows, volcanic breccias (including avalanche and mass-flow deposits) and rare, but relatively widespread, bedded, air-fall tuffs, characterized by rim-type accretionary lapilli.

The volcanic rocks exposed at Tower Peak and Camp Hill (Botany Bay) have also been dated in the interval *c.* 169–167 Ma, but are compositionally and volcanologically very different from the Mapple Formation rocks and their correlatives. Basaltic and intermediate compositions are very rare in the Mapple Formation, but the successions at Tower Peak and Camp Hill are dominated by andesitic lavas and agglomerates, which reach a maximum thickness of *c.* 750 m. Interestingly, the Botany Bay Group sedimentary rocks that underlie the andesitic lavas and agglomerates at Camp Hill and Tower Peak both contain volcanic horizons that are more typical of Mapple Formation volcanic rocks (e.g. rhyolitic crystal tuffs, accretionary lapilli tuffs), which are also similar to the sequences exposed at Hope Bay, Joinville Island and Cape Disappointment.

The basalts and basaltic andesites described from Jason Peninsula are not considered to be directly related to the andesites from Camp Hill and Tower Peak, which have intermediate compositions and are often garnet-bearing. However, both occurrences indicate that there was some mafic–intermediate magmatism associated with the predominantly silicic volcanism of the east coast of the Antarctic Peninsula. A similar scenario exists in the Chon Aike province of Patagonia (Pankhurst *et al.* 1998) where many of the described formations are characterized by silicic volcanism, with the exception of the Lonco Trapial and Bajo Pobre formations, which are dominated by andesite and basaltic andesite lavas, agglomerates, and breccias. The intermediate lavas from Patagonia have been dated at a younger episode (*c.* 152 Ma: Pankhurst *et al.* 2000) than the adjacent silicic volcanic rocks, which have Early–Middle Jurassic ages.

The paucity of basaltic rocks during the main phase of silicic activity at *c.* 170 Ma (Pankhurst *et al.* 2000) is explained by the petrogenetic model of Riley *et al.* (2001), which describes the development of an impermeable rhyolite trap at the base of the continental crust that prevents the migration of basaltic melts to the surface. This impermeable trap is argued to develop following the basaltic underplating associated with the arrival of a mantle plume (perhaps as early as 190 Ma: Riley *et al.* 2005); therefore any basaltic melt escaping to the surface would be more likely to occur early in the history of the province.

Palmer Land Volcanic Group

The Palmer Land Volcanic Group, as defined by Hunter *et al.* (2006) and redefined by Smellie (2020), includes the Sweeney Formation, the Mount Poster Formation, and also the Brennecke and Hjort formations further north.

An intracaldera setting for the Mount Poster Formation is the dominant geological feature, with the rhyolites interpreted to be sourced from several centres, which have resulted in a sequence of overlapping or nested calderas forming the present-day elongate outcrop pattern. The basaltic and sedimentary rocks of the Sweeney Formation would have been emplaced outside the main caldera succession in a terrestrial environment, characterized by lakes and conifer woods (Hunter and Cantrill 2006). Thin, outflow ignimbrite sheets have been identified outside the main caldera succession (Riley and Leat 1999), and are associated with the Sweeney Formation basaltic–sedimentary succession indicating that the Mount Poster and Sweeney formations were coeval.

The Brennecke Formation metavolcanic rocks are less well defined based on field observations in comparison to the Mount Poster Formation and a maximum thickness is difficult to determine (Wever and Storey 1992). The Brennecke Formation rhyolites are interpreted to be broadly contemporaneous with the basaltic Hjort Formation, although the main pulse of basaltic magmatism (180–178 Ma) post-dates the Brennecke Formation silicic volcanic rocks (184 Ma), in contrast to the relationship identified in the Graham Land Volcanic Group.

Tectonic and depositional setting

The extensive silicic volcanism (0.5×10^6 km^3: Pankhurst et al. 2000) of the Chon Aike province, extending from Patagonia into West Antarctica (Antarctic Peninsula and Thurston Island), has been termed a silicic large igneous province (LIP) (Bryan and Ferrari 2013). The silicic magmatism is dominated by subaerial volcanism, although co-magmatic granitoid plutonism has also been recognized.

The emplacement of the Chon Aike province throughout the Jurassic follows the pattern of 'flare-up' magmatism (Ducea and Barton 2007), with three clear peaks of magmatic activity followed by distinct lulls. The eruption of the silicic Chon Aike province coincided with rifting in West Gondwana during the early stages of supercontinent break-up (Storey et al. 2013). The early stages of silicic magmatism also overlapped with the emplacement of the Karoo and Ferrar continental flood basalt provinces at c. 183 Ma (Burgess et al. 2015; Svensen et al. 2012). The silicic LIP developed along the proto-Pacific Cordilleran margin of West Gondwana, and Pankhurst et al. (1998) identified a shift in rhyolite chemistry from 'within plate' during the Early Jurassic to more 'arc-like' by the Late Jurassic.

The preferred petrogenetic model is of Early Jurassic (V1) silicic magmatism being directly related to Gondwana extension, triggering the melting of Grenvillian-age fusible lower crust. The Early Jurassic volcanism of the southern Antarctic Peninsula is probably related to extension in the Weddell Sea (Fig. 4), although subduction was ongoing in the Antarctic Peninsula (Riley et al. 2017a) and South America (Rapela et al. 2005). Middle Jurassic (V2) magmatism of Graham Land and central Patagonia is also related to rifting and extension associated with Gondwana break-up, exploiting highly fusible lower crust, whilst Late Jurassic (V3) magmatism widespread along the Andean Cordillera is directly related to subduction processes.

Evidence for calderas as the primary source edifice for large-volume ignimbrites have often proved difficult to identify in the field (Bryan et al. 2002), and the Chon Aike province is no exception. Field evidence from Patagonia (Aragón et al. 1996) and the Antarctic Peninsula (Riley et al. 2001) identifies caldera centres, whilst Seltz et al. (2018) have determined eruption timescales from the Chon Aike silicic volcanic rocks and suggest short residence times in mid-crustal magma reservoirs, consistent with a caldera setting. Caldera complexes on the Antarctic Peninsula are likely to be controlled by crustal architecture and the orientation of extensional structures, leading to elongate caldera morphology as seen at the Mount Poster Formation. The source edifices would be akin to graben-type features and lead to significant localized thickening of pyroclastic deposits.

Acknowledgements Simon Abrahams, Adam Clark, Bradley Morrell, John Sweeney, Catrin Thomas, James Wake, Tom Weston and the air operations staff at Rothera Base are thanked for their field support over the last 23 years. This manuscript has benefited from the helpful comments of John Smellie, Gerhard Wörner and an anonymous reviewer.

Author contributions TRR: investigation (lead), methodology (lead), writing – original draft (lead), writing – review & editing (lead); **PTL**: writing – review & editing (supporting).

Funding This study is part of the British Antarctic Survey Polar Science for Planet Earth programme, funded by the Natural Environmental Research Council.

Data availability The datasets generated during and/or analysed during the current study are available in the UK Polar Data Centre repository, www.bas.ac.uk/data/uk-pdc/.

References

Aitkenhead, N. 1975. *The Geology of the Duse Bay–Larsen Inlet Area, North-East Graham Land (with Particular Reference to the Trinity Peninsula Series)*. British Antarctic Survey Scientific Reports, **51**.

Aragón, E., Rodriguez, A.M.I. and Benialgo, A. 1996. A calderas field at the Marifil Formation, new volcanogenic interpretation, Norpatagonian Massif, Argentina. *Journal of South American Earth Sciences*, **9**, 321–328, https://doi.org/10.1016/S0895-9811(96)00017-X

Bacon, C.R., Foster, H.L. and Smith, J.G. 1990 Rhyolitic calderas of the Yukon–Tanana Terrane, east central Alaska – volcanic remnants of a mid-Cretaceous magmatic arc. *Journal of Geophysical Research*, **95**, 21 451–21 461, https://doi.org/10.1029/JB09 5iB13p21451

Barbeau, D.L., Olivero, E.B., Swanson-Hysell, N.L., Zahid, K.M., Murray, K.E. and Gehrels, G.E. 2009. Detrital-zircon geochronology of the eastern Magallenes foreland basin: Implications for Eocene kinematics of the northern Scotia Arc and Drake Passage. *Earth and Planetary Science Letters*, **284**, 489–503, https://doi.org/10.1016/j.epsl.2009.05.014

Barbeau, D.L., Davis, J.T., Murray, K.E., Valencia, V., Gehrels, G.E., Zahid, K.M. and Gombosi, D.J. 2010. Detrital-zircon geochronology of the metasedimentary rocks of north-western Graham Land. *Antarctic Science*, **22**, 65–65, https://doi.org/10.1017/S095410200999054X

Birkenmajer, K. 1992. Trinity Peninsula Group (Permo-Triassic?) at Hope Bay, Antarctic Peninsula. *Polish Polar Research*, **13**, 215–240.

Birkenmajer, K. 1993. Jurassic terrestrial clastics (Mount Flora Formation) at Hope Bay, Trinity Peninsula (West Antarctica). *Bulletin of the Polish Academy of Sciences, Earth Sciences*, **41**, 23–38.

Bradshaw, J.D., Vaughan, A.P.M., Millar, I.L., Flowerdew, M.J., Trouw, R.A.J., Fanning, C.M. and Whitehouse, M.J. 2012. Permo-Carboniferous conglomerates in the Trinity Peninsula Group at View Point, Antarctic Peninsula: sedimentology, geochronology and isotope evidence for provenance and tectonic setting in Gondwana. *Geological Magazine*, **149**, 626–644, https://doi.org/10.1017/S001675681100080X

Brown, R.J., Branney, M.J., Maher, C. and Davila Harris, P. 2010. Origin of accretionary lapilli within ground hugging density currents: Evidence from pyroclastic couplets on Tenerife. *Geological Society of America Bulletin*, **122**, 305–320, https://doi.org/10.1130/B26449.1

Bryan, S.E. and Ferrari, L. 2013. Large igneous provinces and silicic large igneous provinces: Progress in our understanding over the last 25 years. *Geological Society of America Bulletin*, **125**, 1053–1078, https://doi.org/10.1130/B30820.1

Bryan, S.E., Riley, T.R., Jerram, D.A., Stephens, C.J. and Leat, P.T. 2002. Silicic volcanism: an undervalued component of large igneous provinces and volcanic rifted margins. *Geological Society of America, Special Paper*, **362**, 97–118, https://doi.org/10.1130/0-8137-2362-0.97

Burgess, S.D., Bowring, S.A., Fleming, T.H. and Elliot, D.H. 2015. High-precision geochronology links the Ferrar large igneous province with early-Jurassic ocean anoxia and biotic crisis. *Earth and Planetary Science Letters*, **415**, 90–99, https://doi.org/10.1016/j.epsl.2015.01.037

Burton-Johnson, A. and Riley, T.R. 2015. Autochthonous v. accreted terrane development of continental margins: a new *in situ* tectonic history of the Antarctic Peninsula. *Journal of the Geological Society, London*, **172**, 832–835, https://doi.org/10.1144/jgs2014-110

del Valle, R.A., Lirio, J.M., Lusky, J.C., Morelli, J.R. and Nunez, H.J. 1997. Jurassic trees at Jason Peninsula, Antarctica. *Antarctic Science*, **9**, 443–444, https://doi.org/10.1017/S0954102097000576

Ducea, M.N. and Barton, M.D. 2007. Igniting flare-up events in Cordilleran arcs. *Geology*, **35**, 1047–1050, https://doi.org/10.1130/G23898A.1

Elliot, D.H. 1967. The geology of Joinville Island. *British Antarctic Survey Bulletin*, **12**, 23–40.

Elliot, D.H. and Gracanin, T.M. 1983. Conglomeratic strata of Mesozoic age at Hope Bay, Antarctic Peninsula. *In*: Oliver, R.L., James, P.R. and Jago, J.B. (eds) *Antarctic Earth Science*. Cambridge University Press, Cambridge, UK, 303–307.

Fanning, C.M. and Laudon, T.S. 1999. Mesozoic volcanism, plutonism and sedimentation in eastern Ellsworth land, West Antarctic. *In*: Skinner, D.N.B. (ed.) *8th International Symposium on Antarctic Earth Sciences, Programme and Abstracts*. Victoria University of Wellington, Wellington, New Zealand, 102.

Farquharson, G.W. 1983. Evolution of Late Mesozoic sedimentary basins in the northern Antarctic Peninsula. *In*: Oliver, R.L., James, P.R. and Jago, J.B. (eds) *Antarctic Earth Science*. Cambridge University Press, 323–327.

Farquharson, G.W. 1984. Late Mesozoic, non-marine conglomeratic sequences of northern Antarctic Peninsula (The Botany Bay Group). *British Antarctic Survey Bulletin*, **65**, 1–32.

Féraud, G., Alric, V., Fornari, M., Bertrand, H. and Haller, M. 1999. $^{40}Ar/^{39}Ar$ dating of the Jurassic volcanic province of Patagonia: migrating magmatism relating to Gondwana break-up and subduction. *Earth and Planetary Science Letters*, **172**, 83–96, https://doi.org/10.1016/S0012-821X(99)00190-9

Fleet, M. 1968. *The Geology of the Oscar II Coast, Graham Land*. British Antarctic Survey Scientific Reports, **59**.

Flowerdew, M.J., Millar, I.L., Vaughan, A.P.M. and Pankhurst, R.J. 2005. Age and tectonic significance of the Lassiter Coast Intrusive Suite, Eastern Ellsworth Land, Antarctic Peninsula. *Antarctic Science*, **17**, 443–452, https://doi.org/10.1017/S0954102005002877

Halle, T.G. 1913. The Mesozoic flora of Graham Land. *Wissenschaftliche Ergebnisse der Schwedischen Südpolar-Expedition, 1901–1903*, **3**(14), 3–124.

Hathway, B. 2000. Continental rift to back-arc basin: Jurassic–Cretaceous stratigraphical and structural evolution of the Larsen Basin, Antarctic Peninsula. *Journal of the Geological Society, London*, **157**, 417–432, https://doi.org/10.1144/jgs.157.2.417

Hervé, F., Lobato, J., Ugalde, I. and Pankhurst, R.J. 1996. The geology of Cape Dubouzet, northern Antarctic Peninsula: continental basement to the Trinity Peninsula Group? *Antarctic Science*, **8**, 407–414, https://doi.org/10.1017/S0954102096000582

Hunter, M.A. and Cantrill, D.J. 2006. A new stratigraphy for the Latady Basin, Antarctic Peninsula: Part 2, Latady Group and basin evolution. *Geological Magazine*, **143**, 797–819, https://doi.org/10.1017/S0016756806002603

Hunter, M.A., Riley, T.R. and Millar, I.L. 2004. Middle Jurassic air fall tuff in the sedimentary Latady Formation, eastern Ellsworth Land. *Antarctic Science*, **16**, 185–190, https://doi.org/10.1017/S0954102004001944

Hunter, M.A., Cantrill, D.J., Flowerdew, M.J. and Millar, I.L. 2005. Middle Jurassic age for the Botany Bay Group: implications for Weddell Sea Basin creation and southern hemisphere biostratigraphy. *Journal of the Geological Society, London*, **162**, 745–748, https://doi.org/10.1144/0016-764905-051

Hunter, M.A., Riley, T.R., Cantrill, D.J., Flowerdew, M.J. and Millar, I.L. 2006. A new stratigraphy for the Latady Basin, Antarctic Peninsula: Part 1, Ellsworth Land Volcanic Group. *Geological Magazine*, **143**, 777–796, https://doi.org/10.1017/S0016756806002597

Laudon, T.S., Thomson, M.R.A., Williams, P.L., Milliken, K.L., Rowley, P.D. and Boyles, J.M. 1983. The Jurassic Latady Formation, southern Antarctic Peninsula. *In*: Oliver, R.L., James, P.R. and Jago, J.B. (eds) *Antarctic Earth Science*. Cambridge University Press, Cambridge, UK, 308–314.

Leat, P.T. and Riley, T.R. 2021. Antarctic Peninsula and South Shetland Islands: volcanology. *Geological Society, London, Memoirs*, **55**, https://doi.org/10.1144/M55-2018-52

Leat, P.T. and Scarrow, J.H. 1994. Central volcanoes as sources for the Antarctic Peninsula Volcanic Group. *Antarctic Science*, **6**, 365–374, https://doi.org/10.1017/S0954102094000568

Lipman, P.W. 1984. The roots of ash-flow calderas in western North America: windows into the tops of granitic batholiths. *Journal of Geophysical Research*, **89**, 8801–8841, https://doi.org/10.1029/JB089iB10p08801

Marsh, A.F. 1968. *Geology of Parts of the Oscar II and Foyn Coasts, Graham Land*. PhD thesis, University of Birmingham, Birmingham, UK.

McPhie, J., Doyle, M. and Allen, R. 1993. *Volcanic Textures. A Guide to the Interpretation of Textures in Volcanic Rocks*. CODES, University of Tasmania, Hobart, Tasmania, Australia.

Millar, I.L., Milne, A.J. and Whitham, A.G. 1990. Implications of Sm–Nd garnet ages for the stratigraphy of northern Graham Land, Antarctic Peninsula. *Zentrablatt für Geologie und Paläeontologie*, **1**, 97–104.

Millar, I.L., Pankhurst, R.J. and Fanning, C.M. 2002. Basement chronology and the Antarctic Peninsula: recurrent magmatism and anatexis in the Palaeozoic Gondwana Margin. *Journal of the Geological Society, London*, **159**, 145–158, https://doi.org/10.1144/0016-764901-020

Pankhurst, R.J. 1982. Rb–Sr geochronology of Graham Land, Antarctica. *Journal of the Geological Society, London*, **139**, 701–711, https://doi.org/10.1144/gsjgs.139.6.0701

Pankhurst, R.J., Leat, P.T., Sruoga, P., Rapela, C.W., Márquez, M., Storey, B.C. and Riley, T.R. 1998. The Chon-Aike silicic igneous province of Patagonia and related rocks in Antarctica: A silicic large igneous province. *Journal of Volcanology and Geothermal Research*, **81**, 113–136, https://doi.org/10.1016/S0377-0273(97)00070-X

Pankhurst, R.J., Riley, T.R., Fanning, C.M. and Kelley, S.P. 2000. Episodic silicic volcanism in Patagonia and the Antarctic Peninsula: chronology of magmatism associated with break-up of Gondwana. *Journal of Petrology*, **41**, 605–625, https://doi.org/10.1093/petrology/41.5.605

Quilty, P.G. 1983. Bajocian bivalves from Ellsworth Land, Antarctica. *New Zealand Journal of Geology and Geophysics*, **26**, 395–418, https://doi.org/10.1080/00288306.1983.10422256

Rapela, C.W., Pankhurst, R.J., Fanning, C.M. and Hervé, F. 2005. Pacific subduction coeval with the Karoo mantle plume: the Early Jurassic Subcordilleran belt of northwestern Patagonia. *Geological Society, London, Special Publications*, **246**, 217–239, https://doi.org/10.1144/GSL.SP.2005.246.01.07

Rees, P.M. 1993. Revised interpretations of Mesozoic palaeogeography and volcanic arc evolution in the northern Antarctic Peninsula region. *Antarctic Science*, **5**, 77–85, https://doi.org/10.1017/S0954102093000100

Riley, T.R. and Leat, P.T. 1999. Large volume silicic volcanism along the proto-Pacific margin of Gondwana: lithological and stratigraphcial investigations from the Antarctic Peninsula. *Geological Magazine*, **136**, 1–16, https://doi.org/10.1017/S0016756899002265

Riley, T.R., Crame, J.A., Thomson, M.R.A. and Cantrill, D.J. 1997. Late Jurassic (Kimmeridgian–Tithonian) macrofossil assemblage from Jason Peninsula, Graham Land: evidence for a significant northward extension of the Latady Formation. *Antarctic Science*, **9**, 434–442, https://doi.org/10.1017/S0954102097000564

Riley, T.R., Leat, P.T., Pankhurst, R.J. and Harris, C. 2001. Origins of large volume rhyolitic volcanism in the Antarctic Peninsula and Patagonia by crustal melting. *Journal of Petrology*, **42**, 1043–1065, https://doi.org/10.1093/petrology/42.6.1043

Riley, T.R., Leat, P.T., Kelley, S.P., Millar, I.L. and Thirlwall, M.F. 2003. Thinning of the Antarctic Peninsula lithosphere through the Mesozoic: evidence from Middle Jurassic basaltic lavas.

Lithos, **67**, 163–179, https://doi.org/10.1016/S0024-4937(02)00266-9

Riley, T.R., Leat, P.T., Curtis, M.L., Millar, I.L. and Fazel, A. 2005. Early–Middle Jurassic Dolerite dykes from Western Dronning Maud Land (Antarctica): Identifying mantle sources in the Karoo Large Igneous Province. *Journal of Petrology*, **46**, 1489–1524, https://doi.org/10.1093/petrology/egi023

Riley, T.R., Flowerdew, M.J., Hunter, M.A. and Whitehouse, M.J. 2010. Middle Jurassic rhyolite volcanism of eastern Graham Land, Antarctic Peninsula: age correlations and stratigraphic relationships. *Geological Magazine*, **147**, 581–595, https://doi.org/10.1017/S0016756809990720

Riley, T.R., Flowerdew, M.J. and Whitehouse, M.J. 2012. U–Pb ion-microprobe zircon geochronology from the basement inliers of eastern Graham Land, Antarctic Peninsula. *Journal of the Geological Society, London*, **169**, 381–393, https://doi.org/10.1144/0016-76492011-142

Riley, T.R., Curtis, M.L., Flowerdew, M.J. and Whitehouse, M.J. 2016. Evolution of the Antarctic Peninsula lithosphere: evidence from Mesozoic mafic rocks. *Lithos*, **244**, 59–73, https://doi.org/10.1016/j.lithos.2015.11.037

Riley, T.R., Flowerdew, M.J., Pankhurst, R.J., Curtis, M.L., Millar, I.L., Fanning, C.M. and Whitehouse, M.J. 2017*a*. Early Jurassic subduction-related magmatism on the Antarctic Peninsula and potential correlation with the Subcordilleran plutonic belt of Patagonia. *Journal of the Geological Society, London*, **174**, 365–376, https://doi.org/10.1144/jgs2016-053

Riley, T.R., Flowerdew, M.J., Pankhurst, R.J., Millar, I.L., Leat, P.T., Fanning, C.M. and Whitehouse, M.J. 2017*b*. A revised geochronology of Thurston Island, West Antarctica and correlations along the proto-Pacific margin of Gondwana. *Antarctic Science*, **29**, 47–60, https://doi.org/10.1017/S0954102016000341

Riley, T.R., Burton-Johnson, A., Flowerdew, M.J. and Whitehouse, M.J. 2018. Episodicity within a mid-Cretaceous magmatic flare-up in West Antarctica: U–Pb ages of the Lassiter Coast intrusive suite, Antarctic Peninsula and correlations along the Gondwana margin. *Geological Society of America Bulletin*, **130**, 1177–1196, https://doi.org/10.1130/B31800.1

Rowley, P.D., Schimdt, D.L. and Williams, P.L. 1982. Mount Poster Formation, southern Antarctic Peninsula and eastern Ellsworth Land. *Antarctic Journal of the United States*, **17**, 38–39.

Saunders, A.D. 1982. Petrology and geochemistry of alkali-basalts from Jason Peninsula, Oscar II Coast, Graham Land. *British Antarctic Survey Bulletin*, **55**, 1–9.

Seltz, S., Putlitz, B., Baumgartner, L., Meibom, A., Escrig, S. and Bouvier, A.-S. 2018. A NanoSIMS investigation recorded in volcanic quartz from the silicic Chon Aike Province (Patagonia). *Frontiers in Earth Science*, **6**, 95, https://doi.org/10.3389/feart.2018.00095

Smellie, J.L. 1991. Middle–Late Jurassic volcanism on the Jason Peninsula, Antarctic Peninsula, and its relationship to the break–up of Gondwana. *In*: Ulbrich, H. and Rocha Campos, A.C. (eds) *Gondwana Seven Proceedings. Papers presented at the Seventh International Gondwana Symposium, Sao Paulo, 1988*. Instituto Geosciências, Universidade de São Paulo, São Paulo, Brazil, 685–699.

Smellie, J.L. 2020. Antarctic Peninsula: geology and dynamic development. *In*: Kleinschmidt, G. (ed.) *Geology of the Antarctic Continent*. Gebrüder Borntraeger Verlagsbuchhandlung, Stuttgart, Germany, 18–86.

Smellie, J.L. and Millar, I.L. 1995. New K–Ar isotopic ages of schists from Nordenskjöld Coast, Antarctic Peninsula: oldest part of the Trinity Peninsula Group. *Antarctic Science*, **7**, 191–196, https://doi.org/10.1017/S0954102095000253

Storey, B.C., Wever, H.E., Rowley, P.D. and Ford, A.B. 1987. Report on Antarctic fieldwork: the geology of the central Black Coast, eastern Palmer Land. *British Antarctic Survey Bulletin*, **77**, 145–155.

Storey, B.C., Alabaster, T., Hole, M.J., Pankhurst, R.J. and Wever, H.E. 1992. Role of subduction plate boundary forces during the initial stages of Gondwana break-up: evidence from the proto-Pacific margin of Antarctica. *Geological Society, London, Special Publications*, **68**, 149–163, https://doi.org/10.1144/GSL.SP.1992.068.01.10

Storey, B.C., Vaughan, A.P.M. and Riley, T.R. 2013. The links between large igneous provinces, continental breakup and environmental change: evidence reviewed from Antarctica. *Earth and Environmental Science Transactions of the Royal Society of Edinburgh*, **104**, 1–14, https://doi.org/10.1017/S175569101300011X

Suárez, M. 1976. Plate tectonic model for southern Antarctic Peninsula and its relation to southern Andes. *Geology*, **4**, 211–214, https://doi.org/10.1130/0091-7613(1976)4<211:PMFSAP>2.0.CO;2

Svensen, H., Corfu, F., Polteau, S., Hammer, Ø. and Planke, S. 2012. Rapid magma emplacement in the Karoo Large Igneous Province. *Earth and Planetary Science Letters*, **325–326**, 1–9, https://doi.org/10.1016/j.epsl.2012.01.015

Thomson, M.R.A. 1983. Late Jurassic ammonites from the Orville Coast, Antarctica. *In*: Oliver, R.L., James, P.R. and Jago, J.B. (eds) *Antarctic Earth Science*. Cambridge University Press, Cambridge, UK, 315–319.

Thomson, M.R.A. and Pankhurst, R.J. 1983. Age of post-Gondwanian calc-alkaline volcanism in the Antarctic Peninsula region. *In*: Oliver, R.L., James, P.R. and Jago, J.B. (eds) *Antarctic Earth Science*. Cambridge University Press, Cambridge, UK, 328–333.

Vaughan, A.P.M. and Storey, B.C. 2000. The eastern Palmer Land shear zone: a new terrane accretion model for the Mesozoic development of the Antarctic Peninsula. *Journal of the Geological Society, London*, **157**, 1243–1256, https://doi.org/10.1144/jgs.157.6.1243

Vaughan, A.P.M., Pankhurst, R.J. and Fanning, C.M. 2002. A mid-Cretaceous age for the Palmer Land event: implications for terrane accretion timing and Gondwana palaeolatitudes. *Journal of the Geological Society, London*, **159**, 113–116, https://doi.org/10.1144/0016-764901-090

Wever, H.E. and Storey, B.C. 1992. Bimodal magmatism in northeast Palmer Land, Antarctic Peninsula: geochemical evidence for a Jurassic ensialic back-arc basin. *Tectonophysics*, **205**, 239–259, https://doi.org/10.1016/0040-1951(92)90429-A

Williams, P.L., Schmidt, D.L., Plummer, C.C. and Brown, L.E. 1972. Geology of the Lassiter Coast area, Antarctic Peninsula: preliminary report. *In*: Adie, R.J. (ed.) *Antarctic Geology and Geophysics*. Universitetsforlaget, Oslo, 143–153.

Wolff, J.A. and Wright, J.V. 1981. Rheomorphism of welded tuffs. *Journal of Volcanology and Geothermal Research*, **10**, 13–34, https://doi.org/10.1016/0377-0273(81)90052-4

Chapter 2.2b

Palmer Land and Graham Land volcanic groups (Antarctic Peninsula): petrology

Teal R. Riley* and Philip T. Leat

British Antarctic Survey, High Cross, Madingley Road, Cambridge CB3 0ET, UK

TRR, 0000-0002-3333-5021; PTL, 0000-0003-3824-8557

*Correspondence: trr@bas.ac.uk

Abstract: Large-volume rhyolitic volcanism along the proto-Pacific margin of Gondwana consists of three major episodes of magmatism or 'flare-ups'. The initial episode (V1) overlaps with the Karoo–Ferrar large igneous provinces at c. 183 Ma. A second (V2) episode was erupted in the interval 171–167 Ma, and a third episode (V3) was emplaced in the interval 157–153 Ma. The magmatic events of the V1 and V2 episodes of the Antarctic Peninsula are reviewed here describing major and trace elements, and isotopic (Sr, Nd, O) data from rhyolitic volcanic rocks and more minor basaltic magmatism. An isotopically uniform intermediate magma developed as a result of anatexis of hydrous mafic lower crust, which can be linked to earlier, arc-related underplating. The subsequent lower-crust partial melts mixed with fractionated mafic underplate, followed by mid-crust storage and homogenization. Early Jurassic (V1) volcanic rocks of the southern Antarctic Peninsula are derived from the isotopically uniform magma, but they have mixed with melts of upper-crustal paragneiss in high-level magma chambers. The V2 rhyolites from the northern Antarctic Peninsula are the result of assimilation and fractional crystallization of the isotopically uniform magma. This process took place in upper-crust magma reservoirs involving crustal assimilants with an isotopic composition akin to that of the magma. A continental margin-arc setting was critical in allowing the development of an hydrous, fusible lower crust. Lower-crustal anatexis was in response to mafic underplating associated with the mantle plume thought to be responsible for the contemporaneous Karoo magmatic province and rifting associated with the initial break-up of Gondwana.

Early–Middle Jurassic volcanic rocks are widespread across the central and eastern Antarctic Peninsula, and have been interpreted as being related to rifting and mantle-plume activity associated with Gondwana break-up and extension in the Weddell Sea (e.g. Pankhurst et al. 2000; Storey et al. 2013). Subaerial silicic volcanic rocks of Early (c. 183 Ma) and Middle Jurassic (c. 170 Ma) age have been described in detail in the previous chapter (Riley and Leat 2021). Both episodes of volcanic activity are correlated with the extensive Chon Aike province of Patagonia (Pankhurst et al. 1998, 2000) and together they have been identified as a silicic large igneous province (LIP) (Pankhurst et al. 1998; Bryan and Ferrari 2013).

The silicic volcanic rocks of eastern Graham Land and SE Palmer Land are dominated by ignimbrites and crystal-lithic tuffs, and their extent and field characteristics are described in Riley and Leat (2021). Rare basaltic rocks are also identified from both the Early and Middle Jurassic volcanic episodes, and are also described in Riley and Leat (2021).

This paper describes the petrology and geochemistry of the Jurassic silicic (and rarer basaltic) volcanic rocks of the Antarctic Peninsula, and discusses their petrogenesis in the context of the Chon Aike province and Gondwana break-up.

Geological setting

The Antarctic Peninsula was interpreted by Suárez (1976) as an autochthonous continental arc of the West Gondwanan margin and was associated with long-lived subduction during the Mesozoic. The identification of a major shear zone in eastern Palmer Land from the southern Antarctic Peninsula lead to the composite geology of the Antarctic Peninsula being reinterpreted (Vaughan and Storey 2000) as a series of para-autochthonous and allochthonous terranes which had accreted onto the Gondwana margin. This hypothesis has been reinterpreted to a simpler, *in situ*, model of continental arc evolution (Burton-Johnson and Riley 2015).

The Middle Jurassic felsic volcanic rocks of the east coast of the Antarctic Peninsula (Riley and Leat 1999) are associated with tonalite, quartz diorite and granodiorite plutonic rocks also of Middle Jurassic age (Pankhurst et al. 2000), and sequences of sedimentary rocks deposited in a terrestrial setting (e.g. Hunter et al. 2005). These Mesozoic formations of the northeastern Antarctic Peninsula unconformably overlie Carboniferous–Triassic metasedimentary rocks of the Trinity Peninsula Group (Barbeau et al. 2010; Bradshaw et al. 2012). The Trinity Peninsula Group has an estimated thickness of c. 5 km, and Hathway (2000) has interpreted the depositional environment as submarine fans along a passive continental margin. The Trinity Peninsula Group overlaps with and, in part, is likely to overlie Ordovician–Permian-age crystalline basement (e.g. Hervé et al. 1996; Bradshaw et al. 2012; Riley et al. 2012). The volcanic successions of eastern Graham Land also overlie or are, in part, interbedded with sedimentary rocks of the Jurassic-age Botany Bay Group (Hunter et al. 2005).

The Mesozoic volcanic sequences of Graham Land have all been subject to low- to medium-grade metamorphism and deformation, possibly during the Palmer Land event (107–103 Ma: Vaughan et al. 2002) or during the Jurassic Peninsula Orogeny (Storey et al. 1987). The Early–Middle Jurassic-age volcanic rocks of Palmer Land include the silicic Brennecke and Mount Poster formations (e.g. Hunter et al. 2006), which are associated with basaltic volcanic units of the Hjort and Swenney formations. The volcanic units are also associated with the widespread shallow-marine sedimentary rocks of the Latady Group (Hunter and Cantrill 2006).

Early Jurassic silicic volcanic rocks (c. 183 Ma) of the southern Antarctic Peninsula (Palmer Land) include the Brennecke and Mount Poster formations (Riley et al. 2001; Hunter et al. 2006), which correlate with the geographically extensive first-stage event (V1) of the Chon Aike province (Pankhurst et al. 2000). These silicic units overlap in age with the voluminous Karoo–Ferrar flood basalt volcanism (Svensen et al. 2012) which was focused towards the continental interior of Gondwana. The silicic rocks are associated with minor basaltic successions (Riley *et al.* 2016) and extensive shallow-marine sedimentary rocks of the Latady Group (Hunter and Cantrill 2006). Middle Jurassic (*c.* 170 Ma) silicic volcanic rocks of the NE Antarctic Peninsula include the Mapple Formation and are associated with the V2 event of the Chon Aike province. These Early and Middle Jurassic volcanic rocks are the subject of this paper.

Granitoids are the most widespread igneous outcrops on the Antarctic Peninsula, identified as individual plutons, composite intrusions and large batholith complexes constructed during Mesozoic–Cenozoic time (Leat *et al.* 1995; Riley *et al.* 2018).

Graham Land Volcanic Group

General geology

The Graham Land Volcanic Group, defined by Riley *et al.* (2010), includes the Mapple Formation of eastern Graham Land and its correlated units at Hope Bay (Kenney Glacier Formation), Joinville Island, and the Churchill and Jason peninsulas. The intermediate volcanic rocks of the Mount Tucker Formation at Botany Bay and Tower Peak, and the basaltic rocks of Jason Peninsula are also included in the Graham Land Volcanic Group (Riley and Leat 2021). The Mapple Formation from the Oscar II Coast has been dated to the interval 173–168 Ma (Pankhurst *et al.* 2000), and similar ages have been obtained for silicic volcanic rocks from Joinville Island (Riley *et al.* 2010).

All of the silicic volcanic units defined by Riley and Leat (1999) and Riley *et al.* (2010) are correlated with the V2 event of the Chon Aike province of Patagonia (Pankhurst *et al.* 2000). These volcanic units of dominantly crystal tuffs were emplaced onto the continental margin of Gondwana, and were derived from the partial melting of subduction-modified crustal underplate and lower crust (Riley *et al.* 2001).

Silicic volcanic rocks of the east coast of the Antarctic Peninsula are temporally associated with Early–Middle Jurassic granitoid plutons (Riley *et al.* 2017) and sequences of terrestrial sedimentary rocks. Jurassic silicic volcanic rocks have been identified at multiple sites across eastern Graham Land (Fig. 1). The petrology and geochemistry of the Mapple Formation silicic volcanic rocks and the Jason Peninsula basaltic rocks are discussed in detail here (summarized in Table 1), as there is little information available for the other units of the Graham Land Volcanic Group (e.g. Kenney Glacier Formation, Hope Bay).

Mapple Formation

Background. The Mapple Formation of the Oscar II Coast is dominated by ignimbrite units and crystal tuffs (*c.* 80%), which display significant variation in terms of their degree of welding and lithic content. The volcanic rocks are almost entirely subaerial, although localized subaqueous emplacement is evident in rare units and may have been in a lacustrine environment (Riley and Leat 1999). Ignimbrite units are typically <20 m in thickness, exceptionally reaching 80 m, and the entire sequence has a maximum observed thickness of *c.* 1 km on the Mapple Glacier, where the Mapple Formation is defined (Riley and Leat 1999). The ignimbrites and crystal tuffs crop out in association with air-fall units, breccias and rare lava flows. The Mapple Formation was probably metamorphosed and deformed during the end-Jurassic Palmer Land compressional event (Kellogg and Rowley 1989) or the mid-Cretaceous deformational event (Vaughan and Storey 2000). The Mapple Formation is metamorphosed up to greenschist facies and was deformed during the Cretaceous-age Palmer Land tectonic event (Vaughan *et al.* 2002). The deformation in eastern Graham was relatively weak. and generated low-angle faults and open folds. Steeply dipping cleavage has developed in epiclastic and many mudflow deposits, and increases in intensity westwards.

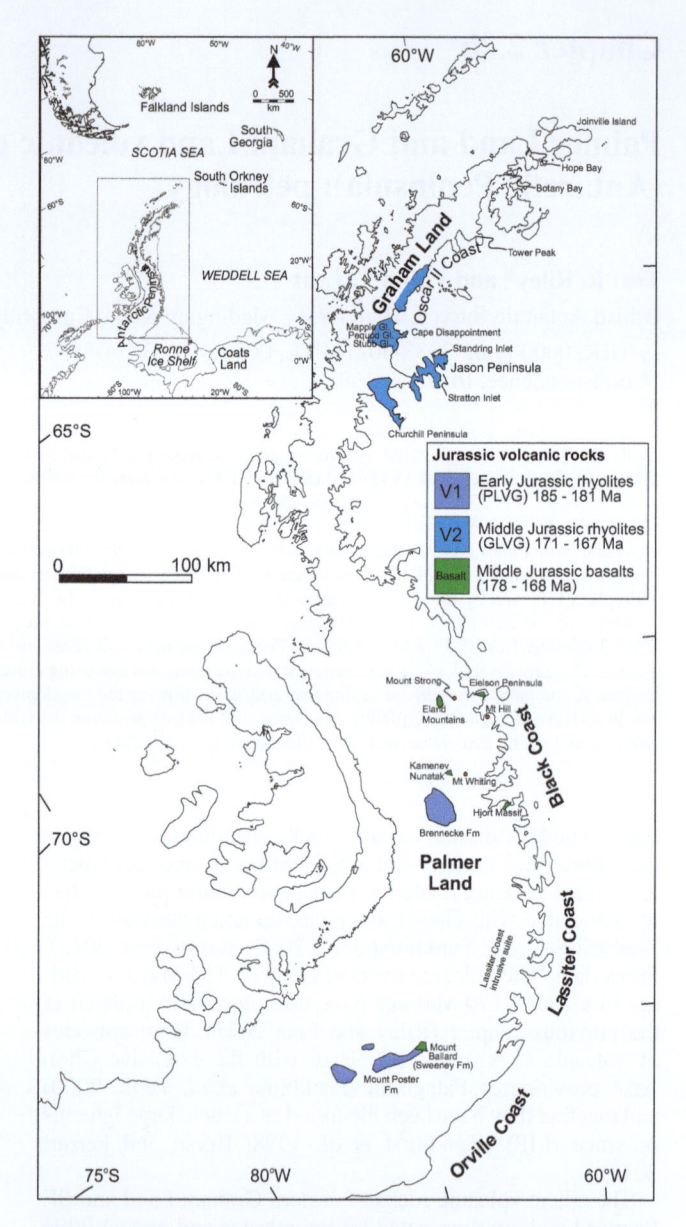

Fig. 1. Map of the Antarctic Peninsula showing the main outcrop area of Jurassic-age silicic volcanic rocks and the key areas of basaltic rock outcrop.

Petrography. The ignimbrite and crystal tuff units are crystal-poor, with a mineral assemblage (in order of abundance) of embayed quartz, K-feldspar, plagioclase, biotite, magnetite, apatite, orthopyroxene, titanite, rutile and zircon. The feldspars are altered, and are typically replaced by calcite, sericite and clay minerals. The development of spherulites (axiolitic and spherical) as a result of high-temperature devitrification of glass is widespread in the rhyolitic ignimbrites.

Major element geochemistry. The Mapple Formation silicic volcanic rocks are predominantly rhyolite–rhyodacite in composition (Fig. 2a) akin to the *c.* 170 Ma Chon Aike Formation of Patagonia, a component of the much wider Chon Aike

Table 1. *Representative geochemical analyses of volcanic rocks from the Graham Land Volcanic Group*

Sample	R.6607.3	R.6612.4	R.6626.1	R.6627.3	R.6632.2	R.6911.3	R.6642.2	R.6646.1	R.6642.6a	R.6642.7	R.6642.9
Formation	Mapple Fm	Mapple Fm	Mapple Fm	Mapple Fm	Mapple Fm	Mapple Fm	Jason Pen	Jason Pen	Jason Pen	Jason Pen	Jason Pen
Latitude (S)	65.533	65.5833	65.6833	65.6833	65.6834	65.4189	65.95	65.95	65.95	65.95	65.95
Longitude (W)	62.433	62.301	62.351	62.351	62.3509	62.4519	61.014	61.014	61.014	61.014	61.014
Major elements (wt%)											
SiO_2	72.04	74.45	68.31	73.79	74.36	72.91	71.34	78.87	53.07	52.74	54.31
TiO_2	0.29	0.17	0.23	0.18	0.26	0.27	0.30	0.08	0.73	0.73	0.88
Al_2O_3	14.19	13.07	15.39	13.6	13.2	13.62	14.50	11.48	15.55	15.45	16.85
Fe_2O_3	2.4	1.7	2.56	1.99	2.34	2.66	2.45	0.38	9.40	9.33	9.20
MnO	0.04	0.04	0.06	0.05	0.04	0.05	0.04	0.00	0.15	0.16	0.18
MgO	0.43	0.4	0.27	0.17	0.51	0.5	0.54	0.00	6.46	6.32	4.43
CaO	1.61	1.1	1.86	1.46	2.51	0.79	2.22	0.00	8.55	8.69	8.72
Na_2O	3.15	4.33	4.16	3.85	4.24	4.8	3.81	0.06	2.75	2.77	2.74
K_2O	4.91	3.82	5.02	4.46	1.89	3.16	3.81	7.96	2.30	2.34	1.88
P_2O_5	0.08	0.06	0.05	0.03	0.05	0.07	0.07	0.01	0.30	0.30	0.30
LOI	0.88	0.53	2.39	0.81	0.71	0.98	0.62	0.84	1.20	1.10	0.65
Total	100.03	99.67	100.29	100.38	100.09	99.81	99.71	99.69	100.46	99.94	100.13
Trace elements (ppm)											
Cr	13	12	9	8	9	16	19	13	148	158	28
Cu	4	5	0	1	2	3	4	0	58	55	21
Ni	2	3	0	1	4	2	4	1	49	50	10
V	34	18	8	7	23	32	31	3	214	226	242
Zn	43	27	58	55	37	24	22	13	71	78	85
Sc							15.4				
Ga	18.0	10.0	19.0	16.0	16.0	13.0	15.0	12.0	17.4	18	20.7
Rb	150	105	161	166	66	100	118.0	346.0	65.4	71.1	51.5
Sr	245	123	340	223	353	126	264.0	24.0	1061	1108.6	1107.2
Y	23.0	21.0	34.0	33.0	28.0	18.0	25.9	20.0	13.9	13.9	16.5
Zr	155.0	108.0	191.0	167.0	165.0	133.0	162.4	72.7	99.9	102.7	107.6
Nb	9.0	8.0	13.0	12.0	10.0	5.0	10.1	14.5	4.5	4.6	5
Ba	955.0	728.0	958.0	816.0	502.0	586.0	1060.0	309.1	766	790	666
La	41.9	26.8	36.8	31.9	41.6	36.9	36.1	19.32	36.8	37.3	26.4
Ce	89.9	52.5	79.8	69.2	82.8	75.7	77.2	43.5	76.7	79.2	60.5
Pr	10.3	6	9.5	8.3	10.2	8.7	9.1	5.2	9.8	10.1	7.8
Nd	39.5	22.2	37.6	33	35.6	32.6	34.7	19.7	40.7	42.4	33.5
Sm	6.2	4.3	7.7	6.6	6.6	5.6	6.2	4.1	6.4	6.6	5.8
Eu	1.14	0.83	1.57	1.09	1.47	1.14	1.06	0.21	1.69	1.78	1.61
Gd	5	3.8	6.8	5.8	6.9	4.5	5.18	3.36	4.4	4.5	4.3

(*Continued*)

Table 1. *Continued*

Sample	R.6607.3	R.6612.4	R.6626.1	R.6627.3	R.6632.2	R.6911.3	R.6642.2	R.6646.1	R.6642.6a	R.6642.7	R.6642.9
Formation	Mapple Fm	Mapple Fm	Mapple Fm	Mapple Fm	Mapple Fm	Mapple Fm	Jason Pen	Jason Pen	Jason Pen	Jason Pen	Jason Pen
Latitude (S)	65.533	65.5833	65.6833	65.6833	65.6834	65.4189	65.95	65.95	65.95	65.95	65.95
Longitude (W)	62.433	62.301	62.351	62.351	62.3509	62.4519	61.014	61.014	61.014	61.014	61.014
Tb	0.69	0.57	1	0.87	0.85	0.65	0.72	0.50	0.52	0.52	0.57
Dy	3.65	3.19	5.85	5.26	4.69	3.56	4.28	3.07	2.55	2.68	2.93
Ho	0.72	0.69	1.2	1.13	1	0.73	0.86	0.65	0.48	0.5	0.54
Er	1.97	2	3.33	3.09	2.72	1.97	2.38	1.94	1.27	1.26	1.52
Tm	0.28	0.32	0.51	0.46	0.44	0.32	0.37	0.32	0.17	0.17	0.21
Yb	1.9	2.1	3.6	3.2	2.7	2.1	2.40	2.32	1.2	1.2	1.4
Lu	0.3	0.34	0.58	0.51	0.41	0.34	0.37	0.36	0.19	0.19	0.23
Hf	3.50	3.00	6.00	4.20	5.00	4.10	4.31	3.19	2.5	2.6	2.8
Ta	0.63	0.95	1.33	0.94	0.83	1.10	1.05	1.80	0.3	0.31	0.29
Pb	18.10	8.70	32.20	20.40	13.90	13.80	11.38	14.83	3.7	6.3	11.1
Th	15.30	17.10	16.50	13.90	11.40	14.10	15.95	23.01	7.6	7.8	6.7
U	2.22	3.06	2.66	3.29	4.70	2.50	2.80	3.06	1.84	1.9	1.28
$^{87}Rb/^{86}Sr$											
$^{87}Sr/^{86}Sr_n$	0.711747	0.7115	0.709689	0.711403	0.707685	0.712328			0.705105	0.705121	0.705598
$^{87}Sr/^{86}Sr_i$	0.70751	0.70559	0.70644	0.7063	0.70644	0.70695			0.704674	0.704673	0.705297
$^{147}Sm/^{144}Nd$	0.1015	0.1187		0.1223	0.1089						
$^{143}Nd/^{144}Nd_n$	0.512413	0.512437		0.512444	0.512396				0.512675	0.512678	0.512577
εNd_i	−2.4	−2.3		−2.2	−2.8				2.9	3.0	0.8

Data sources: Mapple Formation (Riley *et al.* 2001); Jason Peninsula (Riley *et al.* unpublished data; Riley *et al.* 2003*a*).

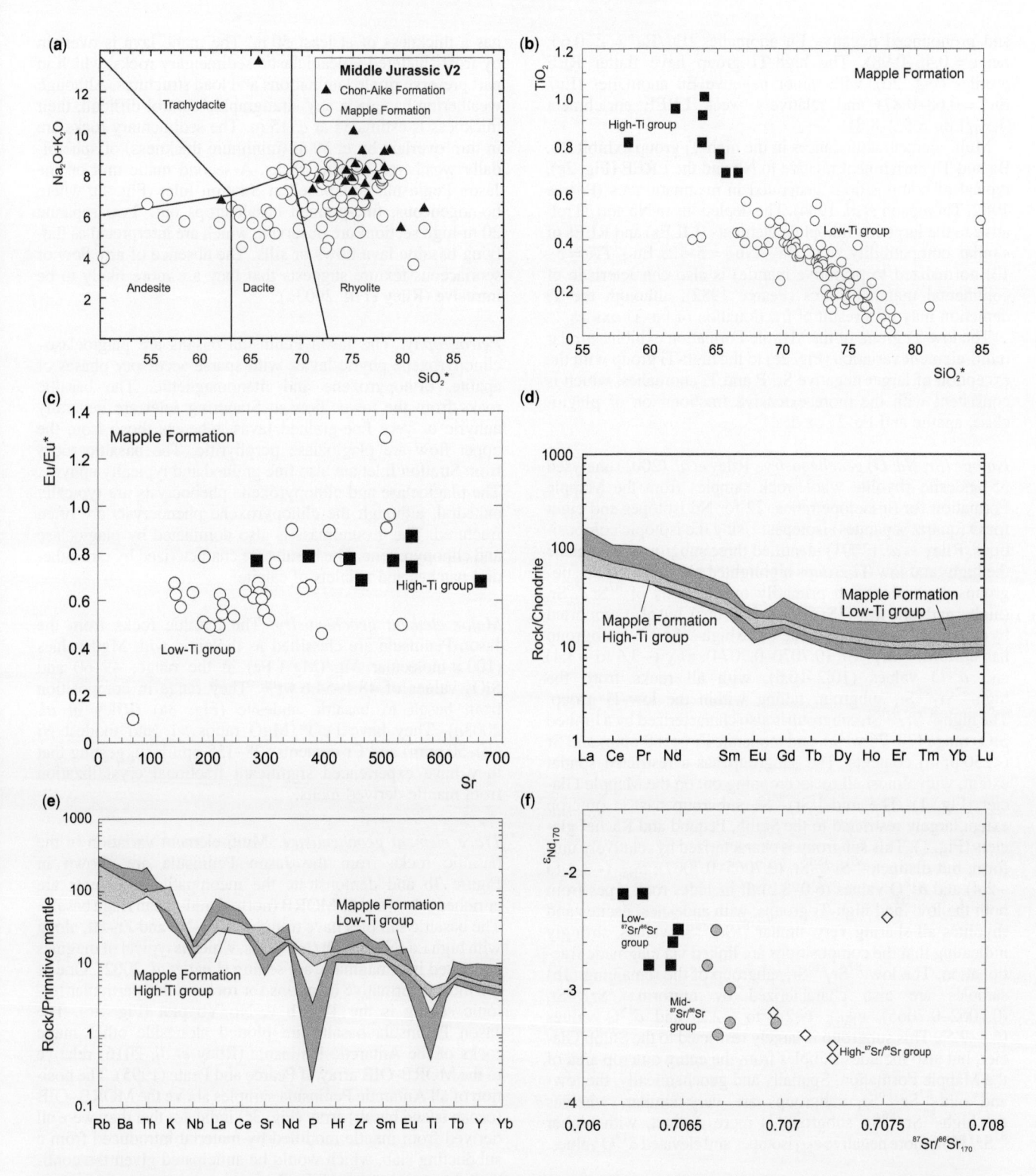

Fig. 2. (a) Total alkali v. SiO$_2$ plot for the Middle Jurassic rocks of the Antarctic Peninsula (Riley *et al.* 2001) and the Chon Aike Formation (Pankhurst *et al.* 1998). SiO$_2$ is recalculated to volatile-free totals of 100% (SiO$_2$*). (b) TiO$_2$ v. SiO$_2$ plots for silicic volcanic rocks from the Mapple Formation (Riley *et al.* 2001). SiO$_2$ is calculated to volatile-free totals of 100% (SiO$_2$*). (c) Eu/Eu* v. Sr plot for volcanic rocks from the Mapple Formation, illustrating the strong control of plagioclase fractionation. (d) Chondrite-normalized (Nakamura 1974) REE diagrams for volcanic rocks from the Mapple Formation (low-Ti) and Mapple Formation (high-Ti). (e) Primitive-mantle-normalized (Sun and McDonough 1989) plots for volcanic rocks from the Mapple Formation (low-Ti) and Mapple Formation (high-Ti). (f) ε_{Nd_i} v. $^{87}Sr/^{86}Sr_i$ plots for volcanic rocks from the the Mapple Formation high-, mid- and low-$^{87}Sr/^{86}Sr_{170}$ subgroups.

province. The SiO$_2$–TiO$_2$ (Fig. 2b) plot illustrates that two groups can be recognized within the Mapple Formation (Riley *et al.* 2001): a low-Ti group (0.02–0.56 wt% TiO$_2$), dominated by dacites and rhyolites (*c.* 64–77 wt% SiO$_2$); and a relatively high-Ti group (0.68–1.00 wt% TiO$_2$), characterized by intermediate compositions (*c.* 62–67 wt% SiO$_2$). The correlative Chon Aike Formation is also dominated by low-Ti compositions.

Trace element geochemistry. In addition to the contrast between the high- and low-Ti groups (Fig. 2b), there are key differences in Sr, Th and Y (Riley *et al.* 2001). Trace element variation within the Mapple Formation, (e.g. Eu/Eu* v. Sr: Fig. 2c), has been affected by plagioclase-dominated fractionation. Rare earth element (REE) abundances in the low-Ti group of the Mapple Formation are generally homogenous (Fig. 2d), with light REE enrichment (La$_N$/Lu$_N$ = 4.0–16.9) and pronounced negative Eu anomalies (Eu/Eu* = c. 0.65; range = 0.46–0.98). The high-Ti group have flatter REE profiles (Fig. 2d), with minor negative Eu anomalies (Eu/Eu* = 0.68–0.87) and relatively weak LREE enrichment (La_N/Lu_N = 5.9–8.8).

Multi-element abundances in the high-Ti group exhibit Rb, Ba and Th enrichment relative to Nb and the LREE (Fig. 2e), typical of compositions generated in magmatic arcs (Pearce 1982; Thompson et al. 1984). The depletions in Nb and Ti relative to the large ion lithophile elements (LILEs) and REEs of similar compatibility (e.g. Ba_N/Nb_N = 5–12; Eu_N/Ti_N = 5–13; normalized to primitive mantle) is also characteristic of continental magmatic arcs (Pearce 1982), although the Ti depletion may be a result of fractionation of Fe–Ti oxides.

The low-Ti group of the Mapple Formation exhibits similar multi-element variation (Fig. 2e) to the high-Ti group with the exception of larger negative Sr, P and Ti anomalies, which is consistent with the more extensive fractionation of plagioclase, apatite and Fe–Ti oxides.

Isotope (Sr, Nd, O) geochemistry. Riley et al. (2001) analysed 55 andesite–rhyolite whole-rock samples from the Mapple Formation for Sr isotope ratios, 22 for Nd isotopes and eight for O (quartz separates) isotopes. Using the isotopic compositions, Riley et al. (2001) identified three subgroups, alongside the high- and low-Ti groups highlighted previously. The subgroups were identified primarily on the basis of $^{87}Sr/^{86}Sr_i$ (high-, mid- and low-$^{87}Sr/^{86}Sr_i$ subgroups), but also supported by ε_{Nd} (Fig. 2f) and $\delta^{18}O$ values. The high-$^{87}Sr/^{86}Sr_i$ subgroup has uniform $^{87}Sr/^{86}Sr_i$ (0.7070–0.7074), ε_{Nd_i} (−3.6 to −3.4) and $\delta^{18}O$ values (10.2–10.6), with all rocks from the high-$^{87}Sr/^{86}Sr_i$ subgroup falling within the low-Ti group. The high-$^{87}Sr/^{86}Sr_i$ subgroup is also characterized by a limited SiO_2 range (70–76 wt%), and moderate Zr (<160 ppm) and Sr (<250 ppm) contents. The subgroup has a restricted spatial extent, with almost all rocks cropping out on the Mapple Glacier (Fig. 1). The mid-$^{87}Sr/^{86}Sr_i$ subgroup has an outcrop extent largely restricted to the Stubb, Pequod and Rachel glaciers (Fig. 1). This subgroup is characterized by relatively uniform, but distinct, $^{87}Sr/^{86}Sr_i$ (0.7065–0.7067), ε_{Nd_i} (−3.4 to −2.4) and $\delta^{18}O$ values (6.0–8.2). It includes rock types from both the low- and high-Ti groups, with andesites, dacites and rhyolites all sharing very similar $^{87}Sr/^{86}Sr_i$ values, strongly indicating that the compositions are linked via magmatic fractionation. The low-$^{87}Sr/^{86}Sr_i$ subgroup of the remaining (16) samples are also characterized by uniform $^{87}Sr/^{86}Sr_i$ (0.7062–0.7065), $\varepsilon_{Nd_{170}}$ (−2.8 to −2.2) and $\delta^{18}O$ values (6.4–9.5). This subgroup is largely restricted to the Stubb Glacier, but also includes samples from the entire outcrop area of the Mapple Formation. Spatially and geochemically, the low- and mid-$^{87}Sr/^{86}Sr_i$ subgroups are very similar, whereas the high-$^{87}Sr/^{86}Sr_i$ subgroup is more distinct, with higher $^{87}Sr/^{86}Sr$, more negative ε_{Nd} isotopes and elevated $\delta^{18}O$ values.

Jason Peninsula basalts

Background. Mafic volcanic rocks have been reported from the Jason Peninsula (Fig. 1) and have been dated ($^{40}Ar/^{39}Ar$) in the interval 175–168 Ma (Riley et al. 2003a). At the eastern margin of Standring Inlet (Fig. 1) three prominent outcrops have been identified: the northernmost outcrop has mafic volcanic rocks and volcaniclastic sedimentary rocks, whilst the other two are outcrops of silicic ignimbrites. No direct field relationships between the silicic and mafic volcanic rocks are observed at Standring Inlet, although, in the absence of observed faulting, they are thought to be at a similar stratigraphic level. The lowest stratigraphic unit is a fine-grained mafic lava flow, which is 'rubbly' in part and has a thickness of at least 50 m. The mafic lava is overlain by frost-shattered volcaniclastic sedimentary rocks, which in part preserve ripple laminations and load structures. Although weathering has made any stratigraphic control difficult, their thickness is estimated at c. 15 m. The sedimentary rocks are in turn overlain by c. 30 m (minimum thickness) of spheroidally weathered basaltic rocks. A second mafic unit on the Jason Peninsula is exposed at Stratton Inlet (Fig. 1), where homogeneous, fine-grained basalt crops out. Two separate 20 m-high sections are observed, which are interpreted as flat-lying basaltic lava flows or sills. The absence of any flow or scoriaceous texture suggests that they are more likely to be intrusive (Riley et al. 2003a).

Petrography. The Jason Peninsula basalts are plagioclase–clinopyroxene-phyric lavas, with sparse accessory phases of apatite, orthopyroxene and titanomagnetite. The basaltic rocks from the lower flow at Standring Inlet are generally aphyric or very fine-grained lavas, whereas those from the upper flow are plagioclase porphyritic. The basaltic rocks from Stratton Inlet are also fine grained and typically aphyric. The plagioclase and clinopyroxene phenocrysts are typically euhedral, although the clinopyroxene phenocrysts are often fractured. The groundmass is also dominated by plagioclase and clinopyroxene. The basalts are characterized by rare anhedral patches and veinlets of calcite.

Major element geochemistry. The basaltic rocks from the Jason Peninsula are classified as tholeiitic with Mg# values (100 × molecular Mg/(Mg + Fe)) in the range, 49–60 and SiO_2 values of 48.1–54.6 wt%. They range in composition from basalt to basaltic andesite (Fig. 3a) (Riley et al. 2003a). They have FeO*/MgO ratios >1, and modest Ni (10–50 ppm) and Cr contents (28–158 ppm), suggesting that they have experienced significant fractional crystallization from mantle-derived melts.

Trace element geochemistry. Multi-element variation in the basaltic rocks from the Jason Peninsula are shown in Figure 3b and demonstrate the incompatible elements are enriched relative to N-MORB (normal mid-ocean ridge basalt). The basaltic samples have troughs at Ta–Nb and Zr–Hf, along with high La/Nb and Th/Nb ratios, which is typical of magmas generated in a magmatic arc setting (Leat et al. 2002). One of the most informative diagrams for rocks in this particular tectonic setting is the Th/Yb v. Nb/Yb plot (Fig. 3c). The Jason Peninsula basalts are plotted alongside other mafic rocks of the Antarctic Peninsula (Riley et al. 2016) relative to the MORB-OIB array of Pearce and Peate (1995). The position of all Antarctic Peninsula samples above the MORB–OIB (ocean island basalt) array (Fig. 3c) indicates that they were all derived from mantle modified by material introduced from a subducting slab, which would be anticipated given the continental margin setting of the Peninsula.

The REE abundances of the Jason Peninsula basalts are strongly enriched in the light REE (LREE) relative to the heavy REE (HREE), with La_N/Yb_N = 6.3–21.0, and with the two Stratton Inlet (Fig. 1) samples being least LREE enriched (Fig. 3d).

Isotope geochemistry. Sr and Nd isotope ratios for the Jason Peninsula basalts initialized to an age of 170 Ma are plotted in Figure 3e alongside mafic dykes from the Oscar II and Black coasts (Leat et al. 2002). Akin to the trace element variations, the basalts from the Jason Peninsula overlap in composition with mafic dykes from the Black Coast, and are distinct to the asthenosphere-like dykes from the Oscar II Coast (Fig. 1). The Jason Peninsula exhibit a similar range in

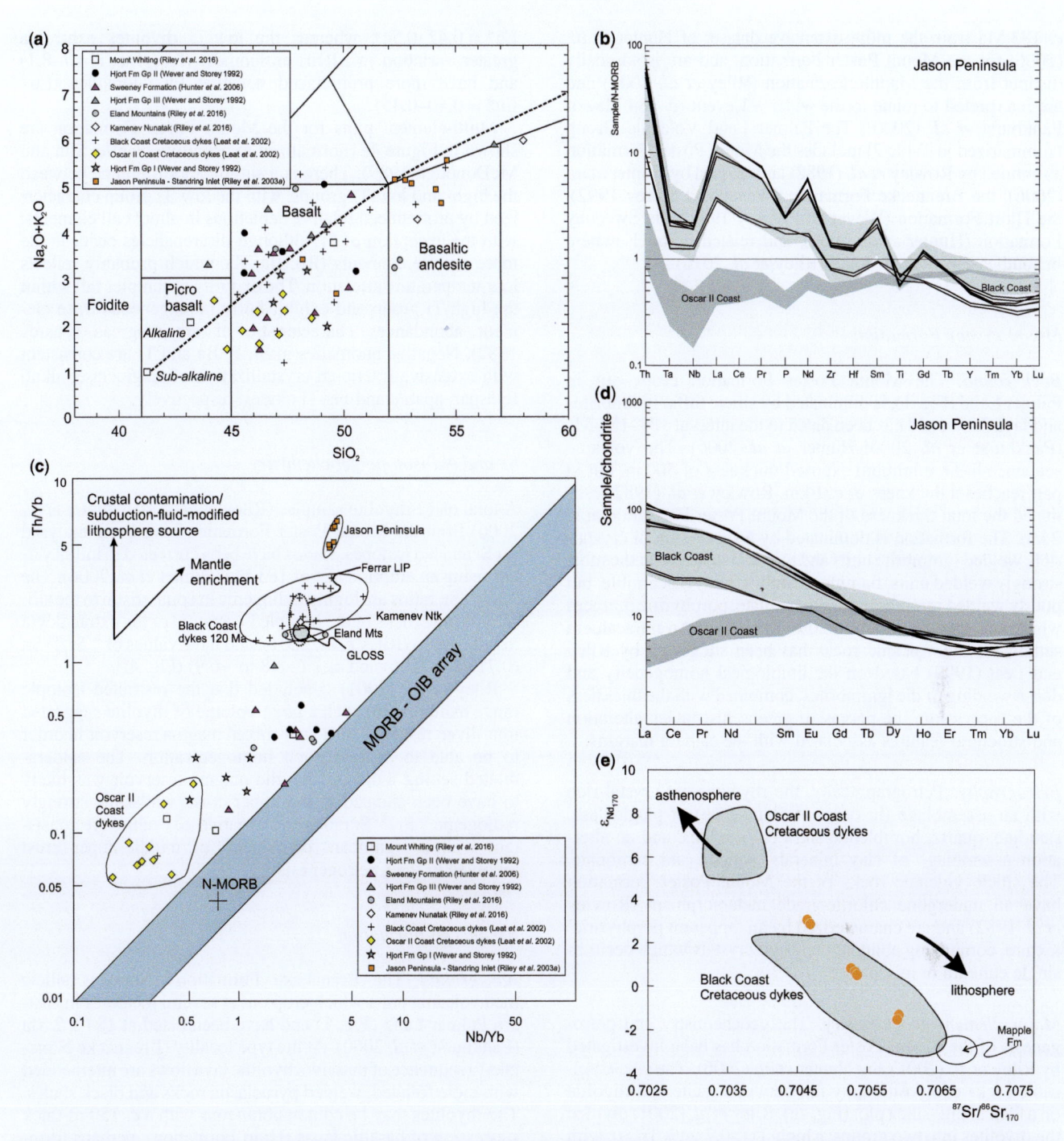

Fig. 3. (a) Total alkalis v. SiO$_2$ diagram (wt%) for the basaltic successions from the east coast of the Antarctic Peninsula. The samples are basalt or basaltic andesite (Riley et al. 2016). (b) MORB-normalized (Sun and McDonough 1989) multi-element abundances in Jason Peninsula basalts (Riley et al. 2003a). The data are compared to mafic dykes from the Oscar II and Black coasts (Leat et al. 2002). (c) Variations in Th/Yb v. Nb/Yb showing the composition of mafic rocks from the east coast of the Antarctic Peninsula (Riley et al. 2016) relative to the MORB–OIB array (Pearce and Peate 1995). The Karoo (Dronning Maud Land) fields are from Riley et al. (2005), average Ferrar is from Molzahn et al. (1996) and GLOSS is average global subducting sediment from Plank and Langmuir (1998). (d) Chondrite-normalized (Nakamura 1974) REE abundances in Jason Peninsula basalts (Riley et al. 2003a). The data are compared to mafic dykes from the Oscar II and Black coasts (Leat et al. 2002). (e) Plot of initial ε_{Nd} v. initial $^{87}Sr/^{86}Sr$ showing the composition of Jason Peninsula basalts (Riley et al. 2003a) relative to the mafic dykes of the Oscar II and Black coasts (Leat et al. 2002), Antarctic Peninsula west coast dykes (Scarrow et al. 1998) and the Mapple Formation rhyolites (Riley et al. 2001).

$^{87}Sr/^{86}Sr$ and ε_{Nd} values to the Black Coast dykes: $^{87}Sr/^{86}Sr_i$ (0.7047–0.7059) and ε_{Nd_i} (−1.6 to +3.0).

Palmer Land Volcanic Group

The Palmer Land Volcanic Group (Smellie 2020; Riley and Leat 2021) replaces the previously defined Ellsworth Land Volcanic Group (Hunter et al. 2006) following the redefined boundaries (UK Antarctic Place Names Committee) for Palmer Land (southern Antarctic Peninsula).

The volcanic rocks of the Palmer Land Volcanic Group (Hunter et al. 2006), which crop out in the southeastern Antarctic Peninsula, are in many respects akin to the Graham Land Volcanic Group defined by Riley et al. (2010). Both volcanic groups are dominated by rhyolitic tuffs forming thick (>1 km), caldera-fed successions. However, the Palmer Land Volcanic Group rocks have an average eruption age of

c. 183 Ma from the more extensive dataset of Hunter et al. (2006) for the Mount Poster Formation, and are isotopically distinct from the Mapple Formation (Riley et al. 2001) and are interpreted to relate to the wider V1 event (c. 184 Ma) of Pankhurst et al. (2000). The Palmer Land Volcanic Group (summarized in Table 2) includes the Mount Poster Formation as defined by Rowley et al. (1982) and revised by Hunter et al. (2006), the Brennecke Formation (Wever and Storey 1992), the Hjort Formation (Wever and Storey 1992), the Sweeney Formation (Hunter et al. 2006), and related units elsewhere in southeastern Palmer Land (Riley et al. 2016).

Mount Poster Formation

Background. The Mount Poster Formation crops out in Palmer Land (Fig. 1), is dominated by silicic tuffs/ignimbrites and lava flows, and has been dated to the interval 189–183 Ma (Pankhurst et al. 2000; Hunter et al. 2006). The volcanic sequence has a minimum exposed thickness of 500 m, but in part reaches a thickness of c. 1 km. Rowley et al. (1982) estimated the total thickness of the Mount Poster Formation at c. 2 km. The formation is dominated by a succession of crystal-rich, welded ignimbrite units and minor lava flows. In the most strongly welded units, fiamme are only weakly discernible, but poorly welded ignimbrites contain oblate, porphyritic pumices which can account for up to 20% of the unit. An intracaldera setting for the volcanic rocks has been suggested by Riley and Leat (1999) based on the lithological homogeneity and dense welding of the ignimbrites, combined with the thickness of the succession, the pervasive greenschist-facies alteration and structural features associated with the caldera margins.

Petrography. Petrographically, the rhyolites are crystal-rich with an assemblage (in order of abundance) of plagioclase, sanidine, quartz, hornblende and Fe–Ti oxides, and an alteration assemblage of clay minerals, sericite and carbonate. The silicic volcanic rocks of the Mount Poster Formation have all undergone chlorite-grade metamorphism (Rowley et al. 1982) and are characterized by an 'apparent porphyritic' texture, comprising abundant feldspar crystals which occur as single euhedra or as clusters.

Major element geochemistry. The geochemistry and petrogenesis of the Mount Poster Formation has been investigated by Riley et al. (2001) and Hunter et al. (2006). The silicic volcanic rocks compositionally fall between dacite and rhyolite on a total alkali–silica plot (Fig. 4a). Riley et al. (2001) divided the rhyolites into two groups: a high-Ti (>0.7 wt% TiO_2), with SiO_2 values of 70–74 wt%; and a low-Ti group (<0.4 wt% TiO_2), at >76 wt% SiO_2 (Fig. 4b). The high-Ti group corresponds to the intracaldera succession, and the low-Ti group represents the extracaldera volcanic rocks.

Trace element geochemistry. The Mount Poster Formation silicic volcanic rocks have significantly higher mean Nb (c. 17 ppm) and Zr (c. 300 ppm) contents relative to the rhyolitic rocks of the c. 170 Ma Mapple Formation (Nb = c. 10 ppm; Zr = c. 180 ppm). The trace element variation of the high-Ti group in the Eu/Eu* v. Sr diagram (Fig. 4c) is interpreted to be a function of plagioclase fractionation. However, the two rhyolites of the low-Ti group have a high Sr (c. 200 ppm) contents but low Eu/Eu* relative to the high-Ti group, which suggests that the two groups are not related by plagioclase fractionation. REE abundances show little variation throughout the Mount Poster Formation (Fig. 4d). The high-Ti group exhibits moderate LREE enrichment (La_N/Lu_N = 6.0–7.8), with negative Eu anomalies (Eu/Eu* = 0.47–0.54), whereas the low-Ti rhyolites exhibit a greater variation in LREE enrichment (La_N/Lu_N = 4.7–8.1) and have more pronounced negative Eu anomalies (Eu/Eu* = 0.40–0.45).

Multi-element plots for the Mount Poster Formation are shown in Figure 4e (normalized to primitive mantle: Sun and McDonough 1989). There is a significant difference between the high- and low-Ti groups, with the low-Ti group characterized by pronounced relative depletions in almost all elements, with the exception of Sr; although discrepancies occur in the more mobile elements (Rb, Ba, K), which probably reflects low-temperature alteration. The majority of samples fall within the high-Ti group and exhibit fairly homogeneous trace element abundances characteristic of arc magmas (Pearce 1982). Negative anomalies in Sr, P, Ba and Ti are consistent with extensive fractional crystallization of plagioclase, alkali feldspar, apatite and Fe–Ti oxides, respectively.

Sr and Nd isotope geochemistry

A total of 12 rhyolite samples (Riley et al. 2001; Hunter et al. 2006) from the Mount Poster Formation have been analysed for Sr and Nd isotopes. Ratios have been corrected to initial values using an eruption age of 183 Ma (Hunter et al. 2006). The $^{87}Sr/^{86}Sr_i$ ratios are highly radiogenic in comparison to the silicic volcanic rocks of the Mapple Formation or the extracaldera low-Ti group (Riley et al. 2001), and have values of $^{87}Sr/^{86}Sr_i$ (0.7180–0.7206) and ε_{Nd_i} (−8.8 to −6.9) (Fig. 4f).

Riley et al. (2001) concluded that the restricted isotopic range exhibited for such a large volume of rhyolite suggested long-lived residence in a well-mixed magma reservoir in order to be able to achieve such homogenization. The caldera-related setting indicates that the magma reservoir was likely to have been situated in the upper crust, whilst the strongly radiogenic $^{87}Sr/^{86}Sr$ ratios and negative ε_{Nd} values are consistent with a significant component of middle-upper crust involved in their petrogenesis.

Brennecke Formation

Background. The Brennecke Formation comprises silicic metavolcanic units which crop out at several localities in eastern Palmer Land (Fig. 1) and have been dated at 184 ± 2 Ma (Pankhurst et al. 2000). At the type locality (Brennecke Nunataks) a sequence of massive rhyolitic lava flows are interbedded with more foliated, welded pyroclastic rocks and black shales. The rhyolites may be contemporaneous with a c. 150 m-thick succession of basaltic lavas (Hjort Formation), perhaps forming a bimodal association (Wever and Storey 1992).

Petrography. The Brennecke Formation metavolcanic silicic rocks are porphyritic dacites, rhyodacites and rhyolites that are plagioclase, quartz and hornblende phyric. The rocks are typically altered, and are characterized by abundant sericite and muscovite and a pervasive greenschist-facies overprint.

Major element geochemistry. The petrogenesis and geochemistry of the Brennecke Formation has been investigated by Wever and Storey (1992), and their findings are summarized here. Compositionally, the silicic volcanic rocks are dacitic–rhyolitic (Fig. 4a) with SiO_2 values in the range 64–76 wt%. They all have TiO_2 values <1 wt%, with the rhyolites having TiO_2 values <0.5 wt%.

Trace element geochemistry. Reported trace element data for the silicic volcanic rocks from the Brennecke Formation is relatively sparse (Wever and Storey 1992). They have REE

Table 2. Representative geochemical analyses of volcanic rocks from the Palmer Land Volcanic Group

Sample	N10-123.1	N10-139.1	N10-178.1	R.6871.3	R.6878.1	R.7102.1	R.4238.1	R.4277.1	R.4197.2	R.8087.1	R.8090.6	R.8109.2
Formation	Hjort Fm	Hjort Fm	Hjort Fm	Mt Poster	Mt Poster	Mt Poster	Brennecke	Brennecke	Brennecke	Sweeney	Sweeney	Sweeney
Latitude (S)	70.6653	70.6373	70.5391	74.9037	75.1795	75.1332	71.9011	72.1984	73.5331	75.1498	75.1083	75.2097
Longitude (W)	62.775	62.6567	62.9912	71.4667	71.4131	69.1833	63.1667	63.5834	64.7504	68.8185	69.2715	70.7962
Major elements (wt%)												
SiO_2	46.69	43.99	48.57	72.06	70.41	75.02	72.43	74.68	68.28	46.24	45.98	46.63
TiO_2	1.34	1.44	1.24	0.84	0.75	0.40	0.43	0.49	0.82	1.50	1.71	1.67
Al_2O_3	15.47	15.84	15.49	11.75	12.94	12.43	11.80	11.23	13.47	16.57	13.42	14.88
Fe_2O_3	12.09	12.21	11.27	4.52	5.06	3.09	5.67	4.40	5.46	10.32	14.26	14.52
MnO	0.28	0.47	0.20	0.07	0.09	0.04	0.09	0.05	0.08	0.15	0.22	0.22
MgO	6.33	7.11	9.60	1.51	1.67	0.71	0.39	0.20	0.87	8.54	9.10	7.89
CaO	5.65	10.56	6.87	1.32	1.42	1.84	1.49	0.57	2.29	9.13	10.28	10.29
Na_2O	4.16	3.17	3.87	2.95	2.05	1.10	3.18	1.54	1.96	2.90	1.82	1.74
K_2O	0.31	0.11	0.03	3.67	5.00	4.00	3.62	5.44	4.58	0.74	0.09	0.40
P_2O_5	0.15	0.16	0.17	0.24	0.20	0.04	0.10	0.09	0.21	0.18	0.17	0.17
LOI	7.49	2.01	2.51	1.17	0.77	1.17	0.80	0.67	1.71	3.36	2.50	1.32
Total	99.97	97.08	99.81	100.10	100.35	99.85	99.73	99.03	99.73	99.63	99.55	99.75
Trace elements (ppm)												
Cr	266	446	143	31	28	28	4	3	23	382	391	175
Cu	26	45	90	6	17	11						
Ni	125	168	77	13	16	9	6	3	15	198	194	114
V	259	301	328	54	78	24	8	9	81			
Zn	86	71	93	69	88	66						
Sc	43.9	40.4	42.7				14.0	15.0	15.0	34.6	37.8	42.4
Ga	16.4	16.9	17.0							15.7	17.8	18.9
Rb	14.6	2.9	24.5	92.5	200.2	203.1	139.0	173.0	196.0	12.1	0.5	12.0
Sr	315.0	291.6	232.8	66.0	130.0	198.0	120.0	135.0	289.0	513.1	118.3	126.0
Y	28.8	28.8	27.6	47.0	48.0	34.0	72.0	51.0	51.0	25.4	33.7	36.2
Zr	38.2	30.3	38.6	284.0	328.0	244.0	386.0	387.0	293.0	94.8	75.0	73.0
Nb	7.3	7.2	3.2	16.0	16.0	15.0	18.0	15.0	16.0	10.7	6.9	7.9
Ba	192.6	55.2	534.2	757.0	875.0	879.0		1219.0		285.0	31.0	122.0
La	8.4	6.9	6.2	42.1	45.6	40.1		38.0		7.8	7.4	7.5
Ce	16.4	15.1	14.0	94.3	97.6	85.5	109.0	113.0	94.0	18.1	17.6	17.6
Pr	3.0	2.5	2.2	12.2	12.0	10.4				2.8	2.7	2.7
Nd	13.5	11.9	10.6	49.9	48.9	40.5	53.6	52.6	44.9	13.5	13.6	13.5
Sm	3.7	3.5	3.0	9.9	9.6	7.4	11.2	11.0	9.5	3.80	4.17	4.13
Eu	1.07	1.22	1.06	1.49	1.58	1.03	2.09	1.36	1.67	1.39	1.41	1.39
Gd	4.53	4.43	3.98	9.30	9.10	6.50				4.67	5.26	5.41

(Continued)

Table 2. *Continued*

Sample	N10-123.1	N10-139.1	N10-178.1	R.6871.3	R.6878.1	R.7102.1	R.4238.1	R.4277.1	R.4197.2	R.8087.1	R.8090.6	R.8109.2
Formation	Hjort Fm	Hjort Fm	Hjort Fm	Mt Poster	Mt Poster	Mt Poster	Brennecke	Brennecke	Brennecke	Sweeney	Sweeney	Sweeney
Latitude (S)	70.6653	70.6373	70.5391	74.9037	75.1795	75.1332	71.9011	72.1984	73.5331	75.1498	75.1083	75.2097
Longitude (W)	62.775	62.6567	62.9912	71.4667	71.4131	69.1833	63.1667	63.5834	64.7504	68.8185	69.2715	70.7962
Tb	0.78	0.77	0.69	1.35	1.36	0.98	1.92	1.59	1.38	0.77	0.92	0.97
Dy	4.92	4.76	4.43	8.00	7.87	5.78				4.61	5.84	6.02
Ho	1.03	1.01	0.96	1.64	1.65	1.19				0.94	1.20	1.29
Er	2.79	2.71	2.69	4.41	4.51	3.33				2.52	3.23	3.51
Tm	0.44	0.42	0.42	0.66	0.68	0.51				0.37	0.50	0.54
Yb	2.64	2.54	2.59	4.20	4.50	3.40	7.09	5.05	4.00	2.34	3.10	3.34
Lu	0.40	0.39	0.41	0.64	0.70	0.51	1.11	0.77	0.63	0.38	0.47	0.51
Hf	1.08	1.00	1.37	6.80	8.00	6.60	10.20	10.20	8.04	2.64	2.37	2.15
Ta	0.48	0.47	0.20	1.12	1.36	1.12	1.41	1.66	1.48	0.71	0.45	0.51
Pb	7.47	1.50	1.82	18.60	25.10	22.80				0.66	1.31	1.68
Th	0.92	0.88	0.82	16.20	18.60	14.90	17.70	19.30	16.50	1.15	1.20	1.20
U	0.17	0.17	0.22	2.88	3.58	3.89				0.40	0.25	0.27
^{87}Rb/^{86}Sr										0.06041	0.02431	0.29151
^{87}Sr/^{86}Sr$_m$				0.731232		0.723727				0.710273	0.705588	0.705656
^{87}Sr/^{86}Sr$_i$				0.71965		0.7156				0.710118	0.705526	0.704910
^{147}Sm/^{144}Nd				0.1229		0.1156	0.128	0.1226	0.1218	0.1742	0.1876	0.1893
^{143}Nd/^{144}Nd$_n$				0.51216		0.512299	0.512206	0.512138	0.512115	0.512848	0.512809	0.512802
εNd$_i$				−7.7		−4.9	−7.1	−7.9	−8.7	4.6	3.5	3.4

Data sources: Hjort Formation (Riley *et al.* 2016); Mount Poster Formation (Riley *et al.* 2001); Brennecke Formation (Wever and Storey 1992); Sweeney Formation (Hunter *et al.* 2006).

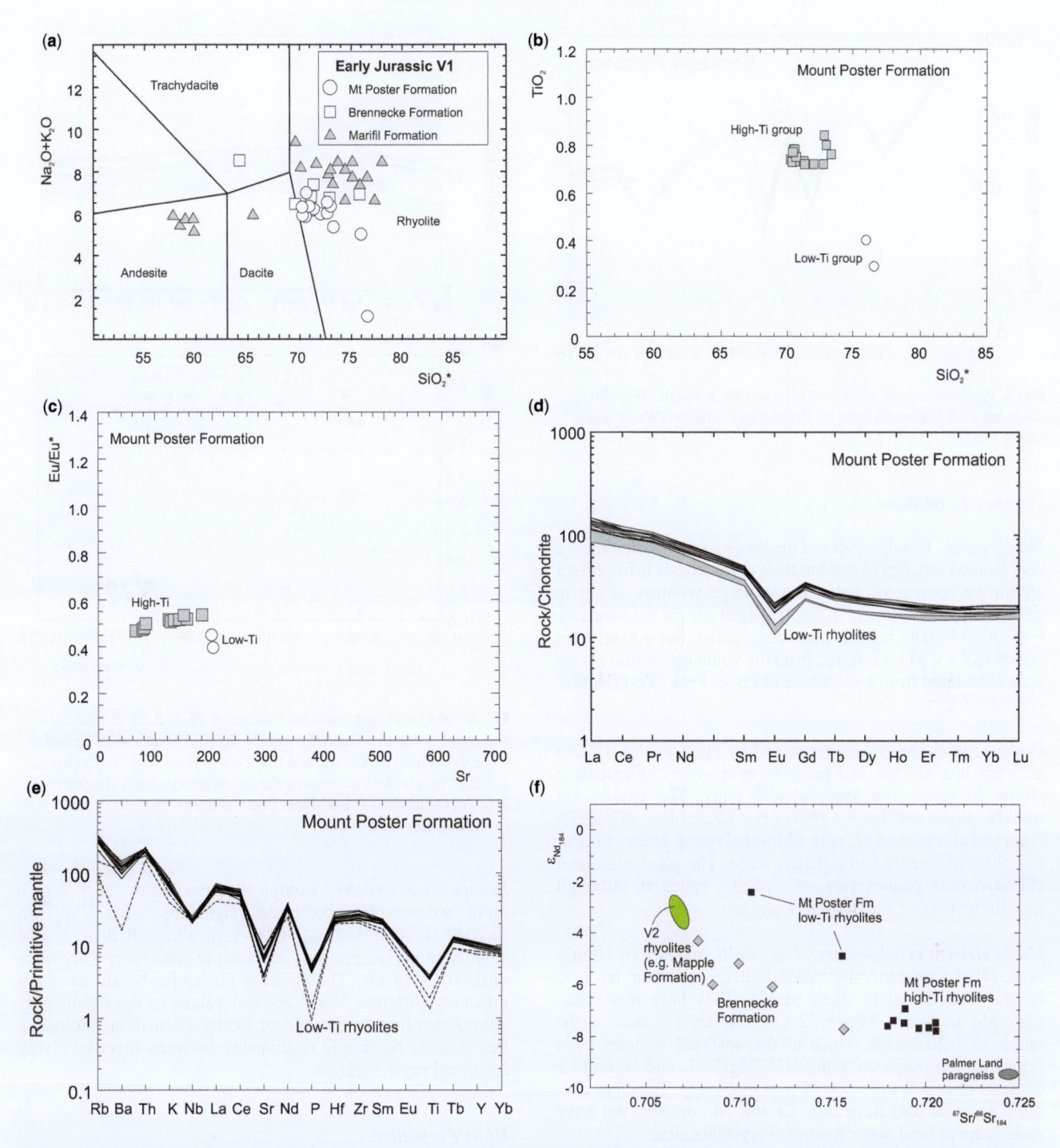

Fig. 4. (a) Total alkali v. SiO$_2$ plot for the Early Jurassic rocks of the Antarctic Peninsula (Riley *et al.* 2001). SiO$_2$ is recalculated to volatile-free totals of 100% (SiO$_2$*). (b) TiO$_2$ v. SiO$_2$ plots for volcanic rocks from the Mount Poster Formation (Riley *et al.* 2001). SiO$_2$ is calculated to volatile-free totals of 100% (SiO$_2$*). (c) Eu/Eu* v. Sr plot for volcanic rocks from the Mount Poster Formation, illustrating the strong control of plagioclase fractionation (Riley *et al.* 2001). (d) Chondrite-normalized (Nakamura 1974) REE diagram for low- and high-Ti silicic volcanic rocks from the Mount Poster Formation (Riley *et al.* 2001). (e) Primitive-mantle-normalized (Sun and McDonough 1989) plot for silicic volcanic rocks from the Mount Poster Formation (high- and low-Ti groups) (Riley *et al.* 2001). (f) ε_{Nd_i} v. $^{87}Sr/^{86}Sr_i$ plot for Mount Poster and Brennecke formation rhyolites (Riley *et al.* 2001). The upper-crustal Palmer Land paragneiss is taken from Wever *et al.* (1994). The Early Jurassic volcanic rocks are shown in comparison to the V2 volcanic rocks of the Mapple Formation (Riley *et al.* 2001).

abundances that are very similar to those of the Mount Poster Formation, with Ce$_N$/Yb$_N$ ratios of 4–6 and a negative Eu anomaly of Eu/Eu* = *c.* 0.4, and also very similar multi-element distributions (Fig. 5) (Wever and Storey 1992; unpublished data).

Isotope geochemistry. Rhyolitic and dacitic volcanic rocks from the Brennecke Formation of eastern Palmer Land (Wever and Storey 1992) have comparable ε_{Nd_i} values (−7.7 to −4.3) and similar $^{87}Sr/^{86}Sr_i$ values (0.7078–0.7157: Riley *et al.* 2001) to the contemporaneous Mount Poster Formation (Fig. 4f). The high-Ti rocks (>0.65% TiO$_2$) of the Brennecke Formation are characterized by more radiogenic $^{87}Sr/^{86}Sr_i$ (0.7100–0.7157) and ε_{Nd_i} (−7.7 to −5.2) in comparison to the low-Ti (>0.5% TiO$_2$) rocks, which have $^{87}Sr/^{86}Sr_i$ (0.7078–0.7086) and ε_{Nd_i} (−6.0 to −4.3).

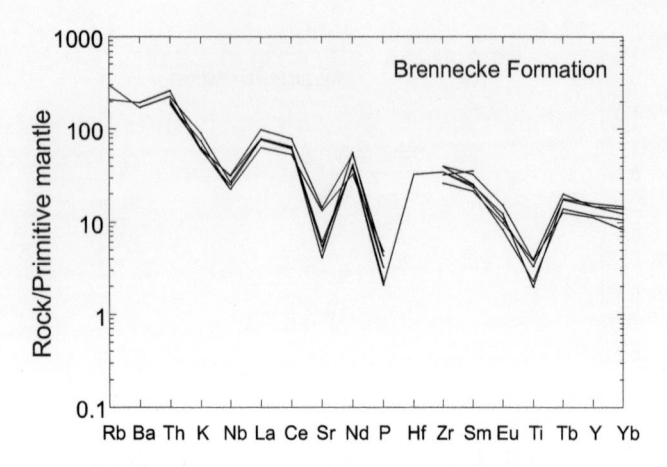

Fig. 5. Primitive-mantle-normalized (Sun and McDonough 1989) plot for silicic metavolcanic rocks from the Brennecke Formation (Wever and Storey 1992).

Sweeney Formation

Background. Basaltic rocks of the Sweeney Formation have a very limited outcrop extent but their petrogenesis is important within the context of generation of large volumes of silicic magmatism at this time. An emplacement age for the Sweeney Formation basalts has not been determined, but a maximum age of 183 ± 4 Ma is inferred from the youngest detrital zircon peak calculated from a sandstone at Potter Peak West (Hunter et al. 2006).

Petrography. The basalts examined by Hunter et al. (2006) are very fine grained and range from aphyric to plagioclase phyric to distinctive amygdaloidal units. The basalts are weakly altered and have a phenocryst assemblage of plagioclase and clinopyroxene, with additional sparse phases of apatite, titanomagnetite and orthopyroxene. The plagioclase and clinopyroxene phenocrysts are typically euhedral, although the clinopyroxene is extensively altered.

Major element geochemistry. The basalts analysed by Hunter et al. (2006) are generally tholeiitic in composition, but also include transitional to alkali basalts. They have moderate–high Mg numbers (Mg# = 52–66) and SiO_2 values in the range 45.9–50.3 wt%. Three of the analysed samples have primitive compositions with FeO*/MgO <1, and Ni and Cr contents of c. 300 and >600 ppm, respectively. The other analysed samples also have high Cr and Ni contents, but have undergone at least some fractional crystallization.

Trace element geochemistry. The analysed samples from the Sweeney Formation are characterized by minor negative Eu anomalies, indicating at least some degree of plagioclase fractionation, and flat REE patterns with $(La/Yb)_N = c.$ 1. The Sweeney Formation basalts are plotted in comparison to the mid-Cretaceous basaltic dykes (Leat et al. 2002) of the Antarctic Peninsula (Fig. 6a), and (with the exception of one sample) are intermediate between the asthenosphere-derived dykes of the Oscar II Coast (NE Antarctic Peninsula) and the lithosphere-derived dykes of the Black Coast (SE Antarctic Peninsula). The Nb/Yb v. Th/Yb plot (Fig. 3c) is a useful discriminant in continental margin arc settings, and supports the interpretation from the REE values that the Sweeney Formation basalts are typically intermediate between asthenosphere- and lithosphere-derived sources and could reflect a back-arc basin setting. Several of the samples plot close to the mantle array, indicating that they have little or no subduction-modified component (Fig. 3c).

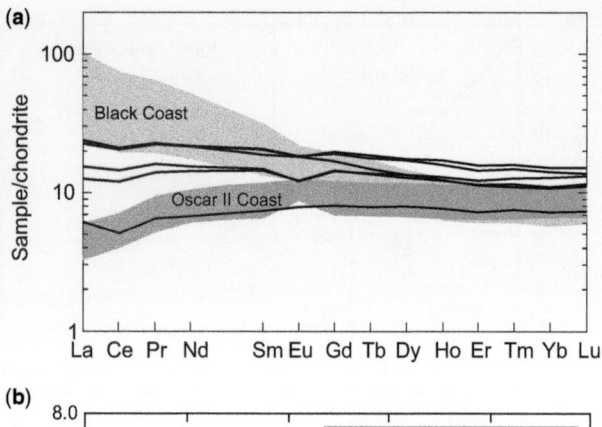

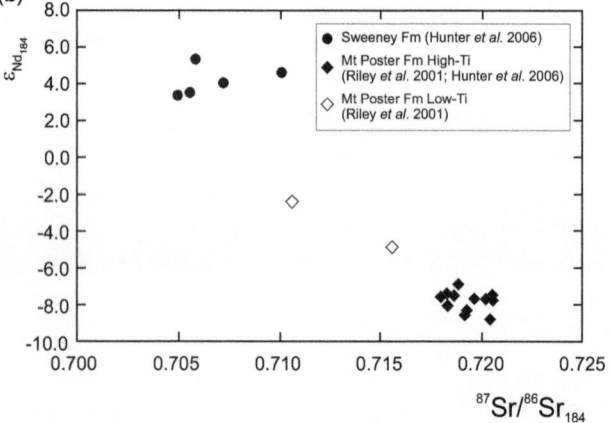

Fig. 6. (a) Chondrite-normalized (Nakamura 1974) REE abundances in Sweeney Formation basalts (Hunter et al. 2006). The data are compared to mafic dykes from the Oscar II and Black coasts (Leat et al. 2002). (b) ε_{Nd_i} v. $^{87}Sr/^{86}Sr_i$ plot for the Sweeney Formation basaltic rocks (Hunter et al. 2006) in comparison to the Mount Poster and Brennecke formation rhyolites (Riley et al. 2001).

Isotope geochemistry. Isotopically, the basalts also have a more asthenosphere-dominated signature, with $^{87}Sr/^{86}Sr_i$ (0.7049–0.7072) and ε_{Nd_i} (3.4–5.3), although there is one slightly more radioge ε_{Nd_i} nic sample with $^{87}Sr/^{86}Sr_i$ value of 0.7101 (Fig. 6b). The Sweeney Formation basalts have significantly different $^{87}Sr/^{86}Sr_i$ and values to the presumably coeval rhyolites of the Mount Poster Formation, indicating that there is no simple relationship between them involving fractional crystallization.

Hjort Formation

Background. Mafic volcanic rocks of the Hjort Formation crop out at several localities along the Black Coast of eastern Palmer Land (Fig. 1) and have been termed greenstones (Wever and Storey 1992). The type locality has been described from the Hjort Massif (Fig. 1), where at least 150 m of amphibolitic greenstones crop out and conformably overlie a westward-dipping sequence of deformed metasedimentary rocks (probably Mount Hill Formation: Wever and Storey 1992).

Amygdaloidal, fine-grained-aphyric basalts occur widely in the Black Coast region (e.g. Eland Mountains, Kamenev Nunatak) and are often associated with metasedimentary rocks of the Mount Hill Formation, and are interpreted to represent isolated units of the Hjort Formation. A probable continuation of these successions is also seen farther south into the Lassiter Coast (Fig. 1) where thin basaltic lava and breccias units are interbedded with clastic sedimentary rocks of the Latady Formation (Riley unpublished field data).

Petrography. The Hjort Formation basaltic rocks are distinctive aphanitic/fine-grained, amygdaloidal metavolcanic rocks, which have, in part, been hornfelsed as a result of extensive Cretaceous magmatism in the area (Riley *et al.* 2018). The amygdales are often quartz- and epidote-filled, particularly at Kamenev Nunataks and at the Eland Mountains. The mineralogy of the mafic rocks is dominated by green hornblende and plagioclase, with minor epidote, titanite and chlorite.

Major element geochemistry. Wever and Storey (1992) defined three basaltic groups from the Hjort Formation, which exhibit a broad geochemical variation. Groups II and III are predominantly alkaline, whereas Group I basalts are subalkaline and tholeiitic in composition (Fig. 3a). Group I basalts are relatively primitive with FeO*/MgO values of c. 1 and Ni contents >100 ppm. The Group II basaltic rocks have MgO contents in the range, 5–9 wt% and Ni contents in the range 46–184 ppm, whilst the group III mafic rocks have MgO contents of 5–7 wt% and low Ni contents of <60 ppm. The lavas from Kamenev Nunatak and Eland Mountains (Fig. 1) are mafic–intermediate in composition and are calc-alkaline.

Trace element geochemistry. The amygdaloidal basaltic lavas of the Eland Mountains (Riley *et al.* 2016) and Kamenev Nunatak (Group III of Wever and Storey 1992; Riley *et al.* 2016) plot alongside the field defined for global subducting sediment (GLOSS: Plank and Langmuir 1998), indicating a significant contribution from continental crust or partial melts of subduction-modified lithosphere (Fig. 3c). Amygdaloidal basalts and mafic greenstones from the Hjort Massif, Mount Hill, Eland Mountains, Mount Strong and Mount Whiting (Fig. 1) are all tholeiitic or transitional to calc-alkaline. They plot close to the MORB–OIB array (Fig. 3c) in the Th/Yb v. Nb/Yb diagram, and range from relatively depleted compositions (Mount Whiting) to compositions that are characteristic of island arc tholeiites and overlap with the amygdaloidal basalts of the marginally older Sweeney Formation of SE Palmer Land (Hunter *et al.* 2006).

Isotope geochemistry. Only Nd isotope data are available for the Hjort Formation basaltic rocks. The Group I and II rocks have ε_{Nd_i} values between −1.2 and 3.7, and are generally akin to the Sweeney Formation basalts, whereas the Group III basalts have ε_{Nd_i} in the range −5.0 to −2.3 and are close in composition to the mid-Cretaceous Black Coast mafic dykes (Leat *et al.* 2002).

Petrogenesis of the Jurassic volcanic rocks

Silicic volcanic rocks

The petrogenetic model for generation of the Jurassic Antarctic Peninsula rhyolites proposed by Riley *et al.* (2001) involves a two-stage model of partial melting of the lower crust and subsequent mixing of this partial melt with fractionates of basaltic (possibly plume related) underplate to generate an andesitic–dacitic magma. These intermediate–silicic melts are interpreted to migrate to upper-crustal magma chambers via a network of dykes where they undergo assimilation and fractional crystallization (AFC) before eruption as rhyolitic ignimbrites and lavas. The composition of the primary basaltic underplate has not previously been possible to determine given the near absence of erupted basaltic material. However the Sweeney Formation basalts, which have Mg numbers of c. 65, and Cr and Ni contents of c. 400 and c. 200 ppm, respectively, could represent approximate compositions of the

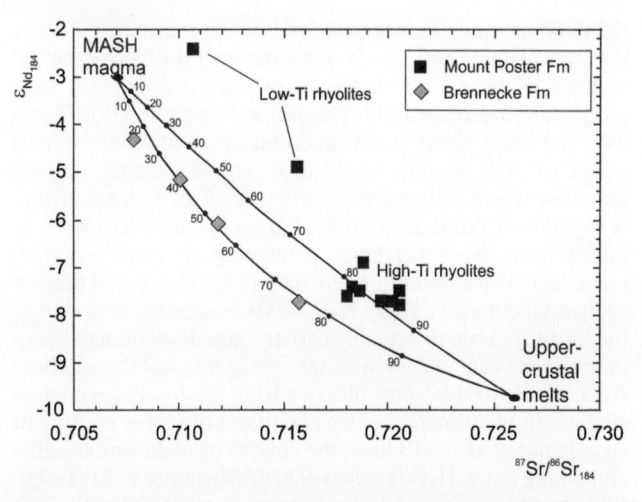

Fig. 7. ε_{Nd_i} v. $^{87}Sr/^{86}Sr_i$ plot for Mount Poster and Brennecke formation rhyolites showing putative mixing curves. The MASH domain magma has $^{87}Sr/^{86}Sr_i = 0.707$, $\varepsilon_{Nd_i} = -3$, Sr = 520 ppm and Nd = 25 ppm (average high-Ti rock from the Mapple Formation). The upper-crustal melt component is taken from Palmer Land paragneiss (Wever *et al.* 1994). The mixing curve for the Mount Poster Formation was calculated using the following parameters for the upper-crustal melt ($^{87}Sr/^{86}Sr_i = 0.7260$, $\varepsilon_{Nd_i} = 9.75$, Sr = 175, Nd = 45 ppm) and for the Brennecke Formation ($^{87}Sr/^{86}Sr_i = 0.7260$, $\varepsilon_{Nd_i} = 9.75$, Sr = 150, Nd = 80 ppm) (Riley *et al.* 2001).

basaltic underplate, although they are not adjudged to be primitive melts.

A mixing model (Fig. 7) for the Mount Poster Formation rhyolites using Sweeney Formation basalt as a mafic end member and Palmer Land paragneiss (Wever *et al.* 1994) as a potential crustal contaminant (Riley *et al.* 2001) provides a plausible model for the composition of the high-Ti (intracaldera) and low-Ti (extracaldera) rhyolites. Although the mixing model illustrates the feasibility that the two end members were important in the petrogenesis of the rhyolites, the more complex two-stage model of Riley *et al.* (2001) using a homogenized end member from a fertile lower crust provides a more satisfactory solution across the broader province.

There are several features of the silicic LIP in the Antarctic Peninsula and Patagonia that are critical in the determination of a petrogenetic model:

- Each formation was erupted in a short time interval, and the entire province was emplaced in three magmatic pulses (Pankhurst *et al.* 2000) or 'flare-ups' of magmatism (cf. Ducea and Barton 2007).
- The focus of magmatism migrated both spatially and temporally, such that the two episodes, V1 (185–182 Ma) and V2 (171–167 Ma), exposed on the Antarctic Peninsula do not overlap in time or geographically. A similar distribution has been identified in Patagonia.
- The volcanic rocks are overwhelmingly silicic (>70 wt% SiO_2). Basaltic rocks are scarce, but some intermediate rocks do occur.
- Uniform Sr and Nd isotope ratios are widespread both spatially and chronologically. This is interpreted as resulting from mixing in the lower crust between fractionated magmas from mafic underplate and partial melts of fertile mafic lower crust (Riley *et al.* 2001). The V1 silicic rocks of the southern Antarctic Peninsula are melts of upper-crustal paragneiss mixed with a minor component of magma from a homogenized fertile source.
- The rhyolites have trace element characteristics of subduction-related magmas inherited from their source, although this is weaker in V1 (185–182 Ma) than V2 (171–167 Ma).

Subduction zones are characterized by the recycling of water into the mantle, and this is a contributory factor for partial melting in the mantle wedge to produce arc magmas (e.g. Stolper and Newman 1992; Tatsumi and Eggins 1995). Therefore, the crust above a subduction-zone setting will, over a period of time, become intruded by hydrous, mantle-derived magmas, which will extend to the rear of the volcanic front. A significant proportion of these magmas are anticipated to intrude into the lower crust, generating an amphibole-rich mafic layer at the point(s) where they freeze because of density or thermal contrasts. Riley et al. (2001) suggested that, during the Early Jurassic or possibly earlier, these lower-crustal conditions were established in eastern Patagonia and the southern Antarctic Peninsula some distance from the arc. Experimental studies allow estimation of the conditions of partial melting of such material at 10–13 kbar, the solidus of mafic amphibolite containing c. 1% H_2O is c. 800–920°C (Rushmer 1991; 1993). Partial melting of amphibolite will generate silicic–intermediate compositions (55–75 wt% SiO_2) at temperatures of 50–1000°C, at 8 kbar, representing partial melting of up to 25% (Rapp and Watson 1995). The partial melting of hydrous, fertile, mafic lower crust could have been initiated in response to extension, occurring without too much additional heat input (Pankhurst and Rapela 1995), or by advection of mantle-derived heat by intrusion of basalt. Lower-crust temperatures in arc environments are thought to be much less than 1000°C (e.g. Tatsumi and Eggins 1995). It is interpreted that crustal melting was initiated by thermal input associated with the underplating of basaltic magma, which may have been related to the peripheral effects of a mantle plume. The mantle plume (Discovery–Shona–Bouvet plume group: Storey et al. 2013) would have been responsible for the eruption of the Karoo lavas. The migration of silicic volcanism from NW to SE in Patagonia and south to north in the Antarctic Peninsula has been attributed by Pankhurst et al. (2000) to the spreading of magmatism away from the plume head, taking into account palaeolocations at the time of the Gondwana break-up. At the onset of intrusion, mafic magmas are interpreted to be trapped at the base of the continental crust due to their density contrast, especially if the magmas have high Mg numbers, in equilibrium with mantle peridotite. Partial melting of the lower crust to produce bodies of intermediate–silicic magma will lead to the reinforcement of this 'barrier', because the mafic melts cannot migrate through the less-dense intermediate–silicic melt body (Huppert and Sparks 1988). At this stage the partial melts of hydrous mafic lower crust can mix with fractionates of the basaltic underplate, leading to the development of an isotopically homogeneous MASH (mixing–assimilation–storage–hybridization) magma. The continued intrusion of mafic melts in the lower crust will pond and lead to continued intermediate–silicic magma generation and subsequent mixing, such that a 'run-away' system can develop whereby the continued intrusion (underplating) of mafic magma leads to increased intermediate–silicic melt which further reduces the possibility of unfractionated mafic melt escaping upwards through the crust.

The emplacement of silicic volcanism at the surface is preceded by the migration of intermediate–silicic composition melts via dykes, which are driven by density contrasts within the crust (e.g. Petford et al. 1994; Weinberg 1999). This process will allow the development of upper-crustal magma reservoirs and subsequent explosive caldera-sourced eruptions (Bryan and Ferrari 2013). The prolonged residence in upper-crustal magma reservoirs permits the interaction between magma and upper-crustal rocks, probably via AFC, causing alteration to the isotopic composition of the MASH domain magma. The cessation of silicic magmatism would have resulted from the reduced thermal input and the slow cooling of the crust. It is interpreted by Riley et al. (2001) that large-volume silicic melt generation would have consumed the most fusible sections of the lower crust, such that there would be no, or limited, spatial overlap of silicic volcanic units. The episodic nature of the magmatism is likely to reflect the thermal migration away from the plume source, and development of pre-conditions for silicic volcanism, which occurs over a few million years prior to rapid eruption of the silicic rocks. Episodic magmatism along the Antarctic Peninsula and elsewhere along the Gondwana margin was repeated during the mid-Cretaceous with three distinct episodes of granitoid emplacement at 125, 115 and 105 Ma (Riley et al. 2018).

Mafic volcanic rocks

The Early–Middle Jurassic mafic volcanic rocks of the eastern Antarctic Peninsula are geochemically varied and have a petrogenesis that is closely related to the continental margin setting of the Antarctic Peninsula. The Nb/Yb v. Th/Yb plot (from Riley et al. 2016) (Fig. 3c) uses trace elements that are considered to have been immobile during both alteration and metamorphism. The fields defined by the basaltic rocks from the Jason Peninsula (Riley et al. 2003a), Sweeney

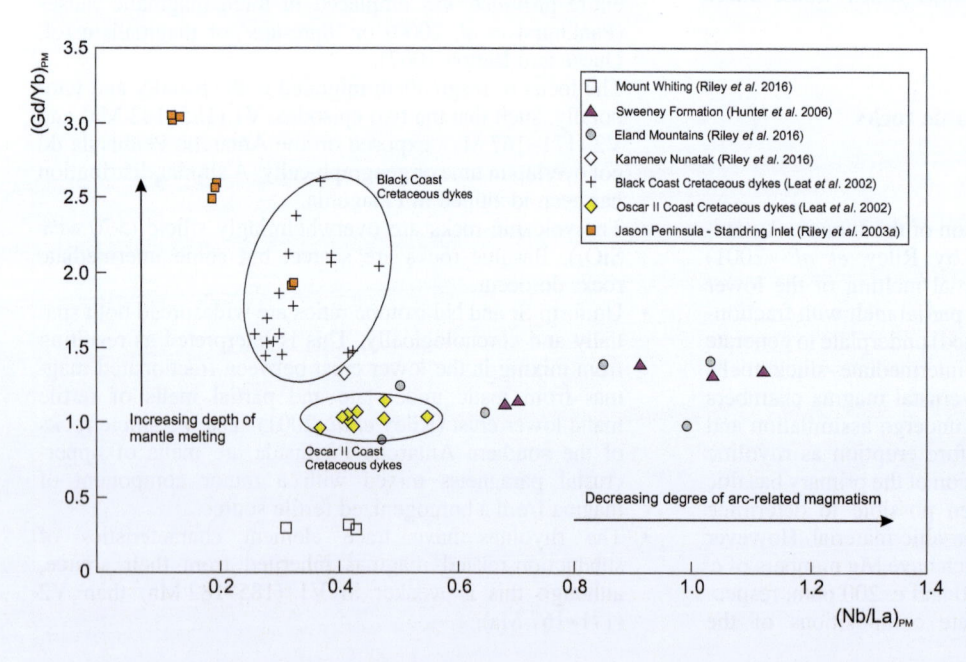

Fig. 8. $(Gd/Yb)_{PM}$ v. $(Nb/La)_{PM}$ with ratios normalized to primitive mantle (PM) values of McDonough and Sun (1995). Increasing $(Gd/Yb)_{PM}$ ratios indicate an increasing depth of mantle melting, and $(Nb/La)_{PM}$ ratios indicate the contribution of arc-related components (Riley et al. 2016).

Formation (Hunter et al. 2006) and several distinct groups of the Hjort Formation of the Black Coast region (Wever and Storey 1992; Riley et al. 2016) are all plotted in Figure 3c. They are plotted relative to the MORB–OIB array (Pearce and Peate 1995) and also to magmatic rocks from the broadly contemporaneous Karoo and Ferrar LIPs of South Africa and East Antarctica (Luttinen and Furnes 2000; Riley et al. 2005, 2006). Also plotted are the Cretaceous mafic dykes of the Antarctic Peninsula (Leat et al. 2002; Vaughan et al. 2012), which represent the arc composition.

The mafic rocks of the Jason Peninsula are calc-alkaline in composition and are characterized by the most incompatible-element-enriched compositions of all the Mesozoic mafic rocks of the Antarctic Peninsula with high Th/Yb and Nb/Yb with respect to MORB and OIB (Fig. 3c). They are characteristic of rocks that have been derived from mature continental margin arcs with a lithospheric mantle source that was enriched in Th as a result of modification by subduction-derived fluids. The outcrops of basaltic rocks from the Jason Peninsula are very minor, limited to flow units of c. 50 m in thickness, and rare dykes and sills. However, they are significant in that they represent the only known mafic magmatism associated with the extensive silicic volcanism (V2: 170 Ma) of the northern Antarctic Peninsula (Riley et al. 2001). The emplacement age of the basalts (175–168 Ma: Riley et al. 2003a) broadly overlap with the Mapple Formation rhyolites (171–168 Ma: Pankhurst et al. 2000) of the Oscar II Coast and their correlatives on the Jason Peninsula. Geochemical data suggest that both the Jason Peninsula basalts and the Mapple Formation rhyolites have subduction-modified compositions. The age of the Jason Peninsula basalts provide evidence that basaltic magmatism was broadly synchronous with the voluminous rhyolitic magmatism, but marginally predating it (Riley et al. 2003a). This suggests that they were erupted just prior to the onset of widespread partial melting of the lower crust, which formed a barrier to basalt traversing the lower crust. Once this barrier formed, the volcanism was entirely silicic, implying that all the mafic magmas were trapped within the crust.

The basaltic rocks of the Sweeney Formation (Hunter et al. 2006) lie close to the MORB–OIB array (Fig. 3c) and are characteristic of island arc tholeiites derived from a mantle source depleted in Nb/Yb but transitional to calc-alkaline rocks, and may have developed in a back-arc basin setting. The Sweeney Formation basalts have significantly different $^{87}Sr/^{86}Sr$ and ε_{Nd} values to the coeval rhyolites of the Mount Poster Formation, indicating that there is no simple relationship between them involving fractional crystallization. The Sweeney Formation basalts with Mg# of 65, and high Cr (c. 400 ppm) and Ni (c. 200 ppm) contents, could represent basaltic underplate.

The lava successions identified from the Black Coast region of Palmer Land are broadly referred to as the Hjort Formation and have been dated as Early Jurassic in age, and show a broad distribution in Figure 3c. A subset of the amygdaloidal basaltic lavas of the Eland Mountains (Riley et al. 2016) and Kamenev Nunatak (Group III of Wever and Storey 1992; Riley et al. 2016) plot close to the field of global subducting sediment (GLOSS: Plank and Langmuir 1998), indicating a significant contribution from continental crust or partial melts of subduction-modified lithosphere. The basaltic rocks from the Eland Mountains and Kamenev Nunataks also overlap with the field of Early Jurassic-age Ferrar LIP dolerite dykes and sills of the Transantarctic Mountains (Riley et al. 2003b, 2006). Two mafic dykes from the Mount Whiting area (Fig. 1) are associated with a nearby gabbroic pluton, which is likely to be mid-Cretaceous in age (Flowerdew et al. 2005). The Mount Whiting dykes are therefore one of the final phases of magmatism in the Black Coast region (Riley et al. 2016).

Table 3. *Summary of Gondwana break-up-related volcanism on the Antarctic Peninsula*

Volcanic Group	Formation (Reference)	Age (Reference)	Field characteristics	Geochemical features	Petrogenesis	Tectonic setting
Graham Land Volcanic Group	Mapple Formation (Riley et al. 2001)	171–167 Ma (Pankhurst et al. 2000; Riley et al. 2010)	Rhyolitic ignimbrites, crystal tuffs and breccia units	Uniform $^{87}Sr/^{86}Sr_i$ (0.7067 ± 0.0005) and ε_{Nd} (−3 to −2) values	Melting of fusible lower crust	Probably rift-related
	Jason Peninsula basalts (Riley et al. 2003a)	175–168 Ma (Riley et al. 2003a)	Homogeneous fine-grained basaltic andesite lava units	SiO_2: 48–55 wt%. Low Mg, Cr, Ni	Incompatible-element-enriched arc basalts	Continental margin extensional setting
Palmer Land Volcanic Group	Mount Poster Formation (Hunter et al. 2006)	189–183 Ma (Pankhurst et al. 2000; Hunter et al. 2006)	Intracaldera rhyodacitic ignimbrites	Highly radiogenic $^{87}Sr/^{86}Sr_i$ (0.7180–0.7206) and (−8.8 to −6.9) values	Plume-influenced lower-crustal melting with upper-crustal contamination	Early-stage Gondwana break-up
	Brennecke Formation (Wever and Storey 1992)			Highly radiogenic $^{87}Sr/^{86}Sr_i$ (0.7100–0.7157) and ε_{Nd} (−7.7 to −5.2) values	Plume-influenced lower-crustal melting with upper-crustal contamination	Early-stage Gondwana break-up
	Hjort Formation (Wever and Storey 1992)	180–177 Ma (Riley et al. 2016)	Fine-grained amygdaloidal basaltic lava units	Basaltic–intermediate, including a subset of more depleted compositions	Lithosphere-derived arc-like basalts and a subset of deeper sourced melts	Continental margin arc setting developing to a back-arc setting
	Sweeney Formation (Hunter et al. 2006)	<183 Ma (Hunter et al. 2006)	Plagioclase-phyric amygdaloidal basaltic lava units	A subset of primitive high Mg, Cr, Ni compositions	Relatively depleted rocks represent a possible sublithospheric source	Developing back-arc basin setting

Amygdaloidal basalts, mafic greenstones and mafic dykes from the Hjort Massif, Mount Hill, Eland Mountains, Mount Strong and Mount Whiting (Fig. 1) are all tholeiitic or transitional to calc-alkaline. They plot close to the MORB–OIB array (Fig. 3c) and range from relatively incompatible-element-depleted compositions (Mount Whiting) to compositions that are characteristic of island arc tholeiiites and overlap with the amygdaloidal basalts of the marginally older Sweeney Formation of SE Palmer Land (Hunter et al. 2006).

The $(Gd/Yb)_{PM}$ v. $(Nb/La)_{PM}$ diagram (Fig. 8) identifies several distinct groups corresponding to the different settings of the basaltic rocks from the eastern Antarctic Peninsula. These ratios (normalized to primitive mantle: McDonough and Sun 1995) are used as a proxy to represent the depth of melting in the mantle $(Gd/Yb)_{PM}$ and also the contribution from the arc lithosphere $(Nb/La)_{PM}$ (Jowitt and Ernst 2013). Those groups with the most enriched 'arc-like' signature are the basaltic rocks from the Jason Peninsula and the mafic dykes from the Black Coast (Fig. 3b), whilst the thick amygdaloidal basaltic successions at Eland Mountains (Riley et al. 2016) and those of the Sweeney Formation (Hunter et al. 2006) have $(Nb/La)_{PM}$ values >0.6, indicating a far less significant contribution from the arc lithosphere.

Summary

Silicic and mafic volcanic rocks of Early–Middle Jurassic age crop out extensively across the eastern Antarctic Peninsula, and are summarized in Table 3. The successions have been correlated geochemically and chronologically to the Chon Aike silicic LIP of Patagonia (Pankhurst et al. 2000; Riley et al. 2001). Mafic rocks are rare in both the Chon Aike province and the Antarctic Peninsula, although isolated outcrops of basalt and basaltic andesite are identified from several localities from the eastern Antarctic Peninsula (e.g. Riley et al. 2016).

A model of rhyolite petrogenesis involves extensive lower-crustal partial melting of a hydrated, melt-fertile, mafic underplate related to long-term suprasubduction processes (Riley et al. 2001). Partial melting of an intermediate, homogenized lower crust is the result of rifting associated with the early stages of Gondwana break-up or plume-related processes associated with the Early Jurassic flood basalt provinces (Storey et al. 2013). The intermediate composition melts migrate to mid-crustal levels and undergo prolonged AFC processes in crustal magma chambers prior to eruption via dyking and caldera collapse. The Early Jurassic silicic volcanic rocks have undergone contamination with strongly radiogenic upper-crustal material, whereas the Middle Jurassic silicic volcanic rocks have a far more homogeneous composition, and are interpreted as melts from a fertile lower-crustal layer and are compositionally indistinct from the Patagonian rhyolites.

The rhyolite petrogenetic model for the Antarctic Peninsula has relevance for the occurrence of basaltic rocks on the Antarctic Peninsula; the ponding of the mafic underplate magma in the lower crust leads to further heating and therefore continued intermediate–felsic melt generation, further reinforcing the barrier to basalt ascent. The intermediate–silicic melt eventually migrates via dyking (cf. Petford et al. 1994) to upper-crustal magma chambers and undergoes AFC, prior to eruption of short-lived ignimbrite forming eruptions. This model satisfactorily explains the geochemistry of the silicic volcanic rocks, their distribution and also the scarcity of basaltic rocks (Riley et al. 2001).

Acknowledgements Simon Abrahams, Malcom Airey, Adam Clark, Bradley Morrell, John Sweeney, Catrin Thomas, James Wake, Tom Weston and the air operations staff at Rothera Base are thanked for their field support over the last 23 years. The authors are grateful to Bob Pankhurst and Michael Flowerdew for helpful reviews of the chapter.

Author contributions TRR: conceptualization (lead), writing – original draft (lead), writing – review & editing (lead); **PTL**: writing – original draft (supporting), writing – review & editing (supporting).

Funding This study is part of the British Antarctic Survey Polar Science for Planet Earth programme, funded by the Natural Environmental Research Council.

Data availability The datasets generated during and/or analysed during the current study are available in the UK Polar Data Centre repository, wwwbas.ac.uk/data/uk-pdc/

References

Barbeau, D.L., Davis, J.T., Murray, K.E., Valencia, V., Gehrels, G.E., Zahid, K.M. and Gombosi, D.J. 2010. Detrital-zircon geochronology of the metasedimentary rocks of north-western Graham Land. *Antarctic Science*, **22**, 65–65, https://doi.org/10.1017/S095410200999054X

Bradshaw, J.D., Vaughan, A.P.M., Millar, I.L., Flowerdew, M.J., Trouw, R.A.J., Fanning, C.M. and Whitehouse, M.J. 2012. Permo-Carboniferous conglomerates in the Trinity Peninsula Group at View Point, Antarctic Peninsula: sedimentology, geochronology and isotope evidence for provenance and tectonic setting in Gondwana. *Geological Magazine*, **149**, 626–644, https://doi.org/10.1017/S001675681100080X

Bryan, S.E. and Ferrari, L . 2013. Large igneous provinces and silicic large igneous provinces: Progress in our understanding over the last 25 years. *Geological Society of America Bulletin*, **125**, 1053–1078, https://doi.org/10.1130/B30820.1

Burton-Johnson, A. and Riley, T.R. 2015. Autochthonous v. accreted terrane development of continental margins: a new *in situ* tectonic history of the Antarctic Peninsula. *Journal of the Geological Society, London*, **172**, 832–835, https://doi.org/10.1144/jgs2014-110

Ducea, M.N. and Barton, M.D. 2007. Igniting flare-up events in Cordilleran arcs. *Geology*, **35**, 1047–1050, https://doi.org/10.1130/G23898A.1

Flowerdew, M.J., Millar, I.L., Vaughan, A.P.M. and Pankhurst, R.J. 2005. Age and tectonic significance of the Lassiter Coast Intrusive Suite, Eastern Ellsworth Land, Antarctic Peninsula. *Antarctic Science*, **17**, 443–452, https://doi.org/10.1017/S0954102005002877

Hathway, B. 2000. Continental rift to back-arc basin: Jurassic–Cretaceous stratigraphical and structural evolution of the Larsen Basin, Antarctic Peninsula. *Journal of the Geological Society, London*, **157**, 417–432, https://doi.org/10.1144/jgs.157.2.417

Hervé, F., Lobato, J., Ugalde, I. and Pankhurst, R.J. 1996. The geology of Cape Dubouzet, northern Antarctic Peninsula: Continental basement to the Trinity Peninsula Group? *Antarctic Science* **8**, 407–414, https://doi.org/10.1017/S0954102096000582

Hunter, M.A. and Cantrill, D.J. 2006. A new stratigraphy for the Latady Basin, Antarctic Peninsula: Part 2, Latady Group and basin evolution. *Geological Magazine*, **143**, 797–819, https://doi.org/10.1017/S0016756806002603

Hunter, M.A., Cantrill, D.J., Flowerdew, M.J. and Millar, I.L. 2005. Middle Jurassic age for the Botany Bay Group: implications for Weddell Sea Basin creation and southern hemisphere biostratigraphy. *Journal of the Geological Society, London*, **162**, 745–748, https://doi.org/10.1144/0016-764905-051

Hunter, M.A., Riley, T.R., Cantrill, D.J., Flowerdew, M.J. and Millar, I.L. 2006. A new stratigraphy for the Latady Basin, Antarctic Peninsula: Part 1, Ellsworth Land Volcanic Group. *Geological*

Magazine, **143**, 777–796, https://doi.org/10.1017/S0016756806002597

Huppert, H.E. and Sparks, R.S.J. 1988. The generation of granitic magmas by intrusion of basalt into continental crust. *Journal of Petrology*, **29**, 599–624, https://doi.org/10.1093/petrology/29.3.599

Jowitt, S.M. and Ernst, R.E. 2013. Geochemical assessment of the metallogenic potential of Proterozoic LIPs of Canada. *Lithos*, **174**, 291–307, https://doi.org/10.1016/j.lithos.2012.03.026

Kellogg, K.S. and Rowley, P.D. 1989. *Structural Geology and Tectonics of the Orville Coast Region, Southern Antarctic Peninsula, Antarctica*. United States Geological Survey Professional Papers, **1498**.

Leat, P.T., Scarrow, J.H. and Millar, I.L. 1995. On the Antarctic Peninsula batholith. *Geological Magazine*, **132**, 399–412, https://doi.org/10.1017/S0016756800021464

Leat, P.T., Riley, T.R., Wareham, C.D., Millar, I.L., Kelley, S.P. and Storey, B.C. 2002. Tectonic setting of primitive magmas in volcanic arcs: an example from the Antarctic Peninsula. *Journal of the Geological Society, London*, **159**, 31–44, https://doi.org/10.1144/0016-764900-132

Luttinen, A.V. and Furnes, H. 2000. Flood basalts of Vestfjella: Jurassic magmatism across an Archaean–Proterozoic lithospheric boundary in Dronning Maud Land, Antarctica. *Journal of Petrology*, **41**, 1271–1305, https://doi.org/10.1093/petrology/41.8.1271

McDonough, W.F. and Sun, S.-S. 1995. The composition of the Earth. *Chemical Geology*, **120**, 223–254, https://doi.org/10.1016/0009-2541(94)00140-4

Molzahn, M., Reisberg, L. and Wörner, G. 1996. Os, Sr, Nd, Pb, O isotope and trace element data from the Ferrar flood basalts, Antarctica: evidence for an enriched subcontinental lithospheric source. *Earth and Planetary Science Letters*, **144**, 529–546, https://doi.org/10.1016/S0012-821X(96)00178-1

Nakamura, N. 1974. Determination of REE, Ba, Fe, Mg, Na and K in carbonaceous and ordinary chondrites. *Geochimica et Cosmochimica Acta*, **38**, 757–773, https://doi.org/10.1016/0016-7037(74)90149-5

Pankhurst, R.J. and Rapela, C.R. 1995. Production of Jurassic rhyolites by anatexis of the lower crust of Patagonia. *Earth and Planetary Science Letters*, **134**, 23–36, https://doi.org/10.1016/0012-821X(95)00103-J

Pankhurst, R.J., Leat, P.T., Sruoga, P., Rapela, C.W., Márquez, M., Storey, B.C. and Riley, T.R. 1998. The Chon-Aike silicic igneous province of Patagonia and related rocks in Antarctica: A silicic large igneous province. *Journal of Volcanology and Geothermal Research*, **81**, 113–136, https://doi.org/10.1016/S0377-0273(97)00070-X

Pankhurst, R.J., Riley, T.R., Fanning, C.M. and Kelley, S.P. 2000. Episodic silicic volcanism in Patagonia and the Antarctic Peninsula: chronology of magmatism associated with break-up of Gondwana. *Journal of Petrology*, **41**, 605–625, https://doi.org/10.1093/petrology/41.5.605

Pearce, J.A. 1982. Trace element characteristics of lavas from destructive plate boundaries. *In*: Thorpe, R.S. (ed.) *Andesites: Orogenic Andesites and Related Rocks*. Wiley, Chichester, UK, 525–548.

Pearce, J.A. and Peate, D.W. 1995. Tectonic implications of the composition of volcanic arc magmas. *Annual Review of Earth and Planetary Sciences*, **23**, 251–285, https://doi.org/10.1146/annurev.ea.23.050195.001343

Petford, N., Lister, J.R. and Kerr, R.C. 1994. The ascent of felsic magmas in dykes. *Lithos*, **32**, 161–168, https://doi.org/10.1016/0024-4937(94)90028-0

Plank, T. and Langmuir, C.H. 1998. The chemical composition of subducting sediment and its consequences for the crust and mantle. *Chemical Geology*, **145**, 325–394, https://doi.org/10.1016/S0009-2541(97)00150-2

Rapp, R.P. and Watson, E.B. 1995. Dehydration melting of metabasalt at 8-32 kbar: implications for continental growth and crust-mantle recycling. *Journal of Petrology*, **36**, 891–932, https://doi.org/10.1093/petrology/36.4.891

Riley, T.R. and Leat, P.T. 1999. Large volume silicic volcanism along the proto-Pacific margin of Gondwana: lithological and stratigraphcial investigations from the Antarctic Peninsula. *Geological Magazine*, **136**, 1–16, https://doi.org/10.1017/S0016756899002265

Riley, T.R. and Leat, P.T. 2021. Palmer Land and Graham Land volcanic groups (Antarctic Peninsula): volcanology. *Geological Society, London, Memoirs*, **55**, https://doi.org/10.1144/M55-2018-36

Riley, T.R., Leat, P.T., Pankhurst, R.J. and Harris, C. 2001. Origins of large volume rhyolitic volcanism in the Antarctic Peninsula and Patagonia by crustal melting. *Journal of Petrology*, **42**, 1043–1065, https://doi.org/10.1093/petrology/42.6.1043

Riley, T.R., Leat, P.T., Kelley, S.P., Millar, I.L. and Thirlwall, M.F. 2003a. Thinning of the Antarctic Peninsula lithosphere through the Mesozoic: evidence from Middle Jurassic basaltic lavas. *Lithos*, **67**, 163–179, https://doi.org/10.1016/S0024-4937(02)00266-9

Riley, T.R., Leat, P.T., Storey, B.C., Parkinson, I.J. and Millar, I.L. 2003b. Ultramafic lamprophyres of the Ferrar large igneous province: evidence for a HIMU mantle component. *Lithos*, **66**, 63–76, https://doi.org/10.1016/S0024-4937(02)00213-X

Riley, T.R., Leat, P.T., Curtis, M.L., Millar, I.L. and Fazel, A. 2005. Early–Middle Jurassic dolerite dykes from western Dronning Maud Land (Antarctica): Identifying mantle sources in the Karoo large igneous province. *Journal of Petrology*, **46**, 1489–1524, https://doi.org/10.1093/petrology/egi023

Riley, T.R., Curtis, M.L., Leat, P.T., Watkeys, M.K., Duncan, R.A., Millar, I.L. and Owens, W.H. 2006. Overlap of Karoo and Ferrar magma types in the KwaZulu–Natal region of South Africa. *Journal of Petrology*, **47**, 541–566, https://doi.org/10.1093/petrology/egi085

Riley, T.R., Flowerdew, M.J., Hunter, M.A. and Whitehouse, M.J. 2010. Middle Jurassic rhyolite volcanism of eastern Graham Land, Antarctic Peninsula: age correlations and stratigraphic relationships. *Geological Magazine*, **147**, 581–595, https://doi.org/10.1017/S0016756809990720

Riley, T.R., Flowerdew, M.J. and Whitehouse, M.J. 2012. U–Pb ion-microprobe zircon geochronology from the basement inliers of eastern Graham Land, Antarctic Peninsula. *Journal of the Geological Society, London*, **169**, 381–393, https://doi.org/10.1144/0016-76492011-142

Riley, T.R., Curtis, M.L., Flowerdew, M.J. and Whitehouse, M.J. 2016. Evolution of the Antarctic Peninsula lithosphere: evidence from Mesozoic mafic rocks. *Lithos*, **244**, 59–73 https://doi.org/10.1016/j.lithos.2015.11.037

Riley, T.R., Flowerdew, M.J., Pankhurst, R.J., Curtis, M.L., Millar, I.L., Fanning, C.M. and Whitehouse, M.J. 2017. Early Jurassic subduction-related magmatism on the Antarctic Peninsula and potential correlation with the Subcordilleran plutonic belt of Patagonia. *Journal of the Geological Society, London*, **174**, 365–376, https://doi.org/10.1144/jgs2016-053

Riley, T.R., Burton-Johnson, A., Flowerdew, M.J. and Whitehouse, M.J. 2018. Episodicity within a mid-Cretaceous magmatic flare-up in West Antarctica: U–Pb ages of the Lassiter Coast intrusive suite, Antarctic Peninsula and correlations along the Gondwana margin. *Geological Society of America Bulletin*, **130**, 1177–1196, https://doi.org/10.1130/B31800.1

Rowley, P.D., Schimdt, D.L. and Williams, P.L. 1982. Mount Poster Formation, southern Antarctic Peninsula and eastern Ellsworth Land. *Antarctic Journal of the United States*, **17**, 38–39.

Rushmer, T. 1991. Partial melting of two amphibolites: contrasting experimental results under fluid-absent conditions. *Contributions to Mineralogy and Petrology*, **107**, 41–59, https://doi.org/10.1007/BF00311184

Rushmer, T. 1993. Experimental high-pressure granulites: Some applications to natural mafic xenolith suites and Archean granulite terranes. *Geology*, **21**, 411–414, https://doi.org/10.1130/0091-7613(1993)021<0411:EHPGSA>2.3.CO;2

Scarrow, J.H., Leat, P.T., Wareham, C.D. and Millar, I.L. 1998. Geochemistry of mafic dykes in the Antarctic Peninsula continental-margin batholith: a record of arc evolution. *Contributions to*

Mineralogy and Petrology, **131**, 289–305, https://doi.org/10.1007/s004100050394

Smellie, J.L. 2020. Antarctic Peninsula – geology and dynamic development. *In*: Kleinschmidt, G. (ed.) *Geology of the Antarctic Continent*. Gebrüder Borntraeger, Stuttgart, Germany, 18–86.

Stolper, E. and Newman, S. 1992. The role of water in the petrogenesis of Mariana trough magmas. *Earth and Planetary Science Letters*, **121**, 293–325, https://doi.org/10.1016/0012-821X(94)90074-4

Storey, B.C., Wever, H.E., Rowley, P.D. and Ford, A.B. 1987. Report on Antarctic fieldwork: the geology of the central Black Coast, eastern Palmer Land. *British Antarctic Survey Bulletin*, **77**, 145–155.

Storey, B.C., Vaughan, A.P.M. and Riley, T.R. 2013. The links between large igneous provinces, continental breakup and environmental change: evidence reviewed from Antarctica. *Earth and Environmental Science Transactions of the Royal Society of Edinburgh*, **104**, 1–14, https://doi.org/10.1017/S175569101300011X

Suárez, M. 1976. Plate tectonic model for southern Antarctic Peninsula and its relation to southern Andes. *Geology*, **4**, 211–214, https://doi.org/10.1130/0091-7613(1976)4<211:PMFSAP>2.0.CO;2

Sun, S.-s. and McDonough, W.F. 1989. Chemical and isotopic systematics of oceanic basalts: implications for mantle composition and processes. *Geological Society, London, Special Publications*, **42**, 313–345, https://doi.org/10.1144/GSL.SP.1989.042.01.19

Svensen, H., Corfu, F., Polteau, S., Hammer, Ø. and Planke, S. 2012. Rapid magma emplacement in the Karoo Large Igneous Province. *Earth and Planetary Science Letters*, **325–326**, 1–9, https://doi.org/10.1016/j.epsl.2012.01.015

Tatsumi, Y. and Eggins, S. 1995. *Subduction Zone Magmatism*. Blackwell Science, Oxford, UK.

Thompson, R.N., Morrison, M.A., Hendry, G.L. and Parry, S.J. 1984. An assessment of the relative roles of crust and mantle in magma genesis: an elemental approach. *Philosophical Transactions of the Royal Society of London A: Mathematical, Physical and Engineering Sciences*, **310**, 549–590.

Vaughan, A.P.M. and Storey, B.C. 2000. The eastern Palmer Land shear zone: a new terrane accretion model for the Mesozoic development of the Antarctic Peninsula. *Journal of the Geological Society, London*, **157**, 1243–1256, https://doi.org/10.1144/jgs.157.6.1243

Vaughan, A.P.M., Pankhurst, R.J. and Fanning, C.M. 2002. A mid-Cretaceous age for the Palmer Land event: implications for terrane accretion timing and Gondwana palaeolatitudes. *Journal of the Geological Society, London*, **159**, 113–116, https://doi.org/10.1144/0016-764901-090

Vaughan, A.P.M., Leat, P.T., Dean, A.A. and Millar, I.L. 2012. Crustal thickening along the West Antarctic Gondwana margin during mid-Cretaceous deformation of the Triassic intra-oceanic Dyer Arc. *Lithos*, **142–143**, 130–147, https://doi.org/10.1016/j.lithos.2012.03.008

Weinberg, R.F. 1999. Mesoscale pervasive felsic magma migration: alternatives to dyking. *Lithos*, **46**, 393–410, https://doi.org/10.1016/S0024-4937(98)00075-9

Wever, H.E. and Storey, B.C. 1992. Bimodal magmatism in northeast Palmer Land, Antarctic Peninsula: Geochemical evidence for a Jurassic ensialic back-arc basin. *Tectonophysics*, **205**, 239–259, https://doi.org/10.1016/0040-1951(92)90429-A

Wever, H.E., Millar, I.L. and Pankhurst, R.J. 1994. Geochronology and radiogenic isotope geology of Mesozoic rocks from eastern Palmer Land, Antarctic Peninsula: crustal anatexis in arc-related granitoid genesis. *Journal of South American Earth Sciences*, **7**, 69–83, https://doi.org/10.1016/0895-9811(94)90035-3

Chapter 2.3

Dronning Maud Land Jurassic volcanism: volcanology and petrology

Arto V. Luttinen

Finnish Museum of Natural History, University of Helsinki, PO Box 44, Jyrängöntie 2, 00014 Helsinki, Finland

0000-0002-3129-0392

arto.luttinen@helsinki.fi

Abstract: The Jurassic igneous rocks of Dronning Maud Land represent Karoo flood basalt magmatism in Antarctica. Fifty years of research has documented systematic differences between magmas associated with the Karoo rift-zone (Vestfjella and Ahlmannryggen) and the rift-shoulder (Sembberget, Kirwanveggen) settings. The 189–182 Ma rift-zone tholeiites were chemically diverse and mainly formed compound-braided flow fields which record several magnetic polarity reversals. In contrast, the c. 181 Ma rift-shoulder tholeiites were chemically uniform and formed thick tabular sheet lavas within a single normal polarity period. The volcanic architecture records a long initial phase of slow eruptions from shield volcanoes in the initial rift and a brief phase of voluminous fissure eruptions flooding the rift shoulder. All of the major magma types in the rift-zone and rift-shoulder settings belong to a Nb-depleted category of Karoo flood basalts and were mainly derived from depleted convective upper mantle by magmatic differentiation. Pyroxenite-rich mantle components may have been significant sources for the most enriched magma types. Geochemical fingerprints of recycled crustal material imply that the Nb-depleted Karoo tholeiites may have been derived from mildly subduction-modified parts of the same overall upper-mantle reservoir which has been associated with the Ferrar tholeiites.

Supplementary material: Geochronological and geochemical data are available at https://doi.org/10.6084/m9.figshare.c.5196613

Jurassic mafic volcanic and intrusive rocks are widespread in southern Africa and across the Transantarctic Mountains, and have been designated as the Karoo and Ferrar continental flood basalt (CFB) provinces, respectively. The formation of both provinces at c. 182 ± 2 Ma (e.g. Encarnación et al. 1996; Jourdan et al. 2007b, 2008; Svensen et al. 2012; Burgess et al. 2015) and the occurrences of contemporaneous flood basalt lavas and intrusions in New Zealand, Australia, Falkland, as well as in Coats Land and western Dronning Maud Land (DML), Antarctica, indicate a great magnitude of igneous activity and a more or less continuous volcanic plateau along the palaeo-Pacific margin of Gondwana before the break-up of the supercontinent (Rex 1967; Hergt et al. 1991; Brewer et al. 1992; Mortimer et al. 1995; Elburg and Soesoo 1999; Hole et al. 2015). Such large igneous provinces (LIPs) constitute perplexing key events of Earth history (e.g. Bryan et al. 2010). Characterizing the chemical and physical variability and emplacement mechanisms of magmas and the timescales of the processes involved is critical for understanding the origin and evolution of LIPs and their environmental impacts.

Geochemical mapping has been utilized to define the compositional characteristics of the Ferrar and Karoo CFBs. Generalizing, the widespread igneous rock types are mainly composed of broadly similar tholeiitic basalt and basaltic andesite (Furnes et al. 1987; Cox 1988; Harris et al. 1990; Elliot et al. 1999) and collectively comprise the Gondwana LIP (Elliot and Fleming 2000). The diversity of isotopic and trace element compositions are indicative of variable magma sources, differentiation histories and plumbing systems across the vast volcanic province, however (e.g. Marsh et al. 1997; Jourdan et al. 2007a, Galerne et al. 2008; Neumann et al. 2011). For instance, the prevalence of different silica contents and initial strontium isotopic ratios in the Karoo and Ferrar tholeiites, first discovered by Faure and Elliot (1971), has been widely interpreted to stem from different principal magma sources in the mantle (e.g. Hergt 2000; Riley et al. 2006). Accordingly, flood basalts and associated intrusions in western DML that were initially included in the Ferrar province were subsequently recognized as Karoo CFBs based on their Sr isotopic composition (Faure et al. 1979). Furthermore, the dykes in Kwa Zulu, southeastern Africa (Riley et al. 2006), the Falkland Islands (Hole et al. 2015), and the Theron Mountains, Coats Land (Leat et al. 2006), and some lavas in Lesotho (Elliot and Fleming 2000) arguably record intercalation of the Karoo and Ferrar province magma types (Fig. 1).

Detailed geochemical research has shown that while the Ferrar magmas were in many respects remarkably uniform low-Ti tholeiites (Elliot et al. 1999; Elliot and Fleming 2018), the Karoo magmas were notably variable low-Ti and high-Ti tholeiites (e.g. Duncan et al. 1984). Advances in geochemical mapping of the Karoo province has led to a plethora of geochemical provincial scenarios and related terminology as the geographical distribution of low-Ti and high-Ti tholeiites has been defined in increasing detail and as the discovery of transitional-Ti tholeiites has led to modified classification schemes (e.g. Erlank et al. 1988; Hergt et al. 1989; Sweeney and Watkeys 1990; Riley et al. 2005; Jourdan et al. 2007a; Luttinen et al. 2010). The presently available data reveal that the Karoo CFBs comprise two contrasting subprovinces: a structurally complex and geochemically very heterogeneous assemblage of low-Ti and high-Ti tholeiites related to a large triple-rift structure; and a contiguous plateau of relatively uniform low-Ti tholeiitic lavas and sills related to major sedimentary basins (Luttinen et al. 2015) (Fig. 1).

Luttinen (2018) recently pointed out that the various low-Ti and high-Ti tholeiites of the Karoo triple rift are characterized by Nb depletion relative to Zr and Y, whereas the monotonous low-Ti tholeiites of the 'plateau association' do not show such a Nb depletion. The abundance of Nb relative to Zr and Y can be readily examined using a logarithmic plot of Zr/Y v. Nb/Y and is quantified by the ΔNb parameter (ΔNb = 1.73 + log(Nb/Y) − 1.94 log(Zr/Y): Fitton et al. 1997). In the classification of Luttinen (2018), the relatively Nb-depleted low-Ti and high-Ti tholeiites associated with the Karoo triple rift constitute South Karoo, and the Nb-undepleted low-Ti tholeiites of the Karoo, Kalahari and Zambezi basins make up North Karoo (Figs 1 & 2). It is important to notice that the new classification into North Karoo and South Karoo illustrated in Figure 1 refers to the Jurassic plate configuration, and is based on the distribution of low-ΔNb and high-ΔNb tholeiites. It is not based on the

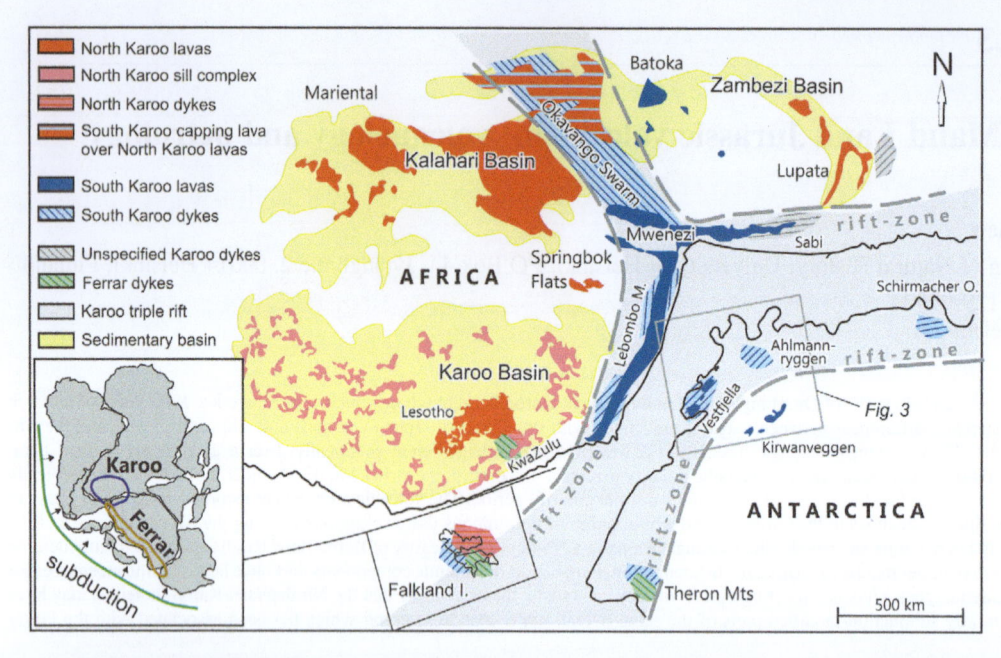

Fig. 1. Distribution of geochemically distinctive North Karoo (red), South Karoo (blue), and Ferrar (green) flood basalts and related intrusive rocks in a schematic Jurassic reconstruction between Africa and Antarctica at c. 180 Ma. The seaward-dipping lava successions (Sabi, Lebombo and Vestfjella) and radiating dyke swarms (Okavango, Sabi, Ahlmannryggen, Vestfjella and Lebombo) define a triple-rift structure (Burke and Dewey 1973). The Karoo Basin CFBs include the Lesotho lava succession and a large dolerite sill complex (e.g. Marsh et al. 1997; Neumann et al. 2011). The division into North Karoo and South Karoo is after Luttinen (2018), and the reconstruction is modified after Lawver et al. (1998). The inset shows the extent of Karoo and Ferrar CFBs (Gondwana LIP) and an active subduction zone along the Gondwana margin (Rapela et al. 2005). The area of Figure 3 is indicated.

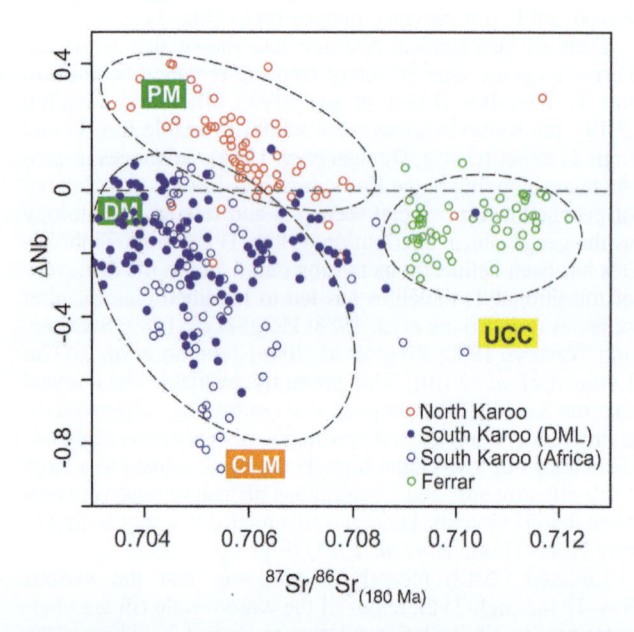

Fig. 2. Geochemical grouping of the Gondwana LIP into three principal clans (i.e. North Karoo (red), South Karoo (Africa, open blue; DML filled blue) and Ferrar (green) subprovinces), based on variations in $^{87}Sr/^{86}Sr$ and ΔNb values ($\Delta Nb = 1.74 + \log(Nb/Y) - 1.92 \log(Zr/Y)$: Fitton et al. 1997). The fields of typical compositions in each subprovince and plausible geochemical major reservoirs are indicated: DM is depleted MORB mantle (Workman and Hart 2005), PM is primitive mantle (Zindler and Hart 1986; McDonough and Sun 1995), CLM is assumed continental lithospheric mantle (Luttinen 2018), UCC is upper continental crust based on Karoo sandstone (Neumann et al. 2011). Geochemical data and references for Karoo are listed in the compilation of Luttinen (2018), and data for Ferrar are from Siders and Elliot (1985), Hergt et al. (1989), Fleming et al. (1992, 1995), Elliot et al. (1995, 1999), Wilhelm and Wörner (1996), Antonini et al. (1999) and Hanemann and Viereck-Götte (2004). Samples from section 82.3 in the Mesa Range (Elliot et al. 1995) have anomalous high Nb contents and were excluded from Ferrar data due to probable analytical error.

conventional low-Ti v. high-Ti classification and does not refer to the original provincial scheme proposed by Cox et al. (1967) for African Karoo.

Overall, the presently available geochemical, geochronological and palaeomagnetic data support the view that the Jurassic Gondwana tholeiites may be best understood using a division into three principal magmatic clans: (1) Ferrar, (2) North Karoo and (3) South Karoo (Table 1). These geographically distinctive subsystems can be geochemically distinguished using strontium isotopic ratios and ΔNb (Fig. 2), and they can be viewed as subprovinces of the greater Gondwana LIP. Importantly, the three subprovinces show affinities to contrasting geochemical reservoirs: depleted mantle and continental lithospheric mantle (South Karoo); primitive mantle (North Karoo); and upper continental crust (Ferrar) (Fig. 2).

This paper presents a synthesis of geological research on flood basalt magmatism in DML over the past five decades. Geochemically, the Jurassic extrusions and intrusions of DML belong to South Karoo (Luttinen 2018) (Figs 1 & 2); and tectonically, the tholeiites of Vestfjella and Ahlmannryggen are associated with the Karoo triple rift, whereas those of Björnnutane, Sembberget and Kirwanveggen can be regarded to represent a rift-shoulder setting (Luttinen et al. 2015) (Figs 1 & 3). Geochemical data from DML manifest profound compositional variability in the Jurassic magmas but, on the other hand, the data also constrain the range of primitive magma compositions and mantle sources involved. Overall, flow-by-flow mapping and sampling of lavas, and comprehensive mapping and sampling of intrusions, record contrasting styles of emplacement and melting conditions for rift and rift-shoulder volcanism and arguably provide the most detailed volcanological and petrological portrait of the southern arm of the Karoo rift zone.

Geological setting

A considerable period of Earth history is recorded in the scattered bedrock exposures of western DML. Over large areas

Table 1. *Characteristic features of North Karoo, South Karoo and the Ferrar subprovinces of the Gondwana LIP*

Subprovince	Age* (Ma)	Polarity[†]	$^{87}Sr/^{86}Sr_{(180\ Ma)}$[‡]	$\varepsilon_{Nd_{180\ Ma}}$[‡]	ΔNb[‡]	MgO[‡] (wt%)	TiO$_2$[‡] (wt%)	Zr/Y[‡]
South Karoo	174–189	R–N–R–N–R–N–R	0.703–0.709	–18 to +9	–	2–25	1–5	3–19
North Karoo	182–183	R–N	0.704–0.708	–5 to +4	+	3–8	1–2	3–6
Ferrar	182–183	N	0.709–0.712	–6 to –2	–	2–10	0.5–2	3–5

*Age data for North Karoo are U–Pb zircon and baddeleyite ages from Svensen et al. (2012); data for South Karoo are $^{40}Ar–^{39}Ar$ plagioclase ages from Duncan et al. (1997), Jourdan et al. (2004, 2005, 2008) and Luttinen et al. (2015); U–Pb zircon ages are from Luttinen et al. (2015); and data for Ferrar are U–Pb zircon ages from Burgess et al. (2015).
[†]Magnetic polarity data for Karoo are from Hargraves et al. (1997) and Moulin et al. (2011); and data for Ferrar are from McIntosh et al. (1982). R and N indicate reverse and normal polarity, respectively.
[‡]Geochemical data for Karoo are from the compilation of Luttinen (2018); and data for Ferrar are from Siders and Elliot (1985), Hergt et al. (1989), Fleming et al. (1992, 1995), Elliot et al. (1995, 1999), Wilhelm and Wörner (1996), Antonini et al. (1999) and Hanemann and Viereck-Götte (2004). Note that data for the Dufek intrusion have not been used due to the strong influence of accumulation processes on the geochemical compositions. $\Delta Nb = 1.74 + \log(Nb/Y) - 1.92 \log(Zr/Y)$, as defined by Fitton et al. (1997).

between Ahlmannryggen–H.U. Sverdrupfjella and Mannefallknausane the bedrock mainly consists of Precambrian basement, which ranges in age from Archean to Mesoproterozoic (Fig. 3).

Archean basement is exposed only on a small outcrop at Annandagstoppane (Fig. 3) (Halpern 1970; Barton et al. 1987), where a leucocratic granite has yielded a crystallization age of 3076 Ma (Marschall et al. 2010). Aeromagnetic data indicate that the Archean crustal domain underlies the sedimentary and volcanogenic sequences of the Mesoproterozoic (c. 1080 Ma) Ritscherflya Supergroup (Wolmarans and Kent 1982) as well as mafic 800–1000 Ma Borgmassivet intrusions (Krynauw et al. 1988, 1991; Moyes et al. 1995) in the Ahlmannryggen region, and its southwestern boundary is estimated to be located under the ice sheet close to northern Vestfjella (Corner 1994) (Fig. 3). The Archean basement and the overlying supracrustal rocks comprise the Grunehogna Craton (Krynauw et al. 1991), which is bounded in the south by the Mesoproterozoic Maud Belt (Groenewald et al. 1995) (Fig. 3). Several lines of evidence indicate that, prior to Jurassic Africa–Antarctica rifting, the Grunehogna Craton was juxtaposed to the Zimbabwe–Kaapvaal Craton of southern Africa and these Archean blocks together formed the Proto-Kalahari Craton (Jacobs et al. 2008).

The Maud Belt and the coeval Namaqua–Natal mobile belt of Africa formed the southern–eastern accretionary rim of the Kalahari Craton (e.g. Jacobs et al. 1993, 2008). The Maud Belt is principally the product of high-grade metamorphism, folding and thrusting of supracrustal successions and syn- to post-tectonic plutonism during the accretion of relatively juvenile crust in the period 1100–1040 Ma (Grantham et al. 1988; Arndt et al. 1991; Moyes et al. 1993; Jacobs 2009). Detailed mapping of the Heimefrontfjella area suggests generation of juvenile crust in a volcanic arc and associated back-arc setting at 1180–1100 Ma, followed by metamorphism, folding and syntectonic granitic plutonism during amalgamation of the Rodinia supercontinent (Hoffman 1991; Groenewald et al. 1995) at 1100–1060 Ma, and subsequent late- and post-tectonic bimodal plutonism at 1050–1040 Ma (Jacobs 2009).

As in southern Africa, the Precambrian basement is overlain by Paleozoic sedimentary rocks at many localitites at Heimefrontfjella and Kirwanveggen (Fig. 3). Most of the outcrops of the sedimentary sequence are small outliers capping the much larger exposures of the basement complex with an angular unconformity (e.g. Juckes 1972). In the Vestfjella region, the basement is unexposed and Paleozoic sedimentary rocks are confined to the Fossilryggen Ridge (Hjelle and Winsnes 1972).

The sedimentary rocks and the basement complex are overlain by Jurassic flood lavas, and are cut by associated sills and dykes. The main exposures of flood lavas are at Vestfjella, north Heimefrontfjella and southern Kirwanveggen (Fig. 3).

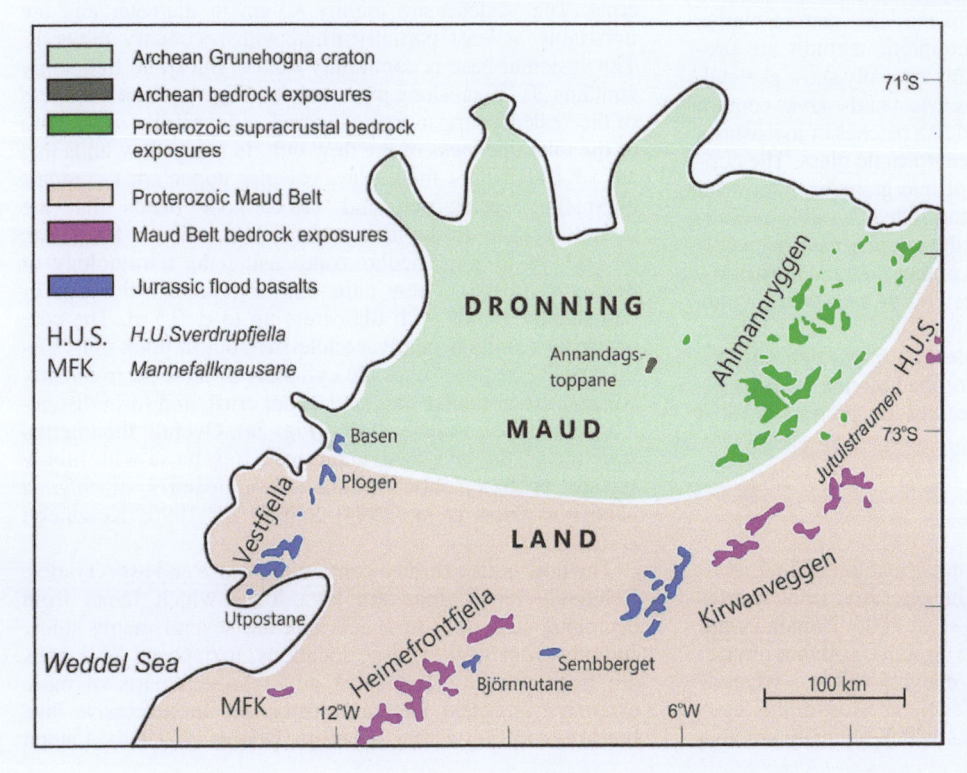

Fig. 3. Distribution of Jurassic flood basalts in western DML. Exposures of Archean and Proterozoic bedrock and the subglacial extent of the Archean Grunehogna Craton and the Proterozoic Maud Belt are indicated.

Intrusive equivalents of the flood basalts are abundant in Vestfjella (Furnes and Mitchell 1978; Luttinen *et al.* 2015), and are quite common in Ahlmannryggen (Harris *et al.* 1991; Riley *et al.* 2005), H.U. Sverdrupfjella (Grantham 1996) and Mannefallkanusane, but rare in Heimefrontfjella (Juckes 1972) and Kirwanveggen (Harris *et al.* 1990; Riley *et al.* 2005). Possible correlates of the intrusions have been reported in the Theron Mountains (Coats Land: Leat *et al.* 2006) and Schirmacher Oasis (central DML: Sushchevskaya *et al.* 2009) (Fig. 1). Moreover, small nepheline–syenite alkaline plutons and correlative dykes at Straumsvola, Tvora and Sistenup nunataks, in the Jutulstraumen area, record silicic Jurassic magmatism in DML (Harris and Grantham 1993; Harris *et al.* 2002; Riley *et al.* 2009). All of the Jurassic flood basalts and most of the intrusions are found within the area of Proterozoic basement, and only the Ahlmannryggen intrusions occur within the Archean craton (Fig. 3).

Petrography

Extrusive rocks

The Jurassic volcanic rocks have rather uniform petrographical features across western DML (Juckes 1972; Faure *et al.* 1979; Furnes *et al.* 1987; Harris *et al.* 1990; Luttinen and Furnes 2000; Luttinen *et al.* 2010): Most of the lavas are porphyritic. Plagioclase is the predominant phenocryst phase and occurs frequently with clinopyroxene or olivine or both. Combined geochemical and petrographical evidence indicates successive crystallization of: (1) olivine and Cr-spinel; (2) plagioclase and olivine; and (3) plagioclase, olivine and clinopyroxene. Phenocrysts are frequently euhedral, but aggregates of subhedral grains are also common and may represent gabbroic autoliths. Typically, olivine is strongly altered to greenish or reddish secondary micaeous material. Plagioclase is rarely unaltered, typically shows alteration to sericite along fractures, and the strongly altered samples record pervasive alteration to saussurite, chlorite, carbonate and/or epidote. In contrast, clinopyroxene is fresh, even in strongly altered samples.

The groundmass consists of plagioclase laths and interstitial clinopyroxene and Fe–Ti oxides. Depending on the relative size of plagioclase and clinopyroxene, the texture varies from ophitic to mesophitic, but subophitic textures are predominant. Opaque Fe–Ti oxide grains typically show skeletal, acicular or dendritic quench textures. Most of the lavas contain interstitial greenish, brownish or reddish patches of mesostasis which represent altered or devitrified volcanic glass. The contents of mesostasis are suggestive of intergranular, intersertal and hyalophitic primary textures, and reflect variable crystallinity of the igneous rocks. Most of the samples have been collected from relatively massive parts of the lavas and intrusions, but many contain at least a few vesicles or dictytaxitic voids filled with secondary minerals. Groundmass olivine has been positively identified in some samples, but it is often difficult to reliably discern totally pseudomorphed olivine from mesostasis or filled vesicles. Apatite needles are present in most samples.

Intrusive rocks

Overall, the intrusive rocks have similar textural and mineralogical characteristics as the lavas (Juckes 1968, 1972; Furnes *et al.* 1982; Grantham 1996; Riley *et al.* 2005; Heinonen and Luttinen 2008; Luttinen *et al.* 2015): picritic (olivine-phyric) dykes and sills are common in many localities, whereas rare picritic lavas are confined to Vestfjella. Most of the intergranular samples are intrusive rocks which typically are also less altered than the volcanic rocks due to higher crystallinity and lower porosity. A few coarse-grained intrusive rock types contain interstitial quartz–feldspathic segregation patches. Two larger intrusions in Vestfjella record the differentiation of phenocryst minerals and the formation of variable gabbroic rock types, most of which display cumulate textures (Vuori and Luttinen 2003; Vuori 2004).

Building blocks of the volcanic edifice

The exposed volcanic and intrusive suites exhibit textural and structural variability indicative of different magma compositions and styles and rates of eruption, and emplacement under a variable crustal stress field. The architecture of the lava successions has not been previously reported in detail. In addition to published papers, the following description utilizes the author's field notes and photographs from Vestfjella, Björnnutane, Sembberget and Kirwanveggen from the period 1989–2018, and field notes provided by Henrik Grind (Vestfjella dykes), Harald Furnes (Vestfjella lavas) and Phil Leat (Sembberget, Kirwanveggen).

Lava flow units, flow fields and feeder fissures

Flow units. Individual cooling units of stalled lava are referred to as flow units. Most of the well-exposed flow units in DML have relatively smooth, billowy or ropy surfaces typical of pāhoehoe (Macdonald 1953) (Fig. 4). Brecciated flow surfaces have been noted in many localities (e.g. Hjelle and Winsnes 1972; Harris *et al.* 1990; Luttinen *et al.* 2010) and some units have brecciated base, but 'a'ā units have not been positively identified. The brecciated material commonly shows characteristic features of pāhoehoe lava, and the lack of clinckery levees and other indications of open channel flows suggests that flow units with brecciated crust and/or base can be regarded as rubbly pāhoehoe (Rowland and Walker 1990; Cashman *et al.* 1999; Bondre *et al.* 2004; Keszthelyi *et al.* 2006).

The flow units typically have a vesicle-rich base and upper crust. The vesicles are mainly <1 cm in diameter and are invariably at least partially filled with secondary minerals. The vesicular base is commonly <0.5 m thick and frequently contains 20–30 cm-long pipe vesicles (Fig. 4c). The thickness of the vesicular upper crust is varied and typically *c.* 30–50% of the total thickness of the flow unit. In many flow units that are several metres thick, the vesicular upper crust contains alternating vesicle-rich and vesicle-poor bands that are broadly parallel to the lava surface. The vesicular bands can be referred to as vesicular zones using the terminology of Self *et al.* (1996). These units may also contain flattened or bell-shaped geoids with diameters up to *c.* 0.5 m. The core of the flow units is nearly vesicle-free, but in thick units distinctive (segregation) vesicle cylinders in the core frequently connect the vesicular base and upper crust, and form distinctive segregation vesicle sheets (Fig. 5a). Overall, the internal structure of the lava flow units, including those with rubbly surface or base, shows characteristic features of *inflated* pāhoehoe (Hon *et al.* 1994; Self *et al.* 1996; Keszthelyi *et al.* 2006).

The flow units exhibit a continuum of size and aspect ratios. Relatively small units are lava lobes which range from decimetre-scale toes to sheets that are several metres thick. Judging from well-exposed locations, juxtaposed flow units are frequently interconnected and represent parts of more extensive branched tongues formed due to successive lava breakouts during active inflation (Hon *et al.* 1994). Lateral

Fig. 4. Structures of Jurassic lavas and intrusions in DML. (a) Hummocky surface of pāhoehoe, Basen (unit #4 in Fig. 8: 73.0402° S, 13.3893° W). (b) Ropy pāhoehoe at Basen (73.0426° S, 13.4069° W). (c) Cross-section of small inflated pāhoehoe flow units belonging to compound-braided flow facies, Basen (73.0224° S, 13.3611° W). (d) Ropy structure within a megavesicle underneath the upper margin of Schievestulen Sill, Heimefrontfjella (74.3310° S, 9.7643° W). The hammer in (a) is 70 cm long; the scale bar in (b) and (d) is 15 cm long; and the scale bar in (c) is 50 cm long. Photographs: A.V. Luttinen.

correlation of the flow units is generally difficult. On cliff faces, the width of the units range from a few metres up to several hundred metres, depending on the cross-section (Fig. 6a). The largest flow units are laterally extensive tabular sheets which are up to several tens of metres or even >100 m thick and up to tens of kilometres wide (Fig. 6b).

Flow fields. Lava flow units that have been emplaced during the same overall eruptive cycle form lava flow fields. They are variably identified based on: (1) clastic interbeds or otherwise well-defined contacts; (2) predominant type of flow units; (3) characteristic colour; and (4) geochemical compositions. Variable vesicularity, jointing, crystallinity, grain size, mineral composition and the overall lava-stacking pattern have a controlling influence on the degree of alteration, the secondary mineral assemblage (e.g. Bevins *et al.* 1991), and the colour variations. Flow fields that represent or have a field appearance of a single tabular lava sheet are *simple* flow fields, whereas *compound* flow fields are made of several intimately overlapping units (Walker 1971). Compound flow fields can be further classified based on the characteristic aspect ratios of the flow units. Apart from a few examples of *tabular-type* compound flow fields made of stacked tabular lava sheets, the compound flow fields are made of intertwining pāhoehoe lobes and represent the *compound-braided* type of Jerram (2002) (Fig. 6).

Intrusive subvolcanic feeders. Most of the intrusive rock types are subvertical dykes. The width of the dykes varies from *c.* 0.2 up to 500 m, but the great majority are 1–3 m wide (Furnes *et al.* 1982; Grind *et al.* 1991; Harris *et al.* 1991; Grantham 1996; Riley *et al.* 2005; Curtis *et al.* 2008; Heinonen and Luttinen 2008; Luttinen *et al.* 2015). Subhorizontal sills 0.2–20 m in thickness have been observed in

Fig. 5. Lava flow contacts at (a) Basen, Vestfjella (73.0385° S, 13.4572° W) and (b) Tunga (Urfjell), Kirwanveggen (73.9192° S, 5.3200° W). In (a) the lava overlies a *c.* 5 m-thick sandstone (the subhorizontally banded top is visible) and shows typical features of inflated pāhoehoe, including inclined pipe vesicles in the lower crust, vesicle cylinders and horizontal vesicle sheets in the massive lava core (unit #2 in Fig. 8). In (b) the lower unit has a flow-top breccia typical of rubbly pāhoehoe with sandstone filling the lava fragments. The upper unit is notably poor in vesicles. The scale bar is 50 cm long. Photographs: A.V Luttinen.

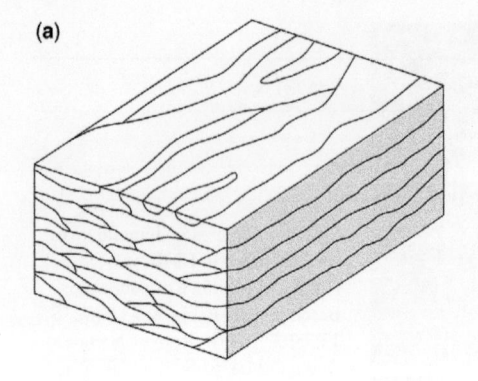

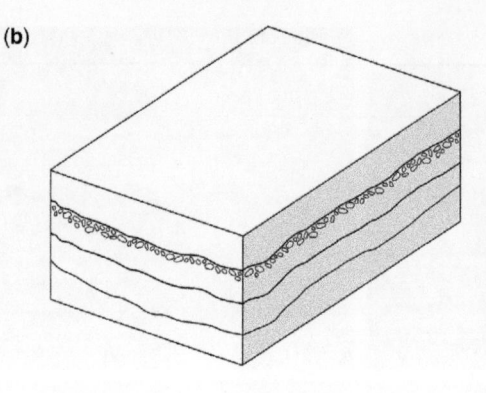

Fig. 6. Schematic representations of two principal lava-flow stacking patterns typical of (**a**) compound-braided facies and (**b**) tabular-type facies of Jerram (2002). Some tabular-type lavas have a brecciated crust and represent rubbly pāhoehoe. Modified after Jerram (2002).

Vestfjella, Mannefallknausane and Heimefrontfjella (Juckes 1968; Hjelle and Winsnes 1972; Luttinen et al. 2015). The dykes and sills exhibit variably developed columnar jointing, many have well-defined chilled margins and some of them contain vesicles indicative of emplacement relatively close to Earth's surface. Judging from geochemical compositions, most of the Vestfjella dykes represent subvolcanic feeders of lava flows that have been removed by erosion (see the subsection 'Intrusive magma types' in the 'Geochemical characteristics' section later in this chapter). The predominant dyke strike directions are broadly parallel to the coast (NNE–SSW in Vestfjella, and ENE–WSW in Ahlmannryggen and H.U. Sverdrupfjella), but SSE–NNW-trending dykes are also abundant, especially in western parts of H.U. Sverdrupfjella. The overall radiating pattern of dyke intrusions in DML and the conjugate African margin is consistent with a major igneous centre in the Karoo triple junction, and studies of anisotropic magnetic susceptibility indicate lateral transport of magmas away from this region (Curtis et al. 2008; Hastie et al. 2014). The roles of mantle-plume activity and pre-existing lithospheric structures in the genesis of the radiating dyke swarms are contested (Jourdan et al. 2004; Curtis et al. 2008; Hastie et al. 2014).

Clastic intercalations

Clastic interbeds are relatively common within the lava successions of DML, but the reports are relatively superficial (Hjelle and Winsnes 1972; Harris et al. 1990; Luttinen and Furnes 2000). Some of the interbeds have been regarded to be pyroclastic and volcaniclastic (Furnes and Mitchell 1978; Peters 1989), but they have not been described in sufficient detail for reliable identification. A report of thick mafic crystal tuffs in Plogen (Peters 1989) turned out to be a misinterpretation. Detailed mapping of the 400 m-thick succession of lavas at Basen revealed four horizons of clastic intercalations (Luttinen 1994). In most places, these are made of fine-grained (clay-size) material and can be referred to as red boles. Some of the fine-grained intercalations are green or yellowish-green or grey-green. The two lower horizons include poorly sorted, but mineralogically mature, sandstone deposits interpreted as redeposited Permian sedimentary strata (Luttinen 1994). The lowermost sandstone attains a maximum thickness of c. 5 m in the SW corner of the nunatak. Similar sandstone interbeds are visible on the NE cliff face of the nearby Plogen nunatak. Occurrences of red and greenish yellow clastic intercalations have been observed at every major nunatak in Vestfjella, but they have not been mapped or sampled in detail. The clastic interbeds in southern Vestfjella are invariably thin (<1 m ?) and appear to be laterally discontinuous lenses.

Red bole intercalations are common in flood basalt successions worldwide, and are considered to represent autochtonous or allochtonous weathering deposits (Shrivastava et al. 2012), or pyroclastic material (Bell et al. 1996). The red boles of DML have not been studied in sufficient detail to draw conclusions concerning possible involvement of pyroclastic material. The report of Juckes (1972) and a visual survey of thin sections representing clastic interbeds at Basen ($n = 15$), Plogen ($n = 1$) and Sembberget ($n = 1$) suggest that most red boles in DML include a significant allochthonous component revealed by angular quartz and microcline clasts. A single geochemical analysis supports a predominantly allochthonous origin (SiO_2 c. 71 wt%; K_2O/Na_2O c. 10; MgO/CaO and $Al_2O_3/(CaO + K_2O + Na_2O)$ c. 1.3: Luttinen 1994). The samples typically contain glassy fragments which probably record entrainment of melt from the overlying lava into unconsolidated sedimentary material. Generalizing, the clastic interbeds in DML appear to indicate pauses in eruptive activity, but the length of dormancy has not been defined.

Volcanic architecture: lava successions and intrusive suites

This section summarizes the physical features and ages of the Jurassic lavas and associated intrusions in western DML. Evidence of excess Ar, Ar-loss, recoil, the lack of plateau ages or insufficient documentation render many of the published dating results doubtful (Zhang et al. 2003; Luttinen et al. 2015). The most reliable U–Pb and $^{40}Ar-^{39}Ar$ mineral plateau ages are listed in Table 2. The $^{40}Ar-^{39}Ar$ ages are reported relative to the FCT (Fish Canyon Tuff) monitor age (28.29 Ma) and the ^{40}K decay constants of Renne et al. (2011).

Vestfjella extrusive–intrusive suite

Volcanic stratigraphy. Vestfjella is a c. 120 km-long range of toothed ridges, clusters of pointed nunataks and isolated small exposures adjacent to the continental margin of East Antarctica (Fig. 3). The Vestfjella range is almost entirely made of Jurassic lavas and intrusions, and minor (<5 m thick) clastic interbeds (Hjelle and Winsnes 1972; Luttinen and Furnes 2000). Fossilryggen represents a unique exposure of Permian sedimentary bedrock that may underlie the lavas in Vestfjella, but the base of the volcanic succession is not exposed (Hjelle and Winsnes 1972; McLoughlin et al. 2005). The lowermost exposures are at c. 200 m elevation (above sea level (asl)). The relatively smooth profile of the slopes and ridges, and the flatness of cliff faces reflect the fact that the Vestfjella succession is mainly composed of relatively thin, superimposed pāhoehoe flow lobes. The thickness of individual flow units is mainly <5 m and the units appear to thin out over distances of <100 m in vertical cross-sections. The thickness of the lava pile exceeds 700 m at Plogen, north Vestfjella (Fig. 7a). The emplacement of the lavas has not been reliably dated due to alteration (Zhang et al. 2003), but the southern exposures are at least 182 myr and possibly up to 189 myr old based on U–Pb zircon and $^{40}Ar-^{39}Ar$ plagioclase dating of cross-cutting intrusions (Luttinen et al. 2015) (Table 2).

Table 2. *Representative age data for Jurassic mafic rocks in Dronning Maud Land*

Location			Rock type*	Method[†]	Age (±2σ)[‡]	Reference
Vestfjella						
Muren West	73° 43.8′ S	15° 06.3′ W	VF1 gabbro	^{206}Pb–^{238}U zrc. ID-TIMS	182 ± 1 Ma	Luttinen *et al.* (2015)
Muren East	73° 43.5′ S	15° 03.0′ W	VF1 gabbro	^{206}Pb–^{238}U zrc. + bad. ID-TIMS	182 ± 1 Ma	Luttinen *et al.* (2015)
Pagodromen	73° 46.3′ S	14° 57.5′ W	VF1 dolerite	^{40}Ar–^{39}Ar plg. plateau	189 ± 2 Ma	Luttinen *et al.* (2015)
Utpostane	73° 55.2′ S	15° 35.6′ W	VF3 dolerite	^{40}Ar–^{39}Ar plg. mini-plateau	187 ± 3 Ma	Luttinen *et al.* (2015)
Kjakebeinet	73° 47.5′ S	14° 50.5′ W	VFD dolerite	^{40}Ar–^{39}Ar plg. plateau	186 ± 2 Ma	Luttinen *et al.* (2015)
Utpostane	73° 53.3′ S	15° 40.9′ W	KW gabbro	^{40}Ar–^{39}Ar plg. plateau	181 ± 1 Ma	Zhang *et al.* (2003)
Kjakebeinet	73° 46.9′ S	14° 52.6′ W	Lamproite	^{40}Ar–^{39}Ar phl. plateau	162 ± 2 Ma	Luttinen *et al.* (2002)
Kirwanveggen						
Lagfjella	74° 05.9′ S	6° 15.3′ W	ARE dolerite	^{40}Ar–^{39}Ar plg. plateau	181 ± 1 Ma	Zhang *et al.* (2003)
Sembberget						
Nunatak	74° 29.4′ S	8° 11.6′ W	KW basalt	^{40}Ar–^{39}Ar plg. plateau	181 ± 1 Ma	This study[§]

*VF1, VF3, VFD, KW and ARE refer to chemically different magma types (see the text).
[†]Mineral abbreviations: zrc., zircon; bad., baddeleyite; plg., plagioclase; phl., phlogopite.
[‡]^{40}Ar/^{39}Ar ages are reported relative to the FCT monitor (28.29 Ma) and the ^{40}K decay constants of Renne *et al.* (2011).
[§]Supplementary Table S2.

The Vestfjella CFB succession is predominantly made of overlapping thin pāhoehoe lobes which show characteristic features of inflated pāhoehoe and comprise compound-braided flow fields. The succession is particularly well exposed on large cliff faces at Basen and Plogen, where different flow fields can be mainly distinguished based on their characteristic colouring (Fig. 7a). Colour variations are quite hard to depict when walking along accessible ridges, and the limited lateral dimension of the flow units also complicates detailed correlation between geochemical samples and the succession of flow fields observed on the cliff faces. Clastic intercalations sporadically sandwiched between flow fields and geochemical variations also help to define several flow fields.

In comparison, laterally extensive, simple sheet lavas are quite rare. The thickness of the sheet lavas range from *c.* 5 to 20 m and they can be followed laterally for up to *c.* 5 km in some locations. Some of the simple lavas have a rubbly flow-top breccia (Fig. 5b). The lava fragments typically have oxidized red surfaces, and the interstitial spaces between scoriaceous blocks are variably filled with clastic material. Some of the relatively thick sheet lavas have poorly developed vertical jointing in the massive lava core, whereas the vesicular

Fig. 7. Occurrences of Jurassic flood basalt lavas and intrusions in (**a**) Vestfjella (Plogen), (**b**) Björnnutane, (**c**) Kirwanveggen (Lagfjella) and (**d**) Heimefrontfjella (Schievestulen) (for localities see Fig. 3). The height of the cliff in (a) is *c.* 400 m. It shows numerous flow fields with variable structures and colours, including possible tumuli. A cross-cutting inclined dyke is *c.* 20 m thick. The sandstone in (b) is *c.* 1 m thick. The lower part of the Björnnutane lava succession is made of compound-braided flow fields, and simple tabular-type units can be seen in the upper part of the photograph. The thick tabular lava sheets in (c) produce a classic trap topography at Kirwanveggen. The 5 m-thick Schievestulen Sill (d) is underlain by *c.* 20 m of Paleozoic sedimentary strata. Photographs: A.V. Luttinen.

upper crust exhibits subhorizontal jointing, which is also seen in small pāhoehoe lobes. Only a few lava flow units have been tentatively classified as 'a'ā lava (Hjelle and Winsnes 1972). Pipe vesicles, round vesicles, geoids and other voids are filled mainly with secondary quartz, calcite, chlorite, prehnite and epidote. Secondary microcline has yielded alteration ^{40}Ar–^{39}Ar ages of 152 and 140 Ma (Zhang *et al.* 2003). In Vestfjella, subsolidus alteration of the lavas and intrusions renders dating of the magmatic events problematic. Notably, variable Cl/K and Ca/K spectra indicate that ^{40}Ar–^{39}Ar plagioclase plateau ages of *c.* 173–150 Ma are more likely to date secondary resetting of the K–Ar isotopic systematics than magma emplacement, even if plagioclase appears to be unaltered in thin sections (Zhang *et al.* 2003; Luttinen *et al.* 2015). Lamproite dykes with an ^{40}Ar–^{39}Ar phlogopite age of 162 Ma record the youngest magmatic event in Vestfjella (Luttinen *et al.* 2002; Table 2) during the purported onset of seafloor spreading between Africa and Antarctica (Jokat *et al.* 2003).

Two nunataks have been mapped in sufficient detail to determine the number of flow fields with reasonable confidence. Colour variations, clastic interbeds and geochemical variations indicate that the >700 m-thick Plogen succession records at least 25 flow fields, most of which belong to the compound-braided type (Fig. 8). The compound-braided lavas generally lack indications of significant breaks in effusive activity and individual compound flow fields may be up to 200 m thick, although most are less than 50 m (Fig. 7a). The nearby Basen succession is estimated to be made of nine compound-braided and three simple flow fields, and up to 5 m-thick clastic interbeds (Luttinen and Furnes 2000) (Fig. 8). A distinctive plagioclase porphyritic lava at nunatak Basen (flow #3) has been correlated with a tabular-type compound flow field of two similarly plagioclase porphyritic overlapping sheets at nunatak Plogen *c.* 25 km away, based on exceptional high-Ti compositions (lava #6) (Luttinen and Furnes 2000). The lava succession dips gently (3°–10°) towards the WSW, and is cut by abundant dykes and faults. On a regional scale, the tilting and faulting of the lavas lead to uncertainty with respect to the total stratigraphic thickness, which has been estimated to be 1000–2500 m (Hjelle and Winsnes 1972; Furnes *et al.* 1987; Peters 1989; Luttinen and Furnes 2000). Flow direction measurements from inclined pipe vesicles (Fig. 5a) show notable variability (Hjelle and Winsnes 1972; Furnes and Mitchell 1978; Luttinen and Furnes 2000), which documents that the tilting is post-volcanic. Chemically different lavas show broadly consistent flow directions (Fig. 8; see also the subsection 'Extrusive magma types' in the 'Geochemical characteristics' section later in this chapter).

Dolerite dykes and sills. The dykes and less abundant sills are here referred to as dolerites regardless of textural and mineralogical variability. Most of the dolerites are subvertical and have a broadly coast-parallel NNE–SSW strike (Furnes *et al.* 1982; Peters 1989; Grind *et al.* 1991; Luttinen *et al.* 2015). Subordinate north–south, NW–SSE and east–west strike directions have been interpreted to be successively younger (Grind *et al.* 1991), but the age relationships have not been confirmed by detailed studies. Plagioclase ^{40}Ar–^{39}Ar dating indicates emplacement of the cross-cutting dolerites in southern Vestfjella during the interval 189–186 Ma (Luttinen *et al.* 2015) (Table 2). The dykes and sills are relatively narrow, typically 1–3 m wide, although some attain widths of several tens of metres (Furnes and Mitchell 1978; Furnes *et al.* 1982; Grind *et al.* 1991; Luttinen *et al.* 2015) (Fig. 4a). Judging from the cross-cutting relationships, some of the dykes are multiple intrusions generated by successive magma injections. The succession shows normal faulting from the centimetre scale to several tens of metres, but the vertical movements appear to have been typically less than 10 m. The main fault direction coincides with the predominant NNE–SSW dyke strike direction and some of the dykes have slickensides at the dyke–wall rock contact, but also within multiple dykes. Some of the dykes record shearing into lensoid bodies (Grind *et al.* 1991).

Gabbroic intrusions. In southern Vestfjella, the lavas are cut by gabbroic intrusive bodies. Two bodies of gabbroic rock types have been interpreted to represent *c.* 500 m-wide dyke-like intrusions at Muren, southern Vestfjella (Luttinen and Vuori 2006). The dykes have indistinguishable U–Pb zircon ages of 182 Ma (Luttinen *et al.* 2015) (Table 2). The East Muren dyke is made of relatively homogeneous gabbro, whereas the West Muren dyke is differentiated and is dominated by olivine cumulates which are overlain by leucogabbros and a capping felsic unit (Vuori and Luttinen 2003; Luttinen and Vuori 2006; Luttinen *et al.* 2015). The Utpostane intrusion was first reported by Juckes (1968) and has been subsequently briefly studied by Hjelle and Winsnes (1972). Plagioclase ^{40}Ar–^{39}Ar dating indicates emplacement of the gabbro at 181 Ma (Zhang *et al.* 2003) (Table 2). Detailed geological mapping has shown that Utpostane is a layered intrusion composed mainly of various gabbroic cumulates (Vuori

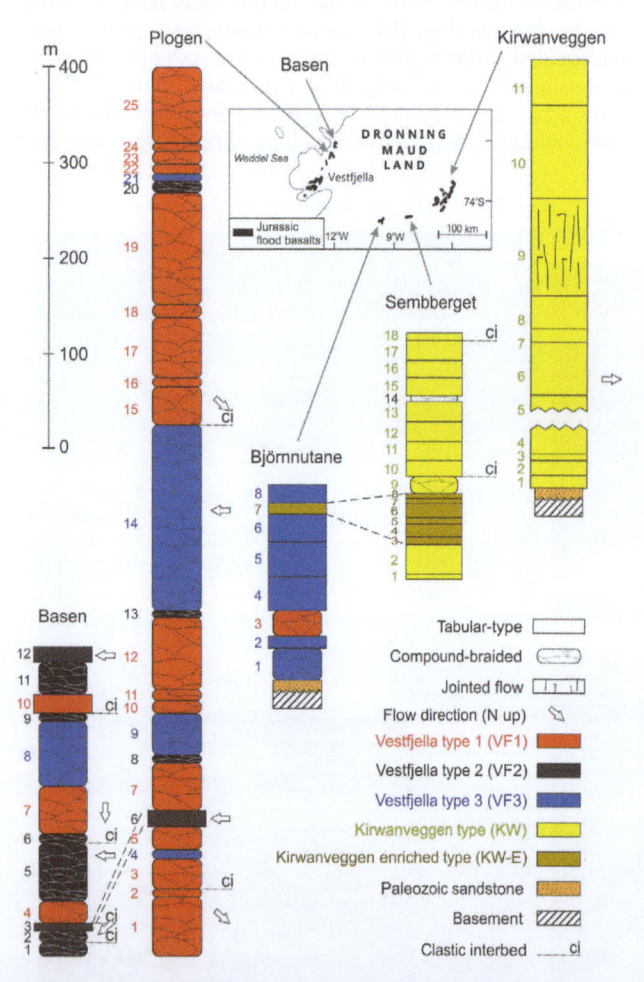

Fig. 8. Stratigraphic columns illustrating the volcanic architecture of Vestfjella (Basen, Plogen), Björnnutane, Sembberget and Kirwanveggen (nunatak B, Lagfjella). For chemical compositions see the subsection 'Extrusive magma types'. Numbers indicate interpreted lava flow fields and dashed lines show possible correlations between different localities. The contact between units #4 and #5 of Kirwanveggen is not exposed, and chemical data for unit #14 of Sembberget are lacking.

and Luttinen 2003; Luttinen and Vuori 2006). The lithological variations and layer structures are suggestive of >3 km thickness for the exposed part of the Utpostane intrusion (Vuori 2004; Vuori and Luttinen 2003).

Björnnutane extrusive suite

Björnnutane is a group of three small nunataks in northern Heimefrontfjella (Fig. 3). The bedrock is principally composed of a lava succession that has a maximum stratigraphic thickness of c. 200 m and records at least eight lava flow fields (Juckes 1972; Luttinen et al. 2010) (Fig. 8). The base of the lava pile is exposed at two localities, at c. 2400 m elevation (asl), where a c. 1 m layer of sandstone is sandwiched between the volcanic succession and the Mesoproterozoic basement (Fig. 7b). The lava succession starts with a c. 35 m-thick compound-braided flow field of lobes that vary in thickness from c. 1 to 10 m and are mainly 2–4 m thick. Measurements of pipe amygdules in the basal crust indicate the flow direction to the NE. The basal lava is overlain by a discontinuous and thin (<1 m) layer of sandstone that fills cracks in the underlying lava crust. The sandstone is succeeded by a c. 10 m-thick, distinctly light-coloured simple flow field, with pipe amygdules indicating the flow direction to the NNW. The next unit is a c. 30 m-thick compound-braided flow field consisting of pāhoehoe lava lobes of up to a few metres thick. The upper part of the succession consists of five relatively thick simple flow fields which range in thickness from c. 10 to 35 m. The contacts between lavas are poorly exposed, probably due to preferable weathering of vesiculated and brecciated lava tops. Vesicles and other cavities typically contain secondary quartz, agate, calcite, scolecite, stilbite and apophyllite (Juckes 1972). Summarizing, the Björnnutane lava succession records a transition from predominantly compound-braided to tabular-type flow fields (Fig. 8).

Kirwanveggen extrusive suite

The nunataks of southern Kirwanveggen are largely composed of lavas (Fig. 3) and exhibit a terraced appearance, trap topography, typical of many flood basalts (Fig. 7c). Plagioclase ^{40}Ar–^{39}Ar dating has yielded plateau ages of 185–182 Ma (Duncan et al. 1997) for samples from the same stratigraphic level (unit #10 in Fig. 8). The reported data do not allow critical evaluation of the ages and the older ages may result from excess Ar (Zhang et al. 2003). The volcanic succession is underlain by Paleozoic sedimentary rocks and Precambrian basement (Aucamp et al. 1972; Wolmarans and Kent 1982) (Fig. 8). The base of the lavas is typically at c. 2100 m elevation (asl). The greatest stratigraphic section is made of 420 m of basalt at 'nunatak B' in the Lagfjella area (Faure et al. 1979; Harris et al. 1990). Field observations are suggestive of shallow (c. 1°–3°), generally SE, dipping of the strata (Faure et al. 1979; Harris et al. 1990), but the irregular lava surfaces render determination of the dip angle and direction difficult.

The Kirwanveggen lava succession is composed almost entirely of simple flow fields that vary in thickness from c. 6 m up to >100 m (Faure et al. 1979; Harris et al. 1990). The lava succession at 'nunatak B' is made of 10 or 11 lava lobes and sheets (Fig. 8). Lava contacts are seldom exposed. Generalizing, the lava units which underlie a distinctive intensively jointed sheet lava are relatively thin (<50 m, frequently c. 6–14 m: Faure et al. 1979), whereas the jointed flow and those atop of it are notably thick (80–100 m). Detailed descriptions of the lavas are lacking, but the lava tops typically are vesicular (Faure et al. 1979) and at least some units have rubbly crust (Fig. 5b). Pipe amygdules at the base are rare, but invariably indicate flow towards the east (Harris et al. 1990). The vesicles and voids are filled mainly with quartz, calcite, agate and zeolite group minerals.

Overall, the Kirwanveggen succession is made of notably thick tabular-type lavas, many of which may be rubbly pāhoehoe. Sandstone intercalations (Aucamp et al. 1972; Harris et al. 1990) and thin clastic interbeds interpreted to represent pyroclastic material ('ash bands') have been reported in the succession (Harris et al. 1990), but the clastic intercalations have not been described in any detail. According to Harris et al. (1990), sparse sandstone intercalations are confined to the lower third of the stratigraphic column, and one such deposit infiltrates a flow-top breccia at the Tunga section in the northernmost part of the basalt exposure of Kirwanveggen (Fig. 5b). Sparse, intrusive Jurassic rocks cut the basement in northern Kirwanveggen (Wolmarans and Kent 1982) and a single dolerite in southern Kirwanveggen, dated at 181 Ma (Zhang et al. 2003; Table 2), shows a chemical affinity to the Ahlmannryggen intrusions (Riley et al. 2005).

Sembberget extrusive suite

Sembberget is an isolated small exposure of basalt lavas between Heimefrontfjella and Kirwanveggen (Fig. 3), and shows a similar trap topography to nunataks in southern Kirwanveggen (Juckes 1972). The lowermost lava unit is at c. 2500 m elevation (asl) and the base of the c. 250 m-thick lava succession is not exposed. The volcanic rocks comprise between 12 and 19 flow fields (Luttinen et al. 2010); the stratigraphic interpretation presented in Figure 8 includes 18 flow fields identified using field observations and geochemical data. Generalizing, the succession is dominated by thick tabular-type simple lavas. The basal lava is only partly exposed, but, based on compositional similarity, it may represent a similar massive simple inflated pāhoehoe lava to the overlying c. 30 m-thick unit which contains vesicular zones and upper crust. It is overlain by a c. 50 m-thick series of six relatively thin (4–14 m) lava sheets which are interpreted to represent discrete eruptions. The overlying c. 16 m-thick compound-braided flow field has a complex, rubbly base and is covered with a thin layer of cross-bedded, red sandstone. The upper part is almost exclusively made of seven massive 18–20 m-thick simple flow fields notably similar to those of Kirwanveggen. The first one directly above the sandstone has a distinct 4 m-thick rubbly base, and the lava contacts, when exposed, show vesicular and variably rubbly upper crusts. A thin compound flow field of strongly altered lobes is sandwiched between massive lava sheets. The capping lava is only partly exposed and is underlain by a red bole. Lava unit #17 has yielded a plagioclase ^{40}Ar–^{39}Ar plateau age of 181 Ma (Fig. 8; Table 2; see also Supplementary Table S1). Flow direction observations are lacking.

Ahlmannryggen–Sverdrupfjella intrusive suite

Flood basalt lavas are not exposed in Ahlmannryggen and H.U. Sverdrupfjella (Fig. 3), but a large number (>500) of mafic and ultramafic dykes and sills have been mapped in the region (Harris et al. 1991; Grantham 1996; Curtis et al. 2008). A significant portion of the dykes may be Proterozoic, however, and the subset of 47 dykes reported by Riley et al. (2005) provide the most reliable insight into the Karoo-age intrusions in the area. Imprecise whole-rock and groundmass ^{40}Ar–^{39}Ar ages point to emplacement at 190–180 Ma (Riley et al. 2005). Most of the dykes have been emplaced along pre-existing joints within the Proterozoic Borgmassivet Intrusive Suite in northern Ahlmannryggen. The predominant strike directions for the dykes and joints are NNE–SSW and ENE–WSW. Oblique jointing has resulted in offsetting segments and en echelon geometries in the Mesozoic dykes. The dykes are generally <5 m wide, but dykes up to 80 m wide have been reported by Riley et al. (2005). Dykes that are over 5 m wide are members of a geochemically distinctive subset (Group 2 of Riley et al. 2005; see the subsection 'Intrusive magma types' later in this chapter). Dykes which were emplaced in a ENE–WSW direction have been interpreted to be the oldest Jurassic intrusions in the area (Riley et al. 2005).

Associated intrusive suites

Schievestolen sill. A single Jurassic intrusion has been reported in the northernmost Heimefrontfjella (Fig. 3). The Schievestolen dolerite sill intrudes Permian sedimentary rocks and is exposed over a distance of c. 2 km (Juckes 1972). A $^{40}Ar–^{39}Ar$ plagioclase correlation age of c. 184 Ma was provided by Brewer et al. (1996), but the result is likely to have been affected by excess Ar. The sill is mostly concordant and has fairly constant thickness of c. 10 m about 30 m above the base of the sediments, but in places the intrusion transgresses the stratification (Fig. 7d). The sill is further typified by columnar jointing, a vesicular upper part and the development of ropy structures within large, flattened megavesicles (Fig. 4d) (cf. Liss et al. 2002). Close to the base, the dolerite contains xenolithic quartz grains and small vesicles.

Mannefallknausane intrusions. Numerous mafic intrusions cross-cut the 1.1 Ga granitoid basement of Mannefallknausane (Fig. 3) (Juckes 1968; Arndt et al. 1991). Most of the intrusions are <5 m-thick subhorizontal sills (dip angles mainly <15°) and usually follow the prominent subhorizontal jointing of the basement, but are sometimes deflected by other joints (Juckes 1968). One sill is 12–20 m thick. Two c. 2 m-wide dykes strike ENE–WSW. None of the intrusions has been dated, but one has a $^{40}Ar–^{39}Ar$ plagioclase spectrum suggestive of Middle Jurassic emplacement (Zhang et al. 2003), and the pristine magmatic mineralogy and geochemical similarities to Vestfjella flood basalts (Heinonen et al. 2010; Luttinen et al. 2015) also lend support for a Jurassic origin.

Geochemical characteristics

Geochemical data for representative samples of Karoo CFBs in DML are listed in Table 3 and the compiled dataset used in this study is given in Supplementary Table S2. For analytical methods, the reader is referred to the original publications.

General features

The flood lavas and related intrusions of DML are subalkaline quartz or olivine normative tholeiites (Harris et al. 1990; Luttinen and Furnes 2000; Riley et al. 2005; Heinonen and Luttinen 2008; Luttinen et al. 2010, 2015). The great majority of them exhibit low-Mg basalt or basaltic andesite compositions. Olivine-rich, high-Mg picrite lavas are sporadically intercalated with basaltic lavas in Vestfjella. Picritic intrusions are relatively common in the dyke swarms at Vestfjella and Ahlmannryggen, and also occur at Mannefallknausane (Riley et al. 2005; Luttinen et al. 2015). The Utpostane layered intrusion is mainly composed of high-Mg cumulate rock types (Vuori and Luttinen 2003).

The extrusive and intrusive rocks are characterized by notably variable compositions (Fig. 9). For instance, MgO ranges from 3 to 28 wt%, SiO_2 from 44 to 57 wt% and TiO_2 from 0.3 to 5.5 wt%. Generalizing, concentrations of most major and minor elements (e.g. TiO_2 and FeO^T) and gabbro-incompatible trace elements (e.g. Zr and Y) correlate negatively with Mg number (molar $MgO/(MgO + 0.9FeO^T)$), whereas compatible elements (e.g. Ni) correlate positively and the trend of Al_2O_3 shows maximum values at a Mg number of c. 0.55 (Fig. 9). While most of the analysed samples comprise broadly uniform trends, many of the intrusive rocks from Ahlmannryggen and Vestfjella have distinctively high contents of FeO^T and incompatible elements (TiO_2, Zr and Y), and low Al_2O_3 at a given Mg number (Fig. 9).

The initial isotopic compositions have been calculated at 180 Ma. The Nd and Sr isotopic ratios show remarkable variations from depleted compositions, indistinguishable from depleted upper mantle (ε_{Nd} = 9; $^{87}Sr/^{86}Sr$ = 0.7035), to variably enriched compositions similar to Precambrian silicic crust (ε_{Nd} down to −18; $^{87}Sr/^{86}Sr$ up to >0.710: Fig. 10a). In general, the isotopic ranges of the lavas and intrusions are similar, but the most isotopically depleted samples represent intrusive rock types. Data on Pb and Os isotopic compositions are limited to rare picritic rocks (Heinonen et al. 2010, 2014). The initial $^{187}Os/^{188}Os$ values vary from 0.124 to 0.143, but mainly record low values typical of primitive and depleted mantle reservoirs (Fig. 10b). The Pb isotopic data are characterized by variable $^{206}Pb/^{204}Pb$ (17.0–18.4), $^{208}Pb/^{204}Pb$ (37.0–38.3) and variable $^{207}Pb/^{204}Pb$ (15.3–15.6) (Fig. 10c, d).

As in African Karoo, the Karoo tholeiites in DML have been divided into normal (low-Ti) and enriched (high-Ti) types based on the incompatible element content. Overall, c. 30% of the analysed rock samples can be described as high-Ti types based on TiO_2 >2 and Zr >200 ppm (Fig. 9), but the enriched types are overrepresented due to duplicate analyses of the same intrusive units (e.g. Riley et al. 2005; Heinonen and Luttinen 2008). The high-Ti types are effectively limited to coastal areas at Vestfjella and Ahlmannryggen, whereas the low-Ti types are found at all localities. Both categories show pronounced variations in mantle-incompatible element and isotopic ratios which facilitates division of the igneous rocks into compositionally different groups often referred to as magma types. The following subsection summarizes the geochemical key features of the magma types using a uniform modified nomenclature. The petrological basis for the grouping is that fractional crystallization of tholeiitic basalt and picrite magmas has only a mild influence on incompatible-element ratios such as Ti/Zr, Ti/P, Nb/La, Nb/Zr and Zr/Y, so that different ratios are likely to indicate different magma plumbing systems.

Extrusive magma types

Vestfjella type 1 (VF1). The northern part of Vestfjella is dominated by low-Ti picrite and basalt (Mg# 0.4–0.8) lavas which are typifid by low Ti/Zr (mainly <80), Ti/P (mainly <14), Nb/La (mainly <0.6) and Zr/Y (mainly 4–6), and high La/Sm (<2) (CT1 of Luttinen and Siivola 1997) (Figs 8 & 11). Compared to other Vestfjella lavas, VF1 lavas have low TiO_2 and FeO^T but high Zr at a given Mg number (Fig. 9). Initial Nd and Sr isotopic compositions are enriched (ε_{Nd} = −18 to −2; $^{87}Sr/^{86}Sr$ = 0.705–0.711) (Fig. 10a). A single picrite lava has initial $^{187}Os/^{188}Os$ (0.130) akin to primitive mantle, notably low $^{206}Pb/^{204}Pb$ (17.0), and low $^{207}Pb/^{204}Pb$ (15.4) and $^{208}Pb/^{204}Pb$ (37.5) values (Fig. 10b–d). Several of the Vestfjella intrusions belong to VF1 (Luttinen et al. 2015) and geochemically similar lavas of Björnnutane (BN2 of Luttinen et al. 2010) can be correlated with this category (Figs 8–11). Some dykes at Ahlmannryggen (AR1: Group 1 of Riley et al. 2005) show broad similarity to VF1.

Vestfjella type 2 (VF2). The Basen section, northernmost Vestfjella, is dominated by basalt and rare picrite lavas (Mg# 0.4–0.7) that are transitional between low-Ti and high-Ti compositions judging from their TiO_2 (mainly >2 wt%) and Zr (<200 ppm) contents (CT2 of Luttinen and Siivola 1997) (Figs 8 & 9). Type VF2 is distinguished from other Vestfjella lavas by its combination of high Ti/Zr (80–120) and Ti/P (mainly 15–20), and low Nb/La (<0.5) values (Fig. 11). Concentrations of TiO_2 and FeO^T are higher than in type VF1 at a given Mg number. Initial Nd and Sr isotopic compositons are variably enriched (ε_{Nd} = –8 to +1; $^{87}Sr/^{86}Sr$ 0.706–0.708) (Fig. 10a). Isotopic data on Os and Pb for VF2 are lacking.

Vestfjella type 3 (VF3). The exposed volcanic bedrock of southern Vestfjella is dominated by VF3 type of low-Ti lavas that effectively lack the strong geochemical crustal signature which typifies other low-Ti lavas of DML (CT3 of Luttinen and Siivola 1997). The VF3 lavas range from picrite to basalt (Mg# 0.4–0.8) and have low Ti/P values (mainly <14) combined with high Ti/Zr (80–150) and low Zr/Y (<5) values, whereas Nb/La is relatively high (mainly 0.5–0.7) and La/Sm low (<2.5) compared to other Vestfjella lavas (Fig. 11). Concentrations of TiO_2 and FeO^T are higher and those of Zr lower than in VF1 at the same Mg numbers (Fig. 9). Isotopic compositions vary from mildly depleted to mildly enriched (ε_{Nd} = –2 to +4; $^{87}Sr/^{86}Sr$ = 0.7035–0.7050) (Fig. 10a). Two VF3 picrites have initial $^{187}Os/^{188}Os$ (c. 0.130) and $^{206}Pb/^{204}Pb$ (c. 17.3) quite similar to those of VF1, but $^{207}Pb/^{204}Pb$ (c. 15.5) and $^{208}Pb/^{204}Pb$ (c. 37.1) are dissimilar (Fig. 10b–d). The Vestfjella dyke swarm is mainly composed of VF3 (Luttinen et al. 2015) and geochemically similar lavas dominate the Björnnutane succession (BN1 of Luttinen et al. 2010) (Figs 8–11).

Kirwanveggen type (KW). The lava successions of Kirwanveggen and Sembberget are composed of geochemically notably uniform basalts (Mg# 0.4–0.6) typified by low TiO_2 and Zr, but high FeO^T at a given Mg number (Harris et al. 1990; Luttinen et al. 2010) (Fig. 9). Their compositions are in many respects intermediate between those of the Vestfjella types VF1–VF3; Ti/Zr are mainly within 70–100, Ti/P is 9–15, La/Sm is 2–3 and Nb/La is mainly <0.6 (Fig. 11). The Kirwanveggen type is readily distinguished from the Vestfjella types due to coupling of low Ti/Zr and Zr/Y (Fig. 11c). In the case of Kirwanveggen lavas, the initial Nd isotopic compositions are chondritic or mildly depleted (ε = 0–3), but the Sembberget lavas also include mildly enriched compositions (ε_{Nd} = –5 to +2) (Fig. 10a). The initial Sr isotopic compositions are mostly mildly enriched, but the Sembberget lavas range to strongly enriched compositions ($^{87}Sr/^{86}Sr$ = 0.7045–0.7055) (Fig. 10a). Isotopic data on Os and Pb for KW are lacking. In Figure 8, the enriched subtype has been distinguished from the main group as Kirwanveggen type E. One simple flow in the Björnnutane section has been correlated with the enriched subtype (Luttinen et al. 2010).

Intrusive magma types

At Vestfjella, the majority of the intrusive, subvolcanic dykes and sills can be correlated with the volcanic magma types (Figs 9–11). Importantly, sampling of the Vestfjella dyke swarm and intrusions in Mannefallkanusane and Ahlmannryggen has also revealed compositional types that lack extrusive equivalents. The geochemical assessment of the Karoo-related intrusions in Ahlmannryggen and H.U. Sverdrupfjella is based on data reported by Riley et al. (2005) and Heinonen et al. (2013, 2014), because the datasets of Harris et al. (1991) and Grantham (1996) lack precise trace element and isotopic data, and also may include Proterozoic dykes (Riley et al. 2005). The intrusive rock types of Ahlmannryggen probably represent feeder fissures of now eroded lavas.

Vestfjella depleted type (VFD). A suite of geochemically quite variable dyke rocks can be collectively designated as Vestfjella depleted type (VFD) based on their depleted Nd isotopic compositions (depleted ferropicrites and low-Nb of Heinonen et al. 2010; CT3-E of Luttinen et al. 2015). Three of the intrusions are found in Mannefallknausane. This magma type shows a notably wide range of basalt and picrite compositions (Mg# 0.3–0.8). The VFD dykes exhibit low-Ti to high-Ti compositions with TiO_2 of 1–5 wt%, Zr of 50–300 ppm and Zr/Y of 4–7, and are typified by high Ti/Zr (90–150), Ti/P (13–18) and Nb/La (0.6–0.9) values, as well as high FeO^T at given Mg numbers (Figs 9 & 11). The initial Nd isotopic compositions vary from strongly to mildly depleted (ε_{Nd} = 2–8) and initial Sr isotopic ratios from depleted to mildly enriched ($^{87}Sr/^{86}Sr$ = 0.7035–0.7055) (Fig. 10a). The initial Os isotopic ratios are low ($^{187}Os/^{188}Os$ = 0.126–0.129) (Fig. 10b). Apart from two crustally contaminated samples that have exceptionally high and low Pb isotopic ratios (Heinonen et al. 2010), the VFD dykes are typified by rather uniform $^{206}Pb/^{204}Pb$ (17.6–17.9), $^{207}Pb/^{204}Pb$ (c. 15.5) and $^{208}Pb/^{204}Pb$ (37.4–37.5) values (Fig. 10c–d).

Vestfjella enriched type (VFE). A small number of dykes show broad geochemical similarities to ocean island basalts (OIBs) and are referred here collctively as Vestfjella enriched type (VFE) (enriched ferropicrites of Heinonen and Luttinen 2008; high-Nb basalts of Luttinen et al. 2015). The VFE dykes have high-Ti basalt and picrite compositions (Mg# 0.4–0.7) with TiO_2 of 2–3 and Zr/Y of c. 6 (Figs 9 & 11). They show overlap with VF3 in a Ti/Zr v. Ti/P plot, but are readily distinguished from all other Vestfjella magma types by their relative enrichment of Nb, with notably high Nb/La at given Ti/Zr and high Nb/Zr (Fig. 11). The initial Nd and Sr isotopic compositions are mainly mildly depleted and cluster close to primitive mantle (ε_{Nd} = –1 to +4; $^{87}Sr/^{86}Sr$ = 0.7035–0.7055) (Fig. 10a). The Pb isotopic ratios are somewhat higher than in most VFD ($^{206}Pb/^{204}Pb$ = c. 17.9, $^{207}Pb/^{204}Pb$ = c. 15.6 and $^{208}Pb/^{204}Pb$ = c. 37.8), but $^{187}Os/^{188}Os$ values are much higher (0.140–0.143) and record the most radiogenic compositions in the Jurassic CFBs of DML (Fig. 10b–d).

Utpostane type (UP). The geochemical compositions of the gabbroic rocks of Utpostane layered intrusion are mainly controlled by accumulation processes (Vuori and Luttinen 2003). Compositions of samples from the capping Zone 1 and a gabbroic apophysis are likely to be least affected by accumulation and provide the best proxy for parental magma compositions (Luttinen and Vuori 2006). Detailed comparison reveals that UP-type magmas are geochemically different from the other magma types of Vestfjella and show many similarities to the Kirwanveggen magma type (Luttinen et al. 2010) with, for example, low Zr/Y (c. 4), intermediate Ti/Zr (c. 90), and enriched Nd and Sr isotopic compositions (ε_{Nd} = –2; $^{87}Sr/^{86}Sr$ = c. 0.7055) (Figs 10a & 11). Isotopic data on Os and Pb for UP are lacking.

Ahlmannryggen type 1 (AR1). A small set of dykes in Ahlmannryggen (Group 1 of Riley et al. 2005) shows affinity to VF1 lavas and dykes of Vestfjella, although the TiO_2 (1.5–2.3 wt%) contents are somewhat higher at given Mg number (0.4–0.6) and straddle the low-Ti–high-Ti boundary (Fig. 9).

Table 3. *Representative elemental and isotopic compositions for Jurassic tholeiites in Dronning Maud Land*

Sample	Z.1801.1[1]	Z.1814.4[1]	Z.1805.1[1]	Z.1810.1[1]	Z.1803.1[1,2]	Z.1816.1[1,2]	Z.1826.1[1,9]	Z.1833.2[1,9]	AL/B7-03[3,4]	AL/WM1e-98[3,4]	AL/B7-98[3,4]	JSH/B006[4]
Magma type[10]	AR1	AR1	AR2	AR2	ARD	ARD	ARE	ARE	VFD	VFD	VFE	VFE
Locality	Ahlmann[11]	Ahlmann[11]	Ahlmann[11]	Ahlmann[11]	Ahlmann[11]	Ahlmann[11]	Ahlmann[11]	Ahlmann[11]	Vestfjella	Vestfjella	Vestfjella	Vestfjella
	73° 09.43′S	72° 02.20′S	72° 13.31′S	72° 16.65′S	72° 08.13′S	72° 03.27′S	71° 59.54′S	72° 02.21′S	73° 02.50′S	73° 43.82′S	73° 01.30′S	73° 01.77′S
	2° 08.18′W	2° 47.80′W	3° 25.02′W	3° 25.43′W	3° 18.25′W	2° 42.74′W	3° 21.64′W	3° 30.34′W	13° 24.00′W	15° 06.17′W	13° 23.93′W	13° 25.17′W
SiO_2 (wt%)	52.43	55.53	51.95	50.47	49.15	48.66	49.33	47.71	47.05	43.65	49.07	44.86
TiO_2	1.58	2.25	2.26	2.34	4.06	3.34	4.39	4.24	1.70	2.07	3.16	3.01
Al_2O_3	12.15	14.59	14.43	14.11	12.02	9.38	8.81	8.42	8.59	8.85	11.29	8.19
FeO^T	11.33	10.06	12.40	13.57	12.96	12.87	11.93	12.37	12.87	15.10	13.69	16.19
MnO	0.20	0.16	0.21	0.21	0.17	0.16	0.15	0.17	0.20	0.20	0.18	0.21
MgO	8.86	5.45	5.59	6.21	8.66	14.73	14.41	15.23	18.57	18.88	10.49	15.54
CaO	10.34	8.66	10.14	9.96	10.46	8.99	8.74	8.83	9.24	9.02	9.37	9.02
Na_2O	2.12	1.92	2.43	2.55	1.84	1.43	1.47	1.91	1.41	1.53	1.76	1.75
K_2O	0.80	1.27	0.36	0.35	0.44	0.21	0.43	0.75	0.23	0.50	0.59	0.87
P_2O_5	0.19	0.12	0.25	0.24	0.24	0.23	0.34	0.38	0.13	0.19	0.39	0.36
LOI	1.67	0.56	1.30	0.50	0.69	1.89	2.05	3.33	1.23	1.58	3.86	2.42
Cr (ppm)	668	76	84	84	479	803	834	918	1132.4	1022	357	579
Ni	294	83	64	66	390	500	666	789	807.3	880	370	660
Sc	24	30	32	33	30	37	34	28	28	28	21	19
V	258	425	327	336	345	375	319	283	321	364	258	285
Cu	95	206	100	97	146	146	140	133	139	171	80	116
Zn	93	124	119	118	144	147	141	101	95	100	132	152
Ga	18	23	20	20	21	20	17	17	15	17	25	20
Ta	0.5	0.6	0.8	0.8	0.7	0.6	1.5	1.8	0.4	0.4	1.2	1.2
Nb	7.2	9.4	11.9	11.9	10.0	8.4	22.0	27.5	5.5	6.0	17.1	18.5
Hf	3.1	4.4	4.1	3.9	7.4	8.1	10.8	12.0	2.2	2.7	4.1	4.0
Zr	122.0	163.0	159.0	153.0	275.0	295.0	413.0	477.0	80.0	104.0	160.0	151.0
Y	27.2	30.1	34.4	34.3	43.7	43.9	38.6	34.8	15.1	16.6	28.2	23.7
Ba	310.0	314.0	115.0	100.0	112.0	38.0	337.0	702.0	55.8	90.5	164.8	225.0
Rb	14.1	12.9	6.5	6.9	11.6	2.4	21.9	33.4	6.7	8.4	7.9	20.8
Sr	248.0	356.0	181.0	213.0	257.0	246.0	584.0	1005.0	242.2	302.8	415.0	520.0
U	0.6	0.5	0.5	0.4	0.3	0.2	0.5	0.9	0.1	0.1	0.5	0.5
Th	1.9	2.4	1.4	1.2	1.2	0.4	2.1	4.5	0.5	0.5	1.9	1.7
Pb	4.2	4.2	11.5	2.0	2.1	0.8	3.0	5.6	1.1	1.3	2.2	2.2
La	14.7	17.5	12.8	12.3	11.0	7.5	27.7	59.8	7.5	7.8	17.9	18.9
Ce	31.8	38.6	30.6	29.9	30.8	24.8	71.2	137.9	20.0	19.8	38.2	43.7
Pr	4.4	5.5	4.6	4.5	5.4	5.0	11.7	20.3	3.0	2.9	5.0	6.1
Nd	18.9	24.5	21.4	21.0	29.3	29.6	55.6	84.5	13.8	14.5	23.5	27.4
Sm	4.5	5.9	5.5	5.5	9.4	9.9	13.0	15.4	3.6	4.3	7.1	7.1
Eu	1.4	1.9	1.8	1.8	3.2	3.4	3.9	4.2	1.3	1.5	2.6	2.5
Gd	5.2	6.4	6.5	6.4	11.0	11.4	12.1	11.7	3.7	4.5	7.5	7.2
Tb	0.9	1.0	1.0	1.0	1.7	1.7	1.7	1.6	0.6	0.7	1.2	1.1
Dy	4.9	5.6	6.0	6.0	8.9	9.1	8.4	7.6	3.4	3.8	6.5	5.9
Ho	1.0	1.1	1.2	1.2	1.6	1.7	1.5	1.3	0.6	0.7	1.1	1.0
Er	2.5	2.8	3.3	3.2	3.9	3.9	3.3	3.0	1.6	1.5	2.5	2.3
Tm	0.4	0.4	0.5	0.5	0.6	0.6	0.5	0.4	0.2	0.2	0.3	0.3
Yb	2.2	2.4	3.1	3.0	3.0	3.0	2.5	2.2	1.1	1.1	1.6	1.4
Lu	0.3	0.4	0.5	0.5	0.4	0.4	0.4	0.3	0.2	0.2	0.2	0.2
$^{87}Rb/^{86}Sr$	0.1637	0.1047	0.1038	0.0936	0.1302	0.0284	0.1084	0.0961	0.06763	0.07992	0.05472	0.11394
$^{87}Sr/^{86}Sr$	0.707608	0.70881	0.704604	0.704802	0.705842	0.703647	0.705652	0.705188	0.703302	0.70320	0.70518	0.70457
$^{87}Sr/^{86}Sr_{180\,Ma}$	0.707189	0.708542	0.704338	0.704563	0.705509	0.703574	0.705374	0.704942	0.703128917	0.70300	0.70504	0.70428
$^{147}Sm/^{144}Nd$	0.1514	0.1479	0.1588	0.1605	0.2019	0.209	0.1481	0.1167	0.1482	0.16330	0.16150	0.14830
$^{143}Nd/^{144}Nd$	0.512271	0.512283	0.512631	0.512682	0.512902	0.513113	0.51266	0.512321	0.513004	0.51301	0.51269	0.51276
$\varepsilon_{Nd_{180Ma}}$	−6.1	−5.8	0.8	1.7	5.1	9.0	1.6	−4.3	8.3	8.0	1.8	3.5
$^{238}U/^{204}Pb$	na	na	na	na	9.90	11.91	10.62	9.28	10.75	7.28	13.85	12.30
$^{232}Th/^{204}Pb$	na	na	na	na	34.84	34.71	45.07	51.13	28.11	24.76	58.28	48.51
$^{206}Pb/^{204}Pb$	na	na	na	na	18.03	18.57	17.61	17.46	18.12	17.83	18.26	18.26
$^{207}Pb/^{204}Pb$	na	na	na	na	15.43	15.51	15.38	15.43	15.50	15.49	15.60	15.54
$^{208}Pb/^{204}Pb$	na	na	na	na	37.80	38.03	37.84	37.86	37.73	37.52	38.28	38.26
$^{206}Pb/^{204}Pb_{180\,Ma}$	na	na	na	na	17.75	18.24	17.31	17.20	17.82	17.62	17.87	17.91
$^{207}Pb/^{204}Pb_{180\,Ma}$	na	na	na	na	15.41	15.49	15.36	15.42	15.49	15.48	15.58	15.53
$^{208}Pb/^{204}Pb_{180\,Ma}$	na	na	na	na	37.49	37.72	37.44	37.40	37.48	37.30	37.76	37.82
Re	na	na	na	na	0.36	0.83	0.28	0.82	0.74	0.79	0.44	0.55
Os	na	na	na	na	0.63	1.31	0.71	1.80	1.22	0.90	0.26	0.15
$^{187}Re/^{188}Os$	na	na	na	na	2.759	3.078	1.936	2.207	2.910	4.271	8.299	17.290
$^{187}Os/^{188}Os$	na	na	na	Na	0.135	0.134	0.144	0.137	0.134	0.141	0.167	0.192
$^{187}Os/^{188}Os_{180\,Ma}$	na	na	na	Na	0.127	0.125	0.138	0.130	0.126	0.128	0.143	0.140

Data sources: [1]Riley *et al.* (2005); [2]Heinonen *et al.* (2014); [3]Heinonen and Luttinen (2008); [4]Heinonen *et al.* (2010); [5]Luttinen and Furnes (2000); [6]Luttinen *et al.* (1998); [7]Luttinen *et al.* (2010); [8]Harris *et al.* (1990); [9]Heinonen, unpublished data; [10]see the text for the definition of magma types; [11]Ahlmann, Ahlmannryggen; [12]Kirwan, Kirwanveggen.

LOI, loss on ignition; na, not available.

307-AVL[4]	VF57-85[5]	B16-AVL[6]	VF72-85[5]	sk229-AVL[4,5]	VF126-85[5]	Z.1610.14[7]	Z.1610.7[7]	255-AVL-92[7]	Z.1610.9[7]	Z.1609.2[7]	Z.1609.26[7]	LAG23[8]	LAG31[8]
VF1	VF1	VF2	VF2	VF3	VF3	VF3	VF3	VF1	VF1	KW	KW	KW	KW
Vestfjella	Vestfjella	Vestfjella	Vestfjella	Vestfjella	Vestfjella	Björnnutane	Björnnutane	Björnnutane	Björnnutane	Sembberget	Sembberget	Kirwan[12]	Kirwan[12]
73° 12.17'S	73° 01.68'S	73° 01.23'S	73° 01.68'S	73° 46.40'S	73° 44.65'S	74° 37.40'S	74° 37.40'S	74° 37.10'S	74° 37.40'S	74° 28.60'S	74° 28.60'S	74° 05.30'S	74° 07.20'S
13° 49.55'W	13° 25.40'W	13° 21.73'W	13° 25.40'W	14° 56.53'W	14° 47.42'W	9° 59.00'W	9° 59.00'W	10° 03.60'W	9° 59.00'W	8° 08.30'W	8° 08.30'W	6° 16.22'W	6° 15.85'W
47.50	53.36	50.79	50.49	48.14	49.51	50.04	47.61	52.28	51.70	51.93	50.40	51.37	50.85
0.50	1.38	1.59	3.08	1.00	1.50	1.70	1.93	1.26	1.44	1.59	1.61	1.44	1.54
7.71	14.97	11.16	13.06	9.49	15.14	14.60	14.92	15.98	15.55	13.95	14.86	14.93	15.43
12.42	9.71	12.36	14.77	13.10	10.72	11.38	12.80	10.14	11.09	12.55	12.87	12.54	12.09
0.18	0.17	0.18	0.17	0.20	0.17	0.17	0.21	0.16	0.17	0.19	0.21	0.19	0.19
26.22	5.99	11.67	5.29	19.13	7.92	7.40	7.53	5.47	5.83	6.24	6.14	5.71	5.92
4.64	10.50	9.87	9.61	7.68	12.57	11.56	12.28	11.81	9.98	10.37	10.53	10.42	10.55
0.66	2.58	1.90	2.39	1.02	2.12	2.48	2.31	2.43	3.25	2.55	2.50	2.65	2.94
0.13	1.07	0.22	0.89	0.10	0.17	0.53	0.16	0.30	0.80	0.47	0.69	0.55	0.31
0.06	0.27	0.12	0.26	0.14	0.19	0.14	0.25	0.17	0.19	0.15	0.19	0.19	0.17
na	1.77	4.22	3.00	4.00	2.29	0.72	2.07	4.79	3.56	0.58	0.11	0.16	0.19
1195	234	702	115	1413	546	255	318	157	175	146	163	96	134
1321	65	328	110	642	170	95	139	77	73	68	64	63	90
17	28	29	26	25	34	29	31	27	28	36	38	33	31
144	277	308	383	223	295	321	324	232	266	288	290	251	278
na	95	150	146	66	95	135	150	74	106	135	114	na	na
na	95	116	140	91	78	88	101	94	94	95	103	na	na
na	20	na	28	15	16	22	20	19	22	21	20	na	na
0.1	0.4	0.2	0.5	0.2	0.4	0.2	0.2	0.2	0.3	0.3	0.3	na	na
1.5	8.1	3.5	6.7	3.1	4.9	3.0	4.0	8.0	4.0	4.0	4.0	4.1	3.0
0.9	3.5	2.6	5.1	1.5	2.0	2.7	2.5	2.4	2.8	3.1	3.0		
41.0	137.0	92.0	173.0	56.0	80.0	101.0	100.0	86.0	110.0	117.0	116.0	105.0	98.0
9.5	28.7	27.0	43.0	15.0	17.0	23.0	24.0	20.0	22.0	29.0	35.0	29.0	28.0
60.0	508.0	79.0	239.0	71.0	140.0	175.0	219.0	282.0	344.0	210.0	190.0	144.0	108.0
10.5	17.0	5.0	18.0	3.0	2.0	12.0	6.0	11.0	20.0	11.0	24.0	10.0	2.5
76.0	349.0	244.0	357.0	125.0	320.0	234.0	330.0	308.0	323.0	188.0	167.0	194.0	227.0
0.1	0.4	0.2	0.7	0.1	0.1	0.2	0.1	0.3	0.4	0.2	0.2	na	na
0.5	1.6	0.6	1.5	0.2	0.2	0.8	0.3	1.4	1.7	1.4	1.2	na	na
na	4.3	1.6	3.5	1.4	0.9	2.1	1.7	3.8	4.1	2.9	2.3	na	na
3.8	20.4	7.0	14.9	5.2	5.7	7.3	7.9	10.6	12.6	9.9	9.3	8.9	7.9
8.5	41.8	16.5	33.8	11.4	13.8	17.8	19.8	23.7	27.7	22.5	21.1	21.0	18.7
na	5.0	na	4.7	1.6	2.0	2.8	3.1	3.3	3.9	3.3	3.3	na	na
5.1	22.0	11.2	22.8	7.8	10.2	14.0	16.0	15.3	17.6	15.7	15.3	12.7	12.6
1.5	5.4	3.7	7.3	2.4	3.2	3.9	4.2	3.7	4.3	4.3	4.2	3.9	3.8
0.6	1.8	1.4	2.5	0.9	1.3	1.4	1.6	1.3	1.4	1.4	1.4	1.3	1.2
1.8	5.6	4.3	8.7	2.8	3.9	4.8	5.0	4.1	4.7	5.2	5.4	4.3	4.1
na	0.9	0.8	1.5	0.5	0.6	0.8	0.8	0.7	0.7	0.9	0.9	na	na
1.9	5.6	na	9.2	3.0	3.7	4.5	4.5	3.8	4.3	5.2	5.6	4.6	4.4
0.4	1.1	na	1.8	0.6	0.7	0.9	0.9	0.7	0.8	1.1	1.2	na	na
na	2.8	na	4.3	1.6	1.8	2.2	2.2	1.9	2.2	2.8	3.2	2.5	2.2
na	0.4	na	0.6	0.2	0.2	0.3	0.3	0.3	0.3	0.4	0.5	na	na
0.8	2.2	2.0	3.3	1.3	1.4	1.8	1.9	1.7	2.0	2.6	3.1	2.6	2.2
0.1	0.3	0.3	0.5	0.2	0.2	0.3	0.3	0.3	0.3	0.4	0.5	0.4	0.3
0.41641	0.14700	0.05346	0.15130	0.06101	0.00900	0.07992	0.00337	0.07805	0.14429	0.087790849	0.3057	na	na
0.70804	0.70703	0.70634	0.70657	0.70509	0.70366	0.70426	0.70409	0.70663	0.70620	0.705677	0.705614869	0.70599	0.70529
0.70697	0.70665	0.70621	0.70619	0.70493	0.70363	0.70405	0.70408	0.70643	0.70583	0.70545232	0.70483253	0.70563	0.70521
0.16684	0.13770	0.17730	0.19330	0.16229	0.19000	0.17834	0.17018	0.15342	0.15097	0.164476116	0.166179672	na	na
0.51206	0.51200	0.51263	0.51252	0.51252	0.51273	0.51278	0.51268	0.51234	0.51232	0.512509822	0.512672815	0.512613	0.512623
−10.7	−11.1	0.3	−2.1	−1.4	2.0	3.2	1.4	−4.9	−5.3	−1.8	1.4	−0.7	0.3
6.74	na	na	na	3.14	na	na	na	na	na	na	na	na	na
45.25	na	na	na	10.71	na	na	na	na	na	na	na	na	na
17.20	na	na	na	17.37	na	na	na	na	na	na	na	na	na
15.44	na	na	na	15.54	na	na	na	na	na	na	na	na	na
37.93	na	na	na	37.16	na	na	na	na	na	na	na	na	na
17.01	na	na	na	17.28	na	na	na	na	na	na	na	na	na
15.43	na	na	na	15.54	na	na	na	na	na	na	na	na	na
37.53	na	na	na	37.06	na	na	na	na	na	na	na	na	na
0.07	na	na	na	0.23	na	na	na	na	na	na	na	na	na
3.02	na	na	na	2.02	na	na	na	na	na	na	na	na	na
0.106	na	na	na	0.545	na	na	na	na	na	na	na	na	na
0.131	na	na	na	0.131	na	na	na	na	na	na	na	na	na
0.130	na	na	na	0.130	na	na	na	na	na	na	na	na	na

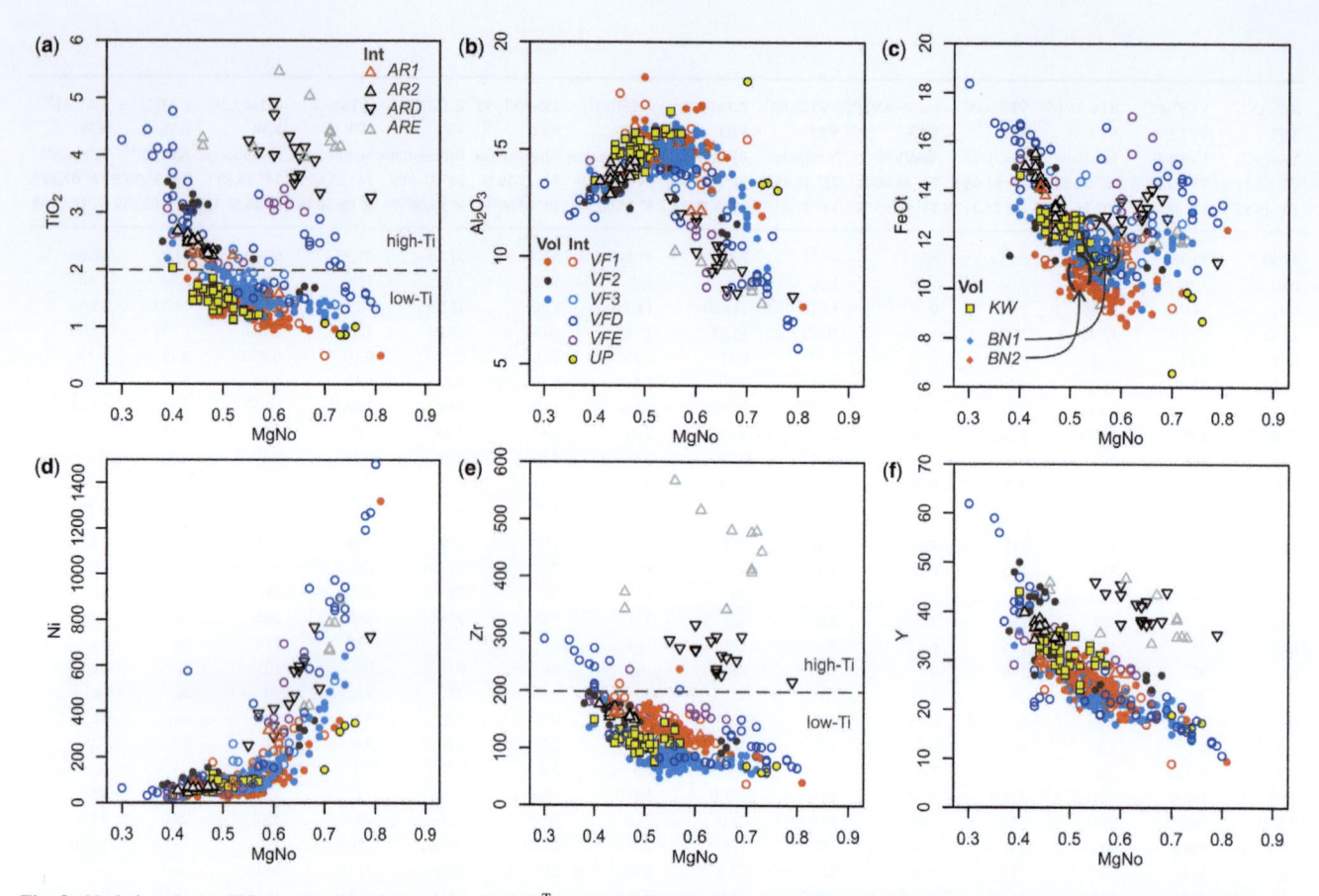

Fig. 9. Variations in (**a**) TiO_2 (wt%), (**b**) Al_2O_3 (wt%), (**c**) FeO^T (wt%), (**d**) Ni (ppm), (**e**) Zr (ppm) and (**f**) Y (ppm) v. Mg number (molar $MgO/(MgO + 0.9FeO^T)$) in the Karoo CFBs of DML. Compositions of different volcanic (Vol) and intrusive (Int) magma types of Vestfjella (VF), Björnnutane (BN), Kirwanveggen and Sembberget (KW), and Ahlmannryggen (AR) are shown with filled and open symbols, respectively (except for Utpostane, UP) (see the text for explanation of magma types). The conventional low-Ti v. high-Ti compositional boundary is shown in (a) and (e) (Erlank *et al.* 1988). The compositions of the Björnnutane lavas (BN1 and BN2) are circled in (c). Geochemical data are from Supplementary Table S2.

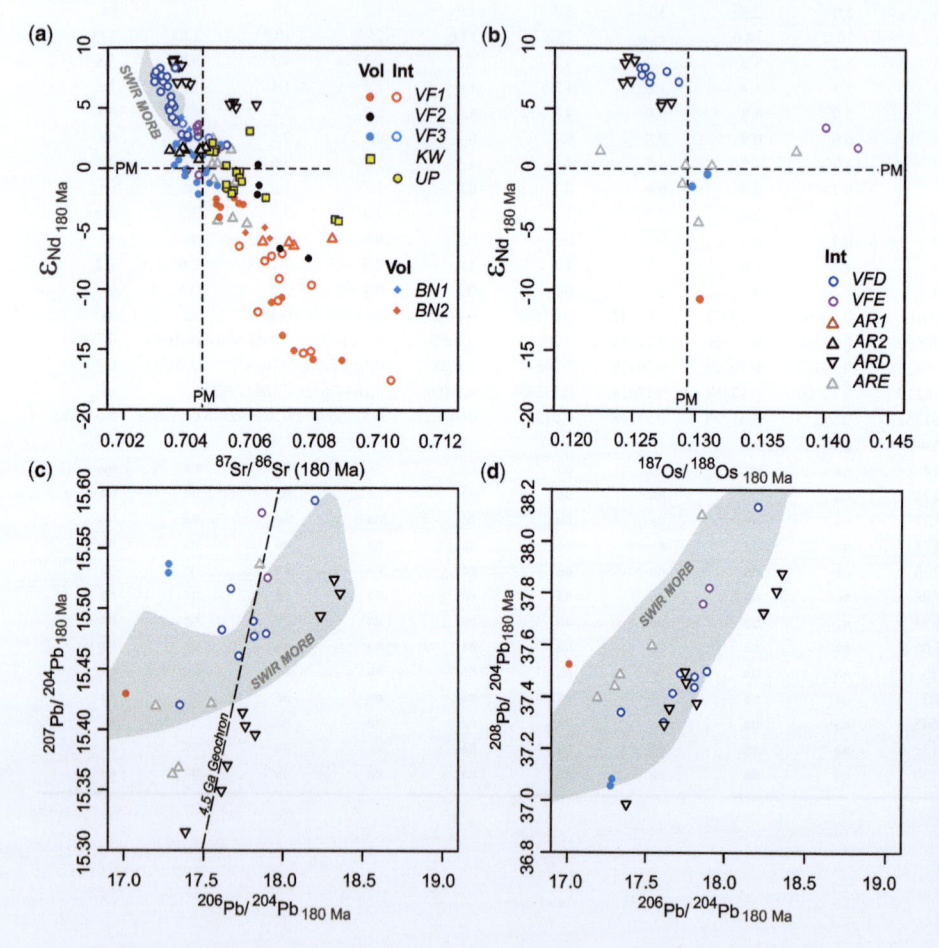

Fig. 10. Variations in initial (**a**) ε_{Nd} v. $^{87}Sr/^{86}Sr$, (**b**) ε_{Nd} v. $^{187}Os/^{188}Os$, (**c**) $^{207}Pb/^{204}Pb$ v. $^{206}Pb/^{204}Pb$ and (**d**) $^{208}Pb/^{204}Pb$ v. $^{206}Pb/^{204}Pb$ in the Karoo CFBs of DML calculated at 180 Ma. Compositions of different volcanic (Vol) and intrusive (Int) magma types are shown with filled and open symbols, respectively (except for Utpostane). Compositions of the SW Indian Ridge (SWIR) MORB (le Roex *et al.* 1983, 1992; Mahoney *et al.* 1992) are shown for comparison in (a), (d) and (e). PM is chondritic primitive mantle (Zindler and Hart 1986; Meisel *et al.* 2001). Geochemical and isotopic data sources and magma type abbreviations are as for Figure 9.

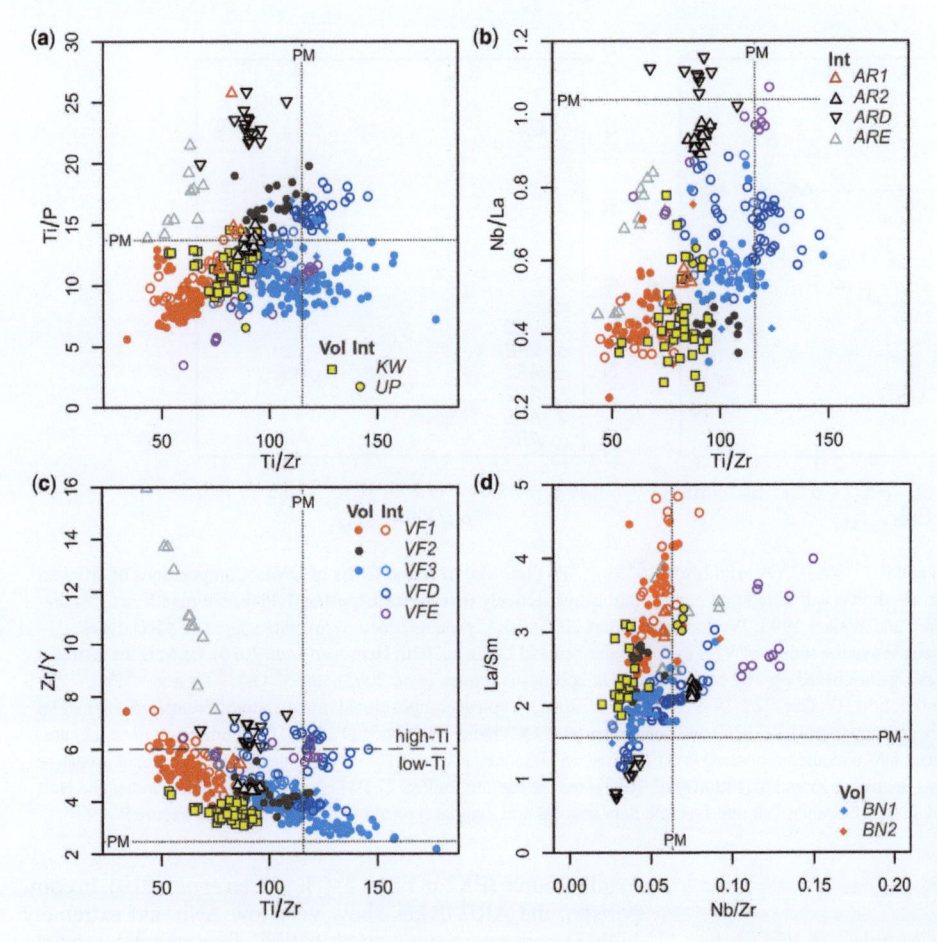

Fig. 11. Geochemical grouping of Karoo CFBs of DML into magma types in (**a**) Ti/P v. Ti/Zr, (**b**) Nb/La v. Ti/Zr, (**c**) Zr/Y v. Ti/Zr and (**d**) La/Sm v. Nb/Zr diagrams. Compositions of different volcanic (Vol) and intrusive (Int) magma types are shown with filled and open symbols, respectively (except for Utpostane). The conventional low-Ti v. high-Ti compositional boundary is shown in (c) (Erlank *et al.* 1988). PM is primitive mantle (McDonough and Sun 1995). Geochemical data sources and magma type abbreviations are as for Figure 9.

Similar to VF1, AR1 is typified by relatively low Ti/Zr (80–90), Ti/P (*c.* 15), Nb/La (*c.* 0.6) and Zr/Y (4–5) values (Fig. 11), as well as enriched initial Nd and Sr isotopic compositions (ε_{Nd} = –6; $^{87}Sr/^{86}Sr$ = 0.706–0.709) (Fig. 10a). Isotopic data on Os and Pb for AR1 are lacking.

Ahlmannryggen type 2 (AR2). Another group of dykes in Ahlmannryggen can be distinguished from other magma types by combined high Nb/La (*c.* 0.9) and low Zr/Y (*c.* 5) (Group 2 of Riley *et al.* 2005). Overall, Ahlmannryggen type AR2 is geochemically quite uniform with low Mg number (0.4–0.5), and intermediate TiO_2 (*c.* 2 wt%), Zr (*c.* 150 ppm), Ti/Zr (90–100) and Ti/P (12–15) values (Figs 9 & 11). Isotopically, these dykes are typified by mildly depleted compositions (ε_{Nd} = 1; $^{87}Sr/^{86}Sr$ = *c.* 0.704) (Fig. 11). The exceptionally wide dykes (5–80 m) in Ahlmannryggen belong to type AR2 (Riley *et al.* 2005). Isotopic data on Os and Pb for AR2 are lacking.

Ahlmannryggen depleted type (ARD). A group of high-Ti basalt and picrite dykes in Ahlmannryggen can be distinguished from other magma types in DML by their high Ti/P (20–25) and Nb/La (>1) values at intermediate Ti/Zr (60–110) (Fig. 11) (Group 3 of Riley *et al.* 2005). The ARD dykes are further characterized by notably high TiO_2 (3–5 wt %), FeO^T (12–14 wt%), Zr (200–300 ppm) and Y (35–45 ppm) at a given Mg number (0.5–0.8) (Fig. 9). High Zr/Y (6–7) underlines the enriched incompatible-element compositions of these dykes, but the strongly incompatible elements are relatively depleted, with, for example, La/Sm and Nb/Zr below primitive mantle values (Fig. 11). The ARD dykes include the isotopically most depleted samples of DML (ε_{Nd} = 5–9; $^{87}Sr/^{86}Sr$ 0.704–0.706) (Fig. 10a). The initial $^{187}Os/^{188}Os$ values are invariably low (0.124–0.128), but $^{206}Pb/^{204}Pb$ (17.4–18.4), $^{208}Pb/^{204}Pb$ (37.0–37.9) and $^{207}Pb/^{204}Pb$ (15.3–15.5) show notable variability (Fig. 10b–d).

Ahlmannryggen enriched type (ARE). Another high-Ti magma type in Ahlmannryggen is distinguished by its very high Zr (>300 ppm) (Group 4 of Riley *et al.* 2005). The ARE dykes resemble the ARD dykes in having exceptionally high TiO_2 (4–5 wt%) and Y (35–45 ppm) contents at a given Mg number (0.45–0.75) (Fig. 9) and plot apart from other magma types at low Ti/Zr (40–70) because of high Ti/P (14–22), Nb/La (0.5–0.9) and Zr/Y (7–16) values (Fig. 11). The initial Nd compositions cluster around primitive mantle (ε_{Nd} = –5 to +2) and those of Sr are mildly enriched ($^{87}Sr/^{86}Sr$ = 0.705–0.706) (Fig. 10a). The initial $^{187}Os/^{188}Os$ (0.122–0.138), $^{206}Pb/^{204}Pb$ (17.2–17.9), $^{207}Pb/^{204}Pb$ (15.4–15.5) and $^{208}Pb/^{204}Pb$ (37.4–38.1) values show notable variability (Fig. 10b–d).

Petrology: mantle sources and differentiation

Some of the intrusive rock types lack geochemical indications of crustal contamination and, therefore, provide the best source of information on the mantle sources (Luttinen *et al.* 1998; Luttinen and Furnes 2000; Riley *et al.* 2005; Heinonen and Luttinen 2008, 2010; Heinonen *et al.* 2010, 2013, 2014, 2015, 2018; Heinonen and Kurz 2015). Detailed isotopic (Nd, Sr, Pb and Os) and chemical studies of picrite samples of VFD, VFE, ARD and ARE type dykes have defined four mantle components that fall into depleted and enriched categories. Tholeiitic compositions indicate a relatively high degree of melting, which means that the ratios of highly incompatible elements in the nearly undifferentiated picrites closely correspond to those of their respective mantle sources.

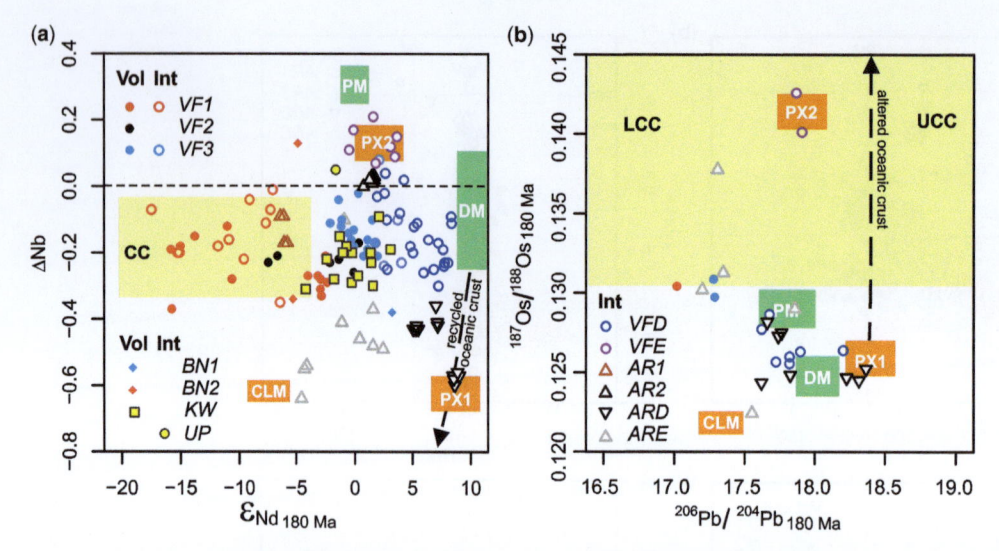

Fig. 12. Variation in (**a**) ΔNb v. ε_{Nd} (180 Ma) and (**b**) $^{187}Os/^{188}Os$ (180 Ma) v. $^{206}Pb/^{204}Pb$ (180 Ma) in Karoo CFBs of DML. Compositions of different volcanic (Vol) and intrusive (Int) magma types are shown with filled and open symbols, respectively (except for Utpostane). Plausible mantle sources are indicated: DM is depleted MORB mantle (Shirey and Walker 1998; Workman and Hart 2005), PX1 is the assumed pyroxenite source of ARD dykes (Heinonen et al. 2013, 2014), PX2 is the assumed pyroxenite source of VFE dykes (Heinonen and Luttinen 2010; Heinonen et al. 2010), CLM is the assumed continental lithospheric mantle with ΔNb and ε_{Nd} values based on 162 Ma Vestfjella lamproites (Luttinen et al. 2002), and $^{187}Os/^{188}Os$ and $^{206}Pb/^{204}Pb$ values are based on southern African peridotite xenoliths (Walker et al. 1989; Simon et al. 2007). Typical compositional range of undifferentiated continental crust (CC) and Archean and Proterozoic continental upper crustal (UCC) and lower crustal (LCC) rocks (Heinonen et al. 2010; Neumann et al. 2011) are shown for comparison. Mixing trajectory between DM and altered oceanic crust is based on Heinonen et al. (2014). ΔNb values indicate depleted (negative ΔNb) and undepleted (positive ΔNb) Nb contents relative to Zr and Y (Fitton et al. 1997) (see the caption for Fig. 2). PM is primitive mantle (Zindler and Hart 1986; McDonough and Sun 1995; Meisel et al. 2001). Geochemical and isotopic data sources and magma type abbreviations are as for Figure 9.

Mantle sources

Depleted mantle sources. The picrite samples of VFD type dykes exhibit Nd, Sr, Pb, Os and He isotopic and chemical characteristics indicative of a depleted upper-mantle peridotite source (DM in Fig. 12) notably similar to that of modern SW Indian Ocean mid-ocean ridge basalt (MORB) (e.g. ε_{Nd} = 8, La/Sm = c. 0.8 and Nb/Zr = c. 0.5 relative to primitive mantle) (Figs 10–12) (Heinonen and Luttinen 2008, 2010; Heinonen et al. 2010, 2018; Heinonen and Kurz 2015; Heinonen and Fusswinkel 2017). Variable contents of fluid-mobile Ba and Sr in the picrites and O isotopic data on olivine phenocrysts point to mild heterogeneity which has been interpreted to result from the addition of subduction-related fluids into sub-Gondwanan convective upper mantle (Heinonen et al. 2018). The ARD picrite dykes reveal another kind of Nd and Sr isotopically (e.g. ε_{Nd} = 9) depleted mantle source that shows strong relative depletion of highly incompatible elements (La/Sm = c. 0.5 and Nb/Zr = c. 0.5 relative to primitive mantle: Fig. 11) but has notably high concentrations of incompatible elements in general (Fig. 9). Their Sr and Os isotopic compositions and whole-rock and olivine chemical characteristics point to derivation from pyroxenite-rich mantle (PX1 in Fig. 12) that may contain recycled altered oceanic crust (Riley et al. 2005; Heinonen et al. 2013, 2014).

Enriched mantle sources. The OIB-like high-Ti VFE and ARE dykes record enriched mantle sources. Both types have a chemical signature of pyroxenite-rich sources with broadly similar mildly depleted Nd and Sr isotopic compositions (ε_{Nd} = 4) (Riley et al. 2005; Heinonen and Luttinen 2008; Heinonen et al. 2010, 2013). A detailed geochemical comparison (e.g. Fig. 11) indicates differences that require generation of the VFE and ARE dykes from distinctive reservoirs, however. Most notably, VFE has unusually high (i.e. primitive mantle-like) ΔNb, and represents the only Nb-undepleted magma type of DML and reveals an enriched (Nb/Zr = 0.10–0.15, La/Sm = 2–4 and $^{187}Os/^{188}Os$ = 0.14: Figs 11 & 12) pyroxenite-rich mantle source (PX2 in Fig. 12) (Heinonen et al. 2010). In comparison, the ARE dykes show very low ΔNb and extremely high Zr contents. Such a combination of geochemical indications of both depletion and enrichment is common in highly alkaline lamproite magmas which are presumably derived from continental lithospheric mantle (Vollmer et al. 1984; Bergman 1987; Luttinen et al. 2002). The most enriched samples of the geochemically variable ARE dykes show particularly strong geochemical affinities to lamproite-type magmas and may record a lithospheric mantle source for DML magmas (CLM in Fig. 12). The less enriched ARE dykes may represent the hybridization of magmas from convective and lithospheric mantle sources (Fig. 13) (Riley et al. 2005). Variations in Os and Pb isotopic compositions point to the additional involvement of lower continental crust in the petrogenesis of the ARE dykes (Fig. 12b).

Differentiation

The most recent petrogenetic studies of Jurassic magmatism in DML suggest that differentiation is largely responsible for the wide compositional ranges in the flood basalt lavas. Depleted upper mantle is the only plausible mantle source which could explain the generation of the predominant flood basalt types by magmatic differentiation processes (Luttinen et al. 2015; Heinonen et al. 2016). The ΔNb parameter (=1.73 + log (Nb/Y) −1.94 log(Zr/Y): Fig. 2) is a particularly useful tracer of parent–daughter relationships as it is not strongly affected by crustal contamination (Fitton et al. 1997; Baksi 2001; Luttinen 2018). For example, the ARD and ARE magma types, both of which have been associated with pyroxenite-rich mantle sources, plot apart from the general compositional trend of DML magmas due to the markedly low ΔNb at a given ε_{Nd}, whereas exceptionally high ΔNb, TiO_2 and $^{187}Os/^{188}Os$ compositions render the VFE magma type an implausible parental magma type (Figs 12 & 13).

In contrast, primitive VFD type dykes consistently define an endmember composition in DML and have been postulated to

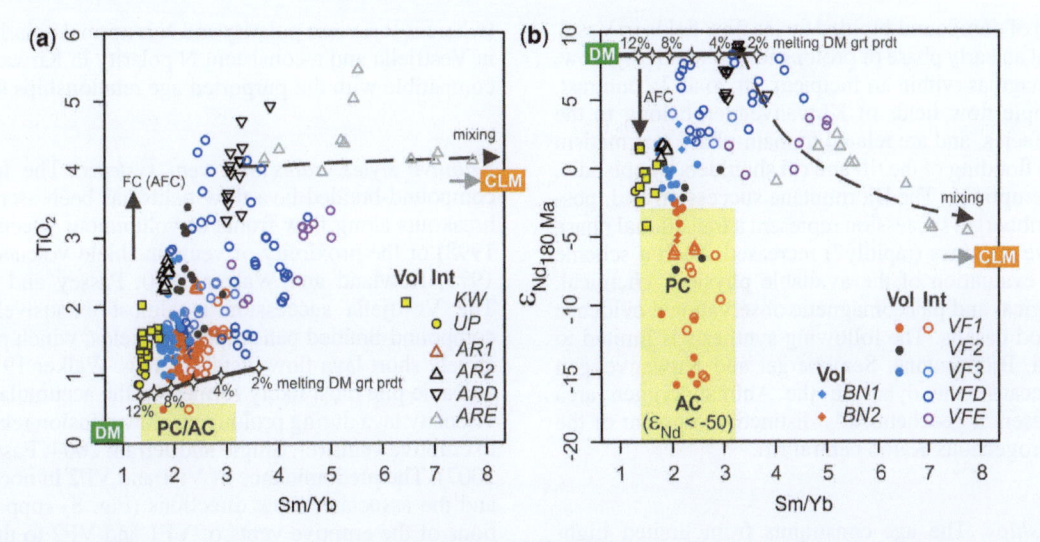

Fig. 13. Variations in (a) TiO$_2$ (wt%) and (b) $\varepsilon_{Nd_{180 Ma}}$ v. Sm/Yb in the Karoo CFBs of DML. Compositions of different volcanic and intrusive magma types are shown with filled and open symbols, respectively (except for Utpostane). Assumed principal mantle sources are indicated: DM is depleted MORB mantle (Workman and Hart 2005) and CLM (Sm/Yb = c. 18) is the assumed continental lithospheric mantle source of 162 Ma Vestfjella lamproites (Luttinen et al. 2002). AC and PC illustrate likely compositions of Archean and Proterozoic upper crustal rock types, respectively, in DML (Luttinen et al. 2010, 2015). Compositions of DM-sourced melts have been modelled using non-modal batch melting equations using a garnet lherzolite source (ol 57%, opx 32%, cpx 6%, grt 5%) and melt mode ol:opx:cpx:grt = 1:1:4:4. The geochemical effect of fractional crystallization and crustal contamination is schematically indicated by the arrow, and a binary mixing model illustrates hybridization between ARD-type magmas (sample Z1816.1; Riley et al. 2005) and CLM-sourced lamproite-like melts. Geochemical and isotopic data sources and magma type abbreviations are as for Figure 9.

record the mantle source of the various flood basalt magma types. In general, quantitative energy-constrained assimilation and fractional crystallization (EC-AFC: Bohrson and Spera 2001) models show that the low-Ti magma types of Vestfjella can be explained quite well by combined fractional crystallization, wall-rock melting, and contamination of VFD-type low-Ti parental magmas with Archean (VF1) and Proterozoic (VF2, VF3) crust (Luttinen et al. 2015; Heinonen et al. 2016) (Fig. 13). An additional lamproite-like contaminant from the continental lithospheric mantle has been associated with some VF3 magmas and the enriched members of VFD dykes. It is worthwhile to note that the VFD dykes record variable degrees of partial melting of depleted mantle, and the parental magmas of the flood basalts represent a relatively high degree of melting (Fig. 13) (Heinonen et al. 2016).

The origin of the Kirwanveggen magma type has not been modelled quantitatively, but its diagnostic features are compatible with the crustal contamination of low-Ti parental magmas that were generated from depleted mantle. Systematically low Zr/Y and Sm/Yb values suggest that the parental magmas of the KW lavas required a somewhat higher degree of melting than those of the Vestfjella lavas (Fig. 13).

Reconstruction of the DML volcanic system

The reconstruction of volcanic systems requires a detailed understanding of the chemical and physical features of the structural components, and the timing of the endogenic and exogenic processes involved. In the case of Jurassic Karoo volcanism, it is crucial to bear in mind that the igneous system is composed of two compositionally different subprovinces, North Karoo and South Karoo (Figs 1 & 2), and that the origin of the two subprovinces may have been fundamentally different with regard to the principal mantle sources and timing of magmatism (Luttinen 2018). Chemical variability within both provinces has been mapped in some detail, whereas the physical volcanological data are rather patchy and the temporal constraints are strongly biased to North Karoo. The high-precision mineral age data (e.g. Zhang et al. 2003; Jourdan et al. 2007b, 2008; Svensen et al. 2012; Luttinen et al. 2015) and palaeomagnetic record (Løvlie 1979; Peters 1989; Hargraves et al. 1997) manifest: (1) rapid province-wide emplacement of the main volume of lavas, possibly within c. 0.5 myr; and (2) continued magmatism in the Karoo triple rift after the province-wide peak; and (3) raises the question as to whether significant volcanism within South Karoo predated the main stage (Table 1). The Jurassic flood basalt exposures of DML are relatively small but provide detailed snapshots of a rift and rift-shoulder volcanic system relevant to the origin and evolution of the compositionally and structurally complex South Karoo CFB succession.

The lava flow facies concept of Jerram and co-workers (e.g. Jerram 2002; Single and Jerram 2004; Jerram and Widdowson 2005) provides a useful conceptual framework for assessing the variable lava-stacking patterns and geochemical compositions which typify the Jurassic volcanism in DML (Figs 8–13).

Structural analysis of the volcanic edifice of three CFB provinces (the Deccan Traps, Paraná-Etendeka and Northern Atlantic Igneous Province) led Jerram and Widdowson (2005) to propose three characteristic phases of flood basalt volcanism: initial, relatively low-volume eruptions are confined by pre-existing lithospheric structures and related topography, and typically produce compound-braided type stacking. The main phase represents a rapid culmination of province-wide igneous activity and is typified by repeated emplacement of large tabular-type flows during episodes of voluminous eruptions. The waning phase characteristically involves localized eruptions and is commonly associated with increasing silica contents. The lava-stacking pattern of the Tertiary flood basalt succession of Skye has been studied in great detail (Single and Jerram 2004) and may exemplify the transition from initial to main phase flood basalt magmatism in the Northern Atlantic Igneous Province (Jerram and Widdowson 2005).

Volcanic architecture of DML flood basalts

It is proposed that the volcanic exposures of DML comprise a flood basalt succession that is structurally notably similar to that described for the Skye flood basalts: the thick accumulation of compound-braided facies flow fields in Vestfjella represent an early phase of prolonged, but relatively slow, effusion of magmas within an incipient rift zone. In contrast, the thick simple flow fields of Kirwanveggen belong to the tabular-type facies, and are related to main-phase magmatism that led to the flooding of the rift and rift shoulders by episodic, voluminous eruptions. The Björnnutane succession and, possibly, the Sembberget succession represent a transitional phase when eruptive volumes (rapidly?) increased. Such a scheme helps in the evaluation of the available physical, chemical, geochronological and palaeomagnetic observational evidence for DML flood basalts. The following synthesis is limited to the Vestfjella, Björnnutane, Sembberget and Kirwanveggen exposures because the dykes in the Ahlmannryggen area seem to represent a geochemically distinctive segment of the laterally heterogeneous Karoo central rift.

Age relationships. The age constraints from limited high-precision results of $^{40}Ar-^{39}Ar$ plagioclase and U–Pb zircon dating, and the available palaeomagnetic data are compatible with a view that Vestfjella represents relatively old, and that Sembberget and Kirwanveggen represent younger, stratigraphic levels of the Jurassic volcanic edifice (Fig. 14; Table 2). Specifically, the age data for Vestfjella are exclusively from the southern exposures, which appear to record the deepest section in DML: the 189–185 Ma ages of the VF1 and VF3 type dolerites are likely to correspond to the age of the lava pile and most of the geochemically similar cross-cutting dykes. The younger ages represent two *c.* 500 m-wide 182 Ma gabbroic dykes and a 181 Ma layered intrusion which intruded a thick accumulated lava pile. Crucially, the Utpostane layered intrusion and the coeval 181 Ma Sembberget lava (Table 2) are geochemically indistinguishable from the Kirwanveggen lavas (Figs 9–13), which suggests rapid emplacement of the KW-type magmas at *c.* 181 Ma (Fig. 14). The palaeomagnetic record of R–N(–R?) polarity (where is R reverse polarity and N is normal polarity) reversals in Vestfjella and a consistent N polarity in Kirwanveggen are compatible with the purported age relationships (Fig. 14).

Eruptive styles, rates and vent systems. The formation of compound-braided lava flow fields has been ascribed to lava breakouts along flow fronts of voluminous sheets (Self *et al.* 1998) or the proximity of vents in shield volcanoes (Walker 1973; Rowland and Walker 1990; Passey and Bell 2007). The Vestfjella succession is almost exclusively made of compound-braided pāhoehoe flow fields, which points to relatively short lava flow distances (e.g. Walker 1973), and the volcanic pile most likely formed by the accumulation of low-viscosity lava during prolonged, slow effusion relatively close to eruptive vents (cf. Single and Jerram 2004; Passey and Bell 2007). The predominance of VF1 and VF2 in north Vestfjella and the associated flow directions (Fig. 8) support the locations of the eruptive vents of VF1 and VF2 to the north and east of Basen and Plogen, respectively. The southern (and possibly central) Vestfjella is principally composed of VF3 (Luttinen and Furnes 2000). The flow directions for VF3 are quite variable, mainly from broadly SW in the southernmost Vestfjella and from the east in northern and central Vestfjella (Hjelle and Winsnes 1972; Furnes *et al.* 1987; Luttinen and Furnes 2000) (Fig. 8), and are generally suggestive of vent systems to the SE of Vestfjella. The intercalated flow fields of VF1–VF3 lavas may record lavas emplaced between three shield volcanoes. In contrast, the notably thick simple flow fields of the tabular-type facies at Kirwanveggen point to high effusion rates (Self *et al.* 1998; Harris and Rowland 2009) and successive flooding of the rift shoulders by large inflated pāhoehoe units from fissure eruptions. The occurrences of rubbly flow-top breccias lend support to high effusion rates (Keszthelyi *et al.* 2006). Evidence of voluminous outpouring of melt and consistent flow direction towards the east (Harris *et al.* 1990) (Fig. 8) can be associated with the opening of

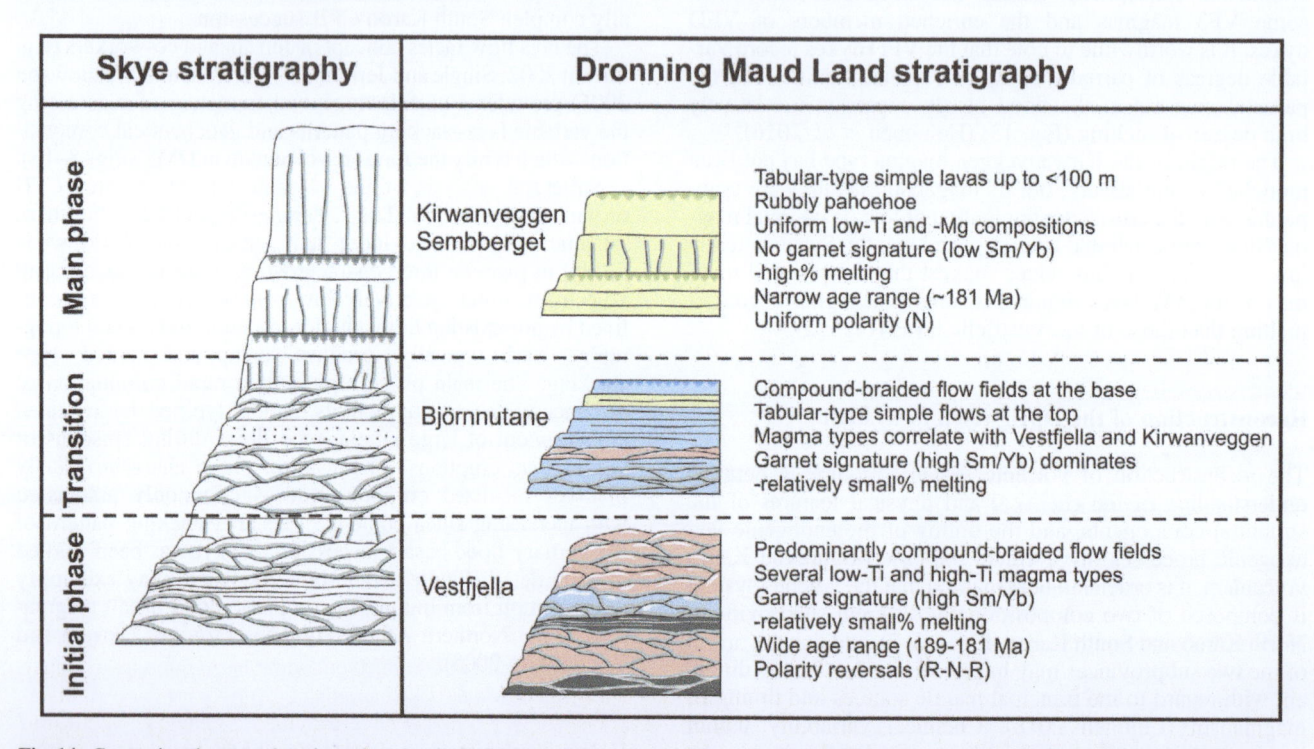

Fig. 14. Comparison between the schematic volcanic stratigraphy of Skye, the Northern Atlantic Volcanic Province and DML Karoo province. The Skye lava succession has been divided into (i) initial phase, (ii) transitional phase and (iii) main phase of eruptive activity (Single and Jerram 2004). The Vestfjella, Björnnutane and Kirwanveggen–Sembberget lava outcrops are ascribed to a similar evolutionary sequence (colours illustrate chemical variations as in Fig. 8). Age data are from Table 2. Palaeomagnetic polarity data are from Løvlie (1979), Peters (1989) and Hargraves *et al.* (1997). The Skye stratigraphy is modified after Single and Jerram (2004).

major fissure vents to the west of Kirwanveggen (i.e. in the general direction of the rift axis) (Fig. 3).

Tectonic setting. The supracrustal strata from the African rifted margin indicate that incipient Jurassic volcanism along the southern arm of the Karoo triple rift occurred in a pre-existing basin (Watkeys 2002), and evidence from the triple junction area suggests that the emplacement of the Jurassic dykes may have been controlled by Precambrian structures (Jourdan *et al.* 2004; Riley *et al.* 2005). Detailed mapping of the African rifted margin in the central Lebombo area led Klausen (2009) to propose that the Karoo rift-zone magmatism was associated with rapid crustal extension, which is also supported by up to 14% dilation recorded by dolerite dykes in southern Vestfjella (Luttinen *et al.* 2015). Consequently, the Vestfjella lavas probably erupted within an actively developing rift system. The lack of intrusions in the Kirwanveggen area is consistent with the drowning of the rift shoulder only during the main phase when voluminous inflated pāhoehoe units were able to flow far from the fissure vents (Walker 1973).

Magma compositions. The geochemical compositions are compatible with physical implications for different eruption rates. Most importantly, the Kirwanveggen-type lavas have low incompatible-element contents at given Mg number and are distinguished by their low Sm/Yb. Relative depletion of the heavy rare earth elements (HREE) has been frequently used as a proxy for mantle melting conditions (e.g. Tegner *et al.* 1998). Specifically, a garnet-bearing mantle source produces relatively high Sm/Yb because the residual garnet is a sink for the HREE. The low Sm/Yb values thus suggest that the mantle source of the Kirwanveggen lavas was poor in garnet either because the melting took place at relatively low pressures where garnet is not stable or because a high degree of melting led to efficient consumption of garnet. Given the low abundances of incompatible elements and the tholeiitic compositions, the degree of melting must have been relatively high. The melting model in Figure 13 suggests that *c.* 12% partial melting of depleted mantle would produce parental magmas with low Sm/Yb typical of Kirwanveggen lavas. Overall, the geochemical features of the Kirwanveggen-type lavas thus imply high melt productivity, which is compatible with high eruption rates indicated by the physical dimensions of the lavas. The homogeneity of the Kirwanveggen lavas, their low Mg numbers and high volumes point to tapping of the magmas from a large magma chamber in which efficient homogenization was possible.

In comparison, the Vestfjella magma types are compositionally quite variable. Some of the magma types have high Sm/Yb due to hybridization with lithospheric mantle material, but the predominant crustally contaminated magma types have probably preserved mantle-derived Sm/Yb quite well. Accordingly, systematically high Sm/Yb values in Vestfjella are compatible with a relatively low degree of melting (Fig. 13). The succession appears to be fed by several plumbing systems typified by long-lived, but rather slow-rate, eruptions and intercalation of flows from closely spaced vents, possibly in juxtaposed shields. Thick tabular-type VF3 lava flows at Björnnutane and wide VF1 gabbro dykes in Vestfjella suggest that the eruption rates may have been increased at the onset of the main phase of magmatism at *c.* 182 Ma (Table 2).

Flood basalts of Dronning Maud Land and the Gondwana LIP

Summarizing, geological mapping of the scattered outcrops of western DML has defined the overall compositional structure of the Antarctic portion of the Karoo continental flood basalt (CFB) province. The DML system is clearly genetically related to geochemically similar basalts and picrites of the Lebombo Monocline at the African rifted margin (Faure *et al.* 1979; Harris *et al.* 1990; Luttinen and Furnes 2000). The compositionally diverse Vestfjella suite probably represents a conjugate margin of the southernmost Lebombo (Luttinen and Furnes 2000; Luttinen *et al.* 2010, 2015), whereas the Jurassic dykes of the Ahlmannryggen–H.U.Sverdrupfjella region show compositional affinities to magma types of the Mwenezi area at the triple junction (Riley *et al.* 2005) (Fig. 1). The compositionally uniform lava suites of Kirwanveggen and Sembberget probably represent a rift-shoulder setting and, apart from the Utpostane layered intrusion, they lack geochemical equivalents elsewhere in the Karoo province.

Geochemical models support the view that the low-Ti flood basalt magmas of DML were mainly generated from depleted upper mantle, and were differentiated into numerous compositional types in physically distinctive plumbing systems largely due to the interaction with different kinds of Archean and Proterozoic crustal wall rocks (Luttinen *et al.* 2015; Heinonen *et al.* 2016). Subordinate high-Ti magma types are confined to the rifted margin, mainly show indications of pyroxenite-rich sources, and are likely to record tapping of enriched components in convective and lithospheric mantle (Heinonen *et al.* 2010, 2013, 2014; Luttinen *et al.* 2015). Recent studies have proposed similar models for the high-Ti magma types at the African rifted margin (Harris *et al.* 2015; Kamenetsky *et al.* 2017).

The general physical architecture of the Jurassic volcanic edifice reveals contrasting lava-flow stacking patterns, and, consequently, different eruption rates and flow distances for the rift and rift-shoulder settings. The age relationships are still ambiguous, but the available geochemical, palaeomagnetic and geochronological data, and the physical architecture of the lava successions are compatible with progressively increased eruption volumes and filling of the Jurassic rift by flood basalts (Fig. 14). Two rift-zone gabbro dykes have been dated at 182 Ma using the U–Pb zircon method (Luttinen *et al.* 2015), whereas most of the published ^{40}Ar–^{39}Ar age data suffer from Ar-loss, excess argon, recoil, use of whole-rock samples or insufficient documentation, and do not yield reliable plateau ages (e.g. Brewer *et al.* 1996; Duncan *et al.* 1997; Zhang *et al.* 2003; Riley *et al.* 2005; Curtis *et al.* 2008; Luttinen *et al.* 2015). Sparse reliable ^{40}Ar–^{39}Ar age data are suggestive of early onset (possibly at *c.* 189 Ma), and a long duration of magmatic–tectonic activity in the rift environment and relatively rapid emplacement of lavas on the rift shoulder (Table 2). Obviously, further geochronological research is required to establish the absolute ages of the different structural components of the DML volcanic system and their relationships to African Karoo. Combined flow-by-flow palaeomagnetic and geochemical sampling and detailed structural analysis has the potential to decipher the evolutionary stages of the DML igneous system (e.g. magma chamber processes, fluctuating magma production rates and eruption volumes, duration of emplacement, and cooling of lavas), many of which may have occurred rapidly in timescales not resolvable using radiometric dating.

The Jurassic flood basalt system of western DML represents a small part of the greater Gondwana large igneous province (LIP), but the lessons learned have province-wide implications. The VFD dykes have provided a detailed insight into a depleted upper-mantle source which may have been the predominant mantle source of the South Karoo subprovince as a whole (Figs 1 & 2) (Ellam and Cox 1991; Luttinen 2018). Importantly, the depleted source was probably variably modified by subducted crustal material: selective enrichment of large ion lithophile elements in the picrites and elevated oxygen isotopic ratios in the olivine phenocrysts indicate that the source material of the low-Ti flood basalts was mildly affected by subduction-related fluids (Heinonen et al. 2018), whereas the various high-Ti compositions may record recycled oceanic crustal components (Heinonen et al. 2010, 2014). The geochemical traces of subduction in the Nb-depleted South Karoo subprovince could stem from long-distance transportation of subducted material from the coeval subduction zone along the Gondwana margin (Rapela et al. 2005; Wang et al. 2015) (inset in Fig. 1) or preservation of an ancient subduction signature generated in Paleozoic or even Proterozoic time (Choi et al. 2019).

Overall, the boundary betweeen the low-ΔNb and high-ΔNb tholeiites (Fig. 1) is likely to reflect plume- and subduction-related large-scale mantle provinciality. The Nb-undepleted North Karoo subprovince, which dominates the African part of the Gondwana LIP, was formed directly above the sub-African large low-shear-velocity anomaly of deep mantle (Torsvik et al. 2010), and its rapid emplacement (Svensen et al. 2012) and geochemical features are compatible with a mantle-plume source (Luttinen 2018). In contrast, the Nb-depleted South Karoo and Ferrar subprovinces (Figs 1 & 2) were probably generated from upper-mantle sources and largely outside the limits of the deep mantle anomaly. However, the South Karoo tholeiites in DML and Africa were formed close to the purported sub-African mantle-plume convection pattern, and the early onset and exceptionally high mantle potential temperatures suggested for magmatism in DML (Heinonen et al. 2015; Luttinen et al. 2015) could stem from the arrival of a mantle plume underneath North Karoo. The voluminous main phase may have been triggered by further plume-related heating of the ambient upper mantle and rapid lithospheric extension.

Despite their different silica contents and Sr isotopic compositions (Faure and Elliot 1971), the Nb-depleted South Karoo and Ferrar subprovinces (Figs 1 & 2) could have been derived from parts of the same overall upper-mantle reservoir that were mildly (South Karoo) and strongly (Ferrar) modified by subducted crustal material (e.g. Hergt et al. 1991; Luttinen 2018). Interestingly, the fundamental petrological boundary in the Gondwana LIP appears to coincide remarkably well with the boundary between East and West Gondwana.

Acknowledgements I thank the editor John Smellie for the invitation to write this review article. I am grateful to Phil Leat, Harald Furnes and Henrik Grind for providing field notes. Fred Jourdan provided the ^{40}Ar–^{39}Ar age data listed in Supplementary Table S1. Katja Bohm helped in the preparaton of the line drawings, and the geochemical plots are based on the GCDkit tool developed and kindly distributed by Vojtěch Janoušek. The crews of Finnish, Swedish, Norwegian, Russian, German, British and South African Antarctic expeditions have facilitated field operations during the period 1989–2018. John Smellie and the reviewers Tom Fleming and Christopher Galerne provided critical and helpful comments which helped to improve the paper.

Author contributions AVL: conceptualization (lead), funding acquisition (lead), writing – original draft (lead), writing – review & editing (lead).

Funding This study has been funded by Academy of Finland (project 305663).

Data availability The ^{40}Ar–^{39}Ar age data generated during the current study are available in Supplementary material Table S1 and the compiled geochemical dataset used in the current study is available in Supplementary material Table S2.

References

Antonini, P., Piccirillo, E.M., Petrini, R., Civetta, L., D'Antonio, M. and Orsi, G. 1999. Enriched mantle – Dupal signature in the genesis of the Jurassic Ferrar tholeiites from Prince Albert Mountains (Victoria Land, Antarctica). *Contributions to Mineralogy and Petrology*, **136**, 1–19, https://doi.org/10.1007/s004100050520

Arndt, N.T., Todt, W., Chauvel, C., Tapfer, M. and Weber, K. 1991. U–Pb zircon age and Nd isotopic composition of granitoids, charnockites and supracrustal rocks from Heimefrontfjella, Antarctica. *Geologisches Rundschau*, **80**, 759–777, https://doi.org/10.1007/BF01803700

Aucamp, A.P.H., Wolmarans, L.G. and Neethling, D.C. 1972. The Urfjell Group, a deformed (?) early Paleozoic sedimentary sequence, Kirwanveggen, western Dronning Maud Land. *In*: Adie, R.J. (ed.) *Antarctica Geology and Geophysics*. Universitetsforlaget, Oslo, 557–561.

Baksi, A.K. 2001. Search for a deep-mantle component in mafic lavas using a Nb–Y–Zr plot. *Canadian Journal of Earth Science*, **38**, 813–824, https://doi.org/10.1139/e00-100

Barton, J.M., Klemdt, R., Allsopp, H.L., Auret, S.H. and Copperthwaite, Y.L. 1987. The geology and geochronology of the Annandagstoppane granite, western Dronning Maud Land, Antarctica. *Contributions to Mineralogy and Petrology*, **97**, 488–496, https://doi.org/10.1007/BF00375326

Bell, B.R., Williamson, I.T., Head, F.E. and Jolley, D.W. 1996. On the origin of a reddened interflow bed within the Palaeocene lava field of north Skye. *Scottish Journal of Geology*, **32**, 117–126, https://doi.org/10.1144/sjg32020117

Bergman, S.C. 1987. Lamproites and other potassium-rich igneous rocks: a review of their occurrence, mineralogy and geochemistry. *Geological Society, London, Special Publications*, **30**, 103–190, https://doi.org/10.1144/GSL.SP.1987.030.01.08

Bevins, R.E., Rowbotham, G. and Robinson, D. 1991. Zeolite to prehnite–pumpellyite facies metamorphism of the late Proterozoic Zig-Zag Dal Basalt Formation, eastern North Greenland. *Lithos*, **27**, 155–165, https://doi.org/10.1016/0024-4937(91)90010-I

Bohrson, W.A. and Spera, F.J. 2001. Energy-constrained open-system magmatic processes II: application of energy-constrained assimilation-fractional crystallization (EC-AFC) model to magmatic systems. *Journal of Petrology*, **42**, 1019–1041, https://doi.org/10.1093/petrology/42.5.1019

Bondre, N.R., Duraiswami, R.A. and Dole, G. 2004. Morphology and emplacement of flows from the Deccan Volcanic Province, India. *Bulletin of Volcanology*, **66**, 29–45, https://doi.org/10.1007/s00445-003-0294-x

Brewer, T.S., Hergt, J.M., Hawkesworth, C.J., Rex, D.C. and Storey, B.C. 1992. Coats Land dolerites and the generation of Antarctic continental flood basalts. *Geological Society, London, Special Publications*, **68**, 185–208, https://doi.org/10.1144/GSL.SP.1992.068.01.12

Brewer, T.S., Rex, D., Guise, P.G. and Hawkesworth, C.J. 1996. Geochronology of Mesozoic tholeiitic magmatism in Antarctica: implications for the development of the failed Weddell Sea rift system. *Geological Society, London, Special Publications*, **108**, 45–62, https://doi.org/10.1144/GSL.SP.1996.108.01.04

Bryan, S.E., Peate, I.U. et al. 2010. The largest volcanic eruptions on Earth. *Earth-Science Reviews*, **102**, 207–229, https://doi.org/10.1016/j.earscirev.2010.07.001

Burgess, S.D., Bowring, S.A., Fleming, T.H. and Elliot, D.H. 2015. High-precision geochronology links the Ferrar large igneous province with early-Jurassic ocean anoxia and biotic crisis. *Earth and Planetary Science Letters*, **415**, 90–99, https://doi.org/10.1016/j.epsl.2015.01.037

Burke, K. and Dewey, J.F. 1973. Plume-generated triple junctions: key indicators in applying plate tectonics to old rocks. *Journal of Geology*, **81**, 406–433, https://doi.org/10.1086/627882

Cashman, K.V., Thornber, C. and Kauahikaua, J.P. 1999. Cooling and crystallization of lava in open channels, and the transition of Pahoehoe Lava to 'A'ā. *Bulletin of Volcanology*, **61**, 306–323, https://doi.org/10.1007/s004450050299

Choi, S.H., Mukasa, S.B., Ravizzac, G., Fleming, T.H., Marsh, B.D. and Bédard, J.H.J. 2019. Fossil subduction zone origin for magmas in the Ferrar Large Igneous Province, Antarctica: Evidence from PGE and Os isotope systematics in the Basement Sill of the McMurdo Dry Valleys. *Earth and Planetary Science Letters*, **506**, 507–519, https://doi.org/10.1016/j.epsl.2018.11.027

Corner, B. 1994. *Geological Evolution of Western Dronning Maud Land within a Gondwana Framework: Geophysics Subprogramme*. Final project report to South African Committee for Antarctic Research (SACAR).

Cox, K.G. 1988. The Karoo Province. *In*: MacDougall, J.D. (ed.) *Continental Flood Basalts*. Kluwer, Boston, MA, 239–271, https://doi.org/10.1007/978-94-015-7805-9_7

Cox, K.G., MacDonald, R. and Hornung, G. 1967. Geochemical and petrographic provinces in the Karoo basalts of southern Africa. *American Mineralogist*, **52**, 1451–1474.

Curtis, M.L., Riley, T.R., Owens, W.H., Leat, P.T. and Duncan, R.A. 2008. The form, distribution and anisotropy of magnetic susceptibility of Jurassic dykes in H.U. Sverdrupfjella, Dronning Maud Land, Antarctica. Implications for dyke swarm emplacement. *Journal of Structural Geology*, **30**, 1429–1447, https://doi.org/10.1016/j.jsg.2008.08.004

Duncan, A.R., Erlank, A.J. and Marsh, J.S. 1984. Geochemistry of the Karoo igneous province. *Geological Society of South Africa Special Publications*, **13**, 355–388.

Duncan, R.A., Hooper, P.R., Rehacek, J., Marsh, J.S. and Duncan, A.R. 1997. The timing and duration of the Karoo igneous event, southern Gondwana. *Journal of Geophysical Research: Solid Earth*, **102**, 18 127–18 138, https://doi.org/10.1029/97JB00972

Elburg, M.A. and Soesoo, A. 1999. Jurassic alkali-rich volcanism in Victoria (Australia): lithospheric v. asthenospheric source. *Journal of African Earth Sciences*, **29**, 269–280, https://doi.org/10.1016/S0899-5362(99)00096-2

Ellam, R.M. and Cox, K.G. 1991. An interpretation of Karoo picrite basalts in terms of interaction between asthenospheric magmas and the mantle lithosphere. *Earth and Planetary Science Letters*, **105**, 330–342, https://doi.org/10.1016/0012-821X(91)90141-4

Elliot, D.H. and Fleming, T.H. 2000. Weddell triple junction: The principal focus of Ferrar and Karoo magmatism during initial breakup of Gondwana. *Geology*, **28**, 539–542, https://doi.org/10.1130/0091-7613(2000)28<539:WTJTPF>2.0.CO;2

Elliot, D.H. and Fleming, T.H. 2018. The Ferrar Large Igneous Province: field and geochemical constraints on supra-crustal (high-level) emplacement of the magmatic system. *Geological Society, London, Special Publications*, **463**, 41–58, https://doi.org/10.1144/SP463.1

Elliot, D.H., Fleming, T.H., Haban, M.A. and Siders, M.A. 1995. Petrology and Mineralogy of the Kirkpatrick Basalt and Ferrar Dolerite, Mesa Range Region, North Victoria Land, Antarctica. *American Geophysical Union Antarctic Research Series*, **67**, 103–141, https://doi.org/10.1002/9781118668207.ch7

Elliot, D.H., Fleming, T.H., Kyle, P.R. and Foland, K.A. 1999. Long-distance transport of magmas in the Jurassic Ferrar Large Igneous Province, Antarctica. *Earth and Planetary Science Letters*, **167**, 89–104, https://doi.org/10.1016/S0012-821X(99)00023-0

Encarnación, J., Fleming, T.H., Elliot, D.H. and Eales, H.V. 1996. Synchronous emplacement of Ferrar and Karoo dolerites and the early breakup of Gondwana. *Geology*, **24**, 535–538, https://doi.org/10.1130/0091-7613(1996)024<0535:SEOFAK>2.3.CO;2

Erlank, A.J., Duncan, A.R., Marsh, J.S., Sweeney, R.J., Hawkesworth, C.J., Milner, R.McG. and Rogers, N.W. 1988. A laterally extensive geochemical discontinuity in the subcontinental Gondwana lithosphere. *In: Proceedings of the Geochemical Evolution of the Continental Crust Conference, Pocos de Caldes, Brazil, 11–16 July, 1988*. International Association of Geochemistry and Cosmochemistry, Columbus, OH, 1–10.

Faure, G. and Elliot, D.H. 1971. Isotope composition of strontium in Mesozoic basalt and dolerite from Dronning Maud Land. *British Antarctic Survey Bulletin*, **25**, 23–27.

Faure, G., Bowman, J.R. and Elliot, D.H. 1979. The initial $^{87}Sr/^{86}Sr$ ratios of the Kirwan volcanics of Dronning Maud Land: comparison with the Kirkpatrick Basalt, Transantarctic Mountains. *Chemical Geology*, **26**, 77–90, https://doi.org/10.1016/0009-2541(79)90031-7

Fitton, J.G., Saunders, A.D., Norry, M.J., Hardarson, B.S. and Taylor, R.N. 1997. Thermal and chemical structure of the Iceland plume. *Earth and Planetary Science Letters*, **153**, 197–208, https://doi.org/10.1016/S0012-821X(97)00170-2

Fleming, T.H., Elliot, D.H., Jones, L.M., Bowman, J.R. and Siders, M.A. 1992. Chemical and isotopic variations in an iron-rich lava flow from the Kirkpatrick Basalt, north Victoria Land, Antarctica: implications for low-temperature alteration. *Contributions to Mineralogy and Petrology*, **111**, 440–457, https://doi.org/10.1007/BF00320900

Fleming, T.H., Foland, K.A. and Elliot, D.H. 1995. Isotopic and chemical constraints on the crustal evolution and source signature of Ferrar magmas, north Victoria Land, Antarctica. *Contributions to Mineralogy and Petrology*, **121**, 217–236, https://doi.org/10.1007/BF02688238

Furnes, H. and Mitchell, J.G. 1978. Age relationships of Mesozoic basalt lava and dykes in Vestfjella, Dronning Maud Land, Antarctica. *Norsk Polarinstitutt Skrifter*, **169**, 45–68.

Furnes, H., Neumann, E.-R. and Sundvoll, B. 1982. Petrology and geochemistry of Jurassic basalt dikes from Vestfjella, Dronning Maud Land, Antarctica. *Lithos*, **15**, 295–304, https://doi.org/10.1016/0024-4937(82)90020-2

Furnes, H., Vad, E., Austrheim, H., Mitchell, J.G. and Garmann, L.B. 1987. Geochemistry of basalt lavas from Vestfjella and adjacent areas, Dronning Maud Land, Antarctica. *Lithos*, **20**, 337–356, https://doi.org/10.1016/0024-4937(87)90015-6

Galerne, C.Y., Neumann, E.-R. and Planke, S. 2008. Emplacement mechanisms of sill complexes: information from the geochemical architecture of the Golden Valley Sill Complex, South Africa. *Journal of Volcanology and Geothermal Research*, **177**, 425–440, https://doi.org/10.1016/j.jvolgeores.2008.06.004

Grantham, G.H. 1996. Aspects of Jurassic magmatism and faulting in western Dronning Maud Land, Antarctica: implications for Gondwana break-up. *Geological Society, London, Special Publications*, **108**, 63–72, https://doi.org/10.1144/GSL.SP.1996.108.01.05

Grantham, G.H., Groenewald, P.B. and Hunter, D.R. 1988. Geology of the northern H. U. Sverdrupfjella, western Dronning Maud Land and implications for Gondwana reconstructions. *South African Journal of Antarctic Research*, **18**, 2–10.

Grind, H., Siivola, J. and Luttinen, A. 1991. A geological overview of the Vestfjella mountains in western Dronning Maud Land; Antarctica. *Antarctic Reports of Finland*, **1**, 11–15.

Groenewald, P.M., Moyes, A.B., Grantham, G.H. and Krynauw, J.R. 1995. East Antarctic crustal evolution: geological constraints and modelling in western Dronning Maud Land. *Precambrian Research*, **75**, 231–250, https://doi.org/10.1016/0301-9268(95)80008-6

Halpern, M. 1970. Rubidium–strontium date of possibly 3 billion years for a granitic rock from Antarctica. *Science*, **169**, 977–978, https://doi.org/10.1126/science.169.3949.977

Hanemann, R. and Viereck-Götte, L. 2004. Geochemistry of Jurassic Ferrar lava flows, sills and dikes sampled during the joint German–Italian Antarctic Expedition 1999–2000. *Terra Antartica*, **11**, 39–54.

Hargraves, R.B., Reháček, J. and Hooper, P.R. 1997. Palaeomagnetsim of the Karoo igneous rocks in southern Africa. *South African Journal of Geology*, **100**, 195–212, https://hdl.handle.net/10520/EJC-929052679

Harris, A.J.L. and Rowland, S.K. 2009. Effusion rate controls on lava flow length and the role of heat loss: a review. *IAVCEI Special Publications*, **12**, 33–51, https://doi.org/10.1144/IAVCEI002.3

Harris, C. and Grantham, G.H. 1993. Geology and petrogenesis of the Straumsvola nepheline syenite complex, Dronning Maud Land, Antarctica. *Geological Magazine*, **130**, 513–532, https://doi.org/10.1017/S0016756800020574

Harris, C., Marsh, J.S., Duncan, A.R. and Erlank, A.J. 1990. The petrogenesis of the Kirwan Basalts of Dronning Maud Land, Antarctica. *Journal of Petrology*, **31**, 341–369, https://doi.org/10.1093/petrology/31.2.341

Harris, C., Watters, B.R. and Groenewald, P.B. 1991. Geochemistry of the Mesozoic regional basic dykes of western Dronning Maud Land, Antarctica. *Contributions to Mineralogy and Petrology*, **107**, 100–111, https://doi.org/10.1007/BF00311188

Harris, C., Johnstone, W.P. and Phillips, D. 2002. Petrogenesis of the Mesozoic Sistefjell syenite intrusion, Dronning Maud Land, Antarctica and surrounding low-$\delta^{18}O$ lavas. *South African Journal of Geology*, **105**, 205–226, https://doi.org/10.2113/1050205

Harris, C., le Roux, P. *et al.* 2015. The oxygen isotope composition of Karoo and Etendeka picrites: High $\delta^{18}O$ mantle or crustal contamination? *Contributions to Mineralogy and Petrology*, **170**, 8, https://doi.org/10.1007/s00410-015-1164-1

Hastie, W.W., Watkeys, M.K. and Aubourg, C. 2014. Magma flow in dyke swarms of the Karoo LIP: Implications for the mantle plume hypothesis. *Gondwana Research*, **25**, 736–755, https://doi.org/10.1016/j.gr.2013.08.010

Heinonen, J.S. and Fusswinkel, T. 2017. High Ni and low Mn/Fe in olivine phenocrysts of the Karoo meimechites do not reflect pyroxenitic mantle sources. *Chemical Geology*, **467**, 134–142, https://doi.org/10.1016/j.chemgeo.2017.08.002

Heinonen, J.S. and Kurz, M.D. 2015. Low-$^3He/^4He$ sublithospheric mantle source for the most magnesian magmas of the Karoo large igneous province. *Earth and Planetary Science Letters*, **426**, 305–315, https://doi.org/10.1016/j.epsl.2015.06.030

Heinonen, J.S. and Luttinen, A.V. 2008. Jurassic dikes of Vestfjella, western Dronning Maud Land, Antarctica: geochemical tracing of ferropicrite sources. *Lithos*, **105**, 347–364, https://doi.org/10.1016/j.lithos.2008.05.010

Heinonen, J.S. and Luttinen, A.V. 2010. Mineral chemical evidence for extremely magnesian subalkaline melts from the Antarctic extension of the Karoo large igneous province. *Mineralogy and Petrology*, **99**, 201–217, https://doi.org/10.1007/s00710-010-0115-9

Heinonen, J.S., Carlson, R.W. and Luttinen, A.V. 2010. Isotopic (Sr, Nd, Pb, and Os) composition of highly magnesian dikes of Vestfjella, western Dronning Maud Land, Antarctica: a key to the origins of the Jurassic Karoo large igneous province? *Chemical Geology*, **277**, 227–244, https://doi.org/10.1016/j.chemgeo.2010.08.004

Heinonen, J.S., Luttinen, A.V., Riley, T.R. and Michallik, R.M. 2013. Mixed pyroxenite–peridotite sources for mafic and ultramafic dikes from the Antarctic segment of the Karoo continental flood basalt province. *Lithos*, **177**, 366–380, https://doi.org/10.1016/j.lithos.2013.05.015

Heinonen, J.S., Carlson, R.W., Riley, T.R., Luttinen, A.V. and Horan, M.F. 2014. Subduction-modified oceanic crust mixed with a depleted mantle reservoir in the sources of the Karoo continental flood basalt province. *Earth and Planetary Science Letters*, **394**, 229–241, https://doi.org/10.1016/j.epsl.2014.03.012

Heinonen, J.S., Jennings, E.S. and Riley, T.R. 2015. Crystallisation temperatures of the most Mg-rich magmas of the Karoo LIP on the basis of Al-in-olivine thermometry. *Chemical Geology*, **411**, 26–35, https://doi.org/10.1016/j.chemgeo.2015.06.015

Heinonen, J.S., Luttinen, A.V. and Bohrson, W.A. 2016. Enriched continental flood basalts from depleted mantle melts: modeling the lithospheric contamination of Karoo lavas from Antarctica. *Contributions to Mineralogy and Petrology*, **171**, 9, https://doi.org/10.1007/s00410-015-1214-8

Heinonen, J.S., Luttinen, A.V. and Whitehouse, M.J. 2018. Enrichment of ^{18}O in the mantle sources of the Antarctic portion of the Karoo large igneous province. *Contributions to Mineralogy and Petrology*, **173**, 21, https://doi.org/10.1007/s00410-018-1447-4

Hergt, J.M. 2000. Comment on: 'Enriched mantle–Dupal signature in the genesis of the Jurassic Ferrar tholeiites from Prince Albert Mountains (Victoria Land, Antarctica)' by Antonini *et al.* (Contributions to Mineralogy and Petrology 136, 1–19). *Contributions to Mineralogy and Petrology*, **139**, 240–244, https://doi.org/10.1007/s004100000130

Hergt, J.M., Chappell, B.W., Faure, G. and Mensing, T.M. 1989. The geochemistry of Jurassic dolerites from Portal Peak, Antarctica. *Contributions to Mineralogy and Petrology*, **102**, 298–305, https://doi.org/10.1007/BF00373722

Hergt, J.M., Peate, D.W. and Hawkesworth, C.J. 1991. The petrogenesis of Mesozoic Gondwana low-Ti flood basalts. *Earth and Planetary Science Letters*, **105**, 134–148, https://doi.org/10.1016/0012-821x(91)90126-3

Hjelle, A. and Winsnes, T. 1972. The sedimentary and volcanic sequence of Vestfjella, Dronning Maud Land. *In*: Adie, R.J. (ed.) *Antarctic Geology and Geophysics*. Universitetsforlaget, Oslo, 539–547.

Hoffman, P.F. 1991. Did the breakout of Laurentia turn Gondwanaland inside out? *Science*, **252**, 1409–1412, https://doi.org/10.1126/science.252.5011.1409

Hole, M.J., Ellam, R.M., Macdonald, D.I.M. and Kelley, S.P. 2015. Gondwana-breakup related magmatism in the Falkland Islands. *Journal of the Geological Society, London*, **173**, 108–126, https://doi.org/10.1144/jgs2015-027

Hon, K., Kauahikaua, J., Denlinger, R. and Mackay, K. 1994. Emplacement and inflation of pahoehoe sheet flows: Observations and measurements of active lava flows on Kilauea, Hawaii. *Geological Society of America Bulletin*, **106**, 351–370, https://doi.org/10.1130/0016-7606(1994)106<0351:EAIOPS>2.3.CO;2

Jacobs, J. 2009. A review of two decades (1986–2008) of geochronological work in Heimefrontfjella, and geotectonic interpretation of western Dronning Maud Land, East Antarctica. *Polarforschung*, **79**, 47–57.

Jacobs, J., Thomas, R.J. and Weber, K. 1993. Accretion and indentation tectonics at the southern edge of the Kaapvaal craton during the Kibaran (Grenville) orogeny. *Geology*, **21**, 203–206, https://doi.org/10.1130/0091-7613(1993)021%3C0203:AAITAT%3E2.3.CO;2

Jacobs, J., Pisarevsky, S., Thomas, R.J. and Becker, T. 2008. The Kalahari Craton during the assembly and dispersal of Rodinia. *Precambrian Research*, **160**, 142–158, https://doi.org/10.1016/j.precamres.2007.04.022

Jerram, D.A. 2002. Volcanology and facies architecture of flood basalts. *Geological Society of America, Special Papers*, **362**, 119–132, https://doi.org/10.1130/0-8137-2362-0.119

Jerram, D.A. and Widdowson, M. 2005. The anatomy of continental flood basalt provinces: geological constraints on the processes and products of flood volcanism. *Lithos*, **79**, 385–405, https://doi.org/10.1016/j.lithos.2004.09.009

Jokat, W., Boebel, T., König, M. and Meyer, U. 2003. Timing and geometry of early Gondwana breakup. *Journal of Geophysical Research: Solid Earth*, **108**, 2428, https://doi.org/10.1029/2002JB001802

Jourdan, F., Féraud, G. *et al.* 2004. The Karoo triple junction questioned: evidence from Jurassic and Proterozoic $^{40}Ar/^{39}Ar$ ages and geochemistry of the giant Okavango dike swarm (Botswana). *Earth and Planetary Science Letters*, **222**, 989–1006, https://doi.org/10.1016/j.epsl.2004.03.017

Jourdan, F., Féraud, G., Bertrand, H., Kampunzu, A.B., Tshoso, G., Watkeys, M.K. and Le Gall, B. 2005. Karoo large igneous province: Brevity, origin, and relation to mass extinction questioned by new $^{40}Ar/^{39}Ar$ age data. *Geology*, **33**, 745–748, https://doi.org/10.1130/G21632.1

Jourdan, F., Bertrand, H., Schaerer, U., Blichert-Toft, J., Féraud, G. and Kampunzu, A.B. 2007a. Major and trace element and Sr, Nd, Hf, and Pb isotope compositions of the Karoo large igneous province, Botswana-Zimbabwe: lithosphere vs mantle plume contribution. *Journal of Petrology*, **48**, 1043–1077, https://doi.org/10.1093/petrology/egm010

Jourdan, F., Féraud, G., Bertrand, H. and Watkeys, M.K. 2007b. From flood basalts to the inception of oceanization: Example from the $^{40}Ar/^{39}Ar$ high-resolution picture of the Karoo large igneous province. *Geochemistry, Geophysics, Geosystems*, **8**, https://doi.org/10.1029/2006GC001392

Jourdan, F., Féraud, G., Bertrand, H., Watkeys, M.K. and Renne, P.R. 2008. The ^{40}Ar/^{39}Ar ages of the sill complex of the Karoo large igneous province: Implications for the Pliensbachian–Toarcian climate change. *Geochemistry, Geophysics, Geosystems*, **9**, Q06009, https://doi.org/10.1029/2008GC001994

Juckes, L.M. 1968. The geology of Mannefallknausane and part of Vestfjella, Dronning Maud Land. *British Antarctic Survey Bulletin*, **18**, 65–78.

Juckes, L.M. 1972. The geology of north-eastern Heimefrontfjella, Dronning Maud Land. *British Antarctic Survey Scientific Reports*, **65**, http://nora.nerc.ac.uk/id/eprint/509222

Kamenetsky, V.S., Maas, R. et al. 2017. Multiple mantle sources of continental magmatism: Insights from 'high-Ti' picrites of Karoo and other large igneous provinces. *Chemical Geology*, **455**, 22–31, https://doi.org/10.1016/j.chemgeo.2016.08.034

Keszthelyi, L., Self, S. and Thordarson, T. 2006. Flood basalts on Earth, Io and Mars. *Journal of the Geological Society, London*, **163**, 253–264, https://doi.org/10.1144/0016-764904-503

Klausen, M.B. 2009. The Lebombo monocline and associate feeder dyke swarm: Diagnostic of a successful and highly volcanic rifted margin? *Tectonophysics*, **468**, 42–62, https://doi.org/10.1016/j.tecto.2008.10.012

Krynauw, J.R., Hunter, D.R. and Wilson, A.H. 1988. Emplacement of sills into wet sediments at Grunehogna, western Dronning Maud Land, Antarctica. *Journal of the Geological Society, London*, **145**, 1019–1032, https://doi.org/10.1144/gsjgs.145.6.1019

Krynauw, J.R., Watters, B.R., Hunter, D.R. and Wilson, A.H. 1991. A review of the field relations, petrology and geochemistry of the Brogmassivet intrusions in the Grunehogna province, western Dronning Maud Land, Antarctica. *In*: Thomson, M.R.A., Crame, J.A. and Thomson, J.W. (eds) *Geological Evolution of Antarctica*. Cambridge University Press, Cambridge, UK, 33–39.

Lawver, L.A., Gahagan, L.M. and Dalziel, I.W.D. 1998. A tight fit-Early Mezozoic Gondawna, a plate reconstruction perspective. *Memoirs of the National Institution of Polar Research, Special Issue*, **53**, 214–229.

Leat, P.T., Luttinen, A.V., Storey, B.C. and Millar, I.L. 2006. Sills of the Theron Mountains, Antarctica: evidence for long distance transport of mafic magmas during Gondwana break-up. *In*: Hanski, E.J., Mertanen, S., Rämö, O.T. and Vuollo, J. (eds) *Dyke Swarms: Time Markers of Crustal Evolution*. Taylor & Francis, Abingdon, UK, 183–199.

le Roex, A.P., Dick, H.J.B., Erlank, A.J., Reid, A.M., Frey, F.A. and Hart, S.R. 1983. Geochemistry, mineralogy and petrogenesis of lavas erupted along the Southwest Indian Ridge between the Bouvet triple junction and 11 degrees east. *Journal of Petrology*, **24**, 267–318, https://doi.org/10.1093/petrology/24.3.267

le Roex, A.P., Dick, H.J.B. and Watkins, R.T. 1992. Petrogenesis of anomalous K-enriched MORB from the Southwest Indian Ridge: 11°53′E to 14°38′E. *Contributions to Mineralogy and Petrology*, **110**, 253–268, https://doi.org/10.1007/BF00310742

Liss, D., Hutton, D.H.W. and Owens, W.H. 2002. Ropy flow structures: A neglected indicator of magma-flow direction in sills and dikes. *Geology*, **30**, 715–718, https://doi.org/10.1130/0091-7613(2002)030<0715:RFSANI>2.0.CO;2

Løvlie, R. 1979. Mesozoic palaeomagnetism in Vestfjella, Dronning Maud Land, East Antarctica. *Geophysical Journal International*, **59**, 529–537, https://doi.org/10.1111/j.1365-246X.1979.tb02571.x

Luttinen, A. 1994. *The Flood Basalt Succession of Basem, Dronning Maud Land, Antarctica*. MSc thesis, University of Helsinki, Helsinki, Finland [in Finnish].

Luttinen, A.V. 2018. Bilateral geochemical asymmetry in the Karoo large igneous province. *Scientific Reports*, **8**, 5223, https://doi.org/10.1038/s41598-018-23661-3

Luttinen, A.V. and Furnes, H. 2000. Flood basalts of Vestfjella: Jurassic magmatism across an Archaean-Proterozoic lithospheric boundary in Dronning Maud Land, Antarctica. *Journal of Petrology*, **41**, 1271–1305, https://doi.org/10.1093/petrology/41.8.1271

Luttinen, A.V. and Siivola, J.U. 1997. Geochemical characteristics of Mesozoic lavas and dikes from Vestfjella, Dronning Maud Land: recognition of three distinct chemical types. *In*: Ricci, C. A. (ed.) *The Antarctic Region: Geological Evolution and Processes, Volume 7*. Terra Antartica, Siena, Italy, 495–503.

Luttinen, A.V. and Vuori, S.K. 2006. Geochemical correlations between Jurassic gabbros and basaltic rocks in Vestfjella, Dronning Maud Land, Antarctica. *In*: Hanski, E., Mertanen, S., Rämö, T. and Vuollo, J. (eds) *Dyke Swarms – Time Markers of Crustal Evolution*. Taylor Francis, London, 201–212.

Luttinen, A.V., Rämö, O.T. and Huhma, H. 1998. Neodymium and strontium isotopic and trace element composition of a Mesozoic CFB suite from Dronning Maud Land, Antarctica: implications for lithosphere and asthenosphere contributions to Karoo magmatism. *Geochimica et Cosmochimica Acta*, **62**, 2701–2714, https://doi.org/10.1016/S0016-7037(98)00184-7

Luttinen, A.V., Zhang, X. and Foland, K.A. 2002. 159 Ma Kjakebeinet lamproites (Dronning Maud Land, Antarctica) and their implications for Gondana breakup processes. *Geological Magazine*, **139**, 525–539, https://doi.org/10.1017/S001675680200674X

Luttinen, A.V., Leat, P.T. and Furnes, H. 2010. Björnnutane and Sembberget basalt lavas and the geochemical provinciality of Karoo magmatism in western Dronning Maud Land, Antarctica. *Journal of Volcanology and Geothermal Research*, **198**, 1–18, https://doi.org/10.1016/j.jvolgeores.2010.07.011

Luttinen, A.V., Heinonen, J.S., Kurhila, M., Jourdan, F., Mänttäri, I., Vuori, S.K. and Huhma, H. 2015. Depleted mantle-sourced CFB magmatism in the Jurassic Africa–Antarctica Rift: petrology and ^{40}Ar/^{39}Ar and U/Pb chronology of the Vestfjella Dyke Swarm, Dronning Maud Land, Antarctica. *Journal of Petrology*, **56**, 919–952, https://doi.org/10.1093/petrology/egv022

Macdonald, G.A. 1953. Pahoehoe, aa, and block lava. *American Journal of Science*, **251**, 169–191, https://doi.org/10.2475/ajs.251.3.169

McIntosh, W.C., Kyle, P.R., Cherry, E.M. and Noltimier, H.C. 1982. Paleomagnetic results from Kirkpatrick Basalt Group, Victoria Land. *Antarctic Journal of the United States*, **17**, 20–22.

Mahoney, J.J., le Roex, A.P., Peng, Z., Fisher, R.L. and Natland, J.H. 1992. Southwestern limits of Indian Ocean ridge mantle and the origin of low ^{206}Pb/^{204}Pb mid-ocean ridge basalt: isotope systematics of the central Southwest Indian Ridge (17°–50°E). *Journal of Geophysical Research: Solid Earth*, **97**, 19 771–19 790, https://doi.org/10.1029/92JB01424

Marschall, H.R., Hawkesworth, C.J., Storey, C.D., Dhuime, B., Leat, P.T., Meyer, H.-P. and Tamm-Buckle, S. 2010. The Annandagstoppane Granite, East Antarctica: evidence for Archaean Intracrustal recycling in the Kaapvaal–Grunehogna Craton from zircon O and Hf isotopes. *Journal of Petrology*, **51**, 2277–2301, https://doi.org/10.1093/petrology/egq057

Marsh, J.S., Hooper, P.R., Rehacek, J., Duncan, R.A. and Duncan, A.R. 1997. Stratigraphy and age of Karoo basalts of Lesotho and implications for correlations within the Karoo Igneous Province. *American Geophysical Union Geophysical Monograph Series*, **100**, 247–272, https://doi.org/10.1029/GM100p0247

McDonough, W.F. and Sun, S.S. 1995. The composition of the Earth. *Chemical Geology*, **120**, 223–253, https://doi.org/10.1016/0009-2541(94)00140-4

McLoughlin, S., Larsson, K. and Lindström, S. 2005. Permian plan macrofossils from Fossilryggen, Vestfjella, Dronning Maud Land. *Antarctic Science*, **17**, 73–86, https://doi.org/10.1017/S0954102005002464

Meisel, T., Walker, R.J., Irving, A.J. and Lorand, J.-P. 2001. Osmium isotopic compositions of mantle xenoliths: a global perspective. *Geochimica et Cosmochimica Acta*, **65**, 1311–1323, https://doi.org/10.1016/S0016-7037(00)00566-4

Mortimer, N., Parkinson, D., Raine, J.I., Adams, C.J., Graham, I.J., Oliver, P.J. and Palmer, K. 1995. Ferrar magmatic province rocks discovered in New Zealand: implications for Mesozoic Gondwana geology. *Geology*, **23**, 185–188, https://doi.org/10.1130/0091-7613(1995)023<0185:FMPRDI>2.3.CO;2

Moulin, M., Fluteau, F. *et al.* 2011. An attempt to constrain the age, duration, and eruptive history of the Karoo flood basalt: Naude's Nek section (South Africa). *Journal of Geophysical Research: Solid Earth*, **116**, B07403, https://doi.org/10.1029/2011JB008210

Moyes, A.B., Barton, J.M.Jr and Groenewald, P.B. 1993. Late Proterozoic to Early Palaeozoic tectonism in Dronning Maud Land, Antarctica: supercontinental fragmentation and amalgamation. *Journal of the Geological Society, London*, **150**, 833–842, https://doi.org/10.1144/gsjgs.150.5.0833

Moyes, A.B., Krynauw, J.R. and Barton, J.M., Jr 1995. The age of the Ristcherflya Supergroup and Borgmassivet intrusions, Dronning Maud Land (Antarctica). *Antarctic Science*, **7**, 87–97, https://doi.org/10.1017/S0954102095000125

Neumann, E., Svensen, H., Galerne, C.Y. and Planke, S. 2011. Multistage Evolution of Dolerites in the Karoo Large Igneous Province, Central South Africa. *Journal of Petrology*, **52**, 959–984, https://doi.org/10.1093/petrology/egr011

Passey, S.R. and Bell, B.R. 2007. Morphologies and emplacement mechanisms of the lava flows of the Faroe Islands Basalt Group, Faroe Islands; NE Atlantic Ocean. *Bulletin of Volcanology*, **70**, 139–156, https://doi.org/10.1007/s00445-007-0125-6

Peters, M. 1989. *Die Vulkanite im westlichen und mittleren Neuschwabenland, Vestfjella und Ahlmannryggen, Antarctica: Petrographie, Geochemie, Geochronologie, Paläomagnetismus, geotektonische Implikationen*. Berichte zur Polarforschung, **61**, https://doi.org/10.2312/BzP_0061_1989

Rapela, C.W., Pankhurst, R.J., Fanning, C.M. and Hervé, F. 2005. Pacific subduction coeval with the Karoo mantle plume: the Early Jurasssic Subcordilleran belt of northwestern Patagonia. *Geological Society, London, Special Publications*, **246**, 217–239, https://doi.org/10.1144/GSL.SP.2005.246.01.07

Renne, P.R., Balco, G., Ludwig, K.R., Mundil, R. and Min, K. 2011. Response to the comment by W.H. Schwarz *et al.* on 'Joint determination of ^{40}K decay constant and ^{40}Ar*/^{40}K for the Fish Canyon sanidine standard, and improved accuracy for ^{40}Ar/^{39}Ar geochronology' by P.R. Renne *et al.* 2010. *Geochimica et Cosmochimica Acta*, **75**, 5097–5100, https://doi.org/10.1016/j.gca.2011.06.021

Rex, D.C. 1967. Age of dolerite from Dronning Maud Land. *British Antarctic Survey Bulletin*, **11**, 101–102.

Riley, T.R., Leat, P.T., Curtis, M.L., Millar, I.L. and Fazel, A. 2005. Early–Middle Jurassic dolerite dykes from western Dronning Maud Land (Antarctica): identifying mantle sources in the Karoo large igneous province. *Journal of Petrology*, **46**, 1489–1524, https://doi.org/10.1093/petrology/egi023

Riley, T.R., Curtis, M.L., Leat, P.T., Watkeys, M.K., Duncan, R.A., Millar, I.L. and Owens, W.H. 2006. Overlap of Karoo and Ferrar magma types in KwaZulu–Natal, South Africa. *Journal of Petrology*, **47**, 541–566, https://doi.org/10.1093/petrology/egi085

Riley, T.R., Curtis, M.L., Leat, P.T. and Millar, I.L. 2009. The geochemistry of Middle Jurassic dykes associated with the Straumsvola–Tvora alkaline plutons, Dronning Maud Land, Antarctica and their association with the Karoo large igneous province. *Mineralogical Magazine*, **73**, 205–226, https://doi.org/10.1180/minmag.2009.073.2.205

Rowland, S.K. and Walker, G.P.L. 1990. Pahoehoe and aa in Hawaii: volumetric flow rate controls the lava sturcture. *Bulletin of Volcanology*, **52**, 615–628, https://doi.org/10.1007/BF00301212

Self, S., Thordarson, Th. *et al.* 1996. A new model for the emplacement of Columbia River basalts as large, inflated pahoeho lava flow fields. *Geophysical Research Letters*, **23**, 2689–2692, https://doi.org/10.1029/96GL02450

Self, S., Keszthelyi, L. and Thordarson, Th. 1998. The importance of pahoehoe. *Annual Review of Earth and Planetary Sciences*, **26**, 81–110, https://doi.org/10.1146/annurev.earth.26.1.81

Shirey, S.B. and Walker, R.J. 1998. The Re–Os isotope system in cosmochemistry and high-temperature geochemistry. *Annual Review of Earth and Planetary Sciences*, **26**, 423–500, https://doi.org/10.1146/annurev.earth.26.1.423

Shrivastava, J.P., Ahmad, M. and Shrivastava, S. 2012. Microstructures and compositional variation in the intra-volcanic bole clayes from the eastern Deccan Volcanic Province: Palaeoenvironmental implications and duration of volcanism. *Journal of the Geological Society of India*, **80**, 177–188, https://doi.org/10.1007/s12594-012-0130-z

Siders, M.A. and Elliot, D.H. 1985. Major and trace element geochemistry of the Kirkpatrick Basalt, Mesa Range, Antarctica. *Earth and Planetary Science Letters*, **72**, 54–64, https://doi.org/10.1016/0012-821X(85)90116-5

Simon, N.S.C., Carlson, R.W., Pearson, D.G. and Davies, G.R. 2007. The origin and evolution of the Kaapvaal cratonic lithospheric mantle. *Journal of Petrology*, **48**, 589–625, https://doi.org/10.1093/petrology/egl074

Single, R.T. and Jerram, D.A. 2004. The 3D facies architecture of flood basalt provinces and their internal heterogeneity: examples from the Palaeogene Skye Lava Field. *Journal of the Geological Society, London*, **161**, 911–926, https://doi.org/10.1144/0016-764903-136

Sushchevskaya, N.M., Belyatsky, B.V., Leichenkov, G.L. and Laiba, A.A. 2009. Evolution of the Karoo-Maud mantle plume in Antarctica and its influence on the magmatism of the early stages of Indian Ocean spreading. *Geochemistry International*, **47**, 1–17, https://doi.org/10.1134/S0016702909010017

Svensen, H., Corfu, F., Polteau, S., Hammer, Ø. and Planke, S. 2012. Rapid magma emplacement in the Karoo Large Igneous Province. *Earth and Planetary Science Letters*, **325–326**, 1–9, https://doi.org/10.1016/j.epsl.2012.01.015

Sweeney, R.J. and Watkeys, M.K. 1990. A possible link between Mesozoic lithospheric architecture and Gondwana flood basalts. *Journal of African Earth Sciences*, **10**, 707–716, https://doi.org/10.1016/0899-5362(90)90037-F

Tegner, C., Lesher, C.E., Larsen, L.M. and Watt, W.S. 1998. Evidence from the rare-earth-element record of mantle melting for cooling of the Tertiary Iceland plume. *Nature*, **395**, 591–549, https://doi.org/10.1038/26956

Torsvik, T.H., Burke, K., Steinberger, B., Webb, S. and Ashwall, L.D. 2010. Diamonds sampled by plumes from the core-mantle boundary. *Nature*, **466**, 352–355, https://doi.org/10.1038/nature09216

Vollmer, R., Ogden, P., Schilling, J.-G., Kingsley, R.H. and Waggoner, D.G. 1984. Nd and Sr isotopes in ultrapotassic volcanic rocks from the Leucite Hills, Wyoming. *Contributions to Mineralogy and Petrology*, **87**, 359–368, https://doi.org/10.1007/BF00381292

Vuori, S.K. 2004. *Petrogenesis of the Jurassic gabbroic intrusions of Vestfjella, Dronning Maud Land, Antarctica*. PhD thesis, University of Helsinki, Helsinki, Finland.

Vuori, S.K. and Luttinen, A.V. 2003. The Jurassic gabbroic intrusions of Utpostane and Muren: insights into Karoo-related plutonism in Dronning Maud Land, Antarctica. *Antarctic Science*, **15**, 283–301, https://doi.org/10.1017/S0954102003001287

Walker, G.P.L. 1971. Compound and simple lava flows and flood basalts. *Bulletin of Volcanology*, **35**, 579–590, https://doi.org/10.1007/BF02596829

Walker, G.P.L. 1973. Lengths of lava flows. *Philosophical Transactions of the Royal Society of London, Series A: Mathematical and Physical Sciences*, **274**, 107–118, https://www.jstor.org/stable/74335

Walker, R.J., Carlson, R.W., Shirey, S.B. and Boyd, F.R. 1989. Os, Sr, Nd, and Pb isotope systematics of Southern African peridotite xenoliths: Implications for the chemical evolution of subcontinental mantle. *Geochimica et Cosmochimica Acta*, **53**, 1583–1595, https://doi.org/10.1016/0016-7037(89)90240-8

Wang, X.-C., Wilde, S., Li, Q. and Yang, Y.-N. 2015. Continental flood basalts derived from the hydrous mantle transition zone. *Nature Communications*, **6**, 7700, https://doi.org/10.1038/ncomms8700

Watkeys, M.K. 2002. Development of the Lebombo rifted volcanic margin of southeast Africa. *Geological Society of America Special Papers*, **362**, 27–46, https://doi.org/10.1130/0-8137-2362-0.27

Wilhelm, S. and Wörner, G. 1996.Crystal size distribution in Jurassic Ferrar flows and sills (Victoria Land, Antarctica): evidence for processes of cooling, nucleation, and crystallisation. *Contributions to Mineralogy and Petrology*, **125**, 1–15, https://doi.org/10.1007/s004100050202

Wolmarans, L.G. and Kent, L.E. (eds). 1982. *Geological Investigations in Western Dronning Maud Land, Antarctica – A Synthesis. South African Journal of Antarctic Research*, Suppl. 2.

Workman, R.K. and Hart, S.R. 2005. Major and trace element composition of the depleted MORB mantle (DMM). *Earth and Planetary Science Letters*, **231**, 53–72, https://doi.org/10.1016/j.epsl.2004.12.005

Zhang, X., Luttinen, A.V., Elliot, D.H., Larsson, K. and Foland, K.A. 2003. Early stages of Gondwana breakup: The $^{40}Ar/^{39}Ar$ geochronology of Jurassic basaltic rocks from western Dronning Maud Land, Antarctica, and implications for the timing of magmatic and hydrothermal events. *Journal of Geophysical Research*, **108**, 2449, https://doi.org/10.1029/2001JB001070

Zindler, A. and Hart, S.R. 1986. Chemical geodynamics. *Annual Review of Earth and Planetary Sciences*, **14**, 493–571, https://doi.org/10.1146/annurev.ea.14.050186.002425

Section 3

Subduction-related volcanism

Flooded crater on Deception Island, an active shield volcano with a large caldera inundated by the sea (seen in the background) in Bransfield Strait, a small, active, ensialic marginal basin. A fissure-sourced block lava has entered the lake at its far end and formed a small lava-fed delta. The lake is 500 m in diameter.

Photograph by A. Geyer

Section 1

Subduction-related volcanism

Chapter 3.1a

Antarctic Peninsula and South Shetland Islands: volcanology

Philip T. Leat[1,2]* and Teal R. Riley[1]

[1]British Antarctic Survey, High Cross, Madingley Road, Cambridge CB3 0ET, UK
[2]School of Geography, Geology and the Environment, University of Leicester, University Road, Leicester LE1 7RH, UK

*Correspondence: ptle@bas.ac.uk

Abstract: The voluminous continental margin volcanic arc of the Antarctic Peninsula is one of the major tectonic features of West Antarctica. It extends from the Trinity Peninsula and the South Shetland Islands in the north to Alexander Island and Palmer Land in the south, a distance of c. 1300 km, and was related to east-directed subduction beneath the continental margin. Thicknesses of exposed volcanic rocks are up to c. 1.5 km, and the terrain is highly dissected by erosion and heavily glacierized. The arc was active from Late Jurassic or Early Cretaceous times until the Early Miocene, a period of climate cooling from subtropical to glacial. The migration of the volcanic axis was towards the trench over time along most of the length of the arc. Early volcanism was commonly submarine but most of the volcanism was subaerial. Basaltic–andesitic stratocones and large silicic composite volcanoes with calderas can be identified. Other rock associations include volcaniclastic fans, distal tuff accumulations, coastal wetlands and glacio-marine eruptions.
 Other groups of volcanic rocks of Jurassic age in Alexander Island comprise accreted oceanic basalts within an accretionary complex and volcanic rocks erupted within a rift basin along the continental margin that apparently predate subduction.

Volcanic rocks are widespread in the Jurassic–Tertiary geological record in the Antarctic Peninsula. For most of this time, the Antarctic Peninsula was an active continental margin, with east-directed subduction of the Pacific oceanic plate taking place along a subduction zone located to the west of, and approximately parallel with, the peninsula (Suárez 1976; Barker 1982; Pankhurst 1982; Tarney et al. 1982). Accretionary complexes developed along the margin (Storey and Garrett 1985; Doubleday and Tranter 1994; Trouw et al. 2000) and belts of plutonic rocks were intruded (Pankhurst 1982; Leat et al. 1995; Riley et al. 2018). Volcanic rocks form a 1300 km-long volcanic arc exposed along the west coast of the peninsula and on adjacent islands, as we shall detail below.

Several groups of volcanic rocks in the Antarctic Peninsula are recognized as distinct from this volcanic arc. Along the east coasts of both Graham Land and Palmer Land, groups of mainly silicic Early–Middle Jurassic volcanic rocks were erupted in the hinterland behind the active margin, and are thought to be more related to continental extension during Gondwana break-up than to proximity to a subduction zone (Riley and Leat 1999, 2021; Pankhurst et al. 2000; Riley et al. 2001). Widespread occurrences of Late Tertiary mafic volcanic rocks are post-subduction in tectonic setting and are described in Hole (2021) and Smellie and Hole (2021). The main volcanic arc is the focus of this chapter, with brief reviews of accretionary and forearc volcanic outcrops. This review covers the Jurassic–Early Tertiary volcanic rocks of the west coasts of Graham Land and Palmer Land, the South Shetland Islands, and Alexander Island.

Before the impact of radiometric dating, fossil evidence for Jurassic volcanism in parts of northern and eastern Graham Land was used to define the term 'Upper Jurassic Volcanic Group' (Adie 1964). Except for obviously recent volcanism, this term was applied from the 1950s to the early 1980s to encompass nearly all volcanic sequences in Graham Land, Palmer Land and the South Shetland Islands (e.g. Hooper 1962; Adie 1964; Barton 1965; Curtis 1966; Davies 1984). However, early radiometric studies indicated that the 'Upper Jurassic Volcanic Group' included rocks that were Cretaceous and Early Tertiary in age (Rex 1976; Pankhurst 1982; Pankhurst and Smellie 1983). To reflect this change in outlook, Thomson (1982a) proposed the name Antarctic Peninsula Volcanic Group to replace Upper Jurassic Volcanic Group and to include all Jurassic and younger volcanic rocks of the volcanic arc. In view of the contrasts between the volcanic rocks of the volcanic arc dominating the west coast of Graham Land and the silicic Jurassic volcanic rocks of the east coast, Riley et al. (2010) separated the latter as forming the Graham Land Volcanic Group. Hunter et al. (2006) allocated similar Jurassic silicic volcanic rocks in eastern Palmer Land into an Ellsworth Land Volcanic Group, now redefined as the Palmer Land Volcanic Group following the redefinition of geographical boundaries in the southern Antarctic Peninsula (Smellie 2020; Riley and Leat 2021). McCarron (1997) separated Late Cretaceous–Early Tertiary volcanic rocks in Alexander Island, which he considered to be forearc in tectonic origin (McCarron and Millar 1997; McCarron and Smellie 1998), from the Antarctic Peninsula Volcanic Group into an Alexander Island Volcanic Group. Late Jurassic–Early Tertiary volcanic rocks in the South Shetland Islands have been included in the Antarctic Peninsula Volcanic Group (Thomson and Pankhurst 1983). Nevertheless, subsequent studies have used group level status for several volcanic and associated sedimentary successions in the archipelago (e.g. Birkenmajer 1989; Hathway and Lomas 1998). In this paper, we use the informal term Antarctic Peninsula volcanic arc to include all Jurassic–Early Tertiary volcanic and associated volcaniclastic rocks thought to be part of the long-lived volcanic arc whose axis lies approximately along the west coast of Graham Land and within western Palmer Land, and including the South Shetland Islands and Alexander Island. The petrology and geochemistry of the arc volcanic rocks is reviewed by Leat and Riley (2021). A compilation of radiometric ages for the volcanic rocks of the arc is presented in Table 1. The geological timescale used is that of Ogg et al. (2008). Smellie (2020) has provided a comprehensive stratigraphical chart showing the names, ages and geographical distribution of geological formations in the Antarctic Peninsula, and it is not reproduced here.

Tectonic development

The Antarctic Peninsula volcanic arc was erupted on continental crust forming the margin of Gondwana (Thomson et al.

Table 1. *Summary of published geochronological ages of volcanic rocks of the Antarctic Peninsula volcanic arc*

Sample	Location	Rock type	Age (Ma)	Method	Reference
South Shetland Islands					
A-378	Destruction Bay Formation, Cape Melville, King George Island	Basaltic tuff	23.6 ± 0.7	K–Ar WR	Birkenmajer et al. (1988)
A-301	Sherratt Bay Formation, Cape Melville, King George Island	Basalt lava	>18	K–Ar	Birkenmajer et al. (1988)
3S5	Three Sisters Point, King George Island	Andesite lava	35.35 ± 0.15	Ar–Ar WR plateau	Pańczyk and Nawrocki (2011)
3S5	Three Sisters Point, King George Island	Andesite lava	29.1 ± 0.2	Ar–Ar plagioclase plateau	Pańczyk and Nawrocki (2011)
TR-8	Turret Point, King George Island	Andesite lava	37.3 ± 0.4	Ar–Ar WR plateau	Pańczyk and Nawrocki (2011)
A-297	Mersey Spit, near Turret Point, King George Island	Basaltic–andesitic lava	34.4 ± 0.5	K–Ar WR	Birkenmajer (1998c)
LR-3	Hennequin Formation, Lions Rump, King George Island	Andesite lava	44.8 ± 0.2	Ar–Ar WR plateau	Pańczyk and Nawrocki (2011)
SK-1	Hennequin Formation, Lions Rump, King George Island	Andesite lava	44.7 ± 0.3	Ar–Ar WR plateau	Pańczyk and Nawrocki (2011)
PR-3	Hennequin Formation, Lions Rump, King George Island	Andesite lava	44.5 ± 0.3	Ar–Ar WR plateau	Pańczyk and Nawrocki (2011)
P.438.1	Lions Rump, King George Island	Andesite lava	42 ± 1	K–Ar	Pankhurst and Smellie (1983); Smellie et al. (1984)
P.2007.2	Low Head Member, King George Island	Basalt lava	23.7 ± 2.0	Ar–Ar WR plateau	Smellie et al. (1998)
P.2007.2	Low Head Member, King George Island	Basalt lava	21.2 ± 2.1	K–Ar WR	Smellie et al. (1998)
P.439.4	Boy Point Formation, King George Island	Dacite lava	22.6 ± 1.7	Ar–Ar WR plateau	Smellie et al. (1998)
P.831.2	Point Hennequin, King George Island	Andesite	45 ± 1	K–Ar	Pankhurst and Smellie (1983); Smellie et al. (1984)
P.831.3	Point Hennequin, King George Island	Andesite	27 ± 1 (R)	K–Ar	Pankhurst and Smellie (1983); Smellie et al. (1984)
P.831.4	Point Hennequin, King George Island	Andesite	32 ± 1 (R)	K–Ar	Pankhurst and Smellie (1983); Smellie et al. (1984)
P.831.5	Point Hennequin, King George Island	Andesite	46 ± 1	K–Ar	Pankhurst and Smellie (1983); Smellie et al. (1984)
P.831.8	Point Hennequin, King George Island	Andesite	47 ± 1	K–Ar	Pankhurst and Smellie (1983); Smellie et al. (1984)
MW-9	Point Hennequin, King George Island	Basaltic andesite	48.7 ± 0.6	Zircon U–Pb	Nawrocki et al. (2011)
MW-7	Point Hennequin, King George Island	Andesite lava	46.7 ± 0.3	Ar–Ar WR plateau	Nawrocki et al. (2011)
MW-7a	Point Hennequin, King George Island	Andesite lava	46.8 ± 0.3	Ar–Ar WR plateau	Nawrocki et al. (2011)
MW-2a	Point Hennequin, King George Island	Andesite lava	43.8 ± 0.3	Ar–Ar WR plateau	Nawrocki et al. (2011)
MW-2	Point Hennequin, King George Island	Andesite lava	46.0 ± 0.3	Ar–Ar WR plateau	Nawrocki et al. (2011)
DL-10	Dufayel Island, Znosko Glacier Formation, King George Island	Andesite lava	47.8 ± 0.5	Zircon U–Pb	Nawrocki et al. (2010)
DL-10	Dufayel Island, Znosko Glacier Formation, King George Island	Andesite lava	45.02 ± 0.2	Ar–Ar WR plateau	Nawrocki et al. (2011)
DL-10a	Dufayel Island, Znosko Glacier Formation, King George Island	Andesite lava	45.50 ± 0.3	Ar–Ar WR plateau	Nawrocki et al. (2011)
PH-3	Point Thomas Formation, King George Island	Basalt lava	48.9 ± 0.7	Zircon U–Pb	Nawrocki et al. (2010)
PL-16	Point Thomas Formation, King George Island	Basalt–andesite lava	44.9 ± 0.7	Ar–Ar WR plateau	Nawrocki et al. (2011)
PL-16a	Point Thomas Formation, King George Island	Basalt–andesite lava	44.6 ± 0.4	Ar–Ar WR plateau	Nawrocki et al. (2011)
PH-3b	Point Thomas Formation, King George Island	Basalt–andesite lava	48.1 ± 0.2	Ar–Ar WR plateau	Nawrocki et al. (2011)
PH-3b plag	Point Thomas Formation, King George Island	Basalt–andesite lava	41.5 ± 0.8	Ar–Ar plagioclase plateau	Nawrocki et al. (2011)
PH-1	Point Thomas Formation, King George Island	Basalt–andesite lava	47.6 ± 0.4	Ar–Ar WR plateau	Nawrocki et al. (2011)
PL-7	Point Thomas Formation, King George Island	Basalt–andesite lava	44.1 ± 1.3	Ar–Ar WR plateau	Nawrocki et al. (2011)
P1-Url	Point Thomas Formation, King George Island	Basalt lava	51.93 ± 0.22	Ar–Ar WR isochron	Spinola et al. (2017)
P1-Lrl	Point Thomas Formation, King George Island	Basalt lava	48.72 ± 0.29	Ar–Ar WR isochron	Spinola et al. (2017)
P2-Lrl	Point Thomas Formation, King George Island	Basalt lava	50.25 ± 0.33	Ar–Ar WR isochron	Spinola et al. (2017)
P2-Url	Point Thomas Formation, King George Island	Basalt lava	42.51 ± 0.20	Ar–Ar WR isochron	Spinola et al. (2017)
P3-Url	Point Thomas Formation, King George Island	Basalt lava	47.89 ± 0.52	Ar–Ar WR isochron	Spinola et al. (2017)
P3-Lrl	Point Thomas Formation, King George Island	Basalt lava	48.71 ± 0.23	Ar–Ar WR isochron	Spinola et al. (2017)
DF-2	Demay Point Formation, King George Island	Trachyte lava	51.6 ± 0.8	Zircon U–Pb	Nawrocki et al. (2010)
DF-25	Demay Point Formation, King George Island	Trachyte lava	53.0 ± 0.7	Zircon U–Pb	Nawrocki et al. (2010)
DF-25	Demay Point Formation, King George Island	Trachyte lava	52.7 ± 0.6	Ar–Ar WR plateau	Nawrocki et al. (2010)
UP-2	Uchatka Point Formation, King George Island	Basalt lava	75.4 ± 0.9	Ar–Ar WR plateau	Nawrocki et al. (2010)
RHAM-17A	Red Hill, Zamek Formation, King George Island	Basalt lava	46.65 ± 2.10	K–Ar WR	Mozer et al. (2015)
RHAM-17	Red Hill, Zamek Formation, King George Island	Basalt lava	42.73 ± 1.47	K–Ar WR	Mozer et al. (2015)

RHAM-16	Red Hill, Zamek Formation, King George Island	Basalt lava	45.45 ± 4.56	K-Ar WR	Mozer et al. (2015)
RHAM-12	Red Hill, Zamek Formation, King George Island	Basalt lava	43.96 ± 2.11	K-Ar WR	Mozer et al. (2015)
RHAM10	Red Hill, Zamek Formation, King George Island	Basalt lava	41.88 ± 1.51	K-Ar WR	Mozer et al. (2015)
RHAM-21	Red Hill, Zamek Formation, King George Island	Basalt lava	48.13 ± 1.56	K-Ar WR	Mozer et al. (2015)
Red Hill-2	Red Hill, Zamek Formation, King George Island	Basalt lava	44.60 ± 3.46	K-Ar WR	Mozer et al. (2015)
Red Hill-1	Red Hill, Zamek Formation, King George Island	Basalt lava	46.14 ± 3.19	K-Ar WR	Mozer et al. (2015)
RHAM-02	Red Hill, Llano Point Formation, King George Island	Basaltic andesite lava	50.49 ± 2.57	K-Ar WR	Mozer et al. (2015)
Red Hill-3	Red Hill, Llano Point Formation, King George Island	Basaltic andesite lava	51.18 ± 2.15	K-Ar WR	Mozer et al. (2015)
PT-2	Llano Point Formation, King George Island	Basalt lava	50.8 ± 1.2	Ar-Ar WR plateau	Nawrocki et al. (2011)
BD-13	Llano Point Formation, King George Island	Basalt lava	52.3 ± 0.5	Ar-Ar WR plateau	Nawrocki et al. (2011)
P.696.2	Potter Peninsula, King George Island	Basaltic andesite	47.6 ± 0.22	Ar-Ar WR	Haase et al. (2012)
P.232.1	Potter Peninsula, King George Island	Basalt lava	44 ± 1	K-Ar	Pankhurst and Smellie (1983); Smellie et al. (1984)
P.760.1	Potter Peninsula, King George Island	Basalt lava	42 ± 1	K-Ar	Smellie et al. (1984)
P.758.1	Potter Peninsula, King George Island	Basaltic andesite	47 ± 1	K-Ar	Pankhurst and Smellie (1983); Smellie et al. (1984)
P.757.2	Potter Peninsula, King George Island	Andesite	48 ± 1	K-Ar	Pankhurst and Smellie (1983); Smellie et al. (1984)
1	Potter Cove, King George Island	Lava	50 ± 2	K-Ar	Watts (1982)
2	Potter Cove, King George Island	Lava	59 ± 2	K-Ar	Watts (1982)
3	Potter Cove, King George Island	Lava	51 ± 3	K-Ar	Watts (1982)
05122402	Weaver Peninsula, King George Island	Basalt	45.77 ± 1.21	Ar-Ar GM isochron	Wang et al. (2009)
05121804	Barton Peninsula, King George Island	Andesite	44.20 ± 1.56	Ar-Ar GM isochron	Wang et al. (2009)
BP04	Barton Peninsula, King George Island	Basaltic andesite	119.5 ± 0.5	Ar-Ar WR plateau	Kim et al. (2000); Zheng et al. (2000)
BP05	Barton Peninsula, King George Island	Basaltic andesite	104.8 ± 5.5	Ar-Ar WR isochron	Kim et al. (2000); Zheng et al. (2000)
BP13	Barton Peninsula, King George Island	Basaltic andesite	61.4 ± 0.9	Ar-Ar plateau	Kim et al. (2000)
BP09	Barton Peninsula, King George Island	Basaltic andesite	73.9 ± 4.4	K-Ar	Kim et al. (2000)
BP12	Barton Peninsula, King George Island	Basaltic andesite	45.9 ± 3.8	K-Ar	Kim et al. (2000)
6	Barton Peninsula, King George Island	Andesite	48.5 ± 4.0	K-Ar WR	Jwa et al. (1992)
7	Barton Peninsula, King George Island	Basaltic andesite	35.5 ± 3.4	K-Ar WR	Jwa et al. (1992)
8	Barton Peninsula, King George Island	Lapilli tuff	44.2 ± 2.4	K-Ar WR	Jwa et al. (1992)
P.1473.5	Marian Cove, King George Island	Andesite lava	46 ± 1	K-Ar	Pankhurst and Smellie (1983); Smellie et al. (1984)
P.1166.1	N Fildes Peninsula, King George Island	Basaltic andesite	56.1 ± 0.3	Ar-Ar WR	Haase et al. (2012)
P.1166.7	North Fildes Peninsula, King George Island	Basalt lava	52 ± 1	K-Ar	Pankhurst and Smellie (1983); Smellie et al. (1984)
P.1147.3	North Fildes Peninsula, King George Island	Basalt lava	48 ± 1	K-Ar	Pankhurst and Smellie (1983); Smellie et al. (1984)
P.1147.4	North Fildes Peninsula, King George Island	Basalt lava	48 ± 1	K-Ar	Pankhurst and Smellie (1983); Smellie et al. (1984)
P.1162.5	North Fildes Peninsula, King George Island	Basalt lava	57 ± 3	K-Ar	Pankhurst and Smellie (1983); Smellie et al. (1984)
P1125.1	North Fildes Peninsula, King George Island	Basaltic andesite lava	43 ± 1	K-Ar	Pankhurst and Smellie (1983); Smellie et al. (1984)
P.1182.1/2	North Fildes Peninsula, King George Island	Dacite lava	42 ± 1	K-Ar	Pankhurst and Smellie (1983); Smellie et al. (1984)
P.1183.2/7	North Fildes Peninsula, King George Island	Dacite lava	46 ± 1	K-Ar	Pankhurst and Smellie (1983); Smellie et al. (1984)
P.608.5b	South Fildes Peninsula, King George Island	Basaltic andesite	55.21 ± 0.49	Ar-Ar WR	Haase et al. (2012)
JH1	South Fildes Peninsula, King George Island	Basalt	55.09 ± 0.25	Ar-Ar GM plateau	Gao et al. (2018)
AB1	South Fildes Peninsula, King George Island	Basalt	56.38 ± 0.02	Ar-Ar GM plateau	Gao et al. (2018)
BH5	South Fildes Peninsula, King George Island	Basalt	54.66 ± 0.56	Ar-Ar GM plateau	Gao et al. (2018)
LH6	South Fildes Peninsula, King George Island	Basalt	52.42 ± 0.19	Ar-Ar GM plateau	Gao et al. (2018)
P.615.1	South Fildes Peninsula, King George Island	Andesite lava	51 ± 1	K-Ar	Pankhurst and Smellie (1983); Smellie et al. (1984)
P.604.1	South Fildes Peninsula, King George Island	Andesite lava	59 ± 2	K-Ar	Pankhurst and Smellie (1983); Smellie et al. (1984)
P.608.5a	South Fildes Peninsula, King George Island	Andesite lava	58 ± 1	K-Ar	Pankhurst and Smellie (1983); Smellie et al. (1984)
P.609.3	South Fildes Peninsula, King George Island	Andesite lava	58 ± 2	K-Ar	Pankhurst and Smellie (1983); Smellie et al. (1984)
P.627.1	South Fildes Peninsula, King George Island	Altered lava	58 ± 1	K-Ar	Pankhurst and Smellie (1983); Smellie et al. (1984)
P.629.1	South Fildes Peninsula, King George Island	Altered lava	31 ± 3 (R)	K-Ar	Pankhurst and Smellie (1983); Smellie et al. (1984)

(Continued)

Table 1. Continued.

Sample	Location	Rock type	Age (Ma)	Method	Reference
A6	South Fildes Peninsula, King George Island	Andesite lava	27 ± 2 (R)	K–Ar	Valencio et al. (1979)
A24	South Fildes Peninsula, King George Island	Andesite lava	61 ± 3	K–Ar	Valencio et al. (1979)
A11	South Fildes Peninsula, King George Island	Andesite lava	88 ± 5	K–Ar	Valencio et al. (1979)
A23	South Fildes Peninsula, King George Island	Andesite lava	110 ± 10	K–Ar	Valencio et al. (1979)
1	Fildes Peninsula, King George Island	Basaltic andesite	59.5 ± 3.0	K–Ar WR	Jwa et al. (1992)
2	Fildes Peninsula, King George Island	Basaltic andesite	61.4 ± 2.4	K–Ar WR	Jwa et al. (1992)
3	Fildes Peninsula, King George Island	Basaltic andesite	56.2 ± 2.7	K–Ar WR	Jwa et al. (1992)
4	Fildes Peninsula, King George Island	Basalt	53.2 ± 2.5	K–Ar WR	Jwa et al. (1992)
5	Fildes Peninsula, King George Island	Basalt	53.2 ± 3.0	K–Ar WR	Jwa et al. (1992)
P.1208.1	Stansbury Peninsula, Nelson Island	Basaltic andesite	55.8 ± 1	Ar–Ar WR	Haase et al. (2012)
P.480.2	Kitchen Point, Robert Island	Basaltic andesite	66.3 ± 23	Ar–Ar WR	Haase et al. (2012)
P.477.1	Kitchen Point, Robert Island	Andesite lava	53 ± 1	K–Ar	Pankhurst and Smellie (1983); Smellie et al. (1984)
P.840.4	Coppermine Peninsula, Robert Island	Basalt lava	82 ± 2	K–Ar	Pankhurst and Smellie (1983); Smellie et al. (1984)
P.840.5	Coppermine Peninsula, Robert Island	Basalt lava	83 ± 3	K–Ar	Pankhurst and Smellie (1983); Smellie et al. (1984)
P.840.6	Coppermine Peninsula, Robert Island	Basalt lava	80 ± 2	K–Ar	Pankhurst and Smellie (1983); Smellie et al. (1984)
P.842.4/6	Coppermine Peninsula, Robert Island	Basalt lava	84 ± 2	K–Ar	Pankhurst and Smellie (1983); Smellie et al. (1984)
P.842.9	Coppermine Peninsula, Robert Island	Basalt lava	82 ± 3	K–Ar	Pankhurst and Smellie (1983); Smellie et al. (1984)
P.1825.7	North of Spark Point, Greenwich Island	Basalt	82.08 ± 0.44	Ar–Ar WR	Haase et al. (2012)
P.1872.5	Mount Bowles Formation, Livingston Island	Basaltic andesite lava	40.5 ± 1.1 (R)	Ar–Ar WR isochron	Willan and Kelley (1999)
105E	Mount Bowles Formation, Livingston Island	Basalt lava	35.0 ± 3.9 (R)	K–Ar	Smellie et al. (1996a)
45D	Mount Bowles Formation, Livingston Island	Andesite lava	44.4 ± 1.3 (R)	K–Ar	Smellie et al. (1996a)
262	Mount Bowles Formation, Livingston Island	Basaltic andesite lava	39.8 ± 1.6 (R)	K–Ar	Smellie et al. (1996a)
P.466.1	Hannah Point, Livingston Island	Basaltic andesite	97.35 ± 0.39	Ar–Ar WR	Haase et al. (2012)
434	Hannah Point, Livingston Island	Basaltic andesite lava	67.5 ± 2.5	K–Ar	Smellie et al. (1996a)
407	Hannah Point, Livingston Island	Basaltic andesite lava	87.9 ± 2.6	K–Ar	Smellie et al. (1996a)
325c3	Cape Shirreff, Livingston Island	Basalt	109.19 ± 0.46	Ar–Ar WR	Haase et al. (2012)
325c3	Cape Shirreff, Livingston Island	Basalt	90.2 ± 5.6	K–Ar	Smellie et al. (1996a)
P.862.4	Start Point, Byers Peninsula, Livingston Island	Basalt	135.17 ± 2.57	Ar–Ar WR	Haase et al. (2012)
P.848	Byers Peninsula, Livingston Island	Rhyolite	111 ± 4	Rb–Sr WR isochron	Pankhurst (1982); Smellie et al. (1984)
P.428.3	Sayer Nunatak, Livingston Island	Clast in vent	74 ± 2	K–Ar	Pankhurst and Smellie (1983)
P.850.8	Eastern Byers Peninsula, Livingston Island	Basalt lava	94 ± 3	K–Ar WR	Pankhurst and Smellie (1983); Smellie et al. (1984)
P.850.11	Eastern Byers Peninsula, Livingston Island	Basalt lava	79 ± 3	K–Ar WR	Pankhurst and Smellie (1983); Smellie et al. (1984)
P.845.1b	Vietor Rock, Byers Peninsula, Livingston Island	Basalt lava	106 ± 4	K–Ar WR	Pankhurst and Smellie (1983); Smellie et al. (1984)
P.845.2c	Vietor Rock, Byers Peninsula, Livingston Island	Basaltic andesite lava	108 ± 4	K–Ar WR	Pankhurst and Smellie (1983); Smellie et al. (1984)
P.845.3a	Vietor Rock, Byers Peninsula, Livingston Island	Basaltic andesite lava	86 ± 3	K–Ar WR	Pankhurst and Smellie (1983); Smellie et al. (1984)
P.845.3b	Vietor Rock, Byers Peninsula, Livingston Island	Basaltic andesite	74 ± 2	K–Ar	Smellie et al. (1984)
P.848.12/14	Chester Cove, Byers Peninsula, Livingston Island	Rhyolite lava	109 ± 4	K–Ar WR	Pankhurst and Smellie (1983); Smellie et al. (1984)
P.417.2	President Head, Snow Island	Dacite	46 ± 2	K–Ar	Pankhurst and Smellie (1983)
Northern Graham Land 63–65°S					
BS.52.3	Tower Island	Basalt	59 ± 2*	K–Ar WR	Rex (1976)
BS.52.1	Tower Island	Basalt	65 ± 2*	K–Ar WR	Rex (1976)
BS.53.2	Tower Island	Basalt	55 ± 2*	K–Ar WR	Rex (1976)
BS.1.2	Two Hummock Island	Basalt	36 ± 2*	K–Ar WR	Rex (1976)
	Cape Spring, Danco Coast	Rhyodacite lava	94 ± 6	K–Ar	Valencio et al. (1979)
	Cuverville Island	Andesitic lava breccia	103.3 ± 1.7	Zircon U–Pb	Zheng et al. (2017)
	Cuverville Island	Andesitic breccia	101.9 ± 1.8	Zircon U–Pb	Zheng et al. (2017)

Central Graham Land 65–67°S
[None]

Southern Graham Land and Adelaide Island 67–69°S

Sample	Locality	Rock type	Age (Ma)	Method	Reference
J6.347.1	Buchia Buttress, Adelaide Island	Silicic crystal tuff	149.5 ± 1.6	Zircon U–Pb	Riley et al. (2011b)
J8.403.1	Milestone Bluff, Adelaide Island	Rhyolitic ignimbrite	113.9 ± 1.2	Zircon U–Pb	Riley et al. (2011b)
J8.20.1s	Reptile Ridge, Adelaide Island	Rhyolitic ignimbrite	67.6 ± 0.7	Zircon U–Pb	Riley et al. (2011b)
R.26.1–R.26.3, R.18.1	Webb Island, east of Adelaide Island	Rhyolite-dacite	67 ± 24	Rb–Sr isochron	Thomson and Pankhurst (1983)
R.2065.1	Mount Liotard, Adelaide Island	Rhyolitic tuff	43.3 ± 0.4 (R)	Ar–Ar WR plateau	Griffiths and Oglethorpe (1998)
R.5155.34	Mount Barré, Adelaide Island	Dacitic tuff	39.5 ± 0.7 (R)	Ar–Ar WR plateau	Griffiths and Oglethorpe (1998)
R.5118.18	Milestone Bluff, Adelaide Island	Dacitic tuff	27.8 ± 0.6 (R)	Ar–Ar WR plateau	Griffiths and Oglethorpe (1998)
R.5210.1	Jenny Island	Andesite	49.6 ± 0.5	Ar–Ar WR plateau	Griffiths and Oglethorpe (1998)
R.5181.2	Dion Islands	Andesite	43.5 ± 1.0	Ar–Ar WR plateau	Griffiths and Oglethorpe (1998)
R.5200.2	Reptile Ridge, Adelaide Island	Rhyolite	57.7 ± 1.1 (R)	Ar–Ar WR plateau	Griffiths and Oglethorpe (1998)
R.5212.3	Piñero Island, E of Adelaide Island	Rhyolite	56.9 ± 0.6	Ar–Ar WR plateau	Griffiths and Oglethorpe (1998)
R.6172.1–R.6172.7	Homing Head, Horseshoe Island	Andesitic–dacitic tuffs and lavas	121 ± 10 Ma	Rb–Sr	Loske et al. (1997)

Northern Palmer Land and Transition Zone 69–71°S

Sample	Locality	Rock type	Age (Ma)	Method	Reference
R.5869.4	East of Crescent Scarp	Andesite lava	107.1 ± 1.7	Zircon U–Pb	Leat et al. (2009)
KG.214	Braddock Nunataks	Andesite	88 ± 4*	K–Ar WR	Rex (1976)

Southern Palmer Land 71–73°S
[None]

Alexander Island

Sample	Locality	Rock type	Age (Ma)	Method	Reference
KG.4411.2A	Elgar Uplands	Andesite lava	55.3 ± 4.4	Ar–Ar plagioclase	McCarron and Millar (1997)
KG.4425.1	Elgar Uplands	Andesite lava	53.3 ± 0.8	Ar–Ar plagioclase	McCarron and Millar (1997)
KG.2011.3	Geode Nunataks	Basaltic andesite lava	41.3 ± 1.0	K–Ar WR	McCarron and Millar (1997)
KG.2011.5	Geode Nunataks	Basaltic andesite lava	59.9 ± 1.4	K–Ar WR	McCarron and Millar (1997)
KG.2011.9	Geode Nunataks	Basaltic andesite lava	56.4 ± 1.6	K–Ar WR	McCarron and Millar (1997)
KG.2015.1	Finlandia Foothills	Basaltic andesite lava	50.2 ± 1.3	K–Ar WR	McCarron and Millar (1997)
KG.2015.1	Finlandia Foothills	Basaltic andesite lava	46.1 ± 7.0	Ar–Ar plagioclase	McCarron and Millar (1997)
KG.2015.4	Finlandia Foothills	Basaltic andesite lava	45.5 ± 1.3	K–Ar WR	McCarron and Millar (1997)
KG.2479, KG.2473	Colbert Mountains	Rhyolite tuffs	62 ± 1	Rb–Sr mineral isochron	Thomson and Pankhurst (1983); McCarron and Millar (1997)
KG.2465.5	Colbert Mountains	Dacite lava	62.7 ± 1.5	K–Ar WR	McCarron and Millar (1997)
KG.2465.7	Colbert Mountains	Dacite lava	58.4 ± 1.4	K–Ar WR	McCarron and Millar (1997)
KG.4356.1	Colbert Mountains	Rhyolitic ignimbrite	64.6 ± 1.0	K–Ar biotite	McCarron and Millar (1997)
KG.4392.1	Colbert Mountains	Rhyolitic ignimbrite	64.9 ± 1.3	K–Ar biotite	McCarron and Millar (1997)
KG.4392.1	Colbert Mountains	Rhyolitic ignimbrite	57.8 ± 1.9	K–Ar amphibole	McCarron and Millar (1997)
KG.4381.3C	Colbert Mountains	Pyroclastic unit	63.3 ± 1.4	Ar–Ar biotite	McCarron and Millar (1997)
KG.4374.1A	Colbert Mountains	Andesite lava	61.7 ± 0.8	Ar–Ar plagioclase	McCarron and Millar (1997)
KG.4375.6	Colbert Mountains	Andesite lava	63.2 ± 1.0	Ar–Ar plagioclase	McCarron and Millar (1997)
KG.4428.1	Monteverdi Peninsula	Andesite lava	79.7 ± 2.5	K–Ar amphibole	McCarron and Millar (1997)

Plutonic and hypabyssal rocks and dates for alteration assemblages are omitted. K–Ar ages published earlier than 1973 are omitted. Some ages in less accessible literature are omitted. (R) indicates that ages can be interpreted as significantly reset. WR, whole rock; GM, groundmass.
*Recalculated to new decay constants (Dalrymple 1979).

1983; Storey and Garrett 1985). The age of the continental crust is as old as Silurian (435–422 Ma) in western Palmer Land (Millar et al. 2002), and Ordovician (487–485 Ma) in eastern Graham Land (Riley et al. 2012). A terrane model for the development of the margin was developed by Vaughan and Storey (2000), Vaughan et al. (2002) and Ferraccioli et al. (2006). In this model, an Eastern Domain consists of autochthonous Gondwana sequences. This is separated from an allochthonous magmatic Central Domain that is thought to have docked in its current position along the Eastern Domain during mid-Cretaceous times (Vaughan et al. 2002; Vaughan et al. 2012a). A Western Domain consisted of accretionary complexes that collided with the margin during Mesozoic times, typified by the LeMay Group of Alexander Island (Burn 1984; Doubleday et al. 1994). More recently, data indicating continuity of lithologies between the Eastern and Central domains have been used to question the allochthonous character of the Central Domain (Burton-Johnson and Riley 2015). Northern Graham Land is underlain by dominantly Carboniferous–Triassic metasediments of the Trinity Peninsula Group (Barbeau et al. 2009; Castillo et al. 2016), which has uncertain relationships in the terrane model. The South Shetland Islands are underlain by the Mesozoic accretionary Scotia Metamorphic Complex and Mesozoic metasediments (Trouw et al. 2000; Hervé et al. 2006).

Seafloor magnetic anomalies on the ocean plate west of the Antarctic Peninsula define a series of ridge–trench collisions that led to the cessation of subduction along the Antarctic Peninsula margin (Barker 1982). As the ocean plate on the west side of the ridge was the same Antarctic Plate as the overriding plate, this led to the extinction of the active margin. The ridge–trench collisions occurred at c. 50 Ma off southern Alexander Island and become younger to the north, and much reduced subduction is still taking place beneath the South Shetland Islands sector (Larter and Barker 1991; Maldonado et al. 1994).

The oldest volcanic rocks that can be assigned to the volcanic arc are Late Jurassic–Early Cretaceous in age (Thomson and Pankhurst 1983; Haase et al. 2012). The plutonic record, largely based on Rb–Sr geochronology, suggests intense magmatism from 142 Ma until the Early Tertiary, preceded by a gap in plutonism from 156 to 142 Ma (Leat et al. 1995). It has been suggested that there was no subduction taking place during this gap (Scarrow et al. 1998; Macdonald et al. 1999; Burton-Johnson and Riley 2015). The earlier tectonic environment is uncertain. A subduction system and volcanic arc were widely thought to have been active almost continuously from Middle Jurassic times or earlier (e.g. Suárez 1976; Pankhurst 1982; Thomson 1982a; Storey and Garrett 1985; Leat et al. 1995). Nevertheless, Jurassic plutons in western Palmer Land and western Graham Land may be related to the extensive Jurassic Chon Aike province of the eastern part of the Antarctic Peninsula (Riley et al. 2017). This province was related to extension during Gondwana break-up rather than subduction (Pankhurst et al. 2000; Riley et al. 2001; Riley and Leat 2021). Detrital zircons from the Trinity Peninsula Group, Miers Bluff Formation and Cape Wallace Beds, all Late Paleozoic–Jurassic metasedimentary units in northern Graham Land and the South Shetland Islands, record a significant peak of Permian age, with a range of c. 320–240 Ma, thought to be sourced from a Permian magmatic province, probably a magmatic arc (Hervé et al. 2006; Barbeau et al. 2009; Castillo et al. 2016). A Jurassic peak in zircon grain ages of c. 180–170 Ma occurs in the Miers Bluff Formation, Livingston Island (Hervé et al. 2006; Castillo et al. 2016). These Jurassic zircons correspond in age to volcanism in the Chon Aike province (Pankhurst et al. 2000) from which they may have been derived (Hervé et al. 2006). The near absence of detrital zircons younger than c. 240 Ma (apart from the 180–170 Ma Chon-Aike-related peak) suggests that any active volcanic arc that existed in the Antarctic Peninsula between the Middle Triassic and Early Cretaceous is not represented in the detrital record. Evidence for Jurassic subduction beneath the Antarctic Peninsula at present remains elusive.

The west coast of the Antarctic Peninsula, notably that of Graham Land and its associated islands, provides abundant evidence for numerous faults with large displacements. The volcanic rocks of the Antarctic Peninsula volcanic arc are commonly deformed by gentle to moderate folds and are cut by faults. These faults control the coastline, fjords, straits and major glacier-filled valleys. Major fault zones approximately parallel to the volcanic arc include the rift that defines the Bransfield Strait (Barker 1982; Birkenmajer 1992), the rift forming George VI Sound between Alexander Island and Palmer Land (Bell and King 1998), and the East Palmer Land Shear Zone, a zone of mid-Cretaceous compression within Palmer Land (Vaughan et al. 2012a, b). Locally significant faults with similar trends have been identified in Pourquoi Pas Island (Matthews 1983a), the Lemaire Channel (Curtis 1966) and in the South Shetland Islands (Smellie et al. 1984; Birkenmajer 1989).

Aeromagnetic data coverage identifies a belt of positive magnetic anomalies c. 70–150 km-wide following the west coast of the Antarctic Peninsula. The source of this so-called Pacific Margin Anomaly (PMA) is modelled as the magmatic arc batholith (Garrett 1990; Johnson 1999; Ferraccioli et al. 2006). Multiple anomaly trends observed in the magnetic fabric suggest a major structural trend approximately parallel to the west coasts of Graham Land and Palmer Land, and a second order set of faults trending approximately NW–SE aligned with flowlines on the subducting plate during Early Tertiary subduction (Johnson and Swain 1995; Johnson 1999).

Occurrence, characteristics and age

We describe the rocks of the Antarctic Peninsula volcanic arc in seven sections; the South Shetland Islands, Alexander Island and subdividing the Antarctic Peninsula into five sectors by latitude (Fig. 1).

South Shetland Islands

The South Shetland Islands (Fig. 2) comprise a 300 km-long archipelago from Smith Island to King George Island (ignoring the northern Elephant Island and Clarence Island on which there are no known volcanic rocks). The major islands are extensively ice covered, and much of the exposure is coastal and on ice-free low-lying areas. There are few extensive ice-free areas, and accurate correlation of lithostratigraphic units is still progressing.

The highest metamorphic grade rocks cropping out in the South Shetland Islands form the Scotia Metamorphic Complex on Smith Island and the Elephant Island group to the NW of King George Island. This complex consists of polydeformed rocks of up to amphibolite and blueschist facies interpreted as an accretionary complex metamorphosed during the Mesozoic (Smellie and Clarkson 1975; Tanner et al. 1982; Hervé et al. 1991; Trouw et al. 2000). A c. 3 km-thick turbiditic sandstone–mudstone sequence on Hurd Peninsula, Livingston Island forms the Miers Bluff Formation (Arche et al. 1992; Smellie et al. 1995). The age of deposition is poorly constrained. Willan et al. (1994) interpreted a Triassic 243 ± 8 Ma Rb–Sr errochron as representing homogenization during deposition and diagenesis. Hervé et al. (2006) found the

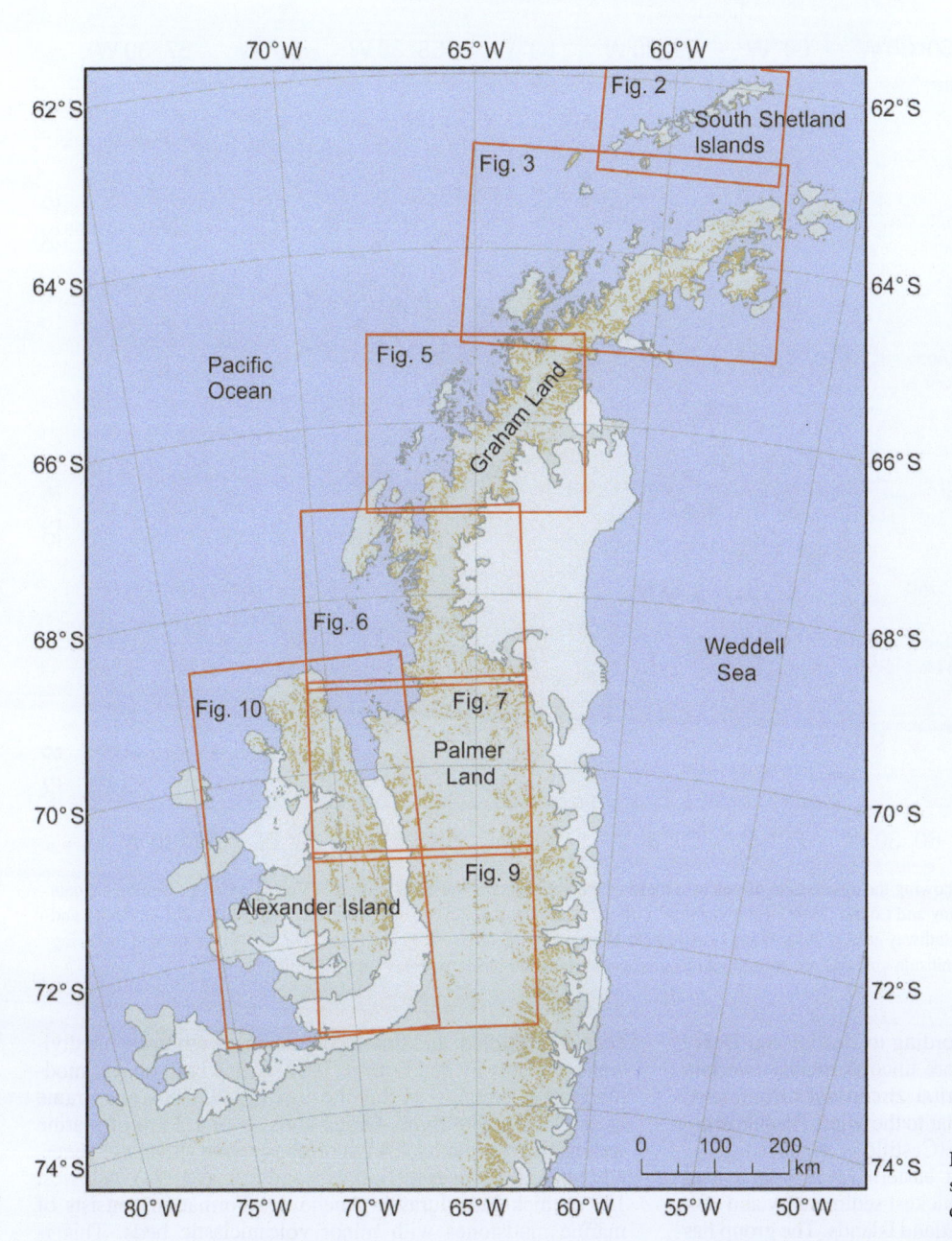

Fig. 1. Map of the Antarctic Peninsula showing areas covered by detailed maps.

youngest U–Pb ages of detrital zircons to be 225–170 Ma in two units of the formation, and provided a minimum age of 137.7 ± 1.4 Ma defined by a dioritic intrusion.

Occurrence. Volcanic and volcaniclastic rocks form most of the outcrop of the archipelago from Snow Island to King George Island (Fig. 2). In the eastern and central parts of the archipelago there are two main stratigraphic units: the Byers Group and the Coppermine Formation. The Byers Formation (Smellie et al. 1984), redefined as the Byers Group, is Late Jurassic–Early Cretaceous in age, includes Early Cretaceous volcanic rocks and forms most of the outcrop in western Livingston Island (Hathway and Lomas 1998). The Late Cretaceous Coppermine Formation (c. 83–78 Ma) crops out mainly on Greenwich and Robert islands (Smellie et al. 1984; Haase et al. 2012). On King George Island and Robert Island, most of the volcanism is Eocene in age, including extensive outcrops on Fildes Peninsula (Gao et al. 2018) and around Admiralty Bay (Nawrocki et al. 2010; Nawrocki et al. 2011). Numerous hypabyssal and subvolcanic intrusions are associated with the Cretaceous–Eocene volcanism. The distribution of ages forms a well-documented general migration of volcanism with time from SW to NE (Pankhurst and Smellie 1983; Smellie et al. 1984; Haase et al. 2012), although, as Haase et al. (2012) noted, the trend becomes less clear when dates of intrusion of hypabyssal bodies are considered. Pliocene–Recent basaltic volcanic deposits overlying the Cretaceous–Eocene volcanic rocks are dealt with in a separate chapter (Smellie 2021).

Geologically, the South Shetland Islands are cut by numerous faults and structure is locally complex, with local tilting and folding (Smellie et al. 1984). Birkenmajer (1989, 1992) proposed that King George Island is composed of four differentially uplifted blocks bounded by trench-parallel faults. Extensional faulting in the South Shetland Islands associated with the formation of the Bransfield Strait is thought to have started at about 26–22 Ma (Birkenmajer 1992), with formation of the deeper Bransfield Rift in the last 4 myr (Barker 1982).

Characteristics. On Low Island (Fig. 3), small, scattered outcrops of marine, volcaniclastic, turbiditic sedimentary rocks forming the Cape Wallace Beds are identified as Late Jurassic in age from macrofauna. Mafic and andesitic lavas have uncertain stratigraphic relationships (Smellie 1980; Thomson

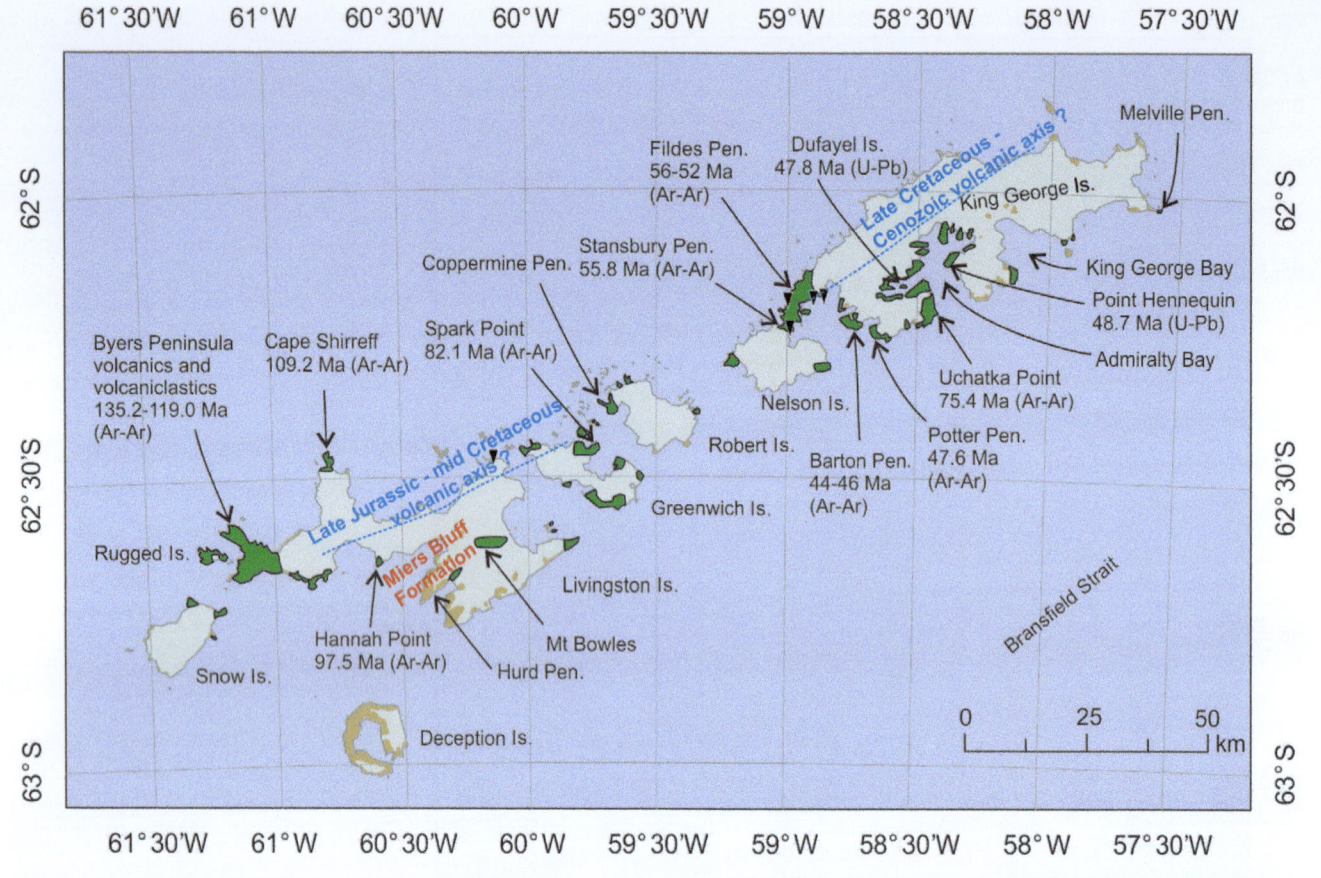

Fig. 2. Map of South Shetland Islands showing the distribution of volcanic rocks of the volcanic arc Antarctic Peninsula Volcanic Group. Geological data after Smellie *et al.* (1984, 1996*a*), Hathway and Lomas (1998), Hathway *et al.* (1999), Machado *et al.* (2005) and Nawrocki *et al.* (2010, 2011). Ar–Ar and U–Pb data are from Haase *et al.* (2012), Hathway *et al.* (1999), Wang *et al.* (2009), Nawrocki *et al.* (2010, 2011) and Gao *et al.* (2018). In this and following maps, volcanic rocks of the Antarctic Peninsula volcanic arc are shown coloured green. Inverted black triangles are vent structures.

1982*b*; Smellie *et al.* 1984). According to Bastias and Hervé (2013), the Cape Wallace Beds are unconformably overlain by the lavas, and they have detrital zircon ages dominated by a Permian (252 Ma) peak, similar to the Miers Bluff Formation and Trinity Peninsula Group (Castillo *et al.* 2016).

The Byers Group, exposed in eastern Livingston Island (Fig. 2), is among the oldest and thickest sedimentary and volcanic successions in the South Shetland Islands. The group has been well studied, and this account uses the stratigraphic divisions of Hathway and Lomas (1998) which built on and modified earlier accounts by Smellie *et al.* (1980, 1984) and Crame *et al.* (1993). The Byers Group consists of 1.3 km of marine sediments overlain by 1.4 km of non-marine strata and generally youngs to the east (Hathway and Lomas 1998). A lowest 120 m-thick Late Jurassic Anchorage Formation consists of marine mudstones with minor volcaniclastic beds. This is

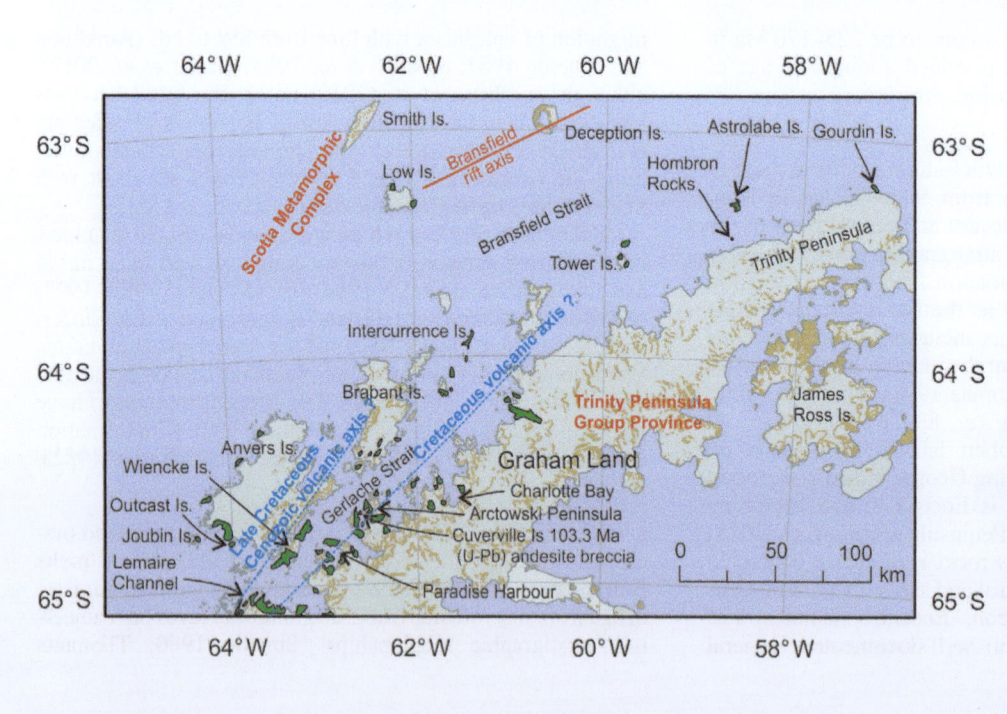

Fig. 3. Map of northern Graham Land 63–65° S showing the distribution of volcanic rocks of the volcanic arc Antarctic Peninsula Volcanic Group. Data are from Hooper (1962), West (1974), Bell (1984), Ringe (1991), Birkenmajer (1995), Smellie *et al.* (1984) and Ryan (2007). Cuverville Island andesite breccias are zircon U–Pb dated at 103.3 ± 1.7 Ma (Zheng *et al.* 2017).

overlain by c. 600 m of sandstone turbidite beds and mudstones of the Early Cretaceous (Berriasian) President Beaches Formation. A 265 m-thick series of volcanic breccias forms the overlying Start Hill Formation (Agglomerate Member of Smellie et al. 1980). These consist of poorly sorted units 2–35 m thick dominated by basaltic sub-angular clasts up to 3 m in diameter interpreted as a marine debris apron on the flanks of a nearby volcano (Smellie et al. 1980; Hathway and Lomas 1998). The Start Hill Formation was Ar–Ar dated at 135.2 ± 2.57 Ma (Valanginian) (Haase et al. 2012). The top of the marine succession is formed by the c. 300 - m-thick shallow-marine Chester Cone Formation which consists of conglomerate and sandstone debris-flow deposits and mudstones of Valanginian age (Crame et al. 1993; Hathway and Lomas 1998). A 1.4 km-thick terrestrial succession, the Cerro Negro Formation, unconformably overlies the marine strata. The deposits include silicic non-welded tuffs and volcaniclastic deposits, silicic welded and non-welded ignimbrites, basaltic tuffs and breccias, and lacustrine and fluviatile deposits (Hathway and Lomas 1998; Hathway et al. 1999). The Byers Group is intruded by many minor igneous intrusions, some of which are sill-like peperitic bodies resulting from intrusion of magma into wet sediment (Smellie et al. 1980; Hathway and Lomas 1998). Ar–Ar age data indicate that the Cerro Negro Formation was deposited during a short interval at c. 120–119 Ma. The hiatus in sedimentation between c. 135 and 120 Ma is interpreted to indicate tectonic uplift followed by deposition of the terrestrial sequence in an intra-arc basin (Hathway and Lomas 1998; Hathway et al. 1999).

Correlatives of the Byers Group are thought to form outcrops on Rugged Island, NE Snow Island and Cape Shirreff, Livingston Island (Smellie et al. 1980, 1984; Hathway and Lomas 1998). On Snow Island, the group comprises dacitic lavas interbedded with silicic tuffs and volcaniclastic deposits (Smellie et al. 1984). The volcanic sequence at Cape Shirreff consists of c. 450 m of mafic lavas and breccias (Smellie et al. 1996a), and was Ar–Ar dated at 109.2 ± 0.46 Ma (Haase et al. 2012).

At Hannah Point, c. 500 m of mafic and dacitic lavas are interbedded with pyroclastic and sedimentary beds (Smellie et al. 1996a). An Ar–Ar date of 97.5 ± 0.39 Ma is consistent with younging of volcanism to the NE (Haase et al. 2012). In southeastern Livingston Island, mafic lavas, breccias and volcaniclastic deposits around Mount Bowles have been strongly altered by hydrothermal activity and are thought to be Cretaceous in age (Willan 1994; Smellie et al. 1996a). This Mount Bowles Formation is inferred from observations of cliff sections to be c. 650 m thick and its volcaniclastic component is thought to consist largely of phreatomagmatic material redeposited by debris flows (Smellie et al. 1995).

Volcanic rocks in northwestern Livingston Island, Greenwich Island and Robert Island are mainly mafic lavas and are grouped into the Coppermine Formation (Smellie et al. 1984). On Coppermine Peninsula, the volcanic rocks are dominated by 5–7 m-thick basaltic lavas with oxidized autobreccias interbedded with coarse-grained pyroclastic air-fall deposits (Smellie et al. 1984). Elsewhere in the formation, mafic lavas are interbedded with coarse volcaniclastic deposits (Smellie et al. 1984). Haase et al. (2012) Ar–Ar dated a Coppermine Formation lava from Greenwich Island at 82.1 ± 0.44 Ma, consistent with K–Ar data for the formation of c. 78–83 Ma (Smellie et al. 1984).

An extensive, gently folded outcrop of volcanic rocks forms Fildes Peninsula, western King George Island. The Fildes Formation (Smellie et al. 1984), or Fildes Peninsula Group (e.g. Barton 1965; Birkenmajer 1989; Gao et al. 2018), has been variously stratigraphically subdivided (Smellie et al. 1984; Birkenmajer 1989; Machado et al. 2005; Gao et al. 2018). According to the recently published system used by Gao et al. (2018), the volcanic sequence is c. 500 m thick with mafic lavas and lava breccias mostly 1–7 m thick (Smellie et al. 1984) forming the lower Jasper Hill and Agate Beach formations, passing up into volcanic breccias, tuffs and sedimentary beds of the Fossil Hill Formation. The sediments contain Eocene fossil plant material and bird footprints (Cao 1992; Shen 1992; Poole et al. 2001; Gao et al. 2018). The overlying Block Hill and Long Hill formations comprise mafic and andesite lavas, breccias, and associated minor intrusions. Debris flows including possible lahar deposits have been identified (Birkenmajer 1989). Gao et al. (2018) used Ar–Ar geochronology to date this succession at 56.4 ± 0.20–52.4 ± 0.42 Ma, consistent with the Ar–Ar dating of Haase et al. (2012) from the same area. A comparable succession forms Stansbury Peninsula (Smellie et al. 1984; Gao et al. 2018) on Nelson Island, which is Ar–Ar dated at 55.8 ± 1 Ma (Haase et al. 2012). In both the Fildes and Stansbury peninsulas, several vents have been described containing mafic agglomerates with bombs, and intruded by irregular dykes, sheets and plugs (Smellie et al. 1984).

At the Barton and Potter peninsulas, King George Island, c. 500 m of mafic–intermediate lavas interbedded with pyroclastic and volcaniclastic deposits have local internal unconformities (Smellie et al. 1984; Tokarski et al. 1987; Birkenmajer 1998a; Wang et al. 2009). Fossil plant fragments are contained within a fine-grained tuffaceous bed at Barton Peninsula (Tokarski et al. 1987). Willan and Armstrong (2002) described debris-flow deposits, lacustrine sediments and welded tuffs including eutaxitic rhyolitic welded tuff, probably an ignimbrite, interbedded with lavas at Barton Peninsula. A c. 1.25 km-diameter caldera may be defined by the arcuate Noel Hill intrusion. Wang et al. (2009) dated two lavas from Barton Peninsula by Ar–Ar geochronology yielding preferred isochron ages of 45.8 ± 1.68 and 44.2 ± 1.56 Ma, similar to the Ar–Ar plateau age of 47.6 ± 0.22 Ma of a mafic lava from Potter Peninsula (Haase et al. 2012). Two volcanic samples yielded split Ar–Ar spectra with high-temperature plateaux at 120.4 ± 1.6 and 119.5 ± 0.5 Ma, a preferred isochron age of 104.8 ± 5.5 Ma, and low-temperature components at 53–52 Ma (Kim et al. 2000; Zheng et al. 2000). The low ages are consistent with resetting during a local intrusion event, and the older ages suggest that an inlier of Cretaceous volcanic rocks may be present.

The numerous outcrops around Admiralty Bay consist mostly of basaltic, andesitic and dacitic lavas interbedded with terrestrial volcaniclastic and sedimentary deposits. Zircon U–Pb and Ar–Ar geochronology indicates that the majority of the volcanism and sedimentation was Early Eocene in age (c. 53–48 Ma: Nawrocki et al. 2010, 2011), and occurred in low-lying, locally lacustrine and wetland environments close to stratocones (Birkenmajer 1989; Mozer 2012, 2013; Warny et al. 2016). Around King George Bay volcanism continued in episodes in Late Eocene times (c. 37–35 Ma: Pańczyk and Nawrocki 2011) and at about the Oligocene–Miocene transition boundary (c. 24–22 Ma), when the volcanism took place in a coastal, glaciated, locally submarine environment (Smellie et al. 1998; Warny et al. 2018). Two of the widespread mafic hypabyssal and subvolcanic intrusions around Admiralty Bay are reliably dated at c. 28–25 Ma (Pańczyk et al. 2009), indicating that Oligo-Miocene volcanic rocks may have been removed by erosion.

Along the NW coast of Admiralty Bay, at least 1100 m of mafic, andesitic and subordinate silicic lavas, tuffs, and volcaniclastic sediments comprises the Martel Inlet Group. Birkenmajer (1989) thought that the sequence represented parts of subaerial stratovolcanoes. The unit has not been reliably dated, with K–Ar data only indicating a Paleogene age (Birkenmajer 1989). Volcanic rocks on the adjacent NE coast of Admiralty Bay are strongly folded mafic lavas, tuffs and breccias more than 540 m thick comprising the Cardoza Cove Group, which also has not been reliably dated (Birkenmajer 1989). A sequence of dominantly andesitic lavas and breccias interbedded with tuffs and conglomerates on Dufayel Island may correlate with part of the Cardoza Cove Group (Birkenmajer 1989; Nawrocki et al. 2011). An andesitic lava from Dufayel Island yielded a zircon U–Pb age of 47.8 ± 0.5 Ma (Middle Eocene) (Nawrocki et al. 2010).

On the headland between Dufayel Island and Bransfield Strait, SE of Admiralty Bay, three volcanic–sedimentary groups have been defined. The Paradise Cove Group crops out around Uchatka Point. An Ar–Ar plateau age of 75.4 ± 0.9 Ma for a mafic lava of the Uchatka Point Formation of this group is the only reliable Late Cretaceous age in the Admiralty Bay area (Nawrocki et al. 2010). The formation consists of mafic lavas associated with terrestrial deposits of the Creeping Stone Formation (Nawrocki et al. 2010, 2011). Zircon U–Pb geochronology has shown that dacitic (or trachydacitic) lavas of the Demay Point Formation are much younger, 53.0 ± 0.7 Ma (Early Eocene) (Nawrocki et al. 2010).

The Demay Point dacites are overlain by mafic lavas of the Baranowski Glacier Group which is distributed c. 5 km north and west of Uchatka Point (Mozer et al. 2015). The group comprises a lower Llano Point Formation consisting of subaerial basaltic andesitic lavas and breccias more than 1100 m thick (Birkenmajer 1989; Mozer et al. 2015), and an upper Zamek Formation up to 150 m thick consisting of subaerial basaltic lavas, breccias, conglomerate, bedded tuffs interbedded with mudstones, siltstones, sandstones, and coals with leaves and other plant remains (Mozer et al. 2015). Two lavas from the Llano Point Formation yielded Ar–Ar plateau ages of 52.3 ± 0.5 and 50.8 ± 1.2 Ma (Nawrocki et al. 2011).

An Ezcurra Inlet Group is exposed on the NW-facing coast of Admiralty Bay east of Dufayel Island. Its lower Arctowski Cove Formation, c. 220 m thick, comprises basaltic–andesitic lavas and breccias, pyroclastic deposits, conglomerates, and mudstones (Birkenmajer 1989; Mozer 2012). The formation contains plant fragments and its upper Petrified Forest Member is rich in fossil wood (Mozer 2012). Its upper Point Thomas Formation, c. 500 m thick, comprises basaltic–dacitic lavas and breccias, tuffs, conglomerates, sandstones, and mudstones. Plant fossils including fossilized wood occur in several localities, and pillow structures at the base of lava flows are interpreted as indicating a shallow lake environment (Mozer 2012). A lava from the upper part of the Point Thomas Formation yielded a zircon U–Pb age of 48.9 ± 0.7 Ma (Nawrocki et al. 2010).

On the east coast of Admiralty Bay, around Point Hennequin, the Point Hennequin Group consists of a lower Viéville Glacier Formation and an upper Mount Wawel Formation, with a total thickness of more than 500 m (Birkenmajer 1989). The formations consist of basaltic andesitic and andesitic lavas, fluvial conglomerate, volcaniclastic breccias, siltstones, and mudstones (Nawrocki et al. 2011; Mozer 2013). Small-scale cross-bedding and leaf imprints occur in the fine-grained deposits in loose blocks, which are also rich in authigenic pyrite associated with the plant material, suggesting deposition in a low-lying coastal area (Mozer 2013). A lava of the Viéville Glacier Formation was dated by zircon U–Pb geochronology at 48.7 ± 0.6 Ma (Early Eocene), similar to two consistent Ar–Ar plateau ages of 46.8 ± 0.3 and 46.7 ± 0.3 Ma from a sample of the same formation (Nawrocki et al. 2011). The Mount Wawel Formation yielded younger Ar–Ar plateau ages of 46.0 ± 0.3 and 43.8 ± 0.3 Ma (Nawrocki et al. 2011).

Outcrops at the SW point of the entrance to King George Bay are volcanic and glacio-marine sediments. A basal group of mafic and andesitic lavas (Mazurek Point Formation) which may correlate with the Hennequin Group is unconformably overlain by diamictites, sandstones and basaltic breccias of the Polonez Cove Formation (Troedson and Smellie 2002). Glacial diamictite deposited by marine-based ice is interpreted to form the lowermost 20 m of the formation (Troedson and Smellie 2002; Warny et al. 2018). The upper 60 m of the formation mostly consists of basaltic breccias and bedded basaltic volcaniclastic sandstones interpreted as having been deposited in a submarine fan and lava-fed delta (Porębski and Gradziński 1990; Troedson and Smellie 2002). Within these breccias, a c. 25 m-thick basaltic lava dome with columnar core and breccia carapace was emplaced in a shallow-marine setting and forms the Low Head Member (Smellie et al. 1998). The formation is overlain by the subaerial dacites lavas and breccia of the c. 30 m-thick Boy Point Formation (Warny et al. 2018). Three samples of the basal Mazurek Point lavas have been Ar–Ar dated, giving consistent plateau ages of 44.8 ± 0.2, 44.7 ± 0.3 and 44.5 ± 0.3 Ma (Pańczyk and Nawrocki 2011). The Low Head lava dome yielded an Ar–Ar plateau age of 23.7 ± 2.0 Ma, and the overlying Boy Point dacites a plateau age of 22.6 ± 1.7 Ma (Smellie et al. 1998). These data imply that the unconformity between the Muzurek Point lavas and the marine diamictites of the Polonez Cove Formation may represent some 22 myr.

Andesite lavas from two localities at the NE point of the entrance to King George Bay are andesite lavas with Ar–Ar plateau ages of 37.3 ± 0.4 and 35.35 ± 0.15 Ma (Pańczyk and Nawrocki 2011), suggesting that they were erupted during the interval of the unconformity above the Muzurek Point lavas.

At Cape Melville, subaerially erupted basaltic lavas more than 60 m thick with a weathered upper surface are overlain by marine and glacio-marine sediments dated at c. 26–24 and 23–21 Ma, respectively (Birkenmajer 1989; Dingle and Lavelle 1998; Troedson and Riding 2002; Warny et al. 2016).

Northern Graham Land 63–65° S

This sector comprises the east coast of Graham Land from Trinity Peninsula to the Danco coast on the eastern side of Gerlache Strait, and islands on both sides of the Gerlache Strait (Fig. 3). The topography generally rises steeply from the coast, with a few low-lying islands (Fig. 4a).

The main outcropping basement in the sector is the Trinity Peninsula Group. These metasediments are turbiditic siliciclastic marine sandstones, mudstones and conglomerates thought to represent Carboniferous–Triassic sedimentation in a marginal basin (Birkenmajer 1994; Castillo et al. 2015). U–Pb ages of detrital zircons form a Permian peak, extending from Late Carboniferous to Triassic, and the source is thought to be a Late Paleozoic magmatic province (Barbeau et al. 2009; Castillo et al. 2016). A single zircon grain dated at 176 Ma suggests that sedimentation might have continued in the Jurassic (Barbeau et al. 2009). The group was deformed and metamorphosed probably during Permo-Triassic times up to greenschist facies (Birkenmajer 1994; Smellie et al. 1996b; Castillo et al. 2015). To the west, the Scotia Metamorphic Complex forms Smith Island.

Volcanic rocks on Anvers Island and Brabant Island are poorly known geochemically. They include alkaline and tholeiitic lavas whose tectonic setting is uncertain (Ringe 1991; Smellie et al. 2006). They are described by Smellie (2021). The opening of the 500 km-long and 100 km-wide extensional Bransfield Strait is thought to have occurred mainly in the Pliocene (c. 4 Ma), and rifted the South Shetland block from the Trinity Peninsula (Barker 1982; Catalán et al. 2013), although this may have continued earlier extension (Birkenmajer 1992). The pronounced bathymetric expression of the rift has a SW

Fig. 4. Photographs of the Antarctic Peninsula volcanic arc. (**a**) Northeast coast of Charlotte Bay, Danco Coast, Graham Land, showing the characteristic topography of the west coast of Graham Land. The dark grey and brown alternations in the cliffs in the foreground are probably volcanic rocks. The elevation rises to the right towards the central plateau of Graham Land. (**b**) The south entrance to the Lemaire Channel, Graham Land. Most of the cliff outcrops are volcanic rocks that are folded and *c.* 1525 m thick (Curtis 1966). The bedding is picked out by snow ledges. (**c**) The volcanic succession near Cape Cloos, Penola Strait, Graham Land. The succession dips to the left, with bedding clearly visible. The succession is described as lavas and lava breccias overlying tuffs and conglomerates (Curtis 1966). (**d**) Southeast coast of Adelaide Island. The dark rocks in the foreground are outcrops of the Late Jurassic Buchia Buttress Formation. The high peak, Mount Liotard, is an eroded remnant of the Late Cretaceous Mount Liotard Formation. The pale rocks in the centre left are a granodiorite intrusion (Riley *et al.* 2011*a*). (**e**) Ignimbrite overlying breccia, intracaldera units of the Zonda Towers caldera, northern Palmer Land. (**f**) Silicic ignimbrite, Zonda Towers, northern Palmer Land.

limit to the east of Low Island (Catalán *et al.* 2013) (Fig. 3), with volcanically active Deception Island approximately overlying the rift axis (Smellie 2021).

Occurrence. Volcanic rocks of the Antarctic Peninsula Volcanic Group crop out along the NW coast of the Trinity Peninsula, on both sides of the Gerlache Strait and around Anvers Island (Fig. 3). Along the Trinity Peninsula, outcrops are relatively small and only described from islands (Bell 1984). Around the Gerlache Strait, outcrops form many steep, often cliff-like outcrops which extend from sea level towards the spine of Graham Land. Many of the volcanic strata have gentle dips, where not deformed by later plutons, as on Wiencke and Astrolabe islands and the Arctowski Peninsula (Hooper 1962; Bell 1984; Birkenmajer 1995). Where the base of the volcanic rocks has been observed, they unconformably overlie the Trinity Peninsula Group.

Characteristics. On Gourdin Island, near the northern point of Trinity Peninsula, banded andesitic–dacitic ignimbrites with flattened shards and lithic fragments up to 6 cm across form most of the outcrop (Bell 1984). In the eastern part of Astrolabe Island, a 230 m-thick sequence of planar bedded andesitic and dacitic tuffs were interpreted as dominantly air-fall tuffs by Bell (1984), and are possibly associated with an ignimbrite cropping out to the south. A distinct basaltic volcaniclastic unit in the NW of the island is probably younger and related to extension of the Bransfield Strait. At Hombron Rocks, the volcanic rocks dominantly consist of tuffs and breccias with angular to subrounded clasts up to 0.5 m in diameter. Bell (1984) thought that the rounding of clasts indicated subaqueous deposition, although there is no further evidence to support this interpretation. Tower Island consists of a granodioritic pluton intruded into volcanic rocks. The volcanic rocks include basaltic–andesitic breccias, coarse- and fine-grained tuffs, and lavas (Bell 1984). Basalts from Tower Island have been dated by K–Ar to fall in the interval 65–55 Ma (Rex 1976).

The sequences of air-fall tuffs, ignimbrites and lavas for Gourdin Island to Tower Island are consistent with an interpretation that they represented accumulations on the eastern flank of the volcanic arc, and rifting of the arc to form the Bransfield Strait occurred along a rear-arc axis.

On Intercurrence Island, a dioritic pluton intrudes similar basaltic–andesitic tuffs and lavas (Bell 1984). A basalt from Two Hummock Island to the SW was dated at 36 ± 2 Ma by K–Ar (Rex 1976).

To the west of the Gerlache Strait, on the western coast of the Arctowski Peninsula, basaltic, basaltic–andesitic and andesitic lavas interbedded with minor breccia and tuff and dacitic–rhyolitic lavas are over 1000 m thick. Volcanic rocks form most of the west coast of Rongé Island and the smaller Cuverville and Danco islands (West 1974). Zheng et al. (2017) reported zircon U–Pb dates for an andesitic lava breccia and an andesitic breccia from Cuverville Island of 103.3 ± 1.7 and 101.9 ± 1.8 Ma, respectively.

West (1974) thought that a granitic pluton that forms the SW coast of the peninsula is older than the volcanic rocks. Rb–Sr dating of this pluton at Neko Harbour yielded an age of 114 ± 11 Ma (Pankhurst 1982). However, Birkenmajer (1995) concluded that the granite intrudes the volcanic rock, and suggested that the latter are probably Early Cretaceous in age.

At Paradise Harbour, c. 2000 m of volcanic rocks unconformably overlie metasediments of the Trinity Peninsula Group (Birkenmajer 1998b; Ryan 2007). Pillow lavas and hyaloclastites are reported from the base of the sequence and no marine fossils have been identified (Birkenmajer 1994). The volcanic rocks consist of mafic lavas and breccias intercalated with minor rhyodacite lavas (Birkenmajer 1998b).

On Wiencke Island, the volcanic rocks are over 1200 m thick, and mainly consist of andesite lavas and tuffs (Hooper 1962). To the south of Anvers Island, the volcanic strata have been deformed and metamorphosed by plutonic intrusions. In the Outcast Islands, the volcanic rocks comprise andesite tuffs, breccias and lavas, and on the Joubin Islands include tuffs, breccias and banded lava-like units, some of which are rhyolitic, interbedded with one or more beds interpreted to be of sedimentary origin (Hooper 1962). The plutons that have contact-metamorphosed boundaries with the Joubin Islands volcanic rocks are geochronologically dated at 21–20 Ma (Parada et al. 1990).

Parada et al. (1990) compiled K–Ar and Rb–Sr geochronological data to indicate that most Danco Coast plutons were intruded at 117–96 Ma, and plutons on Wiencke and Anvers islands were intruded later at 68–20 Ma. This is consistent with Late Cretaceous and Early Cenozoic zircon U–Pb dates on plutonic rocks from the Gerlache Strait area (Tangeman et al. 1996; Ryan 2007; Zheng et al. 2018). The age distribution in the plutonism has been used to suggest a westward shift in the axis of the magmatic arc during the Late Cretaceous (Parada et al. 1990; Birkenmajer 1994; Ryan 2007; Zheng et al. 2018).

Central Graham Land 65–67° S

The sector comprises the Graham and Loubet coasts along the west of Graham Land, the northern tip of Adelaide Island and groups of islands along the coast from the Biscoe Islands to the Argentine Islands (Fig. 5). The west coast of Graham Land is formed by steep descents from the c. 2000 m Bruce Plateau along the axis of Graham Land to the coast, with many heavily crevassed valley glaciers, and exposures are poorly accessible. Most of the coast is formed by ice cliffs, and many of the islands are largely ice covered. Consequently, both the inland and island geology is generally poorly known.

To the east of the central and southern parts of this sector, the Target Hill crustal block (Vaughan and Storey 2000; Riley et al. 2012) forms pre-volcanic arc basement (Fig. 5). This crustal block dominantly consists of high-grade rocks recording Ordovician–Jurassic plutonism (Riley et al. 2012), with cover consisting of Jurassic siliciclastic sediments and Middle Jurassic silicic volcanic rocks of the Chon Aike province (Pankhurst et al. 1998; Riley and Leat 1999; Riley et al. 2001). To the north, the basement consists of Trinity Peninsula Group metasediments which extend northwards from Lahille Island (Curtis 1966; British Antarctic Survey 1981a). The nature of the basement to the west of these domains is undefined, as no pre-Jurassic rocks have been proved, and the oldest known plutonic intrusion is a tonalite on Rigsby Island that has been U–Pb zircon dated at 156.0 ± 0.9 Ma (Jordan et al. 2014). The basement may consist of correlatives of the Western Zone of the Central Domain identified in Palmer Land (Ferraccioli et al. 2006; Riley et al. 2011b).

Occurrence. In the northern part of the sector, volcanic rocks crop out on both sides of the Lemaire Channel–Penola Strait between Cape Renard and the Argentine Islands. Around Penola Strait the volcanic rocks are folded, and marker horizons occurring on both sides of Lemaire Channel (Fig. 4b) suggest that the strait marks a fault with a downthrow of c. 900 m to the west (Curtis 1966). The volcanic rocks are intruded by several of the gabbroic–granitic plutonic intrusions in the area. Volcanic rocks form most of the eastern islands of the Argentine Islands and are intruded by plutonic rocks (Elliot 1964). In the southern part of the sector, volcanic rocks form scattered outcrops along the Loubet Coast and outlying islands, where they are near horizontal or have gentle dips, and are locally intruded by plutonic rocks (Goldring 1962).

Volcanic rocks are widespread in the Bisco Islands (Smellie et al. 1985) and form most of the northern part of the Arrowsmith Peninsula (Moyes and Hamer 1984) and the northern part of Adelaide Island (Riley et al. 2011a).

Characteristics. A thickness of 1525 m of volcanic rocks was documented to the east of the Penola Strait by Curtis (1966), where the succession was described as consisting of c. 300 m of andesite lavas overlain by 1220 m of pyroclastic and volcaniclastic rocks. In a separate succession to the north at Cape Cloos (Fig. 4c), lavas overlie volcaniclastic strata (Curtis 1966). The 37 m-thick sequence consists of 10 andesite lavas individually ranging from 1 m to more than 15 m thick, and some have associated breccias. The volcaniclastic rocks are well bedded with current bedding, and are interbedded with weakly bedded, poorly sorted conglomerates

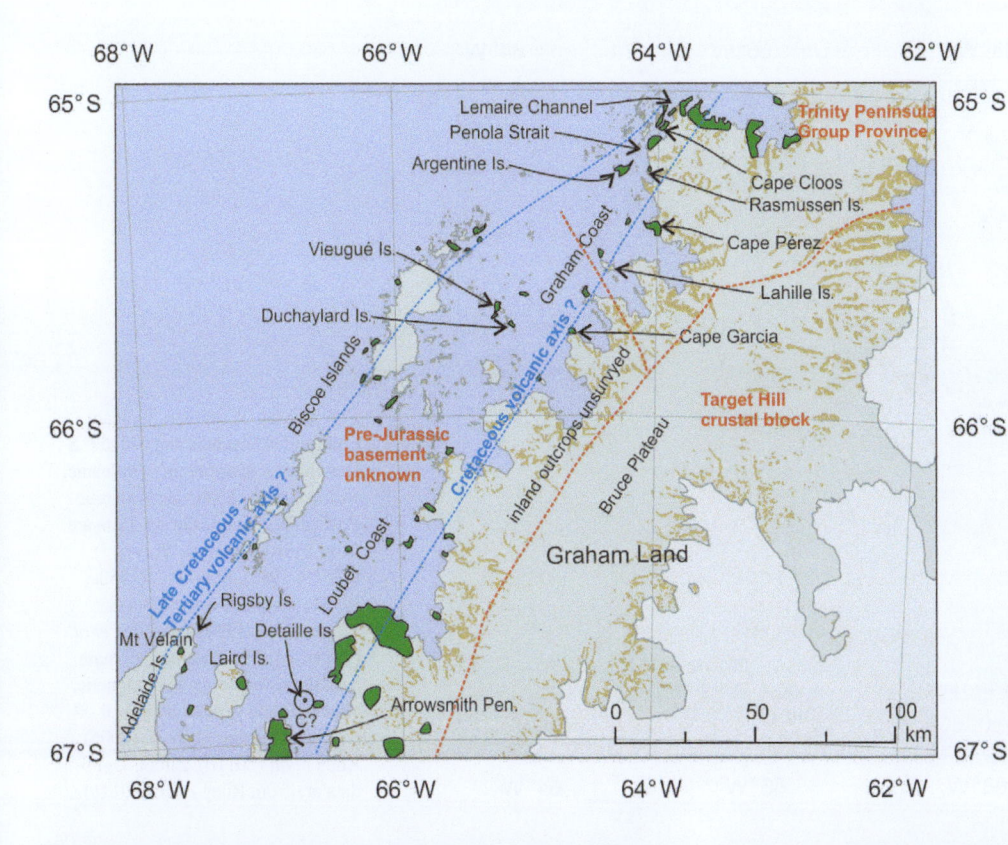

Fig. 5. Map of central Graham Land 65–67° S showing the distribution of volcanic rocks of the volcanic arc Antarctic Peninsula Volcanic Group. Data are from Goldring (1962), Elliot (1964), Curtis (1966), British Antarctic Survey (1981a), Smellie et al. (1985), Moyes et al. (1994) and Riley et al. (2011a). The basement domain boundaries are based on aeromagnetic, structural and geochronological data after Ferraccioli et al. (2006) and Riley et al. (2011b). Circle C is a possible caldera.

containing rounded polymict volcanic clasts (Curtis 1966). A fine- to medium-grained pink granite on Rasmussen Island intruding the volcanic rocks (Curtis 1966) has been Rb–Sr dated at 128 ± 3 Ma (Pankhurst 1982) and zircon U–Pb dated at 117.0 ± 0.8 Ma (Tangeman et al. 1996), suggesting that the volcanism is Early Cretaceous or older. A nearshore marine environment for the current-bedded tuffs and conglomerates was suggested by Curtis (1966), although the absence of any observations of marine sediments or marine fossils would suggest that a terrestrial setting may be more likely.

The volcanic rocks in the Argentine Islands comprise andesitic lavas and poorly bedded dacitic tuffs and breccias (Elliot 1964). The volcanic rocks are intruded by plutonic rocks, one gabbro–granodiorite of which has been Rb–Sr dated at 55 ± 3 Ma (Pankhurst 1982).

To the south, along the Graham Coast, similar sequences of andesite lavas interbedded with volcanic breccias and tuffs, and generally limited in outcrop size and accessibility, are described by Curtis (1966). The thickest described sequence is a c. 300 m thickness of andesite lavas and tuffs at Cape Pérez. At Cape Garcia, a more varied succession of andesitic lava, andesitic breccia, lithic tuff and welded tuff, possibly an ignimbrite, is described. Dacites are only recorded to the west of Graham Coast where they are interbedded with andesites on Duchaylard and Vieugué islands.

The scattered outcrops of volcanic rocks along the Loubet Coast are mostly andesitic but range from basalt to dacites and rhyolite, as described by Goldring (1962). They are mostly gently dipping, and a thickness of volcanic rocks of 1500–1800 m was estimated based on the heights of coastal mountains. According to Goldring (1962), some localities comprise mostly andesitic pyroclastic rocks with subsidiary lavas, while others are dominated by many similar lavas. Several localities comprising crystal, lithic and lapilli tuffs were recorded, ranging from mafic to silicic in composition. Dacites and rhyolites are commonly flow-banded, and at least one has 'elongate lenticles', which may be fiamme.

In the Biscoe Islands, massive, polymict agglomerates and tuffs interbedded with subaqueous sediments are dominant. Massive, brecciated and rare pillowed lavas are interbedded with the sediments. Welded and non-welded dacitic ignimbrites also occur (Smellie et al. 1985). The massive agglomeritic units are interpreted as likely to be submarine debris-flow deposits, and the setting may be a forearc or intra-arc basin (Smellie et al. 1985).

Goldring (1962) described a 'vent agglomerate', which, from his geological map, appears to be a breccia complex that forms all of Detaille Island and outlying islets. The complex consists of a wide range of andesitic, dacitic and basaltic lava, and fragmental clasts. Dacitic bodies in the breccia are described as flow-banded or tuffaceous, range up to 100 m across and have brecciated margins or are 'squeezed into' the breccia. The size of the complex (at least 1.7–1.9 km in diameter), the wide range of clasts and the presence of dacitic layers which may be ignimbritic are indicative that this complex can be tentatively reinterpreted as an intracaldera collapse breccia.

Volcanic rocks at Mount Vélain, in northern Adelaide Island, are 300 m thick and comprise at least 300 m of basaltic andesite and andesite breccias and lavas interbedded with minor, generally <2 m-thick, coarse-grained volcaniclastic sandstones (Riley et al. 2011b). The rocks are correlated with the Bond Nunatak Formation in central Adelaide Island, which is thought to be Late Cretaceous in age (c. 75–67 Ma), and possibly also with volcanic outcrops on Liard Island (Riley et al. 2011b). No base to the volcanic strata has been observed in the sector.

Southern Graham Land and Adelaide Island 67–69° S

The northern part of the sector is dominated by a wide, mountainous, glaciated terrain forming Adelaide Island, the Arrowsmith Peninsula, and Blaiklock, Pourquoi Pas and Horseshoe islands (Fig. 6). In the southern part of the sector, the Avery Plateau along the spine of Graham Land falls away steeply towards Marguerite Bay, with valleys occupied by major glaciers. Several small, isolated islands are mostly situated close to the coast.

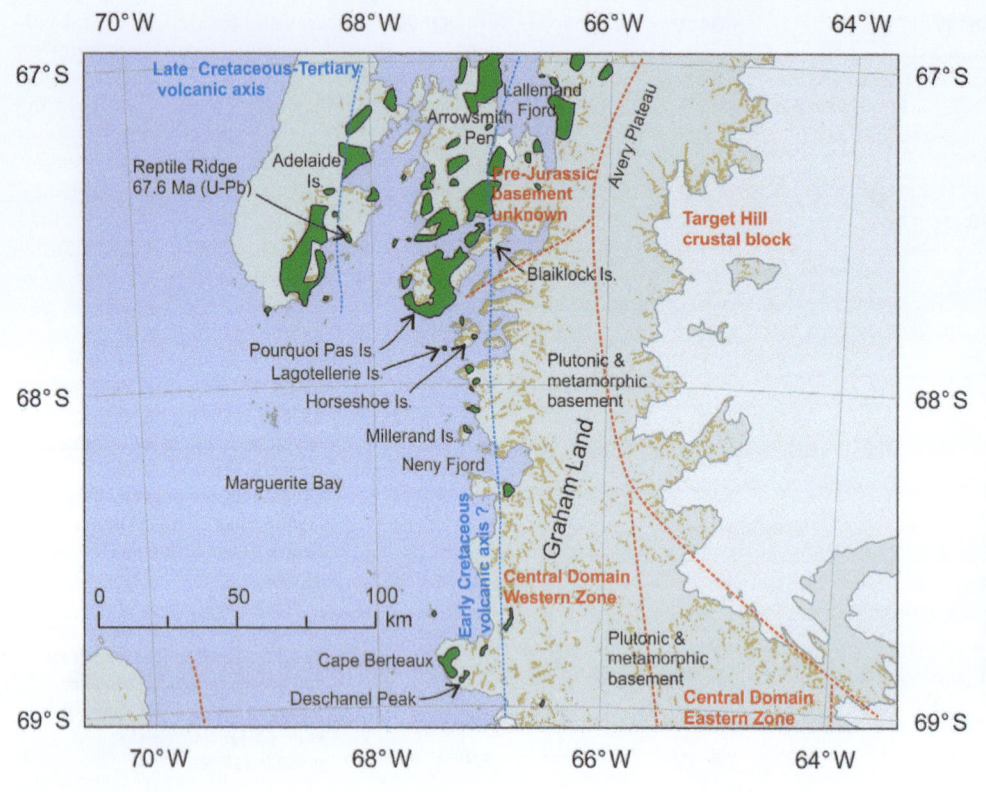

Fig. 6. Map of southern Graham Land and Adelaide Island 67°–69° S showing the distribution of volcanic rocks of the volcanic arc Antarctic Peninsula Volcanic Group. Data are from British Antarctic Survey (1981a, 1982), Matthews (1983a, b), Moyes and Hamer (1984), Moyes et al. (1994) and Riley et al. (2011a, b). The Basement domain boundaries based on aeromagnetic, structural and geochronological data after Ferraccioli et al. (2006), Riley et al. (2011b). Zircon U–Pb data are from Riley et al. (2011b).

High-grade metamorphic and plutonic rocks of the Target Hill crustal block form the basement to the east of the sector. The west coast of Graham Land is included in the Western Zone of the Central Domain in the terrain analysis of Ferraccioli et al. (2006). There is no direct evidence of basement older than Mesozoic in the Domain in this sector. The oldest Rb–Sr dated plutons are Jurassic, 168 ± 13 Ma from the Arrowsmith Peninsula and 162 ± 6 Ma from Horseshoe Island. Gneisses at Horseshoe Island and in the south of the sector have been zircon U–Pb dated at 206 ± 4 and 183.0 ± 2.1 Ma, respectively (Millar et al. 2002; Leat et al. 2009). Most plutons are Cretaceous (Pankhurst 1982; Moyes and Pankhurst 1994; Leat et al. 2009). Zircon U–Pb analysis of plutonic samples from Stonington Island, Neny Fjord, by Tangeman et al. (1996) confirmed Cretaceous ages of crystallization of 125–111 Ma, and U–Pb concordia intercepts in the Stonington Island samples are interpreted as inherited components with ages of 242 and 266 Ma (Late Paleozoic–Mesozoic). Evidence for Paleozoic sources also comes from Horseshoe Island, where granitic clasts in conglomerates have yielded a U–Pb intercept of 431 Ma (Tangeman et al. 1996), and crystallization ages of 445 and 419 Ma (Loske et al. 1997). Similar Paleozoic ages have been obtained for plutonic rocks from the Target Hill crustal block (Riley et al. 2012), and it is uncertain whether the clasts were transported to the Horseshoe Island locality by rivers or whether such basement exists west of the Target Hill crustal block.

Occurrence. There are extensive outcrops of volcanic rocks of the Antarctic Peninsula Volcanic Group in the northern part of the sector, notably on Adelaide and Pourquoi Pas islands and the Arrowsmith Peninsula (Smellie et al. 1994; Riley et al. 2011a, b). The best-documented area is Adelaide Island, where 2–3 km of volcanic and sedimentary rock form successions ranging from Late Jurassic to Tertiary in age. The volcanic rocks are gently folded and dips of strata are generally up to 20°. There are many plutonic intrusions, some of which intrude the volcanic strata. To the south of Horseshoe Island, volcanic outcrops continue as isolated, relatively small outcrops along the west coast of southern Graham Land.

Characteristics. Adelaide Island (Fig. 4d) is dominated by volcanic and sedimentary units of Late Jurassic–Early Tertiary age which were divided into several stratigraphic units by Dewar (1970) and Griffiths and Oglethorpe (1998), and described as formations by Riley et al. (2011a, b). Gabbroic–granitic plutons intruding the volcanic and sedimentary rocks are dated as Tertiary in age (60–44 Ma: Pankhurst 1982; Griffiths and Oglethorpe 1998; Riley et al. 2011b).

The oldest Buchia Buttress Formation is up to 830 m thick, and crops out in the central and southern part of the island (Riley et al. 2011b). It is a volcaniclastic unit, consisting of volcanic breccias, silicic tuffs, volcaniclastic sandstones and conglomerates. The sandstones are cross-bedded and contain a Late Jurassic marine molluscan fauna considered to be Kimmeridgian–Tithonian in age (Thomson and Griffiths 1994). A silicic tuff within the formation yielded a zircon U–Pb concordia age of 149.5 ± 1.6 Ma (Riley et al. 2011b). The formation is thought to represent forearc sedimentation and has been compared to the contemporaneous Himalia Ridge Formation in Alexander Island (Riley et al. 2011b).

The Milestone Bluff Formation is at least 1.5 km thick and is restricted to southern Adelaide Island. It is dominated by turbiditic sandstones and poorly sorted, mostly clast-supported conglomerates containing clasts up to 0.8 m in diameter, interpreted to have been deposited by debris flows. Belemnite and bivalve fossils occur in sandstones and siltstones (Thomson and Griffiths 1994). A zircon U–Pb concordia age of 113.9 ± 1.2 Ma was obtained for a silicic ignimbrite from a silicic tuff unit within the formation (Riley et al. 2011b). The formation is thought to represent deposition in a marine fan in a forearc environment (Griffiths and Oglethorpe 1998; Riley et al. 2011b).

The Mount Liotard Formation is at least 1800 m thick, crops out in southern Adelaide Island, and consists of lavas of basaltic andesitic and andesitic composition associated with lava breccias and rare hyaloclastite beds. Where determined, individual lavas are typically 30–40 m thick. Rare sandstone and mudstone beds contain angiosperm leaf fossils of Cretaceous or Tertiary age (Thomson and Griffiths 1994). The Bond Nunatak Formation comprises a 300–600 m thickness of andesitic and basaltic–andesitic lavas and breccias, volcaniclastic sandstones, and conglomerates, and extends over much of eastern and northern Adelaide Island. The subaerial Reptile Ridge Formation crops out in the southeastern part of the island, and consists of 250–300 m of silicic tuffs, ignimbrites and rare lavas (Riley et al. 2011b). These three formations are thought to be of similar age. A rhyolitic ignimbrite from Reptile Ridge yielded a zircon U–Pb concordia age of 67.6 ± 0.7 Ma (Riley et al. 2011b), and the three formations are thought to have erupted at c. 75–67 Ma. The Mount Liotard and Bond Nunatak mafic–andesitic lava sequences can be interpreted as dominantly subaerial stratocone sequences, and the Reptile Ridge Formation as an approximately contemporaneous silicic centre.

Extensive volcanic rocks on the Arrowsmith Peninsula are intruded by gabbro–granite plutons (Moyes and Hamer 1984). The volcanic rocks are at least 400 m thick, and mostly mafic and rarer andesitic lavas with associated breccias. There are minor silicic tuffs including ignimbrites, and coarse-grained, cross-bedded volcaniclastic beds up to 10 m thick which contain fossil wood (Moyes and Hamer 1984; Smellie et al. 1994). There are no reliable ages for the volcanism on the Arrowsmith Peninsula.

In Blaiklock Island, a volcanic sequence, also undated, forms the eastern part of the island and is intruded by a granite (Moyes and Hamer 1984). The sequence is subaerial, and consists of andesitic and rhyolitic lavas and breccias interbedded with tuffs and weathered horizons described as lateritic (British Antarctic Survey 1981a).

Volcanic rocks form the majority of outcrops on Pourquoi Pas Island where near-horizontal volcanic strata are over 1500 m thick (Matthews 1983a). The sequence is dominated by breccias and tuffs, with relatively few lavas. In the west of the island, the volcanic sequence unconformably overlies dioritic plutonic rocks, the basal bed being a polymict conglomerate including clasts up to 0.5 m in diameter (Matthews 1983a). The overlying volcanic succession generally comprises thick, poorly stratified, fragmental rocks, and basaltic, andesitic and rhyolitic lavas. The presence of a vent agglomerate and the abundance of poorly stratified, volcanic breccias have been interpreted to suggest a volcanic centre on the west of the island. Locally, lavas have well-developed boles, and a subaerial environment is inferred for the volcanism (Matthews 1983a). There are no radiometric ages for the volcanic rocks.

Volcanic rocks form a minor part of Horseshoe Island, forming several isolated outcrops either intruded by or in fault contact with plutonic rocks (Matthews 1983b). Breccias, tuffs and relatively minor lavas are of intermediate–silicic composition and a probable agglomerate-filled vent cutting granite occurs at Trifid Peak in the southern part of the island (Matthews 1983b). In the north of the island at Homing Head, a series of andesitic–dacitic tuffs and lavas has been dated at 121 ± 10 Ma by Rb–Sr geochronology (Loske et al. 1997), suggesting an Early Cretaceous age for the volcanism. A separate sequence of variably deformed mudstones and siltstones locally containing conglomerates with plutonic and volcanic clasts up to 0.5 m in diameter is not seen in contact with the volcanic rocks (Matthews 1983b). Two plutonic boulders from a conglomerate yielded concordant zircon U–Pb ages of 445–419 Ma, and a third gave a poorly defined U–Pb age of c. 225 Ma, constraining the sedimentation age to younger than Permian (Loske et al. 1997).

On Lagotellerie Island, 3 km SW of Horseshoe Island, a volcaniclastic and volcanic sequence c. 100 m thick is in faulted contact with plutonic rocks (Matthews 1983a). The sequence is dominated by breccias and bedded tuffs, and volcaniclastic sandstones. Plant remains, mainly cycadophytes and conifers, are poorly preserved in the volcaniclastic sandstones but no age has been determined (Thomson and Griffiths 1994).

There are few published descriptions of volcanic rocks in the sector south of Horseshoe Island. Around Millerand Island and Neny Fjord, volcanic breccias and tuffs, including welded tuffs, are interbedded with andesitic, dacitic and rhyolitic lavas, and a vent agglomerate occurs in northern Millerand Island (British Antarctic Survey 1981a). Volcanic rocks intruded by granite form the headland at Cape Berteaux. The volcanic succession is c. 800 m thick and comprises mafic, andesitic and silicic lavas interbedded with lithified pyroclastic deposits (unpublished information of Philip Leat). The granite is one of a group of Late Cretaceous high-level granites in the Cape Berteaux–Fleming Glacier area (Leat et al. 2009). To the SE, at Deschanel Peak, over 500 m of volcanic rocks comprise lavas and pyroclastic rocks. The sequence is subaerial and includes compound basaltic lavas, dacitic–rhyolitic lavas, breccias and silicic ignimbrites.

Northern Palmer Land and Transition Zone 69–71° S

This sector includes northern Palmer Land and the transitional zone into southernmost Graham Land (Fig. 7). Volcanic rocks are widespread, forming many nunataks and larger peaks. Significant outcrops of the volcanic rocks include the prominent ridge in northern Palmer Land from Cape Jeremy to the southern flank of the Fleming Glacier, the succession at Carse Point, Orion Massif and neighbouring nunataks, Braddock Nunataks, and a sequence at Mount Charity.

The basement to the volcanic arc in NW Palmer Land is formed by the Western Zone of the Central Domain in the terrane model (Ferraccioli et al. 2006; Vaughan et al. 2012a). The Western Zone contrasts with the Eastern Zone in having a more positive long-wavelength aeromagnetic anomaly, indicating more mafic crust (Ferraccioli et al. 2006). This zone consists of Paleozoic–Early Mesozoic gneisses intruded by plutonic rocks. Two samples of migmatized orthogneiss from Mount Eissenger contain zircon cores U–Pb dated at 435 ± 8 and 422 ± 18 Ma (Silurian) interpreted as ages of the protolith, with zircon rim ages being dated at 225 ± 5 Ma (Millar et al. 2002). Otherwise, gneiss protoliths in the terrane have been dated at c. 253–223 Ma (Permian–Triassic) by zircon U–Pb geochronology (Millar et al. 2002). The oldest granite at Mount Charity in the Eastern Zone is zircon U–Pb dated at 259 ± 5 (Millar et al. 2002).

Occurrence. Graham Land south of 69° S forms what has been called the Transition Zone (Wyeth 1977) between Graham Land and Palmer Land in which the plateau along the spine of the Antarctic Peninsula broadens, glaciers become wider and the slope down to the coast is less precipitous (Fig. 7). This part of western Graham Land is dominated by plutonic rocks, with scattered volcanic outcrops, and is difficult to access. Volcanic rocks occur NW of Mount Castro, east of Triune Peaks and in the Bristly Peaks range (Leat et al. 2009).

In northernmost Palmer Land, volcanic rocks form a chain of peaks and nunataks along the south side of the Fleming Glacier, which marks the transition from Graham Land to Palmer Land. On the Rhyolite Islands, within George VI Sound, dacitic–rhyolitic lavas unconformably overlie gneisses, and on the immediately adjacent mainland a similar unconformity is exposed (Davies 1984). The lower slopes and surrounding nunataks at Mount Edgell are dominantly volcanic – the upper parts are ice covered or inaccessible – and this peak,

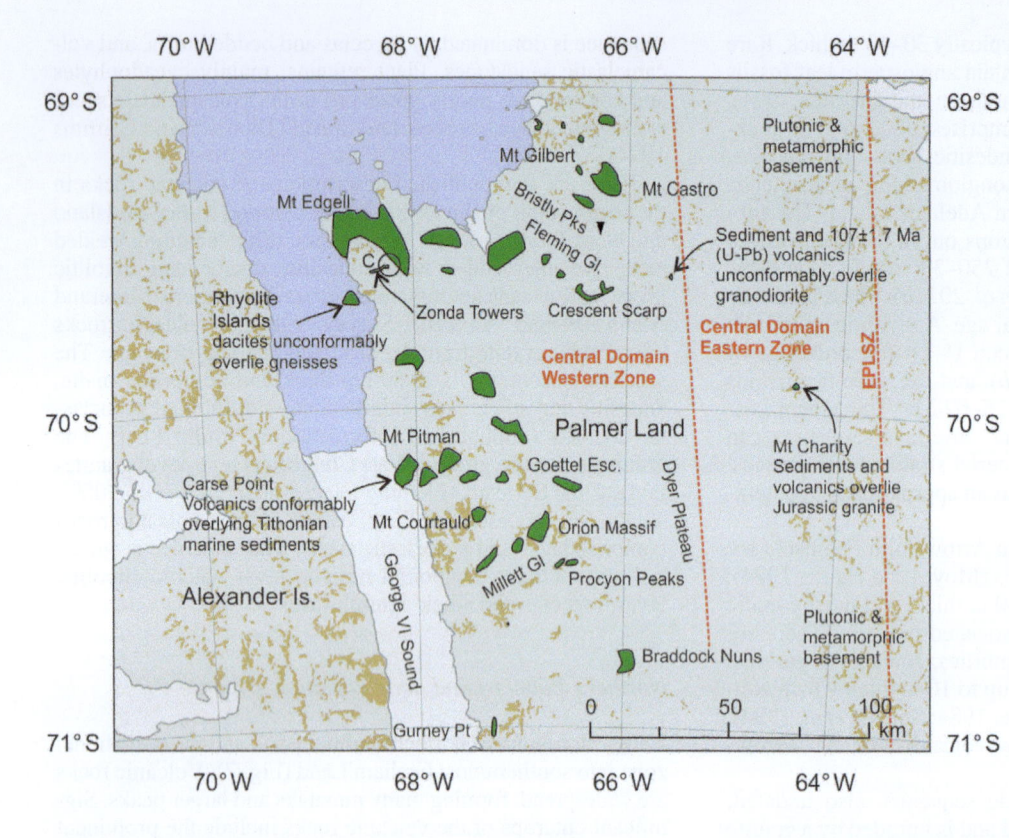

Fig. 7. Map of Northern Palmer Land and Transition Zone 69°–71° S showing distribution of volcanic rocks of the volcanic arc Antarctic Peninsula Volcanic Group. Data from British Antarctic Survey (1982), Ayling (1984), Smith (1987), Leat and Scarrow (1994) and Leat et al. (2009). The basement domain boundaries are based on aeromagnetic, structural and geochronological data after Ferraccioli et al. (2006) and Vaughan et al. (2012a, b). East Palmer Land Shear Zone (EPLSZ) is a compressional zone between the continental interior Eastern Domain and the outboard Central Domain, which culminated in the mid-Cretaceous (Vaughan et al. 2002; Ferraccioli et al. 2006; Vaughan et al. 2012a, b). Zircon U–Pb age data are from Leat et al. (2009). Circle C is a caldera.

at 1676 m the highest peak in NW Palmer Land, appears to be an eroded composite, central volcano (Leat and Scarrow 1994). Zonda Towers, to the south, is interpreted to expose the margin of a silicic caldera (Leat and Scarrow 1994). To the east, the volcanic outcrops continue through Relay Hills, Garcie Peaks and Crescent Scarp. At the easternmost outcrop of the volcanic rocks, a volcanic and sedimentary succession unconformably overlies a granodiorite (Davies 1984; Leat et al. 2009). Tectonic deformation of these volcanic rocks is weak, with dips up to 45° suggesting localized tilting and no folding has been recognized (Leat and Scarrow 1994).

Volcanic rocks at Carse Point form prominent ridges projecting towards George VI Sound, with Late Jurassic marine sediments underlying the volcanic rocks at the base of the ridges (Thomson 1975). Inland, to the east and NE, volcanic rocks dominantly form a series of peaks extending some 25 km to around Mount Pitman (British Antarctic Survey 1982). Further inland and SE of Carse Point, and extending to the Dyer Plateau, volcanic rocks form many nunataks and peaks, including at Mount Courtauld and the Goettel Escarpment. The largest outcrop is at Orion Massif (British Antarctic Survey 1982; Harrison and Piercy 1990) but better described outcrops are SE of Orion Massif at Procyon Peaks (Smith 1987). Volcanic rocks form a large part of the outcrop in Braddock Nunataks where folding of the volcanic strata has been attributed to deformation during intrusion of a later granodiorite (Ayling 1984).

An undeformed volcanic sequence cropping out high (c. 2100 m) on the flanks of Mount Charity overlies sandstones and conglomerates which unconformably overlie a granite (Davies 1984). In view of the position of Mount Charity east of the Dyer Plateau, these represent preservation of distal strata related to the volcanic arc.

Characteristics. To the NW of Mount Castro, a group of crags and nunataks extending to Mount Gilbert expose volcanic and volcaniclastic rocks. The rocks are dominantly volcaniclastic, and include conglomerates, breccias, crystal tuffs, ignimbrites and basaltic–dacitic lavas. On a SW ridge of Mount Castro, volcanic breccias are intruded by a granite. Volcanic outcrops in Bristly Peaks are dominated by volcaniclastics, and also ignimbrites and silicic mafic and rhyolitic lavas, locally intruded by granodiorites. A vent complex with ignimbritic infill has been identified in the southern part of the range (unpublished information of Philip Leat).

In the Rhyolite Islands and adjacent mainland, dacites and rhyolites unconformably overlie metamorphic rocks. The lavas are overlain by lapilli tuffs and coarse, poorly stratified volcanic breccias which extend towards Mouth Edgell (Davies 1984). The accessible outcrops around Mount Edgell comprise dominant volcanic rocks, c. 1.5 km thick, and plutonic rocks and sills. The volcanic rocks are mostly coarse-grained volcaniclastic rocks, lavas and ignimbrites (Leat and Scarrow 1994). The basaltic, andesitic and rhyolitic lavas are up to c. 50 m thick and locally autobrecciated. Silicic, welded and non-welded ignimbrites are up to 100 m thick. Most of the volcaniclastic deposits are poorly sorted, matrix-supported and unstratified, and contain volcanic and minor plutonic clasts. They are thought to be debris-flow, possibly lahar, deposits. The Mount Edgell area was interpreted as a large, subaerial composite volcano built by compositionally varied but largely silicic lavas and pyroclastic deposits, and which formed a silicic topographically high source area for the abundant debris-flow deposits (Leat and Scarrow 1994).

At Zonda Towers, a 160 m-thick silicic ignimbritic unit is interpreted as an intracaldera deposit (Leat and Scarrow 1994). In the eastern part of the outcrop, a single cooling unit comprises a basal breccia (Fig. 4e), ignimbrite (Fig. 4f) and a 75–100 m-thick megabreccia. The megabreccia contains lithic blocks up to 25 m in diameter in an at least partly welded ignimbritic matrix, and is interpreted to have formed by landsliding of a caldera wall during ignimbrite eruption. The western part of the outcrop consists of lavas, ignimbrites up to c. 6 m thick and volcanic breccias, interpreted as an extracaldera succession.

The nunataks from Garcie Peaks to Crescent Scarp are dominated by mafic and intermediate lavas, volcanic breccias and conglomerates. To the east, an unconformity is exposed where a granodiorite is overlain by coarse-grained, poorly sorted beds containing volcanic and plutonic clasts up to 1 m in diameter. These are overlain by a 10 m-thick andesitic lava and 4 m of well-bedded, silicic units containing accretionary lapilli. The andesite lava yielded a zircon U–Pb age of 107.1 ± 1.7 Ma interpreted as the eruption age (Leat *et al.* 2009). This age determination confirms a mid-Cretaceous age, at least locally, for volcanism in NW Palmer Land, overlapping with granitoid emplacement associated with the Lassiter Coast Intrusive Suite (Riley *et al.* 2018).

At Carse Point, nearly 900 m of volcanic rocks overlie fossiliferous Late Jurassic (Tithonian) marine mudstones and sandstones (Thomson 1975). The conformable contact is at an elevation of about 100–200 m and the strata have gentle dips of 5°–20° (Smith 1987). The sandstones are poorly sorted, and at least one is graded (Smith 1987). The volcanic rocks include mafic and intermediate lavas, and andesitic crystal tuffs. A conglomerate bed containing plutonic, sandstone and mudstone clasts is interbedded with mudstones near the base of the sequence, and a breccia containing clasts of sandstone, andesitic tuff and mudstones overlies the lower lavas. The sandstones and coarser beds may be the product of submarine density flows (Smith 1987), suggesting that the whole sequence is submarine. Thomson and Pankhurst (1983) suggested that the volcanic rocks overlying the marine beds may be at least in part Early Cretaceous in age.

The volcanic sequence at Procyon Peaks, south of the Millett Glacier, consists of over 400 m of intermediate and silicic lavas and tuffs forming a ridge and intruded by a granite (Smith 1987). West of the granite, coarse-grained, weakly bedded silicic lithic tuffs containing glassy lapilli dominate the sequence. These are interbedded with intermediate lavas and crystal tuffs which were thought to be reworked (Smith 1987). East of the granite, a different sequence consists of lavas cut by a vent structure with an andesitic plug and vent agglomerate (Smith 1987).

The volcanic rocks at Braddock Nunataks crop out on several nunataks, and include basaltic and andesitic lavas and andesitic–rhyolitic tuffs. Ayling (1984) suggested that basaltic lavas interbedded with bedded andesitic tuffs form the base of the exposed succession. These are overlain by andesitic tuffs and breccias interbedded with rhyolites. The tuffs range from fine-grained, well-bedded units to coarse-grained, polymict volcanic breccias. The rhyolites include flow-banded lavas and vitroclastic units (Ayling 1984) which are possibly ignimbrites. Ayling (1984) described three vent structures up to 15 m across filled with agglomerates, and interpreted the volcanic structure of Braddock Nunataks as an eroded, subaerial, composite cone. Rex (1976) dated an andesite lava from near the northern vent agglomerate at 88 ± 4 Ma by whole-rock K–Ar. This age is difficult to reconcile with intrusion of the volcanic rocks by a granodiorite pluton as dated plutons in the area are older than Late Cretaceous (Harrison and Piercy 1990; Millar *et al.* 2002).

Volcanic rocks at Mount Charity consist mainly of bedded tuffs containing pumices and shards. They overlie *c.* 10 m of poorly sorted, poorly consolidated, polymict conglomerates, sandstones and siltstones thought to have been derived from local plutonic and metavolcanic sources (Davies 1984). The conglomerates contain rounded clasts up to 1 m in diameter and wind-faceted pebbles, indicating a subaerial environment (Scarrow *et al.* 1996). The sediments unconformably overlie a granite dated at 168 ± 1 Ma (Middle Jurassic) by whole-rock Rb–Sr geochronology (Scarrow *et al.* 1996).

Although most of the volcanic outcrops in the sector are radiometrically undated, there are sufficient data to indicate a significant range in eruption ages from Late Jurassic to mid- or Late Cretaceous. The Tithonian marine sediments at Carse Point pass conformably into a volcanic sequence which is probably Late Jurassic at its base and probably, in view of its thickness, extends into the Early Cretaceous (Thomson and Pankhurst 1983). The volcanic rocks in this sector are weakly deformed, and any tectonic folding is gentle. In view of the absence of significant deformation, Harrison and Piercy (1990) suggested a Cretaceous age for most of the volcanism, consistent with the only zircon U–Pb age for a volcanic rock from the sector (Leat *et al.* 2009).

Southern Palmer Land 71–73°S

The character of the volcanism as a volcanic arc becomes less obvious in this sector. There are five significant areas where volcanic rocks which may represent the volcanic arc occur: Gurney Point, Batterbee Mountains, Gutenko Mountains, Seward Mountains and Journal Peaks (Figs 8 and 9). The basement in western Palmer Land in this sector is poorly known. Based on aerogeophysical data, the sector is underlain by the Western and Eastern zones of the Central Domain of the Antarctic Peninsula; both zones are interpreted as parts of a composite magmatic arc terrain (Ferraccioli *et al.* 2006). The Eastern Zone of the Central Domain is separated from the Eastern Domain by the East Palmer Land Shear Zone, a linear belt of compressive deformation during mid-Cretaceous times (Vaughan *et al.* 2012*a*, *b*). A foliated granodiorite from the Batterbee Mountains was dated by zircon U–Pb geochronology at 184 ± 2 Ma, and may be part of the *c.* 187–182 Ma volcanic and plutonic Chon-Aike episode that is widespread in the Antarctic Peninsula (Riley *et al.* 2017; Riley and Leat 2021).

Fig. 8. Photographs of the Antarctic Peninsula volcanic arc. (**a**) Columnar jointed rhyolitic ignimbrites from Mount Mumford, Gutenko Mountains, Palmer Land. (**b**) Mass-flow deposit with granitoid, volcanic and metasedimentary clasts, Mount Mumford.

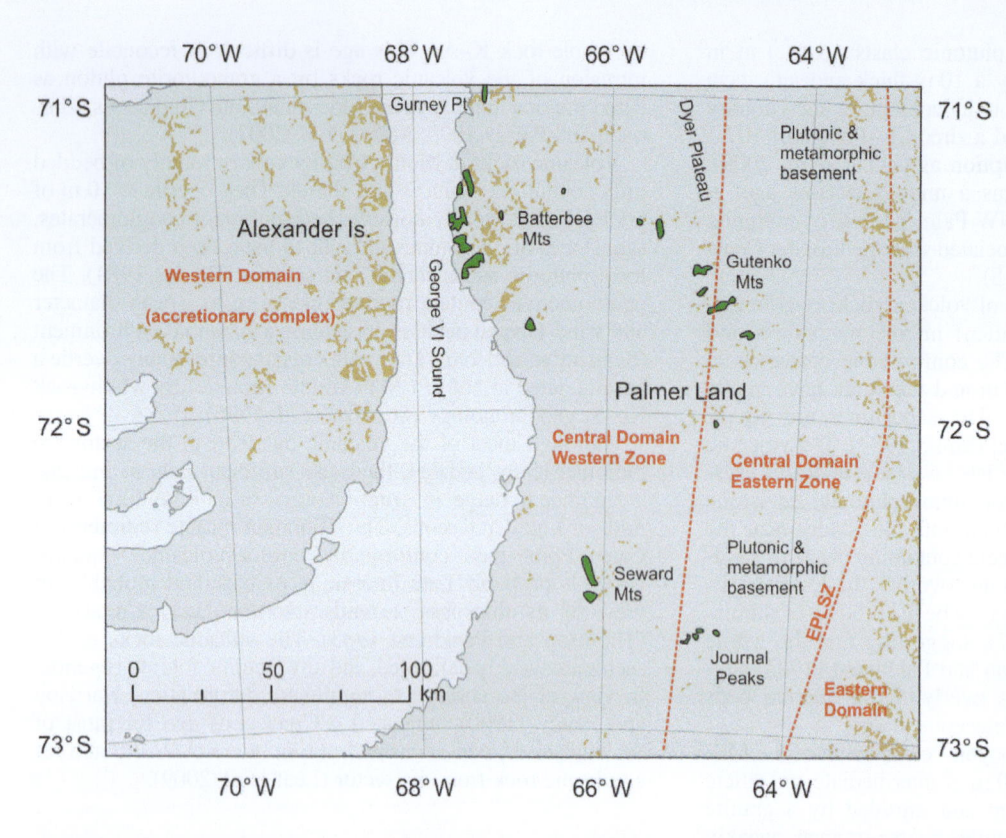

Fig. 9. Map of southern Palmer Land 71–73° S showing the distribution of volcanic rocks of the volcanic arc Antarctic Peninsula Volcanic Group. Data are from Singleton (1979), British Antarctic Survey (1982) and Ayling (1984). The basement domain boundaries are based on aeromagnetic, structural and geochronological data after Ferraccioli *et al.* (2006) and Vaughan *et al.* (2012*a*). EPLSZ, East Palmer Land Shear Zone (see Fig. 7 for the explanation).

Occurrence. Volcanic rocks along the west coast of Palmer Land are included in the Antarctic Peninsula volcanic arc, although their ages and geochemical compositions are poorly known. Some of the outcrops may correlate with the Palmer Land Volcanic Group (Smellie 2020; Riley and Leat 2021), although this distinction cannot be made currently. At Gurney Point, volcanic rocks have limited outcrop, form a north–south-trending ridge and are probably intruded by a quartz-diorite intrusion (Smith 1987). More extensive volcanic rocks form most of the western outcrops of the Batterbee Mountains around Swine Hill, and are gently folded (Ayling 1984). Towards the centre of Palmer Land, volcanic rocks form several peaks and nunataks in the Gutenko Mountains, forming the edge of the Dyer Plateau. Although undated, they have been classified as Mesozoic volcanic rocks (British Antarctic Survey 1982). To the south, volcanic rocks form the western nunataks of the Seward Mountains, and are not seen in contact with plutonic rocks that form higher ridges to the east (Singleton 1979). These volcanic rocks are also gently folded. Outcrops at Journal Peaks are also poorly studied and mapped as Mesozoic volcanic rocks (British Antarctic Survey 1982) but recent mapping (unpublished information of Teal Riley) interpreted the crystal tuffs and ignimbrites to be more closely related to the Early Jurassic Brennecke Formation. This was due to their pervasive greenschist-facies metamorphism and rare interbedded shaley units, which is a common characteristic of the Brennecke Formation of eastern Palmer Land (Wever and Storey 1992).

Characteristics. The volcanic succession at Gurney Points consists of mafic lavas overlain by 170 m of flinty tuffs thought to be thermally metamorphosed by an intrusion (Smith 1987). Volcanic rocks forming several ridges in the Batterbee Mountains are dominantly rhyolitic, consisting of lavas and associated volcaniclastic deposits containing subangular blocks of rhyolite lava (Ayling 1984). Ayling (1984) thought that the succession changes upwards from submarine to subaerial. However, as no associated marine sediments are known, a lacustrine setting is possible. Two separate vent agglomerates have been described from the Batterbee Mountains, and Ayling (1984) suggested that there was a volcanic centre in the western part of the mountain range, near the main rhyolite outcrops at Swine Hill. A zircon U–Pb age of 184 ± 2 Ma for a foliated granodiorite from the Batterbee Mountains (Riley *et al.* 2017) suggests that the area was within the zone affected by widespread Early Jurassic silicic volcanism but it is uncertain what relationship, if any, this granodiorite has to the rhyolites.

The volcanic succession in the Seward Mountains consists of pyroclastic rocks and mafic–intermediate lavas. In the northern outcrops, the pyroclastic rocks are bedded purple and green crystal tuffs. In the southern outcrops, a succession of eutaxitic or parataxitic ignimbrites interbedded with lavas is described (Singleton 1979). At one locality, Singleton (1979) described a tuff containing polymict lithic and crystal fragments intruding lavas as both pipe- and sill-like structures. In the Gutenko Mountains, at least 600 m of rhyolitic ignimbrites, crystal-lithic tuffs and mass-flow deposits crop out (Fig. 8). The ignimbrite units are often strongly welded and occasionally preserve secondary flow fabrics and form columnar jointed units up to 110 m in thickness (TR, unpublished information). Unequivocal volcanic centres have not been identified in the Gutenko Mountains area but, given the nature of the volcanic succession, a caldera setting is likely to be combined with possible flank/lava-dome collapse deposits.

None of the volcanic rocks in this sector has been radiometrically dated and there are no known fossiliferous sediments. The plutonic Early Jurassic U–Pb age from the Batterbee Mountains confirms silicic magmatism at least locally within the sector, along with the Journal Peaks volcanic rocks, which have been likened to the Early Jurassic Brennecke Formation. The folding of the volcanic rocks in the Seward Mountains and Batterbee Mountains suggests that these predate the mid-Cretaceous compressional deformation observed in eastern Palmer Land (Vaughan *et al.* 2012*a*). It is therefore likely that most of the volcanism in the sector was Early Jurassic–mid-Cretaceous but this is tentative and further dating is required.

Alexander Island

Alexander Island (Fig. 10) is situated on the western, forearc side of the Antarctic Peninsula. The island is dominated by an accretionary complex which forms the LeMay Group (Burn 1984). The LeMay Group consists of trench-slope and trench-fill sediments, accreted ocean-floor lavas, pelagic sedimentary rocks, and mélange belts (Burn 1984; Doubleday and Tranter 1994; Doubleday et al. 1994). The LeMay Group is unconformably overlain in the east of Alexander Island by the Fossil Bluff Group, a Middle Jurassic–Early Cretaceous forearc basin sequence of clastic sedimentary rocks (Doubleday et al. 1993; Moncrieff and Kelly 1993).

Occurrence. There are four volcanic groups of Jurassic–Recent age in Alexander Island: accreted oceanic basalts within the LeMay Group; volcanic rocks within the forearc basin Fossil Bluff Group; a Late Cretaceous–Early Tertiary volcanic arc group; and Late Tertiary (c. 7 Ma)–Recent post-subduction basalts. The Late Tertiary–Recent volcanism is described in another chapter (Smellie and Hole 2021).

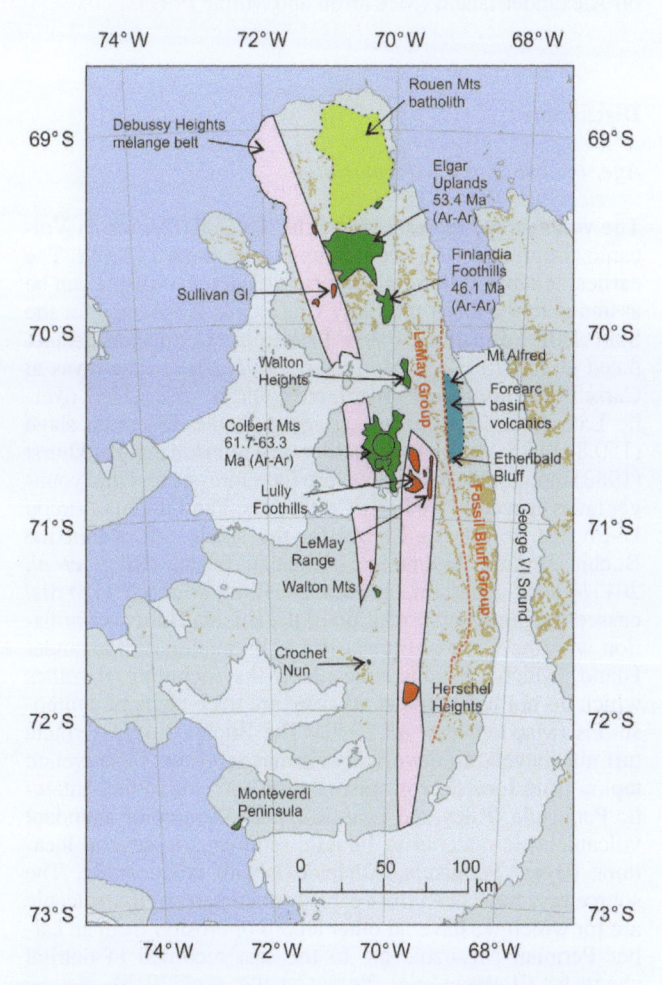

Fig. 10. Map of Alexander Island showing the distribution of Jurassic–Early Tertiary volcanic rocks. Data are from British Antarctic Survey (1981b), Burn (1984), Doubleday et al. (1994), McCarron (1997), McCarron and Millar (1997) and Macdonald et al. (1999). Ar–Ar data are from McCarron and Millar (1997). The LeMay Group, forming the basement to the island, is an accretionary complex. The Fossil Bluff Group, exposed in the east of the island, is a forearc basin fill. The accretionary complex includes two belts containing mélanges and oceanic lavas (red-coloured outcrops), the Debussy Heights mélange belt (Nell 1990), and another trending south from the Lully Foothills and LeMay Range (Doubleday and Tranter 1994). Circle C is a caldera.

Characteristics. Accreted oceanic basalts occur within two belts within the LeMay Group accretionary complex of Alexander Island (Fig. 10). These belts are characterized by oceanic basaltic rocks, cherts and mélanges, and are associated with gravity anomaly highs (Nell 1990; Doubleday and Tranter 1994).

The western Debussy Heights mélange belt dips to the east and is a west-directed thrust zone at least 18 km wide (Nell 1990). Oceanic basalts occur within this belt in the NW of the island and in the Walton Mountains (Burn 1984). At Sullivan Glacier, sequences of basaltic pillow lavas, pillow breccias, volcaniclastics, shales and cherts occur in thrust slices within the mélange belt. Pillows are up to 0.5 m in diameter and interpillow fill is basaltic hyaloclastite (Doubleday et al. 1994). The cherts contain a radiolarian assemblage of approximately late Tithonian–Barremian age (150–125 Ma) (Holdsworth and Nell 1992). The basalts and cherts are in fault contact with greywackes interpreted as trench-fill turbidites deposited over the oceanic rocks as they approached the subduction zone (Nell 1990; Holdsworth and Nell 1992; Doubleday et al. 1994). In the south of the belt, basaltic pillow lavas are associated with cherts and cataclasite (Burn 1984).

The eastern belt includes basaltic rocks at Lully Foothills, NW LeMay Range and Herschel Heights. Basaltic volcanic rocks crop out over an area of c. 24 × 10 km in the Lully Foothills. They are weakly deformed, and consist of interbedded pillowed and massive lavas, basaltic breccias, hyaloclastites, and volcaniclastic material (Doubleday et al. 1994). Pillow lavas are up to 30 m thick, with 0.5–1 m-diameter pillows. Basaltic breccias are up to tens of metres thick. The volcaniclastic material is reworked basaltic tuff with graded bedding, bioturbation, scours and slumps. An Early Jurassic marine fossil assemblage in the volcaniclastic deposits includes gastropods, ammonites, bivalves and plant fossils assigned to the Sinemurian (Thomson and Tranter 1986). The presence of volcanic bombs and reworking of the tuffs indicates shallow-water depths, and the unit is interpreted as a seamount or ocean island that was accreted into the accretionary prism (Doubleday and Tranter 1994; Doubleday et al. 1994). Basaltic rocks in the NW LeMay Range are separated from the Lully Foothills outcrops by the Quinault Pass mélange belt, a major structural boundary within the accretionary prism (Doubleday and Tranter 1992). Basaltic rocks in the NW LeMay Range occur within thrust slivers, and consist of pillowed lavas up to 50 m thick, massive lavas, volcaniclastics and cherts (Doubleday et al. 1994). At Herschel Heights, tectonically deformed massive and pillow lavas are associated with pelagic cherts, several metres thick, and hyaloclastites (Doubleday and Tranter 1992, 1994; Doubleday et al. 1994). Sandstones thought to be arc-derived are tectonically interleaved with the igneous rocks and interpreted as trench-fill sediments.

Volcanic rocks occur within the Fossil Bluff Group between Mount Alfred and the Ethelbald Bluff–Ablation Point area, a distance of c. 50 km (Fig. 10). Elliott (1974) described at Ablation Point a group of basic lavas, breccias, crystal tuffs and agglomerates in a 'disturbed zone' with unclear stratigraphic relationships to the sedimentary strata, and an upper 'lava' (possibly a sill) interbedded with sedimentary strata of the Fossil Bluff Group. Macdonald et al. (1999) described sills and lavas low in the Himalia Ridge Formation (Callovian–earliest Valanginian, 164–140 Ma). The volcanic rocks include silicic tuffs, lavas, and sills and mafic sills. Mafic sills are c. 1–2 m thick and many have complex margins with the sediment, including pillowed structures and mixing of sediment into the sill, suggesting that the sills intruded unconsolidated sediment (Macdonald et al. 1999). Silicic rocks include lavas with pillow structures and rhyolitic and dacitic sills, and volcaniclastic deposits sourced from silicic tephra and siliciclastic material. The volcanic rocks are thought to be Oxfordian–Kimmeridgian in age. The Himalia Ridge Formation is thought to have been deposited during the formation of the Fossil Bluff Group basin on the earlier trench slope (Doubleday et al. 1993). Macdonald et al. (1999) suggested that rifting of the forearc contributed to basin formation and generated the volcanism possibly at a time when subduction was not taking place.

The Early Tertiary volcanic arc sequences form a roughly central north–south zone in Alexander Island extending from the Elgar Uplands in the north to the Monteverdi Peninsula in the south (Fig. 10). McCarron (1997) included all of these volcanic rocks within an Alexander Island Volcanic Group. Significant outcrops of the volcanic rocks occur in the Elgar Uplands, Finlandia Foothills and Colbert Mountains.

All of the volcanic arc rocks unconformably overlie the LeMay Group metasediments, are interpreted to have been subaerial and are locally overlain by late Tertiary–Recent volcanic rocks (Burn 1981; McCarron 1997). The volcanic outcrops in southern Alexander Island are limited by erosion and ice cover, and their maximum observed thickness is c. 250 m (McCarron 1997). In the north, outcrops are more extensive and Burn (1981) estimated that the volcanic rocks exceed 1500 m in thickness, similar to the thickness of 1400 m suggested by McCarron (1997). The age of eruption of the volcanic group is Late Cretaceous–Paleogene (80–46 Ma) based on K–Ar and Ar–Ar dating, with ages decreasing from south to north (McCarron and Millar 1997).

At the Monteverdi Peninsula, the southernmost outcrop, a c. 20–30 m-high cliff exposes basaltic andesite lavas and volcaniclastic rocks which form the Monteverdi Formation (McCarron 1997). Hornblende from a lava has been K–Ar dated at 79.9 ± 2.5 Ma (McCarron and Millar 1997). At Crochet Nunatak, massive beds of tuff, volcanic breccias and andesitic lava crop out on isolated nunataks (Burn 1981), and comprise the Staccato Formation (McCarron 1997). At Walton Peaks, a c. 90 m-thick basal sequence unconformably overlies the LeMay Group and comprises volcaniclastic conglomerate interbedded with two lavas. These are overlain by c. 150 m of inaccessible lavas interbedded with volcaniclastic deposits (Burn 1981; McCarron 1997).

In the Colbert Mountains, volcanic rocks are up to 1400 m thick, and are dominated by dacitic and rhyolitic ignimbrites with associated basaltic–andesitic to rhyolitic lavas, tuffs and volcaniclastic deposits (Burn 1981; McCarron 1997; McCarron and Smellie 1998). Several groups of ignimbrites and lavas, designated as members (McCarron 1997; McCarron and Millar 1997), are separated by unconformities. Welded and non-welded ignimbrites are present, and individual ignimbrites are up to 200 m thick. The thickness of ignimbrites, their distribution, unconformities with relief of c. 50 m and minor lacustrine sediments indicate that the Colbert Mountain may be a caldera or group of overlapping calderas (McCarron and Millar 1997). No caldera ring faults or collapse megabreccias have been identified. McCarron and Millar (1997) suggested that the caldera may have formed by downsag or that the caldera margin facies are no longer preserved. Three Ar–Ar ages on samples from the Colbert Mountains gave ages of 63.3 ± 1.4–61.7 ± 0.8 Ma (McCarron and Millar 1997).

The volcanic rocks in the Elgar Uplands (Elgar Formation) comprise up to c. 1000 m of basaltic–andesitic, andesitic and dacitic lavas interbedded with conglomerates, breccias and volcaniclastic and pyroclastic deposits (Burn 1981; McCarron 1997; McCarron and Smellie 1998). Lavas and fragmental rocks are present in approximately equal proportions (Burn 1981; McCarron and Millar 1997). Lavas are 1–30 m thick, with some of the thicker examples possibly being compound, and red brecciated bases are observed (Burn 1981). Some volcaniclastic sandstones contain abundant small plant fragments (McCarron 1997), whilst some are interpreted as fluvial and volcaniclastic sandstone deposits up to 60 m thick deposited in a lacustrine environment (Burn 1981). Internal unconformities occur within the succession, there is evidence for local downsagging during deposition and the formation may have accumulated in a topographical depression (McCarron 1997; McCarron and Millar 1997). Two andesite lavas from the formation yielded a preferred weighted mean Ar–Ar age of 53.4 ± 0.8 Ma (McCarron and Millar 1997). The age is similar to the 53 ± 3 Ma zircon U–Pb age for the Rouen Mountains diorite–granite batholith (McCarron and Millar 1997), which may be a plutonic equivalent of the Elgar Formation.

In the Finlandia Foothills, basaltic–andesitic lavas are interbedded with volcaniclastic deposits (Burn 1981; McCarron 1997; McCarron and Smellie 1998). In the southern part of the Finlandia Formation, more than 120 m of subhorizontal, polymict conglomerates are overlain by thin andesite lavas. In the northern part, the outcrop comprises moderately dipping lavas interbedded with minor volcaniclastic deposits (McCarron 1997; McCarron and Millar 1997). Deposition is thought to have filled a graben formed by regional extension (McCarron 1997). A lava was Ar–Ar dated at 46.1 ± 7.0 Ma, and the Finlandia lavas represent the youngest known arc volcanism on Alexander Island (McCarron and Millar 1997).

Discussion

Age, thickness and axis migration

The volcanic arc represented by the Antarctic Peninsula Volcanic Group and associated groups spans some 112 myr. The earliest reliable radiometric age for volcanic rocks that can be assigned to the arc is the Early Cretaceous Ar–Ar age for the Start Hill Formation, Byers Peninsula, Livingston Island, dated at 135.20 ± 2.57 Ma (Haase et al. 2012). The lavas at Carse Point, northern Palmer Land could be older. They overlie Late Jurassic sediments assigned to the Tithonian stage (150.8–145.5 Ma) on fossil evidence. Thomson and Pankhurst (1983) thought the lower lavas are Late Jurassic and that younger lavas upsection are Early Cretaceous. The Tithonian zircon U–Pb age of 149.5 ± 1.6 Ma for the silicic tuff within the Buchia Buttress Formation, Adelaide Island (Riley et al. 2011b) is of uncertain significance. Riley et al. (2011b) discussed the possible correlation of the Buchia Buttress Formation with the forearc Himalia Ridge Formation of Alexander Island, which contains volcanic rocks including rhyolites which do not have typical volcanic arc trace element compositions (Macdonald et al. 1999). The Buchia Buttress silicic tuff may have a similar origin or it may represent far-travelled tephra from Jurassic volcanism on the east side of the Antarctic Peninsula (Riley and Leat 2021). The source of abundant volcaniclastic material in Jurassic sediments in several locations (Byers Peninsula, Adelaide Island) is uncertain. The source may have been from early-formed parts of the volcanic arc for which we have no other record or erosion from an earlier Permian–Mesozoic arc to the east recorded in detrital chemistry (Barbeau et al. 2009; Castillo et al. 2016).

All other reliable radiometric and palaeontological ages for the Antarctic Peninsula volcanic arc are Cretaceous or Early Tertiary, which together confirm that this is the age of eruption of the volcanic arc. There is one younger exception, the Early Miocene age of the Boy Point dacites, King George Island which is Ar–Ar dated at 22.61 ± 1.7 Ma, and the underlying Low Head lava dome probably just predates the Miocene age at 23.7 ± 2.0 Ma (Smellie et al. 1998). Similar U–Pb ages of 22.9 ± 0.2 Ma for a diorite intrusion of the Sillard Islands to the north of Adelaide Island (Jordan et al. 2014) and K–Ar ages of 24.0 and 21–20 Ma from Brabant and Anvers islands, respectively (Parada et al. 1990; Ringe 1991), have been obtained. These are the most recent known representatives of the volcanic arc. The likeliest age range of the volcanic arc is therefore c. 135–23 Ma based on available evidence.

Outcrops of the volcanic rocks of the volcanic arc are scattered and non-continuous, resulting from differential subsidence and uplift and erosion, cover by ice, and intrusion by plutonic rocks. Nevertheless, thicknesses of the volcanic rocks estimated in many locations are over 1 km. The thickest sequences occur on Livingston Island, where the 1.4 km-thick subaerial volcanic and volcaniclastic succession at the Byers Peninsula overlies 1.3 km of marine sediments (Hathway and Lomas 1998) and the 2–3 km-thick Late Jurassic–Tertiary sequence on Adelaide Island (Riley et al. 2011b). The thicknesses of interpreted remnants of stratocones range up to 1.1 km for the Martel Inlet Group, King George Island (Birkenmajer 1989), 1.8 km for the Mount Liotard Formation, Adelaide Island (Riley et al. 2011b) and 1.5 km at Mount Edgell (Leat and Scarrow 1994). The 1.2 km-thick sequence of andesite lavas and tuffs at Wiencke Island (Hooper 1962) and the 1.5 km-thick lava, tuff and volcaniclastic sequence at Penola Strait, central Graham Land (Curtis 1966) may be remnants of parts of stratocones. The thickest recorded volcanic sequence is the 2 km-thick volcanic sequence at Paradise Harbour (Birkenmajer 1998b) but it is not known if this is a remnant of a single volcano or several overlapping centres, possibly of different ages. These thicknesses suggest that the arc comprises significant accumulations of volcanic products with thicknesses comparable to the heights of modern volcanic arc volcanoes.

There are strong indications from the admittedly fragmentary data for a westward (towards the trench) migration of volcanism with time. In the northern Graham Land sector, K–Ar, Rb–Sr and zircon U–Pb geochronology on plutonic rocks of the volcanic arc suggest magmatism at 131–80 Ma along the east coast of the Gerlache Strait, 54–46 Ma around Wiencke Island and mostly 34–20 Ma on Anvers Island (Pankhurst 1982; Parada et al. 1990; Birkenmajer 1994; Tangeman et al. 1996; Ryan 2007; Zheng et al. 2018). This westward migration trend is consistent with Ar–Ar data for mafic dykes (Ryan 2007). Insufficient data are available to test for the migration of volcanic centres but the zircon U–Pb ages of 103–101 Ma of lavas from Cuverville Island (Zheng et al. 2017) are consistent with the older plutonic ages east of the Gerlache Strait. A significant westward shift in the volcanic axis of c. 25 km between c. 80 and 54 Ma can be inferred (Fig. 3), with possible further westward migration. In the central Graham Land sector, there are few radiometric ages. However, the proximity of plutons dated at 117 Ma at Rasmussen Island and 55 Ma in the Argentine Islands (Pankhurst 1982; Tangeman et al. 1996) suggests little westward arc migration (Fig. 5). The main volcanic sequences in Adelaide Island are Late Cretaceous in age (75–67 Ma: Riley et al. 2011b). These represent volcanoes built on the Late Jurassic–Early Cretaceous Buchia Buttress and Milestone Bluff formations thought to represent forearc volcaniclastic fans derived from an arc situated to the east (Griffiths and Oglethorpe 1998; Riley et al. 2011b; Jordan et al. 2014). The age of volcanism on Arrowsmith Island immediately to the east is unknown but volcanic rocks to the south at Horseshoe Island have been Rb–Sr dated at 121 Ma (Loske et al. 1997). If this was the location of the Early Cretaceous volcanic axis, a westward shift in the arc axis of c. 60 km between c. 121 and 75 Ma is implied (Fig. 6). In Palmer Land, the arc volcanism is thought to be latest Jurassic and Cretaceous in age (Thomson and Pankhurst 1983; Harrison and Piercy 1990). This is consistent with the evidence from Mount Charity, where the volcanic rocks must be younger than 168 Ma (Scarrow et al. 1996), and the 107 Ma zircon age for a lava in northern Palmer Land (Leat et al. 2009). The earliest age for volcanic arc volcanism in Alexander Island is the K–Ar date of 79.9 ± 2.5 at the Monteverdi Peninsula in the south of the island (McCarron and Millar 1997). The axis of volcanism clearly shifted from Palmer Land to a new axis in Alexander Island, a distance of c. 100–150 km between 107 and 80 Ma.

In the South Shetland Islands, volcanism of the Byers Peninsula, Coppermine Formation and on King George Island ranges in age from 135 to 23 Ma, and all appear to have erupted along more or less the same axis (Fig. 2). There is therefore no evidence for a significant shift in the axis of volcanism. It is not clear whether the absence of arc migration is linked to the eventual rifting of the South Shetland Islands from northern Graham Land. In both the South Shetland Islands and Alexander Island, radiometric ages of volcanic rocks decrease from south to north, and this has been attributed to the northward migration of magmatism with time in these sectors (Smellie et al. 1984; Birkenmajer 1994; McCarron and Millar 1997) or the northward migration of the cessation of magmatism (Pankhurst and Smellie 1983). However, as the trend in the South Shetland Islands is unclear when dykes are considered (Haase et al. 2012), the trend might mainly be a result of erosion level. In Alexander Island, where volcanic outcrops in the southern part of the islands are very limited by ice cover and erosion, apparent migration may also result mainly from erosion level.

In summary, in all sectors from northern Graham Land to Palmer Land there was a westward shift in volcanism that occurred during the Cretaceous, and the new axis was active by 80–75 Ma in Alexander Island and Adelaide Island. In northern Graham Land, the shift may have been later, possibly 54 Ma. The distance the axis shifted generally decreased to the north, and may not have occurred in the South Shetland Islands.

Palaeoenvironments of eruption

The volcanic arc was formed during the period in which the Antarctic Peninsula climate changed from warm Cretaceous greenhouse to cold, glaciated Neogene icehouse. Much of the evidence for changing environments is preserved in the fossil record of plants: importantly, wood, leaves and pollen are well preserved in several important sites around the Antarctic Peninsula, including Alexander Island, the James Ross Island area and King George Island (Poole and Cantrill 2006; Francis et al. 2008). Data from fossil leaves and growth rings in wood indicate warming during the Early Cretaceous and a subtropical, humid climate during the mid–Late Cretaceous with widespread conifer forest (Francis and Poole 2002; Francis et al. 2008a). The climate cooled during the Late Cretaceous and into the Paleogene, at which time cool–warm conifer and angiosperm forest occupied coastal lowlands and subalpine–alpine heath on volcanic peaks (Bowman et al. 2014). A warming phase from the Late Paleocene into the Early Eocene was followed by climate cooling during the Eocene when cool temperate forests were present at low altitudes in the South Shetland Islands (Poole et al. 2001; Mozer 2013). Widespread glaciation in Antarctica is thought to have developed during the Late Eocene–Early Oligocene, although analysis of ice-rafted debris indicates that the Antarctic Peninsula was not glaciated at c. 33 Ma (Carter et al. 2017). The earliest glaciation in the South Shetland Islands is thought to have been during the Late Oligocene (Troedson and Smellie 2002; Warny et al. 2018). The generalized trend is for climate cooling from the Cretaceous to the Miocene, with some reversals. Forest cover on the Antarctic Peninsula was extensive during the Cretaceous but became more restricted to coastal lowlands from the Late Cretaceous onwards and vegetation became sparse by the Oligocene (Francis et al. 2008). Ice caps on volcanic peaks and permafrost in the Antarctic Peninsula may have been present by the latest Cretaceous (Bowman et al. 2014), and extensive glaciation in the Antarctic Peninsula did not occur until the latest Oligocene.

Volcanological associations

The volcanological characteristics of sectors of the Antarctic Peninsula, South Shetland Islands and Alexander Island as detailed above is used here to propose several distinct, recurring volcanological associations within the volcanic arc of the Antarctic Peninsula Volcanic Group. It should be noted that comparisons of lithologies and associations are made difficult because observations were made over many decades from the 1950s to the 2010s. In particular, volcanological processes and deposits were poorly understood before the 1970s and such early field observations were poor by recent standards. Many of the described localities cannot be assigned to an association because of the limited size of accessible outcrops, the mixed composition of the deposits or because of the overgeneralized descriptions.

Volcanic vent structures are an important feature of many localities, indicating at least local eruptions. They are 6–500 m across, contain brecciated monomict lava and bomb-like clasts, are commonly intruded by complex dyke-like intrusions usually of mafic or intermediate composition, and more rarely a tuff-like or lava-like silicic intrusive phase. Examples include basaltic vents in the Fildes and Stansbury peninsulas, South Shetland Islands, where plugs and sill-like bodies are commonly present, ranging up to c. 0.7 km in diameter (Smellie et al. 1984). On the west coast of Pourquoi Pas Island, a vertical contact of vent agglomerate with older volcanic rocks is observed, and many near-vertical mafic dykes nearby, interpreted as an eruptive centre (Matthews 1983a). Three small vents 6.5–15 m across were described from Braddock Nunataks, containing polymict breccias with an andesitic matrix intruded in one case by a 10 m-thick sill (Ayling 1984). A vent at least 50 m across at Procyon Peaks, northern Palmer Land, consists of vent agglomerate intruded by an andesitic plug (Smith 1987). A silicic vent complex with ignimbritic infill was observed south of Bristly Peaks, northern Palmer Land. Other vents have been interpreted on Millerand Island, southern Graham Land (British Antarctic Survey 1981a) and in the Argentine Islands (Elliot 1964).

Marine sediment–lava association. At three widely separated locations, marine sediments are associated with volcanic rocks. All the sequences record local instances of the initiation of the volcanic arc, with an initial marine setting in Late Jurassic–Early Cretaceous times and later emergence of the arc during Cretaceous times. At Carse Point, Palmer Land, Late Jurassic marine mudstones and sandstones are directly overlain by mafic lavas. There is no indication from the accounts that the lavas were submarine, suggesting that the lava sequence was emplaced into shallow water and quickly built a subaerial platform. On Livingston Island, the Start Hill Formation was formed by redeposition of basaltic debris onto a submarine slope, and overlies a Late Jurassic–Early Cretaceous turbiditic sandstone and mudstone sequence. The basaltic deposits comprise thick, poorly sorted breccias derived from a nearby volcano (Smellie et al. 1980). They are overlain by a further 300 m of marine deposits of the shallow-marine Chester Cone Formation before the sequence became subaerial (Hathway and Lomas 1998). On Adelaide Island, the Late Jurassic marine Buchia Buttress sediments are overlain by the Cretaceous Milestone Bluff Formation, which is at least partly marine and is dominated by deposition of volcaniclastic material by density currents on a marine fan (Riley et al. 2011b). By Late Cretaceous times, the volcanism is interpreted to be fully terrestrial, although shallow submarine volcanism occurred again in the Late Oligocene–Miocene on King George Island (Smellie et al. 1998; Troedson and Smellie 2002).

Composite volcano/caldera association. This association formed proximal to calderas or large composite volcanoes. Calderas have been inferred at four locations. At Zonda Towers, northern Palmer Land, a well-exposed 160 m-thick silicic ignimbrite and megabreccia unit is interpreted as an intracaldera deposit (Leat and Scarrow 1994). Only one margin of the caldera is observed, so the caldera diameter is unknown. The 'vent agglomerate' at Detaille Island is at least 1.7–1.9 km in diameter, with no identified margins, and consists of large volcanic blocks and flow-banded dacites that may be ignimbritic (Goldring 1962). The complex is considerably larger than a monogenetic vent, and can be interpreted as a caldera infill. The Colbert Mountains, Alexander Island, are dominated by dacitic and rhyolitic welded and non-welded ignimbrites, and are interpreted as a caldera or group of overlapping calderas (McCarron and Millar 1997). Ignimbrites are up to 200 m thick, and columnar jointing indicates that they form single cooling units. Groups of ignimbrites are separated by unconformities indicating topographical relief resulting from erosion or collapse. Associated lavas and tuffs are mafic–rhyolitic, and volcaniclastic deposits occur within the succession (Burn 1981; McCarron 1997). No ring faults or caldera margins have been identified, possibly indicating downsag calderas (McCarron and Millar 1997). A possible 1.25 km-diameter caldera at Noel Hill on Barton Peninsula, King George Island, is defined by an arcuate diorite–granodiorite intrusion possibly within a ring fault associated with a volcanic series including debris-flow deposits, lavas and welded tuffs, including one eutaxitic welded tuff (Willan and Armstrong 2002). At all four locations, calderas are associated with silicic ignimbrite and within the main axis of contemporaneous volcanism.

At other locations, proximity to large, relatively steep-sided volcanoes erupting a range of mafic, andesitic and silicic compositions is indicated. Successions include lavas, ignimbrites, breccias and coarse-grained volcaniclastic deposits with significant proportions of dacites and rhyolites or both. The 1500 m-thick volcanic succession at Mount Edgell includes ignimbrites up to 100 m thick and abundant coarse deposits interpreted to be debris flows on the flanks of a large composite volcano (Leat and Scarrow 1994). The 1100 m-thick Martel Inlet Group comprises mafic–rhyolitic lavas, pyroclastic and volcaniclastic deposits, and is interpreted as remnants of several stratocone complexes (Birkenmajer 1989). At Pourquoi Pas Island, several features of the c. 1500 m-thick volcanic succession (Matthews 1983a) suggest that it may represent a composite, originally steep-sided, composite volcano (Smellie et al. 1994). No well-bedded sedimentary deposits have been observed, and the rocks are dominated by poorly bedded fragmental rocks with relatively few lavas. Mafic–silicic rocks are present, although their relative abundances are unknown. Smellie et al. (1994) suggested that the deposits resemble those of vulcanian eruptions. At Braddock Nunataks, northern Palmer Land, similar successions comprising tuffaceous rocks and basaltic–rhyolitic lavas were interpreted by Ayling (1984) as relicts of a composite cone. The tuffaceous rocks are probably largely of volcaniclastic origin, and contain abundant lithic rhyolitic and andesitic clasts.

Distal tuff association. Some localities west of the main volcanic axis record a possible distal accumulation of tuffs. At

Mount Charity, northern Palmer Land, tuffs of probable airfall origin are interbedded with polymict conglomerates, sandstones and siltstones interpreted as having a local, non-volcanic origin (Davies 1984; Scarrow et al. 1996). The location is more than 100 km east of the Palmer Land volcanic axis, suggesting a distal air-fall origin for the volcanic material. At Gourdin Island, northern Graham Land, ignimbrites are associated with well-bedded andesitic and dacitic tuffs interpreted as air-fall tuffs (Bell 1984). Situated on the eastern fringe of the South Shetland volcanic axis before opening of the Bransfield Strait, these may represent distal ignimbrite outflows interbedded with air-fall tuffs.

Basalt–andesite stratocone association. This association is interpreted as representing large polygenetic cones consisting of relatively monotonous basaltic–andesitic lavas, autobreccias and interbedded volcaniclastic deposits. In this paper, they are distinguished from composite volcanoes by their more monotonous compositions, and the dominance of basaltic and andesitic lavas. Fluviatile deposits are minor. Significant proportions of silicic rocks and polymict conglomerates and breccias are absent. The Mount Liotard Formation, Adelaide Island, is over 1800 m thick, and consists of basaltic andesite and andesite lavas, with rare hyaloclastic interbeds (Riley et al. 2011b). Similar deposits, but with greater volcaniclastic content, form the associated Bond Nunatak Formation (Riley et al. 2011b). The Fildes Group, King George Island, comprises deposits consistent with a stratocone origin (Birkenmajer 1989). Mafic lavas are dominant and interbedded with volcaniclastic deposits including debris-flow deposits. Fine-grained sediments with bird footprints and abundant fossil wood (Smellie et al. 1984) suggest fluvial or lacustrine environments on the low-angle lower slopes of the stratocone. The sequences dominated by andesitic lavas from central Graham Land, such as Cape Pérez (Curtis 1966), may belong to this association, as may the c. 1200-m-thick andesite lavas and tuffs on Wiencke Island (Hooper 1962), and the 1100 m-thick basaltic andesites of the Llano Formation, King George Island (Birkenmajer 1989; Mozer et al. 2015).

Volcaniclastic fan association. This association is interpreted as having formed as accumulations of volcaniclastic debris on the lower flanks of volcanoes or in topographical depressions near volcanic sources. Lavas of variable compositions within the deposits are not obviously related to one centre. The southern part of the Finlandia Formation, Alexander Island, is dominated by coarse-grained polymict deposits deposited in a topographical depression, possibly a graben (McCarron and Millar 1997). Volcaniclastic deposits comprise approximately half the volume of the Elgar Formation, Alexander Island, including massive tuffs, breccias, conglomerates and volcaniclastic sandstones (Burn 1981; McCarron and Millar 1997). The sequence fills a basin some 40 km across, with evidence for syndepositional faulting according to McCarron and Millar (1997), and rare lacustrine deposits are also present (Burn 1981). The sequence of polymict breccias, bedded tuffs and volcaniclastic sandstones at Lagotellerie Island, Graham Land, contains coarse beds up to 10 m thick, possibly debris-flow deposits, and fine-grained beds containing plant fragments (Matthews 1983a), suggesting deposition in a low-angle debris-fan environment. The successions around Mount Castro, southern Graham Land, which are dominated by volcaniclastic deposits interbedded with ignimbrites and lavas are similar. In the Byers Group, Livingston Island, the Early Cretaceous conglomerates and sandstones deposited by debris flows interbedded with mudstones (Hathway and Lomas 1998) is a marine example of the association.

Coastal wetland–lava association. This association includes some Eocene formations in King George Island interpreted to have accumulated in coastal lacustrine or boggy environments forming the lower slopes or volcanoes. The Point Thomas Formation consists of basaltic–andesitic lavas interbedded with conglomerates, sandstones and mudstones with coals. Lavas form pillows into underlying sediments, which have abundant plant material. The environment is interpreted as a low-lying, coastal wetland periodically invaded by lava (Mozer 2012, 2013). The Point Hennequin Group consists of andesitic lavas associated with loose blocks of coarse- to fine-grained volcaniclastic deposits which contain abundant plant remains. The sediments are cross-bedded and are interpreted as having been deposited in a low-lying freshwater environment (Nawrocki et al. 2011; Mozer 2013). The Zamek Formation is dominated by lavas interbedded with conglomerates and well-bedded sandstones, siltstones and mudstones containing plant remains and coals up to 5 cm thick, and may be similar (Mozer et al. 2015).

Glacio-marine–lava association. The Low Head basaltic dome within the Polonez Cove Formation, King George Island, was intruded onto the seafloor during the latest Oligocene and overlies glacio-marine diamictites (Smellie et al. 1998; Warny et al. 2018). Mobilization of the sediment by heat from the lava mobilized the unconsolidated sediment, which formed intrusions into cracks within the lava dome (Smellie et al. 1998). Basaltic lava-fed deltas are also present, interbedded with marine sediments deposited after the contraction of the Polonez ice sheet (Troedson and Smellie 2002). Such an association of basaltic volcanism with glacial conditions occurred only in the last stages of volcanism in the magmatic arc.

Conclusions

The west coast and islands of the Antarctic Peninsula contain an abundance of Jurassic–Early Tertiary volcanic rocks. The dominant terrain is a volcanic arc that extended from the South Shetland Islands through western Graham Land to western Palmer Land and Alexander Island. Volcanic associations within this arc are varied, and include large composite volcanoes associated with calderas and ignimbrites, basaltic–andesitic stratocones, volcaniclastic fans, distal tuff accumulations, low-lying coastal settings, and glacio-marine eruptions. Initial volcanism was submarine at several locations, and most of the subsequent volcanism was subaerial. The age of initiation of the volcanic arc is uncertain. The earliest known representatives of the arc are Late Jurassic or Early Cretaceous in age, and there is no evidence that the arc was active during most of the Jurassic period. Volcanism continued from initiation to termination of the arc volcanism in Early Miocene times with no known hiatus. The axis of volcanism migrated to the west (towards the trench) in most parts of the arc during the Cretaceous but no such migration appears to have occurred in the South Shetlands sector.

Late Jurassic volcanism in the Fossil Bluff Group of Alexander Island appears to predate the initiation of the volcanic arc and to have been erupted in a rifting basin on the continental margin. Fragments of ocean floor and seamounts occur in the accretionary complex forming the LeMay Group of Alexander Island.

Acknowledgements We thank Jo Rae of the British Antarctic Survey Archives for helping with access to field data and photographs. We are grateful to John Smellie and an anonymous reviewer for constructive comments on the paper.

Author contributions **PTL**: investigation (equal), methodology (lead), writing – original draft (lead), writing – review & editing (lead); **TRR**: funding acquisition (lead), investigation (equal), project administration (lead), resources (lead), writing – original draft (supporting), writing – review & editing (equal).

Funding This research received no specific grant from any funding agency in the public, commercial, or not-for-profit sectors.

Data availability The geochronological dataset generated during this study is reproduced in full in Table 1 The review is based on published work cited in the paper. No other datasets were generated during this study.

References

Adie, R.J. 1964. Geological history. *In*: Priestley, R.E., Adie, R.J. and Robin, G de Q. (eds) *Antarctic Research*. Butterworths, London, 118–162.

Arche, A., Lopez-Martinez, J. and Martinez de Pison, E. 1992. Sedimentology of the Miers Bluff Formation, Livingston Islands, South Shetland Islands. *In:* Yoshida, Y., Kaminuma, K. and Shiraishi, K. (eds) *Recent Progress in Antarctic Earth Science*. Terra Scientific (TERRAPUB), Tokyo, 357–362.

Ayling, M.E. 1984. *The Geology of Parts of Central West Palmer Land*. British Antarctic Survey Scientific Reports, **105**.

Barbeau, D.L., Davis, J.T., Murray, K.E., Valencia, V., Gehrels, G.E., Zahid, K.M. and Gombosi, D.J. 2009. Detrital-zircon geochronology of the metasedimentary rocks of north-western Graham Land. *Antarctic Science*, **22**, 65–78, https://doi.org/10.1017/S095410200999054X

Barker, P.F. 1982. The Cenozoic subduction history of the Pacific margin of the Antarctic Peninsula: ridge crest–trench interactions. *Journal of the Geological Society, London*, **139**, 787–801, https://doi.org/10.1144/gsjgs.139.6.0787

Barton, C.M. 1965. *The Geology of the South Shetland Islands III. The Stratigraphy of King George Island*. British Antarctic Survey Scientific Reports, **44**.

Bastias, J. and Hervé, F. 2013. The Cape Wallace Beds: a Permian: detritus turbidite unit at Low Island, South Shetland Islands. *Bolletino di Geofisica Teorica e Applicata*, **54**(Suppl. 2), 312–314.

Bell, A.C. and King, E.C. 1998. New seismic data support Cenozoic rifting in George VI Sound, Antarctic Peninsula. *Geophysical Journal International*, **134**, 889–902.

Bell, C.M. 1984. The geology of islands in southern Bransfield Strait, Antarctic Peninsula. *British Antarctic Survey Bulletin*, **63**, 41–55.

Birkenmajer, K. 1989. A guide to Tertiary geochronology of King George Island, West Antarctica. *Polish Polar Research*, **10**, 555–579.

Birkenmajer, K. 1992. Evolution of the Bransfield Basin and rift, West Antarcica. *In*: Yoshida, Y., Kaminuma, K. and Shiraishi, K. (eds) *Recent Progress in Antarctic Earth Science*. Terra Scientific (TERRAPUB), Tokyo, 405–410.

Birkenmajer, K. 1994. Evolution of the Pacific margin of the northern Antarctic Peninsula: an overview. *Geologische Rundschau*, **83**, 309–321.

Birkenmajer, K. 1995. Geology of Gerlache Strait, West Antarctica. I. Arctowski Peninsula. *Polish Polar Research*, **16**, 47–60.

Birkenmajer, K. 1998a. Geological structure of Barton and Weaver Peninsula, Maxwell Bay, King George Island (South Shetland Islands, West Antarctica). *Bulletin of the Polish Academy of Sciences, Earth Sciences*, **46**, 193–209.

Birkenmajer, K. 1998b. Geological research of the Polish Geodynamic Expeditions to West Antarctica, 1984–1991: Antarctic Peninsula and adjacent islands. *Polish Polar Research*, **19**, 125–142.

Birkenmajer, K. 1998c. Geology of Tertiary volcanic rocks at Turret Point and Mersey Spit, King George Island (South Shetland Islands, West Antarctica). *Bulletin of the Polish Academy of Sciences, Earth Sciences*, **46**, 223–234.

Birkenmajer, K., Soliani, E. and Kawashita, K. 1988. Early Miocene K–Ar age of volcanic basement of the Melville glaciation deposits, King George Island, West Antarctica. *Bulletin of the Polish Academy of Sciences, Earth Sciences*, **36**, 25–34.

Bowman, V.C., Francis, J.E., Askin, R.A., Riding, J.B. and Swindles, G.T. 2014. Latest Cretaceous–earliest Paleogene vegetation and climate change at high southern latitudes: palynological evidence from Seymore Island, Antarctic Peninsula. *Palaeogeography, Palaeoclimatology, Palaeoecology*, **408**, 26–47, https://doi.org/10.1016/j.palaeo.2014.04.018

British Antarctic Survey 1981a. *British Antarctic Territory Geological Map. Sheet 3. Southern Graham Land, 1:500 000*. British Antarctic Survey, Cambridge, UK.

British Antarctic Survey 1981b. *British Antarctic Territory Geological Map. Sheet 4. Alexander Island, 1:500 000*. British Antarctic Survey, Cambridge, UK.

British Antarctic Survey 1982. *British Antarctic Territory Geological Map. Sheet 5. Northern Palmer Land, 1:500 000*. British Antarctic Survey, Cambridge, UK.

Burn, R.W. 1981. Early Tertiary calc-alkaline volcanism on Alexander Island. *British Antarctic Survey Bulletin*, **53**, 175–193.

Burn, R.W. 1984. *The Geology of the LeMay Group, Alexander Island*. British Antarctic Survey Scientific Reports, **109**.

Burton-Johnson, A. and Riley, T.R. 2015. Autochthonous,v. accreted terrane development of continental margins: a revised *in situ* tectonic history of the Antarctic Peninsula. *Journal of the Geological Society, London*, **172**, 822–835, https://doi.org/10.1144/jgs2014-110

Cao, L. 1992. Late Cretaceous and Eocene palynofloras from Fildes Peninsula, King George Island (South Shetland Islands), Antarctica. *In*: Yoshida, Y., Kaminuma, K. and Shiraishi, K. (eds) *Recent Progress in Antarctic Earth Science*. Terra Scientific (TERRAPUB), Tokyo, 363–369.

Carter, A., Riley, T.R., Hillenbrand, C.-D. and Rittner, M. 2017. Widespread Antarctic glaciation during the Late Eocene. *Earth and Planetary Letters*, **458**, 49–57, https://doi.org/10.1016/j.epsl.2016.10.045

Castillo, P., Lacassie, J.P., Augustsson, C. and Hervé, F. 2015. Petrography and geochemistry of the Carboniferous–Triassic Trinity Peninsula Group, West Antarctica: implications for provenance and tectonic setting. *Geological Magazine*, **152**, 575–588, https://doi.org/10.1017/S0016756814000454

Castillo, P., Fanning, C.M., Hervé, F. and Lacassie, J.P. 2016. Characterisation and tracing of Permian magmatism in the southwestern segment of the Gondwana margin: U–Pb age, Lu–Hf and O isotopic compositions of detrital zircons from metasedimentary complexes of northern Antarctic Peninsula and western Patagonia. *Gondwana Research*, **36**, 1–13, https://doi.org/10.1016/j.gr.2015.07.014

Catalán, M., Galindo-Zaldivar, J., Davila, J.M., Martos, Y.M., Maldonado, A., Gambôa, L. and Schreider, A.A. 2013. Initial stages of ocean spreading in the Bransfield Rift from magnetic and gravity data analysis. *Tectonophysics*, **585**, 102–112, https://doi.org/10.1016/j.tecto.2012.09.016

Crame, J.A., Pirrie, D., Crampton, J.S. and Duane, A.M. 1993. Stratigraphy and regional significance of the Upper Jurassic–Lower Cretaceous Byers Group, Livingston Island, Antarctica. *Journal of the Geological Society, London*, **150**, 1075–1087, https://doi.org/10.1144/gsjgs.150.6.1075

Curtis, R. 1966. *The Petrology of the Graham Coast, Graham Land*. British Antarctic Survey Scientific Reports, **50**.

Dalrymple, G.B. 1979. Critical tables for conversion of K–Ar ages from old to new constants. *Geology*, **7**, 558–560, https://doi.org/10.1130/0091-7613(1979)7<558:CTFCOK>2.0.CO;2

Davies, T.G. 1984. *The Geology of Part of Northern Palmer Land*. British Antarctic Survey Scientific Reports, **103**.

Dewar, G.J. 1970. *The Geology of Adelaide Island*. British Antarctic Survey Scientific Reports, **57**.

Dingle, R.V. and Lavelle, M. 1998. Antarctic Peninsula cryosphere: Early Oligocene (c. 30 Ma) initiation and a revised glacial chronology. *Journal of the Geological Society, London*, **155**, 433–437, https://doi.org/10.1144/gsjgs.155.3.0433

Doubleday, P.A. and Tranter, T.H. 1992. Modes of formation and accretion of oceanic material in the Mesozoic fore-arc of central and southern Alexander Island, Antarctica: a summary. *In*: Yoshida, Y., Kaminuma, K. and Shiraishi, K (eds) *Recent Progress in Antarctic Earth Science*. Terra Scientific (TERRAPUB), Tokyo, 377–382.

Doubleday, P.A. and Tranter, T.H. 1994. Deformation mechanism paths for oceanic rocks during subduction and accretion: the Mesozoic forearc of Alexander Island, Antarctica. *Journal of the Geological Society, London*, **151**, 543–554, https://doi.org/10.1144/gsjgs.151.3.0543

Doubleday, P.A., Macdonald, D.I.M. and Nell, P.A.R. 1993. Sedimentology and structure of the trench-slope to forearc basin transition in the Mesozoic of Alexander Island, Antarctica. *Geological Magazine*, **130**, 737–754, https://doi.org/10.1017/S0016756800023128

Doubleday, P.A., Leat, P.T., Alabaster, T., Nell, P.A.R. and Tranter, T.H. 1994. Allochthonous oceanic basalts within the Mesozoic accretionary complex of Alexander Island, Antarctica: remnants of proto-Pacific oceanic crust. *Journal of the Geological Society, London*, **151**, 65–78, https://doi.org/10.1144/gsjgs.151.1.0065

Elliot, D.H. 1964. *The Petrology of the Argentine Islands*. British Antarctic Survey Scientific Reports, **41**.

Elliott, M.H. 1974. Stratigraphy and sedimentary petrology of the Ablation Point area, Alexander Island. *British Antarctic Survey Bulletin*, **39**, 87–113.

Ferraccioli, F., Jones, P.C., Vaughan, A.P.M. and Leat, P.T. 2006. New aerogeophysical view of the Antarctic Peninsula: more pieces, less puzzle. *Geophysical Research Letters*, **33**, L05310, https://doi.org/10.1029/2005GL024636

Francis, J.E. and Poole, I. 2002. Cretaceous and early Tertiary climates of Antarctica: evidence from fossil wood. *Palaeogeography, Palaeoclimatology, Palaeoecology*, **182**, 47–64, https://doi.org/10.1016/S0031-0182(01)00452-7

Francis, J.E., Ashworth, A. *et al*. 2008. 100 million years of Antarctic climate evolution: evidence from fossil plants. *United States Geological Survey Open-File Report*, **2007-1047**, 19–27, https://doi.org/10.3133/ofr20071047kp03

Gao, L., Zhao, Y., Yang, Z., Liu, J., Liu, X., Zhang, S.-H. and Pei, J. 2018. New paleomagnetic and ^{40}Ar/^{39}Ar geochronological results for the South Shetland Islands, West Antarctica, and their tectonic implications. *Journal of Geophysical Research: Solid Earth*, **123**, 4–30, https://doi.org/10.1002/2017JB014677

Garrett, S.W. 1990. Interpretation of reconnaissance gravity and aeromagnetic surveys of the Antarctic Peninsula. *Journal of Geophysical Research*, **95**, 6759–6777, https://doi.org/10.1029/JB095iB05p06759

Goldring, D.C. 1962. *The Geology of the Loubet Coast, Graham Land*. British Antarctic Survey Scientific Reports, **36**.

Griffiths, C.J. and Oglethorpe, R.D.J. 1998. The stratigraphy and geochronology of Adelaide Island. *Antarctic Science*, **10**, 462–475, https://doi.org/10.1017/S095410209800056X

Haase, K.M., Beier, C., Fretzdorff, S., Smellie, J.L. and Garbe-Schönberg, D. 2012. Magmatic evolution of the South Shetland Islands, Antarctica, and implications for continental crust formation. *Contributions to Mineralogy and Petrology*, **163**, 1103–1119, https://doi.org/10.1007/s00410-012-0719-7

Harrison, S.M. and Piercy, B.A. 1990. The evolution of the Antarctic Peninsula magmatic arc: Evidence from northwestern Palmer Land. *Geological Society of America Special Papers*, **241**, 9–25, https://doi.org/10.1130/SPE241-p9

Hathway, B. and Lomas, S.A. 1998. The upper Jurassic–Lower Cretaceous Byers Group, South Shetland Islands, Antarctica: revised stratigraphy and regional correlations. *Cretaceous Research*, **19**, 43–67, https://doi.org/10.1006/cres.1997.0095

Hathway, B., Duane, A.M., Cantrill, D.J. and Kelley, S.P. 1999. ^{40}Ar/^{39}Ar geochronology and palynology of the Cerro Negro Formation, South Shetland Islands, Antarctica: A new radiometric tie for Cretaceous terrestrial biostratigraphy in the Southern Hemisphere. *Australian Journal of Earth Sciences*, **46**, 593–606, https://doi.org/10.1046/j.1440-0952.1999.00727.x

Hervé, F., Loske, W., Miller, H. and Pankhurst, R.J. 1991. Chronology of provenance, deposition and metamorphism of deformed fore-arc sequences, southern Scotia arc. *In*: Thomson, M.R.A., Crame, J.A. and Thomson, J.W. (eds) *Geological Evolution of Antarctica*. Cambridge University Press, Cambridge, UK, 429–435.

Hervé, F., Faúndez, V., Brix, M. and Fanning, M. 2006. Jurassic sedimentation of the Miers Bluff Formation, Livingston Island, Antarctica: evidence from SHRIMP U–Pb ages of detrital and plutonic zircons. *Antarctic Science*, **18**, 229–238, https://doi.org/10.1017/S0954102006000277

Holdsworth, B.K. and Nell, P.A.R. 1992. Mesozoic radiolarian faunas from the Antarctic Peninsula: age, tectonic and palaeoceanographic significance. *Journal of the Geological Society, London*, **149**, 1003–1020, https://doi.org/10.1144/gsjgs.149.6.1003

Hole, M.J. 2021. Antarctic Peninsula: petrology. *Geological Society, London, Memoirs*, **55**, https://doi.org/10.1144/M55-2018-40

Hooper, P.R. 1962. *The Petrology of Anvers Island and Adjacent Islands*. Falkland Islands Dependencies Survey Scientific Reports, **34**.

Hunter, M.A., Riley, T.R., Cantrill, D.J., Flowerdew, M.J. and Millar, I.L. 2006. A new stratigraphy for the Latady Basin, Antarctic Peninsula: part 1, Ellsworth Land Volcanic Group. *Geological Magazine*, **143**, 777–796, https://doi.org/10.1017/S0016756806002597

Johnson, A.C. 1999. Interpretation of new aeromagnetic anomaly data from the central Antarctic Peninsula. *Journal of Geophysical Research*, **104**, 5031–5046, https://doi.org/10.1029/1998JB900073

Johnson, A.C. and Swain, C.J. 1995. Further evidence of fracture-zone induced tectonic segmentation of the Antarctic Peninsula from detailed aeromagnetic anomalies. *Geophysical Research Letters*, **22**, 1917–1920, https://doi.org/10.1029/95GL00812

Jordan, T.A., Neale, R.F. *et al*. 2014. Structure and evolution of Cenozoic arc magmatism on the Antarctic Peninsula; a high resolution aeromagnetic perspective. *Geophysical Journal International*, **198**, 1758–1774, https://doi.org/10.1093/gji/ggu233

Jwa, Y.-J., Park, B.-K. and Kim, Y. 1992. Geochronology and geochemistry of the igneous rocks from Barton and Fildes Peninsulas, King George Island: a review. *In*: Yoshida, Y., Kaminuma, K. and Shiraishi, K. (eds) *Recent Progress in Antarctic Earth Science*. Terra Scientific (TERRAPUB), Tokyo, 439–442.

Kim, H., Lee, J.I., Choe, M.Y., Cho, M., Zheng, X., Sang, H. and Qiu, J. 2000. Geochronologic evidence for Early Cretaceous volcanic activity on Barton Peninsula, King George Island, Antarctica. *Polar Research*, **19**, 251–260, https://doi.org/10.3402/polar.v19i2.6549

Larter, R.D. and Barker, P.F. 1991. Effects of ridge crest–trench interaction on Antarctic–Phoenix spreading: forces on a young subducting plate. *Journal of Geophysical Research*, **96**, 19 583–19 607, https://doi.org/10.1029/91JB02053

Leat, P.T. and Riley, T.R. 2021. Antarctic Peninsula and South Shetland Islands: petrology. *Geological Society, London, Memoirs*, **55**, https://doi.org/10.1144/M55-2018-68

Leat, P.T. and Scarrow, J.H. 1994. Central volcanoes as sources for the Antarctic Peninsula Volcanic Group. *Antarctic Science*, **6**, 365–774, https://doi.org/10.1017/S0954102094000568

Leat, P.T., Scarrow, J.H. and Millar, I.L. 1995. On the Antarctic Peninsula batholith. *Geological Magazine*, **132**, 399–4127, https://doi.org/10.1017/S0016756800021464

Leat, P.T., Flowerdew, M.J., Riley, T.R., Whitehouse, M.J., Scarrow, J.H. and Millar, I.L. 2009. Zircon U–Pb dating of Mesozoic volcanic and tectonic events in northwest Palmer Land and southwest Graham Land, Antarctica. *Antarctic Science*, **21**, 633–641, https://doi.org/10.1017/S0954102009990320

Loske, W., Hervé, F., Miller, H. and Pankhurst, R.J. 1997. Rb–Sr and U–Pb studies of the pre-Andean and Andean magmatism in the Horseshoe Island area, Marguerite Bay (Antarctic Peninsula).

In: Ricci, C.A. (ed.) *The Antarctic Region: Geological Evolution and Processes*. Terra Antartica, Siena, Italy, 353–360.

Macdonald, D.I.M., Leat, P.T., Doubleday, P.A. and Kelly, S.R.A. 1999. On the origin of fore-arc basins: new evidence from Alexander Island, Antarctica. *Terra Nova*, **11**, 186–193, https://doi.org/10.1046/j.1365-3121.1999.00244.x

Machado, A., Chemale, F. Jr., Conceição, R.V., Kawaskita, K., Morata, D., Oteiza, O. and Van Schmus, W.R. 2005. Modeling of subduction components in the genesis of the meso-Cenozoic igneous rocks from the South Shetland arc, Antarctica. *Lithos*, **82**, 435–453, https://doi.org/10.1016/j.lithos.2004.09.026

Maldonado, A., Larter, R.D. and Aldaya, F. 1994. Forearc tectonic evolution of the South Shetland margin, Antarctic Peninsula. *Tectonics*, **13**, 1345–1370, https://doi.org/10.1029/94TC01352

Matthews, D.W. 1983*a*. The geology of Pourquoi Pas Island, northern Marguerite Bay, Graham Land. *British Antarctic Survey Bulletin*, **52**, 1–20.

Matthews, D.W. 1983*b*. The geology of Horseshoe and Lagotellerie islands, Marguerite Bay, Graham Land. *British Antarctic Survey Bulletin*, **52**, 125–154.

McCarron, J.J. 1997. A unifying lithostratigraphy of late Cretaceous–early Tertiary fore-arc volcanic sequences on Alexander Island, Antarctica. *Antarctic Science*, **9**, 209–220, https://doi.org/10.1017/S0954102097000266

McCarron, J.J. and Millar, I.L. 1997. The age and stratigraphy of fore-arc magmatism on Alexander Island, Antarctica. *Geological Magazine*, **134**, 507–522, https://doi.org/10.1017/S0016756897007437

McCarron, J.J. and Smellie, J.L. 1998. Tectonic implications of fore-arc magmatism and generation of high-magnesian andesites: Alexander Island, Antarctica. *Journal of the Geological Society, London*, **155**, 269–280, https://doi.org/10.1144/gsjgs.155.2.0269

Millar, I.L., Pankhurst, R.J. and Fanning, C.M. 2002. Basement chronology of the Antarctic Peninsula: recurrent magmatism and anataxis in the Palaeozoic Gondwana margin. *Journal of the Geological Society, London*, **159**, 145–157, https://doi.org/10.1144/0016-764901-020

Moncrieff, A.C.M. and Kelly, S.R.A. 1993. Lithostratigraphy of the uppermost Fossil Bluff Group (Early Cretaceous) of Alexander Island, Antarctica: history of an Albian regression. *Cretaceous Research*, **14**, 1–15, https://doi.org/10.1006/cres.1993.1001

Moyes, A.B. and Hamer, R.D. 1984. The geology of the Arrowsmith Peninsula and Blaiklock Island, Graham Land, Antarctica. *British Antarctic Survey Bulletin*, **65**, 41–55.

Moyes, A.B. and Pankhurst, R.J. 1994. Andean Intrusive Suite. *In*: Moyes, A.B., Willan, C.F.H. *et al.* (eds) *Geological Map of Adelaide Island to Foyn Coast, 1:250 000, with Supplementary Text*. BAS GEOMAP Series Sheet 3. British Antarctic Survey, Cambridge, UK, 10–18.

Moyes, A.B., Willan, C.F.H. *et al.* (eds). 1994. *Geological Map of Adelaide Island to Foyn Coast, 1:250 000, with Supplementary Text*. BAS GEOMAP Series Sheet 3. British Antarctic Survey, Cambridge, UK

Mozer, A. 2012. Pre-glacial sedimentary facies of the Thomas Point Formation (Eocene) at Cytadela, Admiralty Bay, King George Island, West Antarctica. *Polish Polar Research*, **33**, 41–62, https://doi.org/10.2478/v10183-012-0002-7

Mozer, A. 2013. Eocene sedimentary facies in a volcanogenic succession on King George Island, South Shetland Islands: a record of pre-ice sheet terrestrial environments in West Antarctica. *Geological Quarterly*, **57**, 385–394, https://doi.org/10.7306/gq.1100

Mozer, A., Pécskay, Z. and Krajewski, K.P. 2015. Eocene age of the Baranowski Glacier Group at Red Hill, King George Island, West Antarctica. *Polish Polar Research*, **36**, 307–324, https://doi.org/10.1515/popore-2015-0022

Nawrocki, J., Pańczyk, M. and Williams, I.S. 2010. Isotopic ages and palaeomagnetism of selected magmatic rocks from King George Island (Antarctic Peninsula). *Journal of the Geological Society, London*, **167**, 1063–1079, https://doi.org/10.1144/0016-76492009-177

Nawrocki, J., Pańczyk, M. and Williams, I.S. 2011. Isotopic ages of selected magmatic rocks from King George Island (West Antarctica) controlled by magnetostratigraphy. *Geological Quarterly*, **55**, 301–322.

Nell, P.A.R. 1990. Deformation in an accretionary melange Alexander Island, Antarctica. *Geological Society, London, Special Publications*, **54**, 405–416, https://doi.org/10.1144/GSL.SP.1990.054.01.37

Ogg, J.G., Ogg, G. and Gradstein, F.M. 2008. *The Concise Geological Time Scale*. Cambridge University Press, Cambridge, UK.

Pańczyk, A. and Nawrocki, J. 2011. Geochronology of selected andesitic lavas from the King George Bay area (SE King George Island). *Geological Quarterly*, **55**, 323–334.

Pańczyk, A., Nawrocki, J. and Williams, I.S. 2009. Isotopic age constraint for the Blue Dyke and Jardine Peak subvertical intrusions of King George Island, West Antarctica. *Polish Polar Research*, **30**, 379–391, https://doi.org/10.4202/ppres.2009.20

Pankhurst, R.J. 1982. Rb–Sr geochronology of Graham Land. *Journal of the Geological Society, London*, **139**, 701–711, https://doi.org/10.1144/gsjgs.139.6.0701

Pankhurst, R.J. and Smellie, J.L. 1983. K–Ar geochronology of the South Shetland Islands, Lesser Antarctica: apparent lateral migration of Jurassic to Quaternary island arc volcanism. *Earth and Planetary Science Letters*, **66**, 214–222, https://doi.org/10.1016/0012-821X(83)90137-1

Pankhurst, R.J., Leat, P.T., Sruoga, P., Rapela, C.W., Márquez, M., Storey, B.C. and Riley, T.R. 1998. The Chon Aike silicic province of Patagonia and related rocks in West Antarctica: a silicic large igneous province. *Journal of Volcanology and Geothermal Research*, **81**, 113–136.

Pankhurst, R.J., Riley, T.R., Fanning, C.M. and Kelly, S.P. 2000. Episodic silicic volcanism in Patagonia and the Antarctic Peninsula: chronology of magmatism associated with break-up of Gondwana. *Journal of Petrology*, **41**, 605–625, https://doi.org/10.1093/petrology/41.5.605

Parada, M.A., Orsini, J.B., Ardila, R., Guerra, R., Munizaga, F. and Berg, K. 1990. The plutonic rocks of the southern Gerlache Strait, Antarctica: geochronology, geochemistry and mineralogy. *In*: *Symposium International Géodynamique Andine: 15–17 Mai 1990, Grenoble, France: Résumés des communications*. ORSTOM, Paris, 293–295.

Poole, I. and Cantrill, D.J. 2006. Cretaceous and Cenozoic vegetation of Antarctica integrating the fossil wood record. *Geological Society, London, Special Publications*, **258**, 63–81, https://doi.org/10.1144/GSL.SP.2006.258.01.05

Poole, I., Hunt, R.J. and Cantrill, D.J. 2001. A fossil wood flora from King George Island: ecological implications for an Antarctic Eocene vegetation. *Annals of Botany*, **88**, 33–54, https://doi.org/10.1006/anbo.2001.1425

Porębski, S.J. and Gradziński, R. 1990. Lava-fed Gilbert-type delta in the Polonez Cove Formation (Lower Oligocene), King George Island, West Antarctica. *International Association of Sedimentology Special Publications*, **10**, 335–350.

Rex, D.C. 1976. Geochronology in relation to the stratigraphy of the Antarctic Peninsula. *British Antarctic Survey Bulletin*, **43**, 49–58.

Riley, T.R. and Leat, P.T. 1999. Large volume silicic volcanism along the proto-Pacific margin of Gondwana: lithological and stratigraphical investigations from the Antarctic Peninsula. *Geological Magazine*, **136**, 1–16, https://doi.org/10.1017/S0016756899002265

Riley, T.R. and Leat, P.T. 2021. Palmer Land and Graham Land volcanic groups (Antarctic Peninsula): volcanology. *Geological Society, London, Memoirs*, **55**, https://doi.org/10.1144/M55-2018-36

Riley, T.R., Leat, P.T., Pankhurst, R.J. and Harris, C. 2001. Origins of large volume rhyolitic volcanism in the Antarctic Peninsula and Patagonia by crustal melting. *Journal of Petrology*, **42**, 1043–1065, https://doi.org/10.1093/petrology/42.6.1043

Riley, T.R., Flowerdew, M.J., Hunter, M.A. and Whitehouse, M.J. 2010. Middle Jurassic rhyolite volcanism of eastern Graham Land, Antarctic Peninsula: age correlations and stratigraphic relationships. *Geological Magazine*, **147**, 581–595, https://doi.org/10.1017/S0016756809990720

Riley, T.R., Flowerdew, M.J. and Haselwimmer, C.E. 2011a. *Geological Map of Adelaide Island, Graham Land, 1:200 000 Scale*. BAS GEOMAP 2 Series Sheet 3. British Antarctic Survey, Cambridge, UK.

Riley, T.R., Flowerdew, M.J. and Whitehouse, M.J. 2011b. Chrono- and lithostratigraphy of a Mesozoic–Tertiary fore- to intra-arc basin: Adelaide Island, Antarctic Peninsula. *Geological Magazine*, **149**, 768–782, https://doi.org/10.1017/S001675681100 1002

Riley, T.R., Flowerdew, M.J. and Whitehouse, M.J. 2012. U–Pb ion-microprobe zircon geochronology from the basement inliers of eastern Graham Land, Antarctic Peninsula. *Journal of the Geological Society, London*, **169**, 381–393, https://doi.org/10.1144/0016-76492011-142

Riley, T.R., Flowerdew, M.J., Pankhurst, R.J., Curtis, M.L., Millar, I.L., Fanning, C.M. and Whitehouse, M.J. 2017. Early Jurassic magmatism in the Antarctic Peninsula and potential correlation with the Subcordilleran plutonic belt of Patagonia. *Journal of the Geological Society, London*, **174**, 365–376, https://doi.org/10.1144/jgs2016-053

Riley, T.R., Burton-Johnson, A., Flowerdew, M.J. and Whitehouse, M.J. 2018. Episodicity within a mid-Cretaceous magmatic flare-up in West Antarctica: U–Pb ages of the Lassiter Coast intrusive suite, Antarctic Peninsula and correlations along the Gondwana margin. *Geological Society of America Bulletin*, **130**, 1177–1196, https://doi.org/10.1130/B31800.1

Ringe, M.J. 1991. Volcanism on Brabant Island, Antarctica. *In*: Thomson, M.R.A., Crame, J.A. and Thomson, J.W. (eds) *Geological Evolution of Antarctica*. Cambridge University Press, Cambridge, UK, 515–519.

Ryan, C.J. 2007. *Mesozoic to Cenozoic Igneous Rocks from North-western Graham Land: Constraints on the Tectonomagmatic Evolution of the Antarctic Peninsula*. PhD thesis, University of Brighton, Brighton, UK.

Scarrow, J.H., Pankhurst, R.J., Leat, P.T. and Vaughan, A.P.M. 1996. Antarctic Peninsula granitoid petrogenesis: a case study from Mount Charity, north-eastern Palmer Land. *Antarctic Science*, **8**, 193–206, https://doi.org/10.1017/S0954102096000260

Scarrow, J.H., Leat, P.T., Wareham, C.D. and Millar, I.L. 1998. Geochemistry of mafic dykes in the Antarctic Peninsula continental-margin batholith: a record of arc evolution. *Contributions to Mineralogy and Petrology*, **131**, 289–305, https://doi.org/10.1007/s004100050394

Shen, Y.B. 1992. Discussion on stratigraphic subdivision and nomenclature in Fildes Peninsula, King George Island, Antarctica. *Antarctic Research*, **4**, 18–26 [in Chinese with English abstract].

Singleton, D.G. 1979. Geology of the Seward Mountains, western Palmer Land. *British Antarctic Survey Bulletin*, **49**, 81–89.

Smellie, J.L. 1980. The geology of Low Island, South Shetland Islands, and Austin rocks. *British Antarctic Survey Bulletin*, **49**, 239–257.

Smellie, J.L. 2020. Antarctic Peninsula – geology and dynamic development. *In*: Kleinschmidt, G. (ed.) *Geology of the Antarctic Continent*. Gebrüder Borntraeger Verlagsbuchhandlung, Stuttgart, Germany, 17–86.

Smellie, J.L. 2021. Bransfield Strait and James Ross Island: volcanology. *Geological Society, London, Memoirs*, **55**, https://doi.org/10.1144/M55-2018-58

Smellie, J.L. and Clarkson, P.D. 1975. Evidence for pre-Jurassic subduction in western Antarctica. *Nature*, **258**, 701–702, https://doi.org/10.1038/258701a0

Smellie, J.L. and Hole, M.J. 2021. Antarctic Peninsula: volcanology. *Geological Society, London, Memoirs*, **55**, https://doi.org/10.1144/M55-2018-59

Smellie, J.L., Davies, R.E.S. and Thomson, M.R.A. 1980. Geology of a Mesozoic intra-arc sequence on Byers Peninsula, Livingston Island, South Shetland Islands. *British Antarctic Survey Bulletin*, **50**, 55–76.

Smellie, J.L., Pankhurst, R.J., Thomson, M.R.A. and Davies, R.E.S. 1984. *The Geology of the South Shetland Islands: VI. Stratigraphy, Geochemistry and Evolution*. British Antarctic Survey Scientific Reports, **87**.

Smellie, J.L., Moyes, A.B., Marsh, P.D. and Thomson, J.W. 1985. Geology of Hugo Island, Quintana Island, Sooty Rock, Betbeder Islands and parts of the Biscoe and Outcast Islands. *British Antarctic Survey Bulletin*, **68**, 91–100.

Smellie, J.L., Griffiths, C.J. and Pankhurst, R.J. 1994. Antarctic Peninsula Volcanic Group. *In*: Moyes, A.B., Willan, C.F.H. et al. (eds) *Geological Map of Adelaide Island to Foyn Coast, 1:250 000, with Supplementary Text*. BAS GEOMAP Series Sheet 3. British Antarctic Survey, Cambridge, UK, 19–24.

Smellie, J.L., Liesa, M., Muñoz, J.A., Sàbat, F., Pallàs, R. and Willan, R.C.R. 1995. Lithostratigraphy of volcanic and sedimentary sequences in central Livingston Island, South Shetland Islands. *Antarctic Science*, **7**, 99–113, https://doi.org/10.1017/S0954 102095000137

Smellie, J.L., Pallàs, R., Sàbat, F. and Zheng, X. 1996a. Age and correlation of volcanism in central Livingston Island, South Shetland Islands: K–Ar and geochemical constraints. *Journal of South American Earth Sciences*, **9**, 265–272, https://doi.org/10.1016/0895-9811(96)00012-0

Smellie, J.L., Roberts, B. and Hirons, S.R. 1996b. Very low- and low-grade metamorphism in the Trinity Peninsula Group (Permo-Triassic) of northern Graham Land, Antarctic Peninsula. *Geological Magazine*, **133**, 583–594, https://doi.org/10.1017/S001675680000786X

Smellie, J.L., Millar, I.L., Rex, D.C. and Butterworth, P.J. 1998. Subaqueous, basaltic lava dome and carapace breccia on King George Island, South Shetland Islands, Antarctica. *Bulletin of Volcanology*, **59**, 245–261, https://doi.org/10.1007/s004450 050189

Smellie, J.L., McIntosh, W.C. and Esser, R. 2006. Eruptive environment of volcanism on Brabant Island: evidence for thin wet-based ice in northern Antarctic Peninsula during the Late Quaternary. *Palaeogeography, Palaeoclimatology, Palaeoecology*, **231**, 233–252, https://doi.org/10.1016/j.palaeo.2005.07.035

Smith, C.G. 1987. *The Geology of Parts of the West Coast of Palmer Land*. British Antarctic Survey Scientific Reports, **112**.

Spinola, D.N., Pi-Puig, T., Solleiro-Rebolledo, E., Egli, M., Sudo, M., Sedov, S. and Kühn, P. 2017. Origin of clay minerals in Early Eocene volcanic paleosols on King George Island, Maritime Antarctica. *Scientific Reports*, **7**, 6368, https://doi.org/10.1038/s41598-017-06617-x

Storey, B.C. and Garrett, S.W. 1985. Crustal growth of the Antarctic Peninsula by accretion, magmatism and extension. *Geological Magazine*, **122**, 5–14, https://doi.org/10.1017/S0016756800 034038

Suárez, M. 1976. Plate-tectonic model for the southern Antarctic Peninsula and its relation to southern Andes. *Geology*, **4**, 211–214, https://doi.org/10.1130/0091-7613(1976)4<211:PMFSAP>2.0.CO;2

Tangeman, J.A., Mukasa, S.B. and Grunow, A.M. 1996. Zircon U–Pb geochronology of plutonic rocks from the Antarctic Peninsula: confirmation of the presence of unexposed Paleozoic crust. *Tectonics*, **15**, 1309–1324, https://doi.org/10.1029/96TC00840

Tanner, P.W.G., Pankhurst, R.J. and Hyden, G. 1982. Radiometric evidence for the age of the subduction complex in the South Orkney and South Shetland islands, West Antarctica. *Journal of the Geological Society, London*, **139**, 683–690, https://doi.org/10.1144/gsjgs.139.6.0683

Tarney, J., Weaver, S.D., Saunders, A.D., Pankhurst, R.J. and Barker, P.F. 1982. Volcanic evolution of the northern Antarctic Peninsula and Scotia arc. *In*: Thorpe, R.S. (ed.) *Andesites*. John Wiley & Sons, Chichester, UK, 371–400.

Thomson, M.R.A. 1975. Upper Jurassic mollusca from Carse Point, Palmer Land. *British Antarctic Survey Bulletin*, **41–42**, 31–42.

Thomson, M.R.A. 1982a. Mesozoic paleogeography of West Antarctica. *In*: Craddock, C. (ed.) *Antarctic Geoscience*. University of Wisconsin Press, Madison, WI, 331–337.

Thomson, M.R.A. 1982b. Late Jurassic fossils from Low Island, South Shetland Islands. *British Antarctic Survey Bulletin*, **56**, 25–35.

Thomson, M.R.A. and Griffiths, C.J. 1994. Palaeontology. *In*: Moyes, A.B., Willan, C.F.H. *et al*. (eds) *Geological Map of Adelaide Island to Foyn Coast, 1:250 000, with Supplementary Text*. BAS GEOMAP Series Sheet 3. British Antarctic Survey, Cambridge, UK, 35–38.

Thomson, M.R.A. and Pankhurst, R.J. 1983. Age of post-Gonwanian calc-alkaline volcanism in the Antarctic Peninsula region. *In*: Oliver, R.L., James, P.R. and Jago, J.B. (eds) *Antarctic Earth Science*. Australian Academy of Science, Canberra, 328–333.

Thomson, M.R.A. and Tranter, T.H. 1986. Early Jurassic fossils from central Alexander Island and their geological setting. *British Antarctic Survey Bulletin*, **70**, 23–39.

Thomson, M.R.A., Pankhurst, R.J. and Clarkson, P.D. 1983. The Antarctic Peninsula – a late Mesozoic–Cenozoic arc (review). *In*: Oliver, R.L., James, P.R. and Jago, J.B. (eds) *Antarctic Earth Science*. Australian Academy of Science, Canberra, 289–294.

Tokarski, A.K., Danowski, W. and Zastawniak, E. 1987. On the age of fossil flora from Barton Peninsula, King George Island, West Antarctica. *Polish Polar Research*, **8**, 293–302.

Troedson, A.L. and Riding, J.B. 2002. Upper Oligocene to lowermost Miocene strata of King George Island, South Shetland Islands, Antarctica: stratigraphy, facies analysis, and implications for the glacial history of the Antarctic Peninsula. *Journal of Sedimentary Research*, **72**, 510–523, https://doi.org/10.1306/110601720510

Troedson, A.L. and Smellie, J.L. 2002. The Polonez Cove Formation of King George Island, Antarctica: stratigraphy, facies and implications for mid-Cenozoic cryosphere development. *Sedimentology*, **49**, 277–301, https://doi.org/10.1046/j.1365-3091.2002.00441.x

Trouw, R.A.J., Passchier, C.W., Valeriano, C.M., Simões, L.S.A., Paciullo, F.V.P. and Riberio, A. 2000. Deformational evolution of a Cretaceous subduction complex: Elephant Island, South Shetland Islands, Antarctica. *Tectonophysics*, **319**, 93–110, https://doi.org/10.1016/S0040-1951(00)00021-4

Valencio, D.A., Mendía, J.E. and Vilas, J.F. 1979. Palaeomagnetism and K–Ar age of Mesozoic and Cenozoic igneous rocks from Antarctica. *Earth and Planetary Science Letters*, **45**, 61–68, https://doi.org/10.1016/0012-821X(79)90107-9

Vaughan, A.P.M. and Storey, B.C. 2000. The eastern Palmer Land shear zone: a new terrane accretion model for the Mesozoic development of the Antarctic Peninsula. *Journal of the Geological Society, London*, **157**, 1243–1256, https://doi.org/10.1144/jgs.157.6.1243

Vaughan, A.P.M., Pankhurst, R.J. and Fanning, C.M. 2002. A mid-Cretaceous age for the Palmer Land event, Antarctic Peninsula: accretion timing and Gondwana palaeolatitudes. *Journal of the Geological Society, London*, **159**, 113–116, https://doi.org/10.1144/0016-764901-090

Vaughan, A.P.M., Leat, P.T., Dean, A.A. and Millar, I.L. 2012a. Crustal thickening along the West Antarctic Gondwana margin during mid-Cretaceous deformation of the Triassic intra-oceanic Dyer Arc. *Lithos*, **142–143**, 130–147, https://doi.org/10.1016/j.lithos.2012.03.008

Vaughan, A.P.M., Storey, C., Kelley, S.P., Barry, T.L. and Curtis, M.L. 2012b. Synkinematic emplacement of Lassiter Coast Intrusive Suite plutons during the Palmer Land event: evidence for mid-Cretaceous sinistral transpression at the Beaumont Glacier in eastern Palmer Land. *Journal of the Geological Society, London*, **169**, 759–771, https://doi.org/10.1144/jgs2011-160

Wang, F., Zheng, X.-S., Lee, J.I.K., Choe, W.H., Evans, N. and Zhu, R.-X. 2009. An ^{40}Ar/^{39}Ar geochronology on a mid-Eocene igneous event on the Barton and Weaver peninsulas: implications for the dynamic setting of the Antarctic Peninsula. *Geochemistry, Geophysics, Geosystems*, **10**, Q12006, https://doi.org/10.1029/2009GC002874

Warny, S., Madison Kymes, C., Askin, R.A., Krajewski, K.P. and Bart, P.J. 2016. Remnants of Antarctic vegetation on King George Island during the early Miocene Melville glaciation. *Palynology*, **40**, 66–82, https://doi.org/10.1080/01916122.2014.999954

Warny, S., Madison Kymes, C., Askin, R., Krajewski, K.P. and Tatur, A. 2018. Terrestrial and marine floral response to latest Eocene and Oligocene events on the Antarctic Peninsula. *Palynology*, **43**, 4–21, https://doi.org/10.1080/01916122.2017.1418444

Watts, D.R. 1982. Potassium–argon ages and paleomagnetic results from King George Island, South Shetland Islands. *In*: Craddock, C. (ed.) *Antarctic Geoscience*. University of Wisconsin Press, Madison, WI, 255–261.

West, S.M. 1974. *The Geology of the Danco Coast, Graham Land*. British Antarctic Survey Scientific Reports, **84**.

Wever, H.E. and Storey, B.C. 1992. Bimodal magmatism in northeast Palmer Land, Antarctic Peninsula: geochemical evidence for a Jurassic ensialic back-arc basin. *Tectonophysics*, **205**, 239–259, https://doi.org/10.1016/0040-1951(92)90429-A

Willan, R.C.R. 1994. Structural setting and timing of hydrothermal veins and breccias on Hurd Peninsula, South Shetland Islands: a possible volcanic-related epithermal system in deformed turbidites. *Geological Magazine*, **131**, 465–483, https://doi.org/10.1017/S0016756800012103

Willan, R.C.R. and Armstrong, D.C. 2002. Successive geothermal, volcanic hydrothermal and contact-metasomatic events in Cenozoic volcanic-arc basalts, South Shetland Islands, Antarctica. *Geological Magazine*, **139**, 209–231, https://doi.org/10.1017/S0016756802006301

Willan, R.C.R. and Kelley, S.P. 1999. Mafic dike swarms in the South Shetland Islands volcanic arc: unravelling multiepisodic magmatism related to subduction and continental rifting. *Journal of Geophysical Research*, **104**, 23 051–23 068, https://doi.org/10.1029/1999JB900180

Willan, R.C.R., Pankhurst, R.J. and Hervé, F. 1994. A probable Early Triassic age for the Miers Bluff Formation, Livingston Island, South Shetland Islands. *Antarctic Science*, **6**, 401–408, https://doi.org/10.1017/S095410209400060X

Wyeth, R.B. 1977. The physiography and significance of the transition zone between Graham Land and Palmer Land. *British Antarctic Survey Bulletin*, **46**, 39–58.

Zheng, G.G., Liu, X.C., Sang, H., Wang, W. and Chen, L.Y. 2017. Mid-Cretaceous volcano-magmatism in the Curverville Island of the Antarctic Peninsula and its tectonic significance: constraints from zircon U–Pb geochronology and Hf isotopic compositions. *Acta Petrologica Sinica*, **33**, 978–992, http://html.rhhz.net/ysxb/20170322.htm [in Chinese with English abstract].

Zheng, G.G., Liu, X.C., Liu, S., Zhang, S.H. and Zhao, Y. 2018. Late Mesozoic–early Cretaceous intermediate–acid intrusive rocks from the Gerlache Strait area, Antarctic Peninsula: Zircon U–Pb geochronology, petrogenesis and tectonic implications. *Lithos*, **312–313**, 204–222, https://doi.org/10.1016/j.lithos.2018.05.008

Zheng, X., Sang, H., Qiu, J., Liu, J., Lee, J.I. and Kim, H. 2000. New discovery of the Early Cretaceous volcanic rocks on the Barton Peninsula, King George Island, Antarctica and its geological significance. *Acta Geologica Sinica*, **74**, 176–182, https://doi.org/10.1111/j.1755-6724.2000.tb00446.x

Chapter 3.1b

Antarctic Peninsula and South Shetland Islands: petrology

Philip T. Leat[1,2]* and Teal R. Riley[1]

[1]British Antarctic Survey, High Cross, Madingley Road, Cambridge CB3 0ET, UK
[2]School of Geography, Geology and the Environment, University of Leicester, University Road, Leicester LE1 7RH, UK

TRR, 0000-0002-3333-5021
*Correspondence: ptle@bas.ac.uk

Abstract: The Antarctic Peninsula contains a record of continental-margin volcanism extending from Jurassic to Recent times. Subduction of the Pacific oceanic lithosphere beneath the continental margin developed after Late Jurassic volcanism in Alexander Island that was related to extension of the continental margin. Mesozoic ocean-floor basalts emplaced within the Alexander Island accretionary complex have compositions derived from Pacific mantle. The Antarctic Peninsula volcanic arc was active from about Early Cretaceous times until the Early Miocene. It was affected by hydrothermal alteration, and by regional and contact metamorphism generally of zeolite to prehnite–pumpellyite facies. Distinct geochemical groups recognized within the volcanic rocks suggest varied magma generation processes related to changes in subduction dynamics. The four groups are: calc-alkaline, high-Mg andesitic, adakitic and high-Zr, the last two being described in this arc for the first time. The dominant calc-alkaline group ranges from primitive mafic magmas to rhyolite, and from low- to high-K in composition, and was generated from a mantle wedge with variable depletion. The high-Mg and adakitic rocks indicate periods of melting of the subducting slab and variable equilibration of the melts with mantle. The high-Zr group is interpreted as peralkaline and may have been related to extension of the arc.

Supplementary material: Representative analyses from the Antarctic Peninsula volcanic arc (Table S1) are available at https://doi.org/10.6084/m9.figshare.c.5192644

The Antarctic Peninsula was the site of east-directed subduction during Mesozoic–Tertiary times, and many elements of the active margin are evident (Suárez 1976; Tarney et al. 1982; Storey and Garrett 1985). Volcanic rocks intruded by plutons form a magmatic arc cropping out along the west coast of Graham Land and on islands to the west, in western Palmer Land, in Alexander Island and in the South Shetland Islands (Saunders et al. 1980; Pankhurst 1982; Tarney et al. 1982; Thomson et al. 1983; Smellie et al. 1984; Birkenmajer 1994; McCarron and Smellie 1998; Ryan 2007; Haase et al. 2012; Zheng et al. 2018). Thicknesses of exposed volcanic successions of the arc are up to c. 1.5 km (Leat and Riley 2021) and the known distribution of the arc volcanic rocks is shown in Figure 1. Thomson (1982) suggested the name 'Antarctic Peninsula Volcanic Group' for the volcanic rocks of the volcanic arc. This term has been widely used, although the group status is uncertain following designation of more local volcanic/sedimentary units as groups (Birkenmajer 1989; Hathway and Lomas 1998; McCarron 1997). To avoid confusion, this paper uses the term Antarctic Peninsula volcanic arc to include all volcanic rocks of the Mesozoic–Early Tertiary volcanic arc along the west coast of the Antarctic Peninsula. We exclude from this paper Jurassic volcanic rocks along the east coast of the Antarctic Peninsula, which are not related to the subduction system (Riley et al. 2001; Riley and Leat 2021). The occurrence and volcanology of the Antarctic Peninsula volcanic arc are described with detailed maps in a companion paper (Leat and Riley 2021). The geological timescale used is that of Ogg et al. (2008).

The age of inception of the volcanic arc is uncertain. Subduction and arc magmatism have been widely thought to have taken place since Middle Jurassic times or earlier (e.g. Suárez 1976; Saunders et al. 1980; Pankhurst 1982; Thomson 1982; Storey and Garrett 1985; Leat et al. 1995). However, the earliest known volcanic rocks belonging to the arc are Late Jurassic–Early Cretaceous in age (Thomson and Pankhurst 1983; Haase et al. 2012), and evidence for arc volcanism before Early Cretaceous times is weak (Leat and Riley 2021). The earliest isotopically dated volcanic rocks belonging to the volcanic arc are Early Cretaceous in age (135 Ma: Haase et al. 2012), and volcanism appears to have been continuous from then until Early Tertiary cessation. The volcanic axis of the volcanic arc shifted to the west (towards the subduction zone) during the Cretaceous. This shift is evident in the Anvers Island area (Ryan 2007; Zheng et al. 2018), and is pronounced in the south where the volcanic arc axis shifted west from Palmer Land to Alexander Island between 107 and 80 Ma (Fig. 1) (Leat and Riley 2021). Formation of the Bransfield Rift, which widened significantly in the last 4 myr (Barker 1982), has rifted the South Shetland Islands from northern Graham Land (Fig. 1). Volcanism ceased at c. 46 Ma on Alexander Island (McCarron and Millar 1997), and cessation of arc volcanism is thought to have migrated northwards. The youngest volcanic rocks assigned to the volcanic arc are the Early Miocene Boy Point Dacites, King George Island, South Shetland Islands, which are Ar–Ar dated at 22.61 ± 1.7 Ma (Smellie et al. 1998).

Two groups of volcanic rocks that are not part of the Antarctic Peninsula volcanic arc crop out in Alexander Island (Fig. 1). One group comprises accreted basalts within deformed metasediments of the LeMay Group on Alexander Island. The LeMay Group is interpreted as an accretionary prism accreted to the western edge of the active continental margin (Burn 1984; Storey and Garrett 1985; Nell 1990), and the basalts are interpreted as slivers of ocean floor and seamounts thrusted into the accretionary complex (Doubleday et al. 1994). The second group comprises volcanic rocks and sills within sedimentary rocks of the Fossil Bluff Group, eastern Alexander Island, which has been interpreted as a forearc basin overlying the LeMay Group (Doubleday et al. 1993). These Jurassic forearc volcanic rocks were erupted during the Late Jurassic, and are thought to be related to lithospheric extension (Macdonald et al. 1999).

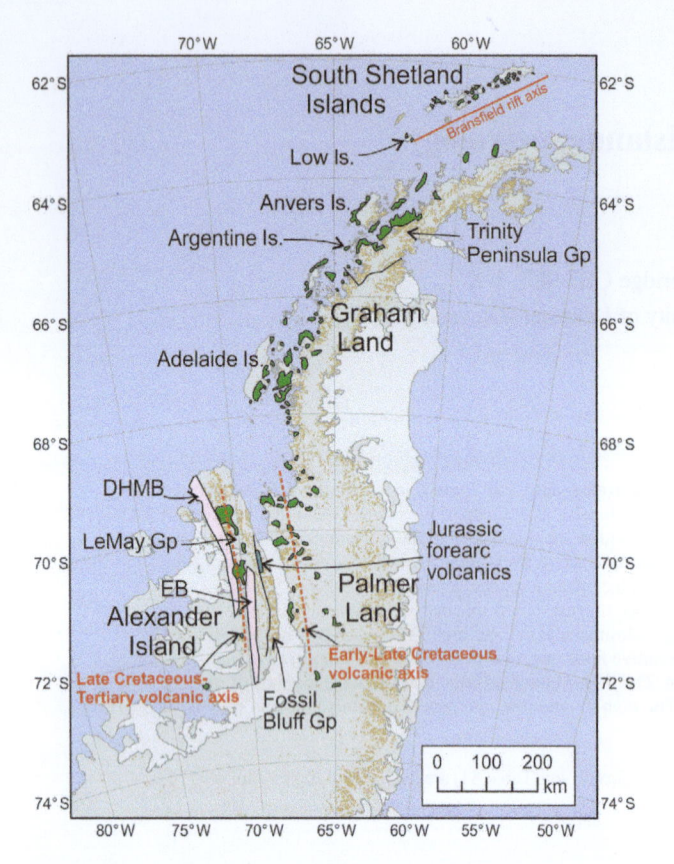

Fig. 1. Map of the Antarctic Peninsula showing the distribution of the Antarctic Peninsula volcanic arc (shaded green). The westward shift in the axis of the arc from Palmer Land to Alexander Island occurred between 107 and 80 Ma. Accreted oceanic basalts within the accretionary LeMay Group form two north–south belts: the western Debussy Heights mélange belt (DHMB) and the eastern belt (EB). Jurassic forearc volcanic rocks of Alexander Island are within the Fossil Bluff Group which overlies the LeMay Group. The Trinity Peninsula Group (metasediments) forms the basement of northern Graham Land. The Bransfield Rift has separated volcanic rocks of the arc in the South Shetland Islands from remnants of the arc in northern Graham Land.

Tectonic and crustal setting

The Antarctic Peninsula volcanic arc was built on continental crust, as well as elements accreted to the continental margin. In western Palmer Land, protolith ages of basement orthogneisses are as old as Silurian (435–422 Ma) according to Millar *et al.* (2002). The oldest known inherited ages for *in situ* rocks from southwestern Graham Land are Late Paleozoic–Mesozoic (Tangeman *et al.* 1996). Northern Graham Land is underlain by metasediments of the Trinity Peninsula Group (Fig. 1), which has dominantly Carboniferous–Triassic sedimentation ages (Barbeau *et al.* 2009; Castillo *et al.* 2016). The South Shetland Islands are underlain by the accretionary Scotia Metamorphic Complex, which was metamorphosed during the Mesozoic (Trouw *et al.* 2000), and Mesozoic metasediments (Willan *et al.* 1994; Hervé *et al.* 2006). The basement to the volcanic arc rocks on Alexander Island is the Mesozoic accretionary LeMay Group (Burn 1984; Doubleday *et al.* 1994).

The cessation of subduction and formation of the present largely inactive ocean–continent margin has been reconstructed from ocean-floor magnetic anomalies (Barker 1982; Larter and Barker 1991). Subduction terminated as successive spreading segments of the Antarctic–Phoenix spreading centre collided with the trench, the collisions migrating northwards over time. Subduction cessation was at *c.* 50 Ma at Alexander Island, *c.* 15 Ma at Adelaide Island and slow subduction is continuing in the South Shetland Islands sector (Maldonado *et al.* 1994). Prior to this Paleogene–Neogene ridge–trench collision, subduction is thought to have been more-or-less continuous since at least the Early Cretaceous, although punctuating episodes of interactions of spreading centres with the subduction zone have been proposed (Alabaster and Storey 1990; Scarrow *et al.* 1997).

Alteration and metamorphism

The Jurassic–Early Tertiary volcanic and associated volcaniclastic rocks of the Antarctic Peninsula and South Shetland Islands have been affected by variable intensities of alteration and metamorphism. Although there are many petrographical descriptions of altered volcanic rocks and assessments of local alteration processes, interpretations of regional variations in metamorphic grade and alteration are largely absent. Alteration of volcanic rocks has resulted from three main processes: hydrothermal processes related to the arc magmatism; contact metamorphism in proximity to plutons; and regional/burial metamorphism. Regional metamorphic grade is generally low in the South Shetland Islands, and effects of hydrothermal alteration are clear (Willan and Armstrong 2002). In Graham Land and Palmer Land, by contrast, hydrothermal alteration and higher grades of overprinting regional metamorphism have proved difficult to disentangle (e.g. West 1974; Davies 1984). Weathering is generally minor, although volcanic rocks are commonly shattered by freeze–thaw action.

Hydrothermal alteration

Several areas of intense hydrothermal alteration of volcanic rocks on the South Shetland Islands and other islands west of Graham Land have been described. In a well-documented example, Hawkes and Littlefair (1981) interpreted alteration zones *c.* 4 km across in volcanic and plutonic rocks in the Argentine Islands as the root zone of a porphyry copper system. A zone of potassic alteration within plutonic rocks forming the western part of the archipelago was interpreted as the core of the system, with associated propylitic alteration of volcanic rocks to the east characterized by widespread alteration of minerals to epidote, sericite, chlorite and carbonate. Magnetite has been introduced to vesicles and groundmasses, and disseminated pyrite is widespread. There are small zones of intense phyllic alteration of volcanic rocks to quartz–muscovite–pyrite assemblages with minor kaolinite, chlorite, epidote and pyrrhotite. Veinlets associated with the system contain quartz, molybdenite, chalcopyrite, pyrite and magnetite. A 13 km-long hydrothermal breccia and vein system on Hurd Peninsula, Livingston Island, in the South Shetland Islands intrudes volcanic rocks as well as metasediments of the Miers Bluff Formation (Willan 1994). It is interpreted as an epithermal system of probably Cretaceous age (Willan 1994, Armstrong and Willan 1996; Willan and Spiro 1996). Along strike in eastern Livingston Island, alteration of volcanic rocks of the Mount Bowles Formation to assemblages including sericite, epidote, chlorite, amphibole and sulfides has been attributed to hydrothermal alteration and contact metamorphism associated with subvolcanic intrusions (Smellie *et al.* 1995). On King George Island, a central zone extending 80 km from the Fildes Peninsula to the NE coast contains numerous examples of hydrothermally altered volcanic rocks, including propylitic alteration (Willan and Armstrong 2002). On the Fildes Peninsula, lavas have been altered by low-temperature hydrothermal fluids (Gao *et al.* 2018). On Barton Peninsula, hydrothermal activity led to the silification of volcanic rocks, which is overprinted by advanced argillic alteration, and the volcanic rocks were also affected by propylitic alteration related to a younger pluton (So *et al.* 1995; Willan and Armstrong 2002).

Polymetallic veining associated with propylitic and phyllic hydrothermal alteration in volcanic rocks has been described from Anvers Island and Adelaide Island (Hawkes 1982), and sphalerite occurs in quartz–epidote veins of probable hydrothermal origin in volcanic rocks on Horseshoe Island (Matthews 1983b). Veins, dominantly filled by calcite, epidote and chlorite, and disseminated pyrite of possible hydrothermal origin, are widely reported (e.g. Goldring 1962; West 1974; Smellie 1980; Matthews 1983a).

Contact metamorphism

Contact metamorphism within the volcanic rocks has been widely described and is associated with intrusion of plutons into the volcanic sequence. Volcanic rocks that have been hornfelsed by intrusions have been described in both Graham Land (Goldring 1962; Hooper 1962; Curtis 1966) and Palmer Land (Ayling 1984; Smith 1987). Smellie *et al.* (1995) suggested albite–hornfels facies volcanic rocks of the Mount Bowes Formation, Livingston Island were contact metamorphosed by unseen subsurface intrusions.

Regional metamorphism

Burial/regional metamorphism in the volcanic rocks is generally zeolite to prehnite–pumpellyite facies, possibly to low-greenschist facies (West 1974; Smellie 1980; Davies 1984; Smellie *et al.* 1984). Typically, plagioclase and ferromagnesian phenocrysts have at least partially been replaced by sericite, calcite, chlorite, epidote, amphibole, biotite, prehnite and Fe–Ti oxides, and commonly the groundmass consists of a fine-grained assemblage of similar replacement minerals, as documented in Graham Land (Goldring 1962; Curtis 1966; West 1974; Matthews 1983a, b; Moyes and Hamer 1984), Adelaide Island (Dewar 1970) and Palmer Land (Singleton 1979; Ayling 1984; Davies 1984; Smith 1987). The regional metamorphic grade in volcanic rocks of the South Shetland Islands is generally very low grade, characterized by alteration of olivine and other phenocrysts, and the presence of zeolite (Smellie *et al.* 1984). On Low Island, the metamorphic grade is at least upper prehnite–pumpellyite facies. In the Eocene volcanic rocks around the southern part of Admiralty Bay, alteration is minor, characterized by alteration of glass, and partial alteration of olivine, orthopyroxene and plagioclase phenocrysts (Smellie *et al.* 1984; Nawrocki *et al.* 2011; Pańczyk and Nawrocki 2011; Mozer *et al.* 2015). In the LeMay Group accretionary complex in Alexander Island, which includes the accreted basalts, the metamorphic grade is zeolite facies to probably as high as low-greenschist facies (Burn 1984; Wendt *et al.* 2008). Metamorphism of the Early Tertiary volcanic rocks of Alexander Island is probably no higher than zeolite facies, and is characterized by extensive alteration of plagioclase to sericite, calcite and epidote. Pyroxenes are commonly altered to biotite, chlorite, calcite and amphibole. Groundmass minerals include calcite, quartz, chlorite and zeolite (Burn 1981). The regional metamorphic grades in Graham Land and Palmer Land suggest temperatures of c. 150–300°C and pressures of up to c. 0.3 GPa, suggesting burial depths of c. 3–8 km, consistent with widespread unroofing of granitic–tonalitic plutons intruded into the volcanic rocks (e.g. Pankhurst 1982; Moyes and Hamer 1984; Smith 1987; Harrison and Piercy 1990).

Geochemical data sources

Geochemical data for the Antarctic Peninsula volcanic arc used in this paper are collated from published sources and unpublished data. The data are restricted to analyses of samples described as volcanic rocks by the collectors (mainly lavas, tuffs, breccias, agglomerates and ignimbrites). Dykes are not included except for three analyses of high-Mg andesite dykes from Alexander Island which are closely related to the nearby volcanic arc (McCarron and Smellie 1998) and provide a more complete dataset on the local arc magmatism. Samples in the dataset are all from volcanic rocks reasonably assumed have erupted more recently than 200 Ma, and are from the West Coast Antarctic Peninsula Volcanic Group, the Alexander Island Volcanic Group and the South Shetland Islands. For plotting the geochemical variation diagrams, we used the datasets of Machado *et al.* (2005) and Haase *et al.* (2012) for the South Shetland Islands, and datasets of Burn (1981) and McCarron and Smellie (1998) for Alexander Island. For Graham Land, the plotted data are from West (1974) and Saunders *et al.* (1980). For Palmer Land, the plotted data are from Davies (1984) and Smith (1987). We also used unpublished data from the British Antarctic Survey Geological Database and unpublished data of P.T. Leat from southeastern Graham Land to supplement the Antarctic Peninsula and Alexander Island datasets. Elemental data are restricted to X-ray fluorescence (XRF) for major and trace elements, inductively coupled plasma mass spectrometry (ICP-MS) data for the South Shetland Islands (Haase *et al.* 2012), and instrumental neutron activation analysis (INAA) data for rare earth elements (REE) and other trace elements for a minority of samples. The XRF data for British Antarctic Survey publications were mostly analysed between the 1960s and 1990 at several UK laboratories, and the INAA data were analysed in the 1980s and 1990s at the Open University, UK. Data were screened for acceptable totals and trace element abundances but variation in analytical accuracy is inevitable. In view of this uncertainty, we restrict the descriptions to broad trends in the data, and to identifying and discussing chemical groups of data.

The volcanic rocks within the forearc basin of the Fossil Bluff Group (Macdonald *et al.* 1999) and accreted basalts from the LeMay Group (Doubleday *et al.* 1994) are described separately. The petrology of Late Cenozoic, post-subduction volcanism is reviewed in separate chapters in this Memoir (Haase and Beier 2021; Hole 2021). All major element data in this paper are recalculated to volatile-free totals of 100%. Mg# is calculated as $100 \times$ molecular Mg/(Mg + Fe).

Accreted oceanic basalts

Accreted basalts occur within two north–south-trending belts within the LeMay Group, which is an accretionary complex forming much of Alexander Island (Burn 1984; Nell 1990; Doubleday and Tranter 1994; Doubleday *et al.* 1994; Leat and Riley 2021). Determination of their accreted, oceanic origin is based on their structural and lithological associations, and immobile trace element compositions. Basalts from the western Debussy Heights mélange belt (Fig. 1) are dated as Late Jurassic–Early Cretaceous in age by radiolarians in associated cherts (Holdsworth and Nell 1992). Samples from Sullivan glacier range within the northern part of the belt are weakly enriched mid-ocean ridge basalt (MORB) to tholeiitic ocean island basalt (OIB) in composition (Doubleday *et al.* 1994). Within the eastern belt (Fig. 1), the Lully Foothills includes both ocean-floor basalts and basaltic and volcaniclastic deposits, indicating shallow-water depths, interpreted as a seamount or ocean island of Early Jurassic age (Tranter

1991; Doubleday et al. 1994). Compositions are MORB-like for the ocean-floor samples, and tholeiitic OIB for the seamount (Doubleday et al. 1994). The basaltic lava–chert association in the NW LeMay Range includes enriched MORB compositions and a depleted MORB sample (Doubleday et al. 1994). The southernmost Herschel Heights locality comprises associations of basaltic lavas with volcaniclastic and chert deposits, and includes both MORB- and OIB-like compositions (Doubleday and Tranter 1992; Doubleday et al. 1994). All of the analysed accreted basalt samples are consistent with the group representing ocean floor with normal and enriched MORB (N-MORB and E-MORB) compositions surmounted by seamounts with E-MORB to OIB compositions. The basalts are interpreted to have been scraped from the subducting Pacific Plate and thrust within the accretionary complex during the subduction that generated the Cretaceous–Tertiary Antarctic Peninsula volcanic arc.

Mahoney et al. (1998) analysed five samples of the N-MORB and E-MORB basalts as representatives of ocean floor derived from Mesozoic Pacific MORB-source mantle for Sr, Nd and Pb isotopes. They demonstrated that they have age-adjusted isotopic compositions similar to present-day Pacific–North Atlantic reservoirs. They concluded that Pacific MORB mantle was isotopically similar to that of today. Barry et al. (2017) identified only one sample from the Lully Foothills as N-MORB, using rigorous trace element criteria, to show that it has a Pacific reservoir composition with respect to Hf and Nd isotopes.

Volcanic rocks within the Fossil Bluff Group

Volcanic rocks and sills occur within the Fossil Bluff Group, western Alexander Island (Leat and Riley 2021). The Fossil Bluff Group is a Middle Jurassic–Early Cretaceous basin-fill sequence in a forearc position in relation to the volcanic arc of western Palmer Land (Doubleday et al. 1993; Moncrieff and Kelly 1993). The volcanic rocks occur in the Himalia Ridge Formation (Callovian–earliest Valanginian: 164–140 Ma) and are thought to be Oxfordian–Kimmeridgian in age (161.2–150.8 Ma: Macdonald et al. 1999). The volcanic rocks are thought not to be geochemically related to the Antarctic Peninsula volcanic arc. They consist of basaltic, dacitic and rhyolitic lavas, tuffs, breccias and sills (Elliott 1974; Macdonald et al. 1999). Igneous mineral phases in the volcanic rocks have been completely replaced during alteration by assemblages dominated by calcite, sericite, chlorite and epidote (Elliott 1974).

The volcanic rocks form a bimodal mafic–silicic group. Three basalts from Tilt Rock analysed by Macdonald et al. (1999) are high-Nb types, with immobile trace element ratios comparable to those of OIB. One basalt from Ethelbald Rock has low Nb typical of arc calc-alkaline magmas. The dacites and rhyolites are calc-alkaline and may be related to the Ethelbald Rock basalt. Macdonald et al. (1999) and Burton-Johnson and Riley (2015) suggested that, because of their OIB-like compositions, the basalts cannot have formed part of the volcanic arc and were, instead, generated during a pause in subduction and were related to lithospheric extension at a time of deepening of the Fossil Bluff Group basin. Similar Jurassic–earliest Cretaceous OIB-like basaltic dykes were identified in western Palmer Land and southwestern Graham Land by Scarrow et al. (1998), who related them to effective cessation of subduction caused by interaction of an oceanic spreading centre with the subduction zone. In view of the evidence that subduction beneath the Antarctic Peninsula may have occurred during Jurassic times, and certainly started again during the Early Cretaceous (Leat and Riley 2021), the Fossil Bluff Group OIB magmas are likely to be related to extension of the forearc continental margin before commencement of the Cretaceous subduction.

The Antarctic Peninsula volcanic arc

Rocks of the Antarctic Peninsula volcanic arc are present in Graham Land, Palmer Land, the South Shetland Islands and Alexander Island (Fig. 1), and are younger than the accreted and forearc volcanic rocks described above. Preserved primary minerals and pseudomorphed grains indicate that basaltic–andesitic lavas of the Antarctic Peninsula volcanic arc contain a normal phenocryst assemblage for subduction-zone magmatism consisting of plagioclase, clinopyroxene, olivine and Fe–Ti oxides ± orthopyroxene ± amphibole ± apatite, with alkali feldspar and biotite phenocrysts appearing in dacitic and rhyolitic lavas and tuffs (e.g. West 1974; Burn 1981; Machado et al. 2005).

In view of the evidence for pervasive alteration of rocks of the Antarctic Peninsula volcanic arc outlined above, classification and interpretation of petrogenesis based on elements that are mobile during low-grade metamorphism and hydrothermal alteration must be treated with caution. In Figure 2, the SiO_2 v. K_2O diagram for classification of volcanic arc rocks is plotted with the Co v. Th analogue (Hastie et al. 2007). In Figure 2b, samples are plotted for which both Co (by XRF or ICP-MS) and Th (by INAA or ICP-MS) are available. This restricts the data to the South Shetland Islands (Machado et al. 2005; Haase et al. 2012) and Alexander Island (McCarron and Smellie 1998). Co–Th data are not available for Graham Land and Palmer Land samples. In Figure 2a, the same data are plotted, with the addition of the full dataset for the Alexander Island Volcanic Group.

In the K_2O v. SiO_2 plot (Fig. 2a), South Shetland Islands samples plot dominantly as basalts and basaltic andesites, and dominantly within the medium-K field, although there is significant overlap with the low-K field, and some scatter into the high-K field. Alexander Island samples have a very different distribution and plot dominantly in the basalt andesite, andesite and dacite/rhyolite fields with significant scatter across both the medium-K and high-K fields. Both Co and Th are immobile at low to moderate degrees of alteration. In the Th v. Co plot (Fig. 2b), Co is used as an inverse index of fractionation, while Th abundances are related to the degree of enrichment of the magma as a result of degree of mantle partial melting and enrichment of the magma source from the subducting slab. As such, it is a proxy for K in the K_2O v. SiO_2 diagram (Hastie et al. 2007). Samples from the South Shetland Islands plot overwhelmingly in the basalt and basaltic andesite–andesite fields, supporting the mafic character of this part of the arc, and consistent with the SiO_2 distribution. The samples plot as dominantly calc-alkaline (medium-K), with a few tholeiitic (low-K) samples, consistent with the observations of Haase et al. (2012). There are fewer low-K samples than suggested by the K_2O v. SiO_2 diagram, consistent with removal of K in some samples during low-temperature alteration. Alexander Island samples are more evenly distributed from basalt to dacite/rhyolite in Figure 2b, and form a higher-Th trend than the South Shetland group. They plot along and slightly above the dividing line between calc-alkaline and high-K to shoshonitic types. There is significantly less scatter in Th compared to K_2O abundances, suggesting significant low-temperature mobilization of K, consistent with the interpretations of McCarron and Smellie (1998).

In view of this evidence for mobility of K in the volcanic arc rocks of Alexander Island and, to a lesser degree, in the South Shetland Islands, it is reasonable to assume that elements which are normally mobile during low-temperature alteration, low-grade metamorphism and hydrothermal alteration of

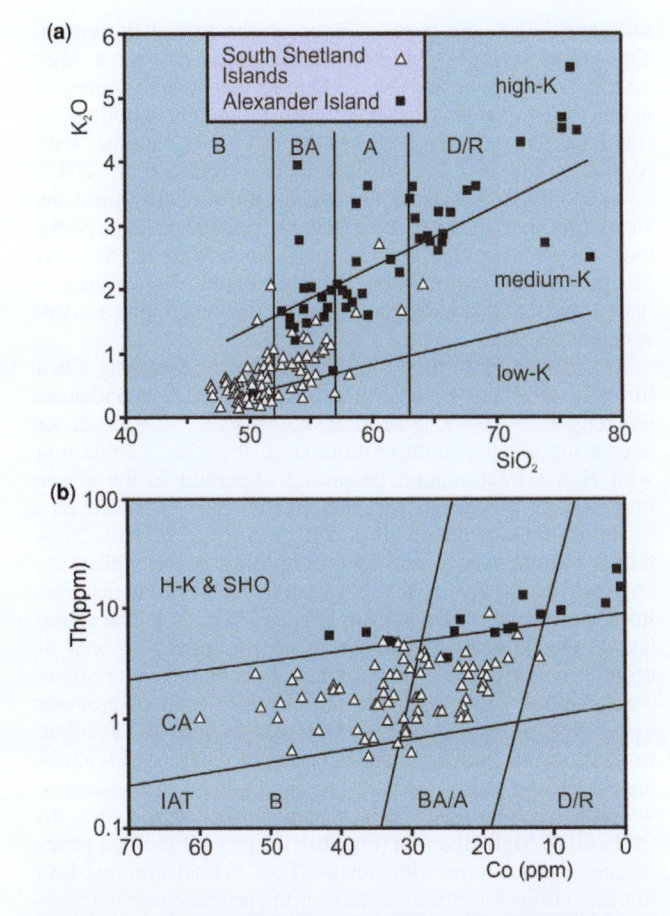

Fig. 2. Plots of SiO$_2$ v. K$_2$O and Th v. Co for Mesozoic and Early Tertiary volcanic arc rocks from the South Shetland Islands and Early Tertiary volcanic arc rocks of the Alexander Island Volcanic Group. The Th v. Co plot is an analogue for the SiO$_2$ v. K$_2$O plot but using immobile elements (Hastie et al. 2007). (**a**) SiO$_2$ v. K$_2$O (wt%, with totals recalculated to 100% volatile-free) with boundaries of compositional fields from Le Maitre (1989). B, basalt; BA basaltic andesite; D/R, dacite and rhyolite. (**b**) Th (ppm) v. Co (ppm) with boundaries of compositional fields from Hastie et al. (2007). B, basalt; BA/A, basaltic andesite and andesite; D/R, dacite and rhyolite; IAT, island arc tholeiite; CA, calc-alkaline; H-K & SHO, high-K and shoshonitic. Data sources – South Shetland Islands: Machado et al. (2005) and Haase et al. (2012); Alexander Island: Burn (1981), McCarron and Smellie (1998) and British Antarctic Survey unpublished data.

volcanic rocks, such as K, Na, Ca, Rb, Ba, Sr and Cs, may not be present in magmatic abundances. This is especially important in the case of Graham Land and Palmer Land, where petrographical and field evidence suggest generally stronger alteration than in the South Shetland Islands and Alexander Island. We therefore concentrate on the more immobile elements when discussing the original composition of the rocks.

Our approach is to attempt to identify distinct geochemical groups within the arc sequences, to indicate the geographical distribution of these groups and to explain the relationship of the groups to the volcanic arc. We distinguish four geochemical groups within the volcanic arc sequence: a dominant, calc-alkaline (including low-K, medium-K and high-K variants) group, representing 'normal' arc magmatism; an adakitic group; a high-Mg andesite group; and a high-Zr group. Of the 262 analyses in the dataset we are using to plot data for the volcanic arc, 191 (73%) belong to the calc-alkaline group, 52 (20%) are adakitic, eight (3%) are high-Mg andesites and 11 (4%) belong to the high-Zr group. The plots that are most important in our identification of these four groups are shown in Figure 3. Additional element variation plots

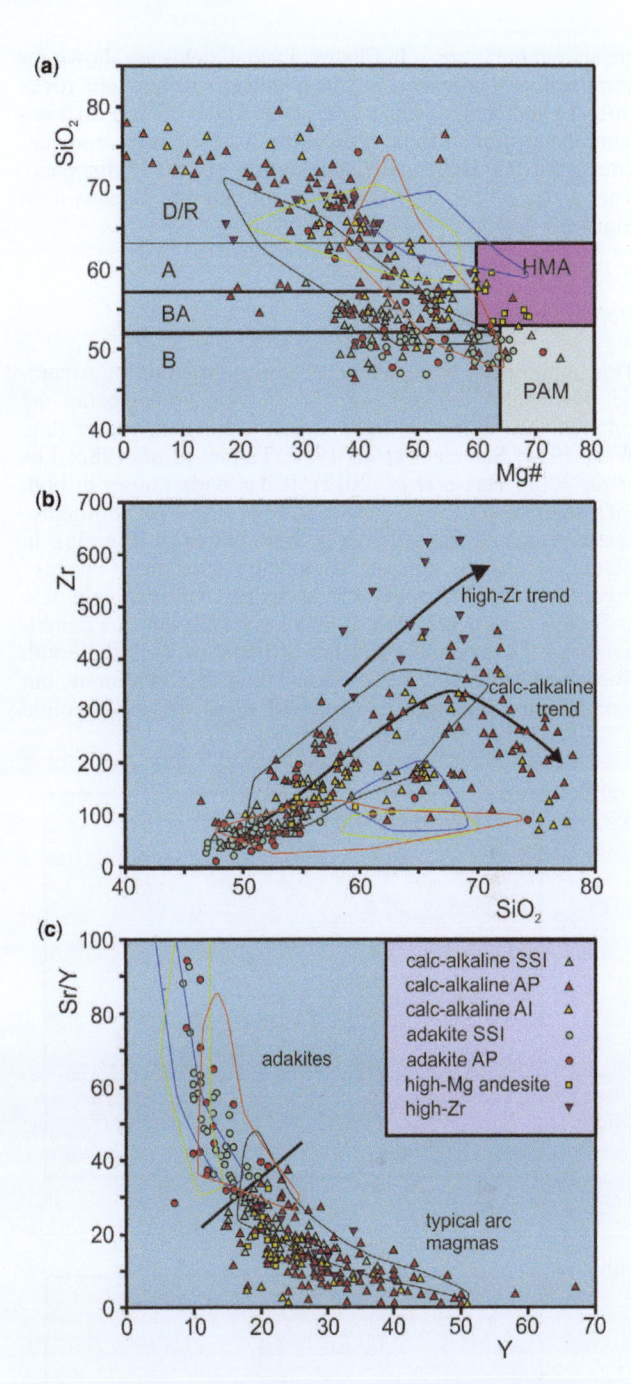

Fig. 3. Plots utilized to distinguish geochemical groups within Mesozoic–Early Tertiary volcanic arc rocks from the Antarctic Peninsula (AP), Alexander Island (AI) and South Shetland Islands (SSI). Major elements are in wt%; trace elements are in ppm. The fields for comparative data are: black, calc-alkaline volcanic arc rocks from the southern volcanic zone of the Andes, Chile (Hickey-Vargas et al. 1989; D'Orazio et al. 2003); blue, adakites from the austral volcanic zone of the Andes, Chile (Stern and Kilian 1996); yellow, adakites from Mindanao Island, the Philippines (Sajona et al. 1993; Sajona et al. 1996); red, adakitic–calc-alkaline rocks from Camiguin Island, the Philippines (Castillo et al. 1999). (**a**) Mg-number (Mg#: molecular Mg/(Mg + Fe)) v. SiO$_2$. The divisions into basalt (B), basaltic andesite (BA), andesite (A) and dacite/rhyolite (D/R) are as in Figure 2. The field for high-Mg andesite (HMA) is Mg# >60 and SiO$_2$ >53 wt%. High-Mg andesites also have Ni>75 ppm and Cr>250 ppm (McCarron and Smellie 1998). Primitive arc magmas (PAM) have Mg# >64, Ni >200 ppm and Cr >400 ppm. (**b**) SiO$_2$ v. Zr showing the distinct high-Zr group in relation to the dominant calc-alkaline series. High-Zr and calc-alkaline trend lines are indicative. (**c**) Y v. Sr/Y plot showing discrimination of adakites from typical volcanic arc rocks (after Atherton and Petford 1993). The dividing line between adakites and typical arc magmas is arbitrary and is drawn from (11, 22) to (26, 45).

are shown in Figure 4. In Figures 3 and 4, fields are shown for comparative purposes for calc-alkaline volcanic arc rocks from the southern volcanic zone of the Andes (Chile), adakites from the austral volcanic zone of the Andes (Chile), and adakites from Mindanao and Camiguin islands (the Philippines). The geographical distribution of the groups is shown in Figures 5 and 6.

Calc-alkaline group

This numerically dominant geochemical group of the Antarctic Peninsula volcanic arc is a typical magmatic arc calc-alkaline series, as noted in many previous papers (e.g. West 1974; Saunders *et al.* 1980; Tarney *et al.* 1982; Lee *et al.* 2008; Haase *et al.* 2012). It has wide ranges in both SiO_2 (from *c.* 47 to >75 wt%) and Mg# (0.7–74), with representatives of all stages between these extremes (Fig. 3a). In Figure 3a there is apparent bimodality with many dacites/rhyolites and basalts/basaltic andesites but relatively few andesites. This distribution is likely to be a result of a combination of the genuine abundance of mafic rocks in the South Shetland Islands and the undersampling of voluminous, but monotonous, andesites in favour of more striking rhyolitic tuffs and lavas, and basaltic lavas, in the rest of the group. The group includes primitive arc magmas that have Mg# >64, Cr >400 ppm and Ni >200 ppm, and which represent mantle-derived melts that have been minimally modified by fractional crystallization. Two primitive arc magmas with Mg# 69.4 and 74.2 were documented by Haase *et al.* (2012) from Lion's Rump, King George Island, probably the Low Head Member locality described by Smellie *et al.* (1998). One sample from the Biscoe Islands has a Mg# of 65.0, and 351 ppm Cr and 144 ppm Ni (British Antarctic Survey unpublished data), and is marginally more fractionated than a primitive arc magma.

Zr abundances in the group increase with increasing silica in mafic and intermediate compositions, with Zr abundances reaching *c.* 300–400 ppm at *c.* 65–70 wt% SiO_2, with Zr decreasing in silicic compositions (Fig. 3b). This is consistent with zircon fractionation becoming important in the silicic magmas, as is normal for calc-alkaline series. There is a lower Zr trend with up to *c.* 250 ppm Zr at 65 wt% SiO_2 that is of uncertain origin. The group plots in the typical arc magmas field in the Sr/Y v. Y diagram that is used to discriminate typical arc magmas from adakites (Fig. 3c). The group has Al_2O_3 abundances that are mostly around 15–17 wt% at up to *c.* 60 wt% SiO_2, with a gradual decline in more silicic compositions (Fig. 4a). This is consistent with plagioclase being a major fractionating phase in basaltic andesitic to silicic magmas. Some basaltic to andesitic samples have Al_2O_3 abundances greater than 17.5 wt% and are high-Al compositions, which is likely to be a result of plagioclase accumulation. In the FeO^T v. MgO diagram (Fig. 4b), the group follows a general calc-alkaline trend with constant FeO^T with declining MgO in mafic compositions, absence of iron-enrichment in intermediate compositions, and declining FeO^T with declining MgO in intermediate–silicic compositions. Some samples have

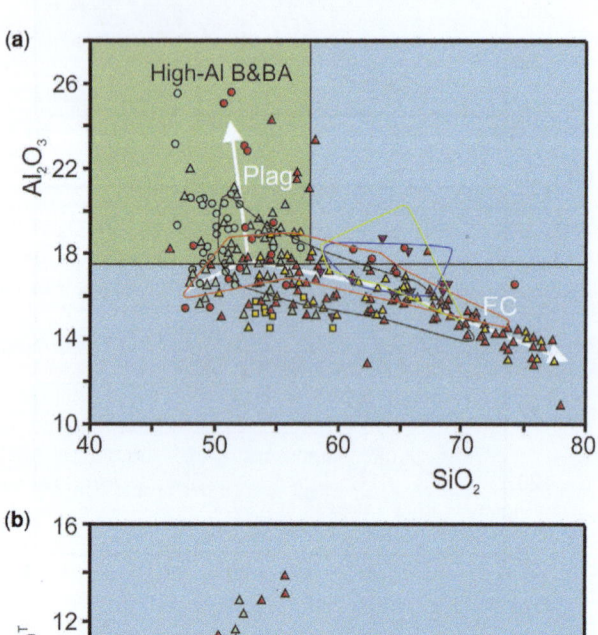

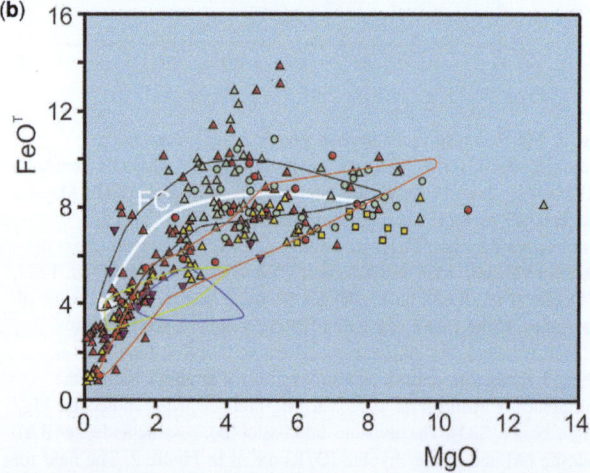

Fig. 4. Major element (wt%) plots for rocks of the Antarctic Peninsula volcanic arc. (**a**) Al_2O_3 v. SiO_2. The typical calc-alkaline fractional crystallization trend (FC) and plagioclase accumulation trends (Plag) are depicted by white arrows. The field for high-Al basalts and basaltic andesites (B&BA) has >17.5 wt% Al_2O_3. (**b**) Total iron as FeO v. MgO. The typical calc-alkaline fractional crystallization trend (white arrow, FC) shows weak iron-enrichment peaking at *c.* 4 wt% MgO and 9 wt% FeO. Symbols and fields for comparative data are as in Figure 3.

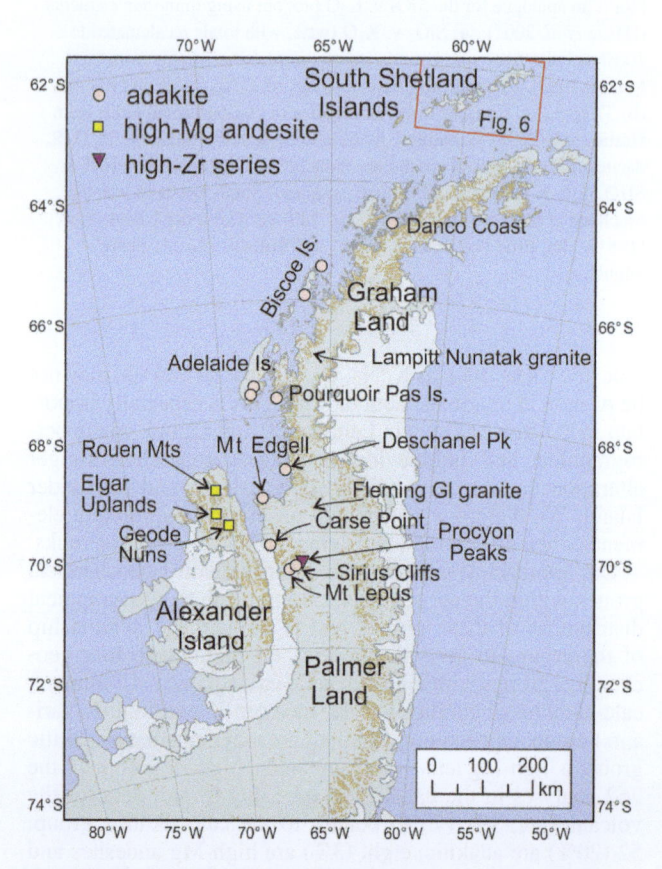

Fig. 5. Map of the Antarctic Peninsula showing the locations of adakitic, high-Mg andesite and high-Zr samples. The calc-alkaline group is widespread and omitted for clarity.

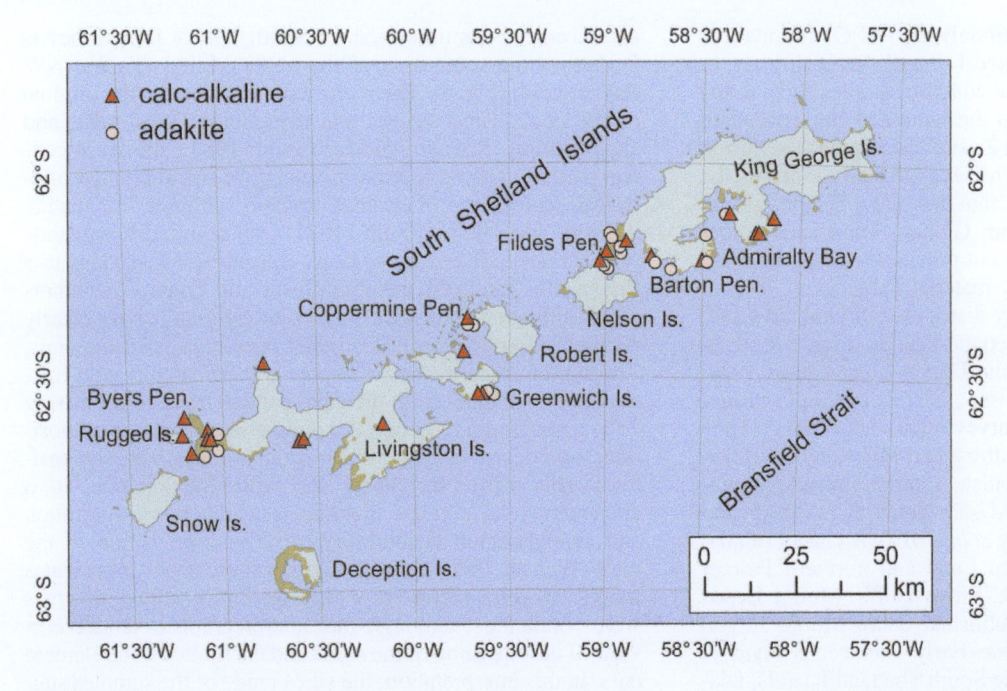

Fig. 6. Map of the South Shetland Islands showing the locations of calc-alkaline and adakitic samples. No high-Mg andesite or high-Zr samples are known from the South Shetland Islands.

>11 wt% FeO^T and plot well above the general trend. The reason for this is uncertain but it may be a result of an accumulation of pyroxene in thick lavas or sills mistakenly sampled as lavas. Incompatible trace element abundances and ratios are consistent with tholeiitic (low-K) and calc-alkaline (medium-K) compositions in the South Shetland Islands, and calc-alkaline to high-K compositions in the Graham Land, Palmer Land and Alexander Island (Saunders et al. 1980; Burn 1981; Saunders and Tarney 1982; Jwa et al. 1992; McCarron and Smellie 1998; Machado et al. 2005; Lee et al. 2008; Haase et al. 2012).

High-Mg andesite

This group has high Si, Mg, Mg# and compatible trace elements relative to 'normal' orogenic andesites, and occurs only on Alexander Island. In this chapter we follow the usage of McCarron and Smellie (1998) with high-Mg andesites with >53 wt% SiO$_2$, Mg# >60 (Fig. 3a), Ni >75 ppm and Cr >250 ppm. The group plots along the calc-alkaline trend in Figure 3b and within the field of typical arc magmas in the Sr/Y v. Y diagram (Fig. 3c). The group has low Al$_2$O$_3$ and FeO^T relative to most calc-alkaline group magmas (Fig. 4). The group has generally low Zr/Y and Zr relative to the associated Alexander Island calc-alkaline samples (Fig. 7), indicating a more depleted source. The high-Mg andesite group has SiO$_2$ 53.4–59.6 wt%, MgO 6.01–9.01 wt%, FeO^T/MgO 0.81–1.13, TiO$_2$ 0.62–0.83 wt%, CaO 7.00–10.26 wt% and Al$_2$O$_3$ 14.5–15.9 wt%, and are compositionally similar to high-Mg andesites of the Setouchi Volcanic Belt, SW Japan, and the western Aleutians (Tatsumi 2006). Incompatible trace element abundances and ratios of the Alexander Island high-Mg andesites are similar to those of the associated normal calc-alkaline to high-K volcanic rocks (McCarron and Smellie 1998). The group occurs only at Elgar Uplands, Geode Nunataks and Rouen Mountains, Alexander Island (Fig. 5). The high-Mg andesites form parts of the same volcanic formations as the calc-alkaline volcanics, and there is no known age difference (McCarron 1997; McCarron and Millar 1997; McCarron and Smellie 1998).

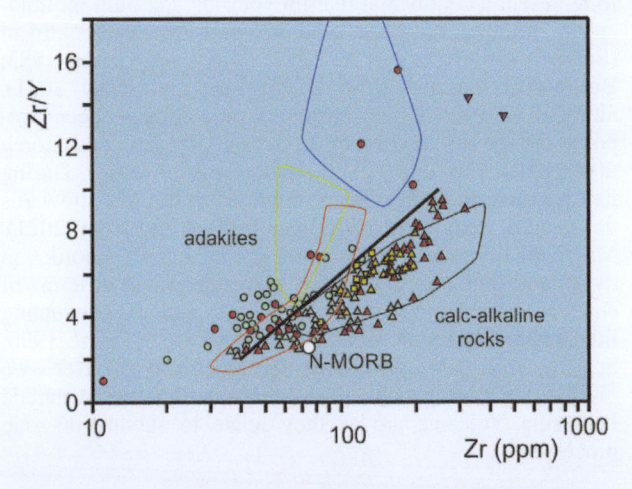

Fig. 7. Zr/Y v. Zr (ppm) plot for mafic rocks of the Antarctic Peninsula volcanic arc. Only samples with <60 wt% SiO$_2$ are plotted. Most data are by XRF. Symbols and fields for comparative data are as in Figure 3. The average N-MORB value is from Sun and McDonough (1989). The line dividing most adakites from most calc-alkaline rocks in the Antarctic Peninsula volcanic arc is from (40, 2) to (250, 10).

Adakitic group

This group is defined by high Sr/Y ratios and low Y values as shown in Figure 3c, following discriminations of adakites from typical arc magmas by Atherton and Petford (1993) and Defant and Drummond (1993). Adakites occur in many volcanic arcs and are characterized by low Y, which is immobile in low-temperature alteration of volcanic rocks, as well as by high Sr/Y, which may be modified during alteration by mobility of Sr. Typical adakites are characterized by low Yb and Sc, and high La/Yb, Sr and Eu, and also have Al$_2$O$_3$ >15 wt%, SiO$_2$ >56 wt% and MgO <3 wt% (Defant and Drummond 1990, 1993; Atherton and Petford 1993).

Adakitic rocks have not, so far as we are aware, previously been identified within the Antarctic Peninsula volcanic arc. The adakitic rocks of the arc have SiO$_2$ abundances of 46.9–74.3 wt% and Mg# 31.5–71.2, ranging from compositions close to those of primitive arc magmas to rhyolites (Fig. 3a). Many of the compositions, notably from the South Shetland Islands, are mafic. The adakitic group has low Zr and Y abundances, lower than typical volcanic arc magmas (Fig. 3b, c).

The group has variable and commonly high Al_2O_3 abundances of 15.4–25.5 wt%, and most have 1–9 wt% MgO (Fig. 4). In the Zr/Y v. Zr plot (Fig. 7), the adakitic samples form a distinct group with higher Zr/Y at the same Zr value compared to the calc-alkaline group. Both Zr and Y are immobile during low-grade metamorphism, and the separate field formed by the group in this diagram confirms that they are a distinct group. Analyses of adakitic rocks from Graham Land and Palmer Land from where adakites have not previously been identified are presented in Supplementary material Table S1.

The adakitic rocks are widely distributed in Graham Land, Palmer Land and the South Shetland Islands (Figs 5 & 6). In Graham Land, they occur on the Danco Coast (West 1974; Saunders *et al.* 1980), in the Biscoe Islands and on Pourquoi Pas Island (British Antarctic Survey unpublished data). They also occur within the Buchia Buttress and Milestone Bluff formations of Adelaide Island (British Antarctic Survey unpublished data) which are zircon U–Pb dated at *c.* 149.5 and *c.* 113.9 Ma, respectively (Riley *et al.* 2011). A cluster of adakitic rocks in southern Graham Land and northern Palmer Land includes Deschanel Peak, Sirius Cliffs, Mount Lepus (British Antarctic Survey unpublished data), Mount Edgell (Davies 1984) and Late Jurassic–Early Cretaceous lavas at Carse Point (Smith 1987). In the South Shetland Islands, adakitic rocks are identified at the Byers Peninsula, where they appear to be lavas and minor intrusions within the Cerro Negro Formation (Machado *et al.* 2005) which is Ar–Ar dated at *c.* 120–119 Ma (Hathway and Lomas 1998). Adakitic rocks on Greenwich Island and the Coppermine Peninsula, Robert Islands (Machado *et al.* 2005; Haase *et al.* 2012) are part of the *c.* 83–78 Ma Coppermine Formation (Smellie *et al.* 1984). Adakitic rocks in the Fildes Peninsula Group are from a succession Ar–Ar dated at *c.* 56–52 Ma (Haase *et al.* 2012; Gao *et al.* 2018) from the Stansbury Peninsula, Nelson Island and the northeastern part of Fildes Peninsula, King George Island (Machado *et al.* 2005; Haase *et al.* 2012). On both theBarton and Potter peninsulas, King George Island, adakitic rocks occur within successions Ar–Ar dated at *c.* 47–44 Ma (Wang *et al.* 2009; Haase *et al.* 2012). Around Admiralty Bay, King George Island, adakitic rocks occur at Point Thomas and Point Hennequin according to the data of Haase *et al.* (2012). The former is likely to be a lava from the Ezcurra Inlet Group, zircon U–Pb dated at 48.9 ± 0.7 Ma (Nawrocki *et al.* 2010). The latter is a lava of the Point Hennequin Group, zircon U–Pb dated at 48.7 ± 0.6 Ma (Nawrocki *et al.* 2011). Within the Baranowski Glacier Group at Red Hill, the upper Zamek Formation consists entirely of adakitic rocks, while the lower Llano Point Formation is non-adakitic, according to the data of Mozer *et al.* (2015). The Llano Formation is Ar–Ar dated at *c.* 52–50.8 Ma (Nawrocki *et al.* 2011) and the Zamek Formation is K–Ar dated at *c.* 46–42 Ma (Mozer *et al.* 2015).

The ages of adakitic rocks in the Antarctic Peninsula range from Late Jurassic or Early Cretaceous to Eocene. In the South Shetland Islands, they are present in three age groups at *c.* 120 Ma (Byers Peninsula), *c.* 83–78 Ma (Coppermine Formation) and *c.* 56–42 Ma (King George Island). In the case of the Baranowski Glacier Group, where there is good stratigraphic control on sampling (Mozer *et al.* 2015), the data suggest that a 46–42 Ma adakitic volcano overlies earlier normal calc-alkaline lavas of the Llano Point Formation.

High-Zr group

This group of 11 samples forms a high-Zr trend in the SiO_2 v. Zr plot (Fig. 3b). The samples have Zr abundances of 438–627 ppm Zr and 58.6–68.9 wt% SiO_2 (see also Supplementary material Table S1). The analyses are from a series of volcanic rocks described by Smith (1987) from Procyon Peaks, on the south side of the upper Millett Glacier, NW Palmer Land (Fig. 6). The rocks were described in the original report as andesitic, dacitic and rhyodacitic lavas, tuffs, and agglomerates. Petrographically (Smith 1987), the lavas contain partially altered feldspar phenocrysts and altered groundmasses consisting of quartz, epidote, chlorite, magnetite, ilmenite and calcite. Smith (1987) also reported hornblende phenocrysts replaced by epidote, chlorite and Fe-oxides in a rhyodacite. Regional metamorphism and contact metamorphism alteration associated with intruding granite have clearly resulted in replacement of original ferromagnesian minerals. The trend formed by these samples of increasing Zr with fractionation is distinct from the calc-alkaline trend of increasing Zr to a maximum of about 300 ppm in intermediate compositions, and decreasing Zr with fractionation in silicic compositions (Fig. 3b). Increasing Zr with fractionation is a characteristic of alkaline magmas in which zircon saturation and crystallization is minimized by the composition of the melt (Watson 1979). It is likely that these rocks represent a mildly alkaline series in which the more silicic magmas were peralkaline, although this interpretation is tentative in view of the alteration of the rocks and the limited trace element data. In this interpretation, the silica range of the samples suggests that they are benmoreites and trachytes. The samples have a wide range in Mg# for a relatively small range in SiO_2 (Fig. 3a), and the lower Mg#, more fractionated, samples in the group have lower Si than the calc-alkaline series, consistent with trachytic compositions. The samples have high Y abundances of 20–40 ppm and all plot within the non-adakite field in the Sr/Y v. Y diagram (Fig. 3c). They have La abundances (measured by XRF) of up to 49 ppm and chondrite-normalized La_N/Y_N (using chondrite values of Sun and McDonough 1989) of up to 11.9 (Smith 1987), indicating that they are moderately light REE (LREE)-enriched. LREE enrichment is a characteristic of peralkaline magmas. The age of the high-Zr volcanism is unknown.

Discussion

The Antarctic Peninsula volcanic arc was active from Late Jurassic or Early Cretaceous times until the Early Miocene, is some 1300 km long and contains successions up to *c.* 1.5 km thick (Leat and Riley 2021). It is widely thought to represent a continental margin volcanic arc, built on lithosphere forming the edge of the Gondwana supercontinent (Suárez 1976; Tarney *et al.* 1982; Storey and Garrett 1985; Burton-Johnson and Riley 2015; Leat and Riley 2021), although alternative allochthonous models have been proposed for the early evolution of the arc (Vaughan and Storey 2000; Ferraccioli *et al.* 2006; Vaughan *et al.* 2012). During its evolution, the volcanic arc migrated to the west (towards the trench) along most of its length (Leat and Riley 2021). Magmatic diversity and structural events, mainly recorded in dykes and plutonic rocks, have been interpreted in terms of changes in subduction dynamics and age of subducting lithosphere (Alabaster and Storey 1990; Storey *et al.* 1996; Scarrow *et al.* 1997; McCarron and Smellie 1998). Here, we discuss the four geochemical groups within the Antarctic Peninsula Volcanic arc as they relate to subduction-zone processes.

Calc-alkaline group

This group ranges from low-K to high-K in composition, and represents typical volcanic arc magmatism. In Figures 3 and 4, the group is similar to the calc-alkaline volcanic arc rocks from

Chile but appears to be more bimodal with respect to SiO_2. In Figure 7, the Antarctic Peninsula group includes samples with lower Zr abundances than the Chile group, consistent with subgroups with low-K compositions.

The generation of the Antarctic Peninsula mafic magmas, including rare primitive magmas, can be assumed to have taken place by partial melting of peridotite in the mantle wedge. The generation of the more silicic magmas, including rhyolites, can be assumed to have involved both fractional crystallization and assimilation and partial melting of crust but insufficient isotopic and trace element data are available to investigate their relative roles. Detailed investigations of mantle source and subduction component compositions have only been carried out for the low-K to medium-K South Shetland Islands sector of the arc. Machado et al. (2005) demonstrated that most South Shetland Islands mafic rocks are products of variable enrichment of a mantle wedge similar to N-MORB source mantle by melts or fluids from subducted sediment, with a possible input from altered subducted oceanic plate. Lee et al. (2008) suggested that the magmas of the South Shetland arc can be modelled by relatively constant subduction input to a heterogeneous mantle wedge similar to a N-MORB source. Haase et al. (2012) modelled the magmas to have been derived from a depleted mantle wedge modified by hydrous fluids from subducted ocean crust and subducted sediments. In Figure 7, lavas of the calc-alkaline group from the Antarctic Peninsula volcanic arc plot overwhelmingly at higher Zr/Y than average N-MORB, suggesting that the mantle wedge was not generally strongly depleted. The lowest Zr abundances and lowest Zr/Y ratios relative to N-MORB are from the South Shetland Islands, consistent with at least local depletion of their sources. Many samples from Alexander Island, Graham Land and Palmer Land have significantly higher Zr/Y than N-MORB, suggesting instead that the mantle source was non-depleted. In the case of Alexander Island, a non-depleted source is consistent with their medium- to high-K compositions (Fig. 2) and other trace element characteristics (McCarron and Smellie 1998).

High-Mg andesite group

The high-Mg andesites were erupted as part of the main Early Tertiary volcanic sequences on Alexander Island. These sequences are volumetrically dominated by the calc-alkaline group. The younger age of this volcanism compared to the youngest known volcanism in western Palmer Land suggests that the volcanic arc axis shifted west from Palmer Land to Alexander Island between 107 and 80 Ma (Leat and Riley 2021). It is questionable whether the Early Tertiary volcanism on Alexander Island tectonically represented a forearc as previously suggested (McCarron and Millar 1997; McCarron and Smellie 1998), as the large volume of the subduction-generated magmatism suggests that this was the main volcanic axis at the time or eruption. Generation of the high-Mg andesitic magmas by partial melting of depleted, hydrous forearc mantle as a result of heating related to subduction of young, hot oceanic lithosphere was proposed by McCarron and Smellie (1998), following models for boninite and high-Mg andesite genesis (e.g. Crawford et al. 1989). In a more recent revision of the magma genesis of high-Mg andesites, Tatsumi (2006) suggested that the examples from the Setouchi Volcanic Belt, southwestern Japan, were generated by partial melting of young (<15 myr old), hot oceanic lithosphere and hydrous equilibration of the melts with mantle. Western Aleutian high-Mg andesites are also interpreted to contain a slab melt component and equilibrated in mantle in hydrous conditions resulting from oblique subduction and a slow subduction path (Yogodzinski et al. 1995). We suggest a similar model of slab partial melting and melt equilibration in the mantle wedge for the Alexander Island high-Mg andesites.

Adakitic group

The characteristics of adakites, including low Y, low HREE, high Sr/Y and high Al_2O_3, are consistent with the involvement of garnet as a residual or fractionating phase. Adakites are minor components of many volcanic arcs including the Aleutians (Kay 1978; Yogodzinski et al. 1995), Cascades (Defant and Drummond 1993), Kamchatka (Yogodzinski et al. 2001), Andes (Atherton and Petford 1993; Stern and Kilian 1996) and others (Martin et al. 2005; Castillo 2012). Several petrogenetic mechanisms have been invoked to explain their origin, including partial melting of eclogite-facies subducting slabs (Defant and Drummond 1990, 1993; Yogodzinski et al. 1995, 2001), partial melting of garnetiferous mafic lower crust (Atherton and Petford 1993), partial melting of mantle wedge peridotite that has been modified by partial melts of subducting slabs (Martin et al. 2005) and by fractional crystallization or assimilation of crust combined with fractionation of normal basaltic arc magmas (Castillo et al. 1999; Petrone and Ferrari 2008). Wareham et al. (1997) thought that Triassic–Cretaceous age plutonic adakites they identified in northern Palmer Land, Antarctic Peninsula, were generated by partial melting of garnetiferous mafic lower crust.

The Antarctic Peninsula adakitic group has low Zr abundances and high Zr/Y, suggesting garnet in the source (Fig. 7). The group has a large range in SiO_2 abundances (Fig. 3a; see also Supplementary material Table S1), and most samples, notably from the South Shetland Islands, have less than the 56 wt% SiO_2 traditionally suggested minimum value for adakites (Defant and Drummond 1990). Adakitic samples from Carse Point, Palmer Land (Smith 1987), the Danco Coast, Graham Land (West 1974; Saunders et al. 1980), and Nelson, Robert and King George islands, South Shetland Islands (Machado et al. 2005; Haase et al. 2012), have a Mg# >60 (Fig. 3a), which suggests that these magmas were derived by melting of mantle, not basaltic slab or lower crust. Martin et al. (2005) and Castillo (2012) described both low-SiO_2 and high-SiO_2 varieties of adakites. High-SiO_2 adakites have compositions consistent with being silicic slab melts. In contrast, low-SiO_2 adakites were interpreted to be melts of a mantle wedge that has consumed partial melts of slabs, and the Antarctic Peninsula adakitic group appears to be of this type. The adakites from the austral volcanic zone of the Chilian Andes shown in Figure 3, 4 and 7 are mostly high-SiO_2 varieties but include a low-SiO_2 (and high Mg#) group from Cook Island (Stern and Kilian (1996). The Mindanao adakites, in the Philippines, include high- and low-Si types (Sajona et al. 1993, 1996). The samples from Camiguin, the Philippines, are transitional between calc-alkaline and low-SiO_2 adakite compositions (Castillo et al. 1999; Castillo 2012). In the Mg# v. SiO_2 plot (Fig. 3a), the Chile and Mindanao adakites are mostly dissimilar to the Antarctic Peninsula examples which have lower SiO_2 contents. However, the field for low-SiO_2 Camiguin samples is a better match for low-SiO_2 Antarctic Peninsula adakites. In Figure 4, the Antarctic Peninsula adakites have a similar distribution of Al and Fe to the other adakite groups. In Figure 7, the adakite groups have higher Zr/Y relative to Zr content than calc-alkaline rocks from Chile and the Antarctic Peninsula, with many of adakites from the Philippines plotting close to low-Zr Antarctic Peninsula examples.

Because of the large age range of the adakites in the Antarctic Peninsula, it is not possible to speculate yet on the tectonic controls on their generation, and more work is needed. Further investigation is required of the Antarctic Peninsula adakitic rocks and their relationship to the high-Mg andesites.

High-Zr group

We interpret the high-Zr group of rocks from Procyon Peaks, Palmer Land, as being a peralkaline series, and suggest that compositions were originally benmoreitic and trachytic. Such compositions are an important minor component in some magmatic arcs. Examples include Mayor Island, Bay of Plenty, New Zealand (Houghton *et al.* 1992), British Columbia (Bevier *et al.* 1979), the D'Entrecasteaux Islands, Papua New Guinea (Stolz *et al.* 1993), Oki-Dogo Island, Sea of Japan (Uto *et al.* 1994), and the Mexico volcanic belt (Nelson and Hegre 1990). In many cases, the peralkaline magmatism is associated with extension within or behind volcanic arcs. The Palmer Land high-Zr volcanism is situated within the general Cretaceous Palmer Land volcanic arc terrain (Thomson *et al.* 1983; Smith 1987; Ferraccioli *et al.* 2006; Leat and Riley 2021). However, neither its absolute age nor its age relative to other elements of the arc is known (Smith 1987).

A peralkaline granite to the south of the Fleming Glacier, northern Palmer Land (Davies 1984), may represent similar magmatism. The granite has not been isotopically dated but is one of a group of vuggy, high-level granites in the transition zone between southern Graham Land and northern Palmer Land that are Rb–Sr dated at 96 ± 2–71 ± 9 Ma (Leat *et al.* 2009). A further undated riebeckite-bearing alkali granite at Lampitt Nunatak, inland of the Loubet Coast (Goldring 1962), is situated in a marginally rear-arc position relative to the distribution of the volcanic arc in southern Graham Land (Fig. 5).

These minor occurrences of peralkaline silicic magmatism may be associated with extension of the arc or may represent a back-arc system magmatism that was emplaced after the westward shift of the axis of volcanism that took place between 107 and 80 Ma in the case of Palmer Land (Leat and Riley 2021). They may be associated with possible Late Cretaceous transpression and intrusion of compositionally varied mafic dykes including OIB-like varieties during the Late Cretaceous (Scarrow *et al.* 1997, 1998).

Conclusions

Most of the Jurassic–Early Tertiary age volcanic rocks of the west coast of the Antarctic Peninsula are related to a subduction zone that was active from Late Jurassic or Early Cretaceous times. Sections of the Antarctic Peninsula volcanic arc that was the major volcanic product of the subduction occur in western Graham Land, western Palmer Land, the South Shetland Islands and Alexander Island.

Mesozoic–Early Tertiary age volcanic rocks of the Antarctic Peninsula have been altered by hydrothermal activity related to the arc magmatism, contact metamorphism in proximity to plutons and regional/burial metamorphism.

Jurassic and Cretaceous basalts within the LeMay Group accretionary complex of Alexander Island have compositions consistent with origins as ocean-floor basalts and seamounts that were scraped from the subducting plate. Compositions include N-MORB, E-MORB and OIB. Isotopic compositions indicate that the magmas were derived from the Pacific mantle reservoir. The basalts were incorporated within the accretionary complex during the subduction of the Pacific Plate that generated the Cretaceous–Tertiary Antarctic Peninsula volcanic arc.

A Late Jurassic bimodal mafic–silicic group of volcanic rocks and sills within the Fossil Bluff Group of Alexander Island include OIB-like basalts. They are interpreted to have been related to extension at the continental margin before initiation of the subduction that produced the Early Cretaceous–Tertiary volcanic arc.

Four geochemical groups are recognized within the Antarctic Peninsula volcanic arc: a calc-alkaline group, a high-Mg andesite group, an adakitic group and a high-Zr group.

A dominant mafic–silicic calc-alkaline group represents typical volcanic arc magmatism. Compositions range from low- to high-K and the mantle wedge was variably depleted. In the South Shetland Island sector, magma genesis by addition of slab-derived components to the mantle wedge source has been modelled.

A high-Mg andesite group is known only in Alexander Island where they occur within a dominantly calc-alkaline succession. They do not have the high Sr/Y ratios of adakites, and may have been generated by partial melting of hot or slowly subducting oceanic lithosphere and hydrous equilibration of the melts with mantle.

An adakitic group is dominated by low-SiO_2 varieties, some of which have a Mg# >60, suggesting derivation from a mantle source, and not by partial melting of basaltic slab or lower crust. They have low Zr abundances and high Zr/Y, suggesting the presence of garnet as a residual phase. They can be interpreted as partial melts of mantle-wedge peridotite that has consumed partial melts of subducting slab.

High-Zr intermediate–silicic volcanic rocks at Procyon Peaks, NW Palmer Land have Zr abundances up to 627 ppm Zr, and are interpreted to have been benmoreites and peralkaline trachytes. They are interpreted to have been erupted during a phase of extension of the arc or in a back-arc position relative to the active volcanic arc.

Acknowledgements We thank Mari Whitelaw for providing access to, and downloading data from, the British Antarctic Survey geochemical database. We thank Kurt Panter, Maria Petrone and an anonymous reviewer for comments on the paper.

Author contributions **PTL**: conceptualization (lead), formal analysis (lead), investigation (lead), writing – original draft (lead); **TRR**: project administration (lead), resources (lead), validation (equal), writing – review & editing (lead).

Funding This research received no specific grant from any funding agency in the public, commercial, or not-for-profit sectors.

Data availability Datasets generated during this review were derived from published work and the British Antarctic Survey geochemical database held by the UK Polar Data Centre https://www.bas.ac.uk/data/uk-pdc/.

References

Alabaster, T. and Storey, B.C. 1990. Antarctic Peninsula continental magnesian andesites: indicators of ridge–trench interaction during Gondwana break-up. *Journal of the Geological Society, London*, **147**, 595–589, https://doi.org/10.1144/gsjgs.147.4.0595

Armstrong, D.C. and Willan, R.C.R. 1996. Orthomagmatic quartz and post-magmatic carbonate veins in a reported porphyry copper deposit, Andean Intrusive Suite, Livingston Island, South

Shetland Islands. *Mineralium Deposita*, **31**, 290–306, https://doi.org/10.1007/BF02280793

Atherton, M.P. and Petford, N. 1993. Generation of sodium-rich magmas from newly underplated basaltic crust. *Nature*, **362**, 144–146, https://doi.org/10.1038/362144a0

Ayling, M.E. 1984. *The Geology of Parts of Central West Palmer Land*. British Antarctic Survey Scientific Reports, **105**.

Barbeau, D.L., Davis, J.T., Murray, K.E., Valencia, V., Gehrels, G.E., Zahid, K.M. and Gombosi, D.J. 2009. Detrital-zircon geochronology of the metasedimentary rocks of north-western Graham Land. *Antarctic Science*, **22**, 65–78, https://doi.org/10.1017/S095410200999054X

Barker, P.F. 1982. The Cenozoic subduction history of the Pacific margin of the Antarctic Peninsula: ridge crest–trench interactions. *Journal of the Geological Society, London*, **139**, 787–801, https://doi.org/10.1144/gsjgs.139.6.0787

Barry, T.L., Davies, J.H. et al. 2017. Whole-mantle convection with tectonic plates preserves long-term global patterns of upper mantle geochemistry. *Scientific Reports*, **7**, 1870, https://doi.org/10.1038/s41598-017-01816-y

Bevier, M.L., Armstrong, M.L. and Souther, J.G. 1979. Miocene peralkaline volcanism in west-central British Columbia – its temporal and plate-tectonics setting. *Geology*, **7**, 389–392, https://doi.org/10.1130/0091-7613(1979)7<389:MPVIWB>2.0.CO;2

Birkenmajer, K. 1989. A guide to Tertiary geochronology of King George Island, West Antarctica. *Polish Polar Research*, **10**, 555–579.

Birkenmajer, K. 1994. Evolution of the Pacific margin of the northern Antarctic Peninsula: an overview. *Geologische Rundschau*, **83**, 309–321.

Burn, R.W. 1981. Early Tertiary calc-alkaline volcanism on Alexander Island. *British Antarctic Survey Bulletin*, **53**, 175–193.

Burn, R.W. 1984. *The Geology of the LeMay Group, Alexander Island*. British Antarctic Survey Scientific Reports, **109**.

Burton-Johnson, A. and Riley, T.R. 2015. Autochthonous v. accreted terrane development of continental margins: a revised *in situ* tectonic history of the Antarctic Peninsula. *Journal of the Geological Society, London*, **172**, 822–835, https://doi.org/10.1144/jgs2014-110

Castillo, P.R. 2012. Adakite petrogenesis. *Lithos*, **134–135**, 304–316, https://doi.org/10.1016/j.lithos.2011.09.013

Castillo, P.R., Janney, P.E. and Solidum, R.U. 1999. Petrology and geochemistry of Camiguin Island, southern Philippines: insights to the source of adakites and other lavas in a complex arc setting. *Contributions to Mineralogy and Petrology*, **134**, 33–51, https://doi.org/10.1007/s004100050467

Castillo, P., Fanning, C.M., Hervé, F. and Lacassie, J.P. 2016. Characterisation and tracing of Permian magmatism in the south-western segment of the Gondwana margin: U–Pb age, Lu–Hf and O isotopic compositions of detrital zircons from metasedimentary complexes of northern Antarctic Peninsula and western Patagonia. *Gondwana Research*, **36**, 1–13, https://doi.org/10.1016/j.gr.2015.07.014

Crawford, A.J., Falloon, T.J. and Green, D.H. 1989. Classification, petrogenesis and tectonic setting of boninites. *In*: Crawford, A.J. (ed.) *Boninites and Related Rocks*. Unwin Hyman, London, 1–49.

Curtis, R. 1966. *The Petrology of the Graham Coast, Graham Land*. British Antarctic Survey Scientific Reports, **50**.

Davies, T.G. 1984. *The Geology of Part of Northern Palmer Land*. British Antarctic Survey Scientific Reports, **103**.

Defant, M.J. and Drummond, M.S. 1990. Derivation of some modern arc magmas by melting of young subducted lithosphere. *Nature*, **347**, 662–665, https://doi.org/10.1038/347662a0

Defant, M.J. and Drummond, M.S. 1993. Mount St. Helens: potential example of the partial melting of the subducted lithosphere in a volcanic arc. *Geology*, **21**, 547–550, https://doi.org/10.1130/0091-7613(1993)021<0547:MSHPEO>2.3.CO;2

Dewar, G.J. 1970. The geology of Adelaide Island. *British Antarctic Survey Scientific Reports*, **57**.

D'Orazio, M., Innocenti, F. et al. 2003. The Quaternary calc-alkaline volcanism of the Patagonian Andes close to the Chile triple junction: geochemistry and petrogenesis of volcanic rocks from the Cay and Maca volcanoes (*c.* 45°S, Chile). *Journal of South American Earth Sciences*, **16**, 219–242, https://doi.org/10.1016/S0895-9811(03)00063-4

Doubleday, P.A. and Tranter, T.H. 1992. Modes of formation and accretion of oceanic material in the Mesozoic fore-arc of central and southern Alexander Island, Antarctica: a summary. *In*: Yoshida, Y., Kaminuma, K. and Shiraishi, K. (eds) *Recent Progress in Antarctic Earth Science*. Terra Scientific (TERRAPUB), Tokyo, 377–382.

Doubleday, P.A. and Tranter, T.H. 1994. Deformation mechanism paths for oceanic rocks during subduction and accretion: the Mesozoic forearc of Alexander Island, Antarctica. *Journal of the Geological Society, London*, **151**, 543–554, https://doi.org/10.1144/gsjgs.151.3.0543

Doubleday, P.A., Macdonald, D.I.M. and Nell, P.A.R. 1993. Sedimentology and structure of the trench-slope to forearc basin transition in the Mesozoic of Alexander Island, Antarctica. *Geological Magazine*, **130**, 737–754, https://doi.org/10.1017/S0016756800023128

Doubleday, P.A., Leat, P.T., Alabaster, T., Nell, P.A.R. and Tranter, T.H. 1994. Allochthonous oceanic basalts within the Mesozoic accretionary complex of Alexander Island, Antarctica: remnants of proto-Pacific oceanic crust. *Journal of the Geological Society, London*, **151**, 65–78, https://doi.org/10.1144/gsjgs.151.1.0065

Elliott, M.H. 1974. Stratigraphy and sedimentary petrology of the Ablation Point area, Alexander Island. *British Antarctic Survey Bulletin*, **39**, 87–113.

Ferraccioli, F., Jones, P.C., Vaughan, A.P.M. and Leat, P.T. 2006. New aerogeophysical view of the Antarctic Peninsula: More pieces, less puzzle. *Geophysical Research Letters*, **33**, L05310, https://doi.org/10.1029/2005GL024636

Gao, L., Zhao, Y., Yang, Z., Liu, J., Liu, X., Zhang, S.-H. and Pei, J. 2018. New paleomagnetic and ^{40}Ar–^{39}Ar geochronological results for the South Shetland Islands, West Antarctica, and their tectonic implications. *Journal of Geophysical Research: Solid Earth*, **123**, 4–30, https://doi.org/10.1002/2017JB014677

Goldring, D.C. 1962. *The Geology of the Loubet Coast, Graham Land*. British Antarctic Survey Scientific Reports, **36**.

Haase, K.M. and Beier, C. 2021. Bransfield Strait and James Ross Island: petrology. *Geological Society, London, Memoirs*, **55**, https://doi.org/10.1144/M55-2018-37

Haase, K.M., Beier, C., Fretzdorff, S., Smellie, J.L. and Garbe-Schönberg, D. 2012. Magmatic evolution of the South Shetland Islands, Antarctica, and implications for continental crust formation. *Contributions to Mineralogy and Petrology*, **163**, 1103–1119, https://doi.org/10.1007/s00410-012-0719-7

Harrison, S.M. and Piercy, B.A. 1990. The evolution of the Antarctic Peninsula magmatic arc: evidence from northwestern Palmer Land. *Geological Society of America Special Papers*, **241**, 9–25, https://doi.org/10.1130/SPE241-p9

Hastie, A.R., Kerr, A.C., Pearce, J.A. and Mitchell, S.F. 2007. Classification of altered volcanic island arc rocks using immobile trace elements: development of the Th–Co discrimination diagram. *Journal of Petrology*, **48**, 2341–2357, https://doi.org/10.1093/petrology/egm062

Hathway, B. and Lomas, S.A. 1998. The upper Jurassic–Lower Cretaceous Byers Group, South Shetland Islands, Antarctica: revised stratigraphy and regional correlations. *Cretaceous Research*, **19**, 43–67, https://doi.org/10.1006/cres.1997.0095

Hawkes, D.D. 1982. Nature and distribution of metalliferous mineralization in the northern Antarctic Peninsula. *Journal of the Geological Society, London*, **139**, 803–809, https://doi.org/10.1144/gsjgs.139.6.0803

Hawkes, D.D. and Littlefair, M.J. 1981. The occurrence of molybdenum, copper, and iron mineralization in the Argentine Islands, West Antarctica. *Economic Geology*, **76**, 898–904, https://doi.org/10.2113/gsecongeo.76.4.898

Hervé, F., Faúndez, V., Brix, M. and Fanning, M. 2006. Jurassic sedimentation of the Miers Bluff Formation, Livingston Island,

Antarctica: evidence from SHRIMP U–Pb ages of detrital and plutonic zircons. *Antarctic Science*, **18**, 229–238, https://doi.org/10.1017/S0954102006000277

Hickey-Vargas, R., Moreno Roa, H., Lopez Escobar, L. and Frey, F.A. 1989. Geochemical variations in Andean basaltic and silicic lavas from the Villarrica–Lanin volcanic chain (39.5° S): an evaluation of source heterogeneity, fractional crystallization and crustal assimilation. *Contributions to Mineralogy and Petrology*, **103**, 361–386, https://doi.org/10.1007/BF00402922

Holdsworth, B.K. and Nell, P.A.R. 1992. Mesozoic radiolarian faunas from the Antarctic Peninsula: age, tectonic and palaeoceanographic significance. *Journal of the Geological Society, London*, **149**, 1003–1020, https://doi.org/10.1144/gsjgs.149.6.1003

Hole, M.J. 2021. Antarctic Peninsula: petrology. *Geological Society, London, Memoirs*, **55**, https://doi.org/10.1144/M55-2018-40

Hooper, P.R. 1962. *The Petrology of Anvers Island and Adjacent Islands*. Falkland Islands Dependencies Survey Scientific Reports, **34**.

Houghton, B.F., Weaver, S.D., Wilson, C.J.N. and Lanphere, M.A. 1992. Evolution of a Quaternary peralkaline volcano: Mayor Island, New Zealand. *Journal of Volcanology and Geothermal Research*, **51**, 217–236, https://doi.org/10.1016/0377-0273(92)90124-V

Jwa, Y.-J., Park, B.-K. and Kim, Y. 1992. Geochronology and geochemistry of the igneous rocks from Barton and Fildes peninsulas, King George Island: a review. *In*: Yoshida, Y., Kaminuma, K. and Shiraishi, K. (eds) *Recent Progress in Antarctic Earth Science*. Terra Scientific (TERRAPUB), Tokyo, 439–442.

Kay, R.W. 1978. Aleutian magnesian andesites: melts from subducted Pacific Ocean crust. *Journal of Volcanology and Geothermal Research*, **4**, 117–132, https://doi.org/10.1016/0377-0273(78)90032-X

Larter, R.D. and Barker, P.F. 1991. Effects of ridge crest–trench interaction on Antarctic–Phoenix spreading: forces on a young subducting plate. *Journal of Geophysical Research*, **96**, 19 583–19 607, https://doi.org/10.1029/91JB02053

Leat, P.T. and Riley, T.R. 2021. Antarctic Peninsula and South Shetland Islands: volcanology. *Geological Society, London, Memoirs*, **55**, https://doi.org/10.1144/M55-2018-52

Leat, P.T., Scarrow, J.H. and Millar, I.L. 1995. On the Antarctic Peninsula batholith. *Geological Magazine*, **132**, 399–4127, https://doi.org/10.1017/S0016756800021464

Leat, P.T., Flowerdew, M.J., Riley, T.R., Whitehouse, M.J., Scarrow, J.H. and Millar, I.L. 2009. Zircon U–Pb dating of Mesozoic volcanic and tectonic events in northwest Palmer Land and southwest Graham Land, Antarctica. *Antarctic Science*, **21**, 633–641, https://doi.org/10.1017/S0954102009990320

Lee, M.J., Lee, J.I., Choe, W.H. and Park, C.-H. 2008. Trace element and isotopic evidence for temporal changes of the mantle sources in the South Shetland Islands, Antarctica. *Geochemical Journal*, **42**, 219–219, https://doi.org/10.2343/geochemj.42.207

Le Maitre, R.W. (ed.) 1989. *A Classification of Igneous Rocks and Glossary of Terms*. Blackwell Scientific, Oxford, UK.

Macdonald, D.I.M., Leat, P.T., Doubleday, P.A. and Kelly, S.R.A. 1999. On the origin of fore-arc basins: new evidence from Alexander Island, Antarctica. *Terra Nova*, **11**, 186–193, https://doi.org/10.1046/j.1365-3121.1999.00244.x

Machado, A., Chemale, F.Jr, Conceição, R.V., Kawaskita, K., Morata, D., Oteíza, O. and Van Schmus, W.R. 2005. Modeling of subduction components in the genesis of the Meso-Cenozoic igneous rocks from the South Shetland arc, Antarctica. *Lithos*, **82**, 435–543, https://doi.org/10.1016/j.lithos.2004.09.026

Mahoney, J.J., Frei, R., Tejada, M.L.G., Mo, X.X., Leat, P.T. and Nähler, T.F. 1998. Tracing the Indian Ocean mantle domain through time: isotopic results from old west Indian, east Tethyan, and south Pacific Seafloor. *Journal of Petrology*, **39**, 1285–1306, https://doi.org/10.1093/petroj/39.7.1285

Maldonado, A., Larter, R.D. and Aldaya, F. 1994. Forearc tectonic evolution of the South Shetland margin, Antarctic Peninsula. *Tectonics*, **13**, 1345–1370, https://doi.org/10.1029/94TC01352

Martin, H., Smithies, R.H., Rapp, R., Moyen, J.-F. and Champion, D. 2005. An overview of adakite, tonalite–trondhjemite–granodiorite (TTG) and sanukitoid: relationships and some implications for crustal evolution. *Lithos*, **79**, 1–24, https://doi.org/10.1016/j.lithos.2004.04.048

Matthews, D.W. 1983*a*. The geology of Pourquoi Pas Island, northern Marguerite Bay, Graham Land. *British Antarctic Survey Bulletin*, **52**, 1–20.

Matthews, D.W. 1983*b*. The geology of Horseshoe and Lagotellerie islands, Marguerite Bay, Graham Land. *British Antarctic Survey Bulletin*, **52**, 125–154.

McCarron, J.J. 1997. A unifying lithostratigraphy of late Cretaceous–early Tertiary fore-arc volcanic sequences on Alexander Island, Antarctica. *Antarctic Science*, **9**, 209–220, https://doi.org/10.1017/S0954102097000266

McCarron, J.J. and Millar, I.L. 1997. The age and stratigraphy of fore-arc magmatism on Alexander Island, Antarctica. *Geological Magazine*, **134**, 507–522, https://doi.org/10.1017/S0016756897007437

McCarron, J.J. and Smellie, J.L. 1998. Tectonic implications of fore-arc magmatism and generation of high-magnesian andesites: Alexander Island, Antarctica. *Journal of the Geological Society, London*, **155**, 269–280, https://doi.org/10.1144/gsjgs.155.2.0269

Millar, I.L., Pankhurst, R.J. and Fanning, C.M. 2002. Basement chronology of the Antarctic Peninsula: recurrent magmatism and anataxis in the Palaeozoic Gondwana margin. *Journal of the Geological Society, London*, **159**, 145–157, https://doi.org/10.1144/0016-764901-020

Moncrieff, A.C.M. and Kelly, S.R.A. 1993. Lithostratigraphy of the uppermost Fossil Bluff Group (Early Cretaceous) of Alexander Island, Antarctica: history of an Albian regression. *Cretaceous Research*, **14**, 1–15, https://doi.org/10.1006/cres.1993.1001

Moyes, A.B. and Hamer, R.D. 1984. The geology of the Arrowsmith Peninsula and Blaiklock Island, Graham Land, Antarctica. *British Antarctic Survey Bulletin*, **65**, 41–55.

Mozer, A., Pécskay, Z. and Krajewski, K.P. 2015. Eocene age of the Baranowski Glacier Group at Red Hill, King George Island, West Antarctica. *Polish Polar Research*, **36**, 307–324, https://doi.org/10.1515/popore-2015-0022

Nawrocki, J., Pańczyk, M. and Williams, I.S. 2010. Isotopic ages and palaeomagnetism of selected magmatic rocks from King George Island (Antarctic Peninsula). *Journal of the Geological Society, London*, **167**, 1063–1079, https://doi.org/10.1144/0016-76492009-177

Nawrocki, J., Pańczyk, M. and Williams, I.S. 2011. Isotopic ages of selected magmatic rocks from King George Island (West Antarctica) controlled by magnetostratigraphy. *Geological Quarterly*, **55**, 301–322, https://gq.pgi.gov.pl/article/view/7721

Nell, P.A.R. 1990. Deformation in an accretionary melange Alexander Island, Antarctica. *Geological Society, London Special Publications*, **54**, 405–416, https://doi.org/10.1144/GSL.SP.1990.054.01.37

Nelson, S.A. and Hegre, J. 1990. Volcán Las Navajas, a Pliocene–Pleistocene trachyte/peralkaline rhyolite volcano in the northwestern Mexican volcanic belt. *Bulletin of Volcanology*, **52**, 186–204, https://doi.org/10.1007/BF00334804

Ogg, J.G., Ogg, G. and Gradstein, F.M. 2008. *The Concise Geological Time Scale*. Cambridge University Press, Cambridge, UK.

Pańczyk, A. and Nawrocki, J. 2011. Geochronology of selected andesitic lavas from the King George Bay area (SE King George Island). *Geological Quarterly*, **55**, 323–334, https://gq.pgi.gov.pl/article/view/7748

Pankhurst, R.J. 1982. Rb–Sr geochronology of Graham Land. *Journal of the Geological Society, London*, **139**, 701–711, https://doi.org/10.1144/gsjgs.139.6.0701

Petrone, C.M. and Ferrari, L. 2008. Quaternary adakite–Nb-enriched basalt association in the western Trans-Mexican volcanic belt: is there any slab melt evidence? *Contributions to Mineralogy and Petrology*, **156**, 73–86, https://doi.org/10.1007/s00410-007-0274-9

Riley, T.R. and Leat, P.T. 2021. Palmer Land and Graham Land volcanic groups (Antarctic Peninsula): volcanology. *Geological Society, London, Memoirs*, **55**, https://doi.org/10.1144/M55-2018-36

Riley, T.R., Leat, P.T., Pankhurst, R.J. and Harris, C. 2001. Origins of large volume rhyolitic volcanism in the Antarctic Peninsula and Patagonia by crustal melting. *Journal of Petrology*, **42**, 1043–1065, https://doi.org/10.1093/petrology/42.6.1043

Riley, T.R., Flowerdew, M.J. and Whitehouse, M.J. 2011. Chrono- and lithostratigraphy of a Mesozoic–Tertiary fore- to intra-arc basin: Adelaide Island, Antarctic Peninsula. *Geological Magazine*, **149**, 768–782, https://doi.org/10.1017/S0016756811001002

Ryan, C.J. 2007. *Mesozoic to Cenozoic Igneous Rocks from Northwestern Graham Land: Constraints on the Tectonomagmatic Evolution of the Antarctic Peninsula*. PhD thesis, University of Brighton, Brighton, UK.

Sajona, F.G., Maury, R.C., Bellon, H., Cotton, J., Defant, M. and Pubellier, M. 1993. Initiation of subduction and generation of slab melts in western and eastern Mindanao, Philippines. *Geology*, **21**, 1007–1010, https://doi.org/10.1130/0091-7613(1993)021<1007:IOSATG>2.3.CO;2

Sajona, F.G., Maury, R.C., Bellon, H., Cotton, J. and Defant, M. 1996. High field strength element enrichment of Pliocene–Pleistocene island arc basalts, Zamboanga Peninsula, western Mindanao (Philippines). *Journal of Petrology*, **37**, 693–726, https://doi.org/10.1093/petrology/37.3.693

Saunders, A.D. and Tarney, J. 1982. Igneous activity in the southern Andes and northern Antarctic Peninsula: a review. *Journal of the Geological Society, London*, **139**, 691–700, https://doi.org/10.1144/gsjgs.139.6.0691

Saunders, A.D., Tarney, J. and Weaver, S.D. 1980. Transverse geochemical variations across the Antarctic Peninsula: implications for the genesis of calc-alkaline magmas. *Earth and Planetary Science Letters*, **46**, 344–360, https://doi.org/10.1016/0012-821X(80)90050-3

Scarrow, J.H., Vaughan, A.P.M. and Leat, P.T. 1997. Ridge–trench collision induced switching of arc tectonics and magma sources: clues from Antarctic Peninsula mafic dykes. *Terra Nova*, **9**, 255–259, https://doi.org/10.1111/j.1365-3121.1997.tb00024.x

Scarrow, J.H., Leat, P.T., Wareham, C.D. and Millar, I.L. 1998. Geochemistry of mafic dykes in the Antarctic Peninsula continental-margin batholith: a record of arc evolution. *Contributions to Mineralogy and Petrology*, **131**, 289–305, https://doi.org/10.1007/s004100050394

Singleton, D.G. 1979. Geology of the Seward Mountains, western Palmer Land. *British Antarctic Survey Bulletin*, **49**, 81–89.

Smellie, J.L. 1980. The geology of Low Island, South Shetland Islands, and Austin rocks. *British Antarctic Survey Bulletin*, **49**, 239–257.

Smellie, J.L., Pankhurst, R.J., Thomson, M.R.A. and Davies, R.E.S. 1984. *The Geology of the South Shetland Islands: VI. Stratigraphy, Geochemistry and Evolution*. British Antarctic Survey Scientific Reports, **87**.

Smellie, J.L., Liesa, M., Muñoz, J.A., Sàbat, F., Pallàs, R. and Willan, R.C.R. 1995. Lithostratigraphy of volcanic and sedimentary sequences in central Livingston Island, South Shetland Islands. *Antarctic Science*, **7**, 99–113, https://doi.org/10.1017/S0954102095000137

Smellie, J.L., Millar, I.L., Rex, D.C. and Butterworth, P.J. 1998. Subaqueous, basaltic lava dome and carapace breccia on King George Island, South Shetland Islands, Antarctica. *Bulletin of Volcanology*, **59**, 245–261, https://doi.org/10.1007/s004450050189

Smith, C.G. 1987. *The Geology of Parts of the West Coast of Palmer Land*. British Antarctic Survey Scientific Reports, **112**.

So, C.S., Yun, S.T. and Park, M.E. 1995. Geochemistry of a fossil hydrothermal system at Barton Peninsula, King George Island. *Antarctic Science*, **7**, 63–72, https://doi.org/10.1017/S0954102095000101

Stern, C.R. and Kilian, R. 1996. Role of the subducted slab, mantle wedge and continental crust in the generation of adakites from the Andean Austral Volcanic Zone. *Contributions to Mineralogy and Petrology*, **123**, 263–281, https://doi.org/10.1007/s004100050155

Stolz, A.J., Davies, G.R., Crawford, A.J. and Smith, I.E.M. 1993. Sr, Nd and Pb isotopic compositions of calc-alkaline and peralkaline silicic volcanics from the D'Entrecasteaux Islands, Papua New Guinea, and their tectonic significance. *Mineralogy and Petrology*, **47**, 103–126, https://doi.org/10.1007/BF01161562

Storey, B.C. and Garrett, S.W. 1985. Crustal growth of the Antarctic Peninsula by accretion, magmatism and extension. *Geological Magazine*, **122**, 5–14, https://doi.org/10.1017/S0016756800034038

Storey, B.C., Vaughan, A.P.M. and Millar, I.L. 1996. Geodynamic evolution of the Antarctic Peninsula during Mesozoic times and its baring on Weddell Sea history. *Geological Society, London, Special Publications*, **108**, 87–103, https://doi.org/10.1144/GSL.SP.1996.108.01.07

Suárez, M. 1976. Plate-tectonic model for the southern Antarctic Peninsula and its relation to southern Andes. *Geology*, **4**, 211–214, https://doi.org/10.1130/0091-7613(1976)4<211:PMFSAP>2.0.CO;2

Sun, S.-s. and McDonough, W.F. 1989. Chemical and isotopic systematics of oceanic basalts: implications for mantle composition and processes. *Geological Society, London, Special Publications*, **42**, 313–345, https://doi.org/10.1144/GSL.SP.1989.042.01.19

Tangeman, J.A., Mukasa, S.B. and Grunow, A.M. 1996. Zircon U–Pb geochronology of plutonic rocks from the Antarctic Peninsula: confirmation of the presence of unexposed Paleozoic crust. *Tectonics*, **15**, 1309–1324, https://doi.org/10.1029/96TC00840

Tarney, J., Weaver, S.D., Saunders, A.D., Pankhurst, R.J. and Barker, P.F. 1982. Volcanic evolution of the northern Antarctic Peninsula and Scotia arc. In: Thorpe, R.S. (ed.) *Andesites: Orogenic Andesites and Related Rocks*. John Wiley & Sons, Chichester, UK, 371–400.

Tatsumi, Y. 2006. High-Mg andesites in the Setouchi volcanic belt, southwestern Japan: analogy to Archean magmatism and continental crust formation? *Annual Review of Earth and Planetary Science*, **34**, 467–499, https://doi.org/10.1146/annurev.earth.34.031405.125014

Thomson, M.R.A. 1982. Mesozoic paleogeography of West Antarctica. In: Craddock, C. (ed.) *Antarctic Geoscience*. University of Wisconsin Press, Madison, WI, 331–337.

Thomson, M.R.A. and Pankhurst, R.J. 1983. Age of post-Gondwanian calc-alkaline volcanism in the Antarctic Peninsula region. In: Oliver, R.L., James, P.R. and Jago, J.B. (eds) *Antarctic Earth Science*. Australian Academy of Science, Canberra, 328–333.

Thomson, M.R.A., Pankhurst, R.J. and Clarkson, P.D. 1983. The Antarctic Peninsula – a late Mesozoic–Cenozoic arc (review). In: Oliver, R.L., James, P.R. and Jago, J.B. (eds) *Antarctic Earth Science*. Australian Academy of Science, Canberra, 289–294.

Tranter, T.H. 1991. Accretion and subduction processes along the Pacific margin of Gondwana, central Alexander Island. In: Thomson, M.R.A., Crame, J.A. and Thomson, J.W. (eds) *Geological Evolution of Antarctica*. Cambridge University Press, Cambridge, UK, 437–441.

Trouw, R.A.J., Passchier, C.W., Valeriano, C.M., Simões, L.S.A., Paciullo, F.V.P. and Riberio, A. 2000. Deformational evolution of a Cretaceous subduction complex: Elephant Island, South Shetland Islands, Antarctica. *Tectonophysics*, **319**, 93–110, https://doi.org/10.1016/S0040-1951(00)00021-4

Uto, K., Takahashi, E., Nakamura, E. and Kaneoka, I. 1994. Geochronology of alkaline volcanism in Oki-Dogo Island, southwest Japan: geochemical evolution of basalts related to the opening of the Japan Sea. *Geochemical Journal*, **28**, 431–449, https://doi.org/10.2343/geochemj.28.431

Vaughan, A.P.M. and Storey, B.C. 2000. The eastern Palmer Land shear zone: a new terrane accretion model for the Mesozoic development of the Antarctic Peninsula. *Journal of the Geological Society, London*, **157**, 1243–1256, https://doi.org/10.1144/jgs.157.6.1243

Vaughan, A.P.M., Leat, P.T., Dean, A.A. and Millar, I.L. 2012. Crustal thickening along the West Antarctic Gondwana margin during mid-Cretaceous deformation of the Triassic intra-oceanic Dyer Arc. *Lithos*, **142–143**, 130–147, https://doi.org/10.1016/j.lithos.2012.03.008

Wang, F., Zheng, X.-S., Lee, J.I.K., Choe, W.H., Evans, N. and Zhu, R.-X. 2009. An ^{40}Ar–^{39}Ar geochronology on a mid-Eocene igneous event on the Barton and Weaver peninsulas: implications for the dynamic setting of the Antarctic Peninsula. *Geochemistry, Geophysics, Geosystems*, **10**, Q12006, https://doi.org/10.1029/2009GC002874

Wareham, C.D., Millar, I.L. and Vaughan, A.P.M. 1997. The generation of sodic granite magmas, western Palmer Land, Antarctic Peninsula. *Contributions to Mineralogy and Petrology*, **128**, 81–96, https://doi.org/10.1007/s004100050295

Watson, E.B. 1979. Zircon fractionation in felsic liquids: experimental results and applications to trace element geochemistry. *Contributions to Mineralogy and Petrology*, **70**, 407–419, https://doi.org/10.1007/BF00371047

Wendt, A.S., Vaughan, A.P.M. and Tate, A. 2008. Metamorphic rocks in the Antarctic Peninsula region. *Geological Magazine*, **145**, 655–676, https://doi.org/10.1017/S0016756808005050

West, S.M. 1974. *The Geology of the Danco Coast, Graham Land*. British Antarctic Survey Scientific Reports, **84**.

Willan, R.C.R. 1994. Structural setting and timing of hydrothermal veins and breccias on Hurd Peninsula, South Shetland Islands: a possible volcanic-related epithermal system in deformed turbidites. *Geological Magazine*, **131**, 465–483, https://doi.org/10.1017/S0016756800012103

Willan, R.C.R. and Armstrong, D.C. 2002. Successive geothermal, volcanic hydrothermal and contact-metasomatic events in Cenozoic volcanic-arc basalts, South Shetland Islands, Antarctica. *Geological Magazine*, **139**, 209–231, https://doi.org/10.1017/S0016756802006301

Willan, R.C.R. and Spiro, B. 1996. Sulphur sources for epithermal and mesothermal veins in Cretaceous–Tertiary magmatic-arc rocks, Livingston Island, South Shetland Islands. *Journal of the Geological Society, London*, **153**, 51–63, https://doi.org/10.1144/gsjgs.153.1.0051

Willan, R.C.R., Pankhurst, R.J. and Hervé, F. 1994. A probable Early Triassic age for the Miers Bluff Formation, Livingston Island, South Shetland Islands. *Antarctic Science*, **6**, 401–408, https://doi.org/10.1017/S095410209400060X

Yogodzinski, G.M., Kay, R.W., Volynets, O.N., Koloskov, A.V. and Kay, S.M. 1995. Magnesian andesite in the western Aleutian Komandorsky region: implications for slab melting and processes in the mantle wedge. *GSA Bulletin*, **107**, 505–519.

Yogodzinski, G.M., Lees, J.M., Churikova, T.G., Dorendorf, F., Wöerner, G. and Volynets, O.N. 2001. Geochemical evidence for the melting of subducting oceanic lithosphere at plate edges. *Nature*, **409**, 500–504, https://doi.org/10.1038/35054039

Zheng, G.G., Liu, X.C., Liu, S., Zhang, S.H. and Zhao, Y. 2018. Late Mesozoic–early Cretaceous intermediate–acid intrusive rocks from the Gerlache Strait area, Antarctic Peninsula: zircon U–Pb geochronology, petrogenesis and tectonic implications. *Lithos*, **312–313**, 204–222, https://doi.org/10.1016/j.lithos.2018.05.008

Chapter 3.2a

Bransfield Strait and James Ross Island: volcanology

John L. Smellie

School of Geography, Geology and the Environment, University of Leicester, University Road, Leicester LE1 7RH, UK

0000-0001-5537-1763

jls55@leicester.ac.uk

Abstract: Following more than 25 years of exploration and research since the last regional appraisal, the number of known subaerially exposed volcanoes in the northern Antarctic Peninsula region has more than trebled, from less than 15 to more than 50, and that total must be increased at least three-fold if seamounts in Bransfield Strait are included. Several volcanoes remain unvisited and there are relatively few detailed studies. The region includes Deception Island, the most prolific active volcano in Antarctica, and Mount Haddington, the largest volcano in Antarctica. The tectonic environment of the volcanism is more variable than elsewhere in Antarctica. Most of the volcanism is related to subduction. It includes very young ensialic marginal basin volcanism (Bransfield Strait), back-arc alkaline volcanism (James Ross Island Volcanic Group) and slab-window-related volcanism (seamount offshore of Anvers Island), as well as volcanism of uncertain origin (Anvers and Brabant islands; small volcanic centres on Livingston and Greenwich islands). Only 'normal' arc volcanism is not clearly represented, possibly because active subduction virtually ceased at *c.* 4 Ma. The eruptive environment for the volcanism varied between subglacial, marine and subaerial but a subglacial setting is prominent, particularly in the James Ross Island Volcanic Group.

The primary focus of this chapter is to provide a comprehensive description of the distribution, stratigraphy and physical volcanology of volcanoes in the northern Antarctic Peninsula region. The volcanism is widely scattered. It includes centres on Brabant and Anvers islands in the south, islands and multiple seamounts in Bransfield Strait and on the South Shetland Islands in the north, and the large volcanic field around James Ross Island in the east (Fig. 1). It also includes a large active volcano at Deception Island, with historical and recently observed eruptions. Evidence for geothermal heat and hydrothermalism in Bransfield Strait, Antarctic Sound and north of James Ross Island also suggests that other volcanoes may still be active. Our knowledge of the volcanoes in the region has increased dramatically in the quarter century that has elapsed since publication of the previous regional summary (LaMasurier *et al.* 1990), and many more volcanic centres are now identified (Table 1). However, for most of the volcanoes, descriptions of the physical volcanology are still relatively rudimentary.

All of the volcanism is broadly linked in some way to subduction but the tectonic setting varies across the region (Smellie 1990*a*). For example, Bransfield Strait is a small ensialic marginal basin. It formed when spreading ceased or decreased to a very slow rate after *c.* 4 or 3 Ma at the offshore Antarctic–Phoenix Ridge, either as a result of slab roll-back at the South Shetland Trench or, more likely, by oblique extension caused by sinistral transcurrent movement between the Antarctic and Scotia plates (Barker 1982; Larter and Barker 1991; Keller *et al.* 1992, 2002; Barker and Austin 1994, 1998; Lawver *et al.* 1996; González-Casado *et al.* 2000; Fretzdorff *et al.* 2004; Solari *et al.* 2008; Košler *et al.* 2009; Hole 2021). Subduction of the Phoenix Plate (also called the Aluk or Drake Plate: Larter 1991; Larter and Barker 1991) continues today at the South Shetland Trench but at a very much reduced rate (Larter 1991). The tectonic relationships of small scattered volcanic centres in the South Shetland Islands (on Livingston, Greenwich and King George islands) are also unclear despite their very young ages (<1 Ma: Smellie *et al.* 1984, 1996; Birkenmajer and Keller 1990). They may be linked in some way to deep-seated fissures and block faulting associated with the virtual cessation of subduction and the imposition of a broad regional extensional regime (cf.

Berrocoso *et al.* 2016). By contrast, extensive sodic alkaline volcanism in the James Ross Island Volcanic Group has been linked to a thermal anomaly beneath a lithospheric thin spot in an arc-rear position (Hole *et al.* 1995; Košler *et al.* 2009; Haase and Beier 2021). Young alkaline and tholeiitic volcanism on Brabant and Anvers islands may be causally related to a SW propagation of extension associated with Bransfield Strait marginal basin but the volcanism is not well studied petrologically and, although two compositional groups are present, the precise tectonic relationship is unclear (Ringe 1991; Smellie *et al.* 2006*b*; unpublished information of J.L. Smellie). Moreover, Bransfield Strait extension may be

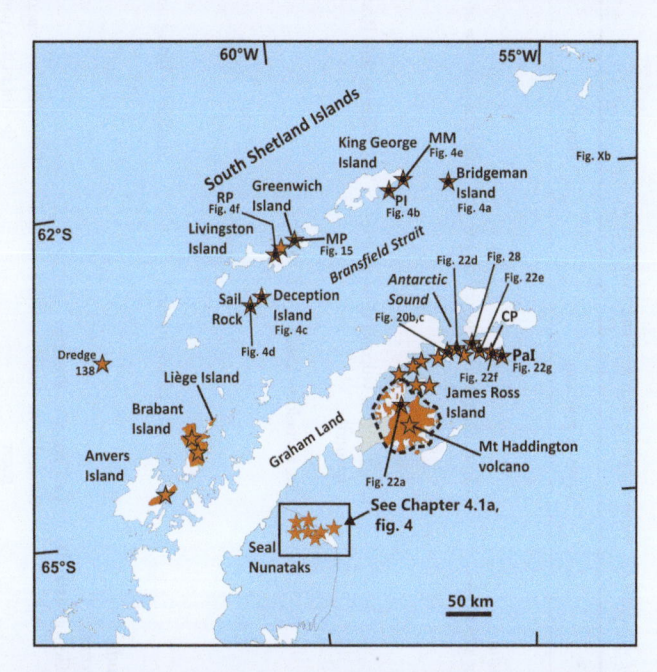

Fig. 1. Location map showing the distribution of volcanoes and volcanic outcrops in the northern Antarctic Peninsula region. The figure numbers of illustrations of selected volcanoes included in this chapter are also shown. Abbreviations: MM, Mount Melville; PI, Penguin Island; MP, Mount Plymouth; RP, Rezen Peak/Knoll; CP, Cape Purvis; PaI, Paulet Island.

From: Smellie, J. L., Panter, K. S. and Geyer, A. (eds) 2021. *Volcanism in Antarctica: 200 Million Years of Subduction, Rifting and Continental Break-up*. Geological Society, London, Memoirs, **55**, 227–284,

First published online January 20, 2021, https://doi.org/10.1144/M55-2018-58

© 2021 The Author(s). Published by The Geological Society of London. All rights reserved.

For permissions: http://www.geolsoc.org.uk/permissions. Publishing disclaimer: www.geolsoc.org.uk/pub_ethics

Table 1. *Inventory of volcanoes in the northern Antarctic Peninsula region*

Name	Type of volcano	Notes	Composition	Key references
Bransfield Strait *Ensialic marginal basin*				
Submarine seamounts and ridges	Pillow mounds and pillow ridges	One occurrence of Pele's hair	Mainly basalts and basaltic andesites; rare silica-poor andesite and rhyolite	Keller *et al.* (1992, 2002); Gràcia *et al.* (1996, 1997); García *et al.* (2011); unpublished information of J.L. Smellie
Bridgeman Island	Large pillow volcano (low-profile mound)	Similar to Sail Rock volcano	Basaltic andesite	González-Ferrán and Katsui (1970); Weaver *et al.* (1979); Keller *et al.* (1992); Aquilina *et al.* (2013)
Deception Island	Large shield volcano		Basalts to trachytes; reports of rhyolites are spurious	González-Ferrán and Katsui (1970); Baker *et al.* (1975); Martí and Baraldo (1990); Smellie (2001); Smellie *et al.* (2002); Martí *et al.* (2013)
Sail Rock	Low-profile pillow mound	Similar to Bridgman Island	Andesite	Keller *et al.* (1992); Barclay *et al.* (2009); Kraus *et al.* (2013)
South Shetland Islands *Tectonic setting uncertain*				
Melville Peak (King George Island)	Tuff cone/stratocone		Basalt and basaltic andesite	Birkenmajer (1982*a, b*)
Penguin Island	Eroded lava shield; scoria cone and maar	Subaerial lavas and scoria cone; one of only two maars known in Antarctica (other on Deception Island 1970)	Alkali basalt	González-Ferrán and Katsui (1970); Birkenmajer (1982*a, b*); Weaver *et al.* (1979); Pańczyk and Nawrocki (2011)
Livingston and Greenwich islands	Numerous small tuff cones, plugs, subaerial lava outcrops	Multiple small volcanic centres	Alkali and tholeiitic basalt	Smellie *et al.* (1995, 1996); Kraus *et al.* (2013); unpublished information of J.L. Smellie
James Ross Island Volcanic Group *Back-arc alkaline volcanism related to a mantle plume rising in a 'thin spot'*				
Cape Purvis (Dundee Island)	Tuya	At least two, possibly three superimposed lava-fed deltas; tuff cone core exposed in hill at north end	Hawaiite	Smellie *et al.* (2006*c*); unpublished information of S. Kraus
Paulet Island	Eroded lava shield, two scoria cones, possible maar	Subaerial lavas and scoria cones	Hawaiite	Baker *et al.* (1973, 1977); Smellie (1990*f*)
Jonassen Island	Tuya?	Glacial origin uncertain but inferred	Unknown	Nelson (1975); Smellie *et al.* (2006*c*); unpublished information of J.L. Smellie
Andersson Island	Tuya	Small tuff cone flank vent present close to Cape Betbeder	Unknown	Nelson (1975); unpublished information of J.L. Smellie
Rosamel Island	Tuff cone	Visited only once (Nelson)	Unknown	Nelson (1975); unpublished information of J.L. Smellie
Jaegyu Knoll	Elongate pillow mound with some tephra?	Submarine volcano, possibly active	Alkali basalt	Lavoie *et al.* (2016)
Brown Bluff	Tuya	One of the first and best-described tuyas	Alkali basalt	Skilling (1994); Smellie and Skilling (1994)
Gamma Hill	Tuya?	Unvisited; Gamma Hill is a likely tuff cone; may have emitted an extensive lava-fed delta that flowed to the south and SW	Unknown	Smellie *et al.* (2013); unpublished information of JL Smellie

Name	Type	Description	Composition	References
Buttress Hill	Tuya?	Unvisited and unconfirmed binocular observation (J.L. Smellie)	Unknown	Ashley (1962); Smellie *et al.* (2013); unpublished information of J.L. Smellie
Unnamed hill SW of Buttress Hill	Tuya?	Little known but possibly emitted an extensive lava-fed delta that flowed south; includes Seven Buttresses; includes a likely tuff cone flank vent at Cone Nunatak	Basalt	Nelson (1975); Smellie *et al.* (2013); unpublished information of J.L. Smellie and I.P. Skilling
Cain Nunatak	Tuff cone	Visited only once (Aitkenhead)	Unknown	Aitkenhead (1975); Nelson (1975)
Abel Nunatak	Tuff cone	Unvisited, binocular observation (Aitkenhead)	Unknown	Aitkenhead (1975); Nelson (1975)
Red Island	Tuya?		Hawaiite	Nelson (1975); Baker *et al.* (1977)
Egg Island	Tuya?		Hawaiite	Nelson (1975); Baker *et al.* (1977)
Vortex Island	Tuff cone	Unvisited and unconfirmed binocular observation (J.L. Smellie)	Unknown	Nelson (1975); unpublished information of J.L. Smellie
Corry Island	Tuya?	The summit crater is essentially uneroded and was used to constrain maximum Plio-Pleistocene ice thicknesses in the region	Unknown	Nelson (1975); Smellie *et al.* (2009); unpublished information of J.L. Smellie
Tail Island	Tuff cone	Initially interpreted as a delta-front sequence; probably subaqueous tuff cone sequence	Unknown	Pirrie and Sykes (1987); unpublished information of J.L. Smellie
Eagle Island	Uncertain	Visited only once (Nelson)	Unknown	Nelson (1975)
Carlson Island	Tuff cone		Unknown	Unpublished information of J.L. Smellie
Mount Haddington	Large shield volcano; the largest volcano in Antarctica	Formed of multiple lava-fed deltas, mainly glacially emplaced; numerous flank and satellite centres (listed below); world's largest polygenetic glaciovolcano	Alkali basalts and hawaiites; mugearite segregation veins in largest intrusion; phonotephrite in one satellite vent (Dobson Dome)	Nelson (1975); Smellie (1987); Smellie *et al.* (2008); Calabozo *et al.* (2015)
Vega Island	Large shield volcano	Formed of multiple lava-fed deltas, mainly glacially emplaced; major source vent possibly at Sandwich Bluff (obscured), and another between Pirrie Col and Cape Gordon (obscured); flank and satellite centres at Mahogany Bluff, unnamed hill SW of Cape Well-met and Devil Island (tuff cones; see below)	Alkali basalt, hawaiite	Nelson (1975); Smellie (1987); Smellie *et al.* (2008, 2013)
Keltie Head, Vega Island	Tuya		Alkali basalt	Smellie *et al.* (2013); unpublished information of J.L. Smellie
Lachman Crags	Volcanic shield?	Extensive marine-emplaced lava-fed delta emitted from source vent near southern Lachman Crags	Alkali basalt, hawaiite	Smellie *et al.* (2008); Nehyba and Nývlt (2015)
Dobson Dome	Tuya	Satellite vent; the tallest and youngest tuya in the JRIVG	Phonotephrite	Smellie *et al.* (2008); unpublished information of J.L. Smellie
Volcan Marina	Scoria cone	Flank vent	Basanite	Strelin and Malagnino (1992); Smellie *et al.* (2013); unpublished information of J.L. Smellie
Volcan Elba	Scoria cone	Flank vent; another possible pyroclastic cone (unvisited) 3 km SW of Volcan Elba on the summit of Rabot Point is postulated from field observations	Hawaiite	Strelin and Malagnino (1992); Smellie *et al.* (2013); unpublished information of J.L. Smellie
Volcan Eugenia	Tuff cone or tuff ring	Marine emplaced or subaerial	Unknown	Strelin and Malagnino (1992); Smellie *et al.* (2013); unpublished information of J.L. Smellie
Terrapin Hill	Tuff cone	Marine-emplaced	Alkali basalt	Smellie *et al.* (2006a)
Tortoise Hill	Tuff cone	Marine-emplaced or subaerial	Basalt	Nelson (1975); Smellie *et al.* (2013); unpublished information of J.L. Smellie

(*Continued*)

Table 1. Continued.

Name	Type of volcano	Notes	Composition	Key references
Lomas Ridge	Tuff cone	Marine-emplaced or subaerial	Basalt	Nelson (1975); Smellie et al. (2013); unpublished information of J.L. Smellie
Lockyer Island	Tuff cone draped by unrelated lava-fed delta (Jefford Point Formation?)	Unvisited and unconfirmed binocular observation (Nelson; J.L. Smellie)	Unknown	Smellie et al. (2013); unpublished information of J.L. Smellie
Sungold Hill	Tuff cone	Marine-emplaced or subaerial	Unknown	Smellie et al. (2013); unpublished information of J.L. Smellie
Persson Island	Tuff cone		Unknown	Nelson (1975); Smellie et al. (2013)
Ineson Glacier/Patalamon Mesa	Tuff cone	Marine-emplaced	Basalt	Williams et al. (2006); Nelson et al. (2008); Smellie et al. (2013); unpublished information of J.L. Smellie
San Fernando Hill	Tuff cone	Marine-emplaced or subaerial	Hawaiite	Nelson (1975); Smellie et al. (2013); unpublished information of J.L. Smellie
Virgin Hill, Seacatch Nunataks	Tuff cone(s)	Marine-emplaced or subaerial; source for prominent tephra beds between deltas at Stickle Ridge and Hambrey Cliffs (Ulu Peninsula)	Hawaiite	Smellie et al. (2013); unpublished information of J.L. Smellie
Flett Buttress	Tuff cone	Unvisited (inaccessible), binocular observation (J.L. Smellie)	Unknown	Smellie et al. (2013); unpublished information of J.L. Smellie
Stoneley Point	Tuff cone	Unvisited and unconfirmed binocular observation (J.L. Smellie)	Unknown	Smellie et al. (2013); unpublished information of J.L. Smellie
Bibby Point	Tuff cone	Marine-emplaced or subaerial	Alkali basalt, hawaiite	Nehyba and Nývlt (2014)
Mahogany Bluff (Vega Island)	Tuff cone	Prominent rising passage zone at west end	Alkali basalt	Smellie et al. (2013); unpublished information of J.L. Smellie
False Island Point (Vega Island)	Tuff cone	Visited only once (Nelson)	Unknown	Nelson (1975); Smellie et al. (2013); unpublished information of J.L. Smellie
Cape Gordon (Vega Island)	Tuff cone	Unvisited and unconfirmed binocular observation (J.L. Smellie)	Unknown	Smellie et al. (2013); unpublished information of J.L. Smellie
Unnamed hill SW of Cape Well-met (Vega Island)	Tuff cone	Unvisited and unconfirmed binocular observation (J.L. Smellie)	Unknown	Smellie et al. (2013); unpublished information of J.L. Smellie
Devil Island	Tuff cone	Unvisited and unconfirmed binocular observation (J.L. Smellie)	Unknown	Smellie et al. (2013); unpublished information of J.L. Smellie
Brabant and Anvers islands				
Tectonic setting unclear				
Brabant Island	Shield volcanoes	At least two large shield volcanoes; much inaccessible	Alkali basalt; basalt, benmoreite	Ringe (1991); Smellie et al. (2006b); unpublished information of J.L. Smellie
Anvers Island	Shield volcano	Largely inaccessible; outcrops grouped within a single large shield volcano	Unknown	Hooper (1962); Smellie et al. (2006b)
Continental shelf				
No-slab-window alkaline volcanism				
Offshore of Anvers Island [DR138]	Small submarine pillow mound		Alkali basalt	Hole and Larter (1993)

propagating either to the NE (Gràcia et al. 1996) or to the SW (Barker and Austin 1998; González-Casado et al. 2000; Christeson et al. 2003; Fretzdorff et al. 2004), illustrating how a consensus still does not exist concerning the precise influence and effects of the dominant tectonic regime. Additionally, the virtual absence of volcanism in the Western Bransfield Basin (Barker and Austin 1998) is at odds with its prominent expression further to the SW at Brabant and Anvers islands. The region also contains a single alkali basalt centre in a submarine trench-proximal position that might be related to post-subduction slab-window formation (Hole and Larter 1993). By far the best-known volcanic centres are Deception Island, a large active volcano in Bransfield Strait, and the long-lived (>12 myr) James Ross Island Volcanic Group to the east of the Antarctic Peninsula, which are particularly well studied and well mapped (e.g. Baker et al. 1973; Sykes 1988; Martí and Baraldo 1990; Birkenmajer 1992; Smellie et al. 2002, 2006a, c, 2008, 2013; Berrocoso et al. 2006; Ben-Zvi et al. 2007; Zandomeneghi et al. 2007; Martí et al. 2013; Nehyba and Nývlt 2014, 2015; Calabozo et al. 2015; see also Geyer et al. 2021). Other centres are much less well known but a few have been the focus of relatively recent volcanological investigations (e.g. Smellie et al. 1995, 1996, 2006b; Kraus et al. 2013; Lavoie et al. 2016).

Stratigraphy and volcanology of the volcanic outcrops

Bransfield Strait submarine seamounts and volcanic ridges

Bransfield Basin is the morphological feature associated with Bransfield Strait and is a composite of three submarine basins. It has associated seismicity and volcanism, high heat flow (49–626 mW m^{-2}) and hydrothermalism typical of regions with active hydrothermal circulation, and other features (including thinned crust, a central positive magnetic anomaly and large negative gravity anomaly), which suggest that it is a young and active rift basin (e.g. Nagihara and Lawver 1989; Lawver et al. 1996; Dählmann et al. 2001; Klinkhammer et al. 2001; Christeson et al. 2003; Somoza et al. 2004; Aquilina et al. 2013, 2014). Only two published isotopic ages exist, for two adjacent seamounts c. 20 km SE of Penguin Island (Fig. 2): 103 ± 35 and 53 ± 36 ka (Fisk 1990). The rift may be opening at 7–10 mm a^{-1} (Dietrich et al. 2001, 2004) but the presence of new oceanic crust is disputed (cf. Larter 1991; Henriet et al. 1992; Lawver et al. 1996; Christeson et al. 2003). The three sub-basins are named the Western, Central and Eastern Bransfield basins. Volcanism is associated with the Central and Eastern Bransfield basins (although the easternmost part of the Eastern Basin is amagmatic: Fig. 2), whereas the Western Bransfield Basin is much shallower than the others and is essentially amagmatic (Gràcia et al. 1996, 1997; Barker and Austin 1998; Fretzdorff et al. 2004). In addition there are several prominent volcanic islands including Deception, Bridgeman and Penguin islands (Smellie 1990a). Six large volcanic edifices protrude from the seafloor in the Central Bransfield Basin and are aligned with the basin axis. In addition, small scattered volcanic cones are particularly common in the Eastern Bransfield Basin. The locations of the larger seamounts in the Central Bransfield Basin might be related to the intersection of two main orthogonal faults systems trending ENE–WSW and NNW–SSE. They may have been constructed in multiple stages (see below), whereas the numerous small centres in Eastern Bransfield Basin are more evenly dispersed (in essence, they form a submarine volcanic field) and had a simpler construction. The variations between the two basins probably relate to their different stages of tectonic development, with the Central Bransfield Basin related to NW–SE extensional faulting and focused active mid-ocean ridge basalt (MORB)-like volcanism related to incipient seafloor spreading (although no true MORB magmas are present: Fretzdorff et al. 2004). By contrast, the Eastern Bransfield Basin may be in an earlier rifting stage dominated mainly by

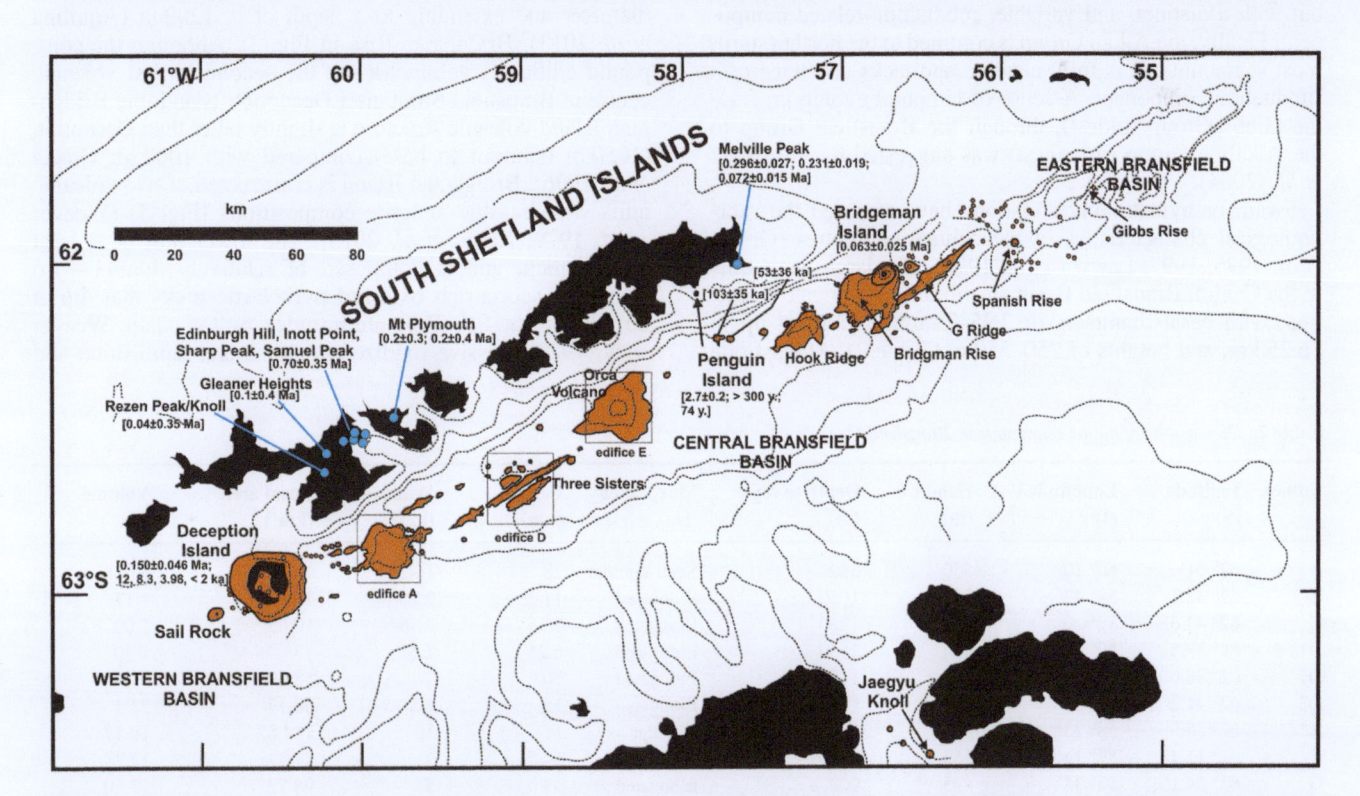

Fig. 2. Map showing the locations of submarine volcanoes and volcanic islands in Bransfield Strait (modified after Gràcia et al. 1996). The locations of terrestrial centres on Livingston, Greenwich and King George islands (blue circles), and the isotopic ages for those and island centres are also shown (ages from Table 3).

NW–SE extension, together with some left-lateral strike-slip faulting.

Magmas erupted in the Bransfield Basin are mainly basaltic but a wide range of compositions is present, from basalt to rhyolite (Keller et al. 1992, 2002; Fretzdorff et al. 2004). Fretzdorff et al. (2004) divided the volcanic centres in the Eastern Basin and northeastern Central Basin into three informally-named compositional groups, comprising:

- The Bransfield Group is the largest group by far, present at G Ridge, Spanish Rise, Bridgeman Rise, Gibbs Rise and Hook Ridge, and including volcanic centres in the Central Basin described by Keller et al. (1992, 2002) (see Fig. 2 for the locations). It is subalkaline (low–medium potassic), has the widest compositional range (basalt to rhyolite) and contains several compositional subgroups.
- The Gibbs Group, found at Gibbs Rise only, is also subalkaline but high potassium, with compositions ranging from basaltic andesite to andesite; the samples are generally more weathered and hydrothermally altered (hence older?) than others in Bransfield Basin.
- An Alkali Group, which includes samples mainly from Spanish Rise, with isolated occurrences at Gibbs Rise, G Ridge and Bridgeman Rise. They are fresh rocks with alkaline compositions ranging from alkali basalt to tephrite. None of the Alkali Group samples were definitely *in situ*. Interestingly, some of the Alkaline Group samples are red and oxidized, indicating that they were erupted subaerially, although they were obtained in deep water (generally >1200 m) and there are no known islands present today.

Although studies by Keller et al. (1992, 2002) showed no systematic variations along-axis, the study by Fretzdorff et al. (2004) demonstrated a somewhat complicated compositional change. They suggested that the Gibbs Group is represented by a single seamount, a bisected cone regarded as a relict of the inactive South Shetland arc. The Bransfield Group is compositionally somewhat like normal-type MORB (N-MORB) but with a distinct, and variable, subduction-related component. Finally, the Alkali Group is confined to the northeasternmost seamounts. It is fully alkaline and lacks any trace of a subduction component. A sense of temporal evolution, from the Gibbs Group (oldest), through the Bransfield Group to the Alkaline Group (youngest) was suggested by Fretzdorff et al. (2004).

Swath bathymetry investigations have revealed the morphological characteristics of the submarine centres (Gràcia et al. 1996, 1997; Lawver et al. 1996). The large seamounts in the Central Bransfield Basin are circular to elongate in outline, with basal diameters of 2.25–7 km but ranging up to 16.25 km, and heights of 250–550 m (Table 2). At least one may have a flat top (Gràcia et al. 1996, edifice A); it also shows evidence for rectilinear faulting and associated graben (Fig. 3). Others show small calderas a few hundreds of metres deep. Several of the edifices have prominent spurs (volcanic ridges) 10–15 km long and up to 350 m high that trend parallel to the NE–SW elongation of the basin (Fig. 3). Both the Central and Eastern Bransfield basins also contain numerous small isolated volcanic cones or mounds with basal diameters up to c. 2.5 km wide and up to 400 m high. However, all but one of the edifices in the Eastern Bransfield Basin are <250 m, so it is dominated by very low profile edifices, and the basin also contains a prominent NE-trending volcanic ridge (G Ridge: Fig. 2). The volcanic edifices in both basins are associated with short-wavelength, high-amplitude positive magnetic anomalies, which suggests that they are formed mainly of pillow lava but at least one, at c. 1200–1500 m depth in the NE Central Bransfield Basin, produced Pele's hair in an explosive eruption of felsic magma (unpublished information of J.L. Smellie and S. Fretzdorff). The total volume represented by seamounts of any type is 87 000 km^3 in the Central Bransfield Basin and 25 000 km^3 in the Eastern Bransfield Basin, consistent with an early rift stage of evolution of the latter.

Bridgeman Island

Nothing significantly new has been discovered about Bridgeman Island since it was examined in the late 1960s and mid 1970s, and published volcanological details are very sparse (González-Ferrán and Katsui 1970; Smellie 1990d). It consists of a cliff-bounded rocky islet rising to 240 m asl (above sea-level) and is just 600–900 m wide (Fig. 4a). The island represents the exposed summit of a submerged volcanic mound measuring 8 × 5 km in basal diameter and rising c. 240 m, which is, in turn, situated on top of a much larger edifice, presumably compound, measuring 10 × 15 km in basal diameter and extending to a depth of c. 1200 m (Aquilina et al. 2013) (Bridgeman Rise in Fig. 2). Although the compound edifice is volumetrically the second largest volcanic centre in Bransfield Strait after Deception Island, the Bridgeman Island volcanic structure is slightly taller than Deception (1050 m (summit to base) compared with 1000 m: Gràcia et al. 1996). Bridgeman Island is constructed of two volcanic units with basaltic andesite compositions (Fig. 5) (Weaver et al. 1979; Kraus et al. 2013). The lower unit is at least 150 m thick, and is composed of relatively thin (1–7 m) lavas and scoria-rich oxidized pyroclastic rocks that dip at 20°–30° to the SE. They are intruded by (or onlap: Weaver et al. 1979) massive oxidized agglomerate, lapillistone and

Table 2. *Dimensions of major seamounts in Bransfield Strait*

Edifice	Latitude (S)	Longitude (W)	Height (m)	Depth to top (m)	Morphology	Length (km)	Width (km)	Basal area (km^2)	Volume (km^3)
A	62° 51'	60° 10'	550	350	Semi-circular	28	16.25	177.69	33.28
B	62° 42.7'	59° 42'	325	1050	Elongated	11.25	2.25	12.13	1.17
C	62° 44.6'	59° 20'	250	1200	Elongated	10.5	2.75	12.05	1.00
D1	62° 38.5'	59° 04'	300	1150	Elongated	6.25	2.5	13	1.30
D2	62° 38.6'	59° 00'	350	1150	Elongated	30	2	53.69	9.40
D3	62° 41.2'	59° 00'	300	1150	Elongated	15	3.5	38.17	5.73
E*	62° 25.5'	58° 23'	550	650	Circular	11	10	114.82	19.17
F	62° 11.4'	57° 15'	550	1150	Circular	14.5	7	51.77	12.73
G	62° 04.1'	56° 35'	475	900	Elongated	33	5	94.72	22.50

*Edifice E corresponds to the ORCA volcano identified by González-Ferrán (1991).
After Gràcia et al. (1996), modified.

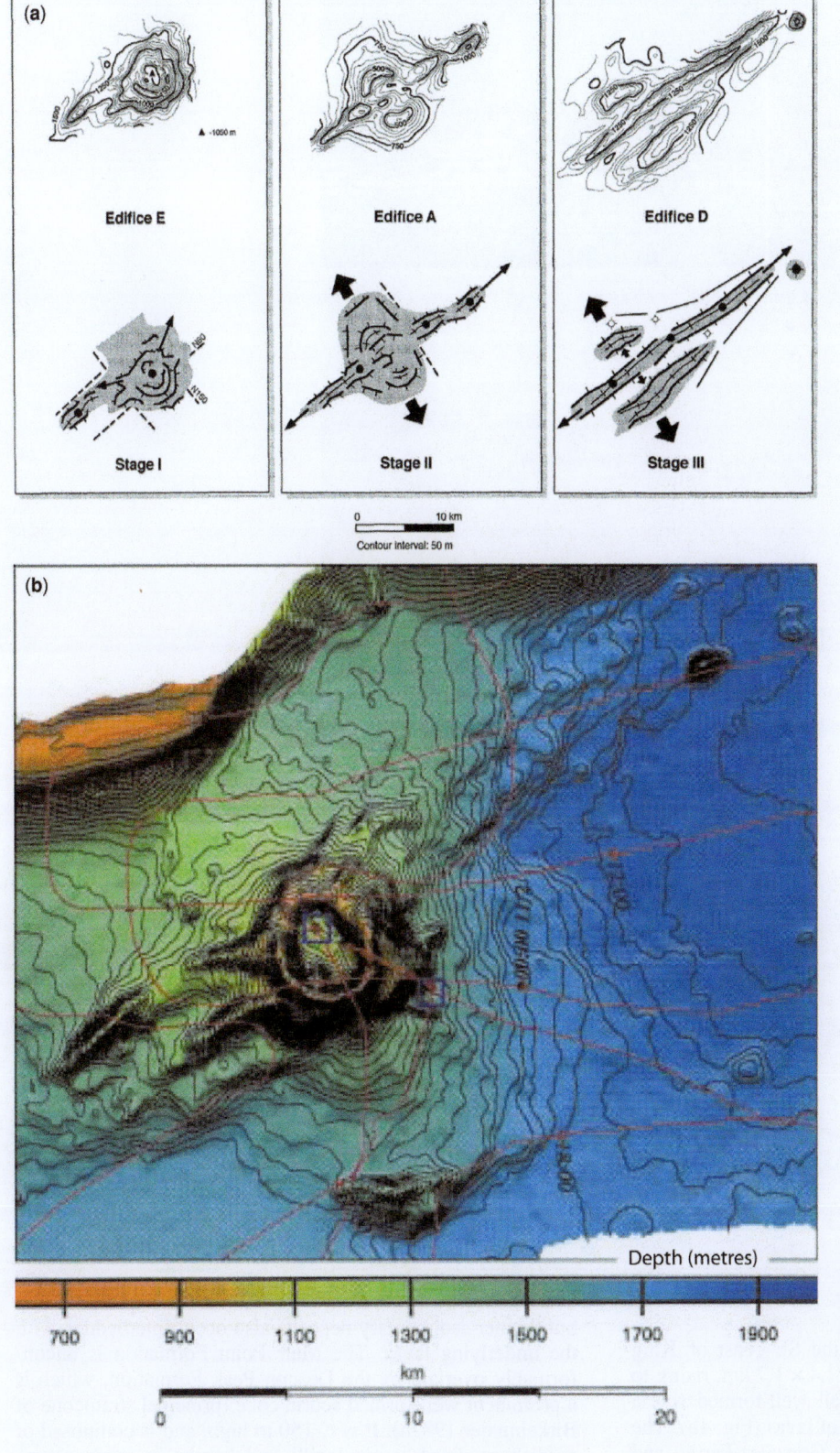

Fig. 3. Morphology of selected submarine volcanoes in Bransfield Strait, showing the variable effect of the regional tectonics. (**a**) Contoured bathymetry and inferred stress directions, after Gràcia *et al.* (1996); the locations of the edifices are shown in Figure 2. (**b**) Swath image to show in greater detail the morphology of edifice E (also called ORCA volcano) with its summit caldera, after Lawver *et al.* (1996).

tuff that may be part of a small scoria cone or vent fill *c.* 500 m in diameter. The lower unit is draped unconformably by an upper unit that thickens to *c.* 150 m northwards, where it cuts down to sea-level through the lower stratigraphical unit and is capped by a summit layer of intensely reddened tephra (Kraus *et al.* 2013). The unconformity is a very uneven surface, with coarse pyroclasts infilling the surface irregularities (Weaver *et al.* 1979). The upper unit is composed of thick (25–30 m) lavas and a greater proportion of interbedded scoria-rich pyroclastic rocks than occur in the lower unit. The report of possible past hydrothermal activity is unverified and no thermal activity exists today (González-Ferrán and Katsui 1970). The volcano now appears to be extinct but Matthies *et al.* (1988) identified Bridgeman Island as the likely source of a marine tephra layer in Bransfield Strait based on a compositional comparison. A K–Ar isotopic age of 63 ± 25 ka was obtained on the upper lava series (Keller *et al.* 1992).

Fig. 4. Compendium of views of selected volcanoes in the northern Antarctic Peninsula region. (**a**) Bridgeman Island, viewed from the NE (image courtesy of Jeronimo López-Martínez). (**b**) Penguin Island, looking WSW. (**c**) Deception Island looking NW. (**d**) Sail Rock. (**e**) Melville Peak, looking ENE. (**f**) Rezen Peak/ Knoll, Livingston Island. Images (d), (e) and (f) are from Kraus *et al.* (2013) (courtesy of Stefan Kraus).

Penguin Island

Penguin Island is situated 2 km off the SE coast of King George Island (Fig. 2). It measures *c.* 1.4 × 1.7 km, rising to 180 m at Deacon Peak, and is a small well-formed scoria cone constructed on top of a platform of lavas (Fig. 4b). The only detailed published studies are by González-Ferrán and Katsui (1970) and Birkenmajer (1982*a*; but see also Smellie 1990*c*) (Fig. 6). Three geological units have been identified. All have mildly alkaline olivine basalt (Weaver *et al.* 1979) or low-potassium calc-alkaline (Pańczyk and Nawrocki 2011) compositions. The Marr Point Formation consists of a basal unit >30 m thick formed of thin (2–10 m) black to grey basaltic 'a'ā lavas that dip gently to the south and increase in thickness in a northeasterly direction (towards the presumed source). The formation is capped by a 10–20 m-thick deposit composed of unconsolidated sand and intercalated gravel, including shingle formed of rounded pebbles, interpreted as beach deposits, which are exposed on the south coast. Similar but thinner sedimentary deposits also occur interbedded with the underlying lavas. The Marr Point Formation is unconformably overlain by the Deacon Peak Formation, which is a prominent well-formed scoria cone (principal stratocone of Birkenmajer 1982*a*). It is *c.* 150 m high, and is composed of reddish oxidized scoria lapilli and bombs in crude coarse beds dipping radially outwards at *c.* 30°–40°; a few thin black basalt lavas may be present within the base of the cone. The principal cone has a crater 350 m wide which contains another, much smaller, fresh-looking scoria cone near its centre that is *c.* 150 m wide and 15–20 m high, and is also formed of oxidized scoria and within which is a short stubby 'a'ā lava. Also present within the main crater on its south side is an isolated stack interpreted by Birkenmajer (1982*a*) as a small plug, together with remnants of radial dykes that are also exposed in a prominent arcuate scar on the west outer flank of the principal cone. However, the stack is formed

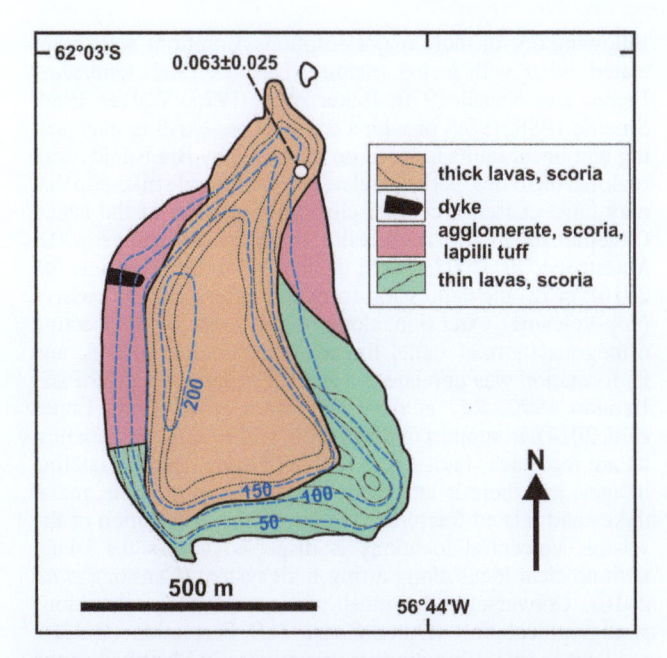

Fig. 5. Geological sketch map of Bridgeman Island (modified after González-Ferrán and Katsui 1970). The location of the dated sample is uncertain (described only as 'upper [lava] series'; Keller et al. 1992) but based on likely access to the geology of the island (author's observation).

of oxidized scoria essentially indistinguishable from that forming the cone itself but much more strongly indurated. The final stage of evolution of Penguin Island consisted of an eruption of a maar on the NE side of the island. The products are subsumed in the Petrel Crater Formation, which is dominated by a well-formed maar crater *c.* 300 m wide with a lake, within which possible fumarolic activity was observed in 1966 (González-Ferrán 1995) but has not been confirmed. The steep crater walls expose a sequence 2–5 m thick composed of thin-bedded lapilli tuffs locally rich in accessory lithic clasts that overlie lavas of the Marr Point Formation (unpublished information of J.L. Smellie). Basalt lava blocks, including several >2 m in diameter, derived from the Marr Point Formation are also liberally strewn over the low-lying ground surrounding the maar. The arcuate scar that affects the west flank of the principal stratocone was interpreted to be a result of a second explosive event similar to, and included within, the Petrel Crater Formation but sourced in a submarine vent west of the island. However, there is no tephra present that might be ascribed to an explosive event at that location and the existence of a submarine crater there, postulated by Birkenmajer (1982*a*; see also Pańczyk and Nawrocki 2011), is unproven. The arcuate scar probably formed simply by local sector collapse following erosional oversteepening of the western side of the island by marine action. Despite the very youthful appearance of the principal scoria cone that dominates the island, Penguin Island may have grown over an extended period of time that included a substantial time gap after the eruption of the basal Marr Point Formation. The basal lavas have yielded a $^{40}Ar/^{39}Ar$ age of 2.7 ± 0.2 Ma (Pańczyk and Nawrocki 2011), whilst the Deacon Peak (principal) cone may be just centuries old (>300 years by the lichenometric method: Birkenmajer 1980). Although a $^{40}Ar/^{39}Ar$ age of 8.8 ± 2.4 Ma was obtained by Kraus (2005) on a dyke cropping out in the western collapse scar, its age must presumably be similar to the Deacon Peak principal scoria cone and it is therefore improbably old (Pańczyk and Nawrocki 2011). The age of the small central scoria cone within the Deacon Peak cone may be *c.* 133 years and the Petrel Crater Formation (maar) *c.* 74 years (both ages relative to AD 1979 and by the lichenometric method: Birkenmajer 1980). In view of the youthful appearance of the Deacon Peak scoria cone, it has been suggested that observations of fumarolic activity attributed to Bridgeman Island in the nineteenth century may correspond to Penguin Island (González-Ferrán and Katsui 1970).

Deception Island

After Mount Erebus, Deception Island is the most active volcano in Antarctica. Compositionally, it ranges between basalt and trachyte but is dominated by basalt and basaltic andesite (Smellie 2002*a*). Interestingly, Deception-sourced marine tephras include rhyolites, which are currently unknown on the island (Moreton and Smellie 1998). In terms of volume of ejecta, Deception Island is the most productive Antarctic volcano, having undergone well-documented recent explosive eruptions in 1967, 1969 and 1970, as well as numerous historical eruptions in the eighteenth, nineteenth and early twentieth centuries (based mainly on englacial tephras: e.g. González-Ferrán et al. 1971; Orheim 1972*a*, *b*; Baker et al. 1975; Pallàs et al. 2001; Smellie 2002*a*, *b*). Additionally, many more eruptions in prehistorical time have been identified from marine, lacustrine, glacial and terrestrial (peat) tephras extending back to >35 ka, at least (e.g. Roobol 1973; Matthies et al. 1988; Björck et al. 1991; Moreton and Smellie 1998; Smellie 1999, 2002*c*; Fretzdorff and Smellie 2002; Hillenbrand et al. 2008; Antoniades et al. 2018). It is also one of the most visited sites in Antarctica, with more than 22 000 tourists annually (for 2018–19; source: http://www.iaato.org). In addition, numerous scientists occupy two summer-only scientific stations (Argentine and Spanish), whilst other scientific stations, by Britain and Chile, were destroyed as a consequence of the 1967–70 volcanic eruptions (Baker et al. 1975). Deception

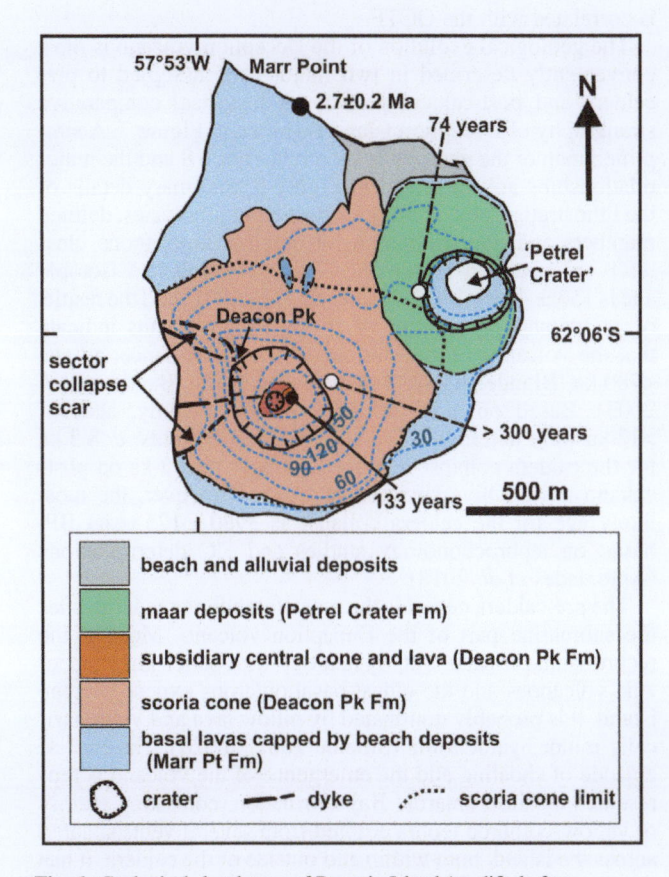

Fig. 6. Geological sketch map of Penguin Island (modified after González-Ferrán and Katsui 1970; Birkenmajer 1982*a*).

Island is situated just off-centre relative to the axis of Bransfield Strait, at the junction between the Western and Central Bransfield basins (Fig. 2). The location of the volcano may be at the intersection of the NE–SW tensional axis of the Bransfield Strait marginal basin and a major orthogonal fault (González-Ferrán 1991; Grad et al. 1992; Maestro et al. 2007). Gravity and magnetic modelling have shown that the volcano sits on top of continental upper crust (Grad et al. 1992; Muñoz-Martín et al. 2005). It is a large volcano with a submerged basal diameter of c. 30 km, rising to 542 m asl at Mount Pond on its eastern side (Smellie 1990a). The island itself is c. 14 km in diameter, horseshoe shaped and with a large central bay (Port Foster), which is entered by a narrow gap on its SE side known as Neptune's Bellows. It sustains two moderate ice caps, at Mount Pond and Mount Kirkwood, and about 57% of the island is covered by snow and ice or ice-cored moraines (Smellie et al. 1997; López-Martínez and Serrano 2002).

Deception Island has a longer history of exploration and scientific research than any other Antarctic volcano and is correspondingly served by a larger database of geological and geophysical publications. It is well studied and described, and is considered in greater detail here than other volcanoes in this chapter. Deception was the site of a historically important expedition to measure Earth's gravity in 1829 (Roobol 1980). It was also the location of a very active whaling station between 1906 and 1931, and the target of numerous scientific expeditions, and there were three all-year-round scientific stations on the island between 1944 and 1967 (Headland 2002). Continued volcanic unrest and the realization that Deception had a restless caldera due to possible volcanic resurgence related to one or more shallow magma chambers has resulted in the two existing stations on the island being summer-only. The evidence for resurgence comprises persistent seismicity, short-lived seismic crises, epicentre resolutions and localized rapid shallowing of the caldera floor beyond normal sedimentation rates (Kowalewski et al. 1990; Ortíz et al. 1992; Cooper et al. 1998; Ibáñez et al. 2003; Carmona et al. 2010). In addition, seismic evidence for hydrothermal venting in Port Foster and metallic element anomalies in the basin-floor sediments, and the presence of numerous fumarolic steam fields situated around the shores of Port Foster, confirm that the volcano should be regarded as active (Rey et al. 2002; Somoza et al. 2004). A swath survey of Port Foster by Barclay et al. (2009) found no evidence for the localized resurgence postulated by Cooper et al. (1998) but the data may still be consistent with a broad region of uplift along the eastern side of the basin. Cooper et al. (1998) also suggested a model of ongoing trapdoor subsidence to explain the SW-dipping seafloor sediment layers identified seismically (Kowalewski et al. 1990). Geophysical surveys suggest that a putative magma chamber with a volume of c. 150 km^3 may be present today, possibly as shallow as 500 m but probably below 1 km, and may extend down to at least 4–5 km (Muñoz-Martín et al. 2005; Berrocoso et al. 2006; Ben-Zvi et al. 2007; Zandomeneghi et al. 2007; Pedrera et al. 2012). Its existence has been challenged on conceptual grounds in favour of several smaller post-caldera magma batches fed by a deeper source that erupted in the numerous small-volume post-caldera volcanic centres (Martí et al. 2013), similar to the idea proposed by Roobol (1980) and Smellie (2002a), and for which there is geochemical support.

Port Foster is surrounded by an annular ring of low hills and ridges (Fig. 4c). Together, they are the morphological expression of a volcanic depression caused by a major volcano–tectonic collapse that formed a large caldera with a diameter of c. 9–11 km (SW–NE and NW–SE, respectively). There are two principal models for the origin of the Deception Island caldera: (1) it formed as a classic example of piston collapse following one or more major eruptions. Eruptions were associated either with a ring fracture (Hawkes 1961; González-Ferrán and Katsui 1970; Baker et al. 1975; Walker 1984; Smellie 1988, 1989) or with a series of pre-existing intersecting rectilinear faults induced on the island by Bransfield Strait regional tectonics, perhaps related to left-lateral strike-slip that may have caused a counter-clockwise rotation of the island (Smellie 2001, 2002a; Smellie and López-Martínez 2002; Maestro et al. 2007; Martí et al. 2013; Berrocoso et al. 2016); or (2) the depression formed progressively by passive (non-volcanic) extension along several sets of intersecting orthogonal normal faults linked to regional extension, and its formation was unrelated to any major eruption (Martí and Baraldo 1990; Rey et al. 1995; Martí et al. 1996; Lopes et al. 2014). In support of the latter hypothesis, the caldera limits are rectilinear fault traces that are clearly seen on satellite images, and there is an absence of arcuate ring faults, radial dykes and related fractures. Moreover, the distribution of the seismic epicentral locations is dispersed across the island with no clear locus along a ring-fault system (Carmona et al. 2010). Conversely, the most widespread and voluminous stratigraphical unit (Outer Coast Tuff Formation (OCTF), see later in this subsection) is unequivocally identified as the caldera-forming deposit (Hawkes 1961; Baker et al. 1975; Smellie 2001, 2002a; Martí et al. 2013). Thus, whilst a strong regional tectonic control on the fault-bounded margins of the caldera is clear, the link to a major voluminous eruption indicates that the caldera formation is primarily volcano–tectonic, formed as a consequence of a major tephra-producing eruption that may have ejected between 30 and 90 km^3 of tephra (c. 20–63 km^3 dense-rock equivalent (DRE): Smellie 2002a; Martí et al. 2013); the smaller estimate is crude and was based principally on the volume of the present caldera topographical depression, whereas the larger estimate, likely to be more accurate, relies on the geophysical identification of a 1.2 km-thick unit subsided within the caldera and which is correlated with the OCTF.

The geological evolution of the Deception volcano is most conveniently described in two major parts assigned to pre-caldera and post-caldera periods. A historical comparative stratigraphy of Deception Island is shown in Figure 7. A simplified map of the geology is shown in Figure 8 and the major relationships are illustrated in Figure 9. Summary details of the lithostratigraphy, including constituent lithofacies, defined members and non-stratigraphical eruptive units (cone clusters), are given in Figure 10. The only published isotopic age is 153 ± 46 ka (by K–Ar: Keller et al. 1992) and the results of palaeomagnetic studies of all the eruptive units indicate that the volcanism has a normal polarity and, hence, an age <780 ka (Blundell 1962; Valencio et al. 1979; Baraldo et al. 2003). Based on a recent palaeomagnetic study, ages of >12 ka were determined for pre-caldera rock units, c. 8.3 ka for the caldera collapse and ages younger than 2 ka on post-caldera units (Oliva-Urcia et al. 2016). However, the most likely age for the caldera collapse is 3980 ± 125 years BP, based on tephrochonology studies and ^{14}C determinations (Antoniades et al. 2018).

The pre-caldera period is the most voluminous as it includes the submarine part of the Deception volcano. Most of the record is inaccessible but, by analogy with better-known oceanic volcanoes and the oldest basal outcrops exposed on the island, it is probably dominated by pillow lava and volumetrically minor hyaloclastite (Smellie 2001, 2002a). The earliest episode of shoaling and the emergence of the volcano is represented by the Fumarole Bay Formation, consisting mainly of yellow-coloured tephra erupted from several vents situated across the island, both within and outside of the caldera. It has also been called the Yellow Scoria Formation by Baraldo and Rinaldi (2000) (Fig. 7) in reference to the dominance of

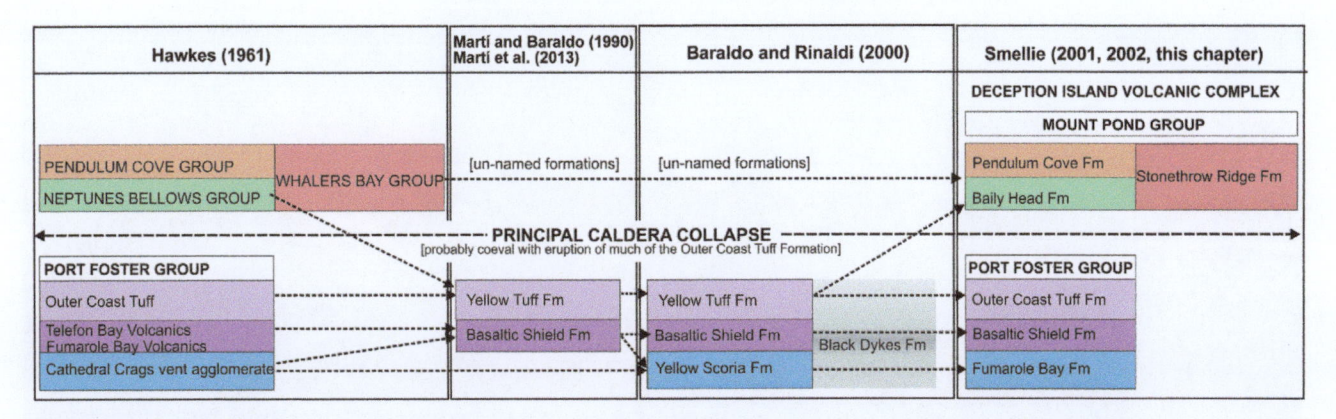

Fig. 7. Comparative stratigraphy of Deception Island.

palagonite-altered (and, hence, yellow) scoria-rich lapillistones. Baraldo and Rinaldi (2000) inferred that their Yellow Scoria Formation may represent an early syn-caldera unit (together with their Black Dykes Formation: Fig. 7) based on an inferred spatial relationship with caldera faults. Several of the locations of the associated likely major eruptive sites coincide with NE-trending caldera faults but others are outside of the caldera (Fig. 8). From a seafloor swath survey, Barclay et al. (2009) noted that the SW–NE-trending structures are dominant in and around Deception, with very little evidence for the cross structures commonly identified by other workers (e.g. Martí et al. 1996; Rey et al. 2002; Maestro et al. 2007). The Fumarole Bay Formation lacks the abundant reworked clasts of bedrock that might be expected in tephras related to caldera collapse (as in the OCTF; see later in this subsection). There is thus no strong reason to infer a genetic relationship to caldera collapse. The deposits of the Yellow Scoria Formation and Fumarole Bay Formation, together with products of a subaerial lava shield, were subsumed into the Basaltic Shield Formation of Martí and Baraldo (1990) and Martí et al. (2013) (Fig. 7) on the basis that the eruption dynamics, products and stratigraphy of the two units do not differ. However, the Fumarole Bay Formation is almost entirely formed of coarse yellow tephra, whilst the Basaltic Shield Formation of Smellie (2001, 2002a) is composed of grey subaerial lavas with minor red and grey Strombolian interbeds. The two units are thus very easily distinguished and their stratigraphical separation is retained here.

The Fumarole Bay Formation exceeds 200 m in exposed thickness and has three named members (from base up): the Lava Lobe Member, Scoria Member and Stratified Lapilli Tuff Member. The Lava Lobe Member is an association of yellow hyaloclastite breccia and irregular amoeboid, lobate and pillow-like masses of lava. It is exposed at only three localities and represents the top of the submarine volcanic pile prior to emergence. It is overlain by the Scoria Member, which consists of massive monomict fine scoria lapillistone and coarse lapilli tuff, mainly yellow coloured and petrologically dissimilar to the underlying Lava Lobe Member (Fig. 11a). The deposits are formed of fine sideromelane scoria (largely palagonite altered and commonly showing blocky shapes with planar fracture surfaces, and a wide range of vesicularity), rare tachylite and coarse tuff, and cowpat and spindle bombs (exceptionally 2 m in diameter) that are often broken. The origin of the Scoria Member is uncertain but there is a clear resemblance to Strombolian deposits (see also Baraldo and Rinaldi 2000; Baraldo et al. 2003). However, the abundance of sideromelane with broken blocky shapes and a wide range of vesicularity, the widespread palagonite alteration, and the lack of fine tuff suggest that it may have formed during episodes of relatively weak explosive activity, perhaps akin to submarine fire fountaining (Smellie 2002a). The gradationally overlying Stratified Lapilli Tuff Member is formed of well-stratified deposits rich in fine tuff, palagonite-altered sideromelane and numerous accessory clasts. The overall characteristics indicate a phreatomagmatic origin, presumably during emergence and construction of a tuff cone edifice, as access of the seawater to the erupting vent became less efficient and magma–water mixing ratios became more optimal for more violent explosivity (e.g. Wohletz 1986; Sohn 1996). No erosional surfaces are known that separate the different centres in the Fumarole Bay Formation (identified by the variable bedding orientations) and they may have been co-eruptive (i.e. erupted during one continuous eruptive

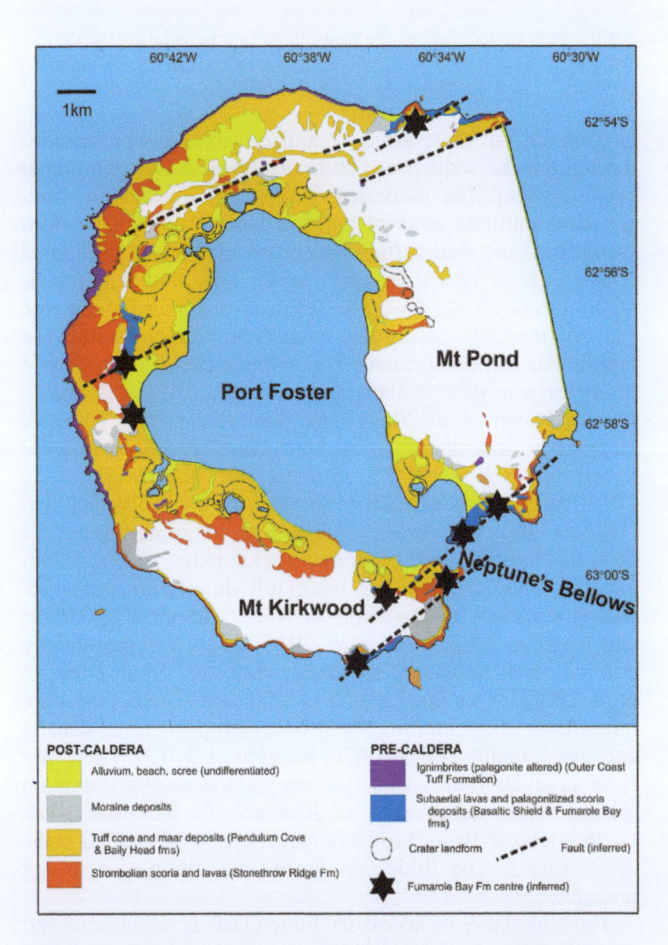

Fig. 8. Simplified geological map of Deception Island (after Smellie 2002a). The locations of Fumarole Bay Formation centres, erupted during the shoaling and emergence phase of the island, are also shown. Note the strong association with NE–SW tectonic fractures on the island.

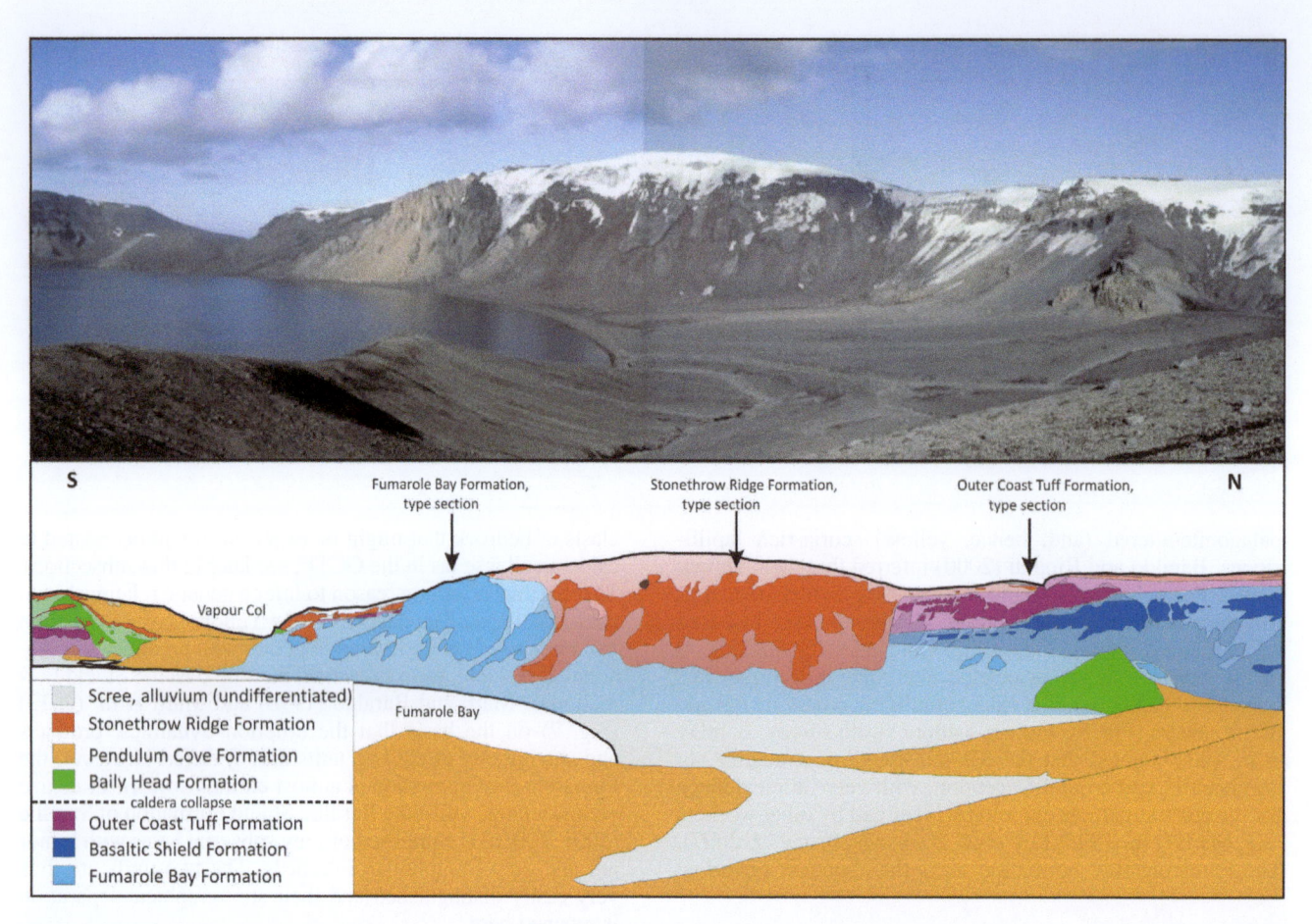

Fig. 9. View and sketch of relationships between the major eruptive units exposed on the inner caldera wall on the west side of Deception Island (modified after Smellie 2002a).

cycle). Conversely, as the centres occur in geographically separate groups (clusters: Fig. 8) not in contact with one another, a more episodic eruptive history involving different groups erupting at different times is possible.

The Basaltic Shield Formation (*sensu* Smellie 2001, 2002*a*) is only exposed on the west side of the island, mainly at Stonethrow Ridge, but also on the western outer coast. It is up to 110 m thick, and is composed of platy jointed sheet-like and pāhoehoe lavas with oxidized surfaces, and dark grey to red scoria lapillistone beds towards the top. Eruption was by subaerial effusion and minor low-energy Hawaiian–Strombolian activity. It formed a small lava shield that probably had an original basal diameter of c. 6–8 km.

The OCTF is the most widely distributed formation on the island. It was also called the Yellow Tuff Formation by Martí and Baraldo (1990) (Fig. 7), and was divided into lower and upper members. However, the upper member recognized by Martí and Baraldo (1990) was subsequently assigned an early post-caldera age based on mapped field relationships (Smellie 2001) and palaeomagnetic results that show a significant time difference between the two units (Baraldo *et al.* 2003). It is therefore called the Outer Coast Tuff Formation (OCTF) based on the name chosen by Hawkes (1961) for its historical precedence (Smellie 2001). Outcrops of the OCTF on the summit of Cathedral Crags were assigned a post-caldera historical age and assigned to a new formation (Cathedral Crags Formation) by Caselli and Agusto (2004) but the correlation is compositionally based and is ambiguous, and is not followed here. The OCTF is 50–70 m thick in most outcrops but locally exceeds 90 m. It is formed of multistorey bedsets individually up to 30 m thick that are continuously traceable several kilometres in the western outer coast cliffs (Fig. 11b). The bedsets fade out laterally and amalgamation is common but there are no major erosional unconformities throughout the sequence. Crude coarse columnar jointing is rarely present. The individual bedsets are dominantly composed of multiple units of massive lapilli tuff 1–15 m thick (usually <3 m). Bed surfaces are often marked by prominent lenses rich in coarse juvenile lapilli and abundant (up to 80%) accessory blocks up to 1.5 m in diameter. Minor thin (<1 m) interbeds of crudely planar or rare cross-stratified lapilli tuffs are also present. The juvenile clasts are blocky to rarely cuspate sideromelane and some abraded or ash coated (*sensu* Brown *et al.* 2012). The sideromelane has basalt to silica-poor andesite compositions (Smellie 2002*a*; unpublished information of J.L. Smellie; see also Martí *et al.* 2013) but there is a small proportion of dacite pumice and obsidian clasts too. Accessory clasts are common throughout beds (up to 35%), and comprise a wide range of Deception-sourced lavas, scoria, yellow lapilli tuff, dolerite and gabbroid, and rare accidental fragments of felsic plutonic rock (Risso and Aparicio 2002; Martí *et al.* 2013) (Fig. 11c), presumably derived from subjacent basement rock (cf. Muñoz-Martín *et al.* 2005). A fine tuff matrix is abundant. The deposits are interpreted as ignimbrites, the products of pyroclastic density currents (Smellie 2001, 2002*a*; Martí *et al.* 2013). They vary from granular fluid based (i.e. the volumetrically dominant massive beds, terminology of Branney and Kokelaar 2002: cf. 'pyroclastic flows') to fully dilute (i.e. the stratified beds: cf. 'surges'). The lithic-rich lenses are interpreted as lag breccias.

The abundance of accessory lithic clasts is remarkable and the proportions of the different accessory clast types vary across the island (Martí *et al.* 2013). Their presence is attributed to incorporation during caldera-collapse episodes, probably during vent widening. The magma chamber may have

Deception Island Volcanic Complex

Mount Pond Group
Post-caldera deposits, the products of phreatomagmatic (Baily Head (early) and Pendulum Cove (late) formations) and magmatic (Stonethrow Ridge Formation) eruptions

Pendulum Cove Formation
Predominantly tuff cone and maar deposits; deposits typically grey and loose or friable; primary landforms well preserved; centres occur in geographically well-defined clusters within caldera although tephra deposits generally blanket the island

Telefon Bay cones*	Deposits of 1967 and 1970 eruptions
Kroner Lake cones*	Three eruptive centres; mainly 19th century
Crimson Hill cones*	At least two large centres; pre-1829
Cross Hill cones*	Multiple centres, pre-1829; includes **White Ash Member**
Vapour Col cones*	Several large centres, pre-1829
Crater Lake cones*	Multiple co-eruptive centres; probably pre-1829
Collins Point cones*	Two centres; includes dacite lava(s)

Stonethrow Ridge Formation
Mainly fissure-erupted, post-caldera Strombolian scoria and lavas (many clastogenic); subdivision mainly based on known or inferred relationship to moraine deposits

Mount Kirkwood Member
Includes products of 1839, 1842 and 1969 eruptions

-------- MORAINES -------- -------- MORAINES --------

Baily Head Formation
Indurated, grey to khaki hydrovolcanic tephra; primary landforms not preserved; centres not restricted to caldera

Degraded, mainly indurated tuff cone deposits

Kendall Terrace Member
At least partly older than Baily Head Formation

-------- CALDERA COLLAPSE --------

Port Foster Group
Pre-caldera deposits, mainly hydrovolcanic tephra, typically indurated and bright yellow; also lavas and Strombolian scoria

Outer Coast Tuff Formation
Indurated yellow lapilli tuffs; particularly conspicuous in outer coast cliffs; multiple thick pyroclastic current deposits

Basaltic Shield Formation
Simple and compound sheet lava flows; Strombolian scoria

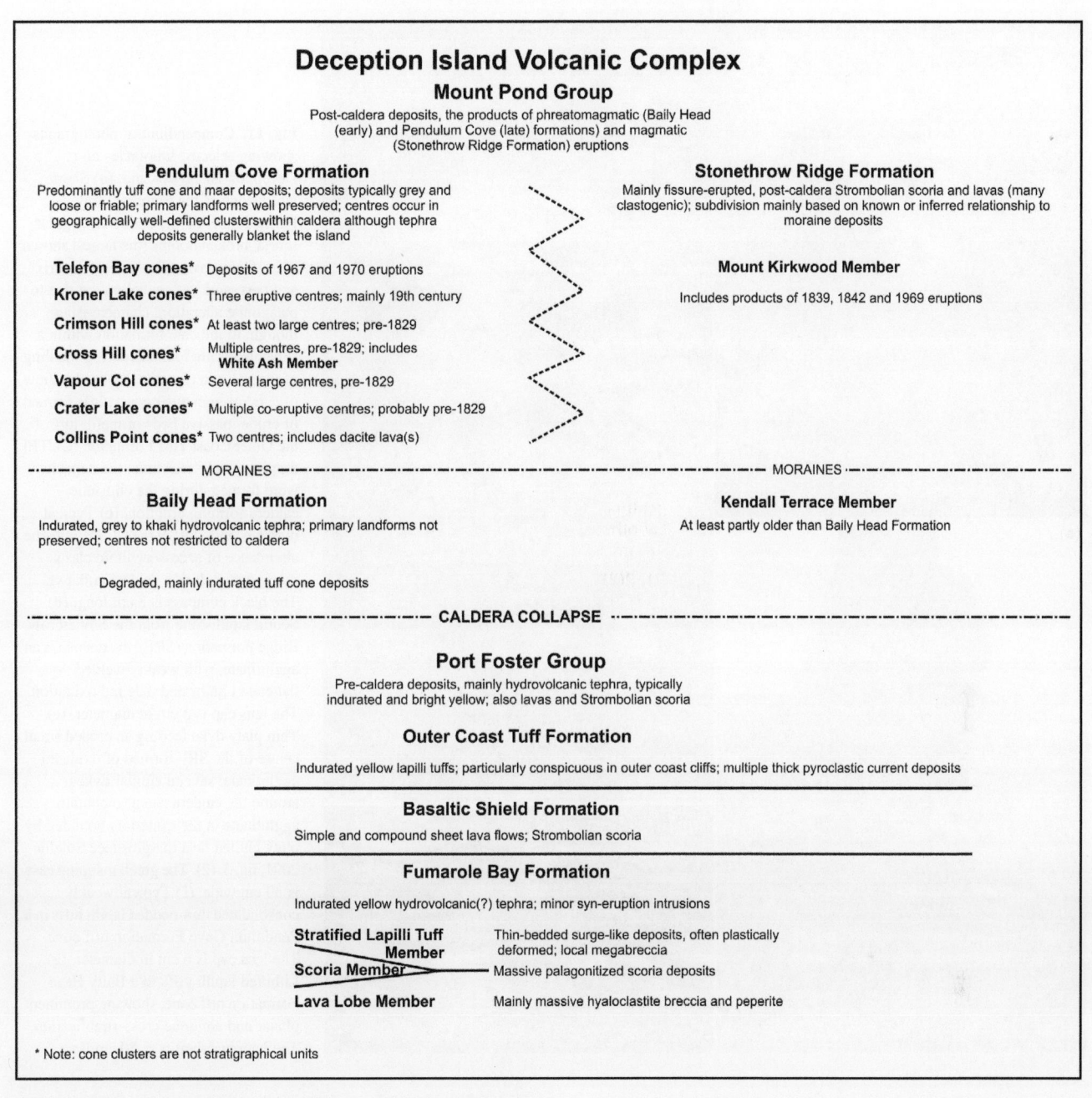

Fumarole Bay Formation
Indurated yellow hydrovolcanic(?) tephra; minor syn-eruption intrusions

Stratified Lapilli Tuff Member	Thin-bedded surge-like deposits, often plastically deformed; local megabreccia
Scoria Member	Massive palagonitized scoria deposits
Lava Lobe Member	Mainly massive hyaloclastite breccia and peperite

* Note: cone clusters are not stratigraphical units

Fig. 10. Summary lithostratigraphy of Deception Island (from Smellie 2001).

---

been situated at a depth of c. 4–6 km (by clinopyroxene–glass geobarometer: Martí et al. 2013). An important trigger for the eruption was probably the incursion into the chamber by a fresh batch of compositionally different mafic magma, causing convective overturn (Smellie et al. 1992; Smellie 2002a). The overall characteristics of the OCTF indicate that the eruption was phreatomagmatic, and the abundance of accessory lithics indicates that the interaction with water probably took place at a shallow level in the conduit. Moreover, the large erupted volume of the formation implies high mass eruption rates and a likelihood that the interacting water was seawater rather than from an aquifer (cf. De Rita et al. 2002). Seawater was also suggested by Martí and Baraldo (1990) and Martí et al. (2013) on the basis of the alteration minerals present. Rapid vent widening and a high discharge rate, together with a high proportion of (cold, dense) lithic fragments and ingress of abundant seawater, probably caused the generation of a low non-convective eruption column consistent with the absence of a Plinian fall layer (Martí and Baraldo 1990;

Martí et al. 1991, 2013). How the water gained and maintained access to the vent is unclear. Martí et al. (2013) suggested the entrance of water via regional faults, melting of an ice cap or sector collapse of the volcano flanks during caldera collapse (e.g. on its SE side). Smellie (2001) and Smellie and López-Martínez (2002) suggested a contemporary sea-level some 200 m higher than present – that is, sufficiently high to surmount the topography surrounding Port Foster. However, the necessity for major sea-level variations, presumably driven by volcanotectonism rather than eustatic effects, seems implausible and there is otherwise no supporting evidence for such a high contemporary sea-level. A possible explanation for how the enclosing caldera rim was surmounted is that infilling of the developing caldera depression with OCTF products more or less matched the space created by subsidence. Thus, the depression was full, at least initially, and the pyroclastic density currents were able to 'spill over' the caldera rim and flow down the outer slopes. Given that the caldera today is a prominent deep depression, it implies

Fig. 11. Compendium of photographs showing selected lithofacies on Deception Island volcano. (**a**) Scoria lapillistone of the Fumarole Bay Formation. Note the abundant angular scoria, broken bombs (the largest shown is *c.* 10 cm across), lack of tuff matrix and pervasive yellow coloration due to palagonite alteration, characteristics thought to indicate formation within a subaqueous fire fountain during shoaling of the volcano (Smellie 2001). (**b**) View of *c.* 60 m-high outer coast cliffs formed of crude massive beds of lapilli tuff of the Outer Coast Tuff Formation (OCTF) dipping gently seawards; the deposits were formed during the climactic caldera-forming eruption. (**c**) Typical massive lapilli tuff of the OCTF; note the abundance of accessory lithic clasts incorporated during caldera collapse. The black compass is 8 cm long. (**d**) Scoria lapillistone from the Stonethrow Ridge Formation (SRF); the deposit is an agglutinate, with weakly welded flattened bombs and pale red oxidation. The lens cap is 6 cm in diameter. (**e**) Thin platy dyke feeding an eroded small centre of the SRF formed of oxidized agglutinate; several similar dykes around the caldera rim also contain agglutinate in their interiors bounded by platy chilled lava margins (see Smellie 2001, fig. 3.12). The green mapping case is 30 cm wide. (**f**) Typical weakly consolidated thin-bedded lapilli tuffs in a Pendulum Cove Formation tuff cone. The lens cap is 6 cm in diameter. (**g**) Lithified lapilli tuffs in a Baily Head Formation tuff cone, showing prominent planar and antidune cross-stratification. The hammer shaft is *c.* 50 cm long.

that subsidence continued well after the OCTF eruption had ceased. However, it remains unclear how seawater was able to continuously interact with the magma throughout the eruption.

The widespread distribution of the OCTF deposits over the whole island is consistent with radial distribution from a collapsing vertical eruption column sourced in a central vent. The presence of 1.2 km of OCTF deposits within the caldera suggests that subsidence began early in the eruption, allowing substantial accumulation within the developing depression, and the absence of internal unconformities suggests that deposition of the OCTF took place over a relatively short period of time probably corresponding to a single event (Smellie 2001; Martí *et al.* 2013). However, the provenance of accessory clasts in the deposits varies with geographical location, suggesting that different caldera-related fractures might have opened at different times (Martí *et al.* 2013) (Fig. 12).

Post-caldera activity is predominantly pyroclastic and the deposits dominate the surface geology (Fig. 8) (Hawkes 1961; Smellie *et al.* 2002). It is divided into three stratigraphical parts: Baily Head Formation, Pendulum Cove Formation and Stonethrow Ridge Formation (Fig. 10). The Stonethrow Ridge Formation has a minimum thickness of 100 m and crops out mainly along the summit ridge encircling Port Foster, where it was erupted from fissures. It consists of red and black scoria, maroon agglutinate, and grey lavas, locally fed by thin dykes (Fig. 11d, e). Many of the lavas are clastogenic and only travelled short distances inside the caldera but several travelled to the outer coast where they formed extensive lava-fed deltas, and at least one travelled to the southern shore of Port Foster and formed a short delta. Two members are defined, an older one (Kendall Terrace Member) which is overlain by glacial moraines (often scoria rich) and forms most of the Stonethrow Ridge Formation, and a younger one (Mount Kirkwood Member) with a much more restricted distribution. The Kendall Terrace Member includes widespread lavas and is locally associated with prominent rows of small craters. It also includes a spectacular 'curtain' of clastogenic lavas that drapes the east face of Stonethrow Ridge (Fig. 9). The Mount Kirkwood Member is one of the better dated geological units, with eruptions observed in 1839–1842 and 1969. One of the lavas associated with the 1839 episode flowed down into Crater Lake where it formed a small lava-fed delta with littoral cones, whilst those in 1842 were associated

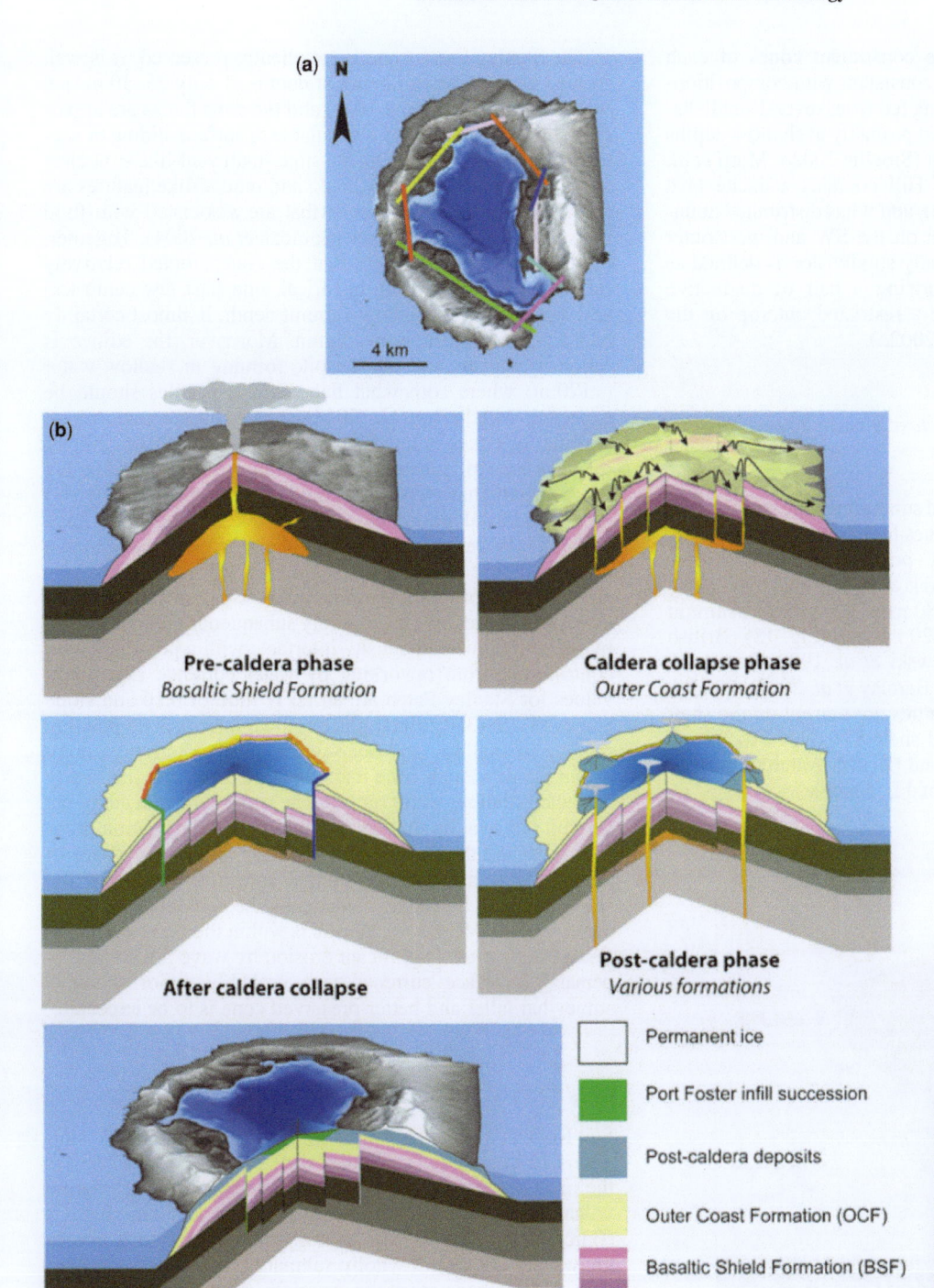

Fig. 12. Schematic diagrams showing the evolution of Deception Island, including eruption of the Outer Coast Tuff Formation during collapse of the Deception Island caldera (from Martí *et al.* 2013). Note that the Basaltic Shield Formation, as depicted here, combines both the Fumarole Bay (shoaling phase) and Basaltic Shield (subaerial phase) formations of Smellie (2001, 2002*a*).

with spectacular fire-fountaining activity. Both episodes formed well-preserved crater rows on the inner slopes below Mount Kirkwood. Deposits and the impact of the 1969 eruption were described by Smellie (2002*b*). Eruptions of the Stonethrow Ridge Formation were entirely subaerial and of Hawaiian and Strombolian type, thought to be due to the position of the erupting fissures on topographical highs away from sources of seawater or groundwater (Baker *et al.* 1975). Although the 1969 eruption was glaciovolcanic, the overlying ice was thin (*c.* 70 m) and did not have any detectable influence on the resulting deposit characteristics (Smellie 2002*b*).

The Baily Head and Pendulum Cove formations are almost entirely pyroclastic, formed of numerous tuff cones up to 180 m high constructed mainly of thinly stratified khaki-grey to yellowish-grey lapilli tuffs (pyroclastic density current deposits: Fig. 11f) (Smellie 2001, 2002*a*). The Baily Head Formation was erupted in early post-caldera time. There are relatively few outcrops and they are widely distributed both inside and outside of the caldera. No primary landforms are preserved, although the locations of the original craters can often be inferred from quaquaversal bedding orientations. Planar stratification is prevalent and is locally associated with dune bedforms, including likely antidunes (Fig. 11g). By contrast, the vents responsible for the Pendulum Cove Formation crop out only within the caldera but their tephra covers most of the island (Figs 8 & 9). They include the centres that erupted in 1967 and 1970, including several water-filled maars (Baker *et al.* 1975; see also Geyer *et al.* 2021). The tuff cones crop out in well-defined clusters. Most of the clusters are geochemically coherent and the constituent cones of each appear to have been co-erupted, consistent with compositionally stratified magma batches being fed from several small discrete magma chambers distributed probably at shallow depths around the margins of Port Foster (Smellie 2002a; Martí et al. 2013). The tuff Cone at Cross Hill contains a dacite lava (unpublished data of J.L. Smellie), and it has distributed abundant dacite pumice and obsidian on the SW and west outer slopes of the cone edifice. The only subdivision is defined as the White Ash Member, comprising a pair of distinctive white dacite ash beds that have a restricted outcrop on the west side of the island (Smellie 2002a).

Submarine volcanic centres in close vicinity to Deception Island

There are at least three unsampled submarine volcanic features in Port Foster. The two smaller ones lack craters and are either degraded pyroclastic cones or possibly elongate pillow mounds. The third (Stanley Patch) is a well-formed pyroclastic cone c. 500 m in basal diameter, 80 m high, and with a summit crater c. 100 m wide and c. 15–20 m deep (Fig. 13) (British Antarctic Survey 1987; Kowalewski et al. 1990; Rey et al. 1992, 2002; Somoza et al. 2004; Barclay et al. 2009). Finally, two small submarine satellite centres are present on the shallow (c. 60 m deep) wave-eroded shelf SW (New Rock) and NE (unnamed) of Deception Island (British Antarctic Survey 1987; unpublished information of J.L. Smellie).

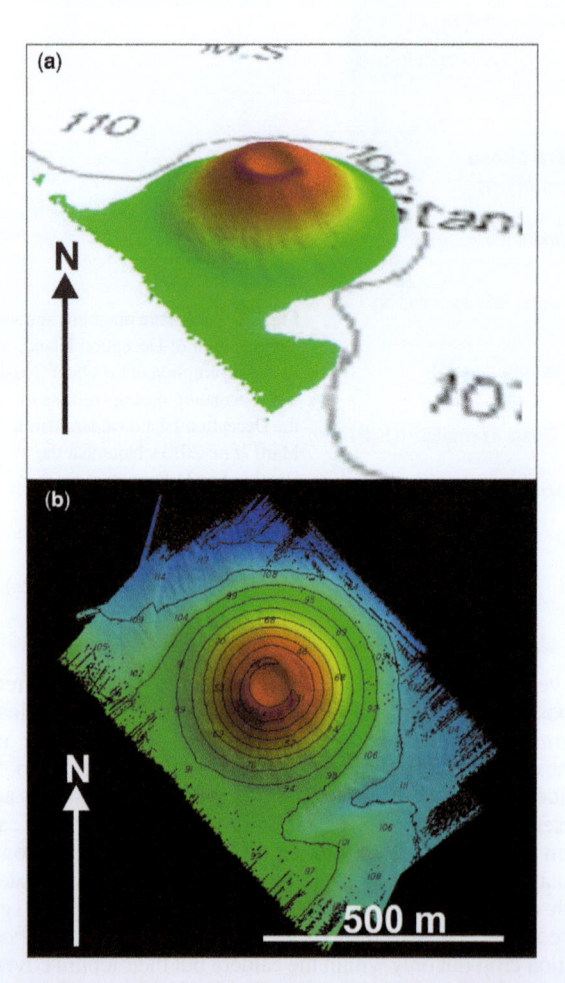

Fig. 13. Swath images of Stanley Patch, a small well-formed pyroclastic cone in Port Foster measuring c. 80 m high and 400–500 m wide; contour interval 10 m. Image: J.L. Smellie unpublished, courtesy of HMS *Endurance*.

The Stanley Patch cone is excellently preserved. It is well formed despite the rim being at depths of only 25–30 m (i.e. well within normal wave base) and the cone flanks are apparently free of indentations attributable to surface sliding or sector collapses. A fissure with possible small vent-like structures may be present in the NW flank, and mound-like features are present to the SE of the cone that are associated with fluid emission in buoyant plumes (Somoza et al. 2004). Together, these observations suggest that the cone erupted relatively recently, probably within historical time (last few centuries) and, because of its shallow summit depth, it almost certainly had a subaerial eruption column. Moreover, the edifice is relatively narrow and tall, despite forming in shallow water (<120 m) where somewhat flatter wider profiles should be expected. Mitchell et al. (2012) showed that a majority of pyroclastic cones erupted at shallow water depths (<300 m: depth to *base* of volcano), broadly similar to those for Stanley Patch, usually have low aspect ratios (i.e. height/width (H/W) <0.2) and low slope angles (<20°); cones in deeper water have higher H/W ratios and flank gradients closer to the stable angle of repose for marine basaltic talus (28°–30°). The overall flattening of shallow-water cones was attributed to the spreading of the erupting, entirely subaqueous column caused by the abrupt air–sea density interface, with a possible lesser contribution from reworking by water currents. Equivalent values for Stanley Patch are an H/W ratio of 0.16 and slope angles of 25°–30°, which place it in an intermediate position relative to shallow- and deep-water cones. The morphology of the Stanley Patch cone may reflect its location, in a highly protected shallow-water setting. The very shallow depth of the summit (25 m at its shallowest) implies that the column probably breached the water surface, at least in the final stages of activity, thus avoiding substantial spreading effects of the sea–air density barrier and enhancing the vertical aggradation of the edifice. Moreover, the site is within the flooded Deception caldera where significant erosion by wave fetch and dispersal by surface currents were probably minor. Thus, a somewhat taller and better-preserved cone is to be expected.

Sail Rock

Sail Rock is a prominent sea stack that rises to 28 m asl, situated c. 15 km SW of Deception Island (Fig. 4d). However, it is the subaerial expression of a large very-low-profile submarine volcanic mound with a basal diameter of c. 5 km (Fig. 2) (Barclay et al. 2009). The geology of the volcano is poorly known but the almost wholly submarine edifice is associated with a prominent magnetic anomaly (Catalán et al. 2013), which implies that it is dominated by pillow lava. By analogy with Deception Island, the pillow lava is probably associated with minor hyaloclastite breccia and possibly a thin capping of lapilli tuffs formed explosively during shoaling but since largely removed by marine erosion and iceberg scouring. Sail Rock itself is composed of alternating brown lavas and reddish-coloured breccias; analysed lavas have basaltic andesite and andesite compositions (Keller et al. 1992; Kraus et al. 2013). There are no isotopic ages. However, the discovery of a submarine tephra layer in the Central Bransfield Basin, with an age probably less than a few hundred years old and compositions unlike Deception but showing some similarity to lava on Sail Rock, suggests that the Sail Rock volcano may be the source (Fretzdorff and Smellie 2002). If true, it implies a very young age for the volcano and that it probably should be considered potentially still active (dormant). However, for such a young age, there is no crater preserved at Sail Rock. An alternative source for the tephra layer, in a hydrothermally active submarine edifice in Central Bransfield Basin, was also suggested by Fretzdorff and Smellie (2002) because of its greater proximity and the presence of ash at the site suggestive of explosive activity, despite its substantial depth (1050 m). However, it is very doubtful whether any tephra could have become subaerially ejected from such a deep volcano.

Young volcanism also forms outcrops at several localities on King George, Greenwich and Livingston islands (Fig. 2).

King George Island

Only a single locality is known on King George Island, at Melville Peak (549 m asl), which consists of the remnants of a small stratocone with a central basaltic plug (Barton 1965; Birkenmajer 1982b; Smellie 1990b). Although largely inaccessible, an excellent cross-section through the cone is well exposed in 450 m-high cliffs caused by marine erosion on its southern side (Fig. 4e). The volcano rests unconformably on a gently undulating erosional surface cut across Oligocene and Miocene sediments (Troedson and Riding 2002). The only detailed description is by Birkenmajer (1982b). Two units of formation status have been identified and were ascribed to two principal phases of eruption. The most important and volumetrically dominant eruptive phase is included in the older Hektor Icefall Formation. It comprises interbedded basalt lavas and pyroclastic deposits with essentially quaquaversal dips of 30°, locally rising to 90°. The pyroclastic deposits were described as alternating yellow, cross-stratified well-cemented 'agglomerates', 'explosion breccias', fine breccias, 'psammitic tuffs or tuff sandstones' and grey lapilli tuffs. They are interbedded with grey to black basalt lavas 5–20 m thick with prominent narrowly spaced (10–30 cm) columnar jointing and autobreccia. Like the pyroclastic rocks, the lavas are locally steep dipping. Birkenmajer (1982b) used a mixture of volcanic and sedimentary terminology to describe the clastic rocks. However, interpreting characteristics such as the steep quaquaversal dips, occurrence within an eroded volcanic cone and intercalation with lavas suggests that the clastic rocks are pyroclastic, and probably mainly lapilli tuffs, tuffs and volcanic breccias. Together with the yellow coloration (palagonite alteration?) and cross-stratification, it is suggested that they formed predominantly as subaerial deposits of fully dilute pyroclastic density currents (*sensu* Branney and Kokelaar 2002). The outcrop is probably a small tuff cone or stratocone. The description of the lavas suggests that they might be 'a'ā lavas and they contain narrow prismatic jointing with varied orientations that may be entablature caused by water cooling. Although the origin of the water involved is unclear, should the presence of 'tuff sandstones' be confirmed it may imply that early emplacement was in a marine setting (and subsequently uplifted), although no pillow lava has been seen. The volcanic succession is intruded by a prominent grey basalt plug c. 300 m in diameter that rises to 549 m asl and shows well-developed vertical prismatic jointing. The later eruptive phase is named the Deacon Peak Formation (after the volcanic feature on Penguin Island, without inferring any time-equivalence) and consists of red bomb-rich basaltic agglomerate and red scoriaceous lava that fill a small vent on the eastern flank of Melville Peak. It was interpreted as a small parasitic vent that erupted after the main eruptive phase and after an intervening unspecified period of erosion, possibly glacial. K–Ar dating has yielded ages of c. 0.296 ± 0.027, 0.231 ± 0.019 and 0.072 ± 0.015 Ma for three lavas from the older stage of the volcano (Hektor Icefall Formation: Birkenmajer and Keller 1990). However, an ash layer in a marine core obtained in Bransfield Strait c. 30 km distant from Melville Peak has a composition that may be comparable with published analyses of Melville Peak lavas (Keller *et al.* 2002, cited in Kraus *et al.* 2013), suggesting Holocene activity despite the absence of any crater landform. The volcanic edifice may have had an original basal diameter at least 3 km. A cone edifice with those dimensions would have surface slopes approximately at the angle of repose, consistent with the internal bedding attitudes and the reconstruction sketched by Birkenmajer (1982b, fig. 7). On that basis, eruption may have been in an unconfined setting (i.e. in the sea or a pluvial lake). A glacially confined setting for the eruptions would result in an edifice with greatly oversteepened flanks due to the tephra beds banking against the confining ice walls (cf. Smellie *et al.* 2006a; Smellie 2009, 2018). A marine setting may thus seem more likely but would imply that the sea-level at 29 or 72 ka was at c. 200 m asl, for which evidence is lacking (cf. John and Sugden 1971). Conversely, eruption was subaerial but with unconstrained access of abundant groundwater (probably sourced in the sea: cf. Sohn 1996; Smellie *et al.* 2016) to the vent. In the latter scenario, which is favoured here, abundant accidental lithic clasts would be expected in the tephras, and may correspond to the numerous fragments of 'consolidated lavas and baked pelitic sediments' described by Birkenmajer (1982b, p. 346). Final eruptive activity at the small vent (Deacon Peak phase) was fully subaerial, when water was presumably excluded from the vent.

Greenwich Island

A single outcrop of young volcanic rock is present on Greenwich Island, and forms Mount Plymouth. It has an isotopic age of 0.2 ± 0.3 Ma (by K–Ar: Smellie *et al.* 1984; Table 3). The outcrop has been studied only at reconnaissance level (Hobbs 1968 and unpublished field notes). The topographical feature is a prominent snow-covered dome-like nunatak rising to 520 m (almost 300 m above the surrounding ice) with prominent rock cliffs on its northern side. It is extensively eroded on all sides and is just a volcano-remnant. The nunatak is c. 600 m wide and its geology is simple (Figs 14 & 15). An alkali olivine basalt mass is exposed at the west end of the cliffs and was interpreted as a plug by Hobbs (1968), who described dominant narrow (15 cm diameter) horizontal columnar joints that change upwards across a nearly abrupt transition to vertical and wider (60 cm) columns. The plug intrudes 'thinly bedded flow agglomerates' (probably lapilli tuffs). The rock face at its SE end, extending almost to the summit, shows thin planar strata that onlap and define a prominent crater rim 'unconformity', suggesting that the outcrop is largely a pyroclastic cone (Fig. 15) (unpublished information of J.L. Smellie). The exposed section reveals a relatively shallow dip of the cone outer slopes (c. 15°), suggestive of a tuff ring. However, there is a hint that the dips may become steeper in the upper half of the exposed section. The position of the crater rim migrates to the east or ESE in the older beds, before possibly becoming near-vertical later (Fig. 15), indicating a widening of the crater during the early eruptive stages followed by rapid vertical aggradation. The stratal relationships suggest that the cone began as a tuff ring and evolved up into a tuff cone. Assuming that the basalt plug was located centrally, the original volcano may have been at least 1–1.2 km in width. The laterally continuous strata suggest eruption in an unconfined (non-ice-confined) setting. Moreover, the elevation of the edifice base (c. 250 m asl) is well above the limit for the highest marine-cut surfaces in the South Shetland Islands (c. 120 m asl: John and Sugden 1971), which suggests that the eruption took place under subaerial conditions, either in an ice-free or ice-poor setting. The eruption was phreatomagmatic.

Table 3. *Summary of published isotopic ages for volcanic rocks in the northern Antarctic Peninsula region*

Sample	Locality	Dated lithologies	Age*	Error*	Method	Reference	Notes
Bransfield Basin							
n.r.	Eastern Seamount	Basalt lava	53 ka	36 ka	K–Ar	Fisk (1990)	
n.r.	Western Seamount	Basalt lava	103 ka	35 ka	K–Ar	Fisk (1990)	
King George Island							
A-406/MP406	Melville Peak	Lava within stratocone sequence	0.296	0.027	K–Ar	Birkenmajer and Keller (1990); Lawver et al. (1995)	
A-372/MP372	Melville Peak	Lava within stratocone sequence	0.231	0.019	K–Ar	Birkenmajer and Keller (1990); Lawver et al. (1995)	
A-375/MP375	Melville Peak	Lava within stratocone sequence	0.072	0.015	K–Ar	Birkenmajer and Keller (1990); Lawver et al. (1995)	
–	Penguin Island	Deacon Peak cone	>300 years		Lichenometric	Birkenmajer (1980)	
–	Penguin Island	Small cone within Deacon Peak cone	c. 133 years		Lichenometric	Birkenmajer (1980)	
–	Penguin Island	Petrel Crater (maar)	c. 74 years		Lichenometric	Birkenmajer (1980)	
PING-3	Penguin Island	Basal lava platform (Marr Point Formation)	2.7	0.2	Ar–Ar	Pańczyk and Nawrocki (2011)	
PI-4	Penguin Island	Dyke cropping out in western slopes of Deacon Peak cone	8.8	2.4	Ar–Ar	Kraus (2005)	Age probably unreliable
Greenwich Island							
P.54.1	Mount Plymouth	Basalt plug	0.2	0.3	K–Ar	Smellie et al. (1984)	
P.55.1	Mount Plymouth	Basalt plug	0.2	0.4	K–Ar	Smellie et al. (1984)	
Livingston Island							
P.51.1	Gleaner Heights	Basalt lava	0.1	0.4	K–Ar	Smellie et al. (1984)	
230	'1.6 km SE of Burdick Peak'	Basalt lava	0.04	0.35	K–Ar	Smellie et al. (2006a, b, c)	Exact locality uncertain; site is possibly Rezen Knoll/Peak
260	1 km south of Samuel Peak	Basalt lava	0.70	0.35	K–Ar	Smellie et al. (2006a, b, c)	
Bridgeman Island							
n.r.	Bridgman Island upper volcanic unit	Lava	63 ka	25 ka	K–Ar	Keller et al. (1992); Lawver et al. (1995)	
Deception Island							
n.r.	Deception Island	'One of the stratigraphically oldest units'	150 ka	46 ka	K–Ar	Keller et al. (1992)	
DI048	Deception Island	Pre-caldera units	105 ka	46 ka	K–Ar	Lawver et al. (1995)	
n.r.	Deception Island	Caldera collapse	>12 ka BP		Palaeomagnetism	Oliva-Urcia et al. (2016)	
n.r.	Deception Island	Post-caldera units	c. 8.3 ka BP		Palaeomagnetism	Oliva-Urcia et al. (2016)	
n.r.	Deception Island		<2. ka BP		Palaeomagnetism	Oliva-Urcia et al. (2016)	
n.r.	Deception Island	Tephra from caldera-forming eruption	3.98 ka BP		^{14}C	Antoniades et al. (2018)	Dated using tephrochronology
Brabant Island							
26203	Mount Parry summit	Basaltic andesite lava	3.07	0.1	K–Ar	Ringe 1991	
26079	Metchnikoff Point	Subaerial capping lava on lava-fed delta	<100 ka	–	K–Ar	Ringe 1991	

DJ.1701.3	Metchnikoff Point	Lava pod in lava-fed delta tuff breccia	0.18	0.09	Ar–Ar	Smellie et al. (2006a, b, c,)	Plateau age
DJ.1701.3	Metchnikoff Point	Lava pod in lava-fed delta tuff breccia	0.1	0.04	Ar–Ar	Smellie et al. (2006a, b, c,)	Isochron age
DJ.1701.4	Metchnikoff Point	Subaerial capping lava on lava-fed delta	0.41	0.33	Ar–Ar	Smellie et al. (2006a, b, c,)	Plateau age
DJ.1701.4	Metchnikoff Point	Subaerial capping lava on lava-fed delta	0.61	0.98	Ar–Ar	Smellie et al. (2006a, b, c,)	Plateau age
DJ.1702.2	'Cairn Point'	Lower sequence lava	0.19	0.07	Ar–Ar	Smellie et al. (2006a, b, c,)	Plateau age
BR.200.23	Claude Point	Upper sequence lava	0.13	0.06	Ar–Ar	Smellie et al. (2006a, b, c,)	Plateau age
BR.200.23	Claude Point	Upper sequence lava	0.02	0.02	Ar–Ar	Smellie et al. (2006a, b, c,)	Isochron age
BR.200.20	Claude Point	Lower sequence lava	0.099	0.067	Ar–Ar	Smellie et al. (2006a, b, c,)	Isochron age
DJ.1703.4#1	'Astrolabe Point'	Lava (=lower sequence at Claude Point)	0.08	0.11	Ar–Ar	Smellie et al. (2006a, b, c,)	Plateau age
DJ.1703.4#2	'Astrolabe Point'	Lava (=lower sequence at Claude Point)	0.08	0.08	Ar–Ar	Smellie et al. (2006a, b, c,)	Plateau age
James Ross Island Volcanic Group							
D.4053.11	Rabot Point, James Ross Island	Olivine basalt	2.2	0.5	K–Ar	Rex (1976)	
D.4053.6	Rabot Point, James Ross Island	Basalt	3.3	0.8	K–Ar	Rex (1976)	
D.4085.2	Limershin Cliffs, James Ross Island	Olivine basalt	3.5	0.5	K–Ar	Rex (1976)	
D.4086.1	Palisade Nunatak, James Ross Island	Dolerite	6.5	0.3	K–Ar	Rex (1976)	
D.4086.1	Palisade Nunatak, James Ross Island	Dolerite	5.4	0.3	K–Ar	Rex (1976)	
D.4096.1	Kipling Mesa, James Ross Island	Basalt	4.6	0.4	K–Ar	Rex (1976)	
D.4097.2	Between Lagrelius Point and Kipling Mesa, James Ross Island	Olivine basalt	2.1	1.0	K–Ar	Rex (1976)	
D.2144.1	Cape Obelisk?, James Ross Island	Basalt	3.0	0.5	K–Ar	Rex (1976)	
D.3753.5	Northern Lachman Crags, James Ross Island	Dolerite	4.6	0.4	K–Ar	Rex (1976)	
D.2166.1	Carlson Island, Prince Gustav Channel	Basalt	2.0	0.5	K–Ar	Rex (1976)	
D.2166.1	Carlson Island, Prince Gustav Channel	Basalt	1.4	0.3	K–Ar	Rex (1976)	
D.2166.1	Carlson Island, Prince Gustav Channel	Basalt	1.4	0.2	K–Ar	Rex (1976)	
D.3711.3	Beak Island, Prince Gustav Channel	Olivine basalt	1.7	0.2	K–Ar	Rex (1976)	
D.3711.3	Beak Island, Prince Gustav Channel	Olivine basalt	2.0	0.2	K–Ar	Rex (1976)	
D.3776.1	Eagle Island, Prince Gustav Channel	Olivine basalt	2.0	0.2	K–Ar	Rex (1976)	
D.3776.1	Eagle Island, Prince Gustav Channel	Olivine basalt	1.7	0.2	K–Ar	Rex (1976)	
D.3771.1	Southern Tail island, Prince Gustav Channel	Basalt	2.7	0.5	K–Ar	Rex (1972)	
D.3787.1	Brown Bluff, Tabarin Peninsula	Basalt	1.1	0.1	K–Ar	Rex (1976)	
D.3787.1	Brown Bluff, Tabarin Peninsula	Basalt	1.1	0.1	K–Ar	Rex (1976)	
27788	North end of Paulet Island	Alkali basalt	0.3	0.1	K–Ar	Baker et al. (1977)	
27831	Seven Buttresses, Tabarin Peninsula, Graham Land	Alkali basalt (hawaiite)	0.9	0.2	K–Ar	Baker et al. (1977)	
27812	Red Island, Prince Gustav Channel	Alkali basalt	1.6	0.2	K–Ar	Baker et al. (1977)	
D.8705.1	Berry Hill, James Ross Island	Basalt lava	4.93	0.27	K–Ar	Sykes (1988)	
D.8707.2	Crame Col, James Ross Island	Basaltic clast in diamict	7.13	0.49	K–Ar	Sykes (1988)	
D.8707.4	Crame Col, James Ross Island	Basalt lava	4.63	0.57	K–Ar	Sykes (1988)	
D.8707.8	Crame Col, James Ross Island	Dolerite sill	5.23	0.57	K–Ar	Sykes (1988)	
D.8711.1	Lookalike Peaks, James Ross Island	Basalt lava	4.44	0.25	K–Ar	Sykes (1988)	
D.8719.3	Akela Col, James Ross Island	Dolerite plug	6.13	0.56	K–Ar	Sykes (1988)	
D.8722.2	Seacatch Nunataks, James Ross Island	Basalt dyke	3.68	0.82	K–Ar	Sykes (1988)	

(Continued)

Table 3. *Continued.*

Sample	Locality	Dated lithologies	Age*	Error*	Method	Reference	Notes
D.8725.4	Donnachie Cliffs, James Ross Island	Basalt lava	1.53	0.18	K–Ar	Sykes (1988)	
D.8727.1	Back Mesa, James Ross Island	Basalt lava	4.68	0.2	K–Ar	Sykes (1988)	
D.8728.1	Palisade Nunatak, James Ross Island	Dolerite laccolith	5.36	0.51	K–Ar	Sykes (1988)	
D.8730.1	Molley Corner, James Ross Island	Basalt lava	5.37	0.35	K–Ar	Sykes (1988)	
D.8739.5	Tumbledown Cliffs, James Ross Island	Basalt pillow	5.45	0.32	K–Ar	Sykes (1988)	
D.8802.7	Virgin Hill, James Ross Island	Basalt lava	5.68	0.31	K–Ar	Sykes (1988)	
D.8814.1	Organpipe Nunatak, James Ross Island	Dolerite laccolith	4.36	0.21	K–Ar	Sykes (1988)	
D.8818.1	Egg Island, Prince Gustav Channel	Basalt lava	2.47	0.38	K–Ar	Sykes (1988)	
D.8820.5	Vertigo Cliffs, Vega Island	Basalt lava	1.27	0.08	K–Ar	Sykes (1988)	
D.8822.1	Flett Buttress, James Ross Island	Basalt lava	5.15	0.33	K–Ar	Sykes (1988)	
D.8826.1	Humps Island	Dolerite plug	6.45	0.6	K–Ar	Sykes (1988)	
D.8831.2	The Naze, James Ross Island	Dolerite laccolith	3.57	0.34	K–Ar	Sykes (1988)	
D.8408.56a	Persson Island	Dolerite sill	4.5	0.5	K–Ar	Smellie *et al.* (1988)	
D.8408.58b	Persson Island	Dolerite sill	5.4	0.7	K–Ar	Smellie *et al.* (1988)	
not given	Fjordo Belen, James Ross Island	Basalt clast in diamict	9.2	0.3	Ar–Ar	Jonkers *et al.* (2002)	
not given	Cape Gage, James Ross Island	Basalt, clast in diamict?	3.1	0.3	Ar–Ar	Jonkers *et al.* (2002)	
n.r.	Cockburn Island	Lava, cap rock	3.65	0.3	K–Ar	Webb and Andreasen (1986)	Lower lava-fed delta
DJ.856.2	Cockburn Island	Lava, cap rock	4.7	0.2	Ar–Ar	Jonkers and Kelley (1998)	Lower lava-fed delta
DJ.856.3	Cockburn Island	Lava, cap rock	4.9	0.4	Ar–Ar	Jonkers and Kelley (1998)	Upper lava-fed delta
DJ.855.1	Cockburn Island	Lava, cap rock	2.9	0.4	Ar–Ar	Jonkers and Kelley (1998)	Upper lava-fed delta
CK3	Cockburn Island	Lava, cap rock?	2.781	0.032	K–Ar	Lawver *et al.* (1995)	
4M18	'Dreadnought' [Sykes Cliffs], James Ross Island	n.r.	6.643	0.102	K–Ar	Lawver *et al.* (1995)	
5M117	'Villar Fabre' [Crisscross Crags], James Ross Island	n.r.	3.94	0.085	K–Ar	Lawver *et al.* (1995)	
B28	The Mesa, NE Seymour Island	Basalt clast in diamict	12.4	0.5	K–Ar	Marenssi *et al.* (2010)	
n.r.	Seymour Island	Dyke	6.8	0.5	K–Ar	Massabie and Morelli (1977)	
n.r.	Seymour Island	Dyke	1.3	n.r.	K–Ar	Massaferro *et al.* (1997)	
DJ.1718.1	Lachman Crags Formation, southern Lachman Crags, James Ross Island	Lava, cap rock	3.95	0.05	Ar–Ar	Kristjánsson *et al.* (2005); Smellie *et al.* (2008)	
DJ.1722.6	Stickle Ridge Formation, at Lookalike Peaks, James Ross Island	Lava in tuff breccia	4.35	0.39	Ar–Ar	Kristjánsson *et al.* (2005); Smellie *et al.* (2008)	Unreliable age?
DJ.1729.8	Patalamon Formation?, at Akela Col, James Ross Island	Lava, cap rock	4.71	0.06	Ar–Ar	Kristjánsson *et al.* (2005); Smellie *et al.* (2008)	Sample may not have been taken *in situ*
DJ.1726.11	Kipling Mesa Formation, at Hambrey Cliffs, James Ross Island	Lava lobe in delta passage zone	4.78	0.07	Ar–Ar	Kristjánsson *et al.* (2005); Smellie *et al.* (2008)	Less reliable age
DJ.1714.5	Johnson Mesa Formation, at Crame Col, James Ross Island	Lava, cap rock	5.04	0.04	Ar–Ar	Kristjánsson *et al.* (2005); Smellie *et al.* (2008)	
DJ.1721.5	Smellie Peak Formation, at Smellie Peak, James Ross Island	Lava pillow in tuff breccia	5.91	0.08	Ar–Ar	Kristjánsson *et al.* (2005); Smellie *et al.* (2008)	Less reliable age
DJ.1721.7	Smellie Peak Formation, at Smellie Peak summit, James Ross Island	Columnar lava	5.14	0.38	Ar–Ar	Kristjánsson *et al.* (2005); Smellie *et al.* (2008)	
DJ.1725.5A	Cape Lachman Formation?, at Hambrey Cliffs, James Ross Island	Lava clast in tuff breccia	5.64	0.25	Ar–Ar	Kristjánsson *et al.* (2005); Smellie *et al.* (2008)	Less reliable age
DJ.1721.11	Lookalike Peaks Formation, at Smellie Peak, James Ross Island	Lava, cap rock	5.89	0.09	Ar–Ar	Kristjánsson *et al.* (2005); Smellie *et al.* (2008)	Less reliable age

Sample	Location	Description	Age (Ma)	±	Method	Reference	Notes
DJ.1722.1	Lookalike Peaks Formation, at Lookalike Peaks, James Ross Island	Lava, cap rock	6.16	0.08	Ar–Ar	Kristjánsson et al. (2005); Smellie et al. (2008)	
DJ.2055.1b	Cape Purvis, Dundee Island	Lava, cap rock	0.132	0.019	Ar–Ar	Smellie et al. (2006c, 2008)	
DJ.2051.4	Brown Bluff, Tabarin Peninsula, Graham Land	Coeval intrusion in tuff breccia	1.22	0.39	Ar–Ar	Smellie et al. (2006c, 2008)	
DJ.436.1	Seven Buttresses, Tabarin Peninsula, Graham Land	Lava, cap rock	1.69	0.03	Ar–Ar	Smellie et al. (2006c, 2008)	
D.8835.1	Keltie Head Formation, at Keltie Head, Vega Island	Lava, cap rock	0.99	0.05	Ar–Ar	Smellie et al. (2008)	
DJ.1711.4	Cape Well-met Formation, Vertigo Cliffs, Vega Island	Lava, cap rock	2.03	0.04	Ar–Ar	Smellie et al. (2008)	
DJ.2052.6	Probably Cape Well-met Formation, at Mahogany Bluff, Vega Island	Lava, cap rock	2.07	0.06	Ar–Ar	Smellie et al. (2008)	Sample not *in situ*
D.8830.1	Cape Well-met Formation, at Cape Lamb, Vega Island	Lava, cap rock	2.09	0.11	Ar–Ar	Smellie et al. (2008)	
DJ.1711.16	Vertigo Cliffs Formation, at Vertigo Cliffs, Vega Island	Lava, caprock	2.67	0.13	Ar–Ar	Smellie et al. (2008)	
D.8838.1	Sandwich Bluff Formation, at Sandwich Bluff, Vega island	Lava, caprock	5.42	0.08	Ar–Ar	Smellie et al. (2008)	
DJ.1733.1	Dobson Dome Formation, at Blyth Spur, James Ross Island	Lava, caprock	0.26	0.34	Ar–Ar	Smellie et al. (2008)	Inferred age <0.08 Ma
DJ.1746.1	Terrapin Hill Formation, Terrapin Hill summit, James Ross Island	Dyke in tuff cone	0.66	0.22	Ar–Ar	Smellie et al. (2008)	
D.8725.4	Donnachie Cliffs Formation, at NW Dobson Dome, James Ross Island	Lava, cap rock	1.15	0.12	Ar–Ar	Smellie et al. (2008)	
DJ.1709.1	Jefford Point Formation at Jefford Point, James Ross Island	Lava, cap rock	1.92	0.09	Ar–Ar	Smellie et al. (2008)	
DJ.1712.2	Jefford Point Formation, at Tortoise Hill, James Ross Island	Lava block in tuff breccia	1.86	0.05	Ar–Ar	Smellie et al. (2008)	
DJ.1710.4	Sungold Hill Formation, at point below Sungold Hill, James Ross Island	Lava, cap rock	2.03	0.07	Ar–Ar	Smellie et al. (2008)	
DJ.1710.10	Sungold Hill Formation, at Sungold Hill, James Ross Island	Lava block in tuff breccia	2.30	0.36	Ar–Ar	Smellie et al. (2008)	
DJ.1740.10a	Terrapin Hill Formation?, Taylor Bluff summit, James Ross Island	Columnar intrusion in tuff cone?	2.03	0.13	Ar–Ar	Smellie et al. (2008)	
DJ.1712.6	Terrapin Hill Formation, at Tortoise Hill, James Ross Island	Bomb in tuff cone	2.23	0.33	Ar–Ar	Smellie et al. (2008)	
DJ.1980.1	Unnamed delta formation, western Cape Gage, James Ross Island	Lava, cap rock	2.23	0.05	Ar–Ar	Smellie et al. (2008)	
DJ.1710.8	Terrapin Hill Formation, at Sungold Hill, James Ross Island	Bomb in tuff cone	2.45	0.09	Ar–Ar	Smellie et al. (2008)	
DJ.1965.4	Forster Cliffs Formation?, at St Rita Point, James Ross Island	Lava, cap rock	2.49	0.02	Ar–Ar	Smellie et al. (2008)	
DJ.1752.3	Forster Cliffs Formation, at west Forster Cliffs, James Ross Island	Lava, cap rock	2.50	0.07	Ar–Ar	Smellie et al. (2008)	
DJ.1734.5	Unnamed delta formation, west of Dreadnought Point, James Ross Island	Lava pillow	2.82	0.66	Ar–Ar	Smellie et al. (2008)	Age of uncertain significance
DJ.1734.11	Unnamed delta formation, west of Dreadnought Point, James Ross Island	Pillow block in tuff breccia	5.65	0.51	Ar–Ar	Smellie et al. (2008)	Age of uncertain significance

(*Continued*)

Table 3. *Continued.*

Sample	Locality	Dated lithologies	Age*	Error*	Method	Reference	Notes
DJ.1734.14	Unnamed delta formation, west of Dreadnought Point, James Ross Island	Pillow block in tuff breccia	5.82	0.56	Ar–Ar	Smellie et al. (2008)	Age of uncertain significance
DJ.1720.2	Unnamed delta formation, west of Dreadnought Point, James Ross Island	Lava block in tuff breccia	2.38	0.41	Ar–Ar	Smellie et al. (2008)	Age of uncertain significance
DJ.2053.1	Forster Cliffs Formation?; unnamed cliff NE of Flett Buttress, James Ross Island	Lava, cap rock	2.68	0.04	Ar–Ar	Smellie et al. (2008)	
DJ.1973.1	Unnamed delta formation, Rhino Corner summit, James Ross Island	Lava, cap rock	3.01	0.07	Ar–Ar	Smellie et al. (2008)	
DJ.1970.1	Jonkers Mesa Formation, at Jonkers Mesa, James Ross Island	Lava, cap rock	3.08	0.15	Ar–Ar	Smellie et al. (2008)	
DJ.1962.11	Jonkers Mesa Formation, at Rhino Corner, James Ross Island	Lava block in tuff breccia	2.85	0.62	Ar–Ar	Smellie et al. (2008)	
DJ.1973.4	Ekelof Point Formation, at Rhino Corner, James Ross Island	Lava, cap rock	3.51	0.07	Ar–Ar	Smellie et al. (2008)	
DJ.1962.8	Ekelof Point Formation, at Rhino Corner, James Ross Island	Lava, cap rock	3.38	0.08	Ar–Ar	Smellie et al. (2008)	
DJ.1977.1	Ekelof Point Formation, between St Rita Point and Rhino Corner, James Ross Island	Lava, cap rock	3.62	0.03	Ar–Ar	Smellie et al. (2008)	
DJ.1976.1	Ekelof Point Formation, at Ekelof Point, James Ross Island	Lava, cap rock	3.41	0.16	Ar–Ar	Smellie et al. (2008)	
DJ.1708.11a	Terrapin Hill Formation, at Lomas Ridge, James Ross Island	Pillow lava in likely tuff cone	3.60	0.29	Ar–Ar	Smellie et al. (2008)	
DJ.1983.1	Hamilton Point Formation, at Hamilton Point, James Ross Island	Lava, cap rock	3.69	0.04	Ar–Ar	Smellie et al. (2008)	
DJ.2070.1	Patalamon Mesa Formation, at Patalamon Mesa, James Ross Island	Lava in tuff breccia	4.64	0.09	Ar–Ar	Smellie et al. (2008)	
DJ.2074.1	Patalamon Mesa Formation, at Patalamon Mesa, James Ross Island	Lava, cap rock	4.53	0.08	Ar–Ar	Smellie et al. (2008)	
DJ.1731.1	Patalamon Mesa Formation, at Davies Dome, James Ross Island	Lava, cap rock	4.74	0.03	Ar–Ar	Smellie et al. (2008)	
DJ.2133.1	Patalamon Mesa Formation, at Crisscross Crags, James Ross Island	Lava, cap rock	4.69	0.25	Ar–Ar	Smellie et al. (2008)	
DJ.2068.1	Unnamed middle delta unit, at Patalamon Mesa, James Ross Island	Subaerial lava	4.70	1.10	Ar–Ar	Smellie et al. (2008)	
DJ.1718.3	Johnson Mesa Formation, at Lachman Crags, James Ross Island	Lava, cap rock	5.08	0.04	Ar–Ar	Smellie et al. (2008)	
DJ.2072.1	Terrapin Hill Formation, at Patalamon Mesa, James Ross Island	Intrusion in upper tuff cone	5.37	0.37	Ar–Ar	Smellie et al. (2008)	
DJ.2111.1	Tumbledown Mesa Formation, at Crisscross Crags, James Ross Island	Lava, cap rock	5.26	0.10	Ar–Ar	Smellie et al. (2008)	
DJ.2083.2	Tumbledown Mesa Formation?, at Back Mesa, James Ross Island	Lava, cap rock	5.17	0.05	Ar–Ar	Smellie et al. (2008)	Sample not *in situ*
D.8739.5	Tumbledown Mesa Formation, at Tumbledown Mesa, James Ross Island	Lava, cap rock	5.01	0.12	Ar–Ar	Smellie et al. (2008)	
DJ.2080.1	Tumbledown Mesa Formation, at Limmershin Cliffs, James Ross Island	Sill in tuff breccia	5.33	0.47	Ar–Ar	Smellie et al. (2008)	

DJ.2062.1	Kipling Mesa Formation, at Kipling Mesa, James Ross Island	Lava, cap rock	5.36	0.05	Ar–Ar	Smellie et al. (2008)	
DJ.1715.1	Cape Lachman Formation?, north Lachman Crags, James Ross Island	Lava/intrusion	5.32	0.16	Ar–Ar	Smellie et al. (2008)	
DJ.1745.2	Unnamed basal delta, at Forster Cliffs, James Ross Island	Lava, cap rock?	5.47	0.11	Ar–Ar	Smellie et al. (2008)	
DJ.1743.1	Dyke cuts unnamed delta, at Blancmange Hill summit, James Ross Island	Dyke in tuff breccia	5.84	0.34	Ar–Ar	Smellie et al. (2008)	
DJ.1744.4	Taylor Bluff Formation, at Blancmange Hill, James Ross Island	Lava, cap rock	5.85	0.03	Ar–Ar	Smellie et al. (2008)	
DJ.1740.8#1	Dyke cuts Taylor Bluff Formation, at Taylor Bluff, James Ross Island	Dyke in tuff breccia	1.67	0.49	Ar–Ar	Smellie et al. (2008)	Probably unreliable age
DJ.1740.8#2	Dyke cuts Taylor Bluff Formation, at Taylor Bluff, James Ross Island	Dyke in tuff breccia	1.94	0.47	Ar–Ar	Smellie et al. (2008)	Probably unreliable age
DJ.2058.1	Unnamed delta formation, at Virgin Hill, James Ross Island	Lava, cap rock	5.87	0.04	Ar–Ar	Smellie et al. (2008)	
DJ.2057.3	Terrapin Hill Formation, SW of Rink Point, James Ross Island	Intrusion in tuff cone	5.90	0.17	Ar–Ar	Smellie et al. (2008)	
DJ.2070.2	Terrapin Hill Formation, at Patalamon Mesa, James Ross Island	Plug in basal tuff cone	6.02	0.12	Ar–Ar	Smellie et al. (2008)	
D.8722.3	Terrapin Hill Formation, at Seacatch Nunataks, James Ross Island	Dyke in tuff cone	5.36	0.28	Ar–Ar	Smellie et al. (2008)	
2SM08	Lookalike Peaks Formation, at Smellie Peak, James Ross Island	Subaerial lava	4.63	0.28	K–Ar	Calabozo et al. (2015)	Conflicts with age published by Kristjánsson et al. (2005) (5.89 ± 0.09 Ma)
SM12	Smellie Peak Formation, at Smellie Peak, James Ross Island	Pillow lava core	4.31	0.26	K–Ar	Calabozo et al. (2015)	Conflicts with age published by Kristjánsson et al. (2005) (5.14 ± 0.38 Ma)
3MAS06	Lookalike Peaks Formation, at Lookalike Peaks, James Ross Island	Lava	5.21	0.32	K–Ar	Calabozo et al. (2015)	Conflicts with age published by Kristjánsson et al. (2005) (6.16 ± 0.08 Ma)
3MAS02	Stickle Ridge Formation, at Lookalike Peaks, James Ross Island	Pillow lava core	4.4	1	K–Ar	Calabozo et al. (2015)	Likely high atmospheric Ar contamination but age agrees with Kristjánsson et al. (2005) (4.35 ± 0.39 Ma)
JR-1	Cape Lachman Formation, at Cape Lachman, James Ross Island	lava pillow	5.85	0.31	Ar–Ar	Nývlt et al. (2011)	Weighted-mean age of 5.67 ± 0.03 Ma recalculated by Smellie et al. (2013), using Nývlt et al. (2011)

*All ages in Ma except where otherwise noted.
n.r., not recorded.

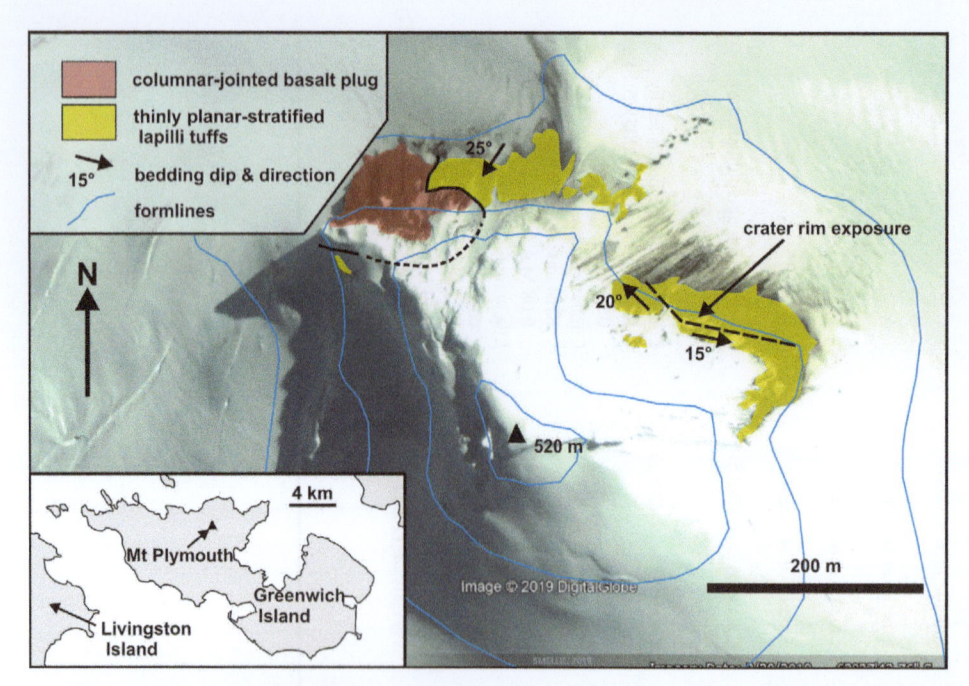

Fig. 14. Geological sketch map of Mount Plymouth, Greenwich Island (South Shetland Islands), based on Hobbs (1968 and unpublished field notes of G.J. Hobbs) and unpublished binocular observations of J.L. Smellie. The outcrop is a small remnant of a tuff cone that may have had an original basal diameter of 1–1.2 km. The formlines are indicative only. Satellite image courtesy of Google Earth; data provider Maxar Technologies 2020.

Livingston Island

Several small scattered volcanic outcrops are also known on Livingston Island and are grouped (together with Mount Plymouth) in the Inott Point Formation (Smellie *et al.* 1995). They consist of the remnants of tuff cones (Inott Point, Sharp Peak, and unnamed nunataks SSW and west of Sharp Peak), plugs (Edinburgh Hill and west of Inott Point), and small outcrops of basalt lavas and associated beds of black and red scoria (at Gleaner Heights; Rezen Knoll or Rezen Peak; ESE of Sharp Peak; summit of Inott Point; and outcrops near Samuel Peak: Smellie *et al.* 1995). Little is known of the Gleaner Heights outcrop other than it includes fresh basalt lava (Smellie *et al.* 1984). Rezen Knoll rises to 433 m, about 130 m above the surrounding ice (Fig. 4f). It was interpreted as a tuya based on the following criteria: an apparent flat-topped morphology, the absence of tephra (neither Strombolian nor phreatomagmatic) and the presence of pillow lava (Kraus *et al.* 2013). Most of the other outcrops are interpreted as relicts of several tuff cones. They consist of stratified lapilli tuff and tuff, the products of fully dilute pyroclastic density currents, but minor massive tuff breccia is also present in outcrops containing a sub-central plug (cf. description of the Melville Peak centre on King George Island, see the earlier 'King George Island' subsection). The tuff breccia is interpreted as a vent facies, whilst the stratified lapilli tuffs represent flank sequences. Although the field evidence is ambiguous, the occurrence of the tuff cone outcrops at elevations mainly above 200 m asl (i.e. above the local limit for marine geomorphological features: John and Sugden 1971), mainly at inland locations and their very young age suggest that they may be glaciovolcanic and formed during eruptions under relatively thin ice (<*c.* 150 m). The lavas and scoria are an association of subaerially erupted lithofacies that includes clastogenic densely welded lenses up to 2.5 m thick; they may have been erupted during interglacial periods of much reduced snow and ice cover. Edinburgh Hill is a spectacularly columnar jointed dolerite plug *c.* 250 m across and 110 m high. Similar columnar plugs occur at Sharp Peak and just west of Inott Point (Smellie *et al.* 1995; Kraus *et al.* 2013).

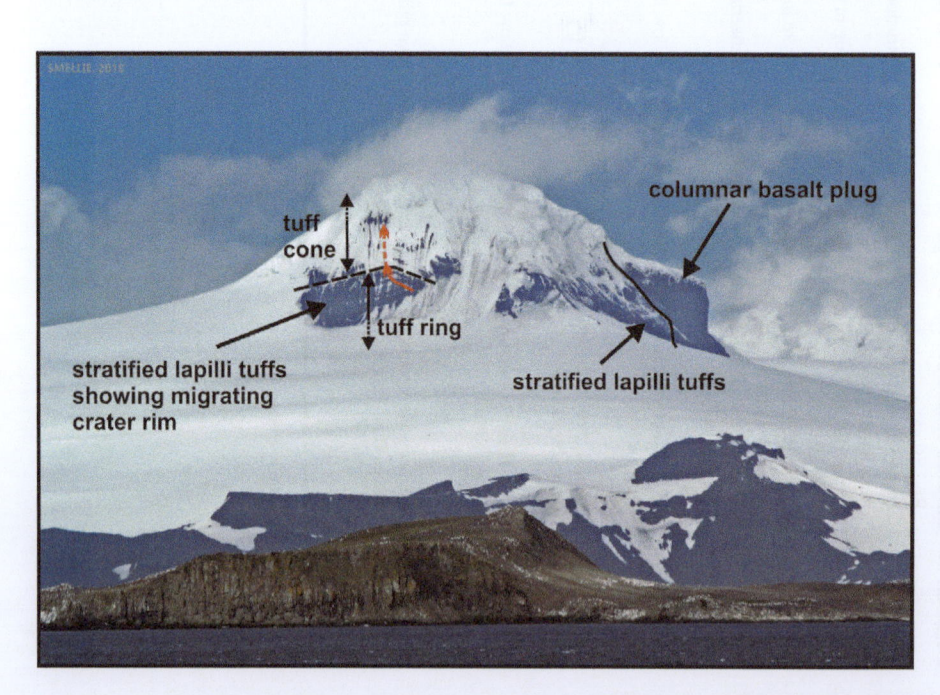

Fig. 15. View of Mount Plymouth, looking SW. A section through a crater rim is well defined by bedding orientations in the left (southeastern) cliff face. The evolution of the crater rim position (red arrows) suggests that the pyroclastic cone evolved from a tuff ring up into a tuff cone. A basaltic plug is exposed at the far right-hand side. The rocks exposed in the lower slopes and foreground belong to the Late Cretaceous Coppermine Formation (Smellie *et al.* 1984).

Other intrusions are uncommon and include an irregular basalt intrusion in lapilli tuffs south of Sharp Peak and a dolerite sill on the tombolo facing Edinburgh Hill. The compositions of the Livingston Island volcanic products are sodic alkaline basalts and hawaiites similar to lavas on Penguin Island but one (Edinburgh Hill) is tholeiitic (Smellie 1990h; Kamenov 2004). The outcrops are poorly dated. Three localities have yielded ages <1 Ma (i.e. 0.70 ± 0.35–0.04 ± 0.35 Ma: effectively too young and poor in potassium to date by the K–Ar method; Table 3) (Smellie et al. 1984, 1996).

Brabant Island and Anvers Island

Brabant and Anvers islands are two of the largest islands off the NW coast of the Antarctic Peninsula (Fig. 1). Both are mountainous and have very difficult access. As a result, neither island is well known geologically, and the knowledge is generally at a reconnaissance level (Hooper 1962; Alarcón et al. 1976; Alfaro and Collao 1985; Smellie 1990e; Ringe 1991; Smellie et al. 2006b). The youngest sequences on both islands are volcanic and the outcrop probably extends to Liège Island (Fig. 1). That on Anvers Island is restricted to the very inaccessible Osterrieth Range, a narrow flat-topped promontory about 1300 m high at the NE end of the island. The volcanic rocks are unconformable on Eocene (51–35 Ma) gabbros, diorites and possibly granodiorite of the Antarctic Peninsula Batholith (Leat et al. 1995). The volcanic rocks are horizontally bedded and >1200 m thick (base unexposed), and have only been accessed at two localities on the east side of the island. Steeper dips of 30°–70° occur at contacts with the plutonic bedrock, which forms the sides of steep-sided valleys. Hooper (1962) identified a major valley trending WSW from Ryswyck Point towards Mount Francais, and a second valley forming a spur curving round the south flank of Mount Francais. The steep valley flanks were tentatively assigned a glacial origin and some of the samples Hooper collected resemble diamicts similar to glacial tills examined on Brabant Island (Smellie et al. 2006b; J.L. Smellie unpublished information). The presence of deep red zones of weathering on some lava surfaces is indicative of subaerial extrusion. The only rocks examined directly were identified as volcaniclastic (tuffs) with basaltic compositions. Hooper (1962) emphasized that the volcanic rocks were very fresh, including unaltered glass and olivine phenocrysts, which contrasts markedly with older volcanic rocks on the island and suggests a very young age but there are no published isotopic ages. Pyroclastic rocks were said to dominate the middle part of the succession, with andesite present towards the top. He suggested a greater petrological affinity with the (subduction-related) volcanism in the South Shetland Islands than with alkaline volcanism in the James Ross Island Volcanic Group (Nelson 1975; Košler et al. 2009). This is confirmed by the phenocryst assemblages, which include lavas with two pyroxenes, similar to calc-alkaline lavas in the South Shetland Islands (Smellie unpublished information; cf. Smellie et al. 1984; Smellie 1990a).

Brabant Island is somewhat better known than Anvers Island following investigations by Alarcón et al. (1976), Alfaro and Collao (1985), Ringe (1991) and Smellie et al. (2006c). The volcanic rocks crop out much more extensively than on Anvers Island and dominate the local geology particularly in the north of the island. Although all workers have divided the volcanism into two sequences, the outcrop distributions and field relationships of the two sequences are very uncertain. Alarcón et al. (1976) suggested that a 'pre-glacial' volcanic sequence ('Bahia Guyou Formation') was present and was described as at least 200 m of continental sediments interbedded with pyroclastic rocks. Sparse gymnosperm and angiosperm spores were obtained by Alfaro and Collao (1985). If the reconnaissance palynological results were confirmed, it would represent an environmentally important discovery. Alarcón et al. (1976) also described a younger sequence of horizontally bedded lavas with maximum thickness of 200–300 m developed at Claude Point ('Bahia Bouquet Formation'). It was said to be unaffected by recent moraines. Ringe (1991) divided outcrops forming the northern part of the island into two effusive sequences with contrasting compositions whose distribution corresponds broadly with the formations described by Alarcón et al. (1976). The older sequence of Ringe (1991) consists of palagonitized lavas and tuffs that are mainly confined to the NW coast but may also be present at Duclaux Point on the north coast; extrusion was thought to have been in a marine or glacial environment, presumably because of the presence of pervasive palagonite alteration, although no reasons were given. By contrast, the younger sequence consisted of horizontally bedded subaerial basalt lavas on the north and east coasts, comprising lavas generally 3–5 m thick with little or no interbedded volcaniclastic rocks. One of the upper-phase lavas, at Metchnikoff Point, yielded an age of <100 ka (by K–Ar). In addition, the high central ridge that rises to 2520 m at Mount Parry is capped by andesite lava (similar to Anvers Island) with an isotopic age of 3.07 ± 0.1 Ma (by K–Ar: Ringe 1991). The volcanic sequences overlie plutons dated as 24.0 ± ±0.7 Ma below Mount Parry and 11.9 ± 0.5 or 8.9 ± 0.4 Ma at Metchnikoff Point (Ringe 1991).

Ringe (1991) suggested that his older sequence comprised sodic alkaline basalts and the younger sequence had 'subduction-related' compositions, mainly basaltic but including andesites and broadly similar compositionally to those in Bransfield Strait volcanic centres. The two compositionally different sequences are also evident in the larger dataset of Smellie (unpublished information) for samples obtained from NW and west coastal outcrops. They also correspond to an older alkaline group and younger group more like Deception Island, similar to the inferences of Ringe (1991).

Smellie et al. (2006b) also identified two groups compositionally like those described by Ringe (1991). $^{40}Ar/^{39}Ar$ ages suggest that the older group is c. 0.19–0.08 Ma and alkali basaltic in composition. The younger group, with mainly Q- and Hy-normative tholeiitic basaltic compositions but also including mugearite, has maximum ages of 0.10–0.02 Ma but most gave ages effectively close to 0 and the younger group was essentially undatable.

The outcrops examined by Smellie et al. (2006b) consist of three contrasting sequence types. Grey coarsely crystalline pāhoehoe lavas, locally with thin strongly oxidized surfaces, form the basal sequence at Claude Point (>150 m thick; base unexposed) and 'Astrolabe Point' nearby. By contrast, the sections at Driencourt Point and 'Cairn Point' consist of multiple sequences individually c. 60–150 m thick separated by erosional unconformities (Fig. 16). They are composed of a basal lithofacies (not always present) comprising weakly stratified gravelly–blocky polymict volcanic breccia or diamict up to 15 m thick with palagonite-altered glassy matrices, or gravelly breccio-conglomerate and faintly planar-laminated sandstone. Overlying lithofacies are composed of lava, often columnar jointed with overthickened entablatures compared to their colonnades, and locally developed hyaloclastite, which alternate with khaki-orange tuff breccia or crudely planar-stratified buff-coloured lapilli tuffs; some of the higher lavas have oxidized surfaces. Finally, the third distinctive sequence occurs at Metchnikoff Point. It is c. 110 m thick and consists of a 40 m-thick capping unit of subaerial pāhoehoe lava that overlies steep (20°–30°) homoclinally dipping tuff breccia beds. Sequence 3 is extensively eroded and

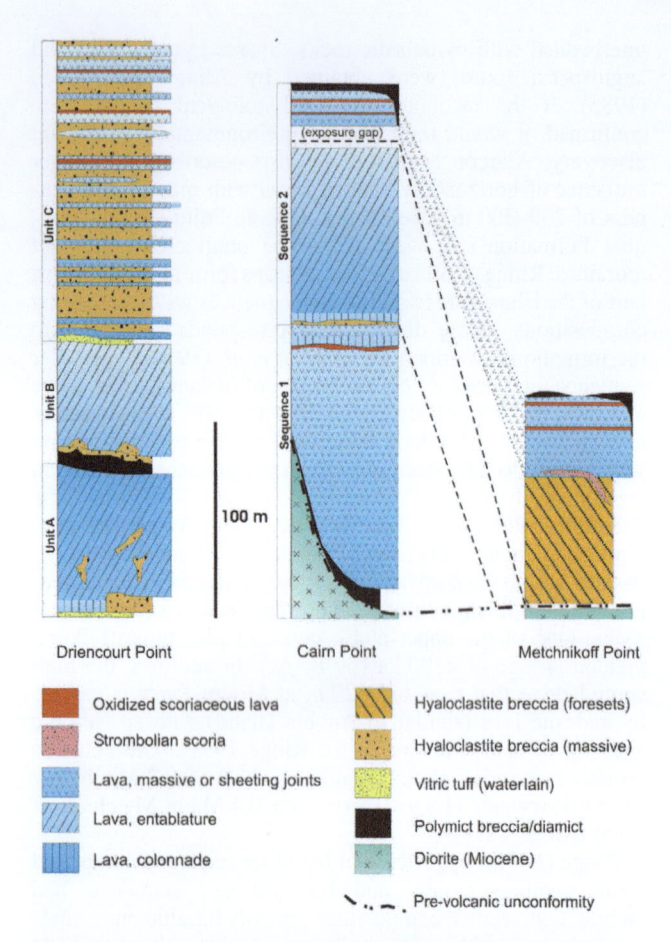

Fig. 16. Vertical profile logs through sequences at several localities on Brabant Island (from Smellie *et al.* 2006*b*). Sections exposed at Driencourt Point and 'Cairn Point' are products of multiple successive glaciovolcanic eruptions of sheet-like sequences (*sensu* Smellie *et al.* 1993), and are characteristic of eruptions under a relatively thin glacial cover. They are dominated by water-cooled mafic lava, with lesser waterlain vitric tuff and local hyaloclastite. Diamicts interpreted as tills overlie erosional surfaces present between each erupted sequence. The Metchnikoff Point sequence is a pāhoehoe lava-fed delta remnant. It is the only one known on Brabant Island so far and has a horizontal passage zone. The associated hyaloclastite breccia forms crude steep-dipping homoclinal foreset beds; the lithofacies is now classified as tuff breccia (Smellie and Edwards 2016). There is a striking compositional similarity between the delta-capping lava and lava at the top of the 'Cairn Point' sequence, suggesting that they might be cogenetic.

overlain unconformably by indurated khaki-brown diamict composed of locally derived volcanic clasts.

Lavas in the first sequence type were erupted under essentially dry subaerial conditions, lacking significant snow or ice. By contrast, lithofacies in the second sequence type contain basal breccia or diamict and conglomerates with facetted and striated clasts, and are interpreted as glacigenic deposits associated with wet-based ice. At 'Cairn Point', the sequence occupies a steep-sided valley with a polished and striated surface carved in plutonic rock. A spectacular erosional surface is also seen at Claude Point, where it undulates with a wavelength of a few hundred metres and has a relief of 20–30 m. The surface is cut in the subaerial lavas of sequence type 1 and it is probably glacial. The volcanic lithofacies in sequence type 2 were interpreted as glaciovolcanic sheet-like sequences (*sensu* Smellie *et al.* 1993; but see Smellie and Edwards 2016 for revised views on sheet-like sequences), with effusion envisaged beneath a relatively thin glacial cover (minimum thicknesses of 60–90 m and unlikely to exceed *c.* 150 m). The glaciovolcanic lithofacies locally pass up into subaerially emplaced thin lava lobes and flows. The lithofacies are consistent with subaerial effusion in the closing effusive stages (i.e. after the local glacial cover was removed by melting caused by emplacement of the earlier lithofacies). Sequence type 3 at Metchnikoff Point is a lava-fed delta, the only one known on Brabant Island, with a horizontal passage-zone elevation of *c.* 70 m. It advanced in a westerly direction. The composition of the upper lava unit is strikingly similar to that of a lava that caps the 'Cairn Point' sequence and may suggest a cogenetic relationship, perhaps as an originally continuous lava flow, with initial eruption high on the volcano and culminating in delta effusion into the sea or through a topographically low ice sheet with a surface elevation of *c.* 110 m.

Both of the groups identified by Smellie *et al.* (2006*b*) contain a mixture of subaerial lava and glaciovolcanic units, and, as the ages are all very young and overlap, it is possible that the two groups are coeval and may be interbedded, implying an alternation of eruptive environments over the past 200 ka, although the age data are insufficient to relate the environments to glacial cycles during the Pleistocene. In addition, the new collection from the younger groups is overwhelmingly dominated by basalts, although there are scarce andesites (Claude Point) and fragments of intermediate-composition lavas occur in the sequence at Driencourt Point (below Mount Parry). By contrast, the lavas at Mount Parry may be much older (3 Ma), if the isotopic age (by K–Ar) is reliable, suggesting that the Mount Parry sequence may be temporally discrete from the coastal outcrops to the NW. No published analyses of the Mount Parry sequence are available.

The combined outcrops on Anvers and Brabant islands, and including small outcrops on Davis and Liège islands (Bell 1984), were reconstructed as at least three large shield volcanoes individually *c.* 40–65 km in length and *c.* 30 km wide (Fig. 17) (Smellie *et al.* 2006*b*). The two edifices on Brabant Island coalesced and they were 1000–1500 m thick with a reconstructed combined volume of *c.* 750 km^3. On the basis of the few published isotopic ages (Table 3), they may be dissimilar in age, with the NW shield significantly younger than that to the SE. The volcanic shield on Anvers Island was similar in thickness to those on Brabant Island but it is now so heavily dissected by glacial erosion and inaccessible that its original volume is much harder to calculate. Its age is wholly unknown. Speculatively, the elongate shapes of the two largest centres may have been caused by eruptions focused along ENE–WSW- and NE–SW-orientated fissures.

Seamount offshore of Anvers Island

Multichannel seismic reflection, magnetic and bathymetric data have identified a broad, low-profile mound ('Dredge 138') at a depth of *c.* 400 m on the seafloor near the shelf edge *c.* 50 km NW of Anvers and Brabant islands (Hole and Larter 1993) (Fig. 1). The feature is approximately circular and about 15 km in diameter, generally rising *c.* 250 m above the surrounding seafloor. It contains at least three distinct peaks that reach *c.* 200–100 m bsl (below sea-level). Two of the peaks are small mounds *c.* 2.4 km apart and with summits about 50 m above the general bathymetrical high. Both are associated with prominent positive magnetic anomalies. The calculated seismic velocities of the mounds are low, which implies that they have a very high overall porosity. This was interpreted as indicating that they may be constructed of loosely packed volcanic rubble. A very young age (post-glacial, probably <12 ka) was also inferred since any overriding ice would have largely removed the volcanic pile. A dredge on one of the mounds recovered fresh vesicular alkali basalt lava whose composition and likely young age imply that they are slab-window related (see

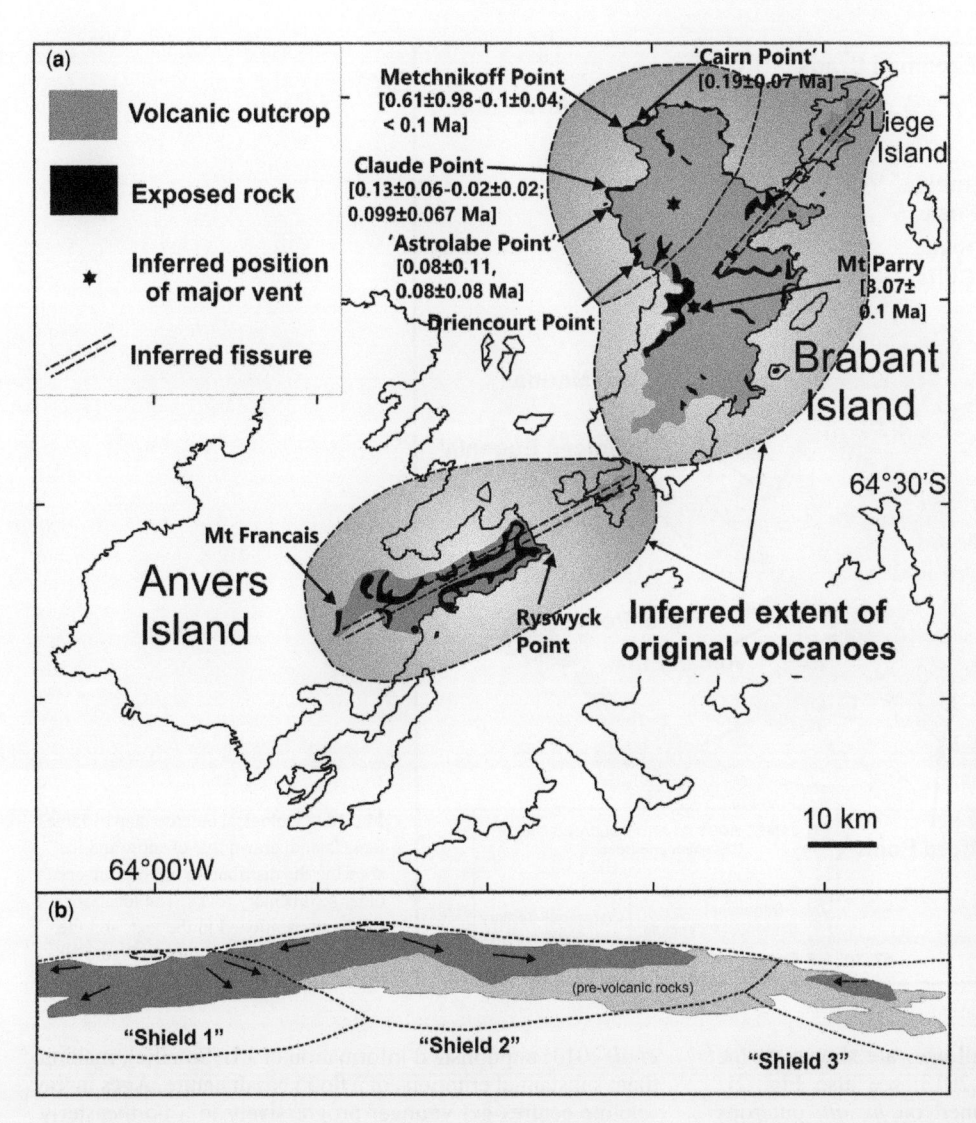

Fig. 17. Geological reconstruction of the three shield volcanoes responsible for Plio-Pleistocene volcanism on Brabant and Anvers islands (from Smellie et al. 2006b). Published isotopic ages are also shown (see Table 3). A perspective view is shown in (**b**), looking SSE. The morphology and lateral extent of the three volcanoes were inferred from a combination of extrapolating bedding orientations, outcrop patterns and topography. However, the reconstruction, and inferred presence of two eruptive fissures, are speculative.

Hole 2021). The calculated combined volume of the mounds is c. 0.16 km³. However, the volcanic mass may be closer to 7 km³ in volume based on a correlation using a prominent seismic reflector at the base of the mounds, which links them and suggests that the volcanic outcrop is at least c. 5.5 km in diameter.

James Ross Island region

James Ross Island is dominated by the basaltic volcanic products of the James Ross Island Volcanic Group (JRIVG), which are spectacularly exposed in coastal cliff sections (Nelson 1975). It has been recently and comprehensively mapped (Smellie et al. 2013; Mlčoch et al. 2015). The JRIVG is the largest Neogene volcanic field in the Antarctic Peninsula region, extending over an area of 7000 km² and with a volume conservatively estimated at >5000 km³ (cf. Smellie et al. 2013). The volcanism includes that which formed Mount Haddington (1630 m), which is a polygenetic shield volcano. The main shield has an elongate shape extending ENE–WSW, possibly caused by eruptions from a similarly orientated fissure for which there is aerogeophysical evidence (Jordan et al. 2009). The subglacial topography suggests that there is no summit caldera on Mount Haddington (unpublished information of J.L. Smellie). It is the largest volcano in Antarctica, 60–80 km wide (including Ulu Peninsula: Fig. 18) and rising to 1630 m. The Mount Haddington volcano also has deep roots extending down to c. 5 km, caused by gravitational subsidence of the volcanic pile into the underlying relatively weak Cretaceous back-arc basin sediments (Fig. 19) (Oehler et al. 2005; Jordan et al. 2009). The gravitational spreading of the Mount Haddington volcanic pile has created an annular ridge around James Ross Island situated at c. 40 km from the summit of Mount Haddington and composed of thrust-faulted weak Cretaceous bedrock (Fig. 20a). As a result, a conservative estimate for the volume of magma contained in the Mount Haddington volcano (excluding Ulu Peninsula) is c. 4500 km³, of which at least 3500 km³ of volcanic products are unexposed below sea-level (cf. Smellie et al. 2013). For comparison, the Mount Erebus stratovolcano on Ross Island in the southern Ross Sea is much higher at 3794 m but has a smaller basal diameter of c. 40 km and a much shallower root, resulting in a smaller volume (c. 2200 km³: Esser et al. 2004; see Smellie and Martin 2021). Other significant volcanic outcrops occur on Vega Island, as islands in Prince Gustav Channel and Antarctic Sound, and on Tabarin Peninsula, Dundee Island and Paulet Island (Figs 21 & 22). They represent multiple satellite centres, most of which were probably erupted subglacially when the coeval ice surfaces were considerably higher than today. Most are monogenetic glaciovolcanic edifices known as tuyas (Baker et al. 1973; Nelson 1975; Skilling 1994; Smellie and Skilling 1994; Smellie et al. 2006c). Conversely, Vega Island is a large multistorey satellite volcanic shield.

The JRIVG is one of the most comprehensively dated volcanic fields in Antarctica, with 125 published isotopic

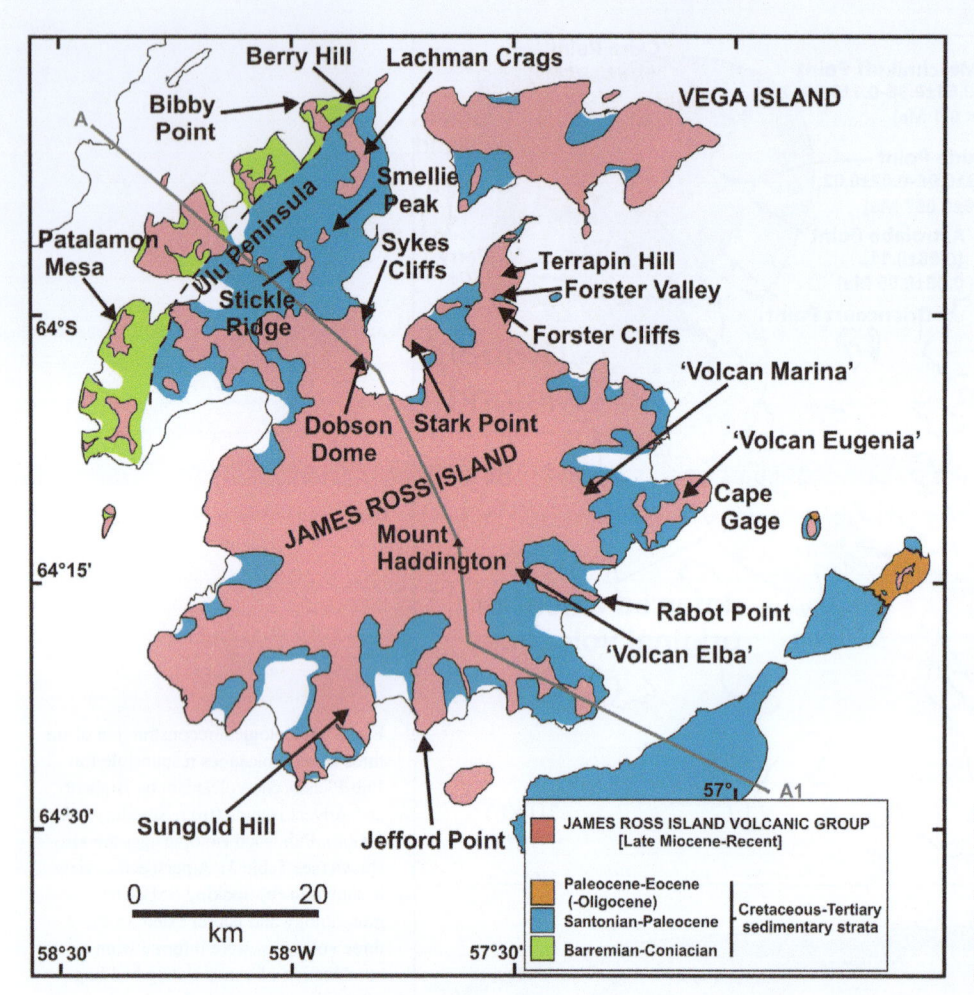

Fig. 18. Geological outcrop map of James Ross Island, drawn free of snow and ice, showing the distribution of volcanic and older sedimentary rocks. The locations of places mentioned in the text, and the line of cross-section shown in Figure 19, are also shown.

ages, including 74 by ^{40}Ar/^{39}Ar (all ages are shown on the map published by Smellie *et al.* 2013; see also Fig. 21; Table 3). The eruptive ages obtained on *in situ* outcrops only extend back to 6.16 ± 0.08 Ma (by ^{40}Ar/^{39}Ar: Smellie *et al.* 2008) but ages determined on JRIVG lava clasts in interbedded diamicts extend the age of the volcanism to 7.13 ± 0.49 Ma (by K–Ar: Sykes 1988), 9.2 ± 0.3 Ma (by ^{40}Ar/^{39}Ar: Jonkers *et al.* 2002) and 12.4 ± 0.5 Ma (by K–Ar: Marenssi *et al.* 2010). At least 50 eruptive episodes have been recognized (Smellie *et al.* 2008). Long periods of inactivity intervened between eruptions, some >100 ka. The individual eruptions were large volume, typically a few tens of km^3, but some may have attained volumes >100 km^3, assuming a source vent for most at Mount Haddington summit (Smellie *et al.* 2013; unpublished information of J.L. Smellie), making them substantial eruptions of a flood basalt nature. Ages in the satellite centres get younger progressively in a northeasterly direction between Prince Gustav Channel and Paulet Island, and are tentatively linked to a northeasterly propagating and opening deep fault system (Smellie *et al.* 2006c). Although a broad age progression is clear in the latter, there are age reversals in the trend, and most of the published ages are very old now and were determined by the K–Ar method (Rex 1972, 1976) so they may not be very accurate. Moreover, the reasons for the presence of the fault system are unclear, and there is no obvious geographical or age relationship with other satellite and flank centres distributed across James Ross Island. The chronology of the volcanic field and the presence

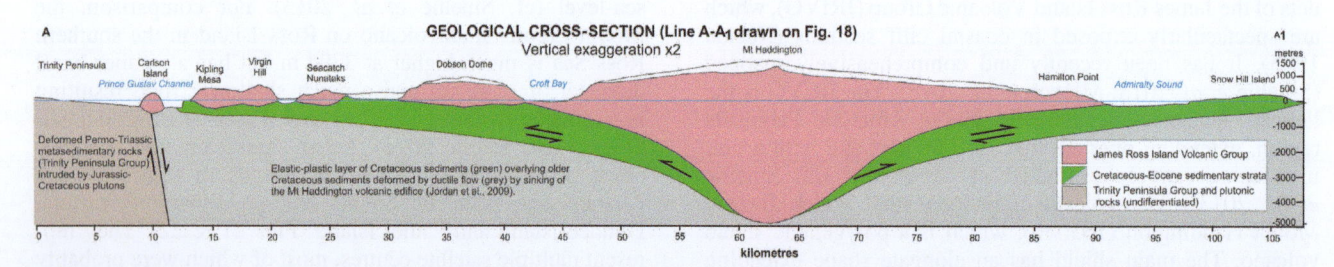

Fig. 19. Schematic cross-section through Mount Haddington based on a gravity survey of James Ross Island by Jordan *et al.* (2009; after Smellie *et al.* (2013), modified; copyright British Antarctic Survey, NERC, all rights reserved). The volcano centre is dominated by low-density clastic rocks, called 'hyaloclastite' by Jordan *et al.* (2009) but a combination of (mainly) tuff breccia and volumetrically lesser pāhoehoe lava together with scattered centres composed of lapilli tuff. The Cretaceous sedimentary basement consists of an upper elasto-plastic 'lid' (coloured green in the figure) deformed by brittle thrusting and related folding associated with the subsidence of the James Ross Island volcano. The deformation may have formed an annular flexural bulge that surrounds James Ross Island and runs through Vega Island, Snow Hill Island and Prince Gustav Channel. The softer basal Cretaceous sediments (coloured grey) were deformed by ductile flow. Note the great thickness of volcanic materials, detected geophysically, forming a 'keel' that has sunk up to 5 km into the weak Cretaceous sedimentary bedrock.

Fig. 20. Photomontage of structural features and representative lithofacies in lava-fed deltas and tuff cones in the James Ross Island Volcanic Group. (**a**) Thrust-faulted Cretaceous marine sedimentary strata on the west coast of James Ross Island, deformed by subsidence and gravitational spreading of the Mount Haddington volcano. The hammer (on the right) is *c.* 30 cm long. (**b**) Massive tuff breccia characteristic of lava-fed delta foreset beds; Sykes Cliffs, northern James Ross Island. The hammer is *c.* 30 cm long. (**c**) Section cut across a lava-fed delta showing flat-lying pāhoehoe capping lavas ('topsets') overlying homoclinal steep-dipping breccia foresets; the section is *c.* 80 m high, Stark Point. Photograph: Ian Skilling. (**d**) View of a prominent large-scale reactivation surface in a 200 m-high lava-fed delta at Stark Point; the structure is caused by a pause in the progradation of the delta with subsequent readvance of the delta front in a different direction, onlapping the earlier delta-front deposits. Photograph: Ian Skilling. (**e**) Annotated view of lava-fed delta at Förster Cliffs, showing a prominent delta-front overlap structure (*sensu* Smellie 2006) comprising left-dipping large-scale breccia foresets (orange) of an earlier delta advance (capping lavas out of view to the right) onlapped by right-dipping brown breccia foresets and overlying flat-lying subaerial pāhoehoe lavas of a younger advance; the younger advance took place with lower coeval (melt)water levels in a glacial setting. (**f**) Orange well-stratified lava-fed delta bottomset beds resting on breccia of an older lava-fed delta; Sykes Cliffs, northern James Ross Island. The notebook is 16 cm long. (**g**) Typical diffusely stratified lapilli tuffs of a tuff cone; Sungold Hill, southern James Ross Island. The pencil is *c.* 16 cm long. (**h**) Syneruptive thin-bedded tuffs emplaced subaqueously and cut by a prominent channel filled by thicker-bedded and massive tuffs; marine-emplaced tuff cone; Terrapin Hill. The hammer is *c.* 70 cm long. All localities are shown in Figure 18.

of at least two small little-eroded scoria cones in eastern James Ross Island ('Volcán Elba', Volcán Marina': Strelin and Malagnino 1992) suggests that volcanism occurred recently and the JRIVG may still be active (Smellie *et al.* 2008, 2013). A third small scoria cone may also be present, 3.5 km SE of 'Volcan Elba' on top of Rabot Point but has not been visited (Smellie *et al.* 2013).

Only two regional-scale geological investigations of the JRIVG have been undertaken, by Nelson (1975) and Smellie *et al.* (2008). Nelson (1975) identified five evolutionary stages that he called eruptive phases. Each phase was associated with lava-fed delta extrusion. His study remains one of the earliest and most detailed studies of lava-fed deltas ever attempted but the study was ultimately flawed in assuming that effusion was into the sea. Thus, the stratigraphical correlations were based on a supposed horizontality of the delta passage zones and that they represented palaeo-sea-levels. Nelson (1975) also recognized, for the first time, that several types

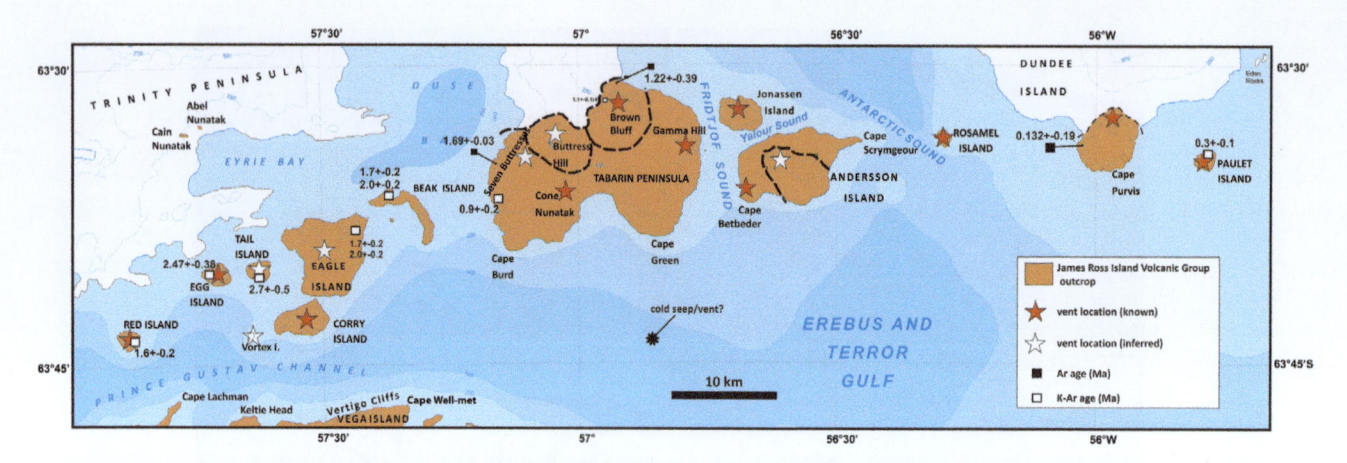

Fig. 21. Map showing the distribution of volcanic centres in Prince Gustav Channel, Tabarin Peninsula and Antarctic Sound (after Smellie *et al.* (2013), modified; copyright British Antarctic Survey, NERC, all rights reserved). Note the ENE-younging of the eruptive ages, the origin of which is uncertain but has been speculatively related to fault propagation (Smellie *et al.* 2006c).

of subaqueous–subaerial transitions occur within lava-fed deltas. He also illustrated a feature that is probably due to a rising water level, which he thought, erroneously, was due to the sea-level varying during the individual eruptive episodes (a hypothesis examined in greater detail by Jones and Nelson 1970; cf. Smellie 2006). Nelson's (1975) study was in press by 1965 but its publication was delayed for a decade. However, it was remarkably prescient and well ahead of its time, and he anticipated many advances made decades later. Isotopic dating by Rex (1972, 1976), Sykes (1988) and Smellie *et al.* (2008) conclusively disproved Nelson's (1975) underlying hypothesis that the deltas could be mapped

Fig. 22. Photomontage of selected satellite volcanic centres in the James Ross Island Volcanic Group. (**a**) Dobson Dome, an 80 ka tuya, looking WSW; it is the largest-known tuya in the Antarctic region, with a thickness of >600 m and a diameter of >5 km. (**b**) Gamma Hill, a small unvisited volcanic centre (probable tuff cone) on the east side of Tabarin Peninsula. (**c**) Brown Buff, Tabarin Peninsula, a prominent well-described tuya (see Skilling 1994 for a description). (**d**) Jonassen Island, a tuya in Antarctic Sound. (**e**) Rosamel Island, a tuff cone in Antarctic Sound. (**f**) Cape Purvis (a tuya; Dundee Island) and Paulet Island. (**g**) Paulet Island, a lava shield capped by two coalesced scoria cones, viewed looking SSW (from Kraus *et al.* 2013; image courtesy of Stefan Kraus).

and correlated by their supposedly horizontal passage zones. Moreover, distinctive features shown by many of the deltas and their alteration characteristics can only be explained by eruption in association with an ice cover, and it is now considered that the deltas are almost all glaciovolcanic (Smellie 2006; Johnson and Smellie 2007; Smellie *et al.* 2008, 2009; Calabozo *et al.* 2015).

Conversely, the laterally extensive horizontal passage zones in a small number of the deltas and the widespread distribution of tephra deposits of most of the tuff cones in the JRIVG can be explained by eruption in a marine setting (Smellie *et al.* 2006*a*, 2008; Calabozo *et al.* 2015). At least one of the tuff cones (at Patalamon Mesa, northwestern James Ross Island) contains marine fossils (Asterozoan external moulds: Bibby 1966; Williams *et al.* 2006; Nelson *et al.* 2008). The constituent pyroclastic beds at Terrapin Hill, which locally extend >5 km from their source vent, were deposited at approximately the angle of repose ($\leq c.$ 10°) and are asymptotic on the Cretaceous bedrock. Features such as these are diagnostic of tephra deposited in an unconfined situation (i.e. subaerial, sea or pluvial lake: Smellie *et al.* 2006*a*). They contrast with the much steeper gradients tuff cone flanks constructed by glaciovolcanic tuff cones, which are far steeper than the stable angle of repose due to banking of tephra beds against enclosing former ice walls, which then melt away (Smellie 2009, 2018). A marine setting for Terrapin Hill is also apparently confirmed by the composition of authigenic zeolites in the tuff cone (Fig. 23).

The deltas that may have been emplaced in a marine setting are that at Lachman Crags (Johnson Mesa Formation of Smellie *et al.* 2013), the distal part of the Jefford Point Formation in southeastern James Ross Island and the Lookalike Peaks Formation in northeastern Ulu Peninsula. Calabozo *et al.* (2015) suggested a marine setting for the Lookalike Peaks Formation based on the very low angle of dip (2°) of the passage zone of the basal delta at Stickle Ridge and Smellie Peak. Mapping and measured GPS-based elevations confirm that the passage-zone elevation dips at *c.* 2–3°, resulting in an elevation drop of at least 70 m over a distance of 1.5 km (unpublished information of J.L. Smellie). Moreover, authigenic phillipsite and chabazite compositions from tuff breccia samples of the Lookalike Formation strongly suggest 'marine' compositions (Fig. 23: cf. Johnson and Smellie 2007). Thus, despite the passage zone not being wholly horizontal, the cumulative evidence available so far suggests that the Lookalike Peaks Formation may also have been marine emplaced. Calabozo *et al.* (2015) also noted that the passage zone in the overlying lava-fed delta (Stickle Ridge Formation of Smellie *et al.* 2013) has a very similar dip of just 3° but suggested that it was glacially emplaced due to evidence for its construction on soft glacial sediments (implying that they were coeval: Smellie *et al.* 2008; Smellie and Edwards 2016; Smellie 2018).

The entire JRIVG is divided into multiple formations (Smellie *et al.* 2013). Twenty-one are lava-fed deltas, whereas all of the tuff cone outcrops are currently grouped together into a single unit (Terrapin Hill Formation), as are all of the relatively uncommon hypabyssal intrusions (Palisade Nunatak Formation). Until now, none of the lithostratigraphical formations has been described individually and the opportunity is taken here to summarize their principal characteristics (Table 4). Although many are not lithologically distinctive compared with neighbouring lava-fed deltas, the deltas are easily correlated where outcrops are continuous or separated by relatively short distances (<5 km), and where passage-zone elevations coincide across the exposure gap. Other specific features of individual deltas used to support correlations

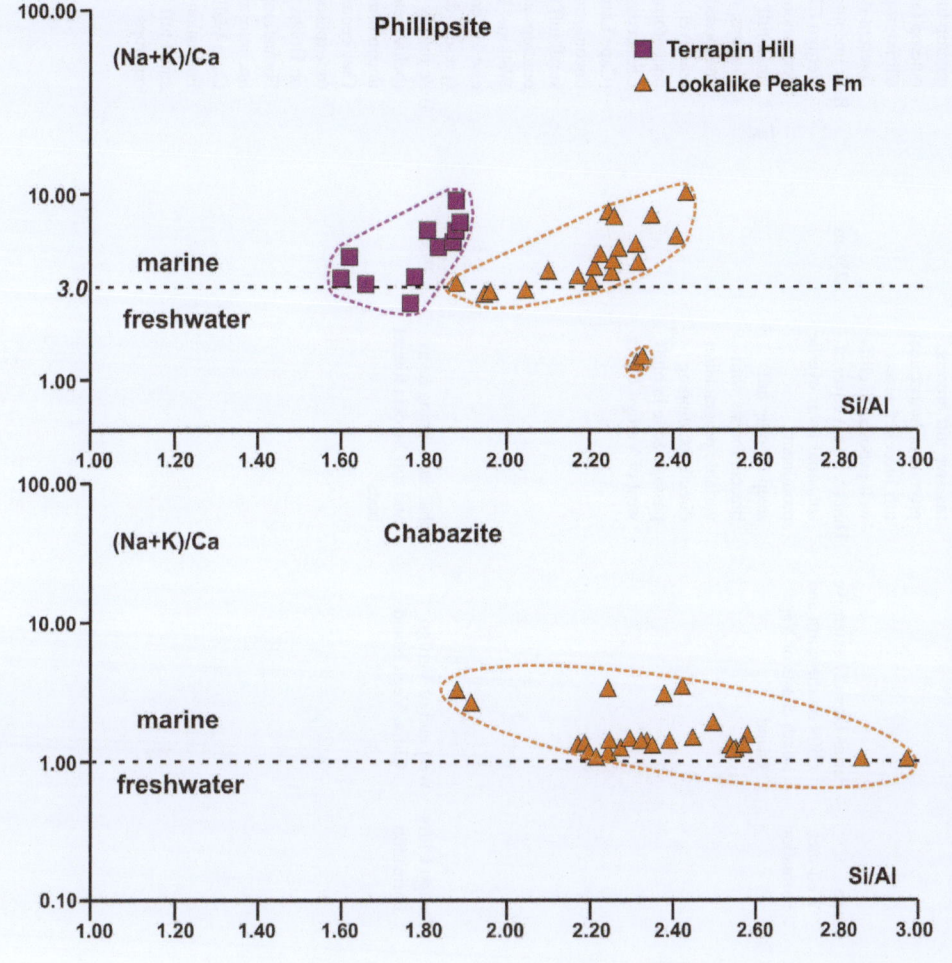

Fig. 23. Discriminant diagrams showing the composition of authigenic zeolites (phillipsite and chabazite) in lapilli tuffs from Terrapin Hill (a marine-erupted tuff cone) and tuff breccias of the Lookalike Peaks Formation (lava-fed delta; unpublished data of J.L. Smellie and J.S. Johnson). The discriminant fields are from Johnson and Smellie (2007). There are no chabazite data for Terrapin Hill (chabazite absent). There are significant caveats in using authigenic zeolite compositions to discriminate eruptive environment owing to the possibility of subsequent compositional overprinting (described by those authors), and the discriminant (Na + K)/Ca value (1.00 for chabazites; 3.00 for phillipsites) being a zone rather than a fixed value. However, the strong association with marine compositions is probably indicative in both examples.

Table 4. *Summary lithostratigraphy of the James Ross Island Volcanic Group (based on Smellie et al. 2013)*

Unit name	Distribution	Principal lithological characteristics	Maximum thickness* (m)	Boundaries	Age (Ma) [and error, 2σ]†	Type section	Notes	Key references
EFFUSIVE UNITS (LAVA-FED DELTAS, MAINLY GLACIOVOLCANIC)								
Vega Island								
Keltie Head Formation	Keltie Head, Vega Island	Subaerial capping lavas contain up to three interbedded units of massive tuff breccia; prominent homoclinal tuff breccia foresets well exposed in cliffs	>200 m	Base mainly unexposed (below sea-level); locally rests on small outcrop tuff cone outcrop (at Waterfall Cliffs); top eroded (present-day surface)	0.99 [0.05]	Keltie Head		Nelson (1975); Smellie et al. (2008, 2013)
Cape Well-met Formation	Forms laterally extensive cliffs on the north and south coasts of Vega Island	Thin (c. 30 m) cover of subaerial lavas overlie prominent steep-dipping tuff breccia foresets with variable progradation directions; dipping passage zone in bluff west of Vertigo Cliffs	>250 m	Base unconformable on Vertigo Cliffs Formation and two small tuff cone outcrops, or is unexposed below sea-level; also rests unconformably on Cretaceous bedrock (Cape Lamb); overlain unconformably by two small tuff cone outcrops and Cape Well-met Formation or else top eroded (present-day surface)	2.03 [0.04]	Cape Well-met	Detailed sketches of variably dipping tuff breccia foresets presented by Skilling (2002); rising and falling passage zone seen in cliffs SW of the west end of Vertigo Cliffs	Nelson (1975); Skilling (2002); Smellie et al. (2008, 2013)
Vertigo Cliffs Formation	West end of Vertigo Cliffs, Vega Island	Subaerial capping lavas and tuff breccia foreset beds	c. 195 m	Base mainly unexposed (below sea-level); also unconformable on Cretaceous bedrock (unexposed) and locally on Hobbs Glacier Formation; overlain unconformably by Cape Well-met Formation and two small tuff cone outcrops	2.67 [0.13]	Vertigo Cliffs, west end	Detailed sketches of variably dipping tuff breccia foresets presented by Skilling (2002)	Nelson (1975); Skilling (2002); Smellie et al. (2008, 2013)

Formation	Location	Description	Thickness (m)	Relations	Age (Ma) [±]	Key localities	Notes	References
Sandwich Bluff Formation	Extensive mainly inland outcrops on Vega Island, including at Leal Bluff, Sandwich Bluff, Pirrie Col and Mahogany Bluff	Subaerial capping lavas and tuff breccia foreset beds	>500	Unconformable on Vertigo Cliffs Formation, Cretaceous bedrock (unexposed) and diverse tuff cone outcrops; base also below sea-level (unexposed); upper boundary eroded (present-day surface)	5.42 [0.08]	Mahogany Bluff, west end	Multiple passage zones seen at western Mahogany Bluff	Nelson (1975); Smellie et al. (2008, 2013)
Western James Ross Island (Ulu Peninsula)								
Dobson Dome Formation	Eastern half of Dobson Dome; principal exposures at Blyth Spur	Subaerial capping lavas and tuff breccia foreset beds	>650	Base unexposed (below sea-level); top eroded / extensively frost shattered	<0.08	Blythe Spur	Subaerial capping lavas largely frost shattered; tuff breccia foreset beds poorly exposed	Nelson (1975); Smellie et al. (2008, 2013)
Donnachie Cliff Formation	Dobson Dome, NW side	Subaerial capping lavas and tuff breccia foreset beds	>250?	Overlies tuff cone outcrop and probably Stickle Ridge Formation	1.15 [0.12]	Unnamed crags NW side of Dobson Dome	Poorly known remote outcrop, difficult access	Nelson (1975); Smellie et al. (2008, 2013)
Lachman Crags Formation	Summit platform of Lachman Crags	Subaerial capping lavas and tuff breccia foreset beds	81	Unconformably overlies Johnson Mesa Formation and small intervening outcrop of Hobbs Glacier Formation; top eroded (present-day surface)	3.95 [0.05]	Southernmost Lachman Crags, summit outcrop	Poorly exposed except in inaccessible rock face on SE side of Lachman Crags	Nelson (1975); Smellie et al. (2008, 2013)
Patalamon Mesa Formation	Mainly SW Ulu Peninsula, at Patalamon Mesa, Back Mesa and northern Crisscross Crags; also flanking Davies Dome	Subaerial capping lavas and tuff breccia foreset beds	270	Unconformable on Kipling Mesa Formation (at Davies Dome) and Tumbledown Formation (SW Ulu Peninsula); locally unconformable on tuff cone outcrops and Hobbs Glacier Formation; top eroded (present-day surface)	4.61 (4.78–4.53)	Patalamon Mesa	Spectacular rising passage zones seen at NW and SW Patalamon Mesa, and rising and falling passage zones at north and west Back Mesa; capping lavas have massive tuff breccia interbeds (cf. Keltie Head Formation)	Nelson (1975); Smellie et al. (2008, 2013)

(Continued)

Table 4. *Continued.*

Unit name	Distribution	Principal lithological characteristics	Maximum thickness* (m)	Boundaries	Age (Ma) [and error, 2σ]†	Type section	Notes	Key references
Johnson Mesa Formation	Northern Ulu Peninsula only, at Lachman Crags, Berry Hill and Johnson Mesa	Subaerial capping lavas and tuff breccia foreset beds	203, possibly 241	Base mainly unconformable on Cretaceous bedrock, locally on Cape Lachman Formation and Hobbs Glacier Formation; overlain unconformably by Lachman Crags Formation, locally by Hobbs Glacier Formation, or is eroded (present-day surface)	5.06 [0.04]	Western Lachman Crags	Subaerial lava cap rock unusually thick (185 m) at southern Lachman Crags; rare marine-emplaced lava-fed delta	Nelson (1975); Smellie *et al.* (2008, 2013); Nehyba and Nývlt (2015)
Smellie Peak Formation	Northern Ulu Peninsula only, at Smellie Peak; inaccessible outcrops at Lookalike Peaks may be capping lavas of Stickle Ridge Formation	Columnar lava of a possible lava lake and tuff breccia foreset beds	c. 200	Unconformable on Lookalike Peaks Formation and locally on Hobbs Glacier Formation and unnamed scoria cone outcrop (at Smellie Peak)	5.91–5.14; younger K-Ar ages of 4.63–4.31, with high errors, were obtained by Calabozo *et al.* (2015)	Smellie Peak, summit	Tuff breccia unit contains pillow lava at Smellie Peak; correlated with the Stickle Ridge Formation by Calabozo *et al.* (2015)	Nelson (1975); Smellie *et al.* (2008, 2013); Calabozo *et al.* (2015)
Tumbledown Cliffs Formation‡	SW Ulu Peninsula, at Tumbledown Mesa, Crisscross Crags, Patalamon Mesa, Back Mesa, Sentinel Buttress and Limmershin Cliffs	Subaerial capping lavas and tuff breccia foreset beds	155	Unconformable on Cretaceous bedrock and locally on Hobbs Glacier Formation and tuff cone outcrops; overlain unconformably by Patalamon Formation, Hobbs Glacier Formation and tuff cone outcrops or is eroded (present-day surface)	5.15 (5.26–5.01)	Tumbledown Mesa		Nelson (1975); Smellie *et al.* (2008, 2013)
Kipling Mesa Formation	NW Ulu Peninsula, at Rink Point, Kipling Mesa, Stonely Point and Hambrey Cliffs	Subaerial capping lavas and tuff breccia foreset beds	125	Mainly unconformable on tuff cone outcrop; also unconformable on Cretaceous bedrock, Hobbs Glacier Formation and Cape Lachman Formation (Davis Dome)	5.36 [0.05]	Rink Point	Interpreted as marine emplaced by Carrizo *et al.* (1998)	Nelson (1975); Carrizo *et al.* (1998); Smellie *et al.* (2008, 2013)

Cape Lachman Formation	North and west Ulu Peninsula, at Cape Lachman, Berry Hill, Johnson Mesa, Hambrey Cliffs and Rink Point	Tuff breccia foreset beds	>100	Unconformable on Cretaceous bedrock, Mendel Formation, Hobbs Glacier Formation and tuff cone outcrops; overlain unconformably by Kipling Mesa Formation, Patalamon Formation, Johnson Mesa Formation and Hobbs Glacier Formation	5.67–5.32	Cape Lachman	Subaerial capping lavas removed by erosion; the formation composed of several geographically separated outcrop remnants in broadly comparable structural positions and isotopic ages but whose correlation is not unequivocally proven	Nelson (1975); Smellie et al. (2008, 2013)
Stickle Ridge Formation	Inland northern Ulu Peninsula only, including Stickle Ridge, Lookalike Peaks and Donnachie Cliff	Tuff breccia foreset beds; pillow lava in tuff breccia at north Lookalike Peak	>150 (Calabozo et al. 2015 suggest 200–300)	Unconformable on tuff cone outcrop (Stickle Ridge), Lookalike Peaks Formation and Hobbs Glacier Formation; overlain unconformably by Smellie Peak Formation and Donnachie Cliff Formation or else top eroded (present-day surface)	<6.16 to >5.91; a younger K–Ar age of 4.4, with very high error, and a preferred age of <4.6, were obtained by Calabozo et al. (2015)	Stickle Ridge	Subaerial capping lavas removed by erosion (Smellie et al. 2013); those indicated on Stickle Ridge outcrop by Calabozo et al. (2015) are tuff breccia foresets of an unrelated overlying delta; Calabozo et al. (2015) correlate the Stickle Ridge and Smellie Peak formations but $^{40}Ar/^{39}Ar$ ages differ	Nelson (1975); Smellie et al. (2008, 2013); Calabozo et al. (2015)
Lookalike Peaks Formation	Restricted to Lookalike Peaks and Smellie Peak only, inland northern Ulu Peninsula	Subaerial capping lavas and tuff breccia foreset beds	173	Unconformable on Cretaceous bedrock and Hobbs Glacier Formation; unconformably overlain by tuff cone and local scoria cone outcrops, Smellie Peak Formation and Hobbs Glacier Formation	6.16 [0.08]; a younger K–Ar age of 5.21, with high error was obtained by Calabozo et al. (2015)	Lookalike Peaks	Subaerial lava cap rock largely eroded to the SE; the passage zone elevation falls 70 m (2°–3° dip) between Smellie Peak and Lookout Peaks; interpreted as marine-emplaced by Calabozo et al. (2015) and possibly supported by authigenic zeolite compositions (see Fig. 13)	Nelson (1975); Smellie et al. (2008, 2013); Calabozo et al. (2015); unpublished information of J.L. Smellie
								Nelson (1975); Smellie et al. (2008, 2013)

(Continued)

Table 4. *Continued.*

Unit name	Distribution	Principal lithological characteristics	Maximum thickness* (m)	Boundaries	Age (Ma) [and error, 2σ]†	Type section	Notes	Key references
Eastern and southern James Ross Island								
Jefford Point Formation	Jefford Point, Hamilton Point and Rabot Point; may be the lava-fed delta that drapes Lockyer Island	Subaerial capping lavas and tuff breccia foreset beds	400 m in Nelson Cliffs?; c. 200 m at coast	Rests on unvisited lava-fed deltas and unconformably on tuff cone outcrops, Hamilton Point Formation and Cretaceous bedrock; top mainly eroded (present-day surface) but locally overlain by scoria cone(s) (e.g. 'Volcan Elba')	1.92–1.86	Jefford Point	Steeply dipping passage zone; lava-fed delta may have commenced effusion subglacially but ultimately became marine-emplaced; compositionally differs from Sungold Hill Formation	Nelson (1975); Smellie *et al.* (2008, 2013)
Sungold Hill Formation	SW James Ross Island, including Sungold Hill and the coastline west from there at least as far as NE of Cape Broms	Subaerial capping lavas and tuff breccia foreset beds	240	Probably unconformable on Cretaceous bedrock, also on tuff cone outcrop at Sungold Hill; top eroded (present-day surface)	2.30 [0.36]	Unnamed headland SE of Sungold Hill	Steeply dipping passage zone; lava-fed delta may have commenced effusion subglacially but ultimately became marine-emplaced; compositionally differs from Jefford Point Formation	Nelson (1975); Smellie *et al.* (2008, 2013)
Forster Cliffs Formation	Widespread, between Flett Buttress and east of Skep Point; possible correlative at St Rita Point (eastern James Ross Island)	Subaerial capping lavas and tuff breccia foreset beds	360–600	Base mainly unexposed (below sea-level); also unconformable on Taylor Bluff Formation, Hobbs Glacier Formation and tuff cone outcrop; top eroded (present-day surface)	2.50 [0.07]	Forster Cliffs	Conspicuous rising passage zone at west Forster Cliffs; interbedded massive tuff breccia within subaerial lava cap rock at St Rita Point	Nelson (1975); Smellie *et al.* (2008, 2013)
Jonkers Mesa Formation	Eastern James Ross Island, at Cape Gage, Jonkers Mesa, Rhino Corner and St Rita Point	Subaerial capping lavas and tuff breccia foreset beds	>146	Unconformable on Ekelof Point Formation, Hobbs Glacier Formation and Cretaceous bedrock; base locally unexposed (below sea-level; at Cape Gage); unconformably overlain by tuff cone outcrop ('Volcan Eugenie') but mostly top eroded (present-day surface)	3.08 [0.15]	Jonkers Mesa		Nelson (1975); Smellie *et al.* (2008, 2013)

Ekelof Point Formation	Eastern James Ross Island, includng wetsern Jonkers Mesa, Ekelof Point, Rhino Corner and St Rita Point; possibly extends SW to Rabot Point	Subaerial capping lavas and tuff breccia foreset beds	202	Unconformable on Cretaceous bedrock and Hobbs Glacier Formation; overlain unconformably by Jonkers Mesa Formation and tuff cone outcrop ('Volcan Eugenie')	3.62–3.41	Ekelof Point	Rising passage zone at St Rita Point	Nelson (1975); Smellie et al. (2008, 2013)
Taylor Bluff Formation	Northern James Ross Island, at Taylor Bluff and Stark Point	Subaerial capping lavas and tuff breccia foreset beds	>350?	Unconformable on Cretaceous bedrock; unconformably overlain by Forster Cliffs Formation	5.85 [0.03]	Taylor Bluff		Nelson (1975); Smellie et al. (2008, 2013)
PYROCLASTIC AND INTRUSIVE UNITS								
Terrapin Hill Formation	Widespread on James Ross Island, including Terrapin Hill, Cape Gage ('Volcan Eugenia'), Tortoise Hill, Lomas Ridge, Sungold Hill, Persson Island, Patalampon Mesa, Back Mesa, San Fernando Hill, Virgin Hill, Seacatch Nunataks and Bibby Point; also Vega Island (e.g. Mahogany Bluff, False Island Point, Cape Gourdon and Devil Island) and islands in Prince Gustav Channel (e.g. Carlson Island, Vortex Island, Eagle Island, Tail Island); Cain and Abel nunataks on eastern Graham Land	Khaki-grey to yellow-orange lapilli tuffs, well stratified; sometimes associated with irregularly shaped dykes	>600 m (Terrapin Hill)	Base either unexposed (below sea-level) or unconformable on Cretaceous bedrock, numerous lava-fed delta formations and Hobbs Glacier Formation; top either eroded (present-day surface) or unconformably overlain by lava-fed delta formations	5.90, 5.37, 3.60, 3.34, 2.23 and 0.66; many outcrops not dated but ages bracketed by associated lava-fed delta formations	Terrapin Hill (but see Notes)	These deposits represent multiple tuff cones, mainly marine-erupted. A few are well-preserved volcanic landforms (e.g. Terrapin Hill; 'Volcan Eugenia'). Could be further subdivided stratigraphically on the basis of differing ages and dispersed locations of the individual volcanic centres	Nelson (1975); Smellie et al. (2008, 2013); Nehyba and Nyvlt (2014)

(*Continued*)

Table 4. Continued.

Unit name	Distribution	Principal lithological characteristics	Maximum thickness* (m)	Boundaries	Age (Ma) [and error, 2σ]†	Type section	Notes	Key references
Palisade Nunatak Formation	Widespread across James Ross Island but volumetrically minor apart from Palisade Nunatak and Organpipe Nunatak (Ulu Peninsula)	Basaltic and doleritic intrusions, many with irregular shapes	The largest outcrop is at Palisade Nunatak; it is at least 300 m thick	Occur mainly within coeval tuff cones (Terrapin Hill Formation); some are 'surface-fed dykes' occurring within coeval lava-fed deltas; two are substantial sills intruding Cretaceous bedrock (Palisade Nunatak and Organpipe Nunatak)	5.90, 5.84, 5.37, 5.36, 5.33, 5.32, 2.03, 1.22 and 0.66; K–Ar ages: 6.45, 6.13, 5.36, 5.23, 3.68 and 3.57	Palisade Nunatak	Relatively few intrusions have been dated, and many occur in tuff cones with which the dykes were probably coeval and provide an age for the tuff cone eruptions; Palisade Nunatak and Organpipe Nunatak are the only large outcrops (columnar-jointed sills)	Nelson (1975); Smellie et al. (2008); K–Ar ages from Sykes (1988)

SEDIMENTARY UNITS

Unit name	Distribution	Principal lithological characteristics	Maximum thickness* (m)	Boundaries	Age (Ma) [and error, 2σ]†	Type section	Notes	Key references
Andreassen Point Formation	Widespread and particularly prominent in northern Ulu Peninsula but also found in most other snow-free areas of James Ross Island	Large jumbled masses of rock formed by the collapse of the steep margins of lava-fed delta units	Individual blocks within each outcrop are tens of metres high and hundreds of metres long	Mainly unconformable on weak Cretaceous bedrock; top eroded (present-day surface)	Mainly post-glacial (Holocene), although rare older examples are known (Pliocene)	Andreassen Point	A distinctive mappable morphostructural unit formed during major post-glacial collapses of the steep margins of the lava-fed deltas	Davies et al. (2013); Smellie et al. (2013)
Weddell Sea Formation	NE Seymour Island only	Unconsolidated massive diamicton with abundant erratics, some fossiliferous, derived from James Ross Island, Cockburn Island, Seymour Island and the Antarctic Peninsula set in a silty clay matrix	4	Unconformable on Eocene–Oligocene sedimentary rocks; top eroded (present-day surface)	Post-Late Pliocene	'La Meseta' plateau, NE Seymour Island	Interpreted as a terrestrial melt-out till	Malagnino et al. (1981); Zinsmeister and de Vries (1983); Gazdzicki et al. (2004)
Cockburn Island Formation	Cockburn Island only	Rusty brown medium to coarse sandstones, pebbly sandstone and conglomerate with abundant fragmented to complete and well-preserved large pectinid shells	>10	Unconformably overlies James Ross Island volcanic rocks (two unnamed lava-fed delta units); top eroded (present-day surface)	4.9–4.7 (mean 4.8) [0.3] by $^{40}Ar/^{39}Ar$; 4.66 [+0.17/−0.24] by $^{87}Sr/^{86}Sr$	Cockburn Island summit platform	Interpreted as a shallow-water (<100 m) marine interglacial sequence	Dingle et al. (1997); Jonkers (1998); Jonkers and Kelley (1998); McArthur et al. (2006)

Hobbs Glacier Formation	Widespread across James Ross Island; also found on Vega Island and NE Seymour Island	Massive diamictite, massive to well-bedded conglomerate and diamict dominate most outcrops; also laminated tuffaceous sandstone; often fossiliferous (fossils reworked and contemporaneous); includes boulder pavements at NE Seymour Island ('La Meseta') and Vega Island (Sandwich Bluff)	150 (typically just a few metres)	Two main stratigraphical occurences: (1) at the base of the James Ross Island Volcanic Group directly and unconformably overlying Cretaceous bedrock; and (2) interbedded with the James Ross Island Volcanic Group; in the latter case, the Hobbs Glacier Formation deposits are frequently coeval with the overlying volcanic unit; occasionally occurs unconformable on Cretaceous but with an eroded top (present-day surface)	Late Miocene and younger (<c. 13 Ma); a supposed Oligocene age (Ivany et al. 2006) is unlikely due to the presence of JRIVG lava clasts (J.L. Smellie unpublished information; cf. Marenssi et al. 2010)	Diamicts interpreted as glaciomarine sediment and basal till; conglomerates are glacigenic debris-flow deposits associated with advancing ice; laminated tuffaceous sandstones were interpreted as delta-front deposits linked to a subaerial braided river plain but may also represent far-travelled toeset beds linked to lava-fed deltas or even ash-fall (±current-reworked) associated with explosive volcanic eruptions	Pirrie et al. (1997); Jonkers et al. (2002); Lirio et al. (2003); Hambrey and Smellie (2006); Ivany et al. (2006); Concheyro et al. (2007); Hambrey et al. (2008); Nelson et al. (2009); Clark et al. (2010); Marenssi et al. (2010); Williams et al. (2010); Pirrie et al. (2011); Calabozo et al. (2015); Rocha-Campos et al. (2017)
Mendel Formation	NE Ulu Peninsula only, between Bibby Point, Berry Hill and Cape Lachman	A wide variety of sediments including massive to stratified diamicts (lodgement, melt-out), fluvial sandstones, glaciomarine debris-flow deposits and diamictites, and laminated marine sandstones and siltstones	>80	Base unexposed, below sea-level and must rest unconformably on Cretaceous bedrock and James Ross Volcanic Group (Cape Lachman Formation); top corresponds to present-day surface but locally covered by modern colluvium and Hobbs Glacier Formation	Maximum age range 6.16–5.23, but probably 5.9–5.4	Interpreted as terrestrial glacigenic and glaciofluvial, glaciomarine. and open-marine sedimentary rocks. Numerous reworked pectinid fossils (ages 20.56–5.80 Ma)	Nyvlt et al. (2011)

*From Smellie et al. (2009, table 1); J.L. Smellie, unpublished field notes.
†^{40}Ar/^{39}Ar ages only (from Smellie et al. 2008) except where otherwise specified.
‡Names modified slightly from Smellie et al. (2013) to maintain nomenclatural consistency.

include: (i) the relative thicknesses of capping subaerial lavas and coeval subaqueous tuff breccia foreset units, which can vary significantly between successive superimposed deltas; (ii) coincident dip directions of homoclinal tuff breccia foresets; (iii) the presence of massive tuff breccia units within lava cap rocks (not present in all of the deltas); (iv) distinctive geometries of the subaqueous–subaerial transitions (cf. Smellie 2006); and (v) similar lava compositions and isotopic ages (by ^{40}Ar/^{39}Ar and K–Ar). Of all the criteria used, lava compositions can be equivocal as the individual formations, which represent large-volume effusive eruptions, appear to have varied compositionally with time during eruption (unpublished information of J.L. Smellie). Isotopic ages can be definitive, particularly when attempting to correlate outcrops that are widely separated and not sufficiently distinctive lithologically. However, high errors may make ages ambiguous and some can be inexplicably unreliable, especially when ages determined by two different methodologies are compared (e.g. conflicting and irreconcilable ages obtained on the Lookalike Peaks and Stickle Ridge formations by Smellie et al. (2008: ^{40}Ar/^{39}Ar) and by Calabozo et al. (2015: K–Ar)).

The spectacular lava-fed deltas have a bipartite division composed of coeval but lithologically contrasting lithosomes. They comprise: (1) an upper unit of essentially flat-lying subaerial pāhoehoe lavas, equivalent to 'topset beds'; and (2) tuff breccia disposed in relatively crude, large-scale, homoclinal dipping beds ('foreset beds': Fig. 20b–d). The subaerial capping lava units are almost always much thinner than the underlying tuff breccias. The most prominent exception occurs at southern Lachman Crags (Johnson Mesa Formation) where the thickness of subaerial capping lavas reaches 185 m and overlies foreset-bedded tuff breccia <<100 m thick (unpublished information of J.L. Smellie). The characteristics are attributed to a vent-proximal situation for that part of the delta (Smellie et al. 2008). The lava-fed deltas range in thickness between c. 50 and >650 m (Smellie et al. 2009; unpublished information of J.L. Smellie). The thickest example forms the prominent eastern spur of Dobson Dome, known as Blyth Spur, which is >600 m (Fig. 22a). The deltas are, in general, superbly exposed in laterally extensive cliffs (Fig. 20c–e) in which numerous environmentally important features are clearly seen, including reactivation surfaces, massive tuff breccia interbeds in the capping lavas (signifying local flooding of the delta top by a rising water level) and a variety of subaerial–subaqueous transitions (Fig. 20e) (Smellie 2006; see also Skilling 2002, figs 2 & 3). In all, the JRIVG contains amongst the best-exposed, accessible and informative multiple lava-fed delta systems in the world, and they have been responsible for several important advances in understanding lava-fed delta emplacement, palaeoenvironmental implications and glaciovolcanism generally (Skilling 1994, 2002; Smellie 2006; Smellie et al. 2008, 2009).

There are few detailed studies focused on the lava-fed deltas in the JRIVG. Skilling (2002) described the lithofacies and structural features of extensive lava-fed deltas on northern Vega Island (Cape Well-met Formation) and at Stark Point (Förster Cliffs Formation). Both deltas are probably glaciovolcanic (Smellie et al. 2013) (Table 4). Important depositional features noted include reactivation surfaces, slump zones, depositional lobes, and chute and scour structures. The study documented the range of lithofacies present, and examined the clast generation and depositional processes. A generic model for the construction of lava-fed deltas in any eruptive environment (lake, sea or englacial) was proposed. Seven coherent lava lithofacies were distinguished, corresponding to the subaerial lava cap-rock lithosome, and eight clastic lithofacies of the subaqueous foreset and related beds. Although Skilling (2002) was describing primary volcaniclastic deposits, he used epiclastic grain-size divisions (cf. White and Houghton 2006). Moreover, although not listed as a lithofacies, Skilling (2002) included products of littoral explosions in his dynamic model, which should produce distinctively vesicular sideromelane pyroclasts that will contrast in morphometry and vesicularity with the (blocky, vesicle-poor) sideromelane created elsewhere in the delta, mainly at the delta brink point (cf. Jurado-Chichay et al. 1996; Smellie et al. 2006b). A few deltas contain large slabs several tens of metres in extent composed of subaerial pāhoehoe lava. Sykes (1989) and Skilling (1994) also identified thick massive beds of polymict tuff breccia distinguished by a wide range of clast types, including pāhoehoe lava and hydrothermally altered lapilli tuffs which were informally called rubble breccias by Sykes (1989). Both the large slabs of lava and the rubble breccias appear to have been emplaced by collapse of the capping lava lithosome at the active delta front.

By contrast, Nehyba and Nývlt (2015) undertook a sedimentological study of three small distal outcrops of the marine-emplaced Johnson Mesa lava-fed delta at northern Lachman Crags and Berry Hill, northern Ulu Peninsula, with a stated focus on delta 'bottomset' beds. Bottomset beds in the JRIVG lava-fed deltas were also described, in less detail, by Hambrey et al. (2008) and Nelson et al. (2009), and they were identified as weak ductile layers implicated in the multiple mechanical collapse events that gave rise to the mainly Holocene-age Andreassen Point Formation, composed of jumbled volcanic megablocks (Oehler et al. 2005; Davies et al. 2013, fig. 10; Smellie et al. 2013). However, Oehler et al. (2005) also suggested that the entire tuff breccia foreset-bedded lithosomes of lava-fed deltas may react ductilely, whilst individual beds are brittle. The bottomset beds examined by Nehyba and Nývlt (2015) typically dip at 7°–15°, which is much lower than the associated overlying tuff breccia foreset beds (30°–35°). Epiclastic terminology was also used for the bottomset lithofacies by Nehyba and Nývlt (2015). However, true bottomsets represent the finer-grained downdip distal terminations of delta foresets. If the associated foresets are volcanic deposits (Smellie and Edwards 2016) then any cogenetic bottomsets are also volcanic, and volcaniclastic terminology should preferably be used as they are also essentially primary volcanic lithofacies, irrespective of the transporting medium (i.e. syneruptive sensu McPhie et al. 1993; White and Houghton 2006; Smellie and Edwards 2016).

The bottomset lithofacies are distinguished by yellow-orange, laminated, fine and coarse tuff, and massive to normally graded very coarse tuff to fine lapilli tuff, using the terminology of White and Houghton (2006) and Smellie and Edwards (2016), which contrast with the associated much coarser and darker khaki-brown to orange-brown tuff breccia foresets (Fig. 20f); interbeds of tuff breccia were also observed within the bottomset strata but were said to be uncommon. Nehyba and Nývlt (2015) suggested that the volcaniclastic bottomset beds were deposited by a variety of processes, including traction currents, debris flows and turbidity currents. The identification of some deposits as products of high-density turbidity currents is unlikely as the clast sizes are too large to be held in turbulent suspension and they were more likely to have been transported in hyperconcentrated flows (Nelson et al. 2009; cf. Mulder and Alexander 2001). Deposit 'proximality' was defined relative to the location of the overlying lava flow (i.e. the brink point, where most of the clasts are generated by thermohydraulic fracturing: Skilling 2002). There are also interbeds of massive to normally graded sandy conglomerate, massive mudstone, polymict pebbly mudstone and muddy pebbly sandstone. The presence in these beds, with rounded to subrounded pebbles to cobbles, polymict compositions and, admittedly rare, Pectinid bivalves and Terebratulid brachiopods suggests that they are epiclastic, and they were interpreted as products of subaqueous debris flows. The mud-rich and pebbly–cobbly beds are probably glacigenic. Some laminated very coarse tuffs to fine lapilli tuffs may also be epiclastic, and may represent deposits of bipartite sediment gravity flows genetically linked to associated sandy pebble conglomerates. In a marine setting with active volcanism it may not be easy to distinguish between rapidly resedimented volcaniclastic deposits (*sensu* McPhie *et al.* 1993) and true erosionally formed epiclastic deposits.

Nehyba and Nývlt (2015) noted that the proportion of volcaniclastic to glacigenic sediments was highly variable between localities. Moreover, they suggested that rafts of glacigenic diamictites and laminated mudstone contained in volcaniclastic beds might indicate overriding of precursor sediments by a lava-fed delta, although the mechanism by which the rafts were incorporated is not clear. An apparent absence of interbedded tuff breccia at one locality and the presence of a sharp contact between the described bottomset beds and overlying foreset beds suggested that a genetic link to a lava-fed delta was not always proven. The outcrops were described within a framework of deposition on the flanks of a volcano, with the deposits subsequently overridden by a lava-fed delta and a time break of unspecified duration intervening. In such a concept, it is unclear if the deposits are then technically lava-fed delta bottomsets in all of the outcrops described by Nehyba and Nývlt (2015) – that is, if there is no genetic link to a delta.

JRIVG tuff cone outcrops

The tuff cone outcrops associated with the JRIVG are strongly eroded and lack any primary landforms apart from two: (i) Terrapin Hill, which is the youngest known tuff cone in the region (0.66 ± 0.22 Ma) and is relatively well formed whilst lacking a summit crater; and (ii) 'Volcán Eugenia', at Cape Gage (eastern James Ross Island: Fig. 18), which preserves a crater and one original flank sandwiched between two lava-fed deltas; the deltas are dated as 3.08 ± 0.15 and 2.23 ± 0.05 Ma (Smellie *et al.* 2008) (Table 3). The tuff cones are constructed of khaki-yellow to orange coloured lapilli tuffs and fewer tuffs, variously thin-bedded, diffusely bedded and diffusely stratified (unpublished information of J.L. Smellie; terminology after Branney and Kokelaar 2002) (Fig. 20g). Less common volcaniclastic lithofacies include massive lapilli tuff, cross-stratified lapilli tuff (including antidunes) and laminated tuff (including rare ripple cross-laminated tuff). Shallow concave-up erosive bases to beds are locally prominent (Fig. 20h). The outcrops may also contain large slabs several tens of metres in extent composed of deformed and brecciated lapilli tuff strata. The lithofacies are characteristic of tuff cones (cf. Sohn and Chough 1992; Sohn 1996), and the combination of rare marine fossils, small-scale ripples and the great lateral extent of the asymptotic, quaquaversal, angle of repose bedding suggests eruption and deposition of the lower parts of the tuff cones in a marine setting (Smellie *et al.* 2008), although a wholly subaerial setting cannot be ruled out for some outcrops (those lacking fossils and ripples; see Smellie 2021). The poorly sorted massive beds may have been deposited from granular fluid-based currents but the bedded or stratified lithofacies were probably deposited mainly from fully dilute pyroclastic density currents (cf. lithofacies dsLT, dbLT and mLT of Branney and Kokelaar 2002; also Smellie and Edwards 2016). The brecciated beds were interpreted to have formed by failure of lithified and locally hydrothermally altered subaqueous and subaerial lithofacies. Skilling (1994) also suggested an origin as dry debris avalanches (e.g. at Brown Bluff) and they were important for assigning a glacial eruptive setting for that edifice (see the following subsection). Only one of the tuff cone outcrops has been investigated in detail, at Bibby Point. Nehyba and Nývlt (2014) recognized 12 lithofacies and three lithofacies associations, the latter comprising deposits of dense and dilute pyroclastic density currents (the dominant association), granular fluid-based pyroclastic density currents (or remobilized slope-failure deposits) and minor 'epiclastic resedimented' deposits formed during quiescent periods. Variations in the distribution of the lithofacies associations were attributed principally to responses to variations in the intensity and type of hydrovolcanic activity (Surtseyan or Taalian) and water availability.

Graham Land mainland and Tabarin Peninsula

The Graham Land mainland, including Tabarin Peninsula, contains numerous satellite centres. Tiny eroded remnants of two tuff cones crop out at Cain and Abel nunataks, on the north flank of the Prince Gustav Channel (Fig. 21). Both have been observed but only one was visited, just once, by Aitkenhead (1975; see also Nelson 1975; Smellie 1990g) and they are not well known. The lapilli tuffs at Cain Nunatak are at least 274 m thick (base unexposed), and they are well stratified, dipping north and east at 35° and 30°, respectively, above a basal more massive section. There are no published details for the unvisited outcrop at Abel Nunatak. Other mainland outcrops of the JRIVG are confined to Tabarin Peninsula, where there are several much larger centres. The two most prominent are Buttress Hill (completely ice covered; almost 300 m high and 3 km wide: Ashley 1962) and Brown Bluff, but subsidiary centres include Gamma Hill and an unnamed hill 4 km SW of Buttress Hill. The latter probably gave rise to the extensive lava-fed delta that is exposed at Seven Buttresses (dated at 1.69 ± 0.03 and 0.9 ± 0.2 Ma; Table 3). The eruptive setting of the Seven Buttresses delta, possibly glacial (Smellie *et al.* 2009), is not well known and requires validation. Cone Nunatak is probably a small tuff cone constructed on the Seven Buttresses lava-fed delta or partially 'drowned' by that delta during its advance. Apart from Brown Bluff, only reconnaissance studies have been carried out on Tabarin Peninsula (Ashley 1962; Nelson 1975).

Gamma Hill is a small but prominent ice dome that rises to *c.* 350 m and is *c.* 1.5 km (NW–SE) to *c.* 3 km (SW–NE) wide (Fig. 22b). The only rock exposures are of khaki-orange stratified rocks, probably lapilli tuffs, which are present on its NE side overlooking Fridtjof Sound. The Gamma Hill centre apparently lacks a capping of subaerial lavas or, if present, the outcrop must be very small and obscured by the ice dome. If lavas were emitted, the centre may have been the source for an extensive mainly ice-covered lava-fed delta that extends to Cape Green, a distance of about 8 km; it is broadly similar in size to the lava-fed delta seen at Seven Buttresses. A source in Brown Bluff is unlikely – the latter erupted beneath a thick ice cover (see below) and could not have emitted a much-lower-elevation lava field. The delta has a surface that dips very gently to the south and SW, implying a horizontal passage zone with an elevation of *c.* <100 m (much lower than Gamma Hill itself). The relationships are probably most consistent with eruption and effusion in a marine rather than a glacial setting, at a time of higher sea-level relative to today (see also Jonassen and Andersson islands below).

Brown Bluff is the most prominent volcanic edifice on Tabarin Peninsula, with a basal diameter of *c.* 7 km and rising to a prominent snow dome with a summit elevation of *c.* 720 m (Fig. 22c). It is an alkali basaltic glaciovolcano that erupted in association with an ice sheet estimated to be >400 m thick (Skilling 1994; Smellie et al. 2009); it has an isotopic age of 1.22 Ma, with high errors (±0.39 Ma: by $^{40}Ar/^{39}Ar$) and there is also a K–Ar age of 1.1 ± 0.1 Ma (Rex 1976; Smellie et al. 2006c, 2008) (Table 3). Support for a glacial setting consists of lithofacies and architectural evidence that suggest at least two lake drainage events, with a drawdown of up to 100 m in each case, followed by lake recharge. There is also an absence of a palaeotopography capable of ponding the water required in the lake. Although there is no published passage-zone elevation for the lava-fed delta eruptive phase, it is c. 350 m asl (Fig. 24). The study of Brown Bluff by Skilling (1994) was the first published detailed description of the lithofacies and architecture of a monogenetic glaciovolcanic edifice of tuya type, and it remains one of the best. However, it is worth noting that apart from the lack of an appropriate palaeotopography capable of impounding a pluvial lake, all the cited evidence for a glacial setting is interpretive and relies mainly on the identification of 'dry' debris avalanche deposits and subaerial lavas abutting supposed subaqueous lapilli tuffs, consistent with a sudden base-level drawdown caused by a jökulhlaup. Because of its accessibility and the excellent exposures in multiple corries, it is also the most detailed published study of any JRIVG volcanic centre and the information is worth reviewing here. Twelve lithofacies were identified, of which nine are volcaniclastic. On the basis of five lithofacies associations, A–E, Skilling (1994; see also Smellie and Skilling 1994) analysed the evolution of the Brown Bluff edifice in terms of two major eruptive episodes corresponding to a subaqueous period (pillow volcano and tuff cone) followed by subaerial eruption (lava-fed delta) and he described five substages of construction (Figs 24–26). Unit A is a pillow volcano stage, which formed a prominent mound >80 m thick dominated by pillow lava in the volcano core, and is similar to that seen in many submarine and glaciovolcanic volcanoes. It is associated with intrusive pillow lava, irregular pillow-margined intrusions, pillow-rich breccia and tuff breccia (terminology of Skilling 1994 reinterpreted following Smellie and Edwards 2016). Towards its top the pillow mound contains several unusual elliptical cavities up to 10 m in diameter with fringing curvicolumnar-jointed to mini-pillowed lava.

Stage A is interpreted as a phase of underwater eruption at high ambient pressures that suppressed explosivity. Two episodes involving pillow lava are recognized: an earlier episode of massive pillow lava characterized by simple continuous effusion (Fig. 26a); and a later episode of pillow lava intrusion that 'bulked up' the earlier mound and was coeval with, and locally intruded, deposits of younger stages C (tuff breccia) and D (lapilli tuff). The mound is locally draped by massive tuff breccia that may represent a slumped deposit sourced in the mound itself or else, less plausibly given that the constituent sideromelane was formed predominantly by non-explosive mechanical disintegration, directly from subaqueous explosive eruptions (Fig. 26b). The formation of the unusual cavities was not interpreted but they may be related to the melting out of large ice blocks engulfed by pillow lava during emplacement in a confining ice vault similar to features seen in Iceland and called ice-block meltout cavities (Skilling 2009; Smellie and Edwards 2016).

Units B and D unconformably overlie unit A, and are dominated by orange-brown stratified lapilli tuffs representing a subaqueous tuff cone eruptive stage (Fig. 26c). They have a combined maximum thickness of only 100 m and the quaquaversal bedding orientations define an original cone said to be as much as 12 km in diameter. However, this estimate, by Skilling (1994), is implausibly large for such a thin sequence erupted from a single vent and confined by ice; it may signify the presence of multiple erupting vents (now obscured) distributed across the summit region, whose co-eruption increased the lateral distribution of the products. Unusually for any glaciovolcanic centre, the proximal summit lithofacies of the tuff cone are also exposed and include a crater fill. They are hydrothermally altered and include unit C, which intervenes between units B and D and is causally related to slope failure, together with a water-cooled lava with well-developed entablature and colonnade that probably represents water-cooled lava ponded within a crater. The stratification is essentially asymptotic on its volcanic substrate, whilst the proximal summit lithofacies are much more gently dipping, a relationship (i.e. upward-shallowing of bedding dips) seen in many other glaciovolcanic tuff cones and attributed to tephra banking against confining ice walls (Smellie and

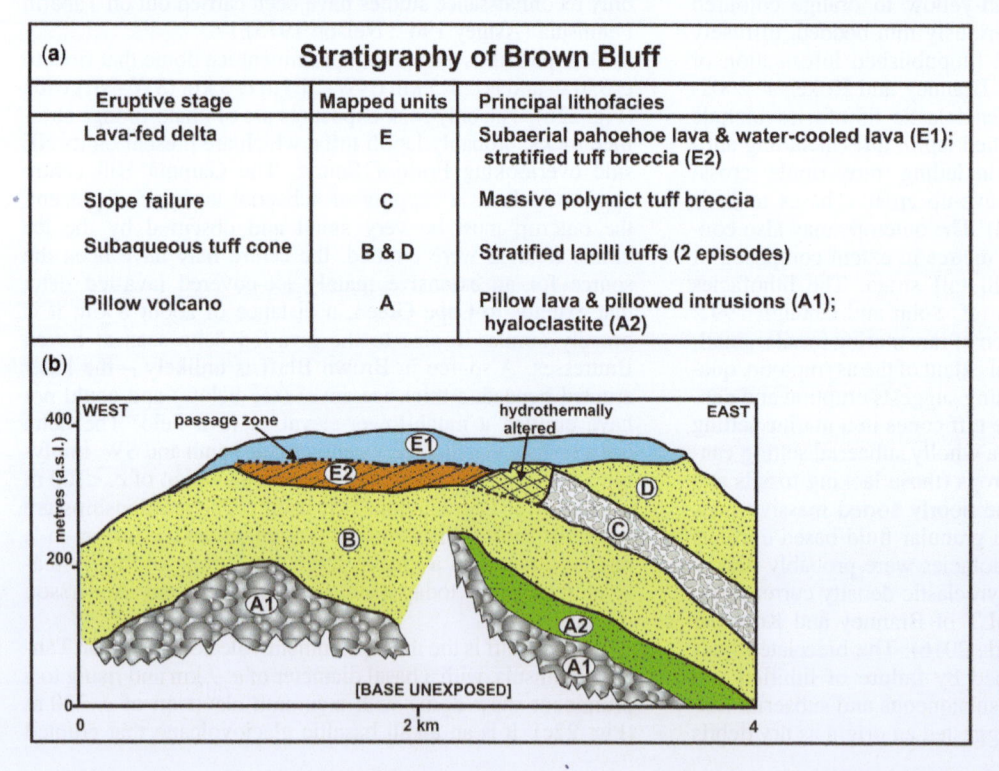

Fig. 24. Stratigraphy of Brown Bluff, summary of the eruptive stages involved in its evolution and the principal lithofacies associated with each stage (modified after Skilling 1994).

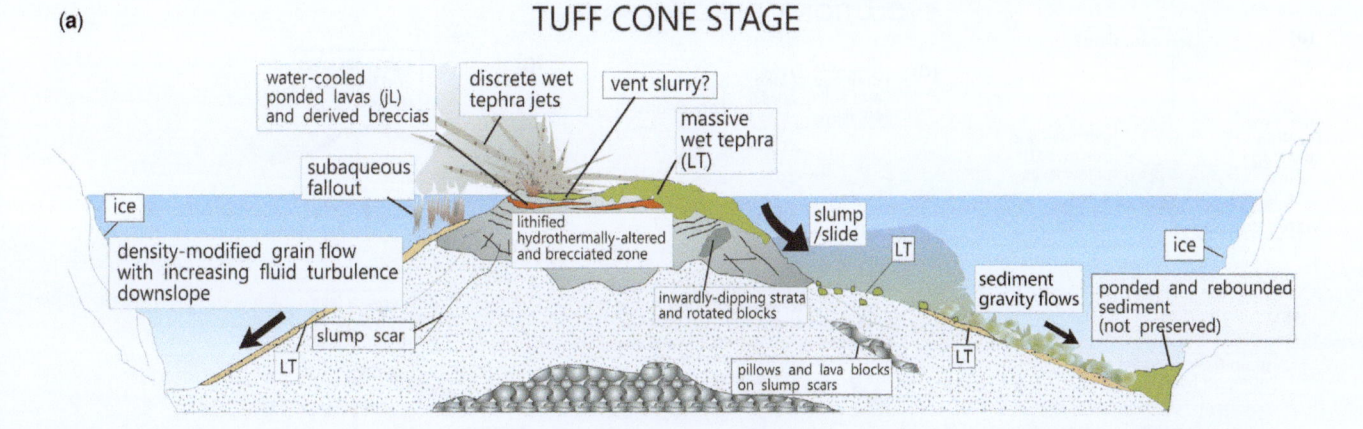

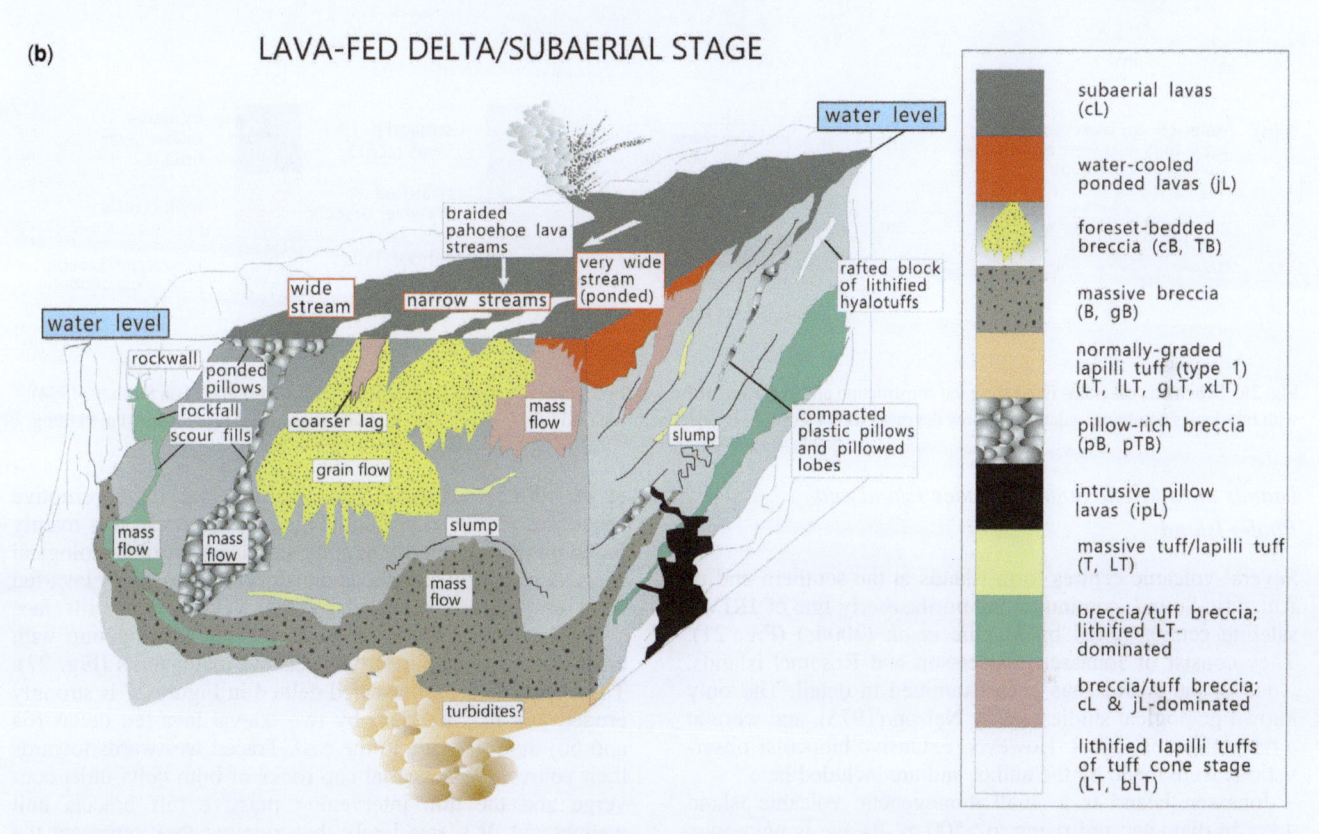

Fig. 25. Interpretive diagrams showing the lithofacies and processes involved in the two major eruptive stages during the construction of Brown Bluff volcano. Both diagrams bear very close scrutiny for the ideas they espouse. Based on Skilling (1994), with lithofacies notations updated to follow the more recent terminology of Smellie and Edwards (2016). The original diagrams were provided by Ian Skilling.

Edwards 2016; Smellie 2018). The cone slope lithofacies were deposited mainly by a variety of sediment gravity flows, probably as a result of slumping or sliding of oversteepened piles of tephra. However, initial transport could have been by direct fallout from subaqueous or subaerial eruptions after overloading the water column with ash and lapilli (the vertical density currents of Manville and Wilson 2004), or they may have been fed directly by wet subaerial pyroclastic density currents that continued to flow underwater. Unit C itself may be related to slope failure with deposition as successive slide events. Uniquely for any glaciovolcano so far described, some of the sediment characteristics of unit C suggest relatively *dry* emplacement as debris avalanche deposits. They were possibly triggered by an englacial lake drainage event (and rapid loss of flank support) that intervened between eruptive episodes B and D (Fig. 26d).

Following recharge of the englacial lake and further tuff cone eruption (Fig. 26e), unit E comprises steeply bedded homoclinal tuff breccia 'foreset' beds and an overlying 50–70 m-thick 'topset' unit of subhorizontally bedded subaerial pāhoehoe lavas that together comprise a lava-fed delta eruptive stage (Fig. 26g). Individual lava lobes extend for up to 15 m down between the breccia foreset beds. The delta lithofacies were emplaced in at least three phases, with units individually up to *c.* 50 m thick that extend laterally for *c.* 400 m. Large slabs of unit E tuff breccia have also slid downslope to be juxtaposed against units B or D. A simple correlation between finer-grained breccia beds being generated by relatively narrow or low-input pāhoehoe streams and the coarser breccias by wider or higher-input lava streams was suggested based on observations of modern lava-fed deltas. Masses of entablature-jointed lavas are also present at the top of unit E, and are interpreted as large-volume lava masses ponded at the waterline, either subaerially and then subsided, or else were emplaced subaqueously and fractured *in situ*. The final stage involved drainage of the englacial lake and draping of subaerial lavas on the flanks of the volcanic edifice (Fig. 20h).

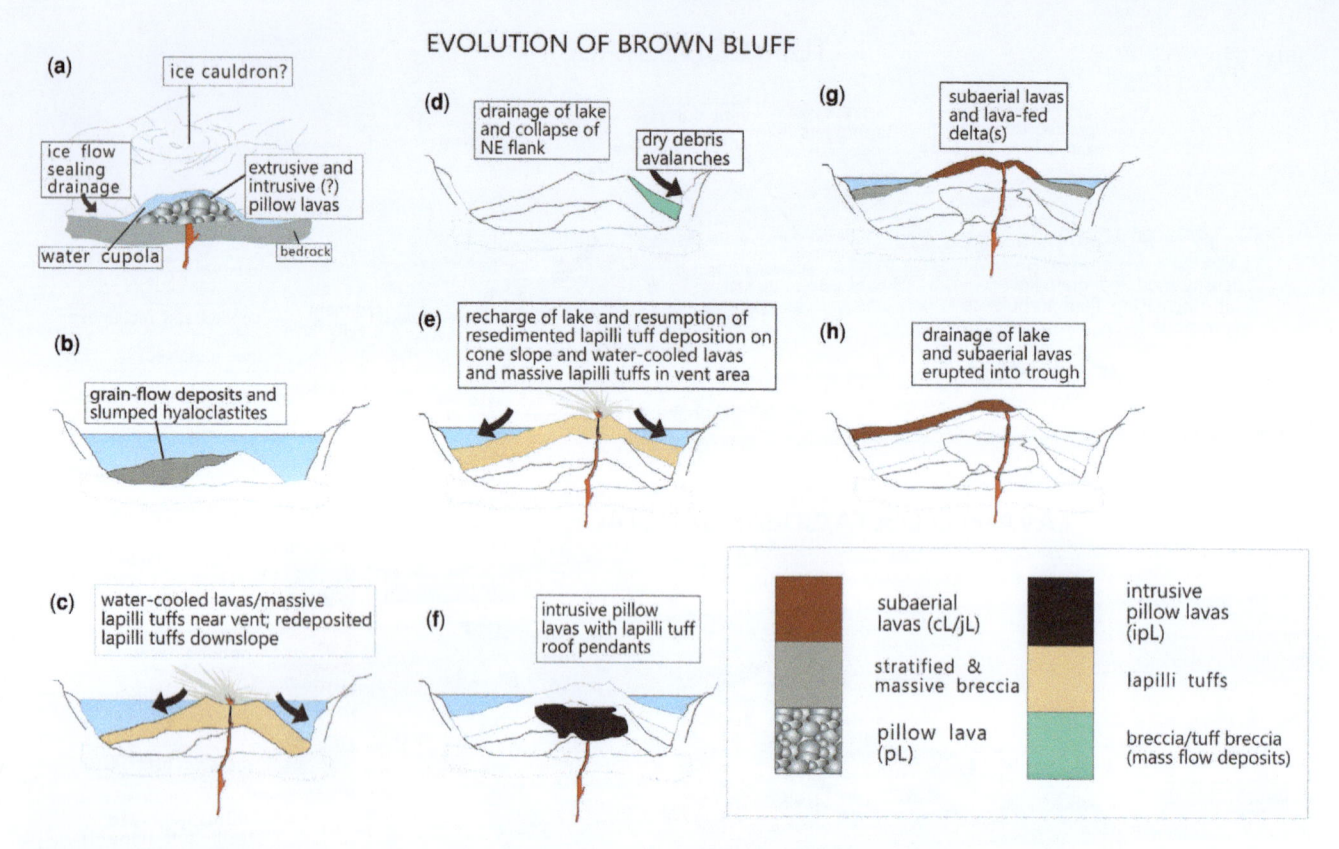

Fig. 26. Schematic diagrams illustrating the construction and evolution of Brown Bluff glaciovolcano. See the text for description. Based on Skilling (1994), with lithofacies notations updated to follow the more recent terminology of Smellie and Edwards (2016). The original diagram was provided by Ian Skilling.

Islands in Antarctic Sound, Dundee Island and Paulet Island

Several volcanic centres form islands at the southern end of Antarctic Sound, continuing the northeasterly line of JRIVG satellite centres noted by Smellie *et al.* (2006*c*) (Fig. 21). They consist of Jonassen, Andersson and Rosamel islands. None of the islands has been examined in detail. The only known geological studies are by Nelson (1975), and were at a reconnaissance level. However, extensive binocular observations were made by the author and are included here.

Jonassen Island is a small monogenetic volcanic island 4 km in diameter and rising to >500 m. Its age is unknown. There are good exposures in cliffs on its north side, where subaerial pāhoehoe lavas overlie a southerly-thickening wedge of tuff breccia foreset beds across a horizontal passage zone at an elevation of *c.* 200 m asl (Fig. 22d). The lavas abut against a steep surface interpreted as a possible cone surface overlap structure, whereas the associated tuff breccia foreset beds depositionally offlap a lower-dipping surface composed of likely stratified lapilli tuffs (see Smellie 2006, fig. 2). The island is covered by a prominent ice dome that masks a shield-like morphology. The sequence was tentatively ascribed a glaciovolcanic origin by Smellie (2006) but reconsideration suggests that the gentle slope of stratification in the tuff cone core and its substantial lateral extent are more consistent with unconfined eruption, and it may be a Surtseyan volcano erupted in the sea. However, the tuff cone and overlying lava-fed delta may not be coeval.

Andersson Island is mainly 9 km in width but it extends by a further 2.2 km at the prominent finger-like easternmost promontory known as Cape Scrymgeour. The island rises to *c.* 432 m asl. Mapping by Nelson (1975) suggested that the island is dominated by a single volcanic edifice, and an associated widespread lava-fed delta erupted from a vent possibly situated near the centre of the north coast. Although Smellie *et al.* (2013) suggested that at least two major eruptive sequences might be present (Fig. 21), that view was mainly based on the present topography and has no strong geological basis. A relatively complicated history of at least four lava-fed delta eruptions is revealed by the well-exposed cliff face *c.* 260 m high on the north side of Cape Scrymgeour, with erosional unconformities between the major units (Fig. 27). The upper surface of lava-fed delta 4 in Figure 27 is strongly eroded and it is overlain by two coeval lava-fed deltas (6a and 6b) that prograde to the east. Traced westwards towards their source, the subaerial cap rocks of both delta units converge and the thin intervening massive tuff breccia unit wedges out. It seems likely that together they represent the rising passage zone of a single lava-fed delta of the type described and illustrated by Smellie (2006, fig. 4). A glacial setting therefore seems likely for delta 6, and the surface separating deltas 6 and 4 is presumably a glacial unconformity. The elevation of the passage zone of lava-fed delta 4 is *c.* 160 m, whilst it rises to a maximum of *c.* 240 m for lava-fed delta 6b. Finally, a tuff cone representing a subsidiary vent is present at the SW corner of the island close to Cape Betbeder (Nelson 1975; unpublished observations by J.L. Smellie). It appears to rest on top of the main Andersson Island lava-fed delta, and its gentle asymptotic flanks suggest eruption in unconfined conditions consistent with either a marine setting and a higher palaeo-sea-level or subaerial eruption. It is conceivable (but unproven) that the main Andersson Island delta extended across Fridtjof Sound to the west side, to include the lava-fed delta exposed there between Gamma Hill and Cape Green. There are no published isotopic ages for Andersson Island.

By contrast, Rosamel Island is a prominent steep-sided eroded tuff cone edifice with a summit elevation of *c.* 416 m constructed on the shallow (*c.* 150 m) southwestern marine shelf of Dundee Island (Fig. 22e). A partly eroded crater was postulated by Nelson (1975) to explain a steep-sided

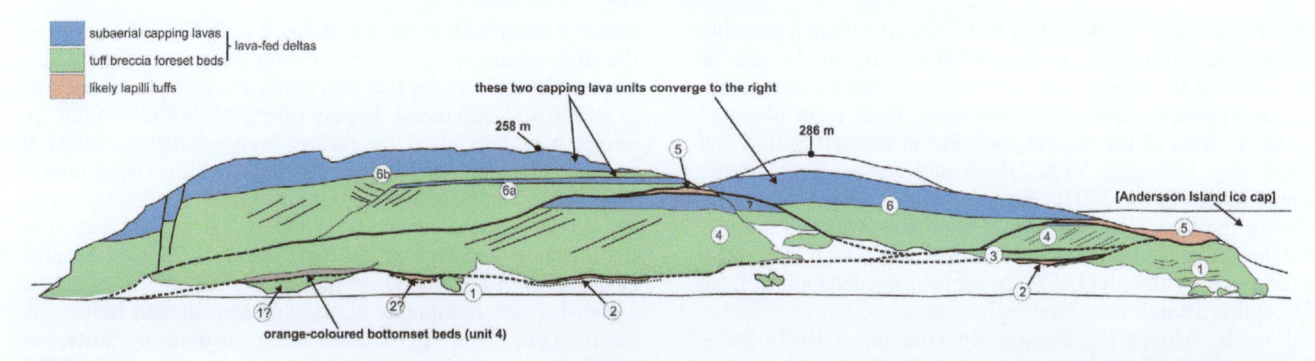

Fig. 27. Sketch showing the geology of Cape Scrymgeour, viewed looking south-southwest. At least six eruptive units are present (numbered), separated by likely glacial erosional surfaces. The field of view is *c.* 2.5 km wide. Spot heights based on BEDMAP, provided by P. Fretwell, British Antarctic Survey.

arcuate embayment on the southeastern side of the island but it is more likely to be a non-volcanic erosional or collapse feature. The age and eruptive setting of Rosamel Island, glacial or marine, are unknown.

A small submarine volcanic edifice named Jaegyu Knoll was discovered in 2004 close to a NW–SE-trending fault(?) scarp on the NE flank of Antarctic Sound, 9 km north of Cape Scrymgeour (Andersson Island: Fig. 28) (Lavoie *et al.* 2016). Jaegyu Knoll has an elongate symmetrical shape aligned NNW–SSE. It measures 3 km in length and is 2 km wide at its base, decreasing to 750 m long and 110 m wide at its summit. The feature rises 700 m above the surrounding seafloor to *c.* 275 m below the ocean surface and has a volume estimated at *c.* 1.5 km^3 of volcanic rock. Initially, it was thought that the edifice showed no obvious signs of glacial scouring and was formed of fresh basalt that was locally devoid of marine colonizing organisms, although the flanks were extensively colonized. Moreover, it was said to be associated with positive temperature anomalies up to 0.052°C, possibly associated with two volcanic vents. Despite the extremely small temperature anomalies, additional observations of discoloured water were used to suggest that the volcano might be active (dormant) (http://volcano.si.edu/reports/reports_additional.cfm?name=JunJaegyu). Subsequently, images showed that although the edifice clearly post-dates the glacially sculpted seafloor, indicating a young (Holocene) age, the summit ridge is draped in volcanic cobbles and pebbles and coarse-grained sediment due to a combination of volcanism, iceberg scouring and strong bottom currents. There is also no crater preserved, although its shallow depth makes it susceptible to erosional scouring. At present, it is impossible to know how recently active it was. Dredged samples were described as glassy and vesicular, and are associated with mounds of rock described as 'agglomerate', suggesting that the summit region, at least, may be at least partly composed of explosively generated tephra. However, it is more likely to be mainly a pillow lava mound. Analysed samples have alkali basalt and 'trachybasalt' (hawaiite?) compositions. Two other small potential volcanic features are also present in Antarctic Sound, 12 and 16 km NW of Jaegyu Knoll, but are undescribed so far (Fig. 28).

Possible hydrothermal vents may also be present 8.5 km south of Cape Green (Fig. 21). Several small mounds were imaged during a multibeam bathymetrical swath survey in 2005–06, on the south side of a prominent seamount of uncertain origin that has a summit at 22 m bsl (R.D. Larter pers. comm. 2017). The mounds are at water depths of *c.* 400–650 m. One of them, rising 25–30 m above the seafloor, showed spurious acoustic returns well above the ambient seafloor on the multibeam bathymetry transects. A 'patch' of different water density was also apparent in the dataset at *c.* 20–60 m above the top of the mound. The water column reflections in the swath data are good evidence that that there was some gas release at the time. As no significant thermal anomalies were recorded, the possibility exists that the activity observed was related to venting either of methane

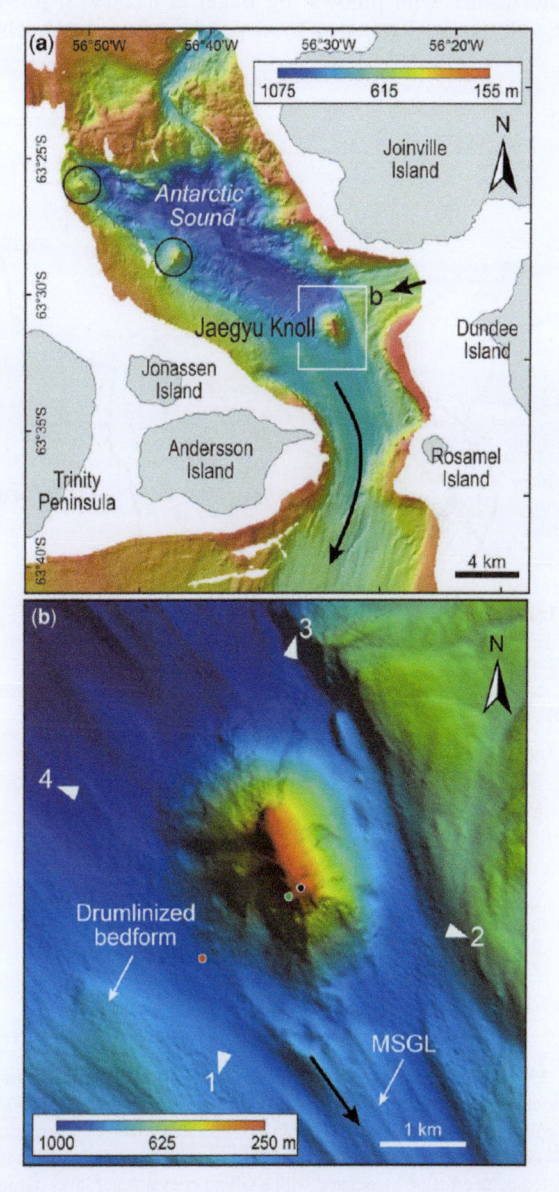

Fig. 28. Swath images showing a young submarine volcano known as Jaegyu Knoll in Antarctic Sound (from Lavoie *et al.* 2016). Two other possible submarine seamounts are also present NW of Jaegyu Knoll (ringed in (a)). MSGL, megascale glacial lineations.

(i.e. a cold seep) or else, given that it occurs within a possibly active volcanic region, hydrothermal activity might also be plausible (R.D. Larter, pers. comm. 2017). Marine tephra layers are known to occur in Holocene sediment cores obtained *c.* 22 km west of the mound, and also in Herbert Sound and Croft Bay (between Vega Island and James Ross Island: Camerlenghi *et al.* 2001; Minzoni *et al.* 2016). The tephras clearly indicate relatively recent explosive volcanic activity but they have not been analysed, so it is unknown if they relate to eruptions within the JRIVG or are far-travelled tephras from Deception Island. Englacial tephras recovered at 349–359 m bsf in the Mount Haddington ice core (*c.* 5040–14 260 ± 500 years BP (1950): Mulvaney *et al.* 2012) have Deception Island compositions (unpublished information of J.L. Smellie).

Cape Purvis is a conspicuous promontory on southernmost Dundee Island (Fig. 21). It is formed by a large volcanic edifice at least 6.5 km in basal diameter (Fig. 29). It has a gently domed ice surface with a general elevation of 300–400 m and a small unnamed peak at its northern end that rises to *c.* 570 m (Fig. 22f). The Cape Purvis volcano was discovered in 2002 (Smellie *et al.* 2006*c*), although a small isolated outcrop of volcanic rock at sea-level on the SW side was described as 'hyaloclastites with pillows' by Baker *et al.* (1973, p. 92). The volcano is dominated by a poorly exposed pāhoehoe lava-fed delta with a passage zone at *c.* 240 m asl that gives the structure its prominent steep outer margin. There are also hints in the topography that two much less conspicuous scarps at *c.* 300 and 380 m asl, largely obscured by the summit ice dome, may be caused by further lava-fed deltas (deltas B and C in Fig. 29) emplaced associated with a rising coeval water level or possibly steps in a single contiguous delta caused by a coeval water level whose elevation dropped in two discrete stages (Fig. 29) (Smellie *et al.* 2006*c*). Smellie *et al.* (2006*b*) postulated on morphological grounds that the earlier-formed lithofacies of the volcano, at and below the northern peak, were pyroclastic and formed during Surtseyan eruptions in a glacial setting – that is, it is a tuya edifice. The peak is now known to be formed of grey and khaki-yellow stratified lapilli tuffs, for which Kraus *et al.* (2013) preferred a subaerial setting but a Surtseyan setting is more consistent with the association with lava-fed deltas. Isotopic dating suggests that the volcano is very young (i.e. 0.132 ± 0.019 Ma by $^{40}Ar/^{39}Ar$: Smellie *et al.* 2006*c*) and contemporaneous ice thicknesses during eruption were much greater than at present (i.e. an ice surface elevation of *c.* 350 m asl: Smellie *et al.* 2009). Lavas in the Cape Purvis volcano are alkali basalts and very rare hawaiites (Smellie *et al.* 2006*c*; Kraus *et al.* 2013). Broadly similar but somewhat more evolved compositions occur in lavas on Paulet Island (Kraus *et al.* 2013).

Paulet Island is a small island situated *c.* 6 km SE of Dundee Island (Fig. 22g). It is *c.* 3.2 km north–south and 2.7 km east–west, and rises to 351 m. Geologically, Paulet and Penguin islands are very similar, both being formed of a basal lava platform, scoria cones and maars. However, there is less published information for Paulet Island. Compositionally, it is mainly formed of hawaiites and is stratigraphically very simple (Fig. 30) (Baker *et al.* 1973; Smellie 1990*f*; Kraus *et al.* 2013). The island is formed of a basal unit of subhorizontal lavas with reddened surfaces that is exposed in coastal cliffs, particularly at the SW end of the island where the sequence

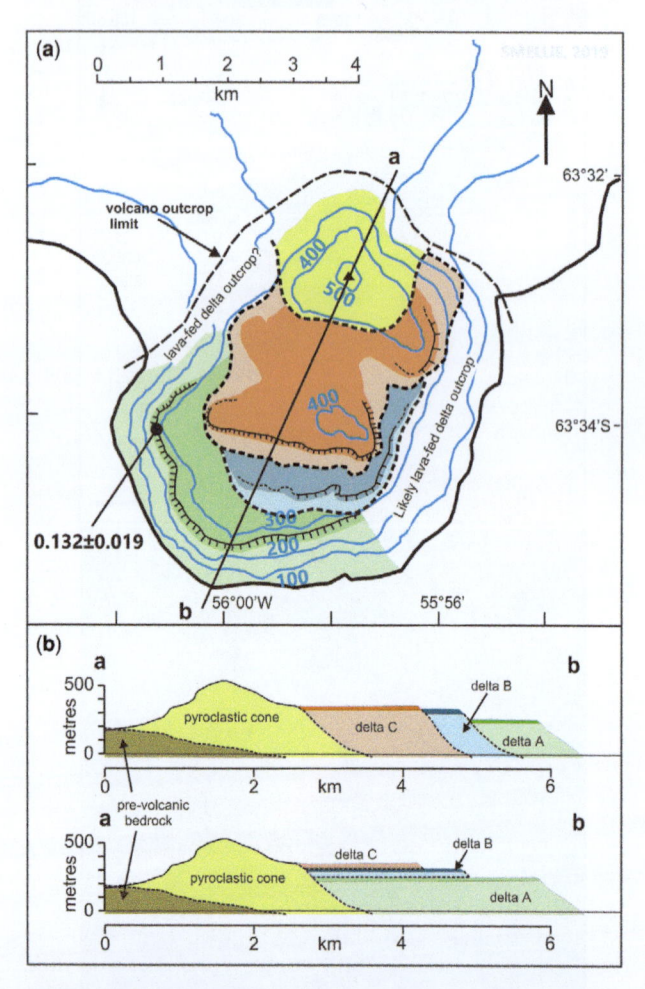

Fig. 29. (a) Geological sketch map of the Cape Purvis volcano, shown without its extensive cover of snow and ice (based on Smellie *et al.* 2006*c*, modified). The cross-sections shown in (b) (with ×2 vertical exaggeration) show two possible explanations for the outcrops and topography – that is, lava-fed delta effusion associated with a falling coeval water level (upper section) and a rising coeval water level (lower section). Exposure of the volcano is relatively poor, and the map and cross-sections, particularly the presence and relationships of deltas B and C, are interpretive. The eruptive environment was glacial. The isotopic age is in Ma (Smellie *et al.* 2006*c*).

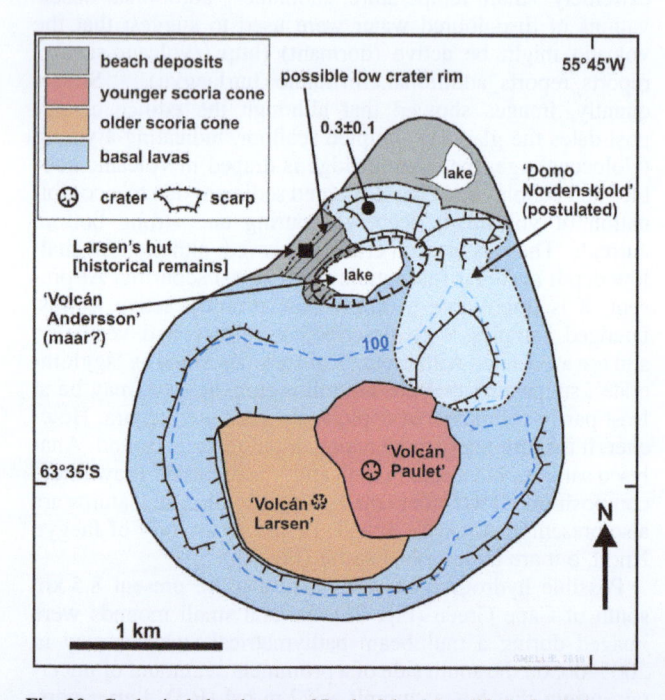

Fig. 30. Geological sketch map of Paulet Island (modified after Baker *et al.* 1973; González-Ferrán 1995). The origin of the raised ground by Larsen's hut, depicted as a possible maar crater rim (diagonal lines ornament), is uncertain but it may be formed of blocky hydrovolcanic ejecta. The depiction of 'Domo Nordenskjold' is included for completeness but its identification is unlikely. The sketched topography is indicative only.

is at least 210 m thick, and in rounded and flat-topped knolls, largely penguin-colonized, at the lower northern end of the island. The latter were included in an outcrop identified as an eroded dome ('Domo Nordenskjold') by González-Ferrán (1995) but the identification seems unlikely. Some individual lavas (e.g. in the coastal crag WSW of Larsen's hut) are very thick, measuring c. 100–125 m (unpublished information of J.L. Smellie). The basal lava sequence is overlain by two steep coalesced pyroclastic cones that form a conical hill c. 150 m high. The older cone ('Volcán Larsen' of González-Ferrán 1995), with a small and very subdued crater, is situated c. 500 m SW of a younger 'main' cone ('Volcán Paulet' of González-Ferrán 1995) that includes the summit of the island and which has a fresher-looking small snow-filled crater surrounded by red scoria. The latter contains agglutinate formed of reddened scoria resting on the basal lava sequence, whilst interbedded similar reddened agglutinate and highly vesicular lavas occur at the summit (unpublished field notes of I.P. Skilling). Kraus et al. (2013) also recorded a small fissure on the west flank of the main cone from which basaltic lavas issued. González-Ferrán (1995) also depicted four small craters (including 'Volcán Larsen' and 'Volcán Paulet') aligned along 080°N but two of the craters are clearly absent in aerial photographs and do not exist. However, a small maar, informally called Volcán Andersson, may be present at the north end of the island (González-Ferrán 1995; see also Kraus et al. 2013). It was postulated to explain a deep oval depression SE of Larsen's hut (Fig. 30). The maar is a pronounced water-filled hollow c. 800 m long (WSW–ENE) and c. 400 m wide (NNW–SSE). It has a low barrier formed of loose angular lava fragments on its west side, although no hydroclastic tephra deposits have been recorded (unpublished observations of J.L. Smellie). A single K–Ar isotopic age of 0.3 ± 0.1 Ma was obtained from the 'Domo Nordenskjold' outcrop at the north end of the island (Fig. 30; Table 3) (Baker et al. 1977). González-Ferrán (1995) regarded the Paulet Island volcano as 'active' based on supposed 'residual high heat' that was presumed responsible for the largely snow-free state of the island but the existence of enhanced geothermal warmth has never been proven. Conversely, the youthful appearance of the younger summit scoria cone suggests that it may be very young (<1 kyr?: Smellie 1990f).

Sedimentary deposits associated with the James Ross Island Volcanic Group

The JRIVG also contains numerous outcrops of glacial and interglacial sedimentary rocks. Although the deposits are not volcanic, they have provided a number of volcanic clasts that were dated isotopically and which have substantially extended the maximum age of the JRIVG (back to 12.4 Ma: Marenssi et al. 2010) (Table 3). Four formations have been defined (Pirrie et al. 1997; Jonkers and Kelley 1998; Jonkers et al. 2002; Nývlt et al. 2011). Their principal characteristics are summarized in Table 4 and they are not described further here but they have proved uniquely important in determining the regional environmental and even climatic conditions during the latest Miocene–Pleistocene period (Jonkers and Kelley 1998; Jonkers et al. 2002; Lirio et al. 2003; Hambrey et al. 2008; Nelson et al. 2009; Clark et al. 2010, 2013; Williams et al. 2010; Smellie 2021).

Effect of James Ross Island Volcanic Group volcanism on Sr isotopic ages

Many of the sediments interbedded with the volcanic rocks were dated using the $^{87}Sr/^{86}Sr$ composition of associated shallow-water bivalve shells (Dingle and Lavelle 1998; McArthur et al. 2006; Smellie et al. 2006a; Nelson et al. 2009; Nývlt et al. 2011). It is suggested here that the association with active volcanism in the JRIVG may have had an important influence on the Sr ages. The published Sr ages are frequently older, by 1–2 myr, than their overlying lava-fed deltas (dated by $^{40}Ar/^{39}Ar$: Troedson and Smellie 2002; Smellie et al. 2006a; Nelson et al. 2009; unpublished information of J.L. Smellie). This is generally explained as the dated shelly material being reworked from older sediment. However, the often excellent preservation of those shells, which, although broken, are often fragile and lack adhering sedimentary rock, implies that the glaciers incorporated unconsolidated sediment during ice advance, suggesting that the sediments, shelly fossils and lava-fed deltas may have been essentially coeval (Nelson et al. 2009; Nývlt et al. 2011). It is unknown how rapidly Neogene sediments may become lithified but the consistent disparity in ages (i.e. fossils older than coeval volcanic rocks) may be significant.

There is an apparent link between flood basalt volcanic eruptions in large igneous provinces (LIPs) and global values of $^{87}Sr/^{86}Sr$ in seawater, which show sharp inflections to slightly lower values following the major eruptive episodes, although the precise relationship is uncertain (McArthur et al. 2001). Effusive eruptions in the JRIVG were also typically voluminous (flood lavas, although not corresponding to a LIP because the time-integrated magma flux was much lower). They may thus have the potential for exerting a local impact on seawater. The lavas have low values of $^{87}Sr/^{86}Sr$ (0.7029–0.7036: Hole et al. 1995; Košler et al. 2009), and meteoric water originating on James Ross Island may equilibrate with the low $^{87}Sr/^{86}Sr$ lava bedrock. This may occur during the (geologically short) cooling period of the lava-fed deltas, during surface runoff in ice-poor interglacials or as groundwater passing through till formed of finely ground-up lavas and highly-reactive glassy tuff breccia during glacials and interglacials. It thus may lower the net $^{87}Sr/^{86}Sr$ value in the surrounding seawater. As a result, the apparent ages of bivalves incorporating the ambient $^{87}Sr/^{86}Sr$ will be older than their true ages. Because of the very low gradient of oceanic $^{87}Sr/^{86}Sr$ in the Pliocene, a lowering of the $^{87}Sr/^{86}Sr$ value in seawater by as little as 0.00 001 will increase the apparent Sr age measured in living bivalves by c. 1 myr (cf. McArthur et al. 2006, fig. 4a). Taken together, it is thus possible that because of the local impact of the volcanism on seawater $^{87}Sr/^{86}Sr$, many of the Sr ages of fossils in sediments associated with the JRIVG are unreliable and the fossils may be somewhat younger than suggested. This is an important but hitherto-undocumented impact of volcanism on understanding the palaeoenvironmental history of the region.

Evidence within the James Ross Island Volcanic Group for higher past sea-levels

The volcanism in the JRIVG also contains widespread, although much less well documented, evidence for significant Mio-Pliocene sea-level variations (Table 5; Fig. 31). Several tuff cone satellite centres currently exposed well above sea-level were erupted in an unconfined, ice-free marine setting, and there are at least three marine-emplaced lava-fed deltas with horizontal passage zone (coeval water level) elevations also well above present datum (Smellie et al. 2006a, 2008; Williams et al. 2006; Nehyba and Nývlt 2014; Calabozo et al. 2015; Smellie 2021). Moreover, marine alteration conditions have been identified using the compositions of authigenic zeolite minerals in several lava-fed deltas, and are again indicative of past sea-level elevations well above modern (Johnson and Smellie 2007). Fossiliferous glaciomarine

Table 5. *Evidence for former high sea-levels in the James Ross Island region*

Locality (all James Ross Island except where indicated)	Locality identifier*	Feature/stratigraphical unit†	Type of outcrop	Age (Ma)	Dating method	Geological feature	Sea-level elevation (m)	Comments	References
Lookalike Peaks, Stickle Ridge, Smellie Peak	1	Lookalike Peak Formation	Volcanic	6.16	$^{40}Ar/^{39}Ar$	Lava-fed delta	480	Laterally extensive horizontal passage zone; authigenic zeolite compositions; marine-emplaced	Smellie et al. (2008); Calabozo et al. (2015)
Johnson Mesa, Lachman Crags, Berry Hill	2	Johnson Mesa Formation	Volcanic	5.04	$^{40}Ar/^{39}Ar$	Lava-fed delta	350	Laterally extensive horizontal passage zone; with wave-eroded top; marine-emplaced	Smellie et al. (2008); Nehyba and Nývlt (2015)
Headland below Sungold Hill and coastal cliffs extending west thereof	3	Sungold Hill Formation	Volcanic	2.03	$^{40}Ar/^{39}Ar$	Lava-fed delta	195	Laterally extensive horizontal passage zone; marine-emplaced	Smellie et al. (2008)
Jefford Point	4	Jefford Point Formation	Volcanic	1.89	$^{40}Ar/^{39}Ar$	Lava-fed delta	195	Laterally extensive horizontal passage zone; marine-emplaced	Smellie et al. (2008)
Terrapin Hill	5	Terrapin Hill Formation at Terrapin Hill	Volcanic	0.66	$^{40}Ar/^{39}Ar$	Tuff cone	c. 200	Laterally continuous strata, including well-bedded sediment gravity-flow deposits up to c. 200 m asl; marine-emplaced	Smellie et al. (2006a, 2008)
Tortoise Hill	6	Terrapin Hill Formation at Tortoise Hill; prominent alteration front at c. 200 m	Volcanic	2.23	$^{40}Ar/^{39}Ar$	Tuff cone	200	Prominent alteration front at c. 200 m resembles possible pyroclastic passage zone (cf. Russell et al. 2013); laterally continuous strata (unconfined); marine-emplaced	Smellie et al. (2008); Russell et al. (2013)
Bibby Point	7	Terrapin Hill Formation at Bibby Point	Volcanic	3.34	$^{40}Ar/^{39}Ar$	Tuff cone	above c. 350	Laterally continuous strata; tuff cone; abundant pillowed intrusions and hyaloclastite breccia indicating water-saturated/unconsolidated, plus local very good stratification; marine-emplaced	Smellie et al. (2008); Nehyba and Nývlt (2014)
Patalamon Mesa, Back Mesa	8	Patalamon Mesa Formation	Volcanic	<4.61	$^{40}Ar/^{39}Ar$	Later marine alteration of glacially-emplaced lava-fed delta	470	Authigenic zeolite compositions indicate marine-related alteration associated with high sea-level elevation; age is a maximum	Johnson and Smellie (2007); Smellie et al. (2008)
St Rita Point, Forster Cliffs	9	Forster Cliffs Formation at St Rita Point	Volcanic	<2.50	$^{40}Ar/^{39}Ar$	Later marine alteration of glacially-emplaced lava-fed delta	285	Authigenic zeolite compositions indicate marine-related alteration associated with high sea-level elevation; age is a maximum	Johnson and Smellie (2007); Smellie et al. (2008)
Forster Cliffs	10	Forster Cliffs Formation at Forster Cliffs	Volcanic	<2.50	$^{40}Ar/^{39}Ar$	Later marine alteration of glacially-emplaced lava-fed delta	200	Authigenic zeolite compositions indicate marine-related alteration associated with high sea-level elevation; age is a maximum	Johnson and Smellie (2007); Smellie et al. (2008)
Jonkers Mesa, Rhino Cliffs	11	Jonkers Mesa Formation	Volcanic	<3.08	$^{40}Ar/^{39}Ar$	Later marine alteration of glacially-emplaced lava-fed delta	365	Authigenic zeolite compositions indicate marine-related alteration associated with high sea-level elevation; age is a maximum	Johnson and Smellie (2007); Smellie et al. (2008)

	Location	Feature	Type	Age (Ma)	Dating method	Description	Elevation (m)	Comments	References
12	Ekelof Point, Rhino Corner	Ekelof Point Formation	Volcanic	<3.52	$^{40}Ar/^{39}Ar$	Later marine alteration of glacially-emplaced lava-fed delta	365	Authigenic zeolite compositions indicate marine-related alteration associated with high sea-level elevation; age is a maximum	Johnson and Smellie (2007); Smellie et al. (2008)
13	Tumbledown Cliffs, Patalamon Mesa, Limmershin Cliffs	Tumbledown Cliffs Formation	Volcanic	<5.15	$^{40}Ar/^{39}Ar$	Later marine alteration of glacially-emplaced lava-fed delta	405	Authigenic zeolite compositions indicate marine-related alteration associated with high sea-level elevation; age is a maximum	Johnson and Smellie (2007); Smellie et al. (2008)
A	Cape Lachman, below Berry Hill	Horizontal erosional surface remnants between Cape Lachman and Berry Hill	Erosional surface	Unknown (Quaternary?)		Raised marine platform	c. 37–39	Covered by erratic-rich glacial drift	Nývlt et al. (2011); Davies et al. (2013); unpublished information of J.L. Smellie
B	Cockburn Island	Erosional surface on top of Cockburn Island	Erosional surface	4.7	$^{87}Sr/^{86}Sr$	Raised marine platform	250	Age is probably a maximum	Jonkers (1998); McArthur et al. (2006)
C	The Mesa, Seymour Island	Horizontal erosional surface on top of The Mesa, Seymour Island	Erosional surface	Quaternary		Raised marine platform	200	Probably the same platform seen at Spath Peninsula, Snow Hill Island	Gazdzicki et al. (2004)
D	Spath Peninsula, Snow Hill Island	Horizontal erosional surface at the SW end of Spath Peninsula, Snow Hill Island	erosional surface	Quaternary		Raised marine platform	c. 150	Probably the same platform seen at the Mesa, Seymour Island	Unpublished observation of J.L. Smellie
E	NE Ulu Peninsula	Mendel Formation	Sedimentary	5.93–5.65	$^{87}Sr/^{86}Sr$	Glaciomarine sediment unit sandwiched between two glacial sedimentary units	>120	Dated by Sr ages on bivalves; maximum current topographical elevation 70 m; minimum palaeodepths >50 m; age by $^{40}Ar/^{39}Ar$ on associated lavas	Nývlt et al. (2011)
F	Stoneley Point	Glaciomarine sediments; near Stoneley Point	Sedimentary	<6.31	$^{87}Sr/^{86}Sr$	Glacigenic debris flow and conglomerates deposited during glacial advances	>120	Depositional age by $^{87}Sr/^{86}Sr$ on included fossil shell material; age is likely a maximum; elevation is a minimum	Nelson et al. (2009); unpublished information of JL Smellie
G	Stickle Ridge	Glaciomarine sediments; Lachman Crags	Sedimentary	<6.16	$^{87}Sr/^{86}Sr$	Glacigenic debris flow and conglomerates deposited during glacial advances	>350	Depositional age by $^{87}Sr/^{86}Sr$ on included fossil shell material; age is likely a maximum; elevation is a minimum	Nelson et al. (2009); unpublished information of J.L. Smellie

(Continued)

Table 5. Continued.

Locality (all James Ross Island except where indicated)	Locality identifier*	Feature/ stratigraphical unit†	Type of outcrop	Age (Ma)	Dating method	Geological feature	Sea-level elevation (m)	Comments	References
Rhino Cliffs	H	Glaciomarine sediments; Rhino Cliffs	Sedimentary	<5.77	$^{87}Sr/^{86}Sr$	Glacigenic debris flow and conglomerates deposited during glacial advances	>350	Depositional age by $^{87}Sr/^{86}Sr$ on included fossil shell material; age is likely to be a maximum; elevation is a minimum	Nelson et al. (2009); unpublished information of J.L. Smellie
Rockfall Valley	I	Glaciomarine sediments; Rockfall Valley	Sedimentary	<5.36	$^{87}Sr/^{86}Sr$	Glacigenic debris flow and conglomerates deposited during glacial advances	>350	Depositional age by $^{87}Sr/^{86}Sr$ on included fossil shell material; age is likely to be a maximum; elevation is a minimum	Nelson et al. (2009); unpublished information of J.L. Smellie
Blancmange Hill	J	Glaciomarine sediments; Blancmange Hill	Sedimentary	<4.89	$^{87}Sr/^{86}Sr$	Glacigenic debris flow and conglomerates deposited during glacial advances	>250	Depositional age by $^{87}Sr/^{86}Sr$ on included fossil shell material; age is likely to be a maximum; elevation is a minimum	Nelson et al. (2009); unpublished information of J.L. Smellie
Jonkers Mesa	K	Glaciomarine sediments; Jonkers Mesa	Sedimentary	<3.08	$^{87}Sr/^{86}Sr$	Glacigenic debris flow and conglomerates deposited during glacial advances	>350	Depositional age by $^{87}Sr/^{86}Sr$ on included fossil shell material; age is likely to be a maximum; elevation is a minimum	Nelson et al. (2009); unpublished information of J.L. Smellie
Pecten Spur	L	Glaciomarine sediments; Pecten Spur	Sedimentary	<3.08	$^{87}Sr/^{86}Sr$	Glacigenic debris flow and conglomerates deposited during glacial advances	>250	Depositional age by $^{87}Sr/^{86}Sr$ on included fossil shell material; age is likely to be a maximum; elevation is a minimum	Nelson et al. (2009); unpublished information of J.L. Smellie

*See Figure 31 for the map showing outcrop locations.
†Volcanic formation names after Smellie et al. (2013).

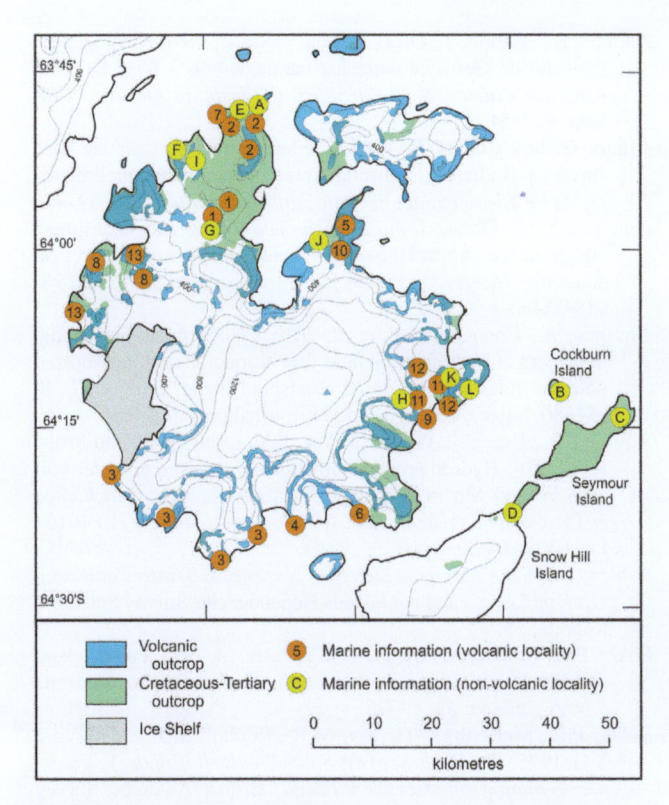

Fig. 31. Map of the James Ross Island region showing localities with evidence of higher past sea-levels. See Table 5 for a description of evidence, ages and sea-level elevations.

suggests that sea-levels may have varied by c. 500 m relative to present datum over the volcanic period, corresponding to a mean uplift rate c. 0.06 mm a^{-1} (Table 5; Fig. 32). Five hundred metres of overall relative sea-level fall is most easily ascribed to progressive uplift of James Ross Island, perhaps related to the effects of the underlying, long-lived, shallow-mantle thermal anomaly ('plume', *sensu lato*) responsible for the volcanism (cf. Hole *et al.* 1995). Despite significant subsidence of the volcanic core of Mount Haddington (Jordan *et al.* 2009), the localities listed in Table 5 are marginal to the main volcanic mass (Fig. 31), where any subsidence effects would probably be limited. A glacio-isostatic effect on sea-levels is also likely but it cannot be separately identified in the data. However, the calculated uplift rate for the Mio-Pliocene is much smaller than during Holocene glacio-isostatic rebound within the area, by 1–2 orders of magnitude (e.g. 0.29–3.91 mm a^{-1} for glacio-isostatic relative sea-level fall during the past c. 7 ka: e.g. Roberts *et al.* 2011). Moreover, glacio-isostatic rebound will be followed by subsidence during each following glacial. The major influence on overall uplift in the region surrounding Mount Haddington is thus probably 'plume'-related.

Summary

This chapter is the most detailed review of the distribution and characteristics of volcanoes in the northern Antarctic Peninsula region. The region, which includes Bransfield Strait and islands on both flanks of the Antarctic Peninsula, contains more volcanoes than any other similar-sized part of Antarctica, conservatively estimated at more than 50 subaerial volcanoes, together with more than 100 submarine seamounts and volcanic ridges. Far more volcanoes are now known than were recognized previously. The age of the volcanism extends back to >12 Ma but most of the volcanoes are Plio-

sediments intercalated between volcanic units are also present, and there are a few high-elevation marine erosional surfaces (e.g. Jonkers 1998; Jonkers *et al.* 2002; Nelson *et al.* 2009; Nývlt *et al.* 2011; Clark *et al.* 2013). The evidence overall

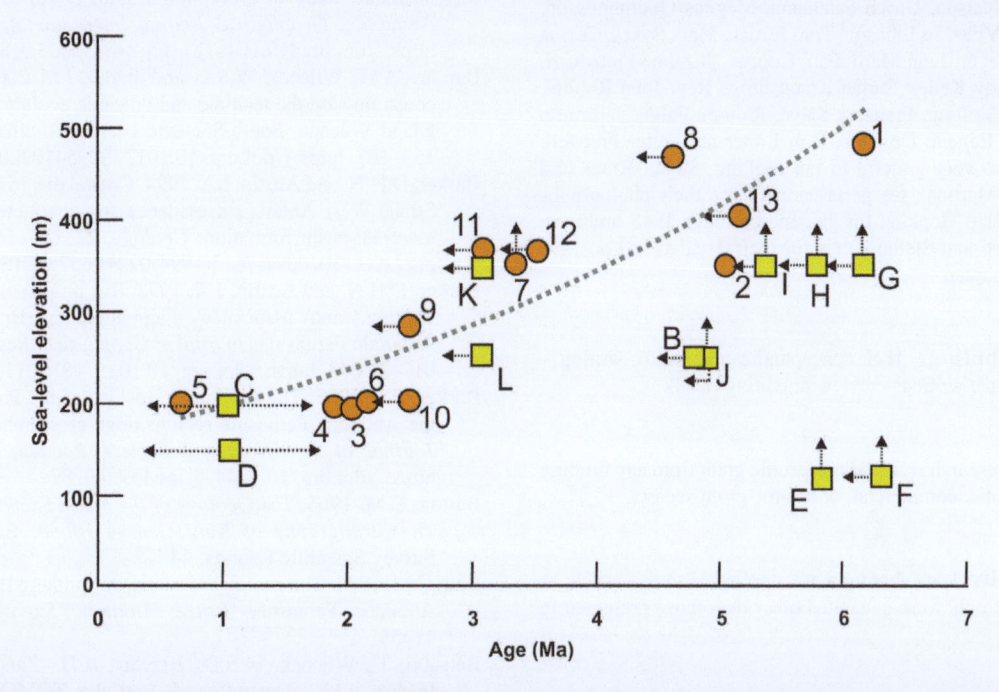

Fig. 32. Diagram showing sea-level elevation (relative to present datum) v. age for volcanic and other features in the James Ross Island region. The curve shown is based on ^{40}Ar/^{39}Ar data for the volcanic localities (numbered orange symbols – numbers refer to localities shown in Fig. 31). Ages for samples from non-volcanic localities (mostly fossils in sediments (see Table 5); yellow symbols) are by ^{87}Sr/^{86}Sr. The errors on Sr ages of Pliocene-age fossils are very high because the reference Pliocene curve used for Sr dating has a very low gradient (McArthur *et al.* 2006). As a result, many Sr ages are merely indicative, and some ages in the JRIVG are of uncertain significance as they do not correspond to any known surface geology (e.g. early Miocene ages reported by Nývlt *et al.* 2011). The arrows indicate the direction(s) in which points may move due to unquantified uncertainties in the data. See Figure 31 for the names of all labelled sample localities.

Pleistocene. They were erupted in a variety of tectonic settings ranging from an ensialic marginal basin (Bransfield Strait) to long-lived back-arc alkaline volcanism associated with mantle diapirism in a crustal thin spot (James Ross Island Volcanic Group) and post-subduction slab-window formation (seamount offshore of Anvers Island). The tectonic setting of some volcanoes is still unclear (e.g. Anvers and Brabant islands; centres on Livingston and Greenwich islands). Many of the volcanoes are either unvisited or have only been subject to minimal investigations at a reconnaissance level; some volcanoes have not been revisited in many decades. Modern volcanological investigations have been focused principally on Deception Island and the James Ross Island Volcanic Group. Deception Island is the most productive active volcano in Antarctica and has been the subject of more scientific investigations than any other Antarctic volcano; and Mount Haddington, on James Ross Island, is the largest Antarctic volcano. The James Ross Island volcanic field may also still be active but is dormant for prolonged periods (often >100 kyr). Our increased knowledge of the James Ross Island Volcanic Group has had a significant impact on our understanding of glaciovolcanism and of the regional Mio-Plio-Pleistocene palaeoenvironmental conditions. As a result, the configuration of the Antarctic Peninsula Ice Sheet is particularly well documented (see Smellie 2021).

Acknowledgements The author is very grateful to the directors and all of the staff of the British Antarctic Survey (BAS), and the captains and complements of HMS *Endurance*, for supporting much of the fieldwork and research on which this chapter is based, spread over very many years. The unflagging support of Mike Thomson, as Geosciences Division Head, was particularly important, as was that of Mike Dinn (BAS Field Operations) and my numerous field assistants (Crispin Day, Rob Smith, Ash Morton, Terry O'Donovan and Asti Taylor). I also gratefully acknowledge fruitful collaborations and numerous discussions, both in the field and in the laboratory, with colleagues too numerous to mention but including (in no order): Mike Hambrey, Anna Nelson, Ulrich Salzmann, Magnus Gudmundsson, Ben van Wyk de Vries, Jo Johnson, Tom Jordan, Mark Sykes, Corina Risso, Adelina Geyer, Joan Martí, Paul Cooper, Suzanne Fretzdorff, Phil Nelson, Randy Keller, Stefan Kraus, Jorge Rey, John Roobol, Peter Baker, Ian Skilling, Francesc Sàbat, Raimon Pallàs, Jeronimo López-Martínez, Eugene Domack, Rob Larter and Peter Fretwell. The author is also very grateful to Ian Skilling, Stefan Kraus and Jeronimo López-Martínez for permission to use their photographs and diagrams; Ieuan Hopkins for guidance through BAS archives; and to Ian Skilling and Stefan Kraus for their detailed and positive reviews of this chapter.

Author contributions JLS: conceptualization (lead), writing – original draft (lead), writing – review & editing (lead).

Funding This research received no specific grant from any funding agency in the public, commercial, or not-for-profit sectors.

Data availability Data sharing is not applicable to this article as no additional datasets were generated other than those presented in this chapter.

References

Aitkenhead, N. 1975. The Geology of the Duse Bay – Larsen Inlet Area, North-East Graham Land (with Particular Reference to the Trinity Peninsula Series). British Antarctic Survey Scientific Reports, **51**.

Alarcón, B., Ambrus, J., Olcay, L. and Vieira, C. 1976. Geología del Estrecho de Gerlache entre los paralelos 648 y 65 8 lat. Sur, Antártica Chilena. *Serie Científica del Instituto Antártico Chileno*, **4**, 7–51.

Alfaro, G. and Collao, S. 1985. Exploracion minera en las islas Anvers y Brabante, Península Antártica: resultados preliminaries. *Serie Científica del Instituto Antártico Chileno*, **30**, 39–47.

Antoniades, D., Giralt, S. *et al.* 2018. The timing and widespread effects of the largest Holocene volcanic eruption in Antarctica. *Scientific Reports*, **8**, 17279, https://doi.org/10.1038/s41598-018-35460-x

Aquilina, A., Connelly, D.P. *et al.* 2013. Geochemical and visual indicators of hydrothermal fluid flow through a sediment-hosted volcanic ridge in the Central Bransfield Basin. *PLoS ONE*, **8**, e54686, https://doi.org/10.1371/journal.pone.0054686

Aquilina, A., Homoky, W.B., Hawkes, J.A., Lyins, T.W. and Mills, R.A. 2014. Hydrothermal sediments are a source of water column Fe and Mn in Bransfield Strait, Antarctica. *Geochimica et Cosmochimica Acta*, **137**, 64–80, https://doi.org/10.1016/j.gca.2014.04.003

Ashley, J. 1962. *A Magnetic Survey of North-East Trinity Peninsula, Graham Land*. Falkland Islands Dependencies Survey Scientific Reports, **35**.

Baker, P.E., González-Ferrán, O. and Vergara, M. 1973. Paulet Island and the James Ross Island Volcanic Group. *British Antarctic Survey Bulletin*, **32**, 89–95.

Baker, P.E., McReith, I., Harvey, M.R., Roobol, M.J. and Davies, T.G. 1975. *The Geology of the South Shetland Islands. V. Volcanic Evolution of Deception Island*. British Antarctic Survey Scientific Reports, **78**.

Baker, P.E., Buckley, F. and Rex, D.C. 1977. Cenozoic volcanism in the Antarctic. *Philosophical Transactions of the Royal Society of London. Series B, Biological Sciences*, **279**, 131–142, http://www.jstor.org/stable/2417758

Baraldo, A. and Rinaldi, C.A. 2000. Stratigraphy and structure of Deception Island, South Shetland Islands, Antarctica. *Journal of South American Earth Sciences*, **13**, 785–796, https://doi.org/10.1016/S0895-9811(00)00060-2

Baraldo, A., Rapalini, A.E., Böhnel, H. and Mena, M. 2003. Paleomagnetic study of Deception Island, South Shetland Islands, Antarctica. *Geophysical Journal International*, **153**, 333–343, https://doi.org/10.1046/j.1365-246X.2003.01881.x

Barclay, A.H., Wilcock, W.S.D. and Ibáñez, J.M. 2009. Bathymetric constraints on the tectonic and volcanic evolution of Deception Island Volcano, South Shetland Islands. *Antarctic Science*, **21**, 153–167, https://doi.org/10.1017/S0954102008001673

Barker, D.H.N. and Austin, J.A. 1994. Crustal diapirism in Bransfield Strait, West Antarctica: evidence for distributed extension in marginal-basin formation. *Geology*, **22**, 657–660, https://doi.org/10.1130/0091-7613(1994)022<0657:CDIBSW>2.3.CO;2

Barker, D.H.N. and Austin, J.A. 1998. Rift propagation, detachment faulting, and associated magmatism in Bransfield Strait, Antarctic Peninsula. *Journal of Geophysical Research*, **103**, 24 017–24 043, https://doi.org/10.1029/98JB01117

Barker, P.F. 1982. The subduction history of the Pacific margin of the Antarctic Peninsula region: ridge crest-trench interactions. *Journal of the Geological Society, London*, **139**, 787–801, https://doi.org/10.1144/gsjgs.139.6.0787

Barton, C.M. 1965. *The Geology of the South Shetland Islands. III. The Stratigraphy of King George Island*. British Antarctic Survey Scientific Reports, **44**.

Bell, C.M. 1984. The geology of islands in southern Bransfield Strait, Antarctic Peninsula. *British Antarctic Survey Bulletin*, **63**, 41–55.

Ben-Zvi, T., Wilcock, W.S.D., Barclay, A.H., Zandomeneghi, D., Ibáñez, J.M., Almendros, J. and the TOMODEC Working Group. 2007. *The P-wave velocity structure of Deception Island, Antarctica, from two-dimensional seismic tomography*. United States Geological Survey Open-File Report, **2007-1047**, Extended Abstract 078.

Berrocoso, M., García-García, A., Martín-Dávila, J., Catalán-Morollón, M., Astiz, M. and Ramírez, M.E. 2006.

Geodynamical studies on Deception Island: DECVOL and GEODEC projects. *In*: Fütterer, D.K., Damaske, D., Kleinschmidt, G., Miller, G. and Tessensohn, F. (eds) *Antarctica: Contributions to Global Earth Sciences*. Springer, Berlin, 283–287.

Berrocoso, M., Fernández-Ros, A., Prates, G., García, A. and Kraus, S. 2016. Geodetic implications on block formation and geodynamic domains in the South Shetland Islands, Antarctic Peninsula. *Tectonophysics*, **666**, 211–219, https://doi.org/10.1016/j.tecto.2015.10.023

Bibby, J.S. 1966. *The Stratigraphy of Part of North-East Graham Land and the James Ross Island Group*. British Antarctic Survey Scientific Reports, **53**.

Birkenmajer, K. 1980. Age of the Penguin Island Volcano, South Shetland Islands (West Antarctica), by the lichenometric method. *Bulletin of the Polish Academy of Sciences*, **27**, 69–76.

Birkenmajer, K. 1982*a*. The Penguin Island volcano, South Shetland Islands (Antarctica): its structure and succession. *Studia Geologica Polonica*, **74**, 155–173.

Birkenmajer, K. 1982*b*. Structural evolution of the Melville Peak Volcano, King George Island (South Shetland Islands, West Antarctica). *Bulletin of the Polish Academy of Sciences*, **24**, 341–351.

Birkenmajer, K. 1992. Volcanic succession at Deception Island, West Antarctic. A revised lithostratigraphic standard. *Studia Geologica Polonica*, **101**, 27–82.

Birkenmajer, K. and Keller, R.A. 1990. Pleistocene age of the Melville Peak volcano, King George Island, West Antarctica, by K–Ar dating. *Bulletin of the Polish Academy of Sciences*, **38**, 17–24.

Björck, S., Sandgren, P. and Zale, R. 1991. Late Holocene tephrochronology of the northern Antarctic Peninsula. *Quaternary Research*, **36**, 322–328, https://doi.org/10.1016/0033-5894(91)90006-Q

Blundell, D.J. 1962. *Palaeomagnetic Investigations in the Falkland Islands Dependencies*. British Antarctic Survey Scientific Reports, **39**.

Branney, M.J. and Kokelaar, P. 2002. *Pyroclastic Density Currents and the Sedimentation of Ignimbrites*. Geological Society, London, Memoirs, **27**, https://doi.org/10.1144/GSL.MEM.2003.027.01.10

British Antarctic Survey 1987. *Report for 1986/1987*. Natural Environment Research Council, Swindon.

Brown, R.J., Bonadonna, C. and Durant, A.J. 2012. A review of volcanic ash aggregation. *Physics and Chemistry of the Earth*, **45–46**, 65–78, https://doi.org/10.1016/j.pce.2011.11.001

Calabozo, F.M., Strelin, J.A., Orihashi, Y., Sumino, H. and Keller, R.A. 2015. Volcano–ice–sea interaction in the Cerro Santa Marta area, northwest James Ross Island, Antarctic Peninsula. *Journal of Volcanology and Geothermal Research*, **297**, 89–108, https://doi.org/10.1016/j.jvolgeores.2015.03.011

Camerlenghi, A., Domack, E. *et al*. 2001. Glacial morphology and post-glacial contourites in northern Prince Gustav Channel (NW Weddell Sea, Antarctica). *Marine Geophysical Researches*, **22**, 417–443, https://doi.org/10.1023/A:1016399616365

Carmona, E., Almendros, J., Peña, J.A. and Ibáñez, J.M. 2010. Characterization of fracture systems using precise array locations of earthquake multiplets: an example at Deception Island volcano, Antarctica. *Journal of Geophysical Research*, **115**, B06309, https://doi.org/10.1029/2009JB006865

Carrizo, H.G., Torielli, C.A., Strelin, J.A. and Muñoz, C.E. 1998. Ambiente eruptivo del Grupo Volcánico Isla James Ross en Riscos Rink, isla James Ross, Antártida. *Revista de la Asociación Geológica Argentina*, **53**, 469–479.

Caselli, A.T. and Agusto, M.R. 2004. Depósitos hudrovolcánicos recientes con indicios de inmiscibilidad magmática en la isla Decepción (Antártida). *Revista de la Asociación Geológica Argentina*, **59**, 495–500.

Catalán, M., Galindo-Zalvidar, J., Davila, J.M., Martos, Y.M., Maldonado, A., Gambôa, L. and Schreider, A.A. 2013. Initial stages of oceanic spreading in the Bransfield Rift from magnetic and gravity data analysis. *Tectonophysics*, **585**, 102–112, https://doi.org/10.1016/j.tecto.2012.09.016

Christeson, G.L., Barker, D.H.N., Austin, J.A. and Dalziel, I.W.D. 2003. Deep crustal structure of Bransfield Strait: Initiation of a back arc basin by rift reactivation and propagation. *Journal of Geophysical Research*, **108**, 2492, https://doi.org/10.1029/2003JB002468

Clark, N., Williams, M. *et al*. 2010. Early Pliocene Weddell Sea seasonality determined from bryozoans. *Stratigraphy*, **7**, 196–206.

Clark, N.A., Williams, M. *et al*. 2013. Fossil proxies of near-shore sea surface temperatures and seasonality from the late Neogene Antarctic shelf. *Naturwissenschaften*, **100**, 699–722, https://doi.org/10.1007/s00114-013-1075-9

Concheyro, A., Salani, F.M., Adamonis, S. and Lirio, J.M. 2007. Los depositos diamictos Cenozoicos de la Cuenca James Ross, Antartida: una synthesis stratigrafica y nuevos hallazgos paleontologicos. *Revista de la Asociacion Geologica Argentina*, **62**, 568–585.

Cooper, A.P.R., Smellie, J.L. and Maylin, J. 1998. Evidence for shallowing and uplift from bathymetric records of Deception Island, Antarctica. *Antarctic Science*, **10**, 455–461, https://doi.org/10.1017/S0954102098000558

Dählmann, A., Wallmann, K. *et al*. 2001. Hot vents in an ice-cold ocean: Indications for phase separation at the southernmost area of hydrothermal activity, Bransfield Strait, Antarctica. *Earth and Planetary Science Letters*, **193**, 381–394, https://doi.org/10.1016/S0012-821X(01)00535-0

Davies, B.J., Glasser, N.F., Carrivick, J.L., Hambrey, M.J., Smellie, J.L. and Nývlt, D. 2013. Landscape evolution and ice-sheet behaviour in a semi-arid polar environment: James Ross Island, NE Antarctic Peninsula. *Geological Society, London, Special Publications*, **381**, 353–395, https://doi.org/10.1144/SP381.1

De Rita, D., Giordano, G., Esposito, A., Fabbri, M. and Rodani, S. 2002. Large volume phreatomagmatic ignimbrites from the Colli Albani volcano (Middle Pleistocene, Italy). *Journal of Volcanology and Geothermal Research*, **118**, 77–98, https://doi.org/10.1016/S0377-0273(02)00251-2

Dietrich, R., Dach, R. *et al*. 2001. ITRF coordinates and plate velocities from repeated GPS campaigns in Antarctica – an analysis based on different individual solutions. *Journal of Geodesy*, **74**, 756–766, https://doi.org/10.1007/s001900000147

Dietrich, R., Rülke, A. *et al*. 2004. Plate kinematics and deformation status of the Antarctic Peninsula based on GPS. *Global and Planetary Change*, **42**, 313–321, https://doi.org/10.1016/j.gloplacha.2003.12.003

Dingle, R.V. and Lavelle, M. 1998. Antarctic Peninsular cryosphere: early Oligocene (*c*. 30 Ma) initiation and a revised glacial chronology. *Journal of the Geological Society, London*, **155**, 433–437, https://doi.org/10.1144/gsjgs.155.3.0433

Dingle, R.V., McArthur, J.M. and Vroon, P. 1997. Oligocene and Pliocene interglacial events in the Antarctic Peninsula dated using strontium isotope stratigraphy. *Journal of the Geological Society, London*, **154**, 257–264, https://doi.org/10.1144/gsjgs.154.2.0257

Esser, R.E., Kyle, P.R. and McIntosh, W.C. 2004. $^{40}Ar/^{39}Ar$ dating and the eruptive history of Mount Erebus, Antarctica: volcano evolution. *Bulletin of Volcanology*, **66**, 671–686, https://doi.org/10.1007/s00445-004-0354-x

Fisk, M.R. 1990. Volcanism in the Bransfield Strait, Antarctica. *Journal of South American Earth Sciences*, **3**, 91–101, https://doi.org/10.1016/0895-9811(90)90022-S

Fretzdorff, S. and Smellie, J.L. 2002. Electron microprobe characterization of ash layers in sediments from the central Bransfield basin (Antarctic Peninsula): evidence for at least two volcanic sources. *Antarctic Science*, **14**, 412–421, https://doi.org/10.1017/S0954102002000214

Fretzdorff, S., Worthington, T.J., Haase, K.M., Hekinian, R., Franz, L., Keller, R.A. and Stoffers, P. 2004. Magmatism in the Bransfield Basin: rifting of the South Shetland Arc? *Journal of Geophysical Research*, **109**, B12208, https://doi.org/10.1029/2004JB003046

García, M., Ercilla, G., Alonso, B., Casas, D. and Dowdeswell, J.A. 2011. Sediment lithofacies, processes and sedimentary models in the Central Bransfield Basin, Antarctic Peninsula, since the

Last Glacial Maximum. *Marine Geology*, **290**, 1–16, https://doi.org/10.1016/j.margeo.2011.10.006

Gazdzicki, A., Tatur, A., Hara, U. and del Valle, R.A. 2004. The Weddell Sea Formation: post-late Pliocene terrestrial glacial deposits on Seymour Island, Antarctic Peninsula. *Polish Polar Research*, **25**, 189–204.

Geyer, A., Pedrazzi, D. *et al.* 2021. Deception Island. *Geological Society, London, Memoirs*, **55**, https://doi.org/10.1144/M55-2018-56

González-Casado, J.M., Giner Robles, J.L. and López-Martínez, J. 2000. Bransfield Basin, Antarctic Peninsula: Not a normal back-arc basin. *Geology*, **28**, 1043–1046, https://doi.org/10.1130/0091-7613(2000)28<1043:BBAPNA>2.0.CO;2

González-Ferrán, O. 1991. The Bransfield Rift and its active volcanism. *In*: Thomson, R.A., Crame, J.A. and Thomson, J.W. (eds) *Geological Evolution of Antarctica*. Cambridge University Press, Cambridge, UK, 505–509.

González-Ferrán, O. 1995. *Volcanes de Chile*. Instituto Geografico Militar, Santiago.

González-Ferrán, O. and Katsui, Y. 1970. Estudio integral del volcanism cenozoico superior de las Islas Shetland del Sur, Antártica. *Series Científica del Instituto Antártico Chileno*, **1**, 123–174.

González-Ferrán, O., Munizaga, F. and Moreno, H. 1971. Sintesis de la evolución volcánica de Isla Decepción y la erupción de 1970. *Instituto Antártico Chileno, Serie Científicas*, **2**, 1–14.

Gràcia, E., Canals, M., Farràn, M., Prieto, M.J., Sorribas, J. and GEBRA Team. 1996. Morphostructural evolution of the Central and Eastern Bransfield Basins (NW Antarctic Peninsula). *Marine Geophysical Researches*, **18**, 429–448, https://doi.org/10.1007/BF00286088

Gràcia, E., Canals, M., Farràn, M.L., Sorribas, J. and Pallàs, R. 1997. Central and eastern Bransfield basins (Antarctica) from high-resolution swath-bathymetry data. *Antarctic Science*, **9**, 168–180, https://doi.org/10.1017/S0954102097000229

Grad, M., Guterch, A. and Šroda, P. 1992. Upper crustal structure of Deception Island area, Bransfield Strait, West Antarctica. *Antarctic Science*, **4**, 469–476, https://doi.org/10.1017/S0954102092000683

Haase, K.M. and Beier, C. 2021. Bransfield Strait and James Ross Island: petrology. *Geological Society, London, Memoirs*, **55**, https://doi.org/10.1144/M55-2018-37

Hambrey, M.J. and Smellie, J.L. 2006. Distribution, lithofacies and environmental context of Neogene glacial sequences on James Ross and Vega islands, Antarctic Peninsula. *Geological Society, London, Special Publications*, **258**, 187–200, https://doi.org/10.1144/GSL.SP.2006.258.01.14

Hambrey, M.J., Smellie, J.L., Nelson, A.E. and Johnson, J.S. 2008. Late Cenozoic glacier–volcano interaction on James Ross Island and adjacent areas, Antarctic Peninsula region. *Geological Society of America Bulletin*, **120**, 709–731, https://doi.org/10.1130/B26242.1

Hawkes, D.D. 1961. *The Geology of the South Shetland Islands. II. The Geology and Petrology of Deception Island*. Falkland Islands Dependencies Survey Scientific Reports, **27**.

Headland, R.K. 2002. Selected chronology of expeditions and historical events at Deception Island. *In*: Smellie, J.L., López-Martínez, J. *et al.* (eds) *Geology and Geomorphology of Deception Island, 1:25 000, with Supplementary Text*. BAS GEOMAP Series Sheets 6-A and 6-B. British Antarctic Survey, Cambridge, UK, 64–67.

Henriet, J.P., Meissner, R., Miller, H. and the GRAPE Team 1992. Active margin processes along the Antarctic Peninsula. *Tectonophysics*, **201**, 229–253, https://doi.org/10.1016/0040-1951(92)90235-X

Hillenbrand, C.-D., Moreton, S.M. *et al.* 2008. Volcanic time-markers for marine isotopic stages 6 and 5 in Southern Ocean sediments and Antarctic ice cores: implications for tephra correlations between palaeoclimatic records. *Quaternary Science Reviews*, **27**, 518–540, https://doi.org/10.1016/j.quascirev.2007.11.009

Hobbs, G.J. 1968. *The Geology of the South Shetland Islands. IV. The Geology of Livingston Island*. British Antarctic Survey Scientific Reports, **47**.

Hole, M.J. 2021. Antarctic Peninsula: petrology. *Geological Society, London, Memoirs*, **55**, https://doi.org/10.1144/M55-2018-40

Hole, M.J. and Larter, R.D. 1993. Trench-proximal volcanism following ridge crest–trench collision along the Antarctic Peninsula. *Tectonics*, **12**, 897–910, https://doi.org/10.1029/93TC00669

Hole, M.J., Saunders, A.D., Rogers, G. and Sykes, M.A. 1995. The relationship between alkaline magmatism, lithospheric extension and slab window formation along continental destructive plate margins. *Geological Society, London, Special Publications*, **81**, 265–285, https://doi.org/10.1144/GSL.SP.1994.081.01.15

Hooper, P.R. 1962. *The Petrology of Anvers Island and Adjacent Islands*. Falkland Islands Dependencies Scientific Reports, **34**.

Ibáñez, J.M., Almendros, J., Carmona, E., Martínez-Arévalo, C. and Abril, M. 2003. The recent seismo-volcanic activity at Deception Island volcano. *Deep-Sea Research II: Topical Studies in Oceanography*, **50**, 1611–1629, https://doi.org/10.1016/S0967-0645(03)00082-1

Ivany, L.C., Van Simaeys, S., Domack, E.W. and Samson, S.D. 2006. Evidence for an earliest Oligocene ice sheet on the Antarctic Peninsula. *Geology*, **34**, 377–380, https://doi.org/10.1130/G22383.1

John, B.S. and Sugden, D.E. 1971. Raised marine features and phases of glaciation in the South Shetland Islands. *British Antarctic Survey Bulletin*, **24**, 45–111.

Johnson, J.S. and Smellie, J.L. 2007. Zeolite compositions as proxies for eruptive palaeoenvironment. *Geochemistry, Geophysics, Geosystems*, **8**, Q03009, https://doi.org/10.1029/2006GC001450

Jones, J.G. and Nelson, P.H.H. 1970. The flow of basalt lava from air into water – its structural expression and stratigraphic significance. *Geological Magazine*, **107**, 13–19, https://doi.org/10.1017/S0016756800054649

Jonkers, H.A. 1998. The Cockburn Island Formation; Late Pliocene interglacial sedimentation in the James Ross Basin, northern Antarctic Peninsula. *Newsletters on Stratigraphy*, **36**, 63–76, https://doi.org/10.1127/nos/36/1998/63

Jonkers, H.A. and Kelley, S.P. 1998. A reassessment of the age of the Cockburn Island Formation, northern Antarctic Peninsula, and its palaeoclimatic implications. *Journal of the Geological Society, London*, **155**, 737–740, https://doi.org/10.1144/gsjgs.155.5.0737

Jonkers, H.A., Lirio, J.M., del Valle, R.A. and Kelley, S.P. 2002. Age and environment of Miocene–Pliocene glaciomarine deposits, James Ross Island, Antarctica. *Geological Magazine*, **139**, 577–594, https://doi.org/10.1017/S0016756802006787

Jordan, T.A., Ferraccioli, F., Jones, P.C., Smellie, J.L., Ghidella, M. and Corr, H. 2009. Airborne gravity reveals interior of Antarctic volcano. *Physics of the Earth and Planetary Interiors*, **175**, 127–136, https://doi.org/10.1016/j.pepi.2009.03.004

Jurado-Chichay, Z., Rowland, S.K. and Walker, G.P.L. 1996. The formation of circular littoral cones from tube-fed pahoehoe: Mauna Loa, Hawaii. *Bulletin of Volcanology*, **57**, 471–482.

Kamenov, B.K. 2004. The olivine basalts from Livingston Island, West Antarctica: petrology and geochemical comparisons. *Bulgarian Academy of Sciences*, **41**, 71–98.

Keller, R.A., Fisk, M.R., White, W.M. and Birkenmajer, K. 1992. Isotopic and trace element constraints on mixing and melting models of marginal basin volcanism, Bransfield Strait, Antarctica. *Earth and Planetary Science Letters*, **111**, 287–303, https://doi.org/10.1016/0012-821X(92)90185-X

Keller, R.A., Fisk, M.R., Smellie, J.L., Strelin, J.A. and Lawver, L.A. 2002. Geochemistry of back arc basin volcanism in Bransfield Strait, Antarctica: subducted contributions and along-axis variations. *Journal of Geophysical Research*, **107**, 2171, https://doi.org/10.1029/2001JB000444

Klinkhammer, G.P., Chin, C.S. *et al.* 2001. Discovery of new hydrothermal vent sites in Bransfield Strait, Antarctica. *Earth and Planetary Science Letters*, **193**, 395–407, https://doi.org/10.1016/S0012-821X(01)00536-2

Košler, J., Magna, T., Mlčoch, B., Mixa, P., Nývlt, D. and Holub, F.V. 2009. Combined Sr, Nd, Pb and Li isotope geochemistry of alkaline lavas from northern James Ross Island (Antarctic Peninsula) and implications for back-arc magma formation. *Chemical Geology*, **258**, 207–218, https://doi.org/10.1016/j.chemgeo.2008.10.006

Kowalewski, W., Rudowski, S. and Zalewski, S.M. 1990. Seismoacoustic studies within flooded part of the caldera of the Deception Island, West Antarctica. *Polish Polar Research*, **11**, 259–266.

Kraus, S. 2005. *Magmatic Dyke Systems of the South Shetland Islands Volcanic Arc (West Antarctica): Reflections of the Geodynamic History*. PhD thesis, University of Munich, Munich, Germany, http://nbnresolving.de/urn:nbn:de:bvb:19-38277

Kraus, S., Kurbatov, A. and Yates, M. 2013. Geochemical signatures of tephras from Quaternary Antarctic Peninsula volcanoes. *Andean Geology*, **40**, 1–40, https://doi.org/10.5027/andgeo V40n1-a01

Kristjánsson, L., Gudmundsson, M.T., Smellie, J.L., McIntosh, W.C. and Esser, R. 2005. Palaeomagnetic, $^{40}Ar/^{39}Ar$, and stratigraphical correlation of Miocene–Pliocene basalts in the Brandy Bay area, James Ross Island, Antarctica. *Antarctic Science*, **17**, 409–417, https://doi.org/10.1017/S0954102005002853

Larter, R.D. 1991. Debate: Preliminary results of seismic reflection investigations and associated geophysical studies in the area of the Antarctic Peninsula. *Antarctic Science*, **3**, 217–222, https://doi.org/10.1017/S0954102091210251

Larter, R.D. and Barker, P.F. 1991. Effects of ridge crest–trench interaction on Antarctic–Phoenix spreading: forces on a young subducting plate. *Journal of Geophysical Research*, **96**, 15 583–19 607, https://doi.org/10.1029/91JB02053

Lavoie, C., Domack, E.W., Heirman, K., Naudts, L. and Brachfeld, S. 2016. A Holocene volcanic knoll within a glacial trough, Antarctic Sound, northern Antarctic Peninsula. *Geological Society, London, Memoirs*, **46**, 125–126, https://doi.org/10.1144/M46.130

Lawver, L.A., Keller, R.A., Fisk, M.R. and Strelin, J.A. 1995. Bransfield Strait, Antarctic Peninsula – active extension behind a dead arc. *In*: Taylor, B. (ed.) *Backarc Basins, Tectonics and Magmatism*. Plenum Press, New York, 315–342.

Lawver, L.A., Sloan, B.J. *et al.* 1996. Distributed, active extension in Bransfield Basin, Antarctic Peninsula: evidence from multibeam bathymetry. *GSA Today*, **6**, 1–6, 16–17.

Leat, P.T., Scarrow, J.H. and Millar, I.L. 1995. On the Antarctic Peninsula batholith. *Geological Magazine*, **132**, 399–412, https://doi.org/10.1017/S0016756800021464

LaMasurier, W.E., Thomson, J.W., Baker, P.E., Kyle, P.R., Rowley, P.D., Smellie, J. and Verwoerd, W.J. (eds). 1990. *Volcanism of the Antarctic Plate and Southern Oceans*. American Geophysical Union Antarctic Research Series, **48**.

Lirio, J.M., Núñez, H.J., Bertels-Psotka, A. and Del Valle, R.A. 2003. Diamictos fosilíferos (Mioceno-Pleistoceno): Formaciones Belén, Gage y Terrapin en la isla James Ross, Antártida. *Revista de la Asociación Geológica Argentina*, **58**, 298–310.

Lopes, F.C., Caselli, A.T., Machado, A. and Barata, M.T. 2014. The development of the Deception Island volcano caldera under control of the Bransfield Basin sinistral strike-slip tectonic regime (NW Antarctica). *Geological Society, London, Special Publications*, **401**, 173–184, https://doi.org/10.1144/SP401.6

López-Martínez, J. and Serrano, E. 2002. Geomorphology. *In*: Smellie, J.L., López-Martínez, J. *et al.* (eds) *Geology and Geomorphology of Deception Island, 1:25 000, with Supplementary Text*. BAS GEOMAP Series Sheets 6-A and 6-B. British Antarctic Survey, Cambridge, UK, 31–39.

Maestro, A., Somoza, L., Rey, J., Martinez-Frias, J. and López-Martínez, J. 2007. Active tectonics, fault patterns, and stress field of Deception Island: a response to oblique convergence between the Pacific and Antarctic plates. *Journal of South American Earth Science*, **23**, 256–268, https://doi.org/10.1016/j.jsames.2006.09.023

Malagnino, E.C., Olivero, E.B., Rinaldi, C.A. and Spikermann, Y.J.P. 1981. Aspectos geomórfologicos de la isla Vicecomodoro Marambio, Antártida. *In*: *VIII Congreso Geológico Argentino, San Luis, Actas II*. Congreso Geologico Argentino, Buenos Aires, 883–896.

Manville, V. and Wilson, C.J.N. 2004. Vertical density currents: a review of their potential role in the deposition of deep-sea ash layers. *Journal of the Geological Society, London*, **161**, 947–958, https://doi.org/10.1144/0016-764903-067

Marenssi, S.A., Casadio, S. and Santillana, S.N. 2010. Record of Late Miocene glacial deposits on Isla Marambio (Seymour Island), Antarctic Peninsula. *Antarctic Science*, **22**, 193–198, https://doi.org/10.1017/S0954102009990629

Martí, J. and Baraldo, A. 1990. Pre-caldera pyroclastic deposits of Deception Island (South Shetland Islands). *Antarctic Science*, **2**, 345–352, https://doi.org/10.1017/S0954102090000475

Martí, J., Diez-Gil, J.L. and Ortiz, R. 1991. Conduction model for the thermal influence of lithic clasts in mixtures of hot gases and ejecta. *Journal of Geophysical Research*, **96**, 21 879–21 885, https://doi.org/10.1029/91JB02149

Martí, J., Vila, J. and Rey, J. 1996. Deception Island (Bransfield Strait, Antarctica: an example of volcanic caldera developed by extensional tectonics. *Geological Society, London, Special Publications*, **110**, 253–265, https://doi.org/10.1144/GSL.SP.1996.110.01.20

Martí, J., Geyer, A. and Aguirre-Diaz, G. 2013. Origin and evolution of the Deception Island caldera (South Shetland Islands, Antarctica). *Bulletin of Volcanology*, **75**, 732, https://doi.org/10.1007/s00445-013-0732-3

Massabie, A.C. and Morelli, J.R. 1977. Buchitas de la Isla Vicecomodoro Marambio, Sector Antartico Argentino. *Revista de la Asociacion Geologica Argentina*, **32**, 44–51.

Massaferro, G.I., Caselli, A.T. and Rovere, E.I. 1997. Petrogénesis de las rocas eruptivas de las Islas Marambio y Cerro Nevado, Península Antártica. *Revista de la Asociación Geológica Argentina*, **52**, 481–490.

Matthies, D., Storzer, D. and Troll, G. 1988. Volcanic ashes in Bransfield Strait sediments: geochemical and stratigraphical investigations (Antarctica). *In*: *Proceedings of the Second International Conference on Natural Glasses, 21–25 September 1987, Prague*. Charles University, Prague, 139–147.

McArthur, J.M., Howarth, R.J. and Bailey, T.R. 2001. Strontium isotope stratigraphy: LOWESS Version 3: Best fit to the marine Sr-isotope curve for 0–509 Ma and accompanying look-up table for deriving numerical age. *Journal of Geology*, **109**, 155–170, https://doi.org/10.1086/319243

McArthur, J.M., Rio, D., Massari, F., Castradori, D., Bailey, T.R., Thirlwall, M. and Houghton, S. 2006. A revised Pliocene record for marine-$^{87}Sr/^{86}Sr$ used to date an interglacial event recorded in the Cockburn Island Formation, Antarctic Peninsula. *Palaeogeography, Palaeoclimatology, Palaeoecology*, **242**, 126–136, https://doi.org/10.1016/j.palaeo.2006.06.004

McPhie, J., Doyle, M. and Allen, R. 1993. *Volcanic Textures: A Guide to the Interpretation of Textures in Volcanic Rocks*. CODES Key Centre, University of Tasmania, Hobart.

Minzoni, R.T., Anderson, J.B., Fernandez, R. and Smith Wellner, J. 2016. Marine record of Holocene climate, ocean, and cryosphere interactions: Herbert Sound, James Ross Island, Antarctica. *Quaternary Science Reviews*, **129**, 239–259, https://doi.org/10.1016/j.quascirev.2015.09.009

Mitchell, N.C., Stretch, R., Oppenheimer, C., Kay, D. and Beier, C. 2012. Cone morphologies associated with shallow water eruptions: east Pico Island, Azores. *Bulletin of Volcanology*, **74**, 2289–2301, https://doi.org/10.1007/s00445-012-0662-5

Mlčoch, B., Nývlt, D. and Mixa, P. 2015. *Geological Map of Northern James Ross Island, 1: 25 000 Scale*. Czech Geological Survey, Brno, Czech Republic.

Moreton, S. and Smellie, J.L. 1998. Identification and correlation of distal tephra layers in deep sea sedimentary cores, Scotia Sea, Antarctica. *Annals of Glaciology*, **27**, 285–289, https://doi.org/10.3189/1998AoG27-1-285-289

Mulder, T. and Alexander, J. 2001. The physical character of subaqueous sedimentary density flows and their deposits. *Sedimentology*, **48**, 269–299, https://doi.org/10.1046/j.1365-3091.2001.00360.x

Mulvaney, R., Abram, N.J. et al. 2012. Recent Antarctic Peninsula warming relative to Holocene climate and ice-shelf history. Nature, **489**, 141–145, https://doi.org/10.1038/nature11391

Muñoz-Martín, A., Catalán, M., Martín-Dávila, J. and Carbó, A. 2005. Upper crustal structure of Deception Island area (Bransfield Strait, Antarctica) from gravity and magnetic modelling. Antarctic Science, **17**, 213–224, https://doi.org/10.1017/S0954102005002622

Nagihara, S. and Lawver, L.A. 1989. Heat flow measurements in the King George Basin, Bransfield Basin. Antarctic Science, **23**, 123–125.

Nehyba, S. and Nývlt, D. 2014. Deposits of pyroclastic mass flows at Bibby Hill (Pliocene, James Ross Island, Antarctica). Czech Polar Reports, **4**, 103–122, https://doi.org/10.5817/CPR2014-2-11

Nehyba, S. and Nývlt, D. 2015. 'Bottomsets' of the lava-fed delta of James Ross Island Volcanic Group, Ulu Peninsula, James Ross Island, Antarctica. Polish Polar Research, **36**, 1–24, https://doi.org/10.1515/popore-2015-0002

Nelson, A.E., Smellie, J.L., Williams, M. and Zalasiewicz, J. 2008. Short Note: Late Miocene marine trace fossils from James Ross Island. Antarctic Science, **20**, 591–592, https://doi.org/10.1017/S0954102008001429

Nelson, A.E., Smellie, J.L. et al. 2009. Neogene glacigenic debris flows on James Ross Island, northern Antarctic Peninsula, and their implications for regional climate history. Quaternary Science Reviews, **28**, 3138–3160, https://doi.org/10.1016/j.quascirev.2009.08.016

Nelson, P.H.H. 1975. *The James Ross Island Volcanic Group of North-East Graham Land*. British Antarctic Survey Scientific Reports, **54**.

Nývlt, D., Kosler, J., Mlčoch, B., MIxa, P., Lisá, L., Bubík, M. and Hendriks, B.W.H. 2011. The Mendel Formation: Evidence for Late Miocene climatic cyclicity at the northern tip of the Antarctic Peninsula. Palaeogeography, Palaeoclimatology, Palaeoecology, **299**, 363–384, https://doi.org/10.1016/j.palaeo.2010.11.017

Oehler, J.-F., van Wyk de Vries, B. and Labazuy, P. 2005. Landslides and spreading of oceanic hot-spot and arc shield volcanoes on Low Strength Layers (LSLs): an analogue modeling approach. Journal of Volcanology and Geothermal Research, **144**, 169–189, https://doi.org/10.1016/j.jvolgeores.2004.11.023

Oliva-Urcia, B., Gil-Peña, I. et al. 2016. Paleomagnetism from Deception Island (South Shetlands archipelago, Antarctica), new insights in the interpretation of the volcanic evolution using a geomagnetic model. International Journal of Earth Sciences, **105**, 1353–1370, https://doi.org/10.1007/s00531-015-1254-3

Orheim, O. 1972a. Volcanic activity on Deception Island, South Shetland Islands. *In*: Adie, R.J. (ed.) *Antarctic Geology and Geophysics*. Universitetsforlaget, Oslo, 117–120.

Orheim, O. 1972b. *A 200-year Record of Glacier Mass Balance at Deception Island, Southwest Atlantic Ocean, and its Bearing on Models of Global Climate Change*. Institute of Polar Studies, Ohio State University Report, **42**.

Ortíz, R., Vila, J. et al. 1992. Geophysical features of Deception Island. *In*: Yoshida, Y., Kaminuma, K. and Shiraishi, K. (eds) *Recent Progress in Antarctic Earth Science*. Terra Scientific (TERRAPUB), Tokyo, 443–448.

Pallàs, R., Smellie, J.L., Casas, J.M. and Calvet, J. 2001. Using tephrochronology to date temperate ice: correlation between ice tephras on Livingston Island and eruptive units on Deception Island volcano (South Shetland Islands, Antarctica). The Holocene, **11**, 149–160, https://doi.org/10.1191/095968301669281809

Pańczyk, M. and Nawrocki, J. 2011. Pliocene age of the oldest basaltic rocks of Penguin Island (South Shetland Islands, northern Antarctic Peninsula). Geological Quarterly, **55**, 335–344.

Pedrera, A., Ruiz-Consta, A. et al. 2012. The fracture system and the melt emplacement beneath the Deception Island active volcano, South Shetland Islands, Antarctica. Antarctic Science, **24**, 173–182, https://doi.org/10.1017/S0954102011000794

Pirrie, D. and Sykes, M.A. 1987. Regional significance of proglacial delta-front, reworked tuffs, James Ross Island area. British Antarctic Survey Bulletin, **77**, 1–12.

Pirrie, D., Crame, J.A., Riding, J.B., Butcher, A.R. and Taylor, P.D. 1997. Miocene glaciomarine sedimentation in the northern Antarctic Peninsula region: the stratigraphy and sedimentology of the Hobbs Glacier Formation, James Ross Island. Geological Magazine, **136**, 745–762, https://doi.org/10.1017/S0016756897007796

Pirrie, D., Jonkers, H.A., Smellie, J.L., Crame, J.A. and McArthur, J.M. 2011. Reworked late Neogene *Austrochlamys andersoni* (Mollusca: Bivalvia) from northern James Ross Island, Antarctica. Antarctic Science, **23**, 180–187, https://doi.org/10.1017/S0954102010000842

Rey, J., Somoza, L. and Hernández-Molina, F.J. 1992. Formas de los sedimentos submarinos superficiales en Puerto Foster, Isla Decepcion, Islas Sherland del Sur. *In*: López-Martínez, J. (ed.) *Geología de la Antártida Occidental. Salamanca, III Congreso Geología de España and VIII Congreso Latinoamericano de Geología, Salamanca, Spain*. Facultad de Ciencias, Universidad de Salamanca, Salamanca, Spain, 163–172.

Rey, J., Somoza, L. and Martínez-Frías, J. 1995. Tectonic, volcanic, and hydrothermal event sequence on Deception Island (Antarctica). Geo-Marine Letters, **15**, 1–8, https://doi.org/10.1007/BF01204491

Rey, J., Maestro, A., Somoza, L. and Smellie, J.L. 2002. Submarine morphology and seismic stratigraphy of Port Foster. *In*: Smellie, J.L., López-Martínez, J. et al. (eds) *Geology and Geomorphology of Deception Island, 1:25 000, with Supplementary Text*. BAS GEOMAP Series Sheets 6-A and 6-B. British Antarctic Survey, Cambridge, UK, 40–46.

Rex, D.C. 1972. K–Ar age determinations on volcanic and associated rocks from the Antarctic Peninsula and Dronning Maud Land. *In*: Adie, R.J. (ed.) *Antarctic Geology and Geophysics*. Universitetsforlaget, Oslo, 133–136.

Rex, D.C. 1976. Geochronology in relation to the stratigraphy of the Antarctic Peninsula. British Antarctic Survey Bulletin, **43**, 49–58.

Ringe, M.J. 1991. Volcanism on Brabant Island, Antarctica. *In*: Thomson, M.R.A., Crame, J.A. and Thomson, J.W. (eds) *Geological Evolution of Antarctica*. Cambridge University Press, Cambridge, UK, 515–519.

Risso, C. and Aparicio, A. 2002. Plutonic xenoliths in Deception Island (Antarctica). Terra Antartica, **9**, 95–99.

Roberts, S.J., Hodgson, D.A. et al. 2011. Geological constraints on glacio-isostatic adjustment models of relative sea-level change during deglaciation of Prince Gustav Channel, Antarctic Peninsula. Quaternary Science Reviews, **30**, 3603–3617, https://doi.org/10.1016/j.quascirev.2011.09.009

Rocha-Campos, A.C., Kuchenbecker, M., Duleba, W., dos Santos, P.R. and Canile, F.M. 2017. A (tidal-marine) boulder pavement in the late Cenozoic of Seymour Island, West Antarctica: contribution to the palaeogeographical and palaeoclimatic evolution of West Antarctica. Antarctic Science, **29**, 555–559, https://doi.org/10.1017/S0954102017000335

Roobol, M.R. 1973. Historic volcanic activity at Deception Island. British Antarctic Survey Bulletin, **32**, 23–30.

Roobol, M.J. 1980. A model for the eruptive mechanism of Deception Island from 1820 to 1970. British Antarctic Survey Bulletin, **49**, 137–156.

Russell, J.K., Edwards, B.R. and Porritt, L.A. 2013. Pyroclastic passage zones in glaciovolcanic sequences. Nature Communications, **4**, 1788, https://doi.org/10.1038/ncomms2829

Skilling, I.P. 1994. Evolution of an englacial volcano: Brown Bluff, Antarctica. Bulletin of Volcanology, **56**, 573–591, https://doi.org/10.1007/BF00302837

Skilling, I.P. 2002. Basaltic pahoehoe lava-fed deltas: large-scale characteristics, clast generation, emplacement processes and environmental discrimination. Geological Society, London, Special Publications, **202**, 91–113, https://doi.org/10.1144/GSL.SP.2002.202.01.06

Skilling, I.P. 2009. Subglacial to emergent volcanism at Hlöðufell, south-west Iceland: A history of ice-confinement. Journal of

Volcanology and Geothermal Research, **185**, 276–289, https://doi.org/10.1016/j.jvolgeores.2009.05.023

Smellie, J.L. 1987. Geochemistry and tectonic setting of alkaline volcanic rocks in the Antarctic Peninsula: a review. *Journal of Volcanology and Geothermal Research*, **32**, 269–285, [https://doi.org/10.1016/0377-0273(87)90048-5]

Smellie, J.L. 1988. Recent observations on the volcanic history of Deception Island, South Shetland Islands. *British Antarctic Survey Bulletin*, **81**, 83–85.

Smellie, J.L. 1989. Deception Island. *American Geophysical Union Field Trip Guidebook*, **T180**, 146–152.

Smellie, J.L. 1990a. Graham Land and South Shetland Islands. Summary. *American Geophysical Union Antarctic Research Series*, **48**, 303–312, https://doi.org/10.1029/AR048p0302

Smellie, J.L. 1990b. Melville Peak. *American Geophysical Union Antarctic Research Series*, **48**, 325–326.

Smellie, J.L. 1990c. Penguin Island. *American Geophysical Union Antarctic Research Series*, **48**, 322–324.

Smellie, J.L. 1990d. Bridgeman Island. *American Geophysical Union Antarctic Research Series*, **48**, 313–315.

Smellie, J.L. 1990e. Anvers and Brabant islands. *American Geophysical Union Antarctic Research Series*, **48**, 334–336.

Smellie, J.L. 1990f. Paulet Island. *American Geophysical Union Antarctic Research Series*, **48**, 337–338.

Smellie, J.L. 1990g. Cain and Abel nunataks, Trinity Peninsula. *American Geophysical Union Antarctic Research Series*, **48**, 344.

Smellie, J.L. 1990h. Northern Livingston and Greenwich islands. *American Geophysical Union Antarctic Research Series*, **48**, 331–333.

Smellie, J.L. 1999. The upper Cenozoic tephra record in the south polar region: a review. *Global and Planetary Change*, **21**, 51–70, https://doi.org/10.1016/S0921-8181(99)00007-7

Smellie, J.L. 2001. Lithostratigraphy and volcanic evolution of Deception Island, South Shetland Islands. *Antarctic Science*, **13**, 188–209, https://doi.org/10.1017/S0954102001000281

Smellie, J.L. 2002a. Geology. In: Smellie, J.L., López-Martínez, J. et al. (eds) *Geology and Geomorphology of Deception Island, 1:25 000, with Supplementary Text*. BAS GEOMAP Series Sheets 6-A and 6-B. British Antarctic Survey, Cambridge, UK, 11–30.

Smellie, J.L. 2002b. The 1969 subglacial eruption on Deception Island (Antarctica): events and processes during an eruption beneath a thin glacier and implications for volcanic hazards. *Geological Society, London, Special Publications*, **202**, 59–79, https://doi.org/10.1144/GSL.SP.2002.202.01.04

Smellie, J.L. 2002c. Chronology of eruptions of Deception Island. In: Smellie, J.L., López-Martínez, J. et al. (eds) *Geology and Geomorphology of Deception Island, 1:25 000, with Supplementary Text*. BAS GEOMAP Series Sheets 6-A and 6-B. British Antarctic Survey, Cambridge, UK, 70–71.

Smellie, J.L. 2006. The relative importance of supraglacial v. subglacial meltwater escape in basaltic subglacial tuya eruptions: an important unresolved conundrum. *Earth-Science Reviews*, **74**, 241–268, https://doi.org/10.1016/j.earscirev.2005.09.004

Smellie, J.L. 2009. Terrestrial sub-ice volcanism: landform morphology, sequence characteristics and environmental influences, and implications for candidate Mars examples. *Geological Society of America Special Papers*, **453**, 55–76.

Smellie, J.L. 2018. Glaciovolcanism – a 21st century proxy for palaeo-ice. In: Menzies, J. and van der Meer, J.J.M. (eds) *Past Glacial Environments (Sediments, Forms and Techniques)*. 2nd edn. Elsevier, Amsterdam, 335–375.

Smellie, J.L. 2021. Antarctic volcanism: volcanology and palaeoenvironmental overview. *Geological Society, London, Memoirs*, **55**, https://doi.org/10.1144/M55-2020-1

Smellie, J.L. and Edwards, B.E. 2016. *Glaciovolcanism on Earth and Mars. Products, Processes and Palaeoenvironmental Significance*. Cambridge University Press, Cambridge, UK.

Smellie, J.L. and López-Martínez, J. 2002. Geological and geomorphological evolution: summary. In: Smellie, J.L., López-Martínez, J. et al. (eds) *Geology and Geomorphology of Deception Island, 1:25 000, with Supplementary Text*. BAS GEOMAP Series Sheets 6-A and 6-B. British Antarctic Survey, Cambridge, UK, 54–57.

Smellie, J.L. and Martin, A.P. 2021. Erebus Volcanic Province: volcanology. *Geological Society, London, Memoirs*, **55**, https://doi.org/10.1144/M55-2018-62

Smellie, J.L. and Skilling, I.P. 1994. Products of subglacial eruptions under different ice thicknesses: two examples from Antarctica. *Sedimentary Geology*, **91**, 115–129, https://doi.org/10.1016/0037-0738(94)90125-2

Smellie, J.L., Pankhurst, R.J., Thomson, M.R.A. and Davies, R.E.S. 1984. *The Geology of the South Shetland Islands: VI. Stratigraphy, Geochemistry and Evolution*. British Antarctic Survey Scientific Reports, **87**.

Smellie, J.L., Pankhurst, R.J., Hole, M.J. and Thomson, J.W. 1988. Age, distribution and eruptive conditions of late Cenozoic alkaline volcanism in the Antarctic Peninsula and eastern Ellsworth Land: review. *British Antarctic Survey Bulletin*, **80**, 21–49.

Smellie, J.L., Hofstetter, A. and Troll, G. 1992. Fluorine and boron geochemistry of an ensialic marginal basin volcano: Deception Island, Bransfield Strait, Antarctica. *Journal of Volcanology and Geothermal Research*, **49**, 255–267, https://doi.org/10.1016/0377-0273(92)90017-8

Smellie, J.L., Hole, M.J. and Nell, P.A.R. 1993. Late Miocene valley-confined subglacial volcanism in northern Alexander Island, Antarctic Peninsula. *Bulletin of Volcanology*, **55**, 273–288, https://doi.org/10.1007/BF00624355

Smellie, J.L., Liesa, M., Muñoz, J.A., Sàbat, F., Pallàs, R. and Willan, R.C.R. 1995. Lithostratigraphy of volcanic and sedimentary sequences in central Livingston Island, South Shetland Islands. *Antarctic Science*, **7**, 99–113, https://doi.org/10.1017/S0954102095000137

Smellie, J.L., Pallàs, R., Sàbat, F. and Zheng, X. 1996. Age and correlation of volcanism in central Livingston Island, South Shetland Islands: K–Ar and geochemical constraints. *Journal of South American Earth Sciences*, **9**, 265–272, https://doi.org/10.1016/0895-9811(96)00012-0

Smellie, J.L., López-Martínez, J., Rey, J. and Serrano, E. 1997. Maps of Deception Island, South Shetland Islands. In: Ricci, C.A. (ed.) *The Antarctic Region: Geological Evolution and Processes*. Terra Antartica, Siena, Italy, 1195–1198.

Smellie, J.L., López-Martínez, J. et al. (eds). 2002. *Geology and Geomorphology of Deception Island, 1:25 000, with Supplementary Text*. BAS GEOMAP Series Sheets 6-A and 6-B. British Antarctic Survey, Cambridge, UK.

Smellie, J.L., McArthur, J.M., McIntosh, W.C. and Esser, R. 2006a. Late Neogene interglacial events in the James Ross Island region, northern Antarctic Peninsula, dated by Ar/Ar and Sr-isotope stratigraphy. *Palaeogeography, Palaeoclimatology, Palaeoecology*, **242**, 169–187, https://doi.org/10.1016/j.palaeo.2006.06.003

Smellie, J.L., McIntosh, W.C. and Esser, R. 2006b. Eruptive environment of volcanism on Brabant Island: Evidence for thin wet-based ice in northern Antarctic Peninsula during the Late Quaternary. *Palaeogeography, Palaeoclimatology, Palaeoecology*, **231**, 233–252, https://doi.org/10.1016/j.palaeo.2005.07.035

Smellie, J.L., McIntosh, W.C., Esser, R. and Fretwell, P. 2006c. The Cape Purvis volcano, Dundee Island (northern Antarctic Peninsula): late Pleistocene age, eruptive processes and implications for a glacial palaeoenvironment. *Antarctic Science*, **18**, 399–408, https://doi.org/10.1017/S0954102006000447

Smellie, J.L., Johnson, J.S., McIntosh, W.C., Esser, R., Gudmundsson, M.T., Hambrey, M.J. and van Wyk de Vries, B. 2008. Six million years of glacial history recorded in the James Ross Island Volcanic Group, Antarctic Peninsula. *Palaeogeography, Palaeoclimatology, Palaeoecology*, **260**, 122–148, https://doi.org/10.1016/j.palaeo.2007.08.011

Smellie, J.L., Haywood, A.M., Hillenbrand, C.-D., Lunt, D.J. and Valdes, P.J. 2009. Nature of the Antarctic Peninsula Ice Sheet during the Pliocene: geological evidence and modelling results compared. *Earth-Science Reviews*, **94**, 79–94, https://doi.org/10.1016/j.earscirev.2009.03.005

Smellie, J.L., Johnson, J.S. and Nelson, A.E. 2013. *Geological Map of James Ross Island. 1. James Ross Island Volcanic Group (1:125 000 Scale)*. BAS GEOMAP 2 Series Sheet 5. British Antarctic Survey, Cambridge, UK, http://nora.nerc.ac.uk/506743/1/BAS%20GEOMAP%202%2C%20sheet%205%20-%20Geological%20map%20of%20James%20Ross%20Island%20-%20I%20-%20James%20Ross%20Island%20volcanic%20group.pdf

Smellie, J.L., Walker, A.J., McGarvie, D.W. and Burgess, R. 2016. Complex circular subsidence structures in tephra deposited on large blocks of ice: Varða tuff cone, Öræfajökull, Iceland. *Bulletin of Volcanology*, **78**, 56, https://doi.org/10.1007/S00445-016-1048-X

Sohn, Y.K. 1996. Hydrovolcanic processes forming basaltic tuff rings and cones on Cheju Island, Korea. *Geological Society of America Bulletin*, **108**, 1199–1211, https://doi.org/10.1130/0016-7606(1996)108<1199:HPFBTR>2.3.CO;2

Sohn, Y.K. and Chough, S.K. 1992. The Ilchulbong tuff cone, Cheju Island, South Korea: depositional processes and evolution of an emergent, Surtseyan-type tuff cone. *Sedimentology*, **39**, 523–544, https://doi.org/10.1111/j.1365-3091.1992.tb02135.x

Solari, M.A., Hervé, F., Martinod, J., Le Roux, J.P., Ramírez, L.E. and Palacios, C. 2008. Geotectonic evolution of the Bransfield Basin, Antarctic Peninsula: insights from analogue models. *Antarctic Science*, **20**, 185–196, https://doi.org/10.1017/S095410200800093X

Somoza, L., Martínez-Frías, J., Smellie, J.L., Rey, J. and Maestro, A. 2004. Evidence for hydrothermal venting and sediment volcanism discharged after recent short-lived volcanic eruptions at Deception Island, Bransfield Strait, Antarctica. *Marine Geology*, **203**, 119–140, https://doi.org/10.1016/S0025-3227(03)00285-8

Strelin, J. and Malagnino, E.C. 1992. Geomorfologia de la Isla James Ross. *In*: Rinaldi, C.A. (ed.) *Geología de la Isla James Ross, Antartida*. Instituto Antartico Argentino, Buenos Aires, 7–36.

Sykes, M.A. 1988. New K–Ar age determinations on the James Ross Island Volcanic Group, north-east Graham Land Antarctica. *British Antarctic Survey Bulletin*, **80**, 51–56.

Sykes, M.A. 1989. *The Petrology and Tectonic Significance of the James Ross Island Volcanic Group, Antarctica*. PhD thesis, University of Nottingham, Nottingham, UK.

Troedson, A.L. and Riding, J.B. 2002. Upper Oligocene to lowermost Miocene strata of King George Island, South Shetland Islands, Antarctica: stratigraphy, facies analysis, and implications for the glacial history of the Antarctic Peninsula. *Journal of Sedimentary Research*, **72**, 510–523, https://doi.org/10.1306/110601720510

Troedson, A.L. and Smellie, J.L. 2002. The Polonez Cove Formation of King George Island, Antarctica: stratigraphy, facies and implications for mid-Cenozoic cryosphere development. *Sedimentology*, **49**, 277–301, https://doi.org/10.1046/j.1365-3091.2002.00441.x

Valencio, D.A., Mendía, J.E. and Vilas, J.F. 1979. Paleomagnetism and K–Ar age of Mesozoic and Cenozoic igneous rocks from Antarctica. *Earth and Planetary Science Letters*, **45**, 61–68, https://doi.org/10.1016/0012-821X(79)90107-9

Walker, G.P.L. 1984. Downsag calderas, ring faults, caldera sizes, and incremental caldera growth. *Journal of Geophysical Research*, **89**, 8407–8416, https://doi.org/10.1029/JB089iB10p08407

Weaver, S.D., Saunders, A.D., Pankhurst, R.J. and Tarney, J. 1979. A geochemical study of magmatism associated with the initial stages of back-arc spreading: the Quaternary volcanics of Bransfield Strait, from South Shetland Islands. *Contributions to Mineralogy and Petrology*, **68**, 151–169, https://doi.org/10.1007/BF00371897

Webb, P.-N. and Andreasen, J.E. 1986. Potassium/argon dating of volcanic material associated with the Pliocene Pecten Conglomerate (Cockburn Island) and Scallop Hill Formation (McMurdo Sound). *Antarctic Journal of the United States*, **21**, 59.

White, J.D.L. and Houghton, B.F. 2006. Primary volcaniclastic rocks. *Geology*, **34**, 677–680, https://doi.org/10.1130/G22346.1

Williams, M., Smellie, J.L., Johnson, J.S. and Blake, D.B. 2006. Late Miocene Asterozoans (Echinodermata) from the James Ross Island Volcanic Group. *Antarctic Science*, **18**, 117–122, https://doi.org/10.1017/S0954102006000113

Williams, M., Nelson, A.E. *et al.* 2010. Sea ice extent and seasonality for the Early Pliocene northern Weddell Sea. *Palaeogeography, Palaeoclimatology, Palaeoecology*, **292**, 306–318, https://doi.org/10.1016/j.palaeo.2010.04.003

Wohletz, K.H. 1986. Explosive magma–water interactions: thermodynamics, explosion mechanisms, and field studies. *Bulletin of Volcanology*, **48**, 245–264, https://doi.org/10.1007/BF01081754

Zandomeneghi, D., Barclay, A.H. *et al.* 2007. Three-dimensional P wave tomography of Deception Island Volcano, South Shetland Islands. *United States Geological Survey Open-File Report*, **2007-1047**, Extended Abstract 025.

Zinsmeister, W.J. and de Vries, T.J. 1983. Quaternary glacial marine deposits on Seymour Island. *Antarctic Journal of the United States*, **18**, 64–65.

Chapter 3.2b

Bransfield Strait and James Ross Island: petrology

Karsten M. Haase* and Christoph Beier

GeoZentrum Nordbayern, Friedrich-Alexander-Universität Erlangen-Nürnberg (FAU), Schlossgarten 5, 91054 Erlangen, Germany

KMH, 0000-0003-4768-5978; CB, 0000-0001-7014-7049

Present addresses: CB, Department of Geosciences and Geography, Research Programme of Geology and Geophysics (GeoHel), University of Helsinki, FIN-00014 Helsinki, Finland

*Correspondence: karsten.haase@fau.de

Abstract: Young volcanic centres of the Bransfield Strait and James Ross Island occur along back-arc extensional structures parallel to the South Shetland island arc. Back-arc extension was caused by slab rollback at the South Shetland Trench during the past 4 myr. The variability of lava compositions along the Bransfield Strait results from varying degrees of mantle depletion and input of a slab component. The mantle underneath the Bransfield Strait is heterogeneous on a scale of approximately tens of kilometres with portions in the mantle wedge not affected by slab fluids. Lavas from James Ross Island east of the Antarctic Peninsula differ in composition from those of the Bransfield Strait in that they are alkaline without evidence for a component from a subducted slab. Alkaline lavas from the volcanic centres east of the Antarctic Peninsula imply variably low degrees of partial melting in the presence of residual garnet, suggesting variable thinning of the lithosphere by extension. Magmas in the Bransfield Strait form by relatively high degrees of melting in the shallow mantle, whereas the magmas some 150 km further east form by low degrees of melting deeper in the mantle, reflecting the diversity of mantle geodynamic processes related to subduction along the South Shetland Trench.

Supplementary material: Analyses of lavas from Bransfield Strait volcanoes (Table S1), and from the James Ross Island Volcanic Group and Antarctic Peninsula post-subduction lavas (Table S2) are available at https://doi.org/10.6084/m9.figshare.c.5188032

Many subduction zones on Earth are associated with extensional regimes of the lithosphere in the upper plate that can have different age, structure and composition. Extensional stress by slab rollback, convection in the mantle wedge or magma formation may split relatively young and thin island arcs but also thick continental lithosphere (e.g. Sleep and Toksöz 1971; Uyeda and Kanamori 1979; Taylor and Karner 1983; Heuret and Lallemand 2005). The extensional features above a subducting plate range from relatively narrow rifts (<2 km) to wide basins (more than tens of kilometres), with newly formed basaltic crust generated at a back-arc spreading centre. Back-arc basins are often volcanically active and the volcanic rocks are typically tholeiitic but alkaline lavas also occur (Taylor and Martinez 2003; Lima et al. 2017). The compositional variability in back-arc basins may result from a combination of magmatic processes along the subducting slab and in the mantle wedge, including compositional heterogeneities both in space and time (e.g. Sinton et al. 2003; Beier et al. 2010). The magmas form largely by adiabatic melting due to extension of the lithosphere but effects of a hydrous slab component on the mantle solidus are frequently observed (Langmuir et al. 2006). The volcanic structures in back-arc basins also show a large variation ranging from central volcanoes along rift zones to volcanic ridges, as observed along spreading axes of the Lau Basin which are similar to those observed at fast-spreading mid-ocean ridges (Martinez and Taylor 2006). The incompatible element variation in back-arc magmas shows transitions from compositions like depleted mid-ocean ridge basalt (MORB) to compositions resembling island arc lavas, and these variations are typically related to the distance of the back-arc to the trench of the subduction zone (e.g. Saunders and Tarney 1979; Pearce and Stern 2006). The Bransfield Strait back-arc basin and the James Ross Island are the youngest volcanic structures of the Antarctic Peninsula, and both are related to subduction and extension of the lithosphere (Gonzalez-Ferran 1985; Garrett and Storey 1987; Hathway 2000; Galindo-Zaldivar et al. 2004). However, whereas the lavas in the Bransfield Strait show variations from MORB-like to arc-like, implying an influence from the subducting slab, the James Ross Island lavas are alkaline basalts typical for continental extensional regimes. The compositional variability of the Antarctic Peninsula lavas shows that extension of continental lithosphere due to subduction can yield different magma types depending on the direct chemical influence of the slab on the mantle and the lithospheric thickness. Here, we review the existing data on the petrology and geochemistry of the lavas from these two regions, and summarize and extend models for the generation of the magmas.

Bransfield Strait

Tectonic setting of the volcanic structures in the Bransfield Strait

The Bransfield Strait formed c. 4 myr ago by rifting of the continental lithosphere of the Antarctic Peninsula which has been the site of subduction of the Phoenix Plate throughout the Mesozoic (Barker 1982; McCarron and Larter 1998). The subduction caused abundant tholeiitic to calc-alkaline magmatic activity on the South Shetland Islands, with ages between c. 150 and 30 Ma (Pankhurst and Smellie 1983; Smellie et al. 1984; Nawrocki et al. 2010; Haase et al. 2012). Spreading of the Phoenix Ridge terminated c. 3.3 myr ago (Livermore et al. 2000). Earthquakes reaching depths of up to 65 km imply that the subduction is continuing at the South Shetland Trench (Robertson Maurice et al. 2003). The subduction zone is bounded in the south by the Hero Fracture Zone (Fig. 1), which also terminates the Bransfield Strait Basin in the south (Barker and Austin 1998). South of the Hero Fracture Zone subduction probably ended between 3 and 5 myr ago because a slab window formed after subduction of the Phoenix Ridge spreading axis (Larter and Barker 1991). North of the Hero Fracture Zone, slab rollback may have been the cause

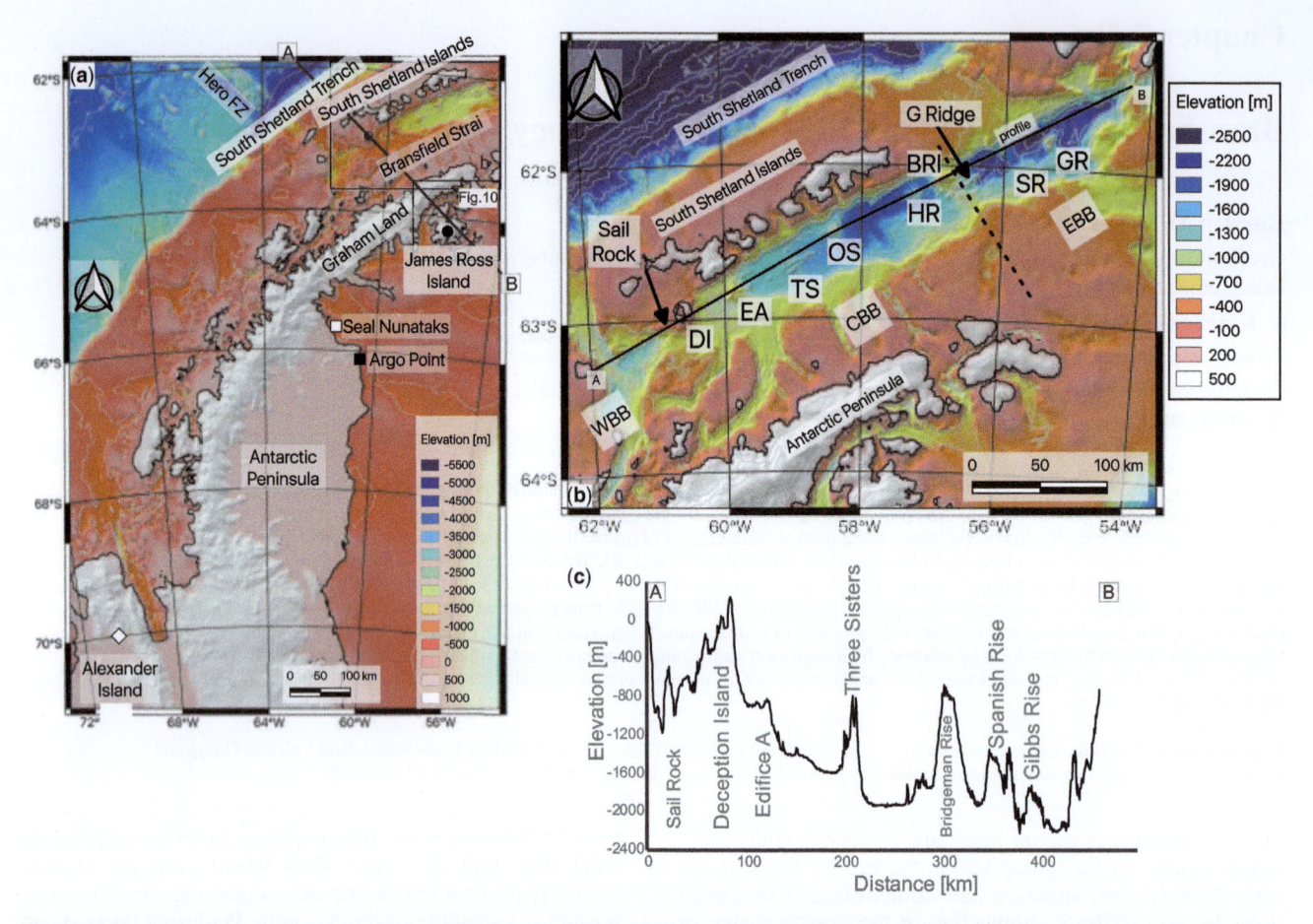

Fig. 1. (**a**) Map of the Antarctic Peninsula with the subduction zone at the South Shetland Trench and the Bransfield Strait back-arc basins, as well as the James Ross Island in the NW. (**b**) Map of the Bransfield Basin with the different volcanic centres along the axis and the distinction between the Eastern (EBB), Central (CBB) and Western Bransfield basins (WBB). The volcanic centres, from west to east, are Sail Rock, Deception Island (DI), Edifice A (EA), Three Sisters (TS), Orca Seamount (OS), Hook Ridge (HR), Bridgeman Rise and Island (BRI), G Ridge, Spanish Rise (SR) and Gibbs Rise (GR). (**c**) Bathymetric profile along the Bransfield Strait showing the different volcanic centres along the axis.

of back-arc extension and formation of the Bransfield Strait (Barker and Austin 1998). The ENE–WSW-striking and 500 km-long Bransfield Strait (Fig. 1) is divided into three basins called the Western, Central and Eastern Bransfield basins (WBB, CBB and EBB, respectively) (Gràcia *et al.* 1996). The CBB and EBB are volcanically active, whereas no evidence for volcanic structures has been found in the WBB. The WBB is relatively shallow and the crust shows features resulting from extensional faulting. Several volcanic structures with ages of 2.7–0.07 Ma occur along the eastern margin of the South Shetland Islands, such as Melville Peak on King George Island and Penguin Island (Pańczyk and Nawrocki 2011).

The crustal thickness decreases from *c.* 25 km beneath the South Shetland Islands to 10–15 km in the CBB (Christeson *et al.* 2003), and gravimetric models suggest the crust thins to 8.5 km along the central volcanic axis (Catalan *et al.* 2013). Back-arc rift propagation from NE to SW along the axis of the Bransfield Strait was suggested on the basis of structural observations in the crust (Barker and Austin 1998; Christeson *et al.* 2003). In contrast, the morphostructural study by Gràcia *et al.* (1996) concluded that the CBB is in the stage of incipient spreading, whereas the EBB is in an earlier, less mature, rifting stage. Magnetic anomalies in the axis of the CBB indicate spreading of basaltic crust and the onset of spreading (Catalan *et al.* 2013; Schreider *et al.* 2015). The opening velocity of the CBB was determined by GPS to be 7 to 9 mm a^{-1}, whereas rifting in the Western Bransfield Basin is slower at 2–3 mm a^{-1} (Dietrich *et al.* 2004; Taylor *et al.* 2008). Seven volcanic structures occur along the CBB from west to east (in Fig. 1b): Sail Rock, Deception Island (DI in Fig. 1b), Edifice A (EA), Three Sisters (TS), Orca Seamount (OS), Hook Ridge (HR), Bridgeman Rise including Bridgeman Island (BRI) and G Ridge (GRI). These volcanic structures represent 10–30 km-long segments that may be fed by magmas from ascending mantle diapirs similar to magmatic segments observed in the northern Red Sea (Barker and Austin 1998). Hook Ridge and Three Sisters are hydrothermally active (Petersen *et al.* 2004). The three volcanic structures of the EBB were named, from west to east, G Ridge, Spanish Rise and Gibbs Rise (Fig. 1).

The submarine structures were sampled during cruises NBP93-1, NBP95-7, JR04, ANT IV and SO155, with the American RVIB *Nathaniel Parker*, the British RV *James Clark Ross*, and the German RV *Polarstern* and RV *Sonne*, respectively. Subaerial volcanoes of Deception Island and Bridgeman Island were studied during several field trips by the British Antarctic Survey, and Polish, Spanish, German and Chilean scientists.

Composition of the magmatic rocks of the Bransfield Strait

Petrology of the volcanic rocks. Lavas from Deception Island range from olivine-bearing basalts via andesites to dacites, and the most abundant mineral phase is plagioclase. The Deception Island lavas are moderately to highly porphyritic and only some of the mafic lavas are aphyric. Olivine and augite are common in the basaltic rocks, whereas plagioclase, orthopyroxene and augite dominate in the andesites and dacites

(Weaver et al. 1979; Martí et al. 2013). In Deception Island lavas, olivine with forsterite (Fo) contents ranging from 36 to 85 occurs in basalts, andesites and dacites (Weaver et al. 1979; Martí et al. 2013). Plagioclase compositions with anorthite (An) contents ranging from 23 to 43 were determined in the Deception Island lavas, and glass compositions range from basaltic to andesitic (Martí et al. 2013). Clinopyroxene–glass barometry suggests crystallization pressures of 150–200 MPa, which corresponds to a depth of 4–6 km (Martí et al. 2013). Most lavas contain augite, whereas orthopyroxene is rare and amphibole is only described from some tonalitic–gabbroic xenoliths on Deception Island (Risso and Aparicio 2002; Martí et al. 2013). Bridgeman Island volcanic rocks are basaltic–andesitic in composition, with plagioclase and augite phenocrysts and typically rare, corroded olivine crystals. Olivine-phyric basalts dominate Penguin Island lavas together with clinopyroxene, whereas plagioclase occurs only in the groundmass (Weaver et al. 1979).

The magmatic rocks recovered in the Bransfield Strait by dredging range from coarse-grained plutonic rocks, such as gabbros and diorites, to volcanic rocks often with glass rims and pillow structures (Keller et al. 2002; Fretzdorff et al. 2004). Because of the proximity to the young magmatic arc of the South Shetland Islands and the young volcanic centres on the Antarctic Peninsula, the rocks have to be selected carefully to rule out ice-drifted material. Fresh and glass-rich lavas were recovered from Hook Ridge and Bridgeman Rise (Fretzdorff et al. 2004). The lavas from the three volcanic structures of the EBB range from fresh to moderately altered rocks. Dolerites were recovered from fault scarps of Gibbs Rise, indicating that deeper parts of the crust may be exposed. Most lavas from the Bransfield Basin are sparsely (<10%) porphyritic, with olivine and plagioclase being the main phenocryst phases. Plagioclase is abundant in all lavas from the Bransfield Strait.

Major element composition of the volcanic rocks and evidence for fractional crystallization. Lavas from the Bransfield Strait volcanic centres are generally tholeiitic (Fig. 2) with relatively low K_2O contents, and range from basalts to rhyolites (Keller et al. 2002; Fretzdorff et al. 2004) (Table 1; see also the full dataset in Supplementary material Table S1). Some lavas from the EBB are alkaline basalts with relatively low SiO_2, resembling those from the young volcanism on the South Shetland Islands (Fig. 3). The samples from most volcanic centres lie along relatively tight compositional trends and, generally, the variation observed within one structure mostly reflects shallow crystal fractionation processes (Weaver et al. 1979; Keller et al. 1991). The lavas from most volcanic structures of the CBB are basaltic–andesitic showing increasing contents of SiO_2, TiO_2, FeO^T and K_2O, with decreasing MgO from 9 to 4 wt% (Fig. 3). Lavas with MgO lower than 4 wt% show decreasing TiO_2 and FeO^T, implying the onset of fractionation of magnetite and ilmenite. In contrast, the lavas from the EBB appear to follow different trends with lower TiO_2 and FeO^T (Fig. 3) reflecting earlier crystallization and fractionation of Fe–Ti oxides than in the CBB magmas (Fisk 1990). Both CaO and Al_2O_3 increase from 9 to 7 wt% MgO and then decrease (Fig. 3), indicating that clinopyroxene and plagioclase begin to crystallize and fractionate in magmas with less than 7 wt% MgO. Lavas with MgO contents >7 wt% appear to lie on an olivine control line. Rhyolitic glasses also have the lowest P_2O_5 contents, indicating fractionation of apatite (Keller et al. 2002). The most evolved lavas appear to occur at the largest volcanic structures of Deception Island and Bridgeman Rise (Fig. 1) suggesting that these volcanoes have stable crustal magma reservoirs, which is supported by geophysical evidence (Ben-Zvi et al. 2009). The gabbroic xenoliths observed on Deception Island are interpreted as cumulates from the magma reservoir beneath the volcano (Risso and Aparicio 2002). Weaver et al. (1979) suggested up to 80% crystallization from the basaltic liquids to form the dacitic melts. The occurrence of tonalitic and quartz-dioritic plutonic xenoliths in the Deception Island magmas (Risso and Aparicio 2002) indicates the presence of evolved intrusive rocks beneath the volcano.

Incompatible element and isotope composition: mantle heterogeneity and slab contribution

In terms of incompatible element composition, the Bransfield Strait lavas show large variations that reflect different magma sources and evolutionary paths (Keller et al. 1991, 2002; Fretzdorff et al. 2004). Lavas from the CBB are relatively homogenous, and show increasing Zr and Ba contents with decreasing MgO (Fig. 4). In contrast, the EBB lavas are variable, and some are enriched in Zr and Ba compared to the CBB rocks, whereas others have relatively low and constant Zr contents with decreasing MgO. This trend of low Zr contents is particularly well defined by the Bridgeman Rise lavas (Fig. 4). The normal MORB (N-MORB)-normalized incompatible element patterns of most lavas from the CBB and EBB show a negative anomaly of Nb and Ta, and positive anomalies for elements like Sr, Pb, K, U and Ba (Fig. 5). Thus, they are typical melts forming above a subducting slab with enrichment of fluid-mobile elements relative to immobile elements (e.g. McCulloch and Gamble 1991). The Nb and Ta contents lower than MORB in some lavas from the CBB and EBB (Fig. 5) may indicate that the mantle beneath some parts of the Bransfield Basin is more depleted than the average upper mantle. Most of the lavas have lower heavy rare earth element (REE) contents than N-MORB which could indicate either extreme depletion of the mantle or residual garnet during partial melting. Some CBB and EBB samples have distinct negative Ti anomalies (Fig. 5) suggesting the fractionation of Ti-rich minerals, most probably oxides as ilmenite or Ti-magnetite. The enrichment of Ba in

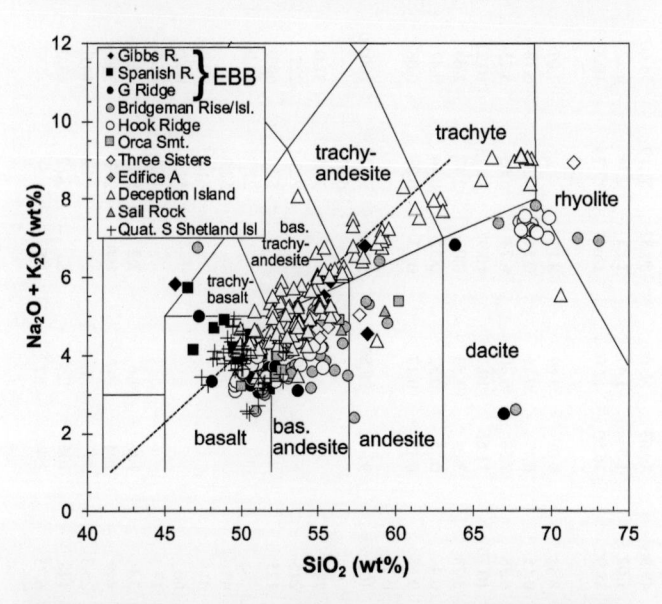

Fig. 2. Total alkalis v. SiO_2 (total alkali–silica (TAS)) diagram after Le Bas et al. (1986) including the division line between alkaline and tholeiitic lavas for Hawaii (Macdonald and Katsura 1964) for volatile-corrected compositions of the Bransfield Strait rocks. Note that the samples consist of both whole rocks and volcanic glasses (Baker et al. 1975; Weaver et al. 1979; Fisk 1990; Keller et al. 1991, 2002; Fretzdorff et al. 2004; Kraus et al. 2013; Martí et al. 2013).

Table 1. *Representative analyses of lavas from Bransfield Strait and Antarctic Peninsula region*

Sample	SR-01	DI-33	D2G	D144.4	D8F	02DR-01wr	BI-05	14DR-01	18DR-01	23DR-12wr	26DR-01	14DR-06	04DR-13	JR1	PAI-02A	CP-05	R.3728.1/2	KG.3612,5	KG.3620,7
Cruise			NBP93-1	JCRoss	NBP93-1	SO155		SO155	SO155	SO155	SO155	SO155	SO155						
Location	Sail Rock	Deception Island	Orca Smt.	Three Sisters, Bransfield St.	Hook Ridge	Bridgeman Rise	Bridgeman Isl.	G Ridge	Spanish Rise	Gibbs Rise	Spanish Rise	G Ridge	Bridgeman Rise	James Ross Island	Paulet Isl.	Cape Purvis	Antarctic Peninsula	Alexander Island	Alexander Island
Rock type		Syn-caldera (Outer Coast Tuff Formation)				Vesicular lava		Basalt			Alkali basalt	Alkali basalt	Alkali basalt	Basalt pillow lava			Seal Nunataks	Hornpipe Heights	Mussorgsky Peaks
Latitude (°S)	63.03					62.12	62.07	62.07	61.88	61.77	61.93	62.07	62.06	63.792	63.5797	63.5833	65.125		
Longitude (°W)	60.95					57.02	56.73	56.40	55.98	55.42	55.88	56.40	56.97	57.810	55.7886	55.9667	60		
Water depth						1377		1330.00	1542.00	2600.00	1902.00	1330.00	771.00						
Reference	1	2	3	3	3	4	1	4	4	4	4	4	4	6	1	1	4	2.5	2.5
																	7	5	5
SiO$_2$	56.53	49.96	49.63	50.06	50.52	51.19	54.60	47.88	51.14	53.46	45.89	47.22	46.26	46.76	46.98	48.22	51.5	45.07	50.67
TiO$_2$	1.07	1.62	1.33	1.47	1.23	0.71	0.72	1.21	0.60	0.78	2.15	1.94	2.30	1.61	2.320	1.804	1.74	3.26	1.86
Al$_2$O$_3$	18.75	18.92	16.50	17.08	16.45	17.74	18.33	17.29	15.44	18.95	15.36	15.00	16.22	14.70	16.62	16.31	14.82	14.49	15.16
Fe$_2$O$_3$						7.57		9.44	7.86	7.04	11.82	12.22	10.32	2.11				1.37	1.24
FeO	6.46	8.42	9.01	8.69	8.15		7.07							8.75	9.73	10.81	10.15	9.19	8.29
MnO	0.12	0.15	0.10	0.16	0.14	0.12	0.13	0.14	0.13	0.11	0.16	0.17	0.14	0.19	0.157	0.172	0.14	0.15	0.14
MgO	3.81	6.38	7.58	6.40	6.73	6.65	5.75	7.78	10.71	2.05	8.92	9.10	7.28	9.93	7.81	9.04	8.32	10.36	7.94
CaO	8.40	10.70	11.66	10.71	10.98	12.13	10.21	12.36	9.46	8.39	8.02	8.74	7.97	9.03	9.78	8.45	8.61	9.30	8.56
Na$_2$O	3.94	3.73	3.03	3.83	3.01	2.47	2.64	2.85	2.78	3.02	4.14	3.74	4.61	2.93	4.11	3.31	3.34	3.89	3.31
K$_2$O	0.70	0.44	0.15	0.32	0.32	0.44	0.47	0.44	0.41	2.04	1.52	1.21	2.00	0.57	1.42	1.27	0.76	2.44	1.53
P$_2$O$_5$	0.22	0.28	0.13	0.17	0.13	0.08	0.08	0.11	0.08	0.24	0.66	0.44	0.93	0.31	1.07	0.59	0.24	0.75	0.38
LOI		0.13																	
Total (wt%)	100.00	100.60	99.12	98.89	97.66	99.33	100.00	99.50	98.63	98.07	99.24	99.78	100.36	99.77	100.00	100.00	99.62	99.87	99.05
Li						5.67		4.98	4.75	11.0	7.26	7.173	6.317	5.9					
Sc	20	28.9				31.8	31.0	32.7	29.4	18.0	19.4	21.125	18.077	25.5	26	26	21.0		
V	164	215				223	227	290	198	201	169	188	166	175	216	189			
Cr	50	125				59.9	60.0	17.8	386	4.7	228	278	171	592	220	250	346	223	294
Co	20					27.1	27.0	33.1	41.0	11.1	46.0	48.9	35.8	47.7	40	43	46.8		
Ni	30	43				52.3		21.4	250		179	171	106	131		90	132	195	108
Cu	20	41				61.7	70.0	52.0	59.9	132.1	46.6	42.9	37.3	45.1	30	50			
Zn	70	66				51.7	80.0	63.1	57.4	58.9	84.5	92.9	74.3		140	130			
Ga	19	17				16.8	18.0			19.1					20	18			
Rb	33	3.1	1.6	4.6	7.2	8.7	11.0	8.5	9.1	54.0	14.4	15.2	16.0	5.1	15	15	11	25	19
Sr	556	449	204	297	190	273	320	275	210	444	758	539	1112	412	975	616	349	838	475
Y	22	25.3	30	31	25	14.8	14.0	19.8	13.0	17.7	22.2	21.5	21.8	24.4	26	24	21	25	22
Zr	164	148	106	124	82	50.2	63.0	76.5	54.2	122.8	230	174	246	145	216	177	118	238	173
Nb	6	6.75	2.4	3.7	1.5	1.5		1.2	0.9	2.8	36.6	31.4	57.7	20.4	44	27	17	68	24

Mo																
Sn																
Sb			0.09	0.24	0.38					0.22						
Cs	0.13					0.2		0.3	0.4							
						0.6		0.7	0.4							
								0.0	0.0							
Ba	133	58	20	71	57	90.1	105.0	54.8	80.2	1.1	3.56	2.48	3.27				
La	13.4	9.46	4.02	5.81	3.92	4.2	6.0	3.7	3.6	1.5	1.80	1.62	1.85				
Ce	30	23.5	12.2	16.1	11.2	10.3	13.6	10.4	8.6	3.5	0.07	0.05	0.08				
Pr	4.02	3.33	2.22	2.51	1.91	1.4	1.9	1.7	1.3	475	0.26	0.28	0.31	96	215	168	66	298	172
Nd	15.9	15.2	10.8	12.8	9.23	6.6	8.4	8.4	6.1	18.3	161	152	304	19.3	39.2	24.1	12.4	40.0	23.9
Sm	3.7	4.14	3.44	3.75	2.9	1.9	2.2	2.7	1.8	40.9	28.7	22.2	44.1	41.5	76.1	48.1	26.5	72.4	40.7
Eu	1.3	1.53	1.3	1.35	1.07	0.7	0.8	1.0	0.7	5.0	58.4	45.2	81.8	5.09	8.41	5.33			
Gd	4	4.54	3.92	4.51	3.35	2.2	2.6	3.3	2.2	20.2	6.94	5.45	9.18	21	30.6	20.4	15.2	35.8	19.8
Tb	0.7	0.8	0.74	0.82	0.61	0.4	0.4	0.6	0.4	4.3	27.9	22.2	34.6	4.85	6.7	4.6	3.92	5.6	5.0
Dy	4.2	4.92	4.67	5.17	3.93	2.6	2.6	3.9	2.5	1.1	5.88	5.00	6.89	1.71	2.37	1.62	1.49	2.4	1.2
Ho	0.9	1.03	1.02	1.11	0.86	0.5	0.6	0.8	0.5	3.7	1.95	1.68	2.19	3.77	6.3	4.5			
Er	2.6	2.74	2.9	3.11	2.53	1.5	1.6	2.3	1.5	0.5	5.70	5.06	6.04	0.64	1	0.7	0.72	1.0	0.8
Tm	0.38	0.39	0.47	0.46	0.39	0.2	0.2	0.3	0.2	3.2	0.85	0.79	0.86	4.73	5.4	4.3			
Yb	2.3	2.35	2.71	2.93	2.39	1.5	1.5	2.2	1.5	0.6	4.86	4.66	4.76	0.98	1.1	0.9			
Lu	0.34	0.38	0.4	0.46	0.34	0.2	0.2	0.3	0.2	1.8	0.92	0.88	0.87	2.88	3	2.5			1.7
Hf	4.3	3.27	2.6	3.1	2.2	1.7	2.0	2.2	1.7	0.3	2.36	2.34	2.24	0.39	0.41	0.35		1.7	0.3
Ta	0.4	0.49	0.19	0.26	0.11	0.1		0.1	0.1	1.8	0.33	0.32	0.30	2.58	2.4	2	1.4	0.3	4.2
W						0.1				0.3	2.04	1.99	1.89	0.38	0.34	0.24	0.26	5.5	1.5
Tl						0.1	0.5	0.1	0.0	3.8	0.30	0.30	0.27	3.13	4.9	4.1	2.98	4.6	
Pb	105	1.91	0.95	2.99	2.27	3.3	5.0	2.0	1.6	0.2	4.87	3.94	5.04	1.06	2.5	1.7	1.26		3.2
Th	2.9	0.75	0.22	0.52	0.58	1.0	1.6	0.7	0.8	0.7	2.26	1.97	3.05	2.47	3.2	2.4	1.82	4.3	
U	0.5	0.26	0.13	0.21	0.41	0.3	0.5	0.2	0.3	0.2	0.044	0.04	0.04	1.81	1.2	0.9			
										14.1	2.304	2.11	3.33	0.62					
										9.0	2.831	2.74	4.74						
										2.4	1.170	0.99	1.74						
$^{87}Sr/^{86}Sr$	0.703965		0.70274	0.703337	0.70326	0.703568		0.703290	0.703507					0.703387	0.703241	0.703340	0.703000		
$^{143}Nd/^{144}Nd$	0.512885		0.51307	0.51306	0.51302	0.512948			0.512944					0.512884	0.512932	0.512888	0.512924		
$^{206}Pb/^{204}Pb$	18.759		18.763	18.723	18.744	18.74	18.733	18.74	18.71					18.855	18.766	18.867	18.976		
$^{207}Pb/^{204}Pb$	15.651		15.564	15.582	15.587	15.58	15.618	15.58	15.58					15.601	15.580	15.620	15.617		
$^{208}Pb/^{204}Pb$	38.671		38.332	38.431	38.473	38.45	38.573	38.45	38.39					38.556	38.420	38.636	38.673		

References: 1, Kraus et al. (2013); 2, Martí et al. (2013); 3, Keller et al. (2002); 4, Fretzdorff et al. (2004); 5, Hole (1988); 6, Kosler et al. (2009); 7, Hole et al. (1993).
LOI, loss on ignition. Major element oxides in wt% and trace elements in ppm.

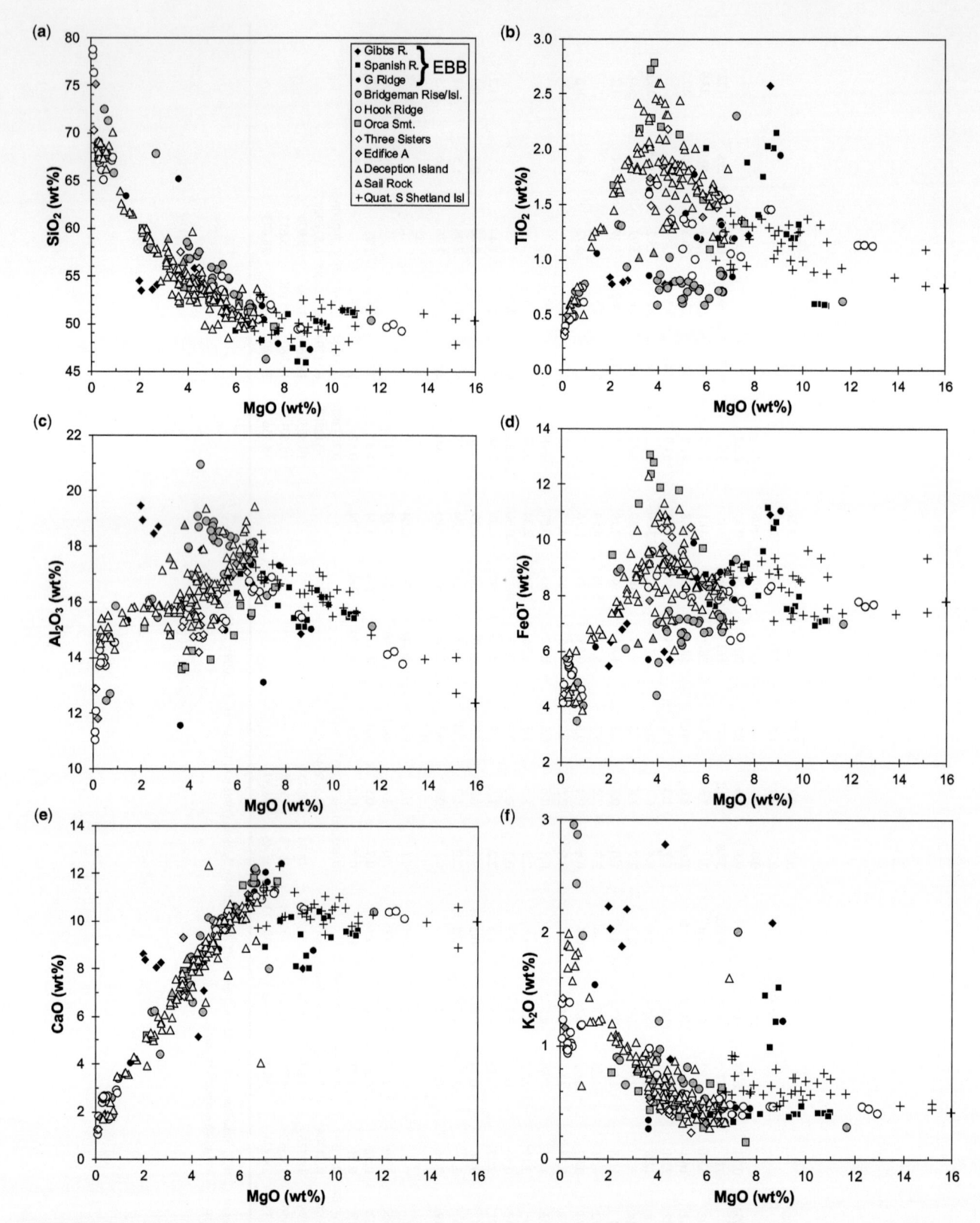

Fig. 3. Variation of the major element composition (**a**) SiO_2, (**b**) TiO_2, (**c**) Al_2O_3, (**d**) FeO^T, (**e**) CaO and (**f**) K_2O v. MgO for the different volcanic centres of the Bransfield Strait and the lavas from the young volcanic centres on the South Shetland Islands. Data sources as in Figure 2.

some EBB rocks resembles that of the alkaline lavas from the South Shetland Islands, implying a similar magma source. Most lavas are enriched in the light REEs (LREEs) with $(Ce/Yb)_N$ (=chondrite-normalized Ce/Yb) values >1 and thus higher than MORB (Fig. 6). The young alkaline rocks from the South Shetland Islands, like Penguin Island, have the highest $(Ce/Yb)_N$, whereas the lavas from the centre of the Bransfield Strait show less fractionation of the LREEs from the heavy REEs (HREEs). Modelling of the REEs of Penguin Island basalt suggests relatively low degrees of partial melting of garnet peridotite, whereas basalts from Deception and Bridgeman Islands formed by 5–20% melting of spinel peridotite (Weaver *et al.* 1979; Keller *et al.* 1991). The REE contents generally increase with decreasing MgO but $(Ce/Yb)_N$ remains constant as observed, for example, in the Deception Island lavas (Fig. 6). The dacitic–rhyolitic rocks show

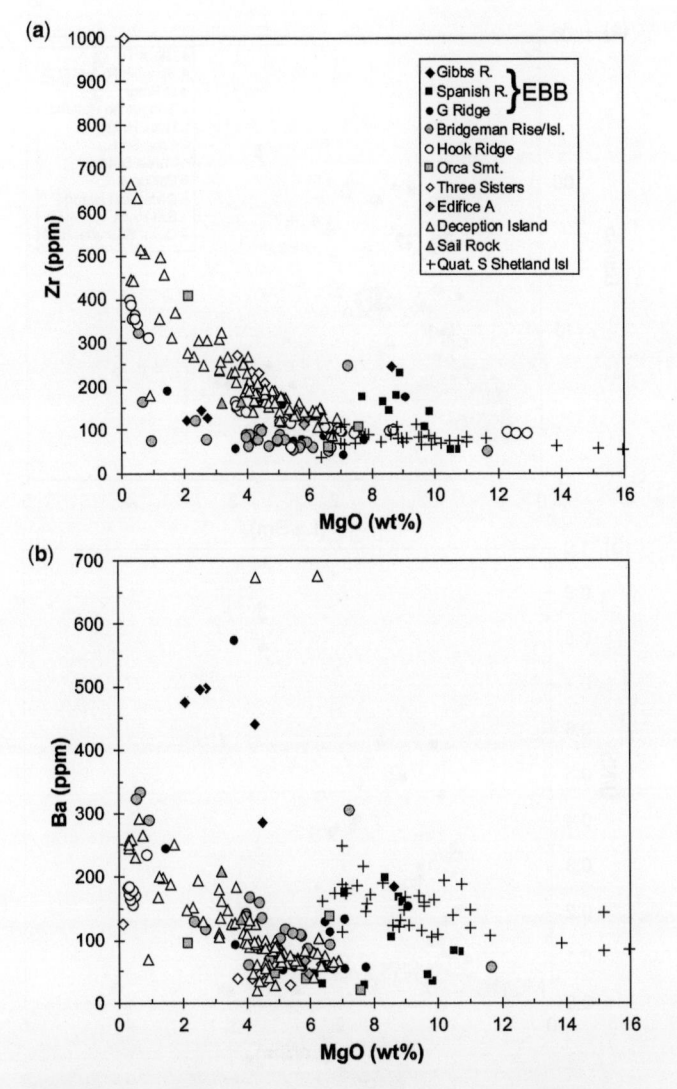

Fig. 4. The concentrations of (**a**) Zr v. MgO and (**b**) Ba v. MgO for the lavas from the Bransfield Strait and the young volcanic centres of the South Shetland islands. Data sources as in Figure 2.

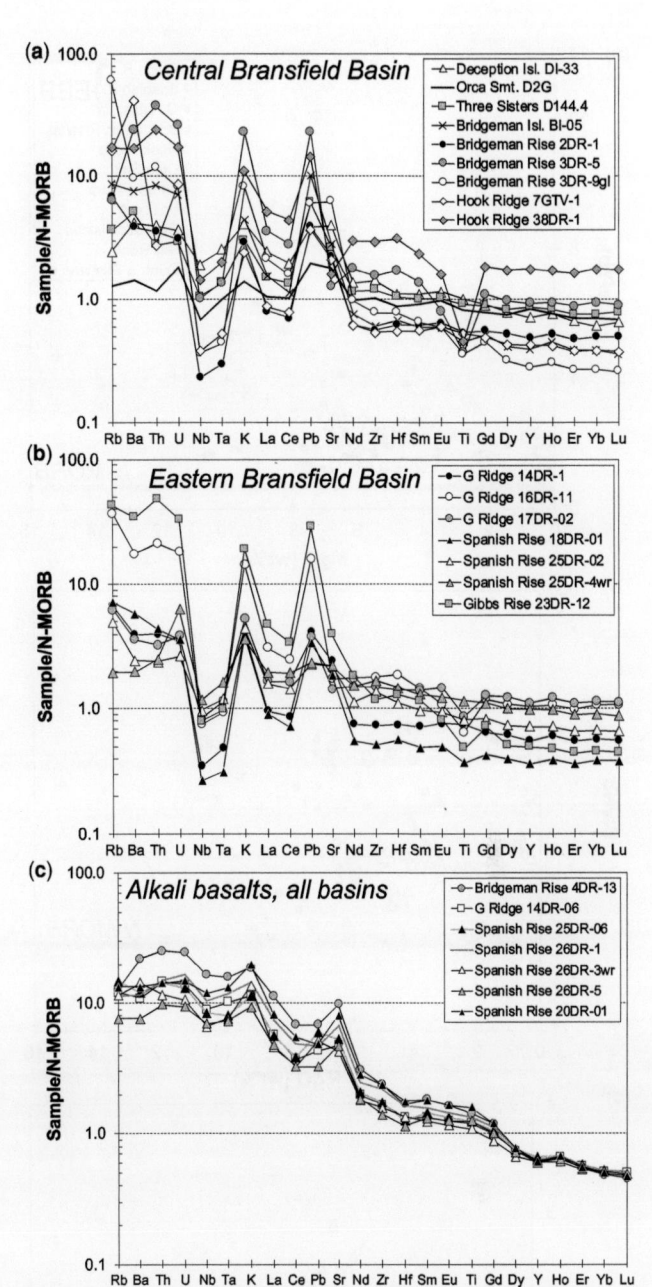

Fig. 5. N-MORB-normalized incompatible element patterns for (**a**) representative lavas from the Central Bransfield Basin (CBB), (**b**) the Eastern Bransfield Basin (EBB) and (**c**) alkaline basalts from Bridgeman Island, G Ridge and Spanish Rise. Data sources as in Figure 2, and N-MORB is from Hofmann (1988).

negative Eu anomalies, reflecting significant plagioclase fractionation (Weaver *et al.* 1979). The Spanish Rise lavas are highly variable in incompatible element ratios and vary from MORB-like to highly enriched in LREEs, Ba and Nb relative to Yb and Nb, respectively (Fig. 7). Several basalts from Spanish Rise in the EBB were found to be depleted in LREEs and to be only slightly enriched in fluid-mobile elements with a Ce/Pb of 20 (Fretzdorff *et al.* 2004). Additionally, several samples recovered from the EBB formed from LREE-enriched mantle but low U/Nb and Ba/Nb (Fig. 7). These enriched basalts resemble those from the young alkaline volcanic centres on the South Shetland Islands, implying that the mantle wedge beneath the Bransfield Strait is heterogeneous on a small scale with enriched portions in typical depleted MORB mantle.

Most Bransfield Strait lavas have higher Ba/Nb and U/Nb than MORB (Fig. 7), indicating the addition of a slab component with high concentrations of the fluid-mobile elements Ba and U. The slab component variably affects the mantle beneath the Bransfield Strait and, for example, Deception Island magmas show much less slab influence than the lavas from the EBB, Bridgeman Rise and Hook Ridge. Keller *et al.* (2002) noticed that the Bransfield lava compositions ranged from close to MORB with high Ce/Pb and low Ba/Nb to those resembling the lavas from the South Shetland island arc with low Ce/Pb and high Ba/Nb. The component with the enrichment of fluid-mobile elements relative to immobile elements also has high $(La/Sm)_N$, implying that the slab component also mobilized the LREEs. The slab enrichment is most clearly observed in the lavas from Gibbs Rise in the easternmost EBB which have the highest Ba/Nb and U/Nb, as well as high $(La/Sm)_N$ (Fig. 7). Because the Gibbs Rise lavas were apparently the oldest and most altered rocks recovered along the Bransfield Strait, Fretzdorff *et al.* (2004) speculated that this structure represents an older volcano related to the South Shetland island arc that was rifted by the opening of the back-arc basin. In contrast, the young lavas of the Spanish Rise further west show little influence of the slab component but have both relatively depleted and enriched mantle sources indicating heterogeneity of the mantle beneath this structure.

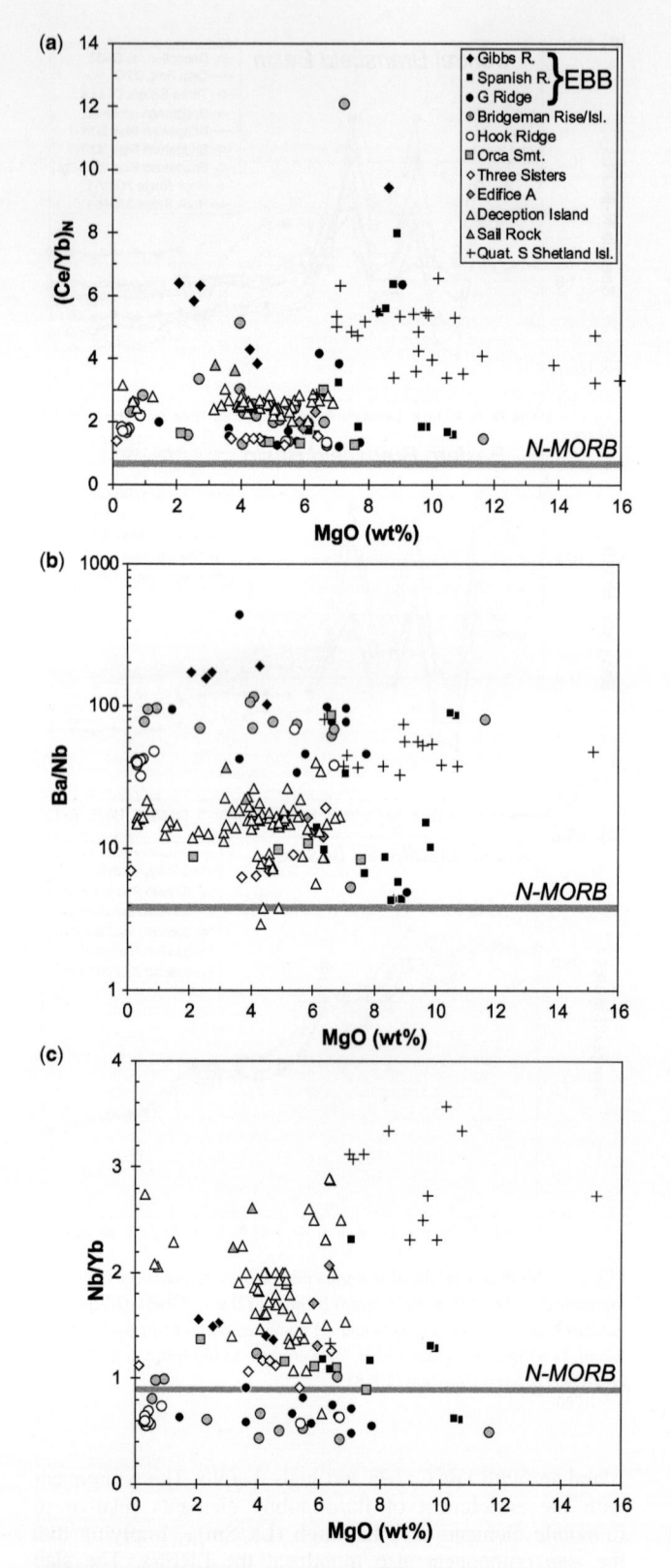

Fig. 6. The variation of incompatible element ratios (a) (Ce/Yb)$_N$ v. MgO, (b) Ba/Nb v. MgO and (c) Nb/Yb v. MgO for the Bransfield Strait lavas and those of the young volcanic centres of the South Shetland islands. Data sources as in Figure 2.

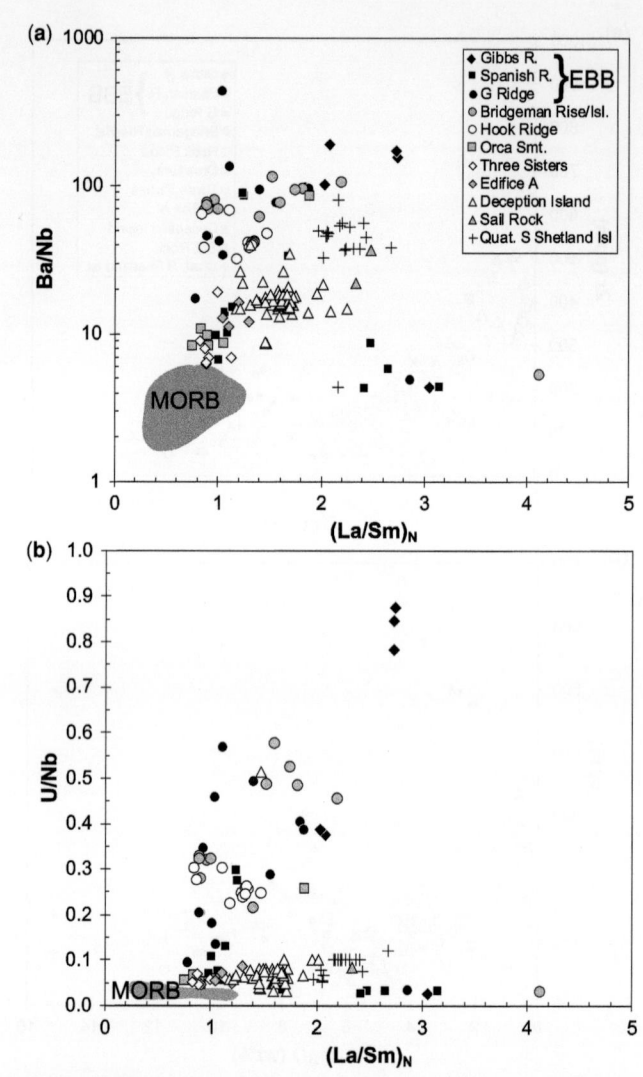

Fig. 7. The variation of (a) Ba/Nb v. (La/Yb)$_N$ and (b) U/Nb v. (La/Sm)$_N$ for the lavas from the Bransfield Strait lavas and those of the young volcanic centres of the South Shetland island. Data sources as in Figure 2, and Pacific MORB data from Bach *et al.* (1994), Wendt *et al.* (1999) and Haase (2002).

The radiogenic isotope ratios support mixing between the upper mantle and up to 3% of a recycled component from the slab (Keller *et al.* 1991; Fretzdorff *et al.* 2004). The component with high $^{87}Sr/^{86}Sr$ and low $^{143}Nd/^{144}Nd$ (Fig. 8) is most likely to represent clastic sedimentary material because it also has higher $^{207}Pb/^{204}Pb$ than MORB (Fretzdorff *et al.* 2004). These authors proposed that the Ce/Pb and Pb isotope systematics imply a slab component with $^{206}Pb/^{204}Pb$ of *c.* 18.8, $^{207}Pb/^{204}Pb$ of *c.* 15.6, and $^{208}Pb/^{204}Pb$ of *c.* 38.5 similar to South Atlantic sediments (Ben Othman *et al.* 1989). In contrast, Keller *et al.* (2002) suggested that the slab component may represent metalliferous sediments because their mixing model requires very high Pb contents commonly not observed in sediments. The high Ba/Nb of >200 probably indicates a hydrous fluid from subducting sediments because these elements would be less fractionated by the partial melting of sediments (Fretzdorff *et al.* 2004). However, this fluid also affects the Nd isotopes and leads to an enrichment in LREEs, as pointed out above, which may indicate that these fluids are critical fluids (Kessel *et al.* 2005).

Whereas most Bransfield Strait lavas indicate a slab component in their source, the Spanish Rise lavas show an enrichment of the LREEs but have low U/Nb and also have high Sr and low Nd isotope ratios (Fig. 7). These samples resemble the alkaline lavas from the South Shetland Islands (Fig. 7), which was previously pointed out by Fretzdorff *et al.* (2004). The high $^{206}Pb/^{204}Pb$ of >18.8 (Fig. 8) indicates a different mantle component with relatively radiogenic Pb isotopes that is comparable to that observed in the alkaline

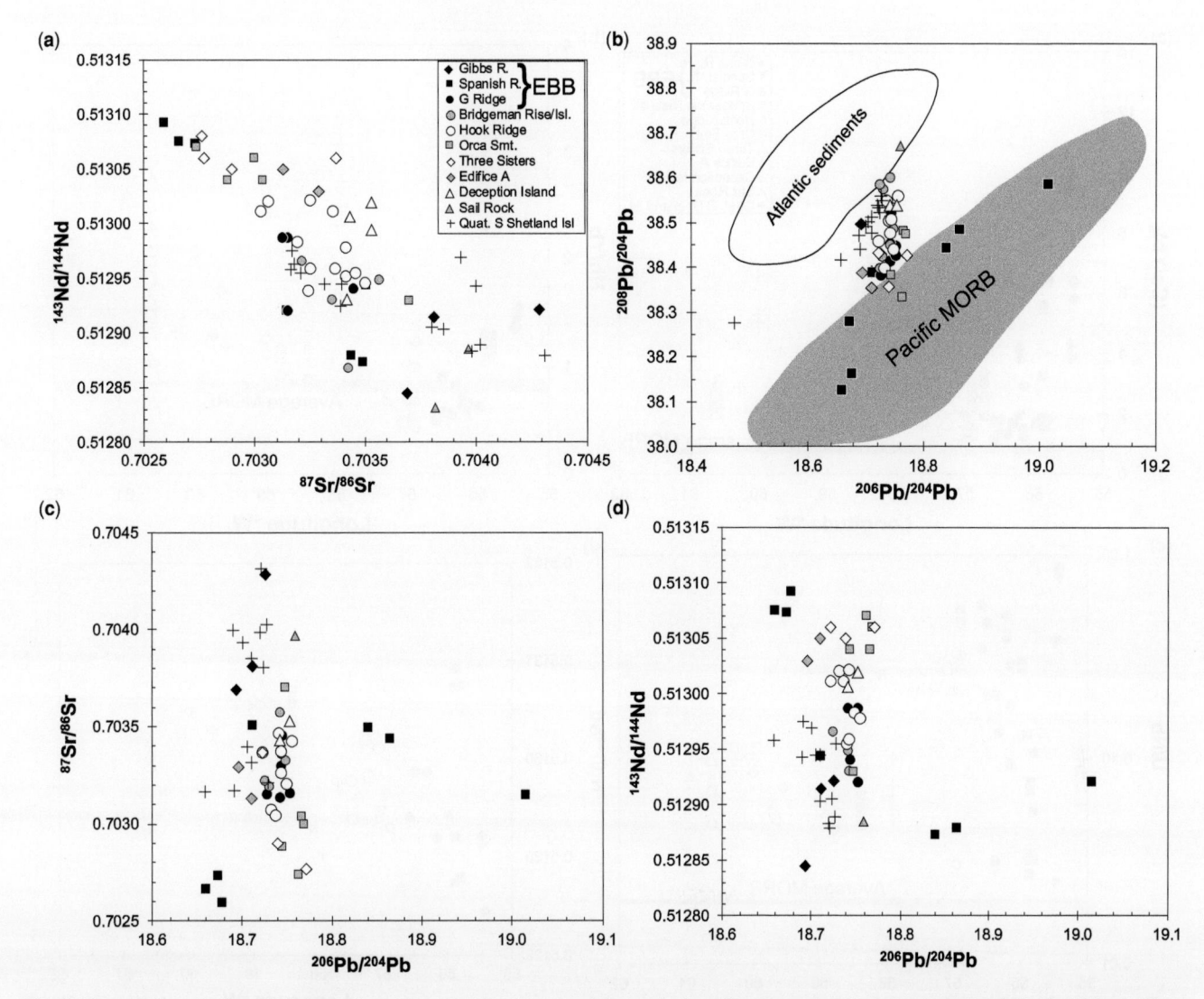

Fig. 8. The variation of radiogenic isotopes for the Bransfield Strait lavas and the basalts from the young volcanic centres on the South Shetland Islands. Data sources as in Figure 6, and Atlantic sediments from Ben Othman et al. (1989) and Hyeong et al. (2011).

basalts of the Antarctic Peninsula (Fretzdorff et al. 2004). The formation of the Antarctic alkali basaltic melts may be related to the opening of slab windows and the ascent of enriched asthenospheric mantle (Hole et al. 1995; Eagles et al. 2009), and such a process could potentially also occur beneath the Bransfield Strait and South Shetland island arc (Fig. 1). The alkali basaltic magmas show distinctly high $(Ce/Yb)_N$, implying low degrees of partial melting of garnet peridotite: that is, deeper than the typical Bransfield Strait basalts (Hole et al. 1995). However, Spanish Rise is too far north from the suggested window in the subducting slab south of the Hero Fracture Zone (Barker and Austin 1998) and the fresh alkali basalts suggest a young formation age, so that the alkaline magmas may have formed by asthenospheric mantle flow around the subducting slab (Fretzdorff et al. 2004) rather than from fluid addition and dehydration melting.

Geographical variation in the mantle sources and magma formation

The incompatible element and radiogenic isotope compositions of the lavas from the EBB and CBB vary considerably (Fig. 9), and indicate that: (1) the mantle source of the lavas between 56° and 60° W is significantly more depleted than those at the terminations of the back-arc basin; and (2) that the EBB magmas contain a higher input from the subducting slab (Keller et al. 2002; Fretzdorff et al. 2004). Lavas from Orca Seamount and Three Sisters in the CBB resemble MORB in terms of many incompatible elements and have high $^{143}Nd/^{144}Nd$ (Fig. 9), whereas the lavas of Deception Island and Sail Rock at the termination of the Bransfield Basin are more enriched in $(Ce/Yb)_N$ and Nb/Yb. The increasing $(Ce/Yb)_N$ and Nb/Yb towards the west at Deception Island possibly correspond to the SW-directed propagation of the Bransfield Strait rift (Barker and Austin 1998), and could reflect decreasing degrees of partial melting or tapping of predominantly enriched mantle sources at the propagating rift tip. Similar variations of magma genesis along rifts propagating into continental lithosphere are known: for example, from the northern Red Sea (Haase et al. 2000). However, the radiogenic isotope compositions of the Deception Island lavas overlap those from Orca Seamount and Three Sisters (Fig. 8), indicating that the mantle sources are roughly similar so that the difference in incompatible element compositions may be largely due to variable degrees of partial melting. Interestingly, the alkaline lavas erupting at the same longitude on the South Shetland Islands at the flanks of the Bransfield Strait are even more incompatible element-enriched in, for example, $(Ce/Yb)_N$ and Nb/Yb (Fig. 9), which probably suggests even lower degrees of partial melting

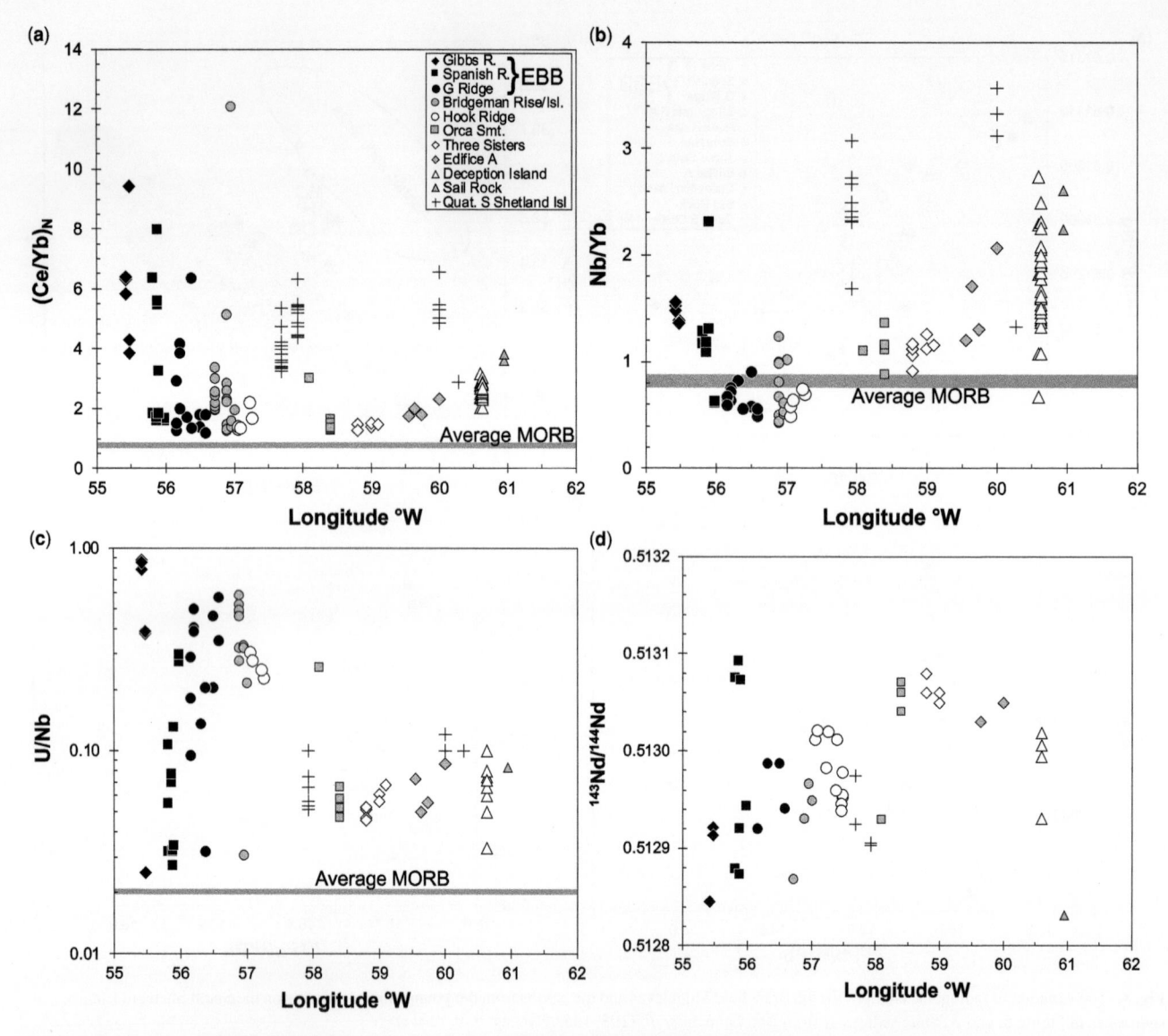

Fig. 9. Variation of incompatible element ratios in the volcanic rocks along the axis of the Bransfield Strait and for the off-axis lavas occurring on the South Shetland Islands. Data sources as in Figure 2.

beneath thicker lithosphere than in the Bransfield Basin rift axis. The EBB lavas show a large range of U/Nb from low ratios close to those of average MORB (c. 0.02: Hofmann et al. 1986) to compositions resembling island arc lavas (Fig. 9c). This range implies that the mantle beneath the EBB is heterogeneous on a small scale, with portions unaffected by slab-derived fluids closely related to metasomatized mantle.

James Ross Island

Tectonic setting of James Ross Island

The James Ross Island Volcanic Group (JRIVG) is part of a belt of young (<7 Ma) volcanic centres along the Larsen Basin (Fig. 10) on the eastern side of the Antarctic Peninsula (Nelson 1975; Gonzalez-Ferran 1985; Smellie 1987). The volcanic occurrences of Seal Nunataks and Argo Point further south along the Antarctic Peninsula may be related to alkaline lavas of Alexander Island with comparable ages (Smellie 1987; Hole et al. 1993). The cause of the volcanism could be either the ascent of asthenospheric mantle through a slab window beneath the eastern Antarctic Peninsula or back-arc extension related to the South Shetland subduction zone (Garrett and Storey 1987; Smellie 1987; Hole et al. 1995; Eagles et al. 2009). Hole et al. (1995) distinguished between syn-subduction alkaline volcanism of James Ross Island, due to extension and thinning of the lithosphere, and post-subduction alkaline volcanism caused by asthenospheric upwelling in a slab window at Seal Nunataks and Alexander Island. The lavas of James Ross Island cover an estimated area of >5000 km^2 (Calabozo et al. 2015) with outcrops also on the neighbouring Vega Island and other small islands (Fig. 10). Radiometric dating of the basalts on James Ross Island suggest that the eruptions occurred between 6.16 and <0.08 Ma, and about the same age range is observed on Vega Island and other outcrops of the JRIVG (Smellie et al. 2008). Two volcanic units of the JRIVG with a thickness of about 400 m occur on Cockburn Island, and basaltic samples from these units yield ^{40}Ar/^{39}Ar ages of 4.9–2.8 Ma (Lawver et al. 1995; Jonkers and Kelley 1998). A potential northeastward age progression from James Ross Island (most volcanism occurred between 6 and 3 Ma) via Prince Gustav Channel (2.0–1.6 Ma) to the Cape Purvis volcanic centre (<0.13 Ma) on Dundee Island was suggested (Hole et al. 1995; Smellie et al. 2006). Hole et al. (1995) related this progression to the opening of the Bransfield Strait, whereas Smellie et al. (2006) interpreted it as a reflection of east-directed propagation of a deep

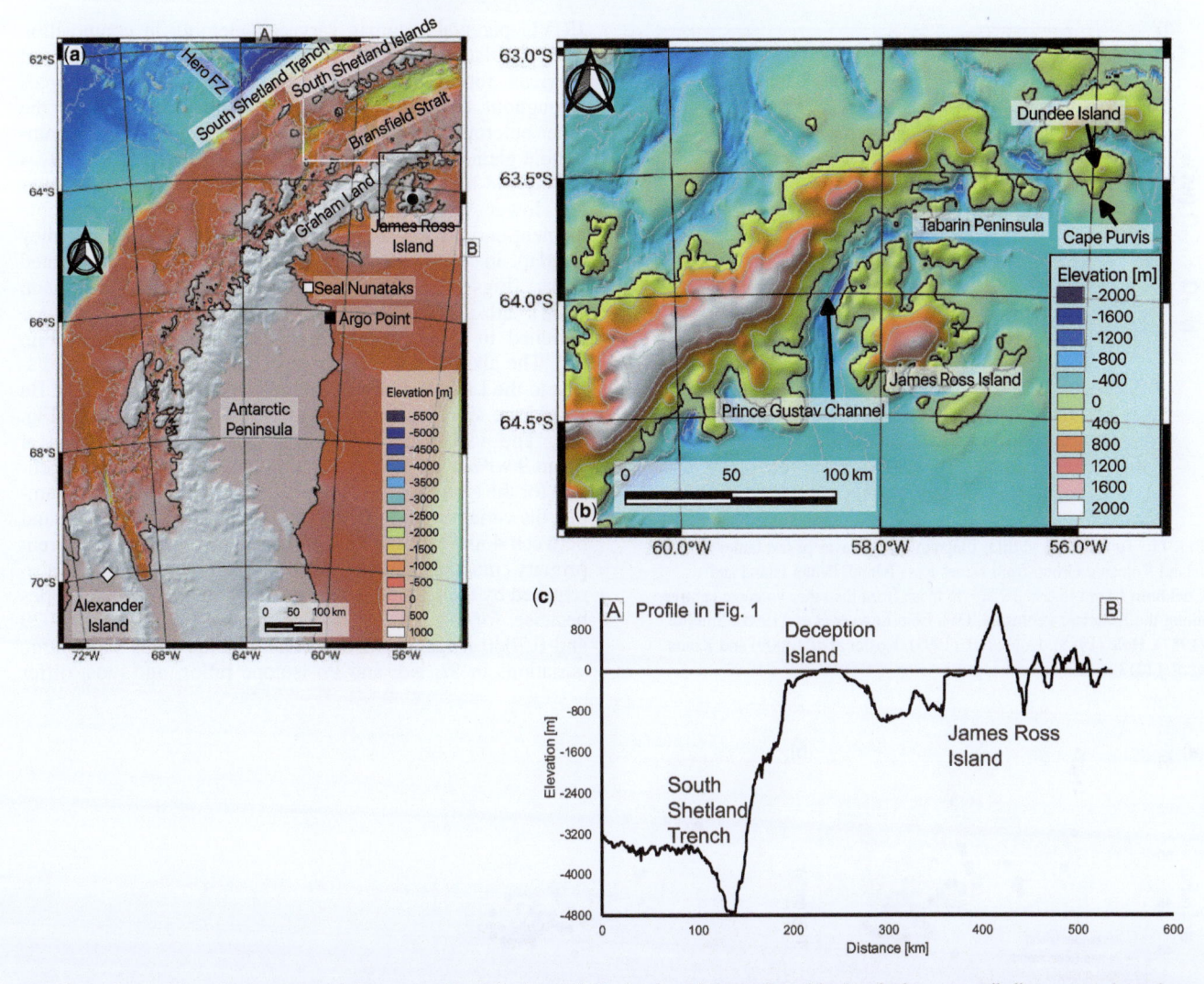

Fig. 10. (**a**) Map of the volcanic centres of the eastern side of the Antarctic Peninsula with James Ross Island and other young alkaline centres shown by symbols. (**b**) Enlarged map of James Ross Island, Dundee Island and Tabarin Peninsula. (**c**) Morphological profile across the subduction zone beneath the Antarctic Peninsula from the South Shetland Trench to James Ross Island.

fault. Interestingly, the apparent northeastward propagation of the JRIVG activity is opposite to the westward propagation of the Bransfield Strait opening.

Composition of the magmatic rocks of James Ross Island

Petrology of the volcanic rocks. At least 50 volcanic eruptions occurred on James Ross Island and formed lava units typically as delta sequences with thicknesses up to 250 m (Smellie *et al.* 2008). The volcanic rocks of James Ross Island range from lava flows to pillow lavas and hyaloclastites, scoria and tuff cones intruded by dykes and plugs (Smellie *et al.* 2008; Kosler *et al.* 2009; Calabozo *et al.* 2015). The lavas typically contain olivine phenocrysts with Cr-spinel inclusions set in a matrix consisting of glass, plagioclase, clinopyroxene, magnetite and ilmenite (Nelson 1975; Smellie 1987; Smellie *et al.* 2006; Kosler *et al.* 2009). Alteration of the olivines and the glass, as well as zeolites in vesicles, is frequently observed and indicate alteration by hydrothermal fluids. The mafic lavas of the JRIVG contain xenoliths of spinel lherzolites, dunites and wehrlites (Smellie 1987), suggesting relatively fast ascent through the lithosphere. The major element variation in the JRIVG possibly indicates high-pressure (9 ± 1.5 kb) fractional crystallization of SiO_2-saturated basaltic magma at the base of the crust (Hole *et al.* 1995).

Geochemical composition of the volcanic rocks. The volcanic rocks of James Ross Island consist of alkali basalts, trachybasalts and trachyandesites that are similar to the lavas from the neighbouring Paulet and Cockburn Islands and Cape Purvis (Fig. 11; Table 1; see also the full dataset in Supplementary material Table S2). In contrast, lavas from Seal Nunataks show more variation in the SiO_2 contents, and tholeiitic basalts, andesites and even basanites may occur. Whereas the JRIVG lavas are relatively homogeneous, the young volcanic samples from Alexander Island show two distinct series: one of alkali basalts to trachybasalts, and the other of tephrites with lower SiO_2 contents (Fig. 11). The JRIVG lavas are generally mafic with MgO contents higher than 4 wt%, and show typical trends of increasing TiO_2 and Al_2O_3 between 10 and 4 wt% MgO (Fig. 12). This implies that plagioclase crystallization and fractionation occurs at MgO contents of <4 wt% in the magmas, and, instead, the fractionating assemblage of the primitive melts consists of olivine and clinopyroxene. The lavas of James Ross Island show a significant variation in incompatible element enrichment, with K_2O ranging from 0.6 to 1.4 wt% and Zr from 120 to 180 ppm at 10 wt% MgO (Fig. 13). The tephrites from Alexander Island have higher TiO_2 and K_2O contents (Fig. 12), as well as other incompatible elements (Fig. 13), than the other alkaline magmas of the Antarctic Peninsula, possibly reflecting lower degrees of partial melting. The

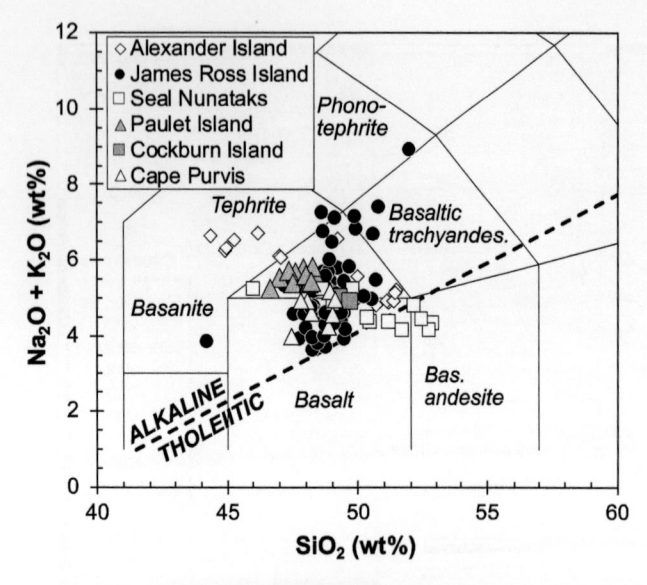

Fig. 11. Total alkalis v. SiO$_2$ diagram for the lavas of the James Ross Island Volcanic Group from James Ross Island, Paulet Island and Cockburn Island in comparison to those from the other volcanic centres along the Antarctic Peninsula. Data from Lawver *et al.* (1995), Smellie (1987), Hole (1988), Hole *et al.* (1995), Kosler *et al.* (2009) and Kraus *et al.* (2013).

JRIVG parental magmas vary considerably in composition but no variation with age is observed, implying that different magmas formed from variably enriched mantle sources throughout the magmatic activity. The samples from the other outcrops of the JRIVG follow distinct trends of incompatible elements like Ba and Zr with the Paulet Island lavas having the highest contents, whereas the Cape Purvis lavas are lower (Fig. 13). Some of the least incompatible element-enriched lavas occur at the Seal Nunataks but they overlap in composition with the JRIVG and associated rocks (Figs 12–14). Compared to N-MORB (Hofmann 1988), all alkaline lavas of the Antarctic Peninsula are enriched in the LREEs and depleted in the HREEs (Fig. 14). The alkaline rocks show enrichment of Nb and Ta relative to the LREEs, whereas the fluid-mobile elements like Ba and U are similarly enriched relative to MORB as Nb (Fig. 14). The variation in the incompatible element ratios at about 9 wt% MgO (Fig. 15) indicates a variable magma genesis for the entire region of the Antarctic Peninsula: for example, the variation in (Ce/Yb)$_N$ between 3 and 6 in JRIVG, and between 4 and 8 in the Seal Nunataks lavas suggest different primary magmas. A variation in magma sources is also required by the significant variation in the radiogenic isotopes because, for example, Sr isotope ratios vary between 0.7029 and 0.7040 (Figs 15c & 16). The alkaline lavas show large variations in Sr, Nd, and Pb isotope ratios, and most differ

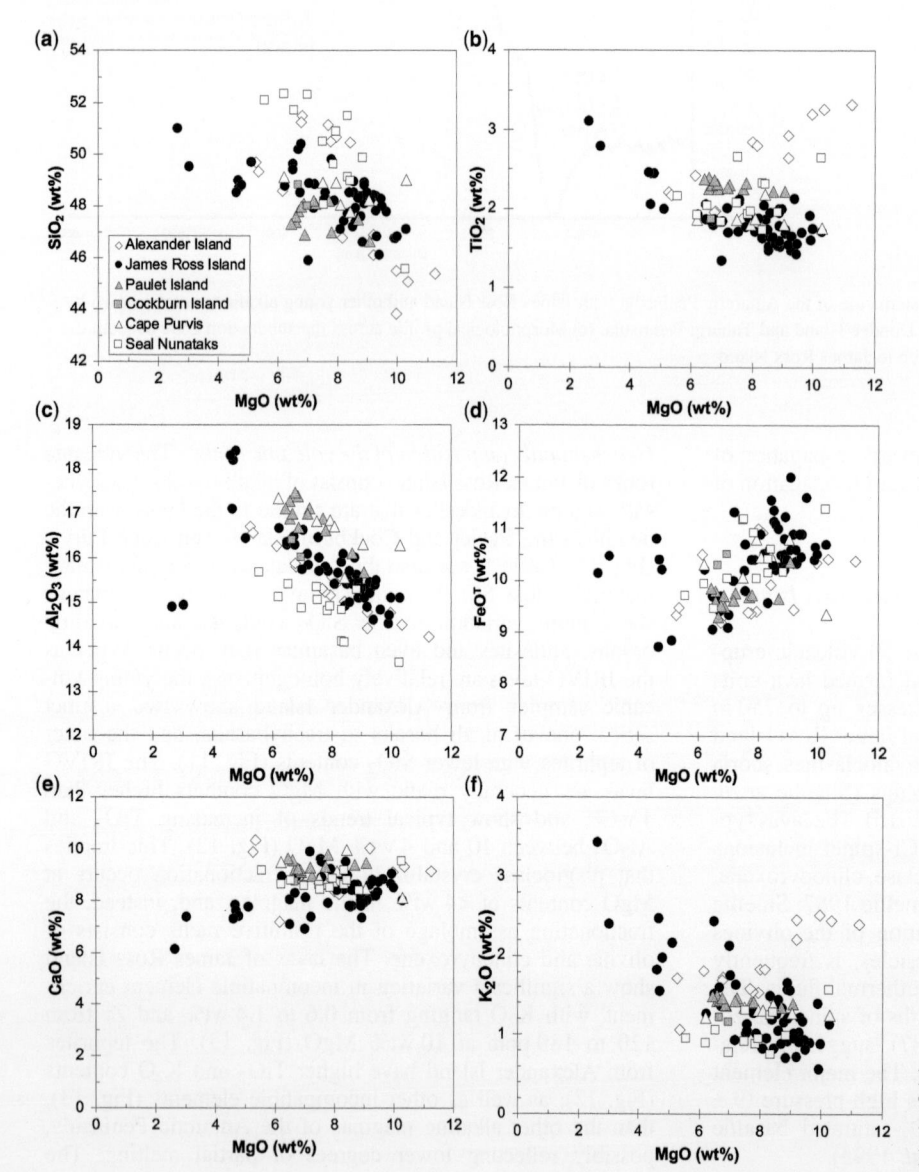

Fig. 12. Major element diagrams (**a**) SiO$_2$, (**b**) TiO$_2$, (**c**) Al$_2$O$_3$, (**d**) FeO^T, (**e**) CaO and (**f**) K$_2$O v. MgO for the lavas of the eastern Antarctic Peninsula. Data sources as in Figure 11.

Variation in partial melting and mantle sources of the alkaline magmas

The lavas of the Antarctic Peninsula erupted through thick continental crust but constant $^{87}Sr/^{86}Sr$ in lavas with 2.6–7.5 wt% MgO suggest that assimilation of crustal rocks did not affect the JRIVG magmas (Fig. 15c). The SiO_2-undersaturated magmas show a strong enrichment in LREE relative to the HREE, implying low degrees (<3%) of partial melting at high pressures (>3 GPa) with residual garnet lherzolite (Hole et al. 1993; Kosler et al. 2009). Relatively low seismic velocities at a depth of 100 km beneath the Antarctic Peninsula were interpreted to reflect the asthenosphere, whereas the lithospheric thickness was estimated to be c. 80 km (An et al. 2015). The most SiO_2-undersaturared lavas with the highest $(Ce/Yb)_N$ occur on Alexander Island, and may reflect a thick lithosphere and very low degrees of partial melting at great depth, whereas the lavas of the Seal Nunataks and JRIVG possibly formed below thinned rifted lithosphere (Garrett and Storey 1987). The low ratios of fluid-mobile elements, like Ba, to immobile elements, like Nb, in the JRIVG lavas (Fig. 14) indicate that the mantle source was not influenced by subduction enrichment (Hole et al. 1993). Thus, although seismic data indicate the presence of the fossil Phoenix Plate beneath the Antarctic Peninsula and the presence of hydrous fluids in the mantle was predicted (An et al. 2015), there is no evidence for the contribution of slab fluids to the young magmatism. The relatively low Sr and high Nd isotope ratios resemble HIMU (high $^{238}U/^{204}Pb = \mu$)-type mantle but the Pb isotopes are too unradiogenic (Fig. 16). Thus, Hole et al. (1993) suggested melting of heterogeneous asthenospheric mantle with little input from lithospheric sources. The hydrous enriched portions of the asthenosphere melt as the mantle flows upwards where the lithosphere is thinned by extension (Hole et al. 1995). More recently, mixing of depleted upper mantle with enriched mantle, possibly enriched mantle type II (EM II), sources was suggested by Kosler et al. (2009) to explain the isotope variation in lavas of James Ross Island. However, the trend of $^{208}Pb/^{204}Pb$ v. $^{206}Pb/^{204}Pb$ of the JRIVG lavas is higher than Pacific MORB and parallel to the MORB trend (Fig. 15). Additionally, the JRIVG lavas have significantly higher $^{87}Sr/^{86}Sr$ at comparable Pb isotope ratios than MORB but they do not show mixing trends towards MORB (Fig. 16c). Thus, MORB mantle may not contribute to the formation of the JRIVG magmas and, rather, they may

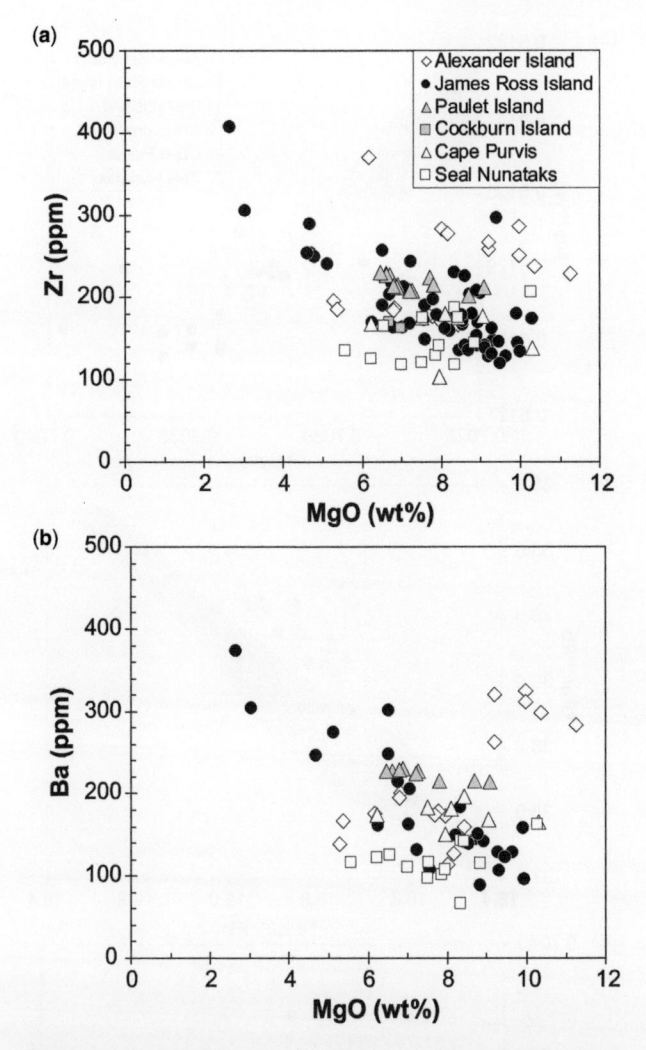

Fig. 13. The trace element variation of (a) Zr v. MgO and (b) Ba v. MgO for the lavas from the eastern Antarctic Peninsula. Data sources as in Figure 11.

considerably from Pacific MORB in having higher Sr and lower Nd isotope ratios and in having higher $^{208}Pb/^{204}Pb$ for a given $^{206}Pb/^{204}Pb$ (Fig. 16).

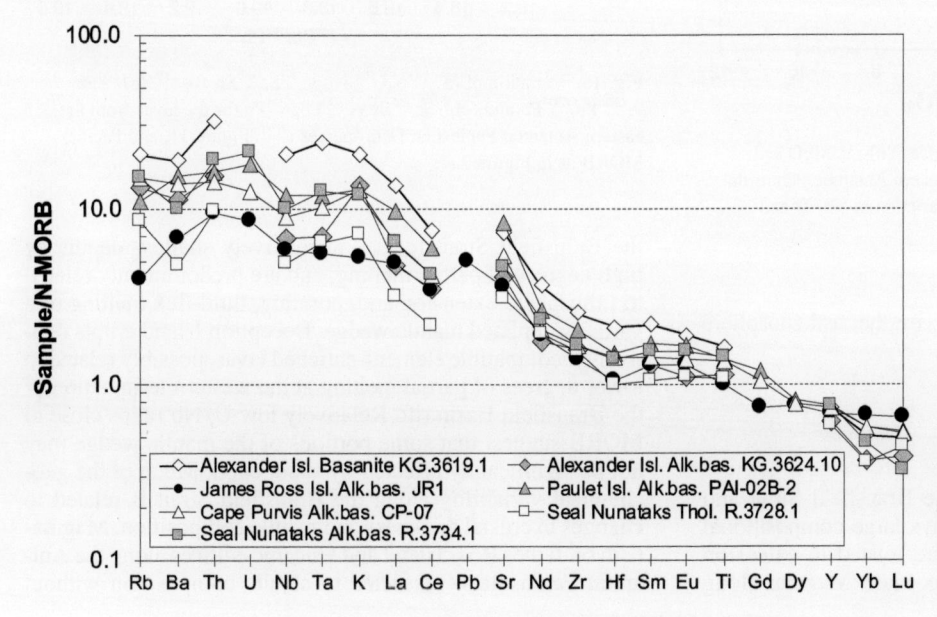

Fig. 14. N-MORB-normalized incompatible element patterns for primitive alkali basalts, basanites and tholeiitic basalts from the Antarctic Peninsula. Data sources as in Figure 11, and N-MORB is from Hofmann (1988).

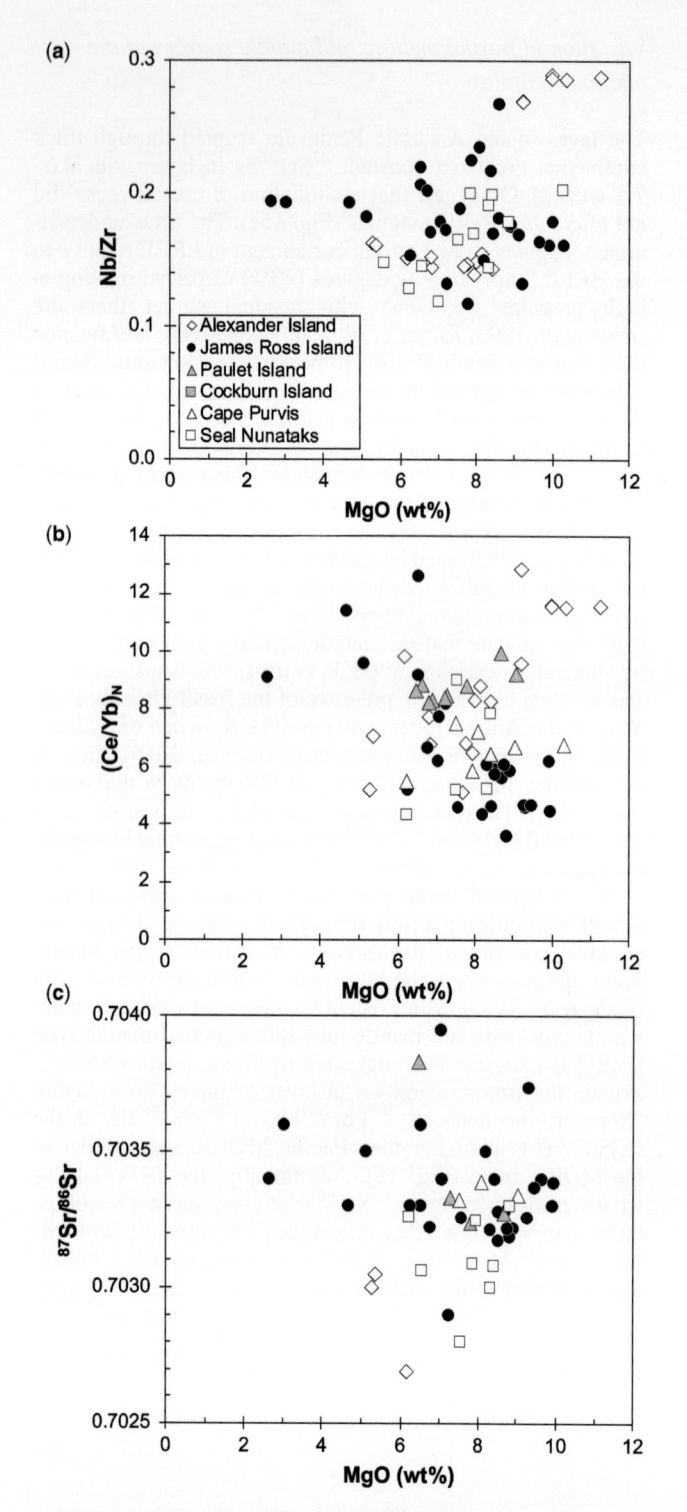

Fig. 15. Variation of (a) Nb/Zr v. MgO, (b) (Ce/Yb)$_N$ v. MgO and (c) ^{87}Sr/^{86}Sr v. MgO for the lavas from the eastern Antarctic Peninsula. Data sources as in Figure 11. Note the large variation in Nb/Zr and (Ce/Yb)$_N$ in the JRIVG.

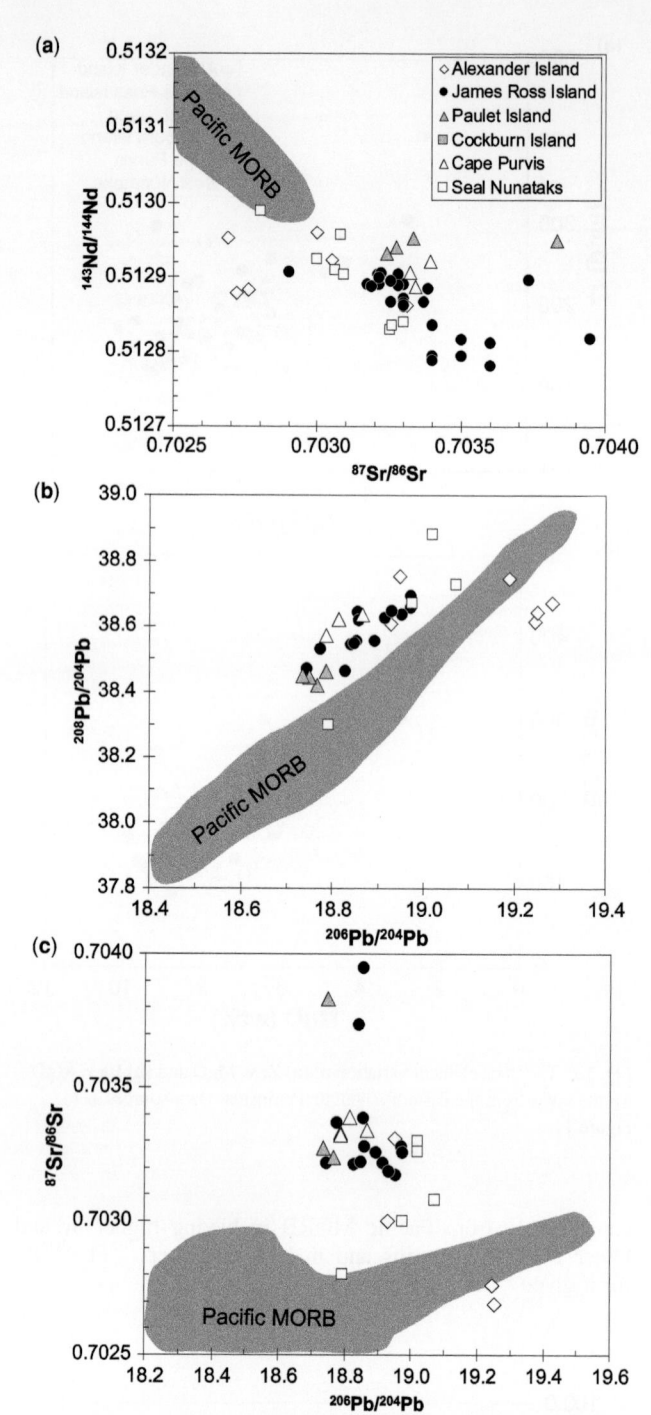

Fig. 16. Variation of (a) ^{143}Nd/^{144}Nd v. ^{87}Sr/^{86}Sr, (b) ^{208}Pb/^{204}Pb v. ^{206}Pb/^{204}Pb and (c) ^{87}Sr/^{86}Sr v. ^{206}Pb/^{204}Pb for the lavas from the eastern Antarctic Peninsula. Data sources as in Figure 11, and Pacific MORB as in Figure 7.

form from enriched heterogeneities in the asthenosphere (Hole *et al.* 1993).

Concluding remarks

Post-subduction magmatism along the Bransfield Strait and east of the Antarctic Peninsula displays large compositional variations, although both structures are related to extension caused by slab rollback of the Phoenix Plate. Magmas along the Bransfield Strait formed at relatively shallow depths by high degrees of partial melting, and are predominantly related to lithosphere extension and, possibly, fluid-flux melting of a variably depleted mantle wedge. Deception Island erupts relatively incompatible element-enriched lavas, possibly related to lower degrees of partial melting at the western termination of the Bransfield Basin rift. Relatively low U/Nb ratios close to MORB suggest that some portions of the mantle wedge may not be significantly affected by subduction. Much of the geochemical variability along the Bransfield Strait is related to changes in crustal thickness and mantle composition. Magmatism on James Ross Island and volcanic edifices along the Antarctic Peninsula is restricted to alkalic composition without subduction zone influence, although the fossil Phoenix Plate apparently lies beneath the region. The alkali basalts to basanites generally display smaller degrees of partial melting (c. 3% as opposed to up to 20% along the Bransfield Strait) of a less heterogeneous mantle. This may result from asthenospheric upwelling along the southern margin of the Bransfield Strait which causes partial melting of enriched heterogeneities in the mantle. The existing data show that the South Shetland subduction zone has a variable influence on magma formation over a distance of c. 150 km.

Acknowledgements We thank M. Hole and P. Leat for constructive reviews that significantly improved the quality of this work.

Author contributions KMH: conceptualization (lead), investigation (equal), supervision (lead), validation (equal), writing – original draft (lead), writing – review & editing (equal); CB: investigation (equal), validation (equal), writing – original draft (equal).

Funding This research received no specific grant from any funding agency in the public, commercial, or not-for-profit sectors.

Data availability All data generated during this study are included in this published article and its supplementary files.

References

An, M., Wiens, D.A. et al. 2015. Temperature, lithosphere–asthenosphere boundary, and heat flux beneath the Antarctic Plate inferred from seismic velocities. *Journal of Geophysical Research*, **120**, 8720–8742, https://doi.org/10.1002/2014JB0 11332

Bach, W., Hegner, E., Erzinger, J. and Satir, M. 1994. Chemical and isotopic variations along the superfast spreading East Pacific Rise from 6 to 30° S. *Contributions to Mineralogy and Petrology*, **116**, 365–380, https://doi.org/10.1007/BF00310905

Baker, P.E., McReath, I., Harvey, M.R., Roobol, M.J. and Davies, T.G. 1975. *The geology of the South Shetland Islands: V. Volcanic evolution of Deception Island*. British Antarctic Survey Scientific Reports, **78**.

Barker, D.H.N. and Austin, J.A. 1998. Rift propagation, detachment faulting, and associated magmatism in Bransfield Strait, Antarctic Peninsula. *Journal of Geophysical Research*, **103**, 24 017–24 043, https://doi.org/10.1029/98JB01117

Barker, P.F. 1982. The Cenozoic subduction history of the Pacific margin of the Antarctic Peninsula: ridge–crust–trench interactions. *Journal of the Geological Society, London*, **139**, 787–801, https://doi.org/10.1144/gsjgs.139.6.0787

Beier, C., Turner, S.P., Sinton, J.M. and Gill, J.B. 2010. Influence of subducted components on back-arc melting dynamics in the Manus Basin. *Geochemistry, Geophysics, Geosystems*, **11**, Q0AC03, https://doi.org/10.1029/2010gc003037

Ben Othman, D., White, W.M. and Patchett, J. 1989. The geochemistry of marine sediments, island arc magma genesis, and crust–mantle recycling. *Earth and Planetary Science Letters*, **94**, 1–21, https://doi.org/10.1016/0012-821X(89)90079-4

Ben-Zvi, T., Wilcock, W.S.D., Barclay, A.H., Zandomeneghi, D., Ibanez, J.M. and Almendros, J. 2009. The P-wave velocity structure of Deception Island, Antarctica, from two-dimensional seismic tomography. *Journal of Volcanology and Geothermal Research*, **180**, 67–80, https://doi.org/10.1016/j.jvolgeores. 2008.11.020

Calabozo, F.M., Strelin, J.A., Orihashi, Y., Sumino, H. and Keller, R.A. 2015. Volcano–ice–sea interaction in the Cerro Santa Marta area, northwest James Ross Island, Antarctic Peninsula. *Journal of Volcanology and Geothermal Research*, **297**, 89–108, https://doi.org/10.1016/j.jvolgeores.2015.03.011

Catalan, M., Galindo-Zaldivar, J., Davila, J.M., Martos, Y.M., Maldonado, A., Gamboa, L. and Schreider, A.A. 2013. Initial stages of oceanic spreading in the Bransfield Rift from magnetic and gravity data analysis. *Tectonophysics*, **585**, 102–112, https://doi.org/10.1016/j.tecto.2012.09.016

Christeson, G.L., Barker, D.H.N., Austin, J.A. and Dalziel, I.W.D. 2003. Deep crustal structure of Bransfield Strait: Initiation of a back arc basin by rift reactivation and propagation. *Journal of Geophysical Research*, **108**, 2492, https://doi.org/10.1029/2003JB002468

Dietrich, R., Rülke, A., Ihde, J., Lindner, K., Niemeier, H.W., Schenke, H.-W. and Seeber, G. 2004. Plate kinematics and deformation status of the Antarctic Peninsula based on GPS. *Global and Planetary Change*, **42**, 313–321, https://doi.org/10.1016/j.gloplacha.2003.12.003

Eagles, G., Gohl, K. and Larter, R.D. 2009. Animated tectonic reconstruction of the Southern Pacific and alkaline volcanism at its convergent margins since Eocene times. *Tectonophysics*, **464**, 21–29, https://doi.org/10.1016/j.tecto.2007.10.005

Fisk, M.R. 1990. Volcanism in the Bransfield Strait, Antarctica. *Journal of South American Earth Sciences*, **3**, 91–101, https://doi.org/10.1016/0895-9811(90)90022-S

Fretzdorff, S., Worthington, T.J., Haase, K.M., Hékinian, R., Franz, L., Keller, R.A. and Stoffers, P. 2004. Magmatism in the Bransfield Basin: Rifting of the South Shetland Arc? *Journal of Geophysical Research*, **109**: https://doi.org/10.1029/2004 JB003046

Galindo-Zaldivar, J., Gamboa, L., Maldonado, A., Nakao, S. and Bochu, Y. 2004. Tectonic development of the Bransfield Basin and its prolongation to the South Scotia Ridge, northern Antarctic Peninsula. *Marine Geology*, **206**, 267–282, https://doi.org/10.1016/j.margeo.2004.02.007

Garrett, S.W. and Storey, B.C. 1987. Lithospheric extension on the Antarctic Peninsula during Cenozoic subduction. *Geological Society, London, Special Publications*, **28**, 419–431, https://doi.org/10.1144/GSL.SP.1987.028.01.26

Gonzalez-Ferran, O. 1985. Volcanic and tectonic evolution of the northern Antarctic Peninsula – Late Cenozoic to recent. *Tectonophysics*, **114**, 389–409, https://doi.org/10.1016/0040-1951 (85)90023-X

Gràcia, E., Canals, M., Farran, M.L., Prieto, M.J. and Sorribas, J. 1996. Morphostructure and evolution of the Central and Eastern Bransfield Basins (NW Antarctic Peninsula). *Marine Geophysical Researches*, **18**, 429–448, https://doi.org/10.1007/BF00286088

Haase, K.M. 2002. Geochemical constraints on magma sources and mixing processes in Easter Microplate MORB (SE Pacific): a case study of plume–ridge interaction. *Chemical Geology*, **182**, 335–355, https://doi.org/10.1016/S0009-2541(01)00327-8

Haase, K.M., Mühe, R. and Stoffers, P. 2000. Magmatism during extension of the lithosphere: geochemical constraints from lavas of the Shaban Deep, northern Red Sea. *Chemical Geology*, **166**, 225–237, https://doi.org/10.1016/S0009-2541(99) 00221-1

Haase, K.M., Beier, C., Fretzdorff, S., Smellie, J.L. and Garbe-Schönberg, D. 2012. Magmatic evolution of the South Shetland Islands, Antarctica, and implications for continental crust formation. *Contributions to Mineralogy and Petrology*, **163**, 1103–1119, https://doi.org/10.1007/s00410-012-0719-7

Hathway, B. 2000. Continental rift to back-arc basin: Jurassic–Cretaceous stratigraphical and structural evolution of the Larsen Basin, Antarctic Peninsula. *Journal of the Geological Society, London*, **157**, 417–432, https://doi.org/10.1144/jgs.157.2.417

Heuret, A. and Lallemand, S. 2005. Plate motions, slab dynamics and back-arc deformation. *Physics of Earth and Planetary Interiors*, **149**, 31–51, https://doi.org/10.1016/j.pepi.2004.08.022

Hofmann, A.W. 1988. Chemical differentiation of the Earth: the relationship between mantle, continental crust, and oceanic crust. *Earth and Planetary Science Letters*, **90**, 297–314, https://doi.org/10.1016/0012-821X(88)90132-X

Hofmann, A.W., Jochum, K.P., Seufert, M. and White, W.M. 1986. Nb and Pb in oceanic basalts: new constraints on mantle evolution. *Earth and Planetary Science Letters*, **79**, 33–45, https://doi.org/10.1016/0012-821X(86)90038-5

Hole, M.J. 1988. Post-subduction alkaline volcanism along the Antarctic Peninsula. *Journal of the Geological Society, London*, **145**, 985–998, https://doi.org/10.1144/gsjgs.145.6.0985

Hole, M.J., Kempton, P.D. and Millar, I.L. 1993. Trace-element and isotopic characteristics of small-degree melts of the asthenosphere: Evidence from the alkalic basalts of the Antarctic Peninsula. *Chemical Geology*, **109**, 51–68, https://doi.org/10.1016/0009-2541(93)90061-M

Hole, M.J., Saunders, A.D., Rogers, G. and Sykes, M.A. 1995. The relationship between alkaline magmatism, lithospheric extension and slab window formation along continental destructive plate margins. *Geological Society, London, Special Publications*, **81**, 265–285, https://doi.org/10.1144/GSL.SP.1994.081.01.15

Hyeong, K., Kim, J., Pettke, T., Yoo, C.M. and Hur, S.-d. 2011. Lead, Nd and Sr isotope records of pelagic dust: Source indication versus the effects of dust extraction procedures and authigenic mineral growth. *Chemical Geology*, **286**, 240–251, https://doi.org/10.1016/j.chemgeo.2011.05.009

Jonkers, H.A. and Kelley, S.P. 1998. A reassessment of the age of the Cockburn Island Formation, northern Antarctic Peninsula, and its palaeoclimatic implications. *Journal of the Geological Society, London*, **155**, 737–740, https://doi.org/10.1144/gsjgs.155.5.0737

Keller, R.A., Fisk, M.R., White, W.M. and Birkenmajer, K. 1991. Isotopic and trace element constraints on mixing and melting models of marginal basin volcanism, Bransfield Strait, Antarctica. *Earth and Planetary Science Letters*, **111**, 287–303, https://doi.org/10.1016/0012-821X(92)90185-X

Keller, R.A., Fisk, M.R., Smellie, J.L., Strelin, J.A., Lawver, L.A. and White, W.M. 2002. Geochemistry of back arc basin volcanism in Bransfield Strait, Antarctica: subducted contributions and along-axis variations. *Journal of Geophysical Research*, **107**, https://doi.org/10.1029/2001JB000444

Kessel, R., Schmidt, M.W., Ulmer, P. and Pettke, T. 2005. Trace element signature of subduction-zone fluids, melts and supercritical liquids at 120–180 km depth. *Nature*, **437**, 724–727, https://doi.org/10.1038/nature03971

Kosler, J., Magna, T., Mlcoch, B., Mixa, P., Nyvlt, D. and Holub, F.V. 2009. Combined Sr, Nd, Pb and Li isotope geochemistry of alkaline lavas from northern James Ross Island (Antarctic Peninsula) and implications for back-arc magma formation. *Chemical Geology*, **258**, 207–218, https://doi.org/10.1016/j.chemgeo.2008.10.006

Kraus, S., Kurbatov, A. and Yates, M. 2013. Geochemical signatures of tephras from Quaternary Antarctic Peninsula volcanoes. *Andean Geology*, **40**, 1–40, https://doi.org/10.5027/andgeoV40n1-a01

Langmuir, C.H., Bézos, A., Escrig, S. and Parman, S.W. 2006. Chemical systematics and hydrous melting of the mantle in back-arc basins. *American Geophysical Union Geophysical Monograph Series*, **166**, 87–146.

Larter, R.D. and Barker, P.F. 1991. Effects of ridge crest–trench interaction on Antarctic–Phoenix spreading: Forces on a young subducting slab. *Journal of Geophysical Research*, **96**, 19 583–19 607, https://doi.org/10.1029/91JB02053

Lawver, L.A., Keller, R.A., Fisk, M.R. and Strelin, J.A. 1995. Bransfield Strait, Antarctic Peninsula active extension behind a dead arc. *In*: Taylor, B. (ed.) *Backarc Basins: Tectonics and Magmatism*. Plenum Press, New York, 315–342.

Le Bas, M.J., Le Maitre, R.W., Streckeisen, A. and Zanettin, B. 1986. A chemical classification of volcanic rocks based on the total alkali–silica diagram. *Journal of Petrology*, **27**, 745–750, https://doi.org/10.1093/petrology/27.3.745

Lima, S.M., Haase, K.M., Beier, C., Regelous, M., Brandl, P.A., Hauff, F. and Krumm, S. 2017. Magmatic evolution and source variations at the Nifonea Ridge (New Hebrides Island Arc). *Journal of Petrology*, **58**, 473–494, https://doi.org/10.1093/petrology/egx023

Livermore, R.A., Balanyá, J.C. *et al.* 2000. Autopsy on a dead spreading center: The Phoenix Ridge, Drake Passage, Antarctica. *Geology*, **28**, 607–610, https://doi.org/10.1130/0091-7613(2000)28<607:AOADSC>2.0.CO;2

Macdonald, G.A. and Katsura, T. 1964. Chemical composition of Hawaiian lavas. *Journal of Petrology*, **5**, 82–133, https://doi.org/10.1093/petrology/5.1.82

Martí, J., Geyer, A. and Aguirre-Diaz, G. 2013. Origin and evolution of the Deception Island caldera (South Shetland Islands, Antarctica). *Bulletin of Volcanology*, **75**, 732, https://doi.org/10.1007/s00445-013-0732-3

Martinez, F. and Taylor, B. 2006. Modes of crustal accretion in back-arc basins: Inferences from the Lau Basin. *American Geophysical Union Geophysical Monograph Series*, **166**, 5–29.

McCarron, J.J. and Larter, R.D. 1998. Late Cretaceous to early Tertiary subduction history of the Antarctic Peninsula. *Journal of the Geological Society, London*, **155**, 255–268, https://doi.org/10.1144/gsjgs.155.2.0255

McCulloch, M.T. and Gamble, J.A. 1991. Geochemical and geodynamical constraints on subduction zone magmatism. *Earth and Planetary Science Letters*, **102**, 358–374, https://doi.org/10.1016/0012-821X(91)90029-H

Nawrocki, J., Pancyk, M. and Williams, I.S. 2010. Isotopic ages and palaeomagnetism of selected magmatic rocks from King George Island (Antarctic Peninsula). *Journal of the Geological Society, London*, **167**, 1063–1079, https://doi.org/10.1144/0016-76492009-177

Nelson, P.H.H. 1975. *The James Ross Island Volcanic Group of North-East Graham Land*. British Antarctic Survey Scientific Reports, **54**.

Pańczyk, M. and Nawrocki, J. 2011. Pliocene age of the oldest basaltic rocks of Penguin Island (South Shetland Islands, northern Antarctic Peninsula). *Geological Quarterly*, **55**, 335–344.

Pankhurst, R.J. and Smellie, J.L. 1983. K–Ar geochronology of the South Shetland Islands, Lesser Antarctica: apparent lateral migration of Jurassic to Quaternary island arc volcanism. *Earth and Planetary Science Letters*, **66**, 214–222, https://doi.org/10.1016/0012-821X(83)90137-1

Pearce, J.A. and Stern, R.J. 2006. Origin of back-arc magmas: Trace element and isotope perspectives. *American Geophysical Union Geophysical Monograph Series*, **166**, 63–86.

Petersen, S., Herzig, P.M., Schwarz-Schampera, U., Hannington, M.D. and Jonasson, I.R. 2004. Hydrothermal precipitates associated with bimodal volcanism in the Central Bransfield Strait, Antarctica. *Mineralium Deposita*, **39**, 358–379, https://doi.org/10.1007/s00126-004-0414-3

Risso, C. and Aparicio, A. 2002. Plutonic xenoliths in Deception Island (Antarctica). *Terra Antartica*, **9**, 95–99.

Robertson Maurice, S.D., Wiens, D.A., Shore, P.J., Vera, E. and Dorman, L.M. 2003. Seismicity and tectonics of the South Shetland Islands and Bransfield Strait from a regional broadband seismograph deployment. *Journal of Geophysical Research*, **108**, 2461, https://doi.org/10.1029/2003JB002416

Saunders, A.D. and Tarney, J. 1979. The geochemistry of basalts from a back-arc spreading centre in the East Scotia Sea. *Geochimica et Cosmochimica Acta*, **43**, 555–572, https://doi.org/10.1016/0016-7037(79)90165-0

Schreider, A.A., Schreider, A.A. *et al.* 2015. Structure of the Bransfield strait crust. *Oceanology*, **55**, 112–123, https://doi.org/10.1134/S0001437014060101

Sinton, J.M., Ford, L.L., Chappell, B. and McCulloch, M.T. 2003. Magma genesis and mantle heterogeneity in the Manus back-arc basin, Papua New Guinea. *Journal of Petrology*, **44**, 159–195, https://doi.org/10.1093/petrology/44.1.159

Sleep, N.H. and Toksöz, M.N. 1971. Evolution of marginal basins. *Nature*, **33**, 548–550, https://doi.org/10.1038/233548a0

Smellie, J.L. 1987. Geochemistry and tectonic setting of alkaline volcanic rocks in the Antarctic Peninsula: a review. *Journal of Volcanology and Geothermal Research*, **32**, 269–285, https://doi.org/10.1016/0377-0273(87)90048-5

Smellie, J.L., Pankhurst, R.J., Thomson, M.R.A. and Davies, R.E.S. 1984. *The Geology of the South Shetland Islands: VI. Stratigraphy, Geochemistry and Evolution*. British Antarctic Survey Scientific Reports, **87**.

Smellie, J.L., McIntosh, W.C., Esser, R. and Fretwell, P. 2006. The Cape Purvis volcano, Dundee Island (northern Antarctic Peninsula): late Pleistocene age, eruptive processes and implications for a glacial palaeoenvironment. *Antarctic Science*, **18**, 399–408, https://doi.org/10.1017/S0954102006000447

Smellie, J.L., Johnson, J.S., McIntosh, W.C., Esser, R., Gudmundsson, M.T., Hambrey, M.J. and van Wyk de Vries, B. 2008. Six million years of glacial history recorded in volcanic lithofacies of the James Ross Island Volcanic Group, Antarctic Peninsula. *Palaeogeography, Palaeoclimatology, Palaeoecology*, **260**, 122–148, https://doi.org/10.1016/j.palaeo.2007.08.011

Taylor, B. and Karner, G.D. 1983. On the evolution of marginal basins. *Reviews of Geophysics and Space Physics*, **21**, 1727–1741, https://doi.org/10.1029/RG021i008p01727

Taylor, B. and Martinez, F. 2003. Back-arc basin basalt systematics. *Earth and Planetary Science Letters*, **210**, 481–497, https://doi.org/10.1016/S0012-821X(03)00167-5

Taylor, F.W., Bevis, M.G. *et al.* 2008. Kinematics and segmentation of the South Shetland Islands–Bransfield basin system, northern Antarctic Peninsula. *Geochemistry, Geophysics, Geosystems*, **9**, Q04035, https://doi.org/10.1029/2007GC001873

Uyeda, S. and Kanamori, H. 1979. Back-arc opening and the mode of subduction. *Journal of Geophysical Research*, **84**, 1049–1061, https://doi.org/10.1029/JB084iB03p01049

Weaver, S.D., Saunders, A.D., Pankhurst, R.J. and Tarney, J. 1979. A geochemical study of magmatism associated with the initial stages of back-arc spreading: The Quaternary volcanics of Bransfield Strait from South Shetland Islands. *Contributions to Mineralogy and Petrology*, **68**, 151–169, https://doi.org/10.1007/BF00371897

Wendt, J.I., Regelous, M., Niu, Y., Hékinian, R. and Collerson, K.D. 1999. Geochemistry of lavas from the Garrett Transform Fault: insights into mantle heterogeneity beneath the eastern Pacific. *Earth and Planetary Science Letters*, **173**, 271–284, https://doi.org/10.1016/S0012-821X(99)00236-8

Section 4

Post-subduction, slab-window volcanism

Spectacular prismatic jointing in a topographically confined, pooled, Late Miocene, slab-window-related basalt lava. The columns define a suppressed basal colonnade and an expanded entablature consistent with surface flooding of the lava during cooling probably in a glacial setting. Mount Pinafore, Alexander Island (Antarctic Peninsula). The prominent basal colonnade, with vertical columns, is *c.* 15 m thick.

Photograph by M. J. Hole

Section 4

Post-subduction, slab-window volcanism

Chapter 4.1a

Antarctic Peninsula: volcanology

John L. Smellie[1]* and Malcolm J. Hole[2]

[1]School of Geography, Geology and the Environment, University of Leicester, University Road, Leicester LE1 7RH, UK

[2]Department of Geology and Petroleum Geology, University of Aberdeen, Meston Building, King's College, Aberdeen AB24 3UE, UK

JLS, 0000-0001-5537-1763; MJH, 0000-0002-1622-4266

*Correspondence: jls55@leicester.ac.uk

Abstract: The Antarctic Peninsula is distinguished by late Neogene volcanic activity related to a series of northerly younging ridge crest–trench collisions and the progressive opening of 'slab windows' in the subjacent mantle. The outcrops were amongst the last to be discovered in the region, with many occurrences not visited until the 1970s and 1980s. The volcanism consists of several monogenetic volcanic fields and small isolated centres. It is sodic alkaline to tholeiitic in composition, and ranges in age between 7.7 Ma and present. No eruptions have been observed (with the possible, but dubious, exception of Seal Nunataks in 1893) but very young isotopic ages for some outcrops suggest that future eruptions are a possibility. The eruptions were overwhelmingly glaciovolcanic and the outcrops have been a major source of information on glaciovolcano construction. They have also been highly influential in advancing our understanding of the configuration of the Plio-Pleistocene Antarctic Peninsula Ice Sheet. However, our knowledge is hindered by a paucity of modern, precise isotopic ages. In particular, there is no obvious relationship between the age of ridge crest–trench collisions and the timing of slab-window volcanism, a puzzle that may only be resolved by new dating.

Post-subduction volcanism occurred in the Antarctic Peninsula as a result of a series of collisions between segments of an oceanic spreading centre and the Antarctic Peninsula trench (Barker 1982) (Fig. 1). The collisions took place at progressively younger times from south to north between c. 50 and 4 Ma, and they caused the locking of the spreading centre at the trench. However, continued subduction of the detached leading plate resulted in the opening up of extensive 'slab windows' that permitted the uprise and decompression melting of fertile mantle into the mantle wedge that formerly intervened between the subducted slab and the overlying continental crust (Hole 1988, 2021; Hole et al. 1991, 1995) (Fig. 2). The eruption of these small-volume melts created a series of extensive monogenetic volcanic fields and small isolated centres scattered along the length of the Antarctic Peninsula, between Seal Nunataks in the north and Snow Nunataks in the south (Smellie et al. 1988; Smellie 1999) (Fig. 3). This contrasts with the formation of a large shield volcano in the James Ross Island Volcanic Group during the same period, which is linked to longer-lived and larger-scale mantle upwelling within an arc-rear 'thin spot' coeval with the final stages of subduction at the South Shetland Trench. However, the relationship between the timing of slab-window formation and the earliest eruptions in each sector is not simple and there is no clear south–north age progression, a feature not yet satisfactorily explained (cf. Barber et al. 1991; Hole et al. 1995) (Fig. 4). The post-subduction mafic magmas consist of two sodic alkaline magmatic series characterized by basanites–phonotephrites and alkali basalts–tholeiites (Smellie 1987; Hole et al. 1995). The outcrops include a variety of volcanic landforms including scoria cones, tuff cones and tuyas, and some outcrops are dominated by dykes. The eruptive environments varied from subaerial to subglacial.

There has only been one new field investigation of the late Neogene volcanic fields described in this chapter since the detailed studies by one of us (M.J. Hole) in 1985–88. That other study took place in December 1994 and during 1997–98, and focused on Seal Nunataks as part of a wider examination of volcanism in the northern Antarctic Peninsula (Veit 2002). Apart from the small size and isolation of the outcrops generally, another major reason for the lack of study is that, in the early 2000s, the Larsen A and B ice shelves collapsed suddenly and catastrophically, leaving only a tiny relict at the nunataks, some of which are now islands. Even during the 1985–88 study the ambient conditions were very difficult, with wet snow and numerous large melt-pools ensuring that normal over-snow travel by skidoo was difficult or impossible. Uniquely in the Antarctic Peninsula, Seal Nunataks was investigated by us using hovercraft logistics, a modus operandi that was very successful in the wet conditions. Access there now will involve intensive dedicated support using helicopters, although minor opportunistic petrological sampling from helicopters has also occurred rarely (e.g. Kraus et al. 2013).

Stratigraphy of the volcanic outcrops

A lithostratigraphy for the post-subduction alkaline volcanic outcrops in the Antarctic Peninsula was published by Smellie (1999) and is broadly followed here but with significant modification. Because the outcrops are composed of laterally discontinuous volcanic rocks erupted from multiple small volcanic centres, they do not easily fit with the rules of formal lithostratigraphy. A simplified revised system is proposed here based on delineating volcanic fields rather than formations. The new stratigraphy broadly mirrors that used in Antarctica for Neogene alkaline volcanic outcrops in Victoria Land (i.e. the McMurdo Volcanic Group: e.g. Kyle 1990), where geographically delineated outcrops have been grouped within volcanic provinces, each containing several volcanic fields (see Smellie and Rocchi 2021 in this Memoir). Table 1 outlines the stratigraphy used here, comprising two volcanic groups (the lithostratigraphic equivalent of volcanic provinces): the Bellingshausen Sea Volcanic Group and the James Ross Island Volcanic Group. Both groups contain multiple volcanic fields.

Most of the Antarctic Peninsula Neogene alkaline volcanic outcrops are described in this chapter; only the Mount Haddington Volcanic Field of the James Ross Island Volcanic Group, which is not post-subduction, is described elsewhere

From: Smellie, J. L., Panter, K. S. and Geyer, A. (eds) 2021. *Volcanism in Antarctica: 200 Million Years of Subduction, Rifting and Continental Break-up*. Geological Society, London, Memoirs, **55**, 305–325,
First published online February 4, 2021, https://doi.org/10.1144/M55-2018-59
© 2021 The Author(s). Published by the Geological Society of London. All rights reserved.
For permissions: http://www.geolsoc.org.uk/permissions. Publishing disclaimer: www.geolsoc.org.uk/pub_ethics

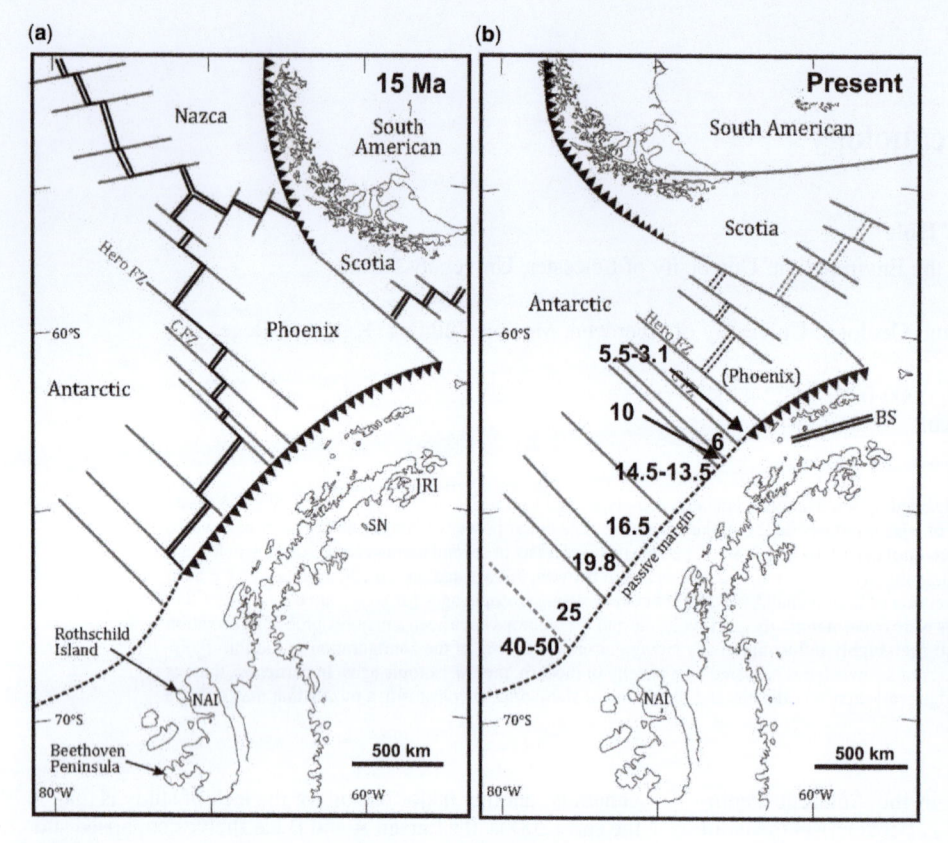

Fig. 1. Sketch maps of the Antarctic Peninsula and southern South America illustrating the tectonic setting of slab-window development along the Antarctic Peninsula between 15 Ma (**a**) and present (**b**), with progressive consumption of a spreading centre northwards with time. Triangle ornament represents active subduction, and double lines represent a spreading-ridge segment. In (b) the collision times for each ridge segment are given on the oceanward margin of the peninsula. Abbreviations: NAI, northern Alexander Island; SN, Seal Nunataks; JRI, James Ross Island; BS, Bransfield Strait. After Hole and Larter (1993) and Hole et al. (1995).

(see Smellie 2021 in this Memoir). The outcrops are mainly contained in the Bellingshausen Sea Volcanic Group. However, outcrops in the Seal Nunataks Volcanic Field are conventionally regarded as part of the James Ross Island Volcanic Group (Fleet 1968; Nelson 1975; Smellie 1999). The Seal Nunataks Volcanic Field is also geographically much closer to the James Ross Island Volcanic Group compared with outcrops of the Bellingshausen Sea Volcanic Group, which are c. 500 km distant (Fig. 3). However, in common with outcrops in the Bellingshausen Volcanic Group, magmas in the Seal Nunataks Volcanic Province probably formed as a result of decompression melting of mantle rising within slab windows (Hole 1988, 1990a). By contrast, James Ross Island Volcanic Group magmas erupted in a back-arc position coeval with subduction at the South Shetland Trench and they are unrelated to a slab window (Hole et al.

1995; see also Haase and Beier 2021; Smellie 2021 in this Memoir). Despite the contrast in tectonic setting, the late Neogene volcanism throughout the Antarctic Peninsula is compositionally indistinguishable, and the lithofacies and eruptive palaeoenvironments are broadly similar (cf. Smellie 2021). Although there may be a case based on tectonic setting and magmagenesis to move the Seal Nunataks Volcanic Field into the Bellingshausen Volcanic Group, at present we retain the distinction.

The outcrops described in this chapter comprise, from north to south: Seal Nunataks, Argo Point, Rothschild Island, Mount Pinafore, Beethoven Peninsula, Snow Nunataks–Rydberg Peninsula–Sims Island and Merrick Mountains (Fig. 3). An additional alkaline volcanic outcrop previously included consists of several lamprophyre (camptonite) dykes at Venus Glacier, eastern Alexander Island (Horne and Thomson 1967) for

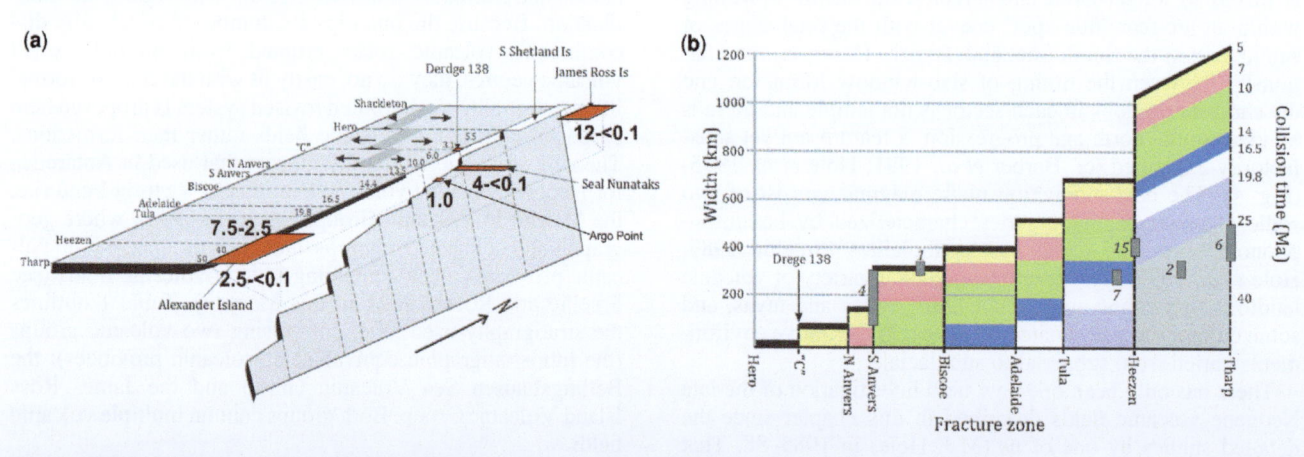

Fig. 2. (**a**) Perspective view of slab-window growth with time, showing the geographical location of the slab-window-related basalts (red-coloured trapezoids with the K–Ar age adjacent). The ages of basalts in each of the outcrops is also given. Note that volcanism in James Ross Island is not related to a slab window and is discussed by Smellie (2021). After Hole et al. (1995). (**b**) Plan view of slab-window development over time along the Antarctic Peninsula. The x-axis represents the palaeotrench and the different ornaments correspond to the amount of slab-window formation associated with each spreading ridge; the solid lines separating the ornaments are 'isochrons' for the slab window as a whole. It is assumed that the subducted slab was planar, and the slab dip did not vary over time. See Hole et al. (1995) for further explanation.

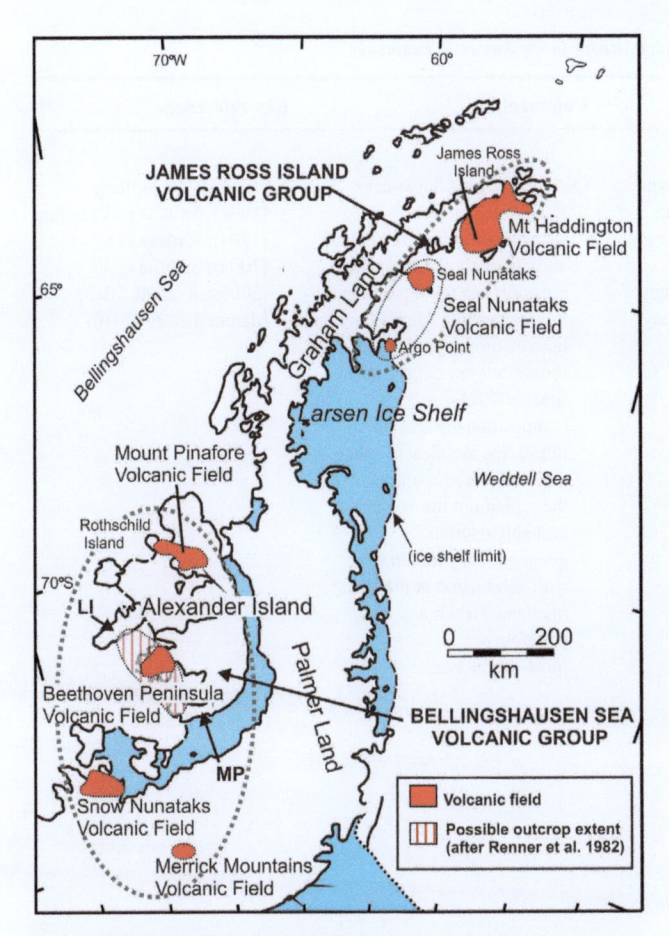

Fig. 3. Map of the Antarctic Peninsula showing the locations of the late Neogene volcanic fields (modified after Smellie 1999). The two major volcanic groups are also shown (grey dashed ellipses). Description of the Mount Haddington Volcanic Field is included in Smellie (2021). Abbreviations: LI, Latady Island; MP, Monteverdi Peninsula.

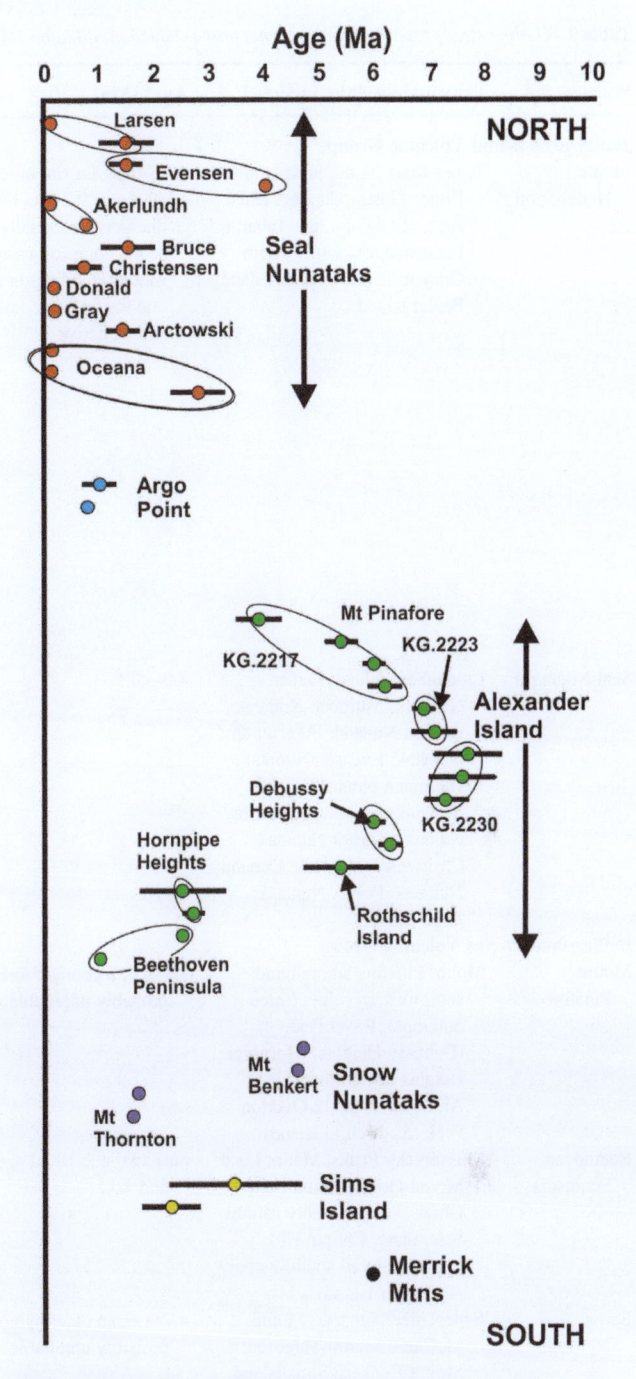

Fig. 4. Diagram showing the age of outcrops of post-subduction volcanism in the Antarctic Peninsula. Error bars are also shown (where known). See the text for data sources. The outcrops are arranged from north to south. Note that there is no obvious progression in ages, despite a postulated link to northerly younging ridge subduction, and associated sequential development of slab windows.

which a mid-Miocene K–Ar age of *c.* 15 Ma has been published (Rex 1970). However, unlike other alkaline volcanic occurrences, the dykes are highly altered and the isotopic age may be substantially in error. Moreover, although alkaline in composition, the Venus Glacier dykes are highly altered, distinctively potassic and relatively evolved (trachybasalt and phonotephrite) with extraordinarily high Sr contents (Rowley and Smellie 1990), and their genesis and affinities to the post-subduction volcanism are highly uncertain. They are therefore excluded from further discussion here. Other potential outcrops consist of isolated dykes with alkaline compositions (e.g. in the Seward Mountains (Palmer Land) and on Adelaide Island: Smellie 1987). Rare alkaline dykes are known to occur throughout the Antarctic Peninsula and, as they appear to be related genetically to the Cretaceous–Tertiary arc terrain (Scarrow *et al.* 1998; Leat and Riley 2021*b*), they do not form part of the post-subduction volcanism discussed in this chapter.

James Ross Island Volcanic Group

Seal Nunataks Volcanic Field

Seal Nunataks is a small cluster of 16 isolated nunataks and islands situated on the east coast of northern Antarctic Peninsula (Figs 5 & 6a, b). Collectively called Seal Islands originally by their discoverer, C.A. Larsen, in 1893, some of the nunataks subsequently became islands following the collapse of the Larsen A ice shelf in 2000 (Pudsey and Evans 2001). Larsen (1894) described eruptions taking place during his visit to Christensen Nunatak and Lindenberg Island, comprising 'funnel-like holes' (fumaroles?) emitting 'very black and thick smoke'. In addition, from observations taken in 1982, González-Ferrán (1983*a*) interpreted the presence of black and red tephra strewn on the ice shelf surrounding Murdoch Nunatak as evidence for recent eruptive activity. A similar argument was used to infer more recent activity (2010: E. Domack quoted in Kraus *et al.* 2013). González-Ferrán (1983*a*) also recorded fumaroles at Dallmann and Murdoch nunataks. However, a visit by one of the authors (M.J. Hole) in January 1988 showed no evidence for fumaroles and

Table 1. *Outline stratigraphy of late Neogene, post-subduction, alkaline volcanic outcrops in the Antarctic Peninsula*

Volcanic field	Principal localities included	Age* (Ma)	Comments	Key references
James Ross Island Volcanic Group				
Mount Haddington	James Ross Island; islands in Prince Gustav Channel and Antarctic Sound; the Tabarin Peninsula and southeastern Graham Land; Dundee Island; Paulet Island	6.25–0.08 (*in situ* outcrops); 12.4 and 9.2 for clasts in diamicts; essentially pristine scoria cones on eastern Mount Haddington suggest that the volcanic field may still be active	Outcrops largely 'layer-cake' (mainly extensive superimposed lava-fed deltas) and, uniquely for Neogene volcanism in the region, lend themselves to a formal lithostratigraphy; 29 rock formations are defined (see Smellie 2021); compositionally and in lithofacies identical to other alkaline volcanic outcrops in the region but the volcanism probably overlaps geographically and in time with subduction at the South Shetland Trench and is therefore not 'post-subduction'	Nelson (1975); Skilling (1994); Smellie and Skilling (1994); Jonkers *et al.* (2002); Smellie *et al.* (2006a, b, 2008, 2013); Marenssi *et al.* (2010)
Seal Nunataks	Lindenberg Island; Larsen Nunatak; Murdoch Nunatak; Donald Nunatak; Akerlundh Nunatak; Evensen Nunatak; Dallmann Nunatak; Bruce Nunatak; Bull Nunatak; Gray Nunatak; Castor Nunatak; Christensen Nunatak; Oceana Nunatak; Pollux Nunatak; Argo Point	4.0–<0.1		del Valle *et al.* (1983); Hole (1990a); Smellie (1999)
Bellingshausen Sea Volcanic Group				
Mount Pinafore	Mount Pinafore summit and southwestern ridges (three outcrops); Ravel Peak (Debussy Heights); Hornpipe Heights (all northern Alexander Island); Overton Peak (Rothschild Island)	7.7–5.4; a younger age of 3.9 is probably unreliable		Care (1980); Smellie *et al.* (1993)
Beethoven Peninsula	Mussorgsky Peaks; Mount Liszt; Mount Grieg; Mount Strauss; Gluck Peak; probably Mount Schumann, Chopin Hill, Mount Lee (all southwestern Alexander Island)	only two published ages: 2.5 and <0.1	Mount Grieg, Schumann, Chopin Hill and Mount Lee outcrops unvisited; good exposures at Mount Grieg observed by binocular	Hole (1990c); Smellie and Hole (1997)
Snow Nunataks	Espenchied Nunatak; Mount McCann; Mount Thornton; Mount Benkert; Sims Island; Rydberg Peninsula (southwestern Palmer Land)	4.7–1.6; an older age of 20 is probably unreliable		O'Neill and Thomson (1985); Rowley and Thomson (1990); Smellie (1999); Hathway (2001); Smellie *et al.* (2009)
Merrick Mountains	Henry Nunataks and outcrop west of Eaton Nunatak (southern Palmer Land)	6		Halpern (1971); Rowley *et al.* (1990); Smellie (1999)

*See Table 2 for isotopic ages.

there are no primary landforms. The observations by Larsen (1894) and González-Ferrán (1983a) are inexplicable, and should probably be discounted. Therefore, the volcanic field at Seal Nunataks is either inactive or extinct (Smellie 1990, 1999). Conversely, the very young isotopic ages (<0.1 Ma: Table 2) obtained at several of the outcrops suggests that eruptions might resume, although all the published ages are by the K–Ar method and are now quite old and should be repeated.

The geology of Seal Nunataks has been described by Fleet (1968), del Valle *et al.* (1983), González-Ferrán (1983a), Hole (1990a), Smellie (1990) and Smellie and Hole (1997), and of Argo Point by Saunders (1982). Published isotopic ages range between *c.* 4 ± 1 and 0.1 Ma (Fig. 5; Table 2). Taken at face value, the ages suggest three broad eruptive episodes: *c.* 3 Ma, *c.* 1.5–1.4 Ma and <0.2 Ma. González-Ferrán (1983b) suggested that the ages are broadly symmetrical and become younger in both directions away from a central axis but the pattern is not well defined. Moreover, the ages were all obtained by the K–Ar method and have high errors (typically 0.3–0.5 myr: del Valle *et al.* 1983) (Table 2). They should be regarded simply as indicative, and the group would benefit substantially from more comprehensive and more precise

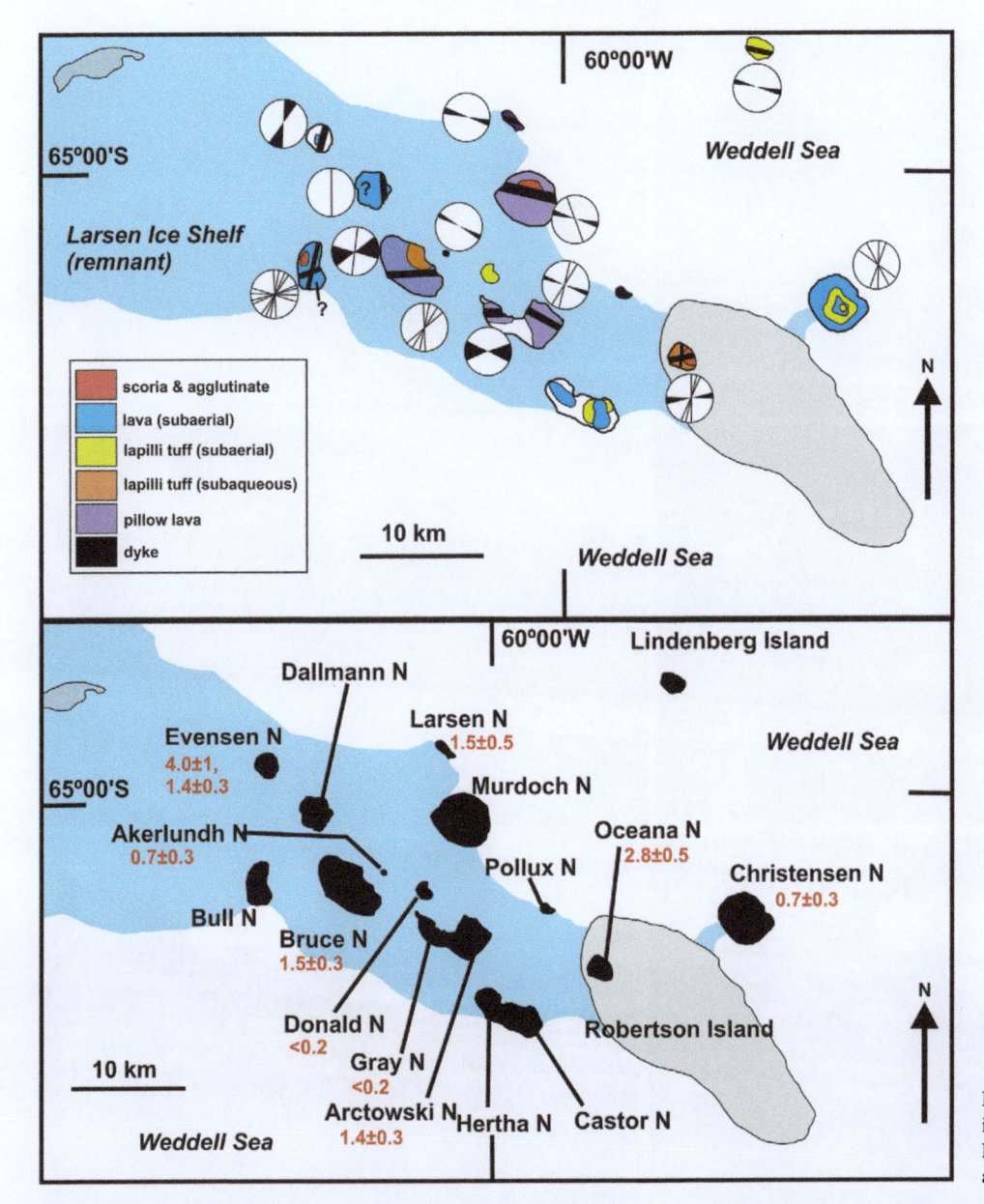

Fig. 5. Maps showing the geology, isotopic ages and dyke trends of Seal Nunataks (modified after Smellie and Hole 1997).

dating. Many of the outcrops in Seal Nunataks are dominated by dykes, whose orientations are more variable than implied by González-Ferrán (1983b). They may describe a broad arcuate pattern that is concave to the north or else they are a result of intrusion along a reticulate system of fractures caused by extension and rifting (Smellie 1990) (Fig. 5). The nunataks are almost all associated with prominent magnetic anomalies associated with pillow lava but also interpreted to reflect deeper pipe-shaped feeders extending to several kilometres rather than fissure-fed dykes (Renner 1980).

The Christensen Nunatak Formation defined by Smellie (1999) was described as mainly subaerially erupted lavas, lapilli tuffs and minor spatter, whilst the Bruce Nunatak Formation was described as some combination of large dykes, pillow lava and subaqueously deposited lapilli tuffs. The distinction was essentially based on eruptive setting, corresponding to either subaerial lithofacies (Christensen Nunatak Formation) or subaqueous lithofacies (Bruce Nunatak Formation). However, unlike the many laterally extensive eruptive units (mainly lava-fed deltas) in the James Ross Island Volcanic Group (Smellie et al. 2013; Smellie 2021), situated <150 km to the NE of Seal Nunataks and overlapping in age, Seal Nunataks is formed of multiple small eruptive centres with a very localized distribution of the erupted products and no meaningful stratigraphical correlations can be made between the centres. There are no known outcrops of lava-fed deltas in the Seal Nunataks Volcanic Field. By contrast, the geographically separate outcrop at Argo Point is distinctive within the volcanic field: it is a well-formed scoria cone.

The principal characteristics of the individual outcrops are now briefly described (see Fig. 5). Although we include relevant descriptions of some of the nunataks by Veit (2002), differences exist between his study and ours, principally in the identification of pillow lava v. pāhoehoe. Although important for identifying the eruptive environment, pillow lava and pāhoehoe can be remarkably similar in appearance and their discrimination can be ambiguous.

Åkerlundh Nunatak. This is the smallest of the Seal Nunataks, just 90 m high (Fig. 6b). It forms an elongate ridge formed around a central 25 m-wide dyke of vesicular olivine–plagioclase-phyric basalt with an age of 0.7 ± 0.3 Ma (del Valle et al. 1983). There are also rare isolated and small outcrops of vesicular plagioclase-phyric pillow lava and poorly bedded black lapillistone.

Arctowski Nunatak. This nunatak rises c. 235 m above the Larsen Ice Shelf and it comprises a 50–70 m-wide dyke

Fig. 6. Photomontage of field photographs of Seal Nunataks and associated lithofacies. (**a**) View of Dalmann Nuntak looking north from Bull Nunatak with a tidal melt-pool in the foreground. Despite the cone-like landform, no primary structure is preserved and the nunatak is dominated by pillow basalt with a central dyke zone. The nunatak rises to 210 m asl. (**b**) View looking SE from Dalmann Nunatak with Åkerlundh Nunatak on the left, and Gray, Hertha and Castor nunataks from front to back, respectively. The largest nunatak in the view (Hertha) rises to 225 m asl. (**c**) Broken slabs and homogenized lapilli tuff at Bruce Nunatak. The structure suggests destabilization and local fluidization of a weakly lithified stratified lapilli tuff sequence. The majority of Bruce Nunatak is composed of pillow basalt. The lens cap (on the centre right) is *c.* 52 mm in diameter. (**d**) Subaerial lava flows at Castor Nunatak overlying subhorizontally bedded yellow-orange lapilli tuffs containing accretionary lapilli. The lapilli tuffs are at approximately the same elevation as pillow lavas at Larsen and Dalmann nunataks, suggesting that eruptions occurred under varied ice thicknesses at different times. The combined thickness of the two prominent volcaniclastic beds beneath the lavas is *c.* 2 m. (**e**) Likely pillow lava at Dalmann Nunatak. The ice axe head is *c.* 35 cm long. (**f**) Dyke zone on the north flank of Larsen Nunatak. The majority of the nunatak is made of pillow lava and the dykes are seen to terminate in the pillow lava pile at the summit of the nunatak. The prominent dyke zone is *c.* 30 m across, perpendicular to strike.

zone orientated $285°N_{mag}$ composed of olivine–plagioclase-phyric basalt. The dyke margins are locally bulbous and there are internal chilled margins suggesting that the dyke is multiple, together with three small (*c.* 35 cm-wide) dykes aligned at *c.* $340°N_{mag}$. Pillow lava with distinctive acicular aggregates of plagioclase phenocrysts is well exposed on the eastern flank of the nunatak. Veit (2002) also recorded 'subaerial rocks' (otherwise undescribed) on the summit of the nunatak. A dyke dated by del Valle *et al.* (1983) yielded an age of 1.4 ± 0.3 Ma.

Bruce Nunatak. Bruce Nunatak comprises three ridges rising to 320 m asl (above sea level) constructed around multiple dykes striking 270, 010 and $221°N_{mag}$. Olivine–plagioclase-phyric pillow lava is conspicuous particularly on the flanks of the ridges, the northern slopes and in the centre. Yellow-orange lapilli tuffs with scattered lava blocks up to 15 cm long and laminated tuffs are exposed in the northern face of the nunatak. The pyroclastic rocks are extensively affected by syndepositional faulting and slumping, including local zones of chaotic blocks in lapilli tuff matrix (Fig. 6c).

Table 2. *Summary of published isotopic ages for post-subduction, slab-window-related volcanic rocks in the Antarctic Peninsula*

Sample	Locality	Dated lithologies	Age (Ma)	Error (Ma)	Error SD	Method	Reference	Notes
Seal Nunataks								
D.4105.1	Akerlundh Nunatak	Lava (dyke?)	<0.1		2σ	K–Ar	Rex (1972, 1976)	Published age modified by Smellie *et al.* (1988)
D.4114.1	Larsen Nunatak	Lava	<0.1		2σ	K–Ar	Rex (1972, 1976)	Published age modified by Smellie *et al.* (1988)
D.727.2	Oceana Nunatak	Lava	<0.1		2σ	K–Ar	Rex (1972, 1976)	Published age modified by Smellie *et al.* (1988)
D.727.3	Oceana Nunatak	Lava	<0.1		2σ	K–Ar	Rex (1972, 1976)	Published age modified by Smellie *et al.* (1988)
41178/101	Akerlundh Nunatak	n.r.	0.7	0.3	n.r.	K–Ar	del Valle *et al.* (1983)	
141178/21	Arctowski Nunatak	n.r.	1.4	0.3	n.r.	K–Ar	del Valle *et al.* (1983)	
41178/102	Bruce Nunatak	n.r.	1.5	0.5	n.r.	K–Ar	del Valle *et al.* (1983)	
	Christensen Nunatak	n.r.	0.7	0.3	n.r.	K–Ar	del Valle *et al.* (1983)	
81178/5	Donald Nunatak	Dyke?	<0.2		n.r.	K–Ar	del Valle *et al.* (1983)	
201178/10	Evensen Nunatak	Dyke?	1.4	0.3	n.r.	K–Ar	del Valle *et al.* (1983)	
201178/11	Evensen Nunatak	Dyke?	4.0	1	n.r.	K–Ar	del Valle *et al.* (1983)	
141178/12	Gray Nunatak	Dyke?	<0.2		n.r.	K–Ar	del Valle *et al.* (1983)	
31178	Larsen Nunatak	n.r.	1.5	0.5	n.r.	K–Ar	del Valle *et al.* (1983)	
1178/10	Oceana Nunatak	n.r.	2.8	0.5	n.r.	K–Ar	del Valle *et al.* (1983)	
Jason Peninsula								
R.217.7	Argo Point	Lava	1.0	0.3	2σ	K–Ar	Smellie *et al.* (1988)	
R.217.7	Argo Point	Lava	0.8	0.1	2σ	K–Ar	Smellie *et al.* (1988)	
R.218.3	*c.* 37 km west of Argo Point	Hematite-coated intrusion	1.6	0.5	2σ	K–Ar	Smellie *et al.* (1988)	Age unreliable; rock probably much older and unrelated to the post-subduction volcanism
R.218.3	*c.* 37 km west of Argo Point	Hematite-coated intrusion	1.3	0.3	2σ	K–Ar	Smellie *et al.* (1988)	Age unreliable; rock probably much older and unrelated to the post-subduction volcanism
Alexander Island								
KG.2217.16	Mount Pinafore area	Lava	3.9	0.4	2σ	K–Ar	Smellie *et al.* (1988)	Age probably unreliable (too young?)
KG.2217.14	Mount Pinafore area	Lava	5.4	0.3	2σ	K–Ar	Smellie *et al.* (1988)	
KG.2217.13	Mount Pinafore area	Lava	6.0	0.2	2σ	K–Ar	Smellie *et al.* (1988)	
KG.2217.13	Mount Pinafore area	Lava	6.2	0.3	2σ	K–Ar	Smellie *et al.* (1988)	
KG.2223.4	Mount Pinafore area	Lava	6.9	0.2	2σ	K–Ar	Smellie *et al.* (1988)	
KG.2223.3	Mount Pinafore area	Lava	7.1	0.4	2σ	K–Ar	Smellie *et al.* (1988)	
KG.2230.1	Mount Pinafore area	Lava	7.7	0.6	2σ	K–Ar	Smellie *et al.* (1988)	
KG.2230.1	Mount Pinafore area	Lava	7.6	0.6	2σ	K–Ar	Smellie *et al.* (1988)	
KG.2230.1	Mount Pinafore area	Lava	7.3	0.4	2σ	K–Ar	Smellie *et al.* (1988)	
KG.2431.5	Mount Pinafore area	Lava	6.0	0.2	2σ	K–Ar	Smellie *et al.* (1988)	
KG.2431.5	Mount Pinafore area	Lava	6.3	0.2	2σ	K–Ar	Smellie *et al.* (1988)	
KG.3619.4	Rothschild Island	Lava	5.4	0.7	2σ	K–Ar	Smellie *et al.* (1988)	
KG.3612.5	Hornpipe Heights	Lava	2.5	0.8	2σ	K–Ar	Smellie *et al.* (1988)	
KG.3608.9	Hornpipe Heights	Lava	2.7	0.2	2σ	K–Ar	Smellie *et al.* (1988)	
n.r.	Mussorgsky Peaks, Beethoven Peninsula	Lava pillow	2.5	n.r.	n.r.	K–Ar	Hole (1990*c*)	No analytical details
n.r.	Gluck Peak, Beethoven Peninsula	Lava pillow	<1	n.r.	n.r.	K–Ar	Hole (1990*c*)	Ages of 2 and 0.68 (±0.97) were reported, without analytical details
Snow Nunataks								
n.r.	Mount McCann	n.r.	20	n.r.	n.r.	K–Ar	Smellie *et al.* (2009)	Unpublished age of J.W. Thomson; no analytical details; age may be unreliable
n.r.	Mount Benkert	n.r.	4.7	n.r.	n.r.	K–Ar	Smellie *et al.* (2009)	Unpublished age of J.W. Thomson; no analytical details
n.r.	Mount Benkert	n.r.	4.6	n.r.	n.r.	K–Ar	Smellie *et al.* (2009)	Unpublished age of J.W. Thomson; no analytical details

(Continued)

Table 2. *Continued.*

Sample	Locality	Dated lithologies	Age (Ma)	Error (Ma)	Error SD	Method	Reference	Notes
n.r.	Mount Thornton	n.r.	1.7	n.r.	n.r.	K–Ar	Smellie *et al.* (2009)	Unpublished age of J.W. Thomson; no analytical details
n.r.	Mount Thornton	n.r.	1.6	n.r.	n.r.	K–Ar	Smellie *et al.* (2009)	Unpublished age of J.W. Thomson; no analytical details
Sims Island								
R.6801.4	South tip of island	Basaltic intrusion	3.46	1.2	n.r.	Ar–Ar	Hathway (2001)	
R.6801.5	South tip of island	Basaltic intrusion	2.3	0.54	n.r.	Ar–Ar	Hathway (2001)	The younger age is regarded as the more reliable
Merrick Mountains								
		Basalt	6	n.r.	n.r.	K–Ar	Halpern (1971)	Reliability uncertain; no analytical details

n.r., not recorded; SD, standard deviation.

Prominent slump planes dip at 23° to the WNW. A dyke dated by del Valle *et al.* (1983) yielded an age of 1.5 ± 0.5 Ma.

Bull Nunatak. This is a conical-shaped nunatak that rises *c.* 175 m above the Larsen Ice Shelf. It is composed of at least two generations of north–south- and east–west-striking olivine-phyric dykes but the majority of the nunatak consists of lava, with pillows up to 1.5 m in diameter. Olivine-phyric agglutinate overlying the lava is exposed locally on the NW side. The lava pillows are slightly to moderately flattened and lack interpillow debris, features that may be more consistent with subaerial pāhoehoe, but the distinction is uncertain and Veit (2002) preferred an origin as pillow lava. Veit (2002) also recorded subaerial lava on the nunatak top, together with bombs and lapilli, and a large xenolith of fossiliferous sedimentary rock of Cretaceous age, which was probably derived from strata similar to those on Robertson Island or James Ross Island.

Castor Nunatak. This nunatak, rising to *c.* 155 m above the Larsen Ice Shelf (Fig. 6b), comprises a snow dome that caps a near-horizontal sequence of lavas and pyroclastic rocks; dykes are absent. The nunatak may preserve a lava-filled crater (Smellie and Hole 1997). The lavas are 'a'ā with bright red oxidized clinkery to locally ropy surfaces. They are locally interbedded with, but mainly overlie, subhorizontally bedded yellow-orange lapilli tuffs with accretionary lapilli, a few thin tuffs, and some breccias in which large channel-like structures are common and conspicuous (Fig. 6d). Veit (2002) considered that the nunatak was constructed from several eruption centres.

Christensen Nunatak. The summit of this nunatak is at *c.* 300 m asl, and it consists of a 20 m-thick lower horizontal columnar jointed lava overlain by *c.* 50 m of yellow-orange pyroclastic rocks and then an upper thin (<3 m) lava. Both lavas have oxidized vesicular upper surfaces and are olivine phyric. The pyroclastic rocks are yellow-orange lapilli tuffs, well bedded and with scattered lava blocks, some with impact structures. There are also a few thin dykes lacking an overall orientation that cut the pyroclastic deposit. There is a published isotopic age of 0.7 ± 0.3 Ma (del Valle *et al.* 1983).

Dallmann Nunatak. This nunatak is *c.* 210 m high and is broadly conical overall (Fig. 6a). A crater that issued a fresh-looking lava flow was sketched and described by González-Ferrán (1983a) but neither the crater nor the lava flow appear to exist. Exposure is very poor and seems to comprise only lava with poorly defined pillow-like structures (Fig. 6e). Although they may indicate pāhoehoe rather than pillow lava, interpretation is equivocal (as on Bull Nunatak, see earlier).

Donald Nunatak. A small nunatak that rises *c.* 100 m above the Larsen Ice Shelf, this outcrop has pyroclastic rocks exposed at its western end. The pyroclastic rocks comprise horizontally bedded yellow-orange and dark-grey lapilli tuffs or lapillistones, tuffs, and breccias with numerous flattened bombs, the latter also occurring as weakly agglutinated layers up to 15 cm thick. The yellow-orange and dark-grey deposits are in subvertical contact with no obvious structural break, suggesting that the yellow coloration is an alteration artefact. A single dyke striking 360°N$_{mag}$ cuts the pyroclastic rocks and is locally brecciated, with mingling of the dyke fragments and yellow lapilli tuff or lapillistone. It has a published K–Ar age of <0.2 Ma (del Valle *et al.* 1983).

Evensen Nunatak. This nunatak is a ridge formed by a grey vesicular olivine-phyric dyke that rises *c.* 160 m above the Larsen Ice Shelf and is *c.* 1 km in length. It has K–Ar ages of 4.0 ± 1 and 1.4 ± 0.3 Ma (del Valle *et al.* 1983). The dyke zone is *c.* 50 m wide and has multiple chilled surfaces, and, hence, is multiple. Small exposures of black vesicular plagioclase-phyric basalt lava commonly with ropy surfaces are present near the nunatak summit and on the north flank.

Gray Nunatak. Gray Nunatak is <2 km long and rises *c.* 100 m above the Larsen Ice Shelf (Fig. 6b). It consists of four east–west-trending en echelon ridges, each formed around a dyke. One of the dykes has a published age of <0.2 Ma (del Valle *et al.* 1983). The dykes are olivine–plagioclase-phyric, and have strikes varying between 233 and 281°N$_{mag}$. There are also rare exposures of pillow lava poorly seen in the scree-covered flanks and, from the abundance of pillow-lava debris; it is likely that the nunatak is dominated by pillow lava. A sill-like outcrop of basalt with conspicuous olivine and plagioclase phenocrysts up to 1 cm long is also present in the SW. Finally, a few very small exposures of yellow-orange lapilli tuff are also present close to the dyke zone, and Veit (2002) suggested that the basal exposures were composed of palagonite breccia [sic].

Hertha Nunatak. This is a small nunatak but it rises *c.* 225 m above the Larsen Ice Shelf (Fig. 6b). It is largely snow covered but its elongate nature suggests that it may have formed around an (unexposed) east–west-striking dyke. A conical hill at the western end is formed of five horizontal olivine–plagioclase-phyric lavas individually up to 7 m thick and with prominent oxidized ropy surfaces. Apart from rare loose blocks of brownish-black lapillistone or lapilli tuff and breccia, no other lithologies were observed by us. By contrast, Veit (2002) suggested that the nunatak is composed of 'palagonitized lapilli' (lapilli tuff?) intruded by dykes that locally fed pillow lava. The conflicting descriptions are currently unresolved.

Larsen Nunatak. Larsen Nunatak is a single ridge <2 km long that rises *c.* 140 m above the Larsen Ice Shelf and is centred around a poorly exposed, east–west-striking olivine–plagioclase-phyric dyke zone (Fig. 6f). It has a K–Ar age of 1.5 ± 0.5 Ma (del Valle *et al.* 1983). The ridge flanks are extensively covered by lava pillows up to 1 m long, whilst Veit (2002) also recorded pāhoehoe lava together with 'subaerial bombs and lapilli'.

Lindenberg Island. Rising *c.* 200 m above its surroundings, this nunatak is an east–west-striking (270°N$_{mag}$) ridge formed by an olivine–plagioclase-phyric dyke There are also rare exposures of yellow-orange lapilli tuff, tuff and breccia at the west end of the ridge.

Murdoch Nunatak. This is the largest nunatak in the group but it is mostly covered by scree. It is *c.* 370 m high (*c.* 320 m above the Larsen Ice Shelf) and has an area of *c.* 4 km^2, with a flat top and steep sides. A central zone of multiple dykes strikes *c.* 320°N$_{mag}$ and other thinner dykes strike 325°N$_{mag}$. Olivine–plagioclase-phyric pillow lava is well exposed on the west side and summit ridge, and there is a series of small exposures of black agglutinate with ropy textures on the north side. Lava bombs litter the surface of the nunatak. Near the summit there are small occurrences of yellow-orange lapilli tuff and tuff, and dark-grey to buff-coloured lapillistones containing fluidal bombs. Veit (2002) also observed subaerial lapilli and bombs, and he identified pāhoehoe in addition to pillow lava.

Oceana Nunatak. This is the only volcanic outcrop on Robertson Island (formed of Late Jurassic sedimentary rocks: Riley *et al.* 1997). It rises to 270 m asl, and consists of an olivine–plagioclase-phyric dyke core (two sets striking 250–260° N$_{mag}$ (main dyke zone) and 348–010°N$_{mag}$). Pyroclastic rocks crop out in a 10 m-high crag at the east end, comprising poorly bedded, massive brownish-grey lapillistone with scattered flattened olivine-phyric bombs. At the west end is a well-bedded association of yellow-orange tuff, lapilli tuff and tuff breccia with rare lava pillows. The western outcrop is also crossed by numerous small-displacement faults reflecting slumping directed towards the NW. A dyke dated by del Valle *et al.* (1983) yielded an age of 2.8 ± 0.5 Ma.

Pollux Nunatak. There is no exposure on this tiny nunatak, which is entirely formed of scree composed of probable dyke fragments.

Argo Point. This locality is situated on the east coast of Jason Peninsula, *c.* 140 km south of Seal Nunataks (Fig. 3). It consists of a small basaltic scoria mound measuring *c.* 300 m in diameter that is breached on its northern side. The mound has numerous bombs and blocks on its surface, and it is associated with a prominent trail of debris extending in a northeasterly direction on the adjacent heavily crevassed ice shelf (Fig. 7). The debris trail is today >4 km long (Saunders 1982 estimated >1 km), and is formed of basalt lava and scoria. It was described by Saunders (1982) as a moraine created by active ice erosion but it also possible that at least some of the debris is a wind tail caused by a combination of strong local winds redistributing loose materials of the scoria cone, followed by northeasterly flow of the ice shelf, as has been observed for similar deposits at Seal Nunataks (González-Ferrán 1983*a*). Inaccessible cliffs beneath the scoria mound

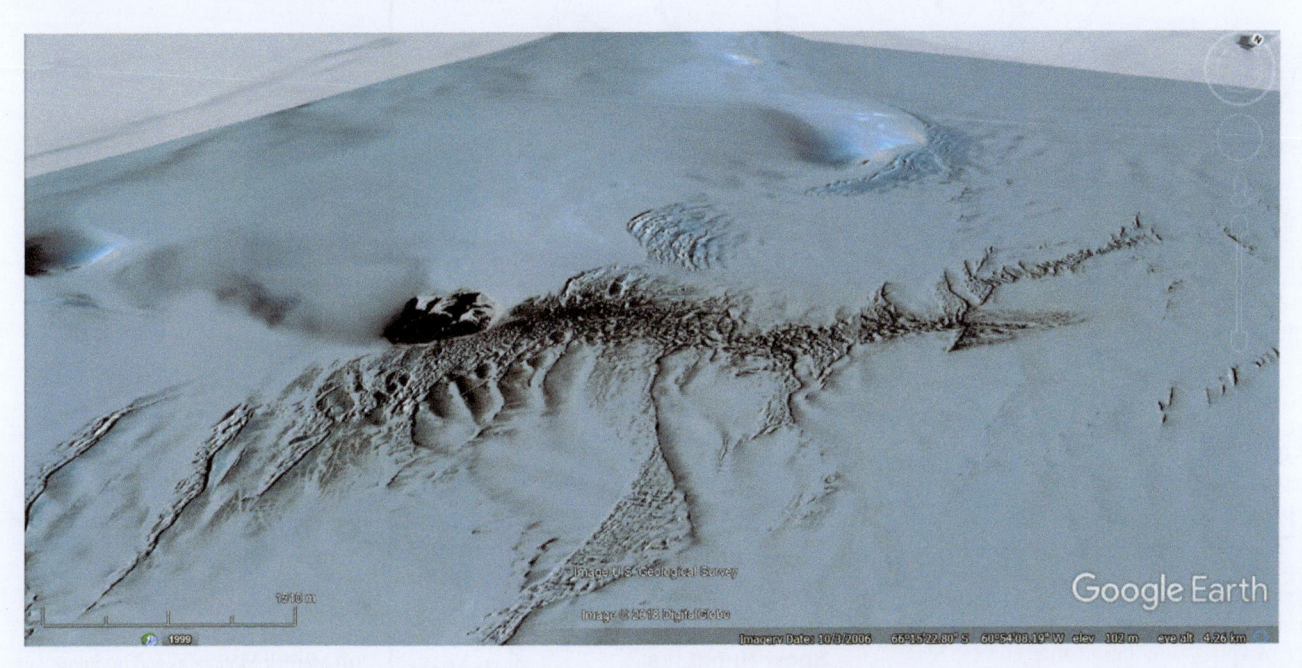

Fig. 7. View of the Argo Point scoria cone, viewed looking NW. Note the prominent debris trail stretching for >4 km NE of the cone, formed by a combination of wind and ice transport. Satellite image courtesy of Google Earth (data provider Maxar Technologies, 2020).

comprise olivine–plagioclase-phyric basalt lavas with ropy textures and scoria (from samples obtained nearby in the moraine: Saunders 1982). The scoria mound has published ages of 1.0 ± 0.3 and 0.8 ± 0.1 Ma (Smellie et al. 1988).

Finally, an isolated outcrop situated on the south side of the Jason Peninsula c. 37 km west of Argo Point also yielded very young K–Ar ages of 1.6 ± 0.5 and 1.3 ± 0.3 Ma (Smellie et al. 1988) (Table 2). However, that outcrop is a hematite-coated intrusion, implying that it is relatively altered, and its composition is tholeiitic, poor in alkalis and low in Nb. It thus more closely resembles the Mesozoic volcanic outcrops that dominate the geology of the Jason Peninsula and which are probably related to extension caused by the migration of a giant plume head associated with Gondwana break-up and triggering widespread bimodal mostly explosive volcanism in the Antarctic Peninsula (Saunders 1982; Smellie 1991; Pankhurst et al. 2000; Riley et al. 2010; see also Riley and Leat 2021 in this Memoir). It therefore does not belong to the very young post-subduction volcanism, and its apparently very young age is currently unexplained.

Bellingshausen Sea Volcanic Group

The Bellingshausen Sea Volcanic group encompasses widely scattered isolated outcrops on Alexander Island and Palmer Land (Fig. 3). They comprise: (1) Rothschild Island; (2) Alexander Island (Mount Pinafore, Debussy Heights, Hornpipe Heights and Beethoven Peninsula); and (2) Palmer Land (Henry Nunataks, Merrick Mountains, Sims Island, Rydberg Peninsula and Snow Nunataks: Fig. 8). The outcrops were first discovered by Bell (1973), with additional outcrops discovered and described by Care (1980), Burn and Thomson (1981), Laudon (1982) and O'Neill and Thomson (1985). All are small and those situated on the flanks of Mount Pinafore (northern Alexander Island) are on high ridges with difficult access (Fig. 9a & b). In addition, Snow Nunataks and outcrops in the Merrick Mountains are remote and have been visited only once (Rowley et al. 1990; Thomson and Kellogg 1990; Thomson and O'Neill 1990). As a result, with a few exceptions (Smellie et al. 1993; Smellie and Hole 1997), most outcrops have been studied only at reconnaissance level.

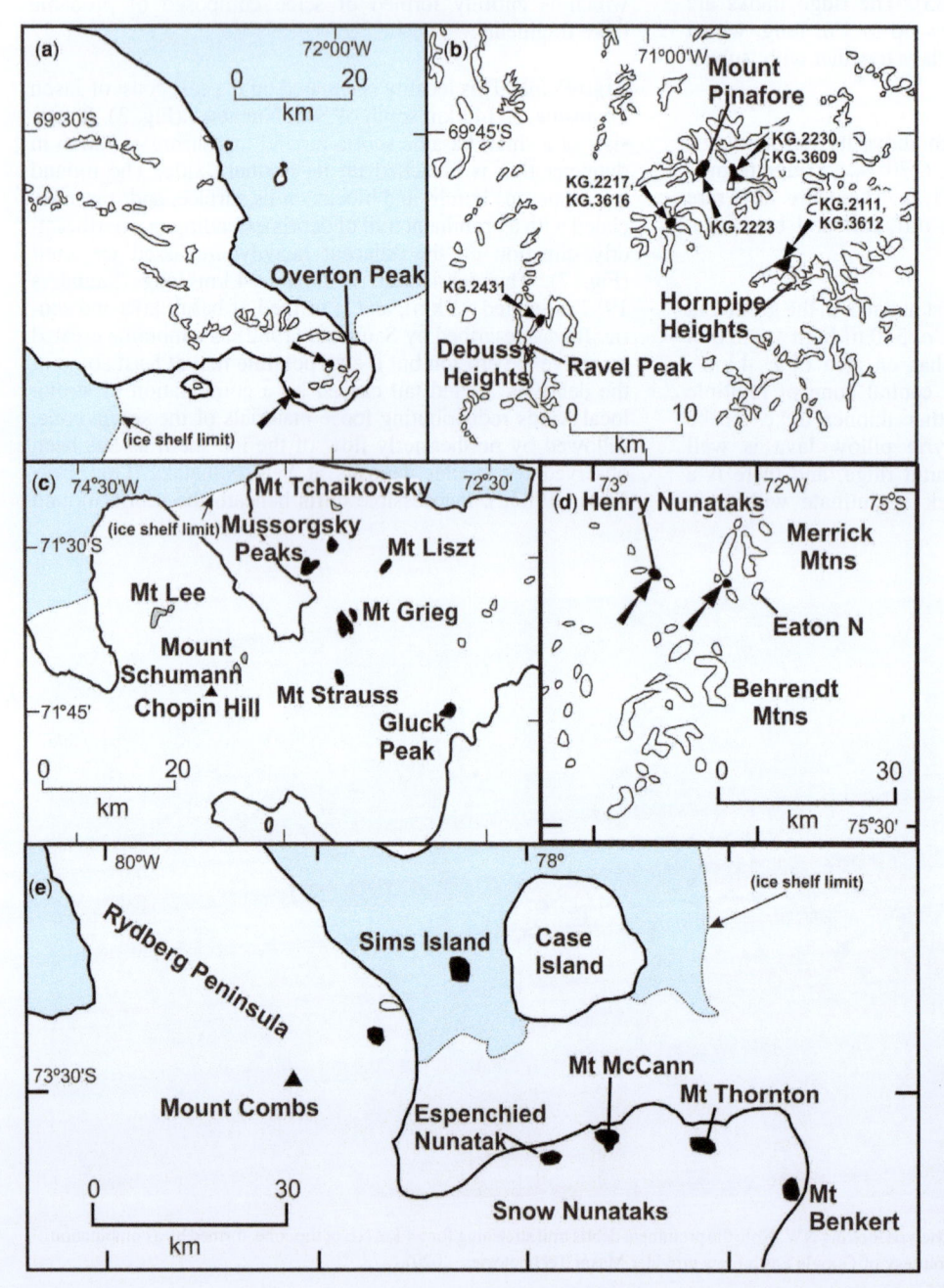

Fig. 8. Maps showing the location of Neogene alkaline volcanic outcrops at (a) Rothschild Island, (b) Mount Pinafore, Alexander Island, (c) Beethoven Peninsula, Alexander Island, (d) Merrick Mountains and (e) Snow Nunataks–Rydberg Peninsula–Sims Island (modified after Smellie 1999).

Fig. 9. Photomontage of field photographs of glaciovolcanic sheet-like sequences near Mount Pinafore and associated lithofacies. (**a**) View looking west towards exposures of volcanic rocks at Mount Pinafore. They rest unconformably (above the yellow line) on steeply dipping accretionary prism metasedimentary rocks (LeMay Group), which make up much of the foreground. Locality KG.2223 (Fig. 10). Image: R.W. Burn. (**b**) View of the 'Twin Peaks' locality on Mount Pinafore (see Fig. 10, locality KG.2217, KG.3616) with the basal unconformity with the underlying LeMay Group indicated (yellow line). Image: P.A.R. Nell. (**c**) Irregular lava lobes and pillowy masses mingled with yellow-orange tuff breccia at the base of a water-chilled lava low in the Mount Pinafore section. Several of the lava masses show prominent chilled margins; the hammer is *c.* 40 cm long. (**d**) Well-bedded reddish-brown volcanic sandstones and fine conglomerates overlying massive to poorly bedded, poorly sorted yellow-orange tuff breccia at the base of the Mount Pinafore section. Locality KG.2223 (Fig. 10). (**e**) A *c.* 75 m-thick pooled tephrite lava flow with a well-developed colonnade and entablature. The sloping apron at the base of the lava flow obscures an outcrop of diamictite overlain by poorly seen volcanic sandstones. The diamictite contains abraded and partly striated cobbles and boulders of greenish volcanic rocks derived from the underlying calc-alkaline lava succession, together with numerous clasts of low-grade metasedimentary rocks of the LeMay Group. The lava contains numerous spinel peridotite xenoliths up to 25 cm in diameter. Locality KG.3609 (Fig. 10). (**f**) Poorly stratified grey diamictite overlain by yellow tuff breccia in irregular contact with overlying coeval tephrite lava. Note the prominent reaction front (yellow) caused by intense palagonitization within the tuff breccia. A small pillow-like lava mass is also present in the tuff breccia at the upper left of the image. Locality KG.3616 (Fig. 10).

Mount Pinafore Volcanic Field

Rothschild Island. This is a large island situated off the NW coast of Alexander Island. It contains two small volcanic outcrops at its southeastern end close to Overton Peak (Care 1980; the Overton Peak Formation of Smellie 1999) (Fig. 8a). No contacts with bedrock are exposed, and Care (1980) speculated that the outcrops may be linked by a NE-striking fault. The southwestern outcrop is a cliff 100 m high. The basal 25 m of the section there comprises irregularly bedded pale-yellow and brown tuffs, and lapillistones showing crude polygonal jointing. Beds are commonly 2–10 cm thick but vary up to *c.* 1 m. Cross-lamination is present. The overlying section is composed of >30 m of dark- and pale-grey lapilli tuffs in beds 2–30 cm thick, often normally graded, and with lapillistones and tuffs towards the top. The northeastern

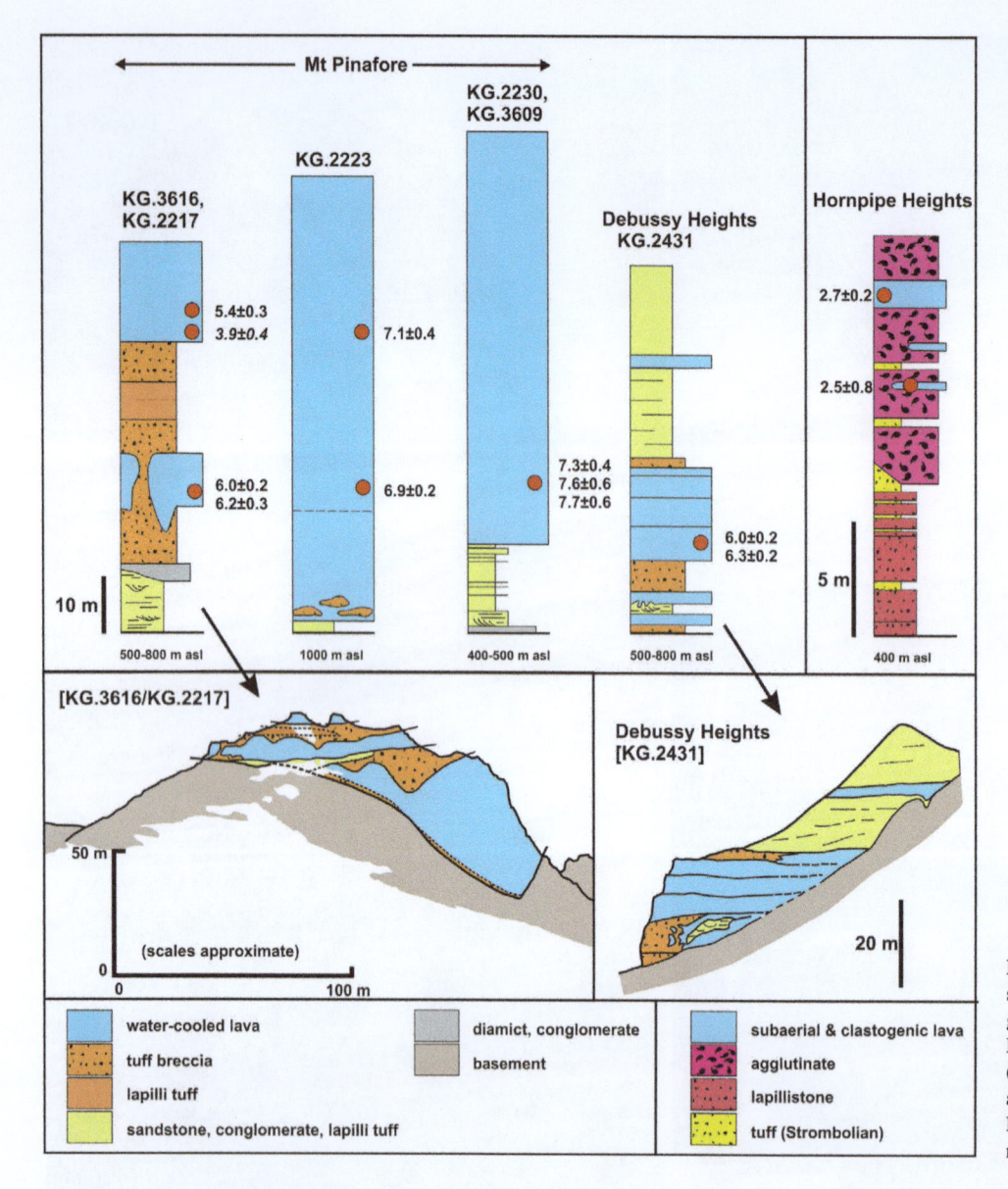

Fig. 10. Vertical profile sections and selected sketched outcrop views of glaciovolcanic sequences in the Mount Pinafore Volcanic Field (after Smellie et al. 1993; Smellie and Skilling 1994); Debussy Heights geology based on field notes of P.A.R. Nell.

outcrops consist of two small rounded hills, which are dominated by basaltic scree including black scoriaceous clinkers; there are also rare exposures of lapilli tuff and tuff similar to the southwestern outcrop described above. Beds are up to 35 cm thick, and they show abundant cross-laminations and channels consistent with subaqueous reworking of explosively generated (phreatomagmatic) tephra, and small-scale faulting. Both of the principal outcrops are intruded by dykes that form a multiple dyke zone 50 m wide with narrow screens of lapilli tuff and tuff in the northeastern outcrop, from which a K–Ar isotopic age of 5.4 ± 0.7 Ma was obtained (Smellie et al. 1988). The two principal outcrop areas are separated by 6 km and they may have been formed by eruptions from two separate small tuff cone centres, an origin that is different to other outcrops in the Mount Pinafore Volcanic Field (see below).

Alexander Island. Mount Pinafore in the Elgar Uplands region of northern Alexander Island includes three localities on Mount Pinafore itself, another situated near Ravel Peak (Debussy Heights) *c.* 15 km SW of the other outcrops, and one at Hornpipe Heights *c.* 10 km to the SSW (Burn and Thomson 1981; Hole and Thomson 1990; Hole 1990*b*) (Fig. 8b). There are two types of outcrop based on the dominant lithofacies, which were used as the basis for separation into two stratigraphical units (formations) by Smellie (1999): the Mount Pinafore Formation, comprising the three outcrops at Mount Pinafore and that at Ravel Peak; and the Hornpipe Heights Formation, found at Hornpipe Heights only (Figs 10 & 11). An additional locality *c.* 13 km NE of Mount Pinafore was shown as a volcanic outcrop of similar age on the map by Hole and Thomson (1990) but was not described. It consists of a plug-like mass *c.* 30 m in diameter and 100 m high formed of massive polymict tuff breccia with numerous blocks of LeMay Group metasediments, Tertiary volcanics and spheroidally weathered dolerite up to 1 m across in a basaltic tuff matrix. The outcrop is capped by a layer of brown well-bedded similarly polymict lapilli tuff, and a fresh-looking olivine dolerite crops out locally at the plug margin and is associated with several thin basaltic dykes. Although it is undated, the locally pervasive deuteric alteration of the outcrop and the likely calc-alkaline composition (based on petrographical characteristics) suggest that the outcrop is not Mio-Pliocene but is probably a vent-fill and part of the compositionally distinctive Early Tertiary subduction-related volcanism that is widespread in Elgar Uplands nearby (Burn 1981; McCarron 1997; McCarron and Millar 1997; McCarron and Smellie 1998; see also Leat and Riley 2021*b* in this Memoir). It is not discussed further here.

The four constituent outcrops at Mount Pinafore and Debussy Heights occupy palaeovalleys cut in bedrock

Fig. 11. Views of the subaerially erupted volcanic sequence and lithofacies at Hornpipe Hts. (**a**) Red (oxidized) agglutinate, scoria and clastogenic lavas draped on a steep underlying slope composed of LeMay Group metasedimentary rocks. (**b**) Aerial view of the Hornpipe Heights outcrop with the dark-coloured apron of red scoriaceous volcaniclastic rocks clearly visible. The volcanic outcrop is *c.* 1 km wide at its base. (**c**) Beautifully preserved reddened basanite agglutinate. Black fine-grained lavas appear to be present towards the base of the sequence but close inspection reveals that they are fine-grained agglutinates with numerous internal chilled margins. Kaersutite megacrysts are common at this locality. The hammer shaft is *c.* 40 cm long.

(deformed metasedimentary strata of the LeMay Group: Burn 1984). The lithofacies can be grouped into two principal associations: a basal epiclastic–volcaniclastic association of volcanic sandstone and conglomerate, and an upper volcanogenic association of lava and hyaloclastite breccia or tuff breccia (Figs 9c–f & 10) (Hole and Thomson 1990; Smellie *et al.* 1993; Smellie and Skilling 1994). Outcrop thicknesses vary from 60 to >80 m. The basal association consists of some combination of beds and lenses of largely massive polymict pale-grey sandy diamictite with abundant abraded non-volcanic (local basement) clasts up to 75 cm across; polymict sandy volcanic pebble conglomerate with crude wavy planar stratification; multistorey beds of polymict gravelly volcanic sandstone with planar stratification and cross-stratification commonly associated with broad shallow channels up to 3 m across (Fig. 9d); and flaggy monomict fine volcanic sandstone or lapilli tuff and tuff. Some of the underlying basement surfaces locally show signs of glacial modification (e.g. striations, ice moulding), and the overlying non-volcanic conglomerates contain rare striated and facetted clasts. The associated sediments are generally yellow-brown in colour due to palagonite alteration of sideromelane. They were sourced in contemporary volcanic materials, mainly sideromelane reworked from unconsolidated lapilli and ash. The blocky variably vesicular sideromelane fragments and abundant ash suggest a phreatomagmatic origin for the volcanism. Because the volcanic-derived deposits are first-cycle sediments, they might also be given primary volcanic names (lapilli tuffs and tuffs: White and Houghton 2006), although the enclosed pebbles are often well rounded. The overlying volcanic lithofacies association is dominated by thick basaltic sheet lavas. They often show spectacular columnar jointing (mainly entablature with much thinner basal and rare upper colonnades), and some are valley-confined (pooled) single lavas up to *c.* 80 m thick (Fig. 9e). The lavas may have pillowed and brecciated lava margins that are in intimate contact with yellow-orange hyaloclastite and tuff breccia (Fig. 9c & f). The latter are massive to crudely and coarsely stratified, and formed of blocky mainly poorly to non-vesicular sideromelane. Two of the outcrops show only a single pair of lithofacies associations consistent with single eruptive events, whereas two others (i.e. those at locality KG.2217/3616 and Ravel Peak) appear to have two or more pairs of associations (Fig. 10). Numerous K–Ar ages have been determined on all four outcrops, and vary between 7.7 ± 0.6 and 3.9 ± 0.4 Ma (Smellie *et al.*

1988). It is likely that the youngest age (3.9 ± 0.4 Ma) is spurious (probably too young) as an older age (5.4 ± 0.3 Ma) was also obtained from the same lava unit. Moreover, the sequence at locality KG. 2217/3616 is petrologically similar throughout, and there is no obvious erosional surface separating the upper and lower lavas there. Thus, it is possible that the 5.4 Ma age obtained on the upper lava may also have been 'younged' slightly (a similar age disparity affects two samples of a single lava at locality KG.2223: Fig. 10; Table 2). Therefore, four potential eruptive episodes are probably reliably dated as *c.* 7.6 and 7.0 Ma, and 6.1 Ma (the last at two different localities), although the possibility exists for a younger episode at 5.4 Ma.

The Hornpipe Heights outcrop is spectacular and has been thoroughly studied (Hole 1990*b*, unpublished; Smellie 1999). The mainly red-coloured (oxidized) lithofacies drape and infill cracks and gullies on the steep north-facing flank of a ridge composed of local non-volcanic basement rocks (LeMay Group metasediments: Figs 10 & 11a, b). The sequence extends up for a few hundred metres, almost the full height of the ridge, but it is just *c.* 20 m in thickness and has a general dip of *c.* 40°, with some beds dipping up to 52°. Close to the base of the sequence, a local lens of massive polymict orthobreccia is present, up to 2.5 m thick and 3 m long. It is composed of abundant crudely parallel slabs and blocks of the local basement (LeMay Group sandstone), together with basalt lava blocks and scoria in a tuffaceous matrix. The rest of the sequence consists of lapillistones with numerous disc-like (cowpat) bombs of olivine basalt. Some beds are reverse graded, and erosion-like surfaces are present which occasionally show channel-like steep-sided profiles. Oxidized agglutinate formed of weakly welded large aerodynamic bombs also occurs (Fig. 11c) and becomes coarser up-sequence. There are also interbedded, yellowish, buff and red-coloured fine lapillistones and tuffs, some of the latter showing asymmetrical ripple-like bedforms with amplitudes up to 5 cm that may be a product of minor sliding due to slope instability together with small-scale syn-depositional faulting. The yellow discoloration affecting some beds is caused by marginal palagonite alteration of sideromelane grains. Several thin (<2 m), platy, grey, olivine- and plagioclase–olivine-phyric highly vesicular lavas, some with poorly developed columnar jointing, occur mainly within the upper part of the sequence; they are probably clastogenic. Others, near the base of the slope, contain 'pillow-like' forms (pāhoehoe toes?). No vent for the sequence has been identified. There are two published K–Ar isotopic ages: 2.7 ± 0.2 and 2.5 ± 0.8 Ma (Smellie et al. 1988).

Beethoven Peninsula Volcanic Field

Beethoven Peninsula in southwestern Alexander Island contains 10 largely snow-covered nunataks and hills scattered over an area of c. 2500 km², of which six of the features have exposed volcanic rock, although all are probably volcanic in origin (Fig. 8c). Only five of the nunataks have been visited (Bell 1973; Hole 1990c; Smellie and Hole 1997). Based on the presence and distribution of strong magnetic anomalies (Renner et al. 1982), the volcanic field at Beethoven Peninsula may be very extensive and it may also underlie much of Monteverdi Peninsula to the SE and Latady Island to the NW (Fig. 3). If true, the combined area of volcanic rock would exceed 7000 km², making it the largest volcanic field in the Antarctic Peninsula south of James Ross Island (Smellie and Hole 1997; Smellie 1999; cf. Smellie et al. 2013). The most accessible and informative outcrop on Beethoven Peninsula is that forming the larger (southwestern) of the two Mussorgsky Peaks (summit elevation c. 500 m asl: Fig. 12a). The inaccessible, but similarly well-exposed, sequence at Mount Grieg (summit elevation of c. 600 m asl) is also informative but has only been viewed at a distance (Fig. 12b). Two lithofacies associations are present, corresponding to lava-fed delta and subaqueous tuff cone lithofacies (Smellie and Hole 1997). Subaqueous lithofacies crop out at Mussorgsky Peaks, Mount Liszt and Mount Strauss, and they form the thick (c. 250–300 m) basal section at Mount Grieg; it is inferred that another four outcrops may be formed of similar lithofacies, at Mount Tchaikovsky, Mount Lee, Mount Schumann and Chopin Hill, but they have not been visited. The subaqueous lithofacies comprise yellow-orange crudely bedded lapilli tuff and lesser thin-bedded tuff (Fig. 12c) containing channel and dewatering structures and displaced slabs of tuff up to 1 m long. Massive chaotic olivine-phyric lava pillows and pillow breccia mingled with yellow lapilli tuff also underlie the volcaniclastic rocks at the eastern Mussorgsky Peaks outcrop. Large and small synsedimentary slump structures (cf. Mount Benkert, below) and faulting are also common and conspicuous (Fig. 12d). Several SW–NE-trending olivine–plagioclase-phyric dykes cut the Mussorgsky Peaks outcrops. The thickest exposed sequence of subaqueous lithofacies is c. 200 m thick at southwestern Mussorgsky Peaks, although the entire sequence may be substantially thicker (up to 500 m, assuming continuity of outcrop down to bedrock: Smellie 1999). Mount Liszt and Mount Strauss are composed of yellow-orange lapilli tuffs similar to the basal Mussorgsky Peaks sequence.

Lithofacies that form lava-fed deltas form a capping sequence up to 100–150 m thick at Mount Grieg and the western Mussorgsky Peak, and it may also include the sequence exposed at Gluck Peak. The sequence at Mount Grieg is inaccessible but it can be studied easily using binoculars. At Mussorgsky Peaks, the lithofacies comprise pillow-fragment breccia and minor pillow lava that form a spectacular thick (100 m) dark-grey unit showing large-scale homoclinal dipping bedding that oversteps the underlying yellow-orange volcaniclastic rocks of the Mussorgsky Peaks Formation (Fig. 12a & e); although flat-topped, the summit of Mussorgsky Peaks is inaccessible and it is unclear if any subaerial lavas are preserved. However, lavas are preserved at Mount Grieg where they form a 50–100 m-thick capping unit of horizontal grey pāhoehoe(?) lavas. Gluck Peak is formed of dark-grey pillow lava and pillow breccia similar to that capping Mussorgsky Peaks together with tuff breccia (Fig. 12f). Moreover, compared to Mussorgsky Peaks, the constituent lava pillows at Gluck Peak are much more highly vesicular and they were fed by dykes similar to relationships seen at Seal Nunataks.

The Beethoven Peninsula outcrops are very poorly dated. K–Ar isotopic ages of 2.5 and 0.68 ± 0.97 Ma were reported for samples from Mussorgsky Peaks and Gluck Peak, respectively, without analytical details (Hole 1990c). The very young age from Gluck Peak (i.e. too young to date by the K–Ar method at the time) suggests that the volcanic field is potentially still active. However, in common with all the other outcrops on the Beethoven Peninsula, Gluck Peak is extensively eroded and there are no primary volcanic landforms, suggesting a much older age. Conversely, thermal waveband satellite imagery of Gluck Peak has revealed elevated heat flow near the summit of Gluck Peak, with inferred temperatures well above background or locally enhanced insolation (Peter Fretwell, British Antarctic Survey, 2016 pers. comm.). The occurrence is most readily explained as enhanced geothermal (volcanic) heat. If confirmed, this would be the first occurrence of geothermally heated warm ground to be discovered in the Antarctic Peninsula region outside of Bransfield Strait (i.e. the Deception Island active volcano and Bransfield Strait seamounts: see Geyer et al. 2021; Leat and Riley 2021a in this Memoir).

Merrick Mountains Volcanic Field

Two very poorly known, small and isolated volcanic outcrops are present at Henry Nunataks and Merrick Mountains in southern Palmer Land (Fig. 8d) (Halpern 1971; Rowley et al. 1990; Smellie 1999). The outcrop at Henry Nunataks (possibly two discrete outcrops, only one visited: Thomson and Kellogg 1990) is composed of c. 100 m of grey aphanitic basalt lavas with rubbly surfaces ('a'ā?) cut by a dyke. The Merrick Mountains outcrop (west of Eaton Nunatak) is extensively frost shattered but is formed of basanite lava breccia with a palagonite-altered glassy matrix (possibly hyaloclastite). It is overlain by thin vesicular to scoriaceous lavas (Laudon 1982; Smellie 1999) for which Halpern (1971) reported a K–Ar age of 6 Ma. However, the age lacks published analytical details and its reliability is uncertain.

Snow Nunataks Volcanic Field

Several volcanic outcrops occur in the Rydberg Peninsula–Sims Island–Snow Nunataks area of southwestern Palmer Land (Renner et al. 1982; O'Neill and Thomson 1985; Rowley and Thomson 1990; Thomson and O'Neill 1990; Hathway 2001) (Fig. 8e). Those at Rydberg Peninsula may include the several-hundred-metre-high and 3.5 km-wide cone-shaped Mount Combs but it is completely snow covered. However, a small exposure of undated subaerial olivine-phyric basalt lavas occurs c. 15 km NE of Mount Combs (Renner et al. 1982; unpublished field notes of R.G.B. Renner 1976). Sims Island is a prominent feature 3 km in length that rises to c. 380 m asl. It is constructed of basal olivine basalt pillow lava overlain by 30–40 m of thick-bedded pillow breccia and 'gravelly volcanic sandstone' (probably lapilli tuff), then more pillow lava that forms the remainder of the pile (Figs 8e & 13). Inaccessible bedded clastic deposits may cap the sequence, and large irregular columnar intrusions are present and especially prominent to base. The Sims Island sequence has published $^{40}Ar/^{39}Ar$ ages of 3.46 ± 1.20 and 2.30 ± 0.54 Ma, of which the latter (with lower errors) is regarded as more reliable (Hathway 2001).

Snow Nunataks are an east–west-trending chain of four volcanic outcrops (O'Neill and Thomson 1985; Thomson and

Fig. 12. Photomontage of field photographs of tuyas on Beethoven Peninsula and associated lithofacies. (**a**) Mussorgsky Peaks (*c.* 630 m asl) viewed from the north from a distance of *c.* 750 m. The lower orange-yellow part of the sequence is entirely volcaniclastic in origin (lapilli tuffs). The black cap rock is composed of basal tuff breccia and pillow lavas possibly overlain by subaerial pāhoehoe lava at the summit. On the extreme left, the poorly seen grey ridge contains a series of east–west-trending multiple dykes. (**b**) View of Mount Grieg (*c.* 600 m asl) looking SE. Like other outcrops on the Beethoven Peninsula, the lower part of the exposed sequence is composed of black pillow lava, hyaloclastite and bedded volcaniclastic rocks. They are capped by pillow basalt, tuff breccia and probably subaerial lavas, corresponding to a typical tuya sequence. Mount Grieg has never been visited due to the impassable surrounding terrain. Mount Strauss is seen in the right-hand background, almost entirely snow-covered. (**c**) Diffusely bedded lapilli tuffs in the lower parts of the Mussorgsky Peaks outcrop. The lapilli tuffs are locally cross-bedded on a metre scale. The lens cap on the extreme left is 52 mm in diameter. (**d**) Diffusely stratified and thin-bedded lapilli tuffs in the lower part of the Mussorgsky Peaks section. The thin dark-brown beds are very-fine-grained palagonite tuff, offset by synsedimentary faulting possibly related to large-scale slumping seen elsewhere on the outcrop. Some palagonite tuff has been mobilized and injected along near-vertical fractures in the centre of the image. The width of the field of view is *c.* 20 m. (**e**) Tuff breccia composed of intact and fragmented lava pillows, and palagonite-altered sideromelane exposed at the base of the black cap rock at Mussorgsky Peaks. The white fragment in the centre of the image is a partially melted xenolith of arkose with a glassy vesicular rim (buchite). The hammer shaft is *c.* 40 cm long. (**f**) Tuff breccia derived from a collapsed pillow-lava pile overlying possible lapilli tuff (yellow) at Gluck Peak (locality KG.3627). Unusually for the Beethoven Peninsula, the pillows are highly vesicular, even in their chilled margins. Plagioclase phenocrysts are also clearly visible in hand specimens and are a rarity on the Beethoven Peninsula. The width of the field of view is *c.* 5 m.

O'Neill 1990). Similar to the Beethoven Peninsula outcrops, two lithofacies associations have been defined based on differing lithofacies characteristics (Smellie 1999). Subaqueously erupted and emplaced basalt pillow lava and orange-brown lapilli tuff crop out at Mount Benkert and Mount Thornton, and form the basal sequence of Mount McCann. They also form the Sims Island outcrop (*c.* 30 km to the NW: Fig. 8e). The proportions of the two main lithofacies vary: pillow lava is 200 m thick at Mount McCann and is capped by just 5 m of lapilli tuff, whereas spectacular exposure at Mount

Espenchied Nunatak is formed of crudely stratified black and reddish-brown lapilli tuff, and tuff breccia intruded by very thin (3 cm) basalt dykes. Isotopic ages (all by K–Ar) range between 20 and 1.6 Ma (unpublished information of J.W. Thomson cited in Smellie *et al.* 2009). There are no published analytical details and the oldest age (from Mount McCann) may be unreliable but Mount Benkert is *c.* 4.6 Ma and Mount Thornton is *c.* 1.6 Ma (Table 2).

Physical volcanology and palaeoenvironmental inferences

Veit (2002) suggested both glacial and marine eruptive settings for the Seal Nunataks centres but without supporting evidence. However, apart from generally small isolated outcrops of subaerial (i.e. non-glacial) lavas or scoria cones (at Argo Point, Hornpipe Heights, Henry Nunataks and Rydberg Peninsula), we suggest that the Antarctic Peninsula post-subduction volcanism appears to have been overwhelmingly glaciovolcanic: that is, erupted in association with an ice sheet (Smellie *et al.* 1988, 1993; Smellie and Skilling 1994; Smellie and Hole 1997). Investigations of the outcrops have been used to reconstruct multiple critical parameters of the mainly Mio-Pliocene Antarctic Peninsula Ice Sheet by Smellie *et al.* (2009). The Antarctic Peninsula (south of James Ross Island: see Smellie 2021 in this Memoir) contains two principal generic types of glaciovolcanic outcrops corresponding to sheet-like sequences and tuyas (Smellie and Edwards 2016, chap. 8). Sheet-like sequences are defined by an association of water-chilled sheet lava and waterlain volcaniclastic deposits that have a laterally extensive sheet- or ribbon-like geometry, and typically rest on glacigenic (till) or epiclastic sediments. The absence of internal erosional or otherwise time-significant unconformities indicates that the two associations form a genetically related multistorey unit constructed during a single eruptive event. Two types have been defined: the Mount Pinafore type, with the Alexander Island occurrences given the status of sequence holotypes by Smellie and Skilling (1994; also Smellie *et al.* 1993); and the Dalsheidi type, based on outcrops in southern Iceland (Smellie 2008). However, it now seems likely that the two 'types' previously distinguished are simply variants occurring in a broad continuum of deposits (Smellie and Edwards 2016, pp. 240–247). Tuyas are the most distinctive

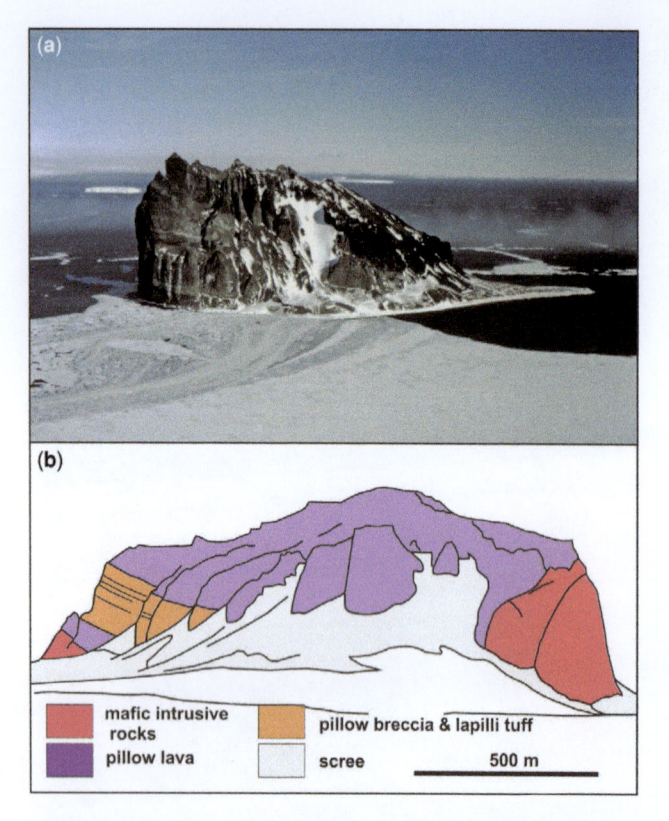

Fig. 13. (a) Photograph of Sims Island (viewed looking west) and (b) geological interpretation (viewed looking south; based on Hathway 2001).

Benkert is formed of at least 350 m of lapilli tuff with minor pillow lava, pillow breccia and thin massive lavas (Fig. 14). The latter outcrop also shows channel structures on a range of scales, including spectacular channels a few hundred metres wide and up to 50 m deep that are probably the two-dimensional traces of slump scars (cf. Smellie 2018). The Mount Thornton sequence is similar to that at Mount Benkert in that it is dominated by massive and thin-bedded lapilli tuffs that show much evidence for slope instability (convolute layering and folding), but it also has poorly exposed pillow lava and blocky lava (lava breccia?) at its base. Subaerially erupted scoriaceous and clinkery basalt rubble and massive vesicular lavas form the upper sequence at Mount McCann, whilst

Fig. 14. Photograph of Mount Benkert, looking north. The sequence is composed of subaqueously erupted and emplaced basalt pillow lava, tuff breccia and orange-brown lapilli tuff. It is crossed by several very large convex-down channel-like structures caused by repeated coeval slumping events (slump-scar surfaces). The cliff is *c.* 350 m high. The red ring encloses a figure on a skidoo. Image: Janet Thomson.

morphological expression of glaciovolcanism, characterized by a flat or gently domed top and steep sides (Mathews 1947; Smellie 2013; Smellie and Edwards 2016, pp. 220–238). The flat or gently domical top is an expression of the construction of a small subaerial shield, whereas the steep flanks are a primary feature formed as a result of the lateral progression of one or more lava-fed deltas, in which the subaerial capping lavas (analogous to delta topset beds) overlie homoclinal, steep-dipping foreset beds of breccia and tuff breccia (formerly called hyaloclastite; see the discussion in Smellie and Edwards 2016, pp. 197–205). The prominent planar structural discontinuity that separates the lavas from tuff breccia is called a passage zone and it is a fossil water level (Smellie 2006). It represents the migrating position of the delta brink point at which the subaerial lava is rapidly cooled by water and mechanically broken, with the glassy and aphanitic lava fragments tumbling down the delta front together with larger-scale delta-front collapses (Smellie and Hole 1997; Skilling 2002).

The sheet-like sequences are confined to northern Alexander Island (i.e. at Mount Pinafore and Ravel Peak). In general, the basal parts of the sequences are composed of fragmental rocks transported and deposited by flowing water. A wet-based glacial (sub-ice) eruptive setting was inferred because of the following observations: (i) the underlying bedrock surface is commonly striated or otherwise ice-moulded and may be overlain by polymict diamict interpreted as till (possibly flow or meltout till); (ii) the associated lavas are very fine grained (aphanitic), indicating rapid chilling; and (iii) the prismatic to hackly (blocky) jointing and rare presence of pillow lava are characteristic of abundant water and associated strong cooling (Smellie and Hole 1997; also Smellie and Edwards 2016, pp. 193–197). Some of the lavas are very thick (up to 80 m), suggesting that they were ponded either in a topographical depression or else by a barrier of coeval ice if the lavas flowing beneath thinner ice in a tributary valley abutted much thicker ice in a major trunk glacier (cf. Lescinsky and Sisson 1998). The basal fragmental beds are mainly traction current deposits laid down under variable flow states, including upper flow-regime conditions. In view of the valley-confined setting and by analogy with the study of a broadly comparable sequence in Iceland by Walker and Blake (1966), formation within a confined ice tunnel was inferred by Smellie et al. (1993). The thickness of tabular cross-stratified units present (typically 40–60 cm) is qualitatively consistent with flow depths of c. 1–5 m, and analogy was made with esker deposits in non-volcanic glacial systems (Smellie and Skilling 1994; but see Smellie and Edwards 2016, p. 176). Although some of the epiclastic deposits are polymict, most are monomict and are formed of reworked and redeposited sideromelane whose blocky shapes and variable vesicularity suggest derivation from unconsolidated phreatomagmatic tephra, probably a tuff cone constructed at an early stage of the glaciovolcanic event. The overlying volcanic lithofacies (sheet lava, locally pillowed and tuff breccia) indicate strong water chilling and it seems likely that the entire sequence was at least intermittently flooded by water during its formation. Although minor hyaloclastite (sensu White and Houghton 2006) may be present locally, where lava was chilled and fragmented in situ in water and against wet tuff breccia, the host volcaniclastic deposit is typically relatively fine grained (lapilli tuff) and shows crude coarse planar stratification characteristic of deposition mainly from hyperconcentrated density flows during subglacial meltwater flood events (Loughlin 2002; Smellie 2008) which, in the Alexander Island examples, were probably tunnel-confined (Smellie et al. 1993). Because the lithofacies in most sheet-like sequences are subaqueous, they only provide a crude minimum estimate for the thickness of coeval ice. Although the associated ice is generally thought to have been relatively thin ($\leq c.$ 150 m: Smellie 2008), consistent with the thinness of the sequences (tens of metres), there are normally no cogenetic subaerial lithofacies that might give a clue to the elevation of the original ice surface.

Most of the other volcanic outcrops represent tuyas in various stages of construction. They make up multiple centres in the Seal Nunataks, southeastern Rothshild Island, Beethoven Peninsula and Snow Nunataks volcanic fields. Those outcrops in the Seal Nunataks and Beethoven Peninsula volcanic fields show different stages in the evolution of tuyas caused by different coeval ice thicknesses and different responses of the glacial hydraulics. They were used to illustrate the varied lithofacies and lithofacies architectures in tuyas, and to erect a general model for tuya construction (Fig. 15) (Smellie and Hole 1997). The outcrops occur in isolation and a glacial setting for the tuya volcanism was inferred mainly because the subaqueous sequences are several hundred metres thick, implying a similar high coeval water elevation and thus ponding of (melt)water by ice. There is no palaeotopography with which a non-glacial (pluvial) lake might have been confined and the elevations of the subaqueous lithofacies are generally too high to be explained by marine construction followed by major regional uplift. Relationships between some of the lithofacies associations are also irreconcilable with a marine setting (see below). The best-exposed examples are at Mussorgsky Peaks and (inaccessible) Mount Grieg on Beethoven Peninsula (Alexander Island), which show the prominent bipartite division into two distinctive lithofacies associations typical of tuyas: that is, a basal succession or lithosome composed of subaqueous, relatively coarse clastic lithofacies deposited from a variety of sediment gravity flows (mainly hyperconcentrated flows) and sector collapses, and a capping sequence of subaerial pāhoehoe lavas and tuff breccia that together comprise a lava-fed delta. The bulk of the clastic lithofacies are crudely bedded relatively coarse lapilli tuffs that probably formed during continuous-uprush episodes of rapid vertical aggradation in a subaqueous tuff cone typical of Surtseyan cone construction; conversely, less common, thinner sequences of thinner-bedded tuffs and finer lapilli tuffs were probably formed during episodes of discrete tephra-jetting activity or during periods of quiescence, allowing redistribution of detritus from unstable volcano flanks (Smellie and Hole 1997; Smellie and Edwards 2016). Snow Nunataks and Sims Island also represent tuyas in various stages of construction and erosion, with many features similar to those seen in the Beethoven Peninsula tuyas (Smellie 1999; Smellie et al. 2009). By contrast, the numerous small outcrops at Seal Nunataks are dominated by multiple dykes; other lithofacies are generally minor apart from pillow lava, of which up to 150 m of vertical thickness is locally exposed above the Larsen Ice Shelf. Additional lithofacies include subaqueously deposited lapilli tuffs showing evidence for slope instability, similar to the basal subaqueous lithofacies at Mussorgsky Peaks. The lithofacies appear to represent the basal lava-dominated (non-explosive) pillow volcano cores of tuyas. The absence of lapilli tuffs associated with explosive hydrovolcanic eruptions at most localities suggests that ambient pressures were relatively high and were sufficient to suppress vesiculation, consistent with either deep-water eruption or substantial contemporaneous ice thicknesses; minimum edifice heights of c. 500–600 m are permissible, assuming the outcrops extend down to the present seafloor. Proving that the outcrops were erupted subglacially is difficult (cf. Veit 2002) but was based on the presence of small patches of agglutinate resting on pillow lava at several localities. Such an association is probably only possible if eruptions were subglacial, beneath a glacial cover that was composed of ice (i.e. with a capping layer

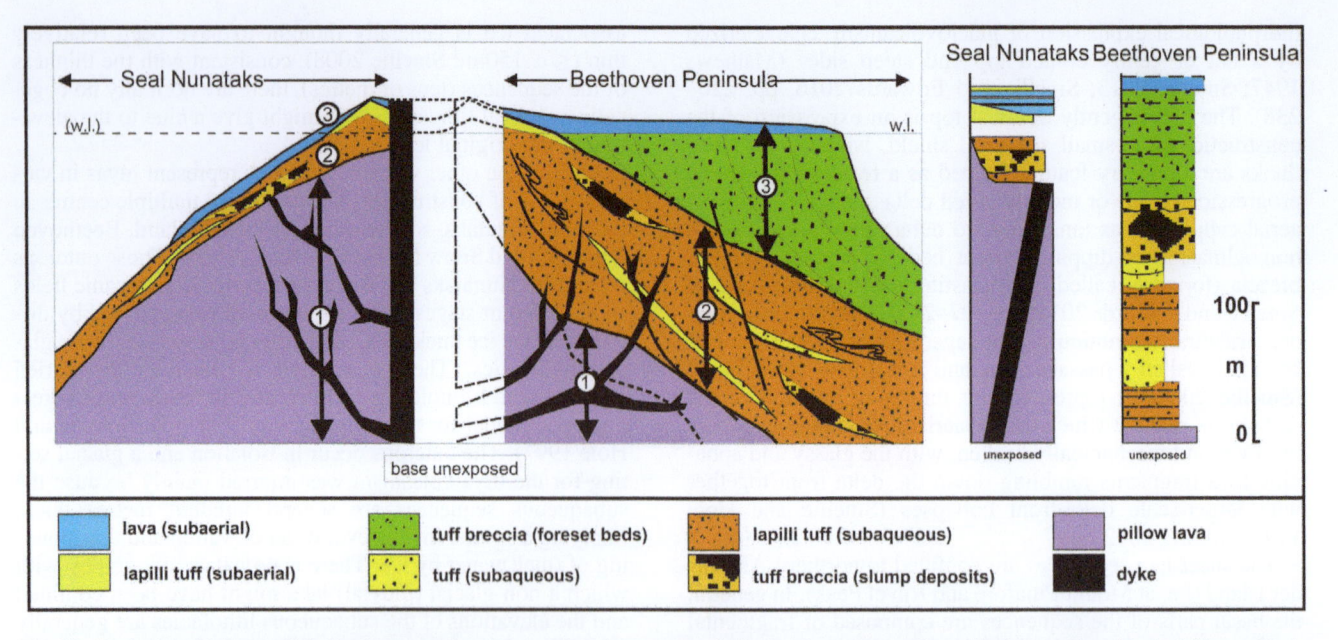

Fig. 15. Vertical profile logs and idealized cross-sections for tuyas based on outcrops in the Beethoven Peninsula and Seal Nunataks volcanic fields; modified after Smellie and Hole (1997). The major edifice-building stages are also numbered: 1, pillow mound; 2, subaqueous tuff cone stage (mainly vertical aggradation; ultimately subaerial); and 3, subaerial stage: lava (Seal Nunataks) or lava-fed delta (Beethoven Peninsula; mainly lateral progradation). The pillow-mound stage is largely unseen at Mussorgsky Peaks but its presence is inferred from minor pillow lava interbedded with lapilli tuff, pillow fragments contained in the lapilli tuffs, and by a 20–30 m-thick basal pillow and breccia mass. Note that Seal Nunataks show no evidence for lava-fed deltas; instead, the coeval (melt)water levels appear to have collapsed, probably by sudden subglacial discharge during jökulhaups, leading to the subaqueous lithofacies (pillow lava, lapilli tuff) becoming draped by subaerial lithofacies (lava, scoria, agglutinate). w.l., water level.

of snow or firn either very thin or absent) and initially within a meltwater-filled englacial vault. Under these conditions, the meltwater in the vault is able to float the surrounding ice, triggering a jökulhlaup, for which there is no evidence preserved (as lithofacies), and draining the vault. Thus, the pillow lava pile became exposed subaerially and the vent dried out, transforming to a magmatic eruption, and resulting in deposition of agglutinate and clastogenic lavas. Such a sequence of events and association of lithofacies cannot occur in a non-glacial (marine) setting, and contemporary ice thicknesses >600 m are therefore implied (Smellie and Hole 1997). The ice was also presumably wet-based ice if it was hydraulically lifted and basal drainage occurred.

There is insufficient known about the small isolated volcanic outcrops at Henry Nunataks, Merrick Mountains and Rydberg Peninsula to be confident of interpreting their eruptive setting, although the eruptions were at least partly subaerial at Rydberg Peninsula and Henry Nunataks. However, the small outcrop in the Merrick Mountains contains hyaloclastite and may be glaciovolcanic (Rowley et al. 1990). Conversely, outcrops at Argo Point (a scoria cone constructed on subaerial lavas) and Hornpipe Heights (dominated by oxidized agglutinate and clastogenic lavas) were fully subaerial and presumably took place in an absence of any significant local ice.

Summary

Late Neogene alkaline volcanic rocks form several monogenetic volcanic fields in the Antarctic Peninsula. Some of the outcrops (Seal Nunataks) will now be difficult to revisit due to contemporaneous ice-shelf collapses and a consequent requirement for intensive helicopter support for field parties. Isotopic dating shows that the volcanism is mainly Mio-Pliocene in age (<7.5 Ma). However, almost all of the published ages were determined by the relatively imprecise K–Ar method and they should be repeated using the $^{40}Ar/^{39}Ar$ method for greater reliability and accuracy, similar to recent investigations of other alkaline volcanic fields in Antarctica (e.g. Smellie 2021; Smellie and Martin 2021; Smellie and Rocchi 2021; Smellie et al. 2021 in this Memoir). The Antarctic Peninsula volcanism is post-subduction, and is thought to be causally related to the rise and decompression melting of mantle through windows in the subducted slab following the sequential (south to north) collision of a segmented spreading centre with the Peninsula trench. However, there is no obvious correlation in timing between the initiation of volcanic activity in the outcrops and cessation of subduction. This is an important problem that still needs to be resolved for a fuller understanding of the genesis of the Neogene post-subduction volcanism.

The outcrops occur as three principal types: rare scoria cones, glaciovolcanic sheet-like sequences and tuyas. There are only two well-exposed examples of scoria cones: at Argo Point and on Alexander Island at Hornpipe Heights. The sequences are relatively simple accumulations of subaerially erupted scoria, agglutinate and clastogenic lavas. Other possible examples include outcrops at Henry Nunataks and Rydberg Peninsula, which are relicts of subaerial lava sequences whose sources and original morphologies are unclear.

Sheet-like sequences are restricted to a comparatively small area surrounding the summit of Mount Pinafore and at Ravel Peak (Debussy Heights) in northern Alexander Island. They include the oldest post-subduction volcanic outcrops in the Antarctic Peninsula south of James Ross Island, extending back to c. 7.5 Ma. Their preservation, the variability of the lithofacies and the generally excellent exposure enabled them to be promoted as sequence holotypes for glaciovolcanic sheet-like eruptions, originally called Mount Pinafore type. Only minimum coeval ice thicknesses can be inferred for the outcrops but the glacial cover was probably thin (several tens of metres to <150 m).

Tuyas are by far the commonest type of post-subduction Neogene volcanic edifice, being characteristic of the Seal Nunataks, Beethoven Peninsula and Snow Nunataks volcanic fields. The tuyas are the glaciovolcanic equivalents of subaqueous to emergent Surtseyan volcanoes. Examples in each of the volcanic fields show different stages in the evolution of tuyas, from the lava-dominated cores of pillow mounds, through explosively generated tephras of a subaqueous tuff cone stage, to capping and pāhoehoe lava-fed deltas that were responsible for laterally extending the edifices. The individual centres were erupted in association with considerably greater ice thicknesses (hundreds of metres) than were associated with the sheet-like sequences, and the lithofacies and architectural features were gathered together in an illustrative general model for tuya construction. Investigations of the post-subduction alkaline volcanism in the Antarctic Peninsula have thus been important in establishing many of the diagnostic characteristics of glaciovolcanic sequences, their styles of eruption and deducing the palaeoenvironmental implications.

Acknowledgements The fieldwork on which this chapter is based was undertaken by M.J. Hole in 1985–88. The authors thank the British Antarctic Survey for originally supporting our project. Andy Saunders is also thanked for additional information on the Argo Point outcrop, and we are grateful to Janet Thomson, Rick Burn and Philip Nell for permission to publish their photographs. Finally, the authors are very grateful for constructive reviews by Corina Risso and Luis Lara, and comments by the editor, Kurt Panter.

Author contributions JLS: conceptualization (lead), formal analysis (lead), investigation (equal), methodology (lead), writing – original draft (lead), writing – review & editing (lead); **MJH**: investigation (equal), writing – review & editing (supporting).

Funding This research received no specific grant from any funding agency in the public, commercial, or not-for-profit sectors.

Data availability Data sharing is not applicable to this article as no additional datasets were generated other than those presented in this chapter.

References

Barber, P.L., Barker, P.F. and Pankhurst, R.J. 1991. Dredged rocks from Powell Basin and the South Orkney Islands. *In*: Thomson, M.R.A., Crame, J.A. and Thomson, J,W. (eds) *Geological Evolution of Antarctica*. Cambridge University Press, Cambridge, 361–367.

Barker, P.F. 1982. The Cenozoic subduction history of the Pacific margin of the Antarctic Peninsula: ridge crest–trench interactions. *Journal of the Geological Society, London*, **139**, 787–801, https://doi.org/10.1144/gsjgs.139.6.0787

Bell, C.M. 1973. The geology of Beethoven Peninsula, south-western Alexander Island. *British Antarctic Survey Bulletin*, **32**, 75–83.

Burn, R.W. 1981. Early Tertiary calc-alkaline volcanism on Alexander Island. *British Antarctic Survey Bulletin*, **53**, 175–193.

Burn, R.W. 1984. *The Geology of the LeMay Group, Alexander Island*. British Antarctic Survey Scientific Reports, **109**.

Burn, R.W. and Thomson, M.R.A. 1981. Late Cenozoic tillites associated with intraglacial volcanic rocks, Lesser Antarctica. *In*: Hambrey, M.J. and Harland, W.B. (eds) *Pre-Pleistocene Tillites: A Record of Earth's Glacial History*. Cambridge University Press, Cambridge, UK, 199–203.

Care, B.W. 1980. The geology of Rothschild Island, north-west Alexander Island. *British Antarctic Survey Bulletin*, **50**, 87–112.

Del Valle, R.A., Fourcade, N.H. and Medina, F.A. 1983. *Interpretacion preliminar de las edades K/Ar y de los analisis quimicos de las rocas volcánicas y de los diques de los nunataks Foca, Antártida*. Contribuciones del Instituto Antártico Argentino, **287**.

Fleet, M. 1968. *The Geology of the Oscar II Coast, Graham Land*. British Antarctic Survey Scientific Reports, **59**.

Geyer, A., Pedrazzi, D. et al. 2021. Deception Island. *Geological Society, London, Memoirs*, **55**, https://doi.org/10.1144/M55-2018-56

González-Ferrán, O. 1983a. The Seal Nunataks: an active volcanic group on the Larsen Ice Shelf, West Antarctica. *In*: Oliver, R.L., James, P.R. and Jago, J.B. (eds) *Antarctic Earth Science*. Australian Academy of Science, Canberra, 334–337.

González-Ferrán, O. 1983b. The Larsen Rift: an active extension fracture in West Antarctica. *In*: Oliver, R.L., James, P.R. and Jago, J.B. (eds) *Antarctic Earth Science*. Australian Academy of Science, Canberra, 344–346.

Haase, K.M. and Beier, C. 2021. Bransfield Strait and James Ross Island: petrology. *Geological Society, London, Memoirs*, **55**, https://doi.org/10.1144/M55-2018-37

Halpern, M. 1971. Evidence for Gondwanaland from a review of West Antarctic radiometric ages. *In*: Quam, L.O. (ed.) *Research in the Antarctic*. American Association for the Advancement of Science, Washington, DC, 717–730.

Hathway, B. 2001. Sims Island: first data from a Pliocene alkaline volcanic centre in eastern Ellsworth Land. *Antarctic Science*, **13**, 87–88, https://doi.org/10.1017/S095410200100013X

Hole, M.J. 1988. Post-subduction alkaline volcanism along the Antarctic Peninsula. *Journal of the Geological Society, London*, **145**, 985–989, https://doi.org/10.1144/gsjgs.145.6.0985

Hole, M.J. 1990a. Geochemical evolution of Pliocene–Recent post-subduction alkalic basalts from Seal Nunataks, Antarctic Peninsula. *Journal of Volcanology and Geothermal Research*, **40**, 149–167, https://doi.org/10.1016/0377-0273(90)90118-Y

Hole, M.J. 1990b. Hornpipe Heights. *American Geophysical Union Antarctic Research Series*, **48**, 271–272.

Hole, M.J. 1990c. Beethoven Peninsula. *American Geophysical Union Antarctic Research Series*, **48**, 273–276.

Hole, M.J. 2021. Antarctic Peninsula: petrology. *Geological Society, London, Memoirs*, **55**, https://doi.org/10.1144/M55-2018-40

Hole, M.J. and Larter, R.D. 1993. Trench-proximal volcanism following ridge crest–trench collision along the Antarctic Peninsula. *Tectonics*, **12**, 897–910, https://doi.org/10.1029/93TC00669

Hole, M.J. and Thomson, J.W. 1990. Mount Pinafore–Debussy Heights. *American Geophysical Union Antarctic Research Series*, **48**, 268–270.

Hole, M.J., Rogers, G., Saunders, A.D. and Storey, M. 1991. The relationship between alkalic volcanism and slab-window formation. *Geology*, **19**, 657–660, https://doi.org/10.1130/0091-7613(1991)019<0657:RBAVAS>2.3.CO;2

Hole, M.J., Saunders, A.D., Rogers, G. and Sykes, M.A. 1995. The relationship between alkaline magmatism, lithospheric extension and slab window formation along continental destructive plate margins. *Geological Society, London, Special Publications*, **81**, 265–285, https://doi.org/10.1144/GSL.SP.1994.081.01.15

Horne, P.R. and Thomson, M.R.A. 1967. Post-Aptian camptonite dykes in south-east Alexander Island. *British Antarctic Survey Bulletin*, **14**, 15–24.

Jonkers, H.A., Lirio, J.M., del Valle, R.A. and Kelley, S.P. 2002. Age and environment of Miocene–Pliocene glaciomarine deposits, James Ross Island, Antarctica. *Geological Magazine*, **139**, 577–594, https://doi.org/10.1017/S0016756802006787

Kraus, S., Kurbatov, A. and Yates, M. 2013. Geochemical signatures of tephras from Quaternary Antarctic Peninsula volcanoes. *Andean Geology*, **40**, 1–40, https://doi.org/10.5027/andgeoV40n1-a01

Kyle, P. 1990. McMurdo Volcanic Group, western Ross Embayment. *American Geophysical Union Antarctic Research Series*, **48**, 19–25.

Larsen, C.A. 1894. The voyage of the Jason to the Antarctic regions. *Geographical Journal*, **4**, 333–344, https://doi.org/10.2307/1773537

Laudon, T.S. 1982. Geochemistry of Mesozoic and Cenozoic igneous rocks, eastern Ellsworth Land. *In*: Craddock, C. (ed.) *Antarctic Geoscience*. University of Wisconsin Press, Madison, WI, 775–785.

Leat, P.T. and Riley, T.R. 2021*a*. Antarctic Peninsula and South Shetland Islands: volcanology. *Geological Society, London, Memoirs*, **55**, https://doi.org/10.1144/M55-2018-52

Leat, P.T. and Riley, T.R. 2021*b*. Antarctic Peninsula and South Shetland Islands: petrology. *Geological Society, London, Memoirs*, **55**, https://doi.org/10.1144/M55-2018-68

Lescinsky, D.T. and Sisson, T.W. 1998. Ridge-forming ice-bounded lava flows at Mount Rainier, Washington. *Geology*, **26**, 351–354, https://doi.org/10.1130/0091-7613(1998)026<0351:RFIBLF>2.3.CO;2

Loughlin, S.C. 2002. Facies analysis of proximal subglacial and proglacial volcaniclastic successions at the Eyjafjallajökull central volcano, southern Iceland. *Geological Society, London, Special Publications*, **202**, 149–178, https://doi.org/10.1144/GSL.SP.2002.202.01.08

Marenssi, S.A., Casadío, S. and Santillana, S.N. 2010. Record of Late Miocene glacial deposits on Isla Marambio (Seymour Island), Antarctic Peninsula. *Antarctic Science*, **22**, 193–198, https://doi.org/10.1017/S0954102009990629

Mathews, W.H. 1947. 'Tuyas': flat-topped volcanoes in northern British Columbia. *American Journal of Science*, **245**, 560–570, https://doi.org/10.2475/ajs.245.9.560

McCarron, J.J. 1997. A unifying lithostratigraphy of late Cretaceous–early Tertiary fore-arc volcanic sequences on Alexander Island, Antarctica. *Antarctic Science*, **9**, 209–220, https://doi.org/10.1017/S0954102097000266

McCarron, J.J. and Millar, I.L. 1997. The age and stratigraphy of fore-arc magmatism on Alexander Island, Antarctica. *Geological Magazine*, **134**, 507–522, https://doi.org/10.1017/S0016756897007437

McCarron, J.J. and Smellie, J.L. 1998. Tectonic implications of fore-arc magmatism and generation of high-magnesian andesites: Alexander Island, Antarctica. *Journal of the Geological Society, London*, **155**, 269–280, https://doi.org/10.1144/gsjgs.155.2.0269

Nelson, P.H.H. 1975. *The James Ross Island Volcanic Group of Northeast Graham Land*. British Antarctic Survey Scientific Reports, **54**.

O'Neill, J.M. and Thomson, J.W. 1985. Tertiary mafic volcanic and volcaniclastic rocks of the English Coast, Antarctica. *Antarctic Journal of the United States*, **20**, 36–38.

Pankhurst, R.J., Riley, T.R., Fanning, C.M. and Kelley, S.P. 2000. Episodic silicic volcanism in Patagonia and the Antarctic Peninsula: chronology of magmatism associated with break-up of Gondwana. *Journal of Petrology*, **41**, 605–625, https://doi.org/10.1093/petrology/41.5.605

Pudsey, C.J. and Evans, J. 2001. First survey of Antarctic sub-ice shelf sediments reveals mid-Holocene ice shelf retreat. *Geology*, **29**, 787–790, https://doi.org/10.1130/0091-7613(2001)029<0787:FSOASI>2.0.CO;2

Renner, R.G.B. 1980. *Gravity and Magnetic Surveys in Graham Land*. British Antarctic Survey Scientific Reports, **77**.

Renner, R.G.B., Dikstra, B.J. and Martin, J.L. 1982. Aeromagnetic surveys over the Antarctic Peninsula. *In*: Craddock, C. (ed.) *Antarctic Geoscience*. University of Wisconsin Press, Madison, WI, 363–370.

Rex, D.C. 1970. Age of a camptonite dyke from south-east Alexander Island. *British Antarctic Survey Bulletin*, **23**, 103.

Rex, D.C. 1972. K–Ar age determinations on volcanic and associated rocks from the Antarctic Peninsula and Dronning Maud Land. *In*: Adie, R.J. (ed.) *Antarctic Geology and Geophysics*. Universitetsforlaget, Oslo, 133–136.

Rex, D.C. 1976. Geochronology in relation to the stratigraphy of the Antarctic Peninsula. *British Antarctic Survey Bulletin*, **43**, 49–58.

Riley, T.R. and Leat, P.T. 2021. Palmer Land and Graham Land volcanic groups (Antarctic Peninsula): volcanology. *Geological Society, London, Memoirs*, **55**, https://doi.org/10.1144/M55-2018-36

Riley, T.R., Crame, J.A., Thomson, M.R.A. and Cantrill, D.J. 1997. Late Jurassic (Kimmeridgian–Tithonian) macrofossil assemblage from Jason Peninsula, Graham Land: evidence for a significant northward extension of the Latady Formation. *Antarctic Science*, **9**, 434–442, https://doi.org/10.1017/S0954102097000564

Riley, T.R., Flowerdew, M.J., Hunter, M.A. and Whitehouse, M.J. 2010. Middle Jurassic rhyolite volcanism of eastern Graham Land, Antarctic Peninsula: age correlations and stratigraphic relationships. *Geological Magazine*, **147**, 581–595, https://doi.org/10.1017/S0016756809990720

Rowley, P.D. and Smellie, J.L. 1990. Southeastern Alexander Island. *American Geophysical Union Antarctic Research Series*, **48**, 277–279.

Rowley, P.D. and Thomson, J.W. 1990. Rydberg Peninsula. *American Geophysical Union Antarctic Research Series*, **48**, 280–282.

Rowley, P.D., Vennum, W.R. and Smellie, J.L. 1990. Merrick Mountains. *American Geophysical Union Antarctic Research Series*, **48**, 296–297.

Saunders, A.D. 1982. Petrology and geochemistry of alkali-basalts from Jason Peninsula, Oscar II Coast, Graham Land. *British Antarctic Survey Bulletin*, **55**, 1–9.

Scarrow, J.H., Leat, P.T., Wareham, C.D. and Millar, I.L. 1998. Geochemistry of mafic dykes in the Antarctic Peninsula continental margin batholith: a record of arc evolution. *Contributions to Mineralogy and Petrology*, **131**, 289–305, https://doi.org/10.1007/s004100050394

Skilling, I.P. 1994. Evolution of an englacial volcano: Brown Bluff, Antarctica. *Bulletin of Volcanology*, **56**, 573–591, https://doi.org/10.1007/BF00302837

Skilling, I.P. 2002. Basaltic pahoehoe lava-fed deltas: large-scale characteristics, clast generation, emplacement processes and environmental discrimination. *Geological Society, London, Special Publications*, **202**, 91–113, https://doi.org/10.1144/GSL.SP.2002.202.01.06

Smellie, J.L. 1987. Geochemistry and tectonic setting of alkaline volcanic rocks in the Antarctic Peninsula: a review. *Journal of Volcanology and Geothermal Research*, **32**, 269–285, https://doi.org/10.1016/0377-0273(87)90048-5

Smellie, J.L. 1990. Seal Nunataks. *American Geophysical Union Antarctic Research Series*, **48**, 349–351.

Smellie, J.L. 1991. Middle–Late Jurassic volcanism on Jason Peninsula, Antarctic Peninsula, and its relationship to the break-up of Gondwana. *In*: Ulbrich, H. and Rocha Campos, A.C. (eds) Gondwana Seven Proceedings. Papers presented at the Seventh International Gondwana Symposium, Sao Paulo, 1988. Instituto de Geociencias, Universidade de Sao Paulo, São Paulo, 685–699.

Smellie, J.L. 1999. Lithostratigraphy of Miocene–Recent, alkaline volcanic fields in the Antarctic Peninsula and eastern Ellsworth Land. *Antarctic Science*, **11**, 362–378, https://doi.org/10.1017/S0954102099000450

Smellie, J.L. 2006. The relative importance of supraglacial v. subglacial meltwater escape in basaltic subglacial tuya eruptions: an important unresolved conundrum. *Earth-Science Reviews*, **74**, 241–268, https://doi.org/10.1016/j.earscirev.2005.09.004

Smellie, J.L. 2008. Basaltic subglacial sheet-like sequences: evidence for two types with different implications for the inferred thickness of associated ice. *Earth-Science Reviews*, **88**, 60–88, https://doi.org/10.1016/j.earscirev.2008.01.004

Smellie, J.L. 2013. Quaternary vulcanism: subglacial landforms. *In*: Elias, S.A. (ed.) *Reference Module in Earth Systems and Environmental Sciences, from The Encyclopedia of Quaternary Science*, 2nd edn. Elsevier, Amsterdam, **1**, 780–802.

Smellie, J.L. 2018. Glaciovolcanism: a 21st century proxy for palaeo-ice. *In*: Menzies, J. and van der Meer, J.J.M. (eds) *Past Glacial Environments*. 2nd edn. Elsevier, Amsterdam, 335–375.

Smellie, J.L. 2021. Antarctic volcanism: volcanology and palaeoenvironmental overview. *Geological Society, London, Memoirs*, **55**, https://doi.org/10.1144/M55-2020-1

Smellie, J.L. and Edwards, B.E. 2016. *Glaciovolcanism on Earth & Mars: Products, Processes and Palaeoenvironmental Significance*. Cambridge University Press, Cambridge, UK.

Smellie, J.L. and Hole, M.J. 1997. Products and processes in Pliocene–Recent, subaqueous to emergent volcanism in the Antarctic Peninsula: examples of englacial Surtseyan volcano construction. *Bulletin of Volcanology*, **58**, 628–646, https://doi.org/10.1007/s004450050167

Smellie, J.L. and Martin, A.P. 2021. Erebus Volcanic Province: volcanology. *Geological Society, London, Memoirs*, **55**, https://doi.org/10.1144/M55-2018-62

Smellie, J.L. and Rocchi, S. 2021. Northern Victoria Land: volcanology. *Geological Society, London, Memoirs*, **55**, https://doi.org/10.1144/M55-2018-60

Smellie, J.L. and Skilling, I.P. 1994. Products of subglacial eruptions under different ice thicknesses: two examples from Antarctica. *Sedimentary Geology*, **91**, 115–129, https://doi.org/10.1016/0037-0738(94)90125-2

Smellie, J.L., Pankhurst, R.J., Hole, M.J. and Thomson, J.W. 1988. Age, distribution and eruptive conditions of late Cenozoic alkaline volcanism in the Antarctic Peninsula and eastern Ellsworth Land. *British Antarctic Survey Bulletin*, **80**, 21–49.

Smellie, J.L., Hole, M.J. and Nell, P.A.R. 1993. Late Miocene valley-confined subglacial volcanism in northern Alexander Island, Antarctic Peninsula. *Bulletin of Volcanology*, **55**, 273–288, https://doi.org/10.1007/BF00624355

Smellie, J.L., McArthur, J.M., McIntosh, W.C. and Esser, R. 2006a. Late Neogene interglacial events in the James Ross Island region, northern Antarctic Peninsula, dated by Ar/Ar and Sr-isotope stratigraphy. *Palaeogeography, Palaeoclimatology, Palaeoecology*, **242**, 169–187, https://doi.org/10.1016/j.palaeo.2006.06.003

Smellie, J.L., McIntosh, W.C., Esser, R. and Fretwell, P. 2006b. The Cape Purvis volcano, Dundee Island (northern Antarctic Peninsula): late Pleistocene age, eruptive processes and implications for a glacial palaeoenvironment. *Antarctic Science*, **18**, 399–408, https://doi.org/10.1017/S0954102006000447

Smellie, J.L., Johnson, J.S., McIntosh, W.C., Esser, R., Gudmundsson, M.T., Hambrey, M.J. and van Wyk de Vries, B. 2008. Six million years of glacial history recorded in the James Ross Island Volcanic Group, Antarctic Peninsula. *Palaeogeography, Palaeoclimatology, Palaeoecology*, **260**, 122–148, https://doi.org/10.1016/j.palaeo.2007.08.011

Smellie, J.L., Haywood, A.M., Hillenbrand, C.-D., Lunt, D.J. and Valdes, P.J. 2009. Nature of the Antarctic Peninsula Ice Sheet during the Pliocene: geological evidence and modelling results compared. *Earth-Science Reviews*, **94**, 79–94, https://doi.org/10.1016/j.earscirev.2009.03.005

Smellie, J.L., Johnson, J.S. and Nelson, A.E. 2013. *Geological Map of James Ross Island. 1. James Ross Island Volcanic Group (1:125 000 Scale)*. BAS GEOMAP 2 Series Sheet 5. British Antarctic Survey, Cambridge, UK, http://nora.nerc.ac.uk/506743/1/BAS%20GEOMAP%202%2C%20sheet%205%20-%20Geological%20map%20of%20James%20Ross%20Island%20-%20I%20-%20James%20Ross%20Island%20volcanic%20group.pdf

Smellie, J.L., Panter, K.S. and Reindel, J. 2021. Mount Early and Sheridan Bluff: volcanology. *Geological Society, London, Memoirs*, **55**, https://doi.org/10.1144/M55-2018-61

Thomson, J.W. and Kellogg, K.S. 1990. Henry Nunataks. *American Geophysical Union Antarctic Research Series*, **48**, 294–295.

Thomson, J.W. and O'Neill, J.M. 1990. Snow Nunataks. *American Geophysical Union Antarctic Research Series*, **48**, 283–285.

Veit, A. 2002. *Volcanology and Geochemistry of Pliocene to Recent Volcanics on Both Sides of the Bransfield-Strait/West Antarctica*. PhD thesis, Ludwig Maximilian University, Munich, Germany [in German].

Walker, G.P.L. and Blake, D.H. 1966. The formation of a palagonite breccia mass beneath a valley glacier in Iceland. *Journal of the Geological Society, London*, **122**, 45–61, https://doi.org/10.1144/gsjgs.122.1.0045

White, J.D.L. and Houghton, B.F. 2006. Primary volcaniclastic rocks. *Geology*, **34**, 677–680, https://doi.org/10.1130/G22346.1

Chapter 4.1b

Antarctic Peninsula: petrology

Malcolm J. Hole

Department of Geology and Petroleum Geology, University of Aberdeen, Meston Building, King's College, Aberdeen AB24 3UE, UK

0000-0002-1622-4266
m.j.hole@abdn.ac.uk

Abstract: Scattered occurrences of Miocene–Recent volcanic rocks of the alkaline intraplate association represent one of the last expressions of magmatism along the Antarctic Peninsula. The volcanic rocks were erupted after the cessation of subduction which stopped following a series of northward-younging ridge crest–trench collisions. Volcanism has been linked to the development of a growing slab window beneath the extinct convergent margin. Geochemically, lavas range from olivine tholeiite through to basanite and tephrite. Previous studies have emphasized the slab-window tectonic setting as key to allowing melting of peridotite in the asthenospheric void caused by the passage of the slab beneath the locus of volcanism. This hypothesis is revisited in the light of more recent petrological research, and an origin from melting of subducted slab-hosted pyroxenite is considered here to be a more viable alternative for their petrogenesis. Because of the simple geometry of ridge subduction, and the well-established chronology of ridge crest–trench collisions, the Antarctic Peninsula remains a key region for understanding the transition from active to passive margin resulting from cessation of subduction. However, there are still some key issues relating to their tectonomagmatic association, and, principally, the poor geochronological control on the volcanic rocks requires urgent attention.

Supplementary material: Analyses of post-subduction, slab-window-related volcanic rocks in the Antarctic Peninsula (S1) and PRIMELT3 primary magma solutions for Antarctic Phoenix Ridge melt inclusions in olivine and plagioclase (S2) are available https://doi.org/10.6084/m9.figshare.c.5203966

Distribution of the volcanic rocks and their tectonic setting

Scattered along the length of the Antarctic Peninsula are exposures of Miocene–Recent volcanic rocks of the intraplate alkaline association (Fig. 1) (Smellie 1987; Hole 1988, 1990a, b, c; Hole and Thomson 1990; Hole et al. 1991, 1993, 1994a; Hole and Larter 1993; Hole and Saunders 1996). Whereas these occurrences are limited in volume, they have great significance because they were erupted after, or nearly synchronous with, the cessation of more than 200 myr of easterly-directed subduction along the Antarctic Peninsula and the formation of slab windows. Subduction ceased because of northeasterly-younging collisions between ridge crests of the Antarctic Phoenix spreading centre and the continental margin, which ultimately resulted in the transformation of the margin from active and convergent to passive. The complex interactions between ridge crests and the trench along the Antarctic Peninsula from 55 Ma until the present have been summarized and illustrated by the animations included in Eagles et al. (2007).

Ridge crest–trench collisions occurred between c. 50 Ma off the coast of Alexander Island and c. 3.1 Ma at the 'C' Fracture Zone (Figs 2–4). At the locus of each collision a ridge crest–trench–transform triple junction formed and subduction stopped at the same time as each collision. The trailing flank of the ridge crest was not subducted and the subducted slab in the collision zone remained attached to the slab to the NE *via* an adjacent fracture zone. Thus, to the NE of each triple junction, the slab continued to sink into the asthenosphere because of continued subduction to the NE. This ultimately resulted in the formation of a slab window. Each collision event destroyed a segment of the Phoenix Plate, such that the continental crust of the Antarctic Peninsula and the seafloor to the west were all part of the Antarctic Plate and consequently represent a passive margin. In the youngest two collision zones, between the North Anvers and Hero fracture zones, there was a slight complication to this simple scenario. The pattern of magnetic anomalies suggests that, starting about 9 myr ago, development of a deformation zone between the 'C' and Hero fracture zones partially de-coupled the part of the leading plate to the SW from the leading plate to the NE. Spreading and subduction continued to the SW of the deformation zone, albeit more slowly than to the NE, until the whole of both ridge-crest

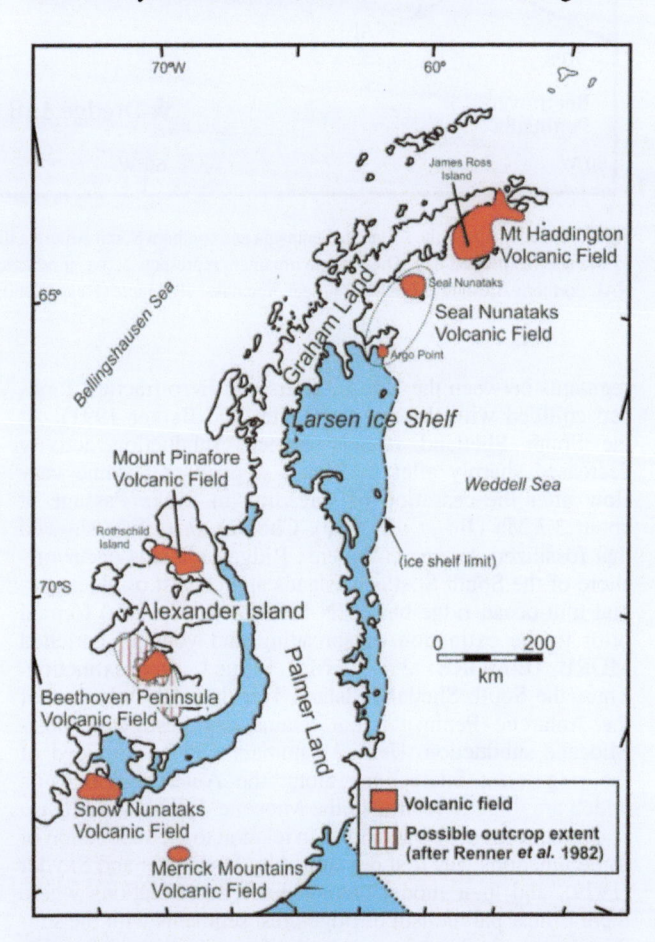

Fig. 1. Map of the Antarctic Peninsula showing the locations of the late Neogene volcanic fields (modified after Smellie 1999).

From: Smellie, J. L., Panter, K. S. and Geyer, A. (eds) 2021. *Volcanism in Antarctica: 200 Million Years of Subduction, Rifting and Continental Break-up.* Geological Society, London, Memoirs, **55**, 327–343,
First published online February 24, 2021, https://doi.org/10.1144/M55-2018-40
© 2021 The Author(s). Published by The Geological Society of London. All rights reserved.
For permissions: http://www.geolsoc.org.uk/permissions. Publishing disclaimer: www.geolsoc.org.uk/pub_ethics

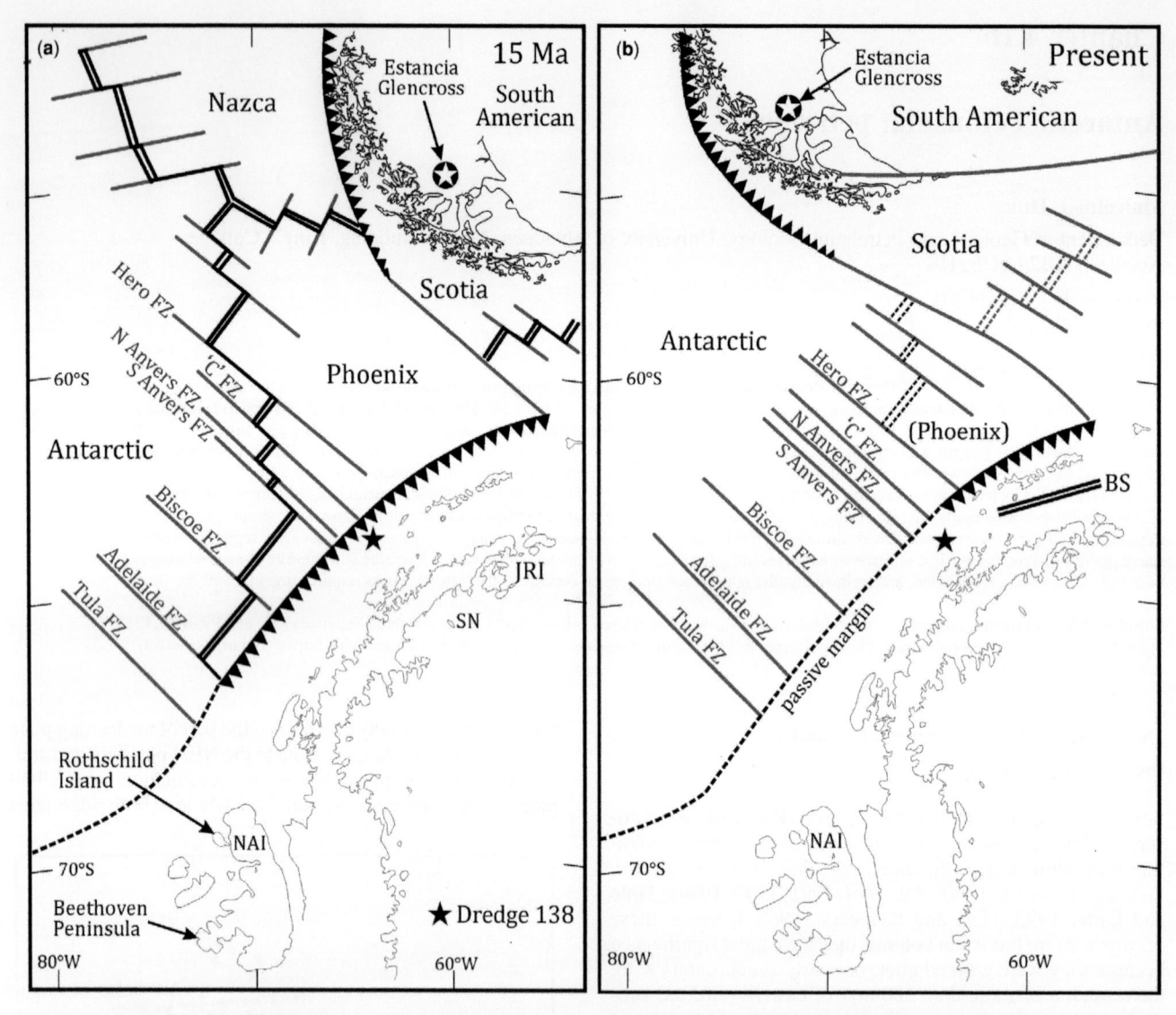

Fig. 2. Sketch maps of the Antarctic Peninsula and southern South America illustrating the tectonic setting and location of the Antarctic–Phoenix Ridge at (**a**) 15 Ma and (**b**) present day. The triangle ornament represents active subduction, and the double lines represent the spreading-ridge segment. Abbreviations; NAI, northern Alexander Island; SN, Seal Nunataks; JRI, James Ross Island; BS, Bransfield Strait. After Hole and Larter (1993).

segments between the North Anvers and Hero fracture zones had collided with the margin (Larter and Barker 1991). At the South Shetland Islands Trench, subduction activity decreased sharply after 6 Ma and stopped or became very slow after the cessation of spreading in Drake Passage at about 3.3 Ma (Jin *et al.* 2009). Choe *et al.* (2007) showed that fossilized Antarctic–Phoenix Ridge segments occur offshore of the South Shetland Islands and consist of older normal mid-ocean ridge basalt (N-MORB: 6.4–3.5 Ma) formed prior to the extinction of spreading and younger enriched MORB (E-MORB: 3.4–1.3 Ma) formed after extinction. Thus, the South Shetlands Island Trench is the only part of the Antarctic Peninsula that retains any vestige of post-Miocene subduction. Here, a summary will be provided of the ridge-crest interactions along the Antarctic Peninsula which are directly related to the Miocene–Recent volcanism.

The concept of slab windows in relation to the subduction of triple junctions was first developed by Dickinson and Snyder (1979), and their model predicted that for situations where there is near parallelism of ridge-crest segments with the subduction trench should result in the formation of a 'zig-zag'-shaped window increasing down-slab in width from the oldest to the youngest collision events. This is what tectonic reconstructions for the Antarctic Peninsula suggest (Fig. 2). The geometry of slab windows is consequently dependent on the convergence angle of the relevant ridge crest; oblique collision results in triangular slab windows, whereas nearly orthogonal collision of a segmented length of ridge crest forms a 'zig-zag'-shaped slab window (Dickinson and Snyder 1979). This hypothesis was developed for the Pacific margin of the Americas (e.g. Storey *et al.* 1989; Thorkelson and Taylor 1989; Hole *et al.* 1991; Thorkelson 1996; Johnson and Thorkelson 2000; D'Orazio *et al.* 2001; Thorkelson and Breitsprecher 2004; Breitsprecher and Thorkelson 2009; Thorkelson *et al.* 2011), demonstrating that the entire region had been underlain by slab windows during various periods throughout the Cenozoic. Consequently, the Antarctic Peninsula is not unique in displaying a relationship between ridge crest–trench collisions and alkaline magmatism but is certainly the best constrained in terms of the seafloor magnetics dataset and accuracy of tectonic reconstructions (e.g. Barker 1982; Eagles *et al.* 2007).

Figure 4, which is adapted from Hole *et al.* (1994a), illustrates the width of individual slab-window segments projected onto the horizontal surface of the continental crust from 15 Ma to the present in relation to the geographical distribution of post-subduction alkalic basalts. Note that some of the collision events were slightly oblique but, for the sake of clarity and

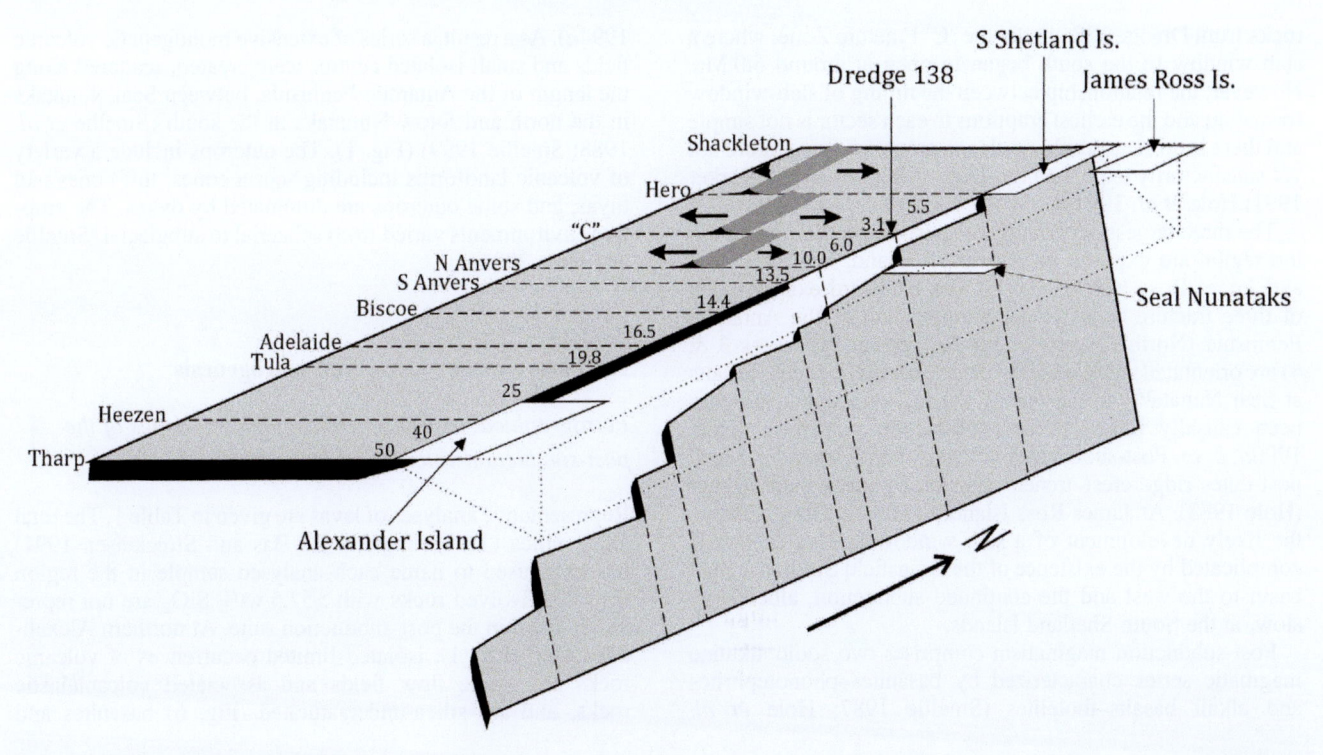

Fig. 3. Schematic diagram illustrating the development of the slab window along the Antarctic Peninsula at the present day. Collision times for each ridge segment are given on the oceanward margin of the peninsula. After Hole *et al.* (1994a).

ease of calculation, we have assumed that all collisions were orthogonal. The *x*-axis represents the palaeotrench and the different ornaments correspond to the amount of slab-window formation associated with each spreading ridge; the solid lines separating the ornaments being 'isochrons' for the slab-window formation. The assumptions that are used in the calculations are that the angle of slab dip was 45°, and the rate of opening of an individual slab-window segment was dependent on the spreading rate of the adjacent ridge crest to the north (Barker 1982; Larter and Barker 1991). It is also assumed that the subducted slab was planar, and the slab dip did not vary over time. Reducing the slab dip to 30° increases the linear width scale from 0–400 km to 0–500 km. The rate of slab-window opening is not constant because spreading took place about a near pole to the SW and spreading rates varied with time. These figures clearly show that trench-proximal volcanic

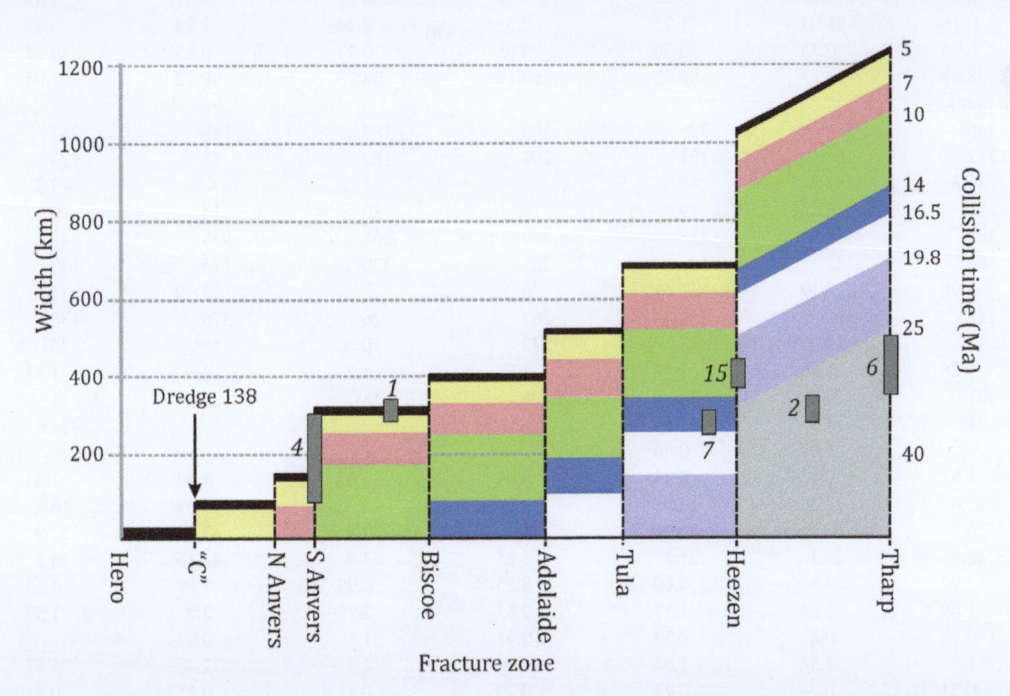

Fig. 4. Plan view of slab-window growth with time, showing the geographical location of the slab-window-related basalts (open symbols with K–Ar age adjacent). The *x*-axis represents the palaeotrench and the different ornaments correspond to the amount of slab-window formation associated with each spreading ridge; the solid lines separating the ornaments being 'isochrons' for the slab window as a whole. The assumptions that are used in the calculations are that the angle of slab dip was *c.* 45° and the rate of opening of an individual slab window segment was dependent on the spreading rate of the adjacent ridge crest to the north (see Barker 1982; Larter and Barker 1991). It is also assumed that the subducted slab was planar, and the slab dip did not vary over time. Reducing the slab dip to 30° increases the linear width scale from 0–1400 km to 0–1650 km. After Hole *et al.* (1994a).

rocks from Dredge 138 overlie the 'C' Fracture Zone, where a slab window to the south began to open at around 6.0 Ma. However, the relationship between the timing of slab-window formation and the earliest eruptions in each sector is not simple and there is no clear south–north age progression, a feature not yet satisfactorily explained (cf. Hole 1988; Larter and Barker 1991; Hole et al. 1994a).

The three largest occurrences of post-subduction basalts in the region are exposed on Alexander Island, Seal Nunataks and James Ross Island (Fig. 1). The landward extrapolation of three fracture zones in the northern part of the Antarctic Peninsula (North Anvers, South Anvers and 'C': Figs 3 & 4) are orientated sub-parallel to the exposures of alkalic basalts at Seal Nunataks, to the rear of the palaeo-arc, and this has been causally linked to post-subduction volcanism (Hole 1990a, b, c). Post-subduction volcanism on Alexander Island post-dates ridge crest–trench collision by more than 40 myr (Hole 1988). At James Ross Island, the relationship between the likely development of a slab window and volcanism is complicated by the existence of the Bransfield Strait marginal basin to the west and the continued subduction, albeit very slow, at the South Shetland Islands.

Post-subduction magmatism comprises two sodic alkaline magmatic series characterized by basanites–phonotephrites and alkali basalts–tholeiites (Smellie 1987; Hole et al. 1994a). As a result, a series of extensive monogenetic volcanic fields and small isolated centres were created, scattered along the length of the Antarctic Peninsula, between Seal Nunataks in the north and Snow Nunataks in the south (Smellie et al. 1988; Smellie 1999) (Fig. 1). The outcrops include a variety of volcanic landforms including scoria cones, tuff cones and tuyas, and some outcrops are dominated by dykes. The eruptive environments varied from subaerial to subglacial (Smellie and Hole 2021).

Petrology, geochemistry and petrogenesis

Petrographical and geochemical characteristics of the post-subduction volcanic rocks

Representative analyses of lavas are given in Table 1. The total alkali–silica (TAS) diagram (Le Bas and Streckeisen 1991) has been used to name each analysed sample in the region (Fig. 5). Evolved rocks with >52.5 wt% SiO_2 are not represented at all in the post-subduction suite. At northern Alexander Island (Fig. 1), isolated limited occurrences of volcanic rocks are single flow fields and associated volcaniclastic rocks, and are silica-undersaturated (Fig. 6) basanites and

Table 1. *Major and trace element geochemistry of selected post-subduction basalts*

Sample #	R.3728.1[1]	R.3710.9[1]	KG3625.2[2]	KG.3619.1[2]	KG.3603.5[2]	KG.3616.5[2]	D.8749.1[3]	DR.138.6[4]
(wt%)								
SiO_2	51.50	48.93	49.51	46.90	46.11	48.60	47.78	50.95
TiO_2	1.74	2.09	1.87	2.75	2.93	2.42	1.86	1.70
Al_2O_3	14.82	14.62	14.83	14.42	14.78	16.27	14.91	17.20
$Fe_2O_3^T$	9.36	11.00	10.21	8.69	9.43	9.58	10.39	7.24
MnO	0.14	0.15	0.15	0.15	0.16	0.17	0.17	0.12
MgO	8.32	8.80	8.90	9.19	9.20	6.16	8.45	7.86
CaO	8.61	8.87	8.58	10.03	8.66	7.18	8.98	8.95
Na_2O	3.34	3.45	3.29	3.88	4.25	4.67	3.34	3.68
K_2O	0.76	1.03	1.23	2.32	2.46	2.24	0.83	1.20
P_2O_5	0.24	0.33	0.36	0.85	0.77	0.83	0.34	0.21
Total	98.83	99.27	98.93	99.18	98.75	98.12	97.05	99.11
Trace elements (in ppm)								
Cr	346	301	326	380	187	188	257	224
Ni	132	156	152	201	152	85	123	129
Co	46.8	49.9	48.9				49.2	35.9
Rb	11	16	19	26	30	14	10	25
Sr	349	436	458	837	868	938	447	446
Ba	66	127	189	263	320	176	140	215
Hf	2.98	3.19	3.61	6.26		7.69	3.1	3.88
Zr	118	140	159	267	326	370	135	165
Nb	17	22	23	72	70	56	23	32
Ta	1.26	1.66	1.46	4.63	4.9	4	1.35	2.18
Y	21	22	24	26	25	26	24	29
Sc	21	24.3	22.9				26.6	25
U		1.05	0.86					1.01
Th	1.82	2.47	2.93	6.03	6.67	5.54	1.85	3.4
La	12.4	17.4	17.5	53.94	43.36	49.37	16.4	21.7
Ce	26.5	37.6	35	91.74	79.5	87.23	32.9	41.9
Nd	15.2	20.7	21.7	41.25	43.4	42.15	16.2	23.4
Sm	3.92	4.82	4.69	8.25	6.99	7.89	4.33	4.8
Eu	1.49	1.64	1.57	2.55	2.72	2.2	1.59	1.64
Tb	0.72	0.8	0.73	0.91	1.1	0.97		0.76
Yb	1.4	1.58	1.64	1.98	2.3	2.46	1.97	2.32
Lu	0.21	0.24	0.25	0.29	0.31	0.35	0.31	0.37

R.3728.1, alkali basalt, Evensen Nunatak, Seal Nunataks; R.3710.9, alkali basalt, Dallman Nuatak, Seal Nunataks; KG3625.2, alkali basalt, Mussorgsky Peaks, Alexander Island; KG.3619.1, basanite, Rothschild Island; KG.3603.5, basanite, Hornpipe Heights, Alexander Island; KG.3616.5, tephrite, Mount Pinafore, Alexander Island; D.8749.1, alkali basalt, James Ross Island; DR.136.6, alkali basalt, trench-proximal dredge.
Data sources (shown following the sample number): [1]Hole (1990a); [2]Hole (1988); [3]Hole et al. (1994a); [4]Hole and Larter (1993).
See also the Supplementary material accompanying this chapter.

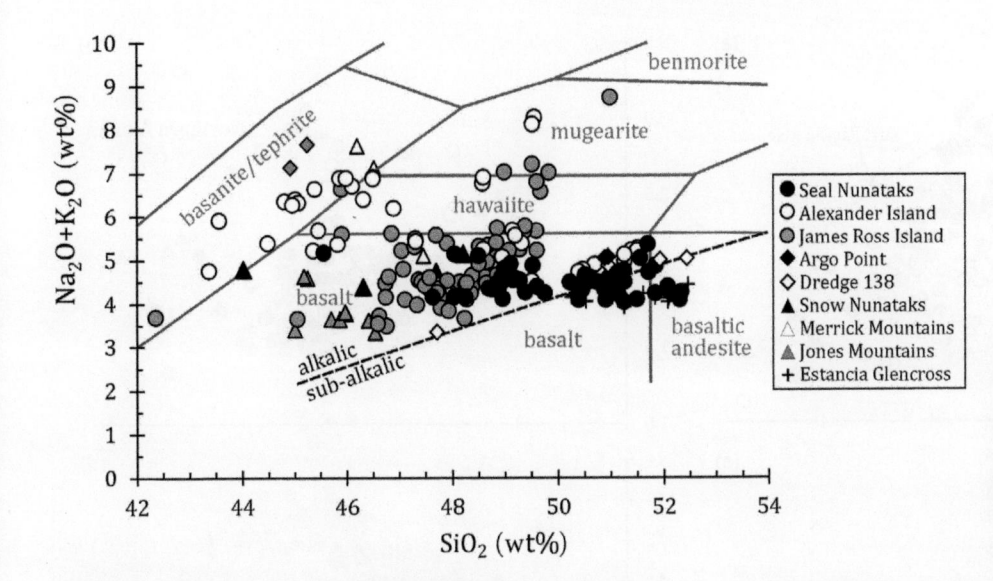

Fig. 5. Total alkali–silica diagram (Le Bas and Streckeisen 1991) used for naming the volcanic rocks. Data sources: Saunders (1982), Hole (1988, 1990a), Smellie (1987), Smellie et al. (1988), Sykes (1988), Hole et al. (1993, 1994a, b), D'Orazio et al. (2001) and Kosler et al. (2009).

tephrites with rare occurrences of hawaiite and mugearite. The subglacial pillow basalts and hyaloclastite sequences of the Beethoven Peninsula are mostly alkali basalt with rare hawaiite at Gluck Peak. Volcanic rocks at James Ross Island are mostly alkali basalt with limited occurrences of hawaiite. At Seal Nunataks (Fig. 1), alkali basalt, subalkalic basalt and basaltic andesite are all represented. On a CIPW-normative tetrahedron (Thompson 1982) the post-subduction basalts span Si-undersaturated, Ne-normative alkali olivine basalts to Si-saturated, Hy-normative olivine tholeiites, with data for some Seal Nunataks samples falling on the Di–Hy join and approaching Si-oversaturation (Fig. 6). Notably, the Hy-normative olivine tholeiites are most abundant at Seal Nunataks and the Beethoven Peninsula, whereas occurrences in northern Alexander Island are predominantly Ne-normative compositions. Also included in Figures 5 and 6 are data for c. 9 Ma volcanic rocks from the Jones Mountains 750 km WSW of Alexander Island (Hole et al. 1993). Whereas these are not specifically part of the Antarctic Peninsula post-subduction volcanic province, they have strong geochemical affinities with lavas from Seal Nunataks and do not appear to be part of the Marie Byrd Land province farther to the south (Hole et al. 1994b; Hole and LeMasurier 1994).

Fine-grained to near-glassy olivine-phyric basalts are the most common petrographical type. Most olivine phenocrysts are euhedral and may be skeletal, suggesting crystallization from the melt rather than accumulation. Plagioclase phenocrysts are rare but are developed in some hawaiites from the Beethoven Peninsula and basaltic andesites from Seal Nunataks. Augite is very rare and is restricted to alkali olivine basalts at Seal Nunataks and occurs in a variolitic texture within the glassy groundmass of the lavas. For Seal Nunataks samples with MgO in the range 7.0–8.6 wt%, olivine varies from $Fo_{74.4}$ to $Fo_{84.3}$, although olivine phenocrysts are generally not in equilibrium with the whole rock and olivine has lower Fo than would be expected for the Mg# of the whole rock for $K_{D[Fe-Mg]}^{Ol/L} = 0.30$ and $FeO/FeO^T = 0.9$ (Putirka 2008). This is best explained by the accumulation of olivine in the most MgO-rich samples, although the volume of olivine accumulation is small. For example, for sample R.3717.1 (8.6 wt% MgO) removal of 10% olivine of the observed phenocryst composition (Fo_{78}) would be needed to provide olivine–liquid equilibrium, and for other Seal Nunataks samples this amount of olivine would be less.

Selected major elements plotted against MgO (wt%) are shown in Figure 7. Covariations between Al_2O_3 and MgO

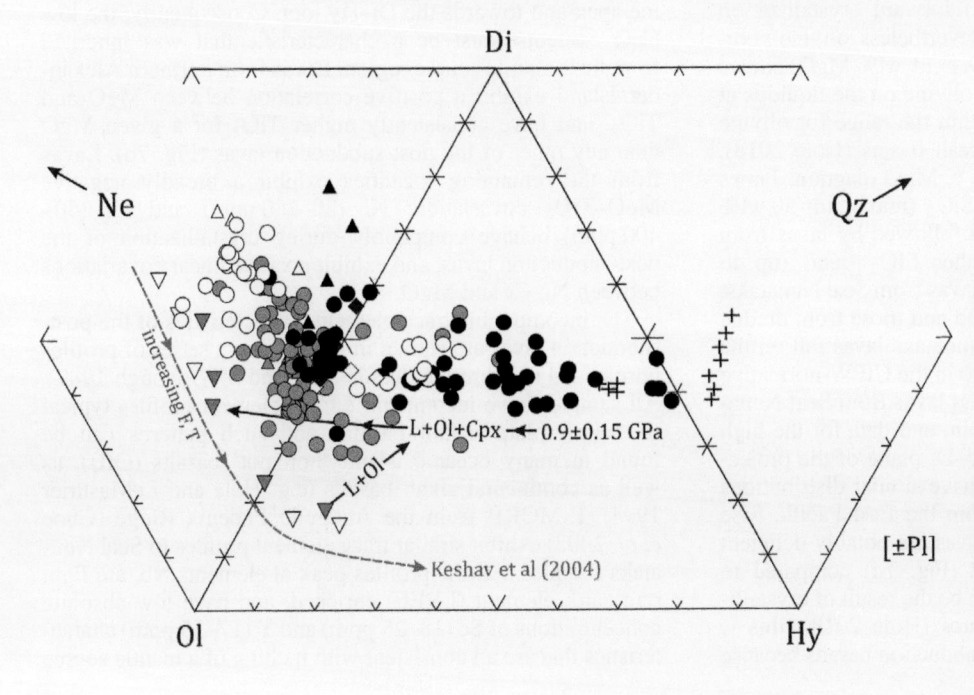

Fig. 6. CIPW-normative tetrahedron projected from or to plagioclase onto the planes Ne–Ol–Di, Ol–Di–Hy and Hy–Di–Qz (Thompson 1982). Inverted triangles are CIPW-norms for experimentally produced melts of pyroxenites 77SL-582 and MixG at 2.0 (grey) and 2.5 GPa (white) with the vector for increasing melting and increasing temperature shown (Keshav et al. 2004). Data for slab-window-related lavas from Estancia Glencross (white triangles) are from D'Orazio et al. (2001). Liquid lines of descent (LLD) along the cotectics L + Ol + Pl and L + Ol + Pl + Cpx are taken from Hole (2018).

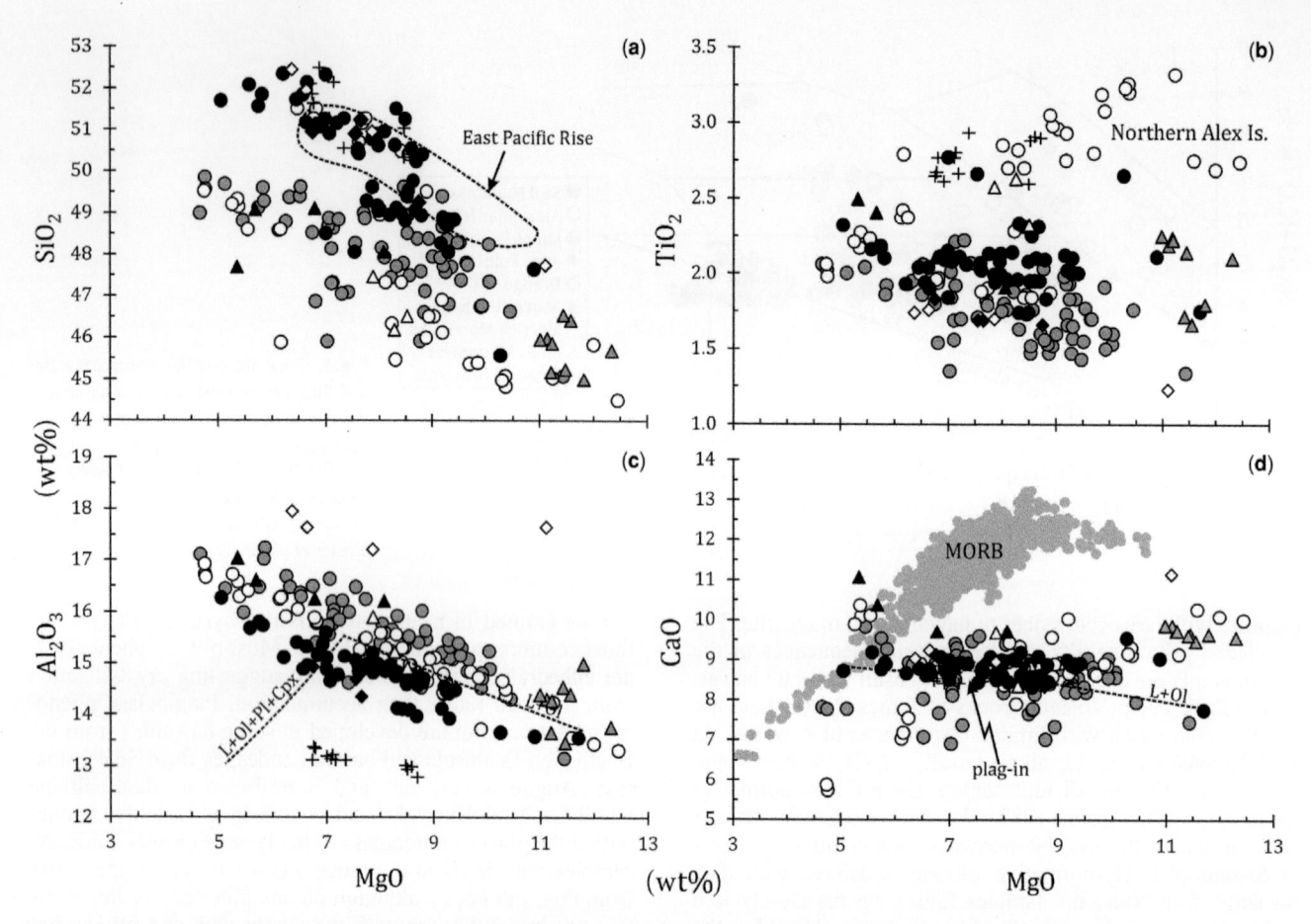

Fig. 7. (a) SiO_2, (b) TiO_2, (c) Al_2O_3 and (d) CaO v. MgO for Antarctic Peninsula and Estancia Glencross post-subduction lavas. In (c) and (d) the LLDs were calculated from the most MgO-rich lava from Seal Nunataks (R.3735.1: 10.27 wt% MgO) using Petrolog3 and assuming crystallization at the FMQ buffer (Danyushevsky and Plechov 2011; Hole 2018). Sample R.3637.7 has a higher MgO content than R.3735.1 and the former has a significantly higher mode of olivine than the latter but this is most likely to be a result of olivine accumulation. The field for the East Pacific Rise MORB in (a) and (d) refers to N-MORB from the Siquieros Fracture Zone with data taken from PetDB (http://www.earthchem.org/petdb/).

(Fig. 7) clearly demonstrate that most of the post-subduction volcanic rocks are related to the olivine liquidus, and there is no major role for plagioclase in their crystallization histories because plagioclase strongly fractionates Al_2O_3 relative to MgO resulting in positive covariations between the two oxides for rocks that have crystallized plagioclase. A Petrolog3 (Danyushevsky and Plechov 2011) forward crystallization model for the highest MgO, but nevertheless olivine-poor, sample from Seal Nunataks (R.3719.1: 11 wt% MgO) shows that plagioclase is predicted to join olivine on the liquidus at c. 7.5 wt% MgO, which is well within the range for olivine tholeiitic compositions from the ocean basins (Hole 2018). Two trends are evident on the SiO_2 v. MgO diagram. James Ross Island lavas represent a low SiO_2 (maximum 50 wt% SiO_2) series of lavas, which is also followed by lavas from northern Alexander Island. A higher SiO_2 trend (up to 52.5 wt% SiO_2) characterizes some lavas from Seal Nunataks, those from southern Alexander Island and those from dredge site 138. A small number of Seal Nunataks lavas fall on the lower SiO_2 trend. This is evident also in the CIPW-normative tetrahedron, and data for the low SiO_2 lavas from Seal Nunataks lavas fall close to the Ol–Di join, and data for the high SiO_2 lavas scatter across the Ol–Hy–Di plane of the projection. Lavas on the high SiO_2 trend have similar distributions of MgO and SiO_2 to N-MORB from the East Pacific Rise (Fig. 7a). All the post-subduction lavas are notably deficient in CaO for a given MgO content (Fig. 7d) compared to MORB. Whereas CaO depletion can be the result of crystallization of augite at elevated pressures (Hole 2018), this is unlikely to be the case for the post-subduction basalts because

they follow olivine control lines in terms of MgO–Al_2O_3 variations, and augite is rare in the lavas. Additionally, the cotectic for crystallization at c. 0.9 GPa results in a vector in the opposite direction to crystallization at near-surface pressure in the CIPW-normative tetrahedron (Fig. 6), and Seal Nunataks lavas exhibit decreasing MgO content away from the olivine apex and towards the Di–Hy join. Consequently, the low CaO contents must be a characteristic that was inherited from their mantle source region. Lavas from northern Alexander Island exhibit a positive correlation between MgO and TiO_2, and have consistently higher TiO_2 for a given MgO than any other of the post-subduction lavas (Fig. 7b). Lavas from the remaining localities exhibit a broadly negative MgO–TiO_2 covariation. Ni (30–250 ppm) and Cr (50–400 ppm) behave compatibly during crystallization of the post-subduction lavas, and exhibit positive linear covariations between Ni, Cr and MgO.

The incompatible trace element characteristics of the post-subduction lavas are shown in Figure 8 as a series of profiles normalized to primitive mantle (Sun and McDonough 1989). All samples have incompatible trace element profiles typical of 'within-plate' alkali basalts, and such patterns can be found in many oceanic island 'hot-spot' basalts (OIB), as well as continental alkali basalts (e.g. Hole and LeMasurier 1994). E-MORB from the Antarctic Phoenix Ridge (Choe et al. 2007) exhibit similar trace element profiles to Seal Nunataks samples. All the profiles peak at elements Nb, are light rare earth element (LREE) enriched, and have low absolute concentrations of Sc (18–25 ppm) and Y (17–25 ppm) characteristics that are all consistent with melting of a mantle source

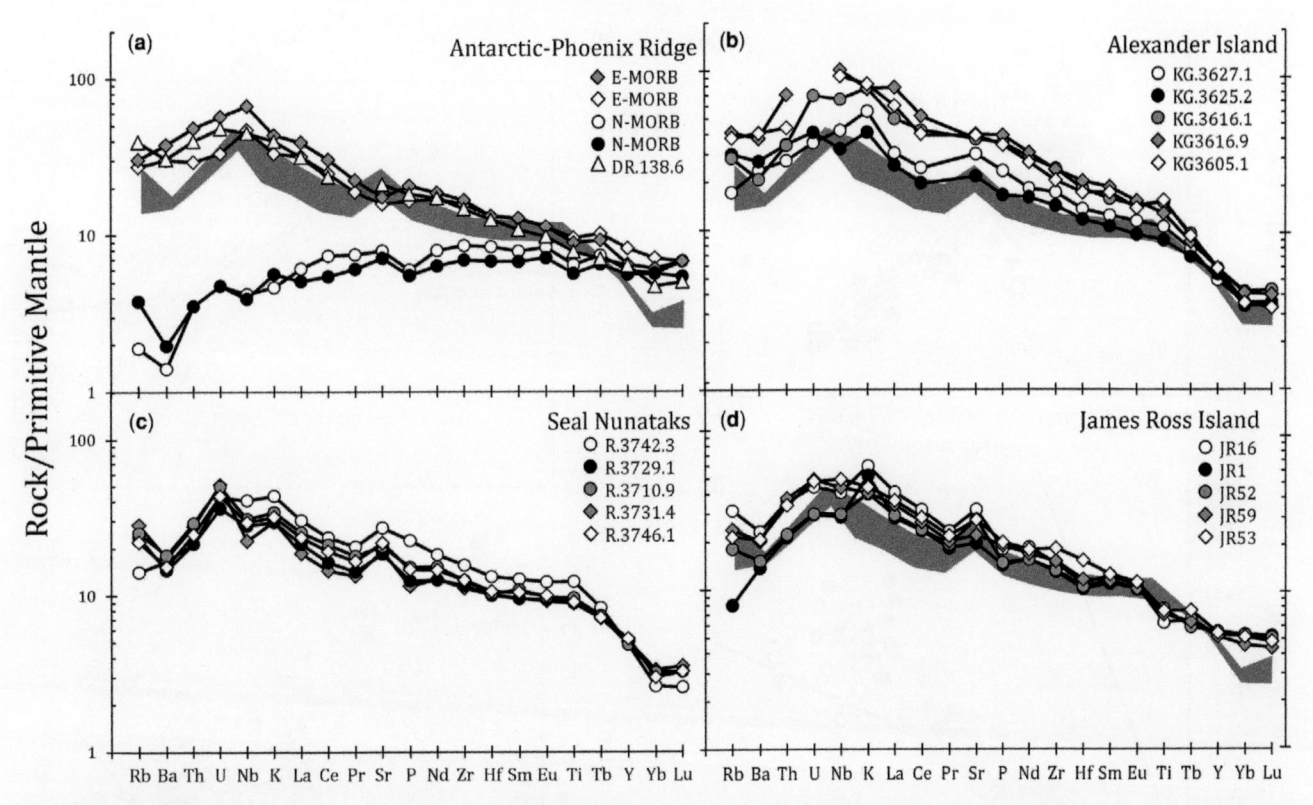

Fig. 8. Primitive mantle-normalized (Sun and McDonough 1989) incompatible trace element diagrams for (a) Antarctic–Phoenix Ridge N- and E-MORB and trench-proximal Dredge 138 lavas; (b) basanite and tephrite from northern Alexander Island; (c) alkali olivine basalts and olivine tholeiites from Seal Nunataks; and (d) alkali basalts from James Ross Island. Data sources: (a) Hole and Larter (1993), Choe *et al.* (2007) and Choi *et al.* (2013); (b) Hole (1988); (c) Hole (1990*a, b, c*); and (d) Kosler *et al.* (2009). The grey shading is the range for the Seal Nunataks lavas.

leaving residual garnet with which the heavy REEs (HREEs), Sc and Y are compatible. Northern Alexander Island basanites exhibit the greatest LREE-enrichment and highest Nb/Zr ($[La/Yb]_N$ c. 19.5; Nb/Zr c. 0.27 for Rothschild Island basanite), and Seal Nunataks Hy-normative olivine tholeiites exhibit the lowest $[La/Yb]_N$ and Nb/Zr (c. 5.3 and c. 0.12, respectively, for Christensen Nunatak). James Ross Island samples exhibit an almost complete overlap with Seal Nunataks samples in Figure 8. Trench-proximal alkali basalts (Fig. 2) (Hole and Larter 1993) similarly fall within the range for Seal Nunataks and overlap in composition with E-MORBs produced at the Antarctic–Phoenix Ridge. Significantly, there is no vestige of a subduction component in any of the basalts: Th/Ta (0.9–1.5) and La/Nb (0.7–1.2) are within the expected ranges for lavas erupted distant in space and time from subduction zones.

The lavas exhibit a restricted range of isotopic compositions (Fig. 9), and differ from N-MORB from the Pacific Ocean in having lower $^{143}Nd/^{144}Nd$ (0.51278–0.51300) for a given $^{87}Sr/^{86}Sr$ (0.7025–0.7040) than N-MORB, and more radiogenic $^{206}Pb/^{204}Pb$ and $^{207}Pb/^{204}Pb$ than N-MORB. The post-subduction basalts have similar Sr and Nd isotopic compositions to so-called HIMU (high $^{238}U/^{204}Pb = \mu$) OIB but a HIMU source is precluded because none of the post-subduction lavas possess the high $^{206}Pb/^{204}Pb$ (>20) that characterizes HIMU OIB (e.g. Palacz and Saunders 1986; Homrighausen *et al.* 2018). Whereas some of the samples from James Ross Island and Seal Nunataks with $^{87}Sr/^{86}Sr$ of c. 0.7034 may have undergone limited interaction with sialic upper crust (Hole *et al.* 1993), they could equally be a feature of the mantle source from which the basalts were derived. In support of the former, glassy 'buchite' xenoliths representing partially melted arkosic sediments occur at Seal Nunataks, James Ross Island and the Beethoven Peninsula but the lack of a trace element signature of such interaction supports the former. Since the available arkosic sedimentary rocks in the area are likely to be Cretaceous back-arc sedimentary rocks, and these were all derived from the Antarctic Peninsula convergent margin, a subduction signature in the trace element compositions of the alkali basalts might be an expected result of crustal contamination but this is not seen. As with the trace element profiles discussed above, there is a strong similarity between the isotopic compositions of E-MORB being erupted at the Antarctic Phoenix Ridge and the post-subduction lavas.

Petrogenesis and relationship to slab-window formation

The lack of any subduction signature in the post-subduction basalts means that the normal subduction-related melting regime involving dehydration melting of the mantle wedge must have ceased before their generation. Therefore, the source of the post-subduction alkali basalts cannot have been from mantle that had been involved in the subduction process. This implies that sub-slab mantle may have been involved in magma genesis (Hole 1990*a*; Hole *et al.* 1993) or that the entire mantle wedge was replenished with pristine peridotite immediately after subduction stopped. In terms of pressure–temperature (*P–T*) conditions, generation of the observed LREE-enrichment would require all melting to occur at >2.7 GPa (Fig. 10a) since this is the minimum pressure at which garnet is a stable phase on the dry peridotite solidus. Since melt is produced by decompression melting, for initial intersection of the dry solidus at c. 2.7 GPa melting would continue into the spinel stability field of mantle peridotite and the garnet signature in the REE would be rapidly lost. Consequently, to preserve the garnet melting signature, all melting would have to occur at pressures substantially >2.7 GPa. This means that the post-subduction basalts should have formed by melting at a mantle temperature that is much

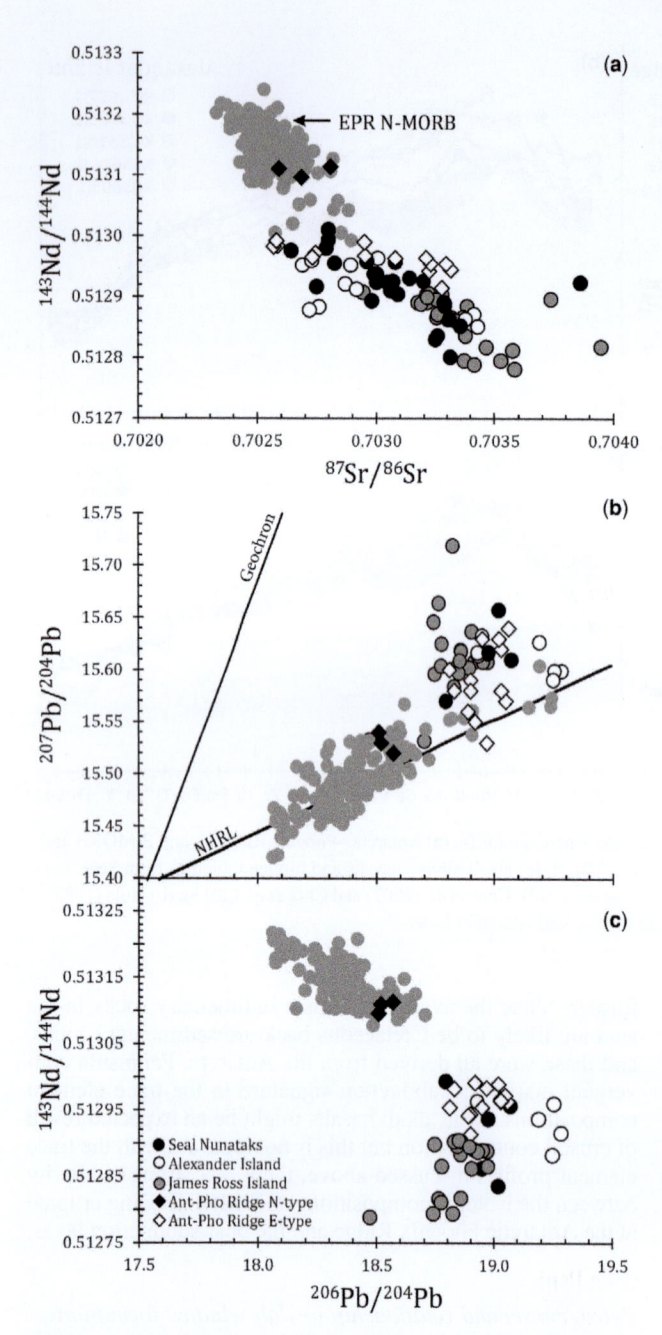

Fig. 9. (a) ^{207}Pb/^{204}Pb and (b) ^{143}Nd/^{144}Nd v. ^{206}Pb/^{204}Pb and (c) ^{143}Nd/^{144}Nd v. ^{87}Sr/^{86}Sr for the post-subduction lavas. Data for Antarctic–Phoenix N- and E-MORB are shown for comparison. Grey dots are the range of isotopic compositions for the present-day East Pacific Rise MORB (data from PetDB: https://www.earthchem.org/petdb, accessed May 2018). Data sources are as for Figure 8 except Seal Nunataks (Hole *et al.* 1993). The Northern Hemisphere Reference Line (NHRL) in (b) is taken from Hart (1988).

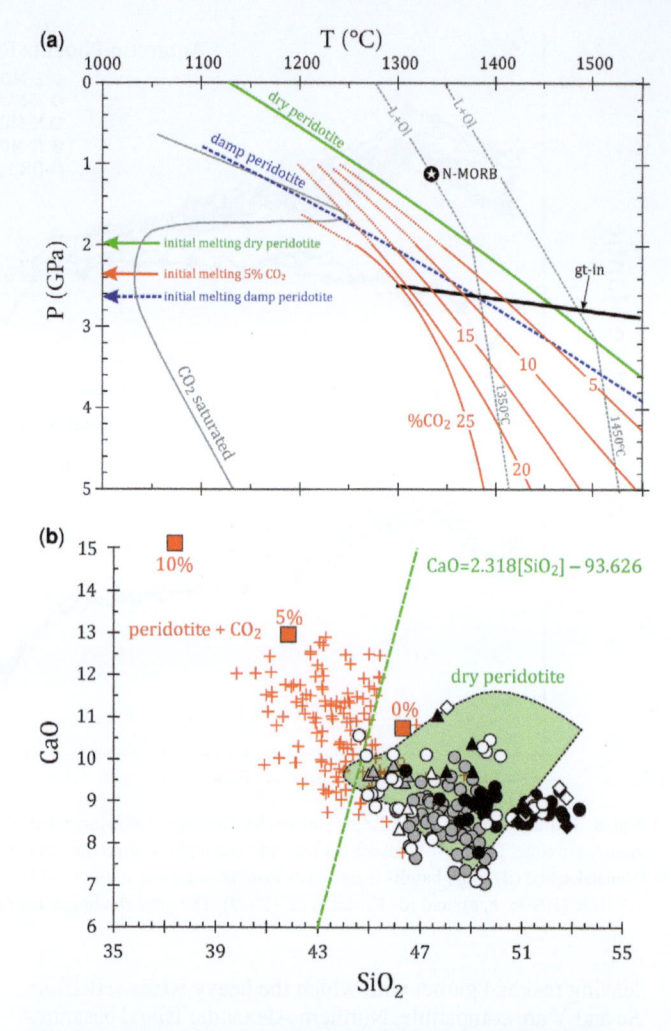

Fig. 10. (a) Pressure–temperature diagram showing the dry peridotite solidus (green curve) of Hirschmann (2000) and 'damp' solidus (blue pecked curve) with 450 ppm H_2O (Sarafian *et al.* 2017), and generalized CO_2-saturated solidus (grey curve: Dasgupta *et al.* 2007). Isopleths for 5–25% CO_2 in dry peridotite were calculated using the polynomial equations of Dasgupta *et al.* (2013). The pecked portions of the curves are extrapolated from the dry peridotite solidus of Hirschmann (2000). The garnet-in (gt-in) contour is taken from McKenzie and O'Nions (1995). The pecked portions of the curves are extrapolated from melting experiments at 2 GPa. The star in the circle is the PRIMELT3 solution for Antarctic–Phoenix Ridge N-MORB, and the grey pecked line is the adiabat for dry peridotite at T_P = 1350°C (Herzberg and Asimow 2015; Hole and Millett 2016). (b) CaO (wt%) v. SiO_2 (wt%) for Antarctic Peninsula post-subduction lavas (symbols as in Fig. 5) and lavas from various volcanoes in the Ross Sea area of the West Antarctic Rift System, which are denoted by red crosses (Wörner *et al.* 1989; Martin *et al.* 2013). The regions for melting of dry peridotite (green) and the dividing line between the compositions of melts from dry and carbonated peridotite are taken from Herzberg and Asimow (2008). The red squares represent the primary magma compositions derived from melting of peridotite + CO_2 with CO_2 (wt%) on the solidus indicated. The red squares relate to the red solidi in (a).

higher, and perhaps up to 1450°C, than that accepted for ambient temperature along continental margins.

The current published model for the generation of the post-subduction lavas requires upwelling of the asthenosphere to fill the incipient void left by the subducting slab after slab-window formation which promoted decompression melting and the ultimate eruption of the alkalic basalts (Hole 1988, 1990*a*, *b*, *c*; Hole and Larter 1993; Hole *et al.* 1993, 1994*a*; D'Orazio *et al.* 2001). Alkalic basalts associated with slab windows are not unique to the Antarctic Peninsula; similar volcanism occurs at San Quintín, Baja California (Storey *et al.* 1989), in central southern British Columbia (Bevier 1983; Thorkelson and Taylor 1989) and in Patagonia (e.g. D'Orazio *et al.* 2001; Breitsprecher and Thorkelson 2009). It has been recognized that there are similarities in the trace element and isotopic compositions of the post-subduction lavas and OIB, and it is generally accepted that the latter are the result of melting at elevated mantle temperatures associated with mantle plumes originating at the core–mantle boundary. However, to produce the post-subduction lavas by this method would require the fortuitous and spontaneous initiation of plumes when subduction ceased; basic geological observations alone also encourage the hypothesis of a genetic link between slab windows and alkalic magmatism (Hole *et al.* 1991). Furthermore, most subduction-related magmatism is generally considered to occur at ambient mantle potential temperature (T_P), with magma generated along the H_2O-saturated peridotite solidus (e.g. Hacker *et al.* 2003; Grove *et al.* 2012), and so there is no reason why a higher than ambient T_P would be expected along the Antarctic Peninsula margin at the time subduction stopped.

If the mantle beneath the Antarctic Peninsula was at an ambient T_P of 1350°C, the difficulty arises that the mantle adiabat does not intersect the dry peridotite solidus to generate melt until *c.* 2 GPa, which is within the spinel stability field of the upper mantle; such a melting regime would not be capable of generating the observed LREE-enrichment in the basalts (Fig. 10a). Conversely, the MgO–SiO_2 covariations (Fig. 7), particularly for the high SiO_2 lineage in lavas at Seal Nunataks and their similarity to N-MORB, is a common manifestation of shallow melting. High-pressure melts formed with garnet on the peridotite solidus are lower in SiO_2 for a given MgO than N-MORB. Herein lies the paradox. If it is assumed that T_P was ambient, then the apparent high-pressure signature in the REE and the apparent low-pressure signature in the major elements in the same lavas need to be explained. D'Orazio *et al.* (2001) recognized the same paradox for the post-subduction alkali basalts of Patagonia and noted that LREE-enrichment required melting at ≥2.7 GPa but SiO_2 contents of up to 51.0 wt% at 8.5 wt% MgO were more akin to those of shallow melting at <1.5 GPa. D'Orazio *et al.* (2001) therefore appealed to a process involving the mixing between shallow harzburgite-derived melt and deeper garnet–peridotite-derived melt to account for this unusual geochemical trait. However, such a model is still unable to account for the intersection of the dry solidus at >2.7 GPa at an ambient T_P of 1350°C.

Further constraints on the mantle temperature in the region can be gained from lavas erupted at the Antarctic Phoenix Ridge. The petrological modelling routine PRIMELT3 (Herzberg and Asimow 2008; 2015) allows estimates of T_P to be made using whole-rock data for lavas that have only fractionated olivine (Herzberg and Asimow 2015; Hole and Millett 2016). None of the post-subduction basalts yield useful temperature information using this method because all are too deficient in CaO at a given MgO content to have been derived from dry peridotite, which in itself is a useful observation. Whole-rock data for the basalts from the 3.5–3.1 Ma Antarctic Phoenix Ridge also do not yield any useful temperature data because they crystallized along the L + Ol + Pl + Cpx cotectic (Choi *et al.* 2013). Melt inclusions trapped in olivine and plagioclase in the Antarctic Phoenix Ridge N-MORB samples are less fractionated parcels of melt than their host whole-rock compositions. Using PRIMELT3, melt inclusions reveal T_P = 1336 ± 21°C (Supplementary material S1), a figure that is within the uncertainty of that determined for the Siqueiros Fracture Zone MORB by the same method (Herzberg and Asimow 2015) and therefore within the temperature range for ambient mantle. This suggests that at *c.* 3 Ma magma generation at the Antarctic–Phoenix Ridge was taking place at ambient mantle temperature. At this time, the slab window had already propagated as far as the Hero Fracture Zone (Fig. 3) and any regional elevation in T_P would therefore be evident in the Antarctic–Phoenix Ridge basalts, which it is not. Indeed, another important observation is that different segments of the Antarctic–Phoenix Ridge produce either N-MORB or E-MORB and the latter is indistinguishable from the post-subduction basalts in terms of incompatible trace elements and isotopic compositions (Figs 8 & 9). Therefore, at the Antarctic–Phoenix Ridge, basalts with similar trace element and isotopic compositions to the post-subduction basalts were formed at ambient T_P. Therefore, the garnet melting signature of the basalts must be due to a factor other than elevated T_P.

Melting of volatile-bearing peridotite

Hole and Saunders (1996) suggested that melting of volatile-enriched, high CO_2–H_2O, peridotite might be the solution to the paradox because the dry peridotite solidus is suppressed by the addition of volatiles such that the mantle adiabat intersects the volatile-bearing solidus at a higher pressure for a given temperature compared to the situation for dry peridotite (Fig. 10a). The presence of kaersutite megacrysts in some of the lavas was cited as evidence for the possible hydrous nature of their mantle source region. However, at the time that this hypothesis was proposed there was little information on the detailed effect of small concentrations of volatiles on the melting behaviour of dry peridotite. More recently, Dasgupta *et al.* (2007, 2013) and Sarafian *et al.* (2017) have performed melting experiments on peridotite in the presence of H_2O (200 and 450 ppm) and CO_2 (5–25 wt%). In both cases, the experiments were for volatile-undersaturated conditions, which differ from the volatile-saturated conditions which occur during subduction-related magmatism. For peridotite plus 450 ppm H_2O, the highest experimentally determined H_2O content (Sarafian *et al.* 2017), the 1350°C adiabat intersects the damp solidus at almost exactly 2.7 GPa and so this is considered an unlikely mechanism for melt generation along the Antarctic Peninsula because most melt would be generated in the presence of spinel not garnet.

Dasgupta *et al.* (2007, 2013) showed that for a pressure range of 2–5 GPa and for 5–25 wt% CO_2, the peridotite solidus is progressively suppressed to lower temperatures at a given pressure (Fig. 10a). For an ambient T_P of 1350°C and *c.* 10 wt% CO_2, the adiabat would intersect the peridotite solidus at >2.7 GPa (Fig. 10a), and this can be considered to be the lower limit of CO_2 required to generate all melting in the presence of garnet. Herzberg and Asimow (2008, 2015) provided a simple plot of CaO v. SiO_2 which serves to discriminate between melt generated in the presence of varying concentrations of CO_2 and dry melting (Fig. 10b). Progressively increasing CO_2 content causes decreases in SiO_2 and increases in CaO compared to dry melting, which also means that melts formed in the presence of CO_2 tend to be strongly Ne-normative. However, none of the post-subduction lavas have CaO and SiO_2 contents that fall within the region reserved for CO_2-rich melting in Figure 10b, even though some are Ne-normative (Fig. 8). For comparison, Cenozoic volcanic rocks from the West Antarctic Rift System (WARS: Wörner *et al.* 1989; Martin *et al.* 2013) are shown in Figure 10b, and WARS lavas with >7.0 wt% MgO are clearly identified as being related to melting in the presence of CO_2. Consequently, neither H_2O-present nor CO_2-present melting can generate lavas that carry a garnet signature and exhibit the observed SiO_2–CaO covariations.

A slab-hosted pyroxenite source for the alkali basalts

It is well established that mantle peridotite is not the only source for alkali basalts (Sobolev *et al.* 2005; Herzberg 2006, 2011) and that the large volumes of olivine tholeiite produced at Hawaii were derived from pyroxenite lithologies within the upper mantle. The pyroxenite that contributes to magmatism at large igneous provinces (LIPs: both oceanic and continental) is generally considered to have its origin in recycled subducted slabs that are later entrained in hot upwelling mantle plumes. On subduction, the transformation of basaltic crust produces quartz or coesite eclogite in the upper mantle and these are termed stage 1 pyroxenites (Herzberg 2011). Cumulates of dunite, troctolite and olivine gabbro will yield olivine pyroxenite lithologies (Ol + Cpx + Gt) in the upper mantle. Stage 1 pyroxenites can therefore be classified into two distinct lithologies; silica-enriched (SE) relating to metamorphism and recycling of basaltic crust, and silica-deficient (SD) relating to recycling of gabbro and other cumulate rocks (Herzberg 2011). Laboratory melting experiments performed on both SD and SE pyroxenite lithologies show that they can generate basaltic melt under a variety of P–T conditions (Lambart et al. 2009, 2012, 2013). Significant to the current discussion is the fact that pyroxenite is likely to melt at a lower temperature for a given pressure than mantle peridotite, such that for a given T_P the pyroxenite solidus is intersected deeper in the mantle than the peridotite solidus (Fig. 11). However, Lambart et al. (2016) showed that the Na, K and Ti content of pyroxenite has a profound effect on melting behaviour, and that some SiO_2-poor pyroxenite, which is also low in total alkalis and TiO_2, may have very similar melting behaviour to mantle peridotite. In general, however, high-SiO_2, alkali-rich pyroxenites will melt at a higher pressure for a given T_P than low SiO_2, alkali-poor pyroxenite. Garnet is stable on the pyroxenite solidus to much lower pressures than for peridotite (c. 1.7 GPa: Kogiso et al. 2003; Lambart et al. 2013), so, regardless of the temperature of the mantle, a large proportion of the melting interval for pyroxenite is in the presence of garnet (Fig. 11) and, additionally, melts derived from SE pyroxenite are SiO_2-rich (Herzberg 2011). Since the ancestral source of pyroxenite lithologies in the upper mantle is subducted slabs, and pyroxenite melts can exhibit LREE-enrichment as well as silica-saturation, the possibility that the post-subduction lavas were derived from pyroxenite resident in the subducted slab is worthy of consideration.

Recognizing a pyroxenite contribution to magmatism

Distinguishing between peridotite- and pyroxenite-derived melts and their related lavas is not a trivial problem. In the upper mantle both lithologies coexist and consequently magmas are generated which are hybrids from the two sources (e.g. Lambart 2017). Additionally, melting of some, particularly SD pyroxenite, can produce magmas that are indistinguishable from melts of peridotite (Hole 2018). Nevertheless, Herzberg and Asimow (2008) showed that at Hawaii, magmas derived by melting of pyroxenite were deficient in CaO for a given MgO content compared to peridotite-derived magmas. This is, in part, due to residual clinopyroxene in pyroxenite source that serves to produce a melting regime in which CaO is compatible ($D_{CaO} >1$), where D_{CaO} is the bulk distribution coefficient for CaO in the melt. Figure 12 shows the CaO and MgO distribution of the Seal Nunataks lavas alongside those for pyroxenite-derived magmas from Hawaii. Tholeiitic lavas from the Hawaiian Scientific Drilling Project (HSDP-2: Rhodes and Vollinger 2004), which are considered to have been derived almost entirely from pyroxenite, all have lower CaO at a given MgO than peridotite-derived magmas, and the peridotite–pyroxenite divide on the diagram is based on those data. Also shown are the positions occupied by near-primary magmas (Sobolev et al. 2005; Herzberg 2006) from Koolau recovered during KSDP (Koolau Scientific Drilling Project), and its Makapuu phase, which are the most extreme examples of pure, near-primary SE pyroxenite melts. Since olivine fractionation produced vectors sub-parallel to the pyroxenite–peridotite divide, the data for the Seal Nunataks lavas shown in Figure 12 clearly have major element compositions that are more deficient in CaO at a given MgO than many Hawaiian SE pyroxenite-derived melts. As discussed earlier, the CaO deficiency in the post-subduction lavas must have been inherited from their source because low-pressure crystallization of olivine does not fractionate MgO relative to CaO. Figure 12b shows covariations between SiO_2 and MgO for the same samples as shown in Figure 12a. There are clear similarities between the low SiO_2 and high SiO_2 lineages, representing the products of melting of SD and SE pyroxenite, respectively (Herzberg 2006), found in the HSDP-2 drill core and those found in the Antarctic Peninsula post-subduction lavas.

Primary or very primitive melts derived from SE pyroxenite (basaltic crust) are recognized because they plot on the SiO_2-rich side of the orthopyroxene–calcium Tschermak's component (Opx–CaTs) join in the molecular projection of the system olivine–orthopyroxene–CaTs–silica (Fig. 13). Conversely, melts derived from SD pyroxenite (cumulates)

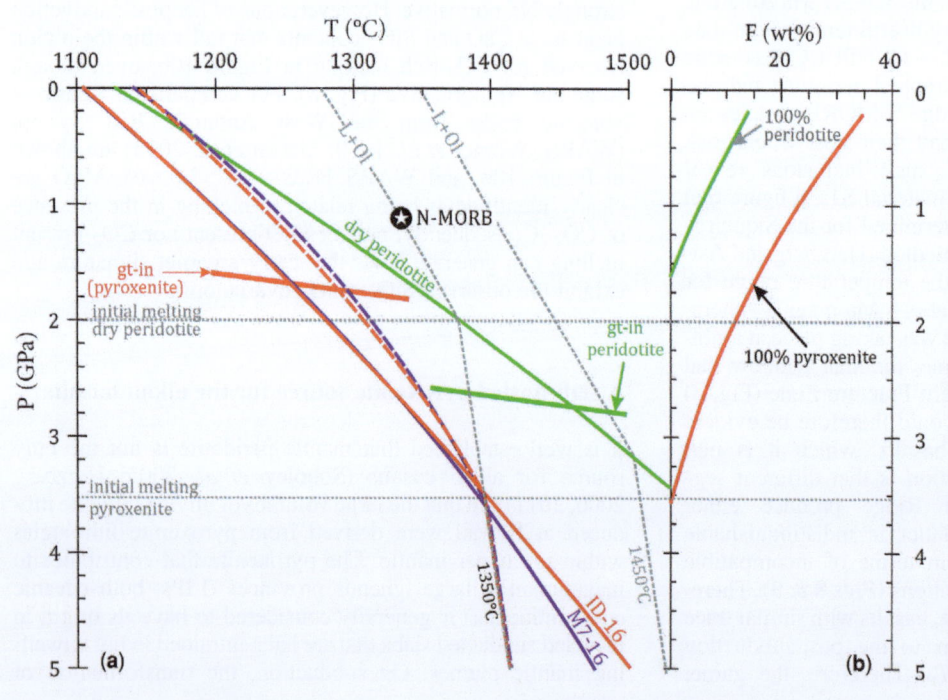

Fig. 11. (a) Pressure–temperature diagram illustrating the differences in melting behaviour for mantle peridotite and pyroxenites ID-16 (SE pyroxenite: Johnston and Draper 1992; Lambart et al. 2016) and M7-16 (SD pyroxenite: Lambart et al. 2009, 2012, 2016), mantle peridotite (Hirschmann 2000), and 'damp' peridotite (Sarafian et al. 2017). The garnet-in (gt-in) for peridotite is from McKenzie and O'Nions (1995) and for pyroxenite from Kogiso et al. (2003). (b) The extent of melting (F (wt%)) v. pressure for peridotite M7-16. ID-16 has an almost identical melt productivity to M7-16. N-MORB is the final pressure of melting at the Antarctic–Phoenix Ridge based on PRIMELT3 solutions for melt inclusions in Choi et al. (2013).

Ol–CaTs–Qz (Fig. 13a) at the predetermined pressure from the CMAS projection. Data for the Seal Nunataks lavas best lends itself to this type of petrological manipulation because there are sufficient data to determine that the lavas are related by olivine crystallization and it is these lavas that have been used here. The CMAS diagram yields a pressure of melting of c. 2.5–3.0 GPa, and by olivine addition in the Ol–CaTs–Qz projection reveals near-primary magmas close to the eutectic at 2.5–3.0 GPa. Most of the model primary magmas are related to an SD parent but some are related to an SE parent (Fig. 13; Table 2), which is consistent with the two lineages of lavas in the MgO–SiO$_2$ diagram (Fig. 12). The important information that this gives us is that the Seal Nunataks lavas could be related by olivine fractionation along a 2.5–3.0 GPa eutectic in equilibrium with Cpx + Gt ± Opx ± Ol: that is, SD or SE pyroxenite.

Pyroxenite melting behaviour

The melting behaviour of pyroxenite compared to peridotite is shown in Figure 11 and has been determined using the Melt–PX routine of Lambart et al. (2016). The pyroxenite starting compositions have been chosen to represent both SD (M7-16) and SE (ID-16) pyroxenites formed by prograde metamorphism of cumulates and basaltic crust, respectively. For both samples there are compositional data from experimentally generated melts (Johnston and Draper 1992; Lambart et al. 2009, 2012, 2016). For a 100% pyroxenite as the source lithology at $T_P = 1350°C$, melting would commence at c. 3.0 GPa for both SD and SE pyroxenites, and garnet would be stable on the pyroxenite solidus. Because M7-16 and ID-16 have similar K_2O, Na_2O and TiO_2 contents, their melting behaviour and melt productivity rates are very similar (Fig. 11). The melt productivity of pyroxenite is much greater than that of peridotite at the same T_P (Fig. 11) and the ability to produce a large range in extents of melting, and, importantly, melts which span a wide range of normative compositions, is available for pyroxenite melting. In Figure 6, the normative composition of experimental melts of SD pyroxenite (Keshav et al. 2004) are shown, and, as expected, the extent of melting increases with increasing temperature, but also melts become less Ne-normative and cross into the Ol–Di–Hy plane at moderate extents of melting. Therefore, melting of subducted slab-derived pyroxenite could explain the major element compositions of the post-subduction lavas without having to appeal to elevated mantle temperatures and/or 'damp' and/or CO_2 present melting. Additionally, the two distinct lineages present in the SiO$_2$ v. MgO diagram (Figs 7a & 12) can be explained by melting of SD (low-SiO$_2$) and SE (high-SiO$_2$) pyroxenite, both of which would be resident in the remnant subducted oceanic crust. This also explains the paradox identified by D'Orazio et al. (2001) that required garnet-present melting at high pressures to explain REE distributions in the lavas, but low pressures of melting to explain the high SiO$_2$ of the *same* lavas at the locus of the slab window in Patagonia.

Using the methods outlined in Lambart (2017), the compositions of accumulated fractional melts generated by melting of basaltic ocean crust have been calculated. For bulk distribution coefficients for trace elements we have used the data of Pertermann and Hirschmann (2003) and Pertermann et al. (2004), and we have assumed a pyroxenite that is close mineralogically to eclogite, containing 25% orthopyroxene, 50% clinopyroxene and 25% garnet (Pertermann and Hirschmann 2003). The relationship between pressure of melting and melt productivity has been derived from Melt–PX, such that for decompression melting of eclogite at 1350°C the melt productivity coefficient is constrained (Fig. 11). For a 'typical' oceanic crust composition, the data of Lambart (2017) have

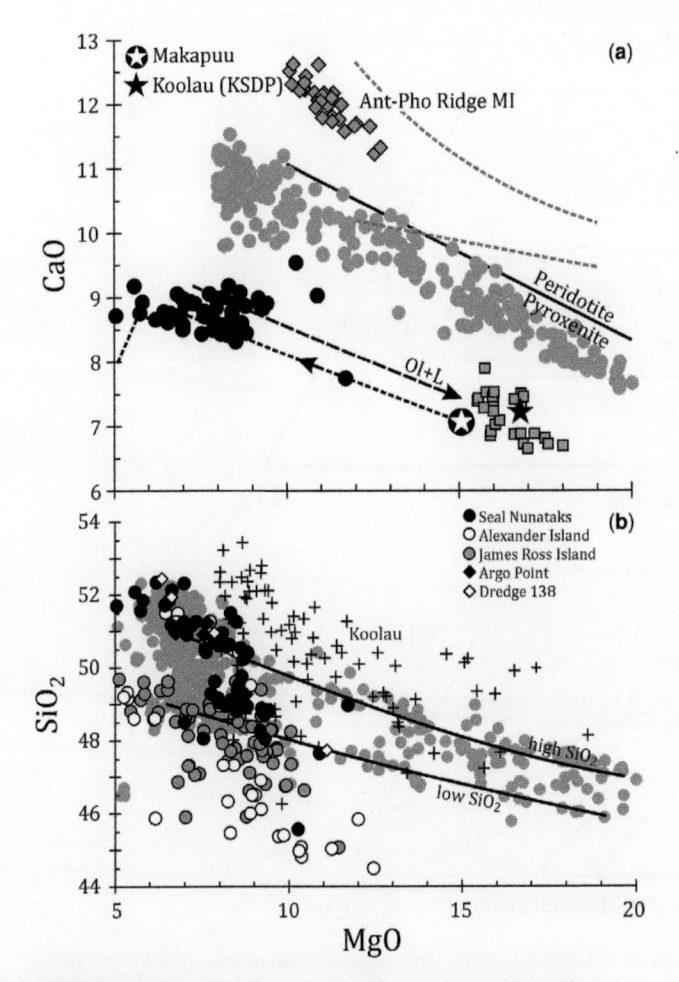

Fig. 12. (a) CaO v. MgO for Seal Nunataks lava (black dots) melt inclusions in Antarctic–Phoenix Ridge MORB (grey diamonds) and pyroxenite-derived lavas from Hawaii (grey dots: HSDP2). Grey squares represent parental, near-primary magmas to Seal Nuntaks lavas calculated by olivine addition and the use of the projections in Figure 13. The star in the circle and black star are primary pyroxenite–derived magmas from Koolau and its Makapuu stage, respectively (Sobolev et al. 2005; Herzberg 2006) with the LLD for L + Ol for Makapuu shown (short pecks line). The pyroxenite–peridotite divide and the curved pecked lines representing the limits for primary magmas derived from mantle peridotite are both from Herzberg and Asimow (2008). (b) SiO$_2$ v. MgO for Antarctic Peninsula post-subduction lavas, HSDP2 (grey dots with no border) and Koolau (crosses, KSDP (Koolau Scientific Drilling Project)). The low- and high-SiO$_2$ trends are from Herzberg (2006).

plot to the SiO$_2$-poor side of the same plane. Since the Opx–CaTs join is a thermal divide at pressures associated with mantle melting (i.e. >c. 1.7 GPa: Kogiso et al. 2003; Herzberg 2011; Lambart et al. 2013), derivative magmas from the melting of high- and low-SiO$_2$ pyroxenite have divergent fractionation histories. At near melting pressures, low-SiO$_2$ pyroxenite may undergo mixing with melts of mantle peridotite to form hybrid magmas, whereas melt derived from high-SiO$_2$ pyroxenite cannot (Sobolev et al. 2005; Herzberg 2011; Lambart 2017). Herzberg (2011) provides a method of calculating the composition of parental or near-primary melts from lavas that have only fractionated olivine, which is the method used here. The CMAS molecular projection (diopside–garnet–enstatite) is pressure sensitive, and so an estimate of pressure of melting can be determined from this diagram assuming olivine is the sole fractionating phase in a melt (Fig. 13b). Once the pressure of melting is determined, the liquid can then be projected along an olivine control line until it intersects the cotectic in the system

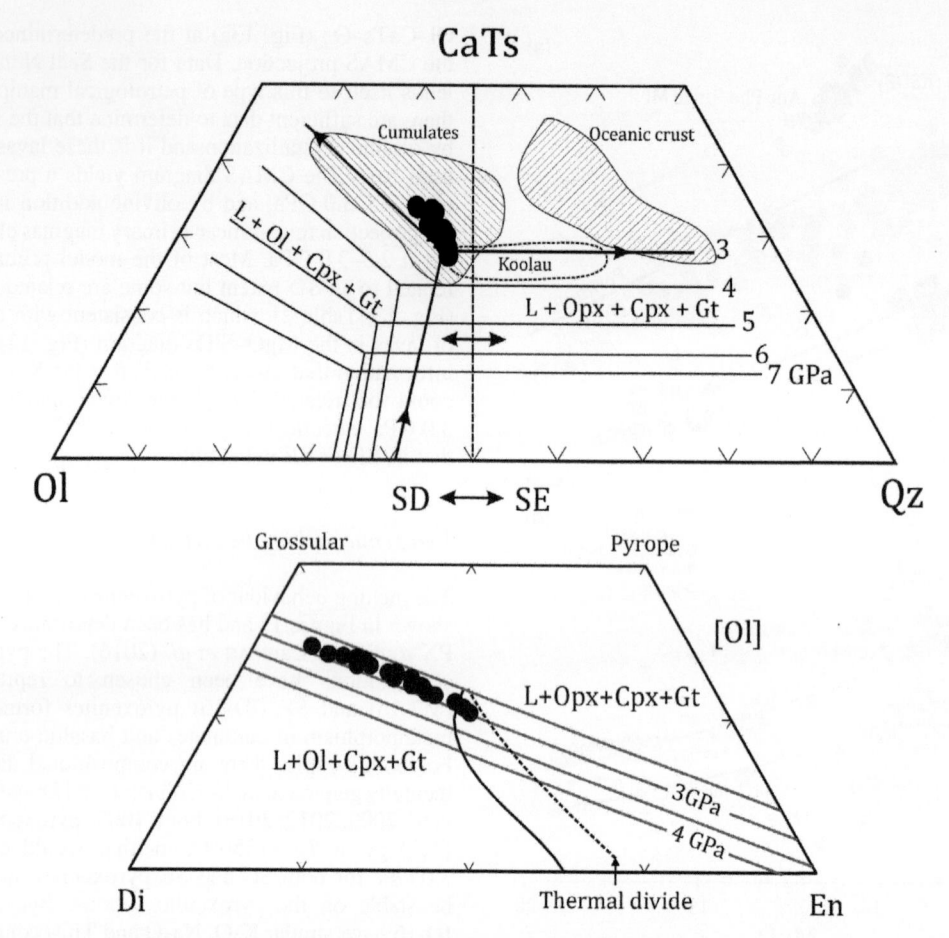

Fig. 13. (**a**) Molecular projection from or towards olivine onto the plane CS–MS–A for Seal Nunataks parental pyroxenite-derived magmas (black dots). This projection is a larger portion of the garnet–pyroxene plane shown in (b) and because the projection is from or towards olivine, olivine fractionation has no effect on the position of individual data points. The black curve separates olivine (to the left) and quartz (to the right) pyroxenite, and the pecked curve represents the thermal divide at magma-generation pressures, such that magmas to the right and left of the divide have divergent fractionation histories. In this projection Seal Nunataks lavas which have crystallized only olivine are related to cotectics at $c.$ 2.5-3.0 GPa. (**b**) Molecular projection from or towards diopside onto the plane olivine–calcium Tschermak's molecule–quartz for Seal Nunataks parental pyroxenite-derived magmas (black dots). Parental compositions were calculated by adding equilibrium olivine to lavas that are related to the olivine liquidus until they reach the cotectic for the pressure indicated by their position in the projection in (a): that is, between 2.5 and 3.0 GPa. The vertical pecked line is the garnet–pyroxene plane, and separates silica-depleted and silica-enriched pyroxenite. Fields for basaltic oceanic crust (which give rise to silica-enriched pyroxenite) and cumulates (which give rise to silica-depleted pyroxenite) are taken from East Pacific Rise MORB glasses and their associated gabbroic cumulates (PetDB database and Herzberg 2011). Projection codes in mol% are from O'Hara (1968):

$$CS(Di) = CaO + 2(Na_2O + K_2O) - 3.333P_2O_5$$
$$MS(Gt) = 2SiO_2 + TiO_2 - (FeO + MnO + MgO) - 2CaO - 8(Na_2O + K_2O) + 6.666P_2O_5$$
$$A(En) = TiO_2 + Al_2O_3 + Na_2O + K_2O$$
$$Ol = (Al_2O_3 + FeO + MnO + MgO) - 0.5(CaO + Na_2O + K_2O) + 1.75P_2O_5$$
$$CaTs = TiO_2 + Al_2O_3 + Na_2O + K_2O$$
$$Qz = SiO_2 + TiO_2(0.5Al_2O_3) - 0.5(FeO + MnO + MgO) - 1.5CaO - 4.5(Na_2O + K_2O) + 5.25P_2O_5.$$

been used. The results of the modelling are shown in Figure 14. As can be seen, using this method we can reproduce all the trace element characteristics of the post-subduction basalts for melting over a pressure interval of 1.0–2.8 GPa and for 5–25% melting, and the similarities with SE pyroxenite-derived tholeiite SR0125-6.25 from Mauna Loa (Rhodes and Vollinger 2004) are clear. Similarly, the composition of the post-subduction alkali basalts from Esperancia Glencross (D'Orazio et al. 2001) are also consistent with such a model (Fig. 8), and the Qz-normative compositions found at that location are most likely to represent large extents of melting coupled with crystallization along the L + Opx + Cpx cotectic at low pressures.

The similarity between the E-MORB and the post-subduction lavas requires some further attention. The role of pyroxenite in the petrogenesis of MORB has been investigated by Lambart et al. (2009), and they noted that, because of their higher melt productivities and lower solidus temperatures, 5% of pyroxenite in the source region may contribute up to 40% of the total melt production at otherwise 'normal' spreading ridges. The key to the contribution of pyroxenite to melting at spreading ridges is the complex solid–solid, liquid–liquid and solid–liquid reactions which take place between decompressing peridotite and pyroxenite (Herzberg 2011). The loss of one component (i.e. peridotite melt) would dramatically reduce melt productivity of the remaining one (i.e. pyroxenite). However, recognizing this contribution from pyroxenite is problematical because of the overwhelming influence of peridotite melting at spreading ridges (Lambart et al. 2009). Choe et al. (2007) showed that the fossilized Antarctic–Phoenix Ridge segments offshore of the South Shetland Islands consist of older N-MORB formed prior to the extinction of

Table 2. *Calculated parental melts for slab-window-related pyroxenite-derived lavas from Seal Nuntaks and Hawaiian pyroxenite parental melt compositions (in wt%)*

	Seal Nuns	HSDP-2	Seal Nuns	HSDP-2	Makapuu	KSDP
	Low-SiO$_2$		High-SiO$_2$			
SiO$_2$	47.69	47.63	49.06	48.6	51.3	50.2
TiO$_2$	1.72	1.85	1.49	2.04	1.5	1.5
Al$_2$O$_3$	12.23	10.54	11.91	9.94	11.3	10.8
Fe$_2$O$_3$	1.70	0.93	1.48	0.87	1.1	1.1
FeO	9.46	10.69	8.69	10.05	9.0	9.3
MnO	0.15	–	0.14	–	–	–
MgO	15.59	17.44	17.15	17.25	15.0	16.8
CaO	7.53	8.79	6.87	9.12	7.0	7.3
Na$_2$O	2.77	1.69	2.59	1.61	2.4	2.0
K$_2$O	0.88	0.28	0.72	0.29	0.4	0.3
P$_2$O$_5$	0.29	0.14	0.21	0.23	0.3	0.3
Mg#	78.5	73.6	74.8	74.6	73.8	75.4

After Sobolev *et al.* (2005) and Herzberg (2006).

spreading and younger E-MORB after extinction. The older N-MORB (3.5–6.4 Ma) occur in the southeastern flank of one segment, and younger E-MORB (3.1–1.4 Ma) comprise a huge seamount at the former ridge axis of a different segment and a large volcanic edifice at the northwestern flank of the same segment. They reasoned that the E-MORBs were formed by melting of enriched veins in the heterogeneous sub-ridge mantle and, when spreading stopped, it was these veins that provided most of the melt. Here, we suggest that the E-MORBs are local melts of pyroxenite and, when spreading stopped, the capacity for melt production by passive upwelling of peridotite melt was reduced and the predominant melt production was from pyroxenite which was already resident in oceanic crust. Upon subduction of half the spreading ridge, the rapid increase in pressure prevented further melting of this young, hot, pyroxenite until it reached *P–T* conditions above its solidus, which it would not do until it reached *c.*

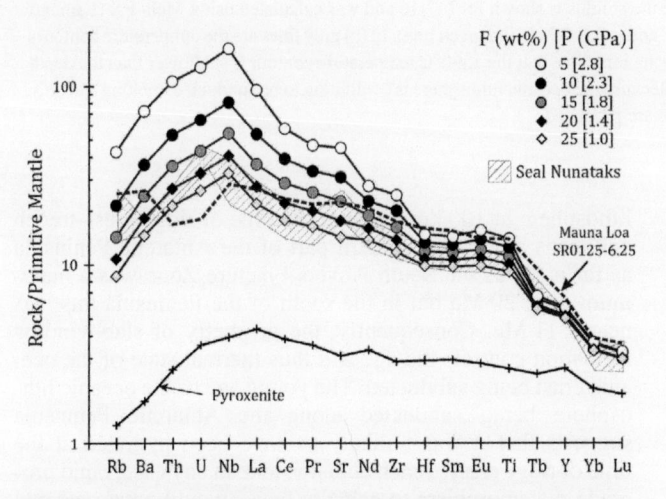

Fig. 14. Model melt compositions based on the trace element content of a typical example of oceanic crust (sample G2: Lambart 2017). The pressure for each model melt and the % melt production is given on the diagram. Model melts represent accumulated melt fractions calculated in increments of 1% melt over the range 5–25% melt using the method described in Lambart (2017). Bulk distribution coefficients for trace elements and the mineralogical composition of pyroxenite were taken from Pertermann and Hirschmann (2003) and Pertermann *et al.* (2004). See the text for the discussion. The Mauna Loa olivine tholeiite is taken from Rhodes and Vollinger (2004).

3.5 GPa or *c.* 100 km (Fig. 11). Using the subduction model for the warm, young slab of Hacker *et al.* (2003), this could be achieved *c.* 300–400 km from the palaeo-trench which in the Antarctic Peninsula would be the location of Seal Nunataks, for example.

Thermal structure before and after slab-window formation

Figure 15 illustrates a possible thermal scenario beneath the Antarctic Peninsula at the instant of ridge crest–trench collision (Fig. 15a) and at the time of passage of the subducted half-ridge beneath Seal Nunataks (Fig. 15b). At the instant of ridge crest–trench collision, subduction-related magmatism might still be active. However, the subducted slab at *c.* 100 km depth would be *c.* 28 myr old (Barker 1982) and whether the relatively young age of the subducted slab affected the magma production rate compared to the earlier history of the Antarctic Peninsula is a subject of conjecture. However, it is possible that more limited alteration to serpentinite in young oceanic crust compared with older oceanic crust may have reduced the flux of volatiles into the mantle wedge (e.g. Grove *et al.* 2012). At this time, the temperature structure beneath the subducted slab is such that the solidus of pyroxenite M7-16 is not intersected and slab melting cannot take place. A few million years after ridge crest–trench collision, the thermal structure below the palaeotrench is not the same as at the time of collision (Fig. 15b). Continued subduction of the half spreading-ridge occurred and therefore the temperature gradient in the mantle beneath the subducted slab must change from the syn-subduction case to that approaching mantle with $T_P = 1350°C$. This thermal re-equilibration would also be aided by the upwelling of asthenosphere to fill the void formed by the passage of the subducted oceanic lithosphere (Hole and Saunders 1996). For $T_P = 1350°C$, the dry solidus for pyroxenite M7-16 would be intersected at 3.5 GPa and 1390°C, producing melt; and for this to occur the 1390°C geotherm must be at <3.5 GPa (Fig. 15b). Because the slab remnant continues to subduct due to its coupling to the active subduction zone north of the North Anvers Fracture Zone (Figs 3 & 4), melting will cease when the slab descends to a depth of *c.* 120 km, equivalent to 3.5 GPa, because the pyroxenite will now be subsolidus. Therefore, the amount of melt that can be generated will be self-limiting, perhaps explaining the short period of post-subduction alkaline magmatism. The lack of a subduction signature in any of the slab-window-related basalts means that melt cannot have been produced from a source that was volatile-enriched during the previous subduction history of the arc. Whereas this thermal regime is appropriate to the tectonic setting of the Antarctic Peninsula, it would not apply in regions where slab windows formed during oblique ridge crest–trench collisions, as will be discussed in the following section.

Hydrous and anhydrous subducted slabs, and their relationship to magmatism

Slab-window-related basalts are not always of the intraplate alkaline association. Thorkelson and Breitsprecher (2004) noted that adakitic magmas, many of which can be classified as high-MgO andesites (e.g. Storey *et al.* 1989), form by partial melting of 'wet' slab crust proximal to the edges of slab windows at depths of 25–90 km, whereas non-adakitic melts of granodiorite to tonalitic composition are generated along plate edges at depths of 5–65 km. Such compositions are not recognized in the Antarctic Peninsula, even in trench-

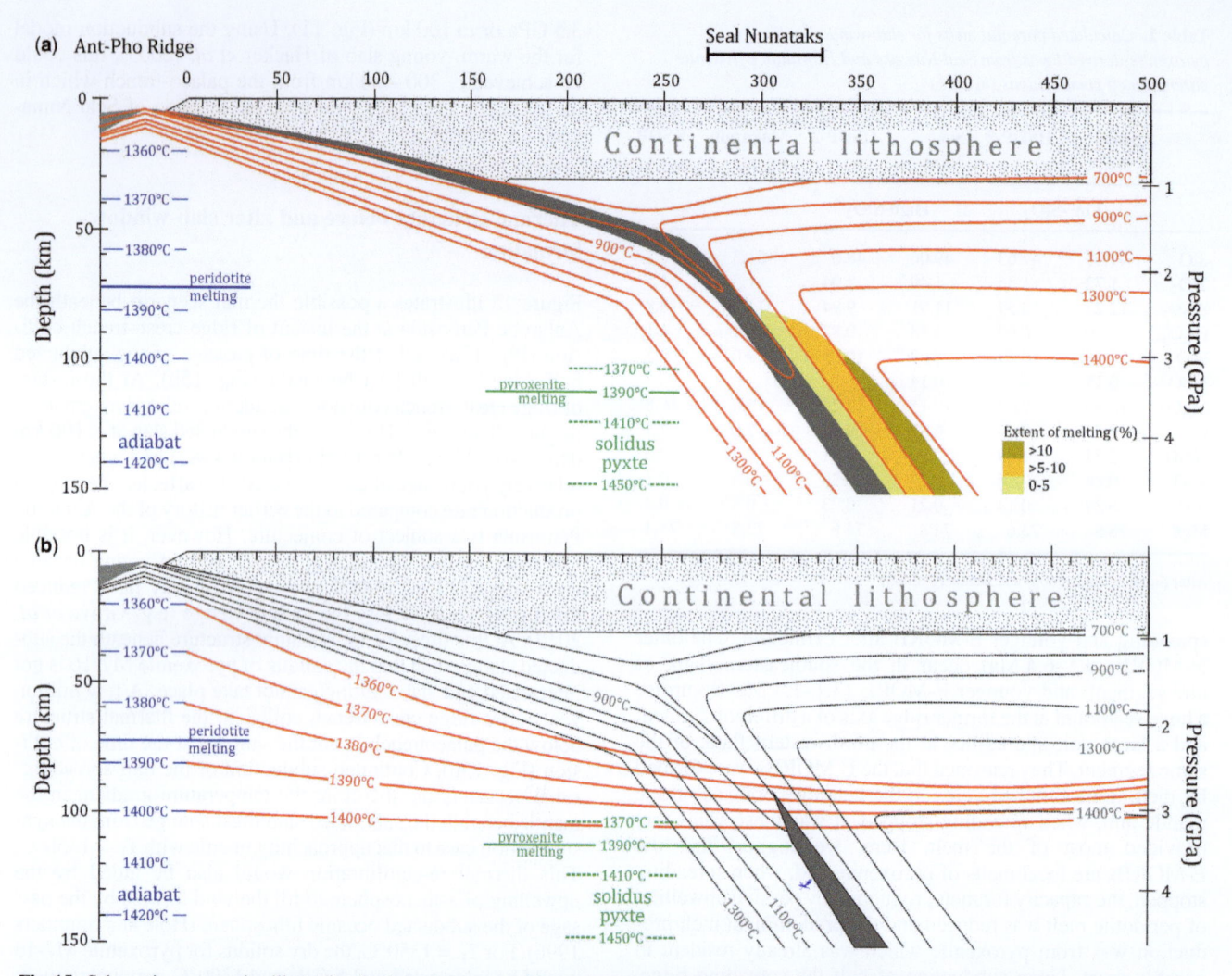

Fig. 15. Schematic cross-section illustrating the thermal structure of the northern Antarctic Peninsula subduction zone in (**a**) at the instant of ridge crest–trench collision and in (**b**) after ridge crest–trench collision and slab-window formation. In (a) note that the oceanic crust thins towards the trench as would be expected for the subduction of a young slab. The temperature contours (red lines) are modified after the 'hot' slab model of Hacker *et al.* (2003), and extents of melting to generate calc-alkaline subduction-related magmas are taken from Grove *et al.* (2012). A 1350°C adiabat at the collided Antarctic–Phoenix Ridge segment is assumed, and the temperature–depth profile is indicated in blue lettering and lines. At this T_P, and for the dry solidus of Hirschmann (2000), peridotite melting would commence at *c.* 2.2 GPa (solid blue line). The pyroxenite solidus is shown for M7-16 and was calculated using Melt–PX (Lambart *et al.* 2016). For $T_P = 1350°C$ pyroxenite melting would commence at *c.* 1390°C and *c.* 3.5 GPa (solid green line). In (b) grey lines are the temperature contours from (a), and red lines are estimated temperature contours after slab-window formation. Note that the 1390°C temperature contour is shallower than the depth required to initiate melting of pyroxenite M7-16 – a requirement of this model. Because the oceanic lithosphere is continuing to be subducted, melting can only occur over a very small temperature interval and therefore small melt fractions are produced.

proximal locations. The Antarctic Peninsula has a unique ridge crest–trench collision history because all collisions were near-orthogonal to the continental margin. At most other locations along the Pacific margin of the America, ridge subduction was strongly oblique, or slab windows formed by ridge crest–trench–transform interactions (Thorkelson *et al.* 2011). In these cases, the oceanic crust forming the edge of the slab window was of variable age and potentially considerably older than any of the oceanic crust being subducted along the Antarctic Peninsula, was probably hydrated, and capable of melting in water-saturated conditions to produce adakite–tonalite magmatism (Thorkelson and Breitsprecher 2004)

In the case of the Antarctic Peninsula, because subduction ceased following orthogonal ridge crest–trench collisions, the oceanic crust beneath the continental margin was both young and hot at the time of collision, and the trailing edge of the slab forming the eastern margin of the slab window was necessarily zero age at the instant of its subduction. Barker (1982) estimated that the age of the subducted oceanic lithosphere at 100 km depth at the time of ridge crest–trench collisions along the northern part of the Antarctic Peninsula as far north as the South Anvers Fracture Zone was a maximum of *c.* 20 Ma but in the south of the Peninsula this was nearer 11 Ma. Consequently, the geometry of slab-window formation controls the age and thus thermal state of the oceanic crust being subducted. The young age of the oceanic lithosphere being subducted along the Antarctic Peninsula suggests that it was unlikely to have been hydrated at the time of ridge crest–trench collision and, in any case, rapid prograde metamorphism to eclogite facies would have rendered the crust anhydrous soon after subduction. In eastern Papua New Guinea, it has been shown that the transformation of oceanic crust into anhydrous coesite-bearing eclogite under ultrahigh–pressure conditions of up to 3.0 GPa and 800°C occurs over the same timescales as plate tectonic processes. Eclogites at D'Entrecasteaux Island, at the boundary of the Woodlark and Australian plates, are dated at 4.33 Ma, meaning that the entire cycle of subduction, prograde metamorphism and exhumation took place over an exceedingly short timescale

(Baldwin et al. 2004). Therefore, there is evidence to support the possibility and probability of transforming young, hot oceanic crust into eclogite over short timescales. Once transformed into garnet eclogite or pyroxenite, the slab was able to contribute near-dry melts.

Remaining conundrums

Whereas melting of slab-hosted pyroxenite provides a very viable alternative to melting at elevated T_P or with CO_2 on the solidus, some aspects of the post-subduction magmatism remain unanswered. For example, the situation at James Ross Island remains rather enigmatic. Here, alkali basalt erupted but this is essentially a back-arc rather than slab-window setting. At the South Shetland Islands, subduction activity decreased sharply after 6 Ma and stopped or became very slow after the cessation of spreading in Drake Passage at about 3.3 Ma (Jin et al. 2009). The present extension in the Bransfield Rift started less than 4 myr ago, and possibly less than 1.5 myr ago, following the demise of the Antarctic–Phoenix spreading centre at about c. 3.5 Ma (Lawver et al. 1995; Keller et al. 2002). One possibility is that once subduction slowed dramatically at the South Shetland Islands Trench, a remnant subducted slab might have been present as far east as the back-arc and it is this that underwent melting to produce the James Ross Island Volcanic Group.

Miocene volcanism at the Jones Mountains (Hole et al. 1994b), 750 km WSW of Alexander Island, shares geochemical affinities with the Antarctic Peninsula occurrences, and do not appear to form part of the larger Marie Byrd Land province of West Antarctica. This being the case, any attempt to explain the Antarctic Peninsula volcanism in relationship to slab-window formation must also account for the origin of the Jones Mountains occurrences. However, Hole et al. (1994b) noted that the Jones Mountains sits astride the projected position of the Undinsev Fracture Zone, which was the locus of the earliest ridge–crest trench collisions for a segment of the Antarctic–Phoenix spreading centre and the continental margin.

Particularly critical issues are the age and timing of post-subduction volcanism in relation to cessation of subduction. Reliable geochronological data are non-existent for many of the key exposures described here, and a concerted campaign for dating these fascinating, but somewhat enigmatic rocks, would further improve our understanding of the relationship between alkalic volcanism and slab-window formation.

Acknowledgements Investigations into alkali basalts in the Antarctic Peninsula were made when M.J. Hole was an employee of the British Antarctic Survey. The number of people who made the fieldwork possible are too numerous to mention but to anyone involved in the 1983–84, 85–86 and 87–88 Austral Summer field seasons, I owe a debt of gratitude. The manuscript was improved by the thoughtful reviews of Kurt Panter and John Gamble. The late Peter Barker first introduced me to the complexities of ridge crest–trench interactions and for that I am grateful.

Author contributions MJH: conceptualization (lead), writing – original draft (lead).

Funding Fieldwork for this project was undertaken whilst the author was a member of the NERC British Antarctic Survey.

Data availability No new data were created during the study.

References

Baldwin, S.L., Monteleone, B.B., Webb, L.E., Fitzgerald, P.G., Grove, M. and Hill, E.J. 2004. Pliocene eclogite exhumation at plate tectonic rates in eastern Papua New Guinea. *Nature*, **431**, 263–267, https://doi.org/10.1038/nature02846

Barker, P.F. 1982. The Cenozoic subduction history of the Pacific margin of the Antarctic Peninsula: ridge crest–trench interactions. *Journal of the Geological Society, London*, **139**, 787–801, https://doi.org/10.1144/gsjgs.139.6.0787

Bevier, M.L. 1983. Implications of chemical and isotopic composition for petrogenesis of Chilcotin group basalts, British Columbia. *Journal of Petrology*, **24**, 207–226, https://doi.org/10.1093/petrology/24.2.207

Breitsprecher, K. and Thorkelson, D.J. 2009. Neogene kinematic history of Nazca–Antarctic–Phoenix slab windows beneath Patagonia and the Antarctic Peninsula. *Tectonophysics*, **464**, 10–20, https://doi.org/10.1016/j.tecto.2008.02.013

Choe, W.-H., Lee, J.-I., Lee, M.-J., Hur, S.-D. and Jin, Y.-K. 2007. Origin of E-MORB in a fossil spreading center: the Antarctic-Phoenix Ridge, Drake Passage, Antarctica. *Geosciences Journal*, **11**, 186–199, https://doi.org/10.1007/BF02913932

Choi, S.-H., Schiano, P., Chen, Y., Devidal, J.-L., Choo, M.-K. and Lee, J.-I. 2013. Melt inclusions in olivine and plagioclase phenocrysts from Antarctic–Phoenix Ridge basalts: Implications for origins of N- and E-type MORB parent magmas. *Journal of Volcanology and Geothermal Research*, **253**, 75–86, https://doi.org/10.1016/j.jvolgeores.2012.12.008

Danyushevsky, L.V. and Plechov, P. 2011. Petrolog3: Integrated software for modeling crystallization processes. *Geochemistry, Geophysics, Geosystems*, **12**, Q07021, https://doi.org/10.1029/2011GC003516

Dasgupta, R., Hirschmann, M.M. and Smith, N.D. 2007. Partial melting experiments on peridotite + CO_2 at 3 GPa and genesis of alkalic ocean island basalts. *Journal of Petrology*, **48**, 2093–2124, https://doi.org/10.1093/petrology/egm053

Dasgupta, R., Mallik, A., Tsuno, K., Withers, A.C., Hirth, G. and Hirschmann, M.M. 2013. Carbon-dioxide-rich silicate melt in the Earth's upper mantle. *Nature*, **493**, 211–216, https://doi.org/10.1038/nature11731

Dickinson, W.R. and Snyder, W.S. 1979. Geometry of subducted slabs related to the San Andreas transform. *Journal of Geology*, **87**, 609–627, https://doi.org/10.1086/628456

D'Orazio, M.D., Agostini, S., Innocenti, F., Haller, M.J., Manetti, P. and Mazzarini, F. 2001 Slab window-related magmatism from southernmost South America: the Late Miocene mafic volcanics from the Estancia Glencross Area (~52°S, Argentina–Chile). *Lithos*, **57**, 67–89, https://doi.org/10.1016/S0024-4937(01)00040-8

Eagles, G., Gohl, K. and Larter, R.D. 2007. High-resolution animated tectonic reconstruction of the South Pacific and West Antarctic Margin. *Geochemistry, Geophysics, Geosystems*, **5**, Q07002, https://doi.org/10.1029/2003GC000657

Grove, T.L., Till, C.B. and Krawczynski, M.J. 2012. The role of H_2O in subduction zone magmatism. *Annual Reviews of Earth and Planetary Sciences*, **40**, 413–439, https://doi.org/10.1146/annurev-earth-042711-105310

Hacker, B.R., Peacock, S.M., Abers, G.A. and Holloway, S.D. 2003. Subduction factory 2. Are intermediate-depth earthquakes in subducting slabs linked to metamorphic dehydration reactions? *Journal of Geophysical Research: Solid Earth*, **108**, 2030, https://doi.org/10.1029/2001JB001129

Hart, S.R. 1988. Heterogeneous mantle domains: signatures, genesis and mixing chronologies. *Earth and Planetary Science Letters*, **90**, 273–296, https://doi.org/10.1016/0012-821X(88)90131-8

Herzberg, C. 2006. Petrology and thermal structure of the Hawaiian plume from Mauna Kea volcano. *Nature*, **444**, 605–609, https://doi.org/10.1038/nature05254

Herzberg, C. 2011. Identification of source lithology in the Hawaiian and Canary Islands: implications for origins. *Journal of Petrology*, **52**, 113–146, https://doi.org/10.1093/petrology/egq075

Herzberg, C. and Asimow, P.D. 2008. Petrology of some oceanic island basalts: PRIMELT2.XLS software for primary magma calculation. *Geochemistry, Geophysics, Geosystems*, **9**, Q09001, https://doi.org/10.1029/2008GC002057

Herzberg, C. and Asimow, P.D. 2015. PRIMELT3 MEGA.XLSM software for primary magma calculation: Peridotite primary magma MgO contents from the liquidus to the solidus. *Geochemistry, Geophysics, Geosystems*, **16**, 563–578, https://doi.org/10.1002/2014GC005631

Hirschmann, M.M. 2000. Mantle solidus: Experimental constraints and the effects of peridotite composition. *Geochemistry, Geophysics, Geosystems*, **1**, 1042, https://doi.org/10.1029/2000GC000070

Hole, M.J. 1988. Post-subduction alkaline volcanism along the Antarctic Peninsula. *Journal of the Geological Society, London*, **145**, 985–989, https://doi.org/10.1144/gsjgs.145.6.0985

Hole, M.J. 1990a. Geochemical evolution of Pliocene–Recent post-subduction alkalic basalts from Seal Nunataks, Antarctic Peninsula. *Journal of Volcanology and Geothermal Research*, **40**, 149–167, https://doi.org/10.1016/0377-0273(90)90118-Y

Hole, M.J. 1990b. Beethoven Peninsula. *Antarctic Research Series*, **48**, 268–270.

Hole, M.J. 1990c. Hornpipe Heights. *Antarctic Research Series*, **48**, 270–272.

Hole, M.J. 2018. Mineralogical and geochemical evidence for polybaric fractional crystallization of continental flood basalts and implications for identification of peridotite and pyroxenite source lithologies. *Earth Science Reviews*, **176**, 51–67, https://doi.org/10.1016/j.earscirev.2017.09.014

Hole, M.J. and Larter, R.D. 1993. Trench-proximal volcanism following ridge crest–trench collision along the Antarctic Peninsula. *Tectonics*, **12**, 897–910, https://doi.org/10.1029/93TC00669

Hole, M.J. and LeMasurier, W.E. 1994. Tectonic controls on the geochemical composition of Cenozoic, mafic alkaline volcanic rocks from West Antarctica. *Contributions to Mineralogy and Petrology*, **117**, 187–202, https://doi.org/10.1007/BF00286842

Hole, M.J. and Millett, J.M. 2016. Controls of mantle potential temperature and lithospheric thickness on magmatism in the North Atlantic Igneous Province. *Journal of Petrology*, **57**, 417–436, https://doi.org/10.1093/petrology/egw014

Hole, M.J. and Saunders, A.D. 1996. The generation of small melt-fractions in truncated melt columns: constraints from magmas erupted above slab windows and implications for MORB genesis. *Mineralogical Magazine*, **60**, 173–189, https://doi.org/10.1180/minmag.1996.060.398.12

Hole, M.J. and Thomson, J.W. 1990. Mount Pinafore–Debussy Heights. *Antarctic Research Series*, **48**, 268–270.

Hole, M.J., Rogers, G., Saunders, A.D. and Storey, M. 1991. Relation between alkalic volcanism and slab-window formation. *Geology*, **19**, 657–660, https://doi.org/10.1130/0091-7613(1991)019<0657:RBAVAS>2.3.CO;2

Hole, M.J., Kempton, P.D. and Millar, I.L. 1993. Trace-element and isotopic characteristics of small-degree melts of the asthenosphere: Evidence from the alkalic basalts of the Antarctic Peninsula. *Chemical Geology*, **109**, 51–68, https://doi.org/10.1016/0009-2541(93)90061-M

Hole, M.J., Saunders, A.D., Rogers, G. and Sykes, M.A. 1994a. The relationship between alkaline magmatism, lithospheric extension and slab window formation along continental destructive plate margins. *Geological Society, London, Special Publications*, **81**, 265–285, https://doi.org/10.1144/GSL.SP.1994.081.01.15

Hole, M.J., Storey, B.C. and LeMasurier, W.E. 1994b. Tectonic setting and geochemistry of Miocene alkalic basalts from the Jones Mountains, West Antarctica. *Antarctic Science*, **6**, 85–92, https://doi.org/10.1017/S0954102094000118

Homrighausen, S., Hoernle, K., Hauff, F., Geldmacher, J., Wartho, J.-A., van den Bogaard, P. and Garbe-Schonberg, D. 2018. Global distribution of the HIMU end member: Formation through Archean plume-lid tectonics. *Earth-Science Reviews*, **182**, 85–101, https://doi.org/10.1016/j.earscirev.2018.04.009

Jin, Y.K., Lee, J., Hong, J.K. and Nam, S.H. 2009. Is subduction ongoing in the South Shetland Trench, Antarctic Peninsula? New constraints from crustal structures of outer trench wall. *Geoscience Journal*, **13**, 59–67, https://doi.org/10.1007/s12303-009-0005-5

Johnson, S.T. and Thorkelson, D.J. 2000. Continental flood basalts: episodic magmatism above long-lived hotspots. *Earth and Planetary Science Letters*, **175**, 247–256, https://doi.org/10.1016/S0012-821X(99)00293-9

Johnston, A.D. and Draper, D.S. 1992. Near-liquidus phase relations of an anhydrous high-magnesia basalt from the Aleutian Islands: Implications for arc magma genesis and ascent. *Journal of Volcanology and Geothermal Research*, **52**, 27–41, https://doi.org/10.1016/0377-0273(92)90131-V

Keller, R.A., Fisk, M.R., Smellie, J.L. and Strelin, J.A. 2002. Geochemistry of back arc basin volcanism in Bransfield Strait, Antarctica: subducted contributions and along-axis variations. *Journal of Geophysical Research: Solid Earth*, **107**, 2171, https://doi.org/10.1029/2001JB000444

Keshav, S., Gudfinnson, G.H., Sen, G. and Fei, Y. 2004. High-pressure melting experiments on garnet clinopyroxenite and the alkalic to tholeiitic transition in ocean-island basalts. *Earth and Planetary Science Letters*, **223**, 356–379, https://doi.org/10.1016/j.epsl.2004.04.029

Kogiso, T, Hirschmann, M.M. and Frost, D.J. 2003. High-pressure partial melting of garnet pyroxenite: possible mafic lithologies in the source of ocean island basalts. *Earth and Planetary Science Letters*, **216**, 603–617, https://doi.org/10.1016/S0012-821X(03)00538-7

Kosler, J., Magna, T., Mlcoch, B., Mixa, P., Nyvlt, D. and Holub, F.V. 2009. Combined Sr, Nd, Pb and Li isotope geochemistry of alkaline lavas from northern James Ross Island (Antarctic Peninsula) and implications for back-arc magma formation. *Chemical Geology*, **258**, 207–218, https://doi.org/10.1016/j.chemgeo.2008.10.006

Lambart, S. 2017. No direct contribution of recycled crust in Icelandic basalts. *Geochemical Perspectives Letters*, **4**, 7–12, https://doi.org/10.7185/geochemlet.1728

Lambart, S., Laporte, D. and Schiano, P. 2009. An experimental study of pyroxenite partial melts at 1 and 1.5 GPa: Implications for the major-element composition of Mid-Ocean Ridge Basalts. *Earth and Planetary Science Letters*, **288**, 335–347, https://doi.org/10.1016/j.epsl.2009.09.038

Lambart, S., Laporte, D., Provost, A. and Schiano, P. 2012. Fate of pyroxenite-derived melts in the peridotitic mantle: Thermodynamic and experimental constraints. *Journal of Petrology*, **53**, 451–476, https://doi.org/10.1093/petrology/egr068

Lambart, S., Laporte, D., Provos, A. and Schiano, P. 2013. Markers of the pyroxenite contribution in the major-element compositions of oceanic basalts: Review of the experimental constraints. *Lithos*, **160–161**, 14–36, https://doi.org/10.1016/j.lithos.2012.11.018

Lambart, S., Baker, M.B. and Stopler, E.M. 2016. The role of pyroxenite in basalt genesis: Melt–PX, a melting parameterization for mantle pyroxenites between 0.9 and 5 GPa. *Journal of Geophysical Research: Solid Earth*, **121**, 5708–5735, https://doi.org/10.1002/2015JB012762

Larter, R.D. and Barker, P.F. 1991. Effects of ridge crest–trench interaction on Antarctic–Pheonix Spreading: Forces on a young subducting plate. *Journal of Geophysical Research*, **96**, 19 583–19 607, https://doi.org/10.1029/91JB02053

Lawver, L.A., Keller, R.A., Fisk, M.R. and Strelin, J.A. 1995. Bransfield Strait, Antarctic Peninsula active extension behind a dead arc. *In*: Taylor, B. (ed.) *Backarc Basins*. Springer, Boston, MA, 315–342.

Le Bas, M.J. and Streckeisen, A.L. 1991. The IUGS systematics of igneous rocks. *Journal of the Geological Society, London*, **148**, 825–833, https://doi.org/10.1144/gsjgs.148.5.0825

Martin, A.P., Cooper, A.F. and Price, R.C. 2013. Petrogenesis of Cenozoic, alkalic volcanic lineages at Mount Moring and their entrained lithospheric mantle xenoliths: lithospheric versus asthenospheric mantle sources. *Geochimica et*

Cosmochimica Acta, **122**, 127–152, https://doi.org/10.1016/j.gca.2013.08.025

McKenzie, D.P. and O'Nions, R.K. 1995. The source region of ocean island basalts. *Journal of Petrology*, **36**, 133–159, https://doi.org/10.1093/petrology/36.1.133

O'Hara, M.J. 1968. The bearing of phase equilibria studies in synthetic and natural systems on the origin of basic and ultrabasic rocks. *Earth-Science Reviews*, **4**, 69–133, https://doi.org/10.1016/0012-8252(68)90147-5

Palacz, Z. and Saunders, A.D. 1986. Coupled trace element and isotope enrichment in the Cook–Austral–Samoa islands, southwestern Pacific. *Earth and Planetary Science Letters*, **79**, 270–280, https://doi.org/10.1016/0012-821X(86)90185-8

Pertermann, M. and Hirschmann, M.M. 2003. Partial melting experiments on a MORB-like pyroxenite between 2 and 3 GPa: Constraints on the presence of pyroxenite in basalt source regions from solidus location and melting rate. *Journal of Geophysical Research: Solid Earth*, **108**, 2125, https://doi.org/10.1029/2000JB000118

Pertermann, M., Hirschmann, M.M., Hametner, D., Günther, D. and Schmidt, M.W. 2004. Experimental determination of trace element partitioning between garnet and silica-rich liquid during anhydrous. *Geochemistry, Geophysics, Geosystems*, **5**, Q05A01, https://doi.org/10.1029/2003GC000638

Putirka, K.D. 2008. Thermometers and barometers for volcanic systems. *Reviews in Mineralogy and Petrology*, **69**, 61–120, https://doi.org/10.2138/rmg.2008.69.3

Renner, R.G.B., Dikstra, B.J. and Martin, J.L. 1982. Aeromagnetic surveys over the Antarctic Peninsula. *In*: Craddock, C. (ed.) *Antarctic Geoscience*. University of Wisconsin Press, Madison, 363–370.

Rhodes, J.M. and Vollinger, M.J. 2004. Composition of basaltic lavas samples by phase-2 of the Hawaii Scientific Drilling Project: geochemical stratigraphy and magma types. *Geochemistry, Geophysics, Geosystems*, **5**, Q03G13, https://doi.org/10.1029/2002GC000434

Sarafian, E., Gaetini, G.A., Hauri, E.H. and Sarafian, A.R. 2017. Experimental constraints on the damp peridotite solidus and oceanic mantle potential temperature. *Science*, **355**, 942–945, https://doi.org/10.1126/science.aaj2165

Saunders, A.D. 1982. Petrology and geochemistry of alkali-basalts from Jason Peninsula, Oscar II Coast, Graham Land. *British Antarctic Survey Bulletin*, **55**, 1–9.

Smellie, J.L. 1987. Geochemistry and tectonic setting of alkaline volcanic rocks in the Antarctic Peninsula: a review. *Journal of Volcanology and Geothermal Research*, **32**, 269–285, https://doi.org/10.1016/0377-0273(87)90048-5

Smellie, J.L. 1999. Lithostratigraphy of Miocene–Recent, alkaline volcanic fields in the Antarctic Peninsula and eastern Ellsworth Land. *Antarctic Science*, **11**, 362–378, https://doi.org/10.1017/S0954102099000450

Smellie, J.L. and Hole, M.J. 2021. Antarctic Peninsula: volcanology. *Geological Society, London, Memoirs*, **55**, https://doi.org/10.1144/M55-2018-59

Smellie, J.L., Pankhurst, R.J., Hole, M.J. and Thomson, J.W. 1988. Age, distribution and eruptive conditions of late Cenozoic alkaline volcanism in the Antarctic Peninsula and eastern Ellsworth Land. *British Antarctic Survey Bulletin*, **80**, 21–49.

Sobolev, A.V., Hofmann, A.W., Sobolev, S.V. and Nikogosian, I.K. 2005. An olivine-free mantle source of Hawaiian shield basalts. *Nature*, **434**, 590–597, https://doi.org/10.1038/nature03411

Storey, M., Rogers, G., Saunders, A.D. and Terrell, D. 1989. San Quintin volcanic field, Baja California, Mexico: 'within-plate' magmatism following ridge subduction. *Terra Nova*, **1**, 195–202, https://doi.org/10.1111/j.1365-3121.1989.tb00352.x

Sykes, M.A. 1988. *The Petrology and Tectonic Significance of the James Ross Island Volcanic Group, Antarctica*. PhD thesis, University of Nottingham, Nottingham, UK.

Sun, S.-s. and McDonough, W.F. 1989. Chemical and isotopic systematics of oceanic basalts: implications for mantle composition and processes. *Geological Society, London, Special Publications*, **42**, 313–345, https://doi.org/10.1144/GSL.SP.1989.042.01.19

Thompson, R.N. 1982. Magmatism of the British Tertiary Volcanic Province. *Scottish Journal of Geology*, **18**, 49–107, https://doi.org/10.1144/sjg18010049

Thorkelson, D.J. 1996. Subduction of diverging plates and the principles of slab window formation. *Tectonophysics*, **255**, 47–63, https://doi.org/10.1016/0040-1951(95)00106-9

Thorkelson, D.J. and Breitsprecher, K. 2004. Partial melting of slab window margins: genesis of adakitic and non-adakitic magmas. *Lithos*, **79**, 24–41, https://doi.org/10.1016/j.lithos.2004.04.049

Thorkelson, D.J. and Taylor, R.P. 1989. Cordilleran slab windows. *Geology*, **17**, 833–836, https://doi.org/10.1130/0091-7613(1989)017<0833:CSW>2.3.CO;2

Thorkelson, D.J., Madsen, J.K. and Sluggett, C.L. 2011. Mantle flow through the Northern Cordilleran slab window revealed by volcanic geochemistry. *Geology*, **39**, 267–270, https://doi.org/10.1130/G31522.1;

Wörner, G., Viereck, L.G., Hertogen, J. and Niephaus, H. 1989. The Mt Melbourne Volcanic Field (Victoria Land Antarctica) II. Geochemistry and magma genesis. *Geologische Jahrbuch*, **E38**, 395–433.

Section 5

Continental extension-related volcanism

View of Mount Melbourne, a small active stratovolcano rising to 2732 m asl situated in northern Victoria Land. Photograph by J. L. Smellie

Section 5

Continental extension-related volcanism

Chapter 5.1a

Northern Victoria Land: volcanology

John L. Smellie[1]* and Sergio Rocchi[2]

[1]School of Geography, Geology and the Environment, University of Leicester, University Road, Leicester LE1 7RH, UK
[2]Dipartimento di Scienze della Terra, Universitá di Pisa, Via Santa Maria 53, I-56126 Pisa, Italy
JLS, 0000-0001-5537-1763; SR, 0000-0001-6868-1939
*Correspondence: jls55@le.ac.uk

Abstract: Neogene volcanism is widespread in northern Victoria Land, and is part of the McMurdo Volcanic Group. It is characterized by multiple coalesced shield volcanoes but includes a few relatively small stratovolcanoes. Two volcanic provinces are defined (Hallett and Melbourne), with nine constituent volcanic fields. Multitudes of tiny monogenetic volcanic centres (mainly scoria cones) are also scattered across the region and are called the Northern Local Suite. The volcanism extends in age between middle Miocene (c. 15 Ma) and present but most is <10 Ma. Two centres may still be active (Mount Melbourne and Mount Rittmann). It is alkaline, varying between basalt (basanite) and trachyte/rhyolite. There are also associated, geographically restricted, alkaline gabbro to granite plutons and dykes (Meander Intrusive Group) with mainly Eocene–Oligocene ages (52–18 Ma). The isotopic compositions of the plutons have been used to infer overall cooling of climate during the Eocene–Oligocene. The volcanic sequences are overwhelmingly glaciovolcanic and are dominated by 'a'ā lava-fed deltas, the first to be described anywhere. They have been a major source of information on Mio-Pliocene glacial conditions and were used to establish that the thermal regime during glacial periods was polythermal, thus necessitating a change in the prevailing paradigm for ice-sheet evolution.

The volcanic rocks described in this chapter crop out on the western margin of the West Antarctic Rift System, where it extends north into the Ross Sea (LeMasurier and Thomson 1990) (Fig. 1). Their origin has been linked to either active (plume-driven) or passive rifting (e.g. Behrendt et al. 1992; LeMasurier and Landis 1996; Storey et al. 1999; Rocchi et al. 2002a, 2003). Recent studies have emphasized a link between the magmatism and Paleozoic tectonic structures in northern Victoria Land, which were reactivated during Southern Ocean spreading between Antarctica and Australia (Salvini et al. 1997). The results of tectonic and seismic tomographic studies by Faccenna et al. (2008) suggest that a low-velocity anomaly, interpreted as a thermal anomaly, underlies the region in which the volcanism occurs. The anomaly was attributed to mantle upwelling at the boundary between the East Antarctic Craton and the West Antarctic Rift System, and it may have continued until relatively recent time. Associated transtensional decompression melting of the subplate mantle was followed by ascent of the melts along the main NW–SE strike-slip fault systems and related north–south fault arrays (Rocchi et al. 2002a, 2003; Vignaroli et al. 2015).

There is no modern regional description of the Neogene volcanic geology of northern Victoria Land, and most published descriptions are now 35–50 years old. The earliest comprehensive account of volcanism in the region was the seminal study by Hamilton (1972), which described outcrops between Coulman Island and Cape Adare; it was also the first study to recognize that much of the volcanism was subglacial and associated with a pre-Quaternary Antarctic Ice Sheet. Kyle and Cole (1974; also Kyle 1990a) placed all the volcanic outcrops in the McMurdo Volcanic Group (following suggestions by Harrington 1958; Nathan and Schulte 1968; Riddolls and Hancox 1968), with further informal subdivision into two major volcanic provinces: the Hallett Volcanic Province, which included northern outcrops between Cape Adare and Coulman Island (McIntosh and Kyle 1990a); and the Melbourne Volcanic Province, for southern outcrops between Malta Plateau and Mount Melbourne (Kyle 1990b) (Fig. 1). In addition, a distinctive geographically restricted group of alkaline plutons and dykes with Eocene–Oligocene ages, called the Meander Intrusive Group, may correspond to the earliest phase of Neogene alkaline magmatism in the region (Müller et al. 1991; Tonarini et al. 1997; Rocchi et al. 2002a, b). Rocchi et al. (2002a) encompassed all the Cenozoic alkaline volcanic and intrusive subvolcanic rocks in Victoria Land, including the Erebus Volcanic Province (see Chapter 5.2a: Smellie and Martin 2021) in the Western Ross Supergroup (Fig. 2).

Early studies of the volcanism in northern Victoria Land quickly divided it into two major types: large polygenetic central volcanoes (both stratovolcanoes and shield volcanoes: called the 'Central Suite' by Nathan and Schulte 1968); and

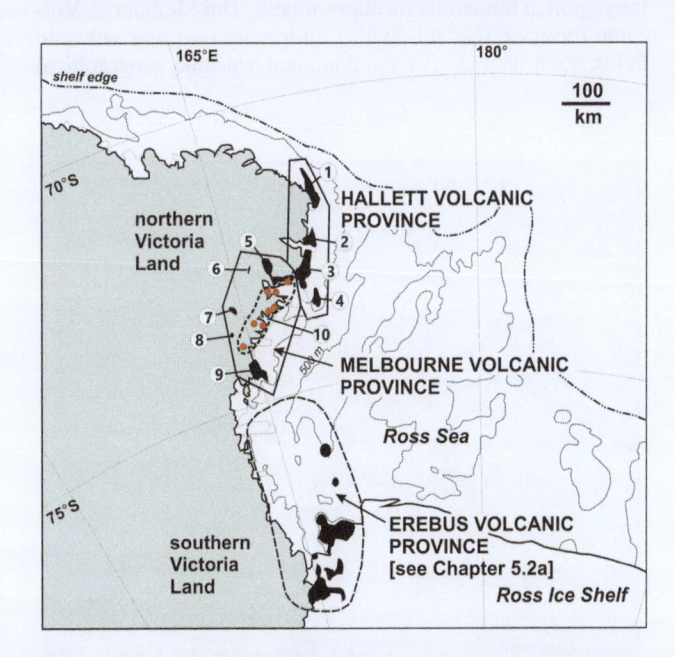

Fig. 1. Map showing the distribution of Neogene volcanic and plutonic outcrops in northern Victoria Land. 1, Adare Volcanic Field; 2, Hallett Volcanic Field; 3, Daniell Volcanic Field; 4, Coulman Volcanic Field; 5, Malta Plateau Volcanic Field; 6, the Pleiades Volcanic Field; 7, Mount Overlord Volcanic Field; 8, Vulcan Hills Volcanic Field (see the text); 9, Mount Melbourne Volcanic Field; 10, Meander Intrusive Group (red circles with dashed field outline).

From: Smellie, J. L., Panter, K. S. and Geyer, A. (eds) 2021. *Volcanism in Antarctica: 200 Million Years of Subduction, Rifting and Continental Break-up*. Geological Society, London, Memoirs, 55, 347–381,
First published online February 19, 2021, https://doi.org/10.1144/M55-2018-60
© 2021 The Author(s). Published by The Geological Society of London. All rights reserved.
For permissions: http://www.geolsoc.org.uk/permissions. Publishing disclaimer: www.geolsoc.org.uk/pub_ethics

WESTERN ROSS SUPERGROUP

McMurdo Volcanic Group
[Alkaline volcanic rocks]

Meander Intrusive Group
[Alkaline plutons & related dykes]

Hallett Volcanic Province
[Large mafic-felsic central volcanoes]
- Adare Peninsula Volcanic Field
- Hallett Peninsula Volcanic Field
- Daniell Peninsula Volcanic Field
- Coulman Island Volcanic Field

Melbourne Volcanic Province
[Large mafic-felsic central volcanoes]
- Mt Melbourne Volcanic Field
- Vulcan Hills Volcanic Field
- Mt Overlord Volcanic Field
- The Pleiades Volcanic Field
- Malta Plateau Volcanic Field

Erebus Volcanic Province
[Mainly large mafic-felsic central volcanoes]
- Ross Island Volcanic Field
- Mt Discovery Volcanic Field
- Mt Morning Volcanic Field
- Terror Rift Volcanic Field
- Southern Local Suite Volcanic Field
  [Products of small isolated mafic eruptions]
- Northern Local Suite Volcanic Field
  [Products of small isolated mafic eruptions]
- Upper Scott Glacier Volcanic Field
  [Mainly products of small isolated mafic eruptions]

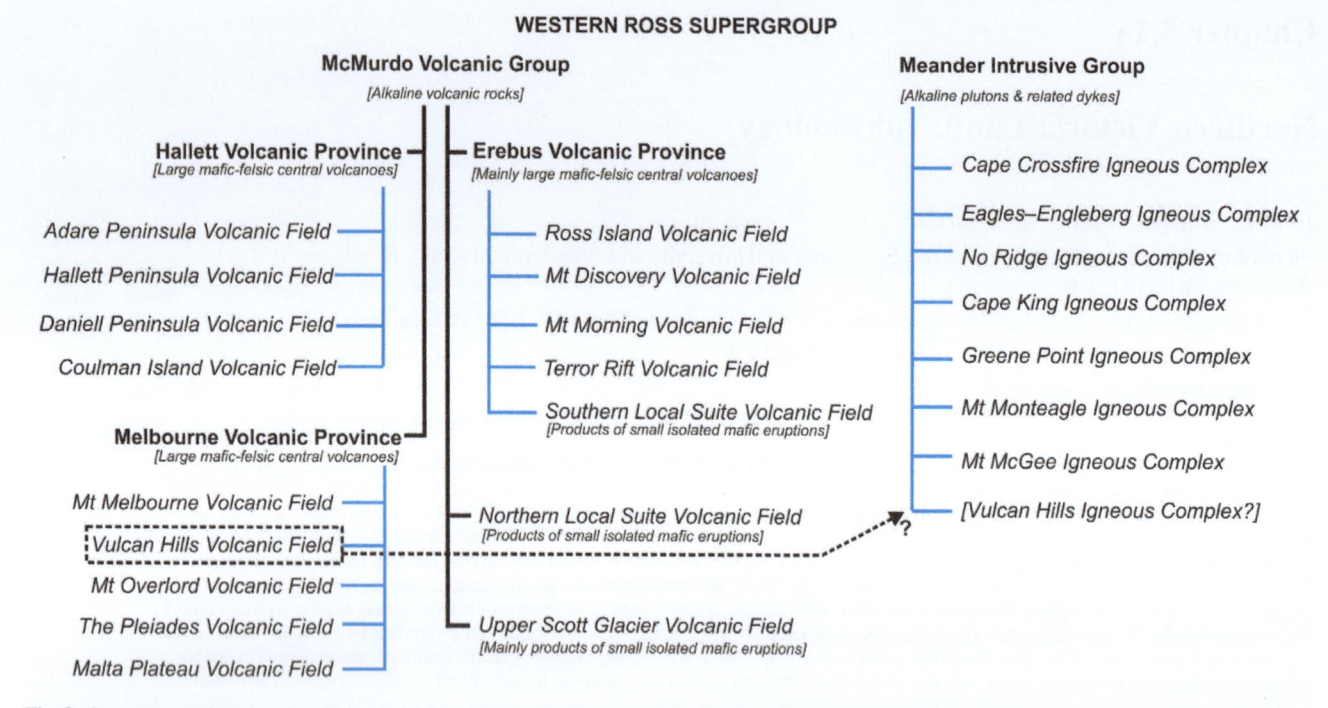

- Cape Crossfire Igneous Complex
- Eagles–Engleberg Igneous Complex
- No Ridge Igneous Complex
- Cape King Igneous Complex
- Greene Point Igneous Complex
- Mt Monteagle Igneous Complex
- Mt McGee Igneous Complex
- [Vulcan Hills Igneous Complex?]

Fig. 2. Summary stratigraphy of the Neogene McMurdo Volcanic Group and subvolcanic Meander Intrusive Group of plutonic outcrops in northern Victoria Land, as used in this chapter. All are incorporated into the Western Ross Supergroup of Rocchi *et al.* (2002*a*). Note that the status of the Vulcan Hills Volcanic Field is uncertain and it may be the local volcanic cover of a geographically associated plutonic outcrop. Both rock units might better be grouped together and renamed the Vulcan Hills Igneous Complex, and assigned to the Meander Intrusive Group (see the text for a further explanation). Details of the Erebus Volcanic Province are included in Chapter 5.2a (Smellie and Martin 2021).

multiple tiny monogenetic centres (the 'Local Suite' of Nathan and Schulte 1968). The large central volcanoes crop out principally in a prominent linear chain along the Ross Sea coast but also occur more rarely inland (Fig. 3), whereas the small monogenetic centres are more widely dispersed throughout the region at numerous locations inland. The Melbourne Volcanic Province was subdivided informally into four volcanic fields, each named after the dominant volcanic geographical feature: Mount Melbourne, Mount Overlord, the Pleiades and Malta Plateau; Vulcan Hills may also qualify as a separate volcanic field but it is small and very poorly known, and it might even be reassigned to the Meander Intrusive Group (see later). Similarly, the Hallett Volcanic Province was divided into Adare Peninsula, Hallett Peninsula, Daniell Peninsula and Coulman Island volcanic fields.

Hamilton (1972) suggested that the large central volcanoes cropping out between Cape Adare and Coulman Island individually showed a broad bipartite association of lithofacies composed of a thick basal sequence of subglacially erupted palagonitized breccias and pillow breccias that were succeeded by subaerial lavas and scoria. The overall stratigraphy of each centre was believed to be consistent with the volcanoes erupting initially beneath a substantial Antarctic Ice Sheet many hundreds of metres thick, with subaerial effusion occurring only in the later stages when the ice sheet was finally penetrated. Unfortunately, Hamilton (1972) used unfamiliar descriptive words, such as marble-cake lava, nodular lava, palagonite breccia and basalt spheroids, in his descriptions that do not correspond closely with more orthodox volcanic terminology as currently recognized. This was a reflection of the complexity of the lithofacies and an absence of published comparable examples, a problem not resolved for several decades. Thus, his seminal study of what have since become known as glaciovolcanic sequences (Smellie 2006) achieved less impact than it deserved, despite being highly advanced for its time. By contrast, a study of the coastal northern Victoria Land outcrops by McIntosh and Gamble (1991) suggested that the sequences were much more complicated than the simple unidirectional subglacial-to-subaerial eruptive scenario proposed by Hamilton (1972). In particular, they highlighted the presence of products of magmatic eruptions at all levels, and the apparent scarcity of evidence for significant glacial interaction. It was suggested that the sequences were erupted within a dominantly *periglacial* environment and that an expanded Antarctic Ice Sheet was absent. These conflicting

Fig. 3. TMA photograph CA103733R0198 looking north from Daniell Peninsula to Adare Peninsula, showing the location of coastal shield volcanoes (shaded yellow). The prominent double arcuate embayment in southern Daniell Peninsula (foreground) may be the site of one or more calderas (see the text).

views were reconciled during the most recent investigations, by Smellie et al. (2011a, b), which established unequivocally that the sequences are overwhelmingly glaciovolcanic. For example, most of the unusual features and lithofacies described with difficulty by Hamilton (1972) almost certainly correspond to chaotic lava-lobe-bearing breccias (*sensu* Smellie and Edwards 2016), which were formed in 'a'ā lava-fed deltas (cf. Smellie et al. 2011a, b, 2013). Moreover, the eruptions took place repeatedly in association with a relatively thin ice cover (<300 m thick) rather than the thick ice sheet envisaged by Hamilton (1972) or the periglacial conditions proposed by McIntosh and Gamble (1991).

For the purposes of description in this chapter, each of the previously informally designated volcanic divisions, together with the Meander Intrusive Group, are raised to full stratigraphical status within the Western Ross Supergroup (Fig. 2). Their principal characteristics are summarized in Table 1. Only the outcrops in northern Victoria Land are included in this chapter; a description of the Erebus Volcanic Province in southern Victoria Land is presented in Chapter 5.2a (Smellie and Martin 2021) and the Upper Scott Glacier Volcanic Field in Chapter 5.3a (Smellie et al. 2021). The distinctions between the volcanic provinces are primarily geographical as the lithofacies erupted in each volcanic field, and the environmental conditions during eruption, were broadly similar and they have comparable alkaline magmatic compositions varying from basanite to phonolite and alkali basalt to trachyte and rhyolite (e.g. Armienti et al. 1991, 2003; Müller et al. 1991; Hornig and Wörner 2003; Nardini et al. 2003; see also Chapter 5.1b: Rocchi and Smellie 2021). They occupy discrete geographical areas that are separated from other volcanic fields by wide areas of basement rocks (Fig. 1). The Hallett Volcanic Province, in particular, is characterized by spectacular high east-facing cliffs, whereas the western flanks of the constituent volcanoes have more gradual slopes terminating in much subdued cliffs. The topographical contrasts suggest that the cliffs facing the Ross Sea were eroded mainly by marine processes, presumably during interglacial periods, although faulting and erosion from glaciers in the Ross Sea may also have played a part. Exposure in the high seaward-facing cliffs is superb but access is only possible working from sea ice, and the cliffs must be visited early in the season when the sea ice is fast (i.e. unbroken). Working on the outcrops, particularly those inland, also requires the extensive use of dedicated helicopters and this has thus far ensured that the region contains some of the least studied volcanoes in Antarctica. With the possible exception of the Meander Intrusive Group (Müller et al. 1991; Rocchi et al. 2002a, b, 2003), all the investigations undertaken so far must be classed as at a reconnaissance level, although some more recent investigations can be regarded as detailed reconnaissance in recognition of the more thorough studies that have been possible at selected localities (e.g. Kyle 1982; Esser and Kyle 2002; Smellie et al. 2011a, 2018). Finally, where possible, the compositional names used here are updated from the original descriptions and are based on the TAS classification (Le Bas et al. 1986).

Stratigraphy and volcanology of the volcanic outcrops

Hallett Volcanic Province

Adare Peninsula Volcanic Field. Adare Peninsula is a narrow finger-like promontory largely covered by snow and ice that extends 72 km NNW from Cape Roget to Cape Adare (Figs 4 & 5a). The peninsula decreases in width from c. 19 km between Cape Roget and Cape McCormick, to c. 12 km flanking 'Downshire Peak' (informal name) and down to <3 km at Cape Adare. The highest point is 2083 m at Downshire Peak. Tall continuous cliffs 600–2000 m high (Downshire Cliffs) are present on the eastern Ross Sea flank. By contrast, the western flank of Adare Peninsula, north of Warning Glacier, slopes down more gently to Robertson Bay and the coast has a low cliff line, typically c. 200 m high, whereas 'Nameless Bluff' (informal name), on the south side of Warning Glacier, rises to c. 1000 m. Adare Peninsula is mainly low lying in the southern half, with little rock exposed except at Cape McCormick and in cliffs >500 m high at Cape Roget. Because of its northernmost location where it is exposed to Southern Ocean waves and the consequent early break-out of the seasonal sea ice, Adare Peninsula is one of the least-visited volcanic areas in northern Victoria Land (Priestley 1923; Hamilton 1972; McIntosh and Gamble 1991; Smellie et al. 2011a, b).

Adare Peninsula is entirely volcanic, with a boundary against basement rocks (Robertson Bay Group metasedimentary rocks (Cambrian); Field and Finlay 1983) passing though Adare Saddle, which connects Robertson Bay in the north and Moubray Glacier in the south. It is formed of the products of at least four large shield volcanoes, descriptively called domes by Hamilton (1972). They are, from north to south: Hansen Peak; Downshire Peak, Nameless Bluff and Cape Roget. In addition, further volcanic outcrops are present inland at Cape Klovstad (also known as Mayr Spur), and at Cape McCormick and the Possession Islands. Compositionally, the Cape Adare Volcanic Field includes basanite, phonolite and trachyte at Cape Adare; basanite and mugearite at Cape Roget; basalt, tephrite and trachyte or phonolite at Nameless Bluff; basanite and basalt at Cape McCormick and Possession Islands; and alkali basalt and possible trachyte at Cape Klovstad (Hamilton 1972; Jordan 1981; McIntosh and Kyle 1990b; Müller et al. 1991; Mortimer et al. 2007; Panter et al. 2018; see Chapter 5.1b: Rocchi and Smellie 2021).

Two discrete elongate topographical domes relating to major volcanic centres dominate on Adare Peninsula (Hamilton 1972). The northernmost has a minimum diameter of c. 23 km but it may originally have been at least 34 km across assuming a symmetrical profile about the highest point (at Hanson Peak; 1256 m above sea level (asl)). The principal vent may have been situated a few kilometres offshore to the east (Hamilton 1972; McIntosh and Gamble 1991) (Fig. 4). The volcano has been considerably reduced by erosion, especially on its east and west flanks, and it has a maximum width today of just 5 km. It has been studied at Cape Adare, where Priestley (1923) described the sequence as northerly-dipping interbedded basalt lava and subordinate breccia. He also described an unusual 'agglomerate' distinguished by abundant basement clasts, including green quartzite, schist and granitoids, together with volcanic lithologies, set in a tuffaceous matrix. Hamilton (1972) considered the latter to have formed by a volcanic eruption through morainic debris at the base of a continental ice sheet. Priestley (1923) also described subaerial basalt lavas and bombs on the summit above Cape Adare. Conversely, unpublished studies by the authors following two short visits in 2006 and 2014 indicated that the sequence exposed in and above the 400 m-high cliffs on the NE tip of Cape Adare is constructed of at least five discrete eruptive sequences (Fig. 6). Each is c. 100 m thick and consists of a basal unit of monomict, coarse lithic breccia with numerous irregular lava masses (lobe hyaloclastite *sensu* Smellie et al. 2011a; lava-lobe-bearing breccia of Smellie and Edwards 2016) (cf. Fig. 7a, b) and an upper unit of multiple thin 'a'ā lavas with oxidized surfaces. Together, they represent products of five 'a'ā lava-fed deltas. Because the passage zones of each delta dip at a shallow angle to the north, they were emplaced in a glacial setting (a passage zone is equivalent to

Table 1. *Stratigraphy and principal characteristics of volcanic provinces of the Western Ross Supergroup in northern Victoria Land**

Volcanic province	Volcanic field	Major constituent volcanoes	Volcano type	Caldera(s)?	Composition	Published ages (Ma)	Comments	Key sources
Hallett	Adare Peninsula	Northern and central Adare Peninsula; several unnamed centres	Several large overlapping shield volcanoes	No	Basanite, basalt, hawaiite, trachyte	3.87 ± 0.03; 3.79 ± 0.03	Outcrops dominated by 'a'ā-lava-fed deltas; age by K–Ar	Hamilton (1972); Müller et al. (1991); J.L Smellie unpublished
		'Nameless Bluff'		Postulated	Basalt, phonolitic tephrite, phonolite, trachyte	11.9–7.12; outlier ages of 13.24 and 1.14 may be unreliable	Five eruptive units identified, four of which are dated (by K–Ar)	Hamilton (1972); Jordan (1981); Müller et al. (1991); J.L Smellie unpublished
		Cape Klovstad/Mayr Spur	Shield volcano covering a tuff cone or tuff ring	No	Basalt	9.7–10.6; mean 10.2	Outcrops very highly eroded; ages by K–Ar	Mortimer et al. (2007); J.L. Smellie unpublished
		Possession Islands	Unknown	Unknown (eroded)	Basanite	Unknown	May be part of Cape Roget centre	Hamilton (1972); McIntosh and Kyle (1990b)
		Cape Roget	Shield volcano	Unknown (eroded)	Basanite	2.21 ± 0.5	Dominated by 'a'ā lava-fed deltas; intruded by Strombolian scoria and lava; age by K–Ar	Hamilton (1972); McIntosh and Kyle (1990b); J.L Smellie unpublished
	Hallett Peninsula	Hallett Peninsula; several unnamed centres	Several large overlapping shield volcanoes	No		7.4–6.6	Outcrops dominated by 'a'ā-lava-fed deltas; ages by Ar–Ar and K–Ar	Hamilton (1972); McIntosh and Kyle (1990c); Smellie et al. (2011a, b)
		Cape Wheatstone ('southern volcano')	Shield volcano	Unknown (eroded)	Uncertain, probably mainly mafic	5.35 ± 0.12	Age by K–Ar	Hamilton (1972); McIntosh and Kyle (1990c); J.L. Smellie unpublished
		Mount Harcourt	Stratovolcano	No	Unknown	Unknown	Never been visited (inaccessible); volcano dissected by Tucker Glacier	Hamilton (1972)
		n/a	n/a – Herschel Tuffaceous Moraine Formation	n/a	n/a	Unknown	Glaciolacustrine and till deposits; major coeval tephra input presumably from unidentified Hallett Peninsula volcano	Harrington et al. (1967); Smellie et al. (2011a, b)
	Daniell Peninsula	Northern Daniell Peninsula, between 'Daniell Summit' and Mount Brewster	Several large overlapping shield volcanoes	Possibly	Alkali basalt to peralkaline trachyte	6.96 (Mount Brewster); 6.7–5.3 (west of Cape Daniell)	Outcrops dominated by 'a'ā lava-fed deltas; possible ice-filled caldera at northern end; ages by Ar–Ar	Hamilton 1972; Nardini et al. (2003); Smellie et al. (2011a, b)
		Southern Daniell Peninsula: Mandible Cirque, Mount Lubbock, Tousled Peak	Mount Lubbock is a stratovolcano; Tousled Peak is a prominent flank vent; Mandible Cirque may be a large shield volcano but morphology much eroded	Large caldera postulated; unconfirmed	Basanite and alkali basalt to peralkaline trachyte and rhyolite	9.866 ± 0.088 (Mount Lubbock); 9.95–8.63 (outcrops surrounding and within Mandible Cirque)	Accessible basal outcrops dominated by 'a'ā lava-fed deltas at all localities except within northern Mandible Cirque (rhyolite domes/coulees); ages by Ar–Ar	Hamilton (1972); Nardini et al. (2003); Smellie et al. (2011a, b)
		Unnamed centre east of Prior Peak	Unknown	Unknown (eroded)	Trachyte (pantelleritic)	11.9 ± 0.6; older age of 12.4 ± 0.4 probably unreliable	Tiny outcrop; volcano very poorly known; age by Ar–Ar	Nardini et al. (2003)

	Name	Type	Caldera	Composition	Age (Ma unless stated)	Comments	References
Coulman Island	Hawkes Heights	Shield volcano	Yes	Basanite, hawaiite, phonotephrite, phonolite, comenditic trachyte	8.59, 7.6, 7.5, 7.4, 7.3 and 7.0	Outcrops dominated by 'a'ā lava-fed deltas; ice-filled caldera 5 km wide at the southern end	Hamilton (1972); Armstrong (1978); Müller et al. (1991); McIntosh and Gamble (1991); Smellie et al. (2011a, b)
Melbourne	Malta Plateau	Unknown	Unknown	Mainly trachyte (comendite), rhyolite (pantellerite); basalt also present	Lavas: 10.7–6.5; dykes: 18.2–14.1	Numerous tiny outcrops, poorly exposed and very inaccessible; ages by K–Ar	Müller et al. (1991); Schmidt-Thomé et al. (1990)
	None identified						
	The Pleiades	Cluster of small scoria cones and at least one lava dome possibly situated along the unexposed rim of an ice-buried caldera	Unknown	Basanite to phonolite; hawaiite to trachyte	Numerous: 0.847–0.826, 0.627; 0.337–0.312, 0.150, 0.093–0.020 and 0.006	Isotopic ages very young and, hence, considered possibly active (dormant); ages by Ar–Ar	Kyle (1982, 1986, 1990a, b, c, d, e); Esser and Kyle (2002)
	Mount Pleiones, Mount Atlas, Tageyte Cone, Aleyone Cone and unnamed others						
	Mount Overlord	Stratovolcano with a small (800 m diameter) summit caldera	Yes	Basanite to phonolite; alkali basalt to trachyte and comendite	7.0–8.3	Rarely visited, very poorly studied volcano; may be constructed on an extensive plateau volcanic field exposed in cliffs overlooking Aviator and Pilot glaciers and at Navigator Nunatak	Nathan and Schulte (1968); Noll (1985); Kyle and Noll (1990)
	Deception Plateau	Ignimbrite plateau?	Unknown	Trachyte and phonolite	14.67 ± 0.3 (Parasite Cone)	Includes Mount Noice and Parasite Cone; age by K–Ar	Noll (1985); Kyle and Noll (1990)
	Mount Rittmann	Unknown (eroded)	Unknown	Unknown	3.97 ± 0.14 Ma; 240–70 ka; possibly active	Tiny, very eroded remnant associated with geothermal heat	Armienti and Tripodo (1991); Bonaccorso et al. (1991); Perchiazzi et al. (1999)
	Vulcan Hills	Unknown (eroded)	Unknown	Basalt (hawaiite) to trachyte; gabbro	Unknown but possibly >10 Ma	Small highly eroded volcanic remnants, rarely visited and very poorly known; little published information	Nathan and Schulte (1968); Kyle (1990a)
	Mount Melbourne	Active (dormant) stratovolcano and numerous small satellite and flank centres	No	Mount Melbourne summit: trachyte and benmoreite; minor alkali basalt	Active volcano; 0.050 and 0.0351 (with very high errors)	Heated ground within small summit crater; ages by Ar–Ar	Wörner and Viereck (1987, 1989); Giordano et al. (2012)
	Cape Washington	Shield volcano	Unknown (eroded)	Tephrite and basanite	2.72 and 1.68	Formed of 'a'ā lava-fed deltas; extensively intruded and covered by Strombolian vents and associated scoria cones; ages by K–Ar; younger age dates younger Strombolian event	Wörner and Viereck (1989); Müller et al. (1991); Giordano et al. (2012)
	Edmonson Point and hinterland to the north and west	Shield volcano	Unknown (eroded)	Hawaiite and benmoreite	0.297 ± 0.055	Very distinctive, formed of megapillows; age by Ar–Ar	Wörner and Viereck (1989); Giordano et al. (2012)

*Excluding numerous small isolated volcanic centres (Northern Local Suite).

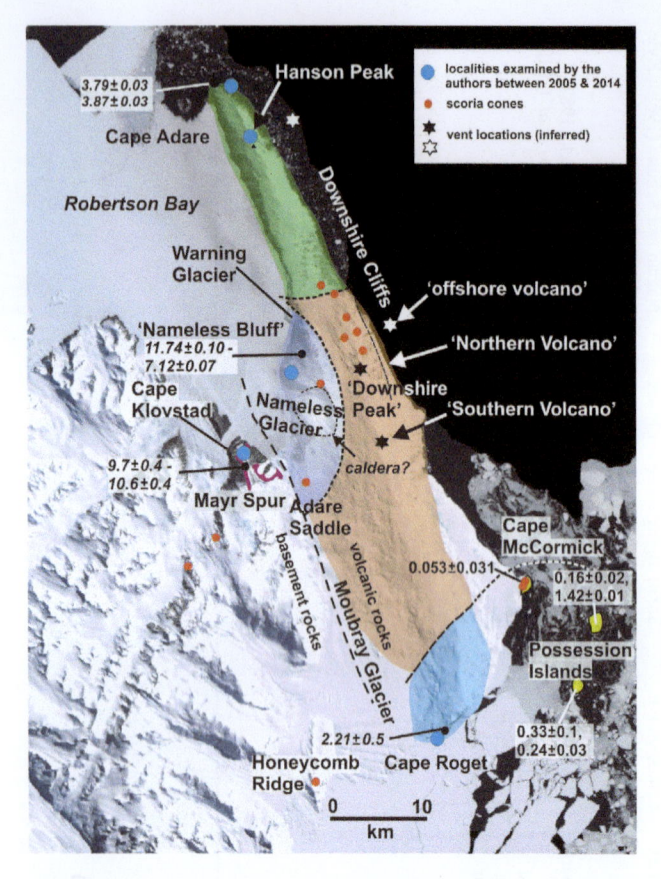

Fig. 4. Simplified geological map of Adare Peninsula Volcanic Field; modified after Smellie *et al.* (2011*b*). The positions and isotopic ages of dated samples are also shown.

a fossil water level; emplacement in a marine or lacustrine environment would have created horizontal passage zones: cf. Smellie *et al.* 2011*b*; Smellie 2018). A small scoria cone exposed in cross-section on the NW corner of Cape Adare was also reported by McIntosh and Kyle (1990*b*). A significant feature of the sequence is the presence of both pristine-looking and eroded surfaces separating the individual eruptive units. Isotopic ages of 3.87 ± 0.03 and 3.79 ± 0.03 Ma (by K–Ar) were obtained on phonolite and phonolitic tephrite lavas at Cape Adare (Müller *et al.* 1991) (Table 2). The pristine state of scoria cones on the summit of Adare Peninsula and elsewhere (e.g. Daniell Peninsula) was used by Hamilton (1972) to suggest that they were Holocene in age and that the underlying volcanoes were still active (unlikely – see Chapter 1.2: Smellie 2021).

The southern volcanic dome extends from Warning Glacier to a shallow col *c.* 17 km north of Cape Roget. It is constructed of the products of at least three coalesced volcanic centres: one situated offshore (labelled 'offshore volcano' in Fig. 4) and the other two erupted from vents situated on the summit axis of the peninsula ('Northern Volcano' and 'Southern Volcano' in Fig. 4). The combined volcanic dome extends at least 36 km in a NNW–SSE direction but is only *c.* 10 km wide for much of its length, broadening to *c.* 15 in the south where it is entirely ice-obscured. To the west, the volcanic mound abuts against Nameless Bluff, which is probably a separate older volcanic centre. The southern volcanic dome is well exposed in Downshire Cliffs. Despite the excellent exposure, the difficult access has ensured that almost nothing is known of its geology. Hamilton (1972) gave the following outline description. Breccias and pillow complexes extend up to *c.* 1800 m and dominate the exposed sequence, with a likely derivation from both the offshore vacanoes and Southern Volcano. They are capped by northward-thickening thin-bedded

Fig. 5. Compendium of views of volcanoes and volcanic outcrops in northern Victoria Land. (**a**) View looking north over 'Nameless Bluff' (foreground) to the narrow northern tip of Adare Peninsula (summit: Hanson Peak); Robertson Bay on the left. (**b**) View looking NE over Mariner Glacier to Malta Plateau (high terrain in the background), showing the paucity of exposed rock. (**c**) Northern tip of Daniell Peninsula, looking south; Cape Daniell on the left; the topography is cut across a crater-like landform with volcanic beds ('a'ā lava-fed deltas) dipping radially away from the centre. (**d**) View of a large basalt plug-like mass in the centre of a tuff cone outcrop at Harrow Peaks, looking north; the outcrop is described in detail by Smellie *et al.* (2018). (**e**) Shield Nunatak, a small tuya, viewed looking NW; the Deep Freeze Range is behind. (**f**) Mount Harcourt (summit: 1571 m asl), a small inaccessible stratovolcano that has never been visited, viewed looking NW from near Cape Wheatstone.

subaerial lavas sourced in the Northern Volcano. Stratification in the breccias is truncated by the lavas, which possibly implies the presence of an erosional hiatus between the two centres. Several small scoria cones are present on the surface of the Northern Volcano. They have well-preserved crater rims but also show varying degrees of erosion attributed to overriding ice. Although subaerial basalt and trachyte lavas were observed near the summit of the Southern Volcano, they were not recorded in the coastal cliffs to the east.

The southernmost volcanic centre in Adare Peninsula is spectacularly exposed in >500 m-high cliffs at Cape Roget. The western upper ice-covered surface of the Cape Roget Volcano may be essentially pristine, with gentle uniform dips towards the west or SW, whereas most of the eastern part of the volcano is absent and has been removed by marine erosion. However, the missing eastern portion may be partially represented by basaltic outcrops at Cape McCormick and the Possession Islands (Possession, Foyn and Heftye islands, and several sea stacks: Hamilton 1972) (Fig. 4). If all the outcrops are related, they suggest a minimum diameter for the original Cape Roget volcano of at least 25 km; its highest point today is 720 m, 10 km NNE of Cape Roget. A lava pillow in palagonite breccia at Cape Roget yielded a K–Ar isotopic age of 2.21 ± 0.5 Ma (Armstrong 1978). However, ages for five basalts from the Possession Islands are significantly younger than those at Cape Roget, varying from 1.42 ± 0.01 to 0.053 ± 0.031 Ma (by Ar–Ar: Panter et al. 2018) (Table 2).

Cape Roget was described by Hamilton (1972) as breccias and pillow complexes possibly overlain by inaccessible subaerial lavas. No generic origin for the lithofacies was suggested but more recent work indicates that the section is composed of several superimposed 'a'ā lava-fed deltas (unpublished information of J.L. Smellie) (Fig. 8). The cliffs extend in a SW–NE direction for c. 5 km and consist of at least four (possibly five) lava-fed delta units similar to those seen at Cape Adare, and, likewise, they are separated by surfaces that are both clearly erosive and possibly pristine but the surfaces were inaccessible and have not been examined closely. An additional lithofacies consists of Strombolian scoria and coeval intrusions that cut across the two basal deltas (unit 3 in Fig. 8). Another feature present in the basal delta is a large mass of stratified lapilli tuffs c. 50 m long and 20 m thick embedded in chaotic lava-lobe-bearing breccia (unit 2a in Fig. 8). The two lower deltas also show prominent passage zones comprising lava-lobe-bearing breccia very rich in lava lobes (Fig. 8). Cape McCormick and the Possession Islands consist of a mixture of massive volcanic breccia, pillow lava and prominent megapillow complexes in sequences 100–250 m thick (Hamilton 1972). The massive breccias lack the 'thin flow units' (lava lobes?) that characterize breccias seen in Adare Peninsula outcrops, whereas megapillows are spectacular and common. The individual megapillow structures are often 50 m across and some probably extend up to several hundred metres in length based on photographs provided by Hamilton (1972). They show prominent columnar jointing, and the descriptions and published photographs resemble features also seen in megapillows in the Melbourne Volcanic Field (see below). Subaerial basalt lava may also be present on the southern part of Foyn Island. Hamilton (1972) postulated that original vents for the Cape Roget Volcano may have been situated in the Possession Islands but their location is not well substantiated. Two well-formed (and, presumably, much younger) scoria cones overlie breccia at Cape McCormick and the presence of a large pile of scoria resting on subaerial lavas on Foyn Island suggests that a scoria cone may also have been present there. Finally, McIntosh and Gamble (1991) indicated that 'Surtseyan deposits' (i.e. presumably stratified lapilli tuffs of a tuff cone) are present on Possession Island. The tuffs contain polished basaltic beach pebbles consistent with a marine eruptive environment.

The Nameless Bluff Volcano was examined by Jordan (1981), who described five discrete eruptive units, designated A–E. Stratigraphically consistent isotopic ages were obtained for the upper three, ranging between 13.24 ± 0.12 and 7.08 ± 0.11 Ma (by K–Ar: Kreuzer et al. 1981) (Table 2); an additional age of 1.14 ± 0.05 Ma (also by K–Ar) for the uppermost unit E is either spurious or it may represent a sample from a scoria cone mapped during our investigation at the SE corner

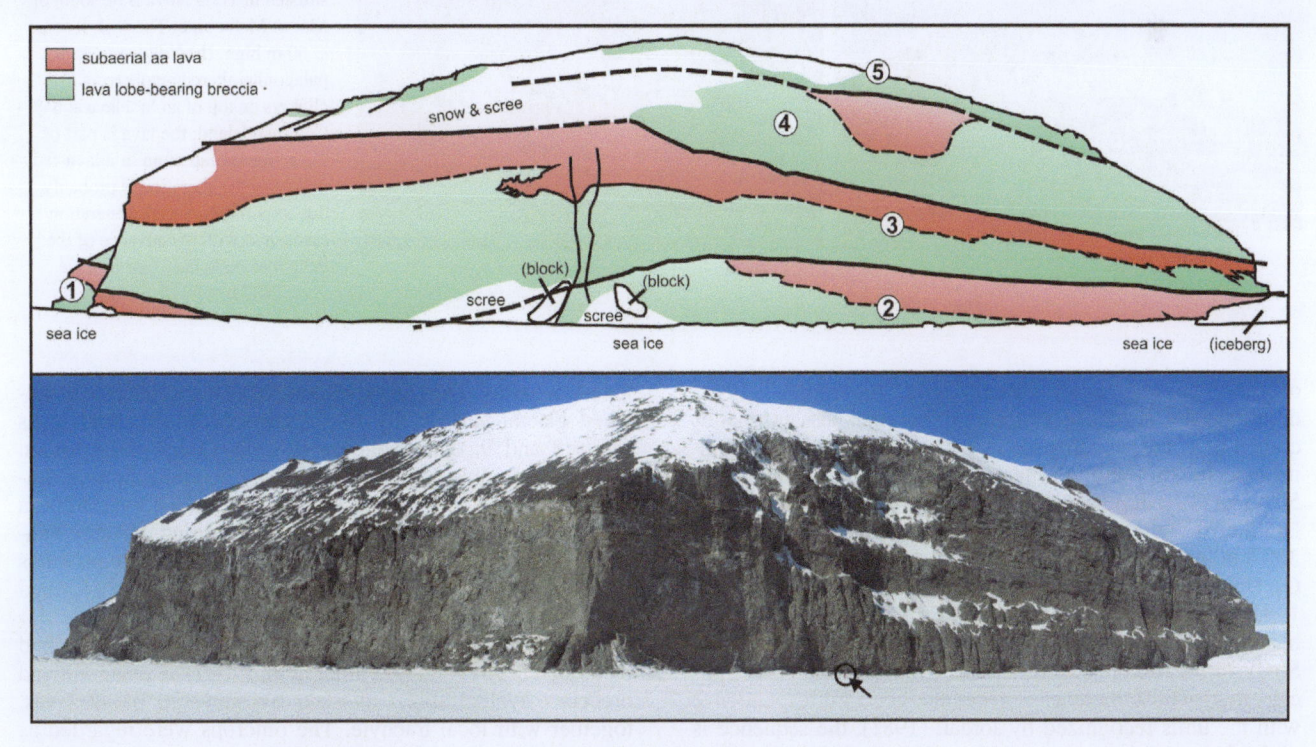

Fig. 6. Photograph and sketch showing the geology of cliffs on the north side of Cape Adare (unpublished information of J.L. Smellie). The section is formed of the products of five 'a'ā lava-fed deltas (numbered in the diagram). The surfaces separating the deltas appear to vary laterally between essentially pristine and glacially eroded. A person is ringed for scale.

Fig. 7. Compendium of photographs showing representative lithofacies in northern Victoria Land volcanic outcrops. (**a**) Lava-lobe-bearing breccia within an 'a'ā lava-fed delta; note the chaotic mixture of highly irregular coherent (water-chilled) lava lobes and lava masses, and enclosing orange-brown-coloured breccia; Cotter Cliffs, Hallett Peninsula. (**b**) Close view of massive lithic breccia, part of a lava-lobe-bearing breccia deposit within an 'a'ā lava-fed delta; Salmon Cliff, Hallett Peninsula. (**c**) Coarse rhyolitic hyaloclastite breccia; 'Mandible Promontory', Daniell Peninsula. (**d**) Grey-green felsic tuff breccia *c.* 1 m thick, probably a block-and-ash deposit related to a felsic dome exposed below (out of photograph), sandwiched between felsic lava-sourced khaki-yellow to pale rust-coloured beds of stratified lapilli tuff; Cotter Cliffs, Hallett Peninsula (see the description by Smellie *et al.* 2011*a*). (**e**) Lapilli tuffs forming the basal outcrop at Cape Klovstad, west of Adare Peninsula; the lapilli tuffs are weakly stratified (not evident in the photograph), contain abundant basement clasts and were formed in a tuff cone or tuff ring. (**f**) Thinly-stratified lapilli tuffs, part of a tuff cone on the coast NE of Baker Rocks, NE of Mount Melbourne. (**g**) Oxidized agglutinate and likely clastogenic basalt lavas at Markham Island, a small volcanic centre with a well-preserved summit crater situated in Terra Nova Bay, south of Mount Melbourne. The rock face is *c.* 60 m high. (**h**) Yellow-coloured, palagonite-altered autobreccia clinkers on top of an 'a'ā lava at SW Coulman Island; the lava is part of the subaerial cap of an 'a'ā lava-fed delta and the alteration is probably due to pervasive steam generation associated with the advance of the delta into meltwater (see Smellie *et al.* 2013).

of the Nameless Bluff outcrop (Fig. 4). Jordan (1981) described the sequence as follows (his thickness estimates, quoted here, are probably too high): a basal chaotic 'palagonite breccia' (his unit A), up to 200 m thick, is overlain by 500 m of alternating basaltic lavas(?), dolerite and intercalated lapilli tuffs with internal angular unconformities (unit B); then 300–400 m of alternating basalt lava, dolerite and scoria (unit C). Unit D is a trachyte lava up to 300 m thick (Kreuzer *et al.* 1981 called it a phonolite) followed by a few tens of metres of black phonolitic tephrite lavas and oxidized scoria (unit E). Studies of a 1000 m section by one of us (J.L. Smellie) indicate that, whilst there is a generally good correspondence with the units recognized by Jordan (1981), the sequence is divisible into at least nine eruptive units (1–9: Fig. 9). Apart from unit 2, which is an epiclastic–volcaniclastic deposit currently of uncertain origin, the basal 880 m consists of six lava-fed deltas. They are overlain, across a striated and polished glacial surface, by two sequences of subaerial lavas (units 8 and 9: corresponding to units D and E of Jordan 1981). A possible caldera bounded by bedding dipping outwards on north and south flanks was identified by Hamilton (1972). It is the only caldera structure identified in Adare Peninsula Volcanic Field. No explosively generated products have been identified and, if it is a caldera, it may have formed by collapse following magma withdrawal.

Finally, volcanic outcrops at Cape Klovstad were described by Jordan (1981) and Mortimer *et al.* (2007) as palagonitized breccia ('hyaloclastite') overlain by subaerial basalt lavas, together with local trachyte. The outcrops were regarded as an extension of the volcanic sequence exposed to the east, flanking Nameless Glacier. Numerous K–Ar isotopic ages by Mortimer *et al.* (2007) for samples from both the basal

Table 2. *Summary of published isotopic ages for large central volcanoes in northern Victoria Land*

Sample	Locality	Dated lithologies	Age*	Error	Error SD	Method	Reference
Hallett Volcanic Province							
Adare Volcanic Field							
A 233D	Cape Roget, Adare Peninsula	Basalt lava pillow	2.21	0.5	1σ	K–Ar	Armstrong (1978)
JO 158	'Unit E', 'Nameless Bluff', Adare Peninsula	Phonolitic tephrite	>9.6?	n.r.	2σ	K–Ar	Müller et al. (1991)
JO 116	'Unit E', 'Nameless Bluff', Adare Peninsula	Phonolitic tephrite	7.08	0.11	2σ	K–Ar	Kreuzer et al. (1981)
JO 115A	'Unit E', 'Nameless Bluff', Adare Peninsula	Phonolitic tephrite	7.35	0.12	2σ	K–Ar	Kreuzer et al. (1981)
JO 148	'Unit E', 'Nameless Bluff', Adare Peninsula	Phonolitic tephrite	1.14	0.05	2σ	K–Ar	Kreuzer et al. (1981)
JO 185	'Unit D', 'Nameless Bluff', Adare Peninsula	Phonolite	8.01	0.07	2σ	K–Ar	Kreuzer et al. (1981)
JO 186	'Unit D', 'Nameless Bluff', Adare Peninsula	Phonolite	7.69	0.05	2σ	K–Ar	Kreuzer et al. (1981)
JO 113C	'Unit D', 'Nameless Bluff', Adare Peninsula	Phonolite	6.77	0.05	2σ	K–Ar	Kreuzer et al. (1981)
JO 119	'Unit D', 'Nameless Bluff', Adare Peninsula	Phonolite	8.28	0.05	2σ	K–Ar	Kreuzer et al. (1981)
JO 141A	'Unit D', 'Nameless Bluff', Adare Peninsula	Phonolite	8.12	0.06	2σ	K–Ar	Kreuzer et al. (1981)
JO 154A	'Unit C', 'Nameless Bluff', Adare Peninsula	Phonolitic tephrite	9.88	0.07	2σ	K–Ar	Kreuzer et al. (1981)
JO 110A	'Unit C', 'Nameless Bluff', Adare Peninsula	Basalt	13.24	0.12	2σ	K–Ar	Kreuzer et al. (1981)
JO 106	'Unit C', 'Nameless Bluff', Adare Peninsula	Basalt	11.9	0.2	2σ	K–Ar	Kreuzer et al. (1981)
JO 107A	'Unit C', 'Nameless Bluff', Adare Peninsula	Andesite	11.74	0.10	2σ	K–Ar	Kreuzer et al. (1981)
JO 187	'Unit B', 'Nameless Bluff', Adare Peninsula	Basalt	11	0.05	2σ	K–Ar	Müller et al. (1991)
JO 173	'Unit B', 'Nameless Bluff', Adare Peninsula	Trachyte	10.9	0.04	2σ	K–Ar	Müller et al. (1991)
TS 4	Cape Adare	Phonolitic tephrite	3.79	0.03	2σ	K–Ar	Müller et al. (1991)
TS 5	Cape Adare	Phonolite	3.87	0.03	2σ	K–Ar	Müller et al. (1991)
Jo 188	Mayr Spur, Cape Klovstad	Olivine basalt	11.6	0.02	2σ	K–Ar	Müller et al. (1991)
Jo 189	Mayr Spur, Cape Klovstad	Olivine basalt	12.2	0.3	2σ	K–Ar	Müller et al. (1991)
Jo 49	Mayr Spur, Cape Klovstad	Olivine basalt	13.8	0.7	2σ	K–Ar	Müller et al. (1991)
Jo 90b	Mayr Spur, Cape Klovstad	Olivine basalt	11	0.2	2σ	K–Ar	Müller et al. (1991)
Jo 80	Mayr Spur, Cape Klovstad	Olivine basalt	11	0.2	2σ	K–Ar	Müller et al. (1991)
P74789	Mayr Spur, Cape Klovstad	'Plug' in lapilli tuff	10.4	0.4	2σ	K–Ar	Mortimer et al. (2007)
P74791	Mayr Spur, Cape Klovstad	'Plug' in lapilli tuff	9.7	0.4	2σ	K–Ar	Mortimer et al. (2007)
P74794	Mayr Spur, Cape Klovstad	'Plug' in lapilli tuff	10.2	0.4	2σ	K–Ar	Mortimer et al. (2007)
P74797	Mayr Spur, Cape Klovstad	'Plug' in lapilli tuff	10.5	0.4	2σ	K–Ar	Mortimer et al. (2007)
P74815	Mayr Spur, Cape Klovstad	Upper lava	10.0	0.4	2σ	K–Ar	Mortimer et al. (2007)
P74822	Mayr Spur, Cape Klovstad	Upper lava	10.1	0.4	2σ	K–Ar	Mortimer et al. (2007)
P74833	Mayr Spur, Cape Klovstad	North dyke	10.6	0.4	2σ	K–Ar	Mortimer et al. (2007)
P74838	Mayr Spur, Cape Klovstad	South dyke	10.0	0.4	2σ	K–Ar	Mortimer et al. (2007)
A227B	'McCormick Island' (Cape McCormick?)	Hawaiite	0.053	0.031	2σ	Ar–Ar	Panter et al. (2018)
HC-SPl4 ('3873')	Possession Island	Tephrite	0.16	0.02	2σ	Ar–Ar	Panter et al. (2018)
NV-3C (5166)	Possession Island	Benmoreite	1.42	0.0	2σ	Ar–Ar	Panter et al. (2018)
NV-4C (5171)	Foyn Island	Hawaiite	0.33	0.1	2σ	Ar–Ar	Panter et al. (2018)
A225A	Foyn Island	Basanite	0.24	0.03	2σ	Ar–Ar	Panter et al. (2018)

(*Continued*)

Table 2. *Continued.*

Sample	Locality	Dated lithologies	Age*	Error	Error SD	Method	Reference	Notes
Hallett Volcanic Field								
A 248	Cape Wheatstone, Hallett Peninsula	Basalt lava pillow	5.35	0.12	1σ	K–Ar	Armstrong (1978)	Age is apparently stratigraphically inconsistent and probably unreliable
A247G	Cotter Cliffs, Hallett Peninsula	Basalt lava pillow	6.4	0.4	1σ	K–Ar	Armstrong (1978)	
T5.25.7	Roberts Cliff, Hallett Peninsula	Basanite lava	7.257	0.050	2σ	Ar–Ar	Smellie et al. (2011b)	
T5.28.5	First bluff north of Salmon Cliff, Hallett Peninsula	Tephriphonolite lava lobe in lobe-hyaloclastite	7.049	0.040	2σ	Ar–Ar	Smellie et al. (2011b)	
T5.30.2	Redcastle Ridge, Hallett Peninsula	Basanite lava	7.572	0.054	2σ	Ar–Ar	Smellie et al. (2011b)	Age probably unreliable; probably not Mount Harcourt sequence, as suggested by Smellie et al. (2011b)
T5.30.5	Redcastle Ridge, Hallett Peninsula	Basanite lava	7.407	0.043	2σ	Ar–Ar	Smellie et al. (2011b)	
Daniell Peninsula Volcanic Field								
MP 15/5	10 km east of Mount Prior	Pantelleritic trachyte (dyke?)	12.4	0.2	1σ	K–Ar	Armienti et al. (2003)	
MP 15/5	10 km east of Mount Prior	Pantelleritic trachyte (dyke?)	11.9	0.3	1σ	Ar–Ar	Nardini et al. (2003)	
MP 10	West of Cape Daniell, Daniell Peninsula	Trachyte lava	6.7	0.2	1σ	K–Ar	Armienti et al. (2003)	
MP 11	West of Cape Daniell, Daniell Peninsula	Benmoreite lava	5.6	0.5	1σ	K–Ar	Armienti et al. (2003)	
MP 12	West of Cape Daniell, Daniell Peninsula	Trachyte lava	5.8	0.1	1σ	K–Ar	Armienti et al. (2003)	
MP 12/1	West of Cape Daniell, Daniell Peninsula	Trachyte lava	5.5	0.2	1σ	Ar–Ar	Nardini et al. (2003)	
MP 12/1	West of Cape Daniell, Daniell Peninsula	Trachyte lava	5.3	0.5	1σ	Ar–Ar	Nardini et al. (2003)	
MP 16	Mount Brewster, Daniell Peninsula	Hawaiite lava	6.9	0.3	1σ	K–Ar	Armienti et al. (2003)	
MP 21	North Mandible Cirque, Daniell Peninsula	Comenditic trachyte	9.5	0.1	1σ	K–Ar	Armienti et al. (2003)	
MP 21/9	North Mandible Cirque, Daniell Peninsula	Comenditic trachyte	9.2	0.2	1σ	Ar–Ar	Nardini et al. (2003)	
70 97	Mount Lubbock–Mandible Cirque, Daniell Peninsula	Alkali basalt	>11		2σ	K–Ar	Müller et al. (1991)	
2702	Mount Lubbock–Cape Jones, Daniell Peninsula	Tephrite	9.85	0.12	2σ	K–Ar	Müller et al. (1991)	
72 80	Mount Lubbock–Cape Jones, Daniell Peninsula	Tephrite	10.05	0.11	2σ	K–Ar	Müller et al. (1991)	
A25/10	'Mandible Bluff', Mandible Cirque, Daniell Peninsula	Comenditic trachyte	11.7	0.3	1σ	Ar–Ar	Nardini et al. (2003)	Age may be slightly too old
SX17/13	South Mandible Cirque, Daniell Peninsula	Mugearite	9.3	0.4	1σ	Ar–Ar	Nardini et al. (2003)	
SX3/15	South Mandible Cirque, Daniell Peninsula	Benmoreite	9.2	0.3	1σ	Ar–Ar	Nardini et al. (2003)	
T29.7	Bluff 7 km west of Cape Daniell, Daniell Peninsula	Mugearite lava lobe in lobe-hyaloclastite	6.338	0.050	2σ	Ar–Ar	Smellie et al. (2011b)	
T5.32.6	Cape Phillips, Daniell Peninsula	Mugearite lava	9.950	0.066	2σ	Ar–Ar	Smellie et al. (2011b)	
T5.2.2	'Mandible Bluff', base, Daniell Peninsula	Rhyolite dyke cutting rhyolite domes, base of sequence	9.659	0.088	2σ	Ar–Ar	Smellie et al. (2011b)	
T5.13.2	'Mandible Bluff', top, Daniell Peninsula	Rhyolite dome, top of sequence	9.683	0.051	2σ	Ar–Ar	Smellie et al. (2011b)	
T5.13.1	'Mandible Bluff', Daniell Peninsula	Hawaiite lava overlying rhyolite sequence	8.863	0.062	2σ	Ar–Ar	Smellie et al. (2011b)	
Coulman Island Volcanic Field								
A220C	West coast, Coulman Island	Lava pillow	7.0	0.5	1σ	K–Ar	Armstrong (1978)	
2706	Cape Anne, Coulman Island	Rhyolite	7.29	0.11	2σ	K–Ar	Müller et al. (1991)	Age conflicts with that obtained on a whole-rock sample at the same locality (8.59 ± 0.14 Ma)

7281	Cape Anne, Coulman Island	Rhyolite	7.33	0.11	2σ	K–Ar	Müller et al. (1991)	Age conflicts with that obtained on a whole-rock sample at the same locality (8.59 ± 0.14 Ma)
MP 33	Cape Wadsworth, Coulman Island	Basanite	7.4	0.3	1σ	K–Ar	Armienti et al. (2003)	
MP 34	Cape Wadsworth, Coulman Island	Basanite	7.6	0.4	1σ	K–Ar	Armienti et al. (2003)	
MP 36	Coulman Island	Phonolite	7.5	0.1	1σ	K–Ar	Armienti et al. (2003)	
MP 37	West side of Hawkes Heights, Coulman Island	Comenditic trachyte	7.3	0.1	1σ	K–Ar	Armienti et al. (2003)	
MP 40	East side of Coulman Island	Comenditic trachyte dyke	7.5	0.1	1σ	K–Ar	Armienti et al. (2003)	

Melbourne Volcanic Province

Malta Plateau Volcanic Field

2704	South Malta Plateau (see Fig. 21)	Tephrite	3.19	0.05	2σ	K–Ar	Müller et al. (1991)	Mean age >3 Ma?
3201	South Malta Plateau (see Fig. 21)	Tephrite	2.85	0.04	2σ	K–Ar	Müller et al. (1991)	Mean age >3 Ma?
2705	South Malta Plateau (see Fig. 21)	Tephrite	3.21	0.05	2σ	K–Ar	Müller et al. (1991)	Mean age 3.22 ± 0.4 Ma
3202	South Malta Plateau (see Fig. 21)	Basalt	3.25	0.06	2σ	K–Ar	Müller et al. (1991)	Mean age 3.22 ± 0.4 Ma
2151	Malta Plateau (see Fig. 21)	Pantellerite	15.13	0.18	2σ	K–Ar	Müller et al. (1991)	Mean age 15.0 ± 0.3 Ma
3199	Malta Plateau (see Fig. 21)	Pantellerite	14.86	0.17	2σ	K–Ar	Müller et al. (1991)	Mean age 15.0 ± 0.4 Ma
2152	Malta Plateau (see Fig. 21)	Comenditic trachyte	10.82	0.13	2σ	K–Ar	Müller et al. (1991)	Mean age 10.88 ± 0.9 Ma
3200	Malta Plateau (see Fig. 21)	Comenditic trachyte	10.93	0.13	2σ	K–Ar	Müller et al. (1991)	Mean age 10.88 ± 0.10 Ma
2153	Malta Plateau (see Fig. 21)	Comenditic trachyte	15.23	0.25	2σ	K–Ar	Müller et al. (1991)	Mean age 15.28 ± 0.19 Ma
3204	Malta Plateau (see Fig. 21)	Comenditic trachyte	15.35	0.31	2σ	K–Ar	Müller et al. (1991)	Mean age 15.28 ± 0.20 Ma
2154	Malta Plateau (see Fig. 21)	Comenditic trachyte	15.01	0.20	2σ	K–Ar	Müller et al. (1991)	Mean age 14.89 ± 0.13 Ma
3213	Malta Plateau (see Fig. 21)	Comenditic trachyte	14.81	0.16	2σ	K–Ar	Müller et al. (1991)	Mean age 14.89 ± 0.13 Ma
2155	Malta Plateau (see Fig. 21)	Comenditic trachyte	14.39	0.27	2σ	K–Ar	Müller et al. (1991)	Mean age 14.45 ± 0.18 Ma
3214	Malta Plateau (see Fig. 21)	Comenditic trachyte	14.51	0.23	2σ	K–Ar	Müller et al. (1991)	Mean age 14.45 ± 0.18 Ma
2156	Malta Plateau (see Fig. 21)	Comenditic trachyte	6.76	0.9	2σ	K–Ar	Müller et al. (1991)	Mean age 6.68 ± 0.17 Ma
3264	Malta Plateau (see Fig. 21)	Comenditic trachyte	6.59	0.9	2σ	K–Ar	Müller et al. (1991)	Mean age 6.68 ± 0.17 Ma
2157	Malta Plateau (see Fig. 21)	Comenditic trachyte, crystal tuff	6.51	0.8	2σ	K–Ar	Müller et al. (1991)	Mean age 6.58 ± 0.14 Ma
3265	Malta Plateau (see Fig. 21)	Comenditic trachyte, crystal tuff	6.65	0.8	2σ	K–Ar	Müller et al. (1991)	Mean age 6.58 ± 0.14 Ma
2158	Malta Plateau (see Fig. 21)	Comenditic trachyte, crystal tuff	8.97	0.11	2σ	K–Ar	Müller et al. (1991)	Mean age 8.95 ± 0.8 Ma
3269	Malta Plateau (see Fig. 21)	Comenditic trachyte, crystal tuff	8.93	0.11	2σ	K–Ar	Müller et al. (1991)	Mean age 8.95 ± 0.8 Ma
2159	Malta Plateau (see Fig. 21)	Pantellerite obsidian	8.96	0.31	2σ	K–Ar	Müller et al. (1991)	Mean age >9 Ma?
2159	Malta Plateau (see Fig. 21)	Pantellerite obsidian	8.42	0.17	2σ	K–Ar	Müller et al. (1991)	Mean age >9 Ma?
2160	Malta Plateau (see Fig. 21)	Comendite	8.30	0.10	2σ	K–Ar	Müller et al. (1991)	Mean age 8.35 ± 0.16 Ma
3285	Malta Plateau (see Fig. 21)	Comendite	8.41	0.10	2σ	K–Ar	Müller et al. (1991)	Mean age 8.35 ± 0.16 Ma
2161	Malta Plateau (see Fig. 21)	Comendite	14.95	0.18	2σ	K–Ar	Müller et al. (1991)	Mean age 14.93 ± 0.13 Ma
3294	Malta Plateau (see Fig. 21)	Comendite	14.90	0.18	2σ	K–Ar	Müller et al. (1991)	Mean age 14.93 ± 0.13 Ma
2162	Malta Plateau (see Fig. 21)	Comenditic trachyte	14.67	0.18	2σ	K–Ar	Müller et al. (1991)	Mean age 14.71 ± 0.13 Ma
3297	Malta Plateau (see Fig. 21)	Comenditic trachyte	14.77	0.19	2σ	K–Ar	Müller et al. (1991)	Mean age 14.71 ± 0.13 Ma
2163	Malta Plateau (see Fig. 21)	Rhyolite	11.8	0.3	2σ	K–Ar	Müller et al. (1991)	Mean age 11.94 ± 0.11 Ma
3218	Malta Plateau (see Fig. 21)	Rhyolite	11.96	0.11	2σ	K–Ar	Müller et al. (1991)	Mean age 11.94 ± 0.11 Ma
MP 15	Between Mount Brewster and Mount Prior (see Fig. 15)	Pantelleritic trachyte dyke	12.4	0.2	1σ	K–Ar	Armienti et al. (2003)	Geographically close to Daniell Peninsula but probably part of Malta Plateau Volcanic Field
MP 15/5	Between Mount Brewster and Mount Prior (see Fig. 15)	Pantelleritic trachyte dyke	11.9	0.6	1σ	Ar–Ar	Nardini et al. (2003)	Geographically close to Daniell Peninsula but probably part of Malta Plateau Volcanic Field

(Continued)

Table 2. *Continued.*

Sample	Locality	Dated lithologies	Age*	Error	Error SD	Method	Reference	Notes
The Pleiades Volcanic Field								
37081	Mount Pleiones	Trachybasalt lava	0.01	0.04	1σ	K–Ar	Armstrong (1978)	(see Kyle 1990c for the sample location)
P15 (25266)	Mount Pleiones, 'cone 1'	Basalt lava	0.04	0.05	1σ	K–Ar	Armstrong (1978)	(see Kyle 1990c for the sample location)
P20 (25271)	Mount Pleiones	Trachyandesite dyke	0.02	0.04	1σ	K–Ar	Armstrong (1978)	(see Kyle 1990c for the sample location)
P56 (25306)	North Pleiones area, eroded cone	Trachyte	0.003	0.014	1σ	K–Ar	Armstrong (1978)	(see Kyle 1990c for the sample location)
25699	The Pleiades (see Fig. 22)	Trachyte	6 ka	6 ka	2σ	Ar–Ar	Esser and Kyle (2002)	
25661	The Pleiades (see Fig. 22)	Tephriphonolite	20 ka	7 ka	2σ	Ar–Ar	Esser and Kyle (2002)	
25665	The Pleiades (see Fig. 22)	Trachyte	38 ka	8 ka	2σ	Ar–Ar	Esser and Kyle (2002)	
25674	The Pleiades (see Fig. 22)	Trachyte	42 ka	4 ka	2σ	Ar–Ar	Esser and Kyle (2002)	
25702	The Pleiades (see Fig. 22)	Trachyte	48 ka	2 ka	2σ	Ar–Ar	Esser and Kyle (2002)	
25666	The Pleiades (see Fig. 22)	Trachyandesite	61 ka	4 ka	2σ	Ar–Ar	Esser and Kyle (2002)	
25667	The Pleiades (see Fig. 22)	Trachyandesite	68 ka	5 ka	2σ	Ar–Ar	Esser and Kyle (2002)	
25687	The Pleiades (see Fig. 22)	Trachyte	93 ka	4 ka	2σ	Ar–Ar	Esser and Kyle (2002)	
25670	The Pleiades (see Fig. 22)	Trachyandesite	150 ka	104 ka	2σ	Ar–Ar	Esser and Kyle (2002)	
25690	The Pleiades (see Fig. 22)	Tephriphonolite	312 ka	6 ka	2σ	Ar–Ar	Esser and Kyle (2002)	
25685	The Pleiades (see Fig. 22)	Phonotephirite	337 ka	37 ka	2σ	Ar–Ar	Esser and Kyle (2002)	
25683	The Pleiades (see Fig. 22)	Phonolite	627 ka	10 ka	2σ	Ar–Ar	Esser and Kyle (2002)	
25680	The Pleiades (see Fig. 22)	Trachyte	826 ka	4 ka	2σ	Ar–Ar	Esser and Kyle (2002)	
25689	The Pleiades (see Fig. 22)	Trachyte	832 ka	56 ka	2σ	Ar–Ar	Esser and Kyle (2002)	
25705	The Pleiades (see Fig. 22)	Trachyte	847 ka	12 ka	2σ	Ar–Ar	Esser and Kyle (2002)	
Mount Overlord Volcanic Field								
35412	Mount Overlord	Trachyandesite	6.8	0.14	1σ	K–Ar	Armstrong (1978)	
35413	Mount Overlord	Trachyandesite	8.1	1.7	1σ	K–Ar	Armstrong (1978)	
37085	Mount Overlord	Trachyandesite	7.2	0.14	1σ	K–Ar	Armstrong (1978)	
81501	Parasite Cone	Basanite clast in 'hyaloclastite'	14.67	0.3	1σ	K–Ar	Kyle (1990b)	
81512B	'Aviator Glacier'	Alkali basalt	7.5	0.14	1σ	K–Ar	Kyle (1990b)	
n.r.	'Pilot Glacier' (Mount Rittman volcano?)	n.r.	3.97	0.14	n.r.	K–Ar	Armienti and Tripodo (1991)	
n.r.	Mount Rittman	n.r.	240 ka	200 ka	n.r.	K–Ar	G. Vita cited in Perchiazzi et al. (1999)	
n.r.	Mount Rittman	n.r.	170 ka	20 ka	n.r.	K–Ar	G. Vita cited in Perchiazzi et al. (1999)	
n.r.	Mount Rittman	n.r.	70 ka	20 ka	n.r.	K–Ar	G. Vita cited in Perchiazzi et al. (1999)	
Mount Melbourne Volcanic Field								
34912	Mount Melbourne summit	Trachyte glass	0.01	0.02	1σ	K–Ar	Armstrong (1978)	
34918	Mount Melbourne summit	Trachyte	0.25	0.06	1σ	K–Ar	Armstrong (1978)	
37175	Mount Melbourne summit	Trachyandesite	0.08	0.015	1σ	K–Ar	Armstrong (1978)	
35425	Baker Rocks	Olivine basalt	0.72	0.1	1σ	K–Ar	Armstrong (1978)	Satellite vent
35422	Baker Rocks	Trachybasalt	0.19	0.04	1σ	K–Ar	Armstrong (1978)	Satellite vent
MM1d	South of Willow Nunatak (Cape Washington?)	Olivine basalt	2.4	0.1	1σ	K–Ar	Armstrong (1978)	Satellite vent

Sample	Locality	Rock type	Age	Error	Sigma	Method	Reference	Notes
MM 31-253	Mount Melbourne summit	Alkali basalt	0.25	0.35	2σ	K–Ar	Müller et al. (1991)	
SN 16-016	Shield Nunatak	Alkali basalt scoria	0.07	0.06	2σ	K–Ar	Müller et al. (1991)	
NR 2803	Shield Nunatak, basal lava	Hawaiite lava	1.61	0.05	2σ	K–Ar	Müller et al. (1991)	
SN 09-009	Shield Nunatak, basal lava	Hawaiite lava	1.65	0.04	2σ	K–Ar	Müller et al. (1991)	
2699	Shield Nunatak, basal lava	Hawaiite lava	3.33	0.12	2σ	K–Ar	Müller et al. (1991)	Satellite vent (tuya)
7083a	Shield Nunatak, basal lava	Tephrite	1.74	0.03	2σ	K–Ar	Müller et al. (1991)	
NE 09-119	North of Edmonson Point	Hawaiite lava	0.9	1.15	2σ	K–Ar	Müller et al. (1991)	Age unreliable?
EP 04-082	Edmonson Point	Benmoreite lava	0.047	0.021	2σ	K–Ar	Müller et al. (1991)	
WR 01-041	'Washington Ridge' (prob. Cape Washington)	Basanite intrusion	2.72	0.17	2σ	K–Ar	Müller et al. (1991)	
WR 02-045	'Washington Ridge' (probably Cape Washington)	Basanite pillow breccia	1.68	0.19	2σ	K–Ar	Müller et al. (1991)	
MB 01a	Oscar Point	Basalt lava	0.71	0.18	2σ	K–Ar	Armienti et al. (1991)	
MB 26a	Shield Nunatak	Basalt lava	0.48	0.24	n.r.	K–Ar	Armienti et al. (1991)	Satellite vent (tuya)
MB 33a	Coast NE of Baker Rocks	Basalt lava	2.96	0.2	n.r.	K–Ar	Armienti et al. (1991)	Satellite vent
MB 33b	Coast NE of Baker Rocks	Basalt lava	2.59	0.11	n.r.	K–Ar	Armienti et al. (1991)	Satellite vent
G77	Mount Melbourne summit crater	Trachytic spatter	50 ka	70 ka	n.r.	Ar–Ar	Giordano et al. (2012)	
G79	Mount Melbourne summit crater	Trachytic pumice	35.1 ka	21.9 ka	2σ	Ar–Ar	Giordano et al. (2012)	Isochron age: 61.6 ± 66.4 ka
G52	Edmonson Point	Basalt lava	111.6 ka	83.8	2σ	Ar–Ar	Giordano et al. (2012)	Isochron age: 37 ± 88 ka
G39	Edmonson Point	Hawaiite lava	297.6 ka	55.4 ka	2σ	Ar–Ar	Giordano et al. (2012)	Isochron age: 32 ± 183 ka
G66	Edmonson Point	Basalt lava	90.7 ka	19.0 ka	2σ	Ar–Ar	Giordano et al. (2012)	Isochron age: 94.4 ± 25.8 ka
G48	Edmonson Point	Trachytic pumice in ignimbrite	118.6 ka	11.6 ka	2σ	Ar–Ar	Giordano et al. (2012)	Isochron age: 103.3 ± 17.3 ka
G02	Edmonson Point	Trachytic pumice in ignimbrite	120.2 ka	12.6 ka	2σ	Ar–Ar	Giordano et al. (2012)	Isochron age: 97.4 ± 33.8 ka
G03	Edmonson Point	Trachytic pumice in ignimbrite	124.3 ka	6.4 ka	2σ	Ar–Ar	Giordano et al. (2012)	Isochron age: 118.2 ± 16.4 ka
G02 + G03	Edmonson Point	Trachytic pumice in ignimbrite	123.6 ka	6.0 ka	2σ	Ar–Ar	Giordano et al. (2012)	Isochron age: 115.2 ± 12.8 ka
G46	Shield Nunatak	Basalt lava	430.5 ka	82.0 ka	2σ	Ar–Ar	Giordano et al. (2012)	Isochron age: 191 ± 199 ka
G49	Markham Island	Basalt lava	415 ka	24 ka	2σ	Ar–Ar	Giordano et al. (2012)	Isochron age: 263 ± 111.7 ka
G83	Harrows Peak (=Harrow Hills)	Basalt lava	744.7 ka	66.2 ka	2σ	Ar–Ar	Giordano et al. (2012)	Isochron age: 669 ± 101 ka
G26	Pinckard Table	Basalt lava	1368 ka	90 ka	2σ	Ar–Ar	Giordano et al. (2012)	Isochron age: 1307 ± 112 ka
MMVFA01	Baker Rocks	'Subaerial lava'	0.33	0.03	n.r.	K–Ar	Lee et al. (2015)	Locality and sample type from Mi Jung Lee (pers. comm.)
MMVFA01	Baker Rocks	'Subaerial lava'	0.28	0.03	n.r.	K–Ar	Lee et al. (2015)	Locality and sample type from Mi Jung Lee (pers. comm.)
MMVFA04	Baker Rocks	'Subaerial lava'	0.2	0.01	n.r.	K–Ar	Lee et al. (2015)	Locality and sample type from Mi Jung Lee (pers. comm.)
MMVFA04	Baker Rocks	'Subaerial lava'	0.2	0.01	n.r.	K–Ar	Lee et al. (2015)	Locality and sample type from Mi Jung Lee (pers. comm.)
MMVFA06	Mount Melbourne summit	'Subaerial lava'	0.07	0.01	n.r.	K–Ar	Lee et al. (2015)	Locality and sample type from Mi Jung Lee (pers. comm.)
MMVFA06	Mount Melbourne summit	'Subaerial lava'	0.12	0.02	n.r.	K–Ar	Lee et al. (2015)	Locality and sample type from Mi Jung Lee (pers. comm.)
MMVFAB01	Willows Nunatak	'Subaerial lava'	1.25	0.09	n.r.	K–Ar	Lee et al. (2015)	Locality and sample type from Mi Jung Lee (pers. comm.)
MMVFB02	East of willows Nunatak	'Subaerial lava'	1.34	0.07	n.r.	K–Ar	Lee et al. (2015)	Locality and sample type from Mi Jung Lee (pers. comm.)
MMVFB03	east of willows Nunatak	'Subaerial lava'	1.31	0.09	n.r.	K–Ar	Lee et al. (2015)	Locality and sample type from Mi Jung Lee (pers. comm.)
MMVFB04	East of willows Nunatak	'Subaerial lava'	1.32	0.07	n.r.	K–Ar	Lee et al. (2015)	Locality and sample type from Mi Jung Lee (pers. comm.)

*All ages in Ma except where otherwise noted; note that the original published ages are given, with no recalculation using modern decay constants.
SD, standard deviation; n.r., not recorded.

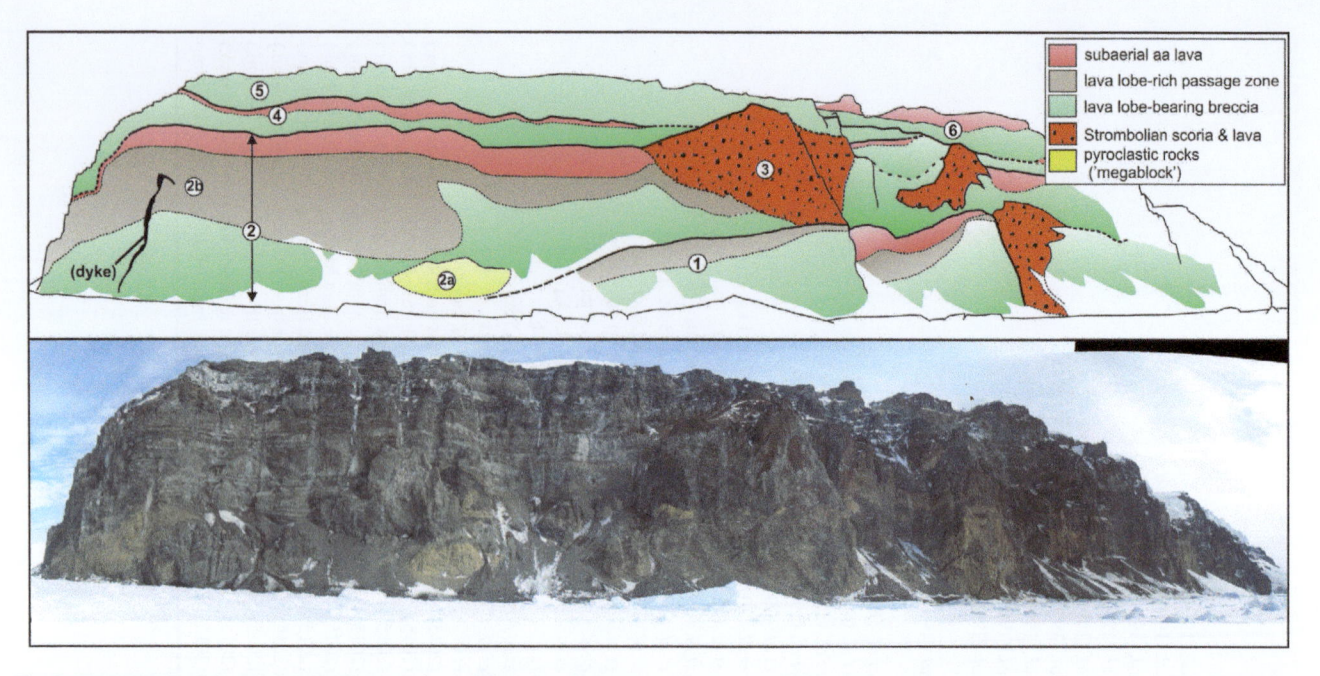

Fig. 8. Photograph and sketch showing the geology of cliffs at Cape Roget on southern Adare Peninsula (unpublished information of J.L. Smellie). The cliffs are >500 m high. See text for description.

volcaniclastic and overlying lava lithofacies range between 10.6 ± 0.4 and 9.7 ± 0.4 Ma, with a mean age of 10.2 Ma. They supersede less-precise K–Ar ages of 13.8 ± 0.7–11.0 ± 0.2 Ma measured by Müller *et al.* (1991) (see Table 2). The narrow range of the ages published by Mortimer *et al.* (2007) indicates that the entire sequence is coeval and cogenetic. A possible subhorizontal passage zone overlying hyaloclastite was mapped at *c.* 700 m asl by Mortimer *et al.* (2007), and those authors proposed a marine eruptive setting based on the isotopic composition of carbonate veins found within the volcanic deposits and presumed to be coeval. The elevation of the passage zone was attributed to a significant post-Miocene uplift of at least 400 m (the elevation of the analysed carbonate veins). A more recent study in December 2014 (unpublished information of J.L. Smellie) indicated that the volcanic outcrops are tiny relicts of a much larger original basaltic shield volcano, and that they are unrelated to the volcanic outcrops exposed further east. They are now preserved as a very thin veneer on a pre-volcanic bluff composed of Cambrian basement rock (Robertson Bay Group metasediments). The basal breccias are not hyaloclastite (*sensu* White and Houghton 2006) but are lapilli tuffs with locally abundant basement lithic clasts (Fig. 7e). They formed explosively in a tuff cone or tuff ring, and there is no passage zone present.

Hallett Peninsula Volcanic Field. Hallett Peninsula is roughly triangular in shape, *c.* 35 km long between Cape Wheatstone in the south and Cape Hallett in the north, and up to 20 km wide between Cape Wheatstone and Football Saddle (Fig. 10). It is one of the best-examined volcanic fields in the region but has few isotopic ages (Table 2). Like Adare Peninsula it is largely snow- and ice-covered, apart from spectacular *c.* 1200–1400 m-high north–south-orientated cliffs on its eastern side (Cotter Cliffs). Rock exposure is also extensive, although much less spectacular, on its west side at Redcastle Ridge, and in low cliffs and steep scree slopes up to 800 m high between Arneb Glacier and Cape Hallett. The peninsula is entirely volcanic, with an unexposed contact with basement rocks in the west (Robertson Bay Group metasediments) extending north from Football Saddle through Edisto Glacier to Edisto Inlet. A further small outcrop, of uncertain age and named 'the Herschel Tuffaceous Moraine' by Harrington *et al.* (1967), is present on the west side of Edisto Inlet. Although it is largely formed of glacial and glaciolacustrine deposits, the detritus is dominated by juvenile sideromelane tephra presumably derived from coeval eruptions of an unidentified Hallett Peninsula volcano (Harrington *et al.* 1967; Smellie *et al.* 2011*a, b*). The volcanic field contains a wide variety of compositions, ranging from basalt and basanite, through mugearite and benmoreite, to phonolite and trachyte (Harrington *et al.* 1967; Hamilton 1972; McIntosh and Kyle 1990*c*; unpublished information of the authors; see also Chapter 5.1b: Rocchi and Smellie 2021).

At least four large shield volcanoes and one stratovolcano (Mount Harcourt) are believed to be present based mainly on binocular mapping of the dip directions of strata (Hamilton 1972) (Fig. 10). Three of the shield volcanoes form a linear almost north–south-aligned outcrop ('Eastern Belt' of Hamilton 1972), whereas two, including Mount Harcourt, are offset to the west by *c.* 8–12 km. Only volcanoes of the Eastern Belt have been examined. Mount Harcourt and the postulated 'Southwestern Volcano' are inaccessible. The lack of prominent summits and the relatively horizontal crest of Quarterdeck Ridge suggest that multiple co-erupting vents were involved in the construction of the combined Hallett volcanic dome, with eruptions probably from fissures and multiple centres distributed along a rift zone (Hamilton 1972; see also Harrington *et al.* 1967). However the different predominant dip directions in large sections of cliff suggest that at least two discrete major eruptive centres were involved, herein called the 'Northern Volcano' and 'Central Volcano', which coalesced to form the Hallett volcanic dome. Conversely, McIntosh and Kyle (1990*c*) suggested that much of the southern part of northern Cotter Cliffs is vent proximal and is composed of a hydrothermally altered assemblage of dykes and tuffs, which they related to a volcanic centre situated at the 1770 m summit of Quarterdeck Ridge. The locus of the postulated vent was placed slightly south of the 'Northern Volcano' identified by Hamilton (1972). McIntosh and Kyle (1990*c*) also suggested that there may have been a second vent situated close offshore of northern Cotter Cliffs, to the NE. The conflicting interpretations cannot be resolved at present and we use the interpretation by Hamilton (1972) here for

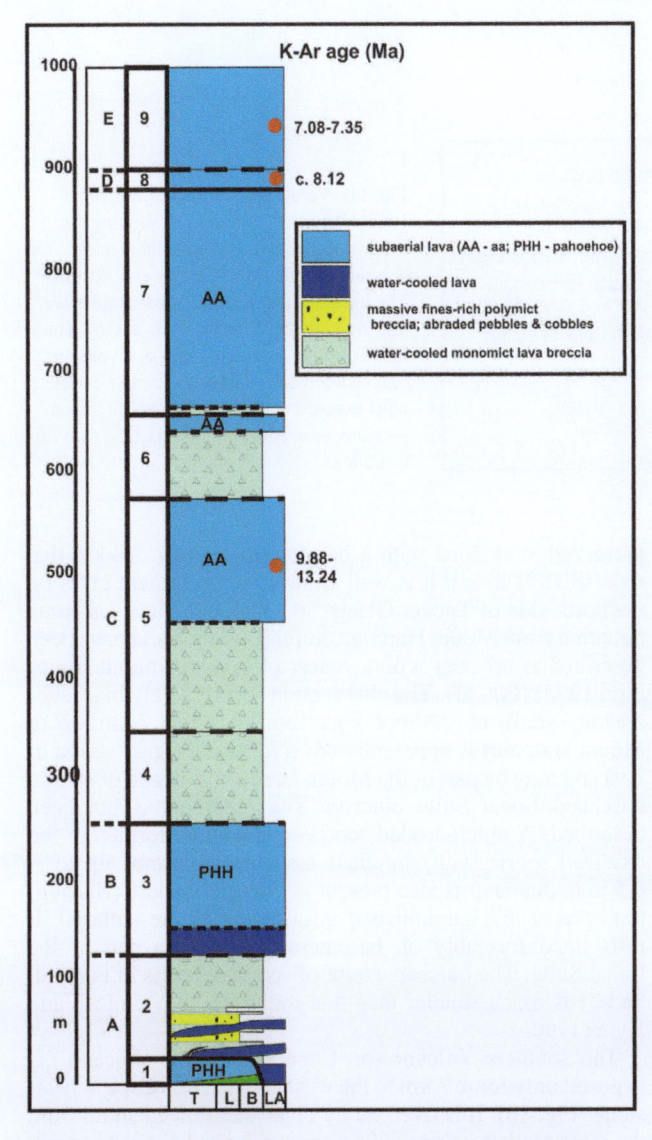

Fig. 9. Simplified stratigraphical column of the volcanic sequence exposed at Nameless Bluff (based on unpublished information of J.L. Smellie). The positions and K–Ar ages of samples dated by Kreuzer et al. (1981) are also shown. Abbreviations: AA, 'a'ā (lava); PHH, pāhoehoe (lava); T, tuff; L, lapilli; B, breccia; LA, lava. Units A–E are after Jordan (1981). Units 1–9 are after J. L. Smellie (unpublished).

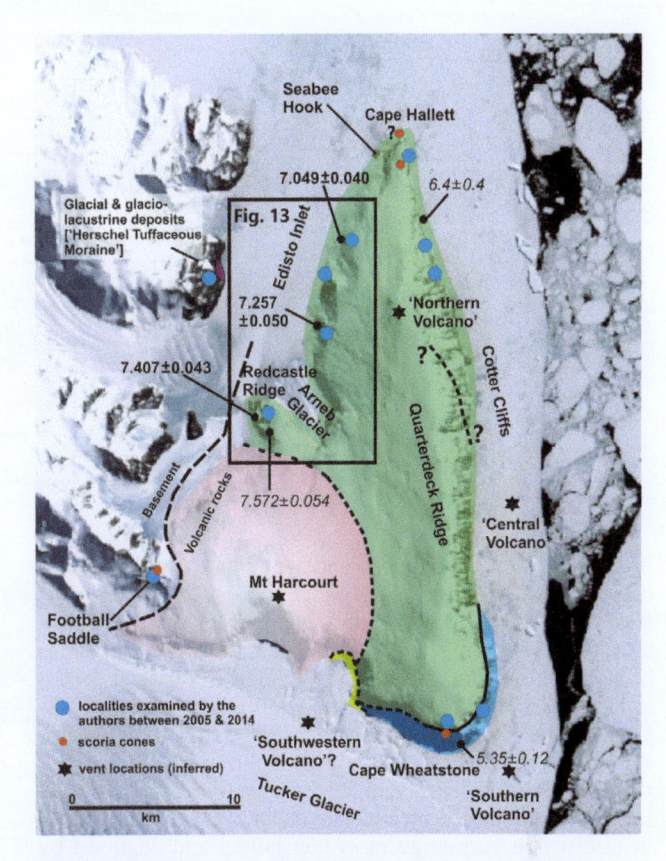

Fig. 10. Simplified geological map of Hallett Peninsula Volcanic Field; modified after Smellie et al. (2011b).

convenience of description, without prejudice to either study. The most detailed recent descriptions of many of the outcrops are presented by Smellie et al. (2011a) and are only briefly summarized here.

Much of the northern half of Hallett Peninsula is formed of the products of the 'Northern Volcano', with its apex situated on Quarterdeck Ridge. It is the best known of all the volcanoes in Hallett Peninsula Volcanic Field principally because of accessible exposures at Redcastle Ridge and along the west coast facing Edisto Inlet (Harrington et al. 1967). Hamilton (1972) mapped a possible contact in Cotter Cliffs with overlying deposits assigned to the 'Central Volcano', with a vent located somewhere to the east in the Ross Sea. However, virtually nothing is known about the deposits of the Central Volcano and it is unclear where the contact extends onto the west flank of the peninsula. We therefore group all the deposits of both volcanoes together here for descriptive purposes. The combined volcanic dome is 32 km in north–south extent, it has a maximum present-day width of 14 km and its highest point, on Quarterdeck Ridge, is 1770 m asl. Hamilton (1972) suggested that the entire height of Cotter Cliffs and the much lower cliffs flanking Edisto Inlet are formed of varied palagonitic breccias, pillow breccias and lavas, with a capping unit of interlayered lavas and scoria several tens of metres thick present in the cliffs on the NW tip of Cape Hallett. Two sections at northern Cotter Cliffs and another at the NE tip of Cape Hallett have been examined recently (Smellie et al. 2011a; unpublished information of the authors) (Figs 11 & 12). They consist mainly of the products of multiple eruptions, each comprising a basal lithosome of chaotic lava-lobe-bearing breccia overlain by subaerial 'a'ā lavas with oxidized surfaces. They represent the 'a'ā lava-fed delta products of repeated glaciovolcanic eruptions, with emplacement through a relatively thin ice-sheet cover that draped the volcano. At least two of the lava-fed deltas in the section at the NE tip of Cape Hallett (Fig. 12) also contain large masses ('megablocks') of stratified lapilli tuffs similar in scale and appearance to that seen at Cape Roget (see above). Surfaces between the lava-fed deltas at Cape Hallett are erosive and, like Cape Roget, there are no sedimentary deposits associated with any surface. The more southerly sections examined at northern Cotter Cliffs also include pumiceous trachytic(?) tuff cone deposits intruded by very thick irregular intrusive basalt sheets, and a trachytic(?) lava dome with an associated block and ash deposit (Fig. 11) (Smellie et al. 2011a, b; unpublished information of the authors). A degraded pyroclastic cone formed of oxidized scoria crops out above Cape Hallett, and two additional scoria cones largely composed of red and black Strombolian tephra have been noted in the cliffs at Cape Hallett (McIntosh and Kyle 1990c); one of the latter occurs at the base of the section shown in Figure 12. It is possible to infer stratigraphical correlations between volcanic units on the west flank of Hallett Peninsula, between Redcastle Ridge and Cape Hallett, over a distance of c. 13 km (Fig. 13). The outcrops are dominated by 'a'ā lava-fed deltas of which, perhaps, five are present together with a

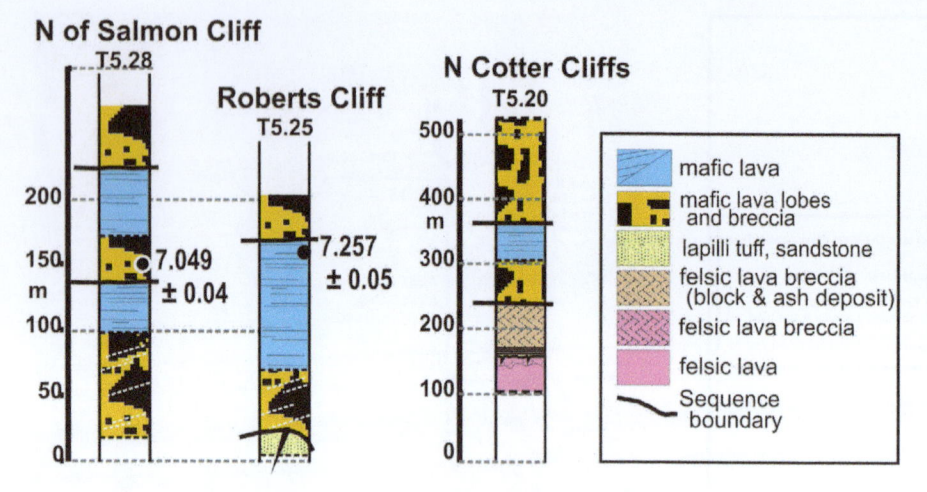

Fig. 11. Vertical profile sections of some Hallett Peninsula outcrops, illustrating the local stratigraphy and typical lithofacies present (modified after Smellie et al. 2011a). The locations and ages of dated samples are also shown (after Smellie et al. 2011b). The lithofacies are dominated by glaciovolcanic 'a'ā lava-fed deltas, whereas the basal felsic units exposed at northern Cotter Cliffs were probably erupted under interglacial conditions.

prominent Strombolian scoria and agglutinate deposit at Redcastle Ridge (Smellie et al. 2011a) (Figs 11 & 13). Although at least one erosional unconformity is present at the latter location, the surfaces of the delta sequences elsewhere appear to be relatively pristine and lack any epiclastic or glacial deposits. The basal exposed unit at Roberts Cliff was called 'Roberts Cliff Tuffite' by Harrington et al. (1967). It is composed of bright yellow deformed stratified lapilli tuffs 30 m thick (Smellie et al. 2011a). It is either an *in situ* older tuff cone relict eroded and buried by a glaciovolcanic 'a'ā lava-fed delta, or else it is a very large mass transported and engulfed by that delta (i.e. a 'megaclast' formed after a sector collapse of the source vent and lateral advection by the delta), similar to those seen in the sections examined at Cape Hallett (Fig. 12) and Cape Wheatstone (Fig. 14 and see below). There are several published isotopic ages for this main phase of Hallett Peninsula volcanism, ranging between 7.257 ± 0.050 and 6.4 ± 0.4 Ma (Armstrong 1978; Smellie et al. 2011b) (Fig. 10).

Strata of a postulated 'Southwestern Volcano' are only known from inaccessible cliff exposures on the north side of Tucker Glacier and comprise breccias that dip to the NE (Hamilton 1972). Mount Harcourt is a prominent well-preserved stratocone with a basal diameter of c. 14 km that rises to 1571 m asl. It is well exposed in prominent cliffs on the north side of Tucker Glacier in which quaquaversal strata radiating from Mount Harcourt summit are seen and have been described as breccias with a veneer of subaerial basalt (Hamilton 1972) (Fig. 5f). The cliff section is completely inaccessible but a gently north-dipping platform situated c. 7 km NW of Mount Harcourt is apparently covered by dark-grey scoria or lava and may be part of the Mount Harcourt volcano or else an unrelated Local Suite outcrop. The outcrop has not been described. A much-eroded scoria cone deposit formed of red oxidized scoria with abundant large dense bombs up to c. 0.5 m in diameter is also present in Football Saddle (Harrington et al. 1967; unpublished information of the authors). It rests unconformably on basement rock and is part of the Local Suite. The outcrop extent of scoria deposits in Football Saddle is much smaller than that indicated by McIntosh and Kyle (1990c).

The Southern Volcano (or 'Cape Wheatstone Volcano') is exposed only for c. 7 km to the west and north of Cape Wheatstone (Fig. 10). It is overlain by lavas that flowed south from the main Hallett volcanic dome and onlapped onto the gently

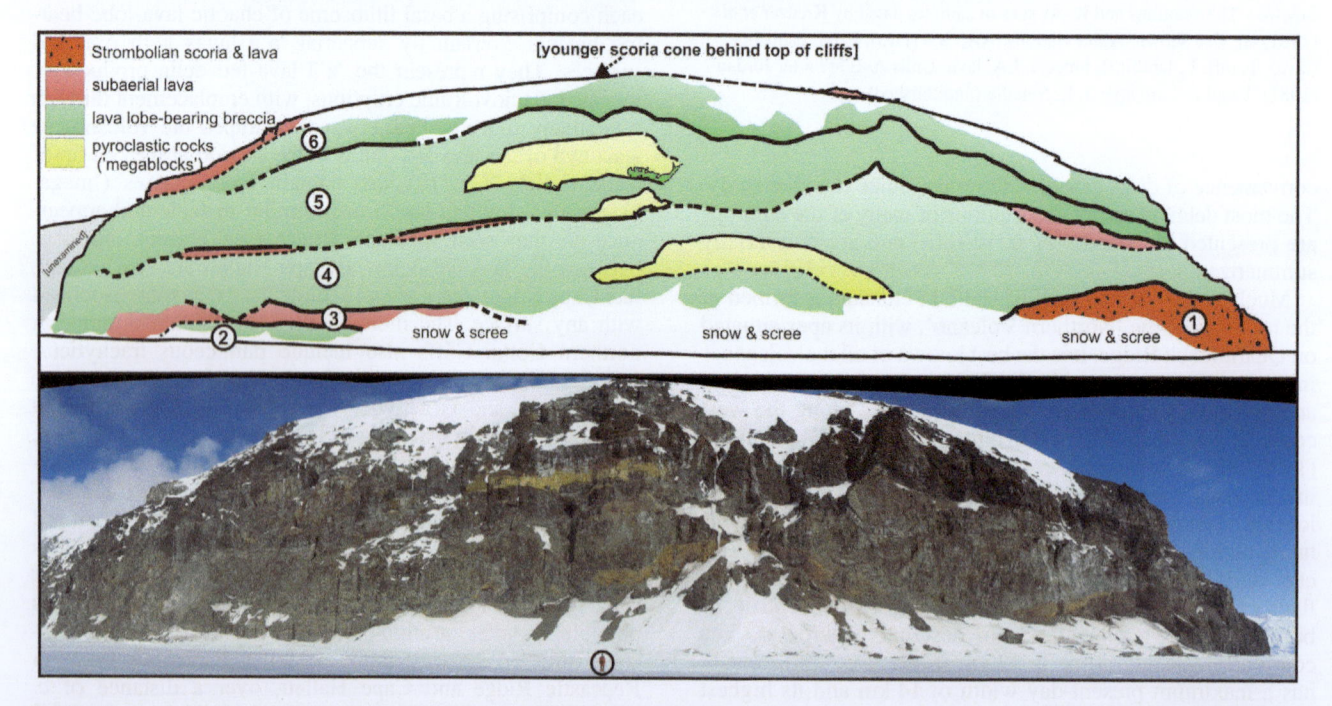

Fig. 12. Photograph and sketch showing the geology of cliffs of NE Hallett Peninsula. See the text for the description. Person (ringed) for scale. Note that the section shown is not that labelled as northern Cotter Cliffs in Figure 11.

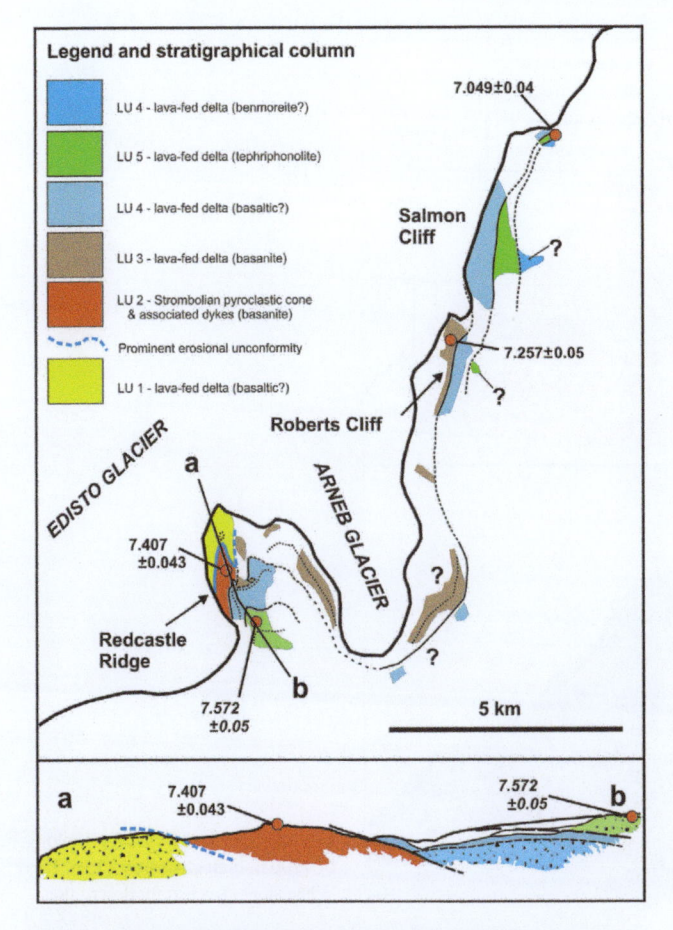

Fig. 13. Geological sketch map of Redcastle Ridge and the west coast of Hallett Peninsula showing the local eruptive units present and their proposed correlation; also shown (below) is a view of the west side of Redcastle Ridge (revised after Smellie *et al.* 2011*a*). The stippled ornament used in the cross-section represents subaqueous lava-lobe-bearing breccia lithofacies. The locations and ages of dated samples (in Ma) are also shown (Smellie *et al.* 2011*b*). Abbreviation: LU, local unit.

dipping upper surface of the Cape Wheatstone Volcano (Hamilton 1972; unpublished information of the authors). At Cape Wheatstone itself, the 1200 m-high cliffs are formed of a chaotic assemblage of coarse breccias and irregular lava masses, with two(?) possible 'a'ā lava-fed delta units present high in the cliff face. The sequence observed by us at Cape Wheatstone resembles that described by McIntosh and Kyle (1990*c*; see also Hamilton 1972) as a vent-proximal assemblage of dykes, hydrothermal alteration and northwesterly-dipping lavas, which they related to a vent situated somewhere to the SE. By contrast, a 1 km section of cliff *c.* 1.5 km NE of Cape Wheatstone reveals a very different sequence (Fig. 14). It consists of a basal 'a'ā lava-fed delta (unit 1 in Fig. 14) dominated by lava-lobe-bearing breccia *c.* 300 m thick. The delta contains two very large masses ('megablocks': units 1a and 1b in Fig. 14), one composed of oxidized Strombolian scoria *c.* 900 m long and 300 m high, which may be one of the two scoria cones indicated by McIntosh and Kyle (1990*c*; also McIntosh and Gamble 1991), and a slightly smaller mass of massive felsic glassy breccia (unpublished information of the authors). The remainder of the sequence, amounting to *c.* 800 or 900 m of the section (unit 2 in Fig. 14), is composed of subaerial 'a'ā lavas. The lavas were observed only by binoculars but they contain multiple uneven surfaces marked by a discontinuous pale-orange coloration, which may represent fragmental deposits that accumulated during time breaks. Finally, a much-eroded pyroclastic cone formed of red oxidized scoria with abundant large bombs (up to 1 m in diameter) is present at the summit of Cape Wheatstone. It rests directly on the Cape Wheatstone sequence and is possibly older than the northerly-derived lavas of the Central Volcano, although the relationships are ambiguous. A single K–Ar isotopic age is available for Cape Wheatstone: 5.35 ± 0.12 Ma (Armstrong 1978) but is stratigraphically inconsistent (i.e. *younger* than supposedly overlying strata to the north).

Daniell Peninsula Volcanic Field. Daniell Peninsula trends NNE–SSW and is 68 km in length. Like other Hallett Coast peninsulas, most of Daniell Peninsula is snow- and ice-covered (Fig. 15). It has two distinct physiographical parts. The northern part is a parallel-sided promontory *c.* 13 km wide that rises to 1910 m asl at an unnamed summit in the north (informally named 'Daniell Summit' here) and 2026 m asl in the south at Mount Brewster. Low inaccessible (ice-capped) cliffs 600 m high form the east flank, whereas the west flank dips more gently down to Whitehall Glacier, where there are few rock outcrops. A prominent cirque at the northern tip, west of Cape Daniell, is flanked by volcanic strata that dip radially to the east and west, leading to the suggestion that it is a breached crater (Hamilton 1972) (Fig. 5c). By contrast, the southern half of Daniell Peninsula is broader and much more deeply eroded. It is roughly 'C'-shaped, concave to the east and with a width of 35 km at its northern end, tapering southwards over a distance of 32 km to Cape Jones. The highest points are 1630 m asl at Mount Lubbock and 1221 m asl at Tousled Peak (Fig. 16). As a result of deep dissection, southern Daniell Peninsula forms a series of narrow dendritic ridges between glaciers; Mandible Cirque on the east side of the peninsula is the most prominent deep embayment. The most extensive rock exposures are at Cape Phillips in the east (cliffs *c.* 600 m high) and a large unnamed headland here informally named 'Mandible Promontory', which rises to 1000 m. Daniell Peninsula is compositionally more varied than other volcanic fields on the Hallett Coast, with sodic alkaline compositions ranging from basanite, hawaiite and benmoreite, to comenditic and pantelleritic trachyte and rhyolite (Armienti *et al.* 2003; Nardini *et al.* 2003; see Chapter 5.1b: Rocchi and Smellie 2021). Uniquely for the Hallett Volcanic Province, the volcanic rocks of Daniell Peninsula are seen to unconformably overlie granodiorite basement at a locality NE of Cape Phillips (Hamilton 1972).

The broad linear ridge that forms northern Daniell Peninsula has an almost uniform summit elevation that rises <200 m southwards over a distance of 25 km. It probably formed from multiple overlapping shield volcanoes along a crestal fissure zone, with final activity focused at Mount Brewster and 'Daniell Summit' (Hamilton 1972). Although Hamilton (1972) suggested the possible presence of an ice-filled caldera on the ridge crest close to Mount Brewster, the satellite imagery now available shows no obvious feature. Conversely, a broad, roughly circular ice-filled feature *c.* 2.5 km in diameter with a nearly flat surface at Daniell Summit is faintly visible on satellite images and speculatively may be a caldera (Smellie *et al.* 2011*a*) (Fig. 15). Hamilton (1972) also postulated that two small breached calderas were responsible for the two deep embayments in the south, including Mandible Cirque, whereas McIntosh and Kyle (1990*d*) indicated only a single caldera encompassing the northernmost embayment. If present, the northern caldera has a maximum diameter of *c.* 12 km and a minimum depth of 1200 m; the southern caldera is *c.* 6 km in diameter. The strongly scalloped margins of both calderas suggest that they have been affected by numerous post-caldera collapses. Identification of the calderas is entirely due to the morphology and no other supporting evidence (such as distinctive explosive deposits and

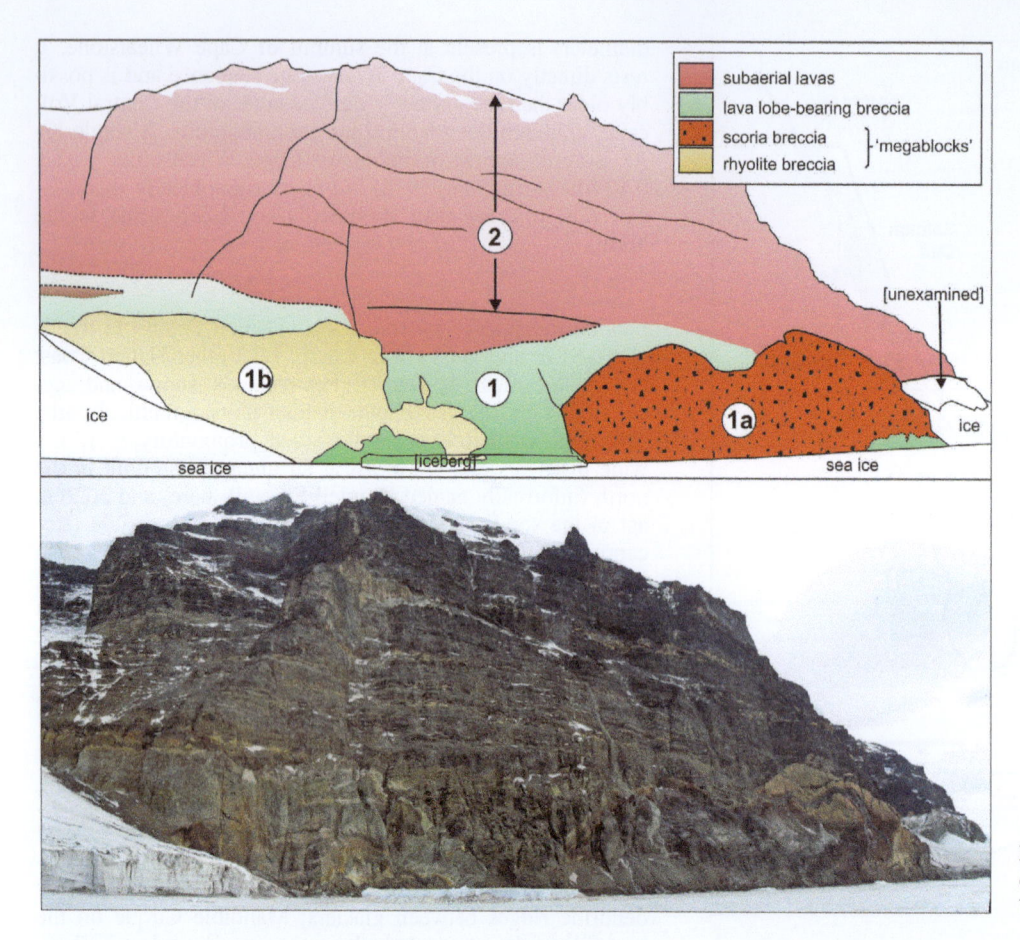

Fig. 14. Sketch showing the geology of 1200 m-high cliffs near Cape Wheatstone. See text for description.

volcanotectonic structures) has been discovered so far. The calderas were omitted from the study by Smellie et al. (2011a, b) but they are included here because the possibility cannot be excluded and they provide a reasonable explanation for the topography. In common with his descriptions of other volcanic fields in the Hallett Coast, Hamilton (1972) described an evolutionary succession comprising initial eruption of breccia and pillow breccia followed by interlayered maroon and grey lavas with sparse 'palagonite lenses'. Multiple well-preserved scoria cones are particularly common scattered along the ridge crest north from Mount Brewster. Phreatomagmatic deposits (tuff cone(s)?) were also recorded by Nardini et al. (2003) on the west flank of the ridge at c. 1000 m. They are distinguished by the presence of abundant clasts of the local basement rocks. A somewhat more complicated history of eruptions was suggested by Nardini et al. (2003). The cliffs west of Cape Daniell were said to comprise basal subaerial lavas overlain by lavas and subaerial pyroclastic rocks lacking evidence for wet conditions (i.e. no pillow lava, hyaloclastite or pillow breccia). Elsewhere, an 'older volcanic edifice' was identified within Mandible Cirque (at Mandible Promontory and the first headland to the west). Unlike the succession west of Cape Daniell, the older edifice was said to be associated with abundant hyaloclastite created during submarine or subglacial eruptions. Although the location of the deposits was unspecified, it probably corresponds to Mandible Promontory where rhyolitic hyaloclastite is found associated with domes or coulées (Smellie et al. 2011a).

Neither of the older published descriptions corresponds closely with results of later studies published by Smellie et al. (2011a, b), which are described briefly here. Localities west of Cape Daniell, and at Cape Phillips and Cape Jones, consist of products of multiple 'a'ā lava-fed deltas with basalt to mugearite compositions (Fig. 17). Those west of Cape Daniell dip radially out on both flanks of the topographical cirque mentioned previously (Fig. 5c). The subaqueous lithosomes in the individual lava-fed deltas are formed of lava-lobe-bearing breccia. They range from 30 to 130 m in thickness, implying associated palaeo-ice thicknesses of c. 120–230 m (Smellie et al. 2011b); from helicopter observations by the authors, these figures are probably typical for the volcanic sequences on much of Daniell Peninsula. The boundaries between sequences appear to be relatively pristine and have experienced only minor erosion. By contrast, the rhyolite domes or coulées at Mandible Promontory are separated by irregular eroded surfaces, including one with a prominent U-shaped profile and tens of metres deep (illustrated by Smellie et al. 2011a, fig. 7e). The basal domes at Mandible Promontory were emplaced subglacially, and are associated with massive glassy hyaloclastite breccia, stratified diamict (likely debris flow deposits), and stratified volcanic sandstones and conglomerate deposited by traction currents and which are probably fluvial (Fig. 18) (Smellie et al. 2011a). They were interpreted as rare glaciovolcanic felsic sheet-like sequences, the first to be described anywhere. By contrast, the upper units at Mandible Promontory are composed of similar-looking foliated rhyolite but lack evidence for subaqueous emplacement and, hence, are probably subaerial. At the north end of the promontory, the subaerial domes are overlain by gently east-dipping hawaiite 'a'ā lavas, mainly thin (few metres) but varying up to c. 50 m thick, with oxidized surfaces and minor lenses of stratified yellow lapilli tuffs. The lavas extend down to a much lower elevation than the tops of the rhyolite domes on their south side, implying that the domes formed an upstanding mass that was simply draped (buried) by the lavas. No evidence was seen that suggested displacement along a caldera fault and similar relationships were observed by binoculars at the unnamed headland 4 km to the NW (unvisited by the authors), with thin lavas apparently draped on trachyte domes/coulées.

Published isotopic ages broadly show an overall northerly younging (Nardini et al. 2003; Smellie et al. 2011b)

migrated north to Mount Brewster (6.9 ± 0.3 Ma) and then to the northern tip of Daniell Peninsula (6.7 ± 0.2–5.3 ± 0.5 Ma).

Coulman Island Volcanic Field. Coulman Island is the southernmost volcanic field in the Hallett Volcanic Province. It measures 34 km north–south and up to 15 km wide, and is roughly triangular in shape with a long northward-tapering snout that declines progressively in elevation northwards from *c.* 1900 to *c.* 600 m asl (Fig. 19). The highest point is 1998 m asl at Hawkes Heights, where there is a prominent ice-filled caldera 6 km wide and at least 700 m deep that is breached by erosion at its SW corner. Although associated caldera-related structures or pyroclastic deposits have not been described, a block of welded ignimbrite has been observed near the summit (G. Giordano pers. comm.); welded ignimbrites are undescribed elsewhere in the Hallett Volcanic Province, although they are present in the Melbourne Volcanic Province (see below). From the morphology of the ice-filled interior of the caldera, it appears that post-caldera volcanism has constructed a small cone near the centre of the structure. The reason for the identification of a major vent near Cape Wadworth by McIntosh and Kyle (1990*e*) is unclear but a minor flank vent is represented by a small dissected hill on top of cliffs on the NW side of the island. Like other volcanic fields in the Hallett Coast region, Coulman Island is presumably formed of the products of multiple overlapping centres with shield-volcano profiles that probably erupted along a fissure or series of fissures and became coalesced. The island is entirely surrounded by ice-capped rock cliffs 400–1800 m high or ice cliffs, and it is one of the hardest areas to access because of ice seracs. Hamilton (1972) suggested a general upward progression from basal massive breccias and pillow breccias into relatively thinly bedded breccias with gentle outward dips of <10°; some bedding dipping gently into the island interior was also noted. Armienti *et al.* (2003) also recorded trachytic lavas and domes, and hawaiitic scoria cones around the caldera rim, which has been locally affected by small collapses. However, McIntosh and Kyle (1990*e*; see also McIntosh and Gamble 1991) described a 1386 m-high composite section at the SW corner of the island (location A in Figs 19 & 20) formed of lavas and related breccias with minor Strombolian tuffs. The presence of slightly pillowed bases to a few of the lavas and associated hyaloclastite suggested limited interaction with snow and ice within a periglacial environment. The section described by McIntosh and Kyle (1990*e*) has not been revisited but the interpretation contrasts with that published for the sequence forming cliffs *c.* 7 km to the north, which is composed of the products of at least three 'a'ā lava-fed deltas; they are described in detail

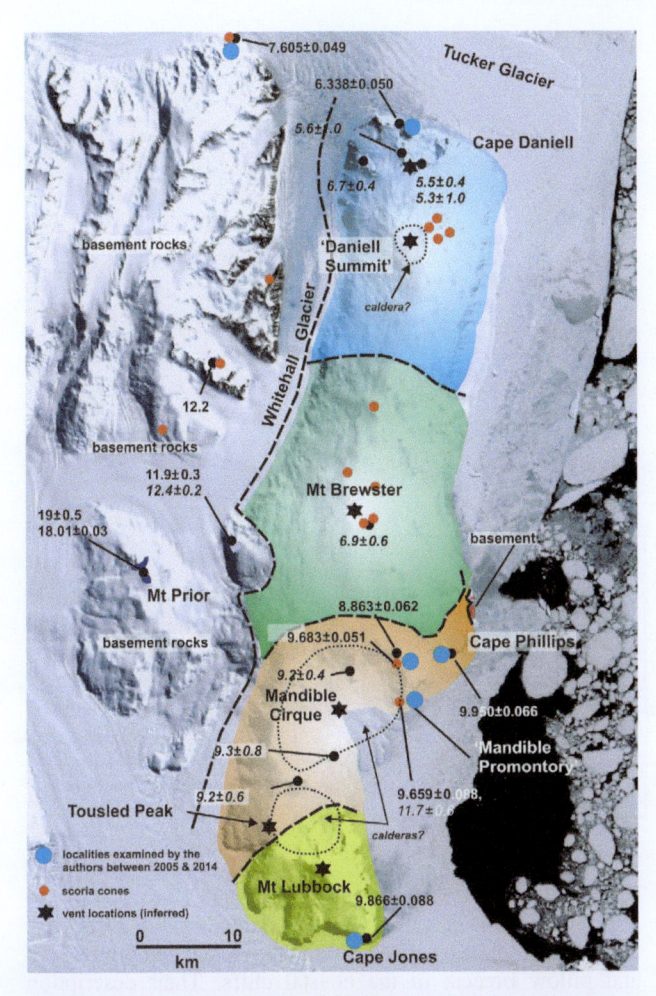

Fig. 15. Simplified geological map of Daniell Peninsula Volcanic Field (modified after Smellie *et al.* 2011*b*). The tiny early Miocene outcrop at Mount Prior comprises dykes that are probably part of the Meander Intrusive Group, whilst the small outcrop *c.* 10 km to the ENE of Mount Prior may be an outlier of Malta Plateau Volcanic Field.

(Fig. 15; Table 2). The earliest activity took place at *c.* 12 Ma in a small and little-known pantelleritic trachyte outcrop (dyke?) *c.* 10 km east of Mount Prior (Nardini *et al.* 2003), which may be an outlier of Malta Plateau volcanism. The next stage occurred at 9.950 ± 0.066–9.2 ± 0.3 Ma, and created the highly dissected compound volcano between Cape Jones and Mandible Promontory. It was a polygenetic volcano, composed of multiple centres that include Mount Lubbock (a stratocone similar in size to Mount Harcourt: Fig. 16) and a smaller centre at Tousled Peak. Activity then

Fig. 16. Google Earth satellite perspective view of the small Mount Lubbock stratocone. Mandible Cirque may be the site of one or more large eroded calderas. Image: Google Earth (2020 Maxar Technologies and US Geological Survey).

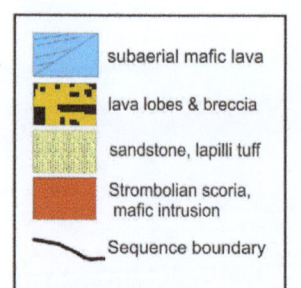

Fig. 17. Vertical profile sections of Daniell Peninsula outcrops, illustrating the local stratigraphy and typical lithofacies present (modified after Smellie *et al.* 2011*a*). The locations and ages of dated samples (in Ma) are also shown (after Smellie *et al.* 2011*b*).

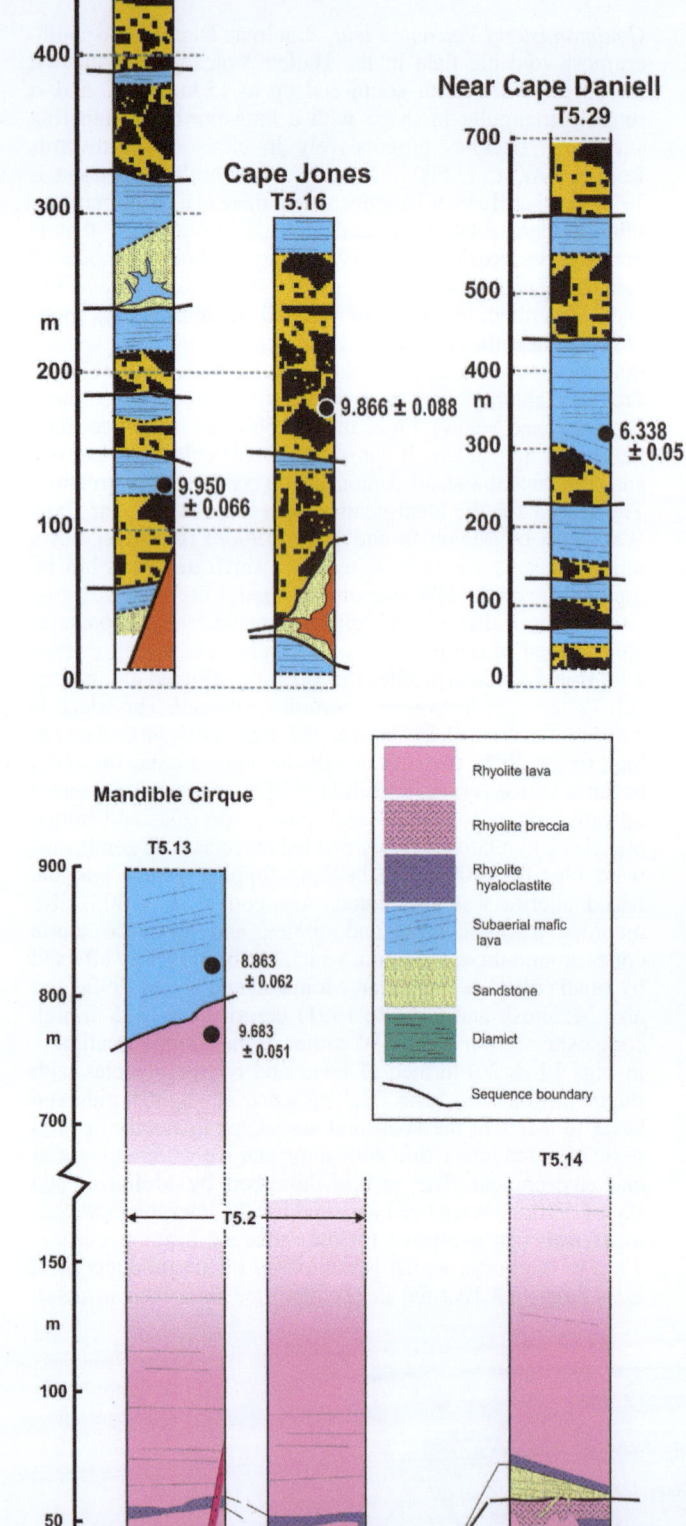

Fig. 18. Vertical profile sections of outcrops at Mandible Cirque, illustrating the local stratigraphy and mainly felsic lithofacies present (modified after Smellie *et al.* 2011*a*). The locations and ages of dated samples (in Ma) are also shown (after Smellie *et al.* 2011*b*).

by Smellie *et al.* (2011*a*) (section B in Figs 19 & 20). Moreover, Armienti *et al.* (2003) described thin basaltic lavas, scoria cones and palagonitized tuffs with intercalated pillow lava and pillow breccia in the coastal cliffs. Their description strongly resembles those of lava-lobe-bearing breccia lithofacies of 'a'ā lava-fed deltas. With the benefit of the more recent interpretations of volcanic sequences in the Hallett Coast (Smellie *et al.* 2011*a*, *b*, 2013), the section described by McIntosh and Gamble (1991) can be more realistically reinterpreted in terms of at least three relatively thick glaciovolcanic 'a'ā lava-fed deltas similar to those seen elsewhere in the region (Fig. 20). The lavas on Coulman Island have basanite, hawaiite, phonotephrite, phonolite and comenditic trachyte compositions (Hamilton 1972; McIntosh and Kyle 1990*e*; Armienti *et al.* 2003). Other than one unverified older age from Cape Anne (8.59 ± 0.14 Ma obtained on sanidine but contrasting with whole-rock ages of 7.33 and 7.29 Ma: Müller *et al.* 1991), all the published isotopic ages are very similar and indicate that eruptions took place within a comparatively narrow interval in the late Miocene (7.6 ± 0.4–7.0 ± 0.5 Ma by K–Ar: Fig. 19).

Melbourne Volcanic Province

Malta Plateau Volcanic Field. Malta Plateau is a high area of relatively gently undulating snow- and ice-covered ground between Borchgrevink and Mariner glaciers (Figs 5b & 21). It measures *c.* 65 km in length (NW–SE) and *c.* 40 km wide, and has a general elevation above 2000 m. Although aeromagnetic surveys suggest that the outcrop of volcanic rock is considerably more extensive than was mapped, and may extend as far as Trafalgar Glacier (cf. Schmidt-Thomé *et al.* 1990; Ferraccioli *et al.* 2009), there is little in the surface geology to support the suggestion. Ferraccioli *et al.* (2009) also inferred that the presence of Malta Plateau volcanics was related to a volcano–tectonic rift zone 80 km west of, and geographically separate from, the rifts in the western Ross Sea (Northern Basin,

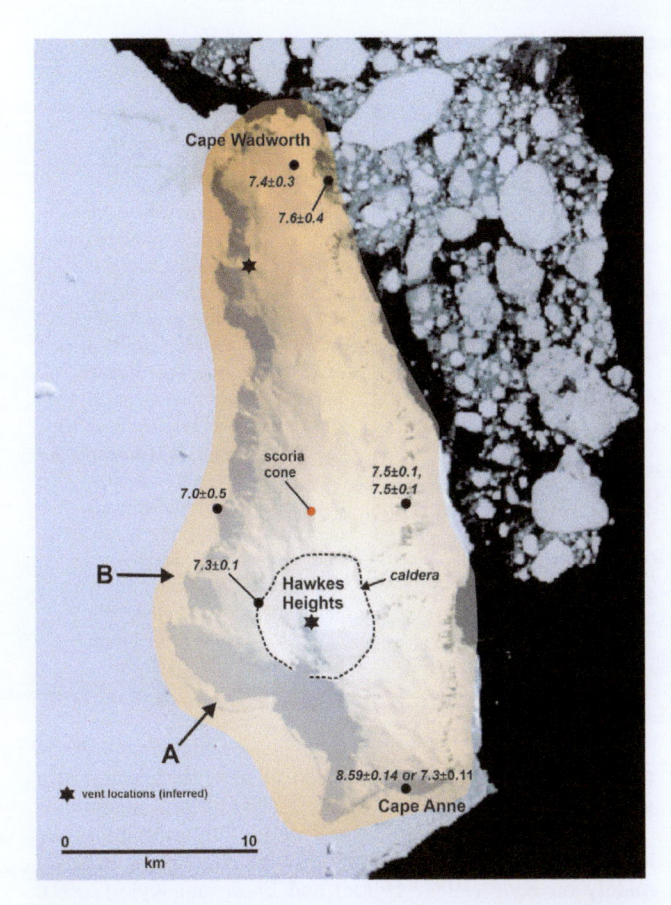

Fig. 19. Simplified geological map of the Coulman Island Volcanic Field showing the locations and ages of dated samples (in Ma: modified after Smellie *et al.* 2011*a*, *b*). A and B refer to logged sections, shown in Figure 20.

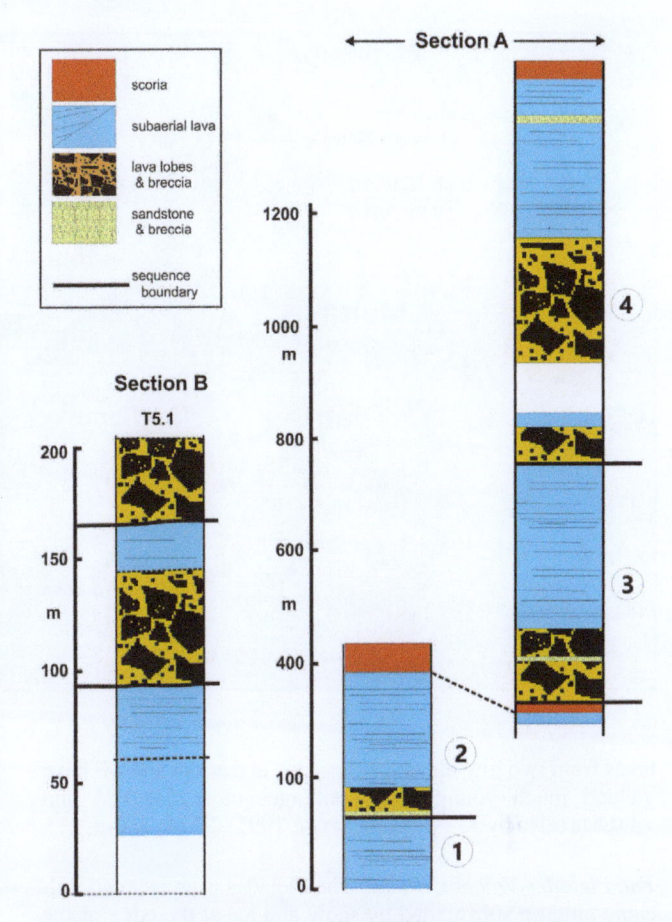

Fig. 20. Vertical profile sections of outcrops on Coulman Island, illustrating the local stratigraphy and typical lithofacies present. Section B is modified after Smellie *et al.* (2011*a*). Section A is a reinterpretation of the section published by McIntosh and Gamble (1991). Note that the lava depicted low in 'unit 4' (Section A) is a thick lava that is succeeded by pillow lava; it is not a unit boundary but it is more likely that the lavas and associated breccias comprise a deposit of lava-lobe-bearing breccia (i.e. the subaqueous lithosome of the unit 4 'a'ā lava-fed delta). The locations of sections A and B are shown in Figure 19.

Victoria Land Basin and Terror Rift: see Chapter 5.2a: Smellie and Martin 2021), that they named Malta Plateau rift. The highest points are Mount Phillips (*c.* 3035 m) and an unnamed summit (*c.* 3080 m) *c.* 7 km SW of Mount Hussey, both composed of volcanic rock. The volcanic rock exposures are small and scattered, and mainly restricted to local peaks and ridges; many are inaccessible, particularly those in the >1000 m-high cliffs and icefalls overlooking Mariner Glacier (Fig. 21). Riddolls and Hancox (1968) speculated that the outcrops may represent the products of a large number of shield volcanoes but it is impossible to judge from the available exposures. A lack of mappable datums also means that outcrops cannot be correlated and the more detailed published maps are based on rock compositions not lithology. The outcrops were originally named Malta Volcanics by Müller *et al.* (1991) to encompass a varied assemblage of dark-grey mafic and pale-coloured felsic lavas (including black obsidian: e.g. at Mount Phillips; personal observation of Laura Crispini), and a wide variety of multi-coloured (pale green, grey, orange and violet) pyroclastic rocks, including lapilli tuffs, welded tuffs, ignimbrites, agglomerates and breccias with maximum thicknesses of between 250 and 400 m. Some of the outcrops are interpreted as tuff rings (Schmidt-Thomé *et al.* 1990). It has also been postulated that Cuneiform Cliffs, at the south end of Malta Plateau, may be part of a caldera (Hornig and Wörner 2003). However, the geology comprises a relatively thin cover of felsic volcanic rocks or horizontal lava flows overlying basement and there is no evidence, other than the arcuate cliff morphology, to suggest that a caldera is present; it is therefore omitted from Figure 21. Moreover, some of the published geology may be unreliable since examination of the SE-trending ridge east of The Ramp by one of us (J.L. Smellie) in 2014 showed it to be formed of basement intruded by dykes and not the felsic volcanics mapped by Schmidt-Thomé *et al.* (1990).

Compositions within Malta Plateau Volcanic Field include basanites but the rocks are mainly alkaline and peralkaline trachytes (comendites) and rhyolites (pantellerites: Schmidt-Thomé *et al.* 1990; Müller *et al.* 1991). Although the Malta Volcanics were distinguished by Müller *et al.* (1991) as a separate compositional group different from volcanism in the Hallett Coast, the subsequent geochemical study by Hornig and Wörner (2003) did not support the petrological distinction. However, there appears to be a greater proportion of felsic rocks, particularly as pyroclastic deposits and including ignimbrites which are rare in the region, so a volcanological distinction seems to exist. Several isotopic ages have been published mainly for felsic extrusive rocks (Müller *et al.* 1991; Hornig and Wörner 2003) (Table 2). They indicate ages of 10.93 ± 0.13 to 6.51 ± 0.8 Ma, although ages determined on associated and presumed coeval peralkaline dykes are significantly older, ranging from $15.35 \pm .31$ to 11.96 ± 0.11 Ma (all ages by K–Ar). Dykes at Mount Prior, which yielded older ages (19.0 ± 0.5–18.1 ± 0.3 Ma) and were included in 'Malta Volcanics' by Müller *et al.* (1991), are probably associated with the Meander Intrusive Group (see below). Conversely, trachytic dykes *c.* 10 km east of Mount Prior, with ages of 12.4 ± 0.2 and 11.9 ± 0.6 Ma (Table 2), may be outliers of Malta Plateau volcanics. Finally, basalt

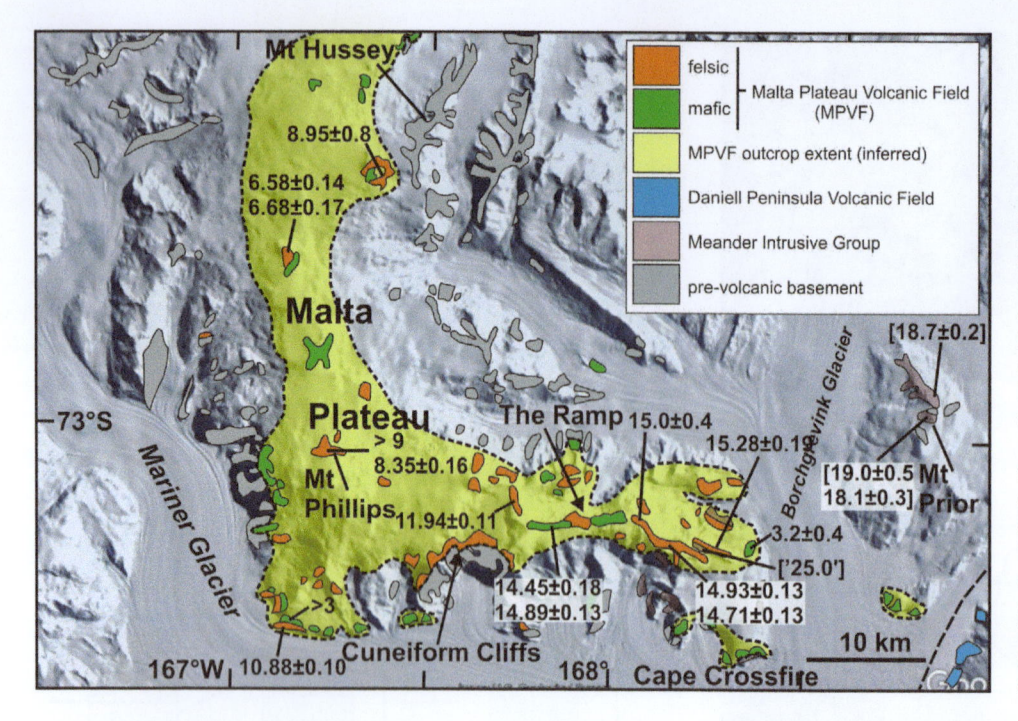

Fig. 21. Geological map of Malta Plateau Volcanic Field draped on a Google Earth satellite image (modified after Schmidt-Thomé et al. 1990; Müller et al. 1991); isotopic ages (in Ma) after Müller et al. (1991). Ages for Meander Intrusive Group outcrops are also shown (in square brackets; note the '25.0' Ma age shown is a mean age for all Meander Intrusive Group outcrops dated by Müller et al. (1991) and may not be very precise). The ages show a hitherto-unrecognized migration, becoming younger clockwise to the NW, between 15 and 6.5 Ma. Image: Google Earth (US Geological Survey).

lavas from two low-elevation outcrops in the SW and SE have yielded much younger apparent ages of 3.23 ± 0.05 and >3.0 Ma (also by K–Ar: Müller et al. 1991) (Table 2; Fig. 21).

The Pleiades Volcanic Field. The Pleiades is an isolated volcanic outcrop surrounded by snow and ice at the edge of the polar plateau overlooking the head of Mariner Glacier, at a general elevation of *c.* 2500 m asl. It is the furthest inland outcrop included in the Melbourne Volcanic Province and is relatively well investigated. The following description is based mainly on Kyle (1982, 1990c) and Esser and Kyle (2002). The outcrop consists of a chain of at least 12 small scoria cones with prominent well-preserved craters and two domes (Taygete Cone and an unnamed feature) that extend *c.* 16 km in a northeasterly direction (Fig. 22). The highest points are Mount Atlas (*c.* 3040 m) and Mount Pleiones (*c.* 3020 m), which together are regarded as a small stratovolcano comprising a composite of several overlapping and nested scoria cones (Esser and Kyle 2002). The broad range of basanite–tephriphonolite and trachyte compositions in the Pleiades, with two major evolutionary petrological lineages present, suggests that the Pleiades outcrops belong to the Central Suite of Nathan and Schulte (1968). The multiple small centres have a curvilinear trend at the edge of a poorly defined snow-covered arcuate feature interpreted as the rim of an older caldera (Kyle 1990c), perhaps *c.* 6 km in diameter, although it is poorly defined topographically (acknowledged by Kyle 1990c) (Fig. 22). The geographical extent of the caldera and its associated volcano are unknown, and there is no convincing magnetic anomaly to suggest an extension of the original volcanic edifice to the SE (Ferraccioli et al. 2009).

The prominent Mount Atlas–Mount Pleiones outcrop rises *c.* 500 m above the surrounding snowfield (Evans Névé) and serves as a typical example of the pyroclastic cones forming the nunataks. It consists of three nested craters variably *c.* 500–1000 m in diameter, formed of benmoreite scoria and numerous large dense bombs, and at least two smaller flank cones on the south side, one of which is composed of hawaiite scoria. Although the three main craters were described as formed by oxidized scoriaceous lavas (Nathan and Schulte 1968; Kyle 1982), like most of the other nunataks they are pyroclastic. Most of the other outcrops to the NE are also pyroclastic cones very similar to the Mount Atlas–Mount Pleiones cluster but they have a variety of basanite, hawaiite and tephriphonolite compositions. By contrast, two of the nunataks (Taygete Cone and the unnamed nunatak *c.* 4 km to the NW) are not pyroclastic. That forming Taygete Cone is a pale-coloured trachyte dome. It is described by Kyle (1982) as showing slickensides on joint surfaces, especially common

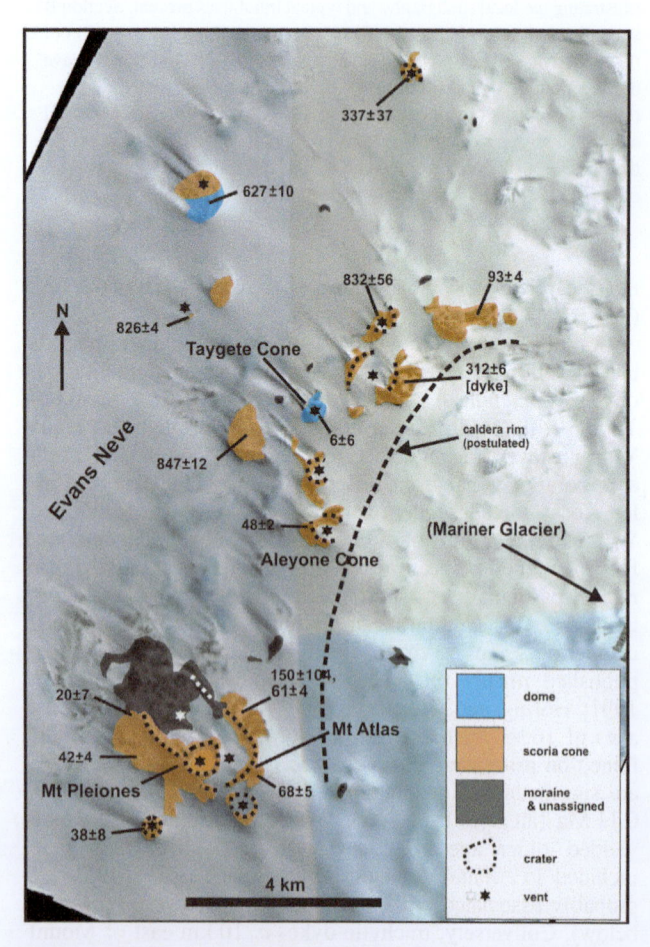

Fig. 22. Geological map of Pleiades Volcanic Field (after Kyle 1990c; isotopic ages (in ka) after Esser and Kyle 2002).

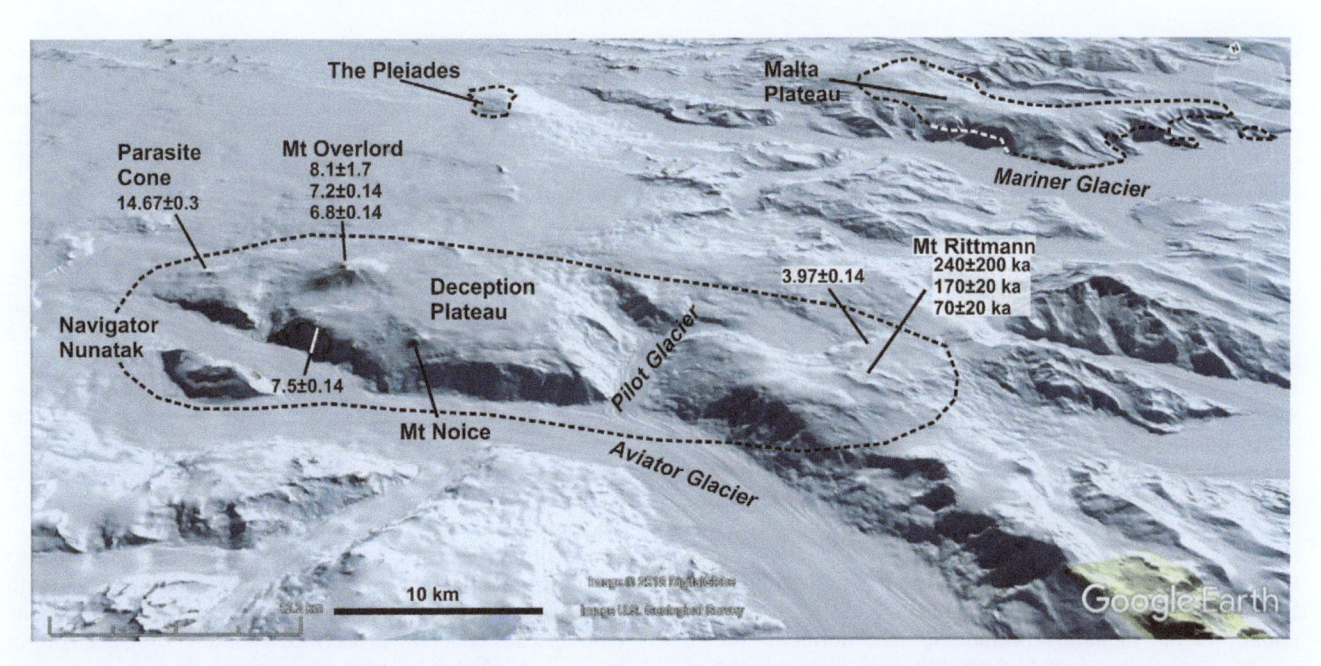

Fig. 23. Google Earth perspective satellite view of the Mount Overlord Volcanic Field, looking north, with places mentioned in the text. Isotopic ages are also shown (in Ma: except for Mount Rittmann which is in ka); the precise location of the dated sample overlooking Pilot Glacier is uncertain. The Pleiades and Malta Plateau are indicated in the background. Image: Google Earth (2018 Digital Globe and US Geological Survey).

on the south side of the outcrop, and weakly developed flow banding consistent with an endogenous origin. The low-lying outcrop situated c. 4 km to the NW of Taygete Cone (the C11 outcrop of Kyle 1982) may also be a dome. It is mainly composed of phonolite but is mantled by basanite lava that forms a semi-circular, possibly crater-like, feature on its east side. Ridges NW of the Mount Atlas–Mount Pleiones cone cluster, originally described by Nathan and Schulte (1968) as the glacially planed-off remnant of an older trachytic volcano, are simply moraines formed entirely of locally derived volcanic clasts, some of which show fine glacial striations (Kyle 1982); further morainic debris crops out within the crater on the NW side of Mount Atlas (Fig. 22) (unpublished observation of J.L. Smellie). Finally, trachyte pumice lapilli are common on several nunataks (Nathan and Schulte 1968; Kyle 1982). They also occur as abundant large broken bombs on the small nunatak c. 1200 m north of Aleyone Cone (outcrop C6 of Kyle 1982), suggesting that it is close to the source vent (unpublished observation of J.L. Smellie).

The presence of several tephra layers recovered in ice cores, with compositions similar to rocks in the Pleiades (or possibly Mount Rittmann), suggests that the Pleiades might have erupted as recently as AD 1254 ± 2 and on several occasions possibly back to c. 123 ka (Narcisi et al. 2001, 2016). Numerous Ar–Ar isotopic ages have also been published for the Pleiades, ranging from 847 ± 12 to 6 ± 6 ka (Esser and Kyle 2002) (Table 2). The youngest age obtained so far was obtained on the trachyte dome at Taygete Cone, which suggested to Kyle (1990c) that the Pleiades should be regarded as dormant. However, Kyle (1982) described five areas of hydrothermal alteration at Taygete Cone, each 3–6 m in diameter and showing a zonal pattern of pervasive centroclinal bleaching, that are present near the summit of the outcrop. Therefore, the reliability of the age may be in doubt due to the alteration possibly affecting the dated sample (acknowledged by Esser and Kyle 2002). The oldest dated outcrops include most of the cones north of Aleyone Cone and east of Taygete Cone, which have very similar ages of 847 ± 12–826 ± 4 ka. Following that activity, eruptions took place at: (1) the small unnamed phonolite dome (627 ± 10 ka); (2) the northernmost outcrop (337 ± 37 ka); and (3) a dyke intruding the cone east of Taygete Cone (312 ± 6 ka) (Fig. 22). A gap of c. 200 kyr then intervened before the next stage of activity, at 93 ± 4 ka in the outcrop c. 4 km east of Taygete Cone. Activity then relocated to the Mount Atlas–Mount Pleiones cone cluster, with dated eruptions at c. 65 ka (ages of 68 ± 5 and 61 ± 4 ka, and an imprecise age of 150 ± 104 ka). The next eruptive event occurred at Aleyone Cone, dated at 48 ± 2 ka, which is almost identical to ages of 42 ± 4–38 ± 8 ka for flank vent(s) on the south side of Mount Pleiones. Activity in the Mount Atlas–Mount Pleiones cone cluster may finally have ceased at 20 ± 7 ka, which is the age obtained on a sample from the northwestern limit of the outcrop.

Mount Overlord Volcanic Field. The Mount Overlord Volcanic Field includes outcrops at Mount Overlord, Parasite Cone and Mount Noice, the western margin of the Deception Plateau, inaccessible cliff exposures overlooking Aviator Glacier, and possibly those overlooking Pilot Glacier, Mount Rittmann and at Navigator Nunatak (Fig. 23). Mount Overlord, discovered by Gair (1964, 1967), is a prominent well-preserved small stratovolcano situated east of Aviator Glacier, on the northern margin of Deception Plateau. It has not been re-examined geologically since January 1982 (Noll 1985; Kyle and Noll 1990). The volcano has a basal diameter of just c. 6 km, rises c. 500 m above the surrounding ice to 3396 m asl and has a prominent ice-filled summit caldera 800 m across. Although largely undissected, the caldera is lower on its eastern side and has shallow breaches there. Nathan and Schulte (1968) tentatively suggested that a low arcuate ridge on the northern flank of Mount Overlord might also be an older 'cone rim' (caldera?) but its status is doubtful (Noll 1985; Kyle and Noll 1990). Because of its relatively fresh appearance, the volcano was initially thought to be very young (a few hundred years: Gair 1967) but subsequent isotopic dating indicated that it has an age of 7.2 ± 0.14–6.8 ± 0.14 Ma (by K–Ar: Armstrong 1978; Kyle and Noll 1990). Several small scoria cones (flank vents) are present on the northwestern flank but rock exposures are few and most of the snow-free ground is described as talus (i.e. scree). However, the volcano-flank gradients are low (c. 15°) and it is probably mainly a surface accumulation, formed more or less *in situ* but with some downslope creep, created by frost action down joints and other fractures, and leading to the mechanical disintegration of the intact rock (i.e. felsenmeer or a block field). One of the cones is located on the northern rim of the caldera; it issued lava(s) that flowed down the outer slopes and also into the caldera (Nathan and Schulte 1968; Noll 1985).

The lithofacies exposed in the western interior caldera wall of Mount Overlord are relatively well exposed (Nathan and Schulte 1968). They consist of mugearite–phonotephrite and trachyte lavas individually >10 m thick, capped by further thinner mugearitic phonotephrite and trachyte lavas. The lower trachyte is pale grey, pumiceous and flow banded, whereas the capping trachyte, which also drapes the caldera rim, is brown to black and glassy, and may also form the summit of Mount Overlord. Elsewhere, the caldera rim on its east side consists of benmoreite and rhyolite (comendite) lavas, and a comenditic welded tuff with large fiammé, which is intensely devitrified and texturally destroyed in thin section, and is interpreted as a fall deposit. The rim on its south and SW sides comprises up to nine discrete lavas, mainly tephriphonolites but including benmoreite. The northern caldera rim exposes an altered trachyte possibly equivalent to that capping the western caldera wall and rim (summit), and a thin trachyte lava; debris of comenditic welded tuff is also present. Nathan and Schulte (1968, citing Watters in Gair 1967) also recorded two specimens of ignimbrite collected loose at Mount Overlord. Ignimbrite is uncommon in Neogene volcanic outcrops across Antarctica and this may be the first recognition of ignimbrite on the continent. Finally, several blocks of alkaline gabbro and sodalite-bearing syenite, up to 30 cm in diameter, were collected from the caldera rim. Their presence may be related to the caldera-forming event and ejection of ignimbrite, although none of these rocks was observed *in situ*.

Although small cones situated on the NW side of Mount Overlord were described as belonging to two types by Noll (1985), the differences appear to be small, with the lower cones slightly more eroded than those higher up. Together, they are formed of dark-brown to black or red (oxidized) scoria with basalt to phonotephrite, mugearite and tephriphonolite–benmoreite compositions. The primary cone shapes and craters are generally well preserved.

Mount Noice is a poorly exposed small nunatak situated at the southwestern edge of Deception Plateau, *c.* 14 km SSE of Mount Overlord (Fig. 23). The summit is at *c.* 2780 m asl, and it rises 100 m above the surrounding snow and ice. It is mainly composed of rhyolite (comendite) lavas, mostly silica-poor (*c.* 69 wt%), that plot close to the boundary with trachyte and are compositionally dissimilar to other lavas found so far in the Mount Overlord Volcanic Field. Obsidian is also present and is more silica-rich (*c.* 73 wt%) and comenditic, similar to comendites on the east caldera rim on Mount Overlord.

There are also several subdued exposures at the western edge of Deception Plateau 7 km SW of Mount Overlord, overlooking Aviator Glacier. Described as 'welded tuff outflow sheets' (Noll 1985), the rocks have trachyte and phonolite compositions, and may be the welded and unwelded tuffs and ignimbrites mentioned by Kyle and Noll (1990), who suggested that similar rocks may also underlie the broad flat snow-covered expanse of Deception Plateau.

Although the felsic rocks are undated, an age of 7.5 ± 0.14 Ma was determined on a basalt overlooking Aviator Glacier and, despite the compositional difference, may provide a minimum age for the Deception Plateau volcanism. The obvious similarity in age with volcanism in Mount Overlord suggests that Mount Overlord may be the source for the Deception Plateau volcanism.

Parasite Cone is an isolated small volcanic centre. It rises to *c.* 2340 m asl and is situated *c.* 14 km NW of Mount Overlord. The nunatak is formed of brown vitroclastic rock, hawaiite in composition, showing local channel-like structures. It also contains basement clasts and is intruded by a dyke. Although described as hyaloclastite, it is probably a lapilli tuff and the outcrop a tuff cone. It was ascribed a possible subglacial origin by Noll (1985) but with no supporting evidence. A juvenile basalt clast from the lapilli tuff yielded an isotopic age of 14.67 ± 0.3 Ma (by K–Ar: Noll 1985), which is the oldest age obtained from the Mount Overlord Volcanic Field (Table 2).

The inclusion within the Mount Overlord Volcanic Field of volcanic outcrops capping basement rocks on Navigator Nunatak is tentative and may only apply to the trachytic outcrop. They have not been described in detail and none of the rocks on Navigator Nunatak are dated. They were observed by Gair (1967), who did not examine them directly but tentatively suggested that they were derived from Mount Overlord, whereas they were regarded by Nathan and Schulte (1968) as three small eruptive centres belonging to the Local Suite. The only published description is by Hornig and Wörner (2003), who noted volcanic rocks overlying basement granodiorite at elevations of *c.* 1400 and 1700 m asl (Fig. 24). In the southern outcrop, a lower section dipping at *c.* 30° is composed of vividly coloured red and white strongly altered platy trachyte and breccia, including a breccia composed of white trachyte scoria with basement clasts; some of the breccia clasts have hawaiite compositions (Hornig and Wörner 2003). A visit in 2005 by the authors found off-white and brownish-green to pale grey-green felsic (trachytic?) rocks with a variable but often steep fissile jointing and foliation that wraps around sparse small cognate(?) enclaves. The outcrops are often pervasively broken, comprising massive fines-poor blocky breccia variably jigsaw-fractured to chaotic. They are monomict, except for a single granitoid block seen. The breccia clasts have a sugary appearance, often with pea-like textures

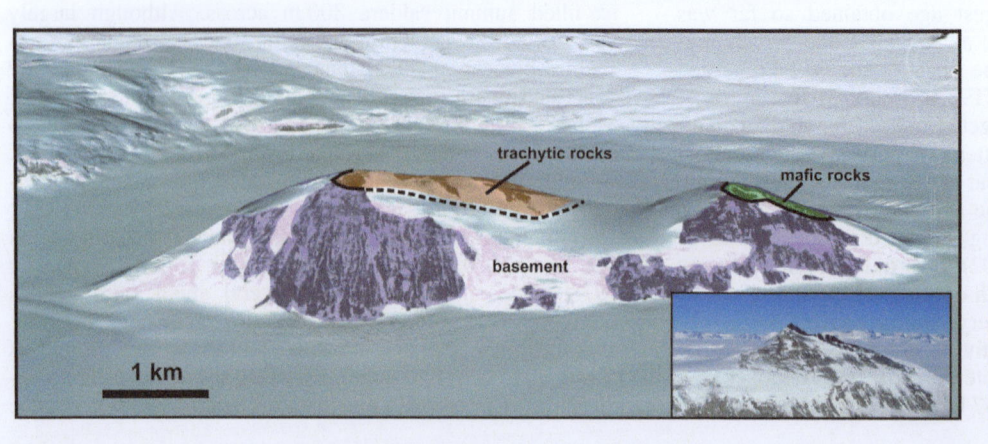

Fig. 24. Google Earth perspective satellite view of Navigator Nunatak looking SW, with the geology indicated. The inset shows the outcrop of dark mafic rocks (mainly pillow lava) draping pale-coloured basement at the northern end of the nunatak. Image: Google Earth (2020 Maxar Technologies and US Geological Survey).

resembling perlite. Both features suggest that they may be formed of pervasively devitrified felsic glass (cf. similar textures in rhyolite at Mandible Promontory described by Smellie et al. 2011a). Although some of the outcrops are apparently dykes, the volcanic country rock may be formed principally of the breccias. The relationships are unclear but the rocks may be brecciated trachytic ignimbrites intruded by trachytic dykes. No bedding was observed and the reasons for the brecciation are unknown. The occurrence of felsic rocks with trachyte compositions, capping basement and showing significant erosion suggests that they may be related to the Deception Plateau sequences and those exposed in the inaccessible cliffs overlooking Navigator Nunatak, which are broadly similar (also suggested by Hornig and Wörner 2003). In this scenario, the trachytic outcrops at Navigator Nunatak are an outlier separated from Deception Plateau by Aviator Glacier. Gair (1967) speculated that the rocks may have travelled across the surface of Aviator Glacier (presumably much thicker at the time) and therefore might be as young as Mount Overlord, which is plausible. If Navigator Nunatak was covered by relatively thick snow when it was draped in ignimbrite (conceivable if the surface elevation of Aviator Glacier was much higher than present), the subsequent uneven collapse of the ignimbrite deposit as the snow melted might be a plausible explanation for the pervasive fracturing and brecciation observed in the ignimbrite, particularly if the underlying basement surface is very uneven. By contrast, the northern summit ridge on Navigator Nunatak is composed of 250 m of dark basanite pillow lava and associated hyaloclastite breccia that includes a layer of blocky breccia with a palagonitized matrix in part and shows several minor faults, said to be volcanotectonic. The mafic sequence was said to be essentially horizontal (Hornig and Wörner 2003). Although unvisited by us, our observations suggest that it drapes a north-dipping surface cut in basement rocks (Fig. 24). The pillowed sequence changes down into subaerial lavas with intercalated epiclastic horizons. The characteristics of the mafic outcrops were thought by Hornig and Wörner (2003) on *a priori* grounds to signify a change from subaerial up into subglacial or subaqueous conditions, consistent with changes in the thickness of a coeval ice sheet. From their characteristics, we consider that the mafic rocks probably formed in one or, perhaps, two small volcanic centres and a time gap must have intervened to have caused the evidence of the contrasting environmental conditions. They may be part of the Local Suite (see below).

Mount Rittmann is also included in the Mount Overlord Volcanic Field (Armienti and Tripodo 1991). It was discovered in 1988–89, and is largely covered by snow and ice. There are only a few small rock exposures, several of which are interpreted as defining a small caldera c. 2 km in diameter that rises only a few tens of metres above the surrounding, almost flat, ice field (Fig. 23). The outcrops are mainly trachytic lavas and hawaiite scoria, and the final eruptive event formed a polymict breccia with numerous clasts of basement granite and metamorphic rocks, together with hydrothermally altered cogenetic 'sub-intrusive' (sic: plutonic?) rocks and phonolite pumice. The polymict nature of the breccia deposit, including plutonic debris and evolved pumice, suggests that might be related to the caldera-forming event. The original outcrop area of the volcano is uncertain although it may extend to the north to include largely inaccessible cliff exposures overlooking Pilot Glacier and in the west overlooking Aviator Glacier. Pilot Glacier exposures show a series of thick grey trachytic sheet lavas with oxidized surfaces overlying orange-coloured planar-stratified probable lapilli tuffs (Armienti and Tripodo 1991; unpublished observations of the authors). The latter were called hyaloclastites by Armienti and Tripodo (1991) but they are probably pyroclastic density current deposits. Some lava has also intruded the lapilli tuffs, pillow-like, and a sample of the subaerial capping lavas has yielded an isotopic age of 3.97 ± 0.14 Ma (by K–Ar: age given by Bonaccorso et al. 1991, attributed to Armienti and Tripodo 1991) (Table 2). K–Ar isotopic ages of 240 ± 200, 170 ± 20 and 70 ± 20 ka were obtained on phonolitic and trachytic lavas at Mount Rittmann (unpublished information of G. Vita cited by Perchiazzi et al. 1999). Considerable interest in the volcano arises from the presence of actively steaming ground caused by weak fumarolic activity in small vents only a few centimetres wide, and ground temperatures of c. 50–63°C at 10 cm depth (Bonaccorso et al. 1991; Bargagli et al. 1996). The heated area occupies a subvertical slope measuring c. 200 m wide by 80 m high. Gases in the fumaroles are only faintly sulfurous and may be composed predominantly of CO_2. The warm ground is consistent with the presence of cooling magma near the surface and, together with the young isotopic ages, suggests that the volcano may still be active. Conversely, no primary landforms are preserved other than a relatively weak caldera-like structure barely visible above the local ice surface, and a present-day landscape that suggests that the volcano has undergone extensive erosion inconsistent with recent activity. However, ice cores have yielded englacial tephra layers with ages between 124 ka and AD 1254 with compositions similar to rocks at Mount Rittmann or possibly the Pleiades (Narcisi et al. 2001, 2016). The published analyses show that the lavas are mainly trachytes with fewer hawaiites, and rare mugearite, benmoreite and phonolite. It is possible that the outcrops overlooking Pilot Glacier that yielded a much older age (3.97 Ma: attributed to Mount Rittman) are, in fact, related to a different centre responsible for some of the plateau ignimbrite volcanism of Deception Plateau (see above), whereas the much younger isotopic ages and fumarolic activity more accurately date the Mount Rittmann volcano.

Vulcan Hills Volcanic Field. Vulcan Hills is a small isolated volcanic and plutonic outcrop situated c. 65 km SSW of Mount Overlord (Fig. 1). It is very poorly known and may have been visited only twice, at reconnaissance level (Nathan and Schulte 1967; Kyle 1990d). The outcrop is described as gently dipping interbedded basaltic (hawaiite) and trachytic lavas overlying Jurassic basalts of the Kirkpatrick Basalt Group. However, although trachyte and hawaiite lavas were confirmed by Kyle (1990d), one of which overlies gabbro, the identification of Jurassic lavas was disputed and they may be gabbroic rocks related to the Vulcan Hills sequence (Kyle 1990d). The outcrop also contains polymict volcanic breccia (G. Giordano pers. comm.). The lavas, gabbroic rocks and associated dykes have a wide range of sodic alkaline compositions ranging from basalt to trachyte, and they are interpreted as the strongly dissected remains of a large volcano. They are generally more altered than other volcanic rocks in the Melbourne Volcanic Province. All but two of the published analyses are for intrusions. There are no published isotopic ages. However, by comparison with Mount Overlord, which is 7 Ma in age and is neither significantly eroded nor altered, the unusually altered and strongly eroded state of the Vulcan Hills outcrops suggests that they may be much older (>10 Ma?: Nathan and Schulte 1968; Kyle 1990d). Assuming that it was originally part of a much larger central volcano, the much-reduced present-day outcrop and the pronounced alteration are in stark contrast with volcanic outcrops <10 Ma in age in the northern Victoria Land region. Speculatively, therefore, Vulcan Hills might contain the sole preserved relict of the volcanic cover associated with the plutonic–hypabyssal activity of the Eocene–Oligocene Meander Intrusive Group, with which it might be correlated (Fig. 2 and see below).

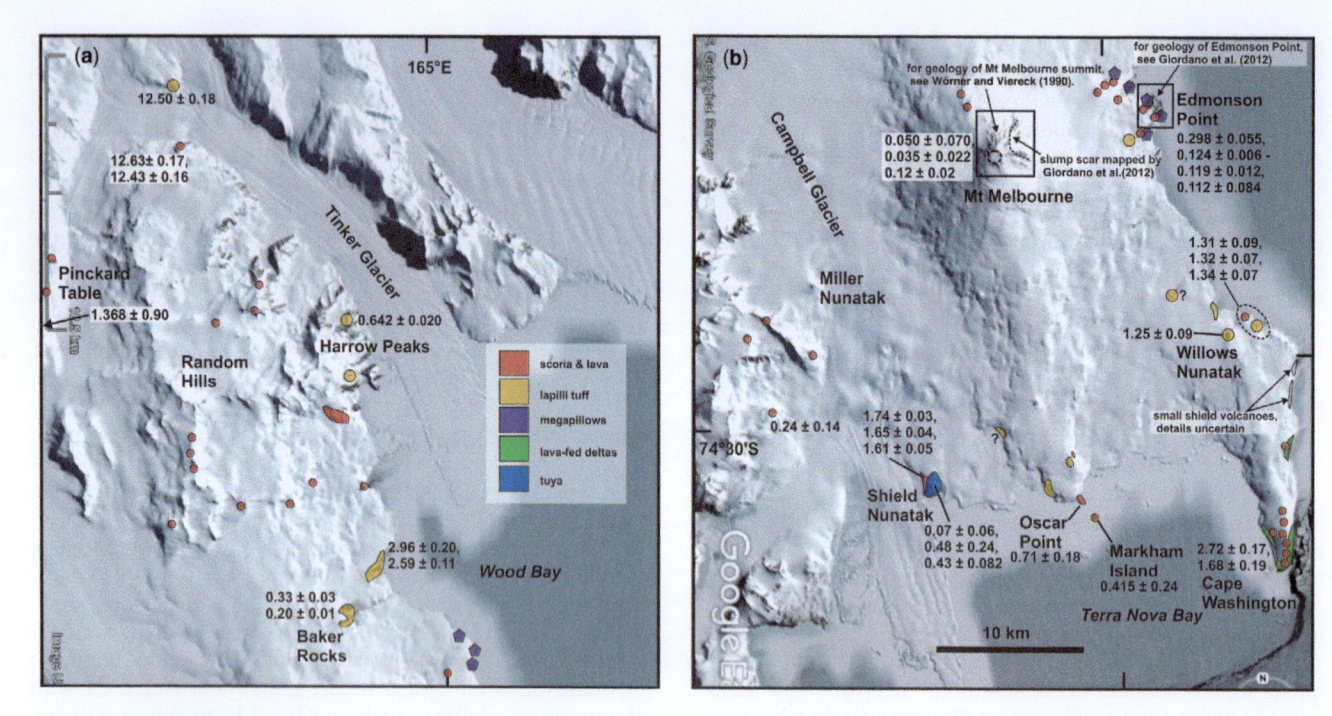

Fig. 25. Simplified geological map of the Mount Melbourne Volcanic Field, based on Armienti *et al.* (1991), Giordano *et al.* (2012) and unpublished information of the authors. The locations and ages of dated samples are also shown (in Ma: see Table 2 for the sources). Image: Google Earth (US Geological Survey).

Mount Melbourne Volcanic Field. The Mount Melbourne Volcanic Field occupies the prominent peninsula extending from Cape Washington north to Tinker Glacier, and flanked to the west and east by Campbell Glacier and Wood Bay, respectively (Fig. 25a, b). It comprises Mount Melbourne, the eroded remnants of two older volcanic shields and numerous small mafic pyroclastic cones (Wörner and Viereck 1989; Giordano *et al.* 2012; unpublished information of the authors). Mount Melbourne is a prominent stratovolcano with a basal diameter of *c.* 21–24 km and summit at 2732 m asl (Fig. 26). It is largely blanketed by snow and ice, except for the summit region from which the rock outcrop extends downslope on the east side to *c.* 1800 m. The very youthful appearance of the volcano, with its undissected flanks, apart from a possible slump scar on the east side (Giordano *et al.* 2012) and a well-formed ice-filled crater *c.* 700 m in diameter (Armienti *et al.* 1991 and Giordano *et al.* 2012 interpreted it as a caldera), suggests that it is very young. This is confirmed by the presence of heated ground and fumaroles both in the crater and 250 m downslope from the summit on its northwestern side, and by the presence of englacial tephra layers found in surrounding ice cliffs (Adamson and Cavaney 1967; Nathan and Schulte 1967; Lyon and Giggenbach 1974; Lyon 1986; Wörner and Viereck 1989). The fumaroles on the northern slope are associated with ice towers 1–6 m in diameter and up to 4 m high, which are similar in appearance and scale to those seen on Mount Erebus (Lyon and Giggenbach 1974; Giggenbach 1976). Air temperatures of 38.5°C have been measured in the voids below the ice towers, and ground temperatures of 59°C. Multiple englacial tephras found in ice cores also suggest that eruptions of Mount Melbourne have been numerous and may extend back to Eemian time at least (123 ka: Narcisi *et al.* 2001, 2016; see also Del Carlo *et al.* 2015). Additionally, although Mount Melbourne has never been observed erupting, an

Fig. 26. Aerial view of the Mount Melbourne Volcanic Field, looking NW. The view shows Cape Washington in the left foreground and the beautiful snow-clad stratocone of Mount Melbourne, which dominates the scenery. The numerous nunataks at lower elevations surrounding Mount Melbourne are the remnants of small monogenetic volcanic centres.

eruption may have occurred as recently as c. AD 1892–1922 based on tephra layers in coastal ice cliffs, and ice tongues of Tinker and Aviator glaciers (Nathan and Schulte 1967; Lyon 1986; Wörner and Viereck 1989). It is therefore regarded as an active volcano (see also Chapter 7.3: Gambino et al. 2021).

Compositionally, the products of Mount Melbourne found in the summit area consist mainly of alkali trachyte and benmoreite, with lesser occurrences of alkali basalt (Wörner et al. 1989; Giordano et al. 2012). The summit region is mainly composed of clusters of trachytic domes and relatively short thick trachytic lavas, which may be rheomorphic effusions from domes situated on the steeper flank surfaces (Wörner et al. 1989; Giordano et al. 2012). There are also two mafic lava fields and prominent well-formed scoria cones, together with a series of three nested craters built of mafic phreatomagmatic tephra (presumably tuff cones) on the southern crater rim. Numerous outcrops of slightly to strongly welded trachytic pumice fall deposits are also present. Some grade into welded agglutinate, and develop crude columnar cooling joints. Petrographical differences between the trachytic fall deposits suggest that they were erupted during different local eruptive events. They include the most recent activity, comprising several layers of trachytic pumice, one of which forms a distinctive black and white unit 5 m thick on the eastern rim of the crater, and basaltic bombs that are scattered around the summit area. Although some other tephra layers show a similar colour variation, the crater-rim deposit can be traced to the east and south where it is found as erosional remnants on the summit and lower flanks, including in the ice cliff NW of Edmonson Point (Wörner and Viereck 1989). Isotopic (Ar–Ar) dating of trachytic spatter and pumice from Mount Melbourne summit gave very imprecise ages of 50 ± 70 and 35.1 ± 21.9 ka (Giordano et al. 2012; see also Armstrong 1978) (Table 2).

Evidence of earlier volcanic-shield edifices is found at two principal locations: (1) the peninsula extending north from Cape Washington; and (2) scattered outcrops at and inland of Edmonson Point and in coastal exposures up to 9 km to the north (Fig. 25a, b). The Cape Washington outcrops consist of a sequence of at least five relatively thin tephrite 'a'ā lava-fed deltas that dip gently to the SW (unpublished information of the authors). They are exposed in 'windows' eroded through spectacular intrusive necks filled by basanite intrusive sheets and oxidized scoria. The younger units also form scoria cones with craters on top of the peninsula, and it appears that they were emplaced as a series of later Strombolian centres that were erupted from a north-trending fissure. Isotopic ages suggest that the older shield formed of lava-fed deltas is late Pliocene in age (2.72 ± 0.17 Ma), whereas the Strombolian centres are early Quaternary (1.68 ± 0.19 Ma; both ages by K–Ar: Müller et al. 1991; sample locations assigned to Cape Washington by Wörner and Viereck 1989) (Table 2). The outcrops associated with Edmonson Point and to the north are very different. They consist of spectacular stacked hawaiite and benmoreite megapillows (Wörner and Viereck 1989; Wörner and Orsi 1990; Giordano et al. 2012). The individual megapillows vary in size from a few metres wide and high to a few tens of metres high and >200 m in length, with well-developed cooling joints that range from colonnade (minor) to entablature and blocky (unpublished observations of the authors). The additional presence of rare ropy lava textures on some of the megapillows, together with minor interstratified pumiceous and hyaloclastite debris with lava pillows, is hard to reconcile in terms of an eruptive setting. Giordano et al. (2012) considered that the megapillows were fed by local dykes and, like Wörner and Viereck (1989), suggested that the megapillows change up into red and black scoria at c. 150 m asl, which Wörner and Viereck (1989) considered to represent a transition from subglacial to subaerial conditions. Giordano et al. (2012) favoured an eruptive setting involving variable water interaction derived from melted snow and ice but neither study provided supporting evidence for a glacial environment. An erosional surface intervenes between the megapillow unit and the overlying subaerial lavas with a prominent planar unconformity (predicted by Giordano et al. 2012) exposed inland c. 4 km NW of Edmonson Point (unpublished observation of the authors). Isotopic dating by Giordano et al. (2012) suggests that the megapillow unit is c. 297.6 ± 55.4 ka in age. The Edmonson Point stratigraphy is also distinguished by the presence of ignimbrites, and a distinctive yellow scoria and tuff cone. The ignimbrites were initially interpreted by Wörner and Viereck (1989) and Wörner and Orsi (1990) as near-vent pumice fall intercalated with ignimbrites and overlain by reworked pumiceous benmoreitic material from different vent(s) situated to the north, and the tuff cone as a complicated tuff ring but Giordano et al. (2012) reinterpreted the outcrops as ignimbrites and tuff cone deposits, respectively. The ignimbrites have yielded $^{40}Ar/^{39}Ar$ ages of 124.3–118.6 ka, with small errors (12.6–6.4 ka: Giordano et al. 2012) (Table 2), which conflicts with its inferred stratigraphical position at the base of the sequence (i.e. older than the megapillow unit). The youngest units at Edmonson Point are mafic 'a'ā lavas with oxidized surfaces and a series of eroded scoria cones dated as 111.6 ± 83.8 ka (Giordano et al. 2012). Most of the published ages for Edmonson Point, and the Mount Melbourne Volcanic Field generally, are very imprecise and should simply be regarded as indicative (Table 2).

All other volcanic outcrops in the Melbourne Volcanic Field consist of numerous small isolated scoria cones, lavas and 'tuff rings' (i.e. tuff cones: cf. Wörner and Viereck 1989; Giordano et al. 2012). Although having the appearance of monogenetic edifices, at least some of the centres are polygenetic (e.g. Shield Nunatak: Wörner and Viereck 1989). Most have alkali basalt compositions but a wide compositional range is present, including basanite–tephrite, hawaiite, mugearite and benmoreite; ages range from 12.63 ± 0.17 to 0.71 ± 0.18 Ma (Wörner and Viereck 1989; Armienti et al. 1991; Giordano et al. 2012; Smellie et al. 2018) (Fig. 25b). The most distinctive outcrop is that at Shield Nunatak (Fig. 5e), a relatively flat-topped steep-sided landform c. 2.2 km wide and rising up to 300 m interpreted by Wörner and Viereck (1987) as a small polygenetic subglacial table mountain (i.e. a tuya) with an alkali basalt composition and published imprecise $^{40}Ar/^{39}Ar$ ages of 480 ± 240–430.5 ± 82 ka (Armienti et al. 1991; Giordano et al. 2012). It is also noteworthy for being the only tuya identified so far in northern Victoria Land. The outcrop rests on a much older lava (1.74 ± 0.03–1.61 ± 0.05 Ma by K–Ar: Müller et al. 1991) (Table 2), whose upper surface shows glacial striations and a thin tillite; the latter features were used as support for a subglacial origin for the nunatak. The well-dated lapilli tuff outcrop at Harrow Peaks is described later (see the following 'Northern Local Suite' subsection).

Northern Local Suite. As described earlier, the products of small isolated eruptions that are not part of the major central volcanoes in the Melbourne Volcanic Province were grouped together by Nathan and Schulte (1968) into their Local Suite, and that name was continued by Kyle (1990e). Smellie et al. (2011a) extended its use to include comparable examples inland all the way north to Adare Peninsula. Figure 27 shows the distribution of the outcrops known so far. However, essentially identical rocks are also widespread in the Erebus Volcanic Province (in the Royal Society Range and Dry Valleys: Wright and Kyle 1990a, b; Wilch et al. 1993; see also Chapter 5.2a: Smellie and Martin 2021). As the latter are

Nathan and Schulte 1968; Hamilton 1972). However, the descriptions are very similar, and most of the outcrops consist of amorphous piles and cones of scoria formed during Strombolian and Hawaiian eruptions. They are composed of black and red (oxidized) scoria, welded maroon agglutinate and bombs up to 2 m long, and subaerial lavas and autoclastic breccias with mafic compositions (unpublished information of the authors). Other related outcrops simply consist of mafic lava. Well-formed craters with negligible erosion characterize several (e.g. scoria cones south of Miller Nunatak, in the Northern Foothills east of Mount Abbott and on the summits on the south side of Murray Glacier). The outcrops unconformably overlie basement rocks but the contact is frequently unexposed. One tiny isolated outcrop in the centre of Gruendler Glacier consists of 'a'ā lava with oxidized autobreccia (unpublished information of the authors). In addition, there are a few outcrops that are different. They comprise the following:

- Tuff cones are extremely rare and the best-described example is a highly degraded outcrop at Harrow Peaks (Fig. 5d) (Wörner and Viereck 1989; Giordano *et al.* 2012), which Smellie *et al.* (2018) interpreted as the remains of a small monogenetic phreatomagmatic centre erupted beneath cold-based ice – the first example of its type to be described anywhere. It has a published age of 0.642 ± 0.020 Ma (by Ar–Ar: Smellie *et al.* 2018). Two other outcrops, said to be similar, were also mentioned by Giordano *et al.* (2012) situated just south of the Harrow Peaks tuff cone and at the confluence of Tinker and Burns glaciers (Fig. 25a). The latter is dated as 12.50 ± 0.18 Ma (by K–Ar: Armienti *et al.* 1991) (Table 3).
- Pillow lava 250 m thick crops out at the north end of Navigator Nunatak. It overlies subaerial lava with epiclastic horizons (Hornig and Wörner 2003). The relationships were interpreted to suggest initially subaerial eruption changing up into subglacial. It is unclear if the outcrop is part of the Northern Local Suite but a correlation is likely based on the mafic composition and gross differences with nearby trachytic volcanic deposits.
- An isolated mafic lava outcrop is present 2 km NW of Crater Cirque, *c.* 30 m above the present surface of Tucker Glacier. It was discovered and briefly described by Harrington *et al.* (1967), and has an age of 7.605 ± 0.049 Ma (Smellie *et al.* 2011b) (Fig. 15; Table 3). The local sequence is described in greater detail by Smellie *et al.* (2011a), and consists of >40 m of entablature-jointed basalt lava overlying a few metres of lapilli tuff or gravelly volcanic sandstone and poorly exposed polymict diamict with abraded and facetted clasts, which together rest on a polished and striated surface cut in basement granodiorite. The sequence was interpreted as a mafic sheet-like sequence (*sensu* Smellie and Edwards 2016) probably erupted below a wet-based palaeo-Tucker Glacier. It is the only example discovered so far in Victoria Land.
- Brownish-orange planar-stratified fine lapilli tuffs are exposed at Eldridge Bluff, a 5 km-long largely inaccessible rock cliff, *c.* 100–150 m high, overlooking Aviator Glacier *c.* 25 km SE of Navigator Nunatak. It is one of several outcrops in the region that have been greatly affected by fracturing related to transtensional tectonics (Vignaroli *et al.* 2015). The lapilli tuffs locally contain numerous blocks and bombs up to 2 m in diameter, which implies that they are close to the source, and they are overlain conformably by horizontal thin dark-grey sheet lavas (unpublished information of the authors). The relationships and lithologies suggest that the outcrop may be part of a small monogenetic basaltic shield volcano with a tuff cone core mantled by subaerial lava. Although viewed mainly from a helicopter, the lithologies,

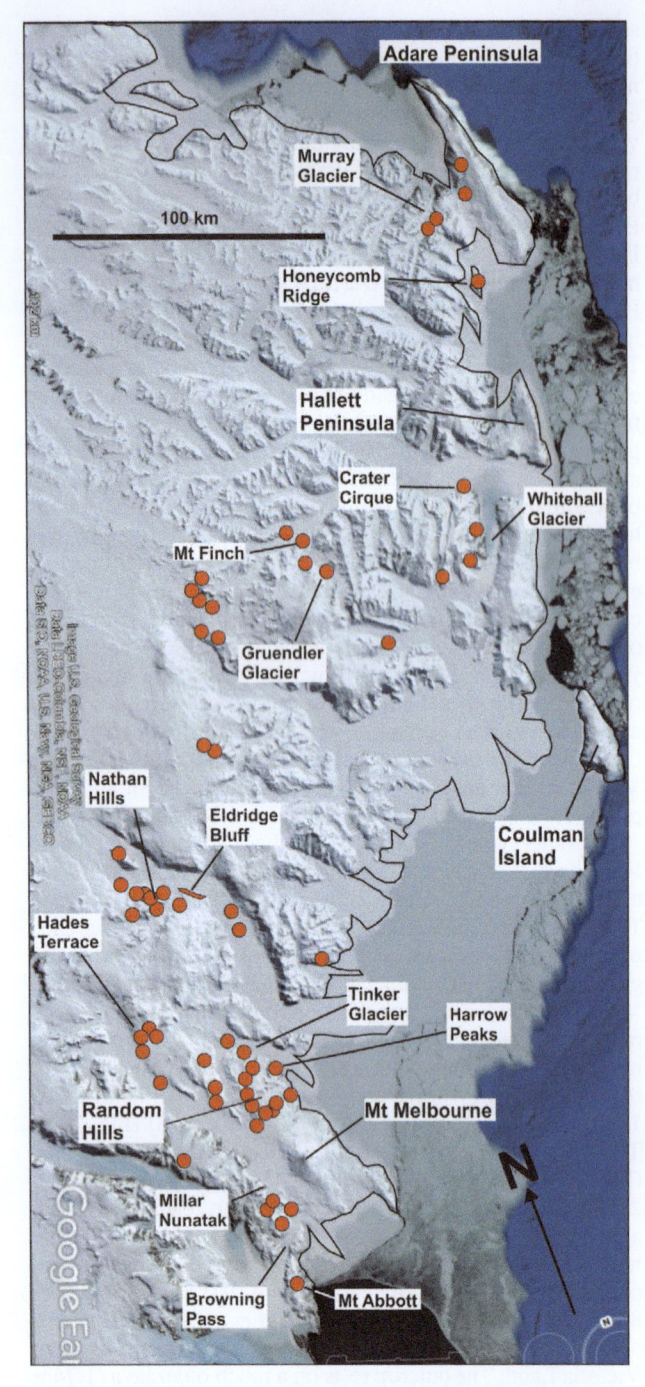

Fig. 27. Satellite image of northern Victoria Land showing the distribution of outcrops assigned to the Northern Local Suite (after Hamilton 1972; Kyle 1990e and references therein; unpublished information of the authors), and localities mentioned in Table 3. Satellite image courtesy of Google Earth (data provider US Geological Survey).

separated from the northern Victoria Land outcrops by *c.* 400 km of coastline lacking any similar outcrops, it would be useful to have a terminology to both relate and distinguish between the two groups. Accordingly, those included in this chapter will be called the Northern Local Suite and those further south in the Erebus Volcanic Province will be called the Southern Local Suite. Both have volcanic field status (see Fig. 2).

Outcrops of the Northern Local Suite were recorded during all of the early geological investigations but they were seldom examined closely (often mapped from helicopters) or described in detail (e.g. Gair 1967; Harrington *et al.* 1967;

Table 3. *Summary of published isotopic ages of small volcanic centres in northern Victoria Land*

Sample	Locality	Age (Ma)	Error	SD	Method	Source	Notes
35421	Hades Terrace	4.3	0.2	1σ	K–Ar	Armstrong (1978)	
35416	Nathan Hills	18	0.7	1σ	K–Ar	Armstrong (1978)	Authors believed the age to be anomalously old
MB 01a	Oscar Point, MMVF[†]	0.71	0.18	n.r.	K–Ar	Armienti *et al.* (1991)	Mount Melbourne satellite centre?
MB 26a	Shield Nunatak (summit sequence), MMVF	0.48	0.24	n.r.	K–Ar	Armienti *et al.* (1991)	Mount Melbourne satellite centre?
MB 65a	Browning Pass (MMVF)	0.24	0.14	n.r.	K–Ar	Armienti *et al.* (1991)	
MB 33a	North of Baker Rocks, MMVF	2.96	0.20	n.r.	K–Ar	Armienti *et al.* (1991)	Mount Melbourne satellite centre?
MB 33b	NE of Baker Rocks, MMVF	2.59	0.11	n.r.	K–Ar	Armienti *et al.* (1991)	Mount Melbourne satellite centre?
MB 61a	Tinker Glacier, MMVF	12.50	0.18	n.r.	K–Ar	Armienti *et al.* (1991)	
MB 61b	Tinker Glacier, MMVF	12.50	0.18	n.r.	K–Ar	Armienti *et al.* (1991)	
MB 63a	Random Hills, MMVF	12.63	0.17	n.r.	K–Ar	Armienti *et al.* (1991)	
MB 62a	Random Hills, MMVF	12.43	0.16	n.r.	K–Ar	Armienti *et al.* (1991)	
MP8	Whitehall Glacier-Martin Hill, DPVF[‡]	12.2	0.3	1σ	K–Ar	Armienti *et al.* (2003)	
G83	Harrow Peaks, MMVF	0.745	0.66	2σ	Ar–Ar	Giordano *et al.* (2012)	
G26	Pinckard Table, MMVF	1.368	0.090	2σ	Ar–Ar	Giordano *et al.* (2012)	
2700	Mount Abbot	0.656	0.015	2σ	K–Ar	Müller *et al.* (1991)	
7095	Mount Abbot	0.599	0.15	2σ	K–Ar	Müller *et al.* (1991)	
MB 65a	Browning Pass	0.24	0.14	n.r.	K–Ar	Armienti *et al.* (1991)	
2703	Mount Finch, HVP[§]	8.46	0.11	2σ	K–Ar	Müller *et al.* (1991)	Mean age >8 Ma?
7250	Mount Finch, HVP	7.42	0.10	2σ	K–Ar	Müller *et al.* (1991)	Mean age >8 Ma?
T5.33.6	South flank of Tucker Glacier, 4.5 km NW of Crater Cirque	7.605	0.049	2σ	Ar–Ar	Smellie *et al.* (2011*b*)	
T5.5.4	Harrow Peaks, MMVF	0.642	0.020	2σ	Ar–Ar	Smellie *et al.* (2018)	

[†]MMVF, Mount Melbourne Volcanic Field.
[‡]DPVF, Daniell Peninsula Volcanic Field.
[§]HVP, Hallett Volcanic Province.

relationships and scale of the outcrop closely resemble those seen at Sheridan Bluff, in the Transantarctic Mountains (Chapter 5.3a: Smellie *et al.* 2021). The surface elevation of *c.* 1800 m asl, *c.* 300 m above the surface of Aviator Glacier, and its location *c.* 67 km inland of Lady Newnes Bay, suggest that it was erupted either in a pluvial lake or subglacially. In common with views of other authors (Gair 1967; Nathan and Schulte 1968), the Eldridge Bluff outcrop is tentatively included in the Northern Local Suite, although it is significantly larger (and thus probably longer lived) than other constituent outcrops.

- A near-vertical section across a basaltic vent cutting basement granodiorite is exposed at Honeycomb Ridge. It comprises an upward-flaring vent-fill 500 m in maximum width and *c.* 200 m high. It contains a core of columnar-jointed basalt flanked by steeply inward-dipping slabs of stratified breccia and tuffs (Harrington *et al.* 1967). The relationships are broadly similar to those seen in a cross-section through a Strombolian pyroclastic cone exposed at Barnes Glacier at the foot of Mount Erebus (Chapter 5.2a: Smellie and Martin 2021).

All outcrops of the Northern Local Suite, for which analyses exist, have basaltic compositions. There are few isotopic ages. They range from 12.63 ± 0.17 to 0.24 ± 0.14 Ma (a single older age of 18 Ma is regarded as unreliable Armstrong 1978) (Table 3). Nathan and Schulte (1968) also suggested that one volcanic outcrop, situated on the south flank of Cosmonaut Glacier, might be very recently formed. The outcrop, which they photographed, was described as 'a small narrow flow extending out on to the glacier ice for more than 100 m' (p. 958). They noted that the lava did not appear to be deformed by movement of the glacier and thus inferred an extremely young age (weeks or a few years?). Examination of the locality using Google Earth satellite imagery shows that the outcrop has changed surprisingly little since it was viewed by Nathan and Schulte (1968). With the resolution now available, the basal outcrop is seen to terminate in a rocky cliff that has been truncated by the glacier. The cliff is *c.* 60 m high and may include rust-coloured fragmental rocks (hyaloclastite or lapilli tuff) as well as an irregular mass of grey lava (interpretation of an image provided by P. R. Kyle). Its age is unknown.

Meander Intrusive Group

The Meander Intrusive Group consists of a suite of at least seven plutonic intrusions and more than 180 hypabyssal dykes that crop out in an area of *c.* 200 km long and 80 km deep in the coastal zone between Campbell and Borchgrevink glaciers (Fig. 28). The following description is based on Müller *et al.* (1991) and Rocchi *et al.* (2002*a, b*, 2003). Isotopic ages (by K–Ar, Ar–Ar and Rb–Sr) have been determined mainly for the dykes but outcrop features of the dykes suggest that many were intruded coeval with the plutons. Apart from an age of 9 ± 1 Ma, which is probably spurious but which was the first identification of Cenozoic plutonic activity on the region (Stump *et al.* 1983), the ages range from 51.6 ± 0.6 to 22.3 ± 4 Ma for the plutons and from 46.7 ± 1.0 to 18.1 ± 0.3 Ma for the dykes (Müller *et al.* 1991; Tonarini *et al.* 1997; Rocchi *et al.* 2002*a, b*; Dallai and Burgess 2011) (Table 4; Figs 21 & 28). No geographical or compositional variations with age have been described but plutons in the SW (between Campbell and Aviator glaciers) may be generally older than those in the NE (Aviator to Borchgrevink glaciers). It is possible that the Vulcan Hills outcrop is part of the Meander Intrusive Group (see above).

From SW to NE, the plutons comprise: Vulcan Hills(?), Mount McGee, Mount Monteagle, Greene Point, Cape King, No Ridge, Eagles–Engberg, Cape Crossfire and Mount Prior igneous complexes. They are generally associated with strong positive magnetic anomalies. The four largest

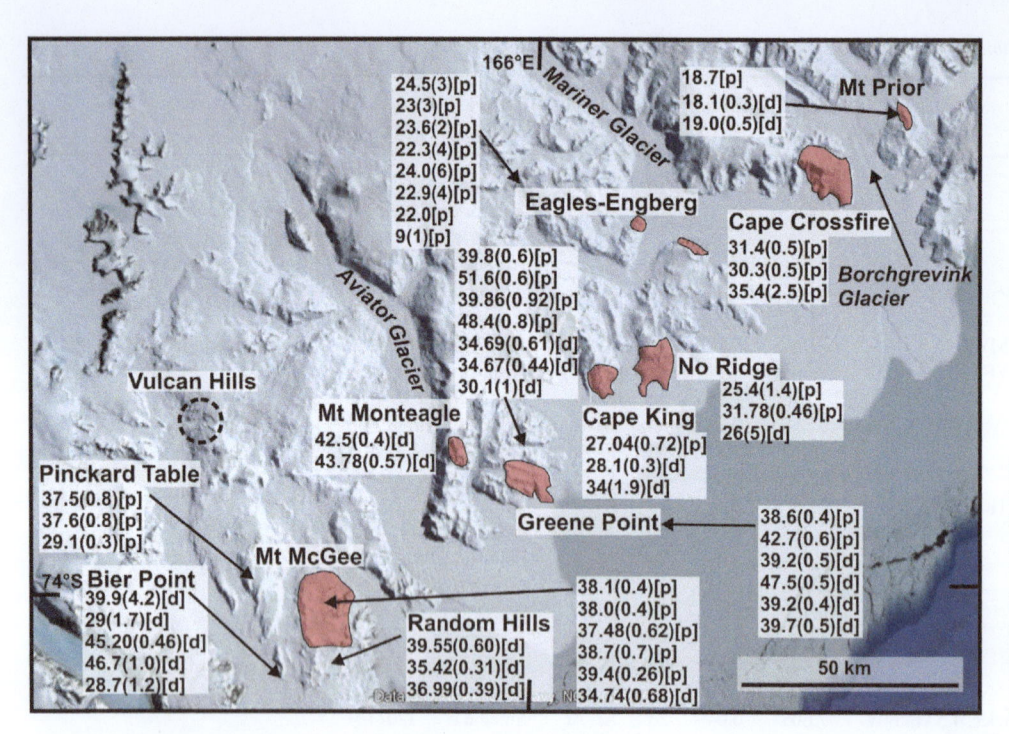

Fig. 28. Simplified geological map showing the distribution of the Meander Intrusive Group (red outcrops: after Rocchi *et al.* 2002*a*). The Vulcan Hills outcrop is also tentatively included (see text for explanation). Isotopic ages (in Ma) are also shown; errors in parentheses (Table 4). Abbreviations: p, pluton; d, dyke.

plutons (Mount McGee, Greene Point, No Ridge and Cape Crossfire) each have areas of *c.* 70–80 km^2. A weak elongation around *c.* N140° E is evident in the individual outcrop patterns but the plutons lack a fabric and are internally isotropic. Compositionally, they are alkaline and vary from gabbro to granite but intermediate compositions are less common and restricted to the McGee Igneous Complex. Compositional zoning is frequently present, as is mafic–felsic interlayering and mingling. The plutons are medium–fine grained and equigranular, and often display a strong variation in grain size. Cumulus textures are a feature of some of the gabbros and diorites.

The associated dykes in the Meander Intrusive Group typically occur as swarms cutting across both the Paleozoic basement and the alkaline plutons. All the dykes are alkaline, ranging between basalt (basanite) and rhyolite comparable with the plutons but often displaced to somewhat more alkaline compositions. They are almost aphyric or poorly porphyritic and display two prominent trends: NNW–SSE and north–south, perhaps reflecting a dextral control on their emplacement (Vignaroli *et al.* 2015). Dyke thicknesses are typically 1–5 m but examples 10–50 m thick are also present. There are no compositional or age differences between dykes intruding basement or the plutons, between the two dominant dyke orientations, or in dyke thicknesses, although the thickest dykes are felsic in composition.

Summary

Volcanism in northern Victoria Land probably commenced in Eocene time, with a major pulse occurring between 48 and 23 Ma (Meander Intrusive Group). The Eocene–Oligocene episode affected at least 200 km of the coastal region and was probably triggered by large-scale plate tectonic effects, in particular the separation of Australia from Antarctica which was initiated in the Eocene (Rocchi *et al.* 2002*a*, 2003). However, the reason for the relatively restricted outcrop area of the Meander Intrusive Group remains enigmatic. The Meander Intrusive Group probably represents the subvolcanic link between formerly extensive volcanism: that is, the magma chambers (plutons) and associated conduits (dykes). The surface expression of the igneous activity is no longer preserved, except possibly at Vulcan Hills. The presence of the associated magma chambers (plutons) at the surface today suggests that significant erosion and exhumation has occurred since Eocene–Oligocene time. The abundance of dykes and the substantial width of many suggest that eruptions were fissure controlled, at least in part, and may reflect a prevailing stress regime characterized by regional strike-slip faulting along NNW–SSE and north–south directions (Salvini *et al.* 1997; Rocchi *et al.* 2002*a*, 2003). It is suggested that the Eocene–Oligocene volcanic landscape probably comprised a variety of large central-vent volcanoes, perhaps a combination of stratovolcanoes associated with the more evolved plutons similar to Mount Melbourne and Mount Overlord, and shield volcanoes for the more mafic plutons and thicker dykes, similar to most of the volcanoes between Coulman Island and Cape Adare. A wide range of small monogenetic centres fed by the many thinner dykes might also be postulated, similar to the Northern Local Suite. The environmental setting of the Meander Intrusive Group is unknown because of the complete removal by erosion of the original Eocene–Oligocene landscape but isotopic analyses of the plutons have been interpreted to suggest a profound climatic cooling that fluctuated significantly during the intrusive period, with intrusion of the younger examples (and, hence, eruption of any associated volcanoes) possibly coinciding with glacial conditions (Dallai and Burgess 2011). The study by Dallai and Burgess (2011) is a rare example that illustrates how terrestrial environmental conditions can be deduced from igneous rock compositions, even with no preserved surficial counterparts.

By contrast, the products of younger (Middle Miocene (*c.* 12 Ma)–present) volcanic activity in northern Victoria Land are widespread and very well exposed, particularly in prominent marine-eroded coastal cliffs up to 2 km high (Hamilton 1972). The outcrops occur in a broad coastal belt *c.* 400 km in length and extending *c.* 150 km inland between Mount Melbourne and Cape Adare (Kyle 1990*a*). They comprise the Hallett and Melbourne volcanic provinces. The two provinces contain a total of at least eight defined volcanic fields, each containing multiple volcanoes. Many of the central volcanoes at inland locations have not been revisited for several decades and our information has not greatly improved as a result. Conversely, many of the coastal examples have undergone

Table 4. *Summary of published isotopic ages of the Meander Intrusive Group*

Sample	Locality	Rock type/mineral	Pluton (p)/Dyke (d)	Age (Ma)	Error (Ma)	Error (SD)	Method	Source	Notes
Mount Prior									
2003	Mount Prior	Granophyric granite	p	>17	–	2σ	K–Ar	Müller *et al.* (1991)	Inferred age c. 18.7 Ma
4414	Mount Prior	Granophyric granite	p	18.7	0.2	2σ	K–Ar	Müller *et al.* (1991)	Inferred age c. 18.7 Ma
2164	Mount Prior	Rhyolite	d	18.50	0.5	2σ	K–Ar	Müller *et al.* (1991)	Mean age 18.1 ± 0.3 Ma
3247	Mount Prior	Rhyolite	d	17.80	0.4	2σ	K–Ar	Müller *et al.* (1991)	Mean age 18.1 ± 0.3 Ma
2165	Mount Prior	Rhyolite	d	18.80	0.40	2σ	K–Ar	Müller *et al.* (1991)	Mean age 19.0 ± 0.5 Ma
3248	Mount Prior	Rhyolite	d	19.20	0.40	2σ	K–Ar	Müller *et al.* (1991)	Mean age 19.0 ± 0.5 Ma
Cape Crossfire									
18.12.96 AR15	Cape Crossfire	Monzogabbro	p	31.4	0.5	1σ	Rb–Sr	Rocchi *et al.* (2002b)	
18.12.96 AR14	Cape Crossfire	Monzodiorite	p	30.3	0.5	1σ	Rb–Sr	Rocchi *et al.* (2002b)	
AR20	Cape Crossfire	Monzodiorite	p	35.4	2.5	2σ	Ar–Ar	Dallai and Burgess (2011)	
Eagles–Engberg									
2364	Engberg Bluff	Peralkaline granite	p	24.4	0.3	2σ	K–Ar	Müller *et al.* (1991)	
1343a	Engberg Bluff	Peralkaline granite	p	24.5	0.3	2σ	K–Ar	Müller *et al.* (1991)	
1994	Eagles Bluff	Granophyric granite	p	26	0.3	2σ	K–Ar	Müller *et al.* (1991)	Mean age <22 Ma
4270	Eagles Bluff	Granophyric granite	p	22.1	0.2	2σ	K–Ar	Müller *et al.* (1991)	Mean age <22 Ma
1996	Eagles Bluff	Granophyric granite	p	23	0.4	2σ	K–Ar	Müller *et al.* (1991)	Mean age 22.85 ± 0.15 Ma
4272	Eagles Bluff	Granophyric granite	p	22.9	0.3	2σ	K–Ar	Müller *et al.* (1991)	Mean age 22.85 ± 0.15 Ma
1997 4273	Eagles Bluff	Granophyric granite	p	25.2	0.3	2σ	K–Ar	Müller *et al.* (1991)	
1998	Eagles Bluff	Arfvedsonite syenite	p	23.7	0.3	2σ	K–Ar	Müller *et al.* (1991)	Mean age 23.6 ± 0.2 Ma
4274	Eagles Bluff	Arfvedsonite syenite	p	23.6	0.3	2σ	K–Ar	Müller *et al.* (1991)	Mean age 23.6 ± 0.2 Ma
1999	Eagles Bluff	Granophyric granite	p	22.60	0.16	2σ	K–Ar	Müller *et al.* (1991)	Mean age c. 23 Ma
4275	Eagles Bluff	Granophyric granite	p	23.04	0.16	2σ	K–Ar	Müller *et al.* (1991)	Mean age c. 23 Ma
2363	Eagles Bluff	Peralkaline granite	p	19.20	1.3	2σ	K–Ar	Müller *et al.* (1991)	Poor age
1342	Eagles Bluff	Peralkaline granite	p	13.70	1.5	2σ	K–Ar	Müller *et al.* (1991)	Poor age
1994	Eagles Bluff	Biotite	p	22	–	2σ	Rb–Sr	Müller *et al.* (1991)	
1994	Eagles Bluff	Biotite	p	21.3	–	2σ	Rb–Sr	Müller *et al.* (1991)	
1996	Eagles Bluff	Biotite	p	22.3	0.4	2σ	Rb–Sr	Müller *et al.* (1991)	
1997	Eagles Bluff	Biotite	p	24	0.6	2σ	Rb–Sr	Müller *et al.* (1991)	
1999	Eagles Bluff	Biotite	p	22.9	0.4	2σ	Rb–Sr	Müller *et al.* (1991)	
No Ridge									
D1	Marinella Camp (part of the No Ridge outcrop)	Syenite	p	25.4	1.4	2σ	Ar–Ar	Dallai and Burgess (2011)	
D2	Marinella Camp (part of the No Ridge outcrop)	Syenite	p	31.78	0.46	2σ	Ar–Ar	Dallai and Burgess (2011)	
2506 1345	No Ridge	Aegirine–quartz trachyte	d	26.0	0.5	2σ	K–Ar	Müller *et al.* (1991)	
2505 1344	No Ridge	Aegirine rhyolitic pneumatolite	?	25.7	0.3	2σ	K–Ar	Müller *et al.* (1991)	
2507 1346	No Ridge	Rhyolitic pneumatolite	?	>28 or <26	–	–	K–Ar	Müller *et al.* (1991)	
2508 1347	No Ridge	Rhyolitic aegirine pneumatolite	?	>25	–	–	K–Ar	Müller *et al.* (1991)	
Cape King									
L1	Cape King	Diorite	p	27.04	0.72	2σ	Ar–Ar	Dallai and Burgess (2011)	
CD260	Cape King	Gabbro	d	34.0	1.9	2σ	Ar–Ar	Dallai and Burgess (2011)	
24-1-86 L4	Cape King	Gabbro	d	28.1	0.3	1σ	Rb–Sr	Tonarini *et al.* (1997)	

(*Continued*)

Table 4. *Continued.*

Sample	Locality	Rock type/mineral	Pluton (p)/Dyke (d)	Age (Ma)	Error (Ma)	Error (SD)	Method	Source	Notes
Greene Point									
21-10-93 A1a	Greene Point	Alkali granite	p	38.6	0.4	1σ	Rb–Sr	Tonarini *et al.* (1997)	
21-10-93 A1e	Greene Point	Monzonite	p	42.7	0.5	1σ	Rb–Sr	Tonarini *et al.* (1997)	
CD47	South Oakley Glacier (just north of Greene Point)	Syenite	p	39.8	0.6	2σ	Ar–Ar	Dallai and Burgess (2011)	
SAX7	Mount Casey to Oakley Glacier (just north of Greene Point)	Syenite	p	51.6	0.6	2σ	Ar–Ar	Dallai and Burgess (2011)	
CD54	NE Mount Knetsch (near Greene Point)	Syenite	p	39.86	0.92	2σ	Ar–Ar	Dallai and Burgess (2011)	
CD64	North Oakley Glacier (just north of Greene Point)	Monzosyenite	p	48.4	0.8	2σ	Ar–Ar	Dallai and Burgess (2011)	
21-10-93 A1d	Greene Point	Gabbro	d	39.2	0.5	1σ	Rb–Sr	Tonarini *et al.* (1997)	
21-10-93 A1c	Greene Point	Gabbro	d	47.5	0.5	1σ	Rb–Sr	Tonarini *et al.* (1997)	
10-11-93 SAX1a	Greene Point	Mugearite	d	39.2	0.4	1σ	Rb–Sr	Tonarini *et al.* (1997)	
10-11-93 SAX8	Greene Point	Benmoreite	d	39.7	0.5	1σ	Rb–Sr	Tonarini *et al.* (1997)	
SAX12	Mount Casey to Oakley Glacier (just north of Greene Point)	Basalt	d	34.69	0.61	2σ	Ar–Ar	Rocchi *et al.* (2002a)	
SAX18	Mount Casey to Oakley Glacier (just north of Greene Point)	Basanite	d	34.67	0.44	2σ	Ar–Ar	Rocchi *et al.* (2002a)	
SAX11	Mount Casey to Oakley Glacier (just north of Greene Point)	Basalt	d	30.1	1	2σ	Ar–Ar	Dallai and Burgess (2011)	
Mount Monteagle									
28-1-91 LZ53	Mount Monteagle	Benmoreite	d	42.5	0.4	1σ	Rb–Sr	Tonarini *et al.* (1997)	
SX2	Mount Monteagle	Tephrite	d	43.78	0.57	2σ	Ar–Ar	Rocchi *et al.* (2002a)	
Mount McGee									
19-1-91 LZ42	Diorite	Mount McGee	p	38.1	0.4	1σ	Rb–Sr	Tonarini *et al.* (1997)	
3-1-87 AP1	Quartz monzonite	Mount McGee	p	38.0	0.4	1σ	Rb–Sr	Tonarini *et al.* (1997)	
CD205	Syenite	Mount McGee	p	37.48	0.62	2σ	Ar–Ar	Dallai and Burgess (2011)	
CD318	Monzonite	Mount McGee	p	38.7	0.7	2σ	Ar–Ar	Dallai and Burgess (2011)	
CD341	Monzonite	Mount McGee	p	39.4	0.26	2σ	Ar–Ar	Dallai and Burgess (2011)	
SX13	Hawaiite	Mount McGee	d	34.74	0.68	2σ	Ar–Ar	Rocchi *et al.* (2002a)	
Random Hills									
AS9	Latite	Random Hills	d	39.55	0.6	2σ	Ar–Ar	Rocchi *et al.* (2002a)	
AS13	Tephrite	Random Hills	d	35.42	0.31	2σ	Ar–Ar	Rocchi *et al.* (2002a)	
AS15	Basanite	Random Hills	d	36.99	0.39	2σ	Ar–Ar	Rocchi *et al.* (2002a)	
Pinckard Table									
CD212	Pinckard Table	Syenite	p	37.6	0.8	2σ	Ar–Ar	Dallai and Burgess (2011)	
CD218	Pinckard Table	Monzosyenite	p	37.5	0.8	2σ	Ar–Ar	Dallai and Burgess (2011)	
17-11-93 SX8	Styx Glacier	Syenite	p	29.1	0.3	1σ	Rb–Sr	Tonarini *et al.* (1997)	
Bier Point									
8.11.93 SA11b	Bier Point	Basalt (kaersutite)	d	39.9	4.2	2σ	Ar–Ar	Rocchi *et al.* (2002a)	
4.11.93 XA2	Bier Point	Hawaiite (kaersutite)	d	45.20	0.46	2σ	Ar–Ar	Rocchi *et al.* (2002a)	
XA9	Bier Point	Basalt	d	29	1.7	2σ	Ar–Ar	Dallai and Burgess (2011)	
XA11	Bier Point	Basalt	d	28.7	1.2	2σ	Ar–Ar	Dallai and Burgess (2011)	

SD, standard deviation.

significant reinvestigation, although logistical constraints have ensured that even the new studies are mainly of a detailed reconnaissance nature. Of necessity, in the newer investigations only a small number of outcrops were selected for detailed study and large areas were only observed using binoculars. Despite these limitations, the studies were carried out over multiple seasons and they have resulted in important advances in our knowledge of the volcanology and eruptive environments.

Most of the large eruptive centres are shield volcanoes, which erupted along coastal fissures, but a few relatively small stratovolcanoes were also constructed, mainly in inland locations (the sole coastal examples are at Mount Harcourt and Mount Lubbock). The age of the volcanism shows no overall geographical progression (Smellie *et al.* 2011*b*). Petrologically, it forms an alkaline suite with compositions ranging from basanite to trachyte and rhyolite (Rocchi and Smellie 2021, and references therein). Two of the central volcanoes are active: Mount Melbourne and, possibly, Mount Rittman. In addition, a large number of small mafic pyroclastic cones were erupted inland, some associated with short lavas, which are grouped within the Northern Local Suite. The age of the latter exactly overlaps with ages of the larger volcanic centres and they are also compositionally comparable, although far fewer geochemical studies of the Northern Local Suite have been conducted. There have been very few studies of the small pyroclastic cones, whereas the larger centres are better investigated. One exception is a small pyroclastic cone at Harrow Peaks (Mount Melbourne Volcanic Field), which consists of a tuff cone erupted under a thin cover of cold-based (frozen-bed) ice (Smellie *et al.* 2018). It is the only example of its type to be described anywhere. From the preserved record, we now know that the large central volcanoes were overwhelmingly effusive, with repeated eruption of glaciovolcanic 'a'ā lava-fed deltas. However, since most of the eruptions took place subglacially, the products of any pyroclastic eruptions would get advected away and the tephra dumped in the sea; thus, completely removing the pyroclastic deposits from the volcanoes themselves and leaving a record biased towards lava flows. Lava-fed deltas are the commonest style of eruption in the Neogene Hallett Coast volcanoes. However, whereas pāhoehoe-lava-fed deltas are common worldwide, the first 'a'ā lava-fed deltas to be described were those in northern Victoria Land (Smellie *et al.* 2011*a*, 2013). They have been used to erect new lithofacies models to examine the processes involved in their formation in a glacial setting and they are now an important new proxy tool used to derive detailed characteristics of the associated palaeo-ice (Smellie *et al.* 2013, 2014). From the new studies, we are also certain that the eruptive setting of the central volcanoes, which was previously only inferred on a weak *a priori* basis (Hamilton 1972; McIntosh and Gamble 1991), was overwhelmingly glacial, with evidence for past ice thicknesses typically <300 m (Smellie *et al.* 2011*a*, *b*). Moreover the variable development of eroded and pristine surfaces between eruptive units is consistent with a coeval glacial regime between eruptions that was mainly polythermal (Smellie *et al.* 2014). The new view of the evolution of the thermal regime is a radical departure from the prevailing paradigm, which was based on a single unidirectional step change from warm- to cold-based ice, necessitating a change in the paradigm.

Acknowledgements The authors are very grateful to each of the Expedition Chiefs and all the personnel at Mario Zuchelli Station for hosting us. We also gratefully thank Samuele Agostini and Valerio Olivetti for participating in the fieldwork in 2014; Laura Crispini, Luigia Di Nicola and Andrea Dini for standing in as field assistants when required; Pietro Armienti, Gianfranco di Vincenzo and Maurizio Gemelli for scientific conversations and input to our publications; Mi Jung Lee for locations of her dated samples in the Mount Melbourne Volcanic Field; Tom Jordan for help with ADMAP information on the Pleiades area; Ray Burgess for help assembling the table of isotopic ages for the Meander Intrusive Group; Warren Hamilton for local information and photographs which helped us plan our field campaigns, especially the first one; Ingrid Kjaarsgard (née Hornig) and Philip Kyle for comments on volcanic outcrops at Navigator Nunatak and Cosmonaut Glacier, respectively; and the pilots and ground crews of Helicopters New Zealand. Conversations with Adam Martin and Kurt Panter about the overall stratigraphy of the McMurdo Volcanic Group are also gratefully acknowledged. Finally, the authors are grateful to Paola Del Carlo and Guido Giordano, and to the editor, Kurt Panter, for their positive critical comments on our chapter.

Author contributions JLS: conceptualization (equal), formal analysis (lead), investigation (equal), methodology (lead), writing – original draft (lead), writing – review & editing (lead); **SR**: conceptualization (equal), formal analysis (supporting), investigation (equal), visualization (equal), writing – review & editing (supporting).

Funding The Italian Antarctic Programme (Programma Nazionale di Ricerche in Antartide (PNRA)) supported all of the studies in northern Victoria Land through successive grants to S. Rocchi; the British Antarctic Survey supported the fieldwork of J.L. Smellie in 2005–06 and his research until 2009.

Data availability Data sharing is not applicable to this article as no additional datasets were generated other than those presented in this chapter.

References

Adamson, R.G. and Cavaney, R.J. 1967. Volcanic debris layers near Mount Melbourne, northern Victoria Land, Antarctica. *New Zealand Journal of Geology and Geophysics*, **10**, 409–421, https://doi.org/10.1080/00288306.1967.10426745

Armienti, P. and Tripodo, A. 1991. Petrography and chemistry of lavas and comagmatic xenoliths of Mount Rittmann, a volcano discovered during the IV Italian expedition in northern Victoria Land (Antarctica). *Memorie della Societa Geologica Italiana*, **46**, 427–451.

Armienti, P., Civetta, L., Innocenti, F., Maetti, P., Tripodo, A., Villari, L. and Vita, G. 1991. New petrological and geochemical data on Mt. Melbourne Volcanic Field, northern Victoria Land, Antarctica. (II Italian Antarctic Expedition). *Memorie della Societa Geologica Italiana*, **46**, 397–424.

Armienti, P., Francalanci, L., Landi, P. and Vita-Scaillet, G. 2003. Age and geochemistry of volcanic rocks from Daniell Peninsula and Coulman Island, Hallett Volcanic Province, Antarctica. *Geologisches Jahrbuch*, **B85**, 409–445.

Armstrong, R.L. 1978. K–Ar dating: Late Cenozoic McMurdo Volcanic Group and Dry Valley glacial history, Victoria Land, Antarctica. *New Zealand Journal of Geology and Geophysics*, **21**, 685–698, https://doi.org/10.1080/00288306.1978.10425199

Bargagli, R., Broady, P.A. and Walton, D.W.H. 1996. Preliminary investigation of the thermal biosystem of Mount Rittmann fumaroles (northern Victoria Land, Antarctica). *Antarctic Science*, **8**, 121–126, https://doi.org/10.1017/S0954102096000181

Behrendt, J.C., LeMasurier, W. and Cooper, A.K. 1992. The West Antarctic Rift System – a propagating rift 'captured' by a mantle plume? *In*: Yoshida, Y., Kaminuma, K. and Shiraishi, K. (eds) *Recent Progress in Antarctic Earth Science*. Terra Scientific (TERRAPUB), Tokyo, 315–322.

Bonaccorso, A., Maione, M., Pertusati, C., Privitera, E. and Ricci, C.A. 1991. Fumarolic activity at Mount Rittmann volcano (northern Victoria Land, Antarctica). *Memorie della Società Geologica Italiana*, **46**, 453–456.

Dallai, L. and Burgess, R. 2011. A record of Antarctic surface temperature between 25 and 50 m.y. ago. *Geology*, **39**, 423–426, https://doi.org/10.1130/G31569.1

Del Carlo, P., Di Roberto, A. et al. 2015. Late Pleistocene-Holocene volcanic activity in northern Victoria Land recorded in Ross Sea (Antarctica) marine sediments. *Bulletin of Volcanology*, **77**, 36, https://doi.org/10:1007/s00445-015-0924-0

Esser, R.P. and Kyle, P.R. 2002. $^{40}Ar/^{39}Ar$ chronology of the McMurdo Volcanic Group at The Pleiades, northern Victoria Land, Antarctica. *Royal Society of New Zealand Bulletin*, **35**, 415–418.

Faccenna, C., Rossetti, F., Becker, T.W., Danesi, S. and Morelli, A. 2008. Recent extension driven by mantle upwelling beneath the Admiralty Mountains (East Antarctica). *Tectonics*, **27**, TC4015, https://doi.org/10.1029/2007TC002197

Ferraccioli, F., Armadillo, E., Zunino, A., Bozzo, E., Rocchi, S. and Armienti, P. 2009. Magmatic and tectonic patterns over the Northern Victoria Land sector of the Transantarctic Mountains from new aeromagnetic imaging. *Tectonophysics*, **478**, 43–61, https://doi.org/10.1016/j.tecto.2008.11.028

Field, B.D. and Finlay, R.H. 1983. The sedimentology of the Robertson Bay Group, northern Victoria Land. In: Oliver, R.L., James, P.R. and Jago, J.B. (eds) *Antarctic Earth Science*. Australian Academy of Science, Canberra, 102–106.

Gair, H.S. 1964. Geology of the upper Rennick, Campbell, and Aviator Glaciers, Victoria Land, Antarctica. In: Adie, R.J. (ed.) *Antarctic Geology*. North Holland, Amsterdam, 188–198.

Gair, H.S. 1967. The geology rom the upper Rennick Glacier to the coast, northern Victoria Land, Antarctica. *New Zealand Journal of Geology and Geophysics*, **10**, 309–344, https://doi.org/10.1080/00288306.1967.10426742

Gambino, S., Armienti, P. et al. 2021. Mount Melbourne and Mount Rittmann. *Geological Society, London, Memoirs*, **55**, https://doi.org/10.1144/M55-2018-43

Giggenbach, W.F. 1976. Geothermal ice caves on Mt Erebus, Ross Island, Antarctica. *New Zealand Journal of Geology and Geophysics*, **19**, 365–372, https://doi.org/10.1080/00288306.1976.10423566

Giordano, G., Lucci, F., Phillips, D., Cozzupoli, D. and Runci, V. 2012. Stratigraphy, geochronology and evolution of the Mt. Melbourne volcanic field (North Victoria land, Antarctica). *Bulletin of Volcanology*, **74**, 1985–2005, https://doi.org/10.1007/s00445-012-0643-8

Hamilton, W. 1972. *The Hallett Volcanic Province, Antarctica*. Geological Survey Professional Paper, **456-C**.

Harrington, H.J. 1958. Nomenclature of rock units in the Ross Sea region, Antarctica. *Nature*, **182**, 290–291, https://doi.org/10.1038/182290a0

Harrington, H.J., Wood, B.L., McKellar, I.C. and Lensen, G.J. 1967. Topography and geology of the Cape Hallett district, Victoria Land, Antarctica. *New Zealand Geological Bulletin*, **80**

Hornig, I. and Wörner, G. 2003. Cenozoic volcanics from Coulman Island, Mandible Cirque, Malta Plateau and Navigator Nunatak, Victoria Land, Antarctica. *Geologisches Jahrbuch*, **B85**, 373–406.

Jordan, H. 1981. Tectonic oservations in the Hallett Volcanic Province, Antartica. *Geologisches Jahrbuch*, **B41**, 111–125.

Kreuzer, H., Höhndorf, A. et al. 1981. K/Ar and Rb/Sr dating of igneous rocks from north Victoria Land, Antarctica. *Geologisches Jahrbuch*, **B41**, 267–273.

Kyle, P.R. 1982. Volcanic geology of The Pleiades, northern Victoria Land, Antarctica. In: Craddock, C. (ed.) *Antarctic Geoscience*. University of Wisconsin Press, Madison, WI, 747–754.

Kyle, P.R. 1986. Mineral chemistry of Late Cenozoic McMurdo Volcanic Group rocks from The Pleiades, northern Victoria Land. *American Geophysical Union, Antarctic Research Series*, **46**, 305–337.

Kyle, P.R. 1990a. McMurdo Volcanic Group, western Ross Embayment. Introduction. *American Geophysical Union Antarctic Research Series*, **48**, 19–25.

Kyle, P.R. 1990b. Melbourne Volcanic Province. *American Geophysical Union Antarctic Research Series*, **48**, 48–52.

Kyle, P.R. 1990c. The Pleiades. *American Geophysical Union Antarctic Research Series*, **48**, 60–64.

Kyle, P.R. 1990d. Vulcan Hills. *American Geophysical Union Antarctic Research Series*, **48**, 69–71.

Kyle, P.R. 1990e. Local Suite basaltic rocks. *American Geophysical Union Antarctic Research Series*, **48**, 79–80.

Kyle, P.R. and Cole, J.W. 1974. Structural control on volcanism in the McMurdo Volcanic Group, Antarctica. *Bulletin Volcanologique*, **38**, 16–25, https://doi.org/10.1007/BF02597798

Kyle, P.R. and Noll, M.R. 1990. Mount Overlord. *American Geophysical Union Antarctic Research Series*, **48**, 65–68.

Le Bas, M.J., Le Maitre, R.W., Streckeisen, A. and Zanettin, B. 1986. A chemical classification of volcanic rocks based on the total alkali–silica diagram. *Journal of Petrology*, **27**, 745–750, https://doi.org/10.1093/petrology/27.3.745

Lee, M.J., Lee, J.I., Kim, T.H., Lee, J. and Nagao, K. 2015. Age, geochemistry and Sr-Nd-Pb isotopic compositions of alkali volcanic rocks from Mt. Melbourne and the western Ross Sea, Antarctica. *Geosciences Journal*, **19**, 681–695, https://doi.org/10.1007/s12303-015-0061-y

LeMasurier, W.E. and Landis, C.A. 1996. Mantle-plume activity recorded by low relief erosion surfaces in West Antarctica and New Zealand. *Geological Society of America Bulletin*, **108**, 1450–1466, https://doi.org/10.1130/0016-7606(1996)108<1450:MPARBL>2.3.CO;2

LeMasurier, W.E. and Thomson, J.W. (eds). 1990. *Volcanoes of the Antarctic Plate and Southern Oceans*. American Geophysical Union Antarctic Research Series, **48**.

Lyon, G.L. 1986. Stable isotope stratigraphy of ice cores and the age of the last eruption at Mt. Melbourne, Antarctica. *New Zealand Journal of Geology and Geophysics*, **29**, 135–138, https://doi.org/10.1080/00288306.1986.10427528

Lyon, G.L. and Giggenbach, W.F. 1974. Geothermal activity in Victoria Land, Antarctica. *New Zealand Journal of Geology and Geophysics*, **17**, 511–521, https://doi.org/10.1080/00288306.1973.10421578

McIntosh, W.C. and Gamble, J.A. 1991. A subaerial eruptive environment for the Hallett Coast volcanoes. In: Thomson, M.R.A., Crame, J.A. and Thomson, J.W. (eds) *Geological Evolution of Antarctica*. Cambridge University Press, Cambridge, UK, 657–661.

McIntosh, W.C. and Kyle, P.R. 1990a. Hallett Volcanic Province. *American Geophysical Union Antarctic Research Series*, **48**, 26–31.

McIntosh, W.C. and Kyle, P.R. 1990b. Adare Peninsula. *American Geophysical Union Antarctic Research Series*, **48**, 32–35.

McIntosh, W.C. and Kyle, P.R. 1990c. Hallett Peninsula. *American Geophysical Union Antarctic Research Series*, **48**, 36–39.

McIntosh, W.C. and Kyle, P.R. 1990d. Daniell Peninsula. *American Geophysical Union Antarctic Research Series*, **48**, 40–42.

McIntosh, W.C. and Kyle, P.R. 1990e. Coulman Island. *American Geophysical Union Antarctic Research Series*, **48**, 43–45.

Mortimer, N., Dunlap, W.J., Isaac, M.J., Sutherland, R.P. and Faure, K. 2007. Basal Adare volcanics, Robertson Bay, north Victoria Land, Antarctica: Late Miocene intraplate basalts of subaqueous origin. *United States Geological Survey Open-File Report*, **2007-1047**, Short Research Paper 045, https://doi.org/10.3133/ofr20071047SRP045

Müller, P., Schmidt-Thomé, M., Tessensohn, F. and Vetter, U. 1991. Cenozoic peralkaline magmatism at the western margin of the Ross Sea, Antarctica. *Memorie della Società Geologica Italiana*, **46**, 315–336.

Narcisi, B., Proposito, M. and Frezzotti, M. 2001. Ice record of a 13th century explosive volcanic eruption in northern Victoria Land, East Antarctica. *Antarctic Science*, **13**, 174–181, https://doi.org/10.1017/S0954102001000268

Narcisi, B., Petit, J.R., Langone, A. and Stenni, B. 2016. A new Eemian record of Antarctic tephra layers retrieved from the Talos Dome ice core (Northern Victoria Land). *Global and Planetary Change*, **137**, 69–78, https://doi.org/10.1016/j.gloplacha.2015.12.016

Nardini, I., Armienti, P., Rocchi, S. and Burgess, R. 2003. ^{40}Ar–^{39}Ar chronology and petrology of the Miocene rift-related volcanism of Daniell Peninsula (northern Victoria Land, Antarctica). *Terra Antartica*, **10**, 39–62.

Nathan, S. and Schulte, F.J. 1967. Recent thermal and volcanic activity on Mount Melbourne, northern Victoria Land, Antarctica. *New Zealand Journal of Geology and Geophysics*, **10**, 422–430, https://doi.org/10.1080/00288306.1967.10426746

Nathan, S. and Schulte, F.J. 1968. Geology and petrology of the Campbell–Aviator Divide, Northern Victoria Land, Antarctica. *New Zealand Journal of Geology and Geophysics*, **11**, 940–975, https://doi.org/10.1080/00288306.1968.10420762

Noll, M.R. 1985. *Mount Overlord, Northern Victoria Land, Antarctica*. MSc thesis, New Mexico Institute of Mining and Technology, Socorro, New Mexico, USA.

Panter, K.S., Castillo, P. et al. 2018. Melt origin across a rifted continental margin: a case for subduction-related metasomatic agents in the lithospheric source of alkaline basalt, NW Ross Sea, Antarctica. *Journal of Petrology*, **59**, 517–558, https://doi.org/10.1093/petrology/egy036

Perchiazzi, N., Folco, L. and Mellini, M. 1999. Volcanic ash bands in the Frontier Mountain and Lichen Hills blue-ice fields, northern Victoria Land. *Antarctic Science*, **11**, 353–361, https://doi.org/10.1017/S0954102099000449

Priestley, R.E. 1923. *British (Terra Nova) Antarctic Expedition, 1910–1913, Physiography (Robertson Bay and Terra Nova Regions)*. Harrison and Sons, London.

Riddolls, B.W. and Hancox, G.T. 1968. The geology of the upper Mariner Glacier region, North Victoria Land, Antarctica. *New Zealand Journal of Geology and Geophysics*, **11**, 881–899, https://doi.org/10.1080/00288306.1968.10420758

Rocchi, S. and Smellie, J.L. 2021. Northern Victoria Land: petrology. *Geological Society, London, Memoirs*, **55**, https://doi.org/10.1144/M55-2019-19

Rocchi, S., Armienti, P., D'Orazio, M., Tonarini, S., Wijbrans, J.R. and Di Vincenzo, G. 2002a. Cenozoic magmatism in the western Ross Embayment: Role of mantle plume v. plate dynamics in the development of the West Antarctic Rift System. *Journal of Geophysical Research: Solid Earth*, **107**, 2195, https://doi.org/10.1029/2001JB000515

Rocchi, S., Fioretti, A.M. and Cavazzini, G. 2002b. Petrography, geochemistry, and geochronology of the Cenozoic Cape Crossfire, Cape King and No Ridge igneous complexes (northern Victoria Land, Antarctica). *Royal Society of New Zealand Bulletin*, **35**, 215–225.

Rocchi, S., Storti, F., Di Vincenzo, G. and Rossetti, F. 2003. Intraplate strike-slip tectonics as an alternative to mantle plume activity for the Cenozoic rift magmatism in the Ross Sea region, Antarctica. *Geological Society, London, Special Publications*, **210**, 145–158, https://doi.org/10.1144/GSL.SP.2003.210.01.09

Salvini, F., Brancolini, G., Busetti, M., Storti, F., Mazzarini, F. and Coren, F. 1997. Cenozoic geodynamics of the Ross Sea region, Antarctica: Crustal extension, intraplate strike-slip faulting, and tectonic inheritance. *Journal of Geophysical Research: Solid Earth*, **102**, 24 669–24 696, https://doi.org/10.1029/97JB01643

Schmidt-Thomé, M., Mueller, P. and Tessensohn, F. 1990. Malta Plateau. *American Geophysical Union Antarctic Research Series*, **48**, 53–59.

Smellie, J.L. 2006. The relative importance of supraglacial v. subglacial meltwater escape in basaltic subglacial tuya eruptions: an important unresolved conundrum. *Earth-Science Reviews*, **74**, 241–268, https://doi.org/10.1016/j.earscirev.2005.09.004

Smellie, J.L. 2018. Glaciovolcanism: A 21st century proxy for palaeo-ice. *In*: Menzies, J. and van der Meer, J.J.M. (eds) *Past Glacial Environments*. 2nd edn. Elsevier, Amsterdam, 335–375.

Smellie, J.L. 2021. Antarctic volcanism: volcanology and palaeoenvironmental overview. *Geological Society, London, Memoirs*, **55**, https://doi.org/10.1144/M55-2020-1

Smellie, J.L. and Edwards, B.E. 2016. *Glaciovolcanism on Earth and Mars: Products, Processes and Palaeoenvironmental Significance*. Cambridge University Press, Cambridge, UK.

Smellie, J.L. and Martin, A.P. 2021. Erebus Volcanic Province: volcanology. *Geological Society, London, Memoirs*, **55**, https://doi.org/10.1144/M55-2018-62

Smellie, J.L., Rocchi, S. and Armienti, P. 2011a. Late Miocene volcanic sequences in northern Victoria Land, Antarctica: products of glaciovolcanic eruptions under different thermal regimes. *Bulletin of Volcanology*, **73**, 1–25, https://doi.org/10.1007/s00445-010-0399-y

Smellie, J.L., Rocchi, S., Gemelli, M., Di Vincenzo, G. and Armienti, P. 2011b. A thin predominantly cold-based Late Miocene East Antarctic ice sheet inferred from glaciovolcanic sequences in northern Victoria Land, Antarctica. *Palaeogeography, Palaeoclimatology, Palaeoecology*, **307**, 129–149, https://doi.org/10.1016/j.palaeo.2011.05.008

Smellie, J.L., Wilch, T. and Rocchi, A. 2013. 'A'ā lava-fed deltas: a new reference tool in paleoenvironmental research. *Geology*, **41**, 403–406, https://doi.org/10.1130/G33631.1

Smellie, J.L., Rocchi, S. et al. 2014. Glaciovolcanic evidence for a polythermal Neogene East Antarctic Ice Sheet. *Geology*, **42**, 39–41, https://doi.org/10.1130/G34787.1

Smellie, J.L., Rocchi, S., Johnson, J.S., Di Vincenzo, G. and Schaefer, J.M. 2018. A tuff cone erupted under frozen-bed ice (northern Victoria Land, Antarctica): linking glaciovolcanic and cosmogenic nuclide data for ice sheet reconstructions. *Bulletin of Volcanology*, **80**, 12, https://doi.org/10.1007/s00445-017-1185-x

Smellie, J.L., Panter, K.S. and Reindel, J. 2021. Mount Early and Sheridan Bluff: volcanology. *Geological Society, London, Memoirs*, **55**, https://doi.org/10.1144/M55-2018-61

Storey, B.C., Leat, P.T., Weaver, S.D., Pankhurst, R.J., Bradshaw, J.D. and Kelley, S. 1999. Mantle plumes and Antarctica–New Zealand rifting: evidence from mid-Cretaceous mafic dykes. *Journal of the Geological Society, London*, **156**, 659–671, https://doi.org/10.1144/gsjgs.156.4.0659

Stump, E., Holloway, J.R., Borg, S.G. and Armstrong, R.L. 1983. Discovery of a Tertiary granite pluton, northern Victoria Land. *Antarctic Journal of the United States*, **18**, 17–18.

Tonarini, S., Rocchi, S., Armienti, P. and Innocenti, F. 1997. Constraints on timing of Ross Sea rifting inferred from Cainozoic intrusions from northern Victoria Land, Antarctica. *In*: Ricci, C.A. (ed.) *The Antarctic Region: Geological Evolution and Processes*. Terra Antartica, Siena, Italy, 511–521.

Vignaroli, G., Balsamo, F., Giordano, G., Rossetti, F. and Storti, F. 2015. Miocene-to-Quaternary oblique rifting signature in the Western Ross Sea from fault patterns in the McMurdo Volcanic Group, north Victoria Land, Antarctica. *Tectonophysics*, **656**, 74–90, https://doi.org/10.1016/j.tecto.2015.05.027

White, J.D.L. and Houghton, B.F. 2006. Primary volcaniclastic rocks. *Geology*, **34**, 677–680, https://doi.org/10.1130/G22346.1

Wilch, T.I., Denton, G.H., Lux, D.R. and McIntosh, W.C. 1993. Limited Pliocene glacier extent and surface uplift in middle Taylor Valley, Antarctica. *Geografiska Annaler*, **75A**, 331–351, https://doi.org/10.1080/04353676.1993.11880399

Wörner, G. and Orsi, G. 1990. Volcanic geology of Edmonson Point, Mt. Melbourne Volcanic Field, north Victoria Land, Antartica. *Polarforschung*, **60**, 84–86.

Wörner, G. and Viereck, L. 1987. Subglacial to emergent volcanism at Shield Nunatak, Mt. Melbourne Volcanic Field, Antarctica. *Polarforschung*, **57**, 27–41.

Wörner, G. and Viereck, L. 1989. The Mt. Melbourne Volcanic Field (Victoria Land, Antarctica) I. Field observations. *Geologisches Jahrbuch*, **E38**, 369–393.

Wörner, G. and Viereck, L. 1990. Mount Melbourne. *American Geophysical Union, Antarctic Research Series*, 48, 72–78.

Wörner, G., Niephaus, H., Hertogen, J. and Viereck, L. 1989. The Mt. Melbourne Volcanic Field (Victoria Land, Antarctica) II. Geochemistry and magma genesis. *Geologisches Jahrbuch*, **E38**, 395–433.

Wright, A.C. and Kyle, P.R. 1990a. Royal Society Range. *American Geophysical Union Antarctic Research Series*, **48**, 131–133.

Wright, A.C. and Kyle, P.R. 1990b. Taylor and Wright valleys. *American Geophysical Union Antarctic Research Series*, **48**, 134–137.

Chapter 5.1b

Northern Victoria Land: petrology

S. Rocchi[1]* and J. L. Smellie[2]

[1]Dipartimento di Scienze della Terra, Università di Pisa, Via Santa Maria 53, I-56126 Pisa, Italy
[2]School of Geography, Geology and the Environment, University of Leicester, University Road, Leicester LE1 7RH, UK

SR, 0000-0001-6868-1939; JLS, 0000-0001-5537-1763
*Correspondence: sergio.rocchi@unipi.it

Abstract: Cenozoic magmatic rocks related to the West Antarctic Rift System crop out right across Antarctica, in Victoria Land, Marie Byrd Land and into Ellsworth Land. Northern Victoria Land, located at the northwestern tip of the western rift shoulder, is unique in hosting the longest record of the rift-related igneous activity: plutonic rocks and cogenetic dyke swarms cover the time span from c. 50 to 20 Ma, and volcanic rocks are recorded from 15 Ma to the present. The origin of the entire igneous suite is debated; nevertheless, the combination of geochemical and isotopic data with the regional tectonic history supports a model with no role for a mantle plume. Amagmatic extension during the Cretaceous generated an autometasomatized mantle source that, during Eocene–present activity, produced magma by small degrees of melting induced by the transtensional activity of translithospheric fault systems. The emplacement of Eocene–Oligocene plutons and dyke swarms was focused along these fault systems. Conversely, the location of the mid-Miocene–present volcanoes is governed by lithospheric necking along the Ross Sea coast for the largest volcanic edifices; while inland, smaller central volcanoes and scoria cones are related to the establishment of magma chambers in thicker crust.

Supplementary material: Published analyses of whole-rock samples from plutons, dykes and volcanic products from northern Victoria Land (Table S1) are available at https://doi.org/10.6084/m9.figshare.c.5198177

Cenozoic magmatism in Antarctica is widespread, and is particularly well represented in the Antarctic Peninsula, Marie Byrd Land and Victoria Land. That in the Antarctic Peninsula is predominantly subduction-related (see Chapter 3.1a: Leat and Riley 2021), whereas igneous activity in Marie Byrd Land, Victoria Land and the Ross Sea is related to the evolution of the West Antarctic Rift System (WARS). The WARS represents a prime example of long-lasting magmatism associated with an extensive area of rifted crust (Behrendt et al. 1991). It is one of the Earth's major active continental extension zones, comparable in scale with the Basin and Range Province of North America and the East African Rift (Trey et al. 1999; LeMasurier 2008). The WARS extends for more than 3000 km across Antarctica, from the Ross Sea to Ellsworth Land (Fig. 1a), and exceeds 10^6 km^2 in area. Associated total extension is approximately 1000–1200 km in the north, and about the half that towards the South Pole. In effect, the WARS is splitting the Antarctic Plate. The plate is stationary, has had a polar position since about 90 Ma and it is almost completely encircled by mid-ocean ridges.

The WARS formed broadly in response to the separation of Antarctica from Australia and New Zealand (Lawver and Gahagan 1994; Siddoway 2008). Regional extension in the WARS is considered to have occurred in at least two distinct phases: a Cretaceous phase of broadly distributed extension between c. 130 and 80 Ma, and an Eocene–Oligocene phase at 46–24 Ma, which was focused on the western margin (Davey et al. 2006; Huerta and Harry 2007; Siddoway 2008; Davey et al. 2016). The latter was also associated with opening of the oceanic Adare Basin and creation of new oceanic crust. Extension of the Adare Basin onto the Antarctic continent gave rise to the Northern Basin, with its subdued magnetic lineations due to burial beneath enhanced sediment accumulation during successive ice-sheet advances and retreats (Davey et al. 2016). Further south, the Victoria Land Basin and its sediment infill also formed during this phase (Henrys et al. 2007). The amount of extension associated with the Eocene–Oligocene phase is believed to be relatively small, perhaps just 150 km (Stock and Cande 2002; Davey and Santis 2006). However, numerical modelling (Huerta and Harry 2007) suggested that pronounced lithospheric necking (stretching resulting in narrow, focused thinning of the lithosphere, with decreasing elevation of the upper surface and rising of the lower boundary) should be associated with the extension that created the Victoria Land Basin. This has been confirmed by petrological studies (Krans 2013; Panter et al. 2018), based on pressure–temperature (P–T) estimates derived from mineral-phase equilibria, which inferred necking below Adare Peninsula. By implication, similar necking is probably also present below other coastal volcanoes in northern Victoria Land.

The Cretaceous episode of extension was amagmatic, probably because extending a lithosphere already thinned during Early Cretaceous back-arc extension was insufficient to cause widespread decompression melting (Bonini et al. 2007; Siddoway 2008). By contrast, significant WARS magmatism occurred in Cenozoic time. It is represented in northern Victoria Land by the emplacement of alkaline plutons at 48–19 Ma, associated synplutonic dykes and dyke swarms (Tonarini et al. 1997; Rocchi et al. 2002a), and by widespread alkaline volcanism between 15 Ma and present (LeMasurier and Thomson 1990).

The magmatism is present in continental areas throughout Victoria Land and Marie Byrd Land, and it also occurs in submerged areas of the Ross Sea (Victoria Land and Adare basins). However, in Marie Byrd Land, the volcanism is distributed across a wide area of thinned crust, while in northern Victoria Land the magmatism is restricted to uplifted thick crust along the border of the Transantarctic Mountains (where necking is likely). The latter represent the roots of the Paleozoic Ross Orogen uplifted during the Mesozoic–Cenozoic activity of the rift. Within this uplifted belt, northern Victoria Land is located in a peculiar position: that is, situated at the northwestern tip of the western WARS shoulder (Fig. 1b). An additional unique aspect of northern Victoria Land relative to elsewhere in the WARS is that it contains the longest record of igneous activity: there, the volcanic record is complemented by the occurrence of cogenetic dyke

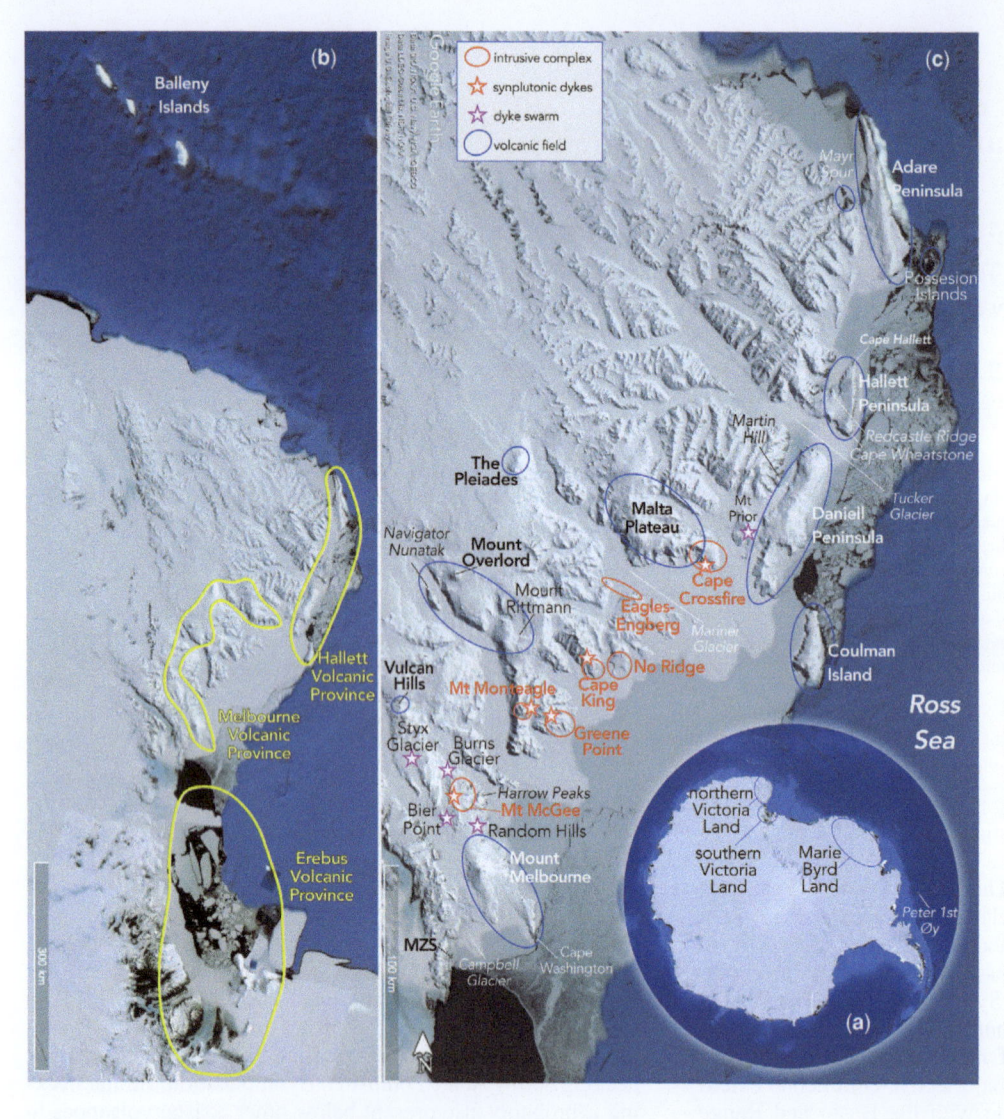

Fig. 1. Location maps. (**a**) Map of Antarctica showing the location of the major areas of Cenozoic igneous outcrops related to the activity of the West Antarctic Rift System (WARS). (**b**) The volcanic provinces of the McMurdo Volcanic Group along the western shoulder of the WARS. (**c**) Map of northern Victoria Land from Cape Adare to Mount Melbourne, showing the occurrence of plutons, dykes and volcanic edifices. Many of the latter are composite volcanic features. Outcrops of Northern Local Suite volcanic rocks are marked in figure 27 of Smellie and Rocchi (2021 – Chapter 5.1a). Base maps from Google Earth, imagery date: 1 January 1999 (accessed 6 April 2019).

swarms and plutons as old as *c.* 50 Ma (Rocchi *et al.* 2002*a*; Dallai and Burgess 2011). The tectonic evolution of this area is well characterized and contains a prominent translithospheric strike-slip fault system directly dated as coeval with the Eocene–Oligocene igneous activity (Di Vincenzo *et al.* 2004). The faulting is also demonstrably active still (Dubbini *et al.* 2010).

Field and petrographical features

The Miocene–Pliocene–Quaternary volcanic rocks erupted within the western Ross Embayment collectively belong to the McMurdo Volcanic Group (Kyle 1990*a*; LeMasurier and Thomson 1990). The occurrence of Eocene–Oligocene plutons and dyke swarms geographically associated with the Miocene–Pliocene–Quaternary volcanic products is a distinguishing feature of northern Victoria Land. The plutonic and hypabyssal (subvolcanic) rocks are called the Meander Intrusive Group (Tonarini *et al.* 1997). As a consequence, the Western Ross Supergroup has been defined to include the plutonic, subvolcanic and volcanic suites of the western Ross Embayment (Rocchi *et al.* 2002*a*).

Plutonic rocks

The plutons of the Meander Intrusive Group are aligned along 200 km of the Ross Sea coast of northern Victoria Land, on the elevated western rift shoulder between Campbell and Mariner glaciers (Fig. 1), and are usually associated with strong positive magnetic anomalies (Müller *et al.* 1991). The largest intrusions, Mount McGee, Greene Point, No Ridge and Cape Crossfire igneous complexes, each cover about 70–80 km^2 (Fig. 1). The intrusions at Mount Monteagle, Cape King and Eagles–Engberg Bluffs are smaller, with the best exposures in steep cliffs. All the intrusions have an isotropic internal fabric, although the shape generally shows a weak elongation (*c.* 1:1.7) striking around N140°E. Compositional zoning is common in the intrusions (Tonarini *et al.* 1997), with mafic and felsic portions sometimes interlayered and/or co-mingled (Rocchi *et al.* 2002*b*). The mafic rocks are medium- to fine-grained equigranular, with gabbro and diorite compositions, and may show evidence of cumulus processes. Syenite rocks are generally equigranular, with grain size strongly variable between intrusions. Intermediate-composition plutons are less common, and mostly occur in the Mount McGee Igneous Complex. Representative textural and mineralogical features are summarized in Table 1.

Subvolcanic rocks

There are at least 180 dykes in the Meander Intrusive Group, and they occur as swarms cutting either Paleozoic basement or Cenozoic intrusive complexes (Rocchi *et al.* 2002*a*). Dykes intruding the plutons have sharp rectilinear contacts, whereas in some instances they show strongly lobate interfaces grading

Table 1. *Synopsis of petrographical features of Cenozoic plutons, dykes and volcanic rocks from northern Victoria Land, the Western Ross Supergroup*

		Notes	Rock type	Texture	Paragenesis	Alteration
Meander Intrusive Group						
Plutonic–subvolcanic complexes	Cape Crossfire Igneous Complex	Layered intrusion	Monzogabbro–monzodiorite	Equigranular, fine–medium grained, hypidiomorphic	Pl, Hbl ± Di ± Bt ± Kfs	Chl after Bt
		Layered intrusion	Alkali feldspar leucomicrogranite	Granophyric	Afs, Qz (Bt)	
	Eagles–Engberg Igneous Complex	Dykes (Mount Prior) Stockwork	Alkali feldspar rhyolite	Granophyric	Afs, Qz, Bt	Hydrothermal alteration
	No Ridge Igneous Complex	Layered intrusion	Alkali feldspar granite	Granophyric	Kfs, Pl, Qz, Bt	
		Unsampled	Gabbro	Fine–medium grained, miarolitic	Afs, Aeg, Arf, Aenig ± Qz	
	Cape King Igneous Complex		Peralkaline syenite	Inequigranular, fine–medium grained	Pl, Bt, Hbl	
		Dyke	Gabbro	Slightly inequigranular, medium grained, hypidiomorphic	Kfs, Pl, Qz, Hbl, Cpx, Ol, Bt	
	Greene Point Igneous Complex		Alkali feldspar syenite			
		Pluton	Gabbro–diorite	Equigranular, medium grained, hypidiomorphic cumulate (scattered biotite)	Pl, Aug, Bt, Ol (rare), Magn, Pl ± biotite	Rare Chl ± Act
		Pluton	Syenite	Equigranular, coarse grained, allotriomorphic	Perthite Kfs, Na-Amph ± Na-Px ± Fa	Minor Kln
				Inequigranular, fine–medium grained, hypidiomorphic	Perthite Kfs, Qz, Na-Px ± Na-Amph	Minor Kln
		Synplutonic dykes	Basanite–tephrite	Aphanitic to slightly porphyritic (<5%)	*pheno*: Krs ± Ol ± Di *gm*: Pl, Ne, Krs, Mag	(Cc, Ep, Chl, Act, Ser)
		Synplutonic dykes	Mugearite to trachyte	Porphyritic (10–20% large phenocrysts) or subaphyric	*pheno*: Plg, Krs ± Fa ± Ano *gm*: Pl, Krs/Bt ± San	(Cc, Ep, Chl, Act, Ser)
		Synplutonic dykes	Peralkaline trachyte	Slightly porphyritic (5%, small phenocrysts)	*pheno*: San *gm*: San, Aeg, Aenigm	Qz + Kfs recrystallized
	Mount Monteagle Igneous Complex	Pluton	Gabbro	Equigranular, medium grained, hypidiomorphic	Pl, Aug, Hyp, Mt, Magn, Ilm	Rare Chl ± Act
		Pluton	Syenite	Equigranular, fine–medium grained, hypidiomorphic, miarolitic	Kfs (perthite), Na-Px ± Aenigm	Minor Kln
		Synplutonic dykes	Basanite–tephrite	Slightly porphyritic (<5%)	*pheno*: Plg ± Ol (rare) ± large Krs *gm*: Plg, Krs/Bt ± Ol ± Ne, Mag	(Cc, Ep, Chl, Act, Ser)
		Synplutonic dykes	Benmoreite	Aphyric, fine grained, to hypocrystalline slightly	*gm*: Plg, Krs/Bt, Hbl, Mag	(Cc, Ep, Chl, Act, Ser)
	Mount McGee Igneous Complex	Pluton	Diorite	Equigranular, medium–coarse grained	Plg, Aug, Bt, Hyp relicts	Minor
		Pluton	Monzonite–syenite	Inequigranular allotriomorphic	Kfs (perthite), Pl (in monzonites), Qz, Hbl, Bt	Minor
		Synplutonic dykes	Basanite–hawaiite–mugearite	Aphyric–subaphyric, sometimes slightly hypocrystalline	*micropheno*: Krs *gm*: Plg ± San, Amph/Krs,Bt, Mag	(Cc, Ep, Chl, Act, Ser)
		Synplutonic dykes	Trachyte	Mainly aphyric–subaphyric	*pheno*: San or Anort	Qz + Kfs recrystallized
Dyke swarms	Random Hills Dyke Swarm	Dykes	Alkali basalt–hawaiite–tephrite	Subaphyric	*pheno*: Ol, Krs *micropheno*: Pl *gm*: Plg, Krs, Magn ± Bt ± San	Cc, Ser ± Chl ± Act
	Bier Point Dyke Swarm	Dykes	Alkali basalt–hawaiite–tephrite	Subaphyric	*pheno*: Ol, Krs *micropheno*: Pl *gm*: Plg, Krs, Magn ± Bt ± San	Cc , Se r ± Ch ± Act
		Dykes	Trachyte–rhyolite	Aphyric–subaphyric	*pheno*: San or Ano ± Qz ± Bt ± Na-Amph	Silicified groundmass

(*Continued*)

Table 1. Continued.

	Notes	Rock type	Texture	Paragenesis	Alteration
McMurdo Volcanic Group					
Hallett Volcanic Province					
Adare Peninsula Volcanic Field	Adare seamounts	Alkali basalt, basanite	Porphyritic, variolitic (Cpx–Plg intergrowth)	*pheno*: Ol (skeletal), Cpx, ± Plg ± Amph	Rare Iddingsite, some amygdales
	Adare Peninsula	Alkali basalt, basanite	Porphyritic (5–25%), vesicular, hypocrystalline–microcrystalline groundmass	*pheno*: Ol, Cpx ± Pl *gm*: Ol, Cpx, Pl, Mag, glass	Bowlingite/s eraentine on olivine, some Amygdales (calcite and clay minerals)
	Possession Islands	Rare trachyte and phonolite	Porphyritic	*pheno*: Ano or San	
		Alkali basalt, basanite	Porphyritic, holo- to hypocrystalline	*pheno*: Ol, Cpx (embayed) ± Pl *gm*: Pl, Mag, Cpx, Ol, Amph ± glass	
Hallett Peninsula Volcanic Field		Hawaiite	Porphyritic–aphyric, with holocrystalline gm, sometimes subophitic	*pheno*: Ol, Aug, Hbl, minor Plg *gm*: Aug, Pl, Ano ± Ol ± Ne	Minor carbonates
		Trachyte	Subaphyric, sometimes microvesicular, sometimes pilotaxitic	*pheno*: Ano, Plg, Ol, Aeg-Aug *gm*: Ano, Pl, Cpx ± Ol, Mag	Iddingsite on Ol
		Phonolite?	Porphyritic holocrystalline	*pheno*: Cpx, San, Ne(?) ± Hbl *gm*: San, Ne(?)	Iddingsite on Ol Analcite, Cal Clay minerals, zeolites
Daniell Peninsula Volcanic Field		Basanite	Poorly porphyritic, with hypocrystalline–microcrystalline gm	*pheno*: Ol, Aug, Mag ± Pl *gm*: Ol, Cpx, Plg, Mag	
		Alkali basalt–hawaiite	Strongly porphyritic	*pheno*: Pl (An$_{50-60}$), Aug, Ol, Mag *gm*: Pl (An$_{35-40}$), Aug, Ol, Mag	
		Mugearite–benmoreite	Microporphyritic	*pheno*: Pl, Cpx, Amph ± Ol *gm*: alk-fs, Aug, Ol, Mag, Ap	
		Trachyte (pantelleritic)	Porphyritic, with holocrystalline to glassy gm	*pheno*: Ano ± Bt ± Ol-Fa ± Na-Amph *gm*: Ano ± Aeg ± Na-Amph ± Sdl	
Coulman Island Volcanic Field		Basalt	Aphyric–porphyritic, microvesicular hypocristalline	*pheno*: Plg, Ol, Cpx *gm*: Pl, Cpx, Ol, glass	
		Trachyte	Porphyritic Pilotaxitic	*pheno*: Ano *gm*: San, Pl, minor Aeg-Aug, Bt	
Northern Local Suite	Inland scoria cones, north of Tucker Glacier	Alkali basalt, basanite	Porphyritic (10–20%), vesicular, hypocrystalline mantle and crustal xenoliths	*pheno*: Ol, Cpx ± Pl	Analcite Rare Iddingsite, minor hydroxides
Melbourne Volcanic Province					
Malta Plateau Volcanic Field		Basanite, hawaiite	Hypocrystalline porphyritic	*pheno*: Ol, Cpx, Plg, Mag ± Ne ± glass *gm*: Ol, Cpx	
		Comendite, pantellerite	Porphyritic	*pheno*: Cpx, Pl, Ol, Fe-Ti oxides, glass *pheno*: Na-Aug, Na-San ± Aen ± Fa	Carbonates hematite, recrystallized groundmass
The Pleiades Volcanic Field		Hawaiite–mugearite	Subaphyric, vesicular	*gm*: Afs, Crs *pheno*: Pl, Cpx, Ol, Mag *gm*: Ol, Pl ± Ne, Mag	
		Benmoreite–trachyte	Subaphyric (rarely porphyritic), poorly vesicular	*pheno*: Ano, ± Bt ± Ol-Fa ± Na-Amph *gm*: Ol, San, Mag	Minor hydroxides

Mount Overlord Volcanic Field		Alkali basalt-basanite	Subaphyric–porphyritic (20%), vesicular (up to 10%), holocrystalline pilotaxitic or hypocrystalline hyalopilitic	*pheno*: Ol, Ti-Aug, Pl (microph.), Amph *gm*: Cpx, Pl, Ol, Mag
	Mount Overlord	Trachyte–minor phonolite	Porphyritic (rarely aphyric), trachytic gm	*pheno*: large San/Ano, Plg ± Cpx ± Fa ± Bt *gm*: San, Pl, Cpx, Mag	Bt + Mag after Cpx/Amph (rare) Spherulite in obsidians
	Mount Rittmann	Alkali basalt-hawaiite	Subaphyric–porphyritic, holocrystalline–hypocrystalline, comagmatic syenitic xenoliths	*pheno*: Pl, Ol, Aug ± Amph relicts *gm*: Pl, Cpx, Ol, Mag	
		Trachyte	Hypocrystalline, trachytic groundmass – glassy pumice shards	*pheno*: San/Ano, Fa, Fe-Aug ± Pl relicts *gm*: Na-Amph, Na-Px	
Vulcan Hills Volcanic Field		Mostly intermediate (?)	Porphyritic with coarse, felsitic gm to aphyric with fine-grained gm	*pheno*: Pl, Ol, Cpx, minor Cpx ± Amph	Strong alteration, carbonates
Mount Melbourne Volcanic Field		Basanite, alkali basalt, hawaiite	Weakly porphyritic	*pheno*: Ol, Ti-Aug, Mag ± Krs megacrysts *gm*: Pl, Ol, Ne, Cpx, Mag ± glass	
		Alkali basalt-hawaiite	Variably porphyritic	*pheno*: Ol, Pl *gm*: Pl, Ol, Cpx, Mag ± glass	
		(Mugearite–benmoreite–) trachyte	Vitrophyric	*pheno*: San/Ano ± Aeg-Aug ± Fa ± Na-Amph *gm*: Pl, Ol, Cpx, Mag ± glass	
Northern Local Suite		Alkali basalt, basanite	Porphyritic, vesicular, hypocrystalline; common mantle and crustal xenoliths	*pheno*: Ol, Cpx *gm*: Ol, Cpx, Pl, Mag ± glass	

Data after observations of the authors and from the literature (Hamilton 1972; Wörner *et al.* 1989; Armienti and Tripodo 1991; Armienti *et al.* 1991; Beccaluva *et al.* 1991; Müller *et al.* 1991; Tonarini *et al.* 1997; Rocchi *et al.* 2002a, b).
Abbreviations: P, plutonic rocks; D, dykes; Act, actinolite; Aenigm, aenigmatite; Amph, amphibole; Ano, anorthoclase; Aug, augite; Bt, biotite; Cc, carbonates; Chl, chlorite; Cpx, clinopyroxene; Di, diopside; Ep, epidote; Fa, fayalitic olivine; Hbl, hornblende; Hyp, hyperstene; Ilm, ilmenite; Kfs, K-feldspar; Krs, kaersutite; Na-Amph, eckermannite/arfvedsonite; Na-Px, aegirinaugite or aegirine; Ne, nepheline; Ol, forsteritic olivine; Mag, magnetite; Pl, plagioclase; Qz, quartz; San, sanidine; Ser, sericite; *pheno*, phenocrysts; *gm*, groundmass; Alteration minerals between parentheses are found in some samples of the group. Mineral abbreviations after Whitney and Evans (2010).

into metre-sized enclave trains, indicating coeval emplacement. Therefore, the geochemical features of the dykes intruding the plutons will be described as part of the plutonic association.

Dykes cutting across the Paleozoic basement at Bier Point, Random Hills and Navigator Nunatak are mostly mafic, whereas felsic, deeply altered dykes crop out in the area between Styx and Burns glaciers and at Navigator Nunatak. At Vulcan Hills, the high elevation of the outcrops (c. 3000 m) has allowed only limited fieldwork, with strongly altered igneous rocks mostly cropping out as dykes (personal observations of S. Rocchi), and the whole appearance of the igneous complex is compatible with the strongly dissected remains of a volcanic edifice (Kyle 1990b). All the dykes in the basement invariably show rectilinear sharp contacts. The mafic dykes usually have thicknesses between 1 and 5 m, and their strikes cluster around north–south or NNW–SSE directions; whereas felsic dykes can reach 50 m in thickness, are deeply weathered and do not show any preferred orientation. The geochemical features of these dyke swarms are described here separately from those of the plutonic and volcanic rocks (Rocchi et al. 2002a).

Rare dykes that occur within volcanic edifices, such as at Mayr Spur (Mortimer et al. 2007) and Martin Hill (west of Daniell Peninsula), are described as part of the volcanic suite. Textural and mineralogical features of all the dyke types are summarized in Table 1.

Volcanic rocks

The volcanic rocks of northern Victoria Land are grouped into the two major provinces: (1) Hallett Volcanic Province, including Adare Peninsula, Hallett Peninsula, Daniell Peninsula and Coulman Island volcanic fields, mainly composite, elongated (c. 25 × 40–75 km) shield volcanoes erupted from fissures; and (2) Melbourne Volcanic Province, including Mount Melbourne (c. 25 × 55 km), Mount Overlord, The Pleiades and Malta Plateau (c. 30 × 50 km) volcanic fields, mainly stratocones with central vents. In addition, there are numerous small volcanic outcrops assigned to the volumetrically minor Northern Local Suite (Fig. 1c), which are mainly scoria cones (Chapter 5.1a: Smellie and Rocchi 2021).

The volcanic products are dominated by lavas, breccia and scoria deposits, according to the eruption style (Smellie et al. 2011a) (Chapter 5.1a: Smellie and Rocchi 2021). Compositionally, they correspond to a moderately alkaline sodic association, with mostly bimodal distribution: basaltic products are dominant (Hamilton 1972), mostly basanites and alkali basalts–hawaiites, along with rare nephelinites; evolved products are mainly trachytes and minor rhyolites, some of which are peralkaline. Most of the samples previously described as tephrites based on their olivine content (lower than basanites) do not actually contain any feldspathoid minerals, and therefore the term tephrite appears misused here (Le Bas et al. 1986). Most primitive samples (Nardini et al. 2009) were extruded as small lava flows, commonly marking the end of explosive activity in monogenetic scoria cones. Products of intermediate composition, such as mugearites and benmoreites, are volumetrically minor. Strongly alkaline compositions are typically very uncommon in northern Victoria Land, and they are limited to phonolites and rare tephrites and foidites. Representative textural and mineralogical features are summarized in Table 1.

In addition, numerous small scattered volcanic centres of the Northern Local Suite, mainly scoria cones, are present inland between Mount Melbourne and Tucker Glacier. They are also present north of Tucker Glacier and are described as a separate subset of the Northern Local Suite in Table 1.

Geochemical features

The geochemical dataset used in this review includes: (i) almost 600 analyses for major elements (47 plutons/synplutonic dykes, 117 dykes and 411 lavas), more than 500 analyses for some trace elements and almost 90 for complete trace elements, and 122 radiogenic isotope analyses in whole-rock samples (volcanic and plutonic/subvolcanic) from northern Victoria Land (see Supplementary Table S1); (ii) c. 900 major element analyses, c. 550 analyses for trace elements and c. 130 for radiogenic isotopes from whole-rock samples of volcanic rocks from southern Victoria Land (excluding englacial tephra and volcanic clasts in onshore/offshore drillcores); and (iii) c. 230 analyses for major elements, c. 180 for trace elements and c. 70 for radiogenic isotopes from whole-rock samples from Marie Byrd Land. Tables 2–5 contain representative published analyses of plutons, dykes and lavas from northern Victoria Land, and also include new Pb isotopic data for selected plutonic rocks and dykes to complement the Pb isotopic data for volcanic rocks.

Major elements

Plutonic rocks mainly have gabbro–syenite bimodal compositions; nevertheless, intermediate compositions like monzonites are quite well represented (Fig. 2b) and a complete differentiation trend is observed. The least evolved intrusive rocks are often Ol–Hy-normative (CIPW-norm) and mildly alkaline, whereas syenites are generally Q-normative and sometimes peralkaline. Most dykes are either mafic or evolved and show a Daly compositional gap typical of alkaline associations (Rocchi et al. 2002a). The entire dyke association is alkaline (Fig. 2b), and is generally Ne-normative.

The lavas have compositions typical of a moderately alkaline association (Fig. 2b), with most mafic examples represented by basanites and alkali-basalts (Wörner et al. 1989; Armienti and Tripodo 1991; Armienti et al. 1991; Beccaluva et al. 1991; Hornig and Wörner 2003; Kim et al. 2018). Tephrites, tephriphonolites, phonotephrites and phonolites are very rare. The few analyses with these compositions, once carefully screened for Na$_2$O content, Na$_2$O/K$_2$O and CaO v. Na$_2$O + K$_2$O, reveal inconsistencies probably linked to chemical modification by alteration processes. The minor group of fresh phonolites does not appear to represent an independent strongly alkaline differention trend. Rather, they may represent products that evolved from trachytes under petrochemical conditions slightly below the critical plane of silica saturation. Trachytes are commonly the most evolved rocks in several volcanic edifices, while rhyolite compositions are reached in others; in some instances, both trachytes and rhyolites reach peralkalinity (Müller et al. 1991). When compared with the other major volcanic regions of the WARS, northern Victoria Land volcanic rocks have a moderately alkaline nature similar to the volcanism in Marie Byrd Land, whereas strongly alkaline products, ranging from tephrite to phonolite, are typically found only in southern Victoria Land (Erebus Volcanic Province: Fig. 2a) (see Chapter 5.2b: Martin et al. 2021).

The compositional features of the lava samples are grouped into three main subsets, linked to the type and location of volcanic edifices, as follows: (i) coastal lavas from the shield volcanoes and stratovolcanoes of Adare, Hallett and Daniell peninsulas and Coulman Island; (ii) inland lavas from the Malta Plateau, Mount Rittmann, The Pleiades and isolated scoria cones of the Northern Local Suite; and (iii) the central edifice of Mount Melbourne, and its parasitic cones and satellite volcanoes. The inland lavas display the most complete differentiation trend, which extends to high-silica rhyolites with normative Q > 30 % (Fig. 2e), and most samples with normative

Table 2. Selected representative major-trace element compositions and radiogenic isotope ratios for Cenozoic igneous rocks from northern Victoria Land – coastal volcanoes

	Hallett Volcanic Province																
Province																	
IVD*	A1					A2			A3			A4			A5		
Unit:	Adare Peninsula		Adare Trough			Hallett Peninsula			Daniell Peninsula			Coulman Island			Possession Islands		
Sample	A240B-1	A210B	A232B	D4-1	D9-1	D12-1	AW82229	VC-9	14-247C	MP8	A223	MP-10	MP34	CI-103	CI-1	MP-37	AW82214	A225A	NV-3B	AW82205
Rock type†	V	V	V	V	V	V	V	V	V	V	V	V	V	V	V	V	V	V	V	V
Reference	Panter et al. (2018)	Panter et al. (2018)	Panter et al. (2018)	Panter et al. (2018)	Panter et al. (2018)	Panter et al. (2018)	Kyle (1990b)	Kyle (1990b)	Hamilton (1972)	Nardini et al. (2009)	Panter et al. (2018)	Nardini et al. (2009)	Nardini et al. (2009)	Rocholl et al. (1995)	Rocholl et al. (1995)	Nardini et al. (2009)	Panter et al. (2018)	Panter et al. (2018)	Aviado et al. (2015)	Panter et al. (2018)
Locality	Cape Adare SW	Cape Roget, SW	Adare Peninsula	Adare Trough	South Adare Basin	South Adare Basin	Hallet Peninsula	Hallet Peninsula	Hallet Peninsula	Martin Hill	West of Daniell Peninsula	Daniell Peninsula	Coulman Island	Coulman Island	Coulman Island	Coulman Island	Possession Islands	Foyn Island	Possession Islands	Foyn Island
Classification	Basanite	Alkali basalt	Hawaiite	Basanite	Tephrite	Basanite	Hawaiite	Mugearite	Comenditic trachyte	Basanite	Alkali basalt	Trachyte	Basanite	Alkali basalt	Benmoreite	Comenditic trachyte	Basanite	Basanite	Basanite	Alkali basalt
Major elements (wt%)																				
SiO₂	42.53	45.65	46.87	43.07	45.26	44.60	47.36	50.14	67.75	44.12	46.31	63.30	42.81	45.08	56.09	61.28	41.64	45.46	44.78	45.57
TiO₂	3.80	2.35	2.87	2.99	3.18	3.06	3.04	2.31	0.24	2.71	2.24	0.52	3.08	3.28	1.02	0.82	3.47	3.15	3.60	2.96
Al₂O₃	14.54	13.42	15.43	12.92	15.18	13.34	16.78	16.03	14.88	14.87	14.48	17.22	14.79	16.77	19.06	16.30	13.07	15.50	15.82	14.74
Fe₂O₃							12.26	11.96	3.77	5.06		2.58	3.19	8.21	3.57	4.01				
FeO	12.66	10.60	11.54	10.20	11.48	11.47			0.31	8.20	11.86	2.02	8.80	3.71	2.81	2.20	12.23	11.22	12.93	11.43
MnO	0.21	0.19	0.21	0.20	0.26	0.21	0.21	0.24	0.14	0.20	0.19	0.20	0.19	0.20	0.25	0.27	0.22	0.20	0.22	0.19
MgO	8.72	13.08	7.14	13.32	6.39	10.70	4.20	2.99	0.02	9.51	9.49	0.47	11.14	4.33	0.92	0.70	9.78	7.74	6.45	9.23
CaO	11.95	9.83	9.84	10.84	9.42	10.02	7.88	6.48	1.01	10.89	11.22	1.75	10.99	9.91	4.06	1.28	12.60	10.58	10.37	10.71
Na₂O	3.47	3.16	3.75	3.64	5.58	4.12	4.40	4.94	5.80	2.52	2.75	6.41	2.80	3.00	6.89	7.73	4.29	4.06	3.65	3.50
K₂O	1.42	1.17	1.61	1.88	2.16	1.54	1.48	2.31	5.25	0.79	0.95	4.86	1.03	1.67	2.88	4.22	1.52	1.37	1.21	1.06
P₂O₅	0.69	0.55	0.73	0.94	1.09	0.93	0.72	1.04	0.01	0.39	0.50	0.18	0.60	0.75	0.24	0.14	1.17	0.71	0.96	0.61
LOI	0.82			1.79	1.73	1.17	1.45	1.34	0.29	0.73		0.48	0.58	2.06	0.89	1.04		0.56		
Trace elements (μg g⁻¹)																				
Sc	29.5	24.7	24.6	27.1	16.4	22.2					32.0	12.1				6.70	27.0	23.0		24.9
V	312		235	233	185	214						2.10		259	16	4.80	296	263	215	278
Cr	305	479	260	540	135	355					7	1		12	1	2	375	238	35	431
Co												1.60		50	44	1.50				
Ni	115		99.0	338	70.9	205						0.67		10	2	0.77	167	117	296	177
Cu	72.0		57.0	45.5	25.3	45.9								18	6		30	58	101	60
Zn	95		100	80.6	136	102								121	113		101	94	87	91
Ga	21		21	15.5	21.3	18.5											21	21		21
Rb	32	26	32	31	40	35				20	39	87	24	37	92	94	47	36	20	26
Sr	800	592	787	861	1216	1007				552	608	210	667	826	1052	63	1034	831	925	684
Y	35.0	26.4	35.0	26.5	33.7	30.3				24.3	40.9	34.6	25.0	35	45	61	38	33	28	30
Zr	263	397	276	292	553	347				158	225	516	204	310	653	889	332	280	217	223
Nb	83.0	67.7	70.0	93.6	155	89.4				40.3	42.4	126.0	67.5	87	206	265	118	82	61	71
Mo										2.22		9.11	2.43			8.75				
Cs	0.40	0.54	0.69								0.31						0.54	0.33		0.29
Ba	422	418	616	467	558	401				251	256	1771	320	404	806	1235	626	396	371	319
La	52.5	43.6	51.8	51.0	88.6	56.5				29.3	32.8	87.4	39.8	51.6	103	136	73.8	54.2	47.0	42.2
Ce	104	83.0	105	95.5	167	112				64.5	69.1	170	81.3	106	200	265	143	108	83.0	85.0

(*Continued*)

Table 2. *Continued.*

Province					Hallett Volcanic Province				
IVD*		A1		A2	A3		A4		A5
Unit:	Adare Peninsula		Adare Trough	Hallett Peninsula	Daniell Peninsula		Coulman Island		Possession Islands

Pr		10.9	10.3	17.1	12.0		8.37	17.6	9.98	12.5	21.0	28.3	61.5	46.4	10.7
Nd	50.0	40.0	45.9	40.6	63.9	48.4	33.0	65.2	40.3	52.3	71.2	97.7	12.01	9.18	46.0	40.5
Sm	9.95	7.54	9.66	7.23	10.8	9.11	6.89	10.4	7.73	9.76	11.12	16.8	3.68	2.82	9.60	8.18
Eu	3.04	2.39	3.44	2.45	3.35	2.90	2.41	3.98	2.63	3.10	3.53	4.36			3.10	2.58
Gd				14.8	24.4	18.0	7.28	10.8	8.25	8.7	8.99	16.8			9.00	
Tb	1.21	1.10	1.13	1.11	1.45	1.30	1.01	1.47	1.06	1.25	1.28	2.44	1.39	1.08	1.20	1.01
Dy		5.19		4.90	6.38	5.74	5.49	7.79	5.46	7.62	8.52	13.1			5.60	
Ho		0.90		0.90	1.14	1.03	0.99	1.42	1.01	1.34	1.69	2.36			0.90	
Er		2.64		1.91	2.50	2.14	2.58	3.62	2.58	3.41	4.49	6.22			2.40	
Tm				0.31	0.41	0.34	0.35	0.58	0.33	0.49	0.72	0.89				
Yb	2.32	2.11	2.33	1.72	2.35	1.94	2.05	3.51	2.10	2.97	4.67	5.84	2.37	2.19	2.20	2.05
Lu	0.33	0.26	0.30	0.28	0.34	0.28	0.30	0.59	0.29	0.66	0.73	0.83	0.31	0.34	0.30	0.26
Hf	6.73	7.48	6.44				4.69	9.92	4.99			16.9	7.57	6.47	5.20	5.41
Ta	5.25	3.90	4.24				2.57	7.99	3.91			14.3	6.89	5.19	4.70	4.51
W							0.61	2.77	0.77			2.88				
Tl																
Pb	2.00	2.45	5.00	2.52	3.44	2.01	1.73	2.06	2.14			9.29	2.00	2.00	1.60	2.00
Th	6.11	5.07	5.53	5.33	6.77	4.81	3.43	12.6	5.22			20.6	8.33	6.64	5.50	4.37
U	1.70	1.68	2.00	1.41	2.20	1.33	1.02	3.70	1.46			6.07	2.80	2.04	1.00	0.90
Time (Ma)[‡]		0.05		3.35	2.86	2.89	12.20	6.70	7.60	8.00	8.00	7.30		0.24		
^{87}Sr/^{86}Sr m	0.702944	0.703458	0.703741	0.702870	0.702900	0.702779	0.703357	0.703836	0.702830	0.702828	0.702902	0.702601	0.702884	0.702970	0.703430	0.702900
^{87}Rb/^{86}Sr		0.128		0.105	0.096	0.0999	0.105	1.198	0.104	0.130	0.253	4.31		0.125		
^{87}Sr/^{86}Sr i	0.702944	0.703458	0.703741	0.702865	0.702896	0.702775	0.703339	0.703722	0.702819	0.702813	0.702873	0.702154	0.702884	0.702970	0.703430	0.702900
^{143}Nd/^{144}Nd m	0.512930	0.512891	0.512874	0.513000	0.512896	0.513001	0.512902	0.512860	0.512958	0.512911	0.512934	0.512941	0.512945	0.512935	0.512820	0.512945
^{147}Sm/^{144}Nd		0.1140		0.1078	0.1018	0.1137	0.1261	0.0964	0.1160	0.1128	0.0944	0.1040		0.1196		
^{143}Nd/^{144}Nd i	0.512930	0.512891	0.512874	0.512998	0.512958	0.512999	0.512892	0.512856	0.512952	0.512905	0.512929	0.512936	0.512945	0.512935	0.512820	0.512945
ε_{Nd} i	5.7	4.9	4.6	7.1	6.3	7.1	5.3	4.4	6.3	5.4	5.9	6.0	6.0	5.8	3.6	6.0
^{206}Pb/^{204}Pb	20.026	19.363	19.432	19.237	20.234	20.111	19.641	19.745		20.143		9.29	19.851	20.044	20.028	20.184
^{207}Pb/^{204}Pb	15.633	15.615	15.642	15.583	15.653	15.625	15.517			15.663			15.620	15.587	15.658	15.653
^{208}Pb/^{204}Pb	39.504	39.202	39.285	38.812	39.715	39.551	39.201			39.600			39.354	39.438	39.916	39.703
^{206}Pb/^{204}Pb i	20.026	19.362	19.432	19.201	20.203	20.080	19.508	19.662		20.056			19.851	20.041	20.028	20.184

Data after observations of the authors and from the literature (Hamilton 1972; Kyle 1982, 1990b; Wörner et al. 1989; Müller et al. 1991; Rocholl et al. 1995; Rocchi et al. 2002a; Hornig and Wörner 2003; Nardini et al. 2009; Aviado et al. 2015; Panter et al. 2018). For unpublished analyses and unpublished Pb isotopic data for symplutonic dykes and dyke swarms, methods are after Nardini et al. (2009).
*IVD, individual volcano description after LeMasurier and Thomson (1990).
[†]Rock type: V, volcanic rock; P, plutonic rock; D, dyke.
[‡]Time: age used to calculate the initial radiogenic isotope ratios; the age is most commonly determined on the sample and, in some cases, is inferred by analogy.
LOI, loss on ignition. ^{87}Sr/^{86}Sr m and ^{143}Nd/^{144}Nd m, measured isotopic ratio; ^{87}Sr/^{86}Sr i and ^{143}Nd/^{144}Nd i, initial isotopic ratio.

Table 3. *Selected representative major-trace element compositions and radiogenic isotope ratios for Cenozoic igneous rocks from northern Victoria Land – inland volcanoes and Mount Melbourne volcano*

Province										Melbourne Volcanic Province									
IVD*	A6			A7				A8				A9	A10				A11	
Unit	Malta Plateau			The Pleiades			Mount Overlord		Mount Rittmann			Vulcan Hills	Mount Melbourne				Local Suite Basaltic Rocks	
Sample	MP24	MA-117	MS3265	MA-100	25682	5.1.2.14 JS1	5.1.2.14 JS12	5.1.2.14 JS13	MB 78	NN004	MB 81A	NN 14	NN 17	NN 24	82050	MB 54	MB 32	MB 26	MB 23	MM 04-173	SAX 20	MB S5b
Rock type†	V	V	V	V	V	V	V	V	V	V	V	V	V	V	D	V	V	V	V	V	V	V
Reference	Nardini et al. (2009)	Rocholl et al. (1995)	Müller et al. (1991)	Rocholl et al. (1995)	Kyle (1982)				Nardini et al. (2009)	Hornig and Wörner (2003)	Armienti and Tripodo (1991)	Armienti and Tripodo (1991)	Armienti and Tripodo (1991)	Armienti and Tripodo (1991)	Kyle (1990b)	Nardini et al. (2009)	Nardini et al. (2009)	Nardini et al. (2009)	Nardini et al. (2009)	Wörner et al. (1989)	Nardini et al. (2009)	Nardini et al. (2009)
Locality	Malta Plateau	Malta Plateau			Cone 11	Mount Pleiones	Aleyone Cone	Aleyone Cone	Nathan Hills–Aviator Glacier	Navigator Nunatak	Mount Noice	Mount Rittmann, calderic rim	Mount Rittmann, calderic rim head	Icebreaker Glacier, Tripodo	Vulcan Hills	Baker Rocks, north	Baker Rocks, north	Shield Nunatak	Edmonson Point, south	Mount Melbourne	Greene Point scoria cone	Browning Pass, cone
Classification	Basanite	Alkali basalt	Comenditic trachyte	Pantellerite	Hawaiite	Benmoreite	Trachyte	Comenditic trachyte	Alkali basalt	Trachyte	Comendite	Trachyte	Mugearite	Comenditic trachyte	Benmoreite	Basanite	Basanite	Basanite	Hawaiite	Trachyte	Basanite	Basanite
Major elements (wt%)																						
SiO₂	43.31	46.84	65.96	72.11	46.11	54.17	63.15	63.64	45.84	66.26	74.88	61.31	50.36	62.10	58.80	44.09	44.68	44.76	49.07	63.10	40.78	43.91
TiO₂	2.80	2.62	0.34	0.32	2.41	1.71	0.36	0.19	2.24	0.20	0.20	0.85	2.47	0.41	1.26	3.01	3.16	2.97	2.70	0.52	3.62	2.76
Al₂O₃	13.87	14.19	15.14	10.15	14.35	17.39	17.16	16.61	16.72	16.69	11.08	16.79	17.63	17.23	15.64	13.92	15.11	14.30	16.29	15.55	11.81	14.30
Fe₂O₃	2.68	3.50	4.66	2.00	3.30	2.45	1.72	1.26	2.06	1.55	2.32	5.26	4.50	2.66	8.20	2.12	2.87	1.69	2.29	2.81	11.00	4.30
FeO	7.94	8.28		3.16	8.35	6.13	3.44	2.52	9.18	0.10	0.90	1.09	6.41	2.68		9.89	9.15	9.97	9.86	3.10	4.60	8.49
MnO	0.18	0.19	0.13	0.13	0.21	0.18	0.16	0.15	0.19	0.05	0.05	0.15	0.20	0.19	0.19	0.19	0.17	0.19	0.25	0.15	0.28	0.20
MgO	12.50	8.38	0.13	0.04	8.94	2.55	0.29	0.12	8.25	0.10	0.07	0.75	3.94	0.18	1.71	12.03	10.31	10.18	3.43	0.21	9.33	10.30
CaO	11.03	10.83	1.12	0.32	10.04	4.80	1.47	0.96	9.67	1.06	0.32	2.33	6.72	0.93	3.66	9.27	8.97	11.44	7.51	1.97	9.23	9.04
Na₂O	2.99	2.62	5.96	5.06	3.84	5.37	6.52	7.31	3.41	6.00	4.81	6.21	5.24	7.60	4.84	2.88	3.25	2.63	4.79	5.58	5.31	2.85
K₂O	1.31	1.14	5.11	4.54	1.48	3.34	5.33	5.09	1.14	5.41	4.33	4.65	1.53	5.18	3.65	1.17	1.16	0.75	2.01	4.68	1.54	1.11
PA	0.56	0.42	0.06	0.01	0.59	0.57	0.08	0.02	0.41	0.05	0.01	0.18	0.59	0.05	0.55	0.51	0.50	0.59	1.12	0.08	1.60	0.53
LOI	0.84	0.44	0.25	0.87		0.33	0.24	0.76	0.89	1.77	1.03	0.43	0.41	0.79	1.08	0.93	0.68	0.54	0.67	0.22	0.90	2.20
Trace elements (μg g⁻¹)																						
Sc			6			8.13	4.16		29.9							31.0	29.7	33.5	16.0	4.6	17.4	26.8
V	302		16	6	185	64.2	0.19		219	7		177	42			165	170	176	87	5	130	163
Cr	349		16	1	317	11.9	1.31		365	1		246	630	12		449	360	632	8	16	369	454
Co	57		9	27		16.7	0.99			26		95	31	52					15	3		
Ni	93		8	86	190	10.2	0.40			2		47	238	701	12				6	4		
Cu	42		12	6	60	17.4	7.68			3		764	65	178	138				38	15		
Zn	108		152	201	97	136	131			99		144			26					153		
Ga						21.3	24.2															
Rb	38	30	236	250	37	98	185		33	203	219		42	113	103	36	34	19	51	152	72	28
Sr	640	531	66	5	745	662	80		603	11	5	246	630	12	537	691	715	648	742	186	1494	684
Y	23	33	59.0	119		33.7	43.2		24.0	46	95	47	31	52	47	30.7	27.7	21.8	39.0	64	48.6	26.3
Zr	194	200	726	1500		471	750		189	779	1392	764	238	701	486	284	269	165	313	719	500	262
Nb	70.6	52	188	257		124	195		53.9	185	170	144	65	178	94	66.7	67.0	44.6	96.0	155	183	67.9
Mo	2.70		12.0			4.72	9.52		2.69							3.63	2.97	2.11			9.65	2.28
Cs						1.77	3.63															

(*Continued*)

Table 3. *Continued.*

Province					Melbourne Volcanic Province											
IVD*	A6			A7				A8		A9		A10			A11	
Unit	Malta Plateau			The Pleiades			Mount Overlord	Mount Rittmann		Vulcan Hills		Mount Melbourne			Local Suite Basaltic Rocks	
Ba	431	337	307	8	416	901	716	358	5	167	686	440	41	361	345	257	669	1047	388	830
La	40.5	32.7	111	172		86.0	121	35.3	89	105	82	42	94	41.9	37.9	30.2	72.1	101	40.2	125
Ce	86.0	67	200	326		159	214	69.0	181	298	175	91	195	87.6	68.6	63.4	139	196	84.4	164
Pr	10.2	8.0		36.5		17.1	21.5	8.20	21.2					10.8	9.64	8.31			10.4	27.6
Nd	39.0	32.0		129		59.7	70.7	31.9	82.1					41.3	38.8	33.5	68.0	76	40.5	104
Sm	7.30	7.08		27.3		9.79	11.1	6.06	14.8					8.51	8.19	7.00	14.2	14.6	8.08	18.9
Eu	2.46	2.13		0.92		2.81	1.76	2.01	4.82					2.72	2.75	2.46	4.90	3.12	2.67	5.71
Gd	7.72	6.8		23.3		7.47	8.05	6.49	12.6					8.65	8.44	7.19	12.8		8.14	17.7
Tb	1.01	1.00		3.76		1.10	1.30	0.93	1.67					1.19	1.10	0.96		1.97	1.05	2.24
Dy	5.11	67.0		22.0		6.38	7.95	5.20	8.92					6.87	6.50	5.01	9.0		5.81	11.8
Ho	0.89	1.10		4.33		1.21	1.56	0.92	1.52					1.21	1.07	0.86		2.3	0.97	1.93
Er	2.25	3.06		11.35		3.42	4.64	2.53	3.98					3.28	2.69	2.22	4.20		2.60	4.96
Tm	0.29	0.40		1.70		0.49	0.70	0.34	0.51					0.42	0.37	0.27			0.32	0.61
Yb	1.87	2.55		11.0		3.16	4.61	2.29	3.06					2.69	2.18	1.56	3.20	5.30	2.09	3.76
Lu	0.26	0.37		1.59		0.48	0.70	0.32	0.34					0.37	0.30	0.21	0.40	0.74	0.27	0.57
Hf	4.62					10.2	15.6	4.20						6.29	6.13	3.97		15.1	5.33	12.8
Ta	4.17		17			7.97	12.8	3.10						3.72	3.63	2.61		10.1	3.52	10.3
W	0.99					1.64	3.88	0.83						1.02	0.85	0.58			0.68	3.06
Pb	2.22		150			9.62	16.7	3.34						2.64	1.32	1.46			1.77	4.97
Th	5.89		33			13.8	26.2	4.65		13				5.92	4.97	3.22		21.8	4.66	16.7
U	1.63		9			3.35	7.03	1.16		12				1.58	1.47	0.95		4.70	1.33	4.93
Time (Ma)‡	10.00	15.00			15.00			18.00						0.30	0.30	0.48			0.24	0.01
⁸⁷Sr/⁸⁶Sr m	0.703160	0.703819		0.734629		0.703738	0.704201	0.704027	0.703376					0.703093	0.702898	0.703291		0.704990	0.702840	0.702989
⁸⁷Rb/⁸⁶Sr	0.171	0.163		143		0.162	1.29	6.70	0.158					*0.151*	*0.138*	*0.085*			0.118	0.139
⁸⁷Sr/⁸⁶Sr i	0.703136	0.703784		0.704167		0.703738	0.704201	0.704027	0.703336					0.703092	0.702897	0.703290		0.704990	0.702840	0.702989
¹⁴³Nd/¹⁴⁴Nd m	0.512876	0.512789		0.512779		0.512821	0.512770	0.512806	0.512883					0.512930	0.512955	0.512953		0.512840	0.512955	0.512979
¹⁴⁷Sm/¹⁴⁴Nd	0.1132	0.1338		0.1279		0.1116	0.0968	0.0950	0.1150					0.1245	0.1277	0.1265			0.1207	0.1095
¹⁴³Nd/¹⁴⁴Nd i	0.512869	0.512776		0.512766		0.512821	0.512770	0.512806	0.512869					0.512930	0.512955	0.512953			0.512955	0.512979
εNd i	4.8	3.1		2.9		3.6	2.6	3.3	5.0					5.7	6.2	6.2			6.2	6.7
²⁰⁵Pb/²⁰⁴Pb	19.647	19.320				19.097	19.065	19.155	19.348					19.450	19.535	19.460			19.558	19.568
²⁰⁷Pb/²⁰⁴Pb	15.567	15.620				15.633	15.635	15.624	15.592					15.591	15.407	15.515			15.571	15.540
²⁰⁸Pb/²⁰⁴Pb	39.280	39.270				39.078	39.091	39.185	39.166					39.210	38.719	38.919			39.086	38.978
²⁰⁶Pb/²⁰⁴Pb i	19.538	19.157				19.097	19.065	19.155	19.152					19.446	19.532	19.455			19.555	19.568

Data after observations of the authors and from the literature (Hamilton 1972; Kyle 1982, 1990*b*; Wörner *et al.* 1989; Müller *et al.* 1991; Rocholl *et al.* 1995; Rocchi *et al.* 2002*a*; Hornig and Wörner 2003; Nardini *et al.* 2009; Aviado *et al.* 2015; Panter *et al.* 2018). For unpublished analyses and unpublished Pb isotopic data for synplutonic dykes and dyke swarms, methods are after Nardini *et al.* (2009).

*IVD, individual volcano description after LeMasurier and Thomson (1990).

†Rock type: V, volcanic rock; P, plutonic rock; D, dyke.

‡Time: age used to calculate the initial radiogenic isotope ratios; the age is most commonly determined on the sample and, in some cases, is inferred by analogy. LOI, loss on ignition. ⁸⁷Sr/⁸⁶Sr m and ¹⁴³Nd/¹⁴⁴Nd m, measured isotopic ratio; ⁸⁷Sr/⁸⁶Sr i and ¹⁴³Nd/¹⁴⁴Nd i, initial isotopic ratio.

Table 4. Selected representative major–trace element compositions and radiogenic isotope ratios for Cenozoic igneous rocks from northern Victoria Land – plutonic complexes and synplutonic dykes

Province							Meander Intrusive Group									
							Plutons and synplutonic dykes									
Unit	No Ridge Igneous Complex		Cape Crossfire Igneous Complex			Cape King Igneous Complex		Greene Point Igneous Complex					Mount Monteagle Igneous Complex		Mount McGee Igneous Complex		
Sample	NO-5	NR2363	AR15	AR14	AR21c	AR1	24.1.86 L4	A1d	SAX17	SAX22	SAX8	SAX2	SAX12	SAX18	SX10b	SX4	LZ53	SX3	SX12	SX13
Rock type*	P	P	P	P	P	P	P	P	D	D	D	D	D	D	P	D	D	D	D	D
Reference	Rocholl et al. (1995)	Müller et al. (1991)	Rocchi et al. (2002b)	Rocchi et al. (2002b)	Rocchi et al. (2002b)	Rocchi et al. (2002b)	Rocchi et al. (2002b)	Rocchi et al. (2002a)	Tonarini et al. (1997)	Rocchi et al. (2002a)	Tonarini et al. (1997)	Tonarini et al. (1997)	Rocchi et al. (2002a, b)	Rocchi et al. (2002a, b)	Rocchi et al. (2002a, b)	Rocchi et al. (2002a, b)	Tonarini et al. (1997)	Tonarini et al. (1997)	Tonarini et al. (1997)	Rocchi et al. (2002a)
Locality	No Ridge	No Ridge	Cape Crossfire	Cape Crossfire	Cape Crossfire	Cape King	Cape King	Greene Point	Greene Point	Greene Point	Greene Point	Greene Point	Greene Point	Greene Point	Mount Monteagle	Mount Monteagle	Mount Monteagle	Mount McGee	Mount McGee	Mount McGee
Classification	Gabbro	Comenditic trachyte	Monzonite	Monzonite	Granite	Trachyte	Tholiitic basalt	Gabbro	Syenite	Basanite	Benmoreite	Comenditic trachyte	K-Trachyte	Basanite	Gabbro	Basanite	Benmoreite	Monzonite	Syenite	Hawaiite
Major elements (wt%)																				
SiO$_2$	45.29	61.78	46.89	53.30	75.80	64.25	48.80	45.44	67.35	42.33	54.03	63.46	45.30	47.20	44.90	42.82	56.95	50.27	62.74	46.94
TiO$_2$	2.45	0.55	3.99	2.11	0.14	0.33	3.31	4.14	0.31	4.48	1.94	0.51	3.62	3.45	5.07	3.73	1.16	2.26	0.79	2.83
Al$_2$O$_3$	19.32	15.37	14.09	16.91	13.22	16.92	15.49	16.02	16.51	15.42	16.79	14.55	14.84	14.41	14.79	13.07	17.47	11.24	15.97	15.16
Fe$_2$O$_3$	5.54	6.13	9.09	6.39	1.12	4.00	3.72	6.66	1.22	3.51	2.26	3.59	4.69	3.84	3.00	6.95	2.27	4.03	2.61	3.28
FeO	4.10		5.10	3.87		0.11	8.87	6.44	1.88	11.00	7.34	4.27	7.75	6.99	11.21	7.28	7.51	13.98	4.24	7.48
MnO	0.15	0.19	0.19	0.16	0.01	0.08	0.18	0.23	0.09	0.22	0.22	0.27	0.19	0.19	0.21	0.18	0.26	0.58	0.19	0.17
MgO	2.99	0.29	5.04	2.65	0.11	0.20	4.52	5.04	0.12	6.15	2.16	0.24	4.70	6.28	7.30	6.88	1.40	2.76	0.58	4.26
CaO	9.48	0.99	8.68	6.22	0.57	1.46	8.45	9.12	0.60	9.08	4.75	0.72	8.25	7.21	8.94	8.08	3.49	5.96	3.05	6.74
Na$_2$O	3.38	7.74	3.25	4.51	3.04	5.27	3.02	3.03	6.97	3.35	6.50	7.79	3.78	4.75	1.98	3.97	4.75	3.05	5.81	4.30
K$_2$O	1.48	5.05	1.85	2.17	5.51	6.74	1.20	0.95	4.39	0.64	2.04	4.02	1.97	2.17	0.67	2.38	2.50	3.07	2.90	1.78
P$_2$O$_5$	0.63	0.10	0.69	0.98	0.02	0.06	0.66	1.85	0.05	1.32	0.76	0.07	1.17	0.89	0.67	0.81	0.89	1.35	0.28	0.85
LOI	3.85	0.59	1.15	0.75	0.47	0.59	1.78	1.07	0.51	2.50	1.19	0.51	3.73	2.62	1.26	3.85	1.34	1.47	0.83	6.22
Trace elements (µg g^{-1})																				
Sc	7.0		32.3	16.2	1.8	11.9	24.5	18.6	4.0	17.5	13.1	4.1	15.4	16.0	25.2	19.0	4.3	26.8	17.1	13.2
V	151		387	114	13.1	3.62	285	168	1.09	232	19.6	0.75	179	184	337	207	2.84	50.4	2.76	138
Cr	9	4	28.6	2.02	28.3	0.81	17.35	1.36		9.62	1.32	0.52	8.46	139	3.65	153		8.31	0.00	33.7
Co	35	20	47.6	19.9	1.45	1.19	35.09	25.7	0.49	40.9	7.49	0.36	31.1	37.9	46.2	40.9	7.51	12.8	1.85	32.00
Ni	13	6	25.0	4.00	2.39	2.57	17.33	8.62	0.73	17.7	3.35	1.46	11.5	122	9.97	117	1.80	11.1	3.06	51.46
Cu	40	23					46.1	42.7	6.32	50.0	21.6		48.2	44.0		72.9	17.8	35.5		
Zn	101	104																		
Ga																				
Rb	36	76	48	56	143	57	32	17	91	5	48	122	60	59	21	67	69	56	66	39
Sr	928	8	532	651	127	61	452	1161	5	1258	876	21	1049	1274	762	843	912	234	493	840
Y	31	43	44.3	30.2	7.28	21.9	38.9	48.1	20.7	32.3	41.0	50.2	43.0	43.1	23.5	32.3	47.4	104	29.7	32.2
Zr	217	343	50	22	94	31	34	157	55	142	257	506	522	626	95	389	173	58	41	373
Nb	44	68	55.4	57.4	7.26	23.7	50.6	66.8	58.7	62.9	90.5	121	108	133	44.0	87.2	126	205	67.2	67.8
Mo		8	4.64	4.44	1.06	12.5	1.90	1.97	5.08	1.26	2.90	9.35	5.38	6.16	1.54	4.54	2.67	18.7	3.93	3.72
Cs			0.79	1.00	1.71	1.18	1.07	0.14	0.64	0.06	1.17	0.55	1.27	5.47	0.52	6.16	0.88	1.73	2.50	1.60
Ba	254	12	511	1330	377	316	234	733	44.4	429	1037	102	690	703	245	453	832	608	946	480

(*Continued*)

Table 4. *Continued.*

Province	No Ridge Igneous Complex	Meander Intrusive Group						
		Plutons and synplutonic dykes						
Unit		Cape Crossfire Igneous Complex		Cape King Igneous Complex		Greene Point Igneous Complex				Mount Monteagle Igneous Complex		Mount McGee Igneous Complex			
La	85	59.4	47.9	71.9	142	65.6	20.6	50.8	67.2	94.6	82.3	96.9	24.8	57.8	115	143	62.3	54.8
Ce	137	119	109	107	262	146	53.4	107.6	142.5	183	167	191	54.5	119	218	320	119	111
Pr		14.1	14.1	9.27	25.9	19.9	6.82	13.6	17.6	22.4	19.7	21.8	7.14	14.5	24.3	43.2	13.8	13.7
Nd		55.9	58.3	26.0	88.6	85.5	27.4	57.7	69.0	82.1	76.1	83.4	30.7	55.7	85.8	171	50.7	53.1
Sm		11.6	10.9	3.07	11.9	17.2	6.07	11.5	13.6	15.6	14.2	14.8	6.75	11.3	14.5	34.2	9.31	10.1
Eu		3.29	3.99	0.68	2.54	6.37	0.59	3.90	5.20	2.15	4.16	4.45	2.34	3.34	3.77	3.69	5.17	3.11
Gd		10.5	9.38	1.51	7.84	12.9	4.30	8.86	11.2	11.2	10.6	11.0	5.44	8.82	11.2	27.1	7.11	8.14
Tb		1.54	1.26	0.22	1.37	2.01	0.80	1.27	1.67	1.87	1.65	1.66	0.84	1.37	1.81	4.14	1.09	1.20
Dy		8.56	6.41	1.13	7.44	10.3	4.36	6.66	8.59	10.3	8.52	8.66	4.63	6.95	9.69	21.3	6.02	6.5
Ho		1.59	1.11	0.24	1.51	1.84	0.83	1.20	1.56	1.93	1.60	1.59	0.87	1.24	1.85	4.14	1.14	1.18
Er		3.88	2.49	0.67	3.67	4.06	2.03	2.76	3.53	4.90	3.84	3.83	2.00	2.82	4.66	9.57	2.77	2.83
Tm		0.55	0.31	0.12	0.51	0.50	0.30	0.35	0.47	0.69	0.53	0.50	0.23	0.39	0.62	1.30	0.37	0.35
Yb		3.21	1.79	0.76	3.08	2.88	1.97	2.06	2.83	5.31	3.24	2.92	1.50	2.20	3.56	7.95	2.37	2.28
Lu		0.44	0.24	0.13	0.42	0.35	0.27	0.30	0.37	0.93	0.45	0.43	0.21	0.26	0.44	1.08	0.38	0.32
Hf		2.16	0.76	2.94	2.06	4.16	2.58	3.73	6.60	15.2	10.9	12.58	2.88	8.72	6.26	2.69	1.71	8.11
Ta		3.82	1.55	1.36	3.50	3.99	4.44	4.12	5.74	7.84	7.11	8.58	3.12	5.92	8.95	13.0	5.32	4.50
W					0.99	0.41	0.46	0.12	0.59	0.39	1.30	1.82	0.35	2.39	1.00	1.35	1.78	0.82
Tl		0.21	0.34	0.38				0.02			0.23	0.17						
Pb	14	3.22	7.64	9.73	6.80	1.61	4.82	1.50	6.31	4.10	5.18	6.18	2.46	5.92	9.32	8.42	7.11	3.30
Th		2.08	0.80	37.3	3.66	2.89	5.52	1.95	6.63	8.98	9.79	12.64	2.30	8.25	13.7	6.32	12.7	6.53
U	11	0.44	0.70	3.79	0.77	0.83	1.29	0.50	1.64	2.99	2.80	3.12	0.47	2.63	1.94	1.64	1.69	1.81
Time (Ma)[†]	25.00	31.40	30.30	31.00	28.10	39.20	40.00	40.00	39.70	39.20	34.69	34.67	45.00	43.78	42.50	38.00	38.00	37.74
^{87}Sr/^{86}Sr m	0.703171	0.704130	0.704337	0.705118	0.705634	0.703610	0.726370	0.703516	0.703714	0.713460	0.703624	0.703280	0.704284	0.703404	0.703640	0.705030	0.704625	0.703330
^{87}Rb/^{86}Sr	0.112	0.262	0.252	3.188	0.206	0.040	35.4	0.011	0.156	16.3	0.166	0.120	0.089	0.231	0.217	0.446	0.396	0.134
^{87}Sr/^{86}Sr i	0.703131	0.704013	0.704228	0.703714	0.705552	0.703494	0.706258	0.703604	0.703626	0.704371	0.703542	0.703221	0.704227	0.703260	0.703509	0.704789	0.704411	0.703258
^{143}Nd/^{144}Nd m	0.512895				0.705552	0.512850	0.512827	0.512850	0.512851	0.512848	0.512860	0.512843	0.512777	0.512893	0.512856	0.512759	0.512774	0.512910
^{147}Sm/^{144}Nd	0.1100				0.1295	0.1219	0.1342	0.1203	0.1195	0.1150	0.1131	0.1077	0.1329	0.1222	0.1021	0.1209	0.1110	0.1155
^{143}Nd/^{143}Nd i	0.512877				0.512659	0.512843	0.512792	0.512819	0.512820	0.512819	0.512834	0.512819	0.512738	0.512858	0.512828	0.512729	0.512746	0.512882
ε_{Nd} i	5.3				1.1	5.0	4.0	4.5	4.5	4.5	4.7	4.4	3.1	5.4	4.8	2.7	3.1	5.7
^{206}Pb/^{204}Pb	19.703								19.212		20.003	20.283			19.459			19.700
^{207}Pb/^{204}Pb	15.577																	
^{208}Pb/^{204}Pb	39.387																	
^{206}Pb/^{204}Pb i	19.594								18.780		19.625	19.906			18.996			19.289

Data after observations of the authors and from the literature (Hamilton 1972; Kyle 1982, 1990b; Wörner *et al.* 1989; Müller *et al.* 1991; Rocholl *et al.* 1995; Rocchi *et al.* 2002a; Hornig and Wörner 2003; Nardini *et al.* 2009; Aviado *et al.* 2015; Panter *et al.* 2018). For unpublished analyses and unpublished Pb isotopic data for synplutonic dykes and dyke swarms, methods are after Nardini *et al.* (2009).

*Rock type: V, volcanic rock; P, plutonic rock; D, dyke.

[†]Time: age used to calculate the initial radiogenic isotope ratios; the age is most commonly determined on the sample and, in some cases, is inferred by analogy. LOI, loss on ignition. ^{87}Sr/^{86}Sr m and ^{143}Nd/^{144}Nd m, measured isotopic ratio; ^{87}Sr/^{86}Sr i and ^{143}Nd/^{144}Nd i, initial isotopic ratio.

Table 5. *Selected representative major-trace element compositions and radiogenic isotope ratios for Cenozoic igneous rocks from northern Victoria Land – dyke swarms*

Province										
			Meander Intrusive Group							
			Dyke swarms							
Unit		Random Hills dyke swarm					Bier Point dyke swarm		Burns dyke swarm	Styx dyke swarm
Sample	1.11.93 AS13	1.11.93 AS1	1.11.93 AS2.	1.11.93 AS16	1.11.93 AS18	1.11.93 AS7	4.11.93 XA14	4.11.93 XA2	11.11.93 SA12	11.11.93 SA13	17.11.93 SX1	17.11.03 SX5
Rock type*	D	D	D	D	D	D	D	D	D	D	D	D
Reference	Rocchi et al. (2002a)	Rocchi et al. (2002a)	Rocchi et al. (2002a)	Rocchi et al. (2002a)	Rocchi et al. (2002a)	Rocchi et al. (2002a)	Rocchi et al. (2002a)	Rocchi et al. (2002a)				
Locality	Random Hills	Random Hills	Random Hills	Random Hills	Random Hills	Random Hills	Bier Point	Bier Point	Burns Glacier	Burns Glacier	Styx Glacier	Styx Glacier
Classification	Tephrite	Alkali basalt	Basanite	Basanite	Tephrite	Comenditic trachyte	Basanite	Hawaiite	Trachyte	Rhyolite	Trachyte	Rhyolite

Major elements (wt%)

SiO_2	44.79	46.47	44.57	43.97	44.91	67.29	42.78	46.08	65.52	75.03	60.77	70.74
TiO_2	3.72	2.77	3.14	3.23	3.45	0.35	4.15	3.43	0.27	0.7	0.32	0.49
Al_2O_3	14.81	16.28	13.71	3.51	14.99	14.44	14.29	14.83	16.01	13.49	18.21	13.16
Fe_2O_3	3.43	1.51	3.66	3.53	3.89	1.47	4.06	2.76	5.70	1.64	4.91	3.12
FeO	8.80	9.61	9.28	9.06	8.62	2.73	9.68	9.80				1.94
MnO	0.16	0.17	0.18	0.17	0.15	0.13	O.16	0.16	0.20	0.02	0.17	0.13
MgO	4.32	8.28	6.16	6.77	4.14	0.64	7.37	4.56	0.26	0.18	0.69	0.22
CaO	7.26	8.95	8.06	8.80	7.83	1.08	8.26	7.24	0.78	0.42	1.29	0.54
Na_2O	4.67	3.44	4.63	4.14	4.94	6.10	3.13	4.27	6.26	4.17	5.41	4.24
K_2O	2.21	1.06	1.69	1.47	1.85	4.84	1.41	1.63	3.20	4.01	5.78	4.69
P_2O	0.82	0.51	1.06	0.91	0.95	0.08	0.85	1.13	0.05	0.02	0.06	0.03
LOI	5.02	0.96	3.85	4.08	4.29	0.85	3.85	4.12	1.75	0.94	2.39	0.70

Trace elements ($\mu g\ g^{-1}$)

Sc	12	22	15	19	12	5	20	13		0.75		
V	164	217	164	199	152	3	245	148		0.27		
Cr	10	241	94	103	7	2	102	20		1.08		
Co	36	50	41	46	33	1	50	36		0.16		
Ni	36	106	95	116	28	2	90	35		0.90		
Cu	43	63	51	58	56	4	70	50		1.50		
Zn												
Ga												
Rb	83	26	51	53	54	117	62	59	98	251	246	212
Sr	1031	620	909	772	962	37	650	905	40	25	46	24
Y	31.3	28	36	32	34	54	33	38	53	119	64	49
Zr	408	204	447	332	449	482	307	474	1280	359	1336	708
Nb	77	54	83	65	90	143	64	81	134	350	221	104
Mo	5.10	3.16	4.90	3.50	3.40	0.61	1.99	4.70		0.31		
Cs	7.70	0.39	2.28	3.90	1.04	2.39	1.61	5.00		1.01		
Ba	596	303	667	404	679	248	347	466	207	34	97	119

(*Continued*)

Table 5. *Continued.*

Province	Meander Intrusive Group					
	Dyke swarms					
Unit		Random Hills dyke swarm			Bier Point dyke swarm	Burns dyke swarm	Styx dyke swarm
La	60	38	67	52	70	194	46	64	115	34	107	50
Ce	123	75	137	107	141	347	93	132	220	84	185	93
Pr	14.8	9.5	16.8	14.4	17.1	36.0	11.9	16.4		11.4		
Nd	60	37-	66	54	66	128	49	66		47		
Sm	11.5	7.5	12.6	10.7	12.7	20.5	10.6	12.8		16		
Eu	3.70	2.34	3.70	3.28	3.80	1.48	3.40	4.10		0.17		
Gd	9.2	6.6	10.3	9.0	10.2	13.4	9.5	10.6		16		
Tb	1.34	1.02	1.54	1.35	1.49	2.09	1.41	1.61		3.30		
Dy	6.5	5.6	7.7	6.8	7.4	10.6	7.3	7.9		19.7		
Ho	1.11	1.07	1.36	1.24	1.31	1.94	1.23	1.41		3.9		
Er	2.59	2.44	3.17	2.73	2.77	5.00	2.92	3.30		10.4		
Tm	0.32	0.35	0.42	0.35	0.36	0.72	0.38	0.44		1.54		
Yb	1.81	2.14	2.19	2.17	2.14	4.10	2.02	2.37		9.00		
Lu	0.26	0.28	0.32	0.27	0.25	0.63	0.29	0.34		1.23		
Hf	9.10	4.80	10.10	7.60	9.80	11.40	7.30	10.3		16.7		
Ta	4.9	3.4	5.5	4.2	6.0	8.6	4.4	5.5		25.6		
W	1.13	0.64	1.12	0.71	0.90	0.50	0.92	1.09		4.10		
Tl	0.78	0.15.	0.24	0.36	0.25	0.43	0.35	0.42		1.64		
Pb	8.67	2.20	5.03	4.13	7.63	7.43	2.87	8.03		44.3		
Th	7.60	4.50	7.90	5.40	8.50	25.1	4.60	7.00		43.0		
U	4.30	1.35	2.38	1.85	2.60	4.10	1.48	2.10		17.1		
Time (Ma)†	35.42	39.55	39.55		36.99		46.72	44.97		46.72		
$^{87}Sr/^{86}Sr$ m			0.703538		0.703570		0.703831					
$^{87}Rb/^{86}Sr$			0.1623		0.162		0.276					
$^{87}Sr/^{86}Sr$ i			0.703507		0.703485		0.703648					
$^{143}Nd/^{144}Nd$ m			0.512907		0.512900		0.512895					
$^{147}Sm/^{144}Nd$			0.1154		0.1163		0.1308					
$^{143}Nd/^{144}Nd$ i			0.512877		0.512872		0.512855					
ε_{Nd} i			5.7		5.5		5.4					
$^{206}Pb/^{204}Pb$	19.333				19.492		18.923					
$^{207}Pb/^{204}Pb$												
$^{208}Pb/^{204}Pb$												
$^{206}Pb/^{204}Pb$ i	18.947				19.090		18.414					

Data after observations of the authors and from the literature (Hamilton 1972; Kyle 1982, 1990b; Wörner et al. 1989; Müller et al. 1991; Rocchi et al. 1995; Rocchi et al. 2002a; Hornig and Wörner 2003; Nardini et al. 2009; Aviado et al. 2015; Panter et al. 2018). For unpublished analyses and unpublished Pb isotopic data for synplutonic dykes and dyke swarms, methods are after Nardini et al. (2009).

*Rock type: V, volcanic rock; P, plutonic rock; D, dyke.

†Time; age used to calculate the initial radiogenic isotope ratios; the age is most commonly determined on the sample and, in some cases, is inferred by analogy. LOI, loss on ignition. $^{87}Sr/^{86}Sr$ m and $^{143}Nd/^{144}Nd$ m, measured isotopic ratio; $^{87}Sr/^{86}Sr$ i and $^{143}Nd/^{144}Nd$ i, initial isotopic ratio.

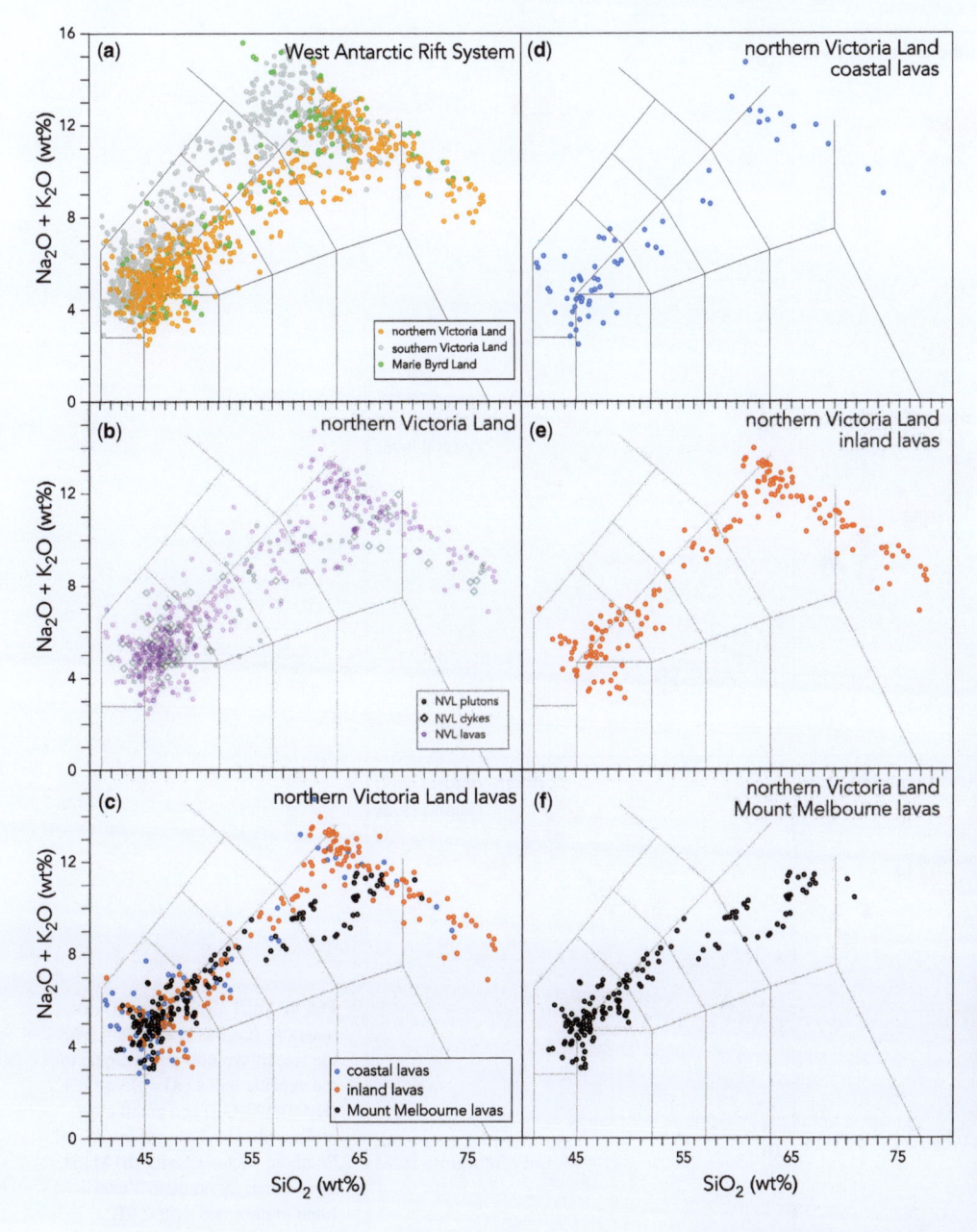

Fig. 2. Total alkali–silica (TAS) diagrams (Le Bas *et al.* 1986) reporting whole-rock samples for (**a**) the whole West Antarctic Rift System (WARS); (**b**) the northern Victoria Land (NVL) plutons, dykes and lavas; and (**c**) the northern Victoria Land lavas, distinguished as coastal, inland and Mount Melbourne lavas – see (**d**)–(**f**) for details. (**d**) Coastal lavas from the large shield volcanoes and stratovolcanoes of Adare, Hallett and Daniell peninsulas and Coulman Island; (**e**) inland lavas from Malta Plateau, Mount Rittmann, The Pleiades and isolated scoria cones of the Northern Local Suite; and (**f**) lavas from Mount Melbourne Volcanic Field.

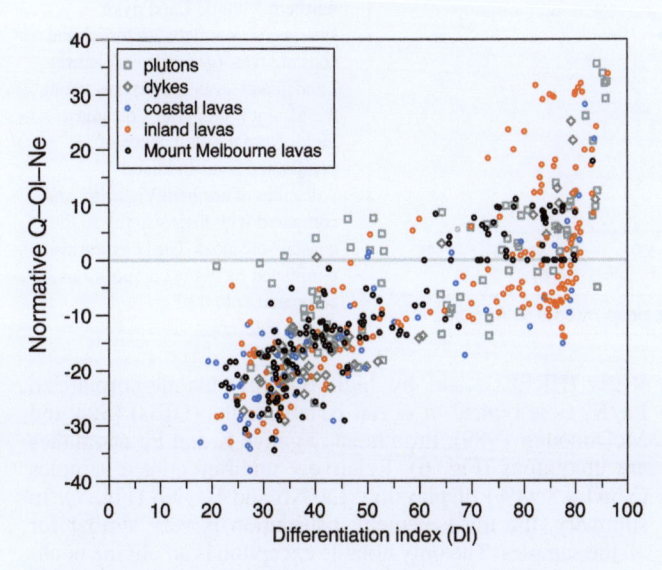

Fig. 3. CIPW normative Q–Ol–Ne v. differentiation index (DI) (Thornton and Tuttle 1960) for all plutons, dykes, coastal lavas, inland lavas and Mount Melbourne lavas of northern Victoria Land.

Q > 10% belong to this group (Fig. 3). Coastal lavas are represented by a lower number of samples and show a similar evolutionary trend, yet reaching less differentiated rhyolites (Fig. 2d); the most undersaturated samples, with normative Ol + Ne > 30%, are from this group. Mount Melbourne lavas only reach trachyte compositions (Fig. 2f). Peralkaline compositions (Fig. 4) are common among trachytes (dominantly comenditic) and rhyolites, with comendites slightly more common than pantellerites (Fig. 5). The plutons also include peralkaline examples (Fig. 4d), while among the dykes, only a single peralkaline trachyte is known (Fig. 4e). Among the evolved lavas, the inland samples almost invariably reach peralkalinity (Fig. 4g), while coastal and Mount Melbourne samples are commonly less evolved and not peralkaline (Fig. 4f, h).

Trace elements

The trace element contents of plutonic rocks are commonly not representative of a melt, and moderate positive Eu anomalies – calculated as $Eu/Eu* = (Eu_{rock}/Eu_{chondrite})/((Sm_{rock}/Sm_{chondrite})(Gd_{rock}/Gd_{chondrite})^{\frac{1}{2}})$ – of up to 1.5 are always present, coupled with the petrographical evidence of

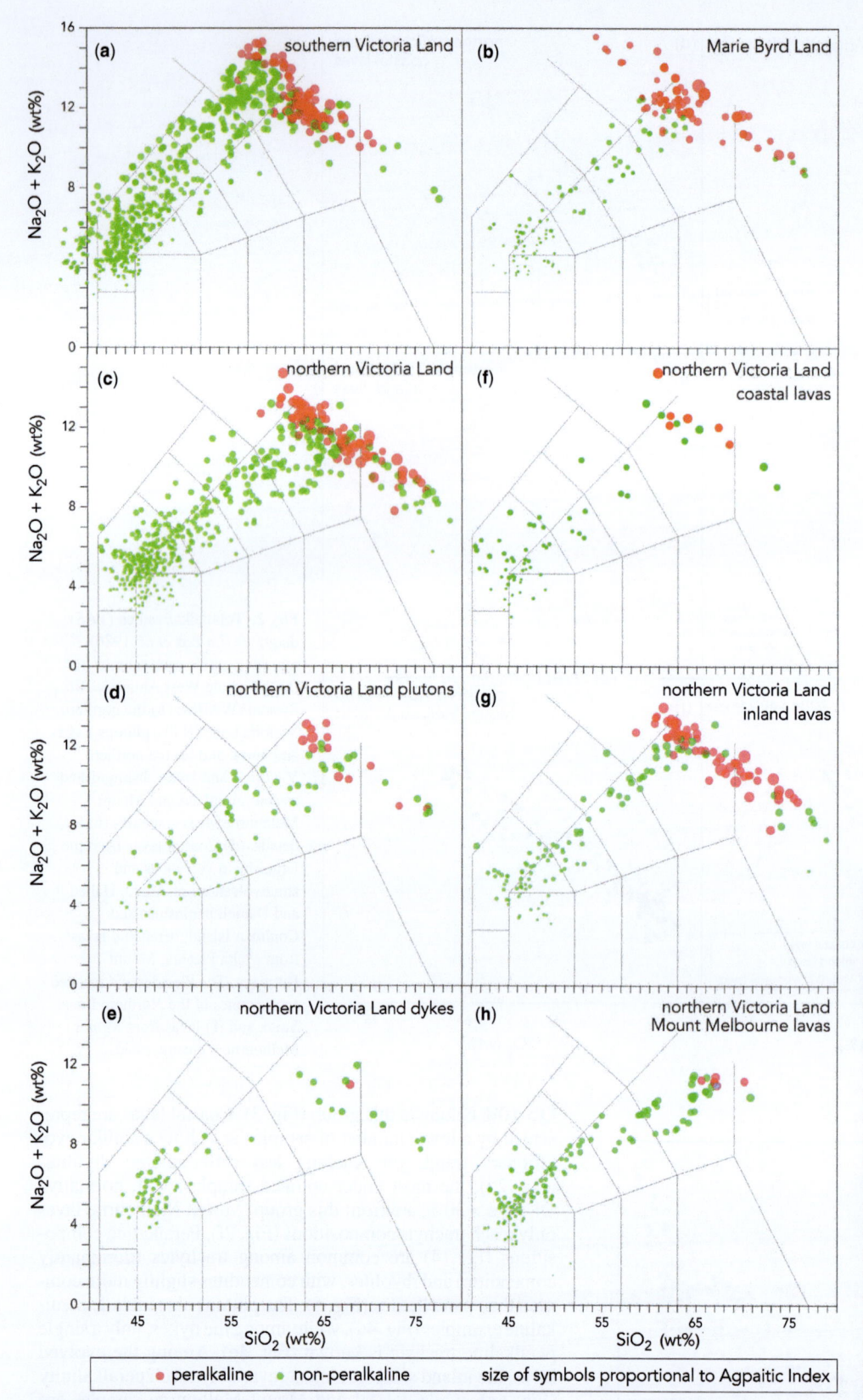

Fig. 4. Total alkali–silica (TAS) diagrams (Le Bas *et al.* 1986) with the size of symbols proportional to the agpaitic index (AI = ([Na$_2$O] + [K$_2$O])/[Al$_2$O$_3$]) and peralkaline rocks with AI > 1 shown in red. (**a**) Southern Victoria Land, (**b**) Marie Byrd Land, (**c**) northern Victoria Land plutons and dykes; (**d**) northern Victoria Land plutons, (**e**) northern Victoria Land dyke swarms, (**f**) northern Victoria Land coastal lavas, (**g**) northern Victoria Land inland lavas and (**h**) lavas from the Mount Melbourne Volcanic Field. Note the abundance of peralkaline lavas in inland volcanoes of northern Victoria Land compared with their scarcity in the coastal volcanoes. The latter are also dominated by mafic compositions compared with the inland volcanoes.

cumulus plagioclase, such as tiling of plagioclase laths. The mafic dykes are always subaphyric or slightly porphyritic; thus, they are suitable representatives of magmatic liquids, similar to lava samples. Therefore, the trace element features are best represented by synplutonic dykes and many of the lavas. Mafic samples from synplutonic dykes, dyke swarms and lavas from any location share the same pattern of incompatible trace elements (Fig. 6), which is dominated by significant fractionation of rare earth elements (REE), high ratios of Nb–Ta to large ion lithophile elements (LILEs) and Y-heavy REEs (HREEs), and by high primitive-mantle-normalized La/K, as is typical of ocean island basalts (OIBs) (Sun and McDonough 1989). Prominent negative K and Pb anomalies are ubiquitous (Fig. 6). Relatively undifferentiated samples (MgO > 8 wt%) display low La/Nb and Ba/Nb (Fig. 7). In summary, the trace element distribution is very similar for all the samples. The only notable exception is an olivine nephelinite (SAX20: from a Northern Local Suite scoria cone located at Greene Point), which is extremely enriched in REE (primitive-mantle-normalized La of c. 200), coupled

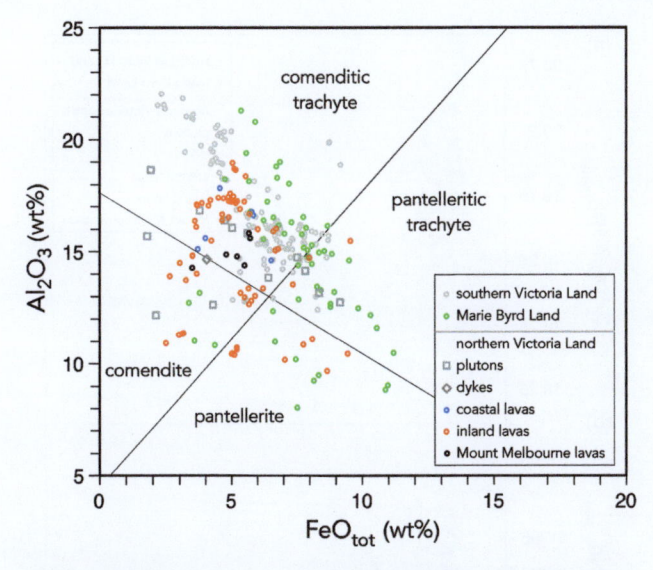

Fig. 5. Whole-rock weight percentage (wt%) Al$_2$O$_3$ v. total FeO (wt%) for peraluminous rocks (i.e. with agpaitic index (AI) > 1).

with the most prominent negative K anomaly of any sample from the region.

Radiogenic isotopes

Analysed primitive lava samples from northern Victoria Land have relatively low ^{87}Sr/^{86}Sr and moderately high ^{143}Nd/^{144}Nd, plotting in the depleted quadrant of the diagram and displaying a negative correlation (Tables 2–5; Fig. 8b). The ^{206}Pb/^{204}Pb isotopic ratios are high but do not reach the typical HIMU (high $\mu = ^{238}$U/^{204}Pb) values recorded in Marie Byrd Land (Fig. 8a).

The ^{87}Sr/^{86}Sr(*t*) ratios for whole-rock samples of plutons and dykes vary from 0.70299 to 0.71297, and ε$_{Nd}$(*t*) is between 2.0 and 6.3 (Tonarini *et al.* 1997; Rocchi *et al.* 2002*a*). However, evolved samples are likely to have been affected by crustal assimilation during fractional crystallization (Tonarini *et al.* 1997), resulting in a shift of isotopic ratios towards the average composition of granitoid basement (Rocchi *et al.* 1998). Samples with high MgO and Sr contents have initial ^{87}Sr/^{86}Sr < 0.7039 and ^{143}Nd/^{144}Nd > 0.51284. Overall, the data plot in the depleted quadrant of the ^{143}Nd/^{144}Nd–^{87}Sr/^{86}Sr diagram and cluster towards slightly more enriched compositions with respect to younger lavas. Nevertheless, this small overall difference in isotopic ratios between the plutonic–subvolcanic association (48–29 Ma) and the volcanic products (<15 Ma) cannot be simply ascribed to the age difference between the two groups. Indeed, if two distinct batches of magmas were extracted from a common source (with Rb/Sr and Sm/Nd typical of OIB sources) at 48 Ma and today, respectively, they should differ by 0.30 ε$_{Nd}$ units and 0.000021 ^{87}Sr/^{86}Sr units. These values are within two–three times the typical error on both Sr and Nd measured isotopic ratios (Tables 2–5), and can therefore be neglected.

It is noticeable that samples with high ^{87}Sr/^{86}Sr are coupled with low MgO (Fig. 8c) and low Sr contents, and are commonly peralkaline, with the highest values of the agpaitic index (AI) (Fig. 8d).

Origins of magmas

The northern Victoria Land Cenozoic igneous rocks depict a petrological framework that can be used to deduce their origins from a specific region of the mantle. Several different models have been proposed for the nature of the source type and its geodynamic significance. These alkaline magmas were initially thought to have originated in an active mantle plume, a fossil plume or a source region metasomatized in the Paleozoic by subduction or in the Mesozoic by extension.

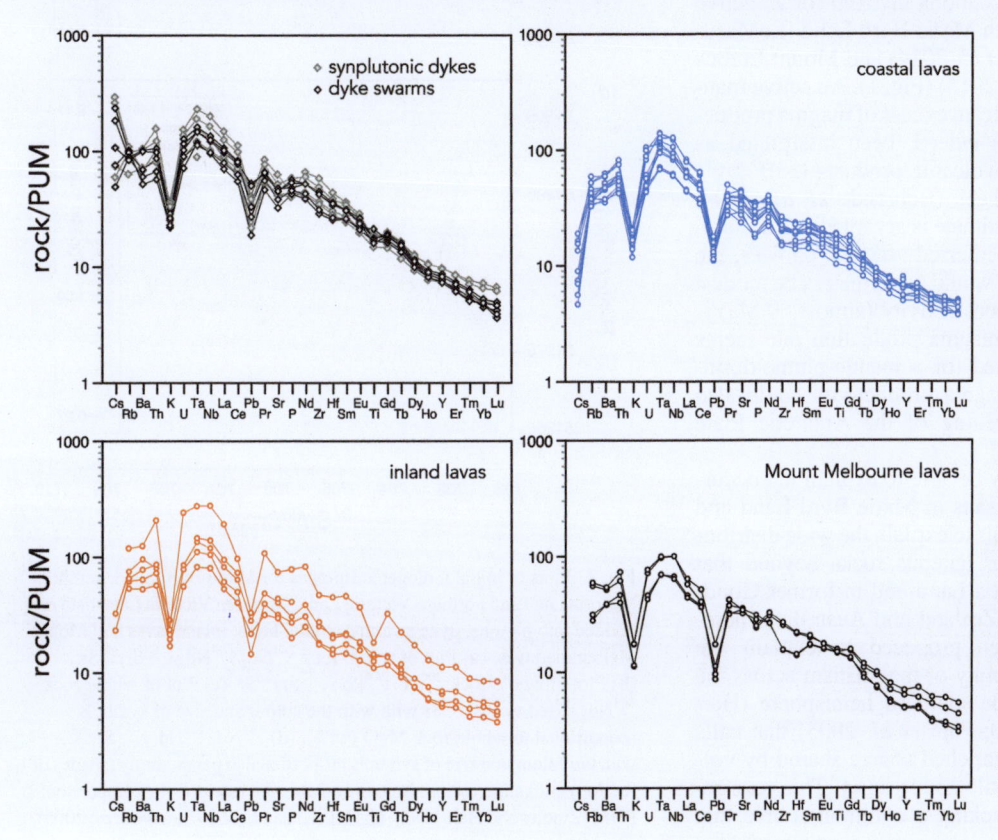

Fig. 6. Trace multi-elemental plots, with concentrations normalized to the primitive undifferentiated mantle (PUM) (McDonough and Sun 1995) for selected samples of: (**a**) dyke swarms (1.11.93 AS1,2,16,18; 4.11.93 XA2,14) (Rocchi *et al.* 2002*a*) along with dykes cross-cutting plutons and synplutonic dykes (10.11.93 SAX12,18, 2.11.93 SX4,13); pluton samples are not reported because their chemistry is usually affected by cumulus effects, mainly of plagioclase (Rocchi *et al.* 2002*a*); (**b**) coastal lavas (P74789, P74794, P74797, P74815, P74822, P74833, P74794, P74833, NV-6E, A240B-1, AW, 82205, A225A, MP8, MP32) (Mortimer *et al.* 2007; Nardini *et al.* 2009; Aviado *et al.* 2015; Panter *et al.* 2018); (**c**) inland lavas (MP24, 5.12.14 JS9, 81510, MB78, SAX20) (Noll 1984; Armienti and Tripodo 1991; Nardini *et al.* 2009); and (**d**) Mount Melbourne lavas (MB26, 27, 32, 54) (Armienti *et al.* 1991; Nardini *et al.* 2009).

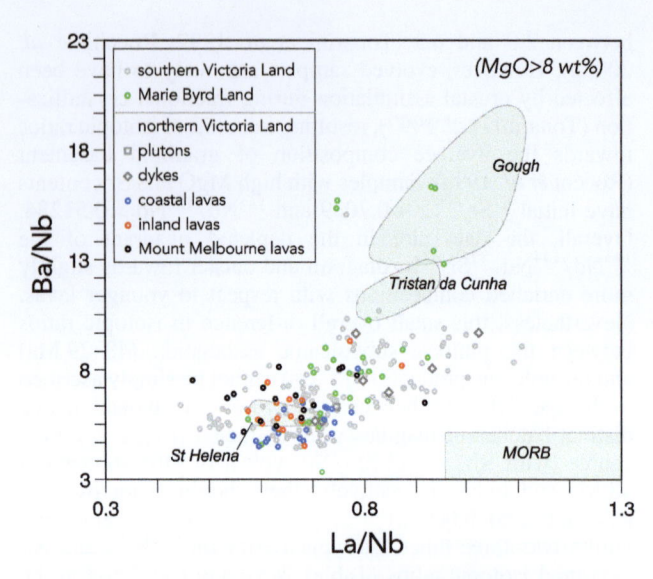

Fig. 7. Ba/Nb v. La/Nb diagram for samples with MgO (wt%) > 8. Reference for primitive mantle, normal MORB (N-MORB), St Helena, Tristan da Cunha, and Gough values (Sun and McDonough 1989).

Mantle-plume scenarios

Across the whole West Antarctic Rift, the magmatic rocks are enriched in incompatible elements, mostly Nb–Ta, and Pb isotope systematics show $^{206}Pb/^{204}Pb$ slightly in excess of 20 (Rocholl et al. 1995; Hart et al. 1997; Panter et al. 2000; Nardini et al. 2009), contributing to the characterization of these rocks as OIB derived from a source with a HIMU tendency. The source of these magmas has to be able to produce such compositions upon partial melting over a wide area encompassing the whole WARS, and probably beyond (Hart et al. 1997; Rocchi et al. 2002a). High $^{206}Pb/^{204}Pb$ ratios are commonly thought to be derived from deep mantle plumes that entrained slab material recycled into the deep mantle over a long time period (1 Ga). The locations invoked for an active mantle plume were beneath both Marie Byrd Land (LeMasurier and Landis 1996; Hansen et al. 2014) and Mount Erebus (Kyle et al. 1992; Phillips et al. 2018) (Fig. 1). An active mantle plume should induce a significant excess of magma production. Aeromagnetic data have, indeed, been interpreted as evidence for large volumes of volcanic products (>10^6 km^3) concealed beneath the ice sheet (Behrendt et al. 1994, 2004). Even if such a volume estimate is accepted and a Cenozoic age is assumed for all the inferred volcanic edifices, the average magma production rate would nevertheless be modest when the long duration of igneous activity (almost 50 Ma) is taken into account. Thus, the magma production rate seems to be much lower than expected for a mantle-plume-dominated scenario, even if low magma productivity could also be related to the stationary setting of the Antarctic Plate (Hole and LeMasurier 1994).

The active plume model may be able to explain the occurrence of Cenozoic OIB-like basalts in Marie Byrd Land and southern Victoria Land but it fails to explain the wide distribution of compositionally similar igneous rocks beyond that area: that is, in northern Victoria Land and in former Gondwana fragments, such as New Zealand and Australia. Therefore, other models have been proposed to explain the large-scale geochemical uniformity of magmatism across the WARS and a wide area in the southern hemisphere (Hart et al. 1997; Panter et al. 2000; Finn et al. 2005) that calls for a uniform, geochemically-enriched source shared by volcanic provinces thousands of kilometres apart. This requirement could be fulfilled by invoking a common source that

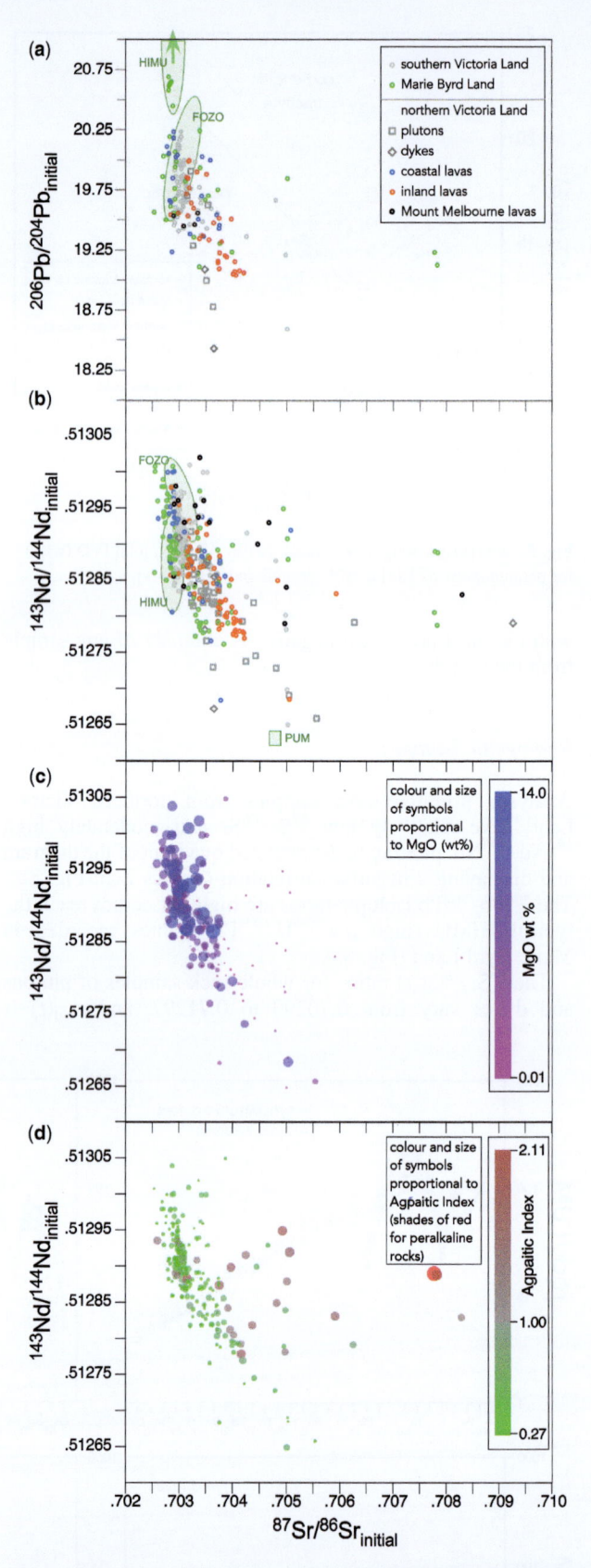

Fig. 8. Plots of initial radiogenic isotopes for Marie Byrd Land, southern Victoria Land and northern Victoria Land. Northern Victoria Land data are divided into plutons, dyke swarms, coastal lavas, inland lavas and Mount Melbourne lavas. (**a**) Plot of whole-rock $^{143}Nd/^{144}Nd$ v. $^{87}Sr/^{86}Sr$. (**b**) Plot of whole-rock $^{206}Pb/^{204}Pb$ v. $^{87}Sr/^{86}Sr$. (**c**) Plot of whole-rock $^{143}Nd/^{144}Nd$ v. $^{87}Sr/^{86}Sr$ with with the colour and size of symbols proportional to whole-rock MgO (wt%). (**d**) $^{143}Nd/^{144}Nd$ v. $^{87}Sr/^{86}Sr$ with the colour and size of symbols proportional to peralkalinity (shades of red for peralkaline rocks). Reference fields are reported for comparison: PUM (Zindler and Hart 1986), and HIMU and FOZO (Stracke et al. 2005).

was generated before the Late Cretaceous Gondwana break-up, when the areas later affected by Cenozoic magmatism were still joined in a single continental block. Such a source would then have been spread out over the southern oceans during Late Cretaceous and Cenozoic continental rifting. A suitable source could consist of a fossil multicomponent plume head, which would have impacted and underplated the Antarctic lithosphere prior to the Late Cretaceous (Hart et al. 1997). This plume head would have had an incubation period when (auto)metasomatism processes produced low-melting-point enriched zones or veins. Such a source would have been suitable for generating magmas: for example, during Cenozoic extensional events in Victoria Land, Marie Byrd Land and other former Gondwana fragments.

On the other hand, the uncommon and geographically-restricted occurrence in Marie Byrd Land of 'true' HIMU basalts (with $^{206}Pb/^{204}Pb$ values up to 20.9, approaching those of OIB HIMU end members) has led to the proposal of a relatively small, HIMU-type mantle plume that impinged before break-up beneath a pre-existing metasomatized layer within the Gondwana lithosphere (Panter et al. 2000).

The timing of plume impingement should not be placed in the Late Cretaceous, when Marie Byrd Land and Zealandia were undergoing separation accompanied by subsidence instead of buoyant uplift (LeMasurier and Landis 1996), although this has been questioned by offshore geophysical investigations (Luyendyk et al. 2001). Furthermore, at that time the rapid separation between Marie Byrd Land and Zealandia is best explained as being a result of the progressive cessation of subduction and of ridge–trench interaction along the Pacific margin of Gondwana (Mukasa and Dalziel 2000). Another possible estimate for the timing of plume impact below the Antarctic lithosphere could be around the Early–Middle Jurassic transition. At that time, tholeiitic magmatism of the Ferrar Large Igneous Province (LIP) affected a 4000 - km-long stripe that crossed the Antarctic continent from northern Victoria Land to Dronning Maud Land (Elliot et al. 1999; Luttinen and Furnes 2000). The Ferrar products have a distribution which partly mimics that of Cenozoic alkaline magmatism along the Transantarctic Mountains in Victoria Land. However, plume activity related to the Ferrar province represents a long-standing matter of debate that is linked to models for Jurassic Gondwana break-up (Hergt et al. 1991; Storey and Alabaster 1991; Antonini et al. 1999; Mukasa 2003; Burgess et al. 2015). Additionally, the Jurassic Ferrar event did not give way to igneous activity in Marie Byrd Land: thus, Victoria Land and Marie Byrd Land are affected by geochemically similar Cenozoic magmatism following different Jurassic histories.

The timing of plume arrival might be reconstructed if a typical hotspot track can be identified. However, the overall distribution of magmatism in the WARS does not show a regional-scale time-progressive track, possibly due to either the absence of a plume or, more simply, to the setting of the Antarctic Plate, which has been stationary since the Late Cretaceous and almost completely encircled by mid-ocean ridges.

A further different role for a fossil plume has been proposed (Rocholl et al. 1995). The temporal geochemical variability of 15 Ma–Recent volcanic rocks from northern Victoria Land may be due to the increasing involvement of progressively deeper sources in a compositionally stratified mantle, where enriched subcontinental lithospheric mantle is underlain by HIMU mantle (originating from a fossil plume head), which in turn overlies depleted asthenospheric mantle. However, stretching in the uppermost layers is required in order to generate melt in the uprisen, formerly deeper, layers, and this conflicts with the minimal amount of Cenozoic extension inferred for the whole Ross Embayment (Lawver and Gahagan 1994). Additionally, the stretching out of deep mantle layers is a questionable process because materials at these depths and temperatures display mainly ductile rheological behaviour and are weakly coupled with the overlying strong lithosphere (Anderson 1995).

Finally, a major tenet of any mantle-plume scenario is the expected isotopic signature of deep material involved in the magma source. In this view, high $^3He/^4He$ ratios (up to 50 R_a, where R_a is the atmospheric He ratio) (Stuart et al. 2003) have generally been used to argue that the deeper mantle is relatively undegassed and preserves its primordial volatile composition compared with the source region of mid-ocean ridge basalts (MORBs), which has a lower time-integrated $^3He/(U + Th)$ ratio (c. 8 R_a) (Graham 2002; Harrison and Ballentine 2005). However, shallower mantle processes should also be considered (Meibom et al. 2003). The $^3He/^4He$ ratios of the northern Victoria Land basalts are c. 6–7 R_a and can be considered as representative of the helium isotope signature of their mantle source (Nardini et al. 2009). These data are similar to lithospheric mantle values (Gautheron and Moreira 2002) and overlap the lower limit for some MORB data (Harrison et al. 2003). Therefore, the available He isotope ratios do not support the involvement of deep undegassed material (i.e. a mantle plume) in the genesis of WARS alkaline basaltic melts, at least not for northern Victoria Land.

Whatever interpretation is proposed for the geochemical data, the main point is that a mantle plume is a physical entity (Ritsema and Allen 2003), not a geochemical reservoir. Therefore, plume occurrence (either active or fossil) cannot be inferred from geochemical features alone. On this basis, the perspective offered by northern Victoria Land studies permits the integration of geochemical data with geological–structural–geochronological data.

Magmatism and local tectonics: geometrical–chronological relationships

Investigations of Cenozoic igneous rocks from northern Victoria Land have contributed a new broader-scale perspective on magma origins, based on several specific geological features such as: (i) the location of northern Victoria Land at the periphery of the Antarctic cratonic lithosphere; (ii) the presence of plutons as old as Middle Eocene (Rocchi et al. 2002a): (iii) outcrops of Eocene–Oligocene dykes with consistent strikes (Rocchi et al. 2002a); and (iv) the occurrence of geochronologically-constrained translithospheric faults (Di Vincenzo et al. 2004). Thus, petrological data from magmas in northern Victoria Land can be compared with both the contexts of local tectonic activity and large-scale geodynamic evolution.

Surface uplift–extension–magmatism and relative timing. Within the WARS, two principal patterns of surface uplift are observed. In Marie Byrd Land, Cenozoic surface uplift has been interpreted to have caused the formation of an elongated dome (LeMasurier and Rex 1989; LeMasurier and Landis 1996; Rocchi et al. 2006); whereas in the western Ross Embayment, Cenozoic uplift created a large-scale linear feature: that is, the Transantarctic Mountains, which are dissected by prominent transverse fault systems (Fig. 9a) (ten Brink and Stern 1992; Salvini et al. 1997). Thus, in northern Victoria Land, the circular symmetry of surface uplift typically associated with mantle plumes is absent.

Besides uplift, extension is also of great significance in defining the origins of magmas. The Ross Embayment underwent extension between the Cretaceous and the Cenozoic. The inferred 150 km of Cenozoic extension in the Adare Trough outside of the Ross Sea (Cande et al. 2000; Cande and Stock 2003; Granot et al. 2010, 2013) is not sufficient to

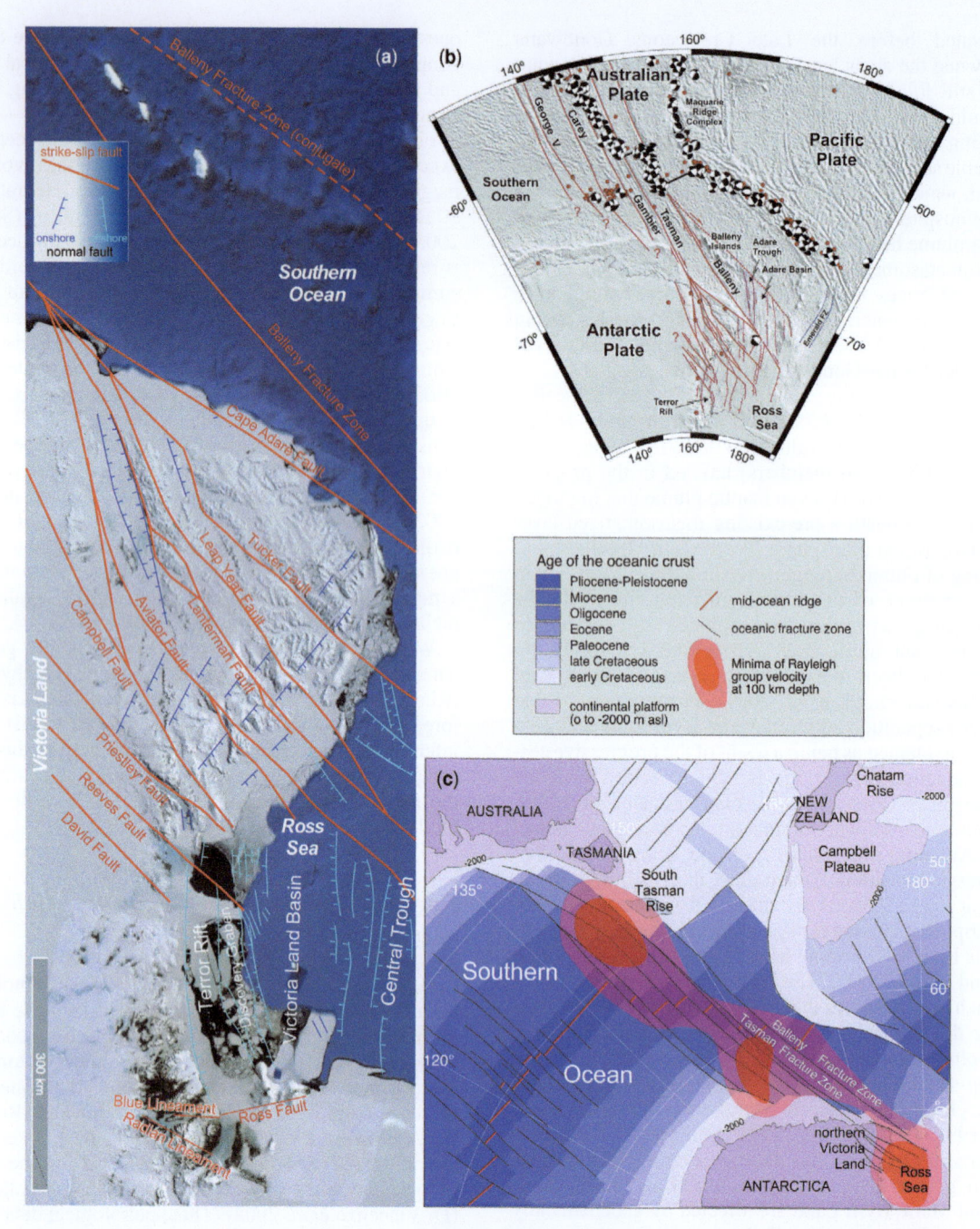

Fig. 9. Geodynamic tectonic framework of Victoria Land. (**a**) Translithospheric strike-slip faults and extensional faults (Salvini *et al.* 1997; Storti *et al.* 2007, 2008). (**b**) Regional geodynamic setting, with oceanic fracture zones and continental translithospheric fault systems (red lines), and earthquake locations (1976–2004: red dots) with their focal mechanisms, modified after Storti *et al.* (2007). (**c**) Map showing the relationships between Victoria Land strike-slip faults and fracture zones present in the Southern Ocean between Antarctica and the Australian passive margin, the conjugate margin to Antarctica (Storti *et al.* 2007). Also shown are the ages of the ocean floor, as well as the anomalous mantle characterized by slow seismic velocities as depicted from seismic tomography, with darkest shades of grey indicating strongest negative anomalies (Danesi and Morelli 2000). Abbreviations: m asl: metres above sea level.

explain the total extension of the area, which was likely to have been between 300 and 500 km (Davey and Brancolini 1995; Fitzgerald and Baldwin 1997; Siddoway *et al.* 2004). The necessity for Cretaceous extension has also been challenged based on the occurrence of sediments of Eocene age overlying basement at the bottom of the Cape Roberts drill holes (Cape Roberts Science Team 2001*a, b*). However, seismic data demonstrate that these sediments are not necessarily the oldest in the Victoria Land Basin (Hamilton *et al.* 2001). Additionally, the site of the Cape Roberts drill holes is very close to the Victoria Land coast, where the recent extension leading to the formation of the Terror Rift and extensive glacial reworking might prevent the preservation of the entire sedimentary basin infill. Therefore, the possibility of Cretaceous extension in the whole rift cannot be excluded. Furthermore, evidence for Early Late Cretaceous tectonism in the eastern Ross Sea is reported by isotopic dating of dredged mylonites (Siddoway *et al.* 2004) and is strengthened by similar cooling ages from apatite-fission tracks obtained for sheared rocks from DSDP (Deep Sea Drilling Program) Site 270 in the central Ross Sea (Fitzgerald and Baldwin 1997). Thus, an important factor that may have affected magma genesis is the occurrence of a significant Late Cretaceous extensional episode (Lawver and Gahagan 1994) some 50 Ma before magmatism started in the Middle Eocene (Tonarini *et al.* 1997; Rocchi *et al.* 2002*a*). For the Cenozoic, in northern

Victoria Land, the dominant type of tectonic activity seems to have been intraplate shearing along strike-slip fault systems (Salvini et al. 1997).

The history of subsidence also helps in placing constraints on the origins of magma. The Cretaceous subsidence was followed by Cenozoic uplift. Conversely, the Cenozoic uplift was mainly in the Paleogene (>c. 25 Ma). Although it overlapped with subsidence in the Victoria Land Basin, subsidence of the latter continued into much younger time and the presence of widespread subaerial volcanic sequences at sea level along the northern Victoria Land coast suggests that uplift did not affect the coastal region after c. 10 Ma, at least (Smellie et al. 2011b). Both scenarios are difficult to reconcile with the simple upwelling history of a plume with a higher than normal mantle potential temperature.

The distribution of igneous activity in space and time is another constraint on the discussion of magma origins. In Marie Byrd Land, perpendicular time-progressive patterns of volcanic activity have been described (LeMasurier and Rex 1989), whereas no radial patterns of igneous activity are observed in Victoria Land. There, different crustal sectors hundreds of kilometres in length display different temporal distributions of igneous activity (Fig. 1b), with no overall evidence for geographical progression (Smellie et al. 2011b). The northern Victoria Land onshore sector is the site of the longest-lasting igneous activity, with emplacement of plutons, dyke swarms and volcanic products from the Middle Eocene to the present. Widespread volcanic activity also occurred in southernmost Victoria Land from the latest Oligocene to the present (McIntosh 2000; see also Chapter 5.2a: Smellie and Martin 2021). Between Mount Erebus and Mount Melbourne, no Cenozoic igneous activity is found onshore (the numerous mafic dykes from this area have proved to be end-Paleozoic in age; Rocchi et al. 2009). Offshore, however, recent volcanic activity is observed that is geometrically linked to the actively extending north–south-elongated Terror Rift (LeMasurier and Thomson 1990; Rilling et al. 2009). Within the northern Victoria Land sector, the magmatism was activated in adjacent crustal blocks separated by strike-slip faults (Fig. 9a). There the dyke swarms have NW–SE or north–south strikes, with coeval emplacement in NW–SE and north–south directions, suggesting a geometrical link with the NW–SE strike-slip fault systems (Rocchi et al. 2002a).

Chronology of tectonic activity. A significant constraint on the chronology of tectonic activity in northern Victoria Land comes from pseudotachylites from the western coast of the Ross Sea. On the Priestley Fault, a NW–SE fault zone with dominant dextral slip (Fig. 9a) (Storti et al. 2001), a cataclastic Paleozoic migmatite is cut by pseudotachylite fault veins and associated injection veins. One of these veins has been dated using $^{40}Ar/^{39}Ar$ laser step heating at 34.11 ± 0.96 Ma, constraining the age of a major episode of coseismic faulting in the area (Di Vincenzo et al. 2004). Thus, the Cenozoic tectonic activity of the fault systems is no longer inferred only from seismic reflection lines but, rather, is dated directly, and the results show that the fault systems were coeval with the emplacement of mafic dykes from nearby areas. In addition, indirect evidence for Middle–Late Eocene fault activity has been found using apatite-fission track thermochronology in the northern segment of the Lanterman Fault (Fig. 9a), on the Southern Ocean side of the right-lateral fault array (Rossetti et al. 2003). The similarity in age between the Priestley Fault pseudotachylite on the Ross Sea coast and movement on the Lanterman Fault attests to the dynamic connection between the Ross Sea and the Southern Ocean during the Cenozoic.

Magmatism and regional plate dynamics: geometrical–chronological relationships

Geometry and timing of igneous activity v. plate dynamics. A tight geometrical link exists between dextral strike-slip systems in northern Victoria Land and Southern Ocean fracture zones (Fig. 9b, c): they share the same orientation and appear to be continuous across the Antarctic continental shelf (Salvini et al. 1997; Rossetti et al. 2003). Thus, the emplacement of the northern Victoria Land dyke swarms is ultimately linked to plate tectonic features. Other examples in Antarctica include the volcanic Balleny Islands, located offshore of northern Victoria Land (Fig. 9) at the termination of the Balleny Fracture Zone at the base of the continental rise, and Peter I Øy (Fig. 1) (Hart et al. 1995; Kipf et al. 2014) on the ocean floor close to the continental rise and adjacent to the formerly active Tharp Oceanic Fracture Zone (Fig. 1). Also in Marie Byrd Land, the location of volcanic activity on time-progressive fractures has been interpreted as evidence for crustal doming above an active plume (LeMasurier and Rex 1989), or the result of anisotropic stress linked to a rapid rotation in the maximum horizontal stress direction from north–south to east–west at approximately 6 Ma as the consequence of major plate reorganization and changes in plate motion along the Pacific–Antarctic Ridge (Paulsen and Wilson 2010). Alternatively, the north–south-trending fractures in Marie Byrd Land could be interpreted as having been activated by differential movement along the Pitman to the Udintsev oceanic fracture zones, which intersect the continental shelf of Marie Byrd Land at a high angle with an approximately north–south orientation (Cande et al. 1995). The extensive ice cover in Marie Byrd Land hampers the possibility of testing the alternative scenarios. The beginning of magmatism in northern Victoria Land is coeval with a global plate reorganization (Lithgow-Bertelloni and Richards 1998; Veevers 2000; Steinberger et al. 2004) and increasing differential movements on Southern Ocean fracture zones (at 47–43 Ma: Cande and Mutter 1982; Richards and Lithgow-Bertelloni 1996). It is therefore likely that the local tectonic framework that governs the emplacement of northern Victoria Land magmas is linked to plate-scale tectonic processes (Rona and Richardson 1978). Also, the onset of magmatism coincides with a process that could trigger differential movement along the strike-slip fault systems, namely differential movement along oceanic fracture zones. Such a process is not paradigmatic in plate tectonics but Cenozoic coseismic fault activity in northern Victoria Land has been demonstrated by the dating of fault-related pseudotachylite, as described above (Di Vincenzo et al. 2004). Onshore, intraplate shearing seems to have been the dominant type of tectonic activity in northern Victoria Land during the Cenozoic (Salvini et al. 1997), suggesting a connection with major transform fracture zones of the Southern Ocean that have developed since the Middle Miocene (Rossetti et al. 2003, 2006) or earlier (Steinberger et al. 2004). Hence, the igneous activity appears most likely to have been governed by top-down processes (i.e. by plate dynamics and strike-slip faulting), not bottom-up processes (i.e. by mantle plumes) (Rona and Richardson 1978; Anderson 2001).

Mantle tomography. Slow wave-speed anomalies in teleseismic tomography images of the mantle are most commonly ascribed to regions hotter than normal, although wetter than normal regions could also explain these anomalies. If they extend to the transition zone, these anomalies can be taken as indicative of deep, hot mantle upwelling. If restricted to the upper c. 200 km, the anomalies can be considered shallower in origin, as in the case of the Yellowstone hotspot (Christiansen et al. 2002). In the Antarctic region, seismic tomography cannot be reliably interpreted deep enough to support or reject the possibility of an active deep mantle plume, although recent tomography models indicate a low seismic-shear-wave velocity anomaly that extends below the mantle transition zone to near 1200 km depth beneath Ross Island (Phillips et al. 2018). Nevertheless, at a regional scale, the zone of slower (hotter?) mantle (Danesi and Morelli 2000) does not show a circular symmetry but, rather, is part of a series of anomalies that extend in a linear zone from the Ross Sea to the oceanic part of both the Antarctic and Australian plates. The anomaly zone is superimposed on a belt of oceanic ridge transform fracture zones, including the Tasman and Balleny fracture zones, that have a cumulative offset of c. 1500 km (Storti et al. 2007) (Fig. 9). The zone of shallow, hot mantle is a linear geodynamic feature >4000 km long (from Tasmania to the Ross Sea) that terminates within the western Ross Sea. A potential source of shallow hot mantle is therefore available for melting in that area.

Importance of northern Victoria Land for understanding continental rifting and related magmatism. Northern Victoria Land offers the opportunity of investigating Cenozoic alkaline magmatism in the WARS, one of Earth's largest continental rift zones, because of the existence of a wealth of geochemical, geochronological, geophysical, and both local and regional structural–tectonic data acquired by multiple investigations over the last few decades and covering a time range of 50 Ma. Interpretations of this unusual combination of multidisciplinary data have already cast doubt on a simple mantle-plume hypothesis as the driving mechanism for creating both the rift itself and all of its magmatism. More complex plume scenarios also appear to be significantly flawed. Thus, hypotheses have evolved that consider the WARS magmatism within the context of melts tapping a metasomatized lithospheric mantle. Discussion has also focused on the nature and timing of that metasomatism, which, from a geodynamic point of view, can plausibly be linked to an early rift extensional stage in the Cretaceous or to long-lasting effects of earlier Paleozoic–Mesozoic subduction, which affected West Antarctica widely.

Metasomatism: constraints on type and age

Determining the nature of any metasomatism that affected the source of WARS magmas requires several constraints. Firstly, the compositional type of metasomatism has to be defined, to demonstrate whether it is most suitably associated with an extensional or subduction process. Secondly, the time interval upon which metasomatism acted upon the mantle magma source is important: it is crucial to determine whether the evolution of *in situ* radiogenic Pb can lead to HIMU-like $^{206}Pb/^{204}Pb$ characteristics in a period of hundred(s) of millions of years ('young' metasomatism') rather than billions of years, as assumed in mantle plume models. Finally, in the case of 'young metasomatism', it also has to be demonstrated whether the Paleozoic or the Cretaceous is the most suitable time for the metasomatism to occur.

Signature of the metasomatic agent: subduction-related v. extension-related. The subduction-related model (Finn et al. 2005) suggested that rapid detachment and sinking of Paleozoic–Mesozoic (c. 500–100 Ma) subducted slabs in the Late Cretaceous induced instabilities along the former Gondwana margin that triggered lateral and vertical flow of warm Pacific mantle. The interaction between the warm mantle and the metasomatized subcontinental lithosphere of this diffuse alkaline magmatic province (DAMP) hypothesis would have concentrated magmatism along zones of weakness. In detail, mantle sources of alkaline magmatism would represent continental lithosphere that hosted amphibole- or phlogopite-rich veins formed by plume- and/or subduction-related metasomatism between 500 and 100 Ma (Panter et al. 2006). Alternatively, the extension-related model suggests that, during the amagmatic Cretaceous phase of rifting, small-degree asthenospheric melts veined the sublithospheric mantle, which, upon lithospheric stretching, replaced the lithospheric mantle and started to age isotopically (Rocchi et al. 2002a, 2003, 2005; Nardini et al. 2009; Panter et al. 2018).

Crucial evidence is therefore the trace element signature of the potential metasomatic agents, found in both magmas and mantle xenoliths. Primitive magmas in northern Victoria Land (and WARS as a whole) are characterized by trace element distributions typical of OIB, such as low LILE/high field strength element (HFSE) ratios (e.g. La/Nb = 0.61–0.76). Alternatively, all types of typically subductable materials, such as primitive undifferentiated mantle (PUM: McDonough and Sun 1995), MORB, global subducting sediment (GLOSS: Plank and Langmuir 1998), arc volcanic and plutonic rocks, have high La/Nb. Indeed, the Jurassic Ferrar LIP emplaced in Victoria Land and inferred to have been generated by subduction-related metasomatism of the mantle prior to extension (Hergt et al. 1989; Hergt et al. 1991) has tholeiitic compositions characterized by high LILE/HFSE ratios (La/Nb = 1.2–1.8). Therefore, the hypothesis of a subduction-related origin for the metasomatism of the magma source, while being able to explain the Pb isotopic signatures of the igneous products across a vast area, has difficulties in reconciling the trace element signature of northern Victoria Land basaltic products with the classic subduction fingerprint such as high LILE/HFSE ratios. A model has been proposed to reconcile subduction metasomatism with low LILE/HFSE ratios in magmas (Stein et al. 1997; Panter et al. 2006), based on the concept that arc magmas tap the upper part of the mantle wedge, enriched in most incompatible elements, while the lowermost part of the wedge, enriched in HFSE, may fossilize and become incorporated into the subcontinental lithosphere, thus preparing the source for the generation of OIB magmas by adiabatic decompression melting during subsequent episodes of lithospheric thinning.

Additional critical evidence for the type of metasomatism is the nature of northern Victoria Land mantle xenoliths. Wehrlite–clinopyroxenite xenoliths contain tephrite–phonotephrite glass domains with La/Nb ratios of <0.6 (Perinelli et al. 2011). Preferential Nb, Zr and Sr enrichments in clinopyroxene associated with high Ti–Fe contents indicate a metasomatic agent corresponding to a mafic alkaline melt (Perinelli et al. 2006). Composite peridotite xenoliths contain glasses with low La/Nb ratios (0.33–0.40) and an overall trace element distribution typical of OIB melts (Fig. 10), and allow modelling of a metasomatic agent (Coltorti et al. 2004) similar to the most enriched nephelinite from northern Victoria Land (Nardini et al. 2009). All the evidence of an alkaline OIB signature of the metasomatic agent, even if potentially generated in a subduction environment, appears to be more simply explained in an extension-related setting.

Age of metasomatism: old v. young. Knowledge of the age(s) of any metasomatic event(s) is essential to constrain the geodynamic time frame of the generation of magma sources. Evidence is provided by radiogenic isotopes, represented by both Pb and Os systematics.

High $^{206}Pb/^{204}Pb$ ratios, typical of HIMU products, are commonly equated to evidence for long-lasting (1 Ga) recycling of crustal material in the deep mantle, followed by upwelling of a hot, buoyant mantle plume. However, high

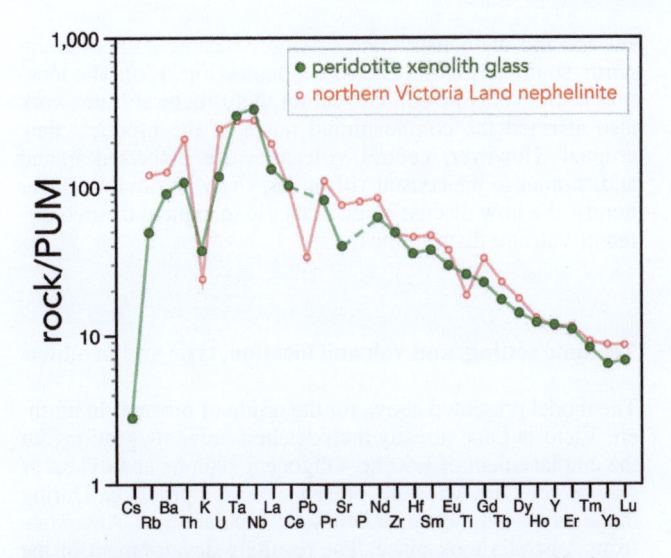

Fig. 10. Trace multi-elemental plot, with concentrations normalized to primitive undifferentiated mantle (PUM) (McDonough and Sun 1995) for glass from northern Victoria Land composite peridotite xenoliths (Coltorti *et al.* 2004) compared with whole-rock composition of the most enriched nephelinite from northern Victoria Land, SAX 20 (Nardini *et al.* 2009).

^{206}Pb/^{204}Pb ratios can be attained by a magma source at comparatively shallow depth in a shorter time interval (100 Ma), provided the source has an adequately high U/Pb ratio (Halliday *et al.* 1995). New Pb isotopic ratios measured for plutonic and subvolcanic rocks in northern Victoria Land extend the Pb isotopic record back in time (Tables 2–5; Fig. 11), and show that older rocks have lower ^{206}Pb/^{204}Pb ratios. This observation suggests that the metasomatic event has to be placed not far back in time with respect to the period of magma genesis. The igneous rocks of any age (plutons, dykes or lavas) with the lowest ^{206}Pb/^{204}Pb ratios can be derived by melting of a common mantle source like FOZO (Stracke *et al.* 2005), which had been affected at *c.* 50 Ma by a metasomatic episode leading to an increase in U/Pb ratios up to 1.13 (corresponding to a value of ^{238}U/^{204}Pb of $\mu = 80$). On the other hand, igneous rocks of any age (plutons, dykes, lavas) with the highest ^{206}Pb/^{204}Pb ratios can be obtained via melting of the same FOZO source that had been affected at *c.* 120 Ma by a metasomatic episode leading to an increase of U/Pb up to 1.80 ($\mu = 128$). Alternatively, if a PREMA source (PREvalent MAntle: Zindler and Hart 1986), rather than FOZO source, is taken as a starting mantle composition, the μ values should be somewhat higher (i.e. 130 and 142). Even acknowledging the slightly lower incompatibility of U with respect to Pb during melting, these U/Pb ratios are not unrealistically high, and have been documented in both northern Victoria Land primitive lavas (Nardini *et al.* 2009) and Zealandia mantle xenoliths (Scott *et al.* 2014; McCoy-West *et al.* 2016). These papers demonstrated that radiogenic Pb signatures can evolve very rapidly in the subcontinental lithospheric mantle, with most metasomatic age estimates younger than 120 Ma. In summary, evidence from Pb isotopes indicates that the incubation time between source metasomatism and melting can be bracketed within the WARS period of amagmatic extension (i.e. from the Early Cretaceous to the Middle Eocene). These results imply that a single event or multiple events that metasomatized the

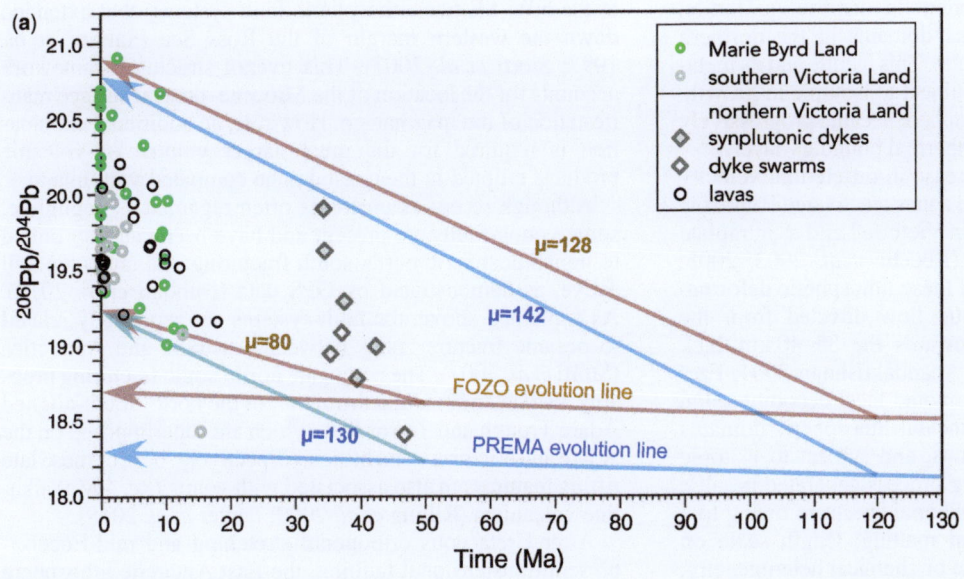

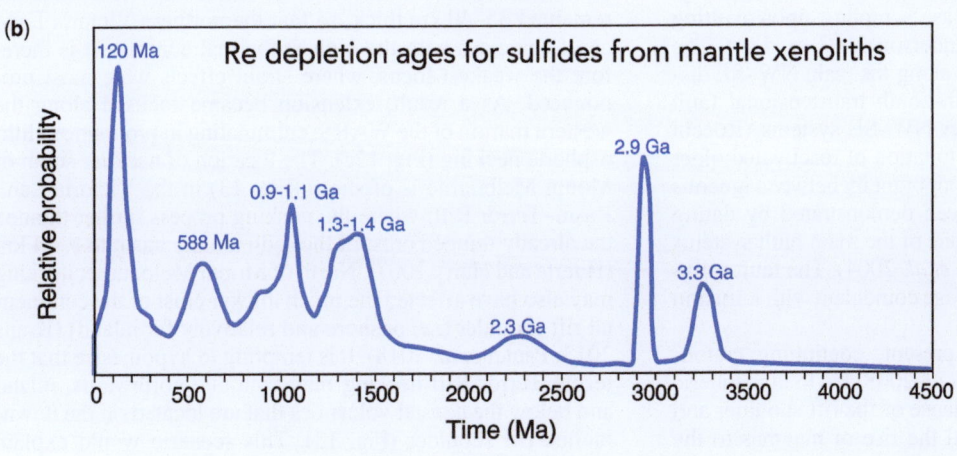

Fig. 11. Constraints for the timing of metasomatism of northern Victoria Land lithospheric mantle. (**a**) ^{206}Pb/^{204}Pb v. time showing that the igneous rocks of various ages (plutons, dykes, lavas) with the lowest ^{206}Pb/^{204}Pb ratios can be derived by the melting of a common mantle source like FOZO, which had been affected at *c.* 50 Ma by a metasomatic episode leading to an increase of U/Pb up to a value of ^{238}U/^{204}Pb of $\mu = 80$; igneous rocks of various ages (plutons, dykes, lavas) with the highest ^{206}Pb/^{204}Pb ratios can be obtained via melting of the same FOZO which had been affected at *c.* 120 Ma by a metasomatic episode leading to an increase of U/Pb up to $\mu = 128$. Alternatively, if a PREMA composition instead of FOZO is taken as a starting mantle composition, the μ values should be somewhat higher (i.e. 130 and 142). (**b**) Relative probability plot of Re depletion model ages calculated for sulfides from mantle xenoliths from northern Victoria Land Cenozoic lava flows (Melchiorre *et al.* 2011).

mantle source, without resulting in magmatism at the surface, may reasonably have occurred during the Cretaceous extension in the Ross Embayment, with the introduction of small-degree partial melts generated by extension-related decompression (Halliday et al. 1995).

Rhenium–Os isotopic analyses of sulfides in mantle peridotite xenoliths provide similar constraints on the evolution of the northern Victoria Land subcontinental lithospheric mantle and its timing of metasomatism. The distribution of Re depletion ages (Fig. 11b) (Melchiorre et al. 2011) indicate events of deviation of the Re–Os system from a chondritic evolution, corresponding either to depletion or metasomatism. This time distribution shows numerous peaks, with the youngest and most significant peak at 120 Ma coinciding with the amagmatic activation of the WARS during the Cretaceous, and for which the best explanation is a link to an extension-related partial melting/metasomatic episode.

Thus, both Pb and Os systematics suggest that the Cretaceous is the most plausible time for the episode of alkaline metasomatism affecting the mantle underlying the Ross Sea.

A comprehensive model for magma origins

A general model for the tectonomagmatic history of the western Ross Embayment is constrained by the lack of both an active mantle plume and minor Cenozoic extension. During the Cretaceous, an early rift phase occurred with orthogonal extension that stretched the crust and the underlying strong lithospheric mantle (Fig. 12a). Lithospheric thinning is likely to have led to the production of very-small-degree partial melts. These were not sufficient to give way to surface magmatism (amagmatic rift phase) but were essential in distributing fertile, enriched, low-melting-point veins/domains in the northern Victoria Land asthenospheric mantle. This mantle and its metasomatic domains started to age, subject to isotopic in-growth. Upon cooling, the veined asthenosphere became progressively incorporated into the lithospheric thermal boundary layer.

In the Middle Eocene, the increase in differential velocity along the Southern Ocean fracture zones reactivated Paleozoic tectonic discontinuities in northern Victoria Land as intraplate dextral strike-slip fault systems (Rocchi et al. 2003, 2005; Storti et al. 2007). The activity of these lithospheric deformation belts, combined with mantle flow directed from the thinned Ross Sea lithosphere towards the 35–40 km-thick Antarctic Craton (Winberry and Anandakrishnan 2004; Faccenna et al. 2008) (Fig. 12b), promoted local decompression melting of the enriched subcontinental lithospheric domains created during the Late Cretaceous, and subject to isotopic in-growth since then. The melting process generated basaltic magmas with a restricted compositional spectrum owing to a combination of degree of partial melting, length scale on which melting took place and size of chemical heterogeneity, a process synthetically defined as 'sampling upon melting and averaging' (Meibom and Anderson 2003, p. 134). The magmas rose and were emplaced along the main NW–SE discontinuities and also along north–south transtensional fault arrays that splay from the master NW–SE systems (Rocchi et al. 2002a), an example of exploitation of reactivated older lithospheric structures. The contemporaneity between igneous activity and fault activity has been demonstrated by dating pseudotachylite associated with one of the main fault systems (the Priestley Fault: Di Vincenzo et al. 2004). The fault activity has been dated to c. 34 Ma, an age coincident with intrusion of nearby dyke swarms.

From the Late Miocene to present, continuing craton-directed mantle flow (Faccenna et al. 2008) led to lithospheric necking (Panter et al. 2018), collapse of the rift shoulder and normal faulting, which promoted the rise of magmas to the surface and the construction of large volcanic edifices along north–south normal-transtensional faults (Fig. 12c). The location of the volcanic edifices within this structural framework also affected the compositional range of the products they erupted. However, central volcanoes are dispersed inland and, similar to the coastal volcanoes, show no obvious linear trends. We now discuss a new scenario to explain the two different volcano distributions.

Tectonic setting, and volcano location, type and products

The model presented above for the origin of magmas in northern Victoria Land already includes first-order suggestions for the emplacement of Eocene–Oligocene plutons and dykes, as well as for Late Miocene–present volcanic products. During the Eocene–Oligocene, accelerated separation of Antarctica from Australia took place. The resulting development of the oceanic transform fracture zones led to the uprise of anomalously hot upper mantle in a linear zone, part of which is still detectable in the western Ross Sea (Fig. 9) (Storti et al. 2007). Magmas generated by decompression melting of the mantle previously metasomatized during the Cretaceous were emplaced as: (i) plutons in NW–SE-elongated pull-apart structures related to the dextral transtensional activity of translithospheric fault systems; and (ii) dykes, also intruded along these faults and their conjugate fractures.

After a magmatic lull of almost 10 Ma, between c. 23 and 15 Ma, magmatic activity started again in northern Victoria Land, with magma generated in the lithospheric mantle rising through late-formed north–south-orientated faults (conjugate to the NW–SE translithospheric fault systems) that extended down the western margin of the Ross Sea (Salvini et al. 1997; Storti et al. 2007). This overall structural framework accounts for the location of the Miocene–present surface manifestation of the magmatism. However, an additional explanation is required for the much larger volume of volcanic products erupted in the coastal zone compared with inland.

Although recent extension is often regarded as negligible, some young faults are present and have been causally linked to translithospheric north–south fracturing; the faults are still active, as demonstrated by GPS data (Dubbini et al. 2010). As explained above, the fault systems are genetically related to oceanic fracture zones between Australia and Antarctica (Storti et al. 2007). The conjugate north–south fracturing probably also determined the formation of the north–south-aligned Adare Trough and Terror Rift, which are superimposed on the Adare and Victoria Land basins, respectively; both of these late rifting features are also associated with young (<c. 4 Ma) alkaline volcanism (Rilling et al. 2009; Panter et al. 2018).

After Cretaceous orthogonal stretching and mid-Eocene–present transtensional faulting, the East Antarctic lithosphere remained 35–40 km thick, so that the northern Victoria Land coast coincides with the highest thermal contrast. It is therefore the weakest locus, where strain effects were most pronounced. As a result, extension became focused along the western margin of the WARS, culminating in pronounced lithospheric necking (Fig. 12c). The location of necking south of Mount Melbourne is offshore (Fig. 13) in the Victoria Land Basin–Terror Rift, where the necking process further thinned the already thinned crust of the sedimentary basin to c. 14 km (Huerta and Harry 2007). North of Mount Melbourne, necking may also have affected the much thicker crust of the continental rift shoulder (i.e. onshore and relatively far inland) (Krans 2013; Panter et al. 2018). It is tempting to hypothesize that the recent Terror Rift necking may continue northwards, inland and below the coastal volcanoes that are located on the down-faulted rift shoulder (Fig. 13). This scenario would explain

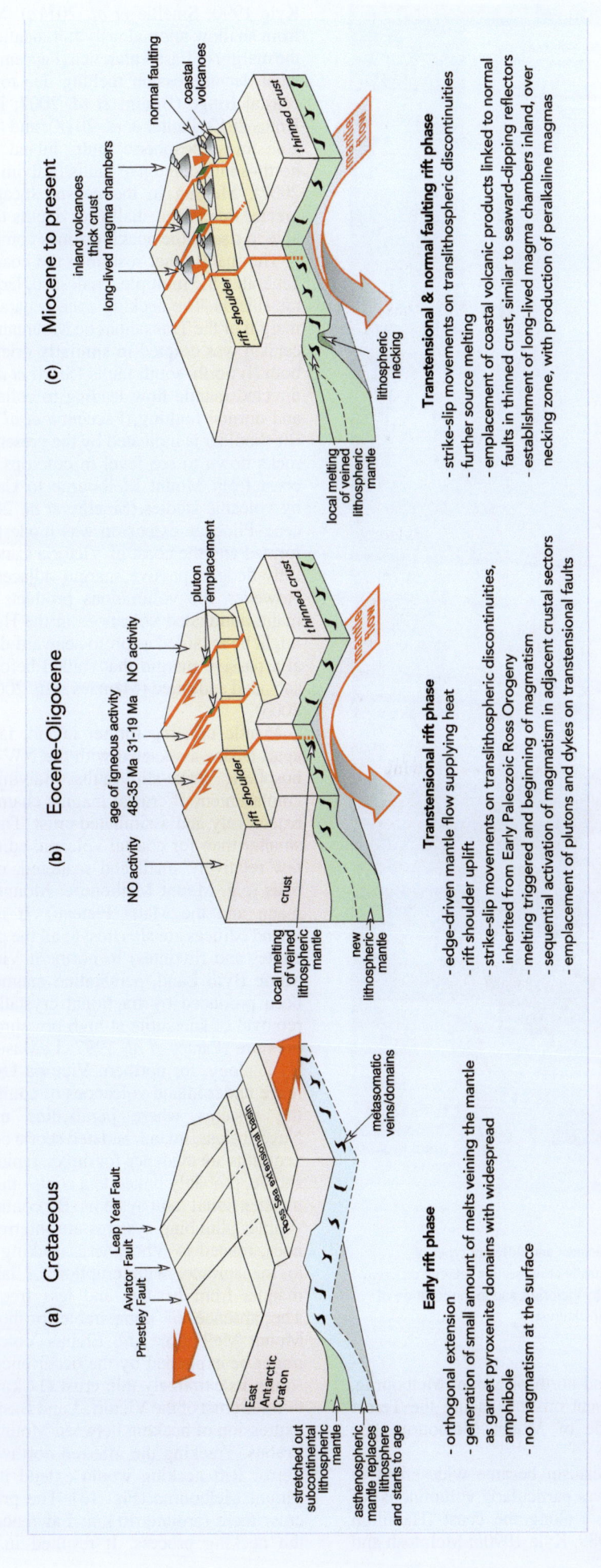

Fig. 12. Three-stage model depicting the tectonomagmatic evolution of northern Victoria Land from (a) the Cretaceous to (b) the Eocene–Oligocene to (c) the Miocene–present.

Kyle 1990; Smellie et al. 2011a). Magmas were generated from shallow anomalously hot mantle that flowed west under the margin of East Antarctica (Faccenna et al. 2008) and underwent decompression melting due to: (1) necking along the coastal fringe (Bonini et al. 2007; Huerta and Harry 2007; Krans 2013; Panter et al. 2018); and (2) reactivation of Paleozoic translithospheric faults inland along with late-formed north–south conjugate faults (Salvini et al. 1997; Storti et al. 2007). Melting in the coastal locations would have been greater due to the shallower depths to which the mantle was able to rise in the necking zones compared to inland.

The mantle that rose into the coast-parallel necking zone generated the multiple, coalesced, large, mainly mafic, volcanic shields. The necking zone is parallel to the thick cratonic margin of the Transantarctic Mountains (Fig. 13), and the volcanism was erupted in similarly orientated fissures linked to both (i) north–south faults (Storti et al. 2007) and (ii) craton-directed mantle flow leading to collapse of the rift shoulder and normal faulting (Faccenna et al. 2008). Collapse of the rift shoulder is indicated by the presence of subaerial volcanic rocks down to sea level in outcrops along the length of the coast from Mount Melbourne to Cape Adare, as indicated by volcanic studies (Smellie et al. 2011a, b). Because Miocene–Pliocene extension was modest, oceanic crust was not formed and the coast of Victoria Land did not develop into a fully-fledged passive margin adjacent to an oceanic basin. However, the voluminous products of the linear chain of mafic-dominated volcanoes in the Hallett Volcanic Province might be regarded as proto-seaward-dipping reflectors formed at a passive margin that stalled before the margin had fully subsided and rifted (Menzies et al. 2002; Direen and Crawford 2003).

Mantle uprising further inland, facilitated by small pull-apart basins associated with the NW–SE continental translithospheric strike-slip faults (Salvini et al. 1997), led to emplacement of crustal magma chambers which fractionated extensively and assimilated crust. The erupted volumes were smaller than for coastal volcanic edifices, and gave rise to a few relatively small and scattered, mafic–felsic stratovolcanoes (e.g. Mount Melbourne, Mount Overlord, Mount Rittmann and the Malta Plateau). It is noticeable that these inland edifices are also host to all the peralkaline products (trachytes and rhyolites) in northern Victoria Land (Fig. 4). In Marie Byrd Land, peralkaline magmas are inferred to have been produced by fractional crystallization of basanite and removal of kaersutite at high pressure and plagioclase at low pressure (Panter et al. 1997; LeMasurier et al. 2011, 2018). By analogy, for northern Victoria Land, we infer the occurrence under inland volcanoes of complex, multi-level plumbing systems where peralkaline magmas are produced. Nevertheless, inland, isolated scoria cones loaded with mantle xenoliths are evidence for direct, rapid magma ascent from the source, probably linked to a connection with translithospheric transtensional fault systems. By contrast to the inland setting, simpler plumbing systems are inferred for the coastal volcanoes, related to lithospheric necking bringing mantle closer to the surface, with eruption of larger volumes of mafic magma from fissures and less frequent felsic derivatives. The absence of comparable onshore volcanism between Mount Melbourne and Erebus Volcanic Province latitudes might be explained by the occurrence of the Terror Rift offshore: its extremely thin crust (14 km), developed in already thinned crust of the Victoria Land Basin, represents the surface expression of necking between Mount Melbourne and Mount Erebus. Tracking the inferred northward continuation of the Terror Rift necking would extend it into the area north of Mount Melbourne (Fig. 13). The presence of much thicker crust there (around 40 km: Lawrence et al. 2006) inhibited the necking process. It resulted in the eruption of much

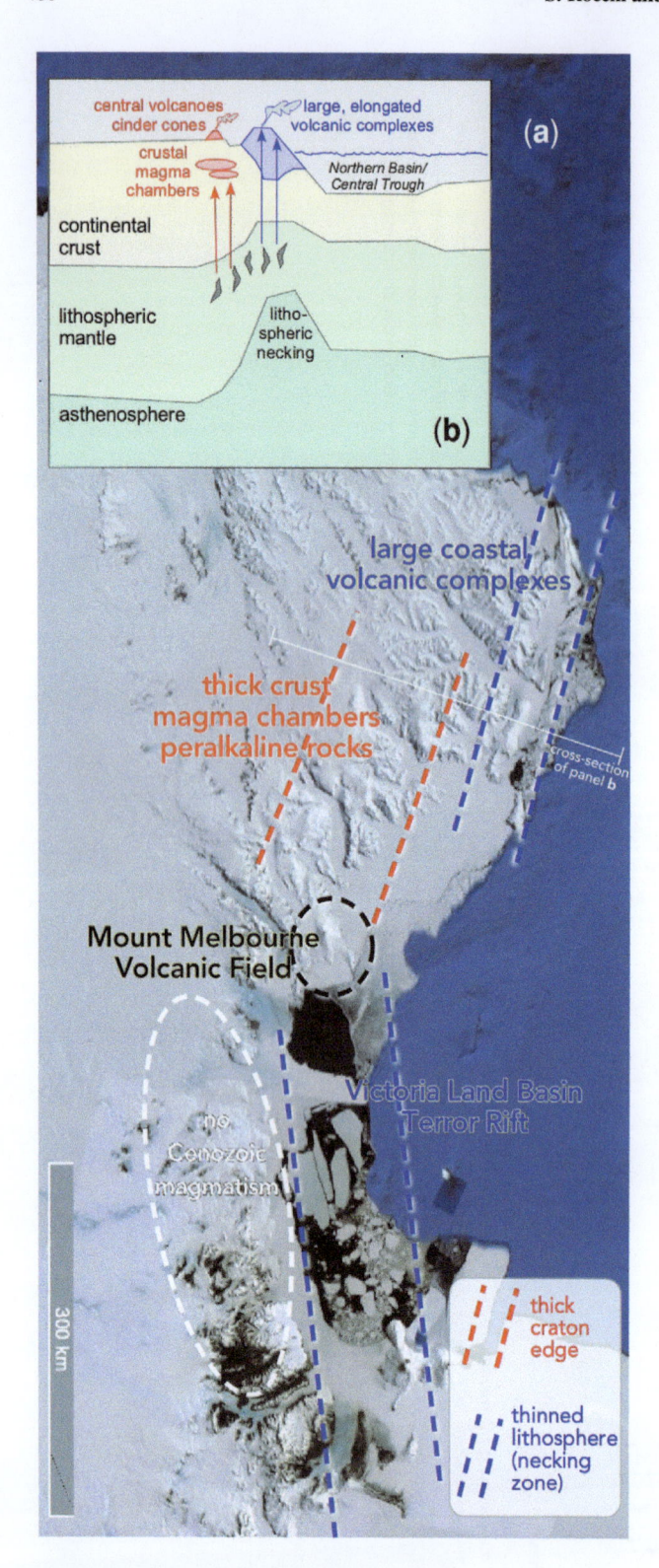

Fig. 13. (a) Location of tectonic structures and different types of volcanoes in Victoria Land. (b) Schematic cross-section relating different volcano plumbing systems in northern Victoria Land to the location of lithospheric necking and 35–40 km-thick crust.

why there are volcanoes inland north of Mount Melbourne, while volcanic activity is absent onshore facing the Terror rift (i.e. between the latitude of Mount Melbourne and Mount Erebus).

From c. 15 to 12 Ma volcanism became widespread in northern Victoria Land, and was particularly voluminous (of the order of at least 10^4 km^3) along the coast (Hamilton 1972; Wörner and Viereck 1989; Kyle 1990b; McIntosh and smaller volumes of volcanic rocks in relatively small stratovolcanoes compared to the fissure-erupted coastal shield volcanoes. These different volcanic expressions north and south of Mount Melbourne, respectively, are possibly linked to the different crustal structure between the northern part, representing the thickened continental magmatic arc of the Early Paleozoic Ross Orogeny, and the southern part, which underwent lithospheric delamination at the end of the same orogeny (Rocchi et al. 1998, 2009, 2011). The reason for the c. 10 Ma lag in time between the Eocene–Oligocene extension, represented magmatically by the Meander Intrusive Group, and the eruption of the northern Victoria Land volcanoes is not yet resolved satisfactorily but it may be a consequence of the time needed for conductive heating at the base of the lithosphere to reach the source solidus (Panter et al. 2018).

Implications for Earth's rift magmatism

Studies on the evolution of continental rift systems and rifted continental margins have led to the development of many hypotheses about their origin. A classic element of controversy is whether rift zones are produced by actively upwelling mantle splitting the continent, possibly along pre-weakened zones, or whether the mantle rises passively as the continents are pulled apart during lithospheric stretching. A major argument used to address this question typically uses the geochemical and isotopic compositions of the rift-related igneous products. Because of the wide multidisciplinary dataset of geochemical, structural and tectonic information now available for magmatism in northern Victoria Land, we are now able to constrain several facets of this discussion. Our results, described in this chapter, and those of ongoing studies should help to resolve many important issues related to the origin of alkaline lavas in intraplate settings, either continental or oceanic.

Acknowledgements The authors wish to thank the base commanders, the helicopter pilots of Helicopters New Zealand and all staff at Mario Zucchelli Station for their unstinting help during all of their field seasons. Thanks are due to the reviewers M. Pompilio and A. Martin, and the editor K. Panter for their very precise work which led to better clarification and discussion of some critical points, as well as cleaning up inconsistencies in the paper.

Author contributions SR: conceptualization (lead), data curation (lead), formal analysis (lead), funding acquisition (lead), investigation (lead), methodology (lead), writing – original draft (lead), writing – review & editing (equal); **JLS**: data curation (supporting), formal analysis (equal), investigation (supporting), writing – original draft (supporting), writing – review & editing (equal).

Funding This work has been supported by the National Research Antarctic Programme of Italy (PNRA18_00037). J.L. Smellie is grateful to the British Antarctic Survey for supporting his participation in fieldwork in northern Victoria Land during 2005–06.

Data availability All data generated or analysed during this study are included in this published article (and its supplementary information files).

References

Anderson, D.L. 1995. Lithosphere, asthenosphere, and perisphere. *Reviews of Geophysics*, **33**, 125–149, https://doi.org/10.1029/94RG02785

Anderson, D.L. 2001. Top-down tectonics? *Science*, **293**, 2016–2018, https://doi.org/10.1126/science.1065448

Antonini, P., Piccirillo, E.M., Petrini, R., Civetta, L., D'Antonio, M. and Orsi, G. 1999. Enriched mantle – Dupal signature in the genesis of the Jurassic Ferrar tholeiites from Prince Albert Mountains (Victoria Land, Antarctica). *Contributions to Mineralogy and Petrology*, **136**, 1–19, https://doi.org/10.1007/s004100050520

Armienti, P. and Tripodo, A. 1991. Petrography and chemistry of lavas and comagmatic xenoliths of Mt. Rittmann, a volcano discovered during the IV Italian expedition in Northern Victoria Land (Antarctica). *Memorie della Società Geologica Italiana*, **46**, 427–451.

Armienti, P., Civetta, L., Innocenti, F., Manetti, P., Tripodo, A., Villari, L. and Vita, G. 1991. New petrological and geochemical data on Mt. Melbourne volcanic field, Northern Victoria Land, Antarctica. (II Italian Antarctic Expedition). *Memorie della Società Geologica Italiana*, **46**, 397–424.

Aviado, K.B., Rilling-Hall, S., Bryce, J.G. and Mukasa, S.B. 2015. Submarine and subaerial lavas in the West Antarctic Rift System: temporal record of shifting magma source components from the lithosphere and asthenosphere. *Geochemistry, Geophysics, Geosystems*, **16**, 4344–4361, https://doi.org/10.1002/2015GC006076

Beccaluva, L., Coltorti, M., Orsi, G., Saccani, E. and Siena, F. 1991. Basanite to tephrite lavas from Melbourne volcanic province, Victoria Land, Antarctica. *Memorie della Società Geologica Italiana*, **46**, 383–395.

Behrendt, J.C., LeMasurier, W.E., Cooper, A.K., Tessensohn, F., Tréhu, A. and Damaske, D. 1991. Geophysical studies of the West Antarctic Rift System. *Tectonics*, **10**, 1257–1273, https://doi.org/10.1029/91TC00868

Behrendt, J.C., Blankenship, D.D., Finn, C.A., Bell, R.E., Sweeney, R.E., Hodge, S.M. and Brozena, J.M. 1994. CASERTZ aeromagnetic data reveal late Cenozoic flood basalts(?) in the West Antarctic rift system. *Geology*, **22**, 527–530, https://doi.org/10.1130/0091-7613(1994)022<0527:CADRLC>2.3.CO;2

Behrendt, J.C., Blankenship, D., Morse, D.L. and Bell, R.E. 2004. Shallow-source aeromagnetic anomalies observed over the West Antarctic Ice Sheet compared with coincident bed topography from radar ice sounding – new evidence for glacial 'removal' of subglacially erupted late Cenozoic rift-related volcanic edifices. *Global and Planetary Change*, **42**, 177–193, https://doi.org/10.1016/j.gloplacha.2003.10.006

Bonini, M., Corti, G., Ventisette, C.D., Manetti, P., Mulugeta, G. and Sokoutis, D. 2007. Modelling the lithospheric rheology control on the Cretaceous rifting in West Antarctica. *Terra Nova*, **19**, 360–366, https://doi.org/10.1111/j.1365-3121.2007.00760.x

Burgess, S.D., Bowring, S.A., Fleming, T.H. and Elliot, D.H. 2015. High-precision geochronology links the Ferrar large igneous province with early-Jurassic ocean anoxia and biotic crisis. *Earth and Planetary Science Letters*, **415**, 90–99, https://doi.org/10.1016/j.epsl.2015.01.037

Cande, S.C. and Mutter, J.C. 1982. A revised identification of the oldest sea-floor spreading anomalies between Autralia and Antarctica. *Earth and Planetary Science Letters*, **58**, 151–160, https://doi.org/10.1016/0012-821X(82)90190-X

Cande, S.C. and Stock, J. 2003. Tertiary seafloor spreading between East and West Antarctica. Presented at the 9th International Symposium on Antarctic Earth Sciences, 8–12 September 2003, Potsdam, Germany.

Cande, S.C., Raymond, C.A., Stock, J. and Haxby, W.F. 1995. Geophysics of the Pitman Fracture Zone and Pacific–Antarctic plate motions during the Cenozoic. *Science*, **270**, 947–953, https://doi.org/10.1126/science.270.5238.947

Cande, S.C., Stock, J.M., Müller, R.D. and Ishihara, T. 2000. Cenozoic motion between East and West Antarctica. *Nature*, **404**, 145–150, https://doi.org/10.1038/35004501

Cape Roberts Science Team. 2001a. Studies from the Cape Roberts Project, Ross Sea, Antarctica. Scientific results from CRP-3 – Part I. *Terra Antartica*, **8**, 119–308.

Cape Roberts Science Team. 2001b. Studies from the Cape Roberts Project, Ross Sea, Antarctica. Scientific results from CRP-3 – Part II. *Terra Antartica*, **8**, 309–622.

Christiansen, R.L., Foulger, G.R. and Evans, J.R. 2002. Upper mantle origin of the Yellowstone hotspot. *Geological Society of America Bulletin*, **114**, 1245–1256, https://doi.org/10.1130/0016-7606(2002)114<1245:UMOOTY>2.0.CO;2

Coltorti, M., Beccaluva, L., Bonadiman, C., Faccini, B., Ntaflos, T. and Siena, F. 2004. Amphibole genesis via metasomatic reaction with clinopyroxene in mantle xenoliths from Victoria Land, Antarctica. *Lithos*, **75**, 115–139, https://doi.org/10.1016/j.lithos.2003.12.021

Dallai, L. and Burgess, R. 2011. A record of Antarctic surface temperature between 25 and 50 m.y. ago. *Geology*, **39**, 423–426, https://doi.org/10.1130/G31569.1

Danesi, S. and Morelli, A. 2000. Group velocity of Rayleigh waves in the Antarctic region. *Physics of the Earth and Planetary Interiors*, **122**, 55–66, https://doi.org/10.1016/S0031-9201(00)00186-2

Davey, F.J. and Brancolini, G. 1995. The late Mesozoic and Cenozoic structural setting of the Ross Sea region. *American Geophysical Union Antarctic Research Series*, **68**, 167–182.

Davey, F.J. and Santis, L.D. 2006. A multi-phase rifting model for the Victoria Land basin, Western Ross Sea. *In*: Fütterer, D.K., Damaske, D., Kleinschmidt, G., Miller, H. and Tessensohn, F. (eds) *Antarctica – Contributions to Global Earth Sciences, Proceedings of the IX International Symposium of Antarctic Earth Sciences, Potsdam, 2003*. Springer, New York, 301–306.

Davey, F.J., Cande, S.C. and Stock, J. 2006. Extension in the western Ross Sea region – links between Adare Basin and Victoria Land Basin. *Geophysical Research Letters*, **33**, L20315, https://doi.org/10.1029/2006GL027383

Davey, F.J., Granot, R., Cande, S.C., Stock, J.M., Selvans, M. and Ferraccioli, F. 2016. Synchronous oceanic spreading and continental rifting in West Antarctica. *Geophysical Research Letters*, **43**, 6162–6169, https://doi.org/10.1002/2016GL069087

Direen, N.G. and Crawford, A.J. 2003. The Tasman Line: where is it, what is it, and is it Australia's Rodinian breakup boundary? *Australian Journal of Earth Sciences*, **50**, 491–502, https://doi.org/10.1046/j.1440-0952.2003.01005.x

Di Vincenzo, G., Rocchi, S., Rossetti, F. and Storti, F. 2004. $^{40}Ar-^{39}Ar$ dating of pseudotachylytes: the effect of clast-hosted extraneous argon in Cenozoic fault-generated friction melts from the West Antarctic Rift System. *Earth and Planetary Science Letters*, **223**, 349–364, https://doi.org/10.1016/j.epsl.2004.04.042

Dubbini, M., Cianfarra, P., Casula, G., Capra, A. and Salvini, F. 2010. Active tectonics in northern Victoria Land (Antarctica) inferred from the integration of GPS data and geologic setting. *Journal of Geophysical Research: Solid Earth*, **115**, B12421, https://doi.org/10.1029/2009JB007123

Elliot, D.H., Fleming, T.H., Kyle, P.R. and Foland, K.A. 1999. Long-distance transport of magmas in the Jurassic Ferrar Large Igneous Province, Antarctica. *Earth and Planetary Science Letters*, **167**, 89–104, https://doi.org/10.1016/S0012-821X(99)00023-0

Faccenna, C., Rossetti, F., Becker, T.W., Danesi, S. and Morelli, A. 2008. Recent extension driven by mantle upwelling beneath the Admiralty Mountains (East Antarctica). *Tectonics*, **27**, TC4015, https://doi.org/10.1029/2007TC002197

Finn, C., Müller, R.D. and Panter, K.S. 2005. A Cenozoic diffuse alkaline magmatic province (DAMP) in the southwest Pacific without rift or plume origin. *Geochemistry, Geophysics, Geosystems*, **6**, Q02005, https://doi.org/10.1029/2004GC000723

Fitzgerald, P. and Baldwin, S. 1997. Detachment fault model for the evolution of the Ross Embayment. *In*: Ricci, C.A. (ed.) *The Antarctic Region: Geological Evolution and Processes*. Terra Antartica, Siena, Italy, 555–564.

Gautheron, C. and Moreira, M. 2002. Helium signature of the subcontinental lithospheric mantle. *Earth and Planetary Science Letters*, **199**, 39–47, https://doi.org/10.1016/S0012-821X(02)00563-0

Graham, D.W. 2002. Noble gas isotope geochemistry of mid-ocean ridge and ocean island basalts: Characterization of mantle source reservoirs. *Reviews in Mineralogy and Geochemistry*, **47**, 247–318.

Granot, R., Cande, S.C., Stock, J.M., Davey, F.J. and Clayton, R.W. 2010. Postspreading rifting in the Adare Basin, Antarctica: Regional tectonic consequences. *Geochemistry, Geophysics, Geosystems*, **11**, Q08005, https://doi.org/10.1029/2010GC003105

Granot, R., Cande, S.C., Stock, J.M. and Damaske, D. 2013. Revised Eocene–Oligocene kinematics for the West Antarctic rift system. *Geophysical Research Letters*, **40**, 279–284, https://doi.org/10.1029/2012GL054181

Halliday, A.N., Lee, D.-C., Tommasini, S., Davies, G.R., Paslick, C.R., Fitton, J.G. and James, D.E. 1995. Incompatible trace elements in OIB and MORB and source enrichment in the sub-oceanic mantle. *Earth and Planetary Science Letters*, **133**, 379–395, https://doi.org/10.1016/0012-821X(95)00097-V

Hamilton, R.J., Luyendik, B.P., Sorlien, C.C. and Bartek, L.R. 2001. Cenozoic tectonics of the Cape Roberts rift basin and the Transantarctic Mountains front, southwestern Ross Sea, Antarctica. *Tectonics*, **20**, 325–342, https://doi.org/10.1029/2000TC001218

Hamilton, W. 1972. *The Hallet Volcanic Province, Antarctica*. United States Geological Survey Professional Papers, **456-C**.

Hansen, S.E., Graw, J.H. et al. 2014. Imaging the Antarctic mantle using adaptively parameterized P-wave tomography: Evidence for heterogeneous structure beneath West Antarctica. *Earth and Planetary Science Letters*, **408**, 66–78, https://doi.org/10.1016/j.epsl.2014.09.043

Harrison, D. and Ballentine, C.J. 2005. Noble gas models of mantle convection and mass reservoir transfer. *American Geophysical Union Geophysical Monograph Series*, **160**, 9–26.

Harrison, D., Leat, P.T., Burnard, P.G., Turner, G., Fretzdorff, S. and Millar, I.L. 2003. Resolving mantle components in oceanic lavas from segment E2 of the East Scotia back-arc ridge, South Sandwich Islands. *Geological Society, London, Special Publications*, **219**, 333–344, https://doi.org/10.1144/GSL.SP.2003.219.01.16

Hart, S., Blundy, J. and Craddock, C. 1995. Cenozoic volcanism in Antarctica: Jones Mountains and Peter I Island. *Geochimica et Cosmochimica Acta*, **59**, 3379–3388, https://doi.org/10.1016/0016-7037(95)00212-I

Hart, S.R., Blusztanjn, J., LeMasurier, W.E. and Rex, D.C. 1997. Hobbs Coast Cenozoic volcanism: Implications for the West Antarctic rift system. *Chemical Geology*, **139**, 223–248, https://doi.org/10.1016/S0009-2541(97)00037-5

Henrys, S., Wilson, T., Whittaker, J.M., Fielding, C., Hall, J. and Naish, T. 2007. Tectonic history of mid-Miocene to present southern Victoria Land Basin, inferred from seismic stratigraphy in McMurdo Sound, Antarctica. *United States Geological Survey Open-File Report*, **2007-1047**, Short Research Paper 049, https://doi.org/10.3133/of2007-1047.srp049

Hergt, J.M., Chappell, B.W., Faure, G. and Mensing, T.M. 1989. The geochemistry of Jurassic dolerites from Portal Peak, Antarctica. *Contributions to Mineralogy and Petrology*, **102**, 298–305, https://doi.org/10.1007/BF00373722

Hergt, J.M., Peate, D.W. and Hawkesworth, C.J. 1991. The petrogenesis of Mesozoic Gondwana low-Ti flood basalts. *Earth and Planetary Science Letters*, **105**, 134–148, https://doi.org/10.1016/0012-821X(91)90126-3

Hole, M.J. and LeMasurier, W.E. 1994. Tectonic controls on the geochemical composition of Cenozoic, mafic alkaline volcanic rocks from West Antarctica. *Contributions to Mineralogy and Petrology*, **117**, 187–202, https://doi.org/10.1007/BF00286842

Hornig, I. and Wörner, G. 2003. Cenozoic volcanics from Coulman Island, Mandible Cirque, Malta Plateau and Navigator Nunatak, Victoria Land, Antarctica. *Geologisches Jahrbuch*, **B85**, 373–406.

Huerta, A.D. and Harry, D.L. 2007. The transition from diffuse to focused extension: Modeled evolution of the West Antarctic Rift system. *Earth and Planetary Science Letters*, **255**, 133–147, https://doi.org/10.1016/j.epsl.2006.12.011

Kim, J., Park, J.-W., Lee, M.J., Lee, J.I. and Kyle, P.R. 2018. Evolution of alkalic magma systems: insight from coeval evolution of sodic and potassic fractionation lineages at the Pleiades volcanic complex, Antarctica. *Journal of Petrology*, **60**, 117–150, https://doi.org/10.1093/petrology/egy108

Kipf, A., Hauff, F. *et al.* 2014. Seamounts off the West Antarctic margin: A case for non-hotspot driven intraplate volcanism. *Gondwana Research*, **25**, 1660–1679, https://doi.org/10.1016/j.gr.2013.06.013

Krans, S.R. 2013. *New Mineral Chemistry and Oxygen Isotopes from Alkaline Basalts in the Northwest Ross Sea, Antarctica: Insights on Magma Genesis across Rifted Continental and Oceanic Lithosphere*. MSc thesis, Bowling Green State University, Bowling Green, Ohio, USA.

Kyle, P.R. 1982. Volcanic geology of The Pleiades, Northern Victoria Land, Antarctica. *In*: Craddock, C. (ed.) *Antarctic Geosciences*. University of Wisconsin Press, Madison, WI, 747–754.

Kyle, P.R. 1990a. McMurdo Volcanic Group, Western Ross Embayment: Introduction. *American Geophysical Union Antarctic Research Series*, **48**, 19–25.

Kyle, P.R. 1990b. Melbourne Volcanic Province. *American Geophysical Union Antarctic Research Series*, **48**, 48–80.

Kyle, P.R., Moore, J.A. and Thirlwall, M.F. 1992. Petrologic evolution of anorthoclase phonolite lavas at Mount Erebus, Ross Island, Antarctica. *Journal of Petrology*, **33**, 849–875, https://doi.org/10.1093/petrology/33.4.849

Lawrence, J.F., Wiens, D.A., Nyblade, A.A., Anandakrishnan, S., Shore, P.J. and Voigt, D. 2006. Crust and upper mantle structure of the Transantarctic Mountains and surrounding regions from receiver functions, surface waves, and gravity: Implications for uplift models. *Geochemistry, Geophysics, Geosystems*, **10**, Q10011, https://doi.org/10.1029/2006GC001282

Lawver, L.A. and Gahagan, L.M. 1994. Constraints on timing of extension in the Ross Sea region. *Terra Antartica*, **1**, 545–552.

Leat, P.T. and Riley, T.R. 2021. Antarctic Peninsula and South Shetland Islands: volcanology. *Geological Society, London, Memoirs*, **55**, https://doi.org/10.1144/M55-2018-52

Le Bas, M.J., Le Maitre, R.W., Streckeisen, A. and Zanettin, B. 1986. A chemical classification of volcanic rocks based on the total alkali–silica diagram. *Journal of Petrology*, **27**, 745–750, https://doi.org/10.1093/petrology/27.3.745

LeMasurier, W.E. 2008. Neogene extension and basin deepening in the West Antarctic rift inferred from comparisons with the East African rift and other analogs. *Geology*, **36**, 247–250, https://doi.org/10.1130/G24363A.1

LeMasurier, W.E. and Rex, D.C. 1989. Evolution of linear volcanic ranges in Marie Byrd Land, Antarctica. *Journal of Geophysical Research*, **94**, 7223–7236, https://doi.org/10.1029/JB094iB06p07223

LeMasurier, W.E. and Landis, C.A. 1996. Mantle-plume activity recorded by low relief erosion surfaces in West Antarctica and New Zealand. *Geological Society of America Bulletin*, **108**, 1450–1466, https://doi.org/10.1130/0016-7606(1996)108<1450:MPARBL>2.3.CO;2

LeMasurier, W.E. and Thomson, J.W. (eds). 1990. *Volcanoes of the Antarctic Plate and Southern Oceans*. American Geophysical Union Antarctic Research Series, **48**.

LeMasurier, W.E., Hi Choi, S., Kawachi, Y., Mukasa, S.B. and Rogers, N.W. 2011. Evolution of pantellerite–trachyte–phonolite volcanoes by fractional crystallization of basanite magma in a continental rift setting, Marie Byrd Land, Antarctica. *Contributions to Mineralogy and Petrology*, **162**, 1165–1179, https://doi.org/10.1007/s00410-011-0646-z

LeMasurier, W., Choi, S.H., Kawachi, Y., Mukasa, S. and Rogers, N. 2018. Dual origins for pantellerites, and other puzzles, at Mount Takahe volcano, Marie Byrd Land, West Antarctica. *Lithos*, **296–299**, 142–162, https://doi.org/10.1016/j.lithos.2017.10.014

Lithgow-Bertelloni, C. and Richards, M.A. 1998. The dynamics of Cenozoic and Mesozoic plate motions. *Reviews of Geophysics*, **36**, 27–78, https://doi.org/10.1029/97RG02282

Luttinen, A.V. and Furnes, H. 2000. Flood basalts of Vestfjella: Jurassic magmatism across the Archean–Proterozoic lithospheric boundary in Dronning Maud Land, Antarctica. *Journal of Petrology*, **41**, 1271–1305, https://doi.org/10.1093/petrology/41.8.1271

Luyendyk, B.P., Sorlien, C.C., Wilson, D.S., Bartek, L.R. and Siddoway, C.S. 2001. Structural and tectonic evolution of the Ross Sea rift in the Cape Colbeck region, Eastern Ross Sea, Antarctica. *Journal of Geophysical Research*, **20**, 933–958, https://doi.org/10.1029/2000TC001260

Martin, A.P., Cooper, A.F., Price, R.C., Kyle, P.R. and Gamble, J.A. 2021. Erebus Volcanic Province: petrology. *Geological Society, London, Memoirs*, **55**, https://doi.org/10.1144/M55-2018-80

McCoy-West, A.J., Bennett, V.C. and Amelin, Y. 2016. Rapid Cenozoic ingrowth of isotopic signatures simulating 'HIMU' in ancient lithospheric mantle: Distinguishing source from process. *Geochimica et Cosmochimica Acta*, **187**, 79–101, https://doi.org/10.1016/j.gca.2016.05.013

McDonough, W.F. and Sun, S.-s. 1995. The composition of the Earth. *Chemical Geology*, **120**, 223–253, https://doi.org/10.1016/0009-2541(94)00140-4

McIntosh, W. 2000. $^{40}Ar/^{39}Ar$ geochronology of tephra and volcanic clasts in CRP-2A, Victoria Land Basin, Antarctica. *Terra Antartica*, **7**, 621–630.

McIntosh, W.C. and Kyle, P.R. 1990. Hallett Volcanic Province. *American Geophysical Union Antarctic Research Series*, **48**, 26–47.

Meibom, A. and Anderson, D.L. 2003. The statistical upper mantle assemblage. *Earth and Planetary Science Letters*, **217**, 123–139, https://doi.org/10.1016/S0012-821X(03)00573-9

Meibom, A., Anderson, D.L., Sleep, N.H., Frei, R., Chamberlain, C.P., Hren, M.T. and Wooden, J.L. 2003. Are high $^{3}He/^{4}He$ ratios in oceanic basalts an indicator of deep-mantle plume components? *Earth and Planetary Science Letters*, **208**, 197–204, https://doi.org/10.1016/S0012-821X(03)00038-4

Melchiorre, M., Coltorti, M., Bonadiman, C., Faccini, B., O'Reilly, S.Y. and Pearson, N.J. 2011. The role of eclogite in the rift-related metasomatism and Cenozoic magmatism of Northern Victoria Land, Antarctica. *Lithos*, **124**, 319–330, https://doi.org/10.1016/j.lithos.2010.11.012

Menzies, M.A., Klemperer, S.L., Ebinger, C.J. and Baker, J. 2002. Characteristics of volcanic rifted margins. *Geological Society of America Special Papers*, **362**, 1–14.

Mortimer, N., Dunlap, W.J., Isaac, M.J., Sutherland, R.P. and Faure, K. 2007. Basal Adare volcanics, Robertson Bay, North Victoria Land, Antarctica: Late Miocene intraplate basalts of subaqueous origin. *United States Geological Survey Open-File Report*, **2007-1047**, Short Research Paper 045.

Mukasa, S.B. 2003. The myth of the Dufek Plume: Nd, Sr, Pb and Os isotopic and trace element data in support of a subduction origin. Presented at the 9th International Symposium on Antarctic Earth Sciences, 8–12 September 2003, Potsdam, Germany.

Mukasa, S.B. and Dalziel, I.W.D. 2000. Marie Byrd Land, West Antarctica: Evolution of Gondwana's Pacific margin constrained by zircon U–Pb geochronology and feldspar common-Pb isotopic compositions. *Geological Society of America Bulletin*, **112**, 611–627, https://doi.org/10.1130/0016-7606(2000)112<611:MBLWAE>2.0.CO;2

Müller, P., Schmidt-Thomé, M., Kreuzer, H., Tessensohn, F. and Vetter, U. 1991. Cenozoic peralkaline magmatism at the western margin of the Ross Sea, Antarctica. *Memorie della Società Geologica Italiana*, **46**, 315–336.

Nardini, I., Armienti, P., Rocchi, S., Dallai, L. and Harrison, D. 2009. Sr–Nd–Pb–He–O isotope and geochemical constraints on the genesis of Cenozoic magmas from the West Antarctic Rift. *Journal of Petrology*, **50**, 1359–1375, https://doi.org/10.1093/petrology/egn082

Noll, M.R. 1984. *Geochemistry and Petrogenesis of the Alkaline Lavas and their Associated Xenoliths, Mt. Overlord, Northern*

Victoria Land, Antarctica. Master thesis, New Mexico Institute of Mininig and Technology, Socorro, New Mexico, USA.

Panter, K.S., Kyle, P.R. and Smellie, J.L. 1997. Petrogenesis of a phonolite–trachyte succession at Mount Sidley, Marie Byrd Land, Antarctica. *Journal of Petrology*, **38**, 1225–1253, https://doi.org/10.1093/petroj/38.9.1225

Panter, K.S., Hart, S.R., Kyle, P.R., Blusztanjn, J. and Wilch, T.I. 2000. Geochemistry of Late Cenozoic basalts from the Crary Mountains: characterization of mantle sources in Marie Byrd Land, Antarctica. *Chemical Geology*, **165**, 215–241, https://doi.org/10.1016/S0009-2541(99)00171-0

Panter, K.S., Blusztajn, J., Hart, S.R., Kyle, P.R., Esser, R. and McIntosh, W.C. 2006. The Origin of HIMU in the SW Pacific: Evidence from Intraplate Volcanism in Southern New Zealand and Subantarctic Islands. *Journal of Petrology*, **47**, 1673–1704, https://doi.org/10.1093/petrology/egl024

Panter, K.S., Castillo, P. *et al.* 2018. Melt origin across a rifted continental margin: a case for subduction-related metasomatic agents in the lithospheric source of alkaline basalt, northwest Ross Sea, Antarctica. *Journal of Petrology*, **59**, 517–558, https://doi.org/10.1093/petrology/egy036

Paulsen, T.S. and Wilson, T.J. 2010. Evolution of Neogene volcanism and stress patterns in the glaciated West Antarctic Rift, Marie Byrd Land, Antarctica. *Journal of the Geological Society, London*, **167**, 401–416, https://doi.org/10.1144/0016-76492009-044

Perinelli, C., Armienti, P. and Dallai, L. 2006. Geochemical and O-isotope constraints on the evolution of lithospheric mantle in the Ross Sea rift area (Antarctica). *Contributions to Mineralogy and Petrology*, **151**, 245–266, https://doi.org/10.1007/s00410-006-0065-8

Perinelli, C., Armienti, P. and Dallai, L. 2011. Thermal evolution of the lithosphere in a rift environment as inferred from the geochemistry of mantle cumulates, northern Victoria Land, Antarctica. *Journal of Petrology*, **52**, 665–690, https://doi.org/10.1093/petrology/egq099

Phillips, E.H., Sims, K.W.W. *et al.* 2018. The nature and evolution of mantle upwelling at Ross Island, Antarctica, with implications for the source of HIMU lavas. *Earth and Planetary Science Letters*, **498**, 38–53, https://doi.org/10.1016/j.epsl.2018.05.049

Plank, T. and Langmuir, C.H. 1998. The chemical composition of subducting sediment and its consequences for the crust and mantle. *Chemical Geology*, **145**, 325–394, https://doi.org/10.1016/S0009-2541(97)00150-2

Richards, M.A. and Lithgow-Bertelloni, C. 1996. Plate motion changes, the Hawaii–Emperor bend, and the apparent success and failure of geodynamic models. *Earth and Planetary Science Letters*, **137**, 19–27, https://doi.org/10.1016/0012-821X(95)00209-U

Rilling, S., Mukasa, S., Wilson, T., Lawver, L. and Hall, C. 2009. New determinations of $^{40}Ar/^{39}Ar$ isotopic ages and flow volumes for Cenozoic volcanism in the Terror Rift, Ross Sea, Antarctica. *Journal of Geophysical Research*, **114**, B12207, https://doi.org/10.1029/2009JB006303

Ritsema, J. and Allen, R.M. 2003. The elusive mantle plume. *Earth and Planetary Science Letters*, **207**, 1–12, https://doi.org/10.1016/S0012-821X(02)01093-2

Rocchi, S., Tonarini, S., Armienti, P., Innocenti, F. and Manetti, P. 1998. Geochemical and isotopic structure of the early Palaeozoic active margin of Gondwana in northern Victoria Land, Antarctica. *Tectonophysics*, **284**, 261–281, https://doi.org/10.1016/S0040-1951(97)00178-9

Rocchi, S., Armienti, P., D'Orazio, M., Tonarini, S., Wijbrans, J. and Di Vincenzo, G. 2002*a*. Cenozoic magmatism in the western Ross Embayment: role of mantle plume v. plate dynamics in the development of the West Antarctic Rift System. *Journal of Geophysical Research*, **107**, 2195, https://doi.org/10.1029/2001JB000515

Rocchi, S., Fioretti, A.M. and Cavazzini, G. 2002*b*. Petrography, geochemistry and geochronology of the Cenozoic Cape Crossfire, Cape King and No Ridge igneous complexes (northern Victoria Land, Antarctica). *Royal Society of New Zealand Bulletin*, **35**, 215–225.

Rocchi, S., Storti, F., Di Vincenzo, G. and Rossetti, F. 2003. Intraplate strike-slip tectonics as an alternative to mantle plume activity for the Cenozoic rift magmatism in the Ross Sea region, Antarctica. *Geological Society, London, Special Publications*, **210**, 145–158, https://doi.org/10.1144/GSL.SP.2003.210.01.09

Rocchi, S., Di Vincenzo, G. and Armienti, P. 2005. No plume, no rift magmatism in the West Antarctic rift. *Geological Society of America Special Papers*, **388**, 435–447.

Rocchi, S., LeMasurier, W.E. and Di Vincenzo, G. 2006. Oligocene to Holocene erosion and glacial history in Marie Byrd Land, West Antarctica, inferred from exhumation of the Dorrel Rock intrusive complex and from volcano morphologies. *Geological Society of America Bulletin*, **118**, 991–1005, https://doi.org/10.1130/B25675.1

Rocchi, S., Di Vincenzo, G., Ghezzo, C. and Nardini, I. 2009. Granite–lamprophyre connection in the latest stages of the Early Paleozoic Ross Orogeny (Victoria Land, Antarctica). *Geological Society of America Bulletin*, **121**, 801–819, https://doi.org/10.1130/B26342.1

Rocchi, S., Bracciali, L., Di Vincenzo, G., Gemelli, M. and Ghezzo, C. 2011. Arc accretion to the early Paleozoic Antarctic margin of Gondwana in Victoria Land. *Gondwana Research*, **19**, 594–607, https://doi.org/10.1016/j.gr.2010.08.001

Rocholl, A., Stein, M., Molzahn, M., Hart, S.R. and Wörner, G. 1995. Geochemical evolution of rift magmas by progressive tapping of a stratified mantle source beneath the Ross Sea Rift, Northern Victoria Land, Antarctica. *Earth and Planetary Science Letters*, **131**, 207–224, https://doi.org/10.1016/0012-821X(95)00024-7

Rona, P.A. and Richardson, E.S. 1978. Early Cenozoic global plate reorganization. *Earth and Planetary Science Letters*, **40**, 1–11, https://doi.org/10.1016/0012-821X(78)90069-9

Rossetti, F., Lisker, F., Storti, F. and Läufer, A. 2003. Tectonic and denudational history of the Rennick Graben (northern Victoria Land): Implications for the evolution of rifting between East and West Antarctica. *Tectonics*, **22**, 1016, https://doi.org/10.1029/2002TC001416

Rossetti, F., Storti, F., Busetti, M., Di Vincenzo, G., Lisker, F., Rocchi, S. and Salvini, F. 2006. Eocene initiation of Ross Sea dextral faulting and implications for East Antarctic neotectonics. *Journal of the Geological Society, London*, **163**, 119–126, https://doi.org/10.1144/0016-764905-005

Salvini, F., Brancolini, G., Busetti, M., Storti, F., Mazzarini, F. and Coren, F. 1997. Cenozoic geodynamics of the Ross Sea region, Antarctica: Crustal extension, intraplate strike-slip faulting, and tectonic inheritance. *Journal of Geophysical Research*, **102**, 24 669–24 696, https://doi.org/10.1029/97JB01643

Scott, J.M., Waight, T.E., van der Meer, Q.H.A., Palin, J.M., Cooper, A.F. and Münker, C. 2014. Metasomatized ancient lithospheric mantle beneath the young Zealandia microcontinent and its role in HIMU-like intraplate magmatism. *Geochemistry, Geophysics, Geosystems*, **15**, 3477–3501, https://doi.org/10.1002/2014GC005300

Siddoway, C.S. 2008. Tectonics of the West Antarctic Rift System: New light on the history and dynamics of distributed intracontinental extension. *United States Geological Survey Open-File Report*, **2007-1047**, 91–114.

Siddoway, C.S., Baldwin, S., Fitzgerald, P., Fanning, C.M. and Luyendik, B.P. 2004. Ross Sea mylonites and the timing of intracontinental extension within the West Antarctic rift system. *Geology*, **32**, 57–60, https://doi.org/10.1130/G20005.1

Smellie, J.L. and Martin, A.P. 2021. Erebus Volcanic Province: volcanology. *Geological Society, London, Memoirs*, **55**, https://doi.org/10.1144/M55-2018-62

Smellie, J.L. and Rocchi, S. 2021. Northern Victoria Land: volcanology. *Geological Society, London, Memoirs*, **55**, https://doi.org/10.1144/M55-2018-60

Smellie, J., Rocchi, S. and Armienti, P. 2011*a*. Late Miocene volcanic sequences in northern Victoria Land, Antarctica: products of glaciovolcanic eruptions under different thermal regimes. *Bulletin* of *Volcanology*, **73**, 1–25, https://doi.org/10.1007/s00445-010-0399-y

Smellie, J.L., Rocchi, S., Gemelli, M., Di Vincenzo, G. and Armienti, P. 2011*b*. A thin predominantly cold-based Late Miocene East Antarctic ice sheet inferred from glaciovolcanic sequences in northern Victoria Land, Antarctica. *Palaeogeography, Palaeoclimatology, Palaeoecology*, **307**, 129–149, https://doi.org/10.1016/j.palaeo.2011.05.008

Stein, M., Navon, O. and Kessel, R. 1997. Chromatographic metasomatism of the Arabian–Nubian lithosphere. *Earth and Planetary Science Letters*, **152**, 75–91, https://doi.org/10.1016/S0012-821X(97)00156-8

Steinberger, B., Sutherland, R. and O'Connell, R.J. 2004. Prediction of Emperor–Hawaii seamount locations from a revised model of global plate motion and mantle flow. *Nature*, **430**, 167–173, https://doi.org/10.1038/nature02660

Stock, J.M. and Cande, S.C. 2002. Tectonic history of Antarctic seafloor in the Antarctic–New Zealand–South Pacific sector: implications for Antarctic continent tectonics. *Royal Society of New Zealand Bulletin*, **35**, 251–260.

Storey, B.C. and Alabaster, T. 1991. Tectonomagmatic controls on Gondwana break-up models: evidence from the proto-Pacific margin of Antarctica. *Tectonics*, **10**, 1274–1288, https://doi.org/10.1029/91TC01122

Storti, F., Rossetti, F. and Salvini, F. 2001. Structural architecture and displacement accomodation mechanisms at the termination of the Priestley Fault, northern Victoria Land, Antarctica. *Tectonophysics*, **341**, 141–161, https://doi.org/10.1016/S0040-1951(01)00198-6

Storti, F., Salvini, F., Rossetti, F. and Phipps Morgan, J. 2007. Intraplate termination of transform faulting within the Antarctic continent. *Earth and Planetary Science Letters*, **260**, 115–126, https://doi.org/10.1016/j.epsl.2007.05.020

Storti, F., Balestrieri, M.L., Balsamo, F. and Rossetti, F. 2008. Structural and thermochronological constraints to the evolution of the West Antarctic Rift System in central Victoria Land. *Tectonics*, **27**, TC4012, https://doi.org/10.1029/2006TC002066

Stracke, A., Hofmann, A.W. and Hart, S.R. 2005. FOZO, HIMU, and the rest of the mantle zoo. *Geochemistry Geophysics Geosystems*, **6**, Q05007, https://doi.org/10.1029/2004GC000824

Stuart, F.M., Lass-Evans, S., Godfrey Fitton, J. and Ellam, R.M. 2003. High ^{3}He/^{4}He ratios in picritic basalts from Baffin Island and the role of a mixed reservoir in mantle plumes. *Nature*, **424**, 57, https://doi.org/10.1038/nature01711

Sun, S.-s. and McDonough, W.F. 1989. Chemical and isotopic systematics of oceanic basalts: implications for mantle composition and processes. *Geological Society, London, Special Publications*, **42**, 313–345, https://doi.org/10.1144/GSL.SP.1989.042.01.19

ten Brink, U. and Stern, T. 1992. Rift flank uplifts and Hinterland Basins: Comparison of the Transantarctic Mountains with the Great Escarpment of southern Africa. *Journal of Geophysical Research: Solid Earth*, **97**, 569–585, https://doi.org/10.1029/91JB02231

Thornton, C.P. and Tuttle, O.F. 1960. Chemistry of igneous rocks: 1. Differentiation Index. *American Journal of Science*, **258**, 664–684, https://doi.org/10.2475/ajs.258.9.664

Tonarini, S., Rocchi, S., Armienti, P. and Innocenti, F. 1997. Constraints on timing of Ross Sea rifting inferred from Cainozoic intrusions from northern Victoria Land, Antarctica. *In*: Ricci, C.A. (ed.) *The Antarctic Region: Geological Evolution and Processes*. Terra Antartica, Siena, Italy, 511–521.

Trey, H., Cooper, A.K., Pellis, G., Della Vedova, B., Cochrane, G., Brancolini, G. and Makris, J. 1999. Transect across the West Antarctic rift system in the Ross Sea, Antarctica. *Tectonophysics*, **301**, 61–74, https://doi.org/10.1016/S0040-1951(98)00155-3

Veevers, J.J. 2000. Change of tectono-stratigraphic regime in the Australian plate during the 99 Ma (mid-Cretaceous) and 43 Ma (mid-Eocene) swerves of the Pacific. *Geology*, **28**, 47–50, https://doi.org/10.1130/0091-7613(2000)28<47:COTRIT>2.0.CO;2

Whitney, D.L. and Evans, B.W. 2010. Abbreviations for names of rock-forming minerals. *American Mineralogist*, **95**, 185–187, https://doi.org/10.2138/am.2010.3371

Winberry, J.P. and Anandakrishnan, S. 2004. Crustal structure of the West Antarctic rift system and Marie Byrd Land hotspot. *Geology*, **32**, 977–980, https://doi.org/10.1130/G20768.1

Wörner, G. and Viereck, L. 1989. The Mt. Melbourne Volcanic Field (Victoria Land, Antarctica). I. Field observations. *Geologisches Jahrbuch*, **E38**, 369–393.

Wörner, G., Vierek, L., Hertoghen, J. and Niephaus, H. 1989. The Mt. Melbourne Volcanic field (Victoria Land, Antarctica). II. Geochemistry and magma genesis. *Geologisches Jahrbuch*, **E38**, 395–433.

Zindler, A. and Hart, S. 1986. Chemical geodynamics. *Annual Review of Earth and Planetary Sciences*, **14**, 493–571, https://doi.org/10.1146/annurev.ea.14.050186.002425

Chapter 5.2a

Erebus Volcanic Province: volcanology

John L. Smellie[1]* and Adam P. Martin[2]

[1]School of Geography, Geology and the Environment, University of Leicester, University Road, Leicester LE1 7RH, UK
[2]GNS Science, Private Bag 1930, Dunedin, New Zealand

JLS, 0000-0001-5537-1763; APM, 0000-0002-4676-8344
*Correspondence: jls55@le.ac.uk

Abstract: The Erebus Volcanic Province is the largest Neogene volcanic province in Antarctica, extending c. 450 km north–south and 170 km wide east–west. It is dominated by large central volcanoes, principally Mount Erebus, Mount Bird, Mount Terror, Mount Discovery and Mount Morning, which have sunk more than 2 km into underlying sedimentary strata. Small submarine volcanoes are also common, as islands and seamounts in the Ross Sea (Terror Rift), and there are many mafic scoria cones (Southern Local Suite) in the Royal Society Range foothills and Dry Valleys. The age of the volcanism ranges between c. 19 Ma and present but most of the volcanism is <5 Ma. It includes active volcanism at Mount Erebus, with its permanent phonolite lava lake. The volcanism is basanite–phonolite/trachyte in composition and there are several alkaline petrological lineages. Many of the volcanoes are pristine, predominantly formed of subaerially erupted products. Conversely, two volcanoes have been deeply eroded. That at Minna Hook is mainly glaciovolcanic, with a record of the ambient mid–late Miocene eruptive environmental conditions. By contrast, Mason Spur is largely composed of pyroclastic density current deposits, which accumulated in a large mid-Miocene caldera that is now partly exhumed.

Supplementary material: Summary of published isotopic ages for volcanic rocks in the Erebus Volcanic Province are available at https://doi.org/10.6084/m9.figshare.c.5332477

Subaerial volcanoes of the Erebus Volcanic Province are conspicuous features of the hinterland flanking McMurdo Sound (Fig. 1). Subaqueous volcanism is also prominent, but less well known, within the Terror Rift, a major submarine graben

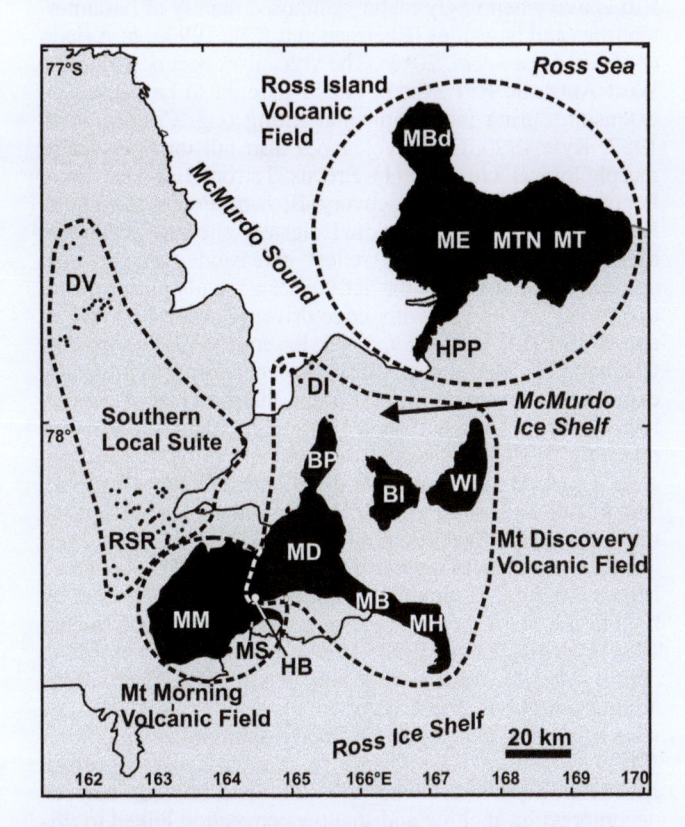

Fig. 1. Distribution of subaerial volcanic fields in the Erebus Volcanic Province (dashed lines). Subaerial volcanic outcrops are coloured black. Abbreviations: BI, Black Island; BP, Brown Peninsula; DI, Dailey Islands; DV, Dry Valleys; HB, Helms Bluff; HPP, Hut Point Peninsula; MB, Minna Bluff; MBd, Mount Bird; MD, Mount Discovery; ME, Mount Erebus; MH, Minna Hook; MM, Mount Morning; MS, Mason Spur; MT, Mount Terror; MTN, Mount Terra Nova; RSR, Royal Society Range; WI, White Island.

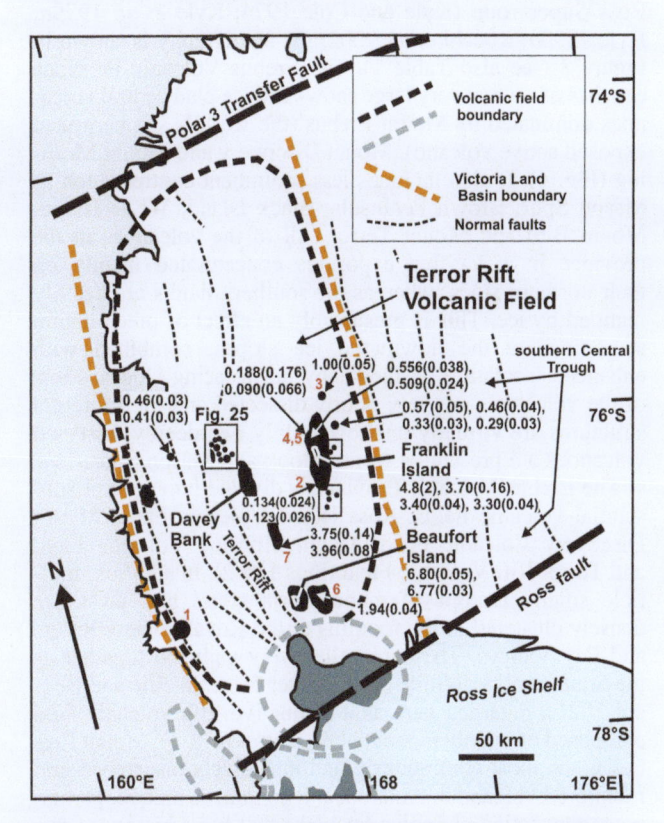

Fig. 2. Distribution of volcanoes (coloured in black) in the Terror Rift Volcanic Field. Subaerial volcanoes south of Terror Rift Volcanic Field are coloured in grey (see Fig. 1). Red numbers indicate localities described in the text. Faults (thick dashed black lines) are from Salvini et al. (1997). The small boxes contain the fields of unusual submarine edifices with pancake-like morphologies described by Lawver et al. (2012). The isotopic ages of dated samples (in Ma) are also shown; errors are shown in parentheses (Supplementary material).

From: Smellie, J. L., Panter, K. S. and Geyer, A. (eds) 2020. *Volcanism in Antarctica: 200 Million Years of Subduction, Rifting and Continental Break-up*. Geological Society, London, Memoirs, **55**, 415–446,
First published online March 24, 2021, https://doi.org/10.1144/M55-2018-62
© 2021 The Author(s). Published by the Geological Society of London. All rights reserved.
For permissions: http://www.geolsoc.org.uk/permissions. Publishing disclaimer: www.geolsoc.org.uk/pub_ethics

WESTERN ROSS SUPERGROUP

McMurdo Volcanic Group
[Alkaline volcanic rocks]

Meander Intrusive Group
[Alkaline plutons & related dykes]

Hallett Volcanic Province
[Large mafic-felsic central volcanoes]
- Adare Peninsula Volcanic Field
- Hallett Peninsula Volcanic Field
- Daniell Peninsula Volcanic Field
- Coulman Island Volcanic Field

Erebus Volcanic Province
[Mainly large mafic-felsic central volcanoes]
- Ross Island Volcanic Field
- Mt Discovery Volcanic Field
- Mt Morning Volcanic Field
- Terror Rift Volcanic Field
- Southern Local Suite Volcanic Field [Products of small isolated mafic eruptions]
- Northern Local Suite Volcanic Field [Products of small isolated mafic eruptions]
- Upper Scott Glacier Volcanic Field [Mainly products of small isolated mafic eruptions]

Melbourne Volcanic Province
[Large mafic-felsic central volcanoes]
- Mt Melbourne Volcanic Field
- Vulcan Hills Volcanic Field
- Mt Overlord Volcanic Field
- The Pleiades Volcanic Field
- Malta Plateau Volcanic Field

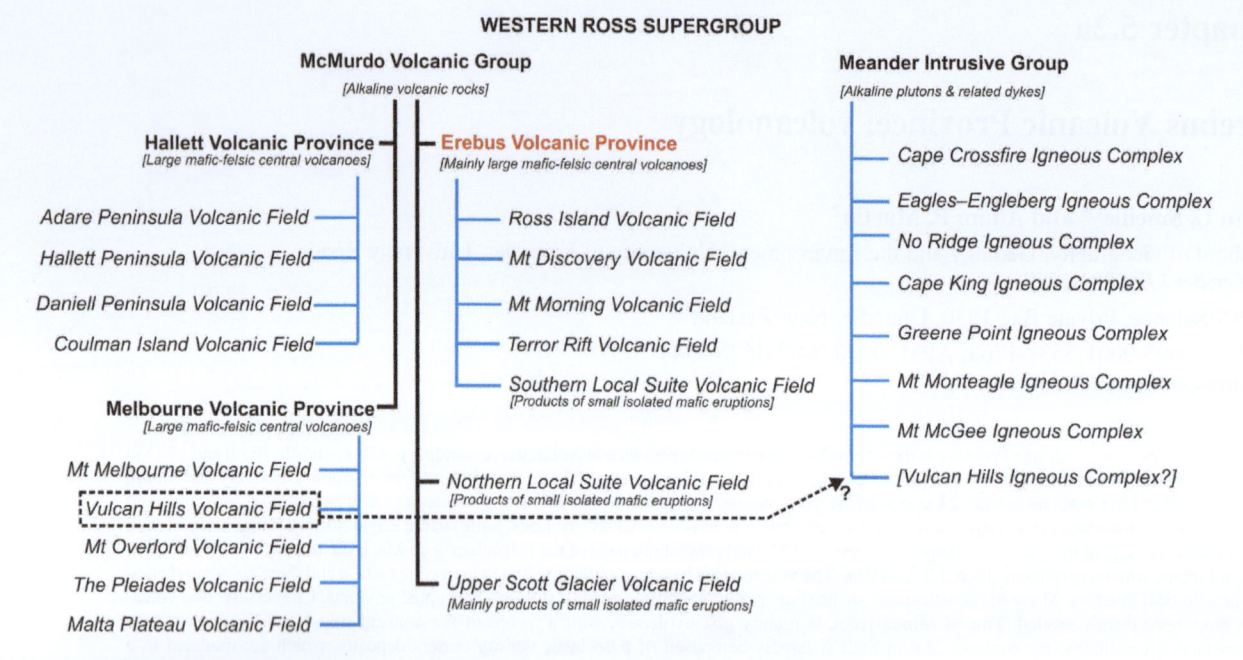

- Cape Crossfire Igneous Complex
- Eagles–Engleberg Igneous Complex
- No Ridge Igneous Complex
- Cape King Igneous Complex
- Greene Point Igneous Complex
- Mt Monteagle Igneous Complex
- Mt McGee Igneous Complex
- [Vulcan Hills Igneous Complex?]

Fig. 3. Summary stratigraphy of the Western Ross Supergroup (after Smellie and Rocchi 2021 – Chapter 5.1a).

that extends north of Mount Erebus almost as far as Mount Melbourne (Fig. 2). All of the volcanism is included in the McMurdo Volcanic Group, which is subsumed in the Western Ross Supergroup (Kyle and Cole 1974; Kyle et al. 1979a; Kyle 1990a; Rocchi et al. 2002). Its stratigraphy is shown in Figure 3 (see also Table 1). The Erebus Volcanic Province consists of several very large snow- and ice-clad central volcanoes dominated by Mount Erebus (the world's southernmost exposed active volcano), Mount Discovery and Mount Morning (Fig. 4). It also includes less prominent centres, such as Mason Spur, Brown Peninsula, Black Island, White Island, Mount Bird and Mount Terror. All of the volcanoes in the province have ice-free exposures concentrated mainly on their northern sides, whereas the southern flanks are heavily mantled by ice. This is presumably an effect of precipitation starvation on the downwind lee slopes, combined with enhanced insolation on the sunny north-facing slopes. Most of the volcanoes are very poorly dissected and their internal structures are virtually unknown. Only two deeply dissected volcanoes are present: at Minna Hook and Mason Spur.

The Erebus Volcanic Province is divided into several substantial volcanic fields: Ross Island Volcanic Field, Mount Discovery Volcanic Field, Mount Morning Volcanic Field and Terror Rift Volcanic Field (Figs 1 & 2). In addition, multiple small outcrops of mafic scoria cones and lavas are densely clustered in the foothills of the Royal Society Range and Dry Valleys. They are collectively gathered together as the Southern Local Suite (see Chapter 5.1a: Smellie and Rocchi 2021), regarded here as a distinctive fifth volcanic field composed of mainly monogenetic centres (Fig. 1). Mount Erebus is the most continuously and intensively researched and monitored volcano in Antarctica. It contains a unique, permanent, convecting phonolite lava lake that has a history of frequent small-scale Strombolian-style eruptions (e.g. Kyle 1994; Rowe et al. 1998; see Chapter 7.2: Sims et al. 2021). Volcanism in the Erebus Volcanic Province is alkaline, with a wide range of basanite/tephrite–phonolite and basanite–trachyte compositions, with mafic compositions predominant and few intermediate compositions. Major evolutionary lineages have been identified: for example, an older Mason Spur Lineage and a younger lineage, which, in the Ross Island Volcanic Field, has been divided into the Dry Valleys Drilling Project (DVDP) Lineage and the Erebus Lineage (Kyle 1981b; Kyle et al. 1992; Martin et al. 2010; see also Chapter 5.2b: Martin et al. 2021). By contrast, volcanism in the Terror Rift is overwhelmingly mafic, composed mainly of basanites, tephrites and hawaiites (Ellerman and Kyle 1990a, b; Aviado et al. 2015; Lee et al. 2015). The volcanism occurs within the West Antarctic Rift System and is thought to be caused by extension during intracontinental rifting (e.g. Cooper et al. 1987; Kyle 1990a). The volcanoes also fall into two major morphological clusters: (1) Erebus–Terror–Bird–Hut Point Peninsula; and (2) Discovery–Brown Peninsula–Minna Bluff–Morning. The individual clusters show a prominent radial symmetry that may reflect extension during crustal doming associated with the activity of a mantle plume, or by updoming associated with edge-driven mantle flow (Kyle and Cole 1974; Kyle et al. 1992; Esser et al. 2004; see also Chapter 5.2b: Martin et al. 2021). Recent seismic tomography experiments have also demonstrated the presence of anomalous mantle with slow P-wave velocities (therefore hot and probably fluid-rich) beneath the volcanic province (Hansen et al. 2014; cf. Phillips et al. 2018 and references therein). The mantle anomaly identified by Hansen et al. (2014) has a much greater diameter than inferred by the petrological studies, and it extends to the north and beneath the Transantarctic Mountains front, a region also known to be characterized by high heat flow (Risk and Hochstein 1974; Decker and Bucher 1982; Blackman et al. 1987). However, it may only extend to depths of 200–300 km and thus would lack deep roots, although this has been contested by Phillips et al. (2018), who suggested that the low-velocity anomaly extends to c. 1200 km beneath Ross Island. In summary, the volcanism has been explained to be a result of either (a) focused decompression-melting and shallow convection linked to rifting and extension of the Terror Rift (Cooper et al. 2007; Hansen et al. 2014; Martin et al. 2015), similar to models for analogous magmatism in northern Victoria Land (e.g. Rocchi et al. 2002, 2003, 2005; Panter et al. 2018); or (2) uprise of a deep-seated mantle plume (Kyle et al. 1992; Phillips et al. 2018; see also Chapters 5.1b (Rocchi and Smellie 2021) and 5.2b (Martin et al. 2021)).

Volcanism in the Erebus Volcanic Province may be particularly voluminous because it occurs at the intersection

Table 1. *Stratigraphy and principal characteristics of the Erebus Volcanic Province (McMurdo Volcanic Group, Western Ross Supergroup)*

Volcanic field	Major constituent volcanoes	Volcano type	Caldera(s)?	Composition	Published ages* (Ma)	Comments	Key references
Ross Island	Mount Erebus	Large compound volcano – shield volcano base, stratovolcano summit	Three calderas identified; oldest at Fang Ridge; two calderas at Erebus summit, the youngest within and much smaller than the main caldera	Mainly hawaiite and phonolite; intermediate compositions (benmoreite and trachyte) also present	Numerous published K–Ar and Ar–Ar ages range between 1.33 Ma and 4 ka; still active (phonolite lava lake)	mafic compositions largely confined to Fang Ridge and northern Mount Erebus	Moore and Kyle (1987, 1990); Kyle (1994); Esser et al. (2004); Harpel et al. (2004); Csatho et al. (2008)
	Mount Bird	Large volcanic shield	No	Mainly basanite and phonolite; rare benmoreite	4.50–3.30		Kyle and Cole (1974); Wright and Kyle (1990d)
	Mount Terror	Large volcanic shield	No	Basanite, phonotephrite, phonolite	1.33–0.80		Cole and Ewart (1968); Wright and Kyle (1990c)
	Mount Terra Nova	Volcanic shield?	No	Basalt, phonotephrite and phonolite	0.80 ± 0.5	Largely obscured by snow and ice; very poorly-exposed – few exposures only at the summit	Wright and Kyle (1990c)
	Hut Point Peninsula	Series of scoria cones, a phonolite dome and tuff cone	No	Basanite–phonolite; intermediate compositions well represented	1.34–0.43	Presumably fissure erupted to give the conspicuous elongation of the peninsula	Cole et al. (1971); Armstrong (1978); Kyle (1981a, b)
Mount Discovery	Mount Discovery	Large compound volcano – shield volcano base, stratovolcano summit	No	Tephriphonolite, phonolite, trachyte; basanite, hawaiite	5.30–1.87; 0.18–0.06	Includes Helms Bluff	Wright-Grassham (1987); Wright and Kyle (1990f)
	Minna Bluff	Unknown	No	Phonolite domes, lavas and basanite lavas	None	Very poor exposure except in McIntosh Cliffs; most recent study still unpublished; probably similar in geology to, and extension of, Minna Hook	Wright-Grassham (1987); Wright and Kyle (1990g); Wilch et al. (2008)
	Minna Hook	Unknown	No	Phonotephrite, tephriphonolite, phonolite, trachyte; rare mugearite, benmoreite	11.8–7.26	The best-exposed volcano in the Erebus Volcanic Province; results of most recent study still largely unpublished	Wright-Grassham (1987); Wright and Kyle (1990g); Fargo (2008); Antibus et al. (2014)
	White Island	Numerous scoria cones and associated lavas; few tuff cones; rare pillow breccia	No	Mostly basanites, few tephrites, rare nephelinite and tephriphonolite	7.65–2.11; 0.2		Cole et al. (1971); Eggers (1979); Wright and Kyle (1990e); Cooper et al. (2007)
	Black Island		No	Basalt, hawaiite, phonolite, trachyte	10.95–3.35		Cole and Ewart (1968); Wright and Kyle (1990e)

(*Continued*)

Table 1. Continued.

Volcanic field	Major constituent volcanoes	Volcano type	Caldera(s)?	Composition	Published ages* (Ma)	Comments	Key references
	Brown Peninsula	Mafic scoria cones and lavas; felsic lavas and domes; lapilli tuff	No	Basanite, hawaiite, mugearite, benmoreite, phonolite	2.70–2.10		Cole and Ewart (1968); Eggers (1979); Kyle et al. (1979a, b); Wright and Kyle (1990e)
	Dailey Islands	Several small scoria cones	n.a.	Basanite	0.77 ± 0.032	Might be part of the Mount Discovery Volcanic Field (i.e. possibly analogous to Hut Point Peninsula volcanics in the Ross Island Volcanic Field)	Wright and Kyle (1990j); Tauxe et al. (2004)
Mount Morning	Mount Morning	Large shield volcano	Yes	Mugearite, trachyte/rhyolite; basanite, tephrite, phonolite	18.7–13.0; 5.01–0.12	Poorly-known, older, early Miocene volcano largely obscured by large late Miocene–Recent shield volcano	Wright-Grassham (1987); Kyle and Muncy (1989); Wright and Kyle (1990h); Martin (2009); Paulsen and Wilson (2009); Martin and Cooper (2010); Martin et al. (2010)
	Mason Spur	Unknown (stratovolcano?)	Large caldera, at least 10 km major diameter	Hawaiite, mugearite, benmoreite, trachyte; basanite, tephrite	12.8–11.40; 6.13; 0.23–0.07	Extensively and deeply eroded, no primary landforms except very young scoria cones; largely formed of caldera-filling pyroclastic density current deposits; results of most recent study still largely unpublished	Wright-Grassham (1987); Wright and Kyle (1990i); Martin (2009); Martin et al. (2010, 2018)
Southern Local Suite	Taylor and Wright valleys (Dry Valleys)	Multiple monogenetic scoria cones and associated short lavas	n.a.	Basanite	4.64–1.50		Wright and Kyle (1990k); Wilch et al. (1993)
	Royal Society Range foothills	Multiple monogenetic scoria cones and associated short lavas	n.a.	Basanite; rare hawaiite	13.80–13.20; 5.70; 2.88–0.27	Some outcrops may be glaciovolcanic: with pillow lava and hyaloclastite, and perched lavas also present	McIver and Gevers (1970); Wright (1980); Wright and Kyle (1990j)
Terror Rift	Beaufort Island	Stratovolcano remnant	No	Tephrite, phonotephrite, tephriphonolite, phonolite	6.80, 6.77		Harrington (1958); Ellerman and Kyle (1990b)
	Franklin Island	Shield volcano remnant	No	Basanite	4.8–3.30		Ellerman and Kyle (1990a)
	Submarine volcanic centres	Mainly monogenetic cones and mounds	One edifice with a possible caldera identified geophysically (unconfirmed)	Basanite	3.96–0.09	Includes at least three clusters of morphologically unusual and enigmatic 'pancake-like' mounds	Behrendt (1990); Rilling et al. (2009); Lawver et al. (2012); Lee et al. (2015)

n.a., not applicable.
*Ages exclude errors; see Supplementary material.

Fig. 4. Compilation of field photographs showing views of selected volcanoes in the Erebus Volcanic Province. (**a**) Mount Erebus viewed looking NE; note the prominent 'shoulders', and conspicuously gentle slopes in the summit region caused by the main Mount Erebus caldera; the present-day small summit cone is also seen; Castle Rock is shown in the lower middle ground. (**b**) View of Mount Bird from the summit of Mount Erebus showing its shield-like profile; Beaufort Island is faintly seen in the background. (**c**) Mount Discovery, viewed looking east, showing the ice-capped, steep stratocone summit. (**d**) Mount Discovery, looking NE from the summit of Mason Spur; Helms Bluff is the dark crag on the left in the middle distance; note the prominent bulge on the right-hand side of Mount Discovery, possibly caused by construction of a large volcano on the SW flank of Mount Discovery and which may have given rise to the Helms Bluff sequence. (**e**) View of Minna Bluff, looking WSW, showing the poor rock exposure and prominent flanking ice shelf moraines; Mount Morning is the sunlit shield volcano in the distance at right. (**f**) View looking NNW along Minna Hook; the cliffs are constructed of multiple glaciovolcanic lava-fed deltas and felsic domes (see Fig. 18). (**g**) Eroded scoria cone and lavas (red and dark grey outcrops) on the SE side of Pyramid Trough (Royal Society Range foothills), part of the Southern Local Suite; Mount Morning towers high in the background. (**h**) Dailey Islands, a series of basanite scoria cones, viewed looking ENE. Photographs by J.L. Smellie, except (c), (g) and (h) (by Dougal Townsend).

between the Terror Rift (and its predecessor, the Victoria Land Basin) and a set of transverse fractures called the Discovery accommodation zone. The volcanoes occur at a prominent east-facing cusp between two major arcuate faults that delineate the margin of the Transantarctic Mountains. Wilson (1999) interpreted the zone as an offset in the broadly north–south-trending rift flank. It is characterized by east–west transverse fractures that were said to be affected by sinistral shear. Wilson (1999) suggested that the voluminous volcanism in the Erebus Volcanic Province was focused within the Discovery accommodation zone, with magma rise facilitated by the transverse fractures. A subsequent study of stress directions based on aligned Plio-Pleistocene scoria cones on Mount Morning by Paulsen and Wilson (2009) further clarified details of the Discovery accommodation zone and its geographical relationship with the Erebus Volcanic Province (Fig. 5). Furthermore, volcanism in the Mount Morning Volcanic Field may also exploit pre-existing, lower-crustal sutures inferred from crustal xenolith studies (Kalamarides *et al.* 1987; Martin *et al.* 2015). A related but somewhat different explanation for a peculiarity of the very elongated outcrop that forms Minna Bluff was suggested by Stump (2002). Minna Bluff is an unusual topographical feature, not dissimilar to Hut Point Peninsula but larger, and its origin is uncertain. It is a volcanic ridge *c.* 1000 m high, 45 km long and 4–6 km wide, which terminates in Minna Hook (Fig. 1). Stump

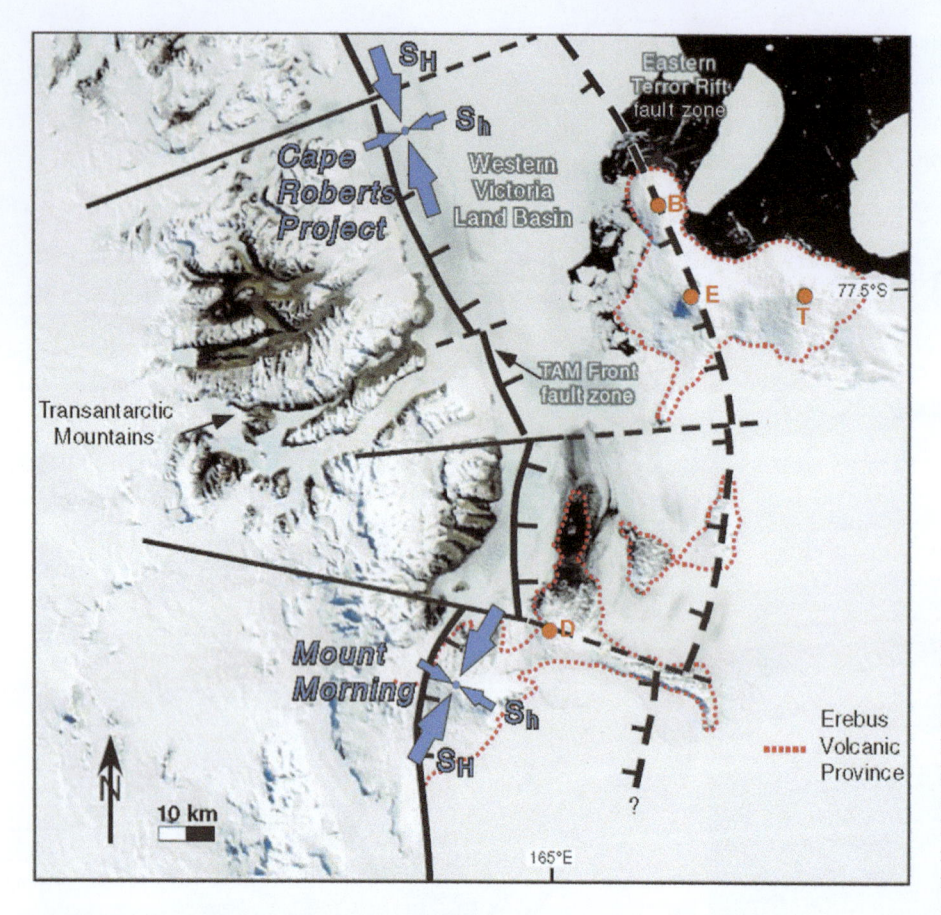

Fig. 5. Pliocene–present stress directions and proposed structural setting of volcanism in the Erebus Volcanic Province (from Paulsen and Wilson 2009). B, Mount Byrd; E, Mount Erebus; T, Mount Terror; D, Mount Discovery; TAM, Transantarctic Mountains. S_H and S_h are maximum and minimum horizontal compressional stress directions, respectively. See Paulsen and Wilson (2009) for further explanation.

(2002) noted that Mount Discovery and Minna Bluff are situated at the intersection of the two major arcuate faults that bound the margin of the Transantarctic Mountains and this may have influenced the unusual shape of the Minna Bluff promontory, especially its 'hooked' termination (Minna Hook). The major north–south fractures are thought to be dextral transtensional structures associated with the formation of the Transantarctic Mountains (Wilson 1995; 1999). Resolution of the forces created by the resulting oblique extension across the arcuate Transantarctic Mountains front may have caused an imbalance in the fault throws with, critically, the throw on the fault to the south being greater than that to the north. Stump (2002) hypothesized that this caused a tear that propagated dextrally (clockwise) into the hanging wall (i.e. West Antarctica), resulting in the creation of a stepped, curved tear fault at Minna Hook. The curved fault then became a major pathway for magmas, resulting in the conspicuous hook-like shape of the Minna Bluff–Minna Hook landform. However, Minna Bluff and Minna Hook have been greatly eroded both by glacial and marine processes, and probably affected by faulting too (Wright-Grassham 1987). The volcanic outcrop was much larger in the Miocene, and its original shape is essentially unknown.

Each of the subaerial volcanic fields identified in the Erebus Volcanic Province is composed of the overlapping products of several large central volcanoes, mainly Pliocene–Holocene in age. The Southern Local Suite is also regarded as a volcanic field. It consists of two geographically discrete, principal outcrop areas of multiple small mafic monogenetic centres (scoria cones and associated lavas: Wright-Grassham 1987; Wright and Kyle 1990a, b) (Fig. 1). Because of the proximity and relatively easy access of the large subaerial volcanoes in the Erebus Volcanic Province to two scientific stations, all of the volcanoes have been examined. As a result, there are far more isotopic ages published for the Erebus Volcanic Province than elsewhere in the region, mainly because of the extraordinary number of published ages for a single centre (Mount Erebus: Supplementary material). A few of the volcanic centres have been studied recently and intensively (Mount Erebus, Minna Hook, Mount Morning, Mason Spur and White Island), although the results of some of those studies are ongoing and have yet to be published in full. By contrast, several other volcanic centres (e.g. Black Island, Brown Peninsula, Mount Discovery, Dailey Islands, Royal Society Range foothills and Dry Valleys) have not been re-examined volcanologically for over a quarter of a century and the published studies should be regarded as reconnaissance. Outcrops of the Southern Local Suite are generally well exposed, and they occur in accessible low-lying snow- and ice-poor terrain. They are individually the products of single short-lived eruptions. By contrast, exposure of the much longer-lived large central volcanoes is typically poor due to extensive snow and ice cover, and a general lack of dissection. Only the final erupted products at any locality (mainly subaerial lavas) are exposed. The most notable exceptions are sequences at Minna Hook and Mason Spur, which are spectacularly displayed in cliff sections 900–1300 m high and extending c. 10 km laterally. Apart from these two examples, knowledge of the internal stratigraphy and construction of the volcanoes is generally poor. Minna Hook and Mason Spur are currently still being investigated, although a summary of some of the results so far is included in this chapter.

Stratigraphy and volcanology of the Erebus Volcanic Province

Ross Island Volcanic Field

The Ross Island Volcanic Field comprises all of Ross Island. It is formed of the products of four large volcanoes (Mount

Erebus, Mount Bird, Mount Terror and Mount Terra Nova) and a low-lying volcanic promontory (Hut Point Peninsula), with a combined erupted volume of *c.* 4520 km^3 (Esser *et al.* 2004). Collectively, these volcanoes show a conspicuous three-spoked radial symmetry dominated at the hub by Mount Erebus, which is the highest volcano in the Erebus Volcanic Province. Mount Erebus is the second largest Neogene volcano in Antarctica (Mount Haddington, James Ross Island (Antarctic Peninsula), is volumetrically larger: cf. Esser *et al.* 2004; Smellie *et al.* 2013; see also Chapter 3.2a: Smellie 2021).

Mount Erebus volcanology. Mount Erebus is the most continuously investigated volcano in Antarctica and, accordingly, it is described here in greater detail than other centres in the volcanic province. It is an active volcano, rises to 3794 m above sea level (asl), and has a basal diameter (above the elevation of the Ross Ice Shelf) of at least 36 km (Fig. 4a). It sits in water 500–700 m deep and geophysical studies have shown that a combination of loading by Ross Island aided by strain-related lithospheric weakening effects of the Terror Rift or the Discovery accommodation zone has caused lithospheric flexure of the underlying Pliocene sediments (Horgan *et al.* 2005). This has resulted in the Ross Island edifice sinking, with its base now at *c.* 2200 m below sea level, and similar effects are seen below the combined Mount Discovery–Mount Morning volcanic massif (Aitken *et al.* 2012). The geology of Mount Erebus has been described in several publications, including Moore and Kyle (1987, 1990), Esser *et al.* (2004), Harpel *et al.* (2004), Csatho *et al.* (2008) and Panter and Winter (2008); maps of many of the better-exposed coastal outcrops are presented by Moore and Kyle (1987). Morphologically, the exposed volcano below *c.* 1600 m asl forms a low shield with slopes of *c.* 9°. It contrasts with the higher part of the edifice, which is a stratocone with slopes of >30° (Esser *et al.* 2004). Rock is exposed primarily in the summit region, mainly above 1600–2000 m on the west, north and NE sides, in several mainly small headlands on the west coast and in low coastal cliffs facing Lewis Bay (Fig. 6). The volcano was constructed in two major stages. The older stage consists of mainly mafic products (predominantly hawaiite but including mugearite and benmoreite) that are exposed principally on the NE upper slopes and are

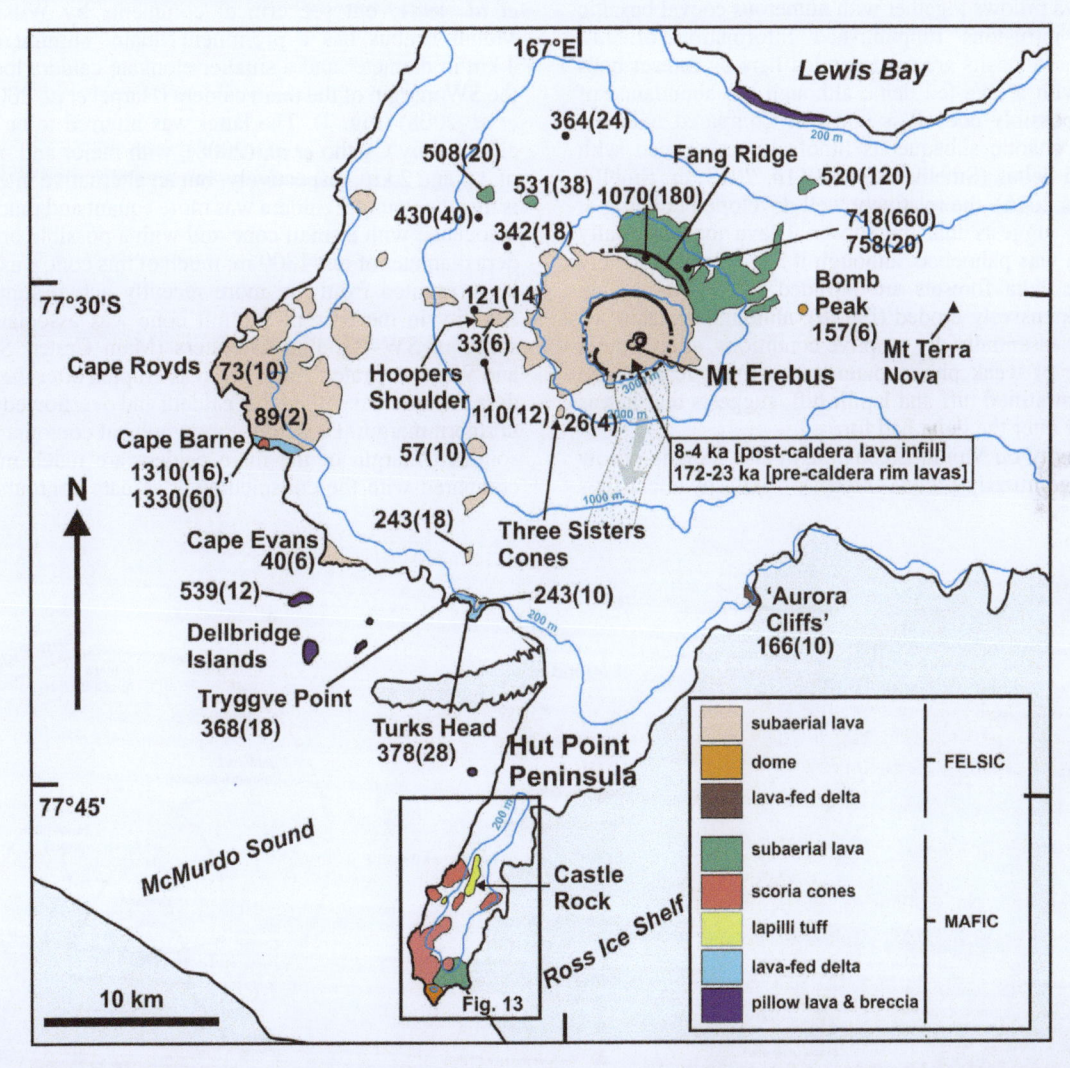

Fig. 6. Geological map of the southwestern part of Ross Island, including Mount Erebus and Hut Point Peninsula (mainly after Harpel *et al.* 2004; additional information is unpublished information of J.L. Smellie). The outcrops shown on this map are normally included in descriptions of 'the Erebus volcano', and we follow that tradition here for convenience of description. However, it is likely that Hut Point Peninsula and its outcrops (see Fig. 13) should be regarded as a separate edifice, admittedly in the early stages of construction as a major satellite centre that may ultimately achieve a similar size and status comparable to Mount Terror and Mount Bird. Isotopic ages are in ka; errors are shown in parentheses (see Supplementary material). The pale-grey stippled area with an arrow on the south side of the Mount Erebus summit represents the sector collapse scar and transport direction, respectively, postulated in the text; see also Figure 7b.

best exposed in Fang Ridge. Fang Ridge is a segment of a caldera rim *c.* 4 km in length and 100–200 m high that is related to an older stratocone with a present-day summit at 3159 m. The 'Fang caldera' or proto-Erebus caldera may have had an original diameter of *c.* 8 km (Csatho *et al.* 2008) (Fig. 7a). These early-formed deposits are dominated by subaerially erupted lavas. The descriptions suggest that they are 'a'ā lavas (cf. Moore and Kyle 1987); 'pyroclastic breccia' interpreted as a lahar deposit is also present at Fang Ridge. Published isotopic ages for the older stage vary from 1070 ± 180 to 342 ± 18 ka (Armstrong 1978; Esser *et al.* 2004) (Supplementary material); the oldest dated rock that unequivocally belongs to the older eruptive stage is a lava at Fang Ridge (1070 ka). However, a basanite outcrop forming a *c.* 100-m-high cliff at Cape Barne is dated as 1311 ± 16 ka; its strongly mafic composition, old age and extensive erosion (it is unconformably overlain by the Cape Royds tephriphonolite lava, or a related lava, dated as 89 ± 4 ka) suggest that it too may be part of the older volcano. The Cape Barne outcrop was previously interpreted as the product of at least three pyroclastic cones (Armstrong 1978; Moore and Kyle 1987), probably on account of the variable, seemingly radiating, bedding orientations (Fig. 8). The outcrop comprises steeply-dipping, homoclinal, monomict breccia and tuff breccia with locally abundant lava pillows together with numerous coeval basaltic hypabyssal intrusions (unpublished information of J.L. Smellie). The deposits are reinterpreted here as foreset beds associated with a lava-fed delta, although the abundance of intrusions (possibly coeval) is unusual. Compared with the much more chaotic subaqueous lithofacies associated with 'a'ā-lava-fed deltas (Smellie *et al.* 2011*a*, 2013; cf. Smellie and Edwards 2016), the relatively well-developed bedding at Cape Barne suggests that the subaerial lava that originally fed the delta was pāhoehoe, although it has been completely eroded. The delta foresets are intruded by a Strombolian vent, also extensively eroded (Fig. 8); although undated, its evidence for essentially dry eruptive conditions, apart from a minor phase of weak phreatomagmatic activity represented by <2 m of stratified tuff and lapilli tuff, suggests that it was erupted long after the delta had formed.

Older activity on Mount Erebus might also include pillow lavas, palagonitized pillow breccias and hyaloclastites exposed in the Dellbridge Islands (Fig. 6). These have been dated at 539 ± 12 ka, consistent with their highly eroded state, and have been interpreted as the remains of volcanic cones (Moore and Kyle 1987), presumably precursor or satellite vents relative to the older 'Fang volcano' stage. Speculatively, considering their coastal exposure and low elevation, and their location on the 'older' north side of the Erebus volcano (Fig. 6), undated palagonitized breccias in the cliffs facing Lewis Bay might also belong to the early eruptive stage. However, they have also been attributed to the Mount Bird volcano by Martin *et al.* (2021 – Chapter 5.2b in this Memoir). The breccias are unconformably overlain by undated benmoreite lavas (A. Wright cited in Moore and Kyle 1987).

By contrast, the younger eruptive stage, which formed the Mount Erebus stratocone, is overwhelmingly dominated by phonolite and tephriphonolite lavas for which isotopic ages of 243 ± 10–4 ± 3 ka have been determined (Esser *et al.* 2004; Harpel *et al.* 2004) (Supplementary material). The lavas are distinguished by abundant, very large, euhedral phenocrysts of anorthoclase, typically several centimetres in diameter (Fig. 9b). The phonolites, initially called 'kenytes', are so distinctive in appearance that their distribution as erratics and in moraines across the volcanic province has been used to infer movements of past ice masses (Kyle 1981*c*; Stuiver *et al.* 1981; but see critical comments by Wilson 2000). Mount Erebus has a prominent 'main' summit caldera *c.* 4 km in diameter and a smaller elongate caldera located near the SW margin of the main caldera (Harpel *et al.* 2004; Csatho *et al.* 2008) (Fig. 7). The latter was inferred to be strikingly elliptical by Csatho *et al.* (2008), with major and minor axes of 3.8 and 2 km, respectively, but an alternative interpretation is that the younger caldera was more equant and much smaller, associated with a small cone and with a possible original caldera diameter of just 1300 m; much of this cone was buried by lavas erupted from the more recently active summit cone. Activity in the present summit cone was associated with a series of SW–NE-aligned craters (Main Crater, Side crater and Western Crater: Fig. 7). Lavas erupted after the main caldera collapse have filled that caldera and overflowed all but the northern margin. However, topographical contours below the southern margin of the main caldera are much more linear compared with the conspicuously arcuate contours on other

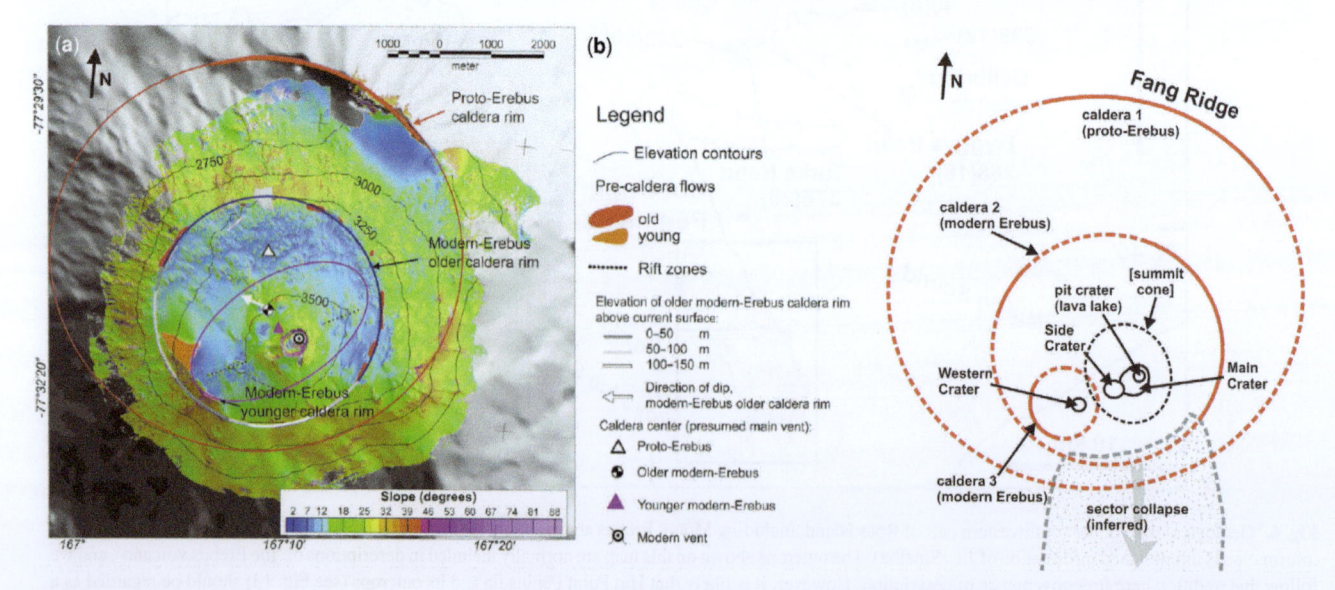

Fig. 7. (a) Map of the Mount Erebus summit region showing topography (hill-shaded digital surface model, illuminated from the NE) and multiple nested calderas (from Csatho *et al.* 2008). (b) Reinterpretation of the shape of summit caldera 3. The location of an inferred zone of sector collapse is also shown, as are the names of the late-formed intra-caldera craters (crater names after Panter and Winter 2008). The dashed red lines are inferred continuations of calderas inferred from extrapolating exposed caldera rims (solid red lines).

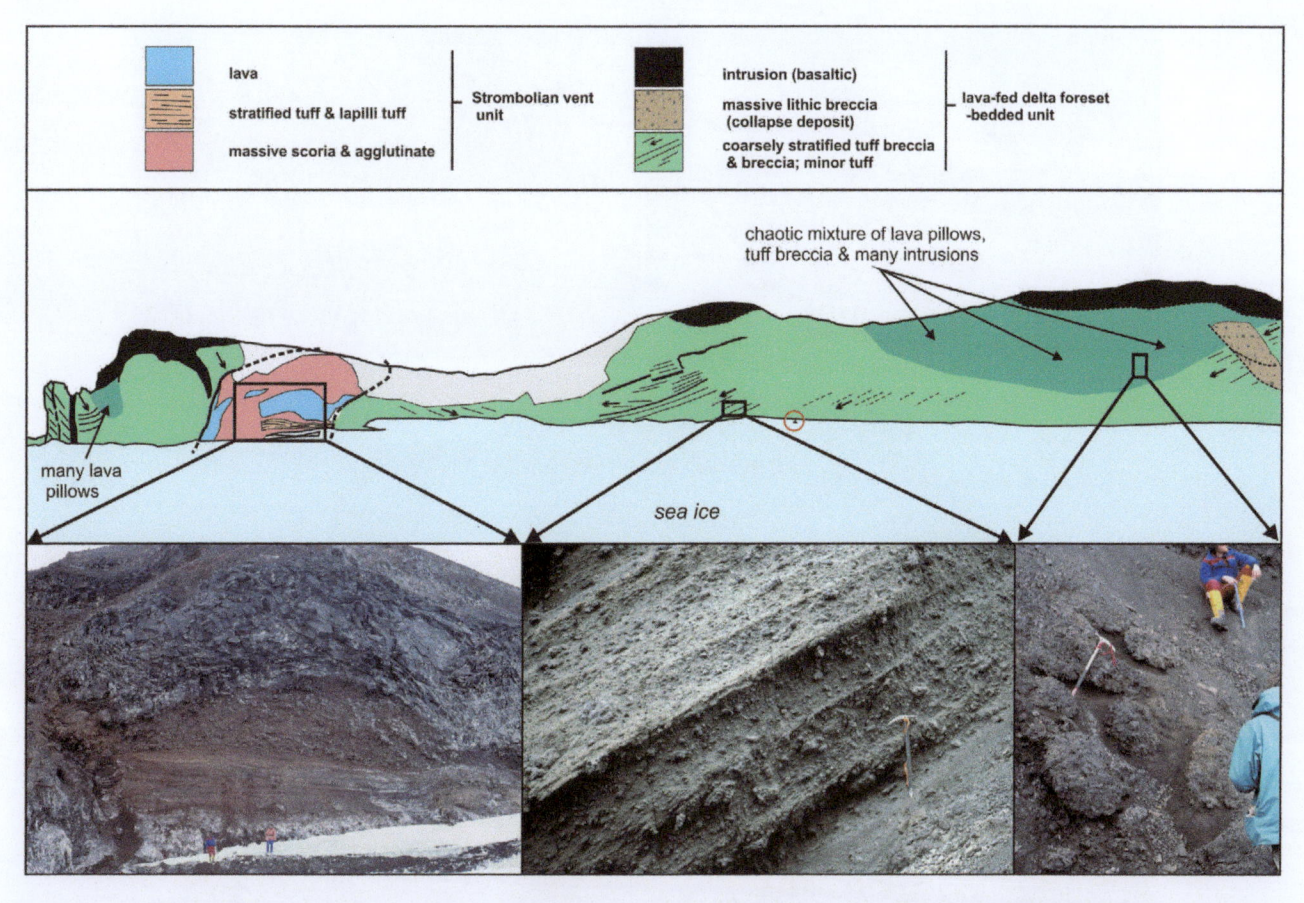

Fig. 8. Geological sketch map of Cape Barne (based on unpublished information of J.L. Smellie) and photographs of the principal lithofacies present. In the main sketch, the red circle encloses a standing person for scale. See the text for the description.

flanks of the edifice, and they coincide with a shallow, slightly curved valley c. 2 km wide on the south flank of the volcano (USGS 1970). This is a feature not previously identified and it is possible evidence for a sector collapse of the southern flank of the volcano prior to activity in the two summit cones; it is partly filled by products of that activity (Figs 6 & 7b). The post-main-caldera activity consisted of numerous small-volume subaerial phonolite and tephriphonolite lavas, of which 10 have been mapped and dated (Harpel et al. 2004; Kelly et al. 2008; Parmelee et al. 2015) (Supplementary material; Fig. 10). Many of the lavas are pāhoehoe, with multiple intertwined lava tubes and ropy surfaces (Fig. 9a). Others have well-developed leveés and are 'a'ā lavas; the latter developed where the underlying slope is steeper, especially on the steep outer (extra-caldera) slopes. The summit region is an uneven, gently-sloping lava plateau that rises to the small asymmetrically positioned, currently active, volcanic cone; the location of the cone is offset to the SE relative to the centre of the main caldera. The lavas have petrologically uniform phonolite compositions and were erupted during a 4 kyr period between 8.50 ± 0.19 and 4.52 ± 0.08 ka, yielding an average eruption rate of $0.01 \text{ km}^3 \text{ ka}^{-1}$ (Parmelee et al. 2015). The summit cone is about 2 km in diameter and rises c. 200 m above the surrounding lavas. It contains two prominent craters: Main Crater, c. 500 m across, which contains an active phonolite lava lake within a small pit crater c. 200 m in diameter; and the smaller Side Crater, 250–350 m in diameter (Fig. 7). Side Crater is inactive but it contains areas of heated ground with temperatures of 40–70°C (J.R. Keys quoted in Panter and Winter 2008). A third, smaller crater (Western Crater), also inactive, snow-filled and c. 180 m in diameter, is present c. 700 m SW of the other two larger craters. All three craters have a linear, presumably fracture-controlled, NE–SW orientation connected with a series of fumarolic ice towers (Panter and Winter 2008) (Fig. 7). The Main and Side craters expose sequences of phonolite lavas and interbedded scoria deposits. Side Crater deposits show signs of weak interaction with external water (Panter and Winter 2008), and similar Strombolian and phreatomagmatic tephras occur on the flanks of Mount Erebus and in the Transantarctic Mountains, some of which are dated (Harpel et al. 2004, 2008; Iverson et al. 2014) (Supplementary material).

The upper slopes of Mount Erebus outside of the main caldera are draped by phonolite and phonotephrite lavas, mainly 20–70 m wide, with prominent leveés; the lavas are highly vesicular to scoriaceous with thick glassy crusts (Moore and Kyle 1987). There are at least six flank cones above 1400 m asl (e.g. at Hoopers Shoulder and Three Sisters Cones: Fig. 6). The cones are constructed of phonolite lavas with short runout distances. Another possible cone, composed of trachyte, occurs c. 1 km east of Cape Barne, whereas Bomb Peak is an endogenous trachyte dome (157 ± 6 ka in age: Esser et al. 2004). The dome is covered by scattered phonolite bombs probably derived from an unexposed flank vent nearby. Lavas forming Cape Evans and Cape Royds, and the lava overlying the basanite at Cape Barne (see above), are formed of anorthite tephriphonolite lavas with ages of 40 ± 6, 73 ± 10 and 89 ± 2 ka, respectively (Esser et al. 2004); the ages of those at Cape Royds and Cape Barne are almost within error and may represent the same event. Although they were emplaced subaerially, it is likely that these lavas advanced into the sea as a substantial lava-fed delta(s) but the coeval subaqueous lithofacies are not exposed (i.e. are below current sea level).

A 200 m-thick cross section through volcanic deposits is exposed at Turks Head and Tryggve Point. The outcrop at

Fig. 9. Compilation of field photographs of lithofacies in the Erebus Volcanic Province. (**a**) Phonolite pāhoehoe lava, Mount Erebus summit; the ice axe in the foreground is for scale. (**b**) 'Kenyte' (phonolite) lava, Mount Erebus summit; the lens cap is 5 cm in diameter. (**c**) View of Helms Bluff showing crude large-scale stratification ('delta foresets') dipping to the left, capped by cogenetic subaerial 'a'ā lavas; together they form a lava-fed delta; the cliff face is c. 400 m high. (**d**) Ignimbrite with numerous ragged pumices, Mason Spur; a pencil is for scale. (**e**) Spatter-rich ignimbrite, Mason Spur (see Martin *et al.* 2018); a pencil is for scale.

Turks Head consists of >150 m of pillow lava, massive hyaloclastite breccia and bedded palagonitized tuff (Wright *et al.* 1983). The beds have northerly dips of 30°, and grade up into a series of 'lava tongues' (lava lobes?). The sequence at Tryggve Point is not as well exposed but comprises disrupted bedded hyaloclastite and broken masses of lava, also dipping at 30° but to the north and west, overlain by massive pillow breccia with pockets of bedded palagonitized tuff and more pillow lava. The breccia/pillow lava outcrops are composed of plagioclase tephriphonolite. The lithofacies are sharply overlain across a 'non-horizontal' (i.e. dipping?) transition by much younger subaerial anorthite tephriphonolite lava. Coeval brecciated dykes and prominent soft-sediment deformation structures are present in both outcrops. The sequences were regarded by Wright *et al.* (1983) and Moore and Kyle (1987) as products of two probably independent eruptions, both showing subaqueous to subaerial transitions; Moore and Kyle (1987) suggested that they were erupted from the same centre that formed the Dellbridge Islands. Dating by Esser *et al.* (2004) indicates that the Tryggve Point and Turks Head outcrops are of similar age (378 ± 28 and 368 ± 18 ka, respectively; the ages are indistinguishable within error), and both outcrops are plagioclase tephriphonolite in composition. The similar lithofacies and indistinguishable ages suggest that the outcrops are probably parts of a single eruptive unit, probably a lava-fed delta, which either flowed into the sea when sea level was higher (or Ross Island has been uplifted), or else the delta melted space in an encircling ice sheet with a surface elevation of c. 150 m asl. The northerly dips of the subaqueous breccias and tuffs suggest that the delta advanced in a northerly direction. The overlying lava (243 ± 10 ka: Esser *et al.* 2004) has developed a distinctive 'hyaloclastite breccia' (tuff breccia?) locally and was sourced from a vent higher on Mount Erebus (Wright *et al.* 1983). The presence of underlying cogenetic hyaloclastite breccia suggests that it, too, might be a lava-fed delta.

Finally, 'Aurora Cliffs' are composed of trachyte dated as 166 ± 10 ka (Fig. 6) (Esser *et al.* 2004). A subglacial to subaerial transition was inferred by Esser *et al.* (2004) but without supporting evidence for a glacial setting and there was no description of the lithologies. The outcrop consists of large-scale homoclinal beds of tuff breccia, blocky breccia, and minor stratified lapilli tuff and tuff that dip steeply to the NW at c. 35°–40° (unpublished information of J.L. Smellie).

Fig. 9. *Continued.* (**f**) Lava-lobe-bearing breccia, Minna Hook; two breccia units are present in the view, separated by a sharp undulating unconformity marked by the yellow dashed line; both breccia deposits represent the subaqueous lithofacies of 'a'ā-lava-fed deltas. (**g**) Close view of lava-lobe-bearing breccia showing a massive lava lobe fractured and spalling into coeval breccia at its margins; Minna Hook; a pencil is for scale. (**h**) Phonolite dome (pale) capped by mafic lavas (dark grey), Minna Hook; the dome is *c.* 200 m high and shows pervasive polygonal jointing. (**i**) Fluidal-textured intrusions intruded into the ignimbrite while it was still unconsolidated, Mason Spur; the rock face is *c.* 25 m high. (**j**) Pale-coloured, crudely-stratified volcaniclastic sediments intercalated between lava-lobe-bearing breccia deposits of two lava-fed deltas, Minna Hook. Photographs by J.L. Smellie except (c) (by Dougal Townsend).

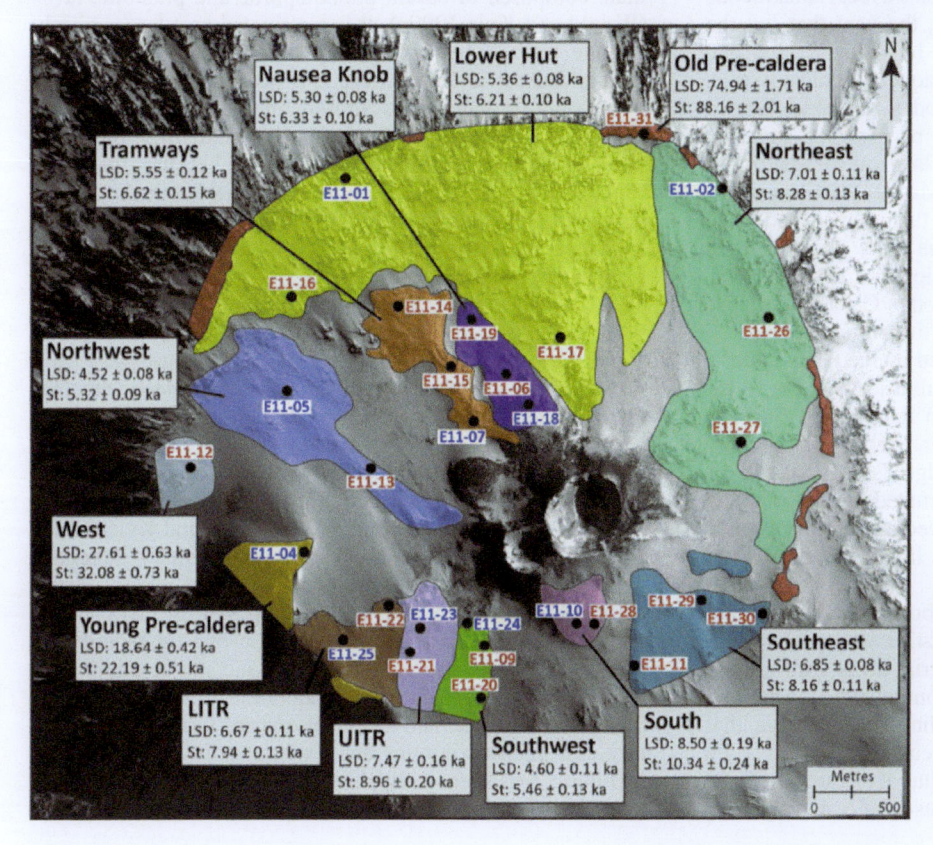

Fig. 10. Mount Erebus summit caldera showing the distribution of mapped phonolite lavas and their eruption ages as determined by cosmogenic exposure dating (after Parmelee *et al.* 2015). Alternative cosmogenic exposure ages are shown for each dated sample, calculated using different scaling models (LSD; St) for cosmogenic production rates (see Parmelee *et al.* 2015 for details).

The coarse beds are massive or have thin reverse-graded bases. They vary from 30 cm to 4 m thick and are dominated by glassy clasts occasionally up to 2 m in diameter. They resemble delta foresets passing up into massive coarsely-jointed lava, whose base, resting on lithic breccia, is lower than the glassy breccias nearby and its elevation becomes lower to the NW. Although the lava also has a trachyte composition, like that forming the glassy breccias, the outcrops are separated by *c.* 50 m of snow. The outcrop architecture is consistent with that of a trachytic lava-fed delta. Although it apparently advanced towards Mount Erebus (similar to that seen at Turks head and Tryggve Point), lava-fed deltas, like sedimentary deltas, often have spreading 'bird's foot' terminations and the progradation direction can thus be very variable locally (e.g. Skilling 2002). The source vent is unknown but a location on Mount Erebus is likely, perhaps an unidentified flank vent. The poorly-exposed upper junction between the breccia and subaerial lava is provisionally interpreted as a passage zone. In an absence of documented effects of very young deformation that has tilted any of the rock outcrops on Mount Erebus, the pronounced dip of the passage zone surface is likely to be a primary feature. It is not explicable as a result of progradation into the sea, which would result in a horizontal surface. A glacial setting is thus more likely (cf. Smellie 2006), consistent with the suggestion of Esser *et al.* (2004).

Mount Erebus evolution. Adapting the summary by Esser *et al.* (2004), the overall evolution of the Mount Erebus edifice is envisaged as occurring in two principal eruptive phases:

(1) The older phase will have consisted of subaqueous activity, mainly mafic pillow lava effusion and associated volcaniclastic rocks, probably mainly hyaloclastite breccias, together with lapilli tuffs erupted during the shoaling phase(s). From the geophysical evidence for crustal loading (Horgan *et al.* 2005; Aitken *et al.* 2012), a thickness of up to *c.* 2 km of these products is possible. The oldest dated rocks that might be attributed to this stage comprise a basanite lava-fed delta at Cape Barne (*c.* 1.310 ± 0.016 Ma), which represents a local subaqueous to subaerial transition. However, the vent that fed the delta must have been subaerial by then and the age must be a minimum for the subaqueous volcanic pile. From the exposures accessible today, the older Erebus edifice was mainly constructed of basanitic–tephritic subaerial lavas, creating a mafic lava shield which was subaerial until *c.* 1.07 ± 0.18 Ma at least (at Fang Ridge: *c.* 2600 m asl). Activity then waned and gave way to fewer or less voluminous more evolved lavas with phonotephrite and tephriphonolite compositions, which constructed a steeper cone. Caldera collapse affected this older edifice some time after *c.* 758 ± 20 or 718 ± 660 ka and is represented by the Fang caldera. The older activity continued until at least 342 ± 18 ka, and included sporadic activity in flank vents. Satellite vents associated with this older Erebus edifice may include centres in the Dellbridge Islands (*c.* 539 ± 12 ka), which were probably erupted from a magma chamber separate from that feeding Mount Erebus (Moore and Kyle 1987) but the lithofacies details are poorly described.
(2) The earliest-dated activity attributed to the younger Mount Erebus edifice began at *c.* 243 ± 10) ka. The younger stage is dominated by subaerial lavas with mainly anorthoclase tephriphonolite compositions. Sporadic flank eruptions occurred, including trachyte lava that formed a small dome at Bomb Peak (157 ± 6 ka) and a glacially-emplaced lava-fed delta at 'Aurora Cliffs' (166 ± 10 ka). Large volumes of anorthoclase tephriphonolite lavas constructed the Erebus summit stratocone, particularly between 121 ± 14 and 90 ± 12 ka, after which (younger than 76 ± 4 ka) the main Erebus caldera collapse took place. Following the main caldera collapse, the cavity became brim full with tephriphonolite lavas. At least two intra-caldera cones were constructed: an earlier one in the SW corner and a second situated to the NE in line with a series of small intra-caldera vents (Harpel *et al.* 2004). The earlier cone underwent a caldera collapse, either forming a highly elliptical caldera (Csatho *et al.* 2008; Kelly *et al.* 2008) or it was more equant and smaller. The phonolite lava products of both cones overflowed and obscured the southern margin of the main Erebus caldera, which may have already undergone a hitherto unrecognized sector collapse prior to their formation (Figs 6 & 7b). Mount Erebus contains one of only two semi-permanent lava lakes known in the Antarctic region (cf. Lachlan-Cope *et al.* 2001; Patrick and Smellie 2013). It occupies a small pit crater within the Main Crater of the younger cone, and is characterized by sporadic but frequent Strombolian and rare phreatic eruptions (Caldwell and Kyle 1994; Dibble *et al.* 1994; see also Chapter 7.2: Sims *et al.* 2021).

Other large volcanic centres on Ross Island. By comparison with Mount Erebus, much less is known about the Mount Bird, Mount Terror and Mount Terra Nova volcanoes. Although a unifying volcanic stratigraphy involving four geographically widespread formations was erected to encompass these volcanoes and those at White Island, Black Island and Brown Peninsula (Cole and Ewart 1968; Cole *et al.* 1971), subsequent isotopic dating has shown that the formations identified are not correlatable (acknowledged by Cole and Ewart 1968; Cole *et al.* 1971) and its use has been abandoned (Kyle *et al.* 1979a). However, it does indicate that eruptions of mafic and felsic lavas have alternated in time. Exposure of rock on Mount Terra Nova (2130 m) is particularly poor, consisting of a few sparse outcrops above 1000 m on the north flank composed of basalt, phonotephrite and phonolite lavas (Wright and Kyle 1990c). It is interpreted as a large, mainly basaltic cone, 22 km long in a north–south direction and 12 km wide east–west, where it is onlapped on both flanks by products erupted from Mount Erebus and Mount Terror. It has a single published isotopic age of 0.8 ± 0.5 Ma (Armstrong 1978). Mount Terror (3262 m) is a basaltic shield volcano, with typical slope gradients of 9°, that forms the eastern termination of Ross Island (Wright and Kyle 1990c). The only extensive ice-free area is a *c.* 10 km-wide swath extending from just west of the summit region to Cape Crozier, a distance of *c.* 25 km; only the coastal area backing Cape Crozier has been mapped geologically (Fig. 11). The volcano has a slightly elliptical basal diameter of 32 km (east–west) and 34 km (north–south). Basanite lavas and pyroclastic breccias erupted from flank vents are associated with basaltic scoria cones (many undated), and endogenous phonolite and phonotephrite domes and cones (originally interpreted as trachytic), for which ages range between 1.75 ± 0.3 and 0.80 ± 0.14 Ma (Cole *et al.* 1971; Armstrong 1978; Wright and Kyle 1990c; Lawrence *et al.* 2009) (Fig. 11; Supplementary material). The coast extending north and west of Cape Crozier consists of numerous small headlands composed of lava, some columnar-jointed, interbedded volcanic breccia (sometimes reddened and possibly autobreccia?) and minor lapilli tuff. Basanitic tuffs with 'trachyte' (phonolite?) clasts are prominent at Cape Crozier and locally elsewhere. Granitic, sedimentary and gneissic erratics are present up to 825 m asl and include a fossiliferous sandstone(?) boulder of Eocene age (Cole *et al.* 1971; Denton and Marchant 2000). Together

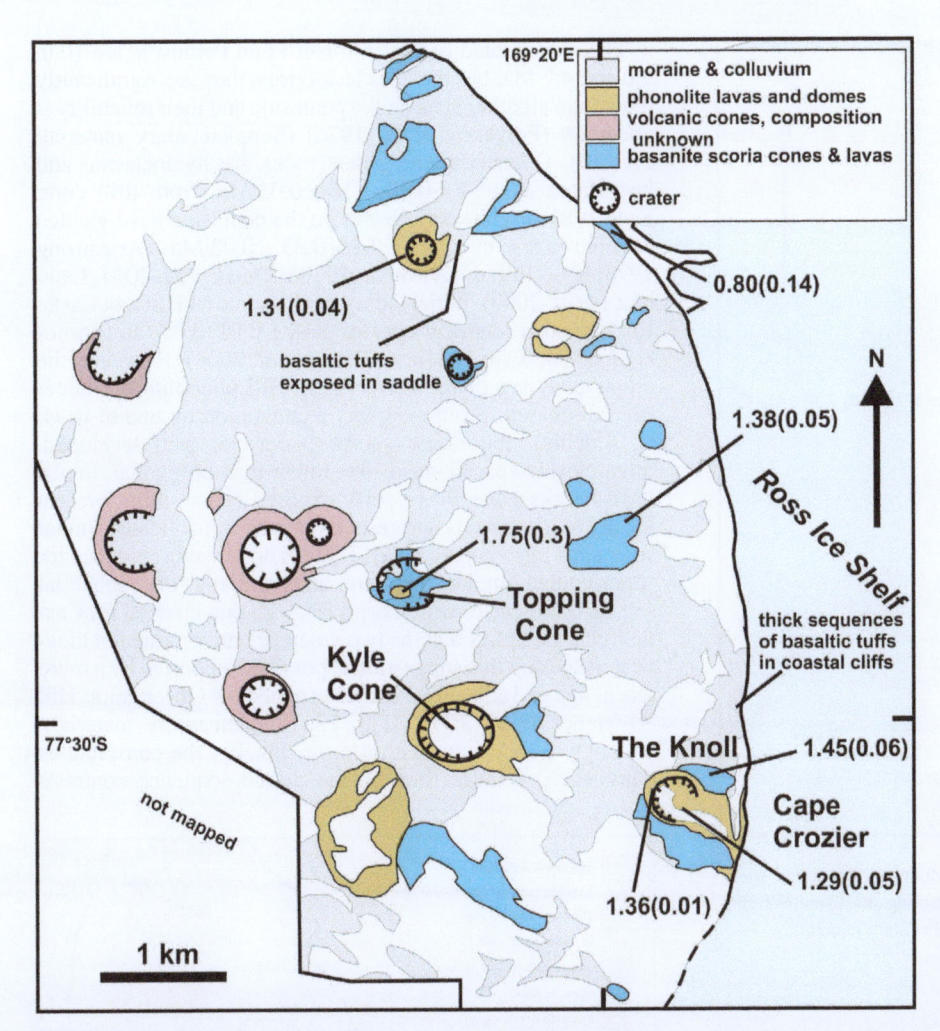

Fig. 11. Geological sketch map of Cape Crozier, Mount Terror volcano (after Cole *et al.* 1971; modified after Denton and Marchant 2000); isotopic ages (in Ma) after Armstrong (1978) and Wright and Kyle (1990c); errors in parentheses (Supplementary material).

with three topographical benches at 150, 200–250 and 300–325 m asl and two raised beaches, the erratics and benches are interpreted to record a formerly expanded Ross Ice Shelf, whilst the raised beaches indicate local isostatic uplift (glacial rebound) of Ross Island (Cole *et al.* 1971).

Mount Bird forms the northern promontory on Ross Island. It is a mainly basaltic shield volcano with a summit at 1800 m asl (Wright and Kyle 1990d) (Fig. 4b). Like the other volcanoes forming Ross Island, exposure is poor and scattered, and only the coastal fringe on the west side has been mapped geologically (Cole and Ewart 1968) (Fig. 12). The east and west coasts are formed of extensive seaward-dipping basaltic (basanite) lavas 1–8 m thick with oxidized scoriaceous autobreccias (presumably 'a'ā); unusually, some of the breccias are distinctively yellow coloured. Although the authors have not visited the outcrops, the presence of yellow autobreccia within subaerial 'a'ā may be evidence for superficial palagonite alteration of the lava clinkers caused by steam activity associated with lava-fed-delta progradation (cf. Smellie *et al.* 2011a, 2013; Smellie and Edwards 2016, p. 198). The lavas are presumed to have been erupted from the main Mount Bird cone. Minor basaltic agglomerate, tuff and hyaloclastite, and 'trachyte' (probably phonolite: Wright and Kyle 1990d) lavas and 'plugs' (domes?) interbedded with basaltic lava are also present on the west coast. Higher flank outcrops comprise numerous poorly-exposed basaltic (basanitic) scoria cones, and phonolite cones and domes (Wright and Kyle 1990d). Published isotopic ages range between 4.5 ± 0.60 and 3.7 ± 0.20 Ma for the main basaltic shield, and between 3.15 ± 0.09 and 3.00 ± 0.15 Ma for a younger phase of phonolitic volcanism (Armstrong 1978) (Supplementary material). Palaeomagnetic analyses of eight lavas from the west coast indicate that they have a reverse polarity consistent with the isotopic ages (P. R. Kyle cited in Wright and Kyle 1990d).

Hut Point Peninsula. Unlike Mount Bird and Mount Terror, the third 'prong' making up the radial symmetry of Ross Island is a topographically low-lying, largely snow- and ice-covered peninsula formed by multiple coalesced mafic scoria cones, a tuff cone and a felsic dome (Fig. 13). Hut Point Peninsula is *c.* 23 km long, mainly <3 km wide and <300 m high, except at Castle Rock (413 m asl: Fig. 4a). It is entirely volcanic, with outcrops exposed mainly at its southwestern termination. It has been described by Cole *et al.* (1971) and Kyle (1981a, c, 1990b). Several of the scoria cones have relatively well-preserved craters consistent with a young age, and that at Crater Hill has issued an extensive lava field, which partly underlies Scott Base and may be the subaerial expression of a lava-fed delta seaward of Scott Base; the lava field is also well exposed in the cliffs on the north side of The Gap and north of Scott Base. Castle Rock and the low ridge extending 1.5 km to the north from there are composed of yellow-orange-coloured stratified lapilli tuffs. They were described as 'autoclastic breccias' by Cole *et al.* (1971), and similarly described rocks (presumably also lapilli tuffs) were said to occur at Boulder Cones, *c.* 1 km SW of Castle Rock (see also Kyle 1981c). Eruption of the Castle Rock deposits may have been subglacial, during a period when the coeval ice was much higher, which contrasts with the 'dry' magmatic eruptions responsible for the abundant scoria cones elsewhere

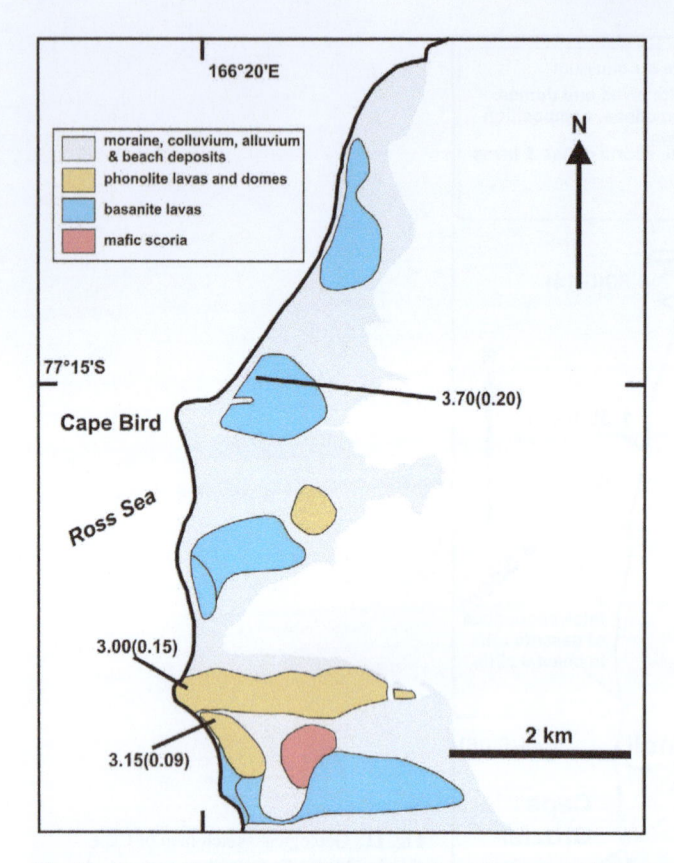

Fig. 12. Geological sketch map of Cape Bird, Mount Bird volcano (after Cole *et al.* 1971). Isotopic ages (in Ma) after Armstrong (1978) and Wright and Kyle (1990*d*); errors in parentheses (Supplementary material).

The oldest dated rocks from Hut Point Peninsula are 10.0, 5.0 and 4.7 Ma, but the ages lack errors, they are significantly older than all other ages on the peninsula and their reliability is uncertain (Polyakov *et al.* 1976) (Supplementary material; Fig. 13). Otherwise, the oldest rocks are hyaloclastite and lava dated as $1.34 \pm 0.23-1.16 \pm 0.03$ Ma (from drill core: Kyle 1981*a*). The scoria cones on the peninsula have yielded isotopic ages of $1.03 \pm 0.10-0.33 \pm 0.02$ Ma (Armstrong 1978; Kyle 1981*a*; Kyle *et al.* 1979*b*; Tauxe *et al.* 2004; Lawrence *et al.* 2009). Initial dating of Observation Hill and Castle Rock yielded identical ages of $1.18 \pm 0.01/0.05$ Ma (Forbes *et al.* 1974; Kyle 1981*a*, *c*; Tauxe *et al.* 2004). However, the eruptive setting of the Observation Hill phonolite has never been discussed. From a cursory examination by one of us (J. L. Smellie), the dome shows pervasive, well-developed, often closely-spaced sheet-like, rather than polygonal, joints. It also lacks glass, has poorly-exposed coarse lithic breccia locally at its margins and rests on weakly consolidated fluvial sediments (also poorly seen). There is no obvious evidence for water-cooled emplacement and an essentially 'dry' subaerial setting is indicated, which contrasts with that inferred for Castle Rock. It implies that the two volcanic outcrops are not identical in age and this suggestion is possibly supported by a more recent age of 1.23 ± 0.02 Ma determined for Observation Hill (Lawrence *et al.* 2009) (Fig. 13; Supplementary material), although the ages are within error. Finally, the conspicuous dominance of lavas forming the drilled sequence contrasts on the peninsula (Kyle 1981*a*, *c*). The presence of east–west-trending striations on the summit of Observation Hill is evidence for higher ice surfaces in the past (Forbes and Ester 1964). The surface rocks are all basanites except for Observation Hill, which is a small endogenous phonolite dome 228 m high and 700 m in diameter. There is also a small 'trachyte' (probably phonolite) lava exposed close to Observation Hill on its north side, and a small outcrop of intermediate-composition (mugearite or benmoreite) lava on the NW side of Observation Hill (Fig. 13). A much broader compositional range, called the DVDP lineage and including phonotephrite and tephriphonolite lavas, was recovered by drilling the Hut Point Peninsula sequence at McMurdo Station (Kyle 1981*a*, *b*; Kyle *et al.* 1992). The cored succession is dominated by lavas with oxidized scoriaceous surfaces (presumably 'a'ā) but it also includes minor red and black airfall scoria beds 5–15 m thick. The drilled succession, together with the exposed scoria cones and their associated lavas (e.g. lava field issued from Crater Hill), were erupted under subaerial conditions. However, the sequence passes down into massive basanite hyaloclastite at least 50 m thick and potentially >200 m. The hyaloclastite, which is not well described and may be either tuff breccia or lapilli tuff (*sensu* White and Houghton 2006), is yellowish and palagonite-altered in the upper 1 m, and was inferred to have formed subaqueously (Kyle 1981*a*). Its surface elevation varies between 76 and 120 m below present sea level between two drilling sites, situated just 500 m apart. The reason for the disparity in elevation is unclear, unless the surface between the hyaloclastite and overlying subaerial lava is a passage zone in a lava-fed delta erupted under a south-dipping ice sheet, but it suggests that Hut Point Peninsula has subsided, or sea level risen, since this sequence was erupted (Kyle 1981*c*).

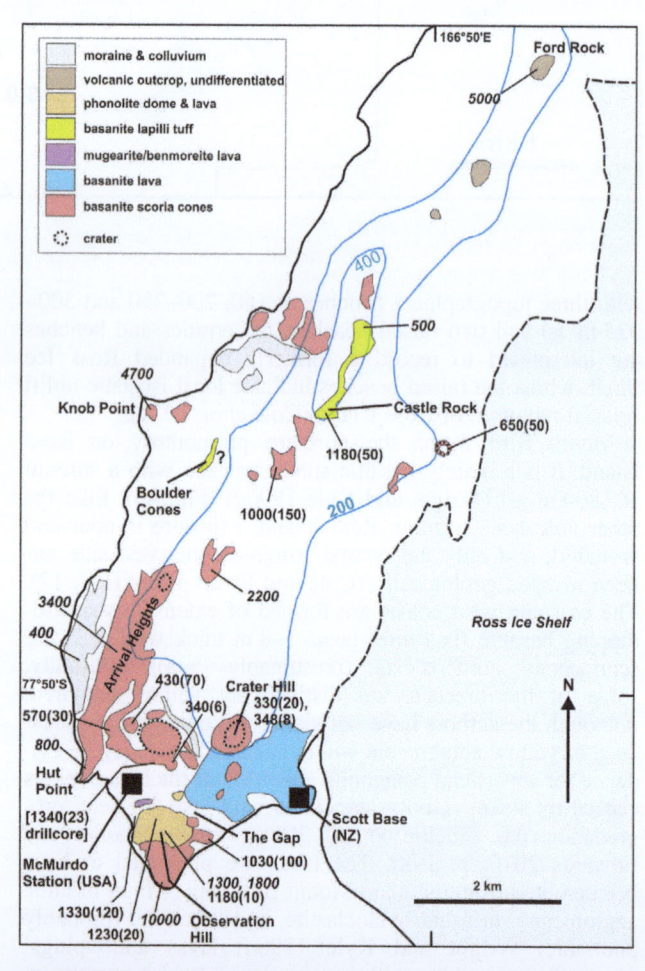

Fig. 13. Geological map of Hut Point Peninsula (modified after Cole *et al.* 1971; Kyle 1981*a*). Isotopic ages (in ka) are after Armstrong (1978), Kyle (1981*a*) and Lawrence *et al.* (2009); errors in parentheses (Supplementary material). Ages in italics and lacking errors are after Polyakov *et al.* (1976). The dashed black line is the approximate boundary between the subaerial extent of Hut Point Peninsula and the Ross Ice Shelf.

with the surface geology, dominated by scoria cones with only minor lavas (apart from the lava field issued from Crater Hill). It may suggest that Hut Point Peninsula is underlain by a substantial volcanic shield in which the final stages of activity were characterized by numerous small-volume eruptions from flank vents represented by scoria cones, a tuff cone and a phonolite dome. This is a similar evolutionary sequence to those more convincingly demonstrated for Mount Bird and Mount Terror, described above, and in northern Victoria Land (Hamilton 1972; see also Chapter 5.1a: Smellie and Rocchi 2021).

Mount Discovery Volcanic Field

Northern outcrops: White Island, Black Island, Brown Peninsula and Dailey Islands. The northern side of the Mount Discovery Volcanic Field is composed of three major volcanic features: White Island, Black Island and Brown Peninsula. All are remnants of shield volcanoes and only one has been examined recently (White Island: Cooper *et al.* 2007). The limited overlap in the published ages (see below) suggests that they are probably the products of at least three unrelated volcanic centres. White Island is 29 km long and up to 14 km wide, tapering to the north, and has a maximum elevation of 762 m. As the name implies, it is largely covered by snow and ice, and rock outcrops are mainly confined to the northern side (Fig. 14). Although initially described as having been constructed by two separate shield volcanoes, the reasoning is unclear (perhaps it refers to the elongate outcrop) and only one volcano may be present on the island. It is dominated by subaerially erupted basanite scoria cones, oxidized agglutinated spatter, and lavas erupted from several vents with a NNE alignment and with isotopic ages of 5.04 ± 0.31–0.17 ± 0.01 Ma; a U/Pb age of 7.65 ± 0.69 Ma on zircon in an anorthosite nodule suggests that the magmatism extends even further back in time (Cole *et al.* 1971; Cooper *et al.* 2007) (Supplementary material). The scoria cones are typically 250–500 m in diameter and many preserve craters, the largest of which (NW of Mount Heine) measures 1.8×0.6 km (north–south and east–west, respectively). Indurated stratified khaki-brown lapilli tuff with abundant accessory and accidental lithic clasts (basanite, gabbro and granulite) forms outcrops at Mount Nipha and Isolation Point, and tuff breccia occurs as large (up to 4 m) loose blocks west of Mount Hayward. The latter show crude bedding, lamination, cross-bedding and grading, and contain lava masses resembling cauliflower bombs and lava pillows, consistent with subaqueous effusion. The Mount Nipha outcrop is interpreted as a tuff ring with a crater 600 m in diameter. Compositions on White Island range from basanite to tephriphonolite. At least five glacial benches are present on the NW and NE flanks, and erratics of granitic basement and calcareous mudstone occur up to 700 m asl; the latter contain a range of marine fossils and are correlated with the Scallop Hill Formation (Pliocene), which crops out *in situ* on Brown Peninsula and probably Black Island (Speden 1962; Eggers 1979). The benches were attributed to different previous elevations of the Ross Ice Shelf by Cole *et al.* (1971), similar to features seen at Cape Crozier (see above), or possibly to overriding by a much expanded Ross Ice Sheet (Cooper *et al.* 2007), but the surface geology does not display any convincing evidence for a glacial eruptive setting.

Black Island is a stratovolcano remnant roughly triangular in shape, with sides *c.* 16–21 km long (Cole and Ewart 1968). It has a maximum elevation of 1041 m asl at Mount Aurora but most of the island is below 600 m asl. Unlike White Island, Black Island is largely snow-free, apart from its southern corner (Fig. 15). Only three published

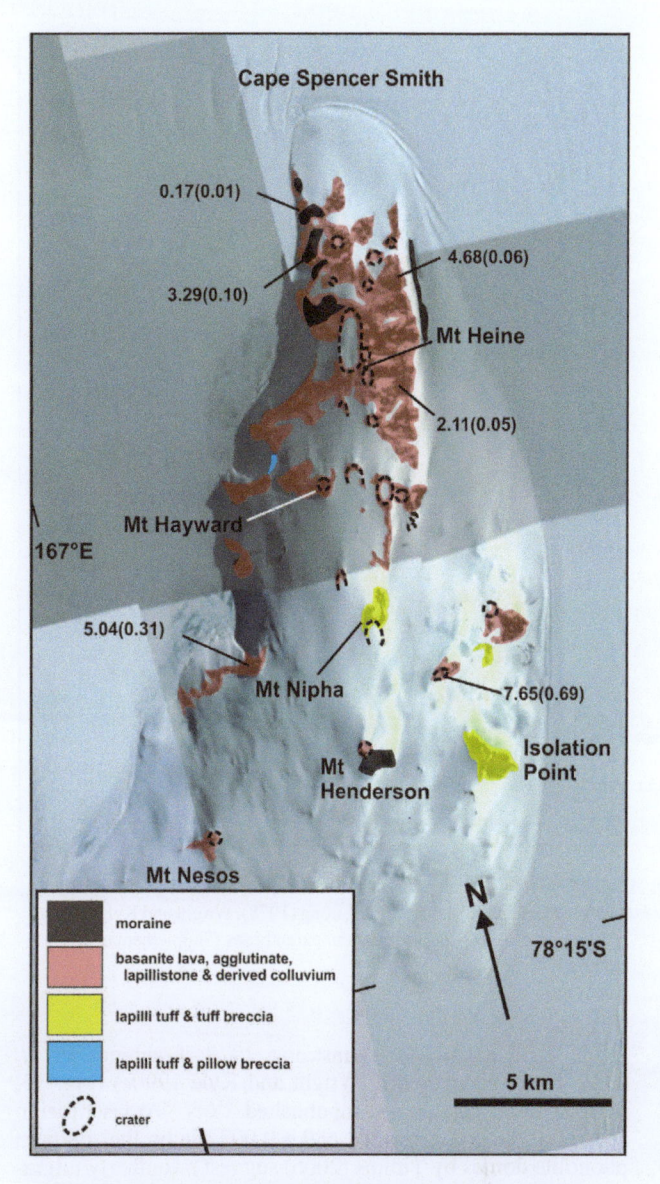

Fig. 14. Geological map of White Island (after Cooper *et al.* 2007). Isotopic ages (in Ma) are also shown; errors in parentheses (Supplementary material).

geochemical analyses are available (Goldich *et al.* 1975). They comprise alkali basalt, 'hawaiite' (mugearite: unpublished data of A.P. Martin) and phonolite compositions, erupted at different times (Cole and Ewart 1968; Wright and Kyle 1990*e*). The surface geology consists of strongly eroded basalt scoria cones and lavas, which may be interbedded with 'trachyte' lavas (probably phonolite: cf. Timms 2006) in the north of the island, and 'trachyte' (phonolite) domes, 'plugs' and lavas, which dominate the central and southern parts of the island. They include Mount Aurora, which may be a major volcanic centre for the felsic units (Wright and Kyle 1990*e*). Two very prominent flat-topped felsic (phonolite) domes at the southern end of the island were thought to be amongst the youngest features on the volcano. From their fresh appearance, they were inferred to be 'post-glacial' (i.e. erupted after the Last Glacial Maximum, when the ice had receded). On the basis of a crudely arcuate outcrop pattern displayed by the eastern scoria cones, Cole and Ewart (1968) speculated that the higher plateau-like centre of the island may mark the site of an old extensive caldera. Published isotopic ages suggest that the northwestern part of Black Island (10.90 ± 0.4 Ma) is older than the centre of the island (9.02

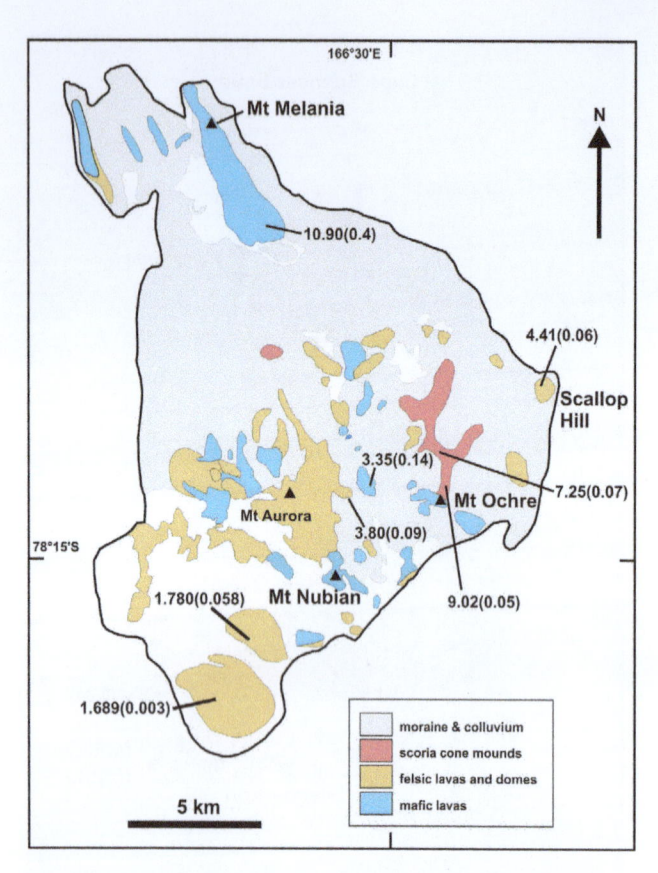

Fig. 15. Geological map of Black Island (after Cole and Ewart 1968). Isotopic ages (in Ma) are after Armstrong (1978), Wright and Kyle (1990e) and Lawrence *et al.* (2009); errors in parentheses (Supplementary material).

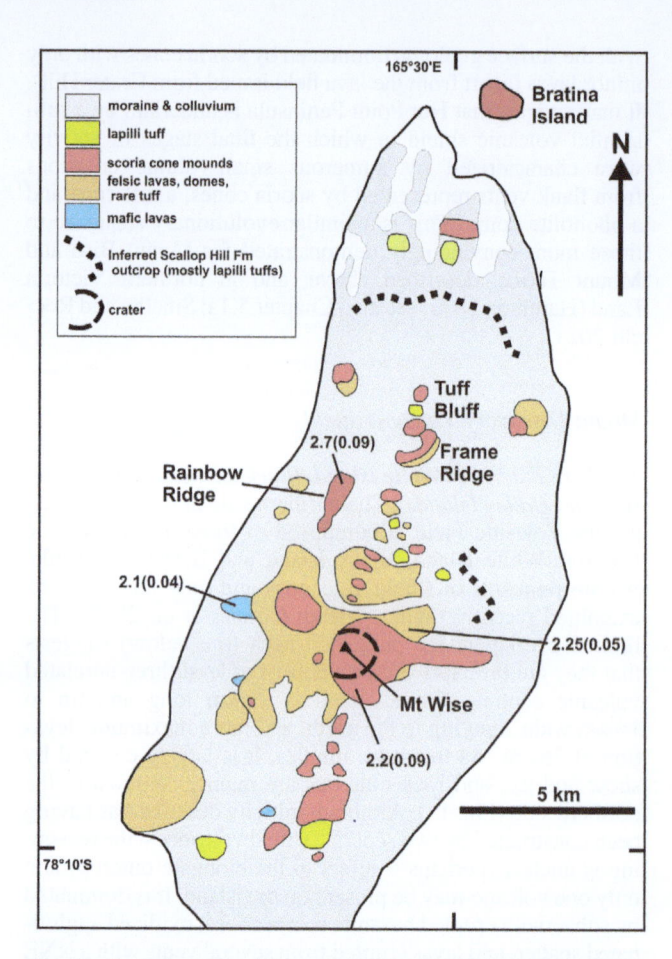

Fig. 16. Geological map of Brown Peninsula (after Cole and Ewart 1968; Eggers 1979). Isotopic ages (in Ma) after Armstrong (1978); errors in parentheses (Supplementary material).

± 0.05–3.35 ± 0.14 Ma: Armstrong 1978; Lawrence *et al.* 2009; P. Webb quoted in Wright and Kyle 1990e) (Fig. 15; Supplementary material). Unpublished ^{40}Ar/^{39}Ar laser fusion ages of 1.780 ± 0.058 and 1.689 ± 0.003 Ma on the southern phonolite domes by Timms (2006) suggest a southerly migration of the volcanism and that the domes are not post-glacial.

Brown Peninsula is essentially snow-free probably because it is situated in the precipitation shadow of Mount Discovery. It is smaller than Black Island and highly elongate, measuring 21 km in length and up to 8 km wide (Fig. 16). The highest elevation is 816 m, at Mount Wise, but most of the peninsula is below 500 m. Geologically, it is similar to Black Island, consisting of mafic and felsic lavas, mafic scoria cones, and felsic domes that were probably erupted from a series of north–south-aligned vents (Cole and Ewart 1968; Kyle *et al.* 1979a; Wright and Kyle 1990e). Most of the outcrops are significantly glacially eroded but the youngest outcrops, near Mount Wise with its shallow summit crater *c.* 1 km in diameter, appear little modified. On that basis, the latter might be 'post-glacial' but, as on Black Island, the suggestion is not supported by isotopic dating (see below). Compositions include basanite, hawaiite, benmoreite and phonolite, and the magmas appear to have been erupted in at least three mafic to felsic cycles (Kyle *et al.* 1979a), although felsic rocks dominate the present-day outcrops. Unlike Black Island, pyroclastic rocks are also present. Cole and Ewart (1968) suggested that they are of limited occurrence and could not be used as datums. However, mapping by Vella (1969; also Eggers 1979) suggested that the distribution of *in situ* outcrops of the distinctive volcaniclastic Scallop Hill Formation, whose type locality is on Black Island, is widespread. It was mapped in a subdued scarp or bench around the east and north sides of the peninsula, and thus might be used as a datum. The Scallop Hill Formation on Brown Peninsula is up to 4 m thick and consists of interbedded yellow-brown 'tuffaceous agglomerate' and brown 'palagonitic agglomerate' (both possibly tuff breccia?), palagonitic tuff and volcanic breccia, in beds *c.* 0.3–2 m thick. An upper 'volcanic breccia' bed is fossiliferous and is a likely source of the Pliocene fossils that have been found on Black Island, White Island and Minna Bluff (Speden 1962). Other lithologies associated with the moraines include tuffaceous sandstones, gravels and conglomerates, many fossiliferous (Speden 1962). The fossiliferous material occurs principally as clasts in moraines. It is unequivocally *in situ* only on Brown Peninsula but a much smaller outcrop on top of Scallop Hill (Black Island) may also be *in situ* (Vella 1969). Although the formation is essentially volcanic (probably phreatomagmatic), it was deposited subaqueously. As the outcrop occurs today at an elevation of *c.* 200–300 m asl (estimated) on Brown Peninsula, and at *c.* 200 m on eastern Black Island, uplift has occurred since the Pliocene (Vella 1969; Eggers 1979). Isotopic ages range from 2.70 ± 0.09 to 2.10 ± 0.40 Ma (Armstrong 1978). None of the dated units is 'post-glacial', despite a lack of glacial erosion on some. Moreover, the ages show little correspondence with the stratigraphy erected by Cole and Ewart (1968), which illustrates the difficulty of correlating laterally discontinuous volcanic units in areas of limited exposure, and is probably the reason why Kyle *et al.* (1979a, b) rejected that stratigraphy.

The Dailey Islands protrude through the northernmost limit of the McMurdo Ice Shelf and as islands in southernmost McMurdo Sound (Figs 1 & 4h). There are six islands in total. Five have geographical names. A sixth island is unnamed but is barely seen emerging from the McMurdo

Ice Shelf to the south. The islands are scattered across an area of c. 20 km². The tallest rises to just 133 m asl but the islands are situated in water depths of c. 200 m (Del Carlo et al. 2009). Most of the islands are <0.5 km in diameter (above ice or water). The largest, West Dailey Island, is 2 km long and 500–800 m wide, and may be a compound edifice, composed of the coalesced products of more than one vent. Very little is published about the geology. Although Wilson (2000) postulated that the islands were 'drift-covered ice pedestals...which may be relicts of a former expanded Koettlitz Glacier grounded in McMurdo Sound', they consist of several scoria cones with basanite compositions from which basement rocks and lower crustal mafic granulites have been obtained (Treves 1967; Berg 1991; Kalamarides and Berg 1991; Del Carlo et al. 2009). The absence of craters and presence of granitic erratics in moraine imply overriding and erosion by ice and thus a likely age of tens of thousands of years, at least. This is confirmed by a published isotopic age of 0.77 ± 0.032 Ma for a dyke on Juergens Island, and the normal magnetic polarity (Brunhes Chron) of samples from two of the islands (McCraw 1967; Mankinen and Cox 1988; Tauxe et al. 2004; Supplementary material).

The Dailey Islands have the appearance of a small monogenetic volcanic field. Broad similarities in eruption style and composition have previously led them to be regarded as analogous to multiple small basaltic centres in the Royal Society Range, situated c. 56 km to the SW (Wright and Kyle 1990j). There are differences, however, in that the Dailey Islands are situated on a prominent short-wavelength magnetic anomaly c. 12 km in diameter (Behrendt et al. 1996; Chiappini et al. 2002; Ferraccioli et al. 2009), which may be an expression of a large underlying pluton. However, there is no means of knowing the age of that pluton, which may be part of the Cambro-Ordovician Granite Harbour Intrusive Complex (Cox et al. 2012) rather than coeval with the Dailey Islands volcanism. The individual scoria cones that form the Southern Local Suite outcrops are also usually small features, typically <60 m high and with basal diameters less than 250 m; larger cones, up to 300 m high, are rare (McIver and Gevers 1970; Wright and Kyle 1990j, k; Wilch et al. 1993). Moreover, because of their terrestrial location, they are formed entirely of 'dry' magmatic products (i.e. scoria, agglutinate and clastogenic lavas). By comparison, the Dailey Islands pyroclastic cones are much larger. They range from 200 to more than 300 m high (measured from the seafloor), and have minimum diameters of 700 m–2 km (above sea level/ice surface). As they erupted either in the sea or under an ice sheet, they are likely to have an internal (unexposed) early-formed cone consisting of explosively-generated (phreatomagmatic) lapilli tuffs possibly resting on a pillow mound, like other marine and glaciovolcanic Surtseyan volcanic edifices elsewhere (e.g. Kokelaar 1983; Smellie and Hole 1997; Smellie 2001; Schopka et al. 2006; Smellie and Edwards 2016). The possibility exists, therefore, that the Dailey Islands are an outlying, poorly-developed part of the Mount Discovery Volcanic Field, stalled at an early evolutionary stage. They are similar in many respects to volcanic centres forming Hut Point Peninsula, in the Ross Island Volcanic Field.

Southern outcrops: Mount Discovery, Minna Bluff, Minna Hook and Helms Bluff. Mount Discovery is a prominent, weakly dissected stratovolcano. It has an elliptical basal diameter (above the surrounding Ross Ice Shelf) of 25 km (NE–SW) and 18 km (NW–SE), and rises to 2681 m asl. It is linked via the long promontory called Minna Bluff to Minna Hook (Fig. 1). Similar to Mount Erebus on Ross Island, Mount Discovery also dominates the centre of a prominent three-fold radial symmetry with radiating 'arms' pointing to Brown Peninsula, Minna Bluff and Mount Morning. Average slopes on the volcano are gentle below c. 1100 m, above which they steepen to c. 30° to form the prominent summit cone (Fig. 4c). Rock exposures are restricted to the northern slopes above 400–600 m; the lower slopes are covered by extensive moraine and colluvium (Anderson et al. 2017), and the south flanks are almost entirely covered by snow and ice. Glacial erosion has affected all the rocks but is only severe below c. 400 m asl (Wright-Grassham 1987). A small ice cap at the summit probably obscures a crater, an interpretation that is supported by the presence of outcrops of unwelded phonolite pumice lapilli nearby, which may be the remains of a pumice cone (Wright and Kyle 1990f). Wright-Grassham (1987) identified two major volcanic stages on Mount Discovery. Stage MD1 is associated with the construction of the main stratovolcano edifice, and consists mainly of tephriphonolite, phonolite, and trachyte lavas, domes, breccias and tuffs, whilst MD2, which dominates the surface geology, incorporates numerous small flank vents that post-date the main volcanic edifice and comprise both basanite–hawaiite scoria cones and associated lavas together with rare phonolite domes (Wright-Grassham (1987; Wright and Kyle 1990f) (Fig. 17). The oldest exposed units of the stratocone-forming period (MD1) comprise steeply-dipping (30–35°) benmoreite lavas, 2–10 m thick, interbedded with a similar proportion of lahar deposits and fluvial volcaniclastic rocks. They are conformably overlain by extensive phonolite lavas 20–100 m thick with minor lenses of lahar deposits, which crop out both in the summit region and near the base of the volcano. The phonolites were erupted from the summit crater and also from small domes, and they include two thin (4–8 m) basal welded agglomerate beds. The numerous flank vents (MD2) comprise scoria cones 50–300 m high, some with agglutinate, and associated lavas 3–10 m thick. Some of the lavas extend several kilometres from their source vent. Glacial erosion has exposed feeder dykes in some cases. Compositions of the effusive rocks are mainly basanite–hawaiite, whilst the domes are phonolitic. Most of the flank vents appear to have erupted from NE-trending fissures but there is also evidence for a weaker NW alignment (Wright-Grassham 1987). Isotopic dating indicates that the exposed phonolitic part of MD1 was formed between 5.46 ± 0.16 and 5.34 ± 0.11 Ma (Armstrong 1978; Wright-Grassham 1987) (Supplementary material). The phonolitic eruptions were followed by mainly basanite to hawaiite eruptions from multiple small flank vents (MD2 stage). Three MD2 units have been dated, with the oldest two yielding ages of 5.19 ± 0.32–4.45 ± 0.26 Ma. The youngest ages obtained so far correspond to a detached phonolite dome low on the SE side of the volcano (1.87 ± 0.43 Ma: Fig. 17), and two ages of 0.18 ± 0.08 and 0.06 ± 0.006 Ma for basaltic rocks at unspecified locations (Tauxe et al. 2004) (Supplementary material).

Helms Bluff, situated at the base of the southern side of the saddle between Mount Morning and Mount Discovery (Figs 1 & 4d), is formed of at least three volcanic units (Wright-Grassham 1987). The description presented here is based on unpublished information of the authors following a half-day visit. At its base is a poorly-exposed subaerial 'a'ā lava succession that overlies massive lava-lobe-bearing breccia (*sensu* Smellie and Edwards 2016). It is overlain by lenses up to c. 1.3 m thick, composed of stratified volcaniclastic sediments (flaggy brownish-grey fine sandstone–pebble conglomerate) and mud-rich diamictite. The sediments are sharply truncated by an erosive surface that shows glacial striations (probably the striated surface recorded by Wright-Grassham 1987), and is overlain by a thick (c. 270–300 m) lava-lobe-bearing basanite breccia which dominates the cliff and has a thinner (c. 100–120 m) capping of 'a'ā lava with oxidized autobreccias. The third volcanic sequence consists of thick grey

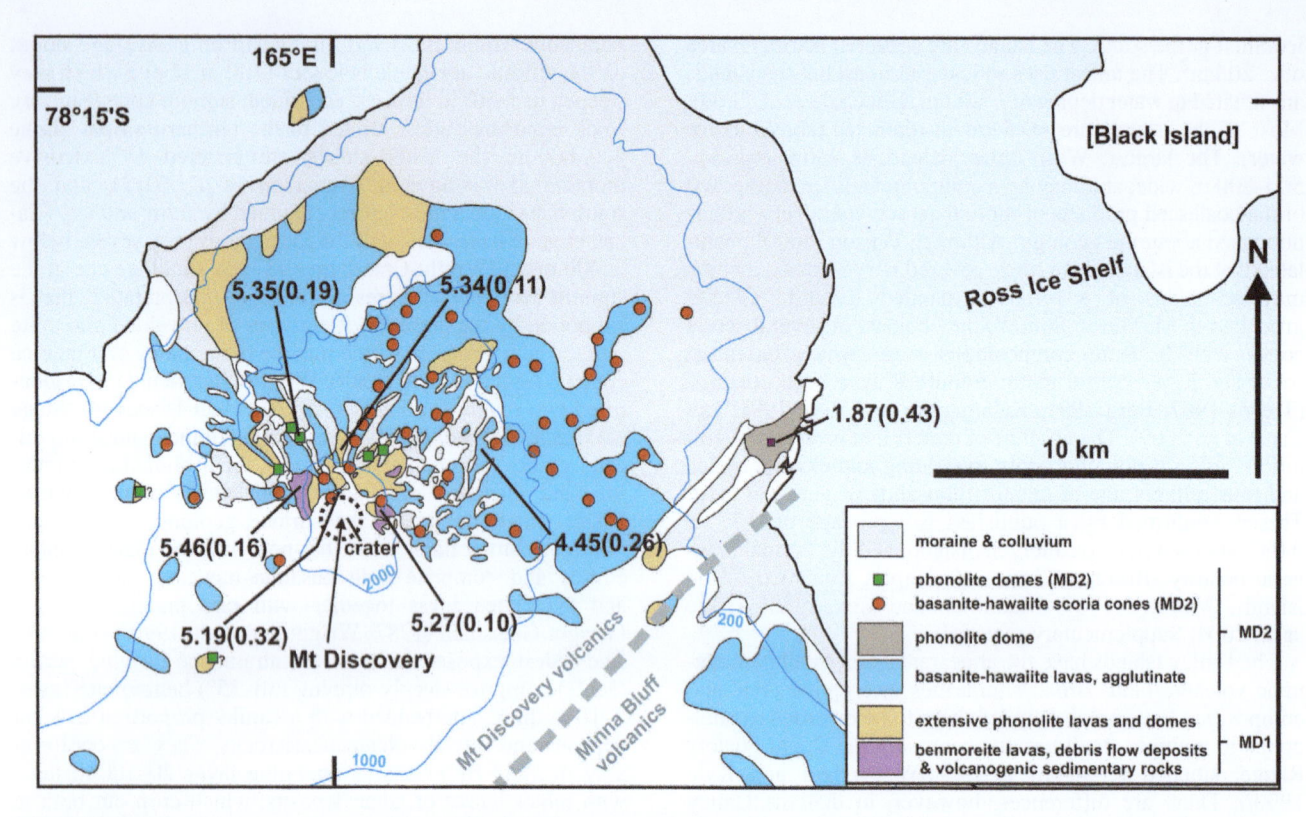

Fig. 17. Geological map of Mount Discovery (after Wright-Grassham 1987; Wright and Kyle 1990f). Isotopic ages (in Ma) are by Wright-Grassham (1987); errors in parentheses (Supplementary material). The bulge in the snow- and ice-covered southern flank (see also Fig. 4e) may be caused by the presence of a subsidiary volcanic centre that may have given rise to volcanic outcrops at Helms Bluff.

subaerial 'a'ā lavas. The basal two volcanic units are interpreted as 'a'ā-lava-fed deltas. Only the middle (thickest) sequence is well exposed. Although mainly chaotic, it also shows a crude homoclinal stratification dipping to the WSW, similar to the much gentler dip of the capping lavas (Fig. 9c). The dips suggest that the delta advanced away from Mount Discovery towards Mount Morning. Moreover, the middle sequence has a published isotopic age of 4.51 ± 0.31 Ma (Wright-Grassham 1987; Martin et al. 2010), which is younger than volcanic units erupted at Mason Spur (apart from late-stage scoria cones), despite the proximity of Helms Bluff to the latter (just 7 km away). By contrast, it overlaps with the age of the MD2 period on Mount Discovery. The weight of the evidence suggests that it may be either (1) a far-travelled lava-fed delta sourced in Mount Discovery; or (2) part of a separate and currently unidentified volcanic centre. In support of the latter suggestion, a large volcanic shield, completely covered by snow and ice, is present on the southern side of Mount Discovery. It forms an asymmetrical 'bulge' at least 10 km in diameter (NW–SE) with a higher general elevation and lower gradients than the corresponding NE flank (Fig. 4d). The high elevation of the base of the Helms Bluff outcrop (c. 800 m asl), together with the presence of diamict and fluvial sediments below a glacially modified surface, are consistent with a terrestrial setting and glacial emplacement for at least some of the volcanism.

Very little is published about the geology of Minna Bluff (excluding Minna Hook; see later), owing to an extensive covering of moraine, colluvium and snow patches (Fig. 4e). It measures 38 km in length, c. 3 km in width and is generally 800 m high, locally reaching 1060 m asl. Rock is only intermittently exposed in small outcrops, apart from on the south side in McIntosh Cliffs, which have difficult access. The only published study is by Wright and Kyle (1990g, based on Wright-Grassham 1987), which suggested that it consists of an older sequence of basanite–alkali trachyte lavas, domes and hyaloclastite, a younger sequence of basanite lavas, and an intermediate, poorly-exposed sequence of phonolite domes and lavas. Numerous small scoria cones are also scattered along the peninsula. However, the sequence described by Wright and Kyle (1990g) is that exposed at Minna Hook, whereas, from recent detailed studies by one of us (J.L. Smellie), the true sequence at Minna Hook is more complicated (see details of Minna Hook below). Wilch et al. (2008) described the sequence at McIntosh Cliffs as dominated by subaerial lava, breccia and vent complexes but lacking the subaqueous lava lithofacies, sedimentary rocks and unconformities that are so prominent at Minna Hook. Wilch et al. (2008) also mapped more than 50 felsic domes and scoria cones on the top of Minna Bluff, together with a single tuff cone or tuff ring in the saddle between Minna Bluff and Mount Discovery. The latter is probably the 'thick sequence of cross-bedded, volcanic conglomerate and tuff' near Minna Saddle recorded by Treves (1967, p. 140). Lithofacies and stratigraphic details of the McIntosh Cliffs sequences have not yet been published and, because Wright-Grassham (1987) and Wright and Kyle (1990g) conflated the geology of Minna Hook and Minna Bluff, it is hard to detach descriptions that apply solely to Minna Bluff. It is therefore not attempted here. There are no published isotopic ages for Minna Bluff. However, of particular interest palaeoenvironmentally is the presence of extensive ice-shelf moraines abutting the north flank of Minna Bluff. They have been an important source of distinctive Eocene and Pliocene fossiliferous debris from which palaeo-ice dynamics have been deduced (Wilson 2000).

Minna Hook forms a spectacular north–south-aligned, possibly fault-bounded (Wright-Grassham 1987), east-facing cliff at the southeastern end of Minna Bluff. The cliffs are c. 12 km long and rise to 1115 m asl, although they are mainly below

700 m. Up to 1000 m of vertical section is available for study via gullies, and exposure is essentially continuous for many hundreds of metres laterally at many localities (Fig. 4f). Wright-Grassham (1987) and Wright and Kyle (1990g) suggested that Minna Hook can be divided into two informal stratigraphical units: MB1 and MB2. The basal unit (MB1) described is >600 m thick and the eruptive setting was said to vary from subaerial up into subaqueous and then subaerial conditions. The sequence began with a trachyte dome, and basanite hyaloclastite breccias and lavas. They are truncated by a prominent unconformity, which is overlain by subaqueously-emplaced lavas and lobe hyaloclastites of basanite to tephriphonolite composition. The unconformity is associated with sedimentary deposits, including volcanic sandstones, debris flow deposits and diamict with facetted clasts interpreted as tillite. It is laterally continuous and locally (in three places) glacially striated, with the striations suggesting a northwesterly ice-flow direction. The subdued topography on the unconformity suggested that it was formed subglacially and the associated glaciation was interpreted as either a palaeo-Ross Ice Shelf or a local ice cap rather than as a valley glaciation. Although the lithofacies characteristics suggested that they were associated with a major, subaerial, fluctuating ice front, a shallow-marine setting was not precluded. The subaqueous sequence is overlain by basanite–tephriphonolite scoria, agglutinate, thin lavas and minor tuff, which extend up to the ridge crest, and were erupted from multiple vents and also crop out at Minna Bluff, corresponding to the subaerial volcanic rocks in McIntosh Cliffs reported by Wilch *et al.* (2008). Unit MB2 consists of phonolite lavas and numerous small domes that crop out along the summit ridge of Minna Hook. They were succeeded by basanite–phonotephrite lavas, scoria and agglutinate, which crop out at the southern end of Minna Hook and more extensively along the crest of Minna Bluff. Isotopic dating reported by Wright-Grassham (1987) and Antibus *et al.* (2014) indicates that the volcanic activity was prolonged and took place between 11.60 ± 0.07 and 5.74 ± 0.15 Ma (Supplementary material).

More recent studies at Minna Hook by one of us (J.L. Smellie) has confirmed many of the broad details described by Wright-Grassham (1987) but the new investigation suggests a considerably more complicated volcanic history that also preserves a detailed palaeoenvironmental record of alternating glacial and interglacial conditions. At least 10 volcanic formations are currently recognized, and unpublished isotopic dating by Fargo (2008) indicates that the volcanism took place between 11.80 ± 0.80 and 8.18 ± 0.10 Ma (Supplementary material). The volcanic units comprise 'a'ā-lava-fed deltas, scoria cones and phonolite domes (Figs 9h & 18) The volcanism was essentially continuous except for gaps of *c.* 800 ka between eruptive stages 3 and 4, and between stages 5 and 6 (as depicted in Fig. 18b), both of which correspond to conspicuous erosional unconformities, with each associated locally with several metres of stratified, gravelly volcaniclastic sediments (Fig. 9j). In addition to the prominent glacial unconformity identified by Wright-Grassham (1987), which is indicated by a bold red line in Figure 18, erosional surfaces are common throughout the sequence; most are probably glacial but associated evidence, such as striations and diamict (tillite), is scarce. The section is dominated by 'a'ā-lava-fed

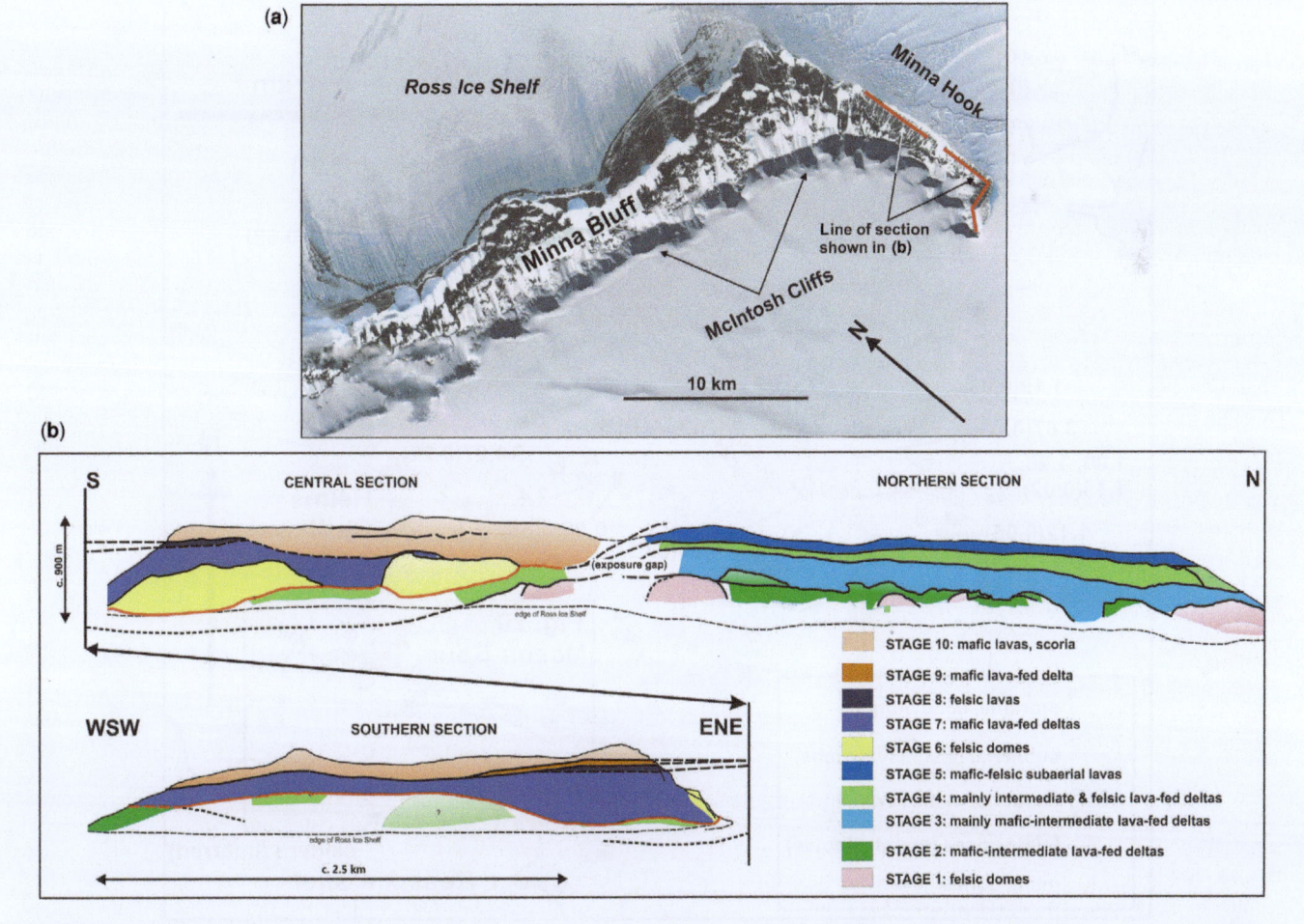

Fig. 18. (**a**) Satellite image showing the location of Minna Bluff, Minna Hook and McIntosh Cliffs. (**b**) Simplified geological sketch of Minna Hook, depicting the products of at least 10 eruptive phases (unpublished information of J.L. Smellie). Most of the boundaries between the units represent periods of erosion. The boundary coloured red is the glacial surface identified by Wright-Grassham (1987).

deltas with a range of mafic–felsic compositions and well-developed subaqueous lithofacies (lava-lobe-rich breccia: *sensu* Smellie and Edwards 2016) (Fig. 9f, g). They have dipping passage zones signifying that they were emplaced in successive ice sheets that draped the Minna Hook edifice (cf. Smellie *et al.* 2011*b*). A glacial environment for the lava-fed deltas is also suggested by the compositions of the authigenic mineral phases (Antibus *et al.* 2014; cf. Johnson and Smellie 2007). The individual lava-fed deltas are relatively thin and indicate coeval ice thicknesses (draping the volcano) of up to *c.* 300 m and mostly <200 m (unpublished information of J.L. Smellie). The subaerial sections are dominated by 'a'ā lavas with local scoria and agglutinate; pāhoehoe lava is very rare. Rock compositions include basanite, phonotephrite, tephriphonolite, phonolite and trachyte, with rare mugearite and benmoreite (Wright-Grassham 1987; Scanlan 2008; unpublished information of K.S. Panter). Contrasting passage-zone dip directions in different eruptive units suggest that the section may be the highly eroded remnant of more than one overlapping, probably shield-like volcanoes, which have now been substantially reduced in size by a combination of faulting (to give the linear outcrop: Wright-Grassham 1987), and glacial and marine erosion.

Mount Morning Volcanic Field

The Mount Morning Volcanic Field includes outcrops at Mount Morning and Mason Spur. Mount Morning is a composite volcanic edifice that was constructed in two major stages. The oldest stage ('palaeo-Mount Morning') is exposed only on the NNE lower slopes, on Riviera Ridge and Gandalf Ridge. The outcrops are the remnants of a volcano that erupted between 18.7 ± 0.3 and 13.0 ± 0.3 Ma (Wright-Grassham 1987; Kyle and Muncy 1989; Wright and Kyle 1990*h*; Martin *et al.* 2010) (Fig. 19; Supplementary material). Very little of the palaeo-Mount Morning volcano is preserved, and its original diameter and morphology (shield or stratovolcano) and environmental eruptive context are unknown.

The palaeo-Mount Morning volcano at Gandalf Ridge is composed of a north-dipping sequence at least 150 m thick, composed of highly disrupted mugearite and trachyte lavas (Martin and Cooper 2010). This sequence has been interpreted as a diamictite, with the degree of disruption increasing in the top 20 m. The sequence is intruded by numerous trachytic–rhyolitic dykes. Multiple generations of faults, striking approximately north–south parallel with the trend of the Transantarctic Mountain Front in this region, cross-cut Gandalf Ridge. A hawaiite dyke, dated at 3.88 ± 0.05 Ma, is offset *c.* 6 m by one of these faults, indicating Plio-Pleistocene movement (Martin and Cooper 2010). The diamictite sequence unconformably overlies *in situ* outcrops of basement rocks composed of granite with screens of metasedimentary schists (Kyle and Muncy 1989) that have been linked to Late Precambrian–Cambrian Ross Orogen rocks (Martin and Cooper 2010). Occurrences of volcanic rocks in contact with basement are rare for any of the large volcanoes in Victoria

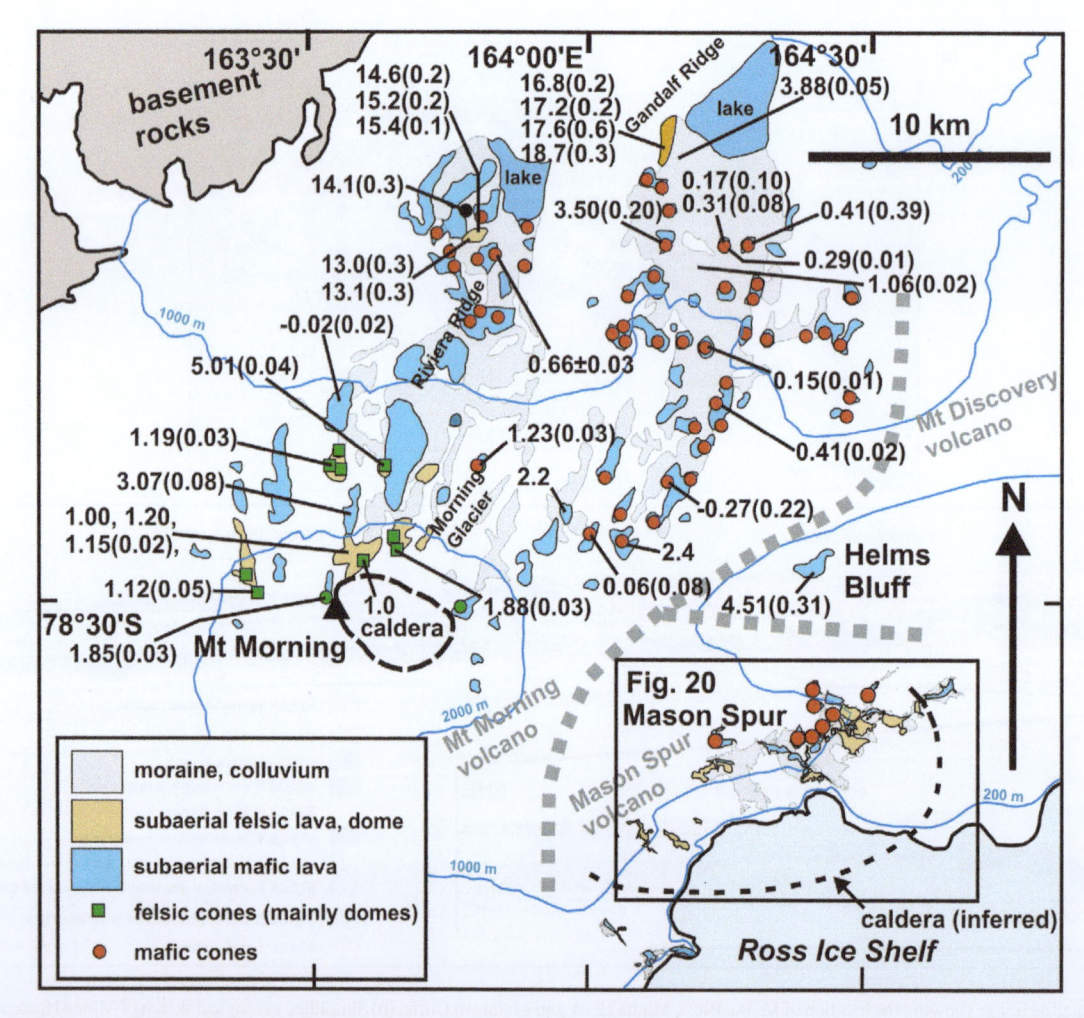

Fig. 19. Geological map of Mount Morning and Mason Spur (after Martin *et al.* 2010; Chapter 5.2b – Martin *et al.* 2021). Isotopic ages (in Ma) are after Wright-Grassham (1987), Kyle and Muncy (1989), Paulsen and Wilson (2009) and Martin *et al.* (2010); errors in parentheses (Supplementary material).

Land as a whole, the only other example being on the Daniell Peninsula in northern Victoria Land (Hamilton 1972; see also Chapter 5.1a: Smellie and Rocchi 2021). Other compositions, including alkali basalt, mugearite and benmoreite, are also present.

At least four, distinctive, eroded remnants of volcanic plugs crop out near present-day sea level on the lower slopes of Riviera Ridge (Treves 1968; Wright-Grassham 1987; Kyle and Muncy 1989; Martin 2009). The southernmost, best-preserved intrusion is over 50 m high and 100 m wide. Large (>1 m) mafic clasts have been entrained. Fluidal structures reminiscent of fiammé are observed and these units have a trachyte composition. Clastic dykes cross-cut the outcrop with irregular, bifurcating morphologies that may be squeezed into cracks or pinch out abruptly. The trachyte plugs are cross-cut by late-stage dykes of both mafic and felsic composition. The dyke margins are sharp and curvi-planar with thermal alteration at some dyke margins ≤ 5 cm wide. No peperite or hyaloclastite has been observed. The dykes are 0.5–2 m across and some bifurcate. They can be traced for the entire length of some plugs, a maximum of 100 m, whilst elsewhere they pinch out. There are two main orientations of the dykes: a NW-striking set (averaging 329°) and a NNE-striking set (averaging 009°). NNE-striking faults offset some of the NNW-striking dykes (Wright-Grassham 1987).

Syenite xenoliths, similar in composition and mineralogy to trachyte lavas of palaeo-Mount Morning, are found in basanite lavas and phonolitic pyroclastic deposits that crop out widely on the much younger Mount Morning shield volcano. Martin *et al.* (2010) inferred from this that palaeo-Mount Morning magmatism extended over a comparable and equivalent area to the modern Mount Morning edifice.

By contrast, Mount Morning itself was constructed during a second, much younger, stage and dominates the local landscape. It is a largely undissected shield volcano with slopes that average 10° (Wright-Grassham 1987). The volcano has a somewhat elliptical shape with a basal diameter of 36 km (SW–NE) and 18 km (NW–SE), and rises to 2723 m asl. It is therefore one of the largest volcanoes in the Erebus Volcanic Province, exceeded only by Mount Erebus and Mount Terror. Like Mount Discovery, rocks are only exposed on the north and NE flanks, and the south-facing slopes are entirely covered by snow and ice. A summit caldera measuring *c*. 4 km in diameter is filled and entirely obscured by ice. It is slightly elliptical with a principal axis aligned NW–SE (Paulsen and Wilson 2009). The caldera rim is slightly lower on its NE side, where it may be breached and is the source for a short glacier (Morning Glacier) that extends down to *c*. 1200 m asl (Fig. 19). Outcrops above *c*. 1500 m asl are mainly phonolitic domes and lavas, and basanite and tephrite lavas; below that elevation only basanites and tephrite lavas are present, together with scoria cones (Wright-Grassham 1987; Wright and Kyle 1990*h*; Martin *et al.* 2010, 2013). Together, they comprise a shield-building phase with ages ranging between 5.01 ± 0.04 Ma and near present day (−0.02 ± 0.02 Ma: Tauxe *et al.* 2004; Paulsen and Wilson 2009; Martin *et al.* 2010) (Supplementary material). Morphologically young scoria cones are generally 50–300 m high. The associated lavas are 'a'ā, typically 2–10 m thick and extend for up to 5 km in length. A study of the distribution and morphology of the scoria cones on Mount Morning by Paulsen and Wilson (2009) showed that they were erupted mainly from NE-trending fissures, with a subordinate northwesterly trend. These trends are related to regional stresses associated with a rift basin margin, consistent with a normal-fault to strike-slip neotectonic regime associated with Terror Rift extension (Fig. 5).

Volcanic products of Mount Morning have partially buried the northwestern flank of the Mason Spur volcano (Fig. 19). Mason Spur is a rugged scarp, aligned WSW–ENE, that is *c*. 10 km long and reveals a discontinuous sequence up to 1000 m thick, exposed in numerous bluffs and ridges. It was initially examined by Wright-Grassham (1987; also Wright and Kyle 1990*i*), who divided the sequence into nine eruptive units labelled MS1–MS9. Wright-Grassham (1987) also produced a detailed geological map, and described a complicated succession of alternating mugearite to trachytic breccias, lavas and domes (units MS1–MS5) which dominate the lower part of the outcrop and contain angular unconformities. An upper sequence (MS6–MS9) was also identified, consisting of trachyte pyroclastic deposits, a dome and lavas, followed by basanite–tephrite scoria cones and lavas. Isotopic dating has shown that, in general, this sequence youngs upwards, was constructed between 12.9 ± 0.1 and 11.4 ± 0.1 Ma, and is cross-cut by numerous generations of dykes (Wright-Grassham 1987; Martin *et al.* 2010) (Supplementary material; Fig. 20). Mafic scoria cones at high elevations have much younger ages of 0.23 ± 0.22 and 0.07 ± 0.08 Ma (Paulsen and Wilson 2009).

Recent unpublished studies by the authors have significantly modified the geological history described by Wright-Grassham (1987). In particular, at least 11 stratigraphical sequences can be recognized, several of which are newly distinguished. This new work combines units MS1–MS5 of Wright-Grassham (1987) into one basal unit, *c*. 800 m thick. It is the infill of a very large caldera with an original basal diameter of *c*. 10 km (Fig. 20). It is largely massive and is composed of trachytic, pyroclastic density current deposits, including breccias and a distinctive and visually spectacular spatter-rich ignimbrite (Martin *et al.* 2018) (Fig. 9d, e). The deposits are pervasively hydrothermally altered and affected by numerous trachyte hypabyssal intrusions, many with fluidal shapes indicating that they were coeval with the pyroclastic material (Fig. 9i). The caldera-filling unit is flanked to the SW and NE by undated extra-caldera mafic lava-fed deltas that are likely to be the oldest-exposed erupted units in the volcano. The caldera fill is overlain by trachyte lava(s). This sequence was then eroded into ravines which became filled by trachytic epiclastic deposits composed of breccio-conglomerate and sandstone capped by white trachyte ignimbrites, measuring *c*. 150 m in total thickness. Above that, the succession consists mainly of basanite lavas showing evidence for both subaerial and subaqueous (possibly subglacial) emplacement but there are also two further trachyte units, a tuff cone and a prominent summit dome *c*. 900 m in diameter. The youngest units present are a series of basanite scoria cones, several with well-preserved craters consistent with the published young isotopic ages (Fig. 20; Supplementary material). Despite inferences that many eruptions at Mason Spur took place during glacial periods (Wright-Grassham 1987; Martin *et al.* 2010), supporting evidence is uncommon and the implicated deposits are volumetrically minor.

Southern Local Suite. The Southern Local Suite is a new stratigraphical term introduced by Smellie and Rocchi (2021 – Chapter 5.1a of this Memoir) for numerous small isolated outcrops of mafic scoria cones clustered principally in two well-defined areas: (a) the southern foothills of the Royal Society Range; and (b) the Taylor and Wright valleys of the Dry Valleys region (Figs 21 & 22).

Outcrops in the Royal Society Range southern foothills have been examined at reconnaissance level by several workers (McIver and Gevers 1970; Wright 1980; Wright-Grassham 1987; Wright and Kyle 1990*j*; Lawrence *et al.* 2009; Cox *et al.* 2012 and references therein). The most informative studies are by Wright (1980) and Wright-Grassham (1987), in which the outcrops are described and geological sketch maps are provided for many. About 50 vents are present in the area,

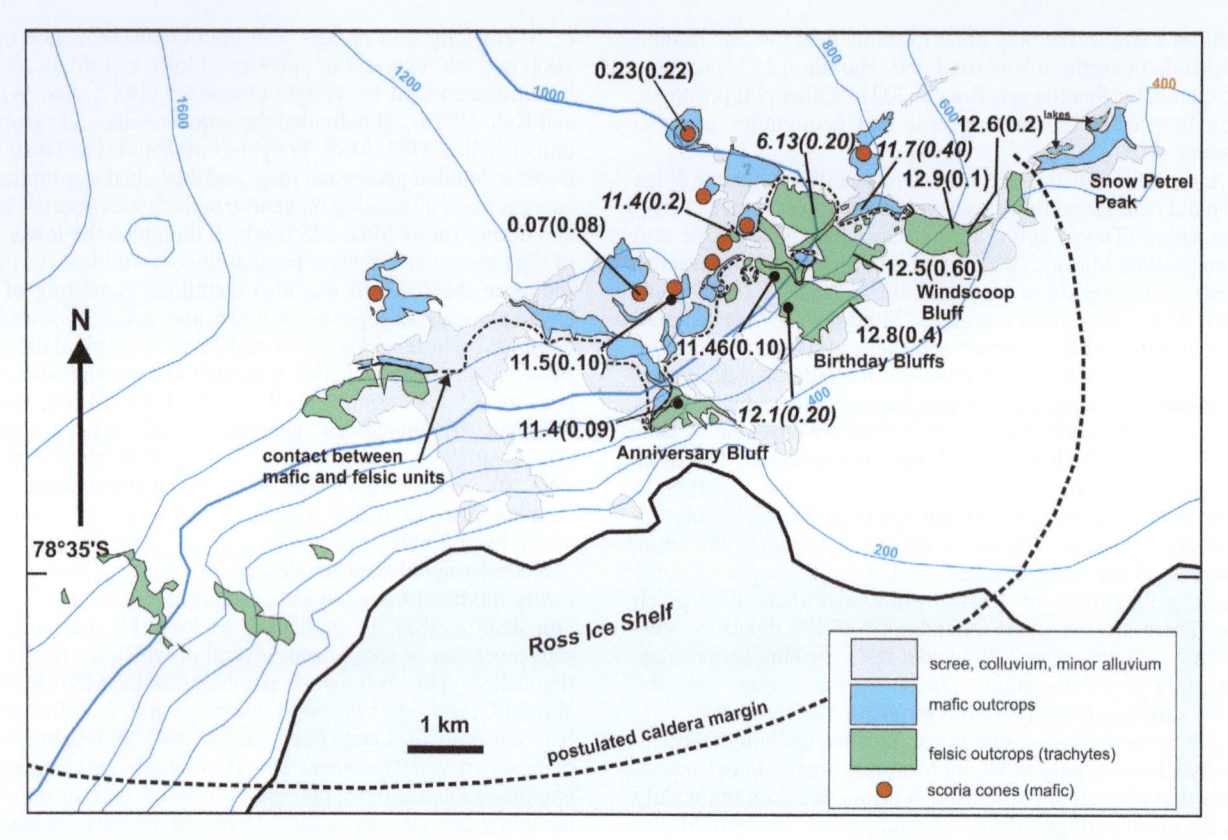

Fig. 20. Simplified geological map of Mason Spur (from unpublished information of J.L. Smellie). Isotopic ages (in Ma) are from Wright-Grassham (1987), Paulsen and Wilson (2009) and Martin *et al.* (2010); errors in parentheses (Supplementary material).

represented overwhelmingly by scoria cones (Figs 4g & 21). Although it would traditionally be regarded as a monogenetic volcanic field, a few of the volcanic cones show evidence for multiple eruptions spaced in time, and representing pre-, syn- and post-glacial activity (i.e. polygenetic: Wright 1980). The majority of the cones were erupted subaerially, and consist of scoria, agglutinate and lavas. Compositionally, they are formed of basanite, with a single occurrence of hawaiite (Miers Valley outcrop: Wright and Kyle 1990*j*) (Fig. 21). Ultramafic xenoliths are present locally, and have been subjected to detailed petrological studies (Gamble and Kyle 1987; Gamble *et al.* 1988; Berg 1991; Kalamarides and Berg 1991). Many outcrops are eroded to their core and are represented by a central basanite plug up to 30 m in diameter with only minor surrounding scoria, and a few are draped by loose till. The outcrops range in size from tiny isolated scoria mounds just 10 m high to large cones up to 300 m high and containing craters up to 1 km in diameter (e.g. Foster Crater: Wright 1980). Most of the larger cones are breached by 'a'ā lava flows. Within the craters, the lavas are thin (typically <1 m) but beyond the cone they are generally 3–10 m thick (see illustrations in Cox *et al.* 2012). They typically extend 0.5–4 km from their source and rarely reach 10 km. In a few cases, lava has rafted a sector of the cone a distance of a few hundred metres. Successions of superimposed lavas are present locally: for example, a 30 m-thick sequence east of the Walcott Glacier terminus, which is composed of at least five lavas. They are inferred to have formed successively by being dammed and banked up against a previously thicker coeval Walcott Glacier (Wright 1980). Wright (1980) also described lavas that crop out as prominent benches on valley walls, ascribed to confinement and diversion of the lavas by glaciers that once occupied those valleys. These may be some of the first published examples of what have come to be called ice-impounded or perched glaciovolcanic lavas (Lescinsky and Sisson 1998; Harder and Russell 2007; Smellie and Edwards 2016). On the basis of sparse evidence, it was also postulated that some outcrops were formed by supraglacial eruptions (i.e. eruption onto a glacier or through a glacier with the edifice constructed on the glacier). The well-formed cone in the Pipecleaner Glacier where it joins the Radian Glacier was cited as an example. However, the feature still exists today and is unmodified since it was photographed almost 40 years ago, which, if the glacier is mobile, makes the interpretation suspect. Other examples include a lava that flowed onto ice and is now preserved as a mantle of debris on bedrock. That example can also be criticized since the preservation of the lava debris would require the coeval ice to be stagnant and to decay *in situ*, which may be unrealistic on a sloping topography. However, examples of lava debris lying on high-level benches in Pyramid Valley, above Trough Lake and up-valley from those locations, suggest that the process may have occurred, although there are no descriptions of the physical characteristics of the debris with which to test the idea (D. Townsend pers. comm.). Rare outcrops show evidence for water interaction, comprising pillow lava, hyaloclastite and likely lapilli tuff, and were presumed to be caused by subglacial eruption. However, no tuyas (i.e. flat-topped, steep-sided glaciovolcanic landforms: cf. Smellie and Edwards 2016) were constructed, probably because of the small volumes of magma extruded and inferred thin coeval ice. Finally, D.N.B. Skinner (cited in Wright 1980) observed a unique occurrence of a dyke that may have been intruded englacially and is now associated with hyaloclastite and pillow lava.

The southern foothills outcrops are extensively dated (Armstrong 1978; Lawrence *et al.* 2009) (Fig. 21; Supplementary material). The ages range mainly between 2.88 ± 0.15 and 0.08 ± 0.13 Ma but at least four outcrops have significantly older ages (13.8 ± 0.20, 13.42 ± 0.18, 13.2 ± 0.40; $12.70 \pm 0.09/12.61 \pm 0.11$ and 5.70 ± 0.15 Ma), indicating that it

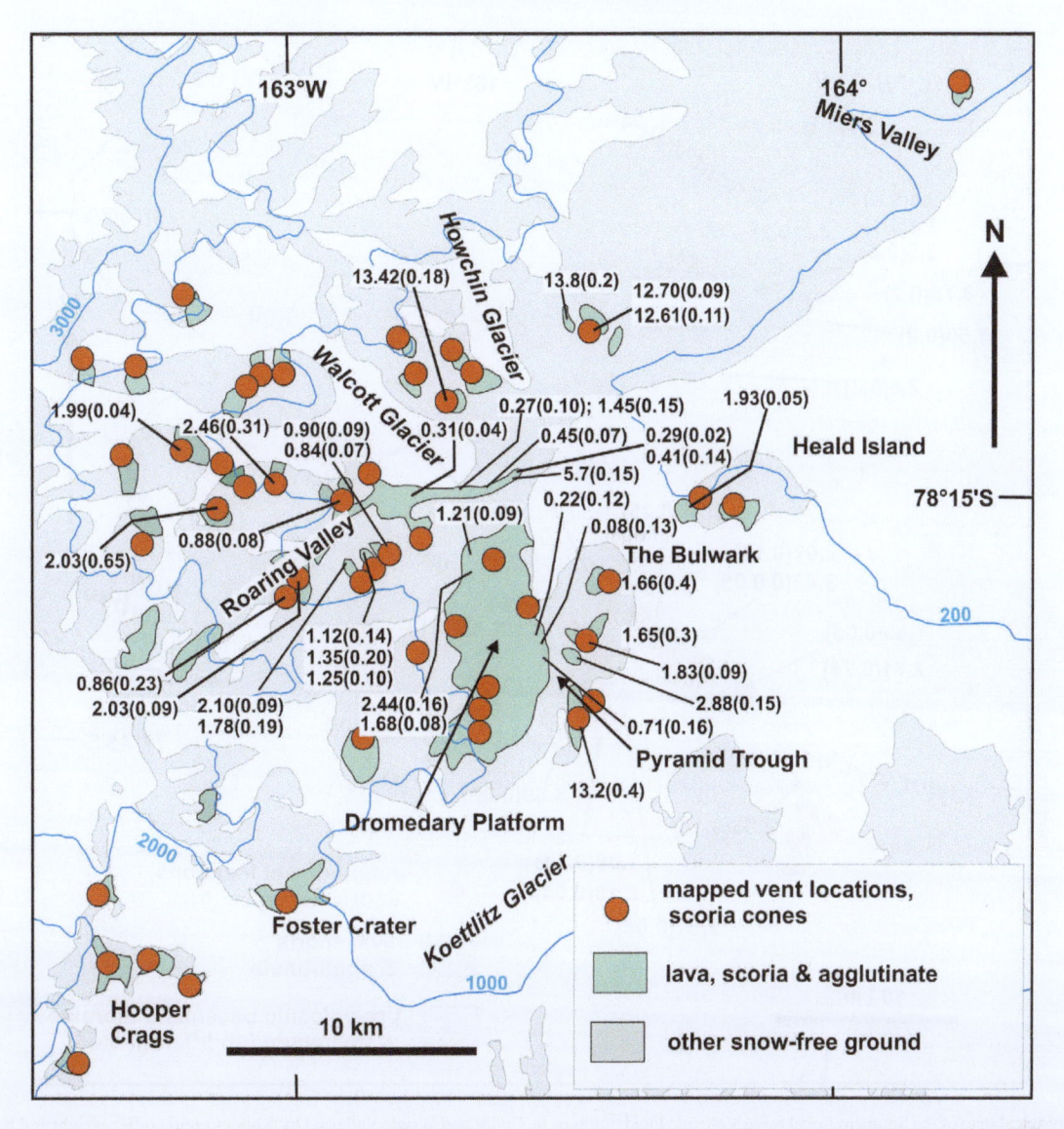

Fig. 21. Geological map of the southern foothillls of the Royal Society Range, showing the distribution of Southern Local Suite Volcanic Field outcrops (after Wright-Grassham 1987). Isotopic ages (in Ma) are after Armstrong (1978) and Lawrence *et al.* (2009); errors in parentheses (Supplementary material).

was a very long-lived volcanic field. On the basis of the youngest ages, Wright and Kyle (1990j) considered that the volcanic field may not be extinct.

Outcrops of the Southern Local Suite in the Dry Valleys region of southern Victoria Land are situated on the valley sides and floors of the Taylor and Wright valleys (Fig. 22). Wright and Kyle (1990k) estimated that there are about 30 scoriaceous basanite mounds up to 60 m high in Taylor Valley, each probably representing an eruptive centre, whilst Wright Valley contains several mafic cones on its south side. Both valleys also have several reported occurrences of windblown and waterlain ash deposits. The outcrops in Wright Valley remain to be mapped in detail and the only systematic study published is for examples in Taylor Valley by Wilch *et al.* (1993), who mapped more than 70 volcanic outcrops. Most of the outcrops consist of eroded cone remnants; pristine cones with well-preserved craters are also present but rare. The outcrops are composed of black scoria, which is red (oxidized) close to the eruptive vents, together with spatter-fed clastogenic lavas (Fig. 23); some deposits were erupted from fissures. All of the eruptions were subaerial and, unlike in the Royal Society Range, there is no evidence for subaqueous or subglacial conditions. Curiously, despite the erosion noted, there is a lack of volcanic debris preserved in the Taylor Valley surface deposits. The Taylor Valley occurrences have been used to provide a robust estimate for the minimum amount of uplift that has taken place in the Dry Valleys crustal block, which was shown to be <300 m since 2.57 Ma (Wilch *et al.* 1993). The locations of the scoria cones, and whether they are overlain by ice today, also indicate the maximum sizes of adjacent glaciers at the time of each eruption (also Fleck *et al.* 1972).

As for outcrops in the southern foothills of the Royal Society Range, multiple isotopic analyses are available for outcrops in the Dry Valleys region (Armstrong 1978; Wilch *et al.* 1993) (Supplementary material). They range from 4.2 ± 0.2 to 2.5 ± 0.3 Ma in Wright Valley, and from 4.64 ± 0.12 to 1.50 ± 0.05 Ma in Taylor Valley. The volcanic field is considered to be extinct (Wright and Kyle 1990k).

Terror Rift Volcanic Field

Volcanism also occurs offshore in the western Ross Sea, associated with what is called the Terror Rift. The rift is a tectonic feature *c.* 70 km in width superimposed on the central Victoria Land Basin, and the Terror Rift Volcanic Field includes all the volcanoes that crop out within the margins of that basin (Fig. 2). The generation of the volcanism may be principally related to decompression associated with extension in the

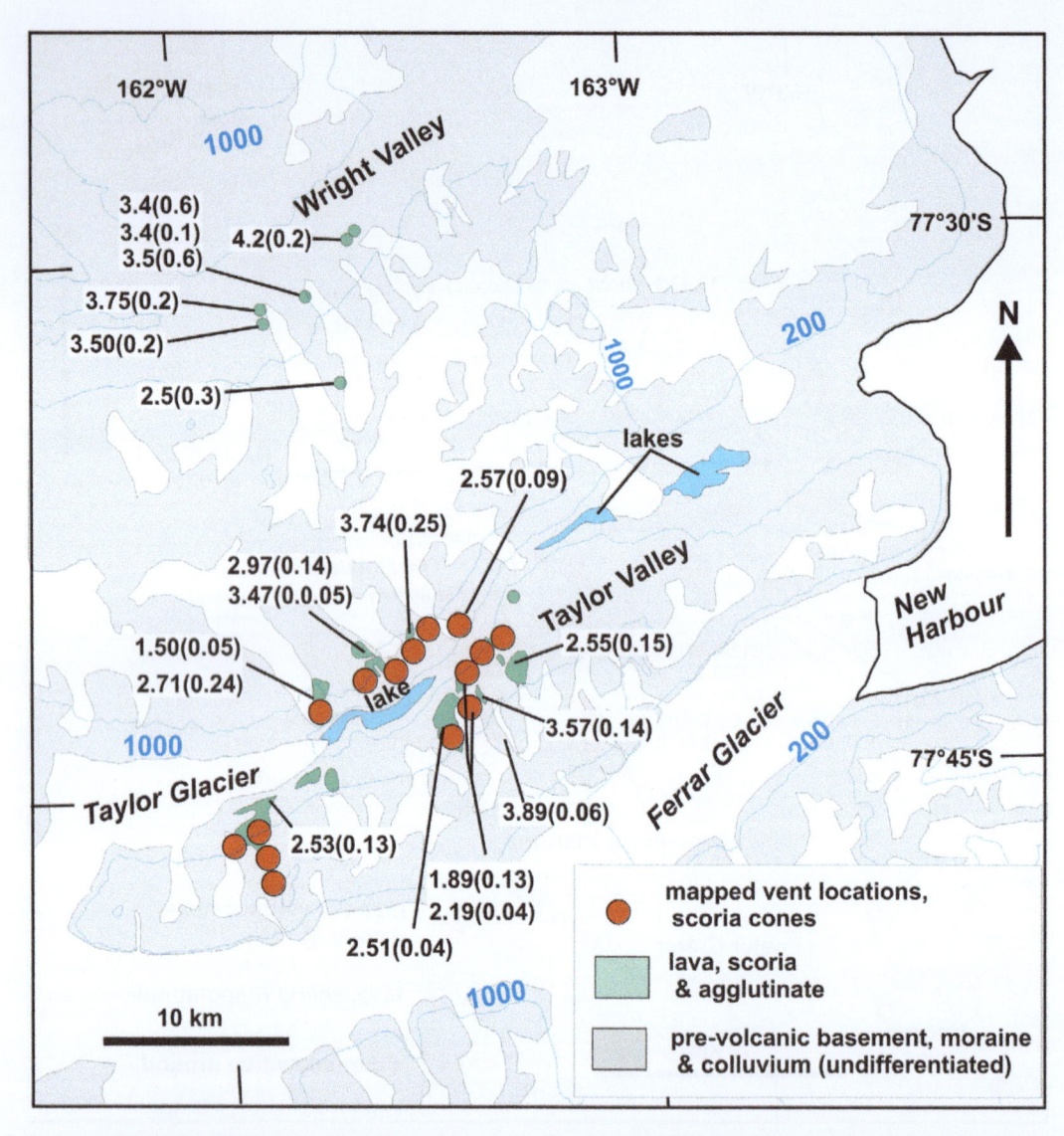

Fig. 22. Geological map of the Southern Local Suite Volcanic Field outcrops in Taylor and Wright valleys, Dry Valleys region (after Wright and Kyle 1990k; Wilch *et al.* 1993). Isotopic ages (in Ma) after those authors and Fleck *et al.* (1972); errors in parentheses (Supplementary material).

rift (Rilling *et al.* 2009). The rift extends for over *c.* 250 km between Mount Erebus and Mount Melbourne. It is truncated in the north by the NE-striking Polar 3 Transfer Fault that dextrally offsets the Victoria Land and Adare basins, and possibly in the south by the Ross Fault, which may also be a transfer fault (Behrendt *et al.* 1996; Chiappini *et al.* 2002; Ferraccioli *et al.* 2009; Davey *et al.* 2016), although other work suggests that it may continue south of the Ross Island Volcanic Field (Johnston *et al.* 2008). The main phase of deformation was probably during the mid–late Miocene, with further less pervasive faulting in the Plio-Pleistocene (Rilling *et al.* 2009; Lawver *et al.* 2012). The rift is characterized by abnormally high heat flow, thin crust and anomalously slow upper-mantle seismic velocities, and is interpreted as the zone of most recent deformation associated with the West Antarctic Rift System (Cooper *et al.* 1987; Della Vedova *et al.* 1997; Salvini *et al.* 1997; Hall *et al.* 2007; Lawver *et al.* 2012; Davey *et al.* 2016). Although the precise kinematics are not well known, the presence of faults cutting the seafloor suggests that faulting and extension may be continuing to present (Salvini and Storti 1999; Rossetti *et al.* 2006; Hall *et al.* 2007).

Only two volcanoes are subaerially exposed in the Terror Rift Volcanic Field: Franklin Island and Beaufort Island. Conversely, there are numerous submarine centres, although the full number and their distribution are unknown and more will undoubtedly be discovered than are shown in Figure 2. Franklin Island is a small island *c.* 12 km long (north–south) and 4 km wide. It rises to 247 m asl and is mostly covered by snow and ice, with exposed rock restricted to cliffs in the south (Bernacchi Head) and along its east coast. It is the subaerial expression of a large, submerged basanite shield volcano, which is at least 20 km long and 8 km wide; it occurs in water depths of *c.* 500–600 m (Lawver *et al.* 2012). In addition to a pronounced north–south elongation, the edifice has much steeper flanks in the lower 200 m. Little is known about the geology. On the basis of a brief visit in 1983, Ellerman and Kyle (1990*a*) described gently-dipping interbedded basanite lavas and yellow tuffs; the lavas are thicker towards the base of the section and bedding attitudes are locally chaotic. Images of the island published online at http://iceblog.puddingbowl.org/archives/2004/01/franklin_island.html show that the tuffs are mainly khaki-grey thinly stratified lapilli tuffs with abundant ultramafic nodules; Ellerman and Kyle (1990*a*) also described cross-bedding. Although it was suggested that the lapilli tuffs dip to the east into the crater of a postulated parasitic cone situated *c.* 1 km north of Bernacchi Head, presumably a tuff cone or tuff ring, no crater structure is obvious in the images available. The images also show that the eastern cliffs are chaotically multicoloured in yellows, reds, browns

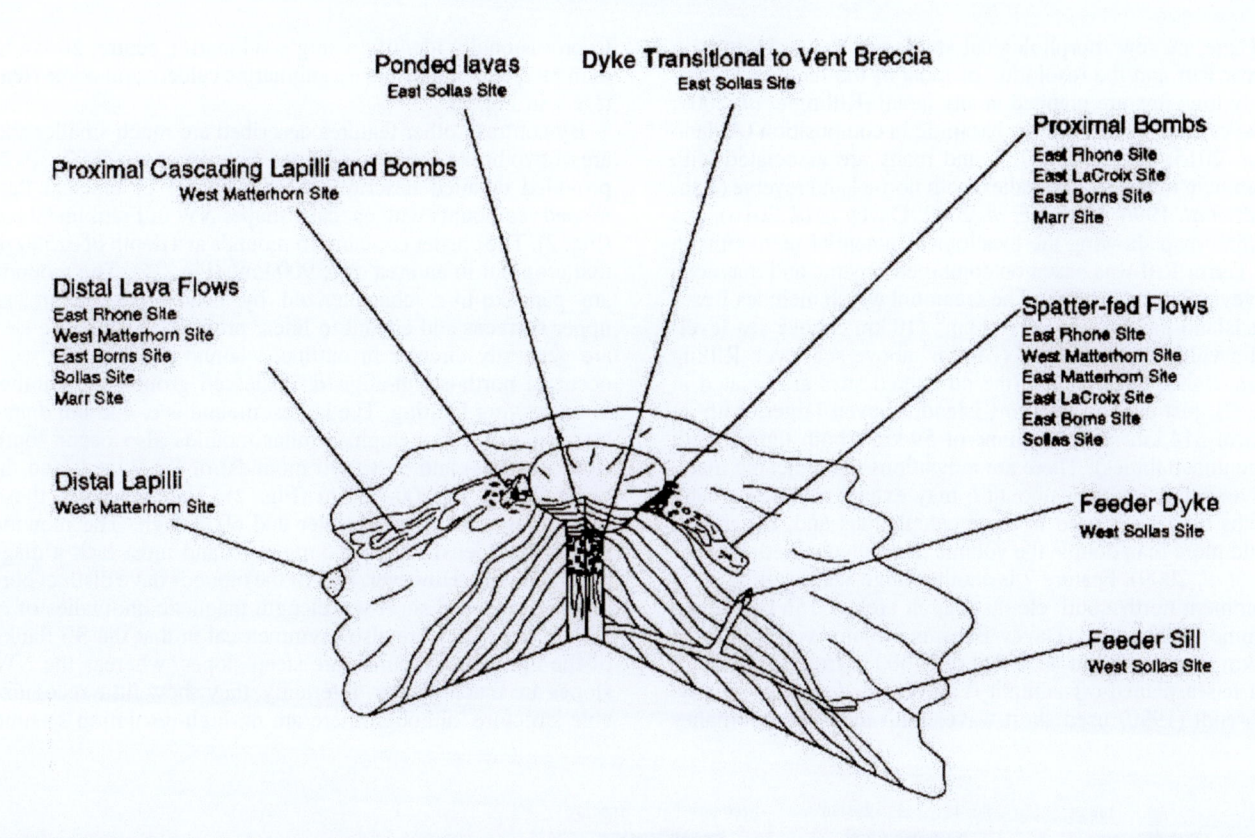

Fig. 23. Generic sketch of a the internal structure and lithofacies of a typical scoria cone from Taylor Valley, Dry Valleys region, showing the main lithofacies present and listing the localities where they are well preserved (after Wilch *et al.* 1993).

and greys, suggesting that they may be affected by hydrothermal alteration, perhaps close to a vent. Bedded scoria or clinkers may overlie the uppermost lavas, and glacial erratics occur up to within 20 m of the summit (Ellerman and Kyle 1990*a*). Overriding and erosion by north-flowing ice might also explain the north–south-elongated morphology of the submarine extension of the volcano. Armstrong (1978) published an imprecise isotopic age of 4.8 ± 2 Ma (by K–Ar), whereas ages of 3.70 ± 0.16–3.30 ± 0.4 Ma by Ar–Ar were reported by Rilling *et al.* (2007, 2009) (Supplementary material).

Beaufort Island is *c.* 4.5 km long and up to 2.5 km wide. It is 771 m high and occurs in water depths of *c.* 500–600 m (Lawver *et al.* 2012). The island, regarded as a stratovolcano remnant (Ellerman and Kyle 1990*b*), is the subaerial expression of a submarine edifice that is much smaller than that at Franklin Island. It is *c.* 12 km in diameter and more or less symmetrical, with Beaufort Island situated on the west side; another larger, wholly submarine edifice is present close by to the east (feature 6 in Fig. 2). The island has a lunate shape which is convex to the east. It is largely snow and ice covered on its western side but prominent arcuate rocky cliffs form the entire east coast; less prominent lower cliffs also occur on the west side (Fig. 24). On the basis of binocular observations, Harrington (1958) suggested that the basal 70 m of the eastern cliffs are composed of variously coloured (grey, brown and red) tuff and 'agglomerate' (breccia?) layers. Bedding appeared almost horizontal but is locally chaotic. The description is very similar to that for Franklin Island. The central part of the cliff was described as poorly-bedded coarse 'agglomerate' (breccia?) that dips to the south at ≥12°. The agglomerate is up to 170 m thick but diminishes to <10 m in a northwesterly direction. It unconformably overlies the basal sequence, and is conformably overlain by pale-coloured tuff and 'agglomerate' (breccia?). The sequence is capped by *c.* 100 m of thin grey blocky lavas (probably 'a'ā) that may also be interbedded with scoria. A small but prominent plug of grey basaltic rock is present at the northwestern end of the island. Numerous dykes are also present and may have caused the hydrothermal alteration suggested by the colours of the host rocks. Ellerman and Kyle (1990*b*) suggested that bedding dips were predominantly to the west and NW, with lavas interbedded throughout the sequence and becoming thinner upwards. Because of the evidence for erosion, Harrington (1958) suggested that the age of the volcano was pre-last glaciation (i.e. >40 ka). However, he believed that the northwestern slopes of the volcano were relatively little modified and implied that the island was probably never completely overridden by ice, ascribing most of the erosion to marine processes. Erratics occur up to *c.* 320 m asl (Denton *et al.* 1975). Erupted compositions include tephrite, phonotephrite, tephriphonolite and phonolite (Ellerman and Kyle 1990*b*; Aviado *et al.* 2015). Isotopic ages of 6.80 ± 0.05 and 6.77 ± 0.03 Ma by Ar–Ar were published by Rilling *et al.* (2009) (Supplementary material).

Fig. 24. View of Beaufort Island, looking east (image: JL Smellie).

There are few morphological studies of the seamounts in Terror Rift and the resolution of most of the features is relatively low; few are mapped in any detail (Rilling et al. 2009; Lawver et al. 2012). All are basanitic in composition (Aviado et al. 2015; Lee et al. 2015) and many are associated with prominent magnetic anomalies, both normal and reverse (Behrendt et al. 1996; Rilling et al. 2009; Davey et al. 2016), and the first map showing the locations of potential seamounts in the Terror Rift was based on combined seismic and magnetic surveys (Behrendt 1990). The seamount which includes Beaufort Island has an area of 169 km^2 (10 km^2 above sea level) and a volume of 42.3 km^3 (2.0 km^3 above sea level: Rilling et al. 2009). Seamount 6 (the numbered sites are located in Fig. 2), just east of Beaufort Island, is even larger, with an area of 314 km^2 and a volume of 59 km^3 (both figures with large uncertainties). There are indications that volcanic material associated with seamount 6 may extend to much greater depths than were used in these calculations and, if verified, could more than double the volume of erupted materials (Rilling et al. 2009). Feature 7 is another large seamount. It has a prominent north–south elongation, an area of 156 km^2 and a volume of 22.3 km^3. Davey Bank is a major volcanic ridge 18 km long and 3 km wide that rises 600 m from the seafloor and has a planed-off summit (Lawver et al. 2012). Finally, Behrendt (1990) used short-wavelength magnetic anomalies to provisionally identify a single submarine centre, 20 km in diameter, which may have a submarine caldera and dome (feature 1 in Fig. 2).

By contrast, other features described are much smaller and are of two broad contrasting types. Lawver et al. (2007, 2012) provided detailed descriptions of a cluster of unusual flat-topped seamounts with pancake shapes NW of Franklin Island (Fig. 2). The cluster contains 15 mounds at a depth of c. 500 m that crop out in an area of c. 900 km^2 (Fig. 25). The mounds are pancake-like, characterized by nearly flat featureless upper surfaces and equant to linear profiles; those in the east are generally circular in outline, whereas those in the west occur in north–south-aligned, coalesced groups sub-parallel to underlying faulting. The largest mound is c. 4 km in diameter but just 100 m high. Similar mounds also occur south (four mounds) and east (two mounds) of Franklin Island, in water depths of 500–650 m (Fig. 2). The largest of those mounds is 2.5 km in diameter and 60 m high. The mounds have a low density (2.2–2.6 mg m^{-3}) and most lack a magnetic anomaly. However, four of the mounds have distinct normal and reversed short-wavelength magnetic anomalies of c. 50–100 nT. They are also asymmetrical in that the SE flanks of the individual mounds are steep slopes, whereas the NW slopes are much gentler. Internally, they show little recognizable structure, although there are no high-resolution seismic

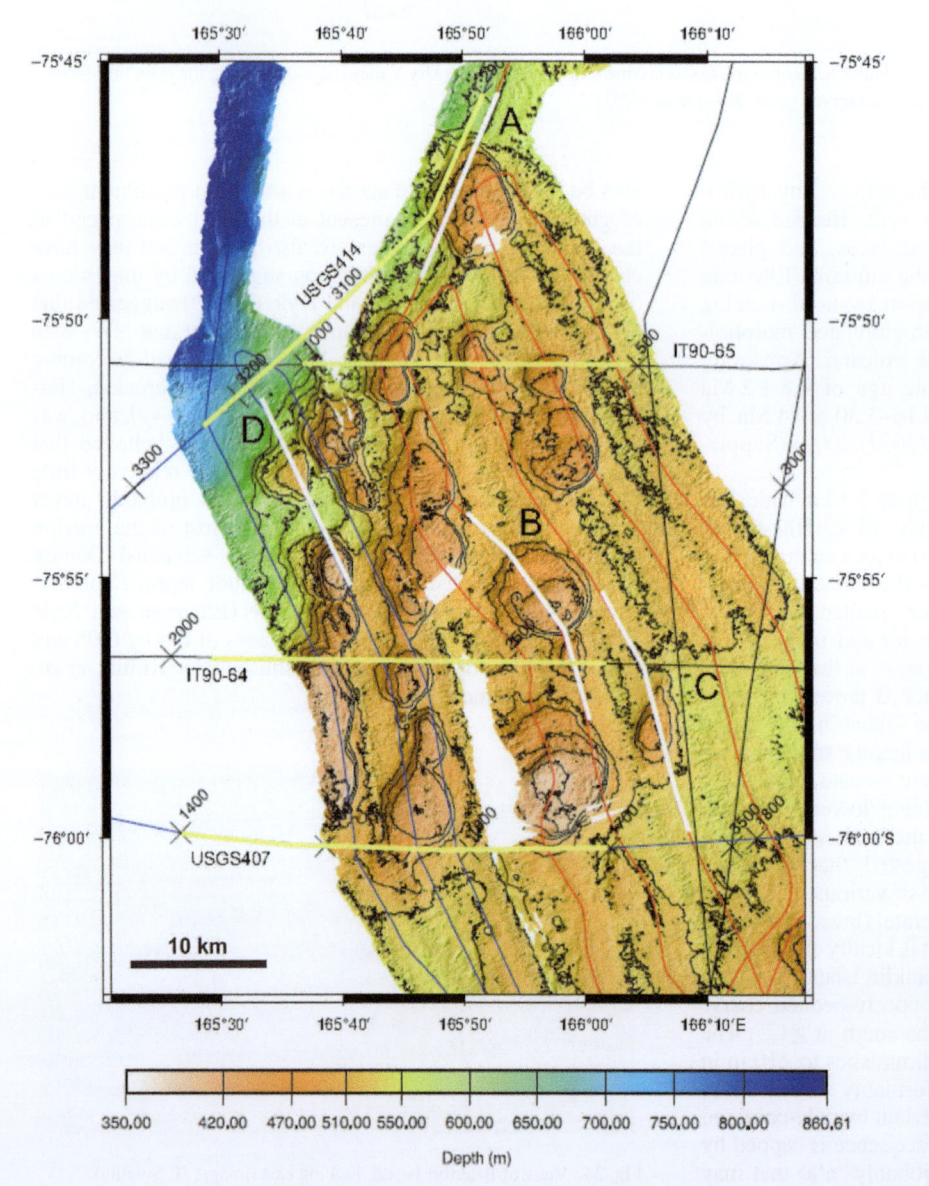

Fig. 25. Swath bathymetry map of central Victoria Land Basin northwest of Davey Bank, showing unusual pancake-shaped, possibly volcanic mounds on the seafloor west of Franklin Island. Contour interval is 25 m. After Lawver et al. 2012. The location of the image is shown in Figure 2.

data available for them. Although an origin as carbonate mounds is possible, the preferred origin for the mounds was as subglacially erupted volcanoes composed mainly of explosively-generated 'hyaloclastite' (i.e. lapilli tuff of White and Houghton 2006). The short-wavelength magnetic anomalies in some of the mounds may represent either a subaerial lava cap or, more plausibly, a pillow-lava basal pile (Lawver et al. 2012). No isotopic ages are available for the mounds but a possible age of c. 800 ka (when the West Antarctic Ice Sheet was much expanded) was suggested. However, there are problems with the glacial interpretation (acknowledged by Lawver et al. 2012). For example, the tops of the mounds would have been under several hundred metres of ice (>c. 400–550 m), whereas explosive hydrovolcanic eruptions are less likely to occur at water depths exceeding 100 m (i.e. c. 110 m of ice); most estimates for the transition to effusive volcanism range between 100 and 200 m (e.g. Zimanowski and Büttner 2003; Schopka et al. 2006). Although explosive eruptions can occur even below ice thicknesses of 500–700 m (Gudmundsson et al. 1997; Schopka et al. 2006), they require special conditions: for example, subglacial meltwater drainage connecting to an ice front and discharging at atmospheric pressures, thus reducing the ambient pressures over the vent – a situation that is hard to envisage for the Terror Rift Volcanic Field examples. In addition, the edifices have extremely low profiles (i.e. they are pancake-shaped), whereas pristine subglacially erupted edifices are characteristically tall (Smellie 2009, 2013; Smellie and Edwards 2016). A possible reason for the low profiles is that the edifices have been largely removed by overriding ice shortly after eruption (cf. Behrendt et al. 2004), which would imply a much older age (with more erosional periods) and also help to explain the combination of steep southerly (stoss) sides, gentler northerly (lee) sides and similar surface elevations above the seafloor; the gentle northerly gradients may be formed by debris eroded from the higher parts and redeposited downstream. Despite the equant shapes of some of the edifices, it might be argued that others look teardrop-shaped as if modified by flowing ice (Fig. 25). Overall, however, the origin of the features remains enigmatic.

Other seamounts are markedly conical. Seamount 2 (Fig. 2) is a small cone-like feature situated at the southern end of the prominent submarine ridge, which also includes Franklin Island. Seamount 3, informally named 'Potter Peak' by Rilling et al. (2009), is a small pyramidal feature elongated in a NNE–SSW direction. It measures c. 3.5 km north–south and 2 km east–west, and rises c. 60 m above the surrounding seafloor, with an undulating summit region. It has an area of 5.24 km^2 and a volume of 0.176 km^3. Seamount 4 is 1.5 km in diameter, with an area of 1.48 km^2 and a volume of 0.069 km^3. It is essentially symmetrical and rises 100 m from the seafloor. Feature 5 is a small cone less than 1 km in diameter. It is associated with a ridge 5 km long and 1 km wide. In general, with their well-preserved volcanic cone morphologies, the conical seamounts are regarded as unlikely candidates for an origin by subglacial eruptions (Rilling et al. 2009). However, like the pancake-shaped mounds, they are also assumed to have been overridden by multiple cycles of grounded ice. Thus, in view of the isotopic ages determined so far (see below), a relatively pristine morphology, in itself, is not a reliable indicator of a post-glacial age (Rilling et al. 2009; cf. e.g. Wright 1980). The smaller seamounts were presumably short-lived (monogenetic), whereas the three much larger edifices may have erupted over a relatively long period of time (polygenetic?). Conversely, for seamount 6, the isotopic age for the top of the feature (1.94 ± 0.04 Ma) combined with its strong positive magnetic anomaly suggests that the seamount might have built up rapidly since the age of the base of the associated normal chron is at 1.95 Ma; a similar argument was proposed for the Beaufort Island seamount (Rilling et al. 2009). With these considerations, minimum average extrusion rates of 6×10^{-3} and 2×10^{-4} km^3 a^{-1}, respectively, were inferred. Isotopic ages obtained on the seamounts vary between 3.96 ± 0.08 and 0.123 ± 0.026 Ma but all but two seamounts are <0.5 Ma (all by Ar–Ar: Rilling et al. 2007, 2009; Lee et al. 2015) (Fig. 2; Supplementary material).

There is no spatial localization of the volcanism nor any clear geographical progression along or across the Terror Rift (Rilling et al. 2009); a similar absence of trends has also been noted for analogous Neogene volcanism in northern Victoria Land (Smellie et al. 2011b). Conversely, there may be a crude northeasterly younging for the initiation of volcanism in the centres between Mount Morning and Ross Island (Martin et al. 2010). The relationship between volcanism in the Erebus Volcanic Province and the opening of the Terror Rift or Victoria Land Basin, either during east–west extension or southerly rift migration, is not obvious. However, the elongation of volcanic ridges and some seamounts show two regional trends, north–south and NNE–SSW, suggesting a structural control on the volcanism. The trends may reflect fissure-fed eruptions controlled by a regional stress field or else they are the orientations of pre-existing faults within the Terror Rift. Rilling et al. (2009) also suggested that there appears to have been an overall increase in the volume of magmatism between the Miocene and present, with the most voluminous volcanic edifices constructed in the Pliocene and Pleistocene. However, it may also be a preservation issue, with greater time available for more prolonged erosion to reduce the older outcrops. It was calculated by Rilling et al. (2009) that the decrease in pressure at the base of the crust required to generate the volcanism need only have been 0.10–0.22 GPa, although an additional heat source (e.g. a plume or mantle upwelling) might also be required unless the mantle source was already modified (high water content or minor amounts of carbonate).

Summary

The Erebus Volcanic Province forms part of the McMurdo Volcanic Group. The predominantly subaerial volcanism is dominated by several large central volcanoes (Mount Erebus, Mount Discovery and Mount Morning); other less conspicuous volcanoes are also present (Black Island, White Island and Brown Peninsula) and some have been significantly reduced in size by erosion (Mason Spur and Minna Hook). Together, they attest to a major pulse of Miocene–present volcanism situated close to the western margin of the West Antarctic Rift System. Mount Erebus has been the subject of nearly continuous investigation since the late 1960s and is also unusual in Antarctica for having at least three nested calderas at its summit. It is the only undoubted active centre in the volcanic field and, uniquely, it contains a convecting phonolite lava lake in its summit crater. There are far more isotopic ages available for the Erebus Volcanic Province than any other volcanic region in Antarctica. The volcanoes typically commenced construction as volcanic shields that then evolved into stratovolcanoes. They are so large that they have deformed the underlying lithosphere by c. 2 km. The large central volcanoes form two major eruptive clusters centred on Mount Erebus and Mount Discovery, each of which shows a prominent tripartite radial symmetry attributed to crustal doming associated either with mantle-plume activity or edge-driven mantle convection. Erupted compositions range from basanite to phonolite and trachyte. The age of the volcanism extends back to 19 Ma in a poorly-exposed 'palaeo-Mount Morning' edifice, and extensive late Miocene volcanism is represented at Mason Spur and Minna Hook (12–8 Ma) but most of the large centres are Plio-Pleistocene (≤5 Ma) in age. Exposure of the volcano interiors is generally very poor, apart from two large very-well-exposed cliff-like outcrops at Minna Hook and Mason Spur. Mason Spur has previously been regarded as either a separate eruptive centre or as part of Mount Morning but here is defined unequivocally as a separate volcanic centre; the eruptive products are mainly caldera-filling ignimbrites, uncommon in Antarctica, and the original caldera was large (c. 10 km in diameter). The central volcanoes are flanked to the NW by a large field of small, mafic, mainly monogenetic volcanoes called the Southern Local Suite, characterized by uniformly basanitic compositions. The constituent centres are scoria cones and small sub-aerial lava fields, and their distribution in the Dry Valleys region has been used to demonstrate limited post-late Pliocene uplift of the Transantarctic Mountains. Volcanism is also distributed widely within the Terror Rift to the north, although the precise number and locations of the constituent centres are not yet well established. It includes two large, polygenetic, basanite–phonolite shield and stratovolcanoes at Franklin Island and Beaufort Island. However, the Terror Rift volcanism is less voluminous compared with other volcanic fields in the Erebus Volcanic Province (except for the Southern Local Suite). It is also mainly submarine, overwhelmingly basanitic in composition, and dominated by small and probably short-lived volcanic centres. The submarine volcanoes include groups of distinctive, small, very low-profile (pancake-shaped) volcanoes that may have erupted beneath a grounded ice sheet, or may have undergone extensive sub-ice-sheet erosion, and whose origin is still enigmatic. The volcanism in the Terror Rift Volcanic Field is mainly ≤1 Ma in age but it extends back to c. 7 Ma in Franklin and Beaufort islands. Curiously, the Terror Rift Volcanic Field volcanism is not associated with volcanism onshore, unlike analogous volcanism further north in northern Victoria Land (see Chapter 5.1a: Smellie and Rocchi 2021).

Acknowledgements The authors are very grateful to the following for their help over the years: Philip Kyle, Thom Wilch, Bill McIntosh, Kurt Panter, Nelia Dunbar, Alan Cooper and Dougal Townsend for stimulating conversations, including during fieldwork; NSF and Antarctica New Zealand for logistical field support at Minna Hook and Mason Spur; Tim Burton, Susan Detweiler, Matt Smith, Matt Windsor and Benji Nicholson for field assistance; Fred Davey with help understanding the Terror Rift and its volcanism; Dougal Townsend for his permission to publish several of his photographs; and Galina Siveter and Nikolai Brilliantov for help with Russian translation. Finally, we are very grateful to Dougal Townsend and Jim Cole for their helpful reviews of our chapter.

Author contributions JLS: conceptualization (lead), formal analysis (lead), investigation (lead), methodology (equal), visualization (lead), writing – original draft (lead), writing – review & editing (lead); **APM:** investigation (equal), writing – review & editing (supporting).

Funding The British Antarctic Survey supported the early fieldwork and research of J.L. Smellie. Projects at Mount Morning and Mason Spur were supported by NZARI Research Project awards and an Antarctica New Zealand (New Zealand Post) scholarship to A.P. Martin.

Data availability Data sharing is not applicable to this article as no additional datasets were generated other than those presented in this chapter and its supplementary files.

References

Aitken, A.R.A., Wilson, G.S., Jordan, T., Tinto, K. and Blakenmore, H. 2012. Flexural controls on late Neogene basin evolution in southern McMurdo Sound, Antarctica. *Global and Planetary Change*, **80–81**, 99–112, https://doi.org/10.1016/j.gloplacha.2011.08.003

Anderson, J.T.H., Wilson, G.S., Fink, D., Lilly, K., Levy, R.H. and Townsend, D. 2017. Reconciling marine and terrestrial evidence for post LGM ice sheet retreat in southern McMurdo Sound, Antarctica. *Quaternary Science Reviews*, **157**, 1–13, https://doi.org/10.1016/j.quascirev.2016.12.007

Antibus, J.V., Panter, K.S. *et al.* and 2014. Alteration of volcaniclastic deposits at Minna Bluff: Geochemical insights on mineralizing environment and climate during the Late Miocene in Antarctica. *Geochemistry, Geophysics, Geosystems*, **15**, 3258–3280, https://doi.org/10.1002/2014GC005422

Armstrong, R.L. 1978. K-Ar dating: Late Cenozoic McMurdo Volcanic Group and dry valley glacial history, Victoria Land, Antarctica. *New Zealand Journal of Geology and Geophysics*, **21**, 685–698, https://doi.org/10.1080/00288306.1978.10425199

Aviado, K.B., Rilling-Hall, S., Bryuce, J.G. and Mukasa, S.B. 2015. Submarine and subaerial lavas in the West Antarctic Rift System: Temporal record of shifting magma source components from the lithosphere and asthenosphere. *Geochemistry, Geophysics, Geosystems*, **16**, 4344–4351, https://doi.org/10.1002/2015GC006076

Behrendt, J.C. 1990. Ross Sea. *American Geophysical Union Antarctic Research Series*, **48**, 89–90.

Behrendt, J.C., Saltus, R., Damaske, D., McCafferty, A., Finn, C.A., Blankenship, D. and Bell, R.E. 1996. Patterns of late Cenozoic volcanic and tectonic activity in the West Antarctic rift System revealed by aeromagnetic surveys. *Tectonics*, **15**, 660–676, https://doi.org/10.1029/95TC03500

Behrendt, J.C., Blankenship, D.D., Morse, D.L. and Bell, R.E. 2004. Shallow-source aeromagnetic anomalies observed over the West Antarctic Ice Sheet compared with coincident bed topography from radar ice sounding – new evidence for glacial 'removal' of subglacially erupted Cenozoic rift-related volcanic edifices. *Global and Planetary Change*, **42**, 177–193, https://doi.org/10.1016/j.gloplacha.2003.10.006

Berg, J.H. 1991. Geology, petrology and tectonic implications of crustal xenoliths in Cenozoic volcanic rocks of southern Victoria Land. *In*: Thomson, M.R.A., Crame, J.A. and Thomson, J.W. (eds) *Geological Evolution of Antarctica*. Cambridge University Press, Cambridge, UK, 311–315.

Blackman, D.K., Von Herzen, R.P. and Lawver, C.A. 1987. Heat flow and tectonics in the western Ross Sea, Antarctica. *Circum-Pacific Council for Energy and Mineral Resources, Earth Science Series*, **5B**, 179–189.

Caldwell, D.A. and Kyle, P.R. 1994. Mineralogy and geochemistry of ejecta erupted from Mount Erebus, Antarctica, between 1972 and 1986. *American Geophysical Union Antarctic Research Series*, **66**, 147–162.

Chiappini, M., Ferraccioli, F., Bozzo, E. and Damaske, D. 2002. Regional compilation and analysis of aeromagnetic anomalies for the Transantarctic Mountains–Ross Sea sector of the Antarctic. *Tectonophysics*, **347**, 121–137, https://doi.org/10.1016/S0040-1951(01)00241-4

Cole, J.W. and Ewart, A. 1968. Contributions to the volcanic geology of the Black Island, Brown Peninsula, and Cape Bird areas, McMurdo Sound, Antarctica. *New Zealand Journal of Geology and Geophysics*, **11**, 793–828, https://doi.org/10.1080/00288306.1968.10420754

Cole, J.W., Kyle, P.R. and Neall, V.E. 1971. Contributions to the geology of Cape Crozier, White Island and Hut Point Peninsula, McMurdo Sound region, Antarctica. *New Zealand Journal of*

Geology and Geophysics, **14**, 528–546, https://doi.org/10.1080/00288306.1971.10421946

Cooper, A.K., Davey, F.J. and Behrendt, J.C. 1987. Seismic stratigraphy and structure of the Victoria Land basin, western Ross Sea, Antarctica. *Circum-Pacific Council for Energy and Mineral Resources, Earth Sciences Series*, **5B**, 27–65.

Cooper, A.F., Adam, L.J., Coulter, R.F., Eby, G.N. and McIntosh, W.C. 2007. Geology, geochronology and geochemistry of a basanite volcano, White Island, Ross Sea, Antarctica. *Journal of Volcanology and Geothermal Research*, **165**, 189–216, https://doi.org/10.1016/j.jvolgeores.2007.06.003

Cox, S.C., Turnbull, I.M., Isaac, M.J., Townsend, D.B. and Smith Lyttle, B. 2012. *Geology of Southern Victoria Land Antarctica*. Institute of Geological and Nuclear Sciences 1:250 000 Geological Map 22. GNS Science, Lower Hutt, New Zealand.

Csatho, B., Schenk, T., Kyle, P., Wilson, T. and Krabill, W.B. 2008. Airborne laser swath mapping of the summit of Erebus volcano, Antarctica: Applications to geological mapping of a volcano. *Journal of Volcanology and Geothermal Research*, **177**, 531–548, https://doi.org/10.1016/j.jvolgeores.2008.08.016

Davey, F.J., Granot, R., Cande, S.C., Stock, J.M., Selvans, M. and Ferraccioli, F. 2016. Synchronous oceanic spreading and continental rifting in West Antarctica. *Geophysical Research Letters*, **43**, 6162–6169, https://doi.org/10.1002/2016GL069087

Decker, E.R. and Bucher, G.J. 1982. Geothermal studies in the Ross Island–Dry Valley region. *In*: Craddock, C. (ed.) *Antarctic Geoscience*. University of Wisconsin Press, Madison, WI, 887–894.

Del Carlo, P., Panter, K.S., Bassett, K., Bracciali, L., Di Vincenzo, G. and Rocchi, S. 2009. The upper lithostratigraphic unit of ANDRILL AND-2A core (Southern McMurdo Sound, Antarctica): Local Pleistocene volcanic sources, paleoenvironmental implications and subsidence in the southern Victoria Land Basin. *Global and Planetary Change*, **69**, 142–161, https://doi.org/10.1016/j.gloplacha.2009.09.002

Della Vedova, B., Pellis, G., Trey, H., Zhang, J., Cooper, A.K. and Makris, J. and the ACRUP Working Group. 1997. Crustal structure of the Transantarctic Mountains, Western Ross Sea. *In*: Ricci, C.A. (ed.) *The Antarctic Region: Geological Evolution and Processes*. Terra Antartica Publication, Siena, Italy, 609–618.

Denton, G.H. and Marchant, D.R. 2000. The geologic basis for a reconstruction of a grounded ice sheet in McMurdo Sound, Antarctica, at the last glacial maximum. *Geografiska Annaler*, **82A**, 167–211, https://doi.org/10.1111/j.0435-3676.2000.00121.x

Denton, G.H., Borns, H.W., Grosswald, M.G., Stuiver, M. and Nichols, R.L. 1975. Glacial history of the Ross Sea. *Antarctic Journal of the United States*, **10**, 160–164.

Dibble, R.R., Kyle, P.R. and Skov, M.J. 1994. Volcanic activity and seismicity of Mount Erebus, 1986–1994. *Antarctic Journal of the United States*, **29**, 11–13.

Eggers, A.J. 1979. Scallop Hill Formation, Brown Peninsula, McMurdo Sound, Antarctica. *New Zealand Journal of Geology and Geophysics*, **22**, 353–361, https://doi.org/10.1080/00288306.1979.10424104

Ellerman, P.J. and Kyle, P.R. 1990a. Franklin Island. *American Geophysical Union Antarctic Research Series*, **48**, 91–93.

Ellerman, P.J. and Kyle, P.R. 1990b. Beaufort Island. *American Geophysical Union Antarctic Research Series*, **48**, 94–96.

Esser, R.E., Kyle, P.R. and McIntosh, W.C. 2004. $^{40}Ar/^{39}Ar$ dating and the eruptive history of Mount Erebus, Antarctica: volcano evolution. *Bulletin of Volcanology*, **66**, 671–686, https://doi.org/10.1007/s00445-004-0354-x

Fargo, A.J. 2008. *$^{40}Ar/^{39}Ar$ Geochronological Analysis of Minna Bluff, Antarctica: Evidence for Past Glacial Events within the Ross Embayment*. MSc thesis, New Mexico Institute of Mining and Technology, Socorro, New Mexico, USA.

Ferraccioli, F., Armadillo, E., Zunino, A., Bozzo, E., Rocchi, S. and Armienti, P. 2009. Magmatic and tectonic patterns over the Northern Victoria Land sector of the Transantarctic Mountains from new aeromagnetic imaging. *Tectonophysics*, **478**, 43–61, https://doi.org/10.1016/j.tecto.2008.11.028

Fleck, R.J., Jones, L.M. and Behling, R.E. 1972. K–Ar dates of the McMurdo volcanics and their relation to the glacial history of Wright Valley. *Antarctic Journal of the United States*, **7**, 244–246.

Forbes, R.B. and Ester, D.W. 1964. Glaciation of Observation Hill, Hutt Point Peninsula, Ross Island, Antarctica. *Journal of Gaciology*, **5**, 87–92, https://doi.org/10.1017/S0022143000028598

Forbes, R.B., Turner, D.L. and Carden, J.R. 1974. Age of trachyte from Ross Island. *Geology*, **2**, 297–298, https://doi.org/10.1130/0091-7613(1974)2<297:AOTFRI>2.0.CO;2

Gamble, J.A. and Kyle, P.R. 1987. The origins of glass and amphibole in spinel–wehrlite xenoliths from Foster Crater, McMurdo Volcanic Group, Antarctica. *Journal of Petrology*, **28**, 755–780, https://doi.org/10.1093/petrology/28.5.755

Gamble, J.A., McGibbon, F., Kyle, P.R., Menzies, M.A. and Kirsch, I. 1988. Metasomatised xenoliths from Foster Crater, Antarctica: Implications for lithospheric structure and processes beneath the Transantarctic Mountain front. *Journal of Petrology*, Special Volume, Issue 1, 109–138, https://doi.org/10.1093/petrology/Special_Volume.1.109

Goldich, S.S., Treves, S.B., Suhr, N.H. and Stuckless, J.S. 1975. Geochemistry of the Cenozoic volcanic rocks of Ross Island and vicinity, Antarctica. *Journal of Geology*, **83**, 415–435, https://doi.org/10.1086/628120

Gudmundsson, M.T., Sigmundsson, F. and Björnsson, H. 1997. Ice–volcano interaction in the 1996 Gjálp eruption, Vatnajökull, Iceland. *Nature*, **389**, 954–957, https://doi.org/10.1038/40122

Hall, J., Wilson, T. and Henrys, S. 2007. Structure of the central Terror Rift, western Ross Sea, Antarctica. *United States Geological Survey Open-File Report*, **2007-1047**, Short Research Paper 108.

Hamilton, W. 1972. *The Hallett Volcanic Province, Antarctica*. Geological Survey Professional Paper, **456-C**.

Hansen, S.E., Graw, J.H. *et al.* 2014. Imaging the Antarctic mantle using adaptively parametrized P-wave tomography: Evidence for heterogeneous structure beneath West Antarctica. *Earth and Planetary Science Letters*, **408**, 66–78, https://doi.org/10.1016/j.epsl.2014.09.043

Harder, M. and Russell, J.K. 2007. Basanite glaciovolcanism at Llangorse Mountain, northern British Columbia, Canada. *Bulletin of Volcanology*, **69**, 329–340, https://doi.org/10.1007/s00445-006-0078-1

Harpel, C.J., Kyle, P.R., Esser, R.P., McIntosh, W.C. and Caldwell, D.A. 2004. $^{40}Ar/^{39}Ar$ dating of the eruptive history of Mount Erebus, Antarctica: summit flows, tephra, and caldera collapse. *Bulletin of Volcanology*, **66**, 687–702, https://doi.org/10.1007/s00445-004-0349-7

Harpel, C.J., Kyle, P.R. and Dunbar, N.W. 2008. Englacial stratigraphy of Erebus volcano, Antarctica. *Journal of Volcanology and Geothermal Research*, **177**, 549–568, https://doi.org/10.1016/j.jvolgeores.2008.06.001

Harrington, H.J. 1958. Beaufort Island, remnant of a Quaternary volcano in the Ross Sea, Antarctica. *New Zealand Journal of Geology and Geophysics*, **1**, 595–603, https://doi.org/10.1080/00288306.1958.10423167

Horgan, H., Naish, T., Bannister, S., Balfour, N. and Wilson, G. 2005. Seismic stratigraphy of the Plio-Pleistocene Ross Island flexural moat-fill: a prognosis for ANDRILL Program drilling beneath McMurdo–Ross Ice Shelf. *Global and Planetary Change*, **45**, 83–97, https://doi.org/10.1016/j.gloplacha.2004.09.014

Iverson, N.A., Kyle, P.R., Dunbar, N.W., McIntosh, W.C. and Pearce, N.J.G. 2014. Eruptive history and magmatic stability of Erebus volcano, Antarctica: Insights from englacial tephra. *Geochemistry, Geophysics, Geosystems*, **15**, 4180–4202, https://doi.org/10.1002/2014GC005435

Johnson, J.S. and Smellie, J.L. 2007. Zeolite compositions as proxies for eruptive paleoenvironment. *Geochemisty, Geophysics, Geosystems*, **8**, Q03009, https://doi.org/10.1029/2006GC001450

Johnston, L., Wilson, G.S., Gorman, A.R., Henrys, S.A., Horgan, H., Clark, R. and Naish, T.R. 2008. Cenozoic basin evolution beneath the southern McMurdo Ice Shelf, Antarctica. *Global and Planetary Change*, **62**, 61–76, https://doi.org/10.1016/j.gloplacha.2007.11.004

Kalamarides, R.I. and Berg, J.H. 1991. Geochemistry and tectonic implications of lower-crustal granulites included in Cenozoic volcanic rocks of southern Victoria Land. *In*: Thomson, M.R.A., Crame, J.A. and Thomson, J.W. (eds) *Geological evolution of Antarctica*. Cambridge University Press, Cambridge, UK, 305–310.

Kalamarides, R.I., Berg, J.H. and Hank, R.A. 1987. Lateral isotopic discontinuity in the lower crust: An example from Antarctica. *Science*, **237**, 1192–1195, https://doi.org/10.1126/science.237.4819.1192

Kelly, P.J., Dunbar, N.W., Kyle, P.R. and McIntosh, W.C. 2008. Refinement of the late Quaternary geologic history of Erebus volcano, Antarctica using $^{40}Ar/^{39}Ar$ and ^{36}Cl age determinations. *Journal of Volcanology and Geothermal Research*, **177**, 569–577, https://doi.org/10.1016/j.jvolgeores.2008.07.018

Kokelaar, B.P. 1983. The mechanism of Surtseyan volcanism. *Journal of the Geological Society, London*, **140**, 939–944, https://doi.org/10.1144/gsjgs.140.6.0939

Kyle, P.R. 1981*a*. Geologic history of Hut Point Peninsula as inferred from DVDP 1, 2 and 3 drillcores and surface mapping. *American Geophysical Union Antarctic Research Series*, **33**, 427–445.

Kyle, P.R. 1981*b*. Mineralogy and geochemistry of a basanite to phonolite sequence at Hut Point Peninsula, Antarctica, based on core from Dry Valley Drilling Project Drillholes 1, 2 and 3. *Journal of Petrology*, **22**, 451–500, https://doi.org/10.1093/petrology/22.4.451

Kyle, P.R. 1981*c*. Glacial history of the McMurdo Sound area as indicated by the distribution and nature of McMurdo Volcanic Group rocks. *American Geophysical Union Antarctic Research Series*, **33**, 403–412.

Kyle, P.R. 1990*a*. McMurdo Volcanic Group, western Ross Embayment. *American Geophysical Union Antarctic Research Series*, **48**, 19–25.

Kyle, P.R. 1990*b*. Hut Point Peninsula. *American Geophysical Union Antarctic Research Series*, **48**, 109–112.

Kyle, P.R. 1994. *Volcanological and Environmental studies of Mount Erebus, Antarctica*. American Geophysical Union Antarctic Research Series, **66**.

Kyle, P.R. and Cole, J.W. 1974. Structural control on volcanism in the McMurdo Volcanic Group, Antarctica. *Bulletin Volcanologique*, **38**, 16–25, https://doi.org/10.1007/BF02597798

Kyle, P.R. and Muncy, H.L. 1989. Geology and geochronology of McMurdo Volcanic Group rocks in the vicinity of Lake Morning, McMurdo Sound, Antarctica. *Antarctic Science*, **1**, 345–350, https://doi.org/10.1017/S0954102089000520

Kyle, P.R., Adams, J. and Rankin, P.C. 1979*a*. Geology and petrology of the McMurdo Volcanic Group at Rainbow Ridge, Brown Peninsula, Antarctica. *Geological Society of America Bulletin*, **90**, 676–688, https://doi.org/10.1130/0016-7606(1979)90<676:GAPOTM>2.0.CO;2

Kyle, P.R., Sutter, J.F. and Treves, S.B. 1979*b*. K/Ar age determinations on drill core from DVDP holes 1 and 2. *Memoirs of the National Institute of Polar Research (Japan)*, **13**, Special Issue, 214–219.

Kyle, P.R., Moore, J.A. and Thirlwall, M.F. 1992. Petrologic evolution of anorthoclase phonolite magmas at Mount Erebus, Ross Island, Antarctica. *Journal of Petrology*, **33**, 849–875, https://doi.org/10.1093/petrology/33.4.849

Lachlan-Cope, T., Smellie, J.L. and Ladkin, R. 2001. Discovery of a recurrent lava lake on Saunders Island (South Sandwich Islands) using AVHRR imagery. *Journal of Volcanology and Geothermal Research*, **112**, 105–116, https://doi.org/10.1016/S0377-0273(01)00237-2

Lawrence, K.P., Tauxe, L., Staudigel, H., Constable, C.G., Koppers, A., McIntosh, W. and Johnson, C.L. 2009. Paleomagnetic field properties at high southern latitude. *Geochemistry, Geophysics, Geosystems*, **10**, Q01005, https://doi.org/10.1029/2008GC002072

Lawver, L.A., Davis, M.B., Wilson, T.J. and Shipboard Scientific Party. 2007. Neotectonic and other features of the Victoria Land Basin, Antarctica, interpreted from multibeam bathymetry data. *United States Geological Survey Open-File Report*, **2007-1047**, Extended Abstract 017.

Lawver, L., Lee, J., Kim, Y. and Davey, F. 2012. Flat-topped mounds in western Ross Sea: Carbonate mounds or subglacial volcanic features? *Geosphere*, **8**, 645–653, https://doi.org/10.1130/GES00766.1

Lee, M.J., Lee, J.I., Kim, T.H., Lee, J. and Nagao, K. 2015. Age, geochemistry and Sr–Nd–Pb isotopic compositions of alkali volcanic rocks from Mt. Melbourne and the western Ross Sea, Antarctica. *Geosciences Journal*, **19**, 681–695, https://doi.org/10.1007/s12303-015-0061-y

Lescinsky, D.T. and Sisson, T.W. 1998. Ridge-forming ice-bounded lava flows at Mount Rainier, Washiington. *Geology*, **26**, 351–354, https://doi.org/10.1130/0091-7613(1998)026<0351:RFIBLF>2.3.CO;2

Mankinen, E.A. and Cox, A. 1988. Paleomagnetic investigation of some volcanic rocks from the McMurdo Volcanic Province, Antarctica. *Journal of Geophysical Research*, **93**, 599–612, https://doi.org/10.1029/JB093iB10p11599

Martin, A.P. 2009. *Mount Morning, Antarctica: Geochemistry, Geochronology, Petrology, Volcanology, and Oxygen Fugacity of the Rifted Antarctic Lithosphere*. PhD thesis, University of Otago, Dunedin, New Zealand.

Martin, A.P. and Cooper, A.F. 2010. Post 3.9 Ma fault activity within the West Antarctic rift system: onshore evidence from Gandalf Ridge, Mount Morning eruptive centre, southern Victoria Land, Antarctica. *Antarctic Science*, **22**, 513–521, https://doi.org/10.1017/S095410201000026X

Martin, A.P., Cooper, A.F. and Dunlap, W.J. 2010. Geochronology of Mount Morning, Antarctica: two-phase evolution of a long-lived trachyte–basanite–phonolite eruptive center. *Bulletin of Volcanology*, **72**, 357–371, https://doi.org/10.1007/s00445-009-0319-1

Martin, A.P., Cooper, A.F. and Price, R.C. 2013. Petrogenesis of Cenozoic, alkalic volcanic lineages at Mount Morning, West Antarctica and their entrained lithospheric mantle xenoliths: Lithospheric v. asthenospheric mantle sources. *Geochimica et Cosmochimica Acta*, **122**, 127–152, https://doi.org/10.1016/j.gca.2013.08.025

Martin, A.P., Cooper, A.F., Price, R.C., Turnbull, R.E. and Roberts, N.M.W. 2015. The petrology, geochronology and significance of Granite Harbour Intrusive Complex xenoliths and outcrop sampled in western McMurdo Sound, Southern Victoria Land, Antarctica. *New Zealand Journal of Geology and Geophysics*, **58**, 33–51, https://doi.org/10.1080/00288306.2014.982660

Martin, A.P., Smellie, J.L., Cooper, A.F. and Townsend, D.B. 2018. Formation of a spatter-rich pyroclastic density current deposit in a Neogene sequence of trachytic–mafic igneous rocks at Mason Spur, Erebus volcanic province, Antarctica. *Bulletin of Volcanology*, **80**, 13, https://doi.org/10.1007/s00445-017-1188-7

Martin, A.P., Cooper, A.F., Price, R.C., Kyle, P.R. and Gamble, J.A. 2021. Erebus Volcanic Province: petrology. *Geological Society, London, Memoirs*, **55**, https://doi.org/10.1144/M55-2018-80

McCraw, J.D. 1967. Soils of Taylor Dry Valley, Victoria Land, Antarctica, with notes on soils from other localities in Victoria Land. *New Zealand Journal of Geology and Geophysics*, **10**, 498–539, https://doi.org/10.1080/00288306.1967.10426754

McIver, J.R. and Gevers, T.W. 1970. Volcanic vents below the Royal Society Range, central Victoria Land, Antarctica. *Transactions of the Geological Society of South Africa*, **73**, 65–88.

Moore, J.A. and Kyle, P.R. 1987. Volcanic geology of Mount Erebus, Ross Island, Antarctica. *Proceedings of the National Institute of Polar Research Symposium on Antarctic Geosciences*, **1**, 48–65.

Moore, J.A. and Kyle, P.R. 1990. Mount Erebus. *American Geophysical Union Antarctic Research Series*, **48**, 103–108.

Panter, K.S. and Winter, B. 2008. Geology of the Side Crater of the Erebus volcano, Antarctica. *Journal of Volcanology and*

Geothermal Research, **177**, 578–588, https://doi.org/10.1016/j.jvolgeores.2008.04.019

Panter, K.S., Castillo, P. et al. 2018. Melt origin across a rifted continental margin: a case for subduction-related metasomatic agents in the lithospheric source of alkaline basalt, NW Ross Sea, Antarctica. *Journal of Petrology*, **59**, 517–558, https://doi.org/10.1093/petrology/egy036

Parmelee, D.E.F., Kyle, P.R., Kurz, M.D., Marrero, S.M. and Phillips, F.M. 2015. A new Holocene eruptive history of Erebus volcano, Antarctica, using cosmogenic ^{3}He and ^{36}Cl exposure ages. *Quaternary Geochronology*, **30**, 114–131, https://doi.org/10.1016/j.quageo.2015.09.001

Patrick, M.R. and Smellie, J.L. 2013. A spaceborne inventory of volcanic activity in Antarctica and southern oceans, 2000–2010. *Antarctic Science*, **25**, 475–500, https://doi.org/10.1017/S0954102013000436

Paulsen, T.S. and Wilson, T.J. 2009. Structure and age of volcanic fissures on Mount Morning: A new constraint on Neogene to contemporary stress in the West Antarctic Rift, southern Victoria Land, Antarctica. *Geological Society of America Bulletin*, **121**, 1071–1088, https://doi.org/10.1130/B26333.1

Phillips, E.H., Sims, K.W.W. et al. 2018. The nature and evolution of mantle upwelling at Ross Island, Antarctica, with implications for the source of HIMU lavas. *Earth and Planetary Science Letters*, **498**, 38–53, https://doi.org/10.1016/j.epsl.2018.05.049

Polyakov, M.M., Krylov, A.Y. and Mazina, T.I. 1976. New data on radiogeochronology of Antarctic Cenozoic vulcanites 160°E–100°W. *Informatsionny byulleten' Sovetskoy antarkticheskoy ekspeditsii*, **93**, 19–26 [in Russian].

Rilling, S.E., Mukasa, S.B., Wilson, T.J. and Lawver, L.A. 2007. ^{40}Ar–^{39}Ar constraints on volcanism and tectonism in the Terror Rift of the Ross Sea, Antarctica. *United States Geological Survey Open-File Report*, **2007-1047**, Short Research Paper 092, https://pubs.usgs.gov/of/2007/1047/srp/srp092/

Rilling, S., Mukasa, S., Wilson, T., Lawver, L. and Hall, C. 2009. New determinations of ^{40}Ar/^{39}Ar isotopic ages and flow volumes for Cenozoic volcanism in the Terror Rift, Ross Sea, Antarctica. *Journal of Geophysical Research*, **114**, B12207, https://doi.org/10.1029/2009JB006303

Risk, G.F. and Hochstein, M.P. 1974. Heat flow at Arrival Heights, Ross Island, Antarctica. *New Zealand Journal of Geology and Geophysics*, **17**, 629–644, https://doi.org/10.1080/00288306.1973.10421586

Rocchi, S. and Smellie, J.L. 2021. Northern Victoria Land: petrology. *Geological Society, London, Memoirs*, **55**, https://doi.org/10.1144/M55-2019-19

Rocchi, S., Armienti, P., D'Orazio, M., Tonarini, S., Wibrans, J.R. and Di Vincenzo, G. 2002. Cenozoic magmatism in the western Ross Embayment: role of mantle plume v. plate dynamics in the development of the West Antarctic Rift System. *Journal of Geophysical Research*, **107**, 2195, https://doi.org/10.1029/2001JB000515

Rocchi, S., Storti, F., Di Vincenzo, G. and Rosetti, F. 2003. Intraplate strike-slip tectonics as an alternative to mantle plume activity for the Cenozoic rift magmatism in the Ross Sea region, Antarctica. *Geological Society, London, Special Publications*, **210**, 145–158, https://doi.org/10.1144/GSL.SP.2003.210.01.09

Rocchi, S., Armienti, P. and Di Vincenzo, G. 2005. No plume, no rift magmatism in the West Antarctic Rift. *Geological Society of America Special Papers*, **388**, 435–447.

Rossetti, F., Storti, F. et al. 2006. Eocene initiation of Ross Sea dextral faulting and implications for East Antarctic neotectonics. *Journal of the Geological Society, London*, **163**, 119–126, https://doi.org/10.1144/0016-764905-005

Rowe, C.A., Aster, R.C., Kyle, P.R., Schlue, J.W. and Dibble, R.R. 1998. Broadband recording of Strombolian explosions and associated very-long-period seismic signals on Mount Erebus Volcano, Ross Island, Antarctica. *Geophysical Research Letters*, **25**, 2297–2300, https://doi.org/10.1029/98GL01622

Salvini, F. and Storti, F. 1999. Cenozoic tectonic lineaments of the Terra Nova Bay region, Ross Embayment, Antarctica. *Global and Planetary Change*, **23**, 129–144, https://doi.org/10.1016/S0921-8181(99)00054-5

Salvini, F., Brancolini, G., Busetti, M., Storti, F., Mazzarini, F. and Coren, F. 1997. Cenozoic geodynamics of the Ross Sea region, Antarctica: Cenozoic geodynamics of the Ross Sea region: Crustal extension, intraplate strike-slip faulting, and tectonic inheritance. *Journal of Geophysical Research*, **102**, 24 669–24 696, https://doi.org/10.1029/97JB01643

Scanlan, M.K. 2008. *Petrology of Inclusion-Rich Lavas at Minna Bluff, McMurdo Sound, Antarctica: Implications for Magma Origin, Differentiation, and Eruption Dynamics*. MSc thesis, Bowling Green State University, Bowling Green, Ohio, USA.

Schopka, H.H., Gudmundsson, M.T. and Tuffen, H. 2006. The formation of Helgafell, southwest Iceland, a monogenetic subglacial hyaloclastite ridge: Sedimentology, hydrology and volcano–ice interaction. *Bulletin of Volcanology*, **152**, 359–377, https://doi.org/10.1016/j.jvolgeores.2005.11.010

Sims, K.W.W., Aster, R. et al. 2021. Mount Erebus. *Geological Society, London, Memoirs*, **55**, https://doi.org/10.1144/M55-2019-8

Skilling, I.P. 2002. Basaltic pahoehoe lava-fed deltas: large-scale characteristics, clast generation, emplacement processes and environmental discrimination. *Geological Society, London, Special Publications*, **202**, 91–113, https://doi.org/10.1144/GSL.SP.2002.202.01.06

Smellie, J.L. 2001. Lithofacies architecture and construction of volcanoes erupted in englacial lakes: Icefall Nunatak, Mount Murphy, eastern Marie Byrd Land, Antarctica. *International Association of Sedimentologists Special Publications*, **30**, 9–34.

Smellie, J.L. 2006. The relative importance of supraglacial versus subglacial meltwater escape in basaltic subglacial tuya eruptions: An important unresolved conundrum. *Earth-Science Reviews*, **74**, 241–268, https://doi.org/10.1016/j.earscirev.2005.09.004

Smellie, J.L. 2009. Terrestrial subice volcanism: Landform morphology, sequence characteristics, environmental influences, and implications for candidate Mars examples. *Geological Society of America Special Papers*, **453**, 55–76, https://doi.org/10.1130/2009.453(05)

Smellie, J.L. 2013. Quaternary vulcanism: subglacial landforms. In: Elias, S.A. (ed.) *The Encyclopedia of Quaternary Science, Volume 1*. 2nd Edn. Elsevier, Amsterdam, 780–802.

Smellie, J.L. 2021. Bransfield Strait and James Ross Island: volcanology. *Geological Society, London, Memoirs*, **55**, https://doi.org/10.1144/M55-2018-58

Smellie, J.L. and Edwards, B.E. 2016. *Glaciovolcanism on Earth and Mars: Products, Processes and Palaeoenvironmental Significance*. Cambridge University Press, Cambridge, UK.

Smellie, J.L. and Hole, M.J. 1997. Products and processes in Pliocene–Recent, subaqueous to emergent volcanism in the Antarctic Peninsula: examples of englacial Surtseyan volcano reconstruction. *Bulletin of Volcanology*, **58**, 628–646, https://doi.org/10.1007/s004450050167

Smellie, J.L. and Rocchi, S. 2021. Northern Victoria Land: volcanology. *Geological Society, London, Memoirs*, **55**, https://doi.org/10.1144/M55-2018-60

Smellie, J.L., Rocchi, S. and Armienti, P. 2011a. Late Miocene volcanic sequences in northern Victoria Land, Antarctica: products of glaciovolcanic eruptions under different thermal regimes. *Bulletin of Volcanology*, **73**, 1–25, https://doi.org/10.1007/s00445-010-0399-y

Smellie, J.L., Rocchi, S., Gemelli, M., Di Vincenzo, G. and Armienti, P. 2011b. Late Miocene East Antarctic ice sheet characteristics deduced from terrestrial glaciovolcanic sequences in northern Victoria Land, Antarctica. *Palaeogeography, Palaeoclimatology, Palaeoecology*, **307**, 129–149, https://doi.org/10.1016/j.palaeo.2011.05.008

Smellie, J.L., Wilch, T.I. and Rocchi, S. 2013. 'A'ā lava-fed deltas: A new reference tool in paleoenvironmental studies. *Geology*, **41**, 403–406, https://doi.org/10.1130/G33631.1

Speden, I.G. 1962. Fossiliferous Quaternary marine deposits in the McMurdo Sound region, Antarctica. *New Zealand Journal of Geology and Geophysics*, **5**, 746–777, https://doi.org/10.1080/00288306.1962.10417636

Stuiver, M., Denton, G.H., Hughes, T.J. and Fastook, J.L. 1981. History of the marine ice sheet in West Antarctica during the last glaciation: A working hypothesis. *In*: Denton, G.H. and Hughes, T.J. (eds) *The Last Great Ice Sheets*. Wiley Interscience, New York, 319–436.

Stump, E. 2002. What put the hook in Minna Bluff and other observations on the Transantarctic Mountains Front. *Royal Society of New Zealand Bulletin*, **35**, 233–237.

Tauxe, L., Gans, P. and Mankinen, E.A. 2004. Paleomagnetism and $^{40}Ar/^{39}Ar$ ages from volcanics extruded during the Matuyama and Brunhes Chrons near McMurdo Sound, Antarctica. *Geochemistry, Geophysics, Geosystems*, **5**, Q06H12, https://doi.org/10.1029/2003GC000656

Timms, C.J. 2006. *Reconstruction of a grounded ice sheet in McMurdo Sound – Evidence from southern Black Island, Antarctica*. MSc thesis, University of Otago, Dunedin, New Zealand.

Treves, S.B. 1967. Volcanic rocks from the Ross Island, Marguerite Bay and Mt. Weaver areas, Antarctica. *Japanese Antarctic Research Expedition Scientific Reports*, Special Issue 1, 136–149.

Treves, S.B. 1968. Volcanic rocks of the Ross Island area. *Antarctic Journal of the United States*, **3**, 108–109.

USGS. 1970. *Ross Island, Antarctica. 1:250 000 Scale Topographical Map ST 57–60/6*. United States Geological Survey, Washington, DC.

Vella, P. 1969. Surficial geological sequence, Black Island and Brown Peninsula, McMurdo Sound, Antarctica. *New Zealand Journal of Geology and Geophysics*, **12**, 761–770, https://doi.org/10.1080/00288306.1969.10431110

White, J.D.L. and Houghton, B.F. 2006. Primary volcaniclastic rocks. *Geology*, **34**, 677–680, https://doi.org/10.1130/G22346.1

Wilch, T.I., Denton, G.H., Lux, D.R. and McIntosh, W.C. 1993. Limited Pliocene glacier extent and surface uplift in middle Taylor Valley, Antarctica. *Geografiska Annaler*, **75A**, 331–351, https://doi.org/10.1080/04353676.1993.11880399

Wilch, T., McIntosh, W.C. *et al*. 2008. Volcanic and glacial geology of the Miocene Minna Bluff Volcanic Complex, Antarctica. Abstract presented at the 2008 AGU Fall Meeting, December 15–19, 2008, San Francisco, California, USA.

Wilson, T.J. 1995. Cenozoic transtension along the Transantarctic Mountains–West Antarctic Rift boundary, southern Victoria Land, Antarctica. *Tectonics*, **14**, 531–545, https://doi.org/10.1029/94TC02441

Wilson, T.J. 1999. Cenozoic structural segmentation of the Transantarctic Mountains rift flank in southern Victoria Land. *Global and Planetary Change*, **23**, 105–127, https://doi.org/10.1016/S0921-8181(99)00053-3

Wilson, G.S. 2000. Glacial geology and origin of fossiliferous-erratic-bearing moraines, southern McMurdo Sound, Antarctica – an alternative ice sheet hypothesis. *American Geophysical Union Antarctic Research Series*, **76**, 19–37, https://doi.org/10.1029/AR076p0019

Wright, A.C. 1980. Landforms of McMurdo Volcanic Group, Southern Foothills of Royal Society Range, Antarctica. *New Zealand Journal of Geology and Geophysics*, **23**, 605–613, https://doi.org/10.1080/00288306.1980.10424132

Wright, A.C., McIntosh, W. and Ellerman, P. 1983. Volcanic geology of Turks Head, Tryggve Point, and Minna Bluff, southern Victoria Land. *Antarctic Journal of the United States*, **18**, 35.

Wright-Grassham, A.C. 1987. *Volcanic Geology, Mineralogy, and Petrogenesis of the Discovery Volcanic Subprovince, Southern Victoria Land, Antarctica*. PhD thesis, New Mexico Institute of Mining and Technology, Socorro, New Mexico, USA.

Wright, A.C. and Kyle, P.R. 1990a. Royal Society Range. *American Geophysical Union Antarctic Research Series*, **48**, 131–133.

Wright, A.C. and Kyle, P.R. 1990b. Taylor and Wright Valleys. *American Geophysical Union Antarctic Research Series*, **48**, 134–137.

Wright, A.C. and Kyle, P.R. 1990c. Mount Terror. *American Geophysical Union Antarctic Research Series*, **48**, 99–102.

Wright, A.C. and Kyle, P.R. 1990d. Mount Bird. *American Geophysical Union Antarctic Research Series*, **48**, 97–98.

Wright, A.C. and Kyle, P.R. 1990e. White Island, Black Island, and Brown Peninsula. *American Geophysical Union Antarctic Research Series*, **48**, 113–116.

Wright, A.C. and Kyle, P.R. 1990f. Mount Discovery. *American Geophysical Union Antarctic Research Series*, **48**, 120–123.

Wright, A.C. and Kyle, P.R. 1990g. Minna Bluff. *American Geophysical Union Antarctic Research Series*, **48**, 117–119.

Wright, A.C. and Kyle, P.R. 1990h. Mount Morning. *American Geophysical Union Antarctic Research Series*, **48**, 124–127.

Wright, A.C. and Kyle, P.R. 1990i. Mason Spur. *American Geophysical Union Antarctic Research Series*, **48**, 128–130.

Wright, A.C. and Kyle, P.R. 1990j. Royal Society Range. *American Geophysical Union Antarctic Research Series*, **48**, 131–133.

Wright, A.C. and Kyle, P.R. 1990k. Taylor and Wright valleys. *American Geophysical Union Antarctic Research Series*, **48**, 134–137.

Zimanowski, B. and Büttner, R. 2003. Phreatomagmatic explosions in subaqueous volcanism. *American Geophysical Union Geophysical Monograph Series*, **140**, 51–60.

Chapter 5.2b

Erebus Volcanic Province: petrology

Adam P. Martin[1]*, Alan F. Cooper[2], Richard C. Price[3], Philip R. Kyle[4] and John A. Gamble[5,6]

[1]GNS Science, Private Bag 1930, Dunedin, New Zealand

[2]Department of Geology, University of Otago, PO Box 56, Dunedin, New Zealand

[3]Science and Engineering, University of Waikato, Hamilton, New Zealand

[4]Department of Earth and Environmental Science, New Mexico Institute of Mining and Technology, Socorro, NM 87801, USA

[5]School of Geography, Environment and Earth Sciences, Victoria University of Wellington, Wellington, New Zealand

[6]School of Biological, Earth and Environmental Science, University College Cork, Distillery Fields, North Mall, Cork, Ireland

APM, 0000-0002-4676-8344; PRK, 0000-0001-6598-8062

*Correspondence: A.Martin@gns.cri.nz

Abstract: Igneous rocks of the Erebus Volcanic Province have been investigated for more than a century but many aspects of petrogenesis remain problematic. Current interpretations are assessed and summarized using a comprehensive dataset of previously published and new geochemical and geochronological data. Igneous rocks, ranging in age from 25 Ma to the present day, are mainly nepheline normative. Compositional variation is largely controlled by fractionation of olivine + clinopyroxene + magnetite/ilmenite + titanite ± kaersutite ± feldspar, with relatively undifferentiated melts being generated by <10% partial melting of a mixed spinel + garnet lherzolite source. Equilibration of radiogenic Sr, Nd, Pb and Hf is consistent with a high time-integrated HIMU *sensu stricto* source component and this is unlikely to be related to subduction of the palaeo-Pacific Plate around 0.5 Ga. Relatively undifferentiated whole-rock chemistry can be modelled to infer complex sources comprising depleted and enriched peridotite, HIMU, eclogite-like and carbonatite-like components. Spatial (west–east) variations in Sr, Nd and Pb isotopic compositions and Ba/Rb and Nb/Ta ratios can be interpreted to indicate increasing involvement of an eclogitic crustal component eastwards. Melting in the region is related to decompression, possibly from edge-driven mantle convection or a mantle plume.

Supplementary material: Whole-rock, clinopyroxene, englacial tephra and marine drill-core volcanic rock and glass chemistry ± isotopes ± chronology from published sources (referenced) and new data (ESM1) are available at https://doi.org/10.6084/m9.figshare.c.5199416

Volcanoes, volcanology and geology were the subjects of the earliest scientific observations in the Ross Sea (Ross 1847), and they formed part of the justification for scientific exploration during the heroic era (*c.* 1900–20). The samples collected in these early studies are still of value and they are part of the dataset used in this chapter (e.g. Prior 1899, 1902, 1907; Thomson 1916; Smith 1954). Igneous rocks were extensively studied as part of the International Geophysical Year programme (the third Polar Year) of 1957–58, and a significant outcome was published descriptions of the petrography of McMurdo Sound igneous rocks (e.g. Harrington 1958*b*, 1965; McCraw 1962). Since the International Geophysical Year, numerous workers have contributed to understanding the petrogenesis of igneous rocks in the McMurdo Sound region, using techniques such as geochronology (e.g. Armstrong 1978; Kelly *et al.* 2008; Martin *et al.* 2010), chemistry and petrology (e.g. Wright-Grassham 1987; Cooper *et al.* 2007), and isotopes (Sun and Hanson 1975; Phillips *et al.* 2018). The most advanced petrological techniques have been applied to igneous rocks from Ross Island (e.g. Kyle *et al.* 1992; Kelly *et al.* 2008; Sims *et al.* 2013; Iacovino *et al.* 2016) and the actively convecting, phonolite lake of Mount Erebus (Kyle *et al.* 1992; Moussallam *et al.* 2013, 2015). Drilling, onshore but mainly offshore, has also contributed considerably to the understanding of McMurdo Sound igneous volcanism, with significant programmes including drilling by the Dry Valley Drilling Project (DVDP), the Cape Roberts Drilling Project (CRP: Cape Roberts Science Team 1999) and the Antarctic Drilling Project (ANDRILL: Pompilio *et al.* 2007; Panter *et al.* 2008; Naish *et al.* 2009). A record of McMurdo Sound volcanic activity is also found in geophysical studies and studies of englacial tephra. This chapter builds on the seminal work presented in the Antarctic Research Series on Antarctic volcanism (Kyle 1990*b*).

The definition of the Erebus Volcanic Province used here closely follows Kyle (1990*b*) as modified by Smellie and Martin (2021; see Fig. 1). It includes Cenozoic-aged igneous rocks erupted and emplaced between Franklin Island and Mason Spur, volcanic islands in the Ross Sea, occurrences in the foothills of the Transantarctic Mountains, tephra in rock, ice core and in sediments, and volcanic rocks from the seafloor (Fig. 1). This chapter will provide a thorough overview of more than 1000 whole-rock and glass analyses, more than 100 isotope analyses on whole rocks and crystals, and more than 100 radiometric dates (Table 1) from the Erebus Volcanic Province. This includes both published (78% of the data) and unpublished analyses. An extensive list of references is provided. Chronology and physical volcanology of the Erebus Volcanic Province are described in Smellie and Martin (2021), and radiometric ages are only used in this chapter to investigate petrogenetic trends. A marine record of Antarctic volcanism from drill cores is also discussed by Di Roberto *et al.* (2021), and only drill cores from the Erebus Volcanic Province (onshore and offshore) are summarized in this chapter. Clinopyroxene chemistry of relatively undifferentiated compositions will be provided where available; 'relatively undifferentiated' is defined here as whole-rock compositions with SiO_2 <55 wt% and MgO >6 wt% (definition after Sprung *et al.* 2007). These analyses are included in the Supplementary material (ESM1), with each analysis linked to a bibliographical reference. An overview of the geological setting is presented, followed by a presentation of a refined subdivision of the province. A petrological review of major volcanic islands and other occurrences in the Transantarctic Mountain

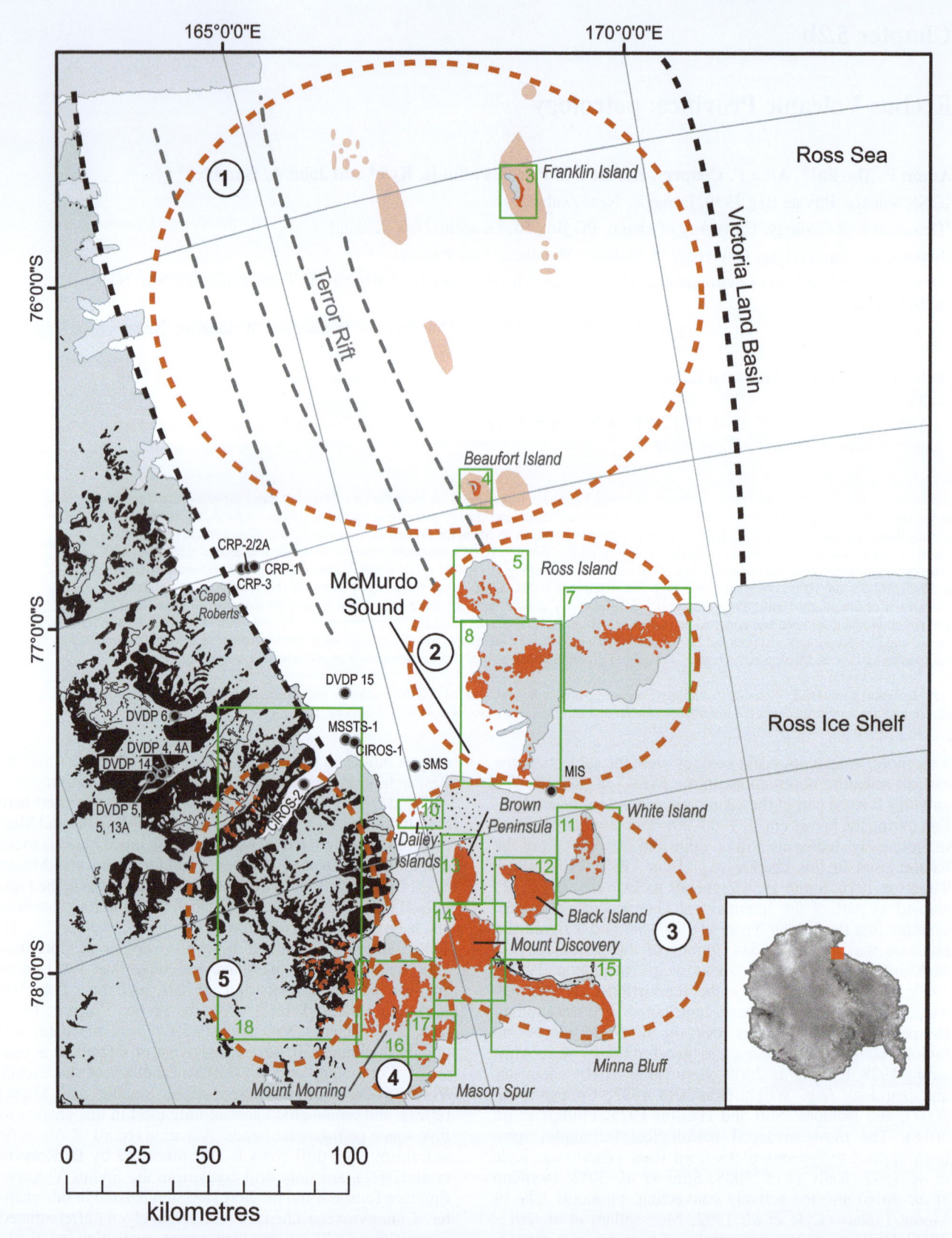

Fig. 1. A map of southern Victoria Land showing the extent of the Erebus Volcanic Province, significant offshore drill-hole locations and key topographical features. The inset map shows the location within Antarctica. Erebus Volcanic Province rock outcrop is shown in red (subaerial) or light red (subaqueous: interpreted from dredge samples and geophysical studies: Lawver *et al.* 2012). Other older rock outcrops are shown in black. The stippled pattern indicates supraglacial till. The geological units here follow the classification of Cox *et al.* (2012). The Victoria Land Basin is located between the dashed black lines, with the western boundary marking the Transantarctic Mountain Front Fault. The positions of the faults associated with the Terror Rift are shown by thin, grey dashed lines (Cooper *et al.* 1987). Offshore drill-hole locations (filled circles) are shown. The green boxes indicate the location of some figures used elsewhere in this study, with the corresponding figure number shown in green text. The red dashed lines indicate the approximate position of volcanic fields and suites within the Erebus Volcanic Province: 1, Terror Rift Volcanic Field; 2, Ross Island Volcanic Field; 3, Mount Discovery Volcanic Field; 4, Mount Morning Volcanic Field; 5, Southern Local Suite. CIROS, Cenozoic Investigations into the western Ross Sea drill-hole collar locations; CRP, Cape Roberts Project; DVDP, Dry Valley Drilling Project drill-hole collar locations; MSSTS, McMurdo Sound Sediment and Tectonic Study drill-hole collar locations; MIS (McMurdo Ice Shelf) and SMS (Southern McMurdo Sound) drill-hole collar locations from ANDRILL (Antarctic Drilling Project). The scale bar is accurate at 78° S.

Table 1. *Summary of the number of analyses and type from each locality or sample type used in this study of the Erebus Volcanic Province*

		Whole-rock chemistry	Isotopes*	Chronology[†]
Terror Rift Volcanic Field	Beaufort Island	3	–	–
	Franklin Island	9	1	3
	Submarine volcanic rocks	17	17	13
	Sub-total	*29*	*18*	*16*
Ross Island Volcanic Field	Cape Crozier	26	5	–
	Holes DVDP 1 and 2	99	11	4
	Hut Point Peninsula	60	17	–
	Mount Bird	33	14	–
	Mount Erebus	107	34	15
	Mount Terror	30	17	–
	Sub-total	*355*	*106*	*19*
Mount Discovery Volcanic Field	Black Island	10	1	–
	Brown Peninsula	8	–	–
	Dailey Islands	1	–	1
	Minna Bluff	39	1	8
	Mount Discovery	49	1	7
	White Island	25	–	4
	Sub-total	*132*	*3*	*20*
Mount Morning Volcanic Field	Mason Spur	110	2	8
	Mount Morning	248	10	22
	Sub-total	*358*	*12*	*30*
Southern Local Suite	–	21	1	–
	Total	895	140	85
Other	Historical Erebus bombs	62	28	28
	Ross Island englacial tephra	37	–	–
	ANDRILL lava	11	–	–
	ANDRILL glass	68	–	24
	Sub-total	*178*	*28*	*52*
	Grand total	1073	168	137

*Where whole-rock chemistry and isotope analysis have been determined on the same sample.
[†]Chronology where the whole-rock chemistry can be associated with the radiometric age with a high degree of certainty (i.e. from the same sample).

foothills, drill core and elsewhere is given along with a review of historical work undertaken in each area. Current lithological maps are provided for many of the major volcanic centres. The whole-rock geochemistry and geothermobarometry of the province as a whole are discussed using the Supplementary material (ESM1) dataset. In the discussion, competing petrogenetic hypotheses are evaluated, and the various mantle sources, age of chemical heterogeneity of the mantle sources, asthenospheric v. lithospheric and types of melting are considered. The final sections of the chapter include suggestions regarding pertinent areas for future research, a summary and a concluding statement.

Geological setting

The earliest, published geological map of the Erebus Volcanic Province rocks is based on notes made by Hartley Ferrar during the 1901–04 British National Antarctic Expedition (Ferrar 1905) and work since then has aimed at adding extra detail, with explanations placed in a modern-science context (Cox *et al.* 2012) (Fig. 2). The region is part of the intracontinental West Antarctic Rift System (Behrendt 1999), with the Transantarctic Mountain Front at the shoulder of the rift. East Antarctica comprises mainly Precambrian cratons (Bentley 1991) compared to West Antarctica, which includes a mosaic of younger crustal blocks (Talarico and Kleinschmidt 2008), with the precise lithological boundary between East and West Antarctica being contentious. The Moho shallows to around 20 km in McMurdo Sound and deepens to around 40 km beneath the Transantarctic Mountains (Bannister *et al.* 2003; An *et al.* 2015). Based on the study of crustal xenoliths, the lower crust beneath the Transantarctic Mountains, Mount Morning and Mason Spur is mainly calc-alkalic, whereas further east, away from the Transantarctic Mountain Front Fault, the lower crust is more alkali–tholeiitic (Berg *et al.* 1985; Kalamarides *et al.* 1987; Martin *et al.* 2015*a*). Subduction beneath the region where the Erebus Volcanic Province now sits last occurred at *c.* 0.5 Ga when the palaeo-Pacific Plate subducted westwards beneath the Gondwana margin (Stump 1995).

Some of the most recent tectonic activity in the West Antarctic Rift System has occurred on faults within the Terror Rift (Cooper *et al.* 1987), with regional correlations suggesting that faulting must be younger than 17 Ma (Fielding *et al.* 2006), and some faults cutting the seafloor may be indications of modern activity (Hall *et al.* 2007). Onshore faulting at Mount Morning occurred post-3.9 Ma (Martin and Cooper 2010). During rifting, the Antarctic Plate is thought to have remained effectively stationary since *c.* 60 Ma (Grindley and Oliver 1983; Torsvik *et al.* 2008). Kyle and Cole (1974) first noted a three-fold radial symmetry of vents around Mount Erebus and Mount Discovery, and suggested these could be related to radial fractures at approximately 120° to each other. They postulated that these fractures resulted from crustal doming, whilst Kyle *et al.* (1992) suggested doming at Ross Island is the result of a mantle plume.

Cenozoic volcanism in the province commenced as recently as 18.7 Ma but tephra in cores, inferred to be sourced within the province, indicate that volcanism is likely to have been initiated around 25 Ma (Kyle and Muncy 1989; McIntosh 2000; Martin *et al.* 2010). Mount Erebus is still volcanically active

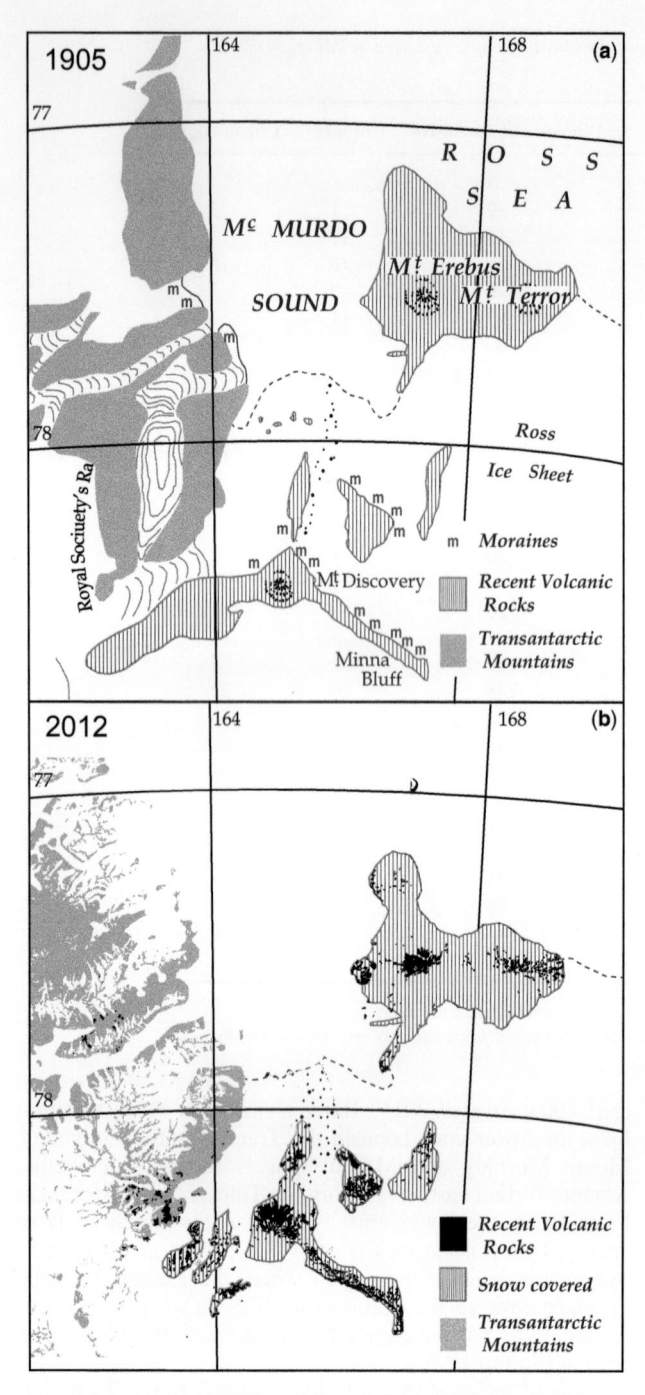

Fig. 2. Geological maps of the Erebus Volcanic Province from two separate eras. (**a**) Possibly the earliest, published geological map of the province, drawn by Ferrar (1905). (**b**) The current state of knowledge (Cox et al. 2012). The shading has been adapted to match the 1905 publication and highlights the quality of the original mapping. Advances in the intervening century have been in understanding petrogenesis of the Erebus Volcanic Province.

today and, based on chronology and abnormally high heat flow, several eruptive centres are considered dormant rather than extinct. Volcanic rock compositions are alkalic, and typically range from alkali basalt and basanite to phonolite and trachyte (Fig. 3). Typical, primitive whole-rock compositions are presented in Table 2. The Erebus Volcanic Province is the southernmost of three provinces that comprise the McMurdo Volcanic Group (Kyle 1990a). To the north is the Melbourne Volcanic Province and north of that is the Hallett Volcanic Province. The McMurdo Volcanic Group occurs along a c. 2000 km-long portion of the western shoulder of the West Antarctic Rift System between Cape Adare and Mount Early. It has long been recognized (e.g. Coombs et al. 1986; Panter et al. 2006) that McMurdo Volcanic Group rocks, including Erebus Volcanic Province rocks, have a unifying character that is ocean island basalt (OIB)- and HIMU-like (high-μ: enriched in ^{206}Pb and ^{208}Pb and relatively depleted in ^{87}Sr/^{86}Sr values). Finn et al. (2005) referred to this as a diffuse alkaline magmatic province (DAMP) encompassing West Antarctica, several sub-Antarctic islands, Zealandia, eastern Australia and Papua New Guinea.

Subdivision of the Erebus Volcanic Province

Harrington (1958b) introduced the term 'McMurdo Volcanics' to describe late Cenozoic volcanic rocks in the Western Ross Sea, including those around Ross Island in McMurdo Sound. Kyle (1990a) recommended formal recognition of the McMurdo Volcanic Group. Cole and Ewart (1968) undertook the first detailed mapping of McMurdo Volcanic Group rocks in McMurdo Sound. They defined four stratigraphic formations on Black Island, which they then extrapolated to several volcanic localities in the region, including Ross Island's Cape Bird (Cole and Ewart 1968), Cape Crozier and the Hut Point Peninsula (Cole et al. 1971). These formations consisted of an early basaltic unit overlain by trachyte (now known from chemical data to be phonolite), then a younger basaltic unit overlain by a younger trachyte (phonolite). It is now also known from geochronology that, although the eruptive sequence of alternating basanite and phonolite is very similar throughout the McMurdo Sound area, it was repeated at different times. Consequently, the formations of Cole and Ewart (1968) have not been widely used in the description of the geology. Instead, Kyle and Cole (1974) used tectonic setting and spatial distribution to assign the occurrences of McMurdo Volcanics (Group) to provinces, including the Erebus Volcanic Province. This system has now endured for more than 40 years.

In this chapter, the Cenozoic volcanic rocks in the Erebus Volcanic Province have been further categorized geographically and petrogenetically, using the thickness and composition of the lithosphere through which they have erupted and the relative clinopyroxene–melt equilibration depth from which primitive magmas originated (Table 3; this study). Systematic patterns in trace element and isotopic ratios across the province record these petrogenetic changes. The groupings are (Fig. 1): (1) Terror Rift Volcanic Field, (2) Ross Island Volcanic Field, (3) Mount Discovery Volcanic Field, (4) Mount Morning Volcanic Field and (5) Southern Local Suite. These names, and the method of defining them, have been applied sympathetically with other studies in the McMurdo Volcanic Group reported in this Memoir (Smellie and Rocchi 2021), and with other studies in Victoria Land (Rocchi et al. 2002).

Petrological overview and previous work

This section describes the history of field visits and subsequent petrological and geochemical research undertaken in each volcanic field. The section is divided geographically, and the volcanic centres are described in a level of detail that reflects the complexity of work undertaken there. Subsequently, volcanic rock from marine drill core and englacial tephra are reviewed. This section builds a holistic picture of the province that allows a regional overview approach to be adopted in the geochemistry and discussion sections below. A summary of the minimum age range and mineral assemblage in relatively undifferentiated rocks of the Erebus Volcanic Province are shown in Table 4. Thin-section photographs of relatively

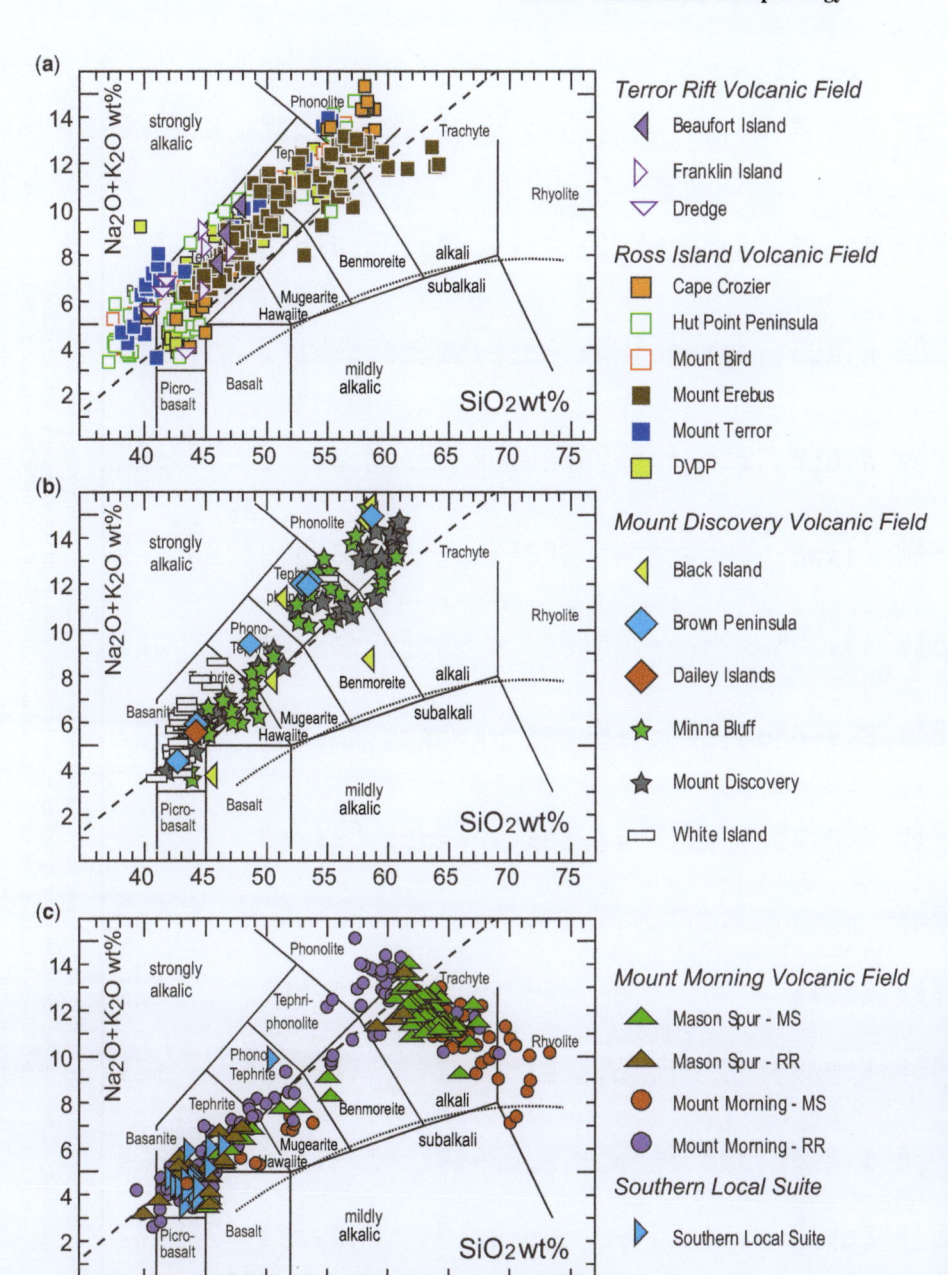

Fig. 3. Total alkali v. silica diagram after Le Bas *et al.* (1986). Data (n = 895; 100%; anhydrous) are shown for the Erebus Volcanic Province whole-rock compositions from the Supplementary material (ESM1). The alkali–subalkali division (dotted line) follows Kuno (1966), and the mildly alkalic–strongly alkalic division (dashed line) follows Saggerson and Williams (1964). (**a**) Data for the Terror Rift Volcanic Field and the Ross Island Volcanic Field. (**b**) Data for the Mount Discovery Volcanic Field. (**c**) Data for the Mount Morning Volcanic Field and Southern Local Suite. MS, Mason Spur Lineage; RR, Riviera Ridge Lineage.

undifferentiated rock types typical of those found in the Erebus Volcanic Province are shown in Figure 4.

Terror Rift Volcanic Field

Franklin Island. Named by James Ross in 1841 for Sir John Franklin, Lieutenant-Governor of Tasmania and Arctic explorer (Ross 1847), Franklin Island is the eroded remnant of a shield volcano. A map of Franklin Island (Fig. 5) shows volcanic rock outcrops and key topographical features. In all map figures (Fig. 5 et seq.), which are adapted from Cox *et al.* (2012), 'basaltic' includes basalts, basanites and intermediate mafic compositions, such as hawaiite and mugearite, contours (grey lines) are in metres, and white areas indicate snow and/or ice cover.

At Franklin Island, lava flows and interbedded tuffs are common, and some dykes occur. A sample collected by the Ross party (Ross 1847) was described by Prior (1899). Subsequent visits in the 1900s (Borchgrevink 1901; Scott 1907), the 1960s (Waterhouse 1965), the 1980s (Ellerman and Kyle 1990*b*) and in 2004 (Rilling *et al.* 2009) collected geological and other samples. Shipborne investigations near Franklin Island were made in 1958 (Brodie 1959), 2004 (Rilling *et al.* 2009) and 2011 (Lawver *et al.* 2012), with several submarine volcanoes being identified (Rilling *et al.* 2009; Lawver *et al.* 2012). The age of emplacement of Franklin Island volcanic rocks is between 3.70 ± 0.05 and 3.30 ± 0.04 Ma (Armstrong 1978; Rilling *et al.* 2009). The four published whole-rock analyses from Franklin Island are basanites (Fig. 3), and mantle nodules and xenocrysts of olivine and clinopyroxene have been noted from this locality (Ellerman and Kyle 1990*b*). Olivine ± plagioclase phenocrysts occur in a fine-grained groundmass of olivine + clinopyroxene + plagioclase + opaque minerals. Glass vesicularity appears to increase upsection (Ellerman and Kyle 1990*b*).

Beaufort Island. In 1841, Captain J.C. Ross named Beaufort Island after the hydrographer, Captain Francis Beaufort (Ross 1847). The island was visited briefly in 1903 but was only described from remote observations (Ferrar 1907; Debenham 1923; Harrington 1958*a*) until New Zealand geologists landed in 1965 (Waterhouse 1965) and others in the 1980s (LeMasurier *et al.* 1983), and samples were collected again in the 2000s

Table 2. Typical whole-rock compositions of Erebus Volcanic Province

Area	Terror Rift Volcanic Field		Ross Island Volcanic Field						Mount Discovery Volcanic Field						Mount Morning Volcanic Field				Southern Local Suite
Location	Beaufort	Franklin	Dredge	Crozier	DVDP	Hut Point	Bird	—	Terror	Black	Brown	Dailey	Minna	Discovery	White	Mason (old)	Mason (young)	Morning (young)	Morning (old)	—
Reference	1	2	3	4	5	6	7	20	9	10	11	12	13	14	15	16	17	18	19	20
Sample No.	1	HC-FR1-2	TRDR02	CCO4	2-105.53	3-179.40	RI115	5	RI17	12	1	2007	AW82113A	AW84738	OU 74785	OU78634	OU78578	OU78710	OU78559	5
TAS	Tephrite	Basanite	Tephrite	Basanite	Basanite	Basanite	Basanite	Basanite	Basanite	Basanite	Basanite	Basanite	Basanite	Basanite	Basanite	Basanite	Basanite	Basanite	Basanite	Basanite
Mg#	50.36	64.16	57.37	50.67	69.43	52.43	48.48	57.42	55.84	64.19	64.57	63.68	61.82	64.47	69.69	54.18	72.87	34.92	60.65	57.42
DI	42.52	35.07	34.57	26.59	23.88	19.88	19.81	32.43	19.50	23.39	25.19	29.84	33.52	26.88	23.51	37.31	21.33	51.47	27.99	32.43
(wt%)																				
SiO_2	45.89	44.83	43.18	40.23	42.02	37.62	37.78	43.60	38.51	45.42	42.71	44.13	45.18	43.31	43.22	45.66	45.39	51.51	44.13	43.60
TiO_2	3.37	2.27	2.98	2.51	4.09	3.71	3.15	3.36	2.85	3.10	4.22	3.26	3.39	3.95	3.27	3.72	2.16	3.02	3.54	3.36
Al_2O_3	16.93	14.03	14.55	13.28	13.02	11.92	11.79	14.47	11.69	13.42	13.34	13.71	15.24	14.51	13.06	14.74	12.62	15.45	14.71	14.47
Fe_2O_3	2.63	2.79	2.69	3.12	1.92	3.45	2.06	2.22	3.27	1.91	2.06	1.95	1.82	1.93	1.89	2.08	1.74	3.14	2.00	2.22
FeO	8.78	9.31	8.96	15.58	9.60	17.23	18.40	11.10	16.36	9.54	10.32	9.73	9.12	9.65	9.44	10.42	8.72	8.96	9.99	11.10
MnO	0.11	0.22	0.17	0.17	0.18	0.15	0.17	0.21	0.16	0.19	0.21	0.19	0.20	0.19	0.19	0.22	0.19	0.27	0.21	0.21
MgO	5.00	9.35	7.30	8.97	12.23	10.65	9.71	8.40	11.60	9.59	10.55	9.56	8.28	9.83	12.18	6.91	13.14	2.70	8.64	8.40
CaO	9.43	10.13	12.10	9.87	11.41	10.20	10.31	10.12	10.59	12.62	11.38	11.27	10.29	11.15	11.65	9.42	12.01	6.98	11.37	10.12
Na_2O	5.10	4.74	5.06	4.12	3.19	2.84	2.93	4.36	2.99	2.63	3.71	3.89	3.57	3.32	3.26	3.94	2.46	4.40	3.63	4.36
K_2O	2.35	1.69	2.02	1.53	1.49	1.48	1.38	1.53	1.25	1.04	0.65	1.69	1.93	1.37	1.15	1.81	1.20	2.41	1.12	1.53
P_2O_5	0.41	0.64	0.98	0.62	0.85	0.75	0.69	0.64	0.73	0.55	0.85	0.62	0.96	0.79	0.69	1.06	0.36	1.15	0.67	0.64
Total	100	100	100	100	100	100	100	100	100	100	100	100	100	100	100	100	100	100	100	100
(ppm)																				
Ba	—	—	767	596	277	406	445	410	494	240	290	—	484	513	—	393	204	276	340	410
Be	—	—	—	—	—	—	—	2.7	—	2.2	2.7	—	—	—	—	—	—	—	—	2.7
Ce	—	—	144	112	109	112	107	bdl	102	bdl	bdl	—	127	119	94	96	52	106	91	bdl
Co	—	—	—	50	—	59	57	54	56	45	56	—	—	—	—	59	63	8	64	54
Cr	—	—	—	465	469	449	384	210	673	510	460	—	288	321	—	218	722	30	363	210
Cs	—	—	0.70	0.59	bdl	0.32	0.35	—	0.39	—	—	—	0.35	0.06	0.30	—	—	—	—	—
Cu	—	54	—	—	—	—	—	50	—	85	41	—	33	43	85	—	—	—	—	50
Dy	—	—	6.8	5.5	5.0	6.4	6.6	—	6.3	—	—	—	—	—	—	4.2	3.8	3.0	5.4	—
Er	—	—	3.3	2.7	2.0	2.9	3.0	—	2.8	—	—	—	3.3	3.2	2.7	bdl	1.8	1.8	2.5	—
Eu	—	—	3.4	2.9	2.5	3.1	3.1	—	2.9	—	—	—	19	20	15	2.7	1.7	2.4	2.4	—
Ga	—	21	—	—	—	—	—	—	—	—	—	—	—	—	—	46	15	223	19	—
Gd	—	—	9.5	7.3	8.4	9.0	8.8	—	8.4	—	—	—	—	—	—	5.81	4.5	3.0	6.9	—
Hf	—	—	6.0	5.9	5.4	7.9	6.7	—	6.0	—	52	—	8.5	7.1	7.2	9.2	3.9	2.7	5.2	—
Ho	—	—	1.2	1.0	1.0	1.2	1.2	—	1.1	—	—	—	—	—	5.8	1.0	0.72	0.56	0.97	—
La	—	—	78	57	68	58	53	54	53	35	—	—	60	55	55	47	26	29	45	54
Lu	—	—	0.40	0.32	—	0.33	0.35	—	0.34	—	—	—	0.41	0.32	0.30	0.82	0.21	0.17	0.29	—
Nb	—	104	108	95	56	81	79	110	75	—	—	—	97	87	60	50	32	24	54	110
Nd	—	—	60	47	—	55	51	4.5	49	—	52	—	63	55	45	43.4	24	13	42	4.5
Ni	—	184	70	209	276	245	206	—	293	150	200	—	127	162	259	170	288	13.8	145	—
Pb	—	2	3.8	3.9	4.3	2.8	2.5	42	2.7	1.6	2.4	—	3.0	5.0	0.95	439	3.7	1.9	3.4	42
Pr	—	9	9.8	13	15	14	13	—	12.6	—	—	—	—	6.1	5.6	10.1	6.0	6.0	5.2	—
Rb	—	59	48	53	30	33	35	—	34	36	24	—	39	37	32	36	22	11	10	—
Sc	—	—	—	27	—	36	32	—	34	—	—	—	23	28	—	19	29	14	29	—
Sm	—	—	11	8.7	10	10	10	—	9.5	—	—	—	11	10	9.1	8.7	5.3	bdl	8.6	—
Sr	—	—	999	966	829	861	730	1000	953	640	880	—	1030	1153	861	999	429	107	794	1000
Ta	—	—	6.7	5.7	0.90	5.3	5.2	—	5.1	—	—	—	5.8	5.5	4.2	2.1	1.7	0.80	3.1	—
Tb	—	35	1.2	1.0	—	1.2	1.2	3.2	1.1	2.7	31	—	1.2	1.2	2.1	1.2	1.4	1.2	1.5	3.2
Th	—	—	9.8	7.5	4.6	6.1	5.6	153	5.7	—	110	—	5.9	6.1	5.6	3.4	3.7	1.9	5.2	153
Tm	—	—	0.40	0.37	—	0.38	0.41	—	0.39	—	—	—	—	—	0.32	bdl	0.30	0.71	0.34	—
U	—	—	2.4	2.5	1.0	1.8	1.2	360	1.4	—	330	—	1.70	2.1	1.5	1.3	6.0	2.0	1.7	360
Y	—	—	—	251	300	288	271.6	240	275	300	280	—	243	317	275	234	22	bdl	305	240
Yb	—	32	32	27	—	32	30	31	28.7	23	31	—	34	30	27	24	18	10	24	31
Zn	—	—	2.7	2.2	1.6	2.4	2.5	3.2	2.4	2.7	3.4	—	2.6	2.2	2.1	bdl	1.9	1.6	2.2	3.2
Zr	—	105	—	—	82	—	—	153	—	126	110	—	89	84	79	231	—	—	—	153
	—	275	266	273	—	363	308	360	266	260	330	—	401	311	249	162	—	99	222	360
$^{87}Sr/^{86}Sr$	—	—	0.703 082	0.702 975	0.702 991	0.703 002	0.703 144	—	0.703 032	0.7032	—	—	—	—	—	0.703 350	—	—	—	—
$^{143}Nd/^{144}Nd$	—	—	0.512 949	0.512 929	0.512 922	0.512 891	—	—	—	—	—	—	—	—	0.512 852	—	—	—	—	—
$^{206}Pb/^{204}Pb$	—	—	19.434	19.5268	19.496	20.1724	19.7321	19.592	19.5684	—	—	—	—	—	—	19.579	—	—	—	19.592
$^{207}Pb/^{204}Pb$	—	—	15.601	15.6146	15.666	15.6663	15.6253	15.602	15.6197	—	—	—	—	—	—	15.645	—	—	—	15.602
$^{208}Pb/^{204}Pb$	—	—	39.020	39.0721	39.096	39.6660	39.4167	39.285	39.0898	—	—	—	—	—	—	39.268	—	—	—	39.285
Nepheline	17.7	17.1	23.2	18.9	14.1	13.0	13.4	15.9	13.7	6.0	11.9	15.4	9.6	11.0	12.8	8.0	7.8	0.0	11.0	15.9
Diopside	22.7	27.5	34.6	26.3	27.4	29.0	21.4	25.0	21.7	30.1	26.6	29.8	20.2	23.8	28.6	18.5	30.0	9.9	25.5	25.0
Olivine	7.3	15.5	8.0	23.2	19.4	29.1	29.1	16.1	30.1	14.8	17.4	14.9	15.1	16.6	19.7	14.5	21.4	2.2	14.8	16.1

Data are taken from the Supplementary material (ESM1). Mineral abundances are CIPW-normative values. TAS: total alkali v. silica diagram classification; bdl, below (lower) detection limit for the method; –, not determined; wt%, weight percentage; ppm, parts per million. Mg#: [Mg/(Mg + Fe^{2+}) × 100] – atomic ratio; Di: differentiation index. Totals are 100% anhydrous. Iron is recalculated based upon Middlemost (1989). See Table 1 and Figure 1 for the location information.

References: 1, Ellerman and Kyle (1990a); 2, Ellerman and Kyle (1990b); 3, Lee et al. (2015); 4, Phillips et al. (2018); 5, Kyle (1981b); 6, Rasmussen et al. (2017); 7, Rasmussen et al. (2017); 8, Kyle (1976); 9, Rasmussen et al. (2017); 10, Goldich et al. (1975); 11, Goldich et al. (1975); 12, Del Carlo et al. (2009); 13, Wright-Grassham (1987); 14, Wright-Grassham (1987); 15, Cooper et al. (2007); 16, Martin et al. (2013); 17, Martin et al. (2013); 18, Martin et al. (2013); 19, Martin et al. (2013); 20, Goldich et al. (1975).

Table 3. *Subdivision of the Erebus Volcanic Province with key eruptive centres and key characteristics, including the nature of the underlying crust and lithosphere, and the estimated depth of clinopyroxene–melt equilibration*

Erebus Volcanic Province		Lower crust	Lithosphere	Clinopyroxene–melt equilibration depth
Terror Rift Volcanic Field		Alkalic–Tholeiitic	Thin (<20 km)	Shallow (c. 1 GPa)*
	Franklin Island			
	Beaufort Island			
	Submarine volcanic rocks			
Ross Island Volcanic Field		Alkalic–Tholeiitic	Thin (<20 km)	Deep (>2 GPa)
	Mount Terror			
	Mount Erebus			
	Mount Bird			
	Hut Point Peninsula			
Mount Discovery Volcanic Field		Alkalic–Tholeiitic	Thin (<20 km)	Shallow (c. 1 GPa)
	Dailey Islands			
	White Island			
	Black Island			
	Brown Peninsula			
	Mount Discovery			
	Minna Bluff			
Mount Morning Volcanic Field		Calc-alkalic crust	Thin (<20 km)	Shallow (c. 1 GPa)
	Mount Morning			
	Mason Spur			
Southern Local Suite		Calc-alkalic crust	Thick (>20 km)	Deep (>2 GPa)

*Not calculated but assumed by comparison with similar data on the normative ternary diagram (Fig. 25).

(Rilling *et al.* 2009). Beaufort Island is the eroded, near-vent remnant of a stratovolcano (Fig. 6). The only ages determined from the island overlap at 6.80 ± 0.05 and 6.77 ± 0.03 Ma (Rilling *et al.* 2009). Less than 5% of Beaufort Island's total volume lies above current sea-level. Lava flows, volcanic breccias and tuffs are cross-cut by numerous dykes which have hydrothermally altered the wall rock. Three analyses from Ellerman and Kyle (1990*a*) and two dated samples

Table 4. *Summary of the mineralogy in relatively undifferentiated rocks from the Erebus Volcanic Province*

Erebus Volcanic Province		Potential age range		Phenocrysts	Groundmass
		From	To		
Terror Rift Volcanic Field					
	Franklin Island	3.70 ± 0.05	3.30 ± 0.04	Ol + Pl	Ol + Cpx + Pl + Opq
	Beaufort Island	6.80 ± 0.05	6.77 ± 0.03	Cpx + Pl + Krs	Cpx + Pl + Krs + Nph ± Ap
	Submarine volcanic rocks	3.73 ± 0.05	0.09	Ol + Cpx	Cpx + Ol + Fe oxides ± Pl
Ross Island Volcanic Field					
	Mount Terror	c. 2.5	0.82 ± 0.14	Ol + Aug	
	Mount Erebus	1.311 ± 0.016	0.040 ± 0.006	Ol + Cpx + Fsp + Ap	Ol + Cpx + Nph + Fsp + Opq + Ap
	Mount Bird	4.62 ± 0.6	3.08 ± 0.15	Pl + Ol + Aug	Pl + Aug + Hbl + Opq
	Hut Point Peninsula	1.68 ± 0.06	0.33 ± 0.02	Ol + Cpx	Ol + Amp + Cpx + Pl + Cr-Spl + titano-Mag + Ilm + Ap + glass
Mount Discovery Volcanic Field					
	Dailey Islands	0.78 ± 0.04	0.76 ± 0.02	Ol + Cpx	Cpx + Pl _ Ol + Mag
	White Island	7.62	0.17	Ol + Di ± Krs	Aug + Mag + Bt + Pl + Nph
	Black Island	10.9 ± 0.4	1.689 ± 0.003	Ol + Aug + Pl + Fe oxides	Pl + Ol + Aug
	Brown Peninsula	2.7	2.2	Ol + Cpx	Pl + Opq + glass
	Mount Discovery	5.46 ± 0.16	0.06 ± 0.006	Ol + Cpx	Pl + Opq + Cpx + Nph
	Minna Bluff	12	4	Ol + Cpx	Pl + Opq + Ap
Mount Morning Volcanic Field					
	Mount Morning – MS	25	13.0 ± 0.3	Pl + Di + Ol	Pl + Cpx + Ilm + Ap
	Mount Morning – RR	5.01 ± 0.04	0.06 ± 0.08	Pl + Cpx + Ol	Pl + Afs + Ne + Opq + Ol + Ap + Cpx
	Mason Spur – MS	12.9 ± 0.1	11.4 ± 0.1	Pl + Di + Ol	Afs + Nph + Pl + Opq + Ap
	Mason Spur – RR	6.13 ± 0.2	0.07 ± 0.08	Pl + Cpx + Ol + Krs	Pl + Afs + Nph + Opq + Ol + Cpx + Ap
Southern Local Suite		13.8 ± 0.20	0.08 ± 0.13	Ol + Cpx ± Pl ± Fe-Ti oxides	Pl + Opq + Nph + Cpx

Mineral abbreviations follow Whitney and Evans (2010). Afs, alkali feldspar; Amp, amphibole; Ap, apatite; Aug, augite; Bt, biotite; Cpx, clinopyroxene; Cr-Spl, chromium spinel; Di, diopside; Fsp, feldspar; Hbl, hornblende; Ilm, ilmenite; Krs, Kaersutite; Mag, magnetite; Nph, nepheline; Ol, olivine; Opq, opaque; Pl, plagioclase. Age ranges are minimum ranges, see text for discussion. Mineralogy is only shown for relatively undifferentiated rock types in the province. MS, Mason Spur Lineage; RR, Rivera Ridge Lineage.

Fig. 4. (a) & (b) Petrographical photographs of relatively undifferentiated rock types typical of the Erebus Volcanic Province. The white bar in each photograph is 5 mm long. The photographs were taken in cross-polarized light. Cpx, clinopyroxene; Fsp, feldspar; Ol, olivine. Porphyritic texture in a fine-grained groundmass.

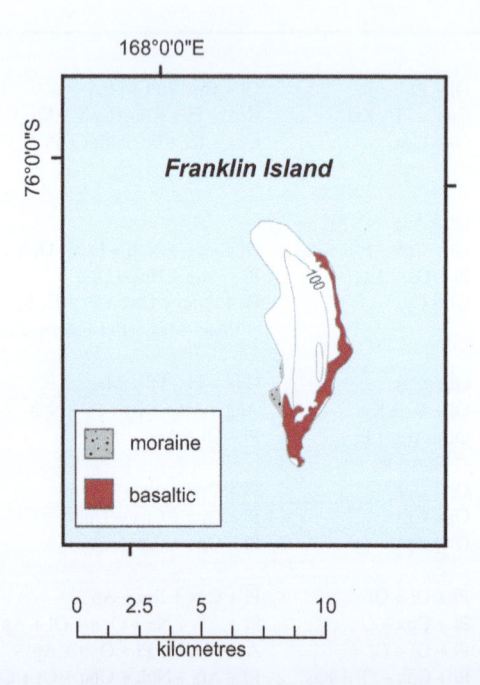

Fig. 5. Map of Franklin Island showing volcanic rock outcrops and key topographical features. In all map figures (Fig. 5 et seq.), 'basaltic' includes basalts, basanite and intermediate compositions, such as hawaiite and mugearite. This grouping reflects the scale of the original mapping, and follows Cox *et al.* (2012) for southern Victoria Land or LeMasurier (2013) for Marie Byrd Land. Contours (grey lines) are in metres, and white areas indicate snow and/or ice cover.

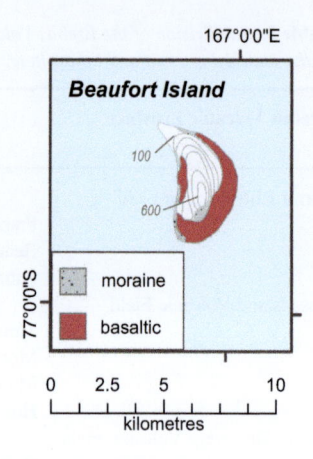

Fig. 6. Map of Beaufort Island showing volcanic rock outcrops and key topographical features.

from Rilling *et al.* (2009, fig. 6) are nepheline-normative and strongly alkali, and they define an apparent lineage between tephrite and phonotephrite (Fig. 3). Phenocrysts of titanaugite + plagioclase + kaersutite are observed with nepheline ± apatite in the groundmass (Ellerman and Kyle 1990*a*). Olivine is conspicuous by its absence. Most rocks are pervasively altered.

Submarine volcanic rocks in the Victoria Land Basin. Geophysical studies indicate that submarine volcanism is widespread in the Ross Sea region where it appears to be closely associated with extensional tectonics (Behrendt *et al.* 1991*a*). The Victoria Land Basin is a broad, sediment-filled half-graben extending for approximately 350 km from Mount Melbourne in the north to Mount Erebus in the south (Cooper *et al.* 1987; Lawver *et al.* 2007). Volcanic rocks with Erebus Volcanic Province affinities have been dredged from the flanks of Franklin Island, in the western Victoria Land Basin, and from the Terror Rift, an extensional trough along the central axis of the basin (Fig. 1).

During a geophysical cruise by the RVIB *Nathaniel B. Palmer* in 2004, samples of volcanic rock were dredged from the north and south flanks of Franklin Island. The samples were basaltic in composition (Fig. 3), and three were dated by the $^{40}Ar/^{39}Ar$ method. Ages ranged from 3.73 ± 0.05 to 3.28 ± 0.04 Ma (Rilling *et al.* 2007). More recently, this suite of dredged samples has been studied by Aviado *et al.* (2015), who documented the petrology and geochemistry of volcanic rocks from seven sea mounts sampled during the 2004 cruise and from subaerial samples from Franklin and Beaufort Islands. The dredged samples are vesicular, weakly porphyritic basanites (Fig. 3) containing phenocrysts of olivine and clinopyroxene in fine-grained groundmasses of clinopyroxene + olivine + iron oxide ± plagioclase. Ultramafic xenoliths are present in some samples and the olivine populations include grains interpreted as xenocrysts.

Samples of Erebus Volcanic Province basaltic rock were also dredged from the Terror Rift during the 2010–11 cruise of the RV *Araon*. Lee *et al.* (2015) carried out geochemical analysis and geochronology on samples from four of these dredge sites, and compared these data with those for the nearby Melbourne Volcanic Province. The dredged volcanic rock samples gave $^{40}Ar/^{39}Ar$ ages ranging from 3.55 ± 0.25 to 2.89 ± 0.18 Ma. Four whole-rock samples were analysed for major element, trace element, and Sr, Nd and Pb isotope composition (Lee *et al.* 2015). The analysed samples have basanite and tephri-basanite compositions (Fig. 3).

Ross Island Volcanic Field

Ross Island comprises the four main eruptive centres of Mount Bird, Mount Terror, Mount Erebus and the Hut Point Peninsula. Mount Terra Nova (named after the ship used by Robert Scott in his 1910–13 British *Terra Nova* Antarctic Expedition) is a 'parasitic' vent that sits on the western flanks of Mount Terror and is included here as part of the Mount Terror eruptive centre. The first sighting of Ross Island in January 1841 by Captain Ross was the beginning of an understanding of the volcanic geology of the region. Ross named mounts Erebus and Terror, after his two tiny sailing ships. Ross noted in his journal on 28 January 1841 'at 4 p.m. Mount Erebus was observed to emit smoke and flames in unusual quantities … and some of the officers believed they could see streams of lava pouring down its side' (Ross 1847, pp. 220–221). At the time, Captain Ross did not identify the volcanoes as part of an island as the Ross Ice Shelf blocked its circumnavigation.

The observations of Captain Ross were followed in the early 1900s by the heroic era of Antarctic exploration. Three British expeditions under the direction of Robert Scott (1901–04 and 1910–13) and Ernest Shackleton (1907–09) made Ross Island their base of operations. All three expeditions included geologists, and the first descriptions of the Ross Island physiography and petrology were made. Notable contributions on the petrology of the volcanic rocks of Ross Island (Ferrar 1907; Prior 1907; Jensen 1916; Smith 1954) were supplemented by descriptions of the physiography, which included many descriptions of the volcanic geology (David and Priestly 1914; Taylor 1922; Debenham 1923).

Mount Bird. Although there are several Adelie penguin rookeries on the western flank of Mount Bird, they are not the origin of the name. Captain Ross named features on the northern extremity of Ross Island after Lieutenant Edward J. Bird who served on the ship HMS *Erebus*. Mount Bird has a classic basaltic shield form with 11° slopes. The summit is 1800 m above sea-level, and the diameter of the volcano at sea-level is 23 km north–south and 18 km east–west, making it weakly ovoid (Fig. 7). The volume of the shield is estimated at 470 km^3 (Esser *et al.* 2004). Rock exposure on the volcano is limited to small scoria cones, a phonolite dome in the summit area, exposures around Lewis Bay to the east and a 10-km-long strip along the western coast in the Cape Bird area (Fig. 7). The timing of eruptions at Mount Bird is poorly constrained from four samples giving ages between 4.62 ± 0.6 and 3.08 ± 0.15 Ma (recalculated to current decay constants for the ^{40}K system: Armstrong 1978). There are marine cliffs bounding Lewis Bay along the eastern coast of Mount Bird which are mostly unexplored and assumed to be basaltic (Fig. 7), except for one area from which several phonolite rock samples were analysed by Phillips *et al.* (2018).

Cape Bird. Smith (1954) reported on several 'trachyte' samples from Cape Bird but very little was known of the volcanic geology at Cape Bird until the work of Cole and Ewart (1968). Recent sampling and geochemical analyses reported in this study (P.R. Kyle), and by Rasmussen *et al.* (2017) and Phillips *et al.* (2018), have provided new insights into the petrology and magmatic evolution and source for the Mount Bird eruptive centre. The geology of Cape Bird consists of basanite lava flows erupted from the main Mount Bird cone overlain by basanite scoria cones and phonolite domes (Cole and Ewart 1968). Lava flows of the main Mount Bird cone form a thick sequence of basanite lava flows which are well exposed along the coastal cliffs. The lava flows vary from about 10 m to less than 1 m in thickness, and typically have oxidized scoriaceous tops. Cole and Ewart (1968) recognized three, older, main Mount Bird cone 'basalt' rock units. The oldest, a 'hornblende

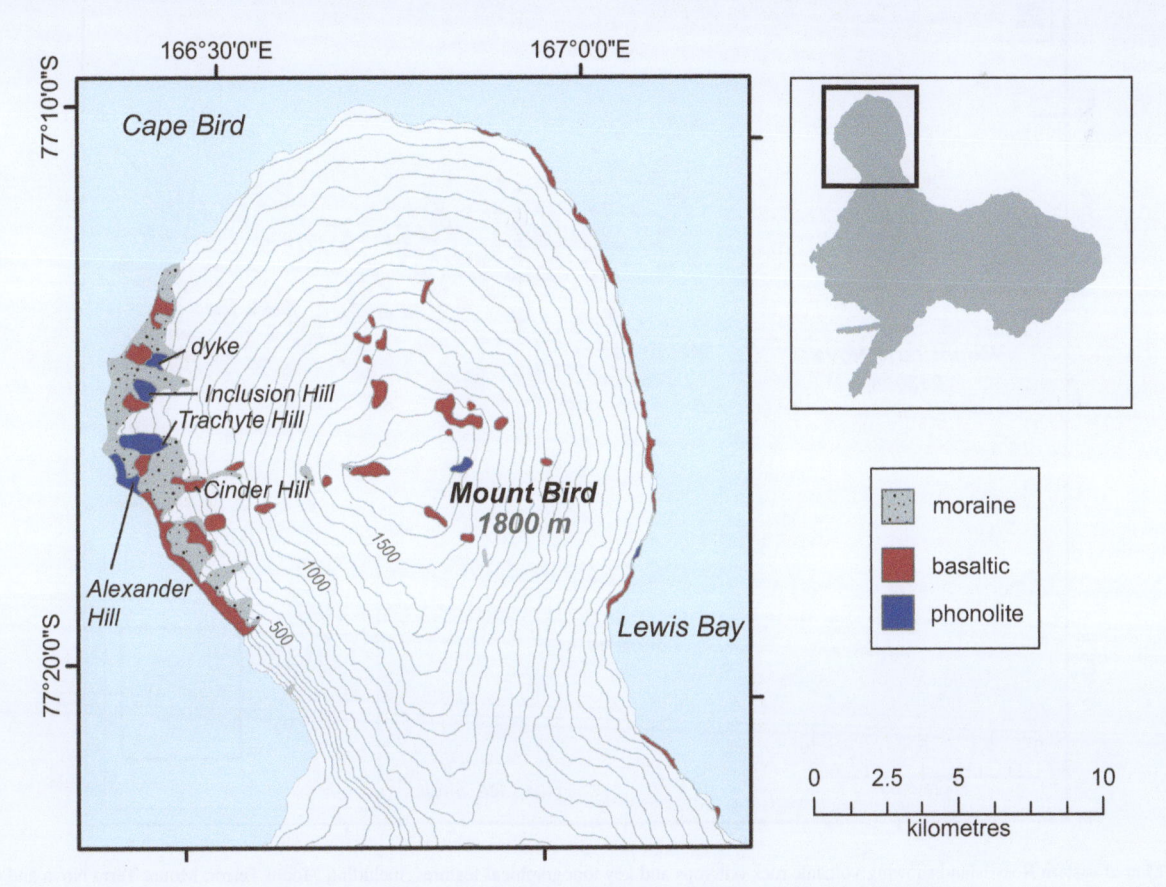

Fig. 7. Map of northwestern Ross Island showing volcanic outcrops and key topographical features, including Mount Bird, Cape Bird and Lewis Bay.

(kaersutite)–augite–plagioclase' basanite, has normal magnetic polarity (Kyle 1976). It is followed by an olivine–augite basanite. This is, in turn, overlain by a magnetically reverse polarized 'olivine–augite–plagioclase' basanite which has a K–Ar age of 3.8 ± 0.2 Ma (Armstrong 1978). Alexander Hill (Fig. 7) consists of the eroded remnants of two complex phonolite–basanite cones and four rock units were recognized by Cole and Ewart (1968). Basanite is interbedded with phonolite and there is no evidence of lava flows with intermediate compositions. A pyroxene phonolite has reversed magnetic polarity (Kyle 1976) and a K–Ar age of 3.23 ± 0.09 Ma (Armstrong 1978). Bright red, scoriaceous, olivine-rich basanite with peridotite xenoliths was erupted at Cinder Hill and this is overlain by black, massive olivine basanite lava flows. Large (no quantitative data are reported) crystals of gem-quality olivine (peridot) are found at Cinder Hill (Wilson *et al.* 1974). At Inclusion Hill, the Cinder Hill igneous rocks are overlain by a phonolite dome. A tephriphonolite dome at Trachyte Hill represents the youngest eruptive event and has a K–Ar age of 3.08 ± 0.15 Ma. A tephriphonolite dyke exposed at the northern end of the Cape Bird exposures is similar in composition to the tephriphonolite at Trachyte Hill.

Lava-flow samples analysed from the Mount Bird eruptive centre have a classic bimodal distribution of compositions comprising mainly basanitic lava flows and scoria cones with later-stage tephriphonolite (Fig. 3) and phonolite domes. There is a Daly gap between 48 and 54 wt% SiO_2 (Fig. 3). Five phonolite vents and a phonolite dyke are known (Fig. 7). Most of the phonolite domes appear to be late in the eruptive sequence. The recent discovery of a phonolite on the Lewis Bay cliffs (Fig. 7) indicates that phonolite may also have been involved in the main cone-building phase of Mount Bird.

Mount Terror. Mount Terror is a 3262 m-high basaltic shield volcano (Wright and Kyle 1990*e*). It is roughly circular in outline, being 35 km north–south and 40 km east–west, with slopes of about 9° (Fig. 8). It has an estimated eruptive volume of about 1700 km^3 (Esser *et al.* 2004). The volcano is mainly snow- and ice-covered but there is a large triangular shaped area of exposed rocks extending from the summit down slope to the east towards the Cape Crozier area (Fig. 8). The summit area and eastern slope consist of numerous basanitic scoria cones, and rare phonotephrite, tephriphonolite and phonolite (Fig. 3) cones and domes, that overlie the main shield. Several samples were collected by early British expeditions from Cape Crozier and these are described by Smith (1954). Phillips *et al.* (2018) sampled some of the cones on the summit and higher parts of the eastern slopes but no detailed geological mapping exists for other areas except around Cape Crozier.

Eruptions at Cape Crozier have been dated at between at least 1.75 ± 0.3 and 0.82 ± 0.14 Ma (Kyle 1976; Armstrong 1978; Lawrence *et al.* 2009). A single sample from Mount Terra Nova (Fig. 8) yielded an imprecise K–Ar age of 0.82 ± 0.5 Ma (Armstrong *et al.* 1968). The eruption age of the main Mount Terror shield is unknown but it must be older than the small phonolite vents at Cape Crozier.

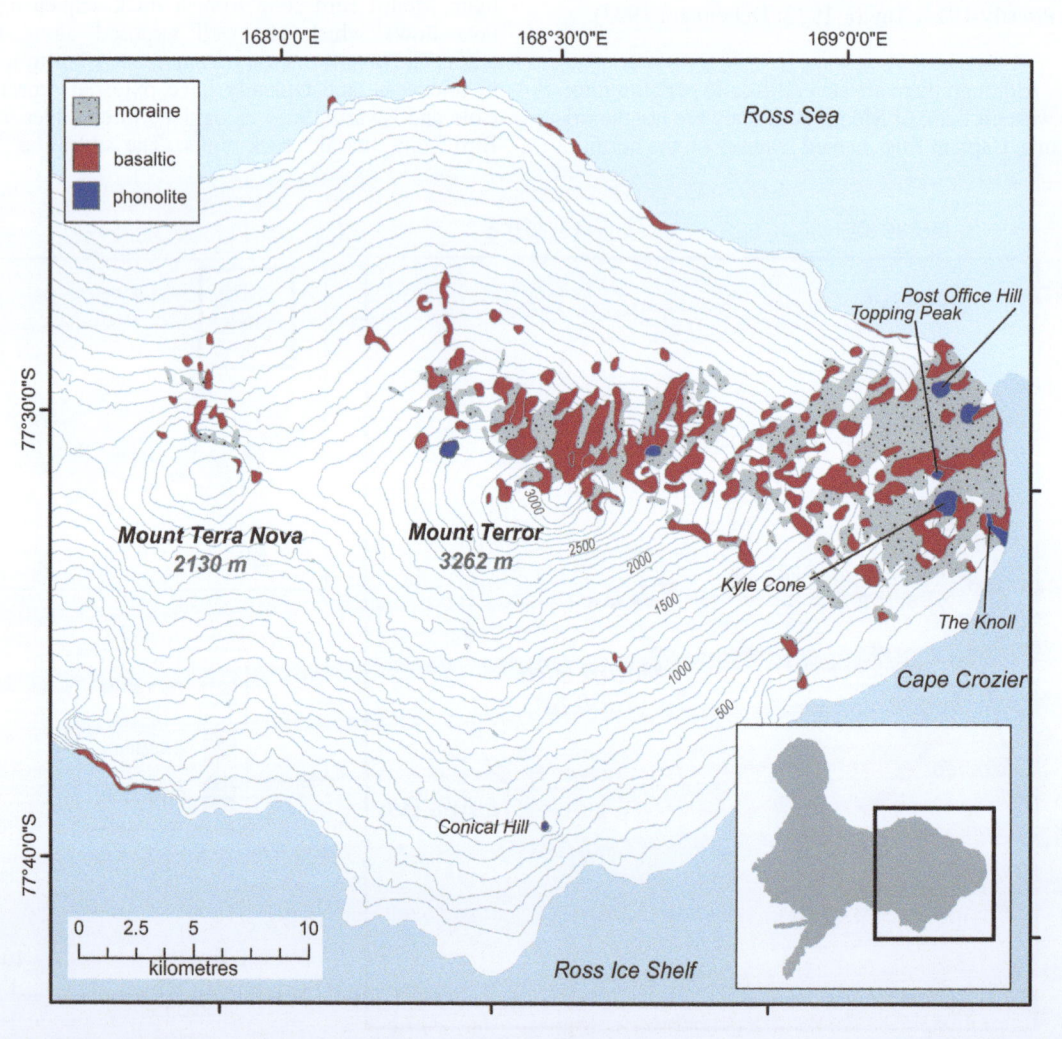

Fig. 8. Map of eastern Ross Island showing volcanic rock outcrops and key topographical features, including Mount Terror, Mount Terra Nova and Cape Crozier.

Considering the size of Mount Terror and that it probably took 1 myr to form, we speculatively suggest it started forming somewhere in the 3–2.5 Ma time period. Compositionally, analysed lava flows from the Mount Terror eruptive centre define a predominantly bimodal distribution with mainly basanitic lava flows and scoria cones and later-stage phonolite domes (Figs 3 & 8). Three lava-flow samples analysed by Phillips et al. (2018) are intermediate in composition (Fig. 3a).

Cape Crozier. The eruptive history of Cape Crozier is representative of the eastern flanks of Mount Terror, and described in detail here and in Figure 8. Cape Crozier extends north–south for over 10 km as an ice-free area along the east margin of Mount Terror (Fig. 8). Cole et al. (1971) described the geology of Cape Crozier and provided a reconnaissance geological map. The oldest lava flows, exposed in coastal cliffs north of The Knoll, comprise olivine–augite basanite lava flows with interbedded volcanic breccia and some tuffs. Lava flows at the base of the exposed sequence originated from the flanks of Mount Terror, and these are commonly overlain by younger lava flows and pyroclastic deposits from small vents at Cape Crozier. Rare dykes indicate local vents for some of the lava flows. On the east flank of Topping Peak, erosion has exposed a section of rocks with compositions of basanite and kaersutite phonolite, the latter were fed by at least four irregular dykes. Post Office Hill is a steep-sided endogenous kaersutite phonolite dome. Using colour and phenocryst variations, Cole et al. (1971) recognized three main rock types. Chemically, the lava flows are very similar (see the Supplementary material (ESM1)). The Knoll and Kyle Cone are steep-sided endogenous domes composed of aegirine–augite phonolite. At Kyle Cone, a kaersutite basanite lava flow underlies the phonolite dome; whereas, at The Knoll, two lava flows of olivine–augite basanite mantle the phonolite. The Knoll basanite lava flows were erupted from a small circular crater near the summit. Five outcrops with a phonolite composition have been identified previously on Cape Crozier but samples from recently identified and analysed domes near the summit of Mount Terror and Conical Hill are confirmed here as having a phonolite composition (Figs 3 & 8). One scoria cone is an alkali basalt, which is a rarity in the Ross Island Volcanic Field.

Mount Erebus. As defined here, the Mount Erebus volcanic centre includes Mount Erebus and the Dellbridge Islands (Tent, Inaccessible, Big Razorback Island and Little Razorback Island) in Erebus Bay (Fig. 9). At 3794 m, Mount Erebus is the highest and largest volcano on Ross Island and dominates the centre of the island. Esser et al. (2004) estimated the volume of the Erebus centre to be about 2200 km^3, making it one of the 20 largest volcanoes in the world. The 'Main Crater' at the summit of Mount Erebus is currently active with a persistent active phonolite lava lake, which has existed for over 45 years (Giggenbach et al. 1973; Kyle et al. 1982; Kyle 1994; Oppenheimer and Kyle 2008).

Anorthoclase-phyric phonolite ('kenyte'). Large (up to 10 cm) rhombic crystals of anorthoclase feldspar occur in Mount Erebus phonolite lava flows. David and Priestly (1914, p. 277) noted 'that the covering of snow became thinner until it almost entirely disappeared, being replaced by a surface formed of crystals of anorthoclase felspar from half an inch to four inches in length'. This rock was originally described as 'kenyte' (e.g. Ferrar 1907), an igneous rock found mainly around Mount Kenya, with a variant, 'Antarctic Kenyte', also proposed for the nepheline phenocryst-free variant on Ross Island (Smith 1954). It is now more properly referred to as anorthoclase-phyric phonolite (e.g. Kyle 1977; Cox et al. 2012). Its significance relates to mapping of the presence or absence of anorthoclase-phyric phonolite erratics to reconstruct grounded ice-sheet flowlines in McMurdo Sound at the last glacial maximum (Vella 1969; Cole et al. 1971; Denton and Marchant 2000; Anderson et al. 2017). On Ross Island, *in situ* anorthoclase-phyric phonolite lava is found in flows between Cape Barne and Cape Royds, as well as in recent ejecta and the convecting phonolitic lake of the Mount Erebus summit (Goldich et al. 1975; Kyle 1977; Kyle et al. 1992). Anorthoclase phenocrysts with the composition $Or_{16}Ab_{63}An_{21}$ (Treves 1967) make up 30–40% of the rock volume (Goldich et al. 1975), and occur in a groundmass of plagioclase, anorthoclase, apatite, opaque mineral and nepheline (Goldich et al. 1975). Anorthoclase-phyric phonolites here are thought to have evolved from basanite parent material as part of the Erebus Lineage (Kyle et al. 1992).

Dellbridge Islands. The earliest study of the field geology and physiography of the Dellbridge Islands was discussed by Debenham (1923). The petrology was summarized by Smith (1954), with Moore and Kyle (1987) describing the four small heavily eroded volcanic islands in more detail. Inaccessible Island (Fig. 9), the most northerly in the group, is composed of many irregular lava flows that dip from 10° to 40° to the north. These are well exposed in coastal cliffs along the south side of the island. The oldest exposed rocks are clinopyroxene–feldspar-phyric phonotephrite (see the Supplementary material (ESM1)). These are overlain by palagonitized breccias with pillow lavas, which are, in turn, overlain by feldspar-phyric phonotephrite lava flows. Kaersutite-bearing lava flows of tephriphonolite and phonolite are interbedded in the sequence. Esser et al. (2004) dated the groundmass from a phonolite at Inaccessible Island as 539 ± 12 ka. The lower part of Tent Island is composed mainly of pillow lavas, lava flows and breccias of clinopyroxene–feldspar-phyric phonotephrite (see the Supplementary material (ESM1)).

Mount Erebus lower slopes. Turks Head and Tryggve Point are two distinct promontories on the SW side of Mount Erebus (Fig. 9). They are composed of hyaloclastite, pillow breccia and palagonitic tuffs. Luckman (1974) described the geology and believed the hyaloclastite rocks to be products of both submarine and subglacial eruptions, and they have yielded an $^{40}Ar/^{39}Ar$ age of 378 ± 28 ka (Esser et al. 2004). A dyke which intrudes the same sequence of hyaloclastites at Tryggve Point has been dated by the $^{40}Ar/^{39}Ar$ method as 368 ± 18 ka (Esser et al. 2004), indistinguishable from the radiometric date of Turks Head. The hyaloclastite comprises porphyritic andesine-rich phonotephrite. Anorthoclase tephriphonolite lava flows from the main cone-building phase of Mount Erebus have a $^{40}Ar/^{39}Ar$ age of 243 ± 18 ka and they overlie the hyaloclastite rocks at Turks Head. The geology of Cape Evans (Fig. 9) was described by numerous early workers and summarized by Smith (1954). Two anorthoclase phonolite lava flows, which reach up to 15 m in thickness (Treves 1962), have been dated by the $^{40}Ar/^{39}Ar$ method at 40 ± 6 ka (Esser et al. 2004) and 55 ± 10 ka (Tauxe et al. 2004).

Fine-grained basanite and phonotephrite volcanic rocks occur along the south coast of Ross Island at Cape Barne (Figs 3 & 9). They have been dated at 1311 ± 16 ka ($^{40}Ar/^{39}Ar$: Esser et al. 2004) and they comprise the oldest eruptive products from the Mount Erebus volcanic centre. The Cape Barne rocks have been interpreted as pyroclastic cones (e.g. Armstrong 1978) but more recent work interprets them as lava-fed deltas intruded by coeval dykes (Smellie and Martin 2021). Erosion has removed over half of the western outcrop, exposing an intrusion which now forms the striking 'Cape Barne Pillar' (Fig. 9). Above Cape Barne is a

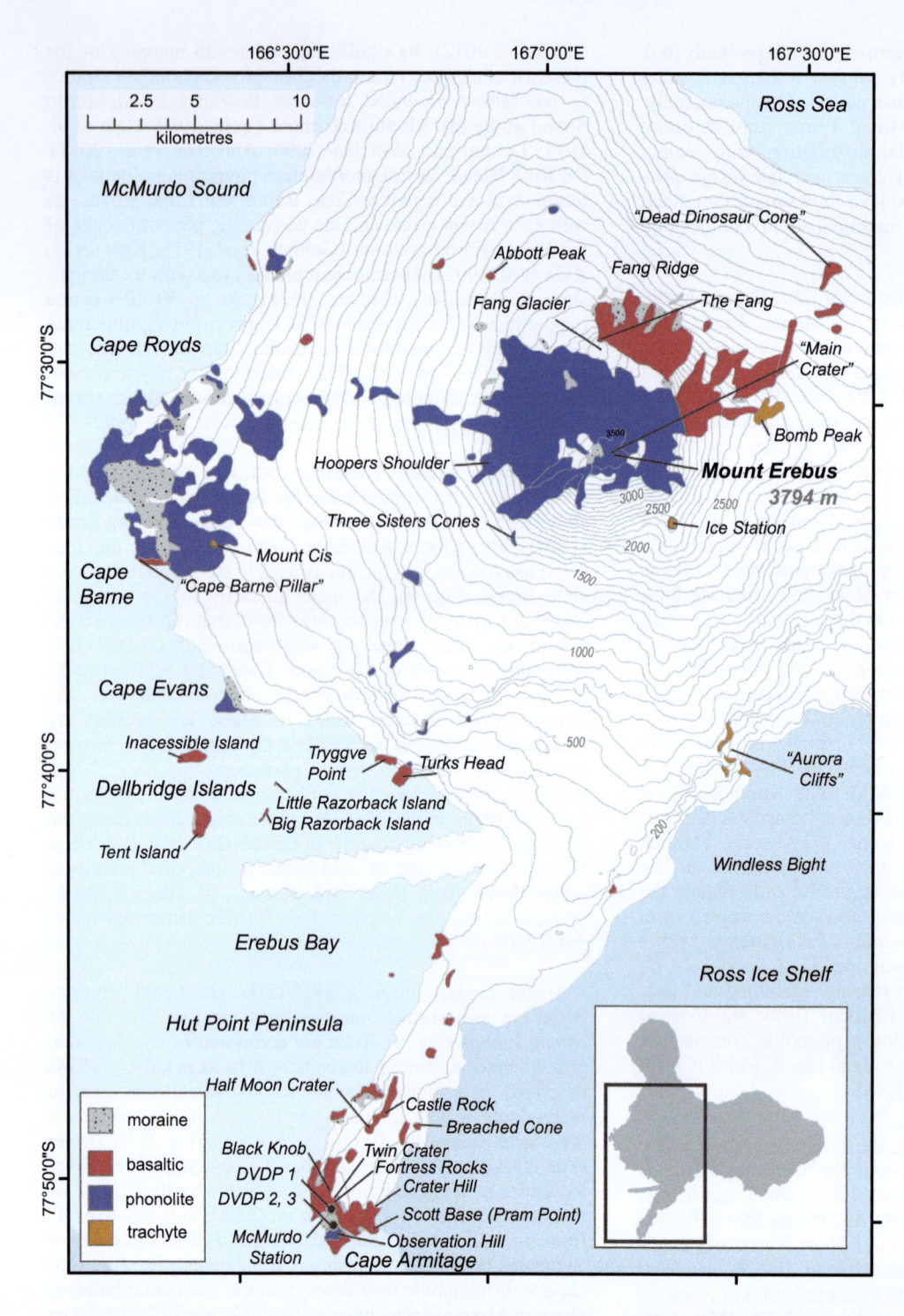

Fig. 9. Map of southwestern Ross Island showing volcanic rock outcrops and key topographical features, including Mount Erebus, Cape Royds, Cape Evans, the Dellbridge Islands and the Hut Point Peninsula, and the locations of the Dry Valley Drilling Project drill holes 1–3.

5 m-high trachyte mound named by the early British explorers as Mount Cis (Fig. 9). This small outcrop has received attention because it contains sandstone xenoliths (Thomson 1916) believed to be from the Beacon Supergroup. The xenoliths indicate that downfaulted rocks, like those in the Transantarctic Mountains, underlie Ross Island and McMurdo Sound. Three anorthoclase tephriphonolite lava flows are exposed along the coast at Cape Royds (Fig. 9). Early work (also summarized by Smith 1954) described the geology and rocks at this location. Many small outcrops on the slopes of Mount Erebus above Cape Royds were described by early workers as cones and potential sources of the lava flows. Most of the outcrops are, however, not cones but small moraine debris mounds, and a source for the lava flows at Cape Royds has not been identified. Until further evidence is found it must be assumed that the lava flows along the coast to the west of Mount Erebus were erupted from vents that are now obscured by the snow and ice cover. Tauxe et al. (2004) reported a whole-rock $^{40}Ar/^{39}Ar$ age of 74 ± 14 ka, and Esser et al. (2004) reported 73 ± 10 ka dated by the $^{40}Ar/^{39}Ar$ method on anorthoclase crystals for a Cape Royds lava flow.

Fang Ridge and Mount Erebus summit. Fang Ridge (the peak, called The Fang, is 3159 m above sea-level) is a prominent feature paralleling the NE slope of Mount Erebus but separated from it by the Fang Glacier (Fig. 9). The north and NE slopes have an average dip of over 45°, and are composed of scree and ribs of rubbly lava flows and pyroclastic rocks. The Fang Glacier side of the ridge has steep, in places vertical, cliffs more than 150 m in height. The lava flows are strongly porphyritic and commonly show well-developed flow banding defined by aligned plagioclase phenocrysts.

The lavas are phonotephrite and tephriphonolite (Fig. 3). Smith (1954) reported the presence of olivine basalt from the lower west end of Fang Ridge but the location of this lava flow has not been found.

The steep (>30°) slopes of Mount Erebus, from 1800 m to the caldera rim at about 3000 m, are made up of numerous (50–100) sinuous, irregular rubbly anorthoclase-phyric phonolitic lava flows. Mount Erebus has five small parasitic vents around its lower flanks. These are Abbott Peak, Hoopers Shoulder, and three endogenous domes that constitute the Three Sisters Cones. All the parasitic vents are composed of black, glassy, porphyritic anorthoclase phonolite which is extremely fresh in appearance. Abbott Peak (1793 m) consists of phonotephrite lava flows that mantle a cone which formed in the main by endogenous growth.

Hut Point Peninsula. This peninsula extends for over 20 km in a SSW direction from Mount Erebus (Fig. 9), indicating that it may be related to a major crustal fracture or weakness radiating from Mount Erebus. Unlike the other Ross Island centres, a major volcanic edifice did not evolve at Hut Point Peninsula. The peninsula is mostly snow- and ice-covered, except for the southern end where the American McMurdo Station and New Zealand Scott Base are situated. The southern tip of the peninsula is made up of an en echelon line of mainly basanite scoria cones and a single endogenous dome of phonolite (Observation Hill). Esser *et al.* (2004) estimated the eruption volume of the peninsula as 82.5 km^3, making it the smallest eruptive centre on Ross Island.

The geology of the Hut Point Peninsula is known from collections made by the early British explorers (e.g. Smith 1954), mapping, for example, by Cole *et al.* (1971) and from onshore drilling of three holes as part of the DVDP. During the period 1973–75, 15 holes were drilled in the Dry Valleys and elsewhere as part of the DVDP, an international research programme involving science agencies from the USA, Japan and New Zealand. There is extensive literature relating to the project, with a large number of papers and reports being published in the 1970s and early 1980s. A comprehensive bibliography was compiled by Rebert (1981). Three of the DVDP holes were cored through volcanic rocks of the Erebus Volcanic Province at the Hut Point Peninsula on Ross Island (Fig. 9); the other holes were largely through lake and glacial sediments.

Hole DVDP 1 cored 40 stratigraphic units and reached a depth of 201 m. Holes DVDP 2 and 3 were drilled 3 m apart on the flank of Observation Hill and are essentially identical. Hole DVDP 2 reached a depth of 179 m, and DVDP 3 sampled 15 stratigraphic units and reached a depth of 381 m (Kyle 1981*b*). The geological evolution of the Hut Point Peninsula can be reconstructed from geological mapping, the DVDP drill cores, palaeomagnetic data and radiometric dating (Kyle 1981*b*). The peninsula has been magmatically active since possibly as early as 1.68 ± 0.06 Ma, continuing to at least 0.33 ± 0.02 Ma. Lithostratigraphic unit 2.4 in the AND-1B (McMurdo Ice Shelf (MIS): discussed in the 'Subsurface volcanic rocks of the Erebus Volcanic Province recovered by offshore drilling' subsection later in this section) core is approximately 12 m thick (Pompilio *et al.* 2007) and is a primary basanite pyroclastic fall deposit. Glass from the unit has an $^{40}Ar/^{39}Ar$ age of 1.68 ± 0.06 Ma (Ross *et al.* 2012*a*). The Hut Point Peninsula is the closest volcanic centre to the AND-1B (MIS) drill core and we suggest the AND-1B (MIS) tephra represents the earliest eruptive phases of the polygenetic Crater Hill on the Hut Point Peninsula, although a source from White Island or other active eruptive centres in the province cannot be excluded. The oldest dated rocks from the peninsula are basanitic (see the Supplementary material (ESM1)) hyaloclastite rocks which are 54 and 214-m-thick in DVDP 1 and DVDP 3, respectively. They have been dated by the K/Ar method at 1.34 ± 0.23 Ma in DVDP 1 and 1.32 ± 0.16 Ma in DVDP 3, yielding ages that overlap within error (Kyle 1981*b*). The hyaloclastite rocks in these drill cores are representative of the submarine pedestal on which the Hut Point Peninsula was erupted (as also seen for Mount Erebus). Lawrence *et al.* (2009) reported an $^{40}Ar/^{39}Ar$ age of a basaltic rock sample collected at McMurdo Station as 1.33 ± 0.12 Ma, which overlaps in time with the eruptive sequence seen in DVDP 1. The ages yielded from the deepest and oldest samples in DVDP 2 and 3 are older than in Observation Hill surface outcrops, where the K/Ar method has yielded a 1.21 ± 0.04 Ma date (Forbes *et al.* 1974), and the $^{40}Ar/^{39}Ar$ method on two different samples has yielded a 1.23 ± 0.02 Ma date (Lawrence *et al.* 2009) and a 1.18 ± 0.01 Ma date (Tauxe *et al.* 2004); this later date from Observation Hill (1.18 ± 0.01 Ma) overlaps with a shallower sample in the DVDP 2 core at 62.38 m dated at 1.16 ± 0.03 Ma (Kyle *et al.* 1979*b*). Volcanic vents at Castle Rock (1.21 ± 0.05 Ma), Half Moon Crater (1.0 ± 0.2 Ma), Cape Armitage (1.03 ± 0.10 Ma) and Breached Cone (0.65 ± 0.05 Ma) were coincident with episodic growth of the peninsula (Kyle *et al.* 1979*b*; Tauxe *et al.* 2004; Lawrence *et al.* 2009). The last eruption at the peninsula was a fissure eruption that formed a sequence of younger lava flows from Crater Hill (including the distinct lava flow forming Pram Point where Scott Base is situated), Twin Crater, Black Knob and Fortress Rocks. Samples dated by the $^{40}Ar/^{39}Ar$ method (Tauxe *et al.* 2004; Lawrence *et al.* 2009) for the fissure eruption yield dates of 0.348 ± 0.008 and 0.33 ± 0.02 Ma (Crater Hill) and 0.34 ± 0.07 Ma (Fortress Rocks), and a K/Ar age of 0.44 ± 0.10 Ma (Kyle *et al.* 1979*b*) for Black Knob.

More than 150 whole-rock analyses of Hut Point Peninsula and DVDP igneous rock samples have been published (Kyle 1976, 1981*b*; Goldich *et al.* 1981; Stuckless *et al.* 1981) (Fig. 3) with new analyses also provided in this study (see the Supplementary material (ESM1)). Surface volcanic rock compositions on the peninsula and lava-flow compositions in DVDP drill core complete a differentiation trend of basanite, phonotephrite, tephra-phonolite to phonolite (Fig. 3a). Several DVDP samples have lower whole-rock Na$_2$O + K$_2$O wt% relative to the majority of DVDP and Hut Point Peninsula compositions (Fig. 3a), and include a benmoreite dyke in DVDP 1 (Kyle 1976). The petrology and possible evolution of these low-alkali lava flows and volcanic rocks have not been examined. Kyle (1981*b*) used electron probe microanalysis (EPM) to determine olivine, clinopyroxene, plagioclase, amphibole, titanomagnetite and ilmenite compositions, as well as those of less common phases such as apatite, sodalite and chromium spinel. Amphibole is mainly kaersutite in composition. The, relatively rare, mineral rhönite occurs in the groundmass of basanites from DVDP 1, 2 and 3 (Kyle and Price 1975). As part of a broader study of Ross Island volcanic rocks, Weiblen *et al.* (1981) analysed clinopyroxene in DVDP core samples where they showed chemical zoning in clinopyroxene phenocrysts may relate to oxidation of magma during ascent or to local reactions. Basanites are porphyritic (15–36% phenocrysts) with phenocrysts of olivine and clinopyroxene in a groundmass comprising olivine or amphibole, clinopyroxene, plagioclase, chrome spinel, titanomagnetite, ilmenite, apatite and glass. Kaersutite is a rare phenocryst phase in basanites. Intermediate rocks are generally less porphyritic (<10% phenocrysts) than basanites, and kaersutite is a more common phenocryst phase. Tephriphonolites and phonotephrites are aphanitic or weakly porphyritic (3–7% phenocrysts) with microphenocrysts of clinopyroxene, kaersutite, plagioclase, iron oxide and apatite.

Mount Discovery Volcanic Field

Dailey Islands. Named after the 1901–04 British National Antarctic Expedition's carpenter (Scott 1907), the Dailey Islands are a group of five, extensively glaciated, volcanic islands that are made up of highly eroded mafic cinder cones and lava flows (McCraw 1967; Treves 1967; Del Carlo *et al.* 2009) (Fig. 10). The two published eruption ages overlap at 0.78 ± 0.04 and 0.76 ± 0.02 Ma (Tauxe *et al.* 2004; Del Carlo *et al.* 2009). A nepheline-normative basanite (Fig. 3) sample from Juergens Island in the group has phenocrysts of olivine + pyroxene in a holocrystalline groundmass of clinopyroxene + plagioclase + olivine + magnetite (Del Carlo *et al.* 2009). The phenocrysts show disequilibrium textures (Del Carlo *et al.* 2009).

White Island. White Island is an extensively glaciated volcanic island (20 × 12 km) in the western Ross Embayment of southern Victoria Land, Antarctica (Fig. 11). It lies approximately 25 km SE of Scott Base (New Zealand), Ross Island, and rises to a height of 741 m above sea-level at Mount Heine in the north, and 762 m at Mount Nipha in the south (Fig. 11). White Island was first visited by Shackleton, Wilson and Ferrar on 19 February 1902, during Scott's British National Antarctic Expedition, and was named for the mantle of snow which covers it (Scott 1907). Rocks collected by this party were described by Smith (1954). In 1969–70, the northern part of the island was mapped by Cole *et al.* (1971) and the geology described as comprising two overlapping olivine–augite basalt shield volcanoes, with subsequent subsidiary cones being constructed on the northern flank. In a study of Cenozoic volcanic rocks from the Ross Island area, Goldich *et al.* (1975, 1981) and Stuckless *et al.* (1981) presented major element and trace element analyses for two basanites: one from the 'younger' unglaciated sequence on Mount Hayward, and the other from the 'older' sequence north of Mount Heine. However, the outcrops occur in geographically separate areas of White Island (Fig. 11), making correlation of the volcanic successions impossible without the use of geochronology.

Large areas in the northern part of the island are essentially snow- and ice-free, and there are exposures of subdued areas of basanitic lava flows and low-rimmed, NNE-elongated craters cut by east–west-striking dykes and plugs. Basanites are the greatly dominant rock type (Figs 3b & 11) and throughout White Island they are olivine–pyroxene phyric, and commonly contain nodules of spinel and plagioclase lherzolite, wehrlite, clinopyroxenite, gabbro, granulite, and hornblendite with megacrysts of olivine, clinopyroxene, kaersutite, spinel and anorthoclase (Cooper *et al.* 2007). At Camp Crater (Fig. 11), a crater filled with a frozen lake has cliff-forming outcrops of nodule- and megacryst-bearing basanite defining a topographically raised rim. Approximately 200 m to the east, a 40 m-wide crater is defined by an inwardly-dipping sheet of kaersutite-phyric tephriphonolite, the most petrogenetically evolved rock found on White Island. The lower part of the Mount Hayward spatter-clast breccia sequence has a unit of coarse breccia interbedded with cross-bedded lapilli tuffs. Basanite clasts (up to 45 cm) are in places rounded with a glassy rind enclosing a more vesicular core, suggesting a water-chilled pillow-like form.

Four samples of groundmass concentrate from three lava flows and a dyke were dated by Ar/Ar techniques and yielded ages of 4.86 ± 0.06–2.11 ± 0.05 Ma (Cooper *et al.* 2007), all considerably older than the 0.17 Ma K–Ar age reported by Kyle (1981*a*) for a basalt collected by Cole *et al.* (1971) from northwestern White Island. An anorthoclasite nodule from basanite at Camp Crater contains inclusions of apatite and zircon. Six zircon grains, dated by laser ablation inductively coupled plasma source mass spectrometry (LA-ICPMS) techniques, define a homogeneous population with a mean $^{206}Pb/^{238}U$ age of 7.62 Ma (MSWD 2.11: Cooper *et al.* 2007). A high-resolution aeromagnetic survey of McMurdo Sound (Wilson *et al.* 2007) showed that the White Island volcanic massif extends a further *c.* 25 km north of White Island itself as a submarine volcanic ridge, which is in close proximity to the AND-1B (MIS) core site. and Talarico and Sandroni (2009) and Di Roberto *et al.* (2010) suggested that clasts from both Minna Bluff and White Island are important glacially-transported volcanic

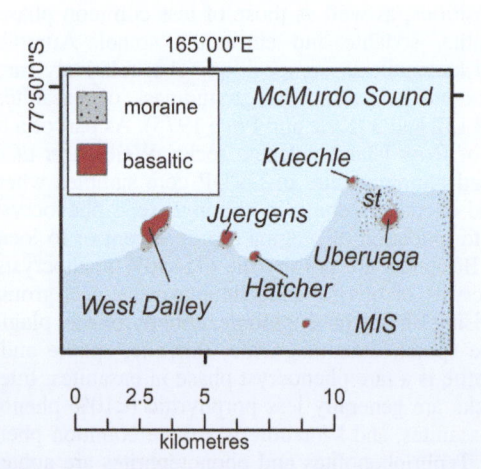

Fig. 10. Map of Dailey Islands showing volcanic rock outcrops and key topographical features. The pattern on the McMurdo Ice Shelf (MIS) indicates supraglacial till (st).

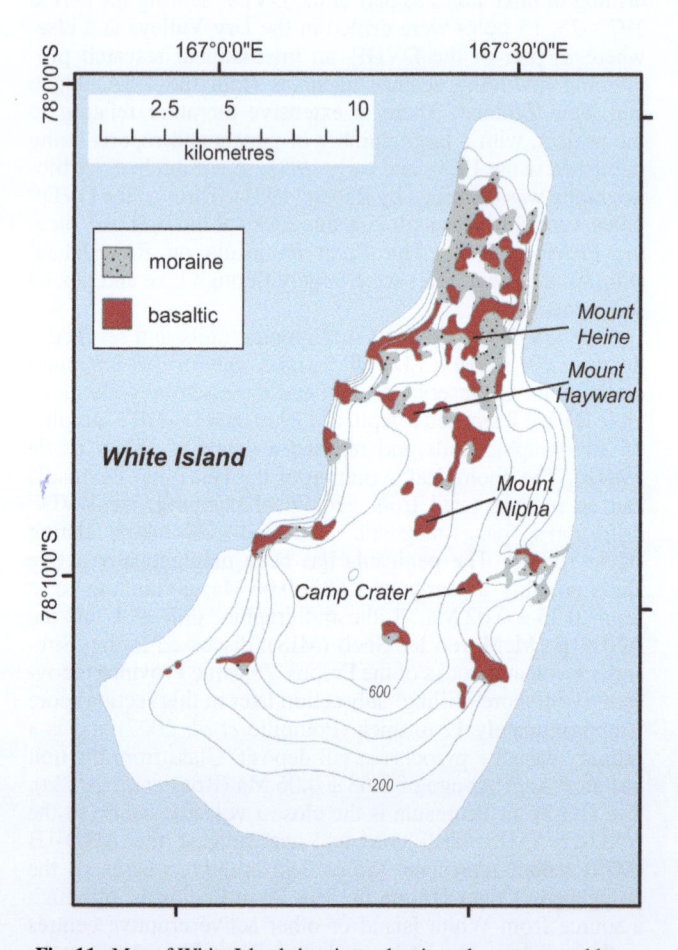

Fig. 11. Map of White Island showing volcanic rock outcrops and key topographical features.

components of the AND-1B (MIS) core. The White Island volcanic history may well be considerably longer than is indicated by dating of onland samples.

Black Island. The island was named during the 1901–04 British National Antarctic Expedition (Scott 1907) because of its mainly snow-free appearance and black volcanic rock. As with the Brown Peninsula, Black Island is believed to have been a stratovolcano (Cole and Ewart 1968), made up predominantly of basanites and basalts (Fig. 12), the latter with varieties characterized by phenocrysts of olivine–augite, plagioclase or hornblende. Trachybasalts and either hornblende- or pyroxene-phyric trachytes make up the more petrologically evolved samples, although Goldich *et al.* (1975) and Timms (2006) described and analysed tephriphonolites and phonolites from the southern part of the island (Figs 3b & 12). Several of the phonolites contain xenocrysts of anorthoclase and 1–20 cm xenoliths of syenite or more plagioclase-rich syenodiorite, interpreted as cumulates. Pyroclastic deposits with clasts ranging in size from 20 cm blocks to fine-grained lapilli and ash are rare.

Although these rock types were assigned formational status by Cole and Ewart (1968), who correlated them with similar lava-flow lithologies on the Brown Peninsula and Cape Bird, different ages indicate that this correlation is invalid (Armstrong 1978). Armstrong (1978) presented the first radiometric (K/Ar) dates from northern and central Black Island, giving ages ranging from 10.9 ± 0.4 and 3.35 ± 0.14 Ma for basalts to a 3.8 ± 0.09 Ma age for trachyte. Timms (2006) obtained $^{40}Ar/^{39}Ar$ laser fusion ages on anorthoclases from two specimens of phonolite of 1.689 ± 0.003 Ma (MSWD 5.8) and 1.780 ± 0.058 Ma (MSWD 7.0). These might indicate a southward migration of eruptive activity during the lifespan of the Black Island eruptive centre.

Brown Peninsula. Named by the British National Antarctic Expedition 1901–04 for the colour of its weathered volcanic lava flows and ash deposits, the Brown Peninsula was first visited by R. Koettlitz in 1902 (Scott 1907), with samples collected during this expedition described by Prior (1907). It was again visited during the British *Terra Nova* Antarctic Expedition 1910–13, and samples collected at this time were described by Smith (1954).

Cole and Ewart (1968) mapped alternating basalt–trachyte eruptive sequences at Black Island, Brown Peninsula and Cape Bird in the McMurdo Sound area, and Vella (1969) mapped glacial moraines at Brown Peninsula. The Brown Peninsula (Fig. 13) consists of a series of north–south- aligned basaltic and felsic eruptive centres erupted both before and after glaciation. A single age determination of 2.7 ± 0.09 Ma on the youngest lava flow at Rainbow Ridge (Fig. 13) postdates glacial erosion (Armstrong 1978). In the Mount Wise area (Fig. 13) three K–Ar age determinations have analytical uncertainties that overlap. The dates indicate that the eruptive events at Mount Wise were only short lived and occurred at about 2.2 Ma (Armstrong 1978).

The basanite rocks contain phenocrysts of olivine and clinopyroxene in a groundmass of plagioclase, opaque oxides and glass (Kyle *et al.* 1979*a*). The eruptive sequence at the Brown Peninsula consists of at least three basaltic (basanite, hawaiite, mugearite) to felsic (benmoreite, phonolite) eruptive cycles. The oldest dated cycles occur at Rainbow Ridge, where two basaltic–felsic eruptive sequences are found.

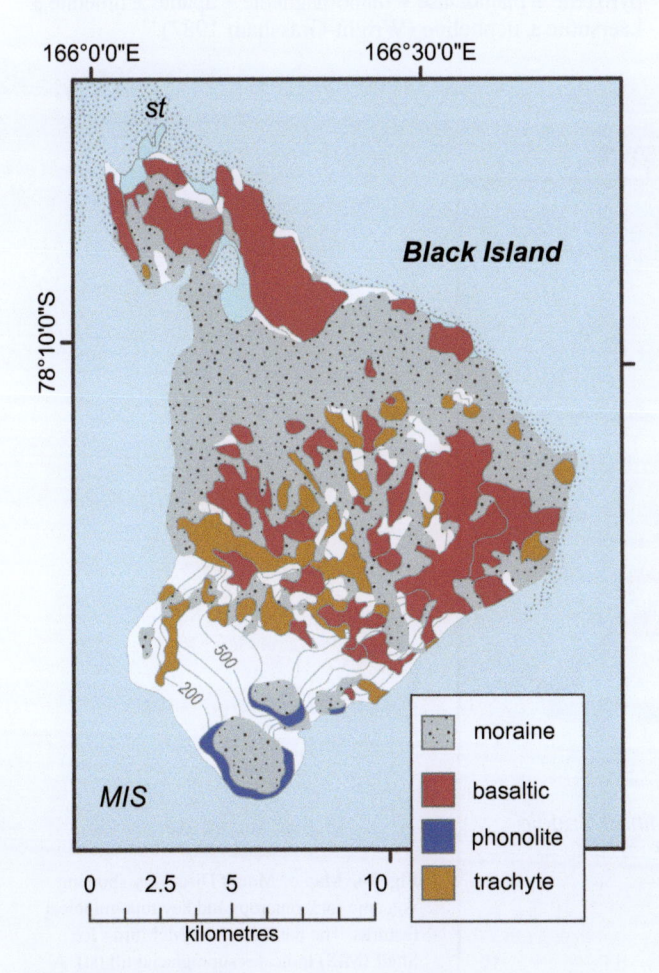

Fig. 12. Map of Black Island showing volcanic rock outcrops and key topographical features. The pattern on the McMurdo Ice Shelf (MIS) indicates supraglacial till (st).

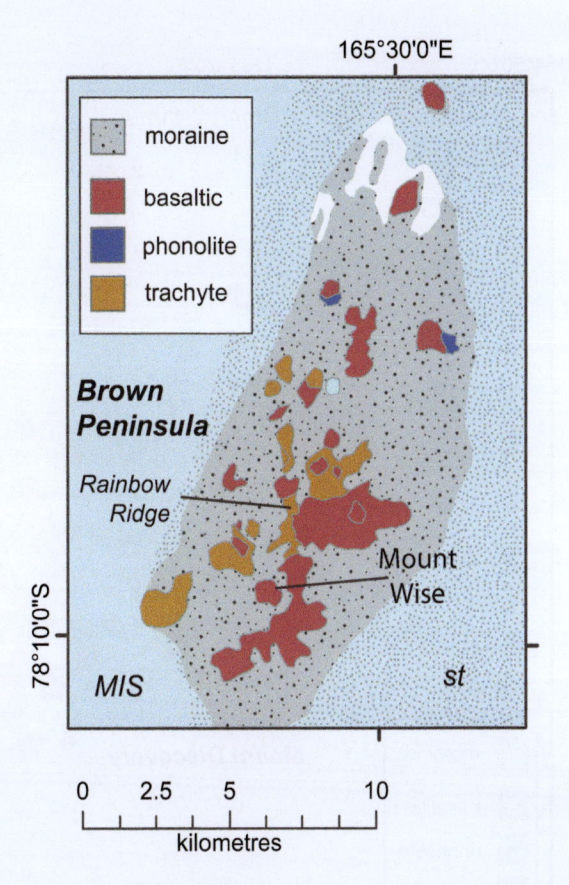

Fig. 13. Map of the Brown Peninsula showing volcanic rock outcrops and key topographical features. The pattern on the McMurdo Ice Shelf (MIS) indicates supraglacial till (st).

These are followed by a younger basaltic–felsic–basaltic eruptive sequence at Mount Wise. The total range in rock types is from basanite, through nepheline hawaiite to nepheline mugearite, nepheline benmoreite and phonolite (Adams 1973; Kyle 1976; Kyle et al. 1979a). The Brown Peninsula nepheline–hawaiite has been modelled to form from basanite by the fractional crystallization of olivine + spinel + clinopyroxene + plagioclase + titanomagnetite + apatite ± ilmenite ± kaersutite (Kyle et al. 1979a). Nepheline–hawaiite to nepheline–benmoreite was modelled by further fractional crystallization of clinopyroxene + kaersutite + titanomagnetite + plagioclase + apatite (Kyle et al. 1979a). At the Brown Peninsula, and many other areas in the Erebus Volcanic Province, small cones and lava flows of kaersutite-bearing and/or clinopyroxene-bearing felsic lava flows (nepheline–benmoreite and phonolite) are common. The cones are small, and most involved the eruption of less than 0.5 km^3 of material (Fig. 13).

Mount Discovery. In 1902, Robert Scott of the British National Antarctic Expedition (1901–04) noted in his diary the first description of Mount Discovery: 'and to the south a peculiar conical mountain ... we named the conical mountain after our ship' (Scott 1907, p. 119). Mount Discovery is a prominent stratovolcano (Global Volcanism Program 2013a) in the Ross Sea (Fig. 14). Rock exposures on Mount Discovery are separated from Mount Morning by the Discovery Glacier, from Minna Bluff by the Minna Saddle and from the Brown Peninsula by Dreary Isthmus (Fig. 14), in a pattern of three-fold radial symmetry (Kyle and Cole 1974).

The volcano was visited once in 1958 by a New Zealand party, with samples collected and described in a doctoral thesis (Kyle 1976). It was visited once again by S.B. Treves between 1960 and 1971 (Goldich et al. 1975). Data for samples collected by Treves have formed part of regional, geochemical, isotopic or chronological overviews of the region (Goldich et al. 1975; Sun and Hanson 1975; Stuckless and Ericksen 1976; Armstrong 1978) but these compilations usually include data for only one Mount Discovery sample. As part of her PhD work, Wright-Grassham (1987) spent 16 days mapping and collecting samples in the only detailed geological study of the mountain. The results from this work are included in a thesis and three short reports (Wright et al. 1984, 1986; Wright and Kyle 1990c).

On Mount Discovery, rock exposures are concentrated on the northern-facing slopes (Fig. 14). Moraine cover is dominant below c. 400 m. Above c. 1100 m the slope angle steepens to 30°. Lava flows, lava domes and pyroclastic deposits are present, and at least 45, morphologically young, cinder cones have been mapped (Wright-Grassham 1987). Ice-mushrooms (personal observations of A.P. Martin) occur on the summit but fumarolic ice towers have not been observed (Wright-Grassham 1987). Volcanism has been dated at between 5.46 ± 0.16 and 1.87 ± 0.43 Ma (Polyakov et al. 1976; Armstrong 1978; Wright-Grassham 1987) but radiometric ages as young as 0.06 ± 0.006 Ma have been reported by Tauxe et al. (2004). The mostly undissected morphology indicates that this is a minimum age range for Mount Discovery.

On a total alkali v. silica diagram, nepheline-normative, strongly alkalic (Saggerson and Williams 1964) rocks from Mount Discovery form a lineage between basanite and phonolite, with rare trachyte (Fig. 3b). This lineage can be modelled by the fractional crystallization of olivine + clinopyroxene + plagioclase + titanomagnetite + apatite ± ilmenite ± kaersutite ± nepheline (Wright-Grassham 1987).

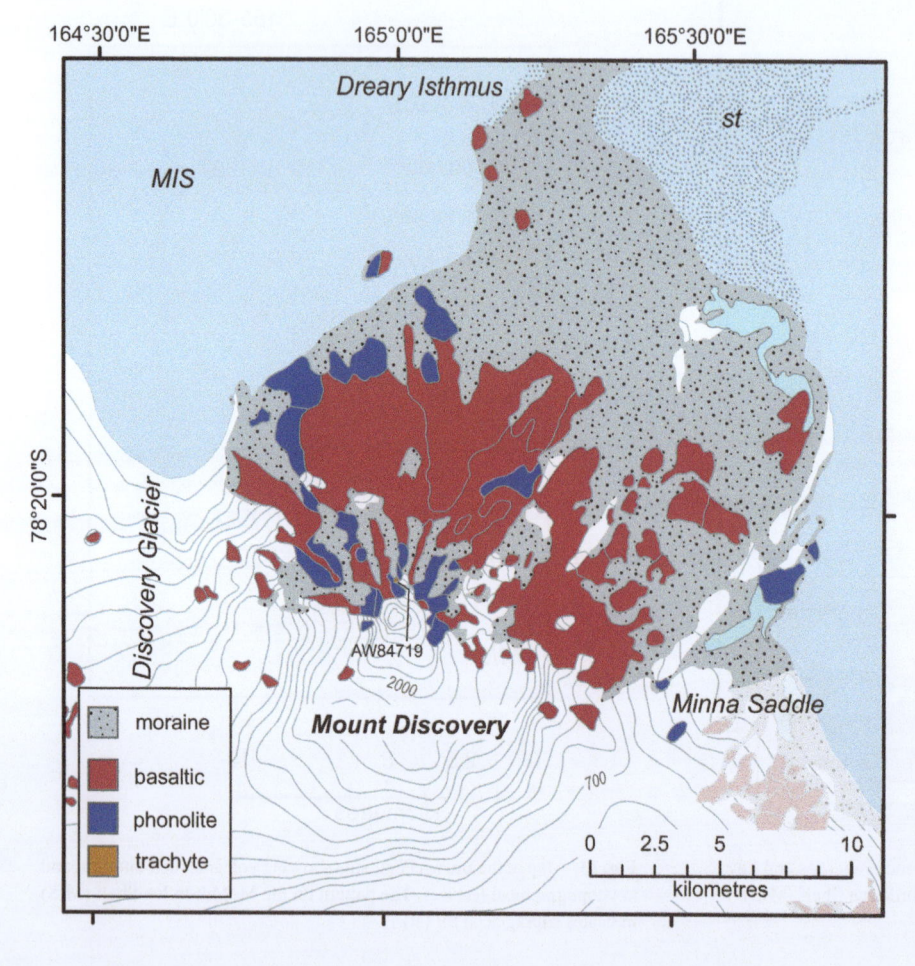

Fig. 14. Map of Mount Discovery showing volcanic rock outcrops and key topographical features. The pattern on the McMurdo Ice Shelf (MIS) indicates supraglacial till (st). A rock outcrop with a trachyte composition is linked to sample AW84719 (see the Supplementary material (ESM1)

Minna Bluff. Known first as 'The Bluff', and later named after the wife of the president of the Royal Geographical Society during the 1901–04 British National Antarctic Expedition (Scott 1907), Minna Bluff is a 45 km-long, southeasterly-trending rock outcrop that ends in a distinct hook (Minna Hook: Fig. 15). Minna Bluff had been visited by S.B. Treeves in the 1960s, with only two of the collected samples being analysed in the 1970s (Goldich *et al.* 1975; Kyle 1976). Wright-Grassham spent 17 days at Minna Bluff in the 1980s, resulting in a thesis and two short publications (Wright *et al.* 1983; Wright-Grassham 1987; Wright and Kyle 1990*d*). In the 2000s, a further two field seasons of research were undertaken, leading to several theses (Scanlan 2008; Fargo 2009; Antibus 2012; Ross 2014; Redner 2016), abstracts and publications (Fargo *et al.* 2008; Panter *et al.* 2011; Wilch *et al.* 2011*a, b*; Ross *et al.* 2012*b*; Antibus *et al.* 2014).

Minna Bluff is made up of coalesced eruptive centres, ranging from small, primitive cinder cones, lava flows and pyroclastic deposits to compositionally evolved domes (Fig. 15). A minimum of four hiatuses separate periods of volcanic activity (Antibus *et al.* 2014). Volcanism occurred between *c.* 12 and 4 Ma (Wright-Grassham 1987; Fargo 2009; Wilch *et al.* 2011*a*; Antibus *et al.* 2014). Eruptions commenced at Minna Hook (12–8 Ma), with phonolite and tephriphonolite compositions preserved (Wilch *et al.* 2011*b*) (Fig. 15). Preserved igneous rocks are more mafic and younger (8–4 Ma) to the NW, along Minna Bluff, relative to compositions at Minna Hook.

On a total alkali v. silica diagram, analysed rocks form a strongly alkalic, nepheline-normative lineage between basanite and phonolite (Fig. 3b). This lineage can be modelled by the fractional crystallization of olivine + clinopyroxene ± kaersutite for basic compositions and olivine + clinopyroxene + plagioclase + magnetite + apatite for more evolved compositions (Panter *et al.* 2011). Kaersutite and feldspar phenocrysts frequently show disequilibrium textures. Megacrysts are common but mantle xenoliths are rare (Panter *et al.* 2011).

Mount Morning Volcanic Field

Mount Morning. Mount Morning is a poorly dissected shield volcano (Global Volcanism Program 2013*b*) forming a volcanic island in the Ross Sea (Fig. 16). It was named during the 1901–04 British National Antarctic Expedition (Ferrar 1905; Fletcher and Bell 1907; Scott 1907) after the relief ship SY *Morning*. Much of the ice-free exposure occurs in two lines on the northern flanks of the mountain, the westerly Riviera Ridge and the easterly Hurricane Ridge, separated by the Vereyken Glacier (Fig. 16). A summit caldera is elongated NW–SE, with axes 4.9 and 4.1 km in length (Paulsen and Wilson 2009). The radius of the volcanic edifice is at least 25 km from the summit caldera to furthest exposed outcrop (Fig. 16), making the volume approximately 1785 km^3, comparable to the volume of Mount Terror. At least 50, morphologically youthful, mafic flank vents occur on the northern slopes (Martin 2009).

Petrological investigations were conducted at a reconnaissance level on Mount Morning in the mid-part of the last century during brief field visits or as part of regional data compilations (e.g. LeMasurier and Wade 1968; Treves 1968, 1977; Kyle and Cole 1974; Goldich *et al.* 1975; Sun and Hanson 1975; Polyakov *et al.* 1976; Armstrong 1978; Stuckless *et al.* 1981; Stuiver and Braziunas 1985). In the 1970s and early 1980s, more detailed, thesis-based investigations were carried out (Kyle 1976; Kyle and Muncy 1978, 1983, 1989; Muncy 1979; Wright *et al.* 1986; Wright-Grassham 1987), and again in the 2000s (Sullivan 2006; van Woerden 2006; Paulsen 2008; Martin 2009). Remote sensing of elongated craters, and dating of young volcanic centres, were also carried out in the 2000s (Paulsen 2002; Csatho *et al.* 2005; Paulsen and Wilson 2007, 2009). Palaeomagnetic studies have also been undertaken at Mount Morning (Mankinen and Cox 1988; Tauxe *et al.* 2004).

Mount Morning varies compositionally between basanite–phonolite and basanite–trachyte. The volcanic rocks are porphyritic with phenocrysts 0.5–4 mm in diameter in a fine

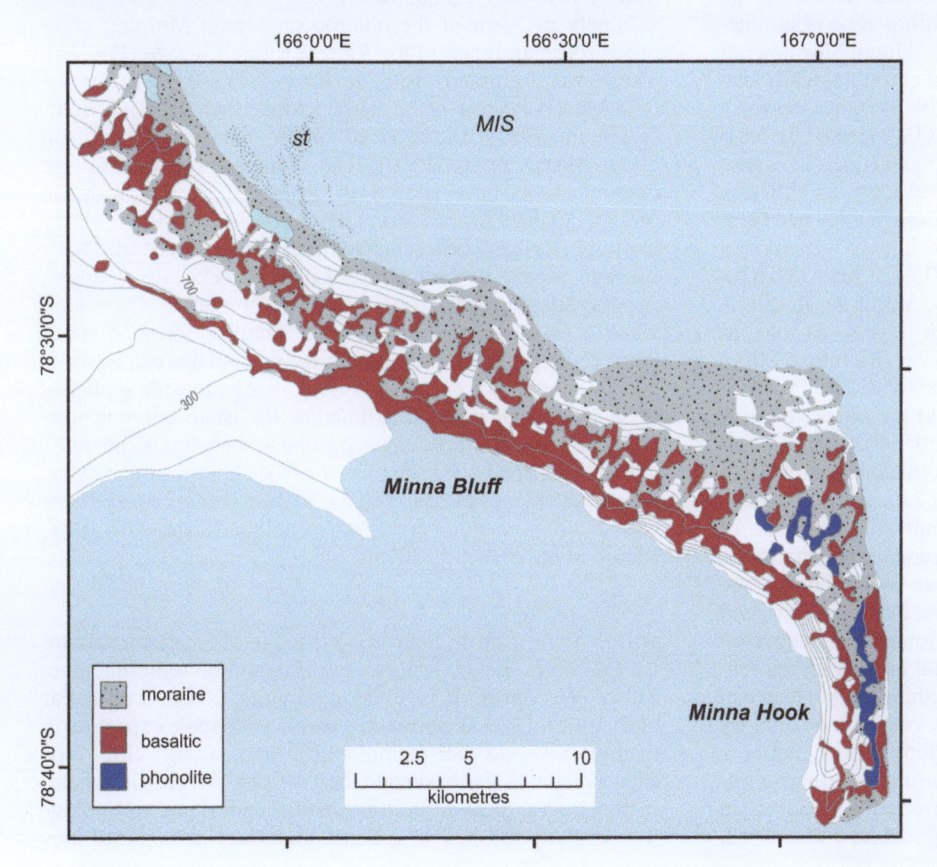

Fig. 15. Map of Minna Bluff showing volcanic rock outcrops and key topographical features. The pattern on the McMurdo Ice Shelf (MIS) indicates supraglacial till (st).

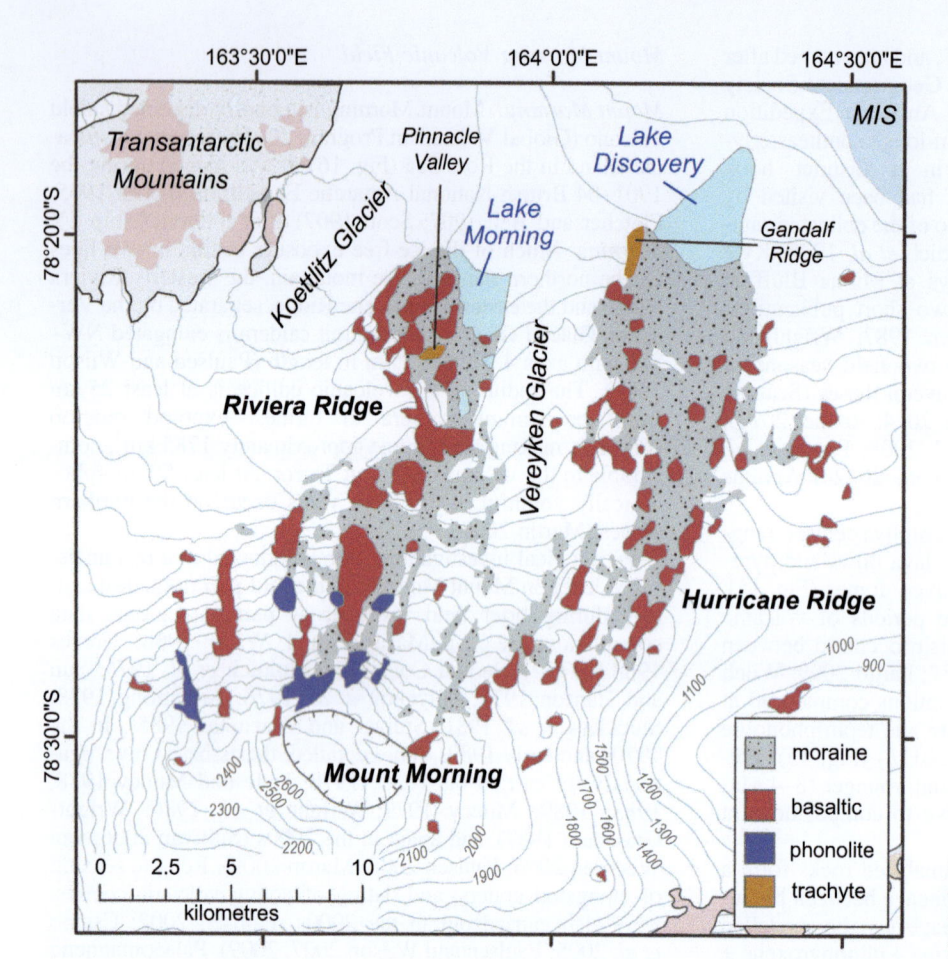

Fig. 16. Map of Mount Morning showing volcanic rock outcrops and key topographical features. The summit caldera is shown by the ticked black line. MIS, McMurdo Ice Shelf.

(0.05 mm) to very fine (<0.05 mm) groundmass. Most volcanic rocks are also poorly vesicular to non-vesicular. Detailed petrographical descriptions are available in a number of theses (Muncy 1979; Wright-Grassham 1987; Sullivan 2006; van Woerden 2006; Martin 2009), and this information is summarized in Martin *et al.* (2013). Volcanism at Mount Morning has been assigned to two geochemically and chronologically distinct lineages (Kyle 1976; Muncy 1979; Wright-Grassham 1987; Wright and Kyle 1990*g*; Martin 2009), termed by Martin *et al.* (2013) the Mason Spur and Riviera Ridge lineages.

Mason Spur Lineage rocks are preserved at two localities on the lower slopes of Mount Morning (Pinnacle Valley and Gandalf Ridge: Fig. 16). Igneous rocks of this lineage were mostly erupted between 18.7 ± 3 and 13.0 ± 0.3 Ma (Wright-Grassham 1987; Kyle and Muncy 1989; Martin *et al.* 2010). Radiometric dates from volcanic ash layers as old as 24.98 Ma in Cape Roberts drill hole 2/2A (McIntosh 2000) are most probably derived from the Erebus Volcanic Province (Smellie 2002) and logically assigned to the oldest, known, eruptive centre, Mount Morning (Smellie 1998). Pinnacle Valley and Gandalf Ridge might be isolated volcanic centres but a more extensive footprint of Mason Spur Lineage magmatism has been proposed based on the widespread occurrence of Mason Spur Lineage-like trachyte and syenite xenoliths across the exposed, volcanic edifice (Martin *et al.* 2010). The lineage includes both quartz- and nepheline-normative rock types, and all analysed rocks are mildly alkalic (following the Saggerson and Williams 1964 definition). On a total alkali v. silica diagram, most data for analysed rocks lie within an array between tephrite and trachyte (Fig. 3c), and this can be modelled by fractional crystallization involving diopside + nepheline ± olivine ± magnetite/ilmenite. A second, quartz-normative lineage of mugearite to trachyte/rhyolite can be modelled in terms of fractional crystallization of plagioclase + diopside ± magnetite/ilmenite ± apatite, with silica oversaturation being related to wall-rock assimilation (Martin *et al.* 2013). Several of the whole-rock compositions are comenditic (Kyle and Muncy 1989; Martin 2009).

Ninety per cent of the outcrop on Mount Morning comprises rocks assigned to the Riviera Ridge Lineage (Fig. 16). These were erupted from at least 5.05 ± 0.04 to 0.06 ± 0.08 Ma (Polyakov *et al.* 1976; Armstrong 1978; Wright-Grassham 1987; Tauxe *et al.* 2004; Paulsen and Wilson 2009; Martin *et al.* 2010). The youngest radiometric age determinations have analytical errors that overlap with the present day, and Martin *et al.* (2010) considered Mount Morning to be dormant rather than extinct. Relative to Mason Spur Lineage rocks, Riviera Ridge Lineage rocks are younger, strongly alkalic and always nepheline-normative. On a total alkali v. silica diagram, analysed samples form an array between basanite and phonolite (Fig. 3c), and this can be modelled by fractional crystallization involving diopside + plagioclase + olivine + magnetite/ilmenite for basic compositions and plagioclase + diopside + aegirine + nepheline + ilmenite/magnetite ± apatite ± kaersutite for felsic compositions (Martin *et al.* 2013). Mafic rocks of the Riviera Ridge Lineage contain abundant mantle and crustal xenoliths (Kyle *et al.* 1987; Martin *et al.* 2014*a*, *b*, 2015*a*, *b*,).

Mason Spur. Mason Spur was named in 1963 after a serving United States Antarctic Research Programme representative, Robert W. Mason. It is a NE-trending, *c.* 10 km-long linear bluff up to 1300 m above sea-level, with rock exposure in mainly SW- to SE-facing cliffs and along cliff tops (Fig. 17). It was first visited in the 1980s by Wright-Grassham, with petrographical descriptions and chemical and chronological results published in a thesis and short communications

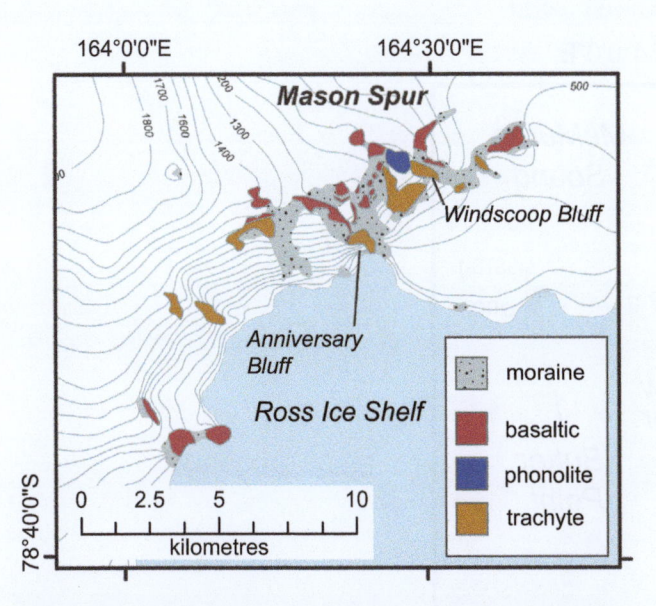

Fig. 17. Map of Mason Spur showing volcanic rock outcrops and key topographical features, adapted from Cox *et al.* (2012) and Martin *et al.* (2018).

(Wright *et al.* 1984, 1986; Wright-Grassham 1987; Wright and Kyle 1990*f*). It was visited by Martin and others in 2005 and again in 2016, resulting in a thesis and publications (Martin 2009; Martin *et al.* 2010, 2013, 2018). Mason Spur is a separate volcanic centre from Mount Morning (cf. Martin *et al.* 2010), as originally proposed by Wright-Grassham (1987) and shown by recent fieldwork. The two lineages identified at Mount Morning are also present at Mason Spur (Wright-Grassham 1987; Martin 2009).

The lowermost, and volumetrically dominant, volcanic suite, referred to here as the Mason Spur Lineage, is part of a complex intra-caldera sequence of pyroclastic deposits and cross-cutting intrusions. These rocks were emplaced between 12.9 ± 0.1 and 11.4 ± 0.1 Ma (Wright-Grassham 1987; Martin *et al.* 2010). The preserved rocks, which form a lineage with compositions between tephrite and trachyte, range from quartz- to nepheline-normative and all are mildly alkalic (Saggerson and Williams 1964) (Fig. 3c). Fractional crystallization can be modelled by either diopside + nepheline ± olivine ± magnetite/ilmenite or plagioclase + diopside ± magnetite/ ilmenite ± apatite for nepheline- or quartz-normative lineages, respectively (Martin *et al.* 2013). Aenigmatite and aegirine augite are present in the groundmass of the most evolved trachyte rocks. Overlying the Mason Spur Lineage sequence are younger rocks (6 Ma to near present day) comparable with the Riviera Ridge Lineage of Mount Morning, and consisting of felsic lava domes and mafic scoria cones. The youngest dated material has ages that overlap with present day, and, like Mount Morning, Mason Spur may be dormant (cf. extinct). These rocks form a lineage between basanite and phonolite that is compatible with fractionation of diopside + plagioclase + olivine + magnetite/ilmenite for basic compositions and plagioclase + diopside + aegirine + nepheline + magnetite/ilmenite ± apatite ± kaersutite for felsic compositions (Martin *et al.* 2013).

Southern Local Suite

Throughout southern Victoria Land, in the foothills of the Transantarctic Mountains (Fig. 18), numerous small basaltic scoria cones pepper the landscape and vary in composition from basanite to alkali basalt. More than 50 small centres, up to 1 km across and 300 m high, are found in the foothills of the Royal Society Range or around the Taylor and Wright valleys, and these vary in age from at least Mid-Miocene to early Pleistocene (Armstrong 1978; Wright 1979*a*, *b*, *c*, *d*; Wright and Kyle 1990*a*, *b*; Tauxe *et al.* 2004; Cox *et al.* 2012), although Holocene ages have been indicated from some unpublished ages obtained by the K–Ar method (Wright and Kyle 1990*a*). The earliest geological investigations were made here during the heroic era (Ferrar 1907; Prior 1907; Mawson 1916; Smith 1954). Further expeditions were made around the time of the International Geophysical Year programme (the third Polar Year), including the Commonwealth Trans-Antarctic Expedition of 1955–58, resulting in the geological maps of Gunn and Warren (1962). The area was again visited by geologists during the early 1960s (Blank *et al.* 1963; Haskell *et al.* 1965), and by geologists from the Geological Survey of New Zealand in the late 1970s (Wright 1979*c*, *d*). Research on the Southern Local Suite rocks has also been undertaken on specific vents, and the xenoliths hosted in the volcanic rocks, at various other times over the past 40 years (e.g. McIver and Gevers 1970; Gamble and Kyle 1987; McGibbon 1991; Wingrove 2005) but a synthesis of the Southern Local Suite has not yet been published.

In the majority of cases, the cones are eroded and the products pyroclastic, with deposits of poorly stratified scoria and agglutinate to rarely preserved hyaloclastite and lava flows (Blank *et al.* 1963; Haskell *et al.* 1965; Skinner *et al.* 1976; Keys *et al.* 1977; Wright 1980; Wright and Kyle 1990*a*, *b*). Accordingly, they can be regarded as monogenetic scoria cones, with relatively short life cycles. In this subsection, a number of occurrences are described that can be considered to be generally representative. The rocks are typically porphyritic, with phenocryst assemblages dominated by olivine and clinopyroxene but with plagioclase ± Fe–Ti oxides and sometimes Ti-amphibole megacrysts (kaersutite). Xenoliths are common, ranging from lithospheric mantle spinel lherzolite, harzburgite and dunite to lower-crustal granulites and metagabbros (McIver and Gevers 1970; Kirsch 1981; Kyle *et al.* 1987). Foster Crater hosts a unique and unusual phlogopite-bearing pyroxenite that varies from coarse-grained granoblastic to fine-grained mylonitic and porphyroclastic in texture (Gamble *et al.* 1988). Shallow crustal granitoids and metasediments, all sintered and partially fused, are also present but less common. The crust throughout this region generally comprises basement of Neoproterozoic–early Cambrian and Ordovician metasedimentary rocks, including quartzites, schists and marbles, intruded by predominantly granitoid plutons of the Granite Harbour Intrusive Complex and unconformably overlain by Permian–Triassic low-grade metasandstone, shales and coal measures of the Beacon Supergroup. These were followed by intrusions and localized lava flows of the Ferrar–Kirkpatrick Large Igneous Province (Cox *et al.* 2012).

Foster Crater. This crater is situated on the northern side of the Koettlitz Glacier, about 110 km south of Ross Island. It consists largely of poorly stratified, highly oxidized agglutinate and scoria, with abundant xenoliths ranging from mantle peridotite, lower-crustal gabbros and pyroxenites (Gamble and Kyle 1987; Gamble *et al.* 1988).

Hooper Crags. Hooper Crags are isolated outcrops towards the head of the Koettlitz Glacier. The exposures are sparse and appear to be remnants of a lava flow that contains xenoliths of peridotite.

Pipecleaner Glacier. Pipecleaner Glacier, Roaring Valley and Radian Glacier are about 10–15 km north of Foster Crater, in a region that was mapped in detail by Worley and others during the Otago University basement study project

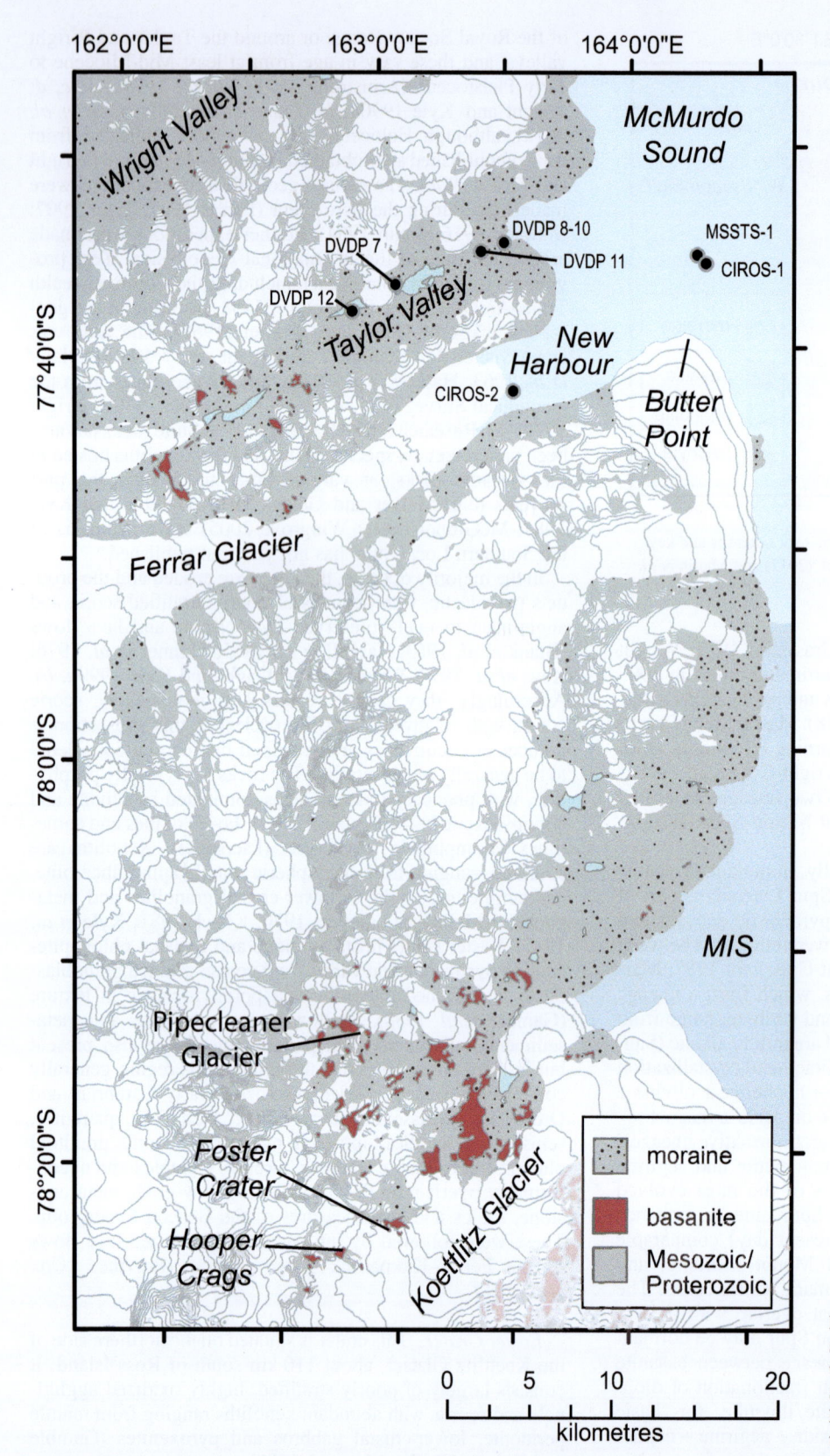

Fig. 18. Map of the Southern Local Suite showing volcanic rock outcrops and key topographical features. MIS, McMurdo Ice Shelf. Mesozoic, rock types older than the Erebus Volcanic Province rocks and part of the Transantarctic Mountains.

(Worley 1992; Worley et al. 1995). In this area there are a number of basaltic outcrops, mostly comprising pyroclastic material but with rare lava flows, both containing abundant peridotite xenoliths.

Subsurface volcanic rocks of the Erebus Volcanic Province recovered by onshore drilling. The DVDP drilled at 14 onshore sites (Figs 1, 9 & 18) between 1973 and 1975. The onshore DVDP (1973) holes 1–3 are discussed in the earlier 'Hut Point Peninsula' subsection. In the Wright Valley, DVDP holes 4, 4A, 5, 5A, 13 and 14 were drilled. No Erebus Volcanic Province material is reported in holes 4, 4A, 5 or 5A (Cartwright et al. 1974a, b) but basaltic fragments, probably sourced from the Erebus Volcanic Province, are reported in DVDP 13 (Mudrey et al. 1975) and DVDP 14 (Chapman-

Smith 1975b). In Victoria Valley, hole DVDP 6 penetrated c. 10 m of sediment and c. 296 m of basement lithologies, with no Erebus Volcanic Material being reported in the sedimentary section (Kurasawa et al. 1974).

Holes DVDP 7–12 were drilled in the Taylor Valley. Drill holes DVDP 8 (c. 156 m deep) and 10 (c. 185 m deep) intersected 185 m of sediment and were sited near the shore of New Harbour (Porter and Beget 1981). Drill hole DVDP 11 (c. 330 m deep) was sited 3 km further inland up Taylor Valley. A major lithological change occurs at c. 154 m in cores 8 and 10, and at c. 185 m in core 11. Above these levels, volcanic clasts of the Erebus Volcanic Province are common but are entirely absent from sections below this interval (Porter and Beget 1981). This lithological change coincides with an unconformity recognized from palaeontological and palaeomagnetic evidence (Porter and Beget 1981). Drill hole DVDP 9 is only c. 39 m deep, and was drilled adjacent to holes 8 and 10 near the head of New Harbour. Drill hole DVDP 9 contained c. 20–30% clasts of Erebus Volcanic Province affinity (Porter and Beget 1981). Drill hole DVDP 7 reached a depth of only 11 m and was drilled some 7 km inland from the mouth of Taylor Valley (Harris and Mudrey 1974). Drill hole DVDP 12 (185 m depth) drilled 165 m of sediments with volcanic lithic fragments recorded throughout the sedimentary section (Chapman-Smith 1975a), these are most likely to be sourced from Erebus Volcanic Province rocks.

Subsurface volcanic rocks of the Erebus Volcanic Province recovered by offshore drilling. Over the past five decades drilling projects have recovered subsurface cores from within the Erebus Volcanic Province and the Ross Sea region (Figs 1 & 19). In chronological order these projects comprise: the McMurdo Sound Sediment and Tectonic Study (MSSTS: 1971); DVDP (1975); Cenozoic Investigations into the western Ross Sea (CIROS: 1980s); the CRP (late 1990s); and ANDRILL (2000s). Cores were also recovered in 1973 during drilling associated with legs 270–273 of the Deep Sea Drilling Project in the Ross Sea but although mention is made of basaltic and doleritic clasts in the cored sediments these are not described in any detail.

McMurdo Sound Sediment and Tectonic Study core MSSTS-1. The stratigraphy and sedimentology of the MSSTS-1 core was described in detail by Barrett and McKelvey (1986). The core was taken 12 km off the Victoria Land coast (Fig. 1) in a water depth of 195 m. Core was recovered to a depth of 227 m, and it comprised a sequence of mudstones and sandstones with intercalated limestone and diamictite units. Examination of the sand fraction of the core indicated that basaltic debris occurs at a depth of 60 m in sediments with an age estimated to be around 21 Ma (Early Miocene) and continues with one short hiatus to the top of the hole. Basaltic pebbles found at a depth of 213 m were described by Gamble et al. (1986) who concluded that they were alkalic basalts with petrological affinities to the volcanic rocks of the Erebus Volcanic Province. Two pebbles dated by the K/Ar method gave ages of 24.3 ± 2 and 13.7 ± 4 Ma but palaeomagnetic and palaeontological information indicates a probable age of 30 Ma for the unit in which the pebbles are contained.

Dry Valley Drilling Project (DVDP). In 1975, DVDP 15 (Fig. 1) was drilled in 122 m water depth with 52% recovery of a total 61.6 m section cored (penetration depth 65 m: Barrett et al. 1976). The top 13 m of the core (unit 1) is fine to coarse sand, with the bottom 13–65 m of the core (unit 2) consisting of fine to medium sand (Barrett and Treves 1981). Petrography shows 65–80% of the sampled intervals are made up of basaltic material derived from the Erebus Volcanic Province (Barrett and Treves 1981). Barrett and Treves (1981) interpreted the recovered sections to be Pliocene–Pleistocene in age.

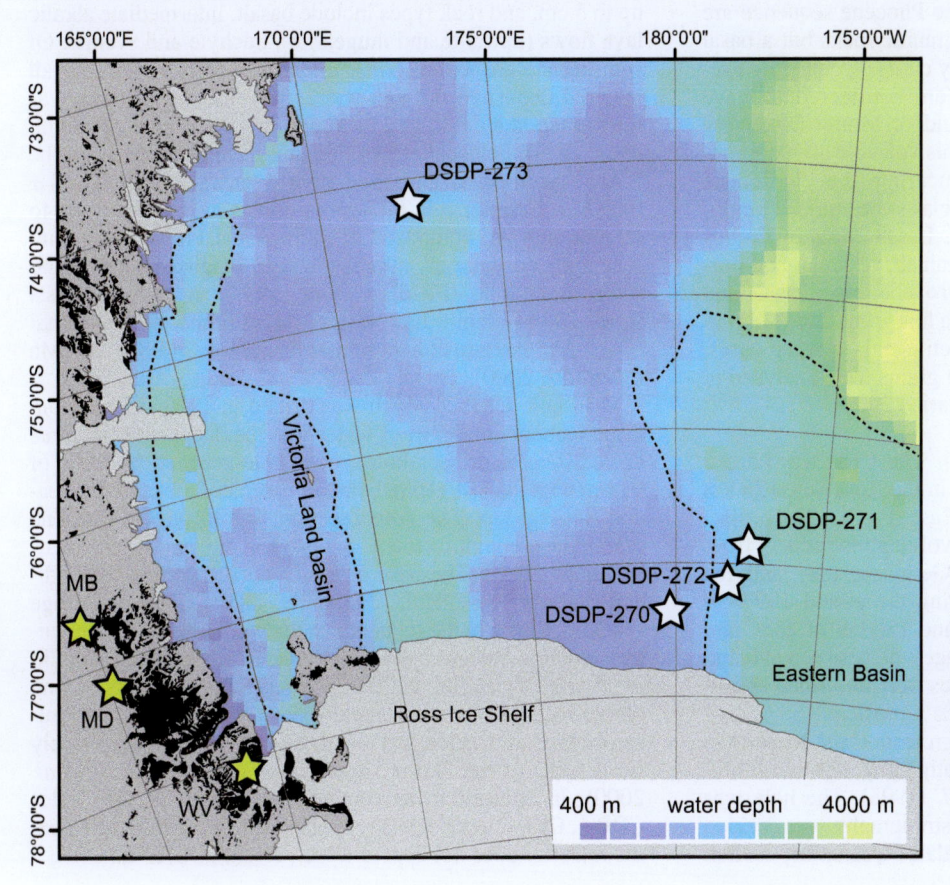

Fig. 19. Regional-scale image of Victoria Land showing the location of englacial tephra deposits (yellow stars) and Deep Sea Drilling Project (DSDP) legs 270–273 drill holes (white stars). The approximate position of the Victoria Land Basin is shown. The bathymetry is shown for reference, with warmer colours indicating shallower water depths (c. 100–4000 m). The filled black polygons are rock outcrops. MD, Mount De Wit; MB, Manhaul Bay; WV, Ward Valley.

Cenozoic Investigations into the western Ross Sea (CIROS). The 'Cenozoic Investigations into the western Ross Sea' (CIROS) Project drilled two holes in New Harbour (Fig. 18) in McMurdo Sound (Fig. 1) in the southern summer seasons of 1984–85 and 1986–87. CIROS-1 was located 12 km off Butter Point, in *c.* 200 m of water. CIROS-2 was located 1.2 km from the snout of the Ferrar Glacier tongue in 211 m of water. The geology of CIROS-1 is comprehensively described in a New Zealand DSIR Bulletin (Barrett 1989). The hole intersected glacial deposits comprising, in order of abundance, diamictites, sandstones, mudstones, and conglomerates and breccias, with the hole terminating at a depth of 702 m in basement gneiss (Hambrey *et al.* 1989). Fossil, palaeomagnetic and isotopic evidence indicates an age for the sequence of Early Oligocene–Early Miocene with two periods of deposition dated at 36–34.5 Ma and *c.* 30.5–22 Ma, respectively (synthesis of Barrett 1989). Dolerite boulders found in the basal units of the cored sequence are from the Ferrar Supergroup (Grapes *et al.* 1989) and grains of this material also occur in the upper 200 m of the section (George 1989).

Basaltic clasts believed to have derived from the Mount Morning and Mount Discovery eruptive centres in the Erebus Volcanic Province occur throughout the cored section and become more abundant near the top (George 1989). These are porphyritic, fine-grained basalts containing phenocrysts of clinopyroxene, plagioclase, olivine and rare amphibole in groundmass consisting of plagioclase, clinopyroxene, iron oxide and glass. There are no whole-rock geochemical data available for CIROS basaltic grains but George (1989) presented clinopyroxenes analyses from basaltic clasts in sediments recovered from the uppermost part of the section.

CIROS-2 cored through 165.5 m of glacial sediments into basement gneiss. Two sequences were identified in the core: a Pliocene lower sequence dominated by diamictites with intercalated mudstones; and a Pleistocene upper sequence of alternating sandstones and diamictites (Barrett and Hambrey 1992). The clast lithologies have been described by Sandroni and Talarico (2006). The clasts in the Pliocene sequence are predominantly Paleozoic basement granitic rocks but a basal 13 m-thick diamictite is dominated by clasts of Cenozoic Erebus Volcanic Province rocks that are considered to have derived from the Mount Morning and/or Mount Discovery eruptive centres. The abundance of this volcanic clast component declines throughout the sequence into the Pleistocene section. Ferrar gabbro and dolerite clasts are minor components occurring throughout the core. Basalt clasts of Erebus Volcanic Province affinity are fine grained, vesicular and porphyritic with phenocrysts of clinopyroxene, plagioclase, and rare amphibole and alkali feldspar in a fine groundmass of plagioclase, clinopyroxene, titanomagnetite and, in some cases, glass. Plagioclase and clinopyroxene grains were analysed in one basalt sample (Sandroni and Talarico 2006).

The Cape Roberts Project (CRP). The CRP was named after its location near Cape Roberts, in the Ross Sea, 125 km NW of McMurdo Station and Scott Base (Fig. 1). The project was an international joint venture involving researchers from New Zealand, Australia, the UK, Germany, Italy and the USA. Three holes were drilled during the period 1997–99 and collectively they recovered around 1500 m of core, and drilled into strata with maximum ages of between 34 and 17 Ma. Initial reports on all three holes were published as collections of papers in the journal *Terra Antartica*.

The CRP-1 hole was located 16 km from Cape Roberts in water 150 m deep. The details of stratigraphy, age and lithology were described by Barrett *et al.* (1998). The hole was drilled to a depth of 148 m and the stratigraphy was divided into two sections: an upper unconsolidated Quaternary section comprising shallow-marine diamictites, sandstones and mudstones dated at 1.8–1.25 Ma (^{40}Ar/^{39}Ar age of 1.2 ± 0.1 Ma at 33 m: McIntosh 1998); and a Miocene section composed of sandstones, siltstones, diamictites and breccias with ages ranging from 22.4 to 17.5 Ma (^{40}Ar/^{39}Ar ages of 19.73 ± 0.86–17.15 ± 0.80 Ma for the interval 114–61 m: McIntosh 1998).

Throughout the core, the larger clasts are granitic and metamorphic rocks derived from the Paleozoic basement or dolerites of the Ferrar Group but in the sand and finer fractions, particularly in the Quaternary section and the upper part of the Miocene sequence, volcanic material is abundant. This component, which includes volcanic rock fragments, volcanic glass and a wide range of alkalic indicator minerals, is from the Erebus Volcanic Province. In the Quaternary section, lithic grains include tuff and lava fragments, and the mineral grains of volcanic provenance include olivine, clinopyroxene and feldspar. Glass is abundant. Olivine has not been identified in the Miocene section but minerals associated with alkalic volcanism are common. These include titanaugite, aegirine, kaersutite and aenigmatite. Brown volcanic glass occurs throughout the Miocene section but it becomes much more abundant in the upper 30–40 m.

Holes CRP-2 and CRP-2A were drilled in 1998, 14.2 km from Cape Roberts, in 178 m of water. CRP-2 was cored to 57 m and CRP-2A to 624 m, with the hole transecting a section with an age range estimated at 33–19 Ma (Fielding *et al.* 1999; Wilson *et al.* 2000). Pliocene and Quaternary sediments in the upper part of the cores are unconsolidated diamictite containing clasts of Paleozoic granitic rocks, Ferrar Dolerite, quartz, feldspar and volcanic rock. Volcanic glass is abundant in the silt and sand fraction of the Pleistocene and Quaternary sections of the cores. The pre-Pliocene section comprises fine- to medium-grained sands, silts and muds, diamictite and conglomerate.

Volcanic clasts and mineral grains derived from the Erebus Volcanic Province are common in the cores, particularly in the upper part of the pre-Pliocene section. The clasts range in size up to 5 cm, and rock types include basalt, intermediate alkalic lava flows (hawaiite and mugearite), trachyte and syenite. Of significance are tephra units, interpreted to represent air-fall material deposited through water. The thickest of these was intersected at a depth of 111 m and is 1.22 m thick (Fielding *et al.* 1999). Feldspars from this unit have been dated by the ^{40}Ar/^{39}Ar method and they give a mean age of 21.44 ± 0.05 Ma. Grain size in the tephra units varies from <1 up to 10 mm, and are composed of pumice and brown glass with scattered crystals of alkali feldspar, aegirine–augite and sodic amphibole (Fielding *et al.* 1999). A rhyolitic clast from 294 m depth has yielded a sanidine single-crystal ^{40}Ar/^{39}Ar laser-fusion radiometric age of 24.98 ± 0.08 Ma (McIntosh 2000).

Drill hole CRP-3, which was drilled in 1999, 12 km from Cape Roberts in 295 m of water was the deepest of the three Cape Roberts project holes. The hole reached a depth of 939 m with the last 116 m being in basement rocks of the Beacon Supergroup. The cored section was through Oligocene diamictites, conglomerates, sandstones and mudstones (Barrett *et al.* 2000). The base of the Cenozoic section, immediately above basement, could be latest Eocene in age (34 Ma). Clasts in the sequence include dolerite, and sedimentary, granitic, metamorphic and volcanic lithologies. Pompilio *et al.* (2001) examined the petrography, mineralogy, and whole-rock and mineral chemistry of volcanic and subvolcanic clasts and concluded that these were derived exclusively from rocks of the Ferrar Supergroup (see also Barrett *et al.* 2000). In contrast to the younger sequences of drill holes CRP-1, CRP-2 and CRP-2A, Erebus Volcanic Province volcanic material does not appear to be present in CRP-3.

The Antarctic Drilling Project (ANDRILL). The Antarctic drilling project (ANDRILL) is a multinational, cooperative research programme involving scientists from Brazil, Germany, Japan, Italy, New Zealand, the Republic of Korea, the UK and the USA. The objective of the project is to build on the outcomes of the Cape Roberts Project and to use similar deep drilling technology to core back in time through the Cenozoic and recent geological record beneath Antarctica. The information is intended to be used primarily to understand palaeoclimate and climate change. The first two holes (Fig. 1), designated the McMurdo Ice Shelf (MIS) and Southern McMurdo Sound (SMS) projects, were completed in the 2006–07 and 2007–08 austral summers, respectively.

Hole MIS (AND-1B) was located in Windless Bight, in McMurdo Sound, 10 km from Scott Base and McMurdo Station, in water 870 m deep. The sedimentology and stratigraphy of the core are described in detail by Krissek *et al.* (2007). Around 1285 m of core were recovered, and this included diamictite, sandstone, mudstone and volcanic ash or tuff. Volcanic material from the Erebus Volcanic Province occurs throughout the cored section, with volcanic clasts representing *c.* 70% of the total clast population (Pompilio *et al.* 2007). Eight volcanic lithostratigraphic units were recognized (Pompilio *et al.* 2007), and clasts and whole-rock samples from these have been dated by the $^{40}Ar/^{39}Ar$ method (Ross *et al.* 2012*a*). Basalt clasts from *c.* 17 m depth have an age of 0.310 ± 0.039 Ma. Near the top of the section (*c.* 85–86 m depth) is a phonolitic pumice, which has been dated at 1.014 ± 0.008 Ma. This has no known correlatives onshore. Basaltic clasts from a sequence of black volcanic sands at 112–145 m give an age of 1.633 ± 0.057 Ma, and this sequence is interpreted to have derived from subaerial Strombolian- or Hawaiian-style eruptions. A thick volcanic sequence in the middle of the core includes a phonolitic lava flow (646–649 m) with an age of 4.800 ± 0.076 Ma. The lava flow and the associated volcanic sequence are interpreted to have come from a submarine vent. Near the base of the sequence is a unit comprising altered tuffs, and a volcanic clast from this part of the core (1280 m depth) has been dated at 13.57 ± 0.13 Ma.

The petrology and geochemistry of clasts in the MIS (AND-1B) core have been described by Pompilio *et al.* (2007), who classified volcanic rocks as mafic (basaltic), intermediate or felsic. Basalts have phenocrysts of olivine and clinopyroxene ± plagioclase. The phenocryst assemblages in intermediate rocks include plagioclase and clinopyroxene ± kaersutite, and in felsic rocks the phenocrysts are K-feldspar ± kaersutite ± aegirine. In all rocks the groundmass is generally glassy, cryptocrystalline or very fine grained. Glasses in samples from throughout the sequence were analysed by EPM (Pompilio *et al.* 2007), and these have compositions varying from basanite, through phonotephrite and tephriphonolite to phonolite and trachyte (classification of Le Bas *et al.* 1986).

The SMS (AND-2A) hole was located in McMurdo Sound, 13 km from the Dailey Islands, 50 km NW of the Hut Point Peninsula on Ross Island (Fig. 1), in water 384 m deep. About 1139 m of core were recovered and the stratigraphy has been subdivided into 14 lithostratigraphic units (Fielding *et al.* 2008) representing 74 glacial marine sequences (Fielding *et al.* 2011). Lithologies recovered included diamictites, conglomerates, breccias, sandstones, mudstones and diatomites, as well as volcanic rocks including lava flows, pyroclastic material and volcanic sedimentary units (Fielding *et al.* 2008). In nine of the 14 lithostratigraphic units, 50% of the clasts are interpreted to be of volcanic origin. Di Roberto *et al.* (2012) studied 27 volcaniclastic units over virtually the full length of the core (37–1139 m) and classified these as pyroclastic fall, resedimented volcanic and volcanic sedimentary deposits.

Di Vincenzo *et al.* (2010) carried out $^{40}Ar/^{39}Ar$ dating on 17 volcanic samples from the core, and 10 of these analyses gave statistically robust ages. Basanite and phonolite clasts from the top of the core (<10 m) gave apparent ages of 0.662 ± 0.042 and 0.124 ± 0.014 Ma, respectively. Samples from the Middle Miocene section, taken between 128 and 358 m, were mainly mafic compositions and they gave ages of between approximately 16 and 11.5 Ma. Most of the samples dated in the Early Miocene section of the core (depths >358 m) were felsic and they gave ages ranging from approximately 20.1 to 16.0 Ma. These ages are in reasonable agreement with those obtained by Nyland *et al.* (2013) for glasses from the 354–765 m section of the core (19.3–15.1 Ma).

The petrology and geochemistry of volcanic clasts are described by Panter *et al.* (2008). Mafic compositions have phenocrysts of clinopyroxene + olivine ± plagioclase in glassy or fine-grained groundmasses of plagioclase and clinopyroxene. In intermediate rock types, the phenocrysts are plagioclase ± clinopyroxene ± amphibole and in felsic compositions the phenocryst assemblage is dominated by K-feldspar.

A number of studies have included analyses of SMS (AND-2A) volcanic rocks and glasses. Panter *et al.* (2008) used X-ray fluorescence (XRF) spectrometric analysis to obtain the whole-rock major element compositions of 20 volcanic clasts from lithostratigraphic units in the upper 760 m of the core. They also reported the results of continuous XRF major element scans through sections of the core and they used EPM to determine the compositions of volcanic glass shards in one of the units (lithostratigraphic unit 10: 649–778 m). Whole-rock compositions vary from mafic (basalt and basanite) through intermediate (phonotephrite, tephriphonolite and trachyandesite) to felsic (phonolite, trachyte and rare rhyolite). Glasses in lithostratigraphic unit 10 are predominantly basanitic with a few phonotephrites and basalts (classification of Le Bas *et al.* 1986).

Del Carlo *et al.* (2009) studied the uppermost 37 m of the core (lithostratigraphic unit 1), which contain the highest proportion of volcanic material of any of the units in the section. Mixed, near primary, volcanic clasts predominate in this lithostratigraphic unit with a minor amount of Paleozoic basement material in the clast population. Volcanic material originated in explosive submarine eruptions. Eleven samples were analysed for whole-rock major element composition and seven of these classify as basanites, with one each of basalt, trachybasalt, phonolite and trachyte. Glasses are predominantly basanitic.

Major and trace element analyses were carried out on SMS (AND-2A) glass samples by Nyland *et al.* (2013). Twenty-four glass samples, spaced through the core and covering the interval between 354 and 765 m, were analysed for major element composition by EPM, and 20 of these were also analysed for trace element composition by LA-ICPMS. The glasses analysed are predominantly basanite, trachybasalt and basalt (SiO_2 in the range 40–52%) but there are also more evolved compositions.

Onshore and englacial tephra in the Erebus Volcanic Province. Tephra has been found across Antarctica in blue ice at the margins of ice sheets, and in englacial settings (Keys *et al.* 1977; Harpel *et al.* 2008; Iverson *et al.* 2014), ice cores (Dunbar and Kurbatov 2011; Dunbar *et al.* 2017; Narcisi *et al.* 2017), in marine sediment cores (Hillenbrand *et al.* 2008; Ross *et al.* 2012*a*) and in outcrop (Keys *et al.* 1977; Cox *et al.* 2012). This subsection discusses onshore and englacial tephra, with marine core tephra discussed in the earlier 'Subsurface volcanic rocks of the Erebus Volcanic Province recovered by offshore drilling' subsection. Tephra from sources both within and external to the Erebus Volcanic Province are found in southern Victoria Land (Cox *et al.* 2012).

In the Transantarctic Mountains, tephra deposits are typically <1 m thick, although an example >1 m thick occurs in Ward Valley (Fig. 19). The latter comprises vesicular glass spherules of tephriphonolitic composition and melanocratic volcanic rock fragments (Cox *et al.* 2012). Other Transantarctic Mountain tephra deposits are either *in situ*, disseminated as ash in colluvium and tills, originally water deposited or redeposited debris flows (Cox *et al.* 2012). In the Transantarctic Mountains the tephra are mostly sourced from eruptive centres in the Erebus Volcanic Province (Hall *et al.* 1993; Marchant *et al.* 1996; Lewis *et al.* 2007, 2008). One sample from Mount De Witt has a phonolitic composition and has been dated at 39 ± 6 ka (Harpel *et al.* 2008) and another from Manhaul Bay is also phonolitic in composition. Both samples are emplaced in the Transantarctic Mountains (Fig. 19) but are inferred to be derived from Mount Erebus (Harpel *et al.* 2008; Iverson *et al.* 2014). Other tephra layers in the Transantarctic Mountains dated by K–Ar and Ar–Ar techniques yield ages between 15.5 and 3.9 Ma (Hall *et al.* 1993; Marchant *et al.* 1996; Lewis *et al.* 2007).

The majority of englacial tephra layers exposed on the flanks of Mount Erebus and Mount Terra Nova have phonolitic compositions indicative of eruption from the Mount Erebus volcano. Anorthoclase crystals in one phonolitic Ross Island tephra at 'Dead Dinosaur Cone' (Fig. 9) were dated by the ^{40}Ar/^{39}Ar method and yielded a preferred plateau age of 40 ± 20 ka (Iverson *et al.* 2014), which overlaps with the Mount De Wit tephra age. Twenty-nine phonolitic tephra layers have glass compositions similar to the matrix glass of bombs erupted from the current lava lake at Mount Erebus, indicating that the major and trace element composition of the magmatic system has remained unchanged for the past *c.* 40 kyr (Iverson *et al.* 2014). Some tephra layers from the Mount Terra Nova summit have a range of chemical compositions including trachybasalt and trachytic. These non-phonolitic tephra layers are correlated with eruptive centres in the Transantarctic Mountains or as far afield as Marie Byrd Land (Iverson *et al.* 2014).

Geochemical overview

Whole rock. A comprehensive table of whole-rock geochemical, isotopic and geochronological data for the Erebus Volcanic Province is included in the Supplementary material (ESM1). In addition, selected data for bombs erupted during historical eruptions from Mount Erebus, englacial tephra, and drill-hole clast and glass compositions are included for reference. The originally reported compositional data are provided with the original references and, where different analysis types have been reported from the same sample across multiple publications, these have been linked for ease of comparison. Each whole-rock major element composition has been recalculated to 100% anhydrous, with Fe_2O_3/FeO ratios recalculated following recommendations by Middlemost (1989). The CIPW wt% normative mineralogy has been calculated, along with differentiation index (normative wt% quartz + orthoclase + albite + nepheline + leucite + kalsilite: Thornton and Tuttle 1960) and whole-rock magnesium number (atomic ratio: $Mg/(Mg + Fe^{2+}) \times 100$). New rock names have been applied following the IUGS classification scheme (Le Maitre *et al.* 2002; Verma and Rivera-Gomez 2013). A summary of the type and location of analyses is shown in Table 1.

Patterns in major and trace element whole-rock data v. wt% MgO are evident: for example, wt% CaO, TiO$_2$ and FeO, CaO/Al$_2$O$_3$ ratios, and ppm Ni are all positively correlated with MgO abundance (Fig. 20), consistent with control by

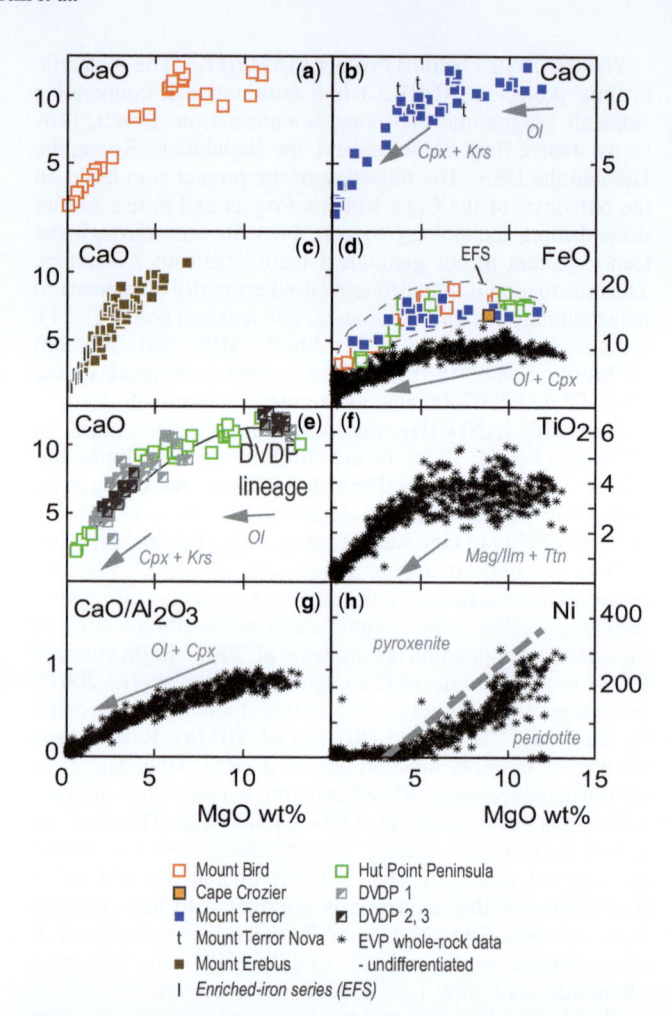

Fig. 20. Major element whole-rock data for the Erebus Volcanic Province (*n* = 895) showing wt% MgO v.: (**a**) wt% CaO Mount Bird data; (**b**) wt% CaO Mount Terror and Mount Terra Nova data; (**c**) wt% CaO Mount Erebus data and data from enriched iron series rocks; (**d**) wt% FeO all Erebus Volcanic Province (asterisks) and enriched iron series data; (**e**) wt% CaO Hut Point Peninsula outcrop data and Dry Valleys Drilling Project (DVDP) data; (**f**) wt% TiO$_2$ all Erebus Volcanic Province (asterisks) data; (**g**) CaO/Al$_2$O$_3$ all Erebus Volcanic Province (asterisks) data; and (**h**) Ni (ppm) all Erebus Volcanic Province (asterisks) data. Fractional crystallization control lines (grey arrows) are shown for various minerals. Cpx, clinopyroxene; Mag/Ilm, magnetite or ilmenite; Ol, olivine; Ttn, titanite. The grey dashed curve in (h) separates pyroxenite-derived melts (field above the curves) from peridotite-derived melts (field below the curves) after Sobolev *et al.* (2005). Data (100%; anhydrous) and references are provided in the Supplementary material (ESM1). Iron has been recalculated based on recommendations of Middlemost (1989).

fractional crystallization. For example, on wt% MgO v. CaO plots for mounts Bird, Terror and Erebus, positive correlation is consistent with clinopyroxene and kaersutite fractionation (Fig. 20a–c). On a wt% MgO v. FeO plot, positive correlation is associated with olivine and clinopyroxene fractionation (Fig. 20d). A plot of whole-rock wt% MgO v. CaO abundance (Fig. 20e) can be interpreted to show the magmatic evolution of the Hut Point Peninsula and DVDP igneous rocks. Deeper samples from the DVDP cores are MgO-rich primitive basanites which are probably close to the parental magmas (possibly derived by partial melting of mantle peridotite). Basanites with 7–11 wt% MgO are absent in the DVDP cores but are well represented by the Hut Point Peninsula samples (Fig. 20e). The highly coherent pattern shown in Figure 20e at whole-rock concentrations >5 wt% MgO is consistent with fractional crystallization of olivine + clinopyroxene; whilst at <5 wt%,

MgO the pattern is consistent with clinopyroxene + kaersutite fractionation. When combined, analyses of the Hut Point Peninsula and DVDP samples define a petrological lineage from basanite to phonolite, which has been named the DVDP lineage (Figs 3 & 20e). Kyle (1981b) modelled the lineage using fractional crystallization mass-balance models based on EMP analyses of olivine, clinopyroxene, kaersutite, opaque oxides, feldspar and apatite. The models showed excellent agreement with the observed whole-rock chemical compositions of the lava flows, and kaersutite was shown to have an important role in the evolution of the DVDP lineage.

Erebus Volcanic Province rocks are predominantly nepheline normative, although some quartz normative rocks are present (Fig. 21a). Quartz-normative compositions are mostly from Mount Morning, with some quartz-normative samples also reported from Mason Spur and Ross Island, and one example from Black Island. The quartz-normative compositions at Mount Morning are mainly from the older (>11.4 Ma) Mason Spur Lineage rocks. All the Erebus Volcanic Province quartz-normative samples are highly evolved with differentiation indices typically >60 (Fig. 21a). There is a wide range of differentiation indices in nepheline-normative rocks (Fig. 21a) reflecting an extensive variation in degree of fractionation, as can be seen on whole-rock total alkali v. silica (TAS) diagrams (Fig. 3). On these TAS diagrams, all Erebus Volcanic Province rocks plot as alkalic. The Mount Morning Volcanic Field rocks are mildly alkali–alkalic, plotting around the divisional line of Saggerson and Williams (1964) (Fig. 3c). The volcanic rocks in the Mount Discovery and Ross Island volcanic fields plot mostly in the strongly alkalic field, with volcanic rocks in the latter having higher $Na_2O + K_2O$ wt% values at any given SiO_2 content, relative to either the Mount Discovery or Mount Morning volcanic fields (Fig. 3). On a cumulative probability plot, the volcanic whole-rock data (excluding bombs, tephra and glass compositions: Fig. 21b) indicate that the bulk of the data (c. 69%) is primitive,

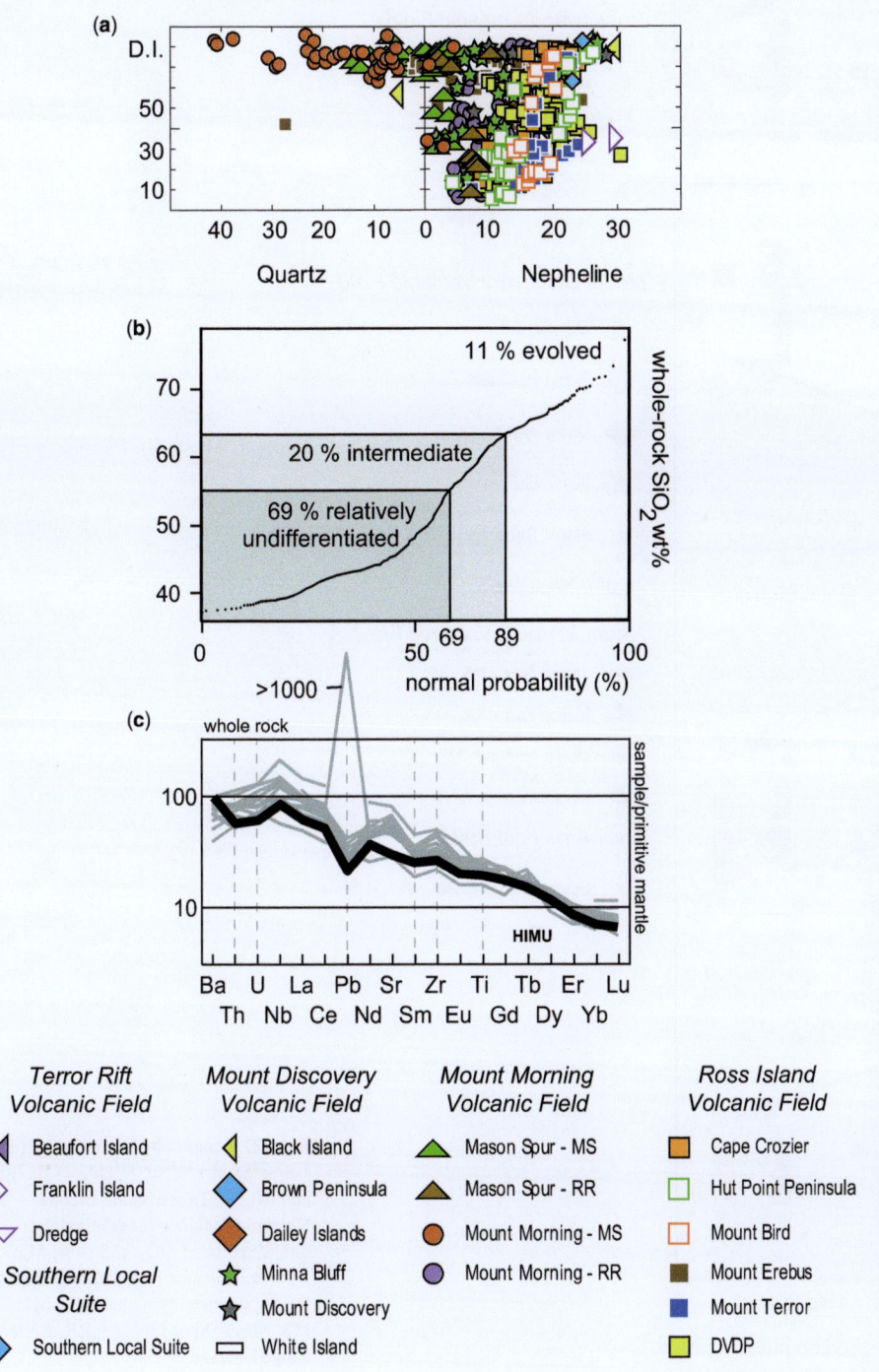

Fig. 21. Normative mineralogy and trace element characteristics of Erebus Volcanic Province rocks. (**a**) A plot of normative quartz or nepheline v. differentiation index (DI: normative wt% quartz + orthoclase + albite + nepheline + leucite + kalsilite: Thornton and Tuttle 1960). (**b**) A cumulative probability plot of volcanic whole-rock SiO_2 (wt%) data with subdivisions of relatively undifferentiated (<55 wt% SiO_2), intermediate (55–63 wt% SiO_2) and evolved (>63 wt% SiO_2) included for comparison. (**c**) Primitive-mantle-normalized extended element plots for selected, relatively undifferentiated, Erebus Volcanic Province whole-rock samples (grey lines). Data from Table 2 have been normalized to the primitive mantle values of McDonough and Sun (1995). The HIMU pattern (black line) is for Mangaia, Austral Islands (sample M-11 of Woodhead 1996). MS, Mason Spur Lineage; RR, Riviera Ridge Lineage.

assuming the primitive cutoff used in this study of 55 wt% SiO_2. Around 20% of the data is of an intermediate composition and only *c.* 11% of the analysed samples are evolved. The distribution of data (Fig. 21b) is not consistent with claims that the province is bimodal with a paucity of intermediate compositions.

Normalized extended element patterns (Fig. 21c) of primitive volcanic rocks are characterized by enrichments in large ion lithophile elements (LILEs) and high field strength elements (HFSEs), and depletions in Pb relative to primitive mantle, which is typical of the HIMU mantle reservoir. One exception from Mason Spur has a strongly positive Pb anomaly, which Martin *et al.* (2013) ascribed to a sedimentary-like component in the source. The mantle-normalized patterns feature moderate U/Th and Zr/Sm fractionation but insignificant variation in Ti relative to Eu or Gd (Fig. 21c). Some isotopic systems and trace element ratios in primitive volcanic rocks of the province can be seen to vary with longitude: for example, $^{87}Sr/^{86}Sr$, Ba/Rb and Nb/Ta values decrease, and $^{143}Nd/^{144}Nd$ values increase, from west to east (Fig. 22).

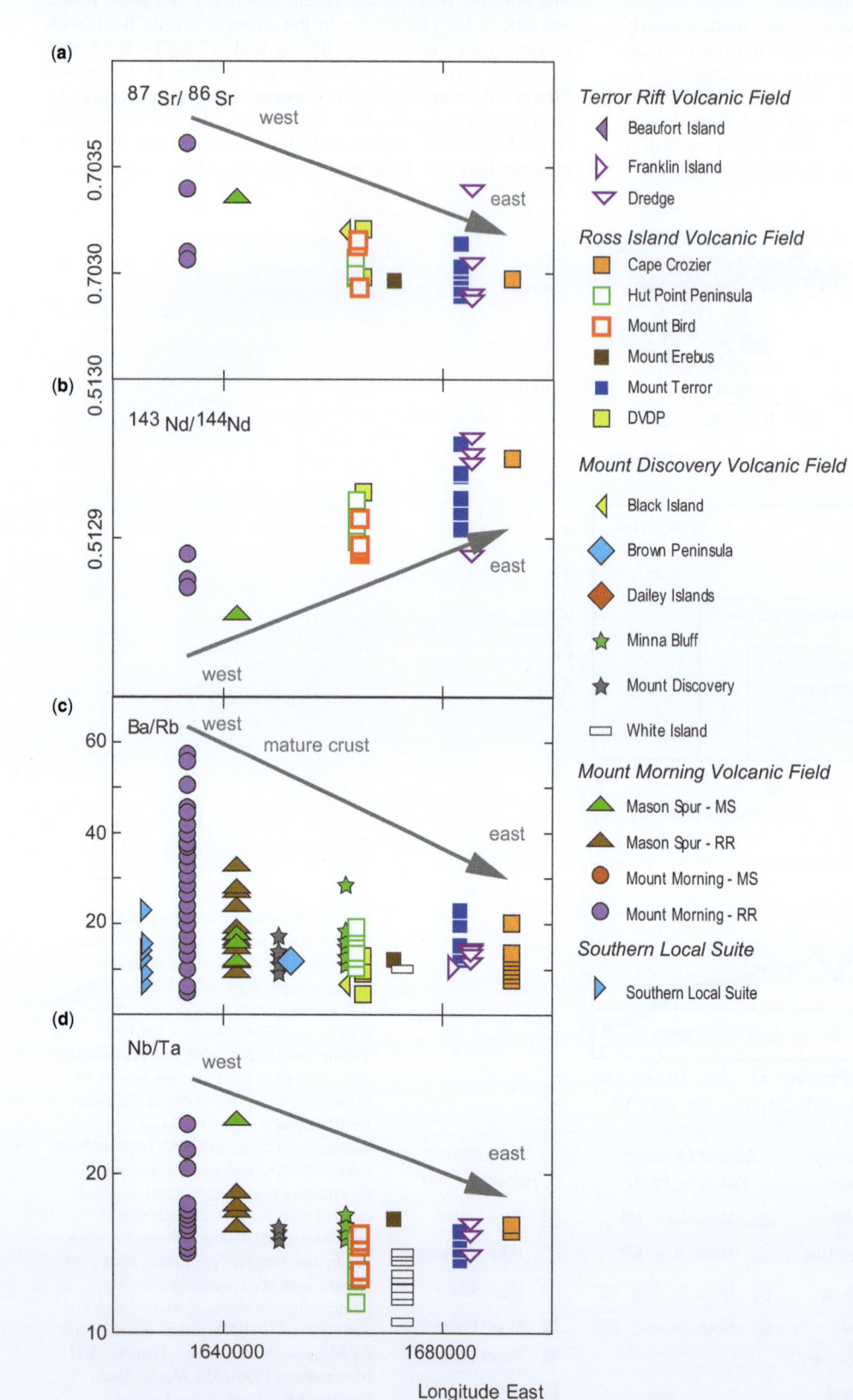

Fig. 22. Longitudinal variation of (**a**) $^{87}Sr/^{86}Sr$, (**b**) $^{143}Nd/^{144}Nd$, (**c**) Ba/Rb and (**d**) Nb/Ta across the Erebus Volcanic Province for relatively undifferentiated SiO_2 <55 wt%; MgO >6 wt%) volcanic rock data included in the Supplementary material (ESM1). MS, Mason Spur Lineage; RR, Riviera Ridge Lineage.

Many trace element ratios in primitive volcanic rocks overlap with the field defined for HIMU, with Ba/Nb ratios that extend towards enriched mantle (Fig. 23a). Strontium and Pb isotopic compositions, however, indicate that the source of primitive melts is too low in radiogenic Pb to overlap with the HIMU field as originally defined (Zindler and Hart 1986), although some data still plot towards enriched mantle (Fig. 23b).

Geothermobarometry. Clinopyroxene–melt equilibration pressures and temperatures were calculated using published and unpublished clinopyroxene data (see the Supplementary material (ESM1)), paired with whole-rock data for relatively undifferentiated volcanic rocks from the Southern Local Suite, Mount Morning, Minna Bluff and Ross Island. The melt compositions used are for rocks considered to represent magmas parental to more evolved compositions; in the case of Mount Morning, at least, these contain mantle xenoliths. The results using the Putirka *et al.* (1996) clinopyroxene–liquid geothermobarometer (Fig. 24) constrain the depths and temperatures at which the pyroxenes equilibrated with melt (Putirka *et al.* 2003). A number of published geothermobarometry equations were tested (Putirka *et al.* 1996, 2003; Putirka 2008), all revealing similar patterns for the province (Fig. 24). Estimates of pressure are broadly correlated with calculated temperatures; the clinopyroxene–melt pairs giving the highest pressures also give the highest temperatures (Fig. 24). The calculated pressures equate to depths greater than or equal to the measured depth to the Moho in the Erebus Volcanic Province (Bannister *et al.* 2003) (Fig. 24). These pressures are interpreted to indicate that the host magmas bypassed magma-staging areas in the crust and, although the possibility of a pause at the Moho boundary cannot be precluded, the time involved was insufficient to allow significant differentiation. This conclusion is consistent with the occurrence of mantle xenoliths. The clinopyroxene geobarometric data are interpreted to reflect variable source melting depths across the province. The highest pressures are associated with the least chemically evolved volcanic rocks in the Southern Local Suite; the pressures indicate shallower depths of equilibration in the Mount Morning and Mount Discovery volcanic fields, and pressures obtained for Ross Island Volcanic Field clinopyroxenes are higher (Fig. 24).

Discussion

Eruptive history and petrogenesis of Mount Erebus. Mount Erebus is the southernmost active volcano in the world, with an extremely long-lived convective lava lake. This has made it one of the best-studied volcanic centres in Antarctica and for this reason a more detailed overview of it is given here. The geological evolution of Mount Erebus was described by

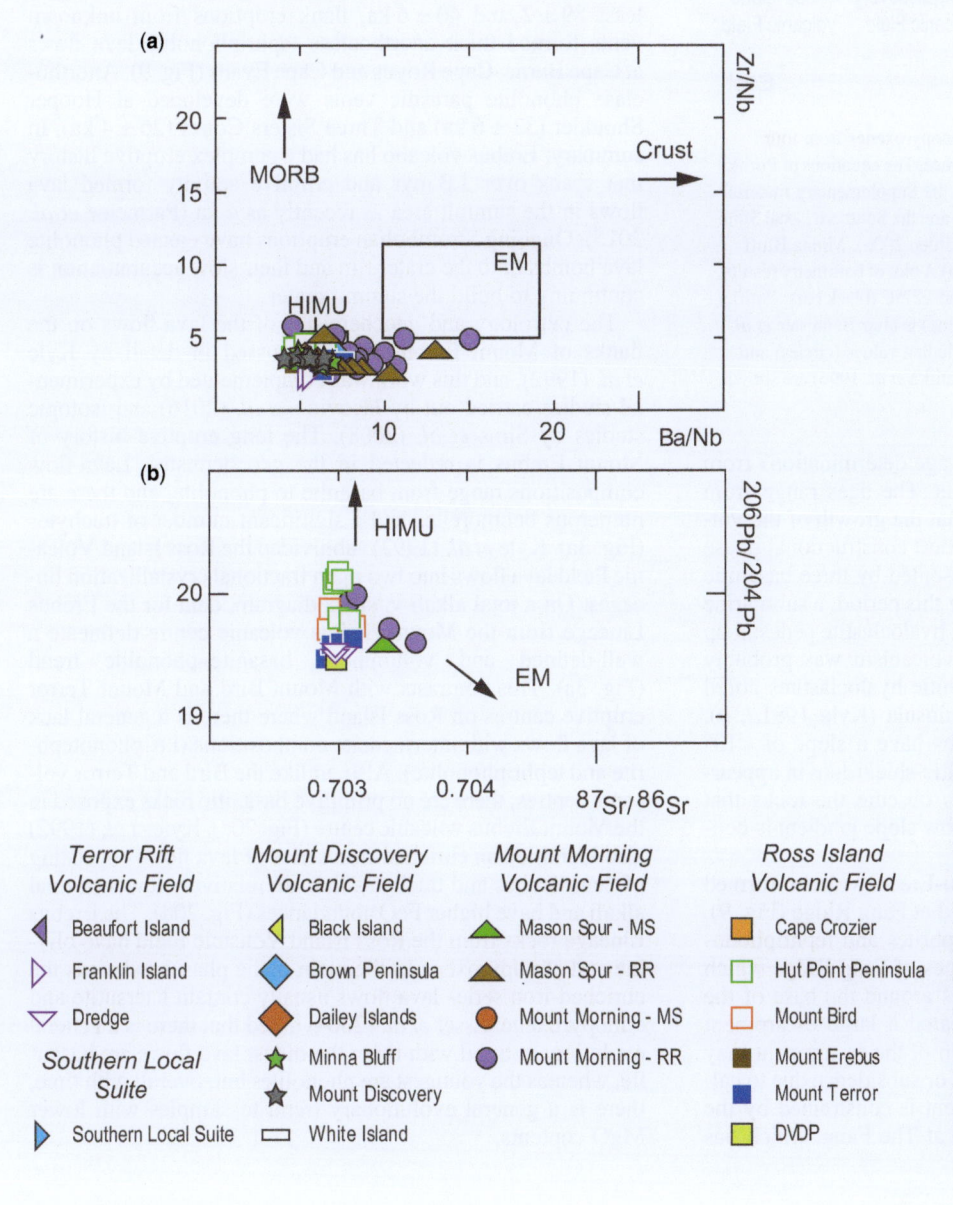

Fig. 23. Erebus Volcanic Province trace element and isotopic compositions for relatively undifferentiated (SiO$_2$ <55 wt%; MgO >6 wt%) volcanic rocks (data from the Supplementary material (ESM1)) compared with the compositions of selected mantle domains. (**a**) A Ba/Nb v. Zr/Nb plot. Compositions of HIMU, enriched mantle (EM), mid-ocean ridge basalt (MORB) and continental crust (crust) are from Weaver (1991). (**b**) A ^{87}Sr/^{86}Sr v. ^{206}Pb/^{204}Pb plot with the direction for HIMU and EM composition taken from Zindler and Hart (1986). MS, Mason Spur Lineage; RR, Riviera Ridge Lineage.

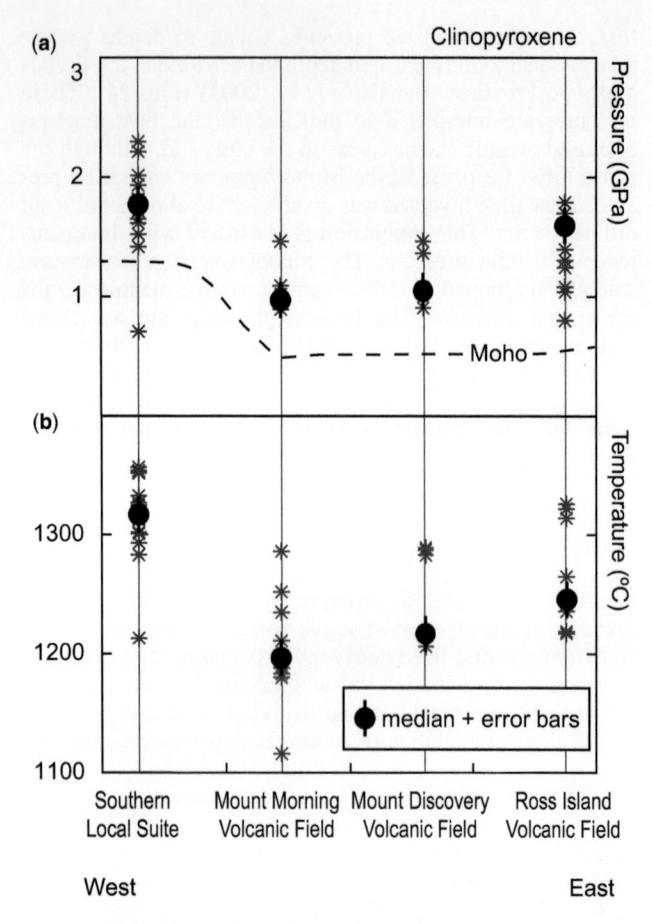

Fig. 24. Geothermobarometry results for clinopyroxenes from four volcanic centres in the Erebus Volcanic Province. The equations of Putirka *et al.* (1996) have been applied to data from the Supplementary material (ESM1). From west to east, the four localities are: the Southern Local Suite (Wingrove 2005), Mount Morning (van Woerden 2006), Minna Bluff (Redner 2016) and Ross Island (this study). (**a**) A plot of barometry results. Median values (circles) and errors estimated at ±27°C (black bars: Putirka *et al.* 1996) are shown. The Moho (dashed line) is after Bannister *et al.* (2003). (**b**) A plot of thermometry results. Median values (circles) and errors estimated at ±0.14 GPa (black bars: Putirka *et al.* 1996) are shown.

Esser *et al.* (2004) using $^{40}Ar/^{39}Ar$ age determinations from 25 sites around the flanks of Erebus. The ages range from 1311 ± 16 to 26 ± 4 ka, indicating that the growth of the volcano took place over 1 myr. The earliest constructional phase was from 1.3 to 1.0 Ma and is represented by three basanitic cones at Cape Barne (Fig. 9). During this period, a submarine phase of volcanism must have built a hyaloclastite pedestal up from the seafloor. The submarine volcanism was probably contemporaneous with similar basanitic hyaloclastites cored by the DVDP at the Hut Point Peninsula (Kyle 1981*a, b*). The lower sections of Mount Erebus have a slope of <10° and they define a profile that is basaltic-shield-like in appearance. Snow and younger lava flows obscure the rocks that comprise most of the shield but the low slope gradient is consistent with these being basaltic.

From 1070 to 718 ka a major proto-Erebus volcano formed and remnants of this are well exposed at Fang Ridge (Fig. 9). The lava flows are mainly phonotephrites and tephriphonolites, and these formed the steep slopes of Fang Ridge which contrast with the lower shield slopes around the base of the mountain. A catastrophic event created a large escarpment which is now Fang Ridge. The origin of the escarpment may have been related to a sector collapse or subsidence due to caldera collapse. The timing of the event is constrained by the stratigraphically youngest lava flow at The Fang, which has an age of 718 ± 66 ka. There is no dated activity from 718 to 539 ka. From 539 to *c.* 250 ka there were eruptions of phonotephrite, tephriphonolite and rare basanite at and surrounding Abbott Peak, at Inaccessible Island and in the Turks Head area (Fig. 9).

The period from *c.* 250 ka to the present was when the modern Mount Erebus volcanic cone ('Main Crater' and summit area: Fig. 9) was built. The younger tephriphonolite and phonolite lava flows of Mount Erebus are characterized by their porphyritic nature and the presence of large anorthoclase phenocrysts (megacrysts). The oldest anorthoclase-phyric tephriphonolite overlies plagioclase-phyric lava flows at Turks Head and is dated at 243 ± 10 ka, which is identical to the age of similar lava flows further upslope on Mount Erebus. These ages give an indication that eruptions were ultimately centred on the modern Erebus cone. There was a major eruptive episode at 160 ka, and this emplaced trachyte rocks on the eastern flank of Mount Erebus with vents formed at Bomb Peak (157 ± 6 ka), Ice Station (159 ± 2 ka) and 'Aurora Cliffs' (166 ± 10 ka: Esser *et al.* 2004; Kelly *et al.* 2008) (Fig. 9). Ice Station at 2730 m high on the upper eastern slopes of Erebus (Fig. 9) indicates that the modern Mount Erebus volcano was well developed by 160 ka. From 121 to 25 ka, anorthoclase-phyric phonolitic lava flows continued to be erupted as parasitic vents and as thick lava flows on the flanks of Mount Erebus volcano. Simultaneous eruptive activity is likely to have occurred in the summit area of the volcano and there were several caldera-forming events. Between at least 89 ± 2 and 40 ± 6 ka, flank eruptions from unknown vents formed thick anorthoclase tephriphonolite lava flows at Cape Barne, Cape Royds and Cape Evans (Fig. 9). Anorthoclase phonolite parasitic vents were developed at Hooper Shoulder (33 ± 6 ka) and Three Sisters Cones (26 ± 4 ka). In summary, Erebus volcano has had a complex eruptive history that spans over 1.3 myr and eruptive activity formed lava flows in the summit area as recently as 4 ka (Parmelee *et al.* 2015). Ongoing Strombolian eruptions have ejected phonolite lava bombs onto the crater rim and their slow accumulation is continuing to build the summit crater.

The petrology and geochemistry of the lava flows on the flanks of Mount Erebus were discussed in detail by Kyle *et al.* (1992), and this work was complemented by experimental studies carried out by Iacovino *et al.* (2016) and isotopic studies by Sims *et al.* (2008). The long eruptive history of Mount Erebus is reflected in the geochemistry. Lava-flow compositions range from basanite to phonolite, and there are numerous benmoreites and a significant number of trachytes (Fig. 3a). Kyle *et al.* (1992) subdivided the Ross Island Volcanic Field lava flows into two main fractional crystallization lineages. On a total alkali v. silica diagram, data for the Erebus Lineage from the Mount Erebus volcanic centre delineate a well-defined and voluminous basanite–phonolite trend (Fig. 3a). This contrasts with Mount Bird and Mount Terror eruptive centres on Ross Island where there is a general lack of lava flows with intermediate compositions (i.e. phonotephrite and tephriphonolite). Also, unlike the Bird and Terror volcanic centres, there are no primitive basanitic rocks exposed in the Mount Erebus volcanic centre (Fig. 20c). Kyle *et al.* (1992) also identified an enriched-iron series of lava flows consisting of benmoreites and trachytes that are relatively lower in total alkali and have higher FeO abundances (Fig. 20d). The Erebus Lineage rocks from the Ross Island Volcanic Field have olivine and clinopyroxene as the main mafic phases, whereas the enriched-iron series lava flows usually contain kaersutite and clinopyroxene. Esser *et al.* (2004) noted that there is a general evolutionary trend with time; the oldest lava flows are basanite, whereas the youngest are phonolites but, overall with time, there is a general evolutionary trend to samples with lower MgO contents.

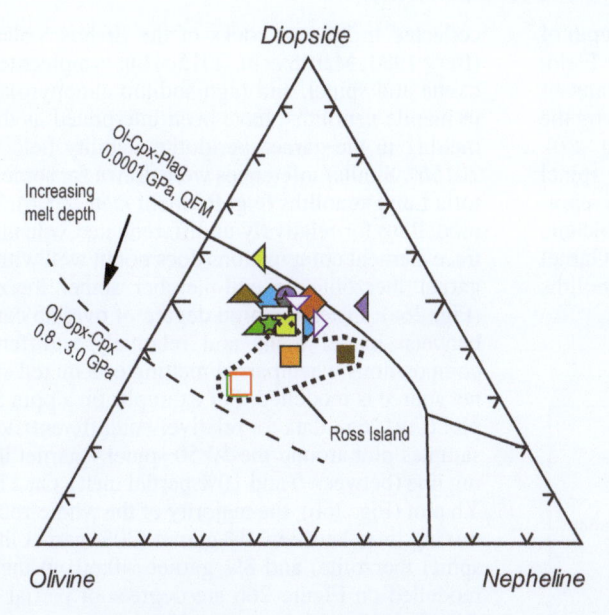

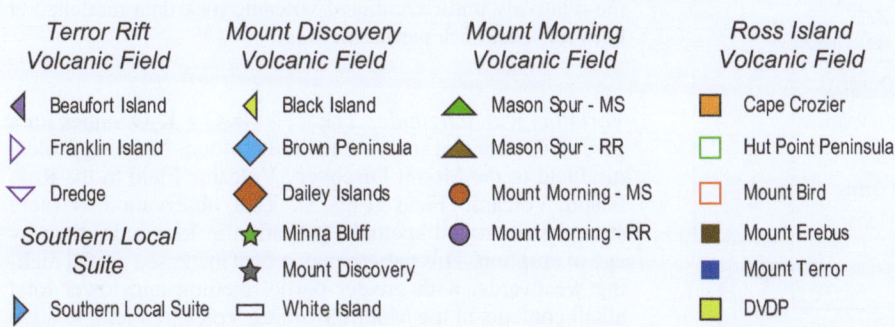

Fig. 25. Normative olivine, diopside and nepheline composition of typical relatively undifferentiated volcanic rocks from the Erebus Volcanic Province (Table 2). Also shown are the 0.0001 GPa and limiting 0.8–3.0 GPa cotectics of Sack *et al.* (1987). Cpx, clinopyroxene; DVDP, Dry Valley Drilling Project; HPP, Hut Point Peninsula; Ol, olivine; Opx, orthopyroxene; Plag, plagioclase; LS, Local Suite; MS, Mason Spur Lineage; RR, Riviera Ridge Lineage.

Assimilation and fractional crystallization processes and lineages. In the following discussion subsections, a regional approach will be used to discuss the petrogenesis of the Erebus Volcanic Province. This is possible because of the data compiled in this study and is thought preferable to discussing each volcanic field individually. For the province as a whole, total alkali v. silica diagrams (Fig. 3), wt% MgO v. major and trace element plots (Fig. 20), normative nepheline content that varies with differentiation index (Fig. 21a) and normative diopside/olivine ratio (not shown) are all consistent with magmatic evolution controlled by fractional crystallization involving olivine + clinopyroxene + magnetite/ilmenite + titanite. Several studies of the petrology of the province have concluded that Ti-amphibole (kaersutite) and feldspar are important fractionating phases (e.g. Kyle 1981b; Kyle *et al.* 1992; Martin *et al.* 2013). Although relatively rare, quartz-normative volcanic rocks have been found in association with otherwise nepheline normative alkalic suites worldwide (e.g. Price and Chappell 1975; Houghton *et al.* 1992; White *et al.* 2006). In Marie Byrd Land, silica-oversaturated trachytic volcanic rocks at Mount Sidley are hypothesized to have formed via wall-rock assimilation during fractional crystallization (Panter *et al.* 1997), and a similar explanation has been used to explain quartz-normative trachyte rocks at Mount Morning (Martin *et al.* 2013).

Several petrological lineages that record crystal fractionation history have been identified in the Erebus Volcanic Province. These include the DVDP (Kyle 1981b) and Erebus lineages (Kyle *et al.* 1992) identified at Ross Island, and the Mason Spur and Riviera Ridge lineages identified in the Mount Morning Volcanic Field (Martin *et al.* 2013). Kyle *et al.* (1992) also identified the enriched-iron series, members of which have significantly higher whole-rock FeO contents at a given wt% MgO content (Fig. 20d) relative to any other whole-rock compositions in the province. The identification and description of these lineages is key to understanding petrogenesis of specific magmatic suites. Variations between lineages represent contrasts in the parameters controlling petrogenesis including pressure, temperature and assimilation histories. The differences between volcanic fields and local suites in this province, however, indicate that applying magmatic lineage nomenclature outside any single field or suite may be problematic. Instead, suites in specific areas should be assigned to particular and unique lineages: for example, the Mount Discovery Volcanic Field lineage and the Terror Rift Volcanic Field lineage (both strongly alkalic, nepheline-normative, basanite–phonolite lineages), and the Southern Local Suite lineage (strongly alkalic, nepheline normative, basanite to phonotephrite). With further study, subtle differences between these various lineages may lead to further subdivision and refinements in nomenclature.

Depth to melting and degree of partial melting. In the diopside–olivine–nepheline experimental phase diagram of Sack *et al.* (1987), a primitive, whole-rock volcanic composition from Black Island plots on the 0.0001 GPa olivine + clinopyroxene + plagioclase cotectic but other compositions plot away from the 0.0001 GPa cotectic towards higher-pressure cotectics (Fig. 25). In this system, the Ross Island Volcanic Field data plot furthest towards the 0.8–3.0 GPa cotectic, indicating that magmas they represent were generated deeper in the mantle relative to other volcanic rocks in the province. This is supported by clinopyroxene-based geothermobarometry, which indicates that melts beneath Ross Island Volcanic Field partially crystallized at greater depths and higher temperatures than those generated beneath either the Mount Morning or Mount Discovery volcanic fields (Fig. 24). The clinopyroxene data from the Southern Local Suite indicate a greater depth of partial crystallization than in the Ross Island Volcanic Field, consistent with the greater depth to the Moho in the Transantarctic Mountains but a conclusion that is not supported by the trend in the diopside–olivine–nepheline projection (Fig. 25).

Spinel peridotite xenoliths and plagioclase-bearing spinel peridotite xenoliths were common cargo in the magmas represented by primitive volcanic rocks of the Erebus Volcanic Province (Kyle *et al.* 1987; Martin *et al.* 2014*a*, *b*). Garnet is never described in the lower-crustal or mantle xenoliths collected in igneous rocks of the Erebus Volcanic Province (Berg 1984; Martin *et al.* 2015*a*) but symplectites of orthopyroxene and spinel, and high-sodium clinopyroxene chemistry in mantle xenoliths, have been interpreted as the products of melting in the garnet–peridotite stability field (Martin *et al.* 2015*b*). Similar inferences were drawn for some northern Victoria Land xenoliths (e.g. Perinelli *et al.* 2006). Trace element modelling for relatively undifferentiated volcanic whole-rock trace element compositions does not fit well with end-member garnet lherzolite or end-member spinel lherzolite melting (Fig. 26a); instead, a high degree of overlap can be achieved between model melts and relatively undifferentiated rock compositions when partial melting of a mixed spinel and garnet source is modelled. For example, on a ppm Sm v. Sm/Yb plot (Fig. 26a), data for relatively undifferentiated whole-rock samples plot around the 50:50 spinel + garnet lherzolite mixing line (between 0 and 10% partial melt). On a La/Yb v. Gd/Yb plot (Fig. 26b), the majority of the whole-rock data plot on mixing lines between 4% garnet (20% garnet lherzolite: 80% spinel lherzolite) and 8% garnet (40:60) in the source. Also modelled on Figure 26b are degrees of partial melting, with the relatively undifferentiated volcanic rock data modelled at between 1 and 5% partial melting.

Variation with longitude. The wt% $Na_2O + K_2O$ values for a given SiO_2 content increase from the Mount Morning Volcanic Field to the Mount Discovery Volcanic Field to the Ross Island Volcanic Field (Fig. 3). This observation is more strongly controlled spatially (variation in longitude) than by age of eruption. This pattern may reflect increased partial melting westwards, with greater partial melting and lower total alkali contents in the Mount Morning Volcanic Field, relative to lower partial melting and higher total alkali abundance in the Ross Island Volcanic Field.

Longitudinal variation in the relatively undifferentiated volcanic rocks of eruptive centres/volcanic fields can also be seen on isotopic (Sr, Nd) and trace element ratio (Ba/Rb, Nb/Ta) plots (Fig. 22). Ratios of Ba/Rb in relatively undifferentiated volcanic rocks are higher in mature lower continental crust, relative to younger lower continental crust and upper continental crust (Stracke *et al.* 2003; Willbold and Stracke 2006). Ratios of Nb/Ta in relatively undifferentiated volcanic rocks can be higher in rocks that have a carbonatite-like component or an eclogite-like component in their sources (e.g. Pfänder *et al.* 2012). Using longitudinal variation of isotopic Sr and Nd (and trace element ratios), Guo *et al.* (2015) argued for a pattern of changing proportions of continental crustal components (including sediment melt and aqueous fluid) in the mantle source of the alkalic magmas they studied. Panter *et al.* (2018) found a pattern of eastward-increasing $^{143}Nd/^{144}Nd$ and eastward-decreasing $^{87}Sr/^{86}Sr$ in McMurdo Volcanic Group rocks in the most northerly part of northern Victoria Land, comparable to relationships seen in the Erebus Volcanic Province (Fig. 22), which they considered to be a function of the thickness and age of the mantle lithosphere. The eastward-decreasing radiogenic Sr, and increasing Nd, in the Erebus Volcanic Province relatively undifferentiated volcanic rocks is consistent with a decreasing proportion of crustal component in the primitive magma source (e.g. derived from fluids from an ancient subducted slab: Fig. 22). The eastwardly change in major and trace element abundances and isotopic compositions can be interpreted to indicate a systematically varying pattern of petrogenesis. This may be related to the involvement of an increasing proportion of mature lower continental crust (Ba/Rb), and/or increasing proportion of carbonatite-like or eclogite-like (Nb/Ta) and/or varying proportions of continental crust-like material (radiogenic Sr and Nd) in the mantle source, possibly

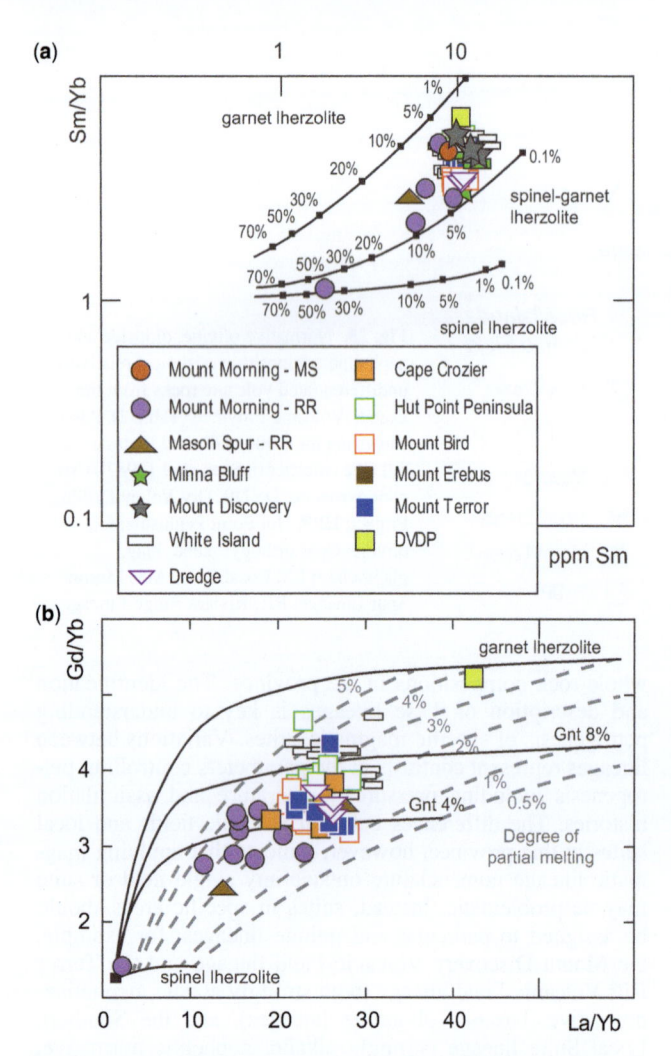

Fig. 26. Trace element plots showing relatively undifferentiated (SiO_2 <55 wt%; MgO >6 wt%) volcanic rock compositions from the Erebus Volcanic Province with various melt curves plotted for comparison. Data are from the Supplementary material (ESM1). (**a**) A ppm Sm v. Sm/Yb plot. Melt curves are for non-modal batch melting (Shaw 1970) for spinel lherzolite with (mode: melt mode) olivine$_{53:6}$ + orthopyroxene$_{27:28}$ + clinopyroxene$_{17:67}$ + spinel$_{3:11}$ (Kinzler 1997); and garnet lherzolite with olivine$_{60:16}$ + orthopyroxene$_{20:16}$ + clinopyroxene$_{10:88}$ + garnet$_{10:9}$ (Walter 1998). The partition coefficients are from McKenzie and O'Nions (1991, 1995), and the diagram follows Aldanmaz *et al.* (2000). (**b**) A plot of Gd/Yb v. La/Yb. Melt curves of accumulated fractional melting for spinel lherzolite with mode olivine$_{46}$ + orthopyroxene$_{28}$ + clinopyroxene$_{18}$ + spinel$_{18}$ and garnet lherzolite with mode olivine$_{54}$ + orthopyroxene$_{17}$ + clinopyroxene$_9$ + spinel$_{20}$, and following Workman *et al.* (2004). Orthopyroxene, clinopyroxene and spinel are assumed to react stoichiometrically to form olivine and garnet. Curves representing 4% garnet in the source (garnet:spinel = 20:80) or 8% garnet in the source (garnet:spinel = 40:60) are plotted for comparison. The partition coefficients are from Halliday *et al.* (1995) and the diagram follows Yokoyama *et al.* (2007). MS, Mason Spur Lineage; RR, Riviera Ridge Lineage.

combined with decreasing partial melting, eastwards (Panter et al. 2018).

Source characteristics. The trace element and isotope compositions of relatively undifferentiated Erebus Volcanic Province basaltic rocks and xenoliths can only be explained in terms of complex mantle sources, and polybaric and variable partial melting events. Several studies have led to the conclusion that the mantle source for the Erebus Volcanic Province relatively undifferentiated magmas is likely to have comprised depleted mantle modified at various times by the addition of enriched mantle, carbonatitic metasomatism and a HIMU-like component (e.g. see Sims et al. 2008; Martin et al. 2013; Aviado et al. 2015; Martin et al. 2015b). It has been suggested that the enriched component may have been compositionally similar to lower continental crust or pelagic sediment and this may have been added during early ancient subduction events (Martin et al. 2015b). The trace element and isotopic composition and mineralogy of potential mantle sources is discussed in the following subsections.

Amphibole, carbonatite and eclogite components. Melts derived as partial melts from an amphibole-bearing mantle source have higher Ba/Rb and lower Rb/Sr ratios than those generated by melting a phlogopite-bearing source (Furman and Graham 1999). Trace element ratios in the Erebus Volcanic Province whole-rock samples (Fig. 27a) indicate that amphibole dominates over phlogopite in the mantle source region; a conclusion that is consistent with trace element modelling (Sun and Hanson 1975) and experimental petrology (Iacovino et al. 2016) for the province.

Primitive-mantle-normalized extended element plots (Fig. 21c) for the Erebus Volcanic Province relatively undifferentiated rocks are characterized by fractionation of U relative to Th and Zr relative to Sm, and some primitive volcanic rocks have low Zr/Sm ratios relative to average continental crust (Fig. 27b). These features have been argued to indicate a carbonatite-like component in the mantle source (Yaxley et al. 1991; Pfänder et al. 2012). Carbonated peridotite and carbonated eclogite/pyroxenite in the mantle sources of relatively undifferentiated volcanic rocks have been evoked by some workers to explain the chemical and trace element relationships in the Erebus Volcanic Province (Martin et al. 2013; Aviado et al. 2015). Cambrian carbonatite dykes have been reported adjacent to the Southern Local Suite (Hall et al. 1995), modal carbonate grains have been reported in mantle

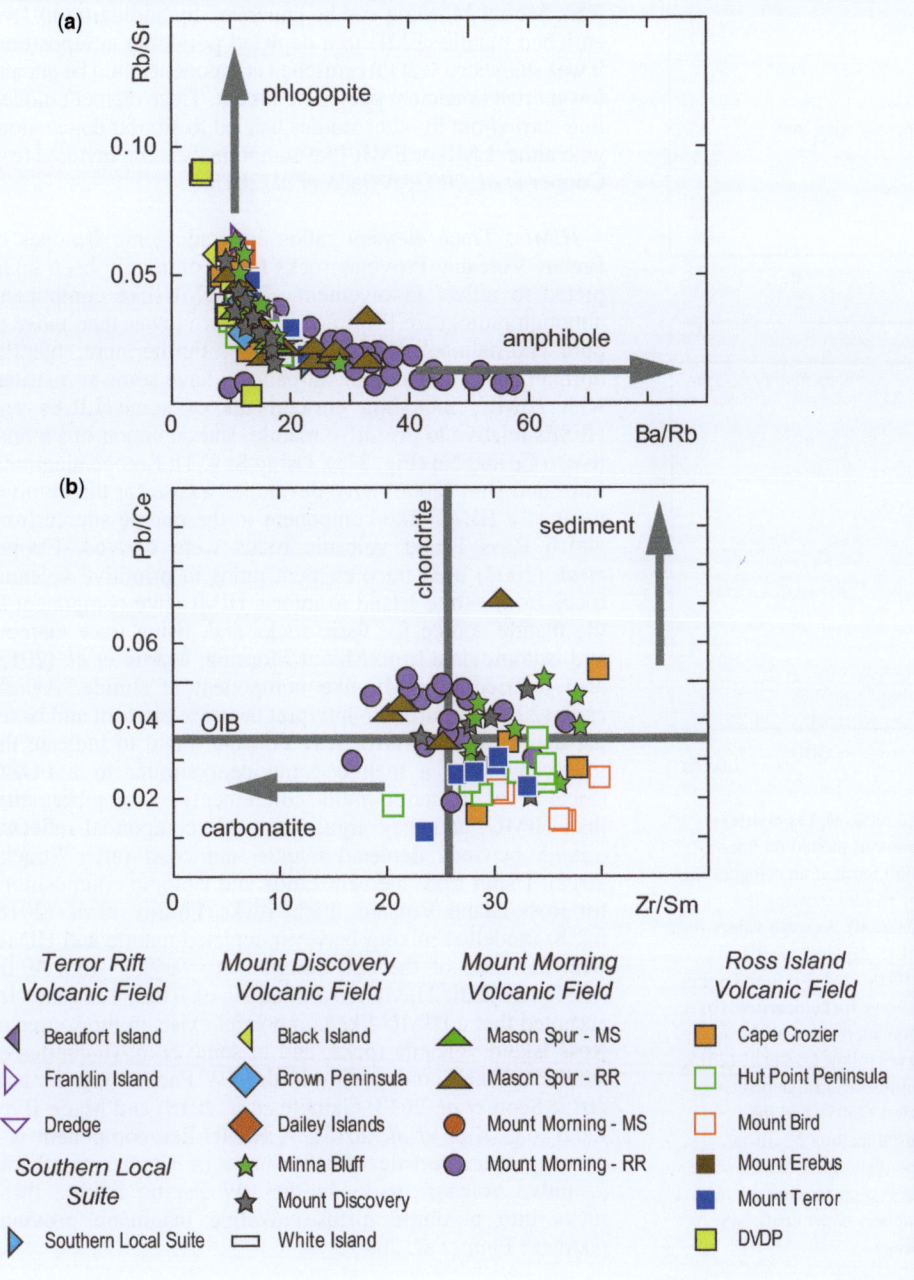

Fig. 27. Trace element ratio diagrams showing relatively undifferentiated (SiO$_2$ <55 wt%; MgO >6 wt%) whole-rock compositions for the Erebus Volcanic Province. The data are available in the Supplementary material (ESM1). (**a**) A Ba/Rb v. Rb/Sr plot showing mantle-source mineralogy that is more strongly amphibole influenced, relative to phlogopite. (**b**) A Zr/Sm v. Pb/Ce plot showing chondritic Zr/Sm (McDonough and Sun 1995) and typical ocean island basalt (OIB) Pb/Ce (Rudnick and Gao 2003). Values below and above these values are indicative of an influence by carbonatite- or sedimentary-like components, respectively. MS, Mason Spur Lineage; RR, Riviera Ridge Lineage.

xenoliths at Mount Morning (Martin *et al.* 2015*b*) and Erebus volcano melts have been modelled to be extremely high in CO_2 (Iacovino *et al.* 2016).

Estimated compositions of Terror Rift Volcanic Field primary melt compositions overlap with experimentally determined major element values of carbonated eclogite (Aviado *et al.* 2015) and an eclogite-like component was used to explain pyroxenite whole-rock trace element ratios at Mount Morning (Martin *et al.* 2015*b*). Ratios La/Nb and Nb/Ta of relatively undifferentiated volcanic rocks from the Erebus Volcanic Province can be used to further test the feasibility of an eclogite-like component in the mantle source. On a La/Nb v. Nb/Ta diagram (Fig. 28), several relatively undifferentiated volcanic rocks from the province plot towards the refractory eclogite reservoir of Rudnick *et al.* (2000), and this trend can also be seen on a Ti v. Ti/Zr plot (not shown). The possible influence of an eclogite component can be further modelled using La/Yb and Dy/Yb ratios (Fig. 28b). On a La/Yb v. Dy/Yb diagram, the trend defined by relatively undifferentiated volcanic rock data from the province is consistent with <15% mixing between eclogite and depleted mantle, with 10–20% eclogite in the source.

Enriched mantle. Trace element and isotopic data for relatively undifferentiated volcanic rocks from the Erebus Volcanic Province have commonly been interpreted to reflect the involvement of an enriched component in the mantle source (e.g. Sims and Hart 2006; Cooper *et al.* 2007; Martin *et al.* 2013; Aviado *et al.* 2015; Phillips *et al.* 2018) (Fig. 23). There is ambiguity about the nature and origin of this material but whole-rock ratios of Ba/Rb >15 (Fig. 22c) and Ba/Nb >9 (Fig. 23a) are similar to those of mature, lower continental crust. This material may have been added to the mantle during ancient subduction events. Modelling carried out by Martin *et al.* (2015*b*) was interpreted to show that peridotite xenoliths from Mount Morning can be generated by addingιup to 15% enriched mantle (EMI) to a depleted peridotite composition; it was suggested that this enriched component could be ancient lower crust or ancient pelagic sediment. Trace element modelling carried out in other studies has led to similar conclusions with either EMI- or EMII-like components being invoked (e.g. Cooper *et al.* 2007; Aviado *et al.* 2015).

HIMU. Trace element ratios and radiogenic isotopes of Erebus Volcanic Province rocks have commonly been interpreted to reflect involvement of a HIMU-like component, although radiogenic Pb isotope ratios are lower than those of pure end-member HIMU (Fig. 23). Furthermore, mantle-normalized extended element patterns have some similarities with HIMU, including enrichments of some LILEs and HFSEs relative to primitive mantle, and depletion of Pb relative to Ce and Nd (Fig. 21c). Using Sr v. Th isotope diagrams, Sims and Hart (2006) have developed a case for the involvement of a HIMU-like component in the mantle source from which Ross Island volcanic rocks were derived. Cooper *et al.* (2007) used trace element ratios in primitive volcanic rocks from White Island to infer a HIMU-like component in the mantle source for these rocks and, using trace element and isotopic data from Mount Morning, Martin *et al.* (2013) also inferred a HIMU-like component at source. Aviado *et al.* (2015) preferred to interpret the trace element and isotopic data from the Terror Rift Volcanic Field to indicate the involvement of a mantle component similar to a FOZO (focal zone – a mantle-plume component) end member rather than HIMU, and they argued that this component reflected mixing between depleted mantle and crust (after Stracke 2012). Using trace element ratios and isotopic compositions for Ross Island Volcanic Field rocks, Phillips *et al.* (2018, fig. 8) modelled mixing between depleted mantle and HIMU with the ratio of the two components varying from 40:60 (depleted mantle:HIMU) to around 90:10. Thus, it is generally accepted that a HIMU-like component exists in the source of Ross Island volcanic rocks, and in some eruptive centres of similar age and composition in the SW Pacific (e.g. Stracke 2012; Scott *et al.* 2013; Gamble *et al.* 2018) and Marie Byrd Land (e.g. Kipf *et al.* 2014). A HIMU-like component is a defining characteristic of the source of Cenozoic, alkalic, primitive volcanic rocks in the SW Pacific linking these rocks into a single diffuse alkaline magmatic province (DAMP: Finn *et al.* 2005).

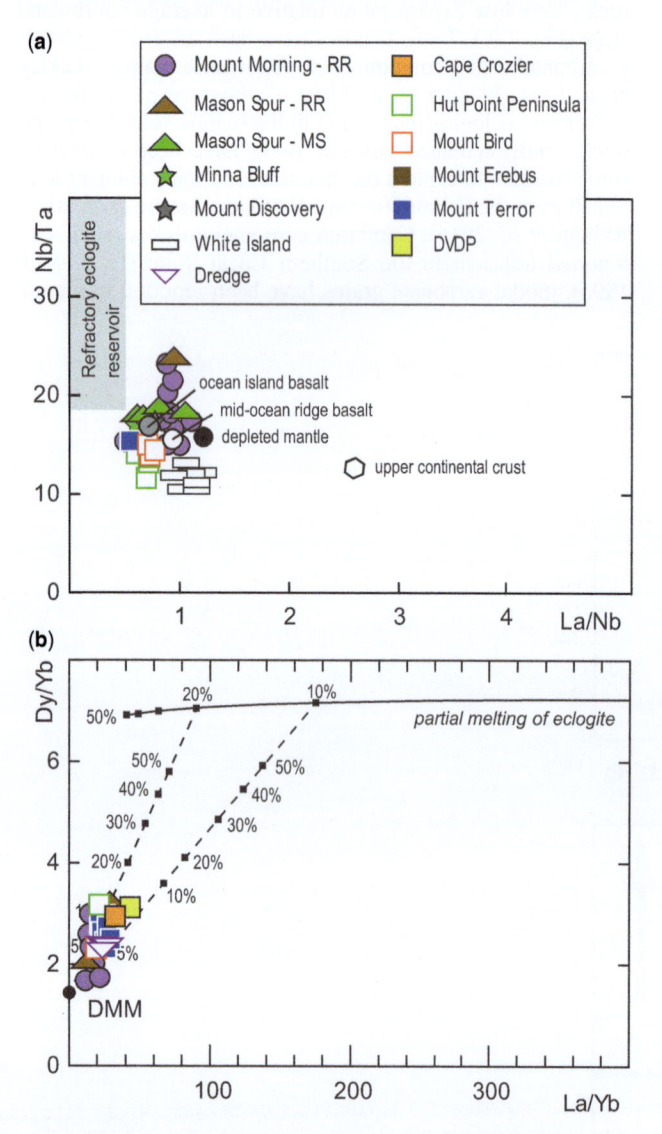

Fig. 28. Relatively undifferentiated (SiO_2 <55 wt%; MgO >6 wt%) Erebus Volcanic Province whole-rock compositions plotted on trace element ratio diagrams to show the potential influence of an eclogite-like component in the source. (**a**) A La/Nb v. Nb/Ta plot showing the refractory eclogite reservoir of Rudnick *et al.* (2000). Average values for upper continental crust (Rudnick and Gao 2014), depleted mantle (Rudnick *et al.* 2000), mid-ocean ridge basalt (Gale *et al.* 2013) and ocean island basalt (Sun and McDonough 1989) are shown for comparison. (**b**) A La/Yb v. Dy/Yb plot. The data and model parameters are available in the Supplementary material (ESM1). The melt curves follow Guo *et al.* (2015, fig. 23 and references therein). The depleted mid-ocean ridge mantle (DMM) composition is from Workman and Hart (2005), and the horizontal melt curve shows the degree of partial melting of crustal eclogite (numbers in per cent). The subvertical melting curves show the proportions of the eclogite-derived melt in the two-component mixture between DMM and eclogite-derived melt (numbers in per cent). MS, Mason Spur Lineage; RR, Riviera Ridge Lineage.

Age of mantle source chemistry and asthenosphere v. lithosphere. For those studying alkalic volcanic provinces, a recurring challenge is the differentiation of lithospheric and asthenospheric sources. On this problem, Herzberg (2011) wrote of the Hawaiian Islands that in mantle peridotite source compositions of volcanic rocks, it is rarely clear whether additional crust-like components are still present as a lithological unit in the source (pyroxenite) or whether only the geochemical signal (fluids or melts) of the recycled crust was imprinted on the source peridotite. To investigate this for the Hawaiian Islands, Sobolev *et al.* (2005) and Herzberg (2011) used major and trace element data to model whether basalt compositions are representative of peridotite- or pyroxenite-derived melts. For example, on a whole-rock wt% MgO v. ppm Ni plot (Fig. 20h) peridotite-derived melts (below the grey dashed curve in Fig. 20h) can be distinguished from pyroxenite-derived melts (above the curve). Most relatively undifferentiated volcanic rocks in the Erebus Volcanic Province have Ni concentrations lower than those expected of pyroxenite-derived melts, indicating that they probably derived from a peridotite mantle source.

The age of the HIMU signature in the source of alkalic rocks continues to be debated. As originally defined, HIMU represented high time-integrated $^{238}U/^{204}Pb$ ratios (Zindler and Hart 1986) with unradiogenic Sr ($^{87}Sr/^{86}Sr$ <0.703) relative to radiogenic Pb ($^{206}Pb/^{204}Pb$ >20.5). Modelling has shown that the appropriate U–Th–Pb ratios in a source will develop HIMU characteristics following extended storage of around 0.5–3.0 Ga (Hofmann and White 1982; Chauvel *et al.* 1992; Stracke *et al.* 2003). More recently, natural and experimental studies have demonstrated that highly radiogenic $^{238}U/^{204}Pb$ can be preserved relatively quickly (<0.5 Ga) through processes involving carbonatization (e.g. Scott *et al.* 2014; McCoy-West *et al.* 2016; van der Meer *et al.* 2017). This carbonatization may occur in the lithosphere or asthenosphere and therefore the age at which highly radiogenic $^{238}U/^{204}Pb$ formed also informs the debate on asthenospheric v. lithospheric sources for primitive magmas.

In the SW Pacific, this debate has been of particular interest because of the widespread dispersal of Cenozoic volcanic rocks with HIMU-like signatures across Zealandia, eastern Australia, Papua New Guinea and West Antarctica (Coombs *et al.* 1986; Finn *et al.* 2005). In Zealandia, isotopic studies have demonstrated disequilibrium between isotopic systems (Sr, Nd and Pb v. Hf), and constructed age isochrons showing Mesozoic ages for high radiogenic Pb are incongruous with the extended times required for *in situ* development of HIMU *sensu stricto* at some localities (e.g. Scott *et al.* 2014; McCoy-West *et al.* 2016).

The Zealandia example differs from that recorded in the Erebus Volcanic Province. The isotopic systems that were in disequilibrium in Zealandia have equilibrated in primitive volcanic rock samples from Ross Island (Sr, Nd, Pb, Hf), which has been interpreted to indicate high time-integrated HIMU in the mantle source (Sims *et al.* 2008; Phillips *et al.* 2018). However, it has also been highlighted that peridotite and pyroxenite mantle xenoliths from the Erebus Volcanic Province have trace element and isotopic ratios that indicate shared characteristics between their source and the source of relatively undifferentiated volcanic rocks in the province, including HIMU-, enriched-mantle-, carbonatite- and eclogite-like components (Martin 2009; Martin *et al.* 2013, 2014*a*, *b*, 2015*b*; Aviado *et al.* 2015). Furthermore, age determinations on pyroxenite xenoliths, intrusions and carbonatite dykes from southern Victoria Land, and eclogite from northern Victoria Land, are around 0.5 Ga (McGibbon 1991; Hall *et al.* 1995; Di Vincenzo *et al.* 1997; Martin *et al.* 2015*a*). This age coincides with the timing of subduction of the palaeo-Pacific margin of Gondwana, leading to the suggestion that the lithospheric mantle was modified by fluids from, or modified by, the subducting plate (Aviado *et al.* 2015; Martin *et al.* 2015*b*). A complication is that the 0.5 Ga time period also overlaps with the minimum time proposed to allow the *in situ* growth of the high time-integrated HIMU signature (Stracke *et al.* 2003). Sims *et al.* (2008) have pointed out that, because of the long (106 Ga) half-life of ^{147}Sm, 0.5 Ga is insufficient time to significantly change the radiogenic $^{143}Nd/^{144}Nd$ values of a vein-infused lithosphere. In northern Victoria Land, Panter *et al.* (2018) explained isotopic changes in Nd with longitude as being caused by the progressive reaction between rising alkalic melt and peridotite in the lithospheric mantle. It is, perhaps, significant that, in a classic paper, Sun and Hanson (1975) determined a two-stage model lead age of 1500 Ma for volcanic rocks from Ross Island. They interpreted this as the time since the development of chemical heterogeneity in the mantle source. Furthermore, $^{143}Nd/^{144}Nd$ is shown to vary with longitude across the Erebus Volcanic Province (Fig. 22b), which, as Sims *et al.* (2008) pointed out, is unlikely to be related to *in situ* processes in periods of time <0.5 Ga.

These observations can be explained by either of two possible hypotheses:

(1) A high time-integrated HIMU signal has developed *in situ* since *c.* 0.5 Ga, with varying radiogenic Nd explained by variable partial melting of the veined lithospheric mantle across the province. The compositions of primitive melts reflect involvement of both lithospheric and asthenospheric sources.
(2) The mantle source is asthenospheric and ancient (much older than 0.5 Ga). The varying radiogenic Nd is explained by variable *in situ* development.

The evidence discussed here, particularly the 1.5 Ga Pb isochron age of volcanic rocks at Ross Island (Sun and Hanson 1975) and the behaviour of TiO_2, FeO, CaO/Al_2O_3 and Ni relative to MgO abundance (Fig. 20), are more consistent with melting of an ancient asthenospheric source. This would support a HIMU mantle source *sensu stricto*. Such a hypothesis would be challenged or strengthened if additional age constraints could be placed on the timing of chemical modifications to the mantle source of relatively undifferentiated volcanic rocks in the province.

Plumes v. decompression melting. A plume-driven model of melting was originally popular for the Erebus Volcanic Province, and Ross Island in particular (e.g. Behrendt *et al.* 1991*b*; Kyle *et al.* 1992; Esser *et al.* 2004), with the three-fold radial symmetry of volcanism on Ross Island, and in volcanic centres about Mount Discovery, being argued to reflect plume-related updoming. Opponents to the Cretaceous plume hypothesis point out the absence of regional uplift (Cooper *et al.* 2007; Martin *et al.* 2013), low magma production rates relative to typical rates associated with plumes (Finn *et al.* 2005), and the size and longevity of mantle pluming required to generate melt simultaneously across >100 km of volcanism in the province (Cooper *et al.* 2007). This model was gradually replaced by the idea of decompression melting promoted by transtensional lithospheric deformation of a mantle metasomatized during a Late Cretaceous amagmatic extensional rift phase (Rocchi *et al.* 2002, 2005). The idea was further developed by Panter *et al.* (2018) for northern Victoria Land, with mantle upwelling possibly related to slab detachment and/or edge-driven mantle flow established at the boundary between the thinned lithosphere of the West Antarctic Rift System and cratonic East Antarctica. Recently, Phillips *et al.* (2018) revived the idea of a Cretaceous mantle plume beneath Ross Island based on three-fold radial symmetry of Ross Island, tomography, $^3He/^4He$ in clinopyroxenes of the DVDP lava flows, isotopic variability between eruptive centres on Ross

Island, and isotopic variability between Ross Island volcanic rock compositions and those in Zealandia.

The cause of contrast in isotopic variability between Zealandia and the Erebus Volcanic Province are, as discussed above, likely to be attributable to differences in the time of *in situ* growth of the HIMU signature in the Erebus Volcanic Province relative to the Mesozoic HIMU-like signature in Zealandia magmatic sources, and it is not diagnostic of a mantle-plume influence. The extensive compilation of data for relatively undifferentiated volcanic rocks of the Erebus Volcanic Province presented here can be used to show that the trace element and isotopic changes seen in the Ross Island Volcanic Field are part of a wider pattern observed across the entire province (e.g. Fig. 22). A plume head, with a >100 km radius centred beneath Ross Island, could explain this element and isotopic variation, as could edge-driven mantle flow. At this width, excess temperatures are modelled to drop significantly (up to four times) at 100 km distance for a plume/edge-driven scenario (Hauri *et al.* 1994, fig. 2b). Median temperatures calculated from clinopyroxene compositions decrease from the Ross Island Volcanic Field, through the Mount Discovery Volcanic Field, to the Mount Morning Volcanic Field, with temperatures rising sharply again in the Southern Local Suite mantle source (Fig. 24). This is contrary to the pattern observed in temperature data calculated for peridotite xenoliths of White Island, Mount Morning and the Southern Local Suite (Martin *et al.* 2014*a*), which increase eastwards. Whole-rock $^{87}Sr/^{86}Sr$, Ba/Rb and Nb/Ta ratios all increase westwards across the province (Fig. 22), whereas $^{143}Nd/^{144}Nd$ ratios decrease. High $^{87}Sr/^{86}Sr$ and relatively low $^{143}Nd/^{144}Nd$ are consistent with an increasingly enriched mantle component in the source melt, whereas increasing Ba/Rb and Nb/Ta ratios indicate increasing involvement of an eclogitic crustal component, eastwards. These patterns are consistent with an increasing input of subduction component eastwards in the province, similar to patterns observed in northern Victoria Land (Panter *et al.* 2018) or southern Tibet (Guo *et al.* 2015). This pattern is opposite to what might be expected if the subduction component is related to the 0.5 Ga subduction of the palaeo-Pacific Plate beneath East Antarctica. These variations could reflect a regional gradient caused by changing dynamics in a plume some 100 km in radius or they could be due to differences in decompression melting associated with uplift and rifting and edge-driven mantle flow. A cross-section down to 350 km-depth across East and West Antarctica in southern Victoria Land, showing shear-wave velocity variations in percentage relative to the Preliminary Reference Earth Model (PREM), corresponds with a sharp discontinuity between cold (fast) mantle beneath the Transantarctic Mountains and cratonic East Antarctica, relative to the warm (slow) anomalies beneath the Ross Sea and West Antarctica (Faccenna *et al.* 2008) (Fig. 29), in agreement with other studies (Watson *et al.* 2006; Martin *et al.* 2014*a*; An *et al.* 2015; Nield *et al.* 2018; Shen *et al.* 2018). This model (Fig. 29) does not have a deep (>150 km) root beneath the Erebus Volcanic Province, which is more consistent with edge-driven mantle flow than mantle pluming (Faccenna *et al.* 2008; Panter *et al.* 2018), although other workers have presented other mantle images which they interpret as evidence of a mantle plume (e.g. Phillips *et al.* 2018, fig. 9). In summary, regional chemical and isotopic changes are consistent with either edge-driven mantle flow or a Cenozoic plume, and geophysical evidence for both has been presented in the literature.

Summary and conclusions

Over 100 years of petrographical research on rocks from the Erebus Volcanic Province has contributed to understanding the petrogenesis of alkalic volcanic rocks in Antarctica and globally. Based upon petrogenetic and geographical discrimination, the province can be subdivided into four volcanic fields and one local suite. The mantle source for relatively undifferentiated volcanic rocks of the province is complex and variable with HIMU, enriched mantle, carbonatitic and eclogitic crustal components all being involved to variable extents. Equilibration of radiogenic Sr, Nd, Pb and Hf isotopic systems is best explained in terms of a high time-integrated HIMU *sensu stricto* component in the mantle source, at least beneath the Ross Island Volcanic Field. One model Pb isotope age, and major and trace element modelling of melting of a peridotitic source are most consistent with an asthenospheric mantle source for the depleted mantle and HIMU components. This is in contrast to some Cenozoic volcanism localities in Zealandia, where a HIMU-like component reflects the relatively young (Mesozoic) development of highly radiogenic Pb. Spatial (west–east) variations in Sr, Nd and Pb isotopic compositions and Ba/Rb and Nb/Ta ratios can be interpreted to indicate increasing involvement of an eclogitic crustal component eastwards, which is the opposite pattern to that expected if this component was derived from subduction of the palaeo-Pacific Plate beneath East Antarctica at *c.* 0.5 Ga. If the Pb isochron ages for Ross Island are applicable across the province, then this increasing eclogitic crustal component would derive from fluids derived from, or modified by, a subducting slab with an age >0.5 Ga: that is, not related to subduction of the palaeo-Pacific Plate at around 0.5 Ga.

The review completed in this study has highlighted areas where additional research could benefit the understanding of alkalic volcanic lineages in continental rift systems. These areas of future research can be summarized as:

- Determining when the chemical heterogeneity was developed in the mantle source at multiple eruptive centres across the province.
- Accurate and precise trace element and isotopic (Sr, Nd, Pb, Hf) measurements performed by a common method across multiple eruptive centres to test isotopic equilibration across the province and to test the eastward migration of trace element ratios further.

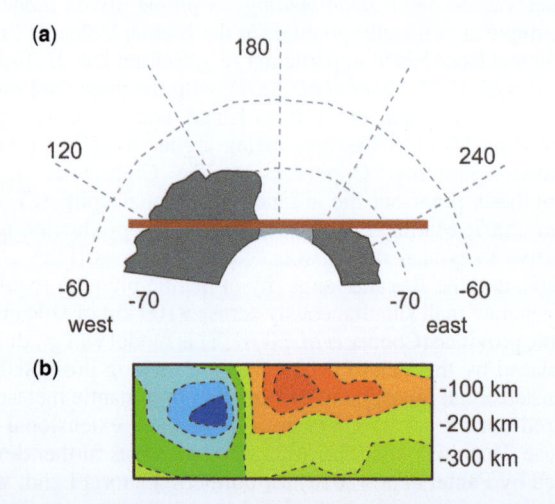

Fig. 29. A location diagram (**a**) and cross-section (**b**) for Antarctica, the latter showing shear-wave variations in model DM01 (Danesi and Morelli 2001), after Faccenna *et al.* (2008). The cross-section colours represent percentage variations with respect to the anisotropic Preliminary Earth Model (PREM), with the darkest blue colour representing 6% faster than PREM and the darkest red colour representing 5% slower than PREM. The discontinuity between fast (blue) and slow (red) anomalies corresponds with the transition from the Transantarctic Mountains into the Ross Sea.

- Analyses that are important but not routinely gathered should be obtained from multiple eruptive centres across the province: for example, halogens (Fl, Cl, Br, I), CO_2, isotopes (Os, Re, Hf) and noble gases (He, Ne, Ar, Kr, Xe).
- Key areas are missing, even basic published petrographical information including submarine volcanic centres, Mount Discovery, the Southern Local Suite and some areas around Ross Island (e.g. Lewis Bay). These areas require fieldwork. Furthermore, the area south of the Mount Morning and Discovery volcanic fields is virtually unknown with respect to Cenozoic volcanism. Geophysical investigations should be made to determine whether Cenozoic volcanic rocks are present beneath the Ross Ice Sheet in these areas.
- More work is needed on the plume v. rifting, extension and upwelling melt models to distinguish one from the other.

Despite the long and detailed study of the petrogenesis of volcanic rocks in the Erebus Volcanic Province, many new and exciting research questions remain, and these should, in the future, provide a new generation of Earth scientists with fruitful areas for further research.

Acknowledgements We thank Kurt Panter for editorial handling and additional review. Sergio Rocchi, Massimo Pompilio and an anonymous reviewer are thanked for thorough reviews.

Author contributions APM: conceptualization (lead), data curation (lead), investigation (equal), writing – original draft (lead), writing – review & editing (equal); **AFC**: investigation (equal), writing – original draft (supporting), writing – review & editing (supporting); **RCP**: investigation (equal), writing – original draft (supporting), writing – review & editing (equal); **PRK**: conceptualization (supporting), investigation (equal), writing – original draft (equal), writing – review & editing (supporting); **JAG**: investigation (equal), writing – original draft (supporting), writing – review & editing (supporting).

Funding Antarctica New Zealand provided financial and logistical support over several field seasons.

Data availability All data generated or analysed during this study are included in this published article (and its supplementary information files).

References

Adams, J. 1973. *Petrology and chemistry of an alkaline cone, McMurdo Sound*. BSc (Hon.) thesis, Victoria University of Wellington, Wellington, New Zealand.

Aldanmaz, E., Pearce, J.A., Thirlwall, M.F. and Mitchell, J.G. 2000. Petrogenetic evolution of late Cenozoic, post-collision volcanism in western Anatolia, Turkey. *Journal of Volcanology and Geothermal Research*, **102**, 67–95, https://doi.org/10.1016/S0377-0273(00)00182-7

An, M., Wiens, D.A. *et al.* 2015. S-velocity model and inferred Moho topography beneath the Antarctic Plate from Rayleigh waves. *Journal of Geophysical Research: Solid Earth*, **120**, 359–383, https://doi.org/10.1002/2014JB011332

Anderson, J.T.H., Wilson, G.S., Fink, D., Lilly, K., Levy, R.H. and Townsend, D. 2017. Reconciling marine and terrestrial evidence for post LGM ice sheet retreat in southern McMurdo Sound, Antarctica. *Quaternary Science Reviews*, **157**, 1–13, https://doi.org/10.1016/j.quascirev.2016.12.007

Antibus, J.V. 2012. *A Petrographic, Geochemical and Isotopic (Sr, O, H and C) Investigation of Alteration Minerals in Volcaniclastic Rocks at Minna Bluff, Antarctica: Petrogenesis and Implications for Paleoenvironmental Conditions*. MSc thesis, Bowling Green State University, Bowling Green, Ohio, USA.

Antibus, J.V., Panter, K.S. *et al.* 2014. Alteration of volcaniclastic deposits at Minna Bluff: Geochemical insights on mineralizing environment and climate during the Late Miocene in Antarctica. *Geochemistry, Geophysics, Geosystems*, **15**, 3258–3280, https://doi.org/10.1002/2014GC005422

Armstrong, R.L. 1978. K–Ar dating: Late Cenozoic McMurdo Volcanic Group and dry valley glacial history, Victoria Land, Antarctica. *New Zealand Journal of Geology and Geophysics*, **21**, 685–698, https://doi.org/10.1080/00288306.1978.10425199

Armstrong, R.L., Hamilton, W. and Denton, G.H. 1968. Glaciation in Taylor Valley, Antarctica, older than 2.7 million years. *Science*, **159**, 187–189, https://doi.org/10.1126/science.159.3811.187

Aviado, K.B., Rilling-Hall, S., Bryce, J.G. and Mukasa, S.B. 2015. Submarine and subaerial lavas in the West Antarctic Rift System: Temporal record of shifting magma source components from the lithosphere and asthenosphere. *Geochemistry, Geophysics, Geosystems*, **16**, 4344–4361, https://doi.org/10.1002/2015GC006076

Bannister, S., Yu, J., Leitner, B. and Kennett, B.L.N. 2003. Variations in crustal structure across the transition from West to East Antarctica, Southern Victoria Land. *Geophysical Journal International*, **155**, 870–884, https://doi.org/10.1111/j.1365-246X.2003.02094.x

Barrett, P.J. 1989. *Antarctic Cenozoic History from the CIROS-1 Drillhole, McMurdo Sound*. DSIR Bulletin, **245**.

Barrett, P.J. and Hambrey, M.J. 1992. Plio-Pleistocene sedimentation in Ferrar Fiord, Antarctica. *Sedimentology*, **39**, 109–123, https://doi.org/10.1111/j.1365-3091.1992.tb01025.x

Barrett, P.J. and McKelvey, B.C. 1986. Stratigraphy. *DSIR Bulletin*, **247**, 9–53.

Barrett, P.J. and Treves, S.B. 1981. Sedimentology and petrology of core from DVDP 15, Western McMurdo Sound. *American Geophysical Union Antarctic Research Series*, **33**, 281–314.

Barrett, P.J., Treves, S.B. *et al.* 1976. *Initial Report on DVDP 15, Western McMurdo Sound, Antarctica*. Dry Valleys Drilling Project Bulletin, **7**.

Barrett, P.J., Fielding, C., Wise, S.W. and the Cape Roberts Science Team. 1998. Initial report on CRP-1, Cape Roberts Project, Antarctica. *Terra Antarctica*, **5**, 187.

Barrett, P.J., Sarti, M., Wise, S. and the Cape Roberts Science Team. 2000. Studies from the Cape Roberts Project, Ross Sea, Antarctica: Initial report on CRP-3. *Terra Antarctica*, **7**, 209.

Behrendt, J.C. 1999. Crustal and lithospheric structure of the West Antarctic Rift System from geophysical investigations – a review. *Global and Planetary Change*, **23**, 25–44, https://doi.org/10.1016/S0921-8181(99)00049-1

Behrendt, J.C., Duerbaum, H.J., Damaske, D., Saltus, R., Bosum, W. and Cooper, A.K. 1991*a*. Extensive volcanism and related tectonism beneath the western Ross Sea continental shelf, Antarctic. Interpretation of an aeromagnetic survey. *In*: Thomson, M.R.A., Crame, J.A. and Thomson, J.W. (eds) *Geological Evolution of Antarctica*. Cambridge University Press, New York, 299–304.

Behrendt, J.C., LeMasurier, W.E., Cooper, A.K., Tessensohn, F., Tréhu, A. and Damaske, D. 1991*b*. Geophysical studies of the West Antarctic Rift System. *Tectonics*, **10**, 1257–1273, https://doi.org/10.1029/91tc00868

Bentley, C.R. 1991. Configuration and structure of the subglacial crust. *In*: Tingey, R.J. (ed.) *The Geology of Antarctica*. Clarendon, Oxford, UK, 335–364.

Berg, J.H. 1984. Crustal inclusions from the Erebus Volcanic Province. *Antarctic Journal of the United States*, **19**, 27.

Berg, J.H., Hank, R.A. and Kalamarides, R.I. 1985. Petrology and geochemistry of inclusions of lower crustal basic granulites from the Erebus Volcanic Province, Antarctica. *Antarctic Journal of the United States*, **20**, 22–23.

Blank, H.R., Cooper, R.A., Wheeler, R.H. and Willis, I.A.G. 1963. Geology of the Koettlitz–Blue Glacier region, Southern Victoria Land, Antarctica. *Transactions of the Royal Society of New Zealand*, **2**, 79–102.

Borchgrevink, C.E. 1901. *First on the Antarctic Continent*. Newnes, London.

Brodie, J.W. 1959. A shallow shelf around Franklin Island in the Ross Sea, Antarctica. *New Zealand Journal of Geology and Geophysics*, **2**, 108–119, https://doi.org/10.1080/00288306.1959.10431316

Cape Roberts Science Team. 1999. Studies from the Cape Roberts Project: initial report on CRP-2/2A, Ross Sea Antarctica – Summary of results. *Terra Antartica*, **6**, 156–169.

Cartwright, K., Treves, S.B. and Torii, T. 1974*a*. Geology of DVDP 4, Lake Vanda, Wright Valley, Antarctica. *Dry Valleys Drilling Project Bulletin*, **3**, 49–74.

Cartwright, K., Treves, S.B. and Torii, T. 1974*b*. Geology of DVDP 5, Don Juan Pond, Wright Valley, Antarctica. *Dry Valleys Drilling Project Bulletin*, **3**, 75–91.

Chapman-Smith, M. 1975*a*. Geologic log of DVDP 12, Lake Leon, Taylor Valley. *Dry Valleys Drilling Project Bulletin*, **5**, 61–70.

Chapman-Smith, M. 1975*b*. Geologic Log of DVDP 14, North Fork Basin, Wright Valley. *Dry Valleys Drilling Project Bulletin*, **5**, 94–99.

Chauvel, C., Hofmann, A.W. and Vidal, P. 1992. HIMU-EM: the French Polynesian connection. *Earth and Planetary Science Letters*, **110**, 99–119, https://doi.org/10.1016/0012-821X(92)90042-T

Cole, J.W. and Ewart, A. 1968. Contributions to the volcanic geology of the Black Island, Brown Peninsula, and Cape Bird areas, Mcmurdo sound, Antarctica. *New Zealand Journal of Geology and Geophysics*, **11**, 793–828, https://doi.org/10.1080/00288306.1968.10420754

Cole, J.W., Kyle, P.R. and Neall, V.E. 1971. Contributions to quaternary geology of Cape Crozier, White Island and Hut Point Peninsula, McMurdo Sound region, Antarctica. *New Zealand Journal of Geology and Geophysics*, **14**, 528–546, https://doi.org/10.1080/00288306.1971.10421946

Coombs, D.S., Cas, R.A., Kawachi, Y., Landis, C.A., McDonough, W.F. and Reay, A. 1986. Cenozoic volcanism in North, East, and Central Otago. *Royal Society of New Zealand Bulletin*, **23**, 278–312.

Cooper, A.K., Davey, F.J. and Behrendt, J. 1987. Seismic stratigraphy and structure of the Victoria Land basin, western Ross Sea, Antarctica. *In*: Cooper, A.K. and Davey, F.J. (eds) *The Antarctic Continental Margin: Geology and Geophysics of the Western Ross Sea*. Circum-Pacific Council Energy and Mineral Resources, Houston, TX, 27–65.

Cooper, A.F., Adam, L.J., Coulter, R.F., Eby, G.N. and McIntosh, W.C. 2007. Geology, geochronology and geochemistry of a basanitic volcano, White Island, Ross Sea, Antarctica. *Journal of Volcanology and Geothermal Research*, **165**, 189–216, https://doi.org/10.1016/j.jvolgeores.2007.06.003

Cox, S.C., Turnbull, I.M., Isaac, M.J., Townsend, D.B. and Smith Lyttle, B. 2012. Geology of southern Victoria Land Antarctica. Institute of Geological and Nuclear Sciences 1:250 000 Geological Map 22. GNS Science, Lower Hutt, New Zealand.

Csatho, B., Schenk, T. *et al.* 2005. Airborne laser scanning for high-resolution mapping of Antarctica. *Eos, Transactions of the American Geophysical Union*, **86**, 237–238, https://doi.org/10.1029/2005EO250002

Danesi, S. and Morelli, A. 2001. Structure of the upper mantle under the Antarctic Plate from surface wave tomography. *Geophysical Research Letters*, **28**, 4395–4398, https://doi.org/10.1029/2001GL013431

David, T.W.E. and Priestly, R.E. 1914. *Glaciology, Physiography, Stratigraphy and Tectonic Geology of south Victoria Land*. British Antarctic Expedition, 1907–09, Reports on the Scientific Investigations, Geology, **1**. W. Heinemann, London.

Debenham, F. 1923. *The Physiography of the Ross Archipelago British Antarctic 'Terra Nova' Expedition, 1910–1913*. Harrison, London.

Del Carlo, P., Panter, K.S., Bassett, K., Bracciali, L., Di Vincenzo, G. and Rocchi, S. 2009. The upper lithostratigraphic unit of ANDRILL AND-2A core (Southern McMurdo Sound, Antarctica): local Pleistocene volcanic sources, paleoenvironmental implications and subsidence in the southern Victoria Land Basin. *Global and Planetary Change*, **69**, 142–161, https://doi.org/10.1016/j.gloplacha.2009.09.002

Denton, G.H. and Marchant, D.R. 2000. The geologic basis for a reconstruction of a grounded ice sheet in McMurdo Sound, Antarctica, at the Last Glacial Maximum. *Geografiska Annaler*, **82A**, 167–211, https://doi.org/10.1111/j.0435-3676.2000.00121.x

Di Roberto, A., Pompilio, M. and Wilch, T.I. 2010. Late Miocene submarine volcanism in ANDRILL AND-1B drill core, Ross Embayment, Antarctica. *Geosphere*, **6**, 524–536, https://doi.org/10.1130/GES00537.1

Di Roberto, A., Del Carlo, P., Rocchi, S. and Panter, K.S. 2012. Early Miocene volcanic activity and paleoenvironment conditions recorded in tephra layers of the AND-2A core (southern McMurdo Sound, Antarctica). *Geosphere*, **8**, 1342–1355, https://doi.org/10.1130/ges00754.1

Di Roberto, A., del Carlo, P. and Pompilio, M. 2021. Marine record of volcanism from drill cores. *Geological Society, London, Memoirs*, **55**, https://doi.org/10.1144/M55-2018-49

Di Vincenzo, G., Palmeri, R., Talarico, F., Andriessen, P.A.M. and Ricci, G.A. 1997. Petrology and geochronology of eclogites from the Lanterman Range, Antarctica. *Journal of Petrology*, **38**, 1391–1417, https://doi.org/10.1093/petroj/38.10.1391

Di Vincenzo, G., Bracciali, L., Del Carlo, P., Panter, K. and Rocchi, S. 2010. ^{40}Ar–^{39}Ar dating of volcanogenic products from the AND-2A core (ANDRILL Southern McMurdo Sound Project, Antarctica): correlations with the Erebus Volcanic Province and implications for the age model of the core. *Bulletin of Volcanology*, **72**, 487–505, https://doi.org/10.1007/s00445-009-0337-z

Dunbar, N.W. and Kurbatov, A.V. 2011. Tephrochronology of the Siple Dome ice core, West Antarctica: correlations and sources. *Quaternary Science Reviews*, **30**, 1602–1614, https://doi.org/10.1016/j.quascirev.2011.03.015

Dunbar, N.W., Iverson, N.A. *et al.* 2017. New Zealand supereruption provides time marker for the Last Glacial Maximum in Antarctica. *Scientific Reports*, **7**, 12238, https://doi.org/10.1038/s41598-017-11758-0

Ellerman, P.J. and Kyle, P.R. 1990*a*. Beaufort Island. *American Geophysical Union Antarctic Research Series*, **48**, 94–96.

Ellerman, P.J. and Kyle, P.R. 1990*b*. Franklin Island. *American Geophysical Union Antarctic Research Series*, **48**, 91–93.

Esser, R.P., Kyle, P.R. and McIntosh, W.C. 2004. ^{40}Ar/^{39}Ar dating of the eruptive history of Mount Erebus, Antarctica: volcano evolution. *Bulletin of Volcanology*, **66**, 671–686, https://doi.org/10.1007/s00445-004-0354-x

Faccenna, C., Rossetti, F., Becker, T.W., Danesi, S. and Morelli, A. 2008. Recent extension driven by mantle upwelling beneath the Admiralty Mountains (East Antarctica). *Tectonics*, **27**, TC4015, https://doi.org/10.1029/2007TC002197

Fargo, A. 2009. *^{40}Ar/^{39}Ar Geochronological Analysis of Minna Bluff, Antarctica: Evidence for Past Glacial Events within the Ross Embayment*. MS thesis, New Mexico Institute of Mining and Technology, Socorro, New Mexico, USA.

Fargo, A., McIntosh, W., Dunbar, N. and Wilch, T. 2008. 40Ar/39Ar geochronology of Minna Bluff, Antarctica: Timing of mid-Miocene glacial erosional events within the Ross Embayment. Abstract presented at the 2008 AGU Fall Meeting, December 15–19, 2008, San Francisco, California, USA.

Ferrar, H.T. 1905. Notes on the physical geography of the Antarctic. *The Geographical Journal*, **25**, 373–382, https://doi.org/10.2307/1776138

Ferrar, H.T. 1907. Report on the field-geology of the region explored during the 'Discovery' Antarctic expedition, 1901–1904. *Natural History*, **1**, 1–100.

Fielding, C.R., Thomson, M.R.A. and the Cape Roberts Science Team. 1999. Studies from the Cape Roberts Project, Ross Sea, Antarctica: initial report on CRP-2/2A. *Terra Antartica*, **6**, 173.

Fielding, C.R., Henrys, S.A. and Wilson, T.J. 2006. Rift history of the western Victoria Land Basin: a new perspective based on integration of cores with seismic reflection data. *In*: Futterer, D.K.,

Damaske, D., Kleinschmidt, G., Miller, H. and Tessensohn, F. (eds) *Antarctica: Contributions to Global Earth Sciences.* Springer, Berlin, 307–316.

Fielding, C.R., Atkins, C.B. *et al.* 2008. Sedimentology and Stratigraphy of the AND-2A Core, ANDRILL Southern McMurdo Sound Project, Antarctica. *Terra Antartica*, **15**, 77–112.

Fielding, C.R., Browne, G.H. *et al.* 2011. Sequence stratigraphy of the ANDRILL AND-2A drillcore, Antarctica: a long-term, ice-proximal record of Early to Mid-Miocene climate, sea-level and glacial dynamism. *Palaeogeography, Palaeoclimatology, Palaeoecology*, **305**, 337–351, https://doi.org/10.1016/j.palaeo.2011.03.026

Finn, C.A., Müller, R.D. and Panter, K.S. 2005. A Cenozoic diffuse alkaline magmatic province (DAMP) in the southwest Pacific without rift or plume origin. *Geochemistry, Geophysics, Geosystems*, **6**, Q02005, https://doi.org/10.1029/2004gc000723

Fletcher, L. and Bell, J.F. (eds). 1907. *National Antarctic Expedition 1901–1904. Natural History. Volume I. Geology (Field-geology: Petrography).* Trustees of the British Museum, London.

Forbes, R.B., Turner, D.L. and Garden, J.R. 1974. Age of trachyte from Ross Island, Antarctica. *Geology*, **2**, 297–298, https://doi.org/10.1130/0091-7613(1974)2<297:AOTFRI>2.0.CO;2

Furman, T. and Graham, D. 1999. Erosion of lithospheric mantle beneath the East African Rift system: geochemical evidence from the Kivu volcanic province. *Lithos*, **48**, 237–262, https://doi.org/10.1016/S0024-4937(99)00031-6

Gale, A., Dalton, C.A., Langmuir, C.H., Su, Y. and Schilling, J.-G. 2013. The mean composition of ocean ridge basalts. *Geochemistry, Geophysics, Geosystems*, **14**, 489–518, https://doi.org/10.1029/2012GC004334

Gamble, J.A. and Kyle, P.R. 1987. The origins of glass and amphibole in spinel–wehrlite xenoliths from Foster Crater, McMurdo Volcanic Group, Antarctica. *Journal of Petrology*, **28**, 755–779, https://doi.org/10.1093/petrology/28.5.755

Gamble, J.A., Barrett, P.J. and Adams, C.J. 1986. Basaltic clasts from Unit 8. *DSIR Bulletin*, **247**, 145–152.

Gamble, J.A., McGibbon, F., Kyle, P.R., Menzies, M. and Kirsch, I. 1988. Metasomatised xenoliths from Foster Crater, Antarctica: Implications for lithospheric structure and process beneath the Transantarctic Mountain front. *Journal of Petrology*, Special Volume, Issue 1, 109–138, https://doi.org/10.1093/petrology/Special_Volume.1.109

Gamble, J.A., Adams, C.J., Morris, P.A., Wysoczanski, R.J., Handler, M. and Timm, C. 2018. The geochemistry and petrogenesis of Carnley Volcano, Auckland Islands, SW Pacific. *New Zealand Journal of Geology and Geophysics*, **61**, 480–497, https://doi.org/10.1080/00288306.2018.1505642

George, A. 1989. Sand provenance. *DSIR Bulletin*, **245**, 159–167.

Giggenbach, W.F., Kyle, P.R. and Lyon, G.L. 1973. Present volcanic activity on Mount Erebus, Ross Island, Antarctica. *Geology*, **1**, 135–136, https://doi.org/10.1130/0091-7613(1973)1<135:PVAOME>2.0.CO;2

Global Volcanism Program. 2013*a*. Discovery (590835). *In*: Venzke, E. (ed.) *Volcanoes of the World, v. 4.6.7.* Smithsonian Institution. Downloaded 06 Apr 2018 (http://volcano.si.edu/volcano.cfm?vn=590835), https://doi.org/10.5479/si.GVP.VOTW4-2013

Global Volcanism Program. 2013*b*. Morning (390017). *In*: Venzke, E. (ed.) *Volcanoes of the World, v. 4.6.7.* Smithsonian Institution. Downloaded 03 Apr 2018 (http://volcano.si.edu/volcano.cfm?vn=390017), https://doi.org/10.5479/si.GVP.VOTW4-2013

Goldich, S., Treves, S., Suhr, N. and Stuckless, J. 1975. Geochemistry of the Cenozoic volcanic rocks of Ross Island and vicinity, Antarctica. *The Journal of Geology*, **83**, 415–435, https://doi.org/10.1086/628120

Goldich, S.S., Stuckless, J.S., Suhr, N.H., Bodkin, J.B. and Wamser, R.C. 1981. Some Trace Element Relationships in the Cenozoic Volcanic Rocks from Ross Island and Vicinity, Antarctica. *American Geophysical Union Antarctic Research Series*, **33**, 215–228.

Grapes, R., Gamble, J. and Palmer, K. 1989. Basal dolerite boulders. *DSIR Bulletin*, **245**, 169–174.

Grindley, G.W. and Oliver, P.J. 1983. Palaeomagnetism of Cretaceous volcanic rocks from Marie Byrd Land, Antarctica. *In*: Oliver, R.L., James, P.R. and Jago, J.B. (eds) *Antarctic Earth Science.* Cambridge University Press, Cambridge, UK, 573–578.

Gunn, B.M. and Warren, G. 1962. *Geology of Victoria Land between the Mawson and Mullock Glaciers, Antarctica.* New Zealand Geological Survey Bulletin, **71**.

Guo, Z., Wilson, M., Zhang, M., Cheng, Z. and Zhang, L. 2015. Post-collisional ultrapotassic mafic magmatism in South Tibet: Products of partial melting of pyroxenite in the mantle wedge induced by roll-back and delamination of the subducted Indian continental lithosphere slab. *Journal of Petrology*, **56**, 1365–1406, https://doi.org/10.1093/petrology/egv040

Hall, B.L., Denton, G.H., Lux, D.R. and Bockheim, J.G. 1993. Late Tertiary Antarctic paleoclimate and ice-sheet dynamics inferred from surficial deposits in Wright Valley. *Geografiska Annaler. Series A, Physical Geography*, **75**(4), 239–267.

Hall, C.E., Cooper, A.F. and Parkinson, D.L. 1995. Early Cambrian carbonatite in Antarctica. *Journal of the Geological Society, London*, **152**, 721–728, https://doi.org/10.1144/gsjgs.152.4.0721

Hall, J., Wilson, T. and Henrys, S. 2007. Structure of the central Terror Rift, western Ross Sea, Antarctica. *United States Geological Survey Open-File Report*, **2007-1047**, Short Research Paper 108.

Halliday, A.N., Lee, D.-C., Tommasini, S., Davies, G.R., Paslick, C.R., Fitton, J.G. and James, D.E. 1995. Incompatible trace elements in OIB and MORB and source enrichment in the suboceanic mantle. *Earth and Planetary Science Letters*, **133**, 379–395, https://doi.org/10.1016/0012-821X(95)00097-V

Hambrey, M.J., Barrett, P.J. and Robinson, P.H. 1989. Stratigraphy. *DSIR Bulletin*, **245**, 23–48.

Harpel, C.J., Kyle, P.R. and Dunbar, N.W. 2008. Englacial tephrostratigraphy of Erebus volcano, Antarctica. *Journal of Volcanology and Geothermal Research*, **177**, 549–568, https://doi.org/10.1016/j.jvolgeores.2008.06.001

Harrington, H.J. 1958*a*. Beaufort Island, remnant of a Quaternary volcano in the Ross Sea, Antarctica. *New Zealand Journal of Geology and Geophysics*, **1**, 595–603, https://doi.org/10.1080/00288306.1958.10423167

Harrington, H.J. 1958*b*. Nomenclature of rock units in the Ross Sea Region, Antarctica. *Nature*, **182**, 290, https://doi.org/10.1038/182290a0

Harrington, H.J. 1965. Geology and morphology of Antarctica. *Monographiae Biologicae*, **15**, 1–71, http://doi-org-443.webvpn.fjmu.edu.cn/10.1007/978-94-015-7204-0_1

Harris, H. and Mudrey, M.G. 1974. Core from Lake Fryxell, DVDP 7, and general geology of Lake Fryxell Area, Taylor Valley. *Dry Valleys Drilling Project Bulletin*, **3**, 109–119.

Haskell, T.R., Kennett, I.P., Prebble, W.M., Smith, G. and Willis, I.A.G. 1965. The geology of the middle and lower Taylor Valley of South Victoria Land, Antarctica. *Transactions of the Royal Society of New Zealand*, **2**, 169–186.

Hauri, E.H., Whitehead, J.A. and Hart, S.R. 1994. Fluid dynamic and geochemical aspects of entrainment in mantle plumes. *Journal of Geophysical Research: Solid Earth*, **99**, 24 275–24 300, https://doi.org/10.1029/94JB01257

Herzberg, C. 2011. Identification of source lithology in the Hawaiian and Canary Islands: implications for origins. *Journal of Petrology*, **52**, 113–146, https://doi.org/10.1093/petrology/egq075

Hillenbrand, C.D., Moreton, S.G. *et al.* 2008. Volcanic time-markers for Marine Isotopic Stages 6 and 5 in Southern Ocean sediments and Antarctic ice cores: implications for tephra correlations between palaeoclimatic records. *Quaternary Science Reviews*, **27**, 518–540, https://doi.org/10.1016/j.quascirev.2007.11.009

Hofmann, A.W. and White, W.M. 1982. Mantle plumes from ancient oceanic crust. *Earth and Planetary Science Letters*, **57**, 421–436, https://doi.org/10.1016/0012-821X(82)90161-3

Houghton, B.F., Weaver, S.D., Wilson, C.J.N. and Lanphere, M.A. 1992. Evolution of a Quaternary peralkaline volcano: Mayor

Island, New Zealand. *Journal of Volcanology and Geothermal Research*, **51**, 217–236, https://doi.org/10.1016/0377-0273(92)90124-V

Iacovino, K., Oppenheimer, C., Scaillet, B. and Kyle, P. 2016. Storage and evolution of mafic and intermediate alkaline magmas beneath Ross Island, Antarctica. *Journal of Petrology*, **57**, 93–118, https://doi.org/10.1093/petrology/egv083

Iverson, N.A., Kyle, P.R., Dunbar, N.W., McIntosh, W.C. and Pearce, N.J.G. 2014. Eruptive history and magmatic stability of Erebus volcano, Antarctica: Insights from englacial tephra. *Geochemistry, Geophysics, Geosystems*, **15**, 4180–4202, https://doi.org/10.1002/2014GC005435

Jensen, H.I. 1916. *Report on the Petrology of the Alkaline Rocks of Mount Erebus, Antarctica*. Heinemann, London.

Kalamarides, R.I., Berg, J.H. and Hank, R.A. 1987. Lateral isotopic discontinuity in the lower crust: an example from Antarctica. *Science*, **237**, 1192–1195, https://doi.org/10.1126/science.237.4819.1192

Kelly, P.J., Kyle, P.R., Dunbar, N.W. and Sims, K.W.W. 2008. Geochemistry and mineralogy of the phonolite lava lake, Erebus volcano, Antarctica: 1972–2004 and comparison with older lavas. *Journal of Volcanology and Geothermal Research*, **177**, 589–605, https://doi.org/10.1016/j.jvolgeores.2007.11.025

Keys, J.R., Anderton, P.W. and Kyle, P.R. 1977. Tephra and debris layers in the Skelton Neve and Kempe Glacier, South Victoria Land, Antarctica. *New Zealand Journal of Geology and Geophysics*, **20**, 971–1002, https://doi.org/10.1080/00288306.1977.10420692

Kinzler, R.J. 1997. Melting of mantle peridotite at pressures approaching the spinel to garnet transition: application to mid-ocean ridge basalt petrogenesis. *Journal of Geophysical Research: Solid Earth*, **102**, 853–874, https://doi.org/10.1029/96jb00988

Kipf, A., Hauff, F. *et al.* 2014. Seamounts off the West Antarctic margin: a case for non-hotspot driven intraplate volcanism. *Gondwana Research*, **25**, 1660–1679, https://doi.org/10.1016/j.gr.2013.06.013

Kirsch, I.D. 1981. *Evidence for Mantle Metasomatism in Ultramafic Inclusions from Foster Crater, Antarctica*. MSc thesis, Ohio State University, Columbus, Ohio, USA.

Krissek, L.A., Browne, G. *et al.* 2007. Sedimentology and stratigraphy of the ANDRILL McMurdo Ice Shelf (AND-1B) core. *United States Geological Survey Open-File Report*, **2007-1047**, Extended Abstract 148.

Kuno, H. 1966. Lateral variation of basalt magma types across continental margins and island arcs. *Bulletin of Volcanology*, **29**, 195–222, https://doi.org/10.1007/BF02597153

Kurasawa, H., Yoshida, Y. and Mudrey, M.G. 1974. Geological log of the Lake Vida core – DVDP 6. *Dry Valleys Drilling Project Bulletin*, **3**, 92–108.

Kyle, P.R. 1976. *Geology, Mineralogy, and Geochemistry of the Late Cenozoic McMurdo Volcanic Group, Victoria Land, Antarctica*. PhD thesis, Victoria University of Wellington, Wellington, New Zealand.

Kyle, P.R. 1977. Mineralogy and glass chemistry of recent volcanic ejecta from Mt Erebus, Ross Island, Antarctica. *New Zealand Journal of Geology and Geophysics*, **20**, 1123–1146, https://doi.org/10.1080/00288306.1977.10420699

Kyle, P.R. 1981a. Glacial history of the McMurdo Sound area as indicated by the distribution and nature of McMurdo Volcanic Group Rocks. *American Geophysical Union Antarctic Research Series*, **33**, 403–412.

Kyle, P.R. 1981b. Mineralogy and geochemistry of a basanite to phonolite sequence at Hut Point Peninsula, Antarctica, based on core from Dry Valley Drilling Project Drillholes 1, 2 and 3. *Journal of Petrology*, **22**, 451–500, https://doi.org/10.1093/petrology/22.4.451

Kyle, P.R. 1990a. McMurdo Volcanic Group, western Ross Embayment: introduction. *American Geophysical Union Antarctic Research Series*, **48**, 18–25.

Kyle, P.R. 1990b. Erebus Volcanic Province summary. *American Geophysical Union Antarctic Research Series*, **48**, 81–88.

Kyle, P.R. (ed.). 1994. *Volcanological and Environmental Studies of Mount Erebus*. American Geophysical Union Antarctic Research Series, **66**.

Kyle, P.R. and Cole, J.W. 1974. Structural control of volcanism in the McMurdo Volcanic Group, Antarctica. *Bulletin Volcanologique*, **38**, 16–25, https://doi.org/10.1007/bf02597798

Kyle, P.R. and Muncy, H.L. 1978. Volcanic geology of the lower slopes of Mount Morning. *Antarctic Journal of the United States*, **13**, 34–36.

Kyle, P.R. and Muncy, H.L. 1983. The geology of the Mid-Miocene McMurdo Volcanic Group at Mount Morning, McMurdo Sound, Antarctica. *In*: Oliver, R.L., James, P.R. and Jago, J.B. (eds) *Antarctic Earth Science, 4th International Symposium on Antarctic Earth Science*. Australian Academy Science, Adelaide, Australia, 675.

Kyle, P.R. and Muncy, H.L. 1989. Geology and geochronology of McMurdo Volcanic Group rocks in the vicinity of Lake Morning, McMurdo Sound, Antarctica. *Antarctic Science*, **1**, 345–350, https://doi.org/10.1017/S0954102089000520

Kyle, P.R. and Price, R.C. 1975. Occurrences of rhönite in alkalic lavas of the McMurdo Volcanic Group, Antarctica and Dunedin volcano, New Zealand. *American Mineralogist*, **60**, 722–725.

Kyle, P.R., Adams, J. and Rankin, P.C. 1979a. Geology and petrology of the McMurdo Volcanic Group at Rainbow Ridge, Brown Peninsula, Antarctica. *Geological Society of America Bulletin*, **90**, 676, https://doi.org/10.1130/0016-7606(1979)90<676:GAPOTM>2.0.CO;2

Kyle, P.R., Sutter, J.F. and Treves, S.B. 1979b. K/Ar age determinations on drill core from DVDP holes 1 and 2. *Memoirs of the National Institute of Polar Research (Japan)*, **13**, Special Issue, 214–219.

Kyle, P.R., Dibble, R.R., Giggenbach, W.F. and Keys, J. 1982. Volcanic activity associated with the anorthoclase phonolite lava lake, Mount Erebus, Antarctica. *In*: Craddock, C. (ed.) *Antarctic Geosciences*. University of Wisconsin Press, Madison, WI, 735–745.

Kyle, P.R., Wright, A.C. and Kirsch, I. 1987. Ultramafic xenoliths in the late Cenozoic McMurdo Volcanic Group, western Ross Sea embayment, Antarctica. *In*: Nixon, P.H. (ed.) *Mantle Xenoliths*. John Wiley & Sons, Chichester, UK, 287–293.

Kyle, P.R., Moore, J.A. and Thirlwall, M.F. 1992. Petrologic Evolution of Anorthoclase Phonolite Lavas at Mount Erebus, Ross Island, Antarctica. *Journal of Petrology*, **33**, 849–875, https://doi.org/10.1093/petrology/33.4.849

Lawrence, K.P., Tauxe, L., Staudigel, H., Constable, C.G., Koppers, A., McIntosh, W. and Johnson, C.L. 2009. Paleomagnetic field properties at high southern latitude. *Geochemistry, Geophysics, Geosystems*, **10**, https://doi.org/10.1029/2008GC002072

Lawver, L.A., Davis, M.B., Wilson, T.J. and Shipboard Scientific Party. 2007. Neotectonic and other features of the Victoria Land Basin, Antarctica, interpreted from multibeam bathymetry data. *United States Geological Survey Open-File Report*, **2007-1047**, Extended Abstract 017.

Lawver, L., Lee, J., Kim, Y. and Davey, F. 2012. Flat-topped mounds in western Ross Sea: Carbonate mounds or subglacial volcanic features? *Geosphere*, **8**, 645–653, https://doi.org/10.1130/GES00766.1

Le Bas, M., Maitre, M.J., Streckeisen, A. and Zanettin, B. 1986. A chemical classification of volcanic rocks based on the total alkali–silica diagram. *Journal of Petrology*, **27**, 745–750, https://doi.org/10.1093/petrology/27.3.745

Lee, M.J., Lee, J.I., Kim, T.H., Lee, J. and Nagao, K. 2015. Age, geochemistry and Sr–Nd–Pb isotopic compositions of alkali volcanic rocks from Mt. Melbourne and the western Ross Sea, Antarctica. *Geosciences Journal*, **19**, 681–695, https://doi.org/10.1007/s12303-015-0061-y

Le Maitre, R.W., Streckeisen, A. *et al.* 2002. *Igneous Rocks. A Classification and Glossary of Terms: Recommendations of the International Union of Geological Sciences Subcommission of the Systematics of Igneous Rocks*, 2nd edn. Cambridge University Press, Cambridge, UK.

LeMasurier, W. 2013. Shield volcanoes of Marie Byrd Land, West Antarctic rift: oceanic island similarities, continental signature, and tectonic controls. *Bulletin of Volcanology*, **75**, 726, https://doi.org/10.1007/s00445-013-0726-1

LeMasurier, W.E. and Wade, A.F. 1968. Fumarolic activity in Marie Byrd Land, Antarctica. *Science*, **162**, 352, https://doi.org/10.1126/science.162.3851.352

LeMasurier, W.E., McIntosh, M.C., Ellerman, P.J. and Wright, A.C. 1983. USCGS Glacier cruise 1, December 1982–January 1983: Reconnaissance of hyaloclastites in the western Ross Sea Region. *Antarctic Journal of the United States*, **18**, 60–61.

Lewis, A., Marchant, D., Ashworth, A., Hemming, S. and Machlus, M. 2007. Major middle Miocene global climate change: Evidence from East Antarctica and the Transantarctic Mountains. *Geological Society of America Bulletin*, **119**, 1449–1461, https://doi.org/10.1130/0016-7606(2007)119[1449:MMMGCC]2.0.CO;2

Lewis, A.R., Marchant, D.R. et al. 2008. Mid-Miocene cooling and the extinction of tundra in continental Antarctica. *Proceedings of the National Academy of Sciences of the United States of America*, **105**, 10 676–10 680, https://doi.org/10.1073/pnas.0802501105

Luckman, P. 1974. *Products of submarine and subglacial volcanism in the McMurdo Sound region, Ross Island, Antarctica*. BSc (Hon.), Victoria University of Wellington, Wellington, New Zealand.

Mankinen, E.A. and Cox, A. 1988. Paleomagnetic investigation of some volcanic rocks from the McMurdo Volcanic Province, Antarctica. *Journal of Geophysical Research*, **93**, 11 599–11 612, https://doi.org/10.1029/JB093iB10p11599

Marchant, D.R., Denton, G.H., Swisher, I.I.I.C.C. and Potter, J.N. 1996. Late Cenozoic Antarctic paleoclimate reconstructed from volcanic ashes in the Dry Valleys region of southern Victoria Land. *Geological Society of America Bulletin*, **108**, 181–194, https://doi.org/10.1130/0016-7606(1996)108<0181:LCAPRF>2.3.CO;2

Martin, A.P. 2009. *Mount Morning, Antarctica: Geochemistry, Geochronology, Petrology, Volcanology, and Oxygen Fugacity of the Rifted Antarctic Lithosphere*. PhD thesis, University of Otago, Dunedin, New Zealand.

Martin, A.P. and Cooper, A.F. 2010. Post 3.9 Ma fault activity within the West Antarctic rift system: onshore evidence from Gandalf Ridge, Mount Morning eruptive centre, southern Victoria Land, Antarctica. *Antarctic Science*, **22**, 513–521, https://doi.org/10.1017/S095410201000026X

Martin, A.P., Cooper, A.F. and Dunlap, W.J. 2010. Geochronology of Mount Morning, Antarctica: Two-phase evolution of a long-lived trachyte–basanite–phonolite eruptive center. *Bulletin of Volcanology*, **72**, 357–371, https://doi.org/10.1007/s00445-009-0319-1

Martin, A.P., Cooper, A.F. and Price, R.C. 2013. Petrogenesis of Cenozoic, alkalic volcanic lineages at Mount Morning, West Antarctica and their entrained lithospheric mantle xenoliths: lithospheric v. asthenospheric mantle sources. *Geochimica et Cosmochimica Acta*, **122**, 127–152, https://doi.org/10.1016/j.gca.2013.08.025

Martin, A.P., Cooper, A.F. and Price, R.C. 2014a. Increased mantle heat flow with on-going rifting of the West Antarctic rift system inferred from characterisation of plagioclase peridotite in the shallow Antarctic mantle. *Lithos*, **190–191**, 173–190, https://doi.org/10.1016/j.lithos.2013.12.012

Martin, A.P., Price, R.C. and Cooper, A.F. 2014b. Constraints on the composition, source and petrogenesis of plagioclase-bearing mantle peridotite. *Earth-Science Reviews*, **138**, 89–101, https://doi.org/10.1016/j.earscirev.2014.08.006

Martin, A.P., Cooper, A.F., Price, R.C., Turnbull, R.E. and Roberts, N.M.W. 2015a. The petrology, geochronology and significance of Granite Harbour Intrusive Complex xenoliths and outcrop sampled in western McMurdo Sound, Southern Victoria Land, Antarctica. *New Zealand Journal of Geology and Geophysics*, **58**, 33–51, https://doi.org/10.1080/00288306.2014.982660

Martin, A.P., Price, R.C., Cooper, A.F. and McCammon, C.A. 2015b. Petrogenesis of the rifted Southern Victoria Land lithospheric mantle, Antarctica, inferred from petrography, geochemistry, thermobarometry and oxybarometry of peridotite and pyroxenite xenoliths from the Mount Morning eruptive centre. *Journal of Petrology*, **56**, 193–226, https://doi.org/10.1093/petrology/egu075

Martin, A.P., Smellie, J.L., Cooper, A.F. and Townsend, D.B. 2018. Formation of a spatter-rich pyroclastic density current deposit in a Neogene sequence of trachytic–mafic igneous rocks at Mason Spur, Erebus volcanic province, Antarctica. *Bulletin of Volcanology*, **80**, 13, https://doi.org/10.1007/s00445-017-1188-7

Mawson, D. 1916. Petrology of rock collections from the mainland of South Victoria Land. British Antarctic Expedition 1907–09. Reports and scientific investigations. *Geology*, **II**, 161–168.

McCoy-West, A.J., Bennett, V.C. and Amelin, Y. 2016. Rapid Cenozoic ingrowth of isotopic signatures simulating 'HIMU' in ancient lithospheric mantle: distinguishing source from process. *Geochimica et Cosmochimica Acta*, **187**, 79–101, https://doi.org/10.1016/j.gca.2016.05.013

McCraw, J.D. 1962. Volcanic Detritus in Taylor valley, Victoria Land, Antarctica. *New Zealand Journal of Geology and Geophysics*, **5**, 740–745, https://doi.org/10.1080/00288306.1962.10417635

McCraw, J.D. 1967. Soils of Taylor Dry Valley, Victoria Land, Antarctica, with notes on soils from other localities in Victoria Land. *New Zealand Journal of Geology and Geophysics*, **10**, 498–539, https://doi.org/10.1080/00288306.1967.10426754

McDonough, W.F. and Sun, S.-S. 1995. The composition of the Earth. *Chemical Geology*, **120**, 223–253, https://doi.org/10.1016/0009-2541(94)00140-4

McGibbon, F.M. 1991. Geochemistry and petrology of ultramafic xenoliths of the Erebus Volcanic Province *In*: Thomson, M.R.A., Crame, J.A. and Thomson, J.W. (eds) *Geological Evolution of Antarctica*. Cambridge University Press, Cambridge, UK, 317–321.

McIntosh, W.C. 1998. ^{40}Ar/^{39}Ar geochronology of volcanic clasts and pumice in CRP_1 core, Cape Roberts, Antarctica. *Terra Antartica*, **5**, 683–690.

McIntosh, W.C. 2000. ^{40}Ar/^{39}Ar geochronology of tephra and volcanic clasts in CRP-2A, Victoria Land Basin, Antarctica. *Terra Antartica*, **7**, 621–630.

McIver, J.R. and Gevers, T.W. 1970. Volcanic vents below the Royal Society Range, Central Victoria Land, Antarctica. *Transactions of the Geological Society of South Africa*, **73**, 65–88.

McKenzie, D.A.N. and O'Nions, R.K. 1991. Partial melt distributions from inversion of rare earth element concentrations. *Journal of Petrology*, **32**, 1021–1091, https://doi.org/10.1093/petrology/32.5.1021

McKenzie, D.A.N. and O'Nions, R.K. 1995. The source regions of ocean island basalts. *Journal of Petrology*, **36**, 133–159, https://doi.org/10.1093/petrology/36.1.133

Middlemost, E.A.K. 1989. Iron oxidation ratios, norms and the classification of volcanic rocks. *Chemical Geology*, **77**, 19–26, https://doi.org/10.1016/0009-2541(89)90011-9

Moore, J. and Kyle, P. 1987. Volcanic geology of Mount Erebus, Ross Island, Antarctica. *Proceedings of the NIPR Symposium on Antarctic Geosciences*, **1**, 48–65.

Moussallam, Y., Oppenheimer, C., Scaillet, B. and Kyle, P.R. 2013. Experimental phase-equilibrium constraints on the phonolite magmatic system of Erebus volcano, Antarctica. *Journal of Petrology*, **54**, 1285–1307, https://doi.org/10.1093/petrology/egt012

Moussallam, Y., Oppenheimer, C. et al. 2015. Megacrystals track magma convection between reservoir and surface. *Earth and Planetary Science Letters*, **413**, 1–12, https://doi.org/10.1016/j.epsl.2014.12.022

Mudrey, M.G., Torii, T. and Harris, H. 1975. Geology of DVDP 13 – Don Juan Pond, Wright Valley, Antarctica. *Dry Valleys Drilling Project Bulletin*, **5**, 78–93.

Muncy, H.L. 1979. *Geologic History and Petrogenesis of Alkaline Volcanic Rocks, Mount Morning, Antarctica*. MSc thesis, Ohio State University, Columbus, Ohio, USA.

Naish, T., Powell, R. *et al.* 2009. Obliquity-paced Pliocene West Antarctic ice sheet oscillations. *Nature*, **458**, 322, https://doi.org/10.1038/nature07867, https://www.nature.com/articles/nature07867#supplementary-information

Narcisi, B., Petit, J.R. and Langone, A. 2017. Last glacial tephra layers in the Talos Dome ice core (peripheral East Antarctic Plateau), with implications for chronostratigraphic correlations and regional volcanic history. *Quaternary Science Reviews*, **165**, 111–126, https://doi.org/10.1016/j.quascirev.2017.04.025

Nield, G.A., Whitehouse, P.L., van der Wal, W., Blank, B., O'Donnell, J.P. and Stuart, G.W. 2018. The impact of lateral variations in lithospheric thickness on glacial isostatic adjustment in West Antarctica. *Geophysical Journal International*, **214**, 811–824, https://doi.org/10.1093/gji/ggy158

Nyland, R.E., Panter, K.S. *et al.* 2013. Volcanic activity and its link to glaciation cycles: Single-grain age and geochemistry of Early to Middle Miocene volcanic glass from ANDRILL AND-2A core, Antarctica. *Journal of Volcanology and Geothermal Research*, **250**, 106–128, https://doi.org/10.1016/j.jvolgeores.2012.11.008

Oppenheimer, C. and Kyle, P.R. 2008. Probing the magma plumbing of Erebus volcano, Antarctica, by open-path FTIR spectroscopy of gas emissions. *Journal of Volcanology and Geothermal Research*, **177**, 743–754, https://doi.org/10.1016/j.jvolgeores.2007.08.022

Panter, K.S., Kyle, P.R. and Smellie, J.L. 1997. Petrogenesis of a phonolite–trachyte succession at Mount Sidley, Marie Byrd Land, Antarctica. *Journal of Petrology*, **38**, 1225–1253, https://doi.org/10.1093/petroj/38.9.1225

Panter, K.S., Blusztajn, J., Hart, S.R., Kyle, P.R., Esser, R. and McIntosh, W.C. 2006. The origin of HIMU in the SW Pacific: Evidence from intraplate volcanism in southern New Zealand and subantarctic islands. *Journal of Petrology*, **47**, 1673–1704, https://doi.org/10.1093/petrology/egl024

Panter, K., Talarico, F. *et al.* 2008. Petrologic and geochemical composition of the AND-2A core, ANDRILL Southern McMurdo Sound Project, Antarctica. *Terra Antartica*, **15**, 147–192.

Panter, K., Dunbar, N.W., Scanlan, M., Wilch, T., Fargo, A. and McIntosh, W. 2011. Petrogenesis of alkaline magmas at Minna Bluff, Antarctica: evidence for multi-stage differentiation and complex mixing processes. Abstract V31F-2590 presented at the AGU 2011 Fall Meeting, December 5–9, 2011, San Francisco, California, USA.

Panter, K.S., Castillo, P. *et al.* 2018. Melt origin across a rifted continental margin: a case for subduction-related metasomatic agents in the lithospheric source of alkaline basalt, northwest Ross Sea, Antarctica. *Journal of Petrology*, **59**, 517–558, https://doi.org/10.1093/petrology/egy036

Parmelee, D.E.F., Kyle, P.R., Kurz, M.D., Marrero, S.M. and Phillips, F.M. 2015. A new Holocene eruptive history of Erebus volcano, Antarctica using cosmogenic 3He and 36Cl exposure ages. *Quaternary Geochronology*, **30**, 114–131, https://doi.org/10.1016/j.quageo.2015.09.001

Paulsen, H.-K. 2008. *A Lithological Cross Section through Mount Morning, Antarctica: A Story Told from Xenolithic Assemblages in a Pyroclastic Deposit*. MSc thesis, University of Otago, Dunedin, New Zealand.

Paulsen, T. 2002. Volcanic cone alignments and the intraplate stress field in the Mount Morning region, South Victoria Land, Antarctica. *Geological Society of America Abstracts with Programs*, **34**, 437.

Paulsen, T.S. and Wilson, T.J. 2007. Elongate summit calderas as Neogene paleostress indicators in Antarctica. *United States Geological Survey Open-File Report*, **2007-1047**, Short Research Paper 072, https://doi.org/10.3133/ofr20071047SRP072

Paulsen, T.S. and Wilson, T.J. 2009. Structure and age of volcanic fissures on Mount Morning: a new constraint on Neogene to contemporary stress in the West Antarctic Rift, southern Victoria Land, Antarctica. *Geological Society of America Bulletin*, **121**, 1071–1088, https://doi.org/10.1130/b26333.1

Perinelli, C., Armienti, P. and Dallai, L. 2006. Geochemical and O-isotope constraints on the evolution of lithospheric mantle in the Ross Sea rift area (Antarctica). *Contributions to Mineralogy and Petrology*, **151**, 245–266, https://doi.org/10.1007/s00410-006-0065-8

Pfänder, J.A., Jung, S., Münker, C., Stracke, A. and Mezger, K. 2012. A possible high Nb/Ta reservoir in the continental lithospheric mantle and consequences on the global Nb budget – Evidence from continental basalts from Central Germany. *Geochimica et Cosmochimica Acta*, **77**, 232–251, https://doi.org/10.1016/j.gca.2011.11.017

Phillips, E.H., Sims, K.W.W. *et al.* 2018. The nature and evolution of mantle upwelling at Ross Island, Antarctica, with implications for the source of HIMU lavas. *Earth and Planetary Science Letters*, **498**, 38–53, https://doi.org/10.1016/j.epsl.2018.05.049

Polyakov, M.M., Krylov, A.Y. and Mazina, T.I. 1976. New data on radiogeochronology of Antarctic Cenozoic vulcanites (in Russian). *Informatsionnyi Biulleten Sovetskoi Antarktich*, **93**, 19–26.

Pompilio, M., Armienti, P. and Tamponi, M. 2001. Petrography, mineral composition and geochemistry of volcanic and subvolcanic rocks of CRP-3, Victoria Land Basin, Antarctica. *Terra Antartica*, **8**, 463–480.

Pompilio, M., Dunbar, N. *et al.* 2007. Petrology and Geochemistry of the AND-1B Core, ANDRILL McMurdo Ice Shelf Project, Antarctica. *Terra Antartica*, **14**, 255–288.

Porter, S.C. and Beget, J.E. 1981. Provenance and depositional environments of Late Cenozoic sediments in permafrost cores from lower Taylor Valley, Antarctica. *American Geophysical Union Antarctic Research Series*, **33**, 351–364.

Price, R.C. and Chappell, B.W. 1975. Fractional crystallisation and the petrology of Dunedin volcano. *Contributions to Mineralogy and Petrology*, **53**, 157–182, https://doi.org/10.1007/bf00372602

Prior, G.T. 1899. Petrographical notes on the rock specimens collected in Antarctic regions during the voyage of H.M.S. Erebus and Terror under Sir James Clark Ross in 1839–43. *Mineralogical Magazine*, **12**, 69–91.

Prior, G.T. 1902. Report on the rock specimens collected by the *Southern Cross* Antarctic Expedition. *In*: Lankester, E.R. and Bell, J. (eds) *Report on the Collections of Natural History made in the Antarctic Regions During the Voyage of the 'Southern Cross'*. British Museum (Natural History), London, 321–332.

Prior, G.T. 1907. Report on the rock specimens collected during the 'Discovery' Antarctic Expedition, 1901–1904. *Natural History*, **1**, 101–160.

Putirka, K., Johnson, M., Kinzler, R., Longhi, J. and Walker, D. 1996. Thermobarometry of mafic igneous rocks based on clinopyroxene–liquid equilibria, 0–30 kbar. *Contributions to Mineralogy and Petrology*, **123**, 92–108, https://doi.org/10.1007/s004100050145

Putirka, K.D. 2008. Thermometers and barometers for volcanic systems. *Reviews in Mineralogy and Geochemistry*, **69**, 61–120, https://doi.org/10.2138/rmg.2008.69.3

Putirka, K.D., Mikaelian, H., Ryerson, F. and Shaw, H. 2003. New clinopyroxene–liquid thermobarometers for mafic, evolved, and volatile-bearing lava compositions, with applications to lavas from Tibet and the Snake River Plain, Idaho. *American Mineralogist*, **88**, 1542–1554, https://doi.org/10.2138/am-2003-1017

Rasmussen, D.J., Kyle, P.R., Wallace, P.J., Sims, K.W., Gaetani, G.A. and Phillips, E.H. 2017. Understanding degassing and transport of CO_2-rich alkalic magmas at Ross Island, Antarctica using olivine-hosted melt inclusions. *Journal of Petrology*, **58**, 841–861, https://doi.org/10.1093/petrology/egx036

Rebert, R. 1981. Bibliography of the Dry Valley Drilling Project. *American Geophysical Union Antarctic Research Series*, **33**, 453–465.

Redner, E.R. 2016. *Magma Mixing and Evolution at Minna Bluff, Antarctica Revealed by Amphibole and Clinopyroxene Analyses.* MSc thesis, Bowling Green State University, Bowling Green, Ohio, USA.

Rilling, S.E., Mukasa, S.B., Wilson, T.J. and Lawver, L. 2007. ^{40}Ar–^{39}Ar Age constraints on volcanism and tectonism in the Terror Rift of the Ross Sea, Antarctica. *United States Geological Survey Open-File Report*, **2007-1047**, Short Research Paper 092, https://pubs.usgs.gov/of/2007/1047/srp/srp092/

Rilling, S., Mukasa, S., Wilson, T., Lawver, L. and Hall, C. 2009. New determinations of ^{40}Ar/^{39}Ar isotopic ages and flow volumes for Cenozoic volcanism in the Terror Rift, Ross Sea, Antarctica. *Journal of Geophysical Research: Solid Earth*, **114**, B12207, https://doi.org/10.1029/2009JB006303

Rocchi, S., Armienti, P., D'Orazio, M., Tonarini, S., Wijbrans, J.R. and Di Vincenzo, G. 2002. Cenozoic magmatism in the western Ross Embayment: role of mantle plume v. plate dynamics in the development of the West Antarctic Rift System. *Journal of Geophysical Research: Solid Earth*, **107**, 2195, https://doi.org/10.1029/2001jb000515

Rocchi, S., Armienti, P. and Di Vincenzo, G. 2005. No plume, no rift magmatism in the West Antarctic Rift. *Geological Society of America Special Papers*, **388**, 435–447, https://doi.org/10.1130/0-8137-2388-4.435

Ross, J. 1847. *A Voyage of Discovery and Research in the Southern and Antarctic Regions, during the Years 1839–43*. John Murray, London.

Ross, J. 2014. *Ar–Ar Geochronology of Southern McMurdo Sound, Antarctica and The Development of Pychron: an Ar–Ar Data Acquisition and Processing Package.* New Mexico Institute of Mining and Technology, Socorro, NM.

Ross, J.I., McIntosh, W.C. and Dunbar, N.W. 2012a. Development of a precise and accurate age–depth model based on ^{40}Ar/^{39}Ar dating of volcanic material in the ANDRILL (1B) drill core, Southern McMurdo Sound, Antarctica. *Global and Planetary Change*, **96–97**, 118–130, https://doi.org/10.1016/j.gloplacha.2012.05.005

Ross, J.I., McIntosh, W.C. and Wilch, T.I. 2012b. Detailed Ar–Ar geochronology of volcanism at Minna Bluff, Antarctica: Two-phased growth and influence on Ross Ice Shelf. Abstract presented at the AGU Fall Meeting, December 3–7, 2012. San Francisco, California, USA.

Rudnick, R.L. and Gao, S. 2003. Composition of the continental crust. In: Rudnick, R.L. (ed.) *Treatise on Geochemistry, Volume 3: The Crust.* Pergamon, Oxford, UK, 1–64.

Rudnick, R.L. and Gao, S. 2014. Composition of the continental crust. In: Holland, H.D. and Turekian, K.K. (eds) *Treatise on Geochemistry.* 2nd edn. Elsevier, Oxford, UK, 1–51.

Rudnick, R.L., Barth, M., Horn, I. and McDonough, W.F. 2000. Rutile-bearing refractory eclogites: missing link between continents and depleted mantle. *Science*, **287**, 278–281, https://doi.org/10.1126/science.287.5451.278

Sack, R.O., Walker, D. and Carmichael, I.S.E. 1987. Experimental petrology of alkalic lavas: constraints on cotectics of multiple saturation in natural basic liquids. *Contributions to Mineralogy and Petrology*, **96**, 1–23, https://doi.org/10.1007/bf00375521

Saggerson, E.P. and Williams, L.A.J. 1964. Ngurumanite from southern Kenya and its bearing on the origin of rocks in the northern Tanganyika Alkaline District. *Journal of Petrology*, **5**, 40–81, https://doi.org/10.1093/petrology/5.1.40

Sandroni, S. and Talarico, F.M. 2006. Analysis of clast lithologies from CIROS-2 core, New Harbour, Antarctica – Implications for ice flow directions during Plio-Pleistocene time. *Palaeogeography, Palaeoclimatology, Palaeoecology*, **231**, 215–232, https://doi.org/10.1016/j.palaeo.2005.07.031

Scanlan, M.K. 2008. *Petrology of Inclusion-Rich Lavas at Minna Bluff, McMurdo Sound, Antarctica: Implications for Magma Origin, Differentiation, and Eruption Dynamics.* MSc thesis, Bowling Green State University, Bowling Green, Ohio, USA.

Scott, J.M., Turnbull, I.M., Auer, A. and Palin, J.M. 2013. The sub-Antarctic Antipodes Volcano: a <0.5 Ma HIMU-like Surtseyan volcanic outpost on the edge of the Campbell Plateau, New Zealand. *New Zealand Journal of Geology and Geophysics*, **56**, 134–153, https://doi.org/10.1080/00288306.2013.802246

Scott, J.M., Waight, T.E., van der Meer, Q.H.A., Palin, J.M., Cooper, A.F. and Münker, C. 2014. Metasomatized ancient lithospheric mantle beneath the young Zealandia microcontinent and its role in HIMU-like intraplate magmatism. *Geochemistry, Geophysics, Geosystems*, **15**, 3477–3501, https://doi.org/10.1002/2014gc005300

Scott, R.F. 1907. *The Voyage of the Discovery.* Charles Scribner, New York.

Shaw, D.M. 1970. Trace element fractionation during anatexis. *Geochimica et Cosmochimica Acta*, **34**, 237–243, https://doi.org/10.1016/0016-7037(70)90009-8

Shen, W., Wiens, D.A. et al. 2018. Seismic evidence for lithospheric foundering beneath the southern Transantarctic Mountains, Antarctica. *Geology*, **46**, 71–74, https://doi.org/10.1130/G39555.1

Sims, K.W.W. and Hart, S.R. 2006. Comparison of Th, Sr, Nd and Pb isotopes in oceanic basalts: implications for mantle heterogeneity and magma genesis. *Earth and Planetary Science Letters*, **245**, 743–761, https://doi.org/10.1016/j.epsl.2006.02.030

Sims, K.W.W., Blichert-Toft, J. et al. 2008. A Sr, Nd, Hf, and Pb isotope perspective on the genesis and long-term evolution of alkaline magmas from Erebus volcano, Antarctica. *Journal of Volcanology and Geothermal Research*, **177**, 606–618, https://doi.org/10.1016/j.jvolgeores.2007.08.006

Sims, K.W.W., Pichat, S. et al. 2013. On the time scales of magma genesis, melt evolution, crystal growth rates and magma degassing in the Erebus volcano magmatic system using the ^{238}U, ^{235}U and ^{232}Th decay series. *Journal of Petrology*, **54**, 235–271, https://doi.org/10.1093/petrology/egs068

Skinner, D.N.B., Waterhouse, B.C., Brehaut, G.M. and Sullivan, K. 1976. *New Zealand Geological Survey Antarctic Expedition 1975–76, Skelton–Koettlitz Glaciers.* New Zealand Geological Survey Report, **DS58**.

Smellie, J.L. 1998. Sand grain detrital modes in CRP-1: provenance variations and influence of Miocene eruptions on the marine record in the McMurdo Sound region. *Terra Antartica*, **5**, 579–587.

Smellie, J.L. 2002. Erosional history of the Transantarctic Mountains deduced from sand grain detrital modes in CRP-2/2A, Victoria Land Basin, Antarctica. *Terra Antartica*, **7**, 545–552.

Smellie, J.L. and Martin, A.P. 2021. Erebus Volcanic Province: volcanology. *Geological Society, London, Memoirs*, **55**, https://doi.org/10.1144/M55-2018-62

Smellie, J.L. and Rocchi, S. 2021. Northern Victoria Land: volcanology. *Geological Society, London, Memoirs*, **55**, https://doi.org/10.1144/M55-2018-60

Smith, W.C. 1954. The volcanic rocks of the Ross archipelago, British 'Terra Nova' Expedition', 1910. In: *Antarctic ('Terra Nova') Expedition, British, 1910: Geology Volume 2: Natural History Report.* British Museum, London, 1–107.

Sobolev, A.V., Hofmann, A.W., Sobolev, S.V. and Nikogosian, I.K. 2005. An olivine-free mantle source of Hawaiian shield basalts. *Nature*, **434**, 590, https://doi.org/10.1038/nature03411, https://www.nature.com/articles/nature03411#supplementary-information

Sprung, P., Schuth, S., Münker, C. and Hoke, L. 2007. Intraplate volcanism in New Zealand: the role of fossil plume material and variable lithospheric properties. *Contributions to Mineralogy and Petrology*, **153**, 669–687, https://doi.org/10.1007/s00410-006-0169-1

Stracke, A. 2012. Earth's heterogeneous mantle: a product of convection-driven interaction between crust and mantle. *Chemical Geology*, **330–331**, 274–299, https://doi.org/10.1016/j.chemgeo.2012.08.007

Stracke, A., Bizimis, M. and Salters, V.J.M. 2003. Recycling oceanic crust: quantitative constraints. *Geochemistry, Geophysics, Geosystems*, **4**, 8003, https://doi.org/10.1029/2001gc000223

Stuckless, J.S. and Ericksen, R.L. 1976. Strontium isotopic geochemistry of the volcanic rocks and associated megacrysts and inclusions from Ross Island and vicinity, Antarctica. *Contributions to*

Mineralogy and Petrology, **58**, 111–126, https://doi.org/10.1007/BF00382180

Stuckless, J.S., Miesch, A.T., Goldich, S.S. and Weiblen, P.W. 1981. A Q-mode factor model for the petrogenesis of the Volcanic Rocks from Ross Island and Vicinity, Antarctica. *American Geophysical Union Antarctic Research Series*, **33**, 257–280.

Stuiver, M. and Braziunas, T. 1985. Compilation of isotopic dates from Antarctica. *Radiocarbon*, **27**, 117–304, https://doi.org/10.1017/S0033822200007037

Stump, E. 1995. *The Ross Orogen of the Transantarctic Mountains*. Cambridge University Press, New York.

Sullivan, R.J. 2006. *The Geology and Geochemistry of Seal Crater, Hurricane Ridge, Mount Morning, Antarctica*. BSc (Hon.) thesis, University of Otago, Dunedin, New Zealand.

Sun, S.S. and Hanson, G.N. 1975. Origin of Ross Island basanitoids and limitations upon the heterogeneity of mantle sources for alkali basalts and nephelinites. *Contributions to Mineralogy and Petrology*, **52**, 77–106, https://doi.org/10.1007/BF00395006

Sun, S.-s. and McDonough, W.F. 1989. Chemical and isotopic systematics of oceanic basalts: implications for mantle composition and processes. *Geological Society, London, Special Publications*, **42**, 313–345, https://doi.org/10.1144/GSL.SP.1989.042.01.19

Talarico, F.M. and Kleinschmidt, G. 2008. The antarctic continent in gondwanaland: A tectonic review and potential research targets for future investigations. *In*: Fabio, F. and Martin, S. (eds) *Antarctic Climate Evolution*. Elsevier, Amsterdam, 257–308.

Talarico, F.M. and Sandroni, S. 2009. Provenance signatures of the Antarctic Ice Sheets in the Ross Embayment during the Late Miocene to Early Pliocene: The ANDRILL AND-1B core record. *Global and Planetary Change*, **69**, 103–123, https://doi.org/10.1016/j.gloplacha.2009.04.007

Tauxe, L., Gans, P. and Mankinen, E.A. 2004. Paleomagnetism and $^{40}Ar/^{39}Ar$ ages from volcanics extruded during the Matuyama and Brunhes Chrons near McMurdo Sound, Antarctica. *Geochemistry, Geophysics, Geosystems*, **5**, Q06H12, https://doi.org/10.1029/2003GC000656

Taylor, T.G. 1922. *The Physiography of the McMurdo Sound and Granite Harbour Region. British Antarctic Terra Nova Expedition, 1910–1913*. Harrison and Sons Ltd, London.

Thomson, J.A. 1916. Report on the inclusions of the volcanic rocks of the Ross Archipelago (with Appendix by F. Cohen). *In*: *Report of the British Antarctic Expedition 1907–1909: Geology Report*. British Museum, London, 129–151.

Thornton, C.P. and Tuttle, O.F. 1960. Chemistry of igneous rocks – [Part] 1, Differentiation index. *American Journal of Science*, **258**, 664–684, https://doi.org/10.2475/ajs.258.9.664

Timms, C.J. 2006. *Reconstruction of a Grounded Ice Sheet in McMurdo Sound – Evidence from Southern Black Island*. MSc thesis, University of Otago, Dunedin, New Zealand.

Torsvik, T.H., Müller, R.D., Van der Voo, R., Steinberger, B. and Gaina, C. 2008. Global plate motion frames: Toward a unified model. *Reviews of Geophysics*, **46**, https://doi.org/10.1029/2007RG000227

Treves, S.B. 1962. The geology of Cape Evans and Cape Royds, Ross Island, Antarctica. *American Geophysical Union Antarctic Research Series*, **33**, 40–46.

Treves, S.B. 1967. Volcanic rocks from the Ross Island, Marguerite Bay and Mt. Weaver areas, Antarctica. *Japanese Antarctic Research Expedition Scientific Reports*, Special Issue 1, 136–149.

Treves, S.B. 1968. Volcanic rocks of the Ross Island area. *Antarctic Journal of the United States*, **3**, 108–109.

Treves, S.B. 1977. Geology of some volcanic rocks from the Ross Island, Mount Morning, and southern Victoria Land areas. *Antarctic Journal of the United States*, **12**, 104–105.

van der Meer, Q.H.A., Waight, T.E., Scott, J.M. and Münker, C. 2017. Variable sources for Cretaceous to recent HIMU and HIMU-like intraplate magmatism in New Zealand. *Earth and Planetary Science Letters*, **469**, 27–41, https://doi.org/10.1016/j.epsl.2017.03.037

van Woerden, T.H. 2006. *Volcanic Geology and Physical Volcanology of Mount Morning, Antarctica*. Masters thesis, University of Waikato, Hamilton, New Zealand.

Vella, P. 1969. Surficial geological sequence, Black Island and Brown Peninsula, McMurdo Sound, Antarctica. *New Zealand Journal of Geology and Geophysics*, **12**, 761–770, https://doi.org/10.1080/00288306.1969.10431110

Verma, S.P. and Rivera-Gomez, M.A. 2013. Computer programs for the classification and nomenclature of igneous rocks. *Episodes*, **36**, 115–124, https://doi.org/10.18814/epiiugs/2013/v36i2/005

Walter, M.J. 1998. Melting of garnet peridotite and the origin of komatiite and depleted lithosphere. *Journal of Petrology*, **39**, 29–60, https://doi.org/10.1093/petroj/39.1.29

Waterhouse, B.C. 1965. *Western Ross Sea–Balleny Islands Expedition: January–March, 1965*. New Zealand Geological Survey.

Watson, T., Nyblade, A. *et al.* 2006. P and S velocity structure of the upper mantle beneath the Transantarctic Mountains, East Antarctic craton, and Ross Sea from travel time tomography. *Geochemistry, Geophysics, Geosystems*, **7**, Q07005, https://doi.org/10.1029/2005gc001238

Weaver, B.L. 1991. Trace element evidence for the origin of ocean-island basalts. *Geology*, **19**, 123–126, https://doi.org/10.1130/0091-7613(1991)019<0123:teefto>2.3.co;2

Weiblen, P.J., Stuckless, J.S., Hunter, W.C., Schulz, K.J. and Mundrey, M.G. 1981. Correlation of clinopyroxene compositions with environment of formation based on data from Ross Island volcanic rocks. *American Geophysical Union Antarctic Research Series*, **33**, 229–246.

White, J.C., Benker, S.C., Ren, M., Urbanczyk, K.M. and Corrick, D.W. 2006. Petrogenesis and tectonic setting of the peralkaline Pine Canyon caldera, Trans-Pecos Texas, USA. *Lithos*, **91**, 74–94, https://doi.org/10.1016/j.lithos.2006.03.015

Whitney, D.L. and Evans, B.W. 2010. Abbreviations for names of rock forming minerals. *American Mineralogist*, **95**, 185–187, https://doi.org/10.2138/am.2010.3371

Wilch, T.I., McIntosh, W. *et al.* 2011a. Two-stage growth of the Late Miocene Minna Bluff Volcanic Complex, Ross Embayment, Antarctica: implications for ice-sheet and volcanic histories. Abstract #V31F-2591 presented at the AGU Fall Meeting, December 5–9, 2011, San Francisco, California, USA.

Wilch, T.I., Panter, K.S. *et al.* 2011b. Miocene evolution of the Minna Bluff Volcanic Complex, Ross Embayment, Antarctica. Abstract PS5.10 2591 presented at the 11th International Symposium on Antarctic Earth Science, 10–16 July 2011, Edinburgh, UK.

Willbold, M. and Stracke, A. 2006. Trace element composition of mantle end-members: implications for recycling of oceanic and upper and lower continental crust. *Geochemistry, Geophysics, Geosystems*, **7**, Q04004, https://doi.org/10.1029/2005gc001005

Wilson, A.T., Hendy, C.H. and Taylor, A.M. 1974. Peridot on Ross Island, Antarctica. *Australian Gemologist*, **12**, 124–125.

Wilson, G., Damaske, D., Möller, H.-D., Tinto, K. and Jordan, T. 2007. The geological evolution of southern McMurdo Sound – new evidence from a high-resolution aeromagnetic survey. *Geophysical Journal International*, **170**, 93–100, https://doi.org/10.1111/j.1365-246X.2007.03395.x

Wilson, G.S., Bohaty, S.M. *et al.* 2000. Chronostratigraphy of CRP-2/2A, Victoria Land Basin, Antarctica. *Terra Antartica*, **7**, 647–654.

Wingrove, D. 2005. *Early mixing in the evolution of alkaline magmas: chemical and oxygen isotope evidence from phenocrysts, Royal Society Range, Antarctica*. MSc thesis, Bowling Green State University, Bowling Green, Ohio, USA.

Woodhead, J.D. 1996. Extreme HIMU in an oceanic setting: the geochemistry of Mangaia Island (Polynesia), and temporal evolution of the Cook–Austral hotspot. *Journal of Volcanology and Geothermal Research*, **72**, 1–19, https://doi.org/10.1016/0377-0273(96)00002-9

Workman, R.K. and Hart, S.R. 2005. Major and trace element composition of the depleted MORB mantle (DMM). *Earth and Planetary Science Letters*, **231**, 53–72, https://doi.org/10.1016/j.epsl.2004.12.005

Workman, R.K., Hart, S.R. *et al.* 2004. Recycled metasomatized lithosphere as the origin of the Enriched Mantle II (EM2) end-member: evidence from the Samoan Volcanic Chain. *Geochemistry, Geophysics, Geosystems*, **5**, Q04008, https://doi.org/10.1029/2003gc000623

Worley, B.A. 1992. *Dismal Geology; A Study of Magmatic and Subsolidus Processes in a Carbonated Alkaline Intrusion, Southern Victoria Land, Antarctica*. MSc thesis, University of Otago, Dunedin, New Zealand.

Worley, B.A., Cooper, A.F. and Hall, C.E. 1995. Petrogenesis of carbonate-bearing nepheline syenites and carbonatites from Southern Victoria Land, Antarctica: origin of carbon and the effects of calcite–graphite equilibrium. *Lithos*, **35**, 183–199, https://doi.org/10.1016/0024-4937(94)00050-C

Wright, A.C. 1979*a*. *McMurdo Volcanics at Foster Crater, Southern Foothills of the Royal Society Range, Central Victoria Land, Antarctica*. New Zealand Geological Survey Internal Report.

Wright, A.C. 1979*b*. *McMurdo Volcanics northwest of Koettlitz Glacier, Antarctica*. New Zealand Geological Survey Internal Report.

Wright, A.C. 1979*c*. *McMurdo Volcanics on Chancellor Ridge, Southern Foothills of the Royal Society Range, Central Victoria Land, Antarctica*. New Zealand Geological Survey Internal Report.

Wright, A.C. 1979*d*. A reconnaissance study of the McMurdo Volcanics northwest of Koettlitz Glacier. *New Zealand Antarctic Record*, **1**, 10–15.

Wright, A.C. 1980. Landforms of McMurdo Volcanic Group, Southern Foothills of Royal Society Range, Antarctica. *New Zealand Journal of Geology and Geophysics*, **23**, 605–613, https://doi.org/10.1080/00288306.1980.10424132

Wright, A.C. and Kyle, P.R. 1990*a*. Royal Society Range. *American Geophysical Union Antarctic Research Series*, **48**, 131–133.

Wright, A.C. and Kyle, P.R. 1990*b*. Taylor and Wright Valleys. *American Geophysical Union Antarctic Research Series*, **48**, 134–135.

Wright, A.C. and Kyle, P.R. 1990*c*. Mount Discovery. *American Geophysical Union Antarctic Research Series*, **48**, 120–123.

Wright, A.C. and Kyle, P.R. 1990*d*. Minna Bluff. *American Geophysical Union Antarctic Research Series*, **48**, 117–119.

Wright, A.C. and Kyle, P.R. 1990*e*. Mount Terror. *American Geophysical Union Antarctic Research Series*, **48**, 99–102.

Wright, A.C. and Kyle, P.R. 1990*f*. Mason Spur. *American Geophysical Union Antarctic Research Series*, **48**, 128–130.

Wright, A.C. and Kyle, P.R. 1990*g*. Mount Morning. *American Geophysical Union Antarctic Research Series*, **48**, 124–126.

Wright, A.C., McIntosh, W.C. and Ellerman, P. 1983. Volcanic geology of Turks Head, Tryggve Point, and Minna Bluff, southern Victoria Land. *Antarctic Journal of the United States*, **18**, 35–36.

Wright, A.C., Kyle, P.R., McIntosh, W.C. and Klich, I. 1984. Geological field investigations of volcanic rocks at Mount Discovery and Mason Spur, McMurdo Sound. *Antarctic Journal of the United States*, **19**, 20–21.

Wright, A.C., Kyle, P.R., More, J.A. and Meeker, K. 1986. Geological investigations of volcanic rocks at Mount Discovery, Mount Morning, and Mason Spur, McMurdo Sound. *Antarctic Journal of the United States*, **21**, 55.

Wright-Grassham, A.C. 1987. *Volcanic Geology, Mineralogy, and Petrogenesis of the Discovery Volcanic Subprovince, Southern Victoria Land, Antarctica*. PhD thesis, New Mexico Institute of Mining and Technology, Socorro, New Mexico, USA.

Yaxley, G.M., Crawford, A.J. and Green, D.H. 1991. Evidence for carbonatite metasomatism in spinel peridotite xenoliths from western Victoria, Australia. *Earth and Planetary Science Letters*, **107**, 305–317, https://doi.org/10.1016/0012-821X(91)90078-V

Yokoyama, T., Aka, F.T., Kusakabe, M. and Nakamura, E. 2007. Plume–lithosphere interaction beneath Mt. Cameroon volcano, West Africa: Constraints from $^{238}U-^{230}Th-^{226}Ra$ and Sr–Nd–Pb isotope systematics. *Geochimica et Cosmochimica Acta*, **71**, 1835–1854, https://doi.org/10.1016/j.gca.2007.01.010

Zindler, A. and Hart, S. 1986. Chemical Geodynamics. *Annual Review of Earth and Planetary Sciences*, **14**, 493–571, https://doi.org/10.1146/annurev.ea.14.050186.002425

Chapter 5.3a

Mount Early and Sheridan Bluff: volcanology

John L. Smellie[1]*, Kurt S. Panter[2] and Jenna Reindel[2]

[1]School of Geography, Geology and the Environment, University of Leicester, University Road, Leicester LE1 7RH, UK

[2]Department of Geology, Bowling Green State University, Overman Hall, Bowling Green, OH 43403, USA

JLS, 0000-0001-5537-1763; KSP, 0000-0002-0990-5880

*Correspondence: jls55@leicester.ac.uk

Abstract: Two small monogenetic volcanoes are exposed at Mount Early and Sheridan Bluff, in the upper reaches of Scott Glacier. In addition, the presence of abundant fresh volcanic detritus in moraines at two other localities suggests further associated volcanism, now obscured by the modern Antarctic ice sheet. One of those occurrences has been attributed to a small subglacial volcano only *c.* 200 km from South Pole, making it the southernmost volcano in the world. All of the volcanic outcrops in the Scott Glacier region are grouped in a newly defined Upper Scott Glacier Volcanic Field, which is part of the McMurdo Volcanic Group (Western Ross Supergroup). The volcanism is early Miocene in age (*c.* 25–16 Ma), and the combination of tholeiitic and alkaline mafic compositions differs from the more voluminous alkaline volcanism in the West Antarctic Rift System. The Mount Early volcano was erupted subglacially, when the contemporary ice was considerably thicker than present. By contrast, lithologies associated with the southernmost volcano, currently covered by 1.5 km of modern ice, indicate that it was erupted when any associated ice was either much thinner or absent. The eruptive setting for Sheridan Bluff is uncertain and is still being investigated.

The only known outcrops of Neogene volcanic rocks exposed in the Transantarctic Mountains south of 79° S consist of two localities in the Queen Maud Mountains at the head of Scott Glacier (Fig. 1). The outcrops are at Mount Early and Sheridan Bluff. They were visited in 1962–63 and briefly described by Doumani and Minshew (1965; see also Treves 1967), followed by a re-examination in 1978–79 (Stump *et al.* 1980, 1990*a, b*). They were visited again by the authors in December 2015. The results of our investigation are still in progress, and only preliminary descriptions and initial interpretations are given here.

Mount Early and Sheridan Bluff are situated *c.* 300 km from South Pole at elevations of 2200–2700 m above sea level (asl). They are included in the McMurdo Volcanic Group (Kyle 1990), although the closest outcrops of the latter are the Erebus Volcanic Province, >1000 km to the NNW (Chapter 5.2a: Smellie and Martin 2021). They are thus part of the Western Ross Supergroup, which was defined by Rocchi *et al.* (2002) and formally redefined by Smellie and Rocchi (2021: Chapter 5.1a) (Fig. 2). Only two small surface outcrops are known. They were previously regarded as the two southernmost volcanoes in the world. However, Licht *et al.* (2018) recorded clasts of basalt lava and hyaloclastite with early Miocene ages in a moraine at Mount Howe, *c.* 40 km south of Mount Early, which may be derived from an even more southerly outcrop, now entirely hidden by ice (Figs 1 & 3). Moreover, a moraine extending from the NE side of Mount Wyatt (*c.* 15 km north of Sheridan Bluff) also contains fragments of likely alkaline volcanic rocks. Both of the morainic occurrences indicate volcanism in addition to that exposed at Mount Early and Sheridan Bluff, and all are included here in a new volcanic field of the McMurdo Volcanic Group (Upper Scott Glacier Volcanic Field), named after the location of its constituent outcrops (Mount Early and Sheridan Bluff) and morainic debris localities (at Mount Howe and Mount Wyatt: Fig. 2). It is therefore possible that other volcanic outcrops will be discovered in the area in future, particularly below the East Antarctic Ice Sheet.

The few published geological accounts available are brief and at a reconnaissance level, or else are hampered by being based only on morainic debris and geophysical interpretation (Doumani and Minshew 1965; Treves 1967; Stump *et al.* 1980, 1990*a, b*; Licht *et al.* 2018). However, the importance of the occurrences far exceeds their small size and scarcity. For example, their presence and characteristics have been used as proof of the existence of an East Antarctic Ice Sheet in early Miocene time; to infer significant uplift of the Transantarctic Mountains by that time; to extend the known geographical extent of the West Antarctic Rift System; and as tacit support for a lithospheric foundering model (cf. Shen *et al.* 2018) that may help to explain uplift of the Transantarctic Mountains. Moreover, the situation of the outcrops, far inland and at a high elevation, is unique compared with the remainder of the McMurdo Volcanic Group, which is exposed in low-elevation coastal locations – see Chapters 5.1a (Smellie and Rocchi 2021) and 5.2a (Smellie and Martin 2021). Environmentally, therefore, the outcrops potentially contain uniquely important information relevant to the interior of the palaeo-East Antarctic Ice Sheet rather than biased towards its margins (cf. Smellie *et al.* 2011, 2014).

Stratigraphy and volcanology of the volcanic outcrops

Mount Early is a small, isolated, conical volcanic nunatak about 1.3 km in longest diameter (north–south) and *c.* 1 km wide (east–west). Its summit is at *c.* 2950 m asl (estimated from a handheld GPS altimeter) and it rises *c.* 450 m above the surrounding ice surface. Rock and extensive scree are exposed mainly on the eastern and northeastern faces, and the other faces are almost wholly covered by snow and ice (Fig. 4a). The outcrop was described by Stump *et al.* (1980, 1990*b*) as a basal shield-like pile of black pillow lava mantled by pillow breccias that grade up into yellow fine-grained 'palagonite breccia'. The latter was described as massive lower down but thinly bedded above; a small dyke was also shown intruding the breccia at the top of the pillow lava pile. It was also suggested that lava capped the southern end of the summit but the outcrop was unvisited. A general comparison was made with subglacially erupted volcanic centres in Iceland (tuyas, which are flat-topped table mountains: called móberg

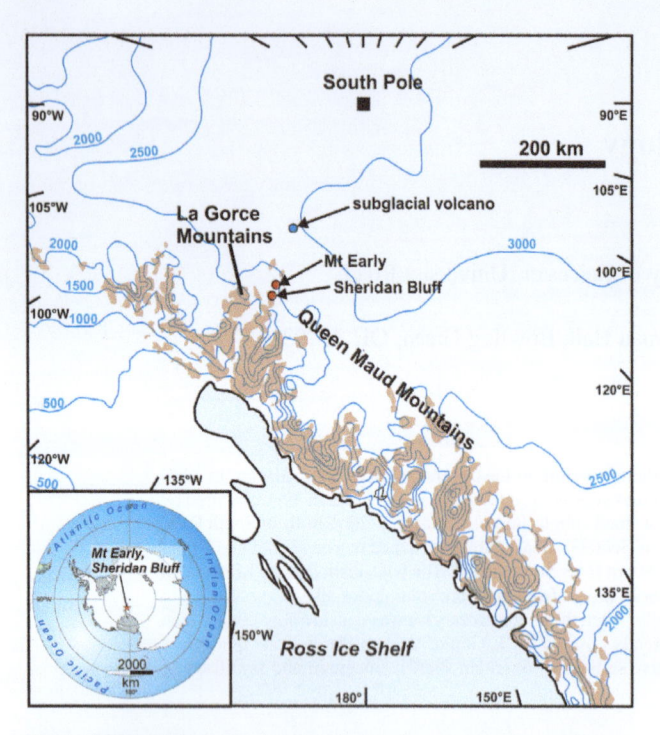

Fig. 1. Topographical map showing the location of volcanoes described in this chapter.

by Stump *et al.* 1980, 1990*b*), and a mean isotopic age of 15.86 ± 0.30 (by K–Ar) was obtained on a duplicated lava sample from the basal pillow lava pile (Table 1). The identification as a glaciovolcanic centre was tentative, however, and was based on: (1) the proximity of Mount Early to Sheridan Bluff, *c.* 18 km to the north, for which a glacial eruptive setting seemed better founded (see below); and (2) the great distance from any standing body of water. Erupted lavas at Mount Early are basaltic (hawaiite and mugearite: Stump *et al.* 1990*b*; S. Hart cited in Licht *et al.* 2018), as confirmed by our new analyses (Panter *et al.* 2021: Chapter 5.3b).

From observations by the authors, we broadly confirm the published description but there are differences in the details. Figure 5 depicts the geology of Mount Early viewed looking WSW. The core of the volcano is composed of a large mound at least 250 m thick (above the surrounding ice surface) composed of massive chaotic pillow lava and hyaloclastite (*sensu* White and Houghton 2006) (Fig. 4b), with a crude sense of dip away from the nunatak shown by lava tubes. The proportion of hyaloclastite increases up and outwards, and the mound is progressively to sharply overlain by orange- to locally grey-coloured finely stratified fine lapilli tuffs (Fig. 4c). The lapilli tuffs lack accidental lithic clasts and they probably form a relatively thin superficial layer tens of metres thick, which drapes the basal pillow lava mound. They show much evidence for syneruptive deformation, including faulting on a variety of scales, and dark-grey hyaloclastite back-injected up into the lapilli tuffs. The tuffs are intruded by irregular and neck-like masses of columnar grey lava. The necks are locally associated with dark-grey agglutinate. A prominent grey-green columnar dyke is present near the crest of the summit ridge but was unvisited. Very rare small granitoid basement erratics are present up to 160 m on the north flank.

Mount Early occurs in isolation in the expanded headward reaches of the Scott Glacier, close to where it merges with the polar plateau. Its base is unexposed. There are no measured ice thicknesses for the surrounding ice but a tentative modelled thickness of *c.* 470 m can be obtained from BEDMAP2 (Fretwell *et al.* 2013; Peter Fretwell, British Antarctic Survey pers. comm.). Combined with the height of the nunatak, this implies that the present Mount Early edifice may be at least 900 m high. The entire nunatak has been affected by erosion, presumably by overriding ice, and the original summit elevation has been reduced by an unknown amount consistent with the absence of a crater. The presence of pillow lava and cogenetic hyaloclastite indicate effusion into water. The coloration and induration of the orange-coloured lapilli tuffs are due to palagonitization of sideromelane, a common and distinctive feature of explosively generated phreatomagmatic deposits. They typically form when magma interacts with water, either groundwater or surface water, resulting in the construction of a

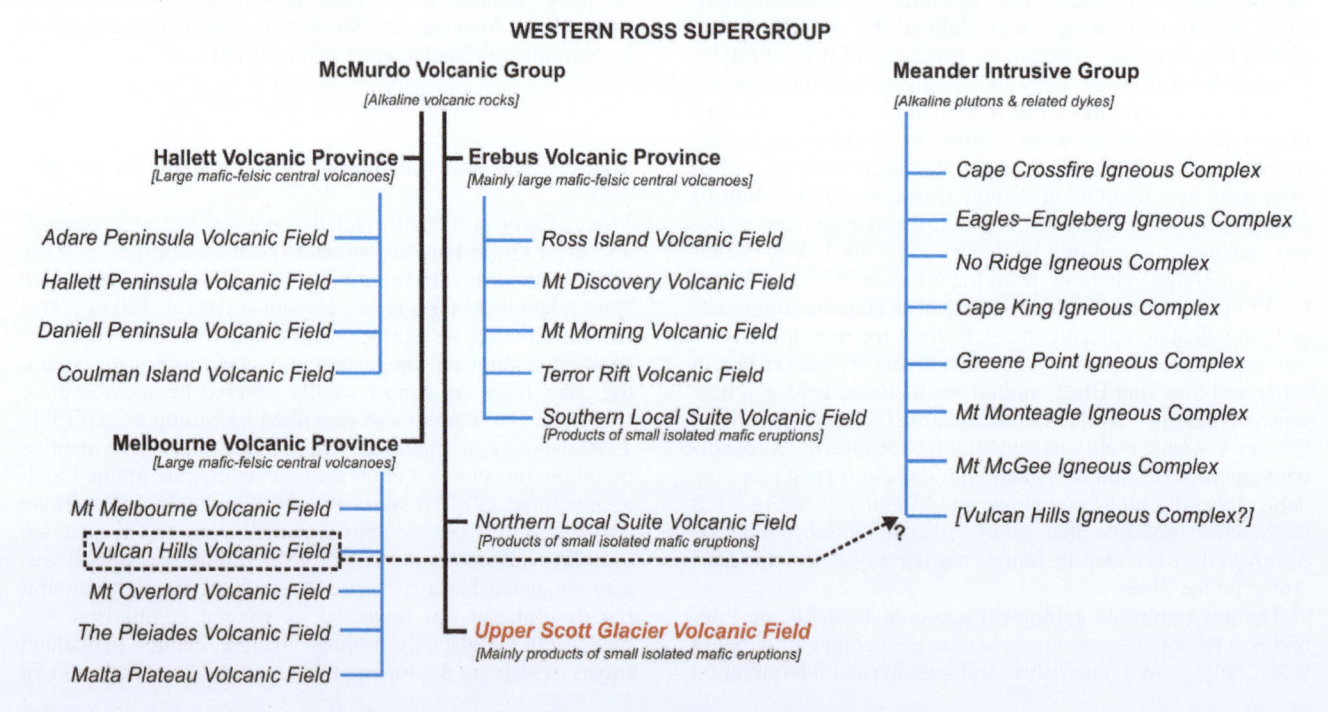

Fig. 2. Summary stratigraphy of the Western Ross Supergroup (after Smellie and Rocchi 2021: Chapter 5.1a).

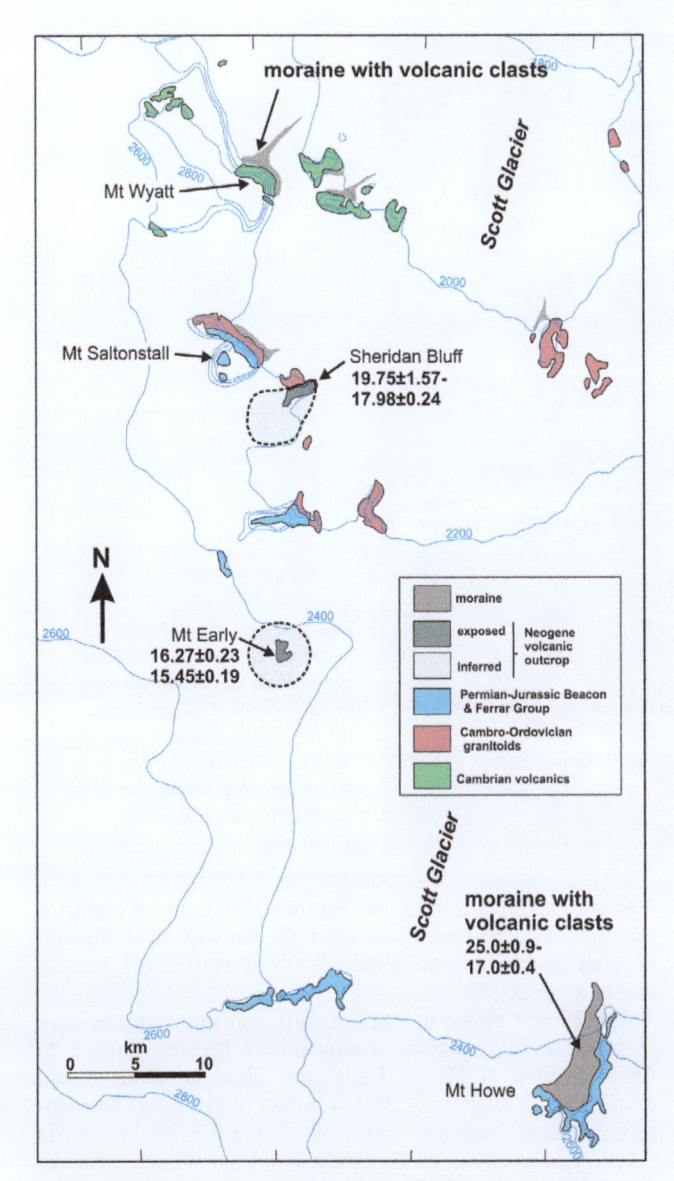

Fig. 3. Geological map of the upper Scott Glacier area, Queen Maud Mountains (Transantarctic Mountains) showing the location and published ages of volcanic outcrops at Mount Early and Sheridan Bluff, and moraines containing volcanic debris. Geological base map from Davis and Blankenship (2005).

tuff ring or tuff cone. The lack of accidental basement clasts and the relatively steep depositional dips in the lapilli tuffs are more consistent with interaction with surface water and formation of a tuff cone, respectively, rather than a tuff ring. Any surface water is unlikely to have been marine in view of the distance inland and high elevation of the outcrop, and a freshwater lake is indicated, either pluvial or glacial. However, the nunatak is surrounded on all sides by a very extensive ice plateau and there is no feasible palaeotopography that might have impounded a pluvial lake (Fig. 6). A glacial eruptive setting and meltwater lake are therefore more likely, consistent with the speculative interpretation of Stump et al. (1980). Thus, the Miocene ice-surface elevation must have been above the nunatak summit: that is, at least 500 m higher than the present-day ice surface.

Sheridan Bluff overlooks Scott Glacier and consists of a steep scree slope capped by a cliff face formed of lava (Figs 3 & 7a). However, although the outcrop is lower than that at Mount Early, it is more extensive and extends c. 3.5 km upslope to the SW, where it reaches a summit elevation of c. 2550 m asl (by handheld Garmin GPS). The difference in elevation between the lowest and highest exposures is c. 370 m, which is substantially greater than the estimate of c. 200 m for the outcrop thickness made by previous workers based only on exposures at Sheridan Bluff itself; however, our estimate assumes a horizontal underlying bedrock surface, which is mostly unexposed and is therefore unverified. The published accounts of the section are very similar and describe a c. 85 m-thick volcaniclastic lower unit overlain by c. 110 m of lavas with red scoriaceous surfaces; between seven and nine lavas were believed to be present (Doumani and Minshew 1965; Stump et al. 1980, 1990a). Three isotopic ages on the capping lavas were determined by K–Ar and average 18.32 ± 0.35 Ma (range 19.21 ± 0.39–17.98 ± 0.24 Ma), whereas two ages determined by $^{40}Ar/^{39}Ar$ yielded more consistent values of 19.75 ± 1.57 and 19.43 ± 0.65 Ma (Stump et al. 1980) (Table 1). The lavas are all mafic (basalt and hawaiite). Uniquely in the West Antarctic Rift System, they include two contrasting types: alkaline basalt and subalkaline tholeiite (Stump et al. 1980, 1990a; Licht et al. 2018; Panter et al. 2021: Chapter 5.3b).

The best-exposed and most complete section occurs at Sheridan Bluff itself (Fig. 8), whereas the higher outcrop to the SW is an undulating surface that dips gently to the NE and is mostly a blockfield of lava fragments, in which patterned ground is locally developed. Smoothed and striated granitoid basement is exposed at the base of the sequence. The basal surface is a prominent hollow a few hundred metres wide that is at least several tens of metres deep. It is overlain by poorly-exposed orange lapilli tuff at least 85 m thick, which is mainly massive but locally shows indistinct planar stratification (Fig. 7b). The basal pillow breccia and hyaloclastite described by Stump et al. (1980, 1990a) were not observed. Only the capping lavas are well exposed in the summit cliff at Sheridan Bluff. They are c. 100 m thick in total, and comprise multiple pāhoehoe lobes with coarsely vesicular oxidized and locally ropy surfaces (Fig. 7c). The highest-elevation outcrop to the SW shows sparse exposures of yellow fine lapilli tuff amongst debris composed of the same. A small proportion of brightly-coloured hydrothermally altered lava blocks are also scattered in the summit area, and there are sparse erratics of granitoid and Beacon sandstone.

An origin as a glaciovolcanic centre was inferred by Stump et al. (1980), principally based on the presence of a glacially modified basal surface and a supposed similarity to Icelandic tuyas (the 'móberg' of those authors). However, although permissive, the presence of a glacially modified surface does not necessarily link that surface with the overlying deposits (cf. Smellie 2018). Secondly, we did not confirm the presence of pillow lava, and the volcaniclastic deposits appear to be lapilli tuffs, with abundant vesicular sideromelane recorded by Treves (1967). Together with a paucity of accidental clasts, the characteristics of the lapilli tuffs suggest an origin by explosive hydrovolcanic eruption rather than the passive mechanical spallation characteristic of hyaloclastite (sensu White and Houghton 2006; see also Smellie and Edwards 2016, table 9.1). Thus, it appears that the Sheridan Bluff outcrop may be a tuff cone draped by a subaerial lava carapace. Based on the uniform NE dip of the overlying lavas and extrapolation of the outcrop extent to the SW, the original edifice had an original basal diameter of c. 6 km. We therefore interpret the outcrop as a remnant of a small mafic shield volcano. The early Miocene eruptive setting is not well understood at present but the lithofacies provide no unequivocal support for a glacial environment and subglacial eruption.

The characteristics of the volcanic debris found in the Mount Howe moraine are vesicular, mildly altered basalt and hyaloclastite. The hyaloclastite contains numerous unaltered, pale-grey, accessory lava fragments. Moreover, the pervasive orange coloration is caused by palagonite alteration of sideromelane, and the rock is relatively fine grained,

Fig. 4. Compilation of field photographs showing Mount Early and selected lithofacies. (**a**) View of the NE face of Mount Early. (**b**) Basaltic lava pillows intruding orange lapilli tuffs; the hammer is c. 40 cm long. (**c**) Diffusely stratified lapilli tuffs; the notebook is 17 cm long.

porous and friable. The alteration and presence of laminated sediment infilling cavities in the rocks suggested to Licht et al. (2018) a subaqueous setting for the volcanism. On overall characteristics, the hyaloclastite samples are probably lapilli tuffs: that is, generated explosively in a hydrovolcanic eruption rather than as hyaloclastite (*sensu* White and Houghton 2006). The volcano that is presumed to have produced the lapilli tuff and associated lavas is currently below ice c. 1500 m thick (see below; Licht et al. 2018). Such a thickness is far too great to permit explosive hydrovolcanic eruptions, which become increasingly unlikely below 100–200 m (e.g. Zimanowski and Büttner 2003), and it confirms that the eruptions took place when the ice was much thinner, or conceivably absent, although water was likely present (Licht et al. 2018).

The Mount Howe morainic debris volcanic samples have tholeiitic–mildly alkaline compositions broadly similar to those erupted at Mount Early and Sheridan Bluff (Licht et al. 2018; Panter et al. 2021: Chapter 5.3b). $^{40}Ar/^{39}Ar$ isotopic dating indicates ages of $20.6 ± 0.3–17.0 ± 0.4$ Ma (Table 1). Moreover, U–Pb dating of detrital zircons extracted from the associated till yielded ages of $25.0 ± 0.9–19.4 ±$

Table 1. *Summary of published isotopic ages for the Upper Scott Glacier Volcanic Field*

Sample	Locality	Dated lithologies	Age* (Ma)	Error	Error SD	Method	References
34	Mount Early	Lava, pillow lava mound	15.45	0.19	n.r.	K–Ar	Stump et al. (1980)
34	Mount Early	lava, pillow lava mound	16.27	0.23	n.r.	K–Ar	Stump et al. (1980)
		Mean age:	15.86	0.30	n.r.	K–Ar	Stump et al. (1980)
30	Sheridan Bluff	Lava	18.54	0.37	n.r.	K–Ar	Stump et al. (1980)
27	Sheridan Bluff	Lava	17.98	0.24	n.r.	K–Ar	Stump et al. (1980)
24	Sheridan Bluff	Lava	18.43	0.23	n.r.	K–Ar	Stump et al. (1980)
		Mean age (for 24, 27 and 30):	18.32	0.35	n.r.	K–Ar	Stump et al. (1980)
22	Sheridan Bluff	Lava, lowest in section	19.21	0.39	n.r.	K–Ar	Stump et al. (1980)
27	Sheridan Bluff	Lava	19.43	0.65	n.r.	Ar–Ar	Stump et al. (1980)
22	Sheridan Bluff	Lava, lowest in section	19.75	1.57	n.r.	Ar–Ar	Stump et al. (1980)
MH-10-7	Mount Howe moraine	Lava clast in moraine	17.5	0.5	n.r.	Ar–Ar	Licht et al. (2018)
MH-10-20C	Mount Howe moraine	Accessory lava in 'hyaloclastite' clast in moraine	17.0	0.4	n.r.	Ar–Ar	Licht et al. (2018)
MH-10-25	Mount Howe moraine	Lava clast in moraine	18.0	0.2	n.r.	Ar–Ar	Licht et al. (2018)
MH-10-26	Mount Howe moraine	Lava clast in moraine	20.6	0.3	n.r.	Ar–Ar	Licht et al. (2018)
MH-35	Mount Howe moraine	Lava clast in moraine	17.3	1.1	n.r.	Ar–Ar	Licht et al. (2018)
1691-11	Mount Howe moraine	Detrital zircon in till	19.4	0.9	n.r.	U–Pb	Licht et al. (2018)
1691-37	Mount Howe moraine	Detrital zircon in till	25.0	0.9	n.r.	U–Pb	Licht et al. (2018)
1691-48	Mount Howe moraine	Detrital zircon in till	23.1	0.5	n.r.	U–Pb	Licht et al. (2018)

*Ages given as published, not recalculated using more recently published decay constants.
n.r., not reported.
SD, standard deviation.

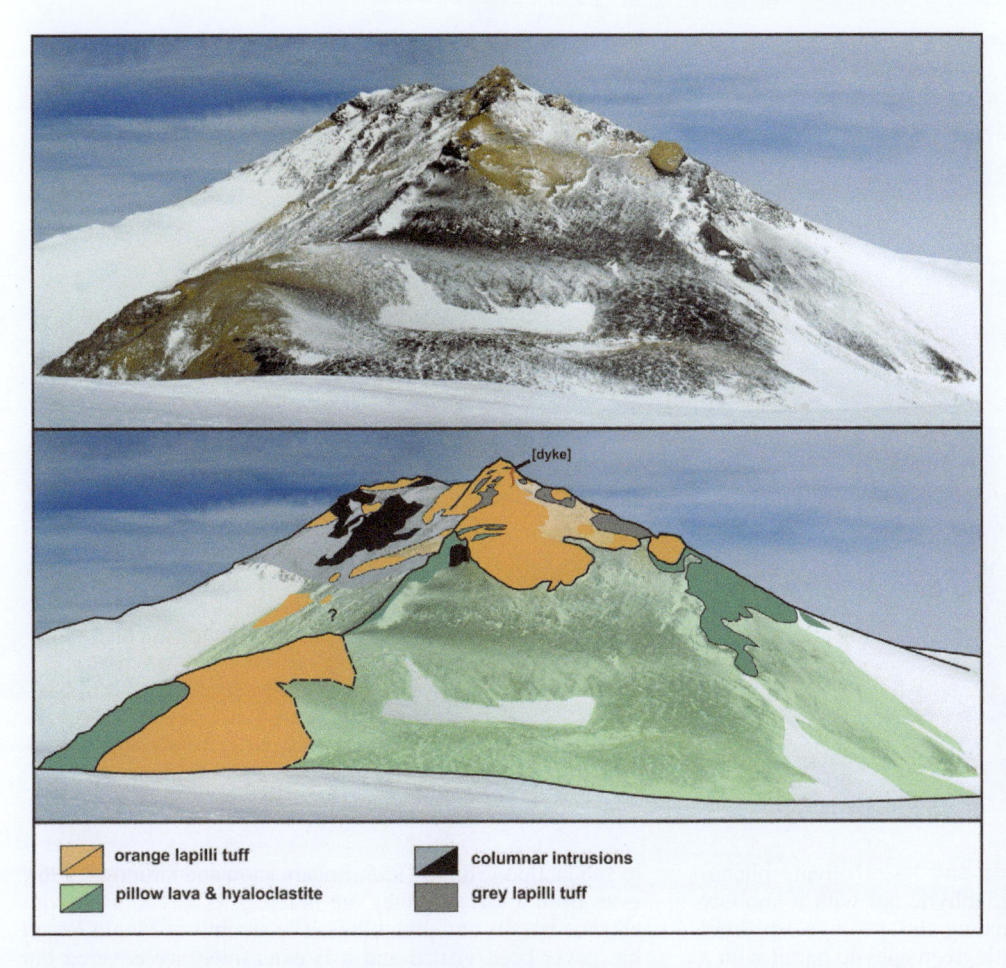

Fig. 5. Geological sketch map of Mount Early, looking south, based on the authors' observations. The nunatak rises *c.* 450 m above the surrounding snowfield. Denser colours indicate exposed rock, less-dense colours indicate outcrop lacking surface exposure. See the text for further description.

0.9 Ma. Since zircons are rare in rocks of basaltic compositions and are more characteristic of felsic volcanism, the older ages may record volcanism additional to that which resulted in the basalt clasts, despite a lack of felsic rock fragments in the moraine. Aeromagnetic data upstream of Mount Howe indicate the presence of a conspicuous high-amplitude negative magnetic anomaly *c.* 7 km in diameter that Licht *et al.* (2018) identified as a plausible source volcano (Fig. 1). The feature has a highly uneven surface but is just *c.* 300–400 m high. It is thus broadly pancake-shaped overall, suggesting that it has undergone significant glacial erosion. On the basis of the negative anomaly of the subglacial feature, the age of the volcanism might be further refined to 17.399–16.807 Ma, which is the age of a magnetic reversal that overlaps with most of the $^{40}Ar/^{39}Ar$ isotopic ages. The location of the postulated volcanic source, *c.* 80 km south of Mount Howe and *c.* 140 km south of Mount Early and Sheridan Bluff, makes it the most southerly volcano in the world currently known.

Volcanic clasts have also been found in the moraine below Mount Wyatt, and are reported here for the first time (Fig. 3). They consist of 'greenish basanite(?); vesicular trachybasalt

Fig. 6. View, looking south, towards Mount Early (on the right). Mount Howe is on the left in the background. Note the wide expanse of ice and an absence of any topography that might have ponded a pluvial lake.

Fig. 7. Compilation of field photographs showing Sheridan Bluff and selected lithofacies. (**a**) View, looking SW, of Sheridan Bluff. The rock face shown is *c.* 185 m high. (**b**) Massive lapilli tuff. (**c**) Pāhoehoe lavas exposed in the cliff capping Sheridan Bluff; the individual lava lobes are *c.* 2–6 m thick; note the prominent oxidized surfaces.

with abundant small plagioclase and lesser olivine phenocrysts; greenish-grey phonolite, aphyric but with a knobbly appearance possibly due to radiating clusters of feldspathoid in the groundmass; compact dark green aphyric basalt with a few nodules of olivine; aphanitic basanite with abundant inclusions [unspecified]; and vesicular trachybasalt with needles of plagioclase and some plagioclase and olivine or pyroxene phenocrysts' (unpublished field notes of M.F. Sheridan; cited by E. Stump in a pers. comm. to J.L. Smellie). Although the compositions of the rocks remain to be verified, the rock assemblage suggests that the clasts were sourced in a hitherto-unknown alkaline volcanic centre, and the wide range of apparent compositions is consistent with a larger (central) volcano than is seen at Mount Early and Sheridan Bluff. The local ice-flow direction indicated by the moraine suggested to Stump (pers. comm.) that the source outcrop might be at the summit of Mount Wyatt or in the bluffs to the NW. Mapped rock at Mount Wyatt consists of Cambrian volcanics (Wyatt Formation: Fig. 3). The Wyatt Formation is predominantly composed of massive pyroclastic deposits, probably ignimbrites, with calc-alkaline compositions related to subduction (Stump 1995; Encarnación and Grunow 1996); even from a distance they are unlikely to be confused with alkaline basalts and phonolites. The summit of Mount Wyatt has never been visited and it is extensively ice covered but no younger volcanic rocks have been described capping the Wyatt Formation in the exposed cliff faces. However, on the basis of the two morainic debris occurrences, it is possible that further Neogene volcanic outcrops exist elsewhere in the upper Scott Glacier region, particularly below the East Antarctic Ice Sheet, and remain to be discovered.

Summary

Small outcrops of Neogene volcanoes are present on the flanks of the upper reaches of the Scott Glacier, in the Queen Maud Mountains (Transantarctic Mountains). They include two well-exposed volcano remnants at Mount Early and Sheridan Bluff, plus two moraine localities below Mount Howe and Mount Wyatt characterized by numerous volcanic clasts. All are included in the new Upper Scott Glacier Volcanic Field

Fig. 8. Geological sketch map of Sheridan Bluff, looking SW, based on the authors' observations. The rock face shown is *c.* 185 m high. Denser colours (lava and granitoid units) indicate exposed rock; less dense colours indicate outcrops lacking surface exposure; individual exposures of lapilli tuff are small and widely scattered, and are not distinguished separately. See the text for description.

(McMurdo Volcanic Group, Western Ross Supergroup). Clasts at Mount Howe have alkaline and tholeiitic compositions and early–mid Miocene isotopic ages similar to those published for Mount Early and Sheridan Bluff, suggesting that they are all petrogenetically related. A plausible source for the Mount Howe debris is a prominent magnetic anomaly situated just 200 km from South Pole, making it the most southerly volcano in the world. Other discoveries of Neogene volcanoes in the same area are likely to be made in the future.

The volcanoes are all quite small, the largest potentially being that south of Mount Howe which may be 7 km in basal diameter. The simple internal structure of the two exposed examples suggests that they are probably monogenetic. In both characteristics it might be argued that they resemble volcanoes of the Southern Local Suite (Erebus Volcanic Province: described in Chapter 5.2a: Smellie and Martin 2021). However, those in the Upper Scott Glacier Volcanic Field are generally taller than those in the Southern Local Suite, with that at Mount Early being the tallest currently known (c. 900 m high). Moreover, volcanoes in the Southern Local Suite are typically small scoria cones, whereas those in the Upper Scott Glacier Volcanic Field have more variable lithofacies, including a pillow mound, lapilli tuffs and subaerial lavas, together with possibly more evolved compositions. In those respects they are most like centres in the Dailey Islands (see Chapter 5.2a: Smellie and Martin 2021). The apparent age of the volcanism varies from c. 20 to 16 Ma, with detrital zircon grains suggesting an older phase at c. 25–23 Ma.

The Mount Early volcano was unequivocally erupted subglacially, beneath an ice sheet substantially thicker than at present, whereas the eruptive setting of the southernmost volcano in the volcanic field, known only from moraine detritus at Mount Howe, was either subaerial or at least there was much less ice than at present. The palaeoenvironment for the Sheridan Bluff volcano is less certain and is still under investigation.

Although the volcanism is currently regarded as part of the West Antarctic Rift System, it occurs 200–350 km inland of the structural margin of the rift and >1000 km from other known surface outcrops of the McMurdo Volcanic Group. The discovery of a prominent linear feature resembling a volcanic ridge under the East Antarctic Ice Sheet, to which the volcanic outcrops are only 50 km distant (Ferraciolli et al. 2011; Fretwell et al. 2013), suggests the possibility that they may be related to that feature or to effects of the interaction between that feature and the West Antarctic Rift System margin (see also Chapter 5.3b: Panter et al. 2021). In this context, the tholeiitic compositions at two of the centres may be particularly significant. Although we now know that tholeiitic basalts occur at Sheridan Bluff and the subglacial volcano south of Mount Howe, tholeiitic compositions are extremely rare elsewhere in the West Antarctic Rift System (see Panter et al. 2021: Chapter 5.3b). This suggests that the outcrops in the Upper Scott Glacier Volcanic Field are petrogenetically related and their origin is distinctive compared with volcanism in the West Antarctic Rift System. The similar petrogenesis also strengthens our inclusion of all the known occurrences within a single volcanic field.

Acknowledgements We are particularly grateful to Ed Stump for his help in providing local geological and logistical information, including field photographs, that helped us to compile our successful grant proposal; the National Science Foundation (NSF) and all the NSF staff at Christchurch, McMurdo Station and the Shackleton Glacier transit camp for their excellent logistical and other support for our project; Tim Burton for his highly professional field assistance, as always; and Peter Fretwell (British Antarctic Survey) for calculating approximate contemporaneous ice thicknesses surrounding Mount Early from BEDMAP2. Finally, the authors are very grateful for constructive comments on our paper by Ed Stump and an anonymous reviewer.

Author contributions JLS: conceptualization (lead), formal analysis (lead), investigation (lead), methodology (lead), visualization (lead), writing – original draft (lead), writing – review & editing (lead); **KSP**: conceptualization (supporting), data curation (lead), funding acquisition (lead), investigation (equal), methodology (supporting), project administration (lead), resources (lead), visualization (supporting), writing – review & editing (supporting); **JR**: investigation (supporting), writing – review & editing (supporting).

Funding The National Science Foundation provided support (with grant NSF PLR 1443576) for this work.

Data availability Data sharing is not applicable to this article as no additional datasets were generated other than those presented in this chapter.

References

Davis, M.B. and Blankenship, D.D. 2005. *Geology of the Scott–Reedy Glaciers Area, Southern Transantarctic Mountains, Antarctica*. Geological Society of America Map and Chart Series, **MCH093**.

Doumani, G.A. and Minshew, V.H. 1965. General geology of the Mount Weaver area, Queen Maud Mountains. *American Geophysical Union Antarctic Research Series*, **6**, 127–139.

Encarnación, J. and Grunow, A. 1996. Changing magmatic and tectonic styles along the paleo-Pacific margin of Gondwana and the onset of early Paleozoic magmatism in Antarctica. *Tectonics*, **15**, 1325–1341, https://doi.org/10.1029/96TC01484

Ferraciolli, F., Finn, C.A., Jordan, T.A., Bell, R.E., Anderson, L.M. and Damaske, D. 2011. East Antarctic rifting triggers uplift of the Gamburtsev Mountains. *Nature*, **479**, 388–392, https://doi.org/10.1038/nature10566

Fretwell, P., Pritchard, H.D. et al. 2013. Bedmap2: improved ice bed, surface and thickness datasets for Antarctica. *The Cryosphere*, **7**, 375–393, https://doi.org/10.5194/tc-7-375-2013

Kyle, P.R. 1990. McMurdo volcanic group, western Ross Embayment. *American Geophysical Union Antarctic Research Series*, **48**, 19–25.

Licht, K.J., Groth, T., Townsend, J.P., Hennessy, A.J., Hemming, S.R., Flood, T.P. and Studinger, M. 2018. Evidence for extending anomalous Miocene volcanism at the edge of the East Antarctic craton. *Geophysical Research Letters*, **45**, 3009–3016, https://doi.org/10.1002/2018GL077237

Panter, K.S., Reindel, J. and Smellie, J.L. 2021. Mount Early and Sheridan Bluff: petrology. *Geological Society, London, Memoirs*, **55**, https://doi.org/10.1144/M55-2019-2

Rocchi, S., Armienti, P., D'Orazio, M., Tonarini, S., Wijbrans, J.R. and Di Vincenzo, G. 2002. Cenozoic magmatism in the western Ross Embayment: role of mantle plume versus plate dynamics in the development of the West Antarctic Rift System. *Journal of Geophysical Research*, **107**, 2195, https://doi.org/10.1029/2001JB000515

Shen, W., Wiens, D.A. et al. 2018. Seismic evidence for lithospheric foundering beneath the southern Transantarctic Mountains, Antarctica. *Geology*, **46**, 71–74, https://doi.org/10.1130/G39555.1

Smellie, J.L. 2018. Glaciovolcanism – a 21st century proxy for palaeo-ice. *In*: Menzies, J. and van der Meer, J.J.M. (eds) *Past Glacial Environments (Sediments, Forms and Techniques)*. 2nd edn. Elsevier, Amsterdam, 335–375.

Smellie, J.L. and Edwards, B.E. 2016. *Glaciovolcanism on Earth and Mars: Products, Processes and Palaeoenvironmental Significance*. Cambridge University Press, Cambridge, UK.

Smellie, J.L. and Martin, A.P. 2021. Erebus Volcanic Province: volcanology. *Geological Society, London, Memoirs*, **55**, https://doi.org/10.1144/M55-2018-62

Smellie, J.L. and Rocchi, S. 2021. Northern Victoria Land: volcanology. *Geological Society, London, Memoirs*, **55**, https://doi.org/10.1144/M55-2018-60

Smellie, J.L., Rocchi, S., Gemelli, M., Di Vincenzo, G. and Armienti, P. 2011 A thin predominantly cold-based Late Miocene East Antarctic ice sheet inferred from glaciovolcanic sequences in northern Victoria Land, Antarctica. *Palaeogeography, Palaeoclimatology, Palaeoecology*, **307**, 129–149, https://doi.org/10.1016/j.palaeo.2011.05.008

Smellie, J.L., Rocchi, S. *et al.* 2014. Glaciovolcanic evidence for a polygenetic Neogene East Antarctic Ice Sheet. *Geology*, **42**, 39–41, https://doi.org/10.1130/G34787.1

Stump, E. 1995. *The Ross Orogen of the Transantarctic Mountains*. Cambridge University Press, Cambridge, UK.

Stump, E., Sheridan, M.F., Borg, S.G. and Sutter, J.F. 1980. Early Miocene subglacial basalts, East Antarctic Ice Sheet, and uplift of the Transantarctic Mountains. *Science*, **207**, 757–759, https://doi.org/10.1126/science.207.4432.757

Stump, E., Borg, S.G. and Sheridan, M.F. 1990a. Sheridan Bluff. *American Geophysical Union Antarctic Research Series*, **48**, 136–137.

Stump, E., Borg, S.G. and Sheridan, M.F. 1990b. Mount Early. *American Geophysical Union Antarctic Research Series*, **48**, 138–139.

Treves, S.B. 1967. Volcanic rocks from the Ross Island, Marguerite Bay and Mt. Weaver areas, Antarctica. *Japanese Antarctic Research Expedition Scientific Reports*, Special Issue 1, 136–149.

White, J.D.L. and Houghton, B.F. 2006. Primary volcaniclastic rocks. *Geology*, **34**, 677–680, https://doi.org/10.1130/G22346.1

Zimanowski, B. and Büttner, R. 2003. Phreatomagmatic explosions in subaqueous volcanism. *American Geophysical Union Geophysical Monograph Series*, **140**, 51–60.

Chapter 5.3b

Mount Early and Sheridan Bluff: petrology

Kurt S. Panter[1]*, Jenna Reindel[1] and John L. Smellie[2]

[1]School of Earth, Environment and Society, Bowling Green State University, Bowling Green, OH 43403, USA
[2]Department of Geology, University of Leicester, Leicester LE1 7RH, UK

KSP, 0000-0002-0990-5880; JLS, 0000-0001-5537-1763
*Correspondence: kpanter@bgsu.edu

Abstract: This study discusses the petrological and geochemical features of two monogenetic Miocene volcanoes, Mount Early and Sheridan Bluff, which are the above-ice expressions of Earth's southernmost volcanic field located at c. 87° S on the East Antarctic Craton. Their geochemistry is compared to basalts from the West Antarctic Rift System to test affiliation and resolve mantle sources and cause of melting beneath East Antarctica. Basaltic lavas and dykes are olivine-phyric and comprise alkaline (hawaiite and mugearite) and subalkaline (tholeiite) types. Trace element abundances and ratios (e.g. La/Yb, Nb/Y, Zr/Y) of alkaline compositions resemble basalts from the West Antarctic rift and ocean islands (OIB), while tholeiites are relatively depleted and approach the concentrations levels of enriched mid-ocean ridge basalt (E-MORB). The magmas evolved by fractional crystallization with contamination by crust; however, neither process can adequately explain the contemporaneous eruption of hawaiite and tholeiite at Sheridan Bluff. Our preferred scenario is that primary magmas of each type were produced by different degrees of partial melting from a compositionally similar mantle source. The nearly simultaneous generation of lower degrees of melting to produce alkaline types and higher degrees of melting forming tholeiite was most likely to have been facilitated by the detachment and dehydration of metasomatized mantle lithosphere.

Supplementary material: Mineral chemistry (olivine, clinopyroxene, plagioclase, titanomagnite, ilmenite, Cr-spinel) is available at https://doi.org/10.6084/m9.figshare.c.5231830

Mount Early and Sheridan Bluff are two basaltic monogenetic volcanoes located at 87° S, 15 km apart, in the upper reaches of Scott Glacier in the southern Transantarctic Mountains (Fig. 1). These Early Miocene volcanoes lie c. 1000 km from any other exposed volcano and c. 175 km inland from the shoulder of the West Antarctic Rift System (WARS), which is the focus of most of the Cenozoic volcanism in Antarctica (LeMasurier and Thomson 1990). Furthermore, apart from Gaussberg (Smellie and Collerson 2021) located on the Wilhelm II Coast (Fig. 1) and several inland volcanoes within the Melbourne Volcanic Province of Northern Victoria Land (Fig. 1), Mount Early and Sheridan Bluff represent rare exposures of Cenozoic volcanism on the East Antarctic Craton. Despite their isolation, they have been included within the McMurdo Volcanic Group (Kyle 1990), which encompass alkaline volcanism along the western margin of the Ross Sea portion of the rift (Fig. 1). Mount Early and Sheridan Bluff are part of the newly designated Upper Scott Glacier Volcanic Field (see fig. 4 in Smellie et al. 2021), which constitutes a much larger area of Miocene volcanism inferred to lie beneath the East Antarctic Ice Sheet. This is based on clasts of alkaline basalt recovered from moraines at Mount Howe (Licht et al. 2018), which is c. 40 km south (upstream) of Mount Early, and from Mount Wyatt (E. Stump pers. comm. 2014), which is c. 15 km north of Sheridan Bluff.

Prior to the December 2015 expedition by the authors, Mount Early and Sheridan Bluff were visited twice before, once in 1962–63 (Doumani and Minshew 1965) and again in 1978–79 (Stump et al. 1980, 1990a, b). Preliminary volcanological information and reconnaissance-level petrological and geochronological data were produced by the later expedition, including some additional major element geochemistry presented in Licht et al. (2018). The Early Miocene age for the volcanism of the Upper Scott Glacier Volcanic Field is constrained based on radiometric dating to between c. 20 and 16 Ma (Stump et al. 1980, 1990a, b; Licht et al. 2018). Our current understanding of the volcanology and stratigraphy of the deposits is described by the authors in Chapter 5.3a (Smellie et al. 2021). The results are important to the reconstruction of palaeoenvironmental conditions at the time of eruption and can provide constraints on the surface elevation of the East Antarctic Ice Sheet in the Early Miocene. In this chapter, the authors focus on the petrology of the deposits. The results of our investigation are still in progress, and only preliminary descriptions and initial interpretations based on mineral chemistry and major and trace elements from whole rocks are provided here.

Our findings confirm that in addition to alkaline basalt compositions there is also olivine tholeiite at Sheridan Bluff, which was originally identified by Stump et al. (1980). This is one of the few occurrences of tholeiite of Cenozoic age, other than several small volcanoes located in the western Marie Byrd Land and western Ellsworth Land provinces (Fosdick and Jones mountains, respectively), and post-subduction tholeiite found on the Antarctic Peninsula (Hole 2021), and thus provides a unique perspective on the genesis of intraplate volcanism in Antarctica. Collectively, the composition of lavas from Mount Early and Sheridan Bluff contain important information that may be associated with fundamental differences between magmas generated beneath and erupted through the East Antarctic Craton and magmas generated beneath and erupted through the younger and thinner lithosphere of West Antarctica. The mechanism for melt production that produced Mount Early and Sheridan Bluff, along with other volcanoes inferred for the Upper Scott Glacier Volcanic Field, is unknown. Beneath these volcanoes lithospheric mantle may not exist, a likely consequence of delamination (Shen et al. 2017); and while the basalts are broadly similar to those erupted within the WARS, there are geochemical differences that shed light on lithospheric and sublithospheric sources in both East and West Antarctica.

Field and petrographical details

At Mount Early a total of 13 basalt samples were collected at different stratigraphic levels for petrological analysis (Table 1). Special care was taken to obtain fresh, unaltered

From: Smellie, J. L., Panter, K. S. and Geyer, A. (eds) 2021. *Volcanism in Antarctica: 200 Million Years of Subduction, Rifting and Continental Break-up*. Geological Society, London, Memoirs, **55**, 499–514,
First published online March 22, 2021, https://doi.org/10.1144/M55-2019-2
© 2021 The Author(s). Published by The Geological Society of London. All rights reserved.
For permissions: http://www.geolsoc.org.uk/permissions. Publishing disclaimer: www.geolsoc.org.uk/pub_ethics

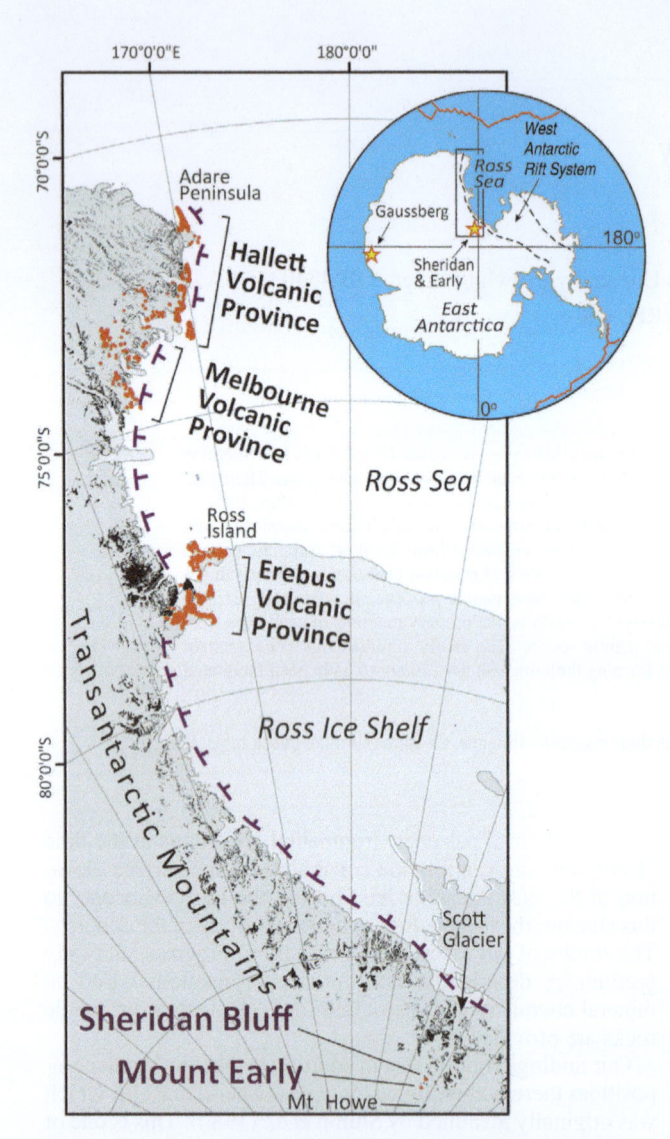

Fig. 1. Location maps. Inset: the locations of Sheridan Bluff, Mount Early and Gaussberg volcanoes in East Antarctica, and the approximate boundary of the West Antarctic Rift System. Detailed map: the western Ross Sea Embayment highlighting the areas of exposed Cenozoic volcanism belonging to the McMurdo Volcanic Group and several of its provinces. Sheridan Bluff and Mount Early lie in the southern Transantarctic Mountains at the head of the Scott Glacier.

interiors of lava flows, lava pillows and dykes. At Sheridan Bluff a total of 10 samples were collected from lavas for petrological study (Table 1). Five separate lava flows were sampled in stratigraphic succession through a *c.* 110 m exposed cliff section on Sheridan Bluff itself. The lowest and stratigraphically oldest lava (SB15-001) lies conformably on top of an orange lapilli tuff, which, in turn, overlies a glacially striated granitoid bedrock surface. Other samples were collected from lava flows exposed on a higher-elevation surface to the SW of the bluff.

The majority of samples are porphyritic to glomeroporphytic (3–15% phenocrysts by volume), with olivine and plagioclase being the most abundant phases with lesser amounts of clinopyroxene (Table 1). Opaque grains are optically identified as magnetite if cubic and ilmenite if tabular (Fig. 2a). Overall, the phenocrysts appear to have been in equilibrium when crystallized (Fig. 2b), although a few phenocrysts (antecrysts?) display resorption textures (Fig. 2c).

Several samples at Sheridan Bluff have subophitic textures with plagioclase laths embedded within or penetrating augite (Fig. 2d). Groundmass varies from holocrystalline to hypohyaline and tachylite (Fig. 2e). Microlites in groundmass consist of opaque oxides, olivine, plagioclase and clinopyroxene, and display pilotaxitic to trachytic textures (Fig. 2f). Samples range from non-vesicular to up to 30% vesicles by volume (Table 1).

Methodology

A total of 8 basalt samples were analysed for mineral chemistry using a Cameca SX-100 electron microprobe at the Robert B. Mitchell Electron Microbeam Analysis Laboratory at the University of Michigan. Core and rim analysis were performed on phenocrysts of olivine, clinopyroxene and plagioclase. Analyses were taken in the middle of grains (cores) and within *c.* 20 μm of their edge (rims). Up to four phenocrysts of each mineral per sample were measured for chemistry. Only cores of opaque oxides were analysed due to their small diameters.

A total of 23 samples were analysed for whole-rock major and trace element concentrations. Each rock was cut, crushed and sieved to <2 cm in diameter pieces, and carefully picked using a binocular microscope to remove fragments that appeared weathered and/or altered. Fused beads for X-ray fluorescence (XRF) analysis were prepared following the method of Johnson *et al.* (1999) by grinding the crushed rock, weighing with di-lithium tetraborate flux with a 2:1 ratio of flux:sample, and fusing at 1000°C in a muffle oven and cooling. The bead produced was reground, refused and polished on a diamond lap for a smooth analysis surface. Analysis was conducted on the ThermoARL Advant'XP + sequential XRF spectrometer at the Peter Hooper GeoAnalytical Laboratory at Washington State University. Analysis of trace elements by inductively coupled plasma mass spectrometry (ICP-MS) utilized the fusion-dissolution method to help to remove unwanted matrix elements. Di-lithium-tetraborate was used as a flux, and the dissolution was by open-vial mixed acid digestion using reagents HNO_3 69–70%, HF 48–52%, $HClO_4$ 67–71% and H_2O_2. The resultant solutions were analysed on an Agilent 7700 ICP-MS at the Peter Hooper GeoAnalytical Laboratory. For this study, 10 major elements and six trace elements (Cr, Ni, Ga, Cu, V and Zn) were measured by XRF and all other trace elements were measured by ICP-MS. For XRF analyses, major elements with concentrations above 5 wt% have standard deviations that are less than 0.07 wt% and XRF trace element concentrations over 10 ppm have standard deviations of ≤5 ppm. For ICP-MS analyses, trace element concentrations over 10 ppm have standard deviations between 0.01 and 3.5 ppm.

Results

Mineral chemistry

Four lavas from Sheridan Bluff and four lavas from Mount Early were analysed for mineral chemistry. The samples were chosen to best represent the compositional spectrum of whole rocks. Whole-rock compositions consist of tholeiite, hawaiite and mugearite (Fig. 3).

Olivine phenocrysts in basalt from Mount Early and Sheridan Bluff are forsteritic (Table 2; Fig. 4). At Sheridan Bluff, olivine phenocyrsts in hawaiite have a lower range in forsterite content (Fo_{58-82}) relative to olivine in tholeiite (Fo_{80-85}); and at Mount Early, olivine phenocryts in mugearite have lower forsterite contents (Fo_{71-78}) relative to hawaiite (Fo_{77-84}). Outermost rims on the majority of olivine phenocyrsts are enriched in Fe relative to cores. Olivine in hawaiite from

Table 1. *Summary of field and petrographical observations, Mount Early and Sheridan Bluff, Antarctica*

Sample	Location	Sample/lithofacies (sequence*)	Rock type	Minerals[†]	Volume %	Texture	Vesicle %
ME15-001	Early	Lava pillow interior	Hawaiite	Ol > Pl	11	Hyh, Glm	<1
ME15-003	Early	Dyke	Mugearite	Pl > Ol > Cpx	15	Hyc	–
ME15-004	Early	Lava pillow interior	Hawaiite	Ol > Pl	5	Hyh	4
ME15-005b	Early	Columnar jointed dyke	Hawaiite	Ol > Pl	6	Hyh, Tac	25
ME15-006	Early	Lava block in lapilli tuff	Hawaiite	Ol > Pl > Cpx	10	Hyc, Tr	15 (large)
ME15-007	Early	Lava block in lapilli tuff	Hawaiite	Ol > Pl	8	Hyc, Tac	<1
ME15-008	Early	Lava pillow interior	Hawaiite	Ol > Pl	8	Hyh, Tac, Glm	<1
ME15-009	Early	Dyke	Mugearite	Pl > Ol > CPx	12	Hc, Glm	<1
ME15-010	Early	Dyke	Mugearite	Pl > Ol > Cpx	12	Hyh, Tac	30
ME15-011	Early	Scoria in lapilli tuff	Mugearite	Pl > Ol > Cpx	10	Tac	30
ME15-012	Early	Dyke	Mugearite	Pl > Ol > Cpx	12	Hc, Glm	10
ME15-013	Early	Lava clast in breccia	Mugearite	Pl > Ol	10	Hc	–
ME15-014	Early	Lava pillow interior	Hawaiite	Ol > Pl	5	Hyh	<1
SB15-001	Sheridan	Lava flow interior (1 – base)	Tholeiite	Ol > Pl	10	Hc	3
SB15-002	Sheridan	Lava flow interior (1 – base)	Tholeiite	Ol > Pl	6	Hyh, Tac	7
SB15-003	Sheridan	Lava flow interior (5 – 'top')	Hawaiite	Pl > Ol	10	Hc	–
SB15-004	Sheridan	Lava flow interior (2 – lower)	Tholeiite	Pl > Cpx > Ol	–	Hc, Grn, Sboph	5
SB15-005	Sheridan	Lava flow interior (3 – middle)	Tholeiite	Ol > Pl > Cpx	–	Hc, Grn	15
SB15-006	Sheridan	Lava flow interior (4 – upper)	Hawaiite	Ol > Pl	6	Hc	8
SB15-007	Sheridan	Lava (float)	Hawaiite	Pl > Ol	15	Hyh	10 (large)
SB15-008	Sheridan	Lava flow interior	Hawaiite	Pl > Ol	10	Hc	20 (large)
SB15-009	Sheridan	Lava flow interior	Tholeiite	Ol > Pl	3	Hc	5 (large)
SB15-010	Sheridan	Lava flow interior	Hawaiite	Ol > Pl	8	Hc, Tr	5 (large)

*Sequence, order of stratigraphic sequence for five lava flows exposed in the cliff section of Sheridan Bluff. Ol, olivine; Pl, plagioclase; Cpx, clinopyroxene.
[†]Abundance of phenocrysts or larger crystals in inequigranular samples. Textures: Glm, glomeroporphyritic; Grn, equigranular; Hc, holocrystalline; Hyc, hypocrystalline; Hyh, hypohyaline; Sboph, subophitic; Tac, tachylite; Tr, trachytic.
Volume % of phenocrysts, larger crystals and vesicles are visual estimates. Large vesicle sizes range from 1 to 7 mm in diameter.

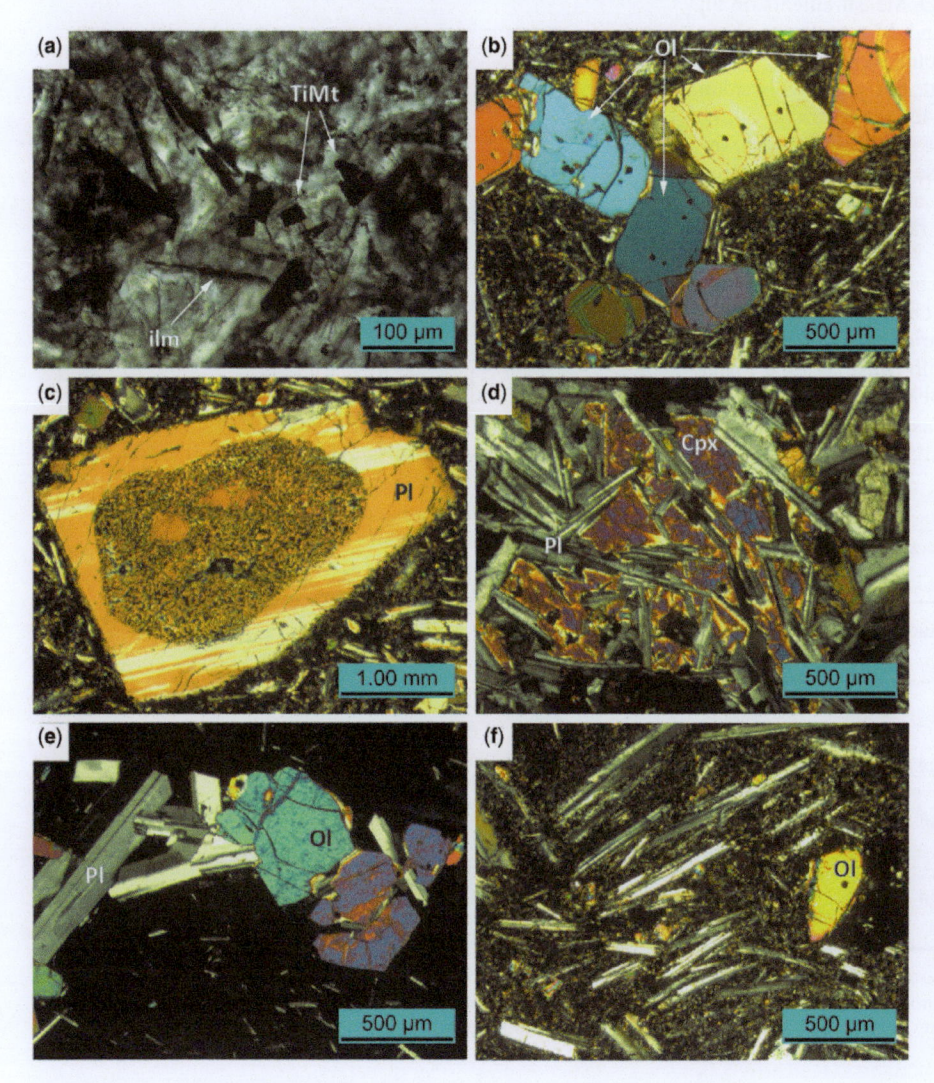

Fig. 2. Representative photomicrographs of Sheridan Bluff and Mount Early lavas and dykes. (**a**) Groundmass crystals of titanomagnetite (Ti-Mag) and ilmenite (Ilm) under plane-polarized light in sample ME15-003. (**b**) A cluster of euhedral to subhedral olivine (Ol) phenocrysts under cross-polarized light (XPL) in sample SB15-010. (**c**) A plagioclase (Pl) antecryst? (cf. Davidson *et al.* 2007) showing a sieved core and resorbed rim under XPL in sample SB15-003. The high first-order birefringence exhibited by plagioclase is due to the sample thickness (*c.* 50 μm). (**d**) Subophitic texture with plagioclase embedded within or penetrating clinopyroxene (Cpx) in sample SB15-004 (XPL). (**e**) A cluster of olivine and plagioclase phenocrysts (glomeroporphyritic texture) within tachylite from sample ME15-005b (XPL). (**f**) trachytic texture in sample SB15-010 (XPL).

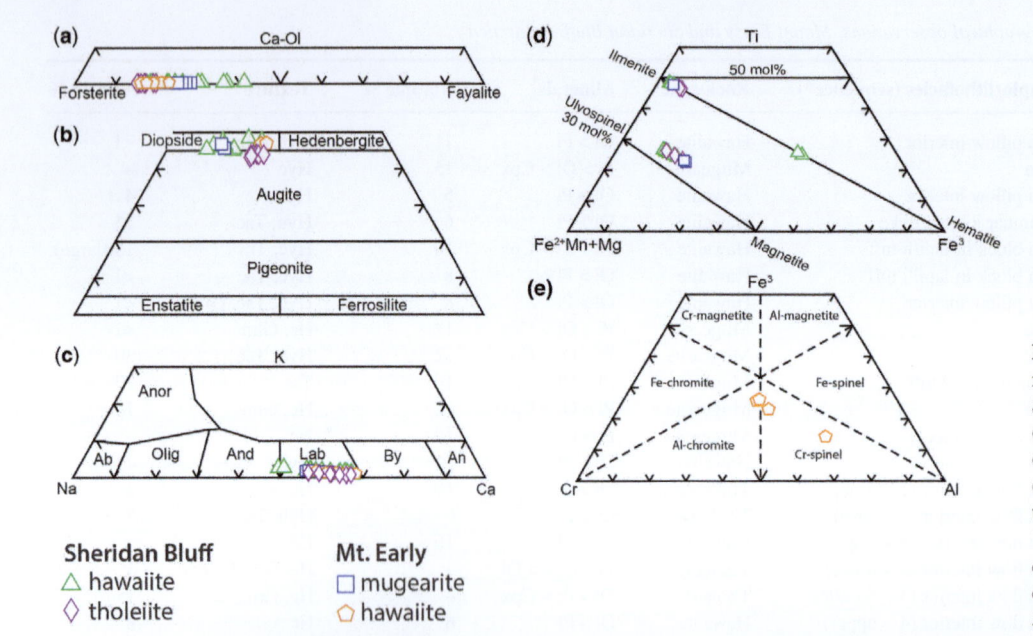

Fig. 3. Chemical classification of minerals in Mount Early and Sheridan Bluff basalts: (**a**) olivine, (**b**) clinopyroxene, (**c**) plagioclase, (**d**) ilmenite and titanomagnetite and (**e**) chromite and Cr-spinel (based on the classification plot of Stevens 1944). Rock names are based on the classification of whole-rock compositions shown in Figures 4 and 5.

Sheridan Bluff show the greatest variability, with rims having fayalite contents as much as 18–23 mol% greater than their cores.

Pyroxene phenocrysts in basalts are calcic (Wo >43 mol%), and range from diopside in hawaiite and mugearite to augite in tholeiite (Table 2; Fig. 4b). The outermost rims on diopside phenocyrsts in hawaiite at Sheridan Bluff are iron enriched relative to their cores by up to 2 wt% FeO. Measurements on all other pyroxenes at Sheridan Bluff and Mount Early show minor differences between cores and outermost rims.

Plagioclase phenocrysts in Mount Early and Sheridan basalts are labradorite (Table 2; Fig. 4c). At Sheridan Bluff, plagioclase in tholeiite overlaps but has higher average anorthite contents (An_{56-67}) than plagioclase measured in hawaiite (An_{48-67}). Plagioclase compositions are similar at Mount Early, with phenocrysts in hawaiite ranging from An_{55} to An_{62} and in mugearite from An_{58} to An_{68}. Outermost rims of plagioclase have a lower anorthite content than their corresponding cores in all basalts. Plagioclase in hawaiite from Sheridan Bluff show the greatest variation in composition between outermost rims and cores, with rims having a lower anorthite content by as much as 16 mol%. The large compositional difference between cores and rims in these samples is in accordance with core–rim measurements of coexisting olivine and diopside as described above.

Spinel group minerals in five samples were analysed and are classified as titaniferous magnetite, ilmenite and Cr-spinel (Table 2; Fig. 4d, e). Titaniferous magnetite and ilmenite coexist in the tholeiite, hawaiite and mugearite. Four grains of Cr-rich spinel were measured in one hawaiite sample (ME15-004) from Mount Early.

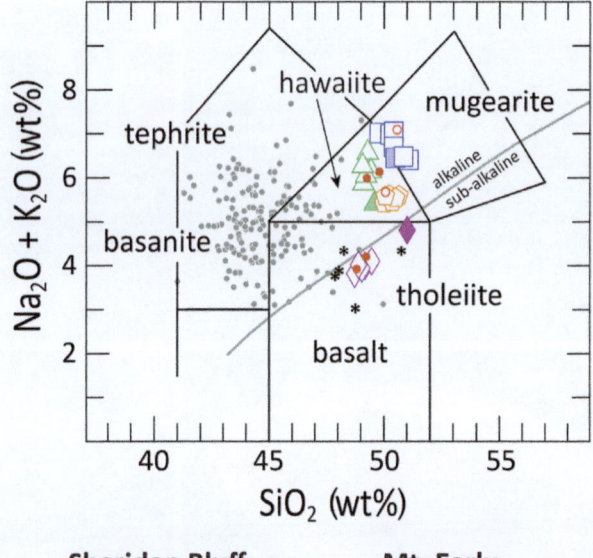

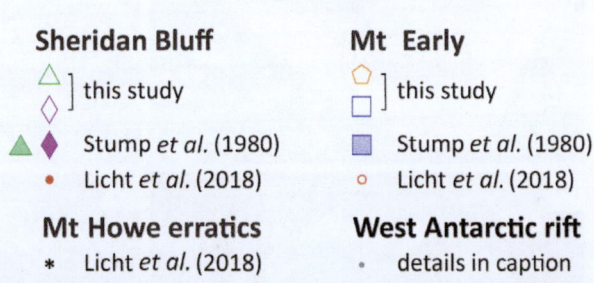

Fig. 4. Total alkali v. SiO_2 (wt%) classification diagram (Le Bas et al. 1986) of basalts from the Upper Scott Glacier Volcanic Field. Also shown are basalts from the West Antarctic Rift System ($n = 171$, MgO ³6 wt%), which includes the Erebus Volcanic Province (see the Table 3 footnote for the references), the Hallett Volcanic Province (Fig. 1) (Nardini et al. 2009; Panter et al. 2018) and the Marie Byrd Land Province (Hart et al. 1997; Panter et al. 1997, 2000). All data are normalized to 100% volatile-free and total Fe.

Table 2. *Summary of composition of major rock-forming minerals at Sheridan Bluff and Mount Early*

Rock type	Location	Olivine	Clinopyroxene	Plagioclase	Spinel group
Tholeiite	Sheridan	Fo_{80-85}	Augite, Mg# 57–63	An_{56-67}	Ti-Mag + Ilm
Hawaiite	Sheridan	Fo_{58-82}	Diopside, Mg# 59–83	An_{48-67}	Ti-Mag + Ilm
Hawaiite	Early	Fo_{77-84}	Diopside, Mg# 57–59	An_{55-62}	Ti-Mag + Ilm + Cr-Spl
Mugearite	Early	Fo_{71-78}	Diopside, Mg# 75–77	An_{58-68}	Ti-Mag + Ilm

Compositional ranges include analyses from both cores and rims. Mg# = 100 × MgO/ MgO + FeO^T. Ti-Mag, titanomagnetite; Ilm, ilmenite; Cr-Spl, chrome-rich spinel. A complete set of mineral analyses is provided in the Supplementary material.

Table 3. *Chemical analyses of whole-rock samples from Sheridan Bluff and Mount Early, Antarctica*

Sample	ME15-001	ME15-003	ME15-004	ME15-005b	ME15-006	ME15-007	ME15-008	ME15-009	ME15-010	ME15-011	ME15-012	ME15-013	ME15-014	SB15-001	SB15-002	SB15-003	SB15-004	SB15-005	SB15-006	SB15-007	SB15-008	SB15-009	SB15-010	*EVP Average (n = 111)
Location	Early	Early	Early	Early	Early	Early	Early	Early	Early	Early	Early	Early	Early	Sheridan	Sheridan	Sheridan	Sheridan	Sheridan	Sheridan	Sheridan	Sheridan	Sheridan	Sheridan	
Rock type	Hawaiite	Mugearite	Hawaiite	Hawaiite	Hawaiite	Hawaiite	Mugearite	Mugearite	Mugearite	Mugearite	Mugearite	Early Hawaiite	Hawaiite	Tholeiite	Tholeiite	Tholeiite	Tholeiite	Hawaiite	Hawaiite	Hawaiite	Hawaiite	Tholeiite	Hawaiite	

Major elements (wt%)

SiO$_2$	48.75	49.39	49.76	49.83	49.39	50.07	49.92	49.77	49.66	49.26	49.61	49.18	49.49	48.58	48.68	48.55	49.37	48.60	48.68	48.64	48.82	48.60	48.63	44.22
TiO$_2$	1.95	2.26	2.00	1.96	1.97	2.00	2.00	2.25	2.10	2.24	2.19	2.21	1.99	1.46	1.53	2.25	1.85	1.54	2.07	2.13	2.07	1.54	2.13	3.41
Al$_2$O$_3$	15.92	17.48	16.10	16.17	16.01	16.10	16.27	17.39	17.17	17.59	17.53	17.06	16.03	15.72	15.95	17.13	15.99	17.01	16.85	17.38	17.17	16.15	17.26	14.27
FeO^T	9.38	9.38	9.51	9.18	9.35	9.32	9.22	8.99	8.73	8.62	8.92	8.63	9.43	10.67	10.47	9.10	10.93	10.00	9.27	8.84	8.99	10.63	9.05	11.52
MnO	0.15	0.14	0.15	0.14	0.14	0.14	0.15	0.14	0.13	0.14	0.14	0.13	0.15	0.17	0.17	0.16	0.18	0.16	0.15	0.15	0.15	0.17	0.15	0.20
MgO	7.01	4.86	6.65	6.50	6.59	6.40	6.46	4.71	4.20	4.27	4.64	4.41	6.95	10.19	8.78	5.09	7.49	7.86	6.49	5.21	6.01	8.72	5.69	9.50
CaO	8.54	7.94	8.80	8.83	8.74	8.77	8.82	8.08	8.38	7.92	8.03	8.06	8.79	8.87	9.22	8.97	9.66	9.78	9.14	9.28	9.32	9.33	9.25	11.06
Na$_2$O	3.61	4.69	3.80	3.75	3.59	3.75	3.70	4.64	4.29	4.65	4.40	4.59	3.77	3.08	3.19	4.44	3.33	3.28	4.15	4.26	4.22	3.25	4.40	3.57
K$_2$O	1.67	2.27	1.75	1.71	1.75	1.78	1.77	2.27	1.94	2.23	2.19	1.68	1.74	0.72	0.76	2.02	0.76	0.62	1.66	1.78	1.77	0.74	1.80	1.44
P$_2$O$_5$	0.53	0.79	0.54	0.53	0.54	0.53	0.54	0.77	0.60	0.78	0.75	0.77	0.54	0.24	0.26	0.78	0.27	0.24	0.65	0.72	0.67	0.26	0.73	0.80
Total	97.50	99.20	99.06	98.60	98.06	98.86	98.85	99.03	97.22	97.71	98.38	96.72	98.87	99.69	99.00	98.47	99.82	99.08	99.11	98.38	99.21	99.39	99.08	100.00
LOI	1.87	0.32	0.30	0.60	0.95	0.57	0.62	0.24	2.26	1.67	1.06	2.33	0.12	0.00	0.84	0.00	0.00	0.42	0.47	0.82	0.00	0.16	0.13	
Mg#	57.1	48.0	55.5	55.8	55.7	55.1	55.5	48.3	46.2	46.9	48.1	47.7	56.8	63.0	59.9	49.9	55.0	58.4	55.5	51.2	54.4	59.4	52.9	59.50

CIPW normative minerals (wt%)

Ne	0.2	5.3	0.8	0.1	1.1	0.4	–	4.5	1.6	4.2	2.8	1.9	1.0	–	–	5.7	–	–	4.2	4.3	4.8	–	5.7	11.2
Hy	–	–	–	–	–	–	0.6	–	–	–	–	–	–	4.8	4.8	–	5.9	3.2	–	–	–	2.8	–	–

Trace element compositions (ppm)

Sc	23.2	17.2	23.8	23.2	23.8	23.5	23.5	18.4	18.3	17.0	17.5	17.9	23.4	27.1	28.1	24.9	33.7	30.1	26.0	25.7	26.8	29.5	26.4	29.0
V	186	188	194	189	193	194	196	192	189	189	184	187	193	179	185	197	218	192	193	198	194	186	193	254
Cr	169	40	161	176	170	181	158	51	54	35	43	42	178	326	291	68	237	230	146	71	112	276	99	356
Ni	100	40	89	93	97	96	85	39	35	32	36	40	100	193	150	29	98	87	60	31	43	147	39	155
Cu	44	34	43	44	44	44	43	34	28	34	31	33	45	56	55	40	57	50	46	37	44	54	42	55
Zn	86	87	87	90	91	86	87	92	93	88	87	92	87	88	87	79	92	80	77	77	80	95	74	91
Ga	20	22	21	19	19	19	19	22	21	22	22	20	20	18	19	20	18	18	19	20	19	19	19	18
Rb	21.9	39.0	31.4	31.6	31.6	24.5	32.0	39.1	33.4	38.9	34.4	13.1	31.3	15.1	21.6	28.6	14.2	9.9	23.1	25.1	24.4	12.9	24.7	34.4
Sr	774	987	727	728	730	762	723	966	842	966	1009	998	712	359	359	839	358	360	782	878	869	380	890	915
Y	24.89	28.11	25.72	25.35	25.43	25.52	25.46	28.19	26.53	28.46	28.09	28.13	25.06	23.76	24.70	32.56	28.12	25.17	28.47	30.91	30.01	27.47	29.91	28.97
Zr	198	263	203	196	200	198	202	258	217	261	255	262	202	124	127	258	146	130	251	266	259	130	267	292
Nb	38.50	65.79	38.85	37.65	37.70	38.64	38.67	62.72	43.23	64.95	62.15	64.35	38.75	13.33	13.74	57.57	15.69	14.05	47.03	53.59	51.72	14.42	53.53	71.87
Cs	0.55	1.07	0.82	0.89	0.92	0.16	0.81	1.20	0.88	1.12	0.75	1.14	0.81	0.37	0.79	0.75	0.26	0.21	0.39	0.64	0.51	0.23	0.70	0.38
Ba	326	484	342	334	345	346	341	472	389	476	476	481	328	135	138	337	141	110	270	319	305	144	313	474
La	33.29	49.81	34.06	33.43	33.68	33.78	34.07	48.52	37.85	49.55	48.52	49.09	33.97	14.84	15.32	47.80	16.29	14.29	38.42	44.65	43.31	16.65	44.28	54.10
Ce	66.75	97.56	68.87	67.50	68.07	68.76	68.60	94.90	75.79	96.67	94.72	96.29	68.14	31.67	32.72	90.36	35.37	30.16	73.44	83.99	80.88	34.06	84.04	106.08
Pr	7.91	10.94	8.17	7.96	8.06	8.09	8.05	10.80	8.92	10.98	10.78	10.82	8.06	4.05	4.15	10.25	4.53	3.80	8.46	9.59	9.28	4.67	9.50	12.19
Nd	30.70	40.83	31.69	31.13	31.60	31.16	31.46	40.00	34.29	40.31	39.83	40.30	31.08	16.85	17.32	38.53	19.18	16.26	32.41	35.97	34.82	19.65	35.71	49.08
Sm	6.43	7.90	6.76	6.55	6.54	6.66	6.55	7.71	6.97	7.88	7.78	7.73	6.67	4.16	4.24	7.77	4.78	4.19	6.70	7.36	7.22	4.91	7.19	9.67
Eu	1.97	2.43	1.99	2.06	2.04	2.02	2.03	2.37	2.14	2.40	2.36	2.35	2.01	1.46	1.50	2.44	1.63	1.52	2.23	2.33	2.32	1.60	2.35	3.04
Gd	5.88	6.86	5.99	5.94	5.97	5.96	6.06	6.79	6.27	6.82	6.75	6.71	5.94	4.42	4.59	7.28	5.18	4.61	6.38	6.79	6.62	5.13	6.77	8.70
Tb	0.90	1.05	0.94	0.93	0.93	0.94	0.94	1.05	0.97	1.08	1.05	1.06	0.94	0.78	0.79	1.15	0.91	0.80	1.03	1.09	1.07	0.87	1.08	1.18
Dy	5.14	5.95	5.38	5.33	5.36	5.35	5.43	5.82	5.49	5.91	5.80	5.94	5.25	4.68	4.84	6.71	5.49	4.97	5.92	6.26	6.10	5.41	6.17	6.26
Ho	1.00	1.12	1.03	1.02	1.01	1.02	1.03	1.10	1.07	1.12	1.10	1.11	1.00	0.96	0.99	1.30	1.13	1.01	1.14	1.24	1.20	1.09	1.19	1.13
Er	2.52	2.86	2.62	2.57	2.56	2.59	2.60	2.83	2.72	2.86	2.80	2.81	2.53	2.58	2.64	3.34	3.01	2.80	2.89	3.14	3.07	2.95	3.10	2.82
Tm	0.35	0.40	0.35	0.35	0.35	0.37	0.35	0.38	0.37	0.39	0.38	0.38	0.35	0.36	0.37	0.46	0.42	0.40	0.41	0.44	0.42	0.42	0.42	0.37
Yb	2.07	2.32	2.09	2.11	2.11	2.12	2.11	2.33	2.22	2.37	2.36	2.35	2.10	2.24	2.35	2.81	2.70	2.42	2.50	2.68	2.60	2.55	2.60	2.18
Lu	0.31	0.36	0.33	0.32	0.31	0.32	0.32	0.35	0.34	0.38	0.35	0.35	0.31	0.34	0.37	0.42	0.42	0.38	0.38	0.40	0.38	0.39	0.39	0.34
Hf	4.55	5.53	4.66	4.54	4.54	4.59	4.64	5.44	4.90	5.49	5.43	5.61	4.54	2.88	3.05	5.47	3.49	3.06	5.11	5.47	5.30	5.43	5.43	6.48
Ta	2.15	3.57	2.20	2.14	2.18	2.17	2.23	3.34	2.38	3.45	3.37	3.43	2.17	0.82	0.82	3.13	0.95	0.86	2.65	2.94	2.87	3.10	2.98	5.04
Pb	5.21	6.32	5.46	5.10	5.48	3.78	5.49	6.80	6.61	6.83	6.84	6.45	5.34	2.66	2.72	5.00	2.66	2.08	3.88	4.59	4.27	0.88	3.94	2.49
Th	3.78	5.17	3.91	3.87	3.89	3.91	3.89	5.09	4.23	5.32	5.25	5.20	3.84	1.65	1.71	5.13	1.90	1.53	4.13	4.85	4.61	1.77	4.74	6.38
U	1.06	1.47	1.08	1.13	1.06	1.12	1.12	1.51	1.23	1.43	1.33	1.52	1.07	0.46	0.89	0.96	0.52	0.46	1.27	1.43	1.37	0.48	1.47	1.68

All major elements were determined by XRF. Trace elements V, Cr, Ni, Cu, Zn and Ga were determined by XRF. All other trace elements were determined by ICP-MS. Total iron as FeO = FeO^T. LOI, loss on ignition. Mg# = 100 × MgO/MgO + FeO^T. Ne, nepheline; Hy, hypersthene.

*EVP is the average of 111 mafic basalts (MgO >7 wt%) from the Erebus Volcanic Province (Fig. 1). Compositions from White Island (Cooper *et al.* 2007) and Minna Bluff (Scanlan 2008; Redner 2016), which are both within the Mount Discovery Volcanic Field; Terror Rift Volcanic Field (Aviado *et al.* 2015); and foothills of the Royal Society Range (Wingrove 2005), which is within the Southern Local Suite (Smellie and Martin 2021).

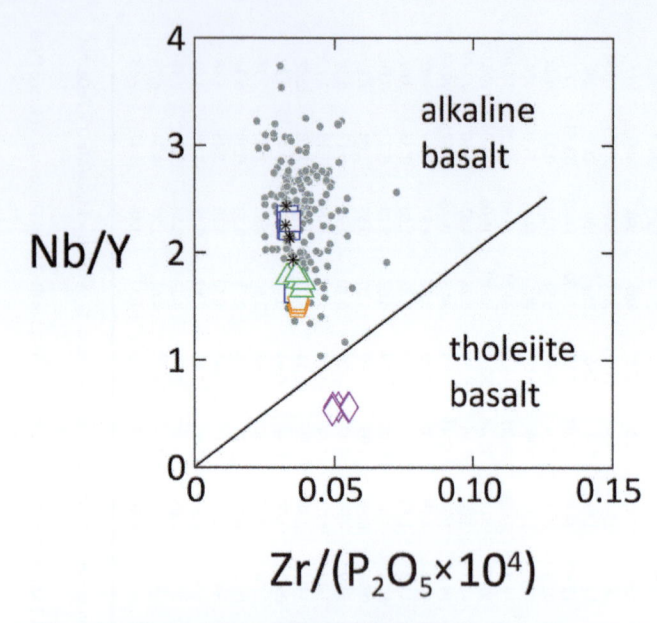

Fig. 5. Discriminant diagram between alkaline and tholeiite basalts (Floyd and Winchester 1975). Symbols and data sources are given in Figure 4.

Whole-rock geochemistry: major and trace elements

Basalts from Mount Early and Sheridan Bluff range in composition from alkaline (up to 5.7 wt% nepheline-normative) to subalkaline (up to 5.9 wt% hypersthene-normative) types (Table 3), and are classified as hawaiite, mugearite and tholeiite (Fig. 4). Previously analysed samples from Mount Early and Sheridan Bluff (Stump *et al.* 1980; Licht *et al.* 2018) are also plotted in Figure 4 along with basaltic clasts collected from the Mount Howe moraine that lies approximately 40 km south of Mount Early (Licht *et al.* 2018). Altogether, the basalts represent all known samples from the Upper Scott Glacier Volcanic Field. Additionally, representative basalts from the WARS are included for comparison. The classification of basalts from Sheridan Bluff as tholeiite is based on their subalkaline nature, overall higher MgO and FeO^T contents and lower trace element concentrations relative to alkaline types. The classification scheme of Floyd and Winchester (1975) is used to further distinguish tholeiite from alkaline types, as well as to highlight their unique geochemistry relative to Cenozoic basalts found within the WARS (Fig. 5).

Basalts from Mount Early and Sheridan Bluff have a broad range in MgO (4.3–10.2 wt%), with a very restricted range in SiO$_2$ (48.7–51.1 wt%: Fig. 4). Overall, the basalts show decreasing concentrations of FeO^T and CaO, and increasing concentrations of TiO$_2$, Al$_2$O$_3$, Na$_2$O, K$_2$O and P$_2$O$_5$, with decreasing MgO content (Fig. 6). The trends are broadly consistent with magmatic evolution by fractional crystallization. Samples from Mount Early and Sheridan Bluff analysed previously (Stump *et al.* 1980; Licht *et al.* 2018) fall within the compositional range as basalts collected in this study. Basalt clasts from the Mount Howe moraine (Licht *et al.* 2018) fall mostly within the same compositional range but have notably lower Na$_2$O concentrations and several samples have elevated CaO concentrations at similar MgO contents. Basalts from the Upper Scott Glacier Volcanic Field overall have lower TiO$_2$ and FeO^T, and higher Al$_2$O$_3$, contents relative to basalt from the WARS. In addition, tholeiite from Sheridan Bluff has lower K$_2$O concentrations relative to all other samples from the Upper Scott Glacier Volcanic Field and the lowest P$_2$O$_5$ concentrations overall. It is interesting to note that the compositional spectrum of tholeiite samples on MgO v. K$_2$O and

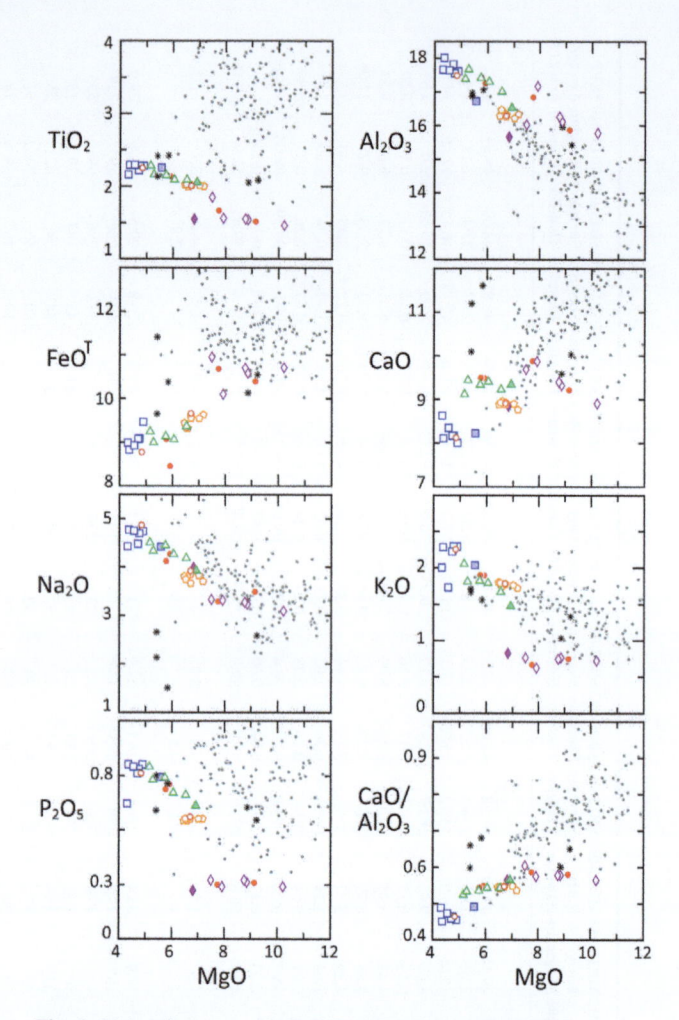

Fig. 6. Major elements v. MgO all in wt% and normalized 100% volatile-free and total iron as FeO = FeO^T. Symbols and data sources are given in Figure 4.

P$_2$O$_5$ plots is relatively flat (Fig. 6), suggesting that the potassium and phosphorus contents were buffered during magma evolution.

Basalts from Mount Early and Sheridan Bluff display an overall decrease in concentration of compatible elements (e.g. Cr, Sc and Ni) and increase in all incompatible elements with decreasing MgO content (Fig. 7). However, for tholeiite, concentrations of incompatible elements (e.g. light REEs (LREEs), Ba, Nb and Sr) remain relatively constant (cf. K$_2$O and P$_2$O$_5$: Fig. 6) over their range of MgO contents (10.2–7.5 wt%). The incompatible trace element concentrations of basalt from Sheridan Bluff and Mount Early fall at or below the concentration levels of basalt from the WARS over the equivalent range of MgO wt% (Fig. 7). However, moderately incompatible elements Sc and Y, heavy REEs (HREEs) Yb and Lu, and compatible elements Ni and Cr have similar concentration levels (Fig. 7). Tholeiites from Sheridan Bluff have the lowest concentrations of LREEs, as well as Ba, Nb, Ta, Sr and Zr contents, overall (Fig. 7). On the Nb/Y v. Th/Yb discriminant diagram of Pearce (2008), the tholeiitic and alkali basalt data are distinctive, with the alkali-rich samples broadly similar to data from the WARS and OIB (Fig. 8b). The tholeiites also display a much lower Zr/Y (*c.* 5) relative to alkaline compositions (*c.* 10: Fig. 8a). The trace element ratios of hawaiite and mugearite from Sheridan Bluff and Mount Early are more similar to basalts within the WARS and oceanic islands (OIB) and for tholeiite are closer to compositions of enriched mid-ocean ridge basalt (E-MORB) (Fig. 8b).

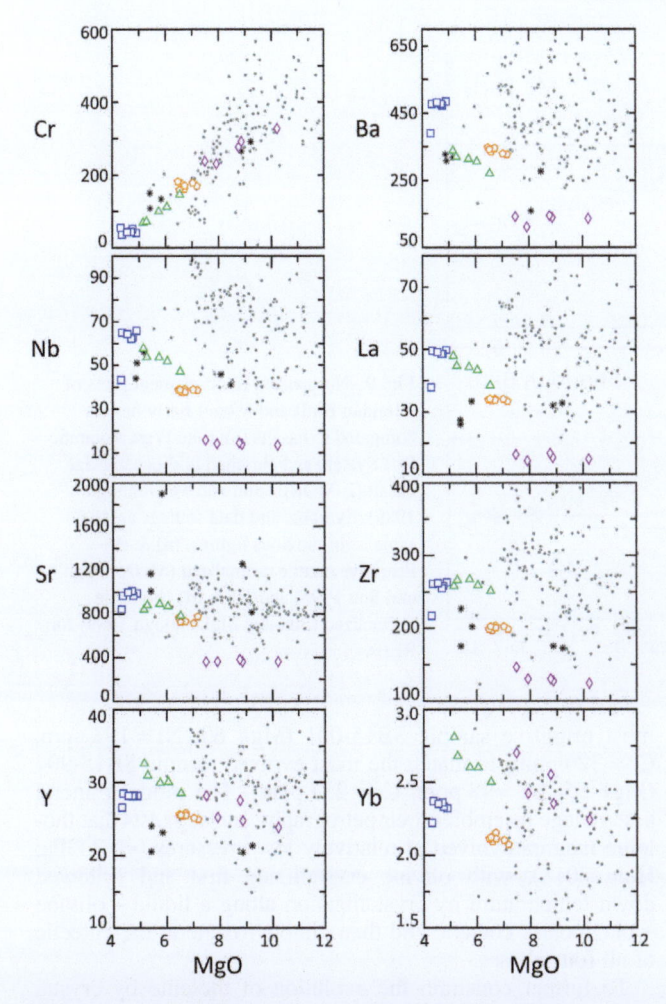

Fig. 7. Selected trace elements (ppm) v. MgO (wt%). Symbols and data sources are given in Figure 4.

All basalt samples from Sheridan Bluff and Mount Early are enriched in incompatible elements relative to primitive mantle (McDonough and Sun 1995), with hawaiite and mugearite samples showing concentration patterns that are very similar to basalt from the WARS (Fig. 9a, b). Tholeiite displays relatively flat patterns with low concentrations relative to all other basalts except for Y, Yb and Lu (Fig. 9b). A common feature on multi-element plots for the majority of basalts from the WARS is that they display prominent negative K- and Pb-anomalies (Panter et al. 2018, 2021; Rocchi and Smellie 2021). In contrast, alkaline basalt from Mount Early and Sheridan Bluff show only minor negative anomalies for Pb, and negative anomalies are mostly absent for K. Tholeiite compositions display neither anomaly (Fig. 9b). Chondrite-normalized REE patterns are smooth (i.e. do not show Eu anomalies) and are LREE-enriched (Fig. 9c, d), with alkaline basalts having La/Yb$_n$ ratios >10 and tholeiites having La/Yb$_n$ ratios c. 4.5. Tholeiite and alkaline (hawaiite) compositions plotted on multi-element diagrams diverge with increasing element incompatibility (i.e. having similar concentrations of HREEs and Y but disparate concentrations for LREEs and Rb, Ba, Th, U and Nb) (Fig. 9b, d). Similar patterns are well documented in other suites of continental intraplate basalts and are often explained as being the result of variable degrees of mantle partial melting from a common source (Jung and Masberg 1998; Lustrino et al. 2002; Boyce et al. 2015). What is particularly notable is that at Sheridan Bluff the occurrence of tholeiite and hawaiite is restricted to a c. 110 m-thick stratigraphic section consisting of between seven and nine lava flows separated by red scoriaceous breccia, with four tholeiite lava flows found in the lower portion of the sequence.

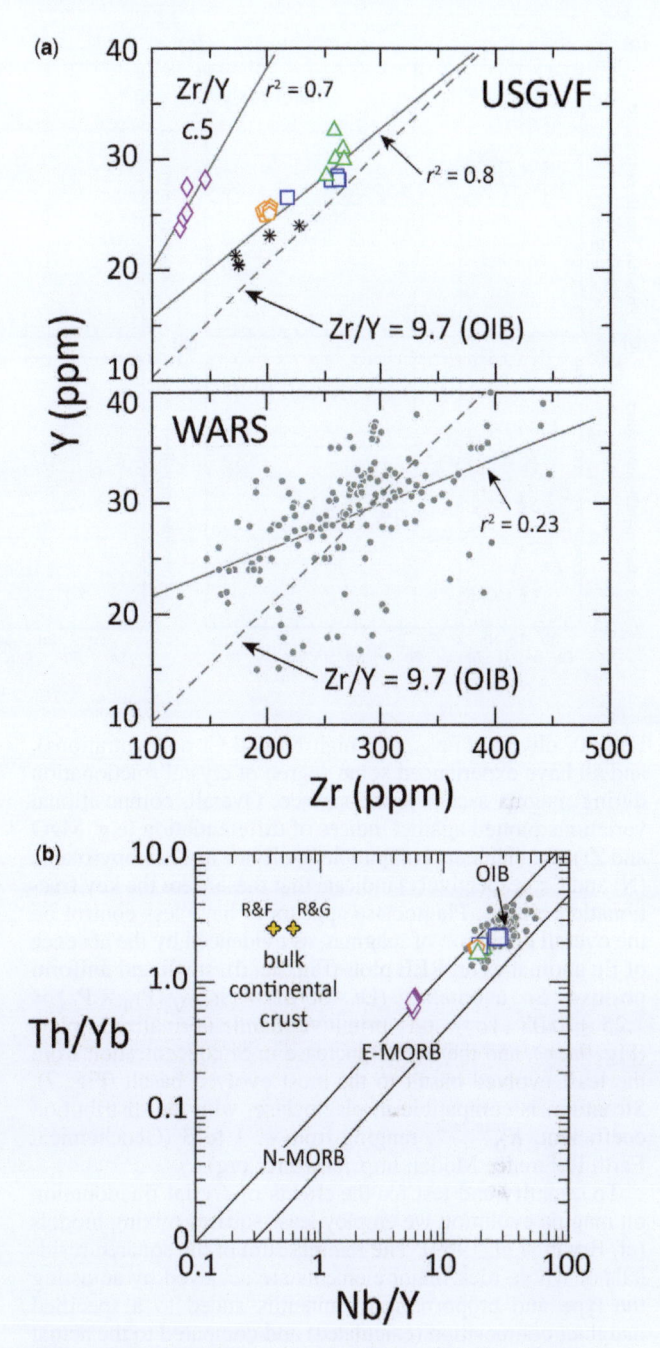

Fig. 8. Selected trace element variation diagrams. (**a**) Zr v. Y plots showing basalts from the Upper Scott Glacier Volcanic Field (USGVF: Sheridan Bluff, Mount Early and Mount Howe erratics) and basalts from the West Antarctic Rift System (WARS). Best-fit (r^2) lines of regression of the data for tholeiite (Sheridan Bluff), alkaline basalts (Sheridan Bluff, Mount Early and Mount Howe) and West Antarctic rift basalts are shown. In addition, the Zr/Y ratios for ocean island basalts (OIB) from Sun and McDonough (1989) are shown for comparison. (**b**) Nb/Y v. Th/Yb (Pearce 2008). Bulk continental crust compositions are from Rudnick and Fountain (R&F) (1995) and Rudnick and Gao (R&G) (2003). See Figure 4 for other symbols and data sources.

Discussion

Fractional crystallization

Basalts from Mount Early and Sheridan Bluff do not represent primary compositions (i.e. whole-rock Mg# 100 × Mg/Mg +

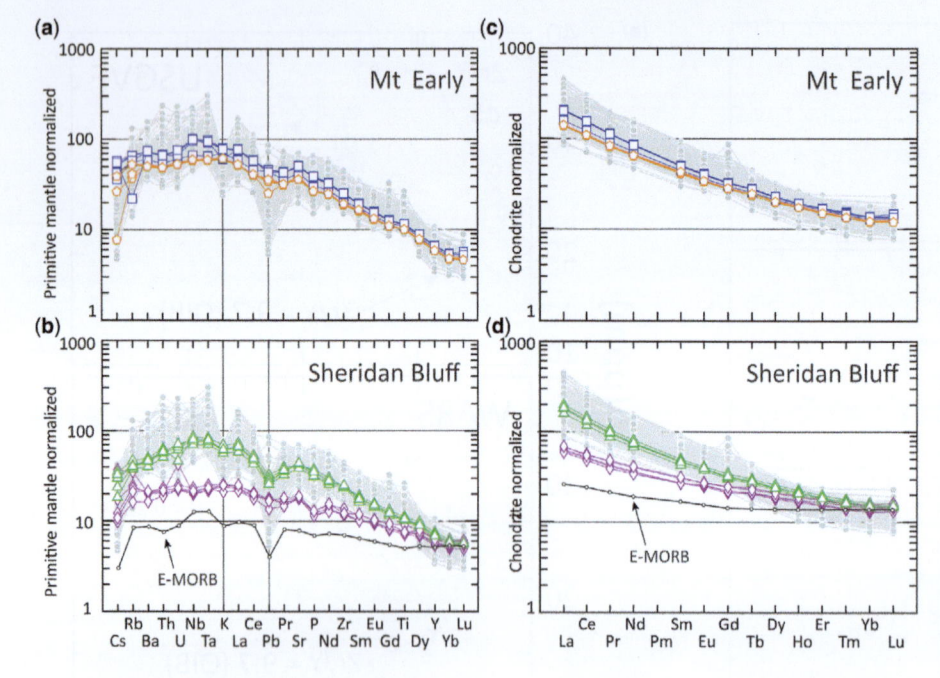

Fig. 9. Normalized multi-element plots of Sheridan Bluff and Mount Early basalts compared to basalts from the West Antarctic Rift System and enriched mid-ocean ridge basalt (E-MORB: Sun and McDonough 1989). Symbols and data sources are the same as in previous figures. (**a**) & (**b**) Primitive mantle normalized (McDonough and Sun 1995) and (**c**) & (**d**) chondrite normalized (Sun and McDonough 1989) for REEs.

Fe >70, olivine = Fo$_{90}$, and high Ni and Cr concentrations), and all have experienced some degree of crystal fractionation during magma ascent to the surface. Overall, compositional variations plotted against indices of differentiation (e.g. MgO and Zr) and elements compatible in olivine and clinopyroxene (Ni and Cr, respectively) indicate that these were the key fractionating phases. Plagioclase appears to have less control on the overall evolution of magmas, as evidenced by the absence of Eu anomalies on REE plots (Fig. 9c, d), small and uniform positive Sr anomalies (i.e. Sr/Sr* = Sr$_n$/$\sqrt{(Pr_n \times P_n)}$ = 1.25 ± 0.07 (1σ)) on primitive-mantle-normalized plots (Fig. 9a, b), and the overall increase in Sr concentration from the least evolved basalt to the most evolved basalt (Fig. 7). Strontium is compatible in plagioclase, with the distribution coefficient, $K_D^{plag/melt}$, ranging from c. 1 to 3 (Geochemical Earth Reference Model: http://earthref.org).

To quantify and test for the effects of crystal fractionation on magma evolution we employ least-squares mixing models (cf. Bryan et al. 1969). The minima sum of the squared residuals of whole-rock major elements are achieved by adjusting the type and proportions of minerals added to a specified daughter composition (calculated) and compared to the actual (measured) values of a specified parent composition. The solutions are used along with mineral partition coefficients (D'Orazio et al. 1998) to determine bulk distribution coefficients (D_i) for trace elements, which allow concentrations of the specified parent to be calculated by the Rayleigh equation and compared to the actual values. First we attempt to model the full compositional spectrum found at each volcano; tholeiite (SB15-001, Mg# 63) to hawaiite (SB15-006, Mg# 50) at Sheridan Bluff, and hawaiite (ME15-001, Mg# 57) to mugearite (ME15-010, Mg# 46) at Mount Early, using a fractionation assemblage of olivine, clinopyroxene, plagioclase and titanomagnetite. The best solution for Mount Early offers a reasonable fit for major elements (Σr^2 <0.1) but a relatively poor match for trace elements ($\Delta_{obs-calc}$ >5%). Results for Sheridan Bluff are poor overall (Σr^2 >1 and $\Delta_{obs-calc}$ >20%). However, when models are applied to more restricted compositional ranges at each volcano, as delimited by basalt type, they offer better approximations of the data. For tholeiite, which show the greatest range in compositions, major elements (Σr^2 <0.1) and the majority of trace elements ($\Delta_{obs-calc}$ <5%) are predicted (Table 4) by removal of 8.6% olivine, 6.4% plagioclase and 0.2% titanomagnetite from the most primitive sample SB15-001 (Mg# 63, Ni = 193 ppm, Cr = 326 ppm) to match the most evolved sample SB15-004 (Mg# 55, Ni = 98 ppm, Cr = 237 ppm). The model mineral assemblage resembles their petrography and suggests that tholeiite magma evolved at relatively low pressures (<0.7 GPa: Hole 2018), with olivine crystallizing first and followed down temperature by crystallization along a liquid + olivine + plagioclase cotectic and then clinopyroxene at the eutectic of all four phases.

To further constrain the evolution of tholeiite by crystal fractionation we use thermodynamically-based model provided by the MELTS algorithm (Ghiorso and Sack 1995). The model results illustrate that the variations in major elements for tholeiite, including the relatively invariant concentrations of K$_2$O and P$_2$O$_5$ with changing MgO content, are well matched by low-pressure (≤0.5 GPa) fractionation of olivine + spinel followed by plagioclase and then by clinopyroxene at temperatures at or below 1160°C from composition SB15-001 (Fig. 10). Two of the models assume a moderate water content (0.5 and 1 wt%). At lower water contents (≤0.25 wt%), the fractionation of clinopyroxene follows plagioclase at much higher temperatures (^{3}1180°C). At high water contents (2 wt%), clinopyroxene does not appear in the fractionation assemblage above 1100°C and plagioclase is absent. The models affirm that alkaline compositions (i.e. hawaiite and mugearite) cannot be derived from tholeiite by fractional crystallization.

Crustal contamination

As mentioned previously, one of the unique geochemical features that distinguish basalts at Sheridan Bluff and Mount Early from basalts associated with the WARS is the absence or subdued negative anomalies for K and Pb on mantle-normalized multi-element plots (Fig. 9a, b). For West Antarctic basalts these anomalies are considered to be inherited from their mantle sources (Hart et al. 1997; Panter et al. 2000, 2018; Rocchi et al. 2002; Nardini et al. 2009; Martin et al. 2013; Aviado et al. 2015; Phillips et al. 2018). Consequently, it must be evaluated whether the anomalies are a result of melting of a different sources or an effect of chemical modification during magma ascent to the surface. Caesium concentrations in mugearite and some hawaiite lavas from Mount Early are

Table 4. *Least squares–Rayleigh distillation models for fractional crystallization of tholeiite to evolved tholeiite*

SB15-001	Measured	Calculated	% difference	Bulk D_i
Major elements (wt%)				
SiO_2	48.73	48.70	0.1	
TiO_2	1.46	1.62	−9.9	
AlO_3	15.77	15.56	1.3	
FeO^T	10.70	10.65	0.5	
MnO	0.17	0.17	0.0	
MgO	10.22	10.25	−0.3	
CaO	8.90	9.09	−2.1	
Na_2O	3.09	3.07	0.7	
K_2O	0.72	0.65	10.8	
P_2O_5	0.24	0.23	4.3	
Σr^2		0.080		
F		0.848		
Trace elements (ppm)				
Sc	27.1	29.9	−9.4	0.27
V	179	190	−5.8	0.16
Cr	326	335	−2.7	3.11
Ni	193	264	−26.9	7.03
Rb	15.1	12.1	24.8	0.04
Sr	359	344	4.4	0.76
Y	23.8	23.9	−0.4	0.01
Zr	124	124	0.0	0.01
Nb	13.33	13.30	0.2	0.02
Cs	0.37	0.22	68.2	0.06
Ba	135	125	8.0	0.24
La	14.84	14.00	6.0	0.06
Ce	31.67	30.20	4.9	0.04
Pr	4.05	3.87	4.7	0.03
Nd	16.85	16.40	2.7	0.03
Sm	4.16	4.07	2.2	0.03
Eu	1.46	1.42	2.8	0.15
Gd	4.42	4.39	0.7	0.00
Tb	0.78	0.77	1.3	0.02
Dy	4.68	4.67	0.2	0.01
Ho	0.96	0.96	0.0	0.01
Er	2.58	2.56	0.8	0.02
Yb	2.24	2.29	−2.2	0.01
Lu	0.34	0.36	−5.6	0.01
Hf	2.88	2.97	−3.0	0.01
Ta	0.82	0.81	1.2	0.02
Th	1.65	1.61	2.5	0.01
U	0.46	0.44	4.5	0.01

Tholeiite (SB15-001) → evolved tholeiite (SB15-004).
SB15-001 = 0.848 × SB15-004 + 0.086 olivine (Fo_{84}) + 0.064 plagioclase (An_{66}) + 0.002 titanomagnetite.
Calculated by least squares after Arth (1976). Mineral chemistry from tholeiite sample SB15-001. Total Fe as FeO^T. Bulk distribution (D_i) is calculated based on phenocryst/matrix partition coefficients from D'Orazio et al. (1998).

elevated (Fig. 9a, b) and may indicate contamination by crust. Additionally, basalts from Mount Early and Sheridan Bluff have low Ce/Pb ratios (15 ± 2, 1σ) relative to basalts from the WARS (>20) and oceanic basalt (25 ± 5: Hofmann et al. 1986), which may also suggest contamination by continental crust (Ce/Pb ≤6: Taylor and McLennan 1985; Rudnick and Fountain 1995; Rudnick and Gao 2003).

In the absence of isotopic data, the possibility of contributions from crust to basaltic magmas is assessed using simple mixing models for major and trace elements. We assume a relatively unfractionated parent, basanite (EVP in Table 3), which represents a calculated average of 111 mafic lavas (c. Mg# 60, Cr = 356 ppm, Ni = 155 ppm) from the Erebus Volcanic Province (Fig. 1). For a contaminant, we select the estimated composition for upper continental crust by Rudnick and Gao (2003). Mixing of 20% upper continental crust with basanite provides a close approximation for mugearite (Fig. 11a), and is able to explain the higher Cs contents and lower Ce/Pb ratios, as well as the reduced negative K-anomalies relative to basalts from the West Antarctica rift (Fig. 9a). Conversely, contamination of tholeiite by crust cannot produce the coexisting hawaiite compositions at Sheridan Bluff (Fig. 11b). Although the range of tholeiite compositions appear to be well matched by models for crystal fractionation (Fig. 10), the primary magma also may have been contaminated by crust. Subtracting 8–10% upper continental crust from the least evolved tholeiite (SB15-001) produces compositions that approach E-MORB (Fig. 11b) (Sun and McDonough 1989), and have Ce/Pb and Nb/U ratios (20–26 and 51–64, respectively) that fall within the range of oceanic basalts (Hofmann et al. 1986; Sims and DePaolo 1997).

Mantle partial melting

We have demonstrated that neither fractional crystallization nor crustal assimilation can explain the intimate association of tholeiite and alkaline compositions at Sheridan Bluff. As

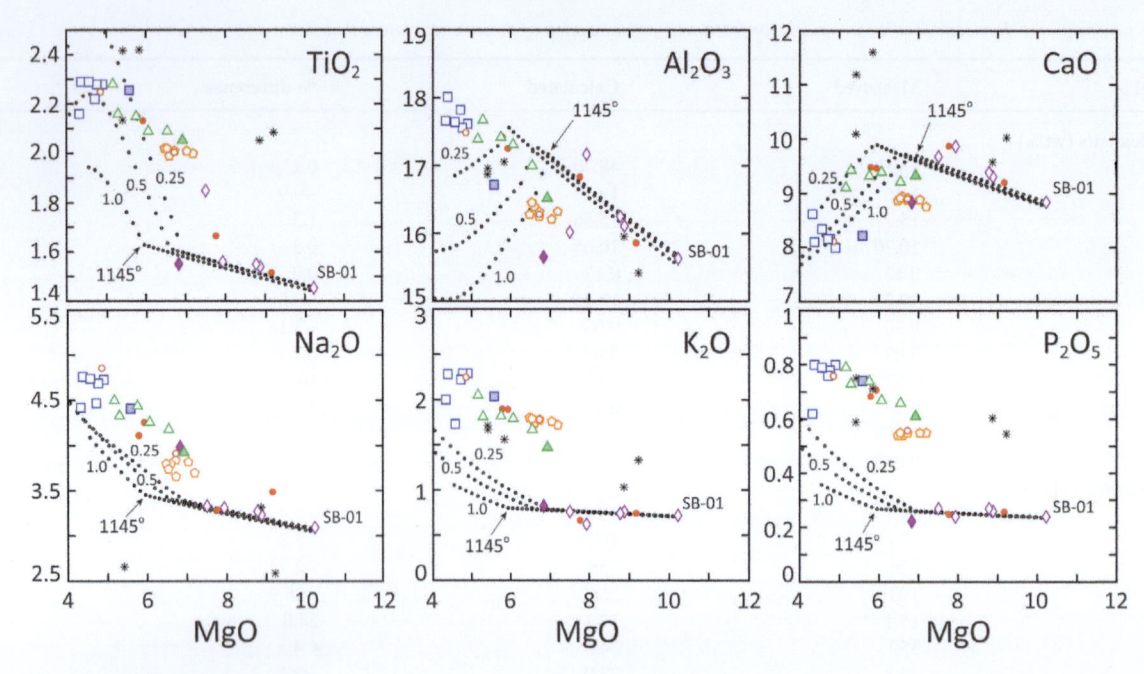

Fig. 10. Model curves for fractional crystallization calculated using the MELTS algorithm (Ghiorso and Sack 1995) are projected onto MgO (wt%) v. TiO_2, Al_2O_3, CaO, Na_2O, K_2O and P_2O_5 (wt%) plots with basalts from the Upper Scott Glacier Volcanic Field. All three models use tholeiite sample SB15-001 (the least fractionated tholeiite) as a starting composition with variable water contents (0.25, 0.5 and 1.0 wt%). Each of the three runs are calculated over a temperature interval of 1325° (temperature above the liquidus) to 1100°C (using steps of 5°C), a pressure interval of 50–500 MPa (using steps of 25 MPa) and a fo_2 buffer = FMQ. The initial conditions were set based on thermobarometric estimates for clinopyroxene–whole rock, olivine–whole rock and magnetite–ilmenite pairs (Reindel 2018). The temperature 1145°C on the 1.0 wt% H_2O fractionation curve marks the step before the inflection caused by the fractionation of plagioclase (at 1140°C) and then clinopyroxene (at 1135°C). Please refer to Figure 4 for symbols and data sources.

stated earlier, changes in the degree of mantle partial melting is often used to explain the coexistence of tholeiite and alkaline magmas (e.g. Jung and Masberg 1998; Wanless *et al.* 2006; Boyce *et al.* 2015; Kocaarslan and Yalçın Ersoy 2018). The hawaiite and tholeiite lavas at Sheridan Bluff have similar Y and HREE contents with diverging, sub-parallel concentrations of highly incompatible elements on normalized multi-element plots (Fig. 9b, d). The geochemical patterns suggest a relationship controlled by variable degrees of partial melting of a compositionally similar source. Yttrium and HREE are compatible in garnet, and their concentrations are buffered during melting if garnet remains in the residue. Thus, low degree melting of a garnet-bearing source yields basalt with high La/Yb ratios. These conditions are commonly invoked for basalt generation within the WARS (Hart *et al.* 1997; Orlando *et al.* 2000; Panter *et al.* 2000, 2018; Nardini *et al.* 2009; Aviado *et al.* 2015). Another mineral that figures prominently in the origin of basalt from the West Antarctic rift is amphibole (±phlogopite). Amphibole is considered to be a metasomatic phase that exists within the mantle lithosphere (Rocchi *et al.* 2002; Nardini *et al.* 2009; Perinelli *et al.* 2011; Martin *et al.* 2013; Aviado *et al.* 2015; Panter *et al.* 2018), and manifests in mafic and ultramafic mantle xenoliths (wehrlite, clinopyroxenites and hornblendites) entrained within basalts (Gamble and Kyle 1987; Gamble *et al.* 1988; Coltorti *et al.* 2004; Perinelli *et al.* 2006, 2011, 2017). Potassium occurs in stoichiometric proportions in pargasite, kaersutite and phlogopite. If these phases are not completely consumed during melting, then K will be retained in the source relative to other incompatible elements (e.g. Ta and La) and thus produce negative K-anomalies on mantle-normalized multi-element plots. We have shown that the assimilation of K-rich upper crustal materials by alkaline magmas may account for the near absence of K-anomalies in some samples (Figs 9a & 11a). However, alternative explanations for the lack of negative anomalies include the possibility that K-bearing phases were totally consumed during melting or that melting occurred at pressures and temperatures outside the stability range of amphibole (>1150°C, >3.7 GPa: Médard *et al.* 2006; Mandler and Grove 2016). Accordingly, for our models we select mineral modes for mantle sources with and without amphibole, and vary their residual proportions.

We have calculated partial melts using a non-modal batch-melting model (Shaw 1970) with source compositions of primitive mantle (PM: McDonough and Sun 1995) and depleted MORB mantle (DMM: Workman and Hart 2005). Starting mineral modes for lherzolite source include olivine (>50%), orthopyroxene and clinopyroxene ± garnet, spinel and amphibole. Garnet lherzolite is paired with PM composition to better represent a deeper source (i.e. asthenosphere or the base of thick continental lithosphere) and spinel lherzolite with DMM composition to better characterize a shallower source (i.e. lithospheric mantle). On REE ratio plots (Fig. 12a, b), model curves for garnet lherzolite and spinel lherzolite without amphibole fail to account for any basalt composition over the modelled range of partial melting (0.5–30%). In contrast, partial melting curves for an amphibole-bearing garnet lherzolite source traverse the compositional field for basalt from the WARS at low melt fractions (<4%), and with small changes to the starting mineral mode and amount of amphibole in the residuum (Fig. 12a, b). Gadolinium is more compatible than Yb in amphibole during mantle melting (McKenzie and O'Nions 1991; Bottazzi *et al.* 1999; Adam and Green 2006); thus, Gd/Yb ratios decrease when higher proportions of amphibole remain in the source. The lower Gd/Yb ratios of Sheridan Bluff and Mount Early basalts can be matched if the proportion of amphibole is increased significantly in both the starting mode (>10%) and residue (>5%) of a PM source. However, considering the evolved compositions for hawaiite and mugearite lavas (<7 wt% MgO, Cr <200 ppm and Ni <100 ppm) and the possible addition of crust, a petrogenesis defined solely by mantle partial melting is untenable.

We therefore combine partial melting models with models for assimilation–fractional crystallization (AFC: DePaolo

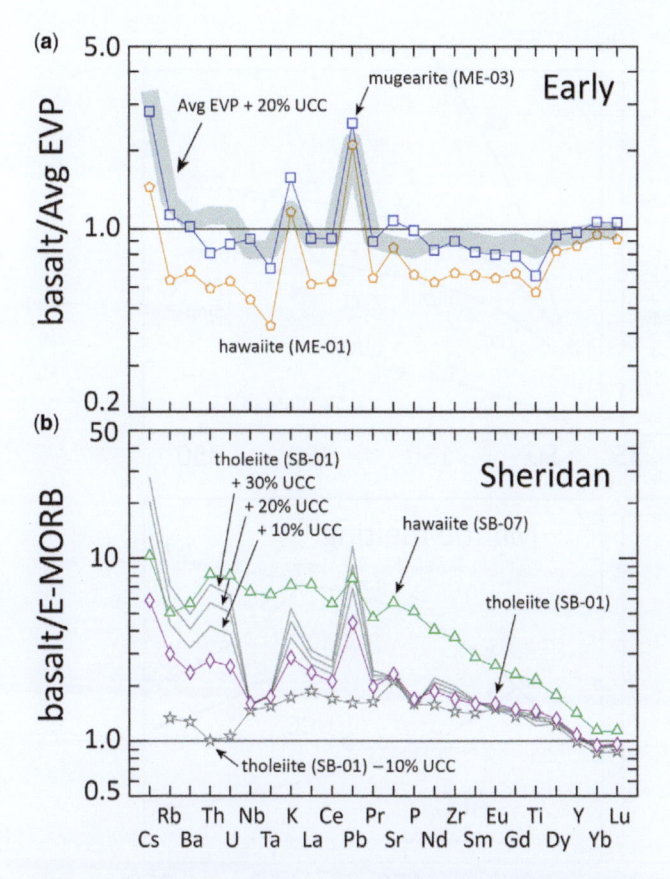

Fig. 11. Normalized multi-element plots of selected Mount Early and Sheridan Bluff basalts, along with calculated simple mixtures of basalt with upper continental crust (UCC) (Rudnick and Gao 2003). (**a**) Mount Early samples (ME-03, ME15-003; ME-01, ME15-001) are normalized to an average of 111 mafic basalts (MgO >7 wt%, c. Cr = 356 ppm, Ni = 155 ppm) from the Erebus Volcanic Province (EVP) (Fig. 1; Table 3). A mixture of average EVP (80%) and 20% UCC provides a close match to Mount Early mugearite. (**b**) Sheridan Bluff samples (SB-01, SB15-001; SB-07, SB15-007) are normalized to values of E-MORB provided by Sun and McDonough (1989). Mixing of tholeiite with 10, 20 and 30% UCC does not successfully emulate the trace element chemistry of coexisting hawaiite lava. Subtracting 10% UCC from tholeiite, however, approaches E-MORB composition.

1981) in Figure 12c & d. Starting from compositions that represent melt fractions of <4 and 20% of the amphibole-bearing garnet lherzolite (PM source), we contaminate with upper continental crust (Rudnick and Gao 2003). The REE ratios for Mount Early and Sheridan Bluff basalts can be predicted with between 80 and 95% liquid remaining by holding the mass assimilation to mass crystallization rate (M_a/M_c) equal to 0.6.

In summary, although our geochemical models for partial melting and AFC processes do not provide unique solutions for basalt petrogenesis, they do allow several fundamental hypotheses to be submitted. First, basalts erupted from Mount Early and Sheridan Bluff were generated by melting of an enriched, hydrous mantle source that contains both amphibole and garnet. Second, tholeiite is produced by higher degrees of mantle partial melting of this source relative to alkaline compositions. Third, magmas were likely to have been contaminated by crustal lithologies as they passed through the thick crust of East Antarctica before erupting at the surface.

Mechanisms for melting beneath East Antarctica

Lithospheric detachment (also known as delamination, foundering or drip) has been proposed as a mechanism for melt generation and eruption of the Upper Scott Glacier Volcanic Field (Shen *et al.* 2017; Licht *et al.* 2018). Evidence to support this mechanism is provided by shear-wave velocity (V_S) models constructed by Heeszel *et al.* (2016) and Shen *et al.* (2017) that reveal a low-velocity zone beneath the southern Transantarctic Mountains at a depth of between 50 and 80 km. The slowest V_S is centred beneath the Mount Early and Sheridan Bluff volcanoes, and is underlain at c. 200 km depth by relatively fast seismic velocities that Shen *et al.* (2017) has interpreted to be foundered lithosphere. The process of detachment and sinking of lithosphere can produce melt under continents by allowing asthenosphere to rise adiabatically and by conductive heating of portions (drips) of lithosphere that become engulfed by asthenosphere (Kay and Kay 1993; Elkins-Tanton 2005, 2007; Furman *et al.* 2016). The mantle that melts, along with the depth and extent of melting, depends on the solidi of the material. Density instability of the lower subcontinental lithosphere can be caused by injections of mafic silicate melts and fluids (Elkins-Tanton 2007; Furman *et al.* 2016). Upon sinking, metasomatized lithosphere will release volatiles into the surrounding mantle to promote wet melting of asthenosphere, in addition to melting of the volatile-rich portions of the lithosphere itself upon heating (Elkins-Tanton 2007; Furman *et al.* 2016). In contrast, mechanisms that have been proposed for volcanism within the WARS have called upon adiabatic partial melting related to *active* mantle plumes (Kyle *et al.* 1992; LeMasurier and Landis 1996; Phillips *et al.* 2018; Martin *et al.* 2021) and *passive* conductive heating of metasomatised lithosphere caused by extension and edge-driven mantle flow (Rocchi *et al.* 2002; Nardini *et al.* 2009; Martin *et al.* 2013, 2021; Kipf *et al.* 2014; Aviado *et al.* 2015; Lee *et al.* 2015; Panter *et al.* 2018). Volcanism triggered by lithospheric detachment, therefore, is uniquely applied to the Upper Scott Glacier Volcanic Field, and is well suited to explain its provenance on the East Antarctic Craton and isolation from other Cenozoic volcanoes. Alternatively, edge-driven convective flow in response to step changes in lithospheric thickness at the cratonic edge may have played a role (cf. Gamble *et al.* 2018). Shen *et al.* (2017) suggested that metasomatism of the continental lithosphere by tectonic events in the Paleozoic (Ross Orogeny: Borg and Stump 1987; Stump 1995) and late Jurassic (rifting and flood magmatism: Elliot and Fleming 2004) may have left a thick and heavy lithospheric lid that foundered when extensional stresses migrated inboard from the developing WARS in the Late Cretaceous–Paleogene. Alternatively, Granot and Dyment (2018) proposed that the motion between East and West Antarctica during the Neogene produced oblique convergence and flexural bending in the southern Transantarctic Mountains to cause the instability. For either case, based on the geochemistry of Mount Early and Sheridan Bluff basalts, we conclude that metasomatic additions to the thick East Antarctic lithosphere played an important role, not only in triggering the detachment but also in melting beneath this region.

Association of alkaline and tholeiite magmas

We have established that mantle partial melting at different degrees is responsible for the occurrence of both tholeiite and alkaline basalt (hawaiite) at Sheridan Bluff. However, why do they occur together at this monogenetic volcano? Based on stratigraphy, tholeiite lavas were erupted before hawaiite and the entire c. 110 m lava sequence was likely to have been deposited over a short time interval, perhaps only a few months or years. One possibility is that the two magma types were stored in the upper crust and, by virtue of contrasting volatile content and density, avoided mixing upon eruption (Moore *et al.* 1995). However considering their average

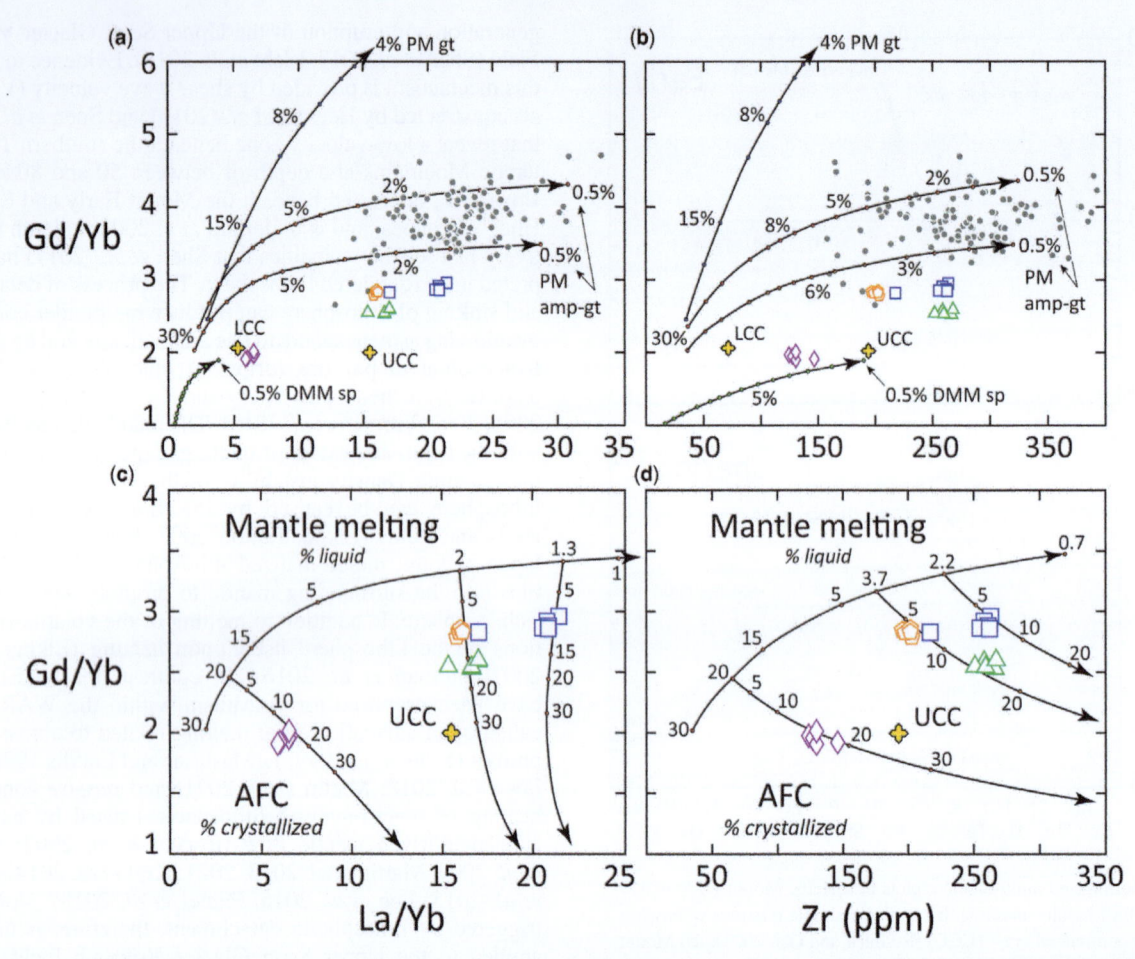

Fig. 12. Non-modal batch melting (Shaw 1970) and assimilation–fractional crystallization (AFC) (DePaolo 1981) models for basalts from Mount Early and Sheridan Bluff. (**a**) & (**b**) Mineral and melt modes for the garnet (gt) and spinel (sp) lherzolite compositions are ol $_{55(0.05)}$ + opx $_{22(0.05)}$ + cpx $_{15(0.81)}$ + gt $_{8(0.09)}$ (Ersoy et al. 2010) and ol $_{53(-0.06)}$ + opx $_{27(0.28)}$ + cpx $_{17(0.67)}$ + sp $_{3(0.11)}$ (Kinzler 1997). Mineral and melt modes for the amphibole (amp)–garnet lherzolite are Ol $_{55(0.05)}$ + opx $_{22(0.05)}$ + cpx $_{15(0.4)}$ + gt $_{3(0.05)}$ + amp $_{5(0.45)}$ (Ersoy et al. 2010), which leaves 2.5% residual amphibole (upper curve), and Ol $_{55(0.05)}$ + opx $_{22(0.05)}$ + cpx $_{15(0.4)}$ + gt $_{4(0.05)}$ + amp $_{4(0.45)}$, which leaves 3.1% amphibole in the residue (lower curve). REE concentrations for primitive mantle (PM) are from McDonough and Sun (1995), and for depleted MORB mantle (DMM) are from Workman and Hart (2005). Partition coefficients used are from Ersoy et al. (2010 and references therein). Numbers on the curves are % liquid generated. Upper and lower continental crust (UCC and LCC, respectively) are from Rudnick and Gao (2003). (**c**) & (**d**) AFC models are calculated starting from different degree partial melt batches obtained from the amphibole–garnet lherzolite melting model (lower curve in a & b). Each melt batch assimilates upper continental crust (UCC) (Rudnick and Gao 2003) with bulk partition coefficients set at $D_{La} = 0.01$, $D_{Yb} = 0.1$, $D_{Gd} = 0.5$ and $D_{Zr} = 0.001$, and rate of assimilation to crystallization (M_a/M_c) set at 0.6 for all models. Numbers on AFC curves are % crystallized.

densities, hawaiite at 2603 kg m^{-3} and tholeiite at 2633 kg m^{-3} (calculated from whole-rock compositions: cf. Bottinga and Weill 1970), and the likelihood that hawaiite magma would have been more volatile-rich (based on incompatible element concentrations: Fig. 9b), it would be expected that from a stratified system the hawaiite would have erupted first.

If lithospheric delamination is the driving mechanism for volcanism in this region, then the two scenarios that seem most likely to explain the coexistence of alkaline and tholeiite basalts at Sheridan Bluff are; (1) that delamination provided a high-relief subplate topography that facilitated simultaneous melting over a range of depths; or (2) that different degrees of melting occurred at approximately the same depth but from a modally heterogeneous source. In schematic models for delamination (i.e. drip magmatism), Elkins-Tanton (2007) and Furman et al. (2016) showed that the devolatilization that causes melting in the asthenosphere would occur above melting within the sinking lithosphere. This could produce simultaneous melt generation at different depths and thus permit a staggered delivery of these different magma types to the surface. In the second scenario, higher degrees of melting may be initiated by a mantle that contains a high proportion of fusible materials (e.g. hornblendite, pyroxenite) and then followed by lower degrees of melting as those materials become exhausted. Again, this provides a possible circumstance that could facilitate a staggered eruption of the different magma types. A thorough assessment of the conditions for mantle melting and the generation of alkaline and tholeiite types at Mount Early and Sheridan Bluff will require constraints on primary melt composition, and careful consideration of thermodynamic models and experimental data. Moreover, in order to distinguish mantle sources from crustal overprint, an evaluation involving isotopic data is also required. These are the objectives of our ongoing study of basalts from the Upper Scott Glacier Volcanic Field.

Conclusions

- Mount Early and Sheridan Bluff, part of the Upper Scott Glacier Volcanic Field, represent rare occurrences of volcanoes exposed in East Antarctica; a craton that covers an area of over 10×10^6 km^2 (>70% of the Antarctic continent) and is almost entirely buried beneath the East Antarctic Ice

Sheet. These Early Miocene volcanoes are distant (c. 1000 km) from other volcanoes and are not associated with the West Antarctic Rift System (WARS), which hosts the majority of Cenozoic volcanism. Our study provides the first detailed geochemical investigation of volcanoes located in East Antarctica, and reveals a complex petrogenesis and a unique set of conditions to explain their origins.

- Basalts (lavas, lava pillows and dykes) range in composition from alkaline to subalkaline, and are classified as hawaiite, mugearite and tholeiite. Major and trace element concentrations for the alkaline basalt association (hawaiite and mugearite) are broadly similar to basalts within the WARS and alkaline basalts from oceanic islands (OIB), while tholeiites approach enriched mid-ocean ridge basalts (E-MORB) in composition.

- Fractional crystallization of olivine, clinopyroxene, plagioclase and titanomagnetite can explain much of the variation in major and trace elements for alkaline compositions. The range in tholeiite compositions is modelled by progressive crystallization of olivine + spinel → plagioclase → clinopyroxene under pressure conditions of less than 500 MPa and water contents ≤1 wt%.

- Contamination of basaltic magmas by upper continental crust may explain why some samples have high Cs contents, low Ce/Pb ratios, and a lack negative K and Pb concentration spikes on mantle-normalized multi-element diagrams. Negative K- and Pb-anomalies are a conspicuous characteristic of basalts from the WARS, and is a common feature of intraplate alkaline basalts found throughout continental fragments of Gondwana (Zealandia, SE Australia: Price *et al.* 2014; Gamble *et al.* 2018). The partial melting of amphibole-rich lithologies in the mantle (metasomatized) is invoked to explain the negative K spikes.

- Neither fractional crystallization nor crustal assimilation can explain the relationship between coexisting tholeiite and alkaline lavas at Sheridan Bluff. Their trace element concentrations suggest an association that is controlled by variations in the degree of partial melting of a compositionally similar mantle source. Melt models suggest that both types were generated by partial fusion of incompatible-trace-element-enriched, hydrous mantle containing garnet and amphibole. Tholeiite was produced by higher degrees of melting of this source relative to alkaline compositions.

- Lithospheric detachment (also known as delamination, foundering or drip) may have initiated melt generation and eruption of the Upper Scott Glacier Volcanic Field. Based on the geochemistry of Mount Early and Sheridan Bluff basalts, we consider that metasomatic additions to the base of a thick East Antarctic lithosphere are likely to have played an key role, not only in triggering the instability leading to detachment but also in controlling the melt production beneath this region.

- A more comprehensive understanding of the role of crustal contamination and possible compositional heterogeneities in mantle sources will be undertaken with the aid of Sr, Nd, Pb, Hf and oxygen isotopes in the ongoing study of Mount Early and Sheridan Bluff basalt.

Acknowledgements We would first like to acknowledge Ed Stump for his help providing local geological and logistical information prior to our 2015 expedition, Tim Burton for his assistance in the field, and all United States Antarctic Program personnel in McMurdo and at the Shackleton Glacier transit camp. Gordon Moore is thanked for his expertise and guidance in mineral analysis, and Susan Krans for tutoring KSP in the use of the MELTS algorithm. We thank Sergio Rocchi and John Gamble for providing insightful comments that improved the quality of this manuscript.

Author contributions KSP: conceptualization (equal), data curation (lead), funding acquisition (lead), supervision (lead), writing – original draft (lead), writing – review & editing (lead); **JR**: conceptualization (supporting), formal analysis (equal), writing – original draft (supporting); **JLS**: conceptualization (equal), investigation (equal), writing – review & editing (supporting).

Funding This work and an assistantship for Reindel (MSc) was provided by a US National Science Foundation grant NSF-PLR 1443576 awarded to KSP.

Data availability All data generated or analysed during this study are included in this published article (and its supplementary information files).

References

Adam, J. and Green, T. 2006. Trace element partitioning between mica- and amphibole-bearing garnet lherzolite and hydrous basanitic melt: 1. Experimental results and the investigation of controls on partitioning behavior. *Contributions to Mineralogy and Petrology*, **152**, 1–17, https://doi.org/10.1007/s00410-006-0085-4

Arth, J.G. 1976. Behavior of trace elements during magmatic processes – a summary of theoretical models and their applications. *Journal of Research of the United States Geological Survey*, **4**, 41–47.

Aviado, K.B., Rilling-Hall, S., Bryce, J.G. and Mukasa, S.B. 2015. Submarine and subaerial lavas in the West Antarctic Rift System: temporal record of shifting magma source components from the lithosphere and asthenosphere. *Geochemistry, Geophysics, Geosystems*, **16**, 4344–4361, https://doi.org/10.1002/2015GC006076

Borg, S.G. and Stump, E. 1987. Paleozoic magmatism and associated tectonic problems of Northern Victoria Land, Antarctica. *American Geophysical Union Geophysical Monograph Series*, **40**, 67–75.

Bottazzi, P., Tiepolo, M., Vannucci, R., Zanetti, A., Brumm, R., Foley, S.F. and Oberti, R. 1999. Distinct site preferences for heavy and light REE in amphibole and the prediction of $^{Amph/L}D_{REE}$. *Contributions to Mineralogy and Petrology*, **137**, 36–45, https://doi.org/10.1007/s004100050580

Bottinga, Y. and Weill, D.F. 1970. Densities of liquid silicate systems calculated from partial molar volumes of oxide components. *American Journal of Science*, **269**, 169–182, https://doi.org/10.2475/ajs.269.2.169

Boyce, J.A., Nicholls, I.A., Keays, R.R. and Hayman, P.C. 2015. Variation in parental magmas of Mt Rouse, a complex polymagmatic volcano in the basaltic intraplate Newer Volcanics Province, southeast Australia. *Contributions to Mineralology and Petrology*, **169**, 11, https://doi.org/10.1007/s00410-015-1106-y

Bryan, W.B., Finger, L.W. and Chayes, F. 1969. Estimating proportions in petrographic mixing equations by least-squares approximation. *Science*, **163**, 926–927, https://doi.org/10.1126/science.163.3870.926

Coltorti, M., Beccaluva, L., Bonadiman, C., Faccini, B., Ntaflos, T. and Siena, F. 2004. Amphibole genesis via metasomatic reaction with clinopyroxene in mantle xenoliths from Victoria Land, Antarctica. *Lithos*, **75**, 115–139, https://doi.org/10.1016/j.lithos.2003.12.021

Cooper, A.F., Adam, L.J., Coulter, R.F., Eby, G.N. and McIntosh, W.C. 2007. Geology, geochronology and geochemistry of a basanitic volcano, White Island, Ross Sea, Antarctica. *Journal of Volcanology and Geothermal Research*, **165**, 189–216, https://doi.org/10.1016/j.jvolgeores.2007.06.003

Davidson, J.P., Morgan, D.J., Charlier, B.L.A., Harlou, R. and Hora, J.M. 2007. Microsampling and isotopic analysis of igneous rocks: implications for the study of magmatic systems. *Annual Reviews of Earth and Planetary Science*, **35**, 273–311, https://doi.org/10.1146/annurev.earth.35.031306.140211

DePaolo, D.J. 1981. Trace element and isotopic effects of combined wallrock assimilation and fractional crystallization. *Earth and Planetary Science Letters*, **53**, 189–202, https://doi.org/10.1016/0012-821X(81)90153-9

D'Orazio, M., Armienti, P. and Cerretini, S. 1998. Phenocryst/matrix trace-element partition coefficients for hawaiite–trachyte lavas from the Ellittico volcanic sequence (Mt. Etna, Sicily, Italy). *Mineralogy and Petrology*, **64**, 65–88, https://doi.org/10.1007/BF01226564

Doumani, G.A. and Minshew, V.H. 1965. General geology of the Mount Weaver area, Queen Maud Mountains. *American Geophysical Union Antarctic Research Series*, **6**, 127–139.

Elkins-Tanton, L.T. 2005. Continental magmatism caused by lithospheric delamination. *Geological Society of America Special Papers*, **388**, 449–461.

Elkins-Tanton, L.T. 2007. Continental magmatism, volatile recycling, and a heterogeneous mantle caused by lithospheric gravitational instabilities. *Journal of Geophysical Research*, **112**, B03405, https://doi.org/10.1029/2005JB004072

Elliot, D.H. and Fleming, T.H. 2004. Occurrence and dispersal of magmas in the Jurassic Ferrar large igneous province, Antarctica. *Gondwana Research*, **7**, 223–237, https://doi.org/10.1016/S1342-937X(05)70322-1

Ersoy, E.Y., Helvacı, C. and Palmer, M.R. 2010. Mantle source characteristics and melting models for the early-middle Miocene mafic volcanism in Western Anatolia: Implications for enrichment processes of mantle lithosphere and origin of K-rich volcanism in post-collisional settings. *Journal of Volcanology and Geothermal Research*, **198**, 112–128, https://doi.org/10.1016/j.jvolgeores.2010.08.014

Floyd, P.A. and Winchester, J.A. 1975. Magma type and tectonic setting discrimination using immobile elements. *Earth and Planetary Science Letters*, **27**, 211–218, https://doi.org/10.1016/0012-821X(75)90031-X

Furman, T., Nelson, W.R. and Elkins-Tanton, L.T. 2016. Evolution of the East African rift: drip magmatism, lithospheric thinning and mafic volcanism. *Geochimica et Cosmochimica Acta*, **185**, 418–434, https://doi.org/10.1016/j.gca.2016.03.024

Gamble, J.A. and Kyle, P.R. 1987. The origins of glass and amphibole in spinel-wehrlite xenoliths from Foster Crater, McMurdo Volcanic Group, Antarctica. *Journal of Petrology*, **28**, 755–779, https://doi.org/10.1093/petrology/28.5.755

Gamble, J.A., McGibbon, F., Kyle, P.R., Menzies, M.A. and Kirsch, I. 1988. Metasomatised xenoliths from Foster Crater, Antarctica: Implications for lithospheric structure and processes beneath the Transantarctic Mountain front. *Journal of Petrology*, Special Volume, Issue 1, 109–138, https://doi.org/10.1093/petrology/Special_Volume.1.109

Gamble, J.A., Adams, C.J., Morris, P.A., Wysoczanski, R.J., Handler, M. and Timm, C. 2018. The geochemistry and petrogenesis of Carnley Volcano, Auckland Islands, SW Pacific. *New Zealand Journal of Geology and Geophysics*, **61**, 480–497, https://doi.org/10.1080/00288306.2018.1505642

Ghiorso, M.S. and Sack, R.O. 1995. Chemical mass transfer in magmatic processes IV. A revised and internally consistent thermodynamic model for the interpolation and extrapolation of liquid–solid equilibria in magmatic systems at elevated temperatures and pressures. *Contributions to Mineralogy and Petrology*, **119**, 197–212, https://doi.org/10.1007/BF00307281

Granot, R. and Dyment, J. 2018. Late Cenozoic unification of East and West Antarctica. *Nature Communications*, **9**, 3189, https://doi.org/10.1038/s41467-018-05270-w

Hart, S.R., Blusztajn, J., LeMasurier, W.E. and Rex, D.C. 1997. Hobbs Coast Cenozoic volcanism: implications for the West Antarctic rift system. *Chemical Geology*, **139**, 223–248, https://doi.org/10.1016/S0009-2541(97)00037-5

Heeszel, D.S., Wiens, D.A. *et al.* 2016. Upper mantle structure of central and West Antarctica from array analysis of Rayleigh wave phase velocities. *Journal of Geophysical Research: Solid Earth*, **121**, 1758–1775, https://doi.org/10.1002/2015JB012616

Hofmann, A.W., Jochum, K.P., Seufert, M. and White, W.M. 1986. Nb and Pb in oceanic basalts: new constraints on mantle evolution. *Earth and Planetary Science Letters*, **79**, 33–45, https://doi.org/10.1016/0012-821X(86)90038-5

Hole, M.J. 2018. Mineralogical and geochemical evidence for polybaric fractional crystallization of continental flood basalts and implications for identification of peridotite and pyroxenite source lithologies. *Earth-Science Reviews*, **176**, 51–67, https://doi.org/10.1016/j.earscirev.2017.09.014

Hole, M.J. 2021. Antarctic Peninsula: petrology. *Geological Society, London, Memoirs*, **55**, https://doi.org/10.1144/M55-2018-40

Johnson, D.M., Hooper, P.R. and Conrey, R.M. 1999. XRF analysis of rocks and minerals for major and trace elements on a single low dilution Li-tetraborate fused bead. *Advances in X-ray Analysis*, **41**, 843–867.

Jung, S. and Masberg, P. 1998. Major- and trace-element systematics and isotope geochemistry of Cenozoic mafic volcanic rocks from the Vogelsberg (central Germany). Constraints on the origin of continental alkaline and tholeiitic basalts and their mantle sources. *Journal of Volcanology and Geothermal Research*, **86**, 151–177, https://doi.org/10.1016/S0377-0273(98)00087-0

Kay, R.W. and Kay, S.M. 1993. Delamination and delamination magmatism. *Tectonophysics*, **219**, 177–189, https://doi.org/10.1016/0040-1951(93)90295-U

Kinzler, R.J. 1997. Melting of mantle peridotite at pressures approaching the spinel to garnet transition: Application to mid-ocean ridge basalt petrogenesis. *Journal of Geophysical Research: Solid Earth*, **102**, 853–874, https://doi.org/10.1029/96JB00988

Kipf, A., Hauff, F. *et al.* 2014. Seamounts off the West Antarctic margin: A case for non-hotspot driven intraplate volcanism. *Gondwana Research*, **25**, 1660–1679, https://doi.org/10.1016/j.gr.2013.06.013

Kocaarslan, A. and Yalçın Ersoy, E. 2018. Petrologic evolution of Miocene-Pliocene mafic volcanism in the Kangal and Gürün basins (Sivas–Malatya), central east Anatolia: evidence for Miocene anorogenic magmas contaminated by continental crust. *Lithos*, **310–311**, 392–408, https://doi.org/10.1016/j.lithos.2018.04.021

Kyle, P.R. 1990. McMurdo Volcanic Group, western Ross Embayment. *American Geophysical Union Antarctic Research Series*, **48**, 19–25.

Kyle, P.R., Moore, J.A. and Thirlwall, M.F. 1992. Petrologic evolution of anorthoclase phonolite lavas at Mount Erebus, Ross Island, Antarctica. *Journal of Petrology*, **33**, 849–875.

Le Bas, M.J., Le Maitre, R.W., Streckeisen, A. and Zanettin, B. 1986. Chemical classification of volcanic rocks based on the total alkali–silica diagram. *Journal of Petrology*, **27**, 745–750, https://doi.org/10.1093/petrology/27.3.745

Lee, M.J., Lee, J.I., Kim, T.H., Lee, J. and Nagao, K. 2015. Age, geochemistry and Sr–Nd–Pb isotopic compositions of alkali volcanic rocks from Mt. Melbourne and the western Ross Sea, Antarctica. *Geosciences Journal*, **19**, 681–695, https://doi.org/10.1007/s12303-015-0061-y

LeMasurier, W.E. and Landis, C.A. 1996. Mantle-plume activity recorded by low-relief erosion surfaces in West Antarctica and New Zealand. *Geological Society of America Bulletin*, **108**, 1450–1466, https://doi.org/10.1130/0016-7606(1996)108<1450:MPARBL>2.3.CO;2

LeMasurier, W.E. and Thomson, J.W. (eds). 1990. *Volcanoes of the Antarctic Plate and Southern Oceans*. American Geophysical Union, Antarctic Research Series, **48**.

Licht, K.J., Groth, T., Townsend, J.P., Hennessy, A.J., Hemming, S.R., Flood, T.P. and Studinger, M. 2018. Evidence for extending anomalous Miocene volcanism at the edge of the East Antarctic craton. *Geophysical Research Letters*, **45**, 3009–3016, https://doi.org/10.1002/2018GL077237

Lustrino, M., Melluso, L. and Morra, V. 2002. The transition from alkaline to tholeiitic magmas: a case study from the

Orosei–Dorgali Pliocene volcanic district NE Sardinia, Italy. *Lithos*, **63**, 83–113, https://doi.org/10.1016/S0024-4937(02)00113-5

Mandler, B.E. and Grove, T.L. 2016. Controls on the stability and composition of amphibole in the Earth's mantle. *Contributions to Mineralogy and Petrology*, **171**, 68, https://doi.org/10.1007/s00410-016-1281-5

Martin, A.P., Cooper, A.F. and Price, R.C. 2013. Petrogenesis of Cenozoic, alkali volcanic lineages at Mount Morning, West Antarctica and their entrained lithospheric mantel xenoliths: lithospheric v. asthenospheric mantle sources. *Geochimica et Cosmochimica Acta*, **122**, 127–152, https://doi.org/10.1016/j.gca.2013.08.025

Martin, A.P., Cooper, A., Price, R., Kyle, P. and Gamble, J. 2021. Erebus Volcanic Province: petrology. *Geological Society, London, Memoirs*, **55**, https://doi.org/10.1144/M55-2018-80

McDonough, W.F. and Sun, S.-s. 1995. The composition of the Earth. *Chemical Geology*, **120**, 223–253, https://doi.org/10.1016/0009-2541(94)00140-4

McKenzie, D. and O'Nions, R.K. 1991. Partial melt distributions from inversion of rare earth element concentrations. *Journal of Petrology*, **32**, 1021–1091, https://doi.org/10.1093/petrology/32.5.1021

Médard, E., Schmidt, M.W., Schiano, P. and Ottolini, L. 2006. Melting of amphibole-bearing wehrlites: an experimental study on the origin of ultra-calcic nepheline-normative melts. *Journal of Petrology*, **47**, 481–504, https://doi.org/10.1093/petrology/egi083

Moore, J.G., Hickson, C.J. and Calk, L.C. 1995. Tholeiitic–alkalic transition at subglacial volcanoes, Tuya region, British Columbia, Canada. *Journal of Geophysical Research: Solid Earth*, **100**, 24 577–24 592, https://doi.org/10.1029/95JB02509

Nardini, I., Armienti, P., Rocchi, S., Dallai, L. and Harrison, D. 2009. Sr–Nd–Pb–He–O isotope and geochemical constraints on the genesis of Cenozoic magmas from the West Antarctic rift. *Journal of Petrology*, **50**, 1359–1375, https://doi.org/10.1093/petrology/egn082

Orlando, A., Conticelli, S., Armienti, P. and Borrini, D. 2000. Experimental study on a basanite from the McMurdo Volcanic Group, Antarctica: inference on its mantle source. *Antarctic Science*, **12**, 105–116, https://doi.org/10.1017/S0954102000000134

Panter, K.S., Kyle, P.R. and Smellie, J.L. 1997. Petrogenesis of a phonolite–trachyte succession at Mount Sidley, Marie Byrd Land, Antarctica. *Journal of Petrology*, **38**, 1225–1253.

Panter, K.S., Hart, S.R., Kyle, P., Blusztajn, J. and Wilch, T. 2000. Geochemistry of Late Cenozoic basalts from the Crary Mountains: characterization of mantle sources in Marie Byrd Land, Antarctica. *Chemical Geology*, **165**, 215–241, https://doi.org/10.1016/S0009-2541(99)00171-0

Panter, K.S., Castillo, P. *et al.* 2018. Melt origin across a rifted continental margin: a case for subduction-related metasomatic agents in the lithospheric source of alkaline basalt, northwest Ross Sea, Antarctica. *Journal of Petrology*, **59**, 517–558, https://doi.org/10.1093/petrology/egy036

Panter, K.S., Wilch, T.I., Smellie, J.L., Kyle, P.R. and McIntosh, W.C. 2021. Marie Byrd Land and Ellsworth Land: petrology. *Geological Society, London, Memoirs*, https://doi.org/10.1144/M55-2019-50

Pearce, J.A. 2008. Geochemical fingerprinting of oceanic basalts with applications to ophiolite classification and the search for Archean oceanic crust. *Lithos*, **100**, 14–48, https://doi.org/10.1016/j.lithos.2007.06.016

Perinelli, C., Armienti, P. and Dallai, L. 2006. Geochemical and O-isotope constraints on the evolution of lithospheric mantle in the Ross Sea rift area (Antarctica). *Contributions to Mineralogy and Petrology*, **151**, 245–266, https://doi.org/10.1007/s00410-006-0065-8

Perinelli, C., Armienti, P. and Dallai, L. 2011. Thermal evolution of the lithosphere in a rift environment as inferred from the geochemistry of mantle cumulates, northern Victoria Land, Antarctica. *Journal of Petrology*, **52**, 665–690, https://doi.org/10.1093/petrology/egq099

Perinelli, C., Gaeta, M. and Armienti, P. 2017. Cumulate xenoliths from Mt. Overlord, northern Victoria Land, Antarctica: A window into high pressure storage and differentiation of mantle-derived basalts. *Lithos*, **268–271**, 225–239, https://doi.org/10.1016/j.lithos.2016.10.027

Phillips, E.H., Sims, K.W.W. *et al.* 2018. The nature and evolution of mantle upwelling at Ross Island, Antarctica, with implications for the source of HIMU lavas. *Earth and Planetary Science Letters*, **498**, 38–53, https://doi.org/10.1016/j.epsl.2018.05.049

Price, R.C., Nicholls, I.A. and Day, A. 2014. Lithospheric influences on magma compositions of late Mesozoic and Cenozoic intraplate basalts (the Older Volcanics) of Victoria, south-eastern Australia. *Lithos*, **206–207**, 179–200, https://doi.org/10.1016/j.lithos.2014.07.027

Redner, E.R. 2016. *Magma Mixing and Evolution at Minna Bluff, Antarctica Revealed by Amphibole and Clinopyroxene Analyses*. MSc thesis, Bowling Green State University, Bowling Green, Ohio, USA.

Reindel, J.L. 2018. *The Origin of Basalt and Cause of Melting Beneath East Antarctica as Revealed by the Southernmost Volcanoes on Earth*. MSc thesis, Bowling Green State University, Bowling Green, Ohio, USA.

Rocchi, S. and Smellie, J.L. 2021. Northern Victoria Land: petrology. *Geological Society, London, Memoirs*, **55**, https://doi.org/10.1144/M55-2019-19

Rocchi, S., Armienti, P., D'Orazio, M., Tonarini, S., Wijbrans, J.R. and Di Vincenzo, G. 2002. Cenozoic magmatism in the western Ross Embayment: role of mantle plume v. plate dynamics in the development of the West Antarctic Rift System. *Journal of Geophysical Research: Solid Earth*, **107**, ECV 5-1–ECV 5-22, https://doi.org/10.1029/2001JB000515

Rudnick, R.L. and Fountain, D.M. 1995. Nature and composition of the continental crust: A lower crustal perspective. *Reviews in Geophysics*, **33**, 267–309, https://doi.org/10.1029/95RG01302

Rudnick, R.L. and Gao, S. 2003. Composition of the continental crust. *In*: Rudnick, R.L. (ed.) *Treatise on Geochemistry, Volume 3: The Crust*. Pergamon, Oxford, UK, 1–64.

Scanlan, M.K. 2008. *Petrology of Inclusion-Rich Lavas at Minna Bluff, McMurdo Sound, Antarctica: Implications for Magma Origin, Differentiation, and Eruption Dynamics*. MSc thesis, Bowling Green State University, Bowling Green, Ohio, USA.

Shaw, D.M. 1970. Trace element fractionation during anatexis. *Geochimica et Cosmochimica Acta*, **34**, 237–243, https://doi.org/10.1016/0016-7037(70)90009-8

Shen, W., Wiens, D.A. *et al.* 2017. Seismic evidence for lithospheric foundering beneath the southern Transantarctic Mountains, Antarctica. *Geology*, **46**, 71–74, https://doi.org/10.1130/G39555.1

Sims, K.W.W. and DePaolo, D.J. 1997. Inferences about mantle magma sources from incompatible element concentration ratios in oceanic basalts. *Geochimica et Cosmochimica Acta*, **61**, 765–784, https://doi.org/10.1016/S0016-7037(96)00372-9

Smellie, J.L. and Collerson, K.D. 2021. Gaussberg: volcanology and petrology. *Geological Society, London, Memoirs*, **55**, https://doi.org/10.1144/M55-2018-85

Smellie, J.L. and Martin, A.P. 2021. Erebus Volcanic Province: volcanology. *Geological Society, London, Memoirs*, **55**, https://doi.org/10.1144/M55-2018-62

Smellie, J.L., Panter, K.S. and Reindel, J. 2021. Mount Early and Sheridan Bluff: volcanology. *Geological Society, London, Memoirs*, **55**, https://doi.org/10.1144/M55-2018-61

Stevens, R.E. 1944. Composition of some chromites of the Western hemisphere. *American Mineralogist*, **29**, 1–34.

Stump, E. 1995. *The Ross Orogen of the Transantarctic Mountains*. Cambridge University Press, Cambridge, UK.

Stump, E., Sheridan, M.F., Borg, S.G. and Sutter, J.F. 1980. Early Miocene subglacial basalts, East Antarctic Ice Sheet, and uplift of the Transantarctic Mountains. *Science*, **207**, 757–759, https://doi.org/10.1126/science.207.4432.757

Stump, E., Borg, S.G. and Sheridan, M.F. 1990*a*. Sheridan Bluff. *American Geophysical Union Antarctic Research Series*, **48**, 136–137.

Stump, E., Borg, S.G. and Sheridan, M.F. 1990b. Mount Early. *American Geophysical Union Antarctic Research Series*, **48**, 138–139.

Sun, S.-s. and McDonough, W.F. 1989. Chemical and isotopic systematics of oceanic basalts: implications for mantle composition and processes. *Geological Society, London, Special Publications*, **42**, 313–345, https://doi.org/10.1144/GSL.SP.1989.042.01.19

Taylor, S.R. and McLennan, S.M. 1985. *The Continental Crust: Its Composition and Evolution. An Examination of the Geochemical Record Preserved in Sedimentary Rocks*. Blackwell Scientific, Oxford, UK.

Wanless, V., Garcia, M., Rhodes, J., Weis, D. and Norman, M.D. 2006. Shield-stage alkalic volcanism on Mauna Loa Volcano, Hawaii. *Journal of Volcanology and Geothermal Research*, **151**, 141–155, https://doi.org/10.1016/j.jvolgeores.2005.07.027

Wingrove, D. 2005. *Early Mixing in the Evolution of Alkaline Magmas: Chemical and Oxygen Isotope Evidence from Phenocrysts, Royal Society Range, Antarctica*. MSc thesis, Bowling Green State University, Bowling Green, Ohio, USA.

Workman, R.K. and Hart, S.R. 2005. Major and trace element composition of the depleted MORB mantle (DMM). *Earth and Planetary Science Letters*, **231**, 53–72, https://doi.org/10.1016/j.epsl.2004.12.005

Chapter 5.4a

Marie Byrd Land and Ellsworth Land: volcanology

T. I. Wilch[1]*, W. C. McIntosh[2] and K. S. Panter[3]

[1]Department of Geology, Albion College, 611 East Porter Street, Albion, MI 49224, USA
[2]New Mexico Bureau of Geology and Mineral Resources, New Mexico Institute of Mining and Technology, 801 Leroy Place, Socorro, NM 87801, USA
[3]School of Earth, Environment and Society, Bowling Green State University, Bowling Green, OH 43403, USA

TIW, 0000-0003-2997-3730
*Correspondence: twilch@albion.edu

Abstract: Nineteen large (2348–4285 m above sea level) central polygenetic alkaline shield-like composite volcanoes and numerous smaller volcanoes in Marie Byrd Land (MBL) and western Ellsworth Land rise above the West Antarctic Ice Sheet (WAIS) and comprise the MBL Volcanic Group (MBLVG). Earliest MBLVG volcanism dates to the latest Eocene (36.6 Ma). Polygenetic volcanism began by the middle Miocene (13.4 Ma) and has continued into the Holocene without major interruptions, producing the central volcanoes with 24 large (2–10 km-diameter) summit calderas and abundant evidence for explosive eruptions in caldera-rim deposits. Rock lithofacies are dominated by basanite and trachyte/phonolite lava and breccia, deposited in both subaerial and ice-contact environments. The chronology of MBLVG volcanism is well constrained by 330 age analyses, including 52 new $^{40}Ar/^{39}Ar$ ages. A volcanic lithofacies record of glaciation provides evidence of local ice-cap glaciation at 29–27 Ma and of widespread WAIS glaciation by 9 Ma. Late Quaternary glaciovolcanic records document WAIS expansions that correlate to eustatic sea-level lowstands (MIS 16, 4 and 2): the WAIS was +500 m at 609 ka at coastal Mount Murphy, and +400 m at 64.7 ka, +400 m at 21.2 ka and +575 m at 17.5 ka at inland Mount Takahe.

Supplementary material: Summary age table with locations (Table S1), Murphy age data (new data only) (Table S2), Takahe age data (new data only) (Table S3) and palaeo-ice history (Table S4) are available at https://doi.org/10.6084/m9.figshare.c.5205362

Marie Byrd Land (MBL) and western Ellsworth Land in West Antarctica have a long history of Cenozoic volcanism defined by 19 polygenetic central volcanoes and numerous smaller volcanic centres exposed above the level of the West Antarctic Ice Sheet (WAIS) (Figs 1 & 2). The central volcanoes include some of the largest composite volcanoes in Antarctica and on Earth. The history is not well known because the volcanoes are remote, mostly snow- and ice-covered, and partially or fully buried by the WAIS. The last major review of the physical volcanology and geochronology of the region was published 30 years ago (LeMasurier and Thomson 1990). This chapter updates that and includes: (1) the establishment of stratigraphic nomenclature for volcanoes on the MBL and Thurston Island tectonic blocks; (2) a general synthesis of regional volcanism; (3) updated summaries of geological histories of individual volcanoes in MBL and western Ellsworth Land; and (4) an updated analysis of the volcanic record of the WAIS. The volcano summaries include general trends in the volcanic histories based on published and unpublished fieldwork, lithofacies descriptions, $^{40}Ar/^{39}Ar$ geochronology data, and an assessment of the state of knowledge of exposed volcanoes.

The MBL Volcanic Group (MBLVG) is formally defined here to include volcanoes on the MBL and Thurston Island tectonic blocks in West Antarctica (from 156° W to 93° W) (Figs 1, 2 & 3). Specifically, the MBLVG is defined to include the MBL Volcanic Province (MBLVP) (Fig. 1) and the Thurston Island Volcanic Province (Fig. 2), the former comprising nine volcanic fields and the latter comprising two volcanic fields. Four of the volcanic fields in the MBLVP consist of central volcanoes that form linear mountain chains and have age trends along the volcano alignments; these fields bear the mountain range names (Flood, Ames and Executive Committee ranges, and Crary Mountains). Each of these volcanic fields consists of three–five large polygenetic central volcanoes. The Eastern MBL Volcanic Field includes three isolated polygenetic central volcanoes and several smaller volcanic centres. The Mount Siple Volcanic Field consists of Mount Siple, the large isolated coastal polygenetic central volcano. The McCuddin Mountains Volcanic Field includes the oldest volcanic centres, all located in central MBL. Finally, the MBLVP includes the Hobbs Coast and Fosdick Mountains volcanic fields, each composed of multiple monogenetic volcanic centres. The much smaller Thurston Island Volcanic Province in western Ellsworth Land includes the Hudson and Jones Mountains volcanic fields, both composed of monogenetic volcanoes. This study focuses more closely on the MBLVP than the Thurston Island Volcanic Province because the MBLVP is much larger and has been studied in greater detail.

The MBLVG volcanoes are exposed as mostly snow-covered nunataks protruding through the marine-based WAIS in West Antarctica. Summit elevations of the large central volcanoes range from 2348 to 4285 m above sea level (asl) and ice-sheet elevations surrounding the volcanoes range from sea level at the coast to more than 2400 m asl at inland locations near the centre of the province. Exposures of the volcano–basement rock contacts are buried at most volcanoes but, where exposed, appear to be higher in elevation towards the centre of the province (LeMasurier and Landis 1996).

The oldest known volcanism in the MBLVG is dated at 36.6 Ma but most of the documented volcanism has occurred since 13.4 Ma (LeMasurier 1990h; Wilch and McIntosh 2000). The $^{40}Ar/^{39}Ar$ ages of volcanic rocks at the large central volcanoes range from 20.46 ± 0.07 Ma (all errors ± 2σ uncertainty) at Mount Flint in the McCuddin Mountains Volcanic Field to 8.3 ± 5.4 ka at Mount Takahe in the Eastern MBL Volcanic Field (Wilch 1997). Mount Berlin in the Flood Range Volcanic Field is an active central volcano with several steaming fumarolic ice towers on the periphery of its summit caldera (LeMasurier and Wade 1968; Wilch et al. 1999). Mounts Takahe and Mount Berlin are the two central volcanoes documented to be active in the Holocene (Wilch et al. 1999), and Mount Waesche in the Executive

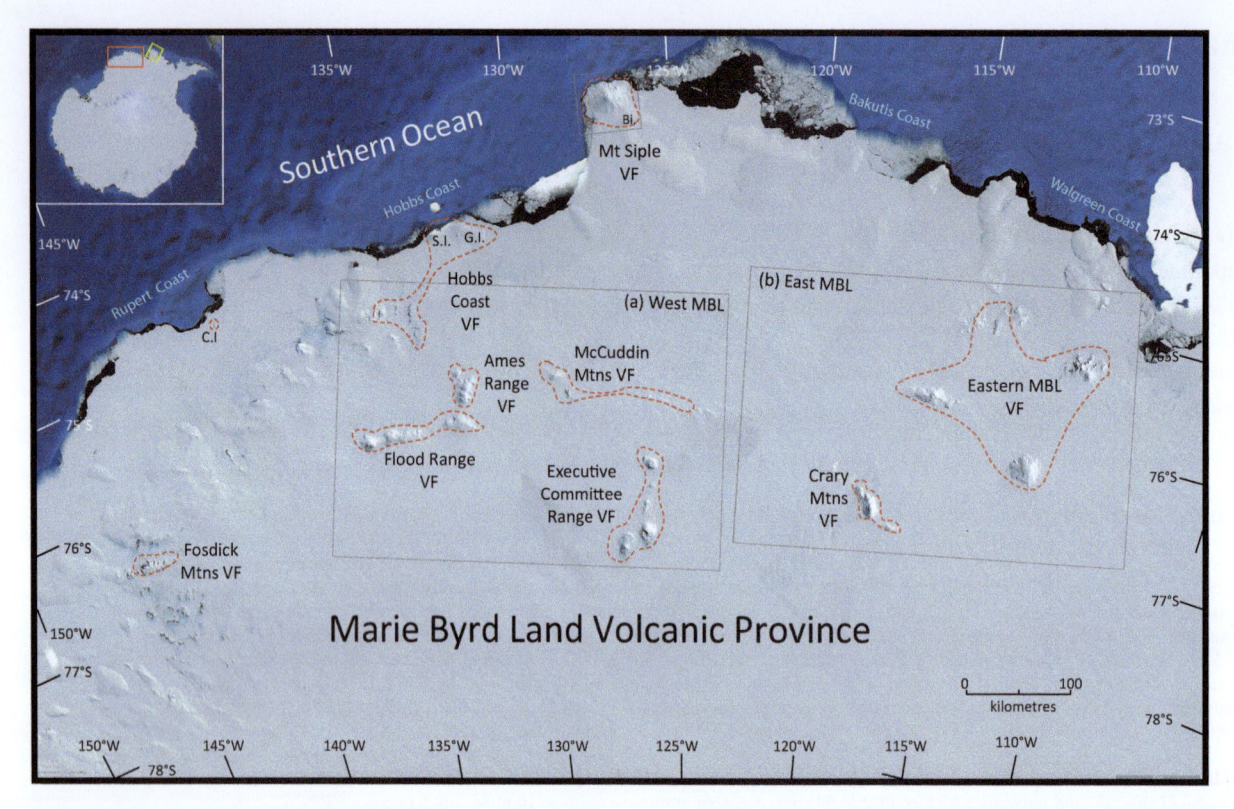

Fig. 1. Map of Marie Byrd Land Volcanic Province (MBLVP), West Antarctica and inset map of Antarctica showing the MBLVP (red box) and Thurston Island Volcanic Province (yellow box, see Fig. 2). Volcanic fields are labelled and volcanoes included in fields are within dashed line polygons and listed in Figure 3. The two large rectangular outlines show the map boundaries of Figures 5a and 5b; an inset map of Mount Siple is also shown in Figure 5a. Image maps of individual central volcanoes or volcanic fields are presented with volcano summaries. Abbreviations: VF, Volcanic Field; USAS Esc., USAS Escarpment; C.I., Cruzen Island; S.I. is Shepard Island; G.I. is Grant Island. The base image maps are derived from Google Earth Pro (image US Geological Survey; Data: SIO, NOAA, US Navy, NGA GEBCO).

Committee Range Volcanic Field may have had eruptions in the Holocene and be considered active because there are abundant englacial tephra layers in blue ice adjacent to the volcano (Dunbar *et al.* 2021). Seismic evidence suggests that an active magmatic intrusive complex exists at 25–40 km below the base of the ice sheet 55 km to the south of Mount Waesche (Lough *et al.* 2013; Quartini *et al.* 2021). Previous workers have noted geographical and temporal patterns of the large central volcanoes, with volcanoes aligned in rectilinear east–west and north–south ranges, and younger volcanoes being located towards the periphery of the province (LeMasurier and Rex 1989, 1991; Paulsen and Wilson 2010).

The volcanoes of the MBLVP are mostly located on the *c.* 1000 × 500 km MBL structural dome on the north flank of the West Antarctic Rift System (WARS) (LeMasurier 2006). The MBL dome is associated with the MBL crustal block, one of four crustal blocks in West Antarctica (Dalziel and Elliot 1982) (Fig. 4). Bedrock elevations on the MBL dome (Fig. 4) are above sea level and rise to an elevation of *c.* 2700 m asl in the centre of the province (LeMasurier 1990*h*, 2006; LeMasurier and Landis 1996; Fretwell *et al.* 2013). The MBL dome has been included by some as part of the WARS based on the presence of horst and graben structures (LeMasurier 1990*g*), although it is more typically described as being situated on the north flank of the WARS (e.g. Paulsen and Wilson 2010). The WARS forms deep marine basins between the southern edge of the MBL dome and the northern front of the Ellsworth Mountains and Transantarctic Mountains (Fig. 4). The WARS extends from eastern MBL and the Amundsen Sea through the Ross Sea Basin to northern Victoria Land, and is one of the major continental rift systems on Earth, similar in size to the US Basin and Range Province and the East African Rift (Tessensohn and Wörner 1991; LeMasurier 2008). Large-scale asymmetrical intracontinental rifting of the nearly stationary Antarctic Plate formed the Ross Sea Basin and deep marine basins beneath the WAIS (Cooper and Davey 1985; Cooper *et al.* 1991; Granot and Dyment 2018). The Ross Sea Basin consists of a series of sediment-filled horst and graben structures. The intracontinental rift zone extends from the Ross Sea to beneath the WAIS (Behrendt *et al.* 1991; 1996). The Byrd Subglacial Basin and Bentley Subglacial Trench form deep submarine east–west-orientated basins that are typically interpreted as downfaulted graben within the WARS. The timing of extension and rifting events are not well constrained in West Antarctica, although it is generally stated that most of extension occurred in the Cretaceous (Siddoway 2008; Jordan *et al.* 2020). Most recognize two rifting episodes: early rifting in the Late Cretaceous; and late rifting that began in the Eocene and intensified in the late Cenozoic (Cooper and Davey 1985; Cooper *et al.* 1991; Jordan *et al.* 2020). Spiegel *et al.* (2016) also identified two episodes of tectonic denudation (presumably associated with uplift and rifting) at 100–60 and 20–0 Ma, based on interpretations of thermochronology data from bedrock outcrops in coastal MBL.

Geophysical data indicate that extended crust in the WARS is as thin as 21 km at the Bentley Subglacial Trench (Winberry and Anandakrishnan 2004). Calculated mean crustal thicknesses beneath the MBL dome from seismic data range from 28 to 33 km, 5–10 km thicker than in the WARS (Chaput *et al.* 2014; Ramirez *et al.* 2017; Shen *et al.* 2018). The presence of the MBL structural dome is attributed to a thermal anomaly beneath the volcanic province rather than isostatic compensation of slightly thicker crust (Winberry and Anandakrishnan 2004). LeMasurier and Landis (1996) interpreted MBL dome elevations increasing towards the centre of the MBLVP as a topographical expression of a mantle plume and suggested that the onset of uplift of the MBL dome

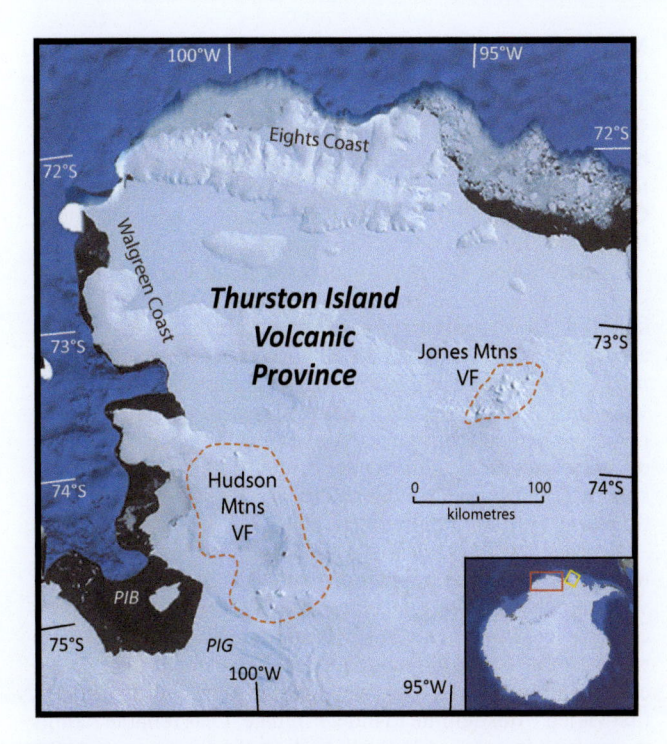

Fig. 2. Map of Thurston Island Volcanic Province, showing the Hudson Mountains and Jones Mountains volcanic fields. Inset map of Antarctica shows the map boundary (yellow box) and the boundary of the MBLVP map (Fig. 1, red box). The base image map is derived from Google Earth Pro. Abbreviations: PIB, Pine Island Bay; PIG, Pine Island Glacier.

Marie Byrd Land Volcanic Group
(Alkaline volcanic rocks)

Marie Byrd Land Volcanic Province
(Mainly large alkaline central volcanoes)

- **Eastern MBL Volcanic Field**
  - Mount Murphy
  - Mount Murphy satellite nunataks
  - Kohler Range
  - Mount Takahe
  - Toney Mountain
- **Crary Mountains Volcanic Field**
  - Mount Rees
  - Mount Steere
  - Mount Frakes
  - Boyd Ridge
- **Mount Siple Volcanic Field**
- **Executive Committee Range Volcanic Field**
  - Mount Hampton
  - Mount Cumming
  - Mount Hartigan
  - Mount Sidley
  - Mount Waesche
- **McCuddin Volcanic Field**
  - Mount Flint
  - Mount Petras
  - USAS Escarpment
- **Ames Range Volcanic Field**
  - Mount Andrus
  - Mount Kosciusko
  - Mount Kaufmann
- **Flood Range Volcanic Field**
  - Mount Bursey
  - Mount Moulton
  - Mount Berlin
- **Hobbs Coast Volcanic Field**
  - Hobbs Coast nunataks
  - Grant Island
  - Shepard Island
  - Cruzen Island
- **Fosdick Mountains Volcanic Field**

Thurston Island Volcanic Province
(Mainly mafic monogenetic volcanoes)

- **Hudson Mountains Volcanic Field**
- **Jones Mountains Volcanic Field**

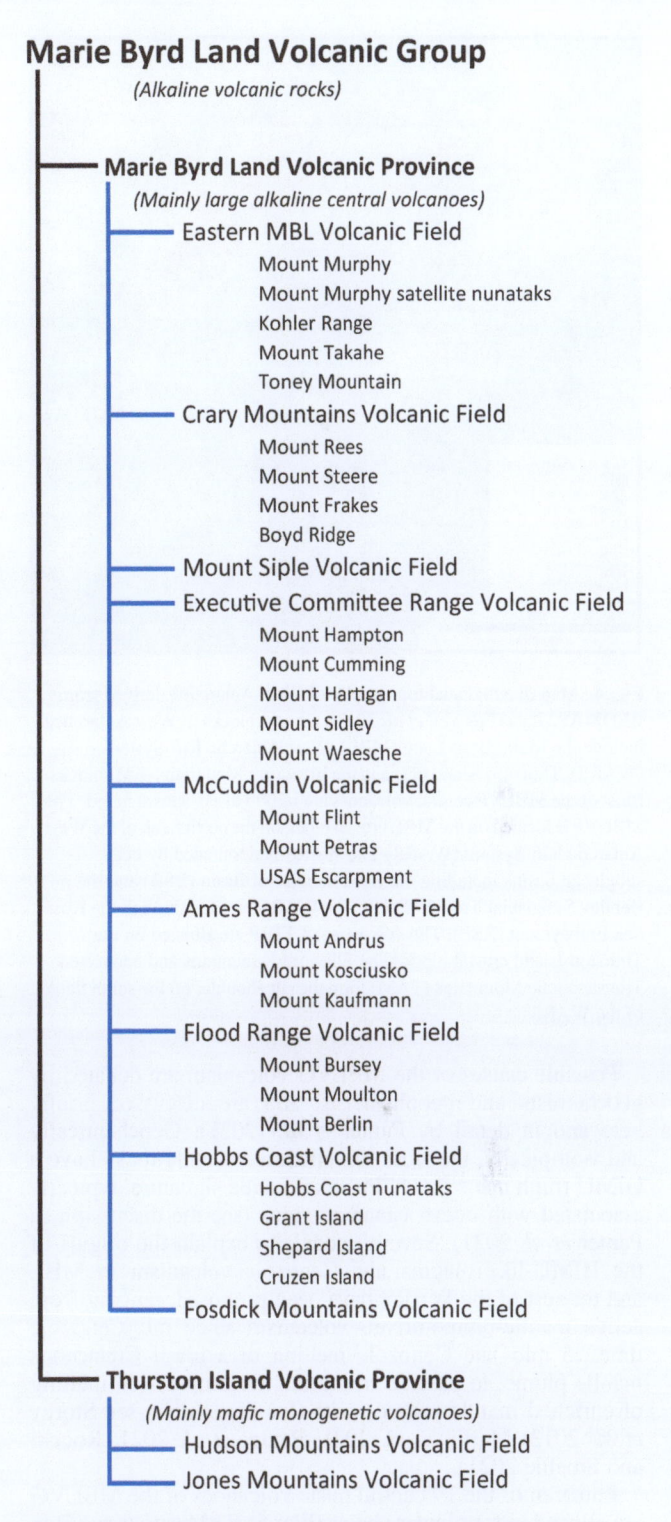

Fig. 3. Stratigraphic nomenclature for the Marie Byrd Land Volcanic Group (MBVG), including volcanic provinces, volcanic fields, and individual central volcanoes and isolated smaller centres.

coincided with an early pulse of volcanism at 29–27 Ma. In contrast, Spiegel *et al.* (2016) noted an absence of tectonic activity between 60 and 20 Ma, with acceleration of tectonic denudation beginning at 20 Ma, based on thermochronological data from MBL bedrock.

There have been significant advances in understanding of the thermal structure of the MBL lithosphere and sublithospheric mantle based on several multi-year geophysics experiments in West Antarctica. Lloyd *et al.* (2015) noted that in West Antarctica the slowest P- and S-wave velocities extend to 200 km below the Executive Committee Range, and Hansen *et al.* (2014) interpreted a low-velocity zone extending down to 800 km beneath MBL using P-wave tomography; both studies support the plume idea. Moreover, a modelling study using ambient seismic noise velocities by Heeszel *et al.* (2016) identified an extensive low-velocity zone down to >200 km centred on the Executive Committee Range that is again consistent with a thermal anomaly and a mantle plume. However, Heeszel *et al.* (2016) also pointed out that this thermal anomaly does not extend to the transition zone in the mid-mantle. Thinning of the transition zone is expected if impacted by elevated temperatures associated with a mantle plume (Reusch *et al.* 2008; Emry *et al.* 2015). In other geophysical studies, Seroussi *et al.* (2017) suggested that a weak mantle plume beneath MBL with a maximum geothermal heat flux of 150 mW m^{-2} is viable, based on modelling and assessment of the hydrology at the base of the WAIS.

In addition to the exposed volcanoes of the MBLVG, there are likely to be many subglacial volcanoes buried beneath the WAIS. Active subglacial volcanoes (Blankenship *et al.* 1993; Behrendt 2013) and an active intrusive magmatic system (Lough *et al.* 2013) have been mapped with confidence beneath the WAIS surface on the basis of geophysics, and Loose *et al.* (2018) inferred active volcanism beneath Pine Island Glacier based on geochemical signatures in meltwater. Extensive aeromagnetic surveys in the WARS show abundant magnetic anomalies, suggesting significant basaltic volcanism (as many as 1000 volcanic centres) beneath the WAIS (e.g. Behrendt 2013). Topographical interpretations of conical subglacial landforms beneath the WAIS, combined with preexisting geophysical and field data, led de Vries *et al.* (2018) to identify 138 subglacial volcanoes in West Antarctica, including 91 previously unknown volcanoes. The inferred subglacial volcanoes are located on the MBL, Thurston Island and WARS tectonic blocks. Although these subglacial volcanoes have not been verified by samples, we suggest that the volcanoes be included in the MBLVG.

Many of these polygenetic volcanoes have two calderas with offsets that reflect the same alignments and age progressions as the central volcanoes (Fig. 5). Linear age progressions of volcanoes elsewhere (e.g. Hawaii, Snake River Plain) have been interpreted as 'plume tracks' recording movement of tectonic plates above mantle plumes or hotspots (Morgan 1972; Courtillot *et al.* 2003) but the discordant directions among the various linear trends in MBL precludes this explanation.

There are two different hypotheses about the origin of the spatial and temporal patterns in the MBLVP. LeMasurier and Rex (1989) recognized a systematic chronological pattern associated with the spatial distribution, with the oldest volcanoes near the centre of the province and progressively younger volcanoes away from the centre. They suggested that the patterns resulted from centrifugal extension and reactivation of relict fracture systems in the brittle crust caused by lithospheric doming associated with the rise of a 550–650 km mantle plume beneath the province.

Paulsen and Wilson (2010) conducted a detailed structural trend analysis of the orientations and ages of the polygenetic volcanic chains, elongate volcano edifices, elongate summit calderas and flank vents in MBL and elsewhere in Antarctica to determine stress directions during volcanism. The shape elongation measurements were assigned reliability ratings that varied from definite to indeterminate. Less reliable ratings were given in cases of low axial ratios of edifices or calderas, snow or ice cover, or erosion of part of the feature. The highest reliability data showed that overall edifice and caldera elongation directions were parallel to overall volcano alignments. The flank vent alignments were deemed mostly unreliable. The stress data combined with all available age K–Ar and ^{40}Ar/^{39}Ar age data showed that Miocene volcanism in MBL occurred along north–south alignments, whereas latest Miocene–Quaternary volcanism (since *c.* 6 Ma) occurred along east–west alignments (Paulsen and Wilson 2010). These data suggested a rapid rotation in the maximum horizontal stress field from north–south to east–west as early as 6 Ma. Paulsen and Wilson (2010) noted that the east–west orientation since 6 Ma is parallel to the absolute motion of the Antarctic Plate. They concluded that the rotation of the stress field was reflected in a change in the spatial pattern of volcanoes, and was driven by a major plate reorganization and a change in plate motion in the Pacific Basin.

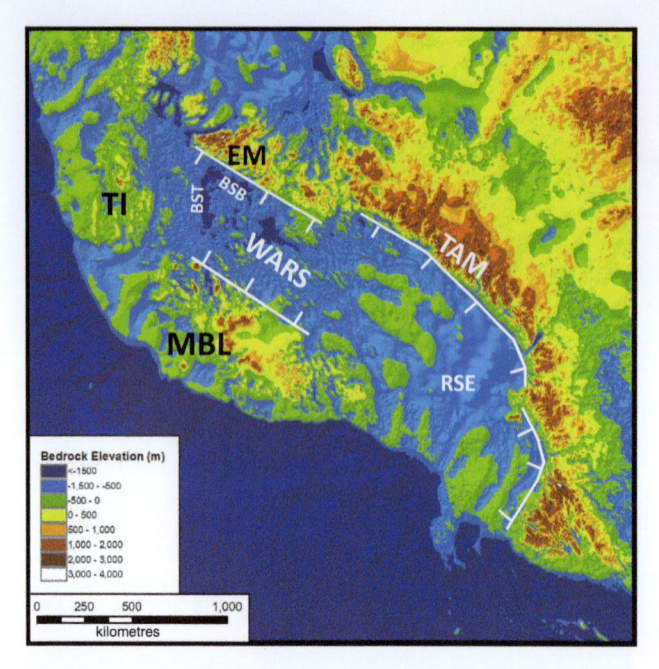

Fig. 4. Map of subglacial topography of West Antarctica derived from BEDMAP2 data (Fretwell *et al.* 2013). Crustal blocks in West Antarctica include the Marie Byrd Land (MBL), West Antarctic Rift System (WARS), Thurston Island (TI) and the Ellsworth Mountains (EM) blocks. Most of the MBLVP central volcanoes are >2000 m asl, shown in red. The MBLVP is located on the MBL crustal block on the north flank of the West Antarctic Rift System (WARS) The WARS is dominated by deep subglacial basins including the Byrd Subglacial Basin (BSB) and the Bentley Subglacial Trough (BST). The WARS extends through the Ross Sea Embayment (RSE). The volcanoes of TIVP are situated on the Thurston Island crustal block. The Ellsworth Mountains and connected Transantarctic Mountains (TAM) form the rift shoulder on the south flank of the WARS.

Possible causes of the MBLVG volcanism are debated by geochemists and geophysicists, and are discussed briefly here and in detail by Panter *et al.* (2021). Geochemically and isotopically, the MBLVG alkaline volcanic rocks have a HIMU (high $\mu = {}^{238}$U/^{204}Pb)-like mantle signature, typically associated with ocean island magmas (see the discussion in Panter *et al.* 2021). Several models to explain the origins of the HIMU-like magma and Cenozoic volcanism in MBL and the rest of the WARS have been proposed, ranging from active mantle-plume-driven volcanism since mid-Cenozoic time, to mid–late Cenozoic melting of a fossil Cretaceous mantle plume, to passive rifting and decompression melting of enriched mantle source rocks (for discussions see Storey *et al.* 2013; Martin *et al.* 2021; Panter *et al.* 2021; Rocchi and Smellie 2021).

Fourteen of the 19 central felsic volcanoes of the MBLVG are aligned in four linear chains (Fig. 5) (LeMasurier and Rex 1989). The linear chains show progressions in the ages of the onset of volcanic activity at the central volcanoes along each chain. In the Executive Committee Range Volcanic Field there is a north–south alignment of five polygenetic volcanoes with southward younging from Mount Hampton (13.4 Ma) to Mount Waesche (2.0 Ma). In the Ames Range Volcanic Field there is a north–south alignment of three polygenetic volcanoes with a northward younging from Mount Andrus (12.7 Ma) to Mount Kauffman (7.1 Ma). In the Crary Mountains Volcanic Field, there is north–south alignment of three polygenetic volcanoes with a northward younging from Mount Rees (9.46 Ma) to Mount Frakes (4.3 Ma). In the Flood Range Volcanic Field, there is an east–west alignment of three polygenetic volcanoes with a westward younging from Mount Bursey (10.1 Ma) to Mount Berlin (0.578 Ma).

Fieldwork

Field research of the MBLVG has required major logistical support because the closest permanent research station is over 1300 km away (Table 1). The US Antarctic Program has provided logistical support for most of the volcano studies, although some field seasons included international teams of researchers and support from other international programmes (e.g. British Antarctic Survey; New Zealand Antarctic Programme). Most of the central volcanoes were discovered during overflights by Admiral Richard Byrd in the second Antarctic Expedition (1933–35) and the United States Service Expedition (1939–41). Scientists participating in several oversnow traverses from 1934 to 1959 made the first geological visits to the volcanoes. Many of the volcanoes and volcanic features are named after the expedition and traverse team members, as summarized in LeMasurier and Thomson (1990). These visits were typically limited to very few outcrops. Expeditions focused on geological mapping and sample collection began in 1960 with an investigation of the Jones Mountains in western Ellsworth Land (Craddock *et al.* 1964). Three major expeditions from 1966 to 1969 focused on understanding the geology of West Antarctica (Wade 1971). The 1967–68 Marie Byrd Land Survey II concentrated

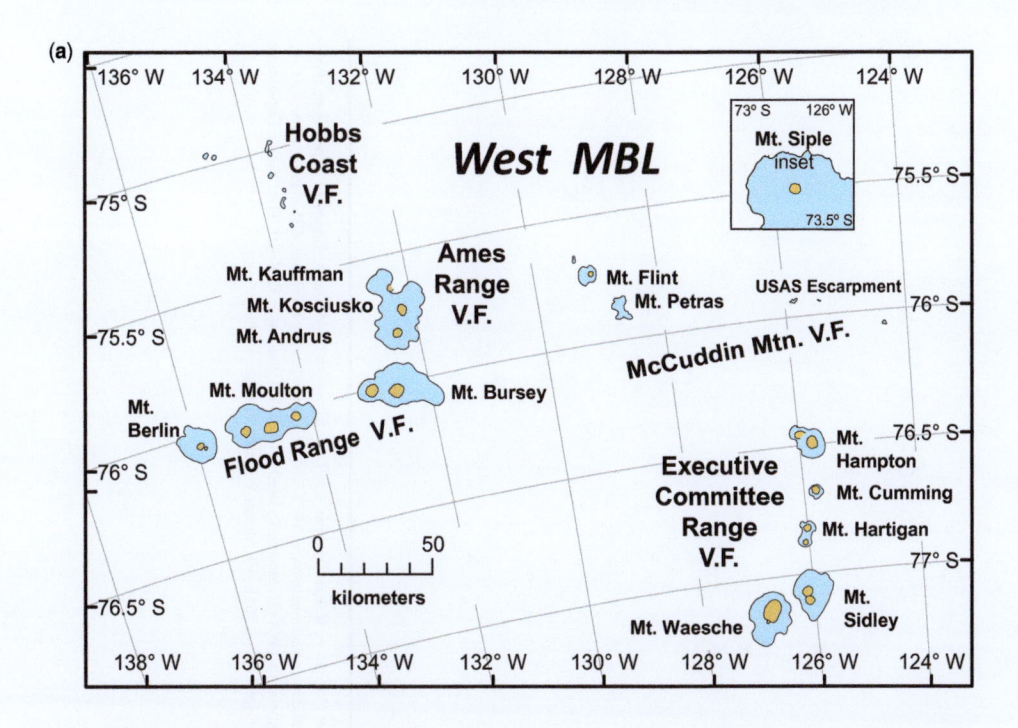

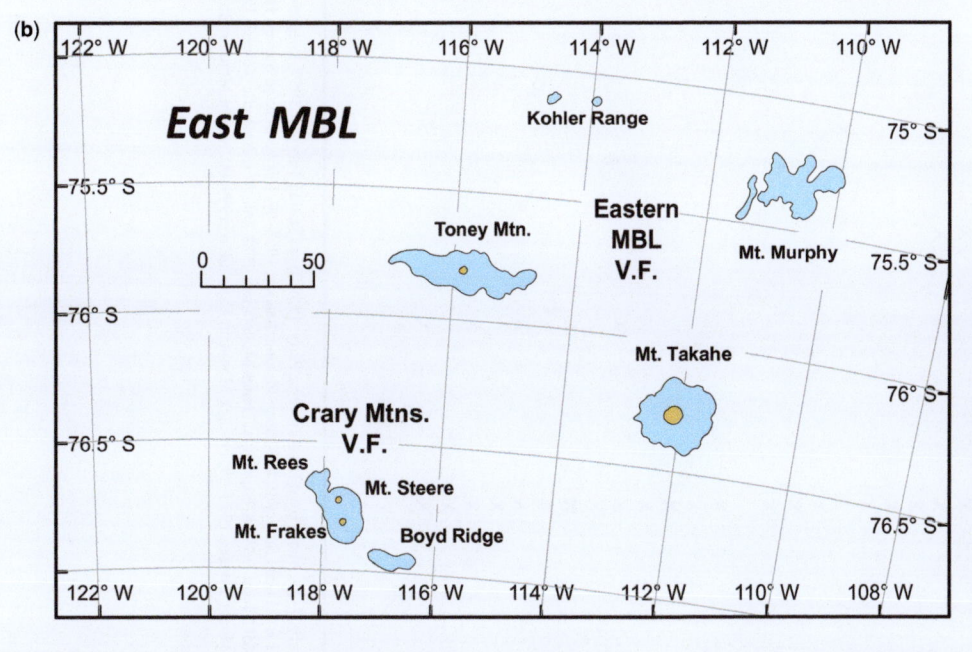

Fig. 5. (a) Map of the western and central MBLVP volcanoes; and (b) a map showing the eastern MBLVP volcanoes. The inset map in (b) shows the Mount Siple Volcanic Field and central volcano. Map boundaries are delineated in Figure 1. Volcanic fields and individual central volcanoes are labelled. The approximate map view extent of the volcanoes at the level of the WAIS is shaded blue, and summit calderas of central polygenetic volcanoes are shaded yellow. Reference map with caldera age data given in Table 2.

on the MBLVP, and all of the central volcanoes and major satellite centres of MBL were visited by helicopter reconnaissance from temporary deep field camps. Wesley LeMasurier led the 1967–68 investigations of the MBL volcanoes that resulted in the first analysis of the nature, extent and significance of the MBL volcanism (e.g. LeMasurier 1972), and laid the foundations for subsequent analysis. The 1968–69 Ellsworth Land Survey included study of the Jones and Hudson mountains in western Ellsworth Land (Wade and Craddock 1968). In 1977–78, another helicopter-based survey was conducted from a single-season deep field camp, focusing on central volcanoes in the Ames and Flood ranges, as well as isolated nunataks along the Hobbs Coast. Results from the 1967–68 and 1977–78 field seasons, and subsequent analyses on the rocks collected, are the basis of most summaries in LeMasurier (1990*h*) and of many subsequent papers. Most volcanoes have been revisited since this early helicopter-based fieldwork, although many outcrops have not been revisited.

The most recent era of fieldwork involving more detailed analysis of selected volcanoes began in 1984–85 at Mount Takahe and Mount Murphy, and was carried out by small autonomous teams transported to the deep field via LC-130 aircraft. These teams accessed outcrops by snowmobiles. Subsequent field seasons in 1989–90 (Executive Committee Range), 1990–91 (Mount Murphy and Mount Hampton), 1992–93 (Crary Mountains), 1993–94 (Ames and Flood ranges, Hobbs Coast Nunataks and McCuddin Mountains), 1997–98 (Mount Waesche), 1998–99 (Mount Takahe), 1999–2000 (Mount Moulton), and 2018–19 (Mount Waesche) used this same autonomous approach augmented by transport to some sites by Twin Otter aircraft. Table 1 lists major expeditions and a fieldwork timetable for each of the central MBL volcanoes and at satellite volcanic centres.

Form and structure of the MBLVP central volcanoes

Most of the 19 central volcanoes have a shield-like morphology, with low-angle (10°–15°) flank slopes and pronounced summit calderas (up to 8 km in diameter) (Fig. 5). Eight of

Table 1. Field expeditions to Marie Byrd Land volcanoes

Volcano	1934	1939–41	1957–58	1959	1959	1960–61	1966–67	1967–68	1968–69	1977–78	1984–85A	1984–85B	1984–85C	1989–90	1990–91	1992–93	1993–94	1998–99	1999–00	2018–19
Mount Murphy								X			X				X					
Mount Takahe			X					X			X							X		
Toney Mountain				X				X												
Mount Steere								X								X				
Mount Frakes								X								X				
Mount Siple												X					X			
Mount Hampton								X							X					
Mount Cumming								X						X						
Mount Hartigan (northern)								X												
Mount Sidley								X						X						
Mount Waesche								X						X						X
Mount Flint								X									X			
Mount Andrus								X		X							X			
Mount Kosciusko								X		X										
Mount Kauffman								X		X							X			
Mount Bursey								X		X							X			
Mount Moulton								X		X							X		X	
Mount Berlin		X						X									X			
Kohler Range								X												
Mount Petras								X		X					X					
Usas Escarpment								X		X							X			
Hobbs Coast (M-P)							X													
Fosdick Mountains	X																			
Jones Mountains					X	X			X				X							
Hudson Mountains					X	X			X				X							

1934, Byrd Antarctic Expedition II; 1939–41, US Antarctic Service Expedition; 1957–58, Marie Byrd Land Traverse Party; 1959, Byrd Station Traverse; 1959, Executive Committee Range Traverse; 1960–61, Ellsworth Land Camp Minnesota Expedition; 1966–67, Marie Byrd Land Survey, Helicopter Reconnaissance; 1967–68, Marie Byrd Land Survey II, Helicopter Reconnaissance; 1968–69, Ellsworth Land Survey, Helicopter Reconnaissance; 1977–78, Helicopter Survey (Western MBL); 1984–85A, Ice Core Tephra Source (Mount Takahe); 1984–85B, Polar Sea Helicopter; 1984–85C, Joint US–UK Expedition (Jones Mountains); 1989–90, West Antarctic Volcano Expedition (WAVE) Season 1; 1990–91, WAVE, Season 2; 1990–91, Ford Range Expedition; 1992–93, WAVE II, Season 1; 1993–94, WAVE II, Season 2; 1998–99, Mount Takahe; 1998–99, Fosdick Mountain Expedition; 1999–00, Mount Moulton Blue-ice Tephra; 2018–19, Mount Waesche Glaciation Study.

the 19 volcanoes have a single summit caldera. Mount Hampton, Mount Waesche and Mount Sidley in the Executive Committee Range each have an older caldera truncated by a younger caldera. Mount Hartigan, Mount Berlin and Mount Bursey each have a summit caldera and multiple non-overlapping older calderas. Mount Moulton has a summit caldera, and a second, and possibly a third, caldera. Mount Rees and Mount Murphy are both deeply dissected and lack preserved summit calderas. If each isolated caldera were considered a distinct central volcano, there would be as many as 25 central volcanoes.

Other features of most of the central volcanoes include a lack of dissection; limited outcrops due to almost complete snow- and ice-cover; lithofacies dominated by lava rather than pyroclastic rocks; and rock compositions having largely bimodal alkaline compositions, dominated by mafic basanite and felsic trachyte and phonolite. LeMasurier (1990*h*) characterized the volcanoes as shield volcanoes, based on low slope angles and the dominance of lavas in outcrop and the flank slopes. Panter *et al.* (1994) characterized Mount Sidley with a large breached summit caldera as a stratovolcano, and early workers (e.g. González-Ferrán and González-Bonorino 1972; LeMasurier 1972) described many of the central volcanoes as stratovolcanoes. In general, the low degree of dissection precludes assessment of the internal structure of most of the volcanoes. Detailed studies of some of the more dissected volcanoes (Panter *et al.* 1994, 1997, 2000; Wilch *et al.* 1999; Wilch and McIntosh 2000) reveal complex polygenetic histories (Panter *et al.* 2021). Caldera wall exposures at many of the central volcanoes include explosively-erupted non-welded and welded pumiceous materials (Wilch *et al.* 1999; authors' personal observations), although most early work described these exposures as lavas. Recognizing the general low angle slopes and also the complex polygenetic evolutionary patterns among the best studied volcanoes, we characterize the central volcanoes as shield-like composite volcanoes. An exception is Mount Sidley, a more typical stratovolcano.

Most of the large central volcanoes and satellite centres in the MBL and Thurston Island volcanic provinces are surrounded and partially buried by the WAIS. Many of the volcanoes and volcanic ranges are significant obstacles to flow of the WAIS, with ice on the upstream sides of the edifices as much as 800 m higher (e.g. Mount Moulton) than ice on the downstream sides. Andrews and LeMasurier (1973) evaluated rates of glacial erosion by comparing the heavily-dissected Mount Murphy and the much less dissected Mount Takahe.

The central volcanoes of the MBLVG are among the largest volcanoes in Antarctica and the world (Table 2). Mount Sidley is the highest Antarctic volcano at 4285 m asl. All of the MBL central volcanoes have summit elevations >2000 m asl and 10 are >3000 m asl. Only nine central volcanoes in Antarctica outside of MBL reach elevations >2000 m asl and only four of those are >3000 m asl. Estimating the size and volume of MBL volcanoes is challenging because most of the volcanoes are buried by the WAIS, and some are likely to extend far below ice and sea level (LeMasurier 1990*h*; LeMasurier 2013). The amount of relief above ice level ranges from just 400 m at Mount Cumming to 2400 m at Toney Mountain. The exception is the island volcano Mount Siple along the Bakutis Coast that is separated from the continent by the Getz Ice Shelf. Mount Siple rises from sea level to about 3100 m asl. The edifice shapes and exposed bases of the central volcanoes vary from circular to very elongate, with exposed basal dimensions from <10 to 70 km in length. The exposed volumes (above ice or sea level) range from 30 km^3 to 1800 km^3, with the largest estimate being Mount Siple at the coast (LeMasurier 1990*h*) (Table 2). Mount Siple would rank third on lists of edifice volumes calculated from digital elevation models (DEMs) of more than 900 shield volcanoes and composite worldwide (Grosse *et al.* 2014; Grosse and Kervyn 2018). Given that many of the MBL volcanoes are likely to extend below ice level, the exposed volumes represent minimum estimates. Toney Mountain is the only volcano in MBL for which there is a geophysically-based depth estimate of the sub-ice volcano–basement contact (Bentley and Clough 1972). The contact is at 3000 m below sea level, suggesting a total relief of the volcano of *c.* 6600 m; with volumes estimates ranging from 2800 to 3613 km^3 (LeMasurier *et al.* 1990*d*; LeMasurier 2013). Given that most of the volcanoes have portions buried beneath the WAIS and that in places the WAIS is grounded far below sea level in deep basins, the calculated exposed volumes represent minimum values. In the absence of good depth data on the volcano–basement contacts, LeMasurier (2013) estimated a likely ratio of felsic to mafic rocks in five of the central volcanoes, and used this to estimate the amount of rock buried beneath the ice-sheet surface. Using this method, calculated volumes were up to 10 times greater than the exposed volumes, with Mount Takahe having the largest estimated volume of 5520 km^3 (LeMasurier 2013). This method assumes that the ratio of mafic to felsic rock is known and that the felsic volume can be estimated from limited outcrops. In this chapter, we use the more conservative estimated volumes based on the volcano volumes above ice level by LeMasurier (1990*h*), realizing that these are minimum volumes and may far underestimate the actual volumes (Table 2).

LeMasurier (1972) proposed a generalized pattern of stratigraphic relationships among most central volcanoes that is characterized by a mafic 'basal succession' overlain by a felsic shield volcano, followed by late-stage post-edifice-building parasitic scoria cones and tuff cones situated on the shield flanks. In this model, the basal succession consists of subglacial to subaerial basaltic lavas and volcaniclastic rocks that form a platform that overlies a flat pre-Cenozoic basement. The basal succession includes many reported hyaloclastite deposits that were interpreted as erupted in subglacial environments (LeMasurier 1972; LeMasurier and Rex 1982, 1983; LeMasurier 1990*h*). The well-exposed Mount Murphy was used as a representative example, and was described as a basaltic shield volcano surmounted by a felsic shield volcano. Although this simple tripartite model of volcanoes built on a flat erosion surface may be accurate in places, detailed studies of the stratigraphy and evolution of some of the more dissected and better-exposed volcanoes indicate more complicated and varied histories, and a less flat erosion surface at many volcanoes (Panter *et al.* 1994; Wilch *et al.* 1999; Wilch and McIntosh 2000, 2002; Smellie 2001). These cases will be addressed in the individual descriptions of volcanoes.

As mentioned, post-edifice-building volcanoes are common in MBL and are superimposed on or are proximal to most of the central volcanoes. The number of preserved and exposed parasitic cones ranges from about two to 10 per central volcano. In a few cases, such as Mount Sidley, these cones are close in age to the main edifice-building interval (<0.4 myr younger); in other cases, such as Mount Cumming and Mount Bursey, the parasitic cones post-date main edifice construction by millions of years (Table 2). The parasitic volcanoes are located on the flanks of the central volcanoes and are typically erosional remnants of mafic scoria cones, characterized by partial crater rims or vent deposits and/or mixtures of welded and non-welded pyroclastic deposits with subordinate clastogenic or massive lavas. Some of the parasitic volcano lavas and tuffs contain large quantities of crustal and mantle xenoliths. Parasitic tuff cones and glaciovolcanic tuyas are less common and are identified by diagnostic lithofacies (Russell *et al.* 2014). Post-edifice-building parasitic intermediate and felsic domes, tuff cones, tuyas, and lavas

Table 2. *MBL Volcanic Province central volcanoes*

Volcanic fields (VF), central volcanoes (1–19), secondary calderas	Latitude (°)	Longitude (°)	Summit elevation* (m asl)	Elevation above ice level (m)	Upstream ice elevation* (m asl)	Downstream ice elevation* (m asl)	Estimated volume[†] (km³)	Calderas	Caldera diameter (km)	Edifice-building age range age ± 2 SD (Ma)	n[‡]	Post-edifice-building age range age ± 2 SD (Ma)	n[§]
Eastern MBL VF													
1. Mount Murphy	−75.3496	−110.7101	2703	200	800	200	580	No caldera		9.46 ± 0.10	23	3.65 ± 0.16	5
										5.18 ± 0.18		0.609 ± 0.027	
2. Mount Takahe	−76.2817	−112.1093	3460	2100	1400	1400	780	Summit caldera	8	0.194 ± 0.006	6	0.105 ± 0.028	2
										0.008 ± 0.005		0.007 ± 0.013	
3. Toney Mountain	−75.8072	−115.8600	3595	2400	1600	1200	550	Summit caldera	3	*9.6 ± 1.0*	*1*	*1.0 ± 0.4*	2
												0.29 ± 0.1	
Crary Mountains VF													
4. Mount Rees	−76.6744	−118.0732	2709	900	2000	1600	400	No caldera	2	9.46 ± 0.24	15	7.01 ± 0.21	2
										7.62 ± 0.06		6.91 ± 0.26	
5. Mount Steere	−76.7250	−117.7812	3558	1800	2000	1600	n.d.	Summit caldera	none	8.66 ± 0.04	30	3.93 ± 0.03	5
										5.81 ± 0.04		0.032 ± 0.010	
6. Mount Frakes	−76.8078	−117.6994	3654	1900	2000	1800	n.d.	Summit caldera	3 × 2.5	4.26 ± 0.05	3	2.17 ± 0.32	3
												1.29 ± 0.03	
Boyd Ridge[¶]	−76.9503	−116.8107	2375	800	1800	1600	n.d.	No caldera	none				
7. Mount Siple VF	−73.4313	−126.7641	3100	3100	0	0	1800	Summit caldera	4.5	0.230 ± 0.008	2	0.746 ± 0.036	2
										0.171 ± 0.005		0.008 ± 0.088	
ECR VF													
8. Mount Hampton	−76.4937	−125.7847	3323	800	2500	2500	70	Summit caldera	6.5 × 5.5	11.43 ± 0.04	5	11.4 ± 1.2	2
										8.6 ± 1.0		10.7 ± 0.8	
Whitney Peak	−76.4577	−125.9767	3003	600				Whitney Peak caldera	indet.	13.36 ± 0.05	4		
9. Mount Cumming	−76.6786	−125.8061	2612	300	2400	2400	na	Summit caldera	4.5 × 3.5	*10.4 ± 1.0*	2	*3.0 ± 0.4*	1
										10.0 ± 1.0			
10. Mount Hartigan	−76.8196	−126.0263	2811	600	n.d.	n.d.	na	Boudette Peak caldera	3.5	8.50 ± 0.66	5		
										6.02 ± 0.50			
Tusing Peak	−76.8717	−126.0742	2652	450				Tusing Peak caldera	3.5	8.36 ± 0.82	3		
										7.57 ± 0.60			
11. Mount Sidley	−77.0570	−126.1346	4181	2200	2600	2000	250	Summit caldera	4.5	4.43 ± 0.06	1	4.37 ± 0.06	3
										4.87 ± 0.06		4.24 ± 0.08	
Weiss Peak	−77.0352	−126.0428	3292	1200				Weiss Peak caldera	2.5	5.77 ± 0.12	5	4.66 ± 0.10	5
												4.51 ± 0.02	
12. Mount Waesche	−77.1686	−126.8938	2920	700	2400	2000	160	Summit caldera	1.5	*1.0 ± 0.2*	5		
										<0.1 ± 0			
Chang Peak	−77.1201	−126.7648						Chang Peak caldera	10	2.01 ± 0.10	3		
										1.09 ± 0.10			
McCuddin Mountains VF													
13. Mount Flint	−75.7223	−129.0590	2695	900	2000	2000	52	Summit caldera	2 × 3	20.46 ± 0.07	3	9.67 ± 0.20	3
Ames Range VF									252				
14. Mount Andrus	−75.8083	−132.3474	2978	1400	1800	1600	115	Summit caldera	4.5	12.71 ± 0.06	5	3.75 ± 0.06	3
										11.18 ± 0.19		<0.1	
15. Mount Kosciusko	−75.7161	−132.2000	2909	1300	1800	1600	107	Summit caldera	3.5 × 5	*9.2 ± 1.3*	1	*11.60 ± 0.80*	3
16. Mount Kauffman	−75.6308	−132.3766	2364	800	1800	1600	30	Summit caldera?	indet.	*7.1 ± 1.5*	1		
Flood Range VF													
17. Mount Bursey	−76.0260	−132.5154	2787	1000	2200	2000	290	Summit caldera	5	6.04 ± 0.48	1	0.49 ± 0.12	1
Koerner Bluff	−76.0078	−132.9713						Koerner Bluff caldera	6 × 8.5	10.12 ± 0.34	3	8.56 ± 0.06	2
										9.31 ± 0.74		0.25 ± 0.03	
18. Mount Moulton	−76.0373	−135.1262	3078	1700	2400	17000	325	Summit caldera	5.5	4.03 ± 0.14	1	1.04 ± 0.04	1
Prahl Crag	−76.0513	−134.6589						Prahl Crag caldera	4.5 × 7	5.95 ± 0.05	1		
Kohler Dome	−76.0459	−134.2753						Kohler Dome caldera?	5	n.d.			
19. Mount Berlin	−76.0547	−135.8655	3478	2100	1900	1400	200	Summit caldera	1.5	0.0279 ± 0.0064	7		
										0.01 ± 0.0053			
Merrem Peak	−76.0378	−135.9588						Merrem Peak Caldera	2.5 × 1	0.578 ± 0.009	16	0.214 ± 0.018	6
										0.143 ± 0.006			

*Elevations from USGS 1:250 000 Antarctic Topographic Map Series.
[†]Estimated volume exposed above the ice sheet, from LeMasurier and Thomson (1990)
[‡] *n* refers to the number of samples included in the edifice-building age range.
[§] *n* refers to the number of samples included in the post-edifice-building age range.
[¶]Boyd Ridge is included in this table because it is likely to be an additional central volcano, although it lacks a caldera and adequate exposure to identify it as such.
Other notes: the '?' in the Caldera column indicates uncertainty about the idenfication of a caldera; indet., indeterminate.
Bracketing ages: italicized ages are conventional K–Ar ages, all other ages are $^{40}Ar/^{39}Ar$ ages; ages between brackets are mostly $^{40}Ar/^{39}Ar$ ages but also include some K–Ar ages.
Central volcanoes (numbered 1–19) are either single centres or coalesced centres, often identified by more than one caldera, indicated by inset.

occur at some locations. Post-edifice-building satellite volcanoes (not on the flanks of, but proximal to, major central volcanoes) and monogenetic volcanoes not proximal to the central volcanoes (e.g. in the Hobbs Coast and Jones and Hudson mountains volcanic fields) range from mostly intact to highly-eroded volcanoes, dominated by scoria cones, with fewer tuyas, tuff cones and lava domes.

Volcanic lithofacies

A wide range of volcanic lithofacies has been documented in the MBLVG. Terminology used to describe the character and interpreted origin of the MBLVG rocks has changed over time, from more general in early studies (pre-1990) to more specific and detailed in later studies. In more recent studies there has been a trend toward more detailed non-genetic descriptions of the rocks (often following suggestions of McPhie et al. 1993) separate from the interpretations of the rock origins and palaeoenvironments (e.g. see Smellie et al. 1993; Wilch and McIntosh 2000, 2002, 2007; Smellie 2001). The more detailed lithofacies approach has led to reinterpretations of some sequences and a refinement in interpretations of others. An emphasis of the MBLVG physical volcanology research has been on differentiating rocks erupted in contact with the atmosphere with no influence of external water from rocks erupted into or in contact with ice or meltwater. Wilch and McIntosh (2000, 2002, 2007) used the general terms 'wet' and 'dry' to differentiate rocks erupted in contact with ice from those erupted subaerially. The ice-contact rocks are very important because they offer the potential to make inferences about the history of the WAIS at the time of eruptions. Smellie and Edwards (2016) discuss the lithofacies approach to describe and interpret ancient volcanic sequences in glacial environments in more detail. This method is particularly powerful because the alkaline rocks are readily datable using K–Ar or $^{40}Ar/^{39}Ar$ techniques.

Here, volcanic lithofacies are subdivided into three classes: primary coherent lithofacies; primary fragmental lithofacies; and secondary epiclastic lithofacies. The coherent lithofacies include lava and dykes (after McPhie et al. 1993). Coherent lava is the most common lithofacies found at the MBLVG outcrops. For this study, coherent lava lithofacies are subdivided into dominant subaerial (also referred to as 'dry') and subordinate subaqueous (or subglacial) (also referred to as 'wet') end members, based on the presence or absence of features (e.g. hyaloclastite breccia) characteristic of interactions with external water.

The subaerial lava lithofacies includes lavas with reddened brecciated bases and pāhoehoe tops or reddened brecciated tops that are interpreted to result from subaerial lava effusion and emplacement, without recognizable water or ice interaction. Many lavas, particularly felsic lavas, lack diagnostic features of subaerial or subaqueous environments. Unless these lavas are associated with other subaqueous deposits, they have been tentatively interpreted as subaerial lavas. Likewise, a common descriptor of MBL lavas in the LeMasurier and Thomson (1990) volume is 'flow rock', which we infer to be subaerial lavas, unless reinterpreted otherwise in subsequent studies. Other textural or morphological descriptors of subaerial lavas (such as compound, sheet, dome, clastogenic) are used when possible. Clastogenic lava, derived from agglutinated pyroclastic spatter, is a common subaerial lithofacies in both mafic and felsic end members, and shows pyroclastic and flowage textures. Intrusive bodies, including dykes, sills and irregular intrusions, are common at more deeply-eroded volcanoes.

The subglacial (or subaqueous) lava lithofacies show evidence of quenching, and are interpreted as water-cooled and generally associated with subaqueous eruptive or depositional environments (see Smellie and Edwards 2016). Two lava lithofacies, pillow lava and blocky or curvi-columnar-jointed sheet lava (see Smellie and Edwards (2016)), are common in several MBL sequences. Two variations of pillow lava lithofacies are recognized: compound nested pillows with minor interpillow hyaloclastite breccia; and lobe hyaloclastites composed of irregular pillow lobes with abundant (>10%) interpillow hyaloclastite breccia. Pillow lava lithofacies are common in MBL at Mount Rees, Mount Steere, Mount Murphy and Mount Takahe but are rare elsewhere. The blocky or curvi-columnar-jointed sheet lava includes slightly glassy, compound and simple lavas and intrusive bodies, with irregular to hackly jointing and rare crude pillow structures. In rare cases, lava apophyses locally deform bedding and exhibit hackly jointing. These apophyses are interpreted as dykes or lava that intruded wet volcaniclastic sediments.

For this study, primary fragmental lithofacies include 'dry' and 'wet' end members to differentiate deposits erupted and emplaced with no interaction between magma or lava and external water from those deposits that show evidence of external water interactions. The principal 'dry' end member is autobreccia. Subaerial autobreccias are ubiquitous in MBL, and are recognized by welding textures, reddening caused by deuteric oxidation and a lack of thick glassy 'quenched' margins. The principal 'wet' end member is hyaloclastite. Use of the term hyaloclastite requires some additional explanation because the term has been confusing in the literature when describing the MBLVG deposits. Early work on MBL volcanoes in LeMasurier and Thomson (1990 and references therein) followed a convention proposed by Fisher and Schmincke (1984) that hyaloclastites include glassy volcaniclastic deposits without regard to the origin of the glassy clasts. In other words, hyaloclastite included deposits formed by cooling contraction granulation, as well as by phreatomagmatic explosions. Hyaloclastites were typically characterized as palagonitized glassy breccia. In general, any palagonitized volcaniclastic deposits were referred to as hyaloclastites and were interpreted as evidence of past ice-sheet expansions. By contrast, following Smellie and Skilling (1994) and Wilch and McIntosh (2000, 2002), differentiated hyaloclastite formed from cooling contraction granulation in contact with water from hyalotuff (Honnorez and Kirst 1975) associated with explosive phreatomagmatic eruptions. In their usage, Wilch and McIntosh (2000, 2002) referred to locally reworked glassy material associated with pillow-lava sequences as reworked hyaloclastite. At several sites (e.g. Mount Takahe and Mount Rees), hyaloclastites are characterized as well stratified, suggesting that they were locally redeposited on underwater slopes by sediment gravity-flow processes. The definition of hyaloclastite proposed by White and Houghton (2006) only includes in situ deposits, whereas reworked deposits, common in lava-fed deltas associated with tuyas, are called breccia or lapilli tuff depending on grain size. In this chapter, we follow Wilch and McIntosh (2000, 2002) with the slightly less-constrained usage of the term hyaloclastite to include locally reworked hyaloclastite. The MBLVG hyaloclastites described here are likely to include peperite, a glassy breccia formed by lava or intrusions quenching and granulating in contact with wet sediment (Skilling et al. 2002). The distinction between hyaloclastite and peperite was not made in the field. In a few instances, it appears that the hyaloclastites and parent pillow lavas were moving downslope together and pillow lobes or fingers were injected up into the hyaloclastite breccia.

Other fragmental lithofacies (breccia, tuff breccia, lapilli tuff and tuff) differentiate deposits based on grain size and do not imply specific emplacement or eruption conditions. These lithofacies are dominated by two types: pyroclastic fall and dilute pyroclastic density current deposits. Pyroclastic fall deposits are products of magmatic or phreatomagmatic explosions and typically form crudely- to well-stratified deposits that mantle topography. A magmatic origin is inferred where pyroclasts are well sorted and exhibit uniform, moderate to high vesicularity and angular to fluidal shapes. Reddening by deuteric oxidation and welding are common features, and are interpreted as indications of close proximity to a sub-aerial vent (Walker and Croasdale 1971). The most common pyroclastic fall deposits are basaltic welded lapilli and bomb-rich outcrops at intact and eroded scoria cones, associated with mildly-explosive Strombolian-style eruptions. We make these interpretations using an ensemble of criteria, understanding the caveat that criteria to distinguish magmatic from phreatomagmatic fragmentation processes are not entirely diagnostic (White and Valentine 2016). A phreatomagmatic origin is inferred by relatively poor sorting, fine grain sizes, variable vesicularity, ash-coated or accretionary lapilli (in basalts), a predominance of sideromelane glass (in basalts) and blocky clast morphology (Wohletz 1983; Fisher and Schmincke 1984). In MBL, dilute turbulent pyroclastic density current deposits are identified by comparison to similar deposits described elsewhere (e.g. Wohletz and Sheridan 1983; Chough and Sohn 1990; Dellino and La Volpe 2000; Douillet et al. 2015) and typically form moderately- to poorly-sorted, planar, cross-stratified or massive beds that thicken and thin laterally. Individual clasts in these dilute pyroclastic density current (surge) deposits resemble those in phreatomagmatic fall deposits, except that they are often much more rounded and abraded due to turbulent lateral transport. In the past, some phreatomagmatic surge and fall deposits have been variably referred to as hyaloclastite (LeMasurier 1990g), hydroclastic tuff (LeMasurier and Rex 1990b) and hyalotuffs (Wilch and McIntosh 2000, 2002). These deposits commonly include signs of wet phreatomagmatic origin followed by sub-aerial deposition, such as ash-coated lapilli and bedding-plane sags caused by ballistic impacts. Pyroclastic density current and fall deposits are commonly interlayered. In western and central MBL, many fall and surge deposits contain clasts that exhibit a continuum of characteristics from phreatomagmatic explosivity (blocky shapes, a wide range of vesicularity) to magmatic explosivity (fluidal to cuspate shapes, high vesicularity).

Explosively-erupted felsic lithofacies are relatively rare compared to mafic counterparts but are, nonetheless, important in the eruptive histories of polygenetic central volcanoes of the MBLVP. Felsic welded pyroclastic fall deposits with variably-flattened fiamme were preserved in many caldera walls of the central volcanoes. At Mount Berlin, these caldera-wall pyroclastic fall deposits are associated with highly-explosive Plinian eruptions; and a similar eruption style is inferred at other calderas. Less commonly, felsic welded ignimbrites have been recognized at Mount Sidley and Mount Berlin.

Secondary epiclastic deposits are less common in MBL and include deposits formed by glacial, mass-flow and fluvial processes. Heterogeneous rock types, subrounded clast shapes, sedimentary structures and stratigraphic context characterize these sedimentary deposits. Despite the intraglacial setting of the MBL volcanic province, interbedded glacial tills/tillites and glacial–erosional unconformities are uncommon, limited to Mount Murphy, the Hobbs Coast Volcanic Field and Mount Aldaz.

Geochronology of MBL volcanoes: K–Ar and $^{40}Ar/^{39}Ar$ dating

Geochronology is an integral part of past studies of the MBLVG. This chapter summarizes published K–Ar and $^{40}Ar/^{39}Ar$ data, and presents unpublished $^{40}Ar/^{39}Ar$ data. Early work on dating MBLVG rocks relied on conventional K–Ar ages (summarized in LeMasurier and Thomson 1990), as well as on a few fission-track ages (Seward et al. 1980; Palais et al. 1988). Conventional K–Ar ages of MBLVG rocks (e.g. in the LeMasurier and Thomson 1990 volume and publications referenced therein) reported uncertainties at the 1σ level (66% confidence level). More recent studies use $^{40}Ar/^{39}Ar$ dating, and several studies presenting $^{40}Ar/^{39}Ar$ data report ages at the 2σ level (e.g. Wilch et al. 1999; Smellie 2001; Wilch and McIntosh 2002). In this paper, analytical 2σ level (95% confidence level) uncertainties are applied to all preferred ages.

In addition to correcting all uncertainties to 2σ, age corrections were applied to accommodate changes in standards and decay constants since the original calculation of ages. For example, the conventional K–Ar results prior to 1989 used a different K–Ar decay constant and the ages have been corrected to the now-accepted constant (Steiger and Jaeger 1977), following the method of Dalrymple (1979). Most of the $^{40}Ar/^{39}Ar$ results used the Fish Canyon sanidine (FCs) standard to monitor the neutron flux during irradiation but used now-outdated FCs ages in the calculation of sample ages. These samples were corrected to the accepted FCs age of Kuiper et al. (2008). Corrections applied to each sample are included in the Supplementary material Table S1.

In this study, $^{40}Ar/^{39}Ar$ ages are preferred over conventional K–Ar ages. Only $^{40}Ar/^{39}Ar$ ages are included in data tables for sites where both methods were applied; K–Ar ages are included for sites that have only been dated by the conventional K–Ar method. The reasons for the preference for $^{40}Ar/^{39}Ar$ ages over K–Ar ages are documented by Wilch (1997) and Wilch and McIntosh (2000, 2002, 2007), and are based on higher precision results, the ability to assess the reliability of results, the ability to date smaller samples including single crystals and the ability to date very young events.

Summaries of MBLVG age data are presented in a series of tables embedded in the volcano summaries. A compilation of more complete age data is available as Supplementary material. In total, results from 330 age analyses (273 $^{40}Ar/^{39}Ar$, 52 K–Ar and five fission track) are presented in this chapter, including 52 previously unpublished $^{40}Ar/^{39}Ar$ ages mostly from Mount Murphy and Mount Takahe. Complete $^{40}Ar/^{39}Ar$ datasets of new data from Mount Murphy and Mount Takahe are included in Supplementary material Tables S2 and S3.

At some localities, multiple samples are correlated to the same eruption event and a total of 198 eruption events are recognized: 80% of the eruption events are associated with the central volcanoes (Fig. 6a) and 20% are associated with minor isolated volcanoes (Fig. 6b). The number of samples and level of detail of the geochronology are quite variable at the MBLVG volcanoes. Well-dated, deeply-dissected volcanoes, including Mount Rees, Mount Steere, Mount Sidley and Mount Murphy, exhibit long-lived, >1 myr, eruption histories.

Summaries of volcanic fields, polygenetic central volcanoes and satellite volcanic centres

The physical volcanology and geochronology of central and satellite volcanoes of the MBLVG is summarized here. The descriptions of the MBLVG volcanoes are organized by volcanic province and volcanic field, as listed in Figure 3. Table 2 summarizes key physical characteristics of the central volcanoes of the MBLVP, as well as age ranges of edifice- and

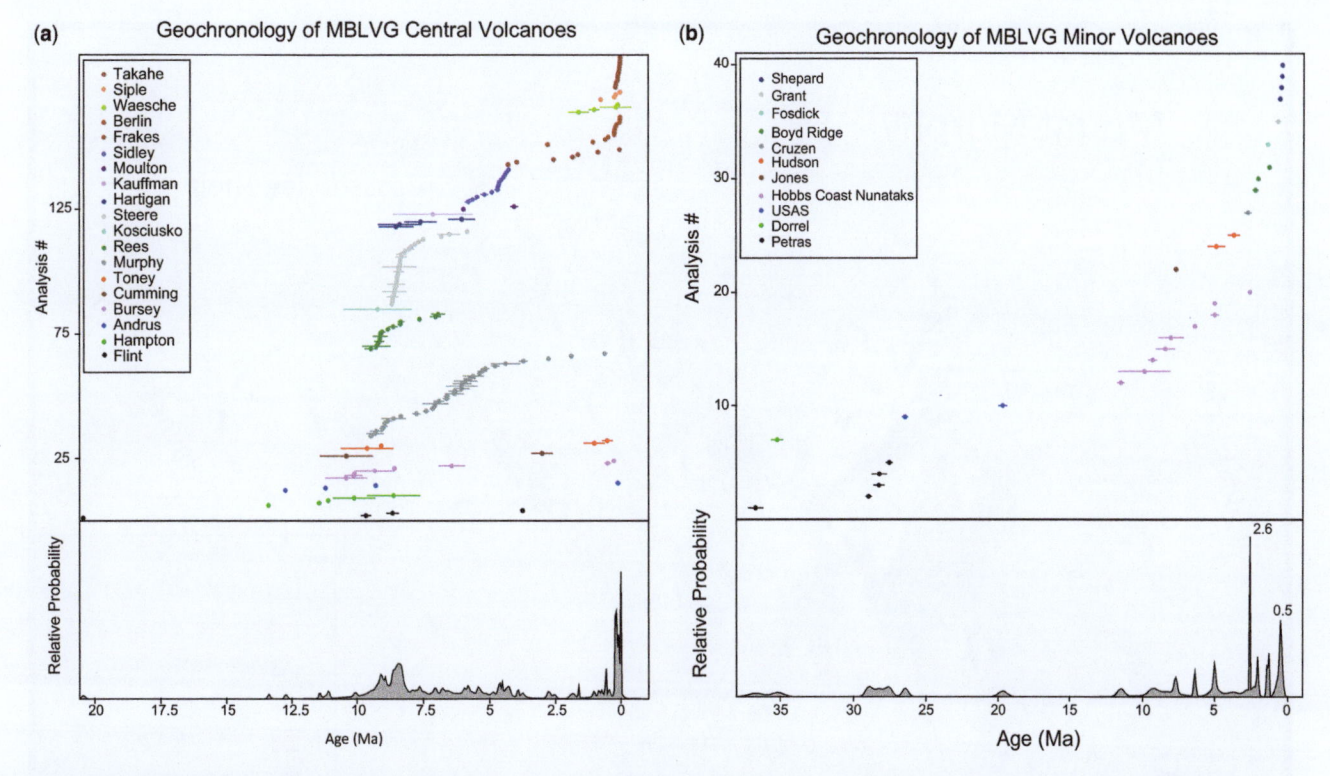

Fig. 6. Geochronology of the MBLVG. (**a**) Central volcano ages. The upper plot shows individual ages of central volcanoes with 2σ (SD) uncertainties. Ages are colour coded by volcano and listed in the legend from youngest to oldest. The lower plot shows relative probability distribution of ages. (**b**) Minor volcano ages. Plots organized similar to that in (a).

post-edifice-building eruptions. Rock compositional types for each volcano and their petrology are provided by Panter *et al.* (2021).

Eastern MBL Volcanic Field

The Eastern MBL Volcanic Field consists of three isolated central volcanoes (Mount Murphy, Mount Takahe and Toney Mountain), satellite centres near Mount Murphy and isolated centres in the Kohler Range.

Mount Murphy. Mount Murphy is a deeply-dissected polygenetic coastal volcano with a summit elevation of 2703 m asl (Fig. 7). Mount Murphy covers an area of about 35 × 45 km, and has an irregular shape with multiple peaks and elongate resistant spurs that are likely to have resulted from, or are enhanced by, glacial erosion. Large cirques are eroded into the volcano and multiple alpine glaciers flow down the volcano flanks (Andrews and LeMasurier 1973). Mount Murphy forms an obstruction to regional ice flow, with inland ice levels at *c.* 800 m asl and coastal ice levels at *c.* 200 m asl. The WAIS abuts the south side of Mount Murphy, and drains around the west and east sides of Mount Murphy through Pope Glacier and Haynes Glacier, respectively. These glaciers flow into the Crosson Ice Shelf located just north of Mount Murphy. Mount Murphy is one of the few MBLVP central volcanoes that lacks a recognizable summit caldera, although some of the tops of cirque headwalls resemble caldera rims.

Most of the work at Mount Murphy has concentrated on the well-exposed SW ridge and on satellite nunataks west of the main edifice. There are many outcrops at Mount Murphy that have not been visited, especially on the eroded spurs, slopes and cliffs away from the west flank. Volcanic rocks on the western flank of Mount Murphy crop out between <400 and 2446 m asl. Several satellite nunataks are exposed as interfluves in Pope Glacier, just west of the main edifice. The lowest rock outcrops at Mount Murphy between *c.* 200 and 400 m asl expose the bedrock underlying the volcano (LeMasurier *et al.* 1990*c*).

LeMasurier *et al.* (1990*c*) characterized Mount Murphy as a basaltic shield surmounted by a much smaller felsic shield, based on 5° dip angles, on slope-forming basanitic lavas as high as 1900 m asl, and a transition to felsic rocks above that elevation. The total original volume of Mount Murphy is estimated at 580 km^3 (LeMasurier *et al.* 1990*c*). The published descriptions of Mount Murphy have largely focused on volcanic sequences along the lower SW flank of Mount Murphy, at Sechrist Peak, and at three satellite nunataks, west and SW of the central volcano (McIntosh *et al.* 1985; LeMasurier *et al.* 1994; Smellie 2001; Wilch and McIntosh 2002) (Fig. 7). Stratigraphic sequences at these localities are composed of intercalated volcanic and glacial deposits that provide evidence of higher palaeo-ice levels during Miocene–Pleistocene times (LeMasurier *et al.* 1994; Wilch and McIntosh 2002). Here, we expand upon these interpretations with new field and ^{40}Ar/^{39}Ar data from additional Mount Murphy localities. In total, 26 new ^{40}Ar/^{39}Ar ages are presented along with descriptions of the rock units (Table 3). A summary of the volcanic geology and geochronology of the main edifice is presented first, followed by a description of the satellite nunataks. The overall evolution was not discussed.

Sechrist Peak–Bucher Peak Ridge. The main edifice that forms the highest and most voluminous part of Mount Murphy is deeply dissected by glacial erosion on the west and south sides, offering exposures of the interior of the volcano. The most complete sequence is exposed along the ridge that extends SW from Bucher Peak (2446 m asl), over Sechrist Peak (1350 m asl) to the base of volcano near 400 m asl (Figs 5 & 6). Compositions within this stratigraphic interval are almost all basanitic, with the exception of one 50 m-thick trachyte lava between 725 and 775 m asl, and several other trachytic lavas near the top of the sequence (1980–2390 m asl).

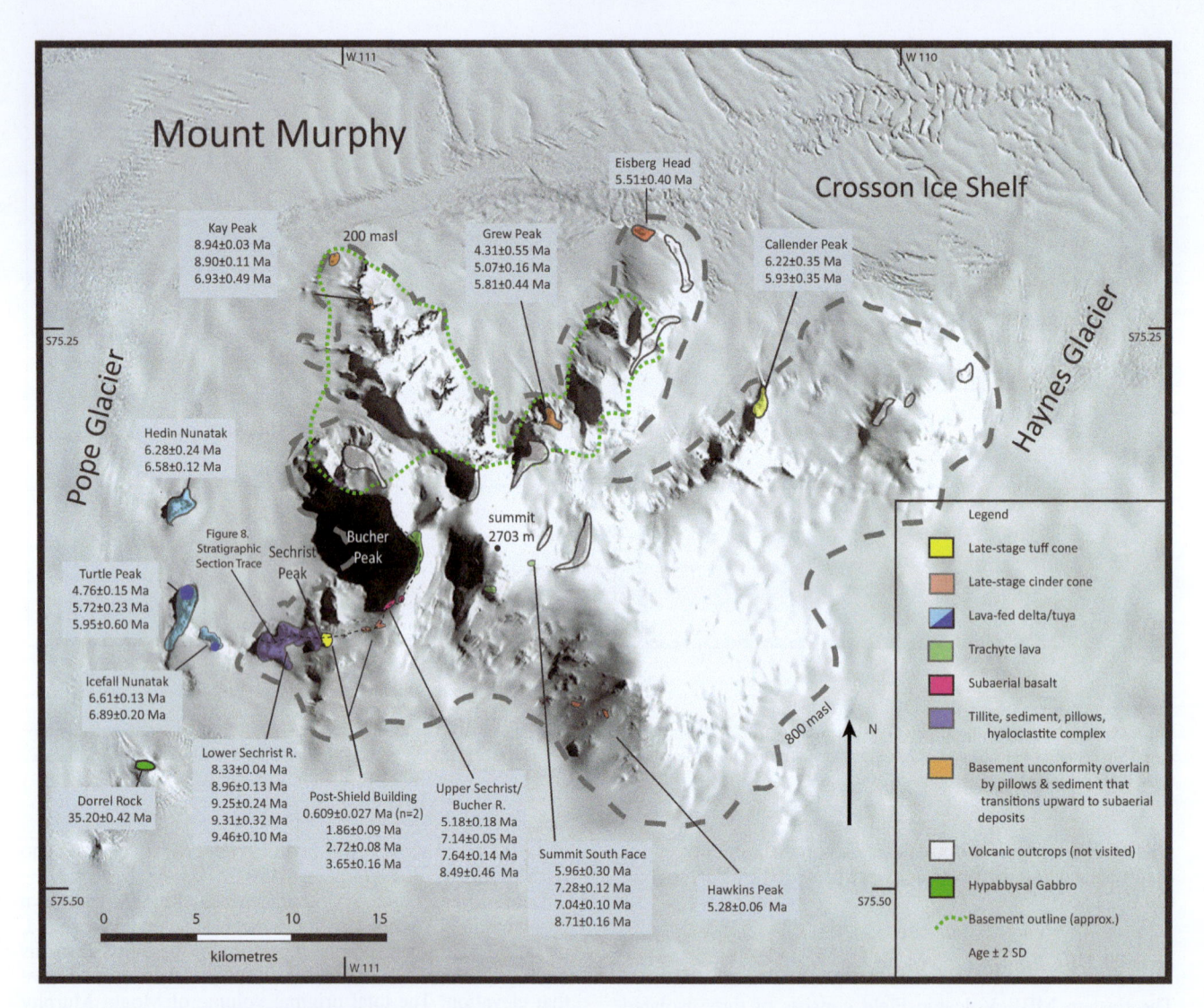

Fig. 7. Geology of Mount Murphy volcano in the Eastern MBL Volcanic Field, with volcanic rock outcrops, lithofacies and ages. The thick dashed line shows the outline of the volcano at the ice-sheet surface. Note the elevation difference between the upstream (800 m asl) and downstream (200 m asl) sides of the volcano. Google Earth Pro image accessed June 2019. Image sources: Landsat from United States Geological Survey and 2019 Digital Globe. The variable resolution image was downloaded from Google Earth Pro and processed in Adobe Photoshop. Processing included conversion to black and white, and adjustment of brightness and contrast to enhance uniformity of surface tones.

^{40}Ar/^{39}Ar ages for this sequence range from 9.46 ± 0.10 Ma for a pillow lava near the base to 5.18 ± 0.18 Ma for a trachytic lava just below Bucher Peak summit. A post-edifice-building Pleistocene tuff cone at Sechrist Peak is superimposed on the sequence; the late-stage tuff cone is considered after the description of the main stratigraphic sequence.

The geology exposed along this SW ridge is summarized in a composite stratigraphic section (Fig. 8: modified and expanded from Wilch and McIntosh 2002). Between 400 and 725 m asl, compositions are basanitic, and the sequence is dominated by pillow lavas, hyaloclastite breccias, bedded hyalotuffs reworked by traction currents, and complexly-jointed subaqueous lavas (Kubbaberg type) (Fig. 9a, b). These lithofacies are interlayered with subordinate amounts of subaerially-erupted hydrothermally-altered or palagonitized Strombolian tuff and welded bomb sequences (Fig. 9c), and columnar-jointed subaerial lavas with pāhoehoe-type fluidal top surfaces and oxidized basal breccias. At least three glacial erosion surfaces occur in the lower sequence, each associated with a polished and striated underlying surface, tillite and/or striated clasts, (Fig. 9d, e) (McIntosh et al. 1985; LeMasurier et al. 1994; Smellie 2001, 2008; LeMasurier 2002; Wilch and McIntosh 2002). Large irregular basanitic water-cooled lavas and intrusive or incursive bodies occur within the section, locally deforming and shearing stratified reworked hyalotuffs (Fig. 9f). Six ^{40}Ar/^{39}Ar ages from this interval range from 9.46 to 8.96 Ma (Wilch and McIntosh 2002). All workers agree that this sequence records complex interactions between glacial ice and a growing polygenetic volcano but there is incomplete agreement about the thickness of the interacting ice. McIntosh et al. (1985) and Wilch and McIntosh (2002) interpreted the interlayered subglacial and subaerial lithofacies as a record of the fluctuations of a regional ice sheet. Smellie (2001, p. 10; 2008) interpreted the lower Murphy sequence as a 'subglacial "sheet-flow type" formed when the slopes of the volcano were mantled by relatively thin "ice" (probably mainly firn and/or snow <100 m thick)'. LeMasurier (2002) described the sequences as a series of complex lava-fed deltas, which have since been interpreted as 'a'ā-lava-fed deltas (Smellie and Edwards 2016). The lower Murphy sequence apparently records eruptive environments that varied from thin, temperate, mantling ice (yielding glacial unconformities, tillites, subaqueous lavas, and fluvially-reworked hyalotuffs and hyaloclastites), to regional ice (yielding lava-fed deltas) and subaerial emergence (yielding welded and non-welded Strombolian deposits and columnar-jointed lavas with oxidized basal breccias). Interpreting regional ice-sheet levels from the lower Sechrist Peak–Bucher Peak Ridge sequence

Table 3. *Mount Murphy and Mount Kohler range summary age table*

Sample ID	Method	Corrected preferred age ± 2 SD (Ma)	Description	Ref.
Mount Murphy				
Murphy Shield Building, Bucher Peak				
W85-060	A	5.18 ± 0.18	North side, lava, 60 m below summit ice cap, c. 2390 m	21
90-72	A	7.14 ± 0.05	West cliff mesa, light-grey lava, 2% plag	21
90-001	A	7.64 ± 0.14	SW ridge, bomb interior, highest cone, trachyte–benmoreite?, 10% xtal, cpx, plag, amph?, porphy lava, spatter/flow, 5% pyx + ol + plag	21
90-005	A	8.49 ± 0.46	SW ridge, lava underlying tillite	21
90-34	A	8.33 ± 0.04	Main cliff, 50 m-thick felsic dome 725–775 m asl	18
90-37	A	8.91 ± 0.20	Main cliff, thin (subaerial?) lava at 685 m	18
90-37	A	8.97 ± 0.15	Main cliff, thin (subaerial?) lava at 685 m	18
90-37	A	**8.95 ± 0.13**	**Mean age (*n* = 2)**	
90-39	A	9.38 ± 0.14	Main cliff, lava assoc. w/hyaloclastite 555 m	18
90-39	A	9.15 ± 0.12	Main cliff, lava assoc. w/hyaloclastite 555m	18
90-39	A	**9.25 ± 0.24**	**Mean age (*n* = 2)**	
90-33	A	9.19 ± 0.09	Main cliff, 1.5 m lava (striated), 527 m	18
90-33	A	9.51 ± 0.12	Main cliff, 1.5 m lava (striated), 527 m	18
90-33	A	**9.31 ± 0.32**	**Mean age (*n* = 2)**	
90-50	A	9.47 ± 0.13	West cliff, pillow 3 m above tillite, 510 m	18
90-50	A	9.44 ± 0.13	West cliff, pillow 3 m above tillite, 510 m	18
90-50	A	**9.46 ± 0.10**	**Mean age (*n* = 2)**	18
Sechrist Peak				
90-69	A	0.607 ± 0.086	West flank	21
90-139	A	0.609 ± 0.028	Dense intrusive interior	21
		0.609 ± 0.027	**Mean age (*n* = 2)**	
Murphy post-shield building, Bucher Peak				
90-007	A	1.86 ± 0.09	Basaltic bomb, lowest cinder cone	21
W85-045	A	2.72 ± 0.08	Basaltic *in situ* flow	21
90-002	A	3.65 ± 0.16	Basaltic flow, middle xeno-bearing cone	21
Murphy shield building, Bucher south face				
90-145	A	5.96 ± 0.30	Dense black bomb interior, high in spatter-fed flow	21
90-143	A	7.28 ± 0.12	Dense spatter-fed lava, 15% pyx + plag + ol	21
90-144	A	7.04 ± 0.10	Green trachtye fragment	21
90-146	A	8.71 ± 0.16	Trachyte, west end of Hawkins–Murphy saddle	21
Kay Peak				
90-123	A	8.94 ± 0.03	Aphanitic vesicular lava	21
90-112	A	8.90 ± 0.11	Second lowest unit, trachytic? lava, porphy w/4% ol + cpx + plag	21
90-115	A	6.93 ± 0.49	Crystal-rich basaltic lava, 15% ol + cpx + plag	21
Grew Peak				
90-127	A	4.31 ± 0.55	Aphyric dyke in epiclast/hyaloclastite	21
90-130	A	5.07 ± 0.16	Aphyric pillow basalt	21
90-131	A	5.81 ± 0.44	Porphy lava, ol? plag, pyx, within hyalo/pillow unit	21
Eisberg Head				
90-133	A	5.51 ± 0.40	J.L. Smellie sample MB58.3B	21
Callendar Ridge				
90-153	A	6.22 ± 0.35	Lower lava, aphyric, coarse plag matrix	21
90-149	A	5.93 ± 0.35	Lowest exposed lava, aphyric	21
Hawkins Peak				
90-148	A	5.28 ± 0.06	Bomb interior, 1% plag	21
Satellite nunataks				
Icefall Nunatak				
90-47	A	6.60 ± 0.13	High subaerial lava	18
90-48	A	6.89 ± 0.20	Lobe in low hyaloclastite	18
Turtle Peak				
90-94	A	4.76 ± 0.15	Very vesicular pāhoehoe top	18
90-92	A	5.72 ± 0.23	Upper pillow	18
90-87	A	5.95 ± 0.60	Lower-flow foot breccia, angular clasts	18
Hedin Nunatak				
90-99	A	6.28 ± 0.24	Subaerial lava, upper tuya	18
90-110	A	6.58 ± 0.12	Lowest tindar	18
Dorrel Rock				
90-105	A	35.50 ± 0.12	Gabbro intrusion	21

(*Continued*)

Table 3. *Continued.*

Sample ID	Method	Corrected preferred age ± 2 SD (Ma)	Description	Ref.
90-105	A	34.83 ± 0.45	Gabbro intrusion	21
90-105	A	34.92 ± 0.12	Gabbro intrusion	21
	A	35.20 ± 0.42	Mean age (*n* = 3)	
60A	A	>34.46 ± 0.22	Gabbro intrusion	13
60A	A	>37.05 ± 0.26	Gabbro intrusion	13
60C	A	33.93 ± 0.24	Dyke	13
60D	A	35.81 ± 0.32	Dyke	13
84	K	10.1 ± 3.4	Thin lava on basement, proximal to tuff breccia	1

Notes: method A is ^{40}Ar/^{39}Ar; method K is K–Ar. Abbreviations: amph, amphibole; cpx, clinopyroxene; hyalo, hyaloclastite; ol, olivine; plag, plagioclase; porphy, porphyritic; pyx, pyroxene.
Ref. code: 1, LeMasurier (1972); 13, Rocchi *et al.* (2006); 18, Wilch and McIntosh (2002); 21, this study. See also Supplementary Material Table S1.

is not straightforward, in part because of uncertainties regarding ice thickness but also because of potential feedback effects on the palaeo-ice-sheet level by the growing Mount Murphy edifice, as further discussed in a subsequent section ('Synthesis of volcanic records of glaciation in the WAIS').

From 725 to 850 m asl, the Sechrist Peak–Bucher Peak Ridge includes subaerial basalt and trachyte–benmoreite lavas (LeMasurier *et al.* 1990c). An approximately 50 m-thick columnar-jointed trachyte lava between 725 and 775 m asl is dated to 8.33 ± 0.04 Ma (Wilch and McIntosh 2002) (Table 3). A basanitic lava near 850 m asl is dated at 7.14 ± 0.05 Ma. From 850 to 1750 m asl, outcrops are sparse and sporadic, and consist of pyroclastic deposits and a few subordinate lavas. Poor exposures and variable dips within this interval hinder determinations of the stratigraphic relationships between these units and the well-exposed, stratigraphically-coherent, lower part of the composite section, described above. Some of the topographically higher units are older than topographically lower units, and other units are significantly younger, apparently overlying eroded older units.

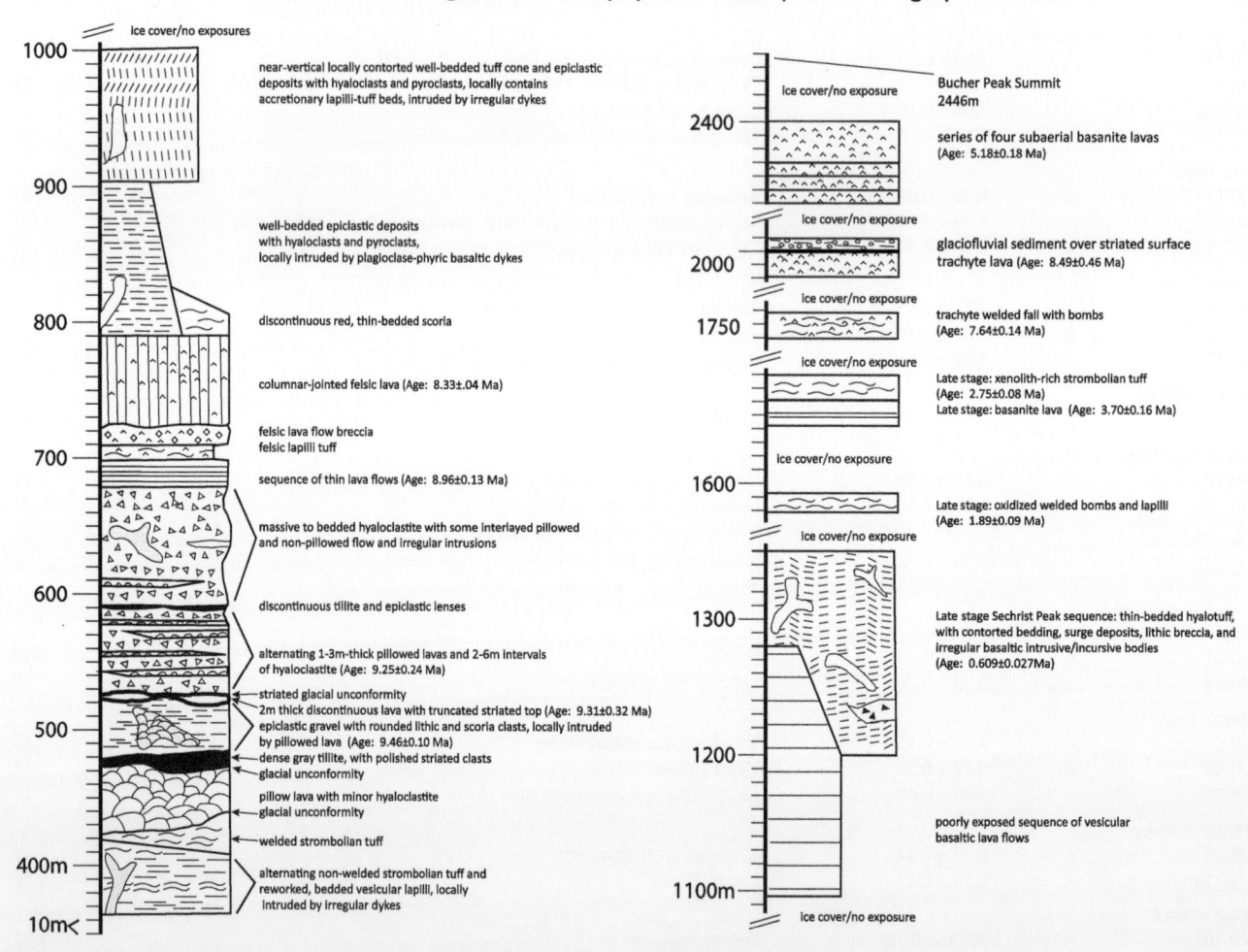

Fig. 8. Composite stratigraphic section of the Bucher–Sechrist Ridge, Mount Murphy, Eastern MBL Volcanic Field. The composite section is based on a traverse shown in Figure 7. The figure is an expanded and updated version of a figure in Wilch and McIntosh (2002).

Fig. 9. Photographs of outcrops and lithofacies at Mount Murphy, Eastern MBL Volcanic Field. (**a**) Sequence near the base of the SW ridge, Mount Murphy, showing diamictite and bedded hyalotuffs, between two striated, polished glacial unconformities. The upper unconformity is overlain by a thick complexly-jointed, water-cooled, Kubbaberg-type lava. The bedded hyalotuffs were reworked and deposited by traction currents, probably water moving downslope in cavities or meltwater channels beneath warm-based, relatively thin, slope-mantling ice. Annotated photograph from Smellie (2008), with permission from Elsevier © 2008. (**b**) Tightly nested pillow lavas near the base the section along the lower SW ridge, Mount Murphy, formed by lava emplaced within meltwater-filled subglacial chambers. (**c**) Hydrothermally altered or palagonitized Strombolian welded bomb sequence, lower SW ridge, Mount Murphy. The flattened and mutually molded forms of the vesicular bombs indicate deposition of molten pyroclasts in a subaerial environment. (**d**) Polished, striated top of subaerially-erupted basanitic lava overlain by younger subaerially-erupted basanitic lava, lower SW ridge, Mount Murphy. This sequence attests to alternating subaerial and subglacial environments.

Some of these younger units are post-edifice-building basanites ranging in age from 3.65 ± 0.16 to 0.61 ± 0.03 Ma. The highest pyroclastic deposit in this interval is a 7.64 ± 0.14 Ma trachytic Strombolian pyroclastic deposit near 1750 m asl, possibly associated with the main edifice-building phase of Mount Murphy.

The uppermost part of the Mount Murphy SW ridge sequence, exposed between 1980 and 2390 m asl, consists of a series six or more basanitic–trachytic subaerial lavas. The lowest lava near 1980 m asl, poorly dated at 8.49 ± 0.46 Ma, has a polished and striated upper surface, overlain by a 1 m-thick sedimentary unit, possibly of glacial origin, composed of dipping (25° downslope) laminated clays containing 0.5–10 cm pebbles; in some cases themselves polished and striated. This laminated deposit contains recycled Neogene microfossils; it is not clear whether these were emplaced during glaciation or later by aeolian processes (LeMasurier et al. 1994). These glacial features suggest erosion and deposition by wet-based local ice on the slopes of Mount Murphy. The uppermost lava, exposed near 2390 m asl and covered by a small summit ice cap at the top of Bucher Peak, is basanitic and dated at 5.18 ± 0.18 Ma.

The main edifice of Mount Murphy has additional very limited outcrops south and east of the summit area between 1600 and 2400 m asl. Basanitic pyroclastic deposits and trachytic lavas exposed in these two areas range in age from 8.71 ± 0.16 to 5.96 ± 0.30 Ma.

Between 1050 and 1350 m asl, along Sechrist Peak–Bucher Peak Ridge, a sequence of deformed tephra-dominated tuya deposits is inset against Murphy edifice-building lavas. Two $^{40}Ar/^{39}Ar$ ages of 0.61 ± 0.09 and 0.61 ± 0.03 Ma indicate that the Sechrist Peak sequence formed several million years after Mount Murphy's main edifice-building period. Much of the sequence consists of thin beds of fine vesicular hyalotuffs, in many cases graded, which may represent volcanically-generated turbidites that accumulated within a subglacial meltwater cavity. These beds are intruded by dykes and irregular intrusive or incursive lava bodies that are locally pillowed at the margins or are brecciated and intermixed with the hyalotuffs. Some of the bedding is strongly deformed, in some cases folded and tilted to vertical or overturned orientations (Fig. 9g). This soft-sediment deformation is probably related to intrusions and slope failures resulting from the changing geometry of the water-filled chamber during and after emplacement. Towards the top of Sechrist Peak, the bedded hyalotuffs contain accretionary lapilli, bomb fragments, reddened Strombolian scoria and local subaerial lavas that suggest the eruptive vent became emergent. The deposits at Sechrist Peak probably erupted below and near the surface of a formerly higher WAIS. The passage

Fig. 9. *Continued.* (**e**) Polished, striated basanitic clast in glacial till, evidence for wet-based glaciation. A 15 cm pencil is for scale. Till overlies a striated unconformity near the base of the lower SW ridge of Mount Murphy. (**f**) Bedded hyalotuffs deformed and sheared by overlying or incursive basanitic water-cooled lavas, lower SW ridge of Mount Murphy. (**g**) Vertical and overturned beds of hyalotuff at Sechrist Peak, Mount Murphy, evidence of slope failure causing soft-sediment deformation during accumulation in a sub-ice-sheet meltwater chamber. (**h**) Glacially-eroded surface at the south end of the upper surface of Turtle Peak. This polished and striated surface with overlying granitic erratic boulders records post-volcanic overriding and erosion by the WAIS.

zone of the Sechrist emergent sequence is not well defined but is located about 500 m above the local current 800 m elevation of the WAIS, just upstream of Sechrist Peak. The currently exposed surfaces of some of the more competent Sechrist lavas are striated and littered with exotic granitic erratics, suggesting glacial overriding subsequent to 0.61 Ma.

Kay Peak. Kay Peak tops a spur that extends to the NNW from the summit of Mount Murphy (Fig. 7). Several outcrops on the eroded SE face of Kay Peak expose a volcanic sequence overlying a smooth, undulating and glacially-striated unconformity eroded into basement metagabbro. Lavas in the sequence at Kay Peak range in age from 8.90 ± 0.11 Ma near the base of the sequence to 6.93 ±0.49 Ma near the top. The basement contact, exposed at elevations as low as 640 m asl, is discontinuously overlain by 0.5 m of crudely-bedded volcanic-rich till containing rounded and striated clasts. The till is in turn overlain by sequence of pillow lavas, massive lavas and pillow hyaloclastites. The lowest lava has well-developed pillows throughout its vertical extent. The upper surface of this pillow lava and most of the overlying massive lavas are eroded, with an undulatory striated top surface. Eroded top surfaces are locally overlain by thin layers of volcanic-rich tillite or crudely-bedded hyaloclastite or reworked hyaloclastic sediments composed of glassy clasts probably derived from the quenching of the overlying lava. Clasts of pre-volcanic basement were not observed in the tillite or bedded sediments. A basanitic lava higher in the sequence has well-developed columnar jointing, and very limited development of pillows and hyaloclastite at the base of the columns. The top of the columnar lava is dramatically glacially sculpted into polished striated knobs with more than 20 m of relief. This well-developed glacial-erosion surface is in turn overlain by sequence of volcaniclastic rocks containing intact volcanic bombs, in turn capped by a deposit of Strombolian bombs and lapilli grading upwards into a black vesicular lava with a locally reddened, oxidized breccia on the upper surface. Minor localized pillows and hyaloclastite at the base of this otherwise subaerial lava is interpreted as evidence of limited interaction with snow or moisture in hollows along the underlying glacial surface. Such 'wet' to 'dry' upward transitions are seen in many lavas at Mount Rees and Mount Steere in the Crary Mountains.

The sequence at Kay Peak is interpreted to have been emplaced in a glacial valley on a bedrock surface adjacent to the slope of developing Mount Murphy. The valley experienced alternating periods of glacial erosion, emplacement of subglacial hyaloclastites and both subglacial and subaerial lavas, and fluvial reworking of hyalotuff and hyaloclastite debris. Similar sequences have been observed in deposits formed beneath valley glaciers in Iceland and elsewhere (Walker and Blake 1966; Smellie *et al.* 1993).

Grew Peak. Grew Peak is located on a spur that extends NNE from the Mount Murphy summit (Fig. 7). Grew Peak, like Kay Peak, exposes a volcanic sequence overlying a pre-volcanic glacial-erosion surface cut into basement metasediments that extends as low as 840 m asl. The glacial unconformity consists of a polished striated surface at the base of a U-shaped valley cut into basement rock, subsequently filled by basanitic lavas and hyaloclastites. The glacial unconformity is directly overlain by 5 m of crudely-bedded, fines-poor basanitic lapilii tuff, containing angular fragments of basanitic lava. The sediment is in turn overlain locally by a massive basanitic lava dated at 5.81 ± 0.44 Ma. This is overlain by a more extensive hyaloclastite sequence containing pillows and lobes of basanitic lava as large as 6 m in length with quenched glassy margins. A pillow from this sequence yielded an age of 5.07 ± 0.16 Ma. Like the sequence at Kay Peak, the sequence at Grew Peak is likely to have formed by lavas flowing down an ice-filled valley on the slope of Mount Murphy. The sequence at Grew Peak is cut by basanitic dykes dated at 4.31 ± 0.55 Ma.

Callender Peak and Ridge. Callender Peak is located along the mostly snow- and ice-covered spur that extends ENE from the summit of Mount Murphy (Fig. 7). Outcrops near Callender Peak and the adjacent ridge extending to the north expose two lavas: a lower aphanitic basanite dated at 6.22 ± 0.35 Ma; and an upper porphyritic basanite dated at 5.93 ± 0.35 Ma. The upper lava contains 15% phenocrysts including plagioclase as large as 2 cm in diameter. Both lavas are vesicular, with brecciated, reddened flow margins indicating subaerial emplacement. The upper surfaces of exposures at Callender Peak are mantled with granitic and gneissic erratics, which indicate overriding by the regional ice sheet.

Hawkins Peak. Hawkins Peak located SE of the summit of Mount Murphy has very limited exposures due to extensive ice cover. Two outcrops were visited. The lower outcrop near 1400 m asl is an agglutinated basaltic pyroclastic deposit containing red-oxidized lapilli and bombs to 1 m in diameter, and abundant ultramafic peridotite and pyroxenite xenoliths. A higher outcrop near 1600 m is a similar deposit of strongly-welded reddened bombs and lapilli with sparse xenoliths. A date from this unit is 5.28 ± 0.06 Ma, suggesting that it represents late edifice building rather than post-edifice-building activity.

Satellite Nunataks near Mount Murphy. There are three complex volcanic nunataks and one hypabyssal plutonic nunatak near Mount Murphy. The three volcanic nunataks are Icefall Nunatak, Hedin Nunatak and Turtle Peak, located 2–5 km west of the base of Mount Murphy on a north–south-orientated glacial interfluve between Pope Glacier to the west and an unnamed glacier to the east (Fig. 7). The plutonic nunatak is Dorrel Rock, located 5 km south of the volcanic nunataks. Of the volcanic nunataks, only Icefall Nunatak has been described in detail (Smellie 2001). General descriptions of Hedin Nunatak and Turtle Peak have been presented in several papers (McIntosh *et al.* 1985; LeMasurier *et al.* 1994; Smellie 2001; Wilch and McIntosh 2002; Smellie and Edwards 2016). The three volcanic nunataks are classified complex tuyas (Russell *et al.* 2014), each consisting of stacked sequences of multiple lava-fed deltas and associated source vents that accumulated in englacial meltwater lakes within a formerly thicker regional ice sheet.

Icefall Nunatak. Smellie (2001) provided a detailed lithofacies analysis and reconstruction of the volcanic history of Icefall Nunatak that includes consideration of the syneruptive ice thickness and hydrological conditions. The volcanic reconstruction is based on detailed mapping, identification and interpretation of 12 volcanic lithofacies and three distinct unconformities exposed in a 200 m-high by 900 m-long bluff section. Smellie (2001) interpreted Icefall Nunatak as a polygenetic basaltic volcanic centre that erupted between 6.89 ± 0.10 and 6.61 ± 0.08 Ma (Smellie 2001). At least five source vents were suggested on the basis of variable phenocryst content or stratigraphic context. Three distinct eruptive stages are differentiated by erosion surfaces and lithofacies assemblages. Deposits associated with Stage I have a maximum thickness of 60 m. Stage I is dominated by a monomict, clast-supported, lithic (juvenile) breccia composed mostly of holocrystalline lapilli to block-size clasts, with less abundant non-vesicular sideromelane clasts. Lenses of blocky-jointed and pillow lava are exposed lower in the section. The lavas and breccia are overlain by two thin (1–6 m) variably scoriaceous lavas separated by 4 m of a massive tuff breccia. The lavas transform laterally into scoriaceous breccia. The uppermost (and youngest) part of the Stage I sequence includes stratified lapilli tuff deposits which contain some highly vesicular clasts. Stage I deposits are inferred to have been emplaced in an ice-contact englacial lake that formed in response to heat from magma and an active eruption. The breccia and blocky-jointed pillow lava are all consistent with subaqueous eruption and deposition. The vesiculation observed in the uppermost volcaniclastic sediments suggests a reduction of hydraulic pressure over the vent and lower water levels. Smellie (2001) interpreted these uppermost deposits as turbidite deposits forming on the slopes of the growing edifice. He further inferred that the ice sheet re-established itself over the Stage I volcano before the onset of Stage II volcanism, although there is no evidence for a significant time break between the two stages.

Deposits associated with Stage II comprise most of the exposed rocks at Icefall Nunatak and have maximum thickness of >150 m. Stage II is interpreted as a subaqueous to emergent tuff cone sequence formed in an englacial lake. The sequence is similar to the middle–late stages of a glaciovolcanic tuya sequence (e.g. Jones 1969, 1970). Stage II is dominated by a wide variety of subaqueous tuff cone lithofacies that include abundant angular, sideromelane-rich, stratified volcaniclastic tuff, lapilli tuff, tuff breccia and breccia. The sequence transitions from a subaqueous tuff cone through a passage zone to subaerial deposition of lava that extends into subaqueous deposition of a lava-fed pillow/hyaloclastite delta. There is evidence of at least two slope-failure events during the construction of the emergent tuff cone sequence. The palaeowater level (a proxy for ice level) is at c. 700 m asl, about 100 m above the current local ice level.

Stage III deposits include thin columnar basalt lavas and thicker entablature overlain by a scoria cone remnant. Stage III is interpreted as emergent volcanism that occurred over thin ice/and or snow.

Turtle Peak and Hedin Nunatak. Turtle Peak and Hedin Nunatak are both stacked sequences of three or more basanitic lava-fed deltas deposits with passage zones that record varying ice levels during their eruption and deposition. Turtle Peak outcrops are larger and better exposed than those at Hedin Nunatak.

Turtle Peak is a flat-topped elongate edifice and measures approximately 2.5 km north–south and 1.5 km east–west. Cliffs as high as 200 m wrap around the east, north and west sides of Turtle Peak, and the flat top is largely free of snow and ice. Together, these outcrops provide excellent three-dimensional exposures of three subhorizontally stacked lava-fed delta sequences and the vent areas for the lower and upper delta sequences. The three lava-fed delta sequences conformably overlie each other without evidence of significant erosion between them (Fig. 10). The lowest sequence consists of gently (7°–12°) south-dipping pillow lava, pillow lava breccia, bedded hyaloclastite and reworked volcaniclastic sediment that formed foreset-bedded breccias as they accumulated in a meltwater chamber below the local ice-sheet level. Dykes and associated faulting and soft-sediment deformation near the north end of Turtle Peak suggest proximity to the eruptive source. Strombolian bombs and lapilli tuff

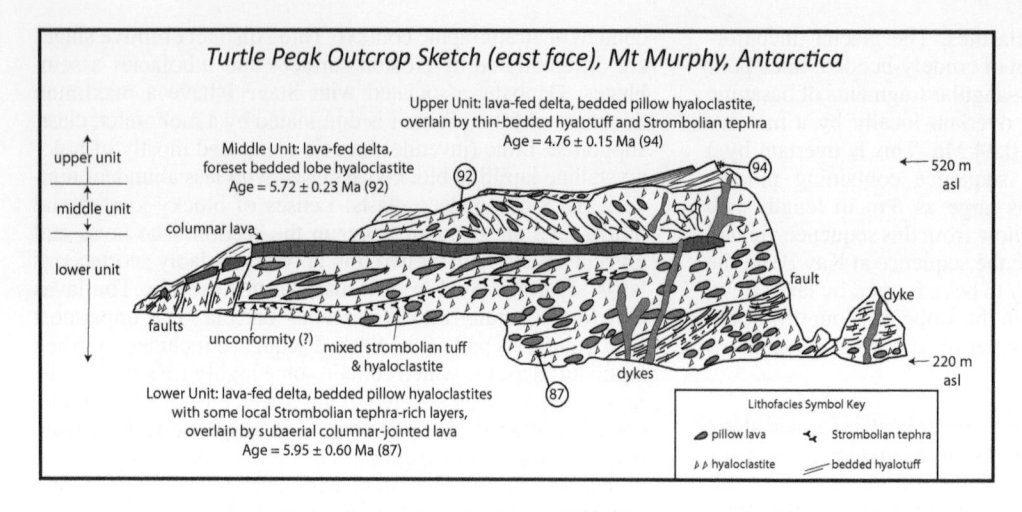

Fig. 10. Field sketch map showing the volcanic stratigraphy of the Turtle Peak satellite nunatak, located just east of Mount Murphy.

intermixed with pillow breccias in this area suggest that the eruptive vent periodically emerged above water level. The gently-dipping breccia sequence is capped by a near-horizontal tabular, columnar-jointed lava, prominently exposed in both the east and west faces of Turtle Peak, which records a transition from subglacial to subaerial conditions either as the pile built above water level or following draining of the meltwater chamber. A $^{40}Ar/^{39}Ar$ age of 5.95 ± 0.60 Ma was obtained from a breccia clast in the lower lava-fed delta. The subaerial lava capping the lower delta sequence is overlain by a similar middle lava-fed delta sequence of pillows, hyaloclastites and reworked volcaniclastic sediments with variable but generally gentle northward dips. This middle pillow hyaloclastite delta sequence closely resembles the lower sequence but lacks a capping subaerial lava. A pillow from the middle sequence yielded a $^{40}Ar/^{39}Ar$ age of 5.72 ± 0.23 Ma, which overlaps within uncertainty with the age of the lower sequence. The middle sequence is overlain by a third lava delta sequence, which again consists of a sequence of pillows, hyaloclastite and reworked volcaniclastic sediments, gently south-dipping and capped at the north end of Turtle Peak by a subaerially-erupted Strombolian scoria cone and pāhoehoe lava that record emergent conditions in the vent area. The cliffs below the vent area expose dykes that appear to have fed the scoria cone and are spatially associated with, and probably caused, extensive deformation of adjacent foreset-bedded pillow breccia and hyaloclastite. The pāhoehoe lava yielded a $^{40}Ar/^{39}Ar$ age of 4.76 ± 0.15 Ma. The prominent subaerial lava capping the lower lava-fed delta at Turtle Peak is exposed as a flat erosional surface at the south end of the upper surface of Turtle Peak. This surface is highly polished, striated and littered with granitic glacial erratics, indicating overriding and erosion by the WAIS some time after activity ceased near 4.8 Ma (Fig. 9h).

Hedin Nunatak, located 4 km north of Turtle Peak, is smaller in area (1 × 1 km) but taller (300 m) than Turtle Peak. The top is ice-covered and outcrops are most extensive on the north side. The sequence at Hedin Nunatak is quite similar to that at Turtle Peak, consisting of at least one source tuya and three lava-fed delta sequences containing foreset-bedded pillow/hyaloclastite breccias, some capped by compound subaerial lavas. $^{40}Ar/^{39}Ar$ ages near the base and top are 6.58 ± 0.12 and 6.28 ± 0.24 Ma, respectively, slightly older than Turtle Peak but similar to the age of Icefall Nunatak, which may have been the source area for some of Hedin Nunatak's lava-fed delta deposits.

Taken together, the sequences of stacked lava-fed deltas and associated source tuyas at Icefall Nunatak, Turtle Peak and Hedin Nunatak provide a record of variably higher icesheet levels during their eruptive activity between 6.5 and 4.7 Ma. Local ice levels were periodically at least 200 m higher than the level of the current ice sheet but determining the regional level of the WAIS from these data is complicated by the position of these nunataks on a glacial interfluve and by their near coastal location, as discussed further below.

Dorrel Rock. Dorrel Rock is an isolated nunatak with a peak elevation of 790 m asl situated about 8 km SW of the base of Mount Murphy and 7 km SSW of Turtle Peak. Dorrel Rock forms an interfluve in the heavily-crevassed and rapidly-descending adjacent north-flowing Pope Glacier. Dorrel Rock, along with Turtle Peak, appears to extend the buttressing effect of Mount Murphy on the WAIS, with upstream ice elevations between 600 m asl and downstream elevations at *c.* 200 m asl.

Dorrel Rock is characterized as an intrusive igneous complex composed of a well-exposed pegmatitic gabbro cut by benmoreite and trachytic dykes (LeMasurier 1990b; Rocchi et al. 2006). The compositions at Dorrel Rock are alkaline and similar to the MBLVG rocks (Rocchi et al. 2006), and the ages are close to the earliest MBLVG volcanism at Mount Petras in the McCuddin Mountains. Biotite in the gabbro was dated by $^{40}Ar/^{39}Ar$ to be >34.2 Ma and a dyke to *c.* 33.5 Ma (Rocchi et al. 2006). New $^{40}Ar/^{39}Ar$ ages on three different mineral phases in the gabbro yielded ages of 35.51 ± 0.12 Ma (hornblende), 34.83 ± 0.45 Ma (biotite) and 34.92 ± 0.12 Ma (potassium feldspar), suggesting an emplacement age of *c.* 35 Ma (Table 3).

Rocchi et al. (2006) estimated that the intrusion was emplaced at a depth of at least 3 km. They argued that most of the exhumation of Dorrel Rock occurred early between 34 and 27 Ma, based on the inference of broad-scale uplift of an MBL structural dome centred at Mount Petras in central MBL. It is interesting to note that the timing of intrusion and uplift proposed at Dorrel Rock coincides with a period of seafloor spreading that formed the Adare Basin (43–26 Ma: Cande et al. 2000; Cande and Stock 2006), as well as the intrusion of similar alkaline magmas in North Victoria Land (Meander Intrusive Group, 48–23 Ma: Rocchi et al. 2002), which altogether signify a widespread tectonomagmatic phase.

Kohler Range. The Kohler Range is situated near the Walgreen Coast, about 90 km west of Mount Murphy and 100 km NE of Toney Mountain. Two volcanic rock patches rest on a bedrock unconformity (LeMasurier 1990b). The rocks are described as a yellow tuff at Morrison Bluff (observed from the air but not visited) and a 'thin veneer' of basalt lava at Leister Peak. Whole-rock geochemical data (XRF analyses) of the basalt sample are provided by LeMasurier (1990b) and there is a K–Ar age of 10.1 ± 3.4 Ma (LeMasurier 1972) (Table 3).

Mount Takahe. Mount Takahe (3460 m asl) is an isolated undissected late Quaternary polygenetic central volcano in eastern Marie Byrd Land, located 80 km SE of Toney Mountain and 90 km SW of Mount Murphy (Figs 5 & 11). The volcano has an 8 km-diameter, snow-filled, circular caldera that is *c.* 2100 m above ice level and a nearly symmetrical circular base, 30 km in diameter. Flank slopes range from 7° to 10°.

Mount Takahe has been studied in detail (McIntosh *et al.* 1985; Palais *et al.* 1988; LeMasurier and Rex 1990*b*; Wilch *et al.* 1999; LeMasurier 2002). Most outcrops on the volcano have been visited. Three groups of outcrops are described here: summit caldera outcrops; lower flank subglacial–subaerial sequences; and lower flank dominantly subaerial sequences. Reconstructions of the volcanic history of Mount Takahe provide key data for WAIS palaeo-ice-level history and for ice-core tephrochronology (McIntosh *et al.* 1985; Palais *et al.* 1988; Wilch *et al.* 1999; LeMasurier 2002). Dating of the young volcanic rocks at Mount Takahe has been challenging. Most K–Ar ages were reported as <0.1 Ma (LeMasurier and Rex 1990*b*). Fission-track analyses of two caldera-rim samples yielded young ages with large uncertainties (Palais *et al.* 1988) (Table 4). $^{40}Ar/^{39}Ar$ dating shows improvement with ages ranging from 194.5 ± 6.3 to 8.3 ± 5.4 ka (Wilch 1997; Wilch *et al.* 1999). New previously unpublished data presented here provide additional improvements in dating precision for selected outcrops (Fig. 11; Table 4).

Outcrops at and near the caldera rim include a 60 m section of welded and non-welded pyroclastic lapilli-tuff deposits, obsidian-bearing bomb-and-block layers, hydrovolcanic tuffs, and lavas (McIntosh *et al.* 1985; Palais *et al.* 1988; Wilch *et al.* 1999). Anorthoclase from a sample of a non-welded obsidian and pumice bomb-and-block layer near the top of this sequence has a $^{40}Ar/^{39}Ar$ age of 8.3 ± 5.4 ka (Wilch *et al.* 1999). This young unit, currently being rapidly eroded, overlies a sequence of lavas and densely-welded pyroclastic deposits with ages ranging from 194.5 ± 6.3 to 94.5 ± 7.9 ka (Wilch *et al.* 1999). These older units indicate that Mount Takahe volcano reached its present elevation by 194.5 ka.

The *c.* 8.3 ka tephra on the Mount Takahe caldera rim has been correlated with 8.2 kyr old tephra layers found in ice cores across West Antarctica, including the Byrd, WAIS Divide and Siple Dome ice cores (Wilch *et al.* 1999; Dunbar *et al.* 2021). The *c.* 8.3 ka eruption was highly dispersive and was likely to have been a Plinian eruption. The Takahe tephra provides an important time stratigraphic horizon in these climate archives (Dunbar *et al.* 2021). Palais *et al.* (1988) interpreted Mount Takahe as the source of many tephra layers in the Byrd ice core. Two Mount Takahe eruption intervals were inferred from the tephra record: alternating subaerial and phreatomagmatic from 30 to 20 ka; and sustained phreatomagmatic eruptions from 20 to 14 ka. These younger eruptions do not appear to be represented in the caldera-rim sequences but are coeval with some late-stage flank eruptions described below.

Volcanic deposits on the lower flanks are younger than the oldest caldera-rim deposits, suggesting that they resulted from post-edifice-building flank eruptions. Three subglacial–subaerial passage-zone sequences associated with these late-stage flank eruptions are preserved near the base of the volcano at Gill Bluff, Möll Spur and Stauffer Bluff (McIntosh *et al.*

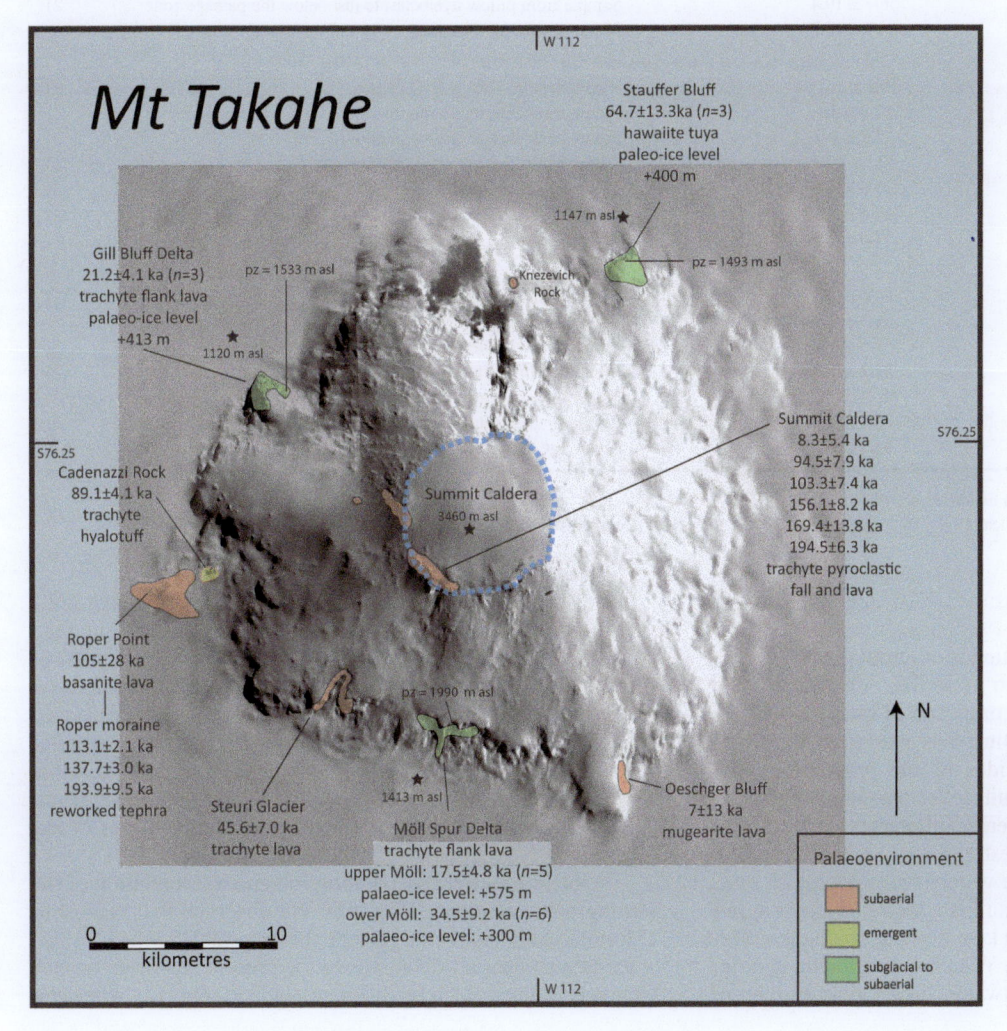

Fig. 11. Satellite image map of Mount Takahe of the Eastern MBL Volcanic field, showing outcrops, lithofacies and ages. Outlines of volcano (gold) and caldera (blue) are approximate. Image source: Google Earth Pro image accessed June 2019; for more information on source and an explanation of image processing see the caption to Figure 7.

Table 4. *Mount Takahe summary age table*

Sample ID	Method	Corrected preferred age ± 2 SD (ka)	Description	Ref.
Summit caldera				
W85009	A	8.3 ± 5.4	Bucher rim (south) pumice and obsidian bombs	20
W85013	A	94.5 ± 7.9	Bucher rim (north) lava beneath welded fall	20
W85015	A	103.3 ± 7.4	Bucher rim (north), welded fall obsidian	20
W85015	F	135 ± 156	Bucher rim (north), welded fall obsidian	15
MT85006	A	156.1 ± 8.2	2835 m outcrop lava overlain by welded fall	20
W85022	A	169.4 ± 13.8	Bucher rim (south), lava	20
W85019	F	175 ± 176	Bucher rim, obsidian in moraine	15
W85011	A	194.5 ± 6.3	Bucher rim (south), 10 m lava over W85-09	20
Post-shield flank outcrops				
Oeschger Bluff				
MT10	A	7.1 ± 13.0	Pāhoehoe lava	16
Roper Point				
MT85-3	A	105.3 ± 28.0	Subaerial lava	16
98-83	A	113.1 ± 2.1	Tephra in moraine	21
98-86	A	137.7 ± 3.0	Tephra in moraine	21
98-87	A	193.9 ± 9.5	Tephra in moraine	21
Cadenazzi Rock				
98-90	A	89.1 ± 4.1	Lava	
Steuri Glacier				
MT85-9	A	45.6 ± 7.0	Subaerial lava	16
Gill Bluff				
98-039	A	26.0 ± 5.1	Lithic clast from debris flow	21
98-017	A	21.5 ± 3.3	Pillow hyaloclastite	21
98-024	A	22.0 ± 3.7	Pillow hyaloclastite	21
98-027	A	20.2 ± 3.7	Debris mixed/pillow hyaloclastite	21
	A	**21.2 ± 4.1**	Mean age (*n* = 3 samples)	
Stauffer Bluff				
98-055	A	69.5 ± 9.6	Pillow hyaloclastite below the passage zone	21
98-058	A	61.3 ± 10.5	Juvenile bomb from accretionary lapilli hyalotuff	21
98-064	A	56.7 ± 19.4	Sample from pillow hyaloclastite just below the passage zone	21
	A	**64.7 ± 13.3**	Mean age (*n* = 3 samples)	
Moll Spur (upper)				
98-100-01	A	19.7 ± 12.1	Pillow hyaloclastite, pillow interior	21
98-100-05	A	22.8 ± 3.1	Pillow hyaloclastite, pillow interior	21
98-100-06	A	17.7 ± 3.9	Pillow hyaloclastite, pillow interior	21
98-100-07	A	20.0 ± 14.7	Pillow hyaloclastite, pillow interior	21
98-100	A	**20.8 ± 4.7**	Mean age (*n* = 4 analyses)	21
98-103	A	16.4 ± 2.8	Matrix-rich pillow hyaloclastite	21
	A	**17.5 ± 4.8**	Mean age (*n* = 2 samples)	21
Moll Spur (lower)				
98-113-1	A	34.3 ± 13.7	Pillow hyaloclastite from the eastern side	21
98-113-2	A	36.3 ± 9.2	Pillow hyaloclastite from the eastern side	21
98-113-4	A	37.6 ± 4.0	Pillow hyaloclastite from the eastern side	21
98-113	A	**37.2 ± 7.0**	Mean age (*n* = 3 analyses)	21
98-110	A	40.7 ± 10.7	Pillow hyaloclastite from the western side	21
98-115	A	28.4 ± 7.5	Pillow hyaloclastite from the eastern side	21
	A	**34.5 ± 9.2**	Mean age (*n* = 3 samples)	

Notes: method A is $^{40}Ar/^{39}Ar$; method F is fission track.
Ref. code: 15, Seward *et al.* (1980); 16, Wilch (1997); 20, Wilch *et al.* (1999); 21, this study. See also Supplementary Material Table S1.

1985; Palais *et al.* 1988; LeMasurier 2002; unpublished data of Wilch and McIntosh).

The Gill Bluff promontory extends out from the NW side of Mount Takahe, where it rises to more than 500 m above the ice-sheet surface. The bluff is composed of trachyte lava and volcaniclastic deposits. Gill Bluff has extensive exposures on the NE- and SW-facing sides of the promontory, and both sides exhibit complex multi-level passage-zone transitions from subglacial to subaerial lithofacies and environments. The passage zone is best exposed on the NE-facing bluff, where it exhibits variable water levels (McIntosh *et al.* 1985; LeMasurier 2002). Gill Bluff is interpreted as a prograding lava delta that formed when lava flowing down the side of Mount Takahe encountered the WAIS and formed a lava-fed delta in an ice-marginal meltwater lake (McIntosh *et al.* 1985). Individual horizontally orientated subaerial lava units can be traced to dipping units of pillow hyaloclastite in delta foresets. LeMasurier (2002) described a rising vertical passage zone at Gill Bluff across which gently-dipping subaerial felsic lavas pass into 20°–25° dipping pillow hyaloclastite foreset beds. LeMasurier (2002) offered two alternative interpretations of the rising passage zone: either it represents a stable ice-sheet level with a fluctuating ice-marginal lake level; or it represents changing ice levels at different times of the eruption.

We offer an alternative description and interpretation of the stratigraphic sequence (Fig. 12). We observed that nearest to Mount Takahe, the passage zone forms a stable horizontal surface at a measured 413 m above the present ice-sheet surface (elevation determined by differential GPS) (Fig. 12). The delta

Fig. 12. Field sketch of a NE-facing outcrop at Gill Bluff, Mount Takahe. Volcanic lithofacies abbreviations: L, subaerial lava; PH, pillow hyaloclastite; DF, debris flow; talus, talus colluvium. Numbers 1–8 designate the sequence of subaerial lavas; numbers 1–7B designated the correlated equivalent sequence of subaqueous deposits in a flow-foot delta. The passage-zone surface is highlighted in red and shows an initial high level at 413 m above current ice level, a drop to low level at c. 200 m above ice level and a rise to c. 400 m above ice level. Composite photograph by T.I. Wilch.

prograded into an ice-marginal meltwater lake for a period until the passage zone dropped steeply by about 200 m in elevation, where it re-established a horizontal surface. This passage-zone fall is attributed to partial draining of the meltwater lake. After another period of stability and delta progradation at this lower lake level, the passage zone exhibits a steep incremental rise, depositing hyaloclastite foresets associated with the rising and prograding topset beds. The rising passage reached about 400 m above present ice level, close to the original lake level. We observed no stratigraphic evidence for any time gaps such as unconformities or changes in weathering, composition or age in the Gill Bluff sequences. Therefore, we infer that the entire bluff sequence was erupted over a short interval. A similar sequence of events in a glaciovolcanic pāhoehoe-lava-fed delta on James Ross Island (Antarctic Peninsula) was described by Smellie (2006) and similarly ascribed to dynamic fluctuations in the surface elevation of a meltwater lake caused by variable subglacial meltwater drainage. The Gill Bluff sequence was known to be young (K–Ar age of <0.1 Ma). Three anorthoclase separate samples with the most ^{40}Ar/^{39}Ar precise ages yielded concordant age spectra with a weighted mean age of 21.2 ± 4.1 ka (Table 4).

The highest horizontal passage zone at +413 m above modern ice level is interpreted as the minimum thickening of the ice-sheet surface at the time of the eruption, c. 21 ka, coincident with the Last Glacial Maximum. The subsequent descending, horizontal and ascending passage zones are interpreted as indicators of partial draining, stability and refilling of an englacial lake during the eruption phase. Deposits at the passage-zone boundary show mixtures of subaqueous and subaerial lithofacies (Fig. 13a). Highly-vesicular welded breccias and massive holocrystalline lavas indicative of subaerial conditions crop out above the passage zone. Pillow lava, bedded hyaloclastite breccia and interbedded debris-flow deposits crop out below and proximal to the passage zone (Fig. 13b, c). These subaqueously emplaced deposits are locally deformed, as evidenced by disrupted bedding, shear surfaces and normal faults. Towards the toe of the delta, glass-rich clastic deposits are finer grained and more uniformly bedded. In places, these distal delta-toe deposits exhibit reverse faulting that may have resulted from pressure due to ice readvance (Fig. 13d).

Möll Spur is a prominent steep ridge composed of trachyte lava and volcaniclastic deposits on the south side of Mount Takahe (Figs 11 & 14). Möll Spur rises up to 800 m above the level of the ice sheet and, similar to Gill Bluff, was formed by late-stage lava eruption on the flank of Mount Takahe. A passage zone from subglacial to subaerial lithofacies occurs at 575 m above the present ice surface at Möll Spur. Below the passage zone, outcrops consist of trachytic pillow lava and palagonitized, matrix- to clast-supported, hyaloclastite breccia deposits (Fig. 13e) (LeMasurier 2002). New ^{40}Ar/^{39}Ar analyses suggest that the Möll Spur sequence was erupted at two different times. Five analyses of three samples from the lowermost pillow and hyaloclastite deposits yield a weighted mean age of 34.5 ± 9.2 ka (Table 4). These subglacially-erupted pillow and lobe hyaloclastite deposits occur from the base of the slope up to about 300 m above ice level. Many of the clastic deposits, especially in the basal sequences, show evidence of syn- or post-depositional deformation, including faulting and slickensided shear surfaces (Fig. 13f). The deformation is likely to have resulted from slumping and collapse during or shortly after deposition into an unstable englacial lake formed between Mount Takahe and the ice sheet. Five ^{40}Ar/^{39}Ar analyses of anorthoclase separated from three samples in the uppermost pillow and hyaloclastite deposits yield a weighted mean age of 17.5 ± 4.8 ka. It is possible that these Möll Spur deposits were part of a more explosive Mount Takahe eruption that has been documented in the WAIS Divide ice core as the '17.7 ka Mount Takahe

Fig. 13. Photographs of outcrops and lithofacies at Mount Takahe, Eastern MBL Volcanic Field. (**a**) View of palagonitized Strombolian breccia just above a falling passage zone at Gill Bluff, Mount Takahe. Unit is underlain by pillow hyaloclastite. Located near the passage zone at +413 m above current ice level. (**b**) View of debris flow and colluvium deposits below the low passage zone (+200 m above ice level) at Gill Bluff, Mount Takahe. Note the steep dips of debris units and sharp shear boundaries between units. (**c**) View of the complex stratigraphy in a SW-facing Gill Bluff sequence at Mount Takahe. Basal tuff breccia is interpreted as remobilized hyaloclastite debris flow. In the middle of the photograph, fine-grained stratified and deformed lapilli tuff and tuff (yellow) are interpreted as remobilized sediment. This is overlain by a 'lobe hyaloclastite' consisting of large irregular-shaped pillow lobes with coarse hyaloclastite breccia. The glassy margins are up to 15 cm thick. (**d**) View of Z-folded, planar-bedded trachytic tuff and lapilli tuff at the distal toe of the Gill Bluff flow-foot delta. (**e**) View of trachyte pillows and hyaloclastite breccia in the upper part of Möll Spur flow-foot delta about 60 m below the passage zone (at +575 m above current ice level). Sample 98-100 was collected from a pillow interior at this location. Four separate anorthoclase mineral aliquots from this sample were $^{40}Ar/^{39}Ar$ dated, yielding a mean age of 20.8 ± 2.8 ka. (**f**) View of a massive shear zone in the central part of the Möll Spur subglacial sequence. (**g**) View of incursive pillows in the complex sediment–hyaloclastite–pillow sequences deposited in an englacial lake during the growth of the parasitic Stauffer Bluff tuya. The location of the photograph is shown in Figure 15. (**h**) View of planar-stratified lapilli tuff with subangular lithic and scoria clasts on the summit of the Stauffer Bluff tuya, Mount Takahe, Eastern MBL Volcanic Field. Note the lithic block sitting in the bedding-plane sag below the snow in the centre of the photograph. Other bedding-plane sags and pyroclastic bombs were observed. The deposit is interpreted as a combination of pyroclastic fall and base surge deposits associated with an emergent phreatomagmatic eruption.

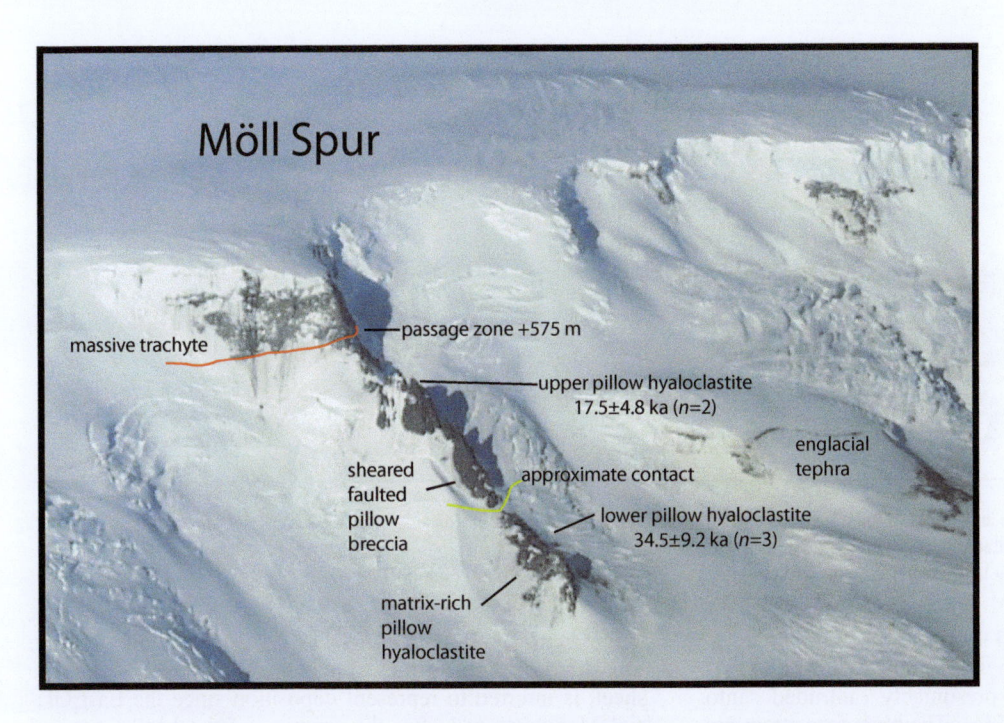

Fig. 14. View of Möll Spur, Mount Takahe, Eastern MBL Volcanic Field, showing subaerial lava at top of the spur with a flow-foot pillow hyaloclastite delta below the passage zone.

Event' (McConnell *et al.* 2017). The dated Möll Spur samples were collected between 300 and 525 m above ice level. The passage zone is at +575 m and is overlain by a massive subaerial lava that forms a prominent cliff section (McIntosh *et al.* 1985; LeMasurier 2002). The Möll Spur sequences place minimum WAIS ice level constraints of >+300 m at *c.* 35 ka and +575 m at *c.* 18 ka. The inferred ice thickening at *c.* 18 ka is coincident with the Last Glacial Maximum.

Stauffer Bluff, situated on the NE flank of Mount Takahe, is composed of hawaiite lava and volcaniclastic deposits; $^{40}Ar/^{39}Ar$ ages of three hawaiite groundmass samples from Stauffer Bluff provide a mean age of 64.7 ± 13.3 ka (Table 4). Stauffer Bluff is part of a 1 km-diameter relatively flat-topped edifice that rises 525 m above the ice-sheet surface (Fig. 15). The Stauffer Bluff edifice is interpreted as a late-stage parasitic tuya with a vent likely to have been centred on the edifice. The bluff face comprises an exposed flank of the tuya, and is composed of stratigraphically complex subaqueously deposited primary and reworked lithofacies (Fig. 16). The dominant subaqueous lithofacies are thick sequences of hawaiite pillow lava dipping outward from the edifice, palagonitized hyaloclastite breccia, and redeposited fine-grained hyalotuff and hyaloclastite sediments (Figs 10 & 13g). Lithofacies also include sills intruded into glassy

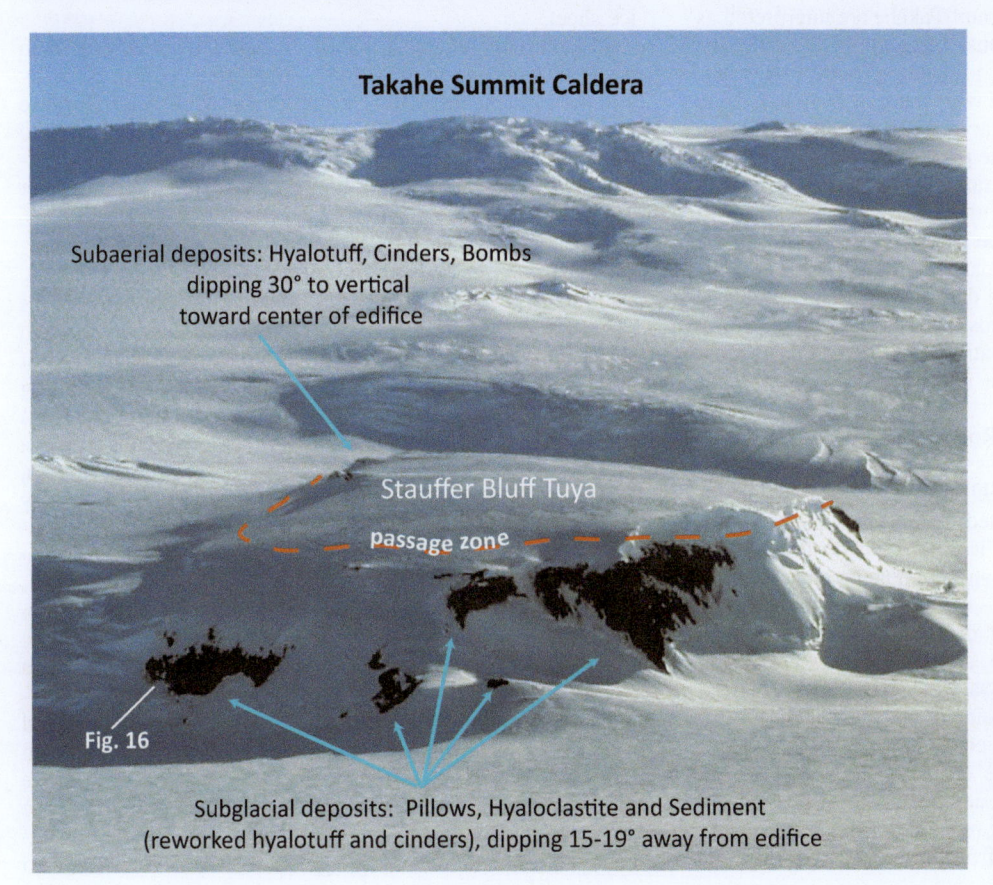

Fig. 15. View of Stauffer Bluff tuya at the NE base of Mount Takahe volcano. The edge of the summit caldera is visible at the top of the photograph. The photograph was taken from an LC-130 airplane by T.I. Wilch.

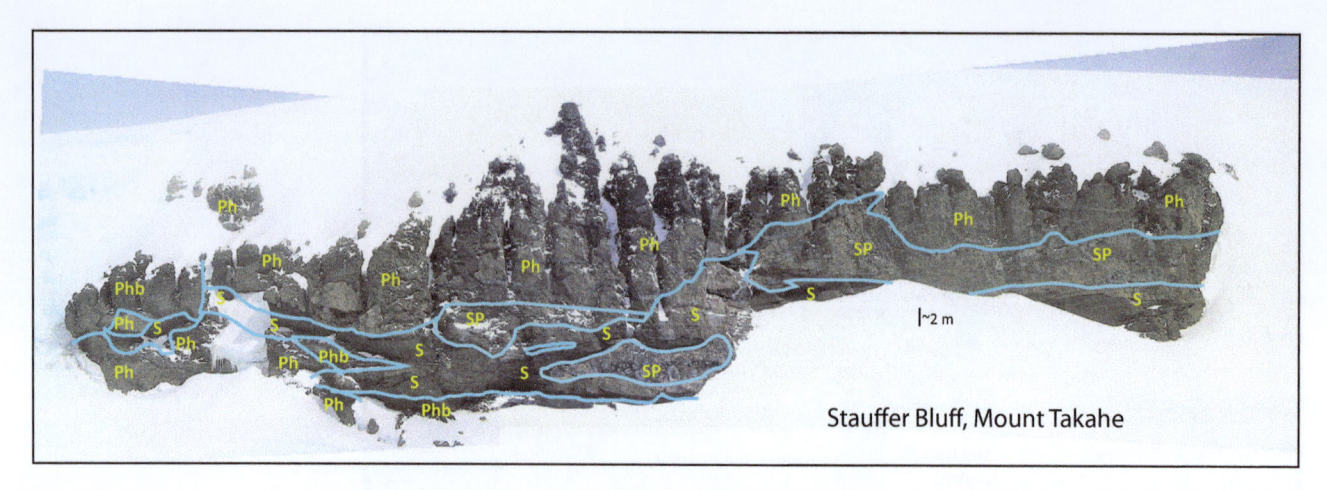

Fig. 16. Field interpretation of outcrop at the base of Stauffer Bluff tuya, Mount Takahe, Eastern MBL Volcanic Field (see Fig. 14). The outcrop is situated below the passage zone. Volcanic lithofacies abbreviations: Ph, pillow with some hyaloclastite; Phb, hyaloclastite with some pillow lava; S, reworked fine-grained hyalotuff sediment (mostly lapilli tuff); SP, pillow lavas intruded into sediment (S). Photograph by T.I. Wilch.

lapilli tuff and beds of reworked scoria lapilli forming peperite. The sills are composed of nested pillows with no hyaloclastite matrix that were presumably intruded into water-saturated sediment (Fig. 13g). The bluff-face strata are interpreted as a lava-fed delta formed in an englacial lake. The passage zone from subaqueous to subaerial conditions is covered by snow and ice but is estimated to be located near the break in slope between the bluff face and the bluff top, 400 m above the present ice level. Outcrops on the top of the bluff consist of horizontally to steeply-bedded, palagonitized lapilli tuff and tuff breccia that are interpreted as phreatomagmatic hyalotuff deposits emplaced in an emergent environment (Fig. 13h). Ash-coated lapilli are common in the lapilli-tuff deposits. Large reddened pyroclastic bombs (up to 1 m in diameter) form bedding-plane sags in the tuff deposits. The Stauffer Bluff sequence places minimum WAIS ice level constraints of >+400 m at 65 ka.

All other flank outcrops at Mount Takahe are interpreted as resulting from subaerial eruptions above the level of the ice sheet and without significant interaction with meltwater. Cadenazzi Rock, located on the west flank of Mount Takahe at 350 m above the ice sheet surface, is a 50 m-high bluff composed of 89.1 ± 4.1 ka mugearite pyroclastic rocks (Table 4). Palagonitized lapilli tuff with rare large (up to 1 m) lithic blocks and pumiceous bombs (up to 15 cm) are exposed in the lower part of the bluff. The upper section of the bluff includes lapilli tuff, rich in ash-coated lapilli with beds of well-sorted achnelith-rich lapillistone deposits. Cadenazzi Rock is interpreted as a pyroclastic deposit that resulted from a mix of magmatic and phreatomagmatic eruptions that formed in a subaerial environment where meltwater had intermittent access to the vent.

Downslope from Cadenazzi Rock, Roper Point consists of eroded outcrops of scoriaceous hawaiite lava, $^{40}Ar/^{39}Ar$ dated to 105 ± 28 ka (Wilch 1997), overlain by glacial till. Roper Point is the oldest dated flank outcrop at Mount Takahe. The till at Roper Point is a poorly-consolidated heterolithic silty gravelly diamict and abundant volcanic erratics. Erratic lithologies are dominated by subaerial lithofacies, including a variety of lava fragments, as well as pumice-rich moderately- to densely-welded pyroclastic rocks. The pyroclastic erratics appear to be derived from welded fall and ignimbrite deposits. The till also includes trachytic pumice lapilli and ash concentrations up to 10 cm thick, with subrounded pumice clasts as large as 2 cm in diameter. The tephra deposits are presumed to be locally derived and are interpreted as reworked. Three $^{40}Ar/^{39}Ar$ dates of anorthoclase separates from different tephra deposits range from 193.9 ± 9.5 ka to 113.1 ± 2.1 ka (Table 4). The till, situated about 270 m above the local level of the ice sheet, is inferred to represent deposition since the Last Glacial Maximum and after the eruption of the Möll Spur and Gill Bluff glaciovolcanic sequences.

On the SW side of Mount Takahe, two outcrops are present near Steuri Glacier (Fig. 11). The eroded rim of a monogenetic basaltic scoria cone is preserved on slopes NW of the glacier, and subaerial trachytic lava and breccia are preserved SE of the glacier. The Steuri Glacier trachyte is about 600 m above present ice level and is $^{40}Ar/^{39}Ar$ dated to 45 ± 7 ka (Wilch 1997). On the SE side of Mount Takahe, a stacked sequence of pāhoehoe basanite lavas is exposed at Oeschger Bluff, about 250 m above the level of the ice sheet. The lava yielded an imprecise $^{40}Ar/^{39}Ar$ age of 7 ± 13 ka (Wilch 1997). On the north side of Mount Takahe, undated trachyte lava is exposed at Knezevitch Rock, less than 200 m above the level of the ice sheet.

In summary, Mount Takahe is a Late Quaternary volcano with a large summit caldera that was constructed by 194.5 ka and the last known eruption was about 8 ka. Because of the Holocene activity, Mount Takahe is still considered to be active. Mount Takahe provides critical data for reconstructing late Pleistocene WAIS ice levels, which are discussed later in this chapter.

Toney Mountain. Toney Mountain is a massive, elongate, east–west-orientated central volcano with basal dimensions of 55×15 km (Figs 1, 2 & 17). The volcano rises c. 2000 m above the WAIS and has a summit peak elevation of 3595 m. Toney Mountain produces a significant damming effect on the north-flowing WAIS. Ice levels on the upstream side of the volcano are 500 m higher than on the downstream side. The summit area of the volcano has a circular caldera, about 3 km in diameter. The edifice appears to be mostly undissected, except at its eastern end. Relatively steep constructional slopes (13°–21°) occur below the caldera on the north and south flanks.

Only five samples from three outcrop localities have been analysed for age (K–Ar, fission track) and geochemistry. Cox Bluff, located c. 15 km east of the summit caldera between 1600 and 1800 m above sea level, exposes a 200-m-thick sequence of Late Miocene (c. 9–10 Ma) subaerial hawaiite lava (LeMasurier et al. 1990d). Zurn Peak, located 7 km NNE of the caldera between 1200 and 1400 m above sea level, exposes trachyte and comendite rhyolite lava. The trachyte yielded a K–Ar age of 1.0 ± 0.4 Ma and the comendite yielded a fission track age of 0.29 ± 0.4 Ma (Seward

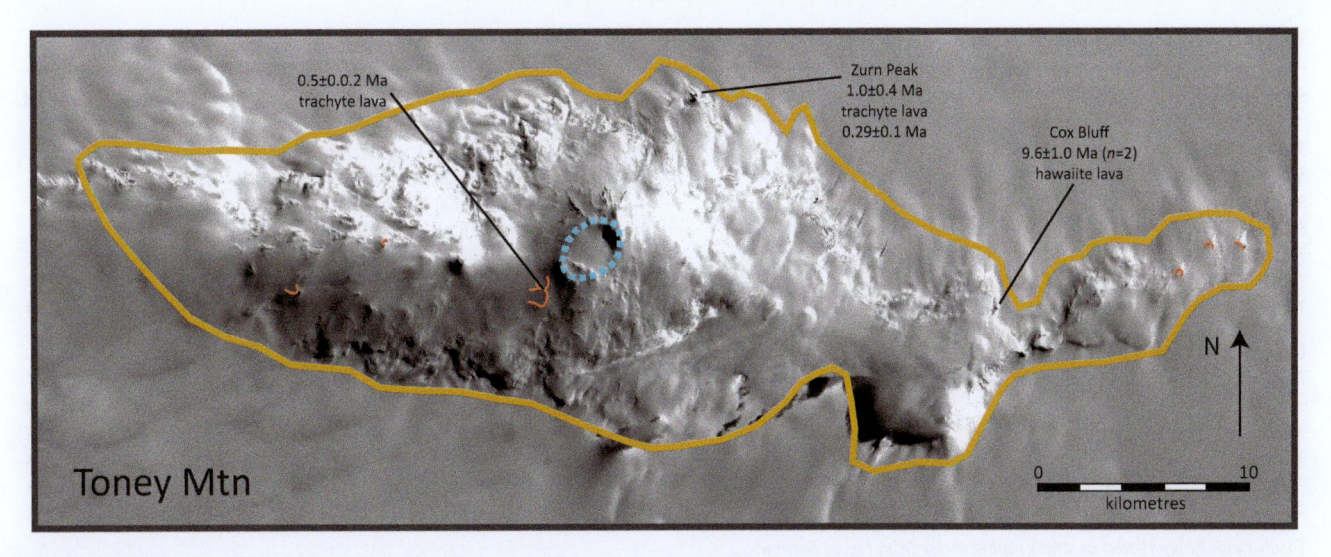

Fig. 17. Satellite image map of studied outcrops of Toney Mountain in the Eastern MBL Volcanic Field. Outlines of volcano (gold) and caldera (blue) are approximate. There are only three analysed outcrops. Image source: Google Earth Pro image accessed June 2019; see the caption to Figure 7 for more information on the source and an explanation of image processing.

et al. 1980; LeMasurier *et al.* 1990*d*) (Table 5). A third (unnamed) outcrop situated 2–3 km SW of the caldera at about 2900 m asl exposes Pleistocene (K–Ar 0.5 ± 0.2 Ma) benmoreite lava (LeMasurier *et al.* 1990*d*). On satellite images, this site appears to be part of a cone rim that is *c.* 0.8 km in diameter (Fig. 17) and is likely to be a parasitic cone. Several other parasitic cones are situated along the long axis of the volcano; some of these cones have been sampled but the outcrops were too altered for analysis (LeMasurier *et al.* 1990*d*). Paulsen and Wilson (2010) noted the alignment of some these cones and included them in their stress-pattern analysis.

Toney Mountain is the only volcano in MBL for which there is a depth estimate of the sub-ice volcano–basement contact. Based on a 1959–60 seismic traverse across the west end of the edifice, the contact occurs at a depth of 3000 m below sea level (Bentley and Clough 1972), suggesting a total relief of the volcano of *c.* 6600 m, and a possible volume ranging from 2800 to 3613 km^3 (LeMasurier *et al.* 1990*d*; LeMasurier 2013).

In summary, Toney Mountain is a major central volcano with a well-defined caldera that appears to have been active in Late Miocene and Pleistocene times. Toney Mountain is the least-studied central polygenetic volcano of the MBLVP and, although it is mostly snow- and ice-covered, there are many outcrops that could be studied to better understand its eruptive history.

Crary Mountains Volcanic Field

The Crary Mountains Volcanic Field consist of three large, coalesced central volcanoes, Mount Rees, Mount Steere and Mount Frakes, and the much smaller Boyd Ridge, which are aligned roughly NW–SE and are less dissected and progressively younger toward the SE (Fig. 18). The Late Miocene Mount Rees and Mount Steere, and Pliocene Mount Frakes, together cover an area of about 33 × 15 km. Boyd Ridge, located about 13 km SE of these central volcanoes, is an east–west-orientated, lower-relief ridge that covers an area of about 13 × 7 km. Mount Steere and Mount Frakes each have >2 km-diameter summit calderas. The Crary Mountains produce a significant damming effect on the NE-flowing WAIS. Ice levels on the upstream SW side of the volcanoes are 200–400 m higher than on the downstream NE side. Deep cirques cut into the east and NE sides of Mount Steere and Mount Rees expose thick stratigraphic sequences and intrusive rocks, making these two volcanoes among the best exposed in Marie Byrd Land.

On the basis of initial observations and four K–Ar ages, the Crary Mountains were characterized as two felsic shield volcanoes (Mount Steere and Mount Frakes) on a platform of basaltic lava and pyroclastic rocks (LeMasurier *et al.* 1990*a*). Subsequent detailed fieldwork, geochemical analysis of 70 samples and ^{40}Ar/^{39}Ar dating of 77 samples (Table 5) provide a more comprehensive record of volcanism in the Crary Mountains (Wilch 1997; Panter *et al.* 2000; Wilch and McIntosh 2002; Chakraborty 2007). This summary of the Crary Mountains volcanoes is drawn mostly from Wilch (1997) and Wilch and McIntosh (2002), with a minor amount of unpublished data.

Mount Rees and Mount Steere. Mount Rees and Mount Steere are overlapping Late Miocene polygenetic central volcanoes (Fig. 18). Given their similar ages, lithofacies and histories, Mount Rees and Mount Steere are presented together in this subsection. Mount Rees is more deeply dissected, lower in elevation (2709 m compared to 3558 m), and slightly older (9.5–6.9 Ma) than Mount Steere (8.7–5.8 Ma). Mount Rees (*c.* 10 × 7 km) has a smaller footprint than Mount Steere (*c.* 9 × 12.5 km). The eroded Mount Rees lacks a summit caldera, whereas Mount Steere has a mostly intact *c.* 2 km-diameter summit caldera. Extensive outcrops of the dissected volcanoes are located on the east flank of Mount Rees and on the north flank of Mount Steere. Outcrops up to 800 m above current ice level have been examined and sampled.

Mount Rees has thick stratigraphic sequences at Trabucco Cliff and Tasch Peak Ridge consisting mostly of mafic–intermediate volcanic rocks with subordinate interlayered felsic lavas (Wilch and McIntosh 2002). The mafic–intermediate rock outcrops (Fig. 19) are characterized by two alternating lithofacies: (1) unbrecciated lavas with oxidized bases (subaerial lithofacies); and (2) palagonitized glassy hyaloclastite breccias and pillow lavas (subaqueous or water-contact lithofacies). In places, oxidized clastogenic lava makes up the subaerial lithofacies. ^{40}Ar/^{39}Ar analyses of five samples from different levels in the Trabucco Cliff sequence agree with the stratigraphic order (from 9.25 ± 0.53 to 9.06 ± 0.06 Ma) but are analytically indistinguishable from one another (Table 5). No glacial unconformities or tillites were observed in either the subaqueous or subaerial lithofacies sequences.

Table 5. *Toney Mountain and Crary Mountains summary age table*

Sample ID	Method	Corrected preferred age ± 2 SD (Ma)	Description	Ref.
Toney Mountain				
80	K	9.1 ± 1.2	Cox Bluff- horizontal lava flows	9
80	K	10.1 ± 1.2	Cox Bluff- horizontal lava flows	9
80	K	**9.6 ± 1.0**	**Mean age (*n* = 2)**	9
76A	K	1.0 ± 0.4	Zurn Peak, trachyte flow rock	9
75	K	0.5 ± 0.2	Unnamed outcrop at 2900 m asl; flow rock	9
76D	F	0.29 ± 0.10	Zurn Peak, comendite	15
Crary Mountains: Mount Rees				
Trabucco Cliff (listed in stratigraphic order from top to bottom)				
92-174	A	9.06 ± 0.06	Lava flow top of section	18
92-175	A	9.06 ± 0.06	Lava flow top of section	18
92-6	A	9.09 ± 0.07	Massive flow, above 92-15	18
92-15	A	9.12 ± 0.04	Lava flow middle of section	18
92-1	A	9.25 ± 0.53	Lobe hyaloclastite	18
Tasch Peak Ridge (listed in stratigraphic order from top to bottom)				
92-36	A	7.62 ± 0.06	Dyke	18
92-38	A	8.32 ± 0.13	2230 m pillow hyaloclastite	18
92-34	A	8.34 ± 0.09	2269 m intrusive hyaloclastite	18
92-41	A	8.61 ± 0.16	Hyaloclastite lobes	18
92-31	A	9.10 ± 0.28	1844 m pillow lobe interior	18
92-28	A	9.19 ± 0.06	1817 m spatter-fed lava	18
92-23	A	9.03 ± 0.12	1739 m pillow lobe	18
92-59	A	9.46 ± 0.24	1643 m lava flow	18
Isolated outcrops				
92-117	A	6.91 ± 0.26	SW of summit; eroded parasitic cone	18
92-114	A	7.01 ± 0.21	Eroded parasitic cone	18
92-112	A	9.14 ± 0.19	North end of Mount Rees, lava-flow, striae (240 m)	18
92-109	A	8.81 ± 0.08	lava-flow, striae (328 m)	18
Crary Mountains: Mount Steere				
Outcrops on west, north and NE sides				
92-118	A	7.64 ± 0.05	Flow-banded lava, west of summit	18
92-95	A	5.81 ± 0.04	Flow-banded lava clast from moraine, base of north side	18
92-64	A	8.16 ± 0.08	Dyke, NE outcrop	18
92-53	A	8.34 ± 0.08	Lava-flow dome, NE outcrop	18
92-93	A	8.35 ± 0.08	Dyke, NE outcrop	18
92-63	A	8.38 ± 0.08	Dyke, NE outcrop	18
92-51	A	8.43 ± 0.06	Lava flow, exposed plug, NE outcrop	18
92-107	A	8.45 ± 0.06	Dyke, 2413 m	18
92-104	A	8.46 ± 0.08	Flow-banded lava, 2278 m	18
92-108	A	8.48 ± 0.06	Flow-banded lava, 2413 m	18
92-91	A	8.57 ± 0.09	Flow-banded lava, east side	18
92-181	A	8.01 ± 0.20	Dyke, fine grained, east side	18
92-182	A	8.63 ± 0.06	Flow-banded lava, east side	18
92-183	A	8.63 ± 0.06	Flow-banded lava, east side	18
92-178	A	8.66 ± 0.04	Flow-banded lava, east side	18
Ridge SE of Lie Cliff (samples listed in stratigraphic order)				
92-169	A	6.49 ± 0.43	SE ridge: 1814 m lava flow	18
92-165	A	7.47 ± 0.07	SE ridge: 1736 m lava, near 92-162	18
92-162	A	6.78 ± 0.05	SE ridge: 1736 m lava, near 92-165	18
Lie Cliff (samples listed in stratigraphic order)				
92-80	A	7.92 ± 0.06	Dyke, 1.5 m wide	18
92-85	A	8.49 ± 0.33	1631 m lava flow	18
92-82	A	8.39 ± 0.21	1600 m hyaloclastite lobe	18
92-79	A	8.54 ± 0.06	Lowest subaerial lava, thin flows	18
92-86	A	8.63 ± 0.23	1558 m lava flow	18
Ridge NW of Lie Cliff (samples listed in stratigraphic order)				
92-89	A	7.78 ± 0.06	Dyke, intrudes entire section	18
92-193	A	8.30 ± 0.18	1798 m feeder dyke, lava	18
92-192	A	8.33 ± 0.07	Feeder dyke, lava	18
92-194	A	8.30 ± 0.22	1753 m glassy breccia	18
92-189	A	8.56 ± 0.46	1747 m glassy lava	18
92-190	A	8.38 ± 0.64	1743 m pillow lava	18
92-186	A	8.51 ± 0.11	1646 m subaerial lava	18

(Continued)

Table 5. *Continued.*

Sample ID	Method	Corrected preferred age ± 2 SD (Ma)	Description	Ref.
Crary Mountains: Mount Frakes				
Morrison Rocks				
92-145	A	1.83 ± 0.10	Subaerial lava	18
92-142	A	1.84 ± 0.05	Subaerial lava	18
92-128	A	2.55 ± 0.06	Subaerial lava	18
92-125	A	2.57 ± 0.09	Subaerial lava	18
92-130	A	3.93 ± 0.03	Subaerial lava	18
92-122	A	4.22 ± 0.05	Lava flow	18
92-121	A	4.24 ± 0.04	Lava flow	18
92-127	A	4.31 ± 0.03	Lava flow	18
		4.26 ± 0.05	**Mean age (*n* = 3)**	
English Rock				
92-151	A	0.032 ± 0.010	Parasitic cone	18
92-151	A	0.035 ± 0.010	Parasitic cone	18
92-151	A	**0.034 ± 0.014**	**Mean age (*n* = 2)**	
92-157	A	0.837 ± 0.079	Parasitic cone	18
92-157	A	0.862 ± 0.036	Parasitic cone	18
92-157	A	**0.858 ± 0.066**	**Mean age (*n* = 2)**	
92-159	A	1.62 ± 0.02	Parasitic cone	18
Crary Mountains: Boyd Ridge				
92-135a	A	2.24 ± 0.19	Runyon Rock: juvenile clast from hyaloclastite	16
92-135b	A	2.00 ± 0.30		16
92-135	A	**2.17 ± 0.32**	**Mean age (*n* = 2)**	
92-134	A	2.05 ± 0.05	Runyon Rock, molded and polished clasts from debris flow	16
92-139	A	1.29 ± 0.03	Subaerial lava from north top side	16

Notes: method A is ^{40}Ar/^{39}Ar; method K is K/Ar; method F is fission track.
Ref. code: 9, LeMasurier *et al.* (1990c); 15, Seward *et al.* (1980); 16, Wilch (1997); 18, Wilch and McIntosh (2002). See also Supplementary Material Table S1.

A short interval of eruptions is consistent with the lack of unconformities in the sections. At Tasch Peak Ridge (Fig. 20), seven ^{40}Ar/^{39}Ar ages are mostly in stratigraphic order and range from 9.46 ± 0.24 to 8.32 ± 0.13 Ma. The mostly intermediate–mafic, alternating lavas and hyaloclastite breccias are cut by a trachytic dyke, ^{40}Ar/^{39}Ar dated to 7.62 ± 0.06 Ma. At both Trabucco Cliff and Tasch Peak Ridge, the contacts between the alternating subaerial and subaqueous lithofacies are dipping conformably with the constructional slopes of the volcano.

The stacked slope-parallel alternating subaerial and subaqueous (meltwater-contact) lithofacies at the base of Mount Rees (and also Mount Steere, described below) comprise slope-forming constructional lithofacies that lack interbedded fluvial deposits, glacial deposits and glacial unconformities. Wilch and McIntosh (2002) interpreted the sloping passage-zone sequence from subaqueous to subaerial lithofacies as being formed by lavas interacting with slope ice and snow to form pillows and hyaloclastites until they built above the level of slope ice and formed subaerial lavas. Furthermore, they interpreted the lack of glacial tills and unconformities as evidence of volcanic interactions with thin or cold-based ice on the slopes of the growing Mount Steere and Mount Rees volcanoes, above the level of the WAIS. The Mount Rees and Mount Steere sequences appear to resemble glaciovolcanic sequences in northern Victoria Land, Antarctica, which have been interpreted similarly by Smellie *et al.* (2011b) as lavas that interacted with a thin cold-based ice field or topography-draping ice. Smellie *et al.* (2011b, p. 135) considered the 'non-horizontal lava-fed delta passage zone(s) orientated parallel to underlying pre-volcanic bedrock slope' to be diagnostic of eruption in a glacial setting and interpreted the northern Victoria Land sequences to be slope parallel 'a'ā-lava-fed hyaloclastite deltas formed in sloping, open-top lava-melted channels in the ice (see also Smellie *et al.* 2013). Although generally similar, the Crary Mountain sequences include both wet to dry and dry to wet transitions, leading Wilch and McIntosh (2002) to suggest that lava flowing through open channels or tunnels in the ice were in some cases resubmerged after becoming emergent. Wilch and McIntosh (2002) concluded that the alternating wet and dry lithofacies, coupled with the sloping passage zones, precludes formation in a water-filled melt-water chamber within a formerly higher WAIS. Smellie and Edwards (2016), using the field descriptions of Wilch and McIntosh (2002), described the Mount Rees and Mount Steere sequences as 'a'ā-lava-fed deltas.

We contend that these alternating lithofacies associated with sloping passage zones are likely to be common features developed during the growth of ice-mantled Antarctic volcanoes. These sequences have not been recognized at other MBL volcanoes, possibly in part due to the lack of dissection and extensive snow- and ice-cover of most MBL volcanoes. Sequences of this type are diagnostic of eruption in an environment mantled with thin (<100 m), cold-based ice and provide only a maximum elevation for syneruptive regional ice sheets.

Two other outcrop areas were sampled at Mount Rees (Wilch and McIntosh 2002). On the NW side of Mount Rees, subaerially-erupted trachyte lava and bombs are exposed in multiple outcrops situated between 200 and 500 m above modern ice levels. Two samples are ^{40}Ar/^{39}Ar dated to 9.14 ± 0.19 and 8.81 ± 0.08 Ma (Table 5). West of Tasch Peak, basanite lapilli and bombs are exposed at three sites situated about 750 m above current ice levels. Bomb interiors from two locations yielded ^{40}Ar/^{39}Ar ages of 7.01 ± 0.21 and 6.91 ± 0.26 Ma. These basanite outcrops are interpreted as remnants of parasitic scoria cones.

The north face of Mount Steere is deeply eroded, with multiple separate outcrops exposing abundant felsic flow-banded lava and breccia, cut by numerous felsic–mafic dykes.

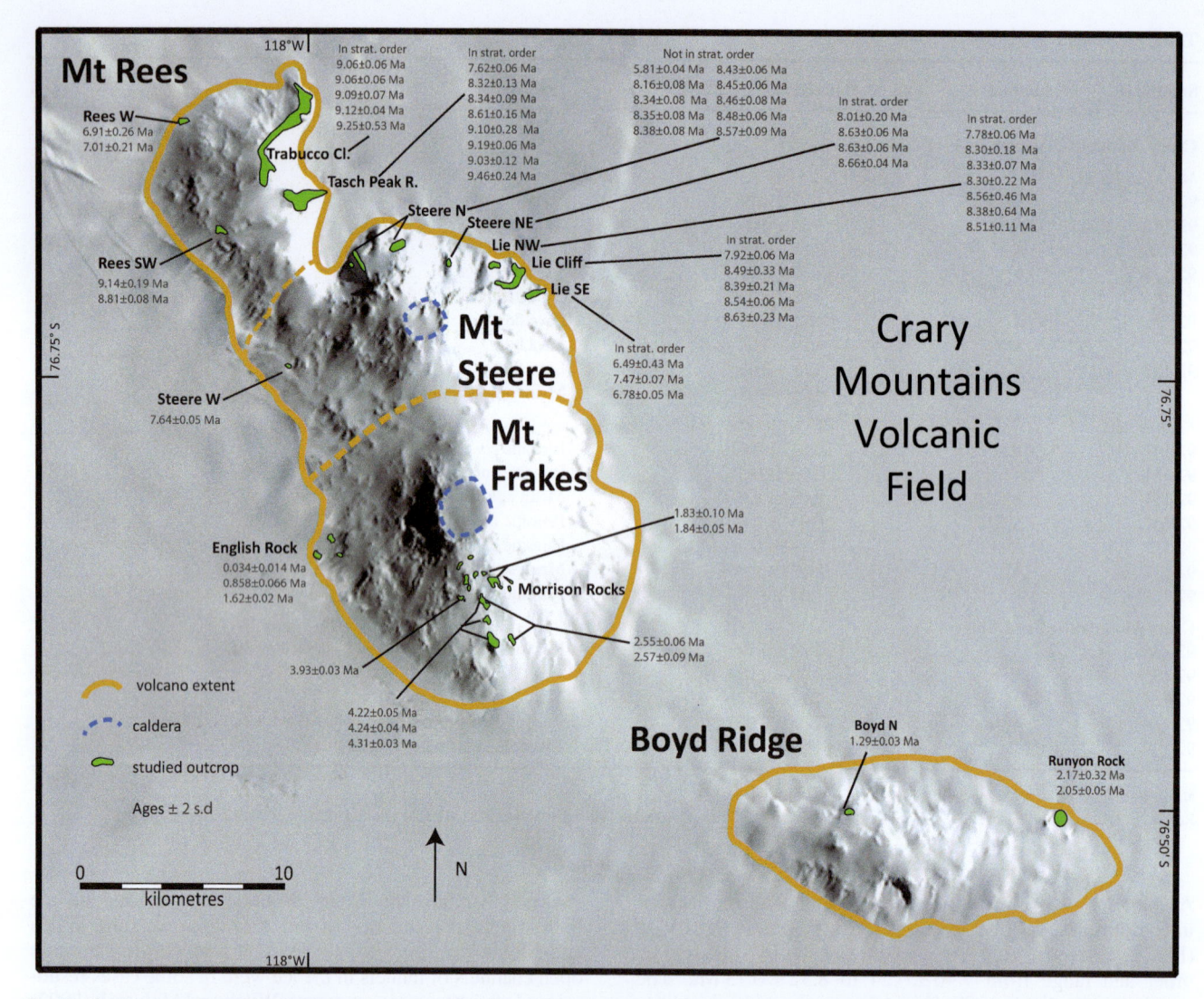

Fig. 18. Satellite image map showing ages and outcrops of the Crary Mountain Volcanic Field. There is a general progression from NW to SE for the four major volcanoes in the Crary Mountains. Both Mount Rees and Mount Steere are deeply dissected and many dykes are exposed. Image source: Google Earth Pro image accessed June 2019; see the caption to Figure 7 for more information on the source and an explanation of image processing.

The oldest dated rocks at Mount Steere are a series of hydrothermally-altered and brecciated, strongly flow-banded trachyte and rhyolite lava, situated between 1830 and 2030 m asl on the lower NE flank of the volcano. Potassium feldspar from three samples date from 8.66 ± 0.04 to 8.63 ± 0.06 Ma (Table 5). The lava sequence is cut by a hydrothermally altered and brecciated phonolite dyke that is dated at 8.01 ± 0.20 Ma. Separate outcrops of trachyte lava yield $^{40}Ar/^{39}Ar$ ages ranging from 8.57 ± 0.09 to 8.43 ± 0.08 Ma. These outcrops are cut by multiple dykes, with ages ranging from 8.45 ± 0.06 to 8.16 ± 0.08 Ma.

On the lower east flank of Mount Steere at Lie Cliff and a ridge to its north, outcrops include a 300 m-thick section of alternating subaqueous and subaerial lithofacies dominated by hawaiite lava and breccia, with subordinate trachyte lava. The sections resemble the alternating slope-forming stratigraphic sequences at Mount Rees and are interpreted in the same way. These sequences were erupted over a short interval that is not differentiated by $^{40}Ar/^{39}Ar$ dating of eight samples, with overlapping ages ranging from 8.63 ± 0.23 to 8.30 ± 0.22 Ma (Table 5). These subaqueous and subaerial sequences are intruded by four trachyte and phonolite dykes with ages ranging from 8.33 ± 0.07 to 7.78 ± 0.06 Ma.

Post-8 Ma eruptions at Mount Steere are represented at three localities. A flow-banded phonolite lava from the west side of Mount Steere is dated to 7.64 ± 0.05 Ma (Table 5). A discontinuous rock ridge SE of Lie Cliff exposes a slightly younger (7.47 ± 0.07–6.49 ± 0.43 Ma) series of dominantly basanite subaerial lava and pyroclastic deposits over a 400 m elevation range. The pyroclastic rocks are mostly subaerial lapilli tuff and bomb-rich agglutinate deposits with some interbedded palagonitized, fine-grained, laminated tuff deposits. Minor trachyte lava occurs in the sequence. This sequence is interpreted as resulting from a series of dry magmatic eruptions with minor intermittent interactions with external water producing phreatomagmatic eruptive phases. The uppermost hydrovolcanic tuff unit at c. 1810 m asl is cut by a glacial unconformity, as evidenced by a polished planar contact with truncated clasts. The units beneath the unconformity are dated to 6.49 ± 0.43 Ma and provide a maximum age of the erosion event. Finally, the youngest rock at Mount Steere is a flow-banded mugearite lava on the lower north flank of Mount Steere, downslope from older lavas and dykes. This subaerially-erupted mugearite lava is dated to 5.81 ± 0.04 Ma and suggests that the eroded Mount Steere slope developed prior to c. 5.8 Ma.

Mount Frakes. Mount Frakes is the least-dissected and highest of three central volcanoes in the Crary Mountains Volcanic Field, with a nearly circular 3 × 2.5 km-diameter summit caldera at 3654 m asl and flank slopes of 11°–15° (Fig. 18). The

Rock. Three outcrops were sampled and dated to 1.62 ± 0.02 Ma, and 858 ± 66 and 34 ± 14 ka (Table 5). The outcrops include clastogenic lava and bombs up to 1.5 m in length. The youngest deposits, situated *c.* 150 m above the level of the ice sheet, limit syneruptive ice-sheet expansion to <150 m above the present ice level at *c.* 33.9 ka.

Other than evidence for minor intermittent water–magma interaction in one outcrop at Morrison Rocks, there is no evidence for glaciovolcanic interactions at Mount Frakes. The overall absence of glaciovolcanic sequences at Mount Frakes may simply reflect the lack of dissection and limited exposure.

Crary Mountains–Boyd Ridge

Boyd Ridge is a low-relief east–west elongate ridge located 13 km SE of Mount Frakes (Fig. 18). At the east end and lee side of Boyd Ridge, Runyon Rock exposes a 130 m-thick stratigraphic section. The base of the section is a 2 m-thick hydrothermally altered phonolite lava breccia. This is overlain by 5 m of massive and poorly-sorted tuff breccia that is dominated by a matrix of angular finely-vesicular hawaiite lapilli and larger (up to 2 m in diameter) phonolite lava clasts that range from angular to very well rounded. Some of the phonolite clasts are polished. The deposit also contains abundant vesicular hawaiite clasts, and some of the clasts and other xenoliths are coated by hawaiite lava. This is interpreted as a debris-flow deposit or, possibly, an explosion breccia emplaced at the time of hawaiite volcanism. The basal tuff breccia grades into a uniform massive to slightly-bedded, matrix-supported, palagonitized hyaloclastite lapilli tuff, with rare larger hawaiite juvenile blocks. The lapilli are dominated by hawaiite clasts. There are no pillow lavas in the hawaiite hyaloclastite sequence. The 130 m-thick hyaloclastite deposit reaches an elevation of 1820 m asl, about 200 m below the upstream ice levels; therefore, in the absence of Boyd Ridge, a hyaloclastite sequence like that at Runyon Rock would be expected to form at the same location today.

Two samples have been dated from the Runyon Rock sequence. A phonolite clast from the debris-flow deposit is $^{40}Ar/^{39}Ar$ dated to 2.05 ± 0.05 Ma, and provides a maximum age for the debris-flow deposit and overlying hyaloclastite. A juvenile hawaiite clast from the hyaloclastite deposit was dated to 2.17 ± 0.32 Ma (Table 5). The maximum age of the phonolite clast and the imprecise eruption age suggest that the hyaloclastite was emplaced between 2.10 and 1.85 Ma. A second outcrop at Boyd Ridge, not identified on USGS topographical maps and located 1 km SW of Runyon Rock, is composed of subaerially erupted phonotephrite lava and lapilli tuff, $^{40}Ar/^{39}Ar$ dated to 1.29 ± 0.03 Ma.

Summary of Crary Mountains. A NW–SE alignment and systematic age progression of the three Crary Mountains polygenetic volcanoes (Mount Rees (9.5–6.9 Ma), Mount Steere (8.7–5.8 Ma) and Mount Frakes (4.3 Ma)) was noted in the structural analysis of MBL volcanoes by Paulsen and Wilson (2010). Younger volcanism occurred at the east–west-aligned Boyd Ridge (*c.* 2–1.3 Ma) and at post-edifice-building volcanoes on the flank of Mount Frakes (3.9–0.03 Ma). The age progression is consistent with the degree of dissection: Mount Rees is much more eroded than Mount Steere, and both of these two Late Miocene volcanoes are more eroded than the Pliocene Mount Frakes. It is likely that the erosion of Mount Rees and Mount Steere occurred early in their histories in the Late Miocene. The fact that the Mount Frakes is undissected suggests that only limited erosion has occurred since early Pliocene time. Numerous erratics are scattered around the flanks of the Crary Mountains. These erratics are

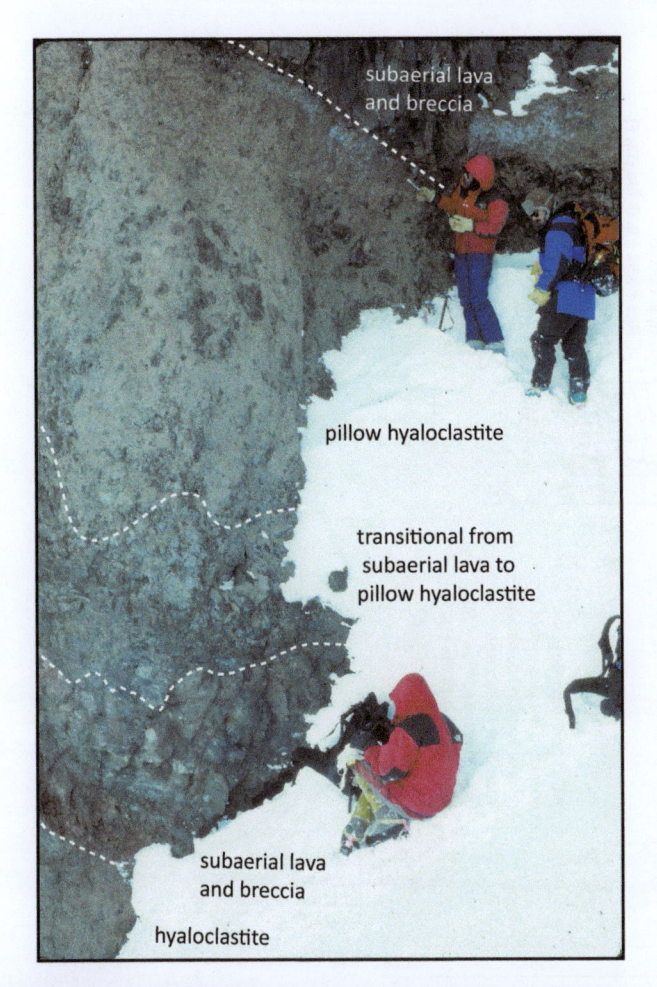

Fig. 19. View of the alternating subaqueous and subaerial lithofacies at Trabucco Cliff, Mount Rees, Crary Mountains Volcanic Field. The layered sequences and passage zones conform to the constructional slopes of the volcanoes. The location of Trabucco Cliff is shown in Figure 18. Adapted and updated from Wilch and McIntosh (2002).

nearly symmetrical base of Mount Frakes at the level of the ice sheet is 13.5 × 14.5 km. Numerous isolated outcrops are exposed at Morrison Rocks on the south flank of the volcano and upslope from English Rocks on its west flank.

Morrison Rocks consists of a series of outcrops on the south slope of Mount Frakes between 2225 and 2990 m asl. The oldest $^{40}Ar/^{39}Ar$ dated rocks (4.26 ± 0.05 Ma: Table 5) are crystal-rich (15–30%) phonolite lavas, with anorthoclase phenocrysts up to 4 cm in length. *In situ* phonolite lava crops out as high as 2550 m asl; phonolite and other felsic erratics occur among more mafic outcrops as high as 2990 m asl. The phonolite outcrops include large lava-flow levees running down the volcano flank. Several outcrops of mafic (basanite and hawaiite) welded Strombolian deposits are also exposed at Morrison Rocks. Spatter ramparts associated with some of these deposits are interpreted as evidence of fissure vents. The mafic rocks have $^{40}Ar/^{39}Ar$ dates of 3.93 ± 0.03, 2.55 ± 0.06 and 1.84 ± 0.05 Ma, and are interpreted as representing post-edifice-building subaerial eruptions. One of the basanite outcrops includes a 6 m-thick section of well-bedded, well-sorted lapilli tuff. Most of the bedding is planar with some cross-bedded layers. This section lacks bedding-plane sags and ash-coated/accretionary lapilli. It is interpreted as a 'dry' turbulent low-density pyroclastic density current deposit that resulted from water–magma interaction during a late-stage low-volume eruption.

Remnants of late-stage basanitic scoria cone deposits also crop out on the western side of Mount Frakes at English

Alternating ice-contact "wet" and subaerial "dry" lithofacies

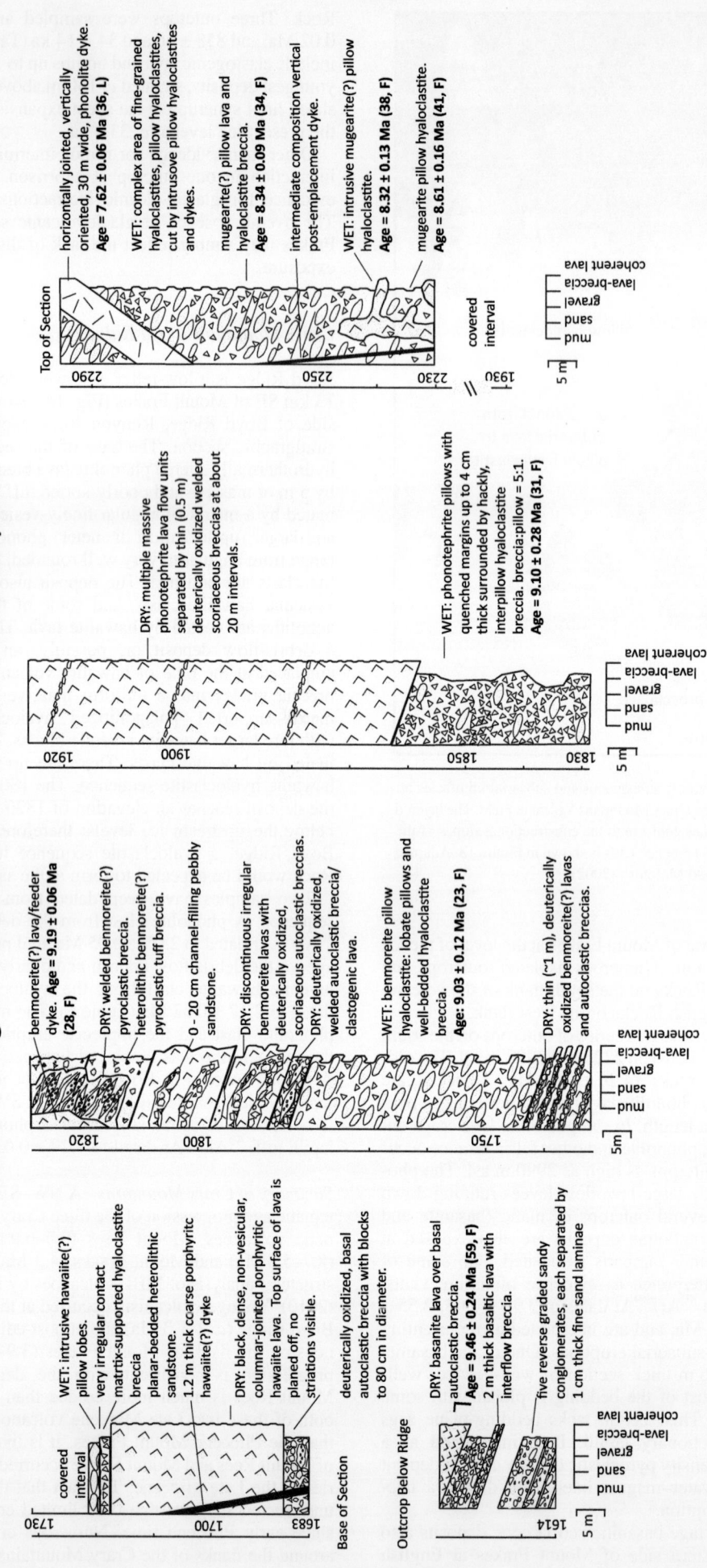

Fig. 20. Stratigraphic section from Tasch Peak Ridge at Mount Rees, Crary Mountains Volcanic Field. Volcanic lithofacies are interpreted as either 'dry' (i.e. subaerial) or 'wet' (i.e. subglacial/ice-contact). Figure adapted from Wilch and McIntosh (2002). The location of Tasch Peak Ridge is shown in Figure 18.

mostly alkaline volcanic lithologies and are likely to result from a combination of alpine/local and ice-sheet/regional glaciation. A few granite erratics occur at Morrison Rocks on the south lower flank of Mount Frakes. An altimeter reading on the highest granite erratic was 2440 m asl, about 600 m above the elevation of the adjacent ice-sheet surface. This granite erratic is interpreted as evidence of former ice-sheet expansion(s) since 1.6 Ma, the age of the youngest volcanic rocks at Morrison Rocks. The lack of glacial till and striated clasts suggests that the ice sheet was likely to have been cold-based.

Mount Rees and Mount Steere are among the more deeply-eroded volcanoes in Marie Byrd Land, and thus offer more complete records of volcanism than do other less-dissected volcanoes. The deep dissection is expressed geomorphologically, and also by the abundance of exposed dykes and the diversity of locally-derived volcanic glacial erratics on the surface. The overlapping ages and bimodal lithology of these volcanoes suggest chemically diverse volcanism since the early stages of the growth of both volcanoes. The alternating subaqueous and subaerial lithofacies at Mount Rees and Mount Steere suggests that during their construction in the Late Miocene (9.5–8.3 Ma) both volcanoes were mantled with snow and ice as they are today. Although extensive work has been completed in the Crary Mountains, many outcrops remain unvisited, especially in the upper elevations at Mount Steere and upper west flank at Mount Frakes.

Mount Siple Volcanic Field

The Mount Siple Volcanic Field consists of one volcano, Mount Siple, an undissected composite volcano on the MBL coast (Fig. 1). The massive volcano rises from sea level to 3100 m asl, with a 4.5 km-diameter ice-filled summit caldera and a base diameter at sea level of 45 × 35 km (Fig. 21). Because the WAIS terminates at the coast, Mount Siple is not buried by the ice sheet and thus has the largest exposed volume of volcanoes in the MBLVP (*c.* 1800 km^3) (LeMasurier and Rex 1990*a*). The volcano is almost completely ice-covered with a few low outcrops along the coast, and limited outcrop near and in the summit caldera wall.

A total of four coastal or lower flank outcrops have been visited (Table 6) (LeMasurier and Rex 1990*a*; Wilch 1997; Wilch *et al.* 1999). LeMasurier and Rex (1990*a*) identified two basaltic tuff cone remnants at Lovill Bluff and at a site 1 km to the south, and a subhorizontal basaltic lava 150 m asl about 4 km north of Lovill Bluff. The sites near Lovill Bluff yielded K–Ar ages with large uncertainties (2.0 ± 1.4 and 1.1 ± 1.0 Ma) and the ages are considered unreliable. The Lovill Bluff outcrop has not been dated reliably: LeMasurier and Rex (1990*a*) reported a <0.1 Ma K–Ar age; and Wilch (1997) ^{40}Ar/^{39}Ar dated a hawaiite clast to 8 ± 88 ka. The Lovill Bluff tuff cone deposits include well-sorted glass-rich lapilli tuff with aphyric hawaiite lava and altered scoria clasts. On the NE side of Mount Siple, at about 320 m asl, a hawaiite tuff cone deposit was dated to 746 ± 36 ka (Wilch 1997). This deposit includes glass-rich, planar-bedded lapilli tuff, significant soft-sediment deformation, ash-coated lapilli and bedding-plane sags.

The summit of Mount Siple was first visited in January 1994. Wilch *et al.* (1999) described a 20+ m-thick densely-welded, fiamme-rich, pyroclastic fall deposit exposed at the highest point of the caldera rim. Three analyses of anorthoclase derived from the trachyte deposit yielded a mean ^{40}Ar/^{39}Ar age of 229.6 ± 7.6 ka. A trachyte lava associated with a subsidiary vent 210 m lower in elevation than the summit crater yielded an age of 171.1 ± 5.4 ka. Although there has been speculation about recent eruptions at Mount Siple (Global Volcanism Program 1988), there is no evidence of this at the summit caldera.

In summary, Mount Siple is a Pleistocene polygenetic central volcano with a 227 ka summit caldera that has been active since at least 746 ka. Parasitic volcanism may be active in the Holocene but existing age data are unreliable. The limited outcrops and observations preclude making more detailed interpretations of the volcanic history. At this stage there is no indication that Mount Siple should be considered 'active'.

The Executive Committee Range Volcanic Field

The Executive Committee Range (ECR) Volcanic Field in central Marie Byrd Land consists of five central volcanoes

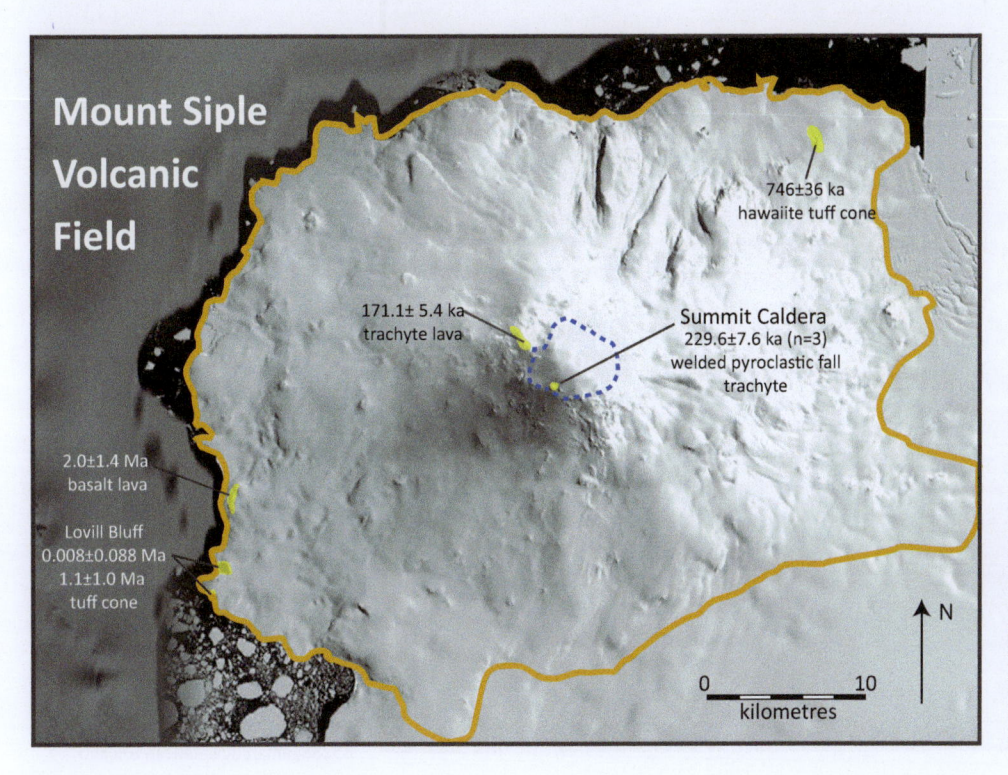

Fig. 21. Satellite image map of Mount Siple volcano showing the approximate volcano outline (orange), caldera rim (dashed blue), studied outcrops (yellow), lithofacies and ages. Mount Siple is an island in the Southern Ocean, separated from the Bakutis Coast by the Getz Ice Shelf. Image source: Google Earth Pro image accessed June 2019; see the caption to Figure 7 for more information on the source and an explanation of image processing.

Table 6. *Mount Siple Volcanic Field summary age table*

Sample ID	Method	Corrected preferred age ± 2 SD (Ma)	Description	Ref.
Mount Siple				
Summit caldera shield building				
93-278	A	0.1711 ± 0.0054	200 m below summit lava	20
93-277	A	0.2329 ± 0.0067	Caldera wall, densely-welded fall, 20 m-thick exposure	20
93-277	A	0.2366 ± 0.0173	Caldera wall, densely-welded fall, 20 m-thick exposure	20
93-277	A	0.2250 ± 0.0069	Caldera wall, densely-welded fall, 20 m-thick exposure	20
93-277	A	**0.2296 ± 0.0076**	**Mean age (*n* = 3)**	20
Flank deposits post-shield building				
93-270	A	0.746 ± 0.036	Lithic clasts in lapilli tuff	16
93-275	A	0.008 ± 0.088	Lithic clasts in lapilli tuff	16
	K	*1.1 ± 1.0*	Tuff cone deposit	8
W83-5	K	*2.0 ± 1.4*	30 m-thick subhorizontal basanite lava	8

Notes: method A is ^{40}Ar/^{39}Ar; method K is K/Ar.
Ref. code: 8, LeMasurier and Rex (1990*a*); 16, Wilch (1997); 20, Wilch *et al.* (1999). See also Supplementary Material Table S1.

that are aligned north–south and progressively young toward the south (Fig. 22). The elevation of the ice-sheet surface is higher at the ECR volcanoes (2000–2600 m asl) than at other MBLVP volcanoes. The diameters of the volcanoes at the level of the ice sheet range from 3 to 15 km. All five of the ECR Volcanic Field volcanoes have well-defined summit calderas, and three of the five have two calderas; caldera diameters range from 2 to 10 km. The volcanoes appear to be deeply buried by the WAIS, in some cases exposing little more than the summit calderas.

Mount Hampton. Mount Hampton, located at the north end of the ECR Volcanic Field, contains an intact slightly NW-elongated 6.5 × 5.5 km summit caldera that rises to 3223 m asl and is 400–700 m above the ice sheet (Fig. 23). The flanks slope from 10° to 20° and the base only extends *c.* 2.5 km beyond the caldera. The remnant of a second, smaller (*c.* 3.5-km-diameter) and older caldera is situated NW of the summit caldera. The remnant caldera has an intact NW rim that includes Whitney Peak. The two calderas are composed of felsic lavas (trachyte and low-silica rhyolite) with superimposed

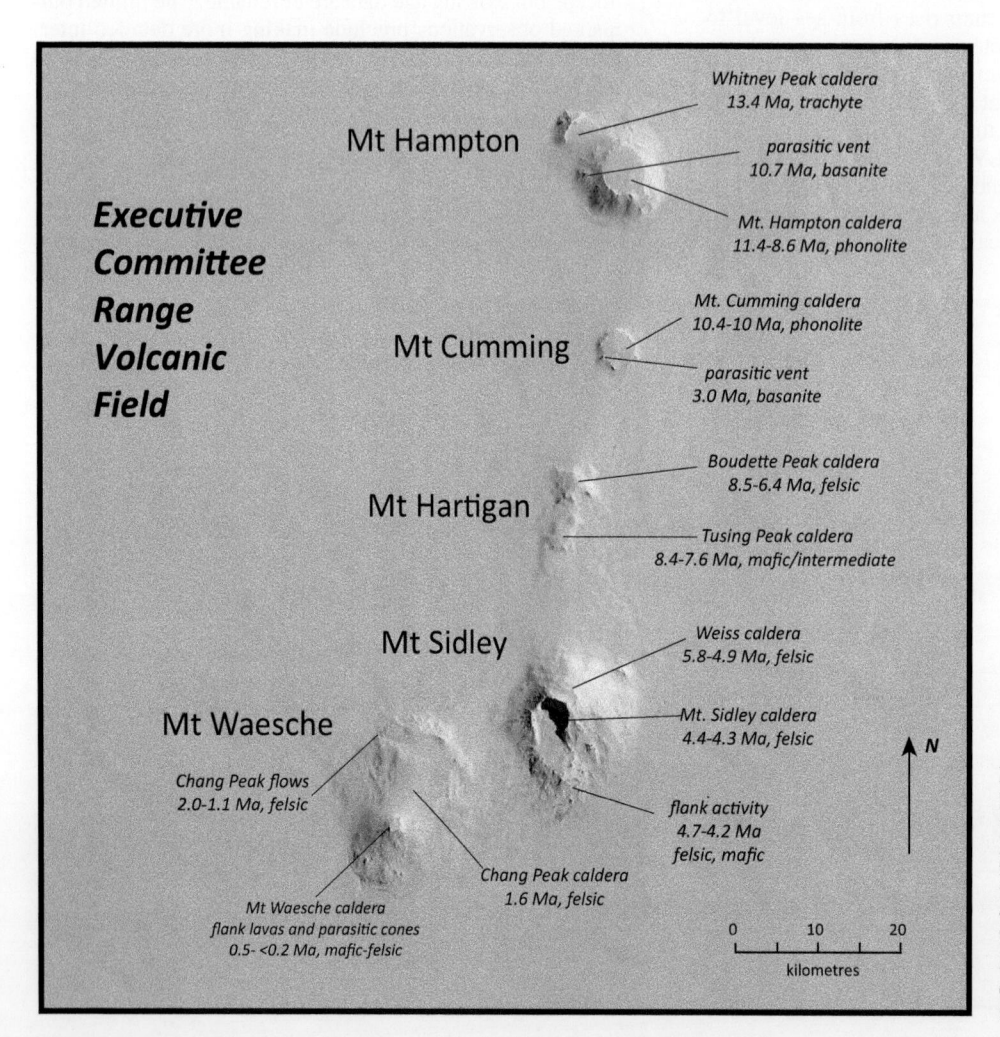

Fig. 22. Satellite image map of the Executive Committee Range (ECR) Volcanic Field. Note the age progression from north to south. NASA Earth Observatory image by Jesse Allen, using Landsat data from the United States Geological Survey. Accessed 21 June 2019 from https://earthobservatory.nasa.gov/images/85238/antarcticas-tallest-volcano

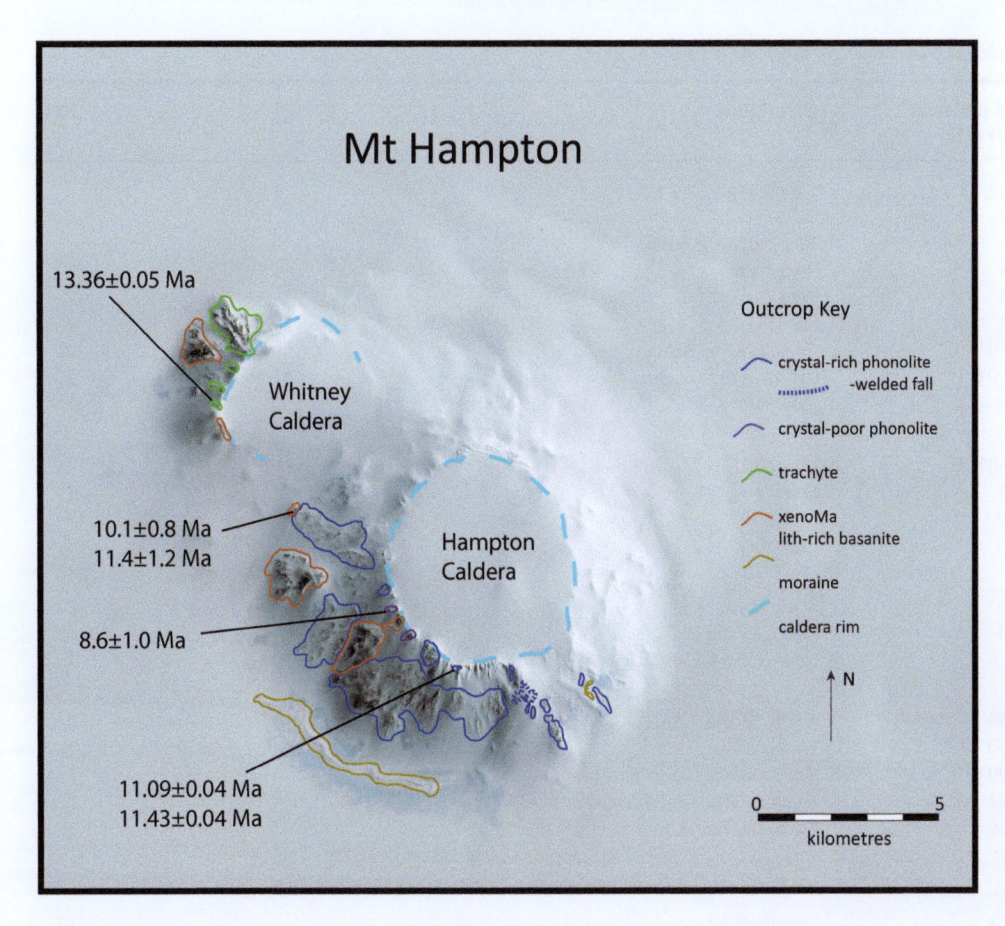

Fig. 23. Simplified geological map of Mount Hampton in the ECR Volcanic Field. Image extracted from NASA Earth Observatory image by Jesse Allen, using Landsat data from the United States Geological Survey. Accessed 21 June 2019 from https://earthobservatory.nasa.gov/images/85238/antarcticas-tallest-volcano

parasitic basanite cones. The basanite deposits include both crustal and mantle xenoliths.

Three rock types are exposed on the older caldera rim and slopes near Whitney Peak (observations of the authors) (Fig. 23). The caldera rim is dominated by a plagioclase-rich (5–25%) trachyte lava, and includes a benmoreite lava (LeMasurier and Kawachi 1990c; observations of the authors). In places, the well-exposed trachyte lava exhibits strong and contorted flow foliations, defined by alternating variably glassy layers. An associated lithic-rich trachyte welded fall, with 5:1 flattening of pumice clasts, crops out locally. The trachyte lava is $^{40}Ar/^{39}Ar$ dated to 13.36 ± 0.05 Ma (Table 7). The third lithology is a late-stage xenolith-bearing basanite that overlies the trachyte lava. Dip direction variations among the basanite outcrops suggest that they constitute multiple parasitic vents. Outcrops include variably-welded lapilli and bombs up to 40 cm, and pyroclastic deposits are locally transitional to spatter-fed, flow-foliated lavas. This unit was not described by LeMasurier and Kawachi (1990c) and has not been dated.

The dominant edifice-building lithology at Mount Hampton is crystal-rich anorthoclase phonolite (kenyte) (LeMasurier and Kawachi 1990c). The crystal-rich phonolite includes both well-exposed lava and welded fall lithofacies. A subordinate but similar lithology is crystal-poor phonolite lava. The stratigraphic relationship of the two phonolite lavas is uncertain. Previously unpublished $^{40}Ar/^{39}Ar$ ages of the crystal-rich phonolite are 11.43 ± 0.04 and 11.09 ± 0.04 Ma; the crystal-poor phonolite is K–Ar dated to 8.6 ± 1.0 Ma (Table 7). The crystal-rich phonolite lava is overlain by a third lithology, a variably-welded xenolith-bearing basanite. Two K–Ar ages for the same basanite sample are 11.4 ± 1.2 and 10.1 ± 0.8 Ma (LeMasurier and Rex 1989). The texture of the basanite varies from non-welded vesicular lapilli tuff and bombs to densely-welded spatter-fed lavas. Bombs are 0.5–1.5 m in diameter. The basanite appears to post-date the erosion of underlying phonolites. However, some spatter-fed basanitic lavas also appear to be cut by the caldera wall, suggesting that this basanite was erupted before caldera collapse.

Glacial moraines are found near both Mount Hampton calderas and contain locally-derived volcanic clasts, including phonolite, trachyte and basanite.

LeMasurier and Wade (1968) and LeMasurier and Kawachi (1990c) noted that the caldera rim has large conical snow and ice mounds, and interpreted these as dormant fumarolic ice towers and indications of recent activity. This interpretation is incorrect and the features are probably snow (rime) mushrooms resulting from wind sculpting (Whiteman and Garibotti 2013). There is no indication of recent volcanic activity or geothermal features.

Mount Cumming. Mount Cumming, located 12 km south of Mount Hampton, is a 3.5 km-diameter circular caldera that protrudes only 200 m above the ice sheet surface (Figs 22 & 24). The exposed base of the volcano extends <1 km beyond the caldera rim. The caldera rim is intact and includes limited outcrops composed of variably foliated pantelleritic trachyte lava, locally underlain by a non-welded to densely-welded fall deposit with flattened glassy fiamme and resorbed cognate xenoliths. Remnants of parasitic basanite scoria cones on the lower slopes are composed of xenolith-rich (e.g. granulite and ultramafic) welded fall deposits transitional to spatter-fed lavas. Conventional K–Ar ages of the trachyte are 10.4 ± 1.0 and 10.0 ± 1.0 Ma, and one basanite cone is dated to 3.0 ± 0.1 Ma (Table 7) (LeMasurier and Rex 1989; LeMasurier and Kawachi 1990b).

Mount Hartigan. Mount Hartigan, located 12 km SSW of Mount Cumming at the centre of the ECR, is made up of two overlapping north–south-aligned volcanoes, each defined by a 3.5 km-diameter circular caldera rim and snow-/ice-filled

Table 7. *Executive Committee Range Volcanic Field summary age table*

Sample ID	Method	Corrected preferred age ± 2 SD (Ma)	Description	Ref.
Mount Hampton				
Hampton caldera				
20D	K	8.6 ± 1.0	Near summit caldera	7
22B	K	10.1 ± 0.8	West flank	7
22D	K	11.4 ± 1.2	West flank	7
25	K	*10.7 ± 0.8*	SE flank	7
90-201	A	11.09 ± 0.04	Crystal-rich phonolite	21
MB-74.2	A	11.43 ± 0.04	Crystal-rich phonolite	21
19C	K	*11.7 ± 1.0*	Flow rock	7
Whitney Peak caldera				
90-174	A	13.36 ± 0.05	Phonolite	21
23A	K	*13.4 ± 1.0*	Flow rock	7
24A	K	*13.7 ± 1.0*	Flow rock	7
23C	K	*13.7 ± 1.0*	Flow rock	7
Mount Cumming				
28	K	10.4 ± 1.0	South flank, lava	7
27	K	10.0 ± 1.0	Near LaVaud Peak, lava	4
26A	K	3.0 ± 0.4	Near Annexstad Peak, parasitic cone	4
Mount Hartigan				
43A	K	6.02 ± 0.50	North caldera, lava	7
45C	K	7.57 ± 0.60	South caldera, lava	7
46B	K	7.86 ± 1.00	South caldera near Mintz, lava	7
48	K	8.36 ± 0.82	South caldera, near Tusing, lava	7
42B	K	8.50 ± 0.66	North caldera, lava	7
Mount Sidley				
Stage IV – post-shield activity				
K168	A	4.24 ± 0.08	Strombolian tephra and lava, parasitiic cone	12
Stage III – formation of the breached Sidley caldera				
K85	A	4.37 ± 0.06	Tuff cone deposits and lava, flank vent	12
K51	A	4.43 ± 0.06	Unwelded ignimbrite, flank vent	12
K137	A	4.31 ± 0.06	Welded fall with fiamme, flank vent	12
Stage II – flank activity				
MB29.4	A	4.59 ± 0.04	Welded fall with abundant lithic fragments	12
K105	A	4.61 ± 0.08	Porphyritic lava and breccia from endogenous dome and commingled lava, flank vent	12
K55	A	4.66 ± 0.10	Porphyritic lava and breccia from endogenous dome and commingled lava, flank vent	12
MB33.3	A	4.64 ± 0.04	Porphyritic lava and breccia from endogenous dome, flank vent	12
MB35.5	A	4.51 ± 0.02	Poorly-welded pyroclastic fall, flank vent	12
Stage I – Weiss caldera formed				
K149	A	4.87 ± 0.06	Vitric and porphyritic lava and basal breccia	12
MB42.3	A	5.15 ± 0.14	Vitric and porphyritic lava and basal breccia	12
K106	A	5.43 ± 0.04	Lava and basal breccia	12
K108	A	5.60 ± 0.14	Porphyritic lava and breccia	12
K68	A	5.77 ± 0.12	Porphyritic lava and breccia	12
Mount Waesche				
Flank deposits				
41A	K	<0.1 ± 0.00	SW flank	7
35A*	K	0.170 ± 0.60	SW flank, cinder cone	7
39A	K	0.200 ± 0.40	SW flank	7
33C	K	1.000 ± 0.20	SW flank	7
	A	0.49 ± 0.02	SW flank, trachyte dome	
Chang Peak caldera				
32A	K	1.6 ± 0.4	Chang Peak, caldera wall	7
32A	F	1.48 ± 0.33	Chang Peak, caldera wall	15
	A	1.09 ± 0.10	Chang Peak, NW flank lava	11
	A	2.01 ± 0.10	Chang Peak, NW flank lava	11

Notes: method A is $^{40}Ar/^{39}Ar$; method K is K/Ar; method F is fission track.
Ref. code: 4, LeMasurier and Kawachi (1990*b*); 7, LeMasurier and Rex (1989); 11, Panter (1995); 12, Panter *et al.* (1994); 15, Seward *et al.* (1980); 21, this study. See also Supplementary Material Table S1.

caldera (Fig. 22). The calderas are less well defined than at Mount Hampton and Mount Cumming. The northern volcano is composed of trachyte, rhyolite and mugearite lava; the southern volcano is composed of hawaiite and mugearite lava (LeMasurier 1990*f*). The volcanoes overlap in age, with two samples from the northern volcano K–Ar dated to

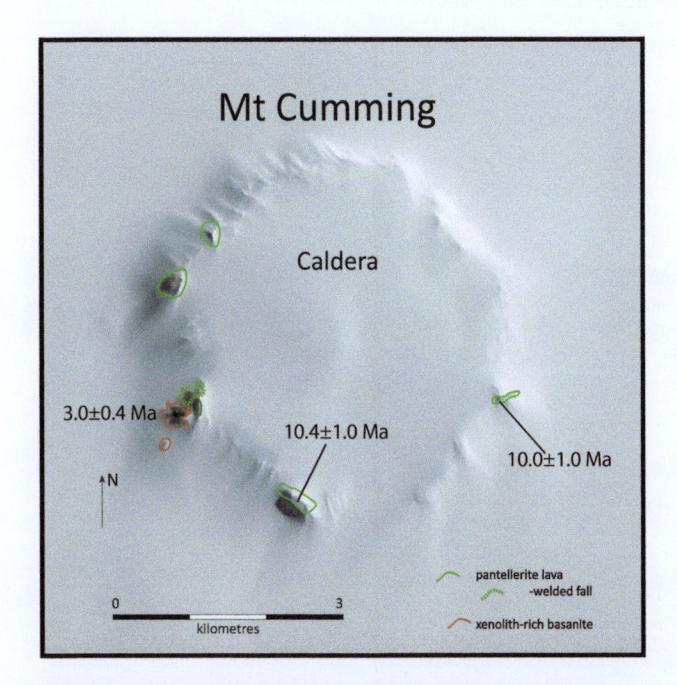

Fig. 24. Simplified geological map of Mount Cumming in the ECR Volcanic Field. Image extracted from NASA Earth Observatory image by Jesse Allen, using Landsat data from the United States Geological Survey. Accessed 21 June 2019 from https://earthobservatory.nasa.gov/images/85238/antarcticas-tallest-volcano

8.50 ± 0.66 and 6.02 ± 0.50 Ma; and three from the southern volcano ranging in age from 8.36 ± 0.82 to 7.57 ± 0.60 Ma (LeMasurier and Rex 1989) (Table 7).

Mount Sidley. Mount Sidley, located 12 km south of Mount Hartigan near the southern end of the ECR Volcanic Field, is one of the most well-exposed and well-studied volcanoes in the MBLVG (Fig. 25). This summary is derived mostly from Panter *et al.* (1994), who published a detailed reconstruction of the history of the volcanic complex, including $^{40}Ar/^{39}Ar$ dating and geochemistry.

Mount Sidley (4285 m asl) is the tallest volcano in Antarctica. The volcano rises 2200 m above the WAIS level and has basal dimensions of 14 × 19 km. Early workers (González-Ferrán and González-Bonorino 1972; LeMasurier 1972) noted two calderas: the older Weiss caldera, a *c.* 2.5 km-diameter, snow- and ice-filled caldera situated north of the summit peak that is truncated by the younger Sidley caldera, a 4.5 km-diameter and >1200 m-deep breached caldera that opens to the south. Parks Glacier occupies the floor of the breached Sidley Caldera and flows from the headwall at the north end of the caldera into the ice sheet SE of the volcano.

Panter *et al.* (1994) described a complex four-stage geological history spanning *c.* 1.5 myr (5.8–4.2 Ma); each stage is defined by one or more geographical shifts of eruptive centres and by geochemical changes of the magma (Fig. 25). Stage I, Sidley Activity, comprised the largest eruptive volume (90% or *c.* 180 km³) of the Sidley massif and resulted in the construction of three successive volcanic edifices, the Byrd (informal name by Panter *et al.* 1994), Weiss and Sidley volcanoes, culminating in the formation of a *c.* 3.5 km-diameter summit caldera. Weiss caldera is the only Stage I caldera preserved mostly intact and is the protocaldera of the later-stage breached caldera. The reconstructed calderas delineate a migration of eruptive centres towards the SW. Phonolite–tephriphonolite lavas dominate these sequences, although minor amounts of pumice-bearing pyroclastic material are interbedded within and between the lava sequences.

Minor pillow lava and hyaloclastite breccia deposits occur locally and are attributed to limited glaciovolcanic interactions. Erosional unconformities separate the volcano sequences. The stratigraphy of Stage I is well dated by $^{40}Ar/^{39}Ar$ geochronology, with ages from 5.77 ± 0.12 Ma at the base of the sequence to 4.87 ± 0.06 Ma at the top of the Sidley volcano sequence.

Stage II, Pirrit Activity, is characterized by a petrological shift to trachyte, as well as a shift in eruption style from the central vent to multiple small monogenetic eruptive centres distributed on the flanks of a largely undissected Mount Sidley. Rocks include thin (5–10 m) lavas associated with small vents, and thick (> 50 m) lavas and carapace breccias associated with endogenous domes. One of the large eruptions is evidenced by a single vent with a thick (70 m) stratigraphic sequence of deposits that includes a basal trachyte/phonolite pyroclastic fall deposit that transitions up into a unit of alternating foliated trachyte lavas and carapace breccias. Stage II lasted from 4.66 ± 0.10 to 4.51 ± 0.02 Ma, and produced *c.* 7% (18 km³) of the erupted volume of Mount Sidley (Panter *et al.* 1994).

Stage III, Doumani Activity, was a short-lived (4.43–4.37 Ma) trachyte pyroclastic eruptive phase that began with emplacement of a lithic-rich welded fall deposit followed closely by non-welded ignimbrite and associated surge deposits and co-ignimbrite breccias (Panter *et al.* 1994). Panter *et al.* (1994) speculated that this explosive phase was coincident with the initial collapse of the present-day breached caldera, which was created by a combination of syneruptive explosive landslide and post-eruptive erosional processes. South of the main Sidley edifice at Doumani Peak, an ignimbrite sequence is dated to 4.43 ± 0.06 Ma and is overlain conformably by a mugearite/benmoreite tuff cone sequence, dated to 4.37 ± 0.06 Ma. This stage comprises an estimated 1% of the erupted volume (*c.* 2.5 km³).

Stage IV is characterized by multiple basanite parasitic scoria cones on the flanks of Mount Sidley. The presence of mantle and lower-crustal xenoliths was interpreted as evidence of rapid ascent rates. A sample from one cone was dated to 4.24 ± 0.08 Ma. This stage comprises <<1% of the erupted volume (*c.* 0.25 km³).

Panter *et al.* (1994) note that magmatic activity at Mount Sidley migrated southwestwards at a rate of 6 cm a^{-1} and follows the same pattern as the ECR Volcanic Field overall. The migration has been interpreted as being due to fracture propagation caused by regional tectonic stresses (LeMasurier and Rex 1989) or related to plate-boundary forces (Paulsen and Wilson 2010). Magma injection into a complex system of conduits and chambers may have aided periodic dilation of pre-existing structures, resulting in the eruption of evolved magmas in discrete pulses (Panter *et al.* 1994).

Mount Waesche. The Mount Waesche massif, located at the south end of the ECR, consists of two coalesced polygenetic volcanoes: the older Chang Peak volcano defined by a NNE-elongated 10 × 6 km-diameter caldera; and the younger Mount Waesche volcano defined by the prominent peak and *c.* 1.5 km-diameter caldera superimposed and centred on the southern rim of Chang Peak caldera (Fig. 22). Outcrops are limited at Chang Peak and include comendite (peralkaline rhyolite) pumice and vitrophyre that is K–Ar dated to 1.6 ± 0.4 Ma (LeMasurier and Rex 1989). Two additional comendite outcrops on the western slope of the caldera have been $^{40}Ar/^{39}Ar$ dated. A flow-banded vitrophyre and lava with abundant spherulites and lithophysae is dated to 2.01 ± 0.10 Ma, and a similar but more phenocryst-rich vitrophyic lava is dated to 1.09 ± 0.10 Ma (Table 7). Thus, the Chang Peak caldera and volcano appear to be early Pleistocene in age.

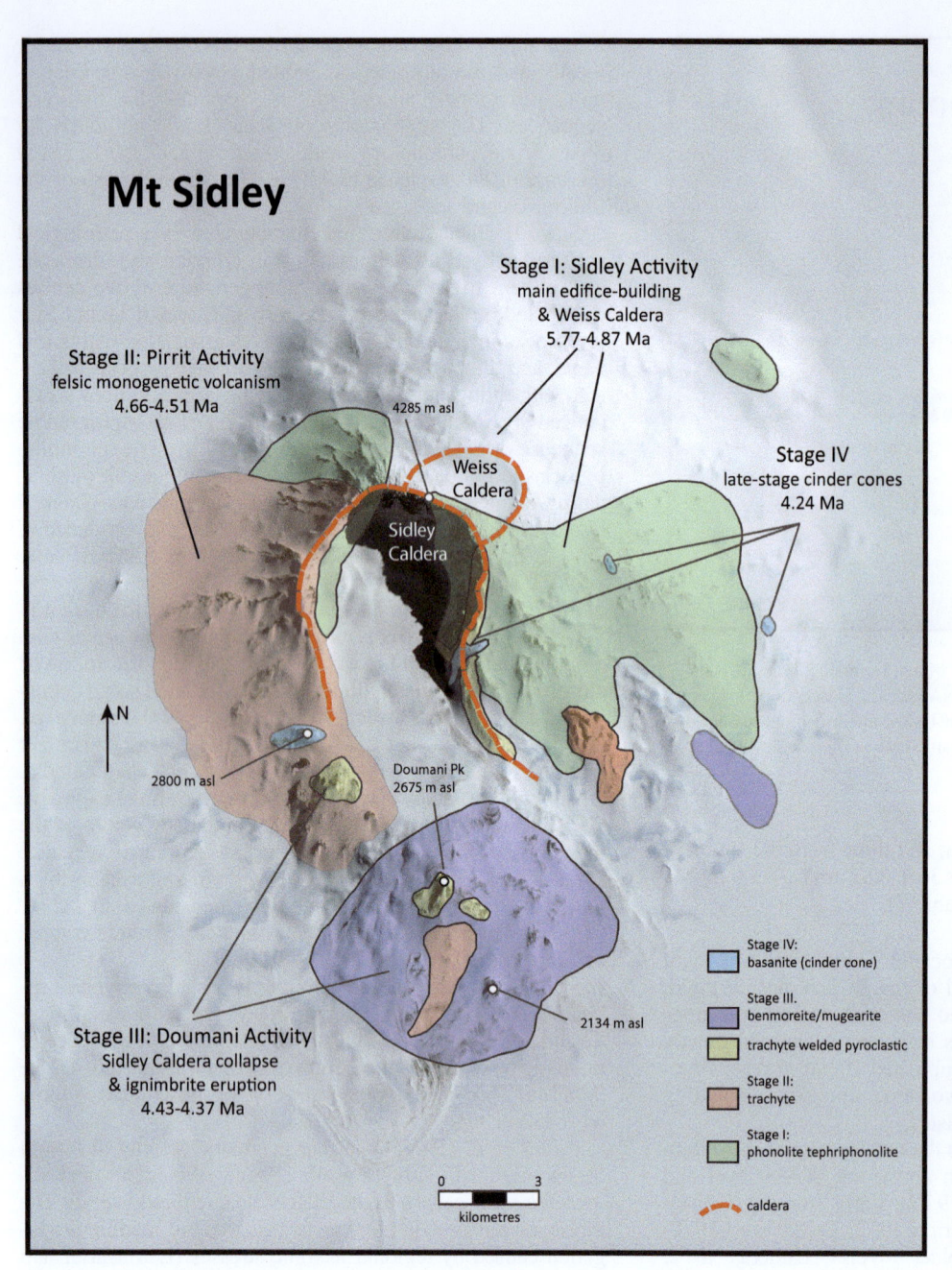

Fig. 25. Simplified geological map on satellite image base map Mount Sidley of the ECR Volcanic Field. Geology adapted from figure 4 in Panter *et al.* (1994); base map extracted from NASA Earth Observatory image by Jesse Allen, using Landsat data from the United States Geological Survey. Accessed June 21, 2019 from https://earthobservatory.nasa.gov/images/85238/antarcticas-tallest-volcano

The younger Mount Waesche volcano has excellent exposures on the SW flank that include a wide range of compositions, from basalts to intermediate types (mugearite, phonotephrite and tephriphonolite) to phonolite and trachyte (Panter *et al.* 2021). Conventional K–Ar ages of hawaiite samples range from 1.0 ± 0.2 to <0.1 Ma (LeMasurier and Rex 1989) (Table 7). Numerous englacial tephra layers are exposed in a blue-ice ablation zone of the ice sheet at the base of the south flank (Dunbar *et al.* 2021).

A geological map and detailed description of the geology are presented by Dunbar *et al.* (2021) and only a summary is given here based on that account. Exposed along the rim of the ice-filled summit caldera of Mount Waesche are deposits of basanite lava, agglutinated lava and welded Strombolian tephra containing xenolith-rich bombs. A heterolithic debris deposit is also found around the rim of the summit caldera and on the upper slopes of Mount Waesche. The debris is composed of fragments of lapilli tuff, green felsic (trachytic?) lavas, gabbro and hypabyssal plutonic lithologies. One large block (*c.* 2 m in diameter) of phonolite contains xenoliths and basalt clasts that are flattened and aligned with reaction rims, all of which suggest that the block is a remnant of a welded pyroclastic fall or flow deposit. The origin of the debris deposit is considered either as a result of glacial deposition or explosive volcanic or a combination of both, and has since been modified by downslope movement and solifluction processes (Smellie *et al.* 1990; Dunbar *et al.* 2021). Ongoing work (2018–19 fieldwork) includes extensive $^{40}Ar/^{39}Ar$ geochronology and detailed mapping.

McCuddin Mountains Volcanic Field

The McCuddin Mountains Volcanic Field is located in central Marie Byrd Land, north of the ECR and east of the Ames Range (Figs 1 & 2). The field includes Mount Flint, a low-relief central volcano, and monogenetic volcanoes at Mount Petras and the USAS Escarpment. Together these represent the oldest known phase of MBLVG volcanism, mostly from 36.58 to 20.46 Ma, followed by Late Miocene and Pliocene (9.67–3.75 Ma) parasitic volcanism at the Mount Flint central volcano (Table 8).

Table 8. *McCuddin Mountains and Ames Range volcanic fields summary age table*

Sample ID	Method	Corrected preferred age ± 2 SD (Ma)	Description	Ref.
USAS Escarpment				
Near Mount Galla				
MB70.1	A	26.40 ± 0.21	Hyalotuff (tuff cone)	16
Mount Aldaz				
90-181	A	19.0 ± 1.4	Mount Aldaz, spatter lava over glacial unconformity and bedrock	16
90-180	A	19.68 ± 0.31	Mount Aldaz, basanitic lava	16
Mount Flint				
93-279	A	9.67 ± 0.20	SW parasitic cone, agglutinate lava	16
93-281	A	3.75 ± 0.06	West upper parasitic cone, agglutinate lava	16
93-284	A	8.66 ± 0.24	West lower parasitic cone agglutinate lava	16
93-288	A	20.45 ± 0.10	Reynolds Ridge, hypabbyssal lava	16
93-291	A	20.48 ± 0.12	Reynolds Ridge, lava, 20–25% feldspar	16
		20.46 ± 0.07	**Mean age (*n* = 2)**	
Mount Petras				
93-323	A	36.71 ± 0.51	SW flank, mugearite lava xenolith	17
93-329	A	27.53 ± 0.23	SW flank, dense haw SW saddlebomb interior	17
93-343	A	28.96 ± 0.22	Aphyric haw lava, 2 m thick	17
93-337	A	28.26 ± 0.38	Near summit, aphyric glassy haw lava, rare xenoliths	17
93-332	A	36.58 ± 0.22	Near summit, massive mugearite lava	17
93-333	A	28.22 ± 0.52	Near summit, dense haw bomb interior	17
Ames Range				
Mount Andrus				
60	K	<0.1	Upper Lind Ridge, parasitic cone	6
41	K	11.3 ± 0.8	Near Rosenburg Glacier	7
44	K	11.6 ± 0.8	Near Rosenburg Glacier parasitic cone	6
93-308	A	9.28 ± 0.06	Lava, parasitic cone	16
93-309	A	12.71 ± 0.06	Pumiceous lava, massive	16
93-311	A	11.19 ± 0.08	Pumiceous lava	16
93-312	A	11.18 ± 0.19	Pumiceous lava	16
Mount Kosciusko				
40B	K	10.00 ± 0.80	Parasitic cone	2
40B	K	8.66 ± 0.70	Parasitic cone	2
	K	**9.20 ± 1.30**	**Mean age (*n* = 2)**	
Mount Kauffman				
67B-8	K	5.9 ± 1.0	Lava	6
AR39B	K	7.6 ± 0.6	Lava	7
	K	**7.1 ± 1.5**	**Mean age (*n* = 2)**	

Notes: method A is $^{40}Ar/^{39}Ar$ method K is K/Ar.
Ref. code: 2, LeMasurier (1990*a*); 6, LeMasurier and Rex (1983); 7, LeMasurier and Rex (1989); 16, Wilch (1997); 17, Wilch and McIntosh (2000). See also Supplementary Material Table S1.

Mount Flint. Mount Flint is a low-relief central volcano that rises to 2695 m, about 900 m above the ice-sheet level (Fig. 26). The edifice is elongate in an east–west direction and is about 10 × 6 km in diameter. The volcano is almost completely covered with snow and ice, with limited outcrops exposed on the west flank. A flat area near the summit edifice suggests a possible *c.* 2 × 3 km-diameter oval caldera. Three outcrops were visited and are interpreted as erosional remnants of parasitic basanite scoria cones (LeMasurier *et al.* 1990*b*; Wilch 1997). The rock includes reddened welded bombs and lapilli, and spatter-fed lava (Wilch 1997). One scoria cone contains abundant mantle xenoliths, whereas two others contain crustal xenoliths. The $^{40}Ar/^{39}Ar$ plateau ages of the three remnant cones are 9.67 ± 0.20, 8.66 ± 0.24 and 3.75 ± 0.06 Ma (Table 8). The post-edifice-building cones provide minimum ages for the Mount Flint edifice. No outcrops have been identified that are part of the underlying (pre-parasitic cone) low-relief edifice.

Reynolds Ridge is a north–south-orientated, 1 km-long, linear, glacially-eroded outcrop ridge situated 5 km NW of the base of Mount Flint. The ridge consists of a 20 m-thick trachyte lava (LeMasurier *et al.* 1990*b*) underlain by a 30 m-thick potassium-feldspar-rich, coarse-grained (crystals up to 4 cm) syenite (i.e. intrusive equivalent of trachyte) (Wilch 1997). Our field observations show that the hypabyssal syenite intrudes into the lava. $^{40}Ar/^{39}Ar$ ages of anorthoclase separates from the syenite hypabyssal intrusion and trachyte lava are 20.45 ± 0.10 and 20.48 ± 0.12 Ma, respectively. Reynolds Ridge records the earliest felsic (trachyte) MBLVG volcanism in the Early Miocene at 20.5 Ma.

The relationship between early Miocene Reynolds Ridge and Mount Flint are uncertain. LeMasurier's (1990*b*) interpretations that Reynolds Ridge is either a dome at the base of Mount Flint or a separate remnant of a trachyte shield volcano are both viable possibilities in the absence of more data. We tentatively assign Reynolds Ridge to the edifice-building phase of Mount Flint (see Table 2).

Mount Petras. Mount Petras is a glacially dissected nunatak (2867 m asl), with about 900 m of relief exposed above the level of the WAIS. Mount Petras is located about 10 km SE of Mount Flint. The eroded spurs that comprise the main nunatak cover an area of about 5 × 8 km.

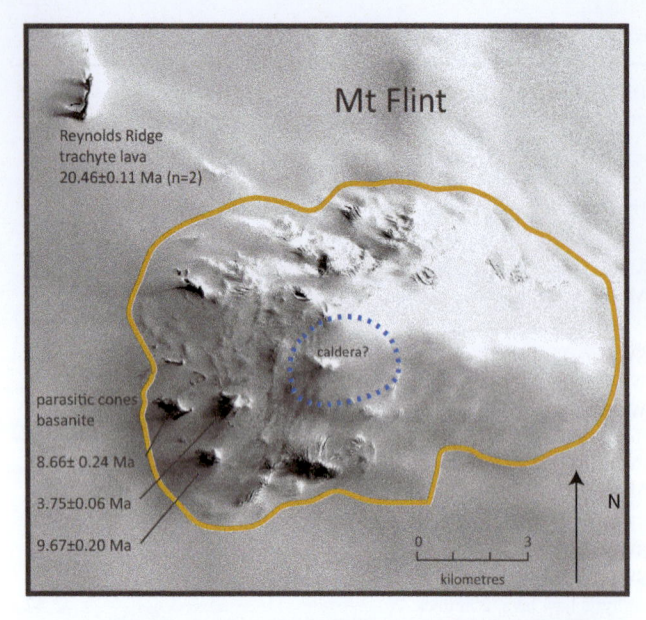

Fig. 26. Satellite image map of Mount Flint volcano of the McCuddin Mountains Volcanic Field, showing the approximate volcano outline (orange), caldera rim (dashed blue), studied outcrops (yellow), lithofacies and ages. Image source: Google Earth Pro image accessed June 2019; see the caption to Figure 7 for more information on source and an explanation of image processing.

Mount Petras is an unusual and important locality because it exposes the oldest known volcanic rocks in the MBLVP and these rocks overlie a currently high-standing non-conformity eroded into Cretaceous rhyodacite basement rocks, K–Ar dated as 82.9 ± 5.7 Ma (LeMasurier and Wade 1976; age adjusted for the decay constant of Steiger and Jaeger 1977). Wilch and McIntosh (2000) significantly revised the original interpretations of Mount Petras summarized in LeMasurier (1990c). The differences in interpretations are important as they have implications for the timing of the onset of MBLVP volcanism, the earliest volcanic records of glaciation, and the nature and significance of pre-volcanic erosion surfaces.

LeMasurier (1990c) interpreted volcanic rocks at Mount Petras as remnants of an Oligocene–early Miocene (25–22 ka) subglacially erupted table mountain overlying a flat, uplifted unconformity surface. The helicopter-supported reconnaissance fieldwork focused on the largest volcanic outcrop at Mount Petras, located on the SW flank. Volcanic rocks there were interpreted as 200 m of subhorizontally-stratified basaltic hyaloclastite, composed of weakly vesicular clasts and lacking any significant subaerial component (LeMasurier 1990c). These interpretations of the deposit characteristics, together with the observation of interbedded rounded basement clasts, led LeMasurier (1990c) to conclude that the volcanic rocks were the remnants of subglacially-erupted sequences and evidence for a Late Oligocene–Early Miocene thick ice sheet in West Antarctica. The original estimates of c. 400 m relief of the erosion surface (LeMasurier and Wade 1976) was revised to <100 m (LeMasurier et al. 1981).

Wilch and McIntosh (2000) used detailed volcanic lithofacies and age data to reach very different conclusions; they identified five eruptive episodes based on ^{40}Ar/^{39}Ar ages, geochemistry, lithofacies analysis and field relationships (Fig. 27). The first stage of volcanism occurred at 36.58 ± 0.22 Ma with an apparently subaerial extrusion of massive mugearite lava (Fig. 27; Table 8). This is the oldest known eruption in the MBLVP. The second stage of eruptions included four pyroclastic hawaiite events from three different vents dated between 28.96 ± 0.22 and 27.53 ± 0.23 Ma. Lithofacies associated with the second stage include welded tuff breccia attributed to a subaerial Strombolian eruption, and stratified lapilli tuff, massive lapilli tuff and ash-coated lapilli tuff attributed to subaerial to shallow-water Surtseyan eruptions. Two of the lapilli tuff units contained rare (<5%) lithic clasts. The lithic clasts are sub-angular to subrounded basement and mugearitic blocks, which are up to 10 cm in diameter and, in some cases, coated with hawaiite lava. No signs of glacial moulding or polish were observed on any of the lithic clasts. Intact and disintegrated pyroclastic bombs and blocks occur as large clasts up to 30 cm in length in some deposits.

Wilch and McIntosh (2000) observed that the outcrops and inferred vents were in contact with basement rocks and that topographical relief on the basement unconformity was >400 m, consistent with the early estimate by LeMasurier and Wade (1976). Wilch and McIntosh (2000) concluded that the 29–27.5 Ma Mount Petras eruptions involved intermittent interaction with water derived from a thin, local ice cap or from snow and ice on the slopes of a relatively high-relief (> 400 m) bedrock nunatak. We believe there is no evidence of extensive ice-sheet glacial conditions, as suggested by LeMasurier (1990c).

The 29–27.5 Ma pyroclastic deposits at Mount Petras provide the oldest terrestrial evidence for glacial ice in MBL but offer no evidence for a thick, continental, ice sheet at that time. The mixed Surtseyan and Strombolian eruptions imply local or intermittent contact with external water, which Wilch and McIntosh (2000) infer resulted from melting of a thin, local ice cap or ice and snow on slopes. The 29–27.5 Ma tuff cone deposits overlie an erosional unconformity, with >400 m of topographical relief. The relatively high-relief pre-volcanic environment is suggestive of ongoing erosion and is inconsistent with interpretations of a regional, low-relief, early Cenozoic West Antarctic Erosion Surface (e.g. Rocchi et al. 2006).

USAS Escarpment. The USAS Escarpment is an east–west-orientated north-facing escarpment located at about 76° S (Fig. 1). It is mostly snow- and ice-covered with five nunataks of basement and/or volcanic rocks exposed at the break in slope along 50 km of the escarpment. Ice on the north side of the escarpment is c. 100–300 lower than ice on the south side. LeMasurier (1990c) noted that the USAS structural escarpment is aligned with Mount Petras. Two volcanic nunataks along the escarpment have been described by LeMasurier (1990c) and unpublished observations of the authors.

At Mount Aldaz, basaltic pyroclastic fall deposits and spatter-fed lavas overlie a felsic hypabyssal intrusive basement outcrop. The top surface of basement rock is smoothed, undulating and striated, and interpreted as a glacial unconformity. Bedrock and volcanic sequence are separated by a 1 m-thickness of strongly deformed laminated sandstone, interpreted as tillite, overlain by a well-sorted, lithic-rich, Strombolian scoria unit grading upwards into a welded bomb deposit, which is overlain by aphyric lava breccias and multiple coherent lavas. The Strombolian deposits are cut by a dyke that appears to have caused local palagonitization in adjacent scoria beds. The volcanic sequence is ^{40}Ar/^{39}Ar dated to 19.68 ± 0.31 Ma (Table 8). The basement rock, currently at about 2300 m asl, was eroded by a wet-based glacier prior to 19.7 Ma. The Mount Aldaz glacial unconformity is more indicative of ice-sheet glaciation than the Mount Petras outcrops, and records the earliest terrestrial evidence for a significant WAIS in the MBLVG. The >19.7 Ma unconformity at Mount Aldaz provides the oldest terrestrial evidence for an ice sheet in MBL.

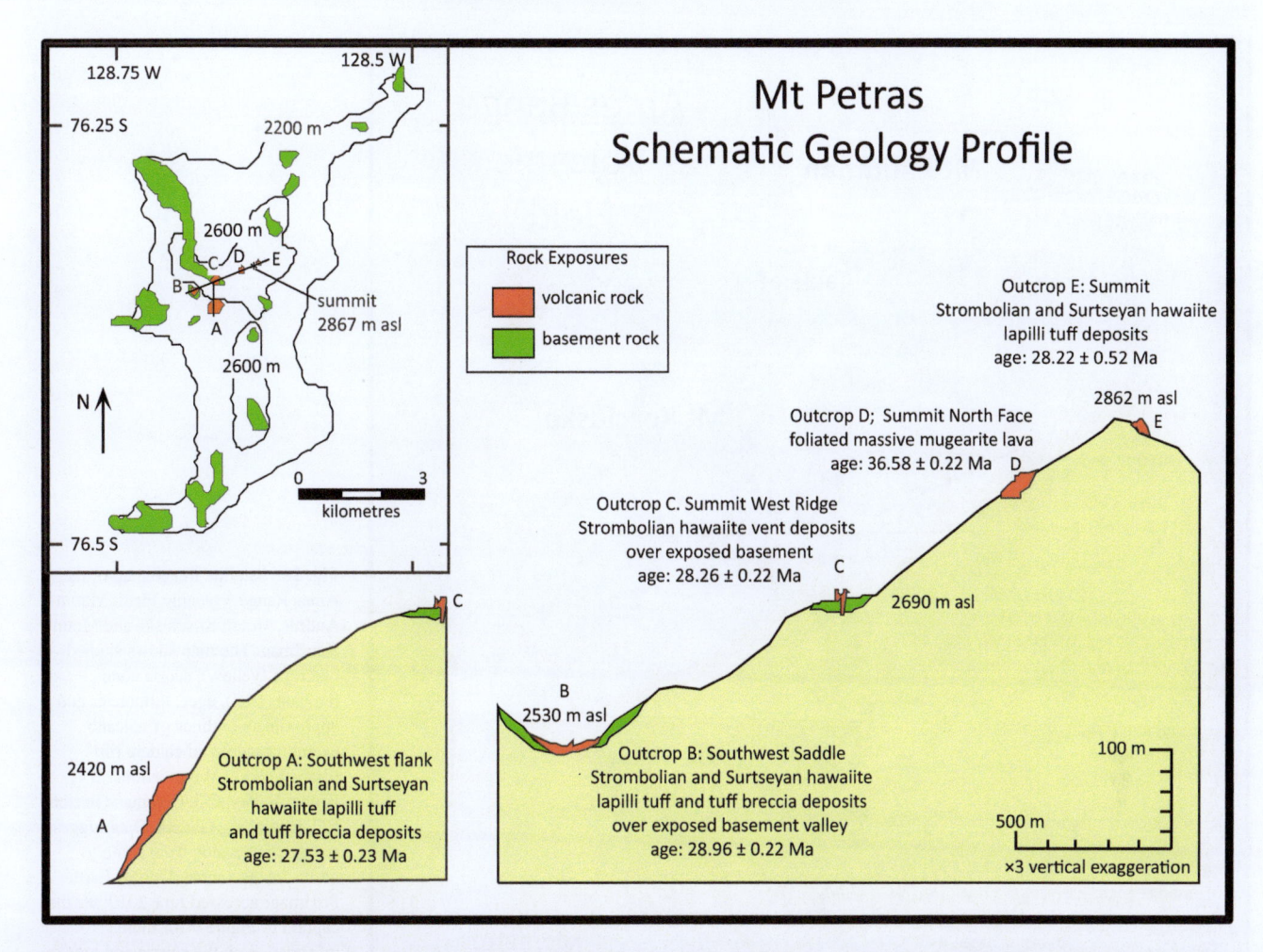

Fig. 27. Topographical profile of the summit area of Mount Petras of the McCuddin Mountains Volcanic Field, showing a schematic cross-section of volcanic outcrops, interpretations and ages. Figure adapted from figure 2 in Wilch and McIntosh (2000). The inset map is derived from the United States Geological Survey McCuddin Mountains 1:250 000 topographical map.

At an unnamed outcrop *c.* 4 km east of Mount Galla, a palagonitized thinly-bedded hawaiite lapilli tuff is dominated by vesicular lapilli with subordinate fracture-bounded non-glassy lithic clasts and rare basement clasts. The bedded lapilli tuff exhibits local channelling and dune bedding. The bedded, palagonitized lapilli tuff is locally underlain by basanitic monomict breccia, possibly a vent-clearing explosion breccia. The sequence is interpreted as a tuff cone remnant and $^{40}Ar/^{39}Ar$ dated to 26.40 ± 0.21 Ma (Table 8).

Ames Range Volcanic Field

The north–south-trending Ames Range Volcanic Field covers an area of about 30 × 15 km and consists of three coalesced central volcanoes: Mount Andrus, Mount Kosciuscko and Mount Kauffman (Fig. 28). The volcanoes are moderately dissected and glaciated with snow- and ice-covered slopes. Based on the limited field observations and sampling, it appears that trachyte lava dominates the Ames Range volcanoes, with a general trend of younger volcanism towards the north. LeMasurier and Rex (1989) noted that the Ames Range is aligned with the Quaternary Shepard Island volcano, 135 km north of Mount Kauffman, and may represent renewed felsic activity along the Ames Range lineament. Paulsen and Wilson (2010) used the ages and alignment of volcanoes in the Ames Range and Koerner Bluff at Mount Bursey in the Flood Range, located 10 km to the south, to establish the timing of a change in regional stress orientation after 6 Ma.

Mount Andrus. Mount Andrus, at the south end of the Ames Range, is about 13 × 13 km in map view and has an estimated volume of 115 km^3 (Fig. 28). The volcano has a partially preserved summit caldera that reaches 2978 m asl and is breached on its west side. Coleman Glacier originates in the caldera, and flows west through the breach and down into the WAIS at about 1600 m asl. There are extensive rock exposures on the south and north sides of Coleman Glacier.

Lind Ridge on the south side of Coleman Glacier has almost continuous outcrop from 2000 to 2800 m asl where it joins the caldera rim. Early reconnaissance reports provided no specific details on exposed rock at Lind Ridge, except that outcrops are dominated by trachyte lava, with minor hydroclastic deposits and no magmatic tuff (LeMasurier 1990*a*). Wilch (1997) provided additional age and geochemical data on four samples from the ridge (Table 8). A trachyte lava near the base of the ridge was dated to 12.71 ± 0.06 Ma; this lava had a massive base and a frothy top that included welded bombs (Wilch 1997). The main ridge outcrops include a welded pyroclastic fall deposit, vertically foliated clastogenic pumiceous trachyte lava and massive trachyte lava. The pumiceous lava and massive lava were $^{40}Ar/^{39}Ar$ dated to 11.19 ± 0.08 and 11.18 ± 0.19 Ma, respectively (Table 8). We interpret the welded trachyte fall and clastogenic lava deposits as products of Strombolian-style eruptions. Two trachyte lava samples from the north flank of Mount Andrus yielded identical K–Ar ages of 11.3 ± 0.8 Ma (LeMasurier 1990*a*).

González-Ferrán and González-Bonorino (1972) mentioned four parasitic scoria cones on the lower west flank of

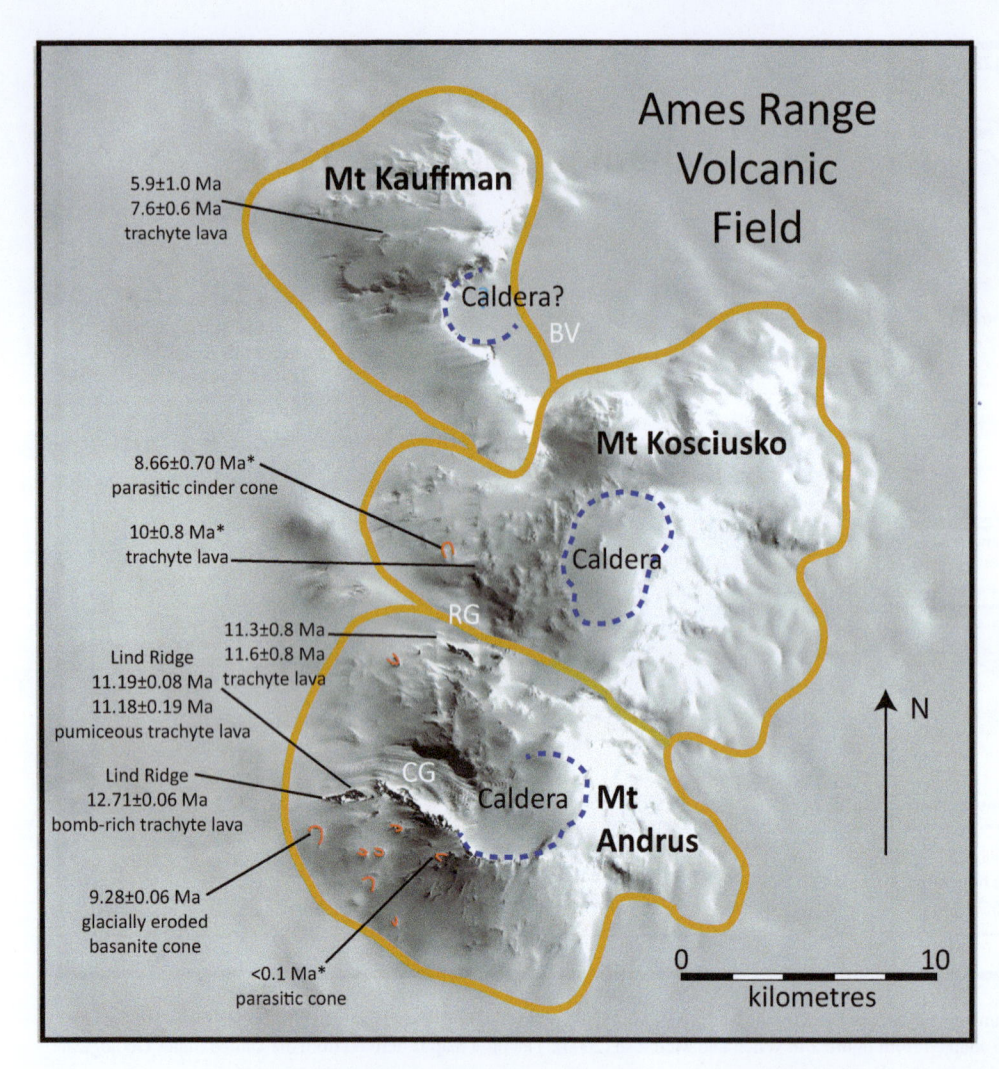

Fig. 28. Satellite image map of the Ames Range Volcanic Field: Mount Andrus, Mount Kosciusko and Mount Kauffman. The map shows studied outcrops (yellow), scoria cone remnants (red), ages, lithofacies and approximate outlines of volcano extent (orange) and caldera rim (dashed blue). Abbreviations: BV, Brown Valley; CG, Coleman Glacier; RG, Rosenberg Glacier. Data suggest an age progression from south to north. Image source: Google Earth Pro image accessed June 2019; see the caption to Figure 7 for more information on the source and an explanation of image processing.

Mount Andrus, noted that three of them appeared aligned north–south and suggested their vent locations are fault-controlled. Analysis of satellite imagery suggests that there are remnants of numerous parasitic scoria cones on the south and west flanks of Mount Andrus (Fig. 28). The cones have partially preserved crater rims that appear to be breached on one side. Narrow linear ridges extending down the slope from a few of the breached rims may be lava-flow levees. Only two parasitic cones have been analysed geochemically and both are basanite (LeMasurier 1990a; Wilch 1997). The lowest elevation outcrop at Mount Andrus is a basanite lava from a glacially eroded scoria cone situated SW of Lind Ridge, with a ^{40}Ar/^{39}Ar age of 9.28 ± 0.06 Ma (Table 8). A parasitic scoria cone located midway up Mount Andrus near Lind Ridge was K–Ar dated but the analysis yielded no radiogenic argon and the reported age was <0.1 Ma (LeMasurier 1990a).

Mount Kosciusko. Mount Kosciusko overlaps with Mount Andrus in the area of Rosenberg Glacier (Fig. 28). It has a summit elevation of 2909 m asl, a c. 3 km-diameter summit caldera and covers an area of about 16 × 16 km. The Mount Kosciusko caldera is about 10 km NNE of the Mount Andrus caldera. Outcrops appear to be limited to one locality north of Rosenberg Glacier, and include older phonolite lav,a K–Ar dated to 10.0 ± 0.8 Ma, and younger parasitic Strombolian scoria cones, K–Ar dated to 10.3 ± 0.8 and 8.66 ± 0.70 Ma, with a mean age of 9.2 ± 1.3 Ma (Table 8) (LeMasurier and Rex 1983; LeMasurier 1990a).

Mount Kauffman. Mount Kauffman, situated at the north end of the Ames Range, is the smallest central volcano and the least studied in the Ames Range (Fig. 28). It is connected to Mount Kosciusko by Gardiner Ridge and has a summit elevation of 2364 m asl. Only one locality has been visited and the rock was described as trachyte lava (LeMasurier 1990a). Two conventional K–Ar dating analyses of one exposure yielded non-overlapping ages of 7.6 ± 0.6 and 5.9 ± 1.0 Ma (Table 8), suggesting that the volcano is Late Miocene in age and the youngest of the Ames Range. Based on helicopter reconnaissance, LeMasurier (1990a) described inaccessible cliff exposures on the SE side as interbedded lavas and fragmental debris, and possibly representing a caldera or an explosion crater.

Flood Range Volcanic Field

The Flood Range Volcanic Field in western Marie Byrd Land (Figs 1 & 5) consists of three east–west-aligned trachytic central shield volcanoes: Mount Bursey (10–6 Ma), Mount Moulton (6–4 Ma) and Mount Berlin (2.8 Ma–active) (Table 9). Mount Bursey, although part of the Flood Range Volcanic Field, forms a solitary edifice located about 20 km east of Mount Moulton and about 10 km south of the north–south-oriented Ames Range Volcanic Field. Mount Moulton and Mount Berlin are separated by a high-elevation (2100 m asl), c. 10 km-wide, saddle. Mount Moulton and Mount Berlin are significant obstacles to the north-flowing WAIS, with surface ice elevations 600–800 m higher on the upstream (south)

Table 9. *Flood Range Volcanic Field summary age table*

Sample ID	Method	Corrected preferred age ± 2 SD (Ma)	Description	Ref.
Mount Bursey				
93-156	A	8.56 ± 0.06	Starbuck crater lava	16
93-158	A	0.25 ± 0.03	Syrstad Rock, bomb, not *in situ*	16
24	K	10.40 ± 0.80	Syrstad Rock, lava	2
93-160	A	10.08 ± 0.06	Koerner Bluff, lava	16
93-161	A	10.12 ± 0.34	Koerner Bluff, bomb	16
29	K	9.31 ± 0.74	Felsic cone on caldera rim	3
28A	K	0.49 ± 0.12	Hutt Peak parasitic cone	3
27	K	6.04 ± 0.48	Heaps Rock, lava	3
Mount Moulton				
93-146	A	1.12 ± 0.15	Gawne Nunatak, bomb	16
93-146	A	1.00 ± 0.10	Gawne Nunatak, bomb	16
	A	1.04 ± 0.04	Gawne Nunatak, bomb	16
93-318	A	5.95 ± 0.05	Prahl Crag, obsidian welded fall	16
2A	F	4.80 ± 0.61	Prahl Crag	15
93-345	A	4.03 ± 0.14	Edward's Spur, platy lava	16
Mount Berlin				
Summit caldera trachyte deposits				
93-16	A	0.0104 ± 0.0053	Fumarolic ice-cave floor trachyte lava	20
93-25	A	0.0184 ± 0.0058	Non-welded pumice fall	20
93-22	A	0.0375 ± 0.0120	Laminated pumice, top of upper welded unit	20
93-22	A	0.0274 ± 0.0031	Laminated pumice, top of upper welded unit	20
93-22	A	0.0255 ± 0.0031	Laminated pumice, top of upper welded unit	20
93-22	A	**0.0268 ± 0.0037**	**Mean age (n = 3 analyses)**	20
93-17	A	0.0253 ± 0.0059	Pumiceous rheomorphic tuff, upper welded unit	20
93-21	A	0.0280 ± 0.0037	Welded fall, below WCM93-22	20
93-21	A	0.0246 ± 0.0029	Welded fall, below WCM93-22	20
93-21	A	**0.0259 ± 0.0039**	**Mean age (n = 2 analyses)**	20
93-23	A	0.0344 ± 0.0130	East side, lower welded trachyte unit	20
93-23	A	0.0259 ± 0.0049	East side, lower welded trachyte unit	20
93-23	A	0.0330 ± 0.0108	East side, lower welded trachyte unit	20
93-23	A	**0.0279 ± 0.0064**	**Mean age (n = 3 analyses)**	20
93-15	A	0.0284 ± 0.0110	Spatter lava with cognate xenoliths	20
93-15	A	0.0241 ± 0.0029	Spatter lava with cognate xenoliths	20
93-15	A	0.0261 ± 0.0042	Spatter lava with cognate xenoliths	20
93-15	A	**0.0249 ± 0.0029**	**Mean age (n = 3 analyses)**	20
	A	**0.0259 ± 0.0020**	**Welded fall deposits (n = 5 samples)**	20
Merrem Peak caldera trachyte				
93-130	A	0.1432 ± 0.0057	1 m-thick pumiceous phonolite fall (non-welded)	20
93-123	A	0.1631 ± 0.0260	Black/yellow fall (SE)	20
93-128	A	0.1862 ± 0.0046	bomb (WNW)	20
93-129	A	0.1763 ± 0.0170	Pumiceous trachyte fall (NE)	20
93-129	A	0.1877 ± 0.0046	Pumiceous fall (NE)	20
93-133	A	0.1843 ± 0.0048	Welded trachyte fall (NE)	20
93-133	A	0.1934 ± 0.0075	Welded trachyte fall (NE)	20
93-133	A	**0.1869 ± 0.0091**	**Mean age (n = 2 analyses)**	20
93-139	A	0.1844 ± 0.0046	Welded clastogenic flow	20
93-140	A	0.182 ± 0.014	Welded clastogenic flow	20
	A	**0.1860 ± 0.0029**	**Welded trachyte fall deposit (n = 5 samples)**	20
Flank trachyte deposits				
93-011	A	0.234 ± 0.011	Near-vent trachyte lava (NE flank)	20
93-152	A	0.231 ± 0.012	Wedemeyer Rock, welded ignmibrite (SE)	20
	A	**0.2328 ± 0.0083**	**Flank trachyte deposits mean age (n = 2 samples)**	20
Mefford Knoll mafic deposits				
93-001	A	0.2142 ± 0.0346	Basanite cinder cone	20
93-001	A	0.1994 ± 0.0487	Basanite cinder cone	20
93-001	A	**0.210 ± 0.031**	**Mean age (n = 2 analyses)**	20
93-004	A	0.215 ± 0.028	Basanite cinder cone	20
93-004	A	0.216 ± 0.036	Basanite cinder cone	20
93-004	A	**0.216 ± 0.022**	**Mean age (n = 2 analyses)**	20
93-008	A	0.210 ± 0.080	Hawaiite flow levee	20
	A	**0.214 ± 0.018**	**Mefford Knoll mafic mean age (n = 3 samples)**	20
Merrem Peak (SW) trachyte lava				
93-134	A	0.456 ± 0.175	Lava	20

(*Continued*)

Table 9. *Continued.*

Sample ID	Method	Corrected preferred age ± 2 SD (Ma)	Description	Ref.
93-127	A	0.460 ± 0.041	Foliated lava	20
	A	**0.460 ± 0.040**	**SW flank trachyte lava 2 (*n* = 2 samples)**	20
93-125	A	0.580 ± 0.015	Foliated lava	20
93-126	A	0.573 ± 0.012	Clastogenic trachyte lava	20
93-126	A	0.571 ± 0.013	Clastogenic trachyte lava	20
93-126	A	**0.572 ± 0.009**	**Mean age (*n* = 2 analyses)**	20
93-137	A	0.583 ± 0.013	Lava	20
93-138	A	0.588 ± 0.040	Clastogenic lava	20
93-135	A	0.594 ± 0.018	Clastogenic phonolite	20
93-135	A	0.588 ± 0.048	Clastogenic phonolite	20
93-135	A	0.560 ± 0.020	Clastogenic phonolite	20
93-135	A	**0.587 ± 0.036**	**Mean age (*n* = 3 analyses)**	20
93-009	A	0.597 ± 0.028	Trachyte flow levee (NW)	20
	A	**0.578 ± 0.009**	**SW flank trachyte lava 1 (*n* = 6 samples)**	20
Brandenberger Bluff trachyte/phonolite tuya				
93-014	A	2.71 ± 0.07	Phonotephrite cinder cone, upslope from bluff	20
93-010	A	2.80 ± 0.11	Phonolite dome lava	20
93-037	A	2.86 ± 0.24	Fractured phonolite lava at bluff base	20
93-053	A	2.76 ± 0.09	Trachyte clast in hyalotuff	20
93-121	A	2.79 ± 0.12	Phonolite clast in hyalotuff	20
93-250	A	2.72 ± 0.13	Trachyte clast in hyalotuff	20
	A	**2.77 ± 0.06**	**Mean age (*n* = 5 samples)**	20

Notes: method A is ^{40}Ar/^{39}Ar; method K is K/Ar; method F is fission track.
Ref. code: 2, LeMasurier (1990*a*); 3, LeMasurier (1990*e*); 15, Seward *et al.* (1980); 16, Wilch (1997); 20, Wilch *et al.* (1999). See also Supplementary Material Table S1.

sides than on the downstream sides of these volcanoes. All three of the Flood Range central volcanoes are covered with snow and ice, and appear to be relatively undissected. Each of the three central volcanoes is a compound polygenetic volcano with two east–west-aligned summit calderas. The preservation of calderas indicates that minimal erosion has occurred, although past overriding by the WAIS at Mount Bursey and Mount Moulton cannot be precluded. Very limited outcrops on Mount Bursey and Mount Moulton provide a glimpse of their histories. Mount Berlin is the best-exposed volcano in the volcanic field and provides the most detailed geological history in the range.

Mount Bursey. Mount Bursey (Fig. 29), located at the east end of the Flood Range Volcanic Field, is a 20 × 30 km ice-covered massif with two east–west-aligned calderas at Hutt Peak and adjacent to Koerner Bluff (LeMasurier 1990*e*). Mount Bursey is interpreted as two coalesced trachytic shield volcanoes (LeMasurier 1990*e*). The Koerner Bluff caldera is situated 300 m lower than the Hutt Peak caldera and is interpreted to be the oldest caldera, ^{40}Ar/^{39}Ar dated to 10.08 ± 0.06 Ma (Table 9). Outcrops associated with the Koerner Bluff caldera include a phonolite lava dome at Koerner Bluff, a felsic crater rim called Starbuck Crater, basaltic rocks at Syrstad Rock, a parasitic felsic cone on the west side of the caldera rim and a parasitic hawaiite cone also on the caldera rim (LeMasurier 1990*e*; Wilch 1997). The phonolite lava dome is flat topped, and is composed entirely of flow-foliated lava with slickensided flow surfaces. A bomb from the hawaiite scoria cone on the caldera rim was

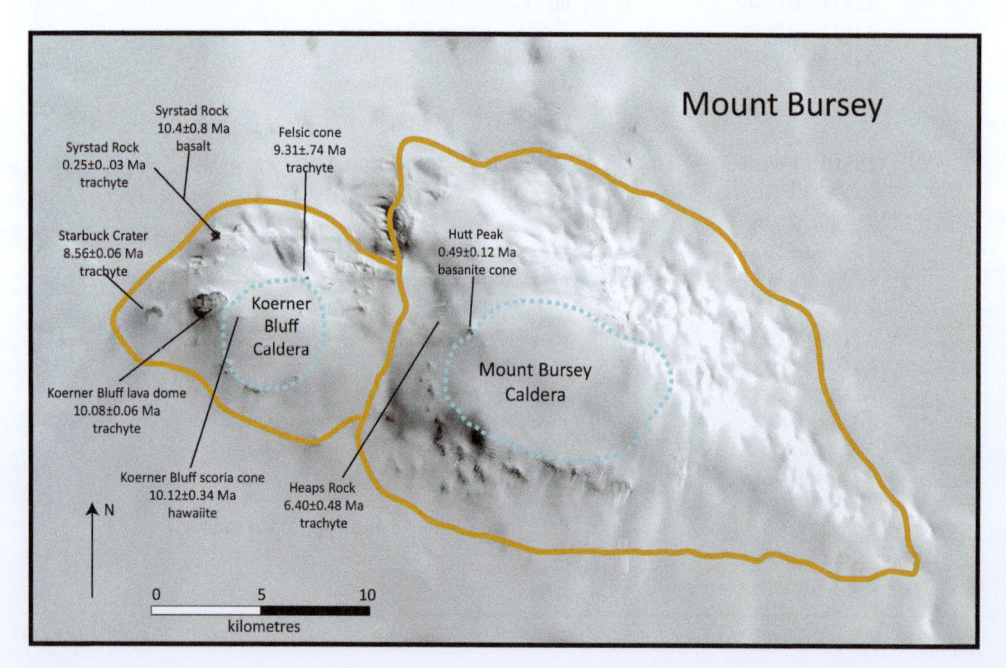

Fig. 29. Satellite image map of the Mount Bursey volcano at the east end of the Flood Range Volcanic Field. The map shows studied outcrops, ages, lithofacies and approximate outlines of volcano extent (orange) and caldera rim (dashed blue). Asterisks indicate conventional ^{40}Ar/^{39}Ar ages. Image source: Google Earth Pro image accessed June 2019; see the caption to Figure 7 for more information on the source and an explanation of image processing.

^{40}Ar/^{39}Ar dated to 10.12 ± 0.34 Ma, consistent with the 10.08 ± 0.06 Ma age for the lava dome. A felsic cone on the north side of the caldera was K–Ar dated to 9.31 ± 0.74, also consistent with the dome age. Starbuck Crater, downslope of Koerner Bluff, is an eroded crater rim with a ^{40}Ar/^{39}Ar age of 8.56 ± 0.06 Ma. The eroded crater is composed of subaerially-erupted trachyte that exhibits welded spatter transitional to clastogenic lava textures. A basaltic sample from Syrstad Rock was dated by conventional K–Ar to 10.4 ± 0.8 Ma. Mugearite pyroclastic rocks also crop out at Syrstad Rock; a mugearite sample has a ^{40}Ar/^{39}Ar date of 0.25 ± 0.03 Ma. The summit caldera at Hutt Peak was dated to 6.04 ± 0.48 Ma, based on a conventional K–Ar date for a sample from Heaps Rock just below the caldera. A mafic cone on the caldera rim of Hutt Peak was K–Ar dated to 0.49 ± 0.12 Ma and interpreted as a late-stage parasitic vent (LeMasurier 1990e).

In summary, edifice-building activity associated with Koerner Bluff caldera occurred from 10.08 to 8.56 Ma, based on high-precision ^{40}Ar/^{39}Ar dating (Table 9). The higher and younger Mount Bursey caldera formed at 6.04 Ma, based on K–Ar dating. Mount Bursey late-stage Pleistocene parasitic eruptions occurred at 0.49 and 0.25 Ma. All of Mount Bursey outcrops appear to have resulted from subaerial eruptions and there is no evidence of ice–magma interactions, although the Starbuck Crater appears to have been overridden by ice at some point after the rocks were erupted. No erratics were observed at Starbuck Crater.

Mount Moulton. Mount Moulton is an almost entirely snow- and ice-covered, 15 × 40 km-long, east–west-orientated massif, with an estimated volume of 325 km^3 (Fig. 30) (LeMasurier and Kawachi 1990d). There are two distinct c. 5–7 km-diameter summit calderas: the eastern caldera associated with the older Prahl Crags, here named the Prahl Crags caldera; and the younger western caldera at Britt Peak, here named the Mount Moulton caldera. There may be a third smaller caldera near Kohler Dome, about 9 km east of the Prahl Crags caldera (González-Ferrán and González-Bonorino 1972), although its morphology is less distinct than the other two. The Mount Moulton calderas are aligned east–west, and the massif and calderas are aligned with Mount Bursey and Mount Berlin, located just to the east and west, respectively. The caldera morphologies become increasingly distinct from east to west, suggesting progressive younging towards the west. Mount Moulton has three known outcrops, Prahl Crags, Edwards Spur at the NW end of the massif, and Gawne Nunatak at the western flank near Wells Saddle which separates Mount Moulton and Mount Berlin.

Prahl Crags consists of a peralkaline rhyolite lava and an associated welded fall deposit, with obsidian fiamme, and is ^{40}Ar/^{39}Ar dated to 5.95 ± 0.05 Ma (Table 9). Prahl Crags is interpreted as a fragment of the eastern caldera rim and represents the early phase of Mount Moulton volcanism. Edwards Spur consists of a rheomorphic welded trachyte fall deposit, with a slightly ropy texture and interlayered with a platy lava (Wilch 1997). Although located near the base of the volcano, Edwards Spur is the only felsic rock proximal to the western caldera. The ^{40}Ar/^{39}Ar age of Edwards Spur is 4.03 ± 0.14 Ma (Table 9) and is inferred to date the growth of the younger caldera. Gawne Nunatak is a remnant of a parasitic hawaiite scoria cone, consisting of weakly-welded pyroclastic lapilli and bomb deposits. The ^{40}Ar/^{39}Ar age of Gawne Nunatak at 1.04 ± 0.04 Ma (Table 9) provides the only constraint on post-edifice-building volcanism. The sampled outcrops at Mount Moulton all exhibit evidence of subaerial eruptions and deposition, with no evidence of glacio-volcanic interactions.

In 1993–94 a very significant finding at Mount Moulton was a sequence of dipping englacial tephra layers exposed in blue ice in the eastern caldera, just north of Prahl Crags (Wilch *et al.* 1999). The englacial tephra layers were sampled in reconnaissance and ^{40}Ar/^{39}Ar dated by Wilch *et al.* (1999), and revisited and studied in detail in 1999–2000 and 2003–4 (Dunbar *et al.* 2008). A total of 48 tephra layers were observed. Most are trachytic in composition with some mafic layers (Dunbar *et al.* 2008). ^{40}Ar/^{39}Ar dating of K-feldspar phenocrysts from coarse-grained pumice-rich trachytic tephra layers yielded eight stratigraphically consistent ages, ranging from 500 to 10 ka. The trachytic tephra layers were erupted from explosive eruptions of Mount Berlin, located 30 km west of the site; the tephra and their chronology are described further by Dunbar *et al.* (2021). A few of the tephra layers are mafic in composition and may have been derived from parasitic vents on the flanks of Mount Berlin or Mount Moulton (Dunbar *et al.* 2008).

Mount Berlin. Mount Berlin is an active volcano with steaming fumaroles and fumarolic ice caves. It is a predominantly trachytic polygenetic composite volcano with a prominent summit caldera and an older subsidiary caldera at Merrem Peak (LeMasurier and Wade 1968; LeMasurier and Kawachi 1990a; Wilch *et al.* 1999) (Figs 31, 32 & 33). Mount Berlin, at the west end of the Flood Range Volcanic Field, has an estimated volume of 125 km^3 (LeMasurier and Kawachi 1990a)

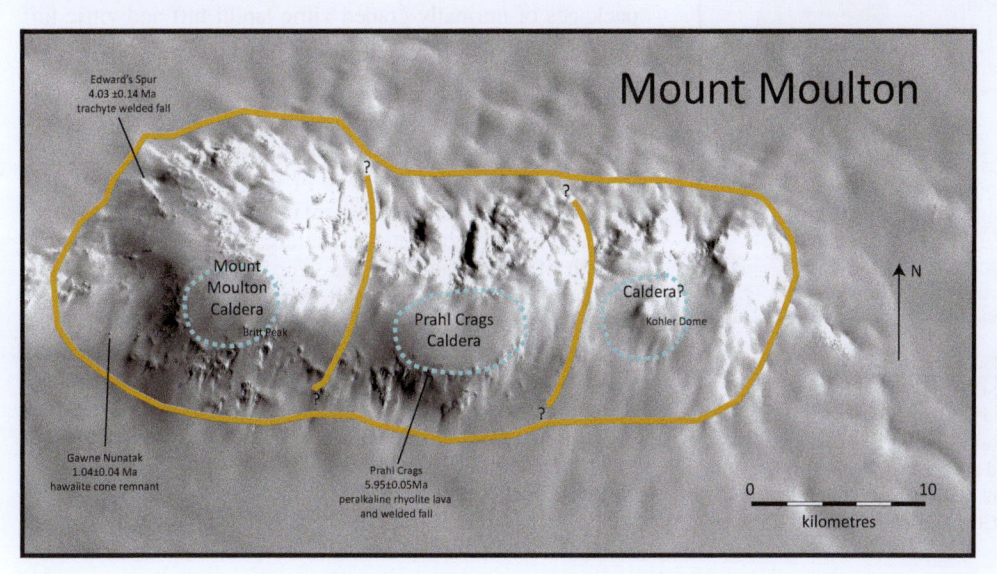

Fig. 30. Satellite image map of the Mount Moulton volcano, located in the centre of the Flood Range Volcanic Field. The map shows studied outcrops, ages, lithofacies and approximate outlines of volcano extent (orange) and caldera rim (dashed blue). The eastern caldera at Kohler Dome is speculative. Possible caldera and degree of erosion suggest that Mount Moulton is composed of three coalesced shield volcanoes that are progressively younger to the west. Image source: Google Earth Pro image accessed June 2019; see the caption to Figure 7 for more information on the source and an explanation of image processing.

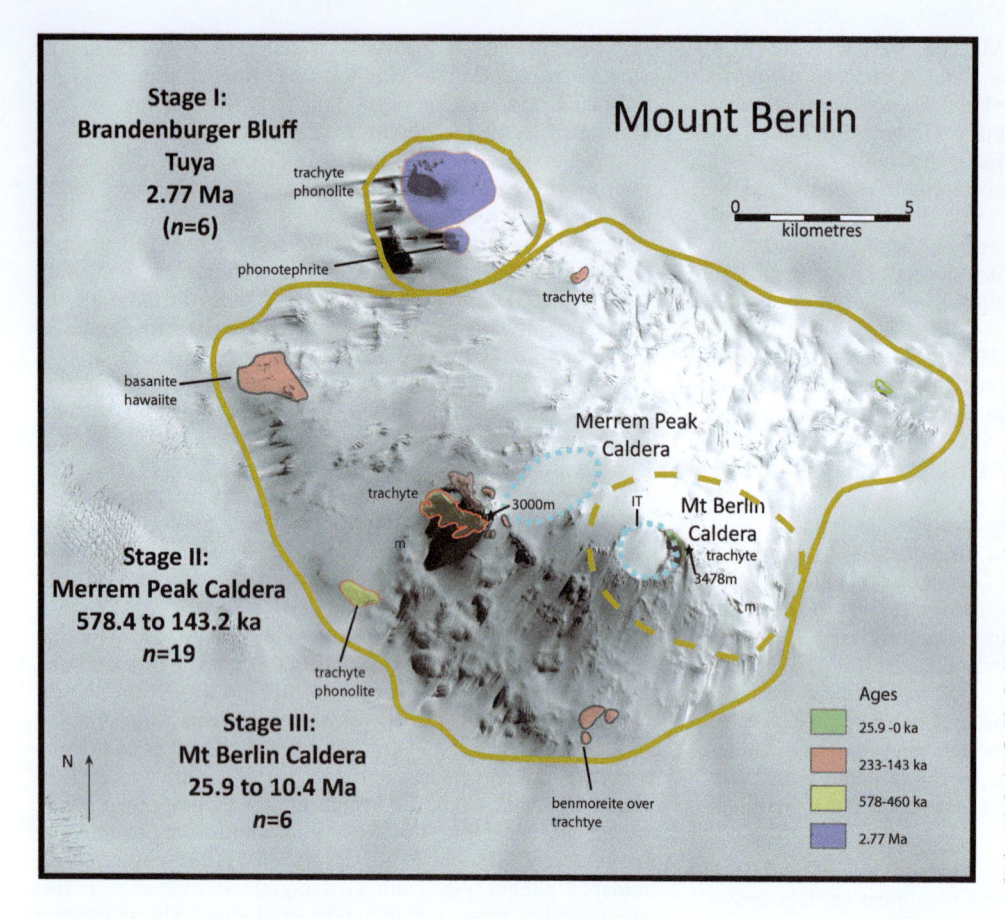

Fig. 31. Satellite image map of Mount Berlin, in the western Flood Range Volcanic Field. The map shows outcrop age ranges and rock types, and approximate outlines of volcano extent (orange) and caldera rim (dashed blue). The three stages of the growth of Mount Berlin are discussed in the text; the number (*n*) of age determinations included in the age ranges of each stage are listed after the ages. In 1993, there were multiple steaming ice towers on the caldera rim; the ice tower visited is marked IT. Letter m designates local moraine. The summits of Merrem Peak (3000 m asl) and Mount Berlin (3478 m asl) are shown with stars. Image source: Google Earth Pro image accessed June 2019; the caption to Figure 7 for more information on the source and an explanation of image processing. Elevations from the Mount Berlin (1973) quadrangle, scale 1:250 000, USGS Reconnaissance Series, Antarctica, United States Geological Survey.

and basal dimensions of 18 × 18 km at the level of the WAIS. Mount Berlin is largely undissected, except for the older north-facing Brandenberger Bluff. The volcanic history is recorded in caldera-rim and flank deposits on the volcano, and in distal englacial tephra deposits at Mount Moulton and in ice cores (see Dunbar *et al.* 2021). Wilch *et al.* (1999) provided a detailed analysis of the volcanic geology and documented three stages in the growth of Mount Berlin, summarized below.

Stage I – Brandenberger Bluff. Brandenberger Bluff, located on the north side of Mount Berlin, is a *c.* 350 m-high Pliocene (2.77 ± 0.06 Ma (*n* = 5): Fig. 31; Table 9) lava and volcaniclastic edifice that exhibits evidence for subglacial, emergent and subaerial glaciovolcanic palaeoenvironments (Wilch *et al.* 1999). The north-facing stratigraphic sequence is about 250 m thick and 1 km wide, and is composed of steeply-dipping (20°–30°), well-stratified, fine-grained trachytic/phonolitic vitric tuff that overlies a highly jointed and brecciated, aphyric, glassy trachyte/phonolite lava (Wilch 1997). The deposits include thick sections of cyclic packages of normally-graded vitric lapilli tuff and vitric tuff and massive fine vitric tuff that are interpreted as volcaniclastic turbidites that formed fan deposits in an ice-marginal lake. The dominance of a fine-grained glassy matrix in the deposits is attributed to intense phreatomagmatic fragmentation during the eruption. The volcaniclastic sediment sequences are interrupted by areas of large (tens of metres in diameter) slide blocks of similar material that are interpreted as collapse breccias formed by gravitational failure of fan deposits during growth of the subaqueous fan. Throughout the stratigraphic sequence both pumice lapilli and blocky, aphyric, glassy lapilli are common. Stratified vitric lapilli tuff at the top of the bluff exhibits shallow-dipping planar beds and cross-beds, and contains abundant ash-coated lapilli, interpreted as subaerial phreatomagmatic pyroclastic density current deposits (Fig. 34a). The margins of the bluff top surface include areas of intense soft-sediment deformation. The Brandenberger Bluff sequence and edifice are interpreted as a felsic variation of a tephra-dominated tuya (Smellie and Edwards 2016), composed of subaqueously-emplaced lava

Fig. 32. Satellite image map of the Mount Berlin summit caldera at the west end of the Flood Range Volcanic Field. Circles locate fumarolic ice towers and steaming vents identified in 1993. The double circle locates the ice tower entrance to the ice cave that was entered, and from which lava on the caldera floor was sampled and dated to 10.3 Ma (Wilch *et al.* 1999). A *c.* 150 m-thick exposure of caldera-wall deposits are highlighted in red. The satellite image was derived from Google Earth in June 2019; image sources: 2019 Digital Globe.

Fig. 33. Photographs of lithofacies and features of Mount Berlin, Flood Range Volcanic Field. (**a**) Lapilli tuff with ash-coated lapilli clasts from near the top of Brandenberger Bluff at Mount Berlin. (**b**) The 143.2 ka non-welded pyroclastic fall breccia and lapilli tuff near the rim of the Merrem Peark caldera. (**c**) The 25.9 ka welded fall deposit in the wall of the Berlin summit caldera. The person is pointing to the 1 m-long flattened pumice bomb. Tephra from this and many other Mount Berlin eruptions was found 30 km away in the summit caldera at Prahl Crags on Mount Moulton, as well as in multiple West Antarctic ice cores. (**d**) A steaming fumarolic ice tower in the Mount Berlin summit caldera. A lava sample from the floor of the ice cave was dated to 10.4 ± 5.4 ka.

and reworked tuff deposits at its base and on its flanks, and subaerially-emplaced phreatomagmatic tuff deposits at the top of the bluff. We infer that this sequence was deposited in an intraglacial lake setting. The abundant deformation evident in the volcaniclastic deposits may have developed as supporting ice walls collapsed. The basal lava and several clasts from within the tuff cone sequences were dated by the $^{40}Ar/^{39}Ar$ method, with a mean age of 2.77 ± 0.06 Ma ($n = 5$) (Table 9). Phonotephritic scoria cone deposits, located upslope and about 100 m in elevation above the top of the bluff, yield an overlapping age of 2.71 ± 0.07 Ma (Table 9) that suggests a single eruptive phase. These subaerially-erupted phonotephrite tuff deposits indicate complete emergence above ice level.

The passage zone from subaqueous to subaerial depositional environments is situated near the top of Brandenberger Bluff, approximately 250 m above the level of the WAIS on the north side of Mount Berlin. Currently, ice-sheet flow is obstructed by Mount Berlin and the elevation of the WAIS on the upstream (south) side of Mount Berlin is at about 1800–2000 m asl, about 400–600 m higher than on the downstream side (Fig. 31). In middle Pliocene time, when the Brandenberger Bluff tuya emerged above ice level, it is likely that the Mount Berlin edifice did not exist and there was probably no local obstruction to ice flow. In the absence of Mount Berlin, local ice elevations at Brandenberger Bluff would be at c. 1800 m asl. Remarkably, the exposed passage zone at Brandenberger Bluff may actually record a lower late Pliocene WAIS level, because of changes in ice-flow patterns after its eruption, caused by growth of the Mount Berlin edifice.

Stage II – Merrem Peak caldera. The second and most voluminous phase of activity at Mount Berlin is characterized by growth of the volcano at Merrem Peak to 3000 m asl and by eruptions from the 2.5 × 1 km-diameter Merrem Peak caldera. The constructional slopes of Merrem Peak volcano average about 13°. The earliest activity associated with this stage of activity is recorded in pumiceous foliated and clastogenic trachyte lava located just west of Merrem Peak, and dated to 578 ± 9 ka ($n = 6$: Table 9). Three subsequent eruptive episodes at Merrem Peak caldera are recorded by proximal deposits west of the caldera, including a trachytic lava dated to 460 ± 40 ka, an 18+ m-thick, densely- to incipiently-welded, trachytic pumiceous pyroclastic breccia dated to 186.0 ± 2.9 ka, and a 1 m-thick, phonolitic, non-welded pumice lapilli layer dated to 143.2 ± 5.7 ka (Fig. 33b). Both of the 186.0 and 143.2 ka fall deposits mantle topography. The youngest (143.2 ka) fall deposit is situated on the rim of the Merrem Peak caldera.

Compositionally-diverse eruptions also occurred during this Merrem Peak caldera interval of volcanism. A pyroclastic vent breccia and a near-vent trachyte lava (234 ± 11 ka) are located on the NE flank of Merrem volcano. On the south flank, a welded trachytic ignimbrite (231 ± 12 ka) contains c. 30% fiamme, with typical aspect ratios of 1:10 (height: length). The ignimbrite may have originated from the caldera. The NE flank trachyte lava and south flank ignimbrite are geochemically identical, with a mean age of 232.8 ± 8.3 ka. A benmoreite lava overlies the ignimbrite and signifies a c. 230 ka or younger effusive eruption. On the NW flank, basanite–hawaiite scoria cone remnants yield a mean age of 214 ± 18 ka.

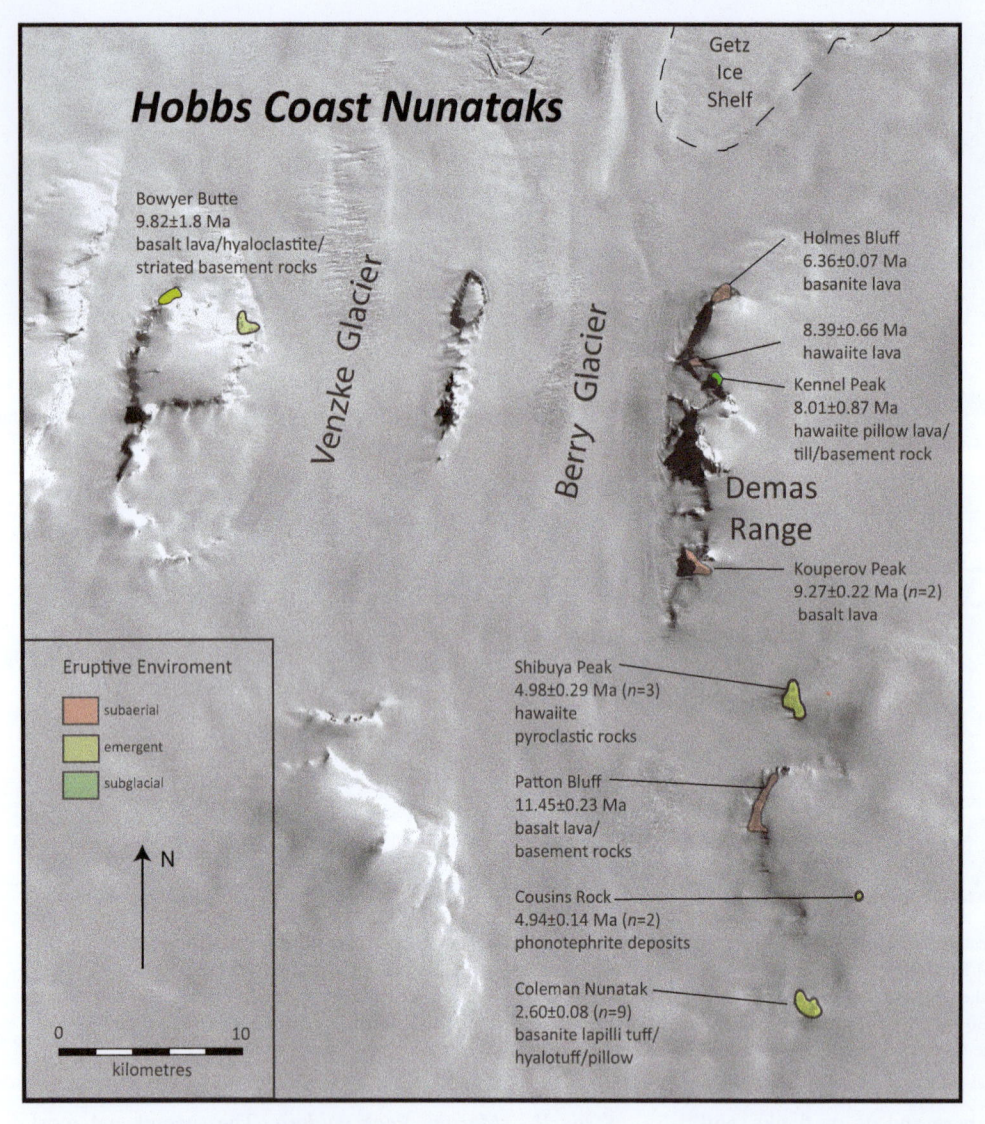

Fig. 34. Satellite image map of the Hobbs Coast nunataks in western MBL. The map shows studied outcrops, ages, lithofacies and inferred eruptive environments. Adapted from Wilch and McIntosh (2007). Image source: Google Earth Pro image accessed June 2019; see the caption to Figure 7 for more information on the source and an explanation of image processing.

Stage III – summit caldera. The final and still-active phase of Mount Berlin volcanism was marked by growth of the volcano by more than 400 m to 3478 m asl and a southeastward shift of the vent area to the 2 km-diameter summit caldera (Figs 31 & 32). The constructional slopes of the volcano above Merrem Peak caldera range from 19° to 30°. Two prominent welded trachytic pyroclastic fall units, totalling more than 150 m in thickness, are exposed in the eastern wall of the summit caldera. These pyroclastic fall deposits are composed of slightly flattened pumiceous bombs and abundant cognate xenoliths. The welding is interpreted as agglutination with minor load-pressure compaction. Recumbent folds seen in one unit suggest that part of the pyroclastic deposit flowed rheomorphically. Ages of five samples from the two lower welded units are analytically indistinguishable and range from 27.9 ± 6.4 to 24.9 ± 2.9 ka, with a mean age of 25.9 ± 2.0 ka (Table 9) (Fig. 33c). These are locally overlain by a >10 m-thick sequence of welded and non-welded pyroclastic fall beds. The fall deposits include a 6 m-thick lithic- and ash-rich explosion breccia containing bomb and lithic clasts up to 50 cm in diameter, and a 2 m-thick densely welded obsidian fall. Anorthoclase phenocrysts from the obsidian yielded a maximum age of 18.4 ± 5.8 ka. These caldera-wall deposits at Mount Berlin were previously interpreted as lava on the basis of reconnaissance investigations (LeMasurier and Kawachi 1990*a*).

Several fumarolic ice towers and steaming vents along the summit caldera rim attest to ongoing geothermal activity (Figs 32 & 33d). One ice tower opens into an underlying ice-cave system more than 70 m long. Lava exposed on the cave floor is dated to 10.4 ± 5.3 ka (Table 9). In 1993, surface temperatures of the cave-floor lava were as high as 12°C. Distal englacial tephra from Mount Berlin have been documented in the ice-filled Mount Moulton caldera, located 30 km east of the summit caldera (Wilch *et al.* 1999; Dunbar *et al.* 2008). On the basis of the size and density of Mount Moulton pumice clasts, and distance from their Mount Berlin source, Wilch *et al.* (1999) calculated eruption column heights ranging from 28 to 40 km, indicative of highly-explosive Plinian eruptions.

Summary. Mount Berlin is a mostly trachytic polygenetic central volcano that was constructed in three stages (Wilch *et al.* 1999). Stage I was the growth of the *c.* 500 m-high, 2.77 myr-old, trachytic Brandenberger Bluff tephra-dominated tuya, in which successive subglacial, phreatomagmatic and subaerial lithofacies record the emergence of the volcano above a palaeo-ice-sheet surface. This volcano formed in an englacial lake, with a passage zone about 250 m above current ice level to the north of the bluff. Stage II is the growth of the Merrem Peak shield volcano with eruptions from the 3000 m asl Merrem Peak caldera from 578 to 143 ka. Most of the estimated volume of Mount Berlin was erupted during construction of the Merrem Peak volcano. This stage was mostly trachytic in composition and was dominated by variably-welded pumiceous pyroclastic rocks. The third and ongoing stage of Mount Berlin volcanism is recorded in explosively-erupted pyroclastic deposits dating back to 25.9 ka and preserved in the summit caldera that is up to 3478 m asl. Englacial tephra found at Mount Moulton and in ice cores records highly-explosive Mount Berlin volcanism during the Merrem Peak and Berlin summit caldera stages (Dunbar *et al.* 2021).

Hobbs Coast Volcanic Field

Several monogenetic, mostly basaltic, volcanic centres located at inland and island sites near the Hobbs Coast in western Marie Byrd Land comprise the Hobbs Coast Volcanic Field (Figs 1 & 34). The inland nunataks include volcanic centres mostly located on the north–south-orientated Demas Range, just east of Berry Glacier; Bowyer Butte is the exception, located 20 km to the west and just west of Venzke Glacier (Fig. 34). Granite bedrock is exposed in contact with or beneath the volcanic rocks at several locations. Basement and volcanic rocks of the inland nunataks exhibit abundant evidence of past glaciations. Wilch and McIntosh (2007) described the lithofacies characteristics, stratigraphic relationships and $^{40}Ar/^{39}Ar$ geochronology of all sites except Bowyer Butte; the summary here is based on this study unless otherwise noted. Ages of volcanism are mostly Late Miocene but range from 11.45 to 2.57 Ma (Table 10). The discussion of the inland volcanic centres is organized according to interpretations of dominant eruptive palaeoenvironments.

Table 10. *Hobbs Coast Volcanic Field summary age table*

Sample ID	Method	Corrected preferred age ± 2 SD (Ma)	Description	Ref.
Shepard Island and nearby islands				
Shepard Island				
93-255	A	0.481 ± 0.049	Lava-flow interior, ponded flows in Mathewson Point tuff cone	16
93-258	A	0.56 ± 0.14	Lava clast from within tuff beds	16
93-265	A	0.41 ± 0.09	Mount Petinos, lithic clast in tuff beds	16
S9F	K	0.43 ± 1.20	Worley Point platy lava	6
Grant Island				
GI11E	K	0.7 ± 0.2	Hyaloclastite from tuff cone	6
Cruzen Island				
CI51C	K	2.75 ± 0.26	Subaerial basalt top of tuya	6
Hobbs Coast Nunataks				
Coleman Nunatak				
93-199	A	2.57 ± 0.06	Dense lava, below upper surge deposit	19
93-185	A	2.64 ± 0.28	Lava, SW moat	19
93-184	A	2.70 ± 0.14	Lava, SW moat	19
93-192	A	2.61 ± 0.20	Lava, SW moat	19
93-188	A	2.49 ± 0.12	Basal lava flow SW moat	19
93-208	A	2.59 ± 0.12	North end, clast within scoriaceous dyke	19
93-207	A	2.62 ± 0.09	North end, dyke	19
93-205	A	2.79 ± 0.27	North end, dyke	19
93-202	A	2.84 ± 0.23	North end, dyke	19
	A	**2.60 ± 0.08**	**Mean (*n* = 9)**	
Cousins Rock				
93-218	A	4.92 ± 0.10	Bomb	19
93-219	A	4.96 ± 0.11	Lava	19
	A	**4.94 ± 0.14**	**Mean (*n* = 2)**	
Shibuya Peak				
93-221	A	5.07 ± 0.14	Bomb	19
93-223	A	5.35 ± 0.20	Lava crusts	19
93-229	A	4.80 ± 0.11	Dyke	19
	A	**4.98 ± 0.29**	**Mean (*n* = 2)**	
Patton Bluff				
93-307	A	11.45 ± 0.23	Lava, 15 m thick	19
Kouperov Peak				
93-303	A	9.24 ± 0.11	South end, lava over unconformity (*n* = 2)	19
93-304	A	9.07 ± 0.57	Lava (*n* = 2)	19
	A	**9.27 ± 0.22**	**Mean (*n* = 2)**	
Kennel Peak				
93-297	A	8.01 ± 0.87	Pillow lobe interior	19
Holmes Bluff				
93-299	A	6.36 ± 0.07	Lava filling valley	19
50	K	8.39 ± 0.66	Subaerial lava at south end, near Kennel Peak	6
Bowyer Butte				
57D	K	9.82 ± 1.80	5 m basalt lava on hyaloclastite over glacially-striated basement	6

Notes: method A is $^{40}Ar/^{39}Ar$; method K is K/Ar.
Ref. code: 6, LeMasurier and Rex (1983); 16, Wilch (1997); 19, Wilch and McIntosh (2007). See also Supplementary Material Table S1.

Subglacial volcanic palaeoenvironments are inferred at two localities: Kennel Peak (Wilch and McIntosh 2007) and Bowyer Butte (LeMasurier 1990d). At Kennel Peak, a c. 125 m-thick section of hawaiitic pillow lavas and interpillow hyaloclastite breccia (8.01 ± 0.87 Ma) unconformably overlies a smooth moulded outcrop of granitic basement rocks (Wilch and McIntosh 2007). A <1 m-thick glacial till occurs locally along the unconformity between the volcanic and basement rocks. The top of the thick pillow hyaloclastite sequence is eroded and overlain by crudely-graded, planar-stratified hyaloclastite lapilli tuff. The pillow lavas and hyaloclastites at Kennel Peak are inferred to have been deposited in a subglacial chamber onto thin basal till and glacially-moulded bedrock. The vesicularity and steep dips of the interbedded pillow lava and hyaloclastite, and the presence of similar age (K–Ar 8.39 ± 0.66 Ma) subaerial lavas nearby at an outcrop just north of Kennel Peak (LeMasurier 1990d), suggest that the pillow hyaloclastite deposits form part of a subaqueous lava-fed delta sequence, possibly associated with subaerial lava effusion. The pillow lava and hyaloclastite sequence extends only c. 25 m above today's ice surface, suggesting that the syneruptive local palaeo-ice level was at least slightly higher than today's ice level (Wilch and McIntosh 2007). At Bowyer Butte, a volcanic sequence of subaerial lava over thin hyaloclastite beds (1–5 m) resting on a striated granite surface has a low precision K–Ar age of 9.82 ± 1.8 Ma (LeMasurier 1990d).

Other volcanic centres along the Berry Glacier show evidence of emergent phreatomagmatic and subaerial eruptions that, in places, deposited volcanic rocks on a previously glaciated bedrock surface. At Patton Bluff, the oldest known volcanic outcrop in the area consists of 11.45 ± 0.23 Ma subaerially-erupted basaltic lava that is situated near, but not in direct contact with, local basement rocks. At Kouperov Peak, two samples with duplicate $^{40}Ar/^{39}Ar$ ages that average 9.27 ± 0.22 Ma consist of subaerially-erupted basaltic lava which unconformably overlies basement rocks. The unbrecciated lava is about 30 m above the present ice surface and the lava–basement contact is poorly exposed but extends over >60 m of topographical relief. The basement rocks are locally striated to within 2 m of the lava, although no striations were observed beneath the lava. The Kouperov Peak locality is tentatively interpreted as a glacial unconformity (older than 9.2 Ma) that is overlain by a subaerial lava.

At Holmes Bluff, a c. 55 m-thick, columnar-jointed, valley-filling 6.36 ± 0.07 Ma (Table 10) lava with a 1.3 m-thick welded and ropy basal breccia overlies a 1.5 m-thick achnelith-rich lapilli tuff and a partially-moulded, slightly-weathered, basement unconformity. The sequence is tentatively interpreted as a basement glacial unconformity (older than 6.4 Ma) that was subsequently exposed and locally weathered before being covered by a thin, subaerially-erupted, pyroclastic fall deposit and a thick, valley-filling, subaerial lava.

Three samples from Shibuya Peak with a mean age of 4.98 ± 0.29 Ma are from an unusual volcaniclastic deposit with abundant exotic granitoid clasts exposed in a >100 m-thick section (Table 10). LeMasurier (1990d) inferred a subglacial eruptive environment based on the presence of bedded basaltic hyaloclastite and interbedded tillite. Wilch and McIntosh (2007) inferred that the volcanic rocks resulted from subaerial Strombolian and phreatomagmatic eruptions, noting the bedding and cross-stratification, the variably vesicular lapilli tuff and tuff breccia, and the mix of sideromelane and tachylite glass clasts. They noted that some of the granitoid cobbles and boulders were encrusted with a thin (1 cm) coat of basaltic lava, and argued that this is consistent with eruption through a tillite but not consistent with the clasts being part of an interbedded glacial tillite. This interpretation provides evidence for a pre-volcanic (>c. 5 Ma) glaciation at this locality. An outcrop of subaerially-erupted phonotephrite and welded agglutinate and slightly reworked phreatomagmatic vitric tuff at Cousins Rock, a small (150 × 150 m) nunatak located near Shibuya Peak, yielded an age of 4.94 ± 0.14 Ma, similar in age to the Shibuya Peak sample.

The 1 × 3 km Coleman Nunatak is one of the largest and the youngest (2.60 ± 0.08 Ma) monogenetic volcanic centres exposed in the Hobbs Coast Volcanic Field (Table 10). Coleman Nunatak has a 75 m-thick, 1 km-wide sequence, well exposed in the wind moat on the south end of the nunatak, which suggests that the volcano is an emergent tuff cone sequence. The base of the sequence includes minor pillow and glassy lava overlain by compound lava with subaerial breccia, and interbedded vitric lapilli tuff and vitric lapillistone units. The vitric lapilli tuff units include bread-crust bombs with bedding-plane sags and ash-coated lapilli, indicative of magmatic and phreatomagmatic fragmentation processes. Discontinuous, erosive, massive, planar and cross-stratified beds are interpreted as turbulent pyroclastic density current deposits (Fisher and Schmincke 1984). The combination of turbulent pyroclastic density current deposits and minor subaqueous lavas is indicative of an emergent intraglacial environment, where interaction with meltwater controlled the explosive phreatomagmatic eruption style and modified the lavas slightly. Cross-cutting exposures of steeply-dipping lapilli tuff and breccia are interpreted as evidence of a vent funnel unconformity overlain by vent slurry lapilli tuff (Sohn and Park 2005; White and Ross 2011).

The inland nunataks of the Hobbs Coast Volcanic Field provide compelling evidence for Miocene and Pliocene ice-sheet glaciation near coastal West Antarctica. The syneruptive thicknesses of the palaeo-ice sheets are difficult to determine because the volcanic rocks and glacial deposits and surfaces are positioned on glacial interfluves. The subglacial lithofacies at two sites and emergent–subaerial lithofacies at the other five sites are generally consistent with regional ice levels similar to the current WAIS at multiple times between 11.5 and 2.6 Ma.

Three small volcanic islands are located offshore and are connected to the WAIS by the Getz Ice Shelf (Fig. 1). Shepard Island and Grant Island are situated 30 km offshore and located 50–70 km NE of Holmes Bluff, the northeasternmost inland nunatak described above. Cruzen Island is situated 30 km offshore along the Rupert Coast, and is located about 240 km west of Shepard Island and 160 km west of Bowyer Butte, the northwesternmost inland nunatak described above.

Shepard Island is a mostly snow- and ice-covered, low-relief, 12 km-diameter volcanic island, with outcrops concentrated along the coast. The island has four main outcrop areas: Mathewson Point, Mount Petinos, Worley Point and Mount Colburn. The rocks are dominated by phreatomagmatic tuff cone sequences erupted between 500 and 400 ka. Our observations confirm and amplify those by LeMasurier (1990g), and show that Mathewson Point is a remnant of a monogenetic basanite tuff cone composed of well-bedded and cross-bedded lapilli tuff. Other features include abundant ash-coated lapilli and small bombs (up to 12 cm) and with bedding-plane sags. The 0.8 km-diameter tuff cone rim at Mathewson Point is well preserved with inward- and outward-dipping beds. A late-stage subaerial lava is ponded in the centre of the tuff cone crater. A lava clast from within the lapilli tuff beds is $^{40}Ar/^{39}Ar$ dated to 0.56 ± 0.14 Ma and the ponded lava is dated to 0.481 ± 0.049 Ma (Table 10). The lapilli tuff and ponded lava are likely to have the same age of about 481 ka.

A c. 500 m-thick sequence of palagonitized hawaiite volcaniclastic deposits and lava extends from sea level to the summit of Mount Petinos. The lapilli tuff and tuff breccia exhibit many of the same textures as the Mathewson Point tuff cone, including planar bedding, accretionary lapilli and

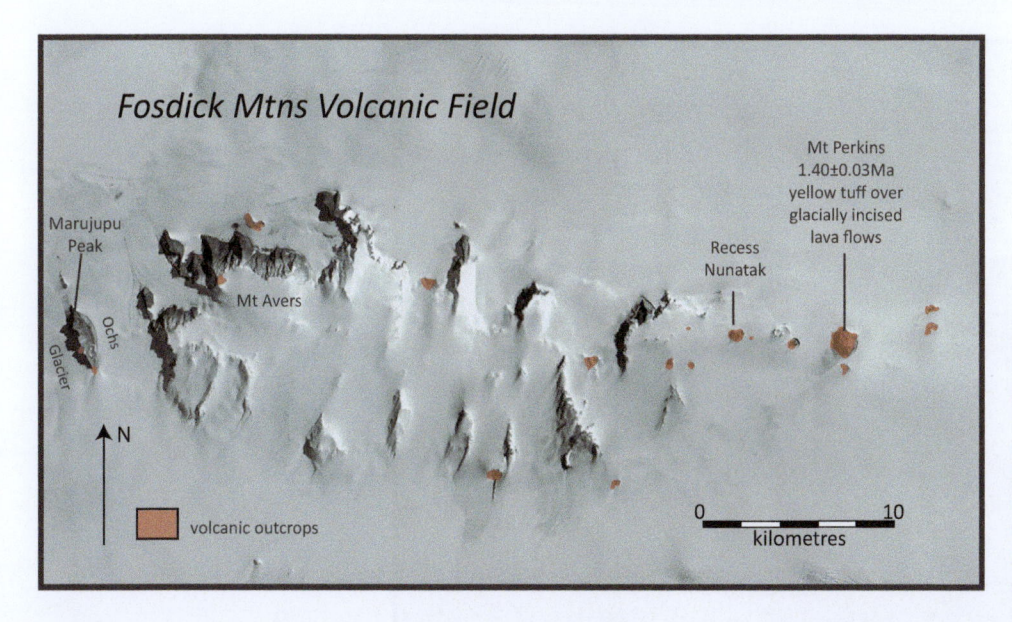

Fig. 35. Satellite image map of volcanoes of the Fosdick Mountains in western Marie Byrd Land. Volcanic outcrop locations are from Gaffney and Siddoway (2007). Image source: Google Earth Pro image accessed June 2019; see the caption to Figure 7 for more information on the source and an explanation of image processing.

bedding-plane sags, suggesting that the sequence resulted from phreatomagmatic tuff cone eruptions. A lithic clast in the lapilli tuff was ^{40}Ar/^{39}Ar dated to 0.41 ± 0.09 Ma (Table 10). Undated trachyte lava is exposed at Mount Colburn, the highest point on Shepard Island. Worley Point consists of trachyte lava, K–Ar dated to 0.43 ± 1.2 Ma (LeMasurier 1990g). The Mount Colburn trachyte may be coeval with a trachyte lava collected at Worley Point.

Grant Island is a mostly snow- and ice-covered, large (35 × 20 km), low-relief island that has a single outcrop situated on Mount Obiglio, the high point (510 m asl) of the island. Mount Obiglio is a small isolated cone composed of palagonitized Strombolian tephra and K–Ar dated to 0.7 ± 0.2 Ma (LeMasurier 1990g).

Cruzen Island is a snow- and ice-capped volcanic island, 1.7 × 2.2 km in diameter. The island is interpreted as a 200 m-high, flat-topped, basalt tuya K–Ar dated to 2.75 ± 0.26 Ma (LeMasurier 1990g). Cliff outcrops expose 50–75 m of hyaloclastite overlain by horizontal subaerial lavas. A passage zone from subglacial to subaerial palaeoenvironment is at least 75 m asl and suggests expanded ice cover in the late Pliocene.

Fosdick Mountains Volcanic Field

The Fosdick Mountains Volcanic Field is located in the Ford Range in far western MBL (Fig. 1). The volcanic field consists of at least 18 separate Quaternary (?) small volcanic centres, many of which are exposed on top of Paleozoic and Mesozoic bedrock (Fenner 1938; LeMasurier and Wade 1990; Gaffney and Siddoway 2007) (Fig. 35). During the course of extensive bedrock and structural geological studies of the Ford Range (Siddoway et al. 2004), many volcanic outcrops were visited and two geochemical studies (Gaffney and Siddoway 2007; Chatzaras et al. 2016) provide information on some of the volcanic geology.

Gaffney and Siddoway (2007) described several of the volcanic outcrops and noted that many volcanic sequences overlie periglacial features and glacial deposits, post-dating glacial event(s). They also noted that many of the rocks preserve delicate features, suggesting they have not been overridden by wet-based glacial ice (Gaffney and Siddoway 2007). In the Ochs Glacier area, volcanic necks and irregular conduits intrude bedrock (Gaffney and Siddoway 2007), including an eroded diatreme near Marujupu Peak (Chatzaras et al. 2016). At Mount Avers, thin vesicular pāhoehoe basanite lavas unconformably overlie the glaciated bedrock summit area (871 m asl) and descend to 610 m asl (C. Siddoway pers. comm.). An array of subvertical, 0.3–1.5 m-wide, dykes exposed over 500 m of vertical elevation cuts through the Mount Avers bedrock and have similar compositions to the summit lavas; the dykes are interpreted as feeder dykes for the summit lava (Gaffney and Siddoway 2007). At Mount Perkins, a c. 200 m sequence of 2–4 m-thick, massive (subaerial?) basanite lavas is capped by lava breccia. The Mount Perkins lava sequence is glacially incised and unconformably overlain by golden-tan-coloured (palagonitized?) tephra layers that dip inwards to a summit depression. The Mount Perkins sequence is tentatively interpreted here as a glacially-eroded subaerial lava sequence overlain by a tuff cone. Recess Nunatak consists of black, vesiculated, vertically-orientated flow-banded, glassy tephra interpreted as a glacially-eroded remnant of a vent complex (Gaffney and Siddoway 2007). Recess Nunatak is the only known site in the Fosdick Mountain Volcanic Field with crustal xenoliths. Many of the nunataks have ultramafic mantle xenoliths, which have been analysed for geochemistry and strain fabrics (Gaffney and Siddoway 2007; Chatzaras et al. 2016).

Two samples from Mount Perkins yielded indistinguishable ^{40}Ar/^{39}Ar ages of 1.41 ± 0.04 and 1.40 ± 0.03 Ma (Table 11); these new ages are considerably younger and more precise than previous K–Ar ages of 4.7 ± 1.0 and 3.5 ± 0.6 Ma (LeMasurier and Wade 1990). Otherwise, rocks of the Fosdick Mountain Volcanic Field are undated.

The Fosdick Mountains Volcanic Field offers the potential for detailed lithofacies and dating studies, similar to those

Table 11. *Fosdick Mountains summary age table*

Sample ID	Method	Corrected preferred age ± 2 SD (Ma)	Description	Ref.
Fosdick Mountains				
28	K	3.5 ± 0.6	Mount Perkins	5
30	K	4.7 ± 1.0	Mount Perkins	5
90207a	A	1.41 ± 0.04	lava, Mount Perkins	21
90207b	A	1.40 ± 0.03	lava, Mount Perkins	21
	A	**1.41 ± 0.03**	**Mean age (*n* = 2)**	21

Notes: method A is ^{40}Ar/^{39}Ar; method K is K/Ar.
Ref. code: 5, LeMasurier and Rex (1982); 21, this study. See also Supplementary Material Table S1.

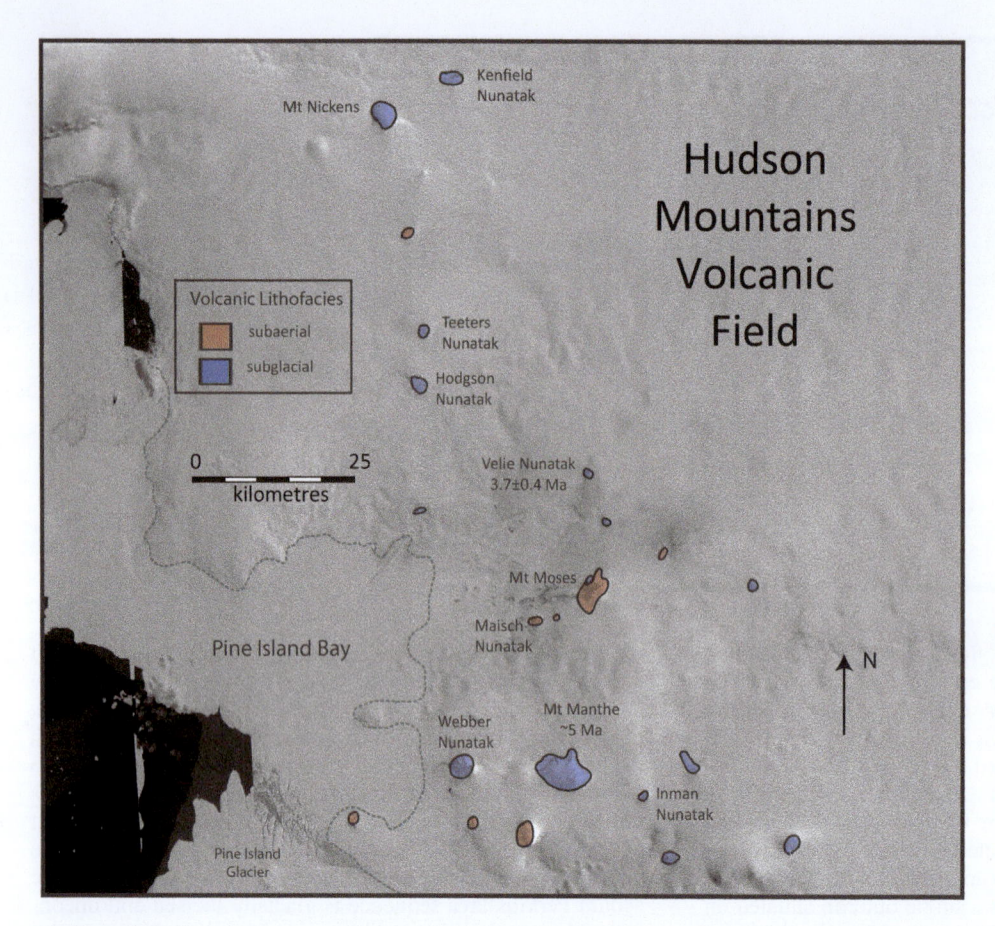

Fig. 36. Satellite image map of volcanoes of the Hudson Mountains Volcanic Field of the Thurston Island Volcanic Province. Volcanoes are designated subaerial or subglacial/subaqueous; adapted from Rowley *et al.* (1990). Image Source: Google Earth Pro image accessed June 2019; see Figure 7 caption for more information on source and an explanation of image processing.

undertaken by Wilch and McIntosh (2007) in the Hobbs Coast Volcanic Field. Such work would be likely to provide additional valuable information on the past extent of the WAIS at the time of the eruptions.

Thurston Island Volcanic Province

The Thurston Island Volcanic Province is located in western Ellsworth Land on the Thurston Island tectonic block, and includes the Hudson Mountains and Jones Mountains volcanic fields (Fig. 2). Many of the outcrops in the Hudson and Jones Mountains volcanic fields have been studied in reconnaissance, mostly in the 1960s (Rowley 1990; Rowley *et al.* 1990); detailed outcrop descriptions and geochronology data are not available.

Hudson Mountains Volcanic Field

The Hudson Mountains Volcanic Field consists of around 20, mostly snow- and ice-covered, volcanic nunataks at 200–750 m asl, located near the Walgreen Coast and just north of Pine Island Glacier in western Ellsworth Land (Fig. 36). No basement rocks are exposed (Rowley *et al.* 1990).

Previous workers suggested there are three, mostly ice-covered, major volcanoes at Teeters Nunatak, Mount Moses and Mount Manthe, which are overlain by smaller parasitic volcanoes (see Rowley *et al.* 1990). Analysed rocks range from tephrite to hawaiite in composition (Rowley *et al.* 1990). Three general lithofacies have been noted: a lower pillow lava and associated hyaloclastite tuff (as much as 60 m thick) observed at Mount Nickens; a middle subhorizontal to

Table 12. *Thurston Island Volcanic Province summary age table*

Sample ID	Method	Corrected preferred age ± 2 SD (Ma)	Description	Ref.
Hudson Mountains Volcanic Field				
Velie Nunatak				
28·3A	K	3.7 ± 0.40	Lava within hyaloclastite tuff sequence	5
Mount Manthe				
42·6A	K	5.0 ± 0.6	Clasts from near the top of a 200 m hyaloclastite section	5
42·5A	K	4.7 ± 0.4		5
42·4A	K	5.1 ± 0.6		5
42		**4.9 ± 0.6**	Mean age (*n* = 3)	
H-6	K	5.6 ± 3.8	Subaerial cap of a 200 m hyaloclastite section	5
H-4	K	8.7 ± 2.0	Clast within hyaloclastite roughly 50 m below the top	5
H-2	K	5.0 ± 3.2	Lava in the middle of a hyaloclastite tuff section	5
Jones Mountains Volcanic Field				
	A	7.68 ± 0.11	Lava above basal pillows and unconformity from a peak west of Forbidden Rocks	14

Notes: method A is $^{40}Ar/^{39}Ar$; method K is K/Ar.
Ref. code: 5, LeMasurier and Rex (1982); 14, Rutford and McIntosh (2007). See also Supplementary Material Table S1.

steeply-dipping stratified hyaloclastite tuff (as much as 350 m thick); and upper subhorizontal (including pāhoehoe) lavas (as much as 60 m thick) observed at Maish Nunatak, Mount Moses and Mount Manthe (see the map in Rowley et al. 1990). The 'hyaloclastite' tuffs are sideromelane rich and interpreted as subglacially erupted, and the upper lava is interpreted as suberially erupted and emplaced (Rowley et al. 1990). Samples from two sites, Mount Manthe and Velie Nunatak, have conventional K–Ar dates (Table 12). The most precise ages for Mount Manthe come from clasts within a hyaloclastite with a weighted mean age of 4.9 ± 0.6 Ma. The age of a lava from within a hyaloclastite at Velie Nunatak is 3.7 ± 0.4 Ma. Late Miocene ages (without errors) were reported by Lopatin and Polyakov (1974) for two other nunataks.

There is great potential for new field and analytical work in the Hudson Mountains. Smellie and Edwards (2016) noted that the edifice shapes and described lithofacies are like those found in tuyas. Our Figure 36 is adapted from the Rowley et al. (1990) map and designates outcrops by lithofacies as either subaerial or subglacial/emergent. We agree that at least some of the subglacial/emergent lithofacies are associated with tuyas. It is likely that a well-documented and dated record of ice–volcanic interactions would provide valuable syneruptive records of past ice-sheet thicknesses.

Jones Mountains Volcanic Field

The Jones Mountains Volcanic Field is located about 150 km NE of the Hudson Mountains Volcanic Field near the Eights Coast in western Ellsworth Land (Fig. 2). The mountains are mostly snow- and ice-covered, and include numerous nunataks that rise up to about 1500 m asl, up to 1000 m higher than the adjacent ice sheet (Fig. 37). The geology of the Jones Mountains has only been mapped in reconnaissance.

The nunataks in the north and east parts of the field consist of Mesozoic basement granitoids at lower elevations, overlain unconformably by alkali basalts on the peaks; in the south only, alkali basalts are exposed (Rowley 1990 after Craddock et al. 1964). There is widespread evidence of glaciation along the unconformity (Craddock et al. 1964; Rutford et al. 1968, 1972). Although thick sections (500–700 m) of basaltic rock are found in outcrop and numerous nunatak outcrops are shown on sketch maps (e.g. Rowley 1990), detailed lithofacies mapping and characterization have not been published.

Near Avalanche Ridge and Pillsbury Tower in the centre of the Jones Mountains, about 500–700 m of Miocene alkali basalt volcanic rocks overlie an unconformity cut into Mesozoic-aged basement rocks. The rocks were described as pillow lavas and nearly horizontal lapilli tuffs by Craddock et al. (1964). Hole et al. (1994) characterized the rocks as massive, highly-vesiculated pillow lava and palagonitized volcaniclastic rocks, subdivided into two units. Each unit has a c. 10 m-thick base composed of variably-stratified, cross-bedded and reworked volcaniclastic tuff and lapilli tuff. These basal units are interpreted as reworked mass flow deposits. The volcanic rocks are interpreted as subglacially erupted. Numerous attempts at K–Ar dating the basal pillow lavas yielded a wide range of ages ranging from 332 to 6 Ma (Table 12), and Rutford et al. (1972) suggested that the most reliable age for Jones Mountains volcanic rocks is between 12 and 7 Ma. The most recent and most reliable determination is a $^{40}Ar/^{39}Ar$ plateau age of 7.68 ± 0.11 Ma (Table 12) from a lava sample taken above the basal sequence from an unnamed nunatak 30 km west of Forbidden Rocks (Rutford and McIntosh 2007). The exact location is uncertain but the basanite had a higher K_2O content than most samples from the Jones Mountains (Rutford et al. 1972).

Lenses of diamictite with striated and faceted erratic clasts are found along the basement unconformity (Rutford et al. 1972; Hole et al. 1994; Rutford and McIntosh 2007). The matrix of the diamictite is composed of volcanic glass and palagonite. Craddock et al. (1964) and subsequent workers have interpreted the diamictite as tillite. The nearly horizontal unconformity extends for 33 km and has a maximum estimated relief of c. 50 m (Rutford and McIntosh 2007). The unconformity is a polished and planed surface with striations and chattermarks (Craddock et al. 1964; Rutford et al. 1972). The unconformity is interpreted as a glacial unconformity that formed coeval with the eruption of the overlying glaciovolcanic sequence and deposition of volcanic-rich diamictite at c. 7.7 Ma (Table 12) (Rutford and McIntosh 2007).

Rutford and McIntosh (2007) noted that the interpretations of the Jones Mountains by Craddock et al. (1964) provided the first well-documented evidence for pre-Quaternary glaciations in Antarctica. There is a need of more detailed fieldwork in the Jones Mountains Volcanic Field, both at the unconformity locations and the several other volcanic nunataks. Establishing a high-resolution chronology will be challenging as the Jones Mountains rocks are largely basaltic with low K_2O contents and glassy textures.

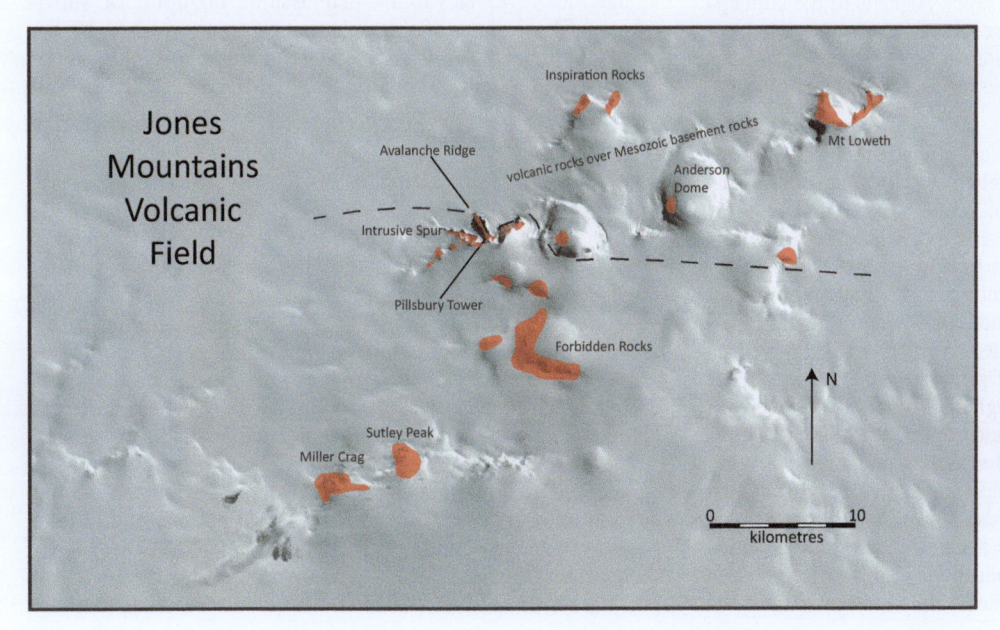

Fig. 37. Satellite image map of volcanic outcrops (red) in the Jones Mountains Volcanic Field of the Thurston Island Volcanic Province. The dashed line separates the southern nunataks that have all volcanic rocks from northern nunataks that are basement rocks overlain by volcanic rocks on peaks. At Avalanche Ridge, the dashed line marks the glacial unconformity between Mesozoic basement and late Cenozoic subglacially emplaced volcanic rocks. Adapted from Rowley (1990), after Craddock et al. (1964). Image source: Google Earth Pro image accessed June 2019; see the caption to Figure 7 for more information on the source and an explanation of image processing.

Synthesis of volcanic records of glaciation in the WAIS

Volcanism and glaciation have been active geological processes in West Antarctica since middle Cenozoic time. As discussed in the individual volcano summaries, the MBLVG volcanic sequences provide snapshot records of eruption and emplacement conditions that can be used to provide syneruptive constraints on ice thickness and palaeoenvironmental conditions (Smellie 2018). A limitation of applying this approach in a currently glaciated area such as West Antarctica is that only records of ice levels that are thicker or equal to today's ice level can be observed. LeMasurier (1972) and LeMasurier and Rex (1982, 1983) reconstructed volcanic records of glaciation in West Antarctica, based on regional reconnaissance fieldwork and K–Ar geochronology. Other studies have provided updated records at selected MBL volcanoes (McIntosh et al. 1991; LeMasurier et al. 1994; Wilch et al. 1999; Smellie 2001; LeMasurier 2002; Wilch and McIntosh 2000, 2002, 2007; LeMasurier and Rocchi 2005). Here we develop a regional synthesis of glaciovolcanism in the MBLVG through time that merges palaeoenvironmental reconstructions and geochronology results.

In the introduction to this paper, we defined the common lithofacies at MBLVG volcanoes and interpreted them as subglacial (or ice-contact), emergent or subaerial to infer the eruptive and depositional palaeoenvironment and the influence of external water in the eruption dynamics. Smellie and Edwards (2016) provided a summary of the lithofacies approach. Detailed lithofacies descriptions and interpretations are a key first step in using the volcanic deposits to reconstruct past ice extent or conditions. In addition to lithofacies interpretations, there are other complexities of the West Antarctic glaciovolcanic environment that are critical to consider in reconstructing the volcanic record of glaciation in West Antarctica (Wilch 1997; Wilch and McIntosh 2000, 2002, 2007). These complexities are explained and addressed below in a conceptual model.

A conceptual model for reconstructing palaeo-ice levels from the volcanic record

Here we review some general concepts and develop a conceptual model that addresses some of the complexities of the glaciovolcanic environment of West Antarctica. A major objective of the model is to differentiate records of local ice level from those of larger WAIS level. Aspects of the conceptual model are illustrated in Figure 38a–f and discussed below. Outcrop examples listed in Figure 38 were introduced in the volcano summaries and are discussed in further detail here.

The traditional approach. Lithofacies analysis and interpretations of palaeoenvironments in relation to local ice conditions provide the foundation of these process-orientated studies. Early compilations of glaciovolcanism in MBL (e.g. LeMasurier and Rex 1982, 1983) relied heavily on general outcrop characteristics, typically interpreting rocks that exhibit any evidence of external water interactions (palagonitization, pillow lava, hyaloclastites, hydroclastic tuff) as evidence of formerly higher ice-sheet levels. If possible, direct evidence of glaciation, such as a tillite or a glacially-incised surface in a volcanic sequence, was used to add confidence to the interpretations. Hyaloclastite and lava-fed deltas and tuyas provide additional evidence of glaciovolcanic interactions (LeMasurier 2002; Smellie and Edwards 2016). Ideally, tuyas or lava-fed deltas preserve passage zones that can provide 'dipstick' measurements of syneruptive palaeo-ice levels (Fig. 38a). Lithofacies analyses and interpretations of glaciovolcanic sequences have become more detailed and nuanced over the past 30 years, with considerable advances in reconstructing the thickness and thermal properties of the glacial environment (Smellie and Edwards 2016). Although detailed lithofacies analysis is critical for understanding palaeoenvironments, an integrative reconstruction of volcanic records of ice-sheet glaciation also requires an understanding of the palaeoenvironments in a regional ice-sheet context.

The uplift caveat. One complication of the volcanic record of the WAIS is that all palaeo-ice-level elevations are relative to modern elevations and do not account for possible uplift or subsidence of volcanic centres (Fig. 38a). This uplift caveat is problematic in many terrestrial ice-sheet reconstructions and is a particular problem for reconstructions of older (pre-Quaternary) glaciations (e.g. Wilch et al. 1993). LeMasurier and Landis (1996) suggested that the progressively higher elevations of the pre-volcanic basement unconformity and older volcanoes towards the centre of the MBLVP are a surface expression of regional structural domal uplift that has been ongoing since inception of volcanism in middle Cenozoic times. Rocchi et al. (2006) suggested that the early Oligocene intrusion at Dorrel Rock near present-day Mount Murphy was rapidly exhumed (500 m Ma^{-1}) and exposed at the surface between 34 and 27 Ma by fluvial and glacial erosion. According to their model, uplift beginning in the centre of the province near Mount Petras has proceeded since that time and has far outpaced erosion. Alternatively, the pre-volcanic unconformity may be a time-transgressive erosion surface that appears to be very flat in some places (e.g. Bowyer Butte) and quite rugged in other places (e.g. Mount Petras exhibits 400 m of relief and Mount Murphy exhibits 900 m of relief: McIntosh et al. 1991; Wilch and McIntosh 2000; Smellie 2001). Although the pattern of higher unconformities and older volcanoes towards the centre of the volcanic province is real, there are limited constraints on the timing and amount of (basement) surface uplift at any location. Wilson et al. (2013) conducted climate–ice sheet modelling experiments to account for calculated ice growth in Antarctica at the Eocene–Oligocene transition, and concluded that the surface topography of West Antarctica may have declined significantly due to glacial erosion and subsidence. Since the Oligocene the topography of West Antarctica has also responded glacioisostatically during glacial–deglacial cycles. In the absence of quantitative geological constraints on uplift or subsidence of MBLVG volcanoes, we propose tentative palaeo-ice-level elevations that assume no uplift or subsidence. The impact of this assumption is diminished by the fact that much of our WAIS reconstruction relies on the younger part of the volcanic history of the MBLVG (since 10 Ma) and the most definitive estimates of past ice level are in late Quaternary sequences.

Volcanoes as ice-flow obstructions. The large polygenetic central volcanoes of the MBLVP are obstructions to regional WAIS ice flow, and produce higher ice levels upstream and lower ice levels downstream (Wilch et al. 1999; Wilch and McIntosh 2002) (Fig. 38b; Table 2). This ice-damming effect can give a false impression of palaeo-ice levels recorded in glaciovolcanic sequences. An example of the ice-damming effect is the coastal volcano, Mount Murphy, where upstream ice levels are 400–600 m higher than downstream ice levels. Elsewhere in MBLVP, many of the central volcanoes have coalesced to form linear ranges. Mount Berlin and Mount Moulton in the Flood Range form a continuous >60 km-long barrier to ice flow, with upstream ice levels 400–800 m higher than downstream ice levels. The tuya at Brandenberger Bluff on the north flank of Mount Berlin is >2 Ma older than the

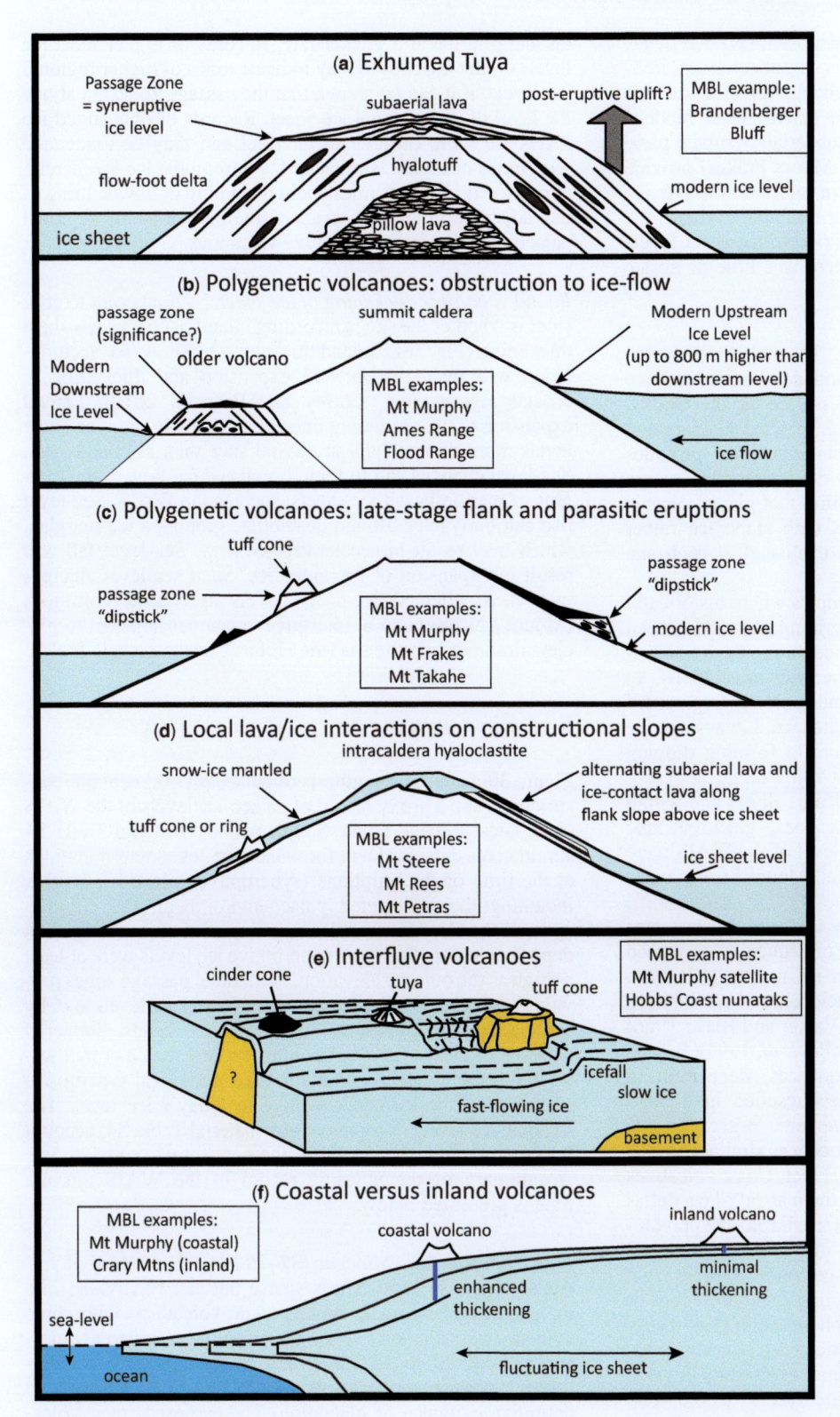

Fig. 38. Cartoons illustrating the key ideas of the conceptual model for interpreting the WAIS history from volcanic sequences. Details are described in the text.

Mount Berlin polygenetic shield volcano and may have formed as an isolated nunatak. The passage zone at 250 m above modern local ice level may have formed when the regional ice-sheet level was similar to or lower than it is today. The more isolated MBLVP central volcanoes produce less significant ice-damming effects: for example, ice-sheet elevations surrounding the isolated inland Mount Takahe polygenetic volcano are <300 m higher on the upstream side.

Early-stage v. late-stage eruptions. Interpretations of early edifice-building sequences can be complicated by issues related to changing ice-flow patterns and ice-sheet levels during volcano construction. At many volcanoes, edifice-building outcrops are not exposed on the lower flanks of the large central volcanoes and problems relating to ice damming can be ignored. In other cases, some of the oldest rocks are exposed at the base of polygenetic volcanoes and palaeo-ice-level interpretations must account for the ice-damming effect (Fig. 38b).

Post-edifice-building flank eruptions and parasitic monogenetic eruptions that occur on pre-existing shields provide less-complicated palaeo-ice-level indicators (Fig. 38c). Presumably, ice-damming effects produced by the pre-existing shield had a similar effect on local ice-flow patterns and elevations as they do today. In flank deposits, lava-fed deltas with passage zones from subaqueous to subaerial environments (e.g. Mount Takahe) provide reliable 'dipstick' records of palaeo-ice levels that can used in glaciological models. Conversely, the lowest elevations of subaerially-erupted parasitic volcanoes (e.g. English Rock at Mount Frakes) provide limits on the maximum elevation of the syneruptive ice sheet at those locations. An important reminder is that these subaerial sequences indicate only maximum ice-sheet levels and may be recording times when there was little or no ice sheet developed in West Antarctica.

Volcanic interactions with slope ice. The volcanoes of the MBLVG are almost completely mantled in snow and ice of variable thickness ranging from no ice at the limited volcanic outcrops to probably >100 m in local glaciers and ice-filled calderas. This mantle of snow and ice provides ample opportunities for glaciovolcanic interactions above the level of the ice sheet. The possibility that glaciovolcanic sequences resulted from interactions with slope ice rather than the regional ice sheet must be evaluated at each site (Fig. 38d).

There are many non-Antarctic examples where historically erupted lavas have been observed overriding ice, cutting open channels into ice and tunnelling into ice on the flanks of volcanoes (see examples in Smellie and Edwards 2016). In these cases, the lavas were quenched and generally fragmented where they came into contact with melted ice. Lava–slope ice interactions offer conditions conducive to forming dipping 'passage zones' where conditions vary from subglacial to subaqueous to subaerial on both open volcanic slopes and within meltwater tunnels. In MBLVP volcanoes, glaciovolcanic interactions with slope ice above the level of the WAIS were dominant processes at Mount Rees and Mount Steere in the Late Miocene.

Models of eruptions beneath valley-confined glaciers provide possible analogues for the type of sequences produced on volcanoes with a moderately thin ice mantle (≤100 m). Eruptions beneath valley-confined glaciers have been inferred from volcanic sequences in Iceland (Walker and Blake 1966) and Alexander Island, Antarctica (Smellie *et al.* 1993). In both Icelandic and Alexander Island sequences, deposition in thermally-excavated subglacial tunnels resulted in a postglacial landform in the form of a volcanogenic 'esker'. Smellie *et al.* (2011*a*, *b*) describe volcanic sequences similar those at Mount Rees and Mount Steere from Hallett Coast volcanoes in North Victoria Land, and interpret them as 'a'ā-lava deltas formed beneath slope-mantling glacial ice. Explosive phreatomagmatic eruptions on volcano slopes may have an associated lava phase and produce dipping passage zones (Smellie *et al.* 1993) or they may simply produce monogenetic tuff cones or rings without an effusive pillow lava phase. Sohn (1996) suggested that phreatomagmatic pyroclastic density current eruptions are typically associated with deeper explosions in weak substrates, where water has only limited access to the vent. Such environments may have been common on the slopes of active Antarctic volcanoes.

Volcanoes on interfluves between glaciers and ice streams. Small volcanoes perched on interfluves between areas of faster flowing ice may exhibit records of changing local ice levels that give false impressions of the magnitude and elevations of regional ice-level variations (Fig. 38e). Examples of interfluve volcanoes include the satellite nunataks at Mount Murphy, and inland volcanic centres in the Hobbs Coast Volcanic Field and Fosdick Mountains Volcanic Field. Regional growth of the ice sheet may not cause higher ice levels at these nunataks if the local ice streams compensate by discharging ice at a greater rate. Records of higher local ice levels on the interfluves may indicate times of higher regional ice levels, if it can be shown that the passage zones lie above the level of the regional ice sheet. Records of unchanged ice levels are more difficult to interpret and may be associated with times of higher, similar or lower regional ice levels relative to today. Uncertainties about the timing of downcutting of the adjacent drainages weaken interpretations of regional palaeo-ice level using interfluve volcanoes.

Inland v. coastal thickening of ice sheet. A final point to consider is whether the site is recording inland or coastal ice-sheet thickening (Fig. 38f). Inland thickening by snow/ice accumulation will result in outward expansion and thickening in coastal regions (e.g. Cuffey and Paterson 2010). Lateral expansion of the grounding line will effectively raise ice-sheet levels more dramatically at coastal sites than at inland sites. Sea-level changes can further complicate ice flow at the margins of marine-based ice sheets, such as the WAIS. Sea-level rise can buoyantly lift and destabilize grounded ice margins, which may result in accelerated ice flow. Sea-level fall will result in expansion of grounded ice. Such sea-level changes may have affected local ice levels at coastal volcanoes (Mount Murphy) and at interfluve volcanoes adjacent to glaciers draining into the sea (the Hobbs Coast Volcanic Field).

A summary of the volcanic record of the WAIS

Figure 39 provides a synthesis of palaeoenvironmental reconstructions and a proxy record of palaeo-ice levels of the WAIS since latest Eocene times. Supplementary material Table S4 summarizes data pertinent for palaeo-ice-level determinations at the time of the eruptions (syneruptive palaeo-ice levels), including the interpreted palaeoenvironment, the eruption age and the elevation relative to today's ice sheet. Subglacial deposits indicate that local syneruptive ice levels were at least as high as the outcrop elevation. Horizontal passage zones provide proxies for local syneruptive englacial lake levels and, by association, for minimum local palaeo-ice levels. Subaerial deposits suggest a 'dry' environment where no ice or meltwater was present and constrain the maximum local syneruptive elevation of the ice sheet relative to today's ice level. The regional ice levels (Supplementary material Table S4) account for complications discussed in the conceptual model above. A summary of the volcanic record of the WAIS through time is presented below.

Latest Eocene–Late Miocene (37–10 Ma). The record of volcanism from 37 to 10 Ma is sparse but can be divided into three intervals: sporadic monogenetic volcanism with some evidence of glaciation (37–20 Ma); a period of apparent quiescence from 20 to 13.4 Ma; and then large-scale caldera-forming polygenetic volcanism from 13.4 to 10 Ma with no definitive evidence of glaciation. The earliest known volcanism in the MBLVG occurred in the late Eocene (36.58 ± 0.22 Ma), when massive mugearite lava was erupted at Mount Petras (Wilch and McIntosh 2000). Mid-Oligocene (29–27 Ma) Surtseyan and Strombolian deposits at Mount Petras provide the first terrestrial evidence for glacial ice in Marie Byrd Land (Wilch and McIntosh 2000). These eroded tuff cone deposits overlie bedrock on an unconformity with over 400 m of relief, suggesting a local hilly or mountainous terrain at 29–27 Ma. Wilch and McIntosh (2000) interpreted the deposits to be products of intermittent contact of magma with external water, probably derived from the melting of a thin, local ice cap or ice and snow on bedrock slopes. Three other MBL volcanic outcrops have been dated to the 36–

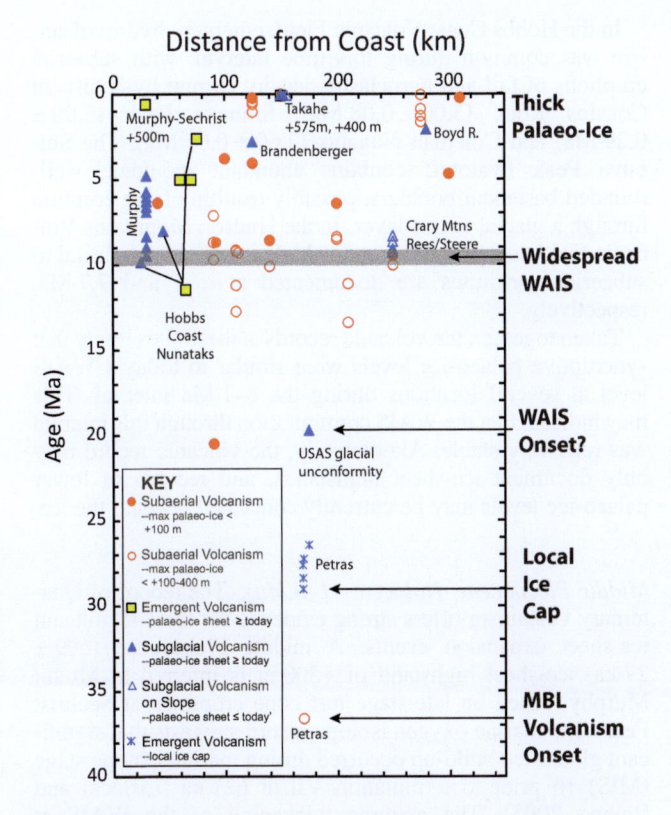

Fig. 39. Volcanic record of the WAIS in Marie Byrd Land with key events listed on the right. A plot of select eruption ages v. distance of the volcano from the coast. Subaerial volcanoes provide maximum syneruptive ice levels. Subaerial volcanoes listed as <100 m indicates that the outcrops are <100 m above the current ice-sheet level; those listed as <400 m include outcrops between 100 and 400 m above the current ice sheet level. Emergent eruptions are associated with Hobbs Coast Nunatak sites, near to present-day ice levels. Palaeo-ice-sheet ≥ today are sites with evidence of subglacial eruptions at or above today's ice level. At most sites, several caveats shown in Figure 37 apply, making it difficult to establish palaeo-ice levels with confidence. Palaeo-ice-level elevations are listed for late-stage eruption sites at Mount Takahe and Mount Murphy. Palaeo-ice sheet ≤ today refer to sites where there is evidence of sloping passage zones inferred to represent a thin mantle of ice on edifice side slopes.

20 Ma interval. A small hyalotuff outcrop situated at the level of the ice sheet near Mount Galla is dated to 26.40 ± 0.21 Ma and provides an additional suggestion of local late Oligocene glaciation, although by no means definitive. The volcanic record indicating Oligocene glaciers in West Antarctica is consistent with other proxy records from around Antarctica (e.g. Anderson 1999; Olivetti *et al.* 2015).

At Reynolds Ridge near the north end of Mount Flint, a 20.20 ± 0.08 Ma subaerial trachyte lava situated at the level of the ice sheet suggests that early Miocene syneruptive ice levels were lower than present levels. It is important to note that this and subsequent examples of subaerial deposits provide only maximum constraints and do not preclude the possibility that MBL was ice-free at the time of the eruptions. Palagonitized Strombolian tuff breccia deposits (19.68 ± 0.31) overlying striated bedrock at Mount Aldaz provide the first definitive indication for >19.7 Ma syneruptive ice levels similar to or lower than they are today.

Several polygenetic volcanoes in the Ames and Executive Committee ranges were apparently formed during the Middle–Late Miocene interval (13.4–10 Ma) interval, all of which are interpreted as consisting of subaerially-erupted, producing mostly felsic lavas and pyroclastic rocks (Fig. 39; see also Supplementary material Table S4). Many of the outcrops are situated near the level of the WAIS and suggest that maximum syneruptive WAIS levels were similar to or lower than today's level (Fig. 38: e.g. Mount Andrus, Mount Hampton and Mount Whitney). A small outcrop of palagonitized hyalotuff at Patton Bluff in the Hobbs Coast nunataks suggests limited hydrovolcanic interactions at 11.32 ± 0.25 Ma but does not provide definitive evidence of palaeo-ice.

Late Miocene (10–8 Ma). The Late Miocene interval from 10 to 8 Ma marks an apparent acceleration in polygenetic volcanism in the MBLVG and provides the first substantial evidence for a widespread WAIS. Late Miocene (10–8 Ma) glaciovolcanic sequences are exposed at the inland Crary Mountains (Mount Steere and Mount Rees), at Kennel Peak and Bowyer Butte near the Hobbs Coast and at coastal Mount Murphy volcano. The Mount Murphy main shield sequence records fluctuating syneruptive (mostly 9.34 ± 0.10–8.84 ± 0.13 Ma) ice–volcano interactions with both thin-ice and regional ice-sheet conditions at different times. The lava-fed delta sequences suggest palaeo-ice levels up to 300 m higher than today's local ice level. Relatively low-elevation striated glacial unconformities and interbedded tillites record fluctuating ice flow across the growing volcano during this interval.

Interpretations of WAIS palaeo-ice levels from the main shield outcrops at Mount Murphy are complicated by three factors. First, the main shield sequences are located on the west side of Mount Murphy, where ice is currently descending from high upstream levels at *c.* 800 m asl to low downstream levels of *c.* 200 m asl. The elevations of these sequences (up to *c.* 700 m asl) are about the same as the elevation of the regional ice sheet on the upstream side. Therefore, these outcrops may record fluctuations of an ice sheet that was smaller than today's ice sheet, assuming that the much smaller Mount Murphy produced a less significant damming effect. Second, because Mount Murphy is at the coast where glaciers are feeding into the Getz Ice Shelf, the local ice configurations may have been very responsive to changes in sea level. Third, the coastal position of Mount Murphy may have facilitated draining of ice-marginal lakes formed by volcanism, so passage-zone sequences may be significantly lower than the palaeo-ice level. By this scenario, the main shield sequences do provide strong evidence for higher than present local ice levels in the Late Miocene. However, the complex setting of Mount Murphy precludes making interpretations about regional palaeo-ice levels based solely on the main shield sequence. The sequence does indicate the presence of an ice sheet but the ice sheet may have been similar in thickness to today's ice sheet.

Coeval glaciovolcanic sequences at Mount Rees and Mount Steere in the Crary Mountains (9.46 ± 0.24–8.30 ± 0.22 Ma) are interpreted as evidence of slope-ice interactions and imply that abundant local slope ice extended to near or below the level of the modern ice sheet. These outcrops are also on the lee or downstream lower-elevation side of the volcanoes. The Mount Rees and Mount Steere reconstructions are consistent with the interpretations of Mount Murphy that suggest lower or unchanged syneruptive WAIS levels in the Late Miocene.

Monogenetic volcanoes at Bowyer Butte (9.82 ± 0.90 Ma by K–Ar: from LeMasurier and Thomson 1990) and Kennel Peak (8.01 ± 0.87 Ma) provide indications of Late Miocene syneruptive palaeo-ice levels that were 50–200 m higher than today's local ice levels (LeMasurier 1990d; Wilch and McIntosh 2007). Interpretations of regional ice-level changes from these local ice levels is complicated by the fact that both sites are situated on interfluves between glaciers that are descending steeply from inland areas to the Getz Ice Shelf. A further complication is that both volcanoes are situated along prominent linear scarps, almost certainly related to faults of unknown age (LeMasurier and Landis 1996). The pillow lava, hyaloclastite breccia, tillite and striated basement rocks do provide additional evidence for widespread glaciation in West Antarctica between 10 and 8 Ma.

During the 10–8 Ma interval of abundant glaciovolcanism, late-stage scoria cones and subaerial lava were erupted at Mount Flint (9.67 ± 0.20 and 8.66 ± 0.24 Ma), Mount Bursey (8.45 ± 0.046 Ma), Mount Andrus (9.28 ± 0.06 Ma) and Kouperov Peak (9.24 ± 0.06 Ma). These subaerially-erupted deposits provide maximum elevation limits on Late Miocene palaeo-ice levels ranging from <50 to ≤300 m above present ice level (Supplementary material Table S4). Poorly-exposed, subaerially-erupted, edifice-building sequences in the Executive Committee Range Volcanic Field (Mount Cumming (10.0 ± 1.0 Ma), Mount Hartigan (8.50 ± 0.66 Ma)) and the Ames Range Volcanic Field (Mount Kosciusko (9.20 ± 1.3 Ma)) provide local limits on syneruptive palaeo-ice levels at <200–250 m above present local ice level (all ages by K–Ar method from LeMasurier and Thomson 1990). Finally, at Kay Peak on Mount Murphy (8.9–6.9 Ma), late-stage glaciovolcanic sequences record interactions between lavas and valley glaciers above the level of the regional ice sheet.

In summary, the record of 10–8 Ma volcanism in MBL provides substantial evidence for a widespread glaciation of West Antarctica by 9 Ma. Assuming no uplift/subsidence, the palaeo-ice levels appear to have been similar to today's level at several times during this interval. The possibility of much higher or lower ice levels during this interval cannot be verified or contradicted by the volcanic record. The volcanic record of a widespread ice sheet in West Antarctica in the Late Miocene is consistent with global cooling and ice-sheet growth recorded in Late Miocene benthic foraminifera records (e.g. Holbourn *et al.* 2013). ANDRILL 1B marine-sediment-core records from the Ross Sea show direct evidence of cyclic glaciation during this interval (Wilson *et al.* 2012).

Latest Miocene–Middle Pleistocene (c. 8–1 Ma). Polygenetic volcanism, late-stage flank eruptions on older volcanoes and isolated monogenetic volcanism persisted through the latest Miocene, Pliocene and into the early Pleistocene. Monogenetic volcanism at Brandenberger Bluff north of Mount Berlin, at the satellite nunataks and on the flanks of Mount Murphy, at several nunataks of the Hobbs Coast, Hudson Mountains, and Jones Mountains volcanic fields provides abundant evidence for glaciovolcanism between 7.8 and 2.6 Ma. Passage zones in stacked lava-delta sequences at Turtle Peak, Hedin Nunatak, Icefall Nunatak near Mount Murphy and in the emergent Brandenberger Bluff tuya at Mount Berlin record local syneruptive palaeo-ice-level elevations as high as 350 m above today's local ice level. The nunataks at Mount Murphy are situated on interfluves at about the same elevation as the main shield outcrops described in the previous subsection. Extracting accurate WAIS syneruptive palaeo-ice-level elevations from these local ice levels is complicated because small changes in WAIS elevations or local erosion or fault displacement at the interfluve margins could have had dramatic effects on local ice level at the interfluve. In addition, ice flow in these localities may have been strongly affected by sea-level fluctuations.

The Brandenberger Bluff tuya (2.77 Ma) also has a passage zone at 250 m above local ice level. However, it is on the lee side of Mount Berlin, about 150 m lower than modern regional ice levels on the upstream side of Mount Berlin, implying a reduced WAIS at 2.77 Ma. This assumes that the regional ice configuration and elevations prior to growth of Mount Berlin were similar to today's configuration.

In the Hobbs Coast Volcanic Field, emergent hydrovolcanism was common during this time interval, with subaerial eruptions of fall and pyroclastic density current hyalotuffs at Cousins Rock (5.00 ± 0.08 Ma), Shibuya Peak (5.00 ± 0.29 Ma) and Coleman Nunatak (2.60 ± 0.01 Ma). The Shibuya Peak hyalotuff contains abundant moulded, well-rounded basement boulders, possibly resulting from eruption through a glacial debris layer. In the Hudson Mountains Volcanic Field and Jones Mountains Volcanic Field subglacial to subaerial transitions are documented at *c.* 5 and 7.7 Ma, respectively.

Taken together, the volcanic records at these sites imply that syneruptive palaeo-ice levels were similar to today's WAIS level at several locations during the 8–1 Ma interval. This may indicate that the WAIS configuration through this interval was relatively stable. Alternatively, the volcanic record may only document ice-sheet highstands, and records of lower palaeo-ice levels may be currently concealed beneath the ice.

Middle Pleistocene–Holocene (1–0 Ma). The record of Quaternary volcanism offers strong evidence for three significant ice-sheet expansion events. A middle Pleistocene (609 ± 27 ka) ice-sheet highstand of +500 m is inferred at Mount Murphy, based on late-stage tuff cone eruptions at Sechrist Peak. The marine oxygen isotope record suggests that significant global-ice build-up occurred during marine isotope stage (MIS) 16 prior to termination VII at 621 ka (Lisiecki and Raymo 2005). The extreme thickening of the WAIS at Mount Murphy may have been driven, at least in part, by sea-level lowering and ice-sheet thickening as the ice sheet expanded onto the exposed continental shelf.

At about the same time (578 ± 9 ka), edifice-building subaerial lava and pyroclastic rocks were erupted from Merrem caldera, on the west flank of Mount Berlin. These rocks imply that syneruptive local ice was at a lower elevation than the lavas. The lowest elevation lava (*c.* 1850 m asl) is situated about 200 m higher than local ice levels but about 200 lower than the upstream level of the WAIS, in an area where present-day ice is descending from high upstream levels of about 2000 m asl to much lower downstream levels of about 1400 m asl. Although it is difficult to place an absolute elevation on the regional palaeo-ice level, it is probable that syneruptive ice-sheet thickening at below Merrem caldera was much less extreme than at Sechrist Peak. Alternatively, these two eruptions may have occurred in different parts of a glacial–interglacial cycle given the age and age uncertainties of the two events.

The Late Quaternary records from Mount Takahe are even more compelling because they show significant ice-sheet thickening at MIS 4 (71 ka) and MIS 2 (18 ka). The Stauffer Bluff tuya, dated to 65 ± 7 ka, forms a table-shaped edifice that rises 525 m above the ice-sheet level on the NE side of Mount Takahe. The highest-elevation, subglacially-emplaced, pillow hyaloclastites, just below the break in slope and inferred passage zone, are situated almost 400 m above the regional ice sheet. The Stauffer Bluff sequence suggests that the WAIS was at least *c.* 400 m higher than today at 65 ka, when global temperatures and sea level were lower and the global-ice volume was higher.

Passage-zone sequences on the lower flanks of Mount Takahe at Gill Bluff and Möll Spur record syneruptive ice-sheet highstands coincident with MIS 2. The post-edifice-building passage zones developed as subaerial lavas flowed into ice-marginal lakes forming lava-fed deltas on the lower flanks of Mount Takahe. The upper horizontal passage zone at Gill Bluff indicates a minimum former ice-sheet level at *c.* 413 m above the present ice-sheet surface at 21.9 ± 2.0 ka. The passage zone at Möll Spur indicates a minimum former ice-sheet level at c. 575 m above the present ice-sheet surface at 18.6 ± 4.4 ka. The lower and older pillow sequences at Möll Spur indicate that ice levels were at least 300 m above present ice level at 36.0 ± 5.0 ka. These post-edifice-building passage-zone sequences at Mount Takahe provide the first glaciovolcanic ice-sheet-level records correlated to the marine isotopic stage timescale.

Limits on the duration of a Last Glacial Maximum ice-sheet highstand at Mount Takahe are provided by lower-elevation (c. 150 m above ice level) subaerial deposits at Oeschger Bluff (7 ± 13 ka) and Steuri Glacier (45 ± 7 ka). Late-stage parasitic scoria cone deposits at English Rock, Mount Frakes (34 ± 14 ka) limit syneruptive ice-sheet expansion to <150 m in the Crary Mountains.

Summary and conclusions

The MBLVG includes two volcanic provinces and 11 geographically distinct volcanic fields, with 19 large central polygenetic alkaline shield-like composite volcanoes and numerous smaller isolated volcanoes that rise above the WAIS. Our understanding of the extent and history of the MBLVG remains fragmentary, largely due to a lack of dissection, extensive local snow and ice cover, and partial or complete burial by the WAIS. We currently have no access the more than 90 hidden sub-ice-sheet volcanoes already documented by geophysical methods (e.g. Behrendt 2013; de Vries et al. 2018). Despite the incomplete record, many details of MBLVG volcanism are known. Key themes have emerged from this research, in some cases reinforcing and in other cases challenging past interpretations.

Documented MBLVG activity dates back to the latest Eocene (c. 37 Ma) and continues to today, with Mount Berlin and Mount Takahe considered active. Early volcanism (37–19 Ma) in the MBLVP was marked by infrequent monogenetic mafic eruptions, preserved in upland areas at Mount Petras in central MBL. The earliest felsic volcanism was the 20 Ma trachyte lava near Mount Flint. Following this, there appears to have been a c. 6 Ma hiatus before the onset of polygenetic volcanism in the middle Miocene (13.4 Ma) at Whitney Peak caldera in the northern ECR. Large-scale polygenetic volcanism has continued since the middle Miocene without major interruptions, producing 19 central volcanoes with 24 preserved large (2–10 km-diameter) summit calderas.

The ubiquity of large summit calderas (24 in the province) is a noteworthy aspect of the MBLVP (LeMasurier 1990h). An anomaly of these calderas is that, in the past, the dominant recognized volcanic outcrops were lavas, with lesser amounts of breccia, including hyaloclastite, and very rare pyroclastic rocks except in late-stage basaltic scoria cones. While lavas dominate, pyroclastic rocks are less rare than previously documented in LeMasurier and Thomson (1990). In this chapter, we presented published (Wilch et al. 1999) and new data documenting explosively-erupted pumiceous welded fall deposits at central volcano caldera-rim outcrops across the MBLVP. These outcrops have been preserved because the pyroclastic deposits are welded. Presumably, the eruptions that generated the caldera-rim welded deposits also generated non-welded tephra that was dispersed across the WAIS and into the Southern Ocean. The two youngest polygenetic central volcanoes, Mount Takahe and Mount Berlin, have Holocene-aged deposits in their caldera walls that have been correlated to tephra in ice cores and englacial tephra layers currently exposed in ablation areas (Dunbar et al. 2021). Although most of the central volcanoes of the MBLVP have shield-like morphologies, they are not simple lava shield volcanoes but have complex explosive and effusive histories.

The extensive application of $^{40}Ar/^{39}Ar$ geochronology has resulted in significant improvements in resolving the history of the volcanoes, especially where detailed analyses have been done. In many cases, the $^{40}Ar/^{39}Ar$ ages are similar to the K–Ar ages but in some cases the ages are significantly different (e.g. Wilch and McIntosh 2007). In almost every instance, $^{40}Ar/^{39}Ar$ ages are at least an order of magnitude more precise than conventional K–Ar ages. The high-precision geochronology can be powerful when tied to detailed fieldwork. For example, Panter et al. (1994) developed a well-dated history of the construction and caldera collapse and breach of Mount Sidley based on detailed field mapping combined with targeted $^{40}Ar/^{39}Ar$ dating. With detailed sampling and $^{40}Ar/^{39}Ar$ dating of multiple 200–800 m-thick stratigraphic sequences in the Crary Mountains and at Mount Murphy, Wilch and McIntosh (2002) compared glaciovolcanic interactions at coastal v. inland sites during the same time interval.

The biggest advance associated with the application of $^{40}Ar/^{39}Ar$ geochronology at MBL volcanoes has been the ability to date very young samples (to Holocene age) (Wilch et al. 1999; Dunbar et al. 2021). This has been critical in reconstructing the history of the youngest volcanoes in the province, especially Mount Berlin and Mount Takahe. Ongoing work at Mount Waesche will complete the first level of detailed work on the Late Pleistocene volcanoes of MBL. $^{40}Ar/^{39}Ar$ dating of young outcrops combined with tephrochronology analyses in blue-ice areas and ice cores has led to fingerprinting and correlating MBL tephra to climate proxy records. Lastly, the high resolution of young $^{40}Ar/^{39}Ar$ ages in glaciovolcanic sequences at Mount Takahe has permitted correlation of interpreted ice levels with Milankovitch-driven marine isotopic cycles.

Detailed lithofacies reconstructions have also advanced our understanding of palaeoenvironments and the eruptive histories of MBL volcanoes. A detailed field and geochronology study of Mount Petras led to major revisions of the earliest history of the province, challenging early interpretations of the glacial and the erosional history of Marie Byrd Land (Wilch and McIntosh 2000). A detailed lithofacies analysis of Icefall Nunatak near Mount Murphy provided an incremental but significant advance in understanding glaciovolcanic processes and how volcanic records can be used to understand local syneruptive ice conditions (Smellie 2001).

Finally, the palaeoenvironmental reconstructions and $^{40}Ar/^{39}Ar$ geochronology of the MBLVG provide snapshot records of the WAIS since the late Eocene, with the following major conclusions:

- The earliest terrestrial indications for glacial ice in West Antarctica are middle Oligocene (29–27 Ma) tuff cone deposits at Mount Petras that suggest the presence of a thin local ice cap or ice and snow on bedrock slopes. Throughout the remainder of MBLVG volcanic record, interactions between volcanism and thin local ice continued to occur, suggesting that significant cover by local snow and ice was common in MBL since the Oligocene.
- The first volcanic evidence for a widespread WAIS is Late Miocene c. 9 Ma glaciovolcanic sequences from across MBL. Two patterns emerge from the record of syneruptive palaeo-ice levels: former WAIS thickening was more extensive at coastal sites (such as Mount Murphy) than at inland sites (such as Mount Rees and Mount Steere); and, currently, the WAIS is in a near-maximum configuration that has existed at many times since 10 Ma but was rarely exceeded.
- A middle Pleistocene ice-sheet highstand of +500 m is inferred at the coastal volcano, Mount Murphy, based on late-stage 609 ± 27 ka tuff cone deposits at Sechrist Peak. This ice-sheet highstand may correspond to global ice

expansion (and eustatic lowering) at MIS 16 in the marine record.

- The Stauffer Bluff tuya at Mount Takahe records minimum syneruptive ice levels at 400 m above today's ice at 64.7 ± 13.3 ka, coincident with a eustatic sea-level lowstand at MIS 4.
- The Gill Bluff and Möll Spur and passage-zone sequences at Mount Takahe indicate minimum syneruptive ice levels at 413 m and c. 575 m above today's ice at 21.2 ± 4.1 and 17.5 ± 4.8 ka, respectively, which is consistent with a eustatic sea-level lowstand at MIS 2.

Acknowledgements We are especially grateful to Phil Kyle, Ian Skilling and John Smellie for detailed and constructive reviews, which improved this chapter significantly. We thank US Navy VXE-6 squadron, Antarctic Support Associates, and Ken Borek Air Ltd for logistical support. We are grateful for field collaborators Nelia Dunbar, John Smellie and John Gamble; and field assistants Paul Rose, Chris Griffiths and Tony Teeling. Most of the $^{40}Ar/^{39}Ar$ geochronology analyses were completed at the New Mexico Geochronological Research Laboratory at New Mexico Tech and we wish to thank the NMGRL staff, particularly Lisa Peters and Matt Heizler, for their assistance. We are grateful to Sean McCuddy and Mike Wagg for additional assistance with lab work. Christine Merritt, Emily Ebaugh and Miriam Wilch provided invaluable technical assistance in preparation of maps and figures for this chapter. This work was supported by National Science Foundation grants DPP-8816342 (WAVE), DPP-9198806 (WAVE II), OPP-9419686, OPP-9725910, and OPP-9814782.

Author contributions TIW: conceptualization (lead), data curation (equal), formal analysis (lead), funding acquisition (equal), investigation (equal), methodology (equal), project administration (lead), writing – original draft (lead), writing – review & editing (equal); **WCM**: conceptualization (supporting), data curation (equal), formal analysis (equal), investigation (equal), methodology (equal), visualization (supporting), writing – original draft (supporting), writing – review & editing (equal); **KSP**: conceptualization (supporting), data curation (equal), formal analysis (equal), investigation (equal), methodology (supporting), project administration (supporting), writing – original draft (supporting), writing – review & editing (equal).

Funding Fieldwork was supported by the National Science Foundation with grants OPP-9725910 and DPP-918806 awarded to W.C. Mcintosh, and grant OPP-9814782 awarded to T.I. Wilch.

Data availability The datasets generated during the current study are available in Supplementary material Tables S1–S4.

References

Anderson, J.B. 1999. *Antarctic Marine Geology*. Cambridge University Press, Cambridge, UK.

Andrews, J.T. and LeMasurier, W.E. 1973. Rates of Quaternary glacial erosion and corrie formation, Marie Byrd Land, Antarctica. *Geology*, **1**, 75–80, https://doi.org/10.1130/0091-7613(1973)1<75:ROQGEA>2.0.CO;2

Behrendt, J.C. 2013. The aeromagnetic method as a tool to identify Cenozoic magmatism in the West Antarctic Rift System beneath the West Antarctic Ice Sheet – A review; Thiel subglacial volcano as possible source of the ash layer in the WAISCORE. *Tectonophysics*, **585**, 124–136, https://doi.org/10.1016/j.tecto.2012.06.035

Behrendt, J.C., LeMasurier, W.E., Cooper, A.K., Tessensohn, F., Trehu, A. and Damaske, D. 1991. Geophysical studies of the West Antarctic rift system. *Tectonics*, **10**, 1257–1273, https://doi.org/10.1029/91tc00868

Behrendt, J.C., Saltus, R., Damaske, D., McCafferty, A., Finn, C.A., Blankenship, D.D. and Bell, R.E. 1996. Patterns of late Cenozoic volcanic and tectonic activity in the West Antarctic rift system revealed by aeromagnetic data. *Tectonics*, **15**, 660–676, https://doi.org/10.1029/95tc03500

Bentley, C.R. and Clough, J.W. 1972. Antarctic subglacial structure from seismic refraction measurements. *In*: Adie, R.J. (ed.) *Antarctic Geology and Geophysics*. Universitetsforlaget, Oslo, 683–691.

Blankenship, D.D., Bell, R.E., Hodge, S.M., Brozena, J.M., Behrendt, J.C. and Finn, C.A. 1993. Active volcanism beneath the West Antarctic ice sheet and implications for ice-sheet stability. *Nature*, **361**, 526–529, https://doi.org/10.1038/361526a0

Cande, S.C. and Stock, J.M. 2006. Constraints on the timing of extension in the Northern Basin, Ross Sea. *In*: Futterer, D.K., Damaske, D., Kleinschmidt, G., Miller, H. and Tessensohn, F. (eds) *Antarctica: Contributions to Global Earth Sciences*. Springer, Berlin, 319–326.

Cande, S.C., Stock, J.M., Müller, R.D. and Ishihara, T. 2000. Cenozoic motion between east and west Antarctica. *Nature*, **404**, 145–150, https://doi.org/10.1038/35004501

Chakraborty, S. 2007. *The Geochemical Evolution of Alkaline magMas from the Crary Mountains, Marie Byrd Land, Antarctica*. MS thesis, Bowling Green State University, Bowling Green, Ohio, USA.

Chaput, J., Aakaster, R.C. et al. 2014. The crustal thickness of West Antarctica. *Journal of Geophysical Research: Solid Earth*, **119**, 378–395, https://doi.org/10.1002/2013JB010642

Chatzaras, V., Kruckenberg, S.C., Cohen, S.M., Medaris, L.G., Jr, Withers, A.C. and Bagley, B. 2016. Axial-type olivine crystallographic preferred orientations: the effect of strain geometry on mantle texture. *Journal of Geophysical Research: Solid Earth*, **121**, 4895–4922, https://doi.org/10.1002/2015jb012628

Chough, S.K. and Sohn, Y.K. 1990. Depositional mechanics and sequences of base surges, Songaksan tuff ring, Cheju Island, Korea. *Sedimentology*, **37**, 1115–1135, https://doi.org/10.1111/j.1365-3091.1990.tb01849.x

Cooper, A.K. and Davey, F.J. 1985. Episodic rifting of phanerozoic rocks in the Victoria Land Basin, Western Ross Sea, Antarctica. *Science*, **229**, 1085–1087, https://doi.org/10.1126/science.229.4718.1085

Cooper, A.K., Davey, F.J. and Hinz, K. 1991. Crustal extension and origin of sedimentary basins beneath the Ross Sea and Ross Ice Shelf, Antarctica. *In*: Crame, J.A., Thomson, J.W. and Thomson, M.R.A. (eds) *Geological Evolution of Antarctica*. Cambridge University Press, Cambridge, UK, 299–304.

Courtillot, V., Davaille, A., Besse, J. and Stock, J. 2003. Three distinct types of hotspots in the Earth's mantle. *Earth and Planetary Science Letters*, **205**, 295–308, https://doi.org/10.1016/S0012-821X(02)01048-8

Craddock, C., Bastien, T.W. and Rutford, R.H. 1964. Geology of the Jones Mountains area. *In*: Adie, R.J. (ed.) *Antarctic Geology*. North-Holland, Amsterdam, 171–187.

Cuffey, K. and Paterson, W.S.B. 2010. *The Physics of Glaciers*. 4th edn. Academic Press, Amsterdam.

Dalrymple, G.B. 1979. Critical tables for conversion of K–Ar ages from old to new decay constants. *Geology*, **7**, 558–560, https://doi.org/10.1130/0091-7613(1979)7<558:CTFCOK>2.0.CO;2

Dalziel, I.W.D. and Elliot, D.H. 1982. West Antarctica – Problem child of Gondwanaland. *Tectonics*, **1**, 3–19, https://doi.org/10.1029/TC001i001p00003

Dellino, P. and La Volpe, L. 2000. Structures and grain size distribution in surge deposits as a tool for modelling the dynamics of dilute pyroclastic density currents at La Fossa di Vulcano (Aeolian Islands, Italy). *Journal of Volcanology and Geothermal Research*, **96**, 57–78, https://doi.org/10.1016/s0377-0273(99)00140-7

de Vries,, M.V., Bingham, R.G. and Hein, A.S. 2018. A new volcanic province: an inventory of subglacial volcanoes in West Antarctica. *Geological Society, London, Special Publications*, **461**, 231–248, https://doi.org/10.1144/SP461.7

Douillet, G.A., Taisne, B., Tsang-Hin-Sun, E., Muller, S.K., Kueppers, U. and Dingwell, D.B. 2015. Syn-eruptive, soft-sediment deformation of deposits from dilute pyroclastic density current: triggers from granular shear, dynamic pore pressure, ballistic impacts and shock waves. *Solid Earth*, **6**, 553–572, https://doi.org/10.5194/se-6-553-2015

Dunbar, N.W., McIntosh, W.C. and Esser, R.P. 2008. Physical setting and tephrochronology of the summit caldera ice record at Mt. Moulton, West Antarctica. *Geological Society of America Bulletin*, **120**, 796–812, https://doi.org/10.1130/b26140.1

Dunbar, N.W., Iverson, N.A., Smellie, J.L., McIntosh, W.C., Zimmerer, M.J. and Kyle, P.R. 2021. Active volcanoes in Marie Byrd Land. *Geological Society, London, Memoirs*, **55**, https://doi.org/10.1144/M55-2019-29

Emry, E.L., Nyblade, A.A. et al. 2015. The mantle transition zone beneath West Antarctica: Seismic evidence for hydration and thermal upwellings. *Geochemistry, Geophysics, Geosystems*, **16**, 40–58, https://doi.org/10.1002/2014gc005588

Fenner, C.N. 1938. Olivine fourchites from Raymond Fosdick Mountains, Antarctica. *Bulletin of the Geological Society of America*, **49**, 367–400, https://doi.org/10.1130/GSAB-49-367

Fisher, R.V. and Schmincke, H. 1984. *Pyroclastic Rocks*. Springer, Berlin.

Fretwell, P., Pritchard, H.D. et al. 2013. Bedmap2: improved ice bed, surface and thickness datasets for Antarctica. *Cryosphere*, **7**, 375–393, https://doi.org/10.5194/tc-7-375-2013

Gaffney, A.M. and Siddoway, C.S. 2007. Heterogeneous sources for Pleistocene lavas of Marie Byrd Land, Antarctica: new data from the SW Pacific diffuse alkaline magmatic province. *United States Geological Survey Open-File Report*, **2007-1047**, Extended Abstract 063, https://pubs.usgs.gov/of/2007/1047/ea/of2007-1047ea063.pdf

Global Volcanism Program. 1988. Report on Siple (Antarctica), Marie Byrd Land, Antarctica. *Scientific Event Alert Network Bulletin*, **13**: 12. Smithsonian Institution, https://doi.org/10.5479/si.GVP.SEAN198812-390025

González-Ferrán, O. and González-Bonorino, F. 1972. The volcanic ranges of Marie Byrd Land between long. 100° and 140°W. *In*: Adie, R.J. (ed.) *Antarctic Geology and Geophysics*. Universitetsforlaget, Oslo, 261–276.

Granot, R. and Dyment, J. 2018. Late Cenozoic unification of East and West Antarctica. *Nature Communications*, **9**, 3189, https://doi.org/10.1038/s41467-018-05270-w

Grosse, P. and Kervyn, M. 2018. Morphometry of terrestrial shield volcanoes. *Geomorphology*, **304**, 1–14, https://doi.org/10.1016/j.geomorph.2017.12.017

Grosse, P., Euillades, P.A., Euillades, L.D. and de Vries, B.V. 2014. A global database of composite volcano morphometry. *Bulletin of Volcanology*, **76**, 16, https://doi.org/10.1007/s00445-013-0784-4

Hansen, S.E., Graw, J.H. et al. 2014. Imaging the Antarctic mantle using adaptively parameterized P-wave tomography: evidence for heterogeneous structure beneath West Antarctica. *Earth and Planetary Science Letters*, **408**, 66–78, https://doi.org/10.1016/j.epsl.2014.09.043

Heeszel, D.S., Wiens, D.A. et al. 2016. Upper mantle structure of central and West Antarctica from array analysis of Rayleigh wave phase velocities. *Journal of Geophysical Research: Solid Earth*, **121**, 1758–1775, https://doi.org/10.1002/2015JB012616

Holbourn, A., Kuhnt, W., Clemens, S., Prell, W. and Andersen, N. 2013. Middle to late Miocene stepwise climate cooling: evidence from a high-resolution deep water isotope curve spanning 8 million years. *Palaeoceanography*, **28**, 688–699, https://doi.org/10.1002/2013pa002538

Hole, M.J., Storey, B.C. and LeMasurier, W.E. 1994. Tectonic setting and geochemistry of Mocene alkaline basalts from the Jones Mountains, West Antarctica. *Antarctic Science*, **6**, 85–92, https://doi.org/10.1017/s0954102094000118

Honnorez, J. and Kirst, P. 1975. Submarine basaltic volcanism: morphometric parameters for discriminating hyaloclastites from hyalotuffs. *Bulletin of Volcanology*, **39**, 1–25, https://doi.org/10.1007/BF02596941

Jones, J.G. 1969. Intraglacial volcanoes of the Laugarvatn region, south-west Iceland – I. *Quarterly Journal of the Geological Society, London*, **124**, 197–211, https://doi.org/10.1144/gsjgs.124.1.0197

Jones, J.G. 1970. Intraglacial volcanoes of the Laugarvatn region, south-west Iceland – I. *Journal of Geology*, **78**, 127–140, https://doi.org/10.1086/627496

Jordan, T.A., Riley, T.R. and Siddoway, C.S. 2020. The geological history and evolution of West Antarctica. *Nature Reviews*, **1**, 117–133, https://doi.org/10.1038/s43017-019-0013-6

Kuiper, K.F., Deino, A., Hilgen, F.J., Krijgsman, W., Renne, P.R. and Wijbrans, J.R. 2008. Synchronizing rock clocks of Earth history. *Science*, **320**, 500–504, https://doi.org/10.1126/science.1154339

LeMasurier, W.E. 1972. Volcanic record of Cenozoic glacial history in Marie Byrd Land. *In*: Adie, R.J. (eds) *Antarctic Geology and Geophysics*. Universitetsforlaget, Oslo, 251–260.

LeMasurier, W.E. 1990*a*. Ames Range. *American Geophysical Union Antarctic Research Series*, **48**, 216–220.

LeMasurier, W.E. 1990*b*. Miocene and older centers, Walgreen Coast. *American Geophysical Union Antarctic Research Series*, **48**, 203–207.

LeMasurier, W.E. 1990*c*. Miocene–Oligocene centers, Mount Petras and USAS Escarpment. *American Geophysical Union Antarctic Research Series*, **48**, 239–243.

LeMasurier, W.E. 1990*d*. Miocene–Pliocene centers, Hobbs Coast. *American Geophysical Union Antarctic Research Series*, **48**, 244–247.

LeMasurier, W.E. 1990*e*. Mount Bursey. *American Geophysical Union Antarctic Research Series*, **48**, 221–224.

LeMasurier, W.E. 1990*f*. Mount Hartigan. *American Geophysical Union Antarctic Research Series*, **48**, 199–202.

LeMasurier, W.E. 1990*g*. Pliocene–Pleistocene centers, Hobbs Coast. *American Geophysical Union Antarctic Research Series*, **48**, 248–250.

LeMasurier, W.E. 1990*h*. Summary. *American Geophysical Union Antarctic Research Series*, **48**, 146–163.

LeMasurier, W.E. 2002. Architecture and evolution of hydrovolcanic deltas in Marie Byrd Land, Antarctica. *Geological Society, London, Special Publications*, **202**, 115–148, https://doi.org/10.1144/gsl.Sp.2002.202.01.07

LeMasurier, W.E. 2006. What supports the Marie Byrd Land dome? An evaluation of potential uplift mechanisms in a continental rift system. *In*: Fütterer, D.K., Damaske, D., Kleinschmidt, G., Miller, H. and Tessensohn, F. (eds) *Antarctica*. Springer, Berlin, 299–302.

LeMasurier, W.E. 2008. Neogene extension and basin deepening in the West Antarctic rift inferred from comparisons with the East African rift and other analogs. *Geology*, **36**: 247–250, https://doi.org/10.1130/G24363A.1

LeMasurier, W.E. 2013. Shield volcanoes of Marie Byrd Land, West Antarctic rift: oceanic island similarities, continental signature, and tectonic controls. *Bulletin of Volcanology*, **75**, https://doi.org/10.1007/s00445-013-0726-1

LeMasurier, W.E. and Kawachi, Y. 1990*a*. Mount Berlin. *American Geophysical Union Antarctic Research Series*, **48**, 229–233.

LeMasurier, W.E. and Kawachi, Y. 1990*b*. Mount Cumming. *American Geophysical Union Antarctic Research Series*, **48**, 195–198.

LeMasurier, W.E. and Kawachi, Y. 1990*c*. Mount Hampton. *American Geophysical Union Antarctic Research Series*, **48**, 189–194.

LeMasurier, W.E. and Kawachi, Y. 1990*d*. Mount Moulton. *American Geophysical Union Antarctic Research Series*, **48**, 225–228.

LeMasurier, W.E. and Landis, C.A. 1996. Mantle-plume activity recorded by low-relief erosion surfaces in West Antarctica and New Zealand. *Geological Society of America Bulletin*, **108**, 1450–1466, https://doi.org/10.1130/0016-7606(1996)108<1450:Mparbl>2.3.Co;2

LeMasurier, W.E. and Rex, D.C. 1982. Volcanic record of Cenozoic glacial history in Marie Byrd Land and western Ellsworth Land: Revised chronology and evalution of tectonic factors. *In*: Craddock, C. (ed.) *Antarctic Geoscience*. University of Wisconsin Press, Madison, WI, 725–734.

LeMasurier, W.E. and Rex, D.C. 1983. Rates of uplift and the scale of ice level instabilities recorded by volcanic rocks in Marie Byrd Land, West Antarctica. *In*: Oliver, R.L., James, P.R. and Jago, J.B. (eds) *Antarctic Earth Sciences*. Australian Academy of Science, Canberra, 660–673.

LeMasurier, W.E. and Rex, D.C. 1989. Evolution of linear volcanic ranges in Marie Byrd Land, West Antarctica. *Journal of Geophysical Research*, **94**, 7223–7236, https://doi.org/10.1029/JB094iB06p07223

LeMasurier, W.E. and Rex, D.C. 1990*a*. Mount Siple. *American Geophysical Union Antarctic Research Series*, **48**, 185–188.

LeMasurier, W.E. and Rex, D.C. 1990*b*. Mount Takahe. *American Geophysical Union Antarctic Research Series*, **48**, 169–174.

LeMasurier, W.E. and Rex, D.R. 1991. The Marie Byrd Land Volcanic Province and its relation to the Cainozoic West Antarctic rift system. *In*: Tingey, R.J. (ed.) *The Geology of Antarctica*. Clarendon Press, Oxford, 249–284.

LeMasurier, W.E. and Rocchi, S. 2005. Terrestrial record of post-Eocene climate history in Marie Byrd Land, West Antarctica. *Geografiska Annaler Series A – Physical Geography*, **87A**, 51–66, https://doi.org/10.1111/j.0435-3676.2005.00244.x

LeMasurier, W.E. and Thomson, J.E. (eds). 1990. *Volcanoes of the Antarctic Plate and Southern Oceans. American Geophysical Union Antarctic Research Series*, **48**, 146–163.

LeMasurier, W.E. and Wade, F.A. 1968. Fumarolic activity in Marie Byrd Land Antarctica. *Science*, **162**, 352, https://doi.org/10.1126/science.162.3851.352

LeMasurier, W.E. and Wade, F.A. 1976. Volcanic history in Marie Byrd Land: implications with regard to southern hemisphere tectonic reconstructions. *In*: González-Ferrán, O. (ed.) *Proceedings of the International Symposium on Andean and Antarctic Volcanology Problems*. IAVCEI, Rome, 398–424.

LeMasurier, W.E. and Wade, F.A. 1990. Fosdick Mountains. *American Geophysical Union Antarctic Research Series*, **48**, 251–253.

LeMasurier, W.E., McIntosh, W.C. and Rex, D.C. 1981. Mid-Tertiary glacial history recorded at Mt. Petras, Marie Byrd Land. *Antarctic Journal of the United States*, **16**, 19–21.

LeMasurier, W.E., Kawachi, Y. and Rex, D.C. 1990*a*. Crary Mountains. *American Geophysical Union Antarctic Research Series*, **48**, 180–184.

LeMasurier, W.E., Kawachi, Y. and Rex, D.C. 1990*b*. Mount Flint–Reynolds Ridge. *American Geophysical Union Antarctic Research Series*, **48**, 212–215.

LeMasurier, W.E., Kawachi, Y. and Rex, D.C. 1990*c*. Mount Murphy. *American Geophysical Union Antarctic Research Series*, **48**, 164–168.

LeMasurier, W.E., Kawachi, Y. and Rex, D.C. 1990*d*. Toney Mountain. *American Geophysical Union Antarctic Research Series*, **48**, 175–179.

LeMasurier, W.E., Harwood, D.M. and Rex, D.C. 1994. Geology of Mount Murphy Volcano: an 8-m.y. history of interaction between a rift volcano and the West Antarctic ice sheet. *Geological Society of America Bulletin*, **106**, 265–280, https://doi.org/10.1130/0016-7606(1994)106<0265:Gommva>2.3.Co;2

Lisiecki, L.E. and Raymo, M.E. 2005. A Pliocene–Pleistocene stack of 57 globally distributed benthic $\delta^{18}O$ records. *Palaeoceanography*, **20**, 17, https://doi.org/10.1029/2004pa001071

Lloyd, A.J., Wiens, D.A. *et al.* 2015. A seismic transect across West Antarctica: evidence for mantle thermal anomalies beneath the Bentley Subglacial Trench and the Marie Byrd Land Dome. *Journal of Geophysical Research: Solid Earth*, **120**, 8439–8460, https://doi.org/10.1002/2015jb012455

Loose, B., Garabato, A.C.N., Schlosser, P., Jenkins, W.J., Vaughan, D. and Heywood, K.J. 2018. Evidence of an active volcanic heat source beneath the Pine Island Glacier. *Nature Communications*, **9**, 9, https://doi.org/10.1038/s41467-018-04421-3

Lopatin, B.G. and Polyakov, M.M. 1974. Geology of the volcanic Hudson Mountains, Walgreen Coast, West Antarctica. *Antarktika*, **13**, 36–51.

Lough, A.C., Wiens, D.A. *et al.* 2013. Seismic detection of an active subglacial magmatic complex in Marie Byrd Land, Antarctica. *Nature Geoscience*, **6**, 1031–1035, https://doi.org/10.1038/ngeo1992

Martin, A.P., Cooper, A.F., Price, R.C., Kyle, P.R. and Gamble, J.A. 2021. Erebus Volcanic Province: petrology. *Geological Society, London, Memoirs*, **55**, https://doi.org/10.1144/M55-2018-80

McConnell, J.R., Burke, A. *et al.* 2017. Synchronous volcanic eruptions and abrupt climate change similar to 17.7 ka plausibly linked by stratospheric ozone depletion. *Proceedings of the National Academy of Sciences of the United States of America*, **114**, 10 035–10 040, https://doi.org/10.1073/pnas.1705595114

McIntosh, W.C., LeMasurier, W.E., Ellerman, P.J. and Dunbar, N.W. 1985. A reinterpretation of glaciovolcanic interaction at Mount Takake and Mount Murphy, Marie Byrd Land, Antarctica. *Antarctic Journal of the United States*, **19**, 57–59.

McIntosh, W.C., Smellie, J.L. and Panter, K.S. 1991. Glaciovolcanic interaction in eastern Marie Byrd Land, Antarctica. *In: Proceedings of the Sixth International Symposium on Antarctic Earth Sciences, Saitama, Japan, 1991*. National Institute of Polar Research, Tokyo, 402.

McPhie, J., Doyle, M. and Allen, R. 1993. *Volcanic Textures: A Guide to the Interpretation of Textures in Volcanic Rocks*. Centre for Ore Deposit and Exploration Studies, University of Tasmania, Hobart, Australia.

Morgan, W.J. 1972. Deep mantle convection plumes and plate motions. *AAPG Bulletin*, **56**, 203–213.

Olivetti, V., Balestrieri, M.L., Rossetti, F., Thomson, S.N., Talarico, F.M. and Zattin, M. 2015. Evidence of a full West Antarctic Ice Sheet back to the early Oligocene: insight from double dating of detrital apatites in Ross Sea sediments. *Terra Nova*, **27**, 238–246, https://doi.org/10.1111/ter.12153

Palais, J.M., Kyle, P.R., McIntosh, W.C. and Seward, D. 1988. Magmatic and phreatomagmatic volcanic activity at Mt. Takahe, West Antarctica, based on tephra layers in the Byrd ice core and field observations at Mt. Takahe. *Journal of Volcanology and Geothermal Research*, **35**, 295–317, https://doi.org/10.1016/0377-0273(88)90025-x

Panter, K.S. 1995. *Geology, Geochemistry and Petrogenesis of the Mt. Sidley Volcano, Marie Byrd Land, Antarctica*. PhD thesis, New Mexico Institute of Mining and Technology, Socorro, New Mexico, USA.

Panter, K.S., McIntosh, W.C. and Smellie, J.L. 1994. Volcanic history of Mt. Sidley, a major alkaline volcano in Marie Byrd Land, Antarctica. *Bulletin of Volcanology*, **56**, 361–376, https://doi.org/10.1007/s004450050045

Panter, K.S., Kyle, P.R. and Smellie, J.L. 1997. Petrogenesis of a phonolite-trachyte succession at Mt. Sidley, Marie Byrd Land, Antarctica. *Journal of Petrology*, **38**, 1225–1253, https://doi.org/10.1093/petrology/38.9.1225

Panter, K.S., Hart, S.R., Kyle, P., Blusztanjn, J. and Wilch, T. 2000. Geochemistry of Late Cenozoic basalts from the Crary Mountains: characterization of mantle sources in Marie Byrd Land, Antarctica. *Chemical Geology*, **165**, 215–241, https://doi.org/10.1016/s0009-2541(99)00171-0

Panter, K.S., Wilch, T.I., Smellie, J.L., Kyle, P.R. and McIntosh, W.C. 2021. Marie Byrd Land and Ellsworth Land: petrology. *Geological Society, London, Memoirs*, **55**, https://doi.org/10.1144/M55-2019-50

Paulsen, T.S. and Wilson, T.J. 2010. Evolution of Neogene volcanism and stress patterns in the glaciated West Antarctic Rift, Marie Byrd Land, Antarctica. *Journal of the Geological Society, London*, **167**, 401–416, https://doi.org/10.1144/0016-76492009-044

Quartini, E., Blankenship, D.D. and Young, D.A. 2021. Active subglacial volcanism in West Antarctica. *Geological Society, London, Memoirs*, **55**, https://doi.org/10.1144/M55-2019-3

Ramirez, C., Nyblade, A. et al. 2017. Crustal structure of the Transantarctic Mountains, Ellsworth Mountains and Marie Byrd Land, Antarctica: constraints on shear wave velocities, Poisson's ratios and Moho depths. *Geophysical Journal International*, **211**, 1328–1340, https://doi.org/10.1093/gji/ggx333

Reusch, A.M., Nyblade, A.A., Benoit, M.H., Wiens, D.A., Anandakrishan, S., Voigt, D. and Shore, P.J. 2008. Mantle transition zone thickness beneath Ross Island, the Transantarctic Mountains, and East Antarctica. *Geophysical Research Letters*, **35**, L12301, https://doi.org/10.1029/2008GL033873

Rocchi, S. and Smellie, J.L. 2021. Northern Victoria Land: petrology. *Geological Society, London, Memoirs*, **55**, https://doi.org/10.1144/M55-2019-19

Rocchi, S., Armienti, P., D'Orazio, M., Tonarini, S., Wijbrans, J.R. and Di Vincenzo, G. 2002. Cenozoic magmatism in the western Ross Embayment: role of mantle plume v. plate dynamics in the development of the West Antarctic Rift System. *Journal of Geophysical Research: Solid Earth*, **107**, 2195, https://doi.org/10.1029/2001jb000515

Rocchi, S., LeMasurier, W.E. and Di Vincenzo, G. 2006. Oligocene to Holocene erosion and glacial history in Marie Byrd Land, West Antarctica, inferred from exhumation of the Dorrel Rock intrusive complex and from volcano morphologies. *Geological Society of America Bulletin*, **118**, 991–1005, https://doi.org/10.1130/b25675.1

Rowley, P.D. 1990. Jones Mountains. *American Geophysical Union Antarctic Research Series*, **48**, 286–288.

Rowley, P.D., Laudon, T.S., La Prade, K.E. and LeMasurier, W.E. 1990. Hudson Mountains. *American Geophysical Union Antarctic Research Series*, **48**, 289–293.

Russell, J.K., Edwards, B.R., Porritt, L. and Ryane, C. 2014. Tuyas: a descriptive genetic classification. *Quaternary Science Reviews*, **87**, 70–81, https://doi.org/10.1016/j.quascirev.2014.01.001

Rutford, R.H. and McIntosh, W.C. 2007. Jones Mountains, Antarctica: Evidence for Tertiary glaciation revisited. United States Geological Survey Open-File Report, **2007-1047**, Extended Abstract 203, https://pubs.usgs.gov/of/2007/1047/ea/of2007-1047ea203.pdf

Rutford, R.H., Craddock, C. and Bastien, T.W. 1968. Late Tertiary glaciation and sea level changes in Antarctica. *Palaeogeography, Palaeoclimatology, Palaeoecology*, **5**, 15–39.

Rutford, R.H., Craddock, C., White, C.M. and Armstrong, R.L. 1972. Tertiary glaciation in the Jones Mountains. *In*: Adie, R.J. (ed.) *Antarctic Geology and Geophysics*. Universitetsforlaget, Oslo, 239–243.

Seroussi, H., Ivins, E.R., Wiens, D.A. and Bondzio, J. 2017. Influence of a West Antarctic mantle plume on ice sheet basal conditions. *Journal of Geophysical Research: Solid Earth*, **122**, 7127–7155, https://doi.org/10.1002/2017jb014423

Seward, D., Kyle, P.R. and LeMasurier, W.E. 1980. Fission track ages of Marie Byrd Land volcanic rocks. *Antarctic Journal of the United States*, **15**, 19.

Shen, W.S., Wiens, D.A. et al. 2018. The crust and upper mantle structure of central and West Antarctica from Bayesian inversion of Rayleigh wave and receiver functions. *Journal of Geophysical Research: Solid Earth*, **123**, 7824–7849, https://doi.org/10.1029/2017jb015346

Siddoway, C.S. 2008. Tectonics of the West Antarctic Rift System: New light on the history and dynamics of distributed intracontinental extension. United States Geological Survey Open-File Report, **2007-1047**, 91–114, https://doi.org/10.3133/ofr20071047KP09

Siddoway, C.S., Baldwin, S.L., Fitzgerald, P.G., Fanning, C.M. and Luyendyk, B.P. 2004. Ross Sea mylonites and the timing of intracontinental extension within the West Antarctic rift system. *Geology*, **32**, 57–60, https://doi.org/10.1130/g20005.1

Skilling, I.P., White, J.D.L. and McPhie, J. 2002. Peperite: a review of magma–sediment mingling. *Journal of Volcanology and Geothermal Research*, **114**, 1–17, https://doi.org/10.1016/s0377-0273(01)00278-5

Smellie, J.L. 2001. Lithofacies architecture and construction of volcanoes erupted in englacial lakes: Icefall Nunatak, Mt. Murphy, eastern Marie Byrd Land, Antarctica. *International Association of Sedimentologists Special Publications*, **30**, 73–98.

Smellie, J.L. 2006. The relative importance of supraglacial v. subglacial meltwater escape in basaltic subglacial tuya eruptions: an important unresolved conundrum. *Earth-Science Reviews*, **74**, 241–268, https://doi.org/10.1016/j.earscirev.2005.09.004

Smellie, J.L. 2008. Basaltic subglacial sheet-like sequences: evidence for two types with different implications for the inferred thickness of associated ice. *Earth-Science Reviews*, **88**, 60–88, https://doi.org/10.1016/j.earscirev.2008.01.004

Smellie, J.L. 2018. Glaciovolcanism – a 21st century proxy for palaeo-ice. *In*: Menzies, J. and van der Meer, J.J.M. (eds) *Past Glacial Environments (Sediments, Forms and Techniques)*. 2nd edn. Elsevier, Amsterdam, 335–375.

Smellie, J.L. and Edwards, B.R. 2016. *Glaciovolcanism on Earth and Mars: Products, Processes and Palaeoenvironmental Significance*. Cambridge University Press, Cambridge, UK.

Smellie, J.L. and Skilling, I.P. 1994. Products of subglacial volcanic eruptions under different ice thicknesses two examples form Antarctica. *Sedimentary Geology*, **91**, 115–129, https://doi.org/10.1016/0037-0738(94)90125-2

Smellie, J.L., McIntosh, W.C., Gamble, J.A. and Panter, K.S. 1990. Preliminary stratigraphy of volcanoes in the Executive Committee Range, central Marie Byrd Land. *Antarctic Science*, **2**, 353–354, https://doi.org/10.1017/S0954102090000487

Smellie, J.L., Hole, M.J. and Nell, P.A.R. 1993. Late Miocene valley-confined subglacial volcanism in northern Alexander Island, Antarctic Peninsula. *Bulletin of Volcanology*, **55**, 273–288, https://doi.org/10.1007/bf00624355

Smellie, J.L., Rocchi, S. and Armienti, P. 2011a. Late Miocene volcanic sequences in northern Victoria Land, Antarctica: products of glaciovolcanic eruptions under different thermal regimes. *Bulletin of Volcanology*, **73**, 1–25, https://doi.org/10.1007/s00445-010-0399-y

Smellie, J.L., Rocchi, S., Gemelli, M., Di Vincenzo, G. and Armienti, P. 2011b. A thin predominantly cold-based Late Miocene East Antarctic ice sheet inferred from glaciovolcanic sequences in northern Victoria Land, Antarctica. *Palaeogeography, Palaeoclimatology, Palaeoecology*, **307**, 129–149, https://doi.org/10.1016/j.palaeo.2011.05.008

Smellie, J.L., Wilch, T.I. and Rocchi, S. 2013. 'A'ā lava-fed deltas: a new reference tool in palaeoenvironmental studies. *Geology*, **41**, 403–406, https://doi.org/10.1130/g33631.1

Sohn, Y.K. 1996. Hydrovolcanic processes forming basaltic tuff rings and cones on Cheju Island, Korea. *Geological Society of America Bulletin*, **108**, 1199–1211.

Sohn, Y.K. and Park, K.H. 2005. Composite tuff ring/cone complexes in Jeju Island, Korea: possible consequences of substrate collapse and vent migration. *Journal of Volcanology and Geothermal Research*, **141**, 157–175, https://doi.org/10.1016/j.jvolgeores.2004.10.003

Spiegel, C., Lindow, J. et al. 2016. Tectonomorphic evolution of Marie Byrd Land – Implications for Cenozoic rifting activity and onset of West Antarctic glaciation. *Global and Planetary Change*, **145**, 98–115, https://doi.org/10.1016/j.gloplacha.2016.08.013

Steiger, R.H. and Jaeger, E. 1977. Subcommision on Geochronology: convention of the use of decay constants in geo- and cosmochronology. *Earth and Planetary Science Letters*, **36**, 359–362.

Storey, B., Vaughan, A. and Riley, T. 2013. The links between large igneous provinces, continental break-up and environmental change: Evidence reviewed from Antarctica. *Earth and Environmental Science Transactions of the Royal Society of Edinburgh*, **104**, 17–30, https://doi.org/10.1017/S175569101300011X

Tessensohn, F. and Wörner, G. 1991. The Ross Sea rift system, Antarctica: structure, evolution, and analogues. *In*: Thomson, M.R.A., Crane, J.A. and Thomson, J.W. (eds) *Geological Evolution of Antarctica*. Cambridge University Press, Cambridge, UK, 273–277.

Wade, F.A. 1971. Marie Byrd Land and Ellsworth Land geologic survey. *Antarctic Journal of the United States*, **6**, 193.

Wade, F.A. and Craddock, C. 1968. The Ellsworth Land Survey. *Antarctic Journal of the United States*, **4**, 92.

Walker, G.P.L. and Blake, D.H. 1966. The formation of a palagonite breccia mass beneath a valley glacier in Iceland. *Quarterly Journal of the Geological Society, London*, **122**, 45–61, https://doi.org/10.1144/gsjgs.122.1.0045

Walker, G.P.L. and Croasdale, R. 1971. Characteristics of some basaltic pyroclastics. *Bulletin of Volcanology*, **35**, 303–317, https://doi.org/10.1007/BF02596957

White, J.D.L. and Houghton, B.F. 2006. Primary volcaniclastic rocks. *Geology*, **34**, 677–680, https://doi.org/10.1130/g22346.1

White, J.D.L. and Ross, P.S. 2011. Maar-diatreme volcanoes: a review. *Journal of Volcanology and Geothermal Research*, **201**, 1–29, https://doi.org/10.1016/j.jvolgeores.2011.01.010

White, J.D.L. and Valentine, G.A. 2016. Magmatic v. phreatomagmatic fragmentation: absence of evidence is not evidence of absence. *Geosphere*, **12**, 1478–1488, https://doi.org/10.1130/ges01337.1

Whiteman, C.D. and Garibotti, R. 2013. Rime mushrooms on mountains: description, formation, and impacts on mountaineering. *Bulletin of the American Meteorological Society*, **94**, 1319–1327, https://doi.org/10.1175/BAMS-D-12-00167.1

Wilch, T.I. 1997. *Volcanic Record of the West Antarctic Ice Sheet in Marie Byrd Land*. PhD thesis, New Mexico Institute of Mining and Technology, Socorro, New Mexico, USA.

Wilch, T.I. and McIntosh, W.C. 2000. Eocene and Oligocene volcanism at Mt. Petras, Marie Byrd Land: implications for middle Cenozoic ice sheet reconstructions in West Antarctica. *Antarctic Science*, **12**, 477–491, https://doi.org/10.1017/S0954102000000560

Wilch, T.I. and McIntosh, W.C. 2002. Lithofacies analysis and $^{40}Ar/^{39}Ar$ geochronology of ice–volcano interactions at Mt. Murphy and the Crary Mountains, Marie Byrd Land, Antarctica. *Geological Society, London, Special Publications*, **202**, 237–253, https://doi.org/10.1144/GSL.SP.2002.202.01.12

Wilch, T.I. and McIntosh, W.C. 2007. Miocene–Pliocene ice–volcano interactions at monogenetic volcanoes near Hobbs Coast, Marie Byrd Land, Antarctica. United States Geological Survey Open-File Report, **2007-1047**, Short Research Paper 074, https://pubs.usgs.gov/of/2007/1047/srp/srp074

Wilch, T.I., Denton, G.H., Lux, D.R. and McIntosh, W.C. 1993. Limited Pliocene glacial extent and surface uplift in Middle Taylor Valley, Antarctica. *Geografiska Annaler*, **75A**, 331–351.

Wilch, T.I., McIntosh, W.C. and Dunbar, N.W. 1999. Late Quaternary volcanic activity in Marie Byrd Land: potential $^{40}Ar/^{39}Ar$-dated time horizons in West Antarctic ice and marine cores. *Geological Society of America Bulletin*, **111**, 1563–1580, https://doi.org/10.1130/0016-7606(1999)111<1563:Lqvaim> 2.3.Co;2

Wilson, D.S., Pollard, D., DeConto, R.M., Jamieson, S.S.R. and Luyendyk, B.P. 2013. Initiation of the West Antarctic Ice Sheet and estimates of total Antarctic ice volume in the earliest Oligocene. *Geophysical Research Letters*, **40**, 4305–4309, https://doi.org/10.1002/grl.50797

Wilson, G.S., Levy, R.H. et al. 2012. Neogene tectonic and climatic evolution of the Western Ross Sea, Antarctica – Chronology of events from the AND-1B drill hole. *Global and Planetary Change*, **96–97**, 189-203, https://doi.org/10.1016/j.glopla cha.2012.05.019

Winberry, J.P. and Anandakrishnan, S. 2004. Crustal structure of the West Antarctic rift system and Marie Byrd Land hotspot. *Geology*, **32**, 977–980, https://doi.org/10.1130/g20768.1

Wohletz, K.H. 1983. Mechanisms of hydrovolcanic pyroclast formation: grain-size, scanning electron microscopy, and experimental studies. *Journal of Volcanology and Geothermal Research*, **16**, 31–63.

Wohletz, K.H. and Sheridan, M.F. 1983. Hydrovolcanic explosions II. Evolution of basaltic tuff rings and tuff cones. *American Journal of Science*, **283**, 385–413.

Chapter 5.4b

Marie Byrd Land and Ellsworth Land: petrology

K. S. Panter[1]*, T. I. Wilch[2], J. L. Smellie[3], P. R. Kyle[4] and W. C. McIntosh[5]

[1]School of Earth, Environment and Society, Bowling Green State University, Bowling Green, OH 43403, USA
[2]Department of Geological Sciences, Albion College, Albion, MI 49224, USA
[3]School of Geography, Geology and the Environment, University of Leicester, Leicester LE1 7RH, UK
[4]Department of Earth and Environmental Sciences, New Mexico Institute of Mining and Technology, Socorro, NM 87801, USA
[5]New Mexico Bureau of Geology and Mineral Resources, New Mexico Institute of Mining and Technology, Socorro, NM 87801, USA

KSP, 0000-0002-0990-5880; PRK, 0000-0001-6598-8062; WCM, 0000-0002-5647-2483
*Correspondence: kpanter@bgsu.edu

Abstract: In Marie Byrd Land and Ellsworth Land 19 large polygenetic volcanoes and numerous smaller centres are exposed above the West Antarctic Ice Sheet along the northern flank of the West Antarctic Rift System. The Cenozoic (36.7 Ma to active) volcanism of the Marie Byrd Land Volcanic Group (MBLVG) encompasses the full spectrum of alkaline series compositions ranging from basalt to intermediate (e.g. mugearite, benmoreite) to phonolite, peralkaline trachyte, rhyolite and rare pantellerite. Differentiation from basalt is described by progressive fractional crystallization; however, to produce silica-oversaturated compositions two mechanisms are proposed: (1) polybaric fractionation with early-stage removal of amphibole at high pressures; and (2) assimilation–fractional crystallization to explain elevated $^{87}Sr/^{86}Sr_i$ ratios. Most basalts are silica-undersaturated and enriched in incompatible trace elements (e.g. La/Yb_N >10), indicating small degrees of partial melting of a garnet-bearing mantle. Mildly silica-undersaturated and rare silica-saturated basalts, including tholeiites, are less enriched (La/Yb_N <10), a result of higher degrees of melting. Trace elements and isotopes (Sr, Nd, Pb) reveal a regional gradient explained by mixing between two mantle components, subduction-modified lithosphere and HIMU-like plume ($^{206}Pb/^{204}Pb$ >20) materials. Geophysical studies indicate a deep thermal anomaly beneath central Marie Byrd Land, suggesting a plume influence on volcanism and tectonism.

Supplementary material: Major element, trace element and isotopic (Sr, Nd, Pb and Hf) compositions of Cenozoic volcanic rocks within the Marie Byrd Land Volcanic Group (Table S1) are available at https://doi.org/10.6084/m9.figshare.c.5233171

Cenozoic volcanism in Marie Byrd Land (hereafter MBL) and the western portion of Ellsworth Land (hereafter wEL) extends for nearly 1500 km along the northern shoulder of the West Antarctic Rift System (WARS), from the Fosdick Mountains in western MBL (c. 76.5° S, c. 145° W) to the Jones Mountains (c. 73.5° S, c. 94° W) near the Eights Coast in wEL (Fig. 1). All of the volcanism is now defined by Wilch et al. (2021) as the MBL Volcanic Group (MBLVG), and consists of the MBL Volcanic Province (MBLVP) and the Thurston Island Volcanic Province (TIVP) (Fig. 1). This region lies between the subduction- to post-subduction-related volcanism of the Antarctic Peninsula to the east and the rift-related volcanism that is exposed in the western Ross Sea over 1000 km to the west (see Hole 2021; Leat and Riley 2021; Martin et al. 2021; Rocchi and Smellie 2021), the latter is collectively known as the McMurdo Volcanic Group (Kyle 1990). The age of volcanism in the MBLVG is comparable to the McMurdo Volcanic Group, with most of the activity occurring since 14 Ma. The rock types are also similar and range in composition from basaltic (e.g. basanite, alkali basalt and hawaiite) to phonolite and trachyte with minor amounts of rhyolite. Overall, the magmatism associated with the broad region (3000 × 1500 km) of thinned lithosphere that comprises the WARS has a relatively consistent spectrum of alkaline compositions emplaced starting at about 48 Ma in North Victoria Land (alkaline plutons of the Meander Intrusive Group: Rocchi et al. 2002) and eruptions starting about 37 Ma in MBL (Wilch et al. 2021).

The WARS was initiated during the final stage of Gondwana break-up in the Cretaceous, which led to the separation of Zealandia and Australia from Antarctica (Lawver and Gahagan 1994; Siddoway 2008; Mortimer et al. 2019; Zundel et al. 2019; Jordan et al. 2020). Prior to continental rifting, the palaeo-Pacific margin of Gondwana was influenced by subduction that may have been nearly continuous since the late Neoproterozoic (Bradshaw 1989; Cawood 2005). The subduction of the Phoenix oceanic plate beneath the eastern margin of Gondwana ceased, abruptly, in the Cretaceous (Bradshaw 1989; Mukasa and Dalziel 2000; Tulloch et al. 2009). The change from a mostly compressional tectonic regime to a tensional one advanced eastwards from the western portion of MBL at c. 105–100 Ma to eastern MBL at c. 100 Ma and then to wEL at c. 95 Ma (Zundel et al. 2019). The change in the stress regime has been explained by oblique subduction of the Pacific–Phoenix spreading centre (Bradshaw 1989), possibly with the aid of a mantle plume that may have weakened the lithosphere and controlled the location of rifting between the crustal blocks of Zealandia and the MBL (Weaver et al. 1994). It has also been proposed that the shutting down of subduction occurred by the mechanism of slab capture (Luyendyk 1995) or by collision with the young and thermally buoyant oceanic Hikurangi Plateau (Davy et al. 2008; Davy 2014; Mortimer et al. 2019). Whatever the cause, rifting led to the final disintegration of Gondwana and the isolation of Antarctica by 83 Ma (Veevers 2012 and references therein). It is estimated that this early rifting phase caused 500–1000 km of crustal extension across the entire Ross Sea between MBL and the Transantarctic Mountains (DiVenere et al. 1994; Luyendyk et al. 1996; Jordan et al. 2020). The early phase of broadly distributed extension transformed into a more focused phase of extension in the Paleogene (Huerta and Harry 2007 and references therein). This younger phase of rifting occurred along the western Ross Sea Embayment and the front of the Transantarctic Mountains resulting in c. 300 km of extension between 80 and 40 Ma (Molnar et al. 1975; Cande et al. 2000; Siddoway et al. 2004). Seafloor

From: Smellie, J. L., Panter, K. S. and Geyer, A. (eds) 2021. *Volcanism in Antarctica: 200 Million Years of Subduction, Rifting and Continental Break-up*. Geological Society, London, Memoirs, **55**, 577–614,
First published online March 9, 2021, https://doi.org/10.1144/M55-2019-50
© 2021 The Author(s). Published by The Geological Society of London. All rights reserved.
For permissions: http://www.geolsoc.org.uk/permissions. Publishing disclaimer: www.geolsoc.org.uk/pub_ethics

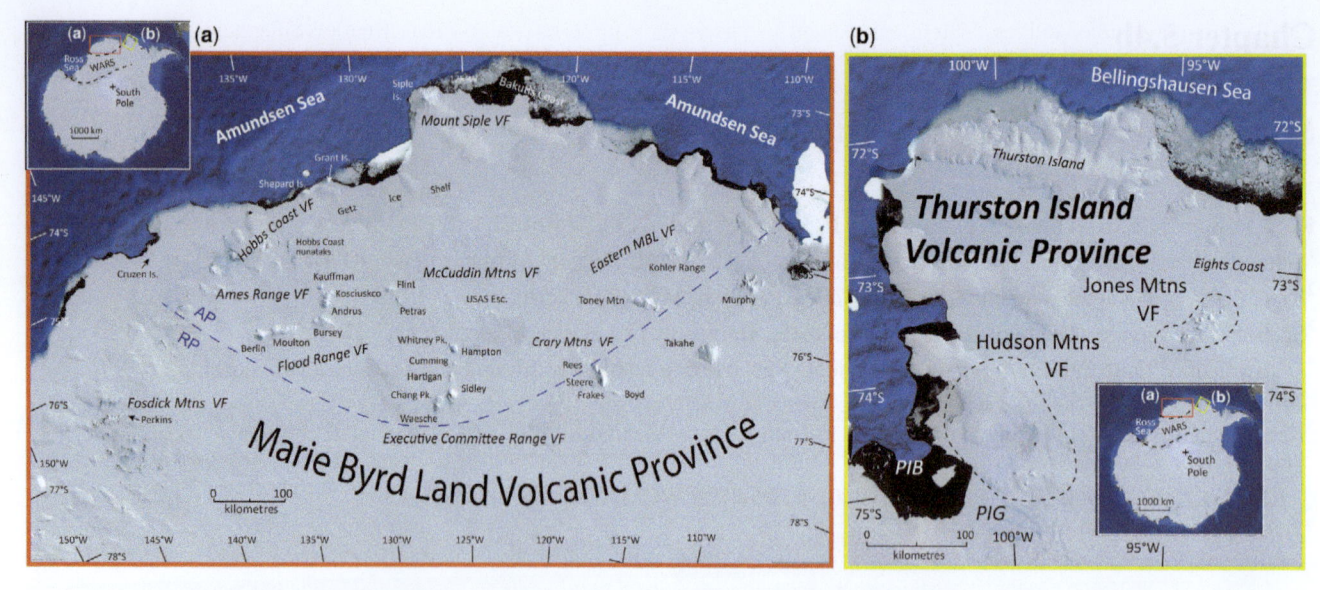

Fig. 1. Map (**a**) of Marie Byrd Land Volcanic Province (MBLVP) and Thurston Island Volcanic Province (TIVP) of western Ellsworth Land; and (**b**) including 11 volcanic fields (*VF*), all of which belong to the Marie Byrd Land Volcanic Group (MBLVG) after Wilch *et al.* (2021). The base maps are derived from Google Earth Pro using US Geological Survey and NASA generated images. Volcanic ranges, central polygenetic volcanoes and satellite volcanic areas are labelled within their respective volcanic fields (*VF*). Inset: West Antarctic Rift System (WARS) and outlines of maps (a) and (b) on the Antarctic continent. USAS Esc. is an abbreviation for the USAS Escarpment, located east of Petras. Mountain and Mountains are abbreviated Mtn and Mtns, respectively. Peak and Island are abbreviated as Pk. and Is., respectively. The blue dashed line in (a) marks the approximate boundary between the Ross Province (RP) and Amundsen Province (AP) as delimited by Jordan *et al.* (2020). In (b), abbreviations PIB and PIG are for Pine Island Bay and Pine Island Glacier, respectively. Satellite image maps of individual volcanoes and ranges are presented in Wilch *et al.* (2021).

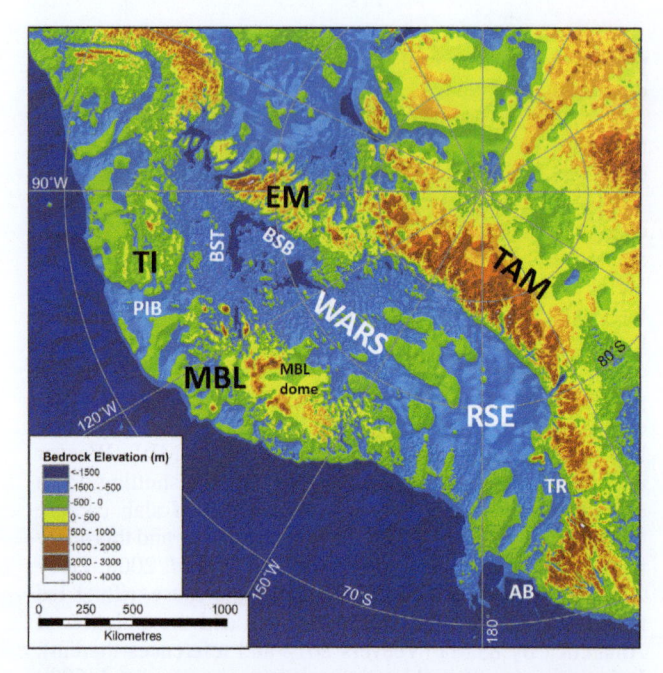

Fig. 2. Map of subglacial topography of West Antarctica derived from BEDMAP2 data (Fretwell *et al.* 2013). Most of the central volcanoes within the Marie Byrd Land Volcanic Province (MBLVP) are >2000 m asl and are shown in red. The whole of the MBLVG is located on the north flank of the West Antarctic Rift System (WARS), which is dominated by deep subglacial basins including the Byrd Subglacial Basin (BSB) and the Bentley Subglacial Trough (BST). The WARS extends through the Ross Sea Embayment (RSE) and includes the Terror Rift (TR) and the oceanic Adare Basin (AB). Crustal blocks in West Antarctica include MBL, WARS, Thurston Island (TI) and the Ellsworth Mountains (EM). The volcanoes of western Ellsworth Land are situated on the Thurston Island crustal block. The Ellsworth Mountain and connected Transantarctic Mountains (TAM) form the rift shoulder on the south flank of the WARS.

spreading in the Adare Basin, northern Ross Sea (Fig. 2), resulted in *c.* 180 km of extension between 43 and 26 Ma (Cande and Stock 2006; Granot *et al.* 2013), with an additional 35 km by 11 Ma (Granot and Dyment 2018). Neogene rifting also produced 10–15 km of extension in the Terror Rift, west-central and southern Ross Sea (Fig. 2). However, according to Granot and Dyment (2018), most of the extension during this period is likely to have occurred in central MBL beneath the West Antarctic Ice Sheet (i.e. *c.* 34 km of extension in the Bentley Subglacial Trench: Fig. 2) and by 11 Ma all motion between East and West Antarctica had ceased.

Cenozoic volcanism within the WARS is comparable to other young continental rift systems such as the Basin and Range–Rio Grande (western USA) and the East Africa Rift (see Panter 2021), and as with these other rift systems the fundamental cause of magmatism is still a matter of debate. Existing models for magmatism within the WARS can be grouped into two fundamental types: those that involve mantle plumes and those that do not. Both types require mantle sources for alkaline magmas that have been enriched in incompatible minor and trace elements relative to primitive mantle in order to explain their geochemical and radiogenic isotope compositions. A mantle plume origin for this signature has been proposed to explain the active Erebus volcano of the McMurdo Volcanic Group in the Ross Sea (Kyle *et al.* 1992; Phillips *et al.* 2018). Additionally, an active mantle plume has been proposed beneath MBL to explain magmatism along with regional doming and volcanic migration patterns (LeMasurier and Rex 1989; Hole and LeMasurier 1994; Hole *et al.* 1994; Hansen *et al.* 2014) in addition to elevated heat flux (Seroussi *et al.* 2017). Larger plumes ('super-plumes') have been invoked to explain Jurassic and Cretaceous continental break-up events and as geochemical reservoirs ('fossil plumes') accreted to the continental lithosphere providing a melt source for Cenozoic volcanism (Lanyon *et al.* 1993; Weaver *et al.* 1994; Rocholl *et al.* 1995; Hart *et al.* 1997; Storey *et al.* 1999; Panter *et al.* 2000; Kipf *et al.*

2014). For non-plume sources, mantle enrichment has been ascribed to metasomatism (Futa and LeMasurier 1983) caused either by small degrees of melting during the early amagmatic phase of rifting prior to continental break-up (Rocchi et al. 2002; Nardini et al. 2009; Rocchi and Smellie 2021) or by melting of slab-derived materials introduced into the upper mantle via subduction (Hart et al. 1995; Finn et al. 2005; Panter et al. 2006, 2018; Gaffney and Siddoway 2007; Martin et al. 2013, 2014; Aviado et al. 2015; LeMasurier et al. 2016). Most of the models that support a metasomatic source for Cenozoic volcanism place the enrichment at the base of the lithosphere. The melting of metasomatized lithosphere to generate alkaline magmatism may have been facilitated by a variety of mechanisms including movement along translithospheric faults (Rocchi et al. 2002; Rocchi and Smellie 2021), conductive heating by regional mantle upwelling (Finn et al. 2005; Panter et al. 2006; Sutherland et al. 2010), edge-driven mantle flow (Faccenna et al. 2008; Panter et al. 2018; Rocchi and Smellie 2021) or by the warming of metasomatized lithosphere that has delaminated and engulfed by asthenosphere (Shen et al. 2018; Licht et al. 2018; Panter et al. 2021).

This chapter provides some background on the geological setting of MBL and wEL, and then gives an overview of the geochemical and isotopic characteristics of the volcanism in this region and within the broader context of magmatism within the WARS. This is followed by more detailed accounts of polygenetic volcanoes that occur within linear ranges, as isolated polygenetic volcanoes, remnants of the earliest volcanism (in the centre of the MBLVP) and, finally, the mostly monogenetic, basaltic volcanic fields that lie on the periphery of the province. We will end the chapter by evaluating province-wide models for volcanism that focus on potential mantle sources and conditions of melting.

Geological setting

The broad tectonic framework for this West Antarctica region consists of two major crustal blocks: MBL and Thurston Island (TI: Fig. 2) (Dalziel and Elliot 1982; Dalziel and Lawver 2001; Jordan et al. 2020). Pankhurst et al. (1998) further divided the MBL block into an eastern half (Amundsen Province) and a western half (Ross Province) based on fundamental differences in crustal affinity (Fig. 1). Western MBL is composed of Lower Paleozoic sedimentary rocks (Swanson Formation) that correlate with the Lachlan Group in Australia and equivalent rocks in western New Zealand. These were subsequently intruded by Devonian–Carboniferous granitoid bodies which are widespread in East Antarctica (Weaver et al. 1991; Korhonen et al. 2010). In contrast, the basement rocks of eastern MBL are devoid of Paleozoic sedimentary strata and include a more extensive series of Late Paleozoic–Mesozoic calc-alkaline granitoid rocks that are considered to be equivalent to those of the Median Tectonic Zone in New Zealand (Pankhurst et al. 1998; Mukasa and Dalziel 2000). The Thurston Island crustal block lies to the east of Pine Island Bay (Fig. 2) and is also composed of Carboniferous–mid-Cretaceous calc-alkaline granitoid with the same continental arc affinity as those in eastern MBL (Leat et al. 1993; Riley et al. 2017). Alkaline plutonism ensued in the eastern portion of the MBL block and the whole of the Thurston Island block in the Late Cretaceous, and this activity delimits the shift from subduction-related to rift-related magmatism prior to final Gondwana break-up and the early development of the WARS (Pankhurst et al. 1993; Weaver et al. 1994; Storey et al. 1999; Siddoway 2008; Zundel et al. 2019). South of the Thurston Island and MBL crustal blocks lie deep subglacial valleys (e.g. Bentley Subglacial Trench and Byrd Subglacial Basin: Fig. 2) that were formed within thinned crust (Winberry and Anandakrishnan 2004; Fretwell et al. 2013; Pappa et al. 2019). These delimit the main axis of the WARS. The rift system in this region is fault-bounded to the south by the Ellsworth–Whitmore Mountains crustal block and the Transantarctic Mountains (Fig. 2). Cenozoic alkaline volcanic rocks are exposed unconformably on top of basement lithologies in eight locations across the MBL block and within the Jones Mountains on the Thurston Island block. The pre-volcanic substrate is considered to have once been part of a widespread low-relief erosion surface that existed across West Antarctica and southern New Zealand prior to continental break-up (LeMasurier and Landis 1996). The West Antarctic erosion surface was disrupted by fault-block uplift that formed a region of high topography known as the Marie Byrd Land Dome (Fig. 2) (Winberry and Anandakrishnan 2004; LeMasurier 2006; Rocchi et al. 2006).

Alkaline volcanism in the MBLVG consists of at least 19 major polygenetic shield-like composite volcanoes along with numerous volcanic fields that have erupted over a period of time beginning in the Late Eocene up to the present (Table 1) (LeMasurier 1990a; Rowley et al. 1990b; Wilch et al. 2021). In the MBLVP, several of the largest polygenetic volcanoes occur as isolated edifices (e.g. Mount Takahe, Mount Murphy and Mount Siple: Fig. 1a) but the majority occur within linear volcanic chains (e.g. Executive Committee Range, Ames Range and Flood Range: Fig. 1a). Pre-existing rift-related structures in the upper crust are considered to have controlled the location and orientation of these volcanic chains (LeMasurier and Rex 1989; Panter et al. 1994; Siddoway 2008; Paulsen and Wilson 2010). Age data demonstrate that the volcanic activity has migrated from the centre of the MBLVP, where the oldest volcanic rocks at Mount Petras and the USAS Escarpment (Fig. 1a) are between 37 and 19 Ma (Wilch 1997; Wilch and McIntosh 2000; Wilch et al. 2021), to the periphery of the province where volcanoes at Mount Takahe, Mount Berlin, Mount Siple and Mount Waesche record eruptions in the Late Quaternary (LeMasurier and Rex 1989; Wilch et al. 1999, 2021; Dunbar et al. 2021). The age distribution along with the ocean island basalt (OIB)-like geochemical signatures of the rocks led LeMasurier and Rex (1989) and Kyle et al. (1991) to propose that the pattern of volcanism in the MBLVP is a result of plume-initiated doming and radial mantle flow. Paulsen and Wilson (2010) proposed an alternative model based on volcano morphology and the distribution of parasitic cones on the flanks of the larger edifices to evaluate past stress conditions. They determined that the pattern of volcanism in MBL is a result of anisotropic stress in two phases: a Middle–Late Miocene phase that formed the roughly north–south alignments of volcanic chains; and a latest Miocene–Pleistocene phase that formed chains orientated roughly east–west. Paulsen and Wilson (2010) conclude that the rapid rotation in the maximum horizontal stress direction from north–south to east–west at c. 6 Ma was the consequence of major plate reorganization and changes in plate motion along the Pacific–Antarctic ridge. It is likely that the migration of volcanism is ongoing, as evidenced by seismic activity at crustal depths of 25–40 km near the southern end of the Executive Committee Range (Lough et al. 2013). Recent subglacial volcanic activity has also been detected to the south of the Marie Byrd Land Dome within the rift (Blankenship et al. 1993; Iverson et al. 2017) and beneath the ice sheet in the Hudson Mountains in wEL (Corr and Vaughan 2008). Furthermore, extensive aeromagnetic surveys in the WARS show abundant magnetic anomalies, suggesting significant basaltic volcanism (as many as 1000 volcanic centres) beneath the West Antarctic Ice Sheet (e.g. Behrendt 2013). Topographical interpretations

Table 1. *Summary of Marie Byrd Land and western Ellsworth Land volcanoes: Marie Byrd Land Volcanic Group (MBLVG)*

Volcanic Provinces, *Volcanic Fields* and Centres within[1]	Latitude[2] (° S)	Longitude[2] (° W)	Age range[3] (Ma)	Estimated volume[4] (km³)	Compositional range[5] Mafic	Intermediate	High total alkali intermediate	Felsic	Total analyses[6]
MBL Volcanic Province									
Fosdick Mountains Volcanic Field									
Avers	−76.48	−145.38	n.d.	n.d.	basanite	–	–	–	2
Perkins	−76.51	−144.15	1.4–4.5	n.d.	alkali basalt	–	–	–	17
Recess	−76.50	−144.37	n.d.	n.d.	basanite	–	–	–	3
location 'F5'	−76.45	−144.92	n.d.	n.d.	tholeiite	–	–	–	1
Flood Range Volcanic Field									
Berlin	−76.00	−135.90	0.0104–2.77	200	bas, haw	mug, ben	phonolite	trachyte	50
Moulton	−76.00	−134.90	1.04–5.95	325	hawaiite	–	phonolite	trac, pan	7
Bursey	−76.00	−132.60	0.25–10.12	290	bas, haw	–	phonolite	trachyte	18
Hobbs Coast Volcanic Field									
Cruzen Island	−74.75	−140.31	2.75	n.d.					1
Shepard Island	−74.40	−132.60	0.41–0.56	n.d.	bas, haw	–	–	trachyte	8
Grant Island	−74.45	−131.84		n.d.	–	latite	–	–	1
Coleman	−75.31	−133.64	2.60	n.d.	bas, teph	–	–	–	21
Cousins	−75.26	−133.52	5.00	n.d.	–	mug, phnt	–	–	3
Shibuya	−75.17	−133.62	5.00	n.d.	hawaiite	–	–	–	3
Holmes	−74.98	−133.71	6.36–8.39	n.d.	bas, akb, haw	–	–	–	4
Kennel	−75.00	−133.76	8.01	n.d.	hawaiite	–	–	–	2
Koupervo	−75.10	−133.78	9.27	n.d.	akb, haw	–	–	trachyte	5
Patton	−75.21	−133.69	11.45	n.d.	alkali basalt	–	–	–	2
Ames Range Volcanic Field									
Andrus	−75.80	−132.30	<0.1–12.71	115	bas, haw	mug, phnt	–	trac, rhy, pan	20
Kosciusko	−75.71	−132.21	9.2	107	–	mugearite	–	trac, pan	2
Kauffman	−75.63	−132.42	7.1	30	–	–	phonolite	trachyte	3
McCuddin Mountains Volcanic Field									
Flint and Reynolds Ridge	−75.70	−129.10	3.75–20.45	n.d.	basanite	–	–	trachyte	7
Petras	−75.87	−128.66	27.53–36.71	52	hawaiite	mugearite	–	–	8
Aldaz and Galla (USAS Escarp.)	−76.00	−124.39	19.0–26.4	n.d.	akb, haw	phonotephrite	–	–	8
Mount Siple VF	−73.40	−126.70	0.008–0.746	1800	akb, haw	–	–	trachyte	8
Executive Committee Range Volcanic Field									
Waesche	−77.17	−126.90	<0.1–1.0	160	akb, teph, haw	mug, phnt, ben	tphn, phn	trachyte	48
Chang Caldera	−77.09	−126.72	1.09–2.01	n.d.	–	–	–	rhyolite	5
Sidley	−77.04	−126.12	4.24–5.77	250	basanite	mug, ben	tphn, phn	trachyte	110
Hartigan	−76.80	−126.00	6.02–8.50	n.d.	hawaiite	mugearite	–	trac, rhy	8
Cumming	−76.68	−125.80	3.0–10.4	n.d.	bas, haw	–	–	trachyte	11
Hampton	−76.49	−125.79	8.6–11.43	70	bas, haw	benmoreite	phonolite	trachyte	25
Whitney	−76.44	−125.05	13.36	n.d.	bas, haw	benmoreite	–	trac, rhy	15
Crary Mountains Volcanic Field									
Rees	−76.65	−118.10	6.91–9.46	n.d.	bas, haw	mug, ben	phonolite	trachyte	12
Steere	−76.72	−117.78	5.81–8.66	n.d.	bas, akb, haw	mug, ben	phonolite	trac, rhy	23
Frakes	−76.81	−117.70	0.034–4.26	n.d.	akb, bas, teph, haw	mug, ben	tphn, phn	trac, rhy	33
Boyd Ridge	−76.94	−116.54	1.29–2.17	n.d.	hawaiite	phonotephrite	phonolite	–	3
Eastern MBL Volcanic Field									
Toney Mountain	−75.80	−115.90	0.5–9.6	550	hawaiite	latite	–	trac, rhy	4
Kohler Range (Leister Peak)	−75.13	−113.90	9.8	n.d.	alkali basalt	–	–	–	1
Mount Takahe	−76.27	−112.00	0.007–0.195	780	akb, bas, teph, haw	mugearite	phonolite	trachyte	103
Mount Murphy	−75.33	−111.38	0.609–9.46	580	akb, bas, teph, haw	mug, ben	phonolite	trachyte	79
west nunataks	−75.35	−111.30	4.76–6.89	n.d.	akb, bas	–	–	–	19
Dorrel Rock[7]	−75.44	−111.38	35.2	n.d.	*basanite*	*benmoreite*	–	*trachyte*	3
Thurston Island Volcanic Province									
Hudson Mountains Volcanic Field	−74.70	−99.30	3.7–5.0	n.d.	akb, thol, haw	–	–	–	7

(*Continued*)

Table 1. *Continued.*

Volcanic Provinces, Volcanic Fields and Centres within[1]	Latitude[2] (° S)	Longitude[2] (° W)	Age range[3] (Ma)	Estimated volume[4] (km³)	Compositional range[5]				Total analyses[6]
					Mafic	Intermediate	High total alkali intermediate	Felsic	
Jones Mountains Volcanic Field	−73.50	−94.10	7.68	n.d.	akb, thol, haw	–	–	–	15
									728

[1]Volcanic Provinces and Volcanic Fields defined within the Marie Byrd Land Volcanic Group (MBLVG) are after Wilch *et al.* (2021).
[2]Latitudes and longitudes were obtained using Google Earth Pro and represent an estimate of the spatial average of each volcanic centre or volcanic field.
[3]Ages based on K–Ar and $^{40}Ar/^{39}Ar$ methods. Following Wilch *et al.* (2021 and references therein), $^{40}Ar/^{39}Ar$ ages are preferred over conventional K–Ar ages, and single mean ages are presented for most monogenetic centres.
[4]Estimated volume exposed above the ice sheet for major volcanic centres is from LeMasurier and Thomson (1990). n.d., not determined.
[5]Mafic compositions (<50 wt% SiO_2 and >4 wt% MgO) = basanite (bsn), alkali basalt (akb), hawaiite (haw), tephrite (teph) and tholeiite (thol). Intermediate compositions (c. 50–58 wt% SiO_2) = mugearite (mug), phonotephrite (phnt), benmoreite (ben) and latite. Intermediate compositions with high total alkalis (>11 wt% $Na_2O + K_2O$) = tephriphonolite (tphn) and phonolite (phn). Felsic compositions (>58 wt% SiO_2) = trachyte (trac), rhyolite (rhy) and pantellerite (pan). Rock names are based on the total alkali v. silica (TAS) diagram of Le Bas *et al.* (1986) with additional criteria of Floyd and Winchester (1975) and Le Maitre (2002).
[6]Totals based on the number of samples measured for major elements.
[7]Dorrel Rock is a nunatak composed of alkaline intrusive rocks (Rocchi *et al.* 2006). Rock names are the extrusive equivalents.

of conical subglacial landforms beneath the West Antarctic Ice Sheet, combined with pre-existing geophysical and field data, led van Wyk de Vries *et al.* (2017) to identify 138 subglacial volcanoes in West Antarctica, including 91 previously unknown volcanoes. The inferred subglacial volcanoes are located on the MBL, Thurston Island and WARS tectonic blocks. Although these subglacial volcanoes have not been verified by samples, Wilch *et al.* (2021) recommend that they be included in the MBLVG.

Sample and data description

In this chapter, we provide published and unpublished geochemical analyses for a total of 728 whole-rock samples from the MBLVG (Table 1; see also Supplementary material Table S1). Subaerial lava flows have been predominantly sampled for geochemistry. However, other lithofacies sampled include pillow lava flows and dykes, and primary juvenile clasts (cf. White and Houghton 2006) were collected from volcaniclastic deposits (e.g. pyroclastic, autoclastic and hyaloclastite). The majority of the samples analysed were collected by expeditions conducted in the 1990s and most of those, including $^{40}Ar/^{39}Ar$ ages, are reported in dissertations and associated publications authored by T. Wilch and K. Panter (see Wilch *et al.* 2021). A significant number of whole-rock analyses (*n* = 185), mostly from Mount Takahe, Mount Murphy, the USAS Escarpment, Mount Cumming and the Whitney Peak–Mount Hampton doublet in the Executive Committee Range, are presented here for the first time. Analyses for 18 basalt samples from the Fosdick Mountains, which were originally reported by Gaffney and Siddoway (2007) but not tabulated, are also provided in Supplementary material Table S1. Other data sources include those within the 1990 AGU Antarctic Research Series (Volume 48) *Volcanoes of the Antarctic Plate and Southern Oceans* (LeMasurier and Thomson 1990 and references therein), and journal articles authored by W.E. LeMasurier, M.J. Hole, S.R. Hart and others since 1990.

In Supplementary material Table S1, full sets of major element oxides are given for all but three samples. Major element concentrations were measured by X-ray fluorescence (XRF) in nearly all cases, except for some taken from LeMasurier and Thomson (1990) which were measured by wet chemistry techniques. Weight percentages (wt%) of FeO and/or Fe_2O_3 are presented as reported in their original data sources; however, iron oxide values are shown exclusively as FeO total (FeO^T) on all plots. On plots, major elements oxides are normalized to 100% volatile-free. Rock names are based on whole-rock compositions and adhere to the classification criteria of Le Maitre (2002). Normative minerals (e.g. nepheline, hypersthene, quartz, diopside, etc.) are presented as wt% based on the criteria of Irvine and Baragar (1971). Minor and trace elements were measured by a variety of instruments and methods that include XRF, instrumental neutron activation analysis (INAA), isotope dilution (ID) and inductively coupled plasma mass spectrometry (ICP-MS). A total of 389 samples have relatively complete trace element datasets with eight of the 17 rare earth elements (REE) measured. A total of 155 samples have complete REE datasets. The number of significant figures for trace element concentrations in parts per million (ppm) presented in Supplementary material Table S1 are as reported in their referenced data sources. In order to facilitate comparisons, we present datasets for each volcano on total alkali v. silica plots (Le Bas *et al.* 1986) and representative compositions on primitive-mantle-normalized multi-element diagrams (also known as spidergrams) using the normalizing values of McDonough and Sun (1995). It is important to be aware of possible pitfalls in comparing concentrations of the same element measured by different analytical techniques. However, we do not exclude any results or attempt to standardize any of the data presented (Supplementary material Table S1), although we identify when compositional differences may be analytical.

The age data (*n* = 258) presented in the text below, and in Table 1 and Supplementary material Table S1, were measured by conventional K–Ar or $^{40}Ar/^{39}Ar$ dating methods, and are compiled from LeMasurier and Rex (1989), LeMasurier and Thomson (1990), Panter *et al.* (1994), Panter (1995), Wilch (1997), Wilch *et al.* (1999, 2021) and Wilch and McIntosh (2000, 2002, 2007). All published and unpublished dates presented below have been recalculated using current accepted constants and standards following Wilch *et al.* (2021). Errors, when presented, are at the 2σ level (95% confidence level).

All known published and unpublished Sr, Nd and Pb isotope analyses are presented in Supplementary material Table S1 ($^{87}Sr/^{86}Sr$, *n* = 170; $^{143}Nd/^{144}Nd$, *n* = 148; $^{206,207,208}Pb/^{204}Pb$, *n* = 76). Measured $^{87}Sr/^{86}Sr$ ratios have been corrected to initial values using available Rb and Sr concentrations and sample age. Some data sources provide only initial $^{87}Sr/^{86}Sr$ ratios (*n* = 25) and therefore the measured

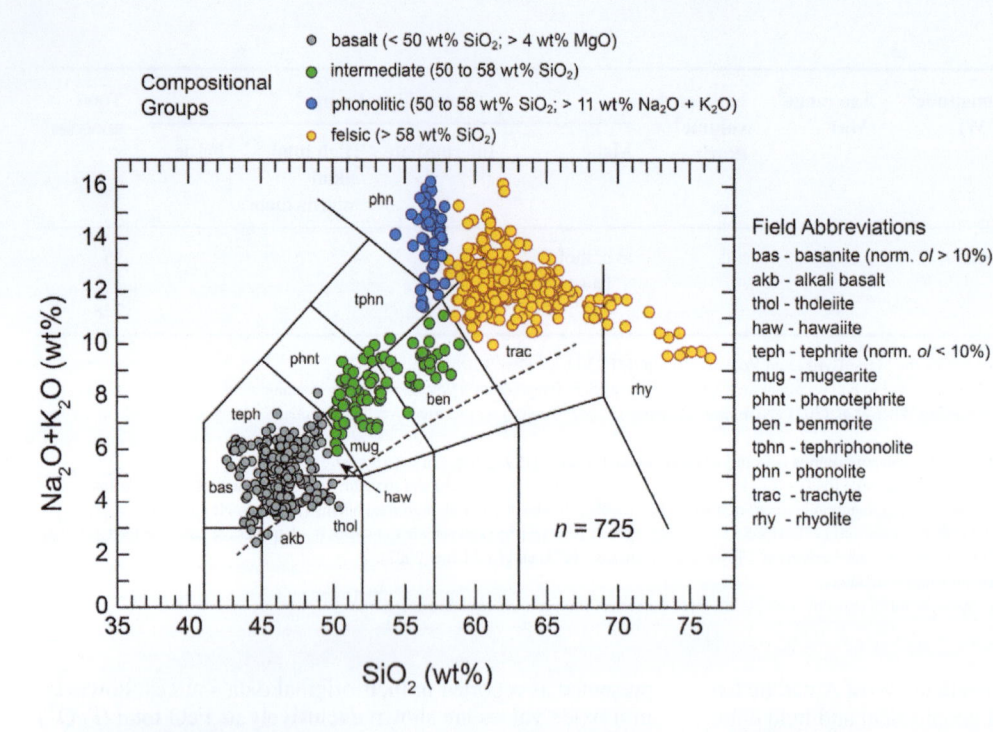

Fig. 3. Total alkali v. SiO$_2$ (TAS) classification (Le Maitre 2002) plot of over 700 volcanic rocks from the MBLVG (Supplementary material Table S1). All analyses shown are the same weight percentage (wt%) and normalized to 100% volatile-free basis. norm. *ol* = CIPW-normative olivine. The line that delimits alkaline from subalkaline composition is from Irvine and Baragar (1971). Compositions are grouped as basalts (<50 wt% SiO$_2$ and >4 wt% MgO), intermediate (50–58 wt% SiO$_2$), phonolitic (50–58 wt% SiO$_2$ and >11 wt% Na$_2$O + K$_2$O) and felsic (>58 wt% SiO$_2$).

values have been back-calculated if the age of the sample is known. Seventy-two analyses are not corrected for in-growth of ^{87}Sr but considering their ^{87}Rb/^{86}Sr ratios and the oldest age from each of their respective volcanoes, the correction on measured ^{87}Sr/^{86}Sr ratios for the majority of these samples is less than 5.0×10^{-5} with a median correction of 8.0×10^{-6}. Four samples from Mount Andrus and two samples from Mount Berlin have been measured for Hf isotopes (LeMasurier *et al.* 2011).

The sample locations listed by longitude and latitude in Table 1 and Supplementary material Table S1 were obtained using Google Earth Pro (https://www.google.com/earth), and represent a spatial average (i.e. roughly centroid) for each major volcano or volcanic field (i.e. the decimal degrees positions do not represent the location of individual samples).

Overview of petrological characteristics

The compositional spectrum of alkaline rocks in MBL and wEL is comparable to continental alkaline volcanism elsewhere in Antarctica (Martin *et al.* 2021; Panter *et al.* 2021; Rocchi and Smellie 2021) and the world; in particular, Cenozoic intraplate alkaline magmatism in Africa (see Baker 1987; Njome and de Wit 2014; Dedzo *et al.* 2019). The volcanic rocks encompass a full range of compositions that vary from mafic types (<50 wt% SiO$_2$ and >4 wt% MgO), basanite, alkali basalt, hawaiite and rare tholeiite, to intermediate types (*c.* 50–58 wt% SiO$_2$), mugearite, benmoreite and rare latite (Na$_2$O–2<K$_2$O (wt%): Le Maitre 2002), including high total alkali types (>11 wt% Na$_2$O + K$_2$O), tephriphonolite and phonolite, to felsic types (>58 wt% SiO$_2$), trachyte, rhyolite and pantellerite (Table 1; Fig. 3).

Mineralogically, phenocrysts and microphenocrysts in basaltic compositions are predominantly olivine, clinopyroxene (diopside, augite) and plagioclase with accessory magnetite and rare amphibole (kaersutite). Felsic compositions contain phenocrysts of clinopyroxene (diopside to hedenbergite), alkali feldspar (anorthoclase), magnetite and less commonly amphibole (arfvedsonite). Accessory minerals include apatite, aenigmatite and rare occurrences of alkaline mica. Anorthoclase is consistently found as a phenocrystic phase in most phonolite (a high-alkali intermediate type) and trachyte rocks, and are up to 6 cm in length in some lava flows at Mount Sidley and the Crary Mountains (authors' observations). These lava flows have been referred to as 'kenytes', and are mineralogically and texturally similar to lava flows on Mount Erebus, southern Ross Sea (Martin *et al.* 2021). Feldspathoids (leucite, nepheline) occur as phenocrysts, microphenocrysts and as groundmass phases in intermediate compositions. Phenocrysts and microphenocrysts of aegirine–augite (high-alkali clinopyroxene) also occur in some trachyte and phonolite rocks. Quartz is an interstitial mineral in rhyolite (LeMasurier and Kawachi 1990*a, b*) but also occurs rarely as phenocrysts in pumice and lava from Chang Peak in the Executive Committee Range (LeMasurier and Kawachi 1990*c*; authors' observations).

Rocks with intermediate SiO$_2$ contents (excluding phonolites) are the least represented, providing a mostly bimodal distribution of compositions (Fig. 3). The dearth of intermediate compositions is often referred to as the 'Daly Gap' (Daly 1925; Clague 1978; Dufek and Bachmann 2010). The bimodal distribution is persistent over the *c.* 30 myr magmatic history of the region (Fig. 4). The majority of samples are

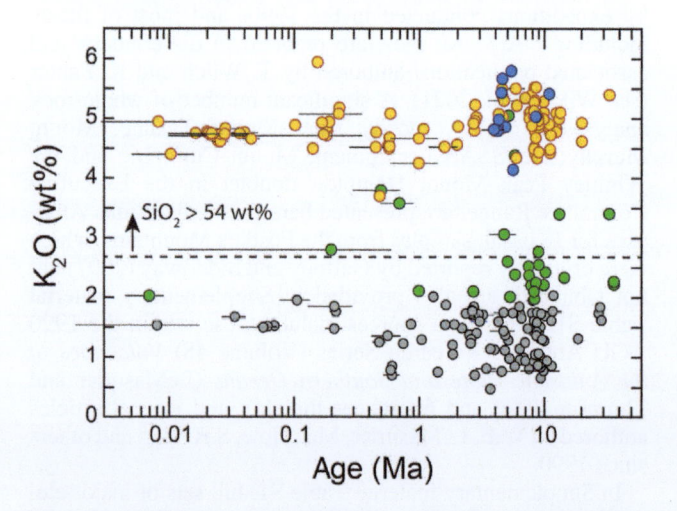

Fig. 4. Age (log Ma) of volcanism (Wilch *et al.* 2021) v. K$_2$O (wt%). The dashed line at 2.6 wt% K$_2$O roughly divides basaltic and less evolved intermediate compositions (≤54 wt% SiO$_2$) from evolved intermediate, phonolitic and felsic compositions. Error bars for age are at 2 standard deviations. Symbols are the same as in Figure 3.

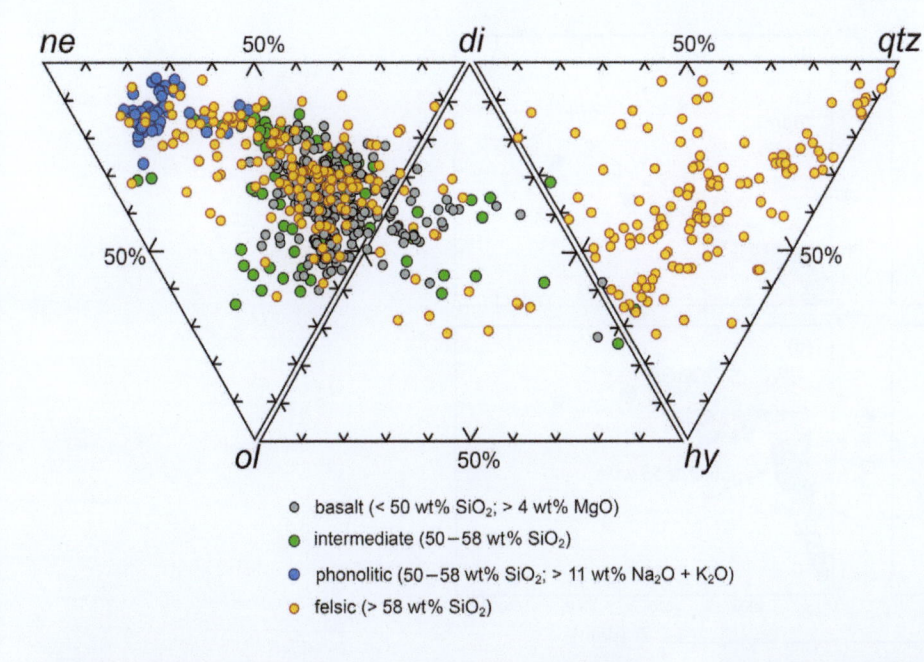

Fig. 5. Plot of normalized (CIPW) compositions (wt%) from MBLVG whole-rock major element data. Silica-undersaturated compositions plot within the nepheline (ne)–olivine (ol)–diopside (di) trangle. Silica-saturated plot within the ol–di–hypersthene (hy) triangle and silica-oversaturated plot within the di–hy–quartz (qtz) triangle. Symbols are the same as in Figure 3.

- basalt (< 50 wt% SiO_2; > 4 wt% MgO)
- intermediate (50–58 wt% SiO_2)
- phonolitic (50–58 wt% SiO_2; > 11 wt% $Na_2O + K_2O$)
- felsic (> 58 wt% SiO_2)

understaturated with respect to silica (i.e. nepheline-normative) but many trachyte samples and all rhyolite samples are silica-oversaturated (i.e. quartz-normative) (Fig. 5). In addition, over 260 samples of phonolites, trachytes and rhyolites are classified as peralkaline (molecular $(Na_2O + K_2O)/Al_2O_3 >1$ and acmite-normative: Le Maitre 2002). Of the 17 samples of peralkaline rhyolite, only four can be subclassified as pantellerite ($SiO_2 \geq 69$ wt% with Al_2O_3 (wt%) less than $1.33 \times FeO^T$ (wt%) + 4.4: Le Maitre 2002), and occur at Mount Moulton, Mount Andrus and Mount Kosciusko (Ames and Flood ranges). Among the mafic rocks, the majority (>95%) of basalts are alkaline. However, some compositions from Mount Murphy and Mount Siple, the Fosdick and Hudson mountains, as well as a few other locations, are transitional to subalkaline.

The entire suite of Cenozoic volcanic rocks from MBL and wEL are evaluated in order to broadly identify petrogenetic relationships and major influences on magma genesis. Variations in major elements with SiO_2 content reveal relatively coherent evolutionary trends of decreasing TiO_2, CaO and MgO, and increasing Na_2O and K_2O, contents with increasing SiO_2 concentrations (Fig. 6). This is largely controlled by early-stage fractionation of mafic minerals (i.e. olivine, clinopyroxene and Ca-plagioclase) and later-stage fractionation of intermediate to felsic minerals (i.e. amphibole, Na-plagioclase, feldspathoid and alkali feldspar). The increase in Al_2O_3 up to about 55 wt% SiO_2 followed by a decrease towards high-Si trachyte and rhyolite compositions (Fig. 6b) is consistent with this mineral fractionation sequence.

Altogether, trace element contents provide broad constraints on magma evolution in MBL and wEL. Zirconium plotted against other high field strength elements (HFSEs: Nb, Th, Hf and Y) and REEs (La, Ce, Nd and Yb) display well-fitted positive correlations, which are consistent with progressive magma evolution from mafic to felsic compositions (Fig. 7a, b). Rubidium, a large ion lithophile element (LILE), also shows a positive correlation with Zr but with significantly more scatter (Fig. 7c) and, along with other LILEs, indicates diverse liquid lines of descent in felsic magmas (Fig. 7d, e). For instance, in rock compositions that have SiO_2 contents of less than 55 wt%, both K_2O and Ba show positive correlations with Zr. However, at higher silica

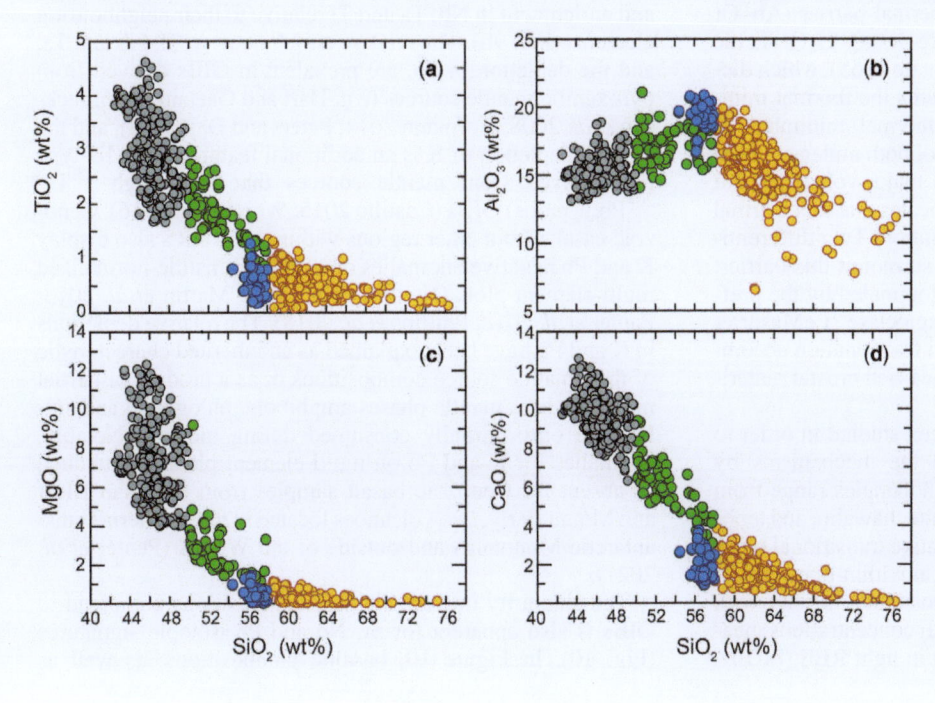

Fig. 6. Major and minor elements (wt%) v. SiO_2 (wt%). Symbols are the same as in Figure 3.

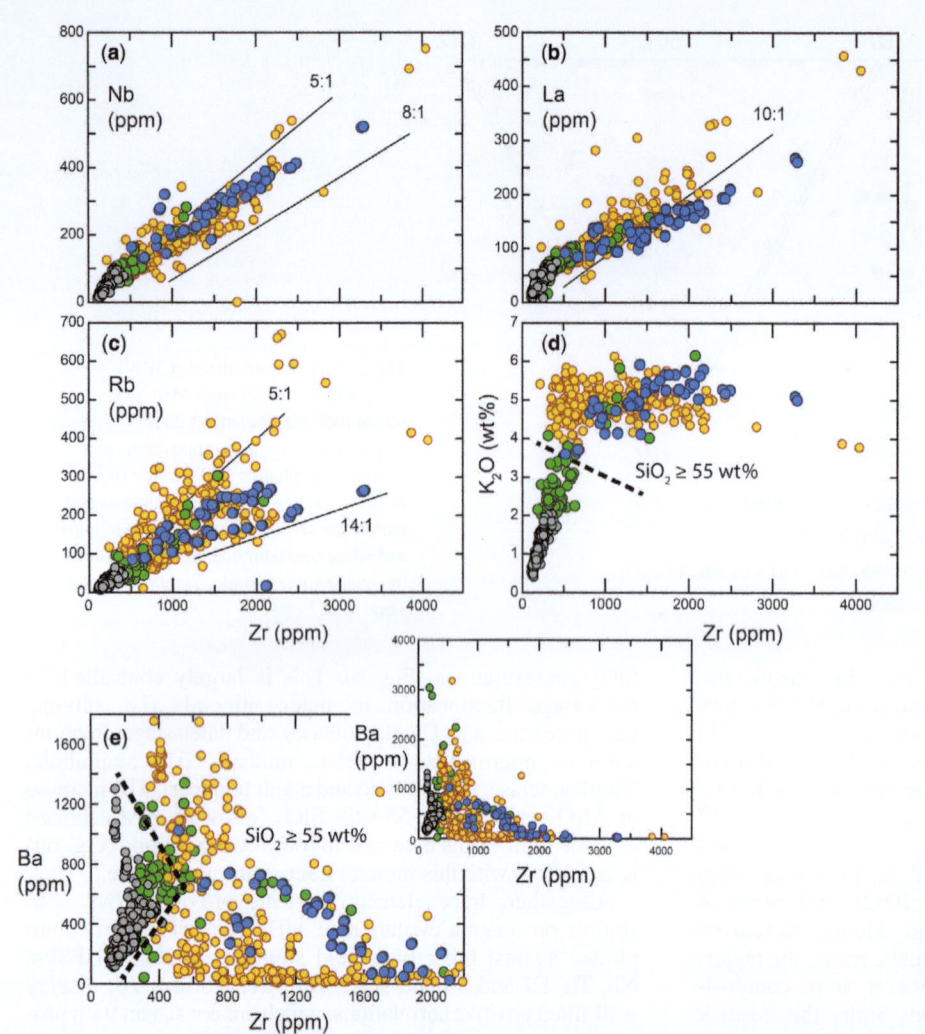

Fig. 7. Zirconium v. selected trace elements (ppm) and K_2O (wt%). In (d) and (e) the dashed line separates intermediate, phonolitic and felsic compositions from less-evolved intermediate and basaltic compositions. Symbols are the same as in Figure 3.

contents the data are dispersed. In particular, Ba contents in samples from a single volcano or volcanic complex can show opposing trends (i.e. decreasing Ba with increasing Zr concentrations, and vice versa). A divergence in compositions is also evident in Figure 3 as delimited by the phonolite v. trachyte–rhyolite lineages. This bifurcation has been classically explained by the presence of a thermal barrier (Ab–Or join) in petrogeny's residua system (Ne–Ks–Q–H_2O: Tuttle and Bowen 1958; Hamilton and MacKenzie 1965), which dictates whether melt will fractionate towards the thermal minimum of silica-oversaturation or the thermal minimum of silica-undersaturation. The coexistence of both undersaturated and oversaturated lineages at nearly all major volcanoes and volcanic complexes in the MBLVG indicates that this thermal barrier is overcome during magma evolution. Two differentiation processes have been proposed to surmount this barrier: (1) high-pressure crystal fractionation dominated by the preferential removal of amphibole over plagioclase (LeMasurier et al. 2003, 2011, 2018); and (2) crystal fractionation accompanied by the assimilation (AFC) of silica-rich crustal materials (Panter et al. 1997).

Basalts from MBL and wEL have been studied in order to characterize their mantle sources and the mechanisms by which melt is generated. The 233 basalt samples range from nepheline-normative alkali basalt, basanite, hawaiite and tephrite to a few ($n = 34$) hypersthene-normative transitional types (Figs 3 & 5). The basalt samples classify as within-plate basalts (cf. Meschede 1986) and have a compositional affinity with OIBs (Fig. 8a, b, after Pearce 2008). REE concentrations measured on 105 samples show enrichment in light REE (LREE) relative to heavy REE (HREE), having chondrite-normalized REE slopes ($[La/Yb]_N$) that range from 6.5 to 26.1 (Fig. 9a). This is broadly consistent with the presence of residual garnet in their mantle source (e.g. Hart et al. 1997; Panter et al. 2000, 2018). On primitive-mantle-normalized multi-element plots the basalts consistently display depletions in Cs, K and Pb, and enrichment in Nb, Ta and Ti relative to their neighbouring elements (Fig. 9b). The relative enrichments in Ti, Nb and Ta, and the depletion in Pb, are prevalent in OIBs derived from pyroxenitic mantle sources (e.g. Hart and Gaetani 2006; Jackson et al. 2008; Hofmann 2014; Peters and Day 2014), and the relative depletion of K is an additional feature of HIMU-type (i.e. derived from mantle sources that have high $^{238}U/^{204}Pb_{t=0}$ ratios) OIBs (Castillo 2015; Weiss et al. 2016). Cenozoic basalts from other regions within the WARS also display K and Pb negative anomalies on primitive-mantle-normalized multi-element plots (Nardini et al. 2009; Martin et al. 2013; Panter et al. 2018; Phillips et al. 2018). The relative depletions in K and Pb have been explained as an inherited characteristic of their mantle source compositions or as a product of partial melting where mantle phases amphibole, phlogopite and sulfide are only partially consumed during melting. Notably, anomalies for K and Pb on multi-element plots are subdued or absent for Cenozoic basalt samples from Sheridan Bluff and Mount Early, two volcanoes located in the southern Transantarctic Mountains and outside of the WARS (Panter et al. 2021).

The similarity in trace element compositions compared to OIBs is also apparent for Sr, Nd and Pb isotopic signatures (Fig. 10). In Figure 10, basaltic compositions, as well as

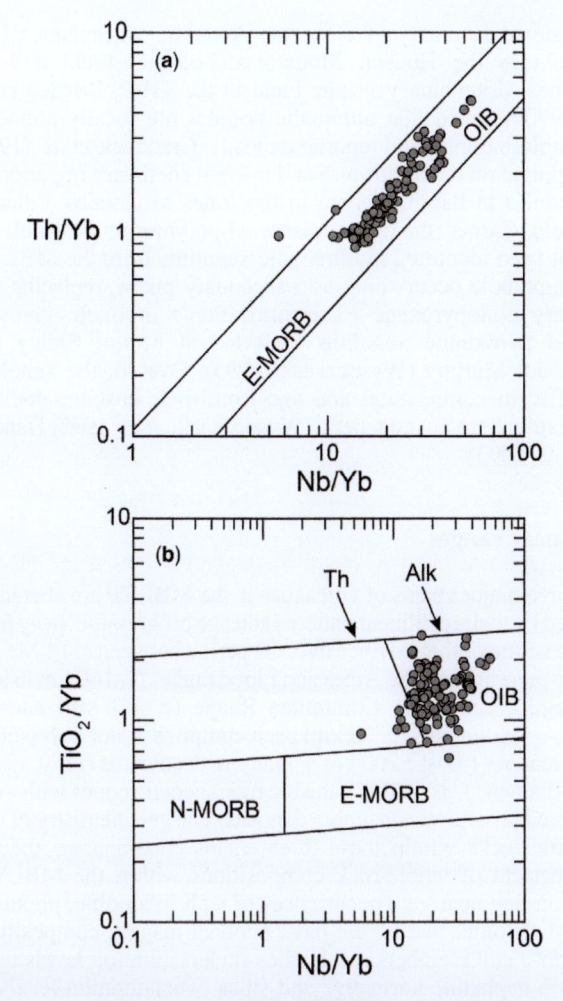

Fig. 8. Plots of (**a**) Nb/Yb v. Th/Yb and (**b**) TiO$_2$/Yb for basalt compositions (see Fig. 3) following the classification of Pearce (2008).

intermediate to phonolitic compositions, in MBL and wEL mostly overlap with the McMurdo Volcanic Group in the western Ross Sea (Rocholl *et al.* 1995; Rocchi *et al.* 2002; Sims *et al.* 2008; Nardini *et al.* 2009; Martin *et al.* 2013, 2021; Aviado *et al.* 2015; Lee *et al.* 2015; Panter *et al.* 2018; Phillips *et al.* 2018; Rocchi and Smellie 2021). Volcanic rocks with the highest silica contents (>58 wt% SiO$_2$) have the widest range in ^{87}Sr/^{86}Sr$_i$ ratios (0.7029–0.7078). Not included in Figure 10a & c are samples with ^{87}Sr/^{86}Sr$_i$ ratios in excess of 0.7100, including three from the Executive Committee Range (two rhyolite samples from Mount Hartigan and a phonolite sample from Mount Hampton: LeMasurier *et al.* 2003) and a rhyolite sample from Mount Andrus within the Ames Range (Fig. 1a). The radiogenic initial ^{87}Sr/^{86}Sr ratio (0.7170) of one of the samples (AR44C: Supplementary material Table S1) was ascribed to magma contamination by upper-crustal materials (LeMasurier *et al.* 2011). Panter *et al.* (1997) also considered slightly elevated ^{87}Sr/^{86}Sr ratios (0.7033–0.7042) of silica-oversaturated trachyte samples from Mount Sidley in the Executive Committee Range (Fig. 1a) to be a consequence of contamination by crust and specifically called for an arc-related calc-alkaline granitoid as the contaminant (i.e. Ford Granodiorite: Weaver *et al.* 1991, 1992). Alternatively, Gaffney and Siddoway (2007) concluded that the slightly elevated ^{87}Sr/^{86}Sr ratios (0.7038–0.7040) measured at Recess Nunatak relative to nearby Mount Perkins and Mount Avers (0.7029–0.7031) in the Fosdick Mountains (Fig. 1a) was not a result of contamination by crust but a product of melting a heterogeneous mantle source.

Isotopes of Pb measured on volcanic rocks from MBL and wEL have restricted ^{207}Pb/^{204}Pb ratios (15.54–15.77) with wider ranges of ^{208}Pb/^{204}Pb (38.47–40.18) and ^{206}Pb/^{204}Pb (19.04–20.93) ratios, the latter of which extends to much higher values than those found within the western Ross Sea (Fig. 10b, d). Basalt and intermediate compositions that have ^{206}Pb/^{204}Pb ratios of >20 are predominantly from two areas within the MBLVP: the Hobbs Coast nunataks (Hart

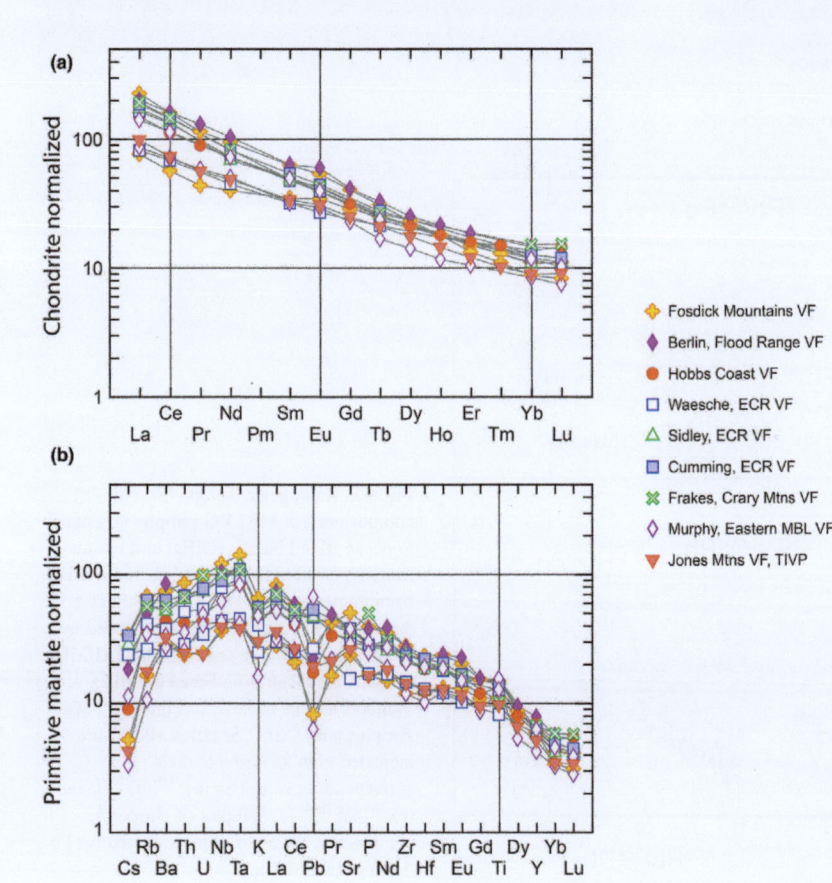

Fig. 9. Normalized multi-element plots of relatively unfractionated basalt compositions (MgO = 8–10 wt%) from volcanic fields (VF) within the MBLVP and the Jones Mountains (Mtns) Volcanic Field within the TIVP. ECR is the Executive Committee Range. (**a**) A plot of REE concentrations normalized to chondrite (Sun and McDonough 1989) and (**b**) normalized to primitive mantle (McDonough and Sun 1995).

et al. 1997) and the Crary Mountains (Panter *et al.* 2000). The isotopic signatures of basalt samples from the WARS coupled with their OIB compositional characteristics have been recognized as part of a much more extensive region of HIMU-like magmatism that spans the now widely scattered continental fragments of East Gondwana (i.e. Zealandia and southeastern Australia) (Finn *et al.* 2005; Panter *et al.* 2006; Sprung *et al.* 2007; Timm *et al.* 2009, 2010; Scott *et al.* 2013, 2014, 2016; McCoy-West *et al.* 2010, 2016; Price *et al.* 2014; Gamble *et al.* 2018; van der Meer *et al.* 2017, 2018), as well as adjacent seamounts and ocean islands (Hoernle *et al.* 2006; Kipf *et al.* 2014; Panter *et al.* 2018; Park *et al.* 2019). The recognition of this 'commonality' has important implications for our understanding of lithospheric and sublithospheric mantle domains, and the possible influence of a superplume on Gondwana break-up (Lanyon *et al.* 1993; Weaver *et al.* 1994; LeMasurier and Landis 1996; Hart *et al.* 1997; Storey *et al.* 1999; Panter *et al.* 2000; Kipf *et al.* 2014; Hoernle *et al.* 2020).

Xenoliths of crustal and mantle origin occur at many locations in MBL and wEL. They are typically found as fragments within basanitic scoria and lava flows from late-stage parasitic cones on the flanks of larger edifices. Collection sites in the MBLVP include those within the Executive Committee Range Volcanic Field (Mount Hampton, Mount Cumming and Mount Sidley), the Fosdick Mountains Volcanic Field (Marujupu Peak, Mount Avers, Demas Bluff, Bird Bluff and Recess Nunatak), the McCuddin Mountains Volcanic Field (Mount Flint, Mount Aldaz and the USAS Escarpment) and Mount Murphy located in the Eastern MBL Volcanic Field (Wysoczanski 1993; Wysoczanski *et al.* 1995; Grapes *et al.* 2003; Handler *et al.* 2003; Chatzaras *et al.* 2016; authors' observations). In the MBLVP, crustal xenoliths include upper-crustal lithologies and lower-crustal granulites, which are predominantly meta-igneous gabbro and norite. Ultramafic mantle xenoliths include pyroxenites and peridotites (e.g. dunite, harzburgite, wehrlite and spinel-bearing lherzolite). In a regional summary of Cenozoic Antarctic volcanism, which includes the Hudson Mountains Volcanic Field and the Jones Mountains Volcanic Field in the TIVP, Rowley *et al.* (1990*b*) stated that ultramafic nodules are locally abundant within basanite and tephrite deposits. Craddock *et al.* (1964) reported on the occurrence and mineral chemistry of peridotite nodules in basaltic tephra in the Jones Mountains Volcanic Field. Garnet (the high pressure Al-polymorph of spinel) has not been identified in ultramafic xenoliths from the MBLVP. Amphibole occurs only as a secondary phase, replacing primary clinopyroxene ('kaersutitization') in mafic granulite and pyroxenite xenoliths collected at Mount Sidley and Mount Murphy (Wysoczanski 1993). Overall, the xenoliths differ in composition and age from their host basalts, and therefore are not cogenetic (Wysoczanski *et al.* 1995; Handler *et al.* 2003).

Linear ranges

Three major chains of volcanoes in the MBLVP are characterized by their rectilinear patterns and age progression away from the centre of the province over the period between *c.* 14 Ma and the present day. The Ames and Flood ranges (*c.* 160 km in total length), Executive Committee Range (*c.* 110 km) and the Crary Mountains (*c.* 70 km) each comprise major polygenetic volcanoes (most >200 km^3), many with calderas (most >2 km in diameter), as well as parasitic monogenetic cones with associated lava flows and tephra deposits. The geochemistry of volcanic rocks within these three chains encompasses the full spectrum of whole-rock compositions within the MBLVG, including nearly all occurrences of tephriphonolite, phonolite and rhyolite: that is, the most evolved magma compositions whose end members reach silica-undersaturation levels up to 29% nepheline-normative and silica-oversaturation levels up to 32% quartz-normative.

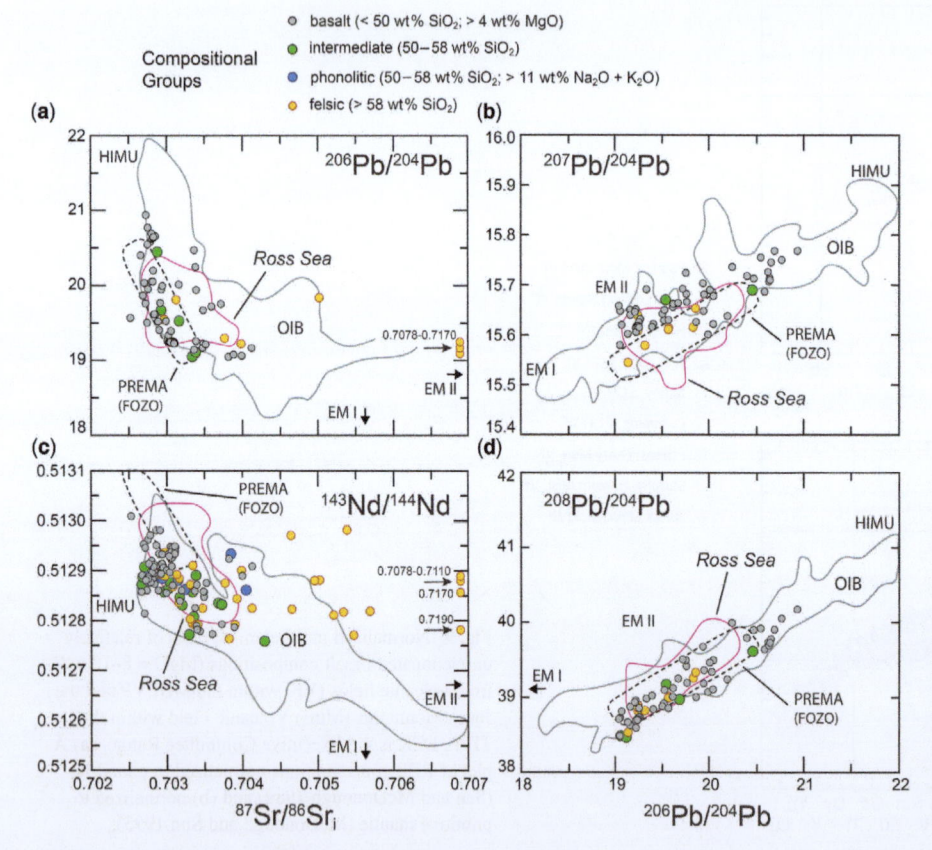

Fig. 10. Radiogenic isotope (Sr, Nd, Pb) compositions of MBLVG samples compared to ocean island basalts (OIBs) and basalts from the western Ross Sea (data and sources from Panter *et al.* 2018). Mantle source end members PREMA (FOZO) as indicated by the black dashed line and regions of HIMU and EM are taken from Stracke (2012). Symbols are the same as in Figure 3. Felsic samples with ^{87}Sr/^{86}Sr ratios >0.707 are indicated with an arrow at their corresponding values for (**a**) ^{206}Pb/^{204}Pb and (**c**) ^{143}Nd/^{144}Nd. Samples of phonolitic composition have not been measured for Pb isotopes (a, b & d).

Executive Committee Range Volcanic Field

The Executive Committee Range Volcanic Field consists of a north–south-aligned range of five major volcanoes (Fig. 1a). Each volcano contains one or more calderas ranging from *c.* 1 to nearly 10 km in diameter at Chang Peak. Mount Sidley (4285 m above sea-level, hereafter abbreviated m asl) has three calderas with the largest and most recent being the result of a catastrophic landslide and paroxysmal eruption that initiated the formation of the *c.* 5 × 7 km and >1200 m-deep Weiss Amphitheater (Panter *et al.* 1994). From north to south, the volcanoes of the Executive Committee Range decrease in age and consist of the coalesced doublet of Whitney Peak and Mount Hampton (13.36–8.6 Ma), Mount Cumming (10.4–10.0 Ma, plus a parasitic cone dated at 3.0 Ma), Mount Hartigan (8.50–6.02 Ma), Mount Sidley (5.77–4.24 Ma), and the Mount Waesche and Chang Peak caldera doublet (1.6–0.17 Ma: Wilch *et al.* 2021).

Basaltic compositions (i.e. basanite, alkali basalt and hawaiite) and trachyte series compositions (i.e. silica-undersaturated to silica-oversaturated) occur at each of the volcanoes. Rhyolite compositions are found at Whitney Peak ('low-silica' rhyolite), Mount Hartigan and Chang Peak (Fig. 11). Phonolitic series compositions (i.e. phonotephrite, tephriphonolite and phonolite) occur only in the larger and better-exposed edifices of Mount Sidley, Mount Waesche and Mount Hampton (Fig. 11). Intermediate compositions (i.e. mugearite, phonotephrite and benmorite) are relatively rare but are best represented at Mount Sidley and Mount Waesche (Fig. 11). Trace and minor element patterns on primitive-mantle-normalized multi-elements plots (Fig. 12) demonstrate the progressive evolution of magmas, showing an overall increase in concentration of highly incompatible elements (e.g. Rb, Th, Nb and La) between mafic and felsic types. Notable is the depletion in Ba, Sr, P, Eu and Ti relative to neighbouring elements that produce pronounced negative anomalies on normalized plots of trachyte and rhyolite with the additional negative anomaly for uranium in phonolite samples from Mount Sidley (Fig. 12). The depletions can be explained by the progressive removal of feldspars and feldspathoids, apatite, and titanomagnetite, which are common phenocryst and microphenocryst phases in these rocks (González-Ferrán and González-Bonorino 1972; LeMasurier 1990*a*, *b*, *c*; LeMasurier and Kawachi 1990*a*, *b*, *c*; Panter *et al.* 1997). At Mount Sidley, the differentiation of the phonolitic series is modelled by the sequential fractionation of diopside, olivine, plagioclase, titanomagnetite, nepheline and/or apatite from basanite to derive 35% mugearite, 25% benmoreite and 20% phonolite as residual liquids (Panter *et al.* 1997). Panter *et al.* (1997) also demonstrated that the removal of a similar mineral assemblage (excluding nepheline) from a parental magma of alkali basalt can explain the production of the silica-undersaturated trachyte series at Mount Sidley and Mount Waesche. For volcanism in the Executive Committee Range, LeMasurier *et al.* (2003) called upon polybaric mineral fractionation within a complex subvolcanic plumbing system that was repeatedly replenished with new basaltic magmas to explain the derivation of phonolite, trachyte and rhyolite compositions with minimal contamination by crust. Complex plumbing beneath Mount Sidley is manifest by inclusion-bearing lavas that represent commingling of basaltic and trachytic magmas during the eruption of a zoned system, as well as magma mixing during an earlier stage of activity that produced hybridized phonolitic compositions (Panter *et al.* 1997). Finally, two

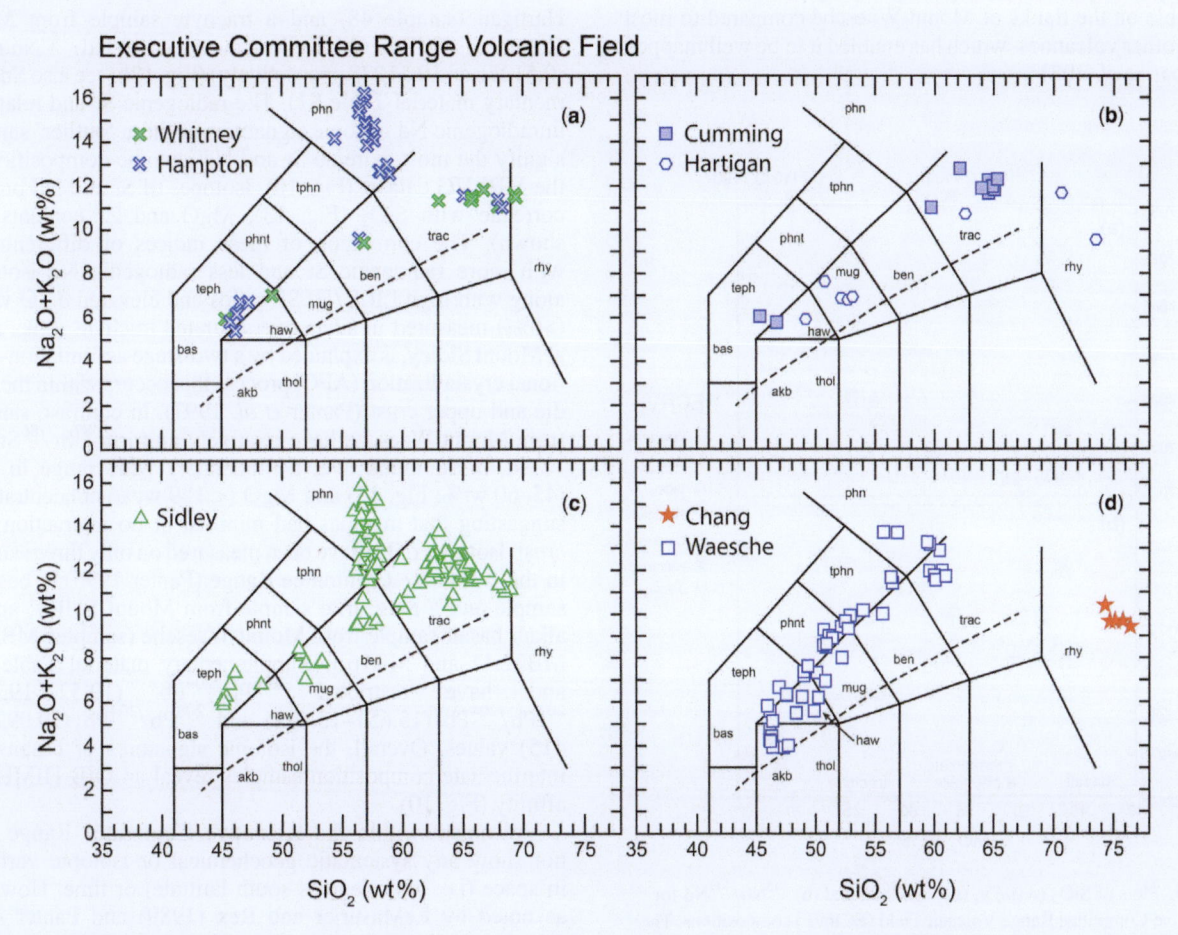

Fig. 11. Total alkali ($Na_2O + K_2O$ (wt%)) v. SiO_2 (wt%) classification plot for compositions from the Executive Committee Range Volcanic Field, MBLVP. Abbreviations for compositional names and the dashed line that delimits alkaline from subalkaline compositions are the same as in Figure 3.

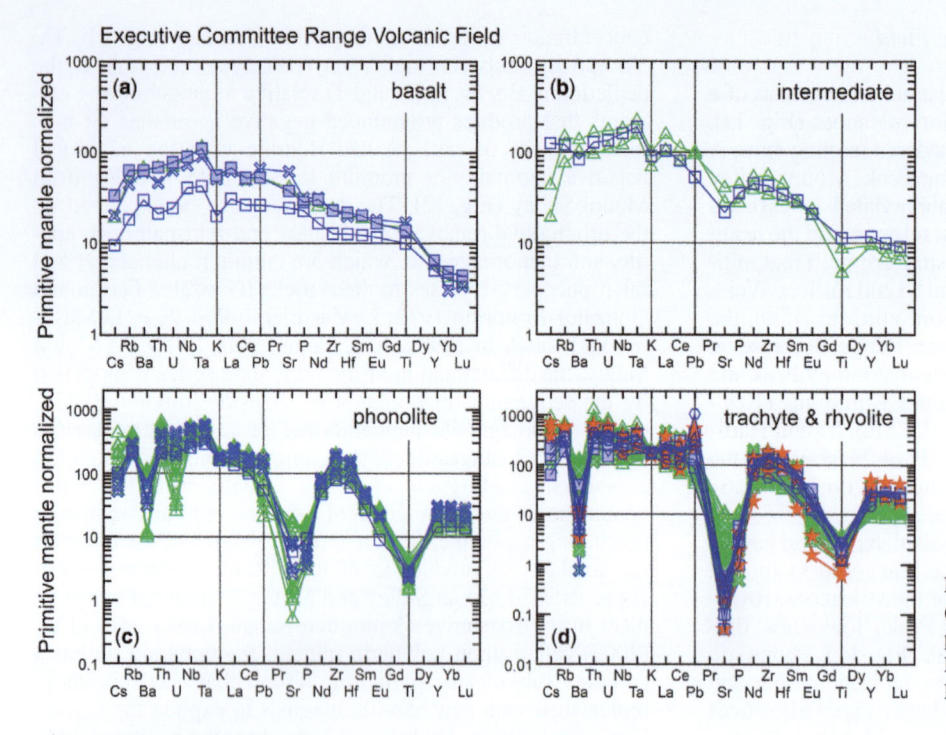

Fig. 12. Primitive-mantle-normalized (McDonough and Sun 1995) multi-element plots of representative compositions from the Executive Committee Range Volcanic Field, MBLVP.

compositional series identified at Mount Waesche may have erupted from the same or geographically very close, yet magmatically isolated, vents (Dunbar et al. 2021), which is consistent with a complicated plumbing system beneath this volcano. Mount Waesche also has the best-represented high-alkaline intermediate series in the MBLVG (cf. Fig. 11). This is possibly a reflection of the generally excellent exposure available on the flanks of Mount Waesche compared to most of the other volcanoes, which has enabled it to be well mapped (Dunbar et al. 2021).

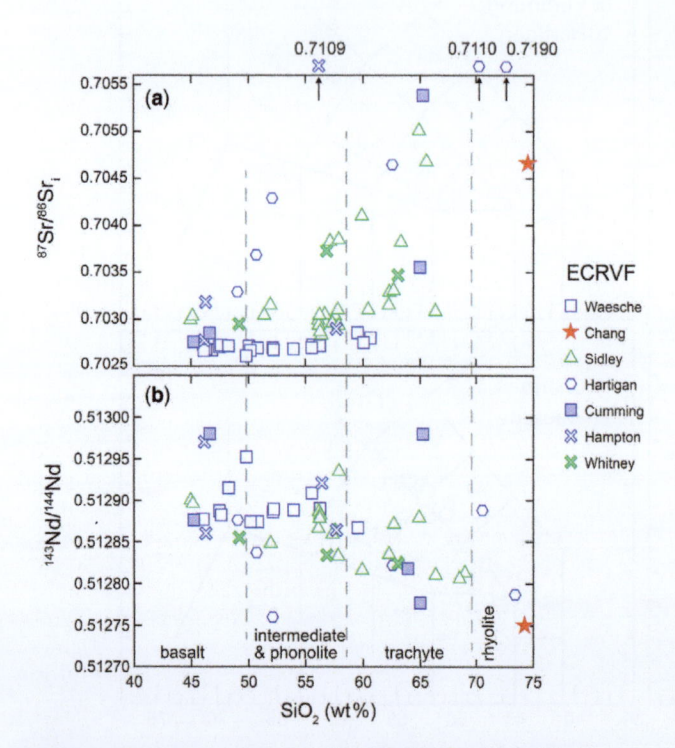

Fig. 13. Plots of SiO$_2$ (wt%) v. (**a**) ^{87}Sr/^{86}Sr$_i$ and (**b**) ^{143}Nd/^{144}Nd for Executive Committee Range Volcanic Field (ECRVF) compositions. The dashed vertical lines delimit the approximate boundaries between rock types as defined in Figure 3.

Isotopic compositions of volcanic rocks within the Executive Committee Range vary and have ^{143}Nd/^{144}Nd ranging from 0.51276 to 0.51298 and most ^{87}Sr/^{86}Sr$_i$ ratios that range from 0.70261 to 0.70568. Two rhyolite samples from Mount Hartigan, along with a phonolite sample from Mount Hampton, have highly radiogenic ^{87}Sr/^{86}Sr$_i$ ratios of 0.71899, 0.71101 and 0.71091, respectively. A mugearite sample from Mount Hartigan (sample 48) and a trachyte sample from Mount Cummings (MB43.6) have low measured ^{143}Nd/^{144}Nd ratios (0.51276 and 0.51278, respectively) (Fig. 13b; see also Supplementary material Table S1). The radiogenic Sr and relatively unradiogenic Nd isotopic signatures of these 'outlier' samples signify the most extreme Sr and Nd isotopic compositions in the MBLVG dataset (Fig. 10). Isotopes of Sr and Nd broadly correlate with SiO$_2$ (Fig. 13), MgO and Zr contents (not shown). The correlation of these indices of differentiation with more radiogenic Sr and less radiogenic Nd isotopes, along with high LILE/HFSE ratios and elevated δ^{18}O values (>6‰) measured in silica-oversaturated trachyte series rocks at Mount Sidley, is explained by a two-stage assimilation–fractional crystallization (AFC) process that occurs within the middle and upper crust (Panter et al. 1997). In contrast, samples from Mount Waesche have relatively uniform ^{87}Sr/^{86}Sr$_i$ and ^{143}Nd/^{144}Nd ratios over a moderately wide range in SiO$_2$ (45–60 wt%: Fig. 13) and MgO (< 1–9 wt%) concentrations, suggesting that magmas had minimal or no interaction with crust. Isotopes of Pb have been measured on only three samples in the Executive Committee Range (Panter 1995): a basanite sample and a mugearite sample from Mount Sidley; and an alkali basalt sample from Mount Waesche (samples MB27.5, MB32.11 and MB16.1: Supplementary material Table S1), and have restricted ^{206}Pb/^{204}Pb (19.524–19.758), ^{207}Pb/^{204}Pb (15.651–15.683) and ^{208}Pb/^{204}Pb (39.099–39.415) values. Overall, the isotopic signatures of basalts and intermediate composition samples reveal an OIB HIMU-like affinity (Fig. 10).

Volcanism within the Executive Committee Range does not show any systematic geochemical or isotopic variation in space (i.e. with degrees south latitude) or time. However, as noted by LeMasurier and Rex (1989) and Panter et al. (1994, 1997), the southward migration of activity from one volcano to the next is coincident with major changes in magma series compositions. For example, the trachyte–rhyolite series at Whitney Peak is followed by the eruption of phonolitic series magmas at Mount Hampton, and then by a trachyte–rhyolite series at Mount Cumming and Mount Hartigan (Fig. 11). Mount Sidley emulates these changes on a smaller scale and over a shorter time span (c. 1.5 myr). The changes were considered by Panter et al. (1997) to be a consequence of variation in degree of mantle partial melting in which lower-degree melts produce basanite that is parental to the phonolite series, whereas higher-degree melts generate alkali basalt that is parental to the trachyte series. Consequently, in this scenario the migration of felsic volcanism is fundamentally linked to a southward displacement of zones of mantle melting. In contrast, LeMasurier and Rex (1989) proposed that the location of mantle melting is nearly random in both space and time, and that the migration of felsic activity is fundamentally controlled by the propagation of relict fractures within the crust in response to regional doming. For Mount Sidley, Panter et al. (1994) suggested that the overall decrease in volume and duration with each successive stage of activity may indicate fracture migration away from magma reservoirs in the crust coupled with the closure of deep fracture systems that cut off melt supply from the mantle.

Ames and Flood ranges

The Ames Range and the Flood Range are two linear volcanic chains in northwestern MBL (Fig. 1a) that are composed of six major edifices, three of which are orientated in an east–west direction (Flood Range Volcanic Field) and three volcanoes orientated roughly in a north–south direction (Ames Range Volcanic Field). Age data reveal that Flood Range volcanism migrated by more than 100 km over a period of c. 10 myr from east to west; Mount Bursey (10.4–6.4 Ma, plus a parasitic cone dated at 0.49 and 0.25 Ma) → Mount Moulton (5.95–4.03 Ma, plus 1.04 Ma parasitic cone) → Mount Berlin (2.77–0.010 Ma). Volcanism in the Ames Range migrated by less than 30 km over c. 5.6 myr from south to north; Mount Andrus (12.71–9.28 Ma, plus a parasitic cone dated at <0.1 Ma) → Mount Kosciusko (9.20 Ma) → Mount Kauffman (7.1 Ma). The overlap in age at about 6 Ma between Mount Bursey in the Flood Range and Mount Kauffman in the Ames Range is considered by Paulsen and Wilson (2010) to be the best constraint on the timing of a province-wide shift in maximum horizontal stress conditions from north–south to east–west.

In the Ames and Flood ranges, magma compositions are most diverse at Mount Berlin and Mount Andrus. The deposits at these volcanoes are better exposed and vary from basalt (basanite and hawaiite) to intermediate compositions (mugearite, phonotephrite and benmoreite) to felsic varieties that include both silica-undersaturated and silica-oversaturated trachyte (Fig. 14). Rocks with phonolite composition are less common, and are limited to deposits found on Mount Berlin and Mount Moulton. Pantellerite compositions (a variety of peralkaline rhyolite) occur at Mount Moulton (WCM93-318), Mount Andrus (AR44-C and AR42-A) and Mount Kosciusko (AR44D2), along with some pantelleritic trachyte compositions (Fig. 15; see also Supplementary material Table S1), but most trachyte–rhyolite series compositions are comenditic (i.e. higher Al and lower Fe contents). In comparison, rocks with a pantellerite composition do not exist within the Executive Committee Range (Fig. 15) nor at any other volcano in the MBLVG.

Trace and minor element concentration patterns for basalt, intermediate and phonolite compositions from the Ames and Flood ranges (Fig. 16) are broadly comparable to those from the Executive Committee Range. However, compositional patterns for trachyte and rhyolite have overall less pronounced negative anomalies for Ba and Sr (Fig. 12), and higher K/Rb ratios (153–633) relative to the Executive Committee Range compositions (61–377). In addition, negative Eu anomalies in trachyte compositions from the Ames and Flood ranges are subdued relative to trachyte from the Executive Committee Range (Fig. 17). These differences are likely to reflect the proportions of Na-plagioclase and alkali feldspar removed. However, the higher relative concentrations of Ba, Sr and Eu in trachyte and rhyolite samples from the Ames and Flood ranges may also be controlled by peralkalinity (Fig. 17), which lowers the partition coefficients of these elements in feldspars (Mahood and Stimac 1990).

LeMasurier et al. (2011) provided a compilation of radiogenic isotope data on volcanic rocks from the Ames and

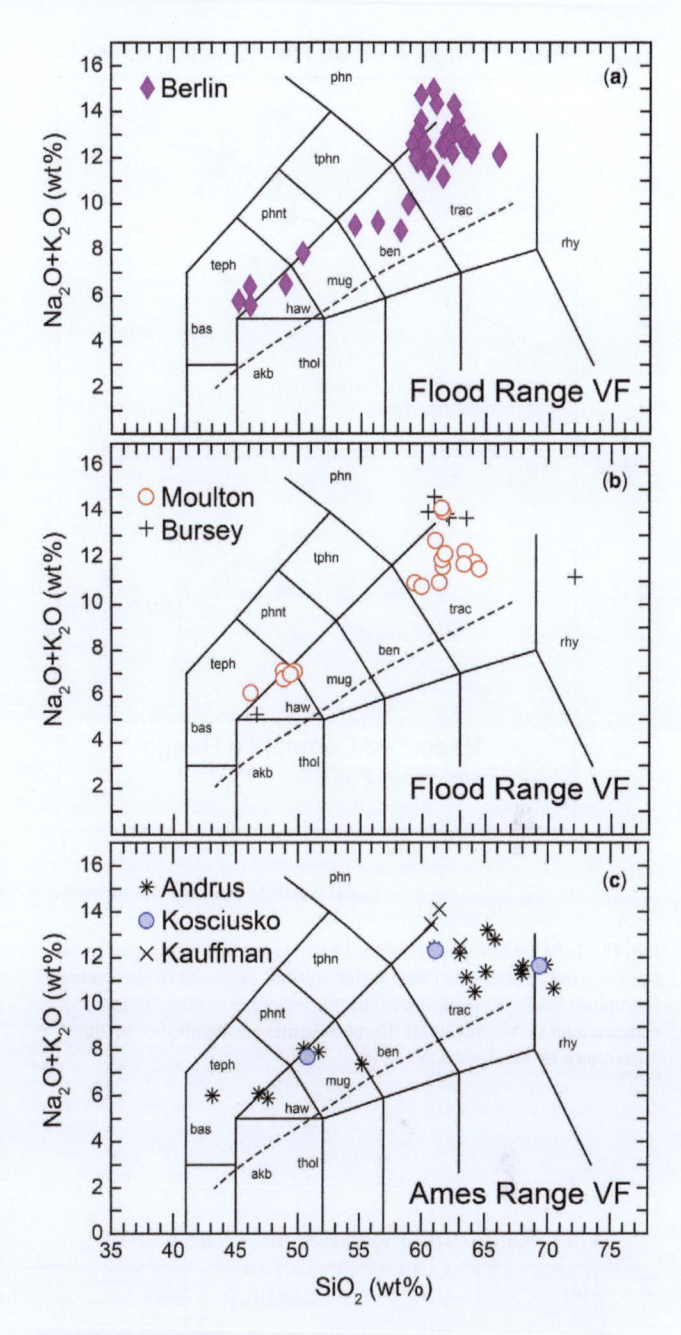

Fig. 14. Total alkali ($Na_2O + K_2O$ (wt%)) v. SiO_2 (wt%) classification plot for compositions from the Ames Range and Flood Range volcanic fields (VF), MBLVP. Abbreviations for compositional names and the dashed line that delimits alkaline from subalkaline compositions are the same as in Figure 3.

Flood ranges. Isotopic compositions mostly fall within a relatively restricted range with ^{143}Nd/^{144}Nd of 0.51284–0.51294 and ^{87}Sr/^{86}Sr$_i$ of 0.70280–0.70502. Measured Pb isotope ratios vary from 19.114 to 20.069 for ^{206}Pb/^{204}Pb, from 15.541 to 15.656 for ^{207}Pb/^{204}Pb and from 38.469 to 39.459 for ^{208}Pb/^{204}Pb ratios. Hafnium isotopes measured on six samples have ^{176}Hf/^{177}Hf ratios that vary from 0.282929 to 0.283010 (Supplementary material Table S1). A pantellerite sample (AR44C, Mount Andrus) and two silica-oversaturated pantelleritic trachyte samples (AR41C and BN31E, Mount Andrus and Mount Berlin, respectively) have elevated ^{87}Sr/^{86}Sr$_i$ ratios (0.70777, 0.70782 and 0.71702, respectively) and relatively unradiogenic ^{206}Pb/^{204}Pb ratios (19.234, 19.131 and 19.232, respectively) that suggest contamination by crust (LeMasurier et al. 2011). Basalt samples from Mount Andrus, Mount Bursey and Mount Berlin have isotopic characteristics (^{87}Sr/^{86}Sr <0.7030, ^{143}Nd/^{144}Nd >0.5129, ^{206}Pb/^{204}Pb values >19.6–20.069) that are OIB HIMU-like, which is consistent with the signatures of basalts found throughout MBL and wEL (Fig. 10).

Volcanic suites with felsic end members that include peralkaline rhyolite (pantellerite) and trachyte (comenditic and pantelleritic) are relatively common in continental rift and oceanic island settings (e.g. Bohrson and Reid 1997; Ren et al. 2006; White et al. 2009; Andreeva and Kovalenko 2011; Rocchi and Smellie 2021). The development of these magma types is often explained by multistage fractional crystallization processes beginning with a mildly alkaline to transitional basalt evolving to trachytic compositions, which, in turn, undergo advanced conditions of crystallization (accumulative $1-F =$ 0.95) dominated by the removal of alkali feldspar, and possibly aided by high halogen contents (Macdonald et al. 2019), to produce pantellerite. Partial melting of mafic alkaline crust followed by fractional crystallization of this magma has also been proposed (Lowenstern and Mahood 1991; Bohrson and Reid 1997). LeMasurier et al. (2011) evaluated the petrogenesis of rocks with pantellerite composition, and their close spatial and temporal association with phonolite and trachyte compositions, within the Ames and Flood ranges. LeMasurier et al. (2011) suggested that removal of kaersutite from basanitic

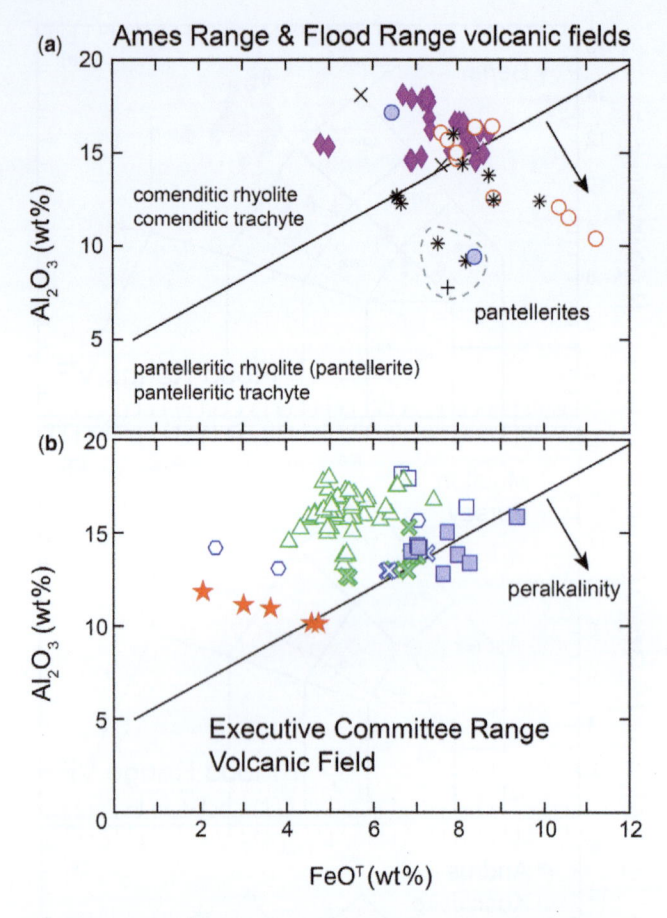

Fig. 15. Total FeO^T v. Al$_2$O$_3$ (wt%) of trachyte and rhyolite compositions from (**a**) Ames Range and Flood Range volcanic fields and (**b**) Executive Committee Range Volcanic Field. The classification of compositional names is after Le Maitre (2002). The pantellerites (i.e. peralkaline rhyolite) are outlined by the dashed line in (a).

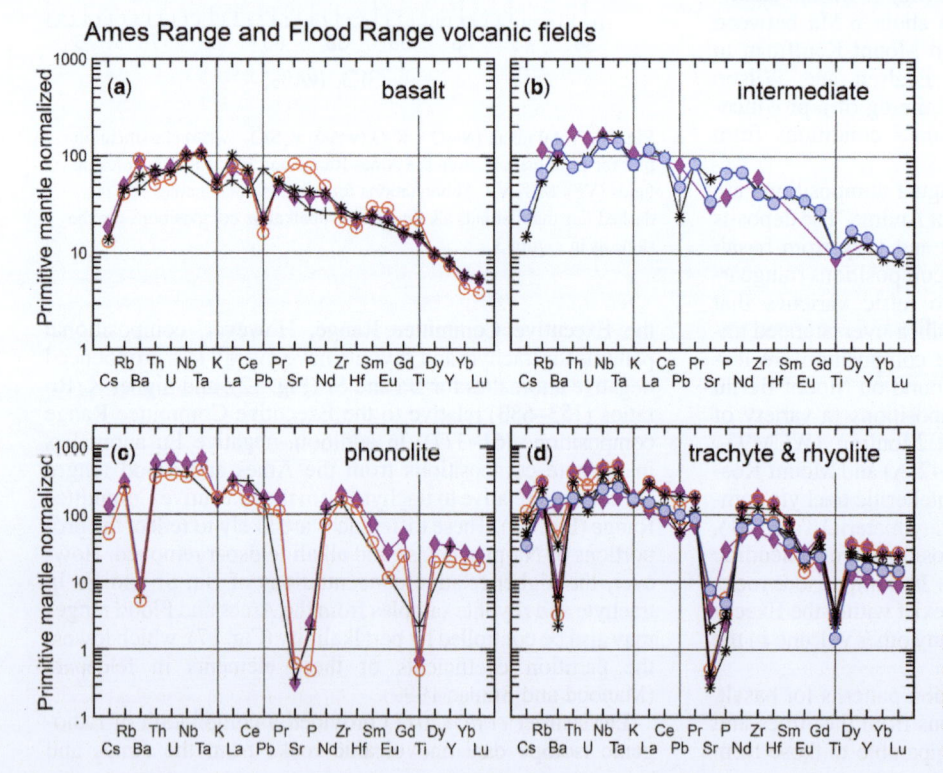

Fig. 16. Primitive-mantle-normalized (McDonough and Sun 1995) multi-element plots of representative compositions from the Ames Range and Flood Range volcanic fields, MBLVP.

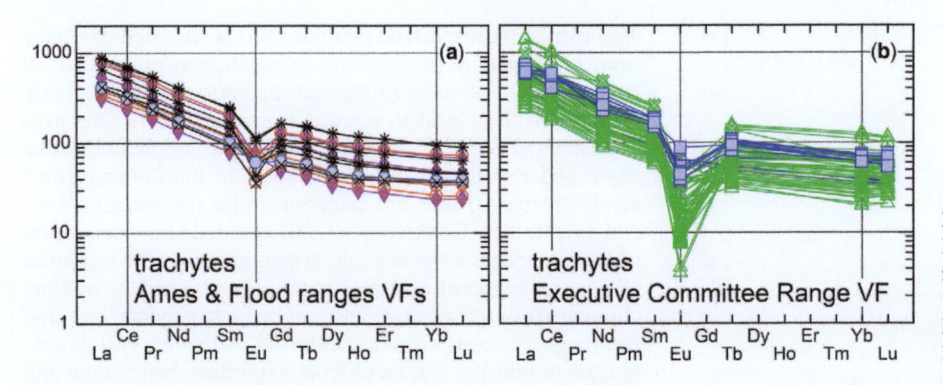

Fig. 17. Chondrite-normalized (Sun and McDonough 1989) REE plots of representative trachytes (SiO_2 = 60–69.8 wt% and Na_2O ≤9 wt%) from (**a**) the Ames and Flood ranges and (**b**) Executive Committee Range, MBLVP.

magma at high pressures (c. 30 km depth) can produce evolved silica-oversaturation liquids with a trachyte composition, and that the prolonged fractionation of the trachytic magmas dominated by the removal of plagioclase at lower pressures in the upper crust can generate pantellerite compositions. Fractionation dominated by the removal of plagioclase from the same or similar basanitic magma at depths of c. 20 km is used to explain the origin of phonolite compositions.

Crary Mountains Volcanic Field

The Crary Mountains Volcanic Field consists of three main coalesced volcanoes aligned roughly north–south over a distance of c. 32 km: Mount Rees, Mount Steere and Mount Frakes (Fig. 1). Approximately 20 km to the SE of these main volcanoes is Boyd Ridge, a c. 15 km east–west- orientated ridge with limited rock exposures. Mount Rees (9.46–6.91 Ma) consists mostly of basaltic through to intermediate to trachyte compositions (Fig. 18). Mount Steere (8.66–5.81 Ma) is deeply dissected and exposes basaltic compositions, as well as rocks with rhyolite, trachyte and phonolite compositions. Mount Frakes (4.26–1.83 Ma, plus parasitic cones at English Rocks (1.62–0.034 Ma); for specific locations in the Crary Mountains Volcanic Field refer to Wilch *et al.* 2021, fig. 18) is the least dissected of the three main volcanoes and is bimodal, consisting of rocks with a phonolite composition and late-stage rocks with a basalt composition (Fig. 18). Boyd Ridge (2.17 Ma) is almost entirely ice covered except for a small scoria cone with a mafic–intermediate composition and a cliff section at Runyon Rock that exposes an extensive sequence of volcaniclastic deposits consisting of phonolite and hawaiite compositions (Wilch *et al.* 2021). The migration of volcanic activity in the Crary Mountains, from NW to SE, occurred at a rate of between 0.7 and 0.8 cm a^{-1} (calculated based on the oldest measured age at each volcano), and is very similar to the migration rate estimated for volcanism in the Executive Committee Range (Panter *et al.* 1994).

Magma compositions show the greatest diversity at Mount Rees and Mount Steere (Fig. 19), which is most likely to be an artefact of better exposures due to greater erosional dissection. Nevertheless, compositions within the Crary Mountains Volcanic Field show an overall decrease in SiO_2 content/silica-saturation with decreasing age of volcanism (Fig. 19 inset). This is not a characteristic of volcanism in either the Executive Committee Range or the Ames and Flood ranges. The compositional trend may also be related to the availability of sampled rock exposures; however, increasing silica-undersaturation with decreasing age is also demonstrated on a smaller scale within volcanic sequences at Mount Steere (8.7–5.8 Ma) and Mount Frakes (4.3–0.03 Ma) (Fig. 19). It is important to note that this relationship does not correspond with magma evolution along any single liquid line of descent. In fact, the

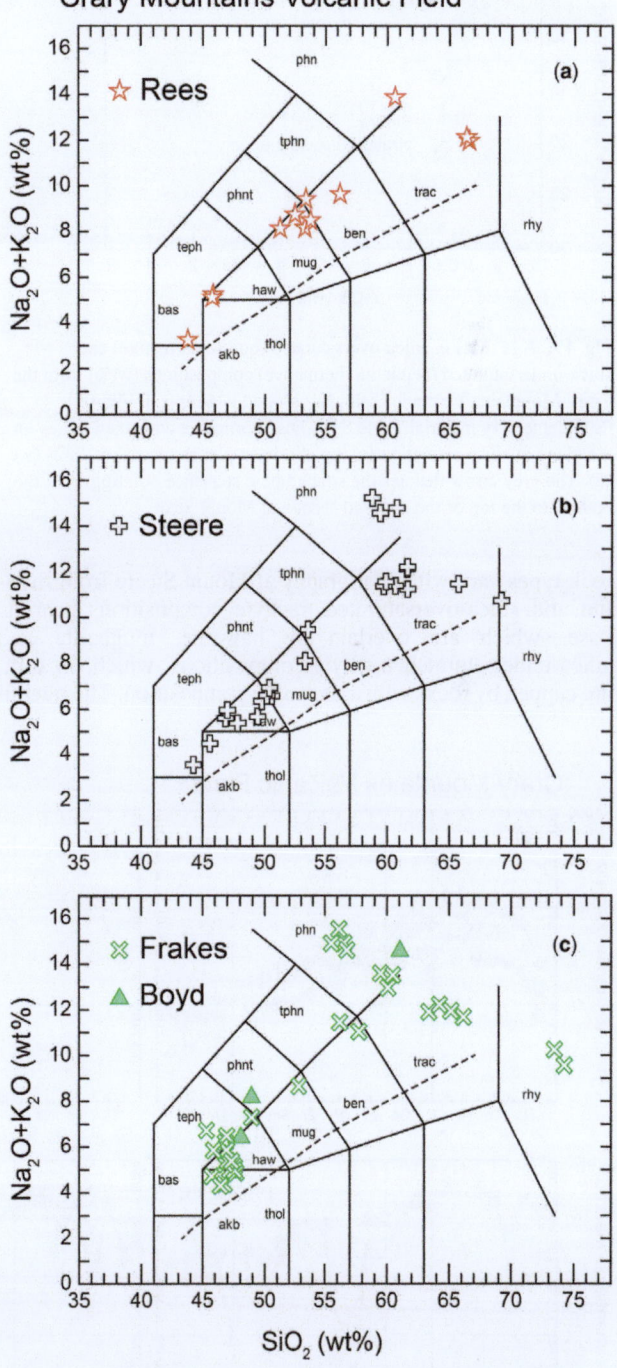

Fig. 18. Total alkali ($Na_2O + K_2O$ (wt%)) v. SiO_2 (wt%) classification plot for compositions from the Crary Mountains Volcanic Field, MBLVP. Abbreviations for compositional names and the dashed line that delimits alkaline from subalkaline compositions are the same as in Figure 3.

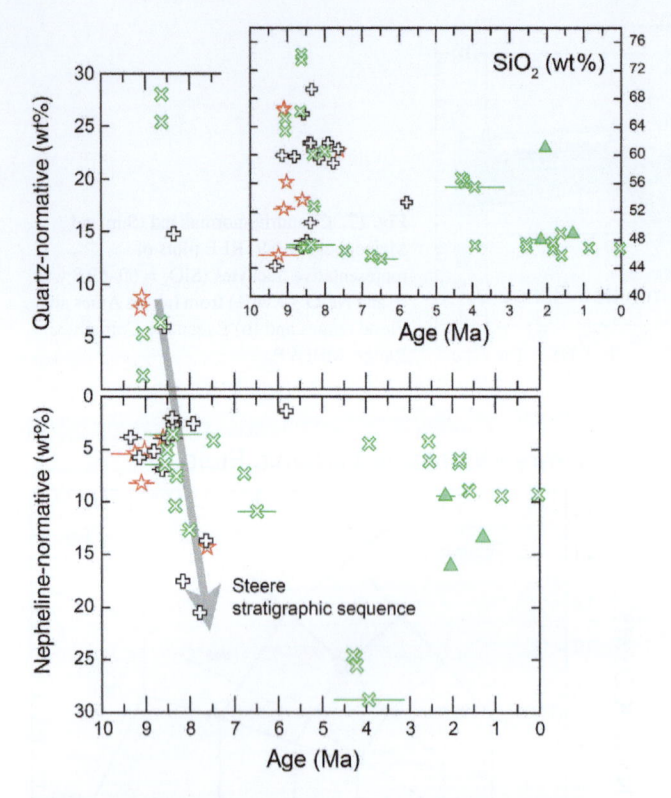

Fig. 19. Age (Ma) v. silica-oversaturated (quartz-normative) and silica-undersaturated (nepheline-normative) compositions (wt%) from the Crary Mountains Volcanic Field. Not plotted is sample TW92-051 (Supplementary material Table S1) whose normative composition lies on the plane of silica-saturation (ol–hy–di). Inset plot shows age v. SiO$_2$ (wt %). The grey arrow denotes the stratigraphic sequence pointing from the bottom to the top of the exposed section at Mount Steer.

rock types vary with stratigraphy at Mount Steer from rhyolite and silica-oversaturated trachyte compositions near its base, which are overlain by hawaiite, mugearite and silica-undersaturated trachyte compositions, which, in turn, are capped by rocks with a phonolite composition. The overall spectrum of these evolved compositions is considered to have been produced by progressive differentiation dominated by crystal fractionation from parental magmas that are variably saturated with respect to silica (Chakraborty 2007). Relative depletions in Ba, Sr, P, Eu and Ti in samples of phonolite, trachyte and rhyolite (Fig. 20c, d) attest to the dominance of alkali feldspar, apatite and magnetite in the fractionating mineral assemblage. Chakraborty (2007) modelled the derivation of a silica-undersaturated trachyte daughter as a 19% residuum of a basanite parent by fractionating the assemblage olivine, clinopyroxene, plagioclase and magnetite to generate hawaiite and mugearite compositions, followed by the removal of apatite, alkali feldspar and amphibole to produce benmoreite and trachyte compositions. The attainment of silica-oversaturated trachyte and rhyolite compositions requires advanced fractionation (accumulative $1-F >92\%$) of this mineral assemblage. However, this produces poor solutions for some trace elements, particularly the REEs. Chakraborty (2007) also modelled assimilation–fractional crystallization involving the contamination of a mugearite magma by granodioritic crust (Weaver et al. 1992) and concluded that this process is likely to have generated the silica-oversaturated magmas. However, a comprehensive evaluation of crustal contamination cannot be undertaken until isotopic data are available.

Basalt compositions from the Crary Mountains Volcanic Field have a narrow range in initial $^{87}Sr/^{86}Sr$ (0.70267–0.70282) and measured $^{143}Nd/^{144}Nd$ (0.51286–0.51298) isotopic ratios, whereas measured Pb isotopic values show much larger variations: $^{206}Pb/^{204}Pb$ 19.95–20.93, $^{207}Pb/^{204}Pb$ 15.68–15.77 and $^{208}Pb/^{204}Pb$ 39.36–40.12. Two samples (TW92059 and TW92135: Supplementary material Table S1) have the highest $^{206}Pb/^{204}Pb$ ratios yet measured within the WARS (>20.7) and, in conjunction with their low $^{87}Sr/^{86}Sr$ ratios, reflect a primitive mantle source containing a HIMU mantle component (Panter et al. 2000). Primitive-mantle-normalized trace element patterns are typical of OIBs in showing the characteristic 'humped' pattern with high relative concentrations of Nb and Ta, and pronounced negative anomalies for K and Pb (Fig. 20a). The magnitude of these anomalies, coupled with the marked depletion in Cs relative to Rb, as well as low Rb concentrations argue against crustal

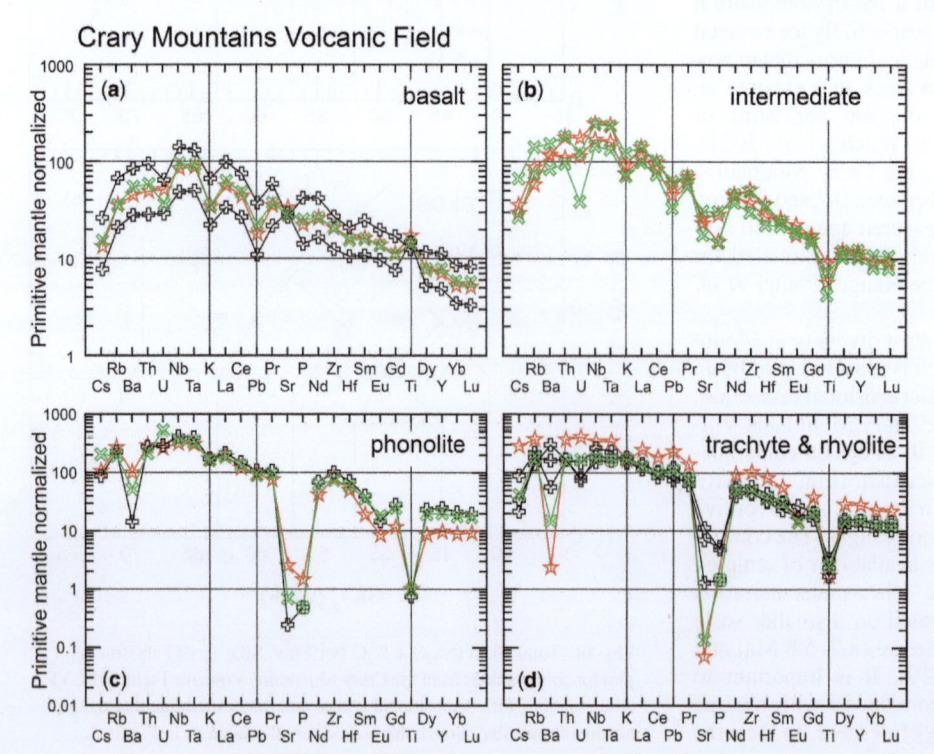

Fig. 20. Primitive-mantle-normalized (McDonough and Sun 1995) multi-element plots of representative compositions from the Crary Mountains Volcanic Field, MBLVP.

contamination, which is also supported by their radiogenic $^{143}Nd/^{144}Nd$ and unradiogenic $^{87}Sr/^{86}Sr$ ratios (Panter et al. 2000). Moderately steep slopes on mantle-normalized REE plots (not shown) for basanite samples (average La/Yb_N = 12.3) and alkali basalt samples (average La/Yb_N = 9.7) indicate that the magmas were derived from a mantle source containing garnet (Panter et al. 2000). In addition, the negative K anomalies shown by basalt samples (Fig. 20a) suggest that amphibole and/or phlogopite must also exist as a residual phase within the source (i.e. metasomatized mantle). Panter et al. (2000) considered small degrees of partial melting between 2 and 3% of a mantle peridotite containing 2.5% amphibole and <1% garnet as being consistent with trace-element variations in the samples with a basalt composition. They concluded that a HIMU-type mantle plume was trapped and stored beneath pre-existing metasomatized lithosphere prior to the Late Cretaceous break-up of New Zealand from West Antarctica. Melting of this layered source in the Late Cenozoic to produce the volcanism at the Crary Mountains may have been facilitated by tectonic stresses imposed by far-field plate-boundary forces, as well as glacial cycling: that is, the loading and unloading of the West Antarctic Ice Sheet (Paulsen and Wilson 2010).

Major isolated volcanic centres

Four major volcanic edifices occur as isolated centres in the MBLVP: Mount Siple, Mount Murphy, Mount Takahe and Toney Mountain (Fig. 1a). In the Eastern MBL Volcanic Field (Wilch et al. 2021), Mount Murphy (c. 590 km^3) is the best-exposed volcano and is comparatively close (<150 km) to the relatively undissected, much more poorly exposed, volcanoes of Mount Takahe (c. 780 km^3) and Toney Mountain (c. 530 km^3). The coastal volcano Mount Siple (c. 1800 km^3) of the Mount Siple Volcanic Field is the northernmost volcanic centre in this region and is almost entirely ice covered. Altogether, the ages of these isolated volcanoes range from Late Miocene (>9 Ma) at Mount Murphy and Toney Mountain to Late Pleistocene (<800 ka) at Mount Takahe and Mount Siple. Ice core tephra and Holocene ages (Palais et al. 1988; Dunbar et al. 2021; Wilch et al. 2021) show that Takahe should be considered active.

Mount Murphy

Mount Murphy (2703 m asl) is a highly dissected Late Miocene–Early Pliocene (9.46–4.37 Ma) polygenetic volcano surmounted by multiple post-edifice-building volcanic sequences (3.70–0.609 Ma). The main volcanic edifice of Mount Murphy overlies Cenozoic and pre-Cenozoic plutonic and hypabyssal igneous rocks (Pankhurst et al. 1998). Immediately to the west of the main edifice are isolated volcanic centres at Icefall Nunatak (6.89–6.61 Ma), Hedin Nunatak (6.58–6.28 Ma) and Turtle Peak (6.03–4.76 Ma), which are described as table-mountain sequences similar to those in Iceland (Wilch et al. 2021 and references therein and locations shown in fig. 7). Dorrel Rock, a nunatak that lies still further to the west, is composed of latest Eocene–earliest Oligocene (c. 35 Ma) alkaline intrusive rocks that have compositional affinity with the MBLVG (Rocchi et al. 2006).

The majority of volcanic rocks are mafic in composition and range from picro-basalt (olivine-phyric) to alkali basalt, basanite and hawaiite (Fig. 21). Basalt compositions range from silica-undersaturated (<1–10% nepheline-normative) to silica-saturated (<1–10% hypersthene-normative and 13–20% diopside-normative) with one basalt (85-33D) being slightly silica-oversaturated (<1% quartz-normative). It is noteworthy that no other basalt from volcanoes found within the linear ranges is silica-saturated. More evolved compositions of mugearite, benmoreite and trachyte occur within the upper portions of the exposed stratigraphy of Mount Murphy. The single phonolite sample (WCM90-144: Fig. 21a; see also Supplementary material Table S1) is a cognate xenolith collected from an agglutinated spatter-fed lava flow below the Mount Murphy summit. The compositional stratigraphy (i.e. basaltic basal succession overlain by intermediate–felsic compositions) exhibited at Mount Murphy was used by LeMasurier (2013) to infer a similar structure for several major undissected and/or ice-covered volcanoes in MBL. For intermediate and felsic compositions (i.e. benmorite and trachyte) only XRF data are available. Nevertheless, several key characteristics distinguish Mount Murphy compositions from those found in the linear volcanic ranges discussed above. For instance, trachyte rocks at Mount Murphy have overall lower concentrations in highly incompatible trace elements Rb, Nb and La relative to trachyte rocks from the well-exposed volcanoes of Mount Steere and Mount Sidley over the same range in SiO_2 (59–63 wt%). On mantle-normalized multi-element plots, negative K anomalies are subdued or absent (Fig. 22), yet the relative depletions of Ba, Sr, P and Ti in trachyte samples suggest that alkali feldspars, apatite and magnetite were controlling phases during late-stage magma evolution.

Basalt samples from Mount Murphy and its satellite volcanic nunataks also have lower Rb contents and have less pronounced positive Nb–Ta anomalies (Fig. 22a) compared to

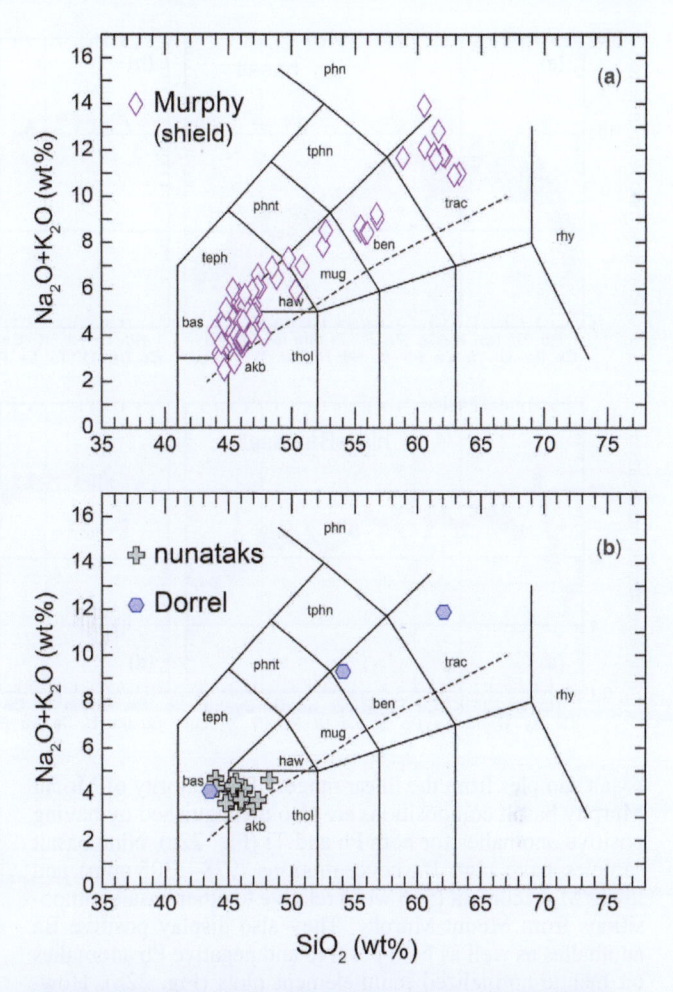

Fig. 21. Total alkali ($Na_2O + K_2O$ (wt%)) v. SiO_2 (wt%) classification plot for compositions from Mount Murphy and satellite nunataks, Eastern MBL Volcanic Field, MBLVP. Abbreviations for compositional names and the dashed line that delimits alkaline from subalkaline compositions are the same as in Figure 3.

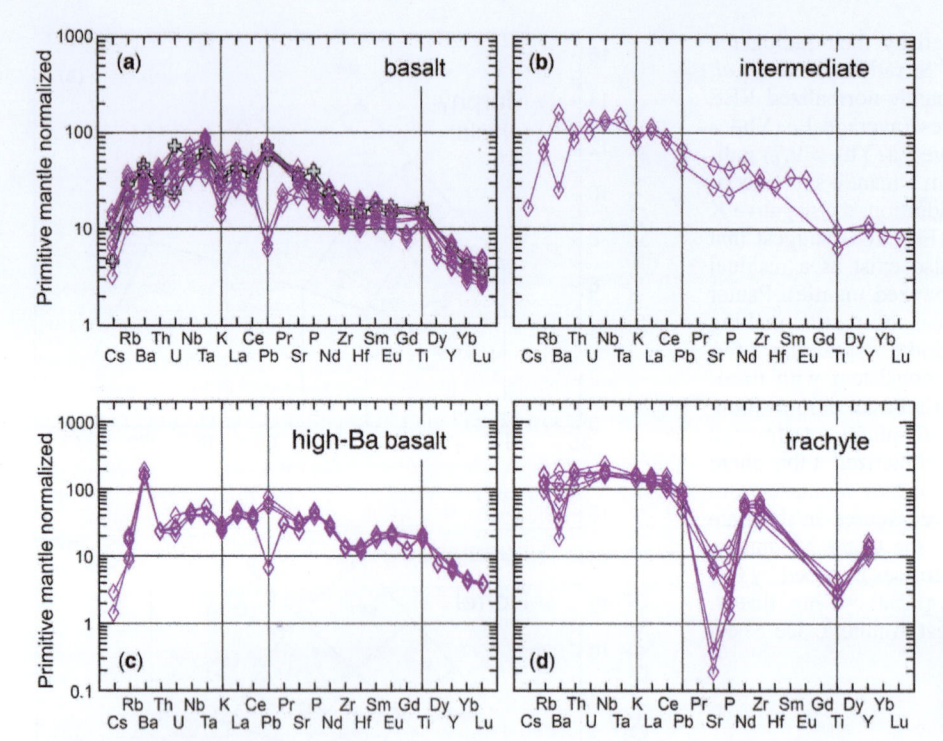

Fig. 22. Primitive-mantle-normalized (McDonough and Sun 1995) multi-element plots of representative compositions from Mount Murphy and satellite nunataks, Eastern MBL Volcanic Field, MBLVP.

basalt samples from the linear ranges. The majority of Mount Murphy basalt compositions are also distinguished by having positive anomalies for both Pb and Ti (Fig. 22a). Nine basalt samples have high Ba concentrations (978–1305 ppm) and lower MgO content (5–6 wt%) relative to other basalt compositions from Mount Murphy. They also display positive Ba anomalies as well as both positive and negative Pb anomalies on mantle-normalized multi-element plots (Fig. 22b). However, it is important to stress that element concentrations displayed in Figure 22 were measured by different techniques (XRF, INAA and ICP-MS). The concentration of Pb in basalts with positive Pb anomalies was measured by XRF, while Pb concentrations for samples that show negative Pb anomalies were measured by ICP-MS. All magmas experienced the effects of olivine and clinopyroxene fractionation but it was most extensive in the high-Ba basalt samples (Fig. 23). Control by clinopyroxene is also indicated by decreasing V content (partition coefficient (K_d) V_{cpx} = 4.8–5.7, basanite: Adam and Green 2006) with increasing Ba concentrations up to c. 600 ppm, as shown in Figure 24a. Fractionation of amphibole from basaltic magmas may have also contributed to this trend (K_d V_{amph} = 3.4–5.7, basanite: Adam and Green 2006). In contrast to other Mount Murphy basalts, the high-Ba basalts have elevated V concentrations and do not display a decrease in V content with increasing Ba concentrations (Fig. 24a). High-Ba basalts also have higher Eu/Eu* ratios (Fig. 24b), which manifest as more pronounced positive Eu anomalies on chondrite-normalized REE plots relative to other basalts (not shown). Overall, TiO_2 concentrations increase with increasing Ba content (Fig. 24c). We suggest that the increase in concentrations of V, Eu and TiO_2 are a consequence of the accumulation of plagioclase, magnetite, amphibole and/or mica; the latter two of which can have high Ba concentrations (Villemant et al. 1981; Latourrette et al. 1995; Arzamastsev et al. 2009). Alternatively, LeMasurier et al. (2016) suggests that the composition of high-Ba basalt is a result of the incorporation of Ba-rich subduction-related materials into the mantle source of Mount Murphy volcanism.

Isotopes of Sr, Nd and Pb have been measured only on two samples (LeMasurier et al. 2016): a high-Ba basalt from the main shield stage at Mount Murphy; and a more mafic basalt (MgO = 11.85 wt%) from the parasitic cone at Sechrist Peak (samples 85-32B and 67–62, respectively: Supplementary material Table S1; for the location of Sechrist Peak refer to Wilch et al. 2021, fig. 7). The high-Ba basalt is more radiogenic in Sr and Pb isotopes ($^{87}Sr/^{86}Sr_i$ = 0.703165, $^{206}Pb/^{204}Pb$ = 20.010, $^{207}Pb/^{204}Pb$ = 15.691 and $^{208}Pb/^{204}Pb$ = 39.463) and less radiogenic in Nd isotopes ($^{143}Nd/^{144}Nd$ = 0.512839) than the younger and more mafic sample ($^{87}Sr/^{86}Sr_i$ = 0.702917, $^{206}Pb/^{204}Pb$ = 19.764, $^{207}Pb/^{204}Pb$ = 15.667, $^{208}Pb/^{204}Pb$ = 39.195 and $^{143}Nd/^{144}Nd$ = 0.512893). The two samples have isotopic signatures that are consistent with the OIB HIMU-like character found in basalts of the MBLVG (Fig. 10).

Mount Takahe

Mount Takahe (3460 m asl) is a large, symmetrical (c. 30 km diameter at the surface of the ice sheet) and non-dissected Late Quaternary shield volcano (Fig. 1a) with a 8 km-diameter

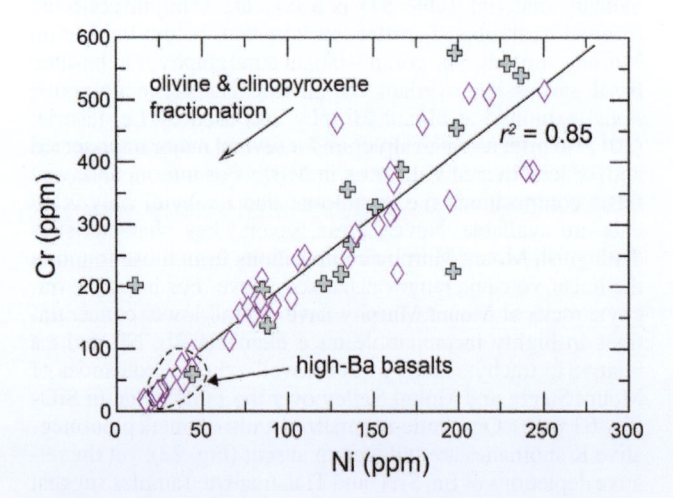

Fig. 23. A plot of Ni (ppm) v. Cr (ppm) for basalt (SiO_2 40–48 wt%, MgO 5–12 wt% and Na_2O ≤6 wt%) samples from Mount Murphy and its satellite nunataks. The trend predicted for the fractional crystallization of olivine and clinopyroxene is denoted by the arrow. High-Ba basalts (see Fig. 22c) are encompassed by the black dashed line.

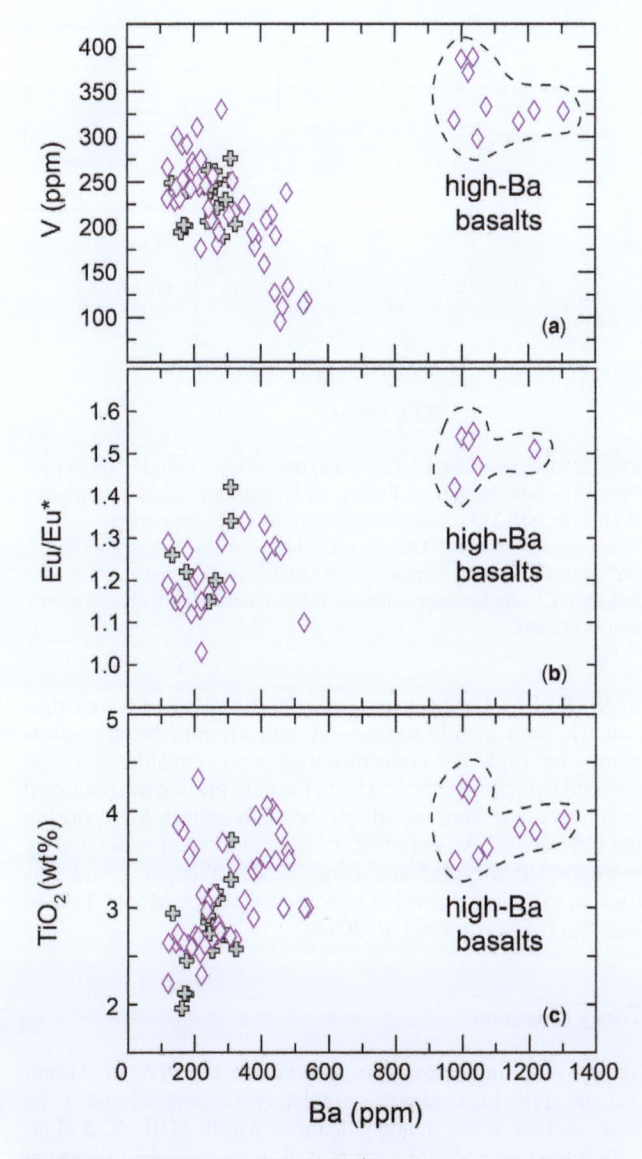

Fig. 24. Plot of Ba (ppm) v. V (ppm), Eu/Eu* and TiO$_2$ (wt%) for basalts from Mount Murphy and satellite nunataks. Eu/Eu* = Eu$_N$/$\sqrt{(Sm_N \times Tb_N)}$, with each element normalized to chondrite (Sun and McDonough 1989). As in Figure 23, the high-Ba basalt samples are encompassed by the black dashed line.

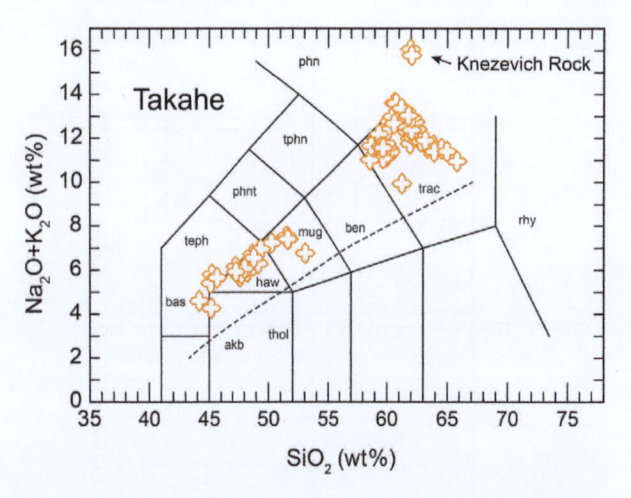

Fig. 25. Total alkali (Na$_2$O + K$_2$O (wt%)) v. SiO$_2$ (wt%) classification plot for compositions from Mount Takahe, Eastern MBL Volcanic Field, MBLVP. Abbreviations for compositional names and the dashed line that delimits alkaline from subalkaline compositions are the same as in Figure 3.

summit caldera. The caldera and the vast majority of the volcano is ice covered. Six samples from on or near the caldera rim yielded ^{40}Ar/^{39}Ar ages from 194.5 to 8.3 ka; post-edifice-building lower-flank outcrops yielded ^{40}Ar/^{39}Ar ages from 105.3 to 7.1 ka (Wilch et al. 1999, 2021).

The compositions of subaerial and subglacial deposits at Mount Takahe are bimodal, consisting of basalt and intermediate types (basanite, hawaiite and mugearite) and trachyte (Fig. 25). The majority of trachyte rocks analysed (n = 67) fall within the classification of pantelleritic trachyte and the rest (n = 7) are comenditic (Le Maitre 2002). Two samples plot marginally within the phonolite field on the total alkali v. silica classification diagram (Fig. 25); W85-15 is weakly nepheline-normative (<1%) and MT85-9 is weakly hypersthene-normative (3.5%). Two other samples, 85-25A and W85-41, collected from a single glassy lava flow (Knezevich Rock, located on the east side of Clausen Glacier, c. 7 km due west of Stauffer Bluff; for locations refer to Wilch et al. 2021, fig. 11) have anomalously high total alkali contents (Fig. 25) dominated by Na$_2$O (c. 12 wt%) and are strongly quartz-normative (c. 16%). They also have lower Al$_2$O$_3$ concentrations, which are less than 1.33FeO^T + 4.4, and, hence, are classified as pantelleritic trachytes (see Le Maitre 2002). LeMasurier et al. (2018) analysed interstitial glass in sample 850-25A and found it to be higher in FeO^T, lower in Al$_2$O$_3$, and slightly lower in SiO$_2$ and K$_2$O relative to whole-rock values. Additionally, the samples from Knezevich Rock are the most enriched in incompatible trace elements (Fig. 26d). To explain the major element composition of sample 85-25A and its higher concentrations of LILEs, LeMasurier et al. (2018) postulated an advanced level of differentiation starting from basanite (accumulative 1−F = 99%) crystallizing within an upper-crustal magma system with the removal of plagioclase and anorthoclase, and minor amounts of clinopyroxene, ilmenite and apatite. This was proposed to be the second stage of evolution for the generation of silica-oversaturated magmas from a silica-undersaturated melt; the first stage being fractional crystallization (c. 60%) dominated by the removal of kaersutite, clinopyroxene and olivine at higher pressures in the upper mantle (LeMasurier et al. 2016, 2018).

A persistent geochemical feature of Mount Takahe basalt, mugearite and low-Si trachyte (SiO$_2$ concentrations between 58 and 61 wt%) samples is the positive anomalies for Ba on mantle-normalized multi-element plots (Fig. 26b, c). Apart from a low-Si trachyte (sample MB43.1) at Mount Cumming in the Executive Committee Range, the Ba concentration in mugearite samples MT85-10 and 85-22C (2892 and 3052 ppm, respectively) from Mount Takahe includes the highest values in the MBLVG (Supplementary material Table S1). Low-Si trachyte and intermediate compositions (i.e. mugearite and benmoreite) with high Ba contents (>1350 ppm) are found at other volcanoes including Mount Hartigan and the Whitney Peak–Mount Hampton doublet in the Executive Committee Range, Mount Berlin in the Flood Range, and Mount Rees in the Crary Mountains. However, what further distinguishes Mount Tahake mugearite and low-Si trachyte from most others is their high Ba/LILE ratios (e.g. high Ba/Ba* = Ba$_N$/$\sqrt{(Rb_N \times Th_N)}$: Fig. 27). Hawaiite rocks from Mount Takahe also show higher Ba/Ba* ratios relative to most other basalt samples with the exception of high-Ba basalt specimens from Mount Murphy (Fig. 27).

Isotopes of Sr and Nd are measured on 15 samples that encompass the compositional spectrum of Mount Takahe deposits and include four basalts analysed for Pb isotopes (LeMasurier et al. 2016, 2018; unpublished data in Supplementary material Table S1). The samples form an array in Sr–Nd isotope space that falls beneath the field for the

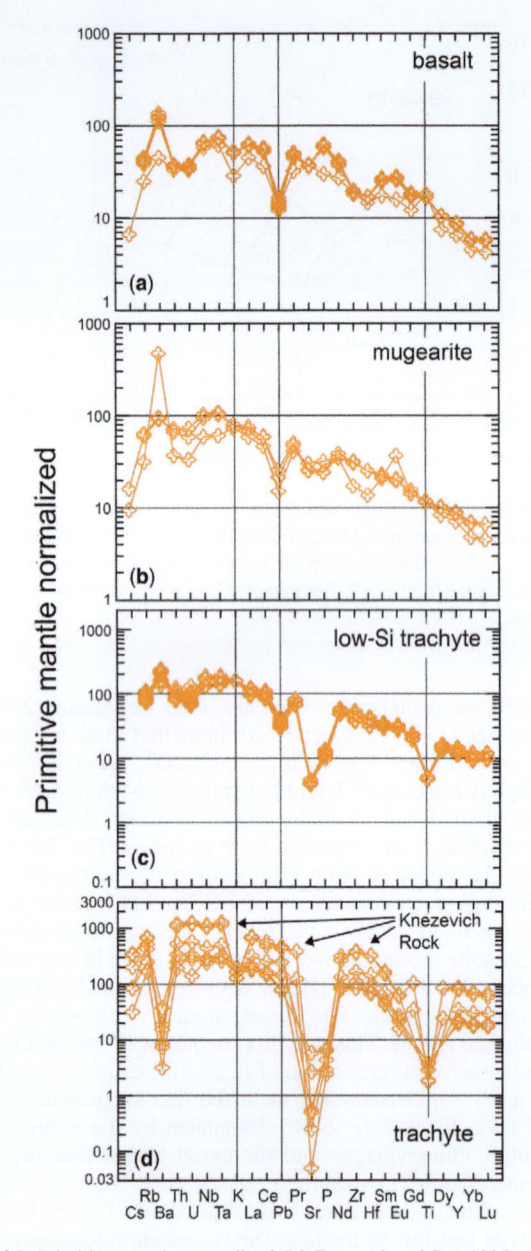

Fig. 26. Primitive-mantle-normalized (McDonough and Sun 1995) multi-element plots of representative compositions from Mount Takahe, Eastern MBL Volcanic Field, MBLVP.

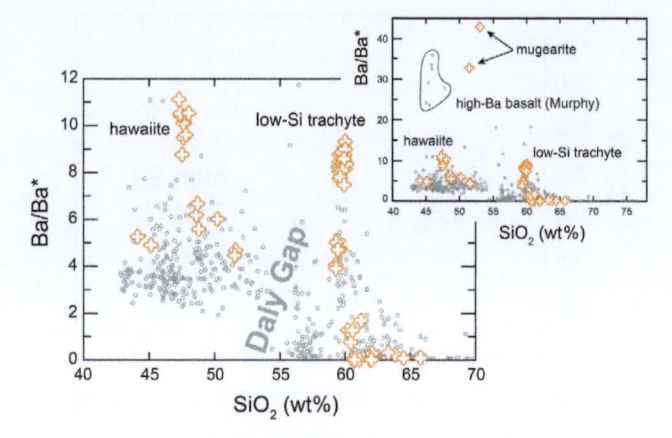

Fig. 27. Diagram of SiO_2 (wt%) v. $Ba/Ba^* = Ba_N/\sqrt{(Rb_N \times Th_N)}$) for Mount Takahe compositions, Eastern MBL Volcanic Field, in comparison to all other MBLVG compositions displayed as grey open circles (Supplementary material Table S1). The inset extends to the higher Ba/Ba* ratios measured in mugearite from Mount Takahe (samples MT85-10 and 85-22C: Supplementary material Table S1) and high-Ba basalt from Mount Murphy.

interpreted by LeMasurier et al. (2016, 2018) to have originated in their mantle source. As with Mount Murphy volcanism, the high Ba concentrations were considered to be sourced from fluids derived from the dehydration of subducted sediments that were introduced beneath eastern MBL during the Cretaceous. An enriched mantle component generated by incorporated recycled sediments (EM1: Weaver 1991) was used to explain the low Nd isotopic values for Mount Takahe samples (LeMasurier et al. 2016).

Toney Mountain

Toney Mountain located *c.* 100 km to the NW of Mount Takahe (Fig. 1a) is almost completely ice covered and is the least studied of the major volcanoes within MBL. The elongate massif extends for over 60 km in an east–west direction and has a single summit caldera (*c.* 3 km in diameter), with Richmond Peak marking the highest point (3595 m asl). The age of Toney Mountain is constrained by four K–Ar whole-rock dates (LeMasurier et al. 1990*a*). Two samples from a lava flow with a hawaiite composition from Cox Bluff near the base of the volcano on its eastern side (for locations refer to Wilch et al. 2021, fig. 17) yielded a mean Late Miocene age of 9.6 Ma. Two other K–Ar dates on intermediate and felsic rocks collected from outcrops near the base of Richmond Peak and on its upper flank yield Pleistocene ages (1.0 ± 0.4 and 0.5 ± 0.2 Ma, respectively).

McMurdo Volcanic Group in the western Ross Sea (Fig. 10) and constitutes the lowest $^{143}Nd/^{144}Nd$ ratios (0.51281–0.51276) over a range of moderately radiogenic $^{87}Sr/^{86}Sr_i$ ratios (0.70330–0.70559) relative to nearly all other deposits in MBL and wEL. Quartz-normative pantelleritic trachyte compositions, including the highly incompatible element-enriched lava flow samples from Knezevich Rock (85-25A), have the most radiogenic Sr values (>0.7038). Lead isotopes for evolved basalt samples (one basanite and three hawaiites: MgO = 4.0–4.9 wt%) are within a comparable range of values to other basalt samples from the MBLVG ($^{206}Pb/^{204}Pb$ = 19.706–20.243, $^{207}Pb/^{204}Pb$ = 15.677–15.737 and $^{208}Pb/^{204}Pb$ = 39.422–39.993) but with much lower Nd (Fig. 28) and higher Sr (not shown) isotopic values. Figure 28 also shows that $^{143}Nd/^{144}Nd$ and $^{206}Pb/^{204}Pb$ ratios of Mount Takahe basalt samples decrease slightly with magma evolution (i.e. increasing Na_2O, SiO_2 and decreasing MgO content), suggesting that MBL crust played a role in the evolution of Mount Takahe magmas.

The anomalously low $^{143}Nd/^{144}Nd$ and uniquely high Ba/LILE ratios measured in Mount Takahe samples were

The geochemistry of Toney Mountain is characterized by four whole-rock XRF analyses, comprising one each for hawaiite, latite, trachyte and rhyolite (Fig. 29; see also Supplementary material Table S1). The latite and trachyte samples are weakly silica-oversaturated (<1% quartz-normative), while the rhyolite is strongly silica-oversaturated (23% quartz-normative) and comenditic. Although the data are few, variations on Harker diagrams (not shown) suggest that the compositions may be related to a single liquid line of descent. Trace and minor element compositions show progressive enrichment in incompatible elements Rb, Th, Nb, Pb, Zr and Y, and progressive depletions in Ba, Sr, P and Ti with magma evolution (Fig. 30). LeMasurier et al. (2003) reported Sr and Nd isotopic ratios measured on two samples: hawaiite (67-80A) and comenditic rhyolite (67-76D). The hawaiite sample has lower $^{87}Sr/^{86}Sr_i$ (0.70270) and lower $^{143}Nd/^{144}Nd$ (0.512814) values relative to the rhyolite sample

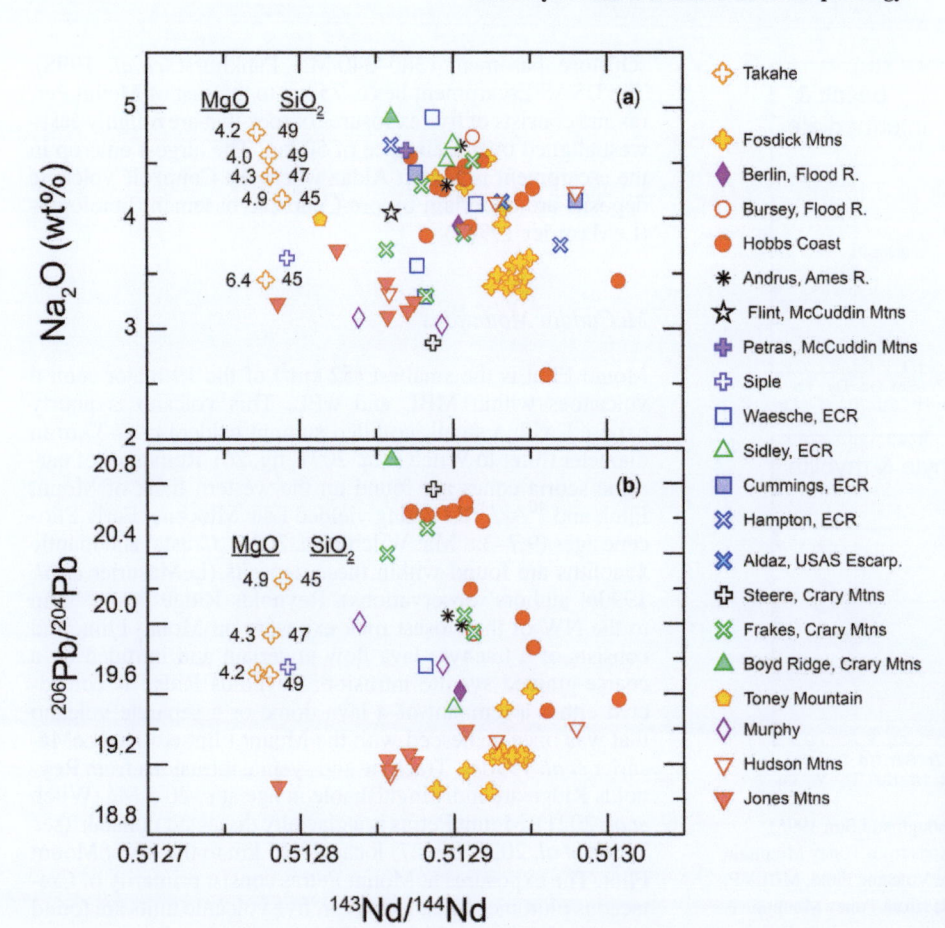

Fig. 28. Plot of measured ^{143}Nd/^{144}Nd ratios v. (**a**) Na$_2$O (wt%) and (**b**) ^{206}Pb/^{204}Pb ratios for basalt samples from Mount Takahe in comparison to basalt samples from other volcanoes in the MBLVG. Abbreviations: Mtns, Mountains; R, Range; ECR, Executive Committee Range. The MgO and SiO$_2$ (wt%) contents of individual Mount Takahe samples are labelled.

(0.70299 and 0.513046, respectively: Supplementary material Table S1). Remarkably, the rhyolite sample has the most radiogenic Nd isotopic signature relative to all other compositions in MBL and wEL. Additionally, the hawaiite sample has the least radiogenic Nd signature of any sample from the MBLVG, except for those with much higher ^{87}Sr/^{86}Sr$_i$ values (>0.7034). It is clear that more data are needed in order to evaluate the origin of these anomalous isotopic signatures, along with additional field and geochemical data to provide a better understanding of the overall petrogenesis of Toney Mountain.

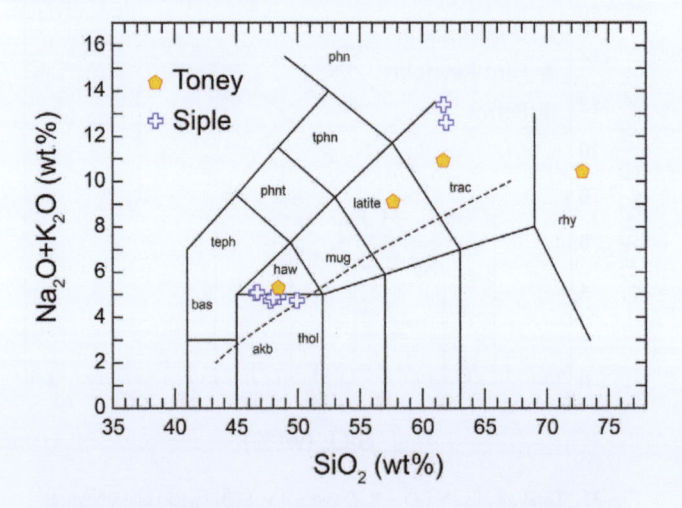

Fig. 29. Total alkali (Na$_2$O + K$_2$O (wt%)) v. SiO$_2$ (wt%) classification plot for compositions from Toney Mountain (Eastern MBL Volcanic Field) and Mount Siple Volcanic Field, MBLVP. Abbreviations for compositional names and the dashed line that delimits alkaline from subalkaline compositions are the same as in Figure 3.

Mount Siple

Mount Siple is located on the northern end of Siple Island and part of the Mount Siple Volcanic Field (Wilch et al. 2021), which lies off the Bakutis Coast of MBL (Fig. 1a). It is similar to Mount Takahe and Toney Mountain in that it is undissected, mostly ice covered and has a single ice-filled summit caldera (c. 4.5 km in diameter: refer to Wilch et al. 2021, fig. 21). The symmetrical cone of Mount Siple rises from sea level to 3110 m in elevation, with a basal diameter of c. 40 km. Trachyte lava flow and pyroclastic deposits near the summit and in the summit crater wall have ^{40}Ar/^{39}Ar dates of 171 and 230 ka, respectively, and a basalt sample collected mid-flank is 746 ka (Wilch et al. 2021). Lower flank and coastal samples of basalt are dated by K–Ar and ^{40}Ar/^{39}Ar methods, and range from 2 Ma to <100 ka, but the ages have high uncertainties (LeMasurier and Rex 1990; Wilch et al. 2021).

Rocks collected from the few available outcrops exposed on Mount Siple are bimodal in composition, and plot within the basalt and trachyte fields (Fig. 29). The six basalt samples show a range from silica-undersaturated hawaiite (sample W83-1, 4% nepheline-normative) to a silica-saturated alkali basalt (sample W83-5C, 4% hypersthene-normative). As noted by LeMasurier and Rex (1990), silica-saturated basalt compositions are relatively uncommon in MBL, and primarily occur at Mount Murphy and in the Fosdick Mountains Volcanic Field (described below). On a primitive-mantle-normalized multi-element plot, Mount Siple alkali basalt and hawaiite samples display very similar concentration patterns apart from titanium (Fig. 30). The alkali basalt sample has <2 wt% TiO$_2$, while the silica-undersaturated basalt samples range from 2.4 to 3.4 wt% TiO$_2$. Overall, the trace element concentrations of basalt from Mount Siple are very similar to the low-Ba basalts from Mount Murphy (Fig. 22a). The two felsic samples from Mount Siple are silica-undersaturated (4 and 8%

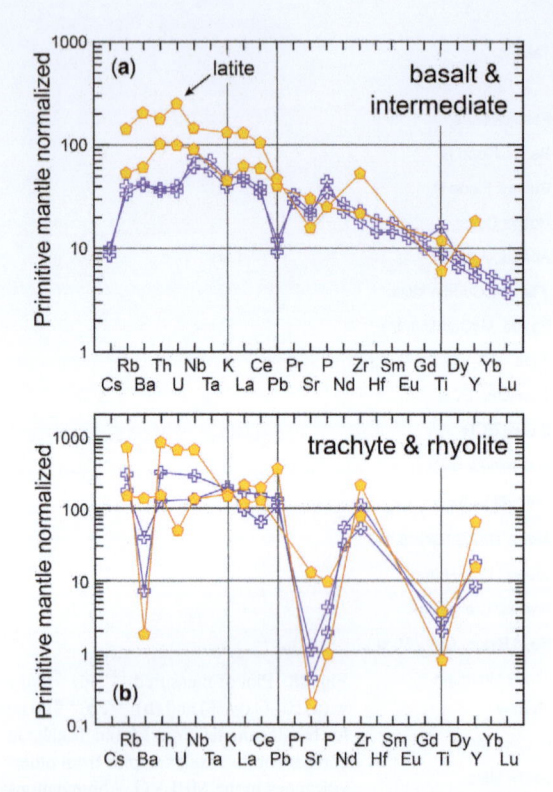

Fig. 30. Primitive-mantle-normalized (McDonough and Sun 1995) multi-element plots of representative compositions from Toney Mountain (Eastern MBL Volcanic Field) and Mount Siple Volcanic Field, MBLVP. Sample 67-75 (Supplementary material Table S1) from Toney Mountain is the only intermediate composition represented and is one of the few samples in the MBLVG classified as latite (Na$_2$O–2<K$_2$O (wt%): Le Maitre 2002).

nepheline-normative), mildly peralkaline and classified as comenditic trachyte (Fig. 29). On a mantle-normalized multi-element plot (Fig. 30) both samples show depletions in Ba, Sr, P and Ti relative to neighbouring elements, which is consistent with advanced fractionation of feldspars, apatite and Fe–Ti oxides.

LeMasurier et al. (2016) reported Sr, Nd and Pb isotopic ratios for two Mount Siple samples: hawaiite (W83-1) and alkali basalt (W83-5C). The alkali basalt has higher ^{143}Nd/^{144}Nd ratios (0.512847) and lower ^{87}Sr/^{86}Sr$_i$ ratios (0.703487) relative to the hawaiite sample (0.512792 and 0.703723, respectively), and is slightly less radiogenic with respect to Pb isotopes (Supplementary material Table S1). The isotopic composition of the hawaiite sample is comparable to several basalt samples from Mount Takahe (Fig. 28). LeMasurier et al. (2016) interpreted the low Nd isotopic ratio of the hawaiite sample, along with low Nd isotopic signatures of basalt samples from Mount Takahe and Mount Murphy, to be a result of partial melting of mantle containing sediments from subduction.

Central MBL: the onset of volcanism

The oldest *in situ* record of Cenozoic volcanic activity (c. 37–20 Ma) in West Antarctica occurs in the McCuddin Mountains Volcanic Field located in central MBL. The McCuddin Mountains consist of two highly dissected and closely spaced edifices, Mount Flint–Reynolds Ridge (refer to fig. 26 in Wilch et al. 2021) and Mount Petras, which lie c. 100 km east of the Ames Range (Fig. 1a). Volcanic rocks at Mount Petras cap a Cretaceous plutonic–hypabyssal igneous complex that was emplaced into a deformed Devonian–Carboniferous schistose basement (380–340 Ma: Pankhurst et al. 1998). The USAS Escarpment lies c. 75 km to the east of Mount Petras and consists of five exposures of rock that are roughly east–west aligned over a distance of 50 km. The largest outcrop in the escarpment is Mount Aldaz where the Cenozoic volcanic deposits are underlain by pre-Cenozoic basement lithologies (LeMasurier 1990d).

McCuddin Mountains

Mount Flint is the smallest (52 km^3) of the 19 major central volcanoes within MBL and wEL. This volcano is poorly exposed with a small ice-filled summit caldera c. 2–3 km in diameter (refer to Wilch et al. 2021, fig. 26). Remnants of parasitic scoria cones are found on the western flank of Mount Flint, and ^{40}Ar/^{39}Ar dating yielded Late Miocene–Early Pliocene ages (9.7–3.8 Ma: Wilch et al. 2021). Crustal and mantle xenoliths are found within these deposits (LeMasurier et al. 1990b; authors' observations). Reynolds Ridge lies c. 5 km to the NW of the closest rock exposure on Mount Flint, and consists of a trachyte lava flow underlain and intruded by a coarse-grained syenite intrusion. Reynolds Ridge is considered either a remnant of a lava dome or a separate volcano that was once coalesced with the Mount Flint edifice (LeMasurier et al. 1990b). Trachyte and syenite intrusions from Reynolds Ridge are indistinguishable in age at c. 20.5 Ma (Wilch et al. 2021). Mount Petras is a glacially dissected nunatak (see Wilch et al. 2021, fig. 27) located <15 km to the SE of Mount Flint. The exposures at Mount Petras consist primarily of Cretaceous plutonic rocks. However, five volcanic units are found on top of a high-relief (c. 400 m) unconformity, and were described and dated by the ^{40}Ar/^{39}Ar method (Wilch and McIntosh 2000). A massive lava flow (c. 25 m thick) sampled near the summit of Mount Petras yields the oldest age of Cenozoic rift-related volcanic activity in Antarctica at 36.71 ± 0.22 Ma. A sequence of basaltic tuff and tuff breccia records activity between 29 and 27 Ma (Wilch and McIntosh 2000; Wilch et al. 2021).

The compositions of six volcaniclastic units on Mount Petras and Mount Flint are silica-undersaturated (6–8% nepheline-normative), and are classified as hawaiite and basanite (Fig. 31; see also Supplementary material Table S1). A glassy

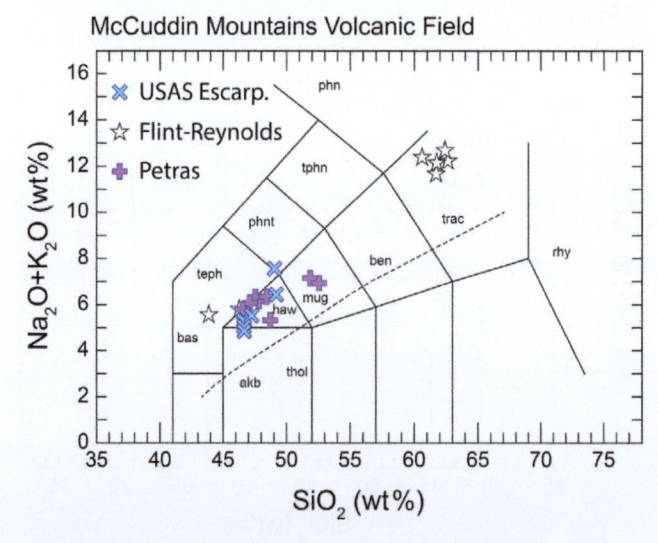

Fig. 31. Total alkalis (Na$_2$O + K$_2$O (wt%)) v. SiO$_2$ (wt%) classification plot for compositions from the McCuddin Mountains Volcanic Field, which includes Mount Flint–Reynolds Ridge and Mount Petras, and the USAS Escarpment (USAS Escp.), MBLVP. Abbreviations for compositional names and the dashed line that delimits alkaline from subalkaline compositions are the same as in Figure 3.

hawaiite lava flow at Mount Petras has a slightly lower total alkali content relative to the other basalt samples and is silica-saturated (16% hypersthene-normative). The massive lava on Mount Petras is classified as mugearite and is silica-saturated (17% hypersthene-normative). A cognate xenolith collected from a volcanoclastic unit with a hawaiite composition is of the same age (within analytical error) and has a nearly identical composition as the mugearite lava flow (Wilch and McIntosh 2000), although with slightly lower SiO_2 and a higher total alkali content (Fig. 31), thus making it less silica-saturated (5% hypersthene-normative) than the lava flow. This compositional difference is likely to reflect minor physicochemical interaction between the silica-undersaturated hawaiite host and the mugearite lava flow during eruption. All samples from Reynolds Ridge plot within the trachyte field shown in Figure 31. The compositions vary from silica-undersaturated (2–6% nepheline-normative) to silica-saturated (2–4% hypersthene-normative). The syenite intrusion (sample WCM93-288: Supplementary material Table S1) is weakly metaluminous (molecular $(Na_2O + K_2O) < Al_2O_3 < (CaO + Na_2O + K_2O)$), while the lava flows are peralkaline and classify as comenditic trachyte.

Minor and trace element compositions of Mount Flint–Reynolds Ridge and Mount Petras, as determined by XRF analysis, are presented for comparison on primitive-mantle-normalized multi-element plots (Fig. 32). Basalt samples display parallel incompatible element-enriched patterns with small negative K anomalies. Mugearite samples from Mount Petras (lava and xenolith) have nearly identical concentrations in all of the elements measured, and show relative depletions in Sr and Ti consistent with the removal of plagioclase, clinopyroxene and Fe–Ti oxides (Fig. 32b). The relative depletions in Ba, Sr, P and Ti shown by three lava-flow samples from Reynolds Ridge on Mount Flint (Fig. 32c) indicate that alkali feldspar and apatite were removed during late-stage differentiation of trachyte magmas. However, two of the samples, WCM93-90 and WCM93-91, have significantly different concentrations in Sr, P and Ba, which suggests that they were either erupted from closely related but separate magma batches or that they are from a single compositionally and/or modally heterogeneous lava flow. The coarse-grained syenite has higher concentrations in Ba, Sr, P and Ti, and lower concentrations of Rb, Th, Nb, Pb, Zr and Y, relative to the trachyte lava-flow samples (Fig. 32c). Given the overall similarity in geochemistry, intimate field relationships and indistinguishable ages, it is possible that the syenite represents a residue of the system that erupted the trachyte lava flow(s).

Hole and LeMasurier (1994) reported Sr and Nd isotopic ratios for two samples in the McCuddin Mountains: a hawaiite sample from Mount Petras (sample PT67D); and a basanite sample from Mount Flint (sample 9D). Initial $^{87}Sr/^{86}Sr$ and $^{143}Nd/^{144}Nd$ ratios are 0.703071 and 0.512870 for hawaiite samples from Mount Petras, and 0.702882 and 0.512860 for basanite samples from Mount Flint. These isotopic values fall within the main cluster of compositions reported for MBL and wEL (Fig. 10).

USAS Escarpment

Of the five nunataks exposed along the USAS Escarpment, only three contain deposits of Cenozoic volcanic rocks (Fig. 1a). At Mount Aldaz, volcaniclastic deposits and clastogenic lava flows overlie a deformed laminated sandstone that, in turn, is underlain by hypabyssal igneous intrusive rocks (Wilch et al. 2021). Mantle xenoliths occur in lava-flow and scoria cone deposits (LeMasurier 1990d; authors'

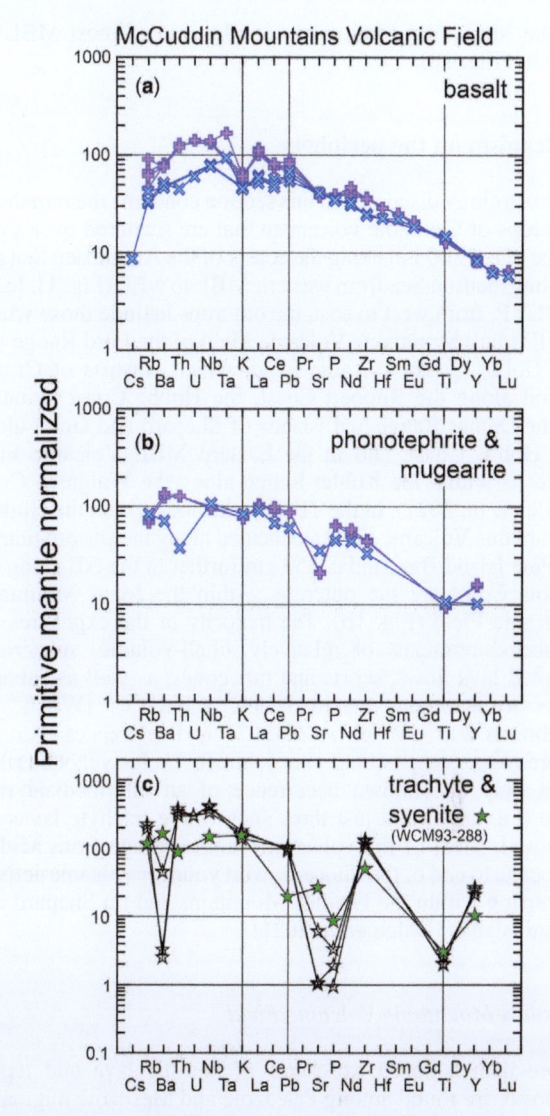

Fig. 32. Primitive-mantle-normalized (McDonough and Sun 1995) multi-element plots of representative compositions from the McCuddin Mountains Volcanic Field, MBLVP. The green-filled star in (c) is a syenite sample WCM93-288 from Reynolds Ridge (Supplementary material Table S1).

observations). The age of the volcanic sequence dated by the $^{40}Ar/^{39}Ar$ method is 19.68 ± 0.31 Ma (Wilch et al. 2021). Volcaniclastic deposits are also present at two other unnamed sites: a small outcrop near the NE flank of Mount Galla, which is described as Strombolian tephra by LeMasurier (1990d); and a more extensive outcrop c. 4 km to the east of Mount Galla. The larger outcrop is considered to be the remains of a tuff cone and yielded a $^{40}Ar/^{39}Ar$ age of 26.40 ± 0.21 Ma (Wilch et al. 2021).

Seven samples from the USAS Escarpment have been analysed for geochemistry by XRF and one of those, sample WCM90-181 (Supplementary material Table S1), for trace elements by INAA. Sample WCM90-181 from Mount Aldaz was also measured for Sr isotopes. All seven samples are silica-undersaturated (1–7% nepheline-normative), and classify as alkali basalt, hawaiite and phonotephrite (Fig. 31). Alkali basalt and hawaiite samples are less enriched in highly incompatible elements relative to Mount Flint and Mount Petras (Fig. 32a) but, overall, have concentrations that are comparable to most other basalts from the MBLVG. The $^{87}Sr/^{86}Sr_i$ value of alkali basalt sample WCM90-181 is relatively unradiogenic (0.702904), and is within the range of the McCuddin Mountains samples and of most MBLVG basalts (Fig. 10).

Volcanism on the periphery

The petrology discussed in this section concerns the remaining outcrops of Cenozoic volcanism that are scattered over a distance of c. 1600 km along the coasts of the Amundsen Sea and Bellingshausen Sea from western MBL to wEL (Fig. 1). In the MBLVP, from west to east, the outcrops include those within the Fosdick Mountains Volcanic Field of the Ford Range and the Hobbs Coast Volcanic Field, which consists of Cruzen Island along the Ruppert Coast, the Hobbs Coast nunataks of the Demas Range and islands of Shepard and Grant along the Hobbs Coast, and in the Eastern MBL Volcanic Field deposits within the Kohler Range along the Walgreen Coast (Wilch et al. 2021). In the TIVP, volcanoes within the Hudson Mountains Volcanic Field are located along the eastern margin of Pine Island Bay, and c. 150 km further to the NE along the Eights Coast are the outcrops within the Jones Mountains Volcanic Field (Fig. 1b). The majority of the exposures are dissected remnants of relatively small-volume, subaerially erupted lava flows, scoria and tuff cones, as well as subaqueous to emergent glaciovolcanic sequences (Wilch and McIntosh 2007; Wilch et al. 2021 and references therein). Moreover, compositions are almost exclusively basaltic, with only one known occurrence of an intermediate rock type (i.e. latite) and just three sites where trachyte lavas are exposed. Most of the volcanism ranges in age from Middle Miocene to end of the Pliocene, with younger volcanic activity occurring within the Fosdick Mountains and on Shepard and Grant islands (Wilch et al. 2021).

Fosdick Mountains Volcanic Field

More than a dozen exposures of basaltic lava and tephra deposits are found among Paleozoic and Mesozoic migmatite gneiss and plutonic rocks that make up the core of the Fosdick Mountains in the westernmost portion of MBL (Siddoway et al. 2004; Gaffney and Siddoway 2007). Gaffney and Siddoway (2007) described four locations where the volcanic rocks host ultramafic xenoliths (Mount Perkins and Mount Avery, Marujupu Peak, and an unnamed outcrop) and one location, Recess Nunatak, where crustal xenoliths occur (refer to Wilch et al. 2021, fig. 35). It is also reported by Luyendyk et al. (1991) and Gaffney and Siddoway (2007) that basaltic necks and dykes, which were the likely feeders of the Cenozoic volcanism, cross-cut basement lithologies in the north-central portion of the range. Lavas from Mount Avery and Mount Perkins are dated by the $^{40}Ar/^{39}Ar$ method, and yield indistinguishable ages of 1.41 ± 0.04 and 1.40 ± 0.03 Ma, respectively (Gaffney and Siddoway 2007; Wilch et al. 2021). The dates contrast with two previous K–Ar ages measured on Mount Perkins basalts at 4.6 and 3.5 Ma (LeMasurier and Wade 1990).

Basalts from the Fosdick Mountains are relatively unfractionated (MgO 7.4–10.3 wt%, Cr 164–363 ppm and Ni 136–263 ppm) and on a total alkali v. SiO₂ classification diagram (Fig. 33) plot in three clusters: (1) strongly silica-undersaturated (12.9–13.9% nepheline-normative) basanite samples from Recess Nunatak; (2) basanite samples collected from Mount Avers which have lower alkali and SiO₂ contents but are also strongly silica-undersaturated (12.6–14.3% nepheline-normative); and (3) alkali basalt samples from Mount Perkins that vary from weakly silica-undersaturated (c. 1% nepheline-normative) to silica-saturated (1–6% hypersthene-normative). The alkali basalt compositions straddle the alkaline to

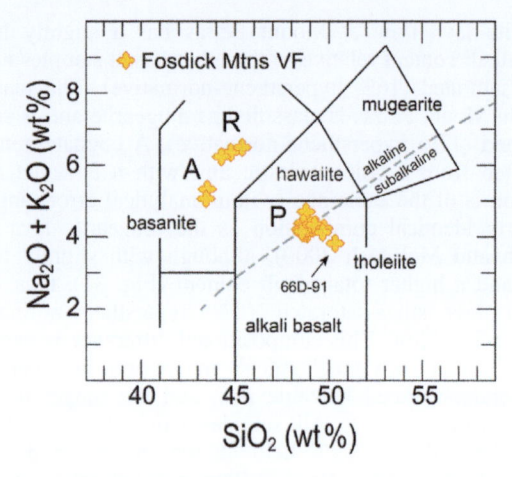

Fig. 33. Total alkali ($Na_2O + K_2O$ (wt%)) v. SiO_2 (wt%) classification plot of basalts from the Fosdick Mountains Volcanic Field (VF). Rock compositions from the Fosdick Mountains are represented by basanites at Recess Nunatak (R) and Mount Avers (A), and alkali basalts at Mount Perkins (P). Tholeiite sample 66D-91 is from location F5 in the central portion of the Fosdick Mountains (see LeMasurier and Wade 1990). The alkaline–subalkaline boundary is from Irvine and Baragar (1971).

subalkaline boundary (Fig. 33). In addition, a basalt sample collected from an unnamed outcrop 'F5' (sample 66D-91: LeMasurier and Wade 1990) (Supplementary material Table S1) is subalkaline (9% hypersthene-normative) and tholeiitic (classification of Floyd and Winchester 1975). It is noteworthy that the only other locations in MBL and wEL where tholeiite occurs are in the Jones Mountains and Hudson Mountains volcanic fields (discussed below), which lie more than 1200 km to the east.

Minor and trace element compositions of basalt samples from the Fosdick Mountains and Cruzen Island are plotted on normalized multi-element diagrams and display overall enriched patterns (Fig. 34). For each of the three volcanic centres in the Fosdick Mountains (Avers, Recess and Perkins) the compositions analysed are relatively homogeneous; however, between centres, the compositional patterns diverge with increasing element incompatibility (i.e. having similar concentrations of HREE, Y and Ti but disparate concentrations for LREE, LILE, and HFS elements Nb, Ta, P and Zr) (Fig. 34). With decreasing total alkalis and silica-undersaturation the basanites from Recess Nunatak are the most enriched in incompatible elements, followed by Mount Avers and then by the alkali basalt samples from Mount Perkins. All compositions show depletions in K and Pb relative to neighbouring elements, and the alkali basalts from Mount Perkins are also depleted in U and Th relative to Ba and Nb (Fig. 34a). Chondrite-normalized REE patterns are smooth (i.e. do not show Eu anomalies) and are LREE-enriched (Fig. 34b), with basanites from Recess Nunatak having La/Yb$_n$ ratios of 25–26, basanite samples from Mount Avers having La/Yb$_n$ ratios c. 19 and alkali basalt samples from Mount Perkins having La/Yb$_n$ ratios of 7–8 (Fig. 35a). The REE patterns are consistent with garnet being a residual phase in the mantle source of these magmas (Gaffney and Siddoway 2007). Based on their broadly comparable minor and trace element patterns, and overall incompatible element abundances, Gaffney and Siddoway (2007) suggested that the basalts from all three volcanoes were derived from compositionally similar sources. Furthermore, the comparable HREE and Y contents with diverging sub-parallel concentrations for the most incompatible elements are similar to patterns displayed by other suites of continental basalts whose origins have been explained by changes in the degree of partial melting from a common source (Jung and Masberg 1998; Lustrino et al. 2002; Boyce et al. 2015;

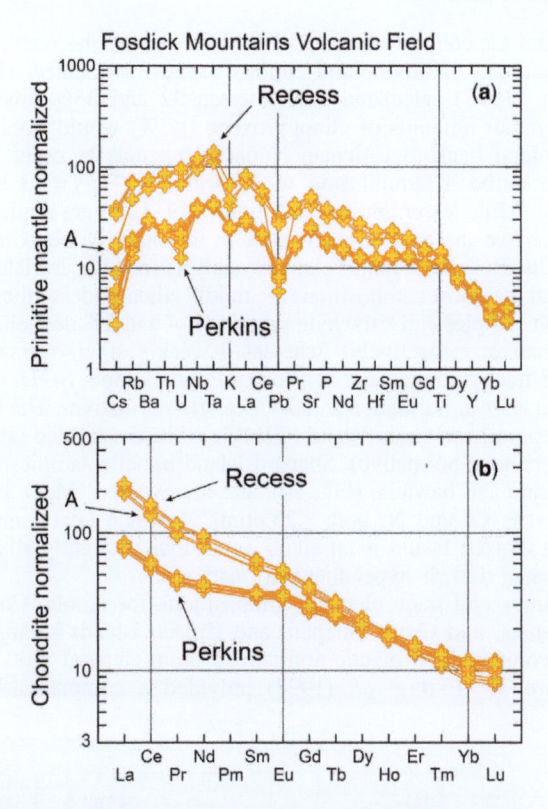

Fig. 34. Normalized multi-element plots of basalts from the Fosdick Mountains Volcanic Field. (**a**) Normalized to primitive mantle (McDonough and Sun 1995) and (**b**) REE normalized to chondrite (Sun and McDonough 1989). The capital A is for Mount Avers basalt samples.

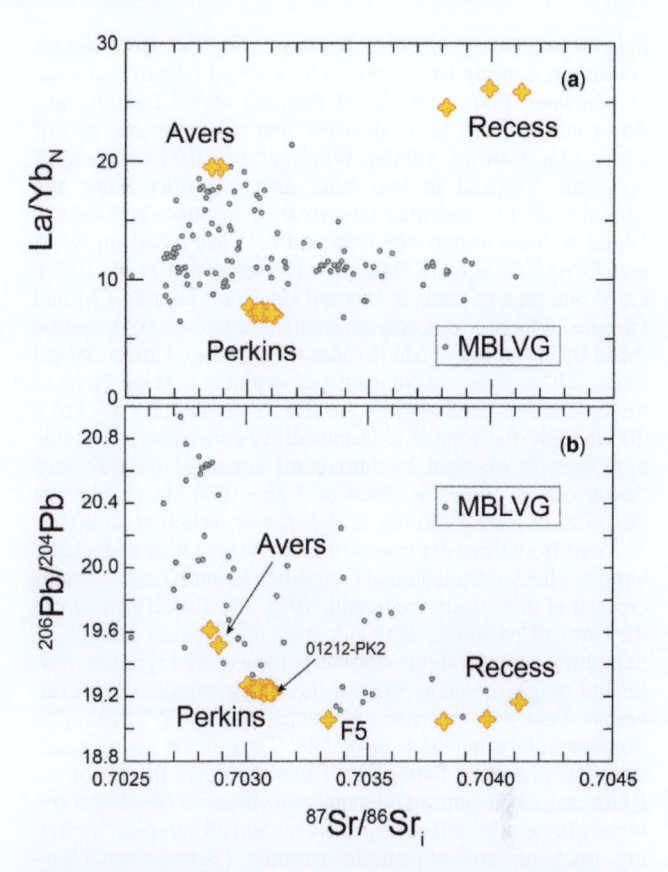

Fig. 35. Initial $^{87}Sr/^{86}Sr$ ratios v. (**a**) La/Yb_N ratios and (**b**) measured $^{206}Pb/^{204}Pb$ ratios for basalt samples from the Fosdick Mountains Volcanic Field in comparison to all other basalts from the MBLVG. Sample 01212-Pk2 was collected from an unnamed outcrop located between Mount Perkins and Recess Nunatak (Gaffney and Siddoway 2007), and tholeiite sample 66D-91 is from location F5 (LeMasurier and Wade 1990).

Panter *et al.* 2021). In this case, basanite samples from Recess Nunatak and Mount Avers, which are the most incompatible-element enriched, would be the result of lower degrees of partial melting relative to the alkali basalts from Mount Perkins.

Gaffney and Siddoway (2007) reported Sr, Nd and Pb isotopic results for samples from Mount Avers ($n = 2$), Recess Nunatak ($n = 3$) and Mount Perkins ($n = 12$), as well as an alkali basalt collected from an unnamed outcrop located between Mount Perkins and Recess Nunatak (sample 01212-Pk2: Supplementary material Table S1). An additional alkali basalt sample from Mount Perkins (sample WCM90-207: Supplementary material Table S1) has been measured for Sr isotopes only (unpublished data). The isotopic compositions of each of the three volcanic centres in the Fosdick Mountains are distinct, as is also seen in the geochemistry (Fig. 35b). Besides having the highest trace element concentrations, Recess Nunatak also has the most radiogenic Sr and lowest $^{206}Pb/^{204}Pb$ ratios (Fig. 35b), as well as the least radiogenic $^{143}Nd/^{144}Nd$ ratios (0.512889–0.512923), in comparison to the two other volcanic centres (0.512925–0.512950). Moreover, basalt samples from Recess Nunatak have the highest La/Yb_N ratios, as well as some of the lowest $^{206}Pb/^{204}Pb$ values, relative to all other MBL and wEL rocks (Fig. 35). Gaffney and Siddoway (2007) ruled out assimilation of crust by Recess Nunatak magmas given the lack of correlation between trace element ratios (those often used to assess contamination e.g. Nb/U, Ce/Pb, Zr/Hf and Nb/Ta) and isotopic ratios. Gaffney and Siddoway (2007) concluded that, from their isotopic characteristics, the basalts from the Fosdick Mountains may have tapped a local mantle source that has not been sampled by other volcanoes in the MBLVP (i.e. not HIMU) and therefore suggested that the mantle beneath MBL is more heterogeneous than previously thought.

Hobbs Coast Volcanic Field

The Hobbs Coast Volcanic Field consists of inland nunataks, and the islands of Grant, Shepard and Cruzen (Fig. 1a). The nunataks of the Hobbs Coast lie inland from the Getz Ice Shelf and include volcanic deposits from eroded monogenetic, mostly basaltic, centres that have a north–south orientation extending over a distance of *c.* 38 km: Holmes Bluff, Kennel Peak, Kouperov Peak, Shibuya Peak, Patton Bluff, Cousins Rock and Coleman Nunatak (for locations refer to Wilch *et al.* 2021, fig. 34). These nunataks compose the Demas Range (Fig. 1a). Volcanic deposits are also found at Bowyer Butte, which lies *c.* 25 km to the west of the Demas Range. The deposits consist of lava flows and volcaniclastic deposits, which in places include lithologies interpreted to have been erupted in subglacial volcanic palaeoenvironments and lie unconformably on glacially sculpted granitoid intrusions (LeMasurier 1990*e*; Wilch and McIntosh 2007) that are middle Cretaceous in age but contain metamorphosed enclaves that range between 400 and 300 Ma (Pankhurst *et al.* 1998). The volcanism is mostly Late Miocene but ranges from 11.45 to 2.60 Ma ($^{40}Ar/^{39}Ar$ ages: Wilch *et al.* 2021). Coleman Nunatak, near the south end of the Demas Range, was interpreted as a long-lived basanitic centre based on a wide range of previous K/Ar ages, 11.7–2.5 Ma (LeMasurier 1990*e*; Hart *et al.* 1997). Detailed sampling ($n = 11$) and high-precision $^{40}Ar/^{39}Ar$ ages suggest that activity at Coleman Nunatak was monogenetic and occurred at 2.60 ± 0.01 Ma (Wilch and McIntosh 2007; Wilch *et al.* 2021).

Grant Island and Shepard Island are located between 60 and 100 km to the NE of the Demas Range and are connected to the Hobbs Coast by the Getz Ice Shelf (Fig. 1a). Both islands are almost entirely ice covered. On Shepard Island, two areas (Mathewson Point and Mount Petinos) reveal basaltic lava flows and volcaniclastic deposits that are remnants of tuff cones (LeMasurier 1990e; Wilch et al. 2021). Trachytic lavas are exposed in two other areas, Worley Point and Mount Colburn, and may be coeval. Volcanism on Shepard Island is dated to between 0.56 and 0.41 Ma based on K/Ar and ^{40}Ar/^{39}Ar ages (LeMasurier 1990e; Wilch et al. 2021). Only one area of rock is exposed on Grant Island at Mount Obiglio, which is a scoria cone of intermediate composition dated by K/Ar at 0.7 Ma (LeMasurier 1990e). Cruzen Island lies c. 230 km west of Shepard Island and c. 230 km NNE of the Fosdick Mountains (Fig. 1a). It is a small (2.7 × 1.7 km × 200 m high) flat-topped volcanic edifice consisting of basaltic hyaloclastite overlain by horizontal subaerial basaltic lava flows, one of which is dated at 2.75 ± 0.26 Ma (K–Ar age from LeMasurier 1990a, b, c, d, e, f, g, h; Wilch et al. 2021).

Apart from three exposures of trachyte lava flow and a latite tephra collected from Mount Obiglio on Grant Island, magmas erupted in this region are basaltic (Fig. 36). Basalt rocks from the Hobbs Coast nunataks range from alkali basalt to basanite to tephrite (<10% olivine-normative) to evolved types, hawaiite and mugearite (Fig. 36). One lava flow sample from Kouperov Peak is trachyte in composition (WCM93-305: Supplementary material Table S1). Two of the alkali basalt samples, one from Patton Bluff and the other from Holmes Bluff, are silica-saturated (samples 48B and 37A: 2 and 6% hypersthene-normative, respectively) and all samples, including trachyte, are silica-undersaturated (2–16% nepheline-normative). It is important to note that all basanite and tephrite samples are from Coleman Nunatak and comprise 20 of the 39 analyses available from the Hobbs Coast (Supplementary material Table S1). Excluding the mugearite and high-silica hawaiite samples, the remainder of the basalt samples has MgO contents that vary from 5.3 to 9.4 wt%, with Ni concentrations of 38–192 ppm and Cr 65–329 ppm. Of these less fractionated basalt samples, those from Coleman Nunatak have the highest FeO^T and TiO$_2$ concentrations, and lower Ni and Cr concentrations. This is likely to be the result of the amount of olivine and clinopyroxene fractionated. Hart et al. (1997) calculated that between 32 and 35% olivine and minor amounts of clinopyroxene (≤3%) would need to be added back to Coleman Nunatak magmas in order for them to be in equilibrium with peridotite (Mg# = 73 and Fo$_{90}$), while lesser amounts of olivine (19–32%) are required to achieve the equilibrium values in the other Hobbs Coast basalts. Results of major element analysis for Shepard Island reveal that most compositions are mildly silica-undersaturated basalt samples and a trachyte sample (1–4 and 2% nepheline-normative, respectively), whereas a weakly silica-oversaturated trachyte is present at Worley Point (sample SI9D: 6% hypersthenes-normative and 0.4% quartz-normative). The latite tephra from Grant Island (GI11C) is silica-saturated (10% hypersthene-normative). Shepard Island basaltic samples are basanite and hawaiite (Fig. 36), and are evolved (MgO 3.7–4.7 wt%, Cr and Ni both ≤25 ppm). A single lava sample from Cruzen Island is an alkali basalt that is slightly silica-saturated (0.02% hypersthene-normative).

Minor and trace element compositions for Hobbs Coast nunataks, and Grant, Shepard and Cruzen islands samples, are compared on mantle-normalized multi-element plots in Figure 37. Hart et al. (1997) provided a comprehensive

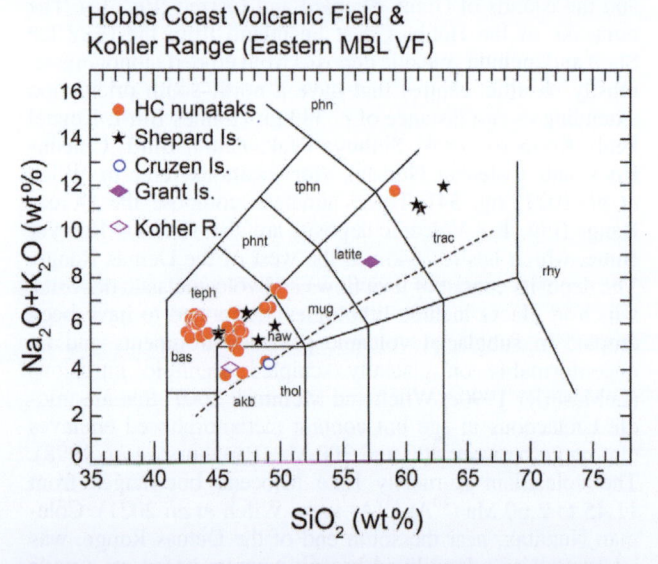

Fig. 36. Total alkali (Na$_2$O + K$_2$O (wt%)) v. SiO$_2$ (wt%) classification plot for compositions from the Hobbs Coast Volcanic Field. HC is Hobbs Coast. The intermediate composition from Mount Obiglio on Grant Island (Is.) is a latite. Also shown is one alkali basalt sample from Leister Peak (unfilled diamond, sample 84: Supplementary material Table S1) in the Kohler Range (R.), Eastern MBL Volcanic Field (VF). Abbreviations for compositional names and the dashed line that delimits alkaline from subalkaline compositions are the same as in Figure 3.

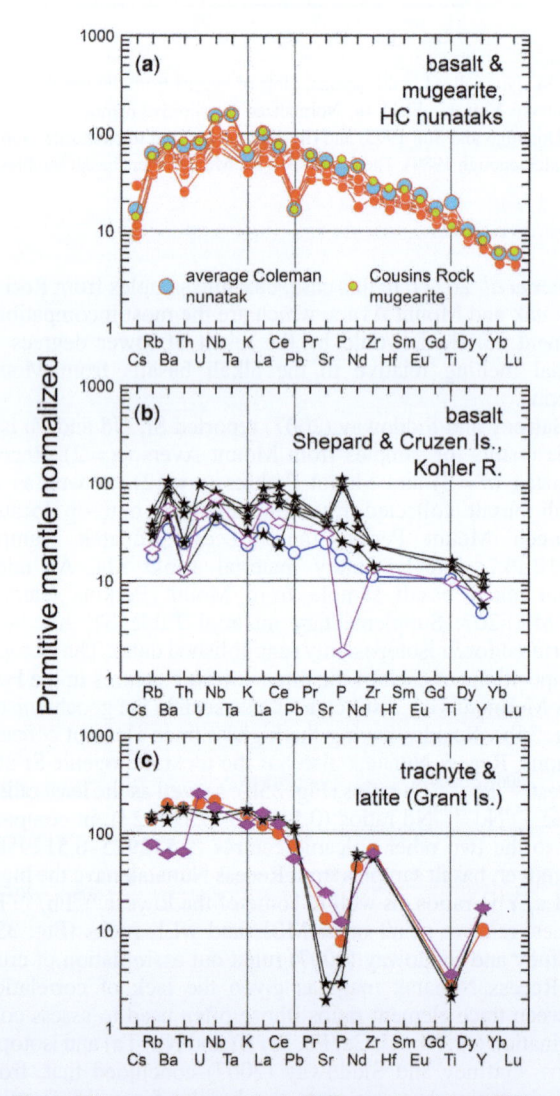

Fig. 37. Primitive-mantle-normalized (McDonough and Sun 1995) multi-element plots of representative compositions from the Hobbs Coast Volcanic Field and Leister Peak in the Kohler Range (R.). Mugearite from Cousins Rock (Hart et al. 1997) and the average of seven samples from Coleman Nunatak are highlighted in (a). Symbols are defined in Figure 36.

geochemical and isotopic study of Hobbs Coast basalt samples, and characterized their concentration patterns as being derived from a homogeneous mantle source tapped by variable degrees of partial melting. They tested and refined this assertion by calculating source concentrations and bulk partition coefficients using a trace element inversion technique for partial melting. Their results indicate that the basalts were derived from 1.6–3.2% partial melting of a garnet and amphibole (±phlogopite)-bearing lherzolite, which was once depleted (harzburgitic) but later enriched in incompatible trace elements. Negative Pb anomalies (Pb measured by ICP-MS: Hart *et al.* 1997) shown by Coleman Nunatak and other basalt samples from the Hobbs Coasts nunataks (Fig. 37a) may be a consequence of a Pb-compatible phase (metal sulfide?) that remained in the mantle during melting. Other samples whose Pb concentrations were measured by XRF display slightly positive Pb anomalies. Whole-rock XRF analyses of five basalt samples from Shepard Island have slightly more enriched patterns relative to the basalts from the Hobbs Coast Nunataks (Fig. 37b), which is expected given their more evolved major element compositions. Analyses obtained by XRF for samples with trachyte compositions (Fig. 37c) have normalized concentration levels similar to low-Si trachyte samples at Mount Takahe (Fig. 26c) and, notably, an absence of negative Ba anomalies (Ba 851–1268 ppm), which suggests that the removal of alkali feldspars (K_d Ba$_{akf}$ = 3.4 for peralkaline trachyte: Larsen 1979) from these magmas was not significant.

Hart *et al.* (1997) also reported Sr, Nd and Pb isotopic results for basalt samples from Hobbs Coast nunataks. Initial $^{87}Sr/^{86}Sr$ ratios of the basalt samples are low (0.702526–0.702858) and measured $^{143}Nd/^{144}Nd$ ratios are high (0.512873–0.513008). Measured Pb isotopic ratios have a large range of values: $^{206}Pb/^{204}Pb$ 19.50–20.69, $^{207}Pb/^{204}Pb$ 15.63–15.77 and $^{208}Pb/^{204}Pb$ 38.86–39.90 (Supplementary material Table S1), which, taken together with Sr and Nd isotopes, form arrays that align between depleted MORB mantle (DMM) and HIMU mantle components. Hart *et al.* (1997) examined the data in 3D isotope space (axes = $^{87}Sr/^{86}Sr$, $^{143}Nd/^{144}Nd$ and $^{206}Pb/^{204}Pb$) along with other basalts from West Antarctica, and observed that the data arrays converge on the region of the FOZO mantle component (Hart *et al.* 1992). The orientation of the 3D array is parallel with the array for basalt compositions from the Crary Mountains, which are also interpreted to be the result of mixing between HIMU and a FOZO-like (lower-μ) mantle component (Panter *et al.* 2000). No isotopic data exist for Shepard, Grant or Cruzen islands.

Hart *et al.* (1997) considered that the volcanism has sampled an isotopically stratified mantle with the HIMU layer underlying a FOZO layer. Through time, the source for Hobbs Coast volcanism was tapped by smaller degrees of melting at greater depths to produce more silica-undersaturated magmas (Fig. 38a) with stronger HIMU signatures (i.e. higher $^{206}Pb/^{204}Pb$ ratios) (Fig. 38b). Hart *et al.* (1997) proposed that the isotopically layered mantle represents a fossilized plume that accreted to the base of the Gondwana lithosphere prior to break-up in the Late Cretaceous. A similar source and melting scenario is used to explain the origins of HIMU-like basalts in the Crary Mountains (Panter *et al.* 2000), although time-progressive compositional changes are not as well defined (Fig. 38a, b).

Kohler Range

The Kohler Range is located near the Walgreen Coast, and lies more than 500 km to the east of the Hobbs Coast and *c.* 100 km to the WNW of Mount Murphy (Fig. 1a). The deposits are

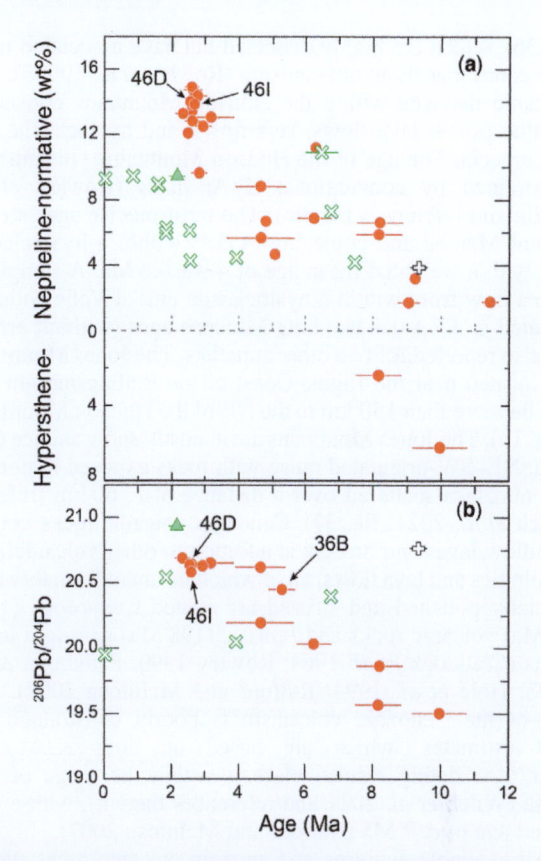

Fig. 38. Age (Ma) v. (**a**) CIPW-normative compositions (nepheline- and hypersthene-normative in wt%) and (**b**) measured $^{206}Pb/^{204}Pb$ ratios for basalt samples, with the exception of sample 36B (mugearite) from the Hobbs Coast nunataks, Hobbs Coast Volcanic Field. Also shown for comparison are basalt samples from the Crary Mountains Volcanic Field (refer to Fig. 18 for the symbols). Dates are from Hart *et al.* (1997) (K–Ar) and Wilch (1997) ($^{40}Ar/^{39}Ar$) with error bars at two standard deviations. The K–Ar dates for Coleman Nunatak samples 46D (7.9 Ma) and 46I (11.7 Ma) from Hart *et al.* (1997) were not used in this plot. Instead, an average of 11 high-precision $^{40}Ar/^{39}Ar$ analyses, yielding an age of 2.60 ± 0.01 Ma, was substituted to provide a better estimate for this monogenetic centre. No other K–Ar dates are modified.

included within the Eastern MBL Volcanic Field (Wilch *et al.* 2021). Two small volcanic outcrops, a yellow tuff and a lava, are exposed and rest on top of Early Permian granitoid intrusions (276 ± 2 Ma: Pankhurst *et al.* 1998). The lava flow is located near Leister Peak and is dated by K/Ar at 10.1 ± 3.4 Ma (LeMasurier 1990*f*). A single geochemical analysis of this lava flow is a mildly silica-undersaturated (4% nepheline-normative) alkali basalt (Fig. 36). The sample is relatively unfractionated (MgO 8.42 wt%, Cr 271 ppm and Ni 151 ppm) but has an unusually low P_2O_5 content (Fig. 37b), which may indicate that it has been altered.

Hudson and Jones mountains: western Ellsworth Land

Both the Hudson and Jones mountains volcanic fields lie within the western portion of the Ellsworth Land Volcanic Province on the Thurston Island crustal block and comprise the TIVP (Fig. 2). The Hudson Mountains are *c.* 400 km east of the Kohler Range, just inland from the eastern coast of Pine Island Bay (Fig. 1b). Over 20 nunataks are scattered in an area of roughly 100 km north–south by 50 km east–west, and are mostly snow and ice covered with no basement lithologies exposed (Rowley *et al.* 1990*a*). Three edifices are described as major volcanoes: Teeters Nunatak, Mount Moses and Mount Manthe (for the locations refer to Wilch *et al.* 2021, fig. 36), which are highly dissected but have parasitic cones that show little erosion (Rowley et al. 1990a). The volcanic deposits within the Hudson Mountains consist of basaltic pillow lava flows, lava flows, and hyaloclastite tuff and breccia. The age of the Hudson Mountains volcanism is constrained by conventional K/Ar dates (Rowley et al. 1990a and references therein). The most precise ages are for Mount Manthe and come from clasts within a hyaloclastite unit with a weighted mean age of 4.9 ± 0.6 Ma. A sample of a lava flow from within a hyaloclastite unit at Velie Nunatak is dated at 3.7 ± 0.4 Ma. Late Miocene ages (without errors) are also reported for two other nunataks. The Jones Mountains are located near the Eights Coast of the Bellingshausen Sea and lie more than 150 km to the NE of the Hudson Mountains (Fig. 1b). The Jones Mountains are a mostly snow and ice covered, NE–SW-orientated range with rocks exposed in numerous nunataks scattered over a distance of c. 60 km (refer to Wilch et al. 2021, fig. 37). Cenozoic volcanic rocks consist of pillow lavas and breccia in addition to other volcaniclastic lithologies and lava flows, all of which lie unconformably atop glacially polished and striated arc-related Cretaceous (100–90 Ma) volcanic rocks and Jurassic (198 Ma) granitoid intrusions (Craddock et al. 1964; Rowley 1990; Pankhurst et al. 1993; Hole et al. 1994; Rutford and McIntosh 2007). The age of the Cenozoic volcanism is poorly constrained but best estimates, which are based on both K/Ar and $^{40}Ar/^{39}Ar$ dating results, place it within the range of 12–7 Ma (Wilch et al. 2021 and references therein), with a preferred age of c. 7 Ma (Rutford and McIntosh 2007).

All available analyses of Cenozoic volcanic rocks within the Hudson and Jones mountains are basaltic in composition. The basalts are classified as hawaiite, alkali basalt and tholeiite (Fig. 39a), and all but one sample (208b: Supplementary material Table S1) is silica-undersaturated (≤11% nepheline-normative). It is notable that of all the basalt samples analysed within the MBLVG, tholeiitic types are found only within these mountain ranges along with the one sample (66D-91) from the Fosdick Mountains Volcanic Field (Fig. 39b). Basalt compositions from the Jones Mountains are more mafic (MgO 9.9–12.3 wt%, Cr 284–440 ppm and Ni 222–320 ppm) than those from the Hudson Mountains (MgO 5.3–9.7 wt%, Cr 104–309 ppm and Ni 35–191 ppm) and when plotted together on Fenner plots (MgO v. other major element oxides) show smooth data arrays, which is consistent with differentiation as a result of progressive fractionation dominated by olivine. Hypothetically, if the basaltic magmas were related to the same system, then the fractionation of clinopyroxene is minimal given that with decreasing MgO content the sample suite shows an increase in V content, which is compatible with clinopyroxene (K_d V_{cpx} = 4.8–5.7 for basanite: Adam and Green 2006), and relatively uniform CaO/Na_2O ratios (average 2.97 ± 0.29, 1 standard deviation).

Minor and trace elements for Hudson Mountains basalt samples are limited to five samples measured by XRF (Rowley et al. 1990a); with another 18 samples from the Jones Mountains, eight of which have trace element concentrations measured by a variety of other techniques (ICP-MS, isotope dilution and INAA: Hole et al. 1994; Hart et al. 1995) (see Supplementary material Table S1). Although more evolved than the Jones Mountains basalt compositions, those from the Hudson Mountains have roughly equivalent LILE/HFSE (e.g. Ba/Nb, K/Zr and Sr/Ti) as well as HFSE ratios (e.g. Zr/Nb, Zr/Y and Nb/Y). On normalized multi-element plots (Fig. 40a, c), basalts from the Jones Mountains display relatively smooth trace element patterns. Their concentration levels are depleted relative to most MBLVP basalt compositions but are remarkably similar to Mount Perkins in the Fosdick Mountains Volcanic Field (Fig. 34), although showing much more subdued negative Pb anomalies. Slightly more enriched compositions (Fig. 40b) have slightly lower Y and HREE concentrations, which, in addition to the relatively steep slope shown on REE plots (Fig. 40c), suggest that the melting of a garnet-bearing mantle source produced the basalts. The Zr/Nb ratios of basalts from the Jones Mountains Volcanic Field, as well as from the Fosdick Mountains Volcanic Field and the Hobbs Coast nunataks of the Hobbs Coast Volcanic Field, vary systematically with La/Yb_N ratios and degree of silica-undersaturation (Fig. 41). While fractional crystallization of clinopyroxene could explain a decrease in Zr/Nb ratios (e.g. K_d Zr_{cpx} = 0.79–1.53 v. K_d Nb_{cpx} = 0.06–0.2 for alkali basalt: Wood and Trigila 2001), it is unlikely given its minimal role in the evolution of these magmas (discussed above), as well as the lack of correlation between Zr/Nb ratios and Cr and MgO contents and negative correlations with V concentrations (not shown). Conversely, the variation in Zr/Nb ratios of basalt samples from Jones Mountains was considered by Hole et al. (1994) to be controlled by partial melting processes where Zr is slightly more compatible than Nb in mantle clinopyroxene (see the partition coefficients of Chazot et al. 1996; Lundstrom et al. 1998; Elkins et al. 2008). The data trends shown by all three basaltic suites in Figure 41 (i.e. decreasing Zr/Nb ratios with increasing silica-undersaturation and LREE-enrichment) therefore would most likely to indicate changing conditions of partial melting from higher to lower degrees, with garnet retained within their mantle sources. This change correlates with time

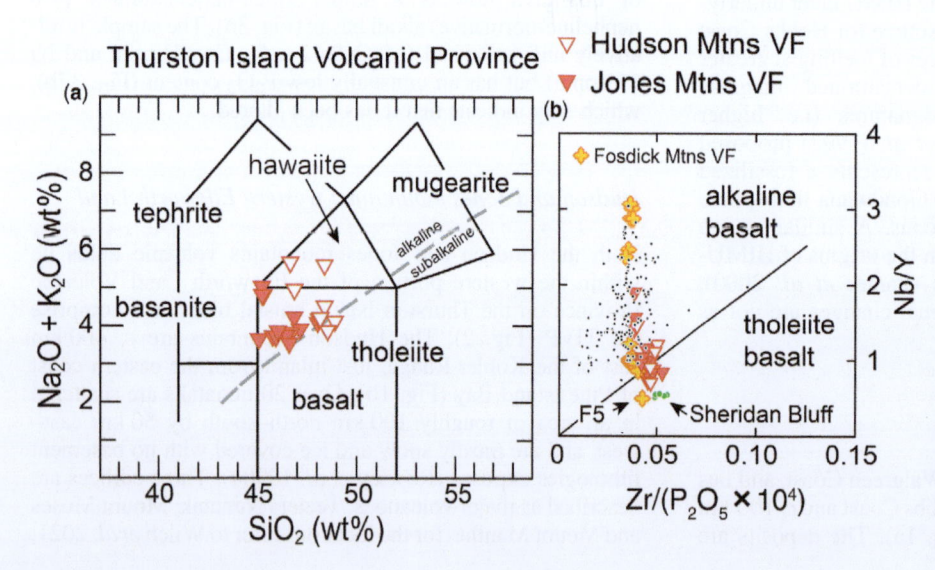

Fig. 39. Basalt classification for basalt compositions from the Hudson and Jones mountains volcanic fields, Thurston Island Volcanic Province (Fig. 1b). (a) Total alkali v. SiO_2 (wt%) plot and (b) the classification plot of Floyd and Winchester (1975) used to distinguish alkaline from subalkaline types. Also plotted in (b) are basalts from the Fosdick Mountains, as well as from the rest of the MBLVG. In addition, tholeiites from Sheridan Bluff (Panter et al. 2021), which is located in the southern Transantarctic Mountains, are also shown for comparison.

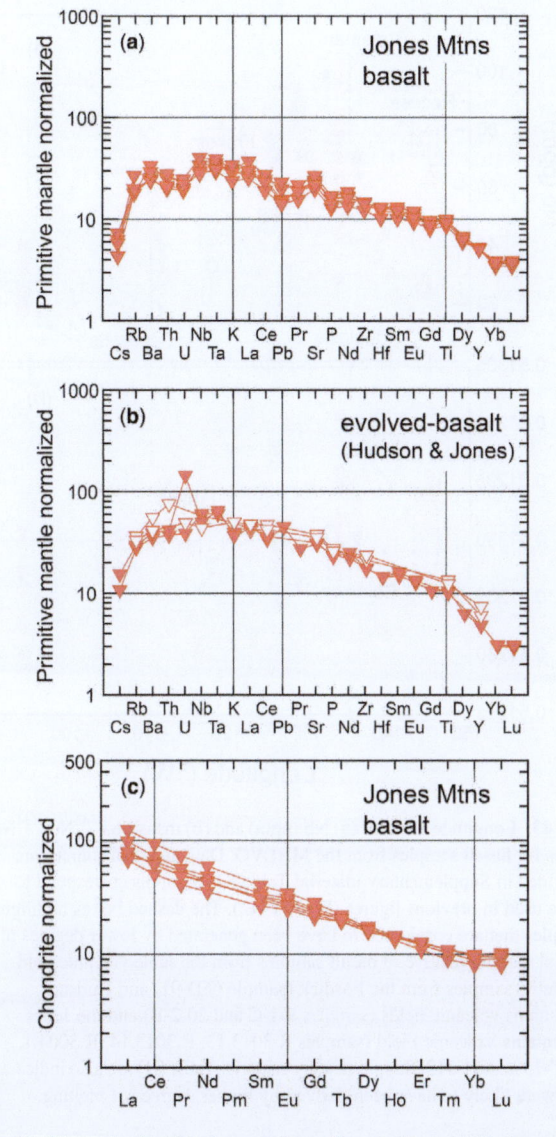

Fig. 40. Normalized multi-element plots of basalts from the Hudson and Jones mountains, Thurston Island Volcanic Province (TIVP). (**a**) & (**b**) Normalized to primitive mantle (McDonough and Sun 1995) and (**c**) REE are normalized to chondrite (Sun and McDonough 1989).

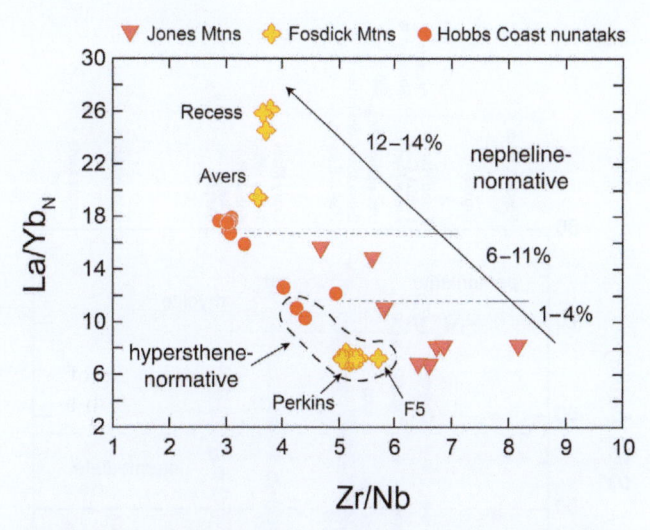

Fig. 41. Plot of Zr/Nb ratios v. La/Yb$_N$ ratios for basalts from the Jones Mountains in comparison to basalts from the Hobbs Coast nunataks (Hobbs Coast Volcanic Field) and the Fosdick Mountains Volcanic Field, MBLVP. Indicated on the diagram are ranges for nepheline-normative compositions (wt%). Dashed outline encompasses samples that are hypersthene-normative.

for basalt samples from the Hobbs Coast nunataks (Hart et al. 1997) (Fig. 38a) but is unknown for the other two suites which lack age constraints.

Hart et al. (1995) reported Sr, Nd and Pb isotopes measured on six samples from the Jones Mountains Volcanic Field, and Hole et al. (1994) provided Sr isotopes for two additional samples, one of which was also measured for Nd isotopes. Strontium, Nd and Pb isotopes for three samples from the Hudson Mountains Volcanic Field are provided by Hart et al. (1997). Compared to most MBLVP basalts, the isotopic signatures of the TIVP basalts extend to more radiogenic Sr values (0.702904–0.703889) over a very broad range in Nd values (0.512786–0.512980) and relatively unradiogenic Pb isotopic ratios (^{206}Pb/^{204}Pb 19.07–19.41, ^{207}Pb/^{204}Pb 15.60–15.69 and ^{208}Pb/^{204}Pb 38.67–39.02) (Supplementary material Table S1). Their Sr and Pb isotopic compositions are, however, similar to basalt samples from the Fosdick Mountains Volcanic Field (Fig. 35). As with the basalt compositions from the Fosdick Mountains, arguments suggested no significant involvement of continental crust based on major and trace element characteristics (Hole et al. 1994; Hart et al. 1995). Moreover, Hart et al. (1995) compared the Jones Mountains basalts with basalts from Peter I Island – a young (≤2 Ma: Prestvik and Duncan 1991; Kipf et al. 2014) volcanic centre, located over 500 km to the north and c. 200 km off the continental shelf on the Antarctic Ocean plate – and concluded that the compositional similarities and overlaps precluded contamination of magmas by continental crust.

The interpretation of the origin of Cenozoic volcanism in the TIVP is debated. Hole et al. (1994) invoked a deep-seated mantle plume ('Marie Byrd Land plume': LeMasurier and Rex 1989; Hole and LeMasurier 1994) that arrived at the base of the lithosphere beneath the central portion of the MBL crustal block at c. 30 Ma and migrated radially outward, flowing beneath the Thurston Island crustal block by c. 12 Ma to trigger melting and volcanism in the Jones Mountains and later in the Hudson Mountains. Hart et al. (1995), on the other hand, implicated both sublithospheric and lithospheric mantle sources for volcanism that were melted by decompression associated with rifting. Hart et al. (1995) further suggested that, prior to volcanism, the continental lithospheric mantle was compositionally modified by subduction during arc volcanism in the Mesozoic. They used this to explain the lower ^{207}Pb/^{204}Pb and Ce/Pb ratios of basalts from the Jones Mountains relative to those from the oceanic Peter I Island.

Origin of volcanism in MBL and wEL

The relatively few detailed petrological studies of volcanoes in MBL and wEL provide important constraints on the petrogenesis of Cenozoic intraplate volcanism in West Antarctica but many issues remain unresolved. Current models are often conflicting with respect to the relative influence of past subduction, rifting and plumes on mantle melt sources and melt generation, as well as regional doming and volcanic migration patterns. Although this chapter is not intended to resolve these disputes, we hope that the extensive compilation of published and unpublished major element, trace element and radiogenic isotope data provided here (see Supplementary material Table S1), in concert with new geophysical findings, will help to stimulate future research in this region, as well as contributing to a better understanding of intraplate continental alkaline volcanism worldwide.

As a primer, we evaluate compositional variations from a geographical point of view. Over a distance of nearly

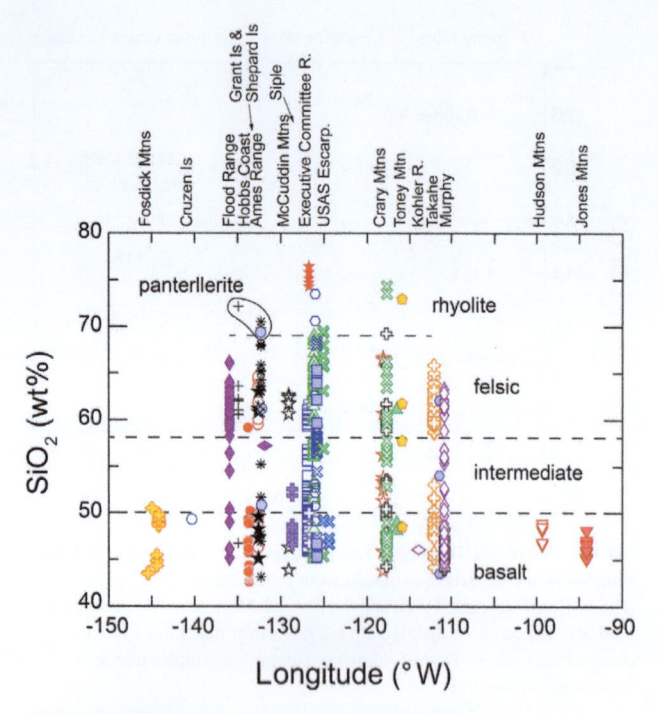

Fig. 42. Longitude (° W) v. SiO$_2$ (wt%) for all MBLVG compositions presented in this study. Data and data sources are provided in Supplementary material Table S1. Compositions are assigned as basalts (<50 wt% SiO$_2$); intermediate (50–58 wt% SiO$_2$), which includes phonolitic samples; and felsic (>58 wt% SiO$_2$), which includes rhyolite (>69 wt% SiO$_2$) and pantellerite (SiO$_2$ ≥69 wt% with Al$_2$O$_3$ (wt%) less than 1.33 × FeO^T (wt%) + 4.4) (Le Maitre 2002).

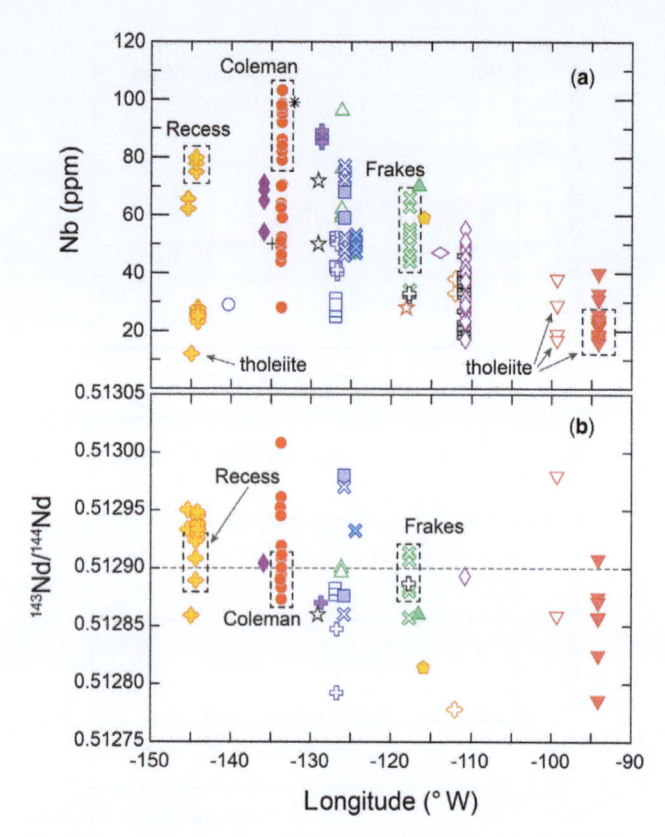

Fig. 43. Longitude (° W) v. (**a**) Nb (ppm) and (**b**) measured ^{143}Nd/^{144}Nd ratios for basalt samples from the MBLVG. Data and data sources are provided in Supplementary material Table S1. Symbols correspond to those used in previous figures (Figs 11–42). The dashed boxes highlight samples that are considered to have been generated by lower degrees of partial melting relative to basalt samples from the same volcanic field. Tholeiite samples from the Fosdick (sample 66D-91) and Hudson mountains volcanic fields (samples 9-1-C and 20-2-B) and the Jones Mountains Volcanic Field (samples R.3013.13, R.3013.14, R.3004.1, 69-C-15 and 61-113: Supplementary material Table S1) are also indicated and were likely to have been derived by higher degrees of melting.

1500 km, from the Fosdick Mountains in western MBL to the Jones Mountains in wEL, Cenozoic volcanism of the MBLVG shows the widest range of major and trace elements concentrations in the middle portion of the region (e.g. Fig. 42) where major volcanic centres are found, including a few whose interiors are well exposed (e.g. Mount Sidley in the Executive Committee Range and Mount Rees in the Crary Mountains). Limiting sample compositions to basalt, region-wide variations can provide insights into mantle processes and sources for the volcanism. From east to west, samples of basalt display a greater range and higher concentrations in incompatible trace elements Nb, Ba and Zr (Fig. 43a). The highest values are in basalts that are considered to have been derived from lower degrees of partial melting relative to other basalt compositions within the same volcanic group (Hart *et al.* 1997; Panter *et al.* 2000; Gaffney and Siddoway 2007). An overall decrease in Zr/Nb ratios from east to west is also observed (not shown). However, other indicators for extent of melting, such as level of silica-undersaturation and La/Yb$_N$ ratios, are not as well correlated. Overall, ^{143}Nd/^{144}Nd ratios exhibit a broad increase from east to west, with values more consistently at or below 0.51290 east of 130° W and the majority of samples at or above 0.51290 west of 130° W (e.g. Hobbs Coast and Fosdick Mountains volcanic fields). However, for the basalt compositions that are identified as derivatives of small degree melts (e.g. from Reese Nunatak, Coleman Nunatak and Mount Frakes), their values are relatively uniform at around 0.51290 (Fig. 43b). Province-wide gradients for Al$_2$O$_3$ and Pb isotopes are also identified, and show decreasing values away from the central portion of the region (Fig. 44). Taken together, the geospatial patterns suggest a gradational change in mantle source compositions and conditions of melting within the region that encompasses the MBLVG. We now will consider these findings within the context of what is already proposed based on much more limited datasets.

Two principal types of mantle reservoirs appear to dominate as sources for volcanism in this region of West Antarctica: (1) HIMU (*sensu lato*) plume(s); and (2) subduction-modified (i.e. metasomatized) depleted (MORB) mantle and/or modified continental lithospheric mantle. On the western (Fosdick Mountains) and eastern (Jones Mountains) extents of this region, melt sources for volcanism have been ascribed to lithosphere metasomatized by subduction (Hart *et al.* 1995; Gaffney and Siddoway 2007). Metasomatized lithosphere is also invoked as the source for volcanoes in the north-central and eastern portion of MBL (i.e. Mount Siple, Mount Murphy and Mount Takahe) (LeMasurier *et al.* 2016). Across the central portion of the region, Panter *et al.* (2000) proposed subduction-metasomatized lithosphere underlain by HIMU plume material for volcanism in the Crary Mountains, while Hart *et al.* (1997) preferred a stratified heterogeneous plume source (FOZO underlain by HIMU) for the Hobbs Coast nunataks. Studies that considered volcanism for the whole of the region are divided into: (1) exclusive melting of plume materials (Lanyon *et al.* 1993; Hole and LeMasurier 1994); and (2) melting of metasomatized sources (Futa and LeMasurier 1983).

Figure 45 shows the geographical distribution of ^{206}Pb/^{204}Pb ratios for basalts. Ratios greater than 20 are restricted to the middle portion of the MBL–wEL region, accompanied by more radiogenic ^{207}Pb/^{204}Pb and ^{208}Pb/^{204}Pb values (Fig. 44a–c), with the highest values

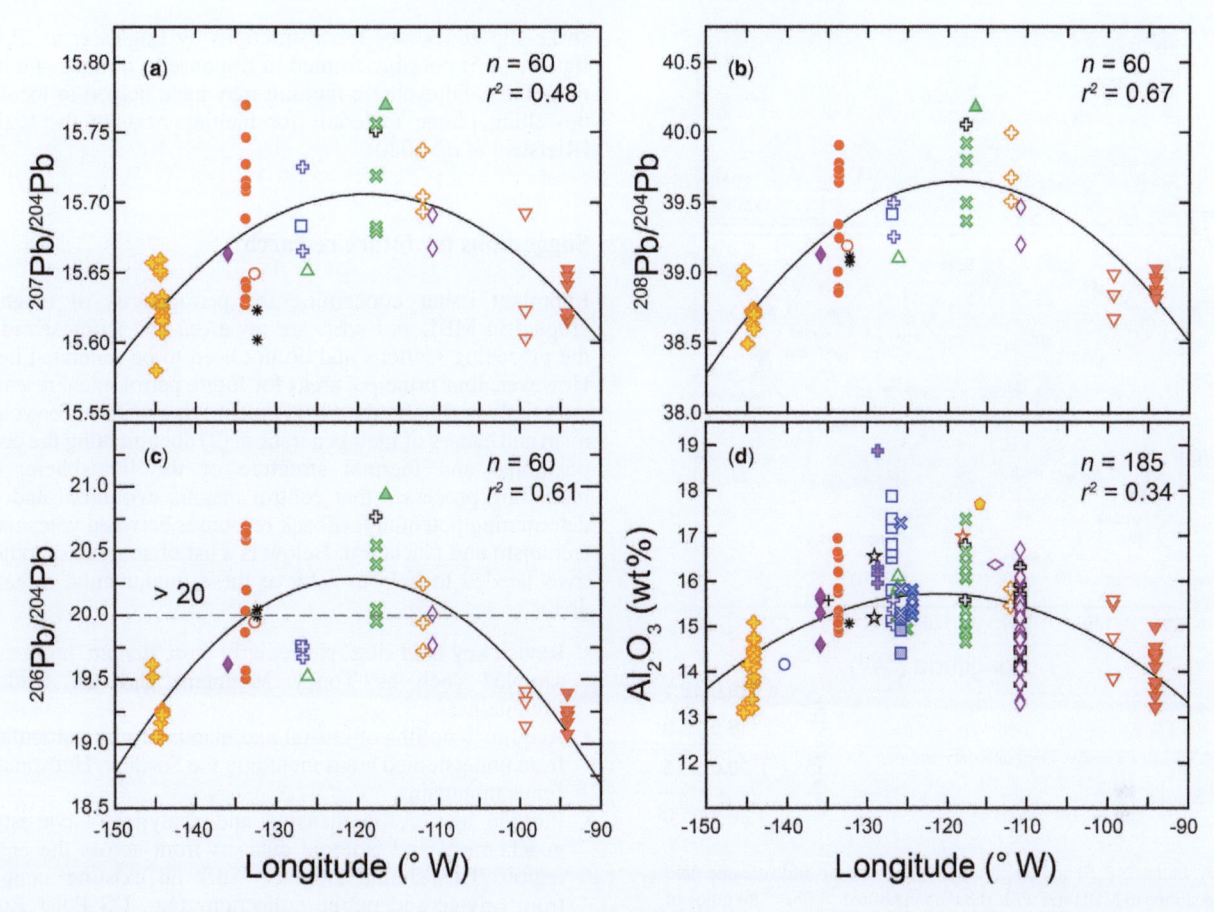

Fig. 44. Longitude (° W) v. Pb isotopic ratios (**a–c**) and Al$_2$O$_3$ (wt%) for MBLVG basalts. Data and data sources are provided in Supplementary material Table S1. The Pb isotope plots include samples with MgO values of 4–12.3 wt%, and the Al$_2$O$_3$ plot in (**d**) includes basalts with MgO values from 6 to 12. 3 wt%. The parabola that best fits the data (quadratic regression), number of samples (*n*) and the coefficient of determination (r^2) are provided on each plot.

measured for Coleman Nunatak (Hobbs Coast) and the Crary Mountains Volcanic Field. The HIMU-like signature of the basalt samples, along with their OIB-like geochemical characteristics, constitute the petrological basis for plume source models.

Other evidence cited in support of a mantle plume includes structural and geophysical data. The volcanoes in the MBLVP are mostly located on the *c.* 1000 × 500 km structural dome (LeMasurier 2006) on the north flank of the WARS (Fig. 2). LeMasurier and Landis (1996) interpreted MBL dome elevations increasing towards the centre of the MBLVP as a topographical expression of a mantle plume. The mean crustal thicknesses beneath the MBL dome are calculated from seismic data to be 28–33 km, which is 5–10 km thicker than in the rest of the WARS (Chaput *et al.* 2014; Ramirez *et al.* 2017; Shen *et al.* 2018). However, the MBL dome is not attributed to isostatic compensation of thicker crust but to a thermal anomaly beneath the volcanic province (Winberry and Anandakrishnan 2004).

More recently, our understanding of the thermal structure of the MBL lithosphere and sublithospheric mantle has been advanced by several multi-year geophysical experiments in West Antarctica. Lloyd *et al.* (2015) noted that in West Antarctica the slowest P- and S-wave velocities extend to 200 km below the Executive Committee Range, and Hansen *et al.* (2014) interpreted a low-velocity zone extending to 800 km beneath MBL using P-wave tomography: both studies support the plume hypothesis. Moreover, a modelling study using ambient seismic noise velocities by Heeszel *et al.* (2016) identified an extensive low-velocity zone extending to >200 km centred on the Executive Committee Range, which is again consistent with a thermal anomaly (i.e. mantle plume). Using a new, continent-scale seismic model, Lloyd *et al.* (2019) resolved slow velocities extending into the transition zone and possibly into the lower mantle beneath MBL. Lloyd *et al.* (2019) also imaged slow seismic velocities at shallow levels (200–250 km) that occur along the length of the West Antarctic coastline from MBL to Ellsworth Land and extending to the Antarctic Peninsula. The shallow slow-velocity anomalies are connected to a broad slow-velocity anomaly that is found at greater depths (*c.* 250 km) beneath oceanic lithosphere that dates to the separation of Zealandia from MBL (*c.* 90 Ma). It is important to note that some researchers have attributed the initiation of seafloor spreading in this region to mantle plume activity (Weaver *et al.* 1994; Storey *et al.* 1999), and this may serve as a cause and source for HIMU-like volcanism forming the Marie Byrd seamounts at 65–56 Ma (Kipf *et al.* 2014), which lie over 300 km to the north of Mount Siple. In other geophysical studies, Seroussi *et al.* (2017) suggested that a weak mantle plume existed beneath MBL with a maximum geothermal heat flux of 150 mW m^{-2}, based on modelling and assessment of the hydrology at the base of the West Antarctic Ice Sheet.

The geophysical evidence for a thermal anomaly beneath central MBL, in concert with the occurrence of high ^{206}Pb/^{204}Pb values in basalts, offers compelling support for a deep HIMU-like plume source for the volcanism. However, the plume source is not consistent with the composition of volcanism in western MBL and wEL (Figs 44 & 45), which, as described above, is considered to be sourced primarily from subduction-modified mantle. A subduction tectonic regime was nearly continuous from the late Neoproterozoic (*c.* 550 Ma) to the Late Cretaceous (*c.* 100 Ma) along the palaeo-Pacific margin of Gondwana (Bradshaw 1989;

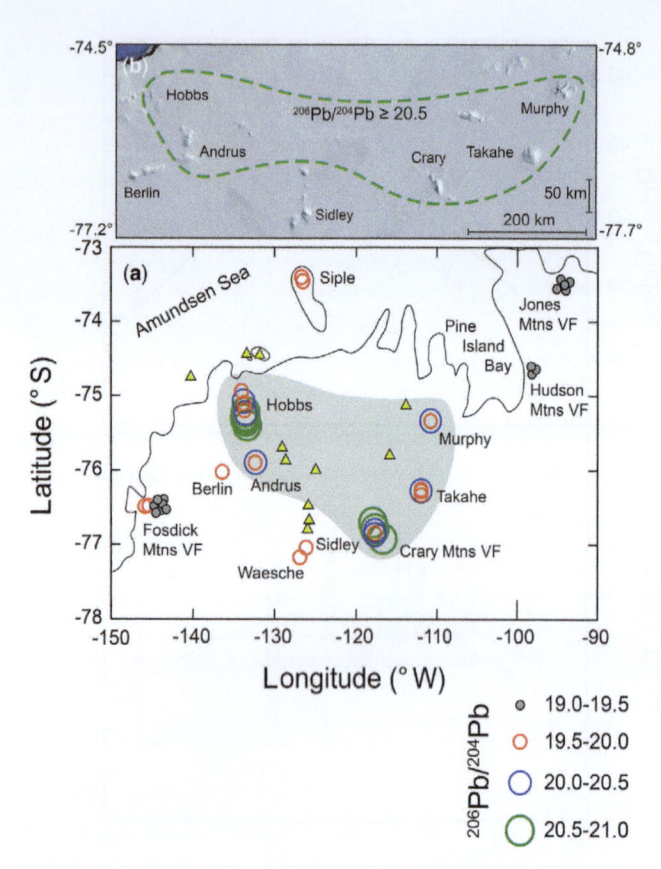

Fig. 45. Latitude (° S) and longitude (° W) of volcano and volcanic field (VF) locations in MBL and wEL showing measured $^{206}Pb/^{204}Pb$ ratios of MBLVG basalt compositions. Data and data sources are provided in Supplementary material Table S1. (**a**) Samples whose values fall within increments of 0.5 are coded by colour and size of circle. Hobbs refers to the nunataks along the Hobbs Coast (Hobbs Coast Volcanic Field). Yellow triangles represent the main volcanoes where Pb isotopes have not been measured. The grey area highlights the volcanoes where basalts have $^{206}Pb/^{204}Pb$ values >20.5. (**b**) (the top panel) is from Google Earth Pro using a US Geological Survey and NASA generated image. The green dashed line encompasses the same volcanoes but offers a less distorted perspective to emphasize the elongate distribution of basalts with the most radiogenic $^{206}Pb/^{204}Pb$ signatures.

Cawood 2005) before the break-up that separated the pieces of the present Zealandia continent (Mortimer et al. 2017) from MBL around 90 Ma (Eagles et al. 2004). Hoernle et al. (2020) proposed that a HIMU plume rose beneath the active convergent margin and flowed along the subducting lithosphere underneath the Zealandia portion of Gondwana where slab tears/windows allowed plume-generated melting to rise and supply Cretaceous volcanism. They also suggested that the compositions of the late Cenozoic HIMU-like volcanism in Zealandia are a result of mixing between plume-derived HIMU-like materials and subduction-modified depleted upper mantle (Hoernle et al. 2020). It is important to re-emphasize here that intraplate alkaline volcanism that has occurred on these now widely scattered continental fragments over the past c. 50 myr share similar compositional characteristics (Finn et al. 2005). An analogous scenario for West Antarctica is proposed here, in which variable mixing between melts from these two source reservoirs (i.e. HIMU-like plume and metasomatized mantle), regulated by the depth and extent of melting, may explain the gradational change in basalt compositions across the MBL–wEL region (Fig. 44). We further speculate that the current known distribution of the most radiogenic Pb signatures found in MBL forms a relatively narrow and broadly linear feature (Fig. 45) which at c. 90 Ma was roughly parallel with several other collinear features that Eagles et al. (2004) suggested were continental strike-slip zones (see reconstructions by Eagles et al. 2004, figs 4 and 5) possibly formed in response to oblique subduction. Trans-lithospheric faulting may have helped to localize upwelling plume materials for melting beneath the region (Riefstahl et al. 2020).

Suggestions for future research

Important issues concerning the petrogenesis of magmas erupted in MBL and wEL are reviewed and summarized in the preceding sections and do not need to be reiterated here. However, four principal areas for future petrological research are briefly highlighted here: (1) identifying sources for volcanism and causes of melt generation; (2) documenting the compositional and thermal structure of the lithosphere; (3) modelling processes that control magma evolution; and (4) determining potential feedback responses between volcanism, tectonism and glaciation. Below is a list of some basic objectives needed to help to achieve these fundamental research goals:

- Revisit key field sites, particularly those that are minimally sampled such as Toney Mountain and the Hudson Mountains.
- Acquire xenoliths of crustal and mantle origin, particularly from understudied areas including the Fosdick, Hudson and Jones mountains.
- Provide new comprehensive and analytically consistent geochemical and isotopic datasets from across the entire region. This should embrace work on existing samples from private and public collections (e.g. US Polar Rock Repository, http://research.bpcrc.osu.edu/rr/). Future comprehensive datasets should also include isotopes of Os and Hf, as well as noble gas (He, Ne, Ar, Kr, Xe) and halogen (Cl, Br, I) analyses which are deficient or non-existent.
- Integrate data from the MBLVG with data gathered from the western Ross Sea (i.e. McMurdo Volcanic Group), as well as from seamounts and volcanism on continental fragments of Gondwana (eastern Australia and Zealandia).
- Initiate petrological collaborations in geophysical programmes (e.g. seismic and magnetotellurics studies).

Acknowledgements We would like to thank our mountaineers Bill Atkinson, Paul Rose, Chris Griffiths and Tony Teeling for keeping us safe during fieldwork. We are grateful to John Gamble and Nelia Dunbar, who played integral roles in fieldwork and research in Marie Byrd Land. Special thanks to Robert Pankhurst, Jerzy Blusztajn and Samuel Mukasa for providing facilities and assistance with isotopic analysis. Antarctic Support Services and VXE-6 aero squadron are acknowledged for critical logistical support in Antarctica. We are grateful to Amy Gaffney and Christine Siddoway for generously providing unpublished analyses of basalts from the Fosdick Mountains. Constructive reviews by Adam Martin and Meritxell Aulinas along with the careful editorial handing of Adelina Geyer are gratefully acknowledged.

Author contributions KSP: conceptualization (lead), writing – original draft (lead), writing – review & editing (lead); **TIW**: conceptualization (equal); **JLS**: conceptualization (equal); **PRK**: conceptualization (equal); **WCM**: conceptualization (equal).

Funding The authors' contributions to this study was supported, in part, by NSF grants DPP-8816342 (WAVE) to P.R. Kyle, DPP-9198806 (WAVE II) to W.I McIntosh (PI) and P.R. Kyle

(Co-PI), and NSF OPP-9419686 to P.R. Kyle (PI) and K.S. Panter (Co-PI).

Data availability All data generated or analysed during this study are included in this published article (and its supplementary information files).

References

Adam, J. and Green, T. 2006. Trace element partitioning between mica- and amphibole-bearing garnet lherzolite and hydrous basanitic melt: 1. Experimental results and the investigation of controls on partitioning behavior. *Contributions to Mineralogy and Petrology*, **152**, 1–17, https://doi.org/10.1007/s00410-006-0085-4

Andreeva, I.A. and Kovalenko, V.I. 2011. Evolution of the trachydacite and pantellerite magmas of the bimodal volcanic association of Dzarta–Khuduk, central Mongolia: investigation of inclusion in minerals. *Petrology*, **19**, 363–385, https://doi.org/10.1134/S0869591111040023

Arzamastsev, A.A., Arzamastseva, L.V., Bea, F. and Montero, P. 2009. Trace elements in minerals as indicators of the evolution of alkaline ultrabasic dike series: LA-ICP-MS data for the magmatic provinces of northeastern Fennoscandia and Germany. *Petrology*, **17**, 46–72, https://doi.org/10.1134/S0869591109010032

Aviado, K.B., Rilling-Hall, S., Bryce, J.G. and Mukasa, S.B. 2015. Submarine and subaerial lavas in the West Antarctic Rift System: Temporal record of shifting magma source components from the lithosphere and asthenosphere. *Geochemistry, Geophysics, Geosystems*, **16**, 4344–4361, https://doi.org/10.1002/2015GC006076

Baker, B.H. 1987. Outline of the petrology of the Kenyan Rift alkaline province. Geological Society, London, Special Publications, **30**, 293–311, http://doi.org/10.1144/GSL.SP.1987.030.01.14

Behrendt, J.C. 2013. The aeromagnetic method as a tool to identify Cenozoic magmatism in the West Antarctic Rift System beneath the West Antarctic Ice Sheet – A review; Thiel subglacial volcano as possible source of the ash layer in the WAISCORE. *Tectonophysics*, **585**, 124–136, https://doi.org/10.1016/j.tecto.2012.06.035

Blankenship, D.D., Bell, R.E., Hodge, S.M., Brozena, J.M., Behrendt, J.C. and Finn, C.A. 1993. Active volcanism beneath the West Antarctic ice sheet and implications for ice sheet stability. *Nature*, **361**, 526–529, https://doi.org/10.1038/361526a0

Bohrson, W.A. and Reid, M.R. 1997. Genesis of silicic peralkaline volcanic rocks in an ocean island setting by crustal melting and open-system processes: Socorro Island, Mexico. *Journal of Petrology*, **38**, 1137–1166, https://doi.org/10.1093/petroj/38.9.1137

Boyce, J.A., Nicholls, I.A., Keays, R.R. and Hayman, P.C. 2015. Variation in parental magmas of Mt. Rouse, a complex polymagmatic volcano in the basaltic intraplate Newer Volcanics Province, southeast Australia. *Contributions to Mineralogy and Petrology*, **169**, 11, https://doi.org/10.1007/s00410-015-1106-y

Bradshaw, J.D. 1989. Cretaceous geotectonic patterns in the New Zealand region. *Tectonics*, **8**, 803–820, https://doi.org/10.1029/TC008i004p00803

Cande, S.C. and Stock, J.M. 2006. Constraints on the timing of extension in the Northern Basin, Ross Sea. *In*: Futterer, D.K., Damaske, D., Kleinschmidt, G., Miller, H. and Tessensohn, F. (eds) *Antarctica: Contributions to Global Earth Sciences*. Springer, Berlin, 319–326.

Cande, S.C., Stock, J.M., Müller, R.D. and Ishihara, T. 2000. Cenozoic motion between east and west Antarctica. *Nature*, **404**, 145–150, https://doi.org/10.1038/35004501

Castillo, P.R. 2015. The recycling of marine carbonates and sources of HIMU and FOZO ocean island basalts. *Lithos*, **216–217**, 254–263, https://doi.org/10.1016/j.lithos.2014.12.005

Cawood, P.A. 2005. Terra Australis Orogen: Rodinia breakup and development of the Pacific and Iapetus margins of Gondwana during the Neoproterozoic and Paleozoic. *Earth-Science Reviews*, **69**, 249–279, https://doi.org/10.1016/j.earscirev.2004.09.001

Chaput, J., Aster, R.C. *et al.* 2014. The crustal thickness of West Antarctica. *Journal of Geophysical Research: Solid Earth*, **119**, 378–395, https://doi.org/10.1002/2013JB010642

Chakraborty, S. 2007. *The Geochemical Evolution of Alkaline Magmas from the Crary Mountains, Marie Byrd Land, Antarctica*. MS thesis, Bowling Green State University, Bowling Green, OH, USA.

Chatzaras, V., Kruckenberg, S.C., Cohen, S.M., Medaris, L.G., Withers, A.C. and Bagley, B. 2016. Axial-type olivine crystallographic preferred orientations: the effect of strain geometry on mantle texture. *Journal of Geophysical Research: Solid Earth*, **121**, 4895–4922, https://doi.org/10.1002/2015JB012628

Chazot, G., Menzies, M.A. and Harte, B. 1996. Determination of partition coefficients between apatite, clinopyroxene, amphibole, and melt in natural spinel lherzolites from Yemen: Implications for wet melting of the lithospheric mantle. *Geochimica et Cosmochimica Acta*, **60**, 423–437, https://doi.org/10.1016/0016-7037(95)00412-2

Clague, D.A. 1978. The oceanic basalt-trachyte association: an explanation of the Daly Gap. *The Journal of Geology*, **86**, 739–743, https://doi.org/10.1086/649740

Corr, H.F.J. and Vaughan, D.G. 2008. A recent volcanic eruption beneath the West Antarctic ice sheet. *Nature Geoscience*, **1**, 122–125, https://doi.org/10.1038/ngeo106

Craddock, C., Bastien, T.W. and Rutford, R.H. 1964. Geology of the Jones Mountains area. *In*: Adie, R.J. (ed.) *Antarctic Geology*. North-Holland, Amsterdam, 171–187.

Daly, R.A. 1925. The geology of Ascension Island. *American Academy of Arts and Sciences Proceedings*, **60**, 1–80.

Dalziel, I.W.D. and Elliot, D.H. 1982. West Antarctica: problem child of Gondwanaland. *Tectonics*, **1**, 3–19, https://doi.org/10.1029/TC001i001p00003

Dalziel, I.W.D. and Lawver, L.S. 2001. The lithospheric setting of the west Antarctic ice sheet. *American Geophysical Union Antarctic Research Series*, **77**, 29–44.

Davy, B. 2014. Rotation and offset of the Gondwana convergent margin in the New Zealand region following Cretaceous jamming of Hikurangi Plateau large igneous province subduction. *Tectonics*, **33**, 1577–1595, https://doi.org/10.1002/2014TC003629

Davy, B., Hoernle, K. and Werner, R. 2008. Hikurangi Plateau: Crustal structure, rifted formation, and Gondwana subduction history. *Geochemistry, Geophysics, Geosystems*, **9**, Q07004, https://doi.org/10.1029/2007GC001855

Dedzo, M.G., Asaah, A.N.E. *et al.* 2019. Petrology and geochemistry of lavas from Gawar, Minawao and Zamay volcanoes of the northern segment of the Cameroon volcanic line (central Africa): constraints on mantle source and geochemical evolution. *Journal of African Earth Sciences*, **153**, 31–41, https://doi.org/10.1016/j.jafrearsci.2019.02.010

DiVenere, V.J., Kent, D.V. and Dalziel, I.W.D. 1994. Mid-Cretaceous paleomagnetic results from Marie Byrd Land, West Antarctica: A test of post-100 Ma relative motion between East and West Antarctica. *Journal of Geophysical Research: Solid Earth*, **99**, 15 115–15 139, https://doi.org/10.1029/94JB00807

Dufek, J. and Bachmann, O. 2010. Quantum magmatism: magmatic compositional gaps generated by melt-crystal dynamics, *Geology*, **38**, 687–690, https://doi.org/10.1130/G30831.1

Dunbar, N.W., Iverson, N.A., Smellie, J.L., McIntosh, W.C., Zimmerer, M.J. and Kyle, P.R. 2021. Active volcanoes in Marie Byrd Land. *Geological Society, London, Memoirs*, **55**, https://doi.org/10.1144/M55-2019-29

Eagles, G., Gohl, K. and Larter, R.D. 2004. High-resolution animated tectonic recon-struction of the South Pacific and West Antarctic margin. *Geochemistry, Geophysics, Geosystems*, **5**, Q07002, https://doi.org/10.1029/2003GC000657

Elkins, L., Gaetani, G. and Sims, K. 2008. Partitioning of U and Th during garnet pyroxenite partial melting: Constraints on the source of alkaline ocean island basalts. *Earth and Planetary*

Science Letters, 265, 270–286, https://doi.org/10.1016/j.epsl.2007.10.034

Faccenna, C., Rossetti, F., Becker, T.W., Danesi, S. and Morelli, A. 2008. Recent extension driven by mantle upwelling beneath the Admiralty Mountains (East Antarctica). Tectonics, 27, TC4015, https://doi.org/10.1029/2007TC002197

Finn, C.A., Müller, R.D. and Panter, K.S. 2005. A Cenozoic diffuse alkaline magmatic province (DAMP) in the southwest Pacific without rift or plume origin. Geochemistry, Geophysics, Geosystems, 6, Q02005, https://doi.org/10.1029/2004GC000723

Floyd, P.A. and Winchester, J.A. 1975. Magma type and tectonic setting discrimination using immobile elements. Earth and Planetary Science Letters, 27, 211–218, https://doi.org/10.1016/0012-821X(75)90031-X

Fretwell, P., Pritchard, H.D. et al. 2013. Bedmap2: improved ice bed, surface and thickness datasets for Antarctica. The Cryosphere, 7, 375–393, https://doi.org/10.5194/tc-7-375-2013

Futa, K. and LeMasurier, W.E. 1983. Nd and Sr isotopic studies on Cenozoic mafic lavas from West Antarctica: another source for continental alkali basalts. Contribution to Mineralogy and Petrology, 83, 38–44, https://doi.org/10.1007/BF00373077

Gaffney, A.M. and Siddoway, C.S. 2007. Heterogeneous sources for Pleistocene lavas of Marie Byrd Land, Antarctica: new data from the SW Pacific diffuse alkaline magmatic province. United States Geological Survey Open-File Report, 2007-1047, Extended Abstract 063.

Gamble, J.A., Adams, C.J., Morris, P.A., Wysoczanski, R.J., Handler, M. and Timm, C. 2018. The geochemistry and petrogenesis of Carnley Volcano, Auckland Islands, SW Pacific. New Zealand Journal of Geology and Geophysics, 61, 480–497, https://doi.org/10.1080/00288306.2018.1505642

González-Ferrán, O. and González-Bonorino, F. 1972. The volcanic ranges of Marie Byrd Land between 100° and 140° W. In: Adie, R.J. (ed.) Antarctic Geology and Geophysics. Universitetsforlaget, Oslo, 261–276.

Granot, R. and Dyment, J. 2018. Late Cenozoic unification of East and West Antarctica. Nature Communications, 9, 3189, https://doi.org/10.1038/s41467-018-05270-w

Granot, R., Cande, S.C., Stock, J.M. and Damaske, D. 2013. Revised Eocene–Oligocene kinematics for the West Antarctic rift system. Geophysical Research Letters, 40, 279–284, https://doi.org/10.1029/2012GL054181

Grapes, R.H., Wysoczanski, R.J. and Hoskin, P.W.O. 2003. Rhönite paragenesis in pyroxenite xenoliths, Mount Sidley volcano, Marie Byrd Land, West Antarctica. Mineralogical Magazine, 67, 639–651, https://doi.org/10.1180/0026461036740123

Hamilton, D.L. and MacKenzie, W.S. 1965. Phase-equilibrium studies in the system $NaAlSiO_4$ (nepheline)–$KAlSiO_4$ (kalsilite)–SiO_2–H_2O. Mineralogical Magazine, 34, 214–231, https://doi.org/10.1180/minmag.1965.034.268.17

Handler, M.R., Wysoczanski, R.J. and Gamble, J.A. 2003. Proterozoic lithosphere in Marie Byrd Land, West Antarctica: Re–Os systematics of spinel peridotite xenoliths. Chemical Geology, 196, 131–145, https://doi.org/10.1016/S0009-2541(02)00410-2

Hansen, S.E., Graw, J.H. et al. 2014. Imaging the Antarctic mantle using adaptively parameterized P-wave tomography: Evidence for heterogeneous structure beneath West Antarctica. Earth and Planetary Science Letters, 408, 66–78, https://doi.org/10.1016/j.eps1.2014.09.043

Hart, S.R. and Gaetani, G.A. 2006. Mantle Pb paradoxes: the sulfide solution. Contributions to Mineralogy and Petrology, 152, 295–308, https://doi.org/10.1007/s00410-006-0108-1

Hart, S.R., Hauri, E.H., Oschmann, L.A. and Whitehead, J.A. 1992. Mantle plumes and entrainment: the isotopic evidence. Science, 256, 517–520, https://doi.org/10.1126/science.256.5056.517

Hart, S.R., Blusztajn, J. and Craddock, C. 1995. Cenozoic volcanism in Antarctica; Jones Mountains and Peter I Island. Geochimica et Cosmochimica Acta, 59, 3379–3388, https://doi.org/10.1016/0016-7037(95)00212-I

Hart, S.R., Blusztajn, J., LeMasurier, W.E. and Rex, D.C. 1997. Hobbs Coast Cenozoic volcanism: Implications for the West Antarctic rift system. Chemical Geology, 139, 223–248, https://doi.org/10.1016/S0009-2541(97)00037-5

Heeszel, D.S., Wiens, D.A. et al. 2016. Upper mantle structure of central and West Antarctica from array analysis of Rayleigh wave phase velocities. Journal of Geophysical Research: Solid Earth, 121, 1758–1775, https://doi.org/10.1002/2015jb012616

Hoernle, K., White, J.D.L. et al. 2006. Cenozoic intraplate volcanism on New Zealand: Upwelling induced by lithospheric removal. Earth and Planetary Science Letters, 248, 350–367, https://doi.org/10.1016/j.epsl.2006.06.001

Hoernle, K., Timm, C. et al. 2020. Late Cretaceous (99–69 Ma) basaltic intraplate volcanism on and around Zealandia: Tracing upper mantle geodynamics from Hikurangi Plateau collision to Gondwana breakup. Earth and Planetary Science Letters, 529, https://doi.org/10.1016/j.epsl.2019.115864

Hofmann, A.W. 2014. Sampling mantle heterogeneity through oceanic basalts: isotopes and trace elements. In: Carlson, R.W. (ed.) Treatise on Geochemistry, Volume 2: The Mantle and the Core. Pergamon, Oxford, UK, 61–101.

Hole, M.J. 2021. Antarctic Peninsula: petrology. Geological Society, London, Memoirs, 55, https://doi.org/10.1144/M55-2018-40

Hole, M.J. and LeMasurier, W.E. 1994. Tectonic controls on the geochemical composition of Cenozoic mafic alkaline volcanic rocks from West Antarctica. Contributions to Mineralogy and Petrology, 117, 187–202, https://doi.org/10.1007/BF00286842

Hole, M.J., Storey, B.C. and LeMasurier, W.E. 1994. Tectonic setting and geochemistry of Miocene alkali basalts from the Jones Mountains, West Antarctica. Antarctic Science, 6, 85–92, https://doi.org/10.1017/S0954102094000118

Huerta, A.D. and Harry, D.L. 2007. The transition from diffuse to focused extension: Modeled evolution of the West Antarctic Rift system. Earth and Planetary Science Letters, 255, 133–147, https://doi.org/10.1016/j.epsl.2006.12.011

Irvine, T.N. and Baragar, W.R.A. 1971. A guide to the chemical classification of the common volcanic rocks. Canadian Journal of Earth Sciences, 8, 523–548, https://doi.org/10.1139/e71-055

Iverson, N.A., Lieb-Lappen, R., Dunbar, N.W., Obbard, R., Kim, E. and Golden, E. 2017. The first physical evidence of subglacial volcanism under the West Antarctic Ice Sheet. Scientific Reports, 7, 11457, https://doi.org/10.1038/s41598-017-11515-3

Jackson, M.G., Hart, S.R., Saal, A.E., Shimizu, N., Kurz, M.D., Blusztajn, J.S. and Skovgaard, A.C. 2008. Globally elevated titanium, tantalum, and niobium (TITAN) in ocean island basalts with high $^3He/^4He$. Geochemistry, Geophysics, Geosystems, 9, Q04027, https://doi.org/10.1029/2007GC001876

Jordan, T.A., Riley, T.R. and Siddoway, C.S. 2020. The geological history and evolution of West Antarctica. Nature Reviews Earth & Environment, 1, 117–133, https://doi.org/10.1038/s43017-019-0013-6

Jung, S. and Masberg, P. 1998. Major- and trace-element systematics and isotope geochemistry of Cenozoic mafic volcanic rocks from the Vogelsberg (central Germany). Constraints on the origin of continental alkaline and tholeiitic basalts and their mantle sources. Journal of Volcanology and Geothermal Research, 86, 151–177, https://doi.org/10.1016/S0377-0273(98)00087-0

Kipf, A., Hauff, F. et al. 2014. Seamounts off the West Antarctic margin: A case for non-hotspot driven intraplate volcanism. Gondwana Research, 25, 1660–1679, https://doi.org/10.1016/j.gr.2013.06.013

Korhonen, F.J., Saito, S., Brown, M., Siddoway, C.S. and Day, J.M.D. 2010. Multiple generations of granite in the Fosdick Mountains, Marie Byrd Land, West Antarctica: implications for polyphaser intracrustal differentiation in a continental margin setting. Journal of Petrology, 51, 627–670, https://doi.org/10.1093/petrology/egp093

Kyle, P.R. 1990. McMurdo Volcanic Group western Ross Embayment. American Geophysical Union Antarctic Research Series, 48, 18–145.

Kyle, P.R., McIntosh, W.C., Panter, K. and Smellie, J. 1991. Is volcanism in Marie Byrd Land related to a mantle plume? Abstract presented at the Sixth International Symposium on Antarctic Earth Sciences, 9–13 September 1991, Saitama, Japan.

Kyle, P.R., Moore, J.A. and Thirlwall, M.F. 1992. Petrologic evolution of anorthoclase phonolite lavas at Mount Erebus, Ross Island, Antarctica. *Journal of Petrology*, **33**, 849–875, https://doi.org/10.1093/petrology/33.4.849

Lanyon, R., Varne, R. and Crawford, A.J. 1993. Tasmanian Tertiary basalts, the Balleny Plume, and opening of the Tasman Sea (southwest Pacific Ocean). *Geology*, **21**, 555–558, https://doi.org/10.1130/0091-7613(1993)021<0555:TTBTBP>2.3.CO;2

Larsen, L.M. 1979. Distribution of REE and other trace-elements between phenocrysts and peralkaline undersaturated magmas, exemplified by rocks from the Gardar Igneous Province, south Greenland. *Lithos*, **12**, 303–315, https://doi.org/10.1016/0024-4937(79)90022-7

Latourrette, T., Hervig, R.L. and Holloway, J.R. 1995. Trace-element partitioning between amphibole, phlogopite, and basanite melt. *Earth and Planetary Science Letters*, **135**, 13–30, https://doi.org/10.1016/0012-821X(95)00146-4

Lawver, L.A. and Gahagan, L.M. 1994. Constraints on the timing of extension in the Ross Sea region. *Terra Antarctica*, **1**, 545–552.

Leat, P.T. and Riley, T.R. 2021. Antarctic Peninsula and South Shetland Islands: petrology. *Geological Society, London, Memoirs*, **55**, https://doi.org/10.1144/M55-2018-68

Leat, P.T., Storey, B.C. and Pankhurst, R.J. 1993. Geochemistry of Palaeozoic–Mesozoic Pacific rim orogenic magmatism, Thurston Island area, West Antarctica. *Antarctic Science*, **5**, 281–296, https://doi.org/10.1017/S0954102093000380

Le Bas, M.J., Le Maitre, R.W., Streckeisen, A. and Zanettin, B. 1986. A chemical classification of volcanic rocks based on the total alkali–silica diagram. *Journal of petrology*, **27**, 745–750, https://doi.org/10.1093/petrology/27.3.745

Lee, M.J., Lee, J.I., Kim, T.H., Lee, J. and Nagao, K. 2015. Age, geochemistry and Sr–Nd–Pb isotopic compositions of alkali volcanic rocks from Mt. Melbourne and the western Ross Sea, Antarctica. *Geosciences Journal*, **19**, 681–695, https://doi.org/10.1007/s12303-015-0061-y

Le Maitre, R.W. 2002. *Igneous Rocks: A Classification and Glossary of Terms: Recommendations of International Union of Geological Sciences Subcommission on the Systematics of Igneous Rocks*. Cambridge University Press, Cambridge, UK.

LeMasurier, W.E. 1990a. Marie Byrd Land. *American Geophysical Union Antarctic Research Series*, **48**, 147–163.

LeMasurier, W.E. 1990b. Mount Hartigan. *American Geophysical Union Antarctic Research Series*, **48**, 199–202.

LeMasurier, W.E. 1990c. Mount Sidley. *American Geophysical Union Antarctic Research Series*, **48**, 203–207.

LeMasurier, W.E. 1990d. Miocene–Oligocene centers, Mount Petras and the USAS Escarpment. *American Geophysical Union Antarctic Research Series*, **48**, 239–243.

LeMasurier, W.E. 1990e. Miocene–Pliocene centers, Hobbs Coast. *American Geophysical Union Antarctic Research Series*, **48**, 244–247.

LeMasurier, W.E. 1990f. Miocene and older centers, Walgreen Coast. *American Geophysical Union Antarctic Research Series*, **48**, 235–238.

LeMasurier, W.E. 1990g. Pliocene–Pleistocene Hobbs Coast volcanoes. *American Geophysical Union Antarctic Research Series*, **48**, 248–250.

LeMasurier, W.E. 1990h. B.12. Ames Range. *American Geophysical Union Antarctic Research Series*, **48**, 216–220.

LeMasurier, W.E. 2006. What supports the Marie Byrd Land Dome? An evaluation of potential uplift mechanisms in a continental rift system. *In*: Fütterer, D.K., Damaske, D., Kleinschmidt, G., Miller, H. and Tessensohn, F. (eds) *Antarctica*. Springer, Berlin, 299–302.

LeMasurier, W.E. 2013. Shield volcanoes of Marie Byrd Land, West Antarctic rift: oceanic island similarities, continental signature, and tectonic controls. *Bulletin of Volcanology*, **75**, 726, https://doi.org/10.1007/s00445-013-0726-1

LeMasurier, W.E. and Kawachi, Y. 1990a. Mount Hampton. *American Geophysical Union Antarctic Research Series*, **48**, 189–194.

LeMasurier, W.E. and Kawachi, Y. 1990b. Mount Cummings. *American Geophysical Union Antarctic Research Series*, **48**, 208–211.

LeMasurier, W.E. and Kawachi, Y. 1990c. Mount Waesche. *American Geophysical Union Antarctic Research Series*, **48**, 195–198.

LeMasurier, W.E. and Landis, C.A. 1996. Mantle-plume activity recorded by low-relief erosion surfaces in West Antarctica and New Zealand. *Geological Society of America Bulletin*, **108**, 1450–1466, https://doi.org/10.1130/0016-7606(1996)108<1450:MPARBL>2.3.CO;2

LeMasurier, W.E. and Rex, D.C. 1989. Evolution of linear volcanic ranges in Marie Byrd Land, West Antarctica. *Journal of Geophysical Research: Solid Earth*, **94**, 7223–7236, https://doi.org/10.1029/JB094iB06p07223

LeMasurier, W.E. and Rex, D.C. 1990. Mount Siple. *American Geophysical Union Antarctic Research Series*, **48**, 185–188.

LeMasurier, W.E. and Thomson, J.W. 1990. Volcanoes of the Antarctic Plate and Southern Oceans. *American Geophysical Union Antarctic Research Series*, **48**.

LeMasurier, W.E. and Wade, F.A. 1990. Fosdick Mountains. *American Geophysical Union Antarctic Research Series*, **48**, 251–252.

LeMasurier, W.E., Kawachi, Y. and Rex, D.C. 1990a. Toney Mountain. *American Geophysical Union Antarctic Research Series*, **48**, 175–179.

LeMasurier, W.E., Kawachi, Y. and Rex, D.C. 1990b. Mount Flint–Reynolds Ridge. *American Geophysical Union Antarctic Research Series*, **48**, 212–215.

LeMasurier, W.E., Futa, K., Hole, M. and Kawachi, Y. 2003. Polybaric evolution of phonolite, trachyte and rhyolite volcanoes in eastern Marie Byrd Land, Antarctica: controls on peralkalinity and silica saturation. *International Geology Review*, **45**, 1055–1099, https://doi.org/10.2747/0020-6814.45.12.1055

LeMasurier, W.E., Choi, S.H., Kawachi, Y., Mukasa, S.B. and Rogers, N.W. 2011. Evolution of pantellerite–trachyte–phonolite volcanoes by fractional crystallization of basanite magma in a continental rift setting, Marie Byrd Land, Antarctica. *Contributions to Mineralogy and Petrology*, **162**, 1175–1199, https://doi.org/10.1007/s00410-011-0646-z

LeMasurier, W.E., Choi, S.H., Hart, S.R., Mukasa, S.B. and Rogers, N.W. 2016. Reconciling the shadow of a subduction signature with rift geochemistry and tectonic environment in eastern Marie Byrd Land, Antarctica. *Lithos*, **260**, 134–153, https://doi.org/10.1016/j.lithos.2016.05.018

LeMasurier, W.E., Choi, S.H., Kawachi, Y., Mukasa, S.B. and Rogers, N.W. 2018. Dual origins for pantellerites and other puzzles, at Mount Takahe volcano, Marie Byrd Land, West Antarctica. *Lithos*, **296–299**, 142–162, https://doi.org/10.1016/j.lithos.2017.10.014

Licht, K.J., Groth, T., Townsend, J.P., Hennessy, A.J., Hemming, S.R., Flood, T.P. and Studinger, M. 2018. Evidence for extending anomalous Miocene volcanism at the edge of the East Antarctic craton. *Geophysical Research Letters*, **45**, 3009–3016, https://doi.org/10.1002/2018GL077237

Lloyd, A.J., Wiens, D.A. et al. 2015. A seismic transect across West Antarctica: Evidence for mantle thermal anomalies beneath the Bentley Subglacial Trench and the Marie Byrd Land Dome. *Journal of Geophysical Research: Solid Earth*, **120**, 8439–8460, https://doi.org/10.1002/2015jb012455

Lloyd, A.J., Wiens, D.A. et al. 2019. Seismic Structure of the Antarctic upper mantle based on adjoint tomography. *Journal of Geophysical Research: Solid Earth*, **125**, https://doi.org/10.1029/2019JB017823

Lough, A.C., Wiens, D.A. et al. 2013. Seismic detection of an active subglacial magmatic complex in Marie Byrd Land, Antarctica. *Nature Geoscience*, **6**, 1031–1035, https://doi.org/10.1038/NGEO1992

Lowenstern, J.B. and Mahood, G.A. 1991. New data on magmatic H_2O contents of pantellerites, with implications for petrogenesis and eruptive dynamics at Pantelleria. *Bulletin of Volcanology*, **54**, 78–83, https://doi.org/10.1007/BF00278208

Lundstrom, C.C., Shaw, H.F., Ryerson, F.J., Williams, Q. and Gill, J. 1998. Crystal chemical control of clinopyroxene-melt partitioning in the Di–Ab–An system: Implications for elemental fractionations in the depleted mantle. *Geochimica et Cosmochimica Acta*, **62**, 2849–2862, https://doi.org/10.1016/S0016-7037(98)00197-5

Lustrino, M., Melluso, L. and Morra, V. 2002. The transition from alkaline to tholeiitic magmas: a case study from the Orosei-Dorgali Pliocene volcanic district NE Sardinia, Italy. *Lithos*, **63**, 83–113, https://doi.org/10.1016/S0024-4937(02)00113-5

Luyendyk, B.P. 1995. Hypothesis for Cretaceous rifting of east Gondwana caused by subducted slab capture. *Geology*, **23**, 373–376, https://doi.org/10.1130/0091-7613(1995)023<0373:HFCROE>2.3.CO;2

Luyendyk, B.P., Richard, S.M., Smith, C.H. and Kimbrough, D.L. 1991. Geological and geophysical exploration in the northern Ford Ranges, Marie Byrd Land, West Antarctica. *In*: Yoshida, Y., Kaminuma, K. and Shiraishi, K. (eds) *Recent Progress in Antarctic Earth Science*, Terra Scienctific (TERRAPUB), Tokyo, 279–288.

Luyendyk, B., Cisowski, S., Smith, C., Richard, S. and Kimbrough, D. 1996. Paleomagnetic study of the northern Ford Ranges, western Marie Byrd Land, West Antarctica: motion between West and East Antarctica. *Tectonics*, **15**, 122–141, https://doi.org/10.1029/95TC02524

Macdonald, R., Baginski, B., Belkin, H.E., White, J.C. and Noble, D.C. 2019. The Gold Flat Tuff, Nevada: insights into the evolution of peralkaline silicic magmas. *Lithos*, **328–329**, 1–13, https://doi.org/10.1016/j.lithos.2019.01.017

Mahood, G.A. and Stimac, J.A. 1990. Trace-element partitioning in pantellerites and trachytes. *Geochimica et Cosmochimica Acta*, **54**, 2257–2276, https://doi.org/10.1016/0016-7037(90)90050-U

Martin, A.P., Cooper, A.F. and Price, R.C. 2013. Petrogenesis of Cenozoic, alkalic volcanic lineages at Mount Morning, West Antarctica and their entrained lithospheric mantle xenoliths: Lithospheric versus asthenospheric mantle sources. *Geochimica et Cosmochimica Acta*, **122**, 127–152, https://doi.org/10.1016/j.gca.2013.08.025

Martin, A.P., Cooper, A.F. and Price, R.C. 2014. Increased mantle heat flow with on-going rifting of the West Antarctic rift system inferred from characterisation of plagioclase peridotite in the shallow Antarctic mantle. *Lithos*, **190–191**, 173–190, https://doi.org/10.1016/j.lithos.2013.12.012

Martin, A.P., Cooper, A.F., Price, R.C., Kyle, P.R. and Gamble, J.A. 2021. Erebus Volcanic Province: petrology. *Geological Society, London, Memoirs*, **55**, https://doi.org/10.1144/M55-2018-80

McCoy-West, A.J., Baker, J.A., Faure, K. and Wysoczanski, R. 2010. Petrogenesis and origins of Mid-Cretaceous continental intraplate volcanism in Marlborough, New Zealand: Implications for the long-lived HIMU magmatic mega-province of the SW Pacific. *Journal of Petrology*, **51**, 2003–2045, https://doi.org/10.1093/petrology/egq046

McCoy-West, A.J., Bennett, V.C. and Amelin, Y. 2016. Rapid Cenozoic ingrowth of isotopic signatures simulating 'HIMU' in ancient lithospheric mantle: distinguishing source from process. *Geochimica et Cosmochimica Acta*, **187**, 79–101, https://doi.org/10.1016/j.gca.2016.05.013

McDonough, W.F. and Sun, S.-s. 1995. The composition of the Earth. *Chemical Geology*, **120**, 223–253, https://doi.org/10.1016/0009-2541(94)00140-4

Meschede, M. 1986. A method of discriminating between different types of mid-ocean ridge basalts and continental tholeiites ith the Nb–Zr–Y diagram. *Chemical Geology*, **56**, 207–208, https://doi.org/10.1016/0009-2541(86)90004-5

Molnar, P., Atwater, T., Mammerickx, J. and Smith, S.M. 1975. Magnetic anomalies, bathymetry and the tectonic evolution of the South Pacific since the Late Cretaceous. *Geophysical Journal International*, **40**, 383–420, https://doi.org/10.1111/j.1365-246X.1975.tb04139.x

Mortimer, N., Campbell, H.J. et al. 2017. Zealandia: Earth's hidden continent. *GSA Today*, **27**, 27–35, https://doi.org/10.1130/GSATG321A.1

Mortimer, N., van den Bogaard, P., Hoernle, K., Timm, C., Gans, P.B., Werner, R. and Riefstahl, F. 2019. Late Cretaceous oceanic plate reorganization and the breakup of Zealandia and Gondwana. *Gondwana Research*, **65**, 31–42, https://doi.org/10.1016/j.gr.2018.07.010

Mukasa, S.B. and Dalziel, I.W.D. 2000. Marie Byrd Land, West Antarctica: Evolution of Gondwana's Pacific margin constrained by zircon U–Pb geochronology and feldspar common-Pb isotopic compositions. *Geological Society of America Bulletin*, **112**, 611–627, https://doi.org/10.1130/0016-7606(2000)112<611:MBLWAE>2.0.CO;2

Nardini, I., Armienti, P., Rocchi, S., Dallai, L. and Harrison, D. 2009. Sr–Nd–Pb–He–O isotope and geochemical constraints on the genesis of Cenozoic magmas from the West Antarctic rift. *Journal of Petrology*, **50**, 1359–1375, https://doi.org/10.1093/petrology/egn082

Njome, M.S. and de Wit, M.J. 2014. The Cameroon Line: Analysis of an intraplate magmatic province transecting both oceanic and continental lithospheres: constraints, controversies and models. *Earth-Science Reviews*, **139**, 168–194, https://doi.org/10.1016/j.earscirev.2014.09.003

Palais, J.M., Kyle, P.R., McIntosh, W.C. and Seward, D. 1988. Magmatic and phreatiomagmatic volcanic activity at Mt. Takahe, West Antarctica, based on tephra layers in the Byrd Ice Core and field observations at Mt. Takahe. *Journal of Volcanology and Geothermal Research*, **35**, 295–317, https://doi.org/10.1016/0377-0273(88)90025-X

Pankhurst, R.J., Millar, I.L., Grunow, A.M. and Storey, B.C. 1993. The pre-Cenozoic magmatic history of the Thurston Island crustal block, West Antarctica. *Journal of Geophysical Research: Solid Earth*, **98**, 11 835–11 849, https://doi.org/10.1029/93JB01157

Pankhurst, R.J., Weaver, S.D., Bradshaw, J.D., Storey, B.C. and Ireland, T.R. 1998. Geochronology and geochemistry of pre-Jurassic superterranes in Marie Byrd Land, Antarctica. *Journal of Geophysical Research*, **103**, 2529–2547, https://doi.org/10.1029/97JB02605

Panter, K.S. 1995. *Geology, Geochemistry and Petrogenesis of the Mount Sidley Volcano, Marie Byrd Land, Antarctica*. PhD thesis, New Mexico Institute of Mining and Technology, Socorro, New Mexico, USA.

Panter, K.S. 2021. Antarctic volcanism: petrology and tectonomagmatic overview. *Geological Society, London, Memoirs*, **55**, https://doi.org/10.1144/M55-2020-10

Panter, K.S., McIntosh, W.C. and Smellie, J.L. 1994. Volcanic history of Mount Sidley, a major alkaline volcano in Marie Byrd Land, Antarctica. *Bulletin of Volcanology*, **56**, 361–376, https://doi.org/10.1007/BF00326462

Panter, K.S., Kyle, P.R. and Smellie, J.L. 1997. Petrogenesis of a phonolite–trachyte succession at Mount Sidley, Marie Byrd Land, Antarctica. *Journal of Petrology*, **38**, 1225–1253, https://doi.org/10.1093/petroj/38.9.1225

Panter, K.S., Hart, S.R., Kyle, P., Blusztanjn, J. and Wilch, T. 2000. Geochemistry of Late Cenozoic basalts from the Crary Mountains: characterization of mantle sources in Marie Byrd Land, Antarctica. *Chemical Geology*, **165**, 215–241, https://doi.org/10.1016/S0009-2541(99)00171-0

Panter, K.S., Blusztajn, J., Hart, S., Kyle, P., Esser, R. and McIntosh, W. 2006. The origin of HIMU in the SW Pacific: Evidence from intraplate volcanism in southern New Zealand and Subantarctic Islands. *Journal of Petrology*, **47**, 1673–1704, https://doi.org/10.1093/petrology/egl024

Panter, K.S., Castillo, P. et al. 2018. Melt origin across a rifted continental margin: a case for subduction-related metasomatic agents in the lithospheric source of alkaline basalt, northwest Ross Sea, Antarctica. *Journal of Petrology*, **59**, 517–558, https://doi.org/10.1093/petrology/egy036

Panter, K.S., Reindel, J. and Smellie, J.L. 2021. Mount Early and Sheridan Bluff: petrology. *Geological Society, London, Memoirs*, **55**, https://doi.org/10.1144/M55-2019-2

Pappa, F., Ebbing, J. and Ferraccioli, F. 2019. Moho depths of Antarctica: comparison of seismic, gravity, and isostatic results. *Geochemistry, Geophysics, Geosystems*, **20**, 1629–1645, https://doi.org/10.1029/2018GC008111

Park, S.-H., Langmuir, C.H. *et al.* 2019. An isotopically distinct Zealandia–Antarctic mantle domain in the southern ocean. *Nature Geoscience*, **12**, 206–214, https://doi.org/10.1038/s41561-018-0292-4

Paulsen, T.S. and Wilson, T.J. 2010. Evolution of Neogene volcanism and stress patterns in the glaciated West Antarctic rift, Marie Byrd Land, Antarctica. *Journal of the Geological Society, London*, **167**, 401–416, https://doi.org/10.1144/0016-76492009-044

Pearce, J.A. 2008. Geochemical fingerprinting of oceanic basalts with applications to ophiolite classification and the search for Archean oceanic crust. *Lithos*, **100**, 14–48, https://doi.org/10.1016/j.lithos.2007.06.016

Peters, B.J. and Day, J.M.D. 2014. Assessment of relative Ti, Ta, and Nb (TITAN) enrichments in ocean island basalts. *Geochemistry, Geophysics, Geosystems*, **15**, 4424–4444, https://doi.org/10.1002/2014GC005506

Phillips, E.H., Sims, K.W.W. *et al.* 2018. The nature and evolution of mantle upwelling at Ross Island, Antarctica, with implications for the source of HIMU lavas. *Earth and Planetary Science Letters*, **498**, 38–53, https://doi.org/10.1016/j.epsl.2018.05.049

Prestvik, T. and Duncan, R.A. 1991. The geology and age of Peter I Øy, Antarctica. *Polar Research*, **9**, 89–98, https://doi.org/10.3402/polar.v9i1.6781

Price, R.C., Nicholls, I.A. and Day, A. 2014. Lithospheric influences on magma compositions of late Mesozoic and Cenozoic intraplate basalts (the Older Volcanics) of Victoria, south-eastern Australia. *Lithos*, **206–207**, 179–200, https://doi.org/10.1016/j.lithos.2014.07.027

Ramirez, C., Nyblade, A. *et al.* 2017. Crustal structure of the Transantarctic Mountains, Ellsworth Mountains and Marie Byrd Land, Antarctica: constraints on shear wave velocities, Poisson's ratios and Moho depths. *Geophysical Journal International*, **211**, 1328–1340, https://doi.org/10.1093/gji/ggx333

Ren, M., Omenda, P.A., Anthony, E.Y., White, J.C., Macdonald, R. and Bailey, D.K. 2006. Application of the QUILF thermobarometer to the peralkaline trachytes and pantellerites of the Eburru volcanic complex, East African Rift, Kenya. *Lithos*, **91**, 109–124, https://doi.org/10.1016/j.lithos.2006.03.011

Riefstahl, F., Gohl, K. *et al.* 2020. Cretaceous intracontinental rifting at the southern Chatham Rise margin and initialisation of seafloor spreading between Zealandia and Antarctica. *Tectonophysics*, **776**, 228–298, https://doi.org/10.1016/j.tecto.2019.228298

Riley, T.R., Flowerdew, M.J., Pankhurst, R.J., Leat, P.T., Millar, I.L., Fanning, C.M. and Whitehouse, M.J. 2017. A revised geochronology of Thurston Island, West Antarctica, and correlations along the proto-Pacific margin of Gondwana. *Antarctic Science*, **29**, 47–60, https://doi.org/10.1017/S0954102016000341

Rocchi, S. and Smellie, J.L. 2021. Northern Victoria Land: petrology. *Geological Society, London, Memoirs*, **55**, https://doi.org/10.1144/M55-2019-19

Rocchi, S., Armienti, P., D'Orazio, M., Tonarini, S., Wijbrans, J.R. and Di Vincenzo, G. 2002. Cenozoic magmatism in the western Ross Embayment: Role of mantle plume v. plate dynamics in the development of the West Antarctic Rift System. *Journal of Geophysical Research: Solid Earth*, **107**, ECV 5-1–ECV 5-22, https://doi.org/10.1029/2001JB000515

Rocchi, S., LeMasurier, W.E. and Di Vincenzo, G. 2006. Oligocene to Holocene erosion and glacial history in Marie Byrd Land, West Antarctica, inferred from exhumation of the Dorrel Rock intrusive complex and from volcano morphologies. *Geological Society of America Bulletin*, **118**, 991–1005, https://doi.org/10.1130/B25675.1

Rocholl, A., Stein, M., Molzahn, M., Hart, S.R. and Wörner, G. 1995. Geochemical evolution of rift magmas by progressive tapping of a stratified mantle source beneath the Ross Sea Rift, Northern Victoria Land, Antarctica. *Earth and Planetary Science Letters*, **131**, 207–224, https://doi.org/10.1016/0012-821X(95)00024-7

Rowley, P.D. 1990. Jones Mountains. American Geophysical Union Antarctic Research Series, **48**, 286–288.

Rowley, P.D., Laudon, T.S., La Prade, K.E. and LeMasurier, W.E. 1990*a*. Hudson Mountains. *American Geophysical Union, Antarctic Research Series*, **48**, 289–293.

Rowley, P.D., Thomson, J.W., Smellie, J.L., Laudon, T.S., La Prade, K.E. and LeMasurier, W.E. 1990*b*. Alexander Island, Palmer Land, and Ellsworth Land. *American Geophysical Union Antarctic Research Series*, **48**, 257–265.

Rutford, R.H. and McIntosh, W.C. 2007. Jones Mountains, Antarctica: evidence for Tertiary glaciation revisited. United States Geological Survey Open-File Report, **2007-1047**, Extended Abstract 203.

Scott, J.M., Turnbull, I.M., Auer, A. and Palin, M. 2013. The sub-Antarctic Antipodes Volcano: a <0.5 Ma HIMU-like Surtseyan volcanic outpost on the edge of the Campbell Plateau. *New Zealand Journal of Geology and Geophysics*, **56**, 134–153, https://doi.org/10.1080/00288306.2013.802246

Scott, J.M., Waight, T.E., van der Meer, Q.H.A., Palin, J.M., Cooper, A.F. and Münker, C. 2014. Metasomatized ancient lithospheric mantle beneath the young Zealandia microcontinent and its role in HIMU-like intraplate magmatism. *Geochemistry, Geophysics, Geosystems*, **15**, 3477–3501, https://doi.org/10.1002/2014GC005300

Scott, J.M., Brenna, M. *et al.* 2016. Peridotitic lithosphere metasomatized by volatile-bearing melts and its association with intraplate alkaline HIMU-like magmatism. *Journal of Petrology*, **57**, 2053–2078, https://doi.org/10.1093/petrology/egw069

Seroussi, H., Ivins, E.R., Wiens, D.A. and Bondzio, J. 2017. Influence of a West Antarctic mantle plume on ice sheet basal conditions. *Journal of Geophysical Research: Solid Earth*, **122**, 7127–7155, https://doi.org/10.1002/2017JB014423.

Shen, W., Wiens, D.A. *et al.* 2018. Seismic evidence for lithospheric foundering beneath the southern Transantarctic Mountains, Antarctica. *Geology*, **46**, 71–74, https://doi.org/10.1130/G39555.1

Siddoway, C.S. 2008. Tectonics of the West Antarctic rift system: new light on the history and dynamics of distributed intracontinental extension. *In*: Cooper, A.K., Raymond, R.C. *et al.* (eds) *Antarctica: A Keystone in a Changing World, Proceedings of the 10th International Symposium on Antarctic Earth Sciences*. The National Academies Press, Washington, DC, 91–114.

Siddoway, C.S., Richard, S.M., Fanning, C.M. and Luyendyke, B.P. 2004. Origin and emplacement of a middle Cretaceous gneiss dome, Fosdick Mountains, West Antarctica. *Geological Society of America Special Papers*, **380**, 267–294, https://doi.org/10.1130/0-8137-2380-9.267

Sims, K.W.W., Blichert-Toft, J. *et al.* 2008. A Sr, Nd, Hf, and Pb isotope perspective on the genesis and long-term evolution of alkaline magmas from Erebus volcano, Antarctica. *Journal of Volcanology and Geothermal Research*, **177**, 606–618, https://doi.org/10.1016/j.jvolgeores.2007.08.006

Sprung, P., Schuth, S., Munker, C. and Hoke, L. 2007. Intraplate volcanism in New Zealand: the role of fossil plume material and variable lithospheric properties. *Contributions to Mineralogy and Petrology*, **153**, 699–687, https://doi.org/10.1007/s00410-006-0169-1

Storey, B.C., Leat, P.T., Weaver, S.D., Pankhurst, R.J., Bradshaw, J.D. and Kelly, S. 1999. Mantle plumes and Antarctica–New Zealand rifting: evidence from mid-Cretaceous mafic dykes. *Journal of the Geological Society, London*, **156**, 659–671, https://doi.org/10.1144/gsjgs.156.4.0659

Stracke, A. 2012. Earth's heterogeneous mantle: a product of convection-driven interaction between crust and mantle. *Chemical Geology*, **330–331**, 274–299, https://doi.org/10.1016/j.chemgeo.2012.08.007

Sun, S.-s. and McDonough, W.F. 1989. Chemical and isotopic systematics of oceanic basalts: implications for mantle composition and processes. *Geological Society, London, Special Publications*, **42**, 313–345, https://doi.org/10.1144/GSL.SP.1989.042.01.19

Sutherland, R., Spasojevic, S. and Gurnis, M. 2010. Mantle upwelling after Gondwana subduction death explains anomalous topography and subsidence histories of eastern New Zealand and

West Antarctica. *Geology*, **38**, 155–158, https://doi.org/10.1130/G30613.1

Timm, C., Hoernle, K., Van den Bogaard, P., Bindeman, I. and Weaver, S. 2009. Geochemical evolution of intraplate volcanism at Banks Peninsula, New Zealand: interaction between asthenosphere and lithospheric melts. *Journal of Petrology*, **50**, 989–1023, https://doi.org/10.1093/petrology/egp029

Timm, C., Hoernle, K. *et al.* 2010. Temporal and geochemical evolution of the Cenozoic intraplate volcanism of Zealandia. *Earth-Science Reviews*, **98**, 38–64, https://doi.org/10.1016/j.earscirev.2009.10.002

Tulloch, A.J., Ramezani, J., Mortimer, N., Mortensen, J., van den Bogaard, P. and Mass, R. 2009. Cretaceous felsic volcanism in New Zealand and Lord Howe Rise (Zealandia) as a precursor to final Gondwana break-up. *Geological Society, London, Special Publications*, **321**, 89–118, https://doi.org/10.1144/SP321.5

Tuttle, O.F. and Bowen, N.L. 1958. Origin of granite in the light of experimental studies in the system NaAlSiO4–KAlSiO4–SiO2–H2O. *Geological Society of America Memoirs*, **74**, 1–153, https://doi.org/10.1130/MEM74-p1

van der Meer, Q.H.A., Waight, T.E., Scott, J.M. and Münker, C. 2017. Variable sources for Cretaceous to recent HIMU and HIMU-like intraplate magmatism in New Zealand. *Earth and Planetary Science Letters*, **469**, 27–41, https://doi.org/10.1016/j.epsl.2017.03.037

van der Meer, Q.H.A., Waight, T.E., Tulloch, A.J., Whitehouse, M.J. and Andersen, T. 2018. Magmatic evolution during the Cretaceous transition from subduction to continental break-up of the eastern Gondwana margin (New Zealand) documented by *in-situ* zircon O–Hf isotopes and bulk-rock Sr–Nd isotopes. *Journal of Petrology*, **59**, 849–880, https://doi.org/10.1093/petrology/egy047

van Wyk de Vries, M.V., Bingham, R.G. and Hein, A.S. 2017. A new volcanic province: an inventory of subglacial volcanoes in West Antarctica. *Geological Society, London, Special Publications*, **461**, 231–248, https://doi.org/10.1144/SP461.7

Veevers, J.J. 2012. Reconstructions before rifting and drifting reveal the geological connections between Antarctica and its conjugates in Gondwanaland. *Earth-Science Reviews*, **111**, 249–318, https://doi.org/10.1016/j.earscirev.2011.11.009

Villemant, B., Jaffrezic, H., Joron, J.L. and Treuil, M. 1981. Distribution coefficients of major and trace-elements – fractional crystallization in the alkali basalt series of Chaine-Des-Puys (Massif Central, France). *Geochimica et Cosmochimica Acta*, **45**, 1997–2016, https://doi.org/10.1016/0016-7037(81)90055-7

Weaver, B.L. 1991. The origin of ocean island basalt end-member compositions: trace element and isotopic constraints. *Earth and Planetary Science Letters*, **104**, 381–397, https://doi.org/10.1016/0012-821X(91)90217-6

Weaver, S.D., Bradshaw, J.D. and Adams, C.J. 1991. Granitoids of the Ford Ranges, Marie Byrd Land, Antarctica. *In*: Thompson, M.R.A., Crame, J.A. and Thomson, J.W. (eds) *Geological Evolution of Antarctica*. Cambridge University Press, Cambridge, UK, 345–351.

Weaver, S.D., Adams, C.J., Pankhurst, R.J. and Gibson, I.L. 1992. Granites of Edward VII Peninsula, Marie Byrd Land: anorogenic magmatism related to Antarctic–New Zealand rifting. *Earth and Environmental Science Transactions of the Royal Society of Edinburgh*, **83**, 281–290, https://doi.org/10.1017/S0263593300007963

Weaver, S.D., Storey, B.C., Pankhurst, R.J., Mukasa, S.B., DiVenere, V.J. and Bradshaw, J.D. 1994. Antarctic–New Zealand rifting and Marie Byrd Land lithospheric magmatism linked to ridge subduction and mantle plume activity. *Geology*, **22**, 811–814, https://doi.org/10.1130/0091-7613(1994)022<0811:ANZRAM>2.3.CO;2

Weiss, Y., Class, C., Goldstein, S.L. and Hanyu, T. 2016. Key new pieces of the HIMU puzzle from olivines and diamond inclusions. *Nature*, **537**, 666–670, https://doi.org/10.1038/nature19113

White, J.C., Parker, D.F. and Ren, M. 2009. The origin of trachyte and pantellerite from Pantelleria, Italy: insights from major element, trace element, and thermodynamic modelling. *Journal of Volcanology and Geothermal Research*, **179**, 33–55, https://doi.org/10.1016/j.jvolgeores.2008.10.007

White, J.D.L. and Houghton, B.F. 2006. Primary volcaniclastic rocks. *Geology*, **34**, 677–680, https://doi.org/10.1130/G22346.1

Wilch, T.I. 1997. *Volcanic Record of the West Antarctic Ice Sheet in Marie Byrd Land*. PhD thesis, New Mexico Institute of Mining and Technology, Socorro, New Mexico, USA.

Wilch, T.I. and McIntosh, W.C. 2000. Eocene and Oligocene volcanism at Mount Petras, Marie Byrd Land: implications for middle Cenozoic ice sheet reconstructions in West Antarctica. *Antarctic Science*, **12**, 477–491, https://doi.org/10.1017/S0954102000000560

Wilch, T.I. and McIntosh, W.C. 2002. Lithofacies analysis and $^{40}Ar/^{39}Ar$ geochronology of ice–volcano interactions at Mt. Murphy and the Crary Mountains, Marie Byrd Land, Antarctica. *Geological Society, London, Special Publications*, **202**, 237–253, https://doi.org/10.1144/GSL.SP.2002.202.01.12

Wilch, T.I. and McIntosh, W.C. 2007. Miocene–Pliocene ice–volcano interactions at monogenetic volcanoes near Hobbs Coast, Marie Byrd Land, Antarctica. United States Geological Survey Open-File Report, **2007-1047**, Short Research Paper 074, https://pubs.usgs.gov/of/2007/1047/srp/srp074/

Wilch, T.I., McIntosh, W.C. and Dunbar, N.W. 1999. Late Quaternary volcanic activity in Marie Byrd Land: potential $^{40}Ar/^{39}Ar$-dated time horizons in West Antarctica ice and marine cores. *Geological Society of America Bulletin*, **111**, 1563–1580, https://doi.org/10.1130/0016-7606(1999)111<1563:LQVAIM>2.3.CO;2

Wilch, T.I., McIntosh, W.C. and Panter, K.S. 2021. Marie Byrd Land and Ellsworth Land: volcanology. *Geological Society, London, Memoirs*, **55**, https://doi.org/10.1144/M55-2019-39

Winberry, P.J. and Anandakrishnan, S. 2004. Crustal structure of the West Antarctic rift system and Marie Byrd Land hotspot. *Geology*, **32**, 977–980, https://doi.org/10.1130/G20768.1

Wood, B. and Trigila, R. 2001. Experimental determination of aluminous clinopyroxene-melt partition coefficients for potassic liquids, with application to the evolution of the Roman province potassic magmas. *Chemical Geology*, **172**, 213–223, https://doi.org/10.1016/S0009-2541(00)00259-X

Wysoczanski, R.J. 1993. *Lithospheric Xenoliths from the Marie Byrd Land Volcanic Province, West Antarctica*. PhD thesis, Victoria University of Wellington, Wellington, New Zealand.

Wysoczanski, R.J., Gamble, J.A., Kyle, P.R. and Thirlwall, M.F. 1995. The petrology of lower crustal xenoliths from the Executive Committee Range, Marie Byrd Land volcanic province, West Antarctica. *Lithos*, **36**, 185–201, https://doi.org/10.1016/0024-4937(95)00017-8

Zundel, M., Spiegel, C., Mehling, A., Lisker, F., Hillenbrand, C.-D., Monien, P. and Klügel, A. 2019. Thurston Island (West Antarctica) between Gondwana subduction and continental separation: a multistage evolution revealed by apatite thermochronology. *Tectonics*, **38**, 878–897, https://doi.org/10.1029/2018TC005150

Chapter 5.5

Gaussberg: volcanology and petrology

J. L. Smellie[1]* and K. D. Collerson[2]

[1]School of Geography, Geology and the Environment, University of Leicester, University Road, Leicester LE1 7RH, UK

[2]School of Earth and Environmental Studies, The University of Queensland, St Lucia, QLD 4072, Australia

JLS, 0000-0001-5537-1763; KDC, 0000-0002-1823-306X

*Correspondence: jls55@le.ac.uk

Abstract: Gaussberg is a nunatak composed of lamproite pillow lava situated on the coast of East Antarctica. It is the most isolated Quaternary volcanic centre in Antarctica but it is important palaeoenvironmentally and petrologically out of all proportion to its small size. The edifice has a likely low, shield-like, morphology *c.* 1200 m high and possibly up to 10 km wide, which is unusually large for a lamproite construct. Gaussberg was erupted subglacially at 56 ± 5 ka, which places it late in the last glacial, close to the peak of marine isotope stage 3. The coeval ice sheet was *c.* 1300 m thick, and *c.* 420 m has been removed from the ice surface since Gaussberg erupted. Lamproite is a rare ultrapotassic mantle-derived magma, and Gaussberg is one of two type examples worldwide. Although traditionally considered as related in some way to the Kerguelen plume, it is more likely that the Gaussberg magma is a product of a separate magmatic event. It is ascribed to the storage and long-term (Gy) isolation of sediment emplaced by subduction in the Transition Zone of the deep mantle, followed by entrainment and subsequent melting in a plume.

Supplementary material: Photomicrographs of representative lamproite lavas and enclaves and published whole-rock analyses of lamproites (Table S1) from Gaussberg are available at https://doi.org/10.6084/m9.figshare.c.5221584

Gaussberg is a small, isolated conical nunatak situated on the coast of East Antarctica at 66° 47′ S, 89° 18′ E, at the east end of West Ice Shelf, between the Davis and Mirny scientific stations (Fig. 1). It was discovered and topographically mapped in 1902 by the Imperial German Antarctic Expedition (Phillipi 1912). Interestingly, a comparison of photographs taken in 1902 and 1997 shows no major difference in the ice-surface elevation surrounding Gaussberg. The nunatak is largely snow-free and is *c.* 1000–1150 m wide, rising to 370 m above sea level (asl) (Figs 2 & 3).

Gaussberg is entirely composed of exceptionally potassic mafic volcanic rock, which is compositionally unusual. It was originally called leucitite because of the presence of leucite phenocrysts but was subsequently renamed lamproite or leucite lamproite to conform with the IUGS classification system (cf. Sheraton and Cundari 1980; Collerson and McCulloch 1983; Tingey *et al.* 1983; Murphy *et al.* 2002; cf. Woolley *et al.* 1996). Lamproites are uncommon worldwide and consist of small intrusive and extrusive bodies, including lavas, necks, pyroclastic cones and diatremes (e.g. Mitchell and Bergman 1991; Prelević *et al.* 2007; Seghedi *et al.* 2007). Gaussberg is the only lamproite occurrence formed of pillow lava and which erupted subglacially (see below). It is also the youngest example on Earth (Murphy *et al.* 2002) and the reasons for its presence are still not well explained.

Gaussberg is situated on the coast of East Antarctica, a passive margin, at the projected southern end of the Kerguelen–Gaussberg Ridge (Fig. 4). The Kerguelen Plateau that forms much of the ridge was constructed by major plume activity extending from the Cretaceous to the Quaternary (e.g. Frey *et al.* 2000; Weis *et al.* 2002; Morgan and Phipps-Morgan 2007; Sushchevskaya *et al.* 2014, 2017). The volcanism at Gaussberg has generally been considered related in some way to the Kerguelen plume. It may have coincided broadly with eruptions on Heard Island and McDonald Islands, both of which are unequivocally part of the Kerguelen plume province (Clarke *et al.* 1983). However, the nature of the relationship between the Kerguelen plume and Gaussberg is uncertain, and a direct connection is unproven. Gaussberg and Kerguelen volcanic products are also compositionally and isotopically distinct (cf. Sheraton and Cundari 1980; Clarke *et al.* 1983; Barling 1990; Nougier and Thomson 1990; Murphy *et al.* 2002; see below). Thus, Gaussberg might represent the products of a different, isolated magmatic event (cf. Sheraton and Cundari 1980; Duncan 1981; Collerson and McCulloch 1983; Tingey *et al.* 1983; Murphy *et al.* 2002; Sushchevskaya *et al.* 2014). Interpretation of the structural setting of Gaussberg is also complicated by the presence and uncertain influence of the 500–700 km-long,

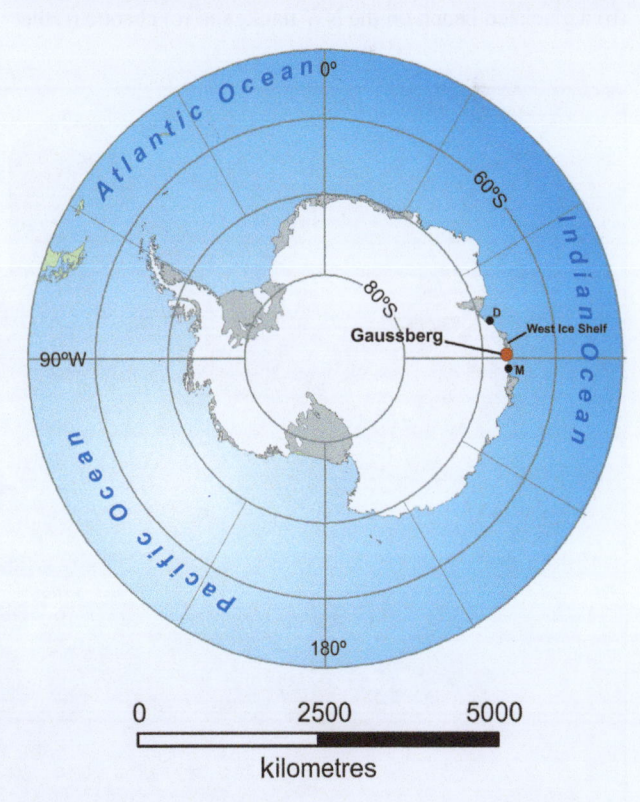

Fig. 1. Map showing the location of Gaussberg in East Antarctica. Abbreviations: D, Davis Station (Australia); M, Mirny Station (Russia).

Fig. 2. View of Gaussberg, looking ESE. Photograph by Darryn Schneider.

60–150 km-wide, poorly-understood Gaussberg rift, which probably formed during Permian–Early Cretaceous large-scale continental extension prior to Gondwana break-up (Golynski and Golynski 2007). Gaussberg sits on top of a segmented horst within that rift.

Tingey et al. (1983) suggested that Gaussberg formed during a single eruptive phase, whereas Murphy et al. (2002) suggested, on limited evidence, that an earlier eroded volcanic construct may also be present, thus splitting the evolution into at least two lamproite eruptive stages. Gaussberg lacks a summit crater, attributed by Tingey et al. (1983) to erosion. Evidence for post-eruptive modification includes the presence of: (a) glacial erratics, striated surfaces and flanking moraines; (b) a glaciated bench on the NW flank; and (c) chaotic (rather than radial) lava pillow orientations, indicative of a larger edifice (Tingey et al. 1983; Tingey 1990; Murphy et al. 2002). Although some glacial erosion is undoubted, the coastal location of Gaussberg suggests that some marine erosion during the Holocene might also have occurred.

Description of Gaussberg: field relationships, lithofacies and age

Gaussberg occurs in isolation. It is surrounded by the East Antarctic Ice Sheet on three sides and flanked to the north by the Southern Ocean. The nunatak is formed entirely of volcanic rock (Fig. 3) and no other geological units are exposed

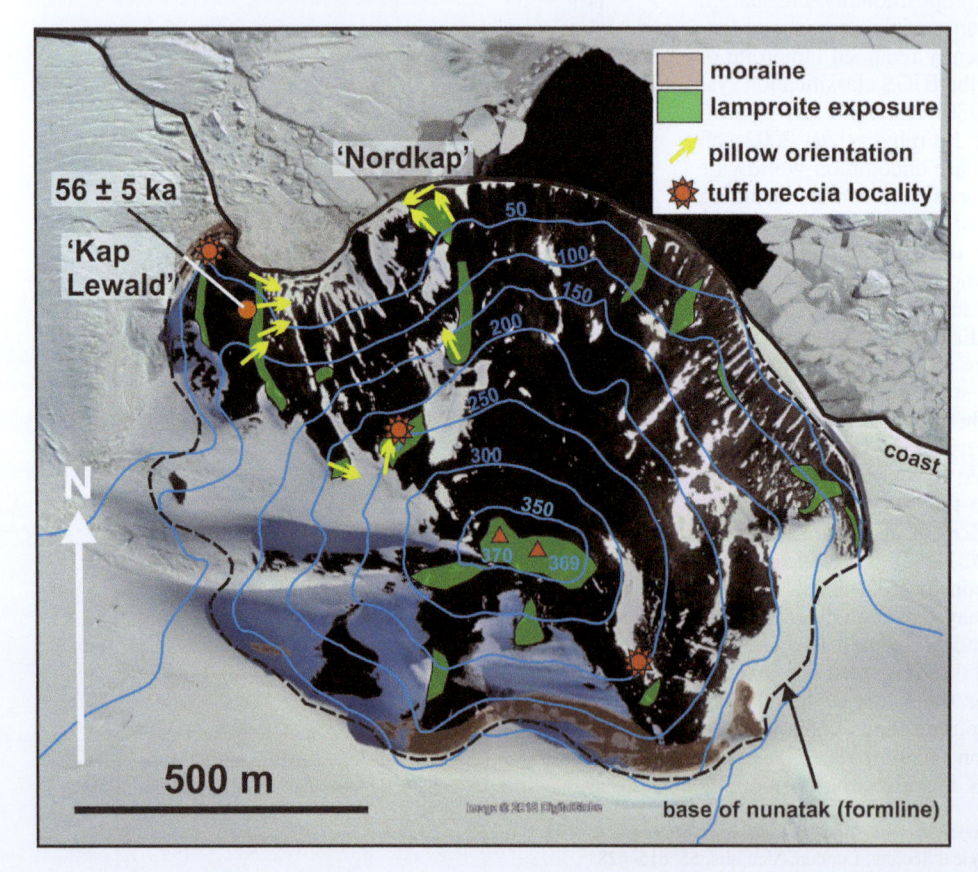

Fig. 3. Topography (after Drygalski 1912) and geological sketch map of Gaussberg. The distribution of exposed lamproite rock is modified after Tingey (1990). Satellite image from Google Earth (2018 DigitalGlobe).

(Tingey et al. 1983). Similar granitic and high-grade metamorphic erratics also occur in flanking moraines (Nockolds 1940), and Tingey et al. (1983) reported striated 'leucitite' (lamproite) on the nunatak.

Gaussberg is composed entirely of lamproite lava, although most of the snow-free surface consists of lamproite scree, much of it glassy or aphanitic. The lamproite is generally fresh, with virtually unaltered small phenocrysts of leucite, olivine and clinopyroxene. It varies from pale grey and holocrystalline, to dark grey and glassy or aphanitic. Subordinate holocrystalline lava is present throughout Gaussberg but it is most prominent in thick, gently inclined, sheets (lavas or sills) 3 m up to a few tens of metres thick that dominate the basal 150 m. They are vesicle-poor and commonly show a likely flow foliation emphasized by dark glass-like joint surfaces. Where present, crude polygonal jointing in the lavas is generally closely spaced at c. 10 cm or less (Fig. 5a). The holocrystalline lava sheets typically form the terrace-like features seen on the flanks of the nunatak. There are also a few ill-defined dykes a few metres wide composed of similar material but generally more vesicular. Crustal and cognate enclaves are abundant, particularly in the basal outcrops although they are also locally common at higher elevations. Pillow-like structures are ubiquitous, closely stacked and generally lacking accompanying interpillow clastic material (Fig. 5b, c). Some of the pillow-like structures were described as resembling entrail pāhoehoe and there are rare ropy textures (Tingey et al. 1983) (Fig. 5d, g). The lava pillows mainly range from 0.1 to 2 m in diameter but often include very large pillows up to 3 m wide and 5 m long. They have fine-grained rims described as thick black glassy selvages variably 2–10 cm wide (Sheraton and Cundari 1980; Collerson and McCulloch 1983). Photographs of the pillows show prominent thick, dark-coloured fine-grained rims (e.g. Tingey 1990, fig. F.8.3) (Fig. 5c, g). However, the published selvage thicknesses are unusually large to be wholly composed of glass. Revised estimates by one of us (KDC) indicate that the thickness of glass is generally just 1–3 cm and the greater thicknesses probably include dark aphanitic lamproite too. The lava pillows are relatively little flattened (high aspect ratios: i.e. the ratio of height to width) but some have irregular or amoeboid shapes. They commonly show tortoise shell and radial joints, and are highly vesicular. The vesicles are larger centrally within the lava pillows, and become flattened towards the pillow rims. Some vesicles contain native sulfur, particularly in the summit area, and many are lined by glass (Vyalov and Sobolev 1959; Sheraton and Cundari 1980; Tingey et al. 1983). Lava tubes are also present and include the 'lava tunnels up to 2 m in diameter' described by Collerson and McCulloch (1983, p. 676). Pillow and tube orientations are highly variable across the outcrop and include some pillows plunging into the slope at 30°–45°, which point towards the summit (Fig. 3). Minor clastic material, described as 'yellow-brown palagonite', is present in interpillow spaces, although Tingey et al. (1983) suggested that 'interstitial sediments' were absent. Sheraton and Cundari (1980) also recorded minor 'thin layers of palagonite tuff', not otherwise described. Tuff breccia is also present. It occurs at three localities, at least, including as numerous blocks in moraine on the NW and south flanks of Gaussberg (Fig. 3). The largest outcrop of tuff breccia measures 20–50 m in area and is at least 30 m thick. Although the margins of the tuff breccias are poorly exposed, the field relationships suggest that they are interbedded with pillow lava. The breccias are monomict, massive, and formed of fragmented and intact lava pillows, with a variable proportion of yellowish-white to orange-coloured, ashy matrix (Fig. 5e–g). Most of the fragments are angular but a few show abrasion. The smaller fragments are dominantly glassy. They are mainly matrix supported and

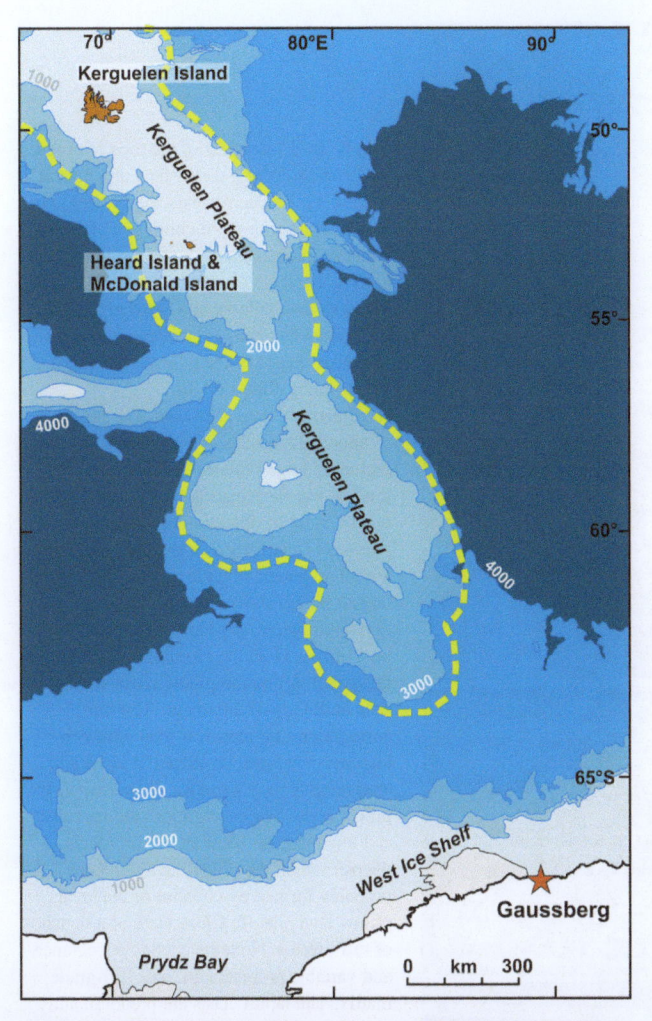

Fig. 4. Map showing the location of Gaussberg relative to the Kerguelen Plateau (yellow dashed line). Gaussberg falls at the projected southern termination of the plateau, formed by activity in the Kerguelen plume, where it might intersect the East Antarctic margin. This has led to suggestions that the two features may be related but the nature of the relationship is still obscure. They are potentially separated by a gap of up to 500 km and it is unclear if they are actually linked.

nearby. However, the presence of a prominent, heavily crevassed, small ice rise, c. 400 m in diameter, c. 2.3 km SW of Gaussberg suggests that other outcrops of volcanic rocks may be present but are obscured by ice. The nearest rock outcrop is at Mount Brown, 150 km away, and is composed of Precambrian basement rocks (felsic orthogneisses, mafic granulites and granultite-facies schists, paragneisses, pyroxenites, and pegmatite: Sheraton and Cundari 1980; Mikhalsky et al. 2015; Liu et al. 2016). Although the field relationships with other geological units are unknown, the lamproite contains numerous subrounded to ovoid crustal xenoliths up to 2 m long (typically <30 cm), which include layered quartz-feldspathic gneiss, granitoids (including rapakivi granite), mafic and intermediate granulite, and metabasaltic rock (Sheraton and Cundari 1980; Collerson and McCulloch 1983; and unpublished information of K.D. Collerson: Fig. 5a). The crustal xenoliths have yielded a variety of Precambrian, Cambrian and Carboniferous U–Pb ages (2000–500 and 320 Ma) obtained on zircons. The xenoliths are probably indicative of the basement on which the nunatak presumably stands, or at least the rocks through which the Gaussberg magma passed prior to eruption (Mikhalsky et al. 2015). Glacial erratics are also scattered on the nunatak, including the summit. They are composed of basement lithologies, including gneiss

Fig. 5. (**a**) Close view of the tops of crude polygonal joints on the surface of a lava at Gaussberg. The narrow spacing of the joints is consistent with rapid chilling. The lava contains a conspicuous crustal xenolith. The knife is *c*. 10 cm in length. (**b**) & (**c**) Close views of lava pillows on Gaussberg. Note the limited flattening (high aspect ratios) and tube-like shapes of the pillows, attributed to buoyancy effects on the subaqueously-emplaced lava. The presence of abundant vesicles in the pillows, including large vesicles, indicates that the magma was vesiculating freely, probably due to some combination of low ambient pressures (relatively thin overlying ice) and a high magmatic volatile content. The largest pillow shown in (b) is *c*. 70 cm in maximum width. (**d**) Fine ropy texture on a lava pillow. Note the prominent 2 cm-thick glassy selvage in the pillow cross-section seen at the top of the image. The fine corrugations are narrower than most ropy textures developed on subaerial lavas and are probably ropy wrinkles, characteristic of pillow lavas. The corrugations are spaced at *c*. 1 cm apart. (**e**) View of a tuff breccia deposit within the pillow lava sequence, probably formed by collapse of the local pillow lava pile. (**f**) Close view of a matrix of tuff breccia. Note the chaotic appearance and variable proportion of pale tuff-grade matrix. The lapilli clasts are predominantly glassy. (**g**) View of tuff breccia showing numerous displaced intact and broken pillows. Note the dark-coloured glassy pillow rims, which are corrugated on the glassy surface of the pillow to the left of the ice axe (cf. d) and the prominent dark glassy rims on other pillows. The ice axe is *c*. 70 cm long. Photographs (b) & (d) are by Darryn Schneider. All others are by Ken Collerson.

the matrix rarely shows fine (millimetre- to centimetre-scale) layering, including possible cross-bedding. Some of the layering is steeply orientated and runs downslope, and it locally wraps around larger clasts. Similar laminated tuff also occurs as angular slabs in the breccia.

There are few isotopic ages (Table 1). K–Ar whole-rock ages of 20 and 9 Ma by Ravich and Krylov (1964) and Soloviev (1972), respectively, lack errors and other analytical data, and are likely to be unreliable. Conversely, Tingey *et al.* (1983) reported two non-overlapping but essentially self-consistent K–Ar ages of 52 ± 3 and 59 ± 2 ka, yielding a mean age of 56 ± 5 ka, which is interpreted as the eruptive age. Conversely, an imprecise fission track age of 26 ± 24 ka was also obtained on 'leucitite' glass by Tingey *et al.* (1983) but it has very high errors and its significance is unknown.

Gaussberg: edifice type and eruptive setting

Gaussberg is almost compositionally invariable (see below), and there are no obvious internal surfaces that might suggest a time break and a polygenetic origin. However, it is a small nunatak, potentially just a minor part of a larger volcanic feature (Murphy *et al.* 2002). The edifice is a pillow mound, which possibly had a relatively prolonged eruptive history. The possibility of multiple eruptive stages separated by time breaks thus cannot be precluded, as has been observed in other pillow complexes (e.g. Edwards *et al.* 2009; Pollock *et al.* 2014).

Vyalov and Sobolev (1959) envisaged a subaerial environment for the volcanism at Gaussberg, whereas Sheraton and Cundari (1980) preferred either a subglacial or submarine setting. The presence of entrail pāhoehoe, ropy lava surfaces and an interpretation of the lava pillows as pāhoehoe toes suggest a subaerial environment but the evidence is ambiguous. While entrail pāhoehoe is a common product of subaerial effusion, it is not confined to that setting and has also been recorded, albeit rarely, in a glaciovolcanic pillow ridge (Edwards *et al.* 2009). The ropy textures may be ropy wrinkles or possibly spreading cracks, which are common in lava pillows and are consistent with the small scale (centimetre spacing) of the corrugations seen at Gaussberg (McPhie *et al.* 1993; Goto and McPhie 2004). If spreading cracks are present they are diagnostic of subaqueous effusion (Goto and McPhie 2012). It is

Table 1. *Summary of published isotopic ages for volcanic rocks at Gaussberg*

Sample	Locality	Dated lithologies*	Age† (Ma)	Error	Error SD	Method	Reference	Notes
K.1152	Gaussberg	Leucite basalt	20	n.r.	n.r.	K–Ar	Ravich and Krylov (1964)	Age probably improbable
n.r.	Gaussberg	Leucite basalt	9	n.r.	n.r.	K–Ar	Soloviev (1972)	Age probably improbable
79-241 (BMR1)	Gaussberg	Leucite concentrates from leucitite	0.052	0.003	1 σ	K–Ar	Tingey *et al.* (1983)	Age is average of three dated separates
79-242 (BMR2)	Gaussberg	Leucite concentrates from leucitite	0.059	0.002	1 σ	K–Ar	Tingey *et al.* (1983)	Age is average of two dated separates
BMR1 and BMR2	Gaussberg	**Mean age**	**0.056**	0.005	1 σ	K–Ar	Tingey *et al.* (1983)	Probably reliable eruption age
n.r.	Gaussberg	Leucitite glass	0.026	0.024	2 σ	Glass fission track	Tingey *et al.* (1983)	Imprecise and probably unreliable eruption age

n.r., not recorded; SD, standard deviation.
*Lithology names have been taken from the original publication, and so do not necessarily comply with modern TAS-based names.
†Ages are given as published, not recalculated using more recently published decay constants.

often difficult to distinguish unambiguously between pāhoehoe toes and lava pillows based on morphology alone. However, photographs of the lavas at Gaussberg show that many of the individual lobes are more or less equant in cross-section, with limited flattening (high aspect ratios), and there are numerous large (>1 m) pillows and tubes. These characteristics are more characteristic of a subaqueous rather than subaerial eruptive setting because of the differing buoyancy and chilling effects of water on the magma compared with air (Deschamps *et al.* 2014; Smellie and Edwards 2016, p. 311). Moreover, the glassy selvages are relatively thick, consistent with water chilling rather than lava cooling subaerially, as is the narrow spacing (≤10 cm) of the crude polygonal jointing (including tortoise shell joints: Grossenbacher and McDuffie 1995; Forbes *et al.* 2014).

A subglacial setting was suggested by Tingey *et al.* (1983) and has been generally accepted. Evidence cited by Tingey *et al.* (1983) for the eruptive palaeoenvironment included the following: (a) the presence of likely pillow lava, signifying subglacial or marine conditions; (b) irregular vesicle shapes and comparatively large vesicle sizes indicating eruption at pressures equivalent to water depths less than *c.* 300 m (by analogy with Icelandic occurrences: e.g. Jones 1969*a*); (c) a young age, falling within the last glacial, when global sea levels would have been lower than today; (d) the absence of a summit crater, said to be consistent with a subglacial setting; and finally (e) the presence of terrace-like features on the nunatak flanks thought to reflect either eruption in a series of pulses (for which supportive evidence was lacking) or else the terraces were of erosional origin and related to changes in the contemporary ice surfaces. On balance, Tingey *et al.* (1983, p. 245) strongly preferred a subglacial eruptive setting with contemporary ice thicknesses 'at least, but not much more than, 370 m'.

The absence of a summit crater is not environment-specific as a crater is unlikely to be present on a volcanic mound formed by passive effusion of pillow lava. The terrace-like features on the flanks of Gaussberg may be an artefact of preferential erosion, which created gently inclined terraces on the massive lava sheets or sills, and steep faces in the more easily eroded overlying pillow lava. The terraces probably do not represent erosional notches related to changes in regional ice levels, as surmised by Tingey *et al.* (1983). Tortoise shell and radial fractures, which are common on Gaussberg lava, are caused by water-induced rapid cooling and contraction of pillow lava (McPhie *et al.* 1993). Although Tingey *et al.* (1983) refuted the presence of hyaloclastite at Gaussberg, the tuff breccias at Gaussberg are comparable with hyaloclastites: that is, formed by thermal contraction and mechanical granulation of water-chilled lava and deposited essentially *in situ* (i.e. *sensu* White and Houghton 2006). However, hyaloclastites are typically volumetrically minor, monomict, fines-poor orthobreccias formed of coarse (often lapilli-size) angular pillow-lava-derived fragments and found essentially *in situ*, mainly in interpillow spaces. By contrast, interpillow spaces between lava pillows at Gaussberg are almost wholly vacant, and the Gaussberg tuff breccias comprise large beds (up to tens of metres in thickness and extent) containing abundant whole and fragmented lava pillows dispersed in massive tuff matrix. We therefore suggest that the tuff breccias formed by the gravitational collapse of locally oversteepened pillow piles and were emplaced as mass flows (cf. Edwards *et al.* 2009; Hungerford *et al.* 2014; Pollock *et al.* 2014). That some of the lava pillows have relatively steep dips (30°–45°) attests to steep original depositional slopes, at least locally. Some of the tuff matrix in the tuff breccias shows layering on millimetre and centimetre scales, including possible rare cross-bedding and steep attitudes suggesting deformation. The layering may have been created initially by reworking of minor interpillow fine hyaloclastite (*sensu stricto*) by traction currents. The layered beds were subsequently broken up and incorporated into the tuff breccias during their formation as gravity-driven mass flows. Finally, the highly variable orientations of lava pillows across the nunatak, including pillows that dip into the slope and point towards the summit, indicate that the nunatak is not simply a slightly degraded volcanic cone. Multiple small centres were probably in eruption during the volcanic period. It may thus be a composite volcano formed from the products of several overlapping pillow mounds, which extended the edifice laterally to create a low-profile shield-like mound. The ice rise to the SW of Gaussberg may be an associated flank vent on the same edifice or else a satellite mound. Alternatively, as no rock is exposed, the rise may simply be a non-volcanic bedrock feature.

A magnetic survey of the area shows several prominent magnetic anomalies within 30 km of Gaussberg, and the nunatak sits on the northern flank of one (Fig. 6) (Golynski *et al.* 2018). The anomalies have higher amplitudes and shorter wavelengths than others in the surrounding region, and their patchy distribution suggests that they are a composite feature and not a uniform geological feature (e.g. an intrusion). Although none appears to be centred on Gaussberg, the data are old (acquired in 1957) and the georeferencing of the survey lines may be slightly imprecise. A pillow mound formed of fresh dense magnetic basalt should be associated with a

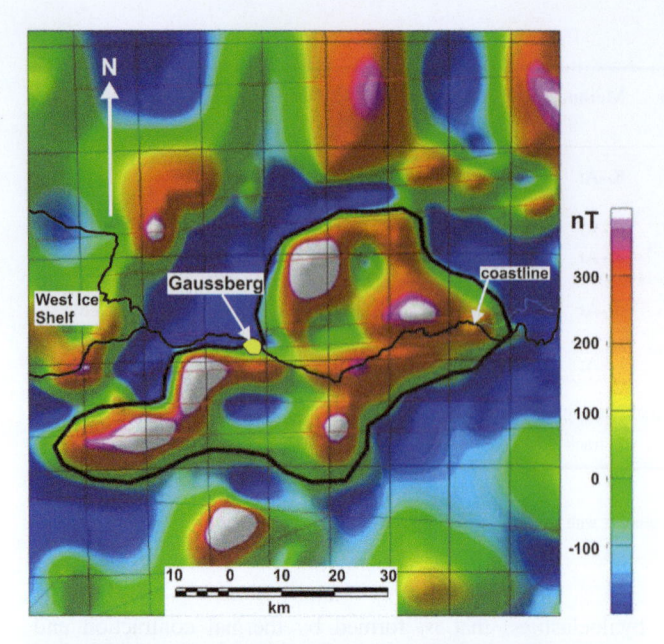

Fig. 6. Map showing magnetic anomalies surrounding Gaussberg (data from ADMAP: Golynski *et al.* 2018). The presence of several small, high-amplitude anomalies within the field, outlined by the thick black line, suggests that they may be sourced in several small satellite vents of which only Gaussberg is currently exposed.

magnetic anomaly, unless the subdued signature is a consequence of a small geophysical contrast with a subsurface body (e.g. gabbro). It is provisionally suggested that the cluster of anomalies may represent several subglacial lamproite mounds similar to Gaussberg. Lamproites on other continents also often occur in clusters of small edifices typically 0.5–2 km in diameter spread over a few tens of kilometres (e.g. Mitchell and Bergman 1991; Prelević *et al.* 2007; Seghedi *et al.* 2007). However, their identification in East Antarctica is uncertain and further investigation is required. The anomalies are *c.* 5–10 km in diameter, which may approximate the width of the Gaussberg volcanic centre, with most of the centre obscured by ice on the south side of the nunatak. Although large compared with most lamproite occurrences, this suggestion is supported by the presence of lamproite debris that dominates moraine at the margin of the Phillipi Glacier, west of Gaussberg, which is consistent with erosion of a larger edifice (Murphy *et al.* 2002) or of additional edifices currently obscured by the ice. Occurrences of lamproite on other continents are typically small mainly subsurface outcrops (e.g. diatremes, hypabyssal intrusions) because of extensive subaerial denudation, which has substantially reduced the original outcrop diameters. Because of the unusually young age of Gaussberg, it (and potentially the associated centres tentatively identified magnetically) may be in a much less eroded state and, hence, larger.

Since the publication by Tingey *et al.* (1983), ancillary environmental evidence for a subglacial setting and glaciovolcanic origin for Gaussberg has become even more compelling. Sea levels at *c.* 56 ka were *c.* 40–90 m below present datum. Although the eruption occurred towards the end of a period of temporarily rising global sea level within the last glacial, contemporary sea levels were still significantly lower than present even at the sea-level peak (probably by a few tens of metres: e.g. Stap *et al.* 2014; Doyle *et al.* 2015; Hansen *et al.* 2016). Moreover, from *c.* 50 ka, global sea level underwent a steady fall until *c.* 20 kyr ago, after which it rose at a more rapid rate to ultimately reach present datum. Consequently, there was never a time during which Gaussberg was submerged by the sea. The changes in global sea level were mirrored by concurrent antithetic changes in ice volume, with the latter driving the observed sea-level changes (Imbrie *et al.* 2011; Doyle *et al.* 2015). Within such a scenario, unless local uplift has occurred since 56 ka, for which there is currently no corroborating regional evidence, eruption of Gaussberg in a subglacial setting seems unequivocal.

Despite extensive volatile exsolution of the Gaussberg magma, phreatomagmatic activity was absent, unless it is represented by the poorly characterized layers of 'palagonite tuff' recorded by other workers. If present, tuff is rare and the overall impression is of passive effusion of a quite strongly vesiculating pillow lava, with growth of abundant large vesicles. However, the possible presence of 'palagonite tuffs' suggests that there might have been brief explosive episodes, possibly phreatomagmatic. Their rarity and thinness, combined with the pervasive coarse vesicularity in the underlying magma, might suggest that the eruption, at least in its final stages, took place close to and occasionally at the transition between effusion and explosivity. Whilst subject to many variables (White *et al.* 2003), the effusive–explosive transition generally takes place at water depths of *c.* 100–200 m, since thermohydraulic explosions are increasingly unlikely in deeper water (or under an equivalent thickness of ice: Jones 1969*b*, 1970; Zimanowski and Büttner 2003). If the palagonite layers do signify explosive activity and they had a subaerial component, it is possible that Gaussberg tephra will also be discovered in ice cores or offshore sediments. Preservation of the latter is probably less likely due to reworking and removal by traction currents and mixing by biological activity. Such tephras would be compositionally highly distinctive and could form valuable time markers for correlation (Kyle and Jezek 1978). However, subaqueous explosivity can be Strombolian even at quite considerable depths (hundreds of metres: Head and Wilson 2003). Moreover, even Strombolian tephras can be yellow coloured due to alteration if erupted under water, and the coloration is not diagnostic of a phreatomagmatic origin (e.g. Pollock *et al.* 2014). Thus, the mode of formation of any tuffs at Gaussberg remains uncertain. However, the absence of thick phreatomagmatic deposits is an indication that explosivity was generally suppressed by the thickness of overlying ice (or meltwater), here estimated, on an *a priori* basis, as ≤100 m above the summit of the nunatak; if the eruptions had occurred under very shallow water (few tens of metres), they probably would have been predominantly phreatomagmatic, as observed in many Surtseyan edifices and glaciovolcanoes (e.g. Jones 1969*b*, 1970; Kokelaar and Durant 1983; Smellie and Hole 1997; Smellie 2001; Schopka *et al.* 2006; Edwards *et al.* 2009). Thus, a comparatively narrow range of potential ice thicknesses is possible, from a few tens of metres to *c.* 100 m *above the summit* of Gaussberg. Using H_2O^+ as a crude estimate for magmatic volatiles (largely H_2O and CO_2: Salvioli-Mariani *et al.* 2004), the relatively high values determined for Gaussberg lava (0.84–2.72 wt%, mean 1.2 wt%: Supplementary material Table S1) implies that the magma was relatively volatile rich. These estimates are therefore likely to be minima and even thicker ice may have been present.

Contemporary grounded-ice thicknesses are *c.* 800 m at the coast flanking Gaussberg (Lythe *et al.* 2001), implying that the Gaussberg volcanic landform might be *c.* 1100–1200 m high, assuming continuity of volcanic outcrop to seafloor. Thus, during the eruption of Gaussberg at *c.* 56 ± 5 ka, the thickness of the coeval ice sheet was probably ≤1300 m, assuming that the ice was grounded. The contemporary ice surface elevation at the coast would therefore have been ≤470 m asl (relative to present datum). Subtracting the general elevation of ice surrounding Gaussberg today (i.e. *c.* 50 m asl, except on its south side where buttressing effects of the nunatak on ice flow have caused the ice surface to rise to *c.* 100 m: Fig. 3) implies that *c.* 420 m of ice thickness has been lost around Gaussberg since it erupted.

The palaeo-ice thickness estimated here for Gaussberg is broadly compatible with published syntheses of ice-sheet thicknesses lost since the Last Glacial Maximum (LGM), which range between 360 and >700 m around coastal East Antarctica (Mackintosh *et al.* 2014). However, most of the geological information on ice-sheet reconstruction in the region comes, naturally, from ice-free coastal oases where rock is exposed and techniques such as cosmogenic nuclide and optically stimulated luminescence dating can be carried out. The oases are very small by comparison with the huge extent of the East Antarctic Ice Sheet and their very existence is predicated on unusual local climatic conditions. Those conditions ensure that they are ice-free today but the reasoning also implies that their glacial histories would have differed from the surrounding ice (e.g. Mackintosh *et al.* 2014). A corollary, not generally acknowledged but which may be important, is that the glacial and deglacial histories of these oases may be atypical and thus unsuited for making regional generalizations. For example, the difference in ice-sheet thickness between LGM and present is 0 m, based on information from Bunger Hills, which is the closest significant ice-free area, located on the coast 500 km east of Gaussberg. The inference that there has been no change in ice thickness since LGM at Bunger Hills contrasts with all other evidence, including that calculated here for Gaussberg, which suggests significant ice-sheet thinning around the East Antarctic coastline. It is also difficult for large-scale ice-sheet modelling to take small-scale local climatic and geographical variations into account satisfactorily. By contrast, glaciovolcanic eruptions record ice-sheet conditions independent of climate. Gaussberg is situated on a coastline that is otherwise rock-free for 1100 km, between Vestfold Hills and Bunger Hills, apart from small outcrops at Mirny Oasis, *c.* 200 km east of Gaussberg. The coastline lacks any other source of information on past glacial conditions. The ice thickness and surface elevations deduced for Gaussberg thus provide a unique 'golden spike' of information created during a geological instant late in the last glacial at 56 ± 5 ka, and free of potential complications caused by local climatic conditions. Moreover, given the evidence for glacial modification of the rock outcrop described by Tingey *et al.* (1983), together with the presence of rounded (abraded) basement erratics scattered on the nunatak, it seems that the local ice overtopped the nunatak and it was wet-based at that time. The palaeoenvironmental importance of Gaussberg is thus out of all proportion to its small physical size.

Petrology of Gaussberg

Lamproite from Gaussberg is a generally dark-coloured rock, almost aphyric but with a small proportion of olivine (Fo_{86-90}), clinopyroxene and leucite phenocrysts, <2 mm and mostly <1 mm in diameter, set in a yellow-brown glassy to aphanitic matrix (Fig. 7a, b; see also the Supplementary material) (Sheraton and Cundari 1980; Sushchevskaya *et al.* 2014). The glass is characterized by abundant quench-textured microlites of leucite, diopside, reddish mica (titaniferous phlogopite) and minor red-brown amphibole (K-richterite: Salvioli-Mariani *et al.* 2004); apatite, ilmenite and chromite are present more rarely. Leucite is the most abundant mineral, and is three times more abundant than either olivine or clinopyroxene. Mineral modes and compositions are given by Sheraton and Cundari (1980) and Salvioli-Mariani *et al.* (2004). In view of the paucity of phenocrysts, whole-rock analyses are considered to approximate magmatic compositions.

Lamproites are relatively primitive, mantle-derived, ultra-potassic volcanic rocks. Gaussberg is recognized as one of only two standard examples (Foley *et al.* 1987). In addition to a distinctive mineralogy, lamproites are characterized by low CaO, Al_2O_3 and Na_2O contents, high Mg-number, and extreme enrichment in incompatible elements, and they carry conspicuous dunite, harzburgite and other ultramafic xenoliths (Foley *et al.* 1987; Woolley *et al.* 1996). At Gaussberg, cognate (cumulate) enclaves up to 10 cm in diameter are present composed of combinations of leucite, clinopyroxene, magnetite, olivine and phlogopite (Fig. 7c, d), together with rare mantle nodules formed of olivine, clinopyroxene and

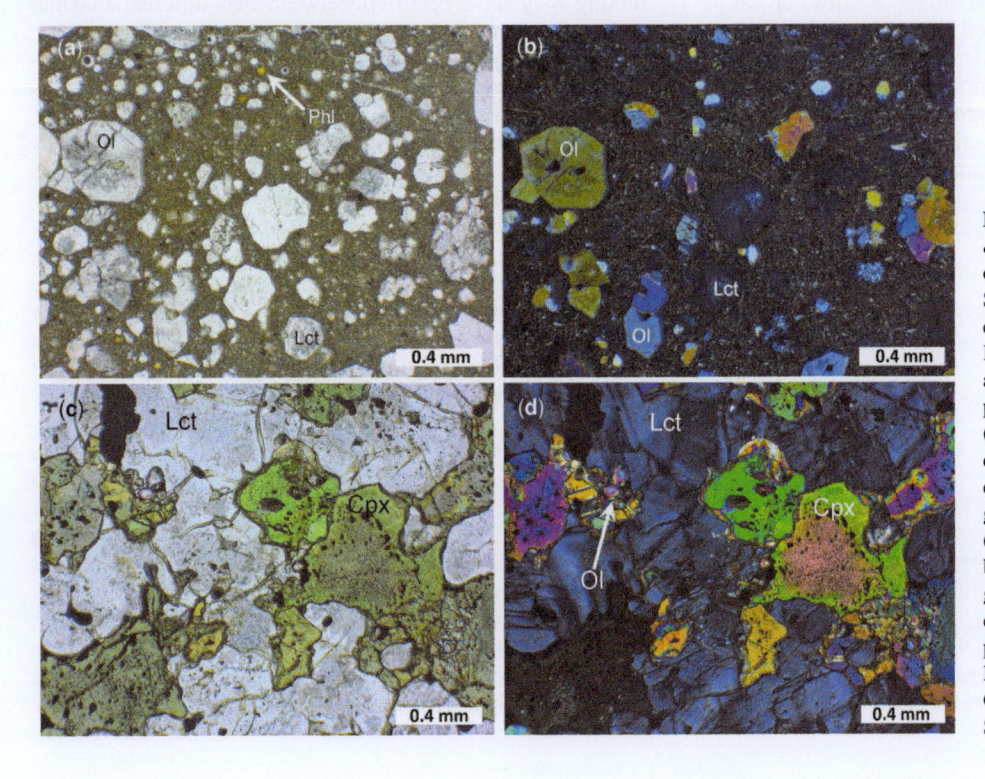

Fig. 7. (**a**) & (**c**) Plane-polarized and (**b**) & (**d**) cross-polarized photomicrographs of an olivine lamproite lava (a & b: Sample KDC-67) and a coarse-grained cognate enclave (c & d: Sample KDC-59) from Gaussberg. The lava has a porphyritic texture defined by euhedral phenocrysts of olivine (Ol) and leucite (Lct) in an aphanitic groundmass. The enclave contains medium- to coarse-grained anhedral crystals of green–pale yellow zoned aegirine augite (Cpx), colourless leucite with low birefringence (Lct), and minor small grains of high-relief very pale green to colourless olivine (Ol). Other photomicrographs of representative lamproite lavas and enclaves from Gaussberg are provided in the Supplementary material.

spinel, and including garnet pyroxenite and garnet harzburgite (Sheraton and Cundari 1980; Collerson and McCulloch 1983; and unpublished information of K.D. Collerson). There are numerous papers discussing various aspects of the petrology of the Gaussberg lamproites, including major and trace elements, mineral analyses, isotopes, and volatiles in melt inclusions (e.g. Sheraton and Cundari 1980; Collerson and McCulloch 1983; Foley et al. 1987; Williams et al. 1992; Murphy et al. 2002; Salvioli-Mariani et al. 2004; Sushchevskaya et al. 2014). This section is summarized from these publications. Selected published whole-rock analyses are given in Supplementary material Table S1 and isotopes in Table 2.

The Gaussberg lamproites are mafic rocks (SiO_2 c. 50 wt%) characterized by very high contents of TiO_2, K_2O, P_2O_5, F, Rb, Sr, Zr, Nb, Ba, La, Ce, Pb, Th, U, and, possibly, As and Sn, whereas values for Al_2O_3, CaO and Na_2O are relatively low. The analysed samples are also strongly olivine- and diopside-normative, and there is significant normative acmite; some samples are slightly leucite-normative. In major and trace element compositions, they are comparable with other occurrences of potassic mafic rocks (see references in Sheraton and Cundari 1980; Murphy et al. 2002). Although some lamproites are associated with kimberlites and diamonds, neither kimberlite nor diamonds have been discovered at Gaussberg. The closest whole-rock compositional comparison is with lamproites from the Leucite Hills (Wyoming) but the slightly more siliceous glass compositions are closer to the quartz-normative ultrapotassic lamproites at West Kimberley (Australia) and Smoky Butte (Montana, USA).

The Gaussberg lavas straddle the outer boundary of the field for tephriphonolites on a total alkali–silica (TAS) diagram (Fig. 8). There is little variation in the analyses, and evidence for fractionation is very limited. Comparison with more differentiated lamproites elsewhere in the world suggests that fractionation or accumulation of leucite and phlogopite, the only two plausible K-rich phases, is unimportant in lamproites generally. Conversely, there is a compositional influence caused by accumulation of K-poor phases such as clinopyroxene (Fig. 9). Thus, part of the (admittedly small) compositional variation at Gaussberg is probably due to minor accumulation of mafic minerals, although the relatively high MgO and Al_2O_3 contents may also be primary features inherited from the source. Analysed glass samples are more silica rich (Supplementary material Table S1), and show that differentiation leads to silica saturation and quartz-normative compositions. Analyses of glass in melt inclusions also indicate that the Gaussberg magma evolved towards peralkaline and more sodic compositions, with K-richterite crystallizing at a late stage, which is reported in many lamproites (Salvioli-Mariani

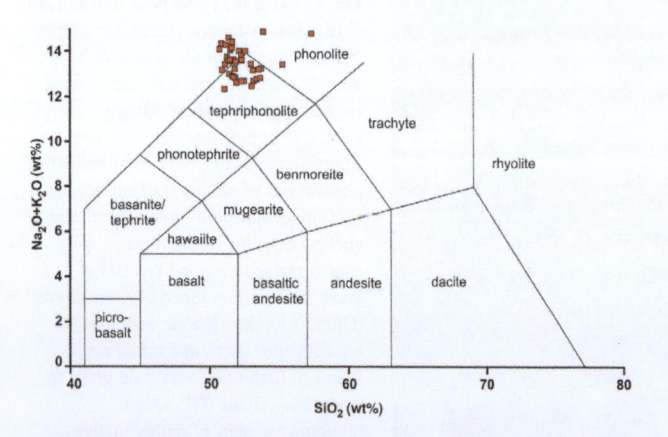

Fig. 8. Total alkali–silica (TAS) diagram (anhydrous values after Le Bas et al. 1986) for Gaussberg lavas, illustrating their alkali-rich nature due to the very high content of K_2O.

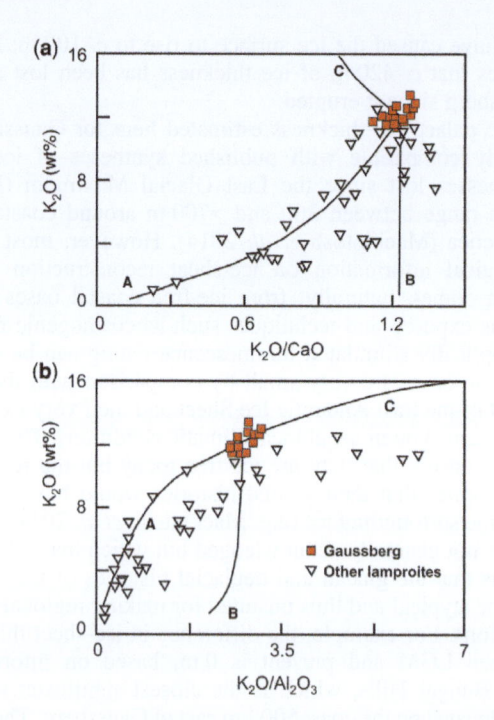

Fig. 9. Diagrams showing (a) K_2O–CaO and (b) K_2O–Al_2O_3 variations in Gaussberg lavas; data for other lamproite occurrences are also shown for comparison (after Murphy et al. 2002). The lines depict indicative fractionation vectors for crystallization of clinopyroxene (A), olivine (B) and leucite (C). Whereas a compositional control involving K-poor phase clinopyroxene and leucite is evident for the other lamproite occurrences, no such influence is shown by lavas at Gaussberg.

et al. 2004). In addition, melt inclusions in leucite and clinopyroxene phenocrysts are more alkaline than the associated groundmass glass, suggesting that the melt inclusions and host crystallized from different batches of magma. It was thus suggested that a pulse of primitive and less alkaline lamproite magma mixed with existing magma in the upper part of the conduit. Based on the composition of clinopyroxene phenocrysts, Sushchevskaya et al. (2014) also postulated that mixing may have occurred between two different magmas during the formation of the lavas.

Cr and Ni contents are moderately high (218–365 and 154–339 ppm, respectively), as are the Mg-numbers (typically 0.69–0.71). They are comparable with little-fractionated, primitive basaltic rocks, although the V contents are relatively low compared with most basaltic rocks. In general, although the range of major and trace element compositions is small, the Gaussberg lamproites show extreme, but irregular, enrichment in incompatible elements, with a notable spike for Pb, and troughs for Nb, Ta and Sr (Fig. 10a). Light REE (LREE) enrichment of c. 800 times chondrite is indicated by the high contents of La and Ce. By contrast, concentrations of Yb are just c. 10 times chondrite, resulting in LREE/heavy REE (HREE) ratios that are very high (Fig. 11). There is no Eu anomaly. The lamproites also have superchondritic Nb/Ta ratios (17.2–20.8), a feature that is extremely rare and distinguishes Gaussberg from most other terrestrial volcanic rocks.

Ratios of the high field strength elements Zr/Hf, as well as Y/Ho, provide useful petrogenetic information for igneous suites. If a geochemical system is characterized by CHArge- and RAdius-Controlled (CHARAC) trace element behaviour, elements with similar charge and ionic radius, such as the twin pairs Y–Ho and Zr–Hf, should display coherent behaviour during crystallization and retain their respective chondritic ratios (Bau 1996). For example, mantle-derived igneous

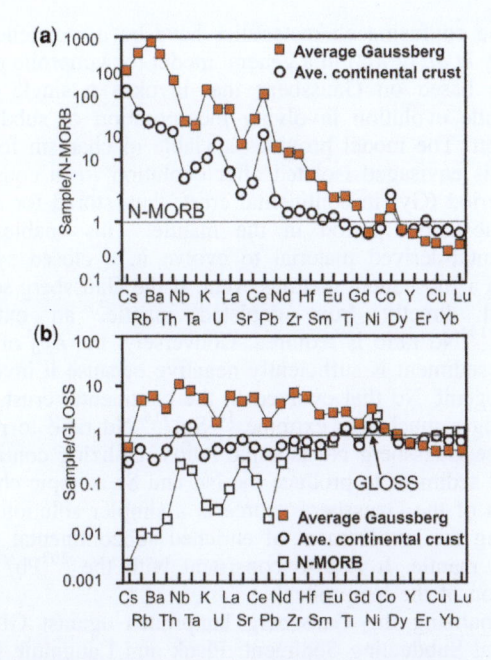

Fig. 10. Spidergrams showing an average Gaussberg lamproite composition normalized against (**a**) N-MORB and (**b**) GLOSS (after Murphy et al. 2002). N-MORB composition of Sun and McDonough (1989). GLOSS, global subducting sediment (Plank and Langmuir 1998). Average continental crust is from Hoffman (1988). See the text for the explanation.

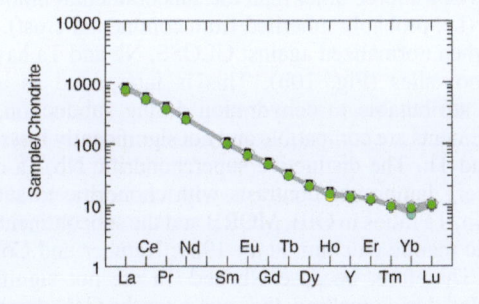

Fig. 11. Chondrite-normalized rare earth element (REE) spidergrams for representative Gaussberg lavas (data in Supplementary material Table S1). Normalizing values of McDonough and Sun (1995). Note the very high LREE/HREE ratios and lack of a Eu anomaly.

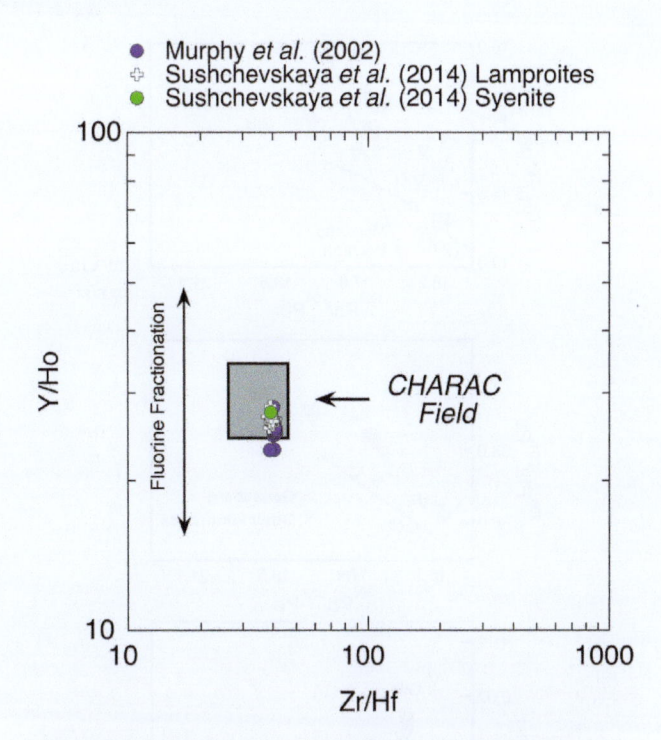

Fig. 12. Covariation between Y/Ho and Zr/Hf ratios in Gaussberg lamproites and in an alkaline syenite xenolith from Gaussberg. These plot within error of each other and indicate derivation from an essentially chondritic reservoir. The alkaline syenite (xenolith) has chondritic Y/Ho = 27.3 and Zr/Hf = 39.3 ratios, and is likely to have been derived from a large alkaline igneous complex linked to the Kerguelen plume that is inferred to underlie the Gaussberg volcanic construct. Several Gaussberg lamproites have slightly subchondritic Y/Ho ratios possibly due to fluoride complexation which fractionates Ho from Y. CHARAC, CHArge and and RAdius-Controlled. See the text for further explanation.

rocks exhibit Y/Ho and Zr/Hf ratios close to chondritic values (i.e. 28 and 38, respectively). These ratios are within error of values exhibited by plume-generated ocean island basalts (OIBs), viz. Y/Ho = 27.7 ± 2.7 and Zr/Hf = 36.6 ± 2.9 (Bau 1996), and values reported from alkaline igneous suites (de Andrade et al. 2002). Y/Ho and Zr/Hf ratios exhibited by samples of Gaussberg lamproites reported in two studies (i.e. Y/Ho = 25.1 ± 0.9; Zr/Hf = 40.8 ± 0.6 (Murphy et al. 2002) and Y/Ho = 26.6 ± 0.6; Zr/Hf = 39.6.8 ± 0.8 (Sushchevskaya et al. 2014)) are within error of each other and indicate derivation from an essentially chondritic reservoir (Fig. 12). The geochemical data reported by Sushchevskaya et al. (2014) also included an analysis of an alkaline syenite (presumably a xenolith). The sample (Sushchevskaya et al. 2014, their sample #464) has chondritic Y/Ho = 27.3 and Zr/Hf = 39.3 ratios that are within error of the ratios for Gaussberg lamproites, suggesting a cogenetic relationship. Thus, the syenite was most likely incorporated during passage of the lamproite magma through a putative large alkaline igneous complex formed by the Kerguelen plume and inferred, from geophysical data, to underlie a wide region, including under the Gaussberg volcanic construct (Sushchevskaya et al. 2014, 2017).

The presence of slightly subchondritic Y/Ho ratios in several Gaussberg lamproites that plot below the CHARAC field in Figure 12 is interpreted to reflect the role of fluoride complexation that facilitates Ho–Y fractionation (Bau and Dulski 1995; Buhn 2008). Evidence for this is not surprising, given the high F content (c. 1 wt%) reported in Gaussberg lamproites by Edgar et al. (1996).

The Gaussberg lavas are considerably more enriched in incompatible elements than average continental crust and significant crustal contamination is not suspected, which would have reduced the level of LREE-enrichment and flattened the normalized REE pattern (Fig. 10a). Anomalies such as the high concentrations of Pb combined with an underabundance of Sr cannot have been created during melting owing to the similar compatibilities of the two elements.

Raman analysis of volatiles in melt inclusions in leucite, olivine, clinopyroxene and apatite phenocrysts showed that H_2O and CO_2 are present. Pure CO_2 is found only in shrinkage bubbles, whereas the melt-inclusion glassy groundmasses contain both CO_2 and up to 0.70 wt% H_2O, suggesting that the two volatile species acted differently during crystallization (Salvioli-Mariani et al. 2004). It was inferred that CO_2 was released during every stage of evolution of the lamproite magma, whereas H_2O was retained for longer in the liquid. CO_2 had a major role at relatively high pressures, which favoured the crystallization of H_2O-poor phenocrysts. With further rise of the magma to shallower levels, water activity increased and was associated with late crystallization of phlogopite and K-richterite.

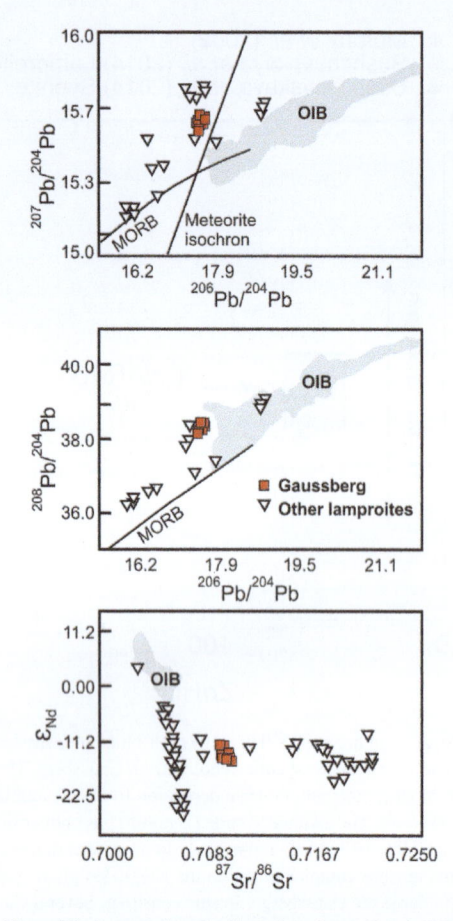

Fig. 13. Pb- and Sr-isotope plots for Gaussberg lavas (data in Table 2) and other lamproite occurrences (modified after Murphy *et al.* 2002). The field for ocean island basalts (OIB: grey shading) is also shown for comparison. Note the very narrow range of compositions for Gaussberg lavas compared with other lamproites and OIB. See the text for further explanation.

Isotopically, $^{87}Sr/^{86}Sr$ ratios are extremely high and they define a narrow range with possibly two populations (0.709185–0.709405 and 0.709532–0.709891: Fig. 13; Table 2). The ratios are within the much broader spectrum seen in other lamproite occurrences but they are significantly higher than values reported for Kerguelen and Heard Island (*c.* 0.704–0.706: Sheraton and Cundari 1980; Clarke *et al.* 1983). Similarly, $^{143}Nd/^{144}Nd$ ratios show a very restricted range (0.511073–0.511170), corresponding to $\varepsilon_{Nd}(0)$ of −14.9 to −13.0. $^{206}Pb/^{204}Pb$, $^{207}Pb/^{204}Pb$ and $^{208}Pb/^{204}Pb$ ratios are also strikingly uniform (17.415–17.609, 15.566–15.630 and 38.145–38.434, respectively). No trends are apparent on plots of $^{207}Pb/^{204}Pb$ v. $^{206}Pb/^{204}Pb$, $^{208}Pb/^{204}Pb$ v. $^{206}Pb/^{204}Pb$ and ε_{Nd} v. $^{87}Sr/^{86}Sr$, indicating a mantle with only a single source rather than multiple potential sources (Fig. 13).

The isotopic composition of Gaussberg lamproites is very unusual for mantle-derived magmas and it places significant constraints on any models for the evolution of the magma source. Indeed, in Pb, Sr and ε_{Nd} space, Gaussberg lamproites plot in a position that is more commonly associated with crustal rocks rather than mantle-derived melts. Yet, the contents of Pb, Sr and Nd indicate that, despite the presence of abundant crustal xenoliths in the lavas, and consistent with the extreme LREE/HREE ratios present (well beyond crustal effects), crustal contamination could have exerted only a subordinate role. Isotopic modelling also shows that if Pb in Gaussberg magma evolved directly from the upper mantle, a multistage history needs to be invoked. But this incurs significant problems.

These confusing characteristics have been reconciled by Murphy *et al.* (2002) in a general model for lamproite petrogenesis based on Gaussberg that invokes a single stage of mantle evolution involving incorporation of subducted sediment. The model provides a viable mechanism for Pb, which is envisaged isolated after evolution for a considerable period (Gy) in continental crust, then stored for a further substantial period in the mantle. This enables the continental-derived material to evolve in a closed system prior to a subsequent melting event. If the Gaussberg source evolved directly from depleted mantle, an extreme $^{147}Sm/^{144}Nd$ ratio is required. Conversely, the ε_{Nd} of subducted sediment is sufficiently negative because it involved unradiogenic Nd that evolved in the continental crust, thus requiring a much less extreme $^{147}Sm/^{144}Nd$ ratio to reproduce the Gaussberg Nd isotopic ratios. Utilizing continent-derived sediment to produce the Nd and Sr isotopic characteristics of the Gaussberg source is a simpler solution than invoking the involvement of enriched subcontinental lithospheric mantle. It is also consistent with the $^{207}Pb/^{204}Pb$ evolution of the lamproites.

Normalizing the Gaussberg lamproites against GLOSS (GLObal Subducting Sediment: Plank and Langmuir 1998) is particularly edifying. It removes the Sr anomaly previously noted but results in a negative Pb anomaly (Fig. 10b). The latter points to a subduction–dehydration influence in the sediment source, since Pb is a highly mobile element and is readily lost from slabs via dehydrating fluids. It is harder to explain the prominent mid-ocean ridge basalt (MORB)-normalized Nb–Ta anomalies but they are regarded as primary features of a source other than the subcontinental lithospheric mantle (i.e. probably inherited from continental crust). Moreover, when normalized against GLOSS, Nb and Ta have positive anomalies (Fig. 10b). This is interpreted as another feature attributable to dehydration during subduction, when both elements are compatible and not significantly lost relative to U and Th. The distinctive superchondritic Nb/Ta ratio in Gaussberg lamproites contrasts with chondritic to subchondritic Nb/Ta ratios in OIB, MORB and the subcontinental lithospheric mantle (Jochum *et al.* 1989; Kamber and Collerson 2000). Therefore, because Nb and Ta are not significantly decoupled during melting, the source for the Gaussberg lamproites must also have had a superchondritic Nb/Ta ratio. Although Nb/Ta ratios of sediments are also low (14.2 in GLOSS: Plank and Langmuir 1998), differential Nb–Ta fractionation during subduction-related dehydration will create variable Nb/Ta ratios, with Ta inferred to be lost to a greater extent than Nb in the sediments that were involved in the source for the Gaussberg magmas. Similarly, the relative depletions of Cs, Th, U and Pb in a GLOSS-normalized spidergram (Fig. 10b) are readily explained by subduction metamorphism, as these elements are particularly mobile in fluids during dehydration (Brenan *et al.* 1994).

Finally, that the concentrations of MgO, Ni and Cr in Gaussberg lamproites are relatively high and the melts are in equilibrium with olivine are convincing evidence for a contribution from a peridotite-derived melt component rather than a source consisting solely of deeply subducted sediment. Melting of eclogitic oceanic crust is also feasible.

The weight of the petrological evidence has thus been interpreted to suggest that the Gaussberg lamproites may have formed by melting of a source which included sediment that was deeply subducted. Based on mantle tomographic studies, it has been inferred that oceanic slabs and some of their subducted sediment cover accumulate at the Transition Zone (*c.* 410–660 km: Simons *et al.* 1999). Under Transition Zone conditions, K is supplied by the continental sedimentary protolith, probably from melting of K-hollandite-bearing assemblages, thus generating K-rich melts (but see a critical comment by

Table 2. *Published isotopic analyses of Gaussberg lavas*

Sample	Reference	^{206}Pb/^{204}Pb	^{207}Pb/^{204}Pb	^{208}Pb/^{204}Pb	^{143}Nd/^{144}Nd	^{87}Sr/^{86}Sr	$\varepsilon_{Nd}(0)$
79-242	1, 4				0.511905	0.70923 ± 7	−14.3
4883	1, 4				0.511915	0.70926 ± 2	−14.1
82-27	1				0.511972	0.70972 ± 4	−13.0
82-30	1				0.511915	0.70974 ± 4	−14.1
4893A	1				0.511910	0.70975 ± 4	−14.2
79-241	1				0.511905	0.70976 ± 8	−14.3
4888	1, 4				0.511920	0.70978 ± 3	−14.0
82-38*	1				0.511874	0.70987 ± 4	−14.9
82-35†	1				0.511910	0.71090 ± 5	−14.2
KC97-1/1	2	17.415	15.594	38.155	0.511894	0.709248	−14.5
KC97-1/2	2	17.550	15.594	38.259	0.511911	0.709742	−14.2
KC97-4A	2	17.526	15.618	38.396	0.511875	0.709816	−14.9
KC97-4B	2	17.557	15.614	38.321	0.511884	0.709759	−14.7
KC97-4C	2	17.509	15.579	38.253	0.511877	0.709708	−14.8
KC97-4D	2	17.499	15.577	38.244	0.511893	0.709741	−14.5
KC97-5c	2	17.527	15.583	38.240	0.511881	0.709796	−14.8
KC97-11	2	17.502	15.580	38.171	0.511913	0.709726	−14.1
KC97-12A	2	17.500	15.581	38.182	0.511907	0.709720	−14.3
KC97-12B	2	17.544	15.630	38.434	0.511915	0.709724	−14.1
KC97-13A	2	17.444	15.566	38.145	0.512013	0.709345	−12.2
KC97-14A	2	17.500	15.629	38.428	0.511870	0.709891	−15.0
KC97-16F	2	17.440	15.579	38.175	0.511999	0.709185	−12.5
KC97-17A	2	17.443	15.582	38.162	0.512015	0.709192	−12.2
KC97-17B	2	17.449	15.618	38.255	0.511915	0.709231	−14.1
KC97-19A	2	17.498	15.590	38.244	0.511927	0.709532	−13.9
KC97-19B	2	17.496	15.585	38.221	0.511917	0.709544	−14.1
KC97-19C	2	17.483	15.579	38.211	0.511909	0.709541	−14.2
KC97-20A	2	17.574	15.626	38.348	0.511942	0.709735	−13.6
KC97-20H	2	17.559	15.614	38.318	0.511922	0.709742	−14.0
KC97-23F	2	17.457	15.593	38.220	0.511922	0.709233	−14.0
KC97-27	2	17.501	15.598	38.241	0.511892	0.709598	−14.6
KC97-28A	2	17.609	15.611	38.331	0.511922	0.709747	−14.0
KC97-28B	2	17.526	15.588	38.251	0.511899	0.709742	−14.4
465	3	17.562	15.625	38.419	0.511898	0.709712	−14.4
457	3	17.508	15.614	38.310	0.511911	0.709294	−14.2
470	3	17.594	15.653	38.546	0.511867	0.709718	−15.0
477	3	17.537	15.622	38.342	0.511906	0.709405	−14.3
472	3	17.611	15.653	38.470	0.511937	0.706915	−13.7
468	3	18.026	15.648	38.468	0.511954	0.709003	−13.3
479	3	18.014	15.644	38.452	0.511913	0.709816	−14.1

*Cognate enclave formed of leucite, clinopyroxene, magnetite and glass.
†Cognate enclave formed of leucite, clinopyroxene and magnetite.
References: 1, Collerson and McCulloch (1983); 2, Murphy *et al.* (2002); 3, Sushchevskaya *et al.* (2014); 4, Williams *et al.* (1992).

Sushchevskaya *et al.* 2014). Migration of the melts to the surface may require entrainment by a plume.

Confirmation of a plume origin using primitive (lower) mantle-normalized ratios

Primitive mantle-normalized Ta/U and Nb/Th ratios using normalizing values from McDonough and Sun (1995) and Lyubetskaya and Korenaga (2007) are shown in Figure 14. This projection, from Niu and Batiza (1997) and Niu *et al.* (1999), shows that plume-derived magmas, formed by melting in lower-mantle upwellings, plot close to unity. Ta/U$_{PMN}$ ratios close to unity indicate the involvement of a lower-mantle plume component in the Gaussberg lamproite magmagenesis. Nb/Th$_{PMN}$ ratios <1 also show the Th-enriched nature of the Gaussberg source previously suggested by Th/U systematics reported by Williams *et al.* (1992). This Th enrichment was caused by the introduction and storage of subducted ancient sediment in the mantle transition zone. The subsequent passage of a rising thermal upwelling (plume) from the primitive lower mantle melted in the transition zone to form the lamproite magma (Murphy *et al.* 2002). The projection (Fig. 14) shows that plume-derived magmas originating from the base of the lower mantle retain a record of Earth's primitive (meteoritic) composition (e.g. Ta/U$_{PMN}$ and Nb/Th$_{PMN}$ ratios close to unity). Magmas contaminated by continental crust, crustal fluids or fluids derived from subducted slabs plot in the lower left-hand quadrant (mantle wedge and slab component), with substantially lower Ta/U$_{PMN}$ and Nb/Th$_{PMN}$ ratios (*c.* 0.01 and 0.1, respectively).

Thus, the evolutionary model proposed for Gaussberg magma involves sediment subduction, long-term (Gy) isolation and storage in the deep mantle (Transition Zone), with subsequent melting following entrainment in a plume. Because of the rarity of lamproites worldwide, it appears that most subducted sediment is not retained at the Transition Zone but gets mixed into the asthenosphere, dispersed and homogenized. The model is potentially applicable to other occurrences of lamproites and can also help to explain certain features of OIB petrogenesis. Thus, although the model is

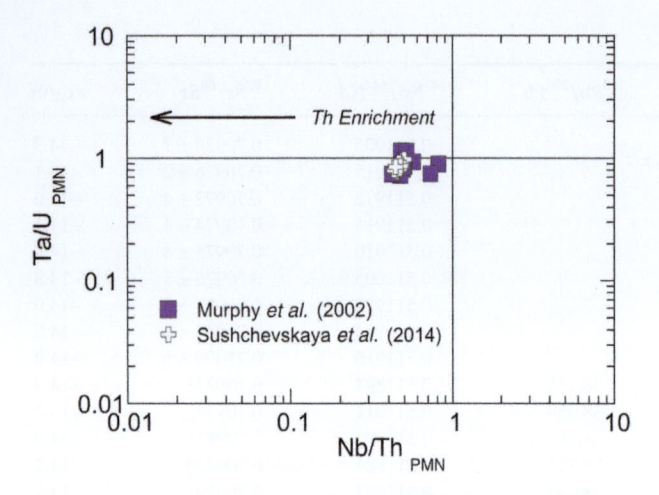

Fig. 14. Ta/U$_{PMN}$ v. Nb/Th$_{PMN}$ ratio plot showing data for Gaussberg lamproites. Ta/U$_{PMN}$ ratios close to unity indicate the involvement of a lower-mantle plume component in Gaussberg lamproite magmagenesis, whilst Nb/Th$_{PMN}$ ratios <1 indicate the Th-enriched nature of the Gaussberg source, as previously shown by Th/U systematics reported by Williams *et al.* (1992). PMN, primitive mantle normalized. See the text for further explanation.

based on a single small outcrop, at Gaussberg, it has a global application. This again illustrates how disproportionately important Gaussberg is, in this case for petrological and chemical geodynamic studies.

Summary

Gaussberg is a small isolated nunatak situated on the coast of East Antarctica at 66° 48′ S, 89° 11′ E. It is the most isolated Quaternary volcanic centre in Antarctica, situated 2500 km from the nearest exposed volcanic outcrops in Victoria Land. The nunatak is formed of lamproite pillow lava. Other lithofacies are rare (i.e. sheet lava or sill, tuff breccia, possible tuff). It represents the summit portion of a larger volcanic edifice, which is a pillow mound with a low shield-like profile, *c.* 1200 m high. Its diameter, possibly up to *c.* 10 km, is unusually large for a lamproite centre but it may be preserved in an unusually pristine (i.e. less eroded) state compared with other occurrences worldwide. Although originally regarded as a little-modified volcanic cone with a now-eroded summit crater, our more detailed examination suggests that it may have had several, probably co-eruptive, vents scattered on its surface, which may have contributed to the width of the edifice. Aerogeophysical surveys tentatively suggest that the volcano may be one of a cluster of several similar small volcanic centres, of which only Gaussberg is exposed. The age of the volcano is 56 ± 5 ka (by K–Ar), which places its construction within a late stage of the last glacial close to the peak of marine isotope stage (MIS) 3. Although early workers found it difficult to distinguish between subaqueously-emplaced lava pillows and subaerial pāhoehoe, Gaussberg was interpreted as a subglacially-erupted volcano using a range of *a priori* reasoning. Since the original descriptions, a glacial eruptive setting has become even more compelling and it is possible to refine the estimate of the thickness of the coeval ice sheet as *c.* 1300 m. A thickness of at least 420 m has been removed from the ice surface since Gaussberg erupted, which is within the range estimated for coastal ice in East Antarctica since LGM. Lamproite is a primitive, ultrapotassic mantle-derived magma that is also very rare worldwide, and Gaussberg is regarded as one of two type examples. Although traditionally regarded as related in some way to the Kerguelen plume, a direct relationship is difficult to prove and it is more likely that the Gaussberg magma is a product of a different, isolated, magmatic event. Because of its rarity, Gaussberg lamproite has been well investigated petrologically. The genesis of the lamproite is ascribed to the storage and long-term (Gy) isolation of sediment emplaced by subduction in the Transition Zone of the deep mantle, followed by entrainment in a plume and subsequent melting. The palaeoenvironmental and petrological importance of Gaussberg is out of all proportion to its small size and isolated location.

Acknowledgements The authors are grateful to the following persons who were very helpful during the writing of this chapter: Tom Jordan for help with extracting and interpreting high-resolution magnetic data from ADMAP; Darryn Schneider for permission to use his photographs; Noel Ward for access to his views of Gaussberg; and to Duanne White for conversations about Gaussberg. The authors are also grateful to David Murphy and an anonymous reviewer for their constructive reviews. Logistics for K.D. Collerson's two visits to Gaussberg were provided by the Australian Antarctic Division.

Author contributions JLS: conceptualization (lead), investigation (equal), methodology (lead), visualization (lead), writing – original draft (lead), writing – review & editing (lead); **KDC**: investigation (equal), writing – original draft (supporting), writing – review & editing (supporting).

Funding Logistical support to KC for the Gaussberg voyage and field work was provided by a grant from the Australian Antarctic Division, which is gratefully acknowledged.

Data availability Data sharing is not applicable to this article as no additional datasets were generated other than those presented in this chapter and its supplementary files.

References

Barling, J. 1990. Heard and McDonald islands. *American Geophysical Union Antarctic Research Series*, **48**, 435–441.

Bau, M. 1996. Controls on the fractionation of isovalent trace elements in magmatic and aqueous systems: Evidence from Y/Ho, Zr/Hf, and lanthanide tetrad effect. *Contributions to Mineralogy and Petrology*, **123**, 323–333, https://doi.org/10.1007/s004100050159

Bau, M. and Dulski, P. 1995. Comparative study of yttrium and rare-earth element behaviours in fluorine-rich hydrothermal fluids. *Contributions to Mineralogy and Petrology*, **119**, 213–223, https://doi.org/10.1007/BF00307282

Brenan, J.M., Shaw, H.F., Phinney, D.L. and Ryerson, F.J. 1994. Rutile–aqueous fluid partitioning of Nb, Ta, Hf, Zr, U and Th – implications for high-field strength element depletions in island-arc basalts. *Earth and Planetary Science Letters*, **128**, 327–339, https://doi.org/10.1016/0012-821X(94)90154-6

Buhn, B. 2008. The role of the volatile phase for REE and Y fractionation in low-silica carbonate magmas: implications from natural carbonatites, Namibia. *Mineralogy and Petrology*, **92**, 453–470, https://doi.org/10.1007/s00710-007-0214-4

Clarke, I., McDougall, I. and Whitford, D.J. 1983. Volcanic evolution of Heard and McDonald islands. *In*: Oliver, R.L., James, P.R. and Jago, J.B. (eds) *Antarctic Earth Science*. Australian Academy of Science, Canberra, 631–635.

Collerson, K.D. and McCulloch, M.T. 1983. Nd and Sr isotope geochemistry of leucite-bearing lavas from Gaussberg, East Antarctica.

In: Oliver, R.L., James, P.R. and Jago, J.B. (eds) *Antarctic Earth Science*. Australian Academy of Science, Canberra, 676–680.

De Andrade, F.R.D., Möller, P. and Dulski, P. 2002. Zr/Hf in carbonatites and alkaline rocks: new data and a re-evaluation. *Revista Brasiliera de Geosciências*, **3**, 361–370, https://doi.org/10.25249/0375-7536.2002323361370

Deschamps, A., Grigne, C., Le Saout, M., Soule, S.A., Allemand, P., Van Vliet Lanoet, B. and Floc'h, F. 2014. Morphology and dynamics of inflated subaqueous basaltic lava flows. *Geochemistry, Geophysics, Geosystems*, **15**, 2128–2150, https://doi.org/10.1002/2014GC005274.

Doyle, T.W., Chivoiu, B. and Enwright, N.M. 2015. *Sea-Level Rise Modelling Handbook: Resource Guide for Coastal Land Managers, Engineers, and Scientists*. United States Geological Survey Professional Paper, **1815**.

Drygalski, E. von 1912. Der Gaussberg, seine Kartierung und seine Formen. *Deutsche Südpolar-Expedition 1901-190, Geographie und Geologie II*, No. 1, 1–46.

Duncan, R.A. 1981. Hotspots in the southern oceans – an absolute frame of reference for motion of the Gondwana continents. *Tectonophysics*, **74**, 29–42, https://doi.org/10.1016/0040-1951(81)90126-8

Edgar, A.D., Pizzolato, L.A. and Sheen, J. 1996. Fluorine in igneous rocks and minerals with emphasis on ultrapotassic mafic and ultramafic magmas and their mantle source regions. *Mineralogical Magazine*, **60**, 243–257, https://doi.org/10.1180/minmag.1996.060.399.01

Edwards, B.R., Skilling, I.P., Cameron, B., Haynes, C., Lloyd, A. and Hungerford, J.H.D. 2009. Evolution of an englacial volcanic ridge: Pillow Ridge tindar, Mount Edziza volcanic complex, NCVP, British Columbia, Canada. *Journal of Volcanology and Geothermal Research*, **185**, 251–275, https://doi.org/10.1016/j.jvolgeores.2008.11.015

Foley, S.F., Venturelli, G., Green, D.H. and Toscani, L. 1987. The ultrapotassic rocks: characteristics, classification, and constraints for petrogenetic models. *Earth-Science Reviews*, **24**, 81–134, https://doi.org/10.1016/0012-8252(87)90001-8

Forbes, A.E.S., Blake, S. and Tuffen, H. 2014. Entablature: fracture types and mechanisms. *Bulletin of Volcanology*, **76**, 820, https://doi.org/10.1007/s00445-014-0820-z

Frey, F.A., Weis, D., Yang, H.-J., Nicolaysen, K., Leyrit, H. and Giret, A. 2000. Temporal geochemical trends in Kerguelen Archipelago basalts: evidence for decreasing magma supply from the Kerguelen Plume. *Chemical Geology*, **164**, 61–80, https://doi.org/10.1016/S0009-2541(99)00144-8

Golynski, A.V., Ferraccioli, F. *et al.* 2018. New magnetic anomaly map of the Antarctic. *Geophysical Research Letters*, **45**, 6437–6449, https://doi.org/10.1029/2018GL078153

Golynski, D.A. and Golynski, A.V. 2007. Gaussberg rift – illusion or reality? *United States Geological Survey Open-File Report*, **2007-1047**, Extended Abstract 168.

Goto, Y. and McPhie, J. 2004. Morphology and propagation styles of Miocene submarine basanite lavas at Stanley, northwestern Tasmania, Australia. *Journal of Volcanology and Geothermal Research*, **130**, 307–328, https://doi.org/10.1016/S0377-0273(03)00311-1

Goto, Y. and McPhie, J. 2012. Morphology and formation of spreading cracks on pillow lavas at Cape Grim, northwestern Tasmania, Australia. *Bulletin of Volcanology*, **74**, 1611–1619, https://doi.org/10.1007/s00445-012-0618-9

Grossenbacher, K.A. and McDuffie, S.M. 1995. Conductive cooling of lava: columnar joint diameter and stria width as functions of cooling rate and thermal gradient. *Journal of Volcanology and Geothermal Research*, **69**, 95–103, https://doi.org/10.1016/0377-0273(95)00032-1

Hansen, J., Sato, M. *et al.* 2016. Ice melt, sea level rise and superstorms: evidence from paleoclimate data, climate modelling, and modern observations that 2°C global warming could be dangerous. *Atmospheric Chemistry and Physics*, **16**, 3761–3812, https://doi.org/10.5194/acp-16-3761-2016

Head, J.W. and Wilson, L. 2003. Deep submarine pyroclastic eruptions: theory and predicted landforms and deposits. *Journal of Volcanology and Geothermal Research*, **121**, 155–193, https://doi.org/10.1016/S0377-0273(02)00425-0

Hoffman, A.W. 1988. Chemical differentiation of the Earth – the relationship between mantle, continental crust, and oceanic crust. *Earth and Planetary Science Letters*, **90**, 297–314, https://doi.org/10.1016/0012-821X(88)90132-X

Hungerford, J.D.G., Edwards, B.R., Skilling, I.P. and Cameron, B.I. 2014. Evolution of a subglacial basaltic lava flow field: Tennena volcanic center, Mount Edziza volcanic complex, British Columbia, Canada. *Journal of Volcanology and Geothermal Research*, **272**, 39–58, https://doi.org/10.1016/j.jvolgeores.2013.09.012

Imbrie, J.Z., Imbrie-Moore, A. and Lisiecki, L.E. 2011. A phase-space model for Pleistocene ice volume. *Earth and Planetary Science Letters*, **307**, 94–102, https://doi.org/10.1016/j.epsl.2011.04.018

Jochum, K.P., McDonough, W.F., Palme, H. and Spettel, B. 1989. Compositional constraints on the continental lithospheric mantle from trace-elements in spinel peridotite xenoliths. *Nature*, **340**, 548–550, https://doi.org/10.1038/340548a0

Jones, J.G. 1969*a*. Pillow lavas as depth indicators. *American Journal of Science*, **267**, 181–195, https://doi.org/10.2475/ajs.267.2.181

Jones, J.G. 1969*b*. Intraglacial volcanoes of the Laugarvatn region, south-west Iceland – I. *Quarterly Journal of the Geological Society, London*, **124**, 197–211, https://doi.org/10.1144/gsjgs.124.1.0197

Jones, J.G. 1970. Intraglacial volcanoes of the Laugarvatn region, south-west Iceland, II. *Journal of Geology*, **78**, 127–140, https://doi.org/10.1086/627496

Kamber, B.S. and Collerson, K.D. 2000. Role of 'hidden' deeply subducted slabs in mantle depletion. *Chemical Geology*, **166**, 241–254, https://doi.org/10.1016/S0009-2541(99)00218-1

Kokelaar, B.P. and Durant, G.P. 1983. The submarine eruption and erosion of Surtla (Surtsey), Iceland. *Journal of Volcanology and Geothermal Research*, **19**, 239–246, https://doi.org/10.1016/0377-0273(83)90112-9

Kyle, P.R. and Jezek, P.A. 1978. Compositions of three tephra layers from Byrd Station ice core, Antarctica. *Journal of Volcanology and Geothermal Research*, **4**, 225–232, https://doi.org/10.1016/0377-0273(78)90014-8

Le Bas, M.J., Le Maitre, R.W., Streckeisen, A. and Zanettin, B. 1986. A chemical classification of volcanic rocks based on the total alkali–silica diagram. *Journal of Petrology*, **27**, 745–750, https://doi.org/10.1093/petrology/27.3.745

Liu, X., Wang, W., Zhao, Y., Liu, J., Chen, H., Cui, Y. and Song, B. 2016. Early Mesoproterozoic arc magmatism followed by early Neoproterozoic granulite facies metamorphism with a near-isobaric cooling path at Mount Brown, Princess Elizabeth Land, East Antarctica. *Precambrian Research*, **284**, 30–48, https://doi.org/10.1016/j.precamres.2016.08.003

Lythe, M.B. and Vaughan, D.G. and the BEDMAP Consortium. 2001. BEDMAP: A new ice thickness and subglacial topographic model of Antarctica. *Journal of Geophysical Research*, **106**, 11 335–11 351, https://doi.org/10.1029/2000JB900449

Lyubetskaya, T. and Korenaga, J. 2007. Chemical composition of the Earth's primitive mantle and its variance. *Journal of Geophysical Research*, **112**, B03211, https://doi.org/10.1029/2005JB004223

Mackintosh, A.N., Verleyen, E. *et al.* 2014. Retreat history of the East Antarctic ice sheet since the last glacial maximum. *Quaternary Science Reviews*, **100**, 10–30, https://doi.org/10.1016/j.quascirev.2013.07.024

McDonough, W.F. and Sun, S.-S. 1995. The composition of the Earth. *Chemical Geology*, **120**, 223–253, https://doi.org/10.1016/0009-2541(94)00140-4

McPhie, J., Doyle, M. and Allen, R. 1993. *Volcanic Textures. A Guide to the Interpretation of Textures in Volcanic Rocks*. Centre for Ore Deposit and Exploration Studies, (CODES), University of Tasmania, Hobart, Australia.

Mikhalsky, E.V., Belyatsky, B.V. et al. 2015. The geological composition of the hidden Wilhelm II Land in East Antarctica: SHRIMP zircon, Nd isotopic and geochemical studies with implications for Proterozoic supercontinent reconstructions. *Precambrian Research*, **258**, 171–185, https://doi.org/10.1016/j.precamres.2014.12.011

Mitchell, R.H. and Bergman, S.C. 1991. Description of lamproite occurrences. *In*: Mitchell, R.H. and Bergman, S.C. (eds) *Petrology of Lamproites*. Plenum Press, New York, 39–102.

Morgan, W.J. and Phipps-Morgan, J. 2007. Plate velocities in the hotspot reference frame. *Geological Society of America Special Papers*, **430**, 65–78, https://doi.org/10.1130/2007.2430(04);

Murphy, D.T., Collerson, K.D. and Kamber, B.S. 2002. Lamproites from Gaussberg, Antarctica: possible transition zone melts of Archaean subducted sediments. *Journal of Petrology*, **43**, 981–1001, https://doi.org/10.1093/petrology/43.6.981

Niu, Y. and Batiza, R. 1997. Trace element evidence from seamounts for recycled oceanic crust in the Eastern Pacific mantle. *Earth and Planetary Science Letters*, **148**, 471–483, https://doi.org/10.1016/S0012-821X(97)00048-4

Niu, Y., Collerson, K.D., Batiza, R., Wendt, I. and Regelous, M. 1999. Origin of enriched-type mid-ocean-ridge basalt at ridges far from mantle plumes: The East Pacific Rise at 11°20′N. *Journal of Geophysical Research*, **104**, 7067–7087, https://doi.org/10.1029/1998JB900037

Nockolds, S.R. 1940. *Petrology of Rocks from Queen Mary Land. Australian Antarctic Expedition, 1911–14.* Scientific Reports Series A, **IV**, Part 2, 15–86.

Nougier, J. and Thomson, J.W. 1990. Îles Kerguelen. *American Geophysical Union Antarctic Research Series*, **48**, 429–434.

Phillipi, E. 1912. Geologische Beschreibung des Gaussbergs. Deutsche Sudpolar Expedition 1901–1903. *Geographie und Geologie*, **11**, 47–71.

Plank, T. and Langmuir, C.H. 1998. The chemical composition of subducting sediment and its consequences for the crust and mantle. *Chemical Geology*, **145**, 325–394, https://doi.org/10.1016/S0009-2541(97)00150-2

Pollock, M., Edwards, B., Hauksdóttir, S., Alcorn, R. and Bowman, L. 2014. Geochemical and lithostratigraphical constraints on the formation of pillow-dominated tindars from Undirhlíðar quarry, Reykjanes Peninsula, southwest Iceland. *Lithos*, **200–201**, 317–333, https://doi.org/10.1016/j.lithos.2014.04.023

Prelević, D., Foley, S.F. and Cvetković, V. 2007. A review of petrogenesis of Mediterranean Tertiary lamproites: A perspective from the Serbian ultrapotassic province. *Geological Society of America Special Papers*, **418**, 113–129.

Ravich, M.G. and Krylov, A.H. 1964. Absolute ages of rocks from East Antarctica. *In*: Adie, R.J. (ed.) *Antarctic Geology*. North Holland, Amsterdam, 578–589.

Salvioli-Mariani, E., Toscani, L. and Bersani, D. 2004. Magmatic evolution of the Gaussberg lamproite (Antarctica): volatile content and glass composition. *Mineralogical Magazine*, **68**, 83–100, https://doi.org/10.1180/0026461046810173

Schopka, H.H., Gudmundsson, M.T. and Tuffen, H. 2006. The formation of Helgafell, southwest Iceland, a monogenetic subglacial hyaloclastite ridge: sedimentology, hydrology and volcano – ice interaction. *Bulletin of Volcanology*, **152**, 359–377, https://doi.org/10.1016/j.jvolgeores.2005.11.010

Seghedi, I., Szakács, A., Hernandez Pacheco, A. and Brändle Matesanz, J.-L. 2007. Miocene lamproite volcanoes in south-eastern Spain – an association of phreatomagmatic and magmatic products. *Journal of Volcanology and Geothermal Research*, **159**, 210–224, https://doi.org/10.1016/j.jvolgeores.2006.06.012

Sheraton, J.W. and Cundari, A. 1980. Leucitites from Gaussberg, Antarctica. *Contributions to Mineralogy and Petrology*, **71**, 417–427, https://doi.org/10.1007/BF00374713

Simons, F.J., Zielhuis, A. and van der Hilst, R.D. 1999. The deep structure of the Australian continent from surface wave tomography. *Lithos*, **48**, 17–43, https://doi.org/10.1016/S0024-4937(99)00041-9

Smellie, J.L. 2001. Lithofacies architecture and construction of volcanoes in englacial lakes: Icefall Nunatak, Mount Murphy, eastern Marie Byrd Land, Antarctica. *International Association of Sedimentologists Special Publication*, **30**, 73–98, https://doi.org/10.1002/9781444304251.ch2

Smellie, J.L. and Edwards, B.R. 2016. *Glaciovolcanism on Earth and Mars: Products, Processes and Palaeoenvironmental Significance*. Cambridge University Press, Cambridge, UK.

Smellie, J.L. and Hole, M.J. 1997. Products and processes in Pliocene–Recent, subaqueous to emergent volcanism in the Antarctic Peninsula: examples of englacial Surtseyan volcano construction. *Bulletin of Volcanology*, **58**, 628–646, https://doi.org/10.1007/s004450050167

Soloviev, D.S. 1972. Platform magmatic formations of East Antarctica. *In*: Adie, R.J. (ed.) *Antarctic Geology and Geophysics*. Universitetsforlaget, Oslo, 531–538.

Stap, L.B., van de Wal, R.S.W., de Boer, B., Bintanja, R. and Lourens, L.J. 2014. Interaction of ice sheets and climate duringteh past 800 000 years. *Climate of the Past*, **10**, 2135–2152, https://doi.org/10.5194/cp-10-2135-2014

Sun, S.-s. and McDonough, W.F. 1989. Chemical and isotopic systematics of oceanic basalts: implications for mantle composition and processes. *Geological Society, London, Special Publications*, **42**, 313–345, https://doi.org/10.1144/GSL.SP.1989.042.01.19

Sushchevskaya, N.M., Migdisova, N.A., Antonov, A.V., Krymsky, R.Sh., Belyatsky, B.V., Kuzmin, D.V. and Bychkova, Ya.V. 2014. Geochemical features of the Quaternary lamproitic lavas of Gaussberg Volcano, East Antarctica: result of the impact of the Kerguelen Plume. *Geochemistry International*, **52**, 1030–1048, https://doi.org/10.1134/S0016702914120106

Sushchevskaya, N.M., Blyatsky, B.V., Dubinin, E.P. and Levchenko, O.V. 2017. Evolution of the Kerguelen Plume and its impact upon the continental and oceanic magmatism of East Antarctica. *Geochemsitry International*, **2017**, 775–791, https://doi.org/10.1134/S0016702917090099

Tingey, R.J. 1990. Gaussberg. *American Geophysical Union Antarctic Research Series*, **48**, 446–448.

Tingey, R.J., MacDougall, I. and Gleadow, A.J.W. 1983. The age and mode of formation of Gaussberg, Antarctica. *Journal of the Geological Society of Australia*, **30**, 241–246, https://doi.org/10.1080/00167618308729251

Vyalov, O.S. and Sobolev, V.S. 1959. Gaussberg, Antarctica (translated from the Russian by L. Drashevska). *International Geological Reviews*, **1**, 30–40, https://doi.org/10.1080/00206815909473430

Weis, D., Frey, F.A. et al. 2002. Trace of the Kerguelen mantle plume: Evidence from seamounts between the Kerguelen Archipelago and Heard Island, Indian Ocean. *Geochemistry, Geophysics, Geosystems*, **3**, https://doi.org/10.1029/2001GC000251

White, J.D.L. and Houghton, B.F. 2006. Primary volcaniclastic rocks. *Geology*, **34**, 677–680, https://doi.org/10.1130/G22346.1

White, J.D.L., Smellie, J.L. and Clague, D. 2003. A deductive outline and topical overview of subaqueous explosive volcanism. *American Geophysical Union Geophysical Monograph Series*, **140**, 1–23.

Williams, R.W., Collerson, K.D., Gill, J.B. and Deniel, C. 1992. High Th/U ratios in subcontinental lithospheric mantle – mass-spectrometric measurements of Th isotopes in Gaussberg lamproites. *Earth and Planetary Science Letters*, **111**, 257–268, https://doi.org/10.1016/0012-821X(92)90183-V

Woolley, A.R., Bergman, S.C., Edgar, A.D., Le Bas, M.J., Mitchell, R.H., Rock, N.M.S. and Scott Smith, B.H. 1996. Classification of lamprophyres, lamproites, kimberlites, and the kalsilitic, melilitic, and leucitic rocks. *The Canadian Mineralogist*, **34**, 175–186.

Zimanowski, B. and Büttner, R. 2003. Phreatomagmatic explosions in subaqueous volcanism. *American Geophysical Union Geophysical Monograph Series*, **140**, 51–60.

Section 6

Tephra record

Coastal ice cliff *c.* 30 m high, showing a prominent dark englacial tephra layer. Although currently unanalysed, the tephra was probably sourced in the active stratovolcano of Mount Melbourne, seen in the background.

Photograph by J. L. Smellie

Chapter 6.1

Marine record of Antarctic volcanism from drill cores

Alessio Di Roberto, Paola Del Carlo and Massimo Pompilio*

Istituto Nazionale di Geofisica e Vulcanologia, Sezione di Pisa, Via della Faggiola 32, 56126 Pisa, Italy

ADR, 0000-0003-1167-8290; PDC, 0000-0001-5506-4579; MP, 0000-0002-0742-0679

*Correspondence: massimo.pompilio@ingv.it

Abstract: We review here data and information on Antarctic volcanism resulting from recent tephrostratigraphic investigations on marine cores. Records include deep drill cores recovered during oceanographic expeditions: DSDP, ODP and IODP drill cores recovered during ice-based and land-based international cooperative drilling programmes DVDP 15, MSSTS-1, CIROS-1 and CIROS-2, DVDP 15, CRP-1, CRP-2/2A and CRP-3, ANDRILL-MIS and ANDRILL-SMS, and shallow gravity and piston cores recovered in the Antarctic and sub-Antarctic oceans. We report on the identification of visible volcaniclastic horizons and, in particular, of primary tephra within the marine sequences. Where available, the results of analyses carried out on these products are presented. The volcanic material identified differs in its nature, composition and emplacement mechanisms. It was derived from different sources on the Antarctic continent and was emplaced over a wide time span.

Marine sediments contain a more complete record of the explosive activity from Antarctic volcanoes and are complementary to those obtained by land-based studies. This record provides important information for volcanological reconstructions including approximate intensities and magnitudes of eruptions, and their duration, age and recurrence, as well as their eruptive dynamics. In addition, characterized tephra layers represent an invaluable chronological tool essential in establishing correlations between different archives and in synchronizing climate records.

Marine sediment sequences from the Southern Ocean surrounding the Antarctic continent may contain tephra produced by explosive eruptions from Antarctic volcanoes (Smellie 1999a, b; Hillenbrand et al. 2008; Del Carlo et al. 2015; Di Roberto et al. 2019, 2020). Marine sediment sequences are often used in preference to the terrestrial realm for the analysis of tephra because the marine environment is characterized by almost continuous sedimentation and few sedimentary disturbances, permitting a more complete record of volcanism. Unfortunately, except for a few cases (e.g. Hillenbrand et al. 2008; Del Carlo et al. 2015; Di Roberto et al. 2019, 2020), marine sediments in Antarctica have not been widely exploited for tephrochronology and volcanology studies, and in some cases only partially (Licht et al. 1996, 1999; Colizza et al. 2003; Xiao et al. 2016). Tephra have a significant potential as time-synchronous marker horizons that are crucial in establishing independent correlations between different geological archives (e.g. marine, terrestrial, lacustrine and ice-core records: Lowe 2011; Riede and Thastrup 2013; Davies 2015; Lowe and Alloway 2015; Ponomareva et al. 2015; Lowe 2016; Di Roberto et al. 2018, 2019, 2020). If fingerprinted using individual glass-shard chemistry and either independently dated or linked to a known and well-dated eruption, tephra layers represent an invaluable chronological tool essential in establishing a correlation between different archives and in synchronizing climate records (Hillenbrand et al. 2008; Di Roberto et al. 2019, 2020). Moreover, tephra layers may provide significant data for volcanological reconstructions, including approximate calculations of the intensity and magnitude of eruptions (e.g. Pyle 2000), their duration, age and recurrence, as well as their eruptive dynamics (e.g. Shane 2000; Lowe 2011, 2016; Albert et al. 2012; Fontijn et al. 2016; Di Roberto et al. 2018, 2019, 2020). This potential is especially effective where the terrestrial record of volcanism is incomplete, somewhat obscured or disturbed owing to erosion (e.g. Fontijn et al. 2014).

In Antarctica, the primary deposits of past eruptive activity are often almost completely covered by thick ice and snow (less than 1% of rock can be studied for volcanological purposes: Burton-Johnson et al. 2016) or may be removed by the ice. In addition, many volcanoes in Antarctica are remote and extremely difficult to study, and the climatic conditions are also robust.

Here we review data and information on Antarctic volcanism derived from marine cores, including drill cores and shallow gravity and piston cores recovered in Antarctic seas. We report on the identification of volcanic rocks, volcaniclastic deposits and, in particular, of primary tephra within the marine sequences, and, where available, the results of the analyses carried out on these products.

Volcanological background

Antarctica contains abundant Neogene–Recent volcanism, including a few active volcanoes (LeMasurier 1990). From the sub-Antarctic South Sandwich Islands, through the Antarctic Peninsula and Marie Byrd Land, and into East Antarctica, a 5000 km-long chain of volcanoes crosses the Antarctic continent (Fig. 1). This volcanism is related to different types of tectonic processes over the past 200 myr (e.g. subduction, rifting and continental break-up, see the introductory chapter to this Memoir: Smellie et al. 2021), and has produced a wide spectrum of volcanic structures, magma compositions and associated deposits.

Volcanic structures range from monogenetic (e.g. Random Hills in Northern Victoria Land) or parasitic scoria cones to large polygenetic central volcanoes (both stratovolcanoes and shield volcanoes like Mount Erebus and Mount Takahe) and calderas (Mount Rittmann).

Magma composition varies from basalt to trachyte–rhyolite (see Chapter 1.3: Panter 2021). Volcanic activity encompasses lava effusion, purely magmatic explosions (from Strombolian to Plinian, including caldera-forming eruptions) and those related to interaction with ice/water. Related products are lava flows (pillow and/or ʻaʻā lava-fed deltas), pyroclastic falls and flow, and hydrovolcanic deposits such as hyaloclastites, tuffs, etc. (see Chapter 1.2: Smellie 2021).

Eruptive activity observed since the beginning of the twentieth century has been scarce. It has occurred on Deception Island (South Shetland Islands) with eruptions between

From: Smellie, J. L., Panter, K. S. and Geyer, A. (eds) 2021. *Volcanism in Antarctica: 200 Million Years of Subduction, Rifting and Continental Break-up*. Geological Society, London, Memoirs, **55**, 631–647,

First published online March 1, 2021, https://doi.org/10.1144/M55-2018-49

© 2021 The Author(s). Published by The Geological Society of London. All rights reserved.

For permissions: http://www.geolsoc.org.uk/permissions. Publishing disclaimer: www.geolsoc.org.uk/pub_ethics

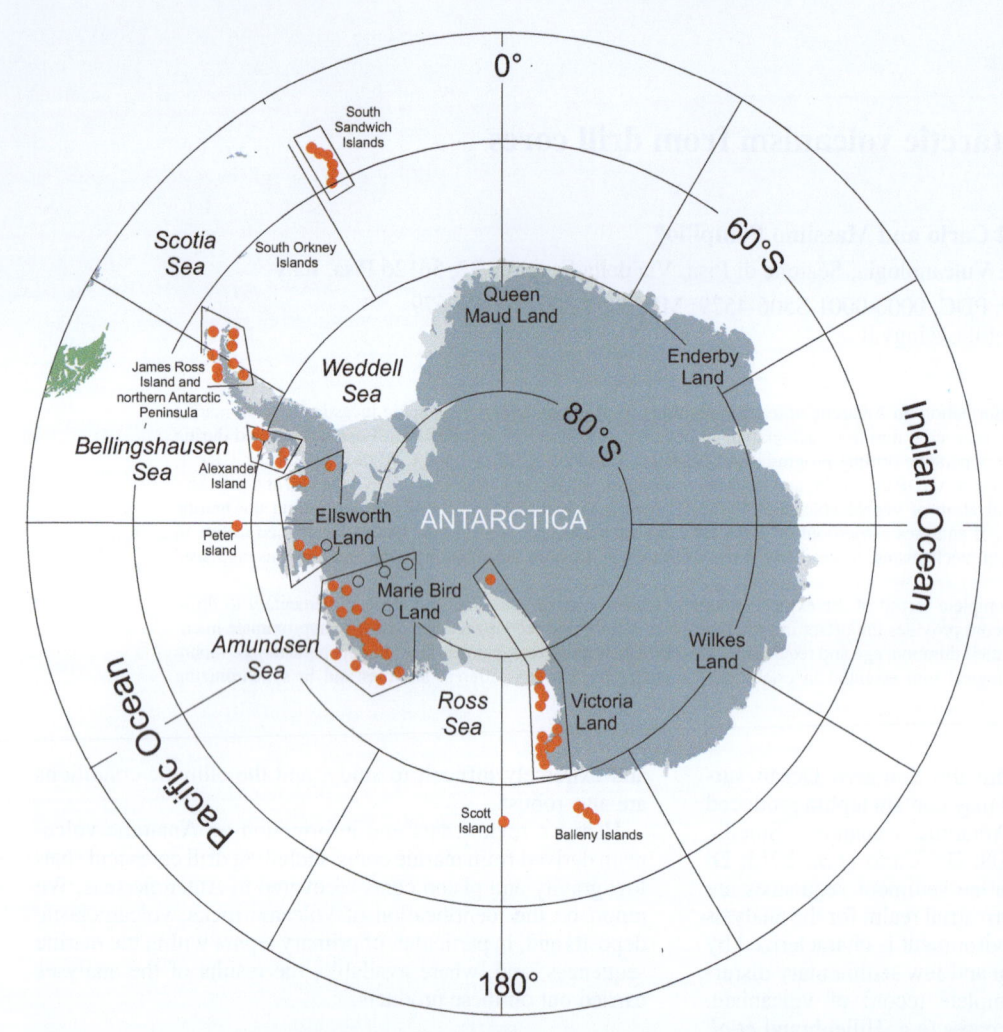

Fig. 1. Map of Antarctica showing the distribution of Neogene (<c. 30 Ma) volcanoes in Antarctica (red dots). The locations of selected potentially active subglacial volcanoes are also shown (open circles). See also van Wyk de Vries et al. (2017). Redrawn from Smellie and Edwards (2016). Base map by courtesy of J.L. Smellie, modified.

1906–10 and 1967–70 (Smellie 2002; Pedrazzi et al. 2014), and at Mount Erebus, where a lava lake in the summit crater has been active for several decades (Iverson et al. 2014). Less well-established eruptions may have occurred on Penguin Island (SE coast of King George Island) and Buckle Island (Balleny archipelago) in the nineteenth century (Global Volcanism Program 2013). More recently, explosive eruptions occurred on Montagu Island in 2002 (Patrick et al. 2005; Global Volcanism Program 2010), at Mount Sourabaya in 2016 (Global Volcanism Program 2017) and at Saunders Island in 2018 (all in the South Sandwich Islands), which produced ash plumes observed by satellites (Patrick and Smellie 2013; Global Volcanism Program 2018).

Nevertheless, several volcanoes are considered to still be active in Antarctica on the basis of the occurrence of fumaroles (as, for instance, at Mount Rittmann in Victoria Land: see Chapters 5.1a (Smellie and Rocchi 2021) and 7.3 (Gambino et al. 2021)) and the presence of Holocene englacial tephra layers (Narcisi et al. 2005, 2006; Iverson et al. 2014; see also Chapters 6.2 (Narcisi and Petit 2021) and 7.4 (Dunbar et al. 2021)). Moreover, large numbers of features that may be volcanoes have recently been identified beneath the West Antarctic Ice Sheet (van Wyk de Vries et al. 2017). Recent discoveries about the presence of subglacial active volcanoes have led to a debate on the volcanic risk in Antarctica (Lough et al. 2013). For example, it has been recognized that large areas at the base of Thwaites Glacier (West Antarctica) are actively melting in response to geothermal flux consistent with rift-associated magma migration and volcanism (Schroeder et al. 2014).

Marine sequences record different types of volcanic activity depending on the distance to the volcanic source. Tephra or cryptotephra represent deposits produced by high-energy explosive activity (sub-Plinian, Plinian or ultra-Plinian) from medio-distal volcanic (Antarctic or extra-Antarctic) or proximal sources. Pillow lavas and associated hyaloclastite usually indicate the occurrence of basaltic submarine/shallow-water eruptions from vents fortuitously close to drill sites. The combination of composition, age, thickness and facies types can make it possible to discriminate between likely volcanic sources (see the discussion in Del Carlo et al. 2018).

Tephras recovered during oceanographic expeditions: DSDP, ODP and IODP

Seafloor sediments and rocks from Antarctica have been extensively drilled during oceanographic expeditions under the auspices of the DSDP, the ODP and their successors the IODP (Table 1). The expeditions were aimed at reconstructing different aspects of the Antarctic continent, including its tectonic and geological history, and the response of the ice cap to past climatic fluctuations. Figure 2 shows the position and details of offshore sites located in Antarctica and the surrounding Southern Ocean.

DSDP Leg 28 (Fig. 2a) explored the area between Australia and the Ross Sea (sites 264, 264A, 265, 266, 267, 268, 269, 270, 271, 272, 273 and 274). At some of the sites abundant volcanic particles dispersed in sediments and volcaniclastic deposits were recovered. Tholeiitic basalt lava, apparently *in situ*, was recovered at four sites: sites 265 and 266 on the south flank of the SE Indian Ridge, at Site 267 in the South Indian Basin, and at Site 274 near Balleny Basin (Fig. 2a;

Table 1. *List and explanation of the initials/acronyms used in the text to identify coring expeditions*

Coring expedition initials/acronym	Full name	Site drilled Rock	Site drilled Soft sediments
DSDP	Deep Sea Drilling Project	X	
ODP	Ocean Drilling Program	X	
IODP	Integrated Ocean Discovery Programs	X	
DVDP	Dry Valleys Drilling Project	X	
MSSTS	McMurdo Sound Sediment and Tectonic Studies	X	
CIROS	Cenozoic Investigation in the Western Ross Sea	X	
CRP	Cape Roberts Project	X	
ANDRILL-MIS	ANtarctic DRILLing projects McMurdo Ice Shelf	X	
ANDRILL-SMS	ANtarctic DRILLing projects Southern McMurdo Sound	X	
TRACERS	TephRochronology and mArker events for the CorrElation of natural archives in the Ross Sea, Antarctica		X

Table 2) (Ford 1975; Hayes et al. 1975). The oldest volcanic rocks were recovered at the southern edge of the Naturaliste Plateau and SE Indian Ridge. These were dated according to sediment biostratigraphy and are Late Cretaceous (Santonian or older)–Early Miocene in age (Hayes et al. 1975). Conversely, Quaternary–Early Oligocene-age volcanic rock detritus has been recovered offshore of Wilkes Land and in the Ross Sea (Hayes et al. 1975).

There have also been several ODP expeditions to Antarctica, including to the Weddell Sea (Leg 113, sites 689–697: Barker et al. 1988, 1990) (Fig. 2b; Table 2) and the sub-Antarctic South Atlantic (Leg 114, sites 698–701: Ciesielski et al. 1991) (Table 2). The aim at both legs was to reconstruct the tectonic and environmental history of Antarctica and the adjacent ocean from the Early Cretaceous through to the Pleistocene. Abundant volcanic material was found interbedded with sediments at most of the drill sites from both legs. It consisted of dispersed volcanic glass shards, ash layers and pumice layers (both altered and unaltered: Hubberten et al. 1991). Basalt lavas were also encountered at one site (Leg 113, Site 690). The volcanic ash showed a wide range of compositions, ranging from basalts to rhyolites and are Quaternary (Leg 114) to Pliocene (legs 113 and 114) and Miocene (Leg 114) in age (Hubberten et al. 1991). According to their potassium/silica ratios, they are classified as island-arc tholeiitic, calc-alkaline and high-potassium calc-alkaline rock series. Twenty-one ash beds from Leg 113, sites 695–697, were sampled close to the South Orkney Islands (Fig. 2b). They are Pliocene in age and calc-alkaline (basaltic andesite–rhyolite) in composition, and have compositions characteristic of subduction-related magmas. Although they were ascribed to volcanic sources in the Antarctic Peninsula (Hubberten et al. 1991), no suitable similar age sources are known (J.L. Smellie pers. comm.).

Conversely, the compositionally variable tephras found in Leg 114 (Site 701) formed three distinct magmatic series. The oldest tephra layers are Early Miocene in age, have a calc-alkaline affinity and were related to the South Shetland Islands arc volcanic activity (Hubberten et al. 1991). Again, however, there are no known similar age sources in the South Shetland Islands (or Antarctic Peninsula: J.L. Smellie pers. comm.). Alternatively, Hubberten et al. (1991) suggested the now-extinct Discovery Arc or Jane Bank, ancestors of the South Sandwich Islands arc, as alternative sources, which are situated much closer to the drill sites.

The Pliocene ash layers can be split into different geochemical fields, possibly indicating different sources. Early and late Pliocene ash with a calc-alkaline affinity is likely to have been derived from the South Shetland Islands, whereas Quaternary layers, with lower potassium contents and displaying island-arc tholeiite affinity, can possibly be correlated to the South Sandwich Islands (Hubberten et al. 1991). Again, there are no known Pliocene calc-alkaline sources in the South Shetland Islands (J.L. Smellie pers. comm.).

Leg 119 was drilled in Prydz Bay (Fig. 2c; Table 2); its location and formation are important for the reconstruction of glaciation in East Antarctica, which in turn is critical to understanding Cenozoic climate and oceanographic development (Barron et al. 1989; O'Brien et al. 2001). Volcanic material recovered consisted of volcanic glass and magmatic minerals (magnetite, hedenbergite, plagioclase and alkali feldspar, apatite with minor biotite, olivine, and augite) mixed with pelagic biogenic sediments. Several discrete volcanic ash layers and dispersed ash were identified in Oligocene–Pleistocene sediments from sites 736, 737, 745 and 746. Analysis of the glass compositions indicated two dominant petrographical rock series: (i) transitional to alkali basalts; and (ii) trachytes with subordinate alkali rhyolites and rhyolites (Morche et al. 1991). Minor dispersed ash of calc-alkaline composition was also identified. The compositions are consistent with derivation from sources in the Kerguelen Islands, Heard Island (Fig. 2c) and possibly other Antarctic locations (e.g. older volcanic structures, eroded or subsided). Dispersed ash of calc-alkaline composition was attributed to sea-ice rafting and is a likely derivation of the South Sandwich island arc (Morche et al. 1991).

Leg 178 (sites 1095–1103: Fig. 2a; Tables 2 and 3) was focused offshore of the Antarctic Peninsula (Barker et al. 1999, 2002) with the objective of recovering high-resolution palaeoclimatic and glacial history records for the past 10 myr, and with a special focus on extracting an ultrahigh-resolution record for the Holocene palaeoclimate and glacial history. Volcanic material recovered consisted of glass shards with a bimodal rhyolite and basalt composition, and ice-rafted debris or clasts in diamictite consisting of basalt, andesite and rhyolite. A 3 cm-thick, possibly primary, fallout ash bed showing normal grading and a sharp basal contact was identified (Barker et al. 1999). This was correlated with products from Deception Island (Hillenbrand et al. 2008). The samples ranged in age from Late Cretaceous to Quaternary (Di Vincenzo et al. 2001).

Leg 183 (Sites 1135–42) was conducted in the Kerguelen Plateau–Broken Ridge area (Fig. 2c; Table 2). The main scientific objectives were to define the chronology and petrogenesis of Kerguelen Plateau–Broken Ridge magmatism and the tectonic history. Volcanic material was obtained from several of the drill sites. At Site 1136, basaltic glass formed 10–15% of the sand fraction; three tholeiitic basalt lavas were found at the base of the core which had an age older than 107–105 Ma. Several basaltic lava flows with interbedded sedimentary and volcaniclastic rock layers were recovered from the lower 147.2 m of the hole at Site 1137. Volcanic grains are dispersed throughout the core. Volcanic material is widespread at Site 1138 and varies from disseminated felsic glass and pumice in sediments with ages of <c. 6.4 Ma, to well-defined felsic tephra with an approximate age of c. 3.8–3.2 Ma, to a suite of tephra with bimodal compositions (basalt and trachyte) and ages of c. 10.6–8.9 Ma. At the base of the hole, a thick succession of basalt tephra was found. A somewhat uniform sequence of tholeiitic to transitional thin basaltic

Fig. 2. Map of the Southern Ocean with DSDP (red dots), ODP (blue dots) and IODP (yellow dots) drill sites shown. See Table 1 for acronyms. Source: IODP (https://iodp.tamu.edu/scienceops/maps.html).

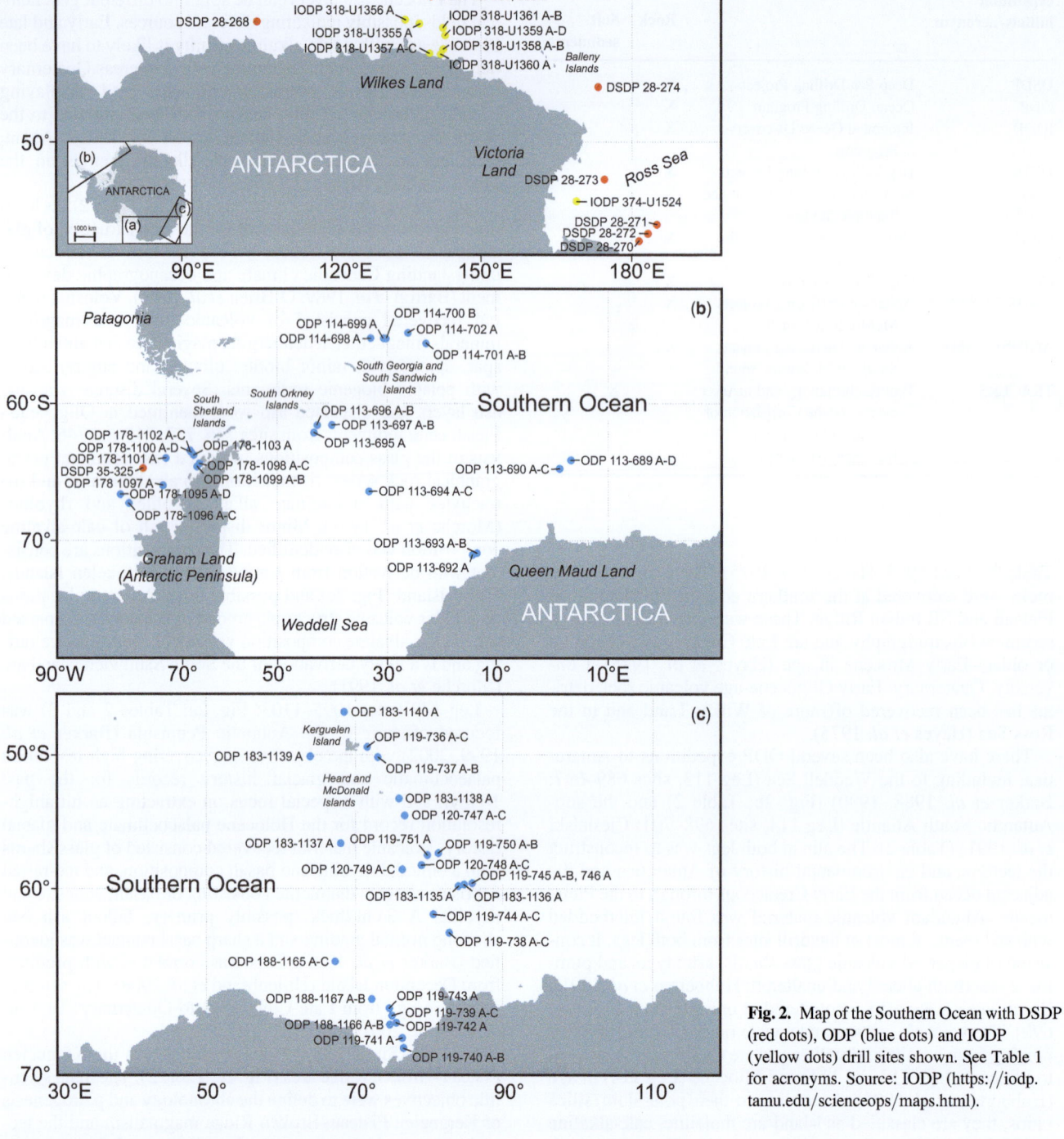

lavas, ranging in thickness from 0.7 to 9.6 m, forms the local basement (Coffin *et al.* 2000).

At Leg 183 drill sites 1139–1140 basaltic and felsic ash occurs as disseminated shards and discrete felsic tephra layers along with variably altered basalt lavas and pillow basalts.

Finally, the IODP performed legs 318, 374 and 379 in Antarctica. Leg 318 was focused on Cenozoic East Antarctic ice-sheet evolution offshore of Wilkes Land (Fig. 2a) but no significant volcanic deposits were identified during the expedition. Vice versa, legs 374 and 379 were carried out in the Ross and Amundsen Seas, respectively, and were focused on the reconstruction of the ice-sheet history and dynamics of those regions. In both legs distinctive primary tephra layers and dispersed volcaniclastic materials were identified (McKay *et al.* 2019).

Tephras recovered during ice-based and land-based drilling programmes: DVDP 15, MSSTS-1, CIROS-1 and CIROS-2, CRP-1, CRP-2/2A and CRP-3, and ANDRILL-MIS and ANDRILL-SMS

Several international cooperative drilling programmes have been designed to recover continuous drill cores from sediments on and close inshore of Antarctica (Fig. 3; Table 2). These programmes were aimed at deriving information on

Table 2. *Details of drilling sites with position, penetration, recovery, type and abundance of volcanic material recovered and reference literature*

Drill project	Leg	Site	Area	Latitude	Longitude	Penetration (m)	Recovery (m)	Volcanic deposits type	Reference*
DSDP	28	265	SE Indian Ridge	53° 32.45' S	109° 56.74' E	462	108	Reworked volcanic materials and of glass; tholeiitic basalt lava flows	1, 2
DSDP	28	266	SE Indian Ridge	56° 24.13' S	110° 06.70' E	384	145.2	Volcanic glass in trace amounts; tholeiitic basalt lava flows	1, 2
DSDP	28	267 (A, B)	Balleny Basin	59° 15.74' S	104° 29.30' E	219.3; 70.5; 323	25.9; 11.6; 53.5	Tholeiitic basalt lava flows	1, 2
DSDP	28	268	Wilkes Land	63° 56.99' S	105° 09.34' E	474.5	65.6	Volcanic pebbles	2
DSDP	28	274	Ross Sea	68° 59.81' S	173° 25.64' E	421	279.1	Traces of volcanic glass; tholeiitic basalt lava flows	1, 2
ODP	113	689 (A, B, C, D)	Weddell Sea	64° 31.01' S	03° 06.00' E	11.8; 297.3; 27.6; 133.8	9.35; 229.44; 20.54; 115.7	Dispersed pyroclastic material and volcanic ash-bearing layers (up to 40 cm thick)	3
ODP	113	690 (A,B,C)	Weddell Sea	65° 9.63' S	1° 12.30' E	9.86; 213.4; 321.2	9.86; 214.59; 185.6	Dispersed volcanic glass and distinct tephra layers. Basalt lava flow	3
ODP	113	691 (A,B,C)	Weddell Sea	70° 44.54' S	13° 48.66' W	0.05; 1.7; 12.7	0.05; 0; 0	Dispersed pyroclastic material	3
ODP	113	692 (A,B)	Weddell Sea	70° 43.48' S	13° 49.21' W	6.7; 97.2	0.65; 29.3	Extensive volcanic layers and lenses (devitrified)	3
ODP	113	693 (A,B)	Weddell Sea	70° 49.89' S	14° 34.41' W	483.9; 167.4	213.5 and 92.2	Layers of volcanic ash/pumice	3
ODP	113	694 (A,B,C)	Weddell Sea	66° 50.83' S	33° 26.79' W	9.8; 179.2; 391.2	9.85; 67.8; 71.7	Dispersed pyroclastic material and volcanic rocks	3
ODP	113	695	Weddell Sea	62° 23.48' S	43° 27.09' W	341.1	254.4	Abundant devitrified volcanic ash layers	3
ODP	113	696 (A,B)	Weddell Sea	61° 50.94' S	42° 55.98' W	103; 645.6	58.3; 156.69	Devitrified volcanic ash and glass	3
ODP	113	697 (A, B)	Weddell Sea	61° 48.63' S	40° 17.27' W	20.9; 322.9	26.6; 188.27	Multiple thin, altered, volcanic ash layers	3
ODP	114	698	South Orkney and South Scotia Ridge	51° 27.51' S	33° 05.96' W	237	52.26	Ash layers	3
ODP	114	699	Subantarctic South Atlantic	51° 32.54' S	30° 40.62' W	518.1	356.52	Dispersed pyroclastic material (basaltic andesite and andesite pyroclastic material, dacitic and rhyolitic)	3
ODP	114	701 (A,B,C)	Subantarctic South Atlantic	51° 59.07' S	23° 12.73' W	74.8	69.66	Multiple thin layers of green, altered ash and dispersed ash	3
ODP	119	736	Kerguelen–Heard Plateau	49° 24.12' S	71° 39.61' E	252.3	147.05	Beds of volcanic ash and lapilli tuffs few centimetres thick; volcanic detritus abundant	4
ODP	119	737 (A, B)	Kerguelen–Heard Plateau	50° 13.67' S	73° 01.97' E	273.2; 715.2	181.64; 298.35	Discrete volcanic ash layers and dispersed ash as volcanic detritus	4
ODP	178	1095 (A, B, C, D)	Antarctic Peninsula	66° 59.13' S	78° 29.24' W	87.30; 570.20; 2.90; 84.60	86.46; 385.75; 2.87; 78.96	Volcanic glass (rhyolite and basalt), volcaniclastic materials and intrusive igneous as ice-rafted debris	5, 6
ODP	178	1096 (A)	Antarctic Peninsula	67° 34.01' S	76° 57.79' W	140.7	118.52	Volcanic glass (rhyolite and basalt), volcaniclastic materials and intrusive igneous as ice-rafted debris	5, 6
ODP	178	1101	Antarctic Peninsula	64° 22.33' S	70° 15.67' W	217.7	215.75	A 3 cm-thick graded bed of volcanic ash with a sharp basal contact	5, 6
ODP	183	1136	Kerguelen Plateau–Broken Ridge	59° 39.10' S	84° 50.09' E	161.4	45	Zeolitic calcareous volcanic clayey sand and fine-grained volcanic sediments; three basalts flows from the lower 30 m of the hole	7

(Continued)

Table 2. Continued.

Drill project	Leg	Site	Area	Latitude	Longitude	Penetration (m)	Recovery (m)	Volcanic deposits type	Reference*
ODP	183	1137	Kerguelen Plateau–Broken Ridge	56° 49.98′ S	68° 5.61′ E	371.2	219.38	Crystal-lithic volcanic siltstone and sandstone, lithic volcanic conglomerate, crystal-vitric tuff and several basaltic lava flows	7
ODP	183	1138	Kerguelen Plateau–Broken Ridge	53° 33.10′ S	75° 58.49′ E	842.7	411.98	Disseminated volcanic ash and pumice lapilli; Discrete primary tephra layers, pumice lithic breccia	7
ODP	183	1139	Kerguelen Plateau–Broken Ridge	50° 11.10′ S	63° 56.19′ E	694.2	356.88	Disseminated ash and a discrete tephra horizon. Altered basalt flows	7
ODP	183	1140	Kerguelen Plateau–Broken Ridge	46° 16.60′ S	68° 29.50′ E	321.9	144.86	A volcanic ash layer, with vesicular basaltic glass shards; scattered sand-sized particles of volcanic material disseminated. Pillow basalts	7
ODP	183	1141	Kerguelen Plateau–Broken Ridge	32° 13.60′ S	97° 7.69′ E	185.6	96.79	Altered and relatively fresh basalts were recovered from the lower 72.1 m of the hole	7
DVDP	15	15	Hut Peninsula, Ross Island	77° 26.24′ S	164° 22.82′ E	64.6	31.8	Dispersed glassy and crystal-rich basaltic fragments	8
MSSTS	1	1	Western McMurdo Sound	77° 33.43′ S	164° 23.22′ E	230	128.75	Abundant volcanoclastic material (<60%) in the upper 60 m and in the lowermost 25 m	9, 10, 11
CIROS	1	1	Butter Point, Ferrar Fjord	77° 34′ 92″ S	164° 29′ 92″ E	702	687	Alkali-rich volcanic rocks of the McMurdo Volcanic Group	12, 13
CIROS	2	2	Western McMurdo Sound, Ferrar Fjord	77° 41.00′ S	163° 32.00′ E	168	117	Volcanic detritus abundant. Tephra layer 30 cm-thick ash interbedded within a stratified diamictite	12, 13
CRP	1	1	Western McMurdo Sound	77° 0.48′ S	163° 46.50′ E	147.69	43.55	Heterogeneous volcanic material from the McMurdo volcanic system (from alkali basalt to trachyte) and the Ferrar Dolerite	14
CRP	2	2	Western McMurdo Sound	77° 0.36′ S	163° 43.14′ E		625	Volcanic detritus throughout the core. Several tephra layers of nearly pure pumice lapilli and ash layers; diffuse alignment of dispersed pumice fragments and glass shards	15
CRP	3	3	Western McMurdo Sound	77° 0.06′ S	163° 38.40′ E	968	939	Volcanics geochemically comparable to Ferrar Supergroup rocks.	16
ANDRILL	1B	MIS	Ross Embayment	77° 53.37′ S	167° 5.36′ E		1284.87	Volcanic material throughout the core. 175 m sequence of volcanic-rich sediments and up to several-decimetre-thick lapilli tuff beds, interbedded with volcanic breccia and a 2.81 m-thick subaqueously emplaced lava flow	17, 18

(Continued)

Table 2. Continued.

Drill project	Leg	Site	Area	Latitude	Longitude	Penetration (m)	Recovery (m)	Volcanic deposits type	Reference*
ANDRILL	2A	SMS	Ross Embayment	77° 45.49' S	165° 16.61' E		1137.84	Topmost 37 m of weakly reworked tephra beds, lava breccias, and ripple cross-laminated volcanic sands. In the second half one pyroclastic fall deposit, several slightly reworked pyroclastic fall deposits and a number of volcanogenic sedimentary deposits occur. Volcanic detritus throughout the core	19, 20

References: 1, Ford (1975); 2, Hayes et al. (1975); 3, Hubberten et al. (1991); 4, Morche et al. (1991); 5, Barker et al. (1999); 6, Barker et al. (2002); 7, Coffin et al. (2000); 8, Barrett and Treves (1981); 9, Barrett (1986); 10, Barrett and McKelvey (1986); 11, Kyle and Muncy (1989); 12, Barrett et al. (1989); 13, Barrett et al. (1992); 14, Armienti et al. (1998); 15, Cape Roberts Science Team (1999); 16, Barrett et al. (2000); 17, Pompilio et al. (2007); 18, Di Roberto et al. (2010); 19, Del Carlo et al. (2009); 20, Di Roberto et al. (2012).

past climate variations and the response of Antarctic ice sheets to projected greenhouse warming, as well as at constraining the geological history of the Antarctic region. In time order, the main projects are DVDP 15, MSSTS-1, CIROS-1 and CIROS-2, CRP-1, CRP-2 and CRP-3, and ANDRILL, comprising the McMurdo Ice Shelf (MIS) and Southern McMurdo Sound (SMS) projects (Tables 1 and 2; Fig. 3). All recovered abundant volcanic material was either in the form of dispersed clasts in sediments or primary deposits.

DVDP 15

The first drill hole (DVDP 15) into the seafloor of McMurdo Sound was attempted in 1975 as part of the Dry Valley Drilling Programme (Barrett et al. 1976). The drilling site was located c. 16 km east of Marble Point on seasonal sea ice (2 m thick) and a water column of 122 m (Fig. 4). This drilling established the feasibility of drilling through seasonal ice, penetrating c. 65 m below seafloor (bsf) and recovering, for the first time, c. 34 m of sediments ranging in age from 50 to 10 Ma (Hayes and Frakes 1975). Volcanic material recovered in the DVDP 15 core consisted of glassy and crystal-rich basaltic fragments, which were attributed to the late Cenozoic McMurdo Volcanic Group outcrops located south of the drill site (Barrett and Treves 1981; see also Chapter 5.2a (Smellie and Martin 2021)).

MSSTS-1

The MSSTS-1 drill hole (227 m) was drilled in November 1979 through a sediment sequence composed mostly of marine glacial sediments and ranging in age from Quaternary to Late Oligocene (Barrett 1986; Harwood et al. 1989). The site was located c. 12 km NE of Butter Point (Barrett 1986) in 195 m of water (Fig. 4). Abundant basaltic volcaniclastic material (<60%) was found in the upper 60 m and lowermost 25 m (Barrett 1986; Barrett and McKelvey 1986; Kyle and Muncy 1989). The age of the drilled sequence was assigned to the Early Miocene–Quaternary and Late Oligocene, respectively, indicating that volcanoes were active in the McMurdo region during those times (Barrett 1986; Kyle and Muncy 1989). Some of the volcanism identified in MMSTS-1 is similar in age to that identified in CIROS-1 (see below).

CIROS-1 and CIROS-2

The CIROS drilling project was conducted to obtain a record of the early history of the Antarctic Ice Sheet and the rise of the

Table 3. Details of piston and gravity cores with sampling area, time interval, volcanic material recovered and reference literature

Project	Site	Time interval	Volcanic deposit type	Reference*	Notes
	Antarctic Peninsula (Bellingshausen and Amundsen seas)	Marine isotope stages 5 and 6	Tepha layers	1	Cores recovered during various British, Italian, German and US cruises including the ODP Leg 178
Tephra Ross TRACERS	Ross Sea (Northern Victoria Land)	137.1–12 ka	Tephra layers	2	Cores recovered during Italian cruises
	Ross Sea (Northern Victoria Land)		Tephra layers	3	Cores recovered during 2016–17 Italian cruise
	Scotia Sea (Southern Ocean)	<274.4 ± 1.5	Tephra layers	4	Sediment cores from R/V Polarstern cruise ANT-XXII/4, ANT-X/5 and ANT-XI/2.
	South Shetland Islands and South Orkney Islands	Marine isotope stages 5 and 6	Tephra layers	5	Tephra analysis of sediments from Midge Lake (South Shetland Islands) and Sombre Lake (South Orkney Islands), Antarctica

References: 1, Hillenbrand et al. (2008); 2, Del Carlo et al. (2015); 3, Di Roberto et al. (2018); 4, Xiao et al. (2016); 5, Hodgson et al. (1998).

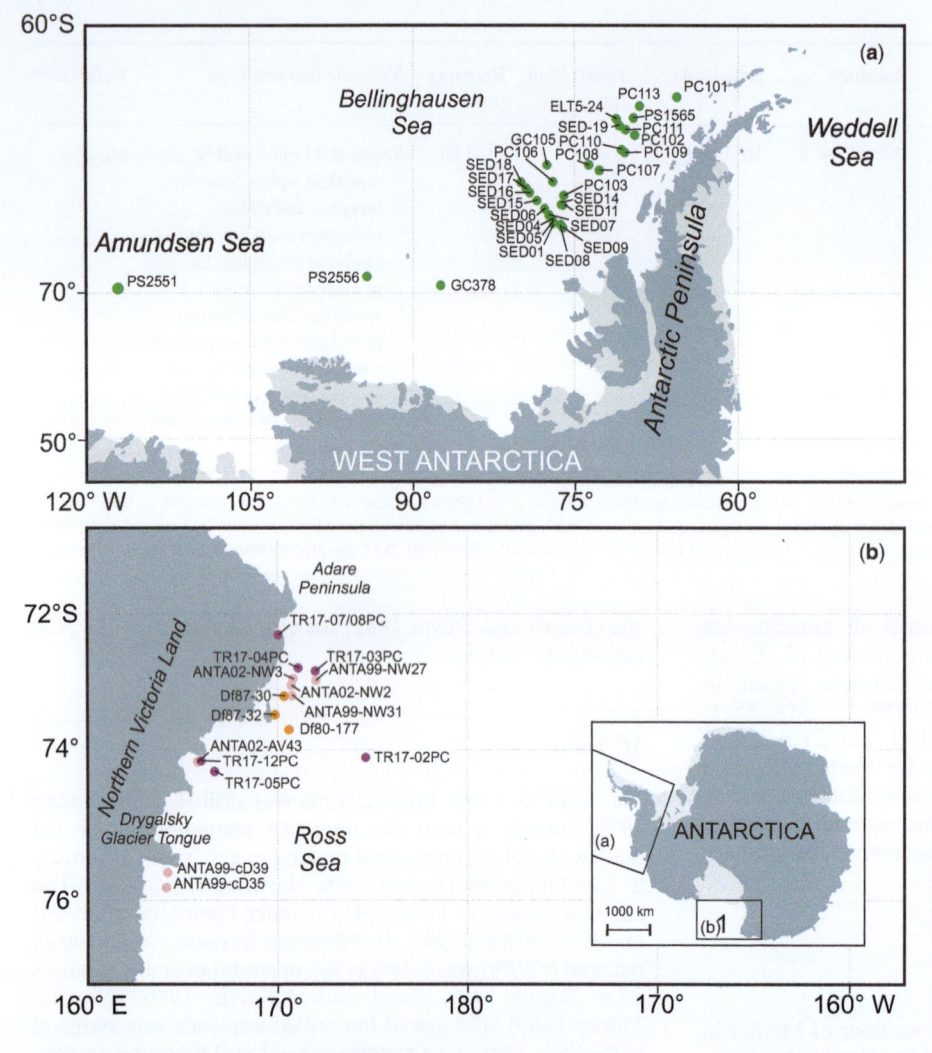

Fig. 3. Maps showing the locations of shallow gravity-piston cores sites where tephra layers have been reported in the Antarctic Peninsula (green dots: Hillenbrand *et al.* 2008) and in the Ross Sea (orange dots: Amrisar *et al.* 1988; Licht *et al.* 1999; pink dots: Colizza *et al.* 2003; Del Carlo *et al.* 2015; magenta dots: Di Roberto *et al.* 2018).

Transantarctic Mountains (Barrett 1989). Two cores were obtained: CIROS-2 in 1984; and CIROS-1 between 1986 and 1987 (Fig. 4).

CIROS-2 was drilled near the axis of the Ferrar Fjord trough in 211 m of water (Barrett 1989; Barrett *et al.* 1989). The core penetrated 168 m of Early Pliocene–Quaternary glaciomarine sediment and reached basement gneiss. Volcanic detritus was relatively abundant throughout the succession, especially as sand-sized grains. It was interpreted as having been derived from the McMurdo Volcanic Group, glacially transported and deposited on the walls of Ferrar Valley and finally blown by wind offshore (Barrett 1989; Barrett *et al.* 1989). A 30 cm-thick ash bed was found interbedded within a stratified diamictite at 125 m bsf. The volcanic ash consisted of abundant pristine shards and micropumices with three distinct compositions: dark-brown phonotephrite, light-brown mugearite and clear benmoreite (Barrett *et al.* 1992). The vitric fraction was found mixed with volcanic minerals (feldspar) and minor detrital, non-volcanic, clasts (<30%). Feldspar crystals provided $^{40}Ar/^{39}Ar$ ages of 2.77 ± 0.03 and 2.91 ± 0.11 Ma, and volcanic glass K–Ar ages of 3.36 ± 0.17 Ma (Barrett *et al.* 1992). 2.9 Ma was considered to be the average eruptive age for the tephra, which may have been derived either from multiple volcanic eruptions or, alternatively, from a single eruption with a compositionally zoned magma chamber.

CIROS-1 was located 12 km off Butter Point in western McMurdo Sound 4 km SE of MSSTS-1 (Barrett *et al.* 1989) (Fig. 5). The CIROS-1 core reached 702 m bsf and had exceptional recovery (*c.* 98%), recovering glaciomarine deposits dating back to the early Oligocene (Barrett *et al.* 1989). In the CIROS-1 drill hole alkali-rich volcanic rocks of the McMurdo Volcanic Group occur throughout the sequence testifying to a continuous influx of volcanic detritus. Volcanic debris is relatively more abundant in the uppermost 300 m of CIROS-1 sequence (Barrett *et al.* 1989). This sequence correlates quite well with the base of MSSTS-1 (Barrett 1987) and is probably around 30 myr old (George 1989).

CRP-1, CRP-2 and CRP-3

The first CRP drill hole (CRP-1) was positioned in 1997 on a seasonal sea-ice platform, about 16 km off Cape Roberts (Fig. 3) (Barrett 1998*a*). It reached a depth of 148 m bsf, and consisted of glacial sediment of Early Miocene and Quaternary age (Barrett 1998*a, b*). The oldest sediment recovered at the bottom of the CRP-1 core is a mudstone dated at 24–22 Ma (early Miocene: Barrett 1998*b*). Analysis of the sand layers widespread throughout the CRP-1 core (Armienti *et al.* 1998) revealed a significant input of compositional heterogeneous volcanic material from the McMurdo Volcanic Group, consisting of non-altered volcanic glasses with alkali basalt to trachyte composition, magmatic minerals and Ferrar dolerite fragments. At 116.55 m bsf, corresponding to an age of 18.4 ± 1.2 Ma (McIntosh 1998), a thin, primary, pumice-rich tephra layer was found (Armienti *et al.* 1998) that was likely to have issued from an explosive, possibly hydromagmatic, eruption (Smellie 1998). This had a peralkaline trachytic composition (Armienti *et al.* 1998; Smellie 1998), which matches somewhat the composition of evolved products from the Erebus and Melbourne Volcanic Province

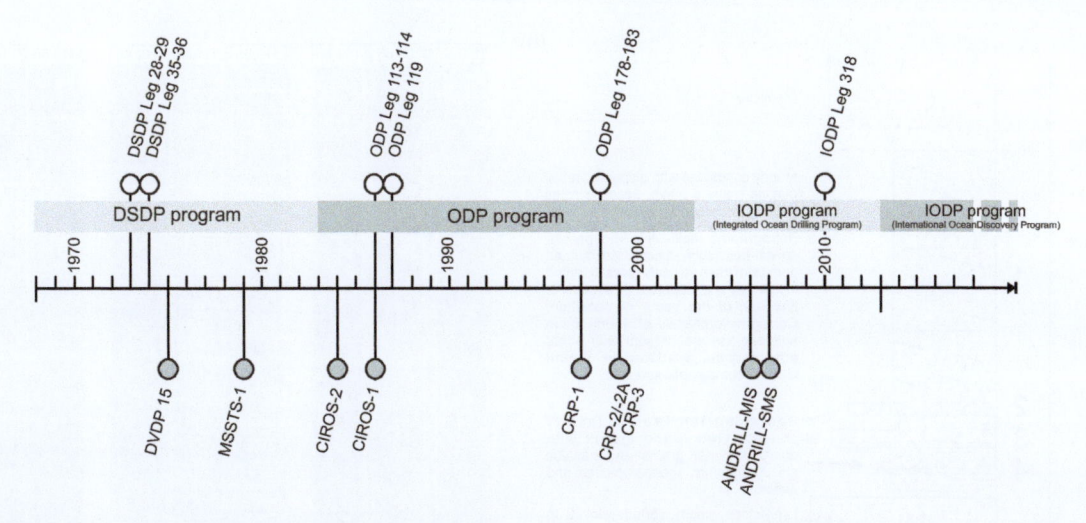

Fig. 4. Time line of the Antarctic drilling campaigns that recovered marine tephra. In the upper part of the figure, ocean drilling expeditions based on a research vessel are shown; while in the lower part the drilling campaigns that took place using an ice platform (sea ice or the Ross Ice Shelf).

products (Armienti *et al.* 1998) but a precise attribution of the source of the ash layers has been uncertain. Concerning the provenance of the volcanic detritus in CRP-1, Smellie (1998) further proposed Mount Morning as the possible source and this attribution was later confirmed by Martin (2009) and Martin *et al.* (2010).

CRP-2A, situated close to CRP-1 (Fig. 5), penetrated down to *c.* 624 m bsf (Cape Roberts Science Team 1999). The upper 27 m of CRP-2A consists of unconsolidated sediment of Pliocene and Quaternary age clearly distinguishable from the underlying sediments that span the early Miocene–early Oligocene (31–19 Ma: Cape Roberts Science Team 1999). Volcanic detritus in CRP-2 and CRP-2A core is abundant throughout and not dissimilar (but less abundant) from that found in CRP-1 core. It consists of clasts deriving from the basement of the Ferrar dolerites and Kirkpatrick basalts and alkaline volcanic rocks from the McMurdo Volcanic Group, including glass and volcanic minerals (Cape Roberts Science Team 1999; Smellie 2000). Detritus influx, at least in the top *c.* 300 m of the core, was attributed to the activity of the McMurdo Volcanic Group (MVG) which would have commenced at *c.* 25 Ma (Smellie 2000). Tephra layers in CRP-2A are quite abundant. Seven main layers were found between 108 and 114.2 m bsf (Cape Roberts Science Team 1999; Smellie 2000) consisting of several-centimetre-thick nearly pure pumice lapilli and ash layers, and a diffuse alignment of dispersed pumice fragments and glass shards (Cape Roberts Science Team 1999; McIntosh 2000). The scarcity of detrital, non-volcanic clasts within these tephra beds suggested a primary origin without significant reworking or resedimentation (Cape Roberts Science Team 1999; McIntosh 2000). The most prominent, up to 1.2 m-thick, tephra layers were found at 111.56–112.78 and 113.86–114.21 m bsf, and their age was determined as 21.44 ± 0.05 Ma (weighted mean age $\pm 2\sigma$: McIntosh 2000). Armienti *et al.* (2001) paid special attention to the tephra layers and to the 1.2 m-thick layer (111.56–112.78 m bsf) for which an evaluation of the height of the associated eruptive column was performed (Fig. 6). Using a model of advection and diffusion of tephra (Armienti *et al.* 1988), and considering in the model two typical size of particles, the height of the eruptive column responsible for the deposition of the tephra was calculated to be less than 17 km; a local source was suggested, possibly within 60 km of the drilling site (Armienti *et al.* 2001). Slightly reworked, less concentrated pumice layers were also found at 193.4 and 280.0 m bsf. These consist of quite rounded pumices with anorthoclase phenocrysts (McIntosh 2000). Although clearly reworked, these layers were interpreted as having been deposited soon after their eruption and thus were considered useful for determining the approximate age of deposition (McIntosh 2000). They yielded ages of 23.98 ± 0.13 and 24.22 ± 0.03 Ma, respectively (McIntosh 2000).

CRP-3 is the deepest stratigraphic drill hole of the CRP project and penetrated approximately 939 m of sediment of early Oligocene (possibly latest Eocene) age (Powell *et al.* 2001) with an exceptional recovery of >97%. Igneous materials form a large fraction of the sediments in CRP-3 core, including a small magmatic body found in the Beacon sandstone of Devonian age (Pompilio *et al.* 2001; Smellie 2001). Data on igneous detritus in CRP-3 point to a provenance from Jurassic

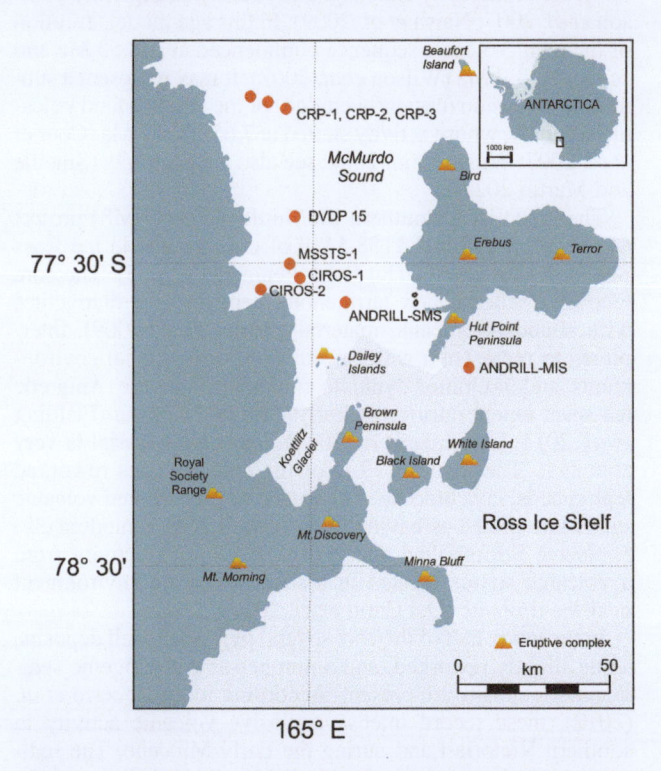

Fig. 5. Map showing the location of the main geographical features in the McMurdo Sound region (southwestern Ross Sea) (the inset shows the location of McMurdo Sound in the map of Antarctica). The positions of DVDP 15, CIROS-1 and CIROS-2, MSSTS-1, CRP-1, CRP-2 and CRP-3, and ANDRILL-MIS and ANDRILL-SMS are shown along with the principal volcanic centres.

640   A. Di Roberto et al.

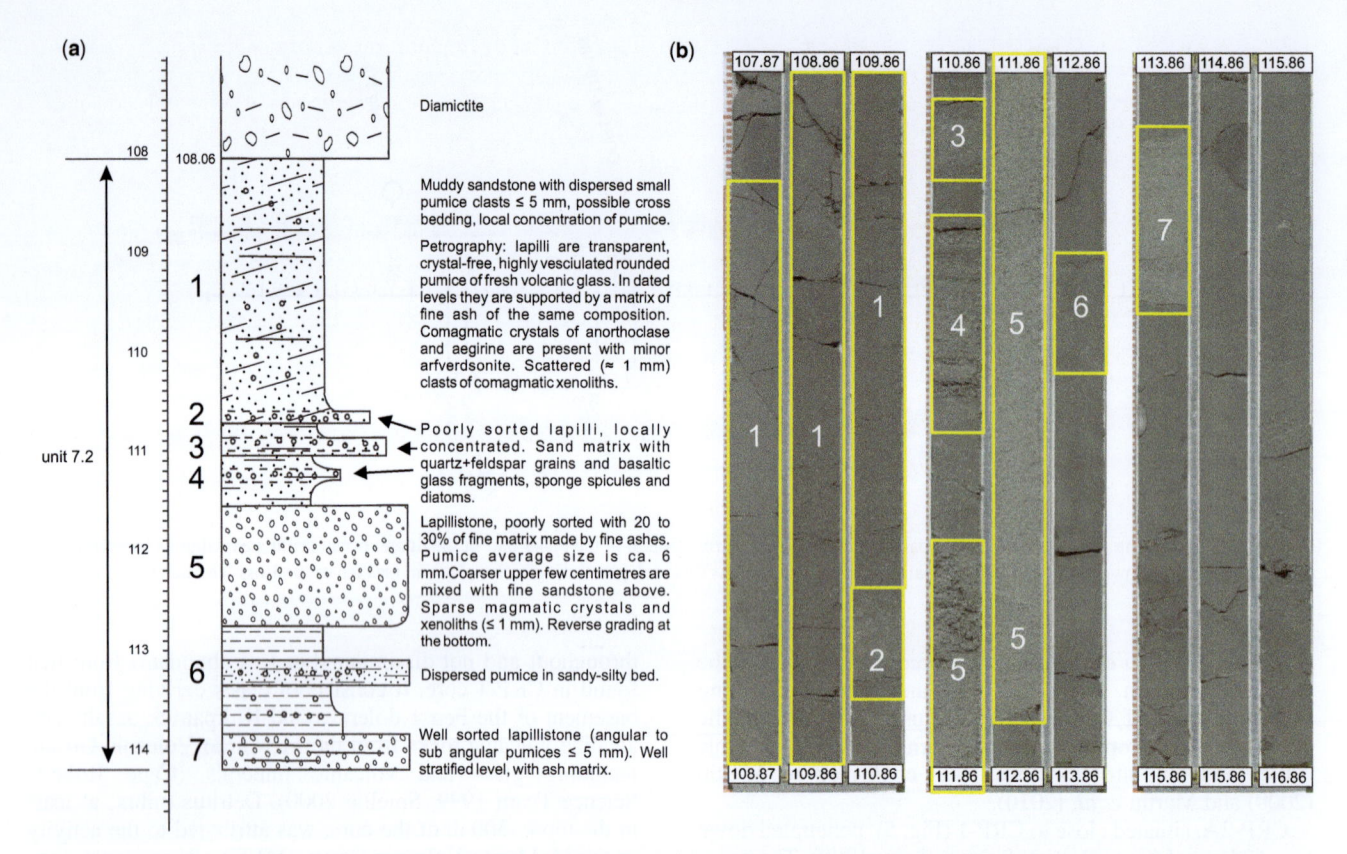

Fig. 6. Representative example of a sedimentary log with several tephra layers present: Unit 7.2 (107.87 and 116.86 m bsf) of the CRP-2 core modified from Armienti et al. (2001) and Fielding and CRP Science Team (2001). Photographs of the logged core are also shown, with the main numbered units identified.

rocks of the Ferrar Supergroup succession and possibly from the underlying Hanson Formation in the central Transantarctic Mountains (Pompilio et al. 2001).

ANDRILL-MIS (AND-1B) and -SMS (AND-2A)

The ANDRILL McMurdo Ice Shelf (MIS) project (AND-1B) recovered a 1285 m-long core consisting of stacked, cyclic glaciomarine sediments interbedded with abundant volcanic deposits (Krissek et al. 2007; Pompilio et al. 2007). This is the longest sedimentary rock core taken from beneath the seafloor under Antarctica's Ross Ice Shelf. The coring site was located on the Ross Ice Shelf about 10 km east of Hut Point Peninsula, Ross Island (Fig. 5). Volcanic material in the AND-1B core consists of dispersed clasts (c. 70% of the total clast count >2 mm are volcanic: Pompilio et al. 2007) and abundant volcanic layers, which occur throughout the core and have thicknesses ranging from millimetres to 175 m (Di Roberto et al. 2010). The thickest and most continuous volcanic sequence occurs between 584.19 and 759.32 m bsf: that is, in a cumulative thickness of c. 175 m (Krissek et al. 2007; Pompilio et al. 2007; Di Roberto et al. 2010). The sequence consists of two main subsequences. The first sequence consists of stacked volcanic-rich mudstone and sandstone, interpreted as having been emplaced by turbidity currents and indicating the abundant influx of volcanic material from active volcanic sources located far from the drill site (Di Roberto et al. 2010). The second, uppermost, subsequence is c. 105 m thick and consists of up to several-decimetre-thick lapilli tuff beds, interbedded with volcanic breccia and a 2.81 m-thick subaqueously emplaced lava (at 648 m bsf) (Fig. 7). This second subsequence was interpreted as being linked to cycles of submarine to emergent volcanic activity that occurred proximal to the drill site (Di Roberto et al. 2010).

A well-constrained chronology has been developed for the AND-1B core from a combination of $^{40}Ar/^{39}Ar$ ages, microfossil biostratigraphy and magnetic polarity stratigraphy (Wilson et al. 2007; Naish et al. 2009). In this age model, eruption of the main volcanic sequence commenced at c. 8.53 Ma and ceased by 4.9 Ma (Wilson et al. 2007). It may represent a submarine extension (currently eroded) of the White Island volcanic complex, whose activity started at 7.65 ± 0.69 Ma (Cooper et al. 2007; Wilson et al. 2007; see also Chapter 5.2a (Smellie and Martin 2021)).

The ANDRILL Southern McMurdo Sound (SMS) project (AND-2A) recovered 1138.54 m of core located in the Ross Sea, c. 50 km NW of Hut Point Peninsula (Fig. 5). The core recovered interbedded terrigenous deposits and diamictites with abundant volcanic material (Panter et al. 2009), interpreted to represent a wide spectrum of depositional environments and multiple, dynamic fluctuations in the Antarctic ice-sheet extent during the Early–Middle Miocene (Fielding et al. 2011; Passchier et al. 2011). Volcanic material is very abundant. The topmost 37 m of the core includes reworked tephra beds, lava breccias and ripple cross-laminated volcanic sands interpreted as having been deposited by a moderately-explosive Strombolian and Hawaiian eruptive activity from a volcanic source located in a shallow-marine environment near the drill site (Del Carlo et al. 2009).

In the lower half of the core several pyroclastic fall deposits, some slightly reworked, and a number of volcanogenic sedimentary deposits are present. According to Di Roberto et al. (2012), these record intense explosive volcanic activity in southern Victoria Land during the Early Miocene. The sedimentological, petrological and geochemical analyses of the deposits provided data about their volcanic sources and eruptions styles, and new insights into their environment of deposition. Di Roberto et al. (2012) recognized two main eruptive styles: moderately-explosive Strombolian–Hawaiian eruptions; and more energetic explosive sub-Plinian and Plinian

Fig. 7. Two examples of tephra occurring in marine cores from the Ross Sea: (a) coarse-ash- to fine-lapilli-grade basaltic tephra bed with a sharp bottom (640.13 m in ANDRILL-SMS); and (b) multiple fine lapilli and pumiceous tephra patches made up of imbricated, subrounded to well-rounded pumice (702.19 m in ANDRILL-SMS core. Photographs courtesy of P. Del Carlo).

eruptions. On the basis of their geochemical composition and age (c. 18 Ma), the proto-Mount Morning–Mount Morning complex, located c. 80 km south of the AND-2A core drill site, was suggested as a possible source for the studied products (Martin 2009; Martin et al. 2010; Di Roberto et al. 2012; see also Chapter 5.2a (Smellie and Martin 2021)). Volcanic detritus and primary tephra layers documented in the CIROS-1, MSSTS-1 and CRP drill cores also indicate volcanic activity in the Erebus Volcanic Province back to 24 Ma (Gamble et al. 1986; Barrett 1987; McIntosh 1998, 2000; Acton et al. 2008; Di Vincenzo et al. 2009).

Tephras recovered by shallow gravity and piston cores

A huge number of shallow cores have been obtained by gravity and piston coring in the Antarctic seas over the past few decades in which primary tephra layers or volcaniclastic deposits are described (Table 3; Fig. 3). Preliminary sedimentological and stratigraphic descriptions for most of these cores has been published but very few studies have attempted to characterize the volcanic materials. In most of the published studies, tephra layers were simply an accidental discovery and they were sampled only rarely. In the following, we broadly follow the geographical subdivision used in a previous review of Cenozoic Antarctic tephra record by Smellie (1999a, b), and we report on gravity and piston cores where tephra layers have been described, sampled and characterized or simply mentioned in the description of sediment sequences.

South Atlantic Ocean

Middle Quaternary–Holocene basaltic and andesitic ash layers a few millimetres to several centimetres thick were found in a set of piston cores taken in the area between the South Sandwich Islands and the Mid-Atlantic Ridge (Ninkovich et al. 1964; Federman et al. 1982; Smellie 1999a, b). According to the texture, mineralogies and glass composition, these deposits are in part primary (airborne) and in part resedimented, and are most probably derived from eruptions in the South Sandwich Islands (Ninkovich et al. 1964; Federman et al. 1982). As a whole, they record a mildly explosive volcanism for the island arc over the last 150 kyr, punctuated by at least one major eruption with basaltic–basalt andesitic composition at 30 ka which produced a visible tephra layer still visible at 700 km downwind from the source.

Bransfield Strait, Scotia Sea and Weddell Sea

Hillenbrand et al. (2008) critically revised a set of 32 sediment cores recovered from the West Antarctic continental slope and rise in the Bellingshausen and Amundsen seas (Table 3; Fig. 3b). Cores were taken during British, Italian, German and US expeditions, and all of them report the occurrence of tephra in the sediment description (Hillenbrand et al. 2008 and references therein). For 13 of 32 cores, geochemical tephra data have been published (Moreton 1999). According to the authors, the cores cover a time interval of approximately 131 kyr, spanning at least marine isotope stages 1–5. An important result of this study was the identification of up to three different macroscopic and widely traceable tephra layers (informally named by the authors as A, B and C) in sediments from most of the analysed sediment cores, with one of them (B) that is traceable in almost all of the studied cores. The geochemical composition of glass, which is peralkaline trachyte, indicates the Marie Byrd Land volcanoes as potential sources for the three tephra layers. These would have been deposited at c. 80, 126 and 132 ka, respectively. Based on a statistical analysis of the chemical composition using the coefficient of similarity (SC: Borchardt et al. 1972) and distance function (D) for concentrations (Perkins et al. 1995), the youngest marine tephra (A) was correlated with a widespread tephra erupted by Mount Berlin at 92.1 ka. This was also found in the ice-core records of EPICA Dome C (Narcisi et al. 2005, 2006) and Dome Fuji (Kohno et al. 2004). The authors also proposed the correlation of Tephra B and Tephra C with tephra erupted from Mount Berlin at c. 118 and 135 ka, respectively, but the correlations are not corroborated statistically. The latter could also correlate with a tephra erupted at c. 136 ka at Mount Berlin found in the ice record of EPICA Dome C (Narcisi et al. 2005, 2006).

Numerous Holocene tephra layers have been found in Bransfield Strait (Baker et al. 1975; Matthies et al. 1990; Björck et al. 1991; Hodgson et al. 1998, 2004; Tatur et al. 1999; Pallàs et al. 2001; Lee et al. 2007; Kraus et al. 2013; Toro et al. 2013; Martínez Cortizas et al. 2014; Liu et al. 2016) including marine sequences from Bransfield Strait (Matthies et al. 1988; Fretzdorff and Smellie 2002) and the Scotia Sea (Moreton and Smellie 1998; Xiao et al. 2016). The ash layers contain basaltic or rhyolite compositions and some layers are bimodal. The compositionally distinctive Deception Island volcano was recognized as the main volcanic source of the Quaternary tephra layers (Kraus et al. 2013; see also Chapter 7.1 (Geyer *et al.* 2021)), including those found in marine records (Moreton and Smellie 1998; Fretzdorff and Smellie 2002; Xiao *et al.* 2016) with only two exceptions. These are represented by a tephra layer on King George Island tentatively linked to Penguin Island (dated 5.5–5 ka BP: Tatur *et al.* 1999; Lee *et al.* 2007) and a relatively recent (few decimetres below the seafloor) andesite ash layer which was not sourced by any known volcano in Antarctica, southern South America or South Sandwich Islands (Fretzdorff and Smellie 2002).

Recently, Xiao *et al.* (2016) found nine tephra beds intercalated within diatomaceous sediments of the Scotia Sea but they included no geochemical data. The ash beds span the period between 274.4 and 7.8 ka (Xiao *et al.* 2016), and Deception Island and South Sandwich Islands have been tentatively suggested as the sources. The attribution to the Deception Island and South Sandwich Islands volcanic sources is based on the fact that the age of Scotia Sea tephra (built on ^{14}C ages, a correlation of magnetic susceptibility to Antarctic ice-core records and diatom stratigraphy) matches that of other tephra beds previously found in the Scotia Sea area (Moreton and Smellie 1998; Moreton 1999: Hillenbrand *et al.* 2008) and whose identifications have a geochemical basis. Similarly, possible correlations of the layers found in the Scotia Sea have also been proposed with the ash layers found in Dome Fuji (Kohno *et al.* 2004) and Vostok ice cores (Basile *et al.* 2001) but, again, these are tentative as they are solely based on age constraints.

Altered tephra layers were found in sediment sequences sampled by piston and gravity cores from the northern Weddell Sea (Barker *et al.* 1988; Pudsey 1992) but no specific studies have been carried out on these samples in terms of age, composition and provenance; thus, their significance as chronostratigraphic markers is unknown.

South Pacific Ocean and Ross Sea

Tephra layers are abundant and widespread in sediment cores of the South Pacific Ocean. Widespread volcanic ash beds were recognized far off the coast of Antarctica in deep-sea sedimentary cores from the SW and south-central Pacific (Huang *et al.* 1973, 1975; Kyle and Seward 1984; Shane and Froggatt 1992). These were geochemically analysed and attributed to several different volcanic sources. Huang *et al.* (1973, 1975) correlated the tephras with a series of volcanic eruptions which possibly occurred in the Balleny Islands during the last 2.45 myr. Later, Kyle and Seward (1984) revised the interpretation and proposed that a source in North Island, New Zealand was more likely. Froggatt *et al.* (1986) also correlated one of the thick volcanic ash layers with New Zealand-derived rhyolite, and in particular to the Mount Curl Tephra which marked the initial eruptive phases of the Whakamaru Ignimbrite, issued from Taupo volcano. However, Froggatt *et al.* (1986) also suggested that most of the ash found in the South Pacific and sub-Antarctic oceans is phonolitic in composition and thus not correlatable with New Zealand eruptions. Finally, the revision of Shane and Froggatt (1992), which was based on a wider geochemical dataset, proved that the ash layers had a composition possibly matching that of products erupted by large stratovolcanoes in Marie Byrd Land and refuting the possibility of extra-Antarctic, New Zealand-derived tephra.

Closer to the Antarctic coasts, there are numerous piston and gravity cores where primary or reworked volcanic deposits have been found. For example, along the coast of northern Victoria Land, in the Ross Sea, tephra layers and volcaniclastic deposits have been widely recognized in sediment cores taken during the USNS *Eltanin* cruises 32–52 (Frakes 1971), during the Operation Deep Freeze cruises 80–87 of USCGC *Glacier* (Amrisar *et al.* 1988) and the R/V *Nathaniel B. Palmer* crises 95–01 (Licht *et al.* 1996). In that framework, Licht *et al.* (1996, 1999) reported the occurrence of distinct tephra layers in the Operation Deep Freeze and *Eltanin* cores from the continental shelf of the Ross Sea, north of Coulman Island. Pumice and glass fragments from two of the cores were analysed for major element glass compositions using an electron microprobe. All of the studied samples were trachytic but the chemical composition of glass from each of the two cores was distinctly different (Licht *et al.* 1999), and was correlated to the Pleiades volcano and Mount Melbourne volcano, respectively. AMS radiocarbon dates on carbonate and organic matter in sediments sandwiching the tephra layers were used to constrain the age of tephra deposition to between 26 520 ± 340 and 22 190 ± 190 years BP (Licht *et al.* 1999).

Additional evidence for thick tephra layers emplaced on the shelf of the western Ross Sea was also reported by Colizza *et al.* (2003), who analysed four marine gravity cores located 50–90 km NE of Coulman Island. These tephra have trachyandesite and trachyte compositions attributed to eruptions in the Mount Melbourne Volcanic Province and possibly to The Pleiades and Mount Rittmann volcanoes. The depositional age of the upper tephra layers (22 ka BP) is constrained by ^{14}C dating of sediment embedding the tephra layers, whereas the ages of the lowermost layers are between 29 and 27 ka BP.

Tephras in the Ross Sea were examined during the ROSSTEPHRA project, funded by the Italian Programma Nazionale di Ricerche in Antartide (PNR-PEA 2010/A2.12). Nine existing gravity cores were reanalysed to identify and compositionally characterize the tephra layers. Several tephra layers have been fully fingerprinted (sedimentological, petrographical and single glass-shard chemistry, ^{40}Ar/^{39}Ar dating of alkali feldspar grains in tephra, and ^{14}C dating of sediments containing tephra). The results indicate that at least five explosive eruptions of middle–high intensity (sub-Plinian–Plinian) occurred from volcanoes in northern Victoria land between *c.* 137 and 12 ka, with three of these possibly sourced in the Mount Melbourne Volcanic Field (Table 3; Fig. 3) (Del Carlo *et al.* 2015).

Most recently, in 2017, the TRACERS project (PNRA-PEA 2016/A3-00055) has continued the approach of the previous ROSSTEPHRA project, recovering eight new piston cores specifically designed for tephrochronological purposes and located on the continental shelf of the Ross Sea, downwind of the main volcanic edifices in northern Victoria Land. This resulted in the discovery and sampling of several macroscopic tephra layers in proximal–medial sites and, for the first time, a long record of distal cryptotephra (Di Roberto *et al.* 2018, 2019, 2020) (Table 3; Fig. 3).

Discussion

A huge number of volcanic rocks, volcaniclastic deposits and primary tephra layers have been found within sediment sequences drilled or cored offshore of Antarctica in the Southern Ocean. The volcanic material, which differs in its nature, composition and emplacement mechanisms, comes from different sources on the Antarctic continent and deposition has occurred over a wide time span.

Despite numerous studies, only a few have included the full characterization of these deposits, comprising textural and geochemical analyses, or study of the volcanic source. For primary fallout tephra layers, this is partially because deposits are often thin, fine grained and cannot be independently dated by the ^{40}Ar/^{39}Ar method because of the scarcity or poor quality of the material. However, rapid advances in analytical methods and dating technology are helping scientists to obtain useful new information on the geochemical composition and age of tephra layers that were previously difficult to analyse.

Where $^{40}Ar/^{39}Ar$ dating is impossible, ^{14}C ages for associated sediments can be used to bracket the ages of the tephras. New ^{14}C techniques, such as ramped pyrolysis (RP), will possibly improve the precision of the conventional ^{14}C method and may be used in samples for which conventional bulk radiocarbon dating of the acid-insoluble organic matter is inadequately informative due to contamination from old allochthonous organic matter (Rosenheim et al. 2008; 2013).

In addition, in order that the tephra layers (either marine or terrestrial) can be used to their full potential, it is necessary to: (i) increase the limited knowledge of the eruptive history of Antarctic volcanoes; and (ii) increase the paucity of textural, mineralogical and geochemical data (major and trace elements) for their erupted products. Without this, it is obviously difficult to make correlations between tephra layers identified in marine sequences and their sources (see Del Carlo et al. 2018).

In the future, the coupling of new techniques for characterizing tephra to improve tephrostratigraphy and tephrochronology in the oceans around Antarctica with an increased knowledge of the possible land-based volcanic sources will substantially increase the value of marine tephra as a powerful chronological tool for the synchronization and correlation of the geological, palaeoclimatic and palaeoenvironmental records in the region. In addition to the tephra delivered to the oceans directly by fall processes, volcanic detritus derived from the erosion, transport and sedimentation of Antarctica's source volcanoes is equally important as it can provide information not only on the potential volcanic sources but can also be used to document the ice conditions in the source regions and provide evidence for the evolving provenance sources (Sandroni and Talarico 2011; Talarico et al. 2012; Bertrand et al. 2014). Similarly, as variations in the clast proportions in Antarctic marine sediments is strongly influenced by changing ice conditions and/or ice-flow patterns, clasts provenance studies can shed light on evolving glacial dynamics. Finally, dust-grade (very fine ash) volcanic particles are also critical indicators of past atmospheric dynamics (Basile et al. 2001; Geyer et al. 2017).

Conclusions

Our review confirms that glaciomarine sediment sequences from the Southern Ocean surrounding the Antarctic continent contain abundant primary tephra and volcaniclastic deposits consisting mainly of tephra produced by explosive eruptions of Antarctic volcanoes.

We emphasize that volcanic material in marine sediment sequences from the Southern Ocean can offer a less disturbed, more complete record of the eruptive activity of Antarctic volcanoes than can be derived from land-based studies.

We also show that when properly characterized, geochemically fingerprinted and dated, tephra layers from the Southern Ocean may represent an invaluable chronological tool essential in establishing independent correlations between different geological archives, such as marine, terrestrial, lacustrine and ice-core records, and providing an accuracy difficult to achieve by other methods. In particular, while volcanic activity registered within englacial tephra might potentially reach a maximum age of just 2.7 Ma (Yan et al. 2017), data derived from marine cores extend the volcanological record back to 26 Ma.

Moreover, we have demonstrated that volcanic deposits in the marine record may provide significant data for volcanological reconstructions, including the intensity and magnitude of eruptions, age and recurrence, as well as the eruptive dynamics. This potential is especially effective where the terrestrial record of volcanism is incomplete, largely obscured by ice or disturbed owing to erosion.

Finally, we highlight that the study of volcanic material within glaciomarine successions also provides important palaeoenvironmental information. This information can be combined and integrated with palaeoecological investigations based on palaeontological and geochemical characterizations of sediments and fossil organisms to reconstruct climate changes and ice-sheet dynamics over time.

Acknowledgements We are grateful to John L. Smellie and two anonymous reviewers for their comprehensive and constructive suggestions.

Author contributions ADR: conceptualization (equal), data curation (equal), investigation (equal), methodology (equal), writing – original draft (lead), writing – review & editing (equal); **PDC**: conceptualization (equal), data curation (equal), investigation (equal), methodology (equal), writing – original draft (equal), writing – review & editing (equal); **MP**: conceptualization (equal), investigation (equal), methodology (equal), supervision (equal), writing – review & editing (equal).

Funding This work was partially funded by the TRACERS (PEA 2016/A3 - 00055: awarded to A. Di Roberto) and ROSSTEPHRA (PEA 2010/A2.12: awarded to M. Pompilio and P. del Carlo) National Antarctic Research Programme projects.

Data availability Part of the datasets generated during and/or analysed during the current study are available in the International Ocean Discovery Program repository, https://iodp.tamu.edu/scienceops/maps.html.

References

Acton, G., Florindo, F. et al. 2008. In: Harwood, D., Florindo, F., Talarico, F. and Levy, R.H. (eds) Studies from the ANDRILL Southern McMurdo Sound Project, Antarctica – Initial Science Report on AND-2A. Terra Antartica, **15**, 193–210.

Albert, P.G., Tomlinson, E.L. et al. 2012. Marine-continental tephra correlations: volcanic glass geochemistry from the Marsili Basin and the Aeolian Islands, Southern Tyrrhenian Sea, Italy. Journal of Volcanology and Geothermal Research, **229**, 74–94, https://doi.org/10.1016/j.jvolgeores.2012.03.009

Amrisar, F.K., Russel, M.D. et al. 1988. The United States Antarctic Research Program in the Western Ross Sea, 1979–1980: The Sediment Descriptions. Sedimentology Research Laboratory Contribution, **53**.

Armienti, P., Macedonio, G. and Pareschi, M.T. 1988. A numerical model for simulation of tephra transport and deposition: applications to May 18, 1980 Mount St. Helens eruption. Journal of Geophysical Research Solid Earth, **93**, 6463–6476.

Armienti, P., Messiga, B. and Vannucci, R. 1998. Sand provenance from major and trace element analyses of bulk rock and sand grains. Terra Antartica, **5**, 589–599.

Armienti, P., Tamponi, M. and Pompilio, M. 2001. Sand provenance from major and trace element analyses of bulk rock and sand grains from CRP2/2A. Terra Antartica, **8**, 2–23.

Baker, P.E., McReath, I., Harvey, M.R., Roobol, M.J. and Davies, T.G. 1975. The Geology of the South Shetland Islands: V. Volcanic evolution of Deception Island. British Antarctic Survey Scientific Reports, **78**.

Barker, P.E., Kennett, J.P. et al. 1988. *Proceedings of the Ocean Drilling Program, Initial Reports, Volume 113*. Ocean Drilling Program, College Station, TX, https://doi.org/10.2973/odp.proc.ir.113.1988

Barker, P.F., Kennett, J.P. et al. 1990. *Proceedings of the Ocean Drilling Program, Scientific Results, Volume 113*. Ocean Drilling Program, College Station, TX, https://doi.org/10.2973/odp.proc.sr.113.1990

Barker, P.F., Camerlenghi, A. et al. 1999. *Proceedings of the Ocean Drilling Program, Initial Reports, Volume 178*. Ocean Drilling Program, College Station, TX, https://doi.org/10.2973/odp.proc.ir.178.101.1999

Barker, P.F., Camerlenghi, A., Acton, G.D. and Ramsay, A.T.S. (eds) 2002. *Proceedings of the Ocean Drilling Program, Scientific Results, Volume 178*. Ocean Drilling Program, College Station, TX, https://doi.org/10.2973/odp.proc.sr.178.2002

Barrett, P.J. (ed.) 1986. Antarctic Cenozoic History from MSSTS-1 Drill Hole, McMurdo Sound. *DSIR Bulletin*, **237**.

Barrett, P.J. 1987. Oligocene sequence cored at CIROS-1, western McMurdo Sound: New Zealand. *Antarctic Record*, **7**, 1–17.

Barrett, P.J. 1989. Antarctic Cenozoic History from the CIROS-1 Drill Hole, McMurdo Sound. *DSIR Bulletin*, **245**.

Barrett, P.J. 1998*a*. Studies from the Cape Roberts Project, Ross Sea, Antarctica. Scientific Report of CRP-1, overview. *Terra Antartica*, **5**, 255–258.

Barrett, P.J. 1998*b*. Chronology of CRP-1. *Terra Antartica*, **5**, 681–682.

Barrett, P.J. and McKelvey, B.C. 1986. Stratigraphy. *DSIR Bulletin*, **237**, 9–52.

Barrett, P.J. and Treves, S.B. 1981. Sedimentology and petrology of core from DVDP 15, western McMurdo Sound. *American Geophysical Union Antarctic Research Series*, **33**, 281–314.

Barrett, P.J., Treves, S.B. et al. 1976. Initial report of DVDP 15, western McMurdo Sound, Antarctica. *Dry Valley Drilling Project Bulletin*, **7**, 1–100

Barrett, P.J., Hambrey, M.J., Harwood, D.M., Pyne, A.R. and Webb, P.N. 1989. Synthesis. *DSIR Bulletin*, **245**, 241–251.

Barrett, P.J., Adams, C.A., McIntosh, W.C., Swisher, C.C.III and Wilson, G.S. 1992. Geochronological evidence supporting Antarctic deglaciation three million years ago. *Nature*, **359**, 816–818, https://doi.org/10.1038/359816a0

Barrett, P.J., Sarti, M. and Wise, S.W. 2000. Studies from the Cape Roberts Project, Ross Sea, Antarctica, Initial Reports on CRP-3. *Terra Antartica*, **7**, 209.

Barron, J., Larsen, B. et al. 1989. *Proceedings of the Ocean Drilling Program, Initial Reports, Volume 119*. Ocean Drilling Program, College Station, TX, https://doi.org/10.2973/odp.proc.ir.119.1989

Basile, I., Petit, J.R., Touron, S., Grousset, F.E. and Barkov, N. 2001. Volcanic layers in Antarctic (Vostok) ice cores: source identification and atmospheric implications. *Journal of Geophysical Research*, **106**, 31 915–31 931, https://doi.org/10.1029/2000JD000102

Bertrand, S., Daga, R., Bedert, R. and Fontijn, K. 2014. Deposition of the 2011–2012 Cordón Caulle tephra (Chile, 40°S) in lake sediments: Implications for tephrochronology and volcanology. *Journal of Geophysical Research: Earth Surface*, **119**, 2555–2573, https://doi.org/10.1002/2014JF003321

Björck, S., Sandgren, P. and Zale, R. 1991. Late Holocene tephrochronology of the northern Antarctic Peninsula. *Quaternary Research*, **36**, 322–328, https://doi.org/10.1016/0033-5894(91)90006-Q

Borchardt, G.A., Aruscavage, P.J. and Millard, H.T.J. 1972. Correlation of the Bishop Ash, a Pleistocene marker bed, using instrumental neutron activation analysis. *Journal of Sedimentary Petrology*, **42**, 301–306.

Burton-Johnson, A., Black, M., Fretwell, P.T. and Kaluza-Gilbert, J. 2016. An automated methodology for differentiating rock from snow, clouds and sea in Antarctica from Landsat 8 imagery: a new rock outcrop map and area estimation for the entire Antarctic continent. *The Cryosphere*, **10**, 1665–1677, https://doi.org/10.5194/tc-10-1665-2016

Cape Roberts Science Team. 1999. Studies from the Cape Roberts Project, Ross Sea, Antarctica: Initial Report on CRP-2/2A. *Terra Antartica*, **6**, 1–173.

Ciesielski, P.F., Kristoffersen, Y. et al. 1991. *Proceedings of the Ocean Drilling Program, Scientific Results, Volume 114*. Ocean Drilling Program, College Station, TX, https://doi.org/10.2973/odp.proc.sr.114.1991

Coffin, M.F., Frey, F.A. et al. 2000. *Proceedings of the Ocean Drilling Program, Initial Reports, Volume 183*. Ocean Drilling Program, College Station, TX, https://doi.org/10.2973/odp.proc.ir.183.2000

Colizza, E., Finocchiaro, F., Marinoni, L., Menegazzo Vitturi, L. and Brambati, A. 2003. Tephra evidence in Marine Sediments from the shelf of the Western Ross Sea. *Terra Antartica*, **8**, 121–126.

Cooper, A.F., Adam, L.J., Coulter, R.F., Eby, G.N. and McIntosh, W.C. 2007. Geology, geochronology and geochemistry of a basanitic volcano, White Island, Ross Sea, Antarctica. *Journal of Volcanology and Geothermal Research*, **165**, 189–216, https://doi.org/10.1016/j.jvolgeores.2007.06.003

Davies, S.M. 2015. Cryptotephras: the revolution in correlation and precision dating. *Journal of Quaternary Sciences*, **30**, 114–130, https://doi.org/10.1002/jqs.2766

Del Carlo, P., Panter, K.S., Bassett, K., Bracciali, L., Di Vincenzo, G. and Rocchi, S. 2009. The upper lithostratigraphic unit of ANDRILL AND-2A core (Southern McMurdo Sound, Antarctica): Local volcanic sources, paleoenvironmental implications and subsidence in the western Victoria Land Basin. *Global and Planetary Change*, **69**, 142–161, https://doi.org/10.1016/j.gloplacha.2009.09.002

Del Carlo, P., Di Roberto, A. et al. 2015 Late Pleistocene–Holocene volcanic activity in northern Victoria Land recorded in Ross Sea (Antarctica) marine sediments. *Bulletin of Volcanology*, **77**, 36, https://doi.org/10.1007/s00445-015-0924-0

Del Carlo, P., Di Roberto, A. et al. 2018 Late Glacial–Holocene tephra from southern Patagonia and Tierra del Fuego (Argentina, Chile): A complete textural and geochemical fingerprinting for distal correlations in the Southern Hemisphere. *Quaternary Science Reviews*, **195**, 153–170, https://doi.org/10.1016/j.quascirev.2018.07.028

Di Roberto, A., Pompilio, M. and Wilch, T.I. 2010. Late Miocene submarine volcanism in ANDRILL AND-1B drill core, Ross Embayment, Antarctica. *Geosphere*, **6**, 524–536, https://doi.org/10.1130/GES00537.1

Di Roberto, A., Del Carlo, P., Rocchi, S. and Panter, K.S. 2012. Early Miocene volcanic activity and paleoenvironment conditions recorded in tephra layers of the AND-2A core (southern McMurdo Sound, Antarctica). *Geosphere*, **8**, https://doi.org/10.1130/GES00754.1

Di Roberto, A., Colizza, E., Del Carlo, P., Gallerani, A. and Giglio, F. 2018. TRACERS project: preliminary results tephrochronology study of the Ross Sea, Antarctica. *Geophysical Research Abstracts*, **20**, EGU2018-14052-1.

Di Roberto, A., Colizza, E., Del Carlo, P., Petrelli, M., Finocchiaro, F. and Kuhn, G. 2019. First marine cryptotephra in Antarctica found in sediments of the western Ross Sea correlates with englacial tephras and climate records. *Scientific Reports*, **10628**, https://doi.org/10.1038/s41598-019-47188-3

Di Roberto, A., Albert, P.G. et al. 2020. Evidence for a large-magnitude Holocene eruption of Mount Rittmann (Antarctica): A volcanological reconstruction using the marine tephra record. *Quaternary Science Reviews*, **250**, 106629, https://doi.org/10.1016/j.quascirev.2020.106629

Di Vincenzo, G., Caburlotto, A. and Camerlenghi, A. 2001. ^{40}Ar–^{39}Ar investigation of volcanic clasts in glacigenic sediments at Sites 1097 and 1103 (ODP Leg 178, Antarctic Peninsula). *Proceeding of the Ocean Drilling Program, Scientific Results*, **178**, 1–26, https://doi.org/10.2973/odp.proc.sr.178.232.2001

Di Vincenzo, G., Bracciali, L., Del Carlo, P., Panter, K. and Rocchi, S. 2009. ^{40}Ar–^{39}Ar dating of volcanogenic products from the AND-2A core (ANDRILL Southern McMurdo Sound Project, Antarctica): correlations with the Erebus Volcanic Province and implications for the age model of the core. *Bulletin of Volcanology*, **72**, 487–505, https://doi.org/10.1007/s00445-009-0337-z

Dunbar, N.W., Iverson, N.A., Smellie, J.L., McIntosh, W.C., Zimmerer, M.J. and Kyle, P.R. 2021. Active volcanoes in Marie Byrd Land. *Geological Society, London, Memoirs*, **55**, https://doi.org/10.1144/M55-2019-29

Federman, A.W., Watldns, N.D. and Sigurdsson, H. 1982. Scotia Arc volcanism recorded in abyssal piston cores downwind from the islands. *In*: Craddock, C. (ed.) *Antarctic Geoscience*. University of Wisconsin Press, Madison, WI, 223–238.

Fielding, C.R. and CRP Science Team 2001. Documentation of sediment core CRP-2A by box images/photos. *PANGAEA*, https://doi.org/10.1594/PANGAEA.58401

Fielding, C.R., Browne, G.H. *et al.* 2011. Sequence stratigraphy of the ANDRILL AND-2A drillcore, Antarctica: A long-term, ice-proximal record of Early to Mid-Miocene climate, sea-level and glacial dynamism. *Palaeogeography, Palaeoclimatology, Palaeoecology*, **305**, 337–351, https://doi.org/10.1016/j.palaeo.2011.03.026

Fontijn, K., Lacgowycz, S.M., Rawson, H.L., Pyle, D.M., Mather, T.A., Naranjo, J.A. and Moreno-Roa, H. 2014. Late Quaternary tephrostratigraphy of southern Chile and Argentina. *Quaternary Science Reviews*, **89**, 70–84, https://doi.org/10.1016/j.quascirev.2014.02.007

Fontijn, K., Rawson, H. *et al.* 2016. Synchronisation of sedimentary records using tephra: A postglacial tephrochronological model for the Chilean Lake District. *Quaternary Science Reviews*, **137**, 234–254, https://doi.org/10.1016/j.quascirev.2016.02.015

Ford, A.B. 1975. Volcanic rocks of Naturaliste Plateau, Eastern Indian Ocean, Site 264, DSDP Leg 28. *Initial Reports of the Deep Sea Drilling Project*, **28**, 821–834, https://doi.org/10.2973/dsdp.proc.28.129.1975

Frakes, L.A. 1971. *USNS Eltanin Core Descriptions Cruises 32 to 45*. Sedimentology Research Laboratory Contribution, Department of Geology, Florida State University, **33**.

Fretzdorff, S. and Smellie, J.L. 2002. Electron microprobe characterization of ash layers in sediments from the central Bransfield basin (Antarctic Peninsula): evidence for at least two volcanic sources. *Antarctic Science*, **14**, 412–421, https://doi.org/10.1017/S0954102002000214

Froggatt, P.C., Nelson, C.S., Carter, L., Griggs, G. and Black, K.P. 1986. An exceptionally large late Quaternary eruption from New Zealand. *Nature*, **319**, 578–582, https://doi.org/10.1038/319578a0

Gambino, S., Armienti, P. *et al.* 2021. Mount Melbourne and Mount Rittmann. *Geological Society, London, Memoirs*, **55**, https://doi.org/10.1144/M55-2018-43

Gamble, J.A., Barrett, P.J. and Adams, C.J. 1986. Basaltic clasts from Unit 8. *DSIR Bulletin*, **237**, 145–152.

George, A. 1989. Sand provenance. *DSIR Bulletin*, **245**, 159–168.

Geyer, A., Martì, A., Giralt, S. and Folch, A. 2017. Potential ash impact from Antarctic volcanoes: Insights from Deception Island's most recent eruption. *Scientific Reports*, **7**, 16534, https://doi.org/10.1038/s41598-017-16630-9

Geyer, A., Pedrazzi, D. *et al.* 2021. Deception Island. *Geological Society, London, Memoirs*, **55**, https://doi.org/10.1144/M55-2018-56

Global Volcanism Program 2010. Report on Montagu Island (United Kingdom). *Bulletin of the Global Volcanism Network*, **35**(9). Smithsonian Institution, https://doi.org/10.5479/si.GVP.BGVN201009-390081

Global Volcanism Program 2013. Volcanoes of the world, v. 4.7.4. *Bulletin of the Global Volcanism Network*, **35**(9). Smithsonian Institution, https://doi.org/10.5479/si.GVP.VOTW4-2013 (downloaded 07 November 2018).

Global Volcanism Program 2017. Report on Bristol Island (United Kingdom). *Bulletin of the Global Volcanism Network*, **42**(9). Smithsonian Institution.

Global Volcanism Program 2018. Report on Saunders (United Kingdom). *In*: Sennert, S.K. (ed.) *Weekly Volcanic Activity Report, 4 July–10 July 2018*. Smithsonian Institution and United States Geological Survey.

Harwood, D.M., Barrett, P.J., Edwards, A.R., Rieck, H.J. and Webb, P.N. 1989. Biostratigraphy and chronology. *DSIR Bulletin*, **245**, 231–239.

Hayes, D.E. and Frakes, L. 1975. General synthesis, Deep Sea Drilling Project Leg 28. *Initial Reports of the Deep Sea Drilling Project*, **28**, 919–942.

Hayes, D.E., Frakes, L.A. *et al.* 1975. *Initial Reports of the Deep Sea Drilling Project, Volume 28*. United States Government Printing Office, Washington, DC.

Hillenbrand, C.-D., Moreton, S.G. *et al.* 2008. Volcanic time-markers for Marine Isotopic Stages 6 and 5 in Southern Ocean sediments and Antarctic ice cores: implications for tephra correlations between palaeoclimatic records. *Quaternary Science Reviews*, **27**, 518–540, https://doi.org/10.1016/j.quascirev.2007.11.009

Hodgson, D.A., Dyson, C.L., Jones, V.J. and Smellie, J.L. 1998. Tephra analysis of sediments from Midge Lake (South Shetland Islands) and Sombre Lake (South Orkney Islands), Antarctica. *Antarctic Science*, **10**, 13–20, https://doi.org/10.1017/S0954102098000030

Hodgson, D.A., Doran, P.T., Roberts, D. and McMinn, A. 2004. Paleolimnological studies from the Antarctic and subantarctic islands. *In*: Pienitz, R., Douglas, M.S.V. and Smol, J.P. (eds) *Long-Term Environmental Change in Arctic and Antarctic Lakes*. Springer, Dordrecht, The Netherlands, 419–474.

Huang, T.C., Watkins, N.D., Shaw, D.M. and Kennett, J.P. 1973. Atmospherically transported volcanic dust in South Pacific deep sea sedimentary cores at distances over 3000 km from the eruptive source. *Earth and Planetary Science Letters*, **20**, 119–124, https://doi.org/10.1016/0012-821X(73)90148-9

Huang, T.C., Watkins, N.D. and Shaw, D.M. 1975. Atmospherically transported volcanic glass in deep-sea sediments: volcanism in sub-Antarctic latitudes of the South Pacific during late Pliocene and Pleistocene time. *Geological Society of America Bulletin*, **86**, 1305–1315, https://doi.org/10.1130/0016-7606(1975)86<1305:ATVGID>2.0.CO;2

Hubberten, H.-W., Morche, W., Westall, F., Fütterer, D.K. and Keller, J. 1991. Geochemical investigations of volcanic ash layers from Southern Atlantic Legs 113 and 114. *Proceedings of the Ocean Drilling Program, Scientific Results*, **114**, 733–749, https://doi.org/10.2973/odp.proc.sr.114.182.1991

Iverson, N.A., Kyle, P.R. *et al.* 2014. Eruptive history and magmatic stability of Erebus volcano, Antarctica: Insights from englacial tephra. *Geochemistry, Geophysics, Geosystems*, **15**, 4180–4202, https://doi.org/10.1002/2014GC005435

Kohno, M., Fujii, Y. and Hirata, T. 2004. Chemical composition of volcanic glasses in visible tephra layers found in a 2503 m deep ice core from Dome Fuji, Antarctica. *Annals of Glaciology*, **39**, 576–584, https://doi.org/10.3189/172756404781813934

Kraus, S., Kurbatov, A. and Yates, M. 2013. Geochemical signatures of tephras from Quaternary Antarctic Peninsula volcanoes. *Andean Geology*, **40**, 1–40, https://doi.org/10.5027/andgeoV40n1-a01

Krissek, L., Browne, G. *et al.* 2007. Sedimentology and stratigraphy of the AND-1B core, ANDRILL McMurdo Ice Shelf Project, Antarctica. *Terra Antartica*, **14**, 185–222.

Kyle, P.R. and Muncy, H.L. 1989. Geology and geochronology of McMurdo Volcanic Group rocks in the vicinity of Lake Morning, McMurdo Sound, Antarctica. *Antarctic Sciences*, **1**, 345–350, https://doi.org/10.1017/S0954102089000520

Kyle, P.R. and Seward, D. 1984. Dispersed rhyolitic tephra from New Zealand in deep-sea sediments of the Southern Ocean. *Geology*, **12**, 487–490, https://doi.org/10.1130/0091-7613(1984)12<487:DRTFNZ>2.0.CO;2

Lee, Y.I., Lim, H.S., Yoon, H.I. and Tatur, A. 2007. Characteristics of tephra in Holocene lake sediments on King George Island, West Antarctica: implications for deglaciation and paleoenvironment. *Quaternary Science Reviews*, **26**, 3167–3178, https://doi.org/10.1016/j.quascirev.2007.09.007

LeMasurier, W.E. 1990. Late Cenozoic volcanism on the Antarctic Plate: An overview. *American Geophysical Union Antarctic Research Series*, **48**, 1–17.

Licht, K.J., Jennings, A.E., Andrews, J.T. and Williams, K.M. 1996. Chronology of late Wisconsin ice retreat from the western Ross Sea, Antarctica. *Geology*, **24**, 223–226, https://doi.org/10.1130/0091-7613(1996)024<0223:COLWIR>2.3.CO;2

Licht, K.J., Dunbar, N.W., Andrews, J.T. and Jennings, A.E. 1999. Distinguishing subglacial till and glacial marine diamictons in the western Ross Sea, Antarctica: Implications for a last glacial maximum grounding line. *Geological Society of America Bulletin*, **111**, 91–103, https://doi.org/10.1130/0016-7606(1999)111<0091:DSTAGM>2.3.CO;2

Lough, A.C., Wiens, D.A. et al. 2013. Seismic detection of an active subglacial magmatic complex in Marie Byrd Land, Antarctica. *Nature Geosciences*, **6**, 1031–1035, https://doi.org/10.1038/NGEO1992

Liu, E.J., Oliva, M. et al. 2016. Expanding the tephrostratigraphical framework for the South Shetland Islands, Antarctica, by combining compositional and textural tephra characterisation. *Sedimentary Geology*, **340**, 49–61, https://doi.org/10.1016/j.sedgeo.2015.08.002

Lowe, D.J. 2011. Tephrochronology and its application: A review. *Quaternary Geochronology*, **6**, 107–153, https://doi.org/10.1016/j.quageo.2010.08.003

Lowe, D.J. 2016. Connecting synchronising and dating with tephras: principles and applications of tephrochronology in Quaternary research. *In*: Vandergoes, M.J., Rogers, K.M., Turnbull, J., Howarth, J., Keller, E. and Cowan, H. (eds) *13th Quaternary Techniques Short Course – Measuring Change and Reconstructing Past Environments*. National Isotope Centre GNS Science, Lower Hutt, New Zealand, 1–31.

Lowe, D. and Alloway, B. 2015. Tephrochronology. *In*: Rink, W. and Thompson, J. (eds) *Encyclopaedia of Scientific Dating Methods*. Springer, Dordrecht, The Netherlands, 783–799.

Martin, A.P. 2009. *Mt. Morning, Antarctica: Geochemistry, Geochronology, Petrology, Volcanology, and Oxygen Fugacity of the Rifted Antarctic lithosphere*. PhD thesis, University of Otago, Dunedin, New Zealand.

Martin, A.P., Cooper, A. and Dunlap, W. 2010, Geochronology of Mt. Morning, Antarctica: Two-phase evolution of a long-lived trachyte–basanite–phonolite eruptive center. *Bulletin of Volcanology*, **72**, 357–371, https://doi.org/10.1007/s00445-009-0319-1

Martínez Cortizas, A., Rozas Muñiz, I., Taboada, T., Toro, M., Granado, I., Giralt, S. and Pla Rabés, S. 2014. Factors controlling the geochemical composition of Limnopolar Lake sediments (Byers Peninsula, Livingston Island, South Shetland Island, Antarctica) during the last ca. 1600 years. *Solid Earth Discussions*, **5**, 651–663, https://doi.org/10.5194/se-5-651-2014

Matthies, D., Storzer, D. and Troll, G. 1988. Volcanic ashes in Bransfield Strait sediments: geochemical and stratigraphical investigations (Antarctica). *Proceedings of the Second International Conference on Natural Glasses*, Prague, 139–147.

Matthies, D., Mausbacher, R. and Storzer, D. 1990. Deception Island tephra: a stratigraphical marker for limnic and marine sediments in Bransfield Strait area, Antarctica. *Zentralblatt fur Geomorphologie und Palaontologie*, **33**, 219–234.

McIntosh, W.C. 1998. ^{40}Ar/^{39}Ar geochronology of volcanic clasts and pumice in CRP-1 core, Cape Roberts, Antarctica. *Terra Antartica*, **5**, 683–690.

McIntosh, W.C. 2000. ^{40}Ar/^{39}Ar geochronology of tephra and volcanic clasts in CRP-2A, Victoria Land Basin, Antarctica. *Terra Antartica*, **7**, 621–630.

McKay, R.M., De Santis, L., Kulhanek, D.K. and the Expedition 374 Scientists 2019. *Proceedings of the International Ocean Discovery Program*, **374**, https://doi.org/10.14379/iodp.proc.374.106.2019

Morche, W., Hubberten, H.-W., Ehrmann, W.U. and Keller, J. 1991. Geochemical investigations of volcanic ash layers from Leg 119, Kerguelen Plateau. *Proceedings of the Ocean Drilling Program, Scientific Results*, **119**, 323–344, https://doi.org/10.2973/odp.proc.sr.119.124.1991

Moreton, S.G. 1999. *Quaternary Tephrochronology of the Scotia Sea and Bellingshausen Sea, Antarctica*. PhD thesis, British Antarctic Survey, Cambridge, UK.

Moreton, S.G. and Smellie, J.L. 1998. Identification and correlation of distal tephra layers in deep sea sediment cores, Scotia Sea, Antarctica. *Annals of Glaciology*, **27**, 285–289, https://doi.org/10.3189/1998AoG27-1-285-289

Naish, T., Powell, R. et al. 2009. Obliquity-paced Pliocene West Antarctic ice sheet oscillations. *Nature*, **458**, 322–328, https://doi.org/10.1038/nature07867

Narcisi, B. and Petit, J.R. 2021. Englacial tephras of East Antarctica. *Geological Society, London, Memoirs*, **55**, https://doi.org/10.1144/M55-2018-86

Narcisi, B., Petit, J.R., Delmonte, B., Basile-Doelsch, I. and Maggi, V. 2005. Characteristics and sources of tephra layers in the EPICA-Dome C ice record (East Antarctica): implications for past atmospheric circulation and ice core stratigraphic correlations. *Earth and Planetary Sciences Letters*, **239**, 253–265, https://doi.org/10.1016/j.epsl.2005.09.005

Narcisi, B., Petit, J.-R. and Tiepolo, M. 2006. A volcanic marker (92 ka) for dating deep east Antarctic ice cores. *Quaternary Sciences Reviews*, **25**, 2682–2687 https://doi.org/10.1016/j.quascirev.2006.07.009

Ninkovich, D., Heezen, B.C., Conolly, J.R. and Burckle, L.H. 1964. South Sandwich tephra in deep-sea sediments. *Deep Sea Research*, **11**, 605–619.

O'Brien, P.E., Cooper, A.K.C. et al. 2001. *Proceedings of the Ocean Drilling Program, Initial Results, Volume 188*. Ocean Drilling Program, College Station, TX, https://doi.org/10.2973/odp.proc.ir.188.2001

Pallàs, R., Smellie, J.L., Casas, J.M. and Calvet, J. 2001. Using tephrochronology to date temperate ice: correlation between ice tephras on Livingston Island and eruptive units on Deception Island volcano (South Shetland Islands, Antarctica). *The Holocene*, **11**, 149–160, https://doi.org/10.1191/095968301669281809

Panter, K.S. 2021. Antarctic volcanism: petrology and tectonomagmatic overview. *Geological Society, London, Memoirs*, **55**, https://doi.org/10.1144/M55-2020-10

Panter, K.S., Talarico, F. et al. 2009. Petrologic and geochemical composition of the AND-2A Core, ANDRILL Southern McMurdo Sound Project, Antarctica. *Terra Antartica*, **15**, 147–192.

Passchier, S., Browne, G. et al. 2011. Early and middle Miocene Antarctic glacial history from the sedimentary facies distribution in the AND-2A drill hole, Ross Sea, Antarctica. *Geological Society American Bulletin*, **123**, 2352, https://doi.org/10.1130/B30334.1

Patrick, M.R. and Smellie, J.L. 2013. A spaceborne inventory of volcanic activity in Antarctica and southern oceans, 2000–2010. *Antarctic Science*, **25**, 475–500, https://doi.org/10.1017/S0954102013000436

Patrick, M.R., Smellie, J.L., Harris, A.J.L., Wright, R., Dean, K., Garbal, I.L. and Pilger, E. 2005. First recorded eruption of Mount Belinda volcano (Montagu Island), South Sandwich Islands. *Bulletin of Volcanology*, **67**, 415–422, https://doi.org/10.1007/s00445-004-0382-6

Pedrazzi, D., Aguirre-Díaz, G., Bartolini, S., Martí, J. and Geyer, A. 2014. The 1970 eruption on Deception Island (Antarctica): eruptive dynamics and implications for volcanic hazards. *Journal of the Geological Society, London*, **171**: 765–778, https://doi.org/10.1144/jgs2014-015

Perkins, M.E., Naish, W.P., Brown, F.H. and Fleck, R.J. 1995. Fallout tuffs of Trapper Creek, Idaho – a record of Miocene explosive volcanism in the Snake River Plain volcanic province. *Geological Society of America Bulletin*, **107**, 1484–1506, https://doi.org/10.1130/0016-7606(1995)107<14 84:FTOTCI>2.3.CO;2

Pompilio, M., Armienti, A. and Tamponi, M. 2001. Petrography, mineral composition and geochemistry of volcanic and subvolcanic rocks of CRP-3, Victoria Land Basin, Antarctica. *Terra Antartica*, **8**, 469–480.

Pompilio, M., Dunbar, N. et al. 2007. Petrology and geochemistry of the AND-1B core, ANDRILL McMurdo Ice Shelf Project, Antarctica. *Terra Antartica*, **14**, 255–288.

Ponomareva, V., Portnyagin, M. and Davies, S.M. 2015. Tephra without borders: Far-reaching clues into past explosive eruptions. *Frontiers in Earth Sciences*, **3**, 83, https://doi.org/10.3389/feart.2015.00083

Powell, R.D., Laird, M.G., Naish, T.R., Fielding, C.R., Krissek, L.A. and Van Der Meer, J.J.M. 2001. Depositional environments for strata cored in CRP-3 (Cape Roberts Project), Victoria Land Basin, Antarctica: palaeoglaciological and palaeoclimatological inferences. *Terra Antartica*, **8**, 207–216.

Pudsey, C.J. 1992. Late quaternary changes in Antarctic bottom water velocity inferred from sediment grain size in the northern Weddell Sea. *Marine Geology*, **107**, 9–33, https://doi.org/10.1016/0025-3227(92)90066-Q

Pyle, D.M. 2000. Sizes of volcanic eruptions. *In*: Sigurdsson, H., Houghton, B.F., McNutt, S.R., Rymer, H. and Stix, J. (eds) *Encyclopedia of Volcanoes*. Academic Press, London.

Riede, F. and Thastrup, M.B. 2013. Tephra tephrochronology and archaeology – a (re-) view from Northern Europe. *Heritage Science*, **1**, 1–17, https://doi.org/10.1186/2050-7445-1-15

Rosenheim, B.E., Day, M.B., Domack, E., Schrum, H., Benthien, A. and Hayes, J.M. 2008. Antarctic sediment chronology by programmed-temperature pyrolysis: methodology and data treatment. *Geochemistry, Geophysics, Geosystems*, **9**, Q04005, https://doi.org/10.1029/2007GC001816

Rosenheim, B.E., Santoro, J.A., Gunter, M. and Domack, E.W. 2013. Improving Antarctic sediment ^{14}C dating using ramped pyrolysis: An example from the Hugo Island Trough. *Radiocarbon*, **55**, 115–126, https://doi.org/10.2458/azu_js_rc.v55i1.16234

Sandroni, S. and Talarico, F.M. 2011. The record of Miocene climatic events in AND-2A drill core (Antarctica): Insights from provenance analyses of basement clasts. *Global and Planetary Change*, **75**, 31–46, https://doi.org/10.1016/j.gloplacha.2010.10.002

Schroeder, D.M., Blankenship, D.D., Young, D.A. and Quartini, E. 2014. Evidence for elevated and spatially variable geothermal flux beneath the West Antarctic ice sheet. *Proceedings of the National Academy of Sciences of the United States of America*, **111**(25), 9070–9072, https://doi.org/10.1073/pnas.1405184111

Shane, P.A. 2000. Tephrochronology: a New Zealand case study. *Earth-Sciences Reviews*, **49**, 223–259, https://doi.org/10.1016/S0012-8252(99)00058-6

Shane, P.A. and Froggatt, P.C. 1992. Composition of widespread volcanic glass in deep-sea sediments of the Southern Pacific Ocean: an Antarctic source inferred. *Bulletin of Volcanology*, **54**, 595–601, https://doi.org/10.1007/BF00569943.

Smellie, J.L. 1998. Sand-grain detrital modes in CRP-1: provenance variations and influence of Miocene eruptions on the marine record in the McMurdo Sound region. *Terra Antartica*, **5**, 579–587.

Smellie, J.L. 1999a. Lithostratigraphy of Miocene–Recent, alkaline volcanic fields in the Antarctic Peninsula and eastern Ellsworth Land. *Antarctic Science*, **11**, 362–378, https://doi.org/10.1017/S0954102099000450

Smellie, J.L. 1999b. The upper Cenozoic tephra record in the south polar region: A review. *Global Planetary Change*, **21**, 51–70, https://doi.org/10.1016/S0921-8181(99)00007-7

Smellie, J.L. 2000. Erosional history of the Transantarctic Mountains deduced from sand grain detrital modes in CRP-2/2A, Ross Sea, Antarctica. *Terra Antartica*, **7**, 545–552.

Smellie, J.L. 2001. History of Oligocene erosion, uplift and unroofing of the Transantarctic Mountains deduced from sandstone detrital modes in CRP-3 drillcore, Victoria Land Basin, Antarctica. *Terra Antartica*, **8**, 481–489.

Smellie, J.L. 2002. Chronology of eruptions of Deception Island. *In*: López-Martínez, J., Smellie, J.L., Thomson, J.W. and Thomson, M.R.A. (eds) *Geology and Geomorphology of Deception Island*. British Antarctic Survey, Cambridge, UK, 70–71.

Smellie, J.L. 2021. Antarctic volcanism: volcanology and palaeoenvironmental overview. *Geological Society, London, Memoirs*, **55**, https://doi.org/10.1144/M55-2020-1

Smellie, J.L. and Edwards, B.R. 2016. *Glaciovolcanism on Earth and Mars: Products, Processes and Palaeoenvironmental Significance*. Cambridge University Press, Cambridge, UK.

Smellie, J.L. and Martin, A.P. 2021. Erebus Volcanic Province: volcanology. *Geological Society, London, Memoirs*, **55**, https://doi.org/10.1144/M55-2018-62

Smellie, J.L. and Rocchi, S. 2021. Northern Victoria Land: volcanology. *Geological Society, London, Memoirs*, **55**, https://doi.org/10.1144/M55-2018-60

Smellie, J.L., Panter, K.S. and Geyer, A. 2021. Introduction to volcanism in Antarctica: 200 million years of subduction, rifting and continental break-up. *Geological Society, London, Memoirs*, **55**, https://doi.org/10.1144/M55-2020-14

Talarico, F.M., McKay, R.M., Powell, R.D., Sandroni, S. and Naish, T. 2012. Late Cenozoic oscillations of Antarctic ice sheets revealed by provenance of basement clasts and grain detrital modes in ANDRILL core AND-1B. *Global and Planetary Change*, **96–97**, 23–40, https://doi.org/10.1016/j.gloplacha.2009.12.002

Tatur, A., del Valle, R. and Barczuk, A. 1999. Discussion on the uniform pattern of Holocene tephrochronology in South Shetland Islands, Antarctica. *Polish Polar Studies*, **26**, 379–389.

Toro, M., Granados, I. et al. 2013. Chronostratigraphy of the sedimentary record of Limnopolar Lake, Byers Peninsula, Livingston Island, Antarctica. *Antarctic Science*, **25**, 198–212, https://doi.org/10.1017/S0954102012000788

van Wyk de Vries, M., Bingham, R.G. and Hein, A.S. 2017. A new volcanic province: an inventory of subglacial volcanoes in West Antarctica. *Geological Society, London, Special Publications*, **461**, 231–248, https://doi.org/10.1144/SP461.7

Wilson, G.S., Levy, R. et al. 2007. Preliminary integrated chronostratigraphy of the AND-1B Core, ANDRILL McMurdo Ice Shelf Project, Antarctica. *Terra Antartica*, **14**, 297–316.

Xiao, W., Frederichs, T., Gersonde, R., Kuhn, G., Esper, O. and Zhang, X. 2016. Constraining the dating of late quaternary marine sediment records from the Scotia Sea (Southern Ocean). *Quaternary Geochronology*, **31**, 97–118, https://doi.org/10.1016/j.quageo.2015.11.003

Yan, Y., Ng, J. et al. 2017. 2.7-million-year-old ice from Allan Hills blue ice areas, East Antarctica reveals climate snapshots since early Pleistocene. *Goldschmidt Abstracts*, **2017**, 4359, https://goldschmidtabstracts.info/abstracts/abstractView?id=2017004920

Chapter 6.2

Englacial tephras of East Antarctica

Biancamaria Narcisi[1]* and Jean Robert Petit[2]

[1]ENEA Centro Ricerche Casaccia, Via Anguillarese 301, Santa Maria di Galeria, I-00123 Rome, Italy
[2]Geosciences de l'Environment (IGE), CNRS, IRD, University Grenoble Alpes, Grenoble F-38000 Grenoble, France

*Correspondence: biancamaria.narcisi@enea.it

Abstract: Driven by successful achievements in recovering high-resolution ice records of climate and atmospheric composition through the Late Quaternary, new ice–tephra sequences from various sites of the East Antarctic Ice Sheet (EAIS) have been studied in the last two decades spanning an age range of a few centuries to 800 kyr. The tephrostratigraphic framework for the inner EAIS, based on ash occurrence in three multi-kilometre-deep ice cores, shows that the South Sandwich Islands represent a major source for tephra, highlighting the major role in the ash dispersal played by clockwise circum-Antarctic atmospheric circulation penetrating the Antarctic continent. Tephra records from the eastern periphery of the EAIS, however, are obviously influenced by explosive activity sourced in nearby Antarctic rift provinces. These tephra inventories have provided a fundamental complement to the near-vent volcanic record, in terms of both frequency/chronology of explosive volcanism and of magma chemical evolution through time. Despite recent progress, current data are still sparse. There is a need for further tephra studies to collect data from unexplored EAIS sectors, along with extending the tephra inventory back in time. Ongoing international palaeoclimatic initiatives of ice-core drilling could represent a significant motivation for the tephra community and for Quaternary Antarctic volcanologists.

Supplementary material: A discussion of methodological and chronological aspects of ice–tephra studies (file S1) and a collection of major element geochemical analyses of EAIS englacial tephra layers (Table S1) are available at https://doi.org/10.6084/m9.figshare.c.5212574

Polar ice sheets represent an extraordinary mine of palaeoclimate archives and geological information, including past volcanism. In addition to a reconstruction of past temperatures deduced from the water isotope composition of snow/ice, the composition of past atmospheres from the entrapped air bubbles, and the origin of past marine and continental aerosols, polar ice contains tephra particles derived from explosive eruptions. The deposition of each layer is essentially instantaneous on a geological timescale. The volcanic material is incorporated into the snow strata and becomes part of the ice sheet. A time series of deposition of englacial tephras can fruitfully expand knowledge of the history of explosive volcanism by complementing the proximal volcanic stratigraphy. This in turn is important for various targets, including understanding the potential relationship between volcanism and climate, and an assessment of volcanic hazards from future eruptions, as well as providing correlatable datums in ice cores that can be used to establish the relative ages of significant episodes of climate change.

Antarctic ice sequences are significant tephra archives. Compared to work in all other depositional settings, studying englacial tephras from Antarctic ice has a number of advantages. Even when faint and very fine-grained, tephra material can be easily isolated from the host ice that in Antarctica is typically characterized by low quantities of ultrafine continental dust. Moreover, due to severe environment conditions, post-depositional chemical alteration of the tephra is minor to absent, permitting reliable characterization of the tephra. Ice is also unique in preserving chemical aerosols associated with volcanic tephra particles, so that through glaciochemical analysis performed along with tephra characterization it is possible to obtain a comprehensive picture of the volcanic emission of individual events. On the other hand, the analysis of Antarctic englacial tephras is challenging due to the typically small amount of available material and the fine size of the volcanic particles. The manipulation and processing of ash samples requires careful dedicated protocols, with equipment set up inside clean laboratories to avoid contamination by extraneous airborne dust. There is also a need for improved procedures to properly embed and polish the sparse material prior to geochemical analysis, as well as specific analytical protocols for the microanalysis of individual tiny particles (see Supplementary material, file S1). Finally, the attribution of an Antarctic englacial tephra to a specific volcano presents difficulties and limitations, especially if it is recovered from the centre of the East Antarctic Ice Sheet (EAIS) at great distances (i.e. several thousands of kilometres) from potential sources. Source identification is typically addressed through comparison with published data on the rock composition and chronostratigraphy of volcanoes located within and around Antarctica; this can be difficult for volcanoes that are remote and hard to access or are largely covered by snow and ice and therefore lack appropriate geological information. In addition, deciphering the source of far-travelled englacial tephras is not straightforward, and requires both multi-elemental geochemical analysis of the tephra composition and consideration of the large number of potential source volcanoes.

Because of a general awareness of the dangers of climate change, much effort has been concentrated in recent years on the study of Antarctic ice as a unique archive of past climate prior to the instrumental period which began in the mid-nineteenth century. Indeed, over the last two decades a number of Antarctic ice records, mainly as ice cores but also from sequences cropping out in more peripheral locations of the EAIS, have been retrieved and analysed, and have provided a wealth of information about Quaternary climate and atmospheric composition through several glacial–interglacial cycles (Brook and Buizert 2018 and references therein).

Taking advantage of the good quality of the sampled ice sequences, and with the advent of suitable analytical techniques for the characterization of fine-grained particles, tephra studies in the EAIS records have been successfully carried out in parallel with palaeoclimate studies. Typical ultimate goals of these studies are both chronostratigraphic and palaeoatmospheric: they allow accurate stratigraphic correlations between distant sites and identify marker horizons useful for constraining glaciological timescales, and also provide information on

From: Smellie, J. L., Panter, K. S. and Geyer, A. (eds) 2021. *Volcanism in Antarctica: 200 Million Years of Subduction, Rifting and Continental Break-up*. Geological Society, London, Memoirs, **55**, 649–664,
First published online February 24, 2021, https://doi.org/10.1144/M55-2018-86
© 2021 The Author(s). Published by The Geological Society of London. All rights reserved.
For permissions: http://www.geolsoc.org.uk/permissions. Publishing disclaimer: www.geolsoc.org.uk/pub_ethics

Table 1. *Location and characteristics of the EAIS ice–tephra sequences considered in this work*

Site	Abbreviation	Location	Elevation (m asl)	Extent of the ice record (ka)
Long ice cores				
EPICA-Dome C	EDC	75° 06′ S, 123° 21′ E	3233	800
Vostok	VK	78° 28′ S, 106° 50′ E	3488	420
Dome Fuji	DF	77° 19′ S, 39° 42′ E	3810	720
EPICA Dronning Maud Land	EDML	75° 00′ S, 0° 04′ E	2892	c. 145
Talos Dome	TALDICE	72° 49′ S, 159° 11′ E	2315	c. 250
Taylor Dome	TD	77° 47′ S, 158° 43′ E	2374	>130
Intermediate and short ice cores				
South Pole	SP	89° 59′ S, 98° 9′ W	c. 2835	c. 50
Law Dome	LD	66° 46′ S, 112° 48′ E	1370	c. 90
Styx Glacier Plateau	Styx G	73° 50′ S, 163° 40′ E	c. 1700	c. 1
GV7	GV7	70° 41′ S, 158° 52′ E	1950	c. 1
Outcrop exposures and blue-ice records				
Allan Hills	AH	c. 75° 43′ S, c. 159° 40′ E	2000	c. 400
Frontier Mountain	FM	72° 59′ S, 160° 20′ E	2800	c. 100
Mount Erebus flanks and blue ice field	ER	c. 77° 5′ S, c. 167° 5′ E	16–1954	c. 70
Nansen Ice Field	NIF	72° 30′–73° S, 23°–25° E	2900–3000	Unknown

pathways complementary to wind-blown dust studies. In addition, the reconstructed tephrostratigraphies contain precious information that could improve our knowledge of past explosive volcanism from a privileged remote viewpoint.

In this chapter, we summarize the most relevant results from englacial tephra studies obtained in the last two decades on snow/ice sequences from the EAIS. Our review covers the last c. 800 kyr (i.e. the last eight glacial cycles). Our literature review is viewed from a primarily palaeovolcanic perspective and is subdivided into three categories based on the characteristics of the ice records: (i) long cores; (ii) intermediate and shallow cores; and (iii) outcrop exposures and blue-ice records. The locations of sites we consider are listed in Table 1 and shown in Figure 1. We also briefly comment on the potential and limitations of tephra studies for ongoing efforts to reconstruct long-term climate–volcanism relationships. A discussion of methodological and chronological aspects is presented in the Supplementary material. In this chapter, the ages of individual tephras use the local ice timescale. Note that ice timescales were developed using various methods and constraints (e.g. Severi *et al.* 2012; Bazin *et al.* 2013). Therefore the ages can be expressed in different ways with respect to the accuracy and the year taken as reference: for example, BP (Before Present, i.e. before 1950), or the date in AD (Anno Domini) or CE (Common Era).

Finally, the reader should refer to the review by Smellie (1999) for a comprehensive summary of studies relating to tephra in East and West Antarctica published prior to 1999.

Description of existing englacial tephra records

Long ice cores

In recent years an array of new ice-core records covering at least the last climatic cycle (c. 130 kyr) has become available from the central EAIS, a region characterized by low accumulation rates and remoteness with respect to potential Antarctic source volcanoes. The inland region has the potential to provide continuous long-term records of tephra layers since surface phenomena, such as mixing of surface snow, erosion and wind scouring, are generally limited. Tephra repositories from such continuous ice cores provide the opportunity to extend the Antarctic ice-core tephra inventory far back in time, thus augmenting our knowledge of Late Quaternary volcanism in the south polar region. Tephra layers within ice-core archives retrieved for palaeoclimatic purposes are typically chronostratigraphically framed within the temperature record for the core and are indirectly dated using *ad hoc* precise glaciological timescales. All of the ice cores, except for TALDICE, almost reach the local bedrock. The collective ice-core stratigraphy extends to c. 800 kyr and the studied tephras are all Pleistocene–Holocene in age.

Dome C. At this site, located in the sector of EAIS facing the Indian Ocean (Fig. 1), an ice core down to 3260 m was drilled as part of the European Project for Ice Coring in Antarctica (EPICA). The stratigraphic continuity of the EPICA Dome C (hereafter EDC) ice-core record above c. 3200 m has yielded an important c. 800 kyr record of multiple climatic cycles, representing the longest continuous ice archive successfully investigated for a wide variety of simultaneously-analysed parameters, including local temperature, greenhouse gases and aerosols of different origins (e.g. Jouzel *et al.* 2007; Lambert *et al.* 2008; Loulergue *et al.* 2008) (Fig. 2).

Thirteen discrete air-fall tephra layers were identified in the last 200 kyr sections primarily through visual inspection of the core (Narcisi *et al.* 2005) (Fig. 2). Quantitative grain-size (Fig. 3) and single-shard major element analyses were carried out by wavelength-dispersive spectrometry (WDS) on each layer (see the Supplementary material). The trace element composition of individual glass shards was also determined for selected ash layers (Narcisi *et al.* 2006, 2008). Based on compositional signatures reflecting the tectonic setting of the volcanic sources (Fig. 4), five low-alkali tholeiitic layers were inferred to originate from activity of the South Sandwich

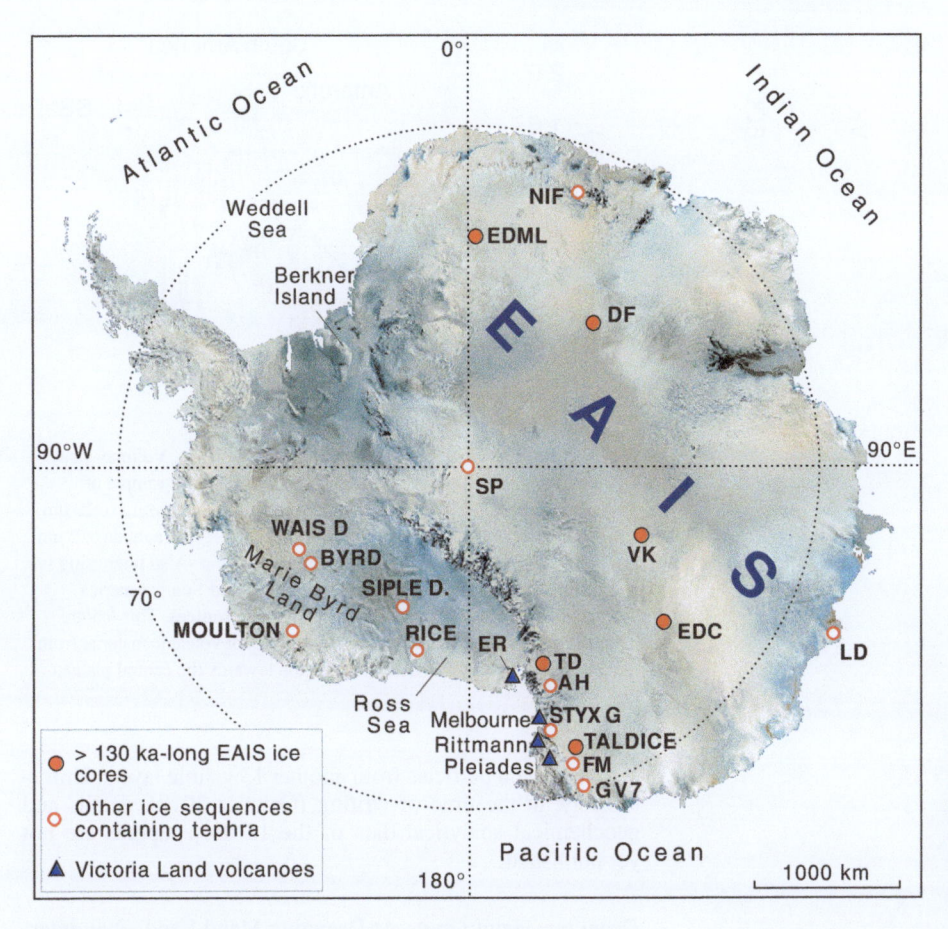

Fig. 1. Satellite image of Antarctica showing the EAIS sites of ice–tephra sequences considered in this work, along with the locations of active Antarctic volcanoes and ice-coring sites in West Antarctica. Abbreviations used are explained in Table 1. ER, Erebus.

Islands (SSI) in the South Atlantic Ocean, two calc-alkaline tephras from South Shetland volcanoes (northern Antarctic Peninsula), a further two calc-alkaline tephras from Andean volcanoes and four showing evolved alkaline compositions from the West Antarctic Rift System (WARS). The downcore inventory of visible tephra layers, investigated by Narcisi et al. (2010b), includes a single layer (c. 360 ka) that was attributed to Antarctic volcanic activity because of its alkali-trachytic composition. The deeper ice cores appear devoid of visually-detectable tephra layers (Fig. 2).

Vostok. The 3623 m-long Vostok ice core (VK, Fig. 1), the deepest ice core ever obtained, spans four climatic cycles over the past 420 kyr (Petit et al. 1999). For a decade prior to the completion of EDC drilling, it represented the longest continuous ice record ever recovered. Systematic studies of discrete tephra layers have been undertaken on the whole ice core (Basile et al. 2001; Narcisi et al. 2010b), representing a substantial advancement over the early investigations by Palais et al. (1989) that had focused on a few layers from the last c. 160 kyr ice record. Basile et al. (2001) characterized 15 visible ash layers spanning in age from c. 213 to 3.5 ka (Fig. 2) using major element analyses of single glass shards (see the Supplementary material), and trace elements, and Sr and Nd isotopic compositions obtained on bulk samples. The large dataset was intended to characterize tectonic conditions for magma genesis in order to facilitate identification of likely volcanic sources (Fig. 4). Through comparison with rock compositions from an inventory of Antarctic and subantarctic volcanoes, nine layers were attributed to activity of the SSI, three from South America, one from the Antarctic Peninsula and two from West Antarctic volcanoes. Basile et al. (2001) also observed that the predominance of atmospheric transport of volcanic material from the southwestern Atlantic did not seem to be dependent on the climatic conditions as it occurred both in glacial and interglacial climatic modes; this pattern is consistent with the continuous advection of continental dust from South America to the Vostok site.

The core record from c. 420 to c. 213 ka is almost devoid of detectable tephras, as only three layers were identified (Narcisi et al. 2010b) (Fig. 2). The tephra at 406 ka displays an evolved alkaline signature typical of Antarctic extension-related volcanism (perhaps Mount Berlin in Marie Byrd Land). The two closely spaced layers at 414 ka show an overlapping composition and were interpreted as being related to the same volcanic horizon that was folded by ice dynamics. The calc-alkaline signature indicates derivation from subduction-related volcanoes in the southern Andes and/or Antarctic Peninsula, although there is no known appropriate and compositionally comparable source in the Antarctic Peninsula region (J.L. Smellie pers. comm.).

Dome Fuji. This site, located in the inner EAIS facing the Atlantic and Indian Ocean sectors (Fig. 1), has been drilled twice, providing an overall climate record of seven glacial cycles (i.e. the last 720 kyr: Dome Fuji Ice Core Project Members 2017 and references therein). Tephra analyses were carried on the first ice core (DF1, 2503 m; 340 ka). Fujii et al. (1999) and Kohno et al. (2004) identified 26 visible layers, spanning in age from 226 to 19 ka (Fig. 2). Individual particles from 21 of these layers were characterized by electron probe microanalysis (major oxides) and one was measured for trace element content using laser ablation inductively coupled plasma mass spectrometry (LA-ICP-MS). Unfortunately, in the publication by Kohno et al. (2004) only plots of major element composition of the studied tephras are presented, supported by a table giving averaged major element abundances of seven layers (Supplementary material). According to the authors, the majority of the layers originate from the SSI, and secondarily from Andes and Antarctic volcanoes

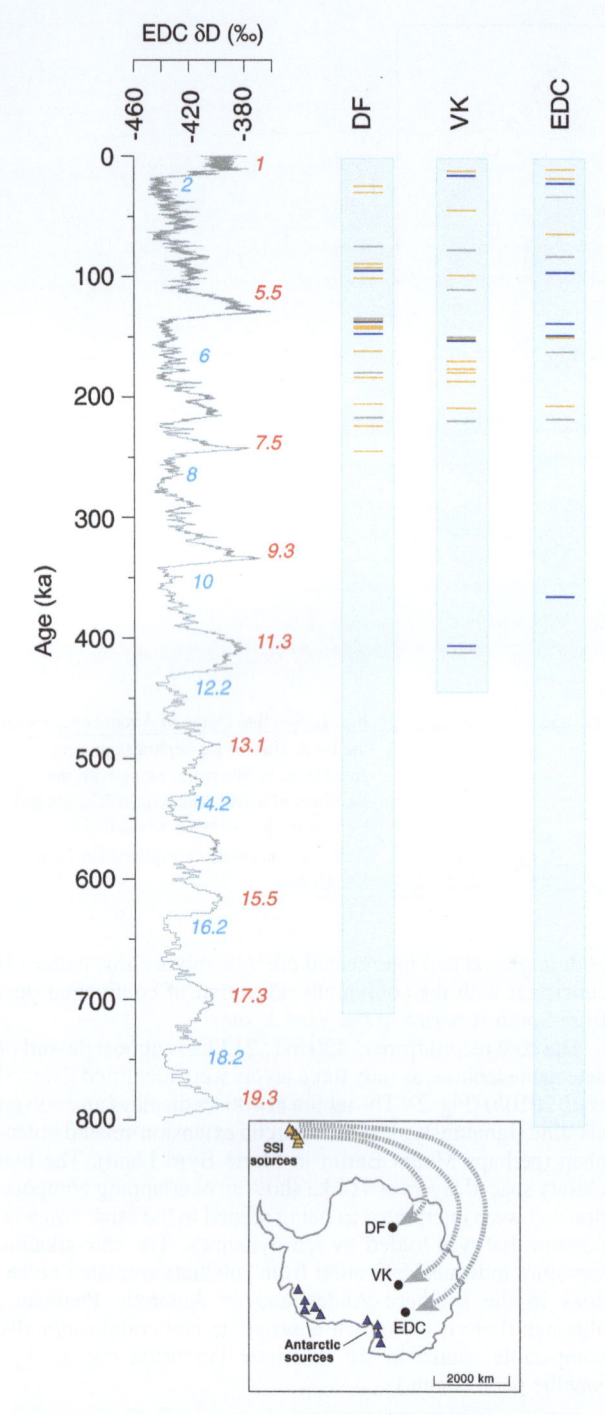

Fig. 2. Tephrostratigraphy for the inner EAIS region based on published tephra studies of DF, VK and EDC ice cores (redrawn and modified from Narcisi *et al.* 2010*b*). South Sandwich Islands (SSI) and Antarctic-sourced tephra layers are represented by orange and blue lines, respectively. Grey lines represent layers from other sources. The identified tephras are plotted alongside the EDC deuterium profile (Jouzel *et al.* 2007), with numbers indicating marine isotope stages (red for interglacial, and blue for glacial periods). Note the preponderance of tephra layers from SSI volcanoes. In the inset map of Antarctica, curved arrows represent schematic atmospheric paths from the southwestern Atlantic penetrating the EAIS. These transport paths match the continuous advection of continental dust from South America to central Antarctica.

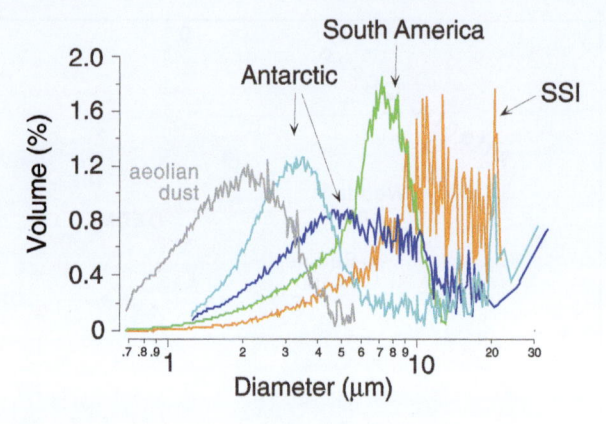

Fig. 3. Results of Coulter counter particle-size analysis. Volume–size distributions of EDC tephra layers compared to a typical sample of background aeolian dust. Modes of tephra samples range from 3 to 20 μm, while aeolian continental dust is typically much finer, with a mode of 2 μm and with particles larger than 5 μm practically absent. Also interesting is that tephras from South Sandwich Islands (SSI) and South America volcanism are coarser than layers from Antarctic sources. This feature indicates fast and efficient atmospheric transport of volcanic material from the southwestern part of the Atlantic region towards the central plateau.

(Fig. 2). Also, the trace element composition of the analysed sample is visually represented (Kohno *et al.* 2004, p. 582, fig. 6) but the data were not provided.

In the ice depth interval between 2500 and 3028 m (depth of DF2, the second ice core), Kohno *et al.* (2007) expected to recover tephra particles from another 13 visible layers. Unfortunately, at the time of writing (October 2018), details and geochemical analytical data of the 13 deep tephras are not yet published.

Dronning Maud Land. At Dronning Maud Land, at an internal site of the EAIS facing the Atlantic sector of the Southern Ocean (Fig. 1), an ice core down to a depth of 2774.15 m was drilled within the EPICA project. With its higher accumulation with respect to Dome C, this core (hereafter named EDML) was intended to provide a detailed climatic record of millennial-scale variability over the entire last glacial cycle (*c*. 145 kyr) and a direct link with the rapid climate oscillations documented in the Greenland Ice Sheet (e.g. EPICA Community Members 2006; Stenni *et al.* 2010). Kohno *et al.* (2005) reported that this core record contains 18 visible tephra layers (Fig. 5a). Seven of these were geochemically analysed for major elements. Unfortunately, to the extent of our knowledge, details of the identified layers and analytical data have not so far been published.

Laluraj *et al.* (2009) investigated five discrete depth intervals from a shallow ice core from the Central Dronning Maud Land region using scanning electron microscopy coupled with X-ray energy dispersive spectroscopy (SED-EDS), and identified volcanic particles showing a major element tholeiitic signature (K_2O 0.15–0.30%). Since this composition is typical of activity of the SSI, these findings confirm that these volcanoes represent a significant tephra contributor, especially in the Atlantic sector of the EAIS.

Talos Dome. This peripheral dome of the EAIS, located at the South Pacific–Ross Sea margin and adjacent to the Victoria Land mountains (Fig. 1), has been the site of ice-core drilling, recovering ice core down to a depth of 1620 m as part of the TALos Dome Ice CorE (TALDICE) European programme. Due to accumulation rates at this site being three–four times greater than those from sites of the inner EAIS, the TALDICE core has provided a high-resolution record of palaeoclimate and atmospheric history back through the previous two interglacial cycles (i.e. *c*. 250 kyr: see http://www.taldice.org for a comprehensive list of papers). Because of its location, relatively proximal (*c*. 200 km) to Quaternary volcanoes

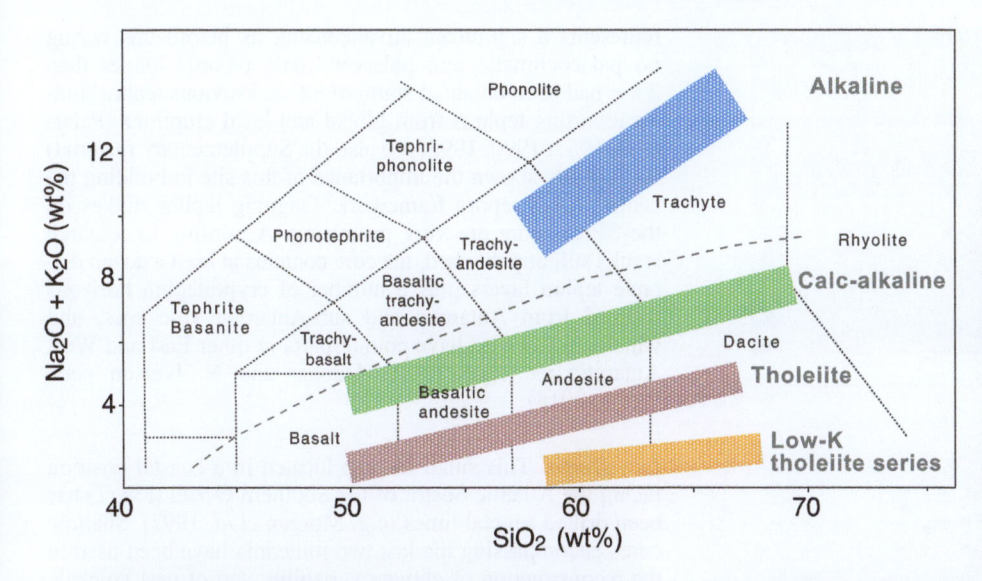

Fig. 4. Schematic chemical classification of ice-core tephra layers from the interior of the EAIS (adapted from Basile *et al.* 2001; and taking into account subsequent published geochemical results for EDC and DF). This schematic pattern provides an indication of the tectonic setting of the volcanic sources, as tholeiitic and calc-alkaline compositions are from subduction-related provinces, and alkaline from within-plate tectonic sources. Note the absence of silicic rhyolitic products that are typical of New Zealand explosive volcanism.

associated with the West Antarctic Rift System and a favourable position for aeolian transport with respect to volcanoes in northern Victoria Land (Scarchilli *et al.* 2011), this ice core also preserves a significant volcanic archive that consists of several tens of tephra layers. Detailed investigations of the layers from the last climatic interglacial sections have been performed, including quantitative grain-size measurements, microscopic inspection of particles, and single glass shard analysis for major and trace elements (Narcisi *et al.* 2001, 2010*a*, 2012, 2016, 2017). The examination of older tephra layers is in progress. Each studied layer was precisely positioned stratigraphically relative to the stable isotope record for the core and the glaciochemical profile. The core was dated using AICC2012, the coherent timescale for four East Antarctic ice cores (Bazin *et al.* 2013; Veres *et al.* 2013), providing a stratigraphic criterion of crucial importance for correlation.

The majority of tephra layers are visually prominent and coarse grained (Fig. 5b), and display an alkaline character (see the Supplementary material), clearly indicating a connection with the nearby Antarctic volcanoes related to extension-related magmatism. Along with distal traces of pyroclastic deposits already mapped at the source (e.g. the so-called Edmonson Point trachytic ignimbrite produced by a Mount Melbourne subaerial Plinian-scale caldera-forming eruption: Giordano *et al.* 2012) (Fig. 6; see also the Supplementary material), several englacial tephras document explosive events not yet recognized in the proximal record. The TALDICE tephra inventory demonstrates the recurrence of local volcanic activity with a regional impact, and represents a significant complement to our knowledge of the volcanic history in northern Victoria Land.

Taylor Dome. A 554 m-deep ice core has been retrieved at this East Antarctic site located in South Victoria Land, *c.* 150 km from the western edge of the Ross Sea (Fig. 1). The climate record spans the complete last glacial–interglacial cycle with good resolution and, although highly compressed, extends back to two or more glacial cycles (Morse *et al.* 2007).

Three tephra layers have been characterized by Dunbar *et al.* (2003), who concluded that the glass composition, basanitic and trachytic, is consistent with derivation from sources on the Antarctic continent (Supplementary material). Based on consistent age and geochemistry, the trachytic ash layer at 79.155 m depth dated at 1321 ± 25 CE has been correlated with a tephra in the West Antarctic Siple Dome ice core.

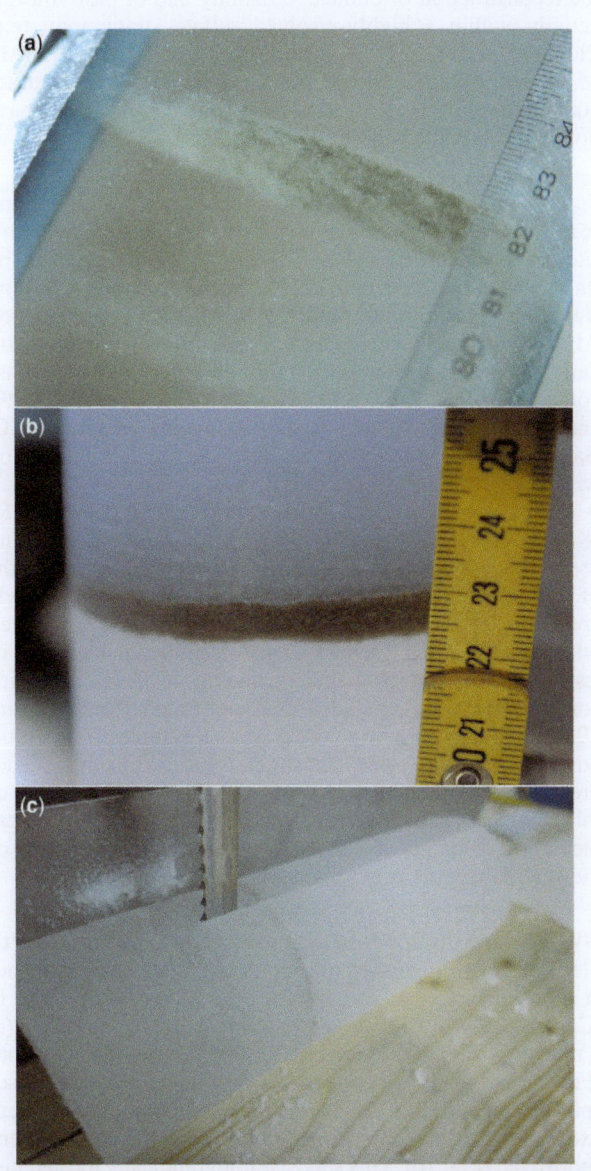

Fig. 5. Visible tephra layers from EAIS ice cores showing the varied macroscopic appeareance. (**a**) Tephra layer at 1489.82 m depth from the EDML ice core (AICC2012: age 48.04 ± 0.96 ka). Photograph: S. Kipfstuhl. (**b**) Tephra layer at 719.78 m depth from the TALDICE ice core (AICC2012: age 13.09 ± 0.34 ka). Photograph: J. Chappellaz. (**c**) Tephra layer at 183.07 m depth in the GV7 ice core (AD 1254 ± 2). The thickness is *c.* 0.2 cm. Photograph: B. Delmonte.

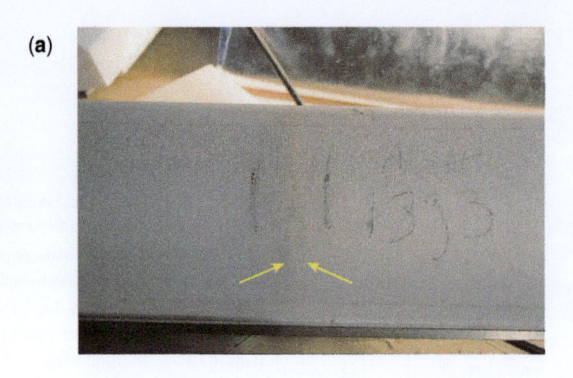

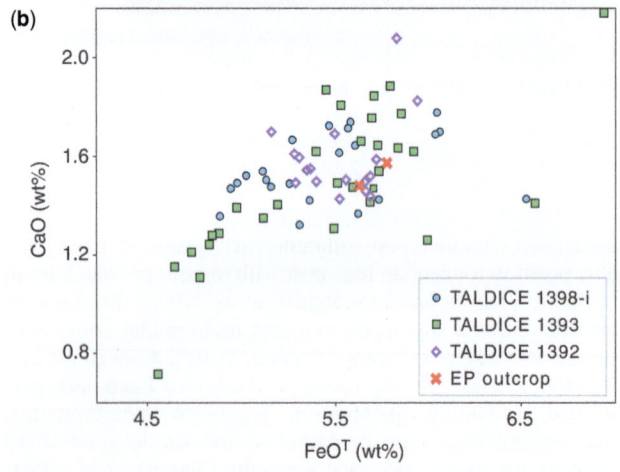

Fig. 6. Correlation between distal englacial tephras and specific proximal volcanic deposits. (**a**) One of the three TALDICE ash layers (AICC21012: 120.0 ± 2.1 ka BP according to the glaciological timescale) represents the distal counterparts of the Edmonson Point trachytic ignimbrite from Mount Melbourne (Giordano *et al.* 2012). The thickness is *c.* 0.7 cm. Photograph: V. Maggi. (**b**) Chemical comparison between single glass shard analyses (total *n* = 69) of the three ice-core tephra layers (Narcisi *et al.* 2016; see also the Supplementary material) and whole-rock analyses of proximal deposits (EP outcrop, Edmonson Point: Giordano *et al.* 2012). Note the compositional similarity, along with the considerable similarity in age, confirm the correlation. Note also that, unlike the bulk chemistry of the proximal samples, the grain-specific approach applied to distal tephras is capable of capturing a wider spectrum of compositions in order to unravel the complex geochemistries inside single horizons.

Intermediate and short ice cores

A number of ice cores, spanning in age from one millennium to a few tens of thousands of years, have been retrieved in the EAIS in recent times. The general purpose of these corings, carried out at relatively high accumulation sites, is to improve our knowledge of natural climatic and environmental variability on decadal to multi-millennial timescales and on a very recent time frame including the transition through to the commencement of significant human activities. Some of these high-resolution palaeoclimate archives have been examined from a tephrostratigraphic perspective. Studying englacial tephras deposited in historical to recent times significantly broadens the monitoring of volcanism and evaluates the present state of volcanoes (Patrick *et al.* 2005; Patrick and Smellie 2013), and enables a more robust assessment of volcanic hazards from future eruptions.

South Pole. Ice-core drilling to 1751 m was completed in January 2016 as part of SPICEcore (South Pole Ice Core) project (http://www.spicecore.org), at a site *c.* 2.7 km from the Admunsen–Scott South Pole Station (Fig. 1). This ice core, expected to provide a continuous record of the past 50 kyr, represents a significant advancement, as before this coring no palaeoclimatic and palaeovolcanic records longer than 2 kyr had been obtained south of 82° S. Previous tephra studies reporting tephras from global and local eruptions (Palais *et al.* 1987, 1990, 1992; see also the Supplementary material) had already shown the importance of this site in building the central EAIS tephra framework. Ongoing tephra studies on the SPICE core are very promising. According to research results still unpublished, the core contains at least a dozen discrete tephra layers plus a number of cryptotephra horizons sourced from Antarctic and sub-Antarctic volcanoes, and which are likely to have counterparts in other East and West Antarctic ice cores (N.W. Dunbar and N. Iverson pers. comm. 2018)

Law Dome. This small ice cap located in a coastal position facing the Atlantic Sector of the Southern Ocean (Fig. 1) has been drilled several times (e.g. Morgan *et al.* 1997). Shallow cores encompassing the last two millennia have been used in the reconstruction of climate variability and of past volcanic activity within a highly-resolved sulfate stratigraphy (e.g. Plummer *et al.* 2012). According to Kurbatov *et al.* (2003), 15 sections of the ice record relating to the last seven centuries contain rhyolitic volcanic glass. The results of this study have not yet been published.

Styx Glacier Plateau. Located in northern Victoria Land some 50 km from the coastline of the Ross Sea (Fig. 1), this site has been drilled twice. The first drilling campaign, by Italy (PNRA) in November 1995, reached a depth of 116 m. Based on a glaciological model, the ice core covers the period AD 1995–1350. Udisti *et al.* (1999) identified a thin (*c.* 1 cm) dark layer estimated to be roughly 500 years old. The glass composition is alkali-trachytic (average SiO_2 of 59.43 wt% and total alkali value ($Na_2O + K_2O$) of 13.83 wt%: C. Barbante pers. comm. 2010), suggesting a possible connection with a volcanic event in northern Victoria Land.

More recently, a 210.5 m ice core was drilled by KOPRI (Korean Polar Research Institute) (Han *et al.* 2015; Yang *et al.* 2018), with three visible tephra layers occurring at depths of 97.01, 99.18 and 165.37 m. Based on ongoing studies (Kyle and Lee 2018), a trachytic tephra within this core is correlated with the AD 1254 ± 2 tephra already identified in a number of East and West Antarctic sites, including the Talos Dome ice core where it was first identified and dated (Narcisi *et al.* 2001). In addition, new analyses of rock samples from northern Victoria Land volcanoes suggest that Rittmann volcano, presently in a quiescent state, is the most likely source of this significant explosive event, with implications for a volcanic hazard from this volcano. The publication of the details of the studied tephras is expected soon.

GV7. During the 2013–14 summer austral expedition, multiple firn/ice drilling down to a depth of 250 m was carried out at this site located at the eastern edge of the EAIS (Fig. 1) (Caiazzo *et al.* 2017). A single macroscopic tephra layer (*c.* 2 mm) was founded embedded in the ice at *c.* 183 m depth (Fig. 5c). It is associated with a prominent sulfate aerosol spike with concentrations comparable with that of the violent extra-Antarctic explosive events recorded in the core but very short-lived (Fig. 7b). This spike is stratigraphically located just below the broad prominent sulfate signal unambiguously related to the well-known AD 1259 event (R. Nardin pers. comm. 2017). We performed geochemical analyses (major elements by WDS electron microprobe and trace elements by LA-ICP-MS) of single glass shards and then compared the results with data from Antarctic rocks and known englacial tephras. The particulate material, up to *c.* 100 μm, is mostly correlated with a tephra layer in the Taylor Dome and Siple Dome ice cores (Dunbar et al. 2003), and to a tephra recently identified both in the RICE ice core (Tuohy 2015) drilled on Roosevelt Island at the northwestern edge of the Ross Ice Shelf (Bertler et al. 2018) and further away in central West Antarctica in the East Antarctic Ice Sheet (WAIS) Divide ice core (Dunbar et al. 2013) (Fig. 7a). Good geochemical matches between our results and published data corroborate this correlation (Fig. 8). According to Narcisi et al. (2001), this tephra could derive from activity of The Pleiades or Mount Rittmann volcanoes. Ongoing studies identify Mount Rittmann volcano, presently in a quiescent state, as the most probable source of this tephra, which also occurs at the Styx Glacier Plateau (Kyle and Lee 2018). The GV7 findings allow enlargement of the known dispersal area of this tephra further north (Fig. 7a) and suggest the likely occurrence in marine sediments from the Ross Sea. Indeed, the available marine records contain pyroclastic fall layers showing alkali-trachytic compositions that were most likely to have been sourced in northern Victoria Land (Del Carlo et al. 2015). However, unfortunately, they are not adequately constrained for chronostratigraphy to attempt a correlation. Interestingly, while widespread dispersal of the tephra occurred across the Pacific sector of the West and East Antarctic ice sheets, no corresponding sulfate peak is present in the EDC record (fig. 3 of Severi et al. 2012). This suggests a decoupling between atmospheric circulation at high-altitude sites of the central EAIS and the periphery of the ice sheet over the last few centuries. This is coherent with evidence arising from dust studies, showing marked differences between high-altitude sites of the central EAIS and marginal locations close to the Transantarctic Mountains (Baccolo et al. 2018). This is the youngest well-documented tephra event involving most of the Pacific sector of Antarctica. Knowledge of such recent explosive activity with widespread tephra dispersal has also implications for hazard management of future eruptions.

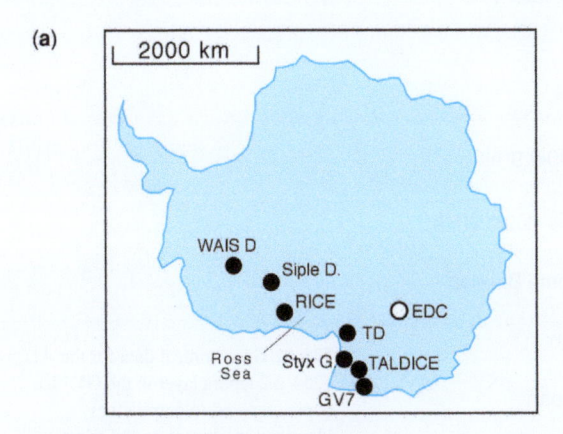

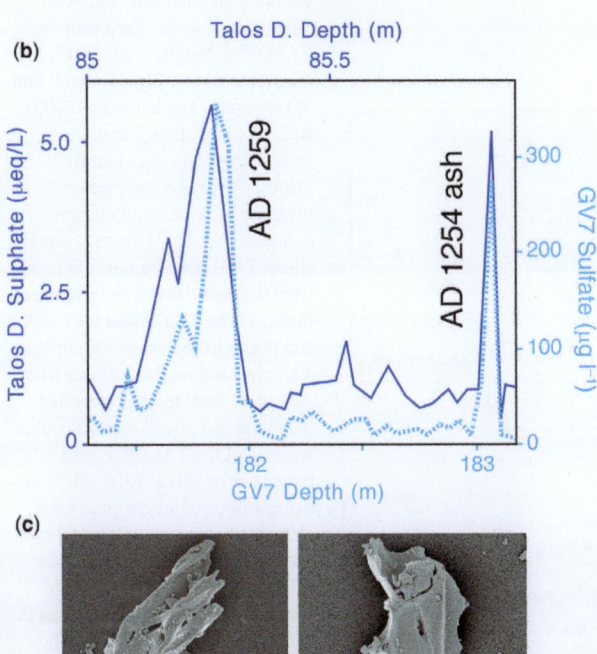

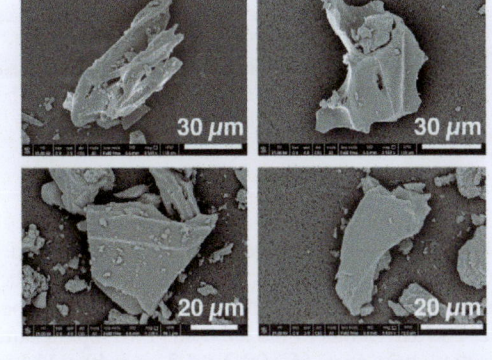

Fig. 7. Details of the AD 1254 ± 2 tephra layer. (**a**) Antarctic map showing the sites of known occurrence of the tephra. Note the inner location of the EDC core where no trace of volcanic deposition is present. (**b**) Volcanic sulfate profile in the GV7 ice core showing the stratigraphic position of the peak associated with the tephra with respect to the AD 1259 signal. The Talos Dome record from a pilot shallow core in which this tephra was identified for the first time is shown for comparison (redrawn from Narcisi et al. 2001). (**c**) Scanning electron photomicrographs of typical volcanic glass particles from the GV7 tephra. Photographs: B. Narcisi and M. Tonelli.

made up of juvenile glass shards (Fig. 7c). Analysis of 48 glass shards indicates that the tephra is alkali-trachytic and phonolitic in composition (Fig. 8; Supplementary material). Both the coarse grain size and geochemical characteristics suggest a volcanic source in northern Victoria Land. Based on chemical (major and trace elements) similarities and the coherency of the stratigraphic position, directly below the AD 1259 volcanic spike, we infer that the GV7 tephra correlates with the volcanic horizon found in the Talos Dome ice where it was dated as AD 1254 ± 2 (Narcisi et al. 2001, 2012). It is also

Outcrop exposures and blue-ice records

Tephra layers cropping out interbedded with snow and ice are common at the periphery of the Antarctic ice sheet (e.g. Dunbar et al. 1995; Pallàs et al. 2001; Moore et al. 2006; Dunbar et al. 2008) (Fig. 9). Although the exposures may be sporadic or spatially limited and discontinuous, these deposits can provide important information for reconstructing past explosive volcanic activity, assuming that the ash layers are preserved within ordered successions and their stratigraphic context is determined. In contrast to ice cores in which the ash samples provide very small samples needing dedicated protocols to enable analysis, surface occurrences typically provide ample material for tephra investigations including, potentially, for radiometric dating (e.g. Wilch et al. 1999).

Among ice–tephra outcrops, those from blue-ice areas are of particular significance. They have long attracted scientists because of their potential to provide easy access to older sections of the Antarctic ice sheet (Sinisalo and Moore 2010). Provided that the internal structure of the ice record is not structurally complicated, a series of samples collected in stratigraphic order can be measured for gases in trapped air, dust and stable isotope composition producing detailed records of past climate conditions and providing a chronostratigraphic framework for the interlayered tephra layers (e.g. Korotkikh et al. 2011; Higgins et al. 2015).

Results of recent studies emphasize the potential of exposed tephra successions in East Antarctica as a significant resource of palaeovolcanic information.

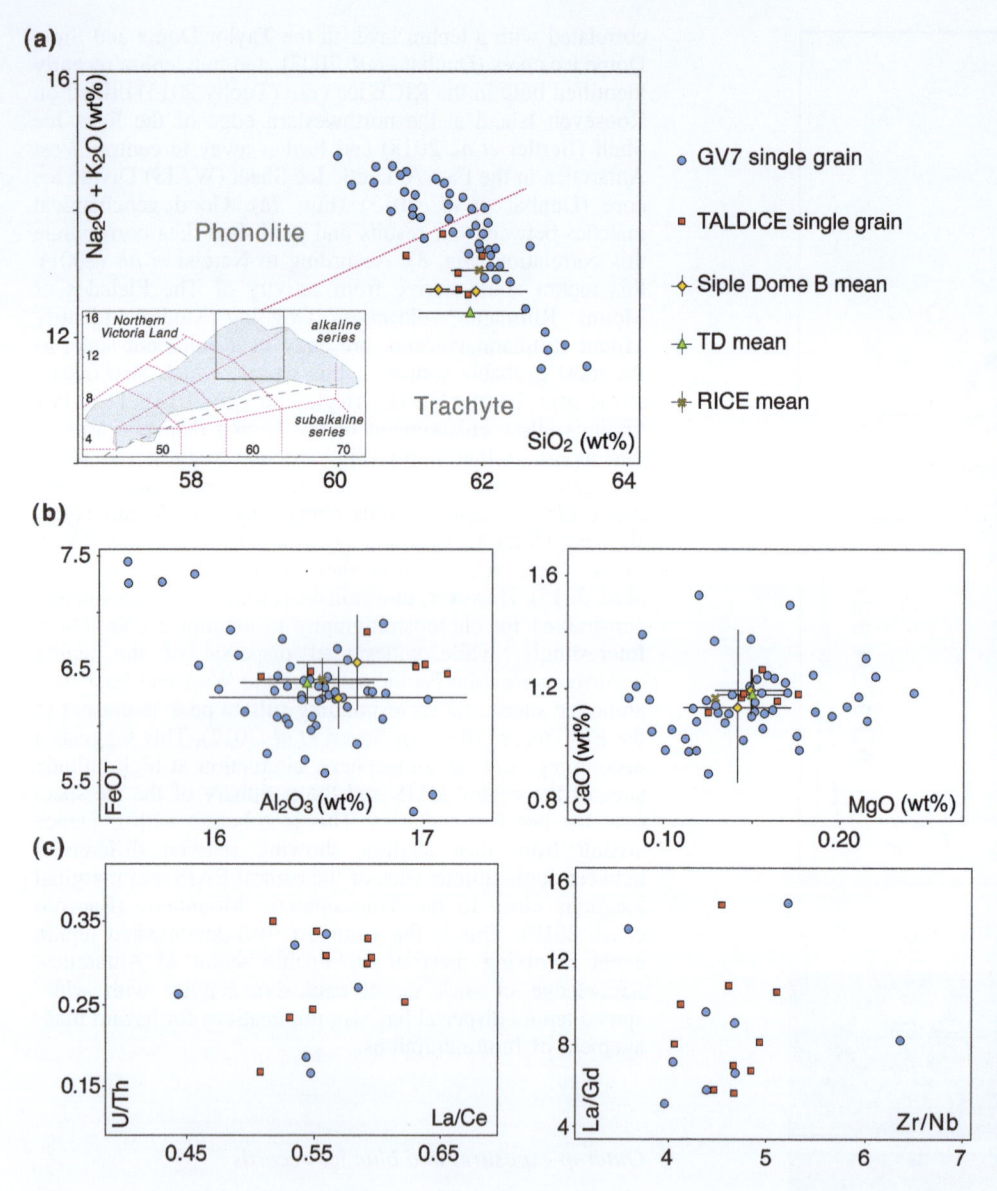

Fig. 8. Geochemical data for the AD 1254 ± 2 tephra layer in the GV7 ice core compared to correlatives identified in other East and West Antarctic ice series. Data sources: TALDICE, Narcisi *et al.* (2012) and the present study; Siple Dome B and TD ice cores, Dunbar *et al.* (2003); RICE ice core, http://antt.tephrochronology.org/I.html?id=AntT-16. (**a**) & (**b**) represent bivariate plots of major element compositions. In (a) the total alkali–silica (TAS) classification diagram is from Rickwood (1989 and references therein). The inset shows the compositional envelope for northern Victoria Land products drawn from data of all local tephras identified within the last glacial–interglacial ice sections of the TALDICE core (Narcisi *et al.* 2012, 2016, 2017). (**c**) Comparisons of single-grain trace element composition of the tephra in the GV7 and TALDICE ice cores determined by LA-ICP-MS (this study; the full dataset is provided in the Supplementary material). The coherent geochemical signature clearly supports the proposed correlation.

Allan Hills. This blue-ice field located near the NW corner of the McMurdo Dry Valleys in South Victoria Land (Fig. 1) is likely to preserve a continuous record through at least marine isotope stage 11 (MIS 11), *c.* 400 kyr (Kehrl *et al.* 2018). It also contains an outstanding tephra record that has not been systematically studied so far to reconstruct the detailed tephrostratigraphy. In an early study, Dunbar *et al.* (1995) presented a map of the ash layers and plots of major and trace element analyses. Basanitic, trachytic and phonolitic compositions were reported for the layers, suggesting an origin in Victoria Land volcanoes and centres in the southern McMurdo Sound area. More recently, Iverson *et al.* (2014) presented results of a few englacial layers used to reconstruct the Mount Erebus eruptive and magmatic history. Iverson (2017) also obtained the first radiometric age (202 ± 7 ka) on a tephra from the Allan Hills site, whose trachytic composition suggests an origin in northern Victoria Land volcanism.

Frontier Mountain. This blue-ice field located within the Transantarctic Mountains at the edge of the EAIS (Fig. 1), along with the nearby Lichen Hills, hosts several tens of englacial tephra layers (Fig. 9b, c). Twenty-two of the largest and best-exposed layers in a *c.* 1150 m succession were studied for their mineralogical, geochemical and morphometrical features (Curzio *et al.* 2008). The major element composition of glass shards is trachytic in most of the layers. This feature, along with the trace element compositions determined in a few layers and with the coarseness of the tephras, indicates derivation from northern Victoria Land volcanoes located within a *c.* 250 km radius of the ice field. Two tephra samples yielded radiometric ages of 100.2 ± 5.3 and 49 ± 11 ka (by $^{40}Ar/^{39}Ar$), suggesting that the succession may represent a *c.* 100 kyr continuous chronicle of explosive volcanic events in northern Victoria Land. However, a number of one-to-one stratigraphic correlations with tephra layers in the TALDICE ice core, *c.* 30 km from Frontier Mountain, suggest that more than half of the exposed blue ice is indeed younger than 15 ka (Narcisi *et al.* 2012, 2017).

Mount Erebus flanks and blue-ice fields. Thorough studies of tephras embedded in blue-ice areas and firn within a radius of a few tens of kilometres from Mount Erebus volcano have provided an invaluable inventory of explosive events that substantially complements the prehistorical eruptive history of this volcano as reconstructed from proximal rock outcrops (see Chapters 5.2a (Smellie and Martin 2021) and 5.2b (Martin *et al.* 2021)).

Harpel *et al.* (2008) presented a micromorphological, mineralogical and geochemical study of tephra samples from 46 individual layers cropping out at the Barne Glacier on the western flank of Erebus. Tephra are mainly from phreatomagmatic eruptions with fewer from Strombolian eruptions.

Fig. 9. Examples of englacial tephras exposed in outcrop. (**a**) Prominent tephra layer in the snow cliff at the Campbell Glacier tongue, close to the Italian 'Mario Zucchelli' station. Photograph: M. Frezzotti. (**b**) & (**c**) Tephra layers cropping out in the Frontier Mountain blue-ice field. Photographs: J. Chappellaz, CNRS/LGGE/IPEV.

Mixed ash morphologies are also common, suggesting a shift from phreatomagmatic activity to magmatic activity as the water source was depleted. Two layers were $^{40}Ar/^{39}Ar$ dated at 71 ± 5 and 39 ± 6 ka. All of the tephra are phonolites with only minor chemical variations, indicating that the Erebus magma composition has been relatively homogeneous throughout the investigated time period (Fig. 10). Two tephra deposited on the EAIS, *c.* 200 km from Erebus, resulted from Plinian and phreatomagmatic eruptions, and show the potential of this volcano to disperse tephra widely onto the Antarctic plateau (Harpel *et al.* 2008).

More recently, Iverson *et al.* (2014) conducted an investigation of tephra layers in blue ice on the flanks of Erebus volcano not sampled before. Thirty-five englacial layers were studied for their particle microscopic characteristics, grain size, and major and trace element composition of individual glass shards. $^{40}Ar/^{39}Ar$ radiometric dating of two samples produced maximum ages of 40 ± 20 and 36 ± 10 ka. The majority (31) of the studied samples are phonolitic with an average SiO_2 of 55.60 wt% and total alkali value ($Na_2O + K_2O$) of 14.48 wt% (Supplementary material). These layers, produced by both magmatic and phreatomagmatic eruptions, are compositionally indistinguishable from the current lava lake at Erebus, indicating that the major and trace element composition of the magmatic system has remained unchanged during the past *c.* 40 kyr regardless of the eruptive style (Fig. 10). The studied ice record has significantly complemented the proximal volcanic stratigraphy, resulting in a more complete reconstruction of the explosive history and magmatic evolution of Erebus volcano.

Nansen Ice Field. This site, located south of the Sør Rondane Mountains in eastern Dronning Maud Land (Fig. 1) and known as one of the richest meteorite traps, contains a series of tephra layers. Several tephra-bearing ice blocks were analysed for geochemistry and the magnetic properties of volcanic ash particles (Oda *et al.* 2016). Both the grain size of the ash particles (*c.* 50 µm) and the low-K basaltic andesitic composition of the volcanic glass suggests an origin from the SSI. Unfortunately, the studied ash samples are of undefined age and stratigraphy, and therefore cannot provide any information about the chronology of past explosive events. These results are nonetheless interesting as they confirm both the importance of the SSI as a source of volcanic ash deposited in the EAIS regions facing the Atlantic sector, and, as we discuss below, the major control exerted by the clockwise atmospheric circulation that affects the transport and dispersal of volcanic ash.

Main characteristics of the East Antarctic tephrostratigraphy

We have seen that within the last two decades considerable progress has been made in the study of Antarctic englacial tephras from different settings and geographical locations (Fig. 1). However, despite recent efforts the data are still scarce in comparison with the vastness of the EAIS and its intrinsic environmental regionality. Note, for example, that the coverage of the interior of the EAIS is made up of a handful of tephra records located several hundred kilometres apart. This limited dataset clashes with the abundance of volcanic aerosol records now available (Sigl *et al.* 2014), which are, however, generally much shorter, less informative of past volcanic sources and of less use for stratigraphic correlative purposes than englacial tephra records. To some extent this

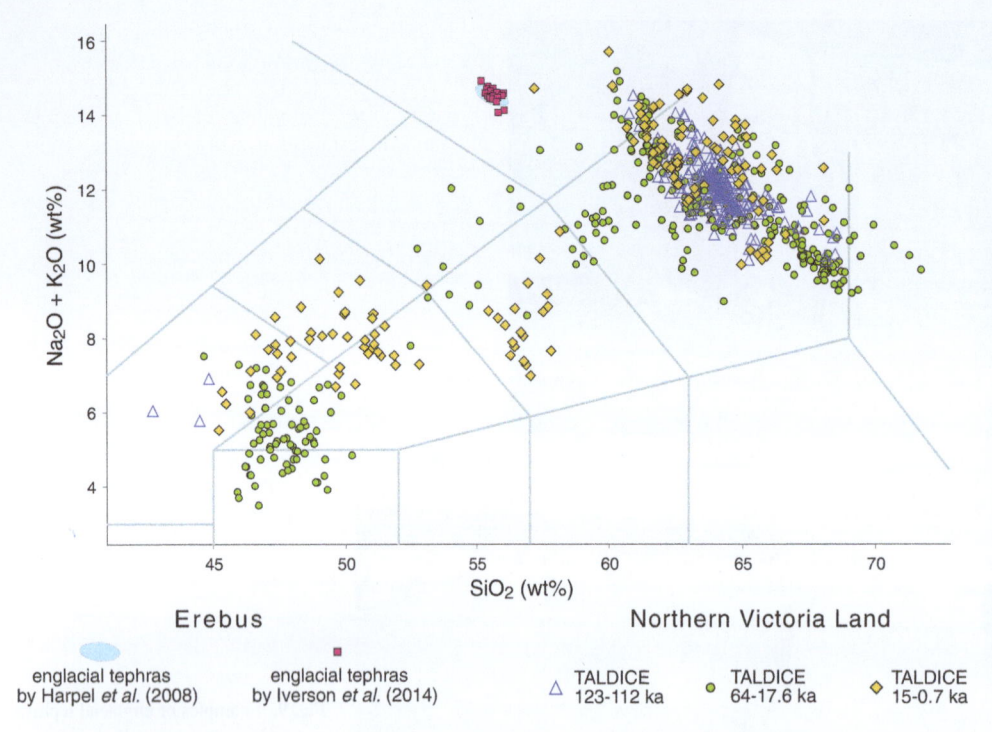

Fig. 10. Major element magma variations of Erebus and northern Victoria Land volcanoes deduced from the compositions of distal englacial tephras (Erebus, various sites from flanks of the volcano: Harpel et al. 2008; Iverson et al. 2014) (northern Victoria Land, TALDICE ice core: Narcisi et al. 2012, 2016, 2017). While the phonolitic magma composition of Erebus volcano has been relatively homogeneous throughout the last few tens of thousands of years, the northern Victoria Land ice-core products show a wide range of compositions and a progressive geochemical shift over the last c. 120 kyr.

paucity reflects the fact that the study of tiny englacial tephras can be challenging and time-consuming, and also requires careful interpretation of the data. Therefore, in spite of the increased opportunity to get good-quality tephra time series, the considerable potential of tephrochronology for successfully correlating widely dispersed palaeoclimate records and for providing information on the eruptive history of source volcanoes is still insufficiently exploited.

Inner EAIS

Based on the available results from deep cores of EDC, VK and DF, all very remotely located with respect to plausible volcanic sources, Narcisi et al. (2010b) developed a tephrostratigraphy for the inner part of the EAIS spanning the last several glacial cycles (Fig. 2). Note that, unfortunately, since the publication of that synthesis very few results from further long cores have become available, and therefore those conclusions are still relevant at the time of writing. First, it is interesting to note that although the ice-core sites are in different geographical positions within the inner EAIS and with respect to Quaternary active volcanoes within and around Antarctica, they show the same order of magnitude of recorded tephra frequency (one–two per 20 kyr over the last two climatic cycles) (Fig. 2). Second, although occurring in different proportions, they all appear to be collectors of tephras from multiple sources with distinct compositions that reflect the different tectonic settings of the volcanic sources (Fig. 4). This distal multi-sourced tephra record includes traces of eruptions from the SSI, possibly Antarctic Peninsula, WARS and South America. Worthy of note is the absence of siliceous rhyolitic compositions (SiO$_2$ >74 wt%) typical of New Zealand volcanoes (e.g. Smith et al. 2005), in spite of their frequency and occurrence of high-magnitude explosive eruptions in the Late Quaternary. Also remarkable is the preponderance of tephra layers from SSI volcanoes (Fig. 2). The latter feature is corroborated by the results on the deeper parts of the DF core, and by findings in the Nansen Ice Field and in shallow Dronning Maud Land core, along with results of investigations of deep-sea sediments in the southern Atlantic Ocean and Weddell Sea (e.g. Hubberten et al. 1991 and references therein; Nielsen et al. 2007). It confirms the previous results by Palais et al. (1987, 1989), who had argued that these sub-Antarctic sources were prolific contributors of compositionally distinctive (tholeiitic) tephras. Some of them appear widely dispersed over large areas of the EAIS, suggesting their generation during major explosive eruptions, evidence for which, however, is virtually absent in the islands themselves (J.L. Smellie pers. comm.). In the Holocene sections, the most widespread englacial tephra sourced in the SSI has an AICC2012 age of 3.60 ± 0.19 ka BP, and has already been identified in EDC and VK ice cores, as well as in an old record from South Pole (so-called 'Vostok tephra' of Palais et al. 1987). A major tephra marker sourced in the SSI is also present in the ice record at 201.75 ± 2.00 ka BP, and is found in EDC, VK and DF cores (Narcisi et al. 2005). It is also noteworthy that the South American volcanoes plausibly represent the most important non-Antarctic contributor that produced widespread detectable englacial tephra layers common to more than one long core (see the Supplementary material). Although Del Carlo et al. (2018), based on geochemical reasoning on the known southern Andes volcanic record, have recently disputed the South American origin of Antarctic englacial tephras, this is strongly confirmed by diverse evidence. Indeed, the observed occurrence of tephra layers in the central EAIS, along with their relative coarseness (Fig. 3), is most easily explained by the major role of the clockwise circum-Antarctic atmospheric circulation penetrating the Antarctic continent. Similar to what has been observed for a recent eruption from the south of Chile whose ash cloud reached the WAIS Divide field camp (Koffman et al. 2017), and in agreement with multi-decadal analysis of atmospheric back trajectories (Narcisi et al. 2012; Neff and Bertler 2015), this circulation has strongly influenced the past transport of volcanic (and continental) dust towards the Atlantic and Indian sectors of the Antarctic plateau (e.g. Delmonte et al. 2008). In this respect, therefore, the inner EAIS sector represents a region that preferentially collects tephras, resulting in an integrated record of past explosive activity in the Atlantic sector. Interestingly, the tephrostratigraphy for the inner Antarctic plateau shows a drastic decrease in detectable tephra before c. 220 ka, with tephra layers even disappearing from c. 800 to 414 ka (i.e. the bottom of EDC core) (Fig. 2). Among various plausible causes that may have determined this continentally significant trend, including artefacts of tephra detection and post-depositional physical processes acting in very deep parts of the ice record and affecting continental particles (Lambert et al. 2008; Traversi et al. 2009), Narcisi et al. (2010b) suggested a decrease in explosive activity for the tephra sources, particularly the SSI.

Although minor compared with tephra derived from South Atlantic volcanism, the interior of the EAIS also preserves traces of Antarctic extension-related eruptions, with plausible volcanic centres situated on both sides of the rift (Fig. 2). For instance, a tephra layer identified in the EDC and Dome Fuji ice cores is correlated by Narcisi et al. (2006) with a pyroclastic layer erupted from Mount Berlin, Marie Byrd Land, and radiometrically dated at 92.5 ± 2.0 and 92.2 ± 0.9 ka (Wilch et al. 1999). A further Antarctic-related tephra showing alkaline-trachyandesitic composition, dated at c. 141 ka, occurs in both the EDC and Dome Fuji ice cores, suggesting widespread deposition at great distances from the source. Also, older EDC (c. 360 ka) and VK (c. 406 ka) tephra layers appear to be related to WARS volcanism (Fig. 2). All of these cases are characterized by remarkable pan-Antarctic continental dispersal of the tephra and indicate significant explosive events, with ash plumes rising high enough in altitude through the very stable Antarctic atmosphere to enter into the stratospheric polar vortex.

Peripheral EAIS

The Antarctic tephrostratigraphy for the margin of the EAIS shows very different features with respect to the inner sector. In particular, the series facing the Ross Sea sector, and collected from areas adjacent to active extension-related volcanoes, show an average frequency of visible tephra layers estimable to be one order of magnitude greater than the tephra frequency in deep cores from the inland EAIS. A similar pattern is shown by West Antarctic tephra sequences recovered in the east Pacific sector and proximal to active Marie Byrd Land volcanoes (e.g. Gow and Meese 2007; Dunbar et al. 2008; Dunbar and Kurbatov 2011; see also Chapter 8.4 (Dunbar et al. 2021)). In these regional depocentres, direct fallout of ash from the troposphere by gravitational settling predominates, with a more limited influence of the continental-scale clockwise circum-Antarctic stratospheric circulation. These tephra sequences are typically rich in visually prominent and coarse-grained layers and appear to be dominated by a local single region producing tephra, at least at the level of macroscopic layers. Although such a plethora of tephra layers with similar signatures within a single ice record could represent a limitation for both widespread and local correlations, it nevertheless has tremendous potential for palaeovolcanic purposes. The two Victoria Land case studies of TALDICE and Erebus tephra series clearly illustrate these benefits (Fig. 10).

The tephra inventory of the Talos Dome ice core is a detailed record of northern Victoria explosive eruptions that occurred during the last glacial–interglacial cycle which substantially increases our palaeovolcanic knowledge of compositional and temporal variations. The TALDICE tephra record, built on a wealth of individual glass shard compositions (Fig. 10), shows that the regional volcanism underwent a progressive compositional shift through time towards less evolved compositions that had not been detected from field studies and related bulk-rock analyses (Giordano et al. 2012). This highly-resolved volcanic record also provides evidence that the northern Victoria Land volcanoes, presently in a quiescent state, were more frequently active than previously recognized from field studies (Narcisi et al. 2019). Moreover, tephra correlations suggest that a few locally originated events were sufficiently powerful to produce ash dispersal to the Ross Sea sector and further away to the WAIS. This is the case, for instance, of the TALDICE trachytic tephra dated at 15.19 ± 0.14 ka, possibly originating from The Pleiades volcanoes (see the Supplementary material), that may equate to a tephra layer in the West Antarctic Siple Dome ice core studied by Dunbar and Kurbatov (2011), c. 1500 km distant from the TALDICE coring site (Narcisi et al. 2012).

The TALDICE record also permits more precise dating of important volcanic events. At Mount Melbourne volcano, the pyoclastic deposit examined in outcrop, comprising three main subunits and emplaced during a Plinian-scale caldera-forming eruption was radiometrically dated at 123.6 ± 6.0 ka (Giordano et al. 2012). The three corresponding tephra layers in the TALDICE ice core (Fig. 6) have more precise glaciological ages of 123.3 ± 2.2, 120.0 ± 2.1 and 119.5 ± 2.1 ka, respectively (Narcisi et al. 2016). Also, lava and scoria samples locally radiometrically dated at 90.7 ± 19.0 ka (Giordano et al. 2012) are represented by a tephritic layer more precisely dated at 116.9 ± 2.0 ka in the TALDICE record (Narcisi et al. 2016).

A few TALDICE layers are situated in particularly interesting stratigraphic positions with respect to the palaeoclimatic and/or glaciological records. The trachytic tephra occurring in the Last Glacial Maximum section at 17.61 ± 0.73 ka (Fig. 11) is especially interesting as it settled during the well-known two–three-centuries-long acidity event detected in ice cores from the EAIS and WAIS (Schwander et al. 2001 and references therein; Sigl et al. 2016). The source volcano for this outstanding event in terms of both duration and intensity has been matter of debate since its first identification in the Byrd ice core (Hammer et al. 1997). La Violette (2005) suggested a possible cosmic origin of this event. More recently, in a study of the EDC core, Vallelonga et al. (2005) inferred an origin from Antarctic volcanism. Indeed, the relative coarseness of this TALDICE tephra (up to c. 40 µm in size) along with its major and trace element composition points to a local source most probably located in Victoria Land. An Antarctic source is further confirmed by an inspection of tephra in ice-core sections associated with the fluoride signal at EPICA Dome C. No distinct volcanic particles were found (S.W. Davies pers. comm. 2013). Therefore, the volcano-derived aerosol sourced somewhere in the West Antarctic Rift System and deposited across distal parts of the inner EAIS was not accompanied by fallout of volcanic particles.

More recently, McConnell et al. (2017) presented the results of a detailed glaciochemical study of this long-lived acidity anomaly in the WAIS Divide ice core, along with ancillary chemical data for ice from Taylor Glacier in the Antarctic Dry Valleys, close to the Ross Ice Shelf. The results indicate that this complex event consisted of several individual pulses. In addition, three of the detected pulses produced tephra, and the tephra particles analysed (c. 10–30 µm) are chemically classified as trachyte, with a composition suggestive of an origin from the Marie Byrd Land volcano of Mount Takahe, located a few hundred kilometres from the coring site.

We point out here that the TALDICE and WAIS Divide tephras, despite both occurring within the acidity anomaly event and classified as trachytes, display a different chemical signature over most of the major elements examined and therefore are genetically unrelated. Therefore, we propose that the long-lived gaseous emission at c. 17.6 ka was produced by coeval activity in different volcanic regions of the West Antarctic Rift System. As such, this example underscores the importance of tephra studies to achieve the correct source linkage for specific events, including the identification of contributions from multiple distinct eruptions (Narcisi et al. 2019). It also indirectly points to limitations in reconstructing past for the 948 m-deep ice core from Berkner Island (Fig. 1) possibly spanning the last glacial–interglacial cycle (Mulvaney et al. 2007). These peripheral records, which are likely to be little influenced by volcanic activity on the Antarctic continent, could fruitfully document past activity of sub-Antarctic volcanoes.

To sum up, due to the marked regionality, at present it is very difficult to develop a comprehensive tephrostratigraphic framework for the peripheral EAIS; however, a few widespread layers already appear common to more than one series. For the last 16 kyr, Narcisi et al. (2012) presented a tephrostratigraphic framework showing key tephras that allow a link between East Antarctic (both peripheral and internal) and West Antarctic ice cores. In addition, we highlight a trachytic layer cropping out in blue ice near Mount Erebus that was correlated by Iverson et al. (2014) with a 28.5 ka tephra sourced from Mount Berlin (Marie Byrd Land), and also identified in the Siple Dome and WAIS Divide ice cores c. 1400 km away (see the Supplementary material).

Long-term climate–volcanism relationships: potential applications of englacial tephra sequences

Volcanism and climate interact. That major volcanic eruptions through the emitted stratospheric aerosols can impact climate on a large scale is widely accepted (e.g. Masson-Delmotte et al. 2013 and references therein). Conversely, there is an ongoing debate over the possible link between volcanism and glaciation on multi-centennial timescales for various geographical contexts including Antarctica (Huybers and Langmuir 2009 and references therein; Watt et al. 2013; Rawson et al. 2016; Cooper et al. 2018). The basic mechanism for regions experiencing deglaciation is that during extreme environmental changes, such as those occurring from glacial to interglacial periods, glacial unloading related to climate change could cause mantle decompression and/or enhanced stress regime and faulting inducing a regional increase in volcanic activity and changes in magma chemistry; however, there are other associated effects producing feedbacks that complicate this simple picture.

It is beyond the scope of the present chapter to discuss this critical topic and draw conclusions on whether Antarctic explosive volcanism may have been triggered by glacial unloading. However, for completeness of our review we report on a few published attempts to use results of Antarctic englacial tephra studies for this purpose. Zielinski (2000) in his review reported that strong evidence exists from ice cores from both polar regions of a relationship between environmental changes associated with rapid climatic fluctuations and an increase of volcanic activity. Nyland et al. (2013) also observed that the physical and geochemical characteristics of glass-rich sediments retrieved from western Ross Sea and sourced from the Erebus Volcanic Province responded to Early–Middle Miocene climate fluctuations. At TALDICE, Narcisi et al. (2016) observed that during the last interglacial period visible tephra layers occur from 123 to 105 ka BP, corresponding to the final stages of the last interglacial and subsequent glacial inception. Tephras are absent in the core sections related to the previous glacial period and to the early phases of the last interglacial. However, Narcisi et al. (2016) concluded that it is unclear whether this pattern represents a causal association between past climate and local volcanism or simple coincidence. Conversely, the last 70 ka visible tephra record of TALDICE shows a minimum of local explosive activity between 20 and 35 ka (i.e. glacial maximum and potential suppression of volcanism), and increased activity back to 65 ka (i.e. pre-glacial maximum

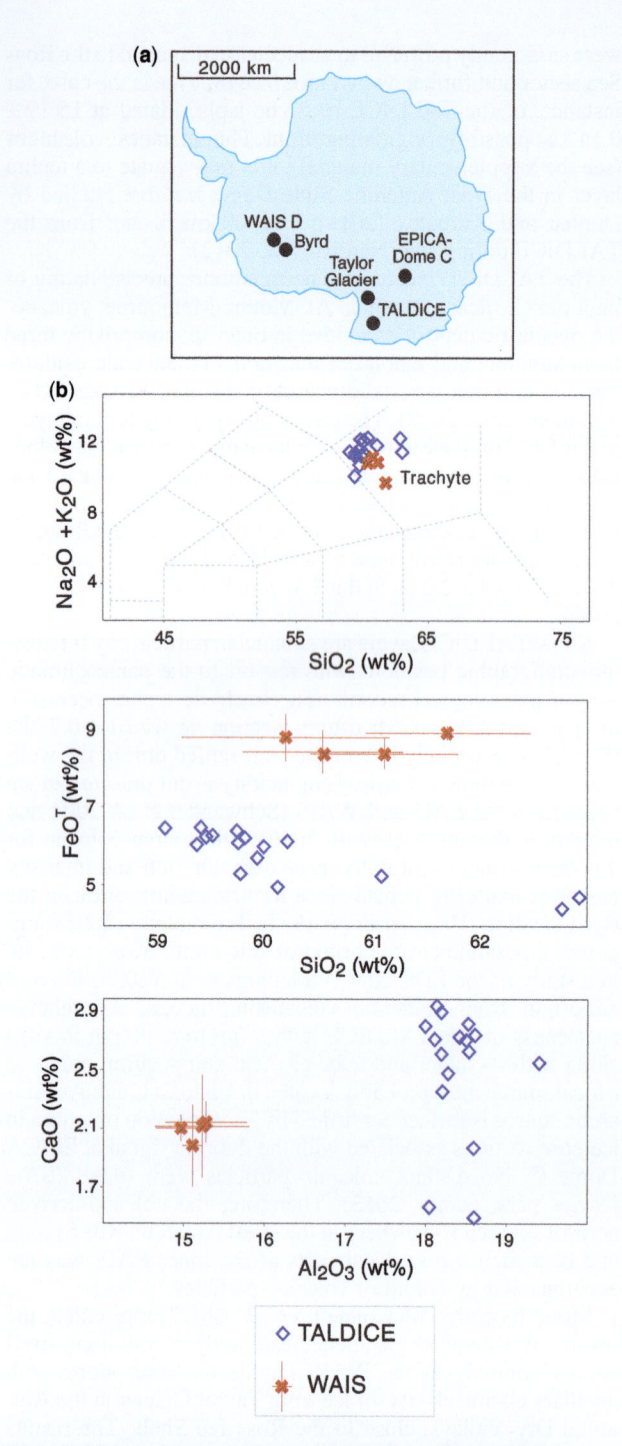

Fig. 11. Features of the well-known long-lasting acidity event at c. 17.6 ka and related ice-core tephras. (**a**) Map showing sites where the long-lasting signal has been identified thus far. (**b**) Biplots comparing major element compositions. TALDICE single-grain data of sample TD822 from Narcisi et al. (2017); West Antarctic Ice Sheet (WAIS) Divide ice-core-averaged data from McConnell et al. (2017). Note the differing compositions which are indicative of at least two eruptive sites contributing tephra to the 17.6 ka 'event'.

volcanic activity merely through volcanic glaciochemical profiles.

Unfortunately, there is still a lack of information on peripheral EAIS tephra series from the Atlantic and Indian Ocean sectors of Antarctica (Fig. 1). Some of them (e.g. the Yamato ice field, Dronning Maud Land) are already known as very promising sites for tephra studies (Nishio et al. 1985) but, unfortunately, have not been systematically investigated thus far. Similarly, to date, no tephra studies have been published and probably somewhat thinner ice: Narcisi et al. 2010a). Tephra records of the West Antarctic Byrd and Siple Dome ice cores (Gow and Meese 2007) show different timings for volcanic deposition: a lower tephra frequency at Talos Dome during the last glacial period corresponds to increased local volcanic input in the two West Antarctic cores. Despite these intriguing observations the current state of knowledge is insufficient to draw any firm conclusions. It is still unclear whether diachronous ice-sheet responses or more complex factors are involved, such as an interplay of climate-related, tectonic and magmatic factors.

It is worth noting that, although glaciochemical profiles are widely established as a reliable tool for the comprehensive reconstruction of past volcanic history (e.g. Sigl et al. 2014), they are unable in themselves to indicate the locations of the volcanoes responsible. Englacial tephras provide a more robust source linkage and are crucial for discriminating volcanic from other signals (Sigl et al. 2015; Narcisi et al. 2019). However, in addition to stratigraphic gaps and the potential presence of unrecognized cryptic tephra, englacial tephra-based volcanic records do not include traces of purely effusive activity; also, mafic tephras, less likely to travel far from source, are likely to be under-represented in ice archives at relatively distal sites. Furthermore, tephra distribution will be controlled by atmospheric and any atypical weather conditions, especially when a long tephra record spanning different palaeoclimatic modes or significant climatic transitions is considered. Thus, temporary and atypical deflections of volcanic plumes are possible and may create a spatial bias in the area(s) of deposition affecting the completeness of the tephra record.

Conversely, volcanic profiles reconstructed using aerosols preserved in the ice strata (e.g. Sigl et al. 2014) are not defect-free. In addition to post-depositional processes affecting the sulfate profile of deep ice-core layers that preclude their use for past volcanic reconstructions prior to 400 ka (Traversi et al. 2009), Antarctic and Greenland ice-core instances of young (<40 ka) tephra layers lack associated aerosol signals in the chemical record (Palais and Kyle 1988; Davies et al. 2008, 2010), thus questioning the completeness of the aerosol record derived by ice cores. In conclusion, investigating the relationship between climate change and any forcing by volcanism is complicated by several factors. There is the potential to reconstruct past event frequencies from englacial tephra sequences and to investigate the response of Antarctic volcanism to climatic variations but tephra data must be treated cautiously.

Conclusions and perspectives

Polar ice is one of the best recorders of explosive volcanic eruptions. Compared to other depositional settings, the study of Antarctic archives of englacial tephras presents a number of advantages. But there are also analytical limitations, mostly related to the small amount of ash and its very fine grain size, that require the use of specialized protocols for extraction and microanalysis and dedicated facilities.

Much work has been done since the preliminary findings of ultrathin microparticles in Antarctic ice by De Angelis et al. (1985). The increased availability of englacial tephra samples mainly recovered in ice cores and from blue ice during palaeoclimatic initiatives has produced considerable advances of Antarctic tephra studies. Publications during the last two decades have provided new insights into distal tephra deposition onto the high Antarctic plateau related to Late Quaternary explosive activity in the Antarctic continent and surrounding southern oceans. Compared with the often limited and inaccessible exposure of near-vent products and the remote locations of volcanoes in the south polar region, englacial EAIS tephras are a remarkable archive that can be used to reconstruct past volcanic activity in the region more completely. Recent Antarctic tephra research has also demonstrated significant volcanological and petrological implications both for volcanological and petrological purposes and for evaluation of volcanic impacts posed by future events.

We also emphasize that, with respect to glaciochemical measurements of volcanic aerosol fallout traditionally used for reconstructions of past volcanic activity, tephra findings have the potential to univocally identify source areas of a given signal. Notable in this respect is the case for the origin of the well-known major acidity event centred at c. 17.6 ka which, before the identification of tephra shed light into source volcanoes, has long been debated. Such robust source identification has consequences for regional palaeovolcanism reconstructions and for a reliable appraisal of the climatic effects of past explosive events. A wealth of continuous glaciochemical volcanic records are already available for Antarctica (Sigl et al. 2014), and an interesting future application of tephra studies could be to examine other well-known prominent signals in order to identify any volcanic particulate material.

In spite of recent achievements and ongoing projects, the potential of the EAIS region as a unique volcanic archive has not been fully exploited thus far. There is still a scarcity of englacial tephra information for periods prior to 800 ka, as well as for sites in sectors facing the Atlantic and Indian EAIS oceans (Fig. 1). It is anticipated that future tephra work will be stimulated by the International Partnership for Ice Core Science (IPICS) initiative, which aims to recover 1.5 myr of ice history by ice coring in inner Antarctica (Fischer et al. 2013). Such a project will clearly provide the opportunity to extend the ice-core tephra record much further far back in time than has been possible until now.

Acknowledgements We are grateful to the Editor J.L. Smellie for inviting us to write this paper. We thank C. Barbante, I. Basile, J. Chappellaz, B. Delmonte, S.W. Davies, M. Frezzotti, N.W. Dunbar, S. Kipfstuhl, P. Kyle, N.A. Iverson and R. Nardin for providing information and materials. We are indebted, in particular, to B. Delmonte for helpful comments on an early draft, and to N.A. Iverson and an unidentified reviewer for constructive suggestions that improved the paper.

We thank M. Tonelli, V. Batanova, V. Magnin and A. Langone for assistance during tephra preparation and microanalytical work within the GV7 project.

This is TALDICE Publication No 53.

Author contributions **BN**: conceptualization (lead), data curation (lead), visualization (lead), writing – original draft (lead); **JRP**: writing – review & editing (equal).

Funding The GV7 study carried out within this work was financially supported by the MIUR (Italian Ministry of University and Research)–PNRA (Italian Antarctic Research Programme) programme through the IPICS-2kyr-It project. This paper contributes to TALos Dome Ice CorE (TALDICE), a joint European programme led by Italy and funded by national contributions from Italy, France, Germany, Switzerland and the UK.

Data availability All data generated or analysed during this study are included in this published article (and its supplementary information files).

References

Baccolo, G., Delmonte, B. et al. 2018. Regionalization of the atmospheric dust cycle on the periphery of the East Antarctic ice sheet since the Last Glacial Maximum. *Geochemistry, Geophysics, Geosystems*, **19**, 3540–3554, https://doi.org/10.1029/2018GC007658

Basile, I., Petit, J.R., Touron, S., Grousset, F.E. and Barkov, N. 2001. Volcanic layers in Antarctic (Vostok) ice cores: source identification and atmospheric implications. *Journal of Geophysical Research*, **106**, 31 915–31 931, https://doi.org/10.1029/2000JD000102

Bazin, L., Landais, A. *et al.* 2013. An optimized multi-proxy, multi-site Antarctic ice and gas orbital chronology (AICC2012): 120–800 ka. *Climate of the Past*, **9**, 1715–1731, https://doi.org/10.5194/cp-9-1715-2013

Bertler, N.A.N., Conway, H. *et al.* 2018. The Ross Sea Dipole – temperature, snow accumulation and sea ice variability in the Ross Sea region, Antarctica, over the past 2700 years. *Climate of the Past*, **14**, 193–214, https://doi.org/10.5194/cp-14-193-2018

Brook, E.J. and Buizert, C. 2018. Antarctic and global climate history viewed from ice cores. *Nature*, **558**, 200–208, https://doi.org/10.1038/s41586-018-0172-5

Caiazzo, L., Baccolo, G. *et al.* 2017. Prominent features in isotopic, chemical and dust stratigraphies from coastal East Antarctic ice sheet (Eastern Wilkes Land). *Chemosphere*, **176**, 273–287, https://doi.org/10.1016/j.chemosphere.2017.02.115

Cooper, C.L., Swindles, G.T., Savov, I.P., Schmidt, A. and Bacon, K.L. 2018. Evaluating the relationship between climate change and volcanism, *Earth-Science Reviews*, **177**, 238–247, https://doi.org/10.1016/j.earscirev.2017.11.009

Curzio, P., Folco, L., Laurenzi, M.A., Mellini, M. and Zeoli, A. 2008. A tephra chronostratigraphic framework for the Frontier Mountain blue-ice field (Northern Victoria Land, Antarctica). *Quaternary Science Reviews*, **27**, 602–620, https://doi.org/10.1016/j.quascirev.2007.11.017

Davies, S.M., Wastegård, S., Rasmussen, T.L., Svensson, A., Johnsen, S.J., Steffensen, J.P. and Andersen, K.K. 2008. Identification of the Fugloyarbanki tephra in the NGRIP ice core: a key tie-point for marine and ice-core sequences during the last glacial period. *Journal of Quaternary Science*, **23**, 409–414, https://doi.org/10.1002/jqs.1182

Davies, S.M., Wastegård, S. *et al.* 2010. Tracing volcanic events in the NGRIP ice-core and synchronising North Atlantic marine records during the last glacial period. *Earth and Planetary Science Letters*, **294**, 69–79, https://doi.org/10.1016/j.epsl.2010.03.004

De Angelis, M., Fehrenbach, L., Jehanno, C. and Maurette, M. 1985. Micrometric-sized volcanic glasses in polar ices and snows. *Nature*, **317**, 52–54, https://doi.org/10.1038/317052a0

Del Carlo, P., Di Roberto, A. *et al.* 2015. Late Pleistocene–Holocene volcanic activity in northern Victoria Land recorded in Ross Sea (Antarctica) marine sediments. *Bulletin of Volcanology*, **77**, 36, https://doi.org/10.1007/s00445-015-0924-0

Del Carlo, P., Di Roberto, A. *et al.* 2018. Late Glacial–Holocene tephra from southern Patagonia and Tierra del Fuego (Argentina, Chile): A complete textural and geochemical fingerprinting for distal correlations in the Southern Hemisphere. *Quaternary Science Reviews*, **195**, 153–170, https://doi.org/10.1016/j.quascirev.2018.07.028

Delmonte, B., Andersson, P.S., Hansson, M., Schöberg, H., Petit, J.R., Basile-Doelsch, I. and Maggi, V. 2008. Aeolian dust in East Antarctica (EPICA-Dome C and Vostok): Provenance during glacial ages over the last 800 kyr. *Geophysical Research Letters*, **35**, L07703, https://doi.org/10.1029/2008GL033382

Dunbar, N.W. and Kurbatov, A.V. 2011. Tephrochronology of the Siple Dome ice core, West Antarctica: correlations and sources. *Quaternary Science Reviews*, **30**, 1602–1614, https://doi.org/10.1016/j.quascirev.2011.03.015

Dunbar, N.W., Kyle, P.R., McIntosh, W.C. and Esser, R.P. 1995. Geochemical composition and stratigraphy of tephra layers in Antarctic blue ice: A new source of glacial tephrochronological data. *Antarctic Journal of the United States*, **30**, 76–78.

Dunbar, N.W., Zielinski, G.A. and Voisins, D.T. 2003. Tephra layers in the Siple Dome and Taylor Dome ice cores, Antarctica: sources and correlations. *Journal of Geophysical Research*, **108**, 2374, https://doi.org/10.1029/2002JB002056

Dunbar, N.W., McIntosh, W.C. and Esser, R.P. 2008. Physical setting and tephrochonology of the summit caldera ice record at Mt. Moulton, West Antarctica. *Geological Society of America Bulletin*, **120**, 796–812, https://doi.org/10.1130/B26140.1

Dunbar, N.W., Iverson, N.A., Kurbatov, A. and McIntosh, W.C. 2013. Continued investigation and correlations of tephra in the WAIS Divide WDC06A ice core. Presented at the WAIS Divide Science Meeting, September 24–25, 2013, La Jolla, California, USA, 37.

Dunbar, N.W., Iverson, N.A., Smellie, J.L., McIntosh, W.C., Zimmerer, M.J. and Kyle, P.R. 2021. Active volcanoes in Marie Byrd Land. *Geological Society, London, Memoirs*, **55**, https://doi.org/10.1144/M55-2019-29

EPICA Community Members. 2006. One-to-one coupling of glacial climate variability in Greenland and Antarctica. *Nature*, **444**, 195–198, https://doi.org/10.1038/nature05301

Fischer, H., Severinghaus, J. *et al.* 2013. Where to find 1.5 million yr old ice for the IPICS 'Oldest-Ice' ice core, *Climate of the Past*, **9**, 2489–2505, https://doi.org/10.5194/cp-9-2489-2013

Fujii, Y., Kohno, M. *et al.* 1999. Tephra layers in the Dome Fuji (Antarctica) deep ice core. *Annals of Glaciology*, **29**, 126–130, https://doi.org/10.3189/172756499781821003

Giordano, G., Lucci, F., Phillips, D., Cozzupoli, D. and Runci, V. 2012. Stratigraphy, geochronology and evolution of the Mt. Melbourne volcanic field (North Victoria Land, Antarctica). *Bulletin of Volcanology*, **74**, 1985–2005, https://doi.org/10.1007/s00445-012-0643-8

Gow, A.J. and Meese, D.A. 2007. The distribution and timing of tephra deposition at Siple Dome, Antarctica: possible climatic and rheologic implications. *Journal of Glaciology*, **53**, 585–596, https://doi.org/10.3189/002214307784409270

Hammer, C.U., Clausen, H.B. and Langway, C.C., Jr. 1997. 50,000 years of recorded global volcanism. *Climate Change*, **35**, 1–15, https://doi.org/10.1023/A:1005344225434

Han, Y., Jun, S.J. *et al.* 2015. Shallow ice-core drilling on Styx glacier, northern Victoria Land, Antarctica in the 2014–2015 summer. *Journal of the Geological Society of Korea*, **51**, 343–355, https://doi.org/10.14770/jgsk.2015.51.3.343

Harpel, C.J., Kyle, P.R. and Dunbar, N.W. 2008. Englacial tephrostratigraphy of Erebus volcano, Antarctica. *Journal of Volcanology and Geothermal Research*, **177**, 549–568, https://doi.org/10.1016/j.jvolgeores.2008.06.001

Higgins, J.A., Kurbatov, A.V. *et al.* 2015. Atmospheric composition 1 million years ago from blue ice in the Allan Hills, Antarctica. *Proceedings of the National Academy of Sciences of the United States of America*, **112**, 6887-6891, https://doi.org/10.1073/pnas.1420232112

Hubberten, H.-W., Morche, W. *et al.* 1991. Geochemical investigations of volcanic ash layers from southern Atlantic Legs 113 and 114. *In*: Ciesielski, P.F., Kristoffersen, Y. *et al.* (eds) *Proceedings of the Ocean Drilling Program, Scientific Results, Volume 114*. Ocean Drilling Program, College Station, TX, 733–749.

Huybers, P. and Langmuir, C. 2009. Feedback between deglaciation, volcanism, and atmospheric CO_2. *Earth and Planetary Science Letters*, **286**, 479–491, https://doi.org/10.1016/j.epsl.2009.07.014

Iverson, N.A. 2017. *Characterization and Correlation of Englacial Tephra from Blue Ice Areas and Ice Cores, Antarctica*. PhD thesis, New Mexico Institute of Mining and Technology, Socorro, New Mexico, USA.

Iverson, N.A., Kyle, P.R., Dunbar, N.W., McIntosh, W.C. and Pearce, N.J.G. 2014. Eruptive history and magmatic stability of Erebus volcano, Antarctica: insights from englacial tephra. *Geochemistry, Geophysics, Geosystems*, **15**, 4180–4202, https://doi.org/10.1002/2014GC005435

Jouzel, J., Masson-Delmotte, V. *et al.* 2007. Orbital and millennial Antarctic climate variability over the past 800,000 years. *Science*, **317**, 793–796, https://doi.org/10.1126/science.1141038

Kehrl, L., Conway, H., Holschuh, N., Campbell, S., Kurbatov, A.V. and Spaulding, N.E. 2018. Evaluating the duration and continuity of potential climate records from the Allan Hills Blue Ice Area, East Antarctica. *Geophysical Research Letters*, **45**, 4096–4104, https://doi.org/10.1029/2018GL077511

Koffman, B.G., Dowd, E.G. *et al.* 2017. Rapid transport of ash and sulfate from the 2011 Puyehue-Cordón Caulle (Chile) eruption to West Antarctica. *Journal of Geophysical Research Atmosphere*, **122**, 8908–8920, https://doi.org/10.1002/2017JD026893

Kohno, M., Fujii, Y. and Hirata, T. 2004. Chemical composition of volcanic glasses in visible tephra layers found in a 2503 m deep ice core from Dome Fuji, Antarctica, *Annals of Glaciology*, **39**, 576–584, https://doi.org/10.3189/172756404781813934

Kohno, M., Kipfstuhl, S., Lambrecht, A., Fuji, Y. and Kronz, A. 2005. Tephra study on the EPICA-DML ice core. *Geophysical Research Abstracts*, **7**, 07573, 1607-7962/gra/EGU05-A-07573.

Kohno, M., Fuji, Y. *et al.* 2007. Tephra study on a 3035.22-m deep ice core from Dome Fuji, Antarctica. Presented at the IUGG XXIV General Assembly, 2–13 July 2007, Perugia, Italy, http://iugg.org/archive/iugg2007perugia/www.iugg2007perugia.it/iuggProc/JM.pdf

Korotkikh, E.V., Mayewski, P.A. *et al.* 2011. The last interglacial as represented in the glaciochemical record from Mount Moulton Blue Ice Area, West Antarctica. *Quaternary Science Reviews*, **30**, 1940–1947, https://doi.org/10.1016/j.quascirev.2011.04.020

Kurbatov, A.V., Dunbar, N.W., Zielinski, G.A., Mayewski, P.A., Curran, M.A., Morgan, V. and van Ommen, T.D. 2003. Evaluation of Tephra Found in the Law Dome Ice Core, East Antarctica. *Eos, Transactions of American Geophysical Union*, **84**, Fall Meeting Supplement, Abstract A31C-56.

Kyle, P. and Lee, L.J. 2018. New occurrences of the widespread 1254 C.E. tephra in Antarctic ice and identification of Rittmann volcano, Antarctica as the eruptive source. *In*: INTAV International Field Conference on Tephrochronology 'Tephra Hunt in Transylvania', Moieciu de Sus, Romania, 24 June–1 July 2018, Book of Abstracts, Romanian Academy and Babes-Bolyai University, Cluj, Romania, 79, http://www.baycer.uni-bayreuth.de/intav2018/en/key_dates/5001/1/16443/INTAV_Programm_final_vers2-2.pdf

Laluraj, C.M., Krishnan, K.P. *et al.* 2009. Origin and characterisation of microparticles in an ice core from the Central Dronning Maud Land, East Antarctica. *Environmental Monitoring and Assessment*, **149**, 377–383, https://doi.org/10.1007/s10661-008-0212-y

Lambert, F., Delmonte, B. *et al.* 2008. Dust–climate couplings over the past 800,000 years from the EPICA Dome C ice core. *Nature*, **452**, 616–619, https://doi.org/10.1038/nature06763

La Violette, P.A. 2005. Solar cycle variations in ice acidity at the end of the Last Ice Age: Possible marker of a climatically significant interstellar dust incursion. *Planetary and Space Sciences*, **53**, 385–393, https://doi.org/10.1016/j.pss.2004.09.020

Loulergue, L., Schilt, A. *et al.* 2008. Orbital and millenial-scale features of atmospheric CH_4 over the last 800,000 years. *Nature*, **453**, 383–386, https://doi.org/10.1038/nature06950

Martin, A.P., Cooper, A.F., Price, R.C., Kyle, P.R. and Gamble, J.A. 2021. Erebus Volcanic Province: petrology. *Geological Society, London, Memoirs*, **55**, https://doi.org/10.1144/M55-2018-80

Masson-Delmotte, V., Schulz, M. *et al.* 2013. Information from paleoclimate archives. *In*: Stocker, T.F., Qin, D. *et al.* (eds) *Climate Change 2013: The Physical Science Basis. Contribution of Working Group I to the Fifth Assessment Report of the Intergovernmental Panel on Climate Change*. Cambridge University Press, Cambridge, UK, 383–464.

McConnell, J.R., Burke, A. *et al.* 2017. Synchronous volcanic eruptions and abrupt climate change ~17.7 ka plausibly linked by stratospheric ozone depletion. *Proceedings of the National Academy of Sciences of the United States of America*, **114**, 10 035–10 040, https://doi.org/10.1073/pnas.1705595114

Moore, J.C., Nishio, F. *et al.* 2006. Interpreting ancient ice in a shallow ice core from the South Yamato (Antarctica) blue ice area using flow modeling and compositional matching to deep ice cores. *Journal of Geophysical Research*, **111**, D16302, https://doi.org/10.1029/2005JD006343

Morgan, V., Wookey, C., Li, J., Van Ommen, T., Skinner, W. and Fitzpatrick, M. 1997. Site information and initial results from deep ice drilling on Law Dome, Antarctica. *Journal of Glaciology*, **43**, 3–10, https://doi.org/10.3189/S0022143000002768

Morse, D.L., Waddington, E.D. and Rasmussen, L.A. 2007. Ice deformation in the vicinity of the ice-core site at Taylor Dome, Antarctica, and a derived accumulation rate history, *Journal of Glaciology*, **53**, 449–460, https://doi.org/10.3189/002214307783258530

Mulvaney, R., Alemany, O. and Possenti, P. 2007. The Berkner Island ice core drilling project. *Annals of Glaciology*, **47**, 115–124, https://doi.org/10.3189/172756407786857758

Narcisi, B., Proposito, M. and Frezzotti, M. 2001. Ice record of a 13th century explosive volcanic eruption in northern Victoria Land, East Antarctica. *Antarctic Science*, **13**, 174–181, https://doi.org/10.1017/S0954102001000268

Narcisi, B., Petit, J.R., Delmonte, B., Basile-Doelsch, I. and Maggi, V. 2005. Characteristics and sources of tephra layers in the EPICA-Dome C ice record (East Antarctica): Implications for past atmospheric circulation and ice core stratigraphic correlations. *Earth and Planetary Science Letters*, **239**, 253–265, https://doi.org/10.1016/j.epsl.2005.09.005

Narcisi, B., Petit, J.R. and Tiepolo, M. 2006. A volcanic marker (92 ka) for dating deep east Antarctic ice cores. *Quaternary Science Reviews*, **25**, 2682–2687, https://doi.org/10.1016/j.quascirev.2006.07.009

Narcisi, B., Petit, J.R., Delmonte, B., Tiepolo, M., Basile-Doelsch, I. and Maggi, V. 2008. Geochemical features of tephra layers in the EPICA-Dome C climatic record (East Antarctica). *Terra Antartica Reports*, **14**, 107–110.

Narcisi, B., Petit, J.R. and Chappellaz, J. 2010*a*. A 70 ka record of explosive eruptions from the TALDICE ice core (Talos Dome, East Antarctic plateau). *Journal of Quaternary Science*, **25**, 844–849, https://doi.org/10.1002/jqs.1427

Narcisi, B., Petit, J.R. and Delmonte, B. 2010*b*. Extended East Antarctic ice core tephrostratigraphy. *Quaternary Science Reviews*, **29**, 21–27, https://doi.org/10.1016/j.quascirev.2009.07.009

Narcisi, B., Petit, J.R., Delmonte, B., Scarchilli, C. and Stenni, B. 2012. A 16,000-yr tephra framework for the Antarctic ice sheet: a contribution from the new Talos Dome core. *Quaternary Science Reviews*, **49**, 52–63, https://doi.org/10.1016/j.quascirev.2012.06.011

Narcisi, B., Petit, J.R., Langone, A. and Stenni, B. 2016. A new Eemian record of Antarctic tephra layers retrieved from the Talos Dome ice core (Northern Victoria Land). *Global and Planetary Change*, **137**, 69–78, https://doi.org/10.1016/j.gloplacha.2015.12.016

Narcisi, B., Petit, J.R. and Langone, A. 2017. Last glacial tephra layers in the Talos Dome ice core (peripheral East Antarctic Plateau), with implications for chronostratigraphic correlations and regional volcanic history. *Quaternary Science Reviews*, **165**, 111–126, https://doi.org/10.1016/j.quascirev.2017.04.025

Narcisi, B., Petit, J.R., Delmonte, B., Batanova, V. and Savarino, J. 2019. Multiple sources for tephra from AD 1259 volcanic signals in Antarctic ice cores. *Quaternary Science Reviews*, **210**, 164–174, https://doi.org/10.1016/j.quascirev.2019.03.005

Neff, P.D. and Bertler, N.A.N. 2015. Trajectory modeling of modern dust transport to the Southern Ocean and Antarctica, *Journal of Geophysical Research Atmosphere*, **120**, 9303–9322, https://doi.org/10.1002/2015JD023304

Nielsen, S.H.H., Hodell, D.A., Kamenov, G., Guilderson, T. and Perfit, M.R. 2007. Origin and significance of ice-rafted detritus in the Atlantic sector of the Southern Ocean. *Geochemistry, Geophysics, Geosystems*, **8**, Q12005, https://doi.org/10.1029/2007GC001618

Nishio, F., Katsushima, T. and Ohmae, H. 1985. Volcanic ash layers in bare ice areas near the Yamato Mountains, Dronning Maud Land and the Allan Hills, Victoria Land, Antarctica. *Annals of Glaciology*, **7**, 34–41, https://doi.org/10.3189/S0260305500005875

Nyland, R.E., Panter, K.S. et al. 2013. Volcanic activity and its link to glaciation cycles: single-grain age and geochemistry of Early to Middle Miocene volcanic glass from ANDRILL AND-2A core, Antarctica. *Journal of Volcanology and Geothermal Research*, **250**, 106–128, https://doi.org/10.1016/j.jvolgeores.2012.11.008

Oda, H., Miyagi, I. et al. 2016. Volcanic ash in bare ice south of Sør Rondane Mountains, Antarctica: geochemistry, rock magnetism and nondestructive magnetic detection with SQUID gradiometer. *Earth, Planets and Space*, **68**, 39, https://doi.org/10.1186/s40623-016-0415-3

Palais, J.M. and Kyle, P.R. 1988. Chemical composition of ice containing tephra layers in the Byrd Station ice core, Antarctica. *Quaternary Research*, **30**, 315–330, https://doi.org/10.1016/0033-5894(88)90007-5

Palais, J.M., Kyle, P.R., Mosley-Thompson, E. and Thomas, E. 1987. Correlation of a 3,200 year old tephra in ice cores from Vostok and South Pole stations, Antarctica. *Geophysical Research Letters*, **14**, 804–807, https://doi.org/10.1029/GL014i008p00804

Palais, J.M., Petit, J.R., Lorius, C. and Korotkevitch, Y.S. 1989. Tephra layers in the Vostok ice core: 160,000 years of southern hemisphere volcanism. *Antarctic Journal of the United States*, **24**, 98–100.

Palais, J.M., Kirchner, S. and Delmas, R.J. 1990. Identification of Some global volcanic horizons by major element analysis of fine ash in Antarctic ice. *Annals of Glaciology*, **14**, 216–220, https://doi.org/10.3189/S0260305500008612

Palais, J.M., Germani, M.S. and Zielinski, G.A. 1992. Interhemispheric transport of volcanic ash from a 1259 A.D. volcanic eruption to the Greenland and Antarctic ice sheets. *Geophysical Research Letters*, **19**, 801–804, https://doi.org/10.1029/92GL00240

Pallàs, R., Smellie, J.L., Casas, J.M. and Calvet, J. 2001. Using tephrochronology to date temperate ice: correlation between ice tephras on Livingston Island and eruptive units on Deception Island volcano (South Shetland Islands, Antarctica). *The Holocene*, **11**, 149–160, https://doi.org/10.1191/095968301669281809

Patrick, M. and Smellie, J. 2013. Synthesis – A spaceborne inventory of volcanic activity in Antarctica and southern oceans, 2000–10. *Antarctic Science*, **25**, 475–500, https://doi.org/10.1017/S0954102013000436

Patrick, M.R., Smellie, J. et al. 2005. First recorded eruption of Mount Belinda volcano (Montagu Island), South Sandwich Islands. *Bulletin of Volcanology*, **67**, 415–422, https://doi.org/10.1007/s00445-004-0382-6

Petit, J.R., Jouzel, J. et al. 1999. Climate and atmospheric history of the past 420,000 years from the Vostok ice core, Antarctica. *Nature*, **399**, 429–436, https://doi.org/10.1038/20859

Plummer, C.T., Curran, M.A.J. et al. 2012. An independently dated 2000-yr volcanic record from Law Dome, East Antarctica, including a new perspective on the dating of the 1450s CE eruption of Kuwae, Vanuatu. *Climate of the Past*, **8**, 1929–1940, https://doi.org/10.5194/cp-8-1929-2012

Rawson, H., Pyle, D.M., Mather, T.A., Smith, V.C., Fontijn, K., Lachowycz, S.M. and Naranjo, J.A. 2016. The magmatic and eruptive response of arc volcanoes to deglaciation: insights from Southern Chile. *Geology*, **44**, 251–254, https://doi.org/10.1130/G37504.1

Rickwood, P.C. 1989. Boundary lines within petrologic diagrams which uses oxides of major and minor elements. *Lithos*, **22**, 247–263, https://doi.org/10.1016/0024-4937(89)90028-5

Scarchilli, C., Frezzotti, M. and Ruti, P.M. 2011. Snow precipitation at four ice core sites in East Antarctica: provenance, seasonality and blocking factors. *Climate Dynamics*, **37**, 2107–2125, https://doi.org/10.1007/s00382-010-0946-4

Schwander, J., Jouzel, J., Hammer, C.U., Petit, J.R., Udisti, R. and Wolff, E. 2001. A tentative chronology for the EPICA Dome Concordia ice core. *Geophysical Research Letters*, **28**, 4243–4246, https://doi.org/10.1029/2000GL011981

Severi, M., Udisti, R., Becagli, S., Stenni, B. and Traversi, R. 2012. Volcanic synchronisation of the EPICA-DC and TALDICE ice cores for the last 42 kyr BP. *Climate of the Past*, **8**, 509–517, https://doi.org/10.5194/cp-8-509-2012

Sigl, M., McConnell, J.R. et al. 2014. Insights from Antarctica on volcanic forcing during the Common Era. *Nature Climate Change*, **4**, 693–697, https://doi.org/10.1038/nclimate2293

Sigl, M., Winstrup, M. et al. 2015. Timing and climate forcing of volcanic eruptions for the past 2,500 years. *Nature*, **523**, 543–549, https://doi.org/10.1038/nature14565

Sigl, M., Fudge, T.J. et al. 2016. The WAIS Divide deep ice core WD2014 chronology – Part 2: annual layer counting (0–31 ka BP). *Climate of the Past*, **12**, 769–786, https://doi.org/10.5194/cp-12-769-2016

Sinisalo, A. and Moore, J.C. 2010. Antarctic blue ice areas - towards extracting palaeoclimate information. *Antarctic Science*, **22**, 99–115, https://doi.org/10.1017/S0954102009990691

Smellie, J.L. 1999. The upper Cenozoic tephra record in the South Polar Region: a review. *Global and Planetary Change*, **21**, 51–70, https://doi.org/10.1016/S0921-8181(99)00007-7

Smellie, J.L. and Martin, A.P. 2021. Erebus Volcanic Province: volcanology. *Geological Society, London, Memoirs*, **55**, https://doi.org/10.1144/M55-2018-62

Smith, V.C., Shane, P. and Nairn, I.A. 2005. Trends in rhyolite geochemistry, mineralogy, and magma storage during the last 50 kyr at Okataina and Taupo volcanic centres, Taupo Volcanic Zone, New Zealand. *Journal of Volcanology and Geothermal Research*, **148**, 372–406, https://doi.org/10.1016/j.jvolgeores.2005.05.005

Stenni, B., Masson-Delmotte, V. et al. 2010. The deuterium excess records of EPICA Dome C and Dronning Maud Land ice cores (East Antarctica). *Quaternary Science Reviews*, **29**, 146–159, https://doi.org/10.1016/j.quascirev.2009.10.009

Traversi, R., Becagli, S. et al. 2009. Sulfate spikes in the deep layers of EPICA-Dome C ice core: Evidence of glaciological artifacts. *Environmental Science & Technology*, **43**, 8737–8743, https://doi.org/10.1021/es901426y

Tuohy, A.J. 2015. *Heavy Metal Pollutants in Snow and Ice from Roosevelt Island, Antarctica*. PhD thesis, Victoria University of Wellington, Wellington, New Zealand, http://researcharchive.vuw.ac.nz/handle/10063/4797

Udisti, R., Barbante, C., Castellano, E., Vermigli, S., Traversi, R., Capodaglio, G. and Piccardi, G. 1999. Chemical characterisation of a volcanic event (about AD 1500) at Styx Glacier plateau, northern Victoria Land, Antarctica. *Annals of Glaciology*, **29**, 113–120, https://doi.org/10.3189/17275649978182 1265

Vallelonga, P., Gabrielli, P., Rosman, K.J.R., Barbante, C. and Boutron, C.F. 2005. A 220 kyr record of Pb isotopes at Dome C Antarctica from analyses of the EPICA ice core. *Geophysical Research Letters*, **32**, L01706, https://doi.org/10.1029/2004GL021449

Veres, D., Bazin, L. et al. 2013. The Antarctic ice core chronology (AICC2012): an optimized multi-parameter and multi-site dating approach for the last 120 thousand years. *Climate of the Past*, **9**, 1733–1748, https://doi.org/10.5194/cp-9-1733-2013

Watt, S.F.L., Pyle, D.M. and Mather, T.A. 2013. The volcanic response to deglaciation: Evidence from glaciated arcs and a reassessment of global eruption records. *Earth-Science Reviews*, **122**, 77–102, https://doi.org/10.1016/j.earscirev.2013.03.007

Wilch, T.I., Mcintosh, W.C. and Dunbar, N.W. 1999. Late Quaternary volcanic activity in Marie Byrd Land: potential ^{40}Ar/^{39}Ar dated time horizons in West Antarctic ice and marine cores. *Geological Society of America Bulletin*, **111**, 1563–1580, https://doi.org/10.1130/0016-7606(1999)111<1563:LQVAIM>2.3.CO;2

Yang, J.-W., Han, Y. et al. 2018. Surface temperature in twentieth century at the Styx Glacier, northern Victoria Land, Antarctica, from borehole thermometry. *Geophysical Research Letters*, **45**, 9834–9842, https://doi.org/10.1029/2018GL078770

Zielinski, G.A. 2000. Use of paleo-records in determining variability within the volcanism–climate system. *Quaternary Science Reviews*, **19**, 417–438, https://doi.org/10.1016/S0277-3791(99)00073-6

Section 7

Active volcanoes

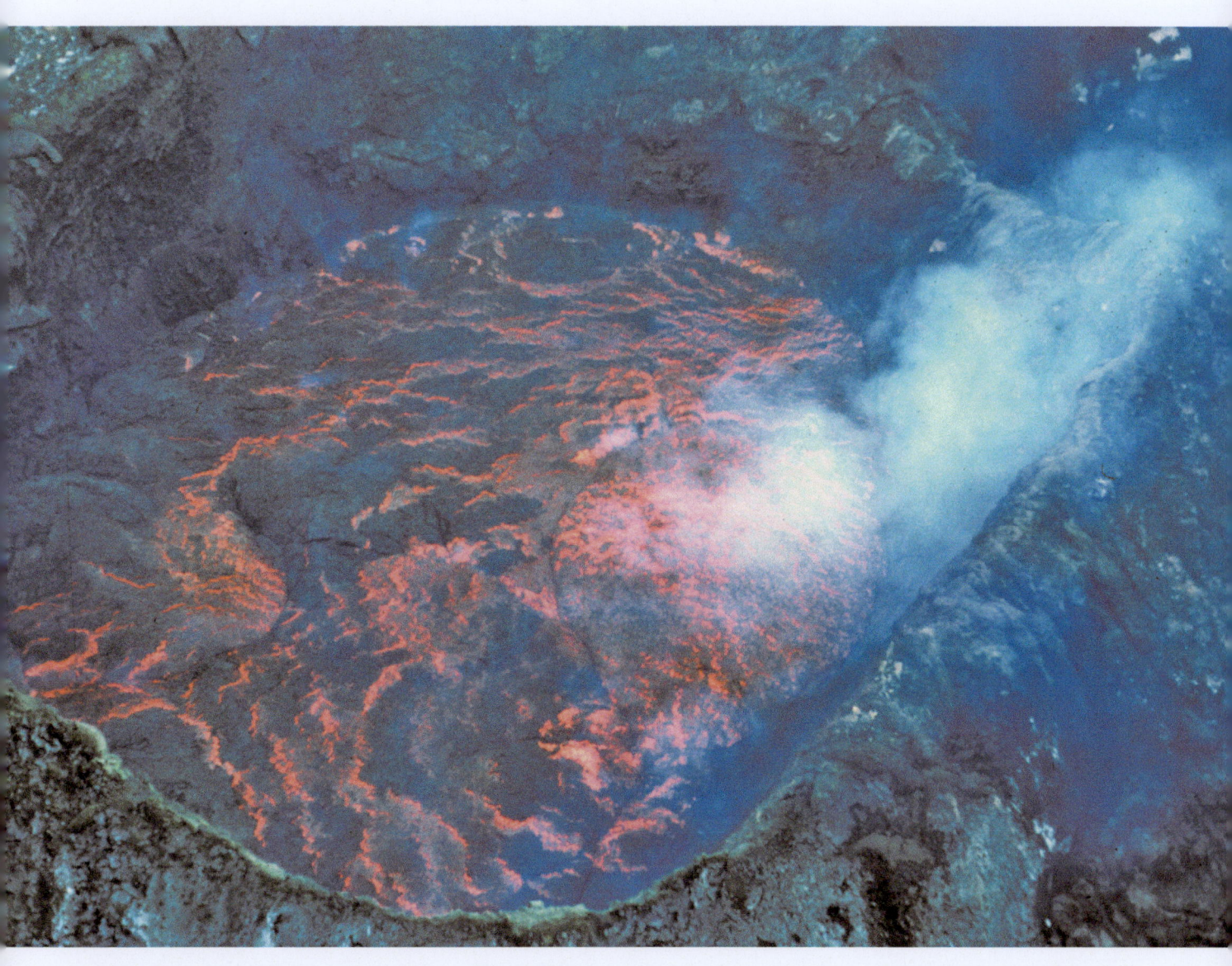

View of lava lake (*c.* 60 m in diameter) in a small pit crater within the post-caldera summit cone on Mount Erebus, southern Victoria Land. The lava lake is phonolite in composition and is an essentially permanent feature.

Photograph by W. C. McIntosh

Chapter 7.1

Deception Island

A. Geyer[1]*, D. Pedrazzi[1], J. Almendros[2], M. Berrocoso[3,4], J. López-Martínez[5], A. Maestro[6], E. Carmona[2], A. M. Álvarez-Valero[7] and A. de Gil[3,4]

[1]Geosciences Barcelona, GEO3BCN - CSIC, Lluis Sole i Sabaris s/n, 08028 Barcelona, Spain
[2]Andalusian Institute of Geophysics, University of Granada, Granada, Spain
[3]Laboratory of Astronomy, Geodesy and Cartography, Department of Mathematics, Faculty of Sciences, University of Cádiz, 11510 Puerto Real (Cádiz), Spain
[4]Instituto Universitario de Investigación Marina (INMAR), Universidad de Cádiz, Campus de Excelencia Internacional del Mar (CEIMAR), Campus Universitario de Puerto Real, 11510 Puerto Real, Cádiz, Spain
[5]Department of Geology and Geochemistry, Faculty of Sciences, University Autónoma of Madrid, 28049 Madrid, Spain
[6]Geological and Mining Institute of Spain, C/Calera, 1, 28760 Tres Cantos, Madrid, Spain
[7]Department of Geology, University of Salamanca, 37008 Salamanca, Spain

AG, 0000-0002-8803-6504; DP, 0000-0002-6869-1325; JA, 0000-0001-5936-6160;
JL-M, 0000-0002-1750-8287; AM, 0000-0002-7474-725X; EC, 0000-0002-4309-4292;
AMA-V, 0000-0001-9707-0168
*Correspondence: ageyer@geo3bcn.csic.es

Abstract: Deception Island (South Shetland Islands) is one of the most active volcanoes in Antarctica, with more than 15 explosive eruptive events registered over the past two centuries. Recent eruptions (1967, 1969 and 1970) and volcanic unrest episodes in 1992, 1999 and 2014–15 demonstrate that the occurrence of future volcanic activity is a valid and pressing concern for scientists, logistic personnel and tourists that are visiting or are working on or near the island. Over the last few decades, intense research activity has been carried out on Deception Island to decipher the origin and evolution of this very complex volcano. To that end, a solid integration of related scientific disciplines, such as tectonics, petrology, geochemistry, geophysics, geomorphology, remote sensing, glaciology, is required. A proper understanding of the island's evolution in the past, and its present state, is essential for improving the efficiency in interpreting monitoring data recorded during volcanic unrest periods and, hence, for future eruption forecasting. In this chapter, we briefly present Deception Island's most relevant tectonic, geomorphological, volcanological and magmatic features, as well as the results obtained from decades of monitoring the island's seismic activity and ground deformation.

Supplementary material: Composition and exact latitude–longitude coordinates of the rock samples behind Figures 3, 12 and 13 are available at https://doi.org/10.6084/m9.figshare.c.5220430

Deception Island, located in the South Shetland Islands, is one of the most active volcanoes in Antarctica with a record of over 30 post-caldera eruptions during the Holocene (e.g. Orheim 1972; Roobol 1982; Smellie 2002c). Since its discovery in 1820, the island's natural harbours of Pendulum Cove and Whalers Bay (Fig. 1) have been actively used during diverse peaks in the commercial exploitation of the Southern Ocean (Roobol 1982; Smellie and López-Martínez 2002a). Between 1905 and 1930, the island served as the shore base for the Antarctic's most important whaling industry and also played a military role during World War II due to its strategic location between the Atlantic and Pacific oceans. This resulted in the construction of a British scientific station, which was occupied from 1944 until its destruction in 1969 (Smellie 2002a, b) (Fig. 2a). Following the British initiative, Argentina and Chile also established scientific bases on the island. The latter was also destroyed and abandoned after the 1967 and 1970 eruptions (Baker et al. 1975) (Fig. 2b). Today, Deception Island hosts the Argentinian and Spanish scientific stations, which operate every year during the Antarctic summer (Fig. 2c); it is also one of the most popular tourist destinations in Antarctica, with over 15 000 visitors per year (figures from the International Association of Antarctica Tour Operators (IAATO) in 2018).

Deception Island's eruptions record from the eighteenth to the twentieth centuries comprise periods of high activity (e.g. 1818–28, 1906–12 and 1967–70) with numerous temporally closely spaced eruptions, followed by decades of dormancy (e.g. 1912–67) (Orheim 1972; Roobol 1980, 1982; Smellie 2002c). Despite the inherent difficulties involved in forecasting future eruptions, the documented historical volcanic activity (since 1820), the recently experienced eruptions (i.e. 1967, 1969 and 1970), and the unrest episodes that occurred in 1992, 1999 (Ibáñez et al. 2003a, 2003b) and 2014–15 (Almendros et al. 2018) confirm that Deception Island is a very active volcano. Consequently, the future occurrence and potential impact of a volcanic event on Deception Island worries the scientists, logistical staff and tourists who visit or stay at (or nearby) the island. To a much greater extent, a future eruption on Deception Island would have more than just an impact at the local and/or regional scale. For example, the region contains more than 20 scientific stations, field camps and refuges within 150 km of the volcano, in the South Shetland Islands and the northern Antarctic Peninsula, many of which are occupied all year round (e.g. Arturo Prat, Chile; Bellingshausen, Russia; Great Wall, China) (figures from the Council of Managers of National Antarctic Programs (COMNAP) in 2019), and are at obvious risk from ash fall and tsunamis should another eruption occur. Moreover, it has been recently demonstrated that volcanic ash emitted from a moderate eruption occurring today on Deception Island could potentially encircle the southern hemisphere, leading to significant economic losses and consequences for global aviation safety (Geyer et al. 2017).

From: Smellie, J. L., Panter, K. S. and Geyer, A. (eds) 2021. *Volcanism in Antarctica: 200 Million Years of Subduction, Rifting and Continental Break-up*. Geological Society, London, Memoirs, **55**, 667–693,
First published online March 16, updated March 31, 2021, https://doi.org/10.1144/M55-2018-56
© 2021 The Author(s). Published by The Geological Society of London. All rights reserved.
For permissions: http://www.geolsoc.org.uk/permissions. Publishing disclaimer: www.geolsoc.org.uk/pub_ethics

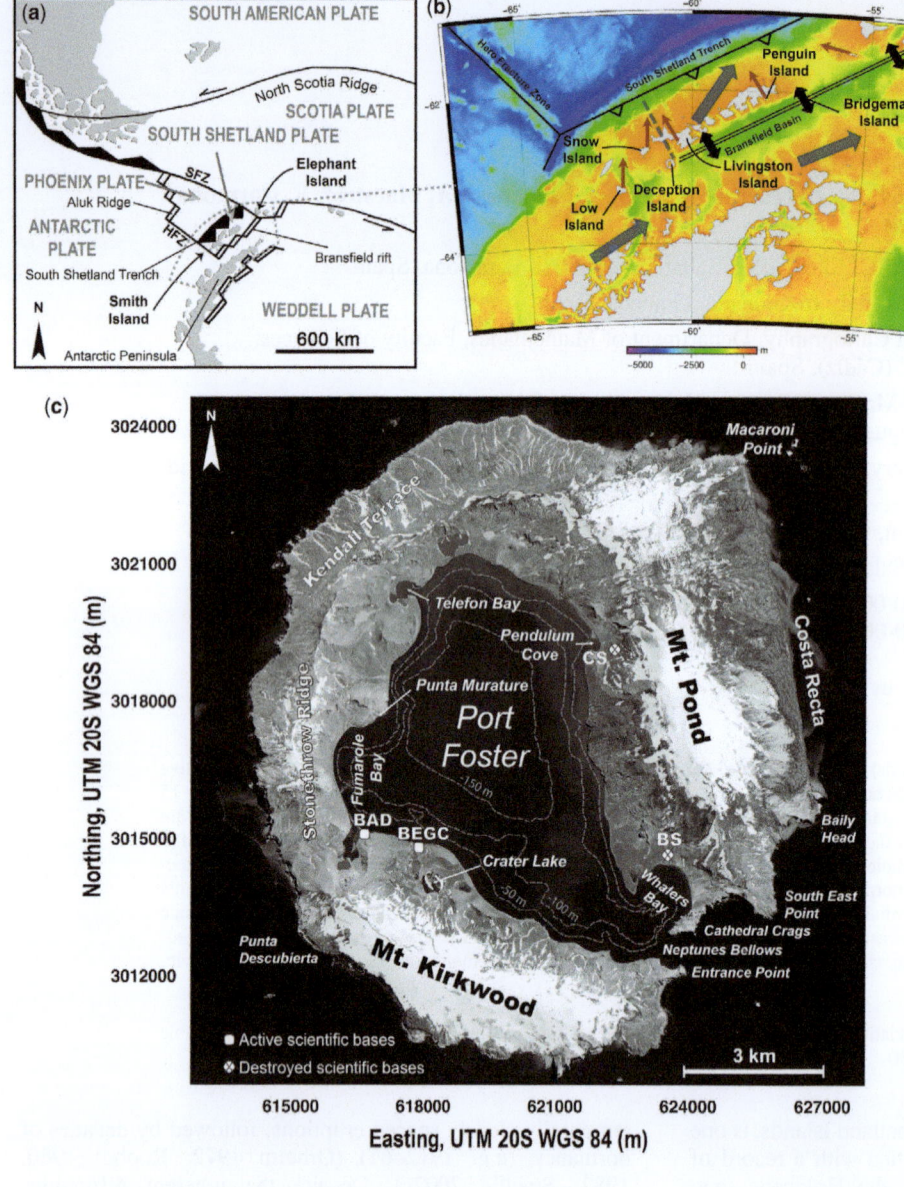

Fig. 1. (a) Simplified regional tectonic map and location of the South Shetland Islands (modified from Martí *et al.* 2013). HFZ, Hero Fracture Zone; SFZ, Shetland Fracture Zone. (b) Proposed geodynamic model for the South Shetland Islands and the Bransfield Basin region. Regions can be distinguished based on their respective geodynamic patterns (see Berrocoso *et al.* 2016 and main text for more details). Horizontal displacement vectors are without scale; thick vectors for the absolute horizontal movement in the ITRF2008 and thin vectors for the regional movement relative to the Antarctic Plate (modified from Berrocoso *et al.* 2016). (c) Deception Island orthophotomap (data obtained from Spatial Data Infrastructure for Deception Island SIMAC: Torrecillas *et al.* 2006) with existing and destroyed scientific stations indicated. BAD, Base Antártica Decepción (Argentinean Scientific base); BEGC, Base Española Gabriel de Castilla (Spanish scientific base); BS, Remains of the British scientific base; CS, Remains of the Chilean scientific base.

During the last few decades, intense research activity has been carried out at Deception Island with the purpose of comprehending the origin and evolution of this very complex volcano. In this chapter, we briefly present Deception Island's most relevant tectonic, geomorphological, volcanological and magmatic features, as well as the results obtained from decades of monitoring the island's seismic activity and ground deformation. A proper understanding of the island's evolution in the past and its present state is essential to efficiently unravelling monitoring data recorded during periods of volcanic unrest. Hence, it is crucial for future eruption forecasting. For further details on the individual topics presented here, the reader is referred to the specific papers mentioned throughout the text.

Geodynamic setting and regional magmatic context

Deception Island is located in Bransfield Basin, between the South Shetland Islands and the Antarctic Peninsula, and is part of an alignment of seamounts and volcanic islands, such as Penguin and Bridgeman islands, arranged parallel to the basin's axis in a NE–SW direction (Fig. 1a, b) (see also Smellie 2021: Chapter 3.2a). Bransfield Basin shows a NNE–SSW trend and is probably a *c.* 4 myr old rift developed in the pre-existing Antarctic Peninsula crust (Barker 1982).

The tectonic evolution of the Bransfield Basin has been explained as related to: (i) the passive subduction of the former Phoenix Plate and slab rollback of the South Shetland Trench (Smellie *et al.* 1984; Maldonado *et al.* 1994; Lawver *et al.* 1995, 1996); (ii) sinistral movement between the Antarctic and Scotia plates causing an oblique extension along the Antarctic Peninsula continental margin (Rey *et al.* 1995; Klepeis and Lawver 1996; Lawver *et al.* 1996; González-Casado *et al.* 2000); or (iii) mechanisms (i) and (ii) occurring simultaneously (Galindo-Zaldívar *et al.* 2004; Maestro *et al.* 2007). The main characteristics of the Bransfield Basin are widespread extension and faulting, the rise of crustal diapirs or domes associated with flower normal-fault structures, and a complicated system of fault-bounded segments across the strike (Barker and Austin 1994). Geophysical evidence also suggests NE–SW propagation of the rift, with initial crustal inflation/doming followed by deflation/subsidence, volcanism and extension along normal faults (Barker and Austin 1994). The Bransfield Rift lies near the critical area that marks the transition from intracontinental rifting to seafloor spreading (Barker and Austin 1994; Lawver *et al.* 1996). Conversely, the seismicity data for the South Shetland Islands area indicates a high number of local earthquakes (e.g. *c.* 150 earthquakes from 1997 to 1999) of relatively low magnitude (body-wave magnitude M_b 2–4) (Robertson *et al.* 2001; Robertson Maurice *et al.* 2003). These events are located in the South

Fig. 2. Remains of (**a**) the British and (**b**) Chilean bases in Whalers Bay and Pendulum Cove, respectively (see Fig. 1b for the exact locations). (**c**) Current Spanish scientific base Gabriel de Castilla (BEGC) (see Fig. 1b for the exact location). Photographs: (a) and (b) A. Villaseñor; (c) J. Galeano.

Shetland Trench with a few earthquakes concentrated in the Bransfield Basin. Although some authors believe that subduction beneath the South Shetland Islands either slowed or stopped with the cessation of spreading at the Aluk Ridge 4 myr ago (e.g. Henriet *et al.* 1992), the existence of many earthquakes at a variety of locations and depths indicates ongoing subduction (e.g. Larter 1991). Earthquakes are located on both the outer rise, where extensional events are typically found, and along the shallow thrust interface. The seismicity of the South Shetland Trench extends to depths of about 50 km (Pelayo and Wiens 1989; Ibáñez *et al.* 1997; Robertson *et al.* 2001) but most events are shallower than 30 km. These seismic results are consistent with GPS data and suggest a subduction rate of 1 cm a^{-1} (Robertson *et al.* 2001). Additionally, the occurrence of some earthquakes concentrated near large seafloor volcanoes along the central rift of the Bransfield Trough suggests that some of the volcanoes may be active.

Recently, horizontal velocities relative to the displacement of the Antarctic Plate obtained by GPS data suggest two different regions in the South Shetlands archipelago (Berrocoso *et al.* 2016) (Fig. 1b). The northeastern region, comprising the area between Livingston and Elephant islands, is horizontally affected by the Bransfield Basin extensional regime, while subsidence observed in the same area may be related to the former Phoenix Plate rollback underneath the South Shetland block (Alfaro *et al.* 2010; Maestro *et al.* 2014). However, the SW region comprising Smith, Low and Snow islands is less affected by the Bransfield Basin extension and the slab rollback at the South Shetland Trench, and is essentially amagmatic (Gràcia *et al.* 1996, 1997). A transition zone between the identified SW and NE regions, comprising Snow, Deception, and Livingston islands has been proposed (Berrocoso *et al.* 2016) (Fig. 1b). The subsidence vectors inferred for these islands have a similar magnitude to those of the northeastern region. SNOW (on Snow Island) and BYER (on Livingston Island) measurement stations give evidence for a NE–SW compression; also inferred in recent years on Deception Island, based on their convergent horizontal movement. Indeed, while the north SNOW velocity direction

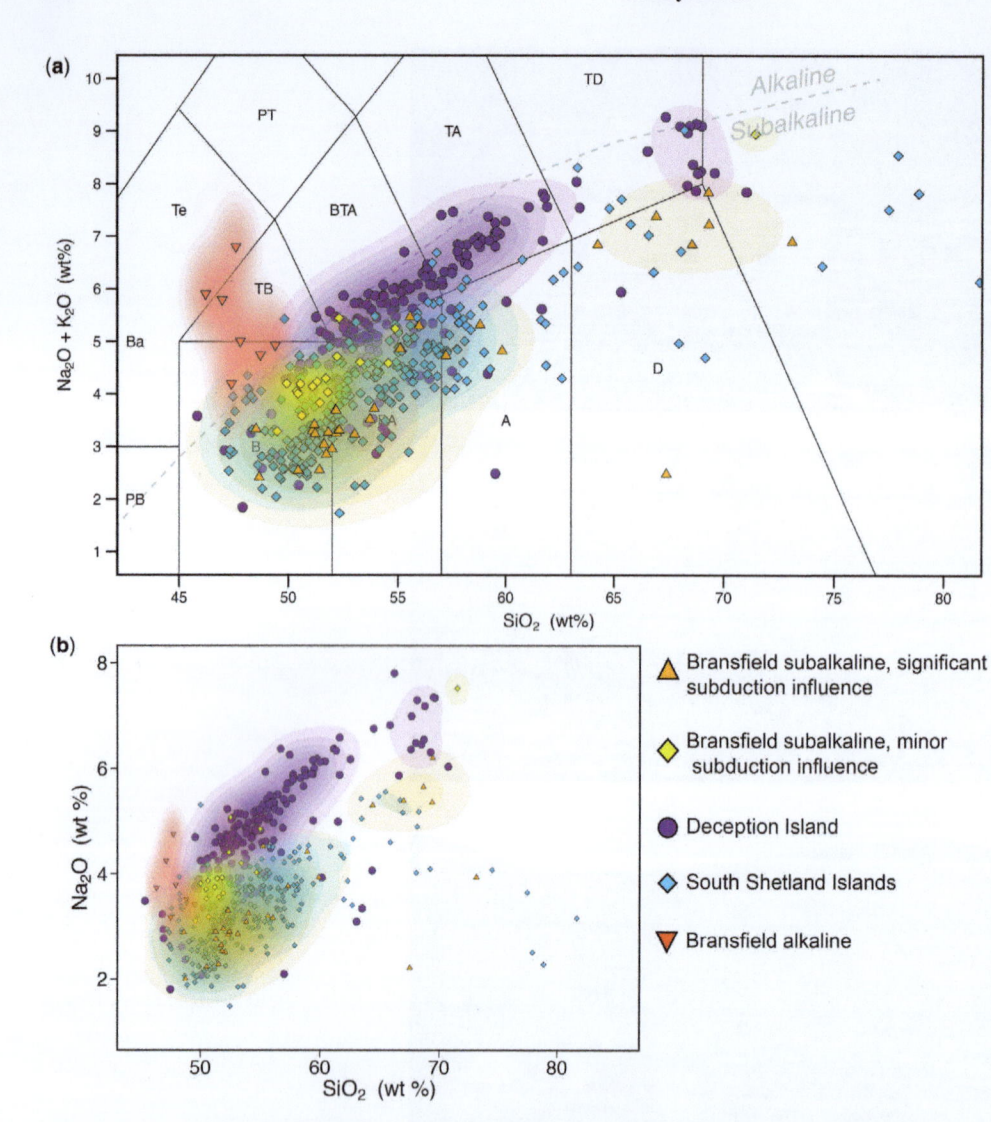

Fig. 3. (a) Total alkalis v. silica diagram (TAS) (Le Bas et al. 1986) and (b) Na$_2$O v. silica diagrams for rock samples along the Bransfield Strait, the South Shetland Islands and Deception Island (modified from Geyer et al. 2019) (see the Supplementary material for details). Major elements normalized to 100% (anhydrous) with Fe recalculated as FeO to Fe$_2$O$_3$ following Middlemost (1989). The black dashed line separates the alkaline and subalkaline fields (Irvine and Baragar 1971).

is similar to that of the southwestern region, the BYER NW velocity direction is similar to that of the northeastern region. On Deception Island, the NW–SE Bransfield Basin extension, the NE–SW compressional regime inferred on Byers Peninsula and the subsidence caused by the South Shetland Trench tectonic regime are all present (Berrocoso et al. 2016) (Fig. 1b). Deception Island and Byers Peninsula may well be closely affected by active faults related to the local compressive state.

This complex regional geodynamics has affected the timing and composition of magmatism in the region (e.g. Košler et al. 2009; Haase et al. 2012; Kraus et al. 2013). Quaternary magmatism in the Bransfield Strait, mostly subalkaline basaltic–basaltic andesitic (Fig. 3a), is strongly connected to rifting and back-arc basin formation, and is concentrated at Deception, Penguin and Bridgeman islands (e.g. Birkenmajer et al. 1990; Hole et al. 1994; Haase et al. 2012). Indeed, the normal magnetic polarity of all Deception Island's exposed rocks indicate that these are younger than 0.78 Ma (Valencio et al. 1979; Baraldo et al. 2003), and K–Ar data (Keller et al. 1992) suggest that most of the subaerial part of the island was built in the last 0.2 myr. Actually, the correlation between the exposed rocks and the tephras found elsewhere in the region suggests that the former are probably even younger than 0.1 Ma (Smellie 2001; Martí et al. 2013).

Despite being contemporary, Deception Island's magmas differ geochemically from the Quaternary South Shetland Islands and Bransfield Rift magmatism (Fig. 3). Rocks on Deception Island range in composition from basalts to trachydacites and rhyolites, and follow a distinctive alkalinity-increasing trend at the upper end of the subalkaline field in the total alkali v. silica diagram (TAS) (Fig. 3a) produced by distinctively high Na$_2$O contents (Fig. 3b) (Hawkes 1961; Smellie 2002c; Kraus et al. 2013), which has enabled tephras sourced at Deception Island to be readily recognized throughout the region, across the Scotia Sea and elsewhere in Antarctica (e.g. Moreton and Smellie 1998; Smellie 1999; Pallas et al. 2001; Liu et al. 2016; Antoniades et al. 2018).

Geomorphology and tectonics

Deception Island is a composite volcano with a basal diameter of 30 km and a summit at Mount Pond (539 m above sea level (asl)), which rises about 1.5 km from the seafloor (Figs 1b & 4). The emerged part of the volcano is a horseshoe-shaped island that is 15 km in diameter, whose central part is occupied by a sea-flooded volcanic collapse caldera (Port Foster) with dimensions of about 6 × 10 km and a maximum water depth of c. 190 m (Figs 1b & 4). In general, Deception Island's morphology is mainly determined by its volcanic activity, as well as the glacial, periglacial and tectonic processes (López-Martínez and Serrano 2002; Maestro et al. 2007) (Fig. 4). Over half of the island is covered by glaciers, partially ice-cored moraines, a diversity of periglacial landforms, such as debris slopes and fans, even slopes, solifluction lobes and flat-floored valleys, and a drainage network due to the relatively high availability of water in summer (López-Martínez

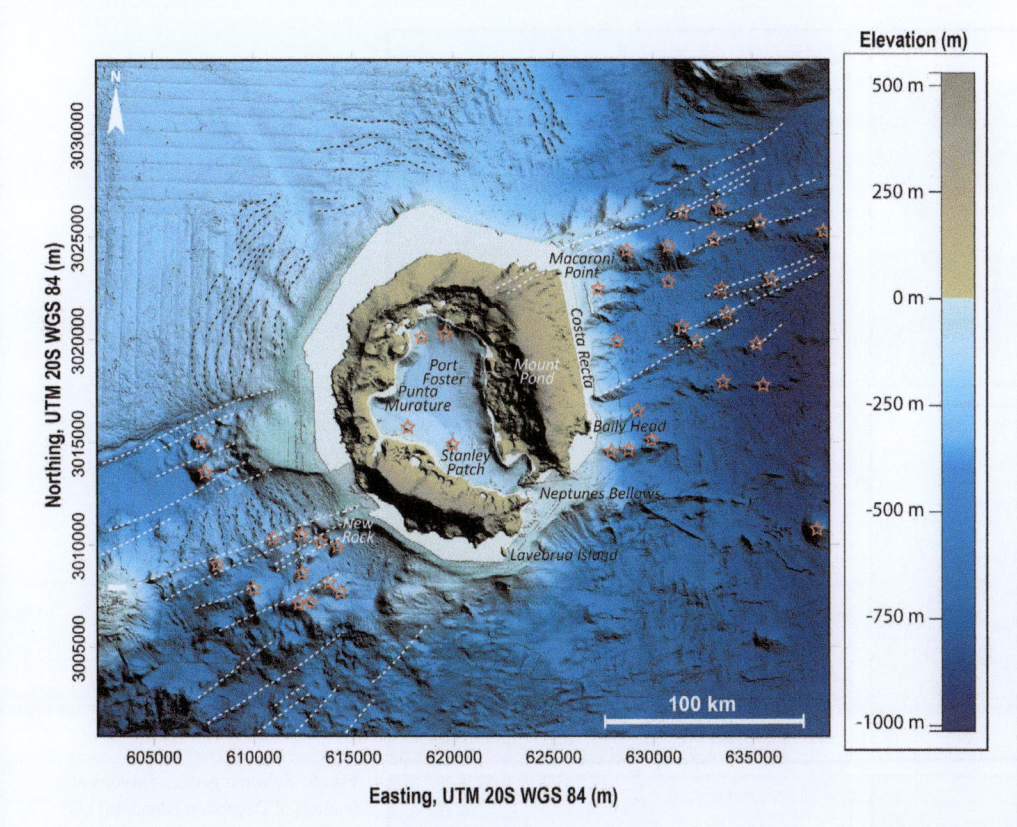

Fig. 4. Digital elevation model of Deception Island and the surrounding seafloor. Some of the most relevant morphostructural features have been highlighted: volcanic cones with a red star; volcanic lineaments and escarpments with discontinuous white lines; and sediment waves related with eruption-fed density flows and slope failures with black discontinuous lines. The bathymetric information was provided by the Navy Hydrographical Institute of Spain (https://armada.defensa.gob.es/ihm).

and Serrano 2002) (Fig. 5). A geomorphological map of the island at 1:25 000 scale shows the distribution of the geomorphological features on land and in Port Foster (López-Martínez et al. 2002a) (see a simplified version in Fig. 6). Permafrost is present on the island, with an active layer normally ranging from 20 cm to more than 1 m in depth, having been detected in recent shallowing of the thaw depth (Ramos et al. 2017).

Submarine morphological features at Port Foster are mainly associated with tectonic instability, volcanism, changes in eustasy and deep-water hydrodynamic processes (e.g. Rey et al. 1997, 2002; Somoza et al. 2004; Barclay et al. 2009) (Fig. 6). Port Foster has been considered to be a half-graben with an active western margin where the slope is rougher, and where slump scars and undulating bottom topography related to slope instability are common. In relation to the instability processes on the basin floor, close to the foot of slope, sedimentary lobes are present and were generated by turbiditic deposition. Moreover, dome and cone volcanic morphologies are present in the southern and western part of Port Foster (Fig. 6). They have an arcuate NNW–SSE alignment; and Stanley Patch (Fig. 4), a large cone with a well-preserved summit crater (see fig. 13 in Smellie 2021: Chapter 3.2a), rises to about 80 m from the seafloor. The main features associated with changes in eustasy are sedimentary progradational wedges, which have an internal prograding sigmoid geometry strongly resembling till deltas, in addition to linear runnels, grooves and channels with U-shaped sections that are perpendicular to the coastline, and which appear to be glacial marks made by moving ice (Kowalewski et al. 1990; Rey et al. 2002) (Fig. 6). Finally, the morphologies associated with hydrodynamic processes are sand waves and small sandbars located along channel margins in Neptunes Bellows.

Regarding Deception Island's tectonic features, mesofractures strike predominantly NNE–SSW, and there are several subordinate maxima striking NE–SW, WSW–ENE to ESE–WNW and NNW–SSE that control the morphology of the island (Figs 6–8). The orientation of these fractures has been compared to Riedel shear fractures related to the development of left-lateral strike-slip motion at the Phoenix and Antarctic plate boundaries (Maestro et al. 2007) (Fig. 8d). This transcurrent movement induced a simple-shear deformation process. En echelon arrays of normal faults developed within the deformation zone, which are oblique to its boundaries; these faults rotated as deformation proceeded (Fig. 8d). It is possible to distinguish two evolutionary stages on the basis of the geometrical and kinematic relationship between the location and the orientation of fractures on Deception Island. These stages relate to an inferred counterclockwise rotation of the island. The three distinct directions of fractures activated during this sequence are consistent with a system of faults with NE–SW, ENE–WSW and NNW–SSE trends. Accordingly, the directions of the fractures are favourably orientated faults with respect to the recent stress field, which we obtain from the fault population analysis carried out on Deception Island; the faults with NNE–SSW and WNW–ESE orientations are mature and unfavourable with respect to the current stress field (Maestro et al. 2007, 2014) (Fig. 8).

The analysis of faults has made it possible to characterize the recent tectonic stress field on Deception Island (Maestro et al. 2007, 2014). The majority of stress tensors correspond to extensional ellipsoids, although a compressional tensor has also been determined. The maximum horizontal stress directions are NW–SE, NNE–SSW and NE–SW, whereas the minimum horizontal stress directions are NE–SW and WNW–ESE to NW–SE. These stress orientations are consistent with: (1) an active subduction along the northern margin of the South Shetland block; and (2) a left-lateral strike-slip zone with simple shear, both associated with the relative movement between the Antarctic and Phoenix plates (Fig. 8).

Volcanic and magmatic evolution: an overview

Deception Island's volcanic and magmatic evolution has been strongly influenced by the development of a large caldera, which collapsed – according to palaeomagnetic data – at about 8.3 ka BC (Oliva-Urcia et al. 2015). However, a more

Fig. 5. Selected geomorphological features of Deception Island: (**a**) cliff on the outer coast of Kendall Terrace; (**b**) crater lake and lava; (**c**) glacier partially covered by pyroclasts; (**d**) drainage system and flat-floored valley on the northern slopes of the caldera rim in the Mount Kirkwood area; (**e**) stream channels cut in unconsolidated pyroclasts; and (**f**) even slopes in the vicinity of Port Foster western coast. Photographs: J. López-Martínez and A. Maestro.

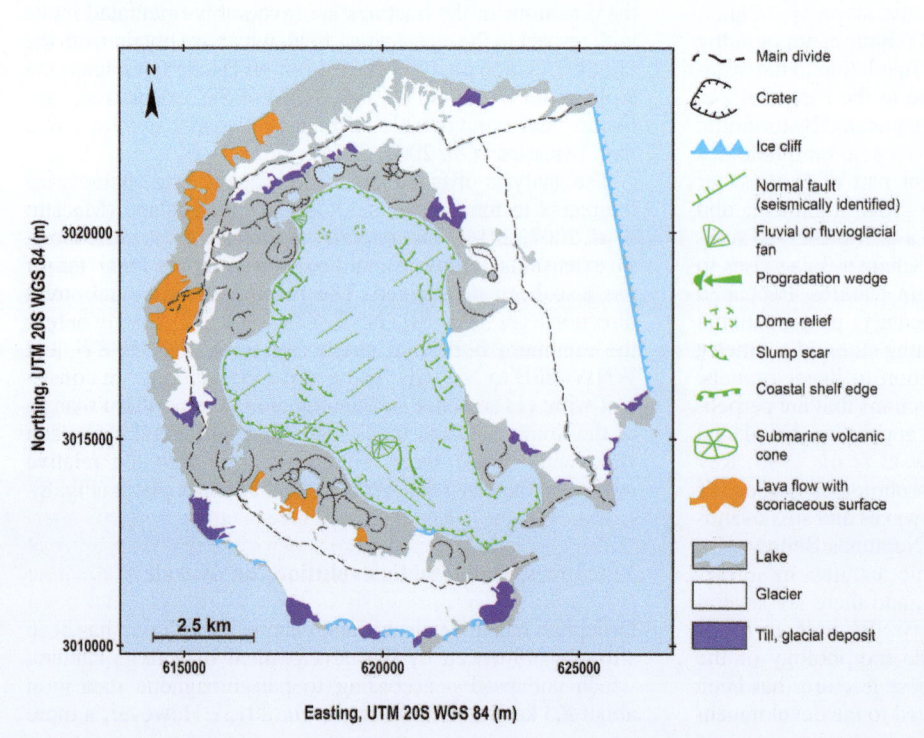

Fig. 6. Simplified geomorphological map of Deception Island (simplified from López-Martínez *et al.* 2002*a*).

Fig. 7. Brittle mesostructures affecting different volcanic units: (**a**) faults and joints at Punta Murature (Pendulum Cove Formation); (**b**) joints at Cathedral Crags (Fumarole Bay Formation); (**c**) normal fault at Punta Murature (Pendulum Cove Formation); (**d**) normal faults close to Irizar Lake (Baily Head Formation); (**e**) normal and reverse faults at Punta Murature (Pendulum Cove Formation); (**f**) reverse faults at Punta Murature (Pendulum Cove Formation); (**g**) basaltic dyke affected by a normal fault close to Irizar Lake (Baily Head Formation); and (**h**) normal faults at Baily Head (Baily Head Formation). The names of the different formations follow the stratigraphy proposed by Smellie (2001), and are assigned according to the geological map by Smellie and López-Martínez (2002c) and field observations by the authors.

recent study of Deception Island caldera-related tephra across Antarctica has placed the age of that event at 3980 ± 125 years BP (Antoniades et al. 2018). During the caldera-forming event about 60 km^3 of magma were extruded (Martí et al. 2013), making Deception Island a medium-sized caldera of comparable dimensions to Krakatau or Santorini (Geyer and Martí 2008). Accordingly, the construction of the island can be separated into three main representative evolutionary stages (Smellie 2001, 2002c; Martí et al. 2013) (Fig. 9a): pre-, syn- and post-caldera.

Traditionally, stratigraphic studies of Deception Island only contemplated whether the rock units pre- or post-dated the formation of the caldera (e.g. Baker and McReath 1971; González-Ferrán et al. 1971; Baker et al. 1975; Smellie 1988, 1989; Martí and Baraldo 1990; Birkenmajer 1992; Baraldo and Rinaldi 2000). By means of a comparison of previously published stratigraphic successions, Smellie (2001) offered a detailed revision and a comprehensive description of Deception Island's lithostratigraphy, formally defining its stratigraphic units (Fig. 9b). A geological map of the island at 1:25 000 scale shows the distribution of the different stratigraphic units (Smellie and López-Martínez 2002c). Martí et al. (2013), following Martí and Baraldo (1990), proposed a more simplified pre-caldera stratigraphy (Fig. 9b), rolling together the three formations units into a single Basaltic Shield Formation (but see Smellie 2021: Chapter 3.2a) (Fig. 9b).

The pre-caldera stage, characterized by the construction of the volcanic lava shield (Fig. 9a-1), is represented by lava flows and different types of eruptive products which resulted from Strombolian-type explosive volcanic activity and hyaloclastic breccias, mainly palagonitized and indurated by alteration processes. Pre-caldera rocks crop out predominantly at Cathedral Crags, in the nearly vertical caldera wall at Fumarole Bay and most conspicuously in the cliffs of the outer coast between Punta Descubierta and Macaroni Point (Figs 10 & 11a). While Smellie (2001) distinguished different discordant units in the pre-caldera sequence (i.e. Fumarole Bay Formation and Basaltic Shield Formation), Martí et al. (2013) suggested

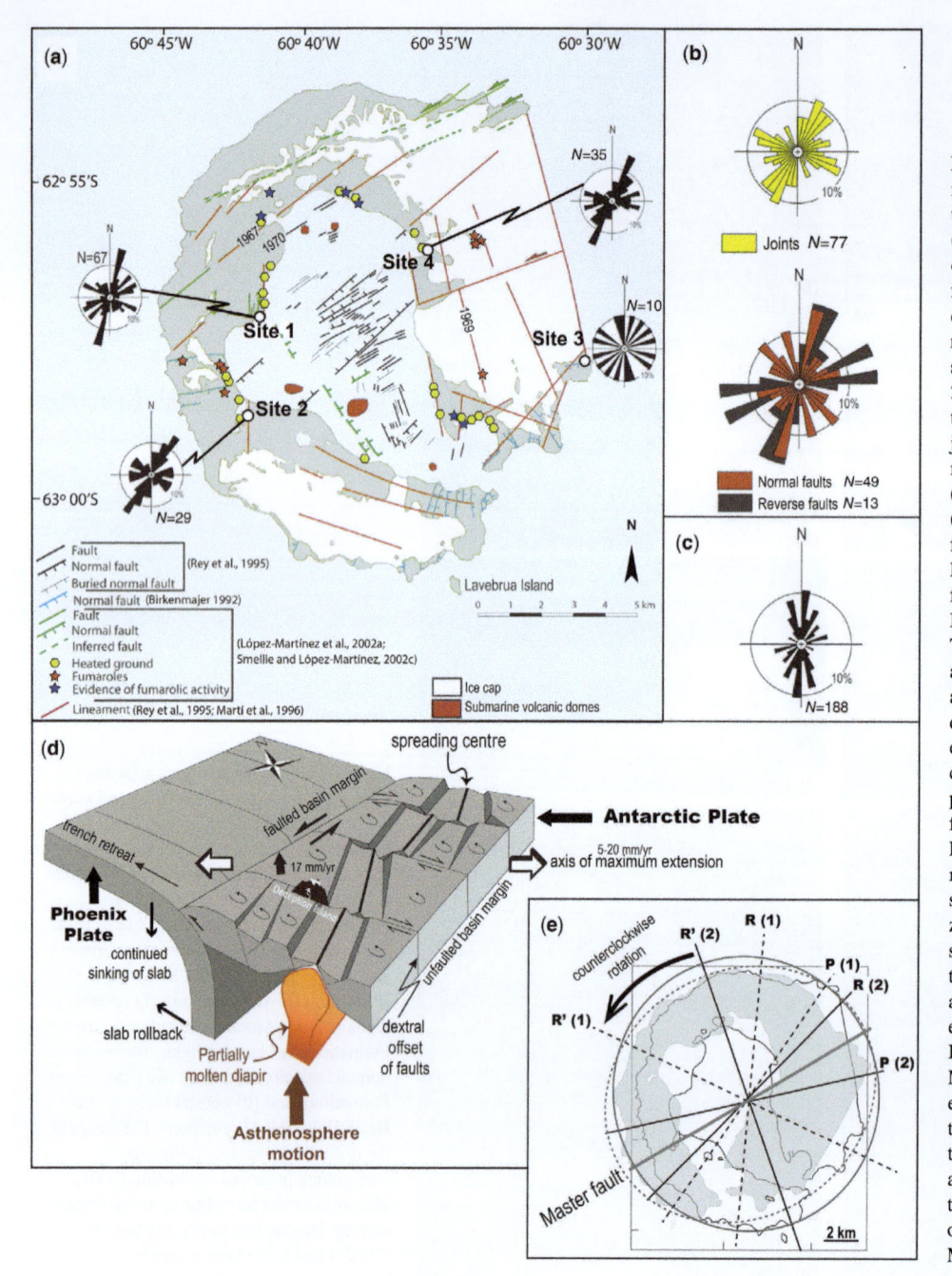

Fig. 8. (a) Simplified geological sketch map of Deception Island indicating the sites studied by Maestro et al. (2007). Rose diagrams show the orientation of fractures (joints and faults) at the outcrop scale (outer circle represents 10%). *N* is the number of fractures measured at each site (modified from Maestro et al. 2007). (b) Rose diagrams indicating the orientation frequency for all joints, normal faults and reverse faults measured by Maestro et al. (2007) (modified from Maestro et al. 2007). (c) Rose diagrams of the orientation frequency for all faults and lineaments mapped in (a) (modified from Maestro et al. 2007). (d) Idealized structure of Bransfield Trough. NW–SE-trending accommodation faults appear to have acted as transfer zones between different rift segments. Both en echelon faults and fault blocks rotate counterclockwise as deformation proceeds. Localized strike-slip faulting occurs within the basin. Large arrows indicate overall sense of motion between the northern and southern margins of the deformation zone. After subduction, the passive sinking of the Phoenix Plate gives rise to trench rollback. Slab rollback and asthenospheric upwelling cause earthquakes and volcanism in Bransfield Trough (modified from Maestro et al. 2007). (e) Sketch of the evolutionary stages of on the basis of the geometrical relationship between the location and orientation of joints and faults. These stages are related through a counterclockwise rotation of Deception Island (modified from Maestro et al. 2007).

grouping all pre-caldera units in the shield-building phase or Basaltic Shield Formation (BSF) (Fig. 9b) given that their eruption dynamics, products and stratigraphy do not differ (but see Smellie 2021: Chapter 3.2a). Geochemically, pre-caldera magmas are among the least evolved beneath Deception Island, with compositions ranging from basaltic to basaltic andesitic and basaltic trachyandesitic (e.g. Aparicio et al. 1997; Smellie 2001; Smellie 2002c; Geyer et al. 2019) (Fig. 12).

The main syn-caldera depositional unit, known as the Outer Coast Tuff Formation (OCTF) (Hawkes 1961; Smellie 2001; Martí et al. 2013), mostly corresponds to massive, lithic-rich, thick deposits of dense pyroclastic density currents (PDC) (Fig. 11b) with a certain number of interbedded diluted PDC units in stratigraphic continuity (Fig. 11c). The OCTF sequence is supposed to have formed in a very short period of time, most probably during a single eruptive event, as suggested by the absence of internal stratigraphic discontinuities (Martí et al. 2013). It has been proposed that the caldera-forming eruption rapidly developed to enormous proportions, causing the continuous collapse of the mixtures of gas and pyroclasts before their development into stable eruption columns (Martí et al. 2013). The OCTF deposits would have emplaced radially from the caldera borders, mantling the pre-caldera units and gradually infilling the caldera depression (Fig. 9a-2). These syn-caldera deposits present a relatively constant average thickness of 50–70 m, and are well exposed on the northern and western coasts, where they form prominent cliffs several tens of metres high (Fig. 11c). The existence of intracaldera OCTF rocks inside Port Foster is also inferred by several seismic reflection profiles and tomographic studies, which provide information on the stratigraphy and tectonics of the collapse infill sequence (e.g. Grad et al. 1992; Rey et al. 1995; Martí et al. 1996; Ben-Zvi et al. 2009; Zandomeneghi et al. 2009; Luzón et al. 2011).

Analysed OCTF samples define two geochemically distinctive groups (Smellie 2002c) (Fig. 13): (1) a main compositional cluster that comprises most of the samples deviating from the principal chemical trend (Magma 1); and (ii) a second cluster within the main Deception Island geochemical trends composed of minor samples with <55 wt% SiO_2 (Magma 2).

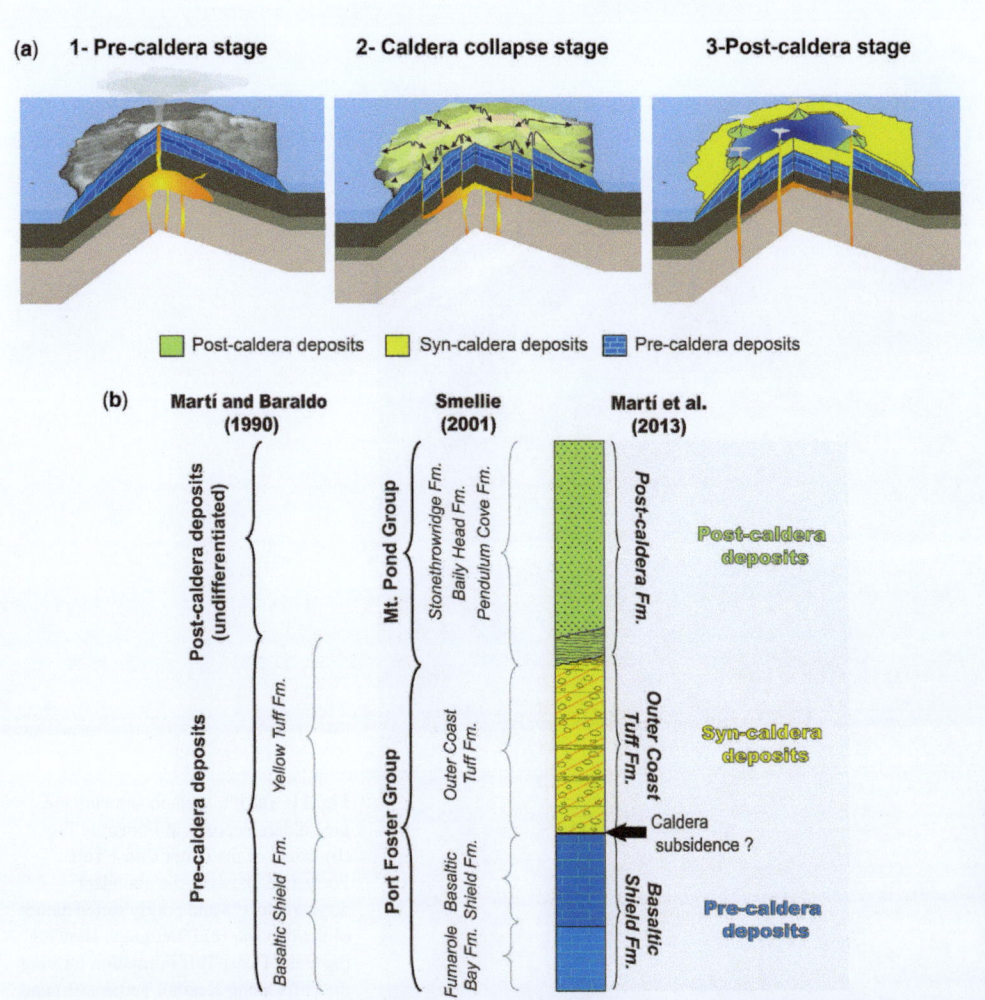

Fig. 9. (a) Simplified sketch illustrating the different stages of Deception Island's evolution (modified from Martí *et al.* 2013). (b) Synthetic stratigraphic section of Deception Island with the divisions proposed by Martí and Baraldo (1990), Smellie (2001) and Martí *et al.* (2013).

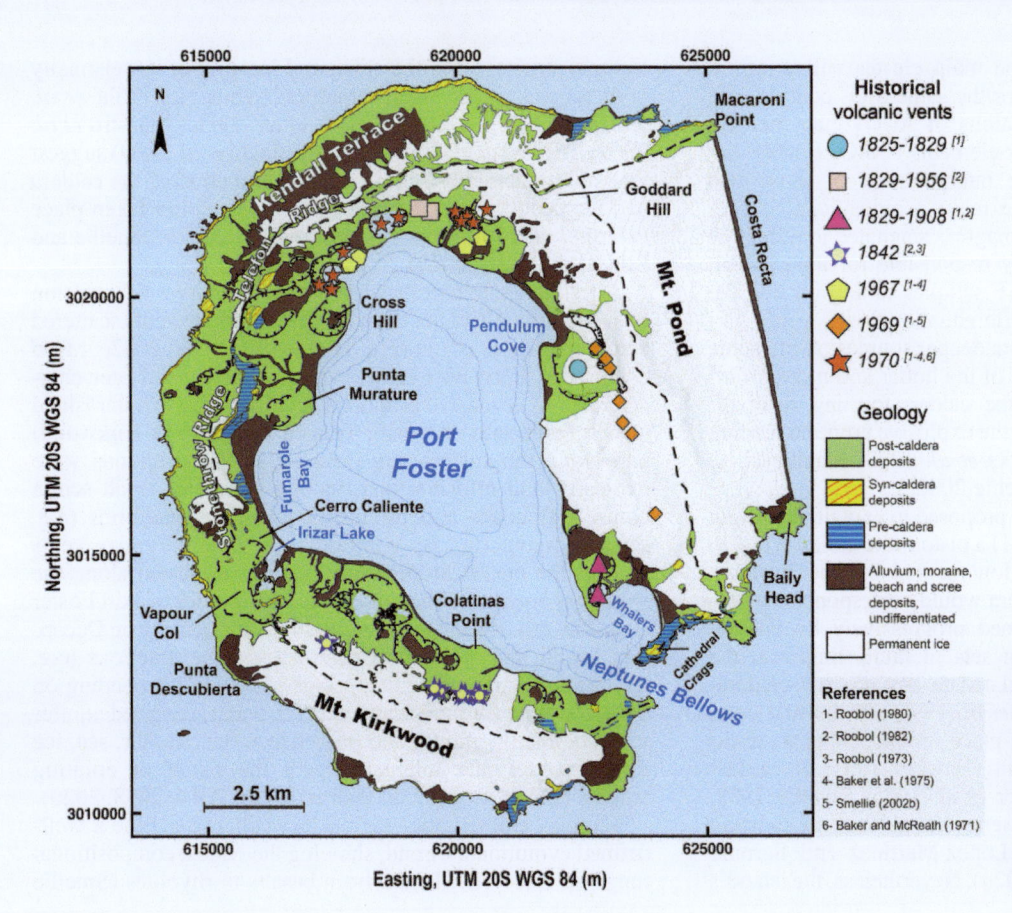

Fig. 10. Simplified geological map of Deception Island (modified from Smellie and López-Martínez 2002c and Martí *et al.* 2013). The shapefiles and digital elevation model were obtained from the SIMAC geodatabase by Torrecillas *et al.* (2006).

Fig. 11. (**a**) Photograph showing the pre-caldera deposits at Fumarole Bay. (**b**) Detail of the Outer Coast Tuff Formation showing the abundant accessory clasts and poorly sorted nature of the deposit. (**c**) Photograph showing the Outer Coast Tuff Formation forming the cliffs along Kendall Terrace. (b) and (c) are modified from Martí *et al.* (2013). Photographs: (a) A.M. Álvarez-Valero; (b) and (c) A. Geyer.

The syn-caldera samples of the main cluster will group at lower TiO_2 and FeO^T values for the same SiO_2 content, and have slightly higher concentrations of Al_2O_3 than the pre- and post-caldera rocks in major elements v. SiO_2 Harker diagrams (Fig. 13b). A possible interpretation of those two magma populations is that the main compositional cluster (i.e. Magma 1) corresponds to magmas stagnated in a shallow magma reservoir being directly responsible for the caldera-forming event. Magma 1 OCTF samples falling outside the main cluster (yet in the main differentiation trend) would correspond to magmas coming from deeper sources. A plausible interpretation is that the arrival of the hotter and more primitive magma Magma 1 into the caldera-forming reservoir (Magma 2) may have triggered the explosive eruption leading to the caldera formation (Sparks *et al.* 1977; Smellie *et al.* 1992; Pallister *et al.* 1996; Smellie 2002*c*).

Two main models have been proposed to explain the origin of Deception Island's caldera: (1) a piston-like collapse due to the massive emptying of a shallow magma chamber during a major eruption; or (2) the caldera would correspond to a volcanic–tectonic depression formed progressively by passive (non-volcanic) extension along sets of faults linked to the regional extension that is unrelated to any specific caldera-forming eruptive event (Martí *et al.* 1996). In the first case, the collapse could have taken place around either a 'traditional' (for the time) ring fault (Hawkes 1961; González-Ferrán and Katsui 1971; Baker *et al.* 1975; Smellie 1988, 1989; Birkenmajer 1992) or a series of regional pre-existing faults (Smellie 2001, 2002*c*; López-Martínez and Serrano 2002; López-Martínez *et al.* 2002*a*). Nevertheless, the island's tectonic structure and the epicentral location of its seismicity with respect to the main structural trends (e.g. Vila *et al.* 1992*b*; Martí *et al.* 1996; Ibáñez *et al.* 2003*a*; Maestro *et al.* 2007; Ben-Zvi *et al.* 2009; Zandomeneghi *et al.* 2009) suggest that strong tectonic control over the formation of the caldera and the post-collapse evolution of the island has taken place (Martí *et al.* 1996; López-Martínez *et al.* 2002*b*; Smellie and López-Martínez 2002*b*).

After the caldera collapse, volcanic activity on Deception Island has mainly come from different eruptive vents scattered across the whole island (Fig. 10) (Smellie 2001, 2002*c*; Martí *et al.* 2013). Early post-caldera volcanic activity has been characterized by extensive eruptions occurring on the outer island slopes (especially Kendall Terrace and Mount Kirkwood) and along the caldera rim (Smellie 2001). Eruptions were explosive and effusive, constructing numerous small scoria cones, tuff cones and prominent lava-delta platforms (e.g. Punta Descubierta). By contrast, most recent eruptive vents (except the one at Stonethrow Ridge) are located along the structural borders of the caldera and the interior of Port Foster Bay (Fig. 10). Recent post-caldera volcanic activity on Deception Island mostly consists of small volume eruptions (e.g. <0.1 km^3) with variable degrees of explosivity depending on the water type (ground water, surface water, confined aquifer v. unconfined), amount and provenance (i.e. aquifer, sea, ice melting, etc.) that interacted with the rising or erupting magma (Baker *et al.* 1975; Pedrazzi *et al.* 2014, 2018, 2020).

Magmas erupted after the caldera collapse outline a well-defined evolutionary trend, showing the widest compositional range on Deception Island from basalts to rhyolites (Smellie

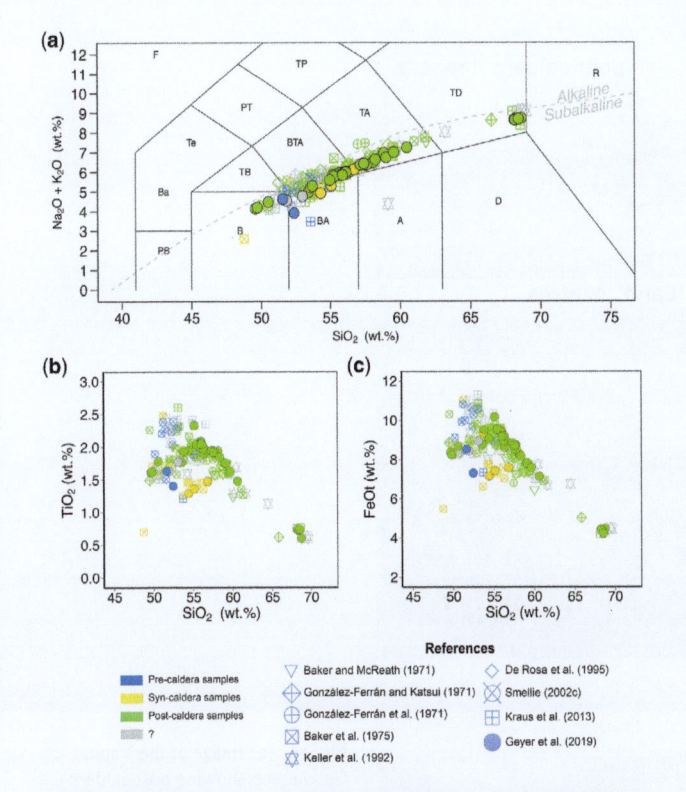

Fig. 12. (a) Total alkalis v. silica diagram (TAS) (Le Bas *et al.* 1986) for Deception Island rock samples (modified from Geyer *et al.* 2019) (see the Supplementary material for details). Major elements normalized to 100% (anhydrous) with Fe recalculated as FeO to Fe_2O_3 following Middlemost (1989). The black dashed line separates the alkaline and subalkaline fields (Irvine and Baragar 1971). (b) TiO_2 and (c) FeO^T v. SiO_2 content Harker diagrams for Deception Island rock samples. Major element compositions have been normalized to 100% in anhydrous base with Fe as FeO^T.

2002c; Geyer *et al.* 2019) (Fig. 12). Magma compositions and pressure–temperature (*P–T*) estimates of juvenile samples from the late post-caldera stage, including historical eruptions, hint that erupted magma can be either directly supplied by the magma accumulation zone at the crust–mantle boundary or by diverse magma batches located at distinct shallow (up to 10 km) depths (e.g. Peccerillo *et al.* 1991; Aparicio *et al.* 1997; Smellie 2001, 2002c; Martí *et al.* 2013; Galé *et al.* 2014; Geyer *et al.* 2019).

Historical volcanic activity

Deception Island's historical volcanism, mostly classified as Volcanic Explosivity Index (VEI) 2–3, has involved small- to moderate-volume monogenetic eruptions (<0.1 km³) with eruptive columns rising up to 10 km in height (Baker *et al.* 1975; Smellie 2001; Bartolini *et al.* 2014; Pedrazzi *et al.* 2014, 2018). However, the common presence of Deception Island tephra in lacustrine cores of neighbouring islands, marine sediments of the Bransfield Strait and Scotia Sea (>800 km distance) (e.g. Hodgson *et al.* 1998; Moreton and Smellie 1998; Fretzdorff and Smellie 2002; Lee *et al.* 2007; Liu *et al.* 2016; Antoniades *et al.* 2018), and even in South Pole ice cores (e.g. Aristarain and Delmas 1998) suggests that some recent post-caldera eruptions may have been significantly higher VEIs than those experienced during historical times. These events would have involved eruptive columns exceeding 20 km in height with much larger volumes of magma (Moreton and Smellie 1998; Smellie 1999; Liu *et al.* 2016).

At Deception Island, such variations in the degree of explosivity can be mostly explained by eventual interactions of the

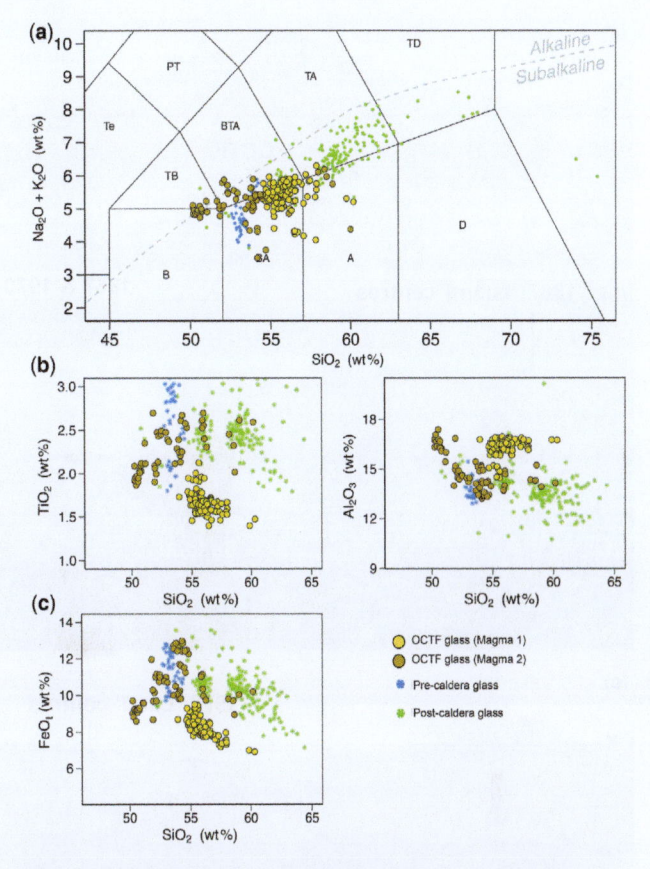

Fig. 13. Glass composition of Deception Island's pre-, post- and syn-caldera (OCTF) juvenile fragments (modified from Antoniades *et al.* 2018) (see Supplementary material for details). (a) Total alkalis v. silica diagram (TAS) (Le Bas *et al.* 1986). Major elements normalized to 100% (anhydrous) with Fe recalculated as FeO to Fe_2O_3 following Middlemost (1989). The black dashed line separates the alkaline and subalkaline fields (Irvine and Baragar 1971). (b) TiO_2 and (c) FeO^T v. SiO_2 content Harker diagrams. Major element compositions have been normalized to 100% in an anhydrous base with Fe as FeO_t. See the supplementary dataset of Geyer *et al.* (2019) for details on composition and exact latitude–longitude coordinates of the rock samples.

rising or erupting magma with seawater, underground aquifers or glacier water (Pedrazzi *et al.* 2014, 2018, 2020). However, in the past, Deception Island has also experienced more evolved and even bimodal eruptions, characterized by the presence of highly explosive dacitic to rhyolitic magmas, such as the ones observed at Cross Hill (Moreton and Smellie 1998; Smellie 2002c; Geyer *et al.* 2019). This suggests that chemical variations between magmas could also be responsible for other eruptive styles (Baker *et al.* 1975; Moreton and Smellie 1998; Smellie 2002a, c; Pedrazzi *et al.* 2018; Geyer *et al.* 2019).

The common occurrence of hydrovolcanic activity episodes on the island is supported by the existence of impact bomb sags underneath blocks and bombs, low vesicularity of juvenile pyroclast, high lithic-rich explosion breccia of fall origin, dilute PDC deposits and palagonitization of glassy groundmass (Baker *et al.* 1975; Smellie 2002a; Pedrazzi *et al.* 2018). Additionally, the presence of abundant fine ash fall and thin PDC deposits might indicate energetic explosive fragmentation due to hydrovolcanic activity with an optimal magma–water energy transfer and/or optimal to deeper subsurface explosion loci of the eruptions (Pedrazzi *et al.* 2018).

During the 1967 and 1970 eruptive episodes, several clustered vents opened simultaneously at the northwestern coast of Port Foster, between Goddard Hill and Cross Hill (Fig. 10). These vents generated diverse types of craters and

Fig. 14. (a) Image of the Vapour Col outcrop showing post-caldera deposits unconformably overlying the syn-caldera OCTF deposits. (b) Panoramic view of several post-caldera volcanic edifices around Post Foster Bay. (c) Photograph of the 1970 eruption 'Land' craters. Photographs: (a) and (b) A.M. Álvarez-Valero; (c) A. Geyer.

cones with contrasting eruptive styles due to contact with water-saturated or icy substrates, or with seawater. Field data from these eruptions suggest alternating magmatic and hydrovolcanic phases with fallout, ballistic blocks and bombs, and subordinated dilute PDCs (Baker *et al.* 1975; Pedrazzi *et al.* 2014, 2018).

The 1967 eruption led to the formation of: (i) a new island (so-called 'island' centres) elongated in a NE–SW direction consisting of three overlapping pyroclastic cones with water-filled craters (Baker *et al.* 1975; Roobol 1980, 1982); and (ii) two roughly circular 'land' centres situated east of the new island (Fig. 10). The 1970 eruption was also characterized by 'land' and 'island' vents (Baker *et al.* 1975; Pedrazzi *et al.* 2014, 2018). The former ones consisted of seven conical edifices located at the foot of Goddard Hill and aligned roughly NW–SE (Figs 10 & 14c). The 'island' centres (six vents) resulted in the formation of a new strip of land about 1700 m long by 400 m wide (Baker *et al.* 1975). On the one hand, the 'land' centres of the 1967 and 1970 eruptions were the result of the interaction between magma and an aquifer inside the OCTF and BSF, with a possible initial first subglacial phase (Fig. 15a). Conversely, 'island' centres started with a submarine eruption in shallow seawater, later evolving into a subaerial phase (Fig. 15b, c) (Pedrazzi *et al.* 2014, 2018).

The February 1969 eruption occurred when a 4 km-long fissure opened beneath glacial ice at Mount Pond (Baker *et al.* 1975; Smellie 2002*b*, *c*) (Fig. 10). During the first eruptive stage, pressurized hot gases may have briefly lifted the glacier, with subglacial melting above each vent and with a subglacial space beneath the lofted glacier that was filled successively with meltwater, eventually overflowing and forming voluminous supraglacial floods that washed the surface of the glacier and destroyed the British station at Whalers Bay. Hydrovolcanic activity was coeval with this early stage (Smellie 2002*b*; Pedrazzi *et al.* 2018). At the last stage, a shift in the location of the eruptive centres is observable but the activity continued and constructed a supraglacial cinder cone at the fissure with only minor meltwater, draining subglacially; the eruption ending up essentially dry (i.e. Hawaiian–Strombolian), although all the preserved deposits appear to be magmatic (i.e. Strombolian) and the early influence of water interaction is inferred rather than proven (Smellie 2002*b*). In any case, an important feature of the eruption was the rapid generation of abundant meltwater due to the melting of the glacier above the eruptive fissures. This meltwater overflowed onto the glacier surface and generated a flood (lahar) that modified the glacier, extended the local coastline, and destroyed the British scientific station and the infrastructures related to the whaling industry at Whalers Bay (Baker *et al.* 1975). The 1969 eruption on Deception Island is still one of the best-described glaciovolcanic eruptions ever observed, owing to an exceptionally prescient descriptive record of the events made by Baker *et al.* (1975; see Smellie 2002*b*).

Unrest episodes and volcano monitoring

Seismic activity

The seismic monitoring of Deception Island volcano began in the 1950s with the deployment of a seismometer at the

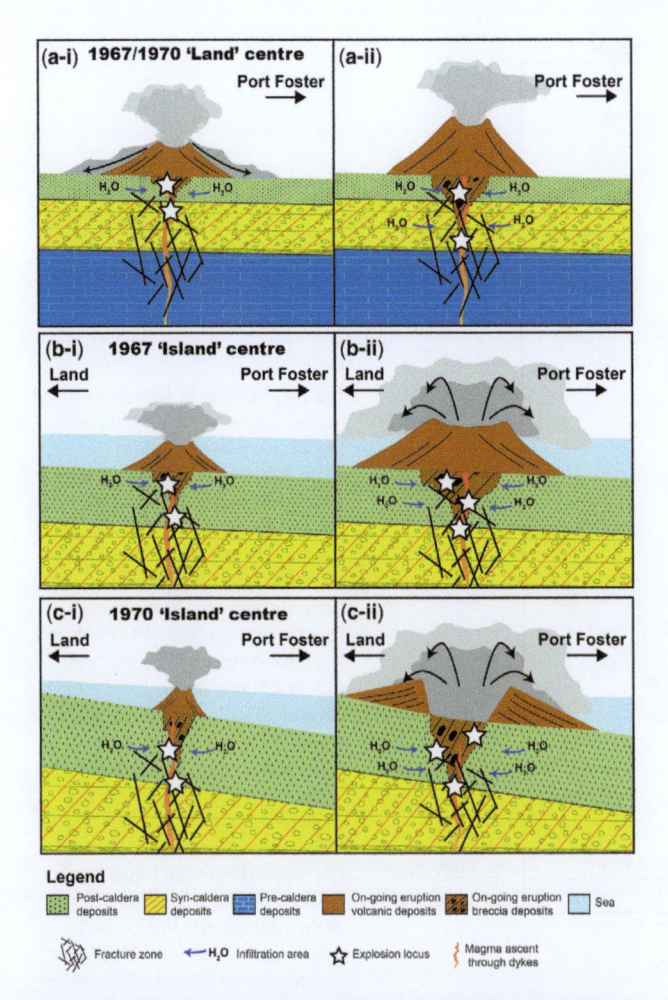

Fig. 15. Sketches (not to scale) illustrating the evolution of the 1967 and 1970 eruptions. (**a-i**) Rise of the magma for the 'land' centres and (**a-ii**) interaction with the post-caldera, syn-caldera and pre-caldera deposits. For the 1967 'island' centres: (**b-i**) interaction of the ascending magma with shallow marine water; and (**b-ii**) interaction with the basement made of syn-caldera deposits. A similar pattern to (b) is also observed for the 1970 eruption 'island' centres in (**c-i**) and (**c-ii**) (modified from Pedrazzi *et al.* 2018).

Argentinian base. In 1965, a second seismometer was installed at the Chilean base (Lorca 1976) and an improvised instrument was used at the British station during the 1969 eruption (Baker *et al.* 1975; Smellie 2002*b*) but all instruments were abandoned once the 1967, 1969 and 1970 eruptions forced the evacuation of these bases. Monitoring of the local seismic activity was re-established in 1986 through field surveys carried out by Argentinian and Spanish researchers during the Antarctic summer (Vila *et al.* 1992*a*, *b*, 1995; Correig *et al.* 1997; Ortiz *et al.* 1997). In 1986–90, seismic data were recorded by a network of five short-period seismic stations deployed around the caldera. In the 1991–93 period, a single three-component seismic station was used at a location near the Argentinian base. In 1994–99, seismic activity was monitored mainly by means of small-aperture seismic arrays (Almendros *et al.* 1997, 1999; Ibáñez *et al.* 2000, 2003*a*). Seismic arrays are dense configurations of seismometers distributed over a small area (compared with the distance to the source). They have applications in volcano seismology: for example, the tracking of continuous signals such as volcanic tremor (Chouet 1996; Almendros *et al.* 1997, 2001, 2007, 2014; Chouet 2003; Zandomeneghi *et al.* 2009; Wasserman 2012). Since 1999 Deception Island has been regularly monitored using a combination of seismic arrays and a seismic network composed of five–six three-component seismometers distributed around Port Foster Bay (Carmona *et al.* 2014) (Fig. 16). Additionally, in 2008 a permanent broadband seismic station was deployed near the Spanish base (Jiménez Morales *et al.* 2017). Apart from these local monitoring efforts, a few temporary experiments have also been carried out in the last few years to understand the seismicity and structure of Deception Island and surrounding areas (e.g. Ibáñez *et al.* 1997; Robertson Maurice *et al.* 2003; Ben-Zvi *et al.* 2009; Zandomeneghi *et al.* 2009; Dziak *et al.* 2010; Luzón *et al.* 2011; Prudencio *et al.* 2015).

Local seismicity at Deception Island volcano includes volcano-tectonic (VT) earthquakes, long-period (LP) events and episodes of volcanic tremor. VT earthquakes are brittle failure earthquakes occurring at faults located within the volcanic edifice as a consequence of stresses related to the internal dynamics of the volcano. They are generally characterized by a double-couple mechanism, the presence of P- and S-phases, an exponentially decaying coda, and broad spectral content reaching frequencies of up to 30 Hz (Chouet 2003; Zobin 2016). Conversely, LP events and volcanic tremor originate from interactions between the solid rock and volcanic or hydrothermal fluids (Chouet 1996, 2003; Zobin 2016). LP events have emergent onsets, a spindle-shaped envelope and a narrow-band, quasi-monochromatic spectral content characterized by narrow peaks in the 0.5–5 Hz band. Volcanic tremor has similar characteristics, although it displays a much longer duration that may reach up to hours or days (McNutt and Nishimura 2008; Zobin 2016). The source mechanism of LP seismicity has been attributed to the resonance of fluid-filled conduits, oscillations of fluid flow, etc. (Chouet 1992; Julian 1994; Konstantinou and Schlindwein 2003). Other types of earthquakes have occasionally been reported at Deception Island volcano, although VT earthquakes and LP seismicity are the dominant types both in terms of number and energy of the seismic events.

In general, seismic data obtained at Deception Island volcano indicate a close relationship between the seismicity and volcanic activity, which emphasizes the usefulness of seismic observations as a monitoring and forecasting tool. For example, the energy and rate of occurrence of the seismic events increased before and during the 1967, 1969 and 1970 eruptions (Baker *et al.* 1975; Newhall and Dzurisin 1988; Smellie 2002*b*). There are reports of large earthquakes felt by the people working at the scientific bases, as well as high-amplitude volcanic tremors, before and during the 1967 and 1969 eruptions (Baker *et al.* 1969; Lorca 1976). There were no eyewitnesses to the 1970 eruption but seismic evidence from nearby seismic stations suggests the occurrence of several large earthquakes with magnitudes of up to 5 in August 1970, coincident in space and time with the eruption (Baker and McReath 1971; Pelayo and Wiens 1989). Years after the 1967–70 eruptions, large earthquakes with magnitudes above 4.5 were detected again near Deception Island: for example, in March 1974 and December 1982 (International Seismological Centre: http://www.isc.ac.uk/). However, the reported source locations have large uncertainties due to the lack of nearby seismic stations. In these cases, we cannot ascertain whether these earthquakes were a consequence of another period of volcanic unrest at Deception Island, or tectonic earthquakes related to extensional dynamics in the Bransfield Rift or subduction in the South Shetland Trench.

Seismic monitoring data recorded since 1986 on periodic surveys during austral summers have allowed further detailed analyses and a better understanding of the local seismic activity on Deception Island. During this period, VT earthquakes and a few volcanic tremor episodes were recorded. Vila *et al.* (1992*a*, *b*) studied the seismic activity from 1986 to 1989 and provided the first epicentral map of the VT activity of Deception Island volcano. The epicentres were distributed

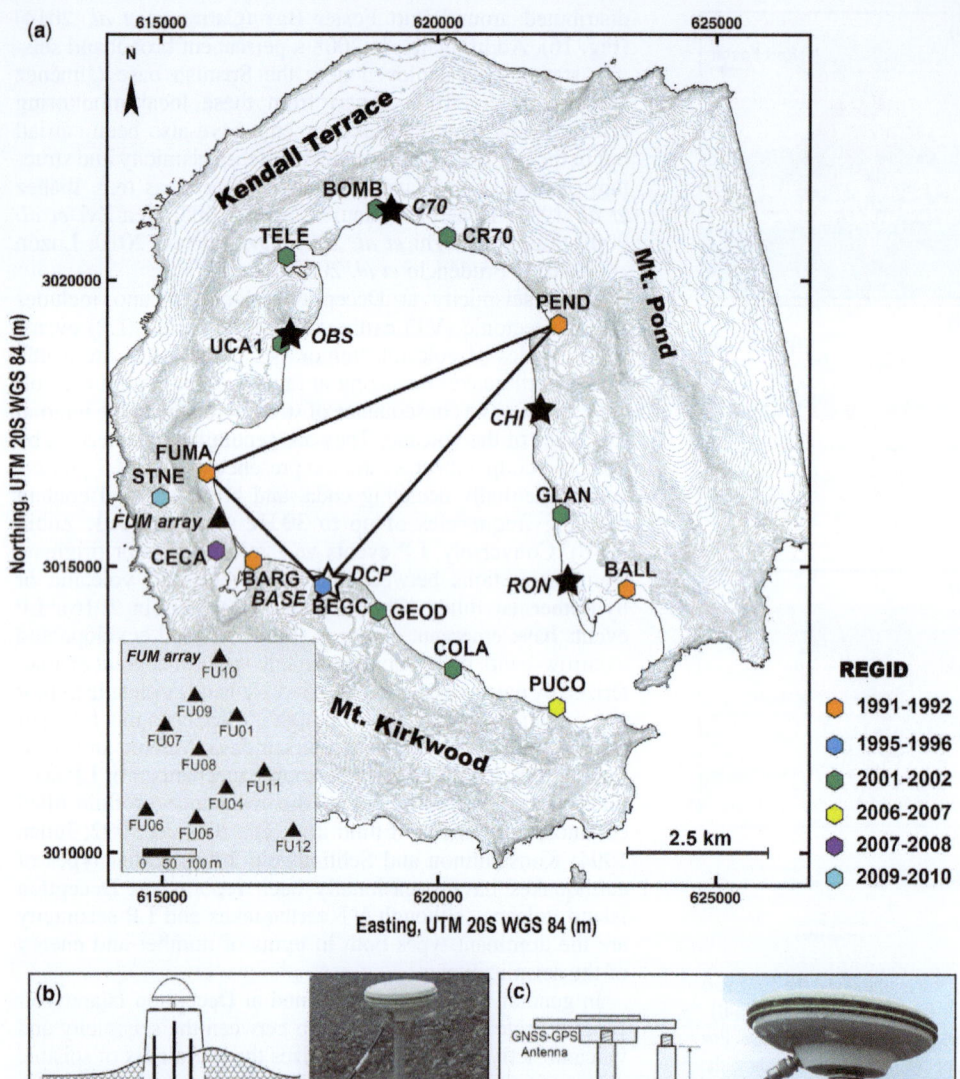

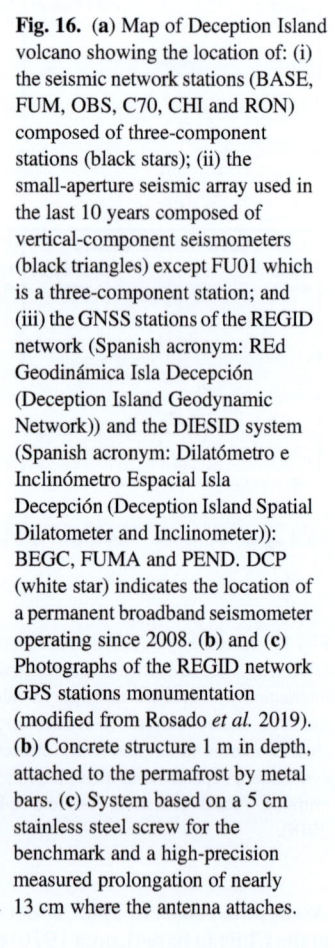

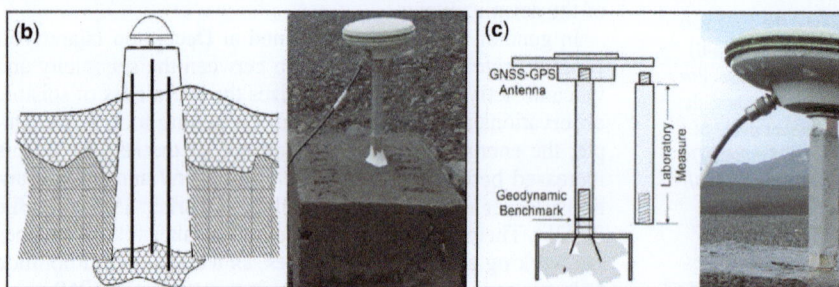

Fig. 16. (a) Map of Deception Island volcano showing the location of: (i) the seismic network stations (BASE, FUM, OBS, C70, CHI and RON) composed of three-component stations (black stars); (ii) the small-aperture seismic array used in the last 10 years composed of vertical-component seismometers (black triangles) except FU01 which is a three-component station; and (iii) the GNSS stations of the REGID network (Spanish acronym: REd Geodinámica Isla Decepción (Deception Island Geodynamic Network)) and the DIESID system (Spanish acronym: Dilatómetro e Inclinómetro Espacial Isla Decepción (Deception Island Spatial Dilatometer and Inclinometer)): BEGC, FUMA and PEND. DCP (white star) indicates the location of a permanent broadband seismometer operating since 2008. (b) and (c) Photographs of the REGID network GPS stations monumentation (modified from Rosado et al. 2019). (b) Concrete structure 1 m in depth, attached to the permafrost by metal bars. (c) System based on a 5 cm stainless steel screw for the benchmark and a high-precision measured prolongation of nearly 13 cm where the antenna attaches.

along a ENE–WSW trend, coincident with one of the dominant fault systems (Martí et al. 1996; Maestro et al. 2007) and the direction of the Bransfield Rift. They also proposed an interpretation for the occurrence of volcanic tremor based on the degasification of an aquifer in contact with deep hot materials. On average, the energy of the earthquakes recorded between 1986 and 1991 was quite low, with magnitudes between 1 and 2.

This situation changed in January 1992, when a significant increase in the number and magnitude of the seismic events was detected (Ortiz et al. 1997). A total of 776 VT earthquakes were recorded in less than 2 months. Some of these earthquakes, and even a few episodes of volcanic tremor, were felt on the island. During this event, the only available seismic instrument was a three-component short-period seismometer deployed near the Argentinian base. Although the epicentral area of the recorded activity could not be determined accurately, indirect evidence shows that the source area was probably located under Fumarole Bay, just 2–3 km from the shoreline. Observations of small gravity irregularities and magnetic anomalies correlated with the successive seismic swarms suggest that a magmatic intrusion took place during this increase of seismic activity (Ortiz et al. 1992, 1997; García et al. 1997). Changes in fumarolic emissions, increased fumarole and groundwater temperatures, and a possible ground deformation process near the Argentinian base (Ortiz et al. 1992), support the hypothesis of magmatic intrusion (Ortiz et al. 1997). After a few weeks, the anomalous level of seismic activity started to decline, and by the end of February it was back to pre-1992 levels.

The introduction of seismic arrays for volcano monitoring in 1994 had a very important effect on our view of the seismicity of Deception Island volcano. Small-aperture seismic arrays allow for the detection and characterization of: (1) continuous signals; and (2) signals with low signal-to-noise ratios. Therefore, they are highly suitable for the analysis of LP seismicity at volcanoes. The application of seismic arrays at Deception Island volcano revealed that LP seismicity was conspicuous (Almendros et al. 1997, 1999; Alguacil et al. 1999; Ibáñez et al. 2000, 2003a; Stich et al. 2011; Carmona et al. 2012; Padrón et al. 2015; Jiménez Morales et al. 2017). Between 1992 and 1999, the seismic activity at Deception Island volcano was clearly dominated by the occurrence of LP events and tremor episodes. Array analyses demonstrated that these events had shallow sources that were spatially related to the hydrothermal features of Deception Island. Therefore, it is very likely that most of the LP activity has a hydrothermal origin (Almendros et al. 1997; Ibáñez et al. 2000). Seismic activity during this period was irregular (Ibáñez et al. 2003b). Of particular note was the increase in the number of LP events during the 1995–96 survey, although in the following surveys (1996–97 and 1997–98) activity decreased considerably.

During the 1998–99 survey, seismic activity increased again with a total of 3643 seismic events recorded during a period of 2 months. Of these, 2072 were VT earthquakes, 1556 were LP events and 15 were hybrid events (Ibáñez et al. 2003a, b). There were also a number of volcanic tremor episodes. The magnitude of these earthquakes ranged from −0.8 to 3.4 (Havskov et al. 2003). The two largest earthquakes in the series, with magnitudes of 2.8 and 3.4, occurred on 11 and 20 January 1999 and were felt by personnel from the Spanish scientific base Gabriel de Castilla. The VT earthquakes were located using array techniques to estimate the propagation direction and apparent velocity of the first P-wave arrivals. A ray-tracing procedure was used in a velocity model obtained by previous studies (Ibáñez et al. 2000; Saccorotti et al. 2001). The distance along the ray path was determined using estimates of the S–P differences. This technique allowed for the location of 863 earthquakes. The locations show that the vast majority of VT events occurred near Fumarole Bay, at depths of between 1 and 4 km. Two alignments can be observed: N45°E and N80°E. The application of precise location techniques developed by Almendros et al. (2004) allowed some of the rupture planes responsible for the VT events to be imaged (Carmona et al. 2010). The strikes for most of these planes trended NW–SE, while a smaller number trended NE–SW. These strikes do not coincide with the dominant fracture directions. The geometry and the position of these planes show that although the 1999 seismicity was influenced by regional tectonics, the origin of the destabilization of the system was a shallow magmatic intrusion that perturbed the regional stress field (Carmona et al. 2010). Additional support for this hypothesis comes from changes in fumarolic emissions (Agusto et al. 2004; Caselli et al. 2004) and variations in the patterns of deformation, from radial extension and uplift to slow compression and subsidence (Fernandez-Ros et al. 2007; Berrocoso et al. 2008). The analysis of the LP events and the volcanic tremor episodes revealed different apparent velocities and azimuths, indicating that the mechanisms causing the LP events and VT earthquakes were different (Ibáñez et al. 2003a, b). These results suggest that the LP activity had a hydrothermal origin and was basically unrelated to the VT series.

Between the 1999 VT earthquake series and 2014, seismic activity was relatively quiet, with occasional peaks of LP activity: for example, during the 2003–04 survey (Ibáñez et al. 2003a; Carmona et al. 2012). This behaviour changed again drastically in September 2014, when a sharp increase in the seismic activity around Deception Island was reported (Almendros et al. 2015, 2018). More than 9000 earthquakes with magnitudes up to 4.6 were recorded over a period of 8 months. The number of events identified at the Deception Island seismic network was an order of magnitude larger than the sum of all earthquakes in the previous 15 years. These earthquakes were located SE of Livingston Island, about 35 km NE of Deception Island, and it is unclear if they had a tectonic or volcanic origin (Almendros et al. 2018). On Deception Island itself, LP seismicity was unusually frequent and intense. Before February 2015, only a few tens of VT earthquakes were detected. They occurred mainly in a region 5–20 km to the SW. However, in mid-February the number and size of VT earthquakes escalated. Moreover, their locations approached Deception Island and, in fact, encompassed the whole volcanic edifice, suggesting a situation of generalized unrest. These observations, in the context of extensive seismo-volcanic activity, led to a change in the volcanic alert system. The Spanish Polar Committee established the volcanic alert level at yellow (enhanced monitoring to corroborate the observed anomalies) for a few days in February 2015 (Almendros et al. 2015). This episode of VT seismicity is similar to the seismic series recorded in 1992 and 1999. A few thousand earthquakes were reported over a period of 5 months, most of them too small to be identified at more than one station. About 400 VT earthquakes were located, with maximum rates of 45 earthquakes per day and a maximum magnitude of 3.2. The similarities between the series suggest that the 2015 VT swarm at Deception Island volcano could also be related to the stress changes induced by a magmatic intrusion at shallow depths, as proposed for the other two episodes (Ortiz et al. 1997). In this case, signs of instability were observed years before the seismic series: for example, the steady increase in soil temperatures at Cerro Caliente (Berrocoso et al. 2018) and the rising number of volcanic tremors (Jiménez Morales et al. 2017). The seismic activity continued at anomalously high levels until at least May 2015 (Almendros et al. 2018).

Given the spatial and temporal coincidence, it is unlikely that the 2014–15 Livingston series and the 2015 Deception VT swarm were unrelated. We propose that the Livingston Island series may have produced a triggering effect on Deception Island volcano. Dynamic stresses associated with the seismic swarm may have induced overpressure in the unstable volcanic system (Manga and Brodsky 2006; Hill and Prejean 2015), leading to a magmatic intrusion that may in turn have triggered the VT swarm. Alternatively, both the Livingston Island earthquakes and the VT swarm could be consequences of a magmatic intrusion at Deception Island. The Livingston Island series would be an example of a precursory distal VT swarm, which seems to be a common feature preceding volcanic eruptions and magma intrusions in long-dormant volcanoes (White and McCausland 2016; Coulon et al. 2017).

All three episodes of enhanced VT activity (1992, 1999 and 2015) have been interpreted to be consequences of magma intrusions. These inferences are based on the seismological characteristics of the earthquakes and the observations of simultaneous volcanological anomalies in gravity and magnetic data, surface deformation, ground and water temperatures, and gas emissions. However, none of the intrusions produced a volcanic eruption. Occurrences of seismic series associated with magma intrusions that do not end in eruptions are not rare in volcanic areas, although perhaps they get less attention from the scientific community (e.g. Poland 2010; Moran et al. 2011). The seismic crises at Deception Island volcano are instances to add to the cases of Akutan (Alaska) in 1996 (Lu et al. 2000), Iwate (Japan) in 1998 (Nishimura and Ueki 2011), Paricutin (Mexico) in 2006 (Gardine et al. 2011) and Harrat Lunayyir (Saudi Arabia) in 2009 (Koulakov et al. 2015), to cite just a few examples. They might constitute clues towards the future behaviour of the volcano (Albert et al. 2016).

As a conclusion of the study of the seismic activity at Deception Island volcano, a clear image seems to emerge. From a seismic point of view, the volcano displays two regimes of seismic activity that correspond to different states of volcanic activity: dormant and restless. The dormant state is most common, and is mainly characterized by the occurrence of shallow, low-energy LP events and volcanic tremor episodes caused by the circulation of fluids in the hydrothermal system (Fig. 17). In this state, some VT and hybrid events may also take place as a response to regional tectonics. The restless state is characterized by a large number of VT earthquakes induced by the destabilizations produced in the stress field by magma intrusions. These earthquakes can be

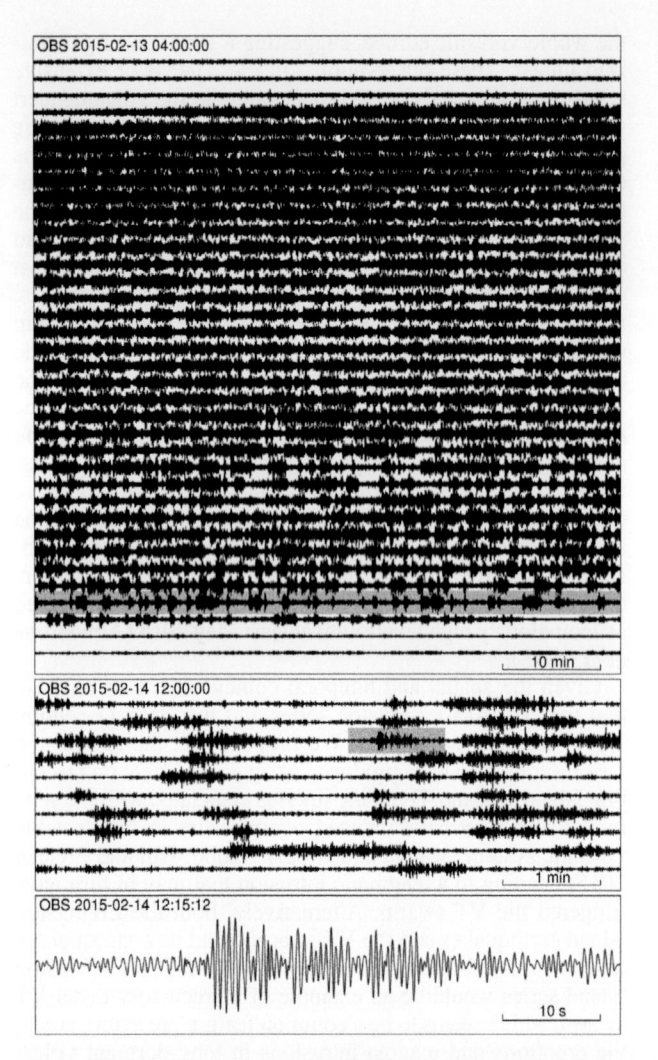

Fig. 17. Example of volcanic tremor and long-period seismicity recorded at Deception Island during a period of high seismo-volcanic activity in February 2015. We show the vertical-component seismogram, filtered in the 1–15 Hz band, for 36 hours (top), 1 hour (middle) and 1 minute (bottom). The periods indicated by a grey background are zoomed in the next plot.

relatively energetic and may even be felt by people on the island. They can be accompanied by LP events and volcanic tremor, large enough to be detected at most stations of the seismic network. The restless state was obviously observed during the volcanic eruptions of 1967–70, and was also reached at least during the seismic crises of 1992, 1999 and 2015 (Fig. 18).

Ground deformation

Deception Island's geodetic reference frame, the REGID (Spanish acronym for REd Geodinámica Isla Decepción: Deception Island Geodynamic Network), has been constructed since 1991–92 throughout several Spanish Antarctic campaigns (Catalán *et al.* 1991; Berrocoso *et al.* 2006*a*, *b*, *c*, *d*, 2008). The REGID network, composed of 15 benchmarks around Deception Island's interior bay (Fig. 16a), was designed with the aim of enabling geodynamic studies based on GNSS–GPS geodetic techniques. All benchmarks are 20 cm in height and are located at the top of an approximately 1 m deep concrete base, which is fastened to the permafrost by metal bars (Fig. 16b). In order to correctly ensure the GPS antenna centre and precisely determine positions relative to the benchmark at each campaign, we designed an accurately measured system based on a 5 cm stainless steel screw for the benchmark and a prolongation nearly 13 cm in length where the antenna is attached (Fig. 16c). For the antenna orientation, washers with a thickness of 1 mm are added until an orientation to the north is reached (Berrocoso *et al.* 2010, 2016; Rosado *et al.* 2019).

From the 2001–02 Antarctic campaign onwards, observations have been consecutively made during every austral summer over periods of 5–6 days. All collected data are processed using the scientific software Bernese v5.0 (Dach *et al.* 2007). The benchmarks' coordinates from 1995 until 2018 were estimated on the basis of 24 h continuous data-gathering, 10° elevation masks and a 30 s sampling rate. Since the software is based on relative positioning, the daily set of baselines is combined in a network adjustment involving every benchmark each day. During the parameter estimation process, carrier-phase double-difference data are used in an ionospheric delay free mode. Tropospheric errors are dealt with using a combination of the *a priori* Saastamoinen model and Neill-mapping functions. Tropospheric parameters are estimated hourly, and ambiguities are resolved for each baseline independently, using the ionosphere-free observable with an *a priori* ionospheric model for determining the widelane ambiguity. The ocean tide loading displacement corrections

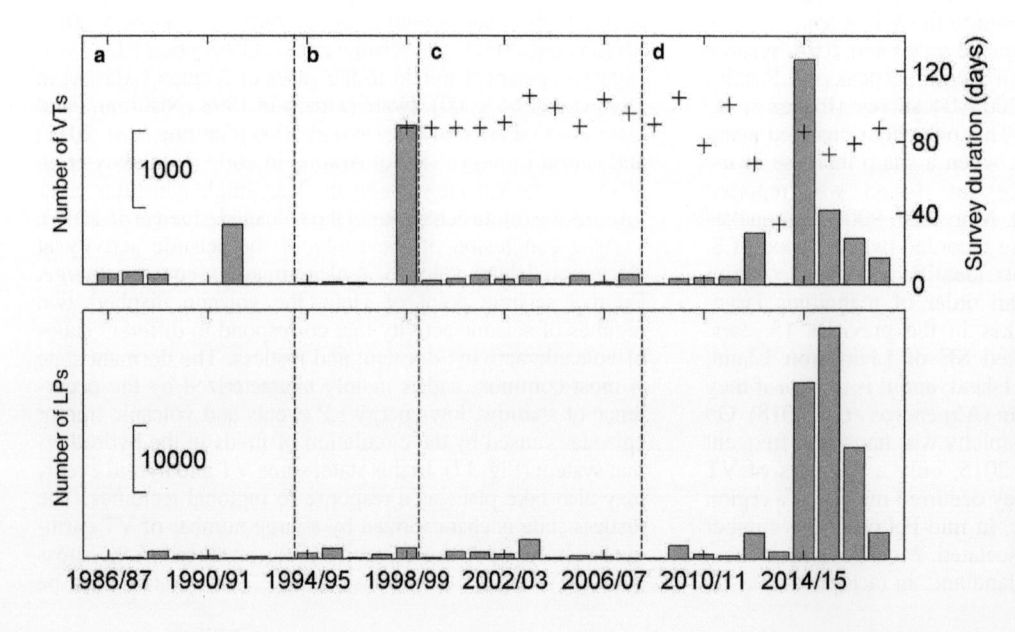

Fig. 18. Histogram of the total number of volcano-tectonic earthquakes and long-period events detected at Deception Island volcano during the austral summer surveys carried out since 1986. Crosses indicate the duration of the surveys in the scale on the top right. The dashed lines separate periods with different instrumental configurations: analogue and digital network, single stations (period a: Vila *et al.* 1992*b*; Ortiz *et al.* 1997); seismic arrays (period b: Almendros *et al.* 1997; Ibáñez *et al.* 2000); seismic network and seismic arrays (periods c and d: Carmona *et al.* 2014).

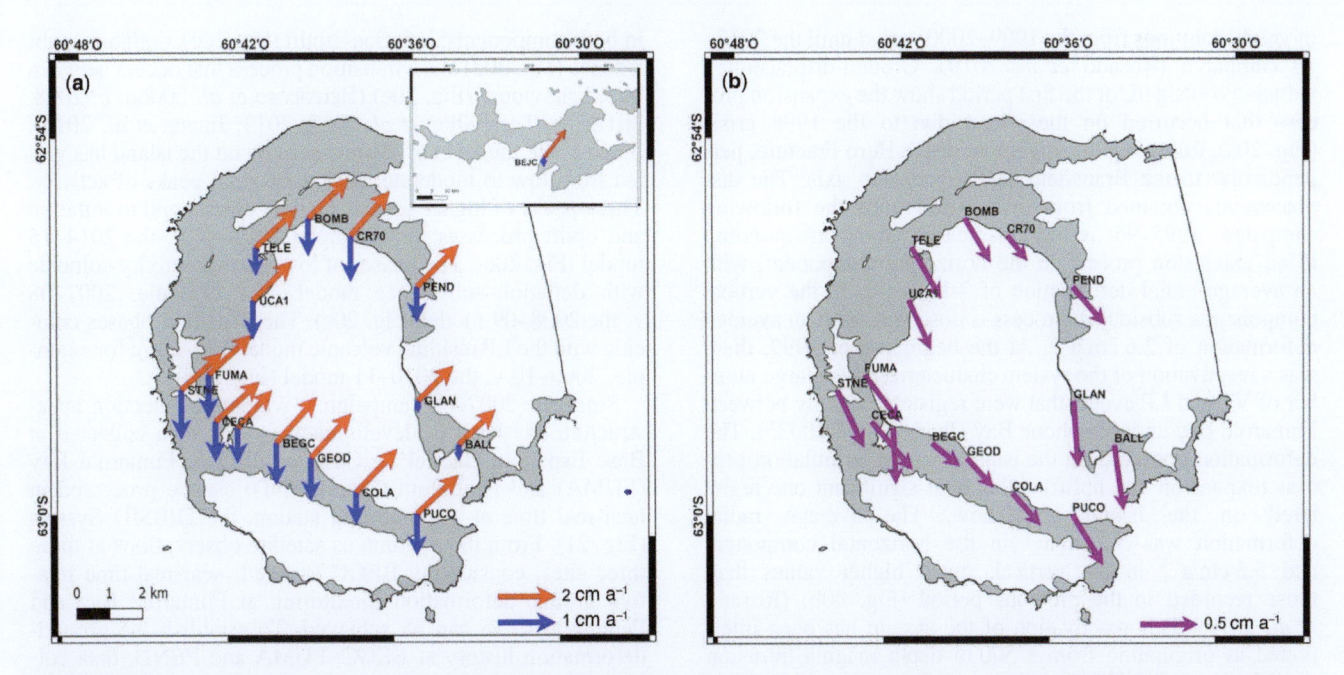

Fig. 19. (a) Horizontal and vertical geodynamic deformation model for the Livingston (tectonic) and Deception (volcano-tectonic) islands. The average horizontal displacement presents a NE orientation and shows a subsidence process due to the subduction process of the Phoenix microplate. Horizontal and vertical displacements are indicated in red and blue arrow, respectively. (b) The residual non-tectonic model for Deception Island was obtained with respect to the Livingston Island stations (modified from Rosado *et al.* 2019).

from Onsala Observatory were also introduced (Berrocoso *et al.* 2010). The normal equations are computed for each daily solution. Finally, for each campaign, the solution is achieved by combining the daily normal equations at the mean epoch of each campaign, using the IGS final orbits and by constraining the movement of one reference benchmark, PALM at Palmer Station (Anvers Island, Antarctica), for the realization of the ITRF2008 reference frame (http://itrf.ensg.ign.fr/) (Altamimi *et al.* 2011). Also, one control benchmark was analysed, OHI2 at O'Higgins Station (Antarctic Peninsula), to validate the inferred velocities against those expected for the ITRF2008 reference frame (Berrocoso *et al.* 2006b, d, 2008, 2016). The above-described process has allowed a 3D VT model for Deception Island to be obtained (Fig. 19) (Rosado *et al.* 2019).

For each campaign, the absolute positioning of high precision in topocentric coordinates (east, north, elevation) is acquired. The analysis of the obtained time series provides the displacement velocity for each of the REGID stations in the regional context of the South Shetland Islands (Berrocoso *et al.* 2016; Rosado *et al.* 2019). In this geodynamic model, we observe an average subsidence of 0.6 cm a^{-1} and a NE horizontal displacement of *c.* 1.95 cm a^{-1} for Livingston Island. Deception Island has the same horizontal geodynamic behaviour as Livingston Island (2 cm a^{-1}), although subsidence is lower (−0.12 cm a^{-1}) due to the island's volcanic activity (Fig. 20a). With respect to the Antarctic Plate, Livingston and Deception islands show a displacement towards the NW of 1.1 and 0.8 cm a^{-1}, respectively. This behaviour is a consequence of the Bransfield Rift expansion and the Phoenix Plate subsidence under the Antarctic Plate (Berrocoso *et al.* 2016; Rosado *et al.* 2019). Figure 20b shows the non-tectonic residual model obtained with respect to Livingston Island. This model represents the LP volcanic activity of Deception Island. The horizontal direction of the stations according to this model is NNW–SSE with a SSW orientation. We obtained the geodynamic model for the volcanic activity by considering the displacement of the reference station BEJ2 located on Livingston Island, which was not influenced by the volcanic activity.

The ground-deformation velocities from the 1991–92 to 2017–18 campaigns have two distinct periods: the period that includes the 1998 crisis; and the continuous time series

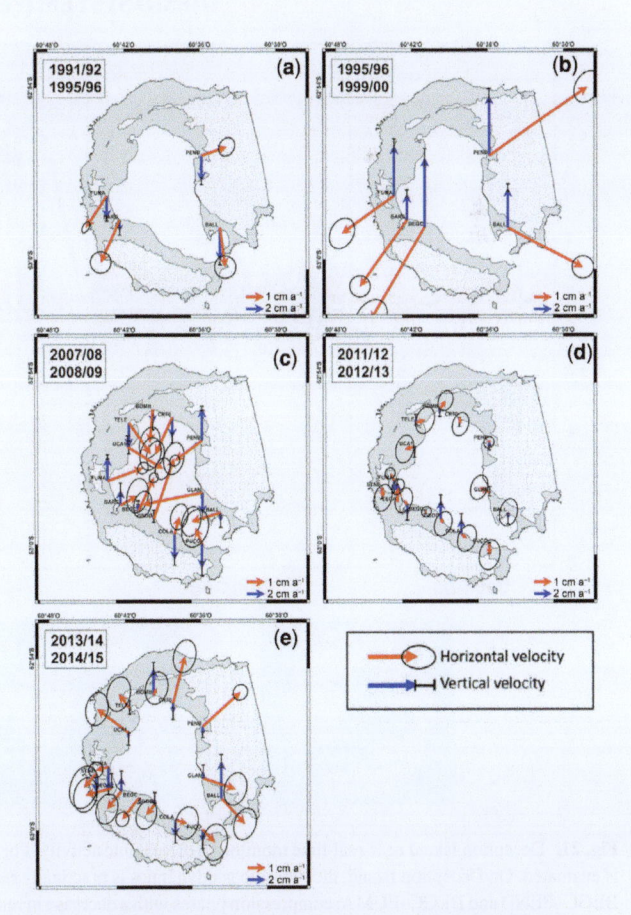

Fig. 20. Displacement models obtained between the campaigns: (a) 1991–92 and 1995–96; (b) 1995–96 and 1999–2000; (c) 2007–08 and 2008–09; (d) 2009–10 and 2010–11; and (e) 2013–14 and 2014–15 (modified from Rosado *et al.* 2019).

of yearly solutions from the 1999–2000 period until the 2017–18 campaign (Rosado et al. 2019). Ground-displacement velocities obtained for the first period show the expansion process that occurred on the island due to the 1998 crisis (Fig. 20a), following the alignment of the Hero Fracture, perpendicular to the Bransfield Strait extension axis. The displacements obtained from that period until the following campaign, 1995–96, reflect this reactivation corresponding to an expansion process in the horizontal component, with an average radial deformation of 2.1 cm a^{-1}. In the vertical component a subsidence process is observed, with an average deformation of 2.6 cm a^{-1}. At the beginning of 1999, there was a reactivation of the system characterized by a large number of VT and LP events that were registered mainly between Fumarole Bay and Telephone Bay (Ibáñez et al. 2003b). The deformation observed on the island showed an inflation process (expansion and uplift) – the most significant one registered on the island until now. The average radial deformation was 5.2 cm a^{-1} in the horizontal component and 5.5 cm a^{-1} in the vertical, much higher values than those recorded in the previous period (Fig. 20b) (Rosado et al. 2019). This reactivation of the system has been interpreted as originating from a 500 m depth magma intrusion that also produced VT earthquakes (Berrocoso et al. 2006d). During the process continuing from 1992 to 1999, deformation acts as a volcanic activity precursor, showing an inflation process in the period 1992–96 which culminated in the 1998 crisis process (Rosado et al. 2019).

Deformation models obtained from the continuous time series 1999–2000 to 2017–18 show three different behaviours in both components: inflation–uplift (Fig. 20c), deflation–subsidence (Fig. 20d) and a transition process that occurs between both behaviours (Fig. 20e) (Berrocoso et al. 2006a, b, 2008, 2012a, b; Torrecillas et al. 2012, 2013; Jigena et al. 2016). Since 1999, the level of seismic activity on the island has varied from low to moderate, with occasional peaks of activity. These peaks of higher seismic activity correspond to inflation and uplift processes: for example, 2013–14 v. the 2014–15 model (Fig. 20c). The phases of low seismic activity coincide with deflation–subsidence models: for example, 2007–08 v. the 2008–09 model (Fig. 20c). The transition phases coincide with the LP residual volcanic model (Fig. 20b): for example, 2009–10 v. the 2010–11 model (Fig. 20d).

Since the 2007–08 campaign, a wireless connection infrastructure has been in development so that data collected at Base Española Gabriel de Castilla (BEGC), Fumarole Bay (FUMA) and Pendulum Cove (PEND) can be processed in near-real time at a monitoring station, the DIESID System (Fig. 21). From the continuous satellite observations at these three sites, considering BEGC as fixed, near-real-time relative ground-deformation monitoring at Fumarole Bay and Pendulum Cove can be achieved. To establish the ground-deformation history at BEGC, FUMA and PEND, data collected during the austral summers from 2001–02 to 2017–18 were processed to attain positioning solutions every 30 min (Fig. 21). The DIESID system was designed to assess inflation and deflation in near-real time, measured with respect to slope-distance and normal-vector magnitude and inclination. The slope distances for BEGC–FUMA and BEGC–PEND are thus computed. In terms of differential relative positioning,

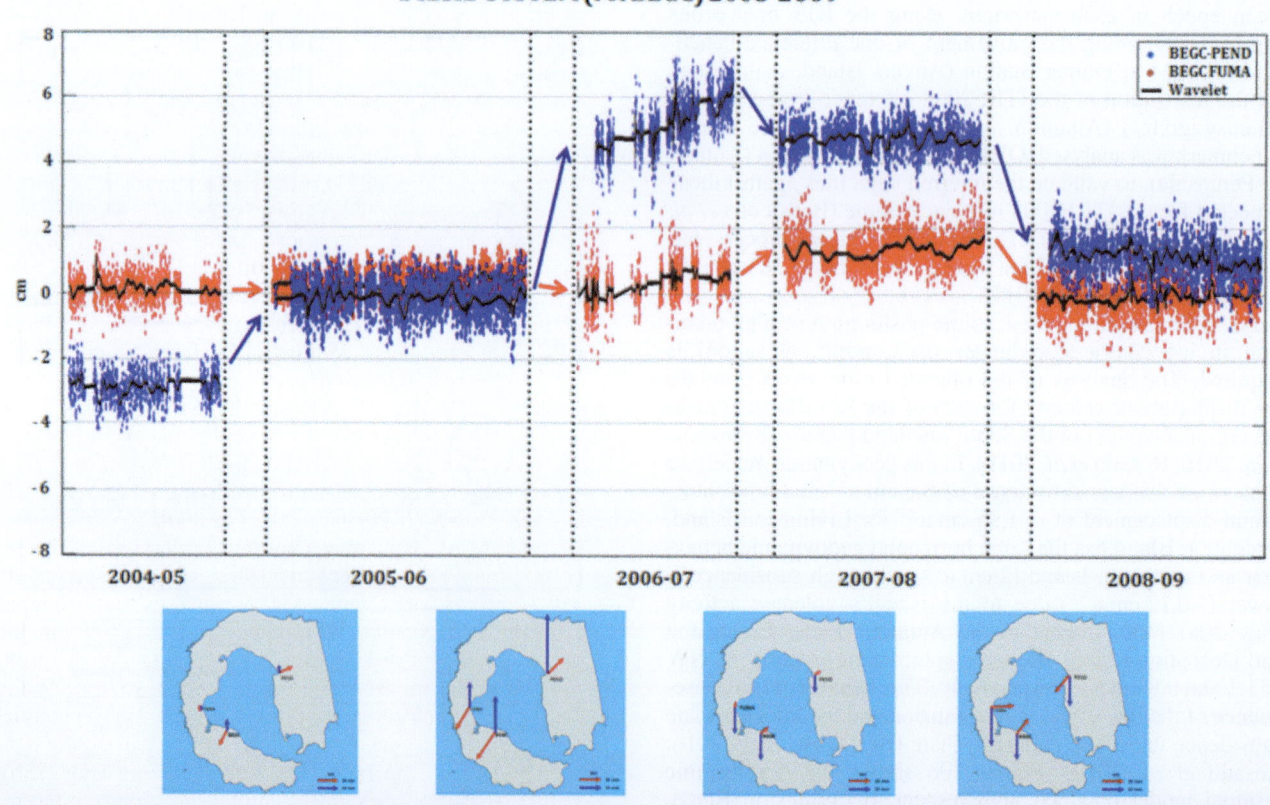

Fig. 21. Deception Island near-real-time monitoring of volcanic activity. For this, the variation in the distance between the BEGC, FUMA and PEND stations is evaluated. On Deception Island, the volcanic geodynamics is practically radial. Thus, expansion phases correspond to an increase in the relative distance of BEGC–PEND and BEGC–FUMA; compression phases with a decrease in relative distance. The analysis of this parameter provides the forecast for the level of volcanic activity. A greater increase in relative distance means more volcanic activity. When the relative distance decreases it implies a relaxation of the volcanic system. Slope–distance time series in near-real time of BEGC–FUMA (red), BEGC–PEND (blue) and wavelet filtering, Coiflet (black) from 2004–05 to 2008–09, and the velocity vectors between austral summers. All through the 2006–07 austral summer campaign the highest BEGC–PEND slope-distance increase was observed.

the reference station is BEGC in near-real-time surveillance, and ground displacements are measured in 30 min–1 h intervals (Berrocoso et al. 2012b; Peci et al. 2012, 2014) (Fig. 21).

Without a stable Internet communication, the near-real-time DIESID system also processes data every 30 min–1 h, with broadcast ephemeris and IERS (International Earth Rotation and Reference Systems Services) Bulletin A pole file, and BEGC as the reference station. The basic ground deformation observables are the slope distances BEGC–FUMA and BEGC–PEND. At Deception, the extension–uplift processes are related to an increase in the slope distances BEGC–FUMA and BEGC–PEND, while compression–subsidence processes are related to their decrease (see Fig. 21).

The noise input of GNSS constellation revolution and Earth gravitational-related effects that affect sub-daily data processing is minimized by wavelet multi-resolution analysis applied to unfiltered data (Peci et al. 2012; Prates et al. 2013a, b; García et al. 2014).

In December 2007, two tide gauges (bottom pressure sensors) were installed within the Deception Island submerged caldera at Colatinas Point (DECMAR) and on Livingston Island at Johnsons Dock (LIVMAR). In addition to an instantaneous sea-level measurement, these tide gauges also record seawater temperature. The tidal constituents and the mean sea levels were determined for both locations (Vidal et al. 2012; Jigena et al. 2015), and the hydrodynamic model for Port Foster was computed (Vidal et al. 2011). In December 2012, a station was installed at Cerro Caliente to record thermometric anomalies at different depths (Peci et al. 2014). Since winter 2012, increased thermal activity in the seawater-filled caldera at Deception Island has been detected, coinciding with the onset of volcano inflation (Fig. 22) (Berrocoso et al. 2018). This thermal activity was manifested in pulses of high water temperature that coincided with ocean tide cycles. The seawater temperature anomalies were detected by the thermometric sensor attached to the tide gauge DECMAR. The detected seawater temperature increase, also observed in soil temperature readings, suggests a rapid and near-simultaneous increase in geothermal activity with the onset of caldera inflation and an increased number of seismic events observed in the following austral summer (Fig. 22).

In summary, Deception Island has shown alternating periods of inflation–deflation for about 6 years since 1999 (Berrocoso et al. 2012b) and similar periods showing an increase–decrease of seismic activity (Carmona et al. 2012; Almendros et al. 2018). Within its complex geodynamic environment, the island's volcanic activity is also influenced by the tectonic extension NW–SE processes occurring in the Bransfield Basin, and alternating NE–SW compression and extension, and NW–SE shear processes, occurring in the fault systems that accommodate the left-lateral component of the South Shetland Trench. These extension processes favour a higher level of hydrothermal activity, producing a fluid temperature increase and magma degassing manifested in the inflation processes. Thus, we see the continued alternating periodic inflation and deflation processes, where deflation occurs with the depressurization of the hydromagmatic system of the Deception Island caldera. The average ground displacement from 2000 to 2018 relative to Livingston shows the NW–SE continuous effect of the Bransfield Basin opening on ground displacements of the geodetic benchmarks at Deception (Rosado et al. 2019). We conclude that the island's deformation is affected mainly by the action of the two dominant fractures: one parallel to the Hero Fracture; and another perpendicular to the previous one and parallel to the Bransfield extension axis. The inflation processes are caused by the Hero Fracture Zone geodynamics, based on the observed NNW–SSE directions. The deflation processes are associated with the island's main fracture activity, in a NE–SW direction, coinciding with the Bransfield Strait axis (Rosado et al. 2019). The correlation between the inflation–uplift processes with the increase in seismic activity, and with the increase in the soil temperature, means that the previous transition models and/or hybrid cases must be considered as precursor characteristics to a possible volcanic crisis on Deception Island (Berrocoso et al. 2016, 2018; Rosado et al. 2019).

Volcanic hazards and hazard assessment

Direct observations of past eruptions indicate that hydrovolcanic activity from maars, tuff rings and tuff cones is the major cause of volcanic hazards at Deception Island (Pedrazzi et al. 2018, 2020). Indeed, due to the potential occurrence of hydrovolcanic activity, even small volume eruptions can be highly explosive in the case of shallow submarine vents or for those located on waterlogged shorelines or beneath the ice caps (Pedrazzi et al. 2018). Therefore, the location of the eruptive vent (in shallow seawater and at onshore ice-free locations) strongly controls the possible hazards to be expected during any future eruption on Deception Island. The latter may mainly include ash fall, ballistic impacts, and extreme temperature and burial effects of subordinate dilute PDCs. In addition, tsunamis triggered by eruptions and slope failures might also be related to Deception Island's future volcanic activity (Smellie 2002a). Hence, Neptunes Bellows, the only exit from (and entry into) Port Foster, may become impassable during an eruptive event, preventing ships from entering/leaving the island's interior bay (Smellie 2002a). Other secondary hazards such as steam fields, fumaroles and ground heating are also common on the island, and are mostly confined to the inside of the caldera along the shores of Port Foster (Baker et al. 1975; Roobol 1982; Smellie 2002c).

Concerning ash fallout, the most common significant hazard, short-term respiratory effects related to ash inhalation including asthma and bronchitis attacks (Horwell and Baxter 2006; Gudmundsson 2011), could affect tourists, scientists and logistics staff on Deception Island and neighbouring islands. On the other side (close to the vent), ash fallout, together with ballistic impact and dilute PDCs, could lead to severe building damage (e.g. the Chilean station at Pendulum Cove) (Fig. 2b), as well as several impacts on humans in the immediate surroundings of the craters. On a more regional scale, due to the strong winds and the low altitude of the tropopause in the area (8–10 km) (Smellie 1999), ash-fall deposits may rapidly disperse over a very wide area far from Deception volcano, as observed in 1970 (Baker et al. 1975), affecting nearby personnel and infrastructures (Pallas et al. 2001; Fretzdorff and Smellie 2002; Liu et al. 2016).

A first attempt to construct a volcanic hazard map of Deception Island was made by Roobol (1982), who focused mainly on assessing the zones threatened by lahars using topographical data and the extent of the ice cap (Fig. 23a). Later, Smellie (2002a), based on the observations of the extent of the products from the historical eruptions of 1842, 1967, 1969 and 1970, produced a summary hazard map identifying potential areas affected by tsunamis, mudflows, dilute pyroclastic density currents, lava flows and/or tephra fall out (Fig. 23b). Recently, Bartolini et al. (2014) provided a volcanic hazard map where they estimated the probability that the different areas may be invaded by lava flows, lahars and/or PDCs (Fig. 23c). Their qualitative hazard map distinguished five hazard levels (from very low to high) and showed that the highest hazard level is confined to the northeastern flanks of Mount Kirkwood, Pendulum Cove and the southeastern slopes of Telefon Ridge (Fig. 23c).

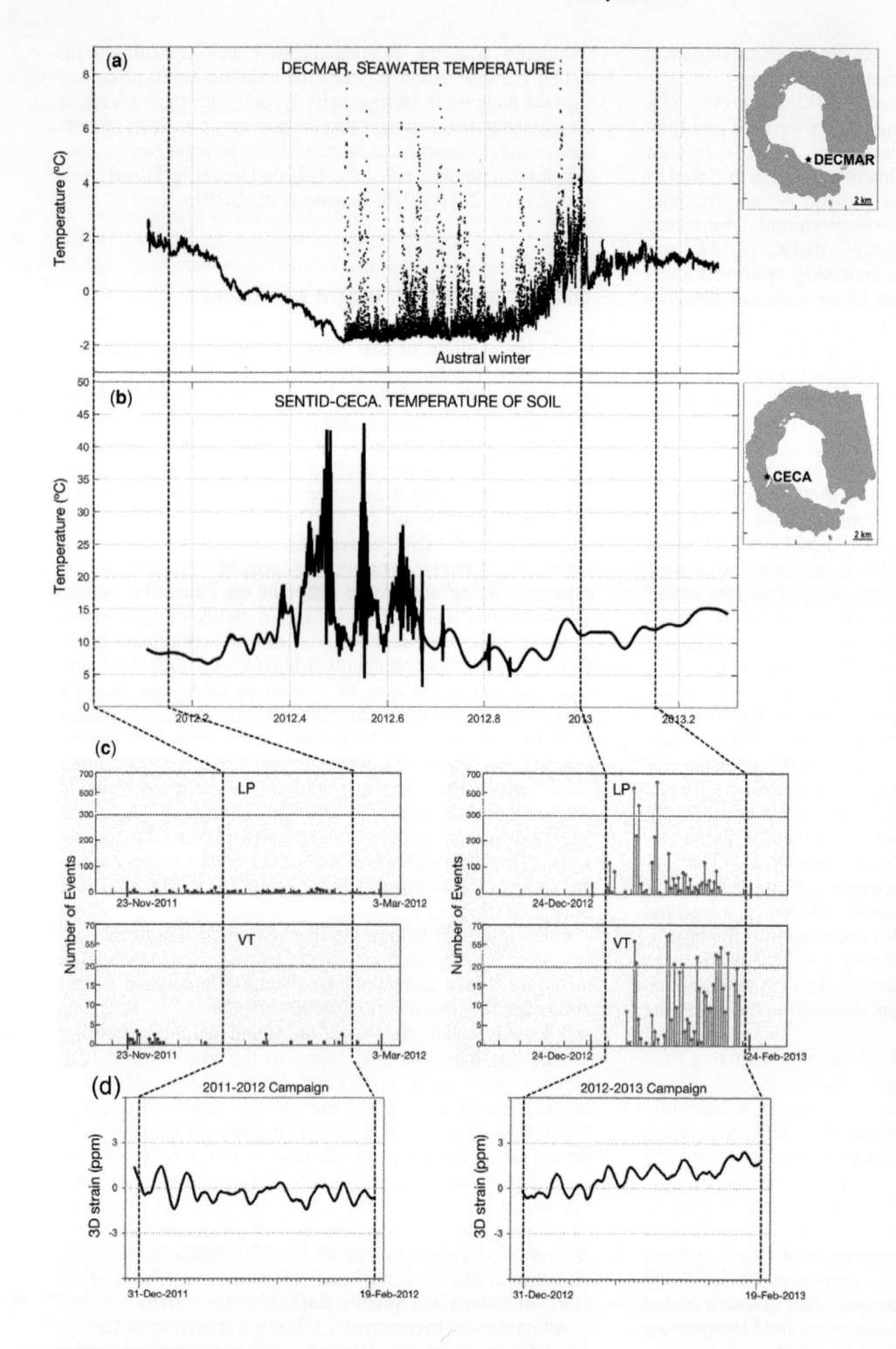

Fig. 22. Multiparameter volcano monitoring dataset for Deception Island. Seawater and soil temperature, and ground deformation from GNSS–GPS was observed by the Laboratory of Astronomy, Geodesy and Cartography of the University of Cadiz, and seismicity was observed by the Andalusian Institute of Geophysics of the University of Granada (adapted from Berrocoso et al. 2018). (**a**) Seawater temperature registered in DECMAR from February 2012 to April 2013. Data peaks correspond to high-temperature anomalies registered by the thermometric sensor attached to the tide gauge (bottom pressure sensor) with amplitudes reaching over 10°C. (**b**) Wavelet analysis of the soil temperature in Cerro Caliente from January 2012 to April 2013. (**c**) The number of long-period (LP) and volcano-tectonic (VT) events recorded in the austral summer campaigns of 2011–12 and 2012–13. (**d**) Normal vector magnitude (strain) in the austral summer campaigns of 2011–12 and 2012–13.

In all three cases, hazard assessment on Deception Island has always been limited by the lack of a complete geological record and full knowledge of the dynamics of post-caldera eruptions. Further research on Deception Island will help to improve the existing hazard maps, a mandatory task for such an active volcano given the increasing number of tourists and scientific expeditions visiting the island and its surroundings.

Concluding remarks and future perspectives

Results of the different investigations carried out on Deception Island during the last decade clearly indicate that it is a very active caldera system, where most recent volcanic activity was dominated by explosive magma–water interaction-driven eruptions. In this sense, even small-volume eruptions may turn into highly explosive ones when located at the waterlogged shorelines of Port Foster Bay, where the rising magma may interact with seawater, the underground aquifer or meltwater from the glaciers. Significant examples are the last eruptive episodes (1967, 1969 and 1970) that destroyed (or severely damaged) the scientific bases operating on the island.

The continuous existence of fumarolic activity and heated ground, the seismic crises of 1992, 1999 and 2015, and the measured ground deformation showing inflation/deflation phases, among other indicators, clearly reveal the volcano continuing magmatic and hydrothermal activity; an eruption in the near future is to be expected. Consequently, Deception Island should be continuously observed and instrumentally

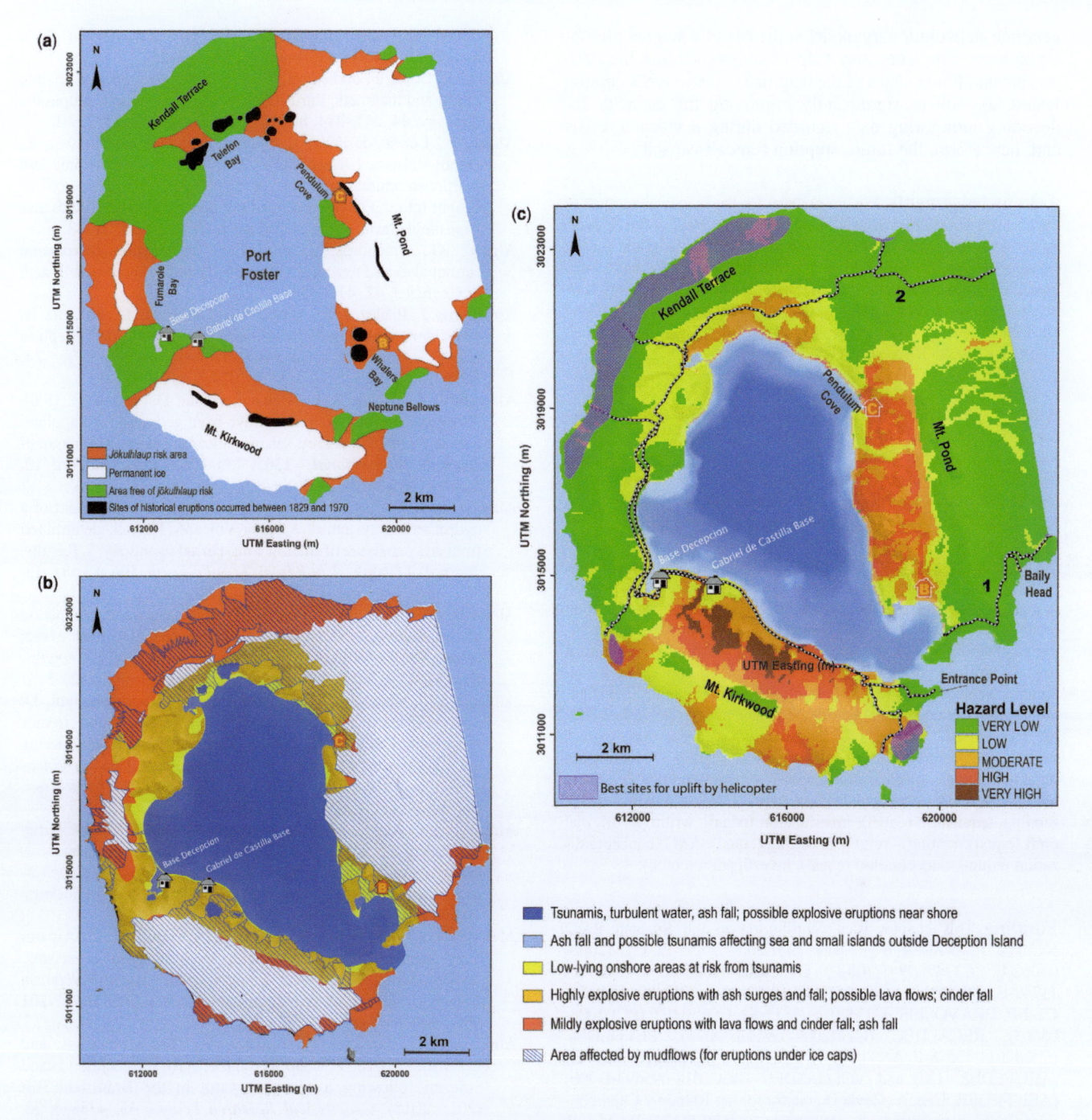

Fig. 23. Hazard maps of Deception Island, after (**a**) Roobol (1982), (**b**) Smellie (2002*a*) and (**c**) Bartolini *et al.* (2014) (modified and corrected from Bartolini *et al.* 2014). In (c) the evacuation routes provided by the Spanish military staff and the best sites for helicopter uplift according to Smellie (2002*a*) are also indicated.

monitored using geochemical and geophysical monitoring methods such as, for example, gas measurements and remote sensing. Indeed, the possibility of acquiring real-time geochemical and geophysical data (separately and in combination) should be contemplated in the future (e.g. Padrón *et al.* 2015; Álvarez-Valero *et al.* 2020). These observables would significantly help to improve our knowledge of the island's ongoing activity and increase the efficiency of deciphering monitoring data recorded during volcanic unrest periods, and, hence, improve the forecasting of upcoming eruptions. For instance, for other active volcanoes (e.g. Tagoro, Canary Islands, Spain), a recent geochemical study of noble gas emissions as sentinels of the magma chamber triggering at depth is revealing promising results for better monitoring of these hazardous volcanoes (Álvarez-Valero *et al.* 2018). In the case of Tagoro's 2011 eruption, the geochemical signal – through noble gases being released at depth – was recognized ahead of the geophysical signal (through the seismic activity).

As said, hazard assessment on Deception Island is biased by the lack of a more complete understanding of the geological evolution and the dynamics of post-caldera eruptions. Further investigations focused on improving available chronological and stratigraphic data for Deception Island may help to refine the current hazard maps, an increasingly urgent task given the growing number of tourists and scientific expeditions visiting the island and its surroundings. Additionally, interdisciplinary and multidisciplinary research activities aimed at understanding the magmatic and volcanic evolution of Deception Island, the nature of the underlying magmatic sources and their relationship to the geodynamic setting are crucial in helping to generate an evolutionary model of the island's magma plumbing system. The latter may help in comprehending the past, present and future states of the magmatic system of Deception Island, as well as significantly improving the capacity for decoding monitoring data recorded during a volcanic crisis and, henceforth, the future eruption forecast capacity.

Acknowledgements We are grateful for the logistical support of the Spanish Polar Program, and to all the military staff of the Spanish Antarctic Base Gabriel de Castilla for their constant help and for the logistic support, without which this research would not have been possible. We thank the editor, John Smellie, and both reviewers Karoly Nemeth and Joan Martí for their constructive comments that have allowed improving a previous version of this manuscript. English editing by Grant George Buffett.

Author contributions AG: conceptualization (lead), data curation (equal), funding acquisition (equal), investigation (equal), writing – original draft (lead); DP: conceptualization (equal), data curation (equal), investigation (equal), writing – original draft (equal), writing – review & editing (equal); JA: conceptualization (equal), data curation (equal), funding acquisition (equal), investigation (equal), writing – original draft (equal), writing – review & editing (equal); MB: conceptualization (equal), data curation (equal), funding acquisition (equal), investigation (equal), visualization (equal), writing – original draft (equal), writing – review & editing (equal); JL-M: conceptualization (equal), data curation (equal), funding acquisition (equal), investigation (equal), writing – original draft (equal), writing – review & editing (equal); AM: conceptualization (equal), data curation (equal), funding acquisition (equal), investigation (equal), writing – original draft (equal), writing – review & editing (equal); EC: conceptualization (equal), data curation (equal), funding acquisition (equal), investigation (equal), visualization (equal), writing – original draft (equal), writing – review & editing (equal); AMÁ-V: conceptualization (equal), data curation (equal), funding acquisition (equal), investigation (equal), writing – original draft (equal), writing – review & editing (equal); AdG: conceptualization (equal), data curation (equal), investigation (equal).

Funding This chapter is a contribution to the Spanish R&D National Plan research projects: CGL2007-28855-E, CTM2008-03062-E, CTM2009-07705-E, CTM2009-08085-E, CTM2010-11740-E, CTM2011-16049-E,CTM2014-57119-R, CTM2014-60451-C2-2-P, BRAVOSEIS (CTM2016-77315-R), CORSHET (POL2006-08663), RECALDEC (CTM2009-05919-E/ANT), PEVOLDEC (CTM2011-13578-E/ANT), POSVOLDEC (CTM2016-79617-P) (AEI/FEDER, UE) and VOLGASDEC (PGC2018-095693-B-I00) (AEI/FEDER, UE). A. Geyer is grateful for her Ramón y Cajal contract (RYC-2012-11024). D. Pedrazzi is grateful for his Beatriu de Pinós (2016 BP 00086) and Juan de la Cierva (IJCI-2016-30482) contracts.

Data availability Relevant data or metadata are available through the Spanish National Polar Data Center, http://hielo.igme.es/index.php/es/. Geochemical data used in Figures 12 and 13 are included in the supplementary dataset of Geyer et al. (2019) and of this published article.

Correction notice The publisher apologizes for the incorrect spelling of the last author's family name. This has been corrected to de Gil.

References

Agusto, M.R., Caselli, A.T. and Dos Santos Afonso, M. 2004. Manifestaciones de piritas framboidales en fumarolas de la Isla Decepción (Antártida): implicancias genéticas. *Revista de la Asociación Geológica Argentina*, **59**, 152–157.

Albert, H., Costa, F. and Martí, J. 2016. Years to weeks of seismic unrest and magmatic intrusions precede monogenetic eruptions. *Geology*, **44**, 211–214, https://doi.org/10.1130/g37239.1

Alfaro, P., López-Martínez, J., Maestro, A., Galindo-Zaldívar, J., Durán-Valsero, J.J. and Cuchí, J.A. 2010. Recent tectonic and morphostructural evolution of Byers Peninsula (Antarctica): insight into the development of the South Shetland Islands and Bransfield Basin. *Journal of Iberian Geology*, **36**, 21–38.

Alguacil, G., Almendros, J.C. et al. 1999. Observations of volcanic earthquakes and tremor at Deception Island – Antarctica. *Annali di Geofisica*, **42**, 417–436.

Almendros, J., Ibáñez, J.M., Alguacil, G., Del Pezzo, E. and Ortiz, R. 1997. Array tracking of the volcanic tremor source at Deception Island, Antarctica. *Geophysical Research Letters*, **24**, 3069–3072, https://doi.org/10.1029/97gl03096

Almendros, J., Ibáñez, J.M., Alguacil, G. and Del Pezzo, E. 1999. Array analysis using circular-wave-front geometry: An application to locate the nearby seismo-volcanic source. *Geophysical Journal International*, **136**, 159–170, https://doi.org/10.1046/j.1365-246X.1999.00699.x

Almendros, J., Chouet, B. and Dawson, P. 2001. Spatial extent of a hydrothermal system at Kilauea Volcano, Hawaii, determined from array analyses of shallow long-period seismicity: 2. Results. *Journal of Geophysical Research: Solid Earth*, **106**, 13 581–13 597, https://doi.org/10.1029/2001jb000309

Almendros, J., Luzón, F. and Posadas, A. 2004. Precise determination of the relative wave propagation parameters of similar events using a small-aperture seismic array. *Pure and Applied Geophysics*, **161**, 1579, https://doi.org/10.1007/s00024-004-2522-5

Almendros, J., Ibáñez, J.M., Carmona, E. and Zandomeneghi, D. 2007. Array analyses of volcanic earthquakes and tremor recorded at Las Cañadas caldera (Tenerife Island, Spain) during the 2004 seismic activation of Teide volcano. *Journal of Volcanology and Geothermal Research*, **160**, 285–299, https://doi.org/10.1016/j.jvolgeores.2006.10.002

Almendros, J., Abella, R., Mora, M.M. and Lesage, P. 2014. Array analysis of the seismic wavefield of long-period events and volcanic tremor at Arenal volcano, Costa Rica. *Journal of Geophysical Research: Solid Earth*, **119**, 5536–5559, https://doi.org/10.1002/2013jb010628

Almendros, J., Carmona, E. et al. 2015. Deception Island (Antartica): sustained deformation and large increase in seismic activity during 2014–2015. *Bulletin of the Global Volcanism Network*, **40**(6), https://doi.org/10.5479/si.GVP.BGVN201506-390030

Almendros, J., Carmona, E., Jiménez, V., Díaz-Moreno, A. and Lorenzo, F. 2018. Volcano-tectonic activity at Deception Island volcano following a seismic swarm in the Bransfield Rift (2014–2015). *Geophysical Research Letters*, **45**, 4788–4798, https://doi.org/10.1029/2018gl077490

Altamimi, Z., Collilieux, X. and Métivier, L. 2011. ITRF2008: an improved solution of the international terrestrial reference frame. *Journal of Geodesy*, **85**, 457–473, https://doi.org/10.1007/s00190-011-0444-4

Álvarez-Valero, A.M., Burgess, R. et al. 2018. Noble gas signals in corals predict submarine volcanic eruptions. *Chemical Geology*, **480**, 28–34, https://doi.org/10.1016/j.chemgeo.2017.05.013

Álvarez-Valero, A.M., Gisbert, G. et al. 2020. δD and δ^{18}O variations of the magmatic system beneath Deception Island volcano (Antarctica): implications for magma ascent and eruption forecasting. *Chemical Geology*, **542**, 119595, https://doi.org/10.1016/j.chemgeo.2020.119595

Antoniades, D., Giralt, S. et al. 2018. The timing and widespread effects of the largest Holocene volcanic eruption in Antarctica. *Scientific Reports*, **8**, 17279, https://doi.org/10.1038/s41598-018-35460-x

Aparicio, A., Menegatti, N., Petrinovic, I., Risso, C. and Viramonte, J.G. 1997. El volcanismo de Isla Decepción (Península Antártida). *Boletín Geológico y Minero*, **108**, 235–258, http://hdl.handle.net/10261/4936

Aristarain, A.J. and Delmas, R.J. 1998. Ice record of a large eruption of Deception Island Volcano (Antarctica) in the XVIIth century. *Journal of Volcanology and Geothermal Research*, **80**, 17–25, https://doi.org/10.1016/s0377-0273(97)00040-1

Baker, P.E. and McReath, I. 1971. 1970 Volcanic Eruption at Deception Island. *Nature Physical Science*, **231**, 5–9, https://doi.org/10.1038/physci231005a0

Baker, P.E., Davies, T.G. and Roobol, M.J. 1969. Volcanic activity at Deception Island in 1967 and 1969. *Nature*, **224**, 553–560, https://doi.org/10.1038/224553a0

Baker, P.E., McReath, I., Harvey, M.R., Roobol, M.J. and Davies, T.G. 1975. *The Geology of the South Shetland Islands: Volcanic Evolution of Deception Island*. British Antarctic Survey Scientific Reports, **78**.

Baraldo, A. and Rinaldi, C.A. 2000. Stratigraphy and structure of Deception Island, South Shetland Islands, Antarctica. *Journal of South American Earth Sciences*, **13**, 785–796, https://doi.org/10.1016/S0895-9811(00)00060-2

Baraldo, A., Rapalini, A.E., Böhnel, H. and Mena, M. 2003. Paleomagnetic study of Deception Island, South Shetland Islands, Antarctica. *Geophysical Journal International*, **153**, 333–343, https://doi.org/10.1046/j.1365-246X.2003.01881.x

Barclay, A.H., Wilcock, W.S.D. and Ibáñez, J.M. 2009. Bathymetric constraints on the tectonic and volcanic evolution of Deception Island Volcano, South Shetland Islands. *Antarctic Science*, **21**, 153–167, https://doi.org/10.1017/S0954102008001673

Barker, D.H.N. and Austin, J.A.J. 1994. Crustal diapirism in Bransfield Strait, West Antarctica: Evidence for distributed extension in marginal-basin formation. *Geology*, **22**, 657–660, https://doi.org/10.1130/0091-7613(1994)022<0657:CDIBSW>2.3.CO;2

Barker, P.F. 1982. The Cenozoic subduction history of the Pacific margin of the Antarctic Peninsula: ridge crest–trench interactions. *Journal of the Geological Society, London*, **139**, 787–801, https://doi.org/10.1144/gsjgs.139.6.0787

Bartolini, S., Geyer, A., Martí, J., Pedrazzi, D. and Aguirre-Díaz, G. 2014. Volcanic hazard on Deception Island (South Shetland Islands, Antarctica). *Journal of Volcanology and Geothermal Research*, **285**, 150–168, https://doi.org/10.1016/j.jvolgeores.2014.08.009

Ben-Zvi, T., Wilcock, W.S.D., Barclay, A.H., Zandomeneghi, D., Ibáñez, J.M. and Almendros, J. 2009. The P-wave velocity structure of Deception Island, Antarctica, from two-dimensional seismic tomography. *Journal of Volcanology and Geothermal Research*, **180**, 67–80, https://doi.org/10.1016/j.jvolgeores.2008.11.020

Berrocoso, M., Fernández-Ros, A. et al. R. 2006a. Geodetic research on Deception Island. *In*: Fütterer, D.K., Kleinschmidt, G., Miller, H. and Tessensohn, F. (eds) *Antarctica: Contributions to Global Earth Sciences*. Springer, Berlin, 391–396.

Berrocoso, M., García-García, A. et al. 2006b. Geodynamical studies on Deception Island: DECVOL and GEODEC Projects. *In*: Fütterer, D.K., Damaske, D., Kleinschmidt, G., Miller, H. and Tessensohn, F. (eds) *Antarctica: Contributions to Global Earth Sciences*. Springer, Berlin, 283–287.

Berrocoso, M., Ramírez, M.E. and Fernández-Ros, A. 2006c. Horizontal deformation models for Deception Island (South Shetland Islands, Antarctica). *In*: Sansó, F. and Gil, A.J. (eds) *Geodetic Deformation Monitoring: From Geophysical to Engineering Roles*. Springer, Berlin, 217–221.

Berrocoso, M., Ramírez, M.E., Fernández-Ros, A. and Jiménez, Y. 2006d. Crustal deformation model in volcanic areas. An application to Deception Island Volcano (South Shetland Islands, Antarctica). *In*: Criado, R., Estep, D., Pérez-García, M.A. and Vigo-Aguiar, J. (eds) *New Trends and Tools in Computacional and Mathematical Methods on Science and Engineering*. Editorial Universidad Rey Juan Carlos I, Madrid, 116–120.

Berrocoso, M., Fernández-Ros, A. et al. 2008. Geodetic research on Deception Island and its environment (South Shetland Islands, Bransfield Sea and Antarctic Peninsula) during Spanish Antarctic campaigns (1987–2007). *In*: Capra, A. and Dietrich, R. (eds) *Geodetic and Geophysical Observations in Antarctica: An Overview in the IPY Perspective*. Springer, Berlin, 97–124.

Berrocoso, M., Carmona, J., Fernández-Ros, A., Pérez-Peña, A., Ortiz, R. and García, A. 2010. Kinematic model for Tenerife Island (Canary Islands, Spain): Geodynamic interpretation in the Nubian plate context. *Journal of African Earth Sciences*, **58**, 721–733, https://doi.org/10.1016/j.jafrearsci.2010.04.007

Berrocoso, M., Prates, G., Fernández-Ros, A. and García, A. 2012a. Normal vector analysis from GNSS–GPS data applied to Deception volcano surface deformation. *Geophysical Journal International*, **190**, 1562–1570, https://doi.org/10.1111/j.1365-246X.2012.05584.x

Berrocoso, M., Torrecillas, C., Jigena, B. and Fernández-Ros, A. 2012b. Determination of geomorphological and volumetric variations in the 1970 land volcanic craters area (Deception Island, Antarctica) from 1968 using historical and current maps, remote sensing and GNSS. *Antarctic Science*, **24**, 367-376, https://doi.org/10.1017/S0954102012000193

Berrocoso, M., Fernández-Ros, A., Prates, G., García, A. and Kraus, S. 2016. Geodetic implications on block formation and geodynamic domains in the South Shetland Islands, Antarctic Peninsula. *Tectonophysics*, **666**, 211–219, https://doi.org/https://doi.org/10.1016/j.tecto.2015.10.023

Berrocoso, M., Prates, G. et al. 2018. Caldera unrest detected with seawater temperature anomalies at Deception Island, Antarctic Peninsula. *Bulletin of Volcanology*, **80**, 41, https://doi.org/10.1007/s00445-018-1216-2

Birkenmajer, K. 1992. Volcanic succession at Deception Island, West Antarctica: a revised lithostratigraphic standard. *Studia Geologica Polonica*, **101**, 27–82.

Birkenmajer, K., Soliani, E. and Kawashita, K. 1990. Reliability of potassium–argon dating of Cretaceous–Tertiary island-arc volcanic suites of King George Island, South Shetland Islands (West Antarctica). *Zentralblatt fur Geologie und Palaöntologie*, **1**, 127–140.

Carmona, E., Almendros, J., Peña, J.A. and Ibáñez, J.M. 2010. Characterization of fracture systems using precise array locations of earthquake multiplets: An example at Deception Island volcano, Antarctica. *Journal of Geophysical Research: Solid Earth*, **115**, B06309, https://doi.org/10.1029/2009jb006865

Carmona, E., Almendros, J., Serrano, I., Stich, D. and Ibáñez, J.M. 2012. Results of seismic monitoring surveys of Deception Island volcano, Antarctica, from 1999–2011. *Antarctic Science*, **24**, 485–499, https://doi.org/10.1017/S0954102012000314

Carmona, E., Almendros, J. et al. 2014. Advances in seismic monitoring at Deception Island volcano (Antarctica) since the International Polar Year. *Annals of Geophysics*, **57**, https://doi.org/10.4401/ag-6378

Caselli, A.T., dos Santos Afonso, M. and Agusto, M.R. 2004. Gases fumarólicos de la isla Decepción (Shetland del Sur, Antártida): Variaciones químicas y depósitos vinculados a la crisis sísmica de 1999. *Revista de la Asociación Geológica Argentina*, **59**, 291–302.

Catalán, M., Berrocoso, M. and García-Solís, M.D. 1991. Forging a South America–Antarctica GPS geodetic link. *GPS World*, **2**, 20–28.

Chouet, B. 1992. A seismic model for the source of long-period events and harmonic tremor. *In*: Gasparini, P., Scarpa, R. and Aki, K. (eds) *Volcanic Seismology, Volume 3*. Springer, Berlin, 133–156.

Chouet, B. 1996. Long-period volcano seismicity: its source and use in eruption forecasting. *Nature*, **380**, 309–316, https://doi.org/10.1038/380309a0

Chouet, B. 2003. Volcano seismology. *Pure and Applied Geophysics*, **160**, 739–788, https://doi.org/10.1007/PL00012556

Correig, A.M., Urquizu, M., Vila, J. and Martí, J. 1997. Analysis of the temporal occurrence of seismicity at Deception Island (Antarctica). A nonlinear approach. *Pure and Applied Geophysics*, **149**, 553–574, https://doi.org/10.1007/s000240050041

Coulon, C.A., Hsieh, P.A., White, R., Lowenstern, J.B. and Ingebritsen, S.E. 2017. Causes of distal volcano-tectonic seismicity inferred from hydrothermal modeling. *Journal of Volcanology and Geothermal Research*, **345**, 98–108, https://doi.org/10.1016/j.jvolgeores.2017.07.011

Dach, R., Hugentobler, U., Fridez, P. and Meindl, M. 2007. *Bernese GPS Software Version 5.0. User Manual*. Astronomical Institute, University of Berne.

De Rosa, R., Mazzuoli, R., Omarini, R., Ventura, G. and Viramonte, J. 1995. A volcanological model for the historical eruptions at Deception Island, Bransfield Strait, Antarctica. *Terra Antartica*, **2**, 95–101.

Dziak, R.P., Park, M., Lee, W.S., Matsumoto, H., Bohnenstiehl, D.R. and Haxel, J.H. 2010. Tectonomagmatic activity and ice dynamics in the Bransfield Strait back-arc basin, Antarctica. *Journal of Geophysical Research: Solid Earth*, **115**, B01102, https://doi.org/10.1029/2009jb006295

Fernandez-Ros, A.M., Berrocoso, M. and Ramirez, M.E. 2007. Volcanic deformation models for Deception Island (South Shetland Islands, Antarctica). *United States Geological Survey Open-File Report*, **2007-1047**, Extended Abstract 094.

Fretzdorff, S. and Smellie, J.L. 2002. Electron microprobe characterization of ash layers in sediments from the central Bransfield basin (Antarctic Peninsula): evidence for at least two volcanic sources. *Antarctic Science*, **14**, 412–421, https://doi.org/10.1017/S0954102002000214

Galé, C., Ubide, T. et al. 2014. Vulcanismo cuaternario de la Isla Decepción (Antártida): una signatura relacionada con la subducción de la Fosa de las Shetland del Sur en el dominio de tras-arco de la Cuenca de Bransfield. *Boletín Geológico y Minero*, **125**, 31–52.

Galindo-Zaldívar, J., Gamboa, L., Maldonado, A., Nakao, S. and Bochu, Y. 2004. Tectonic development of the Bransfield Basin and its prolongation to the South Scotia Ridge, northern Antarctic Peninsula. *Marine Geology*, **206**, 267–282, https://doi.org/10.1016/j.margeo.2004.02.007

García, A., Blanco, I., Torta, J.M., Astiz, M.M., Ibáñez, J. and Ortiz, R. 1997. A search for the volcanomagnetic signal at Deception volcano (South Shetland I., Antarctica). *Annals of Geophysics*, **40**, 319–327, https://doi.org/10.4401/ag-3914

García, A., Berrocoso, M., Marrero, J.M., Fernández-Ros, A., Prates, G., De la Cruz-Reyna, S. and Ortiz, R. 2014. Volcanic alert system (VAS) developed during the 2011–2014 El Hierro (Canary Islands) volcanic process. *Bulletin of Volcanology*, **76**, 825, https://doi.org/10.1007/s00445-014-0825-7

Gardine, M., West, M.E. and Cox, T. 2011. Dike emplacement near Parícutin volcano, Mexico in 2006. *Bulletin of Volcanology*, **73**, 123–132, https://doi.org/10.1007/s00445-010-0437-9

Geyer, A. and Martí, J. 2008. The new worldwide collapse caldera database (CCDB): A tool for studying and understanding caldera processes. *Journal of Volcanology and Geothermal Research*, **175**, 334–354, https://doi.org/10.1016/j.jvolgeores.2008.03.017

Geyer, A., Martí, A., Giralt, S. and Folch, A. 2017. Potential ash impact from Antarctic volcanoes: Insights from Deception Island's most recent eruption. *Scientific Reports*, **7**, 16534, https://doi.org/10.1038/s41598-017-16630-9

Geyer, A., Álvarez-Valero, A.M., Gisbert, G., Aulinas, M., Hernández-Barreña, D., Lobo, A. and Martí, J. 2019. Deciphering the evolution of Deception Island's magmatic system. *Scientific Reports*, **9**, 373, https://doi.org/10.1038/s41598-018-36188-4

González-Casado, J.M., Robles, J.G.L. and López-Martínez, J. 2000. Bransfield Basin, Antarctic Peninsula: Not a normal backarc basin. *Geology*, **28**, 1043–1046, https://doi.org/10.1130/0091-7613(2000)28<1043:Bbapna>2.0.Co;2

González-Ferrán, O. and Katsui, Y. 1971. Estudio integral del volcanismo cenozoico superior de las Islas Shetland del Sur, Antartica. *Serie científica Instituto Antártico Chileno*, **22**, 123–174.

González-Ferrán, O., Munizaga, F. and Moreno, R.H. 1971. 1970 eruption at Deception island: Distribution and chemical features of ejected materials. *Antarctic Journal of the United States*, **6**, 87–89.

Gràcia, E., Canals, M., Lí Farràn, M., José Prieto, M., Sorribas, J. and Team, G. 1996. Morphostructure and evolution of the central and Eastern Bransfield Basins (NW Antarctic Peninsula). *Marine Geophysical Researches*, **18**, 429–448, https://doi.org/10.1007/BF00286088

Gràcia, E., Canals, M., Farràn, M.L., Sorribas, J. and Pallàs, R. 1997. Central and eastern Bransfield basins (Antarctica) from high-resolution swath-bathymetry data. *Antarctic Science*, **9**, 168–180, https://doi.org/10.1017/S0954102097000229

Grad, M., Guterch, A. and Sroda, P. 1992. Upper crustal structure of Deception Island area, Bransfield Strait, West Antarctica. *Antarctic Science*, **4**, 469–476, https://doi.org/10.1017/S0954102092000683

Gudmundsson, G. 2011. Respiratory health effects of volcanic ash with special reference to Iceland. A review. *The Clinical Respiratory Journal*, **5**, 2–9, https://doi.org/10.1111/j.1752-699X.2010.00231.x

Haase, K.M., Beier, C., Fretzdorff, S., Smellie, J.L. and Garbe-Schönberg, D. 2012. Magmatic evolution of the South Shetland Islands, Antarctica, and implications for continental crust formation. *Contributions to Mineralogy and Petrology*, **163**, 1103–1119, https://doi.org/10.1007/s00410-012-0719-7

Havskov, J., Peña, J.A., Ibáñez, J.M., Ottemöller, L. and Martínez-Arévalo, C. 2003. Magnitude scales for very local earthquakes. Application for Deception Island Volcano (Antarctica). *Journal of Volcanology and Geothermal Research*, **128**, 115–133, https://doi.org/10.1016/S0377-0273(03)00250-6

Hawkes, D.D. 1961. *The Geology of the South Shetland Islands: II. The Geology and Petrology of Deception Island*. Falkland Islands Dependencies Survey Scientific Reports, 27.

Henriet, J.P., Meissner, R., Miller, H. and The Grape Team 1992. Active margin processes along the Antarctic Peninsula. *Tectonophysics*, **201**, 229–253, https://doi.org/10.1016/0040-1951(92)90235-X

Hill, D. and Prejean, S. 2015. Dynamic triggering. *Treatise on Geophysics*, **4**, 273–304.

Hodgson, D.A., Dyson, C.L., Jones, V.J. and Smellie, J.L. 1998. Tephra analysis of sediments from Midge Lake (South Shetland Islands) and Sombre Lake (South Orkney Islands), Antarctica. *Antarctic Science*, **10**, 13–20, https://doi.org/10.1017/S0954102098000030

Hole, M.J., Saunders, A.D., Rogers, G. and Sykes, M.A. 1994. The relationship between alkaline magmatism, lithospheric extension and slab window formation along continental destructive plate margins. *Geological Society, London, Special Publications*, **81**, 265–285, https://doi.org/10.1144/GSL.SP.1994.081.01.15

Horwell, C.J. and Baxter, P.J. 2006. The respiratory health hazards of volcanic ash: a review for volcanic risk mitigation. *Bulletin of Volcanology*, **69**, 1–24, https://doi.org/10.1007/s00445-006-0052-y

Ibáñez, J.M., Morales, J., Alguacil, G., Almendros, J., Oritz, R. and Del Pezzo, E. 1997. Intermediate-focus earthquakes under South Shetland Islands (Antarctica). *Geophysical Research Letters*, **24**, 531–534, https://doi.org/10.1029/97gl00314

Ibáñez, J.M., Pezzo, E.D., Almendros, J., La Rocca, M., Alguacil, G., Ortiz, R. and García, A. 2000. Seismovolcanic signals at Deception Island volcano, Antarctica: Wave field analysis and source modeling. *Journal of Geophysical Research: Solid Earth*, **105**, 13 905–13 931, https://doi.org/10.1029/2000jb900013

Ibáñez, J.M., Almendros, J., Carmona, E., Martínez-Arévalo, C. and Abril, M. 2003a. The recent seismo-volcanic activity at Deception Island volcano. *Deep Sea Research Part II: Topical Studies in Oceanography*, **50**, 1611–1629, https://doi.org/10.1016/S0967-0645(03)00082-1

Ibáñez, J.M., Carmona, E., Almendros, J., Saccorotti, G., Del Pezzo, E., Abril, M. and Ortiz, R. 2003b. The 1998–1999 seismic series at Deception Island volcano, Antarctica. *Journal of Volcanology and Geothermal Research*, **128**, 65–88, https://doi.org/10.1016/S0377-0273(03)00247-6

Irvine, T.N. and Baragar, W.R.A. 1971. A guide to the chemical classification of the common volcanic rocks. *Canadian Journal of Earth Sciences*, **8**, 523–548, https://doi.org/10.1139/e71-055

Jigena, B., Vidal, J. and Berrocoso, M. 2015. Determination of the mean sea level at Deception and Livingston islands, Antarctica. *Antarctic Science*, **27**, 101–102, https://doi.org/10.1017/S0954102014000595

Jigena, B., Berrocoso, M., Torrecillas, C., Vidal, J., Barbero, I. and Fernandez-Ros, A. 2016. Determination of an experimental geoid at Deception Island, South Shetland Islands, Antarctica. *Antarctic Science*, **28**, 277–292, https://doi.org/10.1017/S0954102015000681

Julian, B.R. 1994. Volcanic tremor: Nonlinear excitation by fluid flow. *Journal of Geophysical Research: Solid Earth*, **99**, 11 859–11 877, https://doi.org/10.1029/93jb03129

Jiménez Morales, V., Almendros, J. and Carmona, E. 2017. Detection of long-duration tremors at Deception Island volcano, Antarctica. *Journal of Volcanology and Geothermal Research*, **347**, 234–249, https://doi.org/10.1016/j.jvolgeores.2017.09.016

Keller, R.A., Fisk, M.R., White, W.M. and Birkenmajer, K. 1992. Isotopic and trace element constraints on mixing and melting models of marginal basin volcanism, Bransfield Strait, Antarctica. *Earth and Planetary Science Letters*, **111**, 287–303, https://doi.org/10.1016/0012-821X(92)90185-X

Klepeis, K.A. and Lawver, L.A. 1996. Tectonics of the Antarctic–Scotia plate boundary near Elephant and Clarence islands, West Antarctica. *Journal of Geophysical Research*, **101**, 20 211–20 231, https://doi.org/10.1029/96JB01510

Konstantinou, K.I. and Schlindwein, V. 2003. Nature, wavefield properties and source mechanism of volcanic tremor: a review. *Journal of Volcanology and Geothermal Research*, **119**, 161–187, https://doi.org/10.1016/S0377-0273(02)00311-6

Košler, J., Magna, T., Mlčoch, B., Mixa, P., Nývlt, D. and Holub, F.V. 2009. Combined Sr, Nd, Pb and Li isotope geochemistry of alkaline lavas from northern James Ross Island (Antarctic Peninsula) and implications for back-arc magma formation. *Chemical Geology*, **258**, 207–218, https://doi.org/10.1016/j.chemgeo.2008.10.006

Koulakov, I., El Khrepy, S., Alarifi, N., Kuznetsov, P. and Kasatkina, E. 2015. Structural cause of a missed eruption in the Harrat Lunayyir basaltic field (Saudi Arabia) in 2009. *Geology*, **43**, 395–398, https://doi.org/10.1130/G36271.1

Kowalewski, W., Rudowski, S. and Zalewski, S.M. 1990. Seismoacoustic studies within flooded part of the caldera of the Deception Island, West Antarctica. *Polish Polar Research*, **11**, 259–266.

Kraus, S., Kurbatov, A. and Yates, M. 2013. Geochemical signatures of tephras from Quaternary Antarctic Peninsula volcanoes. *Andean Geology*, **40**, 1–40, https://doi.org/10.5027/andgeoV40n1-a01

Larter, R.D. 1991. Preliminary results of seismic reflection investigations and associated geophysical studies in the area of the Antarctic Peninsula: Discussion. *Antarctic Science*, **3**, 217–220, https://doi.org/10.1017/S0954102091210251

Lawver, L.A., Keller, R.A., Fisk, M.R. and Strelin, J. 1995. Bransfield Strait, Antarctic Peninsula: Active extension behind a dead arc. *In*: Taylor, B. (ed.) *Back-arc Basins: Tectonics and Magmatism*. Plenum, New York, 315–342.

Lawver, L.A., Sloan, B.J. *et al.* 1996. Distributed, active extension in Bransfield Basin, Antarctic Peninsula: Evidence from multibeam bathymetry. *GSA Today*, **6**, 1–6.

Le Bas, M.J., Le Maitre, R.W., Streckeisen, A., Zanettin, B. and IUGS Subcommission on the Systematics of Igneous Rocks. 1986. A chemical classification of volcanic rocks based on the total alkali–silica diagram. *Journal of Petrology*, **27**, 745–750, https://doi.org/10.1093/petrology/27.3.745

Lee, Y.I., Lim, H.S., Yoon, H.I. and Tatur, A. 2007. Characteristics of tephra in Holocene lake sediments on King George Island, West Antarctica: implications for deglaciation and paleoenvironment. *Quaternary Science Reviews*, **26**, 3167–3178, https://doi.org/10.1016/j.quascirev.2007.09.007

Liu, E.J., Oliva, M. *et al.* 2016. Expanding the tephrostratigraphical framework for the South Shetland Islands, Antarctica, by combining compositional and textural tephra characterisation. *Sedimentary Geology*, **340**, 49–61, https://doi.org/10.1016/j.sedgeo.2015.08.002

López-Martínez, J. and Serrano, E. 2002. Geomorphology. *In*: López-Martínez, J., Smellie, J.L., Thomson, J.W. and Thomson, M.R.A. (eds) *Geology and Geomorphology of Deception Island, 1:25 000*. BAS GEOMAP Series Sheets 6-A and 6-B. British Antarctic Survey, Cambridge, UK, 31–39.

López-Martínez, J., Serrano, E., Rey, J. and Smellie, J.L. 2002*a*. Geomorphological Map of Deception Island, 1:25 000. *In*: López-Martínez, J., Smellie, J.L., Thomson, J.W. and Thomson, M.R.A. (eds) *Geology and Geomorphology of Deception Island, 1:25 000*. BAS GEOMAP Series Sheet 6-B. British Antarctic Survey, Cambridge, UK.

López-Martínez, J., Smellie, J.L., Thomson, J.W. and Thomson, M.R.A. (eds) 2002*b*. *Geology and Geomorphology of Deception Island, 1:25 000*. BAS GEOMAP Series Sheets 6-A and 6-B. British Antarctic Survey, Cambridge, UK.

Lorca, E. 1976. Deception Island: seismic activity prior to the eruption of 1967. *In*: González-Ferrán, O. (ed.) *Proceedings of the Symposium on Andean andAntarctic Volcanology Problems (Santiago, Chile, September 1974)*. IAVCEI, Naples, Italy, 632–645.

Lu, Z., Wicks, C.Jr, Power, J.A. and Dzurisin, D. 2000. Ground deformation associated with the March 1996 earthquake swarm at Akutan volcano, Alaska, revealed by satellite radar interferometry. *Journal of Geophysical Research: Solid Earth*, **105**, 21 483–21 495, https://doi.org/10.1029/2000jb900200

Luzón, F., Almendros, J. and García-Jerez, A. 2011. Shallow structure of Deception Island, Antarctica, from correlations of ambient seismic noise on a set of dense seismic arrays. *Geophysical Journal International*, **185**, 737–748, https://doi.org/10.1111/j.1365-246X.2011.04962.x

Maestro, A., Somoza, L., Rey, J., Martínez-Frías, J. and López-Martínez, J. 2007. Active tectonics, fault patterns, and stress field of Deception Island: A response to oblique convergence between the Pacific and Antarctic plates. *Journal of South American Earth Sciences*, **23**, 256–268, https://doi.org/10.1016/j.jsames.2006.09.023

Maestro, A., López-Martínez, J., Galindo-Zaldívar, J., Bohoyo, F. and Mink, S. 2014. Evolution of the stress field in the southern Scotia Arc from the late Mesozoic to the present-day. *Global and Planetary Change*, **123**, 269–297, https://doi.org/10.1016/j.gloplacha.2014.07.023

Maldonado, A., Larter, R.D. and Aldaya, F. 1994. Forearc tectonic evolution of the South Shetland Margin, Antarctic Peninsula. *Tectonics*, **13**, 1345–1370, https://doi.org/10.1029/94tc01352

Manga, M. and Brodsky, E. 2006. Seismic triggering of eruptions in the far field: Volcanoes and geysers. *Annual Review of Earth and Planetary Sciences*, **34**, 263–291, https://doi.org/10.1146/annurev.earth.34.031405.125125

Martí, J. and Baraldo, A. 1990. Pre-caldera pyroclastic deposits of Deception Island (South Shetland Islands). *Antarctic Science*, **2**, 345–352, https://doi.org/10.1017/S0954102090000475

Martí, J., Vila, J. and Rey, J. 1996. Deception Island (Bransfield Strait, Antarctica): an example of a volcanic caldera developed by extensional tectonics. *Geological Society, London, Special Publications*, **110**, 253–265, https://doi.org/10.1144/GSL.SP.1996.110.01.20

Martí, J., Geyer, A. and Aguirre-Diaz, G. 2013. Origin and evolution of the Deception Island caldera (South Shetland Islands, Antarctica). *Bulletin of Volcanology*, **75**, 1–18, https://doi.org/10.1007/s00445-013-0732-3

McNutt, S.R. and Nishimura, T. 2008. Volcanic tremor during eruptions: Temporal characteristics, scaling and constraints on conduit size and processes. *Journal of Volcanology and Geothermal Research*, **178**, 10–18, https://doi.org/10.1016/j.jvolgeores.2008.03.010

Middlemost, E.A.K. 1989. Iron oxidation ratios, norms and the classification of volcanic rocks. *Chemical Geology*, **77**, 19–26, https://doi.org/https://doi.org/10.1016/0009-2541(89)90011-9

Moran, S., Newhall, C. and Roman, D. 2011. Failed magmatic eruptions: late-stage cessation of magma ascent. *Bulletin of Volcanology*, **73**, 115–122, https://doi.org/10.1007/s00445-010-0444-x

Moreton, S.G. and Smellie, J.L. 1998. Identification and correlation of distal tephra layers in deep-sea sediment cores, Scotia Sea, Antarctica. *Annals of Glaciology*, **27**, 285–289, https://doi.org/10.3189/1998AoG27-1-285-289

Newhall, C.G. and Dzurisin, D. 1988. *Historical Unrest at Large Calderas of the World*. United States Geological Survey Bulletin, **1855**.

Nishimura, T. and Ueki, S. 2011. Seismicity and magma supply rate of the 1998 failed eruption at Iwate volcano, Japan. *Bulletin of Volcanology*, **73**, 133–142, https://doi.org/10.1007/s00445-010-0438-8

Oliva-Urcia, B., Gil-Peña, I. *et al.* 2015. Paleomagnetism from Deception Island (South Shetlands archipelago, Antarctica), new insights into the interpretation of the volcanic evolution using a geomagnetic model. *International Journal of Earth Sciences*, **105**, 1353–1370, https://doi.org/10.1007/s00531-015-1254-3

Orheim, O. 1972. *A 200-Year Record of Glacier Mass Balance at Deception Island, Southwest Atlantic Ocean, and its Bearing on Models of Global Climate Change*. Institute of Polar Studies, Ohio State University.

Ortiz, R., Vila, J. *et al.* 1992. Geophysical features of Deception Island. *In*: Yoshida, Y., Kaminuma, K. and Shiraishi, K. (eds) *Recent Progress in Antarctic Earth Science*. Terra Scientific (TERRAPUB), Tokyo, 443–448.

Ortiz, R., García, A. *et al.* 1997. Monitoring of the volcanic activity of Deception Island, South Shetland Islands, Antarctica (1986–1995). *In*: Ricci, C.A. (ed.) *The Antarctic Region: Geological Evolution and Processes*. Terra Antartica, Siena, Italy, 1071–1076.

Padrón, E., Hernández, P.A. *et al.* 2015. Geochemical evidence of different sources of long-period seismic events at Deception volcano, South Shetland Islands, Antarctica. *Antarctic Science*, **27**, 557–565, https://doi.org/10.1017/S0954102015000346

Pallas, R., Smellie, J.L., Casas, J.M. and Calvet, J. 2001. Using tephrochronology to date temperate ice: correlation between ice tephras on Livingston Island and eruptive units on Deception Island volcano (South Shetland Islands, Antarctica). *The Holocene*, **11**, 149–160, https://doi.org/10.1191/095968301669281809

Pallister, J.S., Hoblitt, R.P., Meeker, G.P., Newhall, C.G., Knight, R.J. and Siems, D.F. 1996. Magma mixing at Pinatubo volcano: petrographic and chemical evidence from the 1991 deposits. *In*: Newhall, C.G. and Punongbayan, R.S. (eds) *Fire and Mud: Eruptions and Lahars of Mount Pinatubo, Philippines*. Philippine Institute of Volcanology and Seismology, Quezon City, The Philippines/University of Washington Press, Seattle, WA, 687–731.

Peccerillo, A., Tripodo, A., Villari, L., Gurrieri, S. and Zimbalatti, E. 1991. Genesis and evolution of volcanism in back-arc areas. A case history, the island of Decepción (Western Antarctica). *Periodico di Mineralogia*, **60**, 29–44.

Peci, L.M., Berrocoso, M., Páez, R., Fernández-Ros, A. and de Gil, A. 2012. IESID: Automatic system for monitoring ground deformation on the Deception Island volcano (Antarctica). *Computers and Geosciences*, **48**, 126–133, https://doi.org/https://doi.org/10.1016/j.cageo.2012.05.004

Peci, L.M., Berrocoso, M., Fernández-Ros, A., García, A., Marrero, J.M. and Ortiz, R. 2014. Embedded ARM system for volcano monitoring in remote areas: Application to the active volcano on Deception Island (Antarctica). *Sensors*, **14**, 672–690, https://doi.org/10.3390/s140100672

Pedrazzi, D., Aguirre-Díaz, G., Bartolini, S., Martí, J. and Geyer, A. 2014. The 1970 eruption on Deception Island (Antarctica): eruptive dynamics and implications for volcanic hazards. *Journal of the Geological Society, London*, **171**, 765–778, https://doi.org/10.1144/jgs2014-015

Pedrazzi, D., Németh, K., Geyer, A., Álvarez-Valero, A.M., Aguirre-Díaz, G. and Bartolini, S. 2018. Historic hydrovolcanism at Deception Island (Antarctica): implications for eruption hazards. *Bulletin of Volcanology*, **80**, 11, https://doi.org/10.1007/s00445-017-1186-9

Pedrazzi, D., Keresztori, G., Lobo, A., Geyer, A. and Calle, J. 2020. Geomorphology of the post-caldera monogenetic volcanoes at Deception Island, Antarctica — Implications for landform recognition and volcanic hazard assessment. *Journal of Volcanology and Geothermal Research*, **402**, 106986, https://doi.org/10.1016/j.jvolgeores.2020.106986

Pelayo, A.M. and Wiens, D.A. 1989. Seismotectonics and relative plate motions in the Scotia Sea region. *Journal of Geophysical Research: Solid Earth*, **94**, 7293–7320, https://doi.org/10.1029/JB094iB06p07293

Poland, M. 2010. Learning to recognize volcanic non-eruptions. *Geology*, **38**, 287–288, https://doi.org/10.1130/focus032010.1

Prates, G., Berrocoso, M., Fernandez-Ros, A. and Garcia, A. 2013a. Enhancement of sub-daily positioning solutions for surface deformation monitoring at Deception volcano (South Shetland Islands, Antarctica). *Bulletin of Volcanology*, **75**, 1–10, https://doi.org/10.1007/s00445-013-0688-3

Prates, G., García, A., Fernández-Ros, A., Marrero, J.M., Ortiz, R. and Berrocoso, M. 2013b. Enhancement of sub-daily positioning solutions for surface deformation surveillance at El Hierro volcano (Canary Islands, Spain). *Bulletin of Volcanology*, **75**, 1–9, https://doi.org/10.1007/s00445-013-0724-3

Prudencio, J., De Siena, L., Ibáñez, J.M., Del Pezzo, E., García-Yeguas, A. and Díaz-Moreno, A. 2015. The 3D attenuation structure of Deception Island (Antarctica). *Surveys in Geophysics*, **36**, 371–390, https://doi.org/10.1007/s10712-015-9322-6

Ramos, M., Vieira, G., de Pablo, M.A., Molina, A., Abramov, A. and Goyanes, G. 2017. Recent shallowing of the thaw depth at Crater Lake, Deception Island, Antarctica (2006–2014). *Catena*, **149**, 519–528, https://doi.org/10.1016/j.catena.2016.07.019

Rey, J., Somoza, L. and Martínez-Frías, J. 1995. Tectonic, volcanic, and hydrothermal event sequence on Deception Island (Antarctica). *Geo-Marine Letters*, **15**, 1–8, https://doi.org/10.1007/bf01204491

Rey, J., Somoza, L., Martínez-Frías, J., Benito, R. and Martín-Alfageme, S. 1997. Deception Island (Antarctica): a new target for exploration of Fe–Mn mineralization? *Geological Society, London, Special Publications*, **119**, 239–251, https://doi.org/10.1144/gsl.sp.1997.119.01.15

Rey, J., Maestro, A., Somoza, L. and Smellie, J.L. 2002. Submarine morphology and seismic stratigraphy of Port Foster. *In*: López-Martínez, J., Smellie, J.L., Thomson, J.W. and Thomson, M.R.A. (eds) *Geology and Geomorphology of Deception Island, 1:25 000*. BAS GEOMAP Series Sheets 6-A and 6-B. British Antarctic Survey, Cambridge, UK, 40–46.

Robertson, S.D., Wiens, D.A., Shore, P.J., Dorman, L., Adaros, R. and Vera, E. 2001. Seismicity and tectonics of the South Shetland Islands Region from a combined land-sea seismograph deployment. *American Geophysical Union, Fall Meeting 2001*, San Francisco, California, Abstract, S12B-0596. 2001AGUFM.S12B0596R.

Robertson Maurice, S.D., Wiens, D.A., Shore, P.J., Vera, E. and Dorman, L.M. 2003. Seismicity and tectonics of the South Shetland Islands and Bransfield Strait from a regional broadband seismograph deployment. *Journal of Geophysical Research: Solid Earth*, **108**, 2461, https://doi.org/10.1029/2003jb002416

Roobol, M.J. 1973. Historic volcanic activity at Deception Island. *British Antarctic Survey Bulletin*, **32**, 23–30.

Roobol, M.J. 1980. A model for the eruptive mechanism of Deception Island from 1820 to 1970. *British Antarctic Survey Bulletin*, **49**, 137–156.

Roobol, M.J. 1982. The volcanic hazard at Deception Island, South Shetland Islands. *British Antarctic Survey Bulletin*, **51**, 237–245.

Rosado, B., Fernández-Ros, A., Berrocoso, M., Prates, G., Gárate, J., de Gil, A. and Geyer, A. 2019. Volcano-tectonic dynamics of Deception Island (Antarctica): 27 years of GPS observations (1991–2018). *Journal of Volcanology and Geothermal Research*, **381**, 57–82, https://doi.org/10.1016/j.jvolgeores.2019.05.009

Saccorotti, G., Almendros, J., Carmona, E., Ibáñez, J.M. and Del Pezzo, E. 2001. Slowness anomalies from two dense seismic arrays at Deception Island Volcano, Antarctica. *Bulletin of the Seismological Society of America*, **91**, 561–571, https://doi.org/10.1785/0120000073

Smellie, J.L. 1988. Recent observations on the volcanic history of Deception Island, South Shetland Islands. *British Antarctic Survey Bulletin*, **81**, 83–85.

Smellie, J.L. 1989. Deception Island. American Geophysical Union Field Trip Guidebook, **T180**, 146–152.

Smellie, J.L. 1999. The upper Cenozoic tephra record in the south polar region: a review. *Global and Planetary Change*, **21**, 51–70, https://doi.org/10.1016/S0921-8181(99)00007-7

Smellie, J.L. 2001. Lithostratigraphy and volcanic evolution of Deception Island, South Shetland Islands. *Antarctic Science*, **13**, 188–209, https://doi.org/10.1017/S0954102001000281

Smellie, J.L. 2002*a*. Volcanic hazard. *In*: López-Martínez, J., Smellie, J.L., Thomson, J.W. and Thomson, M.R.A. (eds) *Geology and Geomorphology of Deception Island, 1:25 000*. BAS GEOMAP Series Sheets 6-A and 6-B. British Antarctic Survey, Cambridge, UK, 47–53.

Smellie, J.L. 2002*b*. The 1969 subglacial eruption on Deception Island (Antarctica): events and processes during an eruption beneath a thin glacier and implications for volcanic hazards. *Geological Society, London, Special Publications*, **202**, 59–79, https://doi.org/10.1144/GSL.SP.2002.202.01.04

Smellie, J.L. 2002*c*. Geology. *In*: López-Martínez, J., Smellie, J.L., Thomson, J.W. and Thomson, M.R.A. (eds) *Geology and Geomorphology of Deception Island, 1:25 000*. BAS GEOMAP Series Sheets 6-A and 6-B. British Antarctic Survey, Cambridge, UK, 11–30.

Smellie, J.L. 2021. Bransfield Strait and James Ross Island: volcanology. *Geological Society, London, Memoirs*, **55**, https://doi.org/10.1144/M55-2018-58

Smellie, J.L. and López-Martínez, J. 2002*a*. Introduction. *In*: López-Martínez, J., Smellie, J.L., Thomson, J.W. and Thomson, M.R.A. (eds) *Geology and Geomorphology of Deception Island, 1:25 000*. BAS GEOMAP Series Sheets 6-A and 6-B. British Antarctic Survey, Cambridge, UK, 1–6.

Smellie, J.L. and López-Martínez, J. 2002*b*. Geological and geomorphological evolution: summary. *In*: López-Martínez, J., Smellie, J.L., Thomson, J.W. and Thomson, M.R.A. (eds) *Geology and Geomorphology of Deception Island, 1:25 000*. BAS GEOMAP Series Sheets 6-A and 6-B. British Antarctic Survey, Cambridge, UK, 54–57.

Smellie, J.L. and López-Martínez, J. 2002*c*. Geological Map of Deception Island, 1:25 000. *In*: López-Martínez, J., Smellie, J.L., Thomson, J.W. and Thomson, M.R.A. (eds) *Geology and Geomorphology of Deception Island, 1:25 000*. BAS GEOMAP Series Sheet 6-A. British Antarctic Survey, Cambridge, UK.

Smellie, J.L., Hofstetter, A. and Troll, G. 1992. Fluorine and boron geochemistry of an ensialic marginal basin volcano: Deception Island, Bransfield Strait, Antarctica. *Journal of Volcanology and Geothermal Research*, **49**, 255–267, https://doi.org/10.1016/0377-0273(92)90017-8

Smellie, J.L., Pankhurst, R.J., Thomson, M.R.A. and Davies, R.E.S. 1984. *The geology of the South Shetland Islands: VI. Stratigraphy, geochemistry and evolution*. British Antarctic Survey Scientific Reports, **87**.

Somoza, L., Martínez-Frías, J., Smellie, J.L., Rey, J. and Maestro, A. 2004. Evidence for hydrothermal venting and sediment volcanism discharged after recent short-lived volcanic eruptions at Deception Island, Bransfield Strait, Antarctica. *Marine Geology*, **203**, 119–140, https://doi.org/10.1016/S0025-3227(03)00285-8

Sparks, S.R.J., Sigurdsson, H. and Wilson, L. 1977. Magma mixing: a mechanism for triggering acid explosive eruptions. *Nature*, **267**, 315–318, https://doi.org/10.1038/267315a0

Stich, D., Almendros, J., Jiménez, V., Mancilla, F. and Carmona, E. 2011. Ocean noise triggering of rhythmic long period events at Deception Island volcano. *Geophysical Research Letters*, **38**, L22307, https://doi.org/10.1029/2011gl049671

Torrecillas, C., Berrocoso, M. and García-García, A. 2006. The multidisciplinary Scientific Information Support System (SIMAC) for Deception Island. *In*: Fütterer, D., Damaske, D., Kleinschmidt, G., Miller, H. and Tessensohn, F. (eds) *Antarctica*. Springer, Berlin, 397–402.

Torrecillas, C., Berrocoso, M., Pérez-López, R. and Torrecillas, M.D. 2012. Determination of volumetric variations and coastal changes due to historical volcanic eruptions using historical maps and remote-sensing at Deception Island (West-Antarctica). *Geomorphology*, **136**, 6–14, https://doi.org/10.1016/j.geomorph.2011.06.017

Torrecillas, C., Berrocoso, M., Felpeto, A., Torrecillas, M.D. and Garcia, A. 2013. Reconstructing palaeo-volcanic geometries using a Geodynamic Regression Model (GRM): Application to Deception Island volcano (South Shetland Islands, Antarctica). *Geomorphology*, **182**, 79–88, https://doi.org/10.1016/j.geomorph.2012.10.032

Valencio, A., Mendía, E. and Vilas, J. 1979. Palaeomagnetism and K–Ar age of Mesozoic and Cenozoic igneous rocks from Antarctica. *Earth and Planetary Science Letters*, **45**, 61–68, https://doi.org/10.1016/0012-821X(79)90107-9

Vidal, J., Berrocoso, M. and Jigena, B. 2011. Hydrodynamic modeling of Port Foster, Deception island (Antarctica). *In*: Tenreiro, J.A., Baleanu, D. and Luo, A.C.J. (eds) *Nonlinear and Complex Dynamics: Applications in Physical, Biological, and Financial Systems*. Springer, Berlin, 193–204.

Vidal, J., Berrocoso, M. and Fernández-Ros, A. 2012. Study of tides and sea levels at Deception and Livingston islands, Antarctica. *Antarctic Science*, **24**, 193–201, https://doi.org/10.1017/S095410201100068X

Vila, J., Martí, J., Ortiz, R., Garcia, A. and Correig, A.M. 1992*a*. Volcanic tremors at Deception Island (South Shetland Islands). *Journal of Volcanology and Geothermal Research*, **53**, 89–102, https://doi.org/10.1016/0377-0273(92)90076-P

Vila, J., Ortiz, R. and Garcia, A. 1992*b*. Seismic activity on Deception Island. *In*: Yoshida, Y., Kaminuma, K. and Shiraischi, K. (eds) *Recent Progress in Antarctic Earth Sciences*. Elsevier, Amsterdam, 449–456.

Vila, J., Correig, A.M. and Martí, J. 1995. Attenuation and source parameters at deception island (South shetland islands, Antarctica). *Pure and Applied Geophysics*, **144**, 229–250, https://doi.org/10.1007/BF00878633

Wasserman, J. 2012. Volcano seismology. *In*: Bormann, P. (ed.) *New Manual of Seismological Observatory Practice*. Deutsches GeoForschungszentrum (GFZ), Potsdam, Germany, 1–77.

White, R. and McCausland, W. 2016. Volcano-tectonic earthquakes: A new tool for estimating intrusive volumes and forecasting eruptions. *Journal of Volcanology and Geothermal Research*, **309**, 139–155, https://doi.org/10.1016/j.jvolgeores.2015.10.020

Zandomeneghi, D., Barclay, A., Almendros, J., Ibañez Godoy, J.M., Wilcock, W.S.D. and Ben-Zvi, T. 2009. Crustal structure of Deception Island volcano from P wave seismic tomography: Tectonic and volcanic implications. *Journal of Geophysical Research*, **114**, B06310, https://doi.org/10.1029/2008jb 006119

Zobin, V.M. 2016. *Introduction to Volcanic Seismology, Volume 6*. 3rd edn. Elsevier, Amsterdam.

Chapter 7.2

Mount Erebus

Kenneth W. W. Sims[1]*, Richard C. Aster[2], Glenn Gaetani[3], Janne Blichert-Toft[4], Erin H. Phillips[1,5], Paul J. Wallace[6], Glen S. Mattioli[7], Dan Rasmussen[8] and Eric S. Boyd[9]

[1]Department of Geology and Geophysics, University of Wyoming, Laramie, WY 82071, USA
[2]Department of Geosciences, Warner College of Natural Resources, Colorado State University, Fort Collins, CO 80523, USA
[3]Department of Geology and Geophysics, Woods Hole Oceanographic Institution, Woods Hole, MA 02543, USA
[4]Laboratoire de Géologie de Lyon, Ecole Normale Supérieure de Lyon, CNRS UMR 5276, Université de Lyon, 46 Allée d'Italie, 69007 Lyon, France
[5]Center for Economic Geology Research, University of Wyoming, Laramie, WY 82071, USA
[6]Department of Earth Sciences, University of Oregon, Eugene, OR 97403, USA
[7]UNAVCO, Inc., 6350 Nautilus Drive, Boulder, CO 80301-5554, USA
[8]Lamont-Doherty Earth Observatory, Columbia University, Palisades, NY 10964, USA
[9]Department of Microbiology and Immunology, Montana State University, Bozeman, MT 59717, USA

KWWS, 0000-0001-6179-6610; RCA, 0000-0002-0821-4906; EHP, 0000-0003-2092-6890; GSM, 0000-0002-9117-7471; DR, 0000-0003-0137-6715; ESB, 0000-0003-4436-5856
*Correspondence: ksims7@uwyo.edu

Abstract: Erebus volcano, Antarctica, is the southernmost active volcano on the globe. Despite its remoteness and harsh conditions, Erebus volcano provides an unprecedented and unique opportunity to study the petrogenesis and evolution, as well as the passive and explosive degassing, of an alkaline magmatic system with a persistently open and magma-filled conduit. In this chapter, we review nearly five decades of scientific research related to Erebus volcano, including geological, geophysical, geochemical and microbiological observations and interpretations. Mount Erebus is truly one of the world's most significant natural volcano laboratories where the lofty scientific goal of studying a volcanic system from mantle to microbe is being realized.

Supplementary material: Average major and trace elements compositions of lavas from Erebus lineage (Table S1) and ages of dated flows on Erebus volcano and Ross Island (Table S2) are available at https://doi.org/10.6084/m9.figshare.c.5227788

Mount Erebus in Antarctica, or Erebus volcano (EV) as the volcanically active part of the mountain is referred to in more recent literature, is the southernmost active volcano on Earth and one of Earth's rare alkaline volcanoes (Figs 1 & 2). Despite its remote location, and the fact that the conditions at or near its summit and upper slopes can be some of the harshest on the planet, Erebus volcano has provided unprecedented and unique opportunities to study the petrogenesis and evolution of alkaline magmas. First, EV's lavas and tephras span a wide compositional range from primitive basanite to evolved anorthoclase-rich phonolite (Kyle *et al.* 1992). Second, EV's regular Strombolian activity has continuously ejected bombs from a persistent lava lake onto the crater rim and beyond, delivering samples of Erebus magma almost continuously from 1972 until the present (Dunbar *et al.* 1994; Sims *et al.* 2008*a*, 2013*b*). Third, the well-dated lava sequence at EV provides a temporal record ranging from 1.3 Ma to the present, providing a comprehensive record of the long-term evolution of the volcano (Esser *et al.* 2004; Harpel *et al.* 2004; Kelly *et al.* 2008*a*, *b*; Sims *et al.* 2008*a*, 2013*b*). Fourth, EV's persistent lava lake represents an open-conduit degassing magmatic system that provides a unique natural laboratory for studying magma degassing in an alkaline system (Giggenbach *et al.* 1973; Zreda-Gostynska *et al.* 1997; Oppenheimer and Kyle 2008*a*, *b*), as well as the behaviour of open-vent volcanoes. Fifth, EV's location on Ross Island *c.* 35 km north of McMurdo Station, which is the main base for all United States Antarctic Program support and scientific activities on the continent, make it logistically accessible for detailed monitoring studies. As such, a remarkable wealth of observational, geophysical, petrological and geochemical data on Erebus volcano now exists which, together, provides a comprehensive and detailed perspective of the petrogenesis and evolution of this unique alkaline volcanic system, as well as the dynamics of its persistent lava lake.

In this review we present an overview of the most salient aspects of this large body of work. The review is divided into three sections. In the first section, we provide an overview of EV's volcanic characteristics and a synopsis of the geophysical, geochemical, and microbiological measurements conducted on Erebus Volcano since 1972. In the second section, we detail how these geological observations, and geophysical and geochemical measurements, address the following overarching topics and associated questions relating to the driving forces, melting process, magmatic evolution, eruptive behaviour and potential hazards of Erebus volcano:

- Erebus volcanic rocks have a unique HIMU (high μ = high $^{238}U/^{204}Pb$) isotopic signature defined by their highly radiogenic Pb isotopic compositions, the origin of which is of long-standing debate. How do geophysical and geochemical measurements inform on the origin of EV's distinctive HIMU-like source and the causes of volcanism at Erebus volcano and Ross Island?
- Erebus volcano is a rare, active alkaline volcanic system that provides an important window into mantle melting processes. What do geochemical measurements of its lavas reveal about the depth of melting and lithology of the mantle source of Mount Erebus?

Fig. 1. Mount Erebus volcano (3794 m elevation). Looking north from Castle Rock (photograph: Sylvain Pichat).

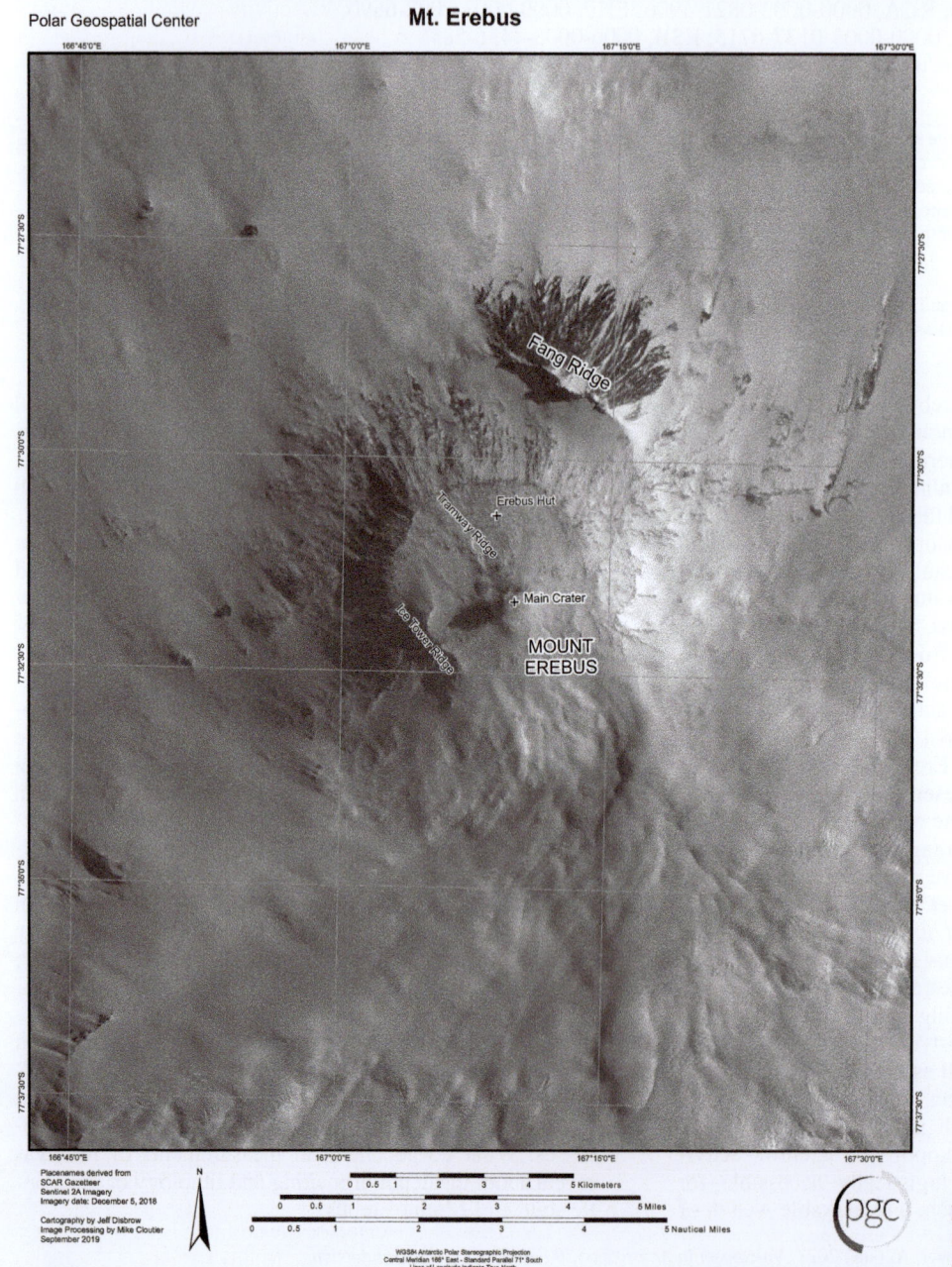

Fig. 2. Satellite image of Mount Erebus (image created by Jeff Dishbrow, Polar Geospatial Center).

- Erebus volcano hosts a unique, long-lived lava lake that allows for direct, long-term study of an open-conduit, alkaline magmatic system. What constraints do geophysical measurements place on the geometry of the EV magmatic plumbing system from the vent down to the lower crust? How do surface gas measurements inform on subsurface processes (magma differentiation and degassing) occurring within the volcanic conduit? What is known about the timescales of magma evolution, crystallization and magma degassing in the Erebus magmatic system?
- The Erebus ice tower fumaroles and hot ground regions host a unique and largely endemic community of microbial life in Antarctica. What is the relationship between EV's magmatic and hydrothermal systems and its biosphere?
- The location of Erebus volcano makes it a threat to both US and international Antarctic science programmes and personnel. How do the observed patterns and changes in magmatic and volcanic activity over both long and short timescales forewarn about potential future eruptive behaviour and help in the assessment of hazards associated with Erebus volcano? Which observations, monitoring and warning systems, and protocols are needed to best forecast short- and long-term volcanic hazards at Erebus volcano and Ross Island?

Finally, in the third section, we provide a summary of the key results.

Geological, geochemical, geophysical and microbiological observations

Geological setting of Erebus volcano

Mount Erebus (Figs 1 & 2) was discovered by Captain James Clark Ross in January 1841:

> [I]t proved to be a mountain twelve thousand four hundred feet of elevation ... emitting flame and smoke in great profusion ... as we drew nearer, its true character became manifest ... I named it 'Mount Erebus'

(Ross 1847, pp. 216, 217).

Mount Erebus is the largest of four volcanic centres forming Ross Island: Mount Erebus (3794 m elevation, 2170 km^3), Mount Terror (3262 m, 1700 km^3), Mount Bird (1800 m, 470 km^3) and Hut Point Peninsula (100 km^3). About 4520 km^3 of volcanic material has been erupted on Ross Island over the last c. 4 myr (Esser et al. 2004; Kelly et al. 2008a). The occurrence, formation and evolution of Erebus volcano can only be fully understood in the context of these other neighbouring volcanoes.

Ross Island is located in the Ross Sea, near the western margin of the Transantarctic Mountains. Ross Island lies at the southern end of the intraplate Terror Rift, which is the locus of the most recent extension within the West Antarctic Rift System (WARS: e.g. Behrendt 1999). Crustal thickness in the immediate Ross Island region ranges from c. 19 to 25 km (Bannister et al. 2003; Finotello et al. 2011; Chaput et al. 2014).

Late Cenozoic, silica-undersaturated, alkaline volcanic rocks erupted on the western margin of the Ross Embayment belong to the McMurdo Volcanic Group (MVG: Kyle (1990a). Ross Island is part of the Erebus Volcanic Province (Kyle 1990a), within the regional McMurdo Volcanic Group of the southern Ross Sea and McMurdo Sound (Kyle 1990b).

Erebus is an active, polygenetic, alkaline composite volcano (Figs 1–6). High-precision $^{40}Ar/^{39}Ar$ dating (Esser et al. 2004; Harpel et al. 2004; Kelly et al. 2008a) has been used to demarcate the volcanic history into three distinct phases (Figs 4 & 5; see also Supplementary Tables S1 and S2): (1) The proto-Erebus shield-building phase (1.3–1.0 Ma), during which basanites were erupted; this phase represents a transition from subaqueous to subaerial activity. Subaqueous activity is characterized by hyaloclastite deposits, while subaerial deposits are distinguished as lava flows with occasional tephra. (2) The proto-Erebus cone-building phase (1.0 Ma–750 ka), dominated by more evolved phonotephrite lavas that largely form the present steep slopes of the volcano. Eruptions were typically effusive with some explosive activity reflected in tephra layers. This phase was terminated by a caldera collapse around 750 ka. (3) The 'modern' Erebus cone-building phase (250 ka–present), which formed the present near-summit edifice (Figs 4–6). Large volumes of anorthoclase-phyric tephriphonolite and phonolite lavas (Fig. 7) were extruded, with minor trachyte also erupted at about 170 ka during the third phase of activity. This phase is characterized by two caldera collapses: an older collapse between 80 and 25 ka, and a younger, smaller collapse, between 25 and 11 ka. Modern Erebus activity has been within the older caldera structure and is dominantly characterized by lava flows, radially distributed around the present crater, although individual eruptive episodes appear to be confined to specific sectors of the volcano with the youngest episode being directed towards the NW. Activity at Mount Erebus during historical times has been characterized by the presence of an active lava lake within the innermost crater, which shows significant variations in elevation. Rising gas bubbles/gas slugs disrupt the lake, with the largest producing Strombolian explosions and the ejection of bombs (Fig. 7), some of which have reached up to 10 m in dimension. The majority of these bombs fall within 500–1000 m of the crater.

Finally, Erebus hydrothermal systems, which are extensively manifest in warm ground and ice-tower fumaroles (Fig. 8), host abundant biodiverse microbial communities (Herbold et al. 2014a, b) comprised of thermophilic Archaea, Bacteria and Eukarya supported by a variety of metabolic functionalities. The habitability of these environments for supporting microbial activity appears to be a direct consequence of continued volcanic activity.

A unique and persistent lava lake

Erebus volcano's persistent, convecting and degassing lava lake of anorthoclase-phyric phonolite magma daylights an open-conduit magma system. This unique magma conduit makes EV an unequalled natural laboratory for studying processes of magma formation, evolution and degassing in an alkaline magmatic system (Figs 6 & 7). The Erebus lava lake was 'first observed directly by a scientific party' in December 1972 (Giggenbach et al. 1973; Kyle et al. 1982); however, it was likely to have been extant for much longer. During the first ascent of Mount Erebus on 10 March 1908 by a party from Shackleton's Nimrod Expedition (Fig. 3), the existence of a lava lake is supported by notes made by T.W. Edgeworth David, the expedition's geologist from the University of Sydney (Shackleton et al. 1909, p. 184):

> After a continuous loud hissing sound, lasting for some minutes, there would come from below a big dull boom, and immediately great globular masses of steam would rush upwards to swell the volume of the snow-white cloud which ever sways over the crater. This phenomenon recurred at intervals during the whole of our stay at the crater. Meanwhile, the air around us was extremely redolent of burning sulphur.

Further, the fact that James Ross reported significant activity, 'emitting flame and smoke in great profusion', suggests that the lava lake also existed in 1841 when Erebus was first discovered.

The EV lava lake is located on the northern floor of the Inner Crater, which sits inside the Main Crater. The main,

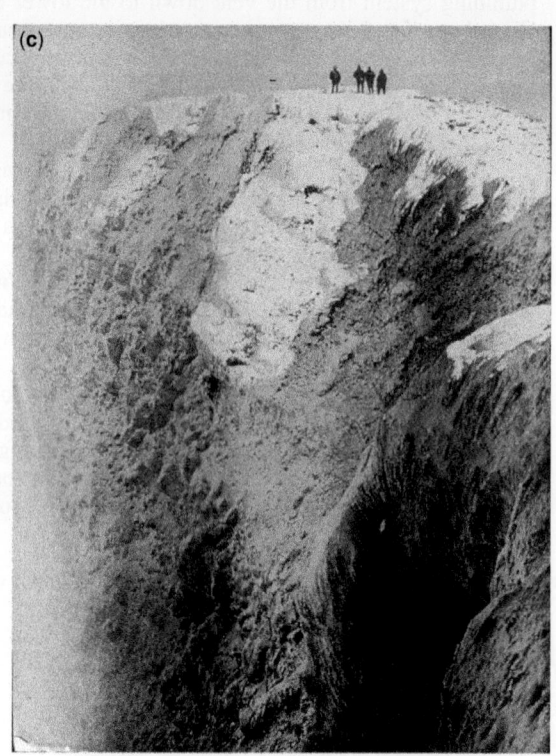

Fig. 3. 'Conquest of Mount Erebus'. Photographs taken during the first ascent of Mount Erebus on 10 March 1908 by a party from Shackleton's Nimrod Expedition (from Shackleton *et al.* 1909). The successful climbing and scientific party included Professor Edgeworth David (the expedition geologist), Sir Douglas Mawson (who took the photos shown here), Dr Alister MacKay, Jameson Adams, Dr Eric Marshall and Phillip Brocklehurst (who did not summit on account of severe frost bite in his toes). Quotes are original photo captions from Shackleton *et al.* (1909). (**a**) 'The (old) crater of Erebus with an older crater in the background'. This perspective is looking approximately north from the rim of the summit plateau and shows the Fang Ridge in the background (e.g. see Fig. 2), which is the remnant of an older caldera. (**b**) 'A remarkable fumarole in the form of a couchant lion'. This fumarolic ice tower is one of several such features on the summit plateau (e.g. see Figs 8 and 18). (**c**) 'The crater of Erebus, 900 feet deep and one-half mile wide'. The summiting Nimrod Expedition members at the Main Crater rim.

prominent, persistent open-conduit lava lake is named Ray's Lava Lake after the New Zealand seismologist Ray Dibble. Adjacent to Ray's Lava Lake are two equally prominent degassing vents, which are referred to as 'Werner's fumarole' and the 'Active vent' (Kyle *et al.* 1982; Oppenheimer and Kyle 2008*a, b*). Werner's fumarole occasionally becomes a small active second lava lake (Csatho *et al.* 2008, Oppenheimer and Kyle 2008*a, b*; Molina *et al.* 2012).

The EV phonolite lava lake is the top of an open convecting magma conduit, which is continuously and quiescently degassing (Figs 1, 4 & 6). All historical eruptive activity has originated from this phonolite lava lake and adjacent vents. Observationally, the EV lava-lake surface shows a steady bidirectional flow of upwelling vesicular magma associated with explosive Strombolian eruptions and downwelling degassed magma. New magma pulses rise to refill the lava lake within about 10 min of an explosive eruption, indicating a shallow magma body with which the lava lake is in 'magma-static equilibrium' (Oppenheimer *et al.* 2009).

Strombolian eruptions (with attendant minor ash) characterize the vast majority of recent eruptive activity (Kyle *et al.* 1982, 1992; Rowe *et al.* 2000; Gerst *et al.* 2008). The explosions are well situated for observations conducted locally and from the crater rim, and have been studied using video (e.g. Dibble *et al.* 2008), seismic (e.g. Aster *et al.* 2008), radar (Gerst *et al.* 2008) and infrasound (e.g. Johnson *et al.* 2008) techniques. In addition to the surface explosions, the eruptions and subsequent refilling of the lava lake produce longer-period seismic signals that arise from a deeper source within the conduit system. The corresponding source centroid for a stack of 293 lava-lake eruptions between January 2005 and April 2006 was located ≤400 m beneath the lava lake and somewhat to the west of the lake centre (Aster *et al.* 2008); however, evidence from the directionality of ejecta and observations of the post-eruptive lava lake indicate that the uppermost tens of metres of the conduit are nearly vertical (Gerst *et al.* 2008).

The style and magnitude of EV's Strombolian eruptions have varied over the years the volcano has been under direct observation. Between 1973 and 1984 there were roughly six small Strombolian eruptions per day. In 1984, their frequency and intensity increased dramatically; then, starting in September 1984, there was an exceptional 4 month period during

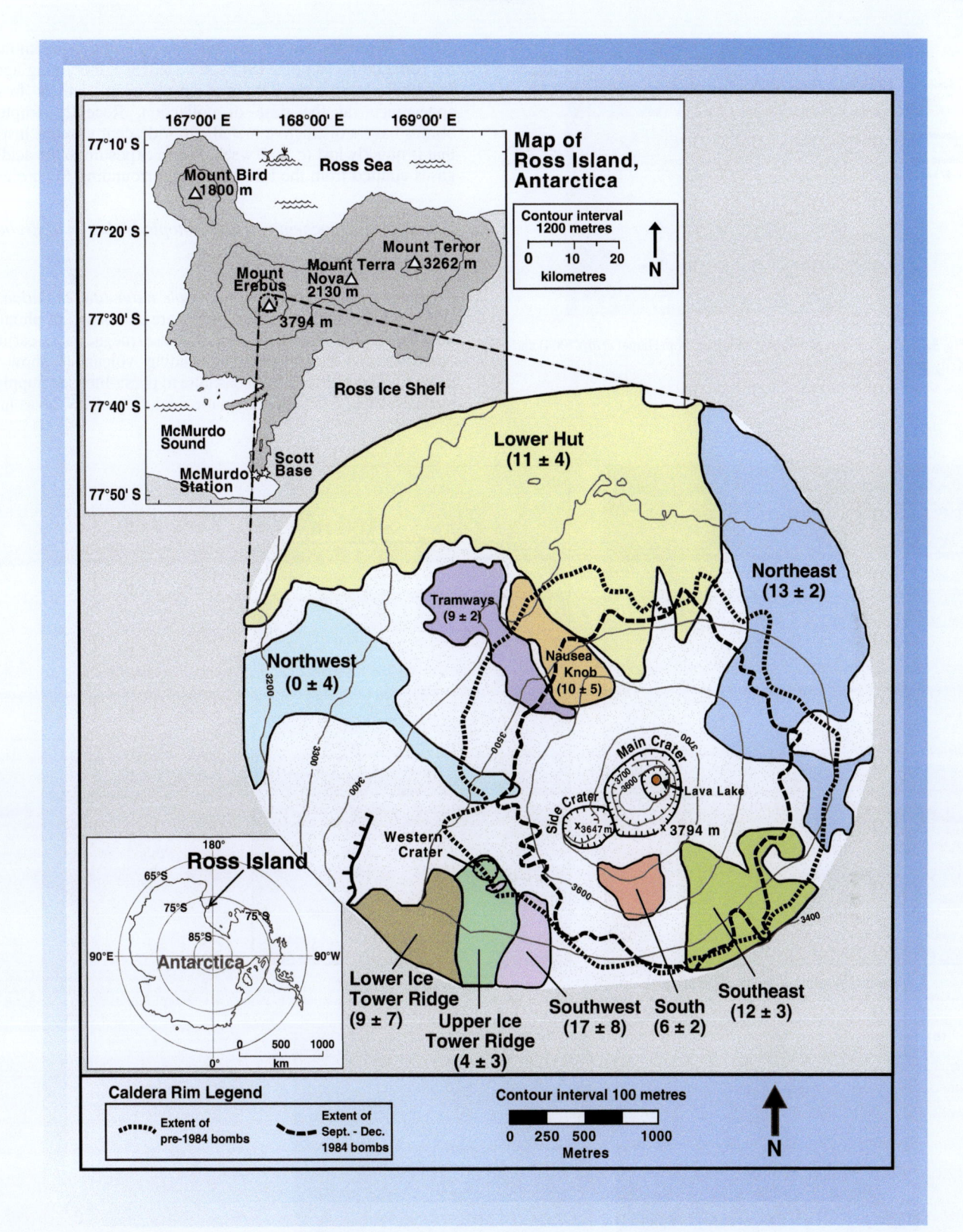

Fig. 4. Map of Mount Erebus showing the location of Mount Erebus, and the location and ^{40}Ar/^{39}Ar ages of its lavas. Map modified from Kelly *et al.* (2008*b*) and Sims *et al.* (2013*b*). Data taken from Esser *et al.* (2004), Harpel *et al.* (2004) and Kelly *et al.* (2008*a*).

which Strombolian eruptions were sending bombs of up to 10 m in diameter more than 2 km from the Erebus summit vent (Dibble *et al.* 1984; Kyle *et al.* 1990, 1992, 1994). From 2001 to 2004 Strombolian eruptions were much less frequent, with a notable hiatus during 2004. Since mid-2005, there has been a return to more frequent Strombolian activity. In October 1993, a powerful gas and steam eruption ejected large lithic bombs and ash beyond the crater rim, and created a new vent at the edge of the inner crater. The exact reasons for these changes in lava-lake eruptive style are not understood explicitly but are likely to be related to processes of gas accumulation at depth in the conduit system feeding the convecting lava lake (Aster *et al.* 2008; Sweeney *et al.* 2008).

The Strombolian eruptions often eject phonolite bombs to and beyond the crater rim where they can be sampled (Fig. 7). These bombs have provided a nearly continuous record of lava-lake magma composition from the early 1970s onwards (Kelly *et al.* 2008*b*; Sims *et al.* 2008*a*,

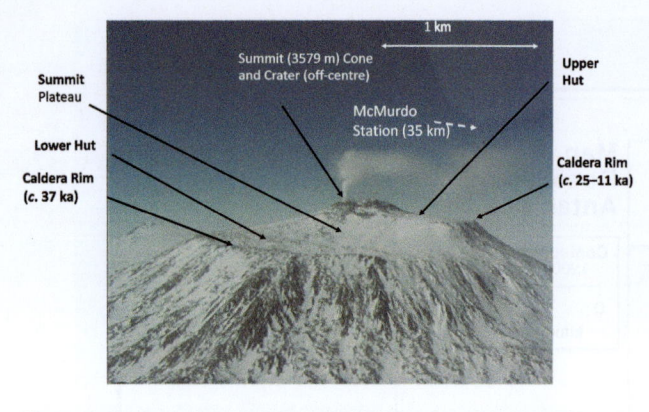

Fig. 5. Erebus summit geography. Modified from Harpel et al. (2004) and Kelly et al. (2008a).

2013b). While the exact dates and times of eruption are known for some bomb samples (Sims et al. 2013b), the young ages for other samples were inferred on the basis of their 'fresh appearance' at the time of collection. Recently erupted bombs have a distinctive metallic to iridescent vitreous luster that is quickly lost (c. 1–2 weeks) upon exposure to the acidic gases emitted from the lava lake and surrounding fumaroles.

Petrological, geochemical, and isotopic evolution of Erebus volcano from source to lava lake

From basanite to phonolite: a simple magmatic evolution. While the most recent EV eruptions are composed of phonolitic lava containing large anorthoclase megacrysts, earlier eruptions and those from surrounding volcanoes show a range of compositions from basanite to phonolite (see Supplementary Table S1). The magmatic evolution of EV lavas has

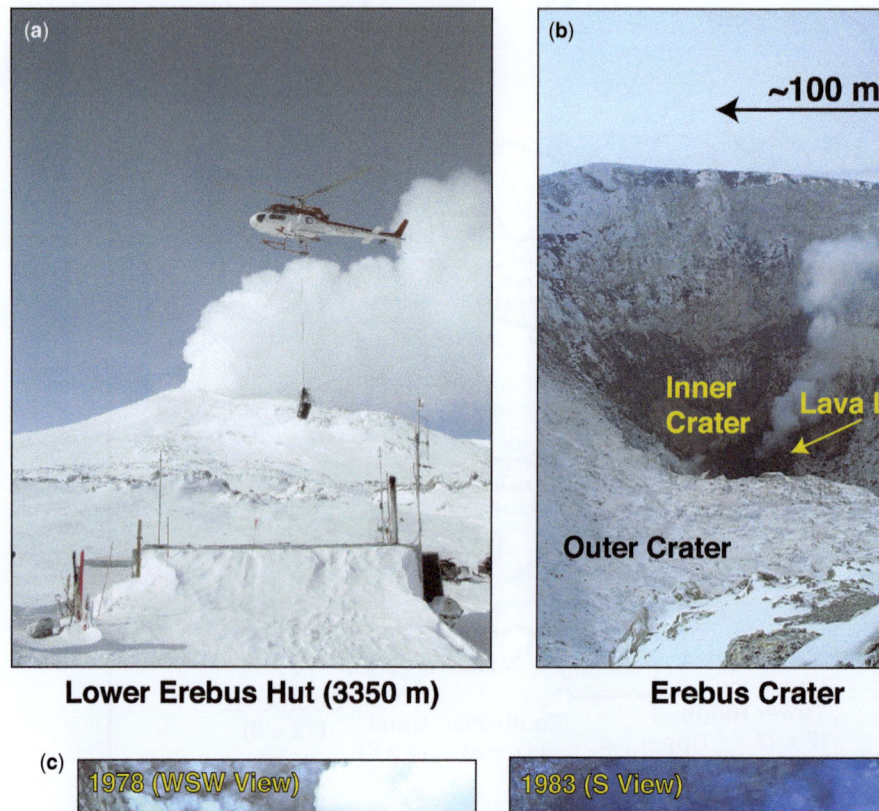

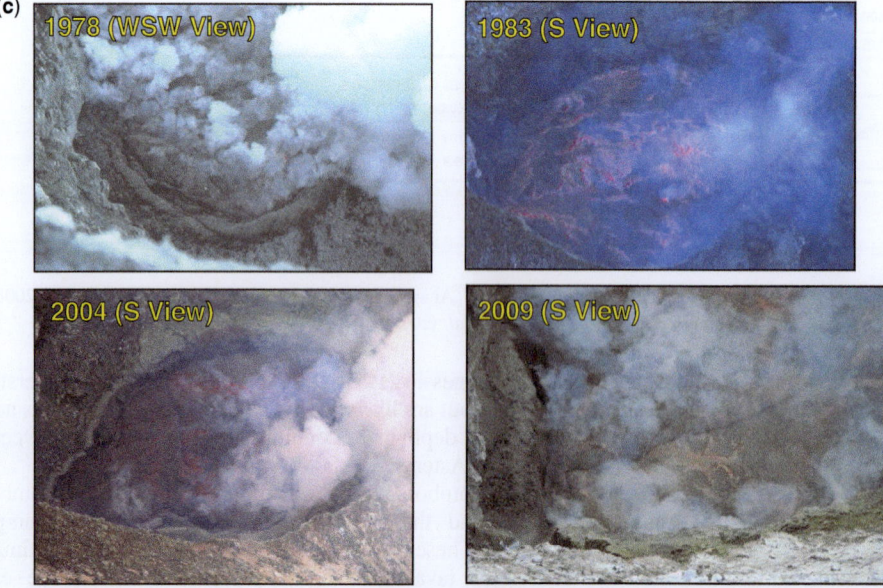

Fig. 6. (**a**) Lower Erebus Hut looking south (Fig. 2). (**b**) Erebus Crater viewed from the Dante robot launch site at the south rim (Wettergreen et al. 1993). (**c**) Photographs spanning 31 years of the long-lived lava lake (approximately 40 m in diameter) looking south to SSW from the crater rim.

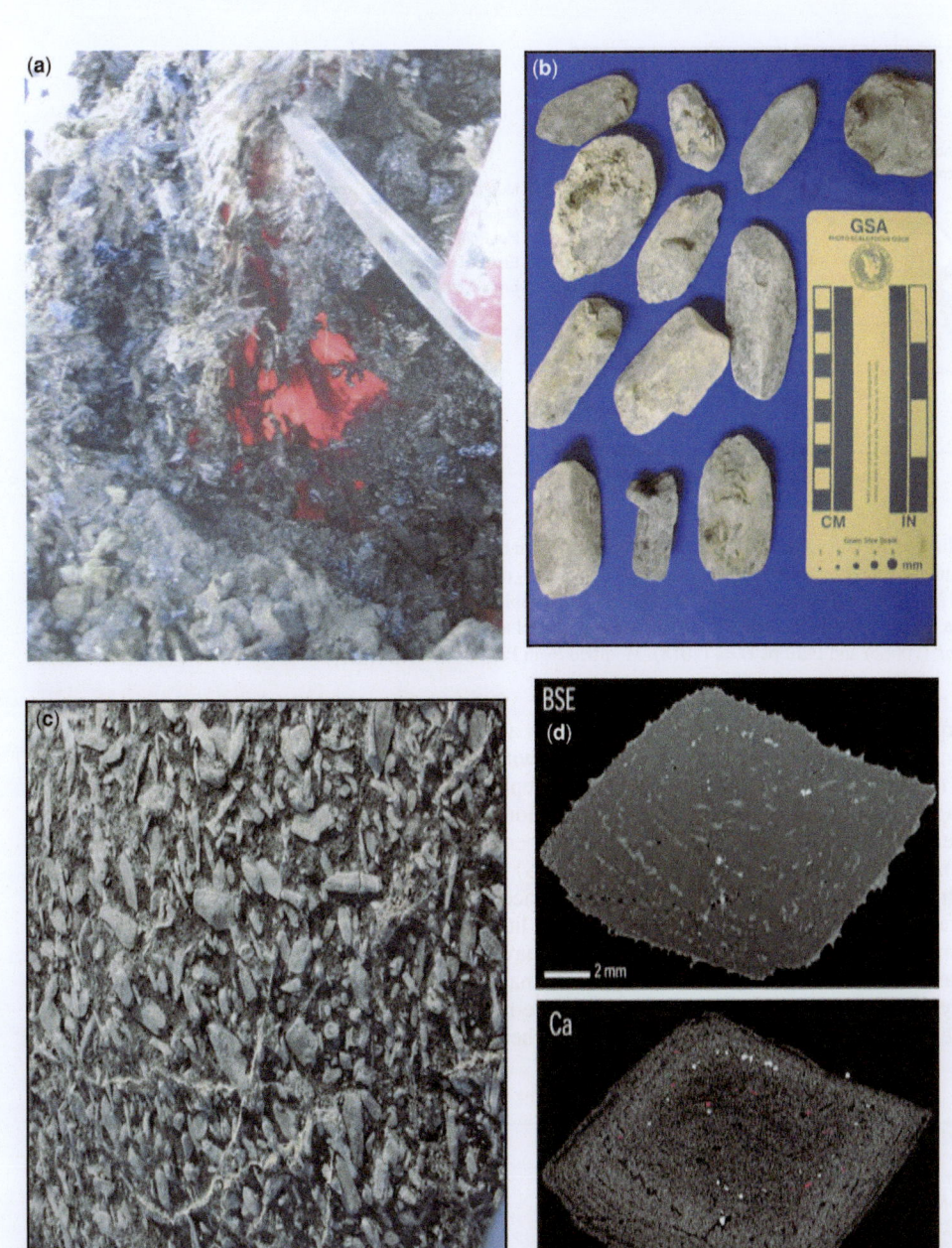

Fig. 7. (a) Molten lava bomb. The ice axe is for scale. (b) Separated anorthoclase megacrystals. (c) Phonolitic lava flow (the scale bar is 4 cm). (d) Backscattered electron (BSE) image and Ca element map of a single anorthoclase crystal. The darkest areas in the BSE image are the anorthoclase host crystal. The abundant irregular areas, slightly brighter than the anorthoclase, are melt inclusions. Other included phases are pyroxene (light-grey phase on the Ca map), apatite (next brightest on BSE, highlighted in red on the Ca map) and magnetite (brightest in BSE, dark on the Ca map). The size of the apatite crystals has been slightly increased in order to improve visibility. SEM image from Sims *et al.* (2013*b*)

Fig. 8. Ice tower showing passive degassing on the Mount Erebus summit flank. The tower is approximately 3 m tall; the circular tube is a collapsed section of the tower that remained intact (photograph: K.W.W. Sims).

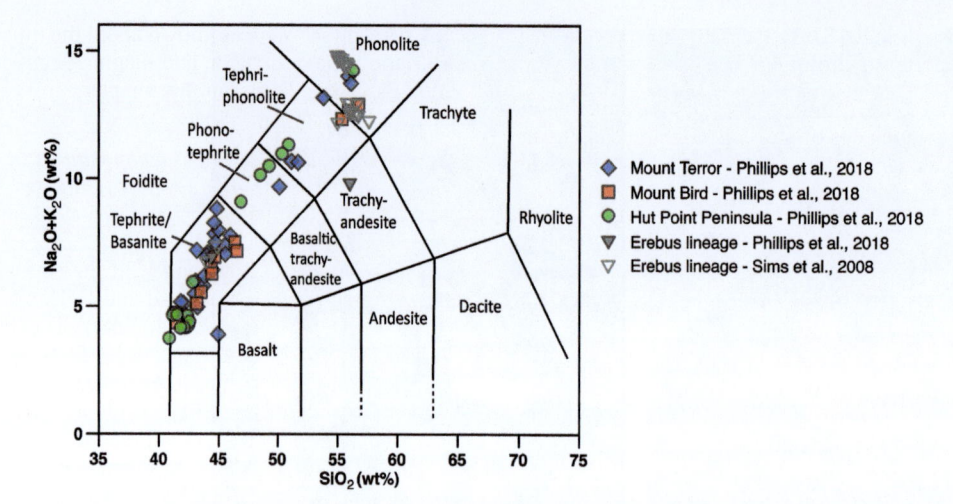

Fig. 9. Total alkalis (wt%) v. silica (wt%) diagram for Ross Island volcanics. Most samples from the peripheral volcanic centres examined in Phillips *et al.* (2018) are basanites. The Mount Terror, Mount Bird and Hut Point Peninsula samples contain SiO_2 ranging from 40.9 to 56.8 wt%, $Na_2O + K_2O$ ranging from 3.74 to 14.7 wt% and MgO ranging from 0.12 to 12.7 wt%, with an average of 6.43 wt%. All samples with MgO less than 1 wt% are either phonolites or plot close to the phonolite–tephriphonolite divide. Major and trace element data for these samples, including both whole-rock analyses and matrix glass analyses, are reported in Kyle *et al.* (1992), Kelly *et al.* (2008b) and Phillips *et al.* (2018). Modified from Phillips *et al.* (2018), with permission from Elsevier.

been studied extensively and is interpreted as a simple coherent magma differentiation trend from mantle-derived basanite to highly evolved phonolite (Fig. 9) (Kyle *et al.* 1992). Lavas on Ross Island show two major magmatic lineages: the Dry Valley Drilling Project (DVDP) lineage (Kyle 1981) and the Erebus lineage (EL: Kyle *et al.* 1992). The DVDP lineage consists predominantly of basanite with minor microporphyritic kaersutite-bearing intermediate differentiates and phonolite. Petrological/geochemical studies undertaken on flank lavas from Mount Terror, Mount Bird and Hut Point Peninsula show the presence of kaersutite (Kyle 1990a; Phillips *et al.* 2018). The EL constitutes a coherent fractionation trend defined by a single liquid line of descent from basanite to phonolite with a complete sequence of intermediate (phonotephrite, tephriphonolite) eruptive products (Kyle *et al.* 1992). Minor volumes of more iron-rich and less silica-undersaturated benmoreite and trachyte, termed the enriched Fe series (EFS), occur as isolated outcrops on the flanks of EV and on adjacent islands in Erebus Bay. The EFS lavas follow a different liquid line of descent, and the trachytes are interpreted to have undergone both shallow crustal assimilation and fractional crystallization during their evolution.

Primitive-mantle-normalized trace element patterns for Erebus and other volcanic rocks from Ross Island are shown in

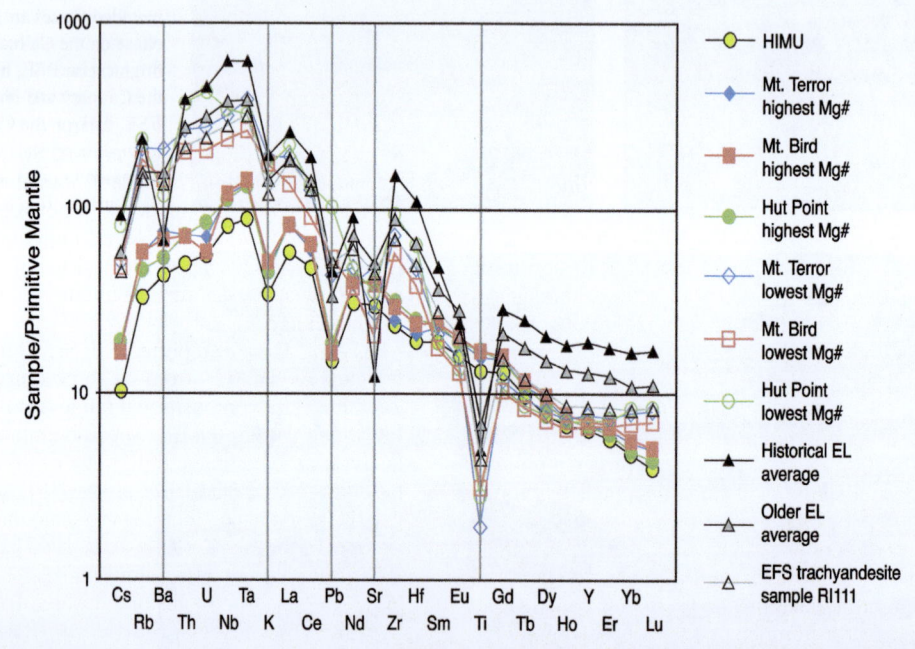

Fig. 10. Primitive-mantle-normalized trace element diagram, showing LREE enrichments in Mount Terror, Mount Bird and Hut Point Peninsula samples, with primitive-mantle-normalized La/Yb ranging from 14.5 to 26.2. Trace element patterns for the highest Mg# samples from the three peripheral volcanic centres are very similar to each other, as well as to the HIMU mantle end member, with negative K and Pb anomalies. HIMU is the mean of 12 samples from St Helena (Willbold and Stracke 2006). Averages for Erebus volcano are for samples reported in Sims *et al.* (2008a). Major and trace element concentrations for these Erebus samples are reported in Kelly *et al.* (2008b) and Kyle *et al.* (1992). Historical Erebus lineage (EL) lavas from Kelly *et al.* (2008b) are for matrix glass, and older EL lavas from Kyle *et al.* (1992) and Kelly *et al.* (2008b) are for whole-rock samples. Primitive- mantle values are from McDonough and Sun (1995). The highest and lowest Mg# samples from Mount Terror, Mount Bird and Hut Point are indicated. Modified from Phillips *et al.* (2018), with permission from Elsevier.

Figure 10 (Phillips *et al.* 2018). These data show several features that have important petrological implications:

- Erebus phonolite glass from historical bombs contains the highest overall abundances of incompatible trace elements.
- The phonolite glass (minus the anorthoclase phenocrysts) is depleted in Ba and Sr because of the relative compatibility of these elements in anorthoclase (Kelly *et al.* 2008*b*; Sims *et al.* 2013*b*).
- Erebus lavas display elevated Nb/La relative to lavas from peripheral Ross Island volcanoes, reflecting variable extents of apatite crystallization, a phase present in Erebus lavas at *c.* 0.5% modal abundance (Kyle 1981; Kyle *et al.* 1992).
- All lavas are light rare earth element (LREE) enriched, suggesting small degrees of partial melting; differences between the highest Mg# basanites and lowest Mg# samples are consistent with crystal fractionation models for the EL (Kyle 1981; Kyle *et al.* 1992).
- The high-Mg# parental basanites from Ross Island lavas and tephras are depleted in heavy REE (HREE) compared to middle REE (MREE) (mean $Dy_N/Yb_N \cong 1.4$), suggesting the presence of residual garnet in their source (Sun and Hanson 1975; 1976; Sims and Hart 2006; Sims *et al.* 2013*b*).
- The radiogenic isotopes and major and trace element abundances and patterns of the high-Mg# parental basanites from Ross Island lavas and tephras closely resemble HIMU basalts from St Helena (Willbold and Stracke 2006), which is the mantle end member that most closely resembles the Erebus and Ross Island lavas (see the subsection 'Sr, Nd, Hf and Pb isotopic compositions of Erebus lavas: a HIMU-like source' below).

Unique mineralogy of EV eruptive products. The Erebus phonolite magma, as represented by erupted bombs, contains matrix glass (*c.* 67 vol%; microlite-free), anorthoclase feldspar (*c.* 30 vol%), titanomagnetite (*c.* 1 vol%), fayalitic olivine (*c.* 0.8 vol%), clinopyroxene (*c.* 0.6 vol%), fluor-apatite (*c.* 0.6%) and occasional blebs of pyrrhotite (Kyle 1977; Kyle *et al.* 1992; Kelly *et al.* 2008*b*). Textural relationships between these phases suggest that the pyroxene, titanomagnetite and olivine crystallized first (Kelly *et al.* 2008*b*). Textures and mineralogical compositions of young lava bombs are consistent with crystallization at low degrees of undercooling in a growth-dominated regime (Kelly *et al.* 2008*b*). Anorthoclase feldspar is the only phase showing compositional zoning.

A unique feature of the Erebus phonolite magma is the presence of abundant large anorthoclase feldspar megacrysts, up to *c.* 10 cm in length (Fig. 7). These crystals are striking because of their size, complex internal zoning and abundance of melt inclusions – up to 30 vol% – as well as for their inclusions of other crystalline phases such as pyroxene, apatite, magnetite and pyrrhotite (Kyle 1977; Dunbar *et al.* 1994; Kelly *et al.* 2008*b*). Large melt inclusions (up to c. 1 mm in length) typically are irregularly shaped (Dunbar *et al.* 1994). However, the anorthoclase megacrysts also contain a population of negative crystal shape inclusions that are 10–40 μm in diameter. Populations of melt inclusions are trapped along what appear to be growth zones in the host crystal, producing a banded appearance in backscattered electron (BSE) microscope images. The anorthoclase megacrysts are complexly compositionally zoned at a number of scales and show evidence for periods of resorption during crystal growth (Dunbar *et al.* 1994; Sumner 2007; Kelly *et al.* 2008*b*). The compositional range of anorthoclase is $An_{10-23}Ab_{62-68}Or_{11-17}$ (Kelly *et al.* 2008*b*). The dominant compositional zoning is defined by high- and low-Ca zones in the crystal, and is attributed to either convective processes (Kelly *et al.* 2008*b*) or boundary-layer effects during crystallization (Sumner 2007). The growth rates of the anorthoclase megacrysts are discussed in detail in the subsection 'What is known about the timescales of magma evolution, crystallization and magma degassing in the Erebus magma system?' later in this chapter.

Sr, Nd, Hf and Pb isotopic compositions of Erebus lavas: a HIMU-like source. The EV lava samples (Figs 11 & 12) have highly radiogenic $^{206}Pb/^{204}Pb$ (Sun and Hanson 1975; Sims and Hart 2006; Sims *et al.* 2008*a*), low $^{87}Sr/^{86}Sr$ (Kyle *et al.* 1992; Sims and Hart 2006; Sims *et al.* 2008*a*), and intermediate $^{143}Nd/^{144}Nd$ and $^{176}Hf/^{177}Hf$ values (Sims and Hart 2006; Sims *et al.* 2008*a*). The highly radiogenic $^{206}Pb/^{204}Pb$ (Sun and Hanson 1975; Sims and Hart 2006; Sims *et al.* 2008*a*) is considered to be the defining isotopic signature for categorizing these lavas as 'HIMU', which is an acronym derived from the inference that lavas with high $^{206}Pb/^{204}Pb$ originate from a source with high time-integrated $^{238}U/^{204}Pb$, which is referred to as mu (μ) in isotope nomenclature. Lavas from St Helena volcano in the Atlantic represent the end-member HIMU mantle component, and the origin of the source of these lavas has been discussed in numerous publications (e.g. Hart 1988; Willbold and Stracke 2006; French and Romanowicz 2015; Weiss *et al.* 2016).

There have been several studies on the isotopic composition of the lavas from Erebus volcano and the other volcanic centres on Ross Island:

- Sun and Hanson (1975) measured Pb isotopes by thermal ionization mass spectrometry (TIMS) on a limited number of basanites and related differentiates from the Erebus Volcanic Province. They were the first to note the unique HIMU-like, or radiogenic, Pb isotopic signature for the McMurdo Volcanic Group (MVG) lavas.
- Stuckless and Ericksen (1976) published low-precision Sr isotopic data on samples from the Erebus Volcanic Province, including a few samples from DVDP cores.
- Kyle *et al.* (1992) measured Sr isotopes on several samples from the EL and EFS suites. They concluded that: (1) the uniform Sr isotopes from basanite to phonolite supported major-element modelling which indicates that the EL resulted from simple fractional crystallization of a parental basanite derived from a homogeneous mantle source; and (2) the high $^{87}Sr/^{86}Sr$ of the trachyte indicates that continental crust was assimilated during evolution of the EFS.
- Sims and Hart (2006) reported Nd, Sr, Th and Pb isotopic compositions of four historical bombs as part of a global study evaluating the relationship between U–Th disequilibria systematics and long-lived radiogenic Nd, Sr and Pb isotope systematics (see the following subsection on 'U- and Th-decay series measurements of Erebus lava bombs of known age').
- Sims *et al.* (2008*a*) reported Nd, Sr, Hf and Pb isotope systematics on samples from all three eruptive phases of EV over the entire compositional range from basanite to phonolite which defines the EL. Specific emphasis was placed on analysing the complete liquid line of descent: this included basanitic DVDP and early phase 1 EL samples, which are used as parental end members for calculating the liquid-line of descent for the EL (Kyle *et al.* 1992); early-stage phonotephrites; dated upper-summit phonolite lavas (Esser *et al.* 2004); and a suite of historical phonolite bombs collected over the past 30 years. Sims *et al.* (2008*a*) came to three conclusions:

1. Isotopic data show a marked distinction between the early-stage basanites and phonotephrites, whose Nd, Hf, Sr and Pb isotope compositions are variable (particularly Pb), and the later, evolved phonolitic lavas and bombs, whose Nd, Hf, Sr and Pb isotope compositions are essentially invariant (see figs 5 & 6 in Sims *et al.* 2008*a*). The observation that the EL lavas define a

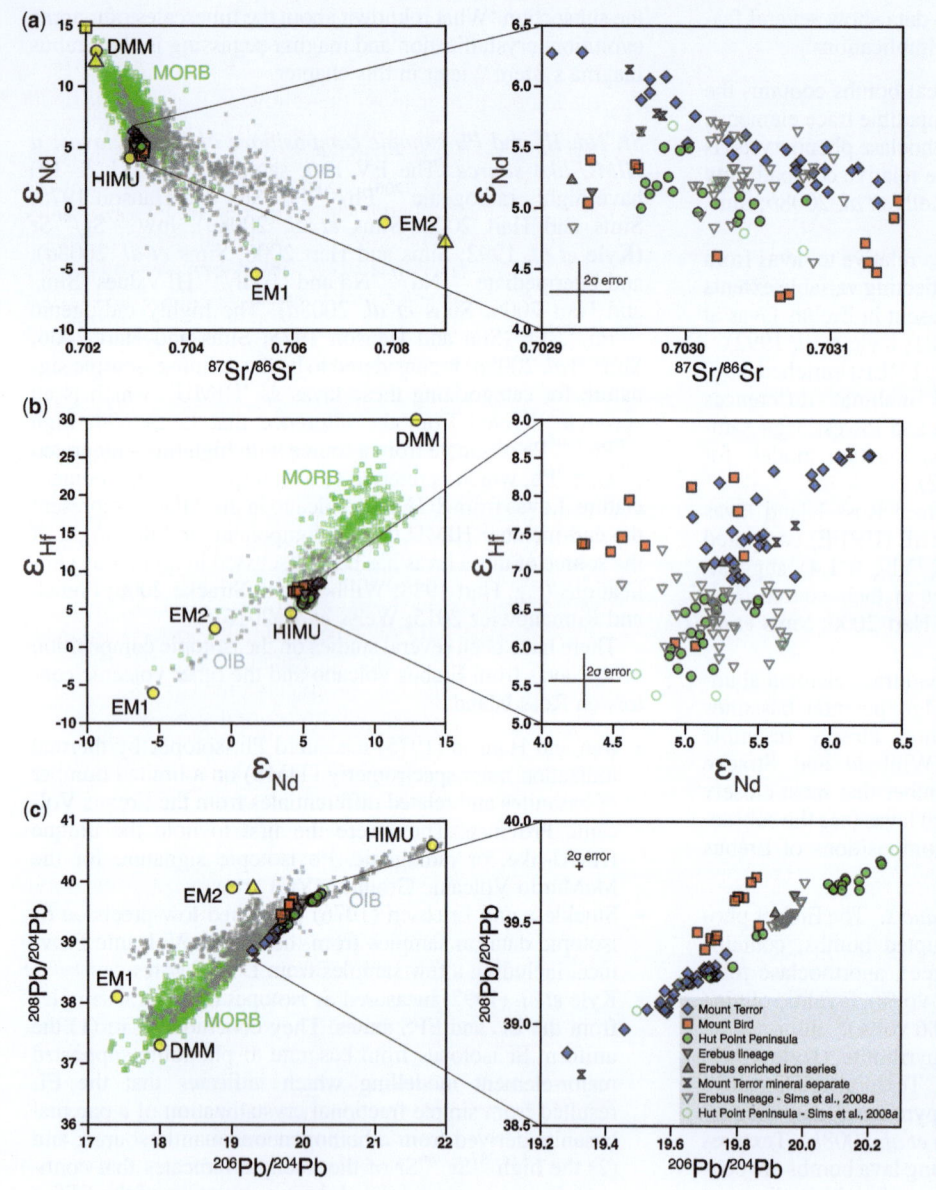

Fig. 11. Isotope plots showing Ross Island rocks in a global context on the left-hand side, including MORB in green and OIB in grey, illustrating the similarity of Ross Island lavas and tephras to OIB and the HIMU mantle end member. Plots on the right-hand side show Ross Island volcanic samples in detail and display the overall greater variability of the peripheral volcanic centres compared to Erebus volcano as well as the general clustering of isotope compositions within each volcanic centre. Erebus lineage samples from Sims *et al.* (2008*a*) exclude enriched iron-series samples, which show clear signs of crustal contamination. (**a**) $^{87}Sr/^{86}Sr$ v. ε_{Nd}. Mantle end members are shown with yellow symbols; depleted MORB mantle (DMM) from Hart *et al.* (1992: circle), Salters and Stracke (2004: square) and Workman and Hart (2005: triangle); HIMU and enriched mantle 1 (EM1) from Hart *et al.* (1992); enriched mantle 2 (EM2) from Hart *et al.* (1992; circle) and Workman *et al.* (2004; triangle). (**b**) ε_{Nd} v. ε_{Hf}. Mantle end members, shown with yellow circles, are inferred for Hf isotope compositions from their Nd isotope compositions in Hart *et al.* (1992). (**c**) $^{206}Pb/^{204}Pb$ v. $^{208}Pb/^{204}Pb$. Mantle end members are shown with yellow symbols; DMM, HIMU and EM1 are from Hart *et al.* (1992: circles); EM2 are from Hart *et al.* (1992: circle) and Workman *et al.* (2004: triangle). After Phillips *et al.* (2018), with permission from Elsevier.

trend of decreasing isotopic variability with increasing extent of differentiation indicates that magma mixing has played a fundamental role in establishing the isotopic and compositional uniformity of the more recent phonolites erupted from EV (Sims *et al.* 2008*a*).

2. In multiple isotope space, the EL lavas lie along a mixing trajectory between the two mantle end members DMM (depleted MORB (mid-ocean ridge basalts) mantle) and HIMU (Sims *et al.* 2008*a*). Because of the position of these mantle components within the 'mantle tetrahedron' they also lie between HIMU and the theoretical mantle component 'FOZO' as defined by Hart *et al.* (1992).

3. All four isotope systems, Sr, Nd, Hf and Pb, are coherent in their mixing relationships with the HIMU end member defined by St Helena, thereby placing important constraints on both the age (Archean or early Proterozoic) and origin of the HIMU-like mantle source of Erebus lavas (see the subsection 'The HIMU-like isotopic signature of Erebus volcano is a well-established, important source characteristic' later in this chapter).

- Phillips *et al.* (2018) measured major and trace element concentrations and Sr, Nd, Pb and Hf isotope compositions of 57 samples from the three peripheral volcanic centres, Mount Terror, Mount Bird, and Hut Point Peninsula, that surround EV on Ross Island (Fig. 11). These samples are mostly older (*c.* 4–0.3 Ma) and mainly basanitic in composition. The Ross Island samples analysed by Phillips *et al.* (2018) have $^{87}Sr/^{86}Sr$ ranging from 0.702907 to 0.703147, ε_{Nd} ranging from +4.3 to +6.3, ε_{Hf} ranging from +5.6 to +8.6 and $^{206}Pb/^{204}Pb$ ranging from 19.3 to 20.2. The Sr, Nd, Hf and Pb isotope compositions for all four volcanoes of Ross Island (Erebus, Terror, Bird and Hut Point Peninsula) fall on a mixing line between the HIMU and DMM mantle end members, with a very small contribution from an EM (enriched mantle) component. Small differences in the isotopic compositions of the four volcanoes, most notably Mount Bird, imply mantle source heterogeneity on a length scale of less than 40 km.

The HIMU-like Pb isotopic signature observed for EL lavas is a pervasive feature of rocks from Antarctica, the sub-Antarctic islands and some New Zealand Cenozoic volcanism as well (Sun and Hanson 1975; Hart *et al.* 1995, 1997; Rocholl *et al.* 1995; Panter *et al.* 2000, 2006; Finn *et al.* 2005; Sims and Hart 2006; Sims *et al.* 2008*a*; Phillips *et al.* 2018). The origin of this HIMU-like signature is a matter of long-standing debate and will be discussed in the subsection 'The HIMU-like isotopic signature of Erebus volcano is a well-established, important source characteristic'.

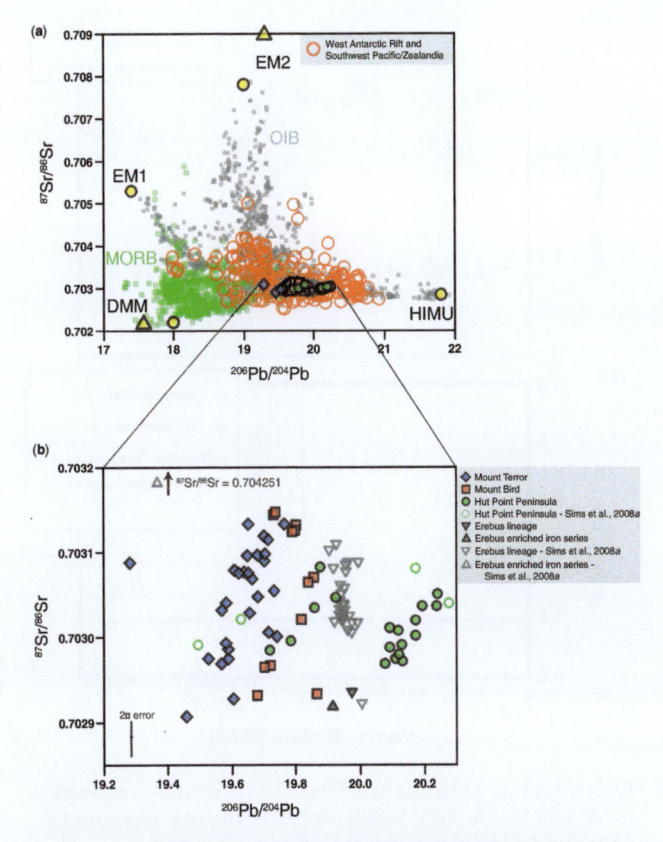

Fig. 12. (a) $^{206}Pb/^{204}Pb$ v. $^{87}Sr/^{86}Sr$ showing the similarity of Ross Island lavas and tephras to those from the West Antarctic Rift and the greater SW Pacific (data from Sun and Hanson 1975; Hart 1988 (Balleny); Hart et al. 1995, 1997; Rocholl et al. 1995; Panter et al. 2000, 2006; Hoernle et al. 2006, 2010 (Hikurangi); Timm et al. 2010; see also Finn et al. 2005). Ross Island volcanics fall on a mixing line between the DMM and HIMU mantle end members, possibly with a small contribution from an EM component. Mantle end members are shown with yellow symbols; depleted MORB mantle (DMM) from Hart et al. (1992: circle) and Workman and Hart (2005: triangle); HIMU and enriched mantle 1 (EM1) from Hart et al. (1992); enriched mantle 2 (EM2) from Hart et al. (1992: circle) and Workman et al. (2004; triangle). (b) Close-up of (a) showing Ross Island samples in detail. After Phillips et al. (2018), with permission from Elsevier.

U- and Th-decay series measurements of Erebus lava bombs of known age. The open-conduit degassing and pyroclastic bombs of known age regularly ejected by Strombolian eruptions at EV make it an ideal system to measure U-series disequilibria to evaluate both the depth of melting and the timescales of magma genesis, melt evolution, crystal growth rates and magma degassing.

U- and Th-decay series nuclides are ideal for determining both the chronology and timescales of the magmatic processes occurring at EV, for three reasons: (1) the elemental characteristics and chemical affinities of the different nuclides make them applicable to a wide range of chemical processes including mantle melting, anorthoclase crystallization and magma degassing; (2) the half-lives of the relevant nuclides are commensurate with the timescales of these processes; and (3) the principles of secular and transient equilibrium (Bateman 1910) constrain the initial relative abundances of parent and daughter nuclides, and thus provide an important and unique constraint on elemental fractionation processes. U-series disequilibria measurements of lava bombs of known age from EV provide important constraints on the timescales of the magmatic processes occurring in the shallow portion of the EV lava lake, namely crystallization rates, magma residence times and degassing rates. U-series measurements also provide constraints on deep melting processes, assuming one can confidently see through disequilibrium effects caused by shallow-level processes.

It is important to note for the following that isotopic abundances are expressed in terms of activity/per unity weight (e.g. disintegrations per min/g of rock: dpm g^{-1}), where the activity is defined as the nuclide's decay constant multiplied by its molar atomic abundance. Activities are denoted by the use of parentheses () around the nuclide in question. This is a convenient formalism making the comparison of parent/daughter activity ratios a direct measure of the extent of elemental fractionation that occurred within a given isotope system.

There have been three studies measuring U-series disequilibria in Erebus lavas:

- Reagan et al. (1992) used $(^{238}U)–(^{230}Th)–(^{226}Ra)$ and $(^{232}Th)–(^{228}Ra)$ disequilibria to obtain internal isochrons to constrain the magma residence times for anorthoclase–glass separates from bombs erupted in 1984 and 1988. Reagan et al. (1992) showed that the anorthoclase crystals were strongly enriched in ^{226}Ra over ^{230}Th, whereas glass separates had ^{226}Ra deficits. Assuming that the mineral/melt partition coefficients for Ra and Ba are the same (i.e. $D_{Ra} = D_{Ba}$, where D is the mineral/melt partition coefficient for each element) during crystal growth, and using two-point $(^{226}Ra)/Ba$ v. $(^{230}Th)/Ba$ isochrons (anorthoclase and glass pairs) for the two samples, calculated ages of 2520 years for the 1984 sample and 2225 years for the 1988 sample were obtained. Using the activity of ^{228}Th as a proxy for ^{228}Ra (which is reasonable given the very short half-life of ^{228}Th), Reagan et al. (1992) also measured significant $(^{228}Th/^{232}Th)$ disequilibrium in the anorthoclase megacrystals (c. 2.16), whereas the glass was in equilibrium with $(^{228}Th)/(^{232}Th)$ of 1.0. Because of the short half-life of ^{228}Ra ($t_{1/2} = 5.77$ years), this result implies that the anorthoclase crystals grew recently and rapidly. As such, Reagan et al. (1992) suggested that the ^{228}Th disequilibria arise from the presence of young rims with ^{228}Ra excesses.
- Subsequently, Sims and Hart (2006) measured Nd, Sr, Th and Pb isotopes on four historical bombs as part of a global study evaluating the relationship between $(^{238}U)–(^{230}Th)$ disequilibria systematics and long-lived radiogenic Nd, Sr and Pb isotope systematics. Their results showed that the $(^{230}Th/^{232}Th)$ and $(^{238}U/^{232}Th)$ of Erebus bombs are intermediate relative to other ocean island and mid-ocean ridge basalts, and fall on the HIMU end member in plots of Pb isotopes v. $(^{230}Th/^{232}Th)$ and $(^{238}U/^{232}Th)$. For the $^{238}U–^{230}Th$ isotope system, Erebus represents the best end-member approximation of the HIMU source because there are no samples young enough for $^{238}U–^{230}Th$ disequilibria studies from St Helena and Mangai–Tubai (HIMU end members). They also noted that the $^{230}Th/^{238}U >1$ in all of the samples required melting in the presence of residual garnet, consistent with fractionations observed for the MREE to HREE.
- Sims et al. (2013b) measured $^{238}U–^{230}Th–^{226}Ra–^{210}Pb–^{210}Po$, $^{232}Th–^{228}Ra–^{228}Th$ and $^{235}U–^{231}Pa–^{227}Ac$ in a suite of 22 phonolite volcanic bombs erupted between 1972 and 2005, and five anorthoclase megacrysts separated from bombs erupted in 1984, 1989, 1993, 2004 and 2005 (Figs 13 & 14). It is noteworthy that these were the first measurements of $^{231}Pa–^{227}Ac$ in volcanic samples, and thus the first dataset for any volcanic system to examine the entire suite of relevant ^{238}U, ^{235}U and ^{232}Th decay series nuclides. As will be discussed in the subsection 'What is known about the timescales of magma evolution, crystallization, and magma degassing in the Erebus magma system?', these results allowed Sims et al.

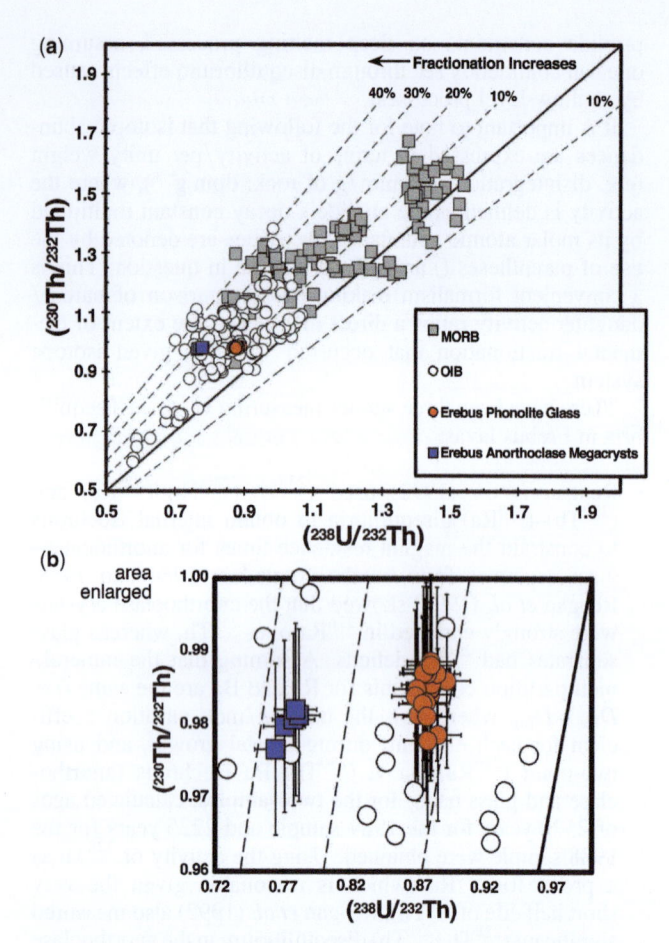

Fig. 13. (a) $^{230}Th/^{232}Th$ v. $^{238}U/^{232}Th$ isochron plot showing Erebus samples compared with the global database for MORB and OIB. (b) Erebus data enlarged. The shift of anorthoclase megacrystals to lower U/Th exceeds what would be predicted from experimental uncertainties and is interpreted to be due to apatite inclusions (see the text for the discussion). After Sims *et al.* (2013*b*).

(2013*b*) to place constraints on both the residence time of anorthoclase in the Erebus magmatic system, and thus crystal growth rates, and the depths and timescales of magma degassing.

Measurements of aerosols and gases from Erebus volcano.
The persistent, open-vent degassing of EV makes it a significant point source of gases and aerosols to the austral polar troposphere. Scientific interest in EV's gaseous emissions are driven by several factors, including: (1) the potential impact of EV's sulfur, halogens and NO_x emissions on the nearly pristine south polar atmospheric environment (Zoller *et al.* 1974; Radke 1982; Zreda-Gostynska *et al.* 1993, 1997; Oppenheimer *et al.* 2005, 2009; Ilyinskaya *et al.* 2010); (2) EV's dynamic open-conduit lava lake and Strombolian activity; (3) the high-altitude and dry climate of EV's crater and the continuously degassing active lava lake which minimize scrubbing of gas emissions and facilitate remote measurements, such as open-path Fourier transform infrared (FTIR) spectroscopy, correlation spectroscopy (COSPEC) and differential optical absorption spectroscopy (DOAS); and (4) EV's location next to the United States Antarctic Program's (USAP) McMurdo Station, which makes access and working conditions logistically feasible and relatively safe, even almost comfortable, when compared to other alkaline volcanoes with persistent lava lakes (e.g. Nyiragongo or Erta Ale). Gas measurements, particularly SO_2 and CO_2 fluxes and ratios, are important for volcano hazard assessment and, as such, should be an integral part of any monitoring programme to evaluate the current state of the EV system and its possible threat to McMurdo Station personnel.

Since the late 1970s, there have been many studies measuring Erebus gas compositions and fluxes. These studies have included: aerosol sampling (Radke 1982; Chuan *et al.* 1986; Zreda-Gostynska *et al.* 1993, 1997; Ilyinskaya *et al.* 2010); ultraviolet spectroscopy for SO_2 flux measurements (Rose *et al.* 1985; Kyle *et al.* 1990, 1994; Sweeney *et al.* 2008; Boichu *et al.* 2010); open-path FTIR spectroscopy of magma degassing (Oppenheimer and Kyle 2008*b*; Oppenheimer *et al.* 2011; Ilanko *et al.* 2015); and infrared analysis of flank CO_2 abundance and flux (Wardell *et al.* 2003; 2004).

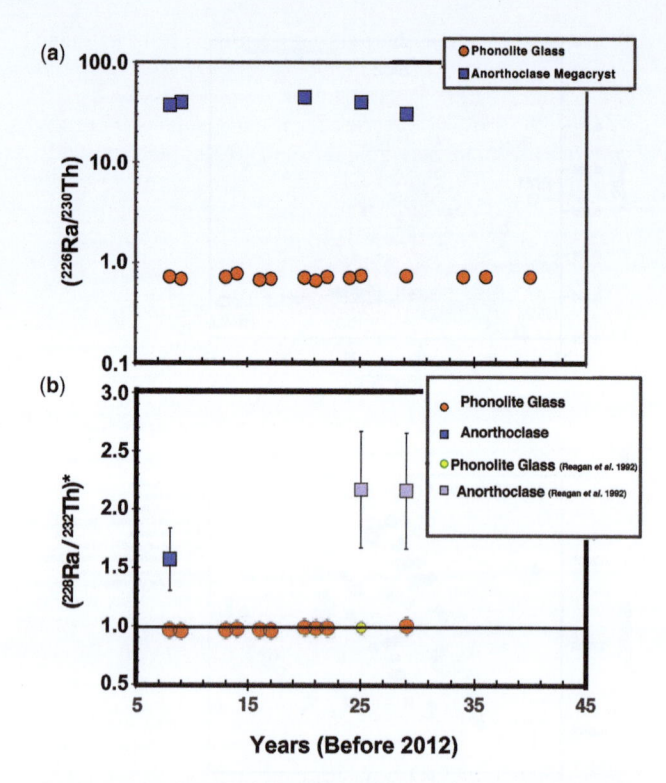

Fig. 14. (a) Time-series data for ($^{226}Ra/^{230}Th$) over the 34 year historical record from 1972 to 2005. Time is since 2012. (b) Time-series data for ^{228}Ra over the 34 year historical record from 1972 to 2005. Time is since 2012. *($^{228}Ra/^{232}Th$) was determined using $^{228}Th/^{232}Th$ measured by alpha spectroscopy as a proxy, as indicated by the asterisk. The half-life of ^{228}Th is 1.9 years, so for this 2005 sample (measured in 2008) the ($^{228}Ra/^{232}Th$) inferred from ($^{228}Th/^{232}Th$) is a minimum value as the system had not yet attained full equilibrium. Reagan *et al.* (1992) data are shown for comparison. After Sims *et al.* (2013*b*).

Filter measurements of Erebus plume aerosols and gases.
Volcanic aerosols are typically collected on filter packs or cascade impactors, and the filters are analysed by scanning electron microscopy, ion chromatography, neutron activation analyses or inductively coupled plasma mass spectrometry (ICP-MS). Kyle *et al.* (1990) showed that the Erebus plume is enriched in the trace metals In, As, Hg, Zn, Au, Se, Co, W, Cs, Mo, Rb, Cu, Na and K and has elevated abundances of halogens (Cl, Fl, Br, etc.) relative to SO_2 and other alkaline volcanoes. In fact, gases from EV have some of the highest halogen/SO_2 ratios of volcanoes globally with HF/SO_2 mass ratios >0.2 and HCl/SO_2 mass ratios >0.5 (Kyle *et al.* 1990; Pyle and Mather 2009; Ilyinskaya *et al.* 2010; Boichu *et al.* 2011).

Aerosol measurements have shown that EV represents a potentially important point source of chemical species to a wide region of the Polar Plateau (Zoller *et al.* 1974; Chuan *et al.* 1986; Meeker *et al.* 1991; Chuan 1994; Zreda-Gostynska *et al.* 1997; Wardell *et al.* 2008; Ilyinskaya *et al.* 2010). As such, the signature and distribution of aerosols from EV have been extensively investigated (Zoller *et al.* 1974; Chuan *et al.* 1986; Meeker *et al.* 1991; Chuan 1994; Ilyinskaya *et al.* 2010). Noteworthy discoveries from early studies were the occurrence of elemental gold in aerosol samples, possibly formed by reduction of Au–Cl complexes by sulfur (Meeker *et al.* 1991), and that aerosol mass was narrowly confined to particles with a diameter of *c.* 0.1 mm (Radke 1982), substantially finer than observed at other volcanoes (e.g. Mather *et al.* 2003; Martin *et al.* 2008). The latter observation is an important constraint on the dispersal of aerosols and led Radke (1982) to conclude that up to 32% of the Antarctic sulfate budget observed at the South Pole could be attributed to Erebus. Zreda-Gostynska *et al.* (1997) calculated emission rates (using SO_2 fluxes from Kyle *et al.* 1990) and enrichment factors (using the method of Zoller *et al.* 1983) for a large suite of trace metals and halogens, and concluded that the similarity of trace element patterns of aerosols collected over the South Pole indicates that the Erebus gas plume provides a significant contribution to the aerosol inventory observed in deep interior Antarctic snow. Lastly, Ilyinskaya *et al.* (2010) measured the chemical composition and size distribution of the Erebus aerosols, focusing on the water-soluble fraction. They showed that: (1) Erebus aerosols are distinct from other volcanic sources in that they are dominated by chloride-bearing particles (over 30% of total mass) and have an unusually high Cl^-/SO_4^{2-} molar ratio of 3.5, similar to early conclusions of Kyle *et al.* (1990); (2) water-soluble aerosols have high concentrations of F^-, Cl^-, Br^- and SO_4^{2-} in a very narrow particle size range (0.1–0.25 mm), this small size fraction is important for aerosol dispersal (Radke 1982); furthermore, the detection of particulate Br^- implies that quiescent Erebus emissions can contribute to tropospheric ozone depletion (Boichu *et al.* 2011); and (3) alkali metal–halide salts (Na, K)(Cl, F) are the most abundant chemical species in the aerosol population. Collectively, aerosol research conducted to date shows that EV's emissions have regional significance, contribute to the Antarctic atmosphere and are preserved in glacio-chemical records.

Ultraviolet spectroscopy for SO_2 flux measurements. The measurement of SO_2 flux from volcanoes is of major importance for both hazard assessment and the evaluation of a volcano's environmental impact. These flux measurements also provide a critical reference value by which to determine emission rates of other chemical species (Kyle *et al.* 1990; Sweeney *et al.* 2008).

SO_2 measurements using UV spectroscopy have evolved significantly over the past several decades (e.g. McGonigle and Oppenheimer 2003) with EV having played a significant early role in the development of this technology (Symonds *et al.* 1985; Kyle and McIntosh 1989; Kyle *et al.* 1990, 1994; Sweeney *et al.* 2008; Boichu *et al.* 2010). Measurements of volcanic SO_2 gas fluxes on Erebus have used scattered light ultraviolet spectroscopy (COSPEC or DOAS). The lower Erebus hut has provided an ideal location from which to profile the plume across its transport direction utilizing a scanning UV zenith-viewing telescope (Kyle and McIntosh 1989; Kyle *et al.* 1990, 1994; Sweeney *et al.* 2008). Importantly, at Erebus volcano, measurement of SO_2 is facilitated by its persistent and degassing lava lake and by its high altitude (3700 m) and resulting dry climate, which help to minimize hydrothermal scrubbing of its emissions and facilitate remote COSPEC and DOAS measurement techniques (Symonds *et al.* 1985; Oppenheimer *et al.* 2011).

New Mexico Tech's Mount Erebus Volcano Observatory (MEVO) conducted decades of COSPEC and DOAS measurements of SO_2 in the Erebus gas plume. Sweeney *et al.* (2008), in a well-considered interpretation of this decade-long effort, assembled 8064 individual SO_2 emission rates measured at EV between 1992 and 2005. The compilation shows a normal distribution skewed slightly towards higher values and ranges from 0.3 Mg/day (0.003 kg s^{-1}) to 224 Mg/day (2.6 kg s^{-1}), with an average of 61 ± 27 Mg/day (0.7 ± 0.3 kg s^{-1}) (Fig. 15). Sweeney *et al.* (2008) also demonstrated fluctuations in SO_2 emission rates occurring on short timescales, with wavelengths varying between *c.* 10 min and 3 h, as well as long-term variations on the timescales of days. Additionally, they noted an increase in average SO_2 emission rates from 39 ± 17 Mg/day (0.5 ± 0.2 kg s^{-1}) in 1996 to *c.* 80 ± 25 Mg/day (0.9 ± 0.3 kg s^{-1}) in 2005 (Sweeney *et al.* 2008). In a follow-up study, Boichu *et al.* (2010) noted that an important source of uncertainty in flux measurements (which is the product of the gas column abundance, integrated across the plume section and the plume transport speed) on Erebus is the estimation of plume speed, which is typically assumed to be the wind speed at or near the gas plume altitude (Boichu *et al.* 2010). To eliminate this source of error in SO_2 flux measurements, Boichu *et al.* (2010) employed a technique using two UV spectrometers equipped with wide-field-of-view telescopes (dual-wide-field-of-view DOAS) that instantaneously collect light from two narrow and parallel entire cross-sections of the plume. This method obviates the need to scan the plume, and also greatly enhances estimates of the plume velocity and calculated gas fluxes (Boichu *et al.* 2010). In a single *c.* 2 h experiment carried out in December 2006, their results showed that SO_2 flux varied from 0.17 to 0.89 ± 0.2 kg s^{-1}, the vertical plume velocity varied from 1 to 2.5 ± 0.1 m s^{-1} and the flux had a cyclicity ranging from about 11 to 24 min.

Two important first-order geochemical observations have been made of sulfur degassing at Erebus volcano:

- The EV average SO_2 emission rate of 61 Mg/day (0.7 kg s^{-1}) is low compared to other persistently active volcanoes (Andres and Kasgnoc 1998), presumably due to the phonolitic magma's evolved nature and subsequent low sulfur content. Basaltic volcanoes, such as Etna (58 kg s^{-1}) and Stromboli (3.5 kg s^{-1}), emit considerably more SO_2 (Caltabiano *et al.* 1994, 2004) than EV. This difference in SO_2 flux is attributed to the formation of EV's phonolite magma, modelled as a 25% residual liquid that formed through crystal fractionation of a mantle-derived parental basanite, during which sulfur is removed from the system by the crystallization of pyrrhotite (Kyle *et al.* 1994).

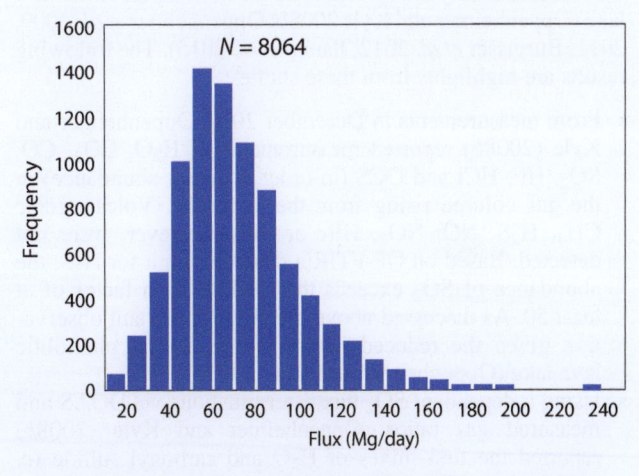

Fig. 15. Histogram of SO_2 emission rates measured between 1992 and 2005. The average is 61 ± 27 Mg/day (0.7 ± 0.3 kg s^{-1}; *N* = 8064). The Gaussian curve has a peak at 60 Mg/day (0.7 kg s^{-1}). Modified from Sweeney *et al.* (2008).

- The phonolitic melts of EV have estimated oxygen fugacities that are 0.5 log units below the quartz–fayalite–magnetite buffer (QFM: Burgisser et al. 2012), making them much reduced. At such oxygen fugacities, the equilibrium gas composition released from the phonolitic magma should have molar H_2S/SO_2 of c. 0.2–1.2 (Burgisser et al. 2012). H_2S has not been detected, however, in the gaseous emissions of EV. Two possible reasons, which could act collectively, have been proposed for the lack of H_2S: (1) H_2S is being oxidized by reaction with O_2 from ambient air at the magma–air interface (Kyle et al. 1990, 1994; Sweeney et al. 2008); or (2) H_2S degasses at greater depth and higher pressure in the magmatic system, and converts to SO_2 during transport to the surface (e.g. de Moor et al. 2013 on the reduced lava lake of Erta Ale volcano, Ethiopia).

Infrared spectroscopy of magma degassing. An important development in the interpretation and understanding of degassing processes at EV has been the introduction of infrared spectroscopy. Infrared spectroscopic techniques now reliably measure CO, CO_2, OCS, H_2O, SO_2, HCl and HF, and have thus provided a wealth of new data for investigating degassing processes in the EV conduit system.

Wardell et al. (2004) reported the first measurements of CO and CO_2 flux from EV. These novel measurements were made by airborne profiling of CO_2 abundances in the plume using an infrared analyser plumbed to an outside inlet attached to the helicopter's nose antenna. CO was measured via open-path Fourier transform infrared spectroscopy (OP-FTIR) using an infrared lamp sited on the crater rim as a thermal source with the FTIR spectrometer located at various distances (c. $\geq$100 m) down the flank. Significant results of this study are:

- The CO_2 flux estimated by airborne profiling of CO_2 abundances (Wardell et al. 2003, 2004) was calculated to be 22.3 $\pm$ 1.0 kg s^{-1} (c. 1900 Mg/day) from the average of 3 years of measurements: 21.4 kg s^{-1} for 18 December 1997, 23.3 kg s^{-1} for 17 December 1999 and 22.2 kg s^{-1} for 16 January 2001.
- Results from the FTIR measurements collected in December 1995 yield a range for the CO/SO_2 molar ratio from 2.51 to 3.28, with an average of 2.96 $\pm$ 10%. Using a corresponding unpublished SO_2 flux from EV as measured by COSPEC of 0.58 kg s^{-1} (50 Mg/day), Wardell et al. (2004) calculated a CO flux of 1.74 kg s^{-1} (150 Mg/day).

All subsequent OP-FTIR measurements have used the Erebus lava lake as the thermal source. The maximum OP-FTIR signal was reasonably interpreted to be the hottest part of the lava lake (Oppenheimer and Kyle 2008b; Oppenheimer et al. 2009, 2011; Burgisser et al. 2012; Ilanko et al. 2015). The following results are highlights from these studies:

- From measurements in December 2004, Oppenheimer and Kyle (2008b) reported measurements of H_2O, CO_2, CO, SO_2, HF, HCl and OCS (in order of molar abundance) in the gas column rising from the lava lake. Volcanogenic CH_4, H_2S, NO, NO_2, HBr or SiF_4, however, were not detected. Based on OP-FTIR's detection limit for H_2S, the abundance of SO_2 exceeds that of H_2S by a factor of at least 50. As discussed above, this is an important observation given the reduced nature of the Erebus phonolitic lava lake (Oppenheimer and Kyle 2008b).
- Using independent SO_2 flux determinations via DOAS and measured gas ratios, Oppenheimer and Kyle (2008b) reported the first fluxes of H_2O and carbonyl sulfide (c. 0.5 Mg day). These SO_2-derived fluxes, given as mass but listed in order of molar abundance, are: H_2O (c. 860 Mg/day), CO_2 (1330 Mg/day), CO (54 Mg/day), SO_2 (74 Mg/day), HCl (21 Mg/day), HF (21 Mg/day) and OCS (c. 0.5 Mg/day) for a total flux of 2360 Mg/day (Oppenheimer and Kyle 2008b). This shows that while H_2O has the highest emission in terms of molar abundance, by mass CO_2 is the dominant component of the Erebus gas plume (Oppenheimer and Kyle 2008b).
- During the December 2004 gas measurements at EV, two lava lakes (the periodic Werner's lava lake and the long-lived Ray's lava lake) were present. The gas compositions of the individual plumes emitted by the two lava lakes are distinct (particularly the H_2O/CO_2 ratio and HF) and point to a more 'evolved' gas being released from Werner's lava lake. This difference was interpreted to suggest that Werner's lava lake is fed by a shallow offshoot of the conduit that supplies Ray's lava lake, or that magma feeding Werner's lava lake is more degassed due to a higher degree of crystallization (Oppenheimer and Kyle 2008b).
- Focusing on the passive degassing regime from the data collected in December 2004, Oppenheimer et al. (2009) identified significant oscillatory variations in gas composition during passive degassing with cycles of 4–15 min (Oppenheimer et al. 2009).
- In December 2005, Oppenheimer et al. (2010) conducted concurrent gas measurements using infrared spectrometers sited both on the crater rim and out to c. 56 km downwind, using a Twin Otter aircraft. Based on wind speeds, the calculated ages of the plume gases sampled ranged from <1 min to as long as 9 h. Three species (CO, OCS and SO_2) were measured from both air and ground. While CO and OCS were conserved in the plume, consistent with their long atmospheric lifetimes, the downwind measurements indicate a SO_2/CO ratio of c. 20% of that observed at the crater rim, suggesting rapid chemical conversion of SO_2 to H_2SO_4.
- Using measurements made in December 2005 during a period when intermittent Strombolian eruptions were occurring in the lava lake, Oppenheimer et al. (2011) found a significant difference in the CO_2/H_2O mass ratio between passive degassing (c. 1.8 by mass) and the eruption of the gas slugs during Strombolian events (c. 4.3 by mass).
- Using the December 2005–January 2006 dataset, Burgisser et al. (2012) noted large differences in the molar CO_2/CO ratios between passive degassing (c. 14.8) and Strombolian eruptions (c. 29 and 65: Fig. 16). The observation that the

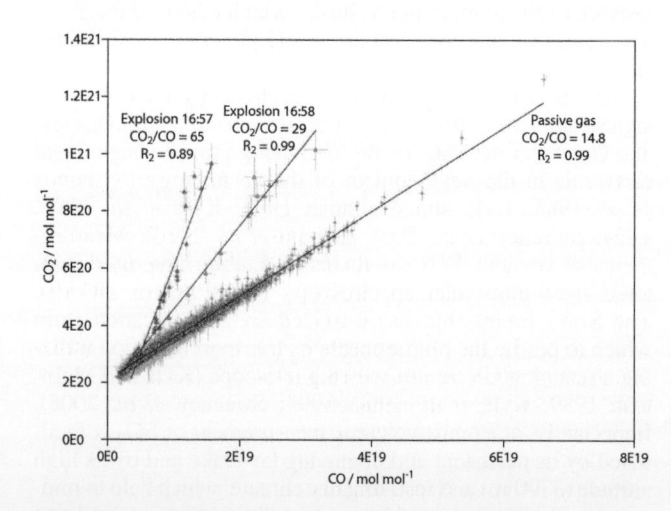

Fig. 16. Example of retrieved CO_2 v. CO immediately following a large Strombolian explosion at 16:57 UT on 13 December 2005 (red triangles) and a smaller explosion just over 1 min later (green triangles) compared with the signature of passive degassing through the lava lake (yellow circles) over the subsequent 2.5 h. Note the much higher CO_2/CO ratio for the explosively released gas but the differing CO_2/CO ratios of the two explosions. Modified from Burgisser et al. (2012).

passive degassing was uniform while the two Strombolian eruptions had very different ratios led to a model wherein adiabatic variances in degassing of phonolitic magma is causing differences in this ratio, which is in stark contrast to Oppenheimer et al. (2011) who called for Strombolian eruption gases to come from a deeper basanitic source.
- Using the OP-FTIR gas measurements from December 2004, Ilanko et al. (2015) use wavelet-based frequency analysis for a time-series gas emission measurement from the lava lake. This frequency analysis reveals a cyclic change in total gas column amount with a period of about 10 min, and a similarly phased cyclic change in proportions of volcanic gases, which they attribute to differences in pressure-dependent solubilities.
- Finally, in addition to the gas emissions from the open-conduit lava lake, degassing also occurs from the flanks of Erebus volcano through warm ground and fumarolic ice towers within and around the summit. On Erebus the fumarolic ice towers allow diffuse degassing to be visually identified (Fig. 8). Measuring the CO_2 concentration, airflow velocity and size of the exit orifice at 43 actively degassing ice towers, Wardell et al. (2004) reported CO_2 fluxes ranging from <0.0001 to 0.034 kg s^{-1}, with small patches of steaming warm ground contributing an additional 0.010 kg s^{-1}. Delta ^{13}C of the highest temperature fumaroles ranged from −4.7 to −2.1‰, suggesting a magmatic origin for the CO_2. The estimated output of flank CO_2 degassing is 0.46 kg s^{-1} (40 Mg/day). Compared with direct airborne measurements of the volcanic plume, passive flank emissions constitute less than 2% of the total volcanic CO_2 budget emitted from Erebus volcano. Subsequent isotopic measurements of these passive flank gases have been shown to be a mixture of magmatic/mantle-derived CO_2 and helium with ambient atmosphere (Fischer et al. 2013).

Melt-inclusion measurements. The study of mineral-hosted melt inclusions can place important constraints on magmatic plumbing systems through measurements of pre-eruptive volatile contents and, with the application of solubility models, determination of the depths at which inclusions were entrapped within growing crystals. Several melt-inclusion studies have targeted EV, as well as the volcanic centres on the periphery of Ross Island (e.g. Dunbar et al. 1994; Eschenbacher 1999; Seaman et al. 2006; Oppenheimer et al. 2011; Rasmussen et al. 2017). Dunbar et al. (1994) measured the concentrations of H_2O (c. 0.15 wt%), F (c. 0.28 wt%) and Cl (c. 0.18 wt%) in phonolitic melt inclusions hosted in clinopyroxene and anorthoclase, as well as the co-existing matrix glass (0.10 ± 0.08 wt% H_2O, 0.30 ± 0.02 wt% F and 0.18 ± 0.01 wt% Cl) from Erebus phonolite bombs. The striking similarity in concentrations among inclusions and matrix glass led them to conclude that crystallization occurred at shallow levels in the magmatic system. They used solubility modelling to estimate melt-inclusion entrapment pressures of less than c. 30 MPa, although the lack of data on CO_2 concentrations makes this a minimum value. Eschenbacher (1999) reported similarly low concentrations of H_2O (0.15 ± 0.04 wt%) in anorthoclase-hosted phonolitic melt inclusions, and found an average CO_2 concentration of 670 ± 120 µg g^{-1}. Volatile contents were also determined for EL olivine-hosted melt inclusions collected from Turk's Head (basanite, phonotephrite, and tephriphonolite), Inaccessible Island (tephriphonolite), and Tent Island (phonotephrite and tephriphonolite). The concentrations of H_2O (0.13 ± 0.05 wt%) and CO_2 (800 ± 500 µg g^{-1}) are similar to those measured in the Erebus inclusions, with the exception of inclusions from Inaccessible Island which contain systematically higher concentrations of H_2O (0.53 ± 0.07 wt%) and CO_2 (2400 ± 700 µg g^{-1}). Here we estimated entrapment pressures from these concentrations using the solubility model of Ghiorso and Gualda (2015), which accurately reproduces results from solubility experiments performed on an Erebus phonotephrite by Iacovino et al. (2013). Our results suggest that the Erebus crystals and those from Turks Head and Tent Island grew at pressures of c. 100 MPa, whereas those collected on Inaccessible Island crystallized at c. 200 MPa. These pressures correspond to crystallization depths of c. 4 and c. 9 km, respectively, for an assumed crustal density of 2.7 g cm^{-3}. These values represent minimum depths if the inclusions contain CO_2-rich vapour bubbles (e.g. Hartley et al. 2014; Moore et al. 2015).

Volatile concentrations in basanitic olivine-hosted melt inclusions from the peripheral volcanic centres – Mount Terror, Mount Bird and Hut Point Peninsula – contrast sharply with those from Mount Erebus. They are associated with the DVDP lineage, which is defined by the presence of kaersutite, and are much richer in H_2O and CO_2. Eschenbacher (1999) measured H_2O and CO_2 in melt inclusions recovered from depths of 283 and 295 m in the DVDP 3 drill core on Hut Point Peninsula. The DVDP 3 drill core inclusions are relatively H_2O- and CO_2-rich (c. 1.5 wt% and c. 5500 µg g^{-1}, respectively), consistent with entrapment pressures of c. 330 MPa (depths of c. 12 km), significantly deeper than inferred for the EL magmas. The CO_2 in vapour bubbles – which can account for the majority of the CO_2 in a melt inclusion – has not been accounted for in these estimates of entrapment pressures. Rasmussen et al. (2017) studied 90 olivine-hosted melt inclusions from nine DVDP lineage basanites from Hut Point Peninsula, Mount Terror and Mount Bird, and took into account the CO_2 sequestered in vapour bubbles, providing the most robust estimates for magma storage depths beneath these volcanic centres. The glasses in these inclusions contain 0.4–2.0 wt% H_2O and 0.2–0.9 wt% CO_2. The total entrapped CO_2 ranges from 0.2 to 1.8 wt%, after correction for vapour bubbles. Entrapment pressures calculated for Mount Terror range from 120 to 730 MPa, with an average of 490 ± 180 MPa, while those for Mount Bird range from 280 to 680 MPa, with an average of 500 ± 140 MPa, and those for Hut Point range from 160 to 650 MPa, with an average of 390 ± 100 MPa. This large range of calculated entrapment pressures below Mount Terror, Mount Bird and Hut Point Peninsula indicate crystallization and storage depths from c. 6 to >19 km (Rasmussen et al. 2017), with this deeper limit approaching the Moho depth for this region (19–27 km: Finotello et al. 2011).

The picture that emerges from these melt-inclusion studies suggests that magma storage beneath EV is relatively shallow (c. 4–9 km), whereas beneath the peripheral eruptive centres it is deep (c. 15–19 km). This has important implications for the volatile contents of parental magmas of the EL and DVDP lineages. It has long been held that magmas parental to both lineages are basanites that are nearly indistinguishable in composition except for their volatile contents, as suggested by the presence of amphibole in the DVDP lineage and its absence in the EL (Kyle et al. 1992). A comparison of the H_2O concentrations in the EL olivine-hosted melt inclusions (c. 0.15 wt%) with those from the DVDP lineage (c. 1.2 wt%) would seem to support this conclusion. However, it must be recognized that while these values may reflect pre-eruptive concentrations, they do not necessarily represent the initial H_2O contents of the parental magmas, as a result of processes that can occur before and after the inclusions were trapped. Melt inclusions trap samples of melt from which the olivine crystal is growing. The pressures of inclusion trapping for the Mount Terror, Mount Bird and Hut Point basanites are in the lower crust, so the H_2O content of parental magmas could have been modified during their transport from the mantle source region into the crust where they began to crystallize.

Because of the relatively high solubility of H_2O in silicate melts, the deep transport stage from mantle to crust is unlikely to have resulted in H_2O loss from the melts but other process such as vapour-saturated fractional crystallization or fluxing with relatively CO_2-rich vapour could potentially have caused H_2O changes (Rasmussen et al. 2017). After entrapment, olivine-hosted melt inclusions are physically isolated from most processes affecting the external magma but rapid diffusion of H^+ through the host olivine causes the inclusions to be susceptible to modification as the external magma degasses (e.g. Chen et al. 2011, 2013; Gaetani et al. 2012; Bucholz et al. 2013). It is probable, therefore, that the low H_2O concentrations that characterize EV lineage olivine-hosted melt inclusions reflect storage at shallow depths rather than low H_2O concentrations at the time of entrapment. The lack of kaersutite in EV lineage lavas reflects their low H_2O contents as they cooled in shallow reservoirs to temperatures at which amphibole becomes stable (c. 1050–1100°C) rather than their initial H_2O contents. A single, H_2O-rich magma was likely to be parental to both lineages.

The very low H_2O contents of phonolitic melt inclusions (Dunbar et al. 1994; Eschenbacher 1999) is interesting in light of the relatively high H_2O of basanite parental magmas. Fractional crystallization of a basanite parent to a highly evolved melt, without any loss of H_2O by degassing, should yield phonolitic melts with relatively high H_2O contents (cf. rhyolitic melts at Hekla volcano formed by extensive crystal fractionation from a basaltic parent have 3.3–6.2 wt% H_2O: Portnyagin et al. 2012). Experimental phase equilibria for Erebus magma compositions constrain the pressures of differentiation from the basanite parent to c. 100–200 MPa (Iacovino et al. 2016), at which pressure H_2O is still relatively soluble. The low H_2O contents of the phonolitic melt inclusions in anorthoclase phenocrysts thus seem to require substantial fluxing of CO_2-rich vapour through the upper-crustal storage and differentiation region (to reduce the dissolved H_2O content; this effect is also confirmed by the phase equilibria). Further modification by H^+ diffusive loss post-entrapment may also have occurred when phonolitic magma with entrained anorthoclase phenocrysts was transported and stored at shallow depths beneath and in the lava lake.

Iacovino et al. (2016) coupled the interpretation of melt-inclusion volatile concentrations and gas emissions from the lava lake using thermodynamic calculations and mixing models. As discussed in the subsection 'How do surface gas measurements inform our knowledge of subsurface processes of magma differentiation and degassing occurring within the volcanic conduit?' later in this chapter, the modelling results support the hypothesis that the surface gas emissions from the EV lava lake are dominated by degassing from the deep basanite but include significant contributions from the phonolitic lava lake.

Melt inclusions can also provide constraints on the chemical and isotopic composition of the magma at the time of crystallization. In some studies, the range of the isotopic compositions among melt inclusions significantly exceeds the isotopic variation in the volcanic suite from which the lava was collected (Saal et al. 1998); in principle, reflecting isotopic heterogeneity in the pre-aggregated melt. To evaluate this possibility for the EV and Ross Island lavas, Sims et al. (2008a) used secondary ionization mass spectrometry (SIMS) to measure the in situ Pb isotope compositions of four anorthoclase-hosted melt inclusions from a 1984 bomb, together with four olivine-hosted melt inclusions from DVDP. Their results showed that the Pb isotopes in the anorthoclase megacrysts are uniform and, within analytical uncertainties, similar to those of the host phonolitic lava. The olivine-hosted melt inclusions, however, show a large range in isotopic composition but this variability is within analytical uncertainty of the host DVDP samples.

Constraints on temperature, viscosity and oxygen fugacity of the Erebus phonolitic magma

Magma temperature. The temperature of the EV phonolitic silicate liquid has not been measured directly in the lava lake. Early indirect estimates of the lava-lake temperature were based on optical pyrometry and mineral thermometry (Kyle 1977), melt-inclusion homogenization (Dunbar et al. 1994), olivine–clinopyroxene thermometry (Caldwell and Kyle 1994), and forward-looking infrared (FLIR: Calkins et al. 2008). All these methods converged on a temperature of roughly 1000°C. Subsequent estimates from phase-equilibrium experiments on EV phonolite (Moussallam et al. 2013) gave a similar value of 925–975°C, whereas thermodynamic modelling of gas redox pairs provides a much higher estimate of 1060–1080°C (Burgisser et al. 2012).

Magma viscosity. Le Losq et al. (2015) compared the explosivity of Vesuvius and EV phonolites, and used experimental data to establish a model of phonolite viscosity that explicitly considers the effects of crystals, bubble fraction, variations of the iron redox state and CO_2 content. In the temperature range of 949–1046°C (the temperature range over which the experiments were conducted), the viscosity of the EV anhydrous and crystal-free melt varies between 10^{10} and 10^{12} Pa. Using reported ranges for temperature (950–1080°C), crystal abundance (30 vol%), H_2O concentrations (c. 0.2 wt%: Oppenheimer et al. 2011; Moussallam et al. 2013) and iron redox state, Le Losq et al. (2015) calculated that the magma viscosity in the upper region of the plumbing system of Erebus ranges between 10^5 and 10^7 Pa, with an uncertainty of, at most, ±0.45 log(Pa). We note that iron is mostly present as Fe^{2+} in the EV phonolite magma ($Fe^{3+}/[Fe^{3+} + Fe^{2+}]$ c. 0.2–0.3: Moussallam et al. 2014), whereas in their experiments, it is mostly present as Fe^{3+} ($Fe^{3+}/[Fe^{3+} + Fe^{2+}]$ >0.9; see the discussion in Le Losq et al. 2015).

Magma oxygen fugacity. Oxygen fugacity (f_{O_2}) was estimated from olivine, titanomagnetite and clinopyroxene mineral compositional data in EV bombs and lavas using the QUILF computer program (Andersen et al. 1993), and an assumed temperature of 1000°C and pressure of 1 bar. Kelly et al. (2008b) obtained a log(f_{O_2}) of −11.86 ± 0.03 (Δlog FMQ − 0.88 ± 0.03) and a silica activity of 0.461 ± 0.007 for magma in the lava lake. The presence of pyrrhotite in EV tephra is consistent with reducing conditions and has been used to constrain the sulfur fugacity in the melt (Kyle 1977). Based on gas measurements and equilibria constraints, Burgisser et al. (2012) confirmed that EV phonolitic melts are very reduced at −0.5Δlog FMQ.

In a set of phase-equilibria experiments, Iacovino et al. (2016) evaluated the P–T–X–f_{O_2} conditions of deep and intermediate magma storage. Given that the natural phase assemblages were reproduced under relatively oxidizing experimental conditions, Iacovino et al. (2016) inferred a relatively oxidizing environment for the deep and intermediate lavas. Basanites have an oxygen fugacity of +1.5Δlog FMQ at 1100°C and 450 MPa (Iacovino et al. 2016), and intermediate tephriphonolites, based on oxide pairs, have an estimated oxygen fugacity of −0.44Δlog FMQ at 1081°C and 300 MPa (Kyle et al. 1992). This is in marked contrast to the reducing conditions (−0.5Δlog FMQ) in the Erebus phonolite 'reservoir' (Moussallam et al. 2013) and (−1.22Δlog FMQ) in the Erebus phonolite lava lake, suggesting upward reduction in redox conditions. This upward-reducing redox trend is confirmed by the systematic analyses of the Fe redox state of a suite of melt inclusions at Erebus, which have conclusively shown a strong reduction trend as pressure decreases (Moussallam et al. 2014). It is posited that the large CO_2 flux proposed as the mechanism for dehydrating Erebus magmas may also contribute to the reduction in oxidation state of the shallow Erebus system by reducing H$_2$O fugacity.

Geophysical studies of Erebus volcano

The open-conduit system of EV has been observed to produce a range of eruptive styles that include ashy vent eruptions, small inner crater lava flows and at least one (1993) phreatic explosion. However, the most common and characteristic type of eruption is the large bubble-bursting lava-lake events. The existence and accessibility of a persistently active volcanic/magmatic system, and its strongly excited elastic wave field in the solid Earth and atmosphere, has made the volcano an attractive target for a range of geophysical investigations utilizing seismic, infrasonic, magnetic, gravity, magnetotelluric and geodetic methods from the earliest days of the modern investigation of EV (e.g. Dibble *et al.* 1984). During the last two decades, a succession of increasingly advanced seismic, photographic/video and infrasonic investigations have revealed increasing levels of understanding and detail regarding the volcano's internal structure and eruption dynamics.

The first seismic and infrasonic observations for monitoring and study of EV initiated in 1974 with a small network of short-period (1 s natural period) vertical-component seismographs, which soon grew into the International Mount Erebus Seismic Study (IMESS) project led by Philip Kyle at New Mexico Tech with geophysical technical assistance from Ray Dibble of Victoria University, Wellington (e.g. Dibble *et al.* 1984). During the early 2000s the IMESS instrumentation was integrated into the Mount Erebus Volcano Observatory (MEVO: Aster *et al.* 2004). MEVO incorporated additional infrasound as well as intermediate-period (30 s natural period) seismomographs and a continuous recording system. The MEVO system functioned until late 2016, when the instruments were removed at the conclusion of funding. Currently, recording of EV near-summit activity is presently being provided by a four-station temporary seismic network maintained by the Incorporated Research Institutions for Seismology (IRIS) Consortium. For many years spanning the IMESS and MEVO projects, seismicity associated with eruptive activity was valuably complemented by video records obtained from successive generations of cameras deployed at the EV crater rim.

Seismic and infrasonic records together provide for highly reliable monitoring and quantification of lava-lake explosions from EV because of their strongly impulsive and strong sonic signatures (Fig. 17). Lava-lake eruptions tend to occur at highly variable intervals, with eruptive rates observed to vary strongly on timescales of weeks to months but can occur in swarms of up to many tens or more strong events per day (e.g. Rowe *et al.* 2000). Seismic and infrasonic signals, corroborated by video, show that many of these events are relatively simple bubble bursts that sometimes span the entire (variably *c.* 5–10 m wide) lava lake. Remarkably for an active volcano, tremor-like signals are rare at EV, and later study revealed that early observed tremor episodes were probably caused by repetitive, stick–slip, tabular-iceberg collisional or grounding events occurring near Ross Island, which are seismically observable in Antarctica up to hundreds of kilometres from their sources (MacAyeal *et al.* 2008; Martin *et al.* 2010) and hydro-acoustically observable as far away as the South Pacific (Talandier *et al.* 2006). In addition, large internal seismic events or swarms are not observed (e.g. Dibble *et al.* 1984). Lack of tremor or other internal seismicity is likely to be a manifestation of the EV's open-vent conditions, which obviates the build-up of sufficient deviatoric stress or magmatic pressures within the system to generate significant and measurable seismicity.

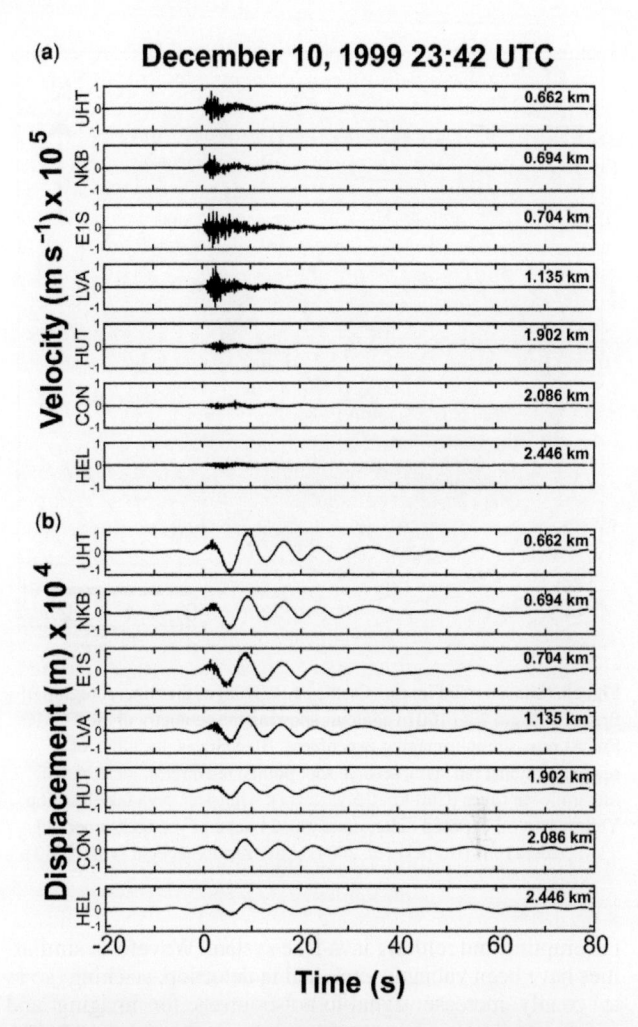

Fig. 17. (**a**) Characteristic vertical-component broadband seismograms of a lava-lake explosion observed at summit seismic stations recorded at distances ranging from 0.7 to 2.4 km from the lava lake. (**b**) Seismograms from (a), prefiltered with a 30 s corner, high-pass, zero-phase, four-pole filter and integrated to displacement. Displacement seismograms are dominated by a VLP signal that begins several seconds prior to the eruption. After Aster *et al.* (2003).

The multi-season seismic and infrasonic networks were augmented by increasingly extensive and focused deployments of portable seismographs in the late 1990s (e.g. Aster *et al.* 2005) supported by the IRIS Portable Array Seismic Studies of the Continental Lithosphere (PASSCAL) Instrument Center. Especially notable portable deployments included the first broadband instruments in December 1996–January 1997, which revealed very-long-period signals (*c.* 8–30 s) associated with lava-lake explosions (Rowe *et al.* 1998), and the large-scale Tomo Erebus imaging-focused deployments of 23 broadband and 100 short-period near-summit stations between 2007 and 2009 (Fig. 18) (Zandomeneghi *et al.* 2013).

Broadband observations of the EV lava-lake eruptive process have been particularly valuable in quantifying and understanding both the explosive decomposition signal of the emerging overpressured gas slug and the subsequent minutes-long response of the magmatic conduit as it re-establishes equilibrium during system refill. A remarkable feature of these events is that the system is nearly reset after this process concludes, so that the next lava-lake eruption may generate nearly identical seismograms and infrasonograms. This quasi-repeating elastic radiation from Erebus's lava-lake explosions is highly unusual for an eruptive source, and reflects the non-destructive and self-reconstructing nature of

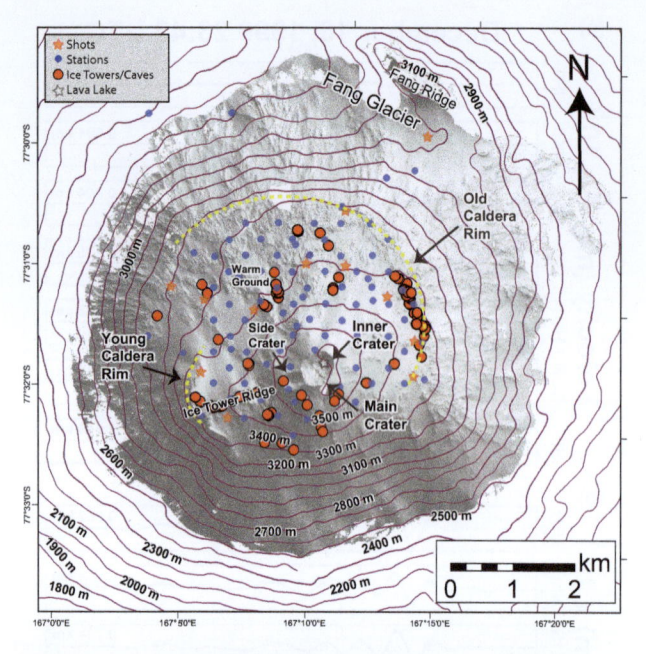

Fig. 18. Shaded relief image (Csatho et al. 2008) of the summit region of Erebus volcano with 100 m contours showing the geometry of the TOMO Erebus near-summit imaging experiment. Blue circles, locations of seismic stations; red stars, seismic shot points; red circles, locations of fumarolic ice caves (Curtis and Kyle 2011); white star, lava-lake location. Yellow curves, exposed older, outer (80–24 ka) and younger, inner (11–9 ka) caldera rims (Harpel et al. 2004). After Zandomeneghi et al. (2013).

the erupting and refilling lava-lake system. Waveform similarities have been valuably exploited in detection, stacking (so as to greatly increase signal-to-noise levels for imaging and source studies) and to reveal temporal changes within the magmatic system.

Lava-lake eruptions are created by the explosive decompression of a large rising gas slug, or slugs, at the surface of the terminal conduit. The eruption mechanism was studied in detail by Gerst et al. (2013), who utilized Doppler radar, seismology, infrasound and video data to constrain the mass and energy balance of the system, and its energy budget as a function of time for kinetic, dissipative, surface, infrasonic and seismic energy terms for 'Type I' eruptions. These observations showed clear lava-lake spanning inflation prior to bursting. A second class of more impulsive 'Type II' eruptions that burst through the lake surface without coherent inflation of the entire lava lake (perhaps because of strength weaknesses at the top of the lake along convective cell boundaries) were noted but not analysed further in this study (see Fig. 19). The modelling in Gerst et al. (2013) invoked a spherically expanding and, subsequently, bursting shell of magma with a basal diameter equal to that of the lava lake. This study, as well as independent infrasonic estimates (Johnson et al. 2008), indicated a bubble overpressure near 400 kPa at the initiation of the eruption, which declines to near 100 kPa at the time of bursting, and typical gas volumes at local temperatures and pressures inside the bubble immediately prior to bursting of 1000–2000 m^3. A commonly observed inflection in bubble-expansion velocity measured by radar was interpreted as splitting of the rising slug upon entering an expanded terminus from a narrower subsurface conduit.

The relatively geometrically simple and impulsive lava-lake eruption process has also made the volcano a target for infrasound study. Infrasound can be particularly diagnostic of surface or near-surface explosive sources because of the strong radiation of such sources into the infrasonic wave field and a relatively simple path function (i.e. Green's function) within the atmosphere, particularly at short ranges. Infrasonic propagation becomes much more complex and subject to temporal variations from atmospheric effects beyond a few kilometres. Infrasonic data have been primarily collected using nearby (<2 km) infrasonic microphones (Dibble et al. 1984; Johnson et al. 2003, 2004, 2008; Johnson and Aster 2005; Jones et al. 2008) but eruptions are also observed regularly at the Comprehensive Test Ban Treaty Organization infrasonic array at Windless Bight on the Ross Ice Shelf (Dabrowa et al. 2014). Although eruptive detection is reliable at Windless Bight, signals must propagate to that site across a distance of c. 25 km and an altitude change of nearly 4000 m, and infrasound propagation can thus be strongly affected by variable atmospheric wind and temperature conditions. Johnson and Aster (2005) noted that the relative unconstrained surface explosion of lava-lake events produced a high and consistent volcano acoustic–seismic elastic wave energy ratio (near 8) relative to that observed at other volcano systems (e.g. Karymsky) with more restricted terminal conduit systems. Johnson et al. (2008) noted that the eruptions were well modelled by a monopole source with a smaller component of dipole acoustic radiation orientated by jetting asymmetries. Jones et al. (2008) further demonstrated the ability of a three-station infrasonic near-summit infrasonic network to monitor activity at multiple inner-crater vents and to locate source locations within the lava lake with accuracies as high as a few metres.

Seismology also has significantly contributed to the present state of knowledge and conceptualization of EV eruptive sources and structure. The volcano is constituted of heterogeneous lava, ash and other features of varying ages with an embedded magmatic system. This creates a complex and seismically contrasting structure that can confound classical seismological analytical methods premised upon direct ray paths and clear seismic phases. In contrast, the path influence at many volcanoes for high-frequency (above c. 1 Hz) seismic signals is commonly very strong, so that multipath, rather than direct-ray path, energy may predominate. At EV, this seismic multipathing is extreme (Chaput et al. 2015a, b), such that seismograms at higher frequencies are dominated by strong scattered waves. This multipathing greatly reduces the amplitude of the initial (P-wave) arrival and creates a highly extended (up to tens of seconds) 'coda' of scattered waves with a cigar-shaped amplitude envelope after even just a few kilometres of propagation (Fig. 20). Strong multipath propagation obscure source information, and icequakes, artificial explosions and eruptions produce grossly similar high-frequency seismograms at EV.

A breakthrough in studying the lava-lake eruption source and plumbing of the upper conduit system was the recognition (initially in Rowe et al. 1998, and subsequently expanded in Aster et al. 2003) of highly repeatable very-long-period (VLP) signals that were not strongly affected by volcano structure, and thus could be analysed to reveal aspects of the conduit response prior to and following eruptions (Fig. 21). EV VLP signals have a period content between approximately 8 and 30 s, are much less affected by the strong structural heterogeneity due to their very long wavelengths, and provide unique constraints on magma- and gas-generated source forces and processes in this period range. VLP signals initiate a few seconds prior to the lava-lake surface explosion during the final expansion of the gas slug and bulging of the lava lake prior to rupture but are predominantly a post-eruptive phenomenon that persists for up to several minutes and ceases after the lava-lake refills from a deeper magmatic reservoir.

Aster et al. (2008) applied moment rate tensor inversion for the VLP source process using stacked similar seismograms from multiple eruptions to reduce microseismic and other noise levels. This inversion determined that VLP sources

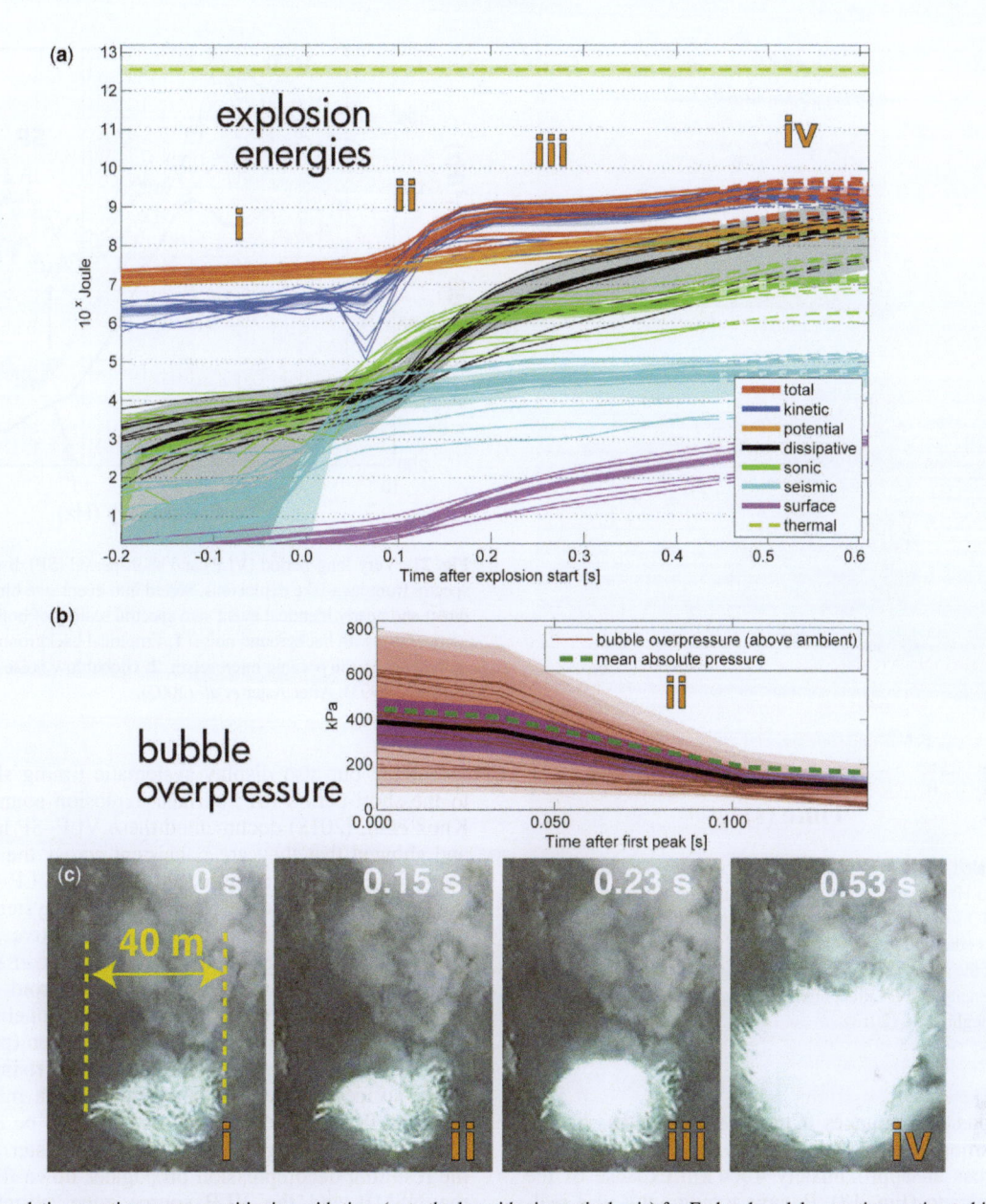

Fig. 19. (a) Cumulative eruptive energy partitioning with time (note the logarithmic vertical axis) for Erebus lava lake eruptions estimated in the eruptive modelling of Gerst et al. (2013) for radially accelerating and bursting gas slug bubbles at the surface of the lava lake (classified therein as 'Type I' eruptions) constrained by Doppler radar observations. Roman numerals refer to eruption episodes (i: pre-emptive; ii: initial bulging; iii: hemispherical bubble expansion; iv: post-bursting). A second identified event type (Type II) was characterized by abrupt explosions that did not exhibit significant lake surface expansion. This behaviour was attributed to premature bubble ruptures due to localized weaknesses in the convecting lava lake. Total energy (red) indicates the sum of all dynamic energy terms (and excludes magmatic ejecta thermal energy shown as a dashed yellow line at top). All lines become dashed after the bubble burst (episode iv), for which energy values have been extrapolated beyond the constraints of the pre-eruptive model. Bold lines show mean estimates for energy terms with standard deviations indicated by shaded regions. (b) Estimated pressure evolution (above ambient and mean absolute) for pre-eruptive expanding gas bubbles at the lava lake surface for a suite of Type I eruptions. Error ranges are shown in light red shade and the average error range is shown in purple shade straddling the mean. (c) Video stills during episodes (i–iv) of a characteristic Type I eruption recorded by a (false colour) infrared sensitive video camera sited at the Main Crater rim. Initial image corresponds to 19 December 2005 at 00:41:01 UTC) and times correspond to the timescale in (a). All figures after Gerst et al. (2013).

had a centroid location approximately 0.5 km NW of the lava lake and at a depth of approximately 0.4 km. The associated couples and forces in the moment rate source were determined as a superposition pressurization, (minor) shear component, and an inclined, approximately north-dipping, single force. The general interpretation is that they represent pressurization and advective (reaction) forces primarily associated with the post-eruptive recovery and return to gravitational equilibrium of the lava-lake system following the eruptive removal of lava-lake mass. The centroid location of the associated forces suggests the presence of an important constrictive region at this location.

A second breakthrough in understanding the shallow eruptive system has been the application of coda-wave autocorrelation (Chaput et al. 2012) to identify and localize scattering from high-impedance contrast interfaces within the volcano. Combining this scattering tomography with P-wave active artificial explosion tomography and a large number of stations deployed in the near-summit region during the Tomo Erebus project, a cogent image of the uppermost c. 2 km of the volcano was obtained (Fig. 18) (Zandomeneghi et al. 2013).

The near-summit model described in Zandomeneghi et al. (2013) is constrained by P-wave arrival times, and by

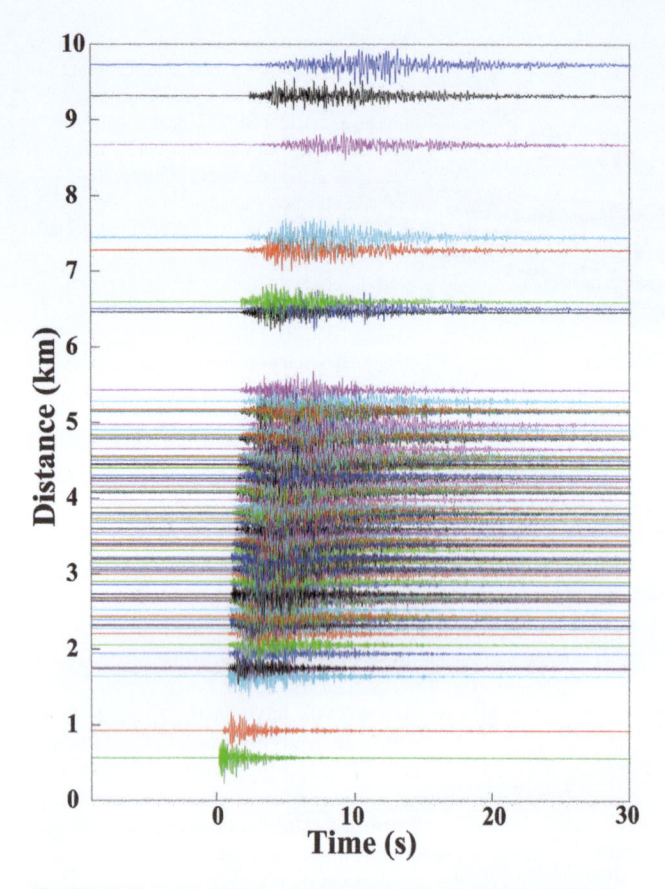

Fig. 20. Self-scaled vertical-component velocity seismogram (high-pass filtered above 0.5 Hz) record section from a shot at Fang Glacier recorded during the TOMO Erebus experiment. Note the rapid development of an extended signal coda (within a few kilometres of the source) and the diminishment of the first (P-wave) arrival amplitude. This reflects the high internal heterogeneity and resultant high-frequency seismic scattering. After Zandomeneghi et al. (2013).

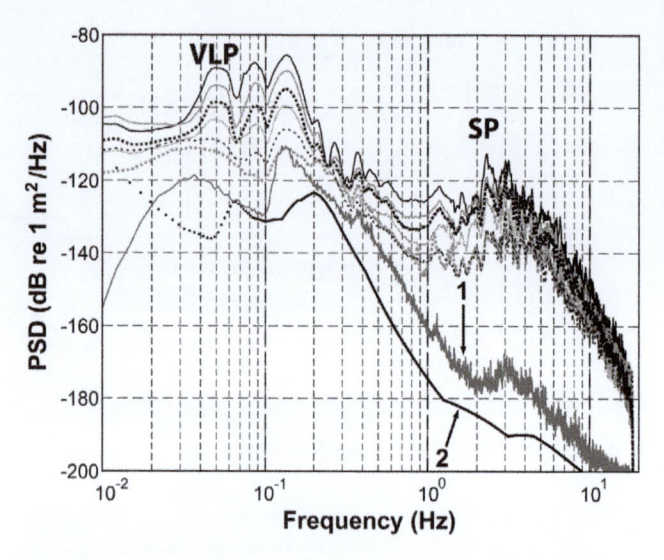

Fig. 21. Very-long-period (VLP) and short-period (SP) displacement spectra from lava-lake explosions, sorted into event size bins, showing direct and nearly identical event size spectral scaling of both signals, and comparison with background noise. **1**, Empirical background noise level, dominated by the oceanic microseism. **2**, Global low noise model of Peterson (1993). After Aster et al. (2003).

autocorrelations of icequakes (Chaput et al. 2015a, b) and eruption seismograms recorded at 91 stations that were deployed across an approximately 4 × 4 km expanse of the uppermost volcano (Fig. 18). P-wave tomography was realized by these instruments coupled with 12 prepared ice borehole shots. A P-wave-tomography code (Toomey et al. 1994) suitable for accommodating the large velocity contrasts expected in magmatic systems was used to model the seismic observations. Prominent features were interpreted to represent regions of fumarolic ice-cave hydrothermal systems, a prominent radial dyke, and a near-summit magma body (NSMB) residing 500–1000 m NW of the lava lake and extending to depths in excess of 500 m below the surface. The NSMB underlies the prominent Tramway Ridge warm-ground region of the summit plateau area of high surface temperatures and gas emissions. The terminal conduit was not resolved in this study, which indicates that it is likely to be narrower than approximately 10 m in its uppermost few hundred metres. If the conduit extends to the nearest edge of the NSMB and the VLP source region, then it would have an inclination of approximately 60° and a length of approximately 500 m, although there is no reason to assume a simple upper-conduit geometry. Indeed, the presence of additional sporadically active vents within the inner crater implies a branching or otherwise complex terminal geometry in at least the uppermost few hundreds of metres below the inner crater (Figs 21–23).

The VLP seismograms associated with eruptions, observed by broadband seismic stations near the summit, are highly repeatable but also display systematic timing shifts relative to the short-period (SP) surface explosion source (Fig. 24). Knox et al. (2018) documented these VLP–SP lag variations and showed that they are consistent across the seismic network (and are thus a source effect). The VLP–SP observations were then used to refine a general system model for the uppermost conduit and lava-lake eruptive system. The essence of the model put forward in Knox et al. (2018) is that the VLP moment tensor source centroid represents a zone of important constriction within the uppermost conduit system. Gas slugs pass through this restriction (perhaps accumulating behind it) aseismically, buoyantly rising to explode at the surface of the lava lake. Once ejecta mass has been removed from the conduit tip (estimated to be c. 8×10^7 kg for conduit-spanning gas slug eruptions: Aster et al. 2008), the resulting decompression propagates down the conduit to the NSMB and the VLP source zone, stimulating rapid (on the order of several minutes) refilling of the system from the deeper reservoir. Knox et al. (2018) investigated several types of elastic processes for facilitating the VLP–SP elastodynamic communication through a magma-filled conduit (Fig. 25). They concluded that an elastic Stoneley (boundary) wave was consistent with VLP–SP timing observations, and that small changes in the effective width of the conduit can explain the observed lag variations, provided that the magma kinematic fluid viscosity is sufficiently large (10^4–10^5 Pa: Dibble et al. 1984; Le Losq et al. 2015) and the conduit is, on average, sufficiently narrow (4–8 m) over several hundreds of metres, consistent with its non-detection in near-summit seismic tomography (Fig. 26).

Seismic imaging and seismogram-analysis-based conclusions of a narrow conduit and the prominent presence of the NSMB just to the NW of the EV crater is also consistent with near-summit gravimetric observations made at a spacing of 100–200 m. Gravimetric lows were observed to correlate with low-seismic-velocity anomalies, including an 8 mGal Bouguer minimum above the NW magma body (Mickus and Kyle 2013, 2014).

EV and Ross Island have also been noted to be promising locales for magnetotelluric (MT) imaging of Earth's conductivity structure, including the imaging of magmatic structures, as a complement to seismology. A recent collection of MT

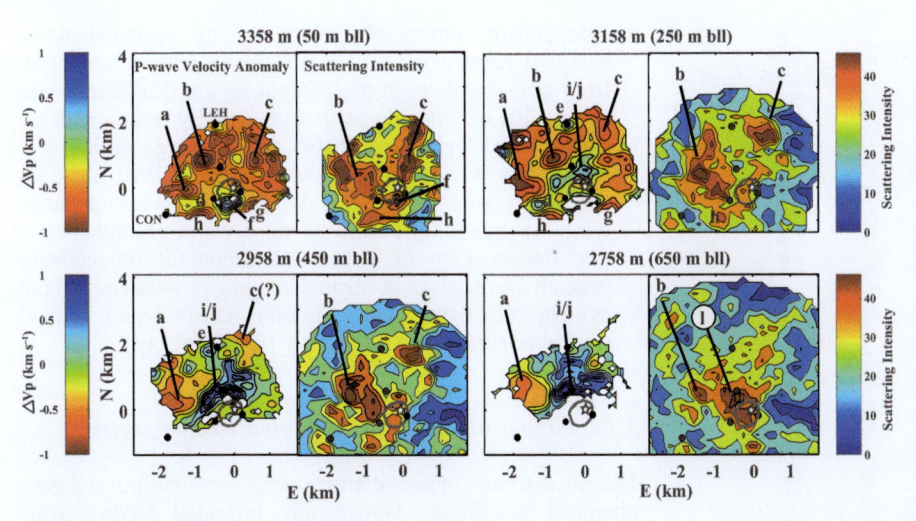

Fig. 22. Four depth slices showing the P-wave velocity model (left panels) and scattering intensity (Chaput *et al.* 2012). Each pair of images is referenced to its depth below the lava lake (bll). Grey contour, crater rim; white star, lava-lake location. Labelling indicates the near-summit features discussed in Zandomeneghi *et al.* (2013). Of particular note is the high scattering and very-low-velocity region (b) centred approximately 1.25 km NW of the crater (indicated by the black contour), which is interpreted as a near-summit magma body (NSMB). The lack of low velocities directly below the crater supports the hypothesis of a thin, inclined or otherwise tortuous, summit conduit that connects the lava lake to the NSMB. After Zandomeneghi *et al.* (2013).

instruments across the summit region and the island is currently being analysed, and early results indicate that these data will provide new insights into presently unimaged magmatic structures in the crust and uppermost mantle (Hill *et al.* 2017; Wannamaker *et al.* 2020).

Eruption-associated geophysical highlights are:

- Lava-lake eruption frequency is highly variable from year to year, suggesting a complex and evolving near-vent plumbing system that affects the formation and delivery of the large gas slugs that drive this eruptive style (Rowe *et al.* 2000; Knox *et al.* 2018).
- Lack of internal transient or tremor seismicity reflects a highly steady-state, open-vent magmatic system and indicates a lack of deviatoric stress or pressure accumulation within the volcano.
- Lava-lake eruptions produce two distinct types of seismic signals from two different processes: a SP (1–8 Hz) signal associated with the surface gas slug explosion and a VLP (8–30 s) signal associated with pre- and post-eruptive gas and magma transport (Rowe *et al.* 1998, 2000; Aster *et al.* 2003).
- The lava-lake eruption system is nearly self-reconstructing, in that the system returns to nearly the same state following lava-lake refill. This situation produces highly repeating infrasonic and seismic signals from eruption to eruption. The passage of the conduit-spanning gas slugs through the uppermost conduit is thus generally non-destructive to the terminal conduit (Aster *et al.* 2003; Knox *et al.* 2018). These highly repetitive signals have proven valuable for improving signal-to-noise in studying the eruptive process and for assessing progressive changes within the volcano.
- Doppler radar, infrasound and seismic integrative study of lava-lake spanning eruptions established an energy budget for the process, and determined that eruptive slugs have typical overpressures between 100 and 600 kPa at the onset of explosion (Gerst *et al.* 2013).
- Moment tensor inversion of VLP post-eruptive signals indicate that the force centroid for these events is about 400 m below the lava lake and offset to the northwestern quadrant of the summit plateau, with an inclined single force component that suggest a northwesterly inclined near-summit conduit (Aster *et al.* 2008).
- Repeating SP and VLP signals from numbers of eruptions show evolving relative timing shifts on the order of ±1 s, consistent with progressive small-scale conduit widening and narrowing occurring on day to week timescales

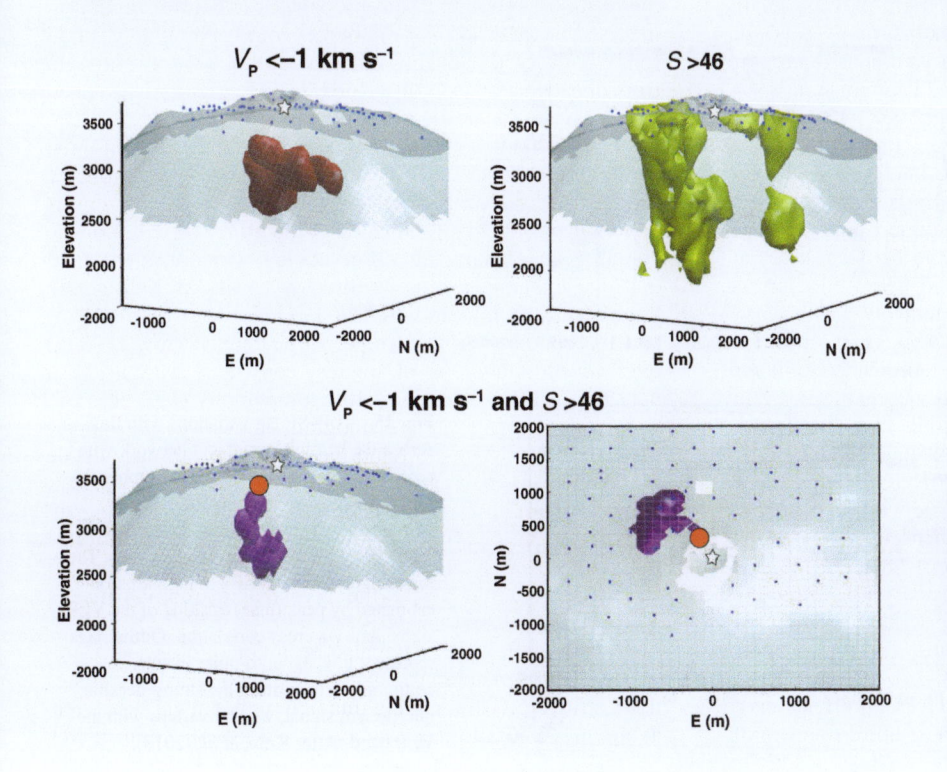

Fig. 23. Isosurfaces showing regions of very low P-wave velocity (top, left) and high scattering coefficient (*S*) (top, right). The intersection of these regions is interpreted as a shallow magma body with a centre *c.* 1 km NW of the crater and lying beneath the Tramway Ridge warm-ground region of the summit plateau. White star, lava-lake location; red sphere, location of the VLP moment tensor centroid from Aster *et al.* (2008). The joint (intersection) volume is shown in the lower panels in cross-section and plan view. After Zandomeneghi *et al.* (2013).

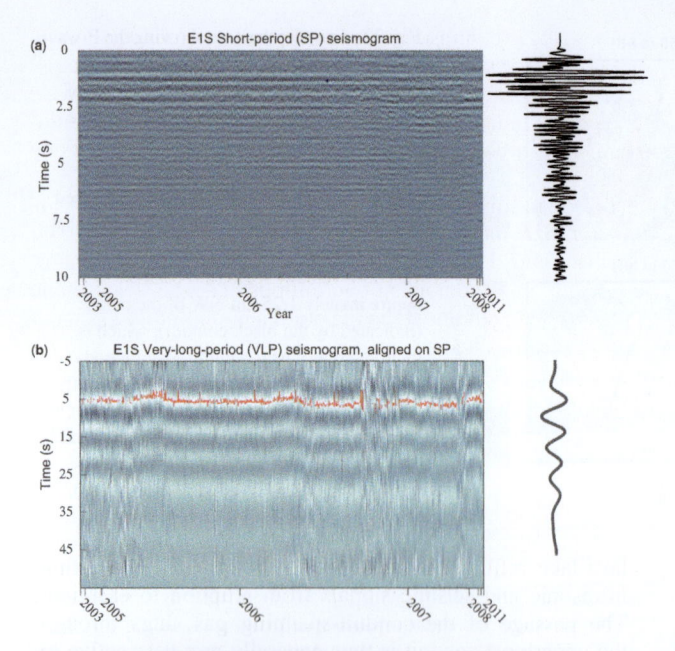

Fig. 24. Temporally progressive VLP–SP lag variations in vertical-component lava-lake eruption seismograms between 2003 and 2011 at near-summit station E1S. (**a**) Correlation-aligned SP signals (1–8 Hz). (**b**) Identically aligned VLP signals (0.03–0.2 Hz) showing systematic timing variations of approximately ±1 s. After Knox *et al.* (2018).

within a near-summit conduit that includes constricted segments with cross-sections as narrow as 4–8 m (Knox *et al.* 2018).

Geophysical imaging highlights are:

- The internal structure of the volcano is extremely seismically and structurally heterogeneous, consistent with lava flows, ash deposits and an intersecting active magmatic system (Chaput *et al.* 2012).

- Independent tomography results using scattered wave energy (Chaput *et al.* 2012, 2015*a*) and P-wave tomography from artificial shots in the summit region (Zandomeneghi *et al.* 2013), corroborated with gravity measurements (Mickus and Kyle 2013, 2014), together suggest the presence of a low-velocity, low-density, NSMB residing beneath the summit plateau, NW of the lava lake and underlying the prominent Tramway Ridge warm-ground area.
- The non-imaging of a prominent conduit immediately beneath the lava lake in seismic imaging provides additional evidence that the terminal conduit is narrow within its final several hundred metres (e.g. <*c*. 10 m in places).

Microbiological studies of the Erebus volcanic system

Despite extremes in temperature, oxygen availability and geochemical conditions, volcanically impacted environments such as hot springs, thermally heated soils and fumaroles host unique, diverse and numerically abundant microbial communities that are distinct in composition from those of non-volcanic areas (Boyd *et al.* 2013; Colman *et al.* 2016; Power *et al.* 2018). Microorganisms inhabiting volcanic environments are often supported by chemical energy (chemosynthesis), rather than light energy (photosynthesis). This is particularly true for microorganisms that inhabit hot springs with temperatures >70°C (Brock 1967; Boyd *et al.* 2010, 2012; Cox *et al.* 2011; Hamilton *et al.* 2012), regardless of their location globally (Brock and Brock 1969; Castenholz 1969). Sources of chemical energy that can support microbial growth in such hot-spring environments come in the form of disequilibrium in oxidation–reduction reactions (Amend and Shock 2001; Shock *et al.* 2010) generated by mixing of reduced volcanic fluids with oxidized atmospheric fluids (Colman *et al.* 2019*a*, *b*). Volcanically influenced fluids are often enriched in reduced compounds such as hydrogen, methane and a variety of sulfur compounds that can serve as electron donors for microbial metabolism, whereas atmospheric-influenced fluids are often enriched in oxidized compounds such as oxygen or nitrate that can serve as electron

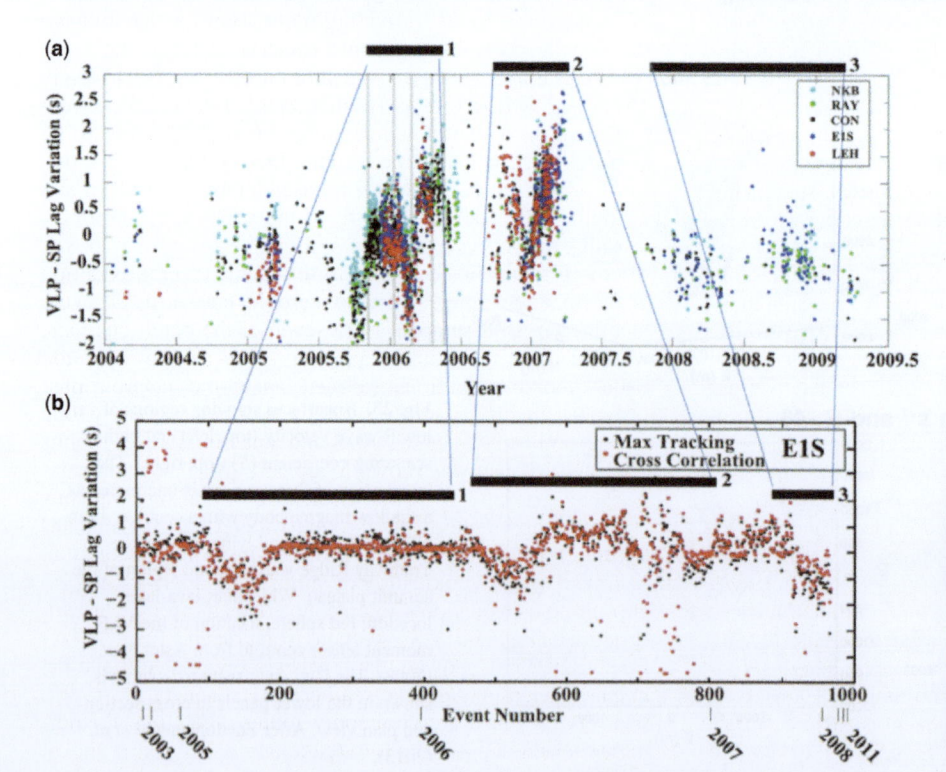

Fig. 25. VLP–SP lag variations with time across the summit broadband network with respect to time (**a**) and by event number (**b**) for high signal-to-noise station E1S. Corresponding time periods of change are represented by the numbered black bars. The lower figure shows VLP–SP lag variation estimated by peak (max) tracking of the VLP signal and via cross-correlation. Outlier lags represent low signal-to-noise events, commonly arising from the strong oceanic microseism signal, which overlaps with the VLP band. After Knox *et al.* (2018).

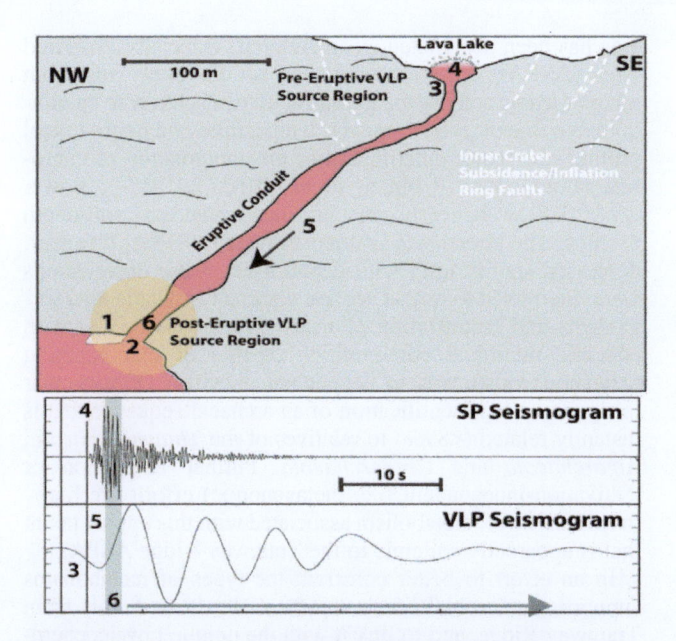

Fig. 26. Conceptualization of lava-lake eruptions and the terminal Erebus conduit system (after Knox *et al.* 2018), based on tomographic results and eruption seismogram interpretation. Eruptive events (1–6) are shown referenced to the SP and VLP seismograms below. **1**, Seismically unobserved pre-eruptive storage in, or transport of, eruptive gas slugs to the NSMB (depicted in the lower left); **2**, buoyant ascent of the gas slug into and through the (inclined and, possibly, geometrically complex) conduit; **3**, gas slug emergence into and bulging of the lava lake; pre-eruptive VLP signal; **4**, surface explosion, mass ejection and SP signal origin; **5**, elastic communication of surface mass removal back into the conduit system; **6**, post-eruptive and prolonged VLP signal (source region indicated) due to oscillatory conduit recharge. Because both the surface (SP) explosion and deeper (VLP) seismic processes and seismograms are similar, correlation or peak tracking produces robust measures of the communication time in process (5). Lengthened communication time is interpreted as indicating increasing construction of the conduit due to internal (i.e. spalling, freezing or thawing of the conduit wall) or external (i.e. collapse of the inner crater) processes.

acceptors. Below, we review existing studies of the EV geothermal system as a unique habitat for microbial life. For a recent review of microbiological geothermal habitats across the Antarctic continent, the reader is referred to Herbold *et al.* (2014b).

Early studies of the EV system by Ugolini and Starkey (Ugolini and Starkey 1966; Ugolini 1967) extended the paradigm of the importance of chemosynthesis in supporting microorganisms in higher-temperature Antarctic hydrothermal systems. Importantly, unlike most other global hydrothermal systems, the EV chemosynthetic communities are largely supplied by water in the form of steam (Broady 1984). In these early studies, two volcanic soils were collected: one from a lower elevation where fumaroles keep snow and ice from accumulating; and the other from near EV's crater. Neither of the soils sampled were found to host extensive macrobiological life, probably due to extreme acidity (pH 2.7 and 4.2) and extensive mineralization of the soils (Ugolini and Starkey 1966). The acidity in these two soils is likely to have developed through the oxygen-dependent oxidation of hydrogen sulfide or native sulfur (Colman *et al.* 2018), the latter of which was enriched in both soil types (Ugolini 1967). Early efforts to characterize microbial communities in these samples used traditional cultivation-based methods, relying on nutrient agar that would allow for the growth and differentiation of bacteria and fungi (Ugolini and Starkey 1966). Cultivatable cells were not detected in the most acidic soil, a finding that was ascribed to the extreme acidic and dry nature of the soils. In contrast, the less acidic soil was found to host up to $c.~10^6$ bacterial cells and $c.~10^4$ fungal cells. Surprisingly, morphological identification suggested that the fungi were apparently strains that are commonplace to non-volcanic environments, including *Penicillium* and *Aspergillus*, among others.

In lower-temperature volcanically influenced soils (<70°C), phototrophic metabolism is possible (Boyd *et al.* 2010, 2012; Cox *et al.* 2011; Hamilton *et al.* 2012), with eukaryotic algae tending to be the dominant phototrophs in acidic systems and cyanobacteria tending to be the dominant phototroph in moderately acidic–alkaline systems (Doemel and Brock 1970; Brock 1973; Hamilton *et al.* 2012). Broady (1984) characterized the diversity of phototrophic organisms in fumarolic soils collected at an elevation of 3500 m on EV. Intriguingly, phototrophic cultivars included both cyanobacteria and chlorophyte algae. Cyanobacteria affiliated with the genera *Mastigocladus*, *Phormidium* and *Lyngbya*, all of which are common in hydrothermal environments (Brock 1967; Castenholz 1969), were also recovered. The dominant algal genus recovered was related to *Chlorella*, which has also been detected in algal mats sampled from other lower-temperature (<30°C) hydrothermal systems (Boyd *et al.* 2009). In a separate study, a phototrophic *Scenedesmus* algal strain was isolated from a microbial mat community that was growing on fumarolic soils near Tramway Ridge (Lesser *et al.* 2002). The ambient temperature of soils where the strain was isolated was 20°C, although the temperature increased to 60.6°C at a depth of only 6 cm, underscoring the extreme thermal gradients in the Tramway Ridge fumarolic field. Intriguingly, this algal strain was shown to be particularly well adapted to stress imposed by ultraviolet light, which would be expected to be intense given the temporal and yearly variation in the UV-attenuating atmospheric ozone in Antarctica (Farman *et al.* 1985). Other studies have revealed that the microbial mats that colonize lower-temperature ($c.~20°C$) geothermally heated soils on Erebus also comprise higher-order photosynthetic plants in the form of the moss *Campylopus pyriformis* (Skotnicki *et al.* 2004).

Carbon-fixing activity of phototrophic bacteria, algae and plants is likely to provide organic forms of carbon (chemosynthate) to support secondary consumers (termed heterotrophs) in such environments, providing motivation for the study of the prevalence of heterotrophic bacteria in fumarolic soils. In higher-temperature soils, the phototrophs *Lyngbya* (Broady 1984) and *Mastigocladus* (Melick *et al.* 1991), which were shown to be active at temperatures up to 59°C and 50°C, respectively, dominated the communities. Soils were sampled for geochemical and cultivation-based analyses (Hudson and Daniel 1988). The temperature of the soils ranged from 37 to 60°C, while the pH of the soils ranged from 3.1 to 8.1. Microorganisms that could catalyse organic carbon oxidation under oxic conditions (termed aerobic heterotrophs) were identified in soils sampled from seven of the eight locations, whereas those that could catalyse organic carbon oxidation under anoxic conditions (termed anaerobic heterotrophs) were identified in six of the eight locations. Intriguingly, one of the soils in which aerobic and anaerobic heterotrophs were not detected was shown to support organisms that catalysed the oxidation of thiosulfate. The organisms used this energy to support inorganic carbon fixation (termed chemoautotrophy) (Hudson and Daniel 1988). Thiosulfate is a common intermediate in the oxidation of sulfur compounds (Xu *et al.* 2000), and its oxidation is often coupled to autotrophic growth (Friedrich *et al.* 2001). Subsequent isolation and characterization of this strain showed that it was closely related to the firmicute bacterium *Bacillus schlegelii* (Hudson *et al.* 1988), relatives of which are commonly found inhabiting geothermally heated soils on Erebus (Nicolaus *et al.* 2000) and in other geothermal habitats such as those in Yellowstone National Park, Wyoming, USA (Belly and Brock 1974). This strain was shown to also be capable of both heterotrophic and autotrophic growth. Under autotrophic conditions, hydrogen could also serve as an electron donor. Other cultivation-based studies by this same group provided evidence supporting the prevalence of *Bacillus* spp. in geothermally heated soils on Erebus, including those that are acidic (Hudson *et al.* 1989). Intriguingly, an extremely halotolerant (growth up to 4.2 M NaCl) thermotolerant (growth up to 40°C) gram-positive *Micrococcus* strain was also isolated from geothermally heated soils collected near Tramway Ridge (Nicolaus *et al.* 1992), which might be attributed to the low water content (low water activity) associated with these environments.

Evidence of thermophilic organisms in geothermally heated soils from these early studies provided impetus to apply next-generation molecular biological techniques, often in combination with cultivation-based approaches, to more comprehensively characterize the composition of microbial communities in geothermally heated soils and in ice caves. Motivation for these more recent studies was likely to have been driven, at least in part, by the recognition of the numerous attributes that EV's volcanically impacted habitats hold as analogues for early Earth and extraterrestrial environments. Attributes that are sought after for such studies include geographical isolation (lower risk of exogenous input of organisms or contaminants) and extremes in pH, salinity, water content and chemistry, as well as the extreme thermal gradients between volcanically impacted environments and their surroundings (Rothschild and Mancinelli 2001). These studies have largely been spearheaded by the laboratories of Ian McDonald and S. Craig Cary out of New Zealand, and include the first molecular characterization of mineralized soils from Tramway Ridge (Soo *et al.* 2009).

Sequencing of 16S rRNA genes and intragenic spacer regions in rRNA operons in DNA extracted from soils along a thermal and chemical gradient extending from an active fumarole at Tramway Ridge confirmed that those sampled near the active fumarole were distinct from those located more distally (Soo *et al.* 2009). Temperature gradients in sampled soils spanned 62°C near the fumarole to <5°C at more distal sites, whereas pH spanned 6.8–7.2 near the fumarole to *c.* 4.0 at more distal sites. This is likely to reflect oxidation of volcanic sulfur at lower temperatures where oxygen can fuel aerobic microbial activity (Colman *et al.* 2018). Mineralized soils sampled near the fumarole vent hosted communities that were distinct from those previously identified in volcanically hosted environments, and many were deeply branching in phylogenetic reconstructions. Many of the bacterial 16S rRNA gene sequences were distantly related to known taxa (85–93% sequence identities) and included a novel lineage that branched between the metabolically diverse and early evolving *Chloroflexi* and the candidate division OP10. Surprisingly, archaeal 16S rRNA gene sequences were closely affiliated with sequences of yet-to-be cultivated crenarchaeotes previously identified in deep subsurface environments. These observations point to the unique biological diversity associated with EV's volcanic environments, which may hold new insight into the metabolic strategies on a volcanically more active early Earth.

In a similar study, biological diversity associated with fumarolic soils from Tramway Ridge was examined along two vertical depth profiles (Herbold *et al.* 2014a). This innovative study showed that soils nearer the atmosphere–soil interface harboured microbial communities that consisted of populations which are more closely related to those identified in other geographically distinct, volcanically impacted environments than those sampled from depth. Since this ice-free area has been designated as an Antarctic Specially Protected Area (ASPA), cross-contamination of this study site from other volcanic activity via human activities is likely to be minimal to non-existent. Moreover, temperature and geochemical profiles, the major determinant of the composition of microbial communities (Colman *et al.* 2016, 2019a, b; Power *et al.* 2018), were nearly constant along the vertical, soil-depth profiles. The abundance of total extractable DNA, however, decreased significantly with depth. Thus, these observations were interpreted to reflect aeolian dispersal of viable microorganisms and colonization of near-surface soils at EV, with endemic microbial communities capable of outcompeting exogenous organisms in deeper soil profiles. Of particular intrigue was the identification of an archaeal sequence that is distantly related (<89%) to relatives of the *Thaumarchaeota*, *Aigarchaeota* and *Crenarchaeota*. Further cultivation or cultivation-independent (i.e. metagenomic) efforts are likely to reveal a novel metabolism associated with this unique taxon that is apparently endemic to the Tramway Ridge ASPA.

In an effort to better constrain the types of metabolisms supporting microbial life in geothermally heated soils from Tramway Ridge, and to link it with the unique physicochemical properties of this particular volcanic environment, Vickers *et al.* (2016) characterized edaphic qualities of soils along three vertical soil samples collected near a fumarole (60–65°C) on Tramway Ridge and subjected soil-associated populations to a battery of physiological analyses. This included an analysis of soil-moisture content and concentrations of total nitrogen, carbon and organic carbon, as well as concentrations of a variety of major and trace elements at 0–2, 2–4 and 4–8 cm vertical depths (Vickers *et al.* 2016). Surprisingly, statistical analyses showed that variations in the location of the soil cores could account for more of the variation in the edaphic attributes of soils than soil depth. This suggests substantial spatial variation in the input and availability of nutrients to support microbial metabolism in these environments, despite their close spatial proximity. Nonetheless, the availability of oxygen at depth in each of the three vertical soil profiles was low relative to atmospheric conditions, suggesting that biological communities inhabiting deeper sediments are consuming oxygen and are likely to be adapted to lower oxygen concentrations (i.e. suboxic conditions). Indeed, phenotypic assays indicate that heterotrophic oxidation of organic compounds by soil communities occurred under suboxic conditions. Like previous studies (Herbold *et al.* 2014a), sequencing of 16S rRNA genes from one of the three soil profiles revealed a community that harboured unique bacterial populations, including sequences affiliated with the GAL35 class of the candidate division OP1, as well as *Chlorflexi*, *Meiothermus* and other unclassifiable sequences. In summary, studies of EV at Tramway Ridge have shown the presence of a microbial habitat that is globally unique among studied hydrothermal systems with unusual qualities including:

- steam as the primary source of water supporting microbial communities (Broady 1984);
- volatiles are the primary source of chemical energy and nutrients supporting chemosynthetic communities in higher-temperature (>60°C) thermal transects (Hudson and Daniel 1988; Vickers *et al.* 2016);
- the presence of deeply branching archaeal and bacterial populations that are apparently endemic to this hydrothermal system (Herbold *et al.* 2014a; Vickers *et al.* 2016). Further work is needed to determine if this unique biodiversity is the consequence of geographical isolation from other thermal areas or if it is due to the unusual physical and chemical conditions at Tramway Ridge.

While most microbiologically focused studies conducted to date on Erebus have focused on geothermally heated soils in surface environments, Tebo *et al.* (2015) recently reported on microbial diversity from three fumarolic ice caves near the summit. These caves are formed by volcanic heat and gas melting overlying ice and snow. Meltwaters can then percolate through the regolith, becoming heated and turning into steam, which, in turn, can be infused by volcanic gases. Over time, these processes can culminate into a cave-like system that provides a moist and relatively warm (>0°C) habitat for microbial life, compared to the cold and dry external Antarctic environment (Curtis and Kyle 2011). Importantly, the overriding ice and snow restrict the input of light, thereby forcing microbial communities to use chemosynthesis to support microbial metabolism. Like fumarolic soils, chemical energy to support microbial metabolism is likely to come in the form of disequilibrium in oxidation–reduction reactions generated through mixing of reduced volcanic gas emissions, which have been shown to contain carbon monoxide (CO) and other reduced gases (Oppenheimer and Kyle 2008*b*; Moussallam *et al.* 2012), with oxidized atmospheric gases.

The three ice caves examined by Tebo *et al.* (2015) included Warren Cave, Harry's Dream and Hubert's Nightmare. Sediments collected from these three caves had temperatures that ranged from 0.1°C in Hubert's Nightmare to 18.5°C in Warren Cave, whereas the pH of sediments varied only slightly from 5.2 to 5.9. Like soils from fumaroles on Tramway Ridge (Vickers *et al.* 2016), the organic carbon contents of sediments collected from the ice caves were low (<126 μg g sediment^{-1}), pointing to the nutrient-limited (termed oligotrophic) nature of these isolated habitats. Nevertheless, the abundance of bacterial 16S rRNA gene templates (a proxy for cell abundance) ranged from 1.6×10^6 to 40×10^6 g sediment^{-1}, which is similar to the abundances of templates and transcripts (indicative of activity) identified in other hydrothermal habitats including a variety of springs in Yellowstone National Park (Colman *et al.* 2016, 2018; Lindsay *et al.* 2018).

The abundance of bacterial cells in sediments from nutrient-limited ice caves prompted additional sequencing of 16S rRNA genes and of the large subunit of the ribulose 1,5-bisphosphate carboxylase/oxygenase gene (*cbbL*) involved in the Calvin cycle of carbon fixation (Tebo *et al.* 2015). Harry's Dream harboured the most diverse bacterial 16S rRNA gene assemblages (archaeal 16S rRNA genes were not detected), including a variety of cyanobacterial phylotypes. This is consistent with the observation that light is only partially excluded from this particular ice-cave system, compared to the other two ice-cave habitats which are completely dark. Surprisingly, however, sediments in Harry's Dream were more limited in total organic carbon than sediments from dark cave systems, the latter of which would preclude input of photosynthetic carbon. Other abundant sequences in Harry's Dream were affiliated with putative hydrogen oxidizers and a variety of novel *Chloroflexi* and *Acidobacteria*; similar novel sequences dominated sediment bacterial communities from Warren Cave. Sediments from Hubert's Nightmare cave, despite being only 50 m from Warren Cave but with significantly colder temperatures than Warren Cave (0.1°C v. 18.5°C), harboured a taxonomically distinct bacterial community that was dominated by a novel member of the *Sphingobacteriales* bacterial order.

The taxonomic composition of *cbbL* genes (encode the large subunit of the primary enzyme involved in the Calvin cycle of CO_2 fixation) recovered from these three environments largely mirrored the taxonomic composition of 16S rRNA genes (Tebo *et al.* 2015). Cyanobacterial- and Chloroflexi-affiliated genes dominated sediment *cbbL* libraries from Harry's Dream, whereas Warren Cave sediment libraries were dominated by Chloroflexi- and Acidobacteria-affiliated *cbbL* genes. *cbbL* libraries from Hubert's Nightmare were dominated by sequences affiliated with Acidobacteria and Bacteroidetes. Similarities in the taxonomic composition of dominant 16S rRNA genes and *cbbL* genes in sediment communities point to the importance of autotrophic carbon fixation and lithogenic energy in supporting microbial communities in dark sub-ice caves on EV.

Volcanic environments have been suggested to represent refugia for organisms, both microbial and macrobial, during periods of extensive glaciation (Fraser *et al.* 2014). The isolated nature of EV, in particular the geothermally impacted caves and soils, makes these latter useful environments for the study of this potential phenomenon. A recent study of ice caves and fumarolic soils near the crater and Tramway Ridge, respectively, focused on the detection and identification of plant and animal DNA using molecular approaches (Fraser *et al.* 2018). This included an examination of recovered 28S rRNA and intragenic spacer-region gene sequences from sediments from Warren Cave and Harry's Dream, in addition to a cave termed '22 Blue'. From this study, a variety of eukaryotic sequences were identified, including those affiliated with mosses, animals and algae. At '22 Blue', a variety of non-human-associated algal and arthropod sequences without representation in current databases were recovered. At Warren Cave, non-human-associated sequences affiliated with a variety of algae and oligochaetes (Earth worms) were recovered, whereas algae were the dominant non-human-associated sequences recovered from environmental DNA at Harry's Dream. It remains unclear if these environments served as refugia for these organisms or if this DNA was exogenously introduced into the system. In support of the former possibility, several of the algal sequences recovered have been previously identified in exposed geothermal sites on other volcanoes and in soils at EV (Broady 1984; Skotnicki *et al.* 2004). In support of the latter possibility, a number of 28S rRNA and intragenic spacer-region gene sequences affiliated with potential human-derived contaminants in Warren Cave sediments were identified, including those affiliated with common wheat and soybean. Further work will be needed to determine the extent to which EV's geothermal habitats have served as refugia during periods of extensive glaciation. Past studies have revealed several unique aspects of EV's ice-cave system, including:

- An isolated, diverse and comparatively abundant microbial community that is supported by chemoautotrophic metabolism. This is confirmed by the recovery of a diverse array of genes involved in the Calvin cycle of CO_2 fixation (Tebo *et al.* 2015). Key questions remain regarding the source of the CO_2 that supports these organisms (volcanic or atmospheric) and what the source of chemical energy is that drives CO_2 fixation. Potential sources include hydrogen and CO, among others.
- EV ice caves are useful analogues to develop life-detection techniques for application to extraterrestrial life-detection missions and for understanding the geobiochemical processes that sustained life during periods of glacial advance.

Driving forces, magmatic evolution, eruptive behaviour and potential hazards of Mount Erebus volcano

The HIMU-like isotopic signature of Erebus volcano is a well-established, important source characteristic

How do current geophysical and geochemical measurements inform about the origin of Erebus volcano's unique HIMU-

like source, and the underlying causes and driving forces behind the volcanism at Erebus volcano and Ross Island?

Hypotheses to explain the Erebus HIMU isotopic signature. The HIMU-like isotopic signature (Fig. 12) observed to be ubiquitous at Ross Island and across West Antarctica and the greater SW Pacific (e.g. Sun and Hanson 1975; Hart 1988; Hart *et al.* 1995, 1997; Rocholl *et al.* 1995; Panter *et al.* 2000, 2006; Rocchi *et al.* 2002; Finn *et al.* 2005; Sims and Hart 2006; Sims *et al.* 2008*a*) has been explained in different ways:

- Ross Island is the manifestation of a localized thermal anomaly in the mantle centred beneath EV (Kyle *et al.* 1992; Storey *et al.* 1999). Noting the large volume of magma erupted over a short period of time (c. 4520 km^3 during only c. 4 myr: Esser *et al.* 2004) and the observation that the volcanic rocks at Ross Island exhibit HIMU-like isotopic signatures akin to St Helena and Tubuai ocean island basalts (OIB), Sun and Hanson (1975) and Kyle *et al.* (1992) suggested that Ross Island, in general, and EV, in particular, are manifestations of an upwelling mantle plume.
- Noting further the widespread nature of the HIMU-like isotopic signature in the Antarctic and SW Pacific (Figs 11 & 12), Hart *et al.* (1997) proposed that a large deep plume with a HIMU signature impinged on the base of the Gondwana lithosphere prior to its break-up at c. 85 Ma and that this large plume head could have been associated with the Ferrar Large Igneous Province at c. 185 Ma.
- The diffuse alkaline magmatic province (DAMP) model of Finn *et al.* (2005) posits that the widespread regional HIMU isotopic signature can be explained alternatively, not by a mantle plume but by melting of a metasomatized lithospheric source. This metasomatism is proposed to have resulted from a sustained Gondwana subduction regime throughout the Paleozoic and into the Mesozoic. Interpretations suggest that slab detachment in the late Cretaceous caused a change in mantle flow. This change in mantle flow was the cause of the onset of Cenezoic magmatism in the region at c. 55 Ma with the metasomatized lithospheric source melting at low melt fractions. Similarly, Panter *et al.* (2006) advocated for widespread metasomatized lithosphere as the primary source for the HIMU signature found in basalts from the Antipodes, Campbell and Chatham islands.

Arguments in support of asthenospheric upwelling beneath Ross Island as the HIMU source. The combination of isotopic and trace element constraints, together with tomographic imaging, strongly indicate that the HIMU source component beneath Antarctica is old (i.e. Archean or early Proterozoic) and deriving from deep-mantle upwelling (Phillips *et al.* 2018).

(1) *Isotope systematics require the HIMU source component to be Archean–Paleoproterozoic*: The third hypothesis, the DAMP model (Finn *et al.* 2005), is readily eliminated for the EV lavas as it is based on the argument that the lithosphere was extensively metasomatized between c. 500 and c. 100 Ma within a subduction regime along the margin of Gondwana, and that this young metasomatized lithosphere later became the source for the HIMU-like signature (Finn *et al.* 2005; Panter *et al.* 2006). As noted by Sims *et al.* (2008*a*) and again by Phillips *et al.* (2018), the DAMP model only considered ^{206}Pb/^{204}Pb. U-enrichment during subduction in the Paleozoic could easily have produced the observed HIMU-like ^{206}Pb/^{204}Pb signature. The observation that all of the radiogenic isotope systems look HIMU-like, however, particularly ^{143}Nd/^{144}Nd, ^{176}Hf/^{177}Hf and ^{207}Pb/^{204}Pb, places explicit temporal constraints on the age of the source. Specifically, it dictates that the <500 myr timescale invoked by DAMP is too short to grow in the HIMU-like Nd and Hf isotopic signatures, as the half-lives of ^{147}Sm (parent of ^{143}Nd) and ^{176}Lu (parent of ^{176}Hf) are 1.06×10^{11} and 3.71×10^{10} years, respectively. Furthermore, ^{207}Pb/^{204}Pb would not be significantly influenced in <500 myr because of the low abundance of ^{235}U relative to ^{238}U on the modern Earth, and as such would not be high if the cause were metasomatism.

Given the relatively recent time window since the proposed metasomatism (<500 myr), the long half-lives of the Sm–Nd and Lu–Hf isotope systems, and the fact that all four isotope systems (Sr, Nd, Hf and Pb) are similar to the St Helena HIMU end member, we argue that it is improbable that the HIMU characteristics of Ross Island represent melting of a relatively young (<500 myr) metasomatically altered lithosphere. Rather, the HIMU mantle end member is consistent with an ancient (Archean–early Proterozoic) recycled source component that was isolated in the lower mantle (Hanyu *et al.* 2011; Weiss *et al.* 2016), thus allowing the development of the characteristic HIMU isotopic signature. Hence, we infer that the Ross Island lavas are derived from a deep-seated, long-lived mantle component that is delivered to the shallow mantle by an upwelling plume.

(2) *Ross Island trace element and isotope systematics are not arc-like*: Phillips *et al.* (2018) presented several lines of evidence that show that the chemical and isotopic compositions of the lavas and tephras from EV and the other Ross Island volcanoes are not arc-like:

- Trace element abundances and patterns of the Ross Island volcanics are not consistent with derivation of these lavas from a metasomatized lithospheric source altered by fluids from subduction processes. In subduction regimes, high field strength elements (HFSEs) are retained in downgoing slabs (Kelemen *et al.* 1993), whereas fluid mobile elements, such as La, are enriched and thus subcontinental lithosphere is expected to be depleted in HFSEs (e.g. Nb) relative to other incompatible elements. As such, ratios like Nb/La are important discriminants of slab-derived fluids related to subduction processes. In arcs, Nb/La is universally <0.5 (Gill 1981). However, as noted in Phillips *et al.* (2018), the Ross Island lavas and tephras, including those from EV, have Nb/La >1.25, indicating little involvement of a subduction-metasomatized lithospheric component (Figs 10 & 27). In contrast to the Ross Island samples, the Mount Somers Volcanic Group (Tappenden 2003), which is often considered the end-member composition for Zealandia magmas generated from subduction-related metasomatically altered lithosphere (Timm *et al.* 2010), and the Schirmacher Oasis minettes from East Antarctica (Hoch *et al.* 2001), all have Nb/La <0.5, suggesting they are either sourced from or have assimilated significant amounts of metasomatized lithospheric mantle (Fig. 27). Accordingly, the Nb/La >> 0.5 of the Ross Island volcanics was interpreted as a contraindication of their being derived by melting or assimilation of a lithospheric source altered by subduction processes (Phillips *et al.* 2018). It was also noted that the Ross Island samples from Mount Bird, Mount Terror and Hut Point Peninsula lie on a mixing curve between DMM and HIMU (Fig. 27), consistent with trends also observed for Sr, Nd, Hf and Pb isotopes (Figs 11 & 12). An interesting feature of the Erebus samples is their high Nb/La relative to lavas from its peripheral volcanoes. Phillips *et al.* (2018) interpreted this increasing Erebus Nb/La as a function of variable extents of apatite fractionation, a phase that is present in EV lavas at approximately 0.5% modal abundance (Kyle 1981; Kyle *et al.* 1992).

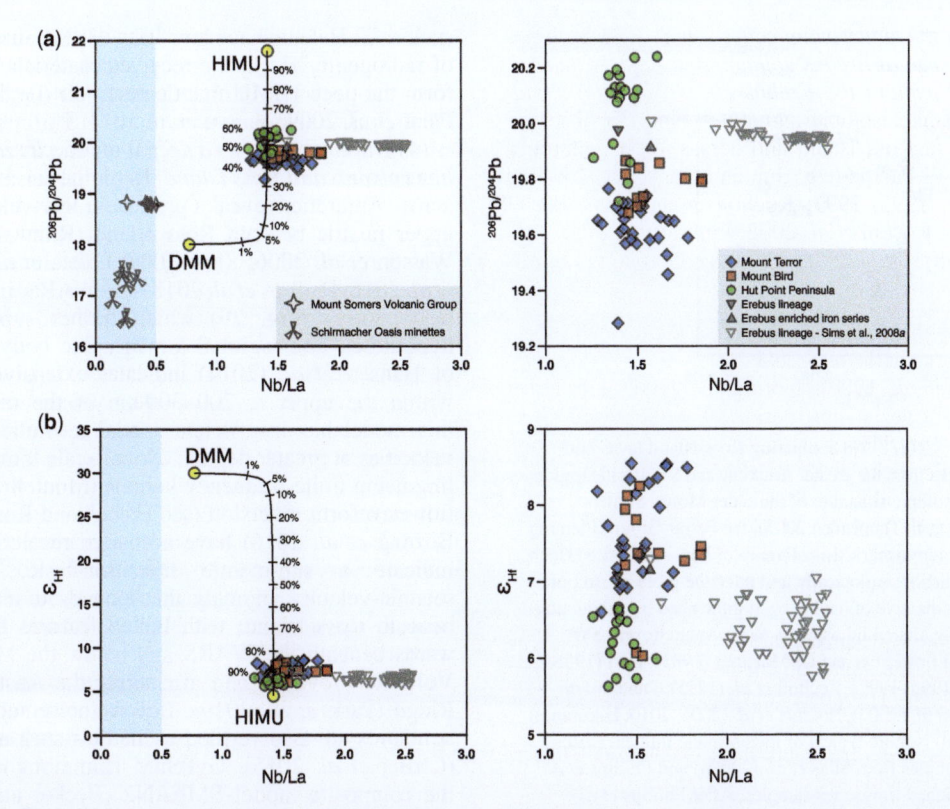

Fig. 27. Nb/La v. (**a**) (^{206}Pb/^{204}Pb) and (**b**) ε_{Hf}, showing the difference between Ross Island volcanic samples (Mount Erebus, Mount Terror, Mount Bird and Hut Point), the Mount Somers Volcanic Group (MSVG: Tappenden 2003) and the Schirmacher Oasis minettes (Hoch *et al.* 2001). Ross Island lavas and tephras lie on a mixing line between the HIMU and DMM mantle end members. Left-hand side panels show Ross Island lavas and tephras in a global context. St Helena (yellow HIMU circle: Willbold and Stracke 2006) represents the HIMU mantle end member, and depleted MORB represents the Earth's depleted mantle end member (yellow DMM circle: Workman and Hart 2005). Also shown are the mixing compositions, with marked increments, between the HIMU and DMM sources. The blue shaded region (Nb/La <0.5) represents Nb/La ratios that represent arc-related processes (Gill 1981). Right-hand side panels are close-ups showing the Ross Island samples in detail. Timm *et al.* (2010) use the MSVG as a possible composition for Zealandia lithosphere enriched by subduction. The MSVG's dissimilarity to Ross Island volcanics indicates a lack of influence from metasomatized subcontinental lithospheric mantle at Ross Island. The data are from Sims *et al.* (2008*a*). After Phillips *et al.* (2018), with permission from Elsevier.

- Long-lived radiogenic isotope compositions (Sr, Nd, Hf and Pb) of the Ross Island lavas and tephras do not resemble what would be expected for a recently metasomatically enriched subcontinental lithospheric source (Phillips *et al.* 2018). Lavas interpreted as lithospheric melts usually have more enriched isotopic signatures (lower ε_{Nd} and ε_{Hf}) than those measured at Ross Island (e.g. Daley and DePaolo 1992; Leat *et al.* 2005). Metasomatic enrichment will typically decrease Sm/Nd and Lu/Hf, leading to lower time-integrated values of ε_{Nd} and ε_{Hf} (Reid *et al.* 2012). In this regard, the older Antarctic samples, such as the minettes of the Schirmacher Oasis (*c.* 455 Ma: Hoch *et al.* 2001) and the Ferrar Group (*c.* 177 Ma: Fleming *et al.* 1995), show much more enriched and variable isotopic signatures (Fig. 28), consistent with melting of metasomatized lithosphere during Paleozoic subduction. It is essential to note that the Ross Island lavas and tephras are similar to the St Helena HIMU mantle source, which is interpreted to be derived from a mid-Atlantic deep-mantle plume.

- The absence of an arc-related metasomatic signature is not only seen in the isotopic and trace element data but also in the ^{238}U–^{230}Th systematics from EV. Arc-related metasomatism would be likely to enrich U over Th (e.g. Reubi *et al.* 2014 and references therein), resulting in high ^{238}U/^{232}Th and high ^{230}Th/^{232}Th values as a result of the decay of ^{238}U to ^{230}Th (over *c.* 500 kyr). Young EV phonolites have relatively low ^{230}Th/^{232}Th compared to global data compilations of MORB and OIB (Sims and Hart 2006; Sims *et al.* 2013*b*), and show ^{238}U/^{232}Th similar to the HIMU end member (Sims and Hart 2006); both observations indicating that the parental basanites were not derived from a source that has been enriched in U relative to Th, as would be expected from arc-related metasomatism.

- Finally, the major element compositions of the volcanic rocks from Ross Island – Mount Erebus, Mount Terror, Mount Bird and Hut Point Peninsula – are neither arc-like nor do they exhibit the anomalous chemistry attributed to melting of a metasomatically altered source. The compositions of Ross Island lavas and tephras are basanitic, or differentiates of parental basanites (e.g. phonolites), and these have long been interpreted to represent small-degree partial melts of mantle peridotite (Green and Ringwood 1967; Green 1973; see the subsection on 'What does Erebus geochemistry indicate about the depth of melting and lithology of its mantle source?'); whereas metasomatic signatures are typically manifest in unusual alkaline lavas such as lamprophyres or minettes (e.g. the minettes of the Schirmacher Oasis in East Antarctica: Hoch *et al.* 2001).

(3) *Helium isotopes for Ross Island are also similar to the St Helena end-member HIMU source*: ^{3}He/^{4}He measured in 11 olivine phenocrysts from hyaloclastites from the DVDP core at Ross Island have R/R$_a$ values (where R/R$_a$ is the ratio of ^{3}He/^{4}He in the sample (R) to ^{3}He/^{4}He in the atmosphere (R$_a$ = 1.384 × 10^{-6})) with an error-weighted mean of 6.91 (Parmelee *et al.* 2015). These values are similar to St Helena (with R/R$_a$ of 5.24–5.97: Hanyu *et al.* 2014). While high ^{3}He/^{4}He is thought to reflect an undegassed primitive mantle source because of the primordial nature of ^{3}He, lavas that represent the HIMU mantle end member at plumes,

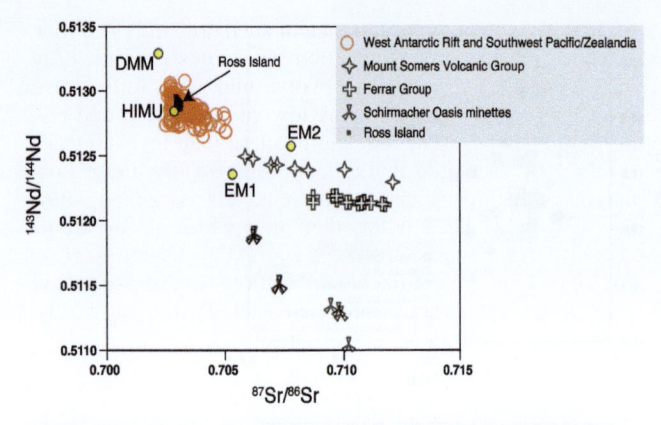

Fig. 28. ^{87}Sr/^{86}Sr v. ^{143}Nd/^{144}Nd comparing Ross Island lavas and tephras to volcanic rocks from the greater Antarctic and SW Pacific region. The highly enriched isotopic signature of the older Mount Somers Volcanic Group (data from Tappenden 2003), the Ferrar Group (Fleming *et al.* 1995) and the Schirmacher Oasis minettes of East Antarctica (Hoch *et al.* 2001) are very distinct isotopically and have the types of isotopic values that are more indicative of having been influenced by subduction-related metasomatically altered lithosphere. West Antarctic and SW Pacific/Zealandia data from Futa and Le Masurier (1983), Hart (1988 : Balleny), Hart *et al.* (1995, 1997), Rocholl *et al.* (1995), Panter *et al.* (2000, 2006), Rocchi *et al.* (2002), Hoernle *et al.* (2006, 2010: Hikurangi) and Timm *et al.* (2010) is more typical of metasomatic influence. Ross Island data include samples from Sims *et al.* (2008*a*) and Phillips *et al.* (2018), excluding enriched iron-series samples. After Phillips *et al.* (2018), with permission from Elsevier.

such as St Helena, have low ^{3}He/^{4}He because of the presence of radiogenic ^{4}He in the recycled materials that presumably form the deep HIMU mantle reservoir (Graham *et al.* 1992; Parai *et al.* 2009; Hanyu *et al.* 2014; Phillips *et al.* 2018).

(4) *The mantle shows a deep low-velocity zone beneath Erebus volcano and Ross Island*: Regional seismic data collected across Antarctica reveal a prominent low-velocity zone in the upper mantle beneath Ross Island (Ritzwoller *et al.* 2001; Watson *et al.* 2006; Zhao 2007; Gupta *et al.* 2009; Heeszel *et al.* 2016; Phillips *et al.* 2018) and WARS, implying elevated temperatures (Fig. 29) which further support the plume hypothesis. The regional tomographic body-wave inversion of Hansen *et al.* (2014) indicates extensive low velocities within the upper *c.* 200–300 km of the mantle. However, this model has insufficient spatial resolution to image low velocities at greater depths. Global-scale tomographic modelling using finite-frequency kernels (Montelli *et al.* 2006) and full-waveform inversion (see French and Romanowicz 2015; Bozdağ *et al.* 2016) have a coarser resolution but strongly indicate a sub-mantle transition-zone, low-shear-wave seismic-velocity anomaly that extends to *c.* 1200 km depth beneath Ross Island, with linked features that extend eastwards beneath the WARS and below the Marie Byrd Land Volcanic Province, and also along the Australian–Antarctic Ridge (Park *et al.* 2019). Low-velocity sub-transition-zone structures are also verified by models such as SGLOBE-rani (Chang *et al.* 2015), GyPSuM (Simmons *et al.* 2010) and the composite model SMEAN2 (Becker and Boschi 2002; Simmons *et al.* 2010; Ritsema *et al.* 2011; Auer *et al.* 2014). These global images, which, as discussed in the earlier subsection on 'Geophysical studies of Erebus Volcano', benefit from vastly expanded recent seismographic data recorded in

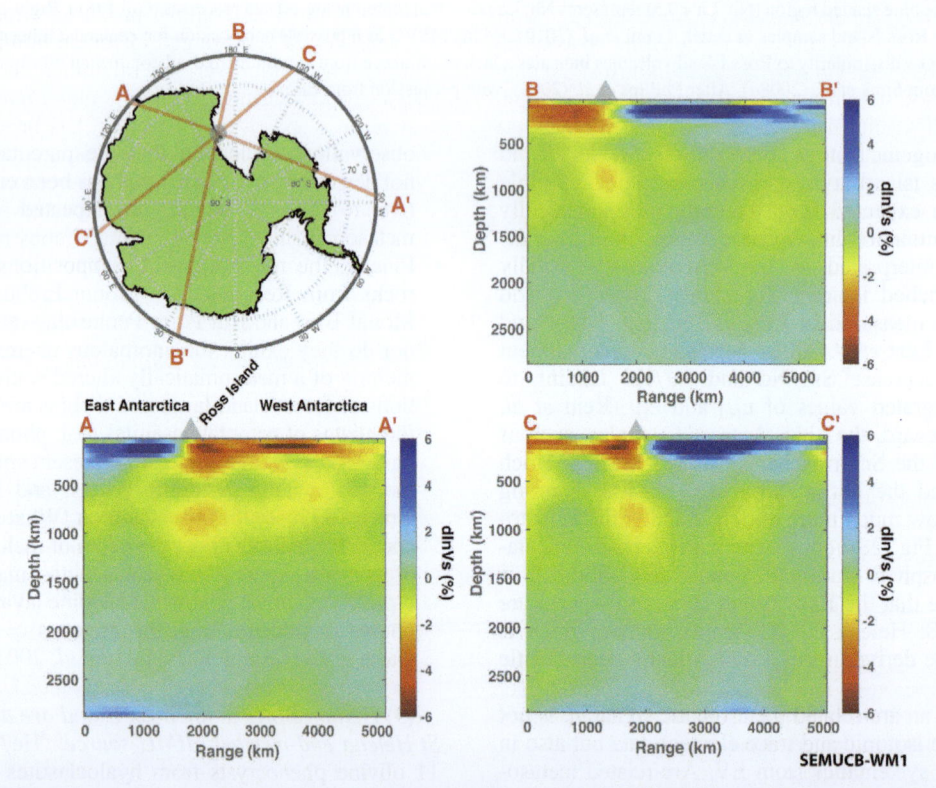

Fig. 29. Seismic tomography transects through Ross Island for three azimuths (upper left) displaying the isotropically averaged seismic shear-wave velocity anomaly from the crust to the core–mantle boundary in global velocity model SEMUCB-WM1 (French and Romanowicz 2015). This and other recent tomography models of the region indicate a low seismic shear-wave velocity anomaly that extends below the mantle transition zone to near 1200 km depth beneath Ross Island, with linked features that extend eastwards beneath the West Antarctic Rift System into the volcanic province of Marie Byrd Land. The colour scale indicates the relative perturbation of shear-wave velocity compared to the corresponding model one-dimensional depth average. SEMUCB-WM1 was calculated using full-waveform inversion of a large dataset of long-period body- (>32 s) and surface-wave (>60 s) global seismograms. Comparable lower-mantle features are also present in the full waveform inversion model GLAD-M15 of Bozdağ *et al.* (2016) and in other global models (see supplement F in Phillips *et al.* 2018). After Phillips *et al.* (2018), with permission from Elsevier.

Antarctica (e.g. Anthony et al. 2015), are consistent with the hypothesis that a disrupted deep-mantle thermal feature continues to drive melting beneath EV and Ross Island.

(5) *A similar deep HIMU-like component farther afield at the Australian–Antarctic Ridge creates a geographical problem*: As recently noted by Park et al. (2019), there is isotopic coherence among the KR-1 and KR-2 segments of the Australian–Antarctic Ridge (AAR), the Ross Island volcanics and the Hikurangi Seamounts, which is consistent with the mixing of ambient mantle with an enriched component with radiogenic Pb, Nd, Hf and Sr isotopic compositions akin to HIMU. Given the geographical setting of the AAR and Hikurangi Seamounts, there are two reasons why this isotopic signature cannot be reasonably explained by a model in which the enriched HIMU component comes from a metasomatically enriched lithospheric mantle resulting from Paleozoic subduction along the Gondwana margin (Finn et al. 2005; Panter et al. 2006). First, the Hikurangi Seamounts were erupted on the oceanic lithosphere subducting along the former Gondwana margin, and, as such, the influence of subcontinental material on the compositions of the Hikurangi volcanics would be minimal. Second, it is physically implausible that this enriched lithospheric material could be influencing mid-ocean ridge volcanism at the AAR, which is 800 km away from both the Zealandia and Antarctica continents. In addition, the influence of any pre-existing lithospheric component would have been eliminated by thermal erosion.

Mount Erebus is a rare, active alkaline volcano that provides an important window into mantle melting processes

What does Erebus geochemistry indicate about the depth of melting and lithology of its mantle source? Key questions remain regarding the petrogenesis of the alkaline lavas of EV and Ross Island. Namely: (1) what is the lithology of the mantle source, peridotite v. pyroxenite, and how do these lithologies and their different fusibilities influence the elemental and isotopic composition of the EV primitive magmas (Sims et al. 2013a); and (2) what role do volatiles such as H_2O and CO_2 play in the petrogenesis and evolution of the EV magmas?

Proposed source lithologies for alkaline lavas include: garnet lherzolite (e.g. Green and Ringwood 1967; Takahashi and Kushiro 1983), carbonated garnet lherzolite (e.g. Hirose 1997; Green and Falloon 1998; Dasgupta et al. 2007), garnet pyroxenite (e.g. Hirschmann et al. 2003; Kogiso et al. 2003), carbonated eclogite (e.g. Dasgupta et al. 2006), eclogite–lherzolite mixtures (e.g. Kogiso et al. 1998) and metasomatic veins (Pilet et al. 2008). Depletion in HREE compared to MREE in lavas and tephras from EV and Ross Island (mean $Dy_N/Yb_N \cong 1.4$: Fig. 10) and the observation that $^{230}Th/^{238}U > 1$ in EL lavas (Fig. 13) require melting in the presence of residual garnet (Sun and Hanson 1976; Sims and Hart 2006; Sims et al. 2013b).

Dasgupta et al. (2007) argued, on the basis of results from partial melting experiments carried out at 30 kbar, that basanites can be generated by up to c. 6% partial melting of a carbonated fertile peridotite. The importance of volatiles to the generation and evolution of Erebus Volcanic Province magmas is also strongly indicated by the presence of up to c. 7300 ppm CO_2 in olivine-hosted melt inclusions from two DVDP basanites (Oppenheimer et al. 2011) and up to c. 1.8 wt% CO_2, after correction for bubbles, in basanites from Mount Terror (Rasmussen et al. 2017). These values are enriched by a factor of c. 30 relative to nominally undegassed MORB melt inclusions from the Siqueiros Fracture Zone (Saal et al. 2002). In addition to elevated CO_2, DVDP and Mount Terror basanite melt inclusions contain up to c. 2 wt% H_2O, enriched by a factor of c. 13 relative to those of Siqueiros MORB. The common occurrence of kaersutite in DVDP lineage lavas is also a strong indication that Erebus Volcanic Province lavas are generated and evolve under the influence of H_2O. Conversely, the extremely low values of CO_2 and H_2O in the Erebus pyroxene and anorthoclase melt inclusions and the *lack* of kaersutite in EL lavas suggest that these lavas may have been generated from a relatively H_2O-poor source that may have been hotter than that which produced the DVDP lineage lavas (Kyle et al. 1992; Iacovino et al. 2016). Perhaps a more likely explanation, as discussed previously, is that EV magmas degassed at shallow depths relative to the basanites from the DVDP lineage.

The simplest interpretation of the EV and Ross Island elemental and isotopic geochemistry (Phillips et al. 2018) is that the Erebus magmas were derived by low-degree partial melting of a source containing residual garnet (i.e. below the garnet–spinel transition zone, c. 80 km depth) and high CO_2. Combined geophysics and geochemistry suggest that the source of these melts is the asthenospheric mantle (Phillips et al. 2018).

The subsequent evolution from basanite to phonolite is recorded in the compositions of lavas and melt inclusions from both the DVDP and EL lavas (Kyle 1977; Kyle et al. 1992; Kelly et al. 2008b; Oppenheimer et al. 2011). The liquid line of descent in these lineages have been experimentally investigated (Moussallam et al. 2013; Iacovino et al. 2016) and thermodynamically modelled (Iacovino 2015). There is also evidence of gases from the basanitic part of the system streaming through to the phonolite magma to be degassed both passively and explosively from the lava lake (Oppenheimer and Kyle 2008b; Oppenheimer et al. 2011; Iacovino 2015).

An important first-order constraint on the recent and current EV magma system comes from the seemingly steady-state composition of its eruptive products, which, as represented in the dated lava flows and historical bombs, have been uniformly phonolitic over several millennia. This observation begs several questions: (1) How much magma is driving the active lava lake at the surface? (2) Is heat (and mass) input from depth (i.e. the mantle) episodic or continuous? (3) Where (i.e. at what depth) does differentiation from primitive basanite to evolved phonolite occur? (4) Which processes control the uniform nature of Mount Erebus phonolitic magmas and where is that uniform character set (at shallow depth, >5 km or deeper)?

Erebus volcano hosts a unique, long-lived lava lake that provides an open-conduit, alkaline magma system natural laboratory

Given the persistent summit lava lake and open magma conduit at EV and the imaged NSMB, two end-member models for the Mount Erebus magmatic system are: (1) full connectivity from a mantle source to the shallow system that feeds the lava lake; and (2) episodic buoyant rise of discrete bodies of mantle-derived basanitic melt which establish transient pre-eruptive crustal reservoirs.

Determining the geometry and physical state of the magmatic plumbing system is essential to understanding how the EV lava lake has remained near its liquidus temperature for decades, probably even millennia, while erupting nearly compositionally uniform magmas over the same timescales. In the following subsection, we examine geophysical and geochemical constraints on the geometry and processes occurring within the Erebus magma system.

What constraints do geophysical measurements place on the geometry of the Erebus magmatic plumbing system from the lava-lake vent to the lower crust? Geophysical imaging of the internal structure of EV provides constraints on models of magma formation, staging, evolution and degassing. A qualitative and quantitative combination of all of these techniques can robustly determine characteristics of the deep system and explain its past and possible future evolution. In the following, we examine the constraints geophysical imaging is providing on the subsurface structure of EV, including its shallow magma conduit system.

Shallow subsurface structure of the upper cone. The surface morphology and geology of the summit cone at EV, as well as its crater walls, indicate spatial and temporal variability in the processes forming the upper volcanic edifice. Evidence includes an uneven distribution of fumaroles and lava flows, a significantly off-centre crater and vent system, and interleaved lava flow, pyroclastic and ice deposits. Temporary dense-array seismic observations of high-frequency scattered waves (c. 1–15 Hz) from regional icequakes, ice borehole shots and eruptions show that this eruptive and morphological heterogeneity is strongly reflected in the seismic wave field (Chaput *et al.* 2014, 2015*a*, *b*).

Shallow magma conduit system (lava lake–1.5 km below the surface). Below the surface expression of the lava lake, the assimilated P-wave tomography and coda-derived autocorrelation scattering imaging (Zandomeneghi *et al.* 2013) confirm the absence of a large centralized vertical conduit below the lava lake (Figs 21 & 22). Instead, imaged areas with low P-wave velocities and high scattering suggest that a near-summit magmatic storage zone exists several hundred metres to the NW of the lava lake at a depth of several hundred metres below the Tramway Ridge warm ground and other well-pronounced hydrothermal features of that sector of the summit plateau. This NSMB resides 500–1000 m NW of the lava lake and extends to depths in excess of 500 m below the surface.

The terminal conduit (i.e. its uppermost few hundred metres above the NSMB) has not been resolved by tomographic study, which indicates that it is likely to be narrower than approximately 10 m in its uppermost few hundred metres. If the conduit extends to the nearest edge of the NSMB, connecting near the VLP source region, then it would have an inclination of approximately 60° and a length of approximately 500 m, although there is no reason to assume a simple upper-conduit geometry. Indeed, the presence of additional sporadically active vents within the inner crater implies a branching or otherwise complex terminal geometry in at least the uppermost few hundreds of metres below the inner crater. This varying near-summit-conduit diameter is also reflected in physical studies of gas slug disruption during lava-lake eruptions (Gerst *et al.* 2013). Although the terminal conduit is evidently too narrow to be resolved in existing tomographic models, it is a source of VLP seismic signal generation (Aster *et al.* 2003), and an average diameter of less than about 10 m is consistent with variations in the timing of these VLP signals with respect to surface explosions (Knox *et al.* 2018). Broadband observations of the lava-lake eruptive process have been particularly valuable in quantifying and understanding both the explosive decomposition signal of the emerging overpressured gas slug and the subsequent minutes-long response of the magmatic conduit as it re-establishes equilibrium during system refill. A remarkable feature of these events is that the system is nearly reset after this process concludes, so that the next lava-lake eruption may generate nearly identical seismograms and infrasonograms. This quasi-repeating elastic radiation from the EV lava-lake explosions reflects the non-destructive and self-reconstructing nature of the erupting and refilling lava-lake system.

Variable terminal conduit size is also reflected in video and geophysical studies of gas slug disruption during lava-lake eruptions (Gerst *et al.* 2013). Broadband observations of the lava-lake eruptive process have been particularly valuable in quantifying and understanding both the explosive decomposition signal of the emerging overpressured gas slug and the subsequent minutes-long response of the magmatic conduit as it re-establishes equilibrium during system refill.

Finally, moment-tensor inversions and spectral analyses of VLP seismic signals associated with lava-lake eruptions demonstrate the existence of significant single-force components in the VLP source region that are consistent with reaction forces generated during post-eruptive lava-lake refill magmatic and gas transport within the shallow conduit system. The predominant single-force component dips NW, providing additional evidence for an inclined conduit with this general orientation (Aster *et al.* 2003, 2008).

Together, these observations suggest that the shallowest (upper few hundred metres to 1.5 km) elements of the conduit system have significantly non-vertical, non-planar geometries, and that a significant feature of this system extends from the lava lake to the VLP source region and NSMB (Fig. 23). The presence of less persistent vents (an intermittent ash vent near the south margin of the inner crater and an intermittent small secondary vent west of the lava lake) furthermore suggests that this system has a multi-threaded geometry across at least the uppermost several hundred metres to support sporadic activity at these secondary vents.

Magma conduit system deeper than 1.5 km. Geophysical investigations on Mount Erebus have focused primarily on imaging the uppermost magma plumbing system (<1.5 km depth), in particular the shallow conduit, but the continuation of the shallow conduit system to greater depth is so far poorly constrained by existing geophysical and geochemical/petrological datasets. Recent reprocessing of dense seismograph deployment data from the Tomo Erebus dataset (Zandomeneghi *et al.* 2013) by Blondel *et al.* (2018) using advanced scattering matrix and adaptive confocal filtering methods has extended seismic imaging to depths near 2.5 km, and suggests additional reservoirs near and below sea level. This model is also consistent with the predominant near-summit magma storage zone (the NSMB also imaged in Chaput *et al.* 2012 and Zandomeneghi *et al.* 2013) lying NW of the lava lake and indicates an inclined conduit system that extends ENE to depths of near 2000 m below sea level with an average dip near 60°. At still greater crustal depths, scattering regions that may represent deeper elements of the magmatic system are visible with a general westward inclination to depths near 6000 m below sea level. This west-dipping deeper trend is also generally consistent with recent, as yet unpublished, MT imaging which shows a generally inclined conduit system below 10 km towards the west and south (Hill *et al.* 2017; Wannamaker *et al.* 2020).

How do surface gas measurements inform our knowledge of subsurface processes of magma differentiation and degassing occurring within the volcanic conduit? The parental lavas of EV are CO_2-rich, silica-undersaturated basanitic magmas, which subsequently evolve via fractional crystallization into intermediate and then phonolitic melts. Simultaneously, as this CO_2-rich silicate melt is evolving from basanite to phonolite, a coexisting C–O–H–S fluid evolves (and is eventually released as gas at the surface). The composition of the fluid is a function of T, P, f_{O_2} and the concentrations of volatiles dissolved in the melt. A discussion and evaluation of relevant solubility parameters for volatiles in basanitic and phonolitic melts can be found in Burgisser *et al.* (2012) and Iacovino *et al.* (2016).

EV's open conduit provides a pathway along which the alkaline magma differentiates and volatiles are exsolved and potentially segregated from the magma. Because the composition of the measured surface gas is the summation of all subsurface processes involving gas exsolution and exchange, deconvolving possible signatures of gasses derived from various depths within the EV system is an important goal that is just beginning to be realized (Iacovino 2015).

Surface gas compositions from the EV lava lake are CO_2 rich (35–40 mol% CO_2) and exhibit a bimodal behaviour linked to eruptive activity. Ambient passive degassing is punctuated by explosive Strombolian degassing associated with large bubble bursts in the lava lake. During these Strombolian eruptions, both CO_2/H_2O and CO_2/CO increase significantly compared to the ambient passive degassing signal. Physically, these Strombolian events are interpreted as eruptions of deeper gas slugs that have accumulated in a shallow magma reservoir and then moved to the surface through the tilted narrow conduit (Aster *et al.* 2003, 2008; Gerst *et al.* 2013; Ilanko *et al.* 2015).

The CO_2-rich nature of gases released by ambient passive degassing at EV have been variably explained by: (1) the complete degassing of a parental basanite melt at *c.* 400 MPa (Oppenheimer and Kyle 2008*b*); (2) differential degassing of a shallow phonolite melt within the lava lake at near-surface pressure (Burgisser *et al.* 2012); or (3) a mixture of both shallow (phonolite) and deep (basanite: 500–800 MPa) degassing (Oppenheimer *et al.* 2011; Iacovino *et al.* 2013, 2016).

Modelling the sources of measured magmatic gases depends on knowing several thermodynamic parameters, including the solubilities of the various gases in the different magma compositions, ranging from basanite to phonolite. Experimental studies have placed tight constraints on the $P–T–f_{O_2}–X_{H_2O}$ conditions of magma differentiation, storage and degassing at various depths throughout the EV plumbing system (Iacovino *et al.* 2013, 2016; Moussallam *et al.* 2013). As a result of poor knowledge of all the required parameters, the solubility models used to calculate the composition of C–O–H–S fluids in equilibrium with sampled silicate melts have come up with very different implications for the EV conduit system.

The highly distinct signatures from explosive degassing (e.g. Fig. 16) are also variably interpreted to be: (1) sourced from deep parental basanite magma (Oppenheimer *et al.* 2009; 2011); or (2) due to different processes occurring entirely with the evolved phonolite (Burgisser *et al.* 2012).

In the third scenario, which is an outgrowth of ideas from the first scenario, the ratio of CO_2/CO tracks redox conditions (Oppenheimer *et al.* 2011), whereas the ratio of CO_2/H_2O relates to the source depth of degassing. In this case, the measured differences in these ratios between quiescent and explosive gas signatures are due to the decompression of two deep, volatile-saturated sources that mixed to various degrees (phonolite at 100–300 MPa and basanite at 500–800 MPa with the basanitic gas signature dominating the explosive signal). The implication for the EV system is that both the deep (mafic) and shallow (phonolitic) portions of the plumbing system are actively degassing (Fig. 30).

In the starkly contrasting second scenario, Burgisser *et al.* (2012) examined whether the difference in the CO_2/CO and CO_2/H_2O ratios between explosive and quiescent degassing can be explained entirely by variations in the adiabatic (isentropic) expansion of the gas bubbles, accompanied by cooling and immediate re-equilibration (Fig. 16). Burgisser *et al.* (2012) interpreted the observed difference in these ratios, particularly CO_2/CO, as indicating that the explosive gases are *not* from a deep basanitic source but, rather, represent gas slugs from the shallow phonolitic source. Essentially, they posit that adiabatic expansion and differences in rates of ascent compared to rates of volatile diffusion between melt and gas are the cause of the observed range of H_2O/CO_2 and CO_2/CO ratios. Their results suggest that the gas slug last equilibrated at temperatures of up to 300°C cooler than those estimated for the lava lake due to rapid gas expansion just prior

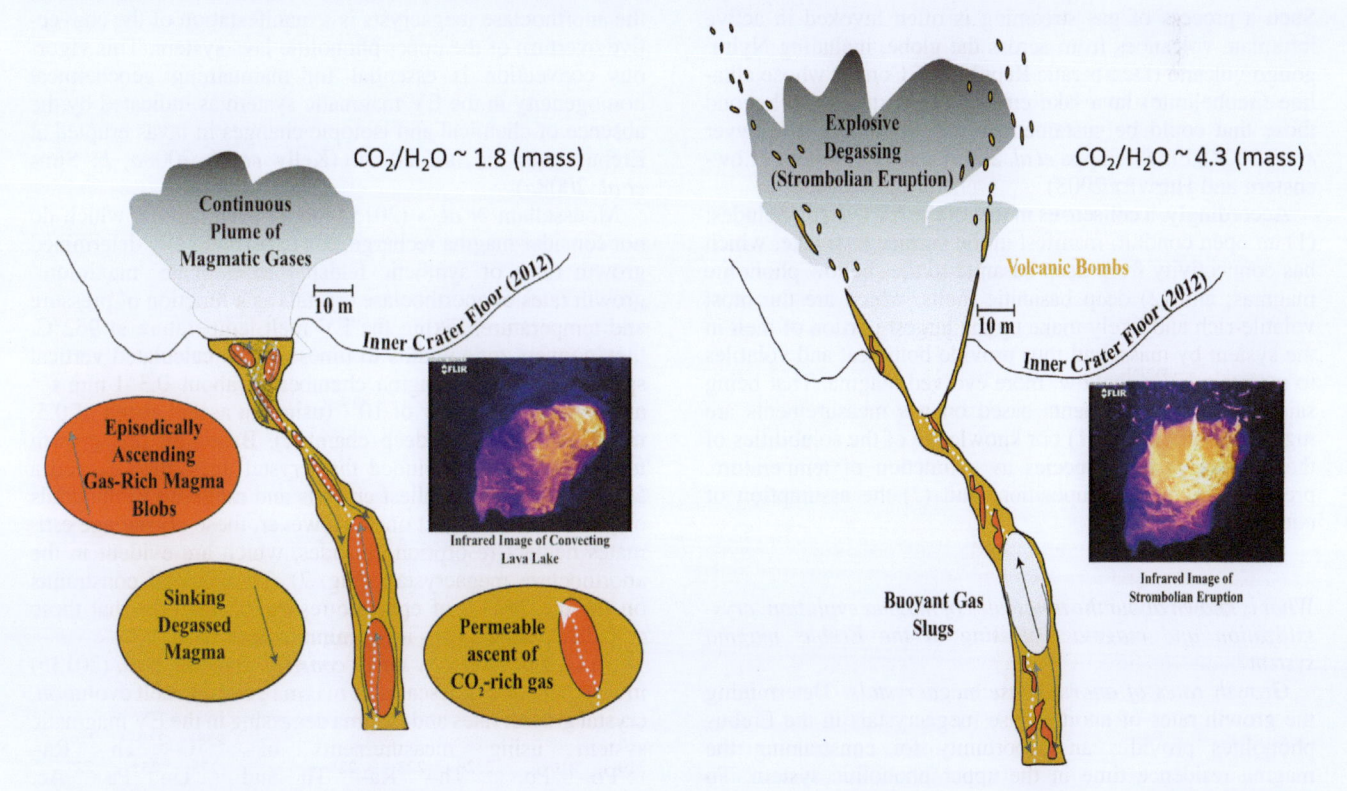

Fig. 30. Schematic image showing how processes and sources of passive (left image) and erupting (right image) degassing result in significantly different CO_2/H_2O (by mass) as determined by Oppenheimer *et al.* (2011).

to burst. In both scenarios, the sporadic periodicity of the explosions is caused by the gas slug migrating through the narrow conduit to the surface.

More recently, Iacovino (2015) again argued for degassing of volatiles from the deeper basanitic magma system. Using measurements of dissolved volatiles recorded in melt inclusions (Oppenheimer et al. 2011), a forward thermodynamic model that computes the composition of a C–O–H–S fluid in equilibrium with melts of different depths (considering the parameters T, P, f_{O_2}, f_{H_2} fH_2 or f_{S_2} and CO_2), and then a mixing model which evaluates all possible mixtures of subsurface fluid that match the surface gas chemistry to within ±2.0 mol%, Iacovino et al. (2016) interpreted the EV lake surface gas emissions to be a mixture dominated by degassing from the deep basanite (at c. 450 MPa) with a lesser, but still substantial, contribution from near-surface fluid in the lava lake.

While the compositions and the ratios of the gases provide important constraints on the source of exsolved EV gases, simple petrological calculations show that the magnitude of CO_2 gas discharge from the EV lava lake can easily be accounted for by gases streaming from deep. Using a CO_2 emission rate of 1000 Mg/day (c. 1300 Mg/day: Oppenheimer and Kyle 2008b) and an initial CO_2 content of 0.6 wt% as measured in melt inclusions (CO_2, 690 ± 125 ppm: Oppenheimer et al. 2011), the CO_2 flux can be accounted for by complete degassing of c. 1 $m^3 s^{-1}$ of parental basanite magma (Oppenheimer et al. 2011). Further, as noted by Oppenheimer et al. (2011), a declining magma flux towards the surface is required to explain the observed CO_2 and H_2O fluxes (Oppenheimer and Kyle 2008b). Given the solubilities and initial concentrations of H_2O and CO_2 in the melt, if all the magma reached very shallow levels in the conduit then the water flux should be significantly higher than observed.

Finally, the suggestion that CO_2 must be streaming up from a deep storage zone of mafic magma in order to explain CO_2-rich surface gas compositions makes physical sense, because the gas phase would be buoyant relative to magma and would help to maintain thermal output from the EV system. Such a process of gas streaming is often invoked in active intraplate volcanoes from across the globe, including Nyiragongo volcano (Democratic Republic of Congo) whose alkaline (nephelinite) lava lake emits CO_2 in quantities beyond those that could be sustained by the erupted lava (Sawyer et al. 2008), Etna (Aiuppa et al. 2006) and Yellowstone (Lowenstern and Hurwitz 2008)

Accordingly, a consensus model of the EV system includes: (1) an open conduit, manifest in the surface lava lake, which has connectivity from deep basanite to the shallow phonolite magmas; and (2) deep basanitic melts, which are the most volatile-rich and likely make up the largest portion of melt in the system by mass, and thus provide both heat and volatiles to maintain EV's shallow, more evolved magma. That being said, the above arguments based on gas measurements are highly dependent on: (1) our knowledge of the solubilities of the various gaseous species as a function of temperature, pressure and melt composition; and (2) the assumption of equilibrium.

What is known about the timescales of magma evolution, crystallization and magma degassing in the Erebus magma system?

Growth rates of anorthoclase megacrysts. Determining the growth rates of anorthoclase megacrysts in the Erebus phonolites provides an opportunity for constraining the magma residence time in the upper phonolitic system. To date, three main approaches have been used to establish the growth rates of Erebus anorthoclase megacrysts: (1) crystal size distributions; (2) crystal speedometry; and (3) U-series measurements.

(1) *Crystal size distributions*: Crystal size distributions of Erebus anorthoclase megacrysts (Dunbar et al. 1994) suggest initial low nucleation rates followed by rapid growth in two or more stages. The initial growth stage produced spongy inclusion-rich cores, which were overgrown by finely laminated rims. The composition of melt inclusions in the anorthoclase is very similar to that of the glass matrix of the samples, indicating that the anorthoclase crystals formed in a fully evolved phonolitic melt. Anorthoclase megacryst size distributions in Erebus bombs show a linear relationship between the natural logarithm of the crystal number density (as a function of linear crystal size) and crystal width, with no accumulation in the larger size classes. This linear relationship suggests that crystallization has been temporally continuous and that accumulation of anorthoclase crystals has not been an important factor (Dunbar et al. 1994). Finally, Dunbar et al. (1994) found that the age of an average EV anorthoclase crystal is 100–300 years. As a caveat, it is critical to note that the presence of resorption/dissolution within some anorthoclase megacrystals suggests that a simple size distribution–age calculation may underestimate the true crystal age.

(2) *Diffusion chronometry*: The first study using oscillatory zoning of anorthoclase feldspar megacrystals erupted from the lava lake of EV was in an unpublished MS thesis (Sumner 2007), who measured Sr and Ba profiles by electron microprobe and laser ablation ICP-MS in five anorthoclase crystals from phonolite bombs, yielding an average growth rate of 1.4×10^{-9} cm s^{-1}. We note, however, that this calculation is significantly influenced by a single outlier, which, when removed, decreases the growth rate to 2.1×10^{-10} cm s^{-1}.

Moussallam et al. (2015) evaluated the origin of the oscillatory zoning in the anorthoclase megacrystals using phase equilibrium, solubility experiments and an analysis of melt inclusion data. Their general conclusions are consistent with other studies, namely that the oscillatory zoning seen in the anorthoclase megacrysts is a manifestation of the convective overturn of the upper phonolitic lava system. This vigorous convection is essential for maintaining geochemical homogeneity in the EV magmatic system as indicated by the absence of chemical and isotopic changes in lavas erupted at Erebus over the last 20 kyr (Kelly et al. 2008a, b; Sims et al. 2008a)

Moussallam et al.'s (2015) model calculations, which do not consider magma recharge, use experimentally determined growth rates of synthetic feldspar to estimate 'maximum' growth rates of anorthoclase crystals as a function of pressure and temperature. Fixing the EV melt temperature at 962°C, they convert zone widths to timescales to calculated vertical speeds within the magma chamber of about 0.3–1 mm s^{-1} and a Raleigh number of 10^{17} (using an ascent speed of 0.5 mm s^{-1} and a 1 km-deep chamber). Based on this growth rate model they concluded that crystal ages must exceed a few years for the smallest crystals and range up to hundreds of years for the largest ones. However, these crystal age estimates neglect resorption episodes, which are evident in the anorthoclase megacrystals (Fig. 7). This lack of constraints on the duration(s) of episodic resorption requires that these calculated crystal ages are minimum ages.

(3) *U- and Th-decay series constraints*: Sims et al. (2013b) investigated the timescales of magma genesis, melt evolution, crystal growth rates and magma degassing in the EV magmatic system using measurements of $^{238}U-^{230}Th-^{226}Ra-^{210}Pb-^{210}Po$, $^{232}Th-^{228}Ra-^{228}Th$ and $^{235}U-^{231}Pa-^{227}Ac$. These different nuclides have starkly contrasting chemistries resulting in significant elemental fractionations during a

variety of magmatic processes, and their different half-lives allow the investigation of timescales ranging from <1 to 10^5 years.

The Sims *et al.* (2013*b*) measurements of ^{238}U–^{230}Th, ^{230}Th–^{226}Ra, ^{226}Ra–^{210}Pb and ^{232}Th–^{228}Ra–^{228}Th disequilibria provide four important observations pertinent to both anorthoclase crystal growth rates and magma residence times in the shallow EV system:

- On a (^{230}Th/^{232}Th) v. (^{238}U/^{232}Th) isochron diagram, both the anorthoclase and phonolite glass show significant (^{230}Th/^{238}U) disequilibria and cluster tightly in two distinct groups forming horizontal two-point isochrons (Fig. 13).
- On a (^{226}Ra)/Ba v. (^{230}Th)/Ba isochron diagram, the anorthoclase and glass lie on opposite sides of the 'equiline' forming steeply inclined two-point isochrons (see fig. 9 in Sims *et al.* 2013*b*).
- (^{210}Pb/^{226}Ra) values are within error of equilibrium for both the anorthoclase and phonolite glass (see fig. 6 in Sims *et al.* 2013*b*).
- (^{228}Ra/^{232}Th) values are within error of equilibrium in the phonolite glasses but (^{228}Ra/^{232}Th) >1 for the anorthoclase (Fig. 14).

Instantaneous crystal formation and chemical isolation with a magma residence time of 10^2–10^3 years is consistent with the first three of these four constraints. The ^{238}U–^{230}Th data form horizontal 'zero-age' isochrons with an age resolution of 5–10 kyr; ^{230}Th–^{226}Ra data form inclined two-point isochrons giving ages of *c.* 2.5 ka for $D_{Ra}/D_{Ba} = 1$ down to hundreds of years for $D_{Ra}/D_{Ba} = 0.1$ (see fig. 10 in Sims *et al.* 2013*b*); and the anorthoclase and glass separates both have equilibrium (^{226}Ra/^{210}Pb) values suggesting they are older than 100 years. The equilibrium (^{227}Ac/^{231}Pa) in the phonolite glasses is also consistent with a magma residence time greater than 100 years, but without measurements of these nuclides in the anorthoclase, these isotopic data have no direct bearing on the timescales of anorthoclase crystallization.

The observation that (^{228}Ra/^{232}Th) significantly exceeds unity in several young anorthoclase separates (1984, 1988 – Reagan *et al.* 1992; 2005 – Sims *et al.* 2013*b*), however, is inconsistent with simple instantaneous crystal growth and requires some portion of the crystal to have grown less than 25 years prior to eruption. This is not surprising as numerous petrographical observations, *in situ* measurements and determinations of crystal size distributions (e.g. Dunbar *et al.* 1994; Sumner 2007; Kelly *et al.* 2008*b*) suggest that anorthoclase crystal growth was episodic and probably continued until eruption. Additionally, there is considerable data indicating that crystallization of magmas occurs over significant periods of time (tens to thousands of years) as heat is transferred away from the magma–wall rock interface (e.g. see Charlier and Zellmer 2000; Vazquez and Reid 2002; Reagan *et al.* 2003, 2008; Turner *et al.* 2003; Hawkesworth *et al.* 2004; Sims *et al.* 2007; Cooper *et al.* 2016; Reubi *et al.* 2017). In light of these observations, it is essential to explicitly consider the timescales of magma differentiation processes in the context of the half-lives of the shorter-lived U- and Th-decay series nuclides (e.g. ^{226}Ra, ^{210}Pb, ^{210}Po, ^{222}Rn, ^{227}Ac, ^{228}Ra, ^{228}Th, etc.). In other words, when the timescale of crystallization is comparable to the half-life of the daughter nuclide of interest (e.g. ^{226}Ra), caution is warranted and a multi-proxy approach to establishing crystal ages is required.

To account for the abundances of all of the short-lived nuclides in the ^{238}U- and ^{232}Th-decay series, including the very short-lived ^{232}Th–^{228}Ra system, in a manner consistent with known mineral abundances, partition coefficients and geochemical variations in the EV system, Sims *et al.* (2013*b*) developed a finite-element, continuous-crystallization model that incorporates ingrowth and decay of the different nuclides in the continuously growing anorthoclase crystals and associated phonolitic melt. This model was used to interrogate a variety of crystallization scenarios, including: (1) closed-system continuous crystallization; (2) variable-rate crystallization; (3) dissolution; and (4) dissolution and recrystallization in open systems with magma recharge (Fig. 31).

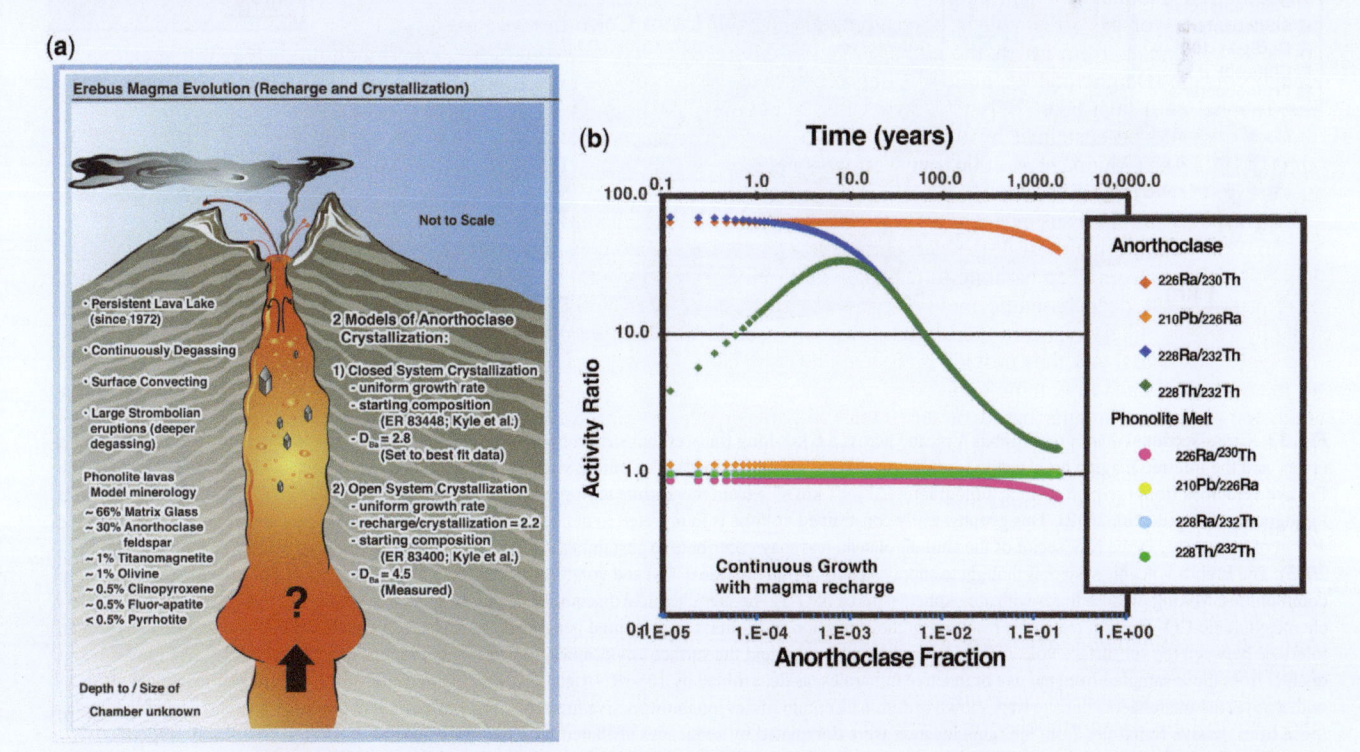

Fig. 31. (**a**) Schematic of the Erebus shallow magmatic system, detailing compositional constraints, model parameters and assumptions. (**b**) Results of finite-element, continuous crystallization model developed to account for all of the abundances of short-lived nuclides in the ^{238}U- and ^{232}Th-decay series in a manner consistent with our understanding of mineral abundances, partition coefficients and geochemical variations in the Erebus system. After Sims *et al.* (2013*b*).

The only way that Sims *et al.* (2013*b*) could replicate the observed Ba and Th concentrations, and the ^{238}U-, ^{235}U- and ^{232}Th-decay series data, was to incorporate magma recharge into the shallow system. Accordingly, the recharge rate had to exceed the crystallization rate. These high recharge rates and degrees of crystallization are close to the values required for steady-state compositions to be attained even for incompatible elements such as Th, implying that the young anorthoclase phonolites from EV have major and trace element compositions that are close to those of a steady-state system (Sims *et al.* 2013*b*). The modelled high recharge rates also imply that the volume of phonolite in the shallow magma chamber and lava-lake system at EV has been growing, which makes sense from a heat-balance point of view, as it explains how a lava lake and shallow magma reservoir can remain liquid (Calkins *et al.* 2008). The result that the shallow magma reservoir within EV is still growing has important implications for hazard assessment (see below).

U-series constraints on the timescales of magma degassing. U-series systematics of the glasses also place constraints on the timescales of magma degassing (Reagan *et al.* 2006; Sims *et al.* 2013*b*; Reubi *et al.* 2014). The observation that (^{210}Pb/^{226}Ra) and (^{227}Ac/^{231}Pa) are in radioactive equilibrium suggests that the residence time of the phonolitic magma is >100 years, consistent with the above calculations based on anorthoclase ages. When the effect of ^{222}Rn degassing on ^{210}Pb/^{226}Ra is considered, the isotopic data indicate that the majority of magma degassing for the phonolitic magma occurred deep and long before eruption. Additionally, for the 2005 lava bomb, whose eruption date (16 December 2005) is known explicitly, ^{210}Po was not completely degassed from the lava at the time of eruption. Incomplete degassing of ^{210}Po is atypical for subaerially erupted lavas (Reagan *et al.* 2006; Sims *et al.* 2008*b*) and suggests the EV shallow magma degasses about 1% of its Po per day.

The Erebus ice tower fumaroles and hot ground host a unique community of chemotrophic life in Antarctica

What is the relationship between EV's magmatic system, hydrothermal system and its biosphere? The near-summit P-wave tomography model of Zandomeneghi *et al.* (2013) describes several prominent features interpreted to represent regions of fumarolic ice-cave hydrothermal systems, a prominent radial dyke, and a NSMB located 500–1000 m NW of the lava lake and extending to depths in excess of 500 m below the surface. The NSMB underlies the prominent Tramway Ridge warm-ground region of the summit plateau, which is an area of high surface temperatures and gas emissions (Fig. 32). Tramway Ridge is also an area that hosts biodiverse and, in some locations, apparently endemic microbial communities (Herbold *et al.* 2014*a, b*) that can comprise thermophilic Archaea, Bacteria and Eukarya supported by a variety of metabolic functionalities. The habitability (Shock and Holland 2007) of these environments appears to be a direct consequence of volcanic activity; in particular, the role that this process has in generating steam, the predominant source of water supporting microbial activity and growth (Broady 1984).

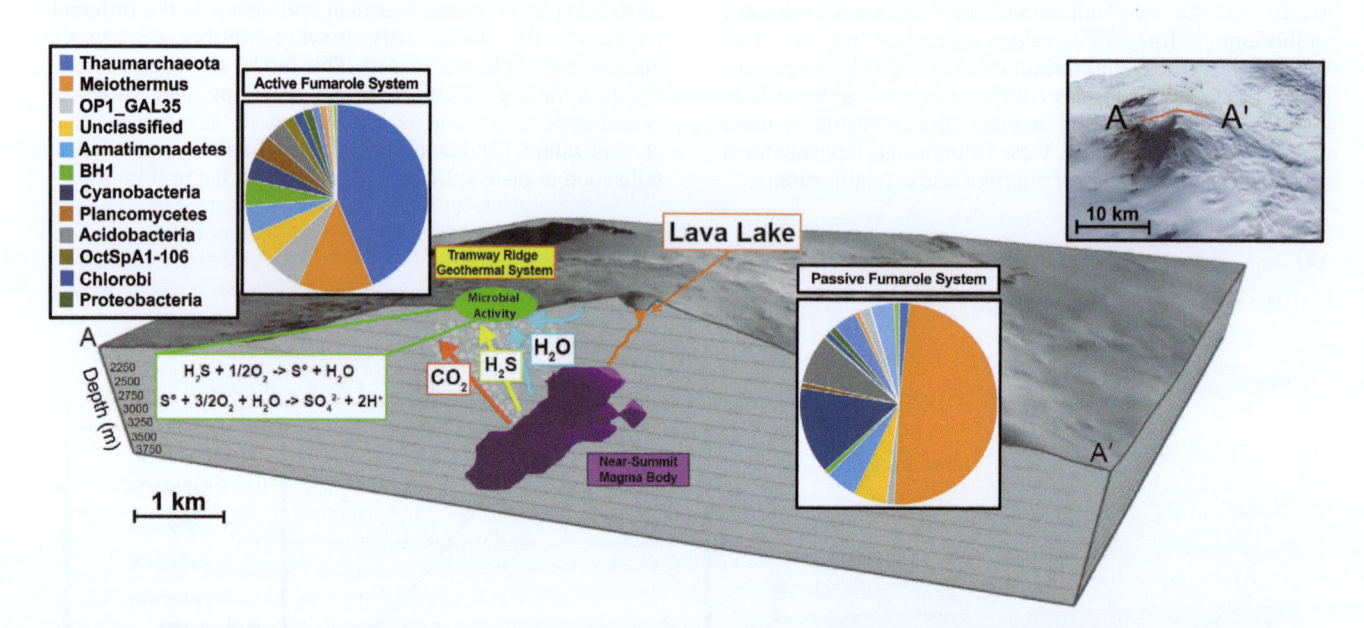

Fig. 32. Cross-sectional diagram of Erebus Volcano across a 6 km-long transect that strikes NW–SE (135°E of N) and goes through the centre of the inner crater, and the inferred magma body centre (see the upper right for the transect). The purple volume corresponds to an isosurface characterized by very low P-wave velocities from ray-path seismic tomography (ΔV_P <1 km s^{-1}) from travel-time tomography and a high scattering coefficient (S >46) from scattering tomography (Chaput *et al.* 2012). This geophysically constrained volume is interpreted to delineate a near-summit magma body that approaches to within *c.* 400 m of the surface in the NW sector of the summit plateau and may contribute to sustaining the Tramway Ridge geothermal system (Zandomeneghi *et al.* 2013). The Erebus volcanic system is thought to supply steam, carbon dioxide (CO$_2$) and volatiles as a source of reductant (e.g. H$_2$S) to near-surface microbial communities. Mixing of these fluids with atmosphere (source of O$_2$) generates chemical disequilibria that can serve as a source of electrons and energy to drive chemosynthetic CO$_2$ fixation reactions, resulting in the formation of biomass (represented generically as CH$_2$O). As such, microbial communities form an interface between the subsurface volcanic and hydrothermal system and the surface environment. Microbial communities sampled from active fumaroles are distinct from those sampled from passive or inactive fumaroles, as determined by 16S rRNA gene sequencing and analysis. For example, sequences affiliated with a novel thaumarchaeota lineage were shown to dominate communities inhabiting active fumarolic systems on Tramway Ridge (top left pie chart), whereas those from passive fumaroles from the same location were dominated by sequences affiliated with Meiothermus (Herbold *et al.* 2014*a, b*). Importantly, sequences affiliated with known thermophiles were more abundant in the active fumarole community. Data adapted from Herbold *et al.* (2014*a, b*). The thin and inclined conduit geometry connecting the near-summit magma body to the lava lake is consistent with that inferred from temporally varying very-long-period and short-period eruptive seismic signals by Knox *et al.* (2018). Image created by Jeff Dishbrow of the Polar Geospatial Center, and Cole Messa and Lisa Kant at the University of Wyoming High Precision Isotope Laboratory using the Reference Elevation Model of Antarctica (Howat *et al.* 2019).

Moreover, the energy metabolism of resident populations is likely to be primarily supported by volatiles generated by volcanic degassing and ensuing interactions between condensed steam, these gases and subsurface minerals.

Further work studying the 'geohydrobiology' of the shallow hydrothermal systems on the summit flanks of EV is necessary to establish the explicit connections between heat and volatile mass transfer from the magma system and their collective role in supporting EV's unique microbiological ecosystem. Such an assessment requires an interdisciplinary approach: shallow geophysical imaging to determine fluid and gas pathways and reservoirs; isotope and elemental geochemistry to determine fluid sources and timescales of water–rock–gas interaction; and cultivation-dependent and cultivation-independent techniques to deduce the redox reactions that support the energy metabolism of the microbial communities. The fumarolic fields on Tramway Ridge and the various fumarolic ice caves present on Erebus provide a unique field laboratory to make direct connections between subsurface processes extending to the mantle and how these integrated processes manifest to generate geochemical conditions conducive to supporting diverse microbial communities. In particular, given the geophysical imaging that links the NSMB and Tramway Ridge, this warm, biologically rich ground could serve as a unique and important site for a long-term ecological study. Clearly, heat and mass transfer from the magma system are strongly coupled to the development and evolution of this microbial community. As such, this connection begs an important question: Do changes in volcanic activity (eruptive style, intensity, gas flux) produce changes in the hydrothermal system (across a variety of timescales, from minutes, associated with abrupt pore pressure changes, to months, associated with long-term changes in volcanic activity), which are then manifest in the microbiological community? If so, monitoring changes in the Tramway Ridge microbial communities (and elsewhere on Mount Erebus) could possibly serve as 'biosensors' or 'bioindicators' of changing volcanic activity.

Erebus volcano's location makes it a threat to both US and international Antarctic science programmes and personnel

Erebus volcano is currently active and sits inspiringly above and 35 miles north of McMurdo Station emitting a gaseous plume that often trails into the horizon (Figs 1–5). McMurdo Station is essential to the US Antarctica programme, and also provides logistical and backup support to many other international programmes (e.g. New Zealand, Italy and Australia).

How do the observed patterns and changes in magmatic and volcanic activity over both long- and short-timescales inform about potential future eruptive behaviour and help in the assessment of EV's hazards? Most of EV's persistent activity consists of frequent Strombolian eruptions. There has been variability in the extent and magnitude of EV's Strombolian eruptions through time. The largest and most significant Strombolian activity observed in recent decades occurred from September 1984 to January 1985, when ashy clouds were visible to altitudes near 6000 m (c. 2 km above the summit) and ejecta were estimated to reach heights of 600 m above the summit (e.g. Smithsonian Institution Global Volcanism Program summary). Events like those occurring in 1984–85, while not likely to disrupt activity in McMurdo, are dangerous and thus would inhibit scientists working on the flanks of Mount Erebus. Variations in the intensity of EV's Strombolian activity could be a function of geometry or roughness of the vent/conduit system, which allows intermittent trapping of larger gas slugs. Volcanic gas data show strong periodicities in fluxes of species that degas from magma at low pressures (H_2O and SO_2), consistent with recharge of the lake by intermittent pulses of magma from shallow depths (Ilanko et al. 2015). Cyclic changes in proportions of other gas species suggest a background gas flux that is decoupled from this very shallow magma degassing. These interpretations are consistent with geometrical complexity in the uppermost conduit system as inferred from seismic and infrasonic studies (Fig. 30) As such, inclination and/or cross-sectional area changes within the shallow magma/gas pathways of EV may play a controlling role in the growth, storage, and eruption size and rate of the large gas slugs driving the current style of lava-lake Strombolian eruptive activity. Alternatively, shallow magma crystallization and/or degassing processes may influence convection and release of gas to a greater degree, although petrological evidence to date does not support this suggestion. Finally, another possibility is that eruption style is dominated by changes in gas/magma supply from the deep system (e.g. through the size and flux of CO_2 slugs from depth), with conduit processes playing a subordinate role in controlling the timing and intensity of each explosive event.

The potential for large accumulated gas slug or even phreatic (ice) or phreatomagmatic eruptions at EV also exists. On 19 October 1993, two substantial gas explosions near the rim of the inner crater created a new sub-crater approximately 80 m in diameter and ejected substantial debris over the crater rim. Thus, a near-term and more likely hazard is a phreatic or phreatomagmatic eruption, similar or larger than the one which occurred in 1993. Such an eruption could be extremely dangerous for anyone working on the volcano's summit or flanks, and could even potentially disrupt air traffic, which is critical for the USAP science support and safety of life operations.

The geological record also indicates numerous periods of more intense activity. Evidence of episodes of greater activity at EV include: (1) an EV edifice that is mantled by numerous lava flows ranging in age from 22 to 2 ka (see Supplementary Table S2); (2) the morphology of the EV edifice which shows evidence of two crater rims inferred to be associated with large eruptions; and (3) englacial tephra layers that have been recognized at a number of locations around Ross Island (Harpel et al. 2004; Harpel et al. 2008; Iverson et al. 2014), as well as several hundred kilometres away from Ross Island in the blue ice of the Transantarctic Mountains (Harpel et al. 2004). Geochemical investigation of these distal tephras reveal that most are compositionally indistinguishable from the modern EV phonolite, suggesting that they are derived from EV's past explosive activity (Harpel et al. 2004, 2008; Iverson et al. 2014).

The probability of a Plinian eruption from the summit, or even an effusive flank eruption, is relatively small. Nevertheless, both geophysical and gas measurements show that the EV magmatic system is likely to be maintained by a mafic magmatic system at depth. Modelled high recharge rates from Sims et al. (2013b) imply that the volume of phonolite in the shallow magma chamber and lava-lake system at EV has been growing. This makes sense from a heat-balance point of view because it explains how a lava lake and shallow magma reservoir can remain liquid (cf. Calkins et al. 2008). Sims et al.'s (2013b) modelled duration of reservoir growth is approximately 2 kyr, which coincides, within error, with the ages of the two youngest lavas from the flanks of EV (the Northwest and Upper Ice Tower Ridge flows: Harpel et al. 2004). This observation has important implications for hazard assessment, and permits speculation that the EV summit magma chamber system began to grow after these eruptive events and is presently amassing magma in advance of another significant lava eruption from a flank vent.

Whatever the short- and long-term hazards, given the nearby proximity of EV to McMurdo Station and its potential to interrupt essential air traffic, understanding volcanic patterns, monitoring signals indicative of the state of volcanic unrest (seismic, deformation, degassing, heat flow, etc.), and forecasting EV's eruptive behaviour is essential.

What information and monitoring are needed to best forecast short- and long-term volcanic hazards at Erebus volcano and Ross Island? Two timescales of information are required to best assess volcanic hazard: (1) long timescales to understand how Erebus volcano has evolved in order to establish patterns of magmatic cyclicity and thus to constrain the potential types of eruptions that could occur; and (2) active and remote volcanic monitoring and interpretation.

Long-term hazard mapping. In volcanology, the past is the key to the future. Establishing magmatic cyclicity is important for understanding the petrogenesis and evolution of a volcano, as well as forecasting its eruption probability. While numerous volcanological and chronological studies have been conducted on the upper flanks of Mount Erebus, there is still significant work to be done for a proper assessment of the long-term volcanic hazards associated with EV, including:

- Better determination of the full dynamic range of EV's eruptive behaviour over time. The assessment of EV's eruptive drivers and styles will require a comprehensive review and characterization of volcanic materials from the proto-Erebus shield- and proto-Erebus cone-building phases to modern deposits.
- Critical evaluation of the inferred link that more explosive periods in EV's history are directly linked to caldera-collapse events. This will require additional high-precision geochronology coupled with stratigraphic, petrological, petrographical and granulometrical analyses of EV's edifice rocks, as well as distal tephra deposits.
- Assessment of whether the overall driving mechanisms and long-term changes in volcanic activity (e.g. Strombolian v. Plinian or caldera forming) are largely determined by the geometry of the deeper magmatic plumbing system, and how this controls magmatic/chemical processing (stagnation, cooling, convection, differentiation, degassing, gas accumulation, etc.) within the system. In this regard, imaging (and periodically reimaging) the mid- to upper-crustal magma plumbing system could greatly enhance our understanding of EV's eruption dynamics. Additionally, joint interpretation of seismic and magnetotelluric (MT) images from the uppermost edifice to the lithospheric scale can further clarify regions of melt or partial melt to constrain their volume, geometry and longer-range connectivity.
- Finally, a systematic geochronological assessment of the entirety of Ross Island volcanism is necessary to better constrain the regional volcanic history and potential hazard, as there is clearly young volcanism on Hut Point Peninsula, the recurrence of which would be extremely problematic for operations and personnel at McMurdo Station.

Revitalize MEVO: a new monitoring program. Monitoring of volcanoes is important for hazard assessment and impact mitigation. For several decades MEVO was led by Phil Kyle at New Mexico Tech. Its observational system functioned until late 2016, when the instruments were removed from the mountain at the conclusion of funding.

At the time of writing (early 2019), the USAP has contracted a small temporary network of seismographs, installed and maintained by the IRIS consortium, to maintain some continuity of data collection on the volcano. USAP is presently considering the installation of a multi-year network of IRIS-supported seismographs and infrasonic sensors to ensure monitoring, and is potentially receptive to science and/or monitoring proposals that can provide a new level of reporting and research on the volcano.

EV continues to be an attractive target for improved monitoring and for the development and application of volcanic studies to further understanding of this remarkable volcano, and as a proving ground for techniques and insights that are generally applicable to active volcanoes. Advances in low-maintenance and high signal-to-noise seismic, infrasonic, structure-from-motion, radar, video, geodetic, and other robust low-power and high-capability instrumentation, recent progress in seismic and MT imaging of the magmatic system of the volcano to sea level and deeper depths, and the continued repetitive seismic illumination of the volcano by quasi-repeating lava-lake eruptions all facilitate the improved monitoring and continued scientific potential of EV. As discussed above, we believe that appropriate and sustained volcano monitoring is essential to ensure the viability and continued safe operation of McMurdo Station for the USA, as well as for leveraging future international research support and collaboration.

Summary: from mantle to microbe

Erebus volcano (EV) is a rare example of an active, intraplate alkaline volcano with an open-conduit system that provides a unique opportunity to study a volcano from its deep source to its surface, or 'from mantle to microbe'. Accordingly, given that this is a review paper, it seems appropriate to conclude by summarizing, as simply as possible, what is currently known about EV and Ross Island, starting from the mantle and moving upward toward the surface.

Mantle source and mantle melting

- The Erebus lineage (EL) is a spectacular alkaline sequence ranging from basanite to phonolite that can be modelled by simple crystal fractionation (Kyle 1981; Kyle *et al.* 1992).
- The parental basanite is derived by low degrees of melting from deep in the mantle. Its mantle source is carbon-rich, as indicated by high CO_2 in olivine-hosted melt inclusions, with garnet being residual during melting, as indicated by HREE depletions and $(^{230}Th/^{238}U) >1$ in EV lavas (Sun and Hanson 1975, 1976; Sims and Hart 2006; Sims *et al.* 2013*b*; Phillips *et al.* 2018).
- The EL lavas, as well as other Ross Island volcanics, have high $^{206}Pb/^{204}Pb$ and mix towards the HIMU mantle end member represented by St Helena (Sun and Hanson 1975, 1976; Sims and Hart 2006; Sims *et al.* 2008*a*; Phillips *et al.* 2018).
- The other long-lived radiogenic isotope systems also mix towards the St Helena HIMU end member. Because of the long half-lives of ^{147}Sm ($t_{1/2} = 108$ Gyr) and ^{176}Lu ($t_{1/2} = 37.1$ Gyr), which are the parents of, respectively, ^{143}Nd and ^{176}Hf, this isotopic coherency requires the age of the HIMU component of the EV and other Ross Island rocks to be old (Archean or early Proterozoic).
- The EV and Ross Island basanitic lavas have trace element characteristics consistent with their being derived in part from a source whose HIMU-like signature is similar to St Helena (Phillips *et al.* 2018). These same characteristics are incompatible with the Erebus and Ross Island volcanics originating from a recently (Paleozoic) metasomatized

lithospheric source related to subduction along the margins of Gondwana.
- A resolvable low-velocity zone, as deep as 1200 km, exists beneath Ross Island, suggesting that a deep, upwelling plume is connected to the shallow, lithosphere-scale magmatic system (Phillips et al. 2018).

Open-system magma conduit

- Seismic methods (P-wave tomography and coda-derived autocorrelation scattering imaging) show no indication of a large centralized vertical conduit below the lava lake (Zandomeneghi et al. 2013). This inability to resolve the EV terminal conduit in tomographic studies, post-eruptive observation of the lava-lake system (e.g. Aster et al. 2003), terminal splitting of eruptive gas slugs (Gerst et al. 2013) and seismogram analysis as indicative of small-scale terminal conduit evolution (Knox et al. 2018) consistently indicate that it is likely to be narrower than c. 10 m, and is geometrically complex and inclined in its uppermost few hundred metres below the lava lake.
- Very low P-wave velocities and high short-period seismic scattering indicate a near-summit magmatic storage zone. This near-summit magma body (NSMB) resides 500 – 1000 m NW of the lava lake and has a top that is less than c. 500 m below the surface. The NSMB is located below the Tramway Ridge warm ground and other well-pronounced hydrothermal features in this sector of the summit plateau.
- Geochemically, EV's open conduit and internal geometry provide pathways along which its alkaline magma differentiates and volatiles exsolve and are released to the atmosphere.
- The EV magma system has been uniformly phonolitic over several millennia, as demonstrated by measurements of its dated lava flows and historical bombs from the summit lava lake (Kelly et al. 2008b; Sims et al. 2008a).
- Combined ^{238}U- and ^{232}Th-decay series data indicate that pyroclasts ejected by Strombolian eruptions at EV have compositions expected for a near-steady-state system, reflecting inmixing of degassed magmas, crystal fractionation and aging (Sims et al. 2013b).
- Gas measurements of CO_2, CO, H_2O and SO_2 taken during both passive degassing and Strombolian eruptions, together with melt-inclusion studies, yield a general consensus model for the EV system that includes the following features: (1) an open conduit that has gas connectivity from the deep basanitic reservoir to the shallow phonolitic lava lake; and (2) deep basanitic melts, which are the most volatile-rich and are likely to make up the largest portion of melt in the system by mass, and provide the heat and volatiles needed to maintain the shallow, more evolved phonolitic lava (Oppenheimer et al. 2011; Iacovino 2015).

Surface expression of magmatic activity

- The most obvious surface manifestation of EV and Ross Island magmatism is its edifice morphology developed as a result of numerous eruptive episodes, which have both deposited lava flows and tephras, and also created large crater rims from past explosive phases (Figs 1–4). In total, about 4520 km^3 of volcanic material has been erupted on Ross Island over the last c. 4 myr (Esser et al. 2004). Currently, the summit of EV is covered with phonolitic lava flows and tephra from lava bombs.
- EV's persistent degassing produces an ever-present volcanic gas plume that represents a potentially important point source of chemical species to the Antarctic atmosphere and subsequent glacio-chemical records (Zoller et al. 1974; Chuan et al. 1986; Meeker et al. 1991; Chuan 1994; Zreda-Gostynska et al. 1997; Ilyinskaya et al. 2010):
  - the average SO_2 emission rates of 61 Mg/day (Sweeney et al. 2008) of the EV plume is low compared to those of other persistently active volcanoes (Andres and Kasgnoc 1998);
  - the EV gas plume is rich in halogens (Cl, Fl, etc.) with some of the highest halogen/SO_2 ratios for volcanoes globally (Kyle et al. 1990; Pyle and Mather 2009; Ilyinskaya et al. 2010);
  - the EV plume is enriched in trace metals of In, As, Hg, Zn, Au, Se, Co, W, Cs, Mo, Rb, Cu, Na and K.
- The final surface manifestation of the EV system is that it hosts biodiverse and endemic microbial communities (Herbold et al. 2014a, b) that can comprise thermophilic Archaea, Bacteria and Eukarya supported by a variety of metabolic functionalities. The habitability of these environments for supporting microbial activity and growth appears to be a direct consequence of EV's sustained thermal and eruptive activity as demonstrated by the spectacular link between the imaged near-surface magma system and the warm, hydrothermally active, ground and the thriving microbial community at Tramway Ridge.

Acknowledgements We thank UNAVCO and the IRIS PASSCAL Instrument Center at New Mexico Tech for facility support and field assistance. The facilities of the IRIS Consortium are supported by the National Science Foundation under Cooperative Agreement EAR-1261681, the NSF Office of Polar Programs and the DOE National Nuclear Security Administration. Computer time was provided through a Blue Waters Innovation Initiative. DEMs were produced using data from DigitalGlobe, Inc. We thank the many individuals and groups at McMurdo and within USAP who make fieldwork possible. Ken Sims and Rick Aster also thank Phil Kyle for introducing us to Mount Erebus volcano. Jake Lowenstern and an anonymous reviewer improved the quality of the manuscript. Cole Messa and Lisa Kant assisted with figures. Finally, we thank the effective and professional editorial handling of editors John Smellie and Adelina Geyer.

Author contributions KWWS: conceptualization (lead), data curation (equal), formal analysis (equal), funding acquisition (equal), investigation (equal), methodology (equal), project administration (lead), supervision (lead), validation (lead), writing – original draft (lead), writing – editing and reviewing (lead); **RA**: conceptualization (supporting), data curation (equal), formal analysis (equal), funding acquisition (equal), investigation (equal), methodology (equal), project administration (supporting), writing – original draft (supporting), writing – review & editing (supporting); **GG**: conceptualization (supporting), writing – original draft (supporting), writing – review & editing (supporting); **JB-T**: writing – review & editing (supporting); **EHP**: visualization (supporting), writing – review & editing (supporting); **PJW**: conceptualization (supporting), writing – review & editing (supporting); **GSM**: conceptualization (supporting), writing – review & editing (supporting); **DR**: formal analysis (supporting); **ESB**: conceptualization (supporting), writing – original draft (supporting), writing – review & editing (supporting).

Funding This work was supported by NSF grants OPP-0126269, OPP-1141167 and OPP-1644013 to K.W.W. Sims; OPP-0229305, ANT-0538414, ANT-0838414 and ANT-1142083 to R. Aster; OPP-1644013 to G. Gaetani; OPP-1644013 to P.J. Wallace; and EAR-1820658 to E.S. Boyd. E.S. Boyd also acknowledges support from NASA grant NNA15BB02A. Geospatial support for this work provided by the Polar Geospatial Center under NSF-OPP awards 1043681 and 1559691. DEMs provided by the Byrd Polar and

Climate Research Center and the Polar Geospatial Center under NSF-OPP awards 1543501, 1810976, 1542736, 1559691, 1043681, 1541332, 0753663, 1548562, 1238993 and NASA award NNX10AN61G.

Data availability The seismic and infrasound data are archived and available at the IRIS Consortium Data Management Center.

References

Aiuppa, A., Federico, C. et al. 2006. Rates of carbon dioxide plume degassing from Mount Etna volcano. *Journal of Geophysical Research*, **111**, B09207, https://doi.org/10.1029/2006JB004307

Amend, J.P. and Shock, E.L. 2001. Energetics of overall metabolic reactions of thermophilic and hyperthermophilic Archaea and bacteria. *FEMS Microbiological Reviews*, **25**, 175–243, https://doi.org/10.1111/j.1574-6976.2001.tb00576.x

Andersen, D.J., Lindsley, D.H. and Davidson, P.M. 1993. QUILF: A pascal program to assess equilibria among Fe–Mg–Mn–Ti oxides, pyroxenes, olivine, and quartz. *Computers & Geosciences*, **19**, 1333–1350, https://doi.org/10.1016/0098-3004(93)90033-2

Andres, R.J. and Kasgnoc, A.D. 1998. A time-averaged inventory of subaerial volcanic sulfur emissions. *Journal of Geophysical Research*, **103**, 25 251–25 262, https://doi.org/10.1029/98JD02091

Anthony, R.E., Aster, R.C. et al. 2015. The seismic noise environment of Antarctica. *Seismological Research Letters*, **86**, 89–100, https://doi.org/10.1785/0220140109

Aster, R., Mah, S. et al. 2003. Very long period oscillations of Mount Erebus Volcano. *Journal of Geophysical Research: Solid Earth*, **108**, 2522, https://doi.org/10.1029/2002JB002101

Aster, R., McIntosh, W. et al. 2004. Real-time data received from Mount Erebus Volcano, Antarctica. *Eos, Transactions of the American Geophysical Union*, **85**, 97–104, https://doi.org/10.1029/2004EO100001

Aster, R., Beaudoin, B., Hole, J., Fouch, M.J., Fowler, J. and James, D. 2005. IRIS PASSCAL program marks 20 years of scientific discovery. *Eos, Transactions of the American Geophysical Union*, **86**, 97–101, https://doi.org/10.1029/2004EO100001

Aster, R., Zandomeneghi, D., Mah, S., McNamara, S., Henderson, D.B., Knox, H. and Jones, K. 2008. Moment tensor inversion of very long period seismic signals from Strombolian eruptions of Erebus Volcano. *Journal of Volcanology and Geothermal Research*, **177**, 635–647, https://doi.org/10.1016/j.jvolgeores.2008.08.013

Auer, L., Boschi, L., Becker, T.W., Nissen-Meyer, T. and Giardini, D. 2014. Savani: A variable-resolution whole-mantle model of anisotropic shear-velocity variations based on multiple datasets. *Journal of Geophysical Research*, **119**, 3006–3034, https://doi.org/10.1002/2013JB010773

Bannister, S., Yu, J., Leitner, B. and Kennett, B.L.N. 2003. Variations in crustal structure across the transition from West to East Antarctica, Southern Victoria Land. *Geophysical Journal International*, **155**, 870–884, https://doi.org/10.1111/j.1365-246X.2003.02094.x

Bateman, H. 1910. The solution of a system of differential equations occurring in the theory of radioactive transformations. *Proceedings of the Cambridge Philosophical Society*, **15**(part V), 423–427.

Becker, T.W. and Boschi, L. 2002. A comparison of tomographic and geodynamic mantle models. *Geochemistry, Geophysics, Geosystems*, **3**, 1003, https://doi.org/10.1029/2001GC000168

Behrendt, J.C. 1999. Crustal and lithospheric structure of the West Antarctic Rift System from geophysical investigations – a review. *Global and Planetary Change*, **23**, 25–44, https://doi.org/10.1016/S0921-8181(99)00049-1

Belly, R.T. and Brock, T.D. 1974. Widespread occurrence of acidophilic strains of *Bacillus coagulans* in hot springs. *Journal of Applied Bacteriology*, **37**, 175–177, https://doi.org/10.1111/j.1365-2672.1974.tb00427.x

Blondel, T., Chaput, J., Derode, A., Campillo, M. and Aubry, A. 2018. Matrix approach of seismic imaging: Application to the Erebus volcano, Antarctica. *Journal of Geophysical Research: Solid Earth*, **123**, 10 936–10 950, https://doi.org/10.1029/2018JB016361

Boichu, M., Oppenheimer, C., Tsanev, V.I. and Kyle, P.R. 2010. High temporal resolution SO_2 flux measurements at Erebus volcano, Antarctica. *Journal of Volcanology and Geothermal Research*, **190**, 325–336, https://doi.org/10.1016/j.jvolgeores.2009.11.020

Boichu, M., Oppenheimer, C., Roberts, T.J., Tsanev, V. and Kyle, P. 2011. On bromine, nitrogen oxides and ozone depletion in the tropospheric plume of Erebus volcano (Antarctica). *Atmospheric Environment*, **45**, 3856–3866, https://doi.org/10.1016/j.atmosenv.2011.03.027

Boyd, E.S., King, S., Tomberlin, J.K., Nordstrom, D.K., Krabbenhoft, D.P., Barkay, T. and Geesey, G.G. 2009. Methylmercury enters an aquatic food web through acidophilic microbial mats in Yellowstone National Park, Wyoming. *Environmental Microbiology*, **11**, 950–959, https://doi.org/10.1111/j.1462-2920.2008.01820.x

Boyd, E.S., Hamilton, T.L., Spear, J.R., Lavin, M. and Peters, J.W. 2010. [FeFe]-hydrogenase in Yellowstone National Park: Evidence for dispersal limitation and phylogenetic niche conservatism. *ISME Journal*, **4**, 1485–1495, https://doi.org/10.1038/ismej.2010.76

Boyd, E.S., Fecteau, K.M., Havig, J.R., Shock, E.L. and Peters, J.W. 2012. Modeling the habitat range of phototrophs in Yellowstone National Park: toward the development of a comprehensive fitness landscape. *Frontiers in Microbiology*, **3**, 221–221, https://doi.org/10.3389/fmicb.2012.00221

Boyd, E., Hamilton, T., Wang, J., He, L. and Zhang, C. 2013. The role of tetraether lipid composition in the adaptation of thermophilic Archaea to acidity. *Frontiers in Microbiology*, **4**, 62, https://doi.org/10.3389/fmicb.2013.00062

Bozdağ, E., Peter, D. et al. 2016. Global adjoint tomography: first-generation model. *Geophysical Journal International*, **207**, 1739–1766, https://doi.org/10.1093/gji/ggw356

Broady, P.A. 1984. Taxonomic and ecological investigations of algae on steam-warmed soil on Mt Erebus, Ross Island, Antarctica. *Phycologia*, **23**, 257–271, https://doi.org/10.2216/i0031-8884-23-3-257.1

Brock, T.D. 1967. Micro-organisms adapted to high temperatures. *Nature*, **214**, 882–885, https://doi.org/10.1038/214882a0

Brock, T.D. 1973. Lower pH limit for the existence of blue-green algae: Evolutionary and ecological implications. *Science*, **179**, 480–483, https://doi.org/10.1126/science.179.4072.480

Brock, T.D. and Brock, M.L. 1969. Effect of light intensity on photosynthesis by thermal algae adapted to natural and reduced sunlight. *Limnology and Oceanography*, **14**, 334–341, https://doi.org/10.4319/lo.1969.14.3.0334

Bucholz, C.E., Gaetani, G.A., Behn, M.D. and Shimizu, N. 2013. Post-entrapment modification of volatiles and oxygen fugacity in olivine-hosted melt inclusions. *Earth and Planetary Science Letters*, **374**, 145–155, https://doi.org/10.1016/j.epsl.2013.05.033

Burgisser, A., Oppenheimer, C., Alletti, M., Kyle, P.R., Scaillet, B. and Carroll, M.R. 2012. Backward tracking of gas chemistry measurements at Erebus volcano. *Geochemistry, Geophysics, Geosystems*, **13**, Q11010, https://doi.org/10.1029/2012GC004243

Caldwell, D. and Kyle, P.R. 1994. Mineralogy and geochemistry of ejecta erupted from Mount Erebus, Antarctica between 1972 and 1986. *American Geophysical Union Antarctic Research Series*, **66**, 147–162, https://doi.org/10.1029/AR066p0147

Calkins, J.A., Oppenheimer, C. and Kyle, P.R. 2008. Ground-based thermal imaging of lava lakes at Mount Erebus Volcano, Antarctica in December 2004. *Journal of Volcanology and Geothermal Research*, **177**, 695–704, https://doi.org/10.1016/j.jvolgeores.2008.02.002

Caltabiano, T., Romano, R. and Budetta, G. 1994. SO$_2$ flux measurements at Mount Etna, Sicily. *Journal of Geophysical Research*, **99**, 12 809–12 819, https://doi.org/10.1029/94JD00224

Caltabiano, T., Burton, M., Giammanco, S., Allard, P., Bruno, N., Murè, F. and Romano, R. 2004. Volcanic gas emissions from the summit craters and flanks of Mt. Etna, 1987–2000. *American Geophysical Union Geophysical Monograph Series*, **143**, 111–128, https://doi.org/10.1029/143GM08

Castenholz, R.W. 1969. Thermophilic blue-green algae and the thermal environment. *Bacteriological Reviews*, **33**, 476–504, https://doi.org/10.1128/MMBR.33.4.476-504.1969

Chang, S.-J., Ferreira, A.M.G., Ritsema, J., van Heijst, H.J. and Woodhouse, J.H. 2015. Joint inversion for global isotropic and radially anisotropic mantle structure including crustal thickness perturbations. *Journal of Geophysical Research*, **120**, 4278–4300, https://doi.org/10.1002/2014JB011824

Chaput, J., Zandomeneghi, D., Aster, R., Knox, H. and Kyle, P. 2012. Imaging of Erebus volcano using body wave seismic interferometry of Strombolian eruption coda. *Geophysical Research Letters*, **39**, L07304, https://doi.org/10.1029/2012GL050956

Chaput, J., Aster, R. et al. 2014. The crustal thickness of West Antarctica. *Journal of Geophysical Research: Solid Earth*, **119**, 378–395, https://doi.org/10.1002/2013JB010642

Chaput, J., Campillo, M., Aster, R., Roux, P., Kyle, P., Knox, H. and Czoski, P. 2015a. Multiple scattering from icequakes at Erebus volcano, Antarctica: Implications for imaging at glaciated volcanoes. *Journal of Geophysical Research*, **120**, 1129–1141, https://doi.org/10.1002/2014JB011278

Chaput, J., Clerc, V., Campillo, M., Roux, P. and Knox, H. 2015b. On the practical convergence of coda-based correlations: A window optimization approach. *Geophysical Journal International*, **204**, 736–747, https://doi.org/10.1093/gji/ggv476

Charlier, B.L.A. and Zellmer, G.F. 2000. Some remarks on U-Th mineral ages from igneous rocks with prolonged crystallisation histories. *Earth and Planetary Science Letters*, **183**, 457–469, https://doi.org/10.1016/S0012-821X(00)00298-3

Chen, Y., Provost, A., Schiano, P. and Cluzel, N. 2011. The rate of water loss from olivine-hosted melt inclusions. *Contributions to Mineralogy and Petrology*, **162**, 625–636, https://doi.org/10.1007/s00410-011-0616-5

Chen, Y., Provost, A., Schiano, P. and Cluzel, N. 2013. Magma ascent rate and initial water concentration inferred from diffusive water loss from olivine-hosted melt inclusions. *Contributions to Mineralogy and Petrology*, **165**, 525–541, https://doi.org/10.1007/s00410-012-0821-x

Chuan, R.L. 1994. Dispersal of volcanic-derived particles from Mount Erebus in the Antarctic atmosphere. *American Geophysical Union Antarctic Research Series*, **66**, 97–102, https://doi.org/10.1029/AR066p0097

Chuan, R.L., Palais, J.M., Rose, W.I. and Kyle, P.R. 1986. Fluxes, size, morphology and composition of particles in the Mt. Erebus volcanic plume, December 1983. *Journal of Atmospheric Chemistry*, **4**, 467–477, https://doi.org/10.1007/BF00053846

Colman, D.R., Feyhl-Buska, J., Robinson, K.J., Fecteau, K.M., Xu, H., Shock, E.L. and Boyd, E.S. 2016. Ecological differentiation in planktonic and sediment-associated chemotrophic microbial populations in Yellowstone hot springs. *FEMS Microbiology Ecology*, **92**, https://doi.org/10.1093/femsec/fiw137

Colman, D.R., Poudel, S., Hamilton, T.L., Havig, J.R., Selensky, M.J., Shock, E.L. and Boyd, E.S. 2018. Geobiological feedbacks and the evolution of thermoacidophiles. *ISME Journal*, **12**, 225–236, https://doi.org/10.1038/ismej.2017.162

Colman, D.R., Lindsay, M.R., Amenabar, M.J. and Boyd, E.S. 2019a. The intersection of geology, geochemistry, and microbiology in continental hydrothermal systems. *Astrobiology*, **19**, https://doi.org/10.1089/ast.2018.2016

Colman, D.R., Lindsay, M.R. and Boyd, E.S. 2019b. Mixing of end-member fluids supports hyperdiverse chemosynthetic hydrothermal communities. *Nature Communications*, **10**, https://doi.org/10.1038/s41467-019-08499-1

Cooper, K., Sims, K.W.W., Eiler, J.M. and Banerjee, N. 2016. Time scales of storage and recycling of crystal mush at Krafla Volcano, Iceland. *Contributions to Mineralogy and Petrology*, **171**, 54, https://doi.org/10.1007/s00410-016-1267-3

Cox, A., Shock, E.L. and Havig, J.R. 2011. The transition to microbial photosynthesis in hot spring ecosystems. *Chemical Geology*, **280**, 344–351, https://doi.org/10.1016/j.chemgeo.2010.11.022

Csatho, B., Schenk, T., Kyle, P., Wilson, T. and Krabill, W.B. 2008. Airborne laser swath mapping of the summit of Erebus volcano, Antarctica: Applications to geological mapping of a volcano. *Journal of Volcanology and Geothermal Research*, **177**, 531–548, https://doi.org/10.1016/j.jvolgeores.2008.08.016

Curtis, A. and Kyle, P. 2011. Geothermal point sources identified in a fumarolic ice cave on Erebus volcano, Antarctica using fiber optic distributed temperature sensing. *Geophysical Research Letters*, **38**, L16802, https://doi.org/10.1029/2011GL048272

Dabrowa, A., Green, D.N., Johnson, J.B., Phillips, J.C. and Rust, A.C. 2014. Comparing near-regional and local measurements of infrasound from Mount Erebus, Antarctica: Implications for monitoring. *Journal of Volcanology and Geothermal Research*, **288**, 46–61, https://doi.org/10.1016/j.jvolgeores.2014.10.001

Daley, E.E. and DePaolo, D.J. 1992. Isotopic evidence for lithospheric thinning during extension: Southeastern Great Basin. *Geology*, **20**, 104–108, https://doi.org/10.1130/0091-7613(1992)020<0104:IEFLTD>2.3.CO;2

Dasgupta, R., Hirschmann, M. and Stalker, K. 2006. Immiscible transition from carbonate-rich to silicate-rich melts in the 3 GPa melting interval of eclogite plus CO$_2$ and genesis of silica-undersaturated ocean island lavas. *Journal of Petrology*, **47**, 647–671, https://doi.org/10.1093/petrology/egi088

Dasgupta, R., Hirschmann, M. and Smith, N. 2007. Partial melting experiments of peridotite + CO$_2$ at 3 GPa and genesis of alkalic ocean island basalts. *Journal of Petrology*, **48**, 2093–2124, https://doi.org/10.1093/petrology/egm053

de Moor, J.M., Fischer, T.P. et al. 2013. Sulfur degassing at Erta Ale (Ethiopia) and Masaya (Nicaragua) volcanoes: Implications for degassing processes and oxygen fugacities of basaltic systems. *Geochemistry, Geophysics, Geosystems*, **14**, 4076–4108, https://doi.org/10.1002/ggge.20255

Dibble, R.R., Kienle, J., Kyle, P.R. and Shibuya, K. 1984. Geophysical studies of Erebus volcano, Antarctica, from 1974–1981. *New Zealand Journal of Geology and Geophysics*, **27**, 425–455, https://doi.org/10.1080/00288306.1984.10422264

Dibble, R.R., Kyle, P.R. and Rowe, C. 2008. Video and seismic observations of Strombolian eruptions at Erebus volcano, Antarctica. *Journal of Volcanology and Geothermal Research*, **177**, 619–634, https://doi.org/10.1016/j.jvolgeores.2008.07.020

Doemel, W.N. and Brock, T.D. 1970. The upper temperature limit of *Cyanidium caldarium*. *Archiv für Mikrobiologie*, **72**, 326–332.

Dunbar, N.W., Cashman, K. and Dupre, R. 1994. Crystallization processes of anorthoclase phenocrysts in the Mount Erebus magmatic system: evidence from crystal composition, crystal size distributions and volatile contents of melt inclusions. *American Geophysical Union Antarctic Research Series*, **66**, 129–146, https://doi.org/10.1029/AR066p0129

Eschenbacher, A.J. 1999. *Open-System Degassing of a Fractionating, Alkaline Magma, Mount Erebus, Ross Island, Antarctica*. MS thesis, New Mexico Institute of Mining and Technology, Socorro, New Mexico, USA.

Esser, R.P., Kyle, P.R. and McIntosh, W.C. 2004. ^{40}Ar/^{39}Ar dating of the eruptive history of Mount Erebus, Antarctica: Volcano evolution. *Bulletin of Volcanology*, **66**, 671–686, https://doi.org/10.1007/s00445-004-0354-x

Farman, J.C., Gardiner, B.G. and Shanklin, J.D. 1985. Large losses of total ozone in Antarctica reveal seasonal ClO$_x$/NO$_x$ interaction. *Nature*, **315**, 207–210, https://doi.org/10.1038/315207a0

Finn, C.A., Muller, R.D. and Panter, K.S. 2005. A Cenozoic diffuse alkaline magmatic province (DAMP) in the southwest Pacific without rift or plume origin. *Geochemistry, Geophysics, Geosystems*, **6**, Q02005, https://doi.org/10.1029/2004GC000723

Finotello, M., Nyblade, A., Julia, J., Wiens, D. and Anandakrishnan, S. 2011. Crustal V_p-V_s ratios and thickness for Ross Island and the Transantarctic Mountain front, Antarctica. *Geophysical Journal International*, **185**, 85–92, https://doi.org/10.1111/j.1365-246X.2011.04946.x

Fischer, T.P., Curtis, A.G., Kyle, P.R. and Sano, Y. 2013. Gas discharges in fumarolic ice caves of Erebus volcano, Antarctica. Presented at the Geological Society of America Fall Meeting, 9–13 December 2013, San Francisco, California, USA.

Fleming, T.H., Foland, K.A. and Elliot, D.H. 1995. Isotopic and chemical constraints on the crustal evolution and source signature of Ferrar magmas, north Victoria Land, Antarctica. *Contributions to Mineralogy and Petrology*, **121**, 217–236, https://doi.org/10.1007/BF02688238

Fraser, C.I., Terauds, A., Smellie, J., Convey, P. and Chown, S.L. 2014. Geothermal activity helps life survive glacial cycles. *Proceedings of the National Academy of Sciences of the United States of America*, **111**, 5634–5639, https://doi.org/10.1073/pnas.1321437111

Fraser, C.I., Connell, L., Lee, C.K. and Cary, S.C. 2018. Evidence of plant and animal communities at exposed and subglacial (cave) geothermal sites in Antarctica. *Polar Biology*, **41**, 417–421, https://doi.org/10.1007/s00300-017-2198-9

French, S.W. and Romanowicz, B. 2015. Broad plumes rooted at the base of the Earth's mantle beneath major hotspots. *Nature*, **525**, 95–99, https://doi.org/10.1038/nature14876

Friedrich, C.G., Rother, D., Bardischewsky, F., Quentmeier, A. and Fischer, J. 2001. Oxidation of reduced inorganic sulfur compounds by bacteria: Emergence of a common mechanism? *Applied and Environmental Microbiology*, **67**, 2873–2882, https://doi.org/10.1128/AEM.67.7.2873-2882.2001

Futa, K. and Le Masurier, W.E. 1983. Nd and Sr isotopic studies on Cenozoic mafic lavas from West Antarctica: Another source for continental alkali basalts. *Contributions to Mineralogy and Petrology*, **83**, 38–44, https://doi.org/10.1007/BF00373077

Gaetani, G.A., O'Leary, J.A., Shimizu, N., Bucholz, C.E. and Newville, M. 2012. Rapid re-equilibration of H_2O and oxygen fugacity in olivine-hosted melt inclusions. *Geology*, **40**, 915–918, https://doi.org/10.1130/G32992.1

Gerst, A., Hort, M., Kyle, P.R. and Voge, M. 2008. 4D velocity of Strombolian eruptions and man-made explosions derived from multiple Doppler radar instruments. *Journal of Volcanology and Geothermal Research*, **177**, 648–660, https://doi.org/10.1016/j.jvolgeores.2008.05.022

Gerst, A., Hort, M., Aster, R.C., Johnson, J.B. and Kyle, P.R. 2013. The first second of volcanic eruptions from the Erebus volcano lava lake, Antarctica – Energies, pressures, seismology, and infrasound. *Journal of Geophysical Research: Solid Earth*, **118**, 3318–3340, https://doi.org/10.1002/jgrb.50234

Ghiorso, M.S. and Gualda, G.A.R. 2015. An H_2O-CO_2 mixed fluid saturation model compatible with rhyolite-MELTS. *Contributions to Mineralogy and Petrology*, **169**, https://doi.org/10.1007/s00410-015-1141-8

Giggenbach, W.F., Kyle, P.R. and Lyon, G. 1973. Present volcanic activity on Mt. Erebus, Ross Island, Antarctica. *Geology*, **1**, 135–136, https://doi.org/10.1130/0091-7613(1973)1<135:PVAOME>2.0.CO;2

Gill, J. 1981. *Orogenic Andesites and Plate Tectonics*. Springer, Berlin, https://doi.org/10.1007/978-3-642-68012-0

Graham, D.W., Humphris, S.E., Jenkins, W.J. and Kurz, M.D. 1992. Helium isotope geochemistry of some volcanic rocks from Saint Helena. *Earth and Planetary Science Letters*, **110**, 121–131, https://doi.org/10.1016/0012-821X(92)90043-U

Green, D.H. 1973. Experimental melting studies on a model upper mantle composition at high pressure under water-saturated and water-undersaturated conditions. *Earth and Planetary Science Letters*, **19**, 37–53, https://doi.org/10.1016/0012-821X(73)90176-3

Green, D.H. and Falloon, T.J. 1998. Pyrolite: A Ringwood concept and its current expression. *In*: Jackson, I. (ed.) *The Earth's Mantle: Composition, Structure, and Evolution*. Cambridge University Press, Cambridge, UK, 311–378.

Green, D.H. and Ringwood, A.E. 1967. The genesis of basaltic magmas. *Contributions to Mineralogy and Petrology*, **15**, 103–190, https://doi.org/10.1007/BF00372052

Gupta, S., Zhao, D. and Rai, S.S. 2009. Seismic imaging of the upper mantle under the Erebus hotspot in Antarctica. *Gondwana Research*, **16**, 109–118, https://doi.org/10.1016/j.gr.2009.01.004

Hamilton, T.L., Vogl, K., Bryant, D.A., Boyd, E.S. and Peters, J.W. 2012. Environmental constraints defining the distribution, composition, and evolution of chlorophototrophs in thermal features of Yellowstone National Park. *Geobiology*, **10**, 236–249, https://doi.org/10.1111/j.1472-4669.2011.00296.x

Hansen, S.E., Graw, J.H. et al. 2014. Imaging the Antarctic mantle using adaptively parameterized P-wave tomography: Evidence for heterogeneous structure beneath West Antarctica. *Earth and Planetary Science Letters*, **408**, 66–78, https://doi.org/10.1016/j.epsl.2014.09.043

Hanyu, T., Tatsumi, Y. et al. 2011. Geochemical characteristics and origin of the HIMU reservoir: A possible mantle plume source in the lower mantle. *Geochemistry, Geophysics, Geosystems*, **12**, Q0AC09, https://doi.org/10.1029/2010GC003252

Hanyu, T., Kawabata, H. et al. 2014. Isotope evolution in the HIMU reservoir beneath St. Helena: Implications for the mantle recycling of U and Th. *Geochimica et Cosmochimica Acta*, **143**, 232–252, https://doi.org/10.1016/j.gca.2014.03.016

Harpel, C.J., Kyle, P.R., Esser, R.P., McIntosh, W.C. and Caldwell, D.A. 2004. $^{40}Ar/^{39}Ar$ dating of the eruptive history of Mount Erebus, Antarctica: summit flows, tephra, and caldera collapse. *Bulletin of Volcanology*, **66**, 687–702, https://doi.org/10.1007/s00445-004-0349-7

Harpel, C.J., Kyle, P.R. and Dunbar, N.W. 2008. Englacial tephrostratigraphy of Erebus volcano, Antarctica. *Journal of Volcanology and Geothermal Research*, **177**, 549–568, https://doi.org/10.1016/j.jvolgeores.2008.06.001

Hart, S.R. 1988. Heterogeneous mantle domains: signatures, genesis and mixing chronologies. *Earth and Planetary Science Letters*, **90**, 273–296, https://doi.org/10.1016/0012-821X(88)90131-8

Hart, S.R., Hauri, E.H., Oschmann, L.A. and Whitehead, J.A. 1992. Mantle plumes and entrainment: isotopic evidence. *Science*, **256**, 517–520, https://doi.org/10.1126/science.256.5056.517

Hart, S.R., Blusztajn, J. and Craddock, C. 1995. Cenozoic volcanism in Antarctica; Jones Mountains and Peter I Island. *Geochimica et Cosmochimica Acta*, **59**, 3379–3388, https://doi.org/10.1016/0016-7037(95)00212-I

Hart, S.R., Blusztajn, J., LeMasurier, W.E. and Rex, D.C. 1997. Hobbs Coast Cenozoic volcanism: implications for the West Antarctic rift system. *Chemical Geology*, **139**, 223–248, https://doi.org/10.1016/S0009-2541(97)00037-5

Hartley, M.E., Maclennan, J., Edmonds, M. and Thordarson, T. 2014. Reconstructing the deep CO_2 degassing behaviour of large basaltic fissure eruptions. *Earth and Planetary Science Letters*, **393**, 120–131, https://doi.org/10.1016/j.epsl.2014.02.031

Hawkesworth, C., George, R., Turner, S. and Zellmer, G. 2004. Timescales of magmatic processes. *Earth and Planetary Science Letters*, **218**, 1–16, https://doi.org/10.1016/S0012-821X(03)00634-4

Heeszel, D.S., Wiens, D. et al. 2016. Upper mantle structure of central and West Antarctica from array analysis of Rayleigh wave phase velocities. *Journal of Geophysical Research: Solid Earth*, **121**, 1758–1775, https://doi.org/10.1002/2015JB012616

Herbold, C.W., Lee, C.K., Mcdonald, I.R. and Cary, S.C. 2014a. Evidence of global-scale aeolian dispersal and endemism in isolated geothermal microbial communities of Antarctica. *Nature Communications*, **5**, 3875, https://doi.org/10.1038/ncomms4875

Herbold, C.W., McDonald, I.R. and Cary, S.C. 2014b. Microbial ecology of geothermal habitats in Antarctica. *In*: Cowan, D.A. (ed.) *Antarctic Terrestrial Microbiology: Physical and*

Biological Properties of Antarctic Soils. Springer, Berlin, 181–215, https://doi.org/10.1007/978-3-642-45213-0_10

Hill, G., Wannamaker, P.E. et al. 2017. Imaging the magmatic system of Erebus volcano, Antarctica using the magnetotelluric method. Presented at the Geological Society of America Fall Meeting, 11–15 December 2017, New Orleans, Loisiana, USA.

Hirose, K. 1997. Partial melt compositions of carbonated peridotite at 3 GPa and role of CO_2 in alkali-basalt magma generation. Geophysical Research Letters, 24, 2837–2840, https://doi.org/10.1029/97GL02956

Hirschmann, M., Kogiso, T., Baker, M.B. and Stolper, E.M. 2003. Alkalic magmas generated by partial melting of garnet pyroxenite. Geology, 31, 481–484, https://doi.org/10.1130/0091-7613(2003)031<0481:AMGBPM>2.0.CO;2

Hoch, M., Rehkämper, M. and Tobschall, H.J. 2001. Sr, Nd, Pb and O isotopes of minettes from Schirmacher Oasis, East Antarctica: A case of mantle metasomatism involving subducted continental material. Journal of Petrology, 42, 1387–1400, https://doi.org/10.1093/petrology/42.7.1387

Hoernle, K., White, J.D.L. et al. 2006. Cenozoic intraplate volcanism on New Zealand: Upwelling induced by lithospheric removal. Earth and Planetary Science Letters, 248, 350–367, https://doi.org/10.1016/j.epsl.2006.06.001

Hoernle, K., Hauff, F. et al. 2010. Age and geochemistry of volcanic rocks from the Hikurangi and Manihiki oceanic Plateaus. Geochimica et Cosmochimica Acta, 74, 7196–7219, https://doi.org/10.1016/j.gca.2010.09.030

Howat, I.M., Porter, C., Smith, B.E., Noh, M.-J. and Morin, P. 2019. The Reference Elevation Model of Antarctica. The Cryosphere, 13, 665–674, https://doi.org/10.5194/tc-13-665-2019

Hudson, J.A. and Daniel, R.M. 1988. Enumeration of thermophilic heterotrophs in geothermally heated soils from Mount Erebus, Ross Island, Antarctica. Applied and Environmental Microbiology, 54, 622–624, https://doi.org/10.1128/AEM.54.2.622-624.1988

Hudson, J.A., Daniel, R.M. and Morgan, H.W. 1988. Isolation of a strain of Bacillus schlegelii from geothermally heated Antarctic soil. FEMS Microbiology Letters, 51, 57–60, https://doi.org/10.1111/j.1574-6968.1988.tb02968.x

Hudson, J.A., Daniel, R.M. and Morgan, H.W. 1989. Acidophilic and thermophilic Bacillus strains from geothermally heated Antarctic soil. FEMS Microbiology Letters, 60, 279–282, https://doi.org/10.1111/j.1574-6968.1989.tb03486.x

Iacovino, K. 2015. Linking subsurface to surface degassing at active volcanoes: A thermodynamic model with applications to Erebus volcano. Earth and Planetary Science Letters, 431, 59–74, https://doi.org/10.1016/j.epsl.2015.09.016

Iacovino, K., Moore, G., Roggensack, K., Oppenheimer, C. and Kyle, P. 2013. H_2O–CO_2 solubility in mafic alkaline magma: applications to volatile sources and degassing behavior at Erebus volcano, Antarctica. Contributions to Mineralogy and Petrology, 166, 845–860, https://doi.org/10.1007/s00410-013-0877-2

Iacovino, K., Oppenheimer, C., Scaillet, B. and Kyle, P. 2016. Storage and evolution of mafic and intermediate alkaline magmas beneath Ross Island. Antarctica. Journal of Petrology, 57, 93–117, https://doi.org/10.1093/petrology/egv083

Ilanko, T., Oppenheimer, C., Burgisser, A. and Kyle, P.R. 2015. Cyclic degassing of Erebus volcano, Antarctica. Bulletin of Volcanology, 77, 56, https://doi.org/10.1007/s00445-015-0941-z

Ilyinskaya, E., Oppenheimer, C., Mather, T.A., Martin, R.S. and Kyle, P.R. 2010. Size-resolved chemical composition of aerosol emitted by Erebus volcano, Antarctica. Geochemistry, Geophysics, Geosystems, 11, Q03017, https://doi.org/10.1029/2009GC002855

Iverson, N.A., Kyle, P.R., Dunbar, N.W., McIntosh, W.C. and Pearce, N.J.G. 2014. Eruptive history and magmatic stability of Erebus volcano, Antarctica: Insights from englacial tephra. Geochemistry, Geophysics, Geosystems, 15, 4180–4202, https://doi.org/10.1002/2014GC005435

Johnson, J.B. and Aster, R. 2005. Relative partitioning of acoustic and seismic energy during Strombolian eruptions. Journal of Volcanology and Geothermal Research, 148, 334–354, https://doi.org/10.1016/j.jvolgeores.2005.05.002

Johnson, J.B., Aster, R.C., Ruiz, M.C., Malone, S.D., McChesney, P.J., Lees, J.M. and Kyle, P.R. 2003. Interpretation and utility of infrasonic records from erupting volcanoes. Journal of Volcanology and Geothermal Research, 121, 15–63, https://doi.org/10.1016/S0377-0273(02)00409-2

Johnson, J.B., Aster, R.C. and Kyle, P.R. 2004. Volcanic eruptions observed with infrasound. Geophysical Research Letters, 31, L14604, https://doi.org/10.1029/2004GL020020

Johnson, J., Aster, R., Jones, K.R., Kyle, P. and McIntosh, W. 2008. Acoustic source characterization of impulsive Strombolian eruptions from the Mount Erebus lava lake. Journal of Volcanology and Geothermal Research, 177, 673–686, https://doi.org/10.1016/j.jvolgeores.2008.06.028

Jones, K.R., Johnson, J.B., Aster, R., Kyle, P.R. and McIntosh, W.C. 2008. Infrasonic tracking of large bubble bursts and ash venting at Erebus Volcano, Antarctica. Journal of Volcanology and Geothermal Research, 177, 661–672, https://doi.org/10.1016/j.jvolgeores.2008.02.001

Kelemen, P.B., Shimizu, N. and Dunn, T. 1993. Relative depletion of niobium in some arc magmas and the continental crust: partitioning of K, Nb, La, and Ce during melt/rock reaction in the upper mantle. Earth and Planetary Science Letters, 120, 111–134, https://doi.org/10.1016/0012-821X(93)90234-Z

Kelly, P.J., Dunbar, N.W., Kyle, P.R. and McIntosh, W.C. 2008a. Refinement of the late Quaternary geologic history of Erebus volcano, Antarctica using $^{40}Ar/^{39}Ar$ and ^{36}Cl age determinations. Journal of Volcanology and Geothermal Research, 177, 569–577, https://doi.org/10.1016/j.jvolgeores.2008.07.018

Kelly, P.J., Kyle, P.R., Dunbar, N.W. and Sims, K.W.W. 2008b. Geochemistry and mineralogy of the phonolite lava lake, Erebus volcano, Antarctica: 1972–2004 and comparison with older lavas. Journal of Volcanology and Geothermal Research, 177, 589–605, https://doi.org/10.1016/j.jvolgeores.2007.11.025

Knox, H.A., Chaput, J.A., Aster, R.C. and Kyle, P.R. 2018. Multiyear shallow conduit changes observed with lava lake eruption seismograms at Erebus volcano, Antarctica. Journal of Geophysical Research: Solid Earth, 123, 3178–3196, https://doi.org/10.1002/2017JB015045

Kogiso, T., Hirose, K. and Takahashi, E. 1998. Melting experiments on homogeneous mixtures of peridotite and basalt: application to the genesis of ocean island basalts. Earth and Planetary Science Letters, 162, 45–61, https://doi.org/10.1016/S0012-821X(98)00156-3

Kogiso, T., Hirschmann, M.M. and Frost, D.J. 2003. High-pressure partial melting of garnet pyroxenite: possible mafic lithologies in the source of ocean island basalts. Earth and Planetary Science Letters, 216, 603–617, https://doi.org/10.1016/S0012-821X(03)00538-7

Kyle, P.R. 1977. Mineralogy and glass chemistry of volcanic ejecta, from Mt. Erebus, Antarctica. New Zealand Journal of Geology and Geophysics, 20, 1123–1146, https://doi.org/10.1080/00288306.1977.10420699

Kyle, P.R. 1981. Mineralogy and geochemistry of a basanite to phonolite sequence at Hut Point Peninsula, Antarctica, based on Core from Dry Valley Drilling Project drillholes 1, 2 and 3. Journal of Petrology, 22, 451–500, https://doi.org/10.1093/petrology/22.4.451

Kyle, P.R. 1990a. McMurdo Volcanic Group Western Ross Embayment: Introduction. American Geophysical Union Antarctic Research Series, 48, 18–25.

Kyle, P.R. 1990b. Erebus volcanic Province: Summary. American Geophysical Union Antarctic Research Series, 48, 81–88.

Kyle, P.R. and McIntosh, W.C. 1989. Automation of a correlation spectrometer for measuring SO_2 emissions. New Mexico Bureau of Mines and Mineral Resources Bulletin, 131, 158.

Kyle, P., Dibble, R., Giggenbach, W. and Keys, J. 1982. Volcanic activity associated with the anorthoclase phonolite lava lake, Mt. Erebus, Antarctica. In: Craddock, C. (ed.) Antarctic Geosciences. University of Wisconsin Press, Madison, WI, 735–745.

Kyle, P.R., Meeker, K. and Finnegan, D. 1990. Emission rates of sulfur dioxide trace gases and metals from Mount Erebus

Antarctica. *Geophysical Research Letters*, **17**, 2125–2128, https://doi.org/10.1029/GL017i012p02125

Kyle, P.R., Moore, J.A. and Thirlwall, M.F. 1992. Petrologic evolution of anorthoclase phonolite lavas at Mount Erebus, Ross Island, Antarctica. *Journal of Petrology*, **33**, 849–875, https://doi.org/10.1093/petrology/33.4.849

Kyle, P.R., Sybeldon, L.M., McIntosh, W.C., Meeker, K. and Symonds, R. 1994. Sulfur dioxide emission rates from Mount Erebus, Antarctica. *American Geophysical Union Antarctic Research Series*, **66**, 69–82, https://doi.org/10.1029/AR066p0069

Leat, P.T., Dean, A.A., Millar, I.L., Kelley, S.P., Vaughan, A.P.M. and Riley, T.R. 2005. Lithospheric mantle domains beneath Antarctica. *Geological Society, London, Special Publications*, **246**, 359–380, https://doi.org/10.1144/GSL.SP.2005.246.01.15

Le Losq, C., Neuville, D., Moretti, R., Kyle, P. and Oppenheimer, C. 2015. Rheology of phonolitic magmas – the case of the Erebus lava lake. *Earth and Planetary Science Letters*, **411**, https://doi.org/10.1016/j.epsl.2014.11.042

Lesser, M.P., Barry, T.M. and Banaszak, A.T. 2002. Effects of UV radiation on a chlorophyte alga (*Scenedesmus* sp.) isolated from the fumarole fields of Mt. Erebus, Antarctica. *Journal of Phycology*, **38**, 473–481, https://doi.org/10.1046/j.1529-8817.2002.01171.x

Lindsay, M.R., Amenabar, M.J. et al. 2018. Subsurface processes influence oxidant availability and chemoautotrophic hydrogen metabolism in Yellowstone hot springs. *Geobiology*, **16**, 674–692, https://doi.org/10.1111/gbi.12308

Lowenstern, J.B. and Hurwitz, S. 2008. Monitoring a supervolcano in repose: Heat and volatile flux at the Yellowstone Caldera. *Elements*, **4**, 35–40, https://doi.org/10.2113/GSELEMENTS.4.1.35

MacAyeal, D.R., Okal, E.A., Aster, R.C. and Bassis, J.N. 2008. Seismic and hydroacoustic tremor generated by colliding icebergs. *Journal of Geophysical Research: Earth Surface*, **113**, F03011, https://doi.org/10.1029/2008JF001005

Martin, R.S., Mather, T.A. et al. 2008. Composition-resolved size distributions of volcanic aerosols in the Mt. Etna plumes. *Journal of Geophysical Research: Atmospheres*, **113**, D17211, https://doi.org/10.1029/2007JD009648

Martin, S., Drucker, R., Aster, R., Davey, F., Okal, E., Scambos, T. and MacAyeal, D. 2010. Kinematic and seismic analysis of giant tabular iceberg breakup at Cape Adare, Antarctica. *Journal of Geophysical Research: Solid Earth*, **115**, B06311, https://doi.org/10.1029/2009JB006700

Mather, T.A., Allen, A.G., Oppenheimer, C., Pyle, D.M. and McGonigle, A.J.S. 2003. Size-resolved characterization of soluble ions in the particles in the tropospheric plume of Masaya Volcano, Nicaragua: Origins and plume processing. *Journal of Atmospheric Chemistry*, **46**, 207–237, https://doi.org/10.1023/A:1026327502060

McDonough, W.F. and Sun, S.S. 1995. The composition of the Earth. *Chemical Geology*, **120**, 223–253, https://doi.org/10.1016/0009-2541(94)00140-4

McGonigle, A.J.S. and Oppenheimer, C. 2003. Optical sensing of volcanic gas and aerosol emissions. *Geological Society, London, Special Publications*, **213**, 149–168, https://doi.org/10.1144/GSL.SP.2003.213.01.09

Meeker, K.A., Chuan, R.L., Kyle, P.R. and Palais, J.M. 1991. Emission of elemental gold particles from Mount Erebus, Ross Island, Antarctica. *Geophysical Research Letters*, **18**, 1405–1408, https://doi.org/10.1029/91GL01928

Melick, D.R., Broady, P.A. and Rowan, K.S. 1991. Morphological and physiological characteristics of a non-heterocystous strain of the cyanobacterium *Mastigocladus laminosus* Cohn from fumarolic soil on Mt Erebus, Antarctica. *Polar Biology*, **11**, 81–89, https://doi.org/10.1007/BF00234270

Mickus, K. and Kyle, P. 2013. Preliminary gravity survey of the Erebus Volcano, Antarctica. Presented at the Geological Society of America Annual Meeting, 27–30 October 2013, Denver, Colorado, USA.

Mickus, K. and Kyle, P. 2014. A detailed gravity survey of the Erebus Volcano, Antarctica. Presented at the Geological Society of America Annual Meeting, 19–22 October 2014, Vancouver, Canada.

Molina, I., Burgisser, A. and Oppenheimer, C. 2012. Numerical simulations of convection in crystal-bearing magmas: a case study of the magmatic system at Erebus, Antarctica. *Journal of Geophysical Research*, **117**, B07209, https://doi.org/10.1029/2011JB008760

Montelli, R., Nolet, G., Dahlen, F.A. and Masters, G. 2006. A catalogue of deep mantle plumes: New results from finite frequency tomography. *Geochemistry, Geophysics, Geosystems*, **7**, Q11007, https://doi.org/10.1029/2006GC001248

Moore, L.R., Gazel, E. et al. 2015. Bubbles matter: An assessment of the contribution of vapor bubbles to melt inclusion volatile budgets. *American Mineralogist*, **100**, 806–823, https://doi.org/10.2138/am-2015-5036

Moussallam, Y., Oppenheimer, C., Aiuppa, A., Giudice, G. and Moussallam, M. 2012. Hydrogen emissions from Erebus volcano, Antarctica. *Bulletin of Volcanology*, **74**, 2109–2120, https://doi.org/10.1007/s00445-012-0649-2

Moussallam, Y., Oppenheimer, C., Scaillet, B. and Kyle, P.R. 2013. Experimental phase-equilibrium constraints on the phonolite magmatic system of Erebus volcano, Antarctica. *Journal of Petrology*, **54**, 1285–1307, https://doi.org/10.1093/petrology/egt012

Moussallam, Y., Oppenheimer, C. et al. 2014. Tracking the changing oxidation state of Erebus magmas, from mantle to surface, driven by magma ascent and degassing. *Earth and Planetary Science Letters*, **393**, 200–209, https://doi.org/10.1016/j.epsl.2014.02.055

Moussallam, Y.C., Oppenheimer, B. et al. 2015. Megacrystals track magma convection between reservoir and surface. *Earth and Planetary Science Letters*, **413**, 1–12, https://doi.org/10.1016/j.epsl.2014.12.022

Nicolaus, B., Marsiglia, F. et al. 1992. Isolation of extremely halotolerant cocci from Antarctica. *FEMS Microbiology Letters*, **99**, 145–149, https://doi.org/10.1111/j.1574-6968.1992.tb05557.x

Nicolaus, B., Lama, L., Esposito, E., Bellitti, M.R., Improta, R., Panico, A. and Gambacorta, A. 2000. Extremophiles in Antarctica. *Italian Journal of Zoology*, **67**, 169–174, https://doi.org/10.1080/11250000009356373

Oppenheimer, C. and Kyle, P. 2008a. Preface: Volcanology of Erebus volcano, Antarctica. *Journal of Volcanology and Geothermal Research*, **177**, v–vii, https://doi.org/10.1016/j.jvolgeores.2008.10.006

Oppenheimer, C. and Kyle, P.R. 2008b. Probing the magma plumbing of Erebus volcano, Antarctica, by open-path FTIR spectroscopy of gas emissions. *Journal of Volcanology and Geothermal Research*, **177**, 743–754, https://doi.org/10.1016/j.jvolgeores.2007.08.022

Oppenheimer, C., Kyle, P.R., Tsanev, V.I., McGonigle, A.J.S., Mather, T.A. and Sweeney, D. 2005. Mt. Erebus, the largest point source of NO_2 in Antarctica. *Atmospheric Environment*, **39**, 6000–6006, https://doi.org/10.1016/j.atmosenv.2005.06.036

Oppenheimer, C., Lomakina, A.S., Kyle, P.R., Kingsbury, N.G. and Boichu, M. 2009. Pulsatory magma supply to a phonolite lava lake. *Earth and Planetary Science Letters*, **284**, 392–398, https://doi.org/10.1016/j.epsl.2009.04.043

Oppenheimer, C., Kyle, P.R. et al. 2010. Atmospheric chemistry of an Antarctic volcanic plume. *Journal of Geophysical Research: Atmospheres*, **115**, D04303, https://doi.org/10.1029/2009JD011910

Oppenheimer, C., Moretti, R., Kyle, P.R., Eschenbacher, A., Lowenstern, J.B., Hervig, R.L. and Dunbar, N.W. 2011. Mantle to surface degassing of alkalic magmas at Erebus volcano, Antarctica. *Earth and Planetary Science Letters*, **306**, 261–271, https://doi.org/10.1016/j.epsl.2011.04.005

Panter, K.S., Hart, S.R., Kyle, P., Blusztanjn, J. and Wilch, T. 2000. Geochemistry of Late Cenozoic basalts from the Crary Mountains: characterization of mantle sources in Marie Byrd Land, Antarctica. *Chemical Geology*, **165**, 215–241, https://doi.org/10.1016/S0009-2541(99)00171-0

Panter, K.S., Blusztajn, J., Hart, S.R., Kyle, P.R., Esser, R. and McIntosh, W.C. 2006. The Origin of HIMU in the SW Pacific: Evidence from intraplate volcanism in southern New Zealand and Subantarctic Islands. *Journal of Petrology*, **47**, 1673–1704, https://doi.org/10.1093/petrology/egl024

Parai, R., Mukhopadhyay, S. and Lassiter, J.C. 2009. New constraints on the HIMU mantle from neon and helium isotopic compositions of basalts from the Cook–Austral Islands. *Earth and Planetary Science Letters*, **277**, 253–261, https://doi.org/10.1016/j.epsl.2008.10.014

Park, S.-H., Langmuir, C.H. et al. 2019. An isotopically distinct Zealandia–Antarctic mantle domain in the Southern Ocean. *Nature Geoscience*, **12**, 206–214, https://doi.org/10.1038/s41561-018-0292-4

Parmelee, D.E.F., Kyle, P.R., Kurz, M.D., Marrero, S.M. and Phillips, F.M. 2015. A new Holocene eruptive history of Erebus Volcano, Antarctica using cosmogenic ^{3}He and ^{36}Cl exposure ages. *Quaternary Geochronology*, **30**, 114–131, https://doi.org/10.1016/j.quageo.2015.09.001

Peterson, J. 1993. *Observations and Modeling of Seismic Background Noise*. United States Geological Survey Open-File Report, **93-322**, https://doi.org/10.3133/ofr93322

Phillips, E.H., Sims, K.W.W. et al. 2018. The nature and evolution of mantle upwelling at Ross Island, Antarctica, with implications for the source of HIMU lavas. *Earth and Planetary Science Letters*, **498**, 38–53, https://doi.org/10.1016/j.epsl.2018.05.049

Pilet, S., Baker, M.B. and Stolper, E.M. 2008. Metasomatized lithosphere and the origin of alkaline lavas. *Science*, **320**, 916–919, https://doi.org/10.1126/science.1156563

Portnyagin, M., Hoernle, K., Storm, S., Mironov, N., van den Bogaard, C. and Botcharnikov, R. 2012. H$_2$O-rich melt inclusions in fayalitic olivine from Hekla volcano: Implications for phase relationships in silicic systems and driving forces of explosive volcanism on Iceland. *Earth and Planetary Science Letters*, **357–358**, 337–346, https://doi.org/10.1016/j.epsl.2012.09.047

Power, J.F., Carere, C.R. et al. 2018. Microbial biogeography of 925 geothermal springs in New Zealand. *Nature Communications*, **9**, 2876, https://doi.org/10.1038/s41467-018-05020-y

Pyle, D.M. and Mather, T.A. 2009. Halogens in igneous processes and their fluxes to the atmosphere and oceans from volcanic activity: a review. *Chemical Geology*, **263**, 110–121, https://doi.org/10.1016/j.chemgeo.2008.11.013

Radke, L.F. 1982. Sulphur and sulphate from Mt Erebus. *Nature*, **299**, 710–712, https://doi.org/10.1038/299710a0

Rasmussen, D.J., Kyle, P.R., Wallace, P.J., Sims, K.W.W., Gaetani, G.A. and Phillips, E.H. 2017. Understanding degassing and transport of CO$_2$-rich alkalic magmas at Ross Island, Antarctica using olivine-hosted melt inclusions. *Journal of Petrology*, **58**, 841–861, https://doi.org/10.1093/petrology/egx036

Reagan, M.K., Volpe, A.M. and Cashman, K.V. 1992. ^{238}U- and ^{232}Th- series chronology of phonolite fractionation at Mt Erebus, Antarctica. *Geochimica et Cosmochimica Acta*, **56**, 1401–1407, https://doi.org/10.1016/0016-7037(92)90071-P

Reagan, M.K., Sims, K.W.W. et al. 2003. Time-scale of differentiation from mafic parents to rhyolite in North American continental arcs. *Journal of Petrology*, **44**, 1703–1726, https://doi.org/10.1093/petrology/egg057

Reagan, M.K., Tepley, F.J.III, Gill, J.B., Wortel, M. and Garrison, J. 2006. Timescales of degassing and crystallization implied by ^{210}Po–^{210}Pb–^{226}Ra disequilibria for andesitic lavas erupted from Arenal volcano. *Journal of Volcanology and Geothermal Research*, **157**, 135–146, https://doi.org/10.1016/j.jvolgeores.2006.03.044

Reagan, M.K., Turner, S., Legg, M.K., Sims, K.W.W. and Hards, V.L. 2008. ^{238}U and ^{232}Th decay series constraints on the timescales of crystal fractionation to produce the phonolite erupted in 2004 near Tristan da Cunha, South Atlantic. *Geochimica et Cosmchimica Acta*, **72**, 4367–4378, https://doi.org/10.1016/j.gca.2008.06.002

Reid, M.R., Bouchet, R.A., Blichert-Toft, J., Levander, A., Liu, K., Miller, M.S. and Ramos, F.C. 2012. Melting under the Colorado Plateau, USA. *Geology*, **40**, 387–390, https://doi.org/10.1130/G32619.1

Reubi, O., Sims, K.W.W. and Bourdon, B. 2014. ^{238}U–^{230}Th equilibrium in arc magmas and implications for the time scales of mantle metasomatism. *Earth and Planetary Science Letters*, **391**, 146–158, https://doi.org/10.1016/j.epsl.2014.01.054

Reubi, O., Scott, S.R. and Sims, K.W.W. 2017. Young U-series crystal ages in andesitic magmas from a hyperactive arc volcano. *Journal of Petrology*, **58**, 261–276, https://doi.org/10.1093/petrology/egx015

Ritsema, J., van Heijst, H.J., Deuss, A. and Woodhouse, J.H. 2011. S40RTS: a degree-40 shear-velocity model for the mantle from new Rayleigh wave dispersion, teleseismic traveltimes, and normal-mode splitting function measurements. *Geophysical Journal International*, **184**, 1223–1236, https://doi.org/10.1111/j.1365-246X.2010.04884.x

Ritzwoller, M.H., Shapiro, N.M., Levshin, A.L. and Leahy, G.M. 2001. Crustal and upper mantle structure beneath Antarctica and surrounding oceans. *Journal of Geophysical Research: Solid Earth*, **106**, 30 645–30 670, https://doi.org/10.1029/2001JB000179

Rocchi, S., Armienti, P., D'Osazio, M., Wijbrans, J. and Di Vincenzo, G. 2002. Cenozoic magmatism in the western Ross Embayment: Role of mantle plume versus plate dynamics in the development of the West Antarctic Rift System. *Journal of Geophysical Research; Solid Earth*, **107**, 2195, https://doi.org/10.1029/2001JB000515

Rocholl, A., Stein, M., Molzahn, M., Hart, S.R. and Worner, G. 1995. Geochemical evolution of rift magmas by progressive tapping of a stratified mantle source beneath the Ross Rift, Antarctica. *Earth and Planetary Science Letters*, **131**, 207–224, https://doi.org/10.1016/0012-821X(95)00024-7

Rose, W.I., Chuan, R.L. and Kyle, P.R. 1985. Rate of sulphur dioxide emission from Erebus volcano, Antarctica, December 1983. *Nature*, **316**, 710–712, https://doi.org/10.1038/316710a0

Ross, J.C. 1847. *A Voyage of Discovery and Research in the Southern Antarctic Regions during the Years 1839–43*. John Murray, London, https://doi.org/10.5962/bhl.title.98449

Rothschild, L.J. and Mancinelli, R.L. 2001. Life in extreme environments. *Nature*, **409**, 1092, https://doi.org/10.1038/35059215

Rowe, C.A., Aster, R.C., Kyle, P.R., Schlue, J.W. and Dibble, R.R. 1998. Broadband recording of Strombolian explosions and associated very-long-period seismic signals on Mount Erebus volcano, Ross Island, Antarctica. *Geophysical Research Letters*, **25**, 2297–2300, https://doi.org/10.1029/98GL01622

Rowe, C.A., Aster, R.C., Kyle, P.R., Dibble, R.R. and Schlue, J.W. 2000. Seismic and acoustic observations at Mount Erebus Volcano, Ross Island, Antarctica, 1994–1998. *Journal of Volcanology and Geothermal Research*, **101**, 105–128, https://doi.org/10.1016/S0377-0273(00)00170-0

Saal, A.E., Hart, S.R., Shimizu, N., Hauri, E.H. and Layne, G.D. 1998. Pb isotopic variability in melt inclusions from oceanic island basalts, Polynesia. *Science*, **282**, 1481–1484, https://doi.org/10.1126/science.282.5393.1481

Saal, A.E., Hauri, E.H., Langmuir, C.H. and Perfit, M.R. 2002. Vapour undersaturation in primitive mid-ocean-ridge basalt and the volatile content of Earth's upper mantle. *Nature*, **419**, 451–455, https://doi.org/10.1038/nature01073

Salters, V.J.M. and Stracke, A. 2004. Composition of the depleted mantle. *Geochemistry, Geophysics, Geosystems*, **5**, Q05B07, https://doi.org/10.1029/2003GC000597

Sawyer, G.M., Carn, S.A., Tsanev, V.I., Oppenheimer, C. and Burton, M. 2008. Investigation into magma degassing at Nyiragongo volcano, Democratic Republic of the Congo. *Geochemistry, Geophysics, Geosystems*, **9**, Q02017, https://doi.org/10.1029/2007GC001829

Seaman, S.J., Dyar, M.D., Marinkovic, N. and Dunbar, N.W. 2006. An FTIR study of hydrogen in anorthoclase and associated melt inclusions. *American Mineralogist*, **91**, 12–20, https://doi.org/10.2138/am.2006.1765

Shackleton, E.H.S., David, T.W.E.S. and Mill, H.R. 1909. *The Heart of the Antarctic: Being the Story of the British Antarctic*

Expedition 1907-1909. J.B. Lippincott, Philadelphia, PA, https://doi.org/10.5962/bhl.title.82322

Shock, E.L. and Holland, M.E. 2007. Quantitative habitability. *Astrobiology*, **7**, 839–851, https://doi.org/10.1089/ast.2007.0137

Shock, E.L., Holland, M., Meyer-Dombard, D.A., Amend, J.P., Osburn, G.R. and Fischer, T.P. 2010. Quantifying inorganic sources of geochemical energy in hydrothermal ecosystems, Yellowstone National Park, USA. *Geochimica et Cosmochimica Acta*, **74**, 4005–4043, https://doi.org/10.1016/j.gca.2009.08.036

Simmons, N.A., Forte, A.M., Boschi, L. and Grand, S.P. 2010. GyPSuM: A joint tomographic model of mantle density and seismic wave speeds. *Journal of Geophysical Research*, **115**, B12310, https://doi.org/10.1029/2010JB007631

Sims, K.W.W. and Hart, S.R. 2006. Comparison of Th, Sr, Nd and Pb isotopes in oceanic basalts: Implications for mantle heterogeneity and magma genesis. *Earth and Planetary Science Letters*, **245**, 743–761, https://doi.org/10.1016/j.epsl.2006.02.030

Sims, K.W.W., Ackert, R.P. Jr, Ramos, F., Sohn, R.A., Murrell, M.T. and DePaolo, D.J. 2007. Determining eruption ages and erosion rates of Quaternary basaltic volcanism from combined U-series disequilibria and cosmogenic exposure ages. *Geology*, **35**, 471–474, https://doi.org/10.1130/G23381A.1

Sims, K.W.W., Blichert-Toft, J. *et al.* 2008a. A Sr, Nd, Hf, and Pb isotope perspective on the genesis and long-term evolution of alkaline magmas from Erebus volcano, Antarctica. *Journal of Volcanology and Geothermal Research*, **177**, 606–618, https://doi.org/10.1016/j.jvolgeores.2007.08.006

Sims, K.W.W., Hart, S.R. *et al.* 2008b. $^{238}U-^{230}Th-^{226}Ra-^{210}Pb-^{210}Po$, $^{232}Th-^{228}Ra$ and $^{235}U-^{231}Pa$ constraints on the ages and petrogenesis of Vailulu and Malumalu Lavas, Samoa. *Geochemistry, Geophysics, Geosystems*, **9**, Q04003, https://doi.org/10.1029/2007GC001651

Sims, K.W.W., Maclennan, J., Blichert-Toft, J., Mervine, E.M., Bluzstajn, J. and Grönvold, K. 2013a. Short length scale mantle heterogeneity beneath Iceland probed by glacial modulation of melting. *Earth and Planetary Science Letters*, **379**, 146–157, https://doi.org/10.1016/j.epsl.2013.07.027

Sims, K.W.W., Pichat, S. *et al.* 2013b. On the time scales of magma genesis, melt evolution, crystal growth rates and magma degassing in the Erebus volcano magmatic system using the ^{238}U, ^{235}U and ^{232}Th decay series. *Journal of Petrology*, **54**, 235–271, https://doi.org/10.1093/petrology/egs068

Skotnicki, M.L., Selkirk, P.M., Broady, P., Adam, K.D. and Ninham, J.A. 2004. Dispersal of the moss *Campylopus pyriformis* on geothermal ground near the summits of Mount Erebus and Mount Melbourne, Victoria Land, Antarctica. *Antarctic Science*, **13**, 280–285, https://doi.org/10.1017/S0954102001000396

Soo, R.M., Wood, S.A., Grzymski, J.J., McDonald, I.R. and Cary, S.C. 2009. Microbial biodiversity of thermophilic communities in hot mineral soils of Tramway Ridge, Mount Erebus, Antarctica. *Environmental Microbiology*, **11**, 715–728, https://doi.org/10.1111/j.1462-2920.2009.01859.x

Storey, B.C., Leat, P.T., Weaver, S.D., Pankhurst, R.J., Bradshaw, J.D. and Kelley, S. 1999. Mantle plumes and Antarctica–New Zealand rifting: evidence from mid-Cretaceous mafic dykes. *Journal of the Geological Society, London*, **156**, 659–671, https://doi.org/10.1144/gsjgs.156.4.0659

Stuckless, J.S. and Ericksen, R.U. 1976. Sr isotope geochemistry of the volcanic rocks and associated megacrysts and inclusions from Ross Island and vicinity, Antarctica. *Contributions to Mineralogy and Petrology*, **58**, 111–126.

Sumner, C.L.K. 2007. *Residence Time Estimates and Controls on Crystallization Patterns for Anorthoclase Phenocrysts in Phonolite Magma, Erebus Volcano, Antarctica*. MS thesis, New Mexico Institute of Mining and Technology, Socorro, New Mexico, USA.

Sun, S.-S. and Hanson, G.N. 1975. Origin of Ross Island basanitoids and limitations upon the heterogeneity of mantle sources for alkali basalts and nephelinites. *Contributions to Mineralogy and Petrology*, **54**, 139–155, https://doi.org/10.1007/BF00372120

Sun, S.-S. and Hanson, G.N. 1976. Rare earth element evidence for differentiation of McMurdo Volcanics, Ross Island, Antarctica. *Contributions to Mineralogy and Petrology*, **54**, 139–155, https://doi.org/10.1007/BF00372120

Sweeney, D., Kyle, P.R. and Oppenheimer, C. 2008. Sulfur dioxide emissions and degassing behavior of Erebus volcano, Antarctica. *Journal of Volcanology and Geothermal Research*, **177**, 725–733, https://doi.org/10.1016/j.jvolgeores.2008.01.024

Symonds, R.B., Kyle, P.R. and Rose, W.I. 1985. SO_2 emission rates and the 1984 activity at Mount Erebus Volcano, Antarctica. *Eos, Transactions of the American Geophysical Union*, **66**, 417.

Takahashi, E. and Kushiro, I. 1983. Melting of dry peridotite at high pressures and basalt magma genesis. *American Mineralogist*, **68**, 859–879.

Talandier, J., Hyvernaud, O., Reymond, D. and Okal, E.A. 2006. Hydroacoustic signals generated by parked and drifting icebergs in the Southern Indian and Pacific Oceans. *Geophysical Journal International*, **165**, 817–834, https://doi.org/10.1111/j.1365-246X.2006.02911.x

Tappenden, V.E. 2003. *Magmatic Response to the Evolving New Zealand Margin of Gondwana during the Mid–Late Cretaceous*. PhD thesis, University of Canterbury, Christchurch, New Zealand.

Tebo, B.M., Davis, R.E., Anitori, R.P., Connell, L.B., Schiffman, P. and Staudigel, H. 2015. Microbial communities in dark oligotrophic volcanic ice cave ecosystems of Mt. Erebus, Antarctica. *Frontiers in Microbiology*, **6**, 179, https://doi.org/10.3389/fmicb.2015.00179

Timm, C., Hoernle, K. *et al.* 2010. Temporal and geochemical evolution of the Cenozoic intraplate volcanism of Zealandia. *Earth-Science Reviews*, **98**, 38–64, https://doi.org/10.1016/j.earscirev.2009.10.002

Toomey, D., Solomon, S. and Purdy, G. 1994. Tomographic imaging of the shallow crustal structure of the East Pacific Rise at 9°30′N. *Journal of Geophysical Research*, **99**, 24 135–24 157, https://doi.org/10.1029/94JB01942

Turner, S.P., George, R.M.M., Jerram, D.A., Carpenter, N. and Hawkesworth, C.J. 2003. Case studies of plagioclase growth and residence times in island arc lavas from Tonga and the Lesser Antilles, and a model to reconcile discordant age information. *Earth and Planetary Science Letters*, **214**, 279–294, https://doi.org/10.1016/S0012-821X(03)00376-5

Ugolini, F.C. 1967. Soils of Mount Erebus, Antarctica. *New Zealand Journal of Geology and Geophysics*, **10**, 431–442, https://doi.org/10.1080/00288306.1967.10426747

Ugolini, F.C. and Starkey, R.L. 1966. Soils and micro-organisms from Mount Erebus, Antarctica. *Nature*, **211**, 440, https://doi.org/10.1038/211440a0

Vazquez, J.A. and Reid, M.R. 2002. Time scales of magma storage and differentiation of voluminous high-silica rhyolites at Yellowstone caldera, Wyoming. *Contributions to Mineralogy and Petrology*, **144**, 274–285, https://doi.org/10.1007/s00410-002-0400-7

Vickers, C.J., Herbold, C.W., Cary, S.C. and Mcdonald, I.R. 2016. Insights into the metabolism of the high temperature microbial community of Tramway Ridge, Mount Erebus, Antarctica. *Antarctic Science*, **28**, 241–249, https://doi.org/10.1017/S095410201500067X

Wannamaker, P., Stodt, J., Hill, G., Maris, V. and Kordy, M. 2020. Thermal Regime, Legacy Structures, Upper Mantle Hydration and Lithospheric-Scale Magmatic Processes of the Antarctic Interior from Regional-Scale Electrical Properties. Presented at the AGU 2020 Fall Meeting, 1–17 December 2020. Earth and Space Science Open Archive, https://doi.org/10.1002/essoar.10505070.1

Wardell, L.J., Kyle, P.R. and Campbell, A.R. 2003. CO_2 emissions from fumarolic ice towers, Mount Erebus volcano, Antarctica. *Geological Society, London, Special Publications*, **213**, 231–246, https://doi.org/10.1144/GSL.SP.2003.213.01.14

Wardell, L.J., Kyle, P.R. and Chaffin, C. 2004. Carbon dioxide and carbon monoxide emission rates from an alkaline intra-plate volcano: Mount Erebus, Antarctica. *Journal of Volcanology and Geothermal Research*, **131**, 109–121, https://doi.org/10.1016/S0377-0273(03)00320-2

Wardell, L.J., Kyle, P.R. and Counce, D. 2008. Volcanic emissions of metals and halogens from White Island (New Zealand) and Erebus volcano (Antarctica) determined using chemical traps. *Journal of Volcanology and Geothermal Research*, **177**, 734–742, https://doi.org/10.1016/j.jvolgeores.2007.07.007

Watson, T., Nyblade, A. *et al.* 2006. P and S velocity structure of the upper mantle beneath the Transantarctic Mountains, East Antarctic craton, and Ross Sea from travel time tomography. *Geochemistry, Geophysics, Geosystems*, **7**, Q07005, https://doi.org/10.1029/2005GC001238

Weiss, Y., Class, C., Goldstein, S.L. and Hanyu, T. 2016. Key new pieces of the HIMU puzzle from olivines and diamond inclusions. *Nature*, **537**, 666–670, https://doi.org/10.1038/nature19113

Wettergreen, D., Thorpe, C. and Whittaker, R. 1993. Exploring Mount Erebus by walking robot. *Robotics and Autonomous Systems*, **11**, 171–185.

Willbold, M. and Stracke, A. 2006. Trace element composition of mantle end-members: Implications for recycling of oceanic and upper and lower continental crust. *Geochemistry, Geophysics, Geosystems*, **7**, Q04004, https://doi.org/10.1029/2005GC001005

Workman, R.K. and Hart, S.R. 2005. Major and trace element composition of the depleted MORB mantle (DMM). *Earth and Planetary Science Letters*, **231**, 53–72, https://doi.org/10.1016/j.epsl.2004.12.005

Workman, R.K., Hart, S.R. *et al.* 2004. Recycled metasomatized lithosphere as the origin of the Enriched Mantle II (EM2) endmember: Evidence from the Samoan Volcanic Chain. *Geochemistry, Geophysics, Geosystems*, **5**, Q04008, https://doi.org/10.1029/2003GC000623

Xu, Y., Schoonen, M.A.A., Nordstrom, D.K., Cunningham, K.M. and Ball, J.W. 2000. Sulfur geochemistry of hydrothermal waters in Yellowstone National Park, Wyoming, USA. II. Formation and decomposition of thiosulfate and polythionate in Cinder Pool. *Journal of Volcanology and Geothermal Research*, **97**, 407–423, https://doi.org/10.1016/S0377-0273(99)00173-0

Zandomeneghi, D., Aster, R., Kyle, P., Barclay, A.H., Chaput, J. and Knox, H. 2013. Internal structure of Erebus volcano, Antarctica imaged by high-resolution active-source seismic tomography and coda interferometry. *Journal of Geophysical Research: Solid Earth*, **118**, 1067–1078, https://doi.org/10.1002/jgrb.50073

Zhao, D. 2007. Seismic images under 60 hotspots: search for mantle plumes. *Gondwana Research*, **12**, 335–355, https://doi.org/10.1016/j.gr.2007.03.001

Zoller, W.H., Gladney, E.S. and Duce, R.A. 1974. Atmospheric concentrations and sources of trace elements at the South Pole. *Science*, **183**, 198–200, https://doi.org/10.1126/science.183.4121.198

Zoller, W.H., Parrington, J.R. and Kotra, J.M.P. 1983. Iridium enrichment in airborne particles from Kilauea volcano. *Science*, **222**, 1118–1121, https://doi.org/10.1126/science.222.4628.1118

Zreda-Gostynska, G., Kyle, P. and Finnegan, D. 1993. Chlorine, fluorine, and sulfur emissions from Mount Erebus, Antarctica and estimated contributions to the Antarctic atmosphere. *Geophysical Research Letters*, **20**, 1959–1962, https://doi.org/10.1029/93GL01879

Zreda-Gostynska, G., Kyle, P.R., Finnegan, D. and Prestbo, K.M. 1997. Volcanic gas emissions from Mount Erebus and their impact on the Antarctic environment. *Journal of Geophysical Research*, **102**, 15 039–15 055, https://doi.org/10.1029/97JB00155

Chapter 7.3

Mount Melbourne and Mount Rittmann

Salvatore Gambino[1]*, Pietro Armienti[2], Andrea Cannata[1,3], Paola Del Carlo[4], Gaetano Giudice[1], Giovanni Giuffrida[5], Marco Liuzzo[5] and Massimo Pompilio[4]

[1]Istituto Nazionale di Geofisica e Vulcanologia, Osservatorio Etneo, Piazza Roma 2, 95123 Catania, Italy
[2]Dipartimento di Scienze della Terra, Università degli Studi di Pisa, Via S. Maria 53, 56126 Pisa, Italy
[3]Dipartimento di Scienze Biologiche, Geologiche e Ambientali-Sezione di Scienze della Terra, Università degli Studi di Catania, Catania, Italy
[4]Istituto Nazionale di Geofisica e Vulcanologia, Sezione di Pisa, C. Battisti 53, 56125 Pisa, Italy
[5]Istituto Nazionale di Geofisica e Vulcanologia, Sezione di Palermo, Via Ugo La Malfa 153, 90146 Palermo, Italy

SG, 0000-0001-8055-3059; GG, 0000-0002-9410-4139; ML, 0000-0002-3099-7505; MP, 0000-0002-0742-0679

*Correspondence: salvatore.gambino@ingv.it

Abstract: Mount Melbourne and Mount Rittmann are quiescent, although potentially explosive, alkaline volcanoes located 100 km apart in Northern Victoria Land quite close to three stations (Mario Zucchelli Station, Gondwana and Jang Bogo). The earliest investigations on Mount Melbourne started at the end of the 1960s; Mount Rittmann was discovered during the 1988–89 Italian campaign and knowledge of it is more limited due to the extensive ice cover. The first geophysical observations at Mount Melbourne were set up in 1988 by the Italian National Antarctic Research Programme (PNRA), which has recently funded new volcanological, geochemical and geophysical investigations on both volcanoes. Mount Melbourne and Mount Rittmann are active, and are characterized by fumaroles that are fed by volcanic fluid; their seismicity shows typical volcano signals, such as long-period events and tremor. Slow deformative phases have been recognized in the Mount Melbourne summit area. Future implementation of monitoring systems would help to improve our knowledge and enable near-real-time data to be acquired in order to track the evolution of these volcanoes. This would prove extremely useful in volcanic risk mitigation, considering that both Mount Melbourne and Mount Rittmann are potentially capable of producing major explosive activity with a possible risk to large and distant communities.

Mount Melbourne (74.35° S, 164.70° E) and Mount Rittmann (73.45° S, 165.50° E) are located in the coastal area of Northern Victoria Land (NVL: Fig. 1), a region where intense volcanic activity has developed since the Oligocene along the western margin of the Ross Sea in Victoria Land and parallel to the Transantarctic Mountains (West Antarctica), forming the McMurdo Volcanic Group (MMVG) (Harrington 1958; Kyle 1990).

NVL is dissected by NW–SE-striking dextral fault systems running through the northern Ross Sea (Fig. 1) which interact with NNE–SSW- to north–south-striking extensional faults. These features are also revealed by seismic profile interpretations (Salvini *et al.* 1997), magnetic anomaly maps (Ferraccioli *et al.* 2009) and GPS measurements (Dubbini *et al.* 2010). Magma ascent has occurred along the main strike-slip fault systems and along the transtensional fault arrays departing from the master faults (Rocchi *et al.* 2005).

In this framework, Mount Melbourne and Mount Rittmann are volcanoes situated to the north of the MMVG, whose activity started at about 4.0 Ma on Mount Rittmann (Armienti and Tripodo 1991) and 2.7 Ma on Mount Melbourne (Wörner and Viereck 1989). They show recent evidence of volcanic activity (Nathan and Schulte 1967; Lyon 1986; Kyle 1990; Bonaccorso *et al.* 1991) and are currently affected by degassing activity characterized by extensive fumarolic fields that are mainly located in the summit areas of the volcanoes. In the second half of the 1980s, the Italian National Antarctic Research Programme (PNRA) began numerous activities in NVL, some of which were focused on investigating and monitoring Mount Melbourne. In particular, between 1988 and 1990, a global positioning system (GPS), tilt and seismic networks were installed on the volcano summit and flanks, and a volcanological observatory was set up (Bonaccorso *et al.* 1997*a*). During the 1988–89 campaign, Mount Rittmann was discovered and studied from a petrographical point of view (Armienti and Tripodo 1991).

Data collected during Mount Melbourne monitoring activities mainly cover the 1990s; however, during the 2016–17 and 2017–18 campaigns, new seismological, geochemical and volcanological research was carried out on both volcanoes in the frame of a recent project (ICE-VOLC).

In this chapter, we focus on the active volcanism of Mount Melbourne and Mount Rittmann, and report on a review of the volcanological investigations and monitoring results achieved over the last 30 years.

Mount Melbourne

Structure, stratigraphy and geochronology

Mount Melbourne is a quiescent stratovolcano (2732 m above sea-level (asl)) located along the western coast of the Ross Sea about 42 km from the Italian Mario Zucchelli Station (MZS) and 33 km from the South Korean Jang Bogo Station (JBS), between Tinker Glacier to the north and Campbell Glacier to the south (Fig. 1). It is largely covered by ice except for some peripheral areas (e.g. Shield Nunatak, Cape Washington, Edmonson Point and Baker Rocks), and numerous scoria cones, lava domes, viscous lava flows and lava fields that are exposed from the summit along the upper flanks (Figs 2 & 3a) (Giordano *et al.* 2012). The volcanic edifice is roughly north–south elongated, the dominant trend of the subglacial and subaerial vents. Its shape is an almost perfect low-angle volcano, with only minor dissection and scarce glacial erosion. As Nathan and Schulte (1967) reported, former writers on Antarctica noted the young appearance of the volcano and compared it with Mount Etna (e.g. Wilson 1966; Ross 1847; Borchgrevink 1901). On the eastern flank, at 2253 m asl, a concave cliff 50–100 m high with a north–south orientation, according to

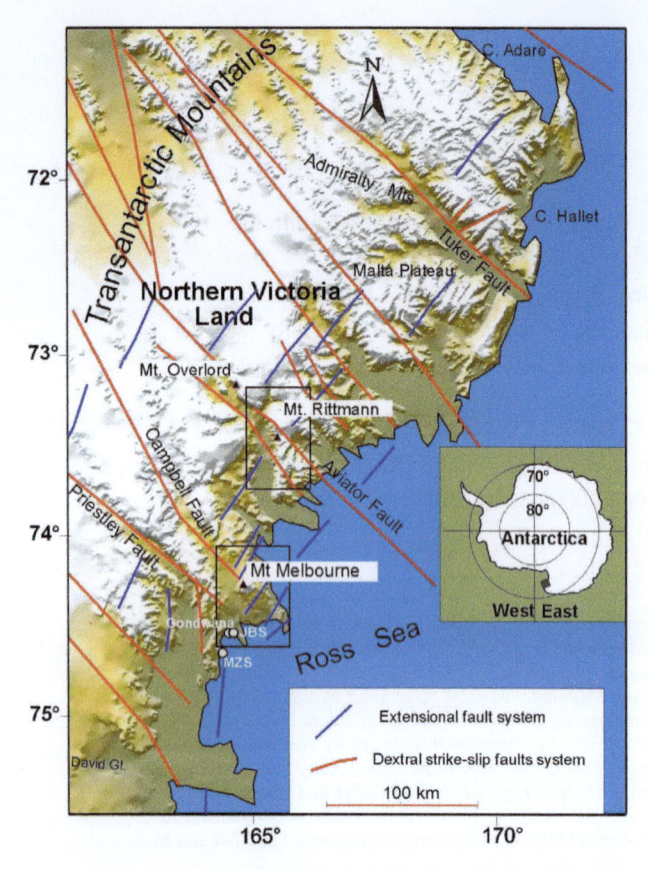

Fig. 1. Sketch map of Northern Victoria Land (tectonic lineaments from Salvini *et al.* 1997 and Vignaroli *et al.* 2015).

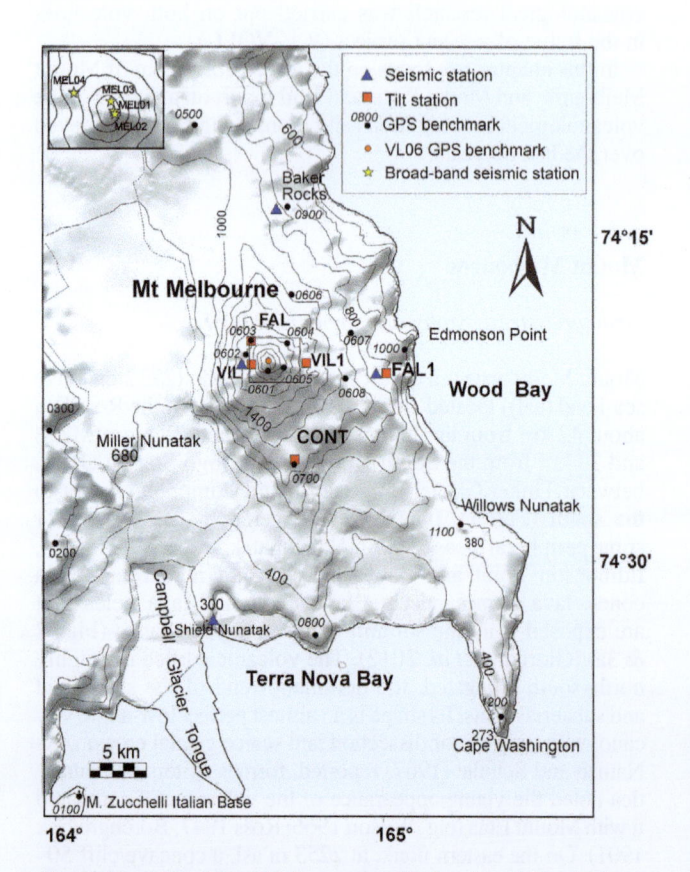

Fig. 2. Map of the Mount Melbourne area. The location of the seismic, GPS and tilt networks are reported. The inset shows the position of mobile broadband seismic stations installed in 2017.

Fig. 3. (**a**) Panoramic view of Mount Melbourne stratovolcano from an area near to MZS (about 40 km from Mount Melbourne summit). (**b**) The northern summit slope of Mount Melbourne covered by meter-sized scoria bombs and pumice lapilli.

Giordano *et al.* (2012), corresponds to an incipient collapse scar and a zone of recent, focused hydrothermal activity.

The summit area is characterized by a *c.* 1 km-wide crater filled by snow and is surrounded by several scoria cones that have produced most of the phonolithic black bombs and scoria (Fig. 3b).

According to Lyon (1986), the most recent volcanic activity of Mount Melbourne occurred between 1862 and 1922, on the basis of the depth of a tephra layer in the ice. Apart from the fumarolic activity at the summit crater, no other type of volcanic activity has been directly observed at least in the last 40 years, bearing in mind that MZS has been a permanent base since 1985 and visited each year during the Antarctic summer season.

The most recent synthesis of the volcanic evolution was presented by Giordano *et al.* (2012), based on stratigraphic, geochemical and age-determination data. According to their results, the first products of Mount Melbourne were mainly alkali basaltic–hawaiitic in composition and monogenetic in style, producing tens of small scoria cones and lava flows scattered over a wide area across the Transantarctic Mountains during the Lower Pleistocene (Random Hills Period). Thereafter, the volcanic activity concentrated in the area of the present-day stratovolcano, where several monogenetic centres show the transition from subglacial/subaqueous to subaerial conditions during the Middle Pleistocene (Shield Nunatak Period). The initial activity of the Mount Melbourne stratovolcano is characterized by a trachytic ignimbrite dated at 123.6 ± 6.0 ka and indicating the formation of a crustal magma chamber (Mount Melbourne Period). A succession of alkali basaltic, hawaiitic, and subordinate benmoreitic lavas and scoria cones were dated at 90.7 ± 19.0 ka, whilst the most recent deposits exposed at the top of Mount Melbourne consist of trachytic–rhyolitic pumice fall deposits, probably produced by Plinian eruptions (Giordano *et al.* 2012). In the framework of the ICE-VOLC project, new geological surveys have revealed the occurrence of englacial volcanic ash layers in the NE flank of the volcano (Cannata *et al.* 2018). Geochemical analyses on glass particles and age determinations (direct or relative) will determine whether they represent deposits of very recent Mount Melbourne volcanic activity.

Geochemistry and petrology. After the exploratory works of Nathan and Schulte (1968) and the first synthesis of Kyle (1990), several papers have described and investigated the petrological and geochemical variability of rocks erupted from Mount Melbourne and peripheral centres (Aviado *et al.* 2015; Armienti *et al.* 1991; Giordano *et al.* 2012; Lee *et al.* 2015; Müller *et al.* 1991; Nardini *et al.* 2009; Perinelli *et al.* 2006, 2012; Wörner *et al.* 1989).

The compositional dataset recovered from the PetDB database (http://www.earthchem.org/petdb) shows that Mount Melbourne volcanics (MMV) form an almost continuous alkaline series, with composition ranging from alkali basalts–basanitoids to trachytes in the total alkali–silica diagram (Le Maitre *et al.* 2002) (Fig. 4). In this series, alkali basalt-hawaiites and basanitoid rocks are the most abundant products (c. 50%) but no real primitive magmas have been found. Less differentiated rocks contain lower-crustal granulite, gabbroic cumulate and mantle nodules comprising spinel lherzolites and harzburgites (Wörner *et al.* 1989; Perinelli *et al.* 2012).

Harker diagrams (Fig. 5) show a negative correlation, without significant gaps, between SiO_2, $Fe_2O_3^T$, MgO and CaO, and a positive correlation between SiO_2, Na_2O and K_2O that is related to the crystal fractionation of phenocryst phases. The positive correlation between SiO_2 and Sr isotopic composition has been interpreted to be the result of assimilation and fractional crystallization (AFC) processes (Armienti *et al.* 1991). A larger scatter in most of the elements is observed in less differentiated rocks (basanite–alkali basalts–hawaiites), indicating the possibility of distinct mantle sources (Wörner *et al.* 1989; Nardini *et al.* 2009).

Detailed petrography of these rocks (Wörner *et al.* 1989; Armienti *et al.* 1991) indicates that basanites are weakly porphyritic, with phenocrysts of olivine and augite, and groundmass made up of plagioclase, olivine, clinopyroxene, Ti-magnetite and sometimes interstitial nepheline.

Alkali basalts are also weakly porphyritic with phenocrysts of olivine and plagioclase, and microlites of augite, olivine, plagioclase and Ti-magnetite immersed in a brown glass. Hawaiites range from subaphyric to strongly porphyritic, and show a phenocryst assemblage of plagioclase and olivine, while clinopyroxene is absent. The same minerals, clinopyroxene, Ti-magnetite and sporadic ilmenite, form the groundmass. More evolved rocks (mugearites, benmoreites and trachytes) show a mineral assemblage dominated by plagioclase and less abundant mafic phases. Sanidine and anorthoclase are present in trachytes. These rocks also show apatite in the groundmass, while in the most evolved terms aegirine–augite, fayalitic olivine and sodic amphiboles are found.

Geochemical monitoring and observations

Early clues to the presence of geothermal activity on Melbourne volcano are dated between the late 1960s and the early 1970s (Nathan and Schulte 1968; Lyon and Giggenbach 1974). However, following these cursory, preliminary surveys, research has been undertaken with the result that very little is known about the volcano. In their exploration of the summit area of Melbourne volcano, Lyon and Giggenbach (1974) noted three different areas where warm ground was uncovered by the otherwise homogeneous ice cover, which, together with the presence of ice towers, evidenced the

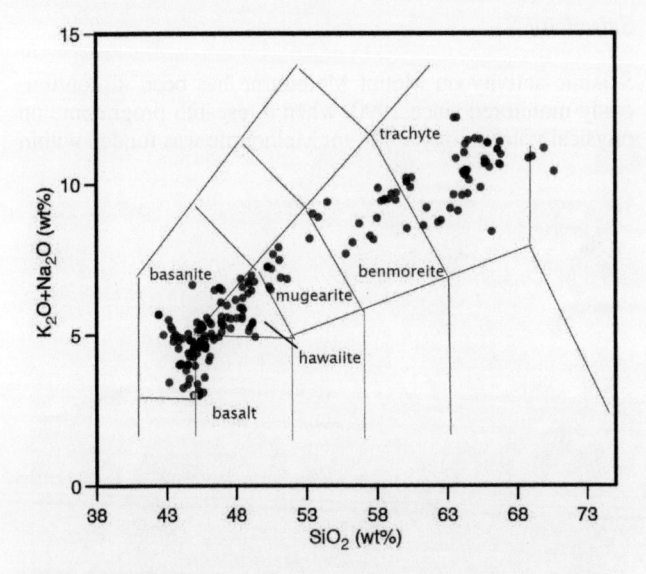

Fig. 4. Total alkali–silica classification diagram (Le Maitre *et al.* 2002) for Mount Melbourne volcanic rocks (dataset retrieved from the PetDB Database at http://www.earthchem.org/petdb). Data sources: Nathan and Schulte (1968), Wörner *et al.* (1989), Armienti *et al.* (1991), Nardini *et al.* (2009), Aviado *et al.* (2015) and Lee *et al.* (2015).

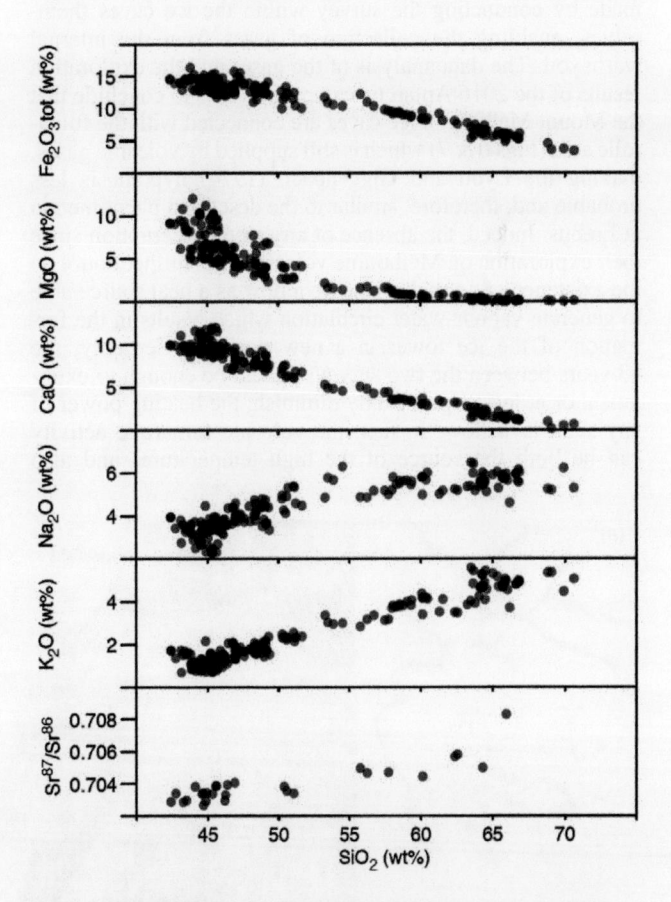

Fig. 5. Harker diagrams for Mount Melbourne volcanic rocks (dataset retrieved from the PetDB Database at http://www.earthchem.org/petdb). Data sources: Nathan and Schulte (1968), Wörner *et al.* (1989), Armienti *et al.* (1991), Nardini *et al.* (2009), Aviado *et al.* (2015) and Lee *et al.* (2015).

presence of geothermal activity. Regarding the ice towers, they concluded that a probable mechanism for their formation was a result of residual heat supplied by extensive lava flows creating a circulation of water vapour from melted snow at the interface between the underlying, still warm lava and the ice sheet. No information has been published relating to the gas composition from the fumarolic area on the south rim of the crater, which is characterized by exposed warm ground not covered by snow or ice, but by the notable presence of mosses.

During an Antarctic exploration carried out on Mount Erebus in the 2000s, Wardell et al. (2003) recognized the presence of a volcanic gas source as the main cause of the genesis of ice caves. In addition, in the same campaign they were able to attempt a first evaluation of the CO_2 efflux emitted by the ice towers on Mount Erebus.

During the Antarctic summer of 2016, a team of INGV researchers within the PNRA programme carried out a survey on the fumarolic area of Mount Melbourne as part of the ICE-VOLC project. The research campaign involved an extensive geochemical investigation of the geothermal area on Mount Melbourne, as well as the mapping and exploring of the ice caves on the summit part of the volcano. During the survey in November 2016, some significant differences were noted in comparison to the description provided by Lyon and Giggenbach (1974) regarding the position of the ice caves, as well as the whole geothermal area (Figs 6 & 7). For instance, the sizes and positions of the ice towers were quite different; and the temperature of the uncovered soil of the Cryptogam area (the yellow area in Fig. 6), corresponding to 'area 2' in the previous work, was different: 33.5° C in our survey and 59°C in the previous ones. Moreover, a step forward from the pioneering work of Lyon and Giggenbach (1974) was made by conducting the survey within the ice caves themselves, enabling the collection of gases from the internal warm soil. The data analysis of the gases and the exploration results of the 2016 Antarctic campaign led us to conclude that the Mount Melbourne ice caves are connected with the fumarolic area (Figs 6 & 7) which is still supplied by volcanic gases, making the Lyon and Giggenbach (1974) hypothesis less probable and, therefore, similar to the described phenomenon at Erebus. Indeed, the absence of any recorded eruption since their exploration on Melbourne volcano casts further doubt on the existence of an underlying lava flow as a heat source able to generate vapour-water circulation which results in the formation of the ice tower in a new position. Certainly, the 43 years between the two surveys should be enough to extinguish, or at least significantly diminish, the heating power of any such lava flow. In fact, the volcanic fumarolic activity can be both the source of the high temperatures and also explain the variation of the position which is a typical phenomenon of fumarolic fields. Mount Melbourne volcano ice caves and ice towers may therefore be considered to be the surface expressions of an established degassing area associated with fumarolic fields similar to those at Erebus (Wardell et al. 2003) or ice caves in similar/matching environmental conditions worldwide: for example, Mount Rainier in the Cascade volcanoes, USA (Zimbelman et al. 2000). In these contexts, ice caves are known to be formed by warm gases and steam escaping from the lava-flow surfaces. These melt the bottom layer of the overlying ice and snow, leaving an underlying cavity together with a typical chimney-like structure that is visible from the external ice field (Fig. 6c). Indeed, the gas geochemistry of the samples collected during the survey show that geothermal activity is still active, and the common gases released are CO_2 and CH_4 in concentrations that are always greater than air concentrations. The isotopic terms of these gases also confirm the volcanic origin (geochemical data are in publication), finally demonstrating a magmatic origin and making the Melbourne volcano still active.

During the 2016 and 2017 expeditions, a team of researchers from the ICE-VOLC project deployed and tested an INGV-designed MultiGAS automatic station coupled with a meteorological station, which allowed the flux of heat, water and CO_2 passing through the median portion of the MC1 ice cave to be measured for 4 days (Fig. 8). In addition, a temperature sensor (Tinytag Plus 2–TGP-4017 working within a temperature range from −40 to +85°C) was installed in a specific section of the cave to monitor the air temperature for 1 year. The preliminary results for this test (Table 1) are not only the first measurements of CO_2 and H_2O fluxes for Mount Melbourne but also demonstrate the feasibility of a surveillance plan for this volcano. Future implementation of the above monitoring system would provide near-real-time data that could recognize the evolution and changes in the state of the volcano – extremely useful for volcanic risk mitigation. In fact, Melbourne volcano lies approximately 40 km NNE of MZS; therefore, any potential unrest in this volcano might also significantly affect the Italian base, as well as the other two nearby scientific bases – Jon Bogo (Korea) and Gondwana (Germany) – suggesting that a geochemical monitoring system is certainly advisable.

Seismicity

Seismic activity on Mount Melbourne has been discontinuously monitored since 1990, when a research programme on physical volcanology at Mount Melbourne was funded within

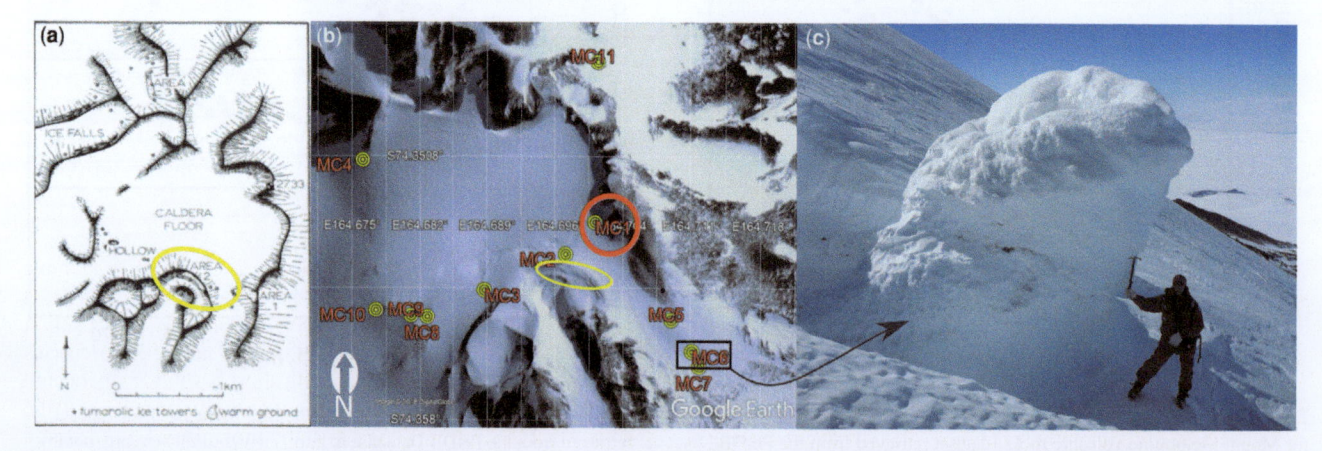

Fig. 6. Location of the ice caves. (**a**) November 2016 survey. (**b**) 1972–73 survey from Lyon and Giggenbach (1974). (**c**) Typical ice tower at the entrance to a cave. The red circle in (b) is the location of the MC1 ice cave also indicated in Figure 7; the yellow ellipses in (b) indicate the Cryptogam area, the main geothermal area uncovered of ice by Mount Melbourne.

In 2010–11, a number of broadband seismic stations were installed on Mount Melbourne by the Korea Polar Research Institute, which studied the structure beneath Mount Melbourne using teleseismic data (Park et al. 2014, 2015).

In 2017, seismic investigation of the Mount Melbourne volcanic activity resumed under the framework of the ICE-VOLC project, funded by the PNRA. In particular, two broadband seismic stations, equipped with Nanometrics Trillium Compact 120s seismometers and sampling ground velocity at a rate of 100 Hz, were temporarily installed at different sites on the summit and flanks of the volcano (Fig. 2).

By putting together all the old and new observations, it can be seen that different kinds of seismic signals have been identified on Mount Melbourne volcano: (i) icequakes, (ii) local seismic events and (iii) tremor.

As for the former, icequakes are defined as coseismic brittle fracture events within the ice (e.g. Podolskiy and Walter 2016) or, more generally, are seismic events associated with ice dynamics. Icequakes, recorded on Mount Melbourne, are generally characterized by short duration (<10 s), high spectral content (>10 Hz), sharp amplitude decrease with distance, and no clear P and S phases (Fig. 9a). On the basis of their features, these events, representing most of the seismic events recorded on Mount Melbourne, are probably associated with ice faulting forming crevassing (e.g. Gambino and Privitera 1994; Walter et al. 2009; Röösli et al. 2014). There is also a second type of icequake recorded on both Mount Melbourne and in the surrounding areas showing higher amplitude, duration of 20–30 s, emergent onset, frequency content below 2 Hz, and clear P and S phases (Fig. 9b) (Gambino and Privitera 1994; Cannata et al. 2017). The sources of these seismic events are located in the David Glacier area (Fig. 1) and are likely to have been generated at the rock–ice interface under the glacier (e.g. Danesi et al. 2007; Zoet et al. 2012; Cannata et al. 2017).

Concerning local seismic events, Gambino and Privitera (1994, 1996) identified seismic events with $M_L<2.0$, characterized by low frequencies (0.7–6.0 Hz), duration longer than 20 s, emergent first arrivals, and S waves that are difficult to identify (Fig. 9c). An analytical solution was only for the main event (at 05:14 on 10 December 1990, $M_L = 1.9$), while 16 other events were located using azimuth obtained from particle motion and the S–P time lapse (Gambino and Privitera 1996). The volcano's eastern flank (Fig. 10) is the source area for all these events; the analytical location showed a depth of 2.8 km, with 4.4 km of error (Gambino and Privitera 1996).

Lastly, in January 2017 a long-lasting tremor-like signal, with frequencies between 2 and 8 Hz, was recorded simultaneously at two summit stations (Fig. 11).

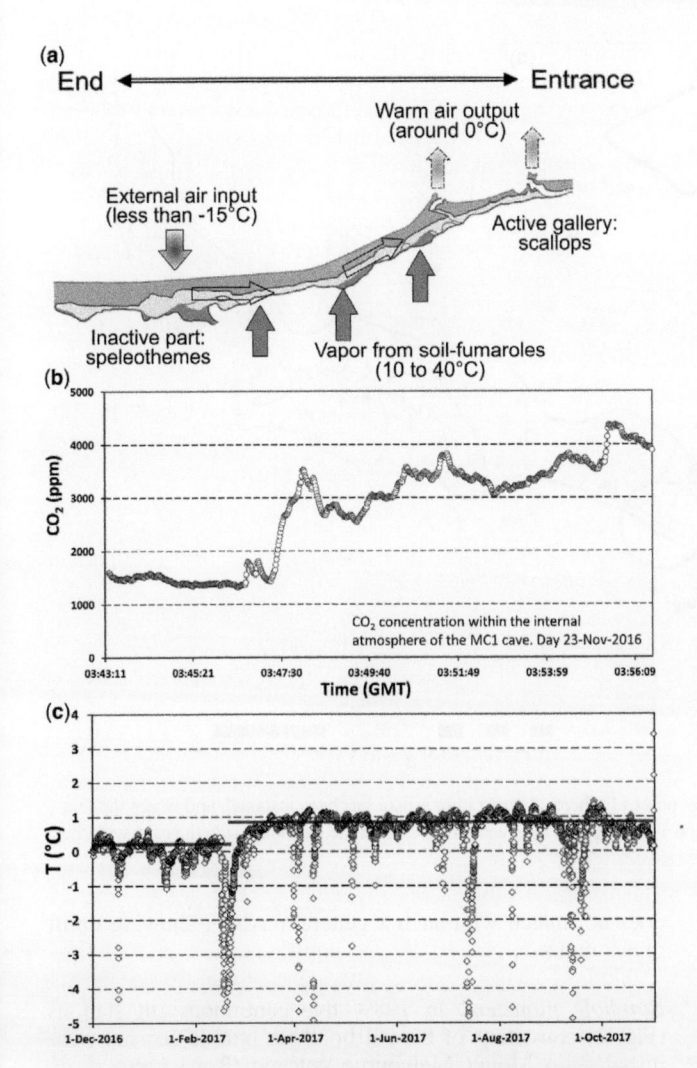

Fig. 7. MC1 ice cave at Mount Melbourne. (a) Schematic profile of the cave. (b) CO_2 molar concentration in the atmosphere inside the cave. (c) Temperature variation between November 2016 and November 2017. The main escape route for the mixed gas circulating inside the cave is through the external chimney-like ice tower (Fig. 6c). Inside the cave, cold air coming in through small fissures or through permeability of snow becomes warmer and enriched in volcanic gases rising up the slope of the volcano, and finally escapes from the ice-tower chimneys. CO_2 concentration inside the cave is not constant. The CO_2 gas emanation coming from the ground is soil-permeability-dependent. However, the CO_2 concentration within the atmosphere of the cave is highly dependent on the accumulation of CO_2 driven by the wind circulation inside the cave. The internal temperature tends to remain fairly stable; however, this equilibrium is likely to be disturbed if the heat flux from the fumaroles within the cave changes as a result of changes in their source. Any variations in air circulation and gas composition within the cave are, therefore, an indication of the possible onset of increased volcanic activity.

the framework of the PNRA (Privitera et al. 1992; Bonaccorso et al. 1997a, b). From 1990 to 1995, a network of four seismic stations (two one-component and two three-component) recorded seismic signals on Mount Melbourne (Fig. 2). These stations, one located close to the summit area and the others at the base of the volcano edifice, were equipped with short-period (0.89–1.1 s) Teledyne Geotech model S-13 geophones, acquiring signals at a sampling rate of 50 Hz. Two stations, located in the line of sight of the Terra Nova Italian base (the former name of the Mario Zucchelli Station), sent the collected data by digital telemetry to the Volcanological Observatory located at MZS, the other two stored the data locally. A total of 4000 events were recorded over 15 months (Bonaccorso et al. 1996).

Ground deformation

Ground deformation at Mount Melbourne has been measured using two techniques: the global positioning system (GPS) and borehole tiltmeters.

GPS. GPS measurements on Mount Melbourne area have been performed since 1990–91 when a geodetic network, set up during the 1988–89 and 1989–90 campaigns, was first surveyed (Capra et al. 1996, 1998). The network comprises 20 benchmarks, eight of which with short baselines are located on the volcanic cone (Fig. 2). It was only surveyed in the 1990s by GPS methodology in static modality. A comparison between 1990–91 and 1993–94 measurements showed no deformation on Mount Melbourne (Capra et al. 1996); however, the results obtained using different software and different

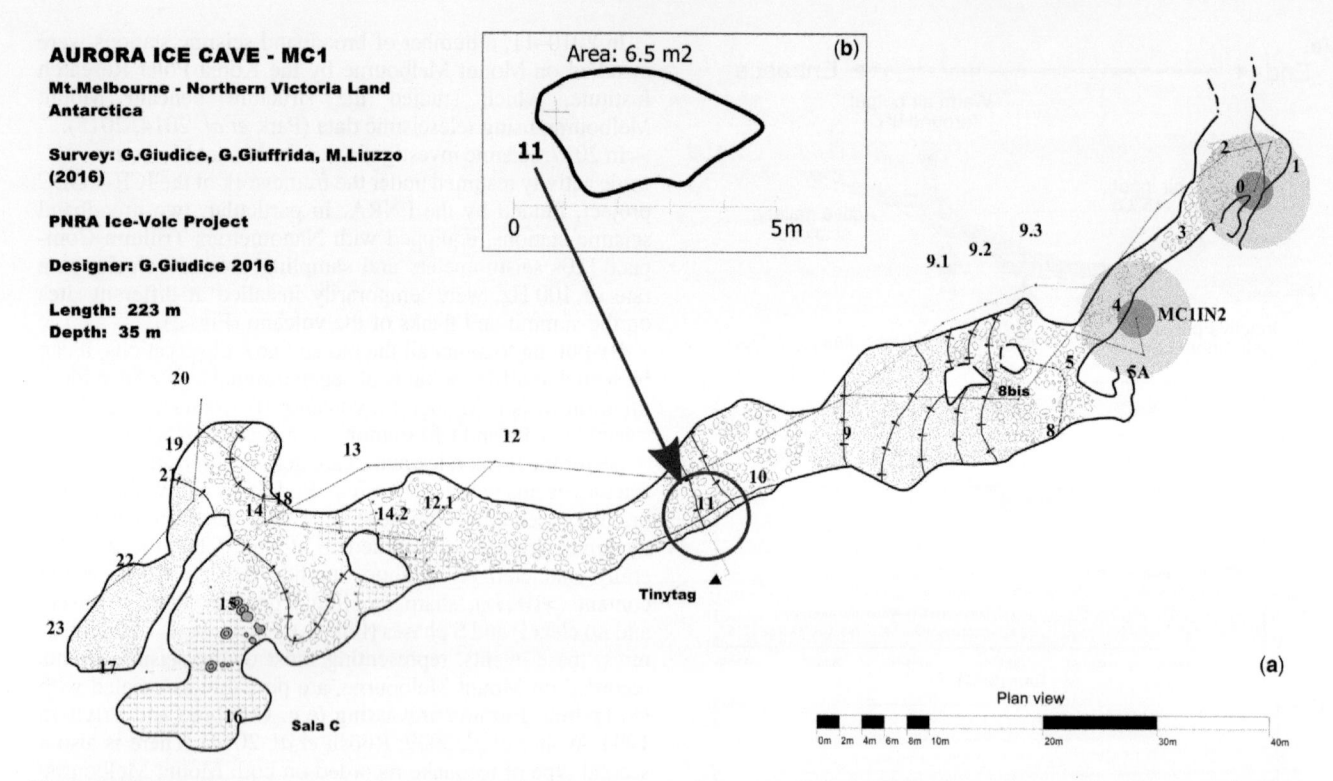

Fig. 8. (a) Topographical map of the MC1 cave and (b) a section of the cave at point 11 where a temperature sensor has been installed, and where the first measurement of the CO_2 and H_2O flux has been carried out (see the text). Section 11 of the cave has been recognized as the location inside the cave with the best physical characteristics for monitoring purposes.

approaches gave significantly different results (Capra *et al.* 1998).

Later, the geodesy group extended the area investigated by GPS observations, leading to the deployment of a new network of 32 benchmarks (VLNDEF), covering the entire Northern Victoria Land area (Dubbini *et al.* 2010), designed for geodynamic purposes. This network includes the VL06 benchmark positioned on the summit area of Mount Melbourne. Uplift rates collected over a more than 15 year span (Zanutta *et al.* 2017) highlighted that the central and southern NVL areas show a general small negative trend (up to -1.3 mm a^{-1}) in the vertical displacements; only VL06, located on the top of Mount Melbourne volcano, does not match with such a general reading, showing uplift of 1.9 mm a^{-1}.

Borehole tiltmeters. In 1989, five continuous tilt stations (Fig. 2) consisting of biaxial borehole bubble sensors were installed on Mount Melbourne volcano (Bonaccorso *et al.* 1995, 1997a).

The sensors, Applied Geomechanics Model 722 with a resolution of 0.1 μrad, were positioned at a depth of *c.* 2.5 m and consisted of two tilt components trending radially (Tilt X) and tangentially (Tilt Y) with respect to the volcano summit (Fig. 12a, b). The boreholes were also equipped with two thermocouples (T1 and T2) at mean depths of 1.0 and 1.9 m, and a

Table 1. *The first measurements of CO_2 and H_2O flux, as well as enthalpy and power, at the MC1 ice cave are shown in bold; the rest of the table indicates the variables used to calculate the above parameters using the psychrometric formulas*

Parameters	Unit	25 November 2016	1 November 2017	2 November 2017	7 November 2017
CO_2 flux	t/day	**0.38**	**0.48**	**0.76**	**0.71**
H_2O flux	t/day	**0.59**	**0.3**	**0.78**	**0.79**
Air flux	t/day	**181.52**	**108.91**	**145.21**	**159.74**
Enthalpy (H)	kJ kg^{-1}	**25.584**	**18.339**	**25.143**	**31.683**
Power	kW	**53.75**	**23.12**	**42.26**	**58.57**
T_{ext}	°C	−16.3	−11	−14.4	−18
RH_{ext}	%	16	15	17	18
H_2O_{ext}	ppm	398	557	670	388
P_{ext}	mbar	690	713	707.7	689.7
CO_{2ext}	ppm	380	512	510	512
T_{int}	°C	0.9	0.2	0.8	1
RH_{int}	%	60	58	73.3	88.9
$CO_{2\ int}$	ppm	1750	3430	3960	3440
H_2O_{int}	ppm	5624	8676	13440	8326
P_{int}	mBar	696	717	708.6	701.6
V_{mean}	m s^{-1}	0.25	0.15	0.2	0.22
Area	m^2	6.5	6.5	6.5	6.5

RH, relative humidity; *P*, pressure; *T*, temperature; *V*, velocity; subscript 'int' refers to a measurement inside the cave; subscript 'ext' refers to a measurement outside the cave.

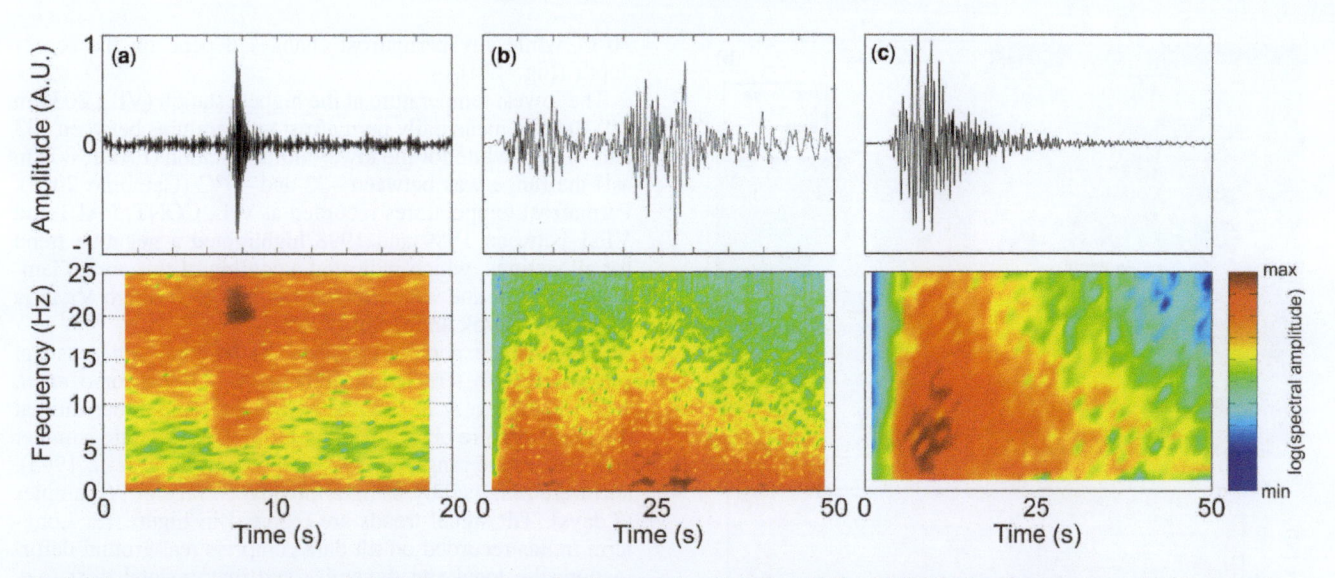

Fig. 9. Waveforms and spectrograms of a high-frequency icequake (**a**), recorded on 21 January 2017 at 13:46:00; a low-frequency icequake (**b**), recorded on 29 November 2016 at 07:33; and a local seismic event (**c**) recorded on 10 December 1990 at 05:14, acquired on Mount Melbourne.

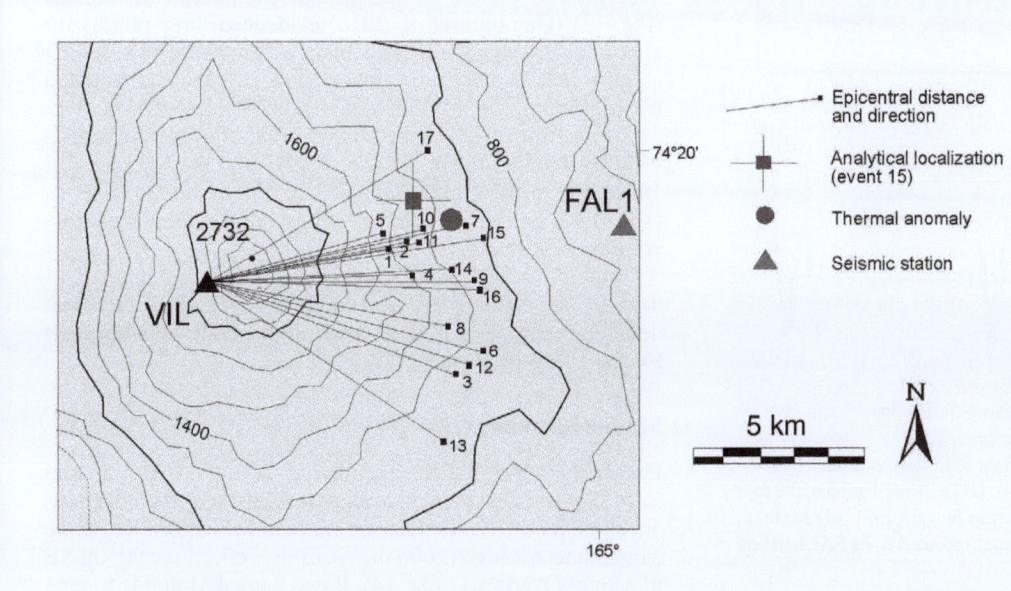

Fig. 10. Distribution of epicentres of local events in Mount Melbourne's summit area. The direction of the seismic rays was obtained from particle motion analysis and the distance from the S–P seismic phase time lapse (Gambino and Privitera 1996). The analytical determination is related to the 10 December 1990 at 05:14 event. The thermal anomaly is referred to by Mazzarini and Salvini (1994).

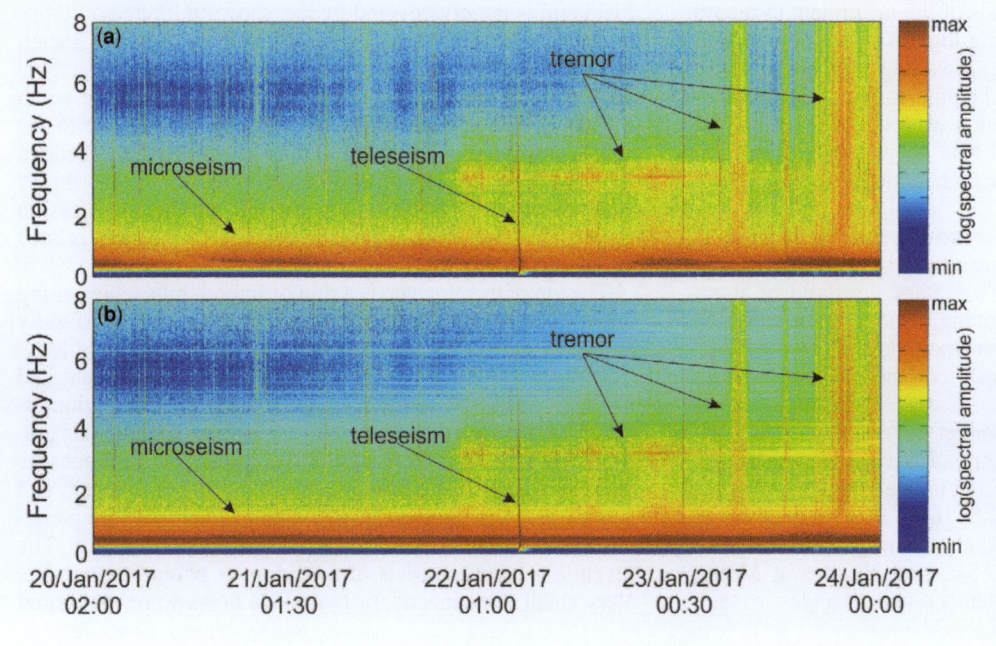

Fig. 11. Spectrograms of the vertical component of the seismic signals recorded by the MEL1 and MEL2 stations during 20–24 January 2017 (see inset of Fig. 2 for information about the location of the stations). Three different types of signals are evidenced: microseism, a teleseism and tremor.

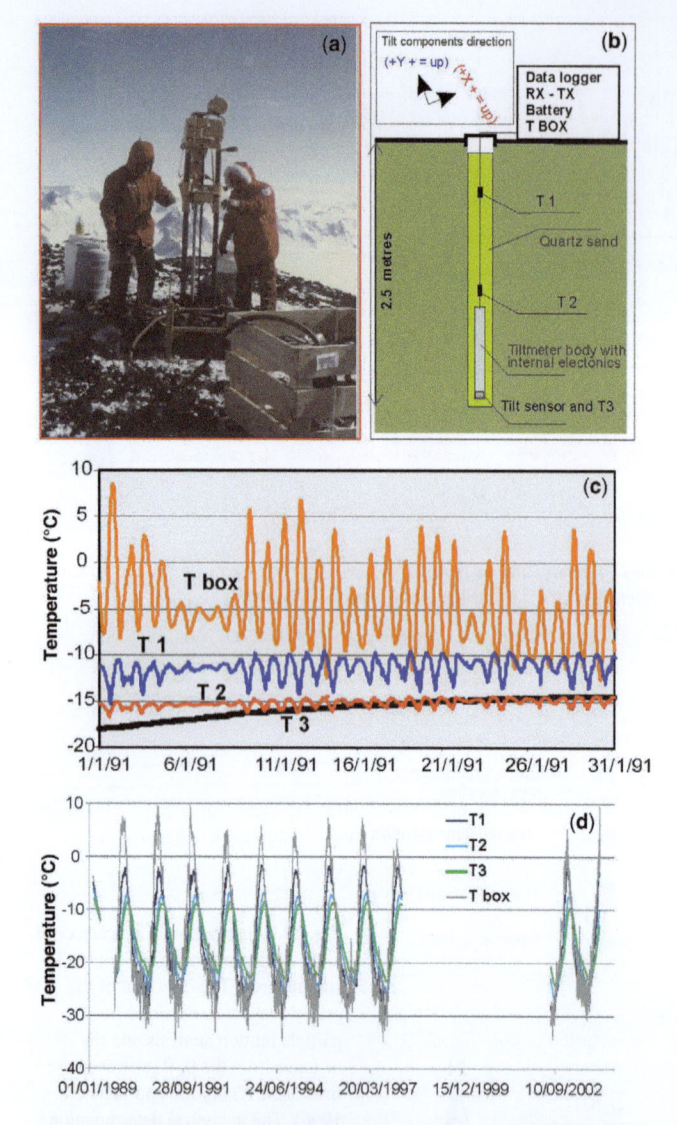

Fig. 12. (a) Photograph of the hole drilling to install the tiltmeter. (b) Scheme of a borehole tiltmeter installation. (c) Daily oscillation of the four temperatures sensors of the CONT station in January 1991. (d) Air and permafrost temperatures in a historical series recorded at the FAL1 station (redrawn from Gambino *et al.* 2016).

tilt temperature sensor at the ends of the instrument to record permafrost temperature. The data loggers were programmed for 48 data/day, with sampling including acquisition of tilt components, air (T-box), ground temperatures and instrumental control parameters. Each station was equipped with solid-state memories that guaranteed data acquisition through the winter months, while high-capacity batteries powered the station (Fig. 12).

The temperature and tilt dataset covers a period from 1989 to 2003. Partial interruptions were caused by data-logger auto-protection that activated under a voltage threshold of about 10 V or by electronic malfunctioning, which is to be expected below −35°C. Some interruptions took place when the external temperature oscillations began to wane and there was no longer any sunlight.

The reduced number of personnel in Antarctica then lessened the ability to acquire data continuously, leading to a complete interruption of the acquisition in late 2003.

Temperatures at different depths show daily fluctuations in summer (Fig. 12c), the amplitude of which lessens with depth and levels out at 2.5 m. Seasonal changes at Mount Melbourne are marked: air temperature changes exceed 30°C, while any permafrost changes depend on the sensor depth (Fig. 12d).

The lowest temperature at the highest station (VIL, 2030 m asl) showed mean daily permafrost temperatures between −33 and −18°C, while for the lower-altitude station (FAL1, 800 m asl) the range was between −22 and −9°C (Gambino 2005). Permafrost temperatures recorded at VIL, CONT, FAL1 and VIL1 between 1989 and 1998 highlighted a negative trend for all stations, which indicated a cooling of this area (Gambino 2005) in line with climatic studies on Northern Victoria Land (e.g. Kwok and Comiso 2002).

Tilt data are generally affected by diurnal and seasonal noise linked to temperature effects (e.g. Bonaccorso *et al.* 1999; Gambino *et al.* 2007). At Mount Melbourne, diurnal fluctuations were observed only during the Antarctic summer and were in the range of 2–12 μrad (Bonaccorso *et al.* 1995), and were removed by using a moving average of 96 samples (2 days). Tilt signal trends are reported in Figure 13. Long-term trends recorded on tilt data comprise real ground deformation, the local site dynamics and instrumental drift (e.g. Kohl and Levine 1993; Anderson *et al.* 2010). The secular marked trends visible at the VIL and CONT (Fig. 13, Phase 2) stations may be linked to instrument drift.

A general picture of the Mount Melbourne tilt signals (Fig. 13) (Gambino *et al.* 2016) evidenced three phases for almost every signal. In particular, a first phase (1), lasting about 1 year, is present on almost all signals and is dominated by instrument/site stabilization effects after installation. Successively, from 1990 to the end of 1997, tilt trends (phase 2) were constant; while at the end of 1997 and the beginning of 1998, the radial components of the three summit stations (VIL, FAL and VIL1) show an almost contemporaneous lowering of their trends (phase 3).

Gambino *et al.* (2016) estimated these changes, in comparison to the stable 1990–97 period, modelling a pressure-deflating source (Fig. 13) at a depth of about −0.8 km (asl) for the 1998–2002 period.

Mount Rittmann

During the fourth Italian expedition to northern Victoria Land in 1988–89, a new volcanic centre was discovered on the eastern shoulder of Aviator Glacier, north of Mount Brabec, in the Mountaineer Range (2600 m, 73.45° S, 165.50° E; 90 km SE of Mount Overlord) (Fig. 14). It was named Mount Rittmann in honour of the celebrated Swiss volcanologist Alfred Rittmann. The volcano belongs to the Mount Overlord Volcanic Field and is largely covered by ice, showing limited outcrops on its summit caldera and on the walls of the Aviator Glacier, WNW of the caldera (Fig. 14b); a few remnants of trachytic lava flows at about 2000 m asl on the borders of the south branch of Pilot Glacier (Fig. 14) can also be ascribed to the activity of Mount Rittmann. Thick lava flows interbedded with hyaloclastites and pillow lavas crop out on the almost vertical cliffs of Pilot Glacier: they fill the large depression north of Mount Rittmann and have been tentatively related to its activity.

The top of this volcano is a ring of smooth hills representing the rims of a caldera about 2 km in diameter and slightly emerging from the surrounding, almost flat, morphology. The outcrops are represented by trachytic lava flows and hawaiitic cinder cones, surrounded by the volcanic agglomerate of the final explosive activity containing xenoliths of granitic and metamorphic rocks of the crystalline basement (*c.* 10% by volume) and of magmatic xenoliths of volcanic and sub-intrusive rocks (*c.* 70% by volume). The latter are generally evolved lavas, often altered by hydrothermal activity. The juvenile fraction consists of a fresh grey phonolitic pumice. Very small remnants of trachytic lava flows were also found

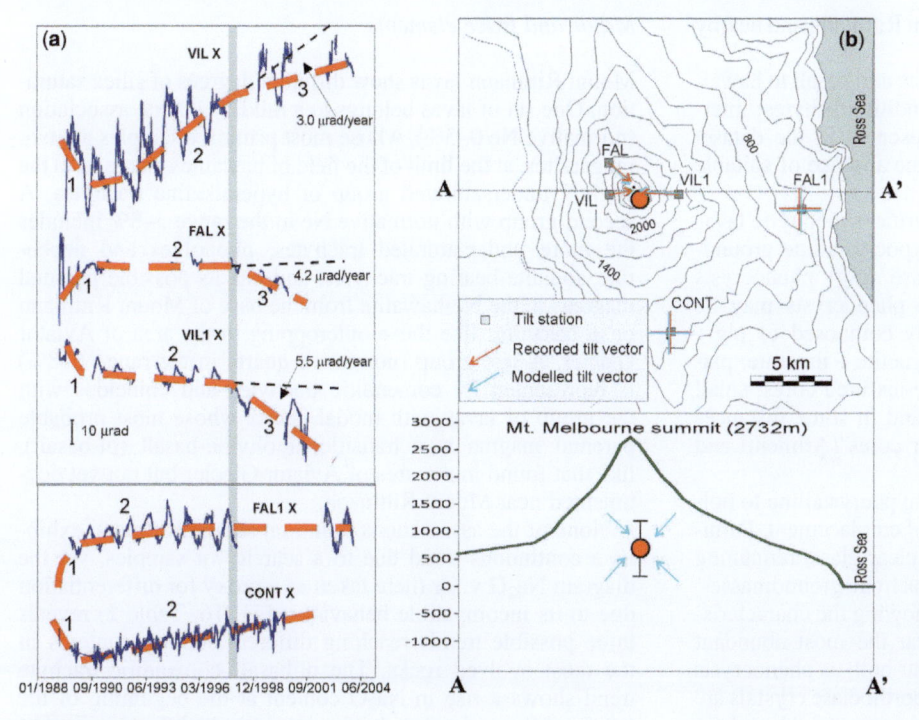

Fig. 13. (a) Radial tilt component at the five stations. Numbers define the three phases discussed in the text. (b) Map and cross-section of the pressure source modelled by Gambino *et al.* (2016).

on the borders of the depressed area that is part of the trimming basin of Pilot Glacier. This outcrop is separated from the caldera of Mount Rittmann by the abundant ice cover of the nearby plateau.

Petrography and chemistry

The petrography and chemistry of lavas and comagmatic xenoliths were studied by Armienti and Tripodo (1991) on a

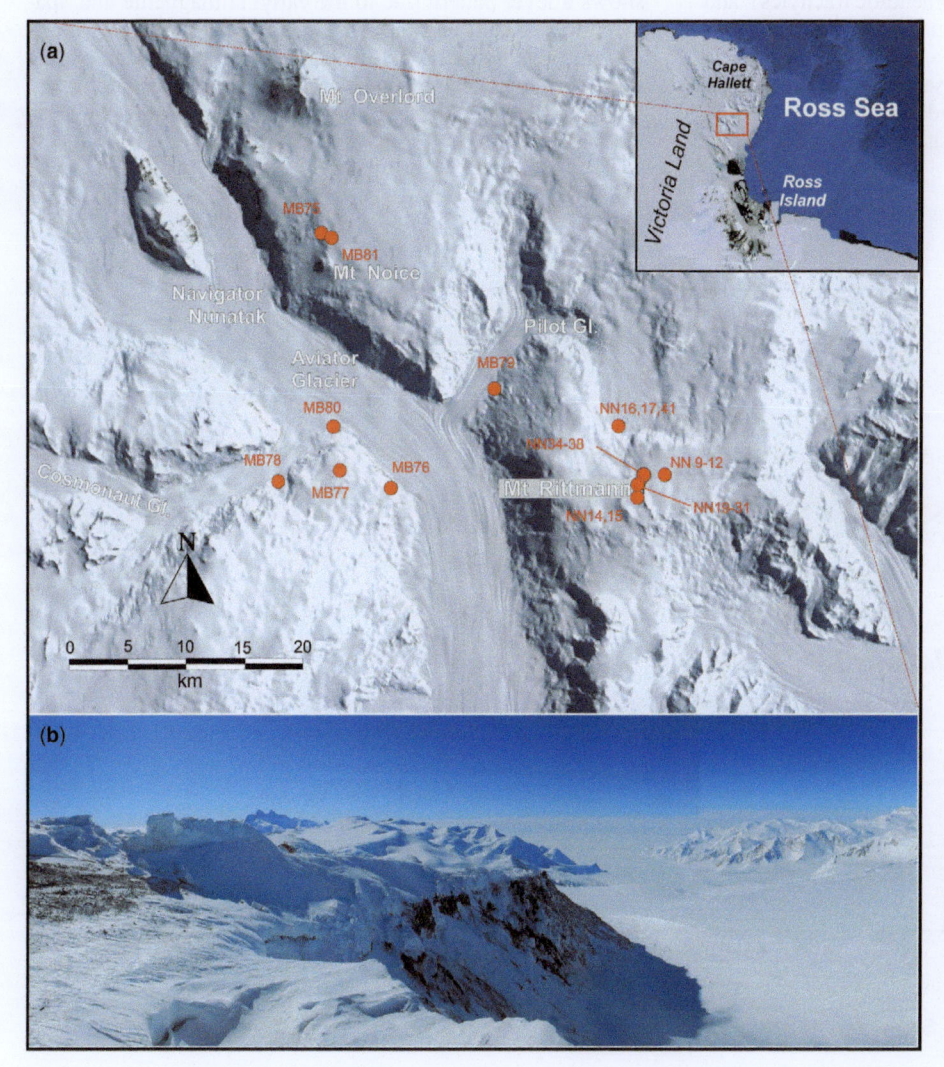

Fig. 14. Satellite image (from Google Earth in 1999) of the Mount Overlord Volcanic Field astride the Aviator Glacier area; the oldest outcrops are the eroded volcanic necks of Navigator Nunatak: subsequent dated activity spans from 14.5 Ma at Parasite Cone to the Holocene age attributed to an isolated vent in the Cosmonaut Glacier (Kyle 1990). Monogenetic cinder cones dot the area, and ice-buried volcanic buildings are highlighted by the morphology of the surface. The ages of Mount Overlord products vary between 8.3 and 7 Ma, while an alkali basalt from the area of Aviator Glacier yielded a K–Ar age of 7.5 Ma (Kyle 1990). To date, no age has been determined for the acidic, mainly pyroclastic, products that underlie Mount Overlord in the Deception Plateau and those that directly lie on the crystalline basement. Armienti and Tripodo (1991) reported a standard K–Ar age of 3.97 ± .14 Ma (MB79) for the lavas cropping out in the Pilot Glacier at the base of Mount Rittmann. This Pliocene age falls between Mount Overlord Miocene activity and the Holocene age of some small centres in the Aviator Glacier area. In red is the position of sampled rocks. The bottom picture shows the fumarolic wall of Mount Rittmann.

total of 36 samples collected from Mount Rittmann and nearby areas (Fig. 14).

Their composition varies from basanite and basalt to hawaiites, hyperalkaline trachytes and comenditic trachytes; intermediate compositions are poorly represented in the dataset due both to their relative scarcity and the absence of suitable rocks for chemical analyses (Fig. 15).

Hawaiites and mugearites are porphyritic–subaphyric lavas with holocrystalline intergranular to hypocrystalline groundmasses. They always show plagioclase (plg) phenocrysts and olivine (ol); clinopyroxene (cpx) phenocrysts may be lacking. The groundmasses are mainly composed of plg + cpx + ol, and contain magnetite or magnetite + ilmenite; plagioclase phenocrysts may exhibit sieve-textured cores; small symplectites of ol + plg are common and in some instances they preserve kaersutite relicts in their cores (Armienti and Tripodo 1991).

Trachytic rock structures vary from hypocrystalline to holocrystalline according to the modality of emplacement. Pumiceous samples are made up of a fresh clear glass; remaining rocks are usually holocrystalline with trachytic groundmasses. Soda-rich sanidine and anorthoclase, showing the characteristic albite–pericline twinning, are by far the most abundant minerals of these rocks where they occur both as phenocrysts and as microlites of the groundmass. Anorthoclase crystals are often patchy zoned with corroded nuclei surrounded by a later rim of a different composition.

The authors divided the trachytic rocks into three groups: a first group showing a tendency to a phonolitic composition in which undersaturation is revealed by the occurrence of modal nepheline and/or sodalite; a second oversaturated group with modal quartz in the groundmasses (comenditic trachytes); and a third set of lavas without feldspatoid or quartz, near saturation condition.

In the trachytic rocks, the plagioclase is rare and only found as rounded phenocrysts in the less undersaturated rocks or in the cores of alkali feldspar phenocrysts.

Major and trace elements

Mount Rittmann lavas show different degrees of silica saturation. One set of lavas belongs to a mildly alkaline association (normative Ne 0–3%), whose most primitive samples are two mugearites, at the limit of the field of hawaiites, parental to the slightly undersaturated group of hyperalkaline trachytes. A second group with normative Ne in the range 3–8% includes the more undersaturated trachytes, phonolites and nepheline/sodalite-bearing trachytes, and has its possible parental magmas in the Ne-hawaiite from the base of Mount Rittmann or in basanites like those outcropping in the area of Aviator Glacier. A last group (normative quartz in the range 0–8%) is represented by comenditic trachytes and coincides with the group of lavas with modal quartz whose most probable parental magma is a transitional olivine-basalt (ol-basalt), like that found in the area of Aviator Glacier but not yet documented near Mount Rittmann.

None of the associations found on Mount Rittmann exhibits a continuous trend due to a scarcity of samples, yet the diagram Na_2O v. Zr (here taken as a proxy for differentiation due to its incompatible behaviour: Fig. 16; Table 2) reveals three possible trends reaching different sodium contents in the most evolved rocks. The ol-basalt–comenditic trachyte trend shows a rise in Na_2O content at the beginning of the trend, while evolved rocks display a marked decrease, probably due to massive anorthoclase fractionation. The other major element variation diagrams do not reveal any significant difference in the behaviour of the three associations: K_2O at the beginning increases abruptly with Zr and then the trend is constant; P_2O_5 initially decreases and then shows a level pattern due to the early Ti-magnetite and apatite fractionation; and the Al_2O_3 content is practically constant and has a slight decrease only in the set of comenditic trachytes, the most evolved rocks of this association, that have undergone a stronger plagioclase fractionation (McDonald 1974).

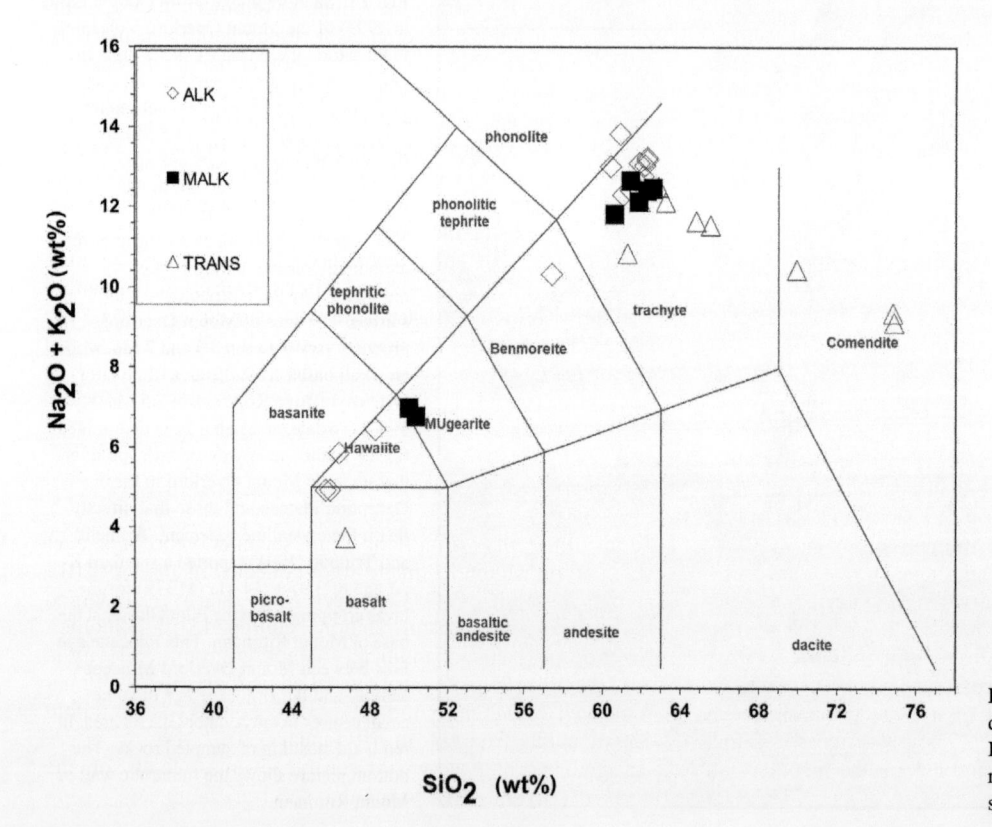

Fig. 15. Total alkali–silica classification diagram of Mount Rittmann rocks (Le Bas et al. 1986). Diamonds, alkaline rocks; black squares, mildly alkaline series; triangles, transitional series.

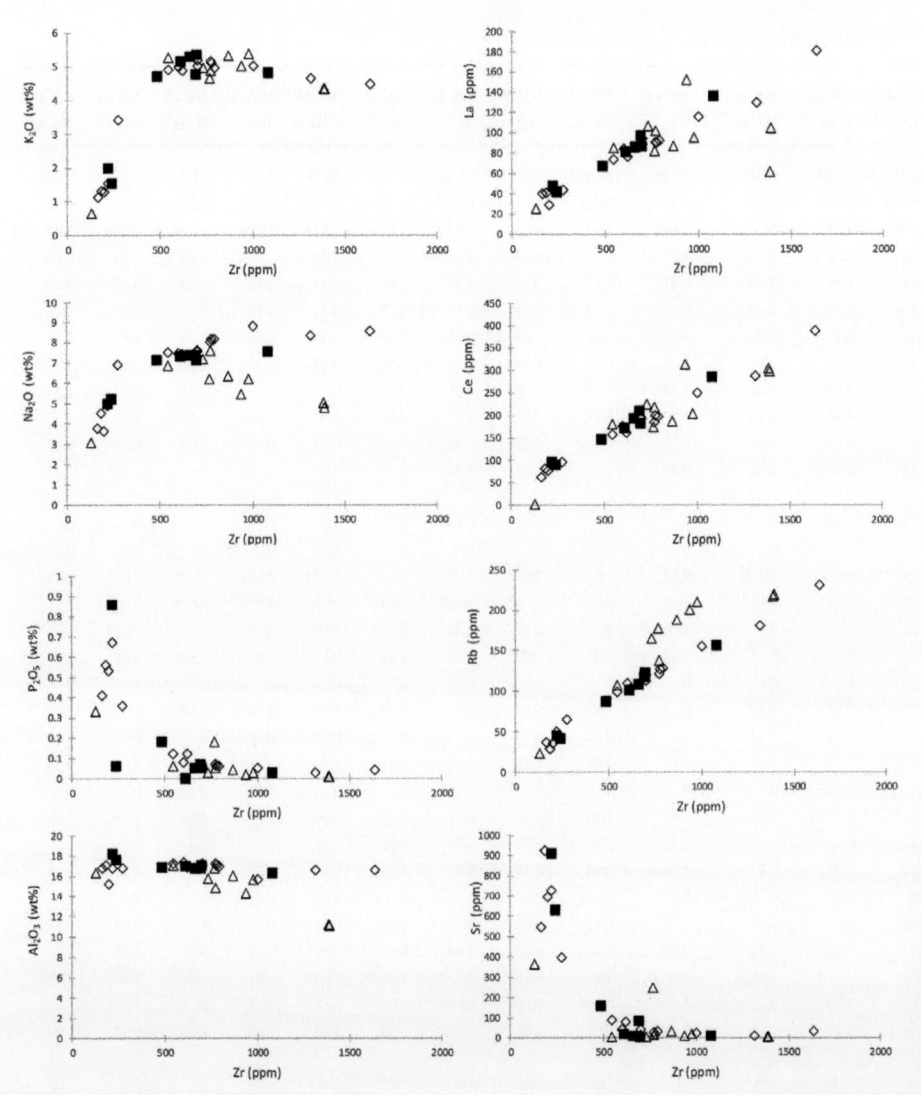

Fig. 16. Selected major and trace element variation diagrams. Zr has been taken as a tracer of degree of evolution due to its incompatible behaviour in this association. Major element variations show the effects of alkali feldspar fractionation and of the early appearance of apatite on the liquid line of descent. The effect of anorthoclase fractionation is evident in the transitional series by the decrease in Al_2O_3 and Na_2O contents for Zr >750 ppm.

In addition, trace element variation diagrams were plotted adopting Zr as the index of differentiation. The variation diagrams in Figure 16 show a generally good positive correlation of La, Ce and Rb with Zr, with an incompatible behaviour only occurring in the less evolved rocks. Comenditic trachytes have a distinctly lower La, Ce and Rb content than the undersaturated trachytes at corresponding Zr concentrations. Sr shows a hyperbolic trend with a sudden decrease in basic and intermediate samples; its values tend to zero in evolved rocks, showing the effects of plagioclase fractionation.

Geochemical monitoring and observations

Due to the recent discovery of Mount Rittmann in late 1988, which is further from the nearest Antarctic bases than Melbourne volcano, very little information has been available and, in particular, no gas geochemistry studies have been made on this volcano. A first attempt to sample gas from the fumaroles was carried out in the summer campaign in 1990 (Bonaccorso et al. 1991); however, it is known that these authors did not have specific equipment to collect fumarole gases and the results were poor quality. Indeed, the results of their analysis unfortunately confirmed that they had merely collected atmospheric air. However, during the survey they were able to provide an account of a wide fumarolic area with more than 20 steaming vents. They also measured the temperature of three of the fumaroles in the geothermal field, finding temperatures ranging between 57 and 65°C.

More recently, during the Antarctic summer of 2016, a survey was carried out by a team of researchers from the ICE-VOLC project. The investigated geothermal field consisted of a steep slope on the east flank of the volcano, uncovered by perennial ice (Fig. 14b). The temperature of the ground measured by a thermal camera (FLIR C2: Fig. 17) showed a wide warm area in which several fumaroles were actively steaming. Further measurements of temperature of the fumaroles, using a thermocouple inserted into the ground at a depth of approximately 50 cm, recorded temperatures ranging between 59 and 72°C (Table 3). In addition, a survey with a portable MultiGAS instrument was carried out along the fumarolic area, with the aim of identifying the presence and the contributions of CO_2, SO_2 and H_2S gas emissions. A track of the survey is shown in Figure 18.

Seismicity

At the time of writing, the only seismic recordings ever acquired on Mount Rittmann were collected in January 2017 as part of the ICE-VOLC project. Such data were recorded by two broadband seismic stations, equipped with Nanometrics Trillium 120s seismometers sampling ground velocity at a rate of 100 Hz, temporarily installed on the rim of the fumarolic wall of Mount Rittmann (73.48° S, 165.63° E), about 130 m apart, from 11 to 19 January 2017 (Fig. 19). We detected approximately 100 seismic events, which, on the basis of several waveform and spectral parameters, were

Table 2. *Major and trace element analyses*

	NN 09 Hy. tr.*	NN 10 Phon.	NN 11 C. tr.	NN 12 Hy. tr.	NN 14 Tr.	NN 15 C. tr.	NN 16A Hy. tr.	NN 17 Mu.	NN 19 Hy. tr.	NN 20 Hy. tr.	NN 21 Tr.	NN 22 Hy.tr.	NN 23 C. tr.	NN 24 Tr.	NN 25 Hy.tr.	NN 26 Mu.	NN 27 Hy. tr.
SiO_2	61.88	60.91	63.26	62.31	61.31	62.89	62.31	50.36	62.6	62.3	61.97	61.44	65.56	62.1	62.6	50.01	61.88
TiO_2	0.5	0.39	0.38	0.39	0.85	0.49	0.42	2.47	0.4.	0.31	0.43	0.61	0.32	0.41	0.43	2.11	0.38
Al_2O_3	17.36	15.68	15.77	17.07	16.79	14.85	17.23	17.63	17.09	16.33	17.12	17.25	15.73	17.23	17.03	18.22	16.8
Fe_2O_3	1.31	4.34	4.42	2.13	5.26	2.41	2.3	4.5	2.22	3.64	2.63	1.11	2.54	2.66	1.98	3.49	2.04
FeO	3.78	2.69	2.03	2.99	1.09	4.51	2.64	6.41	2.48	2.47	2.37	3.89	1.33	2.68	3.22	5.12	3.29
MnO	0.17	0.22	0.22	0.19	0.15	0.24	0.18	0.2	0.19	0.21	0.18	0.17	0.11	0.19	0.19	0.17	0.19
MgO	0.54	0.21	0.15	0.25	0.75	0.2	0.31	3.94	0.26	0.13	0.25	0.79	0.17	0.18	0.32	2.9	0.29
CaO	1.1	1.06	1.01	0.9	2.33	1.23	1.02	6.72	0.99	0.89	1	1.56	0.98	0.93	1.09	7.59	1.3
Na_2O	7.45	8.82	7.19	8.16	6.21	7.63	8.03	5.24	7.12	7.68	6.88	7.52	6.19	7.6	7.32	4.99	7.37
K_2O	5	5.02	4.95	5.09	4.65	4.88	5.16	1.53	5.34	4,83	5.27	4.91	5.37	5.18	5.15	1.99	4.77
P_2O_5	0.08	0.05	0.03	0.06	0.18	0.05	0.07	0.59	0 05	0.03	0.06	0.12	0.03	0.05	0,07	0.86	0.07
LOI	0.83	0.61	0.6	0.46	0.43	0.62	0.32	0.41	1.27	1.38	1.84	0.63	1.67	0.79	0.6	2.55	1.6
CIPW norms																	
Q	0	0	2.14	0	3.13	0.76	0	0	0	0	0	0	7.9	0	0	0	0
C	0	0	0	0	0	0	0	0	0	0	0	0	0	0	0	0	0
Or	29.55	29.66	29.25	30.08	27.48	28.84	30.49	9.04	31.55	28.54	31.14	29.01	31.73	30.61	30.43	11.76	28.19
Ab	53.75	42.31	53.55	50.23	52.54	49.21	49.27	40.81	55.73	55.22	57.52	50.91	51.01	52.76	54.78	37.32	55.72
An	0	0	0	0	4.21	0	0	20.07	0	0	0.27	0	0	0	0	21.44	0
Ne	4.17	5.63	0	5	0	0	5.76	1.91	1.32	1.02	0.38	5.68	0	3.8	2.24	2.65	2.24
Le	0	0	0	0	0	0	0	0	0	0	0	0	0	0	0	0	0
Ac	1.39	12.56	6.41	6.16	0	6.97	6.65	0	1.82	6.2	0	1.97	1.2	3.98	2.66	0	2.21
Ns	0	1.79	0	0.6	0	1.73	0.12	0	0	0	0	0	0	0	0	0	0
DI	4.17	4.33	3.98	3.57	4.03	5.1	4.01	7.6	3.97	3.71	3.68	5.97	1.6	3.75	4.31	8.56	5.23
Hy	0	0	0	0	0	5.73	0	0	0	0	0	0	0	0	0	0	0
Ol	3.71	2.26	0	3.02	0	0	2.42	7.58	1.16	1.25	0.41	3.78	0.00.	1.55	2.46	4.65	2.06
Mt	1.2	0	3.19	0	1.54	0	0	6.52	2.3	2.03	3.81	0.62	3.08	1.86	1.54	5.06	1.85
Il	0.95	0.74	0.72	0.74	1.61	0.93	0.8	4.69	0.76	0.59	0.82	1.16	0.61	0.78	0.82	4.01	0.72
hm	0	0	0	0	4.2	0	0	0	0	0	0	0	0	0	0	0	0
Ap	0.19	0.12	0.07	0.14	0.43	0.12	0.17	71.4	0.12	0.07	0.14	0.28	0.07	0.12	0.17	2.04	0.17
Aq	0.83	0.61	0.6	0.46	0.43	0.62	0.32	0.41	1.27	1.38	1.84	0.63	1.67	0.79	0.6	2.55	1.6
Tot	99.91	100.01	99.91	99.86	99.6	100.01	100.01	170.03	100	98.63	100.01	100.01	98.87	100	100.01	100.04	99.99
DI	88.66	91.94	91.35	92.07	83.16	87.51	92.29	51.76	90.43	90.98	89.04	87.57	91.84	91.16	90.11	51.74	88.35
Agp.lnd	1.02	1.27	1.09	1.11	0.91	1.2	1.09	0.58	1.02	1.08	0.99	1.03	1.02	1.05	1.03	0.57	1.03
Trace elements																	
Ce	176	250	225	199	176	219	185	91	183	286	181	157	204	195	173	96	210
Ba	123	76	9	42	686	146	87	440	16	61	54	179	77	41	195	648	480
La	84	116	106	91	82	102	91	42	87	136	85	74	95	94	81	48	97
Nd	151	245	193	181	144	182	176	66	164	242	153	130	195	178	149	62	166
Zr	602	1000	728	775	764	772	768	238	695	1083	546	543	976	701	608	220	691
Y	47	80	67	56	47	65	65	31	61	83	43	43	65	52	50	27	56
Sr	43	21	5	14	246	13	28	630	6	9	6	89	18	12	17	912	86
Rb	110	155	165	127	177	138	121	42	122	156	107	99	210	113	101	45	118

*C., comenditic; Tr., trachyte; C. tr., comenditic trachyte; Hy. tr., hyperalkaline trachyte; Phon., phonolite, Mu., Mugearite, Ne-hw., Ne-hawaiite; Ol-bas., olivine-basalt; Sy, syenite.
From Armienti and Tripodo (1991).

classified into two different classes: (i) high-frequency events and (ii) low-frequency events.

The first class included events characterized by frequencies higher than 8–10 Hz, short duration (<10 s) and a sharp decrease in amplitude with distance (Fig. 20c, d). The low-frequency events showed a spectral content lower than 8–10 Hz, a duration of about 10 s, emergent onset, and no clear P and S phases (Fig. 20a, b). In December 2017, a permanent integrated seismic and infrasonic station was installed (Contrafatto et al. 2018; Fig. 19).

Discussion

Tomographic studies highlight a slow velocity anomaly, between Ross Island and Victoria Land, reaching a depth of 250 km and interpreted as a broad hot mantle anomaly (e.g. Morelli and Danesi 2004; Hansen et al. 2014). Such a deep hot anomaly is consistent with the geochemical models of volcanism in the Ross area, suggesting a deep magmatic source (e.g. Rocchi et al. 2002).

Park et al. (2015) evidenced a low-velocity region 80 km under Mount Melbourne that was attributed to a c. 300°C thermal anomaly.

Therefore, the magmatic activity of Mount Melbourne and Mount Rittmann is related to the mantle anomaly, and comprises effusive eruption of poorly evolved alkali basalts and hawaiites and explosive products from evolved magmas, such as the trachytes producing Plinian eruptions (Armienti and Tripodo 1991; Giordano et al. 2012).

Geochemical investigations of Mount Melbourne revealed that ice caves on the summit part of the volcano had a different position with respect to the description made by Lyon and Giggenbach (1974). As described in the subsection on 'Geochemical monitoring and observations', volcanic fumarolic activity can be both the source

	NN 28	NN 29	NN 30	NN 31	NN 34	NN 35	NN 37	NN 38	NN 41	MB 75	MB 76	MB 77	MB 78	MB 79	MB 80	MB 81	MB 81A
	C. tr.	Hy. tr.	Hy. tr.	Hy. tr.	Sy.	Hy. tr.	Hy. tr.	Hy. tr.	Tr.	C.	Ol-bas.	Bas.	Ne-hw.	C.	Ne-hw.	C.	C.
SiO_2	64.84	61.08	61.89	61.89	57.38	62.19	61.43	60.44	60.64	69.93	46.75	45.74	45.96	48.29	46.42	74.82	74.88
TiO_2	0.31	0.59	0.49	0.43	1.31	0.26	0.4	0.23	0.67	0.23	1.92	2.87	2.2	2.5	3.27	0.21	0.2
Al_2O_3	16.03	16.94	17.02	16.97	16.79	16.62	16.82	16.65	16.89	14.26	16.32	15.24	16.77	16.8	17.07	11.17	11.08
Fe_2O_3	2.17	2.39	2.78	3.21	1.76	2.73	2.68	4.59	2.85	1.7	2.48	3.01	1.69	3.57	2.69	1.49	2.32
FeO	1.63	3.04	2.61	2.29	5.91	2.86	2.54	1.25	2.65	1.5	8.35	8.18	9.18	6.89	8.71	1.79	0.9
MnO	0.07	0.18	0.19	0.19	0.19	0.18	0.18	0.2	0.17	0.1	0.18	0.18	0.19	0.19	0.2	0.05	0.05
MgO	0.73	0.68	0.49	0.37	1.71	0.19	0.2	0.23	0.6	0.12	7.56	8.82	8.25	5.67	5.12	0.06	0.07
CaO	1.04	1.48	1.2	1.11	3.5	0.83	0.89	0.76	1.8	0.66	9.46	9.6	9.54	8.24	8.13	0.33	0.32
Na_2O	6.34	7.43	7.58	8.16	6.9	8.58	7.36	8.38	7.12	5.45	3.08	3.63	3.8	4.9	4.54	5.02	4.81
K_2O	5.31	4.88	6.01	4.95	3.43	4.48	5.3	4.65	4.71	5.01	0.65	1.28	1.12	1.55	1.31	4.35	4.33
P_2O_5	0.04	0.12	0.06	0.06	0.36	0.04	0.05	0.03	0.18	0.02	0.33	0.53	0.41	0.67	0.56	0.01	0.01
LOI	1.49	1.19	0.69	0.37	0.76	1.03	2.15	2.59	1.7	1.02	2.93	0.92	0.89	0.72	1.99	0.7	1.03
CIPW norms																	
Q	5.21	0	0	0	0	0	0	0	0	16.28	0	0	0	0	0	31.18	31.84
C	0	0	0	0	0	0	0	0	0	0	0	0	0	0	0	0	0
Or	31.38	28.84	29.6	29.25	20.27	26.47	31.32	27.48	27.83	29.6	3.84	7.56	6.62	9.16	7.74	25.7	25.59
Ab	52.89	52.07	52.61	48.56	49.11	52.69	51.09	48.48	55.51	45.45	26.06	17.16	17.51	26.44	25.79	33.23	32.88
An	0	0	0	0	4.71	0	0	0	0.22	0	28.79	21.51	25.39	19.27	22.33	0	0
Ne	0	4.27	3.82	6.05	5.03	4.25	3.2	6.1	2.56	0	0	7.34	7.93	8.14	6.84	0	0
Le	0	0	0	0	0	0	0	0	0	0	0	0	0	0	0	0	0
Ac	0.67	2.56	3.95	8.21	0	7.9	4.64	9.83	0	0.58	0	0	0	0	0	1.2	1.14
Ns	0	0	0	0	0	0.72	n.00	0	0	0	0	0	0	0	0	1.83	1.52
DI	3.98	5.57	4.75	4.41	8.69	3.39	3.57	3.04	6.27	2.77	13.07	18.39	15.69	14.16	11.84	1.39	1.35
Hy	0.89	0	0	0	0	0	0	0	0	3.38	0.42	0	0	0	0	4.22	4.06
Ol	0	1.92	1.48	1.65	5.57	2.95	1.59	0.25	0.07	0	18.35	18.19	18.76	13.65	13.72	0	0
Mt	2.81	2.18	2.05	0.54	2.55	0	1.56	1.73	4.13	0.29	2.03	2.09	2.05	1.94	2.13	0	0
Il	0.59	1.12	0.93	0.82	2.49	0.49	0.76	0.44	1.27	0.44	3.65	5.45	4.18	4.75	6.21	0.4	0.38
hm	0	0	0	0	0	0	0	0	0	0	0	0	0	0	0	0	0
Ap	0.09	0.28	0.14	0.14	0.85	0.09	0.12	0.07	0.43	0.05	0.78	1.26	0.97	1.59	1.33	0.02	0.02
Aq	1.49	1.19	0.69	0.37	0.76	1.03	2.15	2.59	1.7	1.02	2.93	0.92	0.89	0.72	1.99	0.7	1.03
Tot	100	100.01	100.02	100.01	100.02	99.99	100	100	99.99	99.87	99.92	99.87	100	99.8	99.92	99.89	99.81
DI	90.15	87.74	89.98	92.07	74.4	92.03	90.26	91.89	85.91	91.92	29.9	32.07	32.06	43.74	40.37	93.15	92.97
Agp.lnd	1-Jan	1.03	1.05	1.11	0.9	1.14	1.06	1.13	1	1.01	0.35	0.48	0.44	0.58	0.52	1.16	1.14
Trace elements																	
Ce	186	161	194	197	96	389	194	287	146	314	3	78	63	96	82	304	298
Ba	177	146	127	73	1337	121	31	0	672	282	236	464	393	438	388	167	167
La	87	77	89	92	44	181	86	130	67	152	25	29	40	48	41	62	105
Nd	162	142	162	182	77	346	164	298	117	199	24	62	51,	65	62	167	170
Zr	868	621	702	790	276	1636	659	1318	485	936	127	199	165	219	182	1387	1392
Y	55	44	54	60	30	111	53	87	41	75	22	24	23	26	28	94	95
Sr	33	81	44	34	398	33	8	8	159	10	366	695	546	366	923	6	5
Rb	188	104	120	128	65	231	108	181	88	201	23	36	37	23	29	217	219

of the high temperature and also explain the variation in the position of the ice caves. The CO_2 and CH_4 concentrations and the isotopic terms of the sampled gases evidence their volcanic origin.

Geophysical investigations at Mount Melbourne comprise ground deformation and seismicity: long-term tilt trends showed coherent changes at the three highest altitude stations (Gambino et al. 2016) in the 1997–2002 period, while GPS vertical data from 2003 to 2015 evidenced that the GPS benchmark, located atop the Mount Melbourne volcano, does not concord with the regional trend as it is representative of a volcanic evolution (Zanutta et al. 2017). These observations suggest the presence of shallow ground deformation sources, the effects of which seem restricted to the Mount Melbourne summit area (Gambino et al. 2016).

Gambino and Privitera (1996) evidenced microseismicity defined as Mount Melbourne 'local seismic events' represented by earthquakes with M_L <2.0, with emerging onsets, localized on the east slope of Mount Melbourne volcano. They formulated two different hypotheses on the source of these events: (i) long-period events related to fluid dynamics inside the plumbing system; and (ii) volcano-tectonic earthquakes due to fracturing processes taking place within the volcano edifice. In both cases, the identification of these events is a clue to the continuous and active internal dynamics of Mount Melbourne volcano.

In January 2017, a long-lasting tremor-like signal was recorded simultaneously at two seismic stations located on the Mount Melbourne summit (close to the fumarolic areas), 200 m from each other (Fig. 11). Since it is impossible to locate the source of this signal (recorded by only two stations), it is hard to define the source nature of this tremor. On the basis of the higher spectral content, we can exclude oceanic-related sources (similar to microseism). Hence, the possible source contenders are: (i) dynamics of

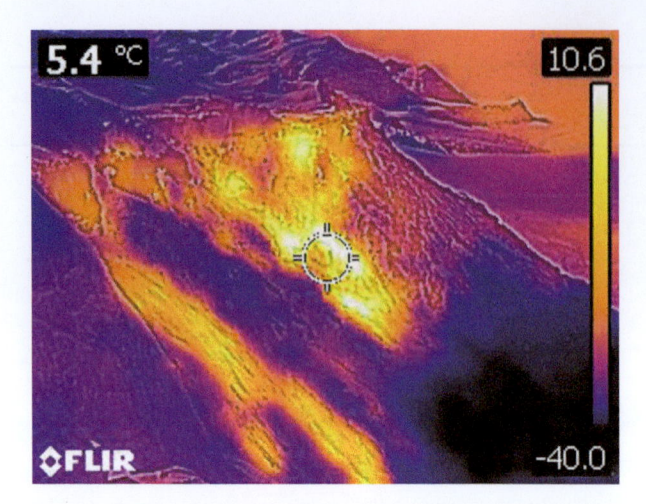

Fig. 17. Thermal image of the fumarolic area at Mount Rittmann. The higher temperature at the soil surface was around 17°C, while in the air was around −10°C. The thermal survey has been useful for focusing on the hotter fumaroles that are more suitable for gas sampling. The following temperature surveys, made with a thermocouple inserted in the ground, revealed temperatures higher than 70°C (see Table 3).

fluids within the volcano plumbing system, and in this case this tremor could be named a volcanic tremor; and (ii) sub-glacial discharge of water melted by the volcanic heat released in the fumarolic areas. Indeed, glacio-hydraulic tremor-like seismic signals have recently been observed in glacial regions (e.g. Bartholomaus *et al.* 2015).

Geochemical and seismological investigations of Mount Rittmann started in 2016. The first results of this work show that, although no SO_2 or H_2S gas species have been detected, the CO_2 and CH_4 concentration was significantly high (Fig. 18), and distinctly higher than the CO_2 and CH_4 contents in the atmosphere. Hence, the geothermal system at Rittmann volcano may still be considered active

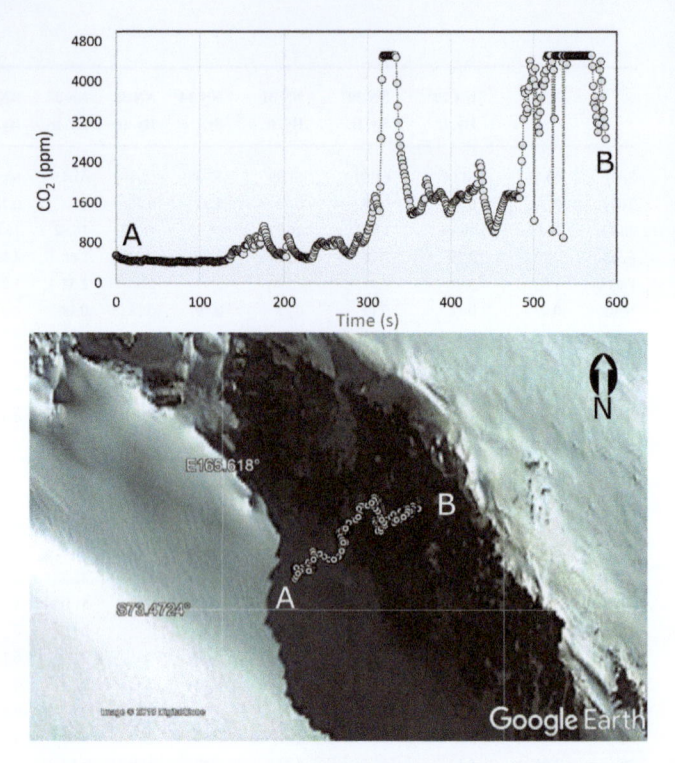

Fig. 18. Track of the MultiGAS (MG) on the fumarolic area of Mount Rittmann. The MG measurement was taken from the air at a distance of around 50 cm from the fumarolic vents. The direction of the track is from A to B, as shown in the image taken from Google Earth. The in-plume CO_2 concentration revealed an increase in concentration as we moved from A to B, which coincides with the higher density of hot fumaroles (see Fig. 17). Interestingly, the CO_2 concentration in the proximity of the fumaroles is distinctly higher than the atmosphere background contents as, in some cases, the CO_2 concentration is even greater than the detection limit of the IR sensor, demonstrating a clear volcanic origin.

Table 3. *External temperatures (T_{ext}) and fumarolic (T_{soil}) temperatures measured at around 50 cm depth on Mount Rittmann*

Date	Sample	T_{ext} (°C)	T_{soil} (°C)
9 December 2016	A RF1 9-12-16	−10	72
9 December 2016	B RF1 9-12-16	−10	72
9 December 2016	A RF1 9-12-16	−10	72
9 December 2016	B RF1 9-12-16	−10	72
9 December 2016	A RF2 9-12-16	−10	66
9 December 2016	B RF2 9-12-16	−10	66
9 December 2016	A RF3 9-12-16	−9	69
9 December 2016	B RF3 9-12-16	−9	69
9 December 2016	A RF3 9-12-16	−9	69
9 December 2016	B RF3 9-12-16	−9	69
9 December 2016	A RF4 9-12-17	−9	69.5
9 December 2016	A RF4 9-12-17	−9	69.5
9 December 2016	B RF4 9-12-17	−9	69.5
9 December 2016	A RF5 9-12-17	−9	71
9 December 2016	B RF5 9-12-17	−9	71
15 December 2016	RF10A 15-12-2016	–	59
15 December 2016	RF10B 15-12-2016	–	59
15 December 2016	RF11A 15-12-2016	–	69
15 December 2016	RF11B 15-12-2016	–	69

During the survey, the atmospheric pressure was around 740 mbar; therefore the temperatures recorded were generally below boiling point.

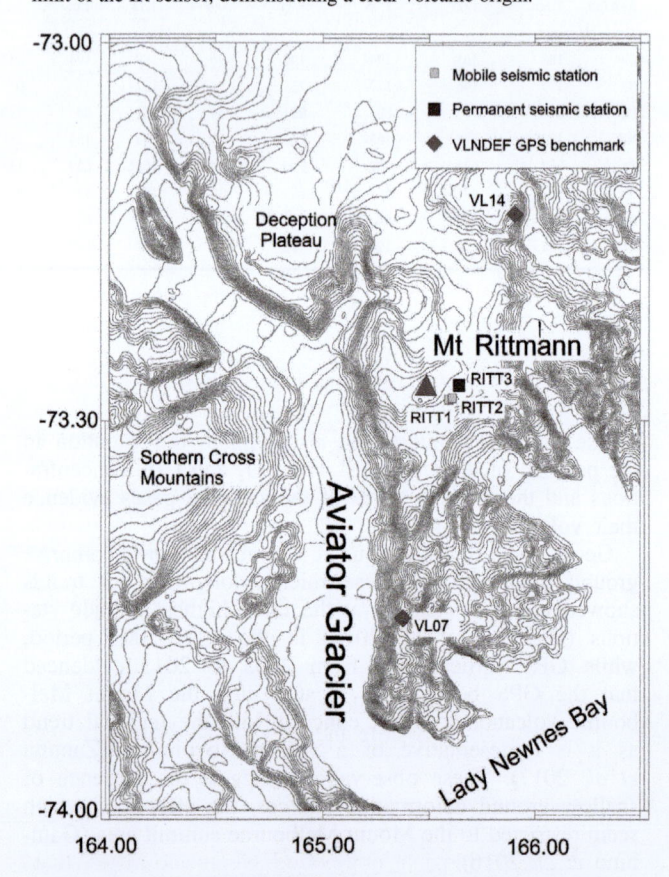

Fig. 19. Map of the Mount Rittmann area showing the location of the mobile (RITT2 and RITT3) and permanent (RITT1) stations installed in 2017.

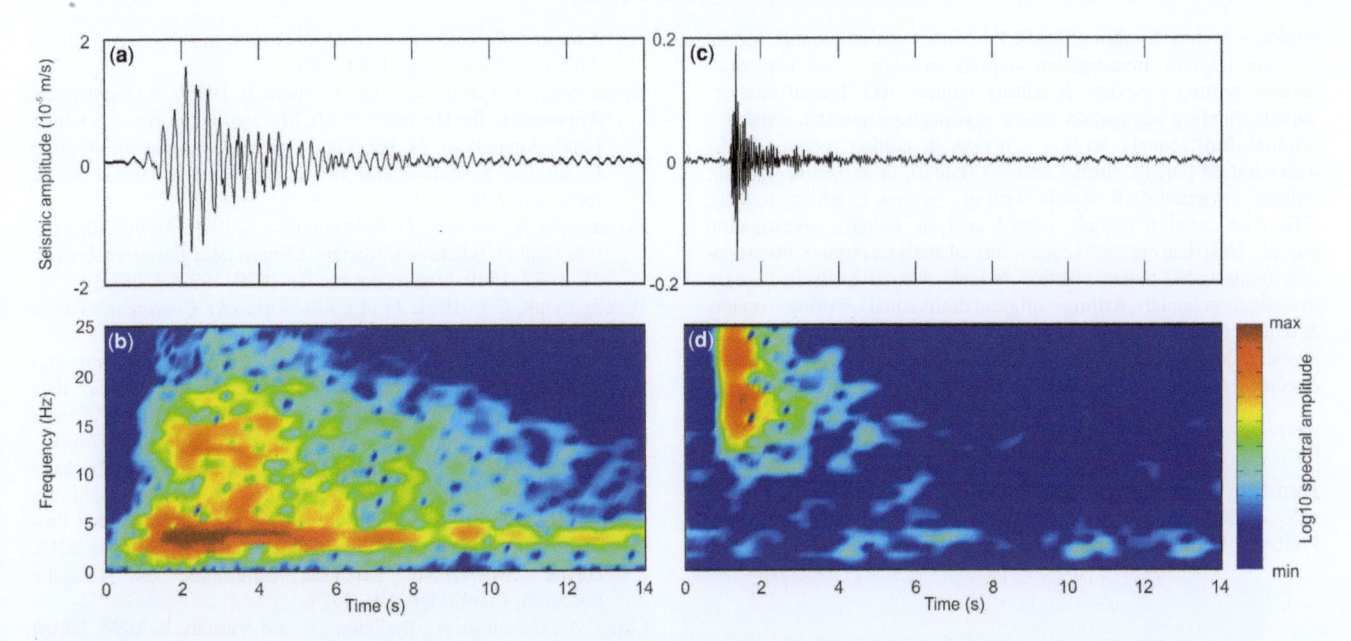

Fig. 20. Examples of waveforms and spectrograms of a low-frequency event (**a** and **b**) recorded on 19 January 2017 at 12:05; and a high-frequency event (**c** and **d**) recorded on 11 January 2017 at 07:09, acquired on Mount Rittmann.

and fed by volcanic fluids. Analysis of 19 sampled fumaroles taken from seven different sites show an enrichment in CO_2 and CH_4 (data still to be published), and denote hydrothermal gas equilibrium in the system H_2O–H_2–CO_2–CO–CH_4. To sum up, even if our data are preliminary, our findings on Mount Rittmann show that magmatic fluids are still feeding the geothermal system and the current geothermal conditions are representative of a late magmatic process.

At the time of writing, we have detected about 100 seismic events on Mount Rittmann that represent the first seismic data recorded on this volcano. On the basis of several waveform and spectral parameters, these seismic events were classified into high- and low-frequency events. The former are characterized by frequencies higher than 8–10 Hz, short duration (<10 s) and a sharp decrease in amplitude with distance (Fig. 20c, d). Similar to the seismic type also identified on Mount Melbourne, these events are probably associated with ice faulting forming crevassing. As for the latter events, they show a spectral content lower than 8–10 Hz, a duration of about 10 s, emergent onset, and no clear P and S phases (Fig. 20a, b). Hence, they resemble the long-period events that on volcanoes are generally associated with fluid dynamics inside the plumbing system (e.g. Chouet and Matoza 2013). However, to shed light on the nature of these low-frequency events, recordings using a greater number of stations will be necessary in order to constrain epicentral coordinates and focal depth. In December 2017, a permanent integrated seismic and infrasonic station was installed.

Conclusions

Data analyses and results collected over the last few decades prove that Mount Melbourne and Mount Rittmann are active and characterized by:

- fumaroles, that at both volcanoes are active and fed by volcanic fluids;
- seismicity, comprising long-period events and tremor that are typical signals of active volcanoes whose sources are related to their internal activity;
- ground deformation, showing evidence of slow deformative phases (inflation/deflation) characterizing the summit area of Mount Melbourne.

Future implementation of a geochemical and geophysical monitoring system (the objective of ongoing and future projects) would help to improve our knowledge and enable the acquisition of near-real-time data that can recognize the evolution and changes in the state of the two volcanoes, all of which would prove highly useful for volcanic risk mitigation. In addition, new geochemical data and dating on the englacial tephra on Mount Melbourne will provide information on the Holocene explosive activity of this volcano.

On Mount Melbourne, the last eruption occurred at some time between 1862 and 1922 (Lyon 1986; Kyle 1990). Its past history and evolution suggests that an intense, potentially Plinian explosive activity in the near future must be considered as a possible scenario, given that the last eruptions were explosive and associated with the most evolved magma compositions (Giordano *et al.* 2012).

Mount Melbourne lies approximately 40 km NNE from MZS, thus any potential unrest of this volcano might significantly affect the Italian base, as well as the other two nearby scientific bases – Jang Bogo (Korea) and Gondwana (Germany). Moreover, as the ash produced by the 2010 Eyjafjallajökull eruption showed (Gislason *et al.* 2011), Mount Melbourne and Mount Rittmann eruptive activity could also be a serious hazard to large and distant communities.

Acknowledgements We are particularly indebted to Prof. Letterio Villari, who set up the first 'Volcanological Observatory of Mount Melbourne' in 1988. We thank Alessandro Bonaccorso, Danilo Contrafatto, Giuseppe Falzone, Angelo Ferro, Gianni Lanzafame, Graziano Larocca, Giuseppe Laudani, Eugenio Privitera, Salvatore Rapisarda, Danilo Reitano and Luciano Scuderi, who all contributed with their commitment and participation to developing the monitoring and research activities in Antarctica. We acknowledge ENEA (Agenzia nazionale per le nuove tecnologie, l'energia e lo sviluppo economico sostenibile) for providing field logistics at MZS Station and CNR (Consiglio Nazionale delle Ricerche) for scientific support.

Author contributions SG: data curation (equal), formal analysis (equal), investigation (equal), writing – original draft (lead), writing – review & editing (lead); **PA**: data curation (equal), formal analysis (equal), investigation (equal), writing – original draft (equal), writing – review & editing (equal); **AC**: formal analysis (equal), funding acquisition (lead), investigation (equal), writing – original draft (equal), writing – review & editing (equal); **PDC**: data curation (equal), formal analysis (equal), investigation (equal), writing – original draft (equal), writing – review & editing (equal); **GG**: data curation (equal), formal analysis (equal), investigation (equal); **GG**: data curation (equal), formal analysis (equal), investigation (equal); **ML**: data curation (equal), formal analysis (equal), investigation (equal), writing – original draft (equal), writing – review & editing (equal); **MP**: data curation (equal), formal analysis (equal), investigation (equal), writing – original draft (equal), writing – review & editing (equal).

Funding This work was supported by a PNRA (Programma Nazionale di Ricerche in Antartide) ICE-VOLC project grant (No. PNRA14_00011) to A. Cannata.

Data availability The datasets generated during and/or analysed during the current study are available from the corresponding author on reasonable request.

References

Anderson, K., Lisowski, M. and Segall, P. 2010. Cyclic ground tilt associated with the 2004–2008 eruption of Mount St. Helens. *Journal of Geophysical Research*, **115**, B11201, https://doi.org/10.1029/2009JB007102

Armienti, P. and Tripodo, A. 1991. Petrography and chemistry of lavas and comagmatic xenoliths of Mt. Rittmann, a volcano discovered during the IV Italian expedition in Northern Victoria Land (Antarctica). *Memorie della Società Geologica Italiana*, **46**, 427–451.

Armienti, P., Civetta, L., Innocenti, F., Manetti, P., Tripodo, S., Villari, L. and Vita, G. 1991. New petrological and geochemical data on Mt. Melbourne Volcanic Field, Northern Victoria Land, Antarctica. (II Italian Antarctic Expedition). *Memorie della Società Geologica Italiana*, **46**, 397–424.

Aviado, K.B., Hall, S.R., Bryce, J.G. and Mukasa, S.B. 2015. Submarine and subaerial lavas in the West Antarctic Rift System: Temporal record of shifting magma source components from the lithosphere and asthenosphere. *Geochemistry, Geophysics, Geosystems*, **16**, 4344–4361, https://doi.org/10.1002/2015GC006076

Bartholomaus, T.C., Amundson, J.M., Walter, J.I., O'Neel, S., West, M.E. and Larsen, C.F. 2015. Subglacial discharge at tidewater glaciers revealed by seismic tremor, *Geophysical Research Letters*, **42**, 6391–6398, https://doi.org/10.1002/2015GL064590

Bonaccorso, A, Mione, M, Pertusati, P.C., Privitera, E and Ricci, C.A. 1991. Fumarolic activity at Mt. Rittmann volcano (Northern Victoria Land, Antarctica). *Memorie della Società Geologica Italiana*, **46**, 453–456.

Bonaccorso, A., Falzone, G., Gambino, S. and Villari, L. 1995. Tilt signals recorded at Mt Melbourne Volcano (Northern Victoria Land, Antarctica) between 1989–94. *Terra Antartica*, **2**, 111–116.

Bonaccorso, A., Gambino, S., Falzone, G. and Privitera, E. 1996. Physics volcanological studies in the activity framework of the Mt. Melbourne Observatory (Northern Victoria Land, Antarctica). *In*: Meloni, A. and Morelli, A. (eds) *Italian Geophysical Observatories in Antarctica*. Editrice Compositori, Bologna, Italy, 67–92.

Bonaccorso, A., Gambino, S., Falzone, G. and Privitera, E. 1997a. The Volcanological Observatory of the Mt. Melbourne (Northern Victoria Land, Antarctica). *In*: Ricci, C.A. (ed.) *The Antarctic Region: Geological Evolution and Processes*. Terra Antartica, Siena, Italy, 1083–1086.

Bonaccorso, A., Gambino, S. and Privitera, E. 1997b. A Geophysical Approach to the Dinamics of Mt. Melbourne (Northern Victoria Land, Antarctica). *In*: Ricci, C.A. (eds) *The Antarctic Region: Geological Evolution and Processes*. Terra Antartica, Siena, Italy, 531–538.

Bonaccorso, A., Falzone, G. and Gambino, S. 1999. An investigation into shallow borehole tiltmeters. *Geophysical Research Letters*, **26**, 1637–1649, https://doi.org/10.1029/1999GL900310

Borchgrevink, C.E. 1901. *First on the Antarctic Continent*. George Newnes, London.

Cannata, A., Larocca, G. *et al.* 2017. Characterization of seismic signals recorded in Tethys Bay, Victoria Land (Antarctica): data from atmosphere–cryosphere–hydrosphere interaction. *Annals of Geophysics*, **60**, S0555, https://doi.org/10.4401/ag-7408

Cannata, A., Contrafatto, D. *et al.* 2018. Melbourne and Rittmann Volcanoes: Results from ICE-VOLC Project. *In*: Schilling Hoyle, A. (ed.) Where the Poles Come Together. Abstract Proceedings of the Open Science Conference, 19 – 23 June 2018, Davos, Switzerland. Scientific Committee on Antarctic Research, Cambridge, UK, 1777.

Capra, A., Gubellini, A., Radicioni, F. and Vittuari, L. 1996. Italian geodetic activities in Antarctica. *In*: Meloni, A. and Morelli, A. (eds) *Italian Geophysical Observatories in Antarctica*. Editrice Compositori, Bologna, Italy, 2–20.

Capra, A., Radicioni, F. and Vittuari, L. 1998. Italian geodetic network as reference frame for geodynamic purposes (Terra Nova Bay–Victoria Land–Antarctica). *In*: Forsberg, R., Feissel, M. and Dietrich, R. (eds) *Geodesy on the Move International*. Association of Geodesy Symposia, **119**, 498–503, https://doi.org/10.1007/978-3-642-72245-5_84

Chouet, B.A. and Matoza, R.S. 2013. A multi-decadal view of seismic methods for detecting precursors of magma movement and eruption. *Joyurnal of Volcanology and Geothermal Research*, **252**, 108–175, https://doi.org/10.1016/j.jvolgeores.2012.11.013

Contrafatto, D., Fasone, R., Ferro, A., Larocca, G., Laudani, G., Rapisarda, S. and Cannata, A. 2018. Design of a seismo-acoustic station for Antarctica. *Review of Scientific Instruments*, **89**(4), 044502, https://doi.org/10.1063/1.5023481

Danesi, S., Bannister, S. and Morelli, A. 2007. Repeating earthquakes from rupture of an asperity under an Antarctic outlet glacier. *Earth and Planetary Science Letters*, **253**, 151–158, https://doi.org/10.1016/j.epsl.2006.10.023

Dubbini, M., Cianfarra, P., Casula, G., Capra, A. and Salvini, F. 2010. Active tectonics in northern Victoria Land (Antarctica) inferred from the integration of GPS data and geologic setting. *Journal of Geophysical Research*, **115**, B12421, https://doi.org/10.1029/2009JB007123

Ferraccioli, F., Armadillo, E., Zunino, A., Bozzo, E., Rocchi, S. and Armenti, P. 2009. Magmatic and tectonic patterns over the Northern Victoria Land sector of the Transantarctic Mountains from new aeromagnetic imaging. *Tectonophysics*, **478**, 43–61, https://doi.org/10.1016/j.tecto.2008.11.028

Gambino, S. 2005. Air and permafrost temperature at Mt. Melbourne (1989–1998). *Antarctic Science*, **17**, 151–152, https://doi.org/10.1017/S095410200500249X

Gambino, S. and Privitera, E. 1994. Characterization of earthquakes recorded by Mt. Melbourne Volcano Seismic Network (Northern Victoria Land, Antarctica). *Terra Antartica*, **1**, 167–172.

Gambino, S. and Privitera, E. 1996. Mt. Melbourne Volcano, Antartica: evidence of seismicity related to volcanic activity. *Pure and Applied Geophysics*, **146**, 305–318, https://doi.org/10.1007/BF00876495

Gambino, S., Campisi, O., Falzone, G., Ferro, A., Guglielmino, F., Laudani, G. and Saraceno, B. 2007. Tilt measurements at Vulcano Island. *Annals of Geophysics*, **50**, 233–247, https://doi.org/10.4401/ag-4419

Gambino, S., Aloisi, M., Falzone, G. and Ferro, A. 2016. Tilt signals at Mt Melbourne Volcano: evidence of a shallow volcanic source. *Polar Research*, **35**, 28269, https://doi.org/10.3402/polar.v35.28269

Giordano, G., Lucci, F., Phillips, D., Cozzupoli, D. and Runci, V. 2012. Stratigraphy, geochronology and evolution of the Mt. Melbourne volcanic field (North Victoria Land, Antarctica). *Bulletin of Volcanology*, **74**, 1985–2005, https://doi.org/10.1007/s00445-012-0643-8

Gislason, S.R., Hassenkam, T. *et al.* 2011. Characterization of Eyjafjallajökull volcanic ash particles and a protocol for rapid risk assessment. *Proceedings of the National Academy of Sciences of the United States of America*, **108**, 7307–7312, https://doi.org/10.1073/pnas.1015053108

Hansen, S.E., Graw, J.H. *et al.* 2014. Imaging the Antarctic mantle using adaptively parameterized P-wave tomography: Evidence for heterogeneous structure beneath West Antarctica. *Earth and Planetary Science Letters*, **408**, 66–78, https://doi.org/10.1016/j.epsl.2014.09.043

Harrington, H.J. 1958. Nomenclature of rock units in the Ross Sea region, Antarctica. *Nature*, **182**, 290–291, https://doi.org/10.1038/182290a0

Kohl, M.L. and J.Levine, , 1993. Measuring low frequency tilts. *Journal of Research of the National Institute of Standards and Technology*, **98**, 191–202, https://doi.org/10.6028/jres.098.014

Kwok, R. and Comiso, J.C. 2002. Spatial patterns of variability in Antarctic surface temperature: connections to the Southern Hemisphere Annular Mode and the Southern Oscillation. *Geophysical Research Letters*, **29**, https://doi.org/10.1029/2002GL015415

Kyle, P.R. 1990. McMurdo Volcanic Group, western Ross Embayment. *American Geophysical Union Antarctic Research Series*, **48**, 19–145.

Le Bas, M.J., Le Maitre, R.W., Streckeisen, A. and Zanettin, B. 1986. A chemical classification of volcanic rocks based on the total alkali-silica diagram. *Journal of Petrology*, **27**, 745–750, https://doi.org/10.1093/petrology/27.3.745

Lee, M.J., Lee, J.I., Kim, T.H., Lee, J. and Nagao, K. 2015. Age, geochemistry and Sr-Nd-Pb isotopic compositions of alkali volcanic rocks from Mt. Melbourne and the western Ross Sea Antarctica. *Geosciences Journal*, **19**, 681, https://doi.org/10.1007/s12303-015-0061-y

Le Maitre, R.W., Streckeisen, A. *et al.* 2002. *Igneous Rocks: A Classification and Glossary of Terms*. Cambridge University Press, Cambridge, UK.

Lyon, G.L. 1986. Stable isotope stratigraphy of ice cores and the age of the last eruption at Mount Melbourne, Antarctica. *New Zealand Journal of Geology and Geophysics*, **29**, 135–138, https://doi.org/10.1080/00288306.1986.10427528

Lyon, G.L. and Giggenbach, W.F. 1974. Geothermal activity in Victoria Land, Antarctica. *New Zealand Journal of Geology and Geophysics*, **17**, 511–521, https://doi.org/10.1080/00288306.1973.10421578

Mazzarini, F. and Salvini, F. 1994. Contribution to geothermal survey by spectral analysis of TM Landsat satellite data in Mt. Melbourne Area, Northern Victoria Land (Antarctica). *Terra Antarctica*, **1**, 104–106.

McDonald, R. 1974. Nomenclature and petrochemistry of the peralkaline oversaturated extrusive rocks. *Bulletin of Volcanology*, **38**, 498–516, https://doi.org/10.1007/BF02596896

Morelli, A. and Danesi, S. 2004. Seismological imaging of the Antarctic continental lithosphere: a review. *Global and Planetary Change*, **42**, 155–165, https://doi.org/10.1016/j.gloplacha.2003.12.005

Müller, P., Schmidt-Thomé, M., Kreuzer, H., Tessensohn, F. and Vetter, U. 1991. Cenozoic peralkaline magmatism at the western margin of the Ross Sea, Antarctica. *Memorie della Società Geologica Italiana*, **46**, 315–336.

Nardini, I, Armienti, P. *et al.* 2009. Sr–Nd–Pb–He–O isotope and geochemical constraints on the genesis of Cenozoic magmas from the West Antarctic Rift. *Journal of Petrology*, **50**, 1359–1375, https://doi.org/10.1093/petrology/egn082

Nathan, S. and Schulte, F.J. 1967. Recent thermal and volcanic activity on Mount Melbourne, Northern Victoria Land, Antarctica. *New Zealand Journal of Geology and Geophysics*, **10**, 422–430, https://doi.org/10.1080/00288306.1967.10426746

Nathan, S. and Schulte, F.J. 1968. Geology and petrology of the Campbell–Aviator divide, Northern Victoria Land, Antarctica. *New Zealand Journal of Geology and Geophysics*, **11**, 940–975, https://doi.org/10.1080/00288306.1968.10420762

Park, Y., Yoo, H.J. *et al.* 2014. Deployment and performance of a broadband seismic network near the New Korean Jang Bogo research station, Terra Nova Bay, East Antarctica. *Seismological Research Letters*, **85**, 1341–1347, https://doi.org/10.1785/0220140107

Park, Y., Yoo, H.J., Lee, Ch-K., Lee, J., Park, H., Kim, J. and Kim, Y. 2015. P-wave velocity structure beneath Mt. Melbourne in northern Victoria Land, Antarctica: Evidence of partial melting and volcanic magma sources. *Earth and Planetary Science Letters*, **432**, 293–299, https://doi.org/10.1016/j.epsl.2015.10.015

Perinelli, C, Armienti, P and Dallai, L. 2006. Geochemical and O-isotope constraints on the evolution of lithospheric mantle in the Ross Sea rift area (Antarctica). *Contributions to Mineralogy and Petrology*, **151**, 245–266, https://doi.org/10.1007/s00410-006-0065-8

Perinelli, C., Andreozzi, G.B. *et al.* 2012. Redox state of subcontinental lithospheric mantle and relationships with metasomatism: insights from spinel peridotites from northern Victoria Land (Antarctica). *Contributions to Mineralogy and Petrology*, **164**, 1053–1067, https://doi.org/10.1007/s00410-012-0788-7

Podolskiy, E.A. and Walter, F. 2016. Cryoseismology. *Reviews of Geophysics*, **54**, 708–758, https://doi.org/10.1002/2016RG000526

Privitera, E., Villari, L. and Gambino, S. , 1992. An approach to the seismicity of Mt. Melbourne Volcano (Northern Victoria Land, Antarctica). *In*: Yoshida, Y., Kaminuma, K. and Shiraishi, K. (eds) *Recent Progress in Antarctic Earth Sciences*. Terra Scientific (TERRAPUB), Tokyo, 499–505.

Rocchi, S., Armienti, P., D'Orazio, M., Tonarini, S., Wijbrans, J.R. and Di Vincenzo, G. 2002. Cenozoic magmatism in the western Ross Embayment: Role of mantle plume versus plate dynamics in the development of the West Antarctic Rift System. *Journal of Geophysical Research*, **107**, 2195, https://doi.org/10.1029/2001JB000515

Rocchi, S., Armienti, P. and Di Vincenzo, G. 2005. No plume, no rift magmatism in the West Antarctic Rift. *Geological Society of America Special Papers*, **388**, 435–447, https://doi.org/10.1130/0-8137-2388-4.435

Röösli, C., Walter, F., Husen, S., Andrews, L.C., Lüthi, M.P., Catania, G.A. and Kissling, E. 2014. Sustained seismic tremors and icequakes detected in the ablation zone of the Greenland ice sheet. *Journal of Glaciology*, **60**, 563–575, https://doi.org/10.3189/2014JoG13J210

Ross, J.C. 1847. *A Voyage of Discovery and Research in the Southern and Antarctic Regions during the Years 1839–43*. 2 vols. John Murray, London.

Salvini, F, Brancolini, G, Busetti, M, Storti, F, Mazzarini, F and Coren, F. 1997. Cenozoic geodynamics of the Ross Sea Region, Antarctica: crustal extension, intraplate strike-slip faulting and tectonic inheritance. *Journal of Geophysical Research*, **102**, 24 669–24 696, https://doi.org/10.1029/97JB01643

Vignaroli, G., Balsamo, F., Giordano, G., Rossetti, F. and Storti, F. 2015. Miocene-to-Quaternary oblique rifting signature in the Western Ross Sea from fault patterns in the McMurdo Volcanic Group, north Victoria Land, Antarctica. *Tectonophysics*, **656**, 74–90, https://doi.org/10.1016/j.tecto.2015.05.027

Walter, F., Clinton, J.F., Deichmann, N., Dreger, D.S., Minson, S.E. and Funk, M. 2009. Moment tensor inversions of icequakes on Gornergletscher, Switzerland. *Bulletin of the Seismological Society of America*, **99**, 852–870, https://doi.org/10.1785/0120080110

Wardell, L.J., Kyle, P.R. and Campbell, A.R. 2003. Carbon dioxide emissions from fumarolic ice towers, Mount Erebus volcano, Antarctica. *Geological Society, London, Special Publications*, **213**, 231–246. https://doi.org/0305-8719/03/515.00

Wilson, E. 1966. *Diary of the 'Discovery' Expedition to the Antarctic Regions, 1901–1904*. Blandford Press, London.

Wörner, G. and Viereck, L. 1989. McMurdo Volcanic Group, western Ross Embayment (Mt. Melbourne). *American Geophysical Union Antarctic Research Series*, **48**, 72–78.

Wörner, G., Viereck, L., Hertoghen, J. and Niephaus, H. 1989. The Mt. Melbourne Volcanic field (Victoria Land, Antarctica), II: geochemistry and magma genesis. *Geologisches Jahrbuch*, **E38**, 395–433.

Zanutta, A., Negusini, M. *et al.* 2017. Monitoring geodynamic activity in the Victoria Land, East Antarctica: Evidence from GNSS measurements. *Journal of Geodynamics*, **110**, 31–42, https://doi.org/10.1016/j.jog.2017.07.008

Zimbelman, D.R., Rye, R.O. and Landis, G.P. 2000. Fumaroles in ice caves on the summit of Mount Rainier preliminary stable isotope, gas, and geochemical studies. *Journal of Volcanogy and Geothermal Research*, **97**, 457–473, https://doi.org/10.1016/S0377-0273(99)00180-8

Zoet, L.K., Anandakrishnan, S., Alley, R.B., Nyblade, A.A. and Wiens, D.A. 2012. Motion of an Antarctic glacier by repeated tidally modulated earthquakes. *Nature Geoscience*, **5**, 623–626, https://doi.org/10.1038/ngeo1555

Chapter 7.4

Active volcanoes in Marie Byrd Land

N. W. Dunbar[1]*, N. A. Iverson[1], J. L. Smellie[2], W. C. McIntosh[1], M. J. Zimmerer[1] and P. R. Kyle[3]

[1]New Mexico Bureau of Geology and Mineral Resources, New Mexico Institute of Mining and Technology, Socorro, NM 87801, USA

[2]School of Geography, Geology and the Environment, University of Leicester, University Road, Leicester LE1 7RH, UK

[3]Earth and Environmental Science Department, New Mexico Institute of Mining and Technology, 801 Leroy Place, Socorro, NM 87801, USA

NWD, 0000-0002-5181-937X; NAI, 0000-0002-7110-5040; JLS, 0000-0001-5537-1763; WCM, 0000-0002-5647-2483; MJZ, 0000-0001-5972-9275; PRK, 0000-0001-6598-8062
*Correspondence: nelia.dunbar@nmt.edu

Abstract: Two volcanoes in Marie Byrd Land, Mount Berlin and Mount Takahe, can be considered active, and a third, Mount Waesche, may be as well; although the chronology of activity is less well constrained. The records of explosive activity of these three volcanoes is well represented through deposits on the volcano flanks and tephra layers found in blue ice areas, as well as by the presence of cryptotephra layers found in West and East Antarctic ice cores. Records of effusive volcanism are found on the volcano flanks but some deposits may be obscured by pervasive glacierization of the edifices. Based on a compilation of tephra depths–ages in ice cores, the activity patterns of Mount Takahe and Mount Berlin are dramatically different. Mount Takahe has erupted infrequently over the past 100 kyr. Mount Berlin, by contrast, has erupted episodically during this time interval, with the number of eruptions being dramatically higher in the time interval between c. 32 and 18 ka. Integration of the Mount Berlin tephra record from ice cores and blue ice areas over a 500 kyr time span reveals a pattern of geochemical evolution related to small batches of partial melt being progressively removed from a single source underlying Mount Berlin.

Supplementary material: Detailed locations, descriptions and geochemistry of Mount Waesche englacial tephra layers, along with selected other samples, and description of Mount Waesche field campaigns (S1); a map of Mount Waesche englacial tephra layers showing sample numbers (S2); and argon geochronology methods and analytical spectra or ideograms (S3) are available at https://doi.org/10.6084/m9.figshare.c.5226158

Following the working definition of 'active volcano' outlined by Simkin and Siebert (2000), any volcano that has erupted during the Holocene (past 10 kyr) can be considered active. Based on this definition, two Marie Byrd Land volcanoes, Mount Berlin and Mount Takahe (Fig. 1), would be considered active. Following a more granular definition of active volcano, in which the time span since the most recent eruption is less than the known quiescence between sequential eruptions, the same two Marie Byrd Land volcanoes would still be considered active. Furthermore, the summit region of Mount Berlin hosts several ice-tower-topped, steaming, fumarolic caves, one of which has a floor temperature of 12°C (Wilch et al. 1999a), which represents clear evidence of volcanic heat in this young crater. West Antarctic volcano Mount Siple was thought to be possibly active due to its undissected appearance (LeMasurier and Rex 1990), as well as a report of a possible volcanic plume based on satellite imagery. However, a visit to the Siple Dome crater in the early 1990s revealed no suggestion of recent activity, and pyroclastic deposits at and near the crater rim yield ages of around 200 ka (Wilch et al. 1999a). Therefore, Mount Siple will not be considered in this chapter. Another West Antarctic volcano, Mount Waesche, hosts well-preserved volcanic deposits, both on the flanks of the volcano and in a large, tephra-bearing, blue ice field on the south side of the volcanic edifice. Mount Waesche does not have the same degree of geochronological constraints as the other two volcanoes but new and published isotopic ages, combined with the youthful appearance of the deposits, indicate that it is very young, as does tephra present in a blue ice field close by below its southern flank and a significant, nearby englacial tephra layer interpreted to be derived from Mount Waesche (Quartini et al. 2021). It will therefore be considered as part of this chapter.

On-site examination of the eruptive history of young volcanoes in Marie Byrd Land is challenging because of the high degree of glaciation in the region. The West Antarctic Ice Sheet (WAIS) in the vicinity of the volcanoes mentioned above is between 1000 and 2000 m thick, completely obscuring the lower halves of the volcanic edifices. Furthermore, the parts of the volcanoes above ice-sheet level are typically glaciated, only revealing a handful of accessible outcrops. The summit craters and calderas, where deposits from the youngest volcanic activity might be expected to be exposed, are generally ice-filled, with only small outcrops around the crater margins. The craters themselves can be difficult to access because of the steep, glaciated, crevassed slopes leading from the lower volcano flanks.

However, the glaciation of the Antarctic continent also provides an unparalleled repository of deposits in ice cores, and blue ice areas from local explosive volcanism, making this part of the Antarctic volcanic record more complete than that in most terrestrial environments. When a volcano on the Antarctic continent erupts explosively, pyroclastic materials are distributed onto the surrounding ice sheet and, in some cases, across large parts of the Antarctic continent (Smellie 1999; Basile et al. 2001; Dunbar et al. 2003, 2007; Narcisi et al. 2005, 2006, 2010b, 2012, 2016; Narcisi 2008; Dunbar and Kurbatov 2011; Iverson 2017; Iverson et al. 2017; DeRoberto et al. 2021; Narcisi and Petit 2021). As snow falls onto the ice sheet, the tephra layers become incorporated into the glacial stratigraphy, in continuity with the ice-sheet layering. Because the silicate tephra is preserved in a non-silicate matrix (ice), vanishingly thin and fine tephra layers (cryptotephra) recovered from ice can be identified, quantitatively analysed and tied to a source volcano. The relatively recent addition of cryptotephra to the field of tephrochronology has expanded the reach of tephra layers as time-

From: Smellie, J. L., Panter, K. S. and Geyer, A. (eds) 2021. *Volcanism in Antarctica: 200 Million Years of Subduction, Rifting and Continental Break-up*. Geological Society, London, Memoirs, 55, 759–783,
First published online April 12, 2021, https://doi.org/10.1144/M55-2019-29
© 2021 The Author(s). Published by The Geological Society of London. All rights reserved.
For permissions: http://www.geolsoc.org.uk/permissions. Publishing disclaimer: www.geolsoc.org.uk/pub_ethics

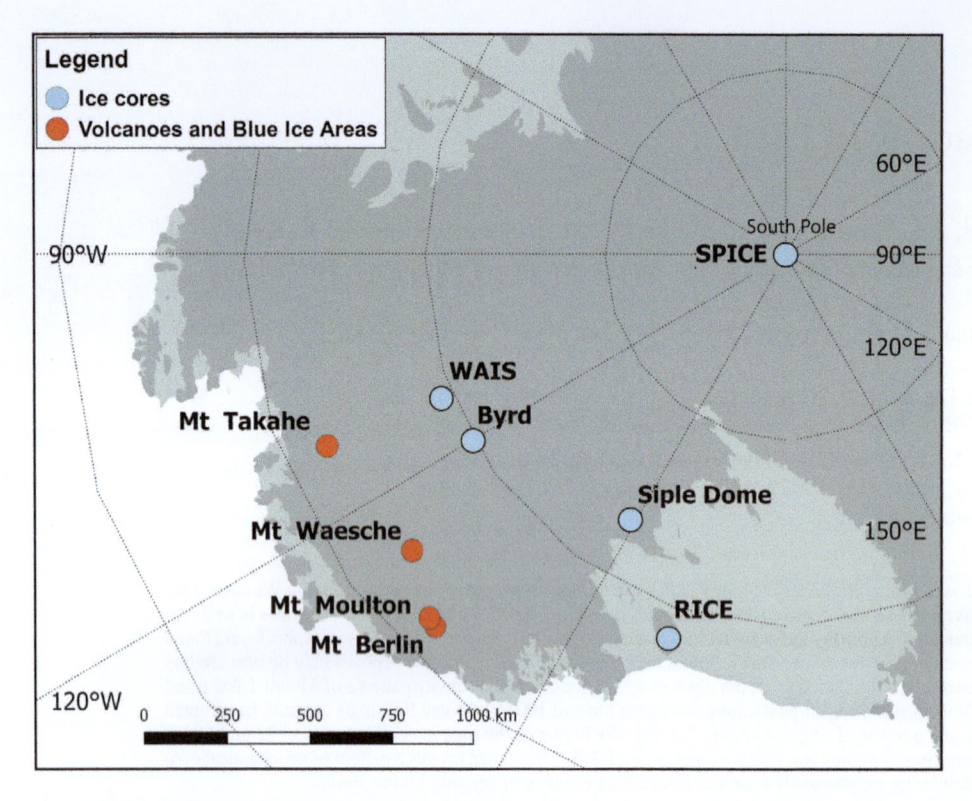

Fig. 1. Map of West Antarctica showing the locations of selected West Antarctic volcanoes (red circles) and ice cores (light blue circles). The dark-grey colour represents the edge of the continent; light grey represents floating ice shelves; and ocean is shown in white.

stratigraphic markers, providing valuable chronology in a range of geoscience studies (Lemieux-Dudon et al. 2010; Davies 2015).

In Marie Byrd Land, many tephra layers are found within the ice sheet, providing an extremely well-resolved record of local explosive volcanism. This record can be sampled in blue ice areas, where deep glacial ice is brought to the surface by ice flow and surface-ablation processes (Cassidy et al. 1977), or in ice cores drilled through the WAIS (e.g. Gow and Williamson 1971; Palais et al. 1988). Two major blue ice sites have been investigated in Marie Byrd Land. The first is the Mount Moulton blue ice site (Wilch et al. 1999a; Dunbar et al. 2008), located in the summit crater of the extinct Mount Moulton volcano (Fig. 1), and which contains tephra erupted from nearby Mount Berlin over the past 550 kyr (Dunbar et al. 2008). The second is the Mount Waesche blue ice site, which is located adjacent to the Mount Waesche volcano (Fig. 1) but, rather than being found in a summit crater, is an integral part of the WAIS. This blue ice area contains englacial tephra layers erupted from Mount Waesche itself, some proximal enough to contain volcanic bombs, but also contains a record of distally erupted tephra as old as 117 ka (Dunbar et al. 2007). The ice core record in Marie Byrd Land includes two inland cores (Byrd core, drilled in 1967–68, and the WAIS Divide core, drilled between 2008 and 2011), and two coastal cores (Siple Dome, drilled between 1997–99, and the recently completed Roosevelt Island core (RICE)) (Fig. 1). All of these ice cores contain abundant tephra, with the richest records preserved at the two inland sites. The ice core repositories also host rare tephra from non-Antarctic explosive volcanism (Cole-Dai et al. 1999; Dunbar et al. 2017; Koffman et al. 2017; Hartman et al. 2019).

In addition to englacial records of volcanism, the marine sediment record can provide a valuable repository for tephra deposits (Smellie 1999). A chapter of this volume (DeRoberto et al. 2021) examines this topic in detail. Marine cores collected in the West Antarctic region are typically dominated by tephra from sub-Antarctic Islands (Moreton and Smellie 1998; Fretzdorff and Smellie 2002), so do not provide substantial insight into Marie Byrd Land volcanism. One exception are cores recovered from the West Antarctic Bellingshausen and Admundsen seas region, which contain tephra layers from Mount Takahe and Mount Berlin, correlated to englacial tephra layers sampled near the source volcanoes (Hillenbrand et al. 2008). As pointed out by Smellie (1999), the marine-sediment-bound tephra record could be studied in more detail and has the potential to yield more information about West Antarctic volcanism.

Below, we present a summary of recent activity of Mount Berlin and Mount Takahe, relying heavily on the englacial tephra records. While examination of the record from this perspective focuses on the explosive activity, at the expense of less explosive eruptions, it is the part of the record that is available for study. Furthermore, the very detailed picture of activity provided through the englacial tephra record allows for a more thorough investigation of eruptive activity and fine-scale geochemical evolution of a single volcanic centre over 500 kyr (Dunbar et al. 2008; Iverson 2017) than is available in almost any other geological setting. Whereas much of the information about Mount Takahe and Mount Berlin has been previously published, information about the Mount Waesche geology and the englacial tephra is new to the literature. Therefore, more detail will be presented for Mount Waesche both in the chapter itself as well as in Supplementary material S1.

Mount Berlin

Physical description and geology

Mount Berlin is a 3478 m-high stratovolcano volcano located at 76° 03′ S and 135° 52′ W (Figs 1 & 2). As outlined by LeMasurier (1990b), Mount Berlin consists of two coalesced volcanoes, Berlin Crater and Merram Peak, each with a summit crater. The 1.4 km-diameter Berlin Crater is the locus of the most recent volcanic activity, whereas the activity from Merram Peak is older than 140 ka (Wilch et al. 1999a). Here, we will focus on the Berlin Crater part of the Mount Berlin complex. Berlin Crater was visited, briefly, in 1967

Fig. 2. Photograph of 3478 m-high Mount Berlin volcano, located in West Antarctica, viewed from nearby Mount Moulton, located to the east. The elevation of the local ice sheet around Mount Berlin is around 1200 m.

(LeMasurier and Wade 1968) and 1977 (LeMasurier and Rex 1982). Then, in 1993–94, an extended field campaign allowed detailed mapping and sampling on the flanks of the volcanic complex, in which every outcrop accessible on foot or by snowmobile was visited (Wilch 1997). The crater was accessed by an overland route, and geological features were examined in more detail than was possible during the earlier visits.

The Berlin Crater is a very well-defined geomorphic feature, with a sharp crater rim (Fig. 3). This alone could suggest that this volcano was recently active but, to further reinforce this conclusion, a steaming fumarolic ice tower in the summit region of Mount Berlin Crater (Fig. 4) was accessed during the 1993–94 field campaign. Entry into the underlying cave system revealed an open passage with a length of 70 m and a cave-floor temperature of 12°C. The lava flow making up the cave floor was dated, using $^{40}Ar/^{39}Ar$ geochronology, at 10.3 ± 5.3 ka (Wilch et al. 1999a). The presence of this icetower system potentially represents a unique example of geothermal activity in Marie Byrd Land.

Outcrops exposed on the walls of Berlin crater consist of complex welded fall deposits, which are likely to be related to significant explosive activity. Two identifiable pyroclastic fall deposits are exposed on the crater rim with a total thickness of around 150 m (Wilch et al. 1999a), and these are likely to have been related to large-scale explosive activity that could have distributed tephra across parts of the WAIS. Two pyroclastic units sampled from the summit area of Berlin crater

Fig. 4. Steaming ice tower in the summit crater on Mount Berlin. The underlying cave, mentioned in the text, was accessed through the entrance at the left base of the ice tower. The tower is roughly 4 m high, and the view is roughly to the north.

yielded ages of 25.5 ± 2.0 and 18.2 ± 5.8 ka (Wilch et al. 1999a).

Although only three young eruptions can be documented through investigation of surface outcrops on Mount Berlin, a wealth of additional information about explosive eruptive activity at Mount Berlin is preserved in the englacial tephra record, accessible through blue ice areas and ice cores. These tephra will be discussed in a later section.

Mount Takahe

Physical description and geology

Mount Takahe is a 3460 m-high shield volcano located at 76° 15′ S and 112° 00′ W (Figs 1 & 5). The volcano, which is 30 km wide at the base, is an isolated feature in the WAIS, and does not form part of a group, or chain, of volcanoes. The symmetrical edifice has an unusual, flat, broad (8 km) snow-filled summit caldera. Unlike some other West Antarctic volcanoes, such as Mount Murphy, Mount Takahe is almost completely undissected, offering limited outcrop to provide insights into the volcano's history.

Investigations of the geology at the 12 outcrop and five moraine localities at Mount Takahe (Wilch 1997) reveal information about the volcano's history, as well as that of the local ice sheet. At three outcrop localities, sequences of deposits indicative of the transition from subaqueous to subaerial volcanism are observed in the lower portions of the outcrops

Fig. 3. Oblique view of the 1.4 km-diameter Mount Berlin Crater, viewed obliquely from the south, showing the sharpness of the crater rim and a shadowed outcrop of pyroclastic rocks.

Fig. 5. Aerial oblique photograph, looking north, of the 3460 m-high Mount Takahe, showing the 7 km-wide summit caldera. The elevation of the local ice sheet around Mount Takahe is around 1200 m.

(McIntosh et al. 1985). The distribution and elevation of the hydroclastic deposits, which consist of pillow lavas, pillow breccias, hyaloclastites and tuff cones, suggest that the ice around Mount Takahe fluctuated through a range of up to c. 400–575 m above the level of the current ice sheet, which, based on argon geochronology, was between 65.5 ± 5.2 and 23.7 ± 5.6 ka (Wilch et al. 1999b). Rocks at nine other outcrops, including around the caldera rim, derive mainly from subaerial volcanic activity and are composed of welded to non-welded pyroclastic deposits, as well as some lavas and hydrovolcanic tuffs (McIntosh et al. 1985). At Bucher Rim, located on the southern part of the caldera, a thick sequence (60 m) includes lavas, welded ash fall tuff, accretionary lapilli tuffs, and obsidian bomb and block units. The observation of these subaerial units at the crater rim of the volcano suggests that the flat-topped shape of the volcano is not related to growth under ice but that that the bulk of the eruption occurred above the level of the ice sheet (McIntosh et al. 1985). However, the undissected nature of the volcano means that only the outer part of the volcano can be observed, leaving the nature of the eruptions that formed the interior unknown. The most recent eruption of Mount Takahe, as dated by $^{40}Ar/^{39}Ar$ geochronology, was dated at 8.2 ± 5.4 ka (Wilch et al. 1999a) but was recently redated at 5.6 ± 0.8 ka (Iverson 2017). Other eruptive events from the crater-rim area have ages of 192.0 ± 6.3, 167.0 ± 14, 154.1 ± 8.2, 102 ± 7.4 and 93.3 ± 7.9 ka (Wilch et al. 1999a).

During fieldwork at Mount Takahe carried out in 1998–99, several tephra layers were found in glacial ice on the flanks of the volcano. These layers contain coarse pumice (up to 1 cm in diameter) and are certainly derived from Mount Takahe. The glass geochemistry of these layers has been analysed by electron microprobe, are all trachytic and fall within a relatively narrow range of geochemical composition (Table 1; see also Supplementary material S1).

Mount Waesche

Physical description and geology

The Mount Waesche massif is the second largest volcano in the Executive Committee Range. It is a compound edifice, composed of two coalesced volcanic shields (Fig. 6). The older larger structure rises to 2920 m above sea-level (asl) at Chang Peak and is referred to here as the Chang Peak volcano. It is at least 17 × 22 km in diameter above the ice. The surrounding ice is at least 2 km thick (Lythe et al. 2001), which implies that the volcano is nearly 5 km high. The Chang Peak volcano is almost entirely covered by snow and ice. It includes a prominent, elongate ice-filled caldera c. 12 km (NE–SW) and 7 km (NW–SE) in diameter, the largest caldera in Marie Byrd Land. Small exposures of rock occur at only four localities. These consist of rhyolite, comprising blocky pumice deposits (Chang Peak) and domes or lavas. Dating indicates ages of 1.6 ± 0.2 Ma for pumice at Chang Peak (LeMasurier and Kawachi 1990), 1.08 ± 0.05 and 1.98 ± 0.05 Ma for two rhyolite lavas on the NW flank, and a 1.60 ± 0.06 Ma age for a rhyolite clast, interpreted to be from Chang Peak, in moraine in ice south of Mount Waesche (Panter 1995: all three by $^{40}Ar/^{39}Ar$) (Fig. 6). The ages are all relatively old and there is no evidence, such as youthful volcanic features or tephra layers recovered in any ice cores, that the Chang Peak volcano is still active. It is therefore not described further.

By contrast, Mount Waesche is a small volcano, in a relatively youthful state of development, situated on the SW flank of the Chang Peak caldera, which it partially infills. It rises to 3292 m asl, almost 500 m above the flat featureless ice field of the Chang Peak caldera. Mount Waesche has a much smaller ice-filled caldera c. 2 km in diameter. Despite being largely undissected, Mount Waesche is one of the best-exposed inland volcanoes in Marie Byrd Land. The SW quadrant of the volcano is largely ice-free, providing excellent exposure and access to outcrops (Figs 6 & 7), although only the more recent eruptive products can be observed and sampled. LeMasurier and Kawachi (1990) suggested that lava compositions are alkali basaltic and hawaiitic but a wider range of compositions is now known to be present (Panter et al. 2021). From a preliminary examination of the geochemistry, the lavas form at least two magmatic lineages comprising: (1) basanites/tephrites–phonolites; and (2) alkali basalts–trachytes. The lineages shared a compositionally similar mantle source but formed by different degrees of melting and have different fractionation histories (unpublished information of J.L. Smellie) (Fig. 8). The outcrops are largely composed of subaerial 'a'ā lavas in which at least four compositional groups have been identified (described later). Each was erupted during a temporally discrete eruptive phase. Few contacts between any groups are exposed, and are limited to three locations on the SW slope where basaltic lavas were observed to directly overlie eroded, more silicic lavas. Three of the lava groups are associated with scoria cones, some of which gave rise to the associated lavas. A distinctive deposit of unconsolidated polymict gravelly volcanic/plutonic breccia is also exposed around the caldera rim and, more extensively, on the south slopes above c. 2100 m asl.

A range of radioisotopic ages have been produced from volcanic rocks erupted from Mount Waesche volcano. Four published K–Ar ages range from very young (0.2 ± 0.2, 0.17 ± 0.3 and <0.1 Ma) to significantly older (1.0 ± 0.1 Ma: LeMasurier 1990a) (see Fig. 7). Unpublished higher-precision $^{40}Ar/^{39}Ar$ ages also exist for a number of samples and are reported here for the first time (see analytical results in Supplementary material S3). The more recently dated materials range from in situ lavas to boulders sampled from moraines. The new ages vary from 0.556 ± 0.824 to 0.168 ± 0.038 Ma, and are discussed, along with the K–Ar ages, below.

Two phases of eruptive activity are identified. The youngest dated sample (<0.1 Ma, by K–Ar) was obtained on a hawaiite lava with 'cognate inclusions' (LeMasurier and Kawachi 1990). The dated sample appears to come from our youngest lava (lava 16 in Fig. 7), which is a hawaiite and part of our Group 4 lavas (Fig. 7). However, more recent examination of this lava did not reveal the 'cognate inclusions' reported in earlier work. Other lavas in Group 4, now dated by $^{40}Ar/^{39}Ar$, have yielded a set of similar ages of 0.168 ±

Table 1. *Average glass compositions, with standard deviations, of englacial tephra layers as analysed by electron microprobe*

	Ice core age* (ka)	Suggested source	P$_2$O$_5$ (wt%)	SiO$_2$ (wt%)	SO$_2$ (wt%)	TiO$_2$ (wt%)	Al$_2$O$_3$ (wt%)	MgO (wt%)	CaO (wt%)	MnO (wt%)	FeO^T (wt%)	Na$_2$O (wt%)	K$_2$O (wt%)	F (wt%)	Cl (wt%)
Mount Waesche blue ice field															
BIT-143		Mount Berlin	0.08	63.24	0.06	0.49	14.59	0.05	1.43	0.31	8.18	6.33	4.87	0.24	0.15
			0.03	1.40	0.03	0.05	0.79	0.06	0.40	0.04	0.32	0.00	0.29	0.09	0.05
BIT-144		Mount Berlin	0.05	63.99	0.06	0.44	13.05	0.00	0.73	0.27	8.15	8.22	4.61	0.17	0.29
			0.03	0.13	0.03	0.02	0.05	0.00	0.05	0.01	0.19	0.00	0.15	0.20	0.03
BIT-145		Mount Waesche	1.78	45.76	0.17	3.58	16.98	4.14	8.46	0.21	11.02	5.64	1.86	0.27	0.10
			0.12	0.50	0.03	0.35	0.62	0.11	0.37	0.06	0.32	0.19	0.04	0.09	0.01
BIT-146		Mount Berlin	0.07	62.50	0.03	0.54	15.62	0.15	1.16	0.26	6.75	7.36	5.18	0.25	0.13
			0.05	0.22	0.01	0.03	0.23	0.07	0.07	0.02	0.25	0.09	0.10	0.13	0.02
BIT-147		Mount Berlin	0.06	62.58	0.05	0.50	15.05	0.11	1.11	0.27	7.09	7.81	5.08	0.14	0.15
			0.02	0.32	0.02	0.05	0.10	0.01	0.03	0.04	0.22	0.00	0.15	0.12	0.02
BIT-148		Mount Berlin	0.06	62.19	0.03	0.52	15.24	0.13	1.10	0.27	6.83	8.00	5.13	0.37	0.14
			0.04	0.30	0.02	0.03	0.17	0.06	0.05	0.04	0.07	0.00	0.06	0.12	0.02
BIT-149		Mount Berlin	0.06	63.61	0.06	0.46	14.08	0.00	1.27	0.31	8.41	6.64	4.78	0.17	0.14
			0.03	0.65	0.03	0.05	0.47	0.00	0.28	0.05	0.35	0.38	0.14	0.09	0.04
BIT-167		Mount Berlin	0.06	61.28	0.07	0.46	14.28	0.02	1.09	0.34	8.73	8.44	4.73	0.36	0.19
			0.03	0.20	0.02	0.05	0.09	0.02	0.04	0.03	0.10	0.00	0.10	0.05	0.01
BIT-168		Mount Berlin	0.07	61.29	0.07	0.46	15.02	0.04	1.14	0.28	8.30	8.07	4.82	0.25	0.23
			0.03	0.29	0.02	0.04	0.20	0.02	0.08	0.04	0.25	0.00	0.11	0.10	0.02
BIT-169		Mount Berlin	0.05	60.98	0.07	0.45	14.96	0.10	0.97	0.32	8.09	8.79	4.81	0.28	0.21
			0.03	0.19	0.02	0.03	0.11	0.02	0.03	0.03	0.24	0.00	0.08	0.13	0.02
BIT-170		Mount Waesche	1.42	49.82	0.10	2.96	16.10	3.08	7.09	0.20	10.97	5.76	2.23	0.21	0.11
			0.20	1.18	0.04	0.36	1.43	0.32	0.38	0.02	0.90	0.00	0.21	0.07	0.01
BIT-171		Mount Berlin	0.05	61.29	0.06	0.48	14.81	0.10	1.03	0.33	8.11	8.42	4.87	0.27	0.21
			0.02	0.11	0.01	0.03	0.06	0.01	0.04	0.02	0.18	0.00	0.06	0.07	0.02
BIT-172		Mount Berlin	0.10	61.34	0.06	0.53	14.51	0.07	1.93	0.32	8.88	7.18	4.84	0.14	0.11
			0.02	0.23	0.02	0.04	0.15	0.02	0.09	0.05	0.10	0.00	0.07	0.06	0.02
BIT-173		Mount Berlin	0.13	61.62	0.08	0.53	14.84	0.09	2.02	0.35	9.09	6.31	4.73	0.13	0.10
			0.03	0.11	0.04	0.02	0.11	0.02	0.06	0.03	0.05	0.00	0.04	0.07	0.02
BIT-174		Mount Waesche	0.84	46.27	0.17	3.45	14.52	4.58	10.13	0.24	13.71	4.49	1.42	0.13	0.06
			0.14	0.21	0.08	0.20	0.79	0.66	0.74	0.04	0.90	0.34	0.18	0.07	0.02
BIT-175		Mount Takahe	0.22	59.97	0.12	0.70	15.42	0.46	2.60	0.35	9.43	6.32	4.22	0.09	0.10
			0.05	0.27	0.02	0.07	0.06	0.04	0.12	0.05	0.15	0.11	0.23	0.06	0.02
BIT-176		Mount Waesche	2.04	46.72	0.25	3.49	13.90	4.28	8.27	0.27	14.49	4.70	1.35	0.21	0.05
			0.17	0.91	0.03	0.16	0.27	0.60	0.43	0.05	0.15	0.29	0.12	0.07	0.01
BIT-177		Mount Waesche	0.95	46.20	0.19	3.50	14.96	3.84	9.53	0.25	13.83	4.89	1.58	0.19	0.06
			0.07	0.76	0.06	0.19	0.98	0.08	0.47	0.04	1.45	0.26	0.27	0.07	0.02
BIT-178		Mount Waesche	2.06	45.25	0.32	3.93	14.88	4.51	9.02	0.29	13.66	4.32	1.44	0.25	0.05
			0.11	0.41	0.04	0.06	0.28	0.10	0.14	0.04	0.23	0.17	0.04	0.10	0.01
BIT-180		Mount Waesche	0.73	45.48	0.16	3.55	13.78	4.84	11.07	0.20	14.93	4.11	1.04	0.07	0.04
			0.04	0.11	0.02	0.03	0.25	0.21	0.14	0.03	0.33	0.21	0.04	0.05	0.01
BIT-182a		Mount Waesche	0.82	46.36	0.14	3.01	15.90	4.20	9.27	0.19	12.83	5.26	1.75	0.18	0.10
			0.10	0.25	0.03	0.07	0.09	0.03	0.08	0.07	0.05	0.16	0.04	0.09	0.01
BIT-182b		Mount Waesche	0.73	46.04	0.19	3.08	15.93	4.33	11.10	0.13	12.07	4.74	1.50	0.11	0.06
			0.22	0.58	0.08	0.53	1.02	0.34	1.86	0.07	1.93	0.47	0.42	0.08	0.02
BIT-182c		Mount Waesche	1.24	47.30	0.18	3.19	14.66	3.63	8.35	0.28	13.18	5.35	2.28	0.30	0.08
			0.05	0.28	0.02	0.15	0.25	0.45	0.37	0.02	0.25	0.21	0.21	0.11	0.02
BIT-183a		Mount Waesche	1.18	46.76	0.14	3.34	15.08	4.00	8.93	0.25	12.54	5.61	1.94	0.16	0.06
			0.19	1.05	0.07	0.29	0.27	0.44	1.18	0.05	1.15	0.53	0.51	0.10	0.04
BIT-183b		Mount Waesche	0.98	45.92	0.20	3.58	15.00	3.92	8.99	0.24	13.67	5.27	1.91	0.23	0.08
			0.13	0.55	0.02	0.23	0.34	0.29	0.48	0.08	0.36	0.38	0.16	0.06	0.02
BIT-195		Mount Waesche	0.70	45.80	0.08	3.23	15.67	5.44	10.15	0.18	12.85	4.29	1.33	0.15	0.05
			0.07	0.37	0.03	0.08	0.15	0.09	0.06	0.08	0.16	0.11	0.08	0.08	0.01
BIT-196		Mount Waesche	0.69	45.75	0.08	3.26	15.93	5.59	9.98	0.16	12.83	4.18	1.30	0.19	0.05
			0.08	0.30	0.03	0.03	0.16	0.35	0.23	0.06	0.11	0.22	0.13	0.12	0.02
BIT-197		Mount Waesche	0.76	46.03	0.06	3.37	15.35	5.43	9.87	0.19	13.13	4.32	1.33	0.11	0.05
			0.07	0.68	0.01	0.28	0.73	0.52	0.13	0.06	0.82	0.21	0.06	0.12	0.01
BIT-198		Mount Waesche	0.70	46.30	0.06	3.14	15.39	5.91	9.74	0.24	12.75	4.33	1.32	0.08	0.04
			0.03	0.00	0.02	0.04	0.03	0.85	0.34	0.08	0.31	0.03	0.08	0.11	0.01
BIT-199		Mount Waesche	0.89	46.42	0.08	3.42	15.40	4.77	9.57	0.21	12.87	4.63	1.53	0.15	0.05
			0.18	0.56	0.02	0.22	0.70	0.61	0.61	0.04	0.26	0.36	0.34	0.12	0.02
BIT-200		Mount Berlin	0.07	61.76	0.06	0.47	15.53	0.09	1.07	0.31	7.31	8.11	4.92	0.17	0.15
			0.03	0.18	0.02	0.03	0.15	0.01	0.04	0.02	0.28	0.00	0.15	0.13	0.02

(*Continued*)

Table 1. *Continued.*

	Ice core age* (ka)	Suggested source	P₂O₅ (wt%)	SiO₂ (wt%)	SO₂ (wt%)	TiO₂ (wt%)	Al₂O₃ (wt%)	MgO (wt%)	CaO (wt%)	MnO (wt%)	FeOᵀ (wt%)	Na₂O (wt%)	K₂O (wt%)	F (wt%)	Cl (wt%)
BIT-201		Mount Berlin	0.05	61.05	0.07	0.41	14.61	0.03	1.00	0.30	8.46	8.92	4.72	0.17	0.22
			0.02	0.14	0.02	0.03	0.13	0.02	0.04	0.02	0.09	0.00	0.07	0.20	0.02
BIT-202		Mount Berlin	0.06	60.30	0.06	0.41	14.69	0.05	1.01	0.32	8.37	9.56	4.75	0.28	0.21
			0.02	0.26	0.02	0.04	0.13	0.01	0.06	0.02	0.22	0.00	0.06	0.11	0.01
BIT-203		Mount Berlin	0.05	60.58	0.04	0.43	14.79	0.04	1.00	0.35	8.71	8.71	4.73	0.34	0.22
			0.03	0.38	0.03	0.03	0.13	0.01	0.06	0.05	0.30	0.00	0.08	0.11	0.02
BIT-204		Mount Berlin	0.04	60.78	0.06	0.46	14.75	0.05	1.05	0.31	8.51	8.79	4.74	0.30	0.19
			0.02	0.25	0.02	0.06	0.09	0.07	0.04	0.02	0.15	0.00	0.10	0.07	0.02
BIT-206		Mount Berlin	0.10	61.29	0.09	0.46	14.75	0.04	1.12	0.34	8.63	8.00	4.81	0.25	0.18
			0.02	0.30	0.01	0.03	0.14	0.02	0.04	0.02	0.16	0.00	0.14	0.05	0.01
BIT-207		Mount Berlin	0.05	61.70	0.08	0.46	14.49	0.06	1.08	0.33	8.47	8.11	4.72	0.27	0.24
			0.02	0.18	0.03	0.03	0.15	0.01	0.03	0.03	0.26	0.00	0.09	0.16	0.04
BIT-208		Mount Berlin	0.08	61.93	0.08	0.52	13.71	0.03	1.12	0.30	8.93	7.86	4.64	0.55	0.27
			0.02	0.47	0.01	0.06	0.12	0.01	0.02	0.02	0.30	0.00	0.07	0.41	0.03
BIT-209		Mount Berlin	0.05	61.30	0.08	0.47	14.10	0.02	1.11	0.34	8.89	8.54	4.68	0.28	0.17
			0.02	0.07	0.02	0.03	0.13	0.01	0.03	0.05	0.12	0.00	0.11	0.03	0.02
BIT-210		Mount Waesche	1.02	46.13	0.21	3.64	13.43	3.62	9.36	0.28	15.94	4.62	1.47	0.20	0.06
			0.13	1.59	0.06	0.50	0.69	0.51	1.02	0.06	1.31	0.73	0.37	0.08	0.03
BIT-211		Mount Berlin	0.07	62.31	0.07	0.44	15.27	0.06	1.05	0.35	8.43	6.57	4.90	0.29	0.20
			0.03	0.19	0.03	0.05	0.10	0.02	0.03	0.03	0.12	0.07	0.08	0.13	0.02
BIT-212		Mount Waesche	1.81	47.12	0.12	3.55	15.27	4.42	7.67	0.21	12.88	4.83	1.75	0.28	0.07
			0.14	0.73	0.03	0.18	0.18	0.22	0.26	0.04	0.41	0.09	0.12	0.05	0.01
BIT-213		Mount Berlin	0.06	61.07	0.04	0.46	14.83	0.04	1.11	0.34	8.59	8.21	4.75	0.36	0.18
			0.04	5.74	0.02	0.08	1.16	0.02	0.31	0.10	1.70	2.33	0.18	0.22	0.08
BIT-214		Mount Waesche	1.75	47.14	0.13	3.52	15.72	4.24	7.71	0.21	12.79	4.69	1.71	0.29	0.07
			0.05	0.12	0.04	0.08	0.18	0.12	0.17	0.03	0.18	0.28	0.04	0.06	0.02
BIT-215		Mount Waesche	0.98	46.34	0.20	3.57	14.50	4.17	9.44	0.26	14.09	4.72	1.53	0.14	0.07
			0.23	0.47	0.03	0.20	0.19	0.23	0.49	0.04	0.44	0.20	0.20	0.04	0.01
BIT-216		Mount Berlin	0.04	67.15	0.02	0.23	16.12	0.02	0.92	0.10	3.54	6.05	5.08	0.50	0.25
			0.04	0.47	0.01	0.03	0.13	0.02	0.11	0.02	0.15	0.00	0.13	0.14	0.02
BIT-217		Mount Waesche	1.79	45.82	0.27	3.88	14.65	4.56	9.09	0.28	13.51	4.40	1.47	0.21	0.07
			0.39	0.61	0.04	0.06	0.24	0.33	0.14	0.05	0.95	0.22	0.10	0.06	0.02
BIT-218		Mount Waesche	0.82	46.21	0.14	3.47	14.80	4.81	10.09	0.23	13.36	4.36	1.39	0.23	0.08
			0.06	0.13	0.06	0.18	0.30	0.47	0.37	0.03	0.74	0.07	0.13	0.10	0.03
BIT-219		Mount Waesche	0.84	44.87	0.22	4.02	15.13	4.85	9.97	0.26	14.27	4.10	1.20	0.20	0.05
			0.10	1.22	0.07	0.25	0.25	0.67	0.53	0.01	0.59	0.47	0.29	0.05	0.03
BIT-220		Mount Berlin	0.08	61.63	0.07	0.43	14.69	0.02	1.15	0.27	8.09	8.20	4.94	0.23	0.22
			0.02	0.20	0.02	0.01	0.09	0.01	0.04	0.03	0.15	0.00	0.10	0.12	0.01
BIT-221		Mount Takahe	0.17	61.20	0.09	0.67	15.16	0.29	2.04	0.32	8.43	6.68	4.73	0.13	0.10
			0.17	61.20	0.09	0.67	15.16	0.29	2.04	0.32	8.43	6.68	4.73	0.13	0.10
BIT-222		Mount Waesche	0.07	62.84	0.04	0.47	15.15	0.03	1.53	0.27	7.42	6.93	4.99	0.16	0.11
			0.05	0.34	0.02	0.07	0.29	0.03	0.10	0.04	0.42	0.00	0.04	0.05	0.01
BIT-224		Mount Berlin	0.11	61.23	0.10	0.55	14.98	0.10	2.01	0.34	9.02	6.57	4.79	0.11	0.10
			0.02	0.30	0.03	0.03	0.27	0.01	0.11	0.08	0.36	0.31	0.10	0.05	0.01
BIT-225		Mount Waesche	0.75	47.47	0.10	3.31	15.61	4.31	9.22	0.19	12.25	4.81	1.85	0.08	0.06
			0.17	1.13	0.04	0.43	0.57	0.23	0.59	0.05	0.49	0.34	0.08	0.05	0.02
BIT-229		Mount Waesche	1.86	45.89	0.27	3.74	14.38	4.89	8.72	0.27	14.28	4.48	1.14	0.04	0.05
			0.37	0.72	0.06	0.18	0.52	0.58	0.66	0.05	0.31	0.36	0.15	0.10	0.02
BIT-230		Mount Takahe	0.16	60.46	0.09	0.87	15.15	0.31	1.67	0.34	8.55	7.17	4.97	0.19	0.09
			0.02	0.30	0.02	0.03	0.12	0.03	0.05	0.04	0.14	0.00	0.07	0.09	0.01
BIT-246		Mount Waesche	0.98	45.91	0.19	3.57	15.09	3.95	9.19	0.27	13.67	5.18	1.74	0.17	0.08
			0.05	0.11	0.00	0.02	0.03	0.15	0.05	0.07	0.24	0.08	0.01	0.05	0.02
BIT-247		Mount Waesche	1.29	48.19	0.18	3.09	15.14	3.43	7.72	0.21	12.73	5.70	2.08	0.15	0.09
			0.08	0.17	0.03	0.07	0.01	0.04	0.05	0.04	0.17	0.11	0.02	0.13	0.02
BIT-248		Mount Waesche	0.94	46.20	0.26	3.43	14.38	4.37	9.06	0.24	14.70	4.87	1.32	0.15	0.07
			0.03	0.39	0.03	0.10	0.48	0.10	0.05	0.06	0.42	0.02	0.05	0.02	0.00
BIT-254		Mount Waesche	0.37	47.83	0.12	2.97	13.68	5.97	10.51	0.21	14.02	3.41	0.82	0.07	0.02
			0.05	0.85	0.10	0.37	0.14	0.86	0.11	0.08	0.41	0.47	0.21	0.08	0.01
BIT-255		Mount Waesche	0.88	46.85	0.17	3.19	14.23	5.05	8.96	0.21	13.84	4.50	1.86	0.17	0.09
			0.17	0.75	0.05	0.18	0.64	1.54	1.08	0.10	1.72	0.42	0.31	0.12	0.03
Mount Takahe englacial tephra															
MTK-083		Mount Takahe	0.08	60.82	0.10	0.54	14.66	0.13	1.18	0.32	8.51	8.38	4.84	0.25	0.18
			0.03	0.72	0.03	0.03	0.59	0.03	0.08	0.06	0.82	0.29	0.17	0.11	0.04

(*Continued*)

Table 1. *Continued.*

	Ice core age* (ka)	Suggested source	P$_2$O$_5$ (wt%)	SiO$_2$ (wt%)	SO$_2$ (wt%)	TiO$_2$ (wt%)	Al$_2$O$_3$ (wt%)	MgO (wt%)	CaO (wt%)	MnO (wt%)	FeO^T (wt%)	Na$_2$O (wt%)	K$_2$O (wt%)	F (wt%)	Cl (wt%)
MTK-084		Mount Takahe	0.09	62.28	0.11	0.59	14.08	0.05	1.17	0.37	8.97	7.10	4.83	0.17	0.21
			0.03	0.92	0.02	0.06	0.18	0.07	0.05	0.03	0.18	1.15	0.04	0.18	0.02
MTK-085		Mount Takahe	0.09	60.02	0.11	0.57	14.05	0.09	1.14	0.33	8.84	9.63	4.73	0.20	0.21
			0.02	1.35	0.03	0.06	0.23	0.06	0.04	0.04	0.35	1.60	0.25	0.16	0.01
MTK-086		Mount Takahe	0.04	61.26	0.07	0.47	15.97	0.17	1.03	0.25	7.13	8.16	5.07	0.23	0.15
			0.04	0.25	0.01	0.03	0.20	0.02	0.02	0.04	0.11	0.07	0.09	0.11	0.02
MTK-087		Mount Takahe	0.07	61.04	0.05	0.53	16.03	0.23	1.12	0.26	7.10	7.85	5.11	0.36	0.13
			0.05	0.35	0.02	0.15	0.30	0.09	0.27	0.05	0.54	0.37	0.12	0.26	0.03
MTK-117		Mount Takahe	0.13	61.11	0.10	0.82	14.95	0.29	1.62	0.26	8.22	7.28	4.99	0.12	0.09
			0.04	0.43	0.03	0.04	0.15	0.13	0.03	0.07	0.16	0.43	0.11	0.12	0.01
MTK-118		Mount Takahe	0.14	61.12	0.13	0.81	14.99	0.27	1.61	0.29	8.24	7.21	5.02	0.07	0.10
			0.04	0.16	0.04	0.02	0.15	0.15	0.03	0.04	0.10	0.16	0.08	0.10	0.01
MTK-119		Mount Takahe	0.15	61.55	0.08	0.80	15.06	0.35	1.59	0.26	8.23	6.77	4.99	0.08	0.09
			0.02	0.39	0.01	0.08	0.10	0.03	0.04	0.06	0.10	0.46	0.06	0.07	0.01
MTK-121		Mount Takahe	0.12	60.59	0.08	0.81	15.36	0.32	1.56	0.28	8.24	7.32	5.05	0.17	0.09
			0.03	0.16	0.02	0.05	0.08	0.02	0.03	0.05	0.13	0.20	0.08	0.08	0.01
WAIS Divide ice core representative tephra compositions															
WDC06A-1589.187	8.077	Mount Takahe	0.17	59.89	0.12	0.84	14.64	0.37	1.83	0.33	8.69	7.39	4.94	0.24	0.12
			0.05	0.77	0.03	0.06	0.17	0.34	0.09	0.03	0.16	0.22	0.11	0.13	0.02
WDC06A-1744.215	9.459	Mount Berlin	0.1	62.54	0.1	0.55	14.15	0.05	1.62	0.31	8.26	6.71	4.64	0.17	
			0.07	0.54	0.03	0.09	0.45	0.03	0.46	0.05	0.72	0.84	0.31	0.14	
WDC06A-2569.205	22.337	Unknown	0.42	58.87	0.05	1.10	16.25	0.98	3.70	0.21	7.87	5.67	4.41	0.31	0.10
			0.19	0.67	0.02	0.11	0.42	0.12	0.46	0.05	0.56	0.48	0.46	0.22	0.03
WDC06A-2758.15	28.21	Mount Berlin	0.08	62.38	0.10	0.50	13.82	0.02	1.30	0.27	8.19	8.05	4.81	0.21	0.28
			0.04	0.81	0.04	0.07	0.65	0.02	0.20	0.06	0.91	0.61	0.21	0.12	0.07
WDC06A-2871.74	32.397	Mount Berlin	0.06	61.55	0.07	0.44	14.17	0.05	1.12	0.33	8.28	8.72	4.77	0.25	0.20
			0.03	0.48	0.02	0.04	0.41	0.01	0.10	0.05	0.61	0.24	0.17	0.11	0.03
WDC06A-3149.138	44.865	Unknown	1.99	45.34	0.3	3.75	14.59	4.72	8.92	0.31	13.74	4.54	1.45	0.29	0.06
			0.11	0.37	0.09	0.08	0.17	0.09	0.15	0.04	0.27	0.20	0.04	0.17	0.01
Other tephra layers referred to in the text															
Byrd Core 788[1]		Mount Takahe	0.17	60.42		0.85	15.44	0.38	1.91		8.52	7.22	4.45		
Siple Dome SDMA9002-9003[2]	8.15	Mount Takahe	0.18	61.55	0.14	0.87	14.86	0.41	1.80	0.33	8.63	6.07	4.99	0.1664	
Mount Moulton BIT-151[3]		Mount Berlin	0.09	60.99	0.06	0.56	15.37	0.10	1.49	0.29	8.27	7.29	5.08	0.25	0.16
Mount Moulton BIT-152[3]		Mount Berlin	0.04	62.57	0.08	0.52	13.88	0.00	1.03	0.26	8.84	7.69	4.69	0.18	0.20
Mount Moulton BIT-159[3]		Mount Berlin	0.03	63.63	0.05	0.43	13.51	0.00	0.84	0.21	8.20	7.78	4.79	0.28	0.26

Notes: Analyses are made using a Cameca SX-100 electon microprobe. Geochemical quantities are in wt%. Analyses are normalized to 100 wt%. Analytical precision, based on replicate analyses of standard reference materials of similar composition to the unknowns, are as follows: SiO$_2$ ± 0.47 wt%, TiO$_2$ ± 0.03 wt%, FeO ± 0.06 wt%, MnO ± 0.06 wt%, MgO ± 0.07 wt%, CaO ± 0.02 wt%, Na$_2$O ± 0.55 wt%, K$_2$O ± 0.27 wt%, P$_2$O5 ± 0.02 wt% and Cl ± 0.07 wt%.

Peak count times of 20 s were used for all elements with the exception of Na (40 s), F (100 s), Cl (40 s) and S (40 s).

Primary calibration standards are: P$_2$O$_5$ and CaO Beeson apatite; SiO$_2$, K$_2$O, Al$_2$O$_3$ orth-1; TiO$_2$ rutile, MgO diopside, MnO MnO, FeO magnetite; Na$_2$O Amelia albite; SO$_2$ barite; Cl scapolite.

Beam sizes used for analysis ranged between 10 and 25 μm, depending on the size of grain available.

Analyses are normalized to 100% analytical totals, as is conventional for tephrochronology research

Full datasets may be accessed in the Supplementary material S1–S3.

*Ice core ages from WDC06A WD2014 time scale. Ages are in ka before 1950.

References: [1]Palais *et al.* (1988); [2]Kurbatov *et al.* (2006); [3]Dunbar *et al.* (2008).

0.038, 0.195 ± 0.011 and 0.195 ± 0.133 Ma, corresponding to our lavas 11, 13 and 14 (Fig. 7). Although the ages are all within error, our mapping suggests that lava 11 might have been extruded prior to an episode of inferred collapse of the Mount Waesche caldera (see below), whereas lavas 13 and 14 are younger than that event. Our lava 5, from Group 2 (Fig. 7), is dated as 0.17 ± 0.3 Ma by K–Ar (LeMasurier 1990*a*). Finally, an age of 0.189 ± 0.009 Ma was measured from a bedrock sample of outcrop protruding through an area mapped as 'mixed lava regolith' (Fig. 7). The ages suggest that a large pulse of young effusive volcanism occurred at Mount Waesche between 0.1 and 0.2 Ma, and that Mount Waesche volcano is undoubtedly very young. It is thus potentially still active. There are no isotopic ages for lavas of Group 3. However, the combination of a lack of significant erosion compared with lavas in Group 1 and Group 2, and the presence of extensive block-field formation (see below), suggests that they may belong to an intermediate volcanic pulse.

Older eruptions are also present. Most were dated using the ^{40}Ar/^{39}Ar method. A trachytic dome in our Group 1 lavas yielded an age of 0.48 ± 0.01 Ma (Panter 1995) (Fig. 7), whereas a sample of mafic lava bedrock yielded an age of

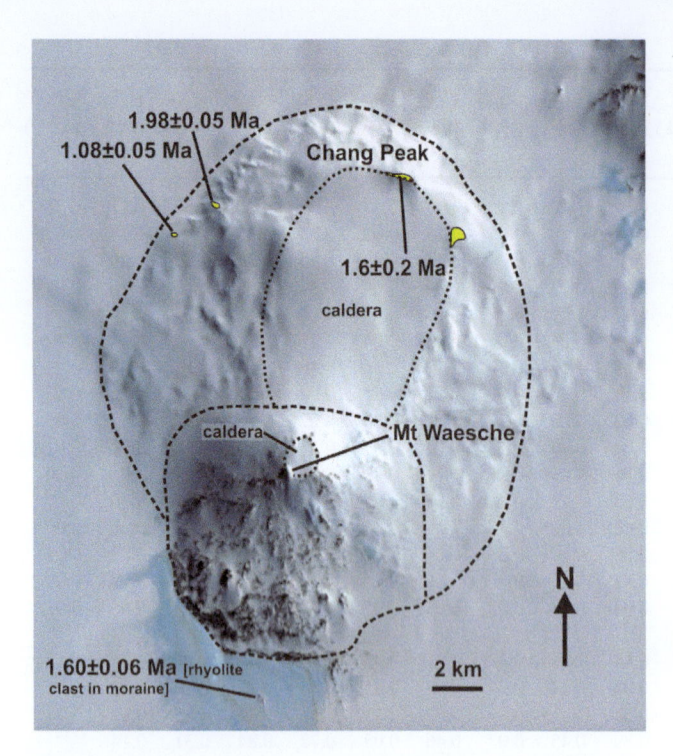

Fig. 6. Google Earth (data provider: US Geological Survey) satellite image showing the morphological outlines of the Chang Peak and Mount Waesche volcanoes (above the 2 km-thick West Antarctic Ice Sheet). Outcrops and isotopic ages (in Ma) for the Chang Peak volcano are also shown (ages after LeMasurier 1990a and Panter 1995). The ages are consistently > 1 Ma and indicate that the Chang Peak volcano is no longer active.

0.335 ± 0.036 Ma (R. Ackert pers. comm. 2019). The latter sample was collected during a field season in the early 1990s and no accurate location is available; hence, we have not included it on our geological map. From a sketch map location, the sample appears to have been collected from somewhere around the SE corner of lava 5 (Fig. 7). However, as noted below, lava 5 is phonolitic, so cannot be the same lava as that dated by Ackert. The location of the sample with the oldest age (1.0 ± 0.1 Ma, dated by K–Ar) appears to correspond to one of the young scoria cones (cone F in Fig. 7), so the accuracy of the age is suspect. This location was resampled during the 2018–19 field season, so new dates will be forthcoming. Other ages, obtained on basaltic lava boulders in moraines on the flanks of Mount Waesche, range from 0.556 ± 0.082 to 0.279 ± 0.028 Ma (Ackert et al. 2013). Together with the ages obtained on in situ rock, it is clear that an older mainly effusive phase of volcanism on Mount Waesche took place mainly between c. 0.5 and 0.3 Ma.

Finally, two ages of 0.2 ± 0.2 Ma (LeMasurier (1990a: K–Ar) and 0.092 ± 0.026 Ma (Panter 1995: $^{40}Ar/^{39}Ar$) were obtained on a prominent vertical englacial tephra layer to the south of Mount Waesche (Fig. 7). The ages of the englacial tephra layers are discussed later in this chapter.

Using previously unpublished geochemical data (see Panter et al. 2021), the oldest lavas (Group 1, lavas 1–3: Fig. 7) are trachytes. Exposures are isolated and largely broken into angular planar-faced blocks. Although three lavas are identified, lavas 2 and 3 are compositionally identical and only a few hundred metres apart; they may be the same unit. The thickest (lava 1) is >30 m thick. The lavas consist of vesicular pale- to mid-grey lava with a local flow foliation. No primary morphology is preserved and the outcrops are extensively glacially eroded. The trachytes were succeeded by effusion of phonolite lavas (Group 2, lavas 4–5). One of the phonolites (lava 5) is the largest lava mapped on Mount Waesche and forms an extensive fan-shaped outcrop on the southernmost slopes; it is locally at least 30 m thick but, like other lavas on Mount Waesche, maximum thicknesses are unknown. The outcrop of lava 4 is very small and it is largely covered by lava 5. It is compositionally almost identical to bombs in scoria cone C, implying derivation of lava 4 from that flank vent. The outcrop of lava 5 wedges out upslope at c. 2300 m asl, suggesting that it may also have issued from a flank vent, now no longer exposed. Both lavas 3 and 4 are distinguished by the presence of abundant ovoid enclaves of medium-grained hypabyssal and gabbroic rock, interpreted to be cognate, typically ≤ 8 cm in diameter but occasionally up to 18 cm. No original lava surfaces are preserved and the lavas are extensively degraded. They comprise loose masses of angular lava fragments (i.e. felsenmeer or block field). Any original autobreccia has been almost completely removed except in rare distinctive arcuate zones containing brown clinkery rock, which probably represent autobreccia infolded by ogives. Erosion of lavas in Group 2 resulted in a very uneven surface, locally with a relief of a few tens of metres that may have deflected lavas of Group 3. Lavas in Group 3 (lavas 6–9) do not show the deep erosional dissection of the older two lava group outcrops, and they preserve a crude lobate landform in one (lava 8 in Fig. 7). However, the lava surfaces are extensively frost shattered and, like Group 2 lavas, they also occur as block fields with relicts of in situ rock, together with faint arcuate solifluction ridges; striated surfaces are rarely present. Compositionally, they are distinguished from other lava groups by their alkali basalt and hawaiite compositions, although more evolved lavas are also rarely present (e.g. mugearite: lava 8). Lavas in the youngest group (Group 4; lavas 10–16) have the widest range of compositions in any group on Mount Waesche (basanite/tephrite to tephriphonolite: Fig. 8). They appear to be the youngest volcanic features on Mount Waesche and form extensive flow fields that dominate the central and northwestern outcrops (Fig. 7). They have prominent levées, and they preserve reddened and brownish autobreccia in places. All the lavas in Group 4 are younger than the polymict gravelly breccia deposit (see below), except for lava 10, which is older (and possibly lava 11, whose stratigraphical position is uncertain).

Scoria cones are associated with lava groups 2, 3 and 4. They are particularly common in Group 4 lavas, and are relatively pristine. By contrast, they are strongly glacially eroded in the older groups, and display well-developed roche moutonnée landforms and smooth surfaces (Fig. 9a). The older scoria cones (A–D in Fig. 7), are aligned NW–SE. They have alkali basalt (lava group 3) and phonolite (lava group 2) compositions and consist of strongly coherent maroon or reddened (oxidized) pyroclastic deposits and grey lava. Clastic textures are ubiquitous, with vesicular bombs, often 10–40 cm in diameter and up to 80 cm, typically showing a moderate to strong flattening due to agglutination; parallel bomb orientation lends a crude stratification to some outcrops. In the phonolite example, more intense flattening has caused a thin foliation that wraps around clasts.

The least eroded, and therefore youngest-appearing, set of scoria cones located on the SW flank of Mount Waesche are aligned along a prominent SW–NE trend. These are identified as 'younger scoria cones' in Figure 7, lettered E–K. All appear to be compositionally uniform and similar to Group 4 lavas, many of which are adjacent to, and probably erupted from, these cones. The cones are composed of crudely stratified well-sorted scoria and bombs, agglutinated in part and locally transitional to lavas. In this respect, these cones are characteristic of proximal facies of typical scoria cones (e.g. Sumner et al. 2005). Large parts of the scoria piles show a reddish colour typical of subaerial, high-temperature oxidation. The agglutinate is locally transitional to dense lavas in which

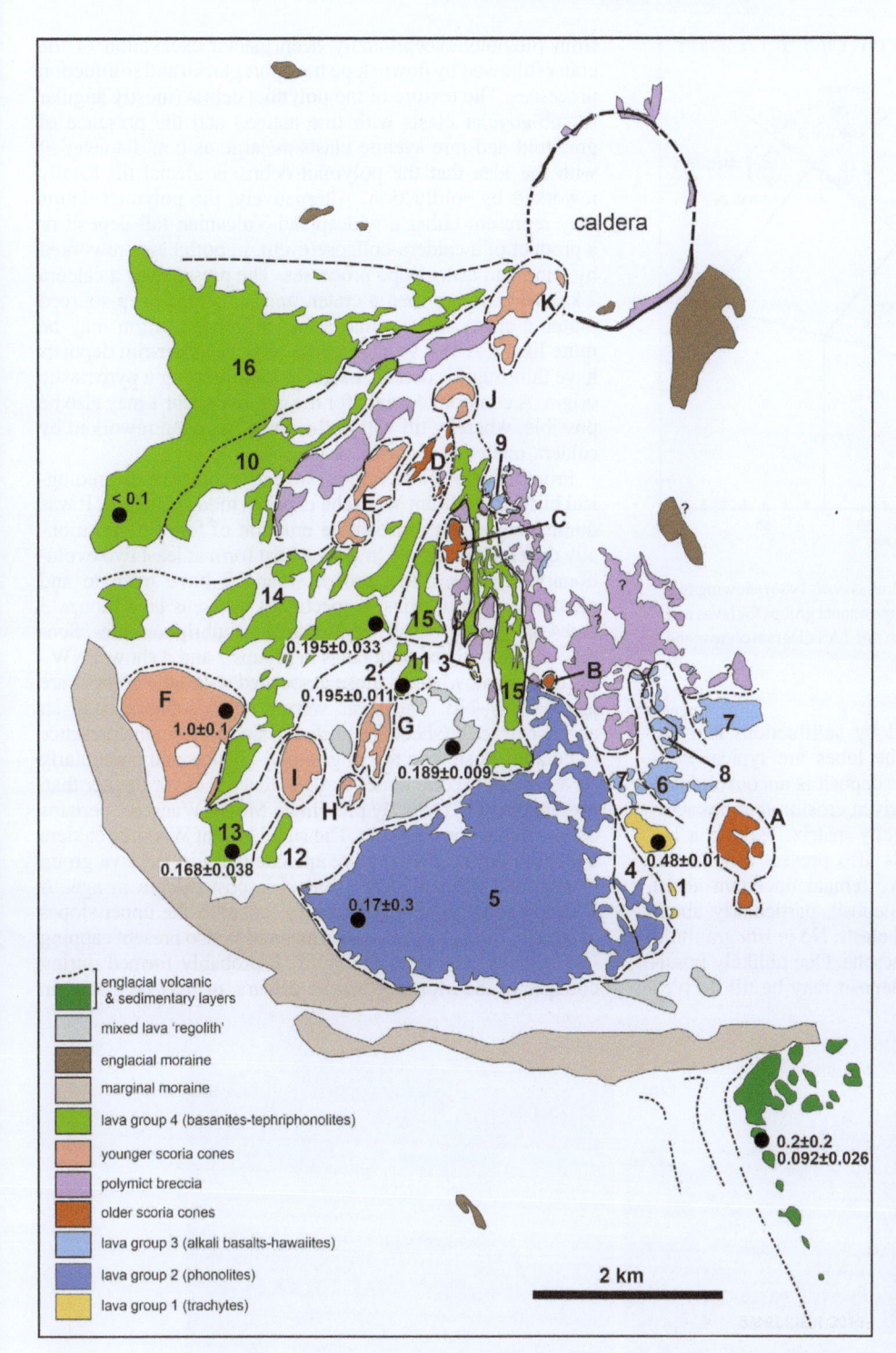

Fig. 7. Geological map of Mount Waesche volcano. Based on mapping by J.L. Smellie, W.C. McIntosh and K.S. Panter. The locations and ages of dated samples (all ages in Ma) are also shown (after LeMasurier 1990a, with sample locations (approximate) from Gonzales-Ferran 1995); and unpublished data of M.J. Zimmerer and W. C. McIntosh. The area in the lower right of the map is the location of the englacial tephra samples, and this area is shown in more detail in Figure 17a & b.

clastic textures are still faintly visible (i.e. they are clastogenic; Sumner *et al.* 2005). The degree of erosion varies from crater to crater, with crater G being nearly pristine (Fig. 9b) and others having lost variable amounts of volume to erosion. Fine, poorly sorted deposits characteristic of hydrovolcanic eruptions are notably lacking from all visited outcrops.

Locally, eroded surfaces of scoria cones and lavas show well-developed erosional lineations (Fig. 9a) that show approximate north–south directions regardless of location or local slope orientation. Although glacial erosion has contributed to the overall shape of outcrops (as described above), we interpret the lineations to be mainly due to wind erosion. They are parallel to the current predominant wind direction, range from parallel to perpendicular to current hillslopes, tend to be on current topographical highs, and preserve fragile ridges and fins which are unlikely to have survived glaciation.

Massive polymict gravelly debris at least a few metres thick crops out widely on the intermediate and upper slopes of Mount Waesche (Figs 7 & 9c); similar material is also present capping Chang Peak. It is distinguished by numerous (15–20%) granitoid blocks, including syenites, typically 20–50 cm in diameter and generally less than 80 cm, although two much larger clasts (2.5 and 7 m in diameter) were observed. Many clasts show rounding, and are set in a matrix of coarse angular granules of lava and granitoid and some sand-sized material. There are also numerous large angular blocks of lava, in some cases as large as 2–5 m in diameter. The deposit is unconsolidated and displays well-formed

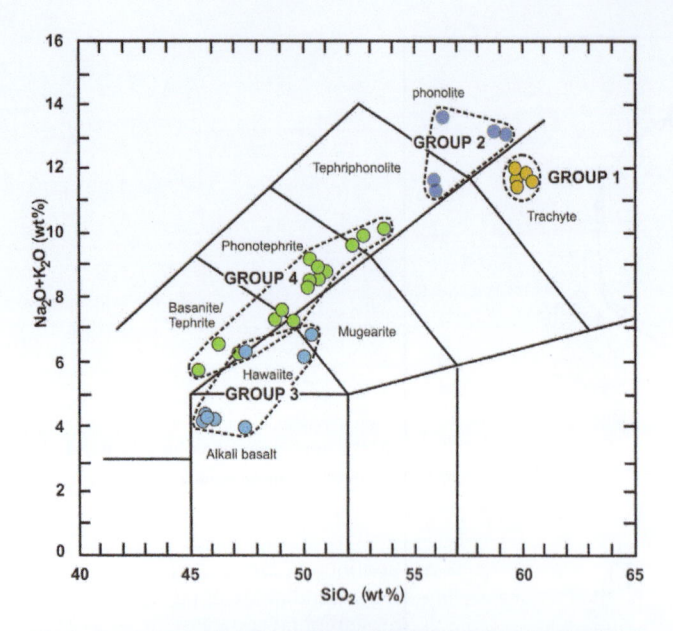

Fig. 8. Total alkali–silica (TAS) diagram (LeBas *et al.* 1986) showing two compositional lineages/series and four compositional groups for lavas and scoria cones on Mount Waesche. Colours on this TAS diagram correspond to those shown on the geological map (Fig. 7).

crescentic lobes, probably formed by solifluction, that are prominent on satellite images. The lobes are typically 6–12 m in width and 2–5 m high. The deposit is unconsolidated and has suffered extensive wind-driven erosion that has winnowed out much of the fine gravelly matrix, leaving a lag deposit of polymict boulders. It is also present around the Waesche caldera rim (Fig. 9d). We remain uncertain about the origin of the polymict debris deposit, particularly about the origin of the abundant granitoid clasts. No *in situ* granitoid outcrops are exposed at Mount Waesche. One unlikely possibility is that the polymict debris deposit may be till derived from previous exceptionally deep glacial excavation of the crater followed by downslope transport glacial and solifluction processes. The texture of the polymict debris (mostly angular to sub-angular clasts with fine matrix) and the presence of granitoid and rare syenite clasts as large as 6 m diameter fit with the idea that the polymict debris is glacial till locally reworked by solifluction. Alternatively, the polymict debris may represent either a widespread Vulcanian fall deposit or a product of a caldera-collapse event, in both cases reworked by wind and downslope processes. The presence of a caldera 2 km wide, rather than a crater, and numerous deep-sourced plutonic clasts suggest that a caldera-related origin may be more likely. A few syenitic clasts in some crater-rim deposits have thin fragile coatings of mafic lava, proving a pyroclastic origin. A compound origin for the polymict debris may also be possible, whereby till within the crater has been reworked by caldera-related pyroclastic processes.

From the above description, we can reconstruct the geological history of Mount Waesche exposed today (Fig. 10). It was dominated by the consecutive eruption of four compositionally different groups of 'a'ā lavas that form at least two evolutionary lineages. Lava groups 1 and 2 have trachyte and phonolite compositions, respectively, whereas lava groups 3 and 4 are mainly basalts and basanites–tephriphonolites. Scoria cones associated with lavas in groups 2 and 3 show a NW–SE alignment, whereas those associated with lava group 4 are aligned SW–NE. The lavas were erupted subaerially in an absence of an ice sheet or significant slope ice but the presence of notable erosion on top of groups 1–3 lavas and particularly on associated scoria cones suggest that ice cover thicker than present day was formerly present on Mount Waesche, perhaps on more than one occasion. The small Mount Waesche caldera collapsed during the eruptive episode that formed lava group 4, although most of lava group 4 is post-caldera in age. A widespread polymict debris deposit mantles the upper slopes of Mount Waesche and similar material is also present capping Chang Peak, 10 km to the NNE. It probably formed during collapse of the Mount Waesche caldera, or may be glacial in

Fig. 9. Field photographs from Mount Waesche. (**a**) Glacially eroded surface cut across agglutinate of scoria cone A (older scoria cones) (clipboard for scale). The glacially polished planated surface shows well-developed striations, probably accentuated by wind-driven erosion. (**b**) Uneroded scoria cone (crater G) (crater is *c.* 200 m in diameter, view is to the south). (**c**) Unconsolidated polymict gravelly debris deposit on upper Mount Waesche slopes. Note the resemblance to till (the hammer is for scale) (**d**) Polymict breccia deposit exposed in the caldera-wall exposures of Mount Waesche (the ice axe is for scale).

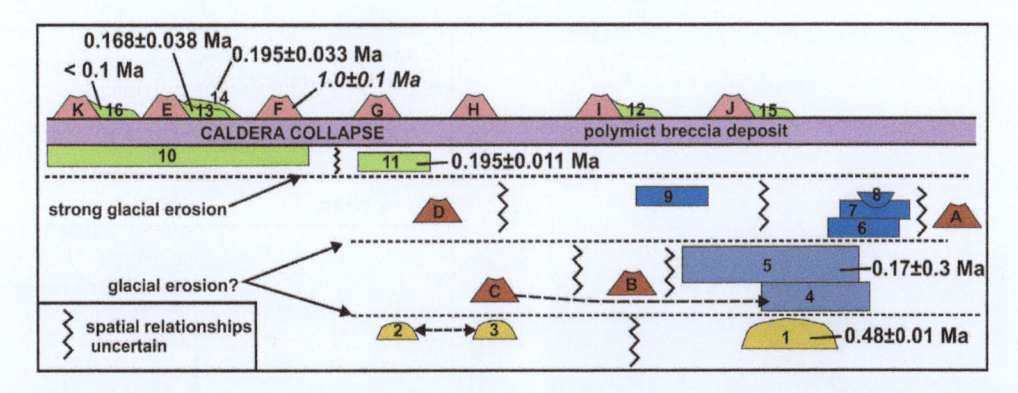

Fig. 10. Schematic diagram showing inferred stratigraphical relationships of the constituent geological units on Mount Waesche with published isotopic (K–Ar) ages (ages after LeMasurier and Kawachi 1990), and unpublished data of M.J. Zimmerer and W.C. McIntosh. The stratigraphical position of lava 11 is uncertain (pre- or post-caldera). Based on strong compositional similarities, lavas 2 and 3 may be equivalent, and scoria cone C may have fed lava 4. Colours and numbered units are as shown in Figure 7.

origin. From the published isotopic dating, all these events took place during the past c. 500 kyr; and many are <200 ka. Mount Waesche should therefore be regarded as potentially still active (dormant).

Blue ice areas

The glaciation of volcanic edifices in West Antarctica results in many eruptive products not being accessible for observation and sampling. However, the glaciation of West Antarctica as a whole provides a repository for products of explosive volcanic events, in the form of englacial tephra layers. In rare cases, local ice dynamics and wind patterns result in deep glacial ice, and associated tephra, being exposed at the surface in stratigraphic section (Cassidy et al. 1977). Two such occurrences in West Antarctica will be discussed below.

Mount Moulton

A summit ice cap of a set of coalesced extinct volcanoes, collectively called Mount Moulton, is located roughly 30 km downwind from Mount Berlin (Fig. 1). During the 1993–94 field season, a field team visited the rock outcrop, Prahl Crags, using Twin Otter aircraft support. This outcrop, interpreted to represent a small exposure of crater-rim pyroclastic rocks, is associated with an extensive blue ice field (Fig. 11). During the visit, a set of englacial tephra layers was recognized, several of which are 10–15 cm thick, containing pumice up to 2 cm in diameter (Fig. 12a). Other tephra layers are thinner, composed of ash (Fig. 12b). A small number of the thickest, coarsest tephra layers were sampled, and then subsequently geochemically analysed and dated using the $^{40}Ar/^{39}Ar$ method (Wilch et al. 1999a). The tephra layers are trachytic, and range in age from 495.4 ± 9.7 to 10.5 ± 2.5 ka (Dunbar et al. 2008), showing that the blue ice section represents a significant time span.

Field campaigns to further examine the Mount Moulton englacial tephra section, and sample not only the tephra layers but local ice through horizontal ice coring and drilling, took place in 1999–2000, and again in 2003–04. During the first field season, the englacial tephra section was thoroughly mapped (Fig. 13) and sampled. Some of these layers were recently redated (Iverson 2017). A total of distinct 48 tephra layers were identified. The tephra-layer geometry is simple and undeformed, resulting in a coherent stratigraphy in which both the record of volcanic eruptions (Dunbar et al. 2008) and the climate record (Popp et al. 2004) have been investigated. Tephra layers in the upper part of the stratigraphic section are continuous, and can be easily traced on the blue ice surface. In the middle–lower part of the stratigraphic section, the thicker tephra layers are exposed as boudins (Fig. 12a), rather than being continuous. The thinner layers can be traced continuously. A set of Mount Moulton tephra layers were recently redated using a high-sensitivity mass spectrometer (Iverson 2017) (Fig. 14), and these new ages should be considered the most accurate and precise ages for this tephra sequence. These ages, from youngest to oldest (all in ka), are: BIT-152, 24.6 ± 0.4; BIT-156, 80 ± 4; BIT-157, 92.3 ± 0.4; BIT-158, 106.3 ± 0.7; BIT-160, 118.8 ± 0.8; BIT-162, 136.9 ± 0.5; and BIT-163, 168 ± 2.

Geochemical analyses and textural observations, carried out by electron microprobe and reported in Dunbar et al. (2008), show that the tephra layers exhibit expanded or bubble-wall textures typical of explosive volcanism (Fig. 15a, b). Furthermore, many of the tephra layers found at Mount Moulton are geochemically similar, and fall within the trachytic field on a total alkali–silica diagram (Le Maitre 1989) (Fig. 16). Several of the tephra layers have less evolved compositions, falling in the basalt, basanite or tephrite fields. The trachytic tephra layers are all interpreted to be from nearby Mount Berlin. The older set of less evolved tephra layers (>500 ka) may be related to proximal Mount Moulton volcanism, although no

Fig. 11. Photograph of the Mount Moulton blue ice field, looking to the west, showing englacial tephra layers as dark lines within the ice. The blue ice area is around 400 m in north–south extent. Many tephra layers are present in the ice field and several have been highlighted with red arrows in the figure.

Fig. 12. Englacial tephra layers at Mount Moulton. (**a**) Very thick tephra layer near the top of the stratigraphic section showing boudins of tephra separated by ice. Mount Berlin can be seen in the background of the photograph. (**b**) The finer, thinner tephra layer is evident as a linear depression in the ice (identified with red arrows). The width of depression is around 20 cm. View is to the west.

direct correlations can be made. Three younger non-trachytes in the section (one that is younger than 10.5 ± 2.5 ka, and two that are around 120 ka) may be related to flank eruptions of

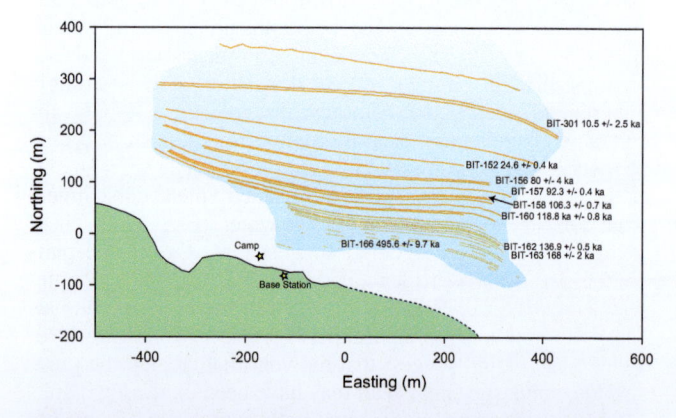

Fig. 13. GPS-produced map of the Mount Moulton tephra section with ice motion and ablation shown. Orange lines represent tephra layers. The area shown in pale blue represents the extent of the exposed blue ice. The areas around the blue ice region are snow-covered. The area shown in green and labelled 'Prahl Crags' in the image represents the approximate extent of the rock outcrop. The tephra-layer ages shown on the map are from Iverson (2017).

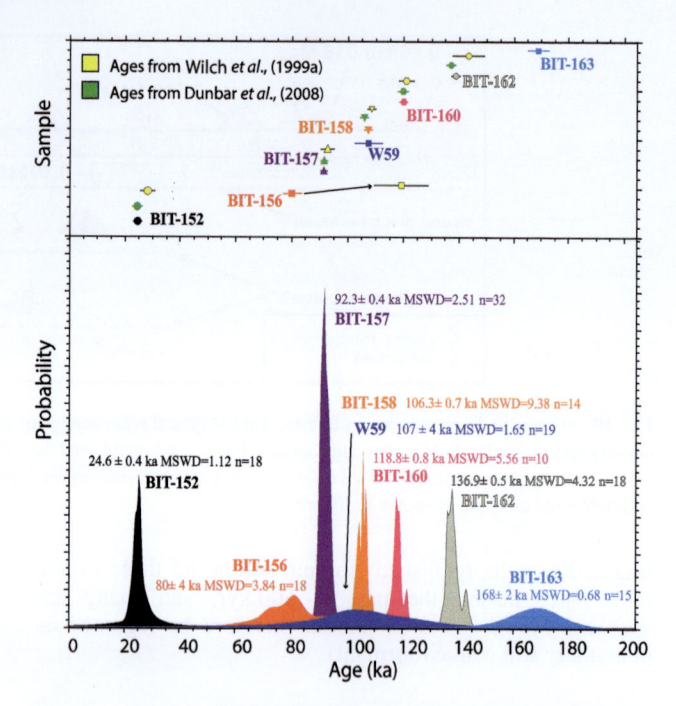

Fig. 14. Comparison of ^{40}Ar/^{39}Ar ages from West Antarctica blue ice areas. The bottom panel shows ideograms (age probability distribution diagram) for the single-feldspar crystal analyses. The top panel is a comparison of different mean ages for each sample. Symbols coloured the same as the probability distribution curve are new ages in this study. Yellow symbols are from Wilch *et al.* (1999*a*) and green symbols are from Dunbar *et al.* (2008). New analyses were measured on the multi-collector ARGUS VI mass spectrometers at NMGRL, Socorro, New Mexico. Uncertainties shown at 2σ. The figure and caption are from Iverson (2017), reprinted with permission from New Mexico Tech.

Mount Berlin, some of which have similar geochemical composition but, again, no direct correlations can be made.

Mount Waesche

A second tephra-bearing blue ice site is located on the south side of Mount Waesche volcano (Fig. 1). The extensive blue ice field, over 8 × 10 km in diameter, is located at between 1900 and 2000 m elevation near Mount Waesche, and contains numerous englacial tephra layers. In contrast to the Mount Moulton site, the Mount Waesche site displays a complex, deformed stratigraphy (Fig. 17a, b). Several field campaigns have focused on studies at Mount Waesche (years of campaigns and focus of work described in Supplementary material S1). The geology was examined during the 1989–90 field season (described in a previous section in this chapter), as well as during a short reconnaissance visit in 1990–91. Mount Waesche was visited again in 1997–98, with the primary focus of examining the englacial tephra layers, and the work from that field season is summarized below.

The englacial tephra layers at Mount Waesche are remarkable for their number and complexity (Fig. 17a, b; see also Supplementary material S2). Some of the tephra layers are coarse (bombs up to 1 m in diameter), dark coloured and mafic in composition. These cannot be far from their volcanic source, are compositionally indistinguishable from lavas on Mount Waesche and are interpreted to be derived from the numerous cinder cones on the volcano flanks. Other tephra layers are thin and fine, greyish-yellow in colour, and trachytic in composition. These are unlikely to be derived from Mount Waesche, and, as will be discussed below, are derived from Mount Berlin or Mount Takahe. Geochemical analyses suggest that some of these distal tephra layers are complexly

Blue Ice tephra | Ice Core tephra

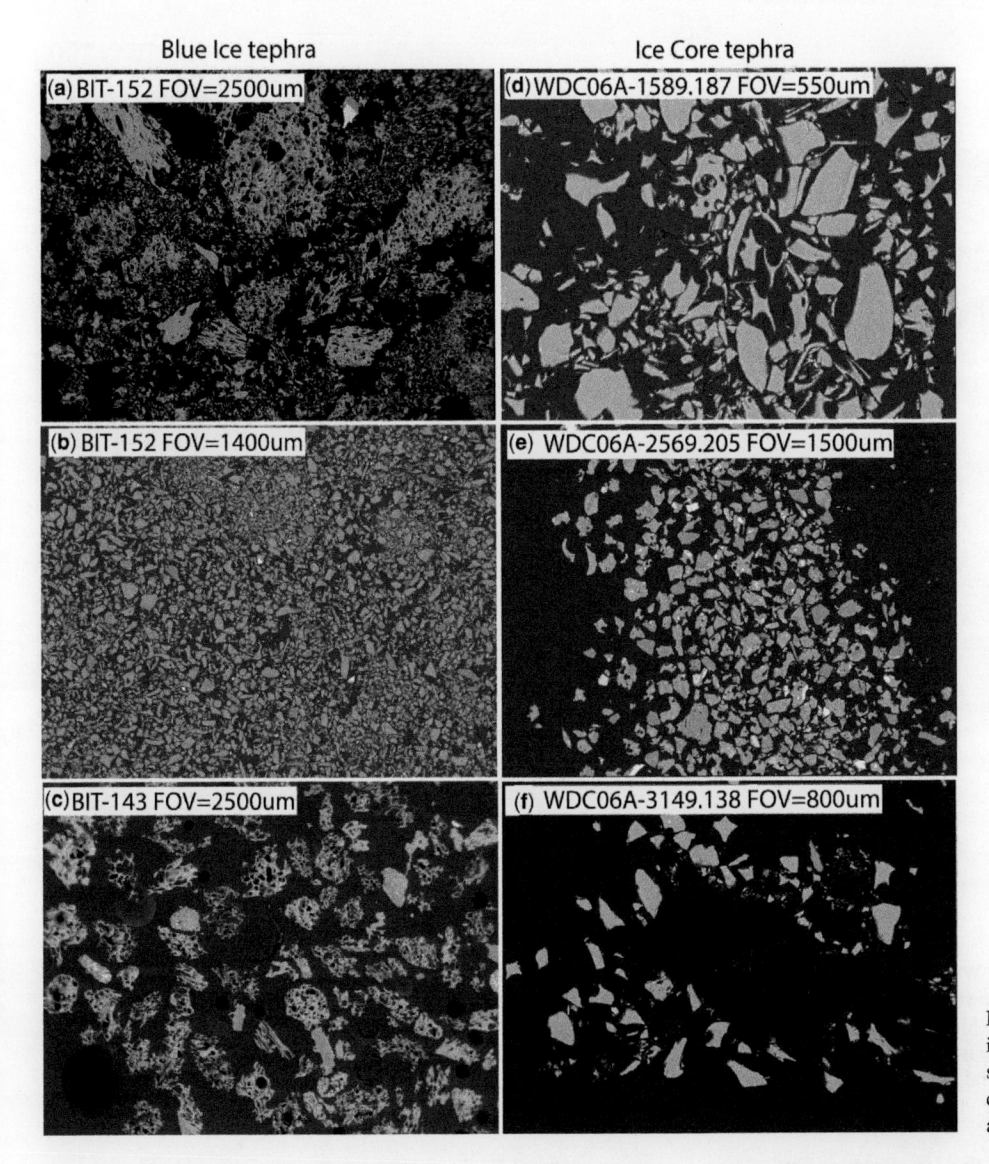

Fig. 15. (a)–(f) Backscattered electron images (BSE) of representative tephra samples from blue ice areas and ice cores. Field of view and sample numbers are shown on the BSE image.

folded, and repeated within the section. Geochemical compositions of glass from Mount Waesche tephra layers are presented in Table 1 and Figure 18, and a backscattered electron (BSE) image of one tephra layer is shown in Figure 15c. The whole-rock geochemical composition of selected, coarser-grained Mount Waesche tephra layers is included in Supplementary material S1.

The largest tephra layer, informally called the 'Great Wall' (BIT-181) (labelled in Fig. 17a), protrudes vertically out of the ice, and contains *in situ* volcanic bombs up to 60 cm in diameter (Fig. 19a, b). The Great Wall is, in places, up to several metres thick in original vertical dimension, and is strongly boudined (Fig. 17a, b). Although the volcanic bombs in the Great Wall indicate subaerial eruption, the deposit also contains a component of granular, non-vesicular lava fragments (Fig. 19c), as well as some finer-grained epiclastic sediment that appears to be transported by water (Fig. 19d). These observations, taken together, suggest that while some parts of the Great Wall were deposited directly from explosive eruptions onto the ice, lava may have also flowed out onto the ice and was granulated by contact between the lava and ice and/or water. We would suggest that the vent that created the pyroclastic/epiclastic Great Wall may have been on the low flanks of Mount Waesche, and created both pyroclastic bombs, but also lava flows that flowed out on the local ice sheet. At one location on the Mount Waesche blue ice field, fragments of an englacial volcanic deposit containing pillow-lava fragments are observed (Fig. 19e). Although this deposit cannot be directly physically linked to the Great Wall, it does reinforce the conclusion that one or more lava flows from

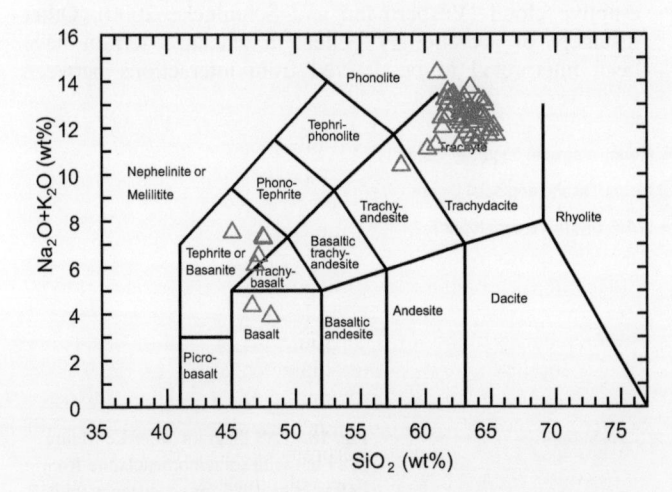

Fig. 16. Total alkali–silica (TAS) diagram (after Le Maitre 1989 but with some nomenclatures from LeBas *et al.* 1986 for consistency with the text) of tephra from all Moulton tephra layers. Each point represents an average glass composition for individual, geochemically homogeneous, tephra layers (from Dunbar *et al.* 2008). Analyses are presented water-free.

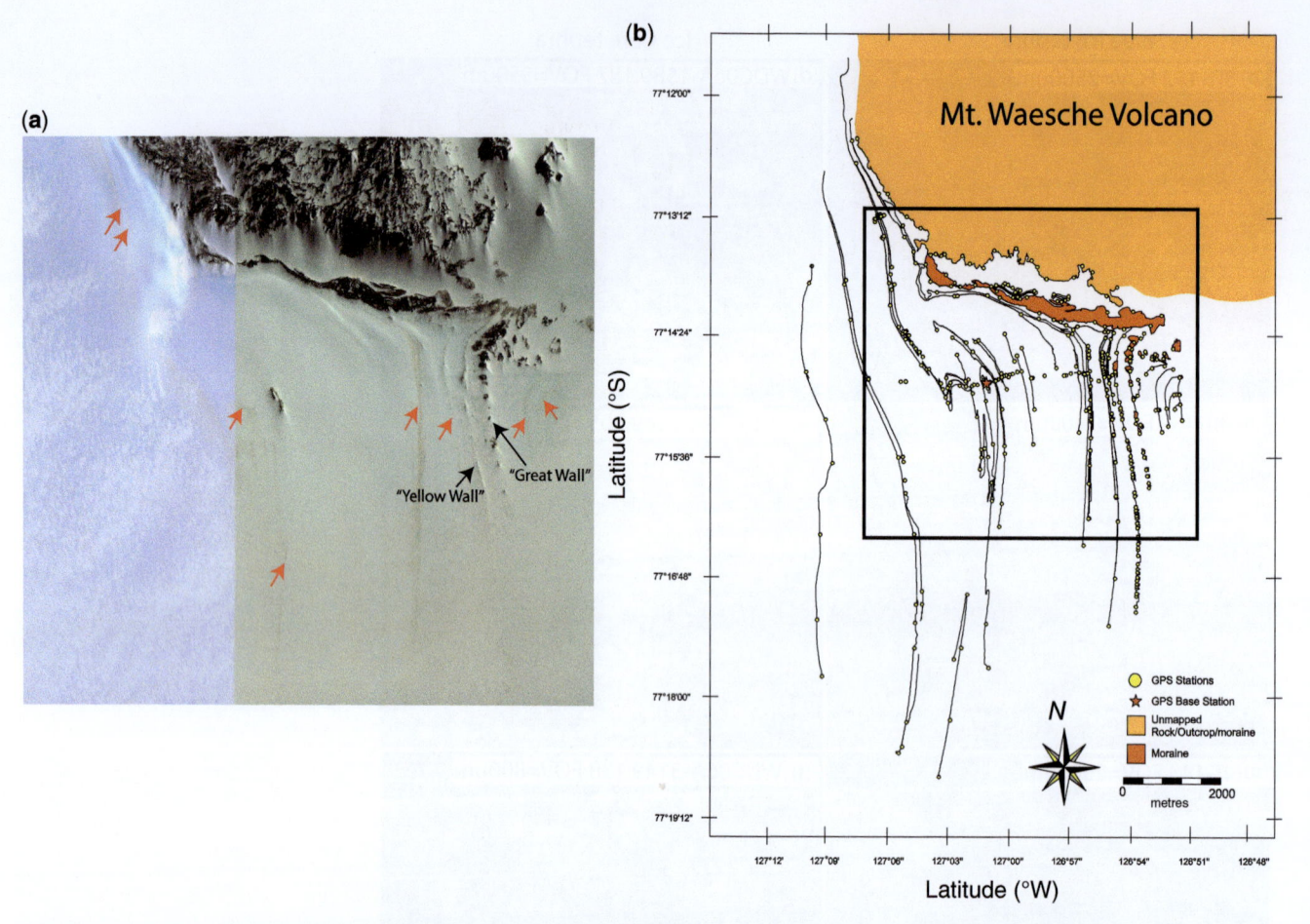

Fig. 17. (**a**) Satellite image of the Mount Waesche blue ice field and englacial tephra layers. The width of the image represents approximately 7 km and the exact scale is noted in (**b**), where a box showing the extent of the satellite image is also shown. Selected tephra layers are highlighted with red arrows, and two are labelled. (**b**) GPS-produced map of the Mount Waesche blue ice area. Dark lines connecting GPS station points present tephra layers. The band of moraine that parallels the flanks of Mount Waesche is a complex morainal deposit, informally called 'The Rock Band'. This is formed mainly by sub-parallel bands of pyroclastic material and lava flows. Several patches of 'moraine' in-line with mapped tephra layers are interpreted as boudins of thick tephra layers in depositional position but rotated by about 90°. See the illustrations in Figure 19.

Mount Waesche flowed out onto the ice sheet, or into water vaults in the ice sheet, possibly created by volcanic heat. Following deposition of Great Wall material onto the ice sheet, the volcanic layer was buried in snow and was incorporated into the ice sheet, later being transported to its current location by ice flow and exhumed.

Upsection from (i.e. west of) the Great Wall is another vertical tephra layer informally called the 'Yellow Wall' (BIT-182) (labelled in Fig. 17a). This unit, if projected back to an original horizontal geometry, is interpreted to be composed of a distinctive, yellow-coloured basal layer, with an overlying black, scoriaceous, horizon (Fig. 20a). The yellow-coloured basal layer of the Yellow Wall everywhere contains abundant accretionary pellets up to 5 mm in diameter (Fig. 20b), indicating that moisture was present within the eruptive cloud (Vespermann and Schmincke 2000). Other instances of accretionary pellets in glaciated terrain have been interpreted to be derived from interactions between

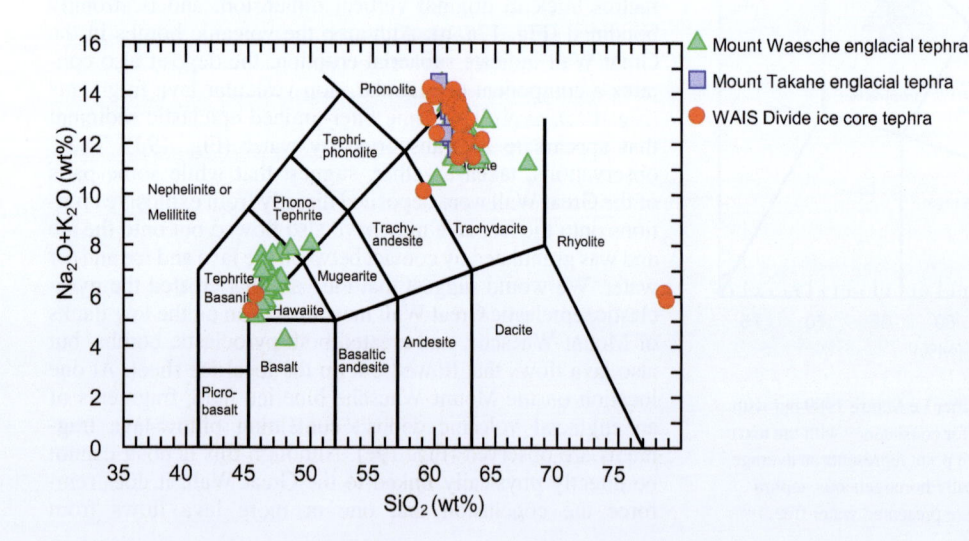

Fig. 18. TAS diagram (after Le Maitre 1989 but with some nomenclature from LeBas *et al.* 1986 for consistency with the text) of average, water-free, glass composition for geochemically homogeneous tephra layers from Mount Waesche, as well as Mount Takahe and the WAIS Divide ice core.

Fig. 19. Photographs of the Mount Waesche vertical englacial tephra layer called 'The Great Wall'. (**a**) Vertical exposure. (**b**) Strombolian bombs in the pyroclastic deposit. (**c**) Granulated lava fragments. (**d**) Reworked pyroclastic material interpreted to be the result of water reworking, possibly related to melting on snow and ice during the eruptive process on the slopes of Mount Waesche and nearby ice sheet during the eruptive process. (**e**) Pillow fragment in a matrix of hyaloclastite breccia. The outcrop is close to the Great Wall but not demonstrably part of the Great Wall deposit. It is found on the 'Rock Band' and may be related to an older englacial tephra.

lava and local ice (Van Eaton et al. 2015). The yellow layer exhibits faint, laminar bedding, and also contains small, black, scoriaceous fragments which are compositionally identical (tephrite–basanite) to those in the black horizon (see samples 182a and 182b in Supplementary material S1). The black scoria in the layer overlying the yellow, accretionary-pellet-bearing layer are up to 5 mm in diameter, along with around 20% free crystals, and with approximately equal amounts of plagioclase and olivine, with rare pyroxene. The scoriaceous texture of the pyroclasts in the black layer resembles pyroclasts found in cinder cones, and is interpreted to be the result of magmatic vesiculation (Cashman et al. 2000), probably resulting from Hawaiian or Strombolian eruption style and scoria cone deposition (Vespermann and Schmincke 2000). The yellow, accretionary-pellet-bearing layer underlies the black layer, suggesting that water was present during the earlier part of the eruption but not during later eruptive stages. Furthermore, with distance from the volcano, the yellow horizon is lost and only the black, cindery horizon is present, suggesting that the yellow layer was only deposited close to the vent, possibly because it was wetter and less buoyant than the scoriaceous black layer. Given the identical chemical composition of scoria in the yellow and black horizons, the simplest explanation is that a single eruption produced both deposits. The scenario that we would suggest for this eruption, following descriptions in Vespermann and Schmincke (2000), would be that the early stages of the eruption were impacted by external water, leading to efficient fragmentation and the development of accretionary pellets. As the eruption progressed, less external water was available, leading to production of the black, scoriaceous, pyroclasts observed in the upper part of the deposit. Another vertical tephra layer, called the 'Little Wall' (BIT-180), which is a black crystal-rich, well-sorted lapilli tuff containing scoriaceous pyroclasts between 3 and 6 mm in diameter is also interpreted to be derived from a vent on the flanks of Mount Waesche.

In addition to the distinctive, vertical tephra walls described above, there are two areas within the Mount Waesche blue ice field that contain sets of closely spaced scoriaceous tephra layers (BIT-184, BIT-185 and BIT-195–BIT-199) with particles up to 5 mm in diameter. One of these scoriaceous bands can be seen as a broad swath of brownish ice in the middle of

Fig. 20. Photographs of the Mount Waesche englacial tephra layer informally called the 'Yellow Wall'. (**a**) Slab of tephra showing the yellow basal layer that contains armoured and accretionary lapilli, and the upper black scoriaceous layer. (**b**) Close-up image of the yellow basal layer, showing accretionary and armoured pellets.

Figure 17a (marked with red arrow). As with the coarser tephra layers described above, these scoriaceous layers are interpreted to be derived from local Mount Waesche eruptions. In contrast to the discrete tephra layers described above, the closely spaced scoriaceous layers with interbedded ice must have been deposited over some time span, although estimating the time elapsed between deposition of the individual layers would be challenging. The appearance and chemical composition (see samples BIT-195–BIT-199) of the closely spaced scoriaceous layers are very similar, suggesting derivation from a continuous eruption that persisted intermittently over some period of time.

The coarse grain size of the set of englacial tephra layers described above, most notably the Great Wall which contains bombs up to 1 m in diameter, suggests that the eruptive source for these layers must have been Mount Waesche, most likely derived from parasitic vents on the flanks of the volcano, described in the earlier 'Physical description and geology' subsection. However, the mechanism of emplacement of these layers is not something for which a modern analogue exists, as far as the authors of this chapter are aware of. The occurrence of the coarse tephra layers within blue ice suggests that deposition occurred onto snow or firn on the ice sheet adjacent to the volcano, possibly on the snowy north side of the volcanic edifice. Over time, the tephra layer would have been snowed over and buried, incorporated into flowing ice moving southwards towards the current blue ice ablation area. We speculate that strong winds coming down the south side of the volcano during the winter season are responsible for ablation, forming a local basin within the ice and creating a flow gradient from the north to the south side of the volcano.

We also speculate that there may be a bedrock high running south from the SE quadrant of Mount Waesche that could result in the ice being rotated, resulting in vertical tephra layers and the acute folding observed in the older tephra layers (Fig. 17a, b). Radar data collected during the 2018–19 field season tentatively supports the presence of this bedrock high (S. Campbell pers. comm. 2019). The radar data, as well as additional geological information that will result from dating samples of tephra layers and potential volcanic sources on the flanks of Mount Waesche collected during the 2018–19 field season, may clarify the dynamics and timescale of local ice flow that transported the tephra layers from their source areas to their current locations.

In addition to the coarse, mafic tephra layers described above, interpreted to be derived from eruptions on the flanks of Mount Waesche, there is a second type of tephra layer (Fig. 21), which is fine and trachytic, and likely to have been derived from Mount Berlin or Mount Takahe (Table 1; Fig. 18; see also Supplementary material S1). A notable aspect of these fine-grained tephra layers is that in some parts of the section, a number of closely spaced layers all have the same geochemical composition. For instance, in the younger part of the Mount Waesche section, samples BIT-167, BIT-169, BIT-171, BIT-201–BIT-204, BIT-206, BIT-207 and BIT-209 are all geochemically indistinguishable (Table 1). As can be seen in the detailed map provided in Supplementary material S2, these layers are all close together or are part of the same set of tephra bands. The interpretation is that these are all the same tephra layer, and that there has been strong deformation and folding, repeating the layer multiple times.

Fig. 21. A fine-grained tephra layer in ice at Mount Waesche.

In some cases, fold noses have been observed in the tephra layers, suggesting structural repetition of the stratigraphic section. Similarly, in the lower part of the stratigraphic section, east of the Great Wall, BIT-146, BIT-147 and BIT-148 all geochemically correlate, suggesting similar strong deformation. Also, BIT-143, BIT-149 and BIT-222 are geochemically indistinguishable. BIT-143 and BIT-149 are apparently downsection of the Great Wall, and BIT-222 is above, suggesting that the lower part of the section may be overturned. The thick, coarse tephra layers at Mount Waesche are folded where they are pushed against the flank of Mount Waesche, and some are broken into boudins but are otherwise relatively undeformed. However, the ice between these competent tephra layers appears to be complexly folded, resulting, in some cases, in multiple repetitions of the same tephra layer within a short distance. Another interpretation would be that each individual tephra band is a single eruptive event, and that the source volcanoes, Mount Berlin and Mount Takahe, produced a number of geochemically indistinguishable tephra deposits. However, a detailed study of tephra layers from Mount Berlin (Dunbar et al. 2008; Iverson 2017) have demonstrated geochemical variability, even for tephra layers formed by eruptions only a few thousand years apart. Therefore, we favour the interpretation of stratigraphic repetition of individual tephra layers.

Several of the fine-grained tephra layers can be geochemically linked to known eruptions from Mount Takahe or Mount Berlin (Fig. 22a–d). A general distinction that can be observed is that Mount Takahe tephra layers typically have a relatively high MgO content (Wilch et al. 1999a; Iverson 2017) (Fig. 22a). These geochemical correlations are described in the paragraphs below.

Three tephra layers (BIT-175, BIT-221 and BIT-230) are geochemically similar to Mount Takahe tephra, sampled both on the volcano (sample W-9 from Wilch et al. 1999a, dated by $^{40}Ar/^{39}Ar$ to 8.2 ± 5.4 ka) and in ice cores (Table 1). One of the stratigraphically highest tephra layers at Mount Waesche (BIT-175) is similar to the Holocene 8.2 ± 5.4 ka from Mount Takahe, although the Ca content of BIT-175 is higher. The tephra layer found at Mount Waesche could come from a different part of the Mount Takahe 8.2 ka eruption, and could therefore have a slightly different chemical composition, as has been observed in other englacial tephra layers in Antarctica (Iverson 2017). Sample BIT-230 is also a close compositional match to the 8.2 ka Mount Takahe tephra layer but is in a blue ice area at Mount Waesche that is not in continuity with the main blue ice area (see the map in Supplementary material S2), so does not provide any chronological constraint.

Other tephra layers at Mount Waesche have been geochemically linked to layers found at Mount Moulton (derived from Mount Berlin), which have been geochemically characterized by Dunbar et al. (2008), and for which additional trace

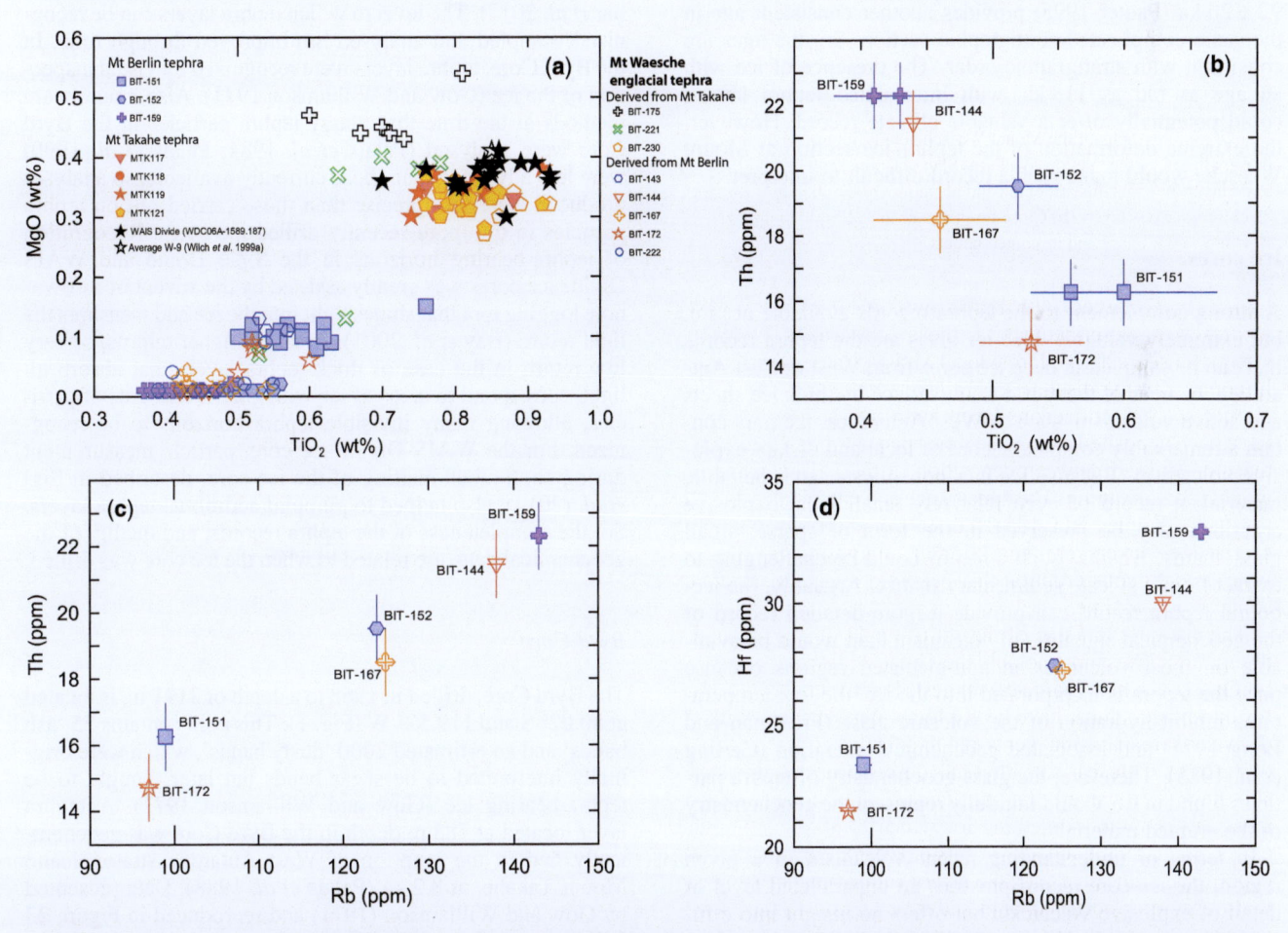

Fig. 22. Geochemical composition tephra samples from Mount Takahe, Mount Waesche and Mount Berlin. Major elements are of individual glass shards, determined by electron microprobe analysis. Trace elements are of bulk tephra samples, determined by instrumental neutron activation analysis (INAA). All data are available in the Supplementary material S1–S3. (a) Geochemistry of individual glass shards from englacial tephra layers derived from Mount Takahe and Mount Berlin, as well as a set of eight tephra layers from Mount Waesche. Analytical uncertainty for TiO_2 is ±0.03 and for MgO is ± 0.07. (b) Average TiO_2 of glass shards (analysed by electron microprobe) plotted against Th analysed by INAA. Uncertainty represented for TiO_2 is the standard deviation of the population of glass shards analysed, and that for Th is the analytical precision. Two separate data points are shown for BIT-151 and BIT-159 because these layers were sampled and analysed twice for major elements. Only one bulk trace element analysis was performed. (c) & (d) Th, Rb and Hf determinations, on bulk tephra samples, by INAA.

element data are presented in Supplementary material S1. The degree of geochemical similarity between many Mount Berlin tephra layers makes direct geochemical correlation based on major elements alone challenging but trace element information can help refine correlations. Although full statistical geochemical correlations are beyond the scope of this chapter, some suggested correlations, based on stratigraphic position and geochemistry, are summarized below. Sample BIT-172 from Mount Waesche is chemically close to BIT-151 from Mount Moulton (Fig. 22b–d), which has an age of between 24.6 ± 0.4 and 10.5 ± 2.5 ka (Iverson 2017). A set of stratigraphically lower samples mentioned above, all correlative with BIT-167, are a close match to BIT-152 from Mount Moulton, which has an age of 24.6 ± 0.4 ka (Fig. 22b–d). Finally, BIT-144, in the lower part of the Mount Waesche section, correlates with BIT-159 at Mount Moulton, which has an age between 118.8 ± 0.8 and 106.0 ± 0.4 ka (Fig. 22b–d). All of these correlations are consistent with the apparent stratigraphic order. Sample locations can be seen plotted onto the map included in the Supplementary material S2. A trachytic tephra layer (W59) found in blue ice at Mount Waesche has been directly dated using $^{40}Ar/^{39}Ar$, and yielded an age of 107 ± 4 ka (Iverson 2017). BIT-222, which is stratigraphically below W59, correlates with BIT-143. BIT-143 is found immediately adjacent to BIT-144, which may be as old as 118.8 ± 0.8 ka. Furthermore, a directly measured $^{40}Ar/^{39}Ar$ age on the Mount-Waesche-derived tephra layer, the 'Great Wall', of 92 ± 26 ka (Panter 1995) provides another consistent age in the same complicated stratigraphic section. So, the ages are consistent with stratigraphic order. The presence of ice with an age as old as 118 ka, with intercalated tephra layers, could potentially offer a valuable climate record. However, the extreme deformation of the tephra–ice section at Mount Waesche would make such a record difficult to interpret.

Ice cores

A strong complement to the tephra records available at rare, but extremely valuable, blue ice areas are the tephra records that can be sampled in deep ice cores from West or East Antarctica. In areas of the world characterized by thick ice sheets and active volcanism, such as West Antarctica, ice may contain a remarkably complete record of local and distant explosive volcanism. Because ice is a non-silicate, and meltable, material, a record of even relatively small local explosive eruptions may be preserved in the form of sparse, small glass shards, whereas such a record could be challenging to extract from a silicate sedimentary matrix. Arguably, the ice-bound tephra record can provide a more detailed record of the geochemical signature of volcanism than would be available on most volcanoes in non-glaciated regions because once the tephra is incorporated into the ice, the low temperatures inhibit hydration of the volcanic glass (Friedman and Long 1976) and associated geochemical alteration (Cerling et al. 1985). Therefore, the glass geochemistry of tephra particles found in ice should faithfully represent the geochemistry of the erupted material.

In terms of understanding recent volcanism in a given region, the ice-core record provides an unparalleled level of detail of explosive volcanism but offers no insight into effusive volcanism, which can typically only be investigated on the source volcano itself. And, as discussed earlier, a complete record of volcanism on West Antarctic source volcanoes is difficult to determine because of local glacierization.

Tephra layers have been detected in a number of deep Antarctic ice cores (Gow and Williamson 1971; Kyle and Jezek 1978; Kyle et al. 1981; Palais et al. 1988; Basile et al. 2001; Dunbar et al. 2003; Narcisi et al. 2005, 2006, 2010a, 2012; Kurbatov et al. 2006; Dunbar and Kurbatov 2011; Iverson 2017; Iverson et al. 2017; Narcisi and Petit 2021). Early research on tephra layers in Antarctic ice cores was hampered by poor knowledge of source eruptions and a lack of quantitative geochemical data to use in making statistical correlations between tephra in different ice cores. Success in tying Antarctic tephra layers to source eruptions has improved in recent years, as a result of a growing body of high-quality analyses (Basile et al. 2001; Dunbar et al. 2003; Narcisi et al. 2005, 2006, 2010b, 2012; Kurbatov et al. 2006; Dunbar and Kurbatov 2011) and broader knowledge of the geochemistry of explosive Antarctic eruptions (Dunbar et al. 2008; Wilch et al. 1999a). A recurring theme in the above work is that West Antarctic volcanoes, particularly Mount Berlin, are the major sources for tephra layers found in both West and East Antarctica, although tephra from East Antarctic volcanoes, such as Mount Melbourne, are also found in West Antarctica.

Four West Antarctic ice cores will be reviewed here. These include the Byrd Core, drilled in 1968, the Siple Dome Core, drilled between 1997 and 1999, the WAIS Divide Core, drilled between 2007 and 2011, and the Roosevelt Island (RICE) Core drilled in 2013. All of these cores contain records of volcanism in the form of visible tephra layers, or the more difficult-to-recognize cryptotephra, which are tephra layers not visibly perceptible but recognizable using other techniques, such as dust logging, particle analysis or association with sulfate peaks in the ice (Dunbar and Kurbatov 2011; Dunbar et al. 2017). The level to which tephra layers can be recognized, sampled and analysed has improved through time. In the Byrd Core, tephra layers were recognized by visual inspection of the ice (Gow and Williamson 1971). Also, microbeam methods at the time that glassy tephra particles in the Byrd Core were analysed (Palais et al. 1984; Palais et al. 1990) were less advanced than those currently available, so analyses produced were less precise than those carried out on tephra particles in the more recently drilled ice cores. Recognition of tephra-bearing horizons in the Siple Dome and WAIS Divide ice cores was greatly assisted by the advent of a downhole logging tool that shines light into the ice and measures the light return (Bay et al. 2001). Areas of higher returns, or very low return in the case of thick tephra layers that absorb all light, correspond to areas of ice with greater amounts of particles, allowing many invisible tephra horizons to be recognized. For the WAIS Divide ice core, particle measurement during continuous melting of the ice core, described in Sigl et al. (2014), also helped to pinpoint additional tephra layers. So, the completeness of the tephra records, and quality of the geochemical data, are related to when the ice core was drilled.

Byrd Core

The Byrd Core, drilled in 1968 to a depth of 2191 m, is located at 80.02° S and 119.52° W (Fig. 1). This core contains 25 'ash bands' and an estimated 2000 'dusty bands', which were originally interpreted to be shear bands but later thought to be tephra-bearing ice (Gow and Williamson 1971). A tephra layer located at 788 m depth in the Byrd Core was geochemically tied to the eruption of West Antarctic stratovolcano Mount Takahe, at 8.2 ka (Palais et al. 1988). Data presented by Gow and Williamson (1971) and reproduced in Figure 23 indicate that few tephra layers are apparent in the upper 1200 m of the core, corresponding to an age of <18 ka. However, between 1400 m (c. 20 ka) and around 1800 m (c. 30 ka) depth in the core, a large number of tephra-bearing horizons are present (Palais et al. 1988). Quantifying the number of tephra layers in the Byrd Core is challenging because not all were geochemically analysed, so some may represent reworked volcanic material rather than primary volcanic

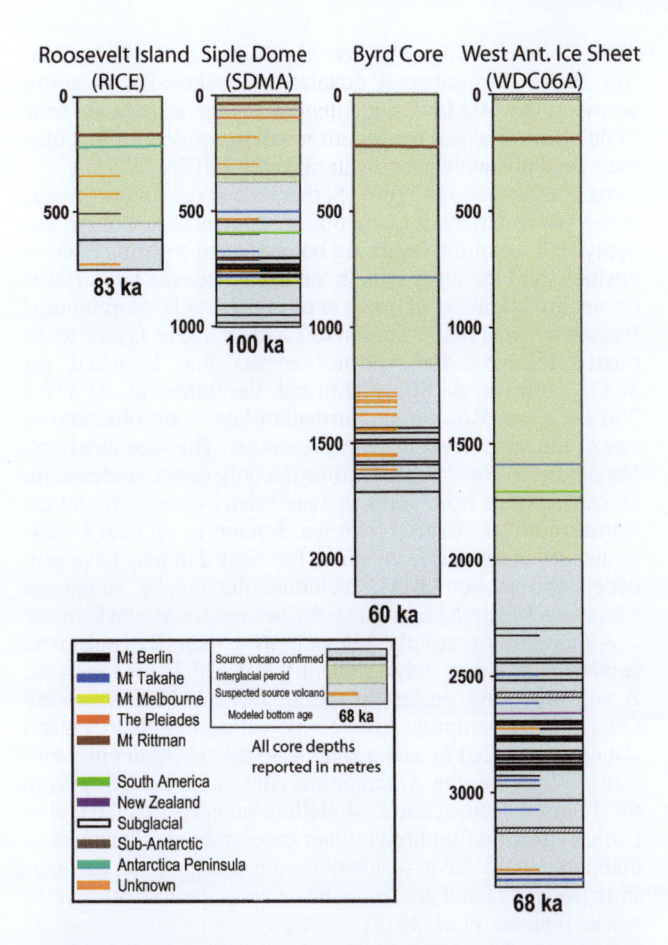

Fig. 23. Representation of ice cores showing horizons identified as tephra layers (after Iverson 2017). Layers are colour coded to represent the source volcano, if one has been identified. The age ranges for the interglacial periods are shown in tan shading.

A total of 36 tephra layers, some visible and some crypto-tephra, were recognized in the deep Siple Dome ice core (Dunbar et al. 2003; Kurbatov et al. 2006; Dunbar and Kurbatov 2011) (Fig. 23). As reported in those papers, most of these layers are derived from explosive volcanism in West Antarctica, with Mount Berlin being a major contributor to the ice-bound tephra record. Twelve of the layers can be statistically correlated to recognized volcanic eruptions based on the geochemical composition of glass shards, and most correlate with known eruptions of the two major West Antarctic stratovolcanoes, Mount Berlin and Mount Takahe. A tephra layer found at a depth of between 503.58 and 503.87 m, corresponding to an age of 8.166 ka before 1950, geochemically correlates to a Mount Takahe explosive eruption at 8.2 ± 5.4 ka, and which is also recognized in the Byrd ice core (Wilch et al. 1999a). A tephra layer at 19.3 ka (SDMA 9058) is geochemically consistent with a Mount Takahe source. Very few tephra layers from non-Antarctic-continent volcanoes were observed but those that were observed were suggested to be derived from South America or sub-Antarctic islands, although no direct ties to known eruptions were made (Dunbar and Kurbatov 2011). In addition to the 12 layers that can be directly tied to a specific eruption, a number of others can be qualitatively correlated to known source volcanoes; so although they do not provide age control points in the SDM-A core, they may provide time-stratigraphic markers that may be recognized in other ice cores. The tephra layers appear to be very widespread across West Antarctica, and have been recognized in East Antarctica and the marine record as well (Hillenbrand et al. 2008; Dunbar and Kurbatov 2011).

Tephra layers in the SDM-A core suggest that Antarctic volcanic activity, mainly from Mount Berlin, was ongoing throughout the roughly 100 kyr history recorded by the core ice, and provide a framework for the timing and frequency of contributions of tephra and sulfate signals to the Antarctic ice sheets by local Antarctic volcanism.

As was observed in the Byrd ice core, the frequency of local volcanism preserved in the SDM-A core does not appear to be uniform through time (Fig. 23). A period of notably higher abundance of tephra layers, and therefore volcanism, occurs between approximately 700–800 m core depth, corresponding to a time interval of between about 35 and 18 ka (Brook et al. 2005). Tephra layers in this time interval from the SDM-A core are derived from Mount Berlin (Dunbar and Kurbatov 2011), suggesting that this volcano underwent a period of high activity during this time interval.

WAIS Divide Core

A deep ice core, drilled at the inland WAIS Divide site (79.28° S, 112.05° W: Fig. 1) and formally called WDC06A, was drilled between the years 2007 and 2011, reaching a final depth of 3405 m, making it the deepest ice core collected in West Antarctica. For simplicity, we will here call the core 'WAIS Divide'. The location of this core, at an area of high accumulation, was chosen in order to provide a high-resolution climate record, and ice at the base of the core is estimated to have an age of 68 ka. The core contains thousands of 'dusty' horizons, recognized through visual observation and downhole optical logging (Fig. 24), and the concentration of these horizons is not uniform throughout the core. Fifty of the sampled dusty horizons were determined to be tephra (Fig. 15c–f), and a number of others are windblown debris which typically consist largely of volcanic material but without a recognizably consistent geochemical composition. There are certainly other primary tephra layers also present but because so many silicate horizons are observed, not all can be geochemically analysed in order to determine whether

layers. Fourteen tephra layers in the Byrd ice core were initially linked to eruptions of Mount Takahe (Kyle et al. 1981). However, later, with the benefit of more geochemical information from source volcanoes, many of these layers were suggested to be derived from Mount Berlin volcano (Wilch et al. 1999a). This interpretation was borne out by geochemical information from tephra layers from the Mount Moulton blue ice field, almost all of which are thought to be from Mount Berlin, and are geochemically more consistent with the tephra composition from the Byrd ice core than are tephra layers from Mount Takahe (Wilch et al. 1999a; Dunbar et al. 2008). Therefore, although one tephra layer in the Byrd ice core is certainly derived from Mount Takahe, most of the tephra layers record extensive volcanic activity from Mount Berlin between about 24 and 18 ka.

Siple Dome Core

Siple Dome is an ice dome located near the coast (81.65° S, 148.81° W) on the WAIS (Fig. 1). Drilling of the deep core occurred between 1996 and 1999. The scientific objectives of this drilling were to obtain a better understanding of the palaeoclimatology and glaciology of West Antarctica (Taylor et al. 2004a), and the deep Siple Dome ice core (SDM-A) reached bedrock at a depth of 1004 m. Depth–age relationships for this core relied on studying annual variability in visual, chemical and electrical properties of the cores (Taylor and Alley 2004; Taylor et al. 2004a, b; Brook et al. 2005).

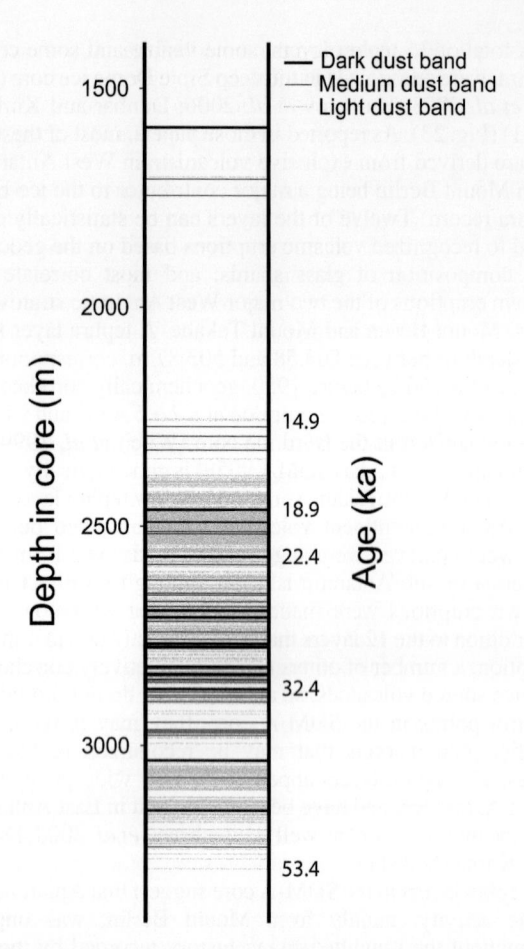

Fig. 24. Stratigraphic column depicting all recognized silicate layers in the WAIS Divide ice core, determined from notes taken by core handlers during core processing. Ages of selected layers are shown, and are determined using timescales presented in Fudge et al. (2013) and Buizert et al. (2015).

or not each exhibits the chemical coherence that identifies the silicate material as a true tephra layer. Although there are abundant dusty layers observed between c. 50 and 15 ka, the strongest concentration is observed between around 35–19 ka, where 29 tephra layers have been identified, and others are also certainly present (Figs 23 & 24) (Iverson 2017).

Many of the tephra layers in the WAIS Divide ice core are trachytic in composition (Iverson 2017), and are found to be geochemically consistent with derivation from Mount Berlin (Iverson 2017). In addition to tephra layers linked to Mount Berlin (Table 1), several tephra layers are geochemically consistent with derivation from Mount Takahe. Notably, a layer at a depth of 1589.187 m (Table 1, 8077 years before 1950) correlates with the Mount Takahe tephra layer recognized in the Byrd ice core (Palais et al. 1988), found also in the Siple Dome ice core (Kurbatov et al. 2006; Dunbar and Kurbatov 2011), and is also similar to a tephra found at the Mount Waesche blue ice site (Table 1, samples BIT-175 and BIT-221). A halogen-rich interval of ice in the WAIS core, spanning 192 years (between 17.556 and 17.748 ka) is interpreted to be related to an extended eruption from Mount Takahe (McConnell et al. 2017), which these authors suggest may have caused ozone depletion and subsequent deglaciation, beginning at 17.7 ka. Tephra from another Mount Takahe eruptive sequence is also recognized in the WAIS Divide ice core at c. 62 ka (Iverson 2017). The geochemical composition of the two tephra layers of this age from Mount Takahe are presented in Table 1 (WDC06A-3373.61 and WDC06A-3376), and are recognizable as Mount Takahe tephra because of their relatively high MgO content (Wilch et al. 1999a; Iverson 2017). A number of detailed geochemical correlations between the WAIS Divide tephra layers and those from Siple Dome core and the Mount Moulton and Mount Waesche blue ice fields have been made (Iverson 2017).

As can be seen in Figure 18, there are several tephra layers in the WAIS Divide ice core that are not trachytic but are less evolved. Two of the layers are basanitic and tephritic in composition, and the other falls in the trachyandesite field. Based on an investigation of grain size and particle morphology, Iverson et al. (2017) interpreted two of these layers to be related to subglacial volcanic events that breached the WAIS, one at 44 865 ± 313 and the other at 22 337 ± 290 cal. years BP. No similar tephra layers are observed in any of the other West Antarctic ice cores. The records of volcanism in ice cores is likely to be the only direct evidence for volcanic events from vents that are briefly emergent, and are subsequently re-covered with ice. Iverson et al. (2017) suggested several possible locations for vents that may have produced two tephra layers, including the nearby subglacial volcanoes Mount Resnik, Mount Thiel and/or Mount Casertz.

Although not strictly relevant to Antarctic volcanism, another significant tephra find in the WAIS Divide ice core is volcanic glass related to the 25 580 ± 258 cal. years BP Oruanui super-eruption from Taupo volcano in New Zealand (Dunbar et al. 2017). Glass shards as large as 30 μm in diameter are found in the Antarctic ice core, 5000 km away from the Taupo volcanic centre. Modelling suggests that the tephra particles from an umbrella cloud thousands of kilometres in diameter could have resulted in the transport of particles from New Zealand to Antarctica over a period of days to weeks (Dunbar et al. 2017).

Roosevelt Island Climate Evolution (RICE) core

The RICE core was drilled in 2013 on Roosevelt Island (79.4° S, 161.7° W) to a depth of 764 m. The location of Roosevelt Island, within the Ross Ice Shelf, makes it a good location to better reconstruct the Antarctic/Southern Ocean climate conditions over the past 70 kyr. Tephra in this core was identified using visual examination of the ice (Kalteyer 2015). Tephra layers were recognized in the RICE core (Kalteyer 2015) and these are shown in Figure 23. Somewhat surprisingly, of the eight tephra layers recognized in this core, only one (found at a depth of 622.89 m) appears to be derived from a West Antarctic volcano. This layer correlates to a tephra layer in the Siple Dome ice core, and is thought to be derived from Mount Berlin (Kalteyer 2015). Other tephra layers in the RICE core are interpreted to be from East Antarctica or sub-Antarctic islands, and a number have been correlated to tephra layers found in other ice cores, including ones at South Pole, Vostok, EPICA-Dome C, Taylor Dome, Talos Dome, Siple Dome and WAIS Divide (Kalteyer 2015). This suggests that the RICE core, geographically closer to East Antarctica than the other cores discussed here, is better placed for deposition of tephra from East Antarctic volcanism than that from West Antarctica. However, another factor may be that because tephra were only identified visually, that other, unrecognized, tephra layers from West Antarctic volcanoes may also be present in RICE ice.

Overview and discussion of recent West Antarctic volcanism

The combination of studies on source volcanoes, along with blue ice areas and ice cores, provides insight into recent volcanism in West Antarctica. Mount Berlin and Mount Takahe both could be considered active, and Mount Waesche certainly exhibits well-preserved cinder cones and young (<200 ka) lava flows on its flanks, as well as mafic pyroclastic deposits preserved in the local blue ice field. On balance, integrating the eruptive records on source volcanoes with those in ice, recent West Antarctic volcanism would appear to be largely explosive but this interpretation may be skewed by the excellent preservation of pyroclastic volcanic material in the WAIS, as well as the possible obscuring of young, effusive eruptions on the glacierized flanks of the three volcanoes mentioned above. Also, we cannot rule out that there may have been subglacial volcanic events in West Antarctica, either that remained completely subglacial, or that breached the ice sheet and produced local pyroclastic deposits that were since buried, and for which there is no longer any visible evidence at the ice sheet surface (Corr and Vaughan 2008; van Wyk de Vries et al. 2017).

Based on observations from blue ice areas and ice cores, Mount Takahe and Mount Berlin have both produced numerous explosive eruptions, generating widespread tephra deposits, over the past 100 kyr. However, the eruptive patterns of the two volcanoes over this time period appear to be significantly different. Mount Takahe appears to have produced a large eruption that distributed tephra widely across West Antarctica at around 8.2 ka, deposits of which are found at the source volcano, in ice cores drilled at Byrd Station, Siple Dome and WAIS Divide ice core sites (Palais et al. 1988; Kurbatov et al. 2006; Dunbar et al. 2007; Iverson 2017), and probably at Mount Waesche. This eruption was large enough to produce a visible tephra layer in a blue ice area and in ice cores, so is likely to have changed the albedo of much of the WAIS for some period of time following the eruption. One other tephra-generating event at <100 ka was recognized at Mount Takahe itself (at 93.3 ± 7.9 ka), and a couple of other tephra layers with chemical compositions consistent with derivation from Mount Takahe have also been recognized in the Mount Waesche blue ice field (Table 1) but the ages are difficult to constrain. So, in summary, the total number of recognized explosive events sourced from Mount Takahe is quite small.

In contrast to the relatively sparse explosive volcanism from Mount Takahe, Mount Berlin appears to have produced a very large number of volcanic events over the past 100 kyr. Gow and Williamson (1971) were able to recognize over 2000 individual ashy dust bands in the Byrd Core, and over 1000 individual dusty bands were noted during optical examination of the WAIS Divide ice core (Fig. 24). The thickest and most visible silicate layers typically have characteristics of primary tephra layers, which include a population of uniformly sized, unabraded and compositionally homogeneous glass shards (Fig. 15c–f; see also Supplementary material S1). However, we would question whether all of the observed dust bands, particularly the finer ones, actually correspond with primary volcanic eruptions. Some of the layers that we have analysed in both the Siple Dome and WAIS Divide ice cores contain glassy shards but do not reveal a geochemically uniform composition. These layers are interpreted to be wind-blown volcanic debris, rather than deposits formed by a volcanic eruption. Investigating all of the more than 1000 individual dusty horizons in the WAIS Divide ice core would be prohibitively time-consuming but we suggest that many, in both the Byrd and WAIS Divide ice cores, are likely to be reworked tephra, rather than primary volcanic deposits. The large number of dusty bands in portions of the Byrd and WAIS Divide ice cores suggests that there was abundant fine volcanic dust available for transport and redeposition during times, as will be discussed below, when Mount Berlin was producing many explosive eruptions.

Despite the interpretation that not all of the silicate layers in the Byrd and WAIS Divide ice cores represent primary fall deposits, volcanism from Mount Berlin appears to have been episodic, rather than evenly spaced through time, over the past 100 kyr. As can been seen in Figures 23 and 24, the number of silicate layers is dramatically higher in the ice aged between about 32 and 18 ka. The geochemistry of analysed tephra layers in this time interval from the Siple Dome and Byrd cores indicates that many of these silicate layers are primary tephra derived from Mount Berlin (Wilch et al. 1999a; Dunbar and Kurbatov 2011), and tephra in this age range from the WAIS Divide Core appear to also be largely derived from Mount Berlin as well (Iverson 2017). Based on this information, and the ice-core chronology, Mount Berlin appears to have produced eruptions roughly every 2 kyr during this time interval, although a number of sets of eruptions are clustered within a few hundred years of each other. This eruptive pattern can be observed in tephra from the Siple Dome ice core (Dunbar and Kurbatov 2011), and is also evident in similar data from the WAIS Divide ice core (Iverson 2017). Looking at a longer timescale, the blue ice tephra record at Mount Moulton indicates that Mount Berlin has been producing tephra-forming eruptions for at least the last 500 kyr (Dunbar et al. 2008) but the episodicity of eruptions observed in the ice core records is not apparent in the longer record, probably because of the lower resolution of the blue ice record compared to that of the ice cores.

Integrating the excellent Mount Berlin tephra record contained in ice cores and the blue ice field at Mount Moulton provides a level of detail of eruptive history for a single volcano that is rarely available. Not only are tephra from many individual pyroclastic eruptions available, in faithful chronological order, in ice cores and the blue ice area but, because of preservation in the ice, the glass shards are perfectly fresh and have not been subject to hydration, which is typically associated with chemical mobility in the glass (Cerling et al. 1985). Geochemical analyses of glass fragments from a large number of individual eruptions have been carried out and average compositions of erupted material, shown in chronological order, are presented in Figure 25. While elements K and Ti appear nearly invariant over the entire 500 kyr time span, Fe and S show a systematic increase with younger age. This pattern was observed in the geochemistry of the Mount Berlin tephra layers found at the Mount Moulton englacial tephra site, and was interpreted by Dunbar et al. (2008) to be related to small batches of partial melt being progressively removed from a single source under Mount Berlin. The geochemical composition of Mount-Berlin-derived tephra layers found in the WAIS Divide ice core, representing more tephra layers in a narrower time range, reinforces the trends observed in the longer Mount Moulton record, so does not require a change in the interpretation presented by Dunbar et al. (2008). The addition of the geochemical compositions of the 27 Mount Berlin tephra layers found in the WAIS Divide ice core to the existing Mount Moulton tephra record provides an additional level of resolution in the geochemical trends through time, providing an excellent opportunity for more detailed chemical and isotopic investigations of the well-preserved volcanic record.

The presence or absence of tephra layers, particularly from Mount Berlin, in a number of ice cores and blue ice areas provides some insight into tephra dispersal patterns around Antarctica. First, the very large number of silicate layers, many likely to be Mount Berlin tephra layers in the Byrd and WAIS Divide ice cores, as well as tephra layers from Mount Berlin present at the Mount Moulton site, and at Mount Waesche, suggest that these sites are all in geographical areas strongly impacted by Mount Berlin tephra-producing eruptions. Mount Berlin tephra layers are also present in the

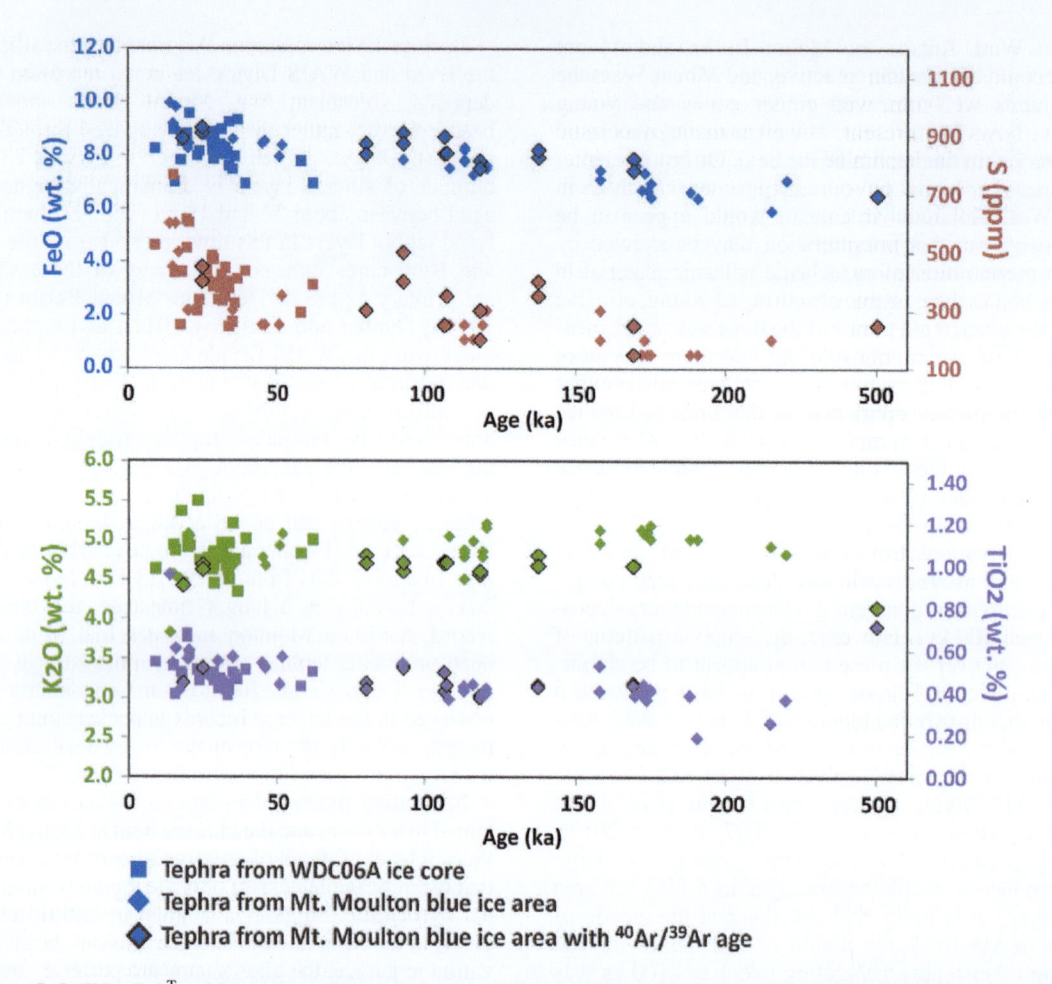

Fig. 25. Average CaO, TiO$_2$, FeO^T and SO$_2$ content of glass shards from the Mount Berlin tephra layers, as analysed by electron microprobe, organized in order of approximate age. Square symbols represent samples collected from the WAIS Divide ice core, and ages are determined using timescale WDC2014. Diamond symbols are samples collected at the Mount Moulton blue ice site. Symbols with a heavy outlines represent samples directly dated using ^{40}Ar/^{39}Ar geochronological analysis (Iverson 2017).

Siple Dome ice core but fewer than appear in the two more inland cores. Interestingly, no Mount Berlin or Mount Takahe tephra layers were found in the RICE core, which, because of its location, would be more influenced by marine air masses than the other core sites mentioned above (Bertler *et al.* 2018). The distribution pattern suggests that the dispersal of tephra from the coastal Mount Berlin was in a clockwise direction and likely to have been controlled by a high-level westerly wind flow around Antarctica, described with a focus on dust transport (Basile *et al.* 2001; Neff and Bertler 2015). This westerly (clockwise) circulation was important in transporting tephra from the Oruanui eruption in New Zealand to West Antarctica, as can be seen particularly in the supplementary material video presented by Dunbar *et al.* (2017). Furthermore, surface katabatic winds are of an appropriate direction to move tephra from Mount Berlin towards the Byrd, WAIS Divide or Siple Dome sites (Parish and Bromwich 2007).

Several East Antarctic ice cores, and one blue ice area, contain tephra layers derived from Mount Berlin, showing that eruptions from West Antarctica have the potential to impact the entire Antarctic continent. Basile *et al.* (2001) reported a West Antarctic tephra layer in the Vostok ice core at a depth of 1996 m, so roughly 141 kyr ago. The geochemistry of this layer is consistent with derivation from Mount Berlin. Kyle *et al.* (1981) reported finding a tephra layer from Mount Takahe in the Dome C ice core. A tephra layer from the Talos Dome ice core (TD1387b-ii) has been geochemically linked to a Mount Berlin eruption (BIT-160), recently redated by ^{40}Ar/^{39}Ar to 118.8 ± 0.8 ka (Iverson 2017). A widespread Mount Berlin tephra (BIT-157), recently redated by ^{40}Ar/^{39}Ar to 92.3 ± 0.4 ka (Iverson 2017), has been identified in the EPICA-Dome C and Dome Fuji ice cores, and therefore may be widely present in East Antarctica (Narcisi *et al.* 2006). Another, deeper, visible tephra layer in EPICA-Dome C (depth of 2631.6 m, age of 358 ka) is also geochemically similar to Mount Berlin tephra but no direct correlation was made (Narcisi *et al.* 2010*b*). Finally, a visible, but faint, tephra layer found in a blue ice site (Terra Nova Saddle, EIT-007) on Ross Island is geochemically indistinguishable from Mount Berlin tephra found at Mount Moulton (BIT-152) and in the WAIS Divide and Siple Dome ice cores (Iverson 2017). As discussed in Iverson (2017), the chronological match between the layers is somewhat complicated but even if the Terra Nova Saddle layer does not correlate to BIT-152, the geochemical match to Mount Berlin more generally is very robust.

In addition to impacting the Antarctic continent, explosive volcanism from West Antarctica has also been recognized in the marine sediment record. Three tephra layers (called A, B and C) were recognized in 28 marine sediment cores collected on the West Antarctic continental margin, mostly near the Antarctic Peninsula (Hillenbrand *et al.* 2008). The youngest of the three tephra layers chemically correlates to a widespread Mount Berlin tephra (BIT-157), recently redated by ^{40}Ar/^{39}Ar to 92.3 ± 0.4 ka (Iverson 2017). As mentioned above, this tephra layer is recognized in the EPICA Dome C and Dome Fuji ice cores. The other two tephra layers are recognized to be geochemically similar to tephra erupted from Mount Berlin, or possibly Mount Takahe, but no direct correlations are offered. These layers contain glass shards with an average diameter of around 100 µm, with individual shards of more than 200 µm in diameter present. The average and maximum shard sizes in the marine tephra layers are significantly larger than almost all shards that have been found in the ice cores (Dunbar *et al.* 2003; Dunbar and Kurbatov 2011; Iverson 2017). Although it is not possible to make 1:1 correlations between grain size of the tephra layers in the marine sediment cores and an equivalent in an ice core, the overall greater coarseness of the marine tephra layers could suggest that the main trajectory of tephra transport from Mount Berlin or Mount Takahe may be more towards the Antarctic Peninsula than onto the continent. Tephra deposited on the WAIS may come from the edges of the tephra cloud, where tephra particles would be expected to be finer, rather than from the central axis of the cloud.

Finally, moving from local to possible global impacts of West Antarctic eruptions, Bay *et al.* (2006) identified a correlation between Antarctic volcanism and abrupt climate changes. These authors suggested several possible mechanisms by which the volcanic eruption and climate change could be linked but do not present a preferred interpretation. However, they noted that the strong correlation they observed is unlikely to be simply coincidental.

Summary

Three volcanoes in West Antarctica have been active in the past 100 kyr and have the potential to produce eruptions again in the future. These are Mount Berlin, Mount Takahe and Mount Waesche; and based on examination of past activity, Mount Berlin and Mount Takahe could generate ash-producing pyroclastic eruptions that could impact much of the Antarctic continent but would have a particularly large impact on West Antarctica, where deposition of volcanic ash could change the albedo of the WAIS and impact air-support operations at nearby field stations. Although volcanic ash could also be deposited in East Antarctica by such eruptions, the quantity and grain size of tephra would be vanishingly small, and therefore would be likely to have little impact on either ice-sheet or human activity. In contrast, an eruption from Mount Waesche would be most likely to form a mafic cinder cone and/or lava flow. An eruption of this style would impact the local environment around Mount Waesche but would be unlikely to have far-reaching effects.

Examination of eruptive activity in West Antarctica based on ice-core investigation also indicates that, over the past 60 kyr, there have been at least two instances of volcanic eruptions that began subglacially and melted through the WAIS to eventually produce subaerially erupted tephra deposits (Iverson *et al.* 2017). Magma movement within the West Antarctic crust has also been detected, using geophysical methods (Lough *et al.* 2013). Therefore, in addition to considering the potential impact of a volcanic eruption from one of the three known volcanoes mentioned above, any assessment of West Antarctic volcanic hazards would also need to take into an account a potential subglacial eruption, which could further destabilize an already unstable WAIS, resulting in global sea-level rise.

Acknowledgements We would like to thank Dr Alessio Di Roberto along with an anonymous reviewer, for providing valuable suggestions that significantly improved the manuscript. Thanks also to Dr Nick Pearce for assistance in trace element analyses of glass presented in this work. We would like to thank the National Science Foundation, Antarctic Support Associates and Raytheon Polar Services also for logistical support while in Antarctica, as well as the Air National Guard and Ken Borek Air for air transportation to and from the field sites from which results are presented here. Dr Kendrick Taylor led the field effort that collected the two ice cores (SDM A and WAIS Divide) from which tephra samples were analysed by authors of this chapter. The National Ice Core Laboratory curated the core. Dr Andrei Kurbatov has also provided valuable collaboration for many aspects of this work.

Author contributions NWD: conceptualization (lead), data curation (equal), funding acquisition (equal), investigation (lead), methodology (equal), project administration (lead), writing – original draft (lead); **NAI**: conceptualization (equal), data curation (equal), investigation (equal), methodology (equal), project administration (supporting), writing – review & editing (supporting); **JLS**: data curation (supporting), funding acquisition (supporting), investigation (supporting), writing – original draft (supporting), writing – review & editing (lead); **WCM**: conceptualization (supporting), data curation (supporting), formal analysis (supporting), funding acquisition (equal), investigation (equal), methodology (supporting), writing – review & editing (supporting); **MJZ**: data curation (supporting), formal analysis (supporting), investigation (supporting), methodology (supporting); **PRK**: conceptualization (supporting), data curation (supporting), funding acquisition (supporting), investigation (supporting), methodology (supporting).

Funding This work was funded by National Science Foundation, with grants to W.C. McIntosh (NSF-OPP-9725910), N.W. Dunbar (NSF-OPP-9814428, NSF-OPP-9615167, NSF-OPP-0230348 and NSF-ANT-1142115) and M.J. Zimmerer (NSF-ANT-1745015). The National Science Foundation Division of Polar Programs also funded the Ice Drilling Program Office (IDPO) and Ice Drilling Design and Operations (IDDO) group for coring activities.

Data availability All data generated or analysed during this study are included in this published article (and its supplementary information files).

References

Ackert, R.P., Putnam, A.E., Mukhopadhyay, S., Pollard, D., DeConto, R.M., Kurz, M.D. and Borns, H.W. 2013. Controls on interior West Antarctic ice sheet elevations; inferences from geologic constraints and ice sheet modeling. *Quaternary Science Reviews*, **65**, 26–38, https://doi.org/10.1016/j.quascirev.2012.12.017

Basile, I., Petit, J.R., Touron, S., Grousset, F.E. and Barkov, N. 2001. Volcanic layers in Antarctic (Vostok) ice cores: Source identification and atmospheric implications. *Journal of Geophysical Research: Atmospheres*, **106**, 31 915–31 931, https://doi.org/10.1029/2000JD000102

Bay, R.C., Price, P.B., Clow, G.D. and Gow, A.J. 2001. Climate logging with a new rapid optical technique at Siple Dome. *Geophysical Research Letters*, **28**, 4635–4638, https://doi.org/10.1029/2001GL013763

Bay, R.C., Bramall, N.E., Price, P.B., Clow, G.D., Hawley, R.L., Udisti, R. and Castellano, E. 2006. Globally synchronous ice core volcanic tracers and abrupt cooling during the last glacial period. *Journal of Geophysical Research: Atmospheres*, **111**, D11108, https://doi.org/10.1029/2005jd006306

Bertler, N.A.N., Conway, H. *et al.* 2018. The Ross Sea Dipole – temperature, snow accumulation and sea ice variability in the Ross Sea region, Antarctica, over the past 2700 years. *Climate of the Past*, **14**, 193–214, https://doi.org/10.5194/cp-14-193-2018

Brook, E.J., White, J.W.C. *et al.* 2005. Timing of millennial-scale climate change at Siple Dome, West Antarctica, during the last glacial period. *Quaternary Science Reviews*, **24**, 1333–1343, https://doi.org/10.1016/j.quascirev.2005.02.002

Buizert, C., Adrian, B. *et al.* 2015. Precise interpolar phasing of abrupt climate change during the last ice age. *Nature (London)*, **520**, 661–665, https://doi.org/10.1038/nature14401

Cashman, K.V., Sturtevant, B., Papale, P. and Navon, O. 2000. *Magmatic Fragmentation*. Academic Press, San Diego, CA.

Cassidy, W.A., Olsen, E. and Yanai, K. 1977. Antarctica: a deep freeze storehouse for meteorites. *Science*, **198**, 727–731, https://doi.org/10.1126/science.198.4318.727

Cerling, T.E., Brown, F.H. and Bowman, J.R. 1985. Low temperature alteration of volcanic glass: Hydration, Na, K, ^{18}O, and Ar mobility. *Chemical Geology*, **52**, 281–293.

Cole-Dai, J., Mosley-Thompson, E. and Qin, D.H. 1999. Evidence of the 1991 Pinatubo volcanic eruption in South Polar snow. *Chinese Science Bulletin*, **44**, 756–760, https://doi.org/10.1007/BF02909720

Corr, H.F.G. and Vaughan, D.G. 2008. A recent volcanic eruption beneath the West Antarctic ice sheet. *Nature Geoscience*, **1**, 122–125, https://doi.org/10.1038/ngeo106

Davies, S.M. 2015. Cryptotephras: the revolution in correlation and precision dating. *Journal of Quaternary Science*, **30**, 114–130, https://doi.org/10.1002/jqs.2766

DeRoberto, A., Del Carlo, P. and Pompilio, M. 2021. Marine record of Antarctic volcanism from drill cores. *Geological Society, London, Memoirs*, **55**, https://doi.org/10.1144/M55-2018-49

Dunbar, N.W. and Kurbatov, A.V. 2011. Tephrochronology of the Siple Dome ice core, West Antarctica: correlations and sources. *Quaternary Science Reviews*, **30**, 1602–1614, https://doi.org/10.1016/j.quascirev.2011.03.015

Dunbar, N., Zielinski, G. and Voisins, D. 2003. Tephra layers in the Siple Dome and Taylor Dome ice cores, Antarctica: Sources and correlations. *Journal of Geophysical Research*, **108**, 2374–2385, https://doi.org/10.1029/2002JB002056

Dunbar, N.W., McIntosh, W.C., Kurbatov, A.V. and Wilch, T.I. 2007. Integrated tephrochronology of the West Antarctic region; implications for a potential tephra record in the West Antarctic ice sheet (WAIS) Divide ice core (peer reviewed extended abstract). *United States Geological Survey Open-File Report*, **2007-1047**, Extended Abstract 179.

Dunbar, N.W., McIntosh, W.C. and Esser, R.P. 2008. Physical setting and tephrochronology of the summit caldera ice record at Mount Moulton, West Antarctica. *Geological Society of America Bulletin*, **120**, 796–812, https://doi.org/10.1130/b26140.1

Dunbar, N.W., Iverson, N.A. *et al.* 2017. New Zealand supereruption provides time marker for the Last Glacial Maximum in Antarctica. *Nature Scientific Reports*, **7**, 12238, https://doi.org/10.1038/s41598-017-11758-0

Fretzdorff, S. and Smellie, J.L. 2002. Electron microprobe characterization of ash layers in sediments from the central Bransfield basin (Antarctic Peninsula): evidence for at least two volcanic sources. *Antarctic Science*, **14**, 412–421, https://doi.org/10.1017/s0954102002000214

Friedman, I. and Long, W. 1976. Hydration rate of obsidian. *Science*, **191**, 347–352, https://doi.org/10.1126/science.191.4225.347

Fudge, T.J., Steig, E.J. *et al.* 2013. Onset of deglacial warming in West Antarctica driven by local orbital forcing. *Nature*, **500**, 440–444, https://doi.org/10.1038/nature12376

Gonzales-Ferran, O. 1995. *Volcanes de Chile*. Instituto Geografico Militar, Santiago.

Gow, A.J. and Williamson, T. 1971. Volcanic ash in the Antarctic ice sheet and its possible climatic implications. *Earth and Planetary Science Letters*, **13**, 210–218, https://doi.org/10.1016/0012-821X(71)90126-9

Hartman, L.H., Kurbatov, A.V. *et al.* 2019. Volcanic glass properties from 1459 CE volcanic event in South Pole ice core dismiss Kuwae caldera as a potential source. *Scientific Reports*, **9**, https://doi.org/10.1038/s41598-019-50939-x

Hillenbrand, C.D., Moreton, S.G. *et al.* 2008. Volcanic time-markers for Marine Isotopic Stages 6 and 5 in Southern Ocean sediments and Antarctic ice cores: implications for tephra correlations between palaeoclimatic records. *Quaternary Science Reviews*, **27**, 518–540, https://doi.org/10.1016/j.quascirev.2007.11.009

Iverson, N.A. 2017. *Antarctica's Englacial Tephra Record: Characterizing and Integrating Tephra from Blue Ice Areas and Ice Cores*. PhD thesis, New Mexico Institute of Mining and Technology, Socorro, New Mexico, USA.

Iverson, N.A., Lieb-Lappen, R., Dunbar, N.W., Obbard, R., Kim, E. and Golden, E. 2017. The first physical evidence of subglacial volcanism under the West Antarctic Ice Sheet. *Nature Scientific Reports*, **7**, 11457, https://doi.org/10.1038/s41598-017-11515-3

Kalteyer, D. 2015. *Tephra in Antarctic Ice Cores*. MS thesis, University of Maine, Orono, Maine, USA.

Koffman, B.G., Dowd, E.G. *et al.* 2017. Rapid transport of ash and sulfate from the 2011 Puyehue–Cordón Caulle (Chile) eruption to West Antarctica. *Journal of Geophysical Research: Atmospheres*, **122**, 8908–8920, https://doi.org/10.1002/2017JD026893

Kurbatov, A.V., Zielinski, G.A., Dunbar, N.W., Mayewski, P.A., Meyerson, E.A., Sneed, S.B. and Taylor, K.C. 2006. A 12 000 year record of explosive volcanism in the Siple Dome Ice Core, West Antarctica. *Journal of Geophysical Research: Atmospheres*, **111**, D12307, https://doi.org/10.1029/2005jd006072

Kyle, P.R. and Jezek, P.A. 1978. Compositions of three tephra layers from the Byrd Station ice core, Antarctica. *Journal of Volcanology and Geothermal Research*, **4**, 225–232, https://doi.org/10.1016/0377-0273(78)90014-8

Kyle, P.R., Jezek, P.A., Mosley-Thompson, E. and Thompson, L.G. 1981. Tephra layers in the Byrd Station ice core, Antarctica, and their climatic importance. *Journal of Volcanology and Geothermal Research*, **11**, 29–39, https://doi.org/10.1016/0377-0273(81)90073-1

LeBas, M.J., LeMaitre, R.W., Streckeisen, A. and Zanettin, B. 1986. A chemical classification of volcanic rocks based on the total alkali silica diagram. *Journal of Petrology*, **27**, 745–750, https://doi.org/10.1093/petrology/27.3.745

Le Maitre, R.W. 1989. *A Classication of Igneous Rocks and Glossary of Terms*. Blackwell, Oxford, UK.

LeMasurier, W.E. 1990a. Mount Sidley. *American Geophysical Union Antarctic Research Series*, **48**, 203–207.

LeMasurier, W.E. 1990b. Marie Byrd Land. *American Geophysical Union Antarctic Research Series*, **48**, 146–163.

LeMasurier, W.E. and Kawachi, Y. 1990. Mount Waesche. *American Geophysical Union Antarctic Research Series*, **48**, 208–211.

LeMasurier, W.E. and Rex, D.C. 1982. Volcanic record of Cenozoic glacial history in Marie Byrd Land and western Ellsworth Land: Revised chronology and evalution of tectonic factors. *In*: Craddock, C. (eds) *Antarctic Geoscience*. University of Wisconsin Press, Madison, WI, 725–734.

LeMasurier, W.E. and Rex, D.C. 1990. Mount Siple. *American Geophysical Union Antarctic Research Series*, **48**, 185–188.

LeMasurier, W.E. and Wade, F.A. 1968. Fumarolic activity in Marie Byrd Land. *Science*, **162**, 352, https://doi.org/10.1126/science.162.3851.352

Lemieux-Dudon, B., Blayo, E. *et al.* 2010. Consistent dating for Antarctic and Greenland ice cores. *Quaternary Science Reviews*, **29**, 8–20, https://doi.org/10.1016/j.quascirev.2009.11.010

Lough, A.C., Wiens, D.A. *et al.* 2013. Seismic detection of an active subglacial magmatic complex in Marie Byrd Land, Antarctica. *Nature Geoscience*, **6**, 1031–1035, https://doi.org/10.1038/ngeo1992

Lythe, M.B., Vaughan, D.G. and Consortium, B. 2001. BEDMAP: A new ice thickness and subglacial topographic model of Antarctica. *Journal of Geophysical Research: Solid Earth*, **106**, 11 335–11 351, https://doi.org/10.1029/2000jb900449

McConnell, J.R., Burke, A. *et al.* 2017. Synchronous volcanic eruptions and abrupt climate change similar to 17.7 ka plausibly linked by stratospheric ozone depletion. *Proceedings of the National Academy of Sciences of the United States of America*, **114**, 10 035–10 040, https://doi.org/10.1073/pnas.1705595114

McIntosh, W.C., LeMasurier, W.E., Ellerman, P.J. and Dunbar, N.W. 1985. A re-interpretation of glacio-volcanic interaction at Mount

Takahe and Mount Murphy, Marie Byrd Land, Antarctica. *Antarctic Journal of the United States*, **20**, 57–59.

Moreton, S.G. and Smellie, J.L. 1998. Identification and correlation of distal tephra layers in deep-sea sediment cores, Scotia Sea, Antarctica. *Annals of Glaciology*, **27**, 285–289, https://doi.org/10.3189/1998AoG27-1-285-289

Narcisi, B. 2008. Geochemical features of tephra layers in the EPICA-Dome C climatic record (East Antarctica). *Terra Antartica Reports*, **14**, 107–110.

Narcisi, B. and Petit, J.R. 2021. Englacial tephras of East Antarctica. *Geological Society, London, Memoirs*, **55**, https://doi.org/10.1144/M55-2018-86

Narcisi, B., Petit, J.R., Delmonte, B., Basile-Doelsch, I. and Maggi, V. 2005. Characteristics and sources of tephra layers in the EPICA-Dome C ice record (East Antarctica); implications for past atmospheric circulation and ice core stratigraphic correlations. *Earth and Planetary Science Letters*, **239**, 253–265, https://doi.org/10.1016/j.epsl.2005.09.005

Narcisi, B., Petit, J.R. and Tiepolo, M. 2006. A volcanic marker (92 ka) for dating deep east Antarctic ice cores. *Quaternary Science Reviews*, **25**, 2682–2687, https://doi.org/10.1016/j.quascirev.2006.07.009

Narcisi, B., Petit, J.R. and Chappellaz, J. 2010*a*. A 70 ka record of explosive eruptions from the TALDICE ice core (Talos Dome, East Antarctic Plateau). *Journal of Quaternary Science*, **25**, 844–849, https://doi.org/10.1002/jqs.1427

Narcisi, B., Petit, J.R. and Delmonte, B. 2010*b*. Extended East Antarctic ice-core tephrostratigraphy. *Quaternary Science Reviews*, **29**, 21–27, https://doi.org/10.1016/j.quascirev.2009.07.009

Narcisi, B., Petit, J.R., Delmonte, B., Scarchilli, C. and Stenni, B. 2012. A 16 000-yr tephra framework for the Antarctic ice sheet: a contribution from the new Talos Dome core. *Quaternary Science Reviews*, **49**, 52–63, https://doi.org/10.1016/j.quascirev.2012.06.011

Narcisi, B., Petit, J.R., Langone, A. and Stenni, B. 2016. A new Eemian record of Antarctic tephra layers retrieved from the Talos Dome ice core (Northern Victoria Land). *Global and Planetary Change*, **137**, 69–78, https://doi.org/10.1016/j.gloplacha.2015.12.016

Neff, P.D. and Bertler, N.A.N. 2015. Trajectory modeling of modern dust transport to the Southern Ocean and Antarctica. *Journal of Geophysical Research: Atmospheres*, **120**, 9303–9322, https://doi.org/10.1002/2015jd023304

Palais, J.M., Kyle, P.R. and Delmas, R. 1984. Detailed studies of tephra layers in the Byrd Station ice core; preliminary results and interpretation. *Antarctic Journal of the United States*, **18**, 109–110.

Palais, J.M., Kyle, P.R., McIntosh, W.C. and Seward, D. 1988. Magmatic and phreatomagmatic volcanic activity at Mt. Takahe, West Antarctica, based on tephra layers in the Byrd Ice Core and field observations at Mt. Takahe. *Journal of Volcanology and Geothermal Research*, **35**, 295–317, https://doi.org/10.1016/0377-0273(88)90025-X

Palais, J.M., Kirchner, S. and Delmas, R. 1990. Identification of some global volcanic horizons by major element analysis of fine ash in Antarctic ice. *Annals of Glaciology*, **14**, 216–220, https://doi.org/10.3189/S0260305500008612

Panter, K.S. 1995. *Geology, Geochemistry and Petrogenesis of the Mount Sidley Volcano, Marie Byrd Land, Antarctica*. PhD thesis, New Mexico Institute of Mining and Technology, Socorro, New Mexico, USA.

Panter, K.S., Wilch, T.I., Smellie, J.L., Kyle, P.R. and McIntosh, W.C. 2021. Marie Byrd Land and Ellsworth Land: petrology. *Geological Society, London, Memoirs*, **55**, https://doi.org/10.1144/M55-2019-50

Parish, T.R. and Bromwich, D.H. 2007. Reexamination of the near-surface airflow over the Antarctic continent and implications on atmospheric circulations at high southern latitudes. *Monthly Weather Review*, **135**, 1961–1973, https://doi.org/10.1175/mwr3374.1

Popp, T.J., Sowers, T., Dunbar, N.W., McIntosh, W.C. and White, J.W. 2004. Radioisotopically dated climate record spanning the last interglacial in ice from Mount Moulton, West Antarctica. *Eos, Transactions of the American Geophysical Union*, **85**, U31A-0015.

Quartini, E., Blankenship, D.D. and Young, D.A. 2021. Active subglacial volcanism in West Antarctica. *Geological Society, London, Memoirs*, **55**, https://doi.org/10.1144/M55-2019-3

Sigl, M., McConnell, J.R. *et al.* 2014. Insights from Antarctica on volcanic forcing during the Common Era. *Nature Climate Change*, **4**, 693–697, https://doi.org/10.1038/nclimate2293

Simkin, T. and Siebert, L. 2000. Earth's volcanoes and eruptions; An overview. *In*: Sigurdsson, H., Houghton, B.F., McNutt, S.R., Rymer, H. and Stix, J. (eds) *The Encyclopedia of Volcanoes*. 2nd edn. Academic Press, San Diego, CA, 239–255, https://doi.org/10.1016/B978-0-12-385938-9.00012-2

Smellie, J.L. 1999. The upper Cenozoic tephra record in the south polar region; a review. *Global and Planetary Change*, **21**, 51–70, https://doi.org/10.1016/S0921-8181(99)00007-7

Sumner, J.M., Grunder, A.L., Blake, S., Matela, R.J., Wolff, J.A. and Russell, J.K. 2005. Spatter. *Journal of Volcanology and Geothermal Research*, **142**, 49–65, https://doi.org/10.1016/j.jvolgeores.2004.10.013

Taylor, K.C. and Alley, R.B. 2004. Two-dimensional electrical stratigraphy of the Siple Dome (Antarctica) ice core. *Journal of Glaciology*, **50**, 231–235, https://doi.org/10.3189/172756504781830033

Taylor, K.C., Alley, R.B. *et al.* 2004*a*. Dating the Siple Dome (Antarctica) ice core by manual and computer interpretation of annual layering. *Journal of Glaciology*, **50**, 453–461, https://doi.org/10.3189/172756504781829864

Taylor, K.C., White, J.W.C. *et al.* 2004*b*. Abrupt climate change around 22 ka on the Siple Coast of Antarctica. *Quaternary Science Reviews*, **23**, 7–15, https://doi.org/10.1016/j.quascirev.2003.09.004

Van Eaton, A.R., Mastin, L.G., Herzog, M., Schwaiger, H.F., Schneider, D.J., Wallace, K.L. and Clarke, A.B. 2015. Hail formation triggers rapid ash aggregation in volcanic plumes. *Nature Communications*, **6**, 7860, https://doi.org/10.1038/ncomms 8860

van Wyk de Vries, M., Bingham, R.G. and Hein, A.S. 2017. A new volcanic province: an inventory of subglacial volcanoes in West Antarctica. *Geological Society, London, Special Publications*, **461**, 231–248, https://doi.org/10.1144/SP461.7

Vespermann, D. and Schmincke, H.-U. 2000. Scoria cones and tuff rings. *In*: Sigurdsson, H., Houghton, B.F., McNutt, S.R., Rymer, H. and Stix, J. (eds) *Encyclopedia of volcanoes*. 2nd edn. Academic Press, San Diego, CA, 683–694.

Wilch, T.I. 1997. *Volcanic Record of the West Antarctic Ice Sheet in Marie Byrd Land*. PhD thesis, New Mexico Institute of Mining and Technology, Socorro, New Mexico, USA.

Wilch, T.I., McIntosh, W.C. and Dunbar, N.W. 1999*a*. Late Quaternary volcanic activity in Marie Byrd Land: Potential ^{40}Ar/^{39}Ar dated time horizons in West Antarctic ice and marine cores. *Geological Society of America Bulletin*, **111**, 1563–1580, https://doi.org/10.1130/0016-7606(1999)111<1563:LQVAIM>2.3.CO;2

Wilch, T.I., McIntosh, W.C., Dunbar, N.W., McCuddy, S.M. and Wagg, M.R. 1999*b*. Higher inland ice levels of the West Antarctic Ice Sheet at Mt. Takahe volcano during Early and late Wisconsinan times. Presented at the Sixth Annual West Antarctic Ice Sheet Workshop, 15–18 September 1999, Sterling, Virginia, USA.

Chapter 7.5

Active subglacial volcanism in West Antarctica

Enrica Quartini[1,2], Donald D. Blankenship[1] and Duncan A. Young[1]*

[1]University of Texas Institute for Geophysics, Jackson School of Geosciences, University of Texas at Austin, Austin, TX 78712, USA

[2]Department of Geological Sciences, Jackson School of Geosciences, University of Texas at Austin, Austin, TX 78712, USA

EQ, 0000-0001-7485-543X; DAY, 0000-0002-6866-8176

Present addresses: EQ, School of Earth and Atmospheric Sciences, Georgia Institute of Technology, Atlanta, GA 30318, USA

*Correspondence: duncan@ig.utexas.edu

Abstract: A combination of aerogeophysics, seismic observations and direct observation from ice cores, and subglacial sampling, has revealed at least 21 sites under the West Antarctic Ice Sheet consistent with active volcanism (where active is defined as volcanism that has interacted with the current manifestation of the West Antarctic Ice Sheet). Coverage of these datasets is heterogeneous, potentially biasing the apparent distribution of these features. Also, the products of volcanic activity under thinner ice characterized by relatively fast flow are more prone to erosion and removal by the ice sheet, and therefore potentially under-represented. Unsurprisingly, the sites of active subglacial volcanism that we have identified often overlap with areas of relatively thick ice and slow ice surface flow, both of which are critical conditions for the preservation of volcanic records. Overall, we find the majority of active subglacial volcanic sites in West Antarctica concentrate strongly along the crustal-thickness gradients bounding the central West Antarctic Rift System, complemented by intra-rift sites associated with the Amundsen Sea–Siple Coast lithospheric transition.

Supplementary material: A list of volcanic sites described in this chapter is available at https://doi.org/10.6084/m9.figshare.c.5226102

Preserved within the depths of the Antarctic ice sheet lies the record of past volcanic eruptions. This record gives us the unique capacity to understand the periodicity of volcanic eruptions at a continental scale, something rarely achieved in other terrestrial environments due to record loss as a result of weathering and limited availability of written documentation.

Importantly, elevated geothermal flux associated with active volcanism can generate basal melt water that contributes to ice-sheet instability by lubricating the base of the ice sheet and causing ice streams to speed up (Pittard *et al.* 2016). Basal water can facilitate sliding at the base of the ice in the area surrounding a heat anomaly and can also sustain subglacial hydraulic systems that lubricate the ice sheet further downstream (Engelhardt 2004b; Vogel and Tulaczyk 2006). Therefore, the effects of geothermal flux on ice-sheet dynamics are not confined to the localized area of elevated heat flux. Through a combination of concentrated and distributed subglacial water systems, melt water produced in regions of high geothermal flux is transported to distal areas (Blankenship *et al.* 1993). Thus, geothermal flux has the potential to indirectly enhance ice flow and associated mass loss in coastal regions through the supply of upstream-produced basal melt water.

In West Antarctica, the West Antarctic Ice Sheet (WAIS) is pinned to a number of subaerial volcanic edifices and subglacial mountain chains distributed throughout the region but destabilized by the deep topographical trough of the West Antarctic Rift System (WARS: Dalziel and Lawver 2001). The WARS is a region of thinned, low-lying continental crust mostly below sea level characterized by local mantle thermal anomalies (Winberry and Anandakrishnan 2004; Chaput *et al.* 2014; Lloyd *et al.* 2015) and locally elevated geothermal flux (Clow *et al.* 2012; Schroeder *et al.* 2014; Fisher *et al.* 2015). WARS first extended in the Cretaceous (Behrendt *et al.* 1991), with subsidence and excavation through the Cenozoic (Wilson and Luyendyk 2009), and subsequent local Neogene topographical enhancement (LeMasurier 2008). Given the importance of geothermal heat flow and water to the stability of the WAIS, the potential threat that volcanism poses motivated extensive efforts to characterize subglacial active volcanism in West Antarctica (Blankenship *et al.* 1993, 2001; Schroeder *et al.* 2014).

Two aerogeophysical indicators have been used to argue for the existence of significant subglacial volcanism in the WARS: a large number of high-frequency (and thus shallow) circular magnetic anomalies observed from airborne magnetic data (Behrendt 1964, 2013; Behrendt *et al.* 1996), and the identification of numerous 'cones' in bed elevation interpolations (van Wyk de Vries *et al.* 2018) derived from airborne radar sounding data (Fretwell *et al.* 2013). However, given the lack of direct sampling, understanding the interactions between active volcanism and the overlying ice sheet is key for understanding if these features are active.

In this chapter we discuss the occurrence of active subglacial volcanism in the context of the WAIS' glaciological regimes, and WARS' geological and topographical structure. We show evidence that active volcanism beneath the WAIS occurs along crustal boundaries and regions of rifted, thinned crust, where local stress regimes allow for the upwelling of magma and hydrothermal fluids (e.g. Maccaferri *et al.* 2014). We also show evidence that active volcanism concentrates around regions of the WAIS that experience large fluctuations in ice thickness during glacial and interglacial cycles. There, changes in ice load can cause decompression melting and magma production in the underlying mantle, similar to what has been proposed for Greenland (Stevens *et al.* 2016).

We concentrate on subglacial active volcanism in West Antarctica due to the abundance of observations in this region. However, the approach to subglacial volcanism identification outlined for West Antarctica can be applied generally to other passive margins in Antarctica to begin a systematic search for similarly active subglacial volcanoes.

WARS geological context for subglacial volcanism

The lithospheric structure of West Antarctica determines key topographical, morphological and thermal basal boundary conditions for the stability of the WAIS. Moreover, the lithosphere's influence on topography and structure may provide insight into the distribution of volcanism, which can occur where stress regimes facilitate the formation and emplacement of shallow-crustal magma chambers. Paulsen and Wilson (2010) examined these factors for West Antarctica's subaerial volcanoes. They found that the age distribution of these volcanoes supports a connection between ice loading and unloading and periods of emplacement; however, the evolution of the distribution of subaerial volcanoes in the highlands was controlled by changes in plate tectonic stresses.

Ice thickness and airborne gravity data reveal that West Antarctica consists of the WARS and five discrete microcontinental blocks characterized by distinct subglacial bed topography, crustal thickness and gravity anomalies (Dalziel and Lawver 2001; Diehl 2008): the Ellsworth–Whitmore Mountains (EWM), Antarctic Peninsula (AP), Thurston Island (TI), eastern Marie Byrd Land (eMBL) and western Marie Byrd Land (wMBL) (Fig. 1b).

Crustal block boundaries are interpreted from gravity anomalies. Bouguer gravity anomalies (which are corrected for the gravity signal of topography) primarily respond to the gravity signal from the crust–mantle interface, and are thus a proxy for crustal thickness. Sharp changes in Bouguer gravity indicate distinct crustal boundaries (Fig. 1b). Gradual changes in the Bouguer anomalies from positive to negative values indicate broad zones of crustal change and are interpreted as partially rifted transitional crust between crustal blocks and the rift zone (Blankenship et al. 2001; Diehl 2008) (Fig. 1b).

Within West Antarctica, the WARS and Marie Byrd Land (MBL, composed of eMBL and wMBL: see Fig. 1b) are two regions identified as volcanically active. The WARS is a region of low bedrock topography and stretched continental crust (Fretwell et al. 2013; Chaput et al. 2014) characterized by a heterogeneous distribution of high geothermal flux (Schroeder et al. 2014; Fisher et al. 2015). Due to the low topography, large portions of central WAIS reside below sea level (up to −2500 m), a highly unstable configuration susceptible to marine ice-sheet instability, which can lead to runaway grounding-line retreat and ice-sheet collapse (Weertman 1974; Schoof 2007).

On the other hand, MBL is an elevated plateau of high bed topography and thicker crust (LeMasurier 2006; Fretwell et al. 2013; Chaput et al. 2014) sustained by a Cenozoic hotspot (LeMasurier and Rex 1989) in the lower mantle beneath MBL, recently imaged by seismic tomography (Accardo et al. 2014; Hansen et al. 2014; An et al. 2015; Emry et al. 2015; Lloyd et al. 2015; Heeszel et al. 2016). Analysis of magnetic anomalies in eMBL shows evidence for multiple stages of tectonic reactivation in the mid-Cretaceous and supports the hypothesis of a hotspot emplacement in the Miocene (Quartini 2018).

Behrendt (1964) found from International Geophysical Year era reconnaissance survey flights that northern West Antarctica is characterized by many high-amplitude, but narrow, magnetic anomalies, consistent with shallow volcanic rocks under the ice. The resolution increased markedly with the 5 km line spacing Corridor Aerogeophysics of the South

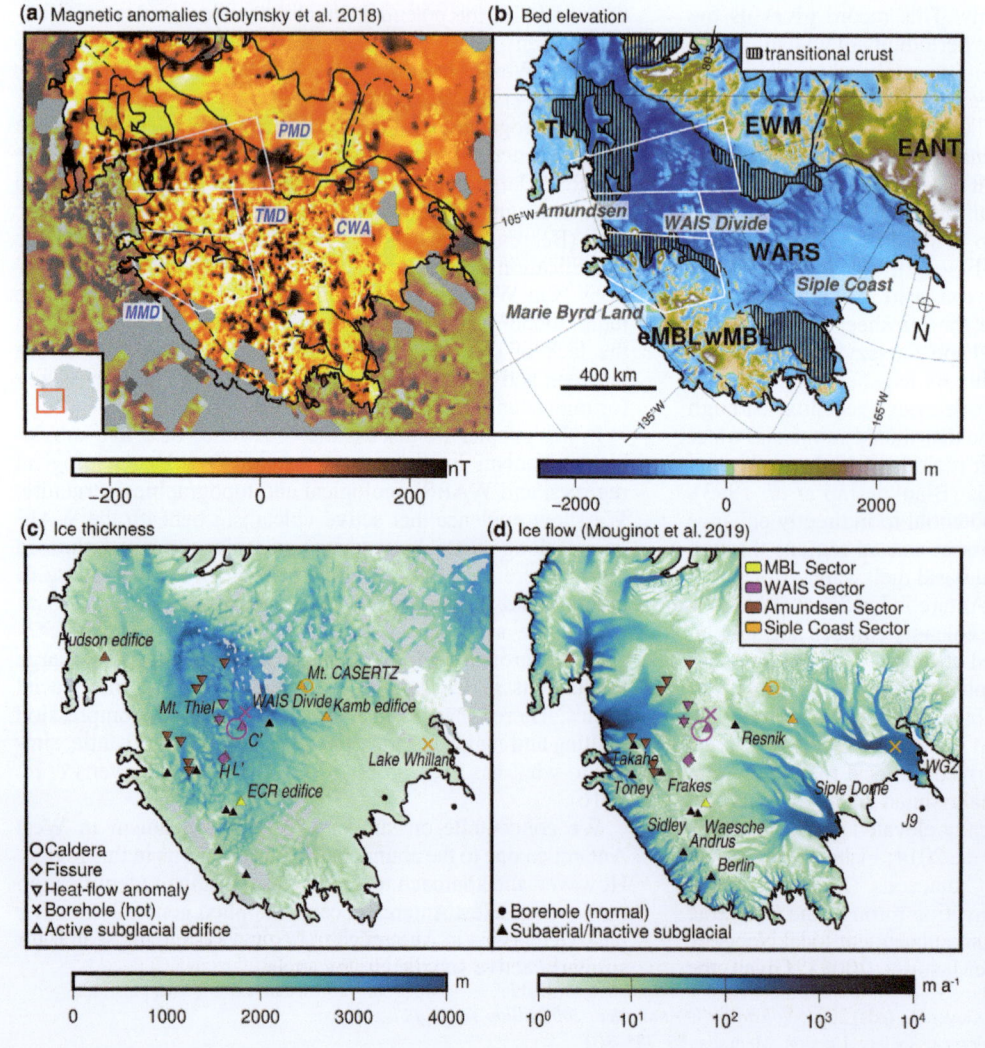

Fig. 1. (a) ADMAP2 magnetic anomalies (Golynsky et al. 2018). White lines outline boundaries between the Pine Island magnetic district (PMD), Thwaites Glacier magnetic district (TMD) and the MBL magnetic district (MMD) (Quartini 2018). Central West Antarctica (CWA) is also shown. (b) Bedrock elevation combining data from Vaughan et al. (2006), Leuschen and Allen (2011), Blankenship et al. (2012) and Young et al. (2017). Thick black lines outline distinct crustal block margins; thin vertical lines indicate transitional crust (Diehl 2008). TI, Thurston Island; EWM, Ellsworth–Whitmore Mountains; eMBL, eastern Marie Byrd Land; wMBL, western Marie Byrd Land; EANT, East Antarctica; WARS, West Antarctic Rift System. Major sectors are shown. Distribution of volcanic features over (c) ice thickness combining data from Vaughan et al. (2006), Leuschen and Allen (2011), Blankenship et al. (2012) and Young et al. (2017); and (d) ice-surface velocity (Mouginot et al. 2019).

Eastern Ross Transect Zone (CASERTZ) surveys of the 1990s (Blankenship *et al.* 2001). The CASERTZ data allowed Behrendt *et al.* (1996) to identify more than 400 circular, small-scale, high-amplitude magnetic anomalies in Central West Antarctica (Fig. 1a), interpreted as small-scale intrusive bodies. In addition, Behrendt *et al.* (1998) found evidence for large annular structures.

The Airborne Geophysical Survey of the Amundsen Embayment (AGASEA) project (Holt *et al.* 2006; Vaughan *et al.* 2006) extended this magnetics coverage north with a 15 km grid. Using these data, Quartini (2018) found a contrasting suite of fabrics reflecting different lithospheric histories. In the WARS, different magnetic fabrics characterize the region adjacent to Pine Island Glacier (the Pine Island Magnetic District or PMD) and the centre of Thwaites Glacier (the Thwaites Magnetic District or TMD). In PMD lies a region of very strong, large, annular anomalies, while in TMD a set of linear magnetic anomalies exists, paralleling similar features identified offshore in the Amundsen Sea and attributed to the mid-Cretaceous break-up of Antarctica (Gohl *et al.* 2013) (Fig. 1a). The circular small-scale magnetic anomalies of Central West Antarctica are not apparent. On the other hand, under eMBL lies a region with suppressed magnetic fabric, with small-scale, low-amplitude magnetic anomalies (the MBL Magnetic District or MMD). The MMD and the TMD are separated by a continuous front of magnetic anomalies that corresponds well with the crustal boundary mapped from gravity by Diehl *et al.* (2008) (Fig. 1a, b). Quartini's (2018) interpretation was that the TMD represented older lithosphere not modified since the Cretaceous surrounded by younger rift and hotspot activity.

Vogel *et al.* (2006) found from dating subglacial samples along the Siple Coast that while there was evidence for petrologically mafic magmatism in West Antarctic, most of the dates obtained were Mesozoic ages, compatible with the primary rifting event. While this result appears to rule out pervasive contemporary volcanism, as suggested by some interpretations of the magnetic data (Behrendt *et al.* 1994), it did not rule out local active volcanism.

WAIS glaciological context for subglacial volcanism

WAIS' ice dynamics and conditions at the bed impact both the type of volcanic products and their record of preservation. Glaciovolcanism has been well studied for ice thicknesses of <800 m (Gudmundsson *et al.* 1997; Smellie 2000) in Iceland. The WAIS has a much larger range in ice thicknesses (and, hence, pressure: Fig. 1c) and ice flow speed (and, hence, erosional capacity: Behrendt *et al.* 1995) (Fig. 1d) relevant to understanding sites of subglacial volcanism. Given the importance of ice–magma interactions to subglacial volcanism, we describe both our approach for defining activity and our framework for the variable response of volcanism to differing ice conditions.

The thermodynamic response of ice to pressure and temperature changes, and the impact of melt water in subglacial environments introduce processes that extend the effect of volcanism both spatially and temporally. For this reason, we consider any subglacial volcanism that affects or has affected the current ice sheet as active. Based on the approximate minimum age of the WAIS' oldest ice retrieved at the WAIS Divide ice-core site (Buizert *et al.* 2015) and the cooling rates of magma chambers (Hawkesworth *et al.* 2000) for compositions observed in West Antarctica (LeMasurier and Rex 1989), we define the time interval of active subglacial volcanism in West Antarctica as the last *c.* 100 kyr.

The main effect of subglacial volcanism is the rapid conversion of ice to water. For very thin ice (<100 m), this process will rapidly breach the ice sheet, and the edifice will evolve to a subaerial style. For ice of the thicknesses seen in Iceland, an englacial vault of melt water can be generated under a depression in the ice surface that will act to trap subglacial water (Smellie 2000; Schmidt *et al.* 2011). Within the vault, the style of volcanism is subaqueous, forming a pile of hyloclastite material and pillow lavas (Gudmundsson *et al.* 1997). Eventually, with sustained activity, this vault can breach the surface, causing a direct transition from a subaqueous mode to a subaerial mode and leading, for basaltic compositions, to flat-topped formations termed 'tuya' or 'table mountains' (Smellie 2006).

At larger ice thicknesses, breaching becomes unlikely. In addition, the overburden pressure of the ice will force the collapse of englacial cavities through ice flow, although this process will be in tension with the hydraulic trapping of water by any surface depression (Fig. 2a). Rapid ice flow will often correspond to inclined surface slopes and significant hydraulic gradients that will dominate water flow, and which will serve to drain englacial vaults and suppress the formation of hyloclastite–pillow lava piles (Fig. 2b). Eruptions under thick ice generate volcanic products typical of high-pressure regimes such as lava sheets and pillow lava flows (Fig. 2c). On the other hand,

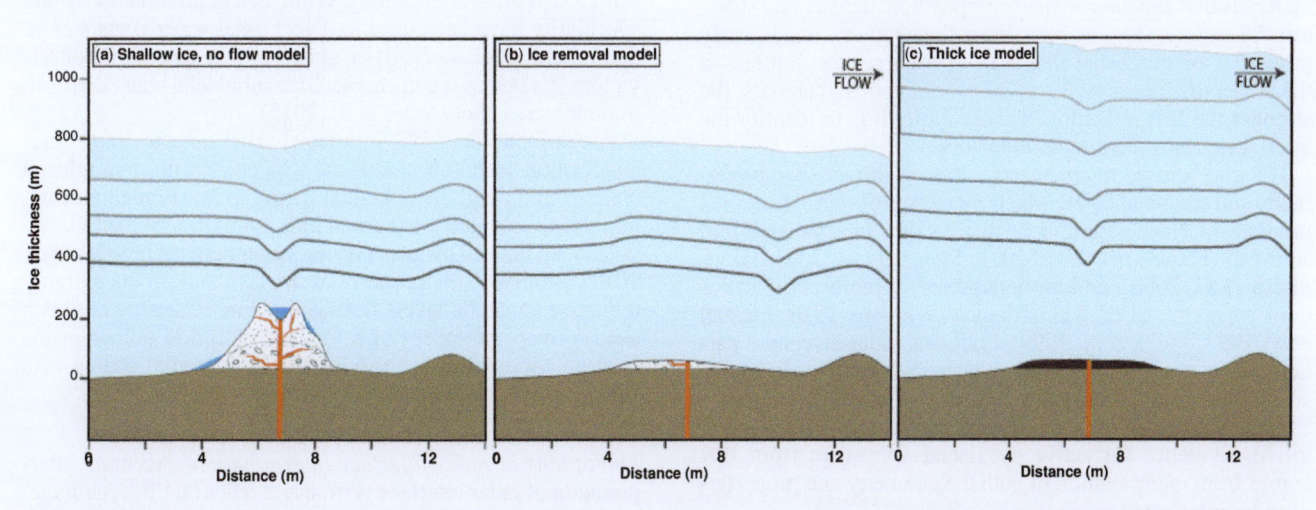

Fig. 2. (**a**) Shallow ice, no flow model: subglacial volcanic edifice produced under shallow ice conditions, composed of poorly consolidated hyaloclastite atop pillow lava (Blankenship *et al.* 1993; Gudmundsson *et al.* 1997) and preserved due to slow to no ice flow. (**b**) Shallow ice, fast-flow model: effect of ice removal due to fast ice flow on a hyloclastite and pillow lava edifice (Nereson *et al.* 1998). (**c**) Thick ice model: removal-resistant basaltic lava flow produced under thick ice sheets.

under shallower ice, volatiles in the magma expand, allowing for faster quenching and breaking up of magma into mixtures of pillow lavas and hyaloclastite deposits.

Regions of fast ice flow are more efficient at removing volcanic deposits at the bed, especially poorly consolidated material such as hyaloclastites (Behrendt et al. 1995). We therefore expect subglacial volcanic products to be more easily removed along tributaries and ice streams, and better preserved along ice divides (Fig. 2).

Techniques and observations

Direct observations of active subglacial volcanism are difficult to obtain due to the thick WAIS cover (>2500 m thick in places). First, thick ice sheets greatly attenuate the subglacial topographical signature that can be directly inferred from the ice surface. Secondly, the single-point nature of ice cores and boreholes make drilling efforts prohibitively expensive and overall inefficient at characterizing subglacial volcanic constructs.

In contrast, indirect observations from geophysical methods provide a more cost-effective way to characterize the internal structure and properties of both the ice sheet and the underlying bed topography. Geophysical surveys have the advantage of covering large areas at varying degrees of spatial resolutions. Some of these methods include, but are not limited to, radar sounding, laser altimetry, potential fields, GPS and passive seismic monitoring.

Identifying active subglacial volcanism requires multiple, independent supporting evidence. While individual techniques can be used to identify plausible subglacial volcano candidates, it is the joint observations across different techniques and measurements that allow one to fully characterize each potential site of active subglacial volcanism. Here we will review some of the most common direct and indirect methods used to investigate active subglacial volcanism in West Antarctica.

Ice-penetrating radar

Ice-penetrating radar (IPR) transmits electromagnetic waves that travel through a medium, and detects changes in the dielectric permittivity of different materials by reflecting and refracting parts of their energy along each interface (Gudmandsen 1971). The dielectric permittivity is a property of materials that depends on their composition, density and structure. Therefore, IPR can both detect the geometry and identify properties of subglacial interfaces. The larger the contrast in the dielectric permittivity of two contiguous materials the stronger the IPR reflection, and the easier it is to identify the interfaces, also called radar reflectors.

IPR can be used to characterize the ice surface, bed topography and englacial layers which are caused by ash and aerosol products of volcanic eruptions that become incorporated into snowfall (Fahnestock et al. 2001; Siegert et al. 2004; Hindmarsh et al. 2006), and are interpreted to represent isochronous events recorded within the ice (Whillans 1976; Jacobel et al. 1993; Fujita et al. 1999). Englacial radar layers are particularly useful as they allow geographical extension of the composition and age–depth record measured in ice cores over hundreds of square kilometres (e.g. Cavitte et al. 2016). Evidence for active subglacial volcanism from IPR comes from interpretation of both the geometry and properties of radar interfaces.

Geometry of radar interfaces. Radar reflectors constrain the shape of the ice surface, bed topography and englacial layers, and can provide evidence for the presence of active subglacial volcanoes in a number of ways. The subglacial bed topography can reveal volcanic constructs based on analysis of the morphology and aspect ratio of the topographical relief (van Wyk de Vries et al. 2018). The internal structure of the ice sheet, revealed by tracing englacial layers, provides a record of changes in surface accumulation rates, ice-flow patterns and basal melt rates in the region. Basal melt rates have been constrained from radar layer drawdown (Fahnestock et al. 2001). In areas of known or negligible basal friction, excessive basal melt production and layer drawdown formations can be indicative of anomalously high subglacial geothermal flux and volcanic activity (Fig. 3c). Excessive basal melt and internal deformation in the ice column associated with active subglacial volcanism can also create a depression in the ice surface sometimes called ice cauldrons (Gudmundsson 1996; Björnsson 2003; Gudmundsson and Högnadóttir 2007), which can be detected by IPR and higher-precision laser altimeters, and can thus be used to infer areas of active subglacial volcanism (Blankenship et al. 1993).

Properties of radar interfaces. The most common properties of interfaces that can be extrapolated from IPR echoes are the reflectivity, roughness and specularity content. While reflectivity is a measure of the strength of a radar reflector, which is proportional to the contrast in dielectric constants between two adjacent materials, roughness and specularity are a measure of the angular spread of IPR echo return energy. Specularity content refers to the ratio of specular over diffuse radar echo return energy, where specular returns identify mirror-like surfaces that reflect waves along the same incident path, and diffuse surfaces scatter the incident wave across a wider range of angles. Water is highly specular, while rock and sediments behave more like diffuse surfaces (Schroeder et al. 2016). Specularity content can be thought of as small (wavelength)-scale roughness, or roughness calculated over a shorter evaluation length. One advantage of specularity content is its insensitivity to attenuation of the radar signal through the ice column, which is the biggest source of uncertainty in reflectivity analyses. The joint interpretation of radar interface character can therefore elucidate the properties and structure of the bounding media.

Liquid water has a much higher dielectric permittivity compared to rock, sediment and ice. Therefore, water at the ice–bed interface can be detected as a strongly reflective, smooth and specular reflector relative to the surroundings (Gudmandsen 1971; Peters et al. 2005). While bed echo reflectivity and specularity have been used to detect basal water (Peters et al. 2005; Schroeder et al. 2014), anisotropy in specularity observations has been used to characterize subglacial water distribution networks (Schroeder et al. 2015).

A combination of IPR subglacial water detection and characterization techniques was used together with a subglacial water-routing and glaciological model to assess melt production due to elevated geothermal flux and map active subglacial volcanism in the Thwaites Glacier catchment (Schroeder et al. 2014). Similarly, the property of ash layers to produce highly reflective englacial layers through volume scattering has been used to map the extent of a volcanic eruption fallout in the Hudson Mountains and to identify the potential source of a subglacial volcanic eruption (Corr and Vaughan 2008).

Limitations in radar interface interpretation. Accurate interpretation of radar interface geometries relies on three main factors: the survey line spacing, the interpolation algorithm, and the radar acquisition system and processing technique. The spacing between survey lines impacts the final resolution of interpolation products such as digital elevation models

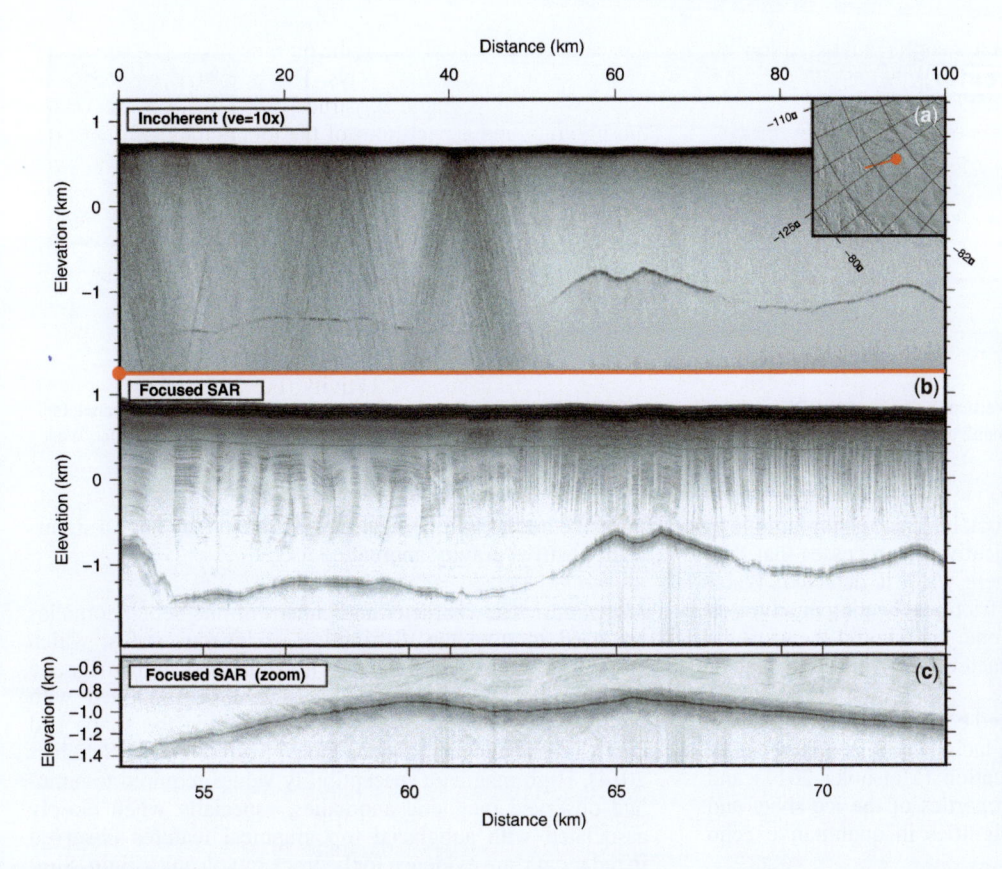

Fig. 3. Comparison between types of radar processing and products. (**a**) Depth-corrected incoherent radargram from CASERTZ (IRE/WCy/Y09a, flown in 1993: Blankenship *et al.* 2001) over the Kamb Ice Stream volcanic edifice (vertical exaggeration is ×10). (**b**) Depth-corrected 2D focused radargram (OND/SJB2/C1D01a) from a reflight on the same line with HiCARS in 2001 (Peters *et al.* 2007 (vertical exaggeration is ×10). (**c**) Zoom-in on the peak of the Kamb Ice Stream volcanic edifice, showing englacial reflectors non-conformable with the bed interface (vertical exaggeration is ×2.5). See Figure 6 for more discussion.

(DEMs). Line spacing is a function of the survey design, usually optimized for the specific science goals of each survey. Exploratory surveys maximize areal coverage at the expense of resolution, while the opposite is true for targeted surveys. Aerogeophysical surveys can have line spacing between a few and tens of kilometres depending on whether the survey was flown with a helicopter or fixed-wing aeroplane, respectively. Ground surveys can achieve the highest level of detail with line spacing that can vary between a few and hundreds of metres but are usually extremely target-specific due to the limit in range of this kind of operation.

The type of radar acquisition system utilized and the subsequent type of processing applied to radar data largely impact the observed geometry of radar interfaces (Fig. 3). The most accurate geometries are rendered by phase-preserving coherent radar acquisition systems (such as High Capability Radar Sounder (HiCARS) (Peters *et al.* 2005) and HiCARS2 (Young *et al.* 2015)), combined with focusing techniques that trace the energy of reflectors to their locations or origin in a radar profile (e.g. Peters *et al.* 2007) (Fig. 3a, b). This processing removes hyperbolae tails in the radar echoes that would otherwise mask subglacial features such as mountains with slopes similar to or greater than the apparent slopes of the hyperbolae tails. The interpolation algorithm applied to generate DEMs of the bed, surface and englacial layer elevation are subject to a number of trade-offs including computational costs, mesh grid sizes, accuracy of results and smoothing techniques. In a rough terrain, the resulting bedrock DEMs often represent smoothed estimates for mountain peaks and valley troughs.

Limitations in both data acquisition and processing techniques can therefore fundamentally impair geometries detected by IPR and all interpolated products derived from it, ultimately invalidating interpretations. One example is the subglacial volcano catalogue compiled by van Wyk de Vries *et al.* (2018), which is based on direct interpretation of the Bedmap2 bed morphology (Fretwell *et al.* 2013) to identify subglacial shield volcanoes. However, the large data acquisition gaps present at the time of Bedmap2 products release (e.g. in MBL) and the use of a smoothing interpolation spline generated a final bed topography DEM that deviates by hundreds of metres from the input IPR data values with significant induced errors in inferred bed slope (Fig. 4).

Without a systematic analysis of independent datasets supporting the volcanogenic nature of characteristic bedrock peaks (such as potential fields, seismic and surface-elevation data), the aspect ratio of erosional features, especially in sediment-rich regions, can easily lead to their misinterpretation as shield volcanoes. Many of the site locations proposed by van Wyk de Vries *et al.* (2018; corrected for a projection error from the coordinates originally reported in table 2 of that paper) lie either in the interfluves or on the centre line of tributaries of the Siple Coast ice streams and may instead be erosional features. Finally, Bedmap2 data from most of van Wyk de Vries *et al.* (2018) targets were constrained either by incoherent profiles collected in the 1990s (which will inherently smooth rough landscapes) or Scott Polar Research Institute (SPRI) point ice thickness data separated by 2 km (and with poor geolocation: Bingham and Siegert 2007). This sparsely sampled and smoothed topography will invalidate many if not most of the sloped/aspect ratios used to infer 'shield volcanoes' by van Wyk de Vries *et al.* (2018).

The interpretation of layer drawdown geometry relies on the existence, persistence and continuity of layers in radar measurements. It is important to note that the interpretation of layer drawdowns is non-unique. A number of processes unrelated to volcanic activity can similarly deform the ice column and englacial layers (Siegert *et al.* 2004). Mechanical deformation due to ice flow over a subglacial obstacle, variations in vertical strain rates (Raymond 1983), ice-sheet-surface slope variations due to changes in surface accumulation rates (Vaughan *et al.* 1999) and changes in ice-flow direction (Siegert *et al.* 2004) deform layers in similar ways. Here we refer to anomalous layer drawdowns as layer geometries which cannot be explained by geo- or ice-dynamic processes other than melting from elevated geothermal heat. Similarly,

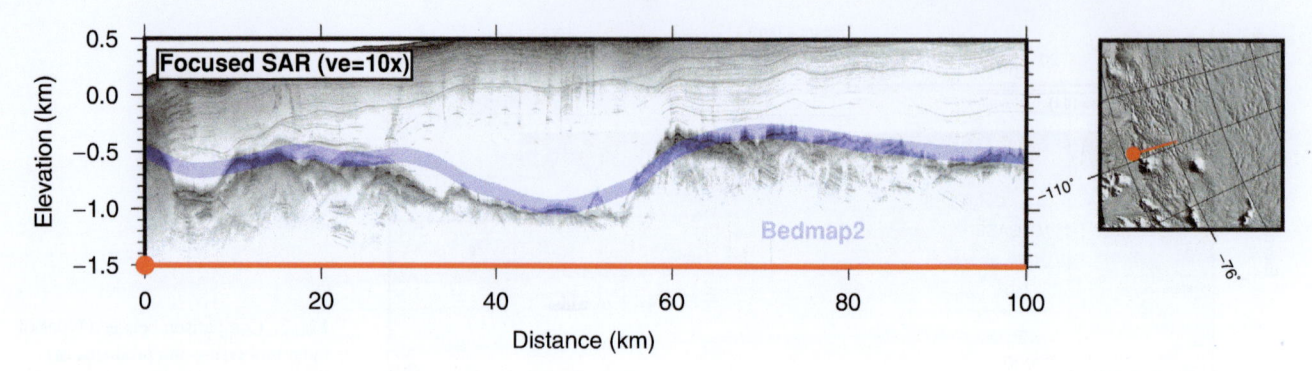

Fig. 4. Portion of depth-corrected and vertically exaggerated 2D focused radargram THW/SJB2/X53d (location shown with a red line in the side panel; the ball indicates the start of the line). Bedmap2 bed-elevation interpolation is overlaid in blue, showing ±67 m uncertainty as determined for Carson Inlet, West Antarctica (Fretwell *et al.* 2013).

the presence of water at the ice–rock interface is non-unique to geothermal heating nor is it indicative of processes that take place *in situ*, at the location where water is detected. Water can be produced locally by basal frictional heating or advected from higher hydropotential regions, and should therefore be representative of processes characteristic of different glaciological/geological settings.

Furthermore, the strength of bed echoes, and thus the interpretation of the presence of subglacial water, is affected by a combination of englacial attenuation (Matsuoka 2011) and the material and geometrical properties of the ice sheet and bed, which can introduce ambiguities in quantitative echo interpretations.

Potential fields

As gravity and magnetic field data smooth as a function of distance from the source, volcanic anomalies related to processes at the ice–rock interface should have relatively high-frequency content compared to deep-crustal processes. Limitations to detection of volcanic constructs from potential field data often arise from the original data acquisition parameters. Both line spacing and source to sensor distance impose limits on the smallest resolvable feature size and require surveys to be designed accordingly.

Gravity. The gravity field can be measured from an aircraft by stable measurement of vertical accelerations, and then subtraction of the acceleration due to the motion of the aircraft. Subtraction of the global reference gravity field predicted at the aircraft's location, and compensation for the Earth' rotation, provides gravity disturbances which are due to deviations in the total mass under the aircraft from a homogeneous model. The spatial variability of the remaining heterogeneous gravity field can provide information on the depth of sources. In Antarctica, the dominant source of gravity disturbances is the ice–rock interface, with the second strongest source being the density contrast between the crust and the mantle. With knowledge of the ice–rock interface, the two sources can be untangled by calculating the Bouguer anomaly. The crustal structure and other spatial density contrasts can be inferred from gravity disturbances, and are commonly used to determine crustal thickness and sediment distribution (e.g. Damiani *et al.* 2014).

The fundamental limit on the resolution of gravity measurements is proportional to the distance between the sensor and the source; given the thickness of ice in Antarctica, this means that gravity is less of a tool for directly inferring the density of volcanic deposits, and more useful for determining the geological context for these features. However, large mafic intrusive bodies within a sedimentary crust may have a significant positive gravity anomaly.

Magnetics. The character and pattern of magnetic anomalies are used to map the distribution of igneous rocks, which have high content in magnetic minerals and high magnetic susceptibility. Magnetic anomalies can be modelled with both forward and inverse methods to define the shape and depth of the source igneous body (Turcotte and Schubert 2014). High magnetic susceptibility values required to simulate observed magnetic anomalies, especially when closely associated with subglacial topographical features observed in radar data, are evidence for bedrock of volcanic origin. Similarly, particular spatial patterns of magnetic anomalies are often good indicators for volcanic origin and activity. As such, circular 'doughnut'-shaped patterns of high-amplitude magnetic anomalies surrounding central magnetic lows mark known volcanically active or young calderas, such as in Yellowstone and the Rocky Mountains area of the USA (Bhattacharyya and Leu 1975; Smith and Braile 1994) (see Fig. 5b).

The magnetization of rocks is a function not only of the presence of magnetic minerals but also temperature. Intrusive and extrusive igneous bodies only consist of a magnetic anomaly if the rocks have cooled below the Curie temperature of magnetite (580°C), whereas hot, molten igneous rock at or above the Curie temperature show no magnetic signature. This leads to the characteristic magnetic 'doughnut'-shaped anomaly seen in active volcanic centres, showing a negligible (close to zero) low-amplitude magnetic signal surrounded by high-amplitude magnetic anomalies.

The 100 kyr limit we impose on the search for active subglacial volcanism helps us to narrow the search for positive magnetic anomalies. Since the last major magnetic reversal was 780 kyr ago (Singer 2014), negative anomalies, if due to negative remnant anomaly, should be older than 1 myr and therefore outside our interval of interest. A brief complete reversal, the Laschamp excursion, occurred only 41 kyr ago during the last glacial (Singer *et al.* 2009). That reversal, though, lasted for only about 440 years and the actual change of polarity lasted for around 250 years.

Joint inversions of magnetic and gravity anomalies are used to test if sources are shared (i.e. a cold, magnetized mafic intrusive) or separated (a hot demagnetized dense body with shallow or offset magnetic sources) and thus test hypotheses for the activity of candidate sites (see the subsection on the 'Kamb Ice Stream subglacial edifice' in the next section).

Seismology

Deep long-period (DLP) earthquakes are a type of volcano–seismic activity identified in many volcanic settings, including

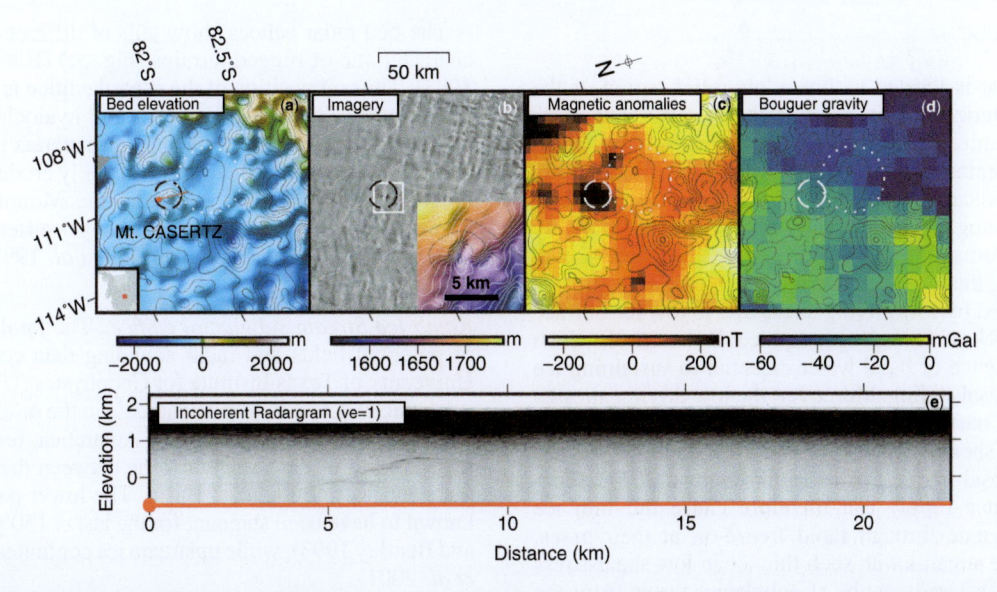

Fig. 5. Observations of Mount CASERTZ (from Blankenship *et al.* 1993). (**a**) Bedrock topography: the red line is part of the profile shown in (e), the ball is the start of line. Mount CASERTZ is the dashed circular shape, the associated caldera is the dotted circle. 100 m bedrock elevation contours are in black for all. (**b**) Imagery from MOA (Scambos *et al.* 2007). The white box shows the inset. The inset shows the ice-surface topography from the Reference Elevation Map of Antarctica (REMA: Howat *et al.* 2019). 10 m surface contours in white. (**c**) Magnetic anomalies from ADMAP2 (Golynsky *et al.* 2018). (**d**) Bouguer gravity anomaly from Scheinert *et al.* (2016). (**e**) Depth-corrected incoherent radargram from the 1991–92 CASERTZ field campaign, part of transect IRE/NEy/Y05a, with no vertical exaggeration. The edifice is at 7 km along track. This transect was flown relatively high at 1000 m above the surface for stable gravity, allowing for significant surface scattering (black band at the top) obscuring englacial layers.

the Aleutian Islands, the Pacific NW of North America, Hawaii and Mount Pinatubo (Power *et al.* 2004; Okubo and Wolfe 2008; Nichols *et al.* 2011). DLP are characterized by deep hypocentres (at or below the brittle–ductile transition zone), low-frequency energy (<5 Hz) and swarm behaviour (Okubo and Wolfe 2008), and are hypothesized to represent the movement of magma and other fluids within volcanic and hydrothermal systems.

Seismic data also reveal variability in seismic-wave velocity and patterns of mantle anisotropy, which are used to image the distribution of thermal anomalies within the mantle and estimate crustal thickness. In West Antarctica, 10 permanent co-located GPS and broadband seismic stations measure relative plate motion and seismicity in central WARS as part of the Antarctic Polar Earth Observing Network (ANET-POLENET). To improve spatial coverage and produce higher-resolution tomography, during the 2010–12 POLENET campaign an additional 13 temporary stations were installed across the WARS between the Whitmore Mountains and Marie Byrd Land (Hansen *et al.* 2014; Emry *et al.* 2015; Lloyd *et al.* 2015; Heeszel *et al.* 2016).

Direct measurements

Direct measurements of geothermal flux via borehole thermometry are extremely difficult to undertake due to inaccessibility of the ice-sheet bed. Ice borehole data coverage is also limited and therefore not representative of large regions. In addition, this type of measurement only provides a minimum bound on geothermal flux as basal melting, if present, would reduce heat conduction into the overlying flowing ice (Engelhardt 2004*b*). Over the last decade, borehole measurements of heat flow have expanded with the WAIS Divide ice-core site and measurements through direct access to the bedrock in the Siple Coast (Begeman *et al.* 2017). The results indicate considerable variability in geothermal heat flux that is consistent with active volcanism (Fisher *et al.* 2015).

In addition, ice cores can sample englacial ash deposits (Palais *et al.* 1988; Iverson *et al.* 2017; McConnell *et al.* 2017) providing a datable indicator for potential subglacial eruptions. Direct measurements thus provide valuable point constraints for geophysical methods.

Active subglacial volcanic sectors

We group active subglacial volcano candidates into sectors based on the portion of the WAIS that is impacted by their activity (Table 1). Available datasets are evaluated for evidence of active volcanism at each candidate site.

Table 1. *Characteristics of volcanic sectors*

Sector	Ice-flow organization (Mouginot *et al.* 2019)	Driving stress (Sergienko *et al.* 2014)	Bed elevation (km) (ice thickness (km)) (Fretwell *et al.* 2013)	Crustal thickness (km) (Chaput *et al.* 2014)
Siple Coast	Fast ice streams	Low	*c.* −0.5 (*c.* 1)	*c.* 30
WAIS Divide	Low to no flow	Low to none	*c.* −1.5 to −0.5 (*c.* 2.5–4)	<25
Marie Byrd Land	Low flow	Low	*c.* −0.5 to +0.5 (*c.* 1–2)	*c.* 30
Amundsen	Fast ice streams	High	*c.* −1.5 (*c.* 2–3)	*c.* 25

These sectors include deep and shallow bedrock, fast and slow flow, and variations in overall crustal thickness. Driving stress is a parameter calculated from ice-surface slope and ice thickness that indicates the degree of coupling of ice to its bed and shear margins.

Siple Coast sector

The Siple Coast is located within a low-relief region of the WARS characterized by relatively thin crust (Chaput *et al.* 2014). It is bounded by the EWM crustal block and the Transantarctic Mountains to the south, and by wMBL to the north.

Subglacial volcanic activity in this region is of particular interest since subglacial melt water production sustained by elevated geothermal flux has the potential to impact ice-stream dynamics along the Siple Coast. The Siple Coast's ice streams are characterized by low driving stress due to low ice-surface slopes and a lack of strong topographical bed control, which makes the presence of basal water essential to sustaining ice flow through basal sliding. Moreover, the low driving stresses of the Siple Coast allow thinning of the ice streams and reduced basal shearing, both of which act to cool the bed and promote basal freezing (Christoffersen *et al.* 2014). Fluctuations in water supply can therefore cause the thin ice streams to stagnate through rapid freeze-on at their bases. To mobilize ice streams with such thin ice in low shear-stress configurations, a large supply of subglacial water from the interior is required.

Mount CASERTZ edifice. Subglacial volcanic activity at Mount CASERTZ was first identified by Blankenship *et al.* (1993) from radar, laser and magnetic data evidence (Fig. 5), and later supported by analysis of englacial ash layers compatible with subglacial eruptions sourced at Mount CASERTZ (Iverson *et al.* 2017). Mount CASERTZ is located NW of the Whitmore Mountains, in the transition zone between the EWM crustal block and the Ross Subglacial Basin, along an ice tributary *c.* 100–200 km upslope of the Siple Coast's area of ice-stream initiation (Blankenship *et al.* 2001) (Fig. 1d).

Mount CASERTZ shows as a cone-shaped subglacial topographical edifice about 6 km across that rises *c.* 650 m above the surroundings (Fig. 5a) (Blankenship *et al.* 1993). Mount CASERTZ stands on the edge of a *c.* 50 km-wide caldera bounded by a rim 100–200 m in height. The cone's steep slopes (*c.* 12°: Fig. 5e) indicate that the edifice was extruded into the ice or possibly under shallow water before the development of the WAIS. The ice-surface elevation at the location of Mount CASERTZ's central edifice shows a 48 m-deep depression (Fig. 5b). Following the regional ice-flow pattern, ice from further upstream enters the surface depression from the north and south, indicating that the large ice-surface anomaly requires a 10–20 W m^{-2} heat source at the base (over two orders of magnitude more than the continental average geothermal flux), melting *c.* 0.07 km^3 of ice each year, to be sustained.

A large-amplitude (*c.* 600 nT), long-wavelength, positive magnetic anomaly *c.* 40–80 km in diameter suggests that both the caldera and central edifice are part of a larger volcanic construct (Blankenship *et al.* 1993) (Fig. 5c). The large magnetic anomaly is associated with a broad uplift of the bedrock topography. The modelled underlying intrusive appears to be several kilometres thick, assuming that the material producing the magnetic anomaly has a magnetic susceptibility 0.10 SI (Behrendt *et al.* 1995). The complex is located on a boundary in crustal thickness, as implied by gradients in the Bouguer anomalies (Blankenship *et al.* 2001) (Fig. 5d).

The bed radar echo strength at the location of Mount CASERTZ is weaker compared to those in areas of both equivalent and larger ice thickness in West Antarctica. The bed echo strength is sensitive to the temperature-dependent dielectric attenuation of radio waves through ice, indicating that the ice above the central edifice is anomalously warm (Blankenship *et al.* 1993).

The bed radar echoes show tails of diffraction hyperbolae characteristic of rugged terrains (Fig. 5e) (Blankenship *et al.* 1993). The composition of the central edifice is therefore consistent with a pile of pillow lava and hyaloclastite material. Unlike a coherent shield, this pile of scatterers is likely poorly consolidated material that would be easily eroded by ice (Behrendt *et al.* 1995) (Fig. 2), indicating that Mount CASERTZ is a recently erupted volcano that has not yet suffered removal by the overlying ice sheet (Blankenship *et al.* 1993).

Kamb Ice Stream subglacial edifice. The joint interpretation of potential fields and radar sounding data collected by the University of Texas Institute for Geophysics (UTIG) indicates a subglacial active volcano candidate in the onset region of the Kamb Ice Stream (KIS) of West Antarctica, few tens of kilometres north of the transition zone between the EWM microplate and Ross Subglacial Basin. The lower parts of KIS are known to have been stagnant for the last *c.* 150 years (Retzlaff and Bentley 1993), while upstream ice continues to flow (Price *et al.* 2001).

The KIS subglacial volcano was first surveyed in 1992 during the CASERTZ aerogeophysical survey with incoherent radar (Blankenship *et al.* 2001) (Fig. 3a) before being partially reflown with HiCARS as part of the Advanced Technology for Radar Sounding of Polar Ice (ATRS) project (Peters *et al.* 2005, 2007). HiCARS-focused data show englacial layers that dip into the edge of a broad conical high in the centre of the ice stream (Fig. 3c). Radar sounding over the KIS onset region shows a bright sub-ice reflection (Peters *et al.* 2007) and high specularity content suggesting the presence of significant water beneath the ice stream (Fig. 6b) (Young *et al.* 2015).

Joint interpretation of the high gravity signal and relatively weak magnetic signature suggests the presence of a subglacial volcano in the area. Magnetic anomalies are relatively weak, narrow and offset from topography (Fig. 6c). This volcanic feature, however, produces a sharp distinctive peak in Bouguer gravity (Fig. 6d) centred on the edifice, likely to be due to a high density contrast between a magma body in the subsurface and the surrounding rocks but a low magnetic signature due to a thin layer of extrusive material just beneath the ice. A single body does not satisfy the gravity, magnetic and topographical constraints, suggesting a deep high-density body too hot to have a magnetic signature, and a carapace of thin cool magnetic material surrounding the edifice with the dipping englacial reflectors.

Subglacial Lake Whillans heat-flow anomaly. Extraordinary high geothermal flux (285 ± 80 mW m^{-2}) was measured at the Subglacial Lake Whillans (SLW) drilling site as part of the Whillans Ice Stream Subglacial Access Research Drilling (WISSARD) project (Fisher *et al.* 2015). The measurement represents the first direct assessment of geothermal flux into the base of the WAIS, obtained with a probe that went *c.* 1 m deep into basal sediments. The value of geothermal flux determined at SLW is significantly higher than the continental average but is representative of areas of active hydrothermal and volcanic activity (e.g. http://www.heatflow.org). A second WISSARD probe 100 km away near the grounding line of Whillians Ice Shelf found a heat flux of 88 ± 7 mW m^{-2} (Begeman *et al.* 2017), while an ice borehole at Siple Dome, where the ice is inferred to be frozen to the bed, yielded a geothermal heat flux of 70 mW m^{-2} (Engelhardt 2004*a*), both consistent with rifted continental crust (Davies 2013).

The highly dynamic hydrological activity observed at the SLW drilling site (Fricker and Padman 2012) is consistent with fluctuations of basal melt water supply generated by elevated geothermal flux. The site location near the convergence between Siple Coast's Whillans and Mercer ice streams makes

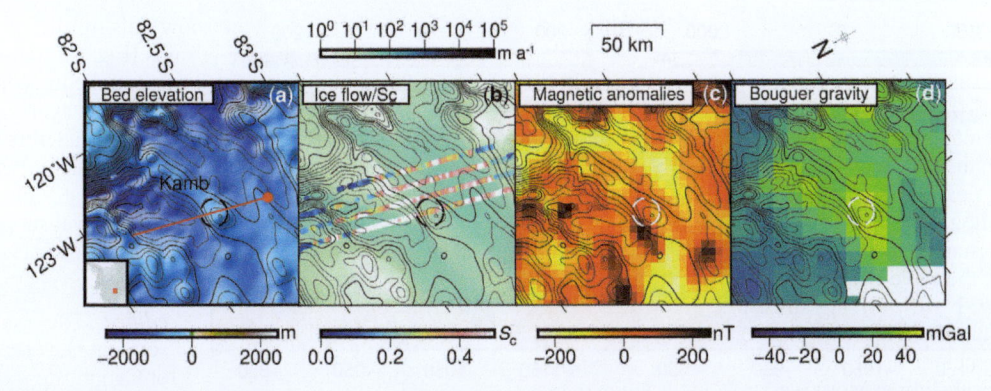

Fig. 6. Volcanic feature at the head of the stalled Kamb Ice Stream. Polar stereographic projected kilometres used as coordinates; bed topography contours in 100 m intervals are shown on all. (a) Bedrock topography; the red line is the line of section shown in Figure 3; the ball indicates the start of the line. (b) Specularity content (S_c: Young et al. 2015) superposed on surface ice-flow velocity (Mouginot et al. 2019). High S_c indicates subglacial water. (c) ADMAP-2 magnetic anomalies (Golynsky et al. 2018). (d) Bouguer gravity anomalies (Scheinert et al. 2016). High values are consistent with either a dense body in the crust or an upwarp of the crust–mantle interface.

changes in the local hydrological configuration particularly relevant to ice-sheet dynamics.

No edifice is observed at Subglacial Lake Whillans (Christianson et al. 2012), instead a minor bedrock depression that is consistent with the high ice velocity in this area (Fig. 1d).

WAIS Divide sector

The WAIS Divide sector is located within the low-lying cradle-shaped topography of the central WARS, up to 2000 m below sea level, characterized by thin crust (Fretwell et al. 2013; Chaput et al. 2014). It is bounded by the EWM crustal block to the south and the eastern MBL crustal block to the north. The WAIS Divide is a region of thick ice and slow to negligible ice flow (Fig. 1c, d). Similar to a continental hydrographical divide, this region separates the WAIS into two catchments: one where the ice flows to the Ross Sea; and one where the ice flows to the Amundsen and Weddell seas. Subglacial volcanic activity in this sector can impact ice organization within those catchments, which can ultimately lead to ice-divide migration.

Detailed shard morphology characterization, and geochemical and micro-CT analyses of two tephra layers from the WAIS Divide ice core suggest a phreatomagmatic origin from a subglacial or close to emergent volcano (Iverson et al. 2017). The tephra layers were erupted from the centre of the WAIS, making Mount CASERTZ one of the potential subglacial volcanoes that sourced the layers.

WAIS ice-core site heat-flow anomaly. The high-precision measurements made in the 3405 m-deep borehole at the WAIS Divide ice-core site (WDS: Fudge et al. 2013) (Fig. 7) represent the first direct measurements of anomalously high subglacial geothermal flux in Antarctica, calculated from vertical strain rates in the ice to 50 m above the estimated bedrock depth where drilling stopped to prevent contamination of the basal hydrology. Site selection for the WAIS Divide ice core had been part of an effort by the ice-coring community to find an ice record extending to the last interglacial in West Antarctica. Criteria for site selection included finding a region with thick ice and slow ice flow to assure recovery of a deep/old and undisturbed/organized ice record (Morse et al. 2002).

However, despite the great ice thickness (>3000 m) and the slow ice flow at WDS, the ice near the bed proved to be relatively young (c. 68 ka) compared to shallower cores drilled in central East Antarctica, which was interpreted to be the result of high accumulation rates (22 cm a^{-1} at present and c. 10 cm a^{-1} during the Last Glacial Maximum) and basal melt (Buizert et al. 2015).

Interpretation of the temperature measurements in the borehole indicates that the ice sheet is melted at the bed at this location and that the basal melting rate is remarkably high, c. 1.5 cm a^{-1}. The inferred geothermal heat flux, using thermal data from the ice sheet and a one-dimensional model of ice dynamics, is estimated to be about 140–220 mW m^{-2}, 4–5 times the continental average. Importantly, the absence of a surface depression at WDS indicates that this high heat flow is likely to be the regional value (horizontal scale c. 30 km), rather than simply a local anomaly. IPR data indicate a very flat ice–bed interface, with no significant local volcanic edifices. WDS is located in a gap between two significant magnetic anomalies (Fig. 7), which is consistent with thermal suppression of magnetic susceptibility.

Central WAIS caldera complex. A series of positive magnetic anomalies between 400 and 1200 nT surrounding a central magnetic low (c. −150 nT) define the rim of what has been inferred to be a large subglacial caldera about 70 km in diameter located in the deep ice region between MBL and the WAIS Divide ice-core site (Figs 1 & 7) (Behrendt et al. 1998). The caldera is surrounded by a low magnetic background (−500 to −300 nT), consistent with a shallow Curie isotherm (Behrendt et al. 1998). The magnetically defined caldera lies on the 'sinuous ridge' subglacial range under the Thwaites Glacier–Siple Coast Divide that was first defined by sparse radar sounding data by Jankowski and Drewry (1981). Additional soundings reveal that this range is highly dissected by deep subglacial valleys (Behrendt et al. 1998; Holt et al. 2006).

Behrendt et al. (1998) inferred from an early analysis of the IPR data that these magnetic anomalies were largely decorrelated from bedrock topography, with the exception of one magnetic anomaly on the SW rim (called 'C' by Behrendt et al. 1998, fig. 3), where current ice thickness is about 3 km. Behrendt et al.'s (1998) observation suggested that the source rock for anomaly C may consist of erosion-resistant pillow lavas, characteristic of eruptions under very thick ice, and supports the interpretation that the caldera complex was recently active under the current, thick WAIS.

However, re-examining the magnetics and radar data, it appears that the original interpretation had a 30 km along-track offset (see Fig. 8 for an updated coregistration).

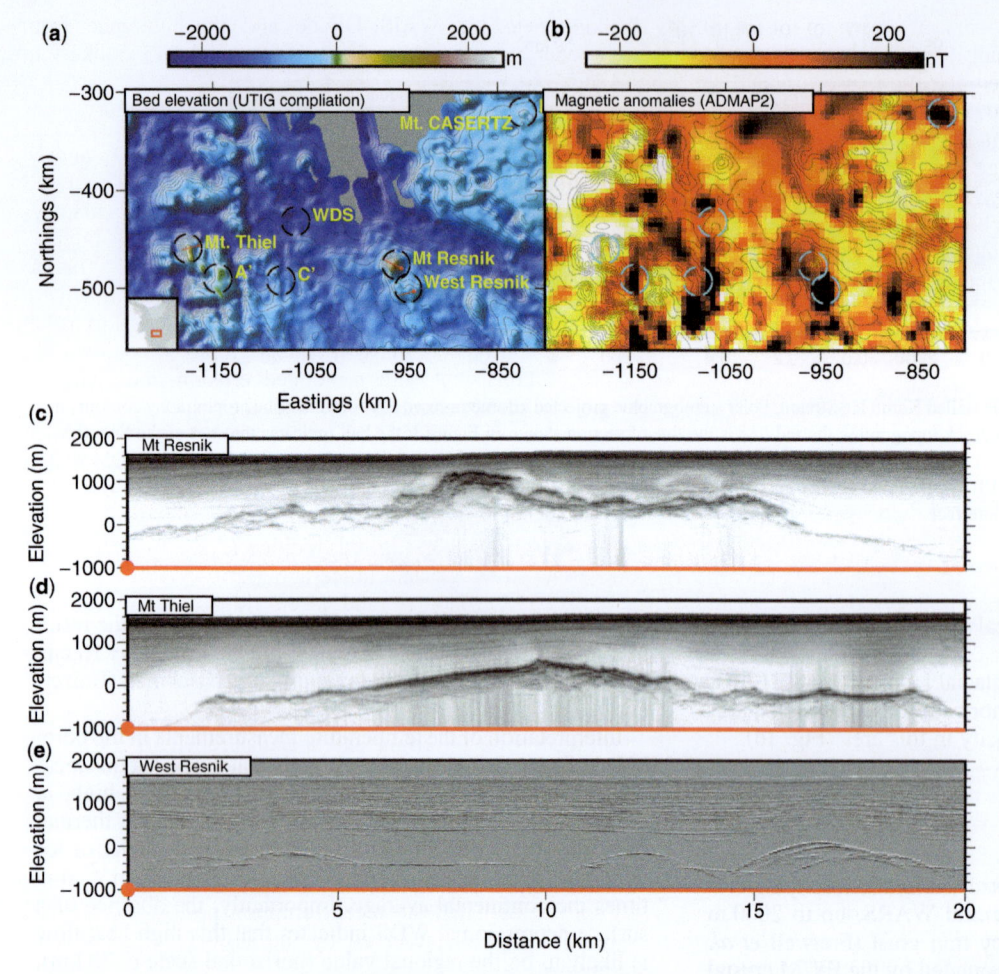

Fig. 7. Map of the central region of the WAIS Divide sector, showing (**a**) bed elevation and the locations of the volcanic features discussed in this section; (**b**) ADMAP-2 magnetic anomalies (Golynsky *et al.* 2018) and radargrams with ×1.5 vertical exaggeration of: (**c**) Mount Resnik (transect OND/SJB2/DVD01a, focused), (**d**) Mount Thiel (transect THW/SJB/Y51a, focused) and (**e**) a low-lying region west of Mount Resnik (transect BSB/Wy/Y19a, incoherent).

Magnetic anomaly C no longer aligns with a deep bedrock peak (C′) but is offset. C′ is surrounded by bright bed returns and overlaid by disturbed englacial layers, while a second, weaker magnetic anomaly (called A by Behrendt *et al.* 1998) aligns closely with a shallower bedrock peak (A′). We suggest that the subglacial mountain C′, offset from anomaly C, may still be warm (and thus have a locally suppressed magnetic anomaly), while A′ may represent an old subaerial volcano composed of solidified lava, which after rebound may have risen above sea level, that is preserved by its proximity to the ice divide.

Mount Resnik edifice. Mount Resnik (informally named after the late astronaut Judith Resnik) is a major subglacial edifice near the Bentley Subglacial Trench (Fig. 7a), which has a negative magnetic anomaly (Behrendt *et al.* 2006) (Fig. 7b). It is unlikely that Mount Resnik was entirely built in the 250 years of the Laschamp event; hence, the negative magnetic anomaly of Mount Resnik dates this volcano at more than 0.8 Ma, and therefore too old and too cold to have had an impact on the current ice sheet. Iverson *et al.* (2017) concluded that Mount Resnik was the most likely source for the subglacial ash layer in the WAIS

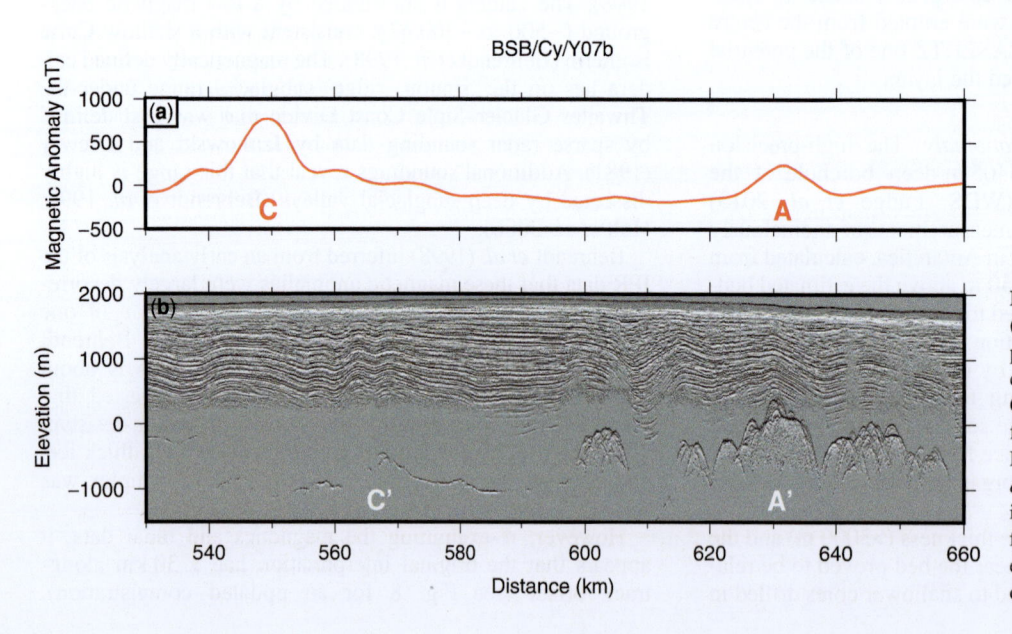

Fig. 8. Updated CASERTZ profile (BSB/Cy/Y07b) across the proposed subglacial caldera complex under the Thwaites–Siple Coast ice divide. Compare with figure 3 of Behrendt *et al.* (1998). (**a**) Magnetic anomaly from (Sweeney *et al.* 1999). (**b**) Depth-corrected incoherent radargram, differentiated for clarity. ×10 vertical exaggeration. Note that these radar data are unfocused.

Divide ice core, based on its shallow depth of burial (c. 300 m). However, the significant amount of erosion apparent on the upper part of the edifice in focused radar data (Fig. 7c) when compared to other edifices under the WAIS also implies that Mount Resnik is not significantly active today. An alternative geomorphic interpretation of Mount Resnik is that it is a 'tuya' or table mountain, formed by a subglacial eruption breaching the surface of the ice; however, additional focused radargrams from additional orientations would be required to make a conclusive identification. Mount Resnik's shallow burial and location in the heart of the main trough of the WARS makes it an ideal target for reconstructing the collapse history of the WAIS through exposure dating methods (Spector et al. 2018).

A site to the west of Mount Resnik does have both a significant magnetic anomaly (Fig. 7b) and a large number of subglacial peaks (Fig. 7e); however, focused data for assessing the morphology of these features are not yet available.

Mount Thiel edifice. Mount Thiel (informally named after the late University of Wisconsin geologist Edward Thiel) was first identified from airborne magnetics, and was interpreted as being composed of erosion-resistant lavas by Behrendt et al. (2002). It has a subglacial topographical relief c. 1800 m high (Fig. 7d), and corresponds to a positive (400 nT) magnetic anomaly situated within the low magnetic background and shallow Curie isotherm surrounding the nearby subglacial caldera complex. Unlike Mount Resnik, the shield of Mount Thiel is largely intact (Fig. 7d). Mount Thiel is located about 100 km distant from the WAIS ice-core drilling site and is a hypothesized source of the prominent volcanic ash layer detected at the WAIS Divide ice core (Behrendt 2013; Iverson et al. 2017). The high bed reflectivity of Mount Thiel, combined with the lack of a catchment for subglacial hydrology, imply heat flows of above 150 mW m^{-2} at this site (Schroeder et al. 2014) (Fig. 1c, d). Given the intact topography, the proximity to the WAIS Divide and the inference from Schroeder et al. (2014), we classify this volcano as likely to be active.

Central WAIS fissure system. Magnetic anomalies L and M identified by Behrendt et al. (1995) have been analysed by Danque (2008) using focused HiCARS radar and magnetics from the AGASEA survey. Danque (2008) inferred that the edifice associated with magnetic anomaly M (hereto referred to as M′) is both older and colder than the edifice associated with magnetic anomaly L (hereto referred to as L′) based on the smooth subglacial topography of edifice M′, which is consistent with prolonged erosion and sediment drape (Behrendt et al. 1995), and the lack of subglacial lakes (discussed below), uniquely associated with edifice L′. Danque (2008) interpreted an additional anomaly, fissure H, which corresponds to a significant downdraw of englacial layers (Fig. 9d).

In radargrams (e.g. Fig. 9d), edifice L′ shows as a broad, dome-shaped subglacial edifice about 250–300 m high with a stairstep morphology and steep ridges on its summit. Coherent radar sounding data over edifice L′ show eight bright basal reflections (200–500 m in diameter) located along the breaks in the slope between the surrounding plains and the dome of edifice L′. These bright spots appear in radar profiles as smooth, bright and specular reflectors located in hydraulic flat regions on the flanks and on the summit of edifice L′, and are interpreted as subglacial lakes. Two of the potential lakes found on the summit of edifice L′ are located in a depression on the SE side of the summit and between the summit ridges, respectively. The steep-sided ridges on edifice L′ could either indicate recent eruption products or an old ridge that is well preserved by the slow flowing ice near the divide.

Even though the ice surface does not dip over edifice L′, the englacial layers have some drawdown over one of these

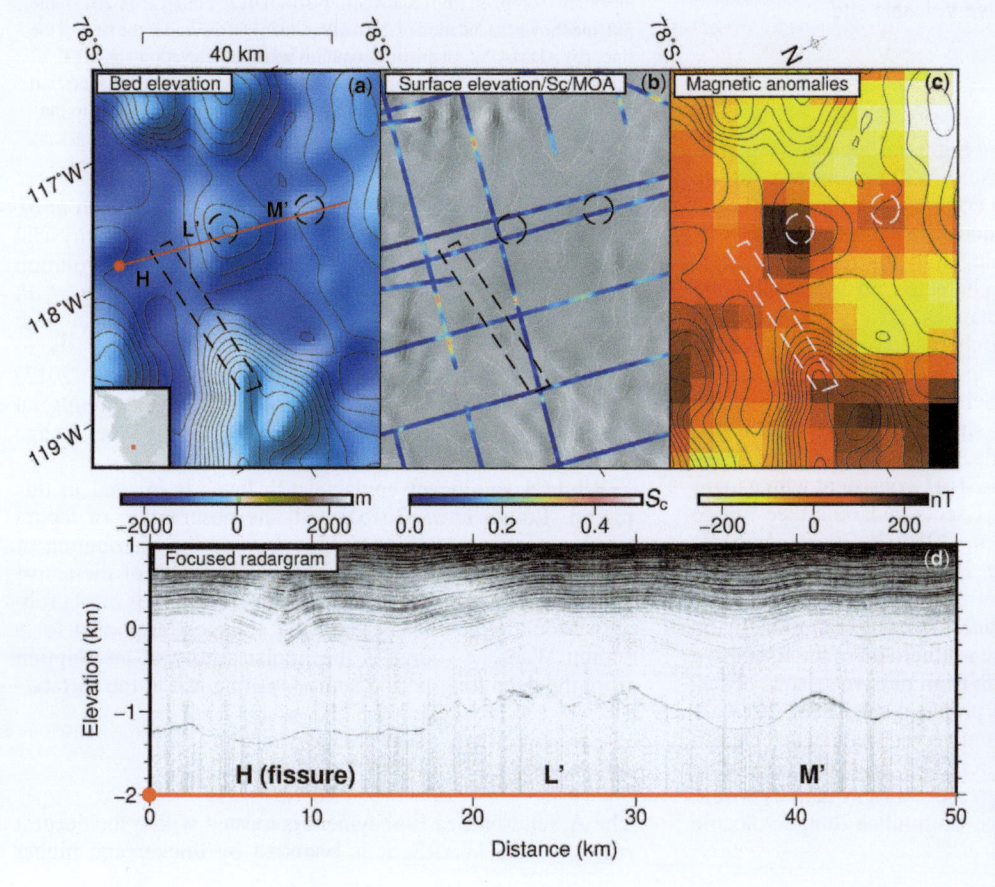

Fig. 9. Map of the H fissure region (dashed rectangle) of the WAIS Divide, showing (**a**) bed elevation, the locations of the volcanic features and the location of the segment of (d) (red line, the ball is the start of the line). (**b**) Specularity content (Schroeder et al. 2013; Young et al. 2015) shown on the MOA image mosaic (Scambos et al. 2007) and 10 m contour lines from the Reference Elevation Map of Antarctica (REMA, white: Howat et al. 2019). A faint dark streak is visible in the dashed rectangle in the MOA imagery and affects REMA contours. (**c**) ADMAP-2 magnetic anomalies with bed-elevation contours (Golynsky et al. 2018). (**d**) Radargram THW/SJB2/X31c showing layer downdraw over H (vertical exaggeration is ×5).

possible lakes, while over the summit of L layers fade out entirely. This might indicate a physical or thermal disturbance over the summit of edifice L' that leads to steep layer slopes dipping >10°, which is the imaging capabilities of the current 2D-focused synthetic-aperture radar (SAR) processing.

A magnetic anomaly c. 300 nT in amplitude is associated with edifice L' (Fig. 9c). The model of the magnetic anomaly is consistent with a large cooling intrusive body hot enough to maintain the subglacial lakes apparent in coherent radar profiles. Because of the position of the lakes with respect to the regional hydraulic potential, it is likely that the lakes are sustained by local sources and not by subglacial water flow.

Edifice M' (Behrendt et al. 1995; Danque 2008) appears in radargrams as a mound situated along an elongated subglacial ridge oriented nearly east–west, adjacent to edifice L'. Edifice M' is associated with a c. 300 nT magnetic anomaly, has no corresponding ice-surface depression or internal layer drawdowns, and lacks evidence for nearby subglacial lakes (Fig. 9). The low topography and smooth sides are consistent with the interpretation that edifice M' is the eroded root of a subglacial volcanic edifice (Behrendt et al. 1995), likely to be older and colder than edifice L'.

The H fissure (Danque 2008) is a large layer drawdown anomaly in the radar englacial layers of the ice sheet located along an area of flat topography adjacent to edifice L. The layer drawdown is identified by a c. 300 m dip in the deepest visible englacial layers that can be traced for 28 km in length. A small, c. 70 m-high and c. 2 km-across topographical relief underlies the layer drawdown for the entire length of the anomaly. Bright, flat and specular echoes on either side of the bed topography relief indicate subglacial lakes 0.5 and 1 km in apparent cross-sectional diameter. The layer drawdown is associated with an ice-surface depression visible in the Mosaic of Antarctica (MOA) surface imagery (Fig. 9b) and a <50 nT magnetic anomaly (Fig. 9c). The presence of a positive but relatively suppressed magnetic signal, basal water and an ice-surface depression in a region of negligible ice flow makes this a likely active volcanic site.

Marie Byrd Land sector

The Marie Byrd Land sector is a region of high bed topography and thicker crust (LeMasurier 2006; Fretwell et al. 2013; Chaput et al. 2014). It is composed of the wMBL and eMBL crustal blocks, and is bounded to the south by the low-lying WARS. It is located hydrologically upstream of the Siple Coast and Amundsen sectors, characterized by fast-flowing ice streams and complex dynamics greatly impacted by changes in subglacial water discharge, which makes volcanism and high geothermal flux in this sector of great importance as a melt water source. Seismic tomography from the broadband seismic POLENET shows evidence for a broad thermal anomaly extending deep into the mantle under Marie Byrd Land (Hansen et al. 2014), consistent with a mantle hotspot centred beneath the Executive Committee Range (Accardo et al. 2014; An et al. 2015; Emry et al. 2015; Lloyd et al. 2015; Heeszel et al. 2016).

Executive Committee Range edifice. Underground magmatic activity has been identified at the southern tip of the Executive Committee Range volcanic chain from passive seismic observations as part of the POLENET project (Lough et al. 2013). A cluster of deep long-period seismic episodes was detected in January–February 2010 and March 2011 beneath a subglacial edifice (Fig. 10) located where present volcanic activity would be expected along the Executive Committee Range volcanic trend north to south migration.

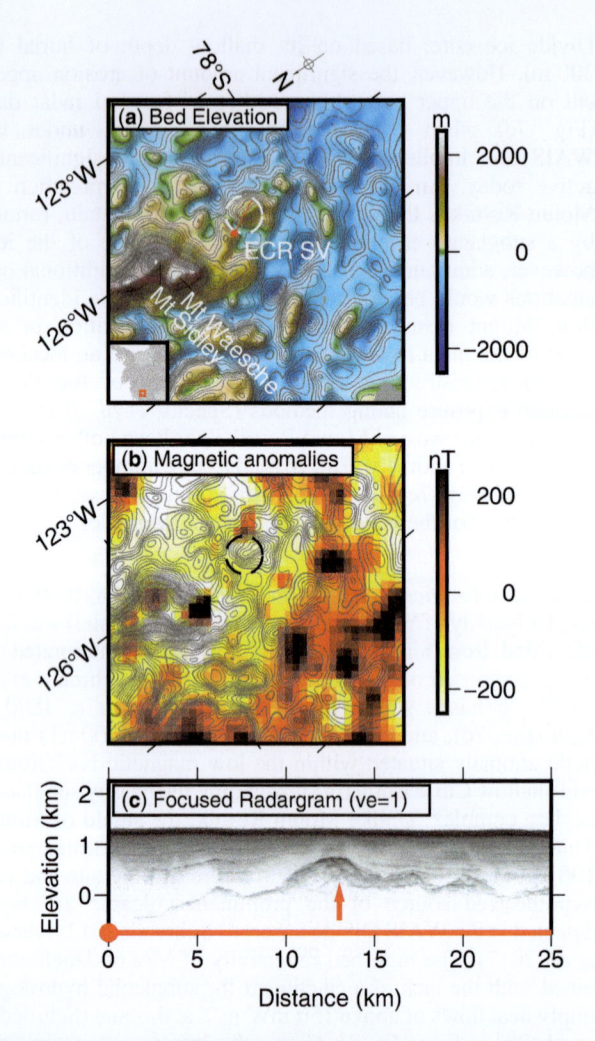

Fig. 10. The Executive Committee Range subglacial volcano (ECR SV), overlying the magma chamber proposed by Lough et al. (2013). (a) Bed elevation (compiled from Blankenship et al. 2012; Young et al. 2017); the red line shows the location of the radargram in (c); the ball is the start of the line. (b) ADMAP-2 magnetic anomalies with bed-elevation contours (Golynsky et al. 2018) showing an offset magnetic anomaly. (c) A portion of focused radargram MBL/MKB2l/Y89a. The red arrow points to the ECR SV.

The active subglacial volcano interpreted by Lough et al. (2013) was directly imaged as part of the Geophysical Investigation of Marie Byrd Land Lithospheric Evolution (GIMBLE) airborne geophysical programme (Young et al. 2017) and corresponds to a subglacial topographical high c. 1000 m above its surroundings (ECR SV in Fig. 10a, c) and a 400 nT magnetic anomaly (Lough et al. 2013) (Fig. 10b). GIMBLE also detected significant subglacial water to the south of the Executive Committee Range (Young et al. 2015).

While a significant englacial ash layer is imaged in this region (Lough et al. 2013), both the observation of tephra bands in a zone of ablating blue ice and the distribution of the ash layer in a wind-orientated streak south of the active subaerial Mount Waesche suggest that the source is most probably Mount Waesche. Perhaps the strongest argument for a Mount Waesche source is the implausibility of an eruption from the deep long-period source venting ash to the surface.

Amundsen Sea Embayment sector

The Amundsen Sea Embayment is located within the deepest region of the WARS. It is bounded by thicker and higher elevation crustal blocks to the south, NE and NW (EWM, eMBL and TI respectively), and by the Siple Coast to the west. Within the Amundsen Sea Embayment's low and landward-sloping bed flow some of the fastest-flowing, most rapidly changing ice streams on Earth currently at risk of rapid collapse (Joughin et al. 2014; Mouginot et al. 2014), which makes this sector of the WAIS a main component in scenarios of rapid deglaciation.

In this sector of the WAIS, geothermal flux can have a large impact on ice-flow organization and ice-stream basal sliding velocity through the production of subglacial melt water. Not only does geothermal flux play a key role on ice-flow velocity; importantly, it can also potentially control ice-flow initiation along ice divides.

Hudson Mountains subglacial edifice. Subglacial volcanic activity near the Hudson Mountains was identified by Corr and Vaughan (2008) as the source of a bright, local englacial layer as imaged in radar data. The strong radar reflection covers an elliptical area of about 23 000 km². The source region for the tephra layer, which has not been overflown with airborne geophysical surveys, was identified using the layer's radar echo strength as a proxy for the tephra thickness and proximity to the volcanic centre, and coincides with a subglacial topographical high close to the Hudson Mountains characterized by a complex pattern of positive magnetic anomalies (Golynsky et al. 2018). The layer depth dates the eruption at 207 BC ± 240 years, which matches strong and previously unattributed conductivity signals measured in two ice cores: Byrd Station and Siple Dome (Hammer et al. 1997; Kurbatov et al. 2006; Corr and Vaughan 2008). Rowley et al. (1986) reported anecdotal evidence of recent volcanic activity, including a report of the 'possible presence of steam' in 1974. However, the report of a possible eruption based on satellite data in 1985 (P.R. Kyle cited by Rowley et al. 1990) is weakly founded and probably should be discounted (J.L. Smellie pers. comm. based on information provided by P.R. Kyle in March 2020).

Thwaites Glacier heat-flow anomalies

Indirect evidence of elevated geothermal flux from airborne radar sounding data and a subglacial hydrological model point to localized heat-flow anomalies of c. 200 mW m^{-2} within the Thwaites Glacier catchment (Schroeder et al. 2014) (Figs 1 & 11). High geothermal flux is postulated along the ice-covered flanks of two subaerial volcanoes active within the last 50 kyr (Mount Takahe (Palais et al. 1988) and Mount Frakes (Wilch and McIntosh 2002)), over subglacial volcano Mount Thiel and possibly associated with two additional circular positive magnetic anomalies associated with subglacial edifices. The spatial distribution of these heat-flow anomalies at the inception of tributaries and fast-flowing ice streams is additional evidence of the potential impact of active volcanism on ice organization and ice-sheet dynamics.

Discussion

We find that active subglacial volcanism identified in this chapter manifests near identified crustal and lithospheric boundary zones. In the Siple Coast sector, volcanism and heat-flow anomalies are mostly associated with the southern edge of the WARS, and within 150 km of the crustal block boundary (Fig. 11) with Mount CASERTZ lying along a sharp crustal-thickness boundary (black line in Fig. 11) marking the edge of a transitional crust block (Figs 5d, 11). Fast-moving ice streams may act to erode volcanic edifices in this sector, biasing the observations. In addition, most of the

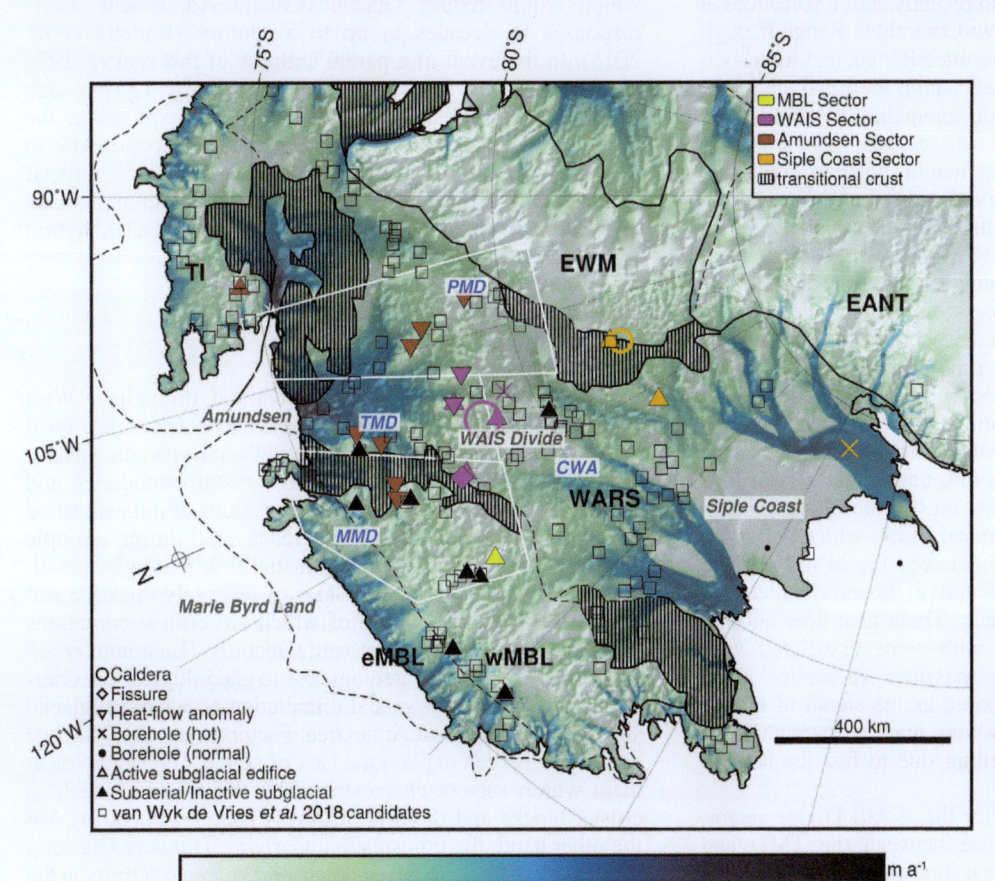

Fig. 11. The distribution of proposed sites of active subglacial volcanism and proposed volcanic edifices over crustal boundaries identified by Diehl et al. (2008) and Quartini (2018). Shading from the bedrock DTM is used in this study; the velocity overlay is from Mouginot et al. (2019). Also shown are candidate volcanoes identified by van Wyk de Vries et al. (2018).

Table 2. *List of active subglacial volcanoes and heat-flow anomalies in West Antarctica described in this study*

Feature	Coordinates	Ice thickness (m)[11]	Bed elevation (m)[11]	Bed elevation rebounded (m)[12]	Crustal thickness (km)[13]	Ice-flow velocity (m a^{-1})[14]
Mount CASERTZ edifice[1]	81.88° S, 111.30° W	1791	−79	518	27.5	10
KIS subglacial edifice[2]	82.00° S, 113.00° W	1816	−295	310	26.5	14
Subglacial Lake Whillans heat-flow anomaly[3]	84.25° S, 153.50° W	799	−667	−400	25.6	350
WAIS Divide heat-flow anomaly[4]	79.47° S, 112.09° W	3441	−1645	−498	22.4	12
L' edifice[5]	78.08° S, 117.96° W	2537	−772	73	23.0	16
H fissure[5]	77.95° S, 118.24° W	2923	−1158	−183	23.1	5
Central WAIS caldera[6]	78.67°S, 114.50°W	2820	−1012	−72	22.5	8
C' edifice[6]	79.12°S, 114.52°W	2540	−752	−51	22.5	8
Mount Thiel edifice[7]	78.42°S, 111.33°W	1425	174	649	22.0	7
ECR edifice[8]	77.65°S, 126.77°W	726	2060	2303	27.6	3
Hudson Mountains edifice[9]	74.67°S, 97.00°W	775	212	470	23.1	2
Thwaites A hotspot[10]	78.23°S, 103.12°W	2220	−611	128	22.6	19
Thwaites B hotspot[10]	77.93°S, 104.52°W	1986	−533	129	22.2	15
Thwaites C heat-flow anomaly[10]	78.77°S, 108.69°W	2333	−598	180	22.3	2
Thwaites D heat-flow anomaly[10]	78.42°S, 111.33°W	1424	175	649	22.1	8
Thwaites E heat-flow anomaly[10]	76.67°S, 115.86°W	3050	−1580	−563	23.2	−*
Thwaites F heat-flow anomaly[10]	76.57°S, 117.08°W	3068	−1525	−503	23.5	−*
Thwaites G heat-flow anomaly[10]	79.22°S, 100.89°W	3000	−1041	−42	23.6	−*
Thwaites H heat-flow anomaly[10]	76.32°S, 110.51°W	2139	−1021	−306	23.0	168
Thwaites I heat-flow anomaly[10]	76.73°S, 112.16°W	2671	−1388	−497	22.7	108
Thwaites J heat-flow anomaly[10]	76.60°S, 123.08°W	2808	−595	340	24.6	11

[1]Blankenship *et al.* (1993); [2]Filina *et al.* (2008); [3]Fisher *et al.* (2015); [4]Clow *et al.* (2012); [5]Danque (2008); [6]Behrendt *et al.* (1998); [7]Behrendt (2013); [8]Lough *et al.* (2013); [9]Corr and Vaughan (2008); [10]Schroeder *et al.* (2014); [11]Fretwell *et al.* (2013); [12]Airy isostatic rebound; [13]Chaput *et al.* (2014); [14]Rignot *et al.* (2014); *no data.

radar data over this region are older incoherent data (Blankenship *et al.* 2001).

The sole identified active subglacial edifice in the Marie Byrd Land sector (Fig. 10) is again within 150 km of a crustal boundary but also lies in line with recently active volcanoes at the southern end of the Executive Committee Range (Lough *et al.* 2013). The Executive Committee Range, and its associated subglacial edifice, lie centred within the MMD, a broad region with suppressed magnetic anomalies surrounded by an intense magnetic front. We suggest that this crust has been modified by an underlying mantle hotspot, consistent with slow mantle velocities observed under the MMD (Hansen *et al.* 2014) and large isostatic anomalies derived from crustal-thickness measurements (Chaput *et al.* 2014).

The WAIS Divide and Amundsen sectors straddle the WARS, and the crustal-thickness boundaries defined by gravity (Fig. 11). Instead, these associations are more controlled by the magnetic crustal provinces defined by Quartini (2018). The Amundsen sector is largely defined by the heat-flow anomalies identified by Schroeder *et al.* (2014); these heat-flow anomalies are separated by the TMD, which was interpreted as old, unmodified Cretaceous lithosphere. Heat-flow anomalies cluster along the eastern edge of the Marie Byrd Land crustal block, which coincides with the well-defined boundary between the MMD and the TMD, as well as the recently active subaerial volcanoes Mount Takahe and Mount Frakes. Three heat-flow anomalies are located in the PMD, with some associated with topography and large circular magnetic anomalies. Care must be taken in this interpretation, as the signal of excess geothermal flux in central Thwaites may be overwhelmed by the substantial frictional melting due to fast ice flow in this region.

Fast ice flow is not a factor for the WAIS Divide sector, which lies in the transition zone between the TMD and CWA. Here the signal of high heat flow and edifice construction is most clearly preserved due to a lack of glacial erosion. Ice thickness varies considerably here. An additional factor is the likely large degree of isostatic rebound in this region (see Table 2 for estimates of Airy isostatic rebound) New GPS measurements show rapid uplift of over 4 cm a^{-1} in ASE, which would reduce Glacial Isostatic Adjustment (GIA) responses to decades to up to a century (Barletta *et al.* 2018). In the event of a partial collapse of this region of the ice sheet in the last interglacial occurring 120 kyr ago (DeConto and Pollard 2016), most of the edifices in the WAIS Divide sector would be expected to rebound close to sea level, leading to shallow submarine volcanic forms. If activity now indicated by high basal echo strengths and layer drawdown extended into the last interglacial, hybrid forms could be observed.

Conclusions

Active subglacial volcanism is identified throughout West Antarctica within different geological and glaciological local contexts. The majority of active subglacial volcanic sites in West Antarctica concentrate along crustal boundaries and within the central WARS, which are regions of thinned, rifted crust that have been tectonically reactivated during multiple stages of the WARS formation (Dalziel 1992). Subglacial volcanic sites also overlap with areas of relatively thick ice and slow ice-surface flow, both of which are critical conditions for the preservation of volcanic records (Behrendt *et al.* 1998). However, other geological and glaciological considerations can explain the spatial distribution of active subglacial sites observed in West Antarctica. Factors such as crustal age and thickness can explain the lack of active subglacial volcanism within sectors characterized by thicker crust, such as crustal blocks and the Siple Coast (Chaput *et al.* 2014). On the other hand, the lithosphere underlying Thwaites Glacier's main trunk is likely to have cooled, and volcanic activity in the area ceased since the last emplacement of thick mafic wedges

(Quartini 2018) during the early mid-Mesozoic stages of WARS formation, similar to those hypothesized offshore (Gohl et al. 2013).

Finally, the glaciological regimes in each sector impose biases to the preservation of subglacial volcanic records which can potentially prevent detection of volcanic activity even where present. Subglacial eruptions under thin ice produce more brittle and less consolidated deposits, such as hyaloclastites, compared to eruptions under the high-pressure conditions of thicker ice columns, where more erosion-resistant pillow lava flows form (Gudmundsson et al. 1997). As a result, the products of volcanic activity under thinner ice columns are more prone to erosion and removal by the ice sheet (Behrendt et al. 1998). This is particularly relevant in regions of fast ice flow, such as the Siple Coast and the main trunk of Thwaites Glacier. While Thwaites Glacier is a region of high driving stresses and thick ice, the Siple Coast ice streams are characterized by thin ice, making this a region particularly prone to loss of subglacial volcanic records.

By altering the ice's thermal structure and generating basal melt water, heterogeneous geothermal flux has the potential to affect the ice dynamics in all sectors of West Antarctica where active subglacial volcanism is observed. Large areas of both West and East Antarctica have been surveyed but not yet searched for evidence of active subglacial volcanism, and it is likely that further discoveries of active subglacial volcanism shall be made.

Acknowledgements The authors thank the anonymous reviewers and the editor for their careful reading of our manuscript and their many insightful comments and suggestions. This is UTIG contribution 3653.

Author contributions EQ: conceptualization (equal), data curation (supporting), formal analysis (lead), visualization (lead), writing – original draft (lead), writing – review & editing (lead); DDB: conceptualization (lead), methodology (equal), project administration (lead), supervision (lead); DAY: conceptualization (supporting), funding acquisition (equal), methodology (supporting), software (supporting), visualization (supporting), writing – original draft (supporting), writing – review & editing (equal).

Funding This work was supported by the G. Unger Veltesen Foundation (D.D. Blankenship), a graduate fellowship from the University of Texas Institute for Geophysics (E. Quartini), and NSF Directorate for Geosciences grant PLR-1043761 (D.A. Young).

Data availability All data (with the exception of the radargrams due to size and complexity) are available at the following archives – ice thickness: https://doi.org/10.15784/601001 (GIMBLE), https://doi.org/10.7265/N5W95730 (AGASEA) and http://www-udc.ig.utexas.edu/external/facilities/aero/data/ (SOAR); gravity: https://doi.org/10.15784/601290 (SOAR); magnetics: https://doi.org/10.1594/PANGAEA.892724 (ADMAP-2); surface elevation: https://www.pgc.umn.edu/data/rema (REMA); surface imagery: https://doi.org/10.7265/N5ZK5DM5 (MOA). Radargrams are available from the authors on reasonable request.

References

Accardo, N.J., Wiens, D.A. et al. 2014. Upper mantle seismic anisotropy beneath the West Antarctic Rift System and surrounding region from shear wave splitting analysis. *Geophysical Journal International*, **198**, 414–429, https://doi.org/10.1093/gji/ggu117

An, M., Wiens, D.A. et al. 2015. Temperature, lithosphere-asthenosphere boundary, and heat flux beneath the Antarctic Plate inferred from seismic velocities. *Journal of Geophysical Research: Solid Earth*, **120**, 8720–8742, https://doi.org/10.1002/2015JB011917

Barletta, V.R., Bevis, M. et al. 2018. Observed rapid bedrock uplift in Amundsen Sea Embayment promotes ice-sheet stability. *Science*, **360**, 6395, 1335–1339, https://doi.org/10.1126/science.aao1447

Begeman, C.B., Tulaczyk, S.M. and Fisher, A.T. 2017. Spatially variable geothermal heat flux in West Antarctica: evidence and implications. *Geophysical Research Letters*, **44**, 9823–9832, https://doi.org/10.1002/2017GL075579

Behrendt, J.C. 1964. Distribution of narrow-width magnetic anomalies in Antarctica. *Science*, **144**, 995–999, https://doi.org/10.1126/science.144.3621.993

Behrendt, J.C. 2013. The aeromagnetic method as a tool to identify Cenozoic magmatism in the West Antarctic Rift System beneath the West Antarctic Ice Sheet – A review; Thiel subglacial volcano as possible source of the ash layer in the WAISCORE. *Tectonophysics*, **585**, 124–136, https://doi.org/10.1016/j.tecto.2012.06.035

Behrendt, J.C., LeMasurier, W.E., Cooper, A.K., Tessensohn, F., Tréhu, A. and Damaske, D. 1991. Geophysical studies of the West Antarctic Rift System. *Tectonics*, **10**, 1257–1273, https://doi.org/10.1029/91TC00868

Behrendt, J.C., Blankenship, D.D., Finn, C.A., Bell, R.E., Sweeney, R.E., Hodge, S.M. and Brozena, J.M. 1994. CASERTZ aeromagnetic data reveal late Cenozoic flood basalts(?) in the West Antarctic rift system. *Geology*, **22**, 527–530, https://doi.org/10.1130/0091-7613(1994)022<0527:CADRLC>2.3.CO;2

Behrendt, J.C., Blankenship, D.D., Damaske, D. and Cooper, A.K. 1995. Glacial removal of late Cenozoic subglacially emplaced volcanic edifices by the West Antarctic ice sheet. *Geology*, **23**, 1111–1114, https://doi.org/10.1130/0091-7613(1995)023<1111:GROLCS>2.3.CO;2

Behrendt, J.C., Saltus, R., Damaske, D., McCafferty, A., Finn, C.A., Blankenship, D. and Bell, R.E. 1996. Patterns of late Cenozoic volcanic and tectonic activity in the West Antarctic rift system revealed by aeromagnetic surveys. *Tectonics*, **15**, 660–676, https://doi.org/10.1029/95TC03500

Behrendt, J.C., Finn, C.A., Blankenship, D. and Bell, R.E. 1998. Aeromagnetic evidence for a volcanic caldera(?) complex beneath the divide of the West Antarctic Ice Sheet. *Geophysical Research Letters*, **25**, 4385–4388, https://doi.org/10.1029/1998GL900101

Behrendt, J.C., Blankenship, D.D., Morse, D.L., Finn, C.A. and Bell, R.E. 2002. Subglacial volcanic features beneath the West Antarctic Ice Sheet interpreted from aeromagnetic and radar ice sounding. *Geological Society, London, Special Publications*, **202**, 337–355, https://doi.org/10.1144/GSL.SP.2002.202.01.17

Behrendt, J.C., Finn, C.A. and Blankenship, D.D. 2006. Examples of models fit to magnetic anomalies observed over subaerial, submarine, and subglacial volcanoes in the West Antarctic Rift System. Abstract V44A-02 presented at the American Geophysical Union Fall Meeting 2006, 11–15 December 2006, San Francisco, California, USA.

Bhattacharyya, B. and Leu, L.-K. 1975. Analysis of magnetic anomalies over Yellowstone National Park: mapping of Curie point isothermal surface for geothermal reconnaissance. *Journal of Geophysical Research*, **80**, 4461–4465, https://doi.org/10.1029/JB080i032p04461

Bingham, R.G. and Siegert, M.J. 2007. Radio-echo sounding over polar ice masses. *Journal of Environmental and Engineering Geophysics*, **12**, 47–62, https://doi.org/10.2113/JEEG12.1.47

Björnsson, H. 2003. Subglacial lakes and jökulhlaups in Iceland. *Global and Planetary Change*, **35**, 255–271, https://doi.org/10.1016/S0921-8181(02)00130-3

Blankenship, D.D., Bell, R.E., Hodge, S.M., Brozena, J.M., Behrendt, J.C. and Finn, C.A. 1993. Active volcanism beneath the

West Antarctic ice sheet and implications for ice-sheet stability. *Nature*, **361**, 526–529, https://doi.org/10.1038/361526a0

Blankenship, D.D., Morse, D. et al. 2001. Geological controls on the initiation of rapid basal motion for West Antarctic Ice Streams: a geophysical perspective including new airborne radar sounding and laser altimetry results. *American Geophysical Union Antarctic Research Series*, **77**, 105–121.

Blankenship, D.D., Young, D.A., Holt, J.W. and Kempf, S.D. 2012. *AGASEA Ice Thickness Profile Data From the Amundsen Sea Embayment, Antarctica*. United States Antarctic Program (USAP) Data Center, Boulder, CO, https://doi.org/10.7265/N5W95730

Buizert, C., Cuffey, K. et al. 2015. The WAIS Divide deep ice core WD2014 chronology – part 1: Methane synchronization (68–31 ka BP) and the gas age–ice age difference. *Climate of the Past*, **11**, 153, https://doi.org/10.5194/cp-11-153-2015

Cavitte, M.G.P., Blankenship, D.D. et al. 2016. Deep radiostratigraphy of the East Antarctic Plateau: connecting the Dome C and Vostok ice core sites. *Journal of Glaciology*, **62**, 323–334, https://doi.org/10.1017/jog.2016.11

Chaput, J., Aster, R.C. et al. 2014. The crustal thickness of West Antarctica. *Journal of Geophysical Research*, **119**, 378–395, https://doi.org/10.1002/2013JB010642

Christianson, K., Jacobel, R.W., Horgan, H.J., Anandakrishnan, S. and Alley, R.B. 2012. Subglacial Lake Whillans – Ice-penetrating radar and GPS observations of a shallow active reservoir beneath a West Antarctic ice stream. *Earth and Planetary Science Letters*, **331–332**, 237–245, https://doi.org/10.1016/j.epsl.2012.03.013

Christoffersen, P., Bougamont, M., Carter, S.P., Fricker, H.A. and Tulaczyk, S. 2014. Significant groundwater contribution to Antarctic ice streams hydrologic budget. *Geophysical Research Letters*, **41**, 2003–2010, https://doi.org/10.1002/2014GL059250

Clow, G., Cuffey, K. and Waddington, E. 2012. High heat-flow beneath the central portion of the West Antarctic Ice Sheet. Abstract C31A-0577 presented at the American Geophysical Union Fall Meeting 2012, 3–7 December 2012, San Francisco, California, USA.

Corr, H.F.J. and Vaughan, D.G. 2008. A recent volcanic eruption beneath the West Antarctic ice sheet. *Nature Geoscience*, **1**, 122–125, https://doi.org/10.1038/ngeo106

Dalziel, I.W.D. 1992. Antarctica: a tale of two supercontinents? *Annual Review of Earth and Planetary Sciences*, **20**, 501–526, https://doi.org/10.1146/annurev.ea.20.050192.002441

Dalziel, I.W.D. and Lawver, L.A. 2001. The lithospheric setting of the West Antarctic Ice Sheet. *American Geophysical Union Antarctic Research Series*, **77**, 13–44.

Damiani, T.M., Jordan, T.A., Ferraccioli, F., Young, D.A. and Blankenship, D.D. 2014. Variable crustal thickness beneath Thwaites Glacier revealed from airborne gravimetry, possible implications for geothermal heat flux in West Antarctica. *Earth and Planetary Science Letters*, **407**, 109–122, https://doi.org/10.1016/j.epsl.2014.09.023

Danque, H.A. 2008. *Subglacial West Antarctic Volcanoes Defined by Aerogeophysical Data and the Potential for Associated Hydrothermal Systems*. Master's thesis, University of Texas at Austin, Austin, Texas, USA.

Davies, J.H. 2013. Global map of solid Earth surface heat flow. *Geochemistry, Geophysics, Geosystems*, **14**, 4608–4622, https://doi.org/10.1002/ggge.20271

DeConto, R.M. and Pollard, D. 2016. Contribution of Antarctica to past and future sea-level rise. *Nature*, **531**, 591–597, https://doi.org/10.1038/nature17145

Diehl, T.M. 2008. *Gravity Analyses for the Crustal Structure and Subglacial Geology of West Antarctica, Particularly Beneath Thwaites Glacier*. Doctoral thesis, University of Texas at Austin, Austin, Texas, USA.

Diehl, T.M., Holt, J.W., Blankenship, D.D., Young, D.A., Jordan, T.A. and Ferraccioli, F. 2008. First airborne gravity results over the Thwaites Glacier catchment, West Antarctica. *Geochemistry, Geophysics, Geosystems*, **8**, Q04011, https://doi.org/10.1029/2007GC001878

Emry, E., Nyblade, A.A. et al. 2015. The mantle transition zone beneath West Antarctica: seismic evidence for hydration and thermal upwellings. *Geochemistry, Geophysics, Geosystems*, **16**, 40–58, https://doi.org/10.1002/2014GC005588

Engelhardt, H. 2004a. Ice temperature and high geothermal flux at Siple Dome, West Antarctica, from borehole measurements. *Journal of Glaciology*, **50**, 251–256, https://doi.org/10.3189/172756504781830105

Engelhardt, H. 2004b. Thermal regime and dynamics of the West Antarctic ice sheet. *Annals of Glaciology*, **39**, 85–92, https://doi.org/10.3189/172756404781814203

Fahnestock, M., Abdalati, W., Joughin, I., Brozena, J. and Gogineni, P. 2001. High geothermal heat flow, basal melt, and the origin of rapid ice flow in central Greenland. *Science*, **294**, 2338–2342, https://doi.org/10.1126/science.1065370

Filina, I.Y., Blankenship, D.D., Thoma, M., Lukin, V.V., Masolov, V.N. and Sen, M.K. 2008. New 3D bathymetry and sediment distribution in Lake Vostok: implication for pre-glacial origin and numerical modeling of the internal processes within the lake. *Earth and Planetary Science Letters*, **276**, 106–114, https://doi.org/10.1016/j.epsl.2008.09.012

Fisher, A.T., Mankoff, K.D., Tulaczyk, S.M., Tyler, S.W. and Foley, N. 2015. High geothermal heat flux measured below the West Antarctic Ice Sheet. *Science Advances*, **1**, https://doi.org/10.1126/sciadv.1500093

Fretwell, P., Pritchard, H.D. et al. 2013. Bedmap2: Improved ice bed, surface and thickness datasets for Antarctica. *The Cryosphere*, **7**, 375–393, https://doi.org/10.5194/tc-7-375-2013

Fricker, H.A. and Padman, L. 2012. Thirty years of elevation change on Antarctic Peninsula ice shelves from multimission satellite radar altimetry. *Journal of Geophysical Research*, **117**, C02026, https://doi.org/10.1029/2011JC007126

Fudge, T.J., Steig, E.J. et al. 2013. Onset of deglacial warming in West Antarctica driven by local orbital forcing. *Nature*, **500**, 440–444, https://doi.org/10.1038/nature12376

Fujita, S., Maeno, H., Uratsuka, S., Furukawa, T., Mae, S., Fujii, Y. and Watanabe, O. 1999. Nature of radio echo layering in the Antarctic ice sheet detected by a two-frequency experiment. *Journal of Geophysical Research*, **104**, 13 049–13 060, https://doi.org/10.1029/1998JB900034

Gohl, K., Denk, A., Eagles, G. and Wobbe, F. 2013. Deciphering tectonic phases of the Amundsen Sea Embayment shelf, West Antarctica, from a magnetic anomaly grid. *Tectonophysics*, **585**, 113–123, https://doi.org/10.1016/j.tecto.2012.06.036

Golynsky, A.V., Ferraccioli, F. et al. 2018. New magnetic anomaly map of the Antarctic. *Geophysical Research Letters*, **45**, 6437–6449, https://doi.org/10.1029/2018GL078153

Gudmandsen, P. 1971. Electromagnetic probing of ice. *In*: Wait, J.R. (ed.) *Electromagnetic Probing in Geophysics*. Golem Press, Boulder, CO, 321–348.

Gudmundsson, M.T. 1996. Ice–volcano interaction at the subglacial Grímsvötn Volcano, Iceland. *In*: Colbeck, S.C. (ed.) *Glaciers, Ice Sheets and Volcanoes: A Tribute to Mark F. Meier*. United States Army Special Report, **96-27**, 34–40.

Gudmundsson, M.T. and Högnadóttir, T. 2007. Volcanic systems and calderas in the Vatnajökull region, central Iceland: Constraints on crustal structure from gravity data. *Journal of Geodynamics*, **43**, 153–169, https://doi.org/10.1016/j.jog.2006.09.015

Gudmundsson, M.T., Sigmundsson, F. and Björnsson, H. 1997. Ice–volcano interaction of the 1996 Gjálp subglacial eruption, vatnajökull, iceland. *Nature*, **389**, 954–957, https://doi.org/10.1038/40122

Hammer, C.U., Clausen, H.B. and Langway, J.C.C. 1997. 50 000 years of recorded global volcanism. *Climatic Change*, **35**, 1–15, https://doi.org/10.1023/A:1005344225434

Hansen, S.E., Graw, J.H. et al. 2014. Imaging the Antarctic mantle using adaptively parameterized P-wave tomography: evidence for heterogeneous structure beneath West Antarctica. *Earth and Planetary Science Letters*, **408**, 66–78, https://doi.org/10.1016/j.epsl.2014.09.043

Hawkesworth, C.J., Blake, S. et al. 2000. Time scales of crystal fractionation in magma chambers – integrating physical, isotopic and geochemical perspectives. *Journal of Petrology*, **41**, 991–1006, https://doi.org/10.1093/petrology/41.7.991

Heeszel, D.S., Wiens, D.A. et al. 2016. Upper mantle structure of central and West Antarctica from array analysis of Rayleigh wave phase velocities. *Journal of Geophysical Research: Solid Earth*, **121**, 1758–1775, https://doi.org/10.1002/2015JB012616

Hindmarsh, R.C., Leysinger Vieli, G.J., Raymond, M.J. and Gudmundsson, G.H. 2006. Draping or overriding: the effect of horizontal stress gradients on internal layer architecture in ice sheets. *Journal of Geophysical Research: Earth Surface*, **111**, F02018, https://doi.org/10.1029/2005JF000309

Holt, J.W., Blankenship, D.D. et al. 2006. New boundary conditions for the West Antarctic ice sheet: Subglacial topography of the Thwaites and Smith Glacier catchments. *Geophysical Research Letters*, **33**, L09502, https://doi.org/10.1029/2005GL025561

Howat, I.M., Porter, C., Smith, B.E., Noh, M.-J. and Morin, P. 2019. The Reference Elevation Model of Antarctica. *The Cryosphere*, **13**, 665–674, https://doi.org/10.5194/tc-13-665-2019

Iverson, N.A., Lieb-Lappen, R., Dunbar, N.W., Obbard, R., Kim, E. and Golden, E. 2017. The first physical evidence of subglacial volcanism under the West Antarctic Ice Sheet. *Scientific Reports*, **7**, 11457, https://doi.org/10.1038/s41598-017-11515-3

Jacobel, R.W., Gades, A.M., Gottschling, D.L., Hodge, S.M. and Wright, D.L. 1993. Interpretation of radar-detected internal layer folding in West Antarctic ice streams. *Journal of Glaciology*, **39**, 528–537, https://doi.org/10.1017/S0022143000016427

Jankowski, E.J. and Drewry, D.J. 1981. The structure of West Antarctica from geophysical studies. *Nature*, **291**, 17–21, https://doi.org/10.1038/291017a0

Joughin, I., Smith, B.E. and Medley, B. 2014. Marine ice sheet collapse potentially under way for the Thwaites Glacier basin, West Antarctica. *Science*, **344**, 735–738, https://doi.org/10.1126/science.1249055

Kurbatov, A.V., Zielinski, G.A., Dunbar, N.W., Mayewski, P.A., Meyerson, E.A., Sneed, S.B. and Taylor, K.C. 2006. A 12 000 year record of explosive volcanism in the Siple Dome Ice Core, West Antarctica. *Journal of Geophysical Research: Atmospheres*, **111**, D12307, https://doi.org/10.1029/2005JD006072

LeMasurier, W.E. 2006. What supports the Marie Byrd Land Dome? An evaluation of potential uplift mechanisms in a continental rift system. *In*: Fütterer, D.K., Damaske, D., Kleinschmidt, G., Miller, H. and Tessensohn, F. (eds) *Antarctica Contributions to Global Earth Sciences*. Springer, Berlin, 299–302, https://doi.org/10.1007/3-540-32934-X_37

LeMasurier, W.E. 2008. Neogene extension and basin deepening in the West Antarctic rift inferred from comparisons with the East African rift and other analogs. *Geology*, **36**, 247–250, https://doi.org/10.1130/G24363A.1

LeMasurier, W.E. and Rex, D.C. 1989. Evolution of linear volcanic ranges in Marie Byrd Land, West Antarctica. *Journal of Geophysical Research: Solid Earth*, **94**, 7223–7236, https://doi.org/10.1029/JB094iB06p07223

Leuschen, C. and Allen, C. 2011. *IceBridge MCoRDS L2 Ice Thickness*. United States Antarctic Program (USAP) Data Center, Boulder, CO, http://nsidc.org/data/irmcr2.html

Lloyd, A.J., Wiens, D.A. et al. 2015. A seismic transect across West Antarctica: evidence for mantle thermal anomalies beneath the Bentley Subglacial Trench and the Marie Byrd Land dome. *Journal of Geophysical Research: Solid Earth*, **120**, 8439–8460, https://doi.org/10.1002/2015JB012455

Lough, A.C., Wiens, D.A. et al. 2013. Seismic detection of an active subglacial magmatic complex in Marie Byrd Land, Antarctica. *Nature Geoscience*, **6**, 1031–1035, https://doi.org/10.1038/ngeo1992

Maccaferri, F., Rivalta, E., Keir, D. and Acocella, V. 2014. Off-rift volcanism in rift zones determined by crustal unloading. *Nature Geoscience*, **7**, 297–300, https://doi.org/10.1038/ngeo2110

Matsuoka, K. 2011. Pitfalls in radar diagnosis of ice-sheet bed conditions: Lessons from englacial attenuation models. *Geophysical Research Letters*, **38**, L05505, https://doi.org/10.1029/2010GL046205

McConnell, J.R., Burke, A. et al. 2017. Synchronous volcanic eruptions and abrupt climate change c. 17.7 ka plausibly linked by stratospheric ozone depletion. *Proceedings of the National Academy of Sciences of the United States of America*, **114**, 10 035–10 040, https://doi.org/10.1073/pnas.1705595114

Morse, D.L., Blankenship, D.D., Waddington, E.D. and Neumann, T.A. 2002. A site for deep ice coring in West Antarctica: results from aerogeophysical surveys and thermo-kinematic modeling. *Annals of Glaciology*, **35**, 36–44, https://doi.org/10.3189/172756402781816636

Mouginot, J., Rignot, E. and Scheuchl, B. 2014. Sustained increase in ice discharge from the Amundsen Sea Embayment, West Antarctica, from 1973 to 2013. *Geophysical Research Letters*, **41**, 1576–1584, https://doi.org/10.1002/2013GL059069

Mouginot, J., Rignot, E. and Scheuchl, B. 2019. Continent-wide, interferometric SAR phase, mapping of Antarctic ice velocity. *Geophysical Research Letters*, **46**, 9710–9718, https://doi.org/10.1029/2019GL083826

Nereson, N.A., Raymond, C.F., Waddington, E.D. and Jacobel, R.W. 1998. Recent migration of Siple Dome ice divide, West Antarctica. *Journal of Glaciology*, **44**, 643–652, https://doi.org/10.1017/S0022143000002148

Nichols, M., Malone, S., Moran, S., Thelen, W. and Vidale, J. 2011. Deep long-period earthquakes beneath Washington and Oregon volcanoes. *Journal of Volcanology and Geothermal Research*, **200**, 116–128, https://doi.org/10.1016/j.jvolgeores.2010.12.005

Okubo, P.G. and Wolfe, C.J. 2008. Swarms of similar long-period earthquakes in the mantle beneath Mauna Loa Volcano. *Journal of Volcanology and Geothermal Research*, **178**, 787–794, https://doi.org/10.1016/j.jvolgeores.2008.09.007

Palais, J.M., Kyle, P.R., McIntosh, W.C. and Seward, D. 1988. Magmatic and phreatomagmatic volcanic activity at Mt. Takahe, West Antarctica, based on tephra layers in the Byrd ice core and field observations at Mt. Takahe. *Journal of Volcanology and Geothermal Research*, **35**, 295–317, https://doi.org/10.1016/0377-0273(88)90025-X

Paulsen, T.S. and Wilson, T.J. 2010. Evolution of neogene volcanism and stress patterns in the glaciated West Antarctic Rift, Marie Byrd Land, Antarctica. *Journal of the Geological Society, London*, **167**, 401–416, https://doi.org/10.1144/0016-7649 2009-044

Peters, M.E., Blankenship, D.D. and Morse, D.L. 2005. Analysis techniques for coherent airborne radar sounding: Application to West Antarctic ice streams. *Journal of Geophysical Research: Solid Earth*, **110**, B06303, https://doi.org/10.1029/2004JB003222

Peters, M.E., Blankenship, D.D., Carter, S.P., Young, D.A., Kempf, S.D. and Holt, J.W. 2007. Along-track focusing of airborne radar sounding data from West Antarctica for improving basal reflection analysis and layer detection. *IEEE Transactions on Geoscience and Remote Sensing*, **45**, 2725–2736, https://doi.org/10.1109/TGRS.2007.897416

Pittard, M.L., Galton-Fenzi, B.K., Roberts, J.L. and Watson, C.S. 2016. Organization of ice flow by localized regions of elevated geothermal heat flux. *Geophysical Research Letters*, **43**, 3342–3350, https://doi.org/10.1002/2016GL068436

Power, J., Stihler, S., White, R. and Moran, S. 2004. Observations of deep long-period (DLP) seismic events beneath Aleutian arc volcanoes; 1989–2002. *Journal of Volcanology and Geothermal Research*, **138**, 243–266, https://doi.org/10.1016/j.jvolgeores.2004.07.005

Price, S.F., Bindschadler, R.A., Hulbe, C.L. and Joughin, I.R. 2001. Post-stagnation behavior in the upstream regions of Ice Stream C, West Antarctica. *Journal of Glaciology*, **47**, 283–294, https://doi.org/10.3189/172756501781832232

Quartini, E. 2018. *The Distribution of Geothermal Flux in West Antarctica*. Doctoral thesis, University of Texas at Austin, Austin, Texas, USA.

Raymond, C.F. 1983. Deformation in the vicinity of ice divides. *Journal of Glaciology*, **29**, 357–373, https://doi.org/10.1017/S0022143000030288

Retzlaff, R. and Bentley, C.R. 1993. Timing of stagnation of Ice Stream C, West Antarctica, from short-pulse radar studies of buried surface crevasses. *Journal of Glaciology*, **39**, 553–561, https://doi.org/10.3189/S0022143000016440

Rignot, E., Mouginot, J., Morlighem, M., Seroussi, H. and Scheuchl, B. 2014. Widespread, rapid grounding line retreat of Pine Island, Thwaites, Smith, and Kohler glaciers, West Antarctica, from 1992 to 2011. *Geophysical Research Letters*, **41**, 3502–3509, https://doi.org/10.1002/2014GL060140

Rowley, P.D., Thomson, J.W., Smellie, J.L., Laudon, T.S., La Prade, K.E. and LeMasurier, W.E. 1986. Alexander Island, Palmer Island, and Ellsworth Land. *American Geophysical Union Antarctic Research Series*, **48**, 256–301, https://doi.org/10.1029/AR048p0256

Rowley, P.D., Laudon, T.S., La Prade, K.E. and LeMasurier, W.E. 1990. Hudson Mountains. *American Geophysical Union Antarctic Research Series*, **48**, 289–293.

Scambos, T.A., Haran, T.M., Fahnestock, M.A., Painter, T.H. and Bohlander, J. 2007. MODIS-based Mosaic of Antarctica (MOA) data sets: Continent-wide surface morphology and snow grain size. *Remote Sensing of Environment*, **111**, 242–257, https://doi.org/10.1016/j.rse.2006.12.020

Scheinert, M., Ferraccioli, F. *et al.* 2016. New Antarctic gravity anomaly grid for enhanced geodetic and geophysical studies in Antarctica. *Geophysical Research Letters*, **43**, 600–610, https://doi.org/10.1002/2015GL067439

Schmidt, B.E., Blankenship, D.D., Patterson, G.W. and Schenk, P.M. 2011. Active formation of 'chaos terrain' over shallow subsurface water on Europa. *Nature*, **479**, 502–505, https://doi.org/10.1038/nature10608

Schoof, C. 2007. Ice sheet grounding line dynamics: steady states, stability, and hysteresis. *Journal of Geophysical Research: Earth Surface*, **112**, F03S28, https://doi.org/10.1029/2006JF000664

Schroeder, D.M., Blankenship, D.D. and Young, D.A. 2013. Evidence for a water system transition beneath Thwaites Glacier, West Antarctica. *Proceedings of the National Academy of Sciences of the United States of America*, **110**, 12 225–12 228, https://doi.org/10.1073/pnas.1302828110

Schroeder, D.M., Blankenship, D.D., Young, D.A. and Quartini, E. 2014. Evidence for elevated and spatially variable geothermal flux beneath the West Antarctic Ice Sheet. *Proceedings of the National Academy of Sciences of the United States of America*, **111**, 9070–9072, https://doi.org/10.1073/pnas.1405184111

Schroeder, D.M., Blankenship, D.D., Raney, R.K. and Grima, C. 2015. Estimating subglacial water geometry using radar bed echo specularity: Application to Thwaites Glacier, West Antarctica. *IEEE Geoscience and Remote Sensing Letters*, **12**, 443–447, https://doi.org/10.1109/LGRS.2014.2337878

Schroeder, D.M., Grima, C. and Blankenship, D.D. 2016. Evidence for variable grounding zone extent and shear margin bed conditions across Thwaites Glacier, West Antarctica. *Geophysics*, **81**, WA35–WA43, https://doi.org/10.1190/geo2015-0122.1

Sergienko, O.V., Creyts, T.T. and Hindmarsh, R.C.A. 2014. Similarity of organized patterns in driving and basal stresses of Antarctic and Greenland ice sheets beneath extensive areas of basal sliding. *Geophysical Research Letters*, **41**, 3925–3932, https://doi.org/10.1002/2014GL059976

Siegert, M.J., Welch, B. *et al.* 2004. Ice flow direction change in interior West Antarctica. *Science*, **305**, 1948–1951, https://doi.org/10.1126/science.1101072

Singer, B.S. 2014. A quaternary geomagnetic instability time scale. *Quaternary Geochronology*, **21**, 29–52, https://doi.org/10.1016/j.quageo.2013.10.003

Singer, B.S., Guillou, H., Jicha, B.R., Laj, C., Kissel, C., Beard, B.L. and Johnson, C.M. 2009. ^{40}Ar/^{39}Ar, K–Ar and ^{230}Th–^{238}U dating of the Laschamp excursion: a radioisotopic tie-point for ice core and climate chronologies. *Earth and Planetary Science Letters*, **286**, 80–88, https://doi.org/10.1016/j.epsl.2009.06.030

Smellie, J. 2000. Subglacial eruptions. *In*: Sigurdsson, H., Houghton, B.F., McNutt, S.R., Rymer, H. and Stix, J. (eds) *The Encyclopedia of Volcanoes*. 2nd edn. Academic Press, San Diego, CA, 403–416.

Smellie, J.L. 2006. The relative importance of supraglacial v. subglacial meltwater escape in basaltic subglacial tuya eruptions: An important unresolved conundrum. *Earth-Science Reviews*, **74**, 241–268, https://doi.org/10.1016/j.earscirev.2005.09.004

Smith, R.B. and Braile, L.W. 1994. The Yellowstone hotspot. *Journal of Volcanology and Geothermal Research*, **61**, 121–187, https://doi.org/10.1016/0377-0273(94)90002-7

Spector, P., Stone, J., Pollard, D., Hillebrand, T., Lewis, C. and Gombiner, J. 2018. West Antarctic sites for subglacial drilling to test for past ice-sheet collapse. *The Cryosphere*, **12**, 2741–2757, https://doi.org/10.5194/tc-12-2741-2018

Stevens, N.T., Parizek, B.R. and Alley, R.B. 2016. Enhancement of volcanism and geothermal heat flux by ice-age cycling: a stress modeling study of Greenland. *Journal of Geophysical Research: Earth Surface*, **121**, 1456–1471, https://doi.org/10.1002/2016JF003855

Sweeney, R.E., Finn, C.A., Blankenship, D.D., Bell, R.E. and Behrendt, J.C. 1999. *Central West Antarctica Aeromagnetic Data: A Web Site for Distribution of Data and Maps*. Online edition. United States Geological Survey Open-File Report, **99-420**, http://pubs.usgs.gov/of/1999/ofr-99-0420/cwantarctica.html

Turcotte, D. and Schubert, G. 2014. *Geodynamics*. Cambridge University Press, Cambridge, UK.

van Wyk de Vries, M., Bingham, R.G. and Hein, A.S. 2018. A new volcanic province: An inventory of subglacial volcanoes in West Antarctica. *Geological Society, London, Special Publications*, **461**, 231–248, https://doi.org/10.1144/SP461.7

Vaughan, D.G., Corr, H.F., Doake, C.S. and Waddington, E.D. 1999. Distortion of isochronous layers in ice revealed by ground-penetrating radar. *Nature*, **398**, 323, https://doi.org/10.1038/18653

Vaughan, D.G., Corr, H.F.J. *et al.* 2006. New boundary conditions for the West Antarctic Ice Sheet: Subglacial topography beneath Pine Island Glacier. *Geophysical Research Letters*, **33**, L09501, https://doi.org/10.1029/2005GL025588

Vogel, S.W. and Tulaczyk, S.M. 2006. Ice-dynamical constraints on the existence and impact of subglacial volcanism on West Antarctic ice sheet stability. *Geophysical Research Letters*, **33**, L23502, https://doi.org/10.1029/2006GL027345

Vogel, S.W., Tulaczyk, S.M., Carter, S.P., Renne, P. and Turrin, B. 2006. Geologic constraints on the existence and distribution of West Antarctic subglacial volcanism. *Geophysical Research Letters*, **33**, L23501, https://doi.org/10.1029/2006GL027344

Weertman, J. 1974. Stability of the junction of an ice sheet and ice shelf. *Journal of Glaciology*, **13**, 3–11, https://doi.org/10.1017/S0022143000023327

Whillans, I.M. 1976. Radio-echo layers and the recent stability of the West Antarctic ice sheet. *Nature*, **264**, 152, https://doi.org/10.1038/264152a0

Wilch, T.I. and McIntosh, W.C. 2002. Lithofacies analysis and ^{40}Ar/^{39}Ar geochronology of ice–volcano interactions at Mt. Murphy and the Crary Mountains, Marie Byrd Land, Antarctica. *Geological Society, London, Special Publications*, **202**, 237–253, https://doi.org/10.1144/GSL.SP.2002.202.01.12

Wilson, D.S. and Luyendyk, B.P. 2009. West Antarctic paleotopography estimated at the Eocene–Oligocene climate transition. *Geophysical Research Letters*, **36**, L16302, https://doi.org/10.1029/2009GL039297

Winberry, J.P. and Anandakrishnan, S. 2004. Crustal structure of the West Antarctic rift system and Marie Byrd Land hotspot. *Geology*, **32**, 977–980, https://doi.org/10.1130/G20768.1

Young, D.A., Schroeder, D.M., Blankenship, D.D., Kempf, S.D. and Quartini, E. 2015. The distribution of basal water between Antarctic subglacial lakes from radar sounding. *Philosophical Transactions of the Royal Society A*, **374**, 1–21, https://doi.org/10.1098/rsta.2014.0297

Young, D.A., Blankenship, D.D., Kempf, S.D., Quartini, E., Muldoon, G.R. and Powell, E.M. 2017. *Ice Thickness and Related Data Over Central Marie Byrd Land, West Antarctica (GIMBLE.GR2HI2)*. United States Antarctic Program (USAP) Data Center, Boulder, CO, https://doi.org/10.15784/601001

Index

Page numbers in *italics* refer to Figures. Page numbers in **bold** refer to Tables.

'a'ā lava 22, 160, 164, 243, 762, 768
'a'ā lava-fed delta 32, 34
    Erebus Volcanic Province 422–427, 431–435
    Victoria Land 349, 352, 353, *354*
        Hallett Volcanic Province *360*, 361–366
        Melbourne Volcanic Province 373–374, 379, 541, 568
accretionary complex 2, 3, 12, 23 26, 33, 185
    Antarctic Peninsula 190, 202–204, 207, 213–215
acidity event *c.* 7.6 ka 659–661
active faulting 384, 449
active mantle plume 45, 406
active rift basin 231
active subduction *328*
    South Shetland Islands 214, 285, *286*, 671
active volcanism, overview 57–68
    hazard, risk and monitoring 67–68
    latest eruptions **55**, 56
    subglacial 571, 785–799
active volcanoes 3–4, 26, 30, 45, 272, 307, 631
    Deception 667–668
    James Ross Island 255
    maps *2, 20, 44, 632, 651, 797*
    Marie Byrd Land Volcanic Province 571, 659
        Mount Berlin 515, *516*
    Mount Melbourne 58–59, *345*
    Mount Rittmann 741, 748–755
    Mount Takahe 60–62, 571, 759–765, 779–781
    Mount Waesche *see* Mount Waesche
    Pleiades (The) 642, 655, 659
    Victoria Land 58, 371, 373, 379
        Mount Erebus 416, 441
adakitic group *217*, 218–222
Adare Basin, seamount *44*, 49
Adare Peninsula Volcanic Field 349, **350**, 352–354, 360, 386
    geochemistry **355**, *397*
    petrology **386**
Adare Trough, extension 383, 401–402
Adelaide Island, arc succession 23–24, 44, *195*, 196–199, 220
    age of volcanics 204–205
    subduction 214
Adelie penguin rookeries 455
aerosols and volcanism 649, 657
    measurement 706–707
    stratosphere, troposphere 60, 660–661
    sulfate spike 654, *655*
agglomerate 128, 129, 135
agglutinate 312, *317*, 322
Ahlmannryggen–Sverdrupfjella intrusive suite 165–166, 174
    geochemistry 167, **168–169**, *170*, 171
air traffic disruption, ash 55, 58, 60, 67, 667, 729–730
air-fall tuff 126, 131, 135, 140, 207
Airborne Geophysical Survey, Amundsen Embayment *786*, 787
Åkerlundh Nunatak 309, *310*
albedo 62

Alexander Island 185, *191*, 221–222
    accretionary complex 202–204, 207, 213–217
    geochemistry 219, **311**, 330–333
    lithofacies 316–318, 320–321
    subglacial eruption 568
    volcanism 23–24, 33
algae 717, 719
Alkali Group 232
alkaline back-arc volcanism 26–28
alkaline magmatism 28–29, 605, *623*, 670, 743
    chemistry 493, *504*, 508, 509–510, 695, 697, 723
    Erebus 416, 429, 449–451, **453**, 457, 465, 480
    Marie Byrd Land 518, 579–593, 595, 597–603
    Mount Early, Sheridan Bluff 491, 496, 499
    post-subduction 327–328, 330–341
    subduction-related 285, 287–299, 305–306, 322
    Victoria Land 349, 379, 383, 388, 399, 404, 406
alkaline volcanism, rift-related 29–33
    petrogenetic model 49
Allan Hills, volcaniclastic rocks 78–82
    blue-ice tephra 656
Ames Range Volcanic Field **522**, 553–554
    geochemistry 589–591
    Miocene ice level 569
Amundsen Sea Embayment *786*
    ice stream flow rate 796–797
amygdaloidal basalt 132–134, 150–151, 154
amygdule pipe 165
Andean-type arc 13
Anderson Island, volcanic rocks 270, *271*
Andes, volcanic ash dispersal 658
andesite 108, 129, 135, 157, 203–207, 238
    geochronology **186–189**, 191–201
ANDRILL, McMurdo Ice Shelf drilling programme **636–637**, *639*, 640–641
anorthoclase crystal/mega- *701*, 706
    growth rate 726–728
    melt inclusions 710, 723
anorthoclase-rich phonolite 695
Antarctic Drilling Project 469
Antarctic Peninsula
    geochemical data **288–289**
    petrology 139–154, 216–222, 327–341
    post-subduction tectonic setting 327–330
    post-subduction volcanism 28–29, 305–323
    subduction 20, 23, 63–64
    volcanism, overview 1–4, 53–66
    volcanology 121–136
Antarctic Peninsula Ice Sheet 27, 29
Antarctic Peninsula Volcanic Group 123, *124*, 185
Antarctic Peninsula, volcanic arc 44, *192, 195*
    age of inception 213
    axis migration 204–205, 207, 213, *214*
    geochronology **186–189**

    metamorphism 214–215
    petrology 216–222, 327–341
    tectonic development 185, 190
    volcanology 206–207, **228–230**, *277, 295*
Antarctic Plate 9, 45, 110, *306*, 449, 516, *668*
    displacement rate 683
    mid-ocean ridges 401
    rotation 518
Antarctic Sound, volcanic centres 270–273
Antarctic Specially Protected Area 58, 59, 66, 68, 718
Antarctic Treaty System 58, 59, 66
Antarctic–Phoenix Ridge 227, 327–328, 332–333, *340*
    mid-ocean ridge basalt 335–338
    spreading centre 341
Antarctica, maps *20*
    subglacial volcanoes *632*
    tectonic *9*
    volcanic rocks *2, 44*
    volcanoes *56*
Anvers Island, volcanic succession 251–252, *253*
    inventory of volcanoes **230**
Ar–Ar age data
    Antarctic Peninsula 193–194, *203*, 204, 220, **244–249**
        post-subduction volcanics **312**, 318–319
    Bransfield Strait 235, 272, 294–298
    East Antarctica, tephra 656–657
    Erebus 454, 457–458, 460–461, 468–470, 474, 697, *699*
    James Ross Island Volcanic Group 251, 254, *266*, 268, **274–275**
    Jurassic 108, 129, *133*, **134**, 174–175
        Dronning Maud Land 162–165
        silicic rocks 122
    Marie Byrd Land 515, 518, 523–526, 571, 582, 595, 597–604
        Pleistocene to recent, volcanics 761–762, 765, 769–770, 775–776, 780
        summary table **522, 527–528**
    Mount Early, Sheridan Bluff 495
    sea level data *277*
    tephra 638, 640, 642, 643
    Victoria Land 369, **375, 377–378, 494**
Ar/Ar laser fusion radiometric age 468
arc magmatism, migration 204–205, 207, 213–214, 589
Archean basement 159, *172*
Arctowski Nunatak 309–310
Argentine Islands 196–197, 205
    porphyry copper, root zone 214
Argo Point, scoria mound 313–314, 322
ash-flow tuff 135
ash, volcanic 55, 58, 62, 64–66
    in marine drill cores 631–643
    plume height 659
    risk 667, 685, *687*, 781
asthenosphere 479, 720
atmosphere, volcanic emissions 87, 707
    ash dispersal 658, 659
Au–Cl complex 707

Aurora Cliffs, lava-fed delta 424–426
Aurora Ice Caves 746
Australasian Antarctic Expedition (1911–14) 75, 95
Australia–Antarctic separation 106, 406
Australian–Antarctic Ridge 723
avalanche deposit 80
aviation disruption 667, 781
Aviator Glacier 748, 749
  ignimbrite deposit 371

back-arc alkaline volcanism 26–28, 34
back-arc basin 13, 285, 286, 306
bacteria 717–718
  in ice caves 719
Baily Head Formation 239, 240–242
Bald Head, volcanic rocks 128
Balleny islands volcanicity 57
basalt 75–87, 215–216, 352, 510, 511
  chemistry 48
  Deception 673, 674
  Erebus **417–418,** 422, 426–427, 450–451, 454–469
  Marie Byrd Land Volcanic Group 577, 579–608
  Mount Early, Sheridan Bluff 493, 499
    geochemistry 504–507, 509
  Victoria Land, north **385–387**
basalt–andesite association 207, 217, 218, 235, 287
basaltic andesite, Ferrar province 75
Basaltic Shield Formation 239, 241, 673–674, 678
basanite 28, 318, 330, 333, 529
  Erebus Volcanic Province 416, **417–418,** 423, 428, 438, 440–442
    petrology 450–451, **452,** 454–466, 469–471, 474–475
  Mount Discovery, Mount Morning 429–436
  Mount Erebus 695, 697, 709
    magma evolution 700–703, 726, 730–731
  Victoria Land 349, 360, 363, 368, 371, 373, 379, **385–387**
basanite agglutinate 317
base station see research stations
basement rocks 159, 172, 434, 493, 617
  melting, HIMU 720
  protolith 196, 197, 199
Basement Sill 93, 97–98, 102, 106, 108, 111, 112
bathymetry 94, 271, 286, 295
  Bransfield Strait 232, 233
  Port Foster 671
  Victoria Land Basin 440
Beacon Supergroup, sedimentary basin 29
Beacon, sills 95, 108
Beaufort Island stratovolcano 415, 438–440, 442
  petrology and age 451, 453–454
  seamount 441
Beethoven Peninsula Volcanic Field 318, 319, 321, 322
  geochemistry 333
  tuya 319, 321
Bellingshausen Sea Volcanic Group 305–306, **308,** 314–320
benmoreite 220, 222
Berkner Island, ice core research 660
bimodal composition 13, 123, 159, 216, 221–222, 521
biological communities/protection 58, 66
bird, fossil footprint 207
Bisco Islands 196, 197

Björnnutane **168–169,** 170–171, 174
  flood basalt 163–164, 165
Black Coast 130, 134
  dykes 145, 150
  magmatism 153–154
Black Island, stratovolcano remnant 429–430
  petrology and age 461
blue ice 655–657
  Marie Byrd Land 769–776
  tephra 759–760, 778–781
bombs, volcanic 237, 373
  Mount Erebus 697, 699–700, 701, 703
    geochemistry 705–706, 709
Botany Bay Group 122, 124–128
  correlation 135
Boyd Ridge **541,** 541, 545
Brabant Island, succession 251–252, 253
  volcano inventory **230**
Braddock Nuntaks 206
Brandenberger Bluff, ice-dam effect 566–567, 570
Brandenberger Bluff, trachyte tuya **556,** 558–560, 566, 570
Bransfield Basin 231
  rifting 668
Bransfield Group 232
Bransfield Strait 20, 25–26, 34
  active volcanism 64–67
  ash dispersal 641–642
  petrology 285–294
  seamounts 46
  volcanic fields 63
  volcanology 231–232, 233, 277
Bransfield Trough, subduction 674
Brennecke Formation 130, 131, 133, 134–136, 139, 151
  geochemistry, petrography 146–150
Bridgeman Island, volcanology 232–233, 234–235
British Antarctic Expeditions 455
  1901–04 95, 461, 462, 463
  1907–09 75, 95, 455
  1910–13 455
British base on Deception, ruins of 669
Broken Ridge, tephra dispersal **635–636**
Brown Bluff, basaltic glaciovolcano 267–270
Brown Peninsula, stratovolcano remnant 429–430
  petrology and age 461–462
Bruce Nunatak Formation, lithofacies 309, 310, 312
bubble bursts, infrasonic signal 711–712, 713
Buchia Buttress tuff 204
Buckle Island, eruption **55,** 632
Bull Nunatak, volcanics 312
Bunger Hills, ice sheet thickness 621
Butcher Peak Ridge, volcanic lithofacies 525–530
Butcher Ridge, igneous complex 109
Byers Group 192–193, 207
Byrd Antarctic Expedition 1928–30 95
Byrd ice core 776–777, 779

Cairn Point, volcanic sequence 252
calc-alkaline composition 13, 449, **453,** 579
  magma 218–222, 285
  South Shetland Islands 251
  tephra 651, 653
caldera 22, 59, 66, 204, 232, 236, 354, 521, 546, 571
  Daniell Summit 363, 365
  Eastern Volcanic Province 533, 538–539
  Marie Byrd Land 518, 519, 524

Miocene 568–569
Mount Erebus 419, 422, 425, 435
Mount Morning 463, 464
Mount Overlord 369, 370, 371
Mount Rittmann 748–749
nested 135, 422, 441
caldera collapse 236–237, 239
  Deception 241, 670–671, 673–676
  Erebus 422–423, 426, 697, 768, 769
caldera complex, subglacial investigation 793
Callender Peak, volcanic lithofacies 531
Camp Crater 460
Camp Hill Formation 128, 135
Cape Alexander, silicification 130
Cape Barne 423
Cape Barne Pillar 457–458
Cape Bird, age and petrology 428, 455–456
Cape Crozier 426, 427
  age and petrology 456–457
Cape Klovstad, volcanics 354
Cape Purvis, volcano 272
Cape Roberts Drilling Project 402, 447, 468
Cape Roget, volcano 353, **357**
Cape Royds, lava flow 458
Cape Scrymgeour, lava-fed delta 271
Cape Wheatstone Volcano (Southern Volcano) 362–363, 364
Carapace Nunatak 78, 80, 85, 86, 98
carbon isotope dating **244,** 642–643, 709
carbon-fixing bacteria 717
carbonatite 476–478, 480
CASERTZ see Corridor Aerogeophysics of the South Eastern Ross Transect Zone
Castor Nunatak, volcanic rocks 310, 312
Cenozoic to Recent, volcanism 697
Cenozoic volcanism
  Erebus 447, 449–450, 499
  Marie Byrd Land 515, 577–579
Cenozoic, tectonomagmatic provinces 44
Central West Antarctic Ice Sheet
  subglacial fissure system 795–796
CH_4 Mount Rittmann 744, 753–755
Chang Peak volcano 762, 766, 767, 768
chemosynthetic CO_2 fixation 728
Chilean scientific base 64, 65
  ruins on Deception 669
chilled margin 97–99, 103–104, 108, 113, 135
Chon Aike, silicic large igneous province 154
  age 126
  Antarctic Peninsula 139–140, 201
  geochemistry 143
  Gondwana break-up 22, 121, 196
  Patagonia 44, 122–123, 135–136
Christensen Nunatak Formation, lithofacies 309, 312
Churchill Peninsula, silicic volcanics 129–130
CIROS, drilling project **636,** 637–638, 639
clastic dyke 80
clastic interbeds, Dronning Maud Land 162–163, 165
climate and volcanism 19, 21–22, 27, 29, 34, 649–650, 660–661
climate change 87, 469
  Cretaceous–Neogene 205–206
  Eocene–Oligocene, cooling 376
climate record, englacial tephra 769, 776–778, 781
  East Antarctica 649, 651, 653–655, 659
CO Mount Rittmann 755
CO_2 emission 60, 731
  measurement 706, 708–709, **746**
  Mount Rittmann 744–745, 746, 753–755

CO₂ fixation 719, 724–726, *728*
coal 87
coastal ice cliff with tephra *629*
collision 214, 305, *306*
  ridge–crest–trench 327–330, 340–341
columnar joints *85*, 162, 353, 526, 530
  Antarctic Peninsula *201*, 202, 206, 317
  Bransfield Strait 238, 243, *250*, 251, *252*
Commonwealth Trans-Antarctic Expedition 465
compound-braided flow facies *162*, 164–165, 173, *174*
compressional tectonics 669–670
conduit (open-), Mount Erebus 695, 700
  degassing 705–710, 725–726, 729, 731
Conquest of Mount Erebus *698*
contact metamorphism 215, 222
continental arc evolution 9, 121
continental rift volcanism 20
convergence rates 26–27
cooling history, Ferrar province 98
Coombs Hills, sill 86
Coombs Hills, volcaniclastic rocks 77–82
Corridor Aerogeophysics of the South Eastern Ross Transect Zone 786–787, *789, 794*
cosmogenic exposure, age *425*, 621
Coulman Island Volcanic Field *347, 348*, **351**, 364–366
  geochemistry *397*
  petrology **386**
  stratigraphy/lithofacies *367*
cracked-lid model *21*
Crary Mountains Volcanic Field *516*, 518, **522**, 539–545, 569
  geochemistry 591–593
  glaciovolcanic sequences 569
Crater Cirque, mafic lava 374
crater lake, Deception *668*
CRP drilling project **636**, *640*
crustal block boundaries and volcanism 110, 797–798
  geophysical investigation 786–787
crustal contamination 109–110, 506–507, 511, 624, 625
crustal doming 416, 441, 449
crustal magma chambers 408
crustal thickness 29, 33–34, 286, 298
  gravity interpretation 23, *28*
  Marie Byrd Land 516, 607
  Ross Island 697
  West Antarctic Ice Sheet 29, 785, **791**, 793
Cruzen Island **561**, 563
crystal tuff 130, 135, 140, 200, 202, 203
cyanobacteria 717, 719

D'Urville Monument, rhyolitic lava 126
dacite 194, 196–197, 200, 203–204, 206–207, 238
  geochemistry 216, *217*, 218
  geochronology, lava **188–189**
dacite–rhyolite 290
Dailey Islands, volcanic field *415*, 429–431, *460*
Dallman volcano 64
Dalmann Nunatak, volcanics *310*, 312
DAMP *see* diffuse alkaline magmatic province
Daniell Peninsula Volcanic Field 388, **389–390**, *397*
Davey Bank 440
Dawson Peak, geochemistry **100–102**, *103*
debris avalanche deposit 268, 269
debris flow 193, 197, 200, 207, *536*, 543

Decepción Base (Spanish) *65*, 66
  eruption protocol 68
Deception Island
  active volcano 26, 46, 66–67, *227*, 298, 667
  caldera 3, 34, *182, 238*
    evolution and collapse 670–671, 676
  crater *182*
  eruptions (1967, 1968, 1970) 677–679, 682, 685–686
  hazard, risk management 67–68, 685–688
  monitoring 678–685
  geochemistry 286–287, 290–291, 293
  historical activity 677–678
  ignimbrite 32
  magmatic evolution 671–677
  stratigraphy 236, *237*, 673–674, *675*
  tectonic setting 668–670
  tephra 631, 633, 641–642
  volcanology 235–242
Deception Plateau *369*
decompression melting 337, 383, 406, 416, 479–480
  ice load 785–786
  slab-window formation 306, 322, 333–334
Deep Sea Drilling Project (DSDP) *467*, 632–634, **635**
deformation model, inflation–deflation 684, 686
degassing, Mount Erebus 695, 697–698, *701*, 703, 705–711
  conduit 723–728
deglaciation 797
delamination 499, 509–510
Dellbridge Islands, petrology 457
delta 240 *see also* lava-fed delta
diamictite 194, 207, 251, 252, 254, 267, *315*, 364
  Erebus 431–434, 468, 469
  Marie Byrd, Ellsworth Land *529*, 565
  marine record 633, 640
diatomaceous sediment 642
diatomite 469
diatreme structure *see* phreatocauldron
diffuse alkaline magmatic province (DAMP) 3, 12, 47, 450, 478, 720
diffusion chronometry 726
distal tuff association 206–207
Dobson Dome, tuya **229, 247**, *256*, **259**, 266
dolerite pegmatite *98*
dolerite sills, dykes 164
  Ferrar 93–96, **100–102**, *103*
Dome C, long ice-core study 650–651, *652–653, 655*, 658–659
Dome Fuji, long ice-core study 651–652, 659
dome, volcanic 126, 352, 360, 361, 373
Dorrel Rock, volcanic lithofacies 532
Drake Passage 26
Drake Plate–Phoenix Plate, subduction 13, 44
Driencourt Point, volcanic sequence 252
drilling hole locations *448*
drilling programme, marine tephrostratigraphy 631
  gravity and piston cores *638*, 641–642
  ocean expeditions 632–641
  acronyms **633**
drilling project, Dry Valleys (DVDP) 416, 447, *458*, 459, 466–472, 474
  offshore **636**, 637, *639*
drilling project, Hut Point Peninsula 428
Dronning Maud Land, long ice core 652
Dronning Maud Land, volcanism 20, 21, 171–173

flood basalts, source 175–176
  geochemistry 166–171
  geology 157–160
  igneous suites 161–166
  lava, flow and feeder fissures 160–162
  petrology 171–173
  stratigraphy *174*
  volcanic system 173–175
Dry Valleys *78*, 86, 93, 108, 111, *415, 438*
  geochemistry 95
  lava and scoria cone *439*, 442
  magmatic lineage 702–703, 709–710, 721, 723
  uplift 437
Dry Valleys Drilling Project *see under* drilling project
Dufek layered intrusion 96–97, 108, 111, 113
  chemistry 99, 100–102, 104, *106*
Dufek Massif *121*
Dundee Island, volcanic centres 270–273
dyke emplacement *111*, 162
dyke feeder *240*
dyke swarm 13, 21, 162, 167, **385**, 388
  age 383, 406
  geochemistry *158*, **396**, *397, 399*, 598
  radiogenic isotopes *400*
  Victoria Land 403
dykes 126, *310, 365*
  Ferrar Large Igneous Province 21, 44, 77, 78, 111
  Meander Intrusive Group 375–376, **396**, *397*

Early Paleozoic Ross Orogeny 409
earthquakes 285, *402*, 684
  Bransfield Basin 669
  Deception 679–682
  Mount Rittmann 753
  volcano-tectonic 66–67
  see also tremor, seismicity
East Antarctic Ice Sheet (EAIS) 29, 32, 34, 497, 499
  possible Miocene age 491
East Antarctic Ice Sheet, ice tephra 649
  dispersal **635–637**
  location and characteristics **650**, *651*
  tephrostratigraphy *652*, 657–661
East Antarctica, basement rocks 9–11, *20*, 44–45
  volcanoes 510–511
Eastern Marie Byrd Land Volcanic Field *61*, 525–539
eclogite 337, 340–341, 476–478, 480
Edinburgh Hill, dolerite plug 250–251
Eland Mountains 134–135
Eldridge Bluff, volcano and tuff cone 374–375
Ellsworth Land Volcanic Group 131, *133*, 145
Ellsworth Land, volcanology 515–572
  active volcano **55**, *56*, 60–63
  age data **580–581**, *582*
  geochemistry 603–605
  petrology 577–608
Ellsworth–Whitmore Mountains 9
Ellsworth, erosion surface 29
englacial lake 269, 535, 538, 560
ensialic marginal basin 26, 227, 278
environmental impact, magmatism 19, 33
epicentres, Mount Melbourne 747
Erebus Volcanic Province 30, 33, *347–348*, 373–374
  active volcano 45, *58*, 59–60
  eruptive centres **453**
  geological setting 48, 449–450, 697

Erebus Volcanic Province (*Continued*)
    magmatic lineage 702–704
    offshore drilling 467–469
    petrology 447–481
    volcanology 415–442
        lithofacies 420–441
        stratigraphy and synopsis **417–418**
    see also Mount Erebus
Erebus volcano, marine tephra source 632, 638, 641
erosion surface 29, 252, *271*, 433
erratics 422, 426–427, 431, *502*, 543, 545, 565
    kenyte 457
    tholeiite 95
eruption history, Marie Byrd Land 605–608
    Eocene–present day 515–572
eruption, geophysical monitoring techniques 711–716
eruption, rate of 30, 34, 86
    Dronning Maud basalts 174
eruptive column, height of 639, 677
Evensen Nunatak, volcanic rocks 312
Executive Committee Range **798**
    active volcanism 60, *61*
    caldera 521, **522**, 545–550, 569
    geochemistry *585*, 586, 587–589
    subglacial volcanoes 569, 796
Executive Committee Volcanic Field *61*, 62
exhumation, Victoria Land 33
expeditions (1934 to 2019) **520**
expeditions and field work, Marie Byrd Land 518–519, **520**
expeditions, early exploration 75, 95, 455, 461–463, 465
extension 285, 454
    Bransfield Basin 685
    Cenozoic 383, 401, 406, 408–409, 416, 481
    Cretaceous 401–402, 406
    and metasomatism 404
extrusion, rate of 441

Fang Glacier, velocity seismogram *714*
Fang Ridge caldera **417**, 422, 426, 458–459, 474, *696*
faults and mafic dykes 403
faults, Deception *673–674*
faults, extensional 231, *742*
felsic dome *419*, 427, 429–430, 432, 465
felsic magmatism 524, 587, 589
felsic shield volcano 518, 521, 539
Ferrar Continental Flood Basalt 157, *158*, **159**, 176
Ferrar Fjord, drill core **636**, 638
Ferrar Group 19, 75, 468
Ferrar Large Igneous Province, petrology 93–113
    alteration, secondary mineralization 106, 108
    geochemistry 95–96, 98–106
    magma origin, transport path 50, 108–112
    petrography 96–98
    sill and dyke distribution 93–95
Ferrar Large Igneous Province, volcanology 10, *44*
    age, nomenclature 75–77
    future study 87
    interbeds 85–86
    lavas and effusive rocks 82–85
    magmatism 46–47, 401
    nomenclature and outcrop 75
    palaeoenvironment 86–87
    sills 19–21, *73*
    vents, eruption rate 86
    volcaniclastic rocks 77–82

field stations, risk from eruptions 781
fissure eruption 459
flare-up 1–3, 22–23, 26, 33, 136
flood basalt 21, 34, *44*, 172
    age, eruption, vents 174–175
    differentiation 172–173
    Dronning Maud Land 157–160
    James Ross Island Volcanic Group 254
    magma and tectonic setting 172–176
Flood Range Volcanic Field 62, *516*, 518, **522, 555–556**
    active volcanism 60, *61*
    geochemistry 589–591
flow banding 124–126, *127*, 130
fluoride 659
foraminifera, global cooling 570
fore-arc basin sequence *203*
forecasting eruptions 681, 685, *686*, 697, 729–730
forest cover, Antarctic Peninsula 205–206
Forrestal Range, oxygen isotope 104, **105**
Fosdick Mountains Volcanic Field 515, *516*, 564, 568
    geochemistry 600–601
    geochronology **563**
Fossil Bluff Group 203, 204, 207, 222
    forearc basin 213, *214*, 216
fossil record, climate change 205–206
fossils 129
    flora 84, 86, 128, 193–194, 199, 203–204, 207
    tree/wood 87, 129, 207
    Neogene 432
    Pliocene 429–430
    shelly fauna 198, 257, 267
    Sr isotopic age 273
Foster Crater 465
fractional crystallization 108–109, *173*, 287, 337, 399, 408
    alkaline magma 584, 590
    basaltic magma 505–506, *508*, *510*, 511
    calc-alkaline *218*, 220–221
    Erebus 463–465, 470–471, 475, 702, 710, 725
    flood basalt magma 151–153
    phonolite 707
Franklin Island *415*, 438–441, *454*
    petrology and age 451
    stratovolcano 442
Frontier Mountain, blue-ice tephra 656, *657*
fumarole activity 235–236, 307, 371–372, 515
    active volcano indication 58–59, 62, 64–67, 632
    biological communities 717–719, *728*
    hazard and risk 685
    ice cave *712*, 761
    ice tower *559*
    Mount Erebus 709, 724
    Mount Rittmann *754*, 755
Fumarole Bay 673, *676*
    earthquake, volcanic tremors 680, 684
Fumarole Bay Formation 236–237, *238, 239*
fungi 717
future research 87
    Erebus Volcanic Province 480–481

gabbroic intrusions 164–165
Gabriel de Castilla, Spanish scientific base 65, 66, 67, *669*
    volcanism warning system 68
Gamma Hill, volcanic rocks 267
gas emission 87, *271*, *687*
    event c.17.6 ka 659
    Mount Erebus 729
    Mount Melbourne 744, *745–746*

gas entrapment pressure 709
gas measurement 697, 706–707, 724–726, 731
gas monitoring 59, 60
    Mount Rittmann 751, 753
Gaussberg volcano *500*
    eruptive setting 618–621
    geology and lithofacies 615–618
    petrology 621–625
    plume origin 625–626
    subglacial volcano 2, 3–4, 33–34, 46
gene sequencing, soil 718–719, *728*
geochemistry 12, 43
    alkaline volcanic rocks 46–50, 518
    Antarctic Peninsula volcanic arc 215–219
    Bransfield Strait 287–294
    Deception Island *670*, 674, 676, **677**
    Dronning Maud Land 162, 166–171
    Ferrar Large Igneous Province 46–48, 76–77, 95–106
    –Karoo 157–158
    Gaussberg 622–625
    Graham Land Volcanic Group 140, **141–142**
    James Ross Island Volcanic Group *257*, 295–298
    lamproite 621–625
    Marie Byrd Land, Ellsworth Land 581–608
    Mount Early, Sheridan Bluff 500–505, *506*, **507**, *508–510*
    Mount Erebus 700–711, 723, 731
        Erebus Volcanic Province **449**, 450–453, 469–473
    Mount Rittmann 748–749, 752
    Palmer Land Volcanic Group 146–151
    post-subduction volcanic rocks 330–335
    Victoria Land 388, **389–396**
geochemistry, tephra 650–661
    Marie Byrd Land **764–765**, *766*, *771–772*, *775*, *780*
geochronology
    Jurassic volcanics *133*, **134**
    Marie Byrd Land volcanics 524, *525*, 526, **527–528**
    West Antarctic Ice Sheet 571–572
geodetic reference framework 66
geomorphology, Deception *672*
geophysical analysis 10, 12, 23, 45
    crustal block boundaries 786–787
    Erebus volcano 711–716, 722–724
        conduit 723–724
    Ferrar intrusives 94
    ice depth estimate 521
    Mount Rittmann 751–755
    subglacial volcanics 788–791
    West Antarctic Rift System 516–517, 579, 581
geophysical experiments, West Antarctica 607
geophysical monitoring, volcanoes 66–67, 755
geothermal heat 318, 744, 751
    microbiological habitat 716–719
    subglacial heat flow 785–786, 788, 791–793, 796–799
geothermobarometry 449, 473
Gibbs Group 232
Gill Bluff, lava-delta deposit 534, *535*
glacial ice, Oligocene 206, 568–569
glacial–interglacial cycles 24, *25*, 660
glaciation and volcanic record 566–571
glaciation, Eocene–Oligocene 205–207
glaciochemical analysis 649, 660, 661, 731
glaciomarine sediments, age **275–276**
glaciovolcanic edifice *see* tuya

glaciovolcanic eruption 678
  Bransfield Strait 250, *252*
  Deception Island 241
glaciovolcanic province 1–3
glaciovolcanic sheet-like facies
    315–316, 320
  holotype 322
glaciovolcanism 27, 29, 32, *632*
  Antarctic Peninsula *315–316*, 320–322
  James Ross Island, Brown Bluff 268,
    269–270
  Marie Byrd Land 570, 571
    palaeo-ice levels 566–569
  Victoria Land 349–366, 379
glass shard 654, *655*, 659
global cooling, Late Miocene 570
global impact, Antarctic eruptions 781
global sea-level rise 620
global warming 28
Gluck Peak, volcanic rocks 318, *319*
golden spike (56±5 ka) 621
Gondwana 9–13
  plate margin 94–95, 110
  reconstruction *22, 76, 121*
  volcanism 19–22, 33
Gondwana break-up 1, 3, 43–48, 577–579
  metasomatism 401
  volcanicity *121*, 123, **153**, 185
Gondwana Large Igneous Province 157, *158*,
    175–176
Gondwana Scientific Base (German) 744, 755
GPS network, Mount Melbourne *742*,
    745–746
Graham Land *63*, 185, *191–192*, 214–215
  tuff cone 267
  volcanic arc succession 194–199, 207
Graham Land Volcanic Group 123–130,
    133, 135, **153**
  petrology 140–145
Grant Island **561**, 563
gravitational tectonics 33, 253, *255*
gravity analysis 236
  Antarctic Peninsula *28*
  subglacial volcanics 790, 792, *793*
  West Antarctic Rift System 786–787
gravity, free-air field *9*
Gray Nunatak, volcanic rocks 312–313
greenschist 126, 134, 140, 194
greenstone 150–151
Greenwich Island 243, *250*
Grew Peak, volcanic lithofacies 530–531
Grosvenor Mountains, red bole 84, *85*, 86
ground deformation
  Deception 682–686
  Mount Rittmann 745–748, 753, 755
  velocity 683–684
Grunehogna Craton 159
GV7 ice-core drilling 654–655

H₂S degassing 708–709
Hallett Peninsula Volcanic Field 360–363
  geochemistry **389–390**, 397
  petrology **386**, 388
Hallett Volcanic Province *30, 347*, 349–366
  petrology **386**
halogen gas 590, 731
Harrow Peaks, tuff cone 374, 379
harzburgite 110, 335, 621
Hawkins Peak, volcanic lithofacies 531
hazard and risk, volcanism 4, 55–57
  Deception 685, *687*
  management 67–68, 655
  McMurdo Station and Scott Base 60
  Mount Erebus 729–730
health risk, volcanic ash 67
heat-flow anomaly **798**

Hedin Nunatak, tuya 531–532
Heimefrontfjella, flood basalt *163–164*
Helms Bluff, subaerial volcanics *415, 424*,
    431–432
Hero Fracture Zone *668*, 684, 685
Hertha Nunatak, volcanic rocks 313
Hf isotope ratio 590, 601
  Mount Erebus 703–704, 720–721,
    725, 730
high field strength elements (HFSE) 472,
    583, 622
  Ferrar province 99, 101, 109–110, 113
  Ross Island 720
high-Mg andesite group 217–219, 221–222
high-Zr group *218*, 220, 222
HIMU (high μ=^{238}U/^{204}Pb) signature 3,
    47–50, 333
  Erebus 450, 471–473, 478–480
  Marie Byrd Land, Ellsworth 518, 586,
    592–593, 603, 606–608
  Mount Erebus 695, *702–704*,
    719–723, 730
  Victoria Land 400–401, 404
Hjort Formation *130*, 134–135, 139,
    146, 153
  geochemistry, petrology 150–151
Hobbs Coast Nunataks *560, 569*
Hobbs Coast Volcanic Field 515, *516*,
    561–563, 568
  geochemistry 601–603
Holocene volcanicity, tephra 55, 632
Honeycomb Ridge basalt vent 375
Hooper Crags 465
Hope Bay, ignimbrite 127–128, 135
hornfels 215
Hornpipe Heights, volcanic rocks *317*, 322
Hubert's Nightmare, cave bacteria 719
Hudson Mountain, subglacial edifice
    786, 797
Hudson Mountains Volcanic Field 62–63,
    515, *517*, 564–565
  geochemistry 603–605
  historic eruption 55
human deaths and volcanism 55
Hut Point volcanic field *415, 421*, 427–429
  age and petrology *458*, 459
  geochemistry 470, 474
hyaloclastite 22, 44, 61, 317, *319*, 321,
    354, 632
  Antarctic Peninsula 126, 133, 196,
    198, 203
  Bransfield Strait 237, 251, *252*
  Erebus 422, 424, 426, 428, 433, 436, 457,
    459, 474
  Ferrar province 85–86, 88
  Marie Byrd and Ellsworth 521, 523–541
    palaeo-environment interpretation 566
  Mount Early and Sheridan Bluff 491–492
  subglacial 762, *787*, 788, 792, 799
hyalotuff *530*, 537
hydrothermal activity 233, 236,
    271–272, 753
  Deception 681, 685–686, 697
  Erebus 728–729, 753
hydrothermal alteration 214–215, 439
hydrothermal metal mineralization 215
hydrous subducted slabs 339–341
hydrovolcanic activity 267, 677–678
hydrovolcanism 67

ice caves, fumarolic 717, 744, *745–746*
ice caves, hydrothermal 728–729, 753
ice core 791
  Marie Byrd Land 776–781
  tephrochronology 533
Ice Core Science Initiative 661

ice cover and magmatism 27–28, 32
ice dome 267
ice sheet 61
  basal melting rate 785, 793
  flow rate *786*, 797, 799
  growth and decay 31
  loading 50, 785–786
  Pleistocene–Holocene expansion 570–571
  pre-Quaternary 347
  West Antarctic Ice Sheet 532
ice shelf, collapse 305, 307, *309*, 322
ice thickness 28–29, 321–323, 379, 537
  and explosive eruption 441
  Gaussberg 620–621, 626
  and glacial volcanism *787*, **791**, 799
  Scott Glacier 492
ice timescales 650
ice tower *701*, 709, *744*, 761
  fumaroles 728–729
ice-contact lithofacies 523, *544*
ice-mushroom 462
ice-penetrating radar, subglacial volcanics
    788–790
ICE-VOLC project 59, 743–745, 751–752
Icefall Nunatak, volcanic lithofacies 531
icequakes 714, 745, *747*
ignimbrite 32, 44, 146, 154, 193, 204,
    206–207, 238
  Graham land, volcanic group 121–130,
    140, 195–199, 202
  Neogene 370, 371, 373, *424*
  Palmer Land 200–202
  Palmer Land Volcanic Group 131–132,
    135, 136
Imperial German Antarctic Expedition
    (1902) 615
Incorporated Research Institutions for
    Seismology 711
India–East Gondwana break-up 44
infrasonic signal 711, 715
Integrated Ocean Discovery Programs
    (IODP) 632, 634
interglacial lake setting 559
interglacial periods *25*, 34
International Geophysical Year (1957–58)
    447, 465, 786
International Mount Erebus Seismic
    Study 711
intracaldera succession 130–132, 134, 135,
    146, **153**
intraplate alkaline magmatism 327
  petrology 46–47
  volcanic 45, 46, 499, 608, 730
intraplate shearing 403
intrusive rocks
  Dronning Maud 160
  geochemistry 167
  volcanic feeders 161–162
island arc, geochemistry 291–294
isostatic effect 277, 798
isotope age *24, 25*, 62, 478–480, 492–493
  Antarctic Peninsula **244–249,
    311–312**, *320*
  Erebus Volcanic Province *415*, 420–422,
    426–439, 441
  Gaussberg lavas 618, **619, 625**
  Hallett Volcanic Province 352
  Marie Byrd Land 762, 765–766
  Melbourne Volcanic Province 367
  Overlord Volcanic Field 369
  Seal Nunataks 308, *309*
  Upper Scott Glacier Volcanic Field **494**
  use in correlation 266
  Victoria Land **355–359**, 375, *376*,
    377–378, 389–396
  volcanic island centres *231*, 236, *237*

isotope composition
  gas, Mount Melbourne 59
  lavas, Mount Erebus 703–705, 710
isotope composition, Jurassic rocks
  Dronning Maud Land 166
  Ferrar province 99
  Graham Land Volcanic Group 144–145
  Palmer Land Volcanic Group 149–151
isotope signature, magma melts 46–49
Italian Expedition (1988–89) 748

Jaegyu Knoll, submarine volcano 271
James Ross Island Volcanic Group 26–28, 46, 63–64, 231
  age 277–278
  alkaline basalt 285, 287–299
  geochemistry 331–333, 341
  historic eruption **55**
  lithology 253–267, 273, **274–276**
  petrology 294–298
  post-subduction volcanology 305–314
  subglacial volcanoes *632*
  tuya 250, 253, *256*, 268, 272
  volcano inventory **228–230**
Jang Bogo Station (South Korea) 741, 745
  volcanic risk 755
Jason Peninsula 314
  basalt 153–154
  geochemistry **141–142**, 144–145
  isotope age **311**
  silicic volcanics 129, 135
joints, Deception *673–674*
Joinville Island, volcanic lithofacies 126–128, 135
jökulhaups 322
Jonassen Island, volcanic rocks *256*, 270, *271*
Jones Mountain Volcanic Field 515, *517*, 565
Jones Mountain, Neogene volcanism 13, 341
  geochemistry 331
Jurassic break-up volcanism 20
juvenile crust 159

K–Ar isotope age 24, 26, 46, 164
  Antarctic Peninsula 129, 196, 201, 204–205, **244–246, 249**
    post-subduction **311–312**, 317, 318
  Brabant and Anvers islands 251–252
  Bransfield Strait 233
  Erebus 456, 459–461, 465, 470, 493
  Gaussberg 618, **619**
  Hallett Volcanic Province 353, 354
  James Ross Island Volcanic Group 254, 266
  King George Island 243
  Marie Byrd Land **763–764**, 766, *768*
    Ellsworth land 518, 582, 597, 600
    volcanics 523–528, 601–604
  Melbourne Volcanic Province 367
  Mount Early, Sheridan Bluff **494**
  Mount Rittmann 749
  Paulet Island 273
  tephra 638
  Victoria Land 352, 360, 363, 369, 371, 373
    summary *361*, **375, 377–378**
kaersutite 335, 408, 590, 595
  Erebus 456, 457, 459–465, 469–471
  melt inclusions 702, 709–710, 723
Kamb Ice Stream, volcanic ediface *786*, 792, *793*
  radar profile *789*
Kamenev Nunataks *140, 145*, 150–153
Karoo Continental Flood Basalt 157–158, *170–173*
  geochemistry 166

Karoo Large Igneous Province 10, *44*, 47
Karoo triple junction 157, *158*, 162, 175
Kay Peak, volcanic lithofacies 530
Kenney Glacier Formation 127–128
kenyte 457, 582
Kerguelen Islands, tephra dispersal 633, **635–636**
Kerguelen Plateau, plume 615, *617, 623*, 626
King George Island 243
  active volcanism 64–65
  adakitic rocks 220
  hydrothermal alteration 214
  stratigraphy 24–25
Kirkpatrick Basalt 75–87
  geochemistry 96
Kirwanveggen, flood lava and intrusives *163–164*, 165, 173–175
  geochemistry 167, **168–169**, *170–171*
Kohler Range, volcanic lithofacies 532
  geochemistry 603
Krakow Icefield Supergroup 24

lacustrine deposit 83, 85–87, 193–194, 204, 207, 360
lahar 66, 67, **79,** 80–82, 193, 678
  risk 685
Lake Whillans, heat flow anomaly *786*, 792–793
lamproite 4, 33–34, 43, 50
  geochemistry 164, 172–173, 621–625
  lithology, and setting 615–617
  magnetic anomaly 620
lamprophyre 46, 96–97, 306
large igneous province (LIP) 1–3, 10, 19–22, 27, 29, 30
large ion lithophile elements (LILE) 101, 144, 398
  Erebus 472, 478
  Marie Byrd, Ellsworth Land 583, 588, 595–596
Larsen A and B ice shelf, collapse 305, 307, *309*
Larsen Nunatak, dykes/volcanics *310*, 313
Lassiter Coast Intrusive Suite 122
Last Glacial Maximum 28, 29, 537–538, 571, 621, 626, 659
Last Interglacial, ice core investigation 793
Latady Group 122, 129, 132, 139–140
Late Quaternary volcanoes 538–539, 571
lava dome 194, 204
lava flow, feeding fissures 160–161, *162*
lava lake, Mount Erebus 4, 59–60, 695–700, 707–708, 731
  geophysical imagining, conduit 723–724
  geophysical monitoring 711–716, *717*
lava-fed delta **274**
  Antarctic Peninsula *182*, 194, 318, 321, *322*
  Bransfield Strait, James Ross Island 252, 266–273, **274**
  Erebus Volcanic Province *419, 423*, 424–427, 432, 457
  James Ross Island 255–257
  Marie Byrd volcanic province 523, 526, 532, 538, 568
  Victoria Land north 354, 373, 379
lava, volume erupted, Ferrar 77
Law Dome, ice-core drilling 654
layered intrusion 96–97, 174–175
LeMay Group 203, 215–216
  accretionary prism 213–215
lherzolite 476, 508–509, 603, 723
lichenometric 235, **244**
LILE *see* large ion lithophile elements
Lindenberg Island, dyke, tuff and breccia 313

LIP *see* large igneous province
lithosphere 297
  chemistry 479
  deformation and volcanoes 441
  delamination 509–511
  necking 406, *407–408*, 409
lithostratigraphy
  Antarctic Peninsula 309, *310, 315*, 316–318, 322
  Bransfield Strait–James Ross Island 227–278
  Deception Island *239*
  James Ross Island Volcanic Group 257–267
  type sections **258–265**
  post-subduction volcanism 305–307, **308**
Livingston Island, earthquakes 681
Livingston Island, ground deformation *683*, 684
Long Period seismicity 679–681, *682*, 684
Lyttleton, Ridge, tuffs 130

maar 65, *237*, 241, 273
  Penguin Island 234
  volcanic hazard 67, 685
maar–diatreme vent complex 3, 21–22
mafic volcanic rocks, petrogenesis 152–153
magma
  chamber 236, 238
  Ferrar 108–113
  mixing *151*, 152, 704
  near-summit body (NSMB) 714, 716, *717*, 724, 728–729
  origin 399–406, *407*, 626
  production rate 400, 479
  source 216, 287, 296–298, 624
magmatic intrusion, seismicity 680–681
magmatism 12–13, 213
  Cenozoic 383
  Deception 670–677
  migrating centres 204–205, 213, *214*, 461
  Mount Erebus 723–724, *727*
  plateau basalt 27
  and tectonics 401–403
  West Antarctic Rift System 579
magnetic anomaly 10, 24, 199, 495, 579, 581
  Antarctic Peninsula 214, 252, 286, 309
  Erebus 440, 441
  Gaussberg 619–620
  subglacial volcanoes 785–787, 790–794
magnetic polarity **159,** 174, 236, 670
  reverse 456, 495
magnetic survey 94, 236, 460
magnetotelluric imaging 714–715, 724
Malta Plateau Volcanic Field *347, 348*
  geochemistry 397
  lithostratigraphy, age **351, 357**, 366–368
  petrology **386**
Mandible Cirque, volcano 364, *365–366*
Mannefalknausane intrusions 166, 167
mantle 10, 12, 28, 33, 408, 480, 626
  anomaly 752
  melting 175, 507–510, 589, 723, 730–731
  melting depth *152*
mantle plume 44–50, 176, 416, 720
  Antarctic Peninsula 152, 335
  Erebus Volcanic Province 441, 479–480
  Gondwana break-up 10–12, 19, *21*, 44–45
  Marie Byrd, Ellsworth Land 516–518, 578, 607
  Victoria Land 400–401
  West Antarctic Rift System 30, 33, 47–50
mantle reservoir 606
mantle source 3, 43, 172, 292–294, 333
  calc-alkaline rocks 218, 221
  Erebus Volcanic Province 449, 476–480

Ferrar dolerites 93, 104, 109–113
  mafic volcanics 153–154, 158
  Marie Byrd volcanic province 579, 584, *586*, 596
  Victoria Land, south 510, 511
mantle tomography 403–404
Mapple Formation 123–126, 153
  correlation 135
  geochemistry, petrography 140–143
Marie Byrd Land 9–13, *20*, 33–34, 50, *121*, 408
  geology 579–581
  seismic profile *31*
  stratigraphy *518, 528, 544*
  tectonic features 403
  thermal anomaly 48
Marie Byrd Land Volcanic Group
  age data **580–581,** *582*
  alkaline volcanism 29–30, 34
  ash dispersal 641
  eruption history 515–572, 598–600, 605–608
  geochemistry *398, 399, 400,* 401
  geochronology 524–528, **534, 540–541, 551, 561, 563**
    Flood Range Volcanic Field **555–556**
  petrology 577–608
  volcano characteristics **522**
Marie Byrd Land Volcanic Province 45, *61, 516,* **580–581**
  central volcanoes **517,** 519–525, 538–539, 542, 545
  isolated edifices 593–598
  linear ranges 586–593
Marie Byrd Land, active volcanoes 60–63, 659, 722, 759–781
  geochemistry **764–765,** *766, 768*
    tephra *771–772, 775, 780*
  heat flow anomaly 798
  magnetic district *786,* 787
  subglacial investigation *632,* 796
marine isotope stage (MIS) 570, 572
  drill core **637,** *652*
  Gaussberg 626
marine platform, age **275**
marine sediment and tephra 206, 760, 780–781
Mario Zucchelli Station (Italian) 741, 744–745, *747*
  volcanic risk 755
Marshall Mountains *77,* 82–84, 87
MASH (mixing–assimilation–storage–hybridization) magma *151,* 152
Mason Spur *415,* 416, **418,** 420, *424*
  geochemistry 471, *472*
  Miocene volcanism 442
  volcano 434–435, *436*
Mason Spur Lineage, petrology 464–465, 471, *472*
mass extinction 21, 22, 23, 29
mass-flow deposits 126, 135, *201,* 202, 619
Matienzo Base (Argentina) 63
Mawson Formation 78, **79,** *80*
McCuddin Mountains Volcanic Field 515, *516,* **522,** 550–551
  geochemistry 598–599
McIntosh Cliffs, subaerial lava 432–433
McMurdo Ice Shelf *415,* 469, 640
McMurdo Sound *639*
  offshore drilling 467
  tephra dispersal **636–637**
McMurdo Station, US base *58,* 59–60, 459, 695, 706, 729–730
McMurdo Volcanic Group 30–31, *347–348,* 384, 450, 491, 697, 741
  isotope composition 703

lithofacies *416,* 441
petrology **386**
Meander Intrusive Group 29, *30,* 347–349, 371, 375–376, 409
  age *365, 368,* **377–378**
  dyke swarm 384, **385,** 397
  geochemical analysis **393–396**
  petrology 384, **385**
  pluton exhumation 33
melange belt 203, 215
Melbourne Volcanic Province *30,* 45, *347–348,* 366–375, 450
  active volcanism 57–59
  geochemical analysis **391–392**
  isotope age **357–358**
  petrology **386–387,** 388
melt-inclusion measurements 709–710
melting model 175, 509–511
  composition and origin 43, 46–50, 473
  depth 475–476
Melville Peak, active volcanism 64–65
Merrem Peak, shield volcano 62
  historic eruption **55,** *56*
  tachytic rocks 555, *558,* 559–560
Merrick Mountain 318, 322
  isotopic age **312**
Mesa Range 86–87
  lavas 103, 106
metal–halide salt 707
metallic element anomaly 236
metamorphic rock 159, 190, 194, 198, 214–215
metasomatism 12, 45, 49, 110, 509
  age 404–406
  Cenozoic 579
  lithosphere 606, 720, 721
  petrological model 47–48
  West Antarctic Rift System 404
meteoric water 103–104
meteorite trap 657
microbial life 697, 728–729
microbiological study 716–719
mid-ocean ridge basalt (MORB) 49, 221, *400, 473*
  accreted basalt 215–216
  Bransfield Strait 285, 287, 290–294, 298, 333–339
  Dronning Maud Land 172
  East Pacific Rise 332–333
  Ferrar province 101, *103,* 109–110
  Gaussberg 624
  James Ross Island 296–297
  Marie Byrd, Ellsworth Land 603, 606
  MORB-OIB array, Antarctic Peninsula *145,* 153
  Mount Early, Sheridan Bluff *505,* 507–511
  Mount Erebus *704–706, 721,* 723
  seamounts 231–232
  South Shetland Islands 328
Milankovitch cyclicity and eruptions 27
mineral analysis
  methodology 500
  Mount Early and Sheridan Bluff **502**
mineralization, alteration 106, 108
mineralogy
  Erebus Volcanic Province **453,** 703
  Marie Byrd Land 582
minerals in vesicles and geodes 84–85
Minna Bluff *415,* 416, **417,** *419,* 420, 431–433
  erratics 460
  petrology and age 463
Minna Hook *425,* 431–434, 442
Miocene volcanics 442, 497, 499, 511
Miocene, ice surface elevation 493
MIS *see* marine isotope stage

mixing, magmas *151,* 152, 704
Moho 28–29, 449, 473, 476
Möll Spur 535, *536–537*
  ice level 571
monitoring
  Deception volcano 668, 678–685, *687,* 751
  gas emission 744, *745*
  Mount Erebus 416, 668
  volcanicity 57, 59–60, 66
  alert scheme 67–68
Montagu Island, explosive eruption 632
moraine *419,* 462, *493,* 496, 497
  ice-cored 236, 670
  Phillipi Glacier 620
MORB *see* mid-ocean ridge basalt
Mount Alexander, volcanics 126, *127*
Mount Andrus 553, *554*
Mount Aurora 429
Mount Benkert, pillow lava *320*
Mount Berlin, active volcano **55,** *56, 61,* 515, *516,* 759–761
  caldera 518, 521, 524, *558,* 560–561
  Eocene–present volcanic activity 62, 571
  ice levels 566–567, 570
  tephra 660, 775–776, *777*
    dispersal 779–781
    glass composition **763–765**
  volcanic history **555,** 557–561
Mount Bird volcanic shield *415,* **417,** *419,* 421, 426–428
  petrology and age **455,** 456
Mount Bumstead, lava *77,* 83, *84*
Mount Bursey volcano *61,* 518, 521, **555,** 556–557
Mount CASERTZ edifice *786,* 792, 797
Mount Cumming, volcanics 546, 547, **548,** *549*
Mount Discovery Volcanic Field *415,* 416, 450
  geochemistry **449,** 471, 475–477
  geothermobarometry 473, *474,* 475
  lithofacies **417,** *419,* 429–434
  petrology and age 460–463
Mount Early and Sheridan Bluff
  geochemistry 500–511
  Miocene basaltic volcanoes 499–500
Mount Early, Neogene volcanics 491–497, 499–511
Mount Early, tuff cone 33
Mount Erebus 421–426
  alkaline volcanic system 695
  blue-ice tephra 656–657, 660
  calderas, nested 441
  englacial tephra 469–470
  eruptive history/overview 473–481
  mantle plume 400
  petrogenesis and evolution 695–731
  petrology and age 457–459
Mount Erebus active volcano 30, 34
  geochemistry 700–711
  geological setting 697
  geophysical study 711–716
  hazard assessment 67
  lava lake 697–700
Mount Erebus Volcano Observatory 711
Mount Falla, lavas *83*
Mount Fazio Chemical Type 76, *85,* 86, 98–106, 109, 112–113
Mount Flint, volcanic lithofacies 551, *552,* 571
Mount Frakes **541,** 542–543
Mount Gran, dolerite plug *83*
Mount Grieg, lithofacies 318, *319,* 321
Mount Haddington, shield volcano 3, *63,* 64
  back-arc volcanism 27–28, 34, 253, *254–255*

Mount Hampton caldera *61*, 521, 546–547, 548
Mount Harcourt stratovolcano *352*, 362, 379
Mount Hartigan, caldera *61*, 521, *546*, 547–549
Mount Howe, moraine debris 494–495
Mount Kauffman, volcanic lithofacies 554
Mount Kirkpatrick, pipe vesicles 84
Mount Kohler, isotopic ages **527–528**
Mount Lubbock stratocone 365
Mount Melbourne active volcano 34, *345*
  recent activity *56*, 58–59
Mount Melbourne quiescent volcano 408
  age 741–743
  geochemistry 642, 743–744, *745*
  ground deformation 745–746
  seismicity 744–745, *747*, 748
Mount Melbourne Volcanic Field *347 348*, **351**, *372*, 373
  geochemistry **358–359**, *397–398*
  petrology **387**
Mount Melbourne, coastal ice cliff with tephra *629*
Mount Morning Volcanic Field *415*, 416, **418**, 434–437, 450, 480
  geochemistry **449**, 471, 476–480
  geothermobarometry 473–475
  petrology and age 463–464
  volcanism 442
Mount Moulton *61*, **555**, 557, 561
  blue ice tephra 769–770, *771*, 780, *780*
  caldera 521
  ice-flow obstruction 566
Mount Murphy volcano 521, 525–531, 568
  geochemistry 593–594
  ice sheet 569, 571
  uplift and exhumation 566
Mount Overlord Volcanic Field *347, 348*
  Aviator Glacier 748, *749*
  isotope age **358**, 369
  lithofacies **351**, 369–371
  petrology **387**
Mount Petras, volcanic lithofacies 551–552, *553*, 568
Mount Pinafore Volcanic Field *303*, 315–318, 320–322
Mount Plymouth, tuff cone 250
Mount Pond Group *239*
Mount Pond, subglacial eruption 66
Mount Poster Volcanic Formation 130, 131–132, *133*, 135, 139, *151*
  geochemistry 146, **147–148**, *149*
Mount Rees volcano 518, 521, 539–545
  glaciovolcanic sequences 569
Mount Resnik, subglacial edifice *786*, 794–795
Mount Rittmann, active volcano 741, 748–755
  ash source 654–655
  fumarole 632, 751
Mount Rittmann, volcano 57–59, *369*, 371
  geochemistry *397*, 749–751, **752–753**
  petrology **387**
Mount Short, pillow lava *86*
Mount Sidley, volcano 30, 45
  geochemistry 597–598
  glaciovolcanic sequences 569
  volcanic lithofacies *546*, **548**, 549, *550*
Mount Siple Volcanic Field 515, *516*, 545
Mount Siple, active volcanism **55**, *56*, 60, *61*
Mount Sourabaya, explosive eruption 632
Mount Steere, lithofacies 539–542, 545, 569
Mount Takahe, active volcano **55**, *56*, 60–62, 759–762, **764–765**
  Eocene–present volcanic activity 571
  tephra dispersal 779–781

Mount Takahe, volcanology 533–538
  geochemistry 594–596
  glaciovolcanic history 570–571, *572*
Mount Terra Nova volcano *415*, 421, 426, *456*
Mount Terror volcano **417**, 421, 426, 697, *699*, 723
  geochemistry *702*, 704, *705*, 709, 720–721
  petrology and age 456–457
Mount Thiel, subglacial edifice *794*, 795, 797, **798**
Mount Tucker, volcanology 129
Mount Waesche, active volcano 55, *56, 61*, 62, *760*, 762–769
  blue ice englacial tephra 759, *760*, 770–776
  caldera 521, *546*, **548**, 549–550
  geochemistry **763–764**, *768*
  tephra dispersal 571, 779–781, 796
Mount Whiting, dykes 153
Mount Wyatt, moraine debris *493*, 495–496
MSSTS-1 drill hole **636**, 637
mudflow risk 687
Murdoch Nunatak, volcanic rocks 313
Mussorgsky Peaks, volcanic rocks 318, *319*, 321

Nameless Bluff volcano *352*, 353–354, **355**
  stratigraphy *361*
Nansen Ice Field, ice-core tephra 657
Navigator Nunatak, pillow lava *370*, 371, 374
Nb isotope composition 176
Nd isotope composition, Erebus 703–704, 720–723
Nd isotope ratio 29, 216, 292, 297, *298*, 333
  Erebus Volcanic Province 454, *472*, 476, 479, 480
  Gaussberg 624
  Jurassic 146, 166–167, 172
    Ferrar province 99, 104–109, 112
  Marie Byrd, Ellsworth Land 581–582, *586*, 588, 590, 592–603
Neogene alkaline volcanic rocks 305–307, **308**, 314
Neogene volcanicity 632
  Transantarctic Mountains 491, 496–497
  Victoria Land *347–348*, 376, 379
  Erebus 415–442
nepheline 462, *471*
  -normative 460, 463–465, 471, 475, 504, *592*, 593, 600
nephelinite 404, *405*
Neptune's Bellows, volcanic risks 685
New Mexico Tech Mount Erebus Volcanic Observatory (MEVO) 707, 711, 730–731–730
Nilsen Plateau, sills *95*, 96, 111, *112*
  chemistry 100
Nimrod Expedition (1908) 697, *698*
Nordenskjöld Formation *124*, 126
North Karoo subprovince *158*, **159**, 173, 176
Northern Local Suite Volcanic Field *348*, 373–376, 379
  petrology **386, 387**
nunataks 62–64

oblique tectonics 48, 227, 608, 668
obsidian 238, 242, 370
Ocean Drilling Program (ODP) 632–634, **635–636**, *639*
ocean island basalt (OIB) 3, 222, 511
  East Pacific Rise 332–334
  Gaussberg 624–625
  Jurassic 109–110, 172
  LeMay Group 215–216

Marie Byrd Land province 12, 579, 584, 586, 592
Mount Erebus 450, *704*, 721
West Antarctic Rift System 30, 47, 49–50, 400
Oceana Nunatak, volcanic rocks 313
oceanic basalt 215–216
ODP *see* Ocean Drilling Program
OIB *see* ocean island basalt
Oligocene glaciers 568–569
olivine basalt 28–29
olivine tholeiite 331–336, *339*
olivine-bearing dolerite *112*
  chemistry 104–106, 109
olivine, chemistry 100, **101**
open-conduit *see* conduit
Operation Deep Freeze, cruise 642
optically stimulated luminescence dating 621
Os osmium isotope 404–406
  Jurassic 106, 109–110, 112, 166–167, 172
Oscar II Coast 123, *124–125*, 126
  dykes *145*, 150
Otway Massif, lava/volcaniclastics 77, 81–83
Outer Coast Tuff Formation 236–240, *241*, 674, *676, 678*
oxide chemistry 100, *101*
oxygen fugacity 106, 109, 708, 710–711
oxygen isotopes, Jurassic 103–106, 109, 176
ozone depletion (c. 17.6 ka) 778
ozone layer 1

P-wave velocity, Mount Erebus 712–715, 724, *728*, 729, 731
Pacific mantle reservoir 216, 222
Pacific Margin Anomaly 24, *25*
Pacific Plate, subduction 9, *306*, 449
pāhoehoe 22, 27, 34, 64, 538
  Antarctic Peninsula 309, 312, 321, 323
  Bransfield Strait, James Ross Island 238, 251, *252, 255*, 272
  Gaussberg 617–619, 626
  Jurassic 84–85, *132*, 133, 135, 160–165, 174
  Victoria Land *361*, 496
palaeo-environment 86–87, 571–572
  glaciovolcanic sequences 31–32, 566
  Paleogene, terrestrial 24
palaeo-ice level, volcanic record 566–569
palaeoclimate 29, 643, 649
palaeomagnetism 10, 236, 238, **244**
palaeosol 22, 83, 84, 86
palagonite 237, 251, 317–319, 457, 492–493
  breccia 354, *361*, 430
  Erebus 422, 424, 427–428, 430
  tuff 617, 620
Palmer Land 185, *191*, 213–217, 219–222
  tectonic event 140
  volcanic arc succession 199–202, 205, 207
Palmer Land Volcanic Group 121, *130*, 131–136, **153**
  petrology 145–151
pantellerite 589–591, 595, 596
parental magma 167, 173, 175, *339*
passive margin 12, 13, 122, 139, 408
Paulet Island, volcano **55**, *56, 63*, 64, *256*, 270–273
Pb isotope composition 216, 297, *298*
  Mount Erebus 703–705, 710, 720–721, 725, 730
Pb isotope ratio 404–406
  Erebus 454, 473, 480
  Gaussberg *624*
  Jurassic 109, 166, 167, 172
  Marie Byrd volcanic province 592–597
    analysis 585, *586*, 588, 601, 603, 606–608
    samples and data 581–582

Pendulum Cove Formation *239, 240,*
   241–242
Penguin Island, volcanology 234–235
   historic eruption **55,** *56,* 632
   scoria cone *63,* 65–66
peperite 80, 193, 523
perched glaciovolcanic lavas 436
peridotite solidus 333–338
periglacial volcanology 348, 365
petrogenesis 171–173
   alkaline volcanism *49*
   mafic volcanics 152–154
   silicic volcanics 151–152
petrology 50
   Bransfield Strait, 285–294
   Dronning Maud Land 160, 171–173
   Ferrar Large Igneous Province 93–113
   Graham Land Volcanic Group 140
   magma evolution 43–50
   Marie Byrd volcanic province 577–608
      samples and data 581–582
      summary and overview **580,** 582–586
   Mount Early, Sheridan Bluff 499–500,
      501
   Palmer Land Volcanic Group 146, 150
pH extremes 718–719
Phoenix Plate 285, 297–299, 577
   subduction 13, 44
   rollback 46, 668–669, *674*
phonolite lava lake 4, 416, 447, *665,* 697–700
phonolite, Mount Erebus 422–435, 439,
   441, 695
   bombs 703–704, 723
   characteristics **417–418**
   chemistry 700–711, 725–730
   lava lake 416, 447, 697–700
   petrology 450, *451,* 455–465, 469–471,
      474–475
phonolitic tephra 655–657, *658*
photomicrograph *454, 655*
   Mount Early and Sheridan Bluff *501*
   olivine lamproite lava 621
phreatocauldron 77, 81, *82,* 88
phreatomagmatic deposits 21–22, 82, 85–87,
   239, 364
   Marie Byrd volcanic province 523–534,
      538, 542, 562
phreatomagmatic eruption 3, 60, 65, 67
   hazard 729
picrite 160, 166–167, 172, 176
pillow lava 22, 33, 46, 242, 268, 632
   Antarctic Peninsula 194, 196, 203,
      309–312, 318–322
   Erebus 422, 424, 436, 457
   Jurassic 80, 85–86, 88, *132,* 133
   Marie Byrd volcanic province *529,*
      615–619
   Mount Early, Sheridan Bluff 492, 493, *494*
   subglacial 787, 792
   Victoria Land, north 353, 366, 373
Pine Island, magnetic district *786,* 787
pipe vesicles *84,* 160–161, 164
Pipecleaner Glacier, volcanic rocks 465–466
plagioclase, chemistry 99, **100,** 502
plant fossils *see under* fossils
plate reconstruction map, Gondwana *11,*
   1, 121
plate tectonic map *668*
plate tectonics and magmatism 403–404
plateau lavas, basalt volcanism 27
platinum group elements 102, 106
Pleiades Volcanic Field 58, 59, *347–348,*
   351, **358,** 368–369
   geochemistry *397*
   petrology **386**
   pyroclastic eruption 32

Pleiades, volcano **55,** *56,* 642, 655, 659
Plinian-type eruption 58, 632, 640–641, 653
   Erebus 60, 729–730, 755
   Marie Byrd, Ellsworth Land 524, 533
plume activity 615, *617,* 707
   Gaussberg lavas 625–626
pluton 23, *30,* 33, 185, 190, 196, 384
   age 205, 251
   geochemistry *399*
   granitoid 122, 123, 126, 130
   Meander Intrusive Group 375–376
   nepheline–syenite 160
   radiogenic isotopes *400*
plutons and dykes **393–394**
   tectonic setting 406–409
polar ice 661
polarity *see* magnetic polarity
Pollux Nunatak, scree 313
polymict debris, caldera collapse 768, *769*
porphyry copper 214
Port Foster Bay *65, 672*
   bathymetry *671*
   caldera collapse *234,* 236
   eruption risk and alert system 68, 686
Port Foster Group *239*
post-glacial volcano 429
post-subduction magmatism 298
   age 340–341
post-subduction volcanism, Antarctic
      Peninsula 305–323
   petrology 327–341
potential field, subglacial volcanics 790
Precambrian domain *110*
precursors of volcanicity 685
pressure–temperature 383, 473, 677
   conditions 333–337, 339
   degassing 725
pressure, crystallization depth 709
pressure, gas bubbles *713*
primitive magma 218, 295, *297,* 333,
   336, 505
   Erebus 405, 450, 470–479
   Gaussberg 625
primitive mantle *172, 400*
Prince Gustav Channel, volcanic centre *256*
prismatic jointing *303*
proto-Weddell Sea 110–111, *112*
protolith age 214
provenance, tephra in ice core *777*
pyroclastic density current 78, 80, 81, 524
   Bransfield Strait, James Ross Island
      238–239, 243, 267, 269
   Deception 674, 677, 678
   hazard 685
pyroclastic deposits 32, 34, 242, 568
pyroxene chemistry 99, **101,** *106,* 502
pyroxenite source, alkali basalts 335–337
   melting behaviour 337–341

Quaternary volcanism 615, 626, 695
Queen Alexandra Range, sills *96*
Queen Maud Mountains 491, *492–493,* 496

radar sounding data
   subglacial volcanoes 785, 788–790,
      792–797
   Thwaites Glacier 793, *794*
radial dykes 234
radiogenic isotopes 292, 293, 399, *400*
radiolarian assemblage 203
radiometric age 76, 294, 447, 459
rapakivi granite 617
rare earth elements (REE) 215, 220–221,
   287, 290–292
   Antarctic Peninsula 332–333, 335–337
   Gaussberg 622–624

Jurassic 102, *103,* 110, 143–145, 150, 175
Marie Byrd volcanic province 583–585,
   591, 592–593, 600, 604
Mount Erebus *702,* 703, 705–706,
   723, 730
Victoria Land 398, 504–505, 508, *510*
Rayleigh equation, calculation 505, **507**
Rb analysis 106, **107,** 108, 596
Rb–Sr isotope age 190, 196–197, 198–199,
   377–378
   silicic rocks 122, 130
Rb/Sr isotope ratio 582
Re isotope in xenolith 406
red bole 162, 165
Redcastle Ridge volcanics 360, *363*
REE *see* rare earth elements
REGID geodetic reference frame 682–683
   network stations *680*
research stations *56,* 60, *63,* 65–67, 667,
   741, *747*
   Deception 235–236
   and field camps 56–57
   risk from volcanic eruption 744, 755
   see also McMurdo Station, US base
rheo-ignimbrite 123–124, *125*
rhönite 459
rhyolite **188–189,** 197, 204, 364
   Deception 670, 677
   geochemistry 216–218
   Graham Land Volcanic Group 123–127,
      129–130
   Palmer Land Volcanic Group 134,
      199–202, 206
   petrogenesis 154
riebeckite granite 222
Riedel shear fractures 671
rifting 11–12, 112, 136, 409, 438, 516
   and magmatism *49,* 404, 577
risk, volcanic eruption 667, 744
Riviera Ridge Lineage 464, 465
Rodinia supercontinent 159
Rodolfo Marsh Martin Aerodrome 64
rollback, subduction 26, 46, 668–669, *674*
Roosevelt Island, ice core 776, 778, 780
Rosamel Island, tuff cone *256,* 270–271
Ross Embayment, tephra dispersal **636**
Ross Fault 438
Ross Ice Shelf *415,* 640
Ross Island 33, 50
   crustal thickness 697
   hazard assessment 730
   magmatism 720–723, 730–731
   seismic profile *31*
   velocity model *722*
Ross Island Volcanic Field *415,* 416, **417,**
   420–429
   geochemistry **449,** 450, 471, 476–480
   geothermobarometry *474,* 475–476
   petrology and age 455–459
Ross Sea 1, 11
   alkaline volcanism 45, *49*
   ash dispersal 659
   metasomatism 406
   offshore drilling 468
   ROSSTEPHRA project 642
   scientific exploration 449
   tectonomagmatic history 406
Ross Sea Basin 516
Ross, Captain, *Terra Nova* Antarctic
   Expedition 1841 455
rotation 9–11, 671, *674*
Rothschild Island, geochemistry 333
Rougier Hill, pegmatite *98*
Royal Society Range *415,* 416, **418,** 420,
   431, 436–437, 465
   volcanic activity **55,** *56*

Salt Rock, volcanics 242, 293
Satellite Vent tuya **359**, 531
Scarab Peak Chemical Type (SPCT) 76, 86, *94*, 97–106, 109, 111–113
Schievestolen sill 166
scientific research stations *see* research stations
scoria *240*, 250, *317*
scoria cone 521, 538
  Antarctic Peninsula *313*, 314, 322
  Bransfield Strait, James Ross Island 234–235, *255*, 273
  Erebus **417–418**, *419*, 427–436, *439*, 455–457, 465
  Marie Byrd Land 766–767, *768*
  Victoria Land, north 352–353, 361, 364, 366, 373
Scotia Plate *668*
Scotia Sea, ash dispersal 641–642
Scott Base, lava-fed delta 427
Scott Base, New Zealand *58*, 59, 459
Scott Glacier 491, 492, *493*, 496
sea ice, fast 349
seafloor spreading 9–11, 13
  initiation 607
sea-ice platform, drill core 638
sea-level change 239, 243, 273–277
  and ice sheet 568, 572
Seal Nunataks 305
  geochemistry 331–333, 335, *337–339*
  isotopic age **311**
  volcanoes **55**, *56*
Seal Nunataks Volcanic Field *63*, 64, 307–314, 321, *322*
seamount 46, 203, 215, 339, 440–441, 454
  geophysical interpretation 252–253
  and volcanic ridges 231–232, *233*
  volcanism 26, *44*, 45, 49, 66
seawater and Sr/Sr age 273
Sechrist Peak, ice sheet 571
Sechrist Peak, volcanic lithofacies 525–530
sediment subduction 624–625
seismic activity 59, 236
  Deception 668–669, 684
    network stations *680*
  monitoring 60, 66–67, 678–682
  Mount Melbourne *742*, 744–745, *747*
  Mount Rittmann 751–753, 755
  subglacial volcanics 790–791
seismic crises [1992, 1999, 2015] 681–682, 686
seismic investigation
  magma conduit 731
  Mount Erebus 711–716
  Mount Erebus, long wave signal 698
  Toney Mountain 539
seismic tomography 30, *31*, 48, 416, *722*
Sembberget extrusive suite *164*, 165, 174–175
  geochemistry **168–169**, *170–171*
serpentinite 339
shear zone *536*
shear-wave velocity *480*, 509, *722*
Shepard Island, volcanic lithofacies **561**, 562
Sheridan Bluff, Neogene volcanics 33, 491–497, 499–511
Shield Nunatak tuya *352*, 373
shield volcano 33, 62–64, 463, 493, 521
  Antarctic Peninsula 252, *253*, 305–306
  geochemistry *397*
  Victoria Land, north 347, *348*, 360, 379, 409
shield-like composite volcano 571, 579
sideromelane 237, 266, 268, *317*, *319*, 492, 493
  tephra 360

silicic magmatism, Jurassic 121–123, 126, 128–131, 134–136, 139, 160
silicic volcanic rocks 185, 193, 200, 202, 203
  petrogenetic model 151–152
  petrography 139–154
silicification 130, 215
sills 13, *111*
  Ferrar dolerite 77, 78
Sims Island, lava and tuff **312**, 319, *320*
  tuya 321
Siple coast, subglacial volcanics 792–793, 797
Siple Dome 759, *786*
  ice core 659, 660–661, 777
  tephra 780
  wind-blown debris 779
Skye, lava stacking pattern 173, *174*
slab 25, 34, 480
  Late Cretaceous detachment 12, 479, 720
  melt chemistry 45, 216, 221, 287, 291–294
slab rollback 3, 64, 227, 285, 298, *674*
  age 26–27
slab window 28–29, 46, 252, 608
  formation 231, 278, 285, 305, *306*, 327–330, 333–335
  volcanism 20, 34, 63
slope gradient, pyroclastic cone 242–243, 257
slope gradient, volcano 431, 458–459, 462, 533, 619
slope ice and volcanic eruption 568
Sm–Nd age 128, 129
Snow Nunatak tuya **311**, 321
Snow Nunataks Volcanic Field 318–320
snow/rime mushroom 547
SO₂ emission rate 60, 706–708, 731
soil temperature, volcanic prediction 681, 685, *686*
soil, fumarolic 717–719
South Atlantic Ocean, ash dispersal 641
South Karoo subprovince *158*, **159**, 173, 175–176
South Orkney, tephra dispersal **635**
South Pacific Ocean, ash dispersal 642
South Pole, aerosol pollution 707
South Pole, ice-core drilling 654, 658
South Sandwich Islands, ash and tephra 641, 658
  subglacial volcanoes *632*
South Scotia Ridge, tephra dispersal **635**
South Shetland Islands *63*, 185, *231*
  age of arc volcanics 205
  calc-alkaline lava 251
  geochemistry 290–293, *294*, 338
  petrology 214–222
  subduction 299, 328–330, 341
    active 45, 285, *286*, 671
  tephra dispersal 633
  volcanic rocks 23–26, 190–194, 207
South Shetland Plate *668*
South Shetland Trench 23–24, 26, 33–34, 44, 227
  subduction 668–669
Southern Local Suite 374, *415*
  geochemistry **449**, 476
  geothermobarometry *474*
  isotope age *437–438*, 442, 450
  petrology and age 465–470
  volcanostratigraphy 416, **418**, 420, 435
Southern McMurdo Sound 640–641
Southern Ocean
  drill sites *634*
  strike-slip faults *402*, 403, 406
Spanish Antarctic programme 682
  scientific base, Deception 65, *669*
spreading centre 23, 45, 286, *306*, 327, 329
  Pacific–Phoenix 577

Sr isotope analysis 216, 292, 297, *298*
  Antarctic Peninsula 146, 149, 176, 333
  Dronning Maud Land 157–159, **159**, 166, 167, 172
  Erebus 454, *472*, 473, 480, 703–705, 720–723
  Ferrar igneous province 99, 104–109, 112
  Gaussberg 624, *624*
  Marie Byrd volcanic province 585, *586*
    isolated centres 588, 590, 592–603
    sampling 581–582
Sr/Sr ratio in alkali basalt 29
Sr/Sr, values in seawater 273, **275–276**, *277*
Stanley Patch, pyroclastic cone 242
Stauffer Bluff tuya *536–538*, 572
  ice sheet level 570–571
steam fields, risk 685
Stonethrow Ridge Formation *238, 239*, 240–241
Storm Peak 77, 83
  geochemistry 100–103
  lava 82, 85–87, 109
stratigraphy/volcanics
  Deception Island 236, *237*
  Marie Byrd Land 521, *528*
  Mount Early, Sheridan Bluff 491–497
  Victoria Land, north 349–366
  Erebus Volcanic Province **417–418**
stratocone 206–207, 235, *365*
  Mount Harcourt 362
stratovolcano 29–30, 45, 57, 59, 193
  Mount Erebus **417–418**, 431
  Mount Melbourne *345, 372, 742*
  Victoria Land 347, 360, 379, 409
stress field 420, 518, 671, 786
striated surface *530*
strike-slip faults *402*, 403, 406, 608
Strombolian-type activity 21, 59–60, 640
  Deception 673
  Erebus, active volcano 695, 698–699, 705
    degassing 706, 708, *725*
    future risk 729–730
  Erebus, Neogene 416, 426, 469, 474
  Marie Byrd Land 524, 526, *529*, 532, *536*, 543, *773*
  tephra 361–362, 620
  Victoria Land 365, 373, 374
Strombolian-type deposits 237–238, 241, 422, *423*
Styx Glacier Plateau ice-core drilling 654, 655
subaerial volcanism *415*, 420, 441
  ice levels 566–570
  lithofacies 523, *544*
subduction 3, 26, 33, 45, 48
  active 23, 45, 285, *286*
  rate of 668–669
  anhydrous slabs 339–341
  Antarctic Peninsula, volcanic arc 213, 214
  fluids 110, 172
  Gaussberg 624–625
  geochemistry 220–221
  and metasomatism 404
  palaeo-Pacific margin 479–480
  processes 12–13, 111–112
subglacial eruptions 20
  Erebus 427, 436, 441
  Gaussberg 615, 619–620, 626
  Marie Byrd, Ellsworth 568–571, *578*, 579, 581
  Mount Early, Sheridan Bluff 497
subglacial lakes 795
subglacial to subaerial transition 570
subglacial volcanism 571, *632*, 781, 785–799
  active volcanoes 791–797

crustal boundary zones 797–799
  palaeoenvironment 517, *518*, 521, 562, 566
  techniques and observations 788–791
submarine volcanism 442, 454, 469, 474
  Port Foster 242
sulfate 654–655, 661
sulfur 106, 617
  degassing, Mount Erebus 707–708
  in microbial communities 717–718
supraglacial eruption 436, *460*
Surtseyan-type deposits 272, 323, 353, 431, 620
Sweeney Formation *130*, 132–135, 139, 146
  geochemistry **147–148**, 153–154

Tabarin Peninsula, volcanic rocks *256*, 267
tabular lava facies 81–82, 84, 162–164, 173–174
Talos Dome ice cores (TALDICE) 780
  tephra and provenance 652–656, *658*, 660–661
Tasch Peak Ridge 541, *544*
Taylor Dome, tephra and provenance 653, 655
tectonic denudation 517
tectonic lineament, Victoria Land *742*
tectonic map *9, 11, 20*
tectonic setting
  Antarctica 9–13
  Bransfield Strait 285–286
  James Ross Island 278, 294–295
  Neogene volcanism 306–307
  silicic volcanics 136, **153**
  subduction 227
  Victoria Land 401–403, 406–409
tectonomagmatic history 43–46
  Early Eocene–Early Miocene 45
  future research priorities 50
  Jurassic–Early Cretaceous 44
  Late Cretaceous–Paleocene 44–45
  Middle Miocene–Holocene 45–46
Telephone Bay, tremor 684
temperature monitoring 748, 751, *754*
temperature of crystallization 101, 480, 710
temperature, seawater 685, *686*
tephra 32, 34, 55, 58–64, 67, 533, *629*, 670
  Antarctic Peninsula region 236–237, 241–243, 257, 272, 360
  chemistry **763–765**
  East Antarctica 649–661
  Erebus 423, 469–470, 729
  Gaussberg 620
  marine record 631–643
  particle size analysis *652*
  source 778–781
  tephrochronology 66, 571, 643
  tephrostratigraphy 650, 657–661
  Victoria Land 369, 371, 372
tephra, Marie Byrd Land 538, 759–762
  blue ice 769–776
  ice core 776–781
tephrite *28*, 352, 416, 435, 439, 586
  Antarctic Peninsula 295, *315*, 330, *333*
Terra Cotta Mountain, sills and dykes *95*
Terra Nova Antarctic Expedition (1841) 455
terrane 9, 13, *24*, 44
Terrapin Hill, tuff cone *255, 257*, 267
terrestrial volcanic centres *231*
Terror Rift 408, 440
  dredge samples 454
Terror Rift Volcanic Field *415*, 416, **418**, 419, 437–442
  geochemistry **449**, **452–453**, 475–478
  petrology and age 450–454
Th-decay series 705–706, 726–728, 731

The Great Wall, tephra *773*
thermal anomaly 11–13, 30, *31*, 227, 271, 277, 347
  Antarctic Peninsula 339, *340*
  Marie Byrd, Ellsworth Land 517, 607
  Ross Island 720
thermometry, borehole 791
Theron Mountains, dolerite sills 93, *94*, 96, 99, 104
tholeiite 28–29, 34, 153, 285, 287
  Dronning Maud Land 157–158, 176
  Ferrar province 94, 99–102, 109
  geochemistry 101–102, 166, **168–169**
    Antarctica Peninsula 215–216, 295, *297*, 330, *333*, 335, 338
  Mount Early and Sheridan Bluff 499, 504–510
tholeiitic tephra 650, 652, *653*, 658
Thomas Rock, hyaloclastite *86*
Thurston Island 13, *20*
  magnetic district *786*, 787
Thurston Island Volcanic Field 515, 564–565, *578*, 603–605
  geochemistry *585*, 600
Thwaites Glacier 799
  geothermal flux 632
  heat flow anomaly 797
  magnetic district *786*, 787
  radar sounding data 793, *794*
tidal gauge 685
till *460, 530*
tilt-meters, Mount Melbourne *742*, 746, 748–749
Toarcian Oceanic Anoxic Event 87
Tomo Erebus Imaging Project 711–713, *714*, 724
Toney Mountain, active volcano 30, **55, 56**, *61*, 62
  geochemistry 596–597
  geophysics 521
  volcanic facies 538–539, **540**
topography, West Antarctica 566
tortoise shell joints 619
tour guides and emergency training 67–68
tourist destinations 66, 667
  Deception Island 235
  risk from volcanoes 56–57
    management 67–68
tower fumarole 697
Tower Peak Formation 129, 135
trace element signature and metasomatism 404
trace elements 48, 215–216, *297, 405*, 702, 730
  Erebus 450, 470, 472–473, *476*, 477–480, 704
  Ferrar province 101–104, 109–110, 112
  Graham Land Volcanic Group **141–142**, 143, 144
  Marie Byrd volcanic province 583–584, 587, 589, 592–593, 599
    Cenozoic outlying volcanic fields 595, 600, 602–605
  Mount Early, Sheridan Bluff **503**, *504, 505*, 506, **507**, 511
  Mount Rittmann 750–751, **752–753**
  Palmer Land Volcanic Group 146–151
  post-subduction basalt **330**
  post-subduction volcanics 332–335, 338, *339*
  Ross Island 720–721
  South Pole 707
  tephra 650–652, 654, 657
  Victoria Land, north 388–399

trace metals 731
TRACERS project, Ross Sea 642
trachyte 220, 222, 702, 748, 750
  Aurora Cliffs, 424–425
  Erebus petrology 450, 461–465, 469–470, 474–475
  Erebus volcanology 416, **417–418**, 423–429, 433, 435, 441
  Marie Byrd, Ellsworth Land 534–536, 577, 580–592, 595–603
  pillows *536*
  Victoria Land **350–351**, 356, **357–359**, 363, 365, 369–371
  petrography 373, 379, **385–387**, 397
Tramway Ridge *696*
  warm ground 714–719
Transantarctic Mountains *21–22, 44*
  extrusive rocks 77, 86
  intrusive rocks 93–113
  mantle 480
  map *9, 20, 500*
transform fracture zones 403
Transition Zone, Graham Land 199–202
Transition Zone, melt 624–626
trap topography 165
trapdoor subsidence 236
tremor, seismicity 711, 715, 745, 753
Trinity Peninsula Group 122, *124*, 126, 139, *214*
triple junction 327–328
triple rift *121*, 157, *158*
troposphere, ozone 1, 707
Tryggve Point 243–244, 246, 457
tsunami risk 667, 685, *687*
tuff breccia **79**, 81–82, 266, 268, *618*, 619
tuff cone **274**, *354*, 521
  Deception Island 237, 242–243
  James Ross Island Volcanic Group 255–257, 267–270
  Mount Plymouth *250*
  Victoria Land 361, 370, 374, 379
  volcanic hazard 67, 685
tuff ring 243, *250*, 685
turbidites, trench fill 203
Turtle Peak, volcanic lithofacies 531–532
tuya, glaciovolcanic edifice 28–29, 32, 566, **359**
  Antarctic Peninsula area 250, 253, *256*, 268, 320–323
  inventory **228–229**
  Beethoven Peninsula *319*, 321
  Brandenberg **556**, 558–560, 566, 570
  Jonassen Island *256*
  Marie Byrd, Ellsworth Land 521, 523, *536*
  age data 531–532, 537, **561**, 563
  Mount Murphy *526*, **527**, 529
  palaeo-environment 566, *567*
  Stauffer Bluff *536*, 537, *538*, 570
  Victoria Land *352*, *359*, 363, 373

U–Pb isotope age 204, 220, **494**
  Dronning Maud Land 162–164, 174
  Graham Land 196–199, 205
  Palmer Land 201, 202
  silicic rocks 122, 126, 128–135
  South Shetland Islands 191, 193–194
U–Pb zircon age, Ferrar rocks **76**, 93
U-decay series, Erebus 705, 726–728, 731
U/Pb isotope ratio, Erebus Volcanic Province 30, 695
ultramafic dykes, sills, nodules 165, 586
ultraviolet spectroscopy 707–708
underplating 135, 151–153
uplift 108, 277, 437
  Cenozoic 33, 403

Upper Scott Glacier Volcanic Field
496–497, 499
  delamination 511
  geochemistry 504–505
  isotope ages **494**
  volcanoes 509–511
USAS Escarpment **551,** 552–553
  geochemistry 599–600
Utpostane layered intrusion 167, 174, 175

Vapour Col *678*
velocity model, Ross Island *722*
velocity of rifting 286
vent 86, *256*, 271, 459
  agglomerate 197, 199, 201, 202, 206
  deposits 80–81, 82
very long period signal (VLP)
  lava source and conduit 712, 714–715, *716–717*, 724
  LP volcanic tremor 679–682, 684
vesicles 84, *132*, 160–162, 165
Vestfjella extrusive–intrusive suite 160–165, *173*, 174–175
  geochemistry 166–167, **168–169,** *170–171*
Victoria Land *20, 44, 347*
  active volcanism 57–60
  ice cover 32
  seismic profile *31*
  sills and dykes 93–113
  uplift and exhumation 33
  volcanism 30–31, 73–88
Victoria Land (north), petrography
  geochemistry 388–399
  magmas and metasomatism 399–406
  magmatism 383–384
  plutons, dykes and volcanic rocks 384–399
  synopsis **385–396**
  tectonics and volcanic setting 406–409
Victoria Land (north), volcanology *345, 347–349*
  lithostratigraphy 349–375
  Meander Intrusive Group 375–376, **377–378**
  satellite image *374*
Victoria Land, south *7, 93, 111, 345, 448*
  subglacial volcanoes *632*
  volcanic fields and suites *347, 415, 448*
viscosity, magma 710
VLP *see* very long period signal
volatiles 620, 709, 723, *728*
  chemosynthetic communities 718–719
  in melt 335, 339, 509–510, 724–726
volcanic arc 22–26
volcanic eruption *see also* eruption and monitoring
volcanic eruption hazard 632
  assessment 67–68, 685–686
  future priorities and monitoring techniques 68
  subglacial eruption, *632*, 781
volcanic eruption protocol
  alert scheme and evacuation plan 67–68
volcanic eruption, prediction 681, 684–686
Volcanic Explosivity Index 677
volcanic fields *347, 415*, 416, 420, *448*
volcanic lithofacies *423–425*
  Antarctic Peninsula Volcanic Group 206–207
  Bransfield Strait and James Ross Island 227, 251, 266–268, **278**
  Ferrar province 77–82
  Marie Byrd 523–524, 571–572
  Mount Early and Sheridan Bluff 491–497
  Victoria Land 349–375, **385–387**, 388
volcanic shield 27, 373, **417–418**, 429, 441
volcanic vents, temperature anomaly 271
volcanism and climate 649–650
volcanism and glaciation 29, 660–661
  ice level record 570–572
  ice-flow obstructions 566–567
  subglacial 57, 60, 62–64, 66, 347–348
volcanism in Antarctica, overview 1–4, 55–68
volcanism, main episodes 22–26, *44, 56, 57, 347*
  alkaline rift volcanism 29–33
  back-arc and Neogene cryosphere 26–28
  Cenozoic 568–569
  Early Jurassic flood lavas and flare-ups 19–22
  Gondwana break-up 10, 12, 13, 33, 50
  Neogene 322, 441
  Pacific margin 22–26
volcano
  building stages *322*
  eruption centre migration 589
  inflation–deflation 67, 684, 686
  inventory, Antarctic Peninsula **228–230**
  location map *492*
  palaeo-environment 320–322
volume of erupted material 30, 77, 521, 538–539, 545, 557
Vostok, long ice core, study 651
Vulcan Hills Volcanic Field *347, 348,* **351,** 371
  age *376*
  petrology **387**

WAIS *see* West Antarctic Ice Sheet
warm ground 709, 714–719, 743–744
WARS *see* West Antarctic Rift System
weathering deposits 162
Weddell Plate *668*
Weddell Sea 113
  ash dispersal 641–642
  extension 136, 139
  geochemistry 104
  tephra dispersal **635**
  triple junction *121*
Werner's fumarole 698
Werner's lava lake 708

West Antarctic Erosion surface 29, 32
West Antarctic Ice Sheet (WAIS) 29–34, 62, 539
  active subglacial volcanism 57
  glaciovolcanic record 566–571, *569*
West Antarctic Ice Sheet Divide, ice core 760, **765,** *772*, 777–781
  subglacial volcanic investigation 793–796
  tephra and wind-blown debris 779–780
West Antarctic Rift System (WARS) 2, 3, 10–13, *20, 44, 47,* 112, 516
  active subglacial volcanism 57, 441, 786–787
  alkaline magmatism 19, 29–33, 47–50, 383–384
  ash dispersal 659
  Cenozoic volcanism, 577–608
  extension–magmatism–uplift 401–403
  geology 449–450
  mantle plume 400–401
  map *20, 44, 500*
  palaeo-ice level 533
  rifting initiation 577
  seismic velocity 438, 722
  subglacial topography *578*
  volcanic rocks 29–33, 45, 491, 497, 499
  geochemistry *502,* 504–509, 511
West Antarctica 9–11, *20,* 44
Western Ross Supergroup *347,* 348–351, *349,* 491
  petrography 384, **385–396**
  stratigraphy *416,* **417–418**
wetland–lava association 207
Whalers Bay, glacial flood *678*
whaling station 236, 667
Whichaway Nunataks *94,* 96, 104
White Island, volcanics 429, 460–461
wind direction, tephra dispersal 780
within-plate magmatism 33, 332–333, 584

xenocryst 451
xenolith 47–50, 287, 521, 531, 543, 593, 748–749
  buchite 333
  crustal 449, 586, 617, *618,* 624
  harzburgite 465
  mantle 405, 408, 463, 473, 476–477, 551, 599
  Os isotope, 406
  peridotite 404–406, 480
  syenite 435, 461, 464, *623*
  ultramafic 436, 454, 600

Young Island, eruption **55**

Zealandia 3, 44, 405, 577, 607–608
  alkaline magmatic province 450
  HIMU signature 479–480
zeolite 85, 108, 215, 273
zircon dating 128–129
Zonda Towers, caldera *195,* 200, 206